COMPUTATIONAL MODELLING OF CONCRETE AND CONCRETE STRUCTURES

Computational Modelling of Concrete and Concrete Structures contains the contributions to the EURO-C 2022 conference (Vienna, Austria, 23–26 May 2022). The papers review and discuss research advancements and assess the applicability and robustness of methods and models for the analysis and design of concrete, fibre-reinforced and prestressed concrete structures, as well as masonry structures. Recent developments include methods of machine learning, novel discretisation methods, probabilistic models, and consideration of a growing number of micro-structural aspects in multi-scale and multi-physics settings. In addition, trends towards the material scale with new fibres and 3D printable concretes, and life-cycle oriented models for ageing and durability of existing and new concrete infrastructure are clearly visible. Overall computational robustness of numerical predictions and mathematical rigour have further increased, accompanied by careful model validation based on respective experimental programmes. The book will serve as an important reference for both academics and professionals, stimulating new research directions in the field of computational modelling of concrete and its application to the analysis of concrete structures.

EURO-C 2022 is the eighth edition of the EURO-C conference series after Innsbruck 1994, Bad Gastein 1998, St. Johann im Pongau 2003, Mayrhofen 2006, Schladming 2010, St. Anton am Arlberg 2014, and Bad Hofgastein 2018. The overarching focus of the conference series is on computational methods and numerical models for the analysis of concrete and concrete structures.

Computational Modelling of Concrete and Concrete Structures

Edited by

Günther Meschke
Ruhr University Bochum, Germany

Bernhard Pichler
Technische Universität Wien, Austria

Jan G. Rots
Delft University of Technology, Netherlands

CRC Press is an imprint of the
Taylor & Francis Group, an **informa** business
A BALKEMA BOOK

CRC Press/Balkema is an imprint of the Taylor & Francis Group, an informa business

Typeset in Times New Roman by MPS Limited, Chennai, India

Library of Congress Cataloging-in-Publication Data

A catalog record has been requested for this book

First published 2022
Published by: CRC Press/Balkema
Schipholweg 107C, 2316 XC Leiden, The Netherlands
e-mail: enquiries@taylorandfrancis.com
www.routledge.com – www.taylorandfrancis.com

ISBN: 978-1-032-32724-2 (Hbk)
ISBN: 978-1-032-32845-4 (Pbk)
ISBN: 978-1-003-31640-4 (eBook)
DOI: 10.1201/9781003316404

Computational Modelling of Concrete and Concrete Structures – Meschke, Pichler & Rots (Eds)
© 2022 Copyright the Editor(s), ISBN: 978-1-032-32724-2

Table of contents

Analysis of concrete structures

Analysis of masonry materials and structures

Constitutive models and computational frameworks

Computational Modelling of Concrete and
Concrete Structures – Meschke, Pichler & Rots (Eds)
© 2022 Copyright the Editor(s), ISBN: 978-1-032-32724-2

Preface

EURO-C 2022 represents the continuation of a series of conferences on computational methods and numerical models for the analysis of concrete and concrete structures.

The Covid-19 pandemic had, unfortunately, an impact on the date and the venue of the current issue of the EURO-C conference series. The main aim of the organizers was to keep the spirit of the previous conferences alive by making every effort to ensure personal interaction by avoiding a hybrid or pure online format. Hence, EURO-C 2022 was postponed from March to May 2022. "Snow", required for the skiing race, one of the traditional elements of the EURO-C conference series, is melted at the end of May, even in the original venue in Obergurgl. The meltwater has flown down the Ötztaler Arche, into the Inn, and finally into the Danube, leading to Vienna. Therefore, the organizers decided to move the conference venue from the Austria alps, exceptionally, to Vienna. Instead of the skiing race, a joint activity on the Danube will take place. Special situations call for exceptional arrangements, but we are very much looking forward to be back in the Austrian alps in 2026.

EURO-C 2022 will take place in Vienna, Austria, from May 23–26, 2022. It is the eighth edition of the EURO-C conference series after Innsbruck 1994, Bad Gastein 1998, St. Johann im Pongau 2003, Mayrhofen 2006, Schladming 2010, St. Anton am Arlberg 2014, and Bad Hofgastein 2018. The series emerged as a joint activity, following early developments in nonlinear Finite Element analysis and softening models for concrete, generated at the time of the ICC 1984 conference in Split, the SCI-C conference in Zell am See, and the two IABSE Concrete Mechanics Colloquia in Delft, 1981 and 1987.

The Proceedings of EURO-C 2022 comprise 6 papers of Plenary Lecturers as well as 85 contributed papers, grouped into 6 sections: (1) Analysis of concrete materials, (2) Analysis of concrete structures, (3) Analysis of masonry materials and structures, (4) Constitutive models and computational frameworks, (5), Durability, coupled, time-dependent, and thermal effects, as well as (6) Safety assessment and design-oriented models. As compared to previous conferences, there are still many contributions on robustness and precision of constitutive models and computational frameworks at the structural scale, for both plain concrete, reinforced concrete and masonry structures. However, trends towards the materials scale with new fibres and 3D printable concretes, multi-scale and multi-physics frameworks, and life-cycle oriented models for ageing and durability of existing and new concrete infrastructure as well as data-driven models are clearly visible.

We are very grateful to the members of the Scientific Advisory Committee for their support and substantial efforts in the reviewing process of over 130 abstracts: Zdenek Bažant, Jan Červenka, Gianluca Cusatis, Guillermo Etse, Dariusz Gawin, Stéphane Grange, Christian Hellmich, Günter Hofstetter, Tony Jefferson, Milan Jirásek, Karin Lundgren, Koichi Maekawa, Chris Pearce, Gilles Pijaudier-Cabot, Ekkehard Ramm, Bert Sluys, Jacek Tejchman, Franz-Josef Ulm, and Yong Yuan.

Prof. Herbert A. Mang, long-standing chairman and currently honorary chairman of the EURO-C conference series as well as key contributor to all previous EURO-C events, celebrated his 80th birthday at the beginning of 2022. We combine our sincerest congratulations with the wish that for many years to come he will maintain in good health his unbroken and extensive scientific activities. Ad multos annos!

We sincerely hope that the EURO-C 2022 Proceedings will serve as a major reference, stimulating new research directions in the field of computational modelling of concrete and its application to the analysis of concrete structures.

Günther Meschke, Bernhard Pichler, Jan Rots, conference chairmen
René de Borst, Herbert Mang, honorary chairmen
Bochum/Vienna/Delft/Sheffield/Vienna, May 2022

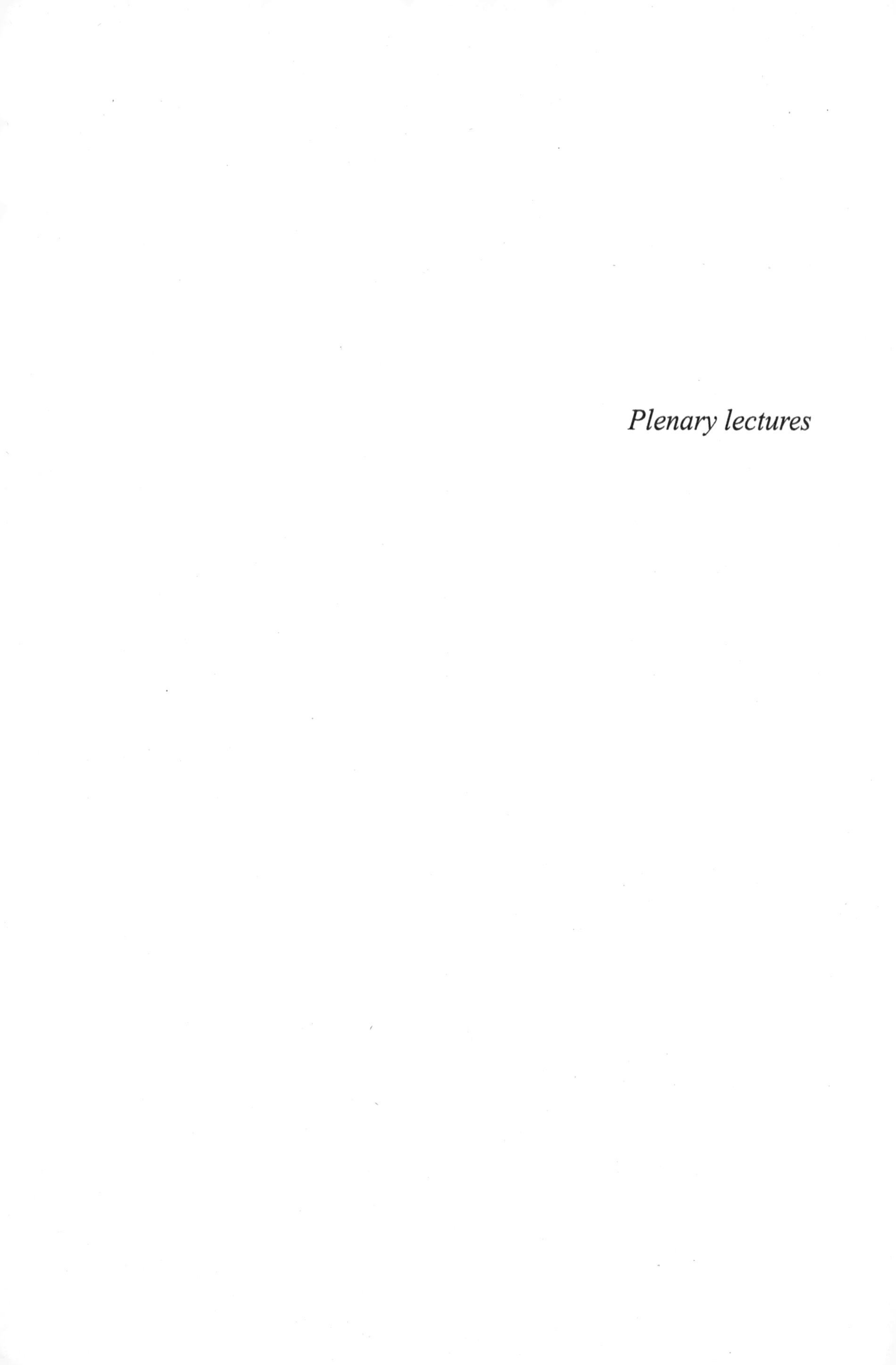

Plenary lectures

Computational Modelling of Concrete and Concrete Structures – Meschke, Pichler & Rots (Eds)
© 2022 Copyright the Author(s), ISBN: 978-1-032-32724-2

Reappraisal of phase-field, peridynamics and other fracture models in light of classical fracture tests and new gap test

Zdeněk P. Bažant* & Hoang T. Nguyen
Northwestern University, Evanston, IL, USA

A. Abdullah Dönmez
Northwestern University, Evanston, IL, USA
Istanbul Technical University, Istanbul, Turkey

ABSTRACT: The newly developed gap test and ten types of classical fracture tests of concrete are used to evaluate the performances of three popular numerical models. The crack band model with microplane damage constitutive model M7 is found to match all the experimental results well. However, the phase-field models show large deviations from the test results, and peridynamic models are even worse. Examination of four recent variants of these models does not change the overall critical appraisal.

1 INTRODUCTION

Recently, a new type of experimental setup, called the gap test [1, 2], has been developed at Northwestern University to reveal in a clear and unambiguous way the effect of crack-parallel stress on the fracture properties of material. Testing specimens of different sizes and applying the size effect method showed that the fracture energy, G_f, and the effective size, c_f, of fracture process zone (FPZ) of concrete depends strongly on level of crack-parallel stresses $\sigma_{xx}(=T)$.

This prediction is confirmed by finite element analysis with the M7 crack band model, which further indicated a strong effect of σ_{zz} and σ_{xz}. The gap test, applied to shale, composites and plastic-hardening metals, to reveal that the crack-parallel stress effects are rather different for different materials. These results shed new light on the validity of numerical models for fracture, such as phase-field (PF) and peridynamics (PD), newly popular in computational mechanics. The gap tests [2] also revealed that the fracture energy of quasibrittle materials, plastic hardening metals and composites depends strongly on the history of crack-parallel stresses (see Figure 1).

2 APPROACH AND MAIN RESULTS

This study uses the new gap test and ten types of classical fracture tests of concrete, most of them previously ignored, to conduct a critical comparison of the phase field (PF) model and peridynamics (PD) with the finite element crack band model (CB) in which the material model is the microplane model M7.

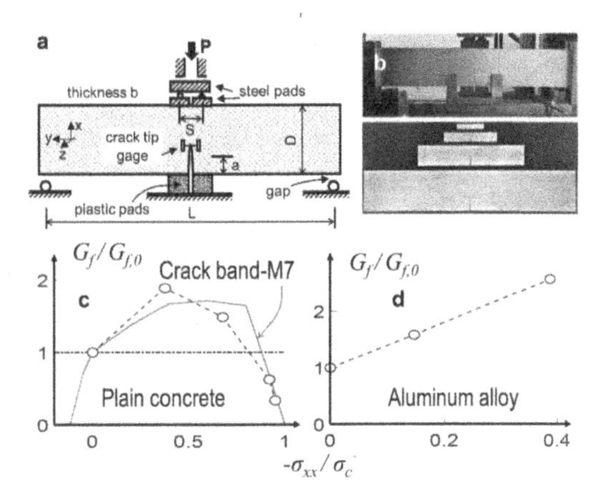

Figure 1. a,b) Setup of the gap test for 2D-geometrically scaled concrete specimens of various sizes. c) Measured and predicted variation of fracture energy G_f with increasing crack-parallel compression; d) the same for polycrystalline metal with millimeter-width yielding zone and micrometer-width fracture process zone.

Optimal fitting of the data by state-of-art phase-field and peridynamics computer programs calibrated by basic material properties reveals severe discrepancies.

Although the phase-field models have certain advantages (being superior for static and dynamic propagation of curved and branching line cracks in perfectly brittle materials obeying LEFM), and could be generalized to different constant (non-varying) levels of crack-parallel stress, they are found incapable of matching the results of the gap test and the classical fracture tests of concrete and rock, provided that the same set of model parameters is used for all the tests conducted on the same material.

*Corresponding Author

DOI 10.1201/9781003316404-1

In these comparisons, the PD, considered as a kind of strongly nonlocal model, is found to disagree with the test data and be even inferior to PF. This reinforces the previous, strictly theoretical, critique of the basic concept of peridynamics [3], both bond- and state-based.

One of the faults of peridynamics is the use of interparticle potential, which is realistic only on the atomic scale. Still another is does not take into account shear-resisted particle rotations (which are what lends LDPM, the lattice particle discrete model, its superior performance). Still another is the unphysical boundary conditions and crack face conditions, along with the problem of unphysical interaction across the fracture process zone (FPZ) softened to various degrees.

The continuum-based finite element crack band model with realistic tensorial damage constitutive law M7 [4, 5] is able to fit the data from all the classical tests and the gap tests closely. The crack band model combined with Grassl's tensorial model and CDPM2 performs in most types of tests almost equally well.

The previously discussed severe limitations of the discrete crack and cohesive crack models are also pointed out. Also, the ubiquity of varying crack-parallel stresses in practical problems and their effects in concrete, shale, fiber composites, plastic-hardening metals and materials on submicrometer scale is emphasized.

3 MODELS AND EXPERIMENTAL DATA USED IN COMPARISONS

Eleven types of experiments on quasibrittle materials (concrete and rock) have been simulated to test the performance of computational models and discussed in the lecture. A few of them are selected here for comments.

- Size effect tests of types 1 and 2 [6]: geometrically scaled specimens with and without notches, subjected to three-point-bend load configuration.
- Compression-torsion fracture tests (mode III) [7]: notched cylindrical specimens subjected to a fixed axial confinement and angle-controlled torque.
- Uniaxial compression fracture tests [2] of cylindrical specimens subjected to uniaxial compressive load with zero or various constant lateral confining pressures rigid confinement.
- Diagonal shear fracture of reinforced concrete (RC) beams [8], reinforced by graded steel bars and subjected to four-point-bend load configuration.
- Gap tests [2] of fracture of notched beams subjected to the loading configuration in Figure 1 and described in Section 1.

Seven computational models are examined in the lecture. They include:

- CB-M7: the crack band model [9] based on the microplane damage constitutive model M7 for concrete [4], as slightly updated in [10] (downloadable codes can be found at http://www.civil.northwestern.edu/people/bazant/m7-coding/m7_cyc_schell_v1.f). The material parameters are optimized for material tests of typical laboratory specimens whose size is close to the size FPZ, or the representative volume of material. This size approximately represents the material characteristic length l_0, which is, of course, kept the same for all specimen sizes.
- CB-Gr: is a tensorial damage constitutive model implemented within the same crack-band finite element framework as CB-M7, except that M7 has been replaced with the concrete constitutive model CDPM2 developed by Grassl et al. [5]. This model is an update of [11] and represents arguably the best plastic-damage constitutive model of concrete formulated in the classical way—in terms of tensors, two loading surfaces in the stress space, and tesorial invariants.
- PF: is the basic phase-field model developed by Francfort and Marigo [12]. Conveniently, this model has been implemented as a user subroutine in Abaqus by Pañeda *et al.* [13].
- PF-Wu: is a phase-field model that is modified to fit better one particular test and is based on the cohesive zone theory of Jiang-Ying Wu [14]. Download both PF models from: https://www.empaneda.com/codes/.
- PD: is an ordinary state-based peridynamic model using a critical stretch with sudden force drop to initiate fracture, developed by Silling [15]. This model has been implemented in the Peridigm [16] code downloadable from the Sandia National Laboratory website.
- PD-Gr: is a state-based non-ordinary (or correspondence-based) peridynamic model, in which Grassl's CDPM2 has been implemented as the constitutive law.
- PDba-Gr: is the same as PD-Gr, except that the deformation gradient needed for the constitutive relation is corrected by Bazilevs et al. according their new bond-associated formulation of peridynamics [17]. Both PD-Gr and PDba-Gr models were implemented as user material subroutines to be used with Peridigm code.

4 CRITICAL COMPARISONS

The size effect on structural strength [18] is salient characteristic of quasibrittle fracture and thus the most important experiment to verify a fracture model. It follows a simple size effect law formulated in 1984 and amply verified for many different quasibrittle materials. This law, whose most important feature is the deviation from the $-1/2$ power law of linear elastic fracture mechanics (LEFM), underlies a simple unambiguous procedure (1990) for measuring the fracture energy and the material characteristic length of quasibrittle materials (even in presence of crack-parallel stresses).

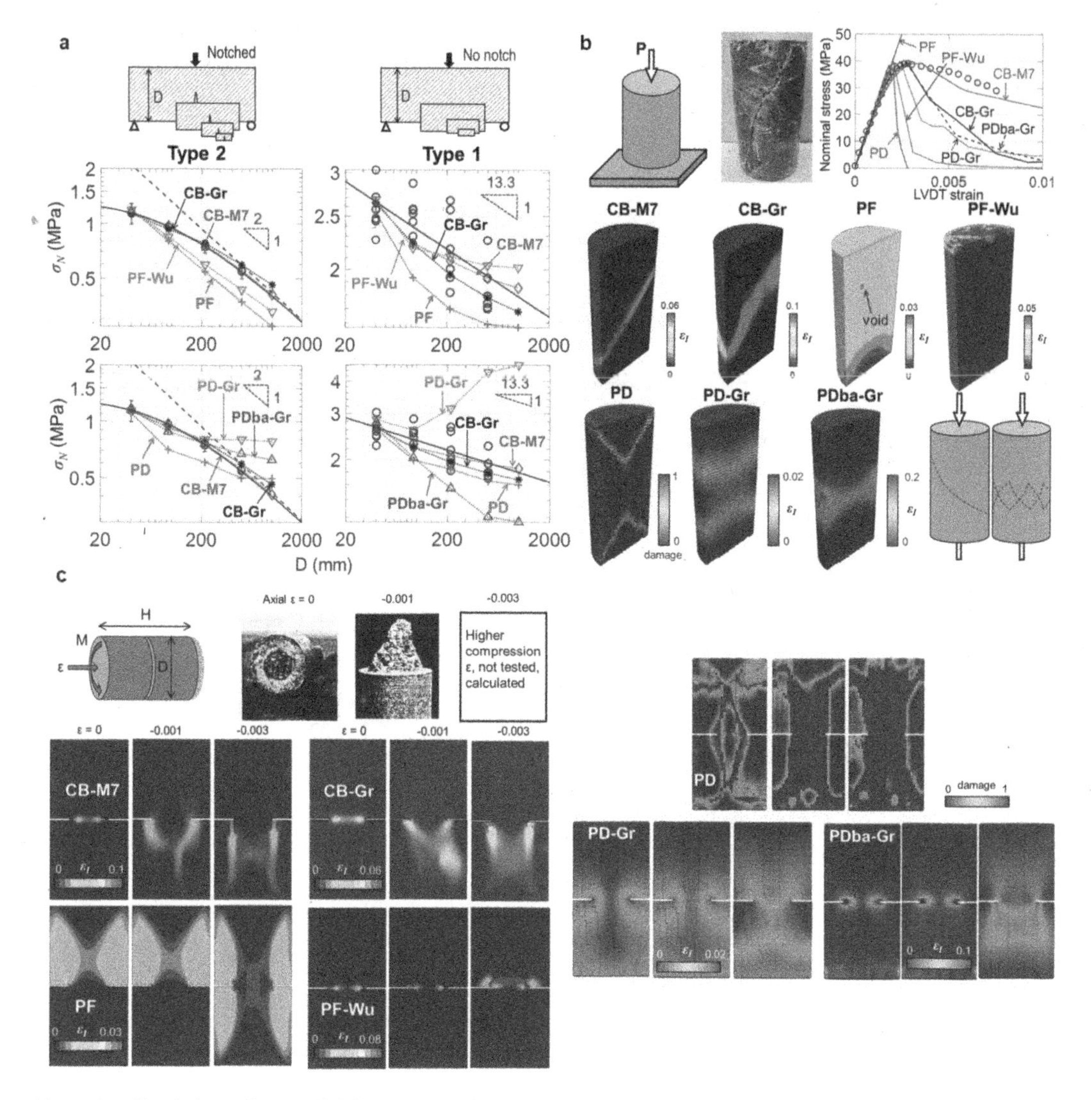

Figure 2. Simulations of a) quasibrittle size effect, b) uniaxial compression fracture, c) mode III shear fracture without and with transverse compression, and d) vertex effect tests (all in concrete).

Figure 2a shows that both PF models result in a power-law behavior in log-log scale. While the slope of the PF model is $-1/2$, which complies with the LEFM, the slope of PF-Wu is different from $-1/2$ which is thermodynamically impossible since it implies a zero-energy flux into the fracture tip. All PD models deviate significantly from the experimental data, and the PD-Gr model even results in an unphysical increase of structural strength. Both CB models yield good results.

Unlike tension, the existence of a discontinuous band of localized strain in concrete could only appear when a material model has the capability of forming frictional or cohesive shear surfaces. Such a capability is absent from both PF models. The same conclusion can be drawn for the basic PD model. Even though the tensorial formulation of Grassl's model allows the emergence of a localized band, only CB implementation of this model shows the presence of such a band. PDba-Gr shows its appearance only vaguely while it is missing completely from the PD-Gr (see Figure 2b).

The transition from a flat to conical and then to distorted cylindrical surfaces of the localized crack when the axially confining strain increases is well captured by both CB models, yet the CB-Gr model produces some secondary diffused cracks. Such a transition is evident in the experimental observation. Neither of the PF models could produce such a transition. Among the PD models, the basic PD model exhibits a rather brittle failure with fragmented pieces which are abruptly released at the peak load but are absent in experiment. The PD-Gr, on the other hand, shows delocalized damage band while the PDba-Gr results in unchanged flat crack surfaces (see Figure 2c).

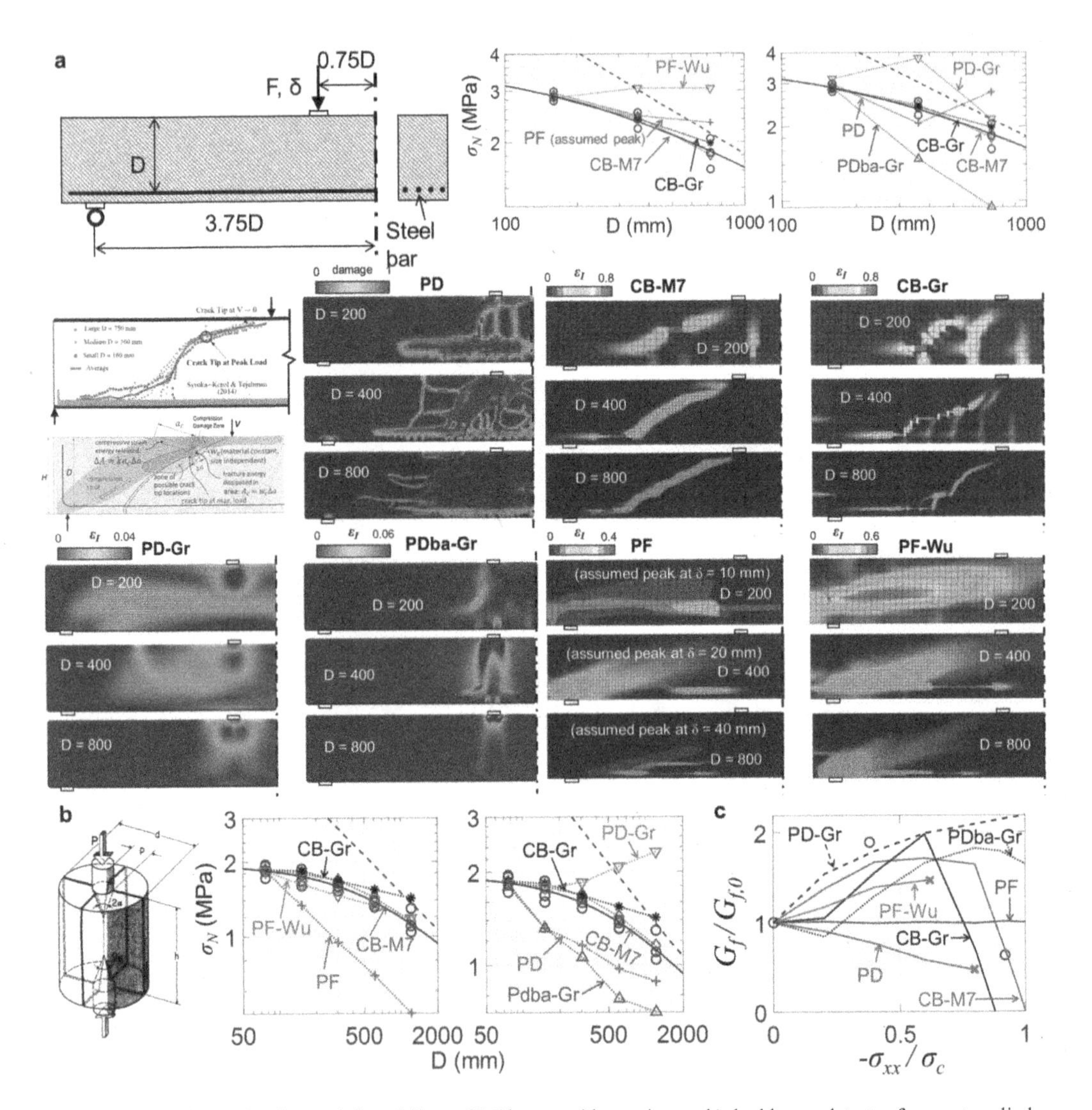

Figure 3. Simulations of a) diagonal shear failure of RC beams without stirrups, b) double punch tests of concrete cylinders, and c) gap test.

The ability to capture diagonal shear of RC beams and the gap test depends on the ability of the model to capture the interaction between components of the stress tensor. Only CB-M7 could reasonably do the job. Though the trend in CB-Gr was reasonable, its prediction of the change in fracture energy could be improved. Other models can capture neither the crack development process nor the peak load corresponding to each structural size (see Figure 3).

5 CLOSING COMMENT

These comparisons document more broadly an unhealthy dichotomy that has recently prevailed between computational mechanics and the concrete testers-designers. The former has relied on minimal selective and insufficient experimental verifications while latter paid insufficient attention to theoretical developments and their critical scrutiny.

ACKNOWLEDGMENT

Funding under NSF Grant CMMI-1439960 to Northwestern University is gratefully acknowledged. A.A.D also thanks for funding from Istanbul Technical University (BAP:42833). Conflict of interest: None.

REFERENCES

[1] Nguyen, H., et al., *New perspective of fracture mechanics inspired by gap test with crack-parallel compression.* Proceedings of the National Academy of Sciences, 2020. **117**(25): p. 14015–14020.

[2] Nguyen, H.T., et al., *Gap test of crack-parallel stress effect on quasibrittle fracture and its consequences.* Journal of Applied Mechanics, 2020. **87**(7): p. 071012.

[3] Bažant, Z.P., et al., *Wave dispersion and basic concepts of peridynamics compared to classical nonlocal damage models.* Journal of Applied Mechanics, 2016. **83**(11).

[4] Caner, F.C. and Z.P. Bažant, *Microplane model M7 for plain concrete. I: Formulation.* Journal of Engineering Mechanics, 2013. **139**(12): p. 1714–1723.

[5] Grassl, P., et al., *CDPM2: A damage-plasticity approach to modelling the failure of concrete.* International Journal of Solids and Structures, 2013. **50**(24): p. 3805–3816.

[6] Hoover, C.G. and Z.P. Bažant, *Comprehensive concrete fracture tests: size effects of types 1 & 2, crack length effect and postpeak.* Engineering Fracture Mechanics, 2013. **110**: p. 281–289.

[7] Bažant, Z.P., P.C. Prat, and M.R. Tabbara, *Antiplane Shear Fracture Tests (Modell).* ACI Materials Journal, 1990.

[8] Syroka-Korol, E. and J. Tejchman, *Experimental investigations of size effect in reinforced concrete beams failing by shear.* Engineering Structures, 2014. **58**: p. 63–78.

[9] Bažant, Z.P. and B.H. Oh, *Crack band theory for fracture of concrete.* Matériaux et construction, 1983. **16**(3): p. 155–177.

[10] Nguyen, H.T., F.C. Caner, and Z.P. Bažant, *Conversion of explicit microplane model with boundaries to a constitutive subroutine for implicit finite element programs.* International Journal for Numerical Methods in Engineering, 2021. **122**(6): p. 1563–1577.

[11] Grassl, P. and M. Jirásek, *Damage-plastic model for concrete failure.* International journal of solids and structures, 2006. **43**(22–23): p. 7166–7196.

[12] Bourdin, B., G.A. Francfort, and J.-J. Marigo, *The variational approach to fracture.* Journal of elasticity, 2008. **91**(1): p. 5–148.

[13] Navidtehrani, Y., C. Betegón, and E. Martínez-Pañeda, *A simple and robust Abaqus implementation of the phase field fracture method.* Applications in Engineering Science, 2021. **6**: p. 100050.

[14] Wu, J.-Y. and Y. Huang, *Comprehensive implementations of phase-field damage models in Abaqus.* Theoretical and Applied Fracture Mechanics, 2020. **106**: p. 102440.

[15] Silling, S.A., et al., *Peridynamic states and constitutive modeling.* Journal of elasticity, 2007. **88**(2): p. 151–184.

[16] Parks, M.L., et al., *Peridigm Users' Guide v1. 0.0.* SAND Report, 2012. **7800**.

[17] Behzadinasab, M., N. Trask, and Y. Bazilevs, *A unified, stable and accurate meshfree framework for peridynamic correspondence modeling—Part I: Core methods.* Journal of Peridynamics and Nonlocal Modeling, 2021. **3**(1): p. 24–45.

[18] Bažant, Z.P. and J. Planas, *Fracture and size effect in concrete and other quasibrittle materials.* 2019: Routledge.

Computational Modelling of Concrete and Concrete Structures – Meschke, Pichler & Rots (Eds)
 ISBN: 978-1-032-32724-2

On the application of nonlinear analysis in the design and assessment of reinforced concrete structures

J. Červenka*, Filip Šmejkal & V. Červenka
Červenka Consulting s.r.o., Prague, Czech Republic

Davide Kurmann
Axpo Power AG, Kernenergie, Switzerland

ABSTRACT: after many years of being only an interesting research topic, nonlinear analysis of reinforced concrete structures is becoming a standard engineering tool that is used in the assessment of existing structures as well as in the design of new structures. This development has been significantly supported by the introduction of new safety formats for nonlinear analysis in the fib model code 2010. Currently the new version of the fib model code 2020 is being finalized. In addition, some of the methods proposed in the fib model code are being implemented in the new version of Eurocodes. The paper provides a summary of the most promising proposals of safety formats for the nonlinear analysis together with the crucial and very important issue of addressing the modelling uncertainty. Some critical and still not fully addressed issues related to the crack band model are discussed and demonstrated using examples of recent blind competition results. Two examples are selected and described in more detail to demonstrate the application of nonlinear analysis, global safety formats and uncertainty treatment in engineering practice. It is concluded that nonlinear modelling represents a very useful engineering tool that can provide better understanding in the structural behavior, expected failure modes and that suitable safety formats, consistent treatment of modelling uncertainties and solid engineering guidelines are needed for their application in practice.

1 INTRODUCTION

The application of finite element method for nonlinear analysis of reinforced concrete structures has been introduced already in the 70's by landmark works of Ngo & Scordelis (1967), Rashid (1968) and Červenka V. & Gerstle (1971). Various material models for concrete and reinforced concrete were developed in 70's, 80's and 90's such as for instance Suidan & Schnobrich (1973), Lin & Scordelis (1975), De Borst (1986), Rots & Blaauwendrad (1989), Pramono & Willam (1989), Etse (1992) or Lee & Fenves (1998). These models are typically based on the finite element method and a concrete material model is formulated as a constitutive model applied at each integration point for the evaluation of internal forces. It was soon discovered that material models with strain softening, if not formulated properly, exhibit severe mesh dependency (De Borst & Rots 1989), and tend to zero energy dissipation if the element size is reduced (Bažant 1976).

The crack band approach was introduced by Bažant and Oh (1983) to remedy the convergence towards zero energy dissipation. A more rigorous solution of the ill-posed nature of the strain softening problem represent nonlocal or higher-order continuum models: such as non-local damage model by Bažant & Pijaudier-Cabot (1987), gradient plasticity model by de Borst & Muhlhaus (1992) or gradient damage model by de Borst et al. (1996). The nonlocal models introduce additional material parameters related to an internal material length scale. Currently these models are mathematically rigorous, but are seldom used in engineering practice or available in commercial finite element codes.

The limitations of the crack band model in practical engineering calculations, namely when it comes to large finite elements or in the presence of reinforcement, were clearly understood already in the work of Bažant and Oh (1983). These limitations were described and treatment was proposed by Červenka (2018) for the cases when large finite element sizes as well as small ones are used in the finite element nonlinear analyses.

Until recently the application of these advanced nonlinear analyses in practice was difficult since it was not compatible in many cases with the existing design codes and engineering practice, which is still mostly based on linear analysis.

The *fib* Model Code 2010 (MC2010) introduced a comprehensive system for the treatment of safety and model uncertainty for structural assessment and design based on nonlinear analysis. On the basis of MC2010,

*Corresponding Author

DOI 10.1201/9781003316404-2

new safety formats are being proposed also in the ongoing revisions of Eurocodes and in the evolution of the new fib Model Code 2020.

The paper summarizes the most prominent safety formats available for nonlinear analysis with stronger focus on the critical treatment of model uncertainty. The important topic of model uncertainty is demonstrated on three recent examples of blind predictions, in which the authors participated.

In the last part, the paper demonstrates the application of nonlinear analysis and the global safety formats to two examples from engineering practice.

2 NONLINEAR ANALYSIS AND CRACK BAND MODEL

As already noted in the introduction, many material models have been developed in the past. It is not the objective of this paper to provide a comprehensive overview and summary of all of them. A comprehensive summary of this topic is available for instance at Jirásek & Bažant (2001). Considering the extremely large variety of existing material models or approaches to nonlinear analysis, it is impossible to provide guidelines or recommendations for their general application in engineering practice. It is therefore evident that very detailed guidelines or approaches for the treatment of for instance the modelling uncertainties can be only specific to certain class of constitutive models or even to a particular material model or even only to a single software.

The examples presented in this paper were calculated using the finite element software ATENA (Červenka 2021), and therefore some of the conclusions are valid only for this software or at most are applicable for the class of models based on smeared crack approach and crack band method. However, the presented safety formats or the general treatment of model uncertainties is applicable for other nonlinear models for concrete structures as well.

The material model used in the examples of this paper is a fracture-plastic model described in more detail in Červenka (1998) and Červenka & Papanikolaou (2008).

The constitutive model formulation assumes small strains, and is based on the strain decomposition into elastic (ε_{ij}^{e}), plastic (ε_{ij}^{p}) and fracture (ε_{ij}^{f}) components. The stress development is described by a rate equation reflecting the progressive damage (concrete cracking) and plastic yielding (concrete crushing):

$$\dot{\sigma}_{ij} = D_{ijkl} \cdot (\dot{\varepsilon}_{kl} - \dot{\varepsilon}_{kl}^{p} - \dot{\varepsilon}_{kl}^{f}) \tag{1}$$

The flow rules govern the evolution of plastic and fracturing strains:

$$\text{Plastic model: } \dot{\varepsilon}_{ij}^{p} = \dot{\lambda}^{p} \cdot m_{ij}^{p}, m_{ij}^{p} = \frac{\partial g^{p}}{\partial \sigma_{ij}} \tag{2}$$

$$\text{Fracture model: } \dot{\varepsilon}_{ij}^{f} = \dot{\lambda}^{f} \cdot m_{ij}^{f}, m_{ij}^{f} = \frac{\partial g^{f}}{\partial \sigma_{ij}} \tag{3}$$

where $\dot{\lambda}^{p}$ is the plastic multiplier rate and g^{p} is the plastic potential function, $\dot{\lambda}^{f}$ is the inelastic fracturing multiplier and g^{f} is the potential defining the direction of inelastic fracturing strains. The multipliers are evaluated from the consistency conditions.

The model of Menetrey & Willam (1995) is used for plasticity of concrete in multiaxial stress state in compression (Figure 1) with nonlinear hardening/softening (Figure 2).

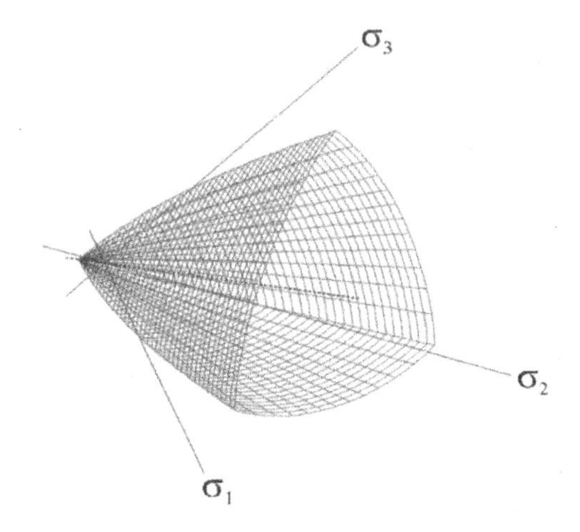

Figure 1. Three-parameter Menetrey & Willam (1995) concrete failure criterion in principal stress frame.

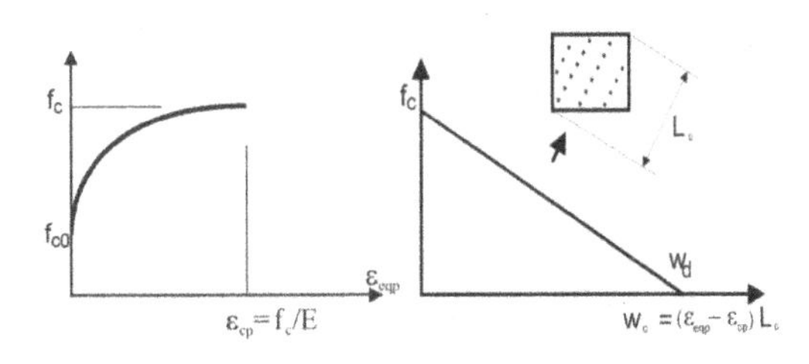

Figure 2. Hardening/softening law for the plasticity model for concrete in compression.

For tensile cracking, Rankine criterion with exponential softening of Hordijk (1991) – see Figure 3 – is used, where w_t stands for the crack width. The crack band approach of Bažant & Oh (1983) is used to relate crack opening displacement to fracturing strains. An analogical approach is used also in compression according to Červenka (2014). The crack band L_t as well as the crush band size L_c are adjusted with regard to the crack orientation approach proposed by Červenka & Margoldová (1995). This method is illustrated in Figure 4 and described by Eq. (4) where the crack angle θ is taken as the average angle between crack direction and element sides.

$$L_t' = \alpha \gamma L_t \text{ and } L_c' = \gamma L_c \tag{4}$$

$$\gamma = 1 + (\gamma_{max} - 1)\frac{\theta}{45}, \ \theta \in \langle 0; 45 \rangle, \ \gamma_{max} = 1.5$$

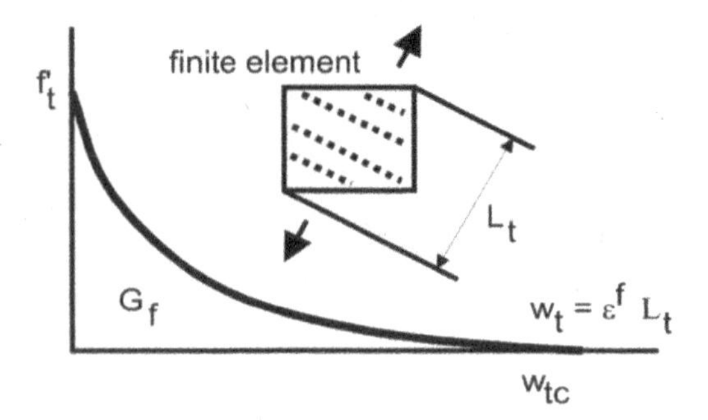

Figure 3. Crack opening law according to Hordijk (1991).

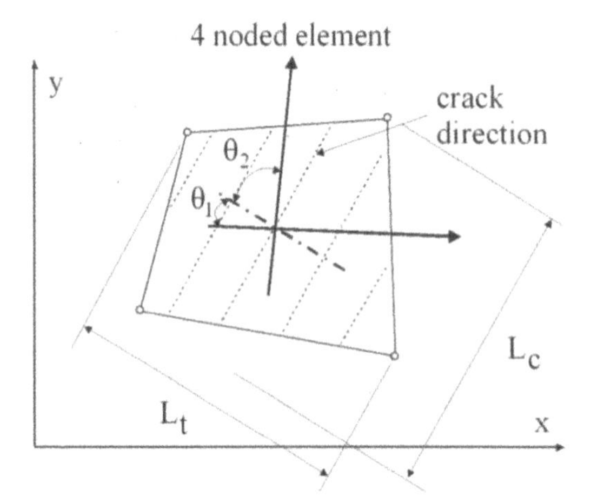

Figure 4. Crack band formulation.

The above formulation controls the strain localization accounting for the mesh size and the crack orientation. Parameter α is introduced to cover the localization effect due to the various element types. according to Slobbe (2013).

Additional important features of cracked reinforced concrete include the reduction of compressive strength, shear stiffness and shear strength degradation, often referred as a shear retention effect. They represent key elements of a successful constitutive model of reinforced concrete (Červenka & Papanikolaou 2008), and are mainly important in problems dominated by shear failure.

Červenka (2018) demonstrated the limits of the crack band approach for modelling of reinforced concrete when very large or very small finite elements are used.

For the case of large finite elements this problem is demonstrated on Figure 5. When very large finite elements are used, the standard assumption of the crack band approach that single crack, i.e. localization zone will develop inside the finite element, is not valid anymore. Due to reinforcement spacing, cover size or reinforcement diameter, cracks will localize at certain distances, which maybe smaller than the used finite element size.

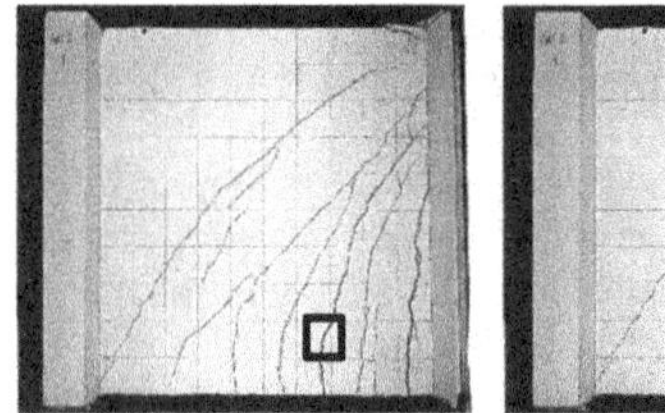

Figure 5. (Left) valid assumption of the crack band and (right) invalid assumption for the large finite element size when more than one crack will localize inside the finite element.

A similar situation may develop in the other extreme when very small finite elements are used as is schematically described in Figure 6. In case of very fine finite element meshes near the reinforcement, artificially stiff response may be obtained at the stage of crack initiation (Figure 7).

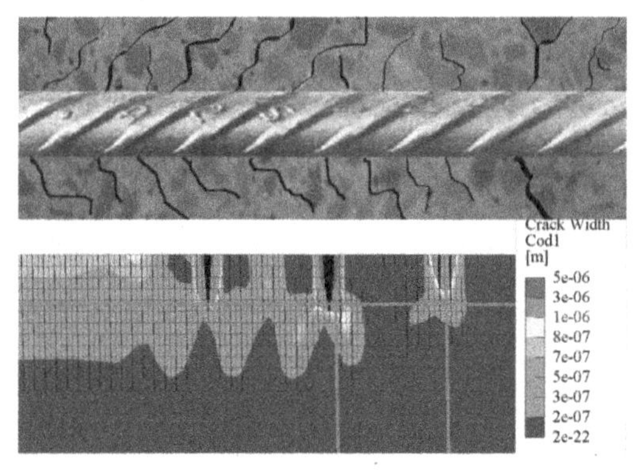

Figure 6. (Top) shows the schematic view of the crack localization near the reinforcement bar influenced by aggregate dimensions and rib spacing, (bottom) shows the incorrect crack initiation if very fine mesh is used.

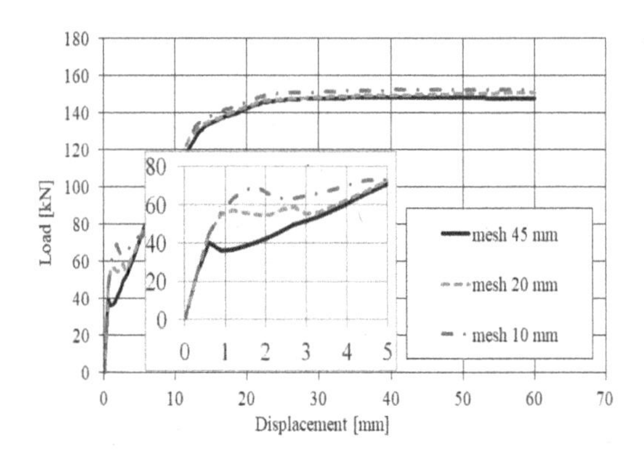

Figure 7. Manifestation of the minimal crack spacing issue in a typical analysis of reinforced concrete element.

If an analogical approach to crack band is used also for the modelling of softening in compression (Figure 2), a similar defective behavior is observed.

Figure 8 shows the failure localization in a typical compression cylinder analysis. Contrary to the direct tensile failure, the crush band cannot localize into a row of single finite elements namely due to the effect of the dilatancy, and the resulting shear band will involve many elements. The crack band approach therefore cannot be used unless the crush band size is much smaller than the size of the finite element. The band size has to be therefore specified as in additional input parameter. The recommended assumption is schematically described in Figure 9 and is equivalent to the

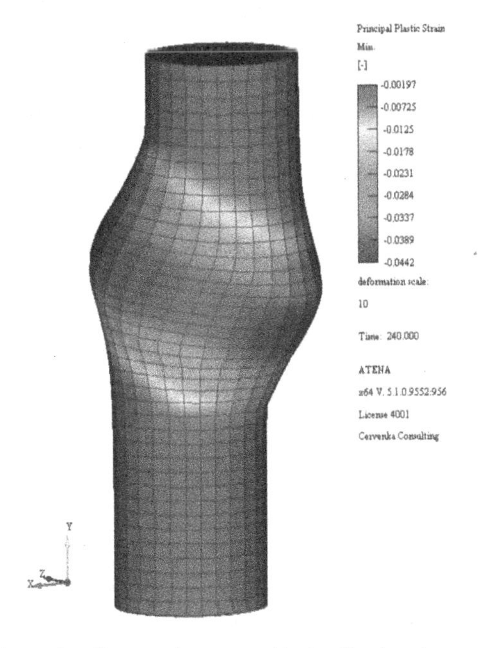

Figure 8. Compression test with localization into an inclined band involving many elements.

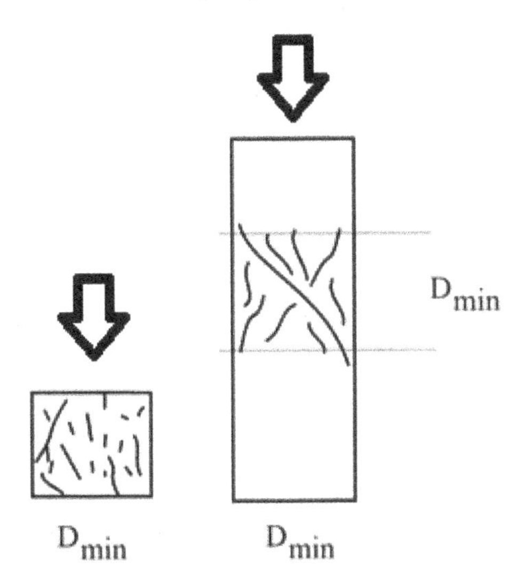

Figure 9. Schematic description of the assumption of the crush band size dependance on the specimen minimal dimension D_{min}.

minimal dimension of the analyzed structural element. The three localization limiters are proposed (Červenka 2018):

$L_{t,\min}$–minimal crack spacing limiter in tension related to aggregate size
$L_{t,\max}$–maximal crack spacing limiter in tension related to reinforcement arrangement
$L_{c,\min}$–minimal crush band limiter in compression related to the minimal size of the compression zone.

3 SAFETY FORMATS FOR NONLINEAR ANALYSIS OF REINFORCED CONCRETE

The design condition is generally formulated as:

$$E_{\mathrm{d}} < R_{\mathrm{d}} \tag{5}$$

where E_{d} represents the design load effect and R_{d} the design resistance. They should include the specified safety margins. For simplicity, the load effect and resistance are considered separately.

The design condition (5) in the standard design practice is applied to critical cross-sections and the load effect is obtained by linear analysis. The inconsistency of this concept is well known as different assumptions are used for the calculation of:

load effects – using typically linear analysis and,
the cross-section resistance – with assumptions of strongly nonlinear material behavior.

In statically indeterminate systems, the section forces may change due to force redistribution at the ultimate limit state. Local safety checks may then be insufficient, and a global safety assessment is needed. Global safety formats combined with nonlinear analysis provide adequate tools for such cases.

The load effect E_{d} in the equation (5) is considered at the global level (typically it represents the total effect or the intensity of the relevant load combination), and analogically the resistance R_{d} is the ultimate load level at failure for the given load combination calculated by the nonlinear analysis.

MC2010 introduces four methods for the global assessment using nonlinear analysis. In this study only two most prominent methods will be treated. i.e. the partial factor (PFM) (Section 7.11.3.4) and GFM-ECoV method (Section 7.11.3.3 of fib model code 2010). The full probabilistic approach on the other hand can be used as a reference solution. These two methods are expected to be included in the new version of Eurocodes, and therefore they will be treated in more detail.

3.1 *Partial safety factor method (PFM)*

This approach is most appealing for practicing engineers as it is a quite natural extension of the current design practice. The resistance R_d is obtained as:

$$R_{\mathrm{d}} = \frac{R\{X_{\mathrm{d}}; a_{\mathrm{nom}}\}}{\gamma_{\mathrm{Rd}}} \tag{6}$$

where:

$R\{-\}$ is the structural resistance by numerical simulation,

X_d is the design value of the material property considering material and geometric uncertainties, but excluding model uncertainty treated separately by γ_{Rd}, and it is estimated as:

$$X_d = \frac{X_k}{\gamma_M}, \quad \gamma_M = \frac{\exp\left(\alpha_R \cdot \beta_{tgt} \cdot V_{RM}\right)}{\mu_{RM}}, \tag{7}$$

a_{nom} is the nominal value of the geometric property, i.e. for instance reinforcement depth or element dimension,

γ_{Rd} is the partial safety factor, which accounts for the model uncertainty, index M where S stands for reinforcement and C for concrete in compression,

α_R is the partial safety factor, which accounts for the model uncertainty, and will be treated in Section 4,

β_{tgt} is sensitivity factor for resistance in MC2010 with the recommended value of 0.8 for a 50-year reference period,

V_{RM} is the coefficient of variation of the resistance estimated for a linear product of resistance parameters as in (8),

μ_{RM} is the bias factor for all uncertainties obtained as a product of the individual bias factors:

$$V_{RM} = \sqrt{\sum_I^n V_I^2} \quad \mu_{RM} = \prod_I^n \mu_I \tag{8}$$

3.2 *GFM-ECoV method – an estimate of the coefficient of variation*

ECoV method originally proposed by Červenka (2008, 2013) is a semi probabilistic approach assuming that the distribution of resistance due to the variability of materials – described by the coefficient of variation V_m - can be estimated from the mean R_m and characteristic value R_k of the resistance. The underlying assumption is that the distribution of the resistance is according to a lognormal distribution, which is however typical for the structural resistance, and it can be expressed as:

$$V_R = \frac{1}{1.65} \ln\left(\frac{R_m}{R_k}\right) \tag{9}$$

Under these assumptions, the global safety factor of the resistance can be calculated as:

$$\gamma_R = \exp\left(\alpha_R \beta\, V_R\right) \cong \exp\left(3.04\, V_R\right) \tag{10}$$

where the typical values for α_R and β are 0.8 and 3.8 respectively leading to 1.12‰ fractile of the design value of resistance, which is calculated as:

$$R_d = \frac{R_m}{\gamma_R\, \gamma_{Rd}} \tag{11}$$

The main task is to estimate the mean and characteristic values R_m and R_k. They can be estimated from two separate nonlinear analyses using mean and characteristic values of the input material parameters, respectively.

The method is quite general and the reliability level β and distribution type can be changed if required. Also the geometric uncertainty can be included (Červenka 2021). It can capture different types of failure and the sensitivity to a random variation of the material parameters is adequately captured in most cases of practical relevance. The slight disadvantage of this method compared to the PFM is the need for two separate non-linear analyses.

4 MODEL UNCERTAINTY

The evaluation of model uncertainty is critical for robust and reliable evaluation of the design resistance as described in Section 3. The model uncertainty should be evaluated by statistical evaluation of the comparison of the model predictions to experimental data. In this respect it should be understood that the obtained partial factors for model uncertainty will be valid only for the investigated material model or simulation software. The model uncertainty is usually defined as the ratio:

$$\theta = R_{exp}/R_{sim} \tag{12}$$

where R_{exp} is the resistance found by an experiment and R_{sim} is the resistance obtained by a numerical simulation. The model uncertainty is considered as an additional random variable with a lognormal distribution.

The experimental resistance is considered as a reference, i.e. true value. Therefore, in order to investigate pure model uncertainties, it is essential to reduce other effects such as aleatory uncertainties to minimum. Material properties of concrete are typically identified by the concrete compressive strength tested on accompanying concrete samples (e.g. cylinders). Other material parameters (elastic modulus, tensile strength, fracture energy, etc.) are usually determined indirectly by formulas available in codes or should be provided as guidelines for a particular model. A random distribution of material properties within the tested structure is also not known. These effects will be therefore included in the model uncertainties.

The experimental data base for the calibration of model uncertainty for a particular material model or modelling approach should include results of experiments relevant to the considered resistance model. Range of parameters in experiments, such as reinforcement, size or concrete strength class, limit also the relevant range of model uncertainty. Failure mode is often also suggested as a classification parameter. However, it may be useful to include more failure modes in one group, since in general a failure mode identification is not unique and straightforward.

For a database containing n samples (experiments) the central moment characteristics of model uncertainty can be estimated for mean μ_θ, standard deviation σ_θ and coefficient of variation V_θ. Considering a log-normal statistical distribution (according to MC2010) a safety factor for model uncertainty can be obtained as:

$$\gamma_{Rd} = \frac{\exp(\alpha_R \beta \times V_\theta)}{\mu_\theta} \tag{13}$$

The values for the sensitivity factor α_R and β are to be selected based on the required reliability levels, and as in Section 3.2 the typical values are 0.8 and 3.8, respectively.

The calibration of the model uncertainty is by its nature dependent on the used constitutive model, its implementation in a particular software or even on the analyst or engineer himself. The human factor can be eliminated or at least addressed by providing modelling guidelines, training and education. Several investigations of model uncertainty for various nonlinear finite element software have been published recently:

Engen (2017) investigated 38 RC members under monotone loading analyzed by several authors. Rather low partial safety factor for the model uncertainty was proposed based on the failure mode in the range 1.02–1.04.

Castaldo (2018) considered 25 structural members from various literature sources including deep beams, shear panels and walls. The investigated members had statically determined static scheme and were tested up to failure with a monotonic incremental loading process. The tests were reproduced by non-linear analysis adopting 9 different modelling hypotheses distinguishing between the software platform and concrete tensile response. A total number of 225 simulations has been performed, and the resulting value of the model uncertainty partial factor was 1.15.

Castaldo (2020) investigated also the model uncertainty for cyclic loading of 17 shear walls with statically determined scheme. 18 different modelling hypothesis were considered and altogether 306 simulations have been performed. This study proposes a model uncertainty factor $\gamma_{Rd} = 1.35$.

Gino (2021) focused on the problem of model uncertainty related to nonlinear analysis of slender RC members, A total number of 40 experiments of concrete columns with slenderness ratio between 15-275 are considered, and the model uncertainty factor γ_{Rd} is proposed in the range 1.15–1.19.

Finally the authors of this paper also performed a similar study in Červenka V. (2018) where 33 RC members were studied with failure modes ranging from ductile modes governed by reinforcement yielding up to brittle failure modes dominated either by tensile fracture or concrete compressive crushing. The study involved slabs and beams with bending or shear or punching failure modes.

The results of this study are summarized in Table 1.

It should be noted that the model uncertainties parameters in Table 1 are valid only for the used software (Červenka 2021) and the constitutive model (Červenka & Papanikolaou 2008).

Table 1. Partial safety factors for model uncertainty (Červenka V. 2018).

Failure type	μ_θ	V_θ	γ_{Rd}
Punching	0.971	0.076	1.16
Shear	0.984	0.067	1.13
Bending	1.072	0.052	1.01
All failure modes	0.979	0.081	1.16

5 VALIDATION AND EXPERIENCE FROM BLIND COMPETITIONS

The nonlinear methods or constitutive models for concrete structures are usually implemented and used as a part of a software tool. It is crucial that the constitutive models as well as the various numerical tools and methods to be used in practical engineering projects are properly tested and validated based on known analytical solutions or best by real material and structural experiments. Such tests and validation are usually part of the software development, and it should contain also the definition of the model uncertainty as described in Section 4.

Since this validation as well as the model uncertainty quantification is performed when the results of the experiments are already known, there is a significant risk that the analyst will adjust certain material or solution parameters to obtain a better fit. It is therefore important to strictly derive all the input material parameters from the known experimental data, i.e. typically concrete compressive strength or reinforcement yield strength. For most material models, more input material parameters are needed then available from the physical tests. It is important that unique formulas and rules are provided how to relate them to available experimental data or otherwise default values should be used.

This is the case also for other parameters necessary for the nonlinear analysis, such as: step size, convergence criteria, arc-length method parameters, finite element size, etc. These should be strictly based on some guidelines or default values.

In this respect, the most crucial test for the robustness of any method or model is the participation in blind robin predictions that are often organized by various research institutions.

The blind robin competitions have certain limitations as well. They often involve a single experiment, therefore they include the influence of both aleatory (inherent uncertainty due to probabilistic variability) and epistemic (missing knowledge) uncertainty mixed together.

The authors participated in the past in various blind competitions, and it is interesting to share the experience of some of the most recent ones to document complications, risks or peculiarities with their evaluation.

The competition organized by Collins (2015) received a lot of attention. It involved a large deep beam with dimensions 4 × 19 m as shown in Figure 10. When the beam was loaded, it failed first on the right side (East) without shear reinforcement. After the failure the right side was strengthened with shear ties, and the beam was reloaded up to the failure on the left side (West), where shear reinforcement was introduced during the beam production.

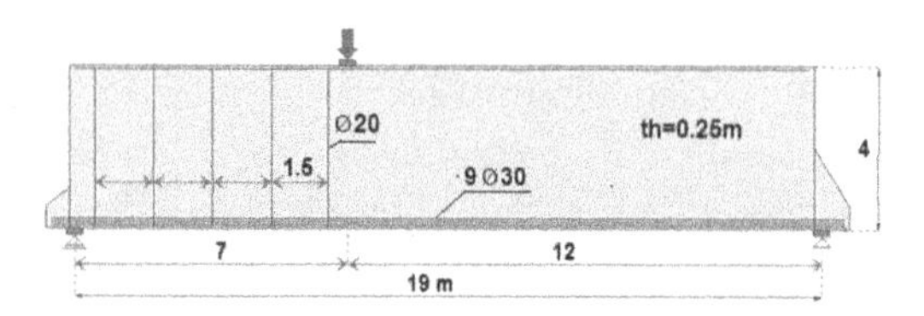

Figure 10. Test specimen geometry for the Toronto beam (Collins 2015).

With this approach the single beam was used to obtain two test results for the shear strength with and without shear reinforcement.

Similar competition was organized recently at the University of California at Berkeley (Moehle & Zhai 2021), Figure 11. Very similar geometry and approach was adopted for testing with the difference that during the first test, i.e. right side (Exp. 1) four longitudinal bottom reinforcement bars were left unbonded inside the beam (see Detail B, Figure 11). After the first test was completed, these bars were injected with grout for the second test, i.e. the left side (Exp. 2).

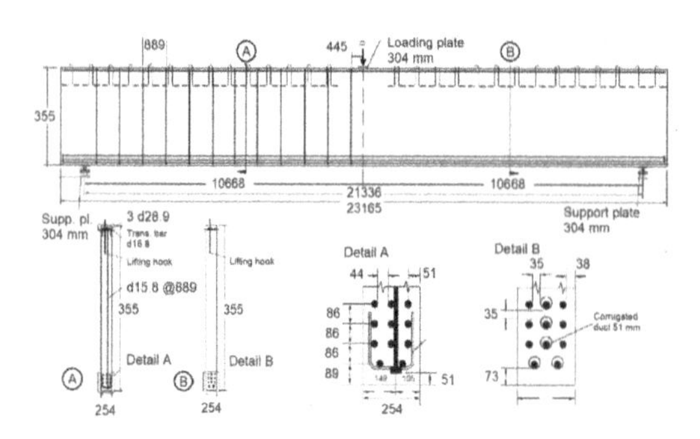

Figure 11. Test specimen geometry (dimensions in mm) for the UC Berkeley beam (Moehle & Zhai 2021).

The material parameters that were used in the prediction simulations for the two competitions are summarized in Table 2. Authors predictions for the two competitions are shown in Figure 12 and Figure 13. The authors predictions for the Toronto competition were the winning predictions. They are indicated by the solid lines in Figure 12.

The predictions for UC Berkeley competition (Figure 13) were not so satisfactory. Especially for the test without shear reinforcement, i.e. the curve labeled "Sim. 1" in Figure 13 has a peak almost 50% lower than in the experiment labeled as "Exp. 1".

Table 2. Material parameters used in the prediction of beam competitions of Toronto and UC Berkeley.

Parameter	Toronto	UC Berkeley
Concrete		
Elastic modulus **E** [MPa]	34129	31 008
Poisson ratio	0.2	0.2
Compressive strength $\mathbf{f_c}$ [MPa]	40.0	30
Tensile strength $\mathbf{f_{ct}}$ [MPa]	3.0	2.3/2.4(*)
Fracture energy $\mathbf{G_F}$ [N/m]	78	100/135(*)
Crushing lim. displ. $\mathbf{w}_d$ [mm]	5	20
Fixed cracks	1.0	0.75/1.0(*)
Strength reduction of cracked concrete r_c^{lim}	0.8	1.0/0.8(*)
Shear factor s_F	50	50/20(*)
Reinforcement		
Elastic modulus **E** [MPa]	200 000	200 000
Yield strength $\mathbf{f}_{ys}$ [MPa]	573/522(**)	830/420(**)
Tensile strength $\mathbf{f_{ts}}$ [MPa]	685/629(**)	1030/620(**)
Limit strain [–]	0.18/0.2(**)	0.025/0.02(**)

(*)The second value indicates the default value normally generated by the software for the given strength class. In Figure 13 the results for these parameters are indicated by the label "default".

(**)The first value is for the shear reinforcement, the second for the longitudinal bars.

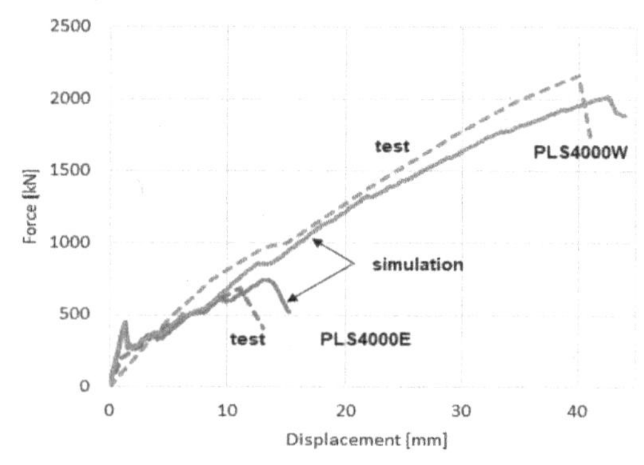

Figure 12. Comparison of Toronto test predictions and experiments.

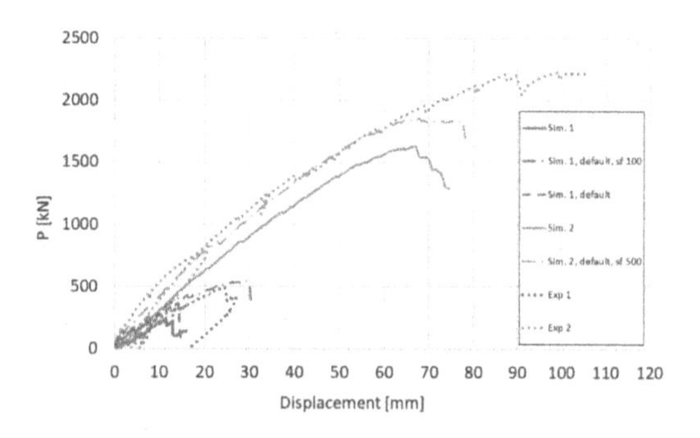

Figure 13. Comparison of UC Berkeley beam predictions and experiments.

It should be however taken into account that the self weight of the beam is significant especially for the test without shear reinforcement. This is actually valid also for the Toronto test.

In UC Berkeley experiment, for instance, the self weight of the beam is about 515 kN. This means that the peak load was in reality underestimated by only about 25% for Exp. 1 and 22% for the Exp. 2. It is also necessary to understand that there was only a single experiment performed in both cases so uncertainty in the material parameters, i.e. aleatory uncertainty was not addressed in the test at all. In both cases, the failure was dominated by concrete. In Exp. 1 the tensile properties played the major role, while in Exp. 2 it was the concrete compressive crushing as well as the steel yielding controlling the peak load. This means that quite high variability of the structural strength can be expected just from the material heterogeneity and variability.

However, the major source of inaccuracy in the UC Berkeley prediction was the consideration of shrinkage (150 μstrains), which by itself reduced the beam strength of the "Sim. 1" by about 25%. This is documented by the curve denoted as "Sim. 1 default", where default material parameters were used without any adjustment and shrinkage was not considered. The error of this analysis was only about 25%, i.e. 12% considering the dead weight. Important parameter for the shear dominated problems is the shear factor parameter s_F(Červenka & Papanikolaou 2008), which controls the shear stiffness of the cracked concrete, which had to be increased to 100 in order to have a good match with the experiment. This analysis is in denoted as "Sim. 1, default, s_F 100" in Figure 13. The default value of this parameter is normally set to 20 to provide conservative results.

The typical crack patterns for the two UC Berkeley experiments are shown in Figure 14, and they are in very good agreement with the observed failure modes, which were reported as diagonal tension failure and flexure-diagonal compression for Exp. 1 and Exp. 2 respectively.

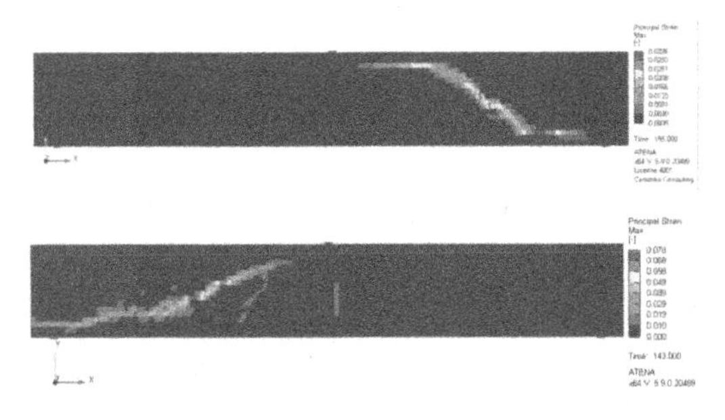

Figure 14. Crack pattern for the UC Berkeley beam predictions.

Gunay & Donald (2021) from UC Berkeley organized another interesting competition in 2021 on the response of reinforced concrete column in cyclic behavior. The results of the competition were announced in December 2021 (see Gunay & Donald 2021).

The load-displacement curves showing the comparison of the predictions with the experiment are shown in Figure 16. The prediction was quite satisfactory even though the initial stiffness of the system is significantly underestimated, however, this can be also due to higher flexibility of the boundary conditions in the experiment itself. Again, it is important to take into consideration that material variability is not addressed by the experiment. The failure mode (see 6) was also in a very good agreement with the experiment, where shear failure in diagonal tension in the middle section was reported.

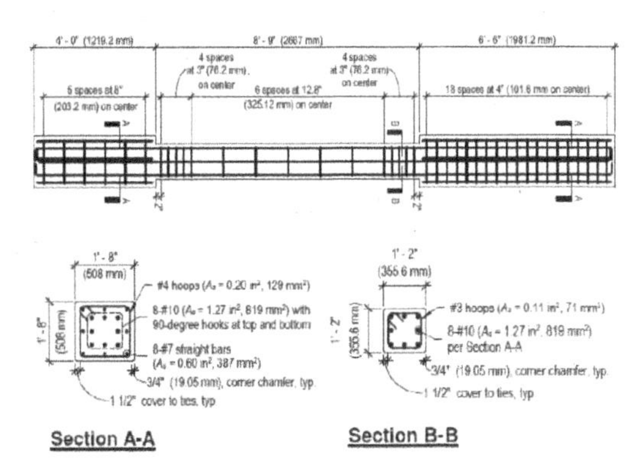

Figure 15. Test specimen geometry for UC Berkeley cyclic column.

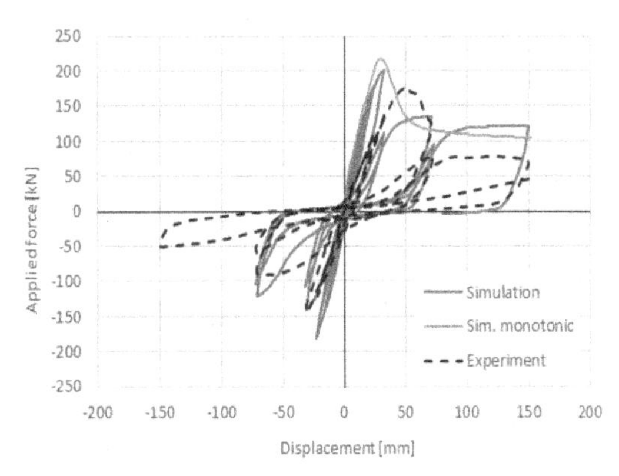

Figure 16. Comparison of UC Berkeley cyclic column experiment and simulation prediction.

6 APPLICATION IN DESIGN AND ASSESSMENT

Two examples of application of nonlinear analysis from engineering practice will be presented in this chapter. The first example is an assessment and design of strengthening for an understrength continuous beam. The second example represents a seismic

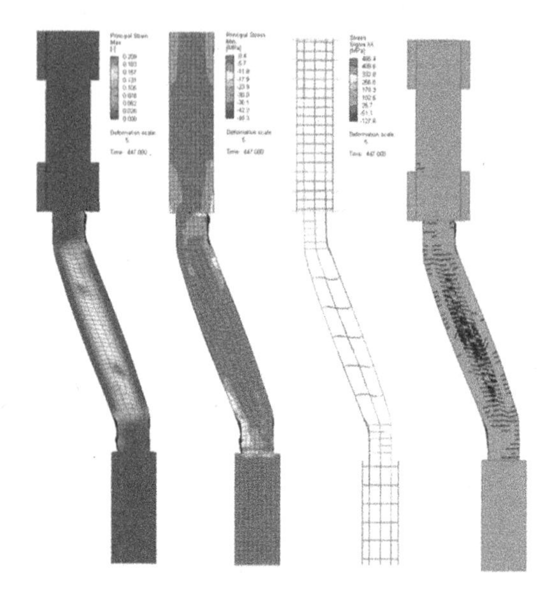

Figure 17. Final failure modes from UC Berkeley cyclic column predictions.

assessment by pushover analysis of existing reinforced concrete building, which is part of a critical infrastructure.

6.1 *Assessment and strengthening design of a continuous beam*

This investigated beam is part of a foundation slab at a food processing and storage plant in southern Bohemia. During the construction an insufficient reinforcement placement was detected in the continuous beams supporting the foundation slab below the storage rooms. The continuous beam is connecting the foundation piles, and is main load bearing element transferring the loads from the structure to the foundation. The problematic area is supposed to be a storage rooms with freezers, and it was critical for the investor that sufficient load carrying capacity is guaranteed.

Figure 18 shows the geometry of the model, which consists of a continuous beam with contributing parts of a slab. The beam dimensions are height 750 mm, thickness 400 mm and 200 mm in its end and middle parts respectively. Figure 19 shows the reinforcement arrangement, which was modelled using embedded truss elements. Based on the construction methods using prefabricated filigran slabs, there is a construction joint between the beam and the slab that was modelled by interface elements as shown in Figure 20. Important aspect of the modelling was to consider this weak connection and its effect on the capacity of the continuous beam. The initial assessment using standard design formulas expected inadequate bending reinforcement above the piles and insufficient shear capacity of the slab-beam connection.

The initial nonlinear analysis confirmed inadequate shear capacity of the continuous beam mainly due to the loss of connection between the slab and the

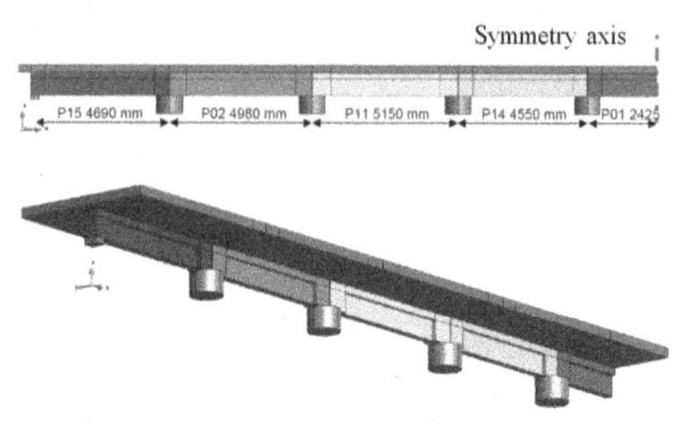

Figure 18. 3D geometric model of the symmetric half of the investigated continuous beam.

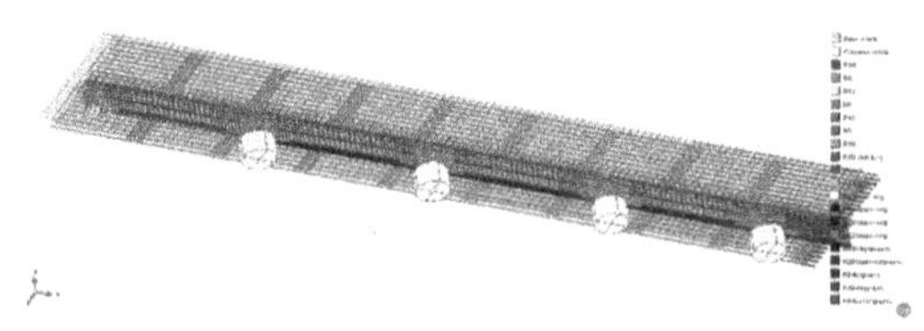

Figure 19. Reinforcement arrangement, label indicates the reinforcement diameter, location and orientation in the model.

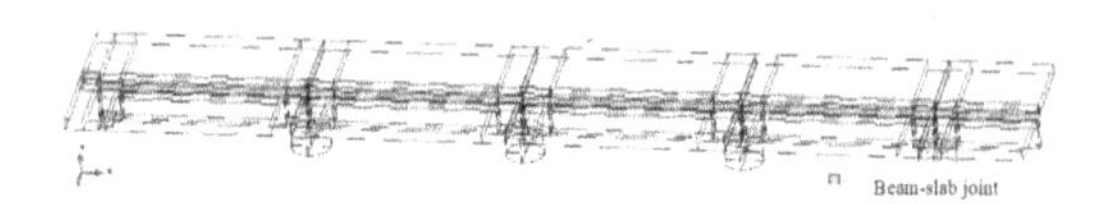

Figure 20. Location of the construction join between the beam and the slab.

beam resulting in their separate behavior without the necessary composite effect.

The calculated load-displacement diagrams are shown in Figure 21. The dashed lines show the structural resistance of the original model without strengthening for two models of the beam-slab connection. Model B considers the contribution of both concrete as well as steel shear studs in the beam-slab interface while model C assumes that due to observed large slab deflection during the construction, the interface is cracked and only steel shear studs are contributing to the interface shear strength.

Figure 21 shows that for the cases without strengthening, only model B provides enough load-carrying capacity. If the required design load level is increased to include the expected model uncertainty (see Chapter 4), none of the models are able to reach this load level.

Figure 22 demonstrates the typical development of the diagonal shear crack during the nonlinear analysis and the crushing of the compressive strut near the right support. From engineering point of view it is interesting to evaluate internal forces (moments, shear) along the beam. Figure 23 shows their evolution at the peak load for the case of model B. It is interesting to compare the value of the shear force at the section where shear failure is observed with the value obtained by Eurocode design formula. The EC2

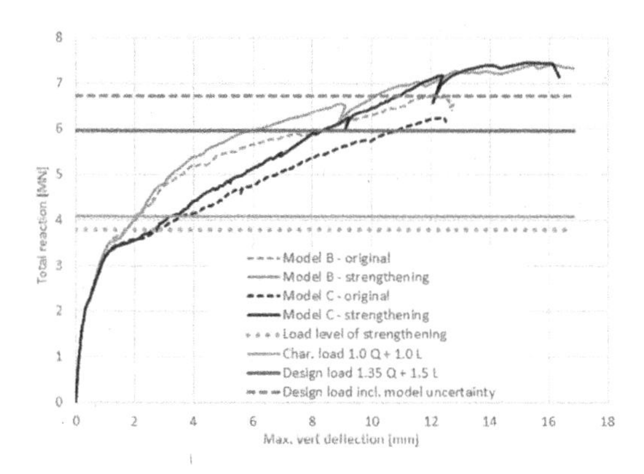

Figure 21. L-D diagram for original and strengthened continuous beam from the first example.

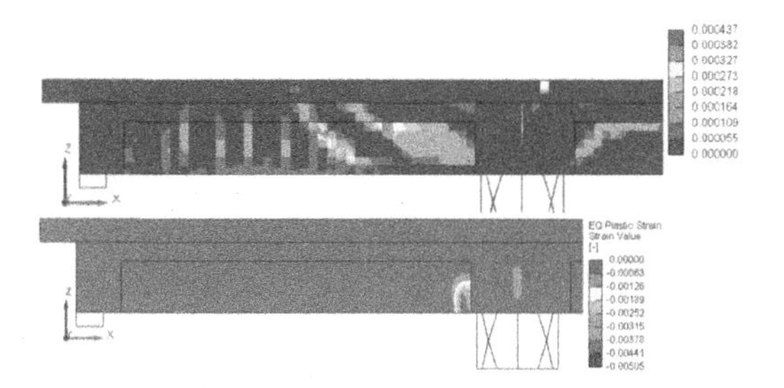

Figure 22. Characteristic development of diagonal shear crack and crushing of the diagonal compressive strut near the right support for the model without strengthening.

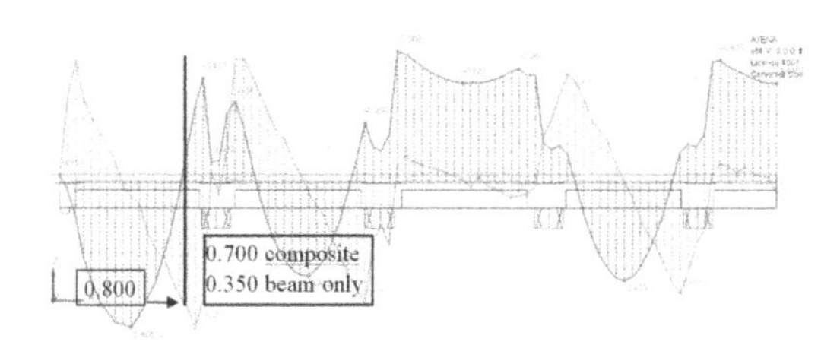

Figure 23. Distribution of moments (MNm) and shear forces (MN) in the model B without strengthening at peak load.

gives shear capacity of 0.350–0.700 MN depending whether we consider only the shear strength of the beam or the whole cross-section including the contributing slab. The analysis shows that structure can carry up to 0.8 MN at failure at this location, which is reasonably close to the EC2 value for the whole section. This indicates that even though model B takes into account the slip between the beam and the slab, the overall shear strength is still provided by the whole section with significant contribution from the slab. Figure 23 also demonstrates the importance of checking the numerical results with analytical formulas whenever possible to verify the reliability of the numerical simulation. In most cases, an exact verification is not possible, but an expected range of reasonable results can be often estimated.

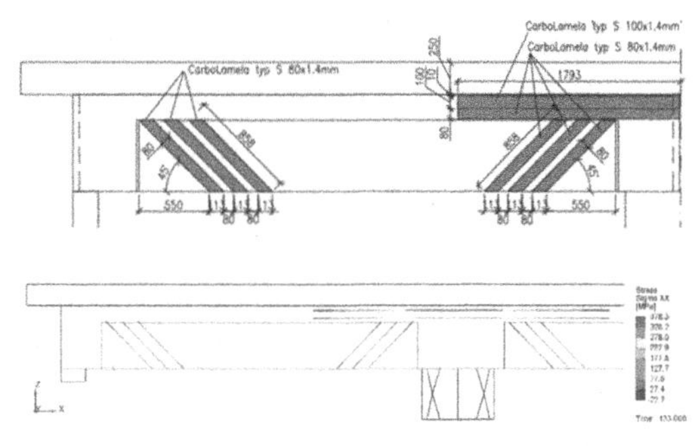

Figure 24. Proposed strengthening and stress carbon lamellas at peak load.

To increase the capacity of the beam, the strengthening was designed and verified by nonlinear analysis. The proposed strengthening arrangement with carbon lamellas is shown in 6.2. The strengthening resulted in the capacity increase by 10%–20% as shown in the l-d diagrams in Figure 21. The impact of the strengthening is not very significant due to the fact that the failure load is governed by crushing of the compressive strut, that can be only partially addressed by the additional lamellas, however, it was enough to address the concerns of the investor.

6.2 *Pushover analysis of a reinforced concrete building*

The investigated reinforced concrete building in the second practical example was constructed in 1960s. The building belongs to a critical infrastructure and it is located in a seismic area. The proof of safety of such buildings requires a sophisticated analysis. The engineers of the building owner developed an analysis method (Kurmann 2013), which inludes certain simplifications, but provides a far more realistic estimation of the seismic load bearing capacity of reinforced concrete structures compared to standard methods. It is based on pushover curves determined by nonlinear analysis and the application of time histories to an equivalent dynamic model of a representative harmonic oscillator (see 7). The dynamic parameters of the oscillator, i.e. model mass and damping are obtained from a soil-structure-interaction analysis. This oscillator is then exposed to a series of time-histories of various earthquakes.

This approach allows the consideration of both uncertainties on the loading side by using different earthquake time histories as well as the uncertainties related to the structural response. These are treated by performing nonlinear structural analyses with mean and characteristic parameters. The resulting realiability of the structure under seismic loading is represented by a fragility curve. The details of this approach are described in more detail in Kurmann (2013). This paper will described only some aspects and results of the nonlinear structural analysis that was used to generate the nonlinear force-deformation curves required by this method.

The geometric model of the building is shown in Figure 25. Two types of reinforcement models were used:

- the critical reinforcement in colums and beams was modelled in a very detailed way using the discrete model, where each reinforcement bar or stirrup is included in the model (Figure 26). In the finite element model these bars are modelled by truss elements embedded inside 3D solid finite elements.
- the reinforcement in the slabs and in most of the walls is modelled by smeared approach, where the reinforcement behavior is included into a composite reinforced concrete constitutive material model (Figure 27).

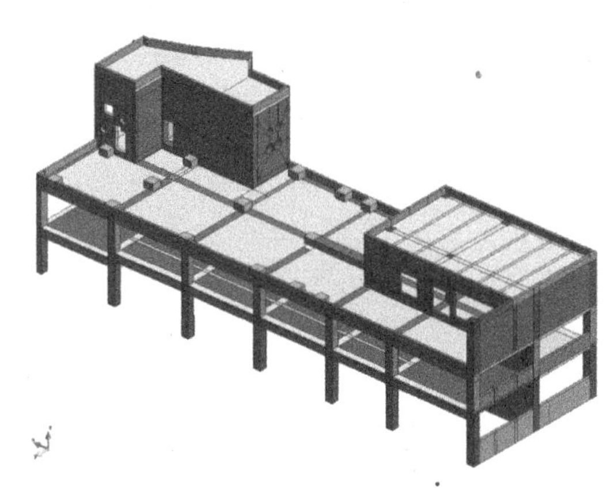

Figure 25. 3D model of the reinforced concrete building from the second example.

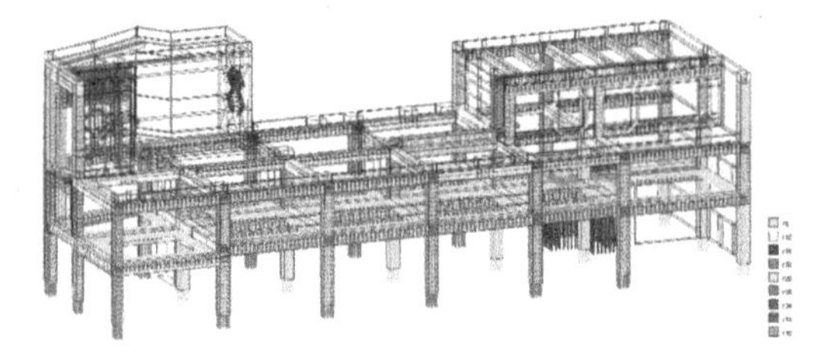

Figure 26. Discrete reinforcement model for the reinforced concrete building example.

Table 3 lists the used material parameters for the characteristic and mean analyses respectively.

The used finite element model is shown in Figure 28. The final mesh was a result of mesh size and element type parametric study.

In the final mesh, columns are modelled using linear isoparametric solid elements. The walls and slabs marked in Figure 27 are modelled by quadratic 3D shell elements. These are special elements described in more detail in Jendele & Červenka (2014) and Červenka (2021). These elements have 3D geometry with 12 nodes with only displacement degrees of freedom, which simplifies their application in combined meshes with standard solid elements. Internally they

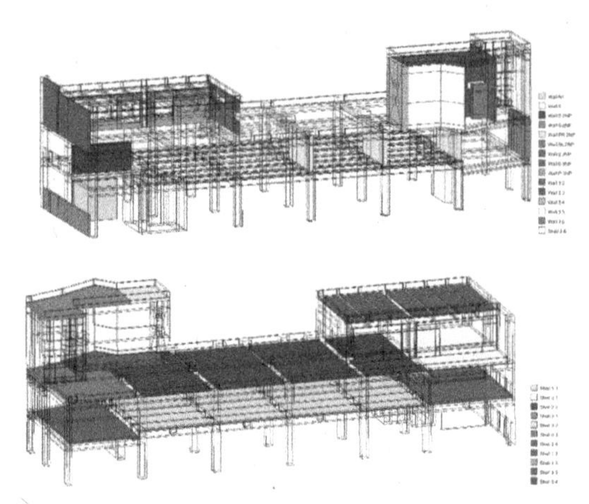

Figure 27. Structural elements with smeared reinforcement model.

Table 3. Summary of material parameters for the reinforced concrete building example.

Concrete	char.		mean	
E [GPa]	36		36	
ν [–]	0.2		0.2	
f_c [MPa]	45		54	
f_t [MPa]	2.7		3.8	
G_f [N/m]	67.5		95	
Steel	Ø ≤ 18		Ø > 18	
E [GPa]	200		200	
f_{ys} [MPa]	380	345	440	370
[–]		0.1		
f_{y2} [MPa]	520		570	

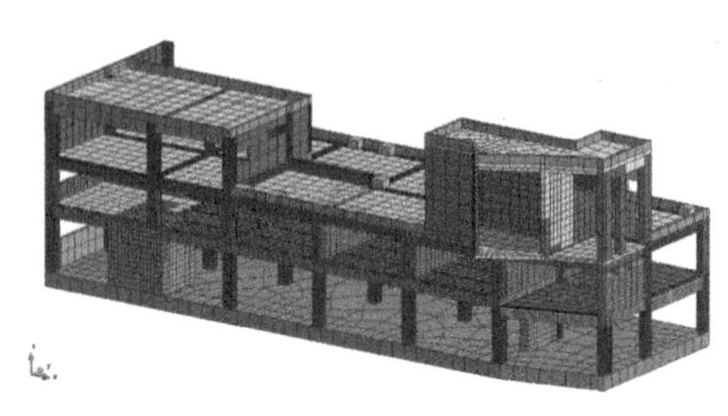

Figure 28. The finite element model for pushover analysis of the reinforced concrete building example.

are however formulated as a layered shell element with displacement and rotational degrees of freedom. This element supports the definition of special internal reinforcement layers, which simplifies the definition of internal reinforcement of the slabs or walls.

Up to 8 nonlinear pushover analyses were performed for characteristic and mean parameters (Table 3) for two directions (x, y) of the pushover forces. For each direction the pushover forces were applied with positive as well as negative orientation. The distribution, i.e. vertical shape, of the pushover forces was determined by a separate soil-structure interaction analysis using SASSI software (Figure 29).

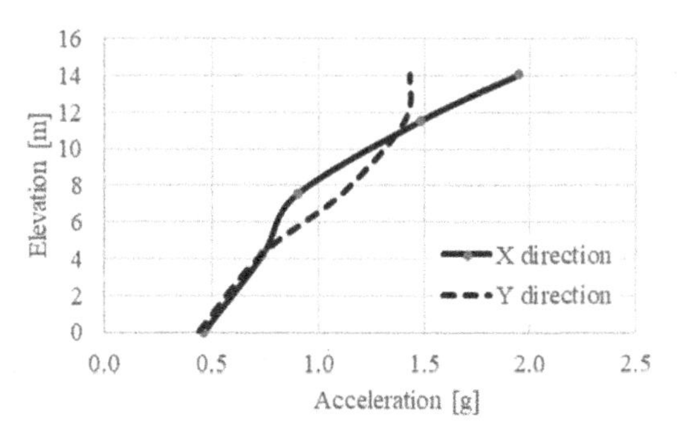

Figure 29. Vertical distribution of accelerations defining the vertical shape of the pushover forces that was used in the nonlinear analysis.

The calculated force-deflection curves are shown in Figure 30 for the two pushover directions, two orientations and two sets of parameters, i.e. mean and characteristic. The pushover curves in Figure 30 show that the structural response is slightly stronger in positive direction, but overall the behavior in both directions and orientations is quite similar. Typical failure modes and crack patterns are shown in Figure 31 for the positive and negative y direction. It shows the deformed shape at failure is slightly unsymmetric. Larger deformations and more damages can be observed on the right side of the figures. This can be attributed to the presence of reinforced concrete walls in both directions on the left side of the building. The stronger response in the negative direction can be attributed to the fact that the shear walls in y-direction are located on the side of the building which is experiencing higher compressive forces when the pushover forces are acting in the negative y-direction.

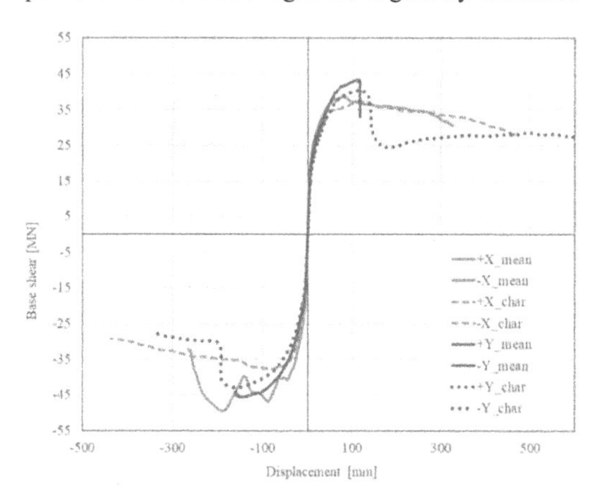

Figure 30. Pushover curves for x,y directions from mean and characteristic analyses for the reinforced concrete building example.

Figure 32 shows the shape and hysteretic response of the equivalent SDOF system that was used in

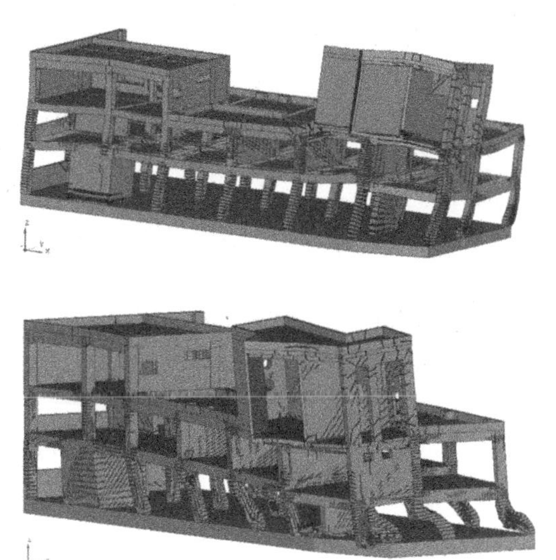

Figure 31. Deformed shape and crack pattern at failure from pushover analysis in -Y direction for the analysis with characteristic parameters.

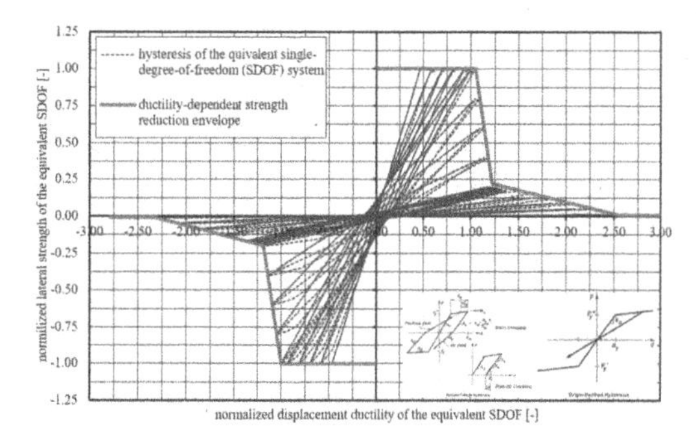

Figure 32. Cyclic behavior of the equivalent SDOF system representing the structural capacity and accounting of a ductility-dependent strength degradation.

the subsequent probabilistic dynamic analyses for the evaluation of the building's fragility curve.

The two practical examples presented in this section provide a very brief overview of typical application cases in the current engineering practice.

The first example covers a situation of a problematic design. Such scenario often occurs in case of very complicated design, which is outside of the scope of standard design methods and tools or if an error or mistake is discovered during the construction or after the completion. This was the case of the first example when a missing reinforcement was discovered during the construction when excessive deformations of the slabs were observed.

The second example comes from the assessment of existing structures, which were originally designed according to older guidelines, and are required to meet the current standards or increased levels of traffic loads, seismic or environmental hazards and risks.

7 CONCLUSIONS

The paper summarizes the most important aspects and issues, at least according to the authors experience, related to the application of nonlinear analysis in engineering practice. At the beginning, typical features of nonlinear models for reinforced concrete analysis by finite element method are discussed namely to provide a consistent background and briefly described the main features of the numerical model used in the examples subsequently presented in the paper.

An important topic is the application and choice of a suitable safety format for nonlinear analysis. Until recently the engineers were left on their own, but the situation is changing now. New safety formats for nonlinear analysis are available in fib model code 2010. They are further enhanced in the new version of the code, i.e. model code 2020 under preparation. CEN committees are working intensively to introduce them into the new generation of Eurocodes.

Another important issue in the application of nonlinear simulation in practice is the treatment of model uncertainties. Approaches are available and significant research effort has been devoted to this topic as discussed in chapter 5.

Blind robin prediction competitions are very useful for the assessment of the robustness and reliability of existing models or software tools. However, they have certain specifics and limits that are demonstrated on three recent prediction competitions. In future competitions, it would be useful, if their organizers make more attempts to separate the various uncertainties involved.

In the last section, two examples of application of nonlinear simulation from engineering practice are presented to illustrate the typical use cases and level of modelling detail that is becoming standard in the nonlinear modeling of reinforced concrete structures.

ACKNOWLEDGEMENTS

This work has been supported by the Czech Science Foundation within the project 20-01781S "Uncertainty modelling in safety formats of concrete structures".

REFERENCES

Bažant, Z.P. 1976. Instability, Ductility and Size Effect in Strain Softening Concrete, J. Engrg. Mech., ASCE, Vol. 102, No. 2, pp. 331–344.

Bažant, Z.P. & Oh, B.H., 1983. Crack band theory for fracture of concrete. *Materials and Structures, RILEM* 16 (3), 155–177.

Bažant, Z.P., Pijaudier-Cabot, G. 1987. Nonlocal continuum damage, localization instability and convergence. Journal of Applied Mechanics, ASME 55 (2), 287–293.

Castaldo, P., Gino, D., Bertagnoli, G., Mancini, G. 2018. Partial safety factor for resistance model uncertainties in 2D non-linear analysis of reinforced concrete structures, Engineering Structures, 176, 746–762. https://doi.org/10.1016/j.engstruct.2018.09.041.

Castaldo, P., Gino, D., Bertagnoli, G., Mancini, G. 2020. Resistance model uncertainty in non-linear finite element analyses of cyclically loaded reinforced concrete systems, Engineering Structures, 211(2020), 110496, https://doi.org/10.1016/j.engstruct.2020.110496

Červenka J. & Papanikolaou V. 2008. Three Dimensional Combined Fracture-Plastic Material Model for Concrete. *Int Journal of Plasticity*. 24:2192–220.

Červenka, J., Červenka, V., Eligehausen, R. 1998. Fracture-Plastic Material Model for Concrete, Application to Analysis of Powder Actuated Anchors, *Proc. FRAMCOS 3*, 1998, pp 1107–1116.

Červenka, J., Červenka, V., Laserna S., 2014, On finite element modelling of compressive failure in brittle materials, Computational Modeling of Concrete Structures. Bicanic et al.(Eds), Euro-C 2014, St. Anton,

Červenka, J., Červenka, V., Laserna, S. 2018. On crack band model in finite element analysis of concrete fracture in engineering practice, Eng. Fract. Mechanics, Vol. 197, pp 27-47, doi.org/10.1016/j.engfracmech.2018.04.010.

Červenka, J., Sýkora, M., Červenka, V. 2021. Uncertainty in Geometry and Modelling in Safety Formats for Nonlinear Analysis of RC Structures – Case Study, fib Symposium 2021, Lisbon, Portugal, June 14–16.

Červenka, V., 2008, Global Safety Format for Nonlinear Calculation of Reinforced Concrete. Beton- und Stahlbetonbau 103, Special Edition, Ernst&Sohn. pp. 37–42.

Červenka, V., Červenka, J. & Jendele, L. 2021. *ATENA Program Documentation, Part 1: Theory, 2021*, Cervenka Consulting s.r.o., www.cervenka.cz

Červenka, V. 2008. Global Safety Format for Nonlinear Calculation of Reinforced Concrete. Beton- und Stahlbetonbau 103, special edition, Ernst & Sohn, pp. 37–42.

Červenka, V. 2013. Reliability-based non-linear analysis according to Model Code 2010, Journal of fib, Structural Concrete 01/2013, pp.19–28

Červenka V, Červenka J, Kadlec L. 2018. Model uncertainties in numerical simulations of reinforced concrete structures. Structural Concrete; 19(6): 2004–16 https://doi.org/10. 1002/suco.201700287.

Červenka, V., Gerstle, K., 1971. Inelastic analysis of reinforced concrete panels. Part I: Theory. Publication I.A.B.S.E. 31(11), 32–45.

Červenka, V., Margoldová, J., 1995, Tension Stiffening Effect in Smeared Crack Model, *Engineering Mechanics*, Stain F(Sture (Eds), Proc. 10th Conf., Boulder, Colorado, pp. 655–658.

Collins, M.P., Bentz, E.C., Quach, P.T. & Proestos, G.T., 2015, Challenge of Predicting the Shear Strength of Very Thick Slabs. Concrete International, V.37, No.11, Nov. 2015, pp 29–37.

de Borst, R. 1986. Non-linear analysis of frictional materials. PhD Thesis, Delft University of Technology, The Netherlands.

de Borst, R., Mühlhaus, H.B. 1992. Gradient dependent plasticity: Formulation and algorithmic aspects. International Journal for Numerical Methods in Engineering 35 (3), 521–539.

de Borst, R., Benallal, A., and Heeres, O.M. 1996. A gradient-enhanced damage approach to fracture. J. de Physique IV, C6, pp. 491–502.

de Borst, R., Rots, J.G. 1989. Occurrence of Spurious Mechanisms in Computations of Strain-Softening Solids, Eng. Computations, Vol. 6, pp. 272–280.

Engen M, Hendriks M., Köhler J, Øverli JA, Åldtstedt E. 2017. A quantification of modelling uncertainty for non-linear finite element analysis of large concrete structures. Structural Safety 2017; 64: 1–8.

Etse, G. 1992. Theoretische und numerische untersuchung zum diffusen und lokalisierten versagen in beton. PhD Thesis, University of Karlsruhe.
Gino D, Castaldo P, Giordano L, Mancini G. 2021. Model uncertainty in nonlinear numerical analyses of slender reinforced concrete members. Structural Concrete. 1–26. https://doi.org/10.1002/suco.202000600
Gunay, S., Donald, E., 2021. Reinforced Concrete Column Blind Prediction Contest, https://peer.berkeley.edu/news-and-events/2021-reinforced-concrete-column-blind-prediction-contest
Jendele, L., Červenka, J., Curvilinear Non-linear Three-Dimensional Isoparametric Layered Shell Elements, Paper 212, Civil-Comp Press, 2014, Proceedings of the Twelfth International Conference on Computational Structures Technology, B.H.V. Topping and P. Iványi, (Editors), Civil-Comp Press, Stirlingshire, Scotland,
Jirásek, M., Bažant, Z.P. 2001. Inelastic Analysis of Structures, John Willey & Sons, LTD, Baffins Lane, Chichester, England, ISBN 0-471-98716-6
Kurmann, D., Proske, D., Červenka, J. 2013. Seismic Fragility of a Reinforced Concrete Structure, Transactions, SMiRT-22, San Francisco, USA, 18-23.8.2013
Lee, J., Fenves, G.L. 1998. Plastic-damage model for cyclic loading of concrete structures. Journal of Engineering Mechanics, ASCE 124 (8), 892–900.
Lin, C.S., and Scordelis, A. 1975. Nonlinear Analysis of RC Shells of General Form, ASCE, J. of Struct. Eng., Vol. 101, No. 3, pp. 152–163.
Menetrey, P. & Willam, K.J. 1995. Triaxial Failure Criterion for Concrete and its Generalization. *ACI Structural Journal*. 1995;92:311-8
Model Code 2010. 2011. *fib* Lausanne, Ernst & Sohn: Switzerland,
Moehle J., Zhai, J. 2021. Thick foundation element blind prediction contest, https://peer.berkeley.edu/news-and-events/2021-thick-foundation-element-blind-prediction-contest
Ngo, D., Scordelis, A.C. 1967. Finite element analysis of reinforced concrete beams, J. Amer. Concr. Inst. 64, pp. 152–163.
Pramono, E., Willam, K.J. 1989. Fracture energy-based plasticity formulation of plain concrete. Journal of Engineering Mechanics, ASCE 115 (6), 1183–1204.
prEN 1992-1-1:2020. 2020. proposal draft, CEN/TC250/SC2/WG1, document N 1057
Rashid, Y.R. 1968. Analysis of prestressed concrete pressure vessels. Nuclear Engineering and Design 7 (4), 334–344.
Rots, J.G., Blaauwendraad, J. 1989. Crack models for concrete : Discrete or smeared ? Fixed, multi-directional or rotating ? Heron 34 (1).
Slobbe, A.T., Hendriks, M.A.N., Rots, J.G., 2013, Systematic assessment of directional mesh bias with periodic boundary conditions: Applied to the crack band model. Engineering Fracture Mechanics, Volume 109, September 2013, Pages 186–208
Suidan, M., Schnobrich, W.C. 1973. Finite Element Analysis of Reinforced Concrete, ASCE, J. of Struct. Div., Vol. 99, No. ST10, pp. 2108–2121

Computational Modelling of Concrete and Concrete Structures – Meschke, Pichler & Rots (Eds)
© 2022 Copyright the Author(s), ISBN: 978-1-032-32724-2

Fluctuation-based fracture mechanics of heterogeneous materials and structures in the semigrand canonical ensemble

F.-J. Ulm*, T. Mulla, T. Vartziotis, A. Attias & M. Botshekan
Department of Civil and Environmental Engineering, Massachusetts Institute of Technology, Cambridge, MA, USA

N. Rahbar
Department of Civil and Environmental Engineering, Worcester Polytechnic Institute, Worcester, MA, USA

ABSTRACT: Fracture mechanics in its most elemental form introduced by Griffith defines the irreversible change of energy between two equilibrium states at constant loading. This particular equilibrium nature permits the use of Monte Carlo sampling for evaluating the energy change introduced by fracture, as well as energy release rate and its critical value, the fracture energy, from fluctuations. This is in short the idea of fluctuation-based fracture mechanics in the semigrand canonical ensemble (SGCMC). Herein, we review recent developments of SGCMC, and show possible applications for heterogeneous materials and structures.

1 INTRODUCTION

1.1 *Gedankenexperiment*

Consider a solid composed of particles subjected to a volume change at constant temperature. The system is further subjected to an external energy source that targets the bonds between particles in the system, akin to a bulk radiation source. At a given energy of this radiation source, denoted by $\Delta\mu$, fracture at the macroscopic level of the sample may occur between two equilibrium states of the system. This transition is defined by the bond potential, $\Delta\mu$, the prescribed volume, V, and temperature, T . In this semigrand canonical ensemble, we measure the ensemble energy average of possible microstates of the system, $\langle\mathcal{U}\rangle$, and the energy fluctuations, as a function of the average number of bonds, $\langle N\rangle$, and their fluctuations. As we repeat the experiment by sweeping possible values for volume changes, stress-strain curves can be traced out for different prescribed bond potentials, $\Delta\mu$.

1.2 *From Griffith's fracture mechanics to fluctuation-based fracture mechanics*

The outlined *Gedankenexperiment* suggested by Mulla et al. 2021 encapsulates the very essence of fluctuation-based fracture mechanics, namely:

1. Fracture mechanics in the classical Griffith sense (Griffith 1920) defines (potential) energy changes between two equilibrium states at constant loading (prescribed forces F^d, prescribed displacements u^d), by means of the energy release rate, $\mathcal{G}$, as thermodynamic driving force of fracture propagation:

$$\mathcal{G} = -\frac{\partial E_{pot}}{\partial \Gamma}|_{F^d,u^d} \leq \mathcal{G}_F \tag{1}$$

where E_{pot} is the potential energy, Γ fracture surface, and $\mathcal{G}_F$ the fracture energy.

2. Yet, two notable differences guide the SGCMC-approach. In contrast to classical fracture mechanics, which samples a single fracture configuration around a notch, we sample by means of Monte Carlo simulations a great deal of possible fracture configurations; so called *microstates*. Each microstate is defined by its number of broken bonds, $N_{b,i} = N_0 - N_i$ (with N_0 = the initial number of bonds; N_i the number of unbroken/active bonds), and the associated bond energy, $\mathcal{U}_i = (\mathcal{U}_0 + \mathcal{U}_\lambda)_i$ (with $\mathcal{U}_0$ the groundstate energy, i.e. the (sum of the) well-depth of the bond potential; and $\mathcal{U}_\lambda$ the (sum of the) elastic energy of the system). Given a sufficient number of sampled bond fracture configurations, SGCMC permits evaluating the change in energy (i.e. dissipation) from the 'heat of bond rupture' (Al-Mulla et al. 2018):

$$q_\lambda = -\frac{\partial\langle\mathcal{U}_\lambda\rangle}{\partial\langle N_b\rangle}|_{\Delta\mu,V,T} = \frac{\mathrm{cov}(\mathcal{U}_\lambda, N)}{\mathrm{var}(N)} \leq -q_0 \tag{2}$$

Herein, $-q_0$ is the bond fracture energy of the considered material or structural system, which derives from the ground state energy activated by bond fracture:

$$-q_0 = \frac{\partial\langle\mathcal{U}_0\rangle}{\partial\langle N_b\rangle}|_{\Delta\mu,V,T} = -\frac{\mathrm{cov}(\mathcal{U}_0, N)}{\mathrm{var}(N)} \tag{3}$$

*Corresponding Author

DOI 10.1201/9781003316404-3

That is, q_λ and $-q_0$ are –for bond rupture– the analogues of respectively the energy release rate, $\mathcal{G} \to q_\lambda$, and the fracture energy, $\mathcal{G}_\mathcal{F} \to -q_0$. Note, however, the difference in dimension: $[\mathcal{G}] = [\mathcal{G}_\mathcal{F}] = MT^{-2}$ vs. $[q_\lambda] = [q_0] = L^2MT^{-2}\text{Mole}^{-1}$. This requires consideration of re-scaling of SGCMC–obtained fracture properties to match with 'classical' Griffith-type fracture properties.

3. The 'new' ensemble quantity in SGCMC-simulations is the external energy potential, $\Delta\mu$, that targets all bonds in the system – akin to a radiation source. Classical fracture simulations correspond to $\Delta\mu = 0$; whereas $\Delta\mu < 0$ may be helpful to simulate dissolution processes, and $\Delta\mu > 0$ bond solidification processes brought to equilibrium. More generally, $\Delta\mu$ permits the development of phase diagrams of fracture that separate, for given values of $\Delta\mu$, the corresponding limit strain of fracture as a first-order phase transition phenomena (Mulla et al. 2021). The coining of this $\Delta\mu VT$ ensemble as *semi*-grand canonical is due to the fact that in contrast to the grand canonical ensemble which targets particles, the semigrand canonical ensemble only targets the bonds between particles in the system. Otherwise said, the mass is conserved.
4. Of critical importance for SGCMC simulations is the (thermodynamic) equilibrium-based nature of the fracture process defined by the thermodynamic ensemble, $\Delta\mu VT$, the rough equivalent of a well-posed boundary value problem in classical continuum mechanics. From the point of view of statistical physics, the well-defined ensemble permits averaging microstates, and hence the evaluation of fracture properties from Eqn. (2) and (3). In return, given this equilibrium nature, out-of-equilibrium fracture situations such as dynamic fracture propagation (see, e.g. Bonamy and Bouchaud 2011), are beyond the (thermodynamic) equilibrium-based focus of Griffith's and SGCMC-based fracture evaluations.

2 SGCMC SIMULATIONS

At the core of the fluctuation-based fracture mechanics approach is the generation of a sufficient number of bond fracture microstates, from which ensemble averages can be derived. This seems on first sight a daunting task. Yet, Monte Carlo simulation techniques provide a solid foundation to effectively address this challenge [for a 'must' read, see (Frenkel & Smit 2002)]. Let's start simple.

2.1 SGCMC acceptance criteria

All boils down to evaluating the probability of acceptance of the addition or removal of a bond. We return to the initial *Gedankenexperiment*, which for purpose of clarity we consider at $\Delta\mu = 0$ (the classical fracture test situation). Our starting configuration is under strain due to the application of displacement boundary conditions (generalized prescribed volume), and temperature, T. In this configuration at a current bond number N, the sum of the bond potential energy is denoted by $\mathcal{U}[o]$, and the probability density is $\mathcal{N}_{\Delta\mu VT}[o] \sim \exp(-\beta\mathcal{U}[o])$, with $\beta^{-1} = (k_BT)$ the Boltzmann energy. We now attempt to either remove (from the active bonds) or add (to the broken bonds) one bond, which entails a change in energy, $\mathcal{U}[o] \to \mathcal{U}[n]$, and a probability density, $\mathcal{N}_{\Delta\mu VT}[n] \sim \exp(-\beta\mathcal{U}[n])$. The probability of acceptance of the trial move from $[o]$ to $[n]$ is thus:

$$acc(o \to n) = \frac{\mathcal{N}[n]}{\mathcal{N}[o]}|_{\Delta\mu=0VT} = \exp(-\beta\Delta\mathcal{U}) < 1 \quad (4)$$

with $\Delta\mathcal{U} = \mathcal{U}[n] - \mathcal{U}[o]$. The extension to considering a non-zero external energy, $\Delta\mu \neq 0$ is straightforward, if we remind us that $\Delta\mu$ targets all bonds in an equiprobable way; so that

$$acc(o \to n) = \exp(\beta(\delta[n]\Delta\mu - \Delta\mathcal{U})) < 1 \quad (5)$$

where $\delta[n] = +1$ if the bond is added and $\delta[n] = -1$ if it is removed.

2.2 Algorithmic realization

We now need to be cognizant of the fact that the addition or removal of a bond from the system entails simultaneous changes of (*i*) the groundstate energy (attached to the bond), (*ii*) the elastic energy stored in the bond; as well as (*iii*) a redistribution of the elastic energy in adjacent bonds. Fortunately, Monte Carlo simulation techniques address this complexity by considering the semigrand canonical ($\Delta\mu VT$) trial moves in a *frozen* particle position configuration, which is updated in the NVT ensemble:

1. *MC trial moves in a frozen particle position configuration:* Each bond $i = 1, N_0$ has an identity $\delta_i = +1$ if ON, and $\delta_i = -1$ if OFF. The MC trial move then consists of (1) choosing a bond at random, $i = \text{ceil}[N_0 * \text{rand}()]$ with $\text{rand}() \sim U_{[0,1]}$ generated from a uniform distribution between 0 and 1; and determining its bond energy,

$$u(i)[o] = u_0(i) + u_\lambda(i) \quad (6)$$

where $u_0(i)$ and $u_\lambda(i)$ stand for the bond's groundstate and elastic energy, respectively. (2) Then toggle the bond's identity,

$$\delta(i)[n] = -1 * \delta(i)[o] \quad (7)$$

Evaluate the trial bond energy, $u(i)[n]$, for the toggled identity, $\delta(i)[n]$, while considering the (frozen) particle position. Finally, the change in energy in the MC trial is $\Delta\mathcal{U} = u(i)[n] - u(i)[o]$, which permits (3) evaluation of the acceptance probability $acc(o \to n)$ according to Eq. (5). Last, (4) the trial is rejected and the bond identity is restored, $\delta(i)[n] = \delta(i)[o]$ if $acc(o \to n) < \text{rand}()$.

2. *Update of particle/node position in the NVT-Ensemble*: For a given microstate of bond identities, $\delta_i(=1, N_0)$, focus of the *NVT*-part is to update the particle positions, $\mathbf{x} = \mathbf{X} + \xi$ (where ξ is the displacement vector. There exists a variety of options for this (canonical) ensemble update in which the number of *particles* (N) is maintained constant (i.e., mass conservation), while displacement boundary conditions (generalized volume change, V) and a temperature (T) are prescribed. These options range from canonical Monte Carlo moves (Frenkel & Smit 2002), to Molecular Dynamics (MD) time integration in the *NVT*-ensemble (Mulla et al. 2021; Villermaux et al. 2021) to (more classical) lattice or finite-element type displacement updates using standard small or large deformation simulation tools (Al-Mulla et al. 2018; Laubie et al. 2017; Wackerfuß 2009).

In Canonical Monte Carlo simulations, trial moves focus on displacing a random particle, while a move is accepted if the change in bond energies, $\mathcal{U}(\mathbf{x}(i)[o]) \rightarrow \mathcal{U}(\mathbf{x}(i)[n])$, satisfies the acceptance criterion [for details, see Frenkel & Smit 2002],

$$acc(o \rightarrow n)_{NVT} = \exp\left(-\beta \Delta \mathcal{U}(\mathbf{x}(i)\right) \leq 1 \tag{8}$$

where $\mathbf{x}(i)[n] = \mathbf{x}(i)[o] + (rand() - 0.5)\Delta\xi\mathbf{1}$.

In contrast, in MD simulations in the *NVT*-ensemble, particle moves are defined by the conservation of linear momentum, while the moves are 'thermalized' by means of thermostats. For instance, in case of a Nosé-Hoover thermostat, the equations of motion are enriched by a mass damping term which evolves in function of the kinetic temperature/energy, $\beta^{-1} = k_B T \sim E_k$ which ensures that the (bonded) particle system attains the prescribed temperature when approaching (static) equilibrium [for a brief 'structural engineering' introduction to thermalization, see Louhghalam et al. 2018]:

$$m(\ddot{\mathbf{x}}_i + \zeta(t)\dot{\mathbf{x}}) = -\nabla_{r_i}(u_\lambda) \tag{9}$$

where the right hand side represents the (bond) forces acting on particle (node) i, while the damping coefficient ζ evolves as:

$$\dot{\zeta} \sim \beta E_k - 1 \tag{10}$$

Herein, $E_k(t)$ is the mean kinetic energy of the particles, and $\beta E_k = T(t)/T_0$ is the kinetic temperature ratio, between the (evolving) kinetic temperature, $T(t)$ and the prescribed (bath) temperature. Such thermostats form an integral part of MD codes, and are readily adopted for structural engineering purposes as well (Louhghalam et al. 2018).

Finally, computational mechanics approaches based on the theorem of minimum potential energy are a a safe (but sometimes computationally expensive) backup for particle (node) position update and related bond energy update in the *NVT* ensemble. At a coarse-grained level of material/structural representation, such energy minima approaches may (eventually) include consideration of continuum-based constitutive equations, that permit access to the elastic energy of bonds understood in a large sense. Given the ubiquity of existing lattice-based or finite-element based minimization procedures, they do not need further development here.

Last, the choice of method for the *NVT* update is often a compromise between computational efficiency and stability. Specifically, while MC, FE and lattice approaches sample stable and (eventually) meta-stable states, MD-approaches equally access unstable particle configurations, incl. rigid body motions.

The algorithm sketch shown in Table 1 provides the elementary structure for a typical SGCMC algorithm, which partitions SGCMC trials of bond activation/de-activation from particle position update in the *NVT* ensemble.

Table 1. Algorithm Sketch for SGCMC fracture simulations. Function BondEnergy($H(i), x$) updates the elastic energy for bond i; and function NVTupdate(H) updates the particle positions and bond energies. [Inspired by Frenkel & Smit 2002].

```
for icycl = 1:ncycle
    if rand()<ratio                                  % SGCMC-Trials
        i = ceil(rand() * N_0)
        % —Store old bond energy
        u_λ(i)[o] = u_λ(i)
        U[o] = (u_0(i) * H(i)[o] + u_λ(i))[o]
        % —Toggle Bond Identity and Calculate Energy
            δ(i) = −δ(i)
            H(i) = 1/2 * (δ(i) + |δ(i)|)
            u_λ(i) = BondEnergy(H(i), x)
            U[n] = (u_0(i) * H(i) + u_λ(i))
        % —Acceptance Criterion
            ΔU = U[n] − U[o]
            acc = exp(β * (δ(i) * Δμ − ΔU))
        % —Reject & Restore
        if acc < rand()
            δ(i) = −δ(i)
            H(i) = 1/2 * (δ(i) + |δ(i)|)
            u_λ(i) = u_λ(i)[o]
        end
    else                                             % NVT Update
        N_act = sum(H)
        [x, u_λ] = NVTupdate(H)
        U_0 = sum(u_0)
        U_λ = sum(u_λ)
    end
end
```

3 APPLICATION

3.1 Reduced units

A convenient way to scale interparticle bond energies in molecular-based simulations is by reduced units. In this unit system, often referred to as LJ units (LJ = Lennard-Jones), bond energies are written in a dimensionless form considering as independent variables physical quantities that define the equilibrium state. These are the groundstate energy, ϵ_0 (i.e. the well-depth of inter-particle potentials), the particle radius

R, and the particle mass m. Derived quantities are the dimensionless energies $\mathcal{U}^* = \mathcal{U}/\epsilon_0$; temperature, $T^* = k_b T/\epsilon_0$; pressure (or stress), $P^* = PR^3/\epsilon_0$; time $t^* = t/(R\sqrt{m/\epsilon_0})$; and their derivatives, such as forces, $\vec{F}^* = -\partial(u^*)/\partial(\vec{r}^*) = \vec{F}(R/\epsilon_0)$; moments, $\vec{M}^* = \vec{F}^* \times \vec{r}^* = \vec{M}/\epsilon^0$, etc. In this reduced unit system, an elastic modulus (e.g. E^*) the energy release rate ($\mathcal{G}^*$), and the fracture energy ($\mathcal{G}_F^*$) [see Eq. (1)] or the fracture toughness ($K_c^* \sim \sqrt{E^* \mathcal{G}_F^*}$) scale as:

$$E^* = E\left(\frac{R^3}{\epsilon_0}\right); \mathcal{G}^* = \mathcal{G}\left(\frac{R^2}{\epsilon_0}\right); K_c^* = K_c\left(\frac{R^{5/2}}{\epsilon_0}\right) \quad (11)$$

A straightforward dimensionless analysis of Eqn. (1), (3) and (11) then reveals that the 'heat of bond rupture', $-q_0^*$ in reduced units obtained from energy fluctuations in the semigrand canonical ensemble [see Eq. 3] provides a direct means to measure the dimensionless fracture energy of the material/ structural system from SGCMC simulations; that is:

$$\mathcal{G}_{\mathcal{F}}{}^* = \mathcal{G}_{\mathcal{F}}\left(\frac{R^2}{\epsilon^0}\right) \sim -q_0^* = -\frac{\mathrm{cov}(U_0^*, N)}{\mathrm{var}(N)} \quad (12)$$

3.2 *Phase diagram of brittle fracture*

The first example deals with the phase diagram of brittle fracture proposed by Mulla et al. 2021. Similar to pressure-temperature phase diagrams of substances, the application of the SGCMC in all its facets permits the development of a phase diagram that delineates a mechanically intact solid from a fractured solid. Herein, the phase line is defined as the pair of critical values of the control variables, i.e. volume $V = V_0(1 + \epsilon_V)$ [with ϵ_V = volume strain] and bath bond potential [$\Delta\mu$], at which the bond energy release rate, q_λ^* [Eq. (2)], equals the 'heat of bond rupture', $-q_0^*$ [Eq. (11)]; as shown in Figure 1. These critical values are developed from simulations of the stress strain diagrams [Figs. 1(a-b)] for different types of 2-pt bond potentials, namely a harmonic potential (in reduced units, $u^* = u/\epsilon_0$):

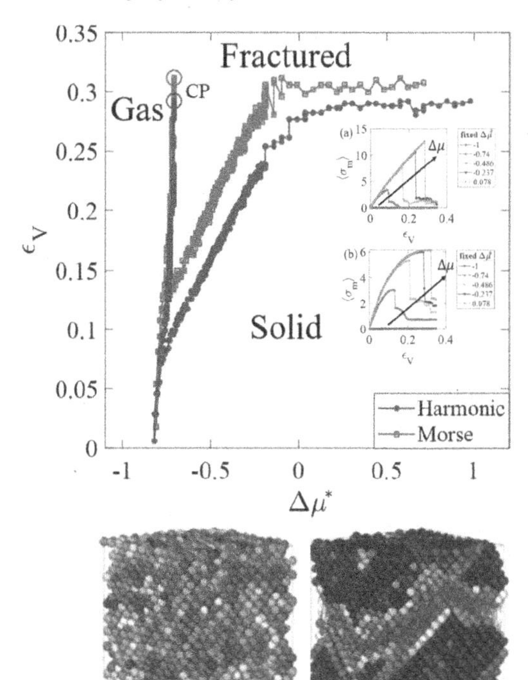

Figure 1. Phase diagram of brittle fracture for a *homogeneous* harmonic/linear and Morse material: ϵ_V is the prescribed volume change; $\Delta\mu$ is the bath potential targeting bonds. The phase line is developed from stress-strain diagrams shown in inset (a) for the harmonic potential, and inset (b) for the Morse potential. [adapted from Mulla et al. 2021].

$$u^* = -1 + u_\lambda^*; \qquad u_\lambda^* = (\lambda/\lambda_c)^2 \quad (13)$$

and a Morse potential:

$$u^* = -1 + u_\lambda^*; \qquad u_\lambda^* = (1 - \exp(-\lambda/\lambda_c))^2 \quad (14)$$

where $\lambda_c = \sqrt{2\epsilon_0/\epsilon_\lambda}$ is the critical dilation for which the harmonic potential is zero, with ϵ_λ the elastic bond energy [for link with continuum elastic moduli for Lattice systems, see e.g. Laubie et al. 2017, and for link with structural mechanics stiffness properties, see Keremides et al. 2018]. Below the phase line shown in Figure 1, $q_\lambda^* < -q_0^*$, there are still elastic energy reserves available in the system to store the externally applied work due to volume change into recoverable energy. In turn, at $q_\lambda^* = -q_0^*$, a first-order phase transition takes place. For the homogeneous system, for which $\mathcal{U}_0^* = \sum u_0/\epsilon^* = -N$, the effective bond fracture energy is:

$$-q_0^* = \frac{\mathrm{cov}(N, N)}{\mathrm{var}(N)} = 1 \quad (15)$$

3.3 *Textured material systems*

The second example here considered is a two-phase material, say A and B phases, with different groundstate energies, ϵ_0^A and ϵ_0^B; so that the total groundstate energy reads:

$$\mathcal{U}_0 = -\epsilon_0^A N_A - \epsilon_0^B N_B \quad (16)$$

where $N_A = N - N_B$ and N_B denote the intact bonds of phase A and B, respectively. In order to evaluate the groundstate energy of the 2-phase composite from Eq. (3), we make use of the covariance and variance of linear expressions and arrive at (Mulla et al. 2022):

$$-q_0 = \epsilon_0^A + (\epsilon_0^B - \epsilon_0^A)S_\beta \quad (17)$$

where:

$$S_\beta = \frac{\mathrm{cov}(N_B, N)}{\mathrm{var}(N)} = \frac{1}{2}\left(1 - \frac{\sigma_{N_A}^2 - \sigma_{N_B}^2}{\sigma_N^2}\right) \quad (18)$$

In reduced units, $-q_0^* = -q_0/\epsilon_0^A$, the SGCMC approach provides a means to evaluate the fracture energy of a two-phase material from the fluctuations of the bond numbers of the two phases, $\sigma_{N_J}^2 = \text{var}(N_J)$ $(J = A, B)$, and the total variance $\sigma_N^2 = \text{var}(N_A + N_B) = \sigma_{N_A}^2 + \sigma_{N_B}^2 + 2\text{cov}(N_A, N_B)$. This is invaluable when evaluating the impact of material texture on fracture properties of heterogeneous materials. As an example, Figure 2 displays simulations results for a 2-D two-phase material system uniaxially strained in a direction inclined w.r.t. the layer axis by an angle θ. Hence, $\theta = 0$ corresponds to a uniaxial strain that activates the layers in parallel [Figure 2(c)], while the system is activated in series for $\theta = \pi/2$. The first case leads to simulation results close to upper bound [Figure 2(a)], for which $S_\beta = f_B$ (with f_B the volume fraction of bonds B); while the latter entails the lower bound, $S_\beta = 0$, dissipating at fracture only groundstate energy in the weaker phase, $\epsilon_0^A < \epsilon_0^B$. Finally, at $\theta = \pi/4$ [Figure 2(e)], the bonds are equally activated, which can be associated with the equiprobable Hill bound, i.e. $S_\beta = f_B/2$ (i.e. the arithmetic mean of upper and lower bound). Yet, simulation results show that the layered system only follows the Hill bound up to a critical volume fraction of $f_B = 1/2$, beyond which S_β drops to zero, and for which the lower bound is recovered. More generally, when plotting S_β vs. θ at constant volume fraction [Figure 2(b)], we recognize that the reinforcing effect of the tougher phase diminishes with increasing load angle θ, and ceases beyond the magic angle, $S_\beta(\theta \geq \arccos(1/\sqrt{3})) = 0$. A detailed analysis of this and other textures can be found in (Mulla et al. 2022).

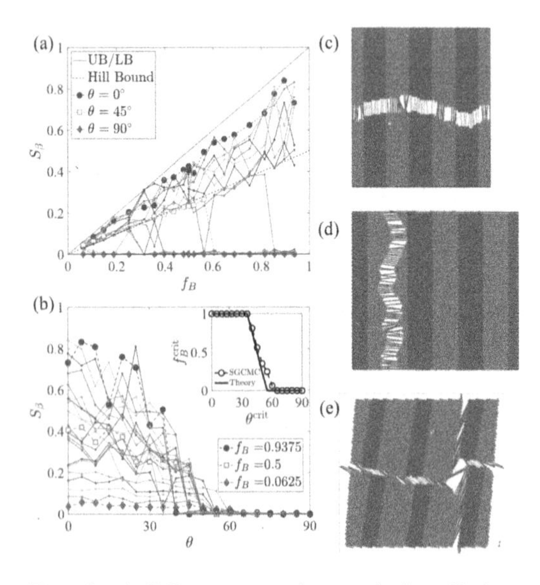

Figure 2. (a–b) Fracture energy homogenization of 2-phase layered composite material, loaded uniaxially in direction (c) $\theta = 0$, (d) $\theta = \pi/2$, and (e) $\theta = \pi/4$. [adapted from Mulla et al. 2022].

3.4 *Scaling relations for structural mechanics applications*

The last example deals with the development of scaling relations of the fracture energy for structural applications. A typical example are fracture hinge relations that can be used for beam-type structural failure analysis (Bažant 2003), coarse-grained Molecular Dynamics structural applications with 3-body (beam) interactions (Keremides et al. 2018; Villermaux et al. 2021) and so on. Focus of our analysis is the calibration of the groundstate and elastic energy for homogeneous and heterogeneous materials, which can be used by Engineers for structural mechanics applications.

Our starting point is an application of the strength scaling in reduced units, applied here to the nominal strength, σ_N, of a structural or material test; i.e.

$$\sigma_N = \sigma_N^* \frac{\epsilon_0}{R^3} \tag{19}$$

where σ_N^* is the nominal structural strength in reduced units, obtained from SGCMC simulations for the reference particle radius $R^* = 1$ and a reference groundstate energy, $\epsilon_0^* = 1$. Therefore, rescaling the reduced nominal strength with actual values of groundstate energy and particle radius, provides a means to determine the actual nominal strength of a specific material or structure. Vice versa, if the experimental and reduced nominal strength are known (the first from experiments, the second from simulations), the ratio σ_N/σ_N^* permits access to ϵ_0/R^3. Similarly, if the fracture energy ratio of the material in 'real' and reduced units, $\mathcal{G}_F/\mathcal{G}_F^*$ is available, it becomes possible to determine the groundstate energy together with the material particle radius from Eqn. (12) and (19):

$$\epsilon_0 = \left(\frac{\sigma_N^*}{\sigma_N}\right)^2 \left(\frac{\mathcal{G}_F}{\mathcal{G}_F^*}\right)^3; \quad R = \left(\frac{\sigma_N^* \mathcal{G}_F}{\sigma_N \mathcal{G}_F^*}\right)^2 \tag{20}$$

This highlights the hybrid experimental-simulation nature of the SGCMC-approach, which provides a clear pathway for fluctuation-based fracture mechanics for the determination of generic values for σ_N^* and $\mathcal{G}_F^* \sim -q_0^*$ from simulations of homogeneous [e.g. Eq. (15)] or heterogeneous [e.g. Eq. (17)] material and structural systems described by interparticle potentials [e.g. Eqn. (13), (14)].

By way of illustration, Figure 3 displays results of SGCMC-simulations of a bending problem. The square material domain is discretized in a hexagonal fashion considering in between particles (radius $R^* = 1$) equidistant bond lengths of $2R^* = 2$, and harmonic bond potentials defined by Eq. (13). Lateral boundary conditions of the form $x^d - X = -Y\omega/2$ and $y^d = Y$ (with $\omega \approx \kappa L^*$ the opening angle, $\kappa =$ curvature) are prescribed; whereas the top and bottom boundary are stress-free. The angle is increased in increments, and for each load angle SGCMC-simulations are carried out until the energy and the bond number reaches equilibrium, at which the probability of bond deletion and bond addition is equal. This SGCMC-equilibrium condition is conveniently achieved by means of the weak stationarity condition, i.e. that the mean and (eventually) the autocovariance of bond number (N), and bond energies ($\mathcal{U}_\iota, \mathcal{U}_\lambda$) do not vary with respect to 'time', set forth by the *NVT* cycles in the Monte Carlo simulations at each constant (displacement) load level (see Table 1). For the *NVT* cycles, a standard truss solver is employed.

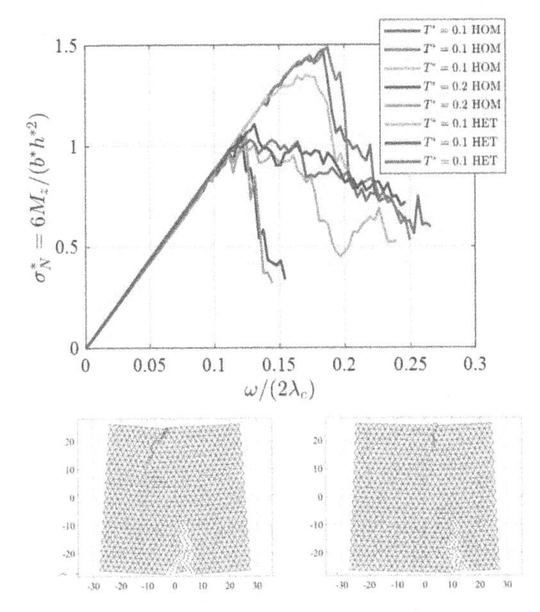

Figure 3. Flexural strength determination using SGCMC approach in reduced units: A square sample (length L^*) is subject to bending deformation (angle $\omega = \kappa L^*$ with $\kappa =$ curvature) at lateral boundaries. The nominal stress σ_N^* is determined from the generated reaction bending moment M_z^*, width $b^* = R$, and height $h^* = L^*$. Displayed are values of σ_N^* of converged SGCMC simulations vs. $\varepsilon_{xy} = \omega/(2\lambda_c)$ (with $\lambda_c = \sqrt{2\epsilon_0/\epsilon_\lambda}$ the critical linear dilation for which the harmonic potential is zero; $\epsilon_0/\epsilon_\lambda = 1/100 =$ groundstate –to– elastic energy ratio). Three simulation samples are considered: Homogeneous sample at $T^* = 0.1$ (3 specimens); homogeneous sample at $T^* = 0.2$ (2 specimens); heterogeneous sample with uniform groundstate energy distribution at $T^* = 0.1$ (3 specimens). The bottom figures show characteristic bond fracture microstates for (left) homogeneous sample and (right) heterogeneous sample ($T^* = 0.1$).

Three SGCMC-simulation experiments are carried out. For reference, the first experiment is carried out for a homogeneous sample at a reference temperature of $T^* = k_B T/\epsilon_0 = 0.1$. The second experiment considers the impact of temperature for a homogeneous sample, by setting the reduced temperature to $T^* = 0.2$. Finally, the last experiment carried out at $T^* = 0.1$ considers a uniform distribution of the groundstate energy in the bonds of mean $E[\epsilon_0^*] = 1$ and variance $\mathrm{var}(\epsilon_0^*) = 1/12$:

$$\epsilon_0^* = 1/2 + \mathcal{U}_{[0,1]} \tag{21}$$

For each experiment, simulations are carried out for 2-3 specimens to gauge the variability of the stress–strain response, as each converged simulation result along the stress curve represents a (bond) fracture microstate.

We observe that temperature significantly reduces the (nominal) flexural strength, $\sigma_N = 6M_z/bh^2$ – in reduced units – from $\sigma_N^*(T^* = 0.1) = 1.5$ to $\sigma_N^*(T^* = 0.2) = 1.0 - 1.1$, whereas the bond fracture energy remains the same [see Eq. (15)]. The apparent increase in brittleness relates to the temperature induced increase in bond energy release rate, q_λ^* [see Eq. (2)]. That is, higher temperatures increase energy/bond fluctuations; whence a lower nominal strength. Note that more realistic simulations of e.g. concrete would need to consider distributions of the elastic energy as well, accessible from e.g. nanoindentation (DeJong & Ulm 2007).

A second observation of interest is the decrease in nominal strength due to the heterogeneous distribution of groundstate energy. Specifically, at same *mean* value of the groundstate energy, the nominal flexural strength decreases from $\sigma_{N,hom}^* = 1.3 - 1.5$ to $\sigma_{N,het}^* = 1.0 - 1.1$, i.e. in similar proportions as observed for a two-fold temperature increase. Yet, the nominal strength for the heterogeneous system plateaus before the stress drops due to localized fracture. The origin of this heterogeneity-induced decrease in nominal strength is quite different. It results from the change in the bond fracture energy from $-q_0^* = 1$ for the homogeneous sample to $-q_0^* = 0.6$ for the considered uniform distribution, Eq. (21). Note clearly that the Griffith-type fracture criterion, $q_\lambda^* \leq -q_0^*$ is never enforced (or checked) in the simulations, but only evaluated during post-processing of the results using Eq. (3). This shows that the SGCMC approach is able to ascertain the 'effective' fracture resistance of heterogeneous materials as an output – not an input, in contrast to the many classical threshold-based computational fracture mechanics approaches, whether local (testing bond strengths, strains, etc.) or global (in the sense of Griffith's criterion, Eq. (1)) [for a discussion, see e.g. Laubie et al. 2017a; 2017b; Wang et al. 2021].

Finally, a comment on fracture modes is in order. As a bond-energy approach toggling bond identities according to the value of the groundstate and elastic energy [i.e., Eq. (4) and Table 1], the SGCMC approach is blind to the tensile or compressive nature of 2-pt bonds. This entails that broken bonds situated in the tension zone of the bending sample in Figure 3 appear as Mode-I opening, whereas those in the compression zone exhibit an apparent Mode-II – type opening. The latter form *de facto* shear fractures due to excessive energy in compression; typically in the softening branch of the stress-strain diagram. More importantly, the key to realistic modeling is the choice of the potentials of mean force (PMF), which for realistic simulations would need to consider the asymmetry of energy in tension and compression beyond harmonic potentials, as captured by e.g. the Morse potential; as well as 3-body and 4-body interactions to accurately reproduce the energy content of heterogeneous materials.

4 CONCLUSIONS

It has long been argued that fracture mechanics of heterogeneous materials and structures cannot be handled satisfactorily with classical averaging rules suitable for elastic and strength problems [see e.g. Dormieux et al. 2006]. This is why we reformulated classical fracture mechanics within the statistical physics framework of fluctuation-based simulations in the semigrand canonical ensemble. The following strengths and limitations define our current understanding of fluctuation-based fracture mechanics:

1. The SGCMC approach is rooted in the equilibrium-nature of fracture processes as defined by Griffith. Out-of-equilibrium situations required e.g. for dynamic fracture propagation would need further refinements. Such refinements may benefit from developments of out-of-equilibrium Monte Carlo simulations techniques applied in glass physics or gelation processes of cement hydration, in which an evolving system during NVT-relaxation is not given enough 'time' to attain complete relaxation (Ioannidou et al. 2016; Masoero et al. 2012).
2. In contrast to classical fracture mechanics theories, there is no need for initial notches or discontinuities to initiate or localize fracture. Instead, it is the acceptance probability that triggers bond fracture initiation and propagation.
3. The key to realistic simulations is an accurate yet efficient way to model the energy content of bonds in terms of both groundstate energy and elastic energy. Compared to classical computational mechanics approaches which only deal with energy variations (i.e. forces, moments,...), the groundstate energy is an active ingredient of the SGCMC approach. It defines the relaxed energy state toward which the system evolves when bonds are held constant. In contrast, in fracture mechanics, it defines the amount of energy that is released when a bond breaks. Based upon this realization, a homogenization rule for fracture energy with heterogeneous groundstate energy is derived based upon groundstate energy fluctuations – as a statistical extension of Griffith's fracture criterion.
4. The elastic energy of bonds is key to capture the interparticle force play, and thus the deformation of the system. While we have considered here only 2-pt interactions, 'bonds' need to be understood in a large sense, encompassing beam and plate bending energy interactions between three and four particles. At even higher coarse-grained levels, bond energies in this large sense can be captured by finite elements. In this case, *NVT* cycles solve classical 'static' or dynamic finite element systems based on energy minimization. Ultimately, such an extension to classical computational approaches as integral part of the SGCMC approach is expected to impact structural mechanics and structural design in that a structural energy release rate and a structural fracture energy of structures is measurable by means of simulations – providing (fracture) strength as output incl. its fluctuation.

ACKNOWLEDGMENT

Research carried out by the CSHUb@MIT with sponsorship provided by the Portland Cement Association (PCA) and the Ready Mixed Concrete Research Education Foundation (RMC EF). The CSHub@ MIT is solely responsible for content.

REFERENCES

Al-Mulla, T., R. J.-M. Pellenq, & F.-J. Ulm (2018). Griffith's postulate: Grand canonical Monte Carlo approach for fracture mechanics of solids. *Engineering Fracture Mechanics 199*, 544–554.

Bažant, Z. P. (2003). Asymptotic matching analysis of scaling of structural failure due to softening hinges. i. theory; ii. implications. *J. Engrg. Mech. 129*, 641–654.

Bonamy, D. & E. Bouchaud (2011). Failure of heterogeneous materials: A dynamic phase transition? *Physics Reports 498*(1), 1–44.

DeJong, M. & F.-J. Ulm (2007). The nanogranular behavior of csh at elevated temperatures (up to 700 c). *Cement and Concrete Research 37*, 1–12.

Dormieux, L., D. Kondo, & F.-J. Ulm (2006). *Microporomechanics*. Chichester, UK: John Wiley & Sons.

Frenkel, D. & B. Smit (2002). *Understanding Molecular Simulations. From Algorithms to Applications, 2nd Edition*. London, UK: Academic Press.

Griffith, A. (1920). The phenomena of rupture and flow in solids. *Phil. Trans. Roy. Soc. A 221*(A 587), 163–198.

Ioannidou, K., K. J. Krakowiak, M. Bauchy, C. G. Hoover, E. Masoero, S. Yip, F.-J. Ulm, P. Levitz, R. J.-M. Pellenq, & E. Del Gado (2016). Mesoscale texture of cement hydrates. *Proceedings of the National Academy of Sciences 113*(8), 2029–2034.

Keremides, K., M. J. A. Qomi, R. J.-M. Pellenq, & F.-J. Ulm (2018). Potential-of-mean-force approach for molecular dynamics–based resilience assessment of structures. *J. Engrg. Mech. 144*, 04018066.

Laubie, H., S. Monfared, F. Radjai, R. Pellenq, & F.-J. Ulm (2017). Effective potentials and elastic properties in the lattice-element method: Isotropy and transverse isotropy. *Journal of Nanomechanics and Micromechanics 7*(3), 04017007.

Laubie, H., F. Radjai, R. Pellenq, & F.-J. Ulm (2017a). A potential-of-mean-force approach for fracture mechanics of heterogeneous materials using the lattice element method. *Journal of the Mechanics and Physics of Solids 105*, 116–130.

Laubie, H., F. Radjai, R. Pellenq, & F.-J. Ulm (2017b). Stress transmission and failure in disordered porous media. *Physical review letters 119*(7), 075501.

Louhghalam, A., R. J. Pellenq, & F.-J. Ulm (2018). Thermalizing and damping in structural dynamics. *Journal of Applied Mechanics 85*, 081001.

Masoero, E., E. Del Gado, R.-M. Pellenq, , F.-J. Ulm, & S. Yip (2012). Nanostructure and nanomechanics of cement: polydisperse colloidal packing. *Phys. Rev. Lett. 109*, 155503.

Mulla, T., S. Moeini, K. Ioannidou, R. J.-M. Pellenq, & F.-J. Ulm (2021). Phase diagram of brittle fracture in the

semi-grand-canonical ensemble. *Phys. Rev. E 103*, 013003.
Mulla, T., R. J.-M. Pellenq, & F.-J. Ulm (2022). Fluctuation-based fracture mechanics of heterogeneous materials in the semi-grand-canonical ensemble. *In Review*.
Villermaux, E., K. Keremidis, N. Vandenberghe, M. Abdolhosseini Qomi, & F.-J. Ulm (2021). Mode coarsening or fracture: Energy transfer mechanisms in dynamic buckling of rods. *Phys. Rev. Lett. 126*, 045501.
Wackerfuß, J. (2009). Molecular mechanics in the context of the finite element method. *Int. J. Numer. Meth. Engng 77*, 969–997.
Wang, X., M. Botshekan, F.-J. Ulm, M. Tootkaboni, & A. Louhghalam (2021). A hybrid potential of mean force approach for simulation of fracture in heterogeneous media. *Computer Methods in Applied Mechanics and Engineering 386*, 114084.

Computational Modelling of Concrete and Concrete Structures – Meschke, Pichler & Rots (Eds)
© 2022 Copyright the Author(s), ISBN: 978-1-032-32724-2

Building resilience and masonry structures: How can computational modelling help?

P.B. Lourenço & M.F. Funari
Department of Civil Engineering, University of Minho, ISISE, Guimarães, Portugal

L.C. Silva
Department of Civil Engineering, Lusófona University, ISISE, Lisbon, Portugal

ABSTRACT: Masonry structures populate European historic centres as both secular and sacred monumental buildings. Several researchers focused on implementing advanced computational strategies to preserve masonry buildings structural integrity. Moreover, today's tools and workflows offer possibilities for assessing vulnerability, simulating scenarios, and reducing vulnerability are opening new perspectives and challenges. This literature review provides a classification of the analytical and numerical modelling strategies for the structural assessment of unreinforced masonry structures. Finally, two-stepped procedures are discussed and suggested as valuable approaches to combine analytical/analytical, numerical/analytical or numerical/numerical methods.

1 INTRODUCTION

Given its important role for economies and societies, the assessment, preventive conservation and maintenance of the historical masonry structures (HMS) continue to stand as major priorities of the overall political strategy at the European level. In this context, the earthquake protection of HMS assumes particular relevance because of the non-negligible seismic vulnerability of this type of ancient building whose tangible and intangible value is further enhanced by the artworks therein located, such as sculptures, paintings and frescos, among others. This means that when a disaster involves historical centres, it is likely that buildings, as well as artworks, are damaged, producing i)a physical loss of artistic and historical materials, ii) an immaterial loss of memory and cultural identity for the people to whom that legacy "belongs", and iii) difficulties in the action of the Civil Protection in assisting the population affected by the disaster (Lourenço 2014).

One of the main issues still not solved in the literature is the indeterminacy in defining the economic value of historical centres containing architectural heritage. Indeed, society tends to define architectural heritage as invaluable, making the problem mathematically indeterminate. However, in order to define a strategy for architectural heritage conservation, one can refer to studies from the National Institute of Building Sciences (US) that show how the investment in mitigation saves six times the amount for damage repair ("prevention pays"). According to this statement, several researchers focused on implementing advanced computational strategies to preserve HMS structural integrity. Moreover, today's tools and workflows offer possibilities for assessing vulnerability, simulating scenarios, and reducing vulnerability are opening new perspectives and challenges.

Because of the need to investigate thousands of buildings, structural engineers often use analytical approaches based on limit analysis, theorems that have the great advantage of being independent of many material properties but inevitably rely on a very simplified material model (Cascini et al. 2020; De Felice & Giannini 2001). Such approaches include force- and displacement-based procedures suitable for a rapid seismic vulnerability assessment. However, force-based approaches do not consider the load-displacement capacity of the structures (Heyman 1966). Moreover, limit analysis-based tools typically neglect the structure's global behaviour, only focusing on assessing a set of local failure mechanisms (D'Ayala & Speranza 2003; Funari et al. 2021; Giuffré 1996).

However, sometimes there is the need to investigate non-linear masonry behaviour deeply, and this may be achieved by adopting sophisticated advanced numerical methodologies, i.e. Finite Element Method (FEM) (Aşıkoğlu et al, 2019; Fortunato et al. 2017) or Disce Element Method (DEM) (Bui et al. 2017; Gonen et al. 2021; Lemos 2007, 2019). Such approaches model the masonry material using different representation scales, i.e. equivalent continuum, macro-blocks or discrete representations. However, despite

DOI 10.1201/9781003316404-4

their reliability, the computational efficiency of the available numerical methods is rarely compatible with the need to have a rigorous real-time post-earthquake assessment (Lourenço & Silva 2020). Hence, researchers are committed in developing alternative modelling approaches and practical tools to decrease the computational cost without losing accuracy.

Several authors have recently proposed two-stepped procedures in order to develop numerical tools for the rapid seismic assessment of historic masonry structures combining analytical/analytical (Funari et al. 2021b), analytical/numerical (D'Altri et al. 2021; Funari et al. 2020), or numerical/numerical approaches (Gams et al. 2017; Mele et al. 2003).

The first attempt to define a workflow based on a two-stepped analysis has been proposed by Mele et al. (2003). In the first step, the structure is analysed in the linear-elastic range using a 3D FE model. Subsequently, a pushover analysis of the single macro-elements is performed. The results obtained through pushover analysis has been compared to the collapse loads derived from limit analysis, proving the ability of the non-linear finite element model to provide reasonable simulations of the actual response of masonry elements.

Betti and Galano (2012) and Cundari et al. (2017) proposed modelling approaches analysing structures with non-linear static or dynamic analysis to detect the most likely collapse mechanisms. The upper bound limit analysis method was applied in the second step to compute the maximum horizontal acceleration that the structure can withstand analytically.

Funari et al. (2020) proposed a two-stepped analysis in which a non-linear static analysis was performed to identify the failure mechanism's geometry. Then, the second step aimed to refine the geometry of the failure mechanism through an optimisation based on limit analysis and genetic algorithm, which explores the research panorama of solutions kinematically compatible.

Recently, D'Altri et al. (2021) proposed a new workflow based on the adaptive limit and pushover analyses. First, limit analysis was used to identify the position of the cracked surfaces by adopting an adaptive NURBS approach. Subsequently, the geometry of the collapse mechanism was imported into FE software to perform a non-linear static analysis simulation by adopting a hardening plasticity model and cohesive-frictional contact-based at the interfaces between macro-blocks.

This study's primary goal is to perform a literature review of the analytical and numerical modelling approaches to assess the structural integrity of HMS. The discussion will be enriched with practical examples of advanced analytical and numerical strategies developed by the Historic Masonry Structures research group (Department of Civil Engineering, University of Minho, ISISE, Guimarães, Portugal). At first, analytical approaches will be discussed. The discussion will only involve force-based analytical formulations.

The subsequent section will be devoted to addressing state of the art numerical methodologies.

Finally, the third section will discuss two-stepped approaches formulated to get an accurate structural response, decreasing the computational demand thanks to the adoption of macro-block representation.

Finally, some meaningful conclusions will be drawn.

2 ANALYTICAL APPROACHES

As aforementioned in the introduction section, several analytical methodologies for the structural assessment of HMS have been proposed to date. These methods can be divided into two main categories:

- Force-equilibrium formulations,
- Displacement-based formulations.

Force-based formulations have been recommended because unreinforced masonry structures are perceived frequently as possessing very limited ductility.

However, several studies have demonstrated that, in one regard, unreinforced masonry walls subjected to dynamic loads can resist accelerations higher than their static strength. In the following subsection, only analytical approaches based on force-based formulations are discussed.

Force-based formulations are set within the theoretical frame of reference of application of plasticity theory for the structural assessment of masonry structures, as proposed for the first time by Heyman in his revolutionary work (Heyman 1966, 1969). According to this, applying the static theorem leads to a lower-bound or safe solution based on equilibrium equations, while applying the kinematic theorem provides an upper-bound multiplier of the collapse load factor. Thus, the solution that satisfies the hypotheses of both theorems, equilibrium, compatibility, and material conditions, is the correct solution and provides the collapse load multiplier for the specific problem.

Once a mechanism is selected, and a set of equilibrated generalised forces and a set of compatible generalised virtual displacements are determined, the work done by the generalised forces in equilibrium with the internal stresses for the given set of generalised virtual displacements is computed. Finding a minimum or a maximum of the resulting equation leads to the optimal solution. In this framework, after the post-seismic damage surveys carried out in the sequence of Irpinia and Syracuse earthquakes in Italy, Giuffrè (1991) presented an original work where he provided an abacus of local failure mechanisms that may be assessed through simple analytical formulations.

Following his pioneering work, some authors have already implemented algorithms able to investigate the most reasonable collapse mechanisms into user-defined routines of analysis that in turn adopted the lower or the upper bound theorem of the limit analysis (Block et al. 2006; D'Ayala & Speranza 2003).

Specifically, the kinematic theorem of limit analysis is a useful tool, and it is the most adopted in formulations recently proposed. One can note that the computation of the load multiplier depends on the macroblocks' geometry that strongly influences the assessment of their structural integrity. Therefore, multiple (theoretically infinite) failure mechanisms need to be considered to evaluate the minimum of the kinematically compatible load multipliers. To address the latter, some researchers proposed using optimisation routines to solve the minimisation problem constrained under specific hypotheses.

Casapulla et al. (2014) proposed a macro-block model coupled with a simplified procedure for predicting the collapse load and the failure mechanism of in-plane loaded masonry walls with non-associative frictional contact interfaces. The same authors (Casapulla & Maione 2018) have recently revisited the previous macro-block approach implementing the frictional resistance computation. Fortunato et al. (2018) developed a numerical procedure for the limit analysis of 2D masonry structures subject to arbitrary loading. Similarly, in the framework of limit analysis methods, other authors have proposed meta-heuristic approaches (i.e. Genetic Algorithms) as a tool to explore the entity of loads associated with considered collapse mechanisms (Funari et al. 2020a, 2020b).

Recently, Turco et al. (2020) developed a novel digital procedure for the assessment of masonry structures embedding the upper bound limit theorem under the hypothesis of no-tension capacity for the masonry material and heuristic optimisation algorithm. Once preselected, the failure mechanism genetic algorithms are employed to search for the failure mode corresponding to the minimum value of the load multiplier that is also statically equilibrated. The workflow is integrated into a computational tool implemented in the visual programming environment offered by Rhinoceros3D+Grasshopper. This is well suited to be confidently used by practitioners, also allowing the user to progress fast. Even though genetic algorithms may require high computational efforts, they make a robust implementation of a multidimensional constrained optimisation problem possible. Furthermore, adopting the upper bound theorem of the limit analysis under the hypothesis of the macro-blocks' discretisation improves the computational efficiency without requiring the detailed knowledge of the mechanical properties of materials and providing a solution of the structural problem showing good accuracy (Figure 1).

3 NUMERICAL APPROACHES

Advanced numerical strategies have been developed in the last few decades. In this framework, sophisticated Finite Element (FE) computational strategies are the ones that deserve more attention from the scientific community. For the masonry field, it is recognisable that two scale levels are of interest when analysing its structural behaviour (Lourenço, 2009; Roca et al., 2010), the macro and the mesoscale as depicted in Figure 2. Again, three main modelling strategies can be put together, namely: i) the direct simulation or the micro-modelling; ii) the macro-modelling; and iii) the multi-scale modelling.

In the micro-modelling approach, both masonry components (units and mortar joints) are explicitly represented. These are certainly capable of well reproducing both in- and out-of-plane orthotropic non-linear behaviour of masonry but are characterised by prohibitive computational cost and is only recommended for limited size structural problems (Adam et al. 2010; Giambanco & Rizzo 200; Lemos 200; Lotfi & Shing 199; Macorini & Izzuddin 2011, 201; Sarhosis et al. 201; Sejnoha et al. 2008. The macro-modelling approaches used fictitious homogeneous anisotropic material to reproduce heterogeneous assemblage of mortar and bricks. The use of closed-form laws to represent the masonry's complex phenomenological behaviour and damage may be cumbersome as it may require a calibration step (usually achieved by thorough experimental campaigns). However, this approach allows studying large-scale structures without the drawbacks exhibited by meso-modelling (Berto et al. 2002; Dhanasekar et al. 198; Paulo B. Lourenço et al. 1997; Roca et al. 2013).

Multi-scale FE (or FE^2) methods are in-between the latter two FE modelling schemes. The framework is being used to investigate composites' response with different natures (Greco et al. 201; Leonetti et al. 201; Spahn et al. 2014; Trovalusci et al. 2015. It typically relies on a meso and macro transition of information and is, therefore, designated as a two-scale or FE^2 approach. Full continuum-based FE^2 approaches result in a good compromise between solution accuracy and computational cost. Nevertheless, these methods still constitute a challenge if one desires to account for the material non-linearity (Geers et al. 2010; Otero et al. 2015).In fact, the constant need of data between the macro-and meso- scales constitute a contentious issue because a new boundary value problem (BVP) must be solved numerically for each load step and in each Gauss integration point. The approach's utility is compromised due to the involved computational time, and thus, full continuum-based FE^2 approaches are seldom used for dynamic or complex structural analysis. An adequate possibility is the use of a two-scale simplified strategy, for instance, by

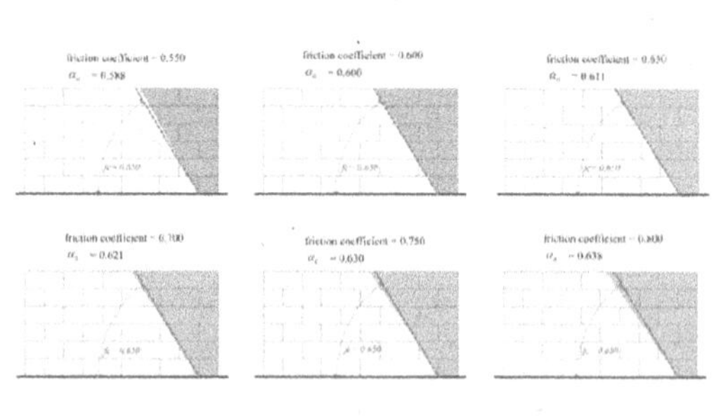

Figure 1. Predicted macro-block geometry by adopting different friction coefficients.

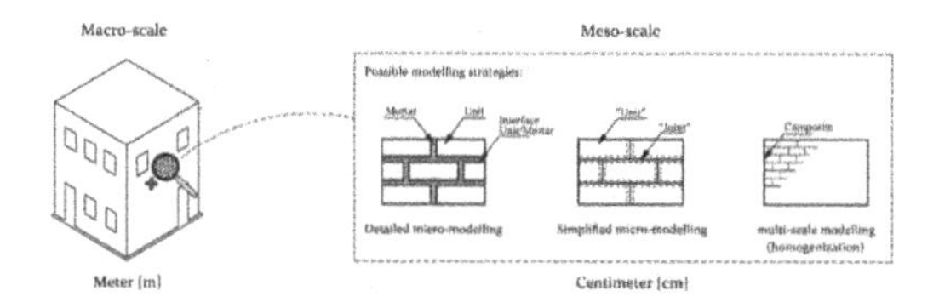

Figure 2. Representation of the three scales considered in the analysis of masonry for this study: macro-scale and meso-scale. Definition of the modelling strategies adopted to represent masonry.

using a kinematic theorem of limit analysis at a macro-level to obtain the homogenised failure surfaces with a minimal computational effort (A. Cecchi & Milani 2008; de Buhan & de Felice 1997; Milani et al. 2006). Yet, the use of discrete FE-based methods at a macro-level seems to be a promising alternative (Casolo & Milani 2010; Milani & Tralli 201; Silva et al. 2017b)

4 TWO-STEPPED APPROACHES

As aforementioned, both analytical and numerical approaches have their own strengths and weaknesses. In particular, micro- or mesoscale approaches requires several input data for the non-linear characterisation of the masonry, making the structural assessment of HMS economically expensive and time-consuming. On the contrary analytical approaches are always based on some hypotheses and may be widely used to study local failure mechanisms but is often not sufficient for a full structural analysis under seismic loads. To this end, new computational modelling strategies are investigating hybrid approaches combining analytical/analytical (Funari et al. 2021b), analytical/numerical (D'Altri et al. 2021; Funari et al. 2020), or numerical/numerical approaches (Gams et al. 2017; Mele et al. 2003). In the following subsections, three of the most recent two-stepped approaches developed are discussed.

4.1 *Visual programming for structural assessment of out-of-plane mechanisms in historic masonry structures*

Funari et al. (2020) proposed a two-stepped procedure for the seismic assessment of HMS. Firstly, digital datasets describing the geometric configuration of historic masonry structures are employed to generate a FE model and investigate possible failure mechanisms automatically. Therefore, a coarse configuration of failure surfaces is detected through the Control Surface Method (CSM), which is generated by interpolating the displacement function obtained during the step-by-step global analysis of all control points (Figure 3).

In the following step of the analysis, structural macroblocks were identified, whereas an upper bound limit analysis approach was employed to estimate the structural capacity of the structure. Genetic Algorithms are

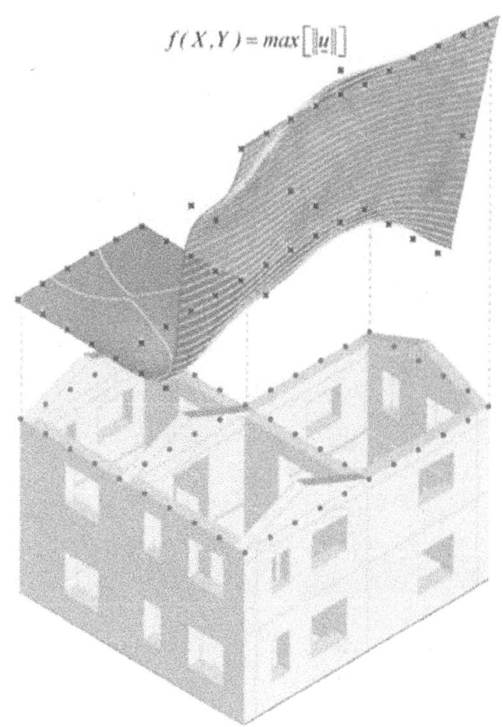

Figure 3. Control surface detected at final load step of the non-linear static analysis.

hence employed to detect the actual failure mode for the structure(Funari et al., 2020).

The procedure was implemented into a visual programming environment, which allows the user to explore all the landscape of possible solutions parametrically (see Figure 4).

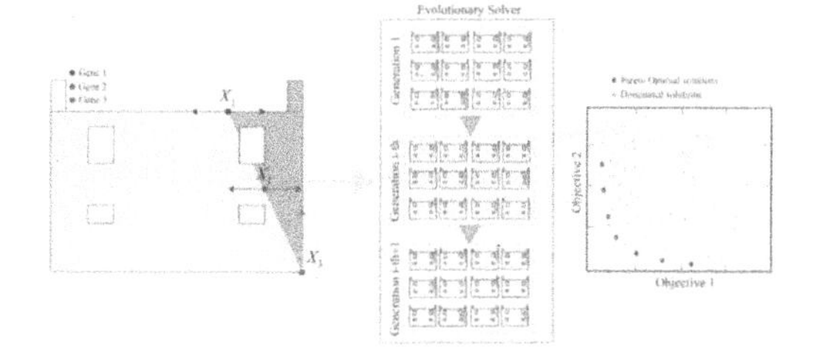

Figure 4. Evolutionary solver: genes definition, generations representation and graphical representation of the genomes.

The approach prosed by Funari et al. (2020) was applied to a benchmark case of study representative of a residential building (Figure 5). The CSM allowed detecting the failure modes of the structure considered, demonstrating its capability to detect the failure mechanism of masonry structures characterised by irregular shape.

Figure 6 shows the optimisation process. During the first generation, the load factor is equal to 0.43. At each succeeding generation, the result tends to converge to

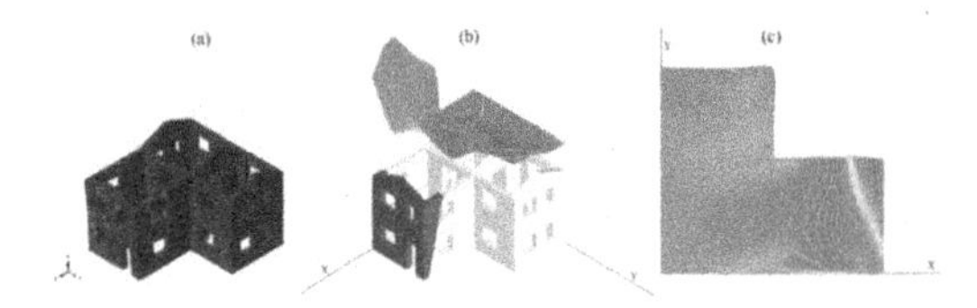

Figure 5. +X direction: (a) damage pattern obtained from the preparatory step (b) CS and identified macro-block (c) Top view of CS.

the value that minimises the load multiplier and simultaneously satisfies the horizontal equilibrium forces. The solving strategy leads to a load multiplier equal to 0.354.

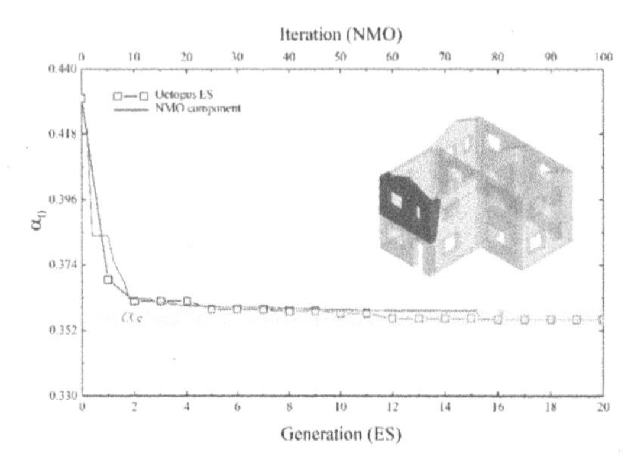

Figure 6. +X direction: Load factor vs generation of the evolutionary solver.

4.2 *A tool for the rapid seismic assessment of historic masonry structures based on limit analysis optimisation and rocking dynamics*

Recently, Funari et al. (2021a) developed a multi-level integrated modelling procedure that uses a combination of upper bound limit analysis and non-linear dynamic (rocking) analysis for the seismic collapse assessment of any user-defined structural configuration. In the first step, parametric modelling of the macro-block geometry is conducted, enabling exploration of the domain of possible solutions using the upper bound method of limit analysis. A heuristic solver based on the Nelder-Mead method is then adopted to refine the geometry of the macro-blocks and search for the minimum value of the load multiplier. Once the macro-block (i.e., collapse mechanism) has been defined, the digital tool then computes the kinematic constants defining the corresponding (rocking) equation of motion, which can be solved for full time-histories. Finally, the structure's response is provided both in terms of the full time-history response and the form of the maximum predicted rotation. An overview of the proposed analysis procedure is reported in Figure 7.

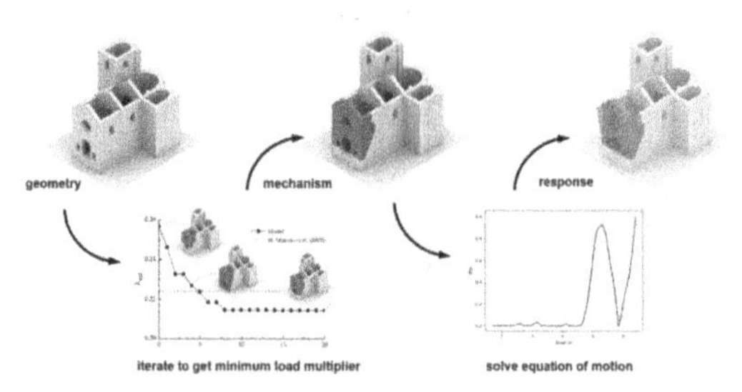

Figure 7. Overview of the proposed multi-level analysis procedure.

As the research work mentioned in the previous section, the methodology was entirely integrated into a user-friendly visual programming environment that allows the user to easily connect data from different sources while clearly understanding the relationships between them thanks to the flowchart-like representation of the different components of the code. The main advantage of this approach with respect to more time-consuming advanced methods of analysis, the proposed method allows the users to perform a seismic assessment of masonry buildings in a rapid and computationally-efficient manner while simultaneously providing more accurate predictions than simplified/code-based methods. Such an approach may be particularly useful for territorial scale vulnerability analysis (e.g., for risk assessment and mitigation in historic city centres) or as post-seismic event response (when the safety and stability of a large number of buildings need to be assessed with limited resources).

4.3 *A concurrent micro/macro FE-model optimised with a limit analysis tool for the assessment of dry-joint masonry structures*

A more detailed description of the failure mechanism may be required in some cases, e.g. using microscale representation. To this end, Funari et al. (2021) presented a numerical framework to accurately describe the in- and out-of-plane failure mechanisms that may affect unreinforced masonry structures. The so-called concurrent approach, firstly presented by Fish (2006), is adopted together with a limit analysis tool. In this regard, the framework has two sequential and coupled steps, in which a limit analysis is conducted first, and a concurrent FE analysis is employed next.

The workflow includes four main tasks, as given in Figure 8, needed to compute the mechanical response of HMS. The first step consists of the geometrical modelling of the structure via an explicit representation of both masonry units and joints (micro-modelling approach). In the second step, masonry units are merged, and its topology is optimised to provide a macro representation. Prone in-plane and/or out-of-plane failure mechanisms are a-priori assigned, and the location of the yielding surfaces is optimised by an Upper bound limit analysis theorem coupled with a heuristic solver. Hence, the third step is conducted, in which an ad-hoc script represents the sub-structure, which is activated by the failure mechanism, through a micro-scale representation. The outer domain, i.e. the rest of the structure that is not involved in the mechanism, keeps a macro and continuous representation. Finally, at the fourth step, the concurrent FE multi-scale model can be used to perform the structural assessment of the structure through a non-linear quasi-static type of analysis and within a FE environment.

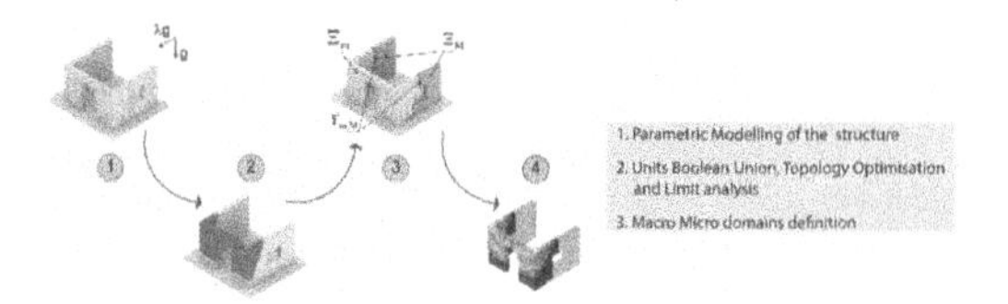

Figure 8. Schematic representation of the proposed procedure.

The authors concluded that the two-step procedure is computational quite attractive, robust, and allows higher levels of accuracy. This is so because it is based on a sequential process in which a continuous transfer of information between scales is not required during the analysis; as observed in classical multi-domain strategies that need activation rules to process the macro-to-micro decomposition (Driesen et al. 2021; Leonetti et al. 2018; Reccia et al., 2018).

5 CONCLUSIONS

The vulnerability assessment and preservation of the HMS need fast, reliable and modern tools. In the last decades, several researchers proposed advanced analysis methods for the preventive assessment of heritage buildings. However, their overall classification is mainly made between numerical and analytical approaches (D'Altri et al., 2020). This paper explored both analytical and numerical methodologies and finally presented a literature review of two-stepped procedures recently developed by the Historic Masonry Structures research group Department of Civil Engineering, University of Minho, ISISE, Guimarães, Portugal).

The following conclusion can be drawn:

1. Because of the need to investigate thousands of buildings, analytical approaches based on force- and displacement-based methodologies have the great advantage of being independent of many material properties and inevitably rely on a very simplified assessment;
2. On the other hand, numerical approaches are typically implemented in the FEM or DEM frameworks. Such approaches model the masonry material using different representation scales, i.e., equivalent continuum, macro-blocks, or discrete representations.
3. In the last decade, the computational efficiency of the available numerical methods has been strongly improved, even though rarely compatible with the need to have a rigorous real-time post-earthquake assessment.
4. The two-stepped procedure represents a valid compromise in terms of both accuracy and computational efficiency.

6 ACKNOWLEDGEMENTS

This study has been partly funded by the STAND4 HERITAGE project (New Standards for Seismic Assessment of Built Cultural Heritage) that has received funding from the European Research Council (ERC) under the European Union's Horizon 2020 research and innovation program (Grant No. 833123) as an Advanced Grant. This work was also partly financed by FCT / MCTES through national funds (PIDDAC) under the R&D Unit Institute for Sustainability and Innovation in Structural Engineering (ISISE), under reference UIDB / 04029/2020. Their support is gratefully acknowledged. However, the opinions and conclusions presented in this paper are those of the authors and do not necessarily reflect the views of the sponsoring organisations.).

REFERENCES

Adam, J. M. et al. (2010) 'Micromodelling of eccentrically loaded brickwork: Study of masonry wallettes', *Engineering Structures*, 32(5), pp. 1244–1251. doi: https://doi.org/10.1016/j.engstruct.2009.12.050.

Aşıkoğlu, A. et al. (2019) 'Effectiveness of seismic retrofitting of a historical masonry structure: Kütahya Kurşunlu Mosque, Turkey', *Bulletin of Earthquake Engineering*, 17(6), pp. 3365–3395. doi: 10.1007/s10518-019-00603-6.

Berto, L. et al. (2002) 'An orthotropic damage model for masonry structures', *International Journal for Numerical Methods in Engineering*, 55(2), pp. 127–157. doi: 10.1002/nme.495.

Betti, M. and Galano, L. (2012) 'Seismic analysis of historic masonry buildings: The Vicarious Palace in Pescia (Italy)', *Buildings*, 2(2), pp. 63–82. doi: 10.3390/buildings2020063.

Block, P., Ciblac, T. and Ochsendorf, J. (2006) 'Real-time limit analysis of vaulted masonry buildings', *Computers & Structures*, 84(29–30), pp. 1841–1852. doi: 10.1016/J.COMPSTRUC.2006.08.002.

de Buhan, P. and de Felice, G. (1997) 'A homogenisation approach to the ultimate strength of brick masonry', *Journal of the Mechanics and Physics of Solids*, 45(7), pp. 1085–1104. doi: 10.1016/S0022-5096(97)00002-1.

Bui, T. T. et al. (2017) 'Discrete element modelling of the in-plane and out-of-plane behaviour of dry-joint masonry wall constructions', *Engineering Structures*, 136 (October), pp. 277–294. doi: 10.1016/j.engstruct.2017.01.020.

Candeias, P. X. et al. (2017) 'Experimental assessment of the out-of-plane performance of masonry buildings through

shaking table tests', *International Journal of Architectural Heritage*, 11(1), pp. 31–58.
Casapulla, C. et al. (2014) '3D macro and micro-block models for limit analysis of out-of-plane loaded masonry walls with non-associative Coulomb friction'. doi: 10.1007/s11012-014-9943-8.
Casapulla, C. and Maione, A. (2018) 'Experimental and Analytical Investigation on the Corner Failure in Masonry Buildings: Interaction between Rocking-Sliding and Horizontal Flexure', *https://doi.org/10.1080/15583058.2018.1529206*, 14(2), pp. 208–220. doi: 10.1080/15583058.2018.1529206.
Cascini, L., Gagliardo, R. and Portioli, F. (2020) 'LiABlock_3D: A Software Tool for Collapse Mechanism Analysis of Historic Masonry Structures', *International Journal of Architectural Heritage*, 14(1), pp. 75–94. doi: 10.1080/15583058.2018.1509155.
Casolo, S. and Milani, G. (2010) 'A simplified homogenization-discrete element model for the non-linear static analysis of masonry walls out-of-plane loaded', *Engineering Structures*, 32(8), pp. 2352–2366. doi: 10.1016/j.engstruct.2010.04.010.
Cecchi, A. and Milani, G. (2008) 'A kinematic FE limit analysis model for thick English bond masonry walls', *International Journal of Solids and Structures*, 45(5), pp. 1302–1331. doi: 10.1016/j.ijsolstr.2007.09.019.
Cundari, G. A., Milani, G. and Failla, G. (2017) 'Seismic vulnerability evaluation of historical masonry churches: Proposal for a general and comprehensive numerical approach to cross-check results', *Engineering Failure Analysis*, 82, pp. 208–228. doi: 10.1016/j.engfailanal.2017.08.013.
D'Ayala, D. and Speranza, E. (2003) 'Definition of Collapse Mechanisms and Seismic Vulnerability of Historic Masonry Buildings', *Earthquake Spectra*, 19(3), pp. 479–509. doi: 10.1193/1.1599896.
Dhanasekar, M., Kleeman, P. and Page, A. (1985) 'The failure of brick masonry under biaxial stresses.', *Proceedings of the Institution of Civil Engineers*, 79(2), pp. 295–313. doi: 10.1680/iicep.1985.992.
Driesen, C., Degée, H. and Vandoren, B. (2021) 'Efficient modeling of masonry failure using a multiscale domain activation approach', *Computers and Structures*, 251, p. 106543. doi: 10.1016/j.compstruc.2021.106543.
De Felice, G. and Giannini, R. (2001) 'Out-of-plane seismic resistance of masonry walls', *Journal of Earthquake Engineering*, 5(2), pp. 253–271. doi: 10.1080/13632460109350394.
Fish, J. (2006) 'Bridging the scales in nano engineering and science', *Journal of Nanoparticle Research*, 8(5), pp. 577–594.
Fortunato, A. et al. (2018) 'Limit analysis of masonry structures with free discontinuities', *Meccanica*, 53(7), pp. 1793–1802.
Fortunato, G., Funari, M. F. and Lonetti, P. (2017) 'Survey and seismic vulnerability assessment of the Baptistery of San Giovanni in Tumba (Italy)', *Journal of Cultural Heritage*. doi: 10.1016/j.culher.2017.01.010.
Funari, Marco Francesco et al. (2020) 'Visual programming for structural assessment of out-of-plane mechanisms in historic masonry structures', *Journal of Building Engineering*, 31. doi: 10.1016/j.jobe.2020.101425.
Funari, M.F. et al. (2020) 'Visual programming for the structural assessment of historic masonry structures', in *REHABEND*.
Funari, M. F. et al. (2021) 'Real-time Structural Stability of Domes through Limit Analysis: Application to St. Peter's Dome', *https://doi.org/10.1080/15583058.2021.1992539*, pp. 1–23. doi: 10.1080/15583058.2021.1992539.
Funari, M. F., Mehrotra, A. and Lourenço, P. B. (2021b) 'A Tool for the Rapid Seismic Assessment of Historic Masonry Structures Based on Limit Analysis Optimisation and Rocking Dynamics', *Applied Sciences*, 11(3), p. 942. doi: 10.3390/app11030942.
Gams, M., Anžlin, A. and Kramar, M. (2017) 'Simulation of Shake Table Tests on Out-of-Plane Masonry Buildings. Part (III): Two-Step FEM Approach', *International Journal of Architectural Heritage*, 11(1), pp. 94–102. doi: 10.1080/15583058.2016.1237589.
Geers, M. G. D., Kouznetsova, V. G. and Brekelmans, W. A. M. (2010) 'Multi-scale computational homogenization: Trends and challenges', *Journal of Computational and Applied Mathematics*, 234(7), pp. 2175–2182. doi: 10.1016/j.cam.2009.08.077.
Giambanco, G., Rizzo, S. and Spallino, R. (2001) 'Numerical analysis of masonry structures via interface models', *Computer Methods in Applied Mechanics and Engineering*, 190(49–50), pp. 6493–6511. doi: 10.1016/S0045-7825(01)00225-0.
Giuffrè, A. (1996) 'A Mechanical Model for Statics and Dynamics of Historical Masonry Buildings', in *Protection of the Architectural Heritage Against Earthquakes*. Springer Vienna, pp. 71–152. doi: 10.1007/978-3-7091-2656-1_4.
Gonen, S. et al. (2021) 'Quasi-static nonlinear seismic assessment of a fourth century A.D. Roman Aqueduct in Istanbul, Turkey', *Heritage*, 4(1), pp. 401–421. doi: 10.3390/heritage4010025.
Greco, F. et al. (2017) 'Multiscale failure analysis of periodic masonry structures with traditional and fiber-reinforced mortar joints', *Composites Part B: Engineering*, 118, pp. 75–95. doi: https://doi.org/10.1016/j.compositesb.2017.03.004.
Heyman, J. (1966) 'The stone skeleton', *International Journal of Solids and Structures*, 2(2), pp. 249–279. doi: 10.1016/0020-7683(66)90018-7.
Heyman, J. (1969) 'The safety of masonry arches', *International Journal of Mechanical Sciences*, 11(4), pp. 363–385. doi: 10.1016/0020-7403(69)90070-8.
Lemos, J. V. (2007) 'Discrete element modeling of masonry structures', *International Journal of Architectural Heritage*, 1(2), pp. 190–213. doi: 10.1080/15583050601176868.
Lemos, J. V. (2019) 'Discrete element modeling of the seismic behavior of masonry construction', *Buildings*, 9(2). doi: 10.3390/buildings9020043.
Leonetti, L., Trovalusci, P. and Cechi, A. (2018) 'A multiscale/multidomain model for the failure analysis of masonry walls: a validation with a combined FEM/DEM approach', *International Journal for Multiscale Computational Engineering*, 16(4), pp. 325–343. doi: 10.1615/IntJMultCompEng.2018026988.
Lotfi, H. R. and Shing, P. B. (1994) 'Interface Model Applied to Fracture of Masonry Structures', *Journal of Structural Engineering*, 120(1), pp. 63–80. doi: 10.1061/(ASCE)0733-9445(1994)120:1(63).
Lourenço, P. B. (2009) 'Recent advances in masonry structures: Micromodelling and homogenisation', in Galvanetto, U. and Aliabadi, M. H. F. (eds) *Multiscale Modeling in Solid Mechanics*. Imperial College Press (Computational and Experimental Methods in Structures), pp. 251–294. doi: 10.1142/p604.
Lourenço, P. B. (2014) 'The ICOMOS methodology for conservation of cultural heritage buildings: Concepts, research and application to case studies', in *REHAB*

2014 – Proceedings of the International Conference on Preservation, Maintenance and Rehabilitation of Historical Buildings and Structures. Green Lines Institute for Sustainable Development, pp. 945–954. doi: 10.14575/gl/rehab2014/095.

Lourenço, P. B., De Borst, R. and Rots, J. G. (1997) 'A plane stress softening plasticity model for orthotropic materials', *International Journal for Numerical Methods in Engineering*, 40(21), pp. 4033–4057. doi: 10.1002/(SICI)1097-0207(19971115)40:21<4033::AID-NME248>3.0.CO;2-0.

Lourenço, P. B. and Silva, L. C. (2020) 'Computational applications in masonry structures: From the meso-scale to the super-large/super-complex', *International Journal for Multiscale Computational Engineering*, 18(1), pp. 1–30. doi: 10.1615/IntJMultCompEng.2020030889.

Macorini, L. and Izzuddin, B. A. (2011) 'A non-linear interface element for 3D mesoscale analysis of brick-masonry structures', *International Journal for Numerical Methods in Engineering*, 85(12), pp. 1584–1608. doi: 10.1002/nme.3046.

Macorini, L. and Izzuddin, B. A. (2013) 'Nonlinear analysis of masonry structures using mesoscale partitioned modelling', *Advances in Engineering Software*, 60–61, pp. 58–69. doi: 10.1016/j.advengsoft.2012.11.008.

Maria D'Altri, A. et al. (2020) 'Modeling Strategies for the Computational Analysis of Unreinforced Masonry Structures: Review and Classification', 27, pp. 1153–1185. doi: 10.1007/s11831-019-09351-x.

Maria D'Altri, A. et al. (2021) 'A two-step automated procedure based on adaptive limit and pushover analyses for the seismic assessment of masonry structures', *Computers and Structures*, 252, p. 106561. doi: 10.1016/j.compstruc.2021.106561.

Mele, E., De Luca, A. and Giordano, A. (2003) 'Modelling and analysis of a basilica under earthquake loading', *Journal of Cultural Heritage*, 4(4), pp. 355–367. doi: 10.1016/j.culher.2003.03.002.

Milani, G., Lourenço, P. B. and Tralli, A. (2006) 'Homogenised limit analysis of masonry walls, Part II: Structural examples', *Computers & Structures*, 84(3–4), pp. 181–195. doi: 10.1016/j.compstruc.2005.09.004.

Milani, G. and Tralli, A. (2011) 'Simple SQP approach for out-of-plane loaded homogenized brickwork panels, accounting for softening', *Computers & Structures*, 89(1–2), pp. 201–215. doi: 10.1016/j.compstruc.2010.09.005.

Otero, F. et al. (2015) 'Numerical homogenization for composite materials analysis. Comparison with other micro mechanical formulations', *Composite Structures*, 122, pp. 405–416. doi: 10.1016/j.compstruct.2014.11.041.

Reccia, E. et al. (2018) 'A multiscale/multidomain model for the failure analysis of masonry walls: a validation with a combined FEM/DEM approach', *International Journal for Multiscale Computational Engineering*, 16(4), pp. 325–343. doi: 10.1615/IntJMultCompEng.2018026988.

Roca, P. et al. (2010) 'Structural Analysis of Masonry Historical Constructions. Classical and Advanced Approaches', *Archives of Computational Methods in Engineering*, 17(3), pp. 299–325. doi: 10.1007/s11831-010-9046-1.

Roca, P. et al. (2013) 'Continuum FE models for the analysis of Mallorca Cathedral', *Engineering Structures*, 46, pp. 653–670. doi: 10.1016/j.engstruct.2012.08.005.

Sarhosis, V., Tsavdaridis, K. and Giannopoulos, I. (2014) 'Discrete Element Modelling (DEM) for Masonry Infilled Steel Frames with Multiple Window Openings Subjected to Lateral Load Variations', *The Open Construction and Building Technology Journal*, 8(1), pp. 93–103. doi: 10.2174/1874836801408010093.

Sejnoha, J. et al. (2008) 'A mesoscopic study on historic masonry', *Structural Engineering & Mechanics*, 30(1), pp. 99–117. doi: 10.12989/sem.2008.30.1.099.

Silva, L. C., Lourenço, P. B. and Milani, G. (2017) 'Nonlinear Discrete Homogenized Model for Out-of-Plane Loaded Masonry Walls', *Journal of Structural Engineering*, 143(9), p. 4017099. doi: 10.1061/(ASCE)ST.1943-541X.0001831.

Silva, L. C., Lourenço, P. B. and Milani, G. (2018) 'Derivation of the out-of-plane behaviour of masonry through homogenization strategies: Micro-scale level', *Computers & Structures*, 209, pp. 30–43. doi: 10.1016/J.COMPSTRUC.2018.08.013.

Spahn, J. et al. (2014) 'A multiscale approach for modeling progressive damage of composite materials using fast Fourier transforms', *Computer Methods in Applied Mechanics and Engineering*, 268, pp. 871–883. doi: 10.1016/j.cma.2013.10.017.

Trovalusci, P. et al. (2015) 'Scale-dependent homogenization of random composites as micropolar continua', *European Journal of Mechanics – A/Solids*, 49, pp. 396–407. doi: https://doi.org/10.1016/j.euromechsol.2014.08.010.

Turco, C. et al. (2020) 'A digital tool based on genetic algorithms and limit analysis for the seismic assessment of historic masonry buildings', in *Procedia Structural Integrity*. doi: 10.1016/j.prostr.2020.10.124.

Computational Modelling of Concrete and Concrete Structures – Meschke, Pichler & Rots (Eds)
 ISBN: 978-1-032-32724-2

The numerical simulation of self-healing cementitious materials

A.D. Jefferson & B.L. Freeman
Cardiff University, Cardiff, UK

ABSTRACT: The behaviour of autonomic self-healing cementitious materials depends on a set of interacting mechanical, chemical and transport processes. A summary is provided of a set of component models developed to simulate these processes along with a description of a linked experimental programme of work. The component models are brought together in a coupled finite element formulation that solves Navier-Stokes and mass-balance equations for healing agent transport and uses elements with embedded strong discontinuities to represent cracks. A compact description is also provided of a new cohesive-zone damage-healing model for discrete concrete cracks. This model simulates evolving curing-fronts within the body of healing-agent using a two-level recursive time-stepping scheme. The formulation naturally accounts for the dependency of the healing response on the crack opening displacement (COD) its rate. The crack-front model is embedded in a damage-healing solution algorithm that addresses simultaneous cracking and healing, as well as re-cracking and re-healing. Validations undertaken of full coupled finite element model are discussed and an illustrative example presented. A new extension is described of the curing-front model that allows healing in wide cracks (i.e. COD>0.5mm) to be simulated. A new parametric study, for a vascular system with cyanoacrylate as the healing agent, is also presented from which a set of graphs are produced that show the expected healing in a crack for a given relative COD and time. The graphs should be useful to researchers working on cementitious self-healing materials.

1 INTRODUCTION

Interest in biomimetic construction materials has grown considerably over the past two decades, as evidenced by a number of recent review articles on the subject (De Belie et al. 2018; Ferrara et al. 2018; Fernandez et al. 2021; Xue et al. 2019). The potential of these materials to greatly improve the durability of future infrastructure has been highlighted by the Royal Society of London in a recent report (The Royal Society 2021).

Concrete is by far the most used construction material, and problems with its durability are well-documented (Gardner et al. 2018). Research to address these durability problems has included the development a range of self-healing cementitious materials (SHCMs). A number of approaches has been taken, and systems developed, to incorporate self-healing capabilities into these materials. The materials have been broadly categorised as (i) autogenic, (ii) microbial and (iii) autonomic (Van Tittelboom & De Bele 2013). It is the latter category that is most relevant to the work reported in this paper. A range of autonomic, or manufactured, healing systems have been developed to store and deliver healing agents to damage sites within structural elements. These include the use of brittle vessels (normally tubes) (Joseph et al. 2010; Minnebo et al. 2017); microcapsules (Kanellopoulos et al. 2017); and interconnected channels or 'vascular networks' (;e Nardi et al. 2020; Li et al. 2020; Shields et al. 2021).

Some research has been undertaken on the development of design and numerical models for these materials, but this has been far more limited that the work on the materials themselves (Mauludin et al. 2019). A review article in 2018, co-written by the presented authors (Jefferson et al. 2018), highlighted that there was no unified coupled model framework for simulating SCHMs. Nor was there a comprehensive experimental data set that provided all of the properties we needed to simulate the transport and mechanical behaviour of any one autonomic SCHM. We therefore embarked on a combined experimental and numerical programme of work to develop a new coupled numerical model for autonomic SHCMs. The work concentrated on understanding and developing a model for cementitious materials with embedded vascular networks, or channels, although the essential model components developed are applicable to a wider range of self-healing materials.

The experimental work from this study is reported in Selvarajoo et al. (2020a and 2020b), the transport component of the model is described in Freeman & Jefferson (2020), a specialised finite element with an embedded strong discontinuity is presented in Freeman et al. (2020) and a new cohesive zone damage-healing model is presented in Jefferson and Freeman (2022). This latter model uses homogenised

 DOI 10.1201/9781003316404-5

damage-healing variables that were developed using the results from a series of simulations undertaken with a multi-ligament model in which the ligament strengths were varied according to a statistical function. Healing is simulated from the interaction of diffuse curing fronts emanating from opposing crack faces. The model naturally allows for the dependency of the degree of healing on the crack opening displacement (COD) and its rate. It also accounts for material that re-damages and re-heals. The model is able to simulate simultaneous and continuous damage and healing with no restrictions on the timings of these events.

The present paper provides an overview of the combined experimental and numerical study. In particular, this article presents a new enhancement to the component of the damage-healing cohesive zone model that allows healing in cracks with larger CODs than those considered in the original study to be simulated accurately. This paper also presents a new study in which the damage-healing model was used to determine the degree of healing expected in cracks of different openings for different healing agents properties.

2 AUTONOMIC VASCULAR SYSTEM, GOVERNING PROCESSES AND MODEL COMPONENTS

Figure 1 shows a schematic of the autonomic self-healing system considered in this study. It comprises

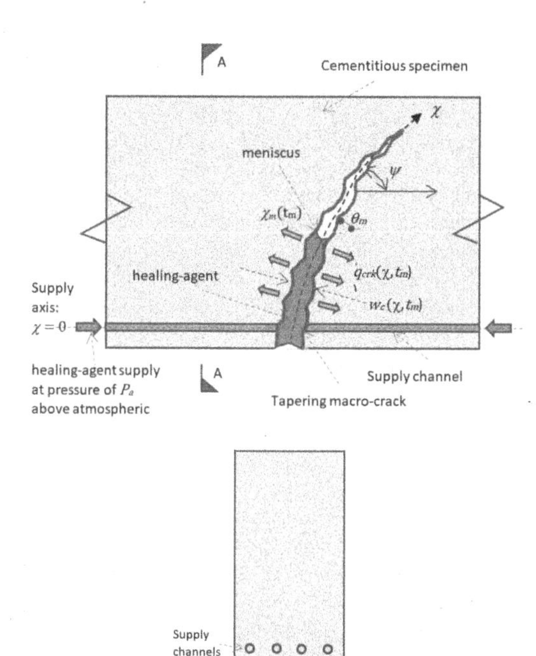

Figure 1. Healing-agent flow in a cementitious specimen with an embedded autonomic healing system.

channels embedded in a cementitious matrix (concrete or mortar) that are used to supply healing-agent to cracks. The healing-agent can be pressurised (with pressure P_a above atmospheric), but also functions without pressurisation, in which case the liquid healing-agent is drawn into a crack by capillary forces alone.

We studied the processes that govern the behaviour of this healing system in a series of experiments (Selvarajoo et al. 2020a, 2020b) and used the evidence to guide the development of our coupled numerical model. All of the tests in this series used concrete specimens and PC20 cyanoacrylate (CA) as the healing-agent. The latter was chosen because it is a relatively fast acting agent that allowed us to study simultaneous cracking–healing processes in tests of modest duration (i.e. 1 to 30 minutes).

The processes considered in the experimental programme include cracking; healing-agent release; healing-agent flow in discrete cracks and within the cementitious matrix; healing-agent curing and its effect on flow properties; mechanical healing resulting from agent curing; and re-cracking and re-healing.

The experiments, associated processes and linked model components are described below.

2.1 *Cracking*

Notched prismatic concrete beams under three-point loading and notched direct tension cube specimens were used to study coupled cracking and healing processes. These testing arrangements were selected because they allowed healing to be studied in specimens with a single discrete crack of known configuration. The applied load – displacement behaviour of the control beams and direct tension specimens (i.e. those without healing-agent) followed the pre-peak hardening/post-peak softening response that is characteristic of this type of specimen. In our finite element (FE) model, concrete cracking is simulated with a cohesive zone damage model applied to an element with embedded strong discontinuity. We chose a strong-discontinuity approach because it provides a precise description of a crack path and opening configuration, both of which are important for properly simulating coupled flow and healing processes. The element is illustrated in Figure 2 and details of its formulation and a linked crack tracking procedure are given in Freeman et al. (2020). The cohesive zone model is described in Jefferson and Freeman (2022)

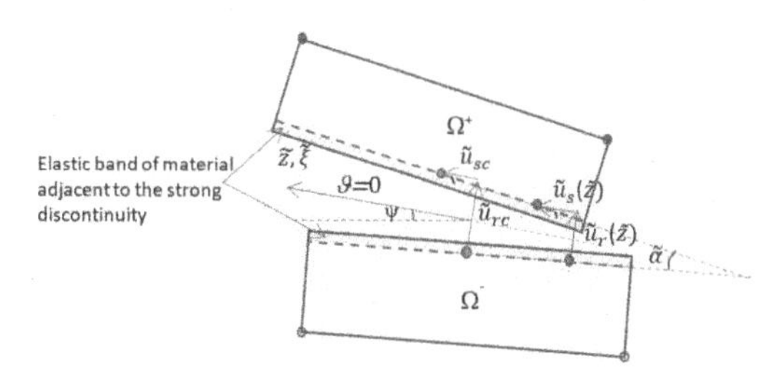

Figure 2. Element with embedded strong discontinuity.

2.2 *Flow of healing-agent in a discrete crack under capillary action and/or external pressure*

Healing-agent is released when a crack breaches one or more capillary channels and reaches a threshold value (w_{c_th}). For the system used in our experiments, this was found to be in the range 20–50 μm.

A set of experiments was undertaken to measure the flow of CA in discrete cracks, the change in viscosity over time and the effect of flow velocity on the dynamic meniscus contact angle. Discrete crack flow was investigated using concrete specimens for a range of natural crack openings (0.1–0.3 mm). The experiments used a high-speed camera and a back-lighting arrangement to capture the meniscus rise behaviour. A range of supply pressures was considered that varied from 0 (atmospheric) to 1 bar (above atmospheric). An illustration of this crack flow process is given in Figure 1. This shows that a crack is defined in terms of a convected coordinate (χ) and its COD ($w_c(\chi, t_m)$) that can vary with the reference time (t_m), measured from the start of loading. The meniscus position is defined by $\chi_m(t_m)$ and the dynamic meniscus contact angle ($\theta_m(\dot{\chi}_m, \theta_{m0})$) depends on the static contact angle (θ_{m0}) and the meniscus speed ($\dot{\chi}_m$). The flow per unit area of healing-agent from a discrete crack into the surrounding matrix is denoted by $q_{crk}(\chi, t_m)$.

We were able to reproduce the observed behaviour with good accuracy using the single fluid Navier-Stokes equations. The resulting expressions for the momentum and mass balance are as follows:

$$\begin{aligned}
&\frac{\partial(\rho\dot{\chi})}{\partial t} + \dot{\chi}\frac{\partial(\rho\dot{\chi})}{\partial \chi} = -\frac{\partial P_{hcrk}}{\partial \chi} + \rho g \sin(\psi) - \eta\dot{\chi} - \rho\dot{\chi} q_{crk} && in\ \Omega_{crk}\\
&\frac{\partial(\rho w_c)}{\partial t} + \frac{\partial(\rho w_c \dot{\chi})}{\partial \chi} + \rho w_c q_{crk} = 0 && in\ \Omega_{crk}\\
&P_{\mathrm{hcrk}} = -P_{c0}(1-\beta_s) + \frac{2\beta_m \dot{\chi}_m}{w_c} && \Gamma_m\\
&P_{\mathrm{hcr}} = P_a && \Gamma_d
\end{aligned} \tag{1}$$

where Ω_{crk}, Γ_m and Γ_d indicate the crack domain, meniscus and the part of the boundary where pressure is prescribed respectively; the superior dot denotes a time derivative, β_m and β_s are meniscus and stick slip material parameters, P_{hcrk} is the healing agent pressure in the crack, ψ is the crack inclination, ρ is the healing agent density and P_c is the capillary pressure; the dependent function for the sink-source term q_{crk} is given in the next sub-section, with the functions for the effective viscous resistance η, dynamic contact angle θ_m and time dependent viscosity μ being given below.

We demonstrated in Freeman & Jefferson (2020) that $\dot{\chi}_m$ in pressurised vascular systems could be sufficiently large for θ_m to depart significantly from its static value (θ_{m0}). In order to investigate this issue, we measured the variation of θ_m with velocity in a series of dynamic flow experiments and found that the observed behaviour was well-matched by the following expression from:

$$\theta_m = arcos(cos(\theta_{m0}) - (cos(\theta_{m0}) + 1)tanh(c_1 C_a^{c_2})) \tag{2}$$

where $C_a = \mu\dot{\chi}_m/\gamma$ is the capillary number, γ is the surface tension and the constants were calibrated to be $c_1 = 1.325$ and $c_2 = 0.350$.

The time-dependent viscosity of CA (PC20) was measured in a series of tests with a bespoke manometer that measured the flow rate in a circular channel within a mortar specimen (Gardner et al. 2014). We found that the chemo-rheological model proposed by Castro & Macosko (1980) matched Gardner et al.'s data with good accuracy. The expression adopted is as follows:

$$\mu = \mu_0 \left(\frac{\varphi_g}{\varphi_g - \varphi}\right)^{n_v} \tag{3}$$

where φ is the degree of cure, μ_0 is the initial viscosity, φ_g is the degree of cure at the gel point and n_v is an exponent which defines the rate of change of μ with φ

2.3 *Flow of healing-agent from the crack into the surrounding cementitious matrix material*

The discrete crack flow tests established that a significant quantity of CA flows from the crack into the surrounding matrix. In order to understand and quantify this process, a series of experiments were conducted that measured the sorption of CA into a concrete specimen through a natural crack surface, with the capillary rise response and weight gain due to CA adsorption both being recorded. We simulated this process using macroscopic balance equations with the capillary pressure in the pores of the continuum, along with the external pressure, as the driving forces. The mass-balance equation for the healing-agent is given below:

$$\begin{aligned}
&\frac{\partial(\overline{\rho_h})}{\partial t} + \nabla \cdot \mathbf{J}_{\mathrm{h}} + \rho q_{mtx} = 0 && in\ \Omega_{mtx}\\
&\vec{\mathbf{n}} \cdot \mathbf{J}_{\mathrm{h}} = q_c && on\ \Gamma_{Nf}
\end{aligned} \tag{4}$$

where Ω_{mtx} defines the cementitious domain, Γ_{Nf} is the part of the boundary where the flux is prescribed and $\vec{\mathbf{n}}$ is a unit vector on the boundary; $\overline{\rho_h} = \rho n S_h$ is the phase averaged density (S_h is the degree of CA saturation), q_{mtx} is a source/sink term and $\mathbf{J}_{\mathrm{h}}$ is the healing-agent flux given by $\mathbf{J}_{\mathrm{h}} = -\rho K_{eff}(S_h)(\nabla P_h - \rho\mathbf{g})$ where $P_h = P_g - P_c$ is the healing-agent pressure and K_{eff} is an effective diffusion coefficient.

The balancing crack boundary flow terms in the discrete crack (q_{crk}) and matrix (q_{mtx}) are given by (2.5), as follows:

$$q_{crk} = \frac{2}{\rho} n\beta_{crk}(P_{hcrk} - P_h) = -q_{mtx} \tag{5}$$

where β_{crk} is a boundary transfer coefficient, n is the matrix porosity, P_h is the matrix pressure, noting that q_{crk} accounts for both crack faces.

2.4 *CA curing and associated healing*

One of the primary concerns of the present work was to understand curing and healing processes in cracks with transient openings. This was a challenge because the healing response is strongly dependent on the COD but prior to this work no information was available on the curing of CA with a concrete substrate or within a concrete crack. To fill this gap and gain a better understanding of the curing process, we undertook a series of experiments aimed at measuring the progression of a curing front in a layer of CA overlying a concrete cube in a sealed container. We then linked the findings from these experiments to a series of mechanical tests on specimens with different CODs during healing (Selvarajoo et al. 2020a).

The experiments showed that CA cures with a diffuse curing front that emanates from the substrate, which -in the case of a crack in a cementitious sample- is the crack wall. The curing front is illustrated below in Figure 3.

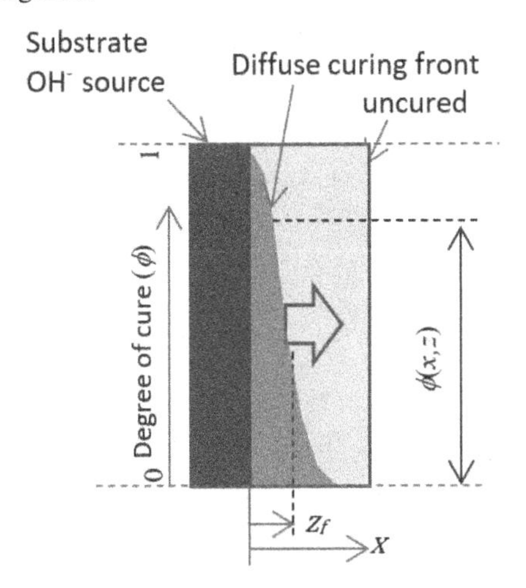

Figure 3. Progression of a diffuse curing front.

The variation of curing was derived by solving an advection diffusion equation (Freeman & Jefferson 2020) and was found to be well-matched by the following equation:

$$\varphi\left(x, z_f\right)=\frac{1}{2}\left(1-\tanh \left(\frac{x-z_f-z_{c2}}{z_{c2}+\sqrt{z_f / z_{c1}}}\right)\right) \tag{6}$$

where x is defined in Figure 3, z_{c1} is the curing front constant and z_{c2} is the wall factor.

The mean position of the curing front is given by the following time-dependent function:

$$z_f\left(t_c\right)=z_{c0}\left(1-e^{-t_c / t}\right) \tag{7}$$

where z_{c0} is the critical curing depth and τ is the curing time parameter.

In a situation where a crack is full of healing agent, only healing occurs and no re-damage or re-healing takes place, the degree of healing was found to be well represented by the degree of cure at the mid-line of a crack (i.e. at $x=w_c/2$) (Jefferson & Freeman 2022). This gives the following basic healing function:

$$h_f=\frac{1}{2}\left(1-\tanh \left(\frac{0.5 w_c-z_f-z_{c2}}{z_{c2}+\sqrt{z_f / z_{c1}}}\right)\right) \tag{8}$$

However, if the amount of healing agent in a crack changes during the healing period and/or healed material re-cracks and re-heals, equations (7) and (8) are no longer directly applicable. To allow for all these scenarios, a new approach was developed that uses a crack propagation variablez that is a convolution integral formed from the product of equation (7) and the rate of change area of curing material. This convolution integral is solved numerically using a two-level recursive scheme. In addition, a set of homogenised healing variables are introduced. The general form of these were derived from the results of a series of simulations with a multi-ligament model that considered the variation of virgin and re-healed material strengths across a representative crack area, along with some thermodynamic considerations.

The variables comprise the virgin healing parameter (h_v), the current healing parameter (h), the effective healed material relative displacement ($\mathbf{u}_h$), the effective healed relative-displacement (ζ_h) and the associated re-damage variable (ω_h). The latter is not a conventional damage variable since it increases with damage but decreases with re-healing.

One of the main findings of the ligament model simulations was that, if the healed material within a representative crack area is plotted against a normalised area variable arranged in order of increasing strength, discrete blocks of healed material develop across the area when the crack is opened continuously during healing. The number and frequency of these blocks tend to increase and decrease respectively as the damage-healing process proceeds.

3 SIMULATING CURING IN WIDE CRACKS

Design serviceability crack limits for concrete structures range from 0.1mm for liquid retaining structures to 0.4mm for structural elements in a humidity-controlled environment. For this range of cracks, the curing front functions (6) and (7) have been shown to be reasonable (Freeman & Jefferson 2020; Jefferson & Freeman 2022); however, there may be occasions when healing in wider cracks needs to be simulated.

When larger bodies of CA cure, thermal convection can become significant, and this provides a second mechanism for transporting OH^- ions (Li et al. 2017). From the data in Tomlinson et al. (2006) and our own data, we conclude that thermal convection becomes significant for layers of CA greater than approximately 0.5 mm. We therefore propose that the following two component forms of curing distribution and curing front position functions are used for crack widths wider 0.5 mm

$$\begin{aligned} z_f\left(t_c\right) &= z_{f1}\left(t_c\right)+z_{f2}\left(t_c\right) \\ &= z_{c0}\left(1-e^{-t_c / \tau}\right)+z_{c02}\left(1-e^{-t_c / \tau_2}\right) \end{aligned} \tag{9}$$

$$\varphi\left(x, z_f\right)=\frac{1}{2}\left(1-\tanh \left(\frac{x-z_f-z_{c2}}{z_{c2}+\sqrt{\frac{z_{f1}}{z_{c1}}+\frac{z_{f2}}{z_{c12}}}}\right)\right) \tag{10}$$

The ability of equations (9) and (10) to represent the curing data of Tomlinson et al. and our own data are illustrated in Figures 4 to 6 respectively. The calibrated constants for considering Tomlinson et al.'s data for the 0.11 mm thick film of CA on glass were $\tau = 120$ s and $z_{c0} = 0.135$ mm, $z_{c02} = 0$. The values of the constants used for the 6 mm thick body of CA overlaying the concrete substrate from our experiments were $\tau = 60$ s, $\tau_2 = 420$ s, $z_{c0} = 0.1$ mm, $z_{c02} = 0.7$ mm, $z_{c1} = 25$ mm, $z_{c12} = 2.5$ mm and $z_{c2} = 0.0001$ mm. Figure 5 uses data from Selvarajoo et al. (2020b) that gave the leading edge of the curing front, which is z_D (a diffusion distance) in front of z_f, where:

$$z_D = \sqrt{z_{f1}/z_{c1} + z_{f2}/z_{c12}} \tag{11}$$

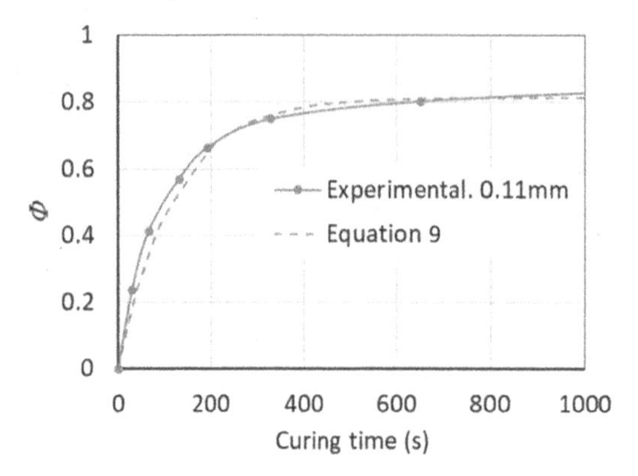

Figure 4. Total degree of cure for 0.11 mm film of CA on glass (Tomlinson et al.).

The comparisons presented in Figures 4 to 6 show that the revised functions can expand the crack widths (thickness of CA layers) to which the curing front equations are applicable.

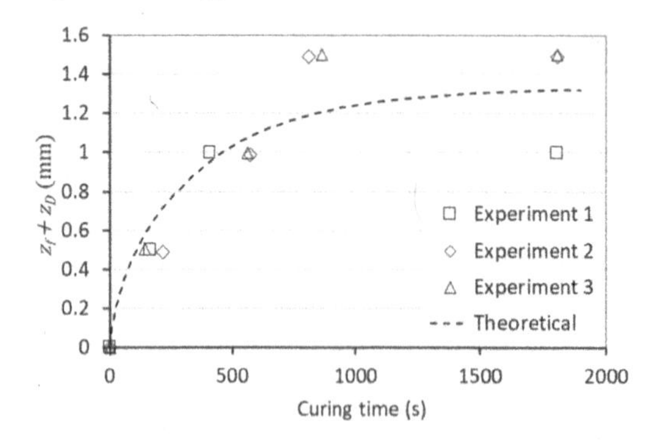

Figure 5. Curing front variable for cementitious substrate (Selvarajoo et al. 2020b).

4 HEALING PARAMETRIC STUDY

The numerical response varies with the value of the material parameters. The role of elasticity, strength and

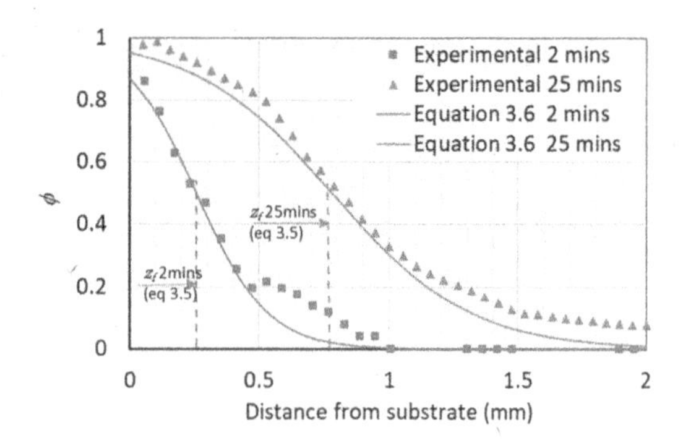

Figure 6. Degree of cure measured from cementitious substrate (Selvarajoo et al. 2020b).

fracture parameters is well understood and these are not considered further here. The new parameters that have the greatest bearing on the healing response are the critical curing front depth (z_{c0}) and the curing time parameter (τ) (Jefferson & Freeman 2022).

In the case that a crack is fully formed, static and filled with healing-agent, the amount of healing depends on the ratios (t_f/τ) and (w_c/z_{c0}), where t_f denotes the healing time. Figure 7 shows how the healing develops over time (relative to τ) for a range of relative CODs. The maximum healing predicted by the model (h_{max}) for a given (w_c/z_{c0}) depends strongly on the value of (w_c/z_{c0}), with less healing predicted for greater relative CODs. 99% of full healing is achieved at between 3 and 6 (t_f/τ), with larger CODs requiring longer to reach this limit. It is evident that there is a time lag for significant healing ($\geq 0.01h_{max}$) to start, which increases with (w_c/z_{c0}) (see the line of $0.01h_{max}$).

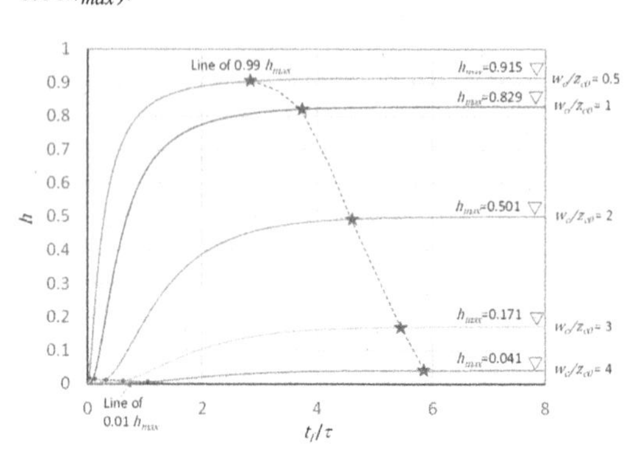

Figure 7. Degree of healing versus the relative healing time for a range of relative CODs.

The relationships shown in Figure 7 should be helpful to researchers selecting a healing agent for a particular design scenario.

5 MODEL VALIDATIONS

The full coupled model has been validated using a range of experimental data. This includes validations of the transport model (Freeman & Jefferson 2020), the element with embedded strong-discontinuity (Freeman et al. 2020) and the full coupled model with the new damage-healing cohesive zone model component (Jefferson & Freeman 2022). In addition, a number of examples are presented in these papers that use data from experimental tests on plain concrete specimens from the literature. The simulations show that the model is able to represent accurately the behaviour of the specimens as measured in the original experiments and then to predict what the behaviour would be if a similar test were undertaken with an embedded healing system. The examples are also used to explore the mesh convergence characteristics of the model. One such example, presented originally in Freeman et al. (2020), concerns an L-shaped specimen undertaken by Winkler et al. (2001), illustrated in Figure 8.

It is assumed that healing agent is supplied from the start of the experiment and the agent is released from the channels when the crack crossing the channel reaches a threshold displacement of 30 μm.

The meshes used for the analysis are shown in Figure 9, and the experimental and predicted load displacement responses of the plain concrete specimen are shown in Figure 10 along with a predicted response of the specimen with the hypothetical embedded healing system.

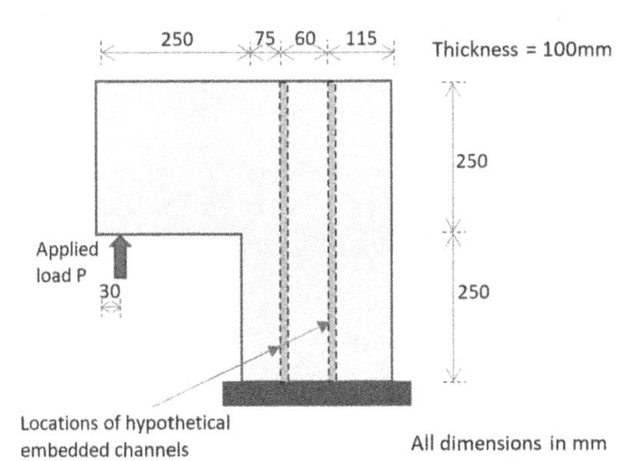

Figure 8. L-Shaped specimen of Winkler et al. with hypothetical embedded channels for the supply of healing agent (Based on Freeman et al. 2020).

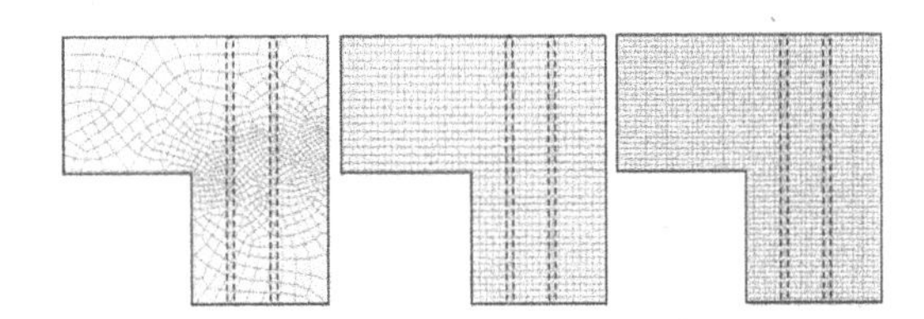

Figure 9. Meshes used for the finite element analysis (based on Freeman et al. 2020).

The degree of healing at a number of load-displacement positions (i.e. i, ii and iii, as indicated in Figure 10) are shown in Figure 11 on mesh extracts.

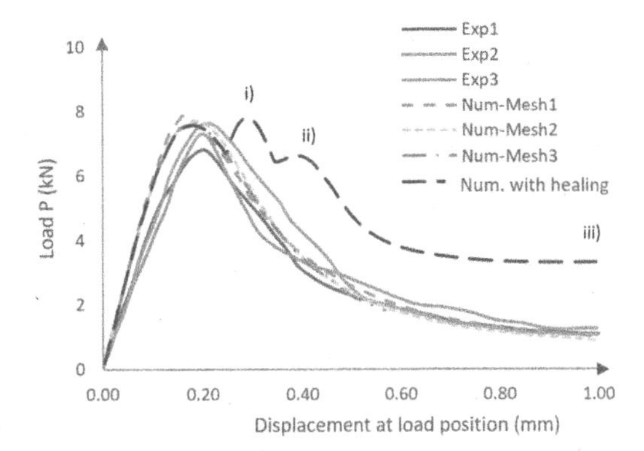

Figure 10. Experimental and numerical responses (based on Freeman et al. 2020).

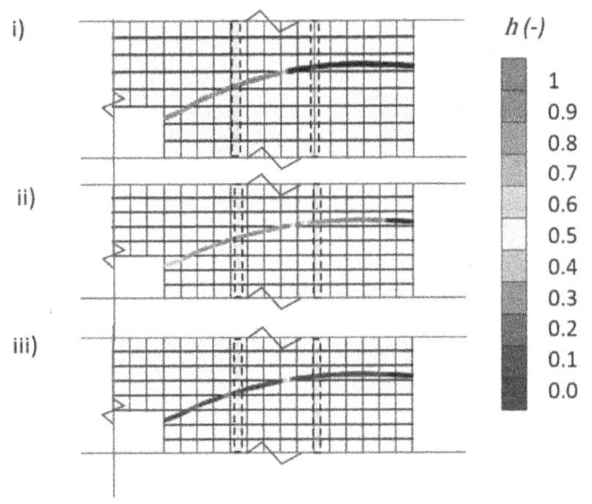

Figure 11. Degree of healing in cracks at selected load-displacement positions shown on a extract from mesh 3 (based on Freeman et al. 2020).

The predicted cracking-healing response has a similar form to those reported in Selvarajoo et al. 2020a, for specimens that have a continuous supply of healing agent from the start of loading. In such a case, the healing and damage rates tend to become equal over time, which results in the plateau in the response. In this case, the response shown in Figure 10 has two peaks, which are believed to be associated with the supply from the two sets of channels, which are breached by the crack at different times.

6 CONCLUDING REMARKS

The behaviour of autonomic self-healing cementitious materials is governed by a set of interacting processes that include, cracking, healing-agent release, healing-agent flow in discrete cracks and within the cementitious matrix, healing-agent curing and mechanical healing. These may be represented by the set of component models described in this paper that were developed using data from a linked experimental investigation. When the model components are coupled together in a finite element formulation, the combined model is able to represent the characteristic behaviour of a vascular self-healing system embedded in concrete specimens with cyanoacrylate as the healing agent. This includes the behaviour of direct tension specimens with different crack opening displacements during healing as well as that of notched beam tests under a range of loading rates and healing conditions.

Mechanical cracking and healing may be simulated effectively using a cohesive zone damage-healing approach linked to a finite element with embedded discontinuity. A Navier-Stokes crack-flow model coupled to a mass balance equation for simulating matrix flow, with degree-of-cure dependent flow properties, provides a sound basis for simulating healing agent transport.

Our healing-agent curing-front model for simulating healing in discrete cracks naturally accounts for COD and COD-rate effects on healing behaviour. The linked two-level recursive scheme, for the evolution of the curing front and the computation of the degree of healing, readily allows for re-cracking and re-healing and can simulate simultaneous damage and healing processes.

The curing front model and its linked parameters are appropriate for simulating healing in serviceability sized cracks in concrete structural elements. A two-part curing front evolution function is required for simulating healing in wider cracks (>0.5 mm).

ACKNOWLEDGEMENTS

Financial support from the UKRI-EPSRC Grant EP/P02081X/1 "Resilient Materials for Life (RM4L)" is gratefully acknowledged.

REFERENCES

Castro JM, Macosko CW (1980) Kinetics and rheolo-gy of typical polyurethane reaction injection mold-ing systems. Society of Plastics Engineers (Tech-nical Papers). 434–438.

De Belie N, Gruyaert E, Al-Tabbaa A, Antonaci P, Baera C, Bajãre D, Darquennes A, Davies R, Ferrara L, Jefferson T, Litina C, Miljevic B, Otlewska A, Ranogajec J, Roig M, Paine K, Pawel L, Serna P, Tulliani JM, Vucetic S, Wang J, Jonkers H. 2018. A review on self-healing concrete for damage management of structures, Adv. Mater. Interfaces 5, 1–28. (doi:10.1002/admi.201800074)

De Nardi C, Gardner DR, Jefferson AD. 2020. Development of 3D printed networks in self-healing concrete. Materials 13(6) 1328

Fernandez CA, Correa M, Nguyen MT. 2021. Progress and challenges in self-healing cementitious materials. J Mater Sci 56, 201–230 (2021). https://doi.org/10.1007/s10853-020-05164-7

Ferrara L, Van Mullem T, Alonso MC, Antonaci P, Borg RP, Cuenca E, Jefferson A, Ng P-L, Peled A, Roig-Flores M, Sanchez M, Schroefl C, Serna P, Snoeck D, Tulliani JM, De Belie N. 2018. Experimental characterization of the self-healing capacity of cement based materials and its effects on the material performance: A state of the art report by COST Action SARCOS WG2. Construction and Building Materials 167, 115–142. (10.1016/j.conbuildmat.2018.01.143)

Freeman BL, Bonilla-Villalba P, Mihai IC, Alnaas WF, Jefferson AD. 2020. A specialised finite element for simulating self-healing quasi-brittle materials. Adv Model Simul Eng Sci 7:32

Freeman BL, Jefferson AD. 2020. The simulation of transport processes in cementitious materials with embedded healing systems. Int J Numer Anal Meth Geomech 44:293–326

Gardner D, Jefferson AD, Hoffman A, Lark R. 2014. Simulation of the capillary flow of an autonomic healing-agent in discrete cracks in cementitious materials. Cement and Concrete Research 58,35–44. (doi:10.1016/j.cemconres.2014.01.005).

Gardner D, Lark R, Jefferson T, Davies R. 2018. A survey on problems encountered in current concrete construction and the potential benefits of self-healing cementitious materials. Case Stud. Constr. Mater. 8, 238–247. (doi:10.1016/j.cscm.2018.02.002)

Jefferson T, Javierre E, Freeman B, Zaoui A, Koenders A, Ferrara L. 2018. Research Progress on Numerical Models for Self-Healing Cementitious Materials. Adv. Mater. Interfaces 5, 1–19. (doi:10.1002/admi.201701378)

Jefferson AD, Freeman BL. 2022. A crack-opening-dependent numerical model for self-healing cementitious materials. Int J Solids Struct. *Under Review (minor revisions)*

Joseph C, Jefferson A, Isaacs B, Lark R, Gardner D. 2010. Experimental investigation of adhesive-based self-healing of cementitious materials. Magazine of Concrete Research 62 (11), 831–843. (doi:/10.1680/macr.2010.62.11.831)

Kanellopoulos A, Giannaros P, Palmer D, Kerr A and Al-Tabbaa A. 2017. Polymeric microcapsules with switchable mechanical properties for self-healing concrete: synthesis, characterisation and proof of concept Smart Materials and Structures Smart Materials and Structures, 26(4)

Li YJ, Barthès-Biesel D, Salsac A V. 2017. Polymerization kinetics of n-butyl cyanoacrylate glues used for vascular embolization. Journal of the Mechanical Behavior of Biomedical Materials 69, 307–317. (doi:10.1016/j.jmbbm.2017.01.003).

Li Z, Souza LRD, Litina C, Markaki A, Al-Tabbaa A 2020. A novel biomimetic design of a 3D vascular structure for self-healing in cementitious materials using Murray's law. Materials & Design. 190:108572

Mauludin LM, Oucif C. 2019. Modeling of Self-Healing Concrete: A Review. J. Appl. Comput. Mech. 5(3), 526–539.(doi: 10.22055/jacm.2017.23665.1167)

Minnebo P, Thierens G, De Valck G, Van Tittelboom K, De Belie N, Van Hemelrijck D, Tsangouri E. 2017. A novel design of autonomously healed concrete: Towards a vascular healing network. Materials 10, 1–23. (doi:10.3390/ma10010049)

Selvarajoo T, Davies RE, Freeman BL, Jefferson AD. 2020a. Mechanical response of a vascular self-healing cementitious material system under varying loading conditions. Construction and Building Materials 254, 119245. (doi: 10.1016/j.conbuildmat.2020.119245)

Selvarajoo T, Davies RE, Gardner DR, Freeman BL, Jefferson AD. 2020b. Characterisation of a vascular self-healing cementitious materials system: flow and curing properties. Construction and Building Materials, 245, 118332. (doi: 10.1016/j.conbuildmat.2020.118332)

Shields Y, De Belie N, Jefferson T, Van Tittelboom K. 2021. A review of vascular networks for self-healing applications. Smart Materials and Structures (10.1088/1361-665X/abf41d)

The Royal Society. 2021. Animate materials. The Royal Society, London, UK

Tomlinson SK, Ghita OR, Hooper RM, Evans KE. 2006. The use of near-infrared spectroscopy for the cure monitoring of an ethyl cyanoacrylate adhesive. Vibrational Spectroscopy. 40(1), 133–141. (doi:10.1016/j.vibspec.2005.07.009)

Van Tittelboom K, De Belie N. 2013. Self-healing in cementitious materials-a review. Materials (Basel), 6(6), 2182-2217. (doi:10.3390/ma6062182).

Xue C, Li W, Li J, Tam VWY, Ye G. 2019. A review study on encapsulation-based self-healing for cementitious materials, Struct. Concr. 20 198–212. (doi:10.1002/suco.201800177)

Winkler B, Hofstetter G, Niederwanger G. Experimental verification of a constitutive model for concrete cracking. 2001. Proceedings of the Institution of Mechanical Engineers Part L Journal of Materials Design and Applications. 215(2):75–86

Computational Modelling of Concrete and Concrete Structures – Meschke, Pichler & Rots (Eds)
 ISBN: 978-1-032-32724-2

Surface and size effects on elasticity and fracture

G. Pijaudier-Cabot, D. Toussaint, M. Pathirage & D. Grégoire
Universite de Pau et des Pays de l'Adour, E2S UPPA, CNRS, TotalEnergies, LFCR, Anglet, France

R. Vermorel
Universite de Pau et des Pays de l'Adour, E2S UPPA, CNRS, TotalEnergies, LFCR, Pau, France

G. Cusatis
Northwestern University, Evanston IL, USA

ABSTRACT: In solid mechanics, size effect is very often observed. This lecture provides a brief overview of two sorts of size effect: size effect on the structural strength observed at the macro-scale and size effect on elasticity and fracture observed at the micro-scale on porous materials. Both size effects are investigated with the same methodology, that is the help of up-scaling techniques : lattice approaches at the meso-to-macro scale and molecular mechanics at the nano-to-micro scale. These two up-scaling techniques provide ways to account for material heterogeneities and for surface effects that are at the origin of these size effects. In the two examples discussed in this contribution, accurate constitutive relations and continuum models at the macro-scale remain a very open issue, without the help of up-scaling. The methodology and results discussed in this paper may enlighten future extended continuum theories.

1 INTRODUCTION

Size effect is a very usual feature in the field of mechanics. Aside from statistical effects, size effect has been related to fracture mechanics over a century ago, when engineers and physicists observed that the apparent strength of a material, as provided by experiments was several orders of magnitude less than that predicted by solid state physics (atomistic considerations). The answer given to this discrepancy was that the material contains defects and that the defect size controls the apparent strength of the material. In linear fracture mechanics (LEFM), the apparent strength is proportional to the toughness of the material and inversely proportional to the size of the defect (crack length) to the power $1/2$.

In concrete and other quasi-brittle materials, size effect on the apparent strength originates from a transition: between strength of materials and LEFM. It is due to the interaction between a fracture process zone of finite size and the size of the structure. This type of size effect – denoted as structural size effect – is very well documented in the literature, see e.g. Bažant & Planas 1998. Cohesive fracture models or regularized continuum models capture this size effect controlled by the size of the internal length entering in these models, that defines also the size (width or length) of the fracture process zone.

Size effect occurs also at a much smaller scale, for instance in microporous materials. Micro-porous materials (with pore sizes under 2nm typically, according to the IUPAC classification) are widely encountered in chemical engineering, novel material design, manufacturing and pharmaceutical industries, and construction. Zeolites, activated carbon, hydrated cement, construction materials and some rocks are among these materials. The behavior of micro-porous materials is the result of the mechanical response of pore walls made of a few layers of atoms. Their elastic response is different from that of the bulk material, and size effects on the elastic response of these pore walls are expected (Liang et al. 2005). Fracture characteristics ought also to be revisited as they are different from those observed in bulk materials (see e.g. Shimada et al. 2015). This kind of size effect is neither due to the defect size nor to the crack length, but to the interaction between the surface of the material and the bulk material. It is a "surface" effect.

These two size effects are investigated with the help of up-scaling techniques. We start by considering structural size effect and focus on lattice modeling; this approach being viewed as a way to up-scale responses from the meso-scale where material heterogeneities are explicitly described to the macro-scale. Then, we consider the atomistic scale and use molecular mechanics to perform the up-scaling to a continuum response. Although the atomistic model is simplistic, it allows to exhibit surface effects, both on elasticity and fracture.

DOI 10.1201/9781003316404-6

2 STRUCTURAL SIZE EFFECT AND MESO-SCALE LATTICE-BASED UP-SCALING

2.1 *Database on size effect experiments*

In the literature, there are very few experimental results dealing with size effect on geometrically similar specimens for the same material with various geometries. To our knowledge, two databases are available, due to Bažant and Hoover (2014) and to Grégoire et al. (2013). In the experiments of Grégoire et al. (2013), three point bending experiments have been performed on specimens made of mortar. Four different sizes and three different geometries are considered: half-notched specimens, short-notched specimens and specimens without any notches (Figure 1).

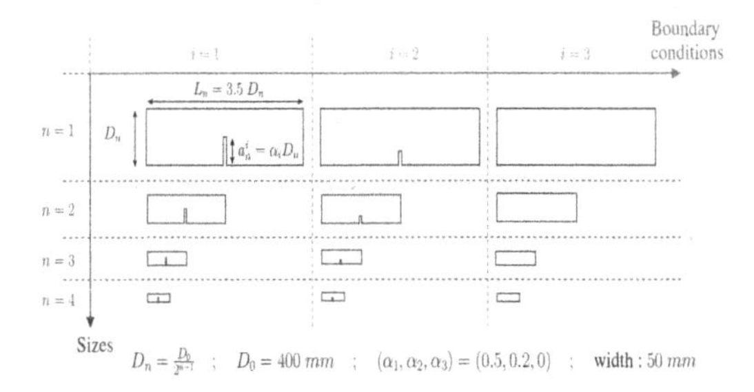

Figure 1. Geometries tested by Grégoire et al. (2013).

The important feature in these experiments is that notched and unnotched specimens, which exhibit very different size effects, have been tested on the same material. Figure 2 shows the size effect on the nominal stress at failure for the three geometries considered. Note that experimental data compare quite well with the universal size effect law proposed by Bažant and Yu (2009).

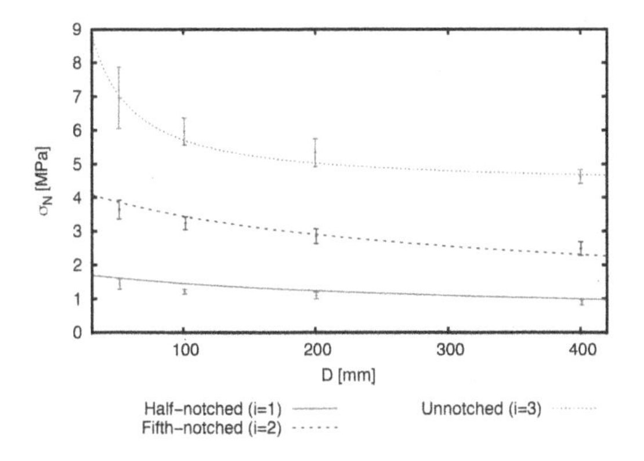

Figure 2. Size effect on the nominal stress obtained experimentally by Grégoire et al. (2013).

2.2 *Lattice modelling of structural size effect*

In a continuum setting, the description of such a structural size effect relies on constitutive relations that are regularized with respect to strain localization due to strain softening. The key ingredient is the internal length which is a model parameter and size effect originates form the ratio of this length to the size of the structure.

Standard non local (e.g. damage) models are capable of capturing it as far as the effect of size on the nominal stress is concerned, this has been documented many times in the literature (see for instance Le Bellego et al. 2003). The challenge, however, is to be able to capture the entire post peak responses, for each size and for each geometry. Grégoire et al. (2013) pointed out this difficulty and also the shortcomings of the standard non local continuum damage model to this respect. Later on, it was also demonstrated that the lattice approach indeed provided a very accurate description on structural size effect with respect to the nominal stress, and also with respect to the description of the entire experimental responses, for notched and unnotched beams of various sizes. In the foregoing, we shall illustrate this result with two different lattice models: the lattice model formed by discrete structural elements (Grassl & Jirasek 2010) and the lattice discrete particle model (Cusatis et al. 2011). For more in-depth considerations about lattice modelling applied to quasi-brittle heterogeneous materials, the reader may refer to the recent review by Bolander at al. (2021).

In the model proposed by Grassl and Jirasek (2010), a heterogeneous material is described by beam elements connecting the nodes of a discretization (Figure 3). The material heterogeneities are described explicitly, provided their size is above a fixed threshold and the interface between the matrix and the inclusions is endowed with a specific response.

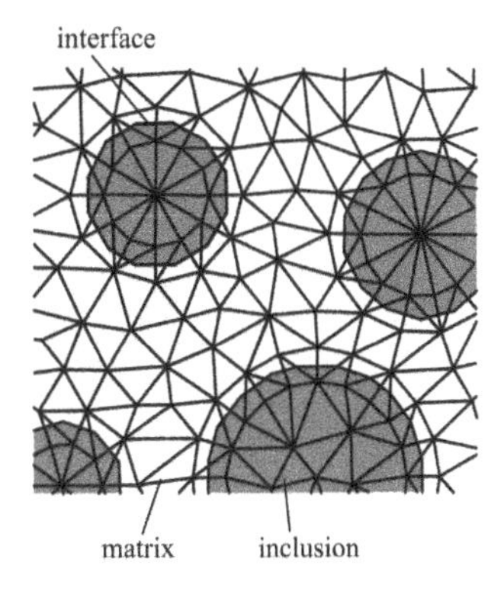

Figure 3. Discrete description of a heterogeneous material (after Grassl et al. 2012).

The mechanical response relies on a continuum damage isotropic model with an elliptical strength envelope (Figure 4). The tensile response is also illustrated on this figure. It is defined by a stress versus crack opening relationship, the crack opening being related to the tensile strain and the length of each element.

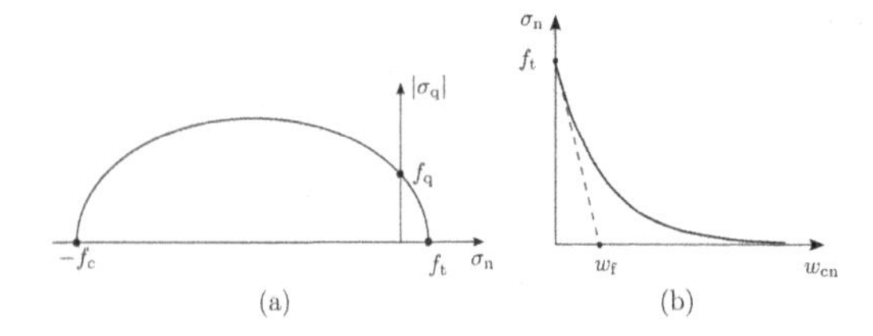

Figure 4. Lattice element response: strength envelope (left) and tensile response (right), after Grassl et al. (2012).

In the model, a meso-scale fracture energy is introduced, defined as the area under the curve in Figure 4b).

Figure 5 shows the results obtained with this lattice approach for the half-notched and unnotched beams. The model parameters entering in the lattice model have been fitted on separate experiments for the elastic constants. The nonlinear response is mainly controlled by the meso-scale tensile strength and fracture energy which are kept the same for all computations.

It can be observed that the lattice model provides a very good description of the various experimental responses, for all sizes and for all geometries.

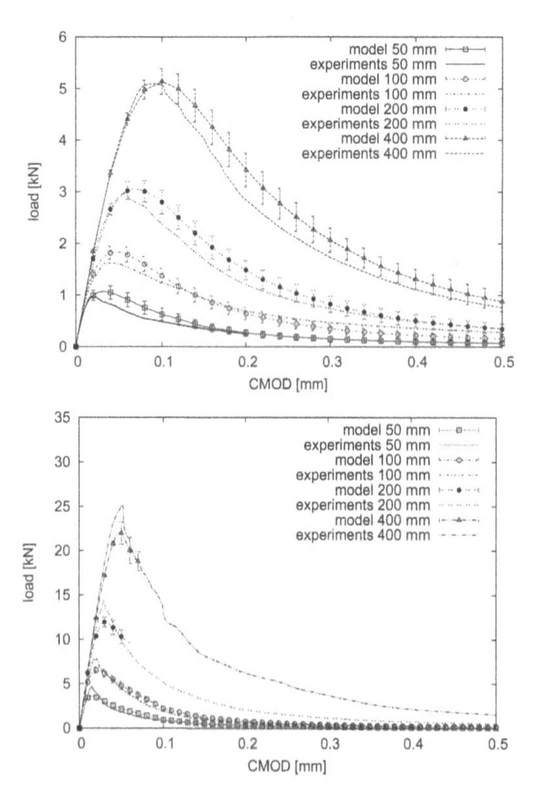

Figure 5. Load – CMOD responses for various sizes and notched (top) and unnotched (bottom) specimens: comparison between experiments and lattice model (after grassl et al. 2012)

The lattice particle discrete model (LPDM) relies on a slightly different approach for constructing the lattice model equivalent to a continuum. Concrete is discretized by placing nodes at the volume centroids of each aggregate particle above a specified size threshold. Elements are defined by the edges of the Delaunay elements constructed from the nodal points (Figure 6). The elements represent the combined actions of the aggregates and intermediary matrix in between the aggregates. The nature of interaction captures tensile failure with strain softening and it has been enriched to account for the effects of triaxial stress conditions. For more details on this model, see e.g. the contribution by Pathirage et al. (2022) in this volume.

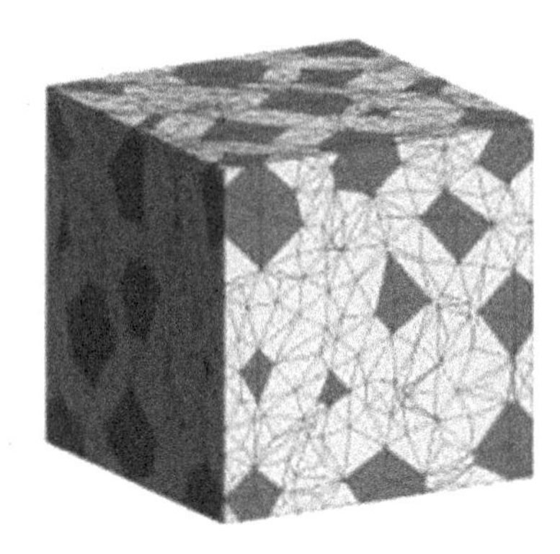

Figure 6. Discrete description of concrete according to LPDM.

Same as for the previous lattice model, the LPDM approach allows a very good description of size effect. Figure 7 is an illustration fort short-notched specimens of 4 various sizes.

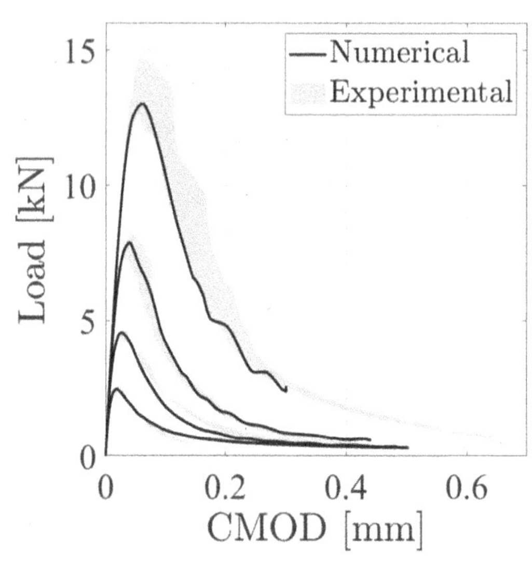

Figure 7. Description of size effect by LPDM for the short-notched specimens (after Pathirage et al. 2022).

In these two modelling approaches, the lattice model lies at the meso-scale. The lattice model can be viewed as an up-scaling technique from the meso-scale to the macro-scale. Heterogeneities on the material are described explicitly and fracture is captured at the meso-scale. It is certainly the combination of these two features that provides, after up-scaling, a broader capability to capture size effect and the complete post-peak responses at the same time. Continuum models at the macro-scale could probably yield similar results, at the price of a greater complexity of the constitutive relations, however. To our knowledge, this has not been completely achieved yet in an entire satisfactory way.

3 SURFACE EFFECT AND MOLECULAR MECHANICS-BASED UP-SCALING

We turn now to a much smaller scale and consider surface and size effects in micro-porous materials. The major common denominator in these materials is that the physics that govern the material response should be considered at the nanometre scale.

Let us look, as an example, at the comparison of the elastic properties as obtained from standard homogenization techniques and experimental data on activated carbon (Perrier 2015): the bulk modulus of a typical activated carbon with micro-porosity of 50% (pore diameters less than 2 nm) is in the range of 0,5 GPa. The bulk modulus of carbon being in the range of 30 GPa, standard homogenization techniques (e.g. the self-consistent method) yield a bulk modulus of the porous material in the range of 4 GPa. There is a gap of one order of magnitude between experiments and modelling.

A simplistic calculation shows that for a porosity of 50% and a pore size in the range of 1nm, the thickness of the solid walls in between pores should be of the order of 1nm, three to four atomic diameters. In such thin walls, atoms experience a modified interaction environment compared to the bulk state. This may explain the above discrepancy. In a broader context, the same phenomenon occurs at the surface of solids, and it induces a surface effect.

At the continuum level, surface effects have been mostly introduced following the concept of surface/interface stress initiated by Gibbs (see e.g. Xia et al. 2011). Surface stresses are qualitatively meant to result from the fact that atoms sitting on the surface of a material do not interact in the same way as in the bulk phase. Hence, forces in the plane of the free surface develop, which do not exist in the bulk state. These forces represent the mechanical effect of the modified interaction environment.

Some homogenization techniques (relating the elasticity of the porous material to that of the skeleton and to the void fraction) introduce such surface stresses. To some extent, these homogenization schemes are similar to the one devised many years ago by Benveniste (1985) in which an imperfect interface is introduced in a composite material (see also Brach et al. 2016). As a consequence, given materials with the same void fraction, the elastic properties depend not only on the shape of the voids but also on their size, as the surface to volume ratio defines the importance of surface forces. Figure 8 shows such an example taken from Duan et al. (2005).

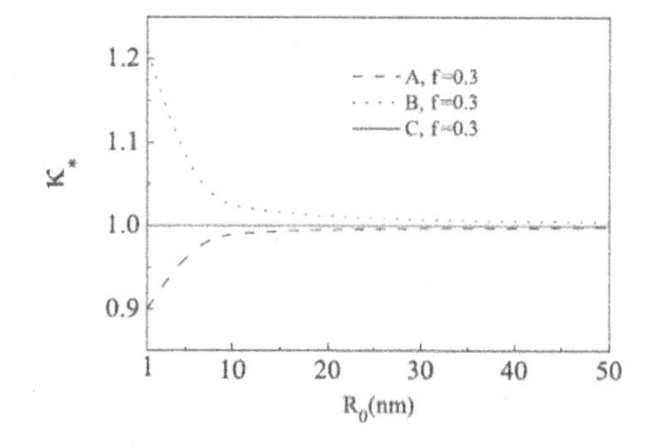

Figure 8. Bulk modulus of a porous material containing spherical voids as a function of the void radii (after Duan et al. 2005).

Connections between surface stresses and the modified energy environment at the atomistic scale depend strongly on the organization of the atoms. We shall consider here the simple model problem of a CFC Lennard-Jones crystal.

3.1 *Description of the model problem*

We are interested in elucidating how the free surface influences the overall mechanical properties (described in terms of elastic moduli, stress distribution and surface energy) of simple model systems when the characteristic size (thickness) of the system is varied. We shall carry out the analysis on a CFC Lennard-Jones crystal as shown below:

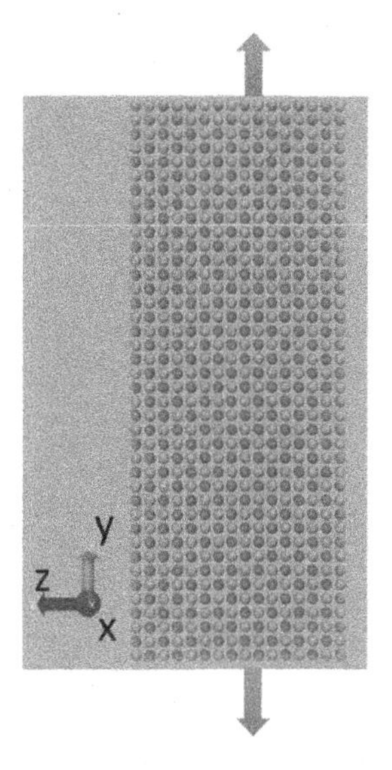

Figure 9. Model CFC structure made of 32 layers of atoms subjected to uniaxial tension.

Periodic boundary conditions are used in the x and y direction so that we are looking at an infinite plate of finite thickness, expressed as a function of the number of layers of atoms. A relative displacement in the y direction is applied so that uniaxial tension is generated in the specimen. Recall that the Lennard Jones potential reads:

$$U(r)=4U_0[\left(\frac{d}{r}\right)^{12}-\left(\frac{d}{r}\right)^{6}] \quad (1)$$

where $U(r)$ is the potential (potential energy) between two atoms lying at distance r (from which the atomic interactions are derived), U_0 and d are two constants. In the foregoing, all the results will be expressed in the Lennard-Jones coordinates, meaning that displacements are divided by d and energies are divided by U_0.

An important issue in such molecular calculations is that, prior to any loading, the system of atoms is indeed at equilibrium. This is performed by minimizing the potential energy of the system, letting the atoms move from an initial arbitrary regular configuration until equilibrium is reached. Minimization is performed in two steps: first the system is considered to be infinite in the z direction with periodic boundary conditions applied in the x, y, z directions and without any load. Upon minimization, we obtain the bulk configuration of the material. Second, periodic boundary conditions are removed in the z direction and a second minimization step is performed. Upon minimization, we have an infinite plate with finite thickness. Then, displacement boundary conditions are applied in the y direction. The relative vertical displacement between the top and bottom boundaries of the periodic cell is increased step by step. At each step, minimization is performed again so that equilibrium is reached for each deformed configuration. It should be underlined that in this simplistic model problem calculations are carried out without any effect of the temperature, meaning that the temperature is 0k. All calculations have been performed with LAMMPS.

3.2 *Global elastic response*

We start the analysis by looking at the overall elastic response of the plates subjected to tension. For each configuration at equilibrium, global quantities can be obtained: the vertical relative displacement divided by the height of the cell provides an average tensile strain ε_{yy}, the induced relative displacements between the cell boundaries in the x direction divided by the cell size provides the average strain ε_{xx} at the boundary, and within the thickness of the plate, relative displacement divided by the thickness yields strain ε_{zz}. The average pressure distribution required to elongate the plate corresponds to the tensile stress σ_{yy}. On average, all other stresses and strain components are zero.

Hence, we may extract from the calculations the average elastic parameters for each value of the thickness of the plate. We should underline that linear elasticity has been recovered at the "macro-scale". In other words, all the responses are linear as a function of the applied relative displacements which remained small.

Figure 10 shows the evolution of the Young's modulus with the number of atomic layers and Figure 11 shows the evolution of the Poisson's ratios defined as:

$$\upsilon_{xy}=-\frac{\varepsilon_{xx}}{\varepsilon_{yy}} \text{ and } \upsilon_{zy}=-\frac{\varepsilon_{zz}}{\varepsilon_{yy}} \quad (2)$$

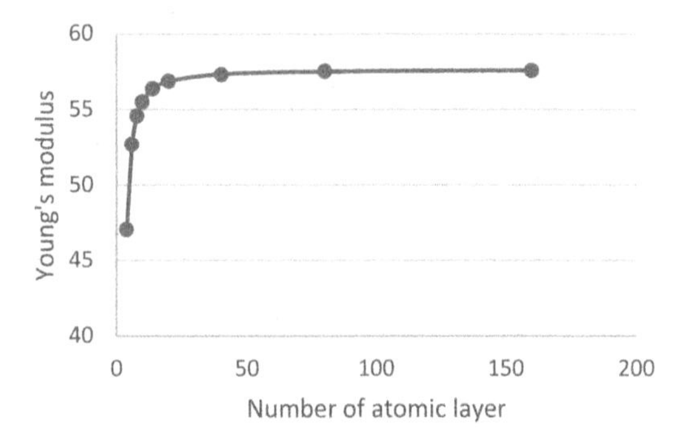

Figure 10. Evolution of the Young's modulus in the y direction as a function of the thickness of the plate.

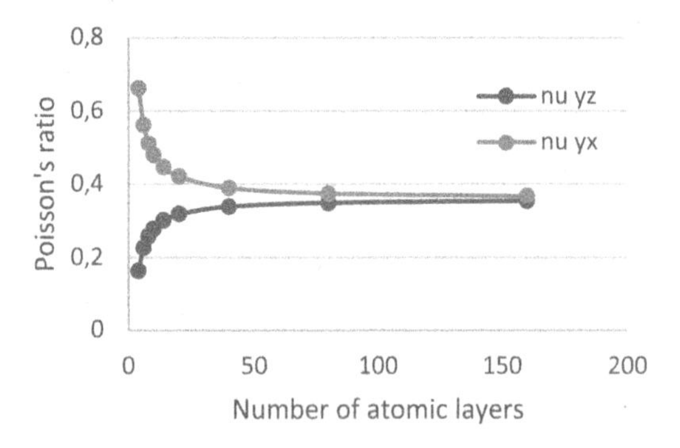

Figure 11. Evolution of the Poisson's ratios as a function of the thickness of the plate.

These two results show that (i) the Young's modulus decreases when the thickness of the plate is decreasing; this trend is consistent with the observations made by Perrier et al. (2015) on porous carbon, and (ii) that the plate is no longer isotropic when the thickness is small. The two Poisson's ratio are different, moreover, their evolution is quite different. It may be surprising to observe that the Poisson's ratio υ_{xy} is larger than 0.5. recall that because the material is no longer isotropic this limit does not necessarily apply.

A complete 3D series of applied displacements on the plate would be required in order to have access to the full average elastic operator. We may use, however, the symmetry of the crystal and infer that x and y directions can be substituted without changing the results and that we should have here a plate that is orthotropic (also because the principal directions of the crystal are the (x,y,z) directions). The important issue is that the elastic constants depend on the thickness of the plate, for thicknesses less than 50 atomic layers typically. In the case of carbon for instance, 50 atomic layers is in the range of 10 nm, which is not necessarily very small.

3.3 *Distribution of stresses within the thickness of the plate*

We look now for the distribution of the stresses within the thickness of the plate in order to better understand surface effects.

A first issue is to define and calculate the stress in this system. Stresses are average quantities and in molecular mechanics calculations they are very often calculated as Virial stresses and that localizes stress quantities on each atom. This works very well for an infinite system, but for a finite system, it yields a bias nearby the boundaries of the solid. This is the reason why we use here the method of planes due to Todd and co-workers (1995) that has been implemented in LAMMPS.

The method of planes uses a very classical definition of the stress, more precisely the stress vector, as the forces acting through a given surface divided by the area of the surface. The solid is cut into two pieces by a plane. The plane is discretized into small subsurfaces, called bins, and the calculation collects the interactions forces between pairs of atoms that cross this surface.

These interacting forces are summed to form a vector whose components are expressed into forces tangent to the bin and forces normal to the bin. Upon dividing these components by the area of the bin, a stress vector is obtained.

The stress distributions that are obtained according to the method of plane may exhibit severe oscillations. There are two reasons for this:

- When the bin size is very small, the sum of interacting forces crossing the bin may oscillate. For instance, if the bin is very small, no interacting forces may cross it. This is the reason why we have taken a bin size in our calculations equal at least to the spacing between two atomic layers, except for the bins near the boundaries where it is divided by 2.
- The plane cut the plate in between atoms and results depend on this location. Consider a series of planes located in the same interatomic spacing but a different distances from atoms. Depending on their position in the interatomic spacing, stresses will vary as interatomic forces may – or not – be accounted for in a specific bin moving within this interatomic spacing. What is observed, however, is that such oscillations are periodic, with a period equal to the interatomic distance. This oscillations have been smoothed by averaging the stress distributions over 10 planes with a spacing equal to the interatomic distance divided by 10.

Let us now have a look at the distribution of the stresses across the thickness of the plate. We shall illustrate the results in the case of a plate with 10 atomic layers. First, let us have a look at the state of stress in the plate prior to any loading.

Figure 12 shows the distribution of the stress components $\sigma_{xx}, \sigma_{xy}, \sigma_{yx}, \sigma_{yy}, \sigma_{xz}$ within the thickness of the plate. We may observe that the in-plane shear σ_{xy} and σ_{yx} are identically zero, that the shear component σ_{xz} oscillates with a small amplitude, and that the two stress components σ_{xx} and σ_{yy} are almost similar (the difference being that the unit cell has not the same number of atoms in the x and y directions). The oscillations of the shear stress may not be regarded as a consistent result. It is most probably generated by discretization effects: amplitudes are decreasing as the thickness of the plate increases. We also checked that the other stress components where small or equal to zero.

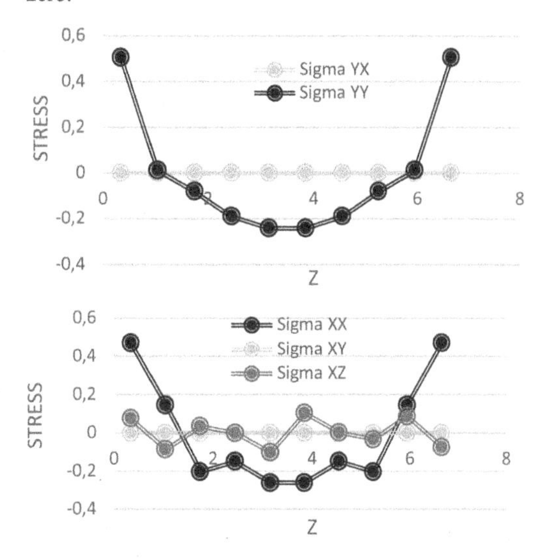

Figure 12. Distributions of stresses within the thickness for an unloaded plate. On the horizontal axis, the depth within the plate is reported (number of the layer times the interatomic spacing in the L-J coordinate system).

Surface effects are most illustrated on the distribution of stresses σ_{xx} and σ_{yy}. Although the plate is not loaded, the surface of the plate experiences tension in both x and y directions. The stress distribution, however, are self-equilibrated, meaning that the integral over the plate thickness is equal to zero.

We may wonder whether this surface effect is dependent on the thickness of the plate or not. Figure 13 shows the distribution of the in-plane stress σ_{yy} for a plate with 40 layers of atoms.

This figure shows that the amplitude of the stress distribution is not the same as in Figure 13 for a smaller thickness. Upon decreasing the thickness from 40 to 10 layers, the maximum stress is multiplied by 2.9. The compressive stress at mid-thickness changes too. We may also observe that the first two atom layers that are at the boundary experience tension only. The others experience compression.

Let us consider now what happens during the tensile loading in the y direction. We are going to compute the

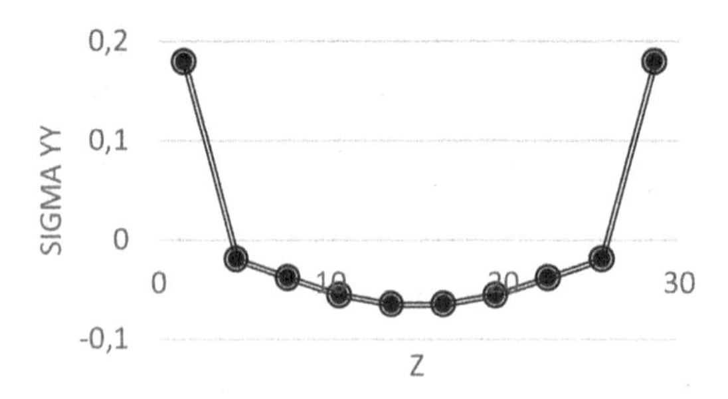

Figure 13. Distribution of the in-plane stress within the thickness of a plate with 40 layers of atoms.

same stress distribution, but for the sake of clarity, we shall illustrate results with differential distributions, that is the distribution under load minus the distribution without any load. Therefore, we will look at the effect of loading and see if it is independent from the initial state of stress or not.

Figure 14 shows the differential distributions of the in-plane stresses σ_{xx} and σ_{yy} for a plate with 10 layers of atoms. Five loading step, with equal relative displacement amplitude have been applied on the system. We may see that (i) the distribution of the in-plane stress σ_{xx} does change, but the difference is only of the order of a few percent; (ii) the in-plane stress σ_{yy} increases at each loading step, but the differential distribution is almost a horizontal line. It means that the stress distributions induced by the applied relative displacement does not exhibit any surface effect, aside from the effect observed initially on the plate without any load. We may conclude also that standard elasticity, that does not account for surface effect, could be superimposed to the initial state of stress that account for the surface effect, both being independent.

With such an assumption, we may now go back to the definition of a surface tension according to Gibbs.

3.4 *Surface tension*

Surface tension arises because the material at the surface does not exhibit the same energy environment than in the bulk. On one hand, one expects that such a quantity is a property of the surface, that does not depend on the size of the bulk phase, in our case the thickness of the plate. On the other hand, our results show that the stress induced nearby the free surface depend on the plate thickness. The conclusion is that our calculations show that the surface tension, as defined by Gibbs, is not independent from the thickness of the plate. It may become a quantity that does not depend on the bulk for plates of sufficiently large thicknesses, meaning that Gibbs' definition holds for sufficiently thick plates.

Extracting the surface tension from our calculation would therefore require running calculation on thicker and thicker plates. We saw also in the previous section that in fact, only the first two layers were experiencing

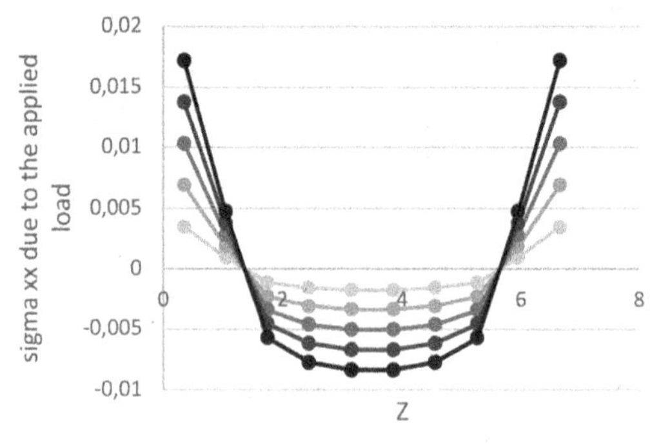

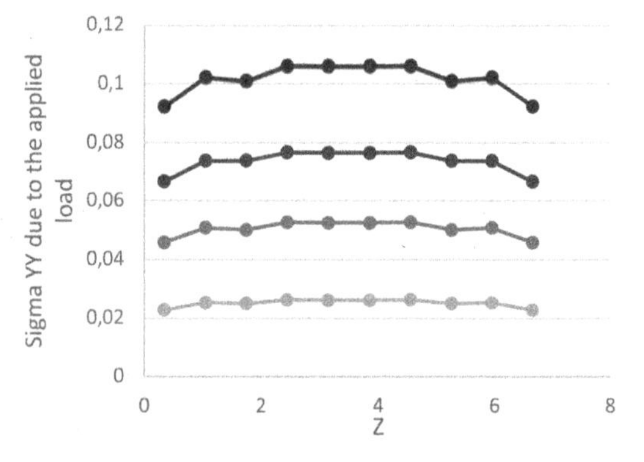

Figure 14. Distribution of differential in-plane stresses σ_{xx} (top) and and σ_{yy} (bottom) upon loading. Step 1 is omitted in the bottom figure.

tension, while the inner layers were experiencing compression. On this basis, we may devise a simplified model in order to extract the surface tension.

We assume that the plate is made of a sandwich of two elastic isotropic plates, perfectly glued to an inner elastic isotropic plate with different mechanical properties (Figure 15). The plate is subjected to a relative displacement at both ends in the y direction.

The thickness of the outer layers is denoted as b times the atomic spacing, and the thickness of the inner layer is a times the atomic spacing. The elastic constants of the inner layer are that of the material in the bulk phase. Numerically, they have been obtained by subjecting a unit cell of crystal to a relative displacement along axis y (periodic boundary conditions in the three direction). The elastic constant of the outer layers should be independent of the bulk phase. In order to obtain those, we have carried out molecular mechanics calculations on plates of thickness equal to $2b$ times the atomic spacing.

In the inner layer we have:

$$\varepsilon_{ij}^{b} = \frac{1+\nu_b}{E_b}\sigma_{ij}^{b} - \frac{\nu_b}{E_b}\sigma_{kk}^{b}\delta_{ij} \tag{3}$$

and in the outer layers we have:

$$\varepsilon_{ij}^{o} = \frac{1+\nu_o}{E_o}\sigma_{ij}^{o} - \frac{\nu_o}{E_0}\sigma_{kk}^{o}\delta_{ij} \tag{4}$$

where subscripts b and o stand for the bulk phase and the overlay (the two outer layers). The inner and outer

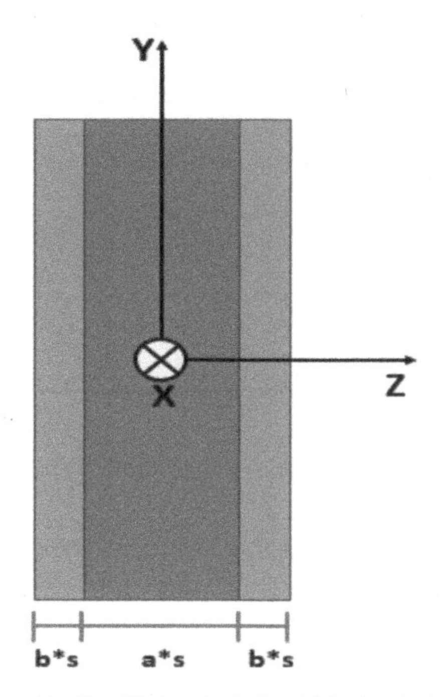

Figure 15. Simplified mechanical model for the calculation of the surface tension.

materials are assumed to be isotropic and elastic and calculations can be performed in order to obtain the stresses and strains in each layer. The in-plane total stresses σ_{yy}^{tot} and σ_{xx}^{tot} read:

$$\begin{aligned} \sigma_{yy}^{tot}(a+2b).s &= \sigma_{yy}^{b}.(a.s) + \sigma_{yy}^{o}(2b.s) \\ \sigma_{xx}^{tot}(a+2b).s &= \sigma_{xx}^{b}.(a.s) + \sigma_{xx}^{o}(2b.s) = 0 \end{aligned} \tag{5}$$

Due to the surface tension, we assume that the outer plates need to be stretched initially, before they are glued onto the inner layer. Due to this stretch, the non-zero in-plane strains are denoted as $\varepsilon_{xx}^{oi} = \varepsilon_{yy}^{oi}$. This will induce a surface stress:

$$\sigma_{xx} = \sigma_{yy} = \frac{E_o}{1-\nu_o}\varepsilon_{xx}^{oi} \tag{6}$$

It is important to remark that this surface stress is independent from the inner plate. It is *not* the actual stress that is calculated once the inner and outer plates are glued and equilibrium is obtained. The actual state of stress in the outer plates will be relaxed and at the same time the inner plate will undergo compression.

Eqs. (3-5) yield the stresses and strains in the inner and outer plates, provided the initial stretching ε^{oi} is known. This initial stretching is obtained by fitting the stress distribution across the thickness in the simplified model and the stress distribution obtained numerically. More precisely, the initial stretching is obtained under the condition that:

$$\int (\sigma_{yy})^2 dz = \left(\sigma_{yy}^{b} * (a*s)\right)^2 + (\sigma_{yy}^{o}(2*b*s))^2 \tag{7}$$

where the integral term in the left hand-side of the equation stands for the integration of the square of the stress obtained numerically according to molecular mechanics.

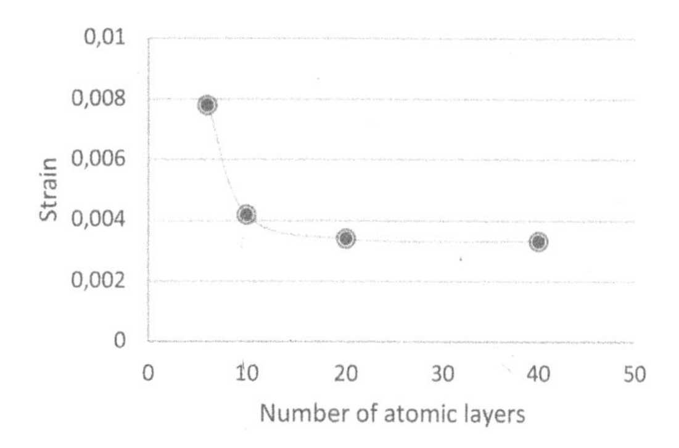

Figure 16. Initial stretch as obtained after fitting for plates of various thicknesses.

Figure 16 shows the fits of this initial stretching for various values of the overall thickness of the plate. We can see that this simplified model provides an initial stretch, equivalent to a surface tension (Eq. 6) that is not constant for very thin plates whereas it becomes almost constant when the plate is sufficiently thick. In this case, an intrinsic surface tension is obtained. It does not depend on the inner layer of material, e.g. on the thickness of the substrate on which the surface is placed. It corresponds indeed to Gibbs' definition of surface tension.

Finally, we may also check that the overall moduli and poisons ratio provided by this simple model are consistent with the numerical results. This is indeed the case.

3.5 *Surface energy*

So far, we have considered surface effects on elasticity. We may obtain also information on surface energy. The surface energy is obtained by comparing the potential energies of the system at equilibrium, when switching to the bulk phase to a finite system. According to the simple molecular mechanics model, the difference should be the energy that is consumed by the creation of a free surface. It is also related to the fracture energy, in its initial definition, meaning that the fracture energy is twice the energy needed to create a crack of unit surface (free surface).

In order to investigate the thickness effect on the surface energy, we shall start from the infinite plate and remove the periodic boundary conditions in the y direction. In this direction, the plate becomes finite – it is a strip of fixed dimension in the y direction and infinite in the x direction. Again, we may consider strips of various thicknesses and look at the influence of this parameter.

Figure 17 shows influence of the thickness of the strip on the surface energy. The thicker the strip, the higher the surface energy. The overall variation, going

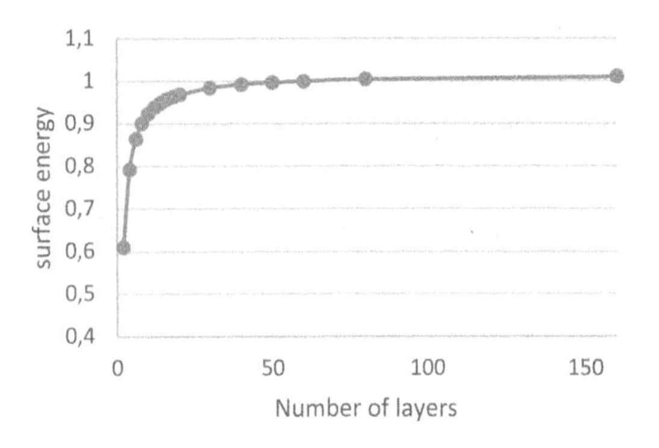

Figure 17. Evolution of the surface energy with the thickness of the plate.

from a thickness of 2 atomic layers to a thickness of 160 atomic layers, is quite large. This growth of the surface energy could have been expected. For very thin plates, many atomic interactions are missing compared to thicker plates. Upon creation of the free surface, these missing interaction do not need to be removed, the plate is easier to separate into a strip and therefore, the surface energy is lower compared to the case of a strip of large thickness.

We may now try to interpret this result in the context of a porous material. For this, we shall consider a porous 2D material containing circular voids. The unit cell of the material is hexagonal, with the void centered in the middle (Figure 18).

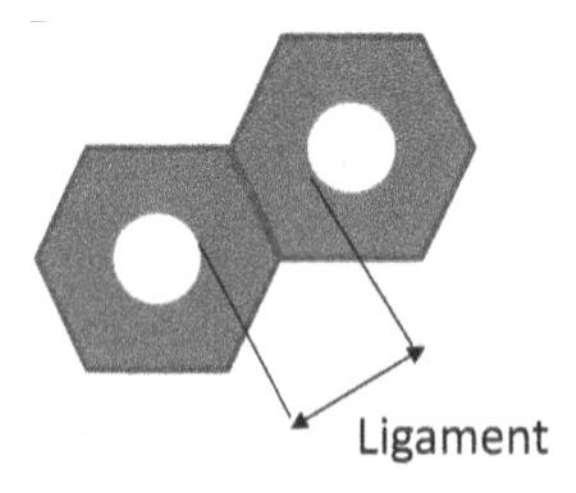

Figure 18. Ligament in a porous material made of hexagonal periodic cells.

We assume that separation occurs in the ligament shown in Figure 18. The crack will run from one void to another propagating in the ligament (which may be a quite rough assumption). For a given porosity and for a given value of the pore radius, which is related to the ligament size, we may now calculate the energy needed to create the crack, keeping in mind that the surface energy is a function of the ligament size as shown in Figure 17.

This calculation simply converts the relationship between the surface energy and the thickness, to a relationship between the surface energy and the ligament size which, for a fixed porosity, is a function of the radius of the pores. We further assume that the fracture energy is twice the energy needed for the creation of a surface of unit area and we plot the evolution of the fracture energy as a function of the pore size, at constant porosity.

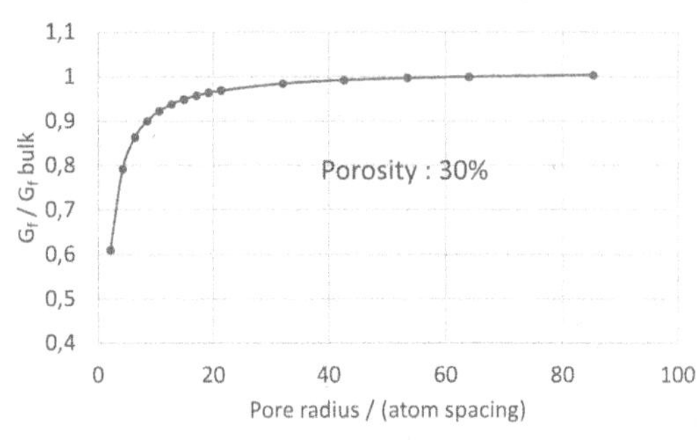

Figure 19. Evolution of the fracture energy of a porous material with the pore size, at constant porosity.

We can see on Figure 19 that the fracture energy grows with the pore size at constant porosity. The same should be expected for the Young's modulus in view of our previous results, it can be obtained with more accurate up-scaling techniques (e.g. a three phase self-consistent method).

4 CONCLUDING REMARKS

In quasi-brittle porous materials, size effect occurs over a wide spectrum of scales. In this contribution two effects operating at very distinct scales have been discussed:

- at the nano-to-micro scale size effect is due to surface effects and the special energy environment that exists near free surfaces. The size effect at stake is related to the ratio of the surface to the volume of the material.
- At the meso scale, it is due to distributed micro-cracking inherited from the presence of heterogeneities in the material. The size effect at stake is related to the ratio between the size of the zone where micro-cracking occurs – a function of the size of the heterogeneities – and the size of the structure.

Both size effects have been investigated with the help of up-scaling techniques. At the structural level, lattice approaches provide a way to account for the size of heterogeneities and therefore enable a quite accurate investigation of structural size effects. Molecular mechanics allow to up-scale the atomistic description to that of a continuum. It should be stressed that the atomistic description of a given material, heterogeneous and amorphous sometimes, may still be looked as very complex task. In many cases, it remains to be achieved.

Nevertheless, simplified nano-scale models provide qualitative trends (see also Vandamme et al. 2015), and limits beyond which existing theories may fail. The results obtained in this paper belong to this category. For instance, the definition of a surface tension, intrinsic to a surface, does not hold very thin solids.

In fact, and in the two situations discussed in this contribution, constitutive relations and models at upper scales, without the help of up-scaling, remain a very open issue. The methodology and results discussed in this contribution may enlighten future extended continuum theories.

ACKNOWLEDGMENTS

Partial financial support from the investissement d'avenir French program (ANR-16-IDEX-0002) under the framework of the E2S UPPA hub Newpores is gratefully acknowledged.

REFERENCES

Bažant, Z.P., Planas, J., (1998). *Fracture and size effect in concrete and other quasi-brittle materials*, CRC press.

Bažant, Z.P., Yu, Q., (2009). Universal size effect law and effect of crack depth on quasi-brittle structure strength, *J. Engrg. Mech. ASCE*, 135:78–84.

Benveniste, J., (1985). The effective mechanical behaviour of composite materials with imperfect contact between constituents, *Mech. Mat.*, 4:197–208.

Bolander, J.E., Elias, J., Cusatis, G., Nagai, K., (2021). Discrete mechanical models of concrete fracture, *Engrg. Fract. Mech.*, 257, 108030.

Brach, S., Dormieux, L., Kondo, D., Vairo, G., (2016). A computational insight into void-size effects on strength properties of nanoporous materials, *Mech. Mat.*, 101:102–117.

Cusatis G, Pelessone D, Mencarelli A., (2011). Lattice Discrete Particle Model (LDPM) for failure behavior of concrete. I: Theory, *Cem Concr Comp.*, 33:881–90.

Duan, H.L., Wang, J., Huang, Z.P., Karihaloo, B.L., (2005). Size-dependent effective elastic constants of solids containing nano-inhomogeneities with interface stress, *J. Mech. Phys. Solids*, 53:1574–1596.

Grassl, P., Grégoire, D., Rojas Solano, L., Pijaudier-Cabot, G., (2012). Meso-scale modelling of the size effect on the fracture process zone of concrete, *Int. J. Solids and Struct.*, 49:1818–1827.

Grassl, P., Jirasek, M., (2010). Meso-scale approach to modelling the fracture process zone of concrete subjected to uniaxial tension, *Int. J. Solids and Struct.*, 48:957–968.

Grégoire, D., Rojas-Solano, L., Pijaudier-Cabot, G., (2013). Failure and size effect of notched and unnotched concrete beams, *Int. J. Num. and Anal. Meths. Geomechanics,* 37:1434–1452.

Hoover, C.G., Bažant, Z.P., (2014). Cohesive crack, size effect, crack band and work-of-fracture models compared to comprehensive concrete fracture tests, *Int. J. Fracture*, 187:133–143.

Le Bellégo, C., Dubé, J.F., Pijaudier-Cabot, G., Gérard, B., (2003). Calibration of non local damage model from size effect tests, *Eur. J. Mech. A/So*lids, 22:33–46.

Liang, H., Upmanyu, M., Huang, H., (2005). Size-dependent elasticity of nanowires: nonlinear effects, *Phys. Rev. B*, 71, 241403(R).

Pathirage, M., Thierry, F., Cusatis, G., Grégoire, D., Pijaudier-Cabot, G., (2022). Numerical modeling of concrete fracturing and size-effect of notched beams, Proc. Euro-C, *this volume*, CRC press.

Perrier, L., (2015). *Couplage entre adsorption et deformation en milieu microporeux*, Ph.D. Dissertation, Université de Pau et des Pays de l'Adour.

Shimada, T., Ouchi, K., Chihara, Y., Kitamura, T., (2015). Breakdown of continuum fracture mechanics at the nanoscale, *Scient. Rep.*, 5, 8596.

Todd, B.D., Evans, D.J., Daivis, P.J., (1995). Pressure tensor for inhomogeneous fluids, *Phys. Rev. E*, 52, 1627.

Vandamme, M., Bažant, Z.P., Keten, S., (2015). Creep of lubricated layered nanoporous solids: application to cementitious materials, *Journal of Nanomechanics and Micromechanics*, 5, 04015002.

Xia, R., Li., X., Qin, Q., Liu, J., Feng, X.-Q., (2011). Surface effect on the mechanical properties of nanoporous materials, *Nanotechnology*, 22, 265714.

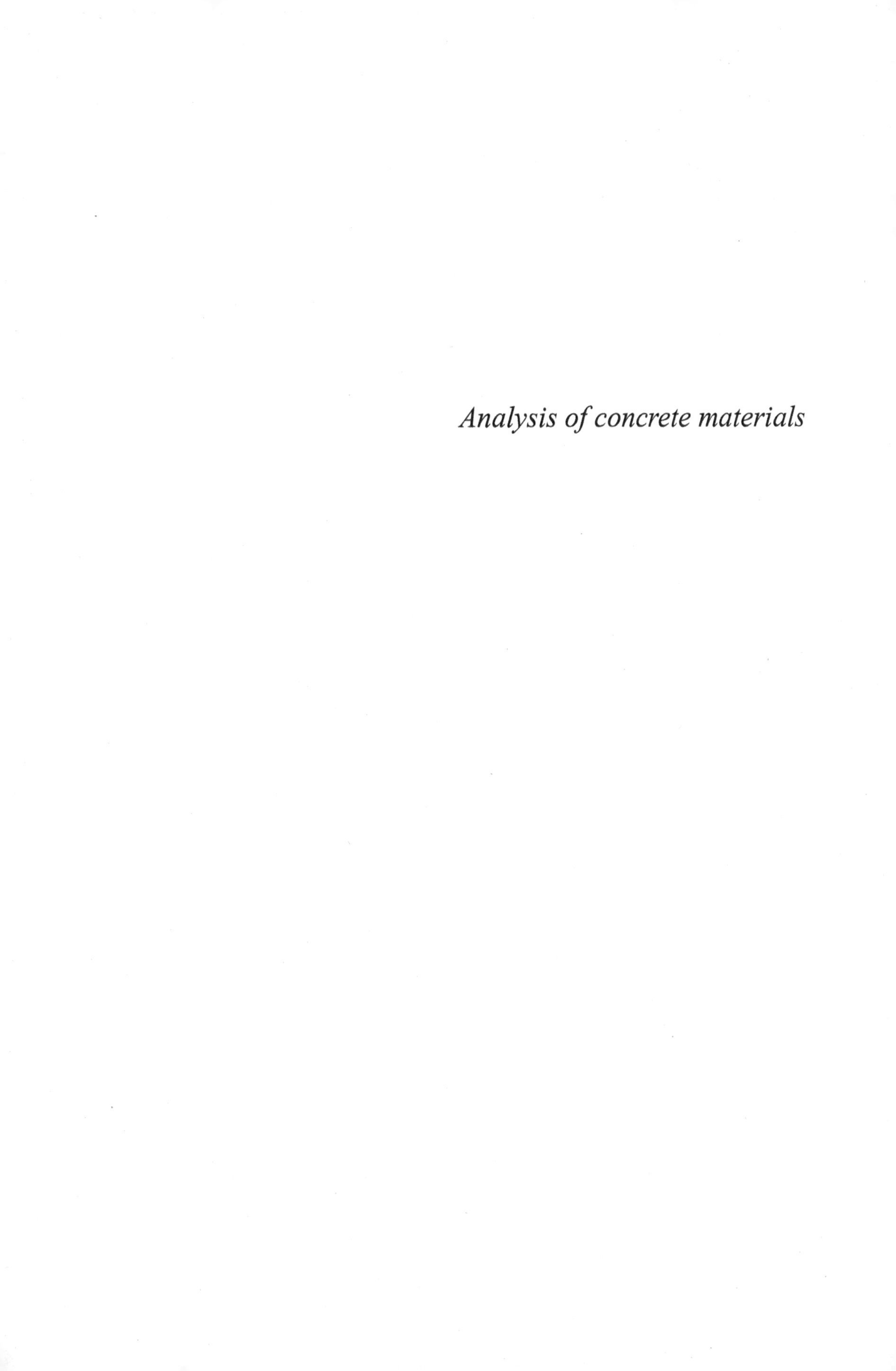

Analysis of concrete materials

Computational Modelling of Concrete and Concrete Structures – Meschke, Pichler & Rots (Eds)
© 2022 Copyright the Author(s), ISBN: 978-1-032-32724-2

Fracture of cement hydrates determined from micro-scratching tests and their modeling

J. Němeček[1], V. Šmilauer & J. Němeček[2]
Faculty of Civil Engineering, Department of Mechanics, Czech Technical University, Prague, Czech Republic
[1]*Orcid 0000-0002-5635-695X*
[2]*Orcid 0000-0002-3565-8182*

R. Čtvrtlík
Joint Laboratory of Optics, Institute of Physics Academy of Sciences, Olomouc, Czech Republic

ABSTRACT: The paper shows both experimental and numerical results for micro-scratching tests applied to main hydration products of cement paste at the scale of 10–100 μm. In the experimental part, micro-scratch tests were conducted along with scanning electron microscopy and acoustic emission measurements to reveal local fracture toughness of individual cement paste constituents. 3-D finite element model of the scratch process utilizing Griffith-type of a fracture-damage model for tension/compression failure was successfully used for replication of the experiments and identification of local cement paste strength. The tensile strength for the outer hydration product was identified as 54 MPa being about 5 times lower compared to FIB-produced micro-cantilevers at 3 μm scale (Němeček, Králík, Šmilauer, Polívka, & Jäger 2016)and about 3.6 times higher than at 500 μm scale (Zhang, Šavija, Figueiredo, & Schlangen 2017). The strong size effect can be attributed to a different number of internal defects in the cement paste microstructure.

1 INTRODUCTION

Initiation of fracture is a localized phenomenon related to small material volume that manifests itself on a higher level as an expanding crack. As proved by Němeček et al. (Němeček, Králík, Šmilauer, Polívka, & Jäger 2016) local tensile strength and fracture energy vary for individual cement paste constituents such as inner and outer products (composed of mainly C-S-H gels), portlandite or clinker at the micrometer scale. The new concept of micro-bending tests has already been introduced and gives access to unique engineering parameters of cement paste at the range from 1 to approximately 100 μm. At this scale, e.g. the inner C-S-H product attains tensile strength around 700 MPa. The high tensile strength of C-S-H becomes substantially reduced by stress concentrations around crystalline inclusions, pores and internal defects as shown in Němeček et al. (Němeček, Šmilauer, Němeček, Kolařík, & Maňák 2018). However, the experimental procedure using micro-bending tests requires specialized lab equipment such as focused ion beam microscope, it is relatively uneasy, lengthy and costly. Gaining larger statistics in a reasonable time is not feasible for most laboratories. The contribution aims to present a novel methodology for fracture testing at the scale of 10-100 μm based on micro-scratching performed with the nanoindenter over a larger representative area of the sample. Multiple scratching experiments are performed in hydrated cements with individual scratch length of several hundreds of micrometers, thus consecutively covering all microscopic phases. While mechanical testing is performed by a nanoindenter which gives information on the horizontal and vertical loads and deflections, position and identification of microscopic phases is provided by the scanning electron microscope (SEM). Along with the tests, acoustic measurements are done for detection of cracking in individual cement paste phases. Fracture toughness is evaluated for microscopic cement paste phases being slightly higher compared to usual macroscopic values. In this contribution, the initiation of the scratching process is modeled by a 3D finite element model with a smeared crack constitutive law combining both tensile and compressive strain softening and the damage mechanics. The experimental data give access to fracture toughness of individual cement paste phases and give the idea about the fluctuation of this quantity at the 10-100 μm scale. With the aid of the numerical model experimental data are matched and the cracking pattern under the indenter tip, otherwise not accessible by microscopic observations, predicted.

DOI 10.1201/9781003316404-7

1.1 *Brief summary of existing experimental data of fracture properties at micro-scale*

Nowadays, micro-scale fracture properties at the scale of 1-100 μm can be accessed only by a few experimental techniques. Focused ion beam milling technique (FIB) (Gianuzzi & Stevie 2005) can be used for fabrication of micro-beams that are loaded by a nanoindenter. Němeček et al. (Němeček, Králík, Šmilauer, Polívka, & Jäger 2016) successfully measured Young's moduli, tensile strength and supremum fracture energies of the outer C-S-H product, inner C-S-H product and CH on micro-cantilevers with triangular cross-section (≈3 μm, length ≈15-20 μm), Table 1.

Table 1. The results from micro-bending and nanoindentation tests reported in Němeček et al. 2016.

	OP	IP	CH
E (GPa)	24.9 ± 1.3	33.6 ± 2.0	39.0 ± 7.1
f_t (MPa)	264 ± 73	700 ± 199	655 ± 258
G_f^{supp} (J/m^2)	4.4 ± 1.9	19.7 ± 3.8	19.9 ± 14.4
K_c (MPa·m$^{1/2}$)	0.33 ± 0.06	0.81 ± 0.08	0.88 ± 0.28

Another way of micro-scale testing of cementitious samples was recently developed by Schlangen et al. (Schlangen, Lukovic, Šavija, & Copuroglu 2015), and Zhang et al. (Zhang, Šavija, Figueiredo, & Schlangen 2017; Zhang et al. 2018). The procedure consists of preparing small-scale cementitious samples: cubes, beams, and cantilevers with a square cross-section and edge length of 100-500 μm with a micro-dicing saw. The produced samples were used for various ranges of experiments e.g.: micro-beam three-point bending test, micro-cantilever static and fatigue tests, compression or split tests. The mechanical experiments are also supported with discrete lattice model, 3D microstructure extracted from X-ray computed tomography (XCT), and SEM images. The measured tensile strength for micro-cantilevers tests lies in the range of 15-25 MPa for 0.4 w/c ratio.

Micro-pillars fabricated by FIB and loaded with a sharp indenter were used to estimate fracture toughness of cement paste value. So far, the only micro-pillar study of cement paste was done by Shahrin and Bobko (Shahrin & Bobko 2019). They investigated micro-pillars in the C-S-H phase (mostly HD C-S-H) with several pillars diameters with volumes of 1.25-38 μm^3. The compressive strength measured was in the range of 181-1145 MPa with a presence of a strong size effect. Also, the authors reported the approximate value of fracture toughness as 0.67 MPa·m$^{1/2}$.

The experiments utilizing FIB are relatively rare, very time consuming and costly which prohibits gaining large statistical data. This disadvantage leads to searching of new approaches that can reliably produce larger measurement sets. Micro-scratching is among such techniques that do not require complicated and laborious procedures and can be done relatively easy and fast.

1.2 *Fracture toughness evaluation from a scratch test*

In standard nanoindetation, the indenter tip is brought to the contact with the specimen and loaded by increasing/decreasing vertical load. Micro-scratching is a technique in which not only vertical but also horizontal force is imposed on the nanoindenter tip. The combination of loads produces a linear scratch in the sample microstructure at the scale depending on the tip geometry and forces applied. Typically the scratch width is about 10 μm and the length is in hundreds of μm which leads to some interactions of phases compared to smaller FIB-produced samples. Thus, evaluation procedures must include reliable phase separation.

For a homogeneous-like material, Akono and Ulm derived an analytical relationship for fracture toughness estimation from the scratch test (Akono 2020; Akono, Randall, & Ulm2012). K_c was derived from the relationship between the horizontal force, F_T and indentation tip geometry and is expressed as

$$K_c = \frac{F_T}{\sqrt{2p(d)A(d)}}, \tag{1}$$

where $A(d)$ is the horizontal load bearing contact area, $p(d)$ is the perimeter and d is the penetration depth. Different indenter geometries can be taken into account considering the shape function

$$f(d) = 2p(d)A(d), \tag{2}$$

which for a spherical probe of radius R is defined as

$$f(d) = \frac{16}{3}\beta\left(\frac{d}{R}\right)d^2R, \tag{3}$$

where β is dimensionless parameter (Akono & Ulm 2012). Since the actual geometry of the indenter varies from the ideal shape, the shape function needs to be calibrated by a suitable approximation as

$$f\left(\frac{d}{R}\right) = R^3\left[\alpha\left(\frac{d}{R}\right)^3 + \delta\left(\frac{d}{R}\right)^2 + \gamma\left(\frac{d}{R}\right)\right], \quad \alpha \geq 0, \delta \geq 0, \gamma \geq 0, \tag{4}$$

where α, δ, and γ are fitting coefficients of the function.

2 EXPERIMENTAL PART

2.1 *Samples preparation and microstructure*

Pure cement paste samples were prepared from Portland cement CEM-I 42.5R with water-to-cement ratio of 0.4. The samples were cast into cylindrical molds with a diameter of 27 mm and a height of 70 mm. The samples were demoulded after one day and put into water for 120 days to achieve a high degree of hydration. Subsequently, the samples were cut into 6 mm

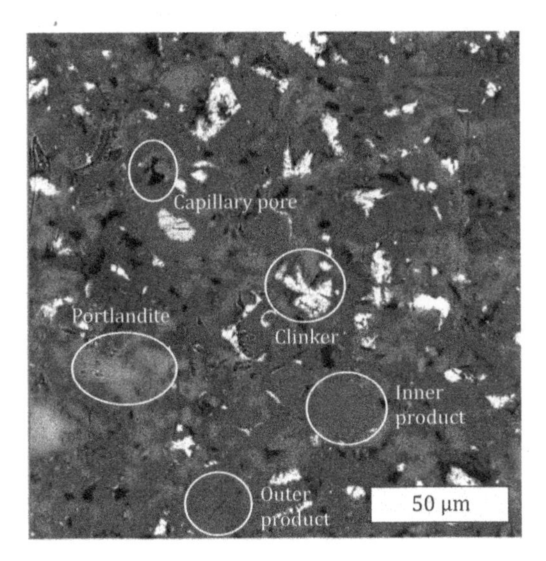

Figure 1. SEM-BSE image of cement paste microstructure.

thick slices, dried in the oven at 50°, and finely polished to achieve a smooth surface according to the procedure described in (Němeček, Lukeš, & Němeček 2020).

The microstructure at the level of cement paste (i.e. below 100 μm) is highly heterogeneous, composed mainly with outer (OP) and inner (IP) products which are rich C-S-H gel mixed with other hydrates, larger crystals of Portlandite (CH), and unreacted grains of clinker (C). Also, the capillary pores are present in the microstructure. The typical situation is shown in Figure 1.

2.1.1 Scratch testing

The scratch tests were performed with the NanoTests Vantage system (Micro Materials) with a spherical diamond tip with the radius of 10 μm. The coefficients for spherical tip with a radius of 10 μm were calibrated using Equation 4 as $\alpha = 298.1$, $\delta = 0$, $\gamma = 0$. The load function was divided into three segments. At first, the vertical force was kept at contact load 0.05 mN (0-50 μm), then in the next segment (50-200 μm) the force was linearly increased up to the maximum vertical force of 25 mN. In the last segment (200-450 μm), the maximum vertical force was kept constant. Also, the tip was moved with a constant horizontal speed of 10 μm/s for all three segments. Thus, during the measurement the penetration depth and horizontal force, F_T are recorded. Since the length of the scratch line is large, the penetration needs to be corrected by sample inclination (measured by pre-scratch scan). Totally 25 scratch tests were performed.

During the scratch test, the acoustic emissions activity was continuously recorded and analyzed using the ZEDO system (Dakel, Prague, Czech Republic). Data were measured in the frequency range of 100–1500 kHz. The sample was fixed using the low-temperature wax on the dedicated AE holder with an inbuild pre-amplifier (Čtvrtlík, et al. 2019).

2.1.2 Results of scratch testing

The wide range of the fracture toughness values from 0.1 to 2 MPa·m$^{1/2}$ were evaluated according to Equation 1. Due to high uncertainties at the beginning of the contact scratch data from ≈0-80 μm were excluded from the analysis. The rest of the scratch was used for K_c evaluation. High fluctuations connected with microstructural variation were encountered. The separation of mechanical response was done into four main phases: OP, IP, CH, and C. The large pores visible on the SEM back-scattered electron (SEM-BSE) images were also excluded from the results. The separation was done manually between two selected points, and the mean value of K_c calculated. Two factors were taken into account during the separation process. First, SEM-BSE images provided a very good overview of phases located in the scratch line (Figure 2a). The second factor was the measured AE signal (Figure 2c), which served as an auxiliary indicator of the fracture process. It was found that AE could reliably record CH cracking during the scratch test. Cracking in other phases was not encountered in the AE signal, even if cracks were observed in SEM-BSE images, which might be the influence of higher AE threshold or acoustic signal damping in disordered C-S-H phases. The results of fracture toughness are summarized in Table 2.

Table 2. The results of fracture toughness from micro-scratch tests.

	OP	IP	CH
K_c (MPa·m$^{1/2}$)	0.34 ± 0.03	0.51 ± 0.05	0.54 ± 0.09
G_f (J/m^2)	5.3 ± 0.9	8.4 ± 1.7	7.1 ± 2.6

The fracture toughness of OP ($K_c = 0.34 \pm 0.03$ MPa·m$^{1/2}$) is with the excellent agreement of un-notched micro-beam bending experiment K_c = 0.33 ± 0.06 MPa·m$^{1/2}$ (Němeček, Králík, Šmilauer, Polívka, & Jäger 2016). Furthermore, the measured values are close to the MD simulation of Bauchy et al. (2015), who evaluated the fracture properties of C-S-H gel as K_c = 0.37 ± 0.03 MPa·m$^{1/2}$. The K_c values for both IP and CH were lower in scratch experiment than for un-notched micro-beam. This is caused by interactions of phases in micro-scratching, since IP and CH zones are relatively small compared to the scratch dimensions. Although, the separation of indents was done to the groups of OP, IP, CH and C, some inevitable interactions appear for phases that occupy volume smaller or closer to the scratch interaction volume. Thus, lower values of K_c in comparison to micro-beams is natural for the smaller phases.

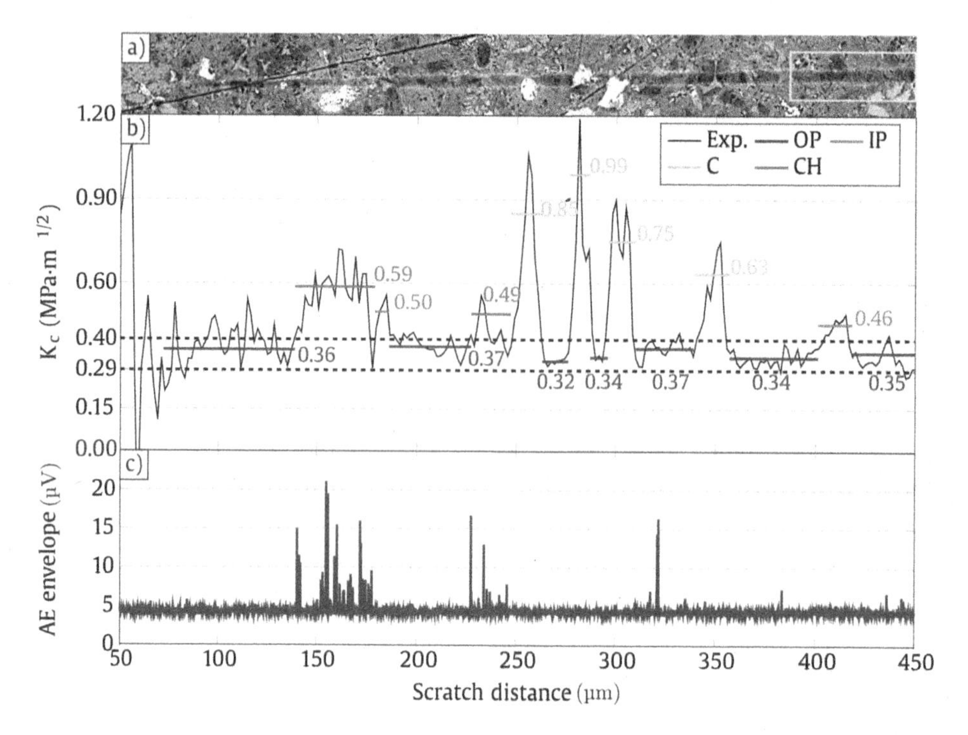

Figure 2. (a) SEM-BSE image of a scratch line, (b) fracture toughness calculated from scratch test with mean values of individual phases, (c) AE signal record.

3 MODELING PART

3.1 *Material model for compressive and tensile failure*

Material model already successfully applied for a combined failure of cement paste e.g. in (Hlobil, Šmilauer, & Chanvillard 2016), (Němeček, Šmilauer, Němeček, & Maňák 2019), (Němeček, Šmilauer, Němeček, Kolařík, & Maňák 2018) was utilized in this work. The main features of the model that combines fracture and damage mechanics will be briefly described here. It uses the concept of damage mechanics where an equivalent uniaxial strain $\tilde{\varepsilon}$ under compression-dominant loading can be derived as (Griffith 1924), (Hlobil, Šmilauer, & Chanvillard 2016)

$$\tilde{\varepsilon}_c = \frac{1}{E} \cdot \frac{-(\sigma_1 - \sigma_3)^2}{8(\sigma_1 + \sigma_3)} \tag{5}$$

where E is the elastic modulus, σ_1 and σ_3 are the maximum positive and negative effective principal stresses of undamaged-like material in uniaxial situations, respectively. An interesting feature of the Griffith model is that the ratio of the uniaxial compressive-to-tensile strength equals to 8, e.g. $|f_c| = 8f_t$. The equivalent strain is evaluated as a maximum of the strain in tension (for $\sigma_1 > 0$ using Rankine criterion) and in compression (using the Griffith criterion, Equation 5) as

$$\tilde{\varepsilon} = \max(\frac{\sigma_1}{E}, \tilde{\varepsilon}_c) \tag{6}$$

Since the damage evolution law has a small effect on the computed macroscopic strength simple linear softening can be assumed. Then, the cohesive law takes the form

$$\sigma = f_t \left(1 - \frac{w}{w_f}\right) \tag{7}$$

where w is a crack opening and w_f is the maximum crack opening at zero stress. According to the formulation of the isotropic damage model, the uniaxial tensile stress obeys the law

$$\sigma = (1 - \omega)E\tilde{\varepsilon} \tag{8}$$

Considering the crack band model (Bažant & Planas 1997) for finite element size of h derived in the direction of the maximum principal strain, objective results, independent on finite element size can be obtained and the damage law formulated as (Jirásek & Bažant 2001)

$$\omega = \left(1 - \frac{\varepsilon_0}{\tilde{\varepsilon}}\right)\left(1 - \frac{hE\varepsilon_0^2}{2G_f}\right)^{-1} \tag{9}$$

with G_f being the mode-I fracture energy and ε_0 the elastic limit strain.

3.2 *Numerical model*

The three-dimensional finite element analysis performed in OOFEM 2.5 software (Patzák 2000) was

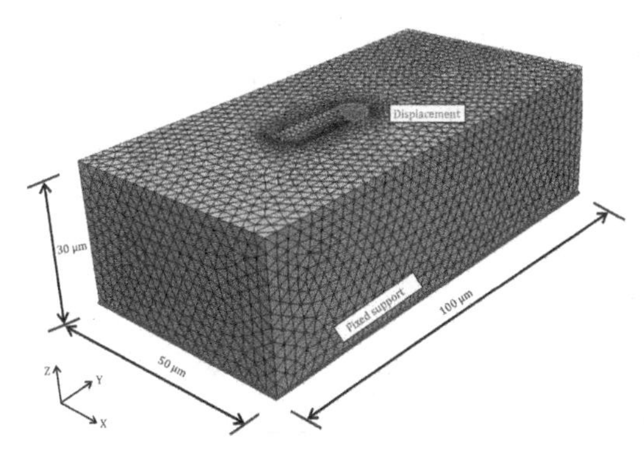

Figure 3. Scheme of the mesh with highlighted boundary conditions.

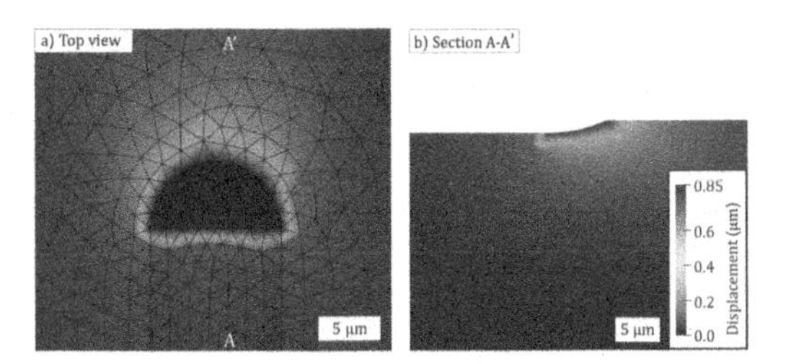

Figure 4. Displacement at the last time step. (a) Top view (b) Section (A-A') view.

used to derive cracking pattern and reproduce the initiation of the cracking process during the scratch experiment. The largest microstructural phase, the OP, was chosen for modeling. The model is schematically shown in Figure 3. The domain (30 × 50 × 100 μm^3) was filled with linear tetrahedra four-node finite elements (LTRSpace), where each node has 3 degrees of freedom corresponding to displacements at each direction. The fixed support was prescribed to all the nodes of the bottom surface. At the upper surface, a trench corresponding to the average penetration of the tip in the OP phase deduced from experiments (900 ± 100 nm) was introduced. The trench also corresponds to the geometry of the tip with radius R=10 µm. Infinitely stiff indenter tip was modeled to be in full contact with the material in lateral y-direction (scratch direction). The analysis was controlled by prescribed displacement imposed on contact part of the sphere in y-direction. Material constants were derived from microindentation experiments using the same tip (R=10 µm) and maximum force of 25 mN performed in OP which yielded Young's modulus of 22 GPa. The Poisson's ratio was assumed as 0.2 (Constantinides & Ulm 2007) and the value of fracture energy was taken 4.4 J/m^2 from the micro-beam experiments (Němeček, Králík, Šmilauer, Polívka, & Jäger 2016). Although the scales are different in micro-bending and micro-scratching experiments, it is assumed that the scaling effect is low in the OP phase. The value of tensile strength was left as a free parameter fitted by the model to find the best match with the record of horizontal force encountered in the experiment (Figure 5a). Since the initial loading stage of the experiment is not reproducible by the current numerical model, only the peak horizontal force was fitted to the mean value of experimental data (5.43 mN) which matches precisely (Figure 5b).

The calculated model response (horizontal load vs. scratch displacement) can be divided into several segments as shown in Figure 5b. Initially, the material behaves elastically until the tensile stress under the tip overcomes the material tensile strength, and the crack is initiated, see Figure 6a. Then, the crack propagates under the tip downwards. The top of such multiple cracks is well visible in SEM images, Figure 7 (the SEM image corresponds to the zone highlighted by yellow rectangle in Figure 2a) and shows the end of the scratch distance 385-450 µm). Fully developed tension crack is formed at end of the second segment of Figure 5 as shown in Figure 6b. The vertical extent of the crack is approximately 2 µm. The horizontal extent of the tension damage zone is approximately 1 µm which corresponds to the crack extent found in SEM images, where typical discrete crack width

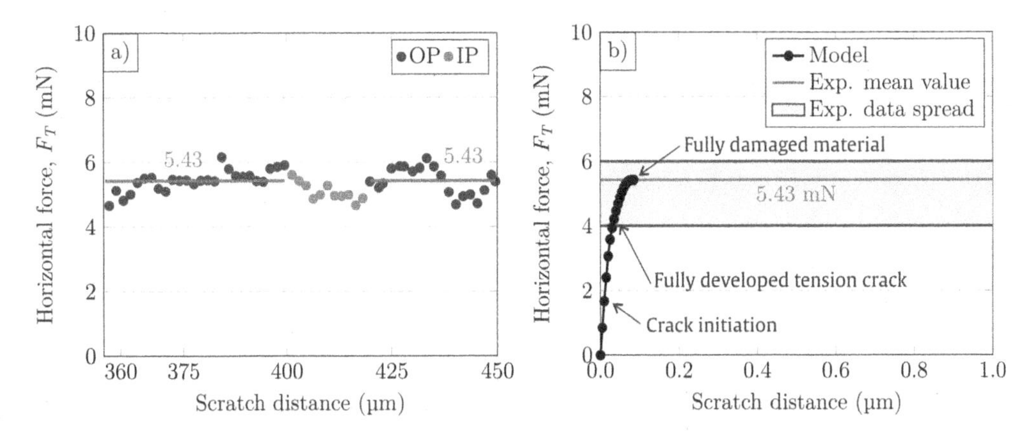

Figure 5. (a) Experimental data corresponding to scratch distance 355-450 µm in Figure 2, (b) calculated model response and experimental mean value of OP.

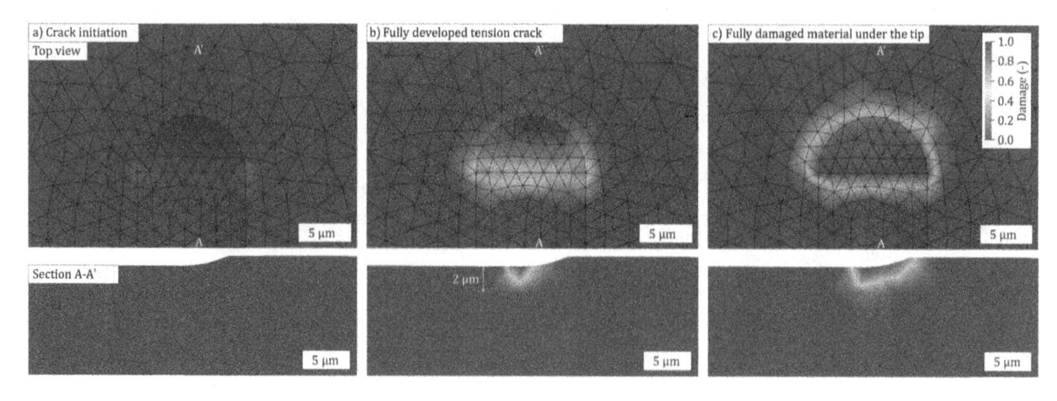

Figure 6. Damage (0-1) propagation in three different states. (a) Crack initiation, (b) fully developed tension crack, (c) fully damaged material under the tip.

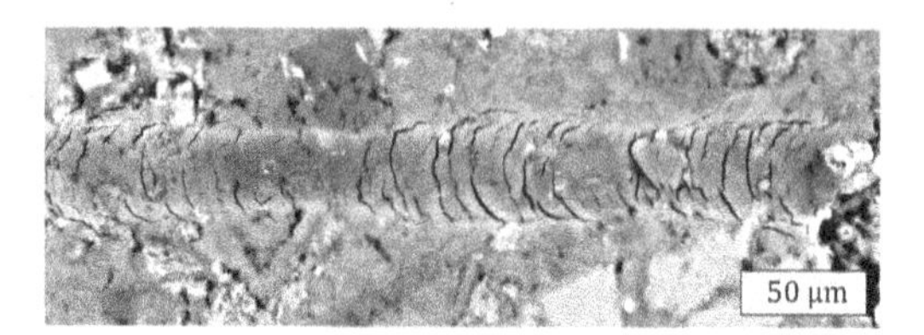

Figure 7. SEM-BSE image with visible cracks corresponding to scratch distance 385-450 μm in Figure 2.

after unloading is 0.1–0.2 μm. In the last segment of Figure 5b, the tension crack under the tip is not evolving anymore, but the compressive damage starts to propagate at the tip front end horizontally. The situation is related to inelastic straining and pushing of the material in the direction of scratch and sidewards, again as observed in SEM images. At the end of the segment, the material under the tip front end is fully damaged in compression (Figure 6c).

The best fit of experimental horizontal force (Figure 5) yielded the cement paste tensile strength of 54 MPa. The value is about 5 times lower than the tensile strength obtained from micro-beam bending experiments (264 ± 73 MPa, (Němeček, Králík, Šmilauer, Polívka, & Jäger 2016)). Such a reduction can be explained by different volumes of the damage zone in both experiments. The tension damage volume for micro-beams derived from their small geometry is about 1 μm^3, while for the scratch test damage zone is much larger, ≈100 μm^3. The damage zone in cement paste contains a variable number of defects at different scales causing the strength scaling. The defects can be in the form of capillary porosity, shrinkage cracks or other inclusions. The number of defects in micro-beam experiments is substantially lower compared to scratch experiment. The scaling of tensile strength in dependence on the defect size was shown for two levels of cement paste in (Němeček, Šmilauer, Němeček, Kolařík, & Maňák 2018), (Němeček, Šmilauer, Němeček, & Maňák 2019). Similarly, the compression strength reduces as the scale of cement paste enlarges (Hlobil, Šmilauer, & Chanvillard 2016).

The cement paste tensile strength of 54 MPa corresponds well with experiments done at similar length scale on micro-cubes/cantilevers with a

Table 3. Summary of cement paste fracture properties obtained at various scales.

Test method	Length scale (μm)	f_t (MPa)	K_c (MPa·m$^{1/2}$)	G_f (J/m^2)	Phase	Reference
Molecular dyn.	0.005		0.37 ± 0.01	1.72±0.29	C-S-H	Bauchy et al. 2015
Micro-pillars	0.5-2		0.67		C-S-H	Shahrin et al. 2019
Micro-cantilevers	3.5	264 ± 73	0.33 ± 0.06	4.4 ± 1.9	OP	Nemecek et al. 2016
		700 ± 199	0.81 ± 0.08	19.7 ± 3.8	IP	Nemecek et al. 2016
		672 ± 370	0.88 ± 0.28	19.9 ± 14.4	CH	Nemecek et al. 2016
Scratch test	10	54	0.34 ± 0.03	5.3 ± 0.9	OP	this study
			0.51 ± 0.05	8.4 1.7	IP	this study
			0.54 ± 0.09	7.1 2.6	CH	this study
Micro-cubes	100	58			OP	Zhang et al. 2016
(multi-scale)		92			IP	Zhang et al. 2016
Scratch test	190		0.65 ± 0.01		All phases	Akono 2020
Micro-beams	500	15.3 ± 2.9			All phases	Zhang et al. 2017

minimum volume of $(100 \times 100 \times 100\ \mu m^3)$ (Zhang, Šavija, Ch. Figueiredo, Lukovic, & Schlangen 2016). In Zhang, Šavija, Xu, & Schlangen (2018) the tensile strength was reproduced from experiments with the aid of XCT scanning and 3D lattice modeling as 58-66 MPa being in excellent agreement with our results. A detailed comparison of the results obtained by several authors at different cement paste scales is summarized in Table 3. Not all data are available for all scales but it is clear that both tensile strength and fracture energy decrease as the testing size increases.

4 CONCLUSIONS

The paper shows application of the micro-scratch test for evaluation of fracture toughness of cement paste at the scale of 10 - 100 μm. The technique was found to be feasible and in line with other available experiments (Table 3). Fracture toughness was assessed for individual micro-scale cement paste phases with the aid of electron microscopy and acoustic emission as $K_c = 0.34 - 0.54\ \text{MPa}{\cdot}\text{m}^{1/2}$ (Table 2). 3-D finite element model was constructed to reproduce the experiments and to identify local tensile strength of main hydration products (OP) in cement paste. As a constitutive law, Griffith-type of a fracture-damage model for tension/compression failure was successfully used. The tensile strength for the outer hydration product was identified as 54 MPa being about 5 times lower compared to FIB-produced micro-cantilevers at 3 μm scale (Němeček, Králík, Šmilauer, Polívka, & Jäger 2016) and about 3.6 times higher than at 500 μm scale (Zhang, Šavija, Figueiredo, & Schlangen 2017). The strong size effect can be attributed to a different number of internal defects in the cement paste microstructure. The scaling effect that can also be deduced from other works (Table 3) was confirmed.

ACKNOWLEDGEMENT

Financial support of the Czech Science Foundation (project 21-11965S) and the Grant Agency of the Czech Technical University in Prague (SGS22/001/OHK1/2T/11) is gratefully acknowledged.

REFERENCES

Akono, A.-T. (2020, 09). Effect of nano-TiO_2 on C–S–H phase distribution within portland cement paste. *Journal of Materials Science 55*.

Akono, A.-T., N. Randall, & F.-J. Ulm (2012). Experimental determination of the fracture toughness via microscratch tests: Application to polymers, ceramics, and metals. *Journal of Materials Research 27*(2), 485–493.

Akono, A.-T. & F.-J. Ulm (2012). Fracture scaling relations for scratch tests of axisymmetric shape. *Journal of the Mechanics and Physics of Solids 60*(3), 379 – 390.

Bauchy, M., H. Laubie, M. A. Qomi, C. Hoover, F.-J. Ulm, & R.-M. Pellenq (2015). Fracture toughness of calcium–silicate–hydrate from molecular dynamics simulations. *Journal of Non-Crystalline Solids 419*(0), 58–64.

Bažant, Z. & J. Planas (1997). *Fracture and size effect in concrete and other quasibrittle materials*, Volume 16. CRC press.

Constantinides, G. & F.-J. Ulm (2007). The nanogranular nature of C–S–H. *Journal of the Mechanics and Physics of Solids 55*(1), 64–90.

Gianuzzi, L. & F. Stevie (2005). *Introduction to Focused Ion Beams. Instrumentation, Theory, Techniques and Practice.* Springer.

Griffith, A. (1924). Theory of rupture. In C. Biezeno and J. Burgers (Eds.), *First International Congress for Applied Mechanics*, Delft, pp. 55–63.

Hlobil, M., V. Šmilauer, & G. Chanvillard (2016). Micromechanical multiscale fracture model for compressive strength of blended cement pastes. *Cement and Concrete Research 83*, 188–202.

Jirásek, M. & Z. Bažant (2001). *Inelastic analysis of structures*. John Wiley & Sons.

Němeček, J., V. Králík, V. Šmilauer, L. Polívka, & A. Jäger (2016). Tensile strength of hydrated cement paste phases assessed by micro-bending tests and nanoindentation. *Cement and Concrete Composites 73*, 164–173.

Němeček, J., J. Lukeš, & J. Němeček (2020). High-speed mechanical mapping of blended cement pastes and its comparison with standard modes of nanoindentation. *Materials Today Communications 23*, 100806.

Němeček, J., V. Šmilauer, J. Němeček, F. Kolařík, & J. Maňák (2018). Fracture properties of cement hydrates determined from mircobending tests and multiscale modeling. In *Computational Modelling of Concrete Structures*, Boca Raton, US. CRC Press.

Němeček, J., V. Šmilauer, J. Němeček, & J. Maňák (2019). Microscale fracture properties of alkali-activated fly ash. In *10th International Conference on Fracture Mechanics of Concrete and Concrete Structures*, Vail, US. IA-FraMCoS.

Patzák, B. (2000). OOFEM home page. http://www.oofem.org.

Schlangen, E., M. Lukovic, B. Šavija, & O. Copuroglu (2015). Nano-indentation testing and modelling of cement paste. In C. H. abd Bernhard Pichler and J. Kollegger (Eds.), *Proceedings of the 10th International Conference on Creep, Shrinkage, and Durability of Concrete and Concrete Structures*, Vienna, Austria. Reston, Virginia: American Society of Civil Engineers.

Shahrin, R. & C. Bobko (2019). Micropillar compression investigation of size effect on microscale strength and failure mechanism of calcium-silicate-hydrates (C-S-H) in cement paste. *Cement and Concrete Research 125*, 105863.

Čtvrtlík, R., J. Tomaštík, L. Václavek, B. Beake, A. Harris, A. Martin, M. Hanák, & P. Abrham (2019). High-resolution acoustic emission monitoring in nanomechanics. *JOM 71*(10), 3358–3367.

Zhang, H., B. Šavija, S. Ch. Figueiredo, M. Lukovic, & E. Schlangen (2016). Microscale testing and modelling of cement paste as basis for multi-scale modelling. *Materials 9*(11).

Zhang, H., B. Šavija, S. C. Figueiredo, & E. Schlangen (2017). Experimentally validated multi-scale modelling scheme of deformation and fracture of cement paste. *Cement and Concrete Research 102*, 175–186.

Zhang, H., B. Šavija, Y. Xu, & E. Schlangen (2018). Size effect on splitting strength of hardened cement paste: Experimental and numerical study. *Cement & Concrete Composites 94*, 264–276.

Computational Modelling of Concrete and Concrete Structures – Meschke, Pichler & Rots (Eds)
© 2022 Copyright the Author(s), ISBN: 978-1-032-32724-2

FE^2 multiscale modelling of chloride ions transport in recycled aggregates concrete

A. Fanara, L. Courard & F. Collin
Urban and Environmental Engineering, University of Liége, Liége, Belgium

ABSTRACT: In the context of climate change, reducing the production of CO_2 emissions and preserving natural resources have proven to be necessary. One way to reach theses objectives is to recycle old concrete members: Recycle Concrete Aggregates (RCA) are aggregates obtained by crushing demolished concrete structures. Those aggregates can substitute the Natural Aggregates (NA) inside the so-called Recycled Aggregates Concrete (RAC). RCA are composed of natural aggregates and adherent mortar paste, the latter increasing the porosity and water absorption of RAC. Furthermore, water is necessary for, and even promotes, the penetration of aggressive ions such as chloride ions, possibly reducing the durability of said concrete.

This paper aims to model the influence of RCA on chloride ions ingress: several experiments have been performed to determine the transfer properties and the chloride ions diffusion coefficients of mortar pastes and concretes produced with NA or 100% RCA. The microstructure of the RCA deeply influences the permeability, the water content distribution and the chloride diffusion. These properties have been included into a numerical model that integrates the microstructural information. A numerical homogenization technique, based on the Finite Element square (FE^2) method, is implemented into a coupled multiscale model of water flows and advection/diffusion of chlorides in saturated concrete, in order to model the complex flow behaviour encountered.

The numerical model developed is compared to existing simple-scale models, using a simple RVE, in order to validate the implementation. The numerical convergence of the developed model is also studied, as far as the numerical cost of the FE square method is expensive.

1 INTRODUCTION

Maintenance and rehabilitation of concrete structures represent a significant and continuously increasing cost. In the vicinity of roads (where de-icing salts are used in winter) and coastal areas, the major cause of degradation of reinforced concrete structures is chloride attacks (Mangat & Molloy 1994; Morga & Marano 2015). Chloride ions leach into the concrete's porous system, reaching the steel rebars where they eventually concentrate. This leads to pitting and loss of section of the reinforcements, decreasing their strength and possibly leading to a structural failure (Angst, Elsener, Larsen, & Vennesland 2009). On the other hand, waste from the construction and demolition sector (C&D Waste) is one of the heaviest and most voluminous waste streams generated (European Commission 2019; Zhao, Courard, Groslambert, Jehin, Léonard, & Xiao 2020). One popular way to reduce the amount of C&DW to be landfilled and simultaneously provide a sustainable source of aggregates for future building materials production is recycling. Recycled Concrete Aggregates (RCA) produced from crushed C&DW as a replacement of Natural Aggregates (NA) is one way to recycle, which has made it a thoroughly studied field of research (Belin, Habert, Thiery, & Roussel 2014; Hussain, Levacher, Quenec'h, Bennabi, & Bouvet 2000; Nagataki, Gokce, Saeki, & Hisada 2004). RCA are coarse particles containing both natural aggregates and residual adherent mortar paste, the latter impairing negatively their properties compared to NA: due to their increased porosity and water absorption, they favour the penetration of water and chloride ions, increasing the diffusivity of Recycled Aggregate Concrete (RAC) (Akbarnezhad, Ong., Tam, & Zhang 2013; Hu, Mao, Xia, Liu, Gao, Yang, & Liu 2018; Rao, Jha, & Misra 2007; Sun, Chen, Xiao, & Liu 2020).

Concrete is a highly heterogeneous material due to its composition: its microstructure is composed of a wide range of components, from nanometre-sized pores to centimetre-sized aggregates (Garboczi & Bentz 1998). Modelling concrete and its entire microstructure is therefore computationally impossible, and often the properties are homogenized over the entire microstructure to obtain mean values.

Nowadays, multiscale modelling and computational homogenization techniques allow to homogenize the concrete's microstructure over a certain scale, and then up-scale it while keeping the computational

DOI 10.1201/9781003316404-8

cost acceptable (Nilenius 2014). The multiscale modelling approach tends to combine the best of both the macroscopic approach and the microscopic approach (Bertrand, Buzzi, Bésuelle, & Collin 2020):

1. Macroscale: the concrete is treated as a homogeneous medium, and the constitutive laws are supposed to represent the whole behaviour of the material. The mixture theory allows to account for multiple phases (e.g. liquid water and water vapour) percolating inside the porous system of the material studied (Bear & Verruijt 1987). This method is easy to implement and allows the use of general properties of concrete, determined experimentally for example. Unfortunately, it means that each modification of the microstructure requires a new experimental campaign to obtain the homogenized properties of the material.
2. Microscale: the whole structure, including the heterogeneities (aggregates, porosity, ...) is directly represented in the model. Each microscopic constituent has its own constitutive equations. Although this increases the precision of the model, its computation cost is too high to be used on metre-sized structures.

The chloride ingress inside RAC being highly dependent on the microstructure of concrete, it is therefore necessary to use a multiscale model for that purpose.

Modelling the advection/diffusion of chloride in the water requires the replacement of the macroscopic phenomenological quantities of interest (e.g. flow measures, pore pressure or gradient of pore pressure) by suitable averages over this RVE. The constitutive equations (Darcy's and Fick's laws among others) are indeed applied only at the microscopic scale and homogenization/localization equations are employed to compute the macroscopic flows based on the pore pressure state at the microscopic scale. It has to be reminded that due to the separation of scale, the diffusion problem is solved under the assumption of steady-state at the microscale. Advective and diffusive transport modes and adsorption of chlorides are to be included in the model.

The homogenization technique used is considered as a numerical homogenization: it is called the unit cell method. This technique is based on the concept of representative volume element (RVE) (Kouznetsova, Brekelmans, & Baaijens 2001). The macroscopic phenomenological equations are replaced by averages over the RVE. The material properties and behaviour at the macroscale are therefore obtained from the modelling of this RVE, volume that contains a detailed model of the microstructure of the material (Bertrand, Buzzi, Bésuelle, & Collin 2020; Kouznetsova, Brekelmans, & Baaijens 2001). In a sense, the RVE is meant to decouple the macrostructure from the microstructure in a computational way (Smit, Brekelmans, & Meijer 1998).

Using this method, the behaviour of the material and its properties are not valid for the whole macroscopic structure, but rather at some macroscopic points where an estimation is obtained through calculations on the RVE assigned to that macroscopic point (Kouznetsova, Brekelmans, & Baaijens 2001).

Each integration point of the discretized homogenized macrostructure is then linked to a RVE and finite element computations are performed separately for each RVE. The macroscopic pressure gradients and mean pressure are then transformed into boundary conditions applied to the RVE, and the macroscopic fluxes are computed by averaging the fluxes obtained for each RVE over their respective volume (Kouznetsova, Brekelmans, & Baaijens 2001; Smit, Brekelmans, & Meijer 1998). This averaging is possible thanks to the periodicity of the microstructure in the vicinity of the integration point.

The method is called FE^2 method because the modelling is achieved by a finite element analysis on both the macroscale and microscale (RVE).

In this work, the microscale is referred to as the mesoscale as it consists of the scale of samples at the laboratory. The Representative Volume Element (RVE) therefore represents the structure of a concrete sample, that is a homogenized mortar paste (with homogenized properties, as for a macroscale solution) as well as impervious aggregates and adherent mortar paste. The ITZ between them could be accounted for through interface elements, but their influence would be difficult to quantify experimentally.

The macroscale, on the other hand, represents a metre-sized civil engineering structure.

At the mesoscale, the parameters used in the equations represent properties of a single phase among the two cited before when the integration point is in that phase. However, this means that for the mortar phase, which is a composite material, the properties used must be effective properties for the composite (Xi & Bazant 1999). At the macroscale, concrete is considered as a single phase composite material and all its properties are therefore averaged effective properties.

The material structure is assumed to be macroscopically homogeneous but microscopically heterogeneous. However, in concrete, the microscopic length scale is still bigger than at the molecular level, allowing the use of continuum mechanics as for the macroscale.

The FE^2 method is therefore a numerical double-scale method based on four consecutive and iterative steps performed on each Gauss point of the mesh until convergence of both scales (Bertrand, Buzzi, Bésuelle, & Collin 2020). For our application, those four steps could be described as follows:

1. From macroscale to mesoscale: the gradients and mean pressures of the macroscale are localised at the mesoscale through boundary conditions;
2. Resolution of the problem based on those boundary conditions at the mesoscale;
3. From mesoscale to macroscale: the fluxes of the mesoscale are homogenised into a unique flux for each Gauss point of the macroscale;
4. Resolution of the boundary value problem at the macroscale.

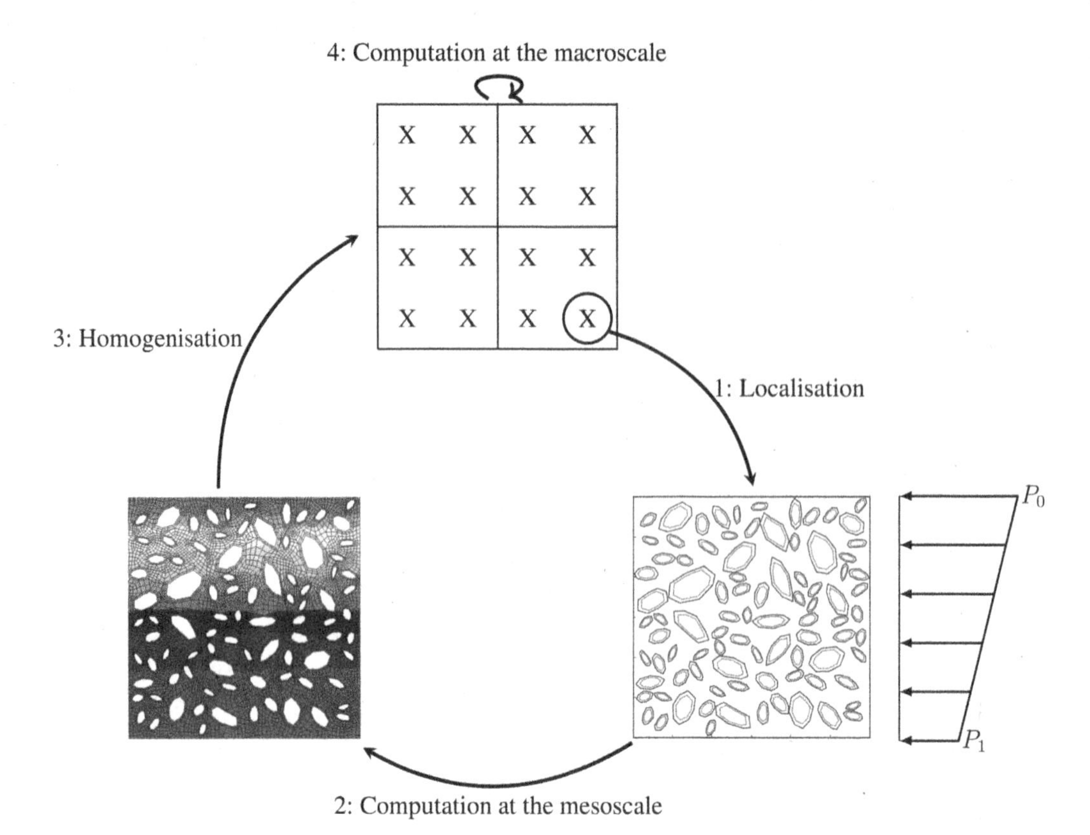

Figure 1. Representation of the iterative process performed on each Gauss point of the mesh during the multiscale computation.

This iterative process is represented in the Figure 1. In the example of an application on an engineering structure, the macroscale would represent its mesh, where four elements are shown, each containing four Gauss points. The conditions applied to this structure (i.e. water pressure and pollutant concentration) and the gradients created by those are first localised in the mesoscale, represented by a slice of concrete multi-centimetres large. Then, based on those gradients and mean values, the boundary value problem is solved at the mesoscale and fluxes are deduced from it. However, all those fluxes must then be homogenised to obtain an unique value for the Gauss point studied. Then, once each Gauss point of each element used to mesh the engineering structure have a macroscopic flux assigned, the boundary value problem is solved at the macroscale.

2 METHODOLOGY

2.1 General multiscale formulation

The start of every multiscale formulation is the splitting of the scalar field ϕ in an additive manner, such that it contains both the macroscale part ϕ^M and the subscale part ϕ^f which contains the fluctuations of the total scalar field (Bertrand, Buzzi, Bésuelle, & Collin 2020; Nilenius 2014):

$$\phi = \phi^M + \phi^f \tag{1}$$

On the boundaries Γ of the RVE, it is assumed that $\phi = \phi^M$ and therefore $\phi^f = 0$.

Following a Taylor expansion, limited to its first order inside the macroscale continuum, the homogenisation follows the assumption that ϕ^M varies linearly within the RVE, yielding the following equation:

$$\phi^M(x, \bar{x}) \approx \bar{\phi}(\bar{x}) + \bar{g}(\bar{x}) \times (x - \bar{x}) \quad \forall x \in \Omega \tag{2}$$

which is represented in the Figure 2 for a 1D RVE. In this formulation, $\bar{x}$ is the center of the RVE and $\bar{g}$ is a gradient defined such that:

$$\bar{g}(\bar{x}) = \text{grad}\,\bar{\phi}\,(\bar{x}) \tag{3}$$

In the subscale, the scalar field is not necessarily continuous and therefore, the higher order terms of the Taylor expansion cannot be neglected. They are then replaced by the fluctuation field, noted ϕ^f and resulting from the variations in the material properties of the RVE:

$$\begin{aligned}\phi(x, \bar{x}) &= \phi^M(x, \bar{x}) + \phi^f(\bar{x}) \\ &= \bar{\phi}(\bar{x}) + \bar{g}(\bar{x}) \times (x - \bar{x}) + \phi^f(\bar{x})\end{aligned} \tag{4}$$

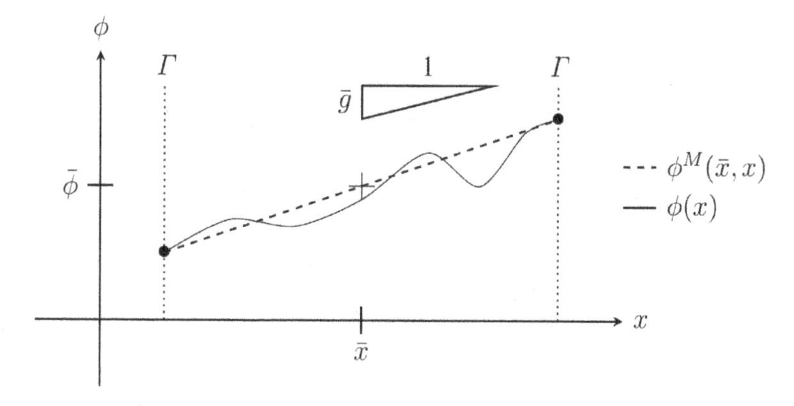

Figure 2. Illustration of the ϕ^M properties and its linear variations within a 1D RVE domain Ω (modified from Nilenius (2014, 2015)).

As the equality between the macroscale-part and the subscale-part of the field is to be true for any point of the macroscale, it follows that:

$$\bar{g}(\bar{x}) \times (x - \bar{x}) + \phi^f(\bar{x}) \ll \bar{\phi}(\bar{x}) \tag{5}$$

which is the concept of separation of scales: the subscale characteristic length l_c^s must be negligible compared to the characteristic fluctuation length L_c^M of the macroscopic field:

$$l_c^s \ll L_c^M \tag{6}$$

If this assumption doesn't hold, then the boundary conditions of the subscale boundary value problem cannot be determined by the local macroscale pressure/concentration gradient, at least not under first-order homogenisation.

The transition from the subscale to the macroscale has an important characteristic that is the transfer of the flux. Indeed, in stationary conditions, the mass balance equation is:

$$\nabla J = 0 \quad \text{in}\ \Omega \tag{7}$$

where J is the flux and Ω the domain where the material heterogeneities are embedded (i.e. the mesoscale). Splitting the scalar field as shown previously, in addition to first order homogenisation, allows to identify the volume average of the flux inside the RVE, that is the macroscale flux $\bar{J}$:

$$\bar{J} = \frac{1}{|\Omega|} \int_{\Omega} J(x)\, d\Omega \tag{8}$$

where Ω denotes the reference domain occupied by the RVE.

2.2 RVE generation

The first step to a multiscale FE^2 study is the generation of the RVE. The RVE is multiphasic: impervious natural aggregates are considered inside a porous mortar matrix. For the case of RCA, an additional mortar gangue is also considered, whose properties are different of the ones of the mortar matrix.

The RVE must be representative of the material studied, and it is therefore essential to use and respect properties related to the concrete: the surface fraction of aggregates, the aspect ratio and the particle size distribution of the aggregates are used in the generation. Each aggregate has a random size, position and orientation following the properties given above. Therefore, it is impossible to generate two times the same RVE.

Then, using an algorithm adapted from the one of Nilenius (Nilenius 2014), a 2D RVE is generated and meshed in 2D by the software GMSH (Geuzaine & Remacle 2009), according to the Frontal-Delaunay algorithm for quads, with a simple recombination algorithm applied to all surfaces, ensuring that all elements are quads.

The size of the RVE is dictated by the maximum aggregate diameter, so as to keep a size of at least 3 times that diameter. An example of a RVE with RCA of 8mm of maximum diameter is represented in the Figure 3.

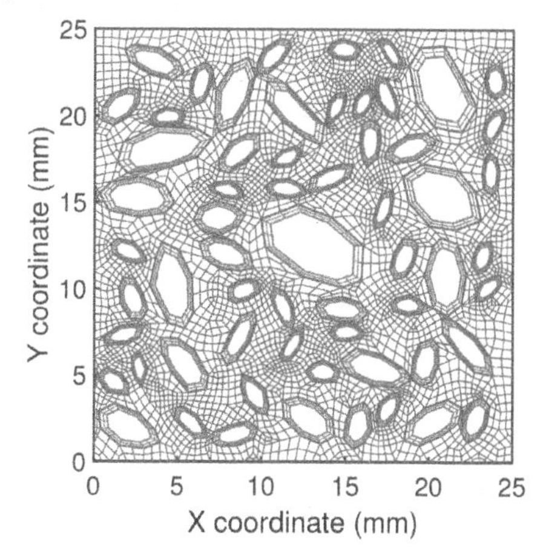

Figure 3. Example of an RVE generated for RAC.

One may see that the mesh is more refined around the aggregates. It is due to the octagonal form of the aggregates that requires many points and elements. Even though the sample is small, its complexity is therefore resulting in a high number of nodes and elements, which is not ideal for the multiscale modelling, but required for precision purposes.

One of the assumptions of this method is that the volume fraction is directly changed into a surface fraction, which is inaccurate because the aggregates are not spherical and therefore, the transformation from a volume to a surface may modify the granulometric curve, which is not accounted for here.

Modelling concrete should be performed with a three dimensional model as it is highly heterogeneous and its 3D porous structure may create preferential path in all directions. However, it requires a greater computational power and the equations solving the problem are also harder to develop.

In this research, the two-dimensional approach is preferred, keeping in mind that other authors, such as (Nilenius 2014), have done the comparison between 2D and 3D, yielding diffusivity coefficients up to 40% higher in 3D than in 2D. This is easily explained by the restriction created in 2D where the flow is required to by-pass the aggregates in the plane, while in 3D, an out-of-plane solution is possible.

Those hypotheses could be corrected by direct modelling methods and inverse modelling: the modelling of the experiments done will allow to verify that the sample is correct and if not, a penalisation will be applied to correct it.

2.3 *Multiscale ingress modelling under saturated conditions*

The first development of the models are performed under saturated conditions. The boundary conditions, i.e. water pressure and pollutant concentration variations, are applied on the macroscale. Gradients are then computed for each Gauss point and transmitted to the mesoscale, as well as the average pressure/concentration at that point. At the mesoscale, each integration point has an assigned value for the water pressure and pollutant concentration, based on the average pressure/concentration and theur respective gradients localized from the macroscale. Once those conditions are applied, the resolution can start.

2.3.1 *Mesoscale water flows under saturated conditions*

The mass balance equation of water, in a fixed and undeformable system, under saturated conditions and under the assumption of steady-state, is:

$$\frac{\partial}{\partial x_i}\left(\rho_w v_i^w\right) = 0 \tag{9}$$

where ρ_w is the water density [kg/m^3] and v_i^w is the fluid flow rate per unit area [m/s].

The Equation 9 represents the mass variation of liquid water inside the porous matrix of concrete. The first factor of the equation, the water density, varies with the internal pressure of the matrix (noted $P_{w,average}$):

$$\rho_w = \rho_{w0} \times \left(1 + \frac{P_{w,average} - P_{w0}}{\chi_w}\right) \tag{10}$$

where ρ_{w0} [kg/m^3] and P_{w0} [Pa] are, respectively, the initial density of liquid water and the initial pressure inside the porous structure. This relation is dependent on the fluid compressibility, noted χ_w [Pa$^-$1] (at 20°C, $1/\chi_w = 5\,10^{-10}$ Pa^{-1}).

The second factor of the Equation 9 is related to the liquid water convection. The Darcy's law is used to describe the movement of a fluid (water) inside a porous medium. Under the hypothesis of a homogeneously permeable medium, and in the absence of gravitational forces, the fluid flux is directly proportional to the gradient of pressure (noted ∇P_w):

$$v_i^w = -\frac{k}{\mu_w}\frac{\partial P_w}{\partial x_i} \tag{11}$$

where k [m^2] is the intrinsic permeability of the porous medium, and μ_w [kg/m.s] is the dynamic viscosity of the fluid.

The stifness matrix, under saturated conditions and for water flows only, is quite simple:

$$K = \begin{bmatrix} \frac{\partial \nabla(\rho_w v_1^w)}{\partial \nabla P_w} & \frac{\partial \nabla(\rho_w v_1^w)}{\partial P_w} \\ \frac{\partial \nabla(\rho_w v_2^w)}{\partial \nabla P_w} & \frac{\partial \nabla(\rho_w v_2^w)}{\partial P_w} \end{bmatrix} = \begin{bmatrix} \frac{k \times \rho_w}{\mu_w} & 0 \\ \frac{k \times \rho_w}{\mu_w} & 0 \end{bmatrix} \tag{12}$$

2.3.2 *Mesoscale pollutant flows under saturated conditions*

The mass balance equation of the pollutant, under saturated conditions, is:

$$\frac{\partial}{\partial x_i}(v_i^c) = 0 \tag{13}$$

where v_i^c is the pollutant flow rate per unit area [m/s].

The pollutant flows are caused by three phenomenon: advection, dispersion and diffusion. The advection is a movement of the pollutant inside the fluid, due to fluid flows. The dispersion is due to the irregularity of the porous system, causing pollutant concentration to vary locally inside the fluid to accommodate for geometrical constraints. Finally, the diffusion is due to a gradient of concentration of the pollutant inside the fluid itself, and is not caused by a fluid flow. The pollutant flow can therefore be calculated according to the following equation:

$$\begin{aligned} v_i^c &= v_i^{\text{advection}} + v_i^{\text{dispersion}} + v_i^{\text{diffusion}} \\ &= C\,u_i - D_{\text{dispersion}}\frac{\partial C}{\partial x_i} - D_{\text{diffusion}}\frac{\partial C}{\partial x_i} \end{aligned} \tag{14}$$

where the dispersion is neglected, and the diffusion coefficient $D_{\text{diffusion}}$, also noted D, is taken as the diffusion coefficient obtained experimentally.

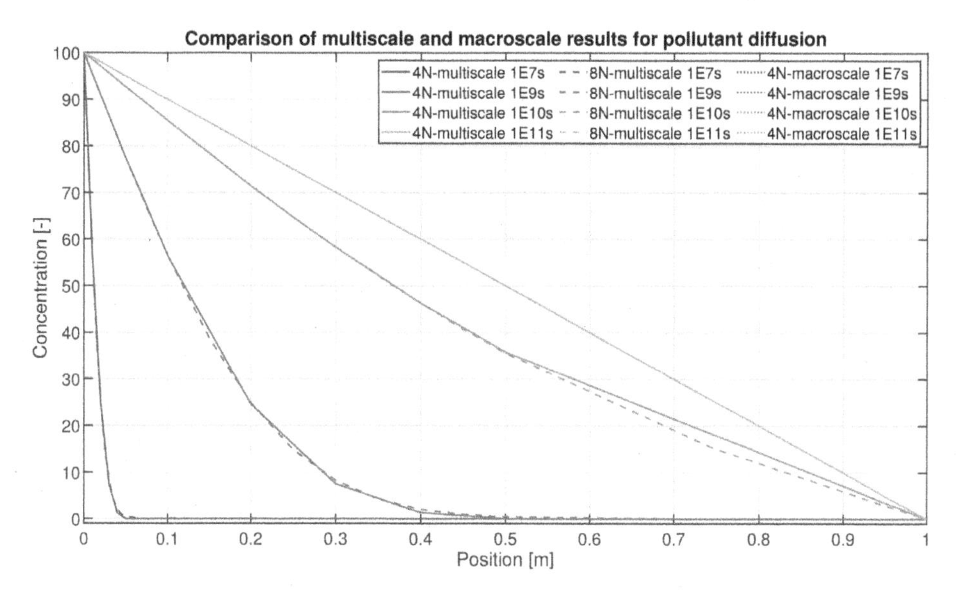

Figure 4. Comparison of the results obtained from a purely 4-node macroscale model and the developed multiscale model (in both 4 and 8-node configurations).

The contribution of the water flows on the pollutant flows are transmitted through the water velocity u_i, taken equal to:

$$u_i = \frac{v_i^w}{\rho_w} \tag{15}$$

The stiffness matrix is obtained by derivation of the nodal fluxes:

$$K_{LK} = \int_V \nabla N_K D \nabla N_L dV - \int_V N_K \underline{u} \nabla N_L dV \tag{16}$$

2.3.3 Homogenized macroscale response

Once the mesoscale fluxes are obtained, they must be homogenized for the macroscale. The fluxes of each integration point are therefore summed up, proportionally to the surface of each integration point.

Once the macroscale has fluxes values for each of its integration point, the forces are computed and the problem is finally solved.

The stiffness matrix of the mesoscale is computed by perturbations: each variable coming from the macroscale (gradient of pressure and mean pressure, gradient of pollutant concentration and mean concentration) are perturbed and the computation is performed at the mesoscale. Then, the results are saved at the macroscale depending on the perturbation applied.

3 RESULTS

The model is fully functional for water flows or diffusion of pollutant. Nonetheless, this paper focuses on the pollutant diffusion only. The results available are therefore:

- Comparison of the multiscale model with a validated macroscale model;
- Comparison of the results for several microstructures: a plain mortar paste, a concrete made from natural aggregates (NAC) and another one from recycled concrete aggregates (RAC);
- Comparison of the results and computation cost for several sizes of RVE.

3.1 Comparison with a validated macroscale model

The multiscale model can easily be validated by comparing the results obtained with the results of an already validated macroscale-only model. The applied conditions and the macroscale meshes used are identically the same for both models, and the microscale RVE consist of a plain material with no aggregates, so that the results can be compared.

The macroscale model is composed of a law for pollutant transport inside porous media (Biver 1992) and its 4-noded elements, while our multiscale model is developed for both 4-node and 8-node elements. The comparison is therefore made for the three possible cases: 4 or 8-node multiscale elements, and 4-node macroscale elements. The results are available in the Figure 4.

The applied conditions are the following: the right border has a fixed concentration of 0, while the left border sees an increase of concentration from 0 to 100, varying linearly from 0s. to 86400s., then kept constant until the end of the simulation.

All the models have the exact same response, except for meshing differences between 4-noded and 8-noded elements. One can therefore assess that the multiscale model is valid and represents the diffusion of pollutant inside a porous medium accurately.

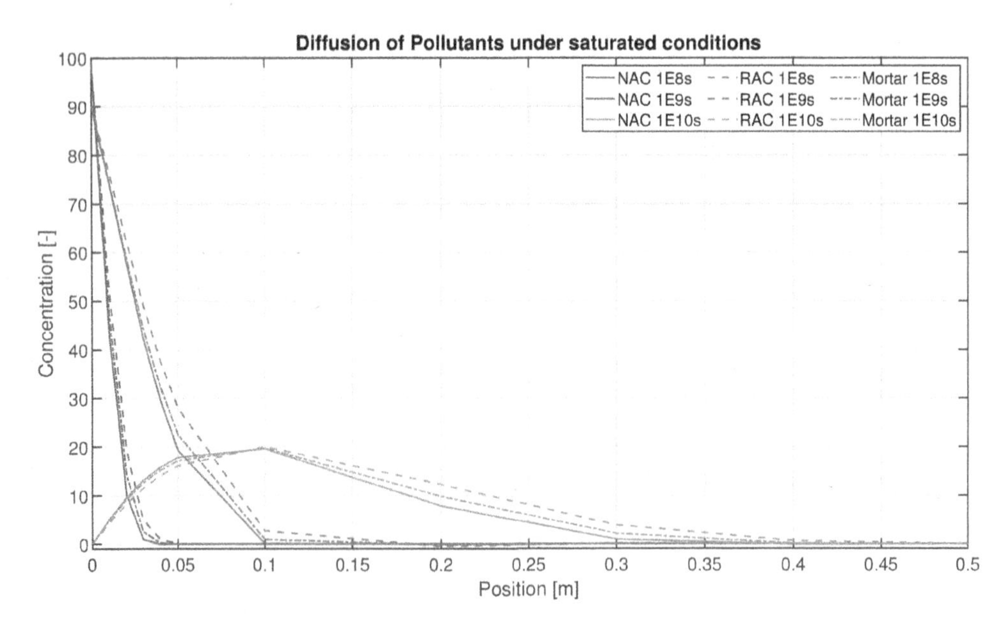

Figure 5. Results obtained for the diffusion of pollutants amongst several types of microstructure: a plain mortar paste, a NAC and a RAC.

3.2 Comparison of several microstructures

For the first application, a 1D diffusion experiment is modelled for different microstructures. The applied conditions on the left border are the following:

- Pollutant concentration of 0% at 0 seconds;
- Pollutant concentration of 100% at 1E5 seconds, and kept constant until 1E7 seconds. The evolution is linear in between 0 and 1E5 seconds;
- Pollutant concentration of 0% at 1E10 seconds, with a linear decrease too.

The microstructure is modelled by a square RVE of 15mm sides[1]. The NAC and RAC microstructures are exactly the same, except for the adherent mortar paste that can be found around the recycled concrete aggregates and that decrease the size of the impervious aggregates inside.

The new mortar paste has a diffusion coefficient of 1E-12 m^2/s while the adherent mortar paste of the RCA has a diffusion coefficient of 5E-12 m^2/s.

The results are available at Figure 5. The first time represented is at 1E8s., when the concentration is already decreasing at the surface. At that time, the RAC has the greater concentration of the three microstructures, followed by the mortar paste and then the NAC. Once the fluxes start going from inside the material towards the exterior surface, the RAC also displays higher exchange rates than the other microstructures, its concentration being the smallest.

This may be surprising as it shows that the small proportion of adherent mortar, whose diffusion coefficient is higher, plays an important role on the diffusion of pollutants. It indeed decreases the total impermeable surface, therefore directly increasing the diffusion capacity of the material. Furthermore, compared to plain mortar paste, the overall diffusion coefficient of the RAC may be higher due to the higher coefficient of the aggregates alone.

Another point worth mentioning is that the two concrete RVE create pollutant fluxes along the y-axis, even though the applied conditions are solely along the x-axis. This is due to the impermeable surfaces that must be by-passed.

3.3 Computational cost dependency

The next results concern the influence of the RVE size. Four RVE of 10mm, two times 15mm and 20mm sides have been used, firstly with NAC microstructure and then with RAC. Using two RVE of the same size is useful to observe whether the random disposition and size of the aggregates impact the overall results of the diffusion experiment or not. The two RVEs of 15mm sides are visible on the Figure 6 for the NAC.

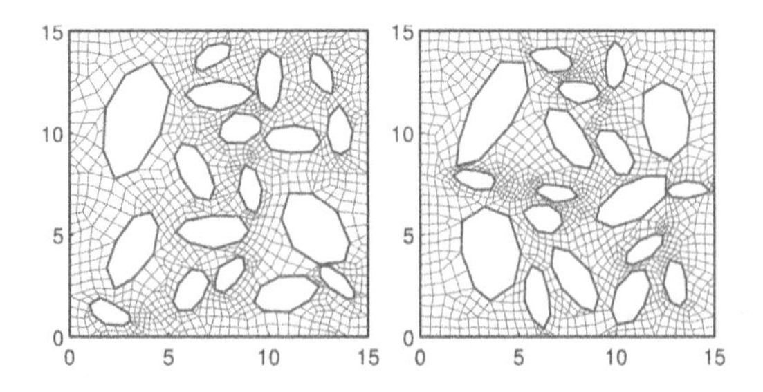

Figure 6. Two RVEs of 15mm side based on the NAC.

[1] A size of 25mm would have been more adequate to respect the $3D_{max}$ rule. However, the computational cost would have been too high for a simple application as this one.

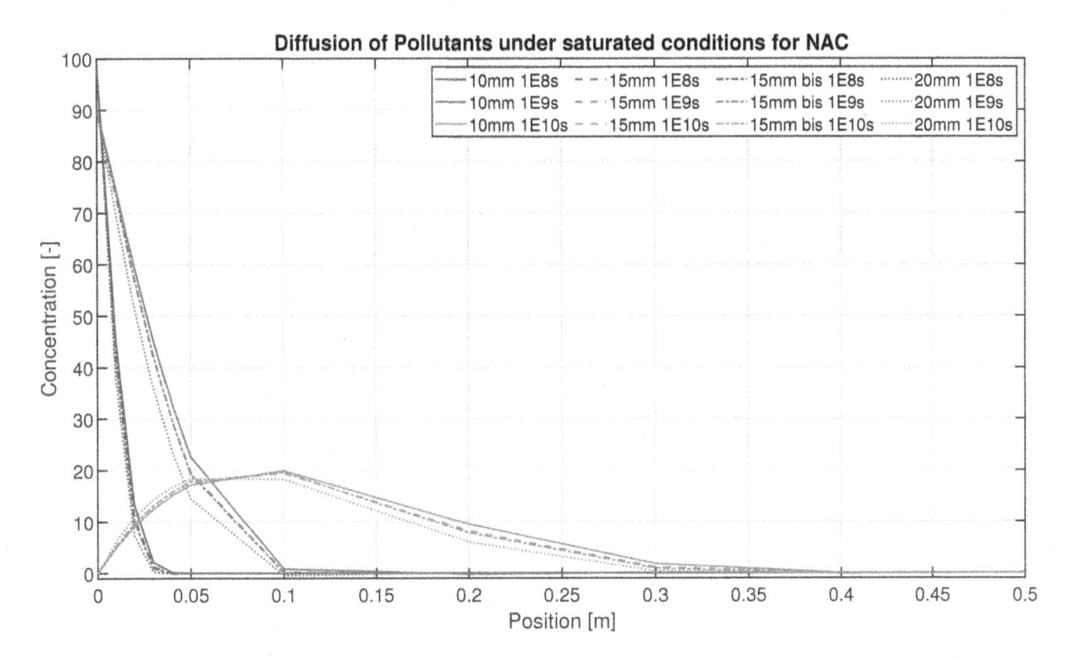

Figure 7. Results obtained for the diffusion of pollutants amongst several sizes of RVE for NAC: 10mm, two times 15mm and 20mm.

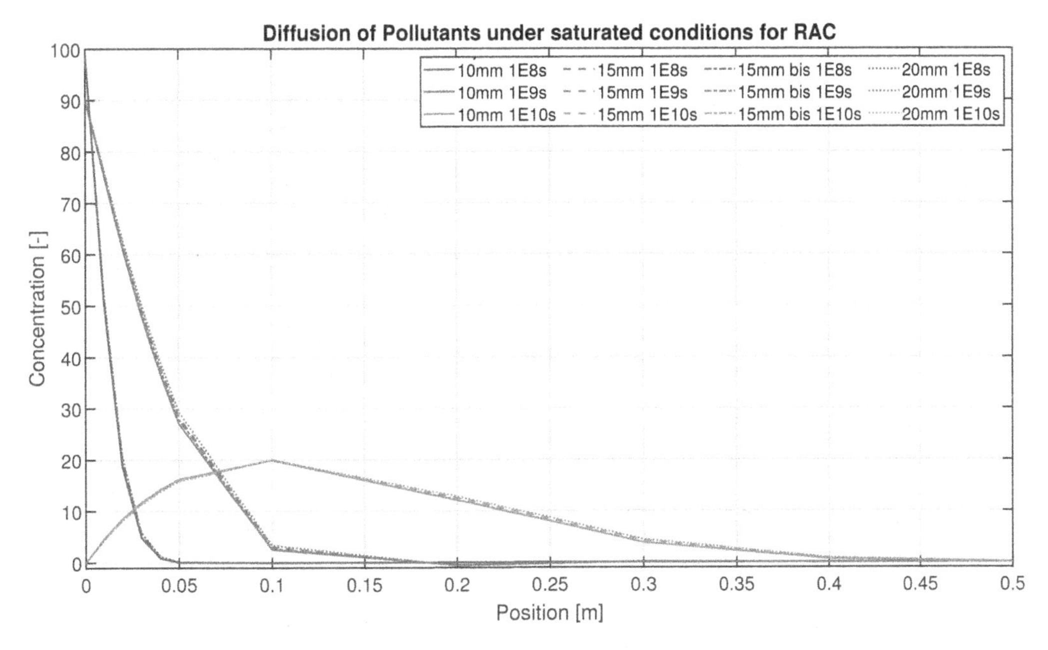

Figure 8. Results obtained for the diffusion of pollutants amongst several sizes of RVE for RAC: 10mm, two times 15mm and 20mm.

Figure 7 represents the evolution of the concentration for the NAC RVE. The first observation is that the two RVE with 15mm side exhibit the same results, which is comforting. Then, one can also observe that the smaller the RVE, the bigger the pollutant fluxes are. This is due to the inability of the RVE generation algorithm to adequately represents the requested granulometric curve when the size of the RVE decreases. There is therefore less aggregates and more mortar paste through which the pollutant can diffuse.

Figure 8 also represents the evolution of the concentration for the same RVE sizes, but for a concrete made from RCA. What is interesting is that the RAC results seem to depend less on the RVE size than the NAC. That may be due to the permeable surface of the RVE being already bigger than for the NAC, leading to fluxes that may not be limited by the permeable surface area but rather by the intrinsic properties of the material.

On Figure 9, representing the computation time with respect to the number of degree of freedom of the RVE, one observes that the computation time is directly linked to both the RVE size and the type of microstructure studied (the RAC is more complex than the NAC and therefore possesses more DOF). Fitting the numerical results with a power equation gives a R-squared value of 0.997 for the following equation:

$$y = 3.46 \times 10^{-12} \times x^{4.793} \tag{17}$$

It is therefore crucial to carefully choose the RVE to be used as it impedes on both the accuracy of the results and the computational time, both being in opposition with each other. In order not to sacrifice one or the other, parallelisation will be necessary.

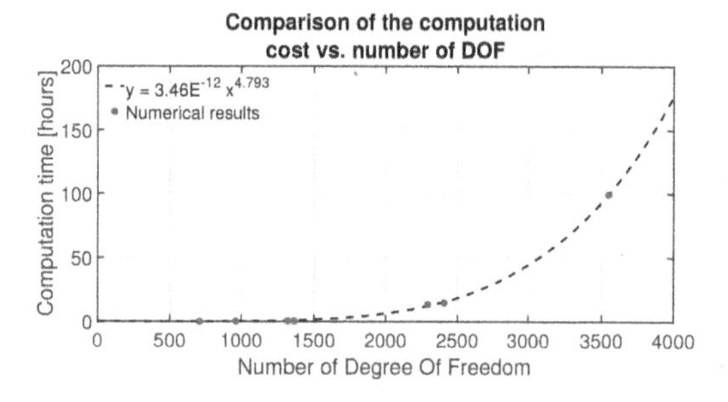

Figure 9. Evolution of the computational cost with respect to the number of Degree of Freedom (DOF) of the problem.

4 CONCLUSION

In conclusion, three observations were made:

1. The multiscale model allows the simulation of a pollutant diffusion inside a porous media, with an accuracy equal to the one of a classical simple-scale model. However, it has the advantage of using intrinsic properties instead of homogenized one, possibly depicting results closer to the reality;
2. The model replicates what has been found experimentally: the greater cement content of the RAC allows a greater chloride diffusion than the NAC;
3. The choice of the RVE has a great influence on the results performed; the aggregates add a certain complexity that increases the computation time while also increasing the accuracy of the results for a porous media such as concrete.

ACKNOWLEDGEMENTS

Funding: This work is supported by the Wallonia regional government (Belgium) in the framework of a FRIA (Fund for Industrial and Agricultural Research) grant.

COMPETING INTERESTS

The authors declare that they have no known competing financial interests or personal relationships that could have appeared to influence the work reported in this paper.

REFERENCES

Akbarnezhad, A., K. C. G. Ong., C. T. Tam, & M. H. Zhang (2013, December). Effects of the Parent Concrete Properties and Crushing Procedure on the Properties of Coarse Recycled Concrete Aggregates. *Journal of Materials in Civil Engineering 25*(12), 1795–1802.

Angst, U., B. Elsener, C. K. Larsen, & Ø. Vennesland (2009). Critical chloride content in reinforced concrete - A review. *Cement and Concrete Research 39*, 1122–1138.

Bear, J. & A. Verruijt (1987). *Modeling Groundwater Flow and Pollution*. D. Reidel Publishing Company.

Belin, P., G. Habert, M. Thiery, & N. Roussel (2014, September). Cement paste content and water absorption of recycled concrete coarse aggregates. *Materials and Structures 47*(9), 1451–1465.

Bertrand, F., O. Buzzi, P. Bésuelle, & F. Collin (2020). Hydro-mechanical modelling of multiphase flowin naturally fractured coalbed using a multiscale approach. *Journal of Natural Gas Science and Engineering 78*, 103303.

Biver, P. (1992). *Phenomenal and Numerical study on the propagation of pollutants*. Ph. D. thesis, University of Liége.

European Commission (2019, August). Construction and Demolition Waste (CDW). https://ec.europa.eu/environment/waste/construction_demolition.htm. Accessed: 28/08/2020.

Garboczi, E. J. & D. P. Bentz (1998). Multiscale Analytical/Numerical Theory of the Diffusivity of Concrete. *Advanced Cement Based Materials 8*, 77–88.

Geuzaine, C. & J.-F. Remacle (2009). Gmsh: a three-dimensional finite element mesh generator with built-in pre- and post-processing facilities. *International Journal for Numerical Methods in Engineering 79*(11), 1309–1331.

Hu, Z., L. Mao, J. Xia, J. Liu, J. Gao, J. Yang, & Q. Liu (2018). Five-phase modelling for effective diffusion coefficient of chlorides in recycled concrete. *Magazine of Concrete Research 70*(11), 583–594.

Hussain, H., D. Levacher, J.-L. Quenec'h, A. Bennabi, & F. Bouvet (2000). Valorisation des aggrégats issus de bétons de démolition dans la fabrication de nouveaux bétons. *Sciences et techniques 19*, 17–22.

Kouznetsova, V., W. A. M. Brekelmans, & F. P. T. Baaijens (2001). An approach to micro-macro modeling of heterogeneous materials. *Computational Mechanics 27*, 37–48.

Mangat, P. S. & B. T. Molloy (1994). Prediction of long term chloride concentration in concrete. *Materials and Structures 27*, 338–346.

Morga, M. & G. C. Marano (2015, June). Chloride Penetration in Circular Concrete Columns. *International Journal of Concrete Structures and Materials 9*(2), 173–183.

Nagataki, S., A. Gokce, T. Saeki, & M. Hisada (2004). Assessment of recycling process induced damage sensitivity of recycled concrete aggregates. *Cement and Concrete Research 34*, 965–971.

Nilenius, F. (2014). *Moisture and Chloride Transport in Concrete - Mesoscale Modelling and Computational Homogenization*. Ph. D. thesis, Chalmers University of Technology, Gothenburg, Sweden.

Rao, A., K. N. Jha, & S. Misra (2007). Use of aggregates from recycled construction and demolition waste in concrete. *Resources, Conservation and Recycling 50*, 71–87.

Smit, R. J. M., W. A. M. Brekelmans, & H. E. H. Meijer (1998). Prediction of the mechanical behavior of nonlinear heterogeneous systems by multi-level finite element modeling. *Computer Methods in Applied Mechanics and Engineering 155*, 181–192.

Sun, C., Q. Chen, J. Xiao, & W. Liu (2020). Utilization of waste concrete recycling materials in self-compacting concrete. *Resources, Conservation & Recycling 161*, 104930.

Xi, Y. & Z. P. Bazant (1999, February). Modeling Chloride Penetration in Saturated Concrete. *Journal of Materials in Civil Engineering 11*(1), 58–65.

Zhao, Z., L. Courard, S. Groslambert, T. Jehin, A. Léonard, & J. Xiao (2020). Use of recycled concrete aggregates from precast block for the production of new building blocks: An industrial scale study. *Resources, Conservation and Recycling 157*, 104786.

Computational Modelling of Concrete and Concrete Structures – Meschke, Pichler & Rots (Eds)
 ISBN: 978-1-032-32724-2

Computational modelling of material behaviour of layered 3D printed concrete

O. Shkundalova & T. Molkens
KU Leuven, campus De Nayer, Sint-Katelijne Waver, Belgium

M. Classen
RWTH University, Institute of Structural Concrete, Aachen, Germany

B. Rossi
KU Leuven, campus De Nayer, Sint-Katelijne Waver, Belgium
New College, University of Oxford, Oxford, UK

ABSTRACT: Based on mechanical test results performed parallel, perpendicular and at an angle to the print direction, an attempt is made to describe the laminar material behaviour of 3D printed concrete. With the resulting material model, several possible mathematical simulation techniques are compared to propose the most suitable design method. In addition to modifying the material properties, equivalent geometrical quantities can also be used to correctly estimate the bending stiffness and resistance with a homogeneous material model. The paper discusses critical aspects of current modelling strategies for 3D printed concrete and highlights possible improvements that are the subject of ongoing research. Together with that, influence, and consequences of uncertainties in the design with extrusion-based concrete are discussed in this paper. The results obtained are analysed in line with the models for conventionally cast concrete, and the discrepancies are addressed in detail.

1 INTRODUCTION

1.1 *State of the art*

Additive manufacturing technologies have become actively used in the construction sector over the past decades (Meurer et al. 2021). Three-dimensional printing of concrete (3DPC) has numerous advantages over conventional production techniques, allowing fast and cost-efficient fabrication of complex geometrical shapes right on the construction site and eliminating the need for formwork, at the same time.

Due to the features of layer-by-layer production process, the mechanical properties of 3D printed concrete differ from conventionally cast concrete owing to a wide variety of different parameters:

- features of production techniques (e.g., layered extrusion, shotcrete, powder-based concrete printing);
- climatic conditions during and after manufacturing;
- material behaviour (flowability of the mixture, faster shrinkage, sensitive interlayer bonding, unstable reaction at elevated temperatures due to hydration processes), etc.

The speed of concrete extrusion, as well as the environmental conditions, such as temperature and humidity, have a big influence on the speed of shrinkage of material, rheological and hardened properties of 3DPC, its quality and material strength. The interval time between subsequent extrusion of layers and the speed of printing can also have a big influence on a final product (Panda et al. 2019). Thus, having a shorter interlayer time, the previous layer may not be hard enough to carry subsequently printed layers. This means that when the next layer of concrete is printed, the bottom layer can be squashed and structural failure of the component may occur during printing.

In July 2020, a first house was completely printed in Belgium as one piece with, at that time, the largest 3D printer in Europe (Figure 1).

Figure 1. The 3D printed house at KampC – Oevel, Belgium.

DOI 10.1201/9781003316404-9

To ensure safe design, stability and durability over the intended service life of the structures, a more thorough investigation into the modelling of 3D printed concrete behaviour is needed. Facing the actual tools used by design engineers, the complex material behaviour should by preference be translated in a straightforward way into more commercial numerical tools as the ones based on the finite element method.

1.2 *Design tools*

Numerical modelling is one of the widely used methods for predicting the structural performance of a material under loads. Though, material models for computer simulation of layered concrete have not yet been sufficiently developed. In this paper, proposals to describe the material behaviour in a straightforward way will be discussed based on proper test results.

Computational modelling of layered 3D printed concrete is a complex endeavour. 3DPC has anisotropic material properties. Material strength varies depending on the raster orientation of the layers. The emphasis is made on explicit modelling of 3DPC material behaviour focusing on a geometrical equivalent description. The first step in the process is to find the proper constitutive law(s) that can de scribe the relation between stresses/strains or forces/displacements. In a second step also failure criteria should be applied to determine the bearing capacity of a bearing element.

The attempt was already made to model structural response of 3DPC at fire (Ni et al. 2021). Several models are proposed for modeling of 3DPC taking into consideration the fresh concrete material strength and avoiding a structural failure during concrete printing process (Nedjar et al. 2021).

In this paper, the accent was made on the use of regular construction software for fast and efficient prediction of structural behaviour available on the market, and which is widely used by specialists in industry. In this case, Diamonds BuildSoft was used.

2 CONSTITUTIVE LAWS, YIELD AND FAILURE CRITERIA

2.1 *Constitutive laws*

For brittle materials subjected to and failing in tension, to which concrete belongs, the maximum applied stress is proportional to the maximum strain. This means, that when designing with conventional concrete based on EC2 recommendations, the tensile strength of concrete would be a determining factor prescribing the strength limit and linear-elastic material response. A distinction should, however, be made between linear and 2D structural elements.

Failure would be expected when the material strength reaches its maximum capacity and can be calculated by a first or second order analysis, assuming a constant geometry of the sample until failure.

When designing with 3D printed concrete, the anisotropic material properties need to be taken into consideration (Liu et al. 2021). The material strength in different direction with respect to the print line can be different due to different printing parameters.

2.1.1 *One dimensional beam element*

The well-known Hook's constitutive law describing the relation between stresses and deformation is defined as Eq. (1):

$$\sigma = \frac{F}{A} = E \cdot \varepsilon \tag{1}$$

where σ is the applied stress, F is the applied force, A is a cross sectional area, E is the Young modulus, and ε represents the strain.

2.1.2 *Orthotropic plate*

Considering the design features of 3D printed concrete, anisotropic material properties should be taken into account. Regarding the elastic properties of the material of the plate, three different planes of symmetry can be considered as the coordinate planes in rectangular coordinates (Stephen et al. 1970). The stress-strain relation in this case can be described as follows:

$$\begin{aligned} \sigma_x &= E'_x \varepsilon_x + E'' \varepsilon_y \\ \sigma_y &= E'_y \varepsilon_y + E'' \varepsilon_x \\ \tau_{xy} &= G \gamma_{xy} \end{aligned} \tag{2}$$

where σ_x, σ_y are normal components of stress parallel to x and y axes; xy is a middle plane of the plate before loading; τ_{xy} is a shear stress component in xy plane; E'_x, E'_y, E'', G are four constants (modulus of elasticity) in different directions and in shear; $\varepsilon_x, \varepsilon_y$ are unit elongations in x and y directions; γ_{xy} is a shear strain component.

When taking into consideration the curvature of the deflection curve equal to d^2w/dx^2, where w is a deflection in a z direction, the following equation for the strain components at a distance z from the middle surface is obtained:

$$\varepsilon_x = -z\frac{\partial^2 w}{\partial x^2}; \varepsilon_y = -z\frac{\partial^2 w}{\partial y^2}; \gamma_{xy} = -2z\frac{\partial^2 w}{\partial x \partial y} \tag{3}$$

The expression for the stress components can be defined as:

$$\sigma_x = -z\left(E'_x\frac{\partial^2 w}{\partial x^2} + E''\frac{\partial^2 w}{\partial y^2}\right)$$
$$\sigma_y = -z\left(E'_x\frac{\partial^2 w}{\partial y^2} + E''\frac{\partial^2 w}{\partial x^2}\right) \quad (4)$$
$$\tau_{xy} = -2Gz\frac{\partial^2 w}{\partial x \partial y}$$

The bending moment of the plate can be calculated as follows:

$$M_x = \int_{-\frac{h}{2}}^{\frac{h}{2}} \sigma_x z dz = -\left(D_x\frac{\partial^2 w}{\partial x^2} + D_1\frac{\partial^2 w}{\partial y^2}\right)$$
$$M_y = \int_{-\frac{h}{2}}^{\frac{h}{2}} \sigma_y z dz = -\left(D_y\frac{\partial^2 w}{\partial y^2} + D_1\frac{\partial^2 w}{\partial x^2}\right) \quad (5)$$
$$M_{xy} = \int_{-\frac{h}{2}}^{\frac{h}{2}} \tau_{xy} z dz = 2D_{xy}\frac{\partial^2 w}{\partial x \partial y}$$

in which:

$$D_x = \frac{E'_x h^3}{12}; D_y = \frac{E'_y h^3}{12}; D_1 = \frac{E'' h^3}{12}; D_{xy} = \frac{G h^3}{12} \quad (6)$$

where D_x, D_y, D_{xy}, D_1, is a flexural rigidity of the plate (Stephen et al. 1970). It is important to notice that the rigidity coefficients are each time the product of a Young modulus and a height.

2.2 *Yield and failure criteria*

Three basic failure criteria were developed in the plane stress states for tensile, compressive failure and a failure in shear.

Based on material strength test results, no yield was observed in flexure and compression, and the 3DPC samples failed in brittle way at maximum load (Feng et al. 2015). The maximum stress criterion is suggested to be used to describe the mechanical behaviour of 3DPC (Jones 1998):

– tension: $\sigma_1 < X_t \quad \sigma_2 < Y_t \quad \sigma_3 < Z_t$
– compression: $\sigma_1 > X_c \quad \sigma_2 > Y_c \quad \sigma_3 > Z_c \quad (7)$
– shear: $\sigma_{12} > S_{12} \quad \sigma_{23} > S_{23}\sigma_{31} > S_{31}$

where $\sigma_1, \sigma_2, \sigma_3$ are maximum principal stresses; X_i, Y_i, Z_i are maximum allowable stresses in different directions; S_{12}, S_{23}, S_{31} are maximum allowable shear stresses in different planes.

The yield criteria for other materials, such as timber and fibre reinforced polymer (FRP) composites were considered instead.

2.2.1 *Analogy based on the criteria for timber*

Although timber can have a non-linear post-peak capacity, timber structures are generally designed in a linear elastic range. The failure in plane stress state for timber structures is described by three basic orthotropic strengths criteria, also known as plasticity conditions (Haasbroek et al. 1994). As timber is a brittle material, and there is almost no ductility in the tensile zone, but linear elastic-plastic behaviour of timber in compression can be assumed. The elastic-plastic behaviour can be described by applying the Hook's low for the elastic part, and the yield function for the plastic part. The typical stress-strain relation for timber can be seen in Figure 2 (Sørensen et al. 2022).

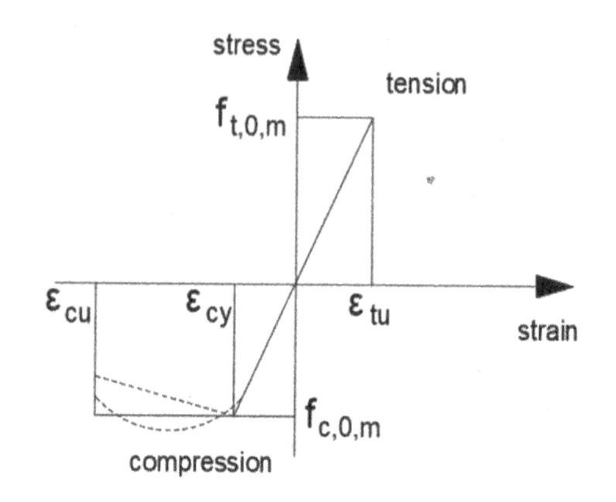

Figure 2. Typical stress-strain curve of timber (Sørensen et al. 2022).

Due to the orthotropic material properties of timber, the compressive stress at an angle to the grain can be calculated as follows (Section 6.2.2, EC5):

$$\sigma_{c,\alpha,d} \leq \frac{f_{c,0,d}}{\frac{f_{c,0,d}}{k_{c,90}\, f_{c,90,d}} sin^2\alpha + cos^2\alpha} \quad (8)$$

where $\sigma_{c,\alpha,d}$ is a compressive stress at an angle α to the grain; $k_{c,90}$ is the factor which takes into consideration the effect of any stresses perpendicular to the grain.

2.2.2 *The ice crushing failure equivalence*

The tensile strength of ice is substantially lower than its compressive strength, which allow us to consider ice as an orthotropic material. Referring to the principal axes of anisotropy, the yield criteria for columnar-grained structure of ice can be described as follows (Chen et al. 1988):

$$f(\sigma_{ij}) = a_1\left[(\sigma_y - \sigma_z)^2 + (\sigma_z - \sigma_x)^2\right] + a_3(\sigma_x - \sigma_y)^2 + a_4(\tau_{yz}^2 + \tau_{zx}^2) + 2(a_1 + 2a_3)\tau_{xy}^2 + a_7(\sigma_x + \sigma_y) + a_9\sigma_z - 1 = 0 \quad (9)$$

where coefficients a_1, a_3, a_7, a_9 are the tensile and compressive strength measurements and can be found as follows:

$$a_1 = \frac{1}{2C_zT_z} \quad a_3 = \frac{1}{C_xT_x} - \frac{1}{2C_zT_z} \quad a_7 = \frac{1}{T_x} - \frac{1}{C_x}$$

$$a_9 = \frac{1}{T_z} - \frac{1}{C_z} \tag{10}$$

where $T_x, T_z, C_x, C_z,$ are the tensile and compressive strength values. The value of a_4 can be determined from the shear tests or compressive tests of the samples with an angle to the vertical direction. Assuming a plane stress conditions, the simplification can be applied:

$$\sigma_z = \tau_{yz} = \tau_{xz} = 0 \tag{11}$$

Having x and y the principal stress directions, the equation takes form (Chen et al. 1988):

$$a_1\left(\sigma_x^2 + \sigma_y^2\right) + a_3\left(\sigma_x - \sigma_y\right)^2 + a_7\left(\sigma_x + \sigma_y\right) = 1 \tag{12}$$

3 TRANSLATING MATERIAL ANISOTROPY

3.1 *Stiffness properties = constitutive laws*

Instead of working with different Young moduli like follows out of Eq. (6), in this paper, it is proposed to adopt the geometry of the section by changing the second moment of area (by the height h). This is needed as most elastic software tools cannot handle differences in Young modulus. However, it is mostly possible to work with ribbed slabs or waffle floors. Material orthotropy is in this way replaced by geometrical orthotropy.

Equal stiffness can be presented by respecting the flexural rigidity EI and adapting I instead of E in both parallel and transverse directions with respect to the print-line. Once cracking occurs in reinforced concrete the second moment of area is significantly reduced. To define the cracked area and by that way the influence on the deformations the cracking moments (M_{cr}) should be properly calculated. Also, here a geometrical equivalent is proposed, by respecting the outcome of the product $M_{cr} = \sigma_{cr} \cdot W$. Direction dependent maximum stresses σ_{cr} will be replaced by different moduli of resistance W.

When using ribbed slabs, differences in height and width of the ribs in each direction can be worked out to fit simultaneously the second moment of area and the resistance modulus in each direction.

3.2 *Material behaviour = failure criteria*

Till so far, the research is limited to 3DPC unreinforced concrete applications. Brittle behaviour of 3DPC is each time observed in the performed test, which will be discussed in this contribution. Reason why, at this moment there is not yet a need for more advanced failure criteria which do account for plasticity. This will be observed when failure in the compression zone can be expected, as for reinforced concrete applications. While, this is the final goal of the research team this field is not yet exploited.

3.3 *Cracking behaviour*

The cracking behaviour of 3D printed concrete significantly differs from that of conventionally cast concrete due to the nature of layer-by-layer extrusion process (Liu et al. 2021). The print layer interfaces and the defects appearing during the printing process lead to different material properties in different directions with respect to the print line (Jenkins et al. 2021). There is insufficient knowledge about the effect on fracture mechanics in weaker interlayer regions (Ven den Heever et al. 2021). By improving the quality of the interlayer interface, the anisotropy can be reduced (Babafemi et al. 2021).

Considering the micro-structure of 3DPC, the pores volume distribution in 3DPC is different compared to conventionally cast concrete and the porosity is primarily aligned with the printing direction (Moini et al. 2021). This will not be usually the case for conventionally cast concrete members with spherical pores randomly distributed in the material (Moini et al. 2021).

There is a large number of factors affecting material strength and structural performance of 3DPC, i.e. speed of concrete extrusion, interlayer time between deposition of the layers, environmental conditions (temperature and humidity in the built chamber). This, in its turn, will influence the quality of the final product, material strength, level of porosities and uncertainties in the material, speed of shrinkage, structural performance under loads.

Together with that, 3DPC has high surface roughness. Concrete slump during the extrusion process is another important parameter to consider. This might have negative influence on bending strength of the produced element. Assuming a good quality of material with low level of porosities, the crack in most cases will be initiated at the surface at the point of geometrical nonlinearity where the stress will reach its extremum.

When choosing a test method it is important to mention that different testing methods can lead to a different results depending on the size of the specimen, support conditions, loading rate, etc (Meurer et al. 2021).

The expected failure modes in 3DPC can be the following (Van den Heever et al. 2021):

- interfacial delamination,
- interface shear-slip,
- intralayer tensile cracking,
- intralayer cracking under compression-shear,
- crushing.

Summarising all mentioned above, crack in 3D printed elements can be initiated due to different reasons: geometrical nonlinearities, inner pores, debonding and lack of fusion between concrete layers leading to the interfacial delamination, which can arise when the interlayer time during printing is too big, and the non-uniform shrinkage takes place.

4 MATERIAL PROPERTIES

4.1 *Material used for 3D concrete printing*

The concrete mixture "Weber 3D 145-2, 3D concrete printing mortar C35/45 – 1 mm – CEM I" was used for the proof of the concepts described in this paper. It is a factory-produced dry mortar made in accordance with EN 206 and available in the market. Properties of concrete mixture can be seen in Table 1.

Table 1. Properties of Weber 3D 145-2 mortar used in the study.

Compressive strength class	C35/45
Largest grain size: D-max	1 mm
Density (28 days, EN 12390-7)	2200 kg/m^3
Compressive strength	>45 MPa

Mechanical material characteristics will be discussed in the next sections for 3DPC tested following different directions and compared with those obtained by casting monolithic beams of equal dimensions as a reference base.

4.2 *Material strength based on test results*

The wall was extruded at ambient temperature with the effective thickness of 40 mm (taken without slumps appearing during concrete extrusion process, Figure 3). The reduced effective thickness is considered for the evaluation of the mechanical properties of the 3DPC and in structural design (Asprone et al 2018).

Figure 3. Wall panel for production of test specimens.

The height of each concrete layer extruded by a 3D printer was 15 mm. The samples for material strength tests were cut out of this wall at longitudinal printing direction, perpendicular and at an angle of 45 degrees to the print-line The samples were cut with the consideration of having a smooth surface without visible connection line between different layers.

For the 3 points bending tests a sample dimension of 40 mm x 40 mm x 160 mm were used, a speed of 0.25 mm/min were used for the displacement control. For the compressive tests, two residual parts of the beam elements with a load surface of 40 mm by 40 mm were used, with a speed of 2.4 kN/s in accordance with EN 196, ISO 679 and EN 12390-5:2009, Figure 4. Preliminary study was based on limited amount of tests.

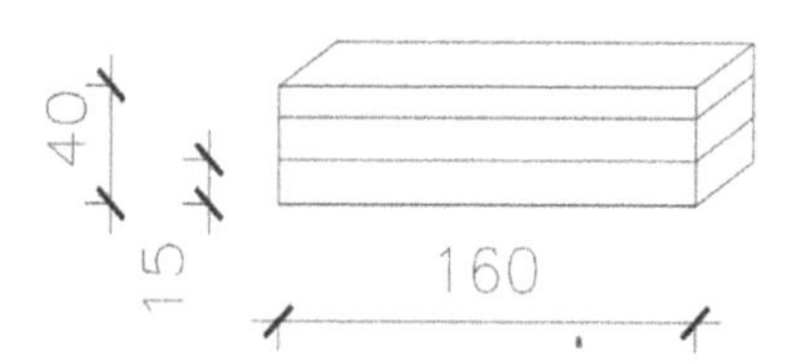

Figure 4. Dimensions of the 3D printed test specimen for flexural tests.

It is worth to mention, that due to the thickness of the printing layer (effective thickness 40 mm + slumps) the water is easily penetrated through connecting surfaces between the layers.

This became clear during sawing operations to prepare the test specimens. The Figure 5 gives a clear representation of the water penetration process. In real life application, this will also happen for the elements subjected to humid environmental conditions and produced with 3D printing technique. This is an important point to consider in the design when choosing the thickness of the printed layer for the real-world application of 3D concrete printing technique.

Figure 5. Water penetration in 3D printed concrete sample.

The samples for material strength tests were cut with the consideration of having as smooth outer surfaces as possible, e.g. without visible connection line between different layers to achieve more accurate results. Out of preliminary tests it was observed that the irregular surface made it very hard to respect an acceptable coefficient of variation when tensile stresses appear perpendicular to the interface.

4.3 *Material behaviour parallel to the layers*

Due to the layered structure of 3D printed element, the material strength is highly dependent on the printing direction. It is already well-known, that bending strength of orthotropic materials, such as timber and fibre reinforced polymer (FRP) composites, is higher in the direction of the fibres. When it comes to 3D printed concrete, a thorough investigation was made, and it was proven that extrusion based concrete has the same tendency: material strength is higher in the direction of the print-line.

In accordance with EN 1990:2002+A 1 :2005 (E), Annex D, the normal and lognormal distribution can be used for the material properties calculation. The JCSS Probabilistic Model Code suggests to use a lognormal distribution for calculation of material properties of concrete. In this study, the lognormal distribution was used for the calculation of the average 3DPC material strength values. Therefore, with the known coefficient of variation V_X calculated with a normal distribution, and the standard deviation s_y applied on the lognormal values of measurements, the s_y was calculated as follows:

$$s_y = \sqrt{\ln(V_X^2 + 1)} \approx V_X \tag{13}$$

The average values of material strength is the exponent of the averaged lognormal values of the measurements.

Table 2 shows the material strength of 3DPC in flexure and compression in longitudinal direction parallel to the layers, the layers were oriented horizontally and vertically as depicted in Figure 6.

Table 2. Mechanical properties of 3DPC in longitudinal direction.

Test	Orientation	Number of spec.	Average strength (MPa)	$s_y \approx V_X$
Sample size: flexural test 40 x 40 x 160 compression tests 40 x 40				
Flexure	Horiz.	10	7.54	0.1
	Vertic.	5	7.1	0.15
Compression	Horiz.	6	42.52	0.08
	Vertic.	9	47.36	0.1

According to the table above, the average flexural strength in longitudinal direction with horizontal and vertical layer orientation has similar values of 7,54 MPa and 7,1 MPa respectively, which is twice

Figure 6. Horizontal and vertical orientation of the 3DPC samples parallel to the layers.

higher than the mean tensile strength for the same class concrete according to EC2 (3.21 MPa for C35/45).

Based on the compressive tests, the compressive strength equals to 42.52 MPa and 47.36 MPa for horizontal and vertical sample orientations respectively. This values are within the safe variation limit to the data provided in product description of the mortar.

4.4 *Material behaviour perpendicular to the layers*

Two series of tests were performed for the samples with transverse raster orientation perpendicular to the payers. Figures 7, 8 represent two variations of surface treatment.

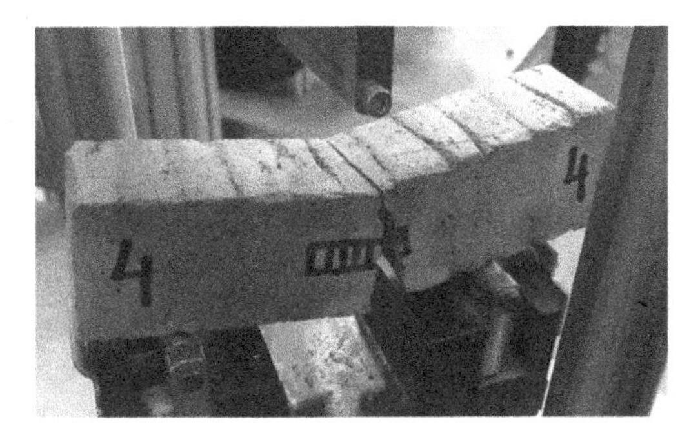

Figure 7. 3-point bending test of the sample with transverse raster orientation (series 1).

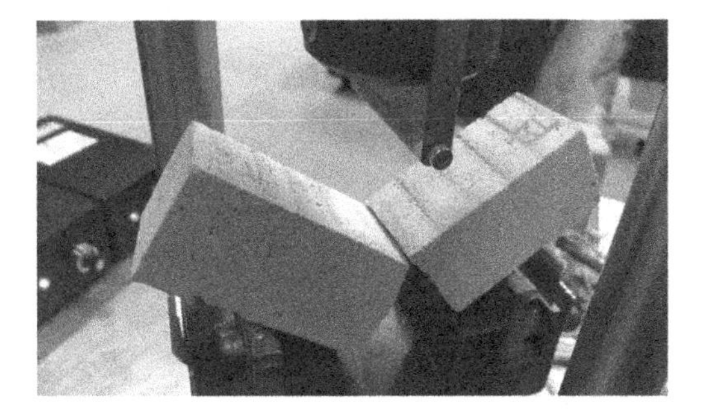

Figure 8. 3-point bending test of the sample with transverse raster orientation (series 2).

It can be observed that the samples from series-1 with dimensions 40 x 40 x 160 had higher surface roughness, and the samples from series-2 with dimensions 35 x 35 x 160 had smooth surface treatment.

All the samples were tested with the load applied along the interlayer interfaces as the strength of the samples highly depend on the bond strength between the concrete layers.

Mechanical properties in transverse raster orientation perpendicular to the layers can be seen in Table 3.

Table 3. Mechanical properties of 3DPC in transverse direction.

Test	Number of specimens	Average strength (MPa)	$s_y \approx V_X$
Sample size: flexural test 40 x 40 x 160 compression tests 40 x 40			
Flexure	3	4.56	0.13
Compression	8	49.16	0.11
Sample size: flexural test 35 x 35 x 160 compression tests 35 x 40			
Flexure	5	7.74	0.04
Compression	9	47.36	0.1

It can be observed that the flexural strength in transverse direction was 4.56 MPa, and compressive strength equals to 49.16 MPa for the samples with higher surface roughness.

For the samples with smooth surfaces, the material strength was 7.74 MPa and 47.36 MPa in flexure and compression respectively.

4.5 *Material behaviour with an angle to the print-line*

The material strength in the direction of 45 degrees was also measured, and the test results in compression and flexure can be seen in Table 4.

Table 4. Mechanical properties of 3DPC in direction of 45 degrees to the print-line.

Test	Number of specimens	Average strength (MPa)	$s_y \approx V_X$
Flexural	2	5.75	0.13
Compressive	4	44.48	0.05

In case of the samples with the orientation of 45 degrees to the layer direction, the crack started at the connection between two layers, Figure 9.

The average flexural strength was 5.75 MPa, and compressive strength was 44.48 MPa.

4.6 *Material strength of the cast specimens*

To be able to compare results, the series of the cast specimens were also tested, produced from the same mortar Weber 3D 145-2 in conventional way in the formwork. The results can be seen in Table 5.

As it can be seen from the table above, the flexural strength of the cast specimens is 7.48 MPa, and the compressive strength is 51.03 MPa.

Figure 9. 3-point bending test of the sample with printing orientation of 45 degrees to the print-line.

Table 5. Mechanical properties of the cast samples from the mortar Weber 3D 145-2.

Test	Number of specimens	Average strength (MPa)	$s_y \approx V_X$
Flexural	6	7.48	0.06
Compressive	12	51.03	0.02

4.7 *Cracking behaviour*

The crack in 3D printed elements was initiated due to different reasons: geometrical non-linearities at outer surfaces, inner pores, debonding. For the cases with transverse layer orientation, the crack initiated at the external surfaces and broke along the interlayer interfaces (Figures 7, 8, 10). This can be caused by initial geometrical imperfections and is also influenced by the bond strength of the interlayer surface.

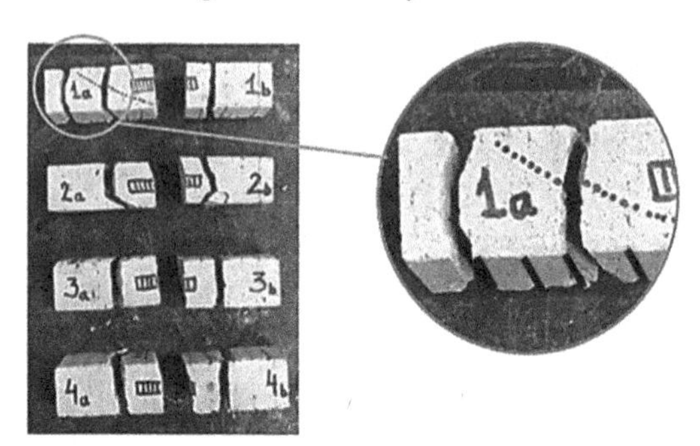

Figure 10. Cracks in 3DPC samples with transverse raster orientation.

Another one important fact is that the 3D printed concrete is not compacted while printing compared to the conventionally cast concrete. This leads to increased number of voids and uncertainties in extrusion based material. This inner voids can be the locations of crack initiation, and it can be seen in Figure 11.

It can be seen, that the crack didn't take a usual shape while breaking in flexure, and had a small inclination. This was caused by a porosity and voids inside the printed material.

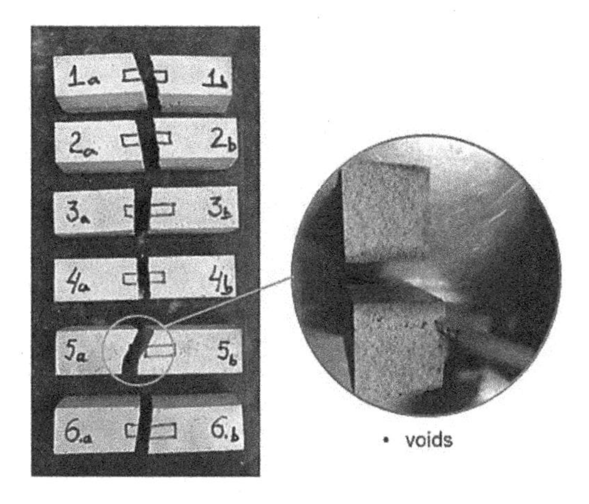

Figure 11. Voids initiating the crack in 3DPC (cast samples).

5 DESIGN EXAMPLE

In order to design material behaviour of 3DPC using conventional construction software, a plate with the dimensions 550 mm x 300 mm x 40 mm was used, with the thickness corresponding to the layer effective thickness during concrete extrusion (40 mm in case of this study). This is due to reduced contact efficiency resulting from the features of 3D printing technology that affect specific layer geometries, which should be taken into account during design (van den Heever et al. 2022).

Based on the material strength test results, the Young's moduli for the 3D printed concrete in different direction to the print-line were calculated. An approximation was used, since the tensile strength for the samples with longitudinal print direction placed horizontally (7.54 MPa) and vertically (7.1 MPa) was similar to that of the samples cast in conventional way (7.48 MPa). The results are as follows:

$$-E_{long} = 14\ GPa$$

$$-E_{transv} = 10.5\ GPa$$

$$-E_{conven} = 14\ GPa$$

where E_{long}, E_{transv}, E_{conven} are the Young's moduli for 3D printed concrete in longitudinal and transversal directions, and for the samples cast conventionally.

It can be observed that the elastic modulus in transversal direction E_{transv} is approximately 75% of that in longitudinal direction E_{long}. Two material models were used:

1) homogeneous isotropic material properties of 3DPC based on mechanical properties of the mortar used;
2) orthotropic model with the transverse stiffness equal to 75% of the longitudinal (main) layer direction.

Generally, there are two ways to estimate flexural rigidity EI of the plate (where E = modulus of elasticity; I = moment of inertia): to modify E in longitudinal and perpendicular printing directions respectively, or by adapting I by changing the geometry of the section in perpendicular direction, in this case by assigning the plate thickness = 36 mm with E = constant in all directions.

At this stage the influence on the resistance modulus is not yet included as the slab part cut out of an unreinforced wall will behave brittle without any post-cracking behaviour. As already described in section 3.1, this is easily feasible without a need of advanced skills.

The plate was modelled with supports at three points, and loaded up to 0.65 kN. Figure 12 represents both design concept for homogeneous isotropic (left) and orthotropic (right) material models respectively.

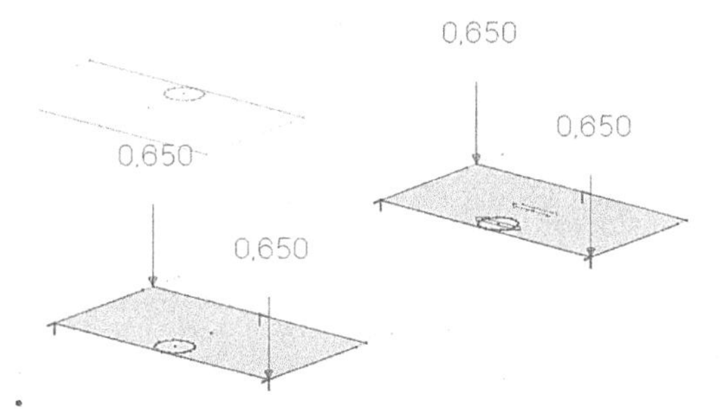

Figure 12. Support and loading conditions of the plate.

The vertical displacement was calculated for both homogeneous and orthotropic material models (Figure 13).

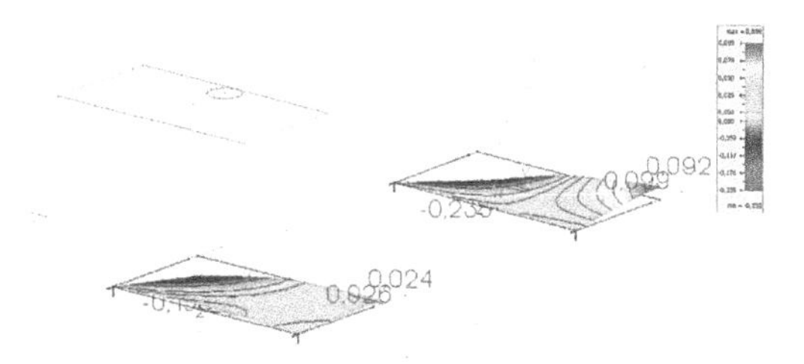

Figure 13. Vertical displacement of the plate.

6 VALIDATION

To validate the results of FE simulations, the real 3d printed concrete plate of the same dimensions 550 mm x 300 mm x 40 mm was tested.

The plate was supported and loaded in the same way as it was previously designed: three point supports and two points of load application.

The vertical displacement was measured at two free corners of the plate: left (L) and right (R). The plate loaded up to 0.65 kN can be seen in Figure 14.

Figure 14. Vertical displacement of the plate.

The loading steps and the values of vertical displacement from the tests and FE simulation for homogeneous and orthotropic material models can be seen in Table 6. Red circles in Figure 14 represents vertical displacement measurement locations.

Table 6. Vertical displacement of free corners of the plate during test and FE simulation (mm).

	L			*R*		
Load (kN)	test	ort. model	hom. model	test	ort. model	hom. model
0.15	−0.01	−0.070	−0.047	0.01	0.021	0.006
0.25	−0.08	−0.117	−0.078	0.015	0.035	0.009
0.35	−0.15	−0.163	−0.109	0.04	0.050	0.013
0.45	−0.23	−0.210	−0.140	0.06	0.064	0.017
0.55	−0.29	−0.256	−0.171	0.08	0.078	0.020
0.65	−0.36	−0.303	−0.202	0.11	0.092	0.024

This study clearly reflects the well-corresponding values of the vertical displacement between tests and FE simulation for the orthotropic material model, which are −0.36 mm and −0.303, respectively, at the maximum applied load on the left corner (L). For the homogeneous material model in the simulation, the vertical displacement was underestimated by about 30% and was −0.202 mm.

A similar situation was observed for the right free corner of the 3DPC plate (R). The lift of the plate during the test corresponds to that in the orthotropic material model, having similar values of 0.11 mm and 0.092 mm, respectively. This was not the case for a model with homogeneous material properties, in which the vertical displacement was underestimated by a factor of 4 and, according to FE simulation, was 0.024 mm.

The vertical displacement at each load step during the test can be seen in Figure 15. A linear increase in vertical displacement in line with the increase in load can be observed.

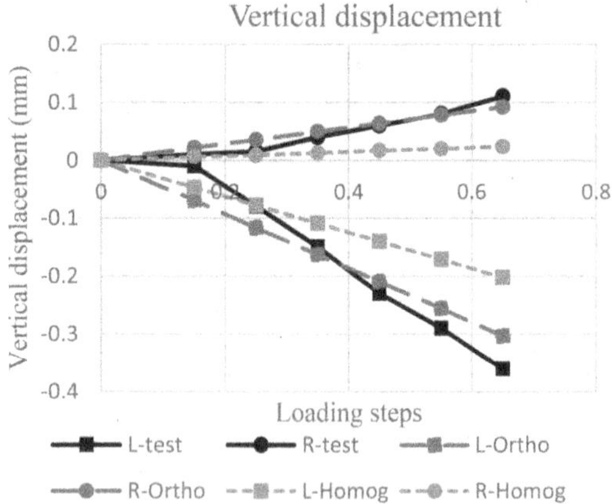

Figure 15. Vertical displacement at left (L) and right (R) corners of the plate at each load step.

7 SUSTAINABILITY

3D printing technology has a bright future, especially when it comes to sustainable construction. New environmental friendly materials can be used for 3D printing, e.i. geopolymers replacing the standard binders, and recycled aggregates. The attempts have been already made, and according to the cost analysis the use of sustainable materials in 3D printing reduces construction time and decreases the energy demand by around 50% compared to the conventional production techniques, as well as reducing the CO2 emission (Munir at el. 2021).

8 CONCLUSIONS

This paper proposes a simple and efficient approach to computational modeling of deformationbehaviour of 3D printed structures prior to failure using conventional construction software that does not require any add-ons for complex material designs.

Material properties of 3DPC were studied empirically:

- the samples oriented parallel to the layers had similar strength to that of conventionally cast samples;
- samples with a transverse raster orientation perpendicular to the printing layers with smooth outer surfaces (series-2) showed results close to the results of samples with a longitudinal orientation and of samples cast conventionally.
- Mechanical properties of the samples in series-1 with higher surface roughness was lower compared to the samples series-2. This is due to the quality of the interlayer interfaces and defects of external surfaces, which were partially removed for the samples in series-2. Having smaller cross section, the samples showed higher mechanical strength.

Taking into account anisotropic material properties of 3D printed concrete, two material models were designed using Diamonds BuildSoft, and the results were verified by tests. This is the first attempt to simulate the pre-peak behaviour of 3DPC with commercial software. It is meant to define the deformation behaviour prior to failure, but does not allow to predict the strength of a 3d printed element.

It can be concluded that material anisotropy for 3DPC can be modelled by adjusting the geometry of the designed element proportionally reducing the geometry of the samples in transversal direction. This approach was verified empirically, and the results obtained correspond well to the material model with the orthotropic material properties. The model with homogeneous material properties showed the results underestimating the vertical deflection at around four times.

ACKNOWLEDGEMENT

- Funding: KU Leuven Impulse Fund IMP/20/018: Development of viable 3D-printing methods for digital fabrication of reinforced concrete.
- Provinciaal Centrum Duurzaam Bouwen & Wonen Kamp C for providing materials for testing.

REFERENCES

Asprone, D., Auricchio, F., Menna, C., & Mercuri, V. (2018). 3D printing of reinforced concrete elements: Technology and design approach. *Construction and Building Materials*, *165*, 218–231. https://doi.org/10.1016/j.conbuildmat.2018.01.018

Babafemi, A. J., Kolawole, J. T., Miah, M. J., Paul, S. C., & Panda, B. (2021). A concise review on interlayer bond strength in 3d concrete printing. In *Sustainability (Switzerland)* (Vol. 13, Issue 13). MDPI AG. https://doi.org/10.3390/su13137137

Chen W.F., Han D.J. (1988). Plasticity for Structural Engineers. New York: Springer-Verlag New-York Inc, 606 pages.

Feng, P., Meng, X., Chen, J. F., & Ye, L. (2015). Mechanical properties of structures 3D printed with cementitious powders. *Construction and Building Materials*, 93, 486-497. https://doi.org/10.1016/j.conbuildmat.2015.05.132

Haasbroek D.F., Pretorius L. (1994). Orthotropic failure criterion for timber. *R&D Journal vol.10*, No1, https://cdn.ymaws.com/www.saimeche.org.za/resource/collection/48DE0CFB-DB04-4952-A201-0AF4845A19C5/Haasbroek_and_Pretorius-1994_09__600_dpi_-_1994__10_1___1-6.pdf

Jenkins, M. C., Brand, A. S., Painter, T. T., & Sherry, S. T. (2021, May). A Preliminary Digital Image Correlation Study of the Anisotropic Mechanical Properties of a 3D-Printed Mortar. *Proceedings of the 6th International Conference on Civil, Structural and Transportation Engineering* (ICCSTE'21). https://doi.org/10.11159/iccste21.146

Jones RM. (1998). Mechanics of composite materials. *CRC press.*

Probabilistic Model Code, Part 3: Material Properties. (2000). JCSS - Joint Committee on Structural Safety https://www.jcss-lc.org/jcss-probabilistic-model-code/

Liu C, Yue S, Zhou C, Sun H, Deng S, Gao F, et al. Anisotropic mechanical properties of extrusion-based 3D printed layered concrete. *Journal of Materials Science*. 2021 Oct 1;56(30):16851–64.

Meurer, M., & Classen, M. (2021). Mechanical properties of hardened 3D printed concretes and mortars-development of a consistent experimental characterization strategy. *Materials, 14 (4)*, 1 – 23. https://doi.org/10.3390/ma14040752

Moini, R., Baghaie, A., Rodriguez, F. B., Zavattieri, P. D., Youngblood, J. P., & Olek, J. (2021). Quantitative microstructural investigation of 3D-printed and cast cement pastes using micro-computed tomography and image analysis. *Cement and Concrete Research, 147*. https://doi.org/10.1016/j.cemconres.2021.106493

Munir, Q., & Kärki, T. (2021). Cost analysis of various factors for geopolymer 3d printing of construction products in factories and on construction sites. *Recycling*, *6*(3). https://doi.org/10.3390/recycling6030060

Nedjar, B. (2021). On a geometrically nonlinear incremental formulation for the modeling of 3D concrete printing. *Mechanics Research Communications, 116*. https://doi.org/10.1016/j.mechrescom.2021.103748

Ni, S., & Gernay, T. (2021). Considerations on computational modeling of concrete structures in fire. *Fire Safety Journal, 120*. https://doi.org/10.1016/j.firesaf.2020.103065

Panda, B., Mohamed, N. A. N., Paul, S. C., Singh, G. V. P. B., Tan, M. J., & Šavija, B. (2019). The effect of material fresh properties and process parameters on buildability and interlayer adhesion of 3D printed concrete. *Materials*, *12*(11). https://doi.org/10.3390/ma12132149

Sørensen, J. D., Rajcic, V., Čizmar, D., Sørensen, J. D., Kirkegaard, P. H., & Rajčić, V. (n.d.). *Robustness analysis of a timber structure with ductile behaviour in compression*. https://www.researchgate.net/publication/266329553

Stephen P., Timoshenko S., Woinowsky-Krieger S. (1989). Theory of plates and shells. *McGraw-Hill international edition. Engineering Mechanics Series.* McGraw-Hill Book Co – Singapore. 28th Printing 1989, 580 pages. ISNB 0-07-Y85820-9

van den Heever, M., Bester, F., Kruger, J., & van Zijl, G. (2021). Mechanical characterisation for numerical simulation of extrusion-based 3D concrete printing. *Journal of Building Engineering, 44*. https://doi.org/10.1016/j.jobe.2021.102944

van den Heever, M., Bester, F., Kruger, J., & van Zijl, G. (2022). Numerical modelling strategies for reinforced 3D concrete printed elements. *Additive Manufacturing*, 50, 102569. https://doi.org/10.1016/j.addma.2021.102569

Computational Modelling of Concrete and Concrete Structures – Meschke, Pichler & Rots (Eds)
 ISBN: 978-1-032-32724-2

Numerical modelling via a coupled discrete approach of the autogenous healing for Fibre-Reinforced Cementitious Composites (FRCCs)

A. Cibelli, G. Di Luzio & L. Ferrara
Department of Civil and Environmental Engineering, Politecnico di Milano, Milan, Italy

ABSTRACT: Aiming to predict long-term performance of advanced cement-based materials and design more durable structures, a reliable modelling of the autogenous healing of cementitious materials is crucial. A discrete model for the regain in terms of water tightness, stiffness and strength induced by the autogenous and/or "žstimulate'ž autogenous healing was recently proposed for ordinary plain concrete. The modelling proposal stemmed from the coupling of two models, namely the Hygro-Thermo-Chemical (HTC) model, on one side, and the Lattice Discrete Particle Model (LDPM), on the other side, resulting in the Multiphysics-Lattice Discrete Particle Model (M-LDPM). Being this approach not customised only for ordinary concrete, but for the whole broad category of cementitious materials, in this paper, its application to Fibre-Reinforced Cementitious Composites is presented. To accurately simulate what has been experimentally observed so far, the mechanical model is updated to also include the self-healing of the *tunnel cracks* at the fibre-matrix interfaces. Therefore, the self-repairing process is modelled to develop on two independent stages: (a) matrix cracks healing, and (b) fibre bridging action restoring. This research activity is part of the modelling tasks framed into the project ReSHEALience, funded from the European Union's Horizon 2020 Research and Innovation Programme.

1 INTRODUCTION

The unavoidable concrete cracking and the ensuing degradation phenomena have encouraged many researchers to increase the efforts in enhancing the comprehension of such processes and the capability of modelling the concrete long-term performance. In this framework, the inherent healing capacity of cement based materials has been gaining an increasing interest by the concrete professional and scientific community. As demonstrated by several authors since its discovery (Snoeck & De Belie 2015), and mainly in the last decades, the self-healing of concrete can lead to a considerable recovery of physical and, in some cases, mechanical properties of damaged concrete.

Through a painstaking literature survey, an unbalanced scientific production clearly stands out. Over the years, an extensive research effort has been placed on the experimental investigation of the self-healing phenomenon, aiming to detect its peculiar features and which techniques were worth being further explored to turn it into a predictable and/or engineered process. On the contrary, few models have been developed to account for the healing-induced effects on both durability performance and mechanical behaviour. As a consequence, in literature there is a limited number of numerical studies on this phenomenon (Aliko-Benítez et al. 2015; Barbero et al. 2005; Chen et al. 2021; Davies & Jefferson 2017; Di Luzio et al. 2018; Hilloulin et al. 2014; Hilloulin et al. 2016; Mergheim & Steinmann 2013; Oucif et al. 2018; Voyiadjis et al. 2011). The majority of them relies on continuum-based approaches, leading to consider the aforementioned effects on the mechanical properties only as a smeared contribution in terms of either stiffness and/or strength regain in the cracked state. Likewise, the impact of the crack self-repairing on durability performance indicators, e.g. permeability, can be simulated only as an overall effect, missing in simulating the local nature of the phenomena, e.g. where the water permeability increases dramatically and restores after healing.

The research activity presented in this paper aims to formulate a discrete model for capturing the mechanical recovery induced by an actual damage healing into which the cracks sealing might eventually evolve.

Building more durable structures in order for concrete to result in a more sustainable material, developing sound models to predict the structural life span of concrete structures, and accounting for durability as a *governing* performance within the design process: these are only three of the many concurrent causes that have made the concrete durability worthwhile deserving an increasing interest by the scientific community. These issues also represent the guidelines of the Horizon 2020 project *ReSHEALience*, in which this work is framed. The project aims to define the concepts of Ultra High Durability Concrete (UHDC) and Durability Assessment-based Design (DAD). The UHDC material concept encompasses advanced cementitious

DOI 10.1201/9781003316404-10

materials which fully exploit their own inherent capacity of autonomously repairing the cracks. To the purpose, supplementary cementitious materials, such as slag and crystalline admixtures, are included into the mixture. In the project *ReSHEALience*, the identification of a quantitative approach to predict long-term performance of concrete structures, even when exposed to extremely aggressive environments, was performed through both laboratory experimental tests and monitoring campaigns on pilot UHDC structures exposed to real exposure conditions, together with the development of numerical models at meso- and macro-scale (Al-Obaidi et al. 2020, 2021; Lo Monte & Ferrara 2020, 2021).

2 RESEARCH BACKGROUND

The modelling proposal stems from the coupling of two models, namely the Hygro-Thermo-Chemical (HTC) model, on one side, and the Lattice Discrete Particle Model (LDPM), on the other side (Di Luzio & Cusatis 2009a, 2009b; Cusatis et al. 2011a, 2011b; Pathirage et al. 2019). The result is the Multiphysics-Lattice Discrete Particle Model (M-LDPM) (Abdellatef et al. 2015; Alnaggar et al. 2017; Cibelli et al. 2022; Yang et al. 2021).

2.1 *Lattice Discrete Particle Model*

In LDPM the geometrical configuration is generated by a trial-and-error random procedure, in which the aggregate particles, whose size distribution derives from a Fuller-type curve, are assumed to have spherical shape and are randomly placed within the volume. Then, zero-radius particles are located along the external surfaces to facilitate the imposition of boundary conditions. Based on the Delaunay tetrahedralisation of the generated system of points, a three-dimensional domain tessellation is carried out, and linear segments, namely tetrahedra edges, are generated to connect all particles centres. The outcome is a system of lattice-connected cells interacting through triangular facets: the mechanical interaction among particles is based on four particle-subsystems (Figure 1a), in which the spheres (nodes) are connected by struts (edges), having cross section (triangular facets) resulting from the volume tessellation (Figure 1b).

In LPDM, rigid body kinematics is employed to describe the deformation of the lattice particle system, and the displacement step $[\![\mathbf{u}_C]\!]$ at the centroid of each facet, C_k (Figure 1b), is used to define the strain measures which read $\varepsilon_N = (n^T[\![\mathbf{u}_C]\!])/l$; $\varepsilon_L = (l^T[\![\mathbf{u}_C]\!])/l$; $\varepsilon_M = (m^T[\![\mathbf{u}_C]\!])/l$, where n, l, m are the unit vectors which identify a local reference system on each facet in normal and shear directions, respectively.

Vectorial constitutive laws are defined at the centroid of each projected facet to describe the mesoscopic stress. In the elastic regime, normal and shear stresses are proportional to the corresponding strains: $\sigma_N = E_N\varepsilon_N$; $\sigma_L = E_T\varepsilon_L$; $\sigma_M = E_T\varepsilon_M$, where the elastic moduli are $E_N = E_0$ and $E_T = \alpha E_0$, in which E_0 is the effective normal modulus and α the shear-normal coupling parameter. One of the unique feature of the LDPM formulation consists of being able to automatically capture the effects of the heterogeneity of the concrete, such as splitting cracks and failure in compression, which can not be achieved by employing the classical theory of elasticity, e.g. see (Cusatis et al. 2011).

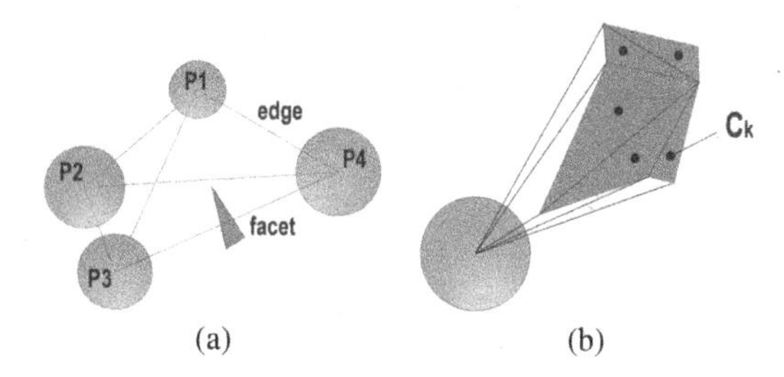

Figure 1. (a) four-particle subsystem; (b) triangular facets.

When in a facet under tension the strain reaches the tensile elastic limit, the meso-scale crack opening is calculated as $w_N = l(\varepsilon_N - \sigma_N/E_N)$; $w_L = l(\varepsilon_L - \sigma_L/E_T)$; $w_M = l(\varepsilon_M - \sigma_M/E_T)$. Then, the crack opening vector associated to each facet is $w_c = w_N n + w_L l + w_M m$, where w_N is the actual opening/closure of the crack, along the direction orthogonal to the facet, while w_L and w_M are two sliding components, catching shear displacements at crack surfaces.

The non-linear behaviour is analysed considering three non-linear meso-scale phenomena: (1) fracture and cohesion, (2) compaction and pore collapse, and (3) friction. For the latter two and further details about the model calibration and validation, the reader can refer to (Cusatis et al. 2011a, 2011b). Hereinafter, for the sake of clarity, the constitutive law for the fracturing behaviour is briefly recalled as the healing effect is therein implemented.

In LDPM the fracture behaviour is modelled by setting damage-type constitutive laws, which stem from the definition of effective strain, $\varepsilon = \sqrt{\varepsilon_N^2 + \alpha\left(\varepsilon_L^2 + \varepsilon_M^2\right)}$, and stress, $\sigma = \sqrt{\sigma_N^2 + \left(\sigma_L^2 + \sigma_M^2\right)/\alpha}$. Then, for tensile loading ($\varepsilon > 0$), the effective mechanical parameters permit to define the following relationships between strain and stress in the local reference systems: $\sigma_N = \varepsilon_N\,(\sigma/\varepsilon)$; $\sigma_L = \alpha\,\varepsilon_L\,(\sigma/\varepsilon)$; $\sigma_M = \alpha\,\varepsilon_M\,(\sigma/\varepsilon)$. The effective stress σ is incrementally elastic ($\dot{\sigma} = E_0\,\dot{\varepsilon}$) and must satisfy the inequality $0 \le \sigma \le \sigma_{bt}(\varepsilon, \omega)$, in which $\sigma_{bt}(\varepsilon, \omega)$ is a yield surface enforced by means of a vertical (at constant strain) return algorithm. The strain-dependent limit can be expressed as

$$\sigma_{bt}(\varepsilon, \omega) = \sigma_0(\omega)\exp\left[-H_0(\omega)\frac{\langle\varepsilon_{max} - \varepsilon_0\rangle}{\sigma_0(\omega)}\right] \quad (1)$$

where the brackets $\langle\cdot\rangle$ are used in Macaulay sense: $\langle x\rangle = \max\{x, 0\}$, and H_0 is the post-peak softening modulus, whose formulation allows for a smooth transition from a softening behaviour under pure tensile stress ($H_0(\omega=\pi/2)=H_t$) to perfectly plastic response under pure shear ($H_0(0)=0$). In fact, the formulation of H_0 reads $H_0(\omega)=H_t(2\omega/\pi)^{n_t}$, with n_t softening exponent.

In Eq. 1, ω is the parameter representing the degree of interaction between shear and normal loading. It is worth noting that ε_{max} is a history-dependent variable, making, on turn, the yield surface a history-dependent exponential function. Therefore, the actual fracture strength is assumed dependent on the actual level of damage. Finally, in Eq. 1 the function $\sigma_0(\omega)$ is the strength limit for the effective stress and is formulated as

$$\sigma_0(\omega)=\sigma_t\frac{-\sin(\omega)+\sqrt{\sin^2(\omega)+4\alpha\cos^2(\omega)/r_{st}^2}}{2\alpha\cos^2(\omega)/r_{st}^2} \quad (2)$$

in which $r_{st}=\sigma_s/\sigma_t$ is the ratio between the shear strength, σ_s (cohesion), and the tensile strength, σ_t.

2.2 *Lattice Discrete Particle Model for FRC (LDPM-F)*

The extension of LDPM to include fibre-reinforcing mechanisms is obtained by inserting straight fibers, in proportion to the volume fraction V_f, with random positions and orientations, into the LDPM geometrical configuration. The geometry of each individual fiber is characterised by the diameter d_f and length L_f. The fibre system is overlapped to the polyhedral cell system, and each facet is paired with its intersecting fibres. At the facet level, the matrix-fibre interaction is described by the bridging forces carried by the fibres crossing the facet, which are activated when the crack opening initiates. In this configuration, equilibrium considerations permit to reasonably assume a parallel coupling between the fibres and the surrounding concrete matrix. Then, the total stresses on each LDPM facet can be computed as $\sigma=\sigma_c+(\sum_{f\in A_c}P_f)/A_c$, where A_c is the facet area, and P_f represents the crack-bridging force for each fibre crossing the given facet.

Since the mechanical interaction between the fibres and the surrounding matrix occurs at a scale smaller than the typical modelling scale of LDPM, the micromechanics governing such interaction is not explicitly simulated in the mesoscopic LDPM numerical framework. The micro-mechanical crack-bridging mechanisms, featuring the bond between the single fibre and the embedding matrix, are implemented into the model within the formulation for computing the bridging force P_f, briefly reported hereinafter as it was published by (Schauffert & Cusatis 2012).

In addition to the above consideration, additional hypothesis are postulated: (i) the contribution of fibres to the equilibrium is negligible in case of either compression stress on the facet or stress not exceeding the elastic limit; (ii) the interaction between adjacent fibres and the effect that adjacent mesoscale cracks are both neglected; (iii) each fibre is assumed to be straight, elastic, with negligible bending stiffness, and non-circular cross sections are simulated through an equivalent diameter, calculated as $d_f=2(A_f/\pi)^{1/2}$ with A_f fibre cross-sectional area.

As proposed by Li et al. (Lin et al. 1999), in LDPM-F the slippage at full debonding v_d is computed as $v_d=(2\tau_0L_e^2)/(E_fd_f)+[(8G_dL_e^2)/(E_fd_f]^{1/2}$, in which L_e is the embedment length, E_f the modulus of elasticity of the fibre, τ_0 the constant value of frictional stress for the portion of the embedded fibre that has debonded, and G_d the bond fracture energy. The parameters τ_0 and G_d govern the debonding stage, modelled as a tunnel-type cracking process (Yang et al. 2008).

During the debonding stage ($v<v_d$), the fibre bridging force is given as (Lin et al. 1999)

$$P(v)=\left[\frac{\pi^2E_fd_f^3}{2}(\tau_0v+G_d)\right]^{1/2} \quad (3)$$

After full debonding ($v>v_d$), the mechanism is entirely frictional and the fibre load results from (Lin et al. 1999)

$$P(v)=P_0\left(1-\frac{v-v_d}{L_e}\right)\left[1+\frac{\beta(v-v_d)}{d_f}\right] \quad (4)$$

where $P_0=\pi L_ed_f\tau_0$, whereas β is the coefficient in charge of shaping the relationship to capture the high variability of the frictional interface nature (Lin & Li 1997). When the friction at the interface does not depend on the slippage, β is set to zero. In case of either slip hardening or slip softening friction, it assumes positive ($\beta>0$) or negative ($\beta<0$) values, respectively.

If the orientations of the embedded and free fibre portions is different, at the point where the fibre exits the matrix and changes orientation, the bearing stress is partially supported by the underlying matrix. When this localised stress field reaches a sufficient intensity, spalling occurs, and the embedment length of the fibre is consequently reduced by a length s_f. Furthermore, when the fibre exits the tunnel crack, the latter shortened because of the spalling, it wraps around the intact matrix. This phenomenon is generally referred to as *snubbing effect*, and it is modelled through the frictional pulley idealisation (Li et al. 1990), which complies with the fibre pull-out model adopted in the LDPM-F model (Yang et al. 2008). The fibre load is updated to account for spalling and snubbing phenomena (see (Schauffert & Cusatis 2012)). The updated value of the fibre load must comply with its rupture strength, then the following relationship must always hold: $\sigma_f=(4P_f)/(\pi d_f^2)\leq\sigma_{u,f}\exp\left(-k_{rup}\varphi_f'\right)$, in which k_{rup} is a material parameter, and $\sigma_{u,f}$ the ultimate tensile strength of the fibre. In case of fibre stress exceeding the corrected value of strength, P_f is set to zero. The exponentional term reflects experimental evidence showing lower rupture loads in single fibre pull-out tests for increasing values of φ'_f (Kanda & Li 1998).

For a generic fibre, with embedment segment orientation n_f, subject to pull-out from both embedment depths due to a crack opening w, and with a spalling length s_f on both sides, the crack-bridging force is given by $P_f = P_f n'_f$, with the crack-bridging segment computed as $||w'|| = 2s_f + v_s + v_l$ and $||n'|| = w'/||w'||$, where $||w'||$ is the vector length, and s_f the slippage reduction due to the matrix spalling. The embedment segments have the relative slippage v_s and v_l, respectively. The pullout resisting forces is then $P_f = P(v_s)\exp\left(k_{sn}\varphi'_f\right) = P(v_l)\exp\left(k_{sn}\varphi'_f\right)$, and on each side must be the same. From the last equality, the relative slippages v_s and v_l can be computed by an interactive procedure in which the compatibility between the bridging segment and the slippages is enforced.

Further details on the constitutive relations of fibres and matrix-fibre interaction as well as on the calibration of the governing parameters can be found in (Schauffert & Cusatis 2012; Schauffert et al. 2012).

3 MESOSCALE HEALING MODEL

The modelling approach relies on the identification of two different levels of damage: (i) matrix and (ii) fibre-matrix interface cracks. The matrix cracks (Figure 2a) are induced by the loads, either mechanical or environmental, and are responsible of the fibres mechanical activation: as long as no cracks intersect a fibre, the latter does not play any role in the structural response. The fibre-matrix interface cracks (Figure 2b) develop during the interface debonding instead, and are hereinafter also referred to as tunnel cracks between the fibre and the surrounding embedding matrix.

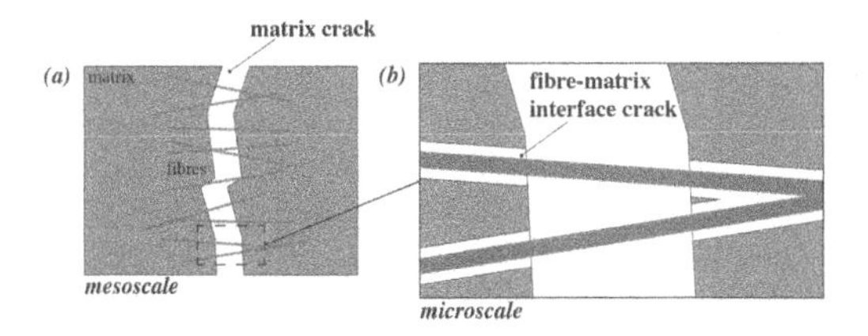

Figure 2. Two levels in the damage modelling: (a) **matrix cracks** at the mesoscale; (b) **fibre-matrix interface cracks** at the microscale.

The self-healing model is in line with the LDPM approach, dealing with matrix and tunnel cracks separately. The autogenous repairing of the former is implemented within the constitutive fracture law at the mesoscale, whereas the effect of healing on the fibres response is taken into account within the calculation of the bridging force carried by the steel reinforcement. This approach stems from the idea for which the recovery of matrix damage and tunnel cracks along fibre-mortar interface affect the material mechanical behaviour differently.

3.1 *Healing characterization*

The healing kinetic law formulated for plain cementitious materials (Di Luzio et al. 2018; Cibelli et al. 2022) presents no limitations in being used for fibre-reinforced composites as well. Following the conceptual differentiation between matrix and tunnel cracks, it can be exploited for capturing the autogenous, and eventually stimulated, healing of the matrix cracks. On the other hand, in order to have two separate internal variables feeding the mechanical model at two different levels, in the improved version of M-LDPM a distinction is made between the normalised healing degree for matrix cracks and that for fibre-matrix interface cracks, λ^m_{sh} and λ^f_{sh} respectively. In the following the formulation emphasising such splitting is reported, with no theoretical differences with respect to the original one (Di Luzio et al. 2018).

The kinetic laws for matrix (superscript m) and tunnel (superscript f) cracks read

$$\dot{\lambda}^m_{sh} = \tilde{A}^m_{sh}\left(1 - \lambda^m_{sh}\right) \tag{5a}$$

$$\dot{\lambda}^f_{sh} = \tilde{A}^f_{sh}\left(1 - \lambda^f_{sh}\right) \tag{5b}$$

in which $\tilde{A}^m_{sh}$ and $\tilde{A}^f_{sh}$, inversely proportional to the reaction characteristic times, are calculated as

$$\tilde{A}^m_{sh} = \tilde{A}^m_{sh0} \cdot f_h(h) \cdot f^m_w(w_c) \cdot e^{\left[-E^m_{sh}/R(1/T - 1/T_{ref})\right]} \tag{6a}$$

$$\tilde{A}^f_{sh} = \tilde{A}^f_{sh0} \cdot f_h(h) \cdot f^f_w(w_c) \cdot e^{\left[-E^f_{sh}/R(1/T - 1/T_{ref})\right]} \tag{6b}$$

where $\tilde{A}^m_{sh.0}$ and $\tilde{A}^f_{sh.0}$, namely the inverse of the reaction characteristic times in standard conditions (RH=100%, $T = T_{ref}$, $w_c = 0$), value

$$\tilde{A}^m_{sh0} = \tilde{A}^m_{sh1}\left(1 - \alpha^{sh0}_c\right)c + \tilde{A}^m_{sh2} \cdot ad \tag{7a}$$

$$\tilde{A}^f_{sh0} = \tilde{A}^f_{sh1}\left(1 - \alpha^{sh0}_c\right)c + \tilde{A}^f_{sh2} \cdot ad \tag{7b}$$

where c and ad are the cement and healing-promoting admixture content, respectively. The material parameters E^m_{sh}, E^f_{sh}, $\tilde{A}^m_{sh1}$, $\tilde{A}^f_{sh1}$, $\tilde{A}^m_{sh2}$, and $\tilde{A}^f_{sh2}$ are calibrated against experimental data, allowing to catch the peculiarities of phenomena occurring at two different scales. Furthermore, the double degree of freedom permits to properly simulate the effect of crack opening, modelled through the coefficient $f_w(w_c)$, on the process evolution. The coefficient $f_h(h)$ accounts for relative humidity and simulates the relevant role played by the moisture supply, making the process proceed or stop whether the healing water-driven reactions are fed or not. In the Eqs. the relative humidity, h, and temperature, T, fields are provided by the HTC model.

3.2 *Healing implementation in LDPM and LDPM-F*

The healing-induced effect on the mechanical response of the cementitious materials involves recovery of post-cracking residual fracture strength. Depending on which cracks are healed, the aforementioned recovery is the result of different physical phenomena. For this reason, the implementation in the mechanical models follows two separate dedicated approaches.

3.2.1 *Matrix cracks*

For matrix cracks, the healing effect is modelled by enforcing a homothetic expansion of the boundary limit curve $\sigma_{bt}(\varepsilon,\omega)$ (Eq. 1), as more pronounced as more the repairing process has developed.

What has been experimentally observed so far is that plain concrete specimens, once loaded, fractured and unloaded, might show a recovery in strength and stiffness if re-loaded after a long enough curing period. It is due to the concurring delayed hydration and carbonation self-healing mechanisms. This partially restores the material continuity, having straightforward consequences on the concrete bulk permeability and its proneness to the attacks of environmental aggressive agents. The effects on the mechanical response, instead, depend on the chemical bounds between the filling products and the crack walls; then, it is not granted that the recovery in water tightness and the regain in strength and stiffness proceed to the same extent. In fact, the crack sealing might not result in an actual concrete healing.

With reference to plain concrete specimens, pre-cracked by means of three-point bending tests up to damage threshold beyond the material linear limit (Eq. 2), the healing effect on fracture behaviour might be measured by carrying out the same fracture tests after varying curing periods. The recorded load-CMOD curve may show reloading branches (1) stiffer than the unloading ones, and (2) crossing the un-healed material boundary curve (Figure 3).

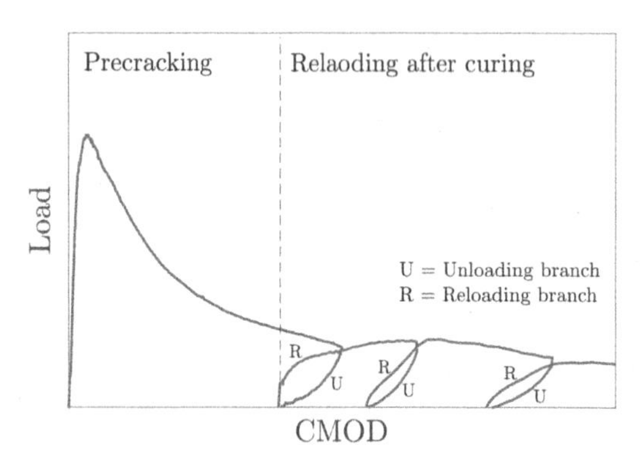

Figure 3. An example of the experimental curves gained in a laboratory campaign to assess the mechanical regain induced by the autogenous healing (Ferrara et al. 2014).

In this work, the modelling strategy adopted aims at preserving the inherent mechanical meaning of the impact due to the healing on the fracture strength, and relies on the homothetic expansion of the boundary curve (Figure 4). The expansion extent is assumed to be proportional to the healing degree λ^m_{sh}, thus capturing the recovery in strength, without varying the

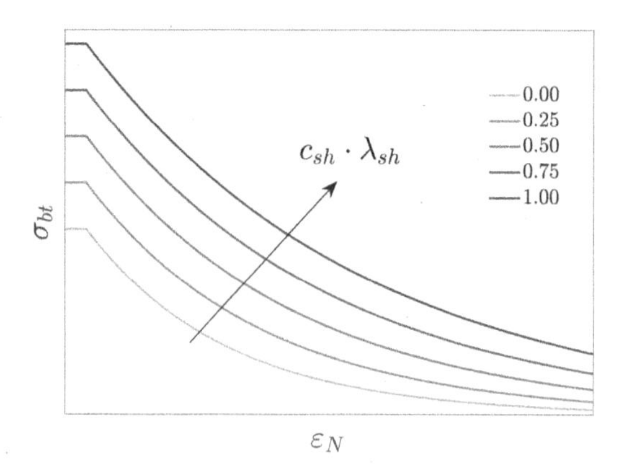

Figure 4. **matrix cracks** - Effect of healing on the boundary curve for the fracturing behaviour.

crack width within the numerical framework. In other words, the boundary expansion is conceived to catch the behaviour described above: the material must be allowed to overcome the strength value reached at the beginning of the unloading branch, for the previously reached value of crack width, if any healing has occurred.

In LDPM, the healing implementation affects the strength limit calculation (Eq. 2), thus, on turn, the limit curve (Eq. 1). The updated version of the healing dependent-constitutive law relevant to the fracture behaviour reads

$$\sigma_0(\omega,\lambda^m_{sh}) = \sigma_0(\omega)\,\left(1 + c_{sh}\cdot\lambda^m_{sh}\right) \tag{8a}$$

$$\sigma_{bt}(\varepsilon,\omega,\lambda^m_{sh}) = \sigma_0(\omega,\lambda^m_{sh})\, e^{\left[-H_0(\omega)\frac{\langle\varepsilon_{max}-\varepsilon_0\rangle}{\sigma_0(\omega,\lambda^m_{sh})}\right]} \tag{8b}$$

In Eq. 8a, c_{sh} is an empirical coefficient governing the impact of crack closure on mechanical strength. It is defined as *healing mechanical impact coefficient*. The parameter c_{sh} depends on several aspects, e.g. curing conditions and mixture composition, therefore, it has to be calibrated experimentally.

Looking at the updated equation of the boundary curve (Eq. 8b), it is important to notice that the healing *plays* an active role as internal variable in both shaping the softening branch and setting the stress limit for the earlier stage of the constitutive law, namely when the maximum strain does not exceed the elastic limit. It is worth emphasising that, though the modelling strategy yields a recovery of both linear and post-peak behavior,the former is never imposed at the mesoscale, being only the limit curve expanded exclusively on those facets which experience cracking and healing.

3.2.2 *Fibre-matrix interface cracks*

With single-fibre pull-out tests, stopped after the first load drop and resumed up to rupture after curing periods featuring different duration and exposure conditions, it has been observed that the healing of the interface cracks does affect the pull-out strength.

Whenever the healing process happens, it yields delayed hydration products and $CaCO_3$ crystals fulfilling the tunnel between the fibre and the surrounding mortar ((Qiu et al. 2019)). This results in a recovery of the interface frictional bond. The phenomenon is implemented in LDPM-F by updating the value of the fibre bridging force $P(v)$ with a coefficient proportional to λ^f_{sh}. The updated constitutive law for the fibre load reads

$$P\left(v, \lambda^f_{sh}\right) = \left(1 + \gamma_{sh} \cdot \lambda^f_{sh}\right) P(v) \leq \alpha \cdot P_0 \tag{9}$$

Referring to a single-fibre pull-out test, in Figure 5 the effect of the tunnel crack self-healing on the mechanical response is qualitatively shown. After the loading and unloading stages (branches L and U), the specimen is exposed to given environmental conditions for a time span long enough to permit the self-healing process to develop. The cured specimen is then reloaded (branch R) up to rupture. Due to the recovered frictional bond, the specimen might experience a recovery in stiffness and strength, to an extent proportional to the degree of completion of the healing process. By means of the device in Eq. 9 LDPM-F is updated to be capable of capturing this experimental evidence. In Figure 5 the updated constitutive law is plotted with reference to increasing self-healing degrees, in the hypothesis of $\gamma_{sh} = 1.00$.

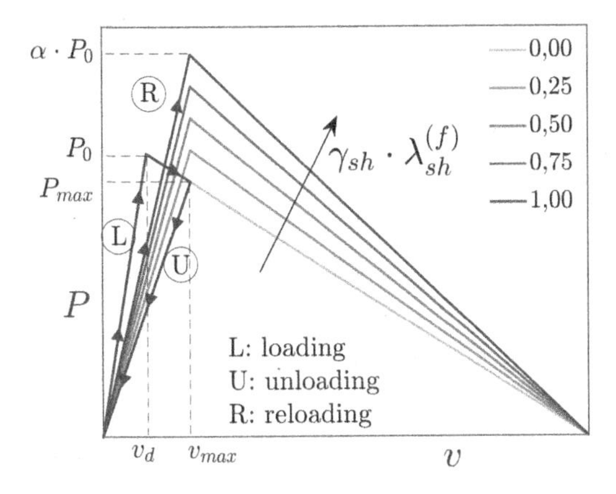

Figure 5. **fibre-matrix interface cracks** - Effect of healing on the fibre load vs. slippage law.

The coefficient γ_{sh} has a physical meaning similar to c_{sh}. It governs the impact that the healing of the tunnel cracks has on the fibres contribution to the mechanical equilibrium. With $\gamma_{sh} = 0$ it is possible to capture the crack sealing, whereas if $\gamma_{sh} \geq 0$ the load carried by the fibre is enhanced thanks to the increased friction along the crack walls. The latter has an upper bound ($\alpha \cdot P_0$) in which the bridging force at full debonding P_0 is either amplified or reduced by the coefficient α. Both γ_{sh} and α are material parameters to calibrate against experimental data. Depending on the composition of the cementitious composites, the technique adopted to engineer the process, the fibres nature, the curing conditions, and the loading regimes the healing might allow to recover either partially or entirely the fibre load bearing capacity. The parameter α sets the maximum achievable level of recovery. Once calibrated experimentally, γ_{sh} must comply with the condition for which, in case of full fulfilment of the tunnel crack:

$$\text{if} \quad \lambda^f_{sh} = 1.00 \Longrightarrow \gamma_{sh} \leq \frac{\alpha \cdot P_0}{P(v)} - 1 \tag{10}$$

4 NUMERICAL SIMULATIONS

4.1 Healing of matrix cracks

The concrete self-healing is expected to affect the meso-scale mechanical response of the material, in tension as much as in shear. The model has been implemented to catch this phenomenon, with the possibility of calibrating the entity of the induced strength recovery by means of the parameter c_{sh}. In order to investigate the model capability of capturing the self-repairing effect on tensile and shear behaviours, the numerical simulations of how two ordinary plain concrete (OPC) specimens behave after being damaged in tension and brought to collapse, after curing, either in pure tension or shear have been executed.

Table 1. Mix composition of the reference concrete (dosages in kg/m^3).

constituent	content
cement	300
water	190
aggregates 5.5-16 mm	1950

The material adopted has been an ordinary plain concrete whose mix composition is presented in Table 1.

Concerning the geometry, the collapse in tension has been investigated for a dogbone specimen, as usual for pure tensile tests, having the dimensions reported in Figure 6a and thickness of 20 mm. These dimensions have been chosen in order to have the narrowest part of the sample larger than the maximum aggregate size of the adopted material, and, at the same time, as smaller as possible to localise there the damage. The other geometrical characteristics have been set accordingly, with the aim of having a sample weak at the midspan, and the parts 70 mm wide covering a portion of the total length as smaller as possible. For the shear failure, instead, a bi-notched prismatic specimen has been used (Figure 6b), having dimensions 100x70x20 mm^3, and the notches 2 mm wide and 25 mm deep. In this case, it has been necessary to avoid a slender sample, as the dog-bone specimen presented above is. In fact, a stocky element presents a larger proneness

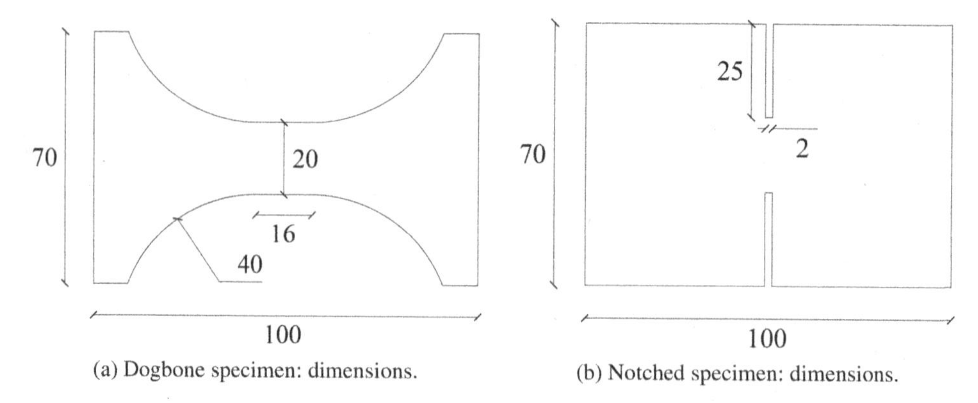

(a) Dogbone specimen: dimensions. (b) Notched specimen: dimensions.

Figure 6. Geometrical dimensions of the simulated specimens in millimetres.

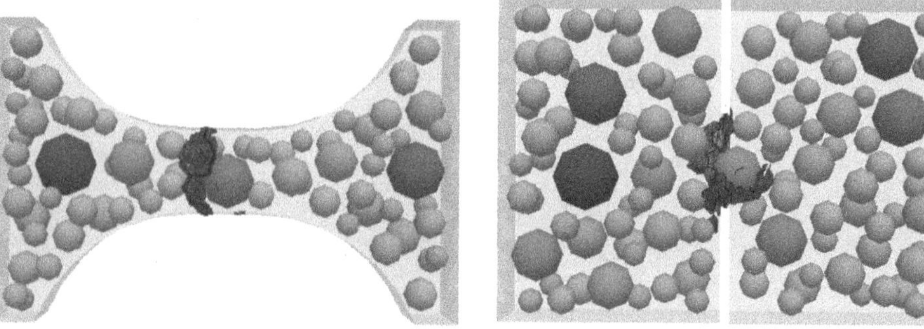

(a) Dogbone specimen: aggregate particles and cracks.

(b) Notched specimen: aggregate particles and cracks.

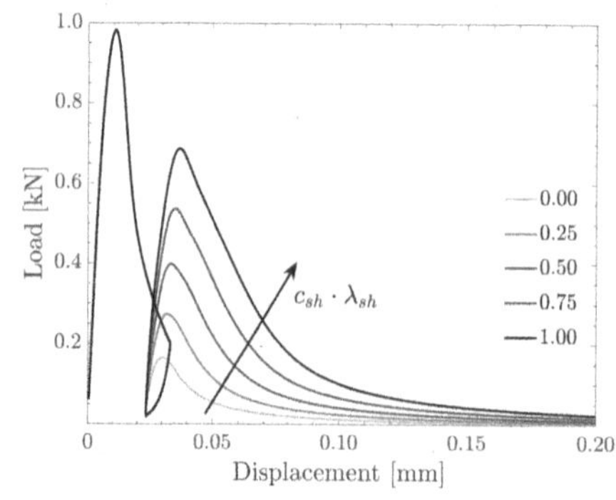

(c) Dogbone specimen undergoing pure tension for precracking and up to failure

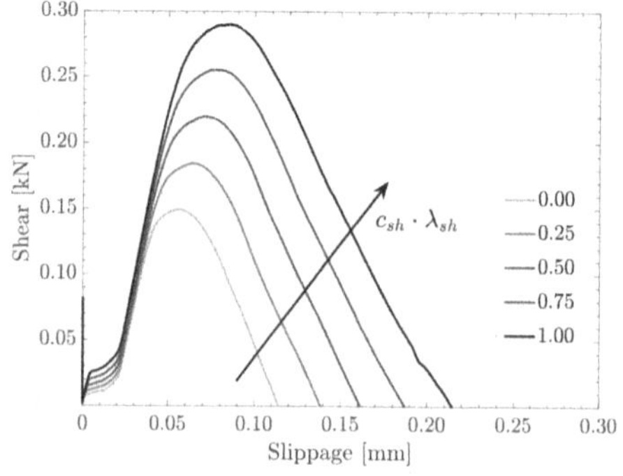

(d) Notched specimen undergoing pure shear up to failure, after being damaged in tension.

Figure 7. LDPM modelled specimens for assessing the influence of healing implementation on (a) tension and (b) shear behaviours.

to shear failure. However, likewise for the investigation in pure tension, it has been necessary to shape the sample in order to have all the mechanical energy channelled into the growth of the fracture at the mid-span, with no dispersion due to multi-cracking scenarios. For this reason, it has been used a bi-notched shape, with narrow and deep notches. It is worth mentioning that also in this case the narrowest sample cross-section has been set in order to have the smallest dimension larger than the maximum aggregate size. The other dimensions have been derived to result in a stocky sample.

Once the samples geometry has been generated, both specimens have been damaged by means of an increasing tensile loading, up showing a single crack roughly 350μm wide (Figures 7a,b). Afterwards, the dog-bone sample has been brought to failure in tension, whereas the bi-notched one in shear. This second stage has been repeated after having imposed increasing value of the normalised healing degree, λ^m_{sh}, ranging from 0.00 to 1.00, and in the hypothesis of having unit healing mechanical impact coefficient, c_{sh}. Then, in Figures 7c,d, the model ability of catching the healing-induced recovery in tensile and shear strength is shown plotting the (e) tensile load vs. displacement and (f) shear load vs. slippage curves.

4.2 *Healing of tunnel cracks*

The dogbone specimen in Figure 6a has been used also for testing the implementation of the tunnel cracks healing, by generating a FRC-based mesh with the same geometry. The concrete composition is reported in Table 2, where it is possible to see that the aggregate size has been reduced in order to have fibres length complying with specimen dimensions and aggregate size: $L_f \geq 3D_{max}$.

Table 2. Mix composition of the reference fibre-reinforced concrete (dosages in kg/m^3).

constituent	content
cement	600
water	200
aggregates 3-6 mm	1518
steel fibres $d_f = 0.22$ mm, $L_f = 20$ mm	0.50% by volume

As for matrix cracks, the purpose of investigating if the healing implementation affects the fibre load-slippage constitutive law as shown in Figure 5 is achieved through a simple set of numerical simulations. The dogbone specimens has been loaded in pure axial tension up to feature a single prominent crack approximately 60μm wide. Then, it has been completely unloaded. After having reached the zero-load condition, the sample has been reloaded up to failure. The reloading stage has been performed by assuming for the tunnel cracks self-healing degree, λ^f_{sh}, increasing fixed values between 0.00 and 1.00, namely 0.00, 0.25, 0.50, 0.75, and 1.00. The numerical simulations have been carried out in two different scenarios: with no matrix cracks healing, $\lambda^m_{sh} = 0.00$, and in the hypothesis of matrix and tunnel cracks healing evolving identically, $\lambda^m_{sh} = \lambda^f_{sh}$.

Firstly, it is important to assess how the model performs at the single fibre-facet intersection, to see if the P-v curve actually evolves as presented in Figure 5. The comparison between the fibre load vs. slippage curves on one of the most damaged LDPM facets obtained with λ^f_{sh} equals to 0.00 and 1.00 are shown in Figure 8b. The effect of healing acts as expected, though the re-loading in presence of healing stops before reaching the ultimate slippage (Figure 8b). In fact, as stands out from Figures 8c,d, in the numerical simulations the specimen experiences a sudden drop in strength, disregarding whether the healing of the matrix cracks is considered or not.

5 CLOSING REMARKS

The healing implementation for both matrix and tunnel cracks show promising capability in capturing the experimental evidence.

The healing of the matrix cracks affects the macroscale response of the two specimens as expected, in tension as much as in shear. With an increasing healing degree, in the hypothesis of $c_{sh} = 1.00$, the material experiences increasing stiffness during the reloading and higher strength. It is worth underlining that the full recovery occurs at the mesoscale, shaping the macroscale behaviour accordingly. The peak load after re-loading, even in case of $c_{sh} \cdot \lambda^m_{sh} = 1.00$ on the damaged facets, is not equals to the peak load of the virgin material. This is in line with laboratory results showing that the hydration outcomes at the crack faces, the main contributors to autogenous cracks healing, have generally lower performance compared to those in bulk cement paste.

With $\lambda^m_{sh} = 0.00$ and increasing λ^f_{sh} ($\gamma_{sh} = 1.00$), the model returns a recovery in stiffness and strength during the re-loading, even though the numerical results do not show a stable re-loading branch when the slippage overcomes the value of the pre-cracking stage. This is likely due to the limited energy redistribution allowed by the specimen geometry, imposed by the necessity of having localised damage. This deduction is justified also by the fibre load vs. slippage curve on the most damaged facets. It is evident that the specimen failure anticipates the full depletion of the load-bearing capacity of the system fibre-matrix.

In case of $\lambda^m_{sh} = \lambda^f_{sh}$ ($\gamma_{sh} = 1.00$), the recovery in stiffness and strength is more pronounced as expected. Also in this condition, the limited energy redistribution due to specimen geometry does not permit to exploit the full material ductility.

The model presented seems to have the potential for capturing phenomenological trends and mechanics standing out from the experimental investigations available in the literature. However, the calibration and validation against laboratory results, currently matter of study, will help in further improving the proposed approach.

ACKNOWLEDGMENTS

The work described in this paper has been performed in the framework of the project ReSHEALience - Rethinking coastal defence and green-energy Service

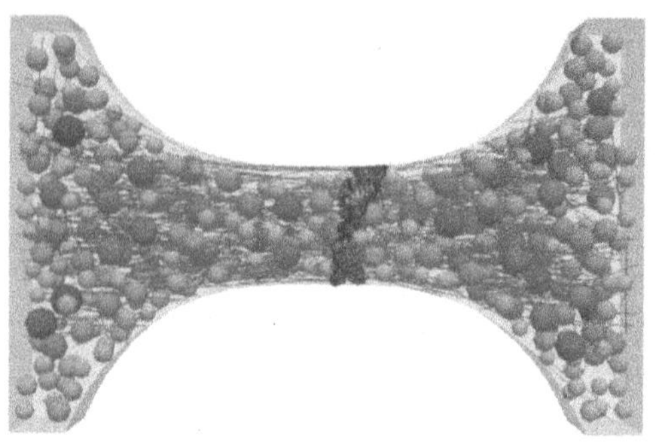

(a) Dogbone FRC specimen: aggregate particles, fibres and cracks.

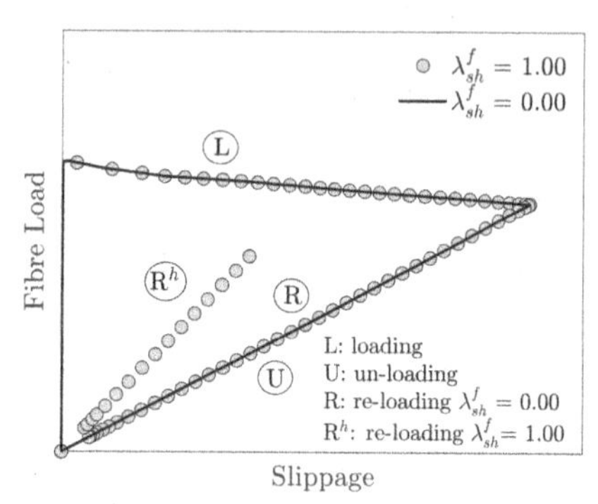

(b) Fibre load vs. slippage curve experienced along one of the most damaged facets.

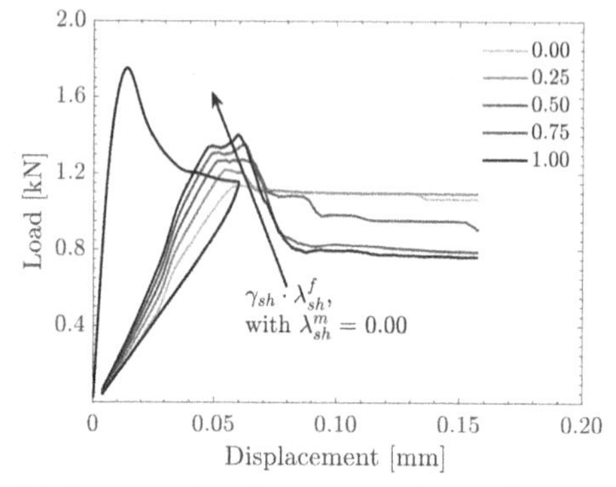

(c) Pure tension for precracking and up to failure with only tunnel cracks healing.

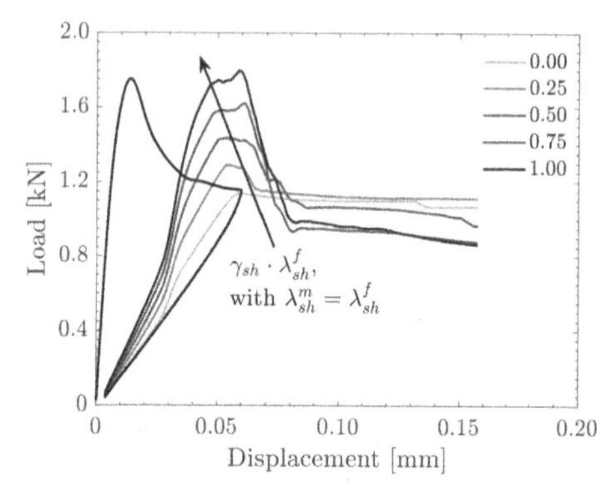

(d) Pure tension for precracking and up to failure with both matrix and tunnel cracks healing.

Figure 8. Influence of healing implementation on FRC dogbone specimen under uniaxial tension load.

infrastructures through enHancEd-durAbiLity high-performance cement-based materials, whose funding the first, fourth and last author gratefully acknowledge. This project has received funding from the European Union Horizon 2020 research and innovation programme under grant agreement No 760824. The information and views set out in this publication do not necessarily reflect the official opinion of the European Commission. Neither the European Union institutions and bodies nor any person acting on their behalf, may be held responsible for the use which may be made of the information contained therein. The numerical analyses have been performed by means of MARS, an explicit dynamic code distributed by ES3 Inc. (Engineering and Software System Solutions), which is gratefully acknowledged.

REFERENCES

Abdellatef, M., M. Alnaggar, G. Boumakis, G. Cusatis, G. Di Luzio, & R. Wendner (2015, September 21–23). Lattice discrete particle modeling for coupled concrete creep and shrinkage using the solidification microprestress theory. In C. Hellmich, B. Pichler, and J. Kollegger (Eds.), *10th International Conference on Mechanics and Physics of Creep, Shrinkage, and Durability of Concrete and Concrete Structures - CONCREEP-10*, Vienna, Austria, pp. 184–193.

Al-Obaidi, S., P. Bamonte, F. Animato, F. Lo Monte, I. Mazzantini, M. Luchini, S. Scalari, & L. Ferrara (2021). Innovative design concept of cooling water tanks/basins in geothermal power plants using ultra-high-performance fiber-reinforced concrete with enhanced durability. *Sustainability 13*(17).

Al-Obaidi, S., P. Bamonte, M. Luchini, I. Mazzantini, & L. Ferrara (2020). Durability-based design of structures made with ultra-high-performance/ultra-high-durability concrete in extremely aggressive scenarios: Application to a geothermal water basin case study. *Infrastructures 5*(11).

Aliko-Benítez, A., M. Doblaré, & J. Sanz-Herrera (2015). Chemical-diffusive modeling of the self-healing behavior in concrete. *International Journal of Solids and Structures 69-70*, 392–402.

Alnaggar, M., G. Di Luzio, & G. Cusatis (2017). Modeling time-dependent behavior of concrete affected by alkali silica reaction in variable environmental conditions. *Materials 10*(5).

Barbero, E. J., F. Greco, & P. Lonetti (2005). Continuum damage-healing mechanics with application to self-healing composites. *International Journal of Damage Mechanics 14*(1), 51–81.

Chen, Q., X. Liu, H. Zhu, J. W. Ju, X. Yongjian, Z. Jiang, & Z. Yan (2021). Continuum damage-healing framework for the hydration induced self-healing of the cementitious composite. *International Journal of Damage Mechanics 30*(5), 681–699.

Cibelli, A., M. Pathirage, L. Ferrara, G. Cusatis, & G. Di Luzio (2022). A discrete numerical model for the effects of crack healing on the behaviour of ordinary plain concrete: Implementation, calibration, and validation. *Engineering Fracture Mechanics in press.*

Cusatis, G., A. Mencarelli, D. Pelessone, & J. Baylot (2011). Lattice discrete particle model (LDPM) for failure behavior of concrete. II: Calibration and validation. *Cement and Concrete Composites 33*(9), 891–905.

Cusatis, G., D. Pelessone, & A. Mencarelli (2011). Lattice discrete particle model (LDPM) for failure behavior of concrete. I: Theory. *Cement and Concrete Composites 33*(9), 881–890.

Davies, R. & A. Jefferson (2017). Micromechanical modelling of self-healing cementitious materials. *International Journal of Solids and Structures 113-114*, 180–191.

Di Luzio, G. & G. Cusatis (2009a). Hygro-thermo-chemical modeling of high-performance concrete. II: Numerical implementation, calibration, and validation. *Cement and Concrete Composites 31*(5), 309–324.

Di Luzio, G. & G. Cusatis (2009b). Hygro-thermo-chemical modeling of high performance concrete. II: Theory. *Cement and Concrete Composites 31*(5), 301–308.

Di Luzio, G., L. Ferrara, & V. Krelani (2018). Numerical modeling of mechanical regain due to self-healing in cement based composites. *Cement and Concrete Composites 86*, 190–205.

Ferrara, L., V. Krelani, & M. Carsana (2014). A "fracture testing" based approach to assess crack healing of concrete with and without crystalline admixtures. *Construction and Building Materials 68*, 535–551.

Hilloulin, B., F. Grondin, M.Matallah, & A. Loukili (2014). Modelling of autogenous healing in ultra high performance concrete. *Cement and Concrete Research 61-62*, 64–70.

Hilloulin, B., D. Hilloulin, F. Grondin, A. Loukili, & N. De Belie (2016). Mechanical regains due to self-healing in cementitious materials: Experimental measurements and micro-mechanical model. *Cement and Concrete Research 80*, 21–32.

Kanda, T. & V. C. Li (1998). Interface property and apparent strength of high-strength hydrophilic fiber in cement matrix. *Journal of Materials in Civil Engineering 10*(1), 5–13.

Li, V., Y. Wang, & S. Backer (1990). Effect of inclining angle, bundling and surface treatment on synthetic fibre pull-out from a cement matrix. *Composites 21*(2), 132–140.

Lin, Z., T. Kanda, & V. C. Li (1999). On interface property characterization and performance of fiber reinforced cementitious composites. *Journal of Concrete Science and Engineering, RILEM 1*, 173–184.

Lin, Z. & V. C. Li (1997). Crack bridging in fiber reinforced cementitious composites with slip-hardening interfaces. *Journal of the Mechanics and Physics of Solids 45*(5), 763–787.

Lo Monte, F. & L. Ferrara (2020). Tensile behaviour identification in ultra-high performance fibre reinforced cementitious composites: indirect tension tests and back analysis of flexural test results. *Materials and Structures 53*(6), 145.

Lo Monte, F. & L. Ferrara (2021). Self-healing characterization of uhpfrcc with crystalline admixture: Experimental assessment via multi-test/multi-parameter approach. *Construction and Building Materials 283*, 122579.

Mergheim, J. & P. Steinmann (2013). Phenomenological modelling of self-healing polymers based on integrated healing agents. *Computational Mechanics 52*(3), 681–692.

Oucif, C., G. Z. Voyiadjis, & T. Rabczuk (2018). Modeling of damage-healing and nonlinear self-healing concrete behavior: Application to coupled and uncoupled self-healing mechanisms. *Theoretical and Applied Fracture Mechanics 96*, 216–230.

Pathirage, M., D. Bentz, G. Di Luzio, E. Masoero, & G. Cusatis (2019). The onix model: a parameter-free multiscale framework for the prediction of self-desiccation in concrete. *Cement and Concrete Composites 103*, 36–48.

Qiu, J., S. He, Q. Wang, H. Su, & E. Yang (2019). Autogenous healing of fiber/matrix interface and its enhancement. *Proc. of the 10st Intern. Conf. on Fracture Mechanics of Concrete and Concrete Structures (FraMCoS-X), G. Pijaudier-Cabot, P. Grassl and C. La Borderie Eds. 24-26 June, Bayonne, France.*

Schauffert, E. A. & G. Cusatis (2012). Lattice discrete particle model for fiber-reinforced concrete. i: Theory. *Journal of Engineering Mechanics 138*(7), 826–833.

Schauffert, E. A., G. Cusatis, D. Pelessone, J. L. O'Daniel, & J. T. Baylot (2012). Lattice discrete particle model for fiber-reinforced concrete. ii: Tensile fracture and multiaxial loading behavior. *Journal of Engineering Mechanics 138*(7), 834–841.

Snoeck, D. & N. De Belie (2015). From straw in bricks to modern use of microfibers in cementitious composites for improved autogenous healing – a review. *Construction and Building Materials 95*, 774–787.

Voyiadjis, G. Z., A. Shojaei, & G. Li (2011). A thermodynamic consistent damage and healing model for self healing materials. *International Journal of Plasticity 27*(7), 1025–1044.

Yang, E.-H., S. Wang, Y. Yang, & V. C. Li (2008). Fiber-bridging constitutive law of engineered cementitious composites. *Journal of Advanced Concrete Technology 6*(1), 181–193.

Yang, L., M. Pathirage, H. Su, M. Alnaggar, G. Di Luzio, & G. Cusatis (2021). Computational modeling of temperature and relative humidity effects on concrete expansion due to alkali–silica reaction. *Cement and Concrete Composites 124*, 104237.

Computational Modelling of Concrete and Concrete Structures – Meschke, Pichler & Rots (Eds)
© 2022 Copyright the Author(s), ISBN: 978-1-032-32724-2

Experimental and computational micromechanics of dental cement paste

P. Dohnalík, B.L.A. Pichler, L. Zelaya-Lainez, O. Lahayne & C. Hellmich
Institute for Mechanics of Materials and Structures, TU Wien (Vienna University of Technology), Vienna, Austria

G. Richard
Research and Development, Septodont, Saint-Maur-des-Fossés, France

ABSTRACT: Biodentine, a cementitious material used in dentistry, significantly outperforms the stiffness and strength properties reached by standard construction cements. This motivates a deeper mechanical analysis, both experimentally and computationally, at different observation scales. At the tens of microns scale, normalized histograms of hardness and modulus, obtained from more than 5000 nanoindentation tests, are suitably represented by the superposition of three lognormal distributions (LNDs) with increasing median values in stiffness and hardness (Dohnalík et al. 2021). The two lower LNDs characterize two variants of calcite-reinforced hydrates, while the highest LND underestimates the elastic properties of the small, very stiff inclusions of unhydrated clinker and zirconium dioxide. In order to quantify the micromechanical interactions of the aforementioned material constituents, the LNDs of the two calcite-reinforced hydrate types enter a self-consistent elastic homogenization scheme, which, at the millimeter scale, is linked to the elastic properties obtained from longitudinal and transverse ultrasonic wave velocities. This reveals the existence of defects, acting as an additional micromechanical phase, and also provides detailed insight into the microstress fluctuations within the key element of Biodentine's mechanics: the calcite-reinforced hydrates.

1 INTRODUCTION

Over the little less than the two decades which have passed since the pioneering works of Constantinides & Ulm (2004), the combination of grid nanoindentation and micromechanical modeling has become a broadly accepted standard for the state-of-the-art characterization of cementitious construction materials (Königsberger et al. 2018; Němeček & Lukeš 2020; Sarris & Constantinides 2013; Vandamme & Ulm 2009). The high level of maturity reached in the aforementioned studies motivates to enter, with very similar methods, the somehow related, still distinct, world of *dental* cement pastes, a fascinating field where both experimental and computational micromechanics play an only minimal role (if any).

The present contribution concisely summarizes very recent experimental results obtained at the Institute for Mechanics of Materials and Structures of TU Wien (Vienna University of Technology, Austria) in cooperation with Septodont, Saint-Maur-le-Fossés, France. The portfolio of the latter company comprises a cementitious product called Biodentine which largely outperforms the mechanical strength properties of ordinary construction cement pastes, by reaching compressive strengths of some 300 MPa (Butt et al. 2014). In this context, the obviously arising question concerns the microstructural and micromechanical features which lie at the origin of this impressive mechanical competence; and an interesting answer to this question comes from evaluating the aforementioned experimental data in the framework of an innovative modeling technique which blurs the border between statistical and deterministic micromechanics.

2 WIDE GRID NANOINDENTATION OF ULTRASMOOTH SURFACES – IDENTIFICATION OF MATERIAL PHASES

It is well known that a sufficiently smooth surface is the key to reliable and informative data obtained from nanoindentation tests (Donnelly et al. 2006; Miller et al. 2008): The smoother the surface, the smaller the minimum indentation depth needed to obtain theoretically meaningful experimental results, and hence, the higher the microstructural resolution at which the material properties can be deciphered.

Aiming at diving as deeply as possible into the origin of the fascinating mechanical properties of Biodentine, particular precautions were taken in order to attain a new level of surface smoothness (Dohnalík et al. 2021): In more detail, the top surface

 DOI 10.1201/9781003316404-11

of a well-hardened Biodentine specimen measuring $12 \times 7 \times 3\,\text{mm}^3$ was first ground by hand. Subsequently, it was polished by means of a PM5 precision polishing machine (Logitech, Scotland), operated for 20 hours at 40 to 50 revolutions per minute, using silicon carbide (SiC) grinding paper with a grain size of 5 microns. This resulted in a root-mean-squared average roughness, which was as low as 18 nm, when averaged over a quadratic test area with 50 microns side length. To the best knowledge of the authors, this has significantly shifted the limits in the field, where customary roughnesses, when averaged over 50 microns, are about 36 nanometers (Miller et al. 2008).

Also in terms of number of indentations making up the testing grid per sample, customary numbers of indentations per sample, typically amounting to some two hundred, were significantly increased in the present study, with more than 5000 tests having been performed on a sample of Biodentine. We also note that the grid was very wide, with its spacing of 70 microns being orders of magnitude larger than the on average 140 nm nanoindentation depth. In this way, it reflects the properties of wide testing area spanning some 5 times 5 mm, while guaranteeing virtually total independence of the individual nanoindentation test results. In this context, the individually obtained hardness and elastic properties refer to representative volume elements with a characteristic length amounting to half of the indentations depths (Jagsch et al. 2020; Königsberger et al. 2021), i.e. to around 70 nm.

The aforementioned, very high number allowed for a quantitative assessment of distribution functions used for the representation of the histograms of hardness and modulus values obtained from the nanoindentation tests: While Gaussian functions have been adopted as the standard choice, the very high number of tests performed on dental cement pastes could be remarkably better described by means of lognormal distributions, more precisely by the superposition of three such probability density functions, see Figures 1 and 2. As explained in further detail in (Dohnalík et al. 2021), the three lognormal distributions refer to cement clinker and zirconium dioxide (with a median modulus of 92.2 GPa and a median hardness of 6.66 GPa), to high-density calcite-reinforced hydrates (with a median modulus of 62.6 GPa and a median hardness of 2.78 GPa), and to less dense calcite-reinforced hydrates (with a median modulus of 45.1 GPa and a median hardness of 1.15 GPa). Moreover, these lognormal distributions, rather than Gaussians, naturally reflect the physical nature of stiffness and hardness, the values of which need to be strictly positive.

However, one must be cautious when evaluating the force-displacement curves of the indents made into the relatively small inclusions made up of the stiffest phases (i.e. of cement clinker and zirconium dioxide - see the light gray and white inclusions in Figure 3) through the standard Oliver-Pharr formulae for infinite halfspaces (Oliver & Pharr 1992). Such an evaluation typically delivers insufficiently high elastic values; and this is because the indenter does not so much probe the stiff, small inclusion, but the latter acts as kind of a larger indenter pressed into the surrounding matrix (with smaller elastic properties than those of the inclusion). This has been explicitly shown by a combination of imaging and nanoindentation, applied to two different types of supplementary cementing materials (Königsberger et al. 2021; Ma et al. 2017). In full accordance with these deliberations, the indentation modulus of clinker has been reported to amount to 125 GPa (Constantinides & Ulm 2007).

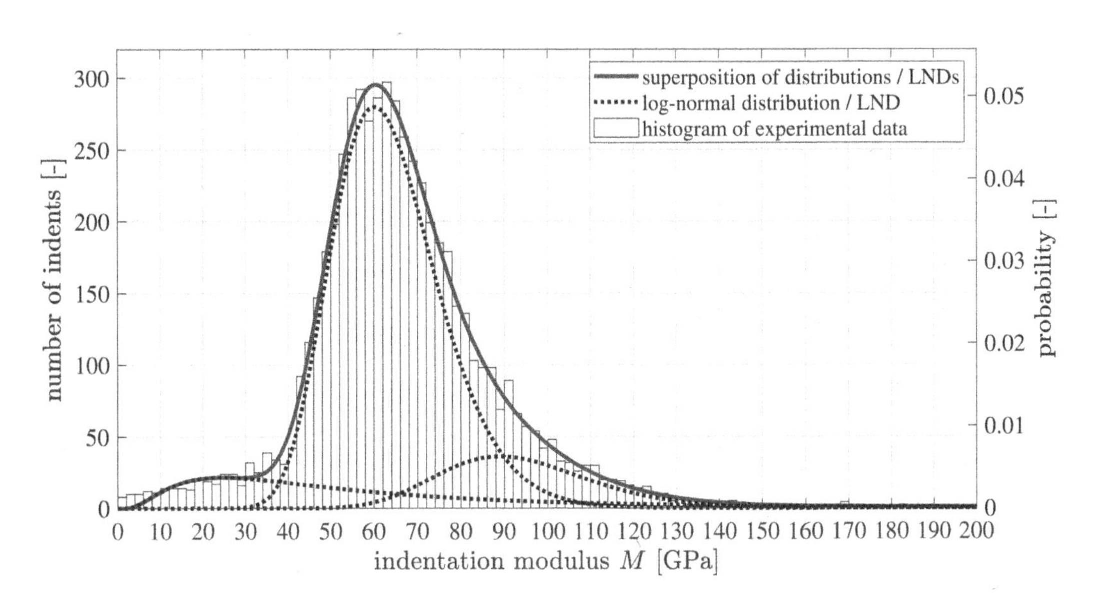

Figure 1. Nanoindentation modulus values obtained from microscopic characterization of Biodentine: histogram and representation in terms of superimposed lognormal probability distributions (LND).

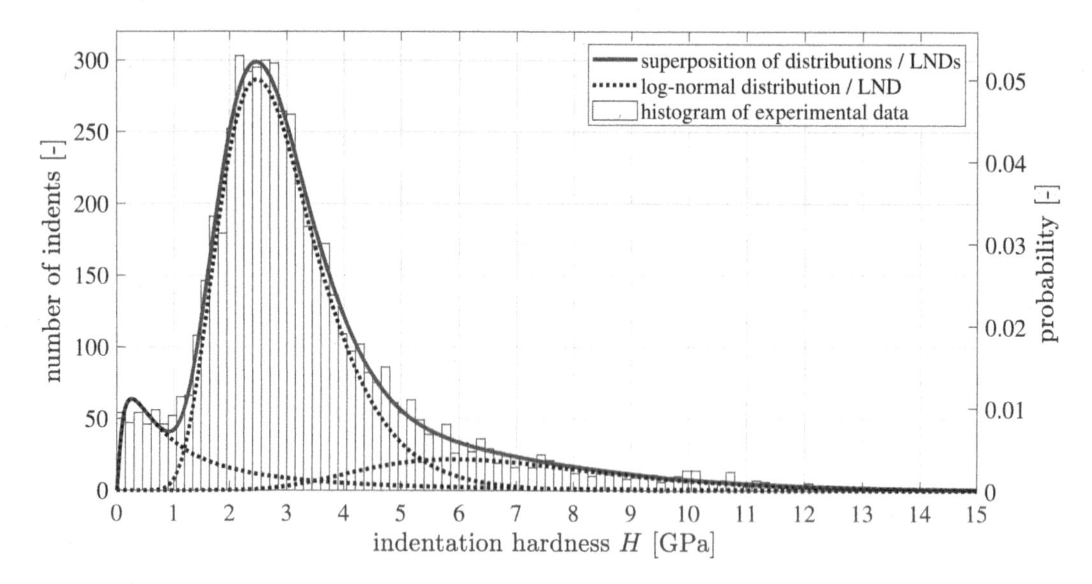

Figure 2. Nanoindentation hardness values obtained from microscopic characterization of Biodentine: histogram and representation in terms of superimposed lognormal probability distributions (LND).

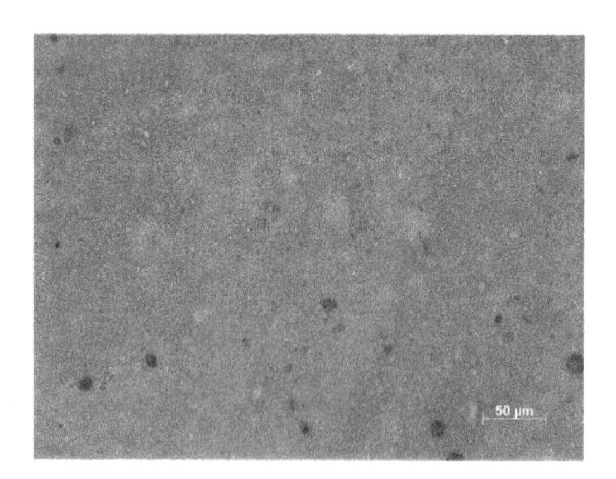

Figure 3. Black-and-white light microscopy image of a polished surface of Biodentine, magnification of 200 fold.

3 ULTRASONIC TESTING – MACROSTIFFNESS CHARACTERIZATION AND DETECTION OF "ZERO-VOLUME" DEFECTS

The non-destructive ultrasonic pulse transmission technique was used to characterize the macroscopic elastic properties of hardened Biodentine. Both longitudinal and transversal waves, with frequencies in the kHz and MHz regime, were sent through the material.

The test setup consisted of a serial arrangement of a pulse generator, a layer of honey (serving as a coupling medium), a plastic foil, the specimen, another plastic foil, another layer of honey, and a pulse detector. The plastic foils protected the sample against contamination of its open porosity with the coupling medium. The specimens, the test setup, and its surrounding environment were conditioned to 37°C.

The wave velocities v of Biodentine are equal to the height b of the tested specimens divided by the time of flight t_f of the ultrasonic pulse through the tested specimen,

$$v = \frac{b}{t_f}\,. \tag{1}$$

t_f cannot be directly measured, but results from the difference of two other time measurements,

$$t_f = t_{tot} - t_d\,, \tag{2}$$

where t_{tot} is the travel time of the pulse from the transducer – through the coupling medium, the plastic foils, *and* the specimen – to the receiver, while the delay time t_d is needed by the pulse to just travel from the generator, through honey and plastic foils (but without specimen), to the receiver.

325 measurements of longitudinal waves were performed at material ages from 7 to 28 days. The central excitation frequencies amounted to 50 kHz, 500 kHz, 1 MHz, 2.25 MHz, 5 MHz, 10 MHz, and 20 MHz. The longitudinal wave velocities were fairly independent of the material age as well as the testing frequency. On average, they amount to 4.977 km/s, see (Dohnalík et al. 2021) and Figure 4.

122 measurements of transversal waves were performed at material ages from 7 to 28 days. The central excitation frequencies amounted to 2.25 MHz and 5 MHz. The transversal wave velocities are also fairly independent of the material age and the testing frequency. On average, they amount to 2.473 km/s, see Figure 5.

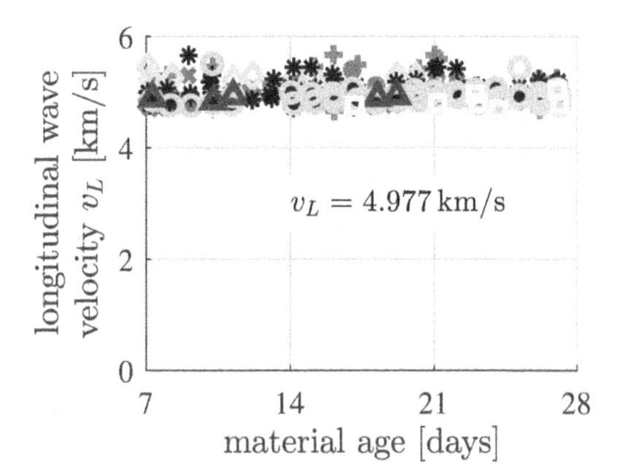

Figure 4. Longitudinal wave velocities sent at different frequencies through cylindrical samples of Biodentine, with of 5 mm diameter and 10 mm height; the mean longitudinal wave velocity is equal to $v_L = 4.977$ km/s; the pink markers correspond to 50 kHz transducers' central frequency, red to 500 kHz, cyan to 1 MHz, black to 2.25 MHz, green to 5 MHz, blue to 10 MHz, and yellow to 20 MHz transducers' central frequency, after (Dohnalík et al. 2021).

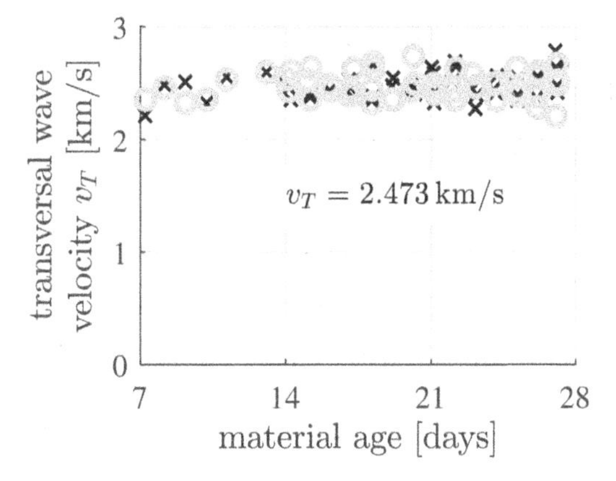

Figure 5. Transversal wave velocities v_T as a function of the material age ranging from 7 to 28 days after production, for different central frequencies: 2.25 MHz (black '×') and 5 MHz (green 'o'); the mean transversal wave velocity amounts to $v_T = 2.473$ km/s.

The principle of separation of scales states that the wavelengths λ must be significantly larger than the size ℓ_{rve} of a representative volume element of the tested material (Kohlhauser & Hellmich 2013; Zaoui 2002), and that ℓ_{rve} must be significantly larger than the characteristic size ℓ_{het} of the microheterogeneities:

$$\lambda \gg \ell_{rve} \gg \ell_{het} \,. \tag{3}$$

Residual clinker grains are the largest microheterogeneities of hardened Biodentine. Their characteristic size amounts to some 4 microns (Dohnalík et al. 2021): $\ell_{het} \approx 4\,\mu\mathrm{m}$. The characteristic size of a representative volume of Biodentine is some three times larger (Drugan & Willis 1996; Pensée & He 2007). Thus, $\ell_{rve} \approx 12\,\mu\mathrm{m}$. This size is to be compared with the wavelenghts of the ultrasonic pulses.

The wavelength is indirectly proportional to the ultrasonic frequency. Therefore, the largest testing frequency yields a lower bound for the wavelengths. As for the longitudinal waves, this lower bound follows as

$$\lambda_L \geq \frac{v_L}{\max f} = \frac{4.977\,\mathrm{km/s}}{20\,\mathrm{MHz}} = 249\,\mu\mathrm{m}\,. \tag{4}$$

As for the transversal waves, it follows as

$$\lambda_T \geq \frac{v_T}{\max f} = \frac{2.473\,\mathrm{km/s}}{5\,\mathrm{MHz}} = 495\,\mu\mathrm{m}\,. \tag{5}$$

Eqs. (4) and (5) underline that the wavelengths were by a factor of 20 (longitudinal waves) and 40 (transversal waves) larger than $\ell_{rve} \approx 12\,\mu\mathrm{m}$. The principle of separation of scales, see Eq. (3), is fulfilled (Kohlhauser & Hellmich 2013; Zaoui 2002). This provides evidence that wave velocities of Figs. 4 and 5 are representative for the homogenized composite Biodentine.

According to the theory of wave propagation through isotropic linear-elastic media, longitudinal and transversal wave velocities, together with the mass density ρ of the tested material, allow for quantifying the bulk modulus, k, and the shear modulus, μ, as (Achenbach 1973; Carcione 2007; Kohlhauser & Hellmich 2012),

$$k = \rho\left(v_L^2 - \frac{4}{3}v_T^2\right), \tag{6}$$

$$\mu = \rho\, v_T^2\,, \tag{7}$$

respectively. Evaluation of Eqs. (6) and (7) based on $\rho = 2.311\,\mathrm{kg/dm^3}$ (Dohnalík et al. 2021) and the wave velocities of Figs. 4 and 5 gives access to constant isotropic elastic properties, namely to a bulk modulus of 38.4 GPa and a shear modulus of 14.1 GPa. These stiffness properties are *lower* than all of the different phase properties identified by nanoindentation. This indicates that another, "zero-volume" material phase is mechanically active at the single-microns scale. This additional phase may be called "microcracks", or more appropriately "micro-defects", with the latter probably arising from non-perfect bonding of the calcite-reinforced hydrate and clinker phases alluded to before.

4 STATISTICAL MICROMECHANICS – DEFECT DENSITY AND MICROSTRESS DISTRIBUTIONS

In order to elucidate the mechanical functioning within a representative volume element of Biodentine, we resort to the theoretical framework of Eshelby problem-based continuum micromechanics (Eshelby 1957; Pensée et al. 2002; Zaoui 2002). However, extending the traditional types of homogenization schemes, we introduce infinitely many spherical hydrate phases with micro-elastic properties following the lognormal probability distributions associated with

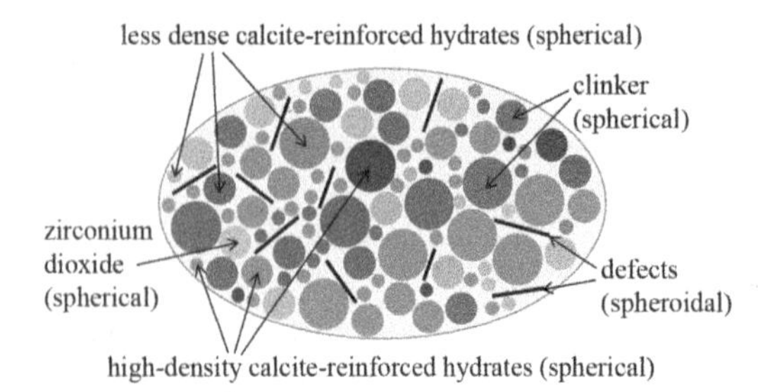

Figure 6. Micromechanical representation of Biodentine ("material organogram"): the two-dimensional sketch shows qualitative properties of a three-dimensional representative volume element.

the high-density and lower-density calcite-reinforced hydrates, as given in Figure 1. The model is completed as seen in Figure 6, by two additional spherical phases representing clinker and zirconium dioxide, and by zero-shear-stiffness, isotropically oriented, infinitely many flat spheroidal phases representing defects (Dormieux et al. 2006). Phase input properties comprise bulk and shear moduli for the non-defect, non-hydrate phases as known from literature sources (Hussey and Wilson 1998; Pichler & Hellmich 2011), and the lognormal distributions of the indentation modulus of the two hydrate phases, whereas the crack or defect density parameter in the sense of Budiansky & O'connell (1976) and Poisson's ratio of the hydrate phases remain *a priori* unknown, and are back-computed from the model predictions for the bulk and shear moduli of the homogenized material, given in more detail in (Dohnalík et al. 2022). The resulting crack (or defect) density parameter amounting to 0.77, underlines the pronounced micromechanical effect of the defects (much more pronounced than e.g. the crack density parameter found in nanoindentation-probed rail steel (Jagsch et al. 2020)); while the Poisson's ratio of calcite-reinforced hydrates, amounting to 0.20, is smaller than the standard value of 0.24 known from construction cements (Constantinides & Ulm 2004).

Having in this way elucidated the microelasticity of Biodentine, the understanding of strength upscaling is a still open topic. As a first steps towards the latter endeavor, it is illustrative to evaluate the concentration components quantifying the relation between the (macro-)stresses subjected to a piece of Biodentine and the (micro-)stresses prevailing in the individual hydrate phases, in terms of probability distributions. As shown by example of the deviatoric portion of the stress concentration tensor (see B_{vol} in Figs. 7 and 8), the corresponding probability distributions are not any more of the lognormal, but of the beta type. Their shape makes simple statements, like one hydrate being more loaded than the other one, obsolete, and given also the distributed nature of the phase-specific hardness properties, indicates a quite complex microstructural interaction pattern which eventually determines the fascinatingly high strength properties at the Biodentine level.

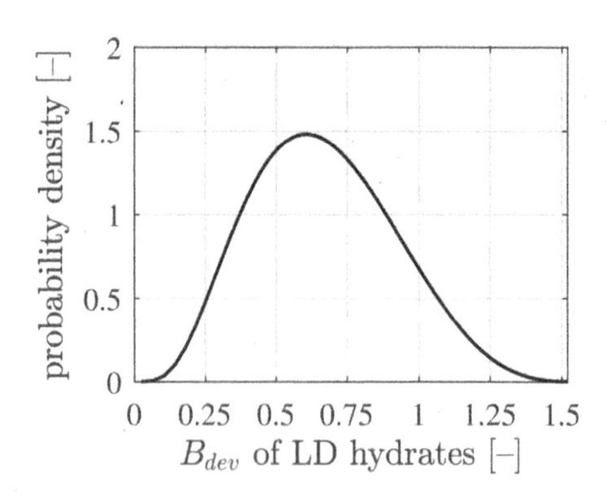

Figure 7. Probability distribution of the deviatoric component of the stress concentration tensors providing the scale-transition from Biodentine to the lower-density calcite-reinforced hydrates.

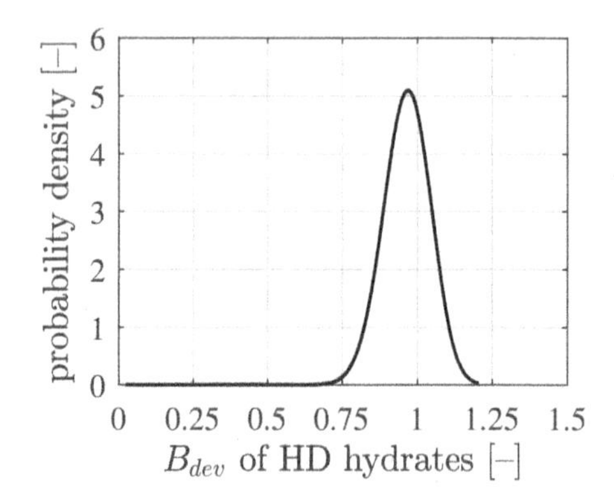

Figure 8. Probability distribution of the deviatoric component of the stress concentration tensors providing the scale-transition from Biodentine to the high-density calcite-reinforced hydrates.

5 CONCLUSIONS

Combining microscopic characterization (grid nanoindentation), macroscopic experiments (ultrasonic pulse velocity testing) with multiscale modeling (Eshelby-problem-based self-consistent scheme) allowed for developing a realistic micromechanics model for hardened dental cement paste "Biodentine". This model provides quantitative insight into microscopic stress fluctuation resulting from corresponding microstructural stiffness distribution.

ACKNOWLEDGEMENTS

This project has received funding from the European Union's Horizon 2020 research and innovation programme under the Marie SkÅ‚odowska-Curie Grant Agreement No. 764691. The authors gratefully acknowledge the support of the ultrasound experiments by Wolfgang Dörner, TU Wien.

REFERENCES

Achenbach, J. (1973). *Wave propagation in Elastic Solids, One-dimensional motion of an elastic continuum.* Elsevier.

Budiansky, B. & R. J. O'connell (1976). Elastic moduli of a cracked solid. *International Journal of Solids and Structures 12*(2), 81–97.

Butt, N., S. Talwar, S. Chaudhry, R. R. Nawal, S. Yadav, & A. Bali (2014). Comparison of physical and mechanical properties of mineral trioxide aggregate and Biodentine. *Indian Journal of Dental Research 25*(6), 692.

Carcione, J. M. (2007). *Wave fields in real media: Wave propagation in anisotropic, anelastic, porous and electromagnetic media.* Elsevier.

Constantinides, G. & F.-J. Ulm (2004). The effect of two types of C-S-H on the elasticity of cement-based materials: Results from nanoindentation and micromechanical modeling. *Cement and Concrete Research 34*(1), 67–80.

Constantinides, G. & F.-J. Ulm (2007). The nanogranular nature of C–S–H. *Journal of the Mechanics and Physics of Solids 55*(1), 64–90.

Dohnalík, P., C. Hellmich, G. Richard, & B. L. Pichler (2022). Stiffness and stress fluctuations in dental cement paste: a continuum micromechanics approach. Manuscript in preparation.

Dohnalík, P., B. L. Pichler, L. Zelaya-Lainez, O. Lahayne, G. Richard, & C. Hellmich (2021). Micromechanics of dental cement paste. *Journal of the Mechanical Behavior of Biomedical Materials 124*, 104863.

Donnelly, E., S. P. Baker, A. L. Boskey, & M. C. van der Meulen (2006). Effects of surface roughness and maximum load on the mechanical properties of cancellous bone measured by nanoindentation. *Journal of Biomedical Materials Research Part A 77*(2), 426–435.

Dormieux, L., D. Kondo, & F.-J. Ulm (2006). *Microporomechanics.* John Wiley & Sons.

Drugan, W. & J. Willis (1996). A micromechanics-based nonlocal constitutive equation and estimates of representative volume element size for elastic composites. *Journal of the Mechanics and Physics of Solids 44*(4), 497–524.

Eshelby, J. D. (1957). The determination of the elastic field of an ellipsoidal inclusion, and related problems. *Proceedings of the royal society of London. Series A. Mathematical and physical sciences 241*(1226), 376–396.

Hussey, R. J. & J. Wilson (1998). *Advanced technical ceramics directory and databook.* Springer Science & Business Media.

Jagsch, V., P. Kuttke, O. Lahayne, L. Zelaya-Lainez, S. Scheiner, & C. Hellmich (2020). Multiscale and multitechnique investigation of the elasticity of grooved rail steel. *Construction and Building Materials 238*, 117768.

Kohlhauser, C. & C. Hellmich (2012). Determination of Poisson's ratios in isotropic, transversely isotropic, and orthotropic materials by means of combined ultrasonic-mechanical testing of normal stiffnesses: Application to metals and wood. *European Journal of Mechanics – A/ Solids 33*, 82–98.

Kohlhauser, C. & C. Hellmich (2013). Ultrasonic contact pulse transmission for elastic wave velocity and stiffness determination: Influence of specimen geometry and porosity. *Engineering Structures 47*, 115–133.

Königsberger, M., M. Hlobil, B. Delsaute, S. Staquet, C. Hellmich, & B. Pichler (2018). Hydrate failure in ITZ governs concrete strength: A micro-to-macro validated engineering mechanics model. *Cement and Concrete Research 103*, 77–94.

Königsberger, M., L. Zelaya-Lainez, O. Lahayne, B. L. Pichler, & C. Hellmich (2021). Nanoindentation-probed Oliver-Pharr half-spaces in alkali-activated slag-fly ash pastes: multimethod identification of microelasticity and hardness. *Mechanics of Advanced Materials and Structures*, 1–12.

Ma, Y., G. Ye, & J. Hu (2017). Micro-mechanical properties of alkali-activated fly ash evaluated by nanoindentation. *Construction and Building Materials 147*, 407–416.

Miller, M., C. Bobko, M. Vandamme, & F.-J. Ulm (2008). Surface roughness criteria for cement paste nanoindentation. *Cement and Concrete Research 38*(4), 467–476.

Němeček, J. & J. Lukeš (2020). High-speed mechanical mapping of blended cement pastes and its comparison with standard modes of nanoindentation. *Materials Today Communications 23*, 100806.

Oliver, W. C. & G. M. Pharr (1992). An improved technique for determining hardness and elastic modulus using load and displacement sensing indentation experiments. *Journal of Materials Research 7*(6), 1564–1583.

Pensée, V. & Q.-C. He (2007). Generalized self-consistent estimation of the apparent isotropic elastic moduli and minimum representative volume element size of heterogeneous media. *International Journal of Solids and Structures 44*(7-8), 2225–2243.

Pensée, V., D. Kondo, & L. Dormieux (2002). Micromechanical analysis of anisotropic damage in brittle materials. *Journal of Engineering Mechanics 128*(8), 889–897.

Pichler, B. & C. Hellmich (2011). Upscaling quasi-brittle strength of cement paste and mortar: A multi-scale engineering mechanics model. *Cement and Concrete Research 41*(5), 467–476.

Sarris, E. & G. Constantinides (2013). Finite element modeling of nanoindentation on c–s–h: Effect of pile-up and contact friction. *Cement and Concrete Composites 36*, 78–84. Special issue: Nanotechnology in Construction.

Vandamme, M. & F.-J. Ulm (2009). Nanogranular origin of concrete creep. *Proceedings of the National Academy of Sciences 106*(26), 10552–10557.

Zaoui, A. (2002). Continuum micromechanics: Survey. *Journal of Engineering Mechanics 128*(8), 808–816.

Computational Modelling of Concrete and Concrete Structures – Meschke, Pichler & Rots (Eds)
© 2022 Copyright the Author(s), ISBN: 978-1-032-32724-2

Autogenous healing in cement: A kinetic Monte Carlo simulation of $CaCO_3$ precipitation

Aleena Alex
Newcastle University, UK

Enrico Masoero
Cardiff University, UK

ABSTRACT: Autogenous healing induced by the dissolution of C-S-H and CH in a cracked cement paste was modelled in this study, at the mesoscale of tens of nanometres. The pore solution contains carbon dioxide (CO_2) resulting in the precipitation of calcium carbonate ($CaCO_3$) into the crack. The simulations were performed using MASKE, a recently developed coarse-grained Kinetic Monte Carlo framework where the molecules of the solid phases are modelled as mechanically interacting particles that can also precipitate and dissolve. The precipitation of $CaCO_3$ molecules was initially observed in tiny gel pores within the C-S-H, but eventually extends completely filling the crack. The mechanical properties of the healed system were also investigated by straining the simulation box, computing the corresponding virial stress, and plotting the resulting stress-strain relationship.

1 INTRODUCTION

1.1 *Self-healing in concrete*

Concrete from Ordinary Portland Cement (OPC) is quasi-brittle construction material which can withstand high compressive load. During its life cycle, microcracks can develop within its structure. Microcracks are too small (<0.1 mm) to significantly affect the mechanical strength of a structure, but they do make the cement matrix more permeable. This increases the likelihood of chemical attack and corrosion of reinforcement, which may ultimately compromise the structural integrity of concrete (Qureshi et al. 2018). Self-healing is a possible solution to this issue.

Self-healing is classified as (1) Autogenous healing: where the materials in concrete continue to react with water in the cement paste forming new hydration product that seal the cracks, or (2) Autonomous healing: where sealing agents (such as bacteria) are encapsulated in the cement matrix and become active when a crack changes its local exposure conditions.

1.2 *Autogenous healing models*

Autogenous healing in cement/concrete can occur in 2 ways: (1) Continuous hydration at the crack where previously anhydrous cement comes in contact with water/moisture. The anhydrous cement reacts with water producing hydration products, such as calcium silicate hydrate (C-S-H) and calcium hydroxide (CH); Another way is (2) the dissolution of C-S-H, CH and other hydration products that have already formed in the cement matrix, releasing ions that react with dissolved carbon dioxide (CO_2) in the pore solution. This process eventually precipitates carbonates that fill the cracks (mainly calcium carbonate, $CaCO_3$, although other alkali ions, such as Na or K, may contribute with other carbonates too, especially in low-calcium cements).

Very few models exist which describe autogenous healing. Some models were proposed to predict the amount of unhydrated cement present in the matrix after hydration based on the water-cement (w/c) ratio and the fineness of cement (He et al. 2007). This is an indirect measure of the self-healing potential of the sample. Traditionally, continuous hydration of residual anhydrous phases was believed to be the governing self-healing mechanism. However, in recent years it has been shown that continued hydration reduces in a few weeks after casting, whereas calcite formation from existing hydration product becomes the main mechanism of healing afterwards (Van Tittelboom & De Belie 2013). Numerical models have been proposed for autogenous healing by both the mechanisms (further hydration and calcite formation). Huang et al. (2013) proposed a reactive transport model for the self-healing of microcracks in cement paste by further hydration. They established the relationship between self-healing efficiency and extra water provided. Hilloulin et al. (2014) reported a combined

DOI 10.1201/9781003316404-12

hydro-chemo-mechanical model to simulate autogenous healing which also predicts the mechanical regain after healing. Chitez and Jefferson (2016) combined the existing approaches to propose a comprehensive mathematical model for early age autogenous healing. The model uses reactive water transport to predict the movement of healing materials (C-S-H and CH) under a thermo-hygro-chemical (THC) framework. This model also predicts early age crack healing due to continuous hydration and is not applicable to long term calcite formation. It was only in recent years that calcite formation induced healing was attempted to be modelled. Ranaivomanana and Benkemoun (2017) proposed a numerical model in which chemical reactions and transport phenomena were modelled for the porous matrix and the crack by diffusion and permeation. .

While hydration induced self-healing has been studied extensively, models on calcite formation induced autogenous self-healing are far and few between. The existing models have their own strength, but also are limited in the mechanistic description of $CaCO_3$ precipitation, and this limits the possibility to use them to explore new solutions. For instance, altering the chemistry of the phases favouring $CaCO_3$ formation as a way to control the rate of the process or even the morphology of the carbonates to optimize healing. This study aims to address calcite induced self-healing and the resulting mechanical regain using a discrete particle-based model.

2 METHODOLOGY

2.1 *Overview*

In this study we have examined autogenous healing by simulating a crack in a paste of C-S-H and CH under the assumption of full hydration. The size of the simulated system is kept very small, with crack with size of 1 nm only. In this way we can use kinetic constants for individual reactions and model these directly from Transition State Theory, without involving additional assumptions to further coarse grain the system (Shvab et al. 2017). Once generated, the crack was filled with water and CO_2 at atmospheric saturation level which was kept constant throughout the simulation. This induces the dissolution of both C-S-H and, most significantly, CH releasing free calcium ions. These ions then react with the CO_2 resulting in the precipitation of $CaCO_3$ and healing the crack.

2.2 *MASKE: A kinetic Monte Carlo framework*

The simulations were performed using MASKE, a recently developed Kinetic Monte Carlo framework (Shvab et al. 2017). In MASKE, the system is discretized representing the mineral phases as agglomerates of nanoparticles which interact via effective potentials (energy as a function of distance) whose spatial derivatives are the interaction forces. The particles can dissolve and precipitate via reaction rates obtained from transition state theory (TST). These rates depend on macroscopic rate constants for the chemical reactions involved, the saturation index of the solution with respect to each reaction and the excess free energy coming from the interactions between particles. This excess free energy is particularly important because it renders the solubility of individual particles dependent on local morphology i.e. the number of interacting neighbours as well as the presence of any local mechanical stress.

The dissolution and precipitation rates for a particle are described by Equations 1 and 2.

$$r_{diss} = \frac{k_B T}{h} V_m \frac{c^*}{\gamma^*} exp\left[-\frac{\Delta G^*}{k_B T}\right] exp\left[\frac{-\Delta U_{diss} - U_{kink}}{k_B T}\right] \tag{1}$$

$$r_{prec} = \frac{k_B T}{h} V_m \frac{c^*}{\gamma^*} exp\left[-\frac{\Delta G^*}{k_B T}\right] \beta_{prec} \tag{2}$$

where kB: Boltzmann constant; T: temperature, K; h: Planck constant; γ^*: activity coefficient; c^*: standard state concentration; V_M: molar volume of the particle; ΔG^*: standard state activation energy for dissolution; β_{prec}: saturation index of the solution. ΔU_{diss} is the change in interaction energy following the dissolution of a particle and ΔU_{kink} is the interaction energy between kink particle and its nearest neighbour. The derivation and physical meaning of these equations and its validation are discussed in detail in earlier works (Coopamootoo & Masoero 2020; Shvab et al. 2017).

2.3 *Simulation steps*

The simulations were carried out in 6 steps:

(1) Starting from an initially empty simulation box, CH crystals were created by agglomerating nanoparticles representing individual CH molecules, until reaching a volume fraction of 28% in the box. This is the theoretical volume fraction of CH in a paste obtained from fully hydrated C_3S if the C-S-H is assumed to have a gel porosity of 34.5% (Masoero et al. 2014).
(2) A spherical agglomerate of randomly closed packed C-S-H particles was created separately and parametrized to get the correct interaction potential parameters for C-S-H. This ensured that the C-S-H grain made of the agglomerated particles is at equilibrium (no dissolution nor precipitation). The equilibrium is reached when the activity product of the surrounding solution coincides with the equilibrium constant of C-S-H (i.e., saturation index ($\beta = 1$)).
(3) Other interaction potentials such as C-S-H/CH, C-S-H/$CaCO_3$ and CH/$CaCO_3$ were analytically determined averaging the interaction strength between pure phases.

(4) Once the interaction potentials were determined, the box was filled to the prescribed volume fraction of solid C-S-H (47%). This leaves 25% gel porosity in the system.
(5) The C-S-H / CH system was then cracked by carving out a 1nm thick slice of particles from the center of the box.
(6) The dissolution-mineralization in and around the crack was finally simulated using MASKE.

These steps are detailed in the sections below.

2.4 *Creating $Ca(OH)_2$ crystals*

This first simulation was to create two crystallites of $Ca(OH)_2$, targeting a desired volume fraction $\eta = 28\%$. This η assumed that the paste was pure C_3S and was fully hydrated, hence leading only to C-S-H gel and $Ca(OH)_2$ as hydration products. The C-S-H gel was assumed to have an internal gel porosity of approximately 34.5%, whereas the $Ca(OH)_2$ was considered be a solid crystal. The volume fractions occupied by the two minerals were then obtained from the stoichiometry and molar volumes (Masoero et al. 2014). It was also assumed that the uncracked paste features no capillary pores.

Two small face centered cubic (FCC) nuclei of $Ca(OH)_2$ were created (Figure 1 (a)) with different orientations, and setting the solution to a high concentration of Ca^{2+} and OH^-, so that further precipitation will occur. Snapshots were saved during the simulation and a configuration with $\eta \approx 28\%$ was chosen (Figure 1(b)).

2.5 *Parameterising amorphous C-S-H for solubility*

The particles in MASKE interact via harmonic potentials as described in Coopamootoo and Masoero (2020).

$$U(r) = \frac{1}{2}k(r - r_0)^2 - \varepsilon_0 \tag{3}$$

where U is the interaction energy, r is the interparticle distance, $k = EA/r_0$, E being the Young's modulus of the particle and $r_0 = \pi D_0^2/4$ and ε_0 is the minimum energy at equilibrium. $\varepsilon_0 = \gamma\Omega/n_{kink}$, where γ is the surface energy, Ω is the surface area of particle and n_{kink} number of neighbors for a kink particle. A kink particle has half the number of neighbors as a bulk particle. For an FCC packed CH n_{kink} is easily determined. However, for amorphous C-S-H it was estimated by a trial-and-error process.

A random close packing (RCP) of C-S-H particles were created separately in another simulation box. This was done using the random space filling algorithm in Masoero & Di Luzio 2020. The equilibrium distance between interacting C-S-H particles was set to be smaller than the actual molecular diameters, to compensate for the porosity of a random close packing and eventually obtain a dense C-S-H with same density as solid C-S-H without pores. See Coopamootoo 2020 paper for a similar approach, although there aimed at producing non-porous C_3S. Attention was paid to zero the average axial stresses on the dense C-S-H system.

An iterative scheme was employed to compute the number of interacting neighbors in the bulk of the dense C-S-H domain, which in turn defines the interaction potential between C-S-H particles: see Coopamootoo and Masoero (2020) for details on how the number of neighboring particles in the bulk and the water-solid interfacial energy of a phase can be used to obtain interaction parameters. When the RCP with consistent interactions was obtained, a spherical grain was carved out and the surrounding solution was set to match the equilibrium constant for C-S-H dissolution. At this stage, the bond energy of C-S-H was tweaked until the correct zero-rate of dissolution/precipitation was obtained.

2.6 *Parameterising amorphous C-S-H/CH*

The relations described in section 2.5 were used to determine the interaction potentials of particles from the same solid phase. However, for two particles

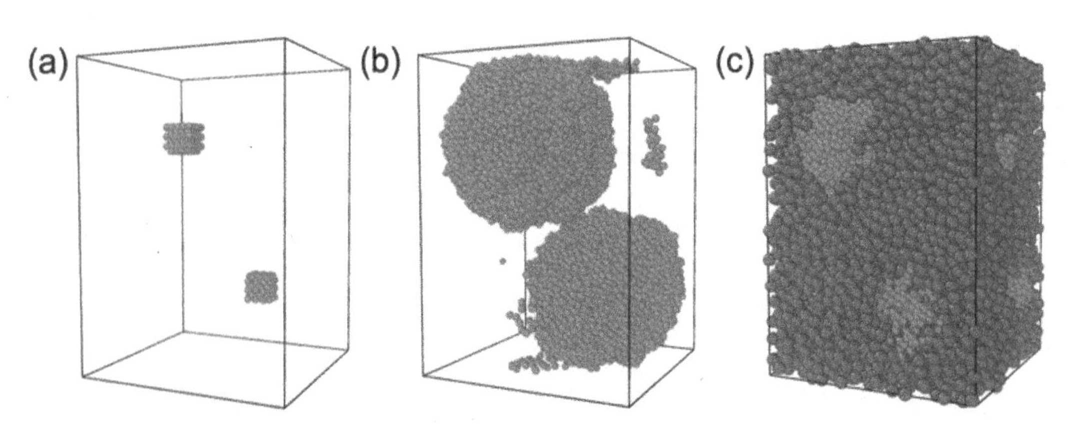

Figure 1. Box size 14x14x10 nm. CH (red) and C-S-H (blue). (a) Two FCC seeds of CH were placed in the box (b) CH precipitation carried out to the desired 28% and (c) C-S-H packed to 47%.

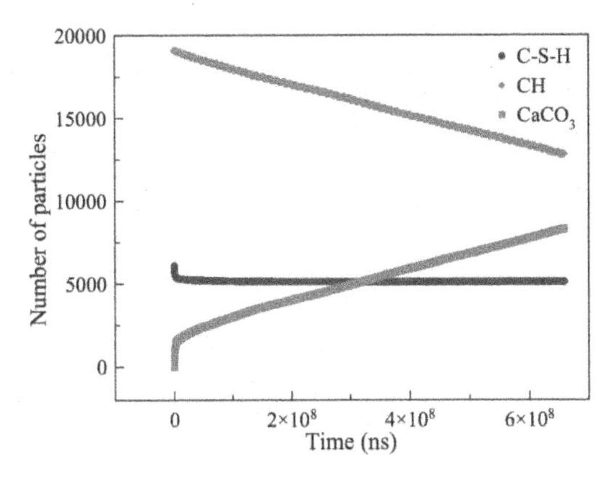

Figure 2. Number of particles vs time as the simulation proceeds. Results are provided for the first 15000 steps.

representing different phases, 1 and 2, such as C-S-H and CH, it was assumed that the minimum interaction energy can be determined by the expression

$$\varepsilon_{12} = \varepsilon_{21} = (k_{12}\gamma_1 + k_{21}\gamma_2)\frac{A_1}{n} \tag{4}$$

where $k_{12} = k_{21}$ was assumed to be on average 0.5. This factor is a measure of individual surface energy contribution towards the interaction potential between surfaces of particles 1 and 2. Detailed atomic scale simulation studies may give a better estimate of k.

2.7 *Packing and relaxing*

Once the interaction potentials between all possible combination of particles were determined, C-S-H was packed to a volume fraction of 47% around the pre-existing CH grains, as presented in Figure 1(c). This leaves 25% of gel pores. This simulation box was then relaxed by energy minimization ensuring that the pressure in all directions was zero. A crack of 1 nm was then created at the centre of the box as shown in Figure 3 (T0).

The concentration of carbonate ions in the implicit solution was fixed to atmospheric saturation levels as determined by Henry's law. Finally, the dissolution-precipitation reactions were run with MASKE and the results were obtained as detailed in the next section.

3 RESULTS AND DISCUSSION

Number of particles of C-S-H, CH and $CaCO_3$ vs time for the first 15000 steps of the simulation is shown in Figure 2. The rates of dissolution of CH and

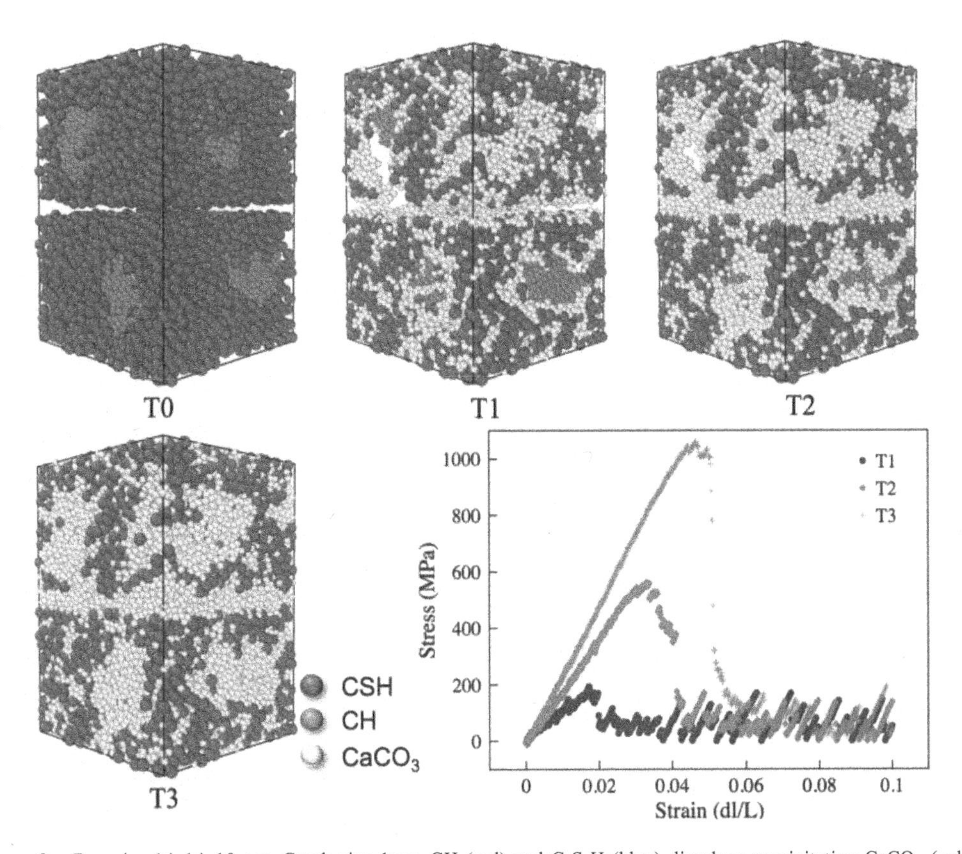

Figure 3. Box size 14x14x10 nm. Crack size 1nm. CH (red) and C-S-H (blue) dissolves precipitating $CaCO_3$ (yellow). Snapshots at 3 consecutive stages (T1, T2, T3) were saved and the evolution of mechanical properties as the crack heals was determined by the stress-strain plot.

precipitation of $CaCO_3$ were found to be constant at this stage. Initially, dissolution of C-S-H particles was observed close to the crack due to the amorphous structure of C-S-H. However, CH dissolution takes over and for the rest of the simulation CH dissolution drives the precipitation of $CaCO_3$. It was observed that initially, the $CaCO_3$ precipitation was predominantly in the gel pores. However, as simulation proceeds precipitation was observed in the crack as well.

3.1 *Mechanical properties*

The crack was entirely bridged by $CaCO_3$ when CH had completely dissolved. The volume occupied by solid particles in the box (with fixed box sizes) increased by 15-20% due to the larger size of $CaCO_3$ compared to CH.

When the crack started to heal (T1 in Figure 3), a snapshot was saved and the box was strained perpendicular to the crack. Stress in the direction of load was determined as the negative of pressure, this latter obtained from the virial method implementation in LAMMPS (Thompson et al. 2022). Stress-strain relationship was plotted as shown in Figure 3. There was a slight regain of mechanical strength ($\sim$100 MPa) at this point. Consequently, on further bridging this regain increases steadily up to $\sim$1000 MPa of ultimate strength (T3 in Figure 2) once the crack was fully bridged.

These strength values are clearly much greater than the tensile strength of cement paste, but one must consider that these simulations are at the nanoscale, hence fracture-inducing defects are very small (consider the scaling of strength with defect size e.g. from Griffith's law). Such high tensile strengths are indeed typically obtained from simulations at the nanoscale on cement hydrates (Lolli et al. 2018).

These simulations show that MASKE is a powerful tool to model the calcite precipitation induced autogenous healing in cementitious systems. Further investigations need to be done on how the chemistry of the solution evolves during the dissolution and precipitation process. Furthermore, all the observations in this work were made at nanoscale. Future research will address the scaling-up to micro and macro scale.

4 CONCLUSIONS

- MASKE was used to simulate dissolution of C-S-H and CH and precipitation of $CaCO_3$ in a model cement paste, leading to autogenous healing.
- A crack of size 1 nm was bridged completely by the precipitated $CaCO_3$. A volume gain of 15–20% was obtained.
- Mechanical properties of the system steadily improved with time.
- A clear strength gain was measured as the crack healed.

ACKNOWLEDGEMENTS

Research funded by the UK Engineering and Physics Research Council, EPSRC, grant EP/S013997/1.

REFERENCES

Chitez, A. S., & Jefferson, A. D. 2016. A coupled thermo-hygro-chemical model for characterising autogenous healing in ordinary cementitious materials. *Cement and Concrete Research 88*: 184–197.

Coopamootoo, K., & Masoero, E. 2020. Simulations of Crystal Dissolution Using Interacting Particles: Prediction of Stress Evolution and Rates at Defects and Application to Tricalcium Silicate. *The Journal of Physical Chemistry C 124*(36): 19603–19615.

He, H., Guo, Z., Stroeven, P., Stroeven, M., & Sluys, L. J. 2007. Self-healing capacity of concrete-computer simulation study of unhydrated cement structure. *Image Analysis & Stereology* 26(3): 137–143.

Hilloulin, B., Grondin, F., Matallah, M., & Loukili, A. 2014. Modelling of autogenous healing in ultra high performance concrete. *Cement and Concrete Research 61*: 64–70.

Huang, H., Ye, G., & Damidot, D. 2013. Characterization and quantification of self-healing behaviors of microcracks due to further hydration in cement paste. *Cement and Concrete Research 52*: 71–81.

Lolli, F., Manzano, H., Provis, J. L., Bignozzi, M. C., & Masoero, E. 2018. Atomistic simulations of geopolymer models: the impact of disorder on structure and mechanics. *ACS applied materials & interfaces 10*(26): 22809–22820.

Masoero, E., & Di Luzio, G. 2020. Nanoparticle simulations of logarithmic creep and microprestress relaxation in concrete and other disordered solids. *Cement and Concrete Research* 137: 106181.

Masoero, E., Thomas, J. J., & Jennings, H. M. 2014. A reaction zone hypothesis for the effects of particle size and water-to-cement ratio on the early hydration kinetics of C_3S. *Journal of the American Ceramic Society* 97(3): 967–975.

Qureshi, T., Kanellopoulos, A. & Al-Tabbaa, A. 2018. Autogenous self-healing of cement with expansive minerals-I: Impact in early age crack healing. *Construction and Building Materials* 192: 768–784.

Ranaivomanana, H., & Benkemoun, N. 2017. Numerical modelling of the healing process induced by carbonation of a single crack in concrete structures: Theoretical formulation and Embedded Finite Element Method implementation. *Finite Elements in Analysis and Design 132*: 42–51.

Shvab, I., Brochard, L., Manzano, H., & Masoero, E. 2017. Precipitation mechanisms of mesoporous nanoparticle aggregates: off-lattice, coarse-grained, kinetic simulations. *Crystal Growth & Design* 17(3): 1316–1327.

Thompson, A. P., Aktulga, H. M., Berger, R., Bolintineanu, D. S., Brown, W. M., Crozier, P. S., ... & Plimpton, S. J. 2022. LAMMPS-a flexible simulation tool for particle-based materials modeling at the atomic, meso, and continuum scales. *Computer Physics Communications 271*: 108171.

Van Tittelboom, K., & De Belie, N. 2013. Self-healing in cementitious materials—A review. *Materials* 6(6): 2182–2217.

Computational Modelling of Concrete and Concrete Structures – Meschke, Pichler & Rots (Eds)
 ISBN: 978-1-032-32724-2

Numerical investigations of discrete crack propagation in Montevideo splitting test using cohesive elements and real concrete micro-structure

B. Kondys, J. Bobiński & I. Marzec
Faculty of Civil and Environmental Engineering, Gdansk University of Technology, Gdansk, Poland

ABSTRACT: The paper is aimed at accurately predicting the discrete fracture process in concrete specimens under complex stress states in two dimensional (2D) simulations. Plain concrete specimens subjected to Montevideo splitting test (MVD) were used for consideration due to non-negligible shear stresses impact in this type of test. In order to reflect the heterogeneous nature of the concrete, the meso-structure of the samples was included in the numerical models. The concrete was modelled as a four-phase material consisting of a cement matrix with air voids, aggregates and Interfacial Transitions Zones (ITZ) between aggregates and cement matrix. The meso-structure was created on the basis of X-ray μCT image of real specimens. The analysis was performed using the finite element method (FEM) with cohesive interface elements in a quasi-static approach carried out by Abaqus. The results of the numerical simulations were compared with the values obtained experimentally in terms of crack patterns and force versus crack mouth opening displacement (CMOD) diagrams.

1 INTRODUCTION

The phenomenon of cracking is particularly important in detailed predictions of the behavior of concrete and reinforced concrete elements in service state, because it is an unavoidable feature and it also determines the key mechanical parameters of concrete such as strength and stiffness (e.g. Tejchman & Bobiński 2013).

The nature of concrete cracking is described as quasi-brittle, meaning that a complete material failure does not occur immediately after the material reaches its tensile strength. For quasi-brittle materials, reaching stresses equal to the tensile strength results in the initiation of a crack. Then the crack begins to propagate due to loading into the creation of a fracture surface until complete dissipation of the elastic strain energy. In other words, the cracked material can still transfer the stress although the weakening of the material by cracking causes a decrease in its strength relatively to the initial value. The behaviour of the material after crack initiation can be described in terms of fracture energy G_F. The fracture energy is a material constant which is defined as the amount of energy required to nucleate a finite unit crack area (Hillerborg et al. 1976).

Among the tests to determine fracture energy, three main groups of test methods can be distinguished: direct uniaxial tension tests (UTT) and indirect experiments based on beam bending i.e. 3-point bending test (3PBT/TPBT) (Hillerborg 1985) or specimen splitting i.e. wedge splitting test (WST) (Brühwiler & Wittmann 1990; Linsbauer & Tschegg 1986). Due to the fact that the direct UTT method is time-consuming and requires special equipment, while the results obtained depend on the interaction of the machine with the specimen, indirect methods TPBT and WST are most frequently used (Löfgren et al. 2005).

In the TPBT, single edge pre-notched concrete beams are tested. During the test, the CMOD and the applied force are measured. Thus, the work required to create the crack can be determined and then the fracture energy can be calculated by dividing by the fracture surface (in perpendicular plane to the tensile stress direction). Main disadvantage of this method is relatively large size of specimens (150 × 150 × 600 mm) which requires a quantity of concrete equivalent to making four standard concrete cubes (150 × 150 × 150 mm).

The main idea of the WST approach is to convert the vertical compressive force into splitting horizontal forces. This is achieved by pressing a rigid wedge into a pre-prepared notch in the cube specimen. Thus, the test requires smaller specimens (150 × 150 × 150 mm) and it can be performed with basic test equipment. An additional advantage of using cube specimens is the ability to test on specimens taken directly from the existing structure or to use specimens prepared for compressive strength testing. An important consideration in WST is to eliminate the effect of wedge-notch interaction on the test results. To reduce friction between the steel wedge and the concrete, special frames with bearing rollers or additional steel spacers and cylinders in contact with the surface of the wedge placed directly in the notch are used (Figure 1A). However, these procedures

DOI 10.1201/9781003316404-13

require additional widening of the notch and as a result, the notch is divided into two parts of different widths.

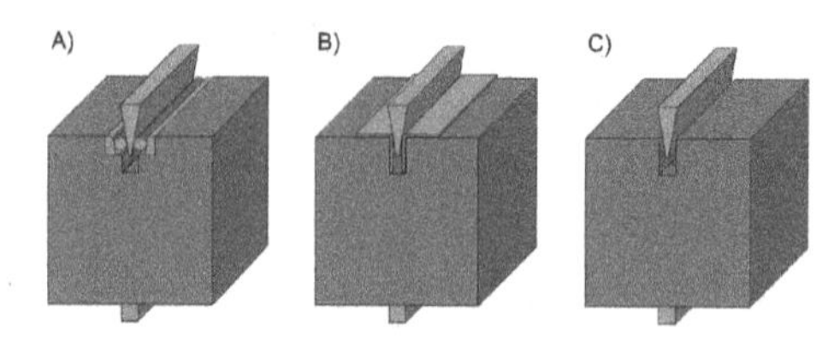

Figure 1. Load application schemes for different types of tests A) WST, B) MVD, C) simplified MVD.

Therefore, efforts are being made to develop even more simplified testing methods such as Montevideo Splitting Test (MVD) (Segura-Castillo et al. 2018) in which additional rollers are omitted, friction is allowed to occur between the wedge and the specimen and the effect of friction is considered in the calculation of the effective splitting force P_{eff}. The authors focused on the effect of friction on the test results and the correlation of the Force-CMOD diagrams of the MVD test with the results obtained from the TPBT according to EN 14651. The fracture energy can be determined from the corresponding F-CMOD diagram, although this quantity was not determined in the paper (Segura-Castillo et al. 2018).

Studies on the crack formation process in materials built of cementitious matrices can be successfully supported by numerical simulations. Computational methods allow quickly and easily create and test various types of specimens in complex stress states that are often difficult and time consuming to perform in laboratory conditions. There are two main methods to simulate cracking process - using continuous models, where the crack is interpreted as a strains band on which microcracks occur (Desmorat et al. 2007; Grassl et al. 2013; Marzec & Bobiński 2019; Menetrey & Willam 1995; Meschke et al. 1998; Skarżyński et al. 2017, 2020) or discontinuous models (Marzec & Bobiński 2022; Moes & Belytschko 2002; Welss & Sluys 2001), where the formation of the actual crack is considered.

Currently, due to high accuracy and wide availability, approaches based on the fictitious crack model (Hillerborg et al. 1976) implemented in the Finite Element Method are particularly interesting. These approaches usually use XFEM or interface elements (IE) which are available in the standard element libraries of widely used calculation software such as Abaqus or ANSYS. They allow for an almost accurate representation of the crack paths obtained in experimental tests especially when the mesoscale of the concrete is sufficient mapping (Trawiński et al. 2016; Zhang et al. 2018).

The topic of modeling crack propagation in quasi-brittle materials at the mesoscale level is currently still dynamically developed especially in the field of multiaxial stress cases.

The paper aims to investigate the applicability of a discrete crack approach using cohesive elements to model concrete mesoscale specimens (taking into account the following phases in concrete: cement matrix, aggregate, interfacial transition zones (ITZ) and air voids) in complex stress states which occur in MVD testing. The successful correlation of simplified experimental testing with advanced numerical methods will allow for a significant acceleration of mesoscale concrete research.

2 EXPERIMENTAL TEST

2.1 *Standard Montevideo splitting test*

As mentioned the MVD is a simplified variant of the Wedge Splitting Test. It allows to obtain meaningful fracture parameters results without the need for additional apparatus and a complex specimen shape to eliminate the influence of vertical forces due to friction between the wedge and the specimen. Another advantage of the simplified variant of the test is the possibility to test specimens of different sizes, which allow simultaneous X-ray μCT imaging of the specimen (Skarżyński & Suchorzewski 2018). The simplification of the procedure has the consequence that more parameters will eventually be taken into account when interpreting the results. When the MVD approach is used the wedge is applied through steel angles (Figure 1B) or directly (Figure 1C) (Segura-Castillo et al. 2018; Skarżyński & Suchorzewski 2018) to the surface of the notch, the vertical stresses components are significant and affect the formation of the crack. As the original authors of the MVD method recommend the use of steel angles in the test, alternative variant with direct steel-concrete contact has been named as "simplified MVD" for the purposes of this work.

2.2 *Simplified Montevideo splitting test*

The experimental simplified Montevideo splitting tests were carried out on cubic specimens with dimensions $70 \times 70 \times 70$ mm supported along its entire length by a steel flat bar. An initial notch was placed in the center of the upper surface of the specimen. The dimensions of the notch were 15 mm high and 5 mm wide.

The concrete mix consisted of cement CEM II/A-LL 42.5R, flying ash, aggregate and water (Table 1). Aggregate was divided into three main fractions i.e.

Table 1. Concrete recipe details.

Concrete components	Concrete mix ($d_{50} = 2$ mm, $d_{max} = 16$ mm)
Cement CEM II/A-LL 42.5R (c)	300 kg/m^3
Sand (0 – 2 mm)	735 kg/m^3
Gravel aggregate (2 – 8 mm)	430 kg/m^3
Gravel aggregate (8 – 16 mm)	665 kg/m^3
Fly ash (a)	70 kg/m^3
Superplasticizer (s)	1,8 kg/m^3
Water (w)	150 kg/m^3

sand with maximum grain size equal 2 mm, gravel with the grain size in the range of 2 mm and 8 mm and gravel with the maximum grain size equal 16 mm.

The sand point was equal to 41% and the water to cement ratio was established at $w/c = 0.50$.

The test consisted in quasi-static pressing of a steel wedge (with an inclination of 10°) into a notch of the specimen by the loading machine Instron 5569 with a constant controlled crack mouth opening displacement (CMOD) rate (Figure 2). Vertical force and CMOD were measured during the experiment.

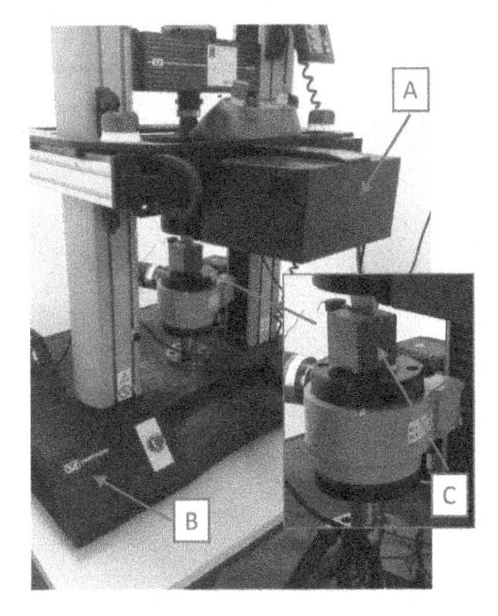

Figure 2. Experimental tests set-up of Montevideo splitting test: A) 1173 Skyscan X-ray micro-tomograph, B) Instron 5569 static machine, C) cubic concrete specimen with notch.

Additionally, μCT imaging by the X-ray micro-tomography Skyscan 1173 (Figure 2) was performed to show the real 3D meso-structure of the specimen and shape of the resulting crack. Finally, the crack pattern was extracted in three different sections of the test specimen: S1-S3 as a reference data to numerical investigation (Figure 3).

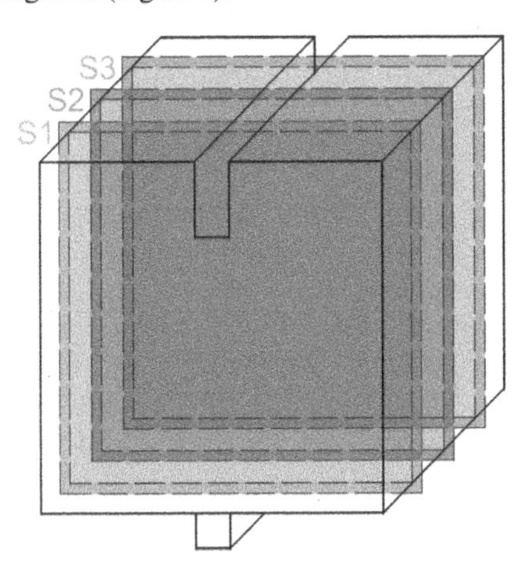

Figure 3. Schematic representation of the scanning sections.

The basic parameters of the hardened concrete i.e. compressive strength f_c, tensile strength f_t and flexural strength $f_{t,flex}$ were tested on additional cube specimens and beams. The measured parameters are listed in Table 2.

Table 2. Experimental mechanical properties of hardened concrete.

Parameter	Type of sample	Number of samples [-]	Average density [kg/m^3]	Average stress [MPa]
Compressive strength - f_c	Cube*	6	2359.2	47.60
Tensile strength - f_t	Cube*	6	2361.1	3.46
Flexural strength - $f_{t,flex}$	Beam**	6	-	3.60

* Cube specimen dimensions 150 × 150 × 150 mm.
** Beam specimen dimensions 600 × 150 × 150 mm.

2.3 *Experimental test results*

The vertical force versus CMOD curve, which is shown in Figure 4, was determined on the basis of experimental studies. Furthermore three μCT scans (Figure 5) showing the crack patterns and meso-structure of the tested specimen were also taken.

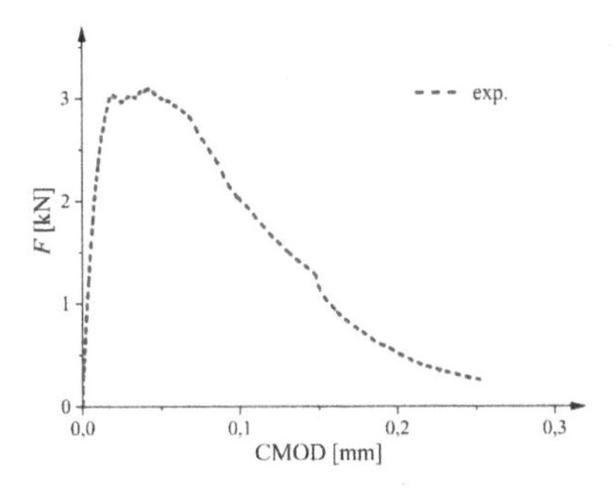

Figure 4. Experimental vertical reaction force vs. CMOD curve.

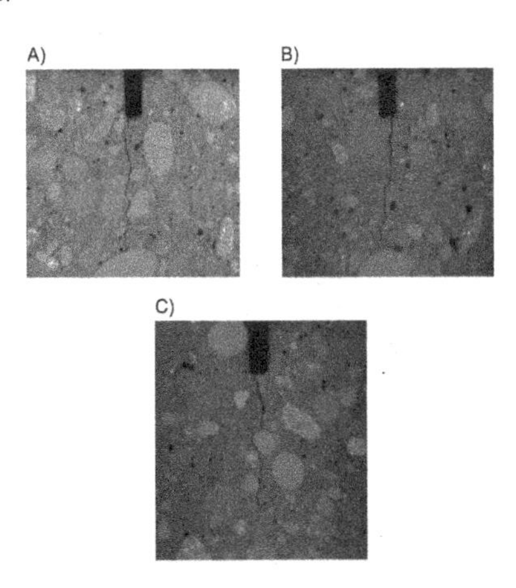

Figure 5. Experimental crack patterns A) section S1, B) section S2, C) section S3.

3 CONSTITUTIVE LAWS

3.1 *Solid elements*

Two-dimensional three-node elements in the plane-stress (CPS3 in Abaqus nomenclature) were used to represent the linear-elastic behaviour of the cement matrix, aggregate, steel support and wedge. Relation between strains ε and stresses σ in solid elements was described by the Hooke's law using two material constants Young's modulus E and Poisson's ratio v:

$$\begin{bmatrix} \Sigma_{11} \\ \varepsilon_{20} \\ \varepsilon_{12} \end{bmatrix} = \frac{1}{E} \begin{bmatrix} 1 & -v & 0 \\ -v & 1 & 0 \\ 0 & 0 & (1-v) \end{bmatrix} \begin{bmatrix} cT_{11} \\ \mathrm{T}_{22} \\ \mathrm{c}_{12} \end{bmatrix} \tag{1}$$

3.2 *Interface cohesive elements*

To take into account the crack initiation and propagation in the analyzed numerical models of the tested specimens, zero-thickness interface 2D elements (COH2D4 in Abaqus nomenclature) with built-in traction-separation law were used. Interfaces can be modelled as elements with zero thickness. This gives the possibility to introduce a strong discontinuity into the continuous model by inserting IE between solid elements.

3.2.1 *Traction-separation law*

Traction-separation law describes the relationship between vectors of cohesive tractions and relative separation displacements. For the 2D problem, this relation can be defined as follows:

$$\begin{bmatrix} t_n \\ t_s \end{bmatrix} = \begin{bmatrix} k_n & 0 \\ 0 & k_s \end{bmatrix} \begin{bmatrix} \delta_n \\ \delta_s \end{bmatrix} \tag{2}$$

where t_n and t_s are normal (mode I) and shear in tangential direction tractions (mode II) respectively, k_n, k_s are interface stiffness in the corresponding directions, whereas δ_n, δ_s are the displacements related to the directions of traction.

Quadratic nominal stress criterion was assumed as the crack initiation criterion:

$$\left\{\frac{\langle t_n \rangle}{t_{n0}}\right\}^2 + \left\{\frac{t_s}{t_{s0}}\right\}^2 = 1 \tag{3}$$

where t_{n0}, t_{s0} are the critical stress in the normal and tangential direction respectively and $\langle\ \rangle$ denotes Macaulay bracket which represents ramp function:

$$\langle t_n \rangle = \begin{cases} 0, & t_n < 0 \\ t_n, & t_n \geq 0 \end{cases} \tag{4}$$

Equation 4 implies that crack initiation cannot occur due to compressive stresses. In order to describe damage under a combination of normal and tangential shear deformations across the interface, the effective relative displacement δ_m was introduced:

$$\delta_m = \sqrt{\langle \delta_n \rangle^2 + \delta_s^2} \tag{5}$$

where symbol $\langle\rangle$ means Macaulay bracket which represents ramp function. It implies that negative displacements shall not be taken into account in the description of the crack propagation:

$$\langle \delta_n \rangle = \begin{cases} 0, & \delta_n < 0 \\ \delta_n, & \delta_n \geq 0 \end{cases} \tag{6}$$

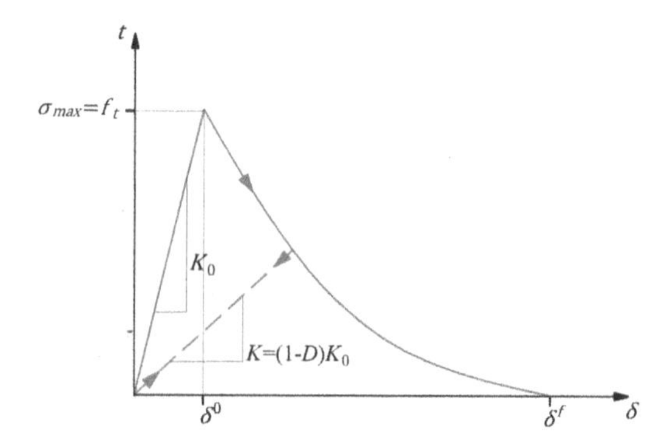

Figure 6. Exponential traction-separation law curve.

When the initiation criterion described by Equation 3 is fulfilled in a particular cohesive element, the stiffness degradation process begins (Figure 6). To describe this effect scalar variable $D$was used:

$$k_n = (1 - D)k_{n0} \tag{7}$$

$$k_s = (1 - D)k_{s0} \tag{8}$$

where k_{n0} and k_{s0} mean initial interface stiffness in normal and tangential direction, respectively. The evolution of the D parameter can be described by different curves. For materials based on cementitious matrices, the bilinear (Petersson 1981) or exponential curve (Barenblatt 1962) are most commonly used. In this paper numerical model with an exponential softening curve defined using effective relative displacements was assumed:

$$D = \begin{cases} 1 - \left\{\frac{\delta_m^0}{\delta_m^{\max}}\right\}\left\{1 - \frac{1-\exp\left[-\alpha\left(\frac{\delta_m^{\max}-\delta_m^0}{\delta_m^f-\delta_m^0}\right)\right]}{1-\exp(-\alpha)}\right\}, \\ \qquad \delta_{\max} \leq \delta_m^f \\ 1,0, \quad \delta_{\max} > \delta_m^f \end{cases} \tag{9}$$

where δ_m^0 means the effective, relative displacement at the crack initiation, δ_m^f is the effective relative displacement at the complete stiffness degradation, $\delta_m^{\max}$ stands for maximum effective relative displacement obtained during the loading history and α is a non-dimensional material parameter that defines the rate of damage evolution.

4 MESH DEFINITION

DXF files containing the shapes and distribution of aggregate grains and air pores were created from the

μCT scans. A finite element mesh containing solid elements was created in Abaqus/CAE on the basis of these DXF drawings. In conventional Cohesive Zone Model (CZM) it is necessary to a priori insert cohesive interfaces in the path of predicted crack propagation. If propagation path cannot be assumed in advance, it is therefore required to place cohesive elements between all the solid elements. Due to the fact that crack propagation through the aggregate usually only occurs in high strength concretes, the possibility of cracks forming within the aggregate grains was excluded. Thus, interfaces elements were created only between the cement matrix elements (CM-CM) and represents ITZs between the aggregate and the matrix (CM-AGG) (Figure 7). A suitably modified program DEIP (Truster 2018) written in MATLAB was used to place the interfaces into the finite element mesh.

Furthermore, in order to simplify the model for reduce time-consuming computing and unwanted numerical issues of the cohesive elements appearing in areas of excessive compressive stress concentrations (the contact points between the specimen and the wedge and the steel flat bar). It was decided to generate cohesive elements only in the central area of the specimen where a crack was most expected to occur (Figure 7). The area containing the cohesive elements was connected to the rest of the model using "elastic" interface elements without a defined cohesive law.

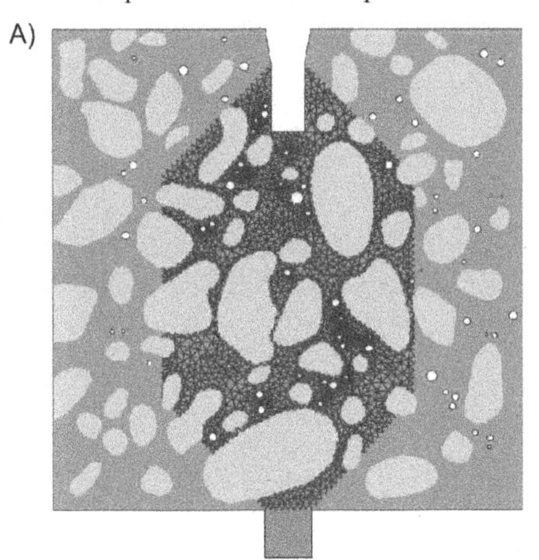

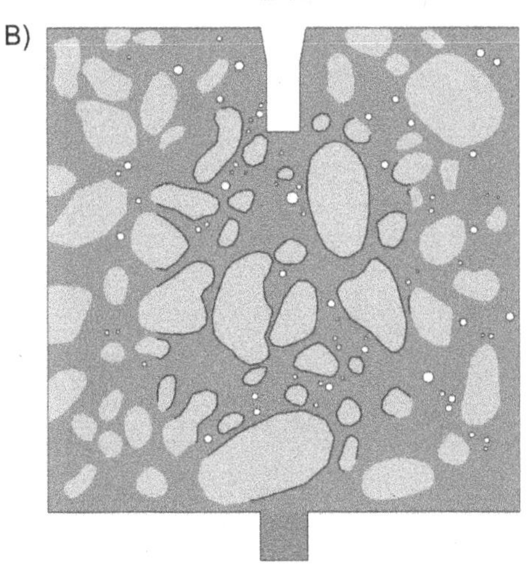

Figure 7. Localization of interfaces in the mesh: A) CM-CM, B) CM-AGG (ITZ) for section S1.

5 FE SIMULATIONS

5.1 *Material parameters*

On the basis of experimental tests it was determined that Young's modulus for concrete specimens was $E_c = 34$ GPa. Then, based on the ratio of the area of the aggregate and the matrix, the Young's modulus was estimated for the aggregate of $E_a = 40$ GPa and the cement matrix of $E_{cm} = 20$ GPa. The Poisson's ratio $\nu = 0.2$ was assumed for all phases of the concrete. For the secondary steel elements (flat bar support and wedge), parameters corresponding to structural steel $E_s = 200$ GPa and $\nu = 0.3$ were assumed.

Different material parameter values were considered for both types of interfaces used. The input parameters of the CM-CM interfaces were taken from the experimental strength results, specifying the tensile strength as $f_{t,CEM} = 3.5$ MPa. The strength of the cohesive elements in the normal and tangential directions was assumed to be the same (Trawiński et al. 2016). Parameters of ITZ interfaces (CM-AGG) were determined based on parametric studies and recommendations from the literature (Xi et al. 2021). The ratio of tensile strength and fracture energy of CM-CM interfaces - $f_{t,CEM}$ to ITZ interfaces $f_{t,ITZ}$ was assumed as 0.5. The initial stiffnesses of the cohesive elements in both directions were assumed to $k_{n0} = k_{s0} = 10^6$ MPa/mm, consistent with the observations of other authors (Trawiński et al. 2016; Wang et al. 2019). The values for were calculated from assumed value of the fracture energy. All material parameters used in the model are listed in Table 3.

5.2 *Numerical results*

2D simulations in plane stress state were performed for three different meshes including different aggregate and air voids distribution based on μCT scans S1–S3. The load was applied by displacement of a linear elastic steel wedge. A contact law with defined tangential penalty friction was assumed between the edges of the wedge and the specimen. A steel-concrete friction coefficient $\mu = 0.5$ was preliminary assumed as a default value. In order to avoid stress concentration at the contact point and to take into account the actual contact area of the wedge with the specimen, a two-sided edge slope was made in the specimen model at 1/3 of the notch height to match the wedge slope (Figure 7).

Finite element size of 1 mm was assumed. As it was shown in other studies (Trawiński et al. 2016), in the case of 2D numerical simulations at the mesoscale level with this FE size there is no mesh dependence on the results.

The result for FE simulation for different section S1–S3 for initial values of material constants are given on Figures 8 and 9. The F-CMOD curves obtained from the numerical calculations remain consistent with the experimental results (Figure 8). The best convergence in terms of force peak was obtained for section S2, the other two sections had a higher peak value but similar post-peak characteristic and total fracture energy. Whereas comparing the crack patterns obtained by the numerical analysis (Figure 9) with the μCT scans from the experiment (Figure 5), the greatest correspondence of the crack pattern was observed for section S3. It was observed that the largest inconsistency, both in the curve and the crack pattern appeared in the case of section S1. It can be clearly noticed that this was caused by the large aggregate grain near the notch (Figure 9A). The aggregate which was surrounded by a weaker ITZ layer has attracted the crack. This is consistent with predictions because cracking is a three-dimensional phenomenon and strongly depends on the location of grains and air voids over the sample volume, whereas two-dimensional simulations consider only three selected sections and can be perturbed by local conditions.

Table 3. Material parameters in FEA.

Solid elements			
Material Parameter	Aggregate	Cement matrix	Steel
E [GPa]	40	20	200
ν [-]	0.2	0.2	0.3
Cohesive elements – displacement approach			
Material Parameter	"Elastic"	Cement matrix (CM–CM)	ITZ (CM–AGG)
k_{n0} [MPa/mm]	10^6	10^6	10^6
$f_{t,n} = f_{t,s}$ [MPa]	-	3.5	1.75
α [-]	-	7.5	7.5
G_F [N/m]	-	70	35

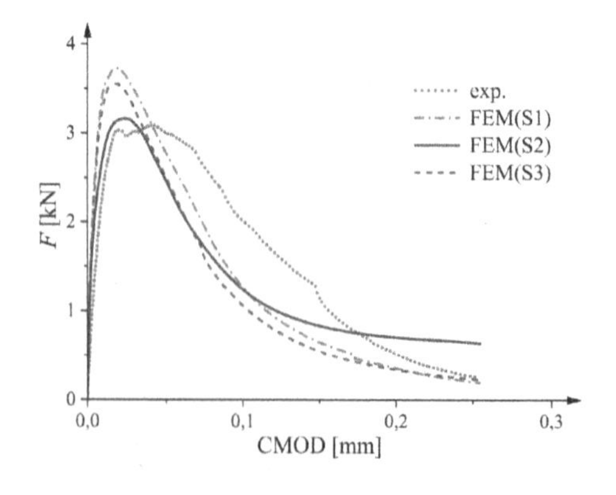

Figure 8. Calculated force vs. CMOD curves for different sections.

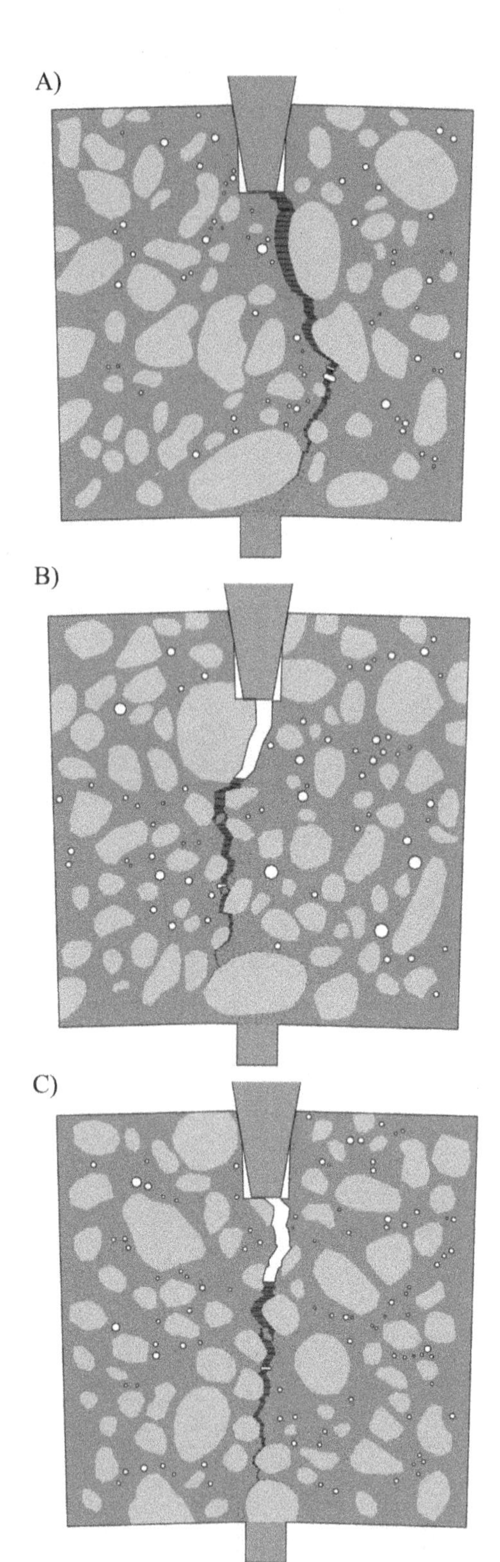

Figure 9. Crack patterns obtain from numerical simulations ($f_{t,ITZ}/f_{t,CEM} = 0.5$, $G_{F,ITZ}/G_{F,CEM} = 0.5$) from: A) section S1, B) section S2, C) section S3.

The results of numerical parametric study as compared to experiments for section S2 are shown on Figures 10–14. The influence of different tensile strength ratio $f_{t,ITZ}/f_{t,CEM}$, fracture energy ratio $G_{F,ITZ}/G_{F,CEM}$ and various friction coefficient was investigated. Increasing the tensile strength and the fracture energy of ITZ relative to the tensile strength and fracture energy of cement matrix leads to evident increase of vertical force (Figures 10 and 11) for section S2.

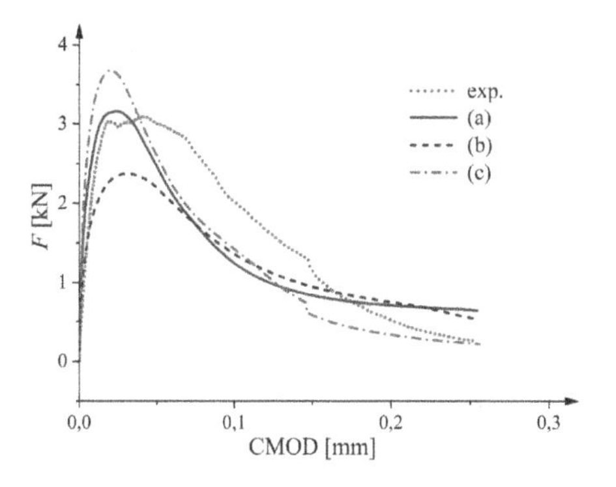

Figure 10. Calculated force vs. CMOD curves for section S2 ($f_{t,CEM} = 3.5$ MPa, $G_{F,CEM} = 70$ N/m, $G_{F,ITZ} = 35$ N/m) for $f_{t,ITZ}$ to $f_{t,CEM}$ ratio equal to: 0.5 (a), 0.75 (b) and 0.25 (c).

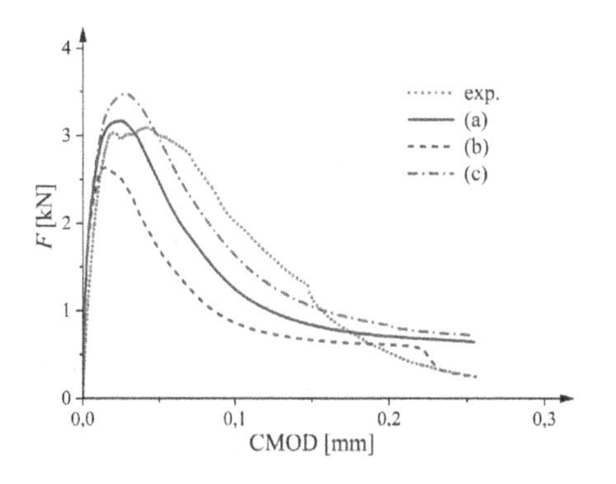

Figure 11. Calculated force vs. CMOD curves section S2 ($f_{t,CEM} = 3.5$ MPa, $f_{t,ITZ} = 1,75$ MPa, $G_{F,CEM} = 70$ N/m) for $G_{F,ITZ}$ to $G_{F,CEM}$ ratio equal to: 0.5 (a), 0.25 (b) and 0.75 (c).

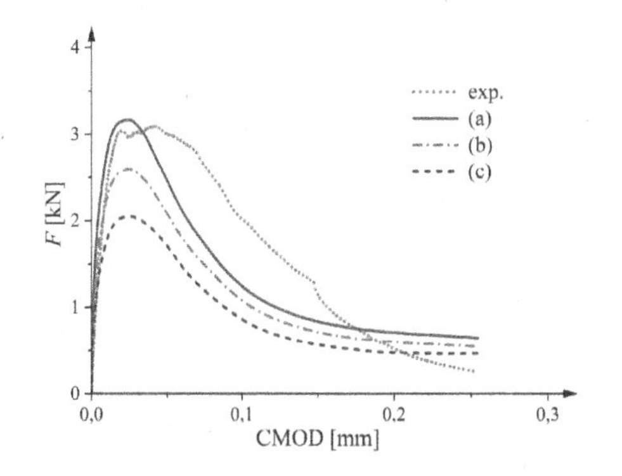

Figure 12. Calculated force vs. CMOD curves for section S2 ($f_{t,ITZ}/f_{t,CEM} = 0.5$, $G_{F,ITZ}/G_{F,CEM} = 0.5$) for different friction coefficient μ: 0.5 (a), 0.4 (b) and 0.3 (c).

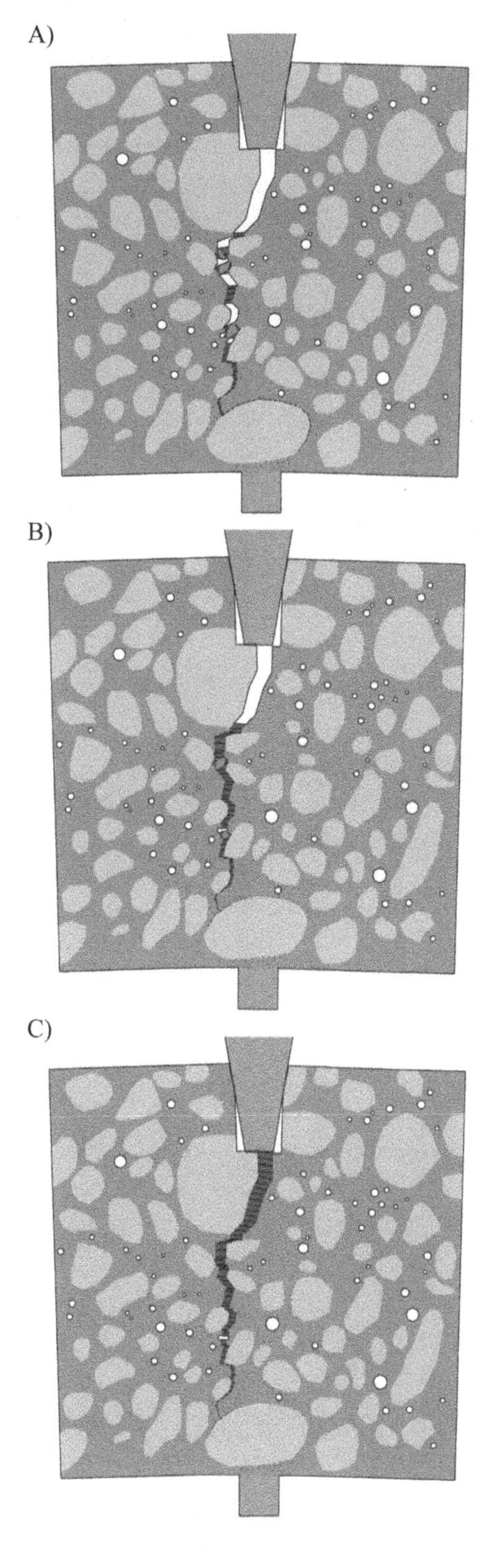

Figure 13. Influence of ITZ fracture energy $G_{F,ITZ}$ on the numerical crack pattern obtain in section S2 ($f_{t,CEM} = 3.5$ MPa, $f_{t,ITZ} = 1,75$ MPa, $G_{F,CEM} = 70$ N/m): A) $G_{F,ITZ}/G_{F,CEM} = 0.25$, *B*) $G_{F,ITZ}/G_{F,CEM} = 0.50$, C) $G_{F,ITZ}/G_{F,CEM} = 0.75$.

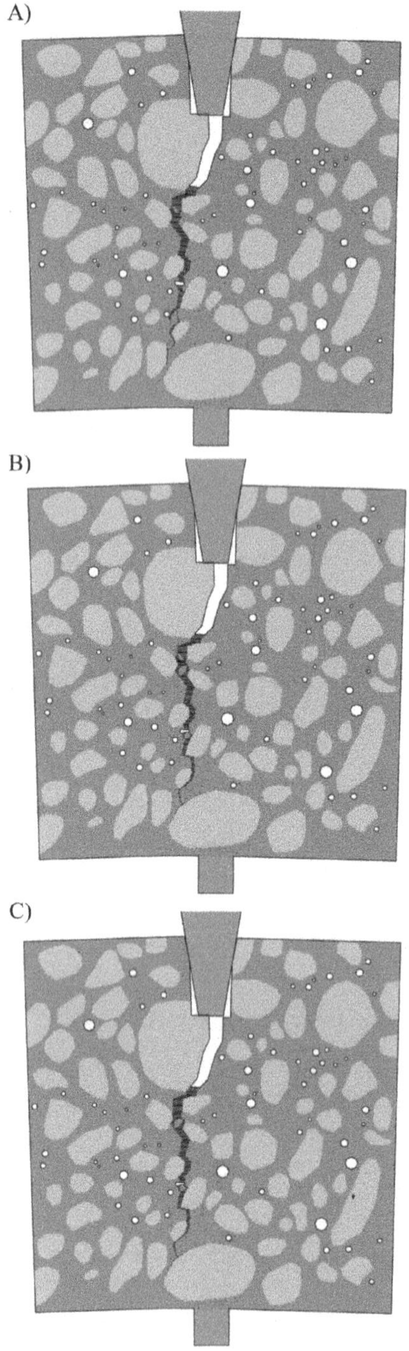

Figure 14. Influence of friction coefficient μ on the numerical crack pattern obtain in section S2 ($f_{t,CEM} = 3.5$ MPa, $f_{t,ITZ} = 1,75$ MPa, $G_{F,CEM} = 70$ N/m, $G_{F,ITZ} = 35$ N/m): A) $\mu = 0.30$, B) $\mu = 0.40$, C) $\mu = 0.50$.

Despite the influence on Force-CMOD curves, the change in tensile strength and fracture energy for the ITZ did not significantly affect the crack pattern (Figure 13), the only difference noted being the number of elements that fully degraded.

As expected, friction has a significant effect on simplified MVD splitting test results, the friction co efficient is strongly correlated with the calculated vertical force in terms of F-CMOD curve response (Figure 12). Simultaneously, reduction of the friction coefficient did not affect the crack path for the section S2 (Figure 14).

Analogous series of FE simulation was performed also for section S1. The analysis of the result of these calculations exhibits the same trends and leads to similar conclusion in case of influence of fracture energy ratio $G_{F,ITZ}/G_{F,CEM}$ and friction coefficient. Interestingly, varying the tensile strength ratio $f_{t,ITZ}/f_{t,CEM}$ had only minor influence of vertical force, in particular for ratio larger then 0.5 (Figure 15). Instead of increase the value of vertical force the response became more ductile with $f_{t,ITZ}/f_{t,CEM} = 0.75$. Moreover, the shape of crack changed significantly with increasing the tensile strength of ITZ (Figure 16).

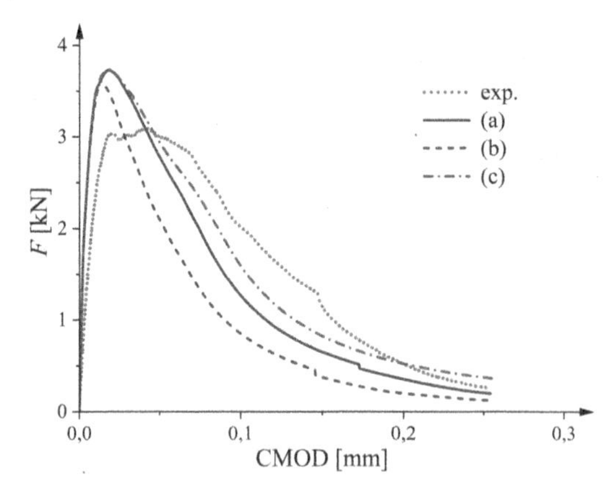

Figure 15. Calculated force vs. CMOD curves section S1 ($f_{t,CEM} = 3.5$ MPa, $f_{t,ITZ} = 1,75$ MPa, $G_{F,CEM} = 70$ N/m) for $G_{F,CEM}$ to $G_{F,ITZ}$ ratio equal to: 0.5 (a), 0.25 (b) and 0.75 (c).

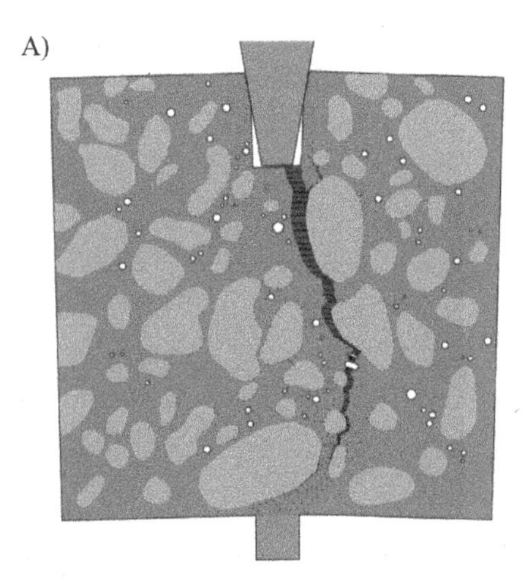

Figure 16. *continued.*

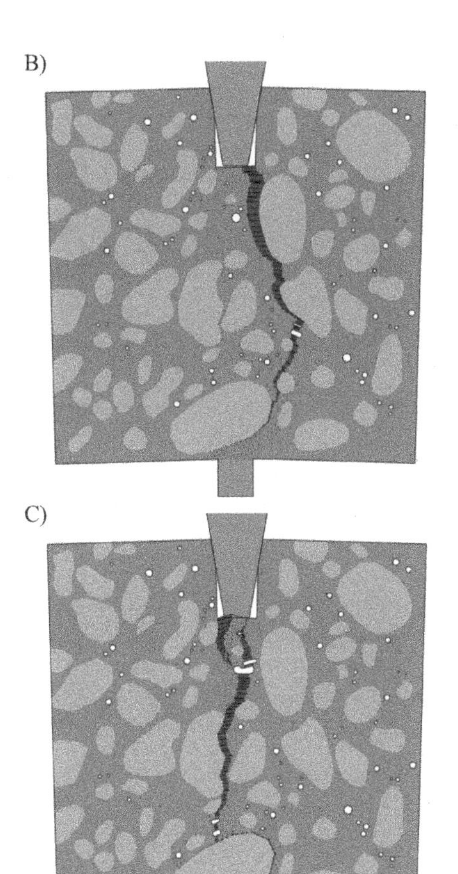

Figure 16. Influence of ITZ tensile strength $f_{t,ITZ}$ on the numerical crack pattern obtain in section S1 ($G_{F,ITZ} = 35$ N/m, $G_{F,CEM} = 70$ N/m): A) $f_{t,ITZ}/f_{t,CEM} = 0.25$, B) $f_{t,ITZ}/f_{t,CEM} = 0.50$, C) $f_{t,ITZ}/f_{t,CEM} = 0.75$.

6 CONCLUSIONS

The following basic conclusions may be derived from our preliminary calculation for simplified MVD splitting test using cohesive elements and real concrete micro-structure:

- Due to the predominant crack length runs through the ITZ interfaces, the *F*-CMOD response strongly depends on the tensile strength and fracture energy of this zone.
- Modification of the ITZ strength and fracture energy may also lead to change of the shape of crack. In particular for very small or very large value of $f_{t,ITZ}/f_{t,CEM}$ and $G_{F,ITZ}$ /$G_{F,CEM}$ ratios respectively.
- Since, the friction coefficient plays a crucial role for obtaining the proper results for MVD splitting test, it is advisable to calibrate the ITZ strength and fracture energy based on separate tests (e.g. simple bending) before calibrating the friction.

The numerical calculations also investigated the possibility of simulating this issue using a quasi-static simulation with explicit integration. However, due to too many variable parameters, including the large effect of specimen inertia, the axial location of the wedge and support, and the definition of the contact, stable and repeatable simulation parameters could not be established at this stage of the study. Nevertheless, this will be the subject of future research.

ACKNOWLEDGMENTS

Calculations were carried out at the Centre of Informatics Tricity Academic Supercomputer & Net-work.

REFERENCES

Barenblatt, G. I. (1962) Mathematical theory of equilibrium cracks in brittle fracture. *Advances in Applied Mechanics VII*: 55–129.

Brühwiler, E. & Wittmann, F.H. (1990) The wedge splitting test, a new method of performing stable fracture mechanics tests. *Engineering Fracture Mechanics* 35 (1–3): 117–125.

Desmorat, R., Gatuingt, F. & Ragueneau, F. (2007) Nonlocal anisotropic damage model and related computational aspects for quasi-brittle materials. *Engineering Fracture Mechanics* 74(10): 1539–1560.

Grassl, P., Xenos, D., Nystrom, U., Rempling, R. & Gylltoft, K. (2013) CDPM2; a damage-plasticity approach to modelling the failure of concrete. *International Journal of Solids and Structures* 50: 3805–3816.

Hillerborg, A., Modéer, M. & Petersson, P.-E. (1976) Analysis of crack formation and crack growth in concrete by means of fracture mechanics and finite elements. *Cement and Concrete Research* 6 (6): 773–781.

Hillerborg, A. (1985) The theoretical basis of a method to determine the fracture energy G_F of concrete. *Materials and Structures* 18: 291–296.

Linsbauer, H. & Tschegg, E. (1986) Fracture energy determination of concrete with cube specimens. *Zement und Beton* 31: 38–40.

Löfgren, I., Stang, H. & Olesen, J.F. (2005) Fracture properties of FRC Determined through Inverse Analysis of Wedge Splitting and Three-Point Bending Tests. *Journal of Advanced Concrete Technology* 3(3): 423–434.

Marzec, I. & Bobiński, J. (2019) On some problems in determining tensile parameters of concrete model from size effect tests. *Polish Maritime Research* 26 (2): 115–125.

Marzec, I. & Bobiński, J. (2022) Quantitative assessment of the influence of tensile softening of concrete in beams under bending by numerical simulations with XFEM and cohesive cracks. *Materials* 15(2), 626.

Menétrey, P. & Willam, K.J. (1995) Triaxial failure criterion for concrete and its generalization. *ACI Structural Journal* 92(3): 311–318.

Meschke, G., Lackner, R. & Mang, H.A. (1998) An anisotropic elastoplastic-damage model for plain concrete. *International Journal for Numerical Methods in Engineering* 42(4):702–727.

Moes, N. & Belytschko, T. (2002) Extended finite element method for cohesive crack growth. *Engineering Fracture Mechanics*, 69(7): 813–833.

Petersson, P. E. (1981) Crack growth and development of fracture zones in plain concrete and similar materials. *Report TVBM* 1006.

Segura-Castillo, L., Monte, R. & Figueiredo, A.D. de (2018) Characterisation of the tensile constitutive behaviour of fibre-reinforced concrete: A new configuration for the Wedge Splitting Test. *Construction and Building Materials* 192: 731–741.

Skarżyński, Ł., Marzec, I. & Tejchman, J. (2017) Experiments and numerical analyses for composite RC-EPS slabs. *Computers and Concrete* 20(6): 689–704.

Skarżyński, Ł. & Suchorzewski, J. (2018) Mechanical and fracture properties of concrete reinforced with recycled and industrial steel fibers using Digital Image Correlation technique and X-ray micro computed tomography. *Construction and Building Materials* 183: 283–299.

Skarżyński, Ł., Marzec, I., Drąg, K. & Tejchman, J. (2020) Numerical analyses of novel prefabricated structural wall panels in residential buildings based on laboratory tests in scale 1:1. *European Journal of Environmental and Civil Engineering* 24(9): 1450–1482.

Tejchman, J. & Bobiński, J. (2013) *Continuous and Discontinuous Modelling of Fracture in Concrete Using FEM.* Berlin-Heidelberg, Springer.

Trawiński, W., Bobiński, J. & Tejchman, J. (2016) Two-dimensional simulations of concrete fracture at aggregate level with cohesive elements based on X-ray μCT images. *Engineering Fracture Mechanics* 168: 204–226.

Truster, T.J. (2018) DEIP, discontinuous element insertion Program — Mesh generation for interfacial finite element modeling. *SoftwareX* 7: 162–170.

Wang, J., Jivkov, A.P., Engelberg, D.L. & Li, Q.M. (2019) Parametric Study of Cohesive ITZ in Meso-scale Concrete Model. *Procedia Structural Integrity* 23: 167–172

Welss, G.N. & Sluys, L.J. (2001) A new method for modelling cohesive cracks using finite elements. *International Journal for Numerical Methods in Engineering* 50(12): 2667–2682.

Xi X., Yin Z., Yang S., Li Ch-Q., Using artificial neural network to predict the fracture properties of the interfacial transition zone of concrete at the meso-scale, *Engineering Fracture Mechanics 242*.

Zhang, C.C., Yang, X.H. & Gao, H. (2018) XFEM Simulation of Pore-Induced Fracture of a Heterogeneous Concrete Beam in Three-Point Bending. *Strength of Materials* 50 (5): 711–723.

Computational Modelling of Concrete and Concrete Structures – Meschke, Pichler & Rots (Eds)
 ISBN: 978-1-032-32724-2

A 3D coupled chemo-mechanical model for simulating transient damage-healing processes in self-healing cementitious materials

B.L. Freeman & A.D. Jefferson
Cardiff University, Cardiff, UK

ABSTRACT: This study presents a 3D coupled chemo-mechanical finite element model for the simulation of the damage-healing behaviour of cementitious materials with embedded vascular networks. The mechanical damage-healing behaviour is described using a cohesive zone model that is implemented into an embedded strong discontinuity element. The mechanical model allows for damage and healing to occur simultaneously, and places no restrictions on the number of damage-healing events. An important feature of the damage-healing model is the inclusion of a healing strain that ensures thermodynamic consistency when healing takes place in non-zero strain conditions. The mechanical model is coupled to a reactive transport model that describes the transport of healing agent to the damage site, along with the chemical reaction governing crack healing. The model considers the reactive transport of healing agent within the discrete cracks, the surrounding cementitious matrix and the embedded vascular network. Richard's equation describes the matrix flow, which is coupled to the mass balance equation combined with Darcy's law for the flow in the discrete cracks and embedded vascular network. For the crack and embedded vascular network flow, a cut finite element framework is employed that allows for discontinuities, such as those found at the fluid interface, internal to the elements. The performance of the model is demonstrated through comparison to data from an experimental investigation undertaken at Cardiff University. The results of the comparison show that the model is able to predict realistic transport of healing agents, as well as being able to represent the damage-healing behaviour with good accuracy.

1 INTRODUCTION

Researchers have shown self-healing systems to be an effective means of mitigating the cracking-related durability problems associated with cementitious materials. A wide range of approaches have been developed, many of which employ embedded healing agents that are transported to damage sites when cracking occurs. Inspired by biological systems, vascular networks represent an effective method of embedding the healing agent into the cementitious matrix. The form of vascular networks utilised ranges from linear channels (Selvarajoo et al. 2020), to mini-vascular networks (De Nardi et al. 2020) and complex 3D biomimetic networks (Li et al. 2020).

Alongside the experimental work on developing self-healing cementitious materials (SHCMs), there has been a great deal of research effort aimed towards the development of numerical models for simulating SHCMs (Di Luzio et al. 2018; Freeman & Jefferson 2022; Granger et al. 2007; Huang & Ye 2016; Koenders, 2012; Oucif et al. 2018; Romero Rodríquez et al. 2019; Zhang & Zhuang 2018;). In spite of the complexity associated with simulating the many interacting physical processes that govern SHCMs, significant progress has been made and research has shown that numerical models can accurately capture various aspects of their behaviour.

The focus of the present study is the simulation of SHCMs with embedded vascular networks. To this end, a 3D coupled chemo-mechanical finite element model is presented. The model comprises a number of components for simulating the interacting physical processes and accounts for transient damage-healing behaviour, healing agent transport and healing agent curing that governs the mechanical healing.

This paper presents an outline description of the model along with an illustrative example concerning a direct tension test on a SHCM with an embedded vascular network.

2 TRANSPORT MODEL

The healing agent transport model comprises flow through the cementitious matrix, discrete cracks and the embedded vascular networks and builds upon that presented in Freeman and Jefferson (2020).

2.1 *Matrix transport*

The transport of the healing agent in the cementitious matrix is governed by Richard's equation, given as:

$$\frac{\partial(\rho n S)}{\partial t} + \nabla \cdot \mathbf{J}_h + Q_{mtx} = 0 \tag{1}$$

DOI 10.1201/9781003316404-14

where ρ is the healing agent density, n is the porosity, S is the degree of saturation, Q_{mtx} represents the flux between the matrix and discrete cracks and $\mathbf{J}_h$ is the healing agent flux, given by Darcy's law:

$$\mathbf{J}_h = -\rho \frac{K_i K_r\,(S)}{\mu}(\nabla P_h - \rho \mathbf{g}) \tag{2}$$

where K_i is the intrinsic permeability of the medium, μ is the dynamic viscosity, $\mathbf{g}$ is the acceleration due to gravity, P_h is the healing agent pressure and K_r is the relative permeability that depends on the degree of saturation according to:

$$K_r(S) = S^{\lambda}\left(1 - \left(1 - S^{\frac{1}{m}}\right)^{m}\right)^{2} \tag{3}$$

where λ accounts for the connectivity and tortuosity of the pores.

The degree of saturation is related to the capillary pressure, P_c, through the moisture retention curve that reads:

$$P_c(S) = a\left(S^{-\frac{1}{m}} - 1\right)^{1-m} \tag{4}$$

where a and m are constants that depend on the medium and $P_c = P_g - P_h$, where P_g is the gas pressure.

Finally the flux between the matrix and discrete cracks is given by:

$$Q_{mtx} = 2n\beta_{crk}\,(P_h - P_{hcrk}) \tag{5}$$

where β_{crk} is a transfer coefficient and P_{hcrk} is the healing agent pressure in the crack.

2.2 *Discrete crack and embedded vascular network transport*

The transport of the healing agent in discrete macro-cracks and embedded vascular networks is governed by the mass balance equation combined with Darcy's law, given as:

$$\mathbf{u} = -\frac{k}{\mu}\,(\nabla P_{hcrk} - \mathbf{u}Q_{mtx} + \rho\mathbf{u}Q_{ch} - \rho\mathbf{g}) \tag{6}$$

$$\frac{\partial\,(\rho A)}{\partial t} + \nabla\cdot(\rho A\mathbf{u}) - AQ_{mtx} + \rho AQ_{ch} = 0 \tag{7}$$

where $\mathbf{u}$ is the healing agent velocity vector, A is the area, k is the permeability and Q_{ch} is the flux between the discrete cracks and embedded vascular channels. For discrete cracks, the area and permeability terms are given as $A = w \times 1$ and $k = w^2/12$, where w is the crack width, whilst for the embedded vascular channels, $A = \pi r^2$ and $k = r^2/8$, where r is the channel radius.

The boundary conditions for (7) are given as:

$$P_{hcrk} = P_{app}\ on\ \Gamma_{app},\ \ P_{hcrk} = P_c(w, \theta_d)\ on\ \Gamma_f \tag{8}$$

where P_{app} and P_c are the pressure applied to the network and capillary pressure in the crack respectively, Γ_{app} and Γ_f are the parts of the boundary to which the applied pressure, and free surface stress balance are applied and θ_d is the dynamic contact angle that is a function of the fluid velocity.

An outline description of the cut finite element framework employed to capture discontinuities associated with the discrete crack transport is presented in Freeman and Jefferson (2021). A full description is the subject of a forthcoming journal paper.

3 MECHANICAL MODEL

The mechanical model describes the damage-healing behaviour using a cohesive zone model that is implemented into an embedded strong discontinuity element. A description of the 2D version of the element is presented in Freeman et al. (2020).

The cohesive zone model is applied to a crack plane that contains a macro-crack and relates the crack-plane traction vector ($\boldsymbol{\tau}_{cp}$) to the relative displacement vector ($\mathbf{d}$) as follows:

$$\boldsymbol{\tau}_{cp} = (1-\omega)\cdot\mathbf{K}:\mathbf{d} + h\cdot\mathbf{K}:(\mathbf{d} - \mathbf{d}_h) \tag{9}$$

where $\mathbf{K}$ is the elastic stiffness, h is the degree of healing, $\mathbf{d}_h$ is the relative displacement at the time of healing and ω is the damage variable that is governed by an exponential softening function (Freeman et al. 2020):

$$\omega = 1 - \frac{d_t}{\varsigma}e^{-c1\frac{\varsigma - d_t}{d_m - d_t}} \tag{10}$$

in which $d_t = f_t/K$, where f_t is the tensile strength of the material, $c_1 = 5$ is a softening constant, d_m is the relative displacement at the end of the softening curve and ς is the damage evolution parameter that depends on the maximum value of the inelastic relative displacements.

In order to account for re-damage and re-healing the evolution of the healing variable is described by:

$$h = h\cdot e^{-\frac{\Delta t}{\tau}} + a\cdot\left(1 - e^{-\frac{\Delta t}{\tau}}\right) \tag{11}$$

in which the relative area of the crack exposed to healing agent reads:

$$a = a + \Delta a_c - \Delta a_{redam} + \Delta a_{rec} \tag{12}$$

where Δa_c is the incremental area of virgin filled crack, Δa_{redam} is the incremental area of re-damaged material and Δa_{rec} is the incremental area of re-filled cracks.

In order to ensure that healing takes place in a stress-free state the following expression is derived from (9) for a healing update with **d** remaining constant:

$$(h+\Delta h)\cdot \mathbf{K}:(\mathbf{d}-\mathbf{d}_h-\Delta \mathbf{d}_h)-h\cdot \mathbf{K}:(\mathbf{d}-\mathbf{d}_h)=0 \quad (13)$$

that is used to derive an update for $(\Delta \mathbf{d}_h)$ as:

$$\Delta \mathbf{d}_h=\frac{\Delta h}{(h+\Delta h)}(\mathbf{d}-\mathbf{d}_h) \quad (14)$$

A full description of the damage-healing cohesive zone model is presented in Jefferson and Freeman (2022).

4 CHEMICAL MODEL

In the present work, the healing agent considered is cyanoacrylate. The chemical model therefore simulates the curing of cyanoacrylate that is a polymerization reaction driven by the transport of moisture into the glue from the substrate or surrounding air. This type of reaction can be simulated as the propagation of a reaction front that is diffuse in nature, and can be described by (Freeman & Jefferson 2020):

$$\phi_x(x,t)=\frac{1}{2}\left(1-\tanh\left(\left(\frac{2}{\sqrt{\pi}}\right)\left(\frac{x-z(t)-z_c}{z_c+\sqrt{\frac{z(t)}{z_{c1}}}}\right)\right)\right) \quad (15)$$

where x denotes the position measured from the crack face, z_c is a wall factor, z_{c1} is a diffusion coefficient and $z(t)$ is the position of the reaction front.

The propagation of the reaction front is described by:

$$z(t)=z_{c0}\left(1-e^{-\frac{t}{\tau}}\right) \quad (16)$$

where z_{c0} is the critical reaction front depth and τ is the curing time parameter.

The effect of the chemical reaction on the mechanical properties is accounted for through the degree of mechanical healing that –in the absence of re-damage- is given as the degree of cure at the center of the crack:

$$h(w,t)=\phi_x\left(\frac{w}{2},t\right) \quad (17)$$

Finally, the effect of the healing agent curing on the transport properties is accounted for using a chemo-rheological model that relates the degree of cure with the associated increase in viscosity as follows (Castro & Macosko 1980):

$$\mu=\mu_i\left(\frac{\phi_g}{\phi_g-\phi}\right)^{nv} \quad (18)$$

where μ_i is the initial viscosity, nv is an exponent, ϕ_g is the degree of cure at the gel point at which a rapid increase in viscosity is observed and ϕ is overall degree of cure across the width of the crack.

5 SOLUTION STRATEGY

Having presented the various model components, the solution strategy is now described. In the present work, the finite element method is used to solve the nonlinear-coupled problem using a staggered solution method. The algorithm employs a sequential coupling procedure that utilises sub-stepping for the transport problem. The nonlinearity of the mechanical model and matrix transport model is dealt with using a Newton-Raphson procedure, whilst a Picard procedure is employed to deal with the nonlinearity of the discrete crack and embedded vascular network transport. Iterative updates are terminated once the L_2 norm of the error meets a specified tolerance.

6 EXAMPLE PROBLEM

To illustrate the performance of the model an example problem is considered. The example concerns the a direct tension test on doubly notched concrete specimens containing an embedded vascular network (Selvarajoo et al. 2020). In the test, the specimen was loaded until a macro-crack opened to a given

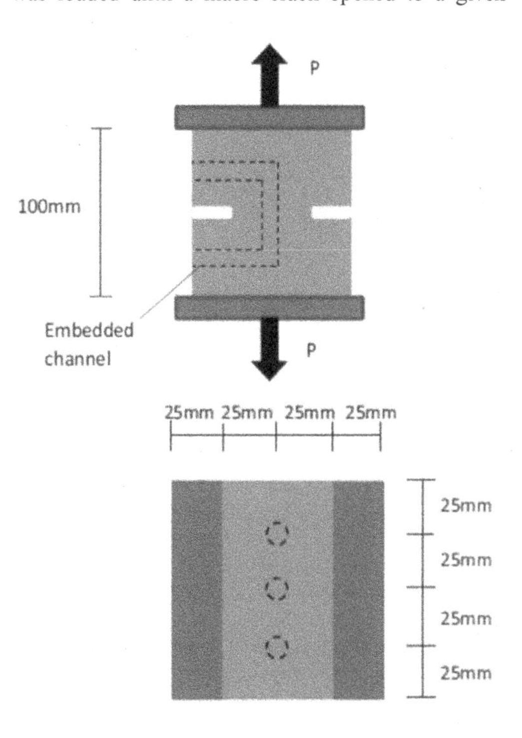

Figure 1. Schematic of test set up elevation (top) and cross-section (bottom) (after Freeman et al. 2020).

crack mouth opening displacement (CMOD). Following this, the healing agent was released and the crack held at the specified CMOD value for a range of healing periods. Finally, the specimen was loaded further until the CMOD reached 0.3 mm. The tests considered crack openings of 0.1 and 0.2 mm and healing periods ranging from 0-1200s.

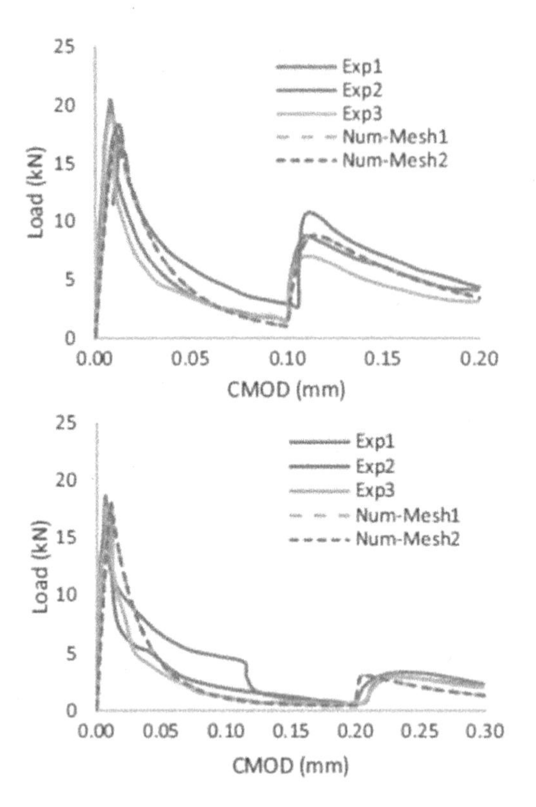

Figure 2. Comparison of load-CMOD curve with experimental data for a healing period of 60 s and crack width of 0.1 mm (top) and 0.2 mm (bottom) (after Freeman et al. 2020).

A schematic of the test set up, including the location of the embedded vascular channels can be seen in Figure 1. It is noted that in this example, following experimental observations, the degree of healing was limited to 85%. This was due to the fact that the healing agent was in constant flux and therefore did not stabilize and cure.

A comparison of the predicted load response with the experimental data can be seen in Figure 2, whilst the prediction of the crack filling for the 0.1mm CMOD can be seen in Figure 3. Finally, the comparison of predicted and experimental crack pattern can be seen in Figure 4. It can be seen from the Figures that the numerical simulations accurately capture the experimental behaviour. In addition, the crack completely filled during the healing period.

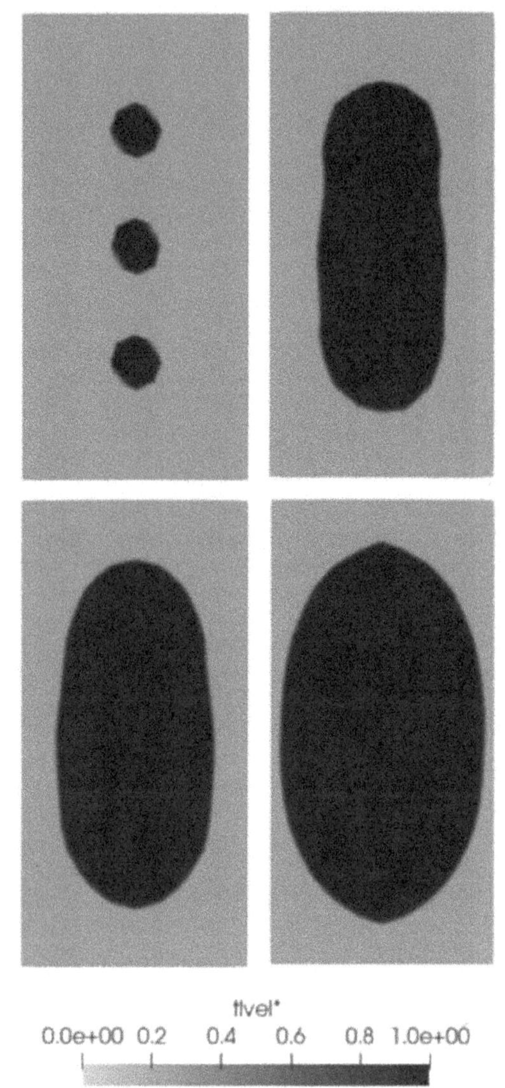

Figure 3. Coverage of 0.1 mm CMOD crack with healing agent as indicated by level set after 0 s (top left), 5.616 s (top right), 6.48 s (bottom left) and 8.64 s (bottom right).

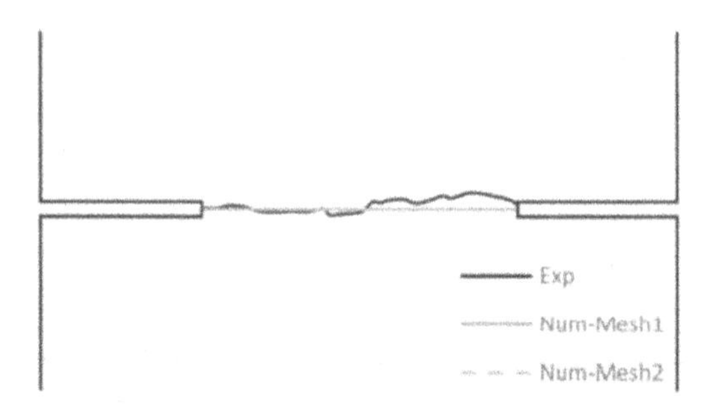

Figure 4. Comparison of numerically computed and experimentally observed crack pattern (after Freeman et al. 2020).

7 CONCLUDING REMARKS

This paper has provided an outline description of a 3D coupled chemo-mechanical model for simulating transient processes in self-healing cementitious materials. The model considers the mechanical damage-healing behaviour, allowing for damage and healing to occur simultaneously, transport of healing agent in the cementitious matrix, discrete macro-cracks and embedded vascular networks and the chemical reaction governing mechanical healing. The performance of the model has been demonstrated through the consideration of a direct tension test and subsequent comparison with experimental data. The results of the comparison showed that the model is capable of predicting realistic transport of healing agents, in addition to accurately capturing the damage-healing behaviour.

ACKNOWLEDGEMENTS

Financial support from the UKRI-EPSRC Grant EP/P02081X/1 "Resilient Materials for Life (RM4L)" is gratefully acknowledged.

REFERENCES

Castro JM, Macosko CW (1980) Kinetics and rheology of typical polyurethane reaction injection molding systems. *Society of Plastics Engineers (Technical Papers)*. 434–438.

De Nardi C, Gardner DR, Jefferson AD (2020) Development if 3d printed networks in self-healing concrete. Materials 13(6) 1328.

Di Luzio G, Ferrara L, Krelani V (2018) Numerical modeling of mechanical regain due to self-healing in cement based composites. Cement Concr Compos 86:190–205.

Freeman BL, Bonilla-Villalba P, Mihai IC, Alnaas WF, Jefferson AD (2020) A specialised finite element for simulating self-healing quasi-brittle materials. Adv Model Simul Eng Sci 7:32.

Freeman BL, Jefferson AD (2020) The simulation of transport processes in cementitious materials with embedded healing systems. Int J Numer Anal Meth Geomech 44:293–326.

Freeman BL, Jefferson AD (2022) Numerical simulation of self-healing cementitious materials. In: Kanellopoulos A., Norambuena-Contreras J. (eds) Self-Healing Construction Materials. Engineering Materials and Processes. Springer, Cham. https://doi.org/10.1007/978-3-030-86880-2_6.

Freeman BL, Jefferson AD (2021) A cutfem approach for simulating coupled 3d matrix- 2d crack plane flow of a healing agent in cementitious materials, Proceedings Resilient Materials 4 Life 2020 (RM4L2020), Maddalena R, Wright-Syed M, (RM4L Eds), pp. 110–114, Cardiff, UK, 20–22 Sep 2021, ISBN 978-1-3999-0832-0.

Granger S, Pijaudier-Cabot G, Loukili A (2007) Mechanical behaviour of self-healed ultra high performance concrete: from experimental evidence to modeling. In: Proceedings of the 6th international conference on fracture mechanics of concrete and concrete structures, vol 3, pp 1827–1834.

Huang H, Ye G (2016) Numerical studies of the effects of water capsules on self-healing efficiency and mechanical properties in cementitious materials. Adv Mater Sci Eng 8271214.

Jefferson AD, Freeman BL (2022) A crack-opening-dependent numerical model for self-healing cementitious materials. Int J Solids Struct. *Under Review*.

Koenders EAB (2012) Modelling the self-healing potential of dissoluble encapsulated cement systems, Final report IOP project SHM08707.

Li Z, Souza LRD, Litina C, Markaki A, Al-Tabbaa A (2020). A novel biomimetic design of a 3D vascular structure for self-healing in cementitious materials using Murray's law. Materials & Design. 190:108572.

Oucif C, Voyiadjis GZ, Rabczuk T (2018) Modeling of damage-healing and nonlinear self healing concrete behaviour: application to coupled and uncoupled self-healing mechanisms.Theoret Appl Fract Mech 96: 216–230.

Romero Rodríquez C, Chaves Figueiredo S, Deprez M, Snoeck D, Schlangen E, Šavija B (2019) Numerical investigation of crack self-sealing in cement-based composites with superabsorbent polymers. Cement Concrete Comp 104:103395.

Selvarajoo T, Davies RE, Freeman BL, Jefferson AD (2020) Mechanical response of a vascular self-healing cementitious material system under varying load conditions. Construct Build Mater 254:119245.

Zhang Y, Zhuang X (2018) A softening-healing law for self-healing quasi-brittle materials: analysing with strong discontinuity embedded approach. Eng Fract Mech 192: 290–306.

Computational Modelling of Concrete and Concrete Structures – Meschke, Pichler & Rots (Eds)
© 2022 Copyright the Author(s), ISBN: 978-1-032-32724-2

Meso-scale simulation of non-uniform steel corrosion induced damage in recycled aggregate ductile concrete

J. Fan, M.J. Bandelt & M.P. Adams
John A. Reif, Jr. Department of Civil and Environmental Engineering, New Jersey Institute of Technology, USA

ABSTRACT: Engineered cementitious composite (ECC) is a ductile construction material with higher damage tolerance compared to conventional concrete. This paper investigates the damage pattern and propagation of reinforced recycled aggregate concrete (RAC) using finite element simulation. Simulations were conducted on RAC members containing brittle matrix and ductile matrix (ECC) subject to uniform and non-uniform corrosion product expansion loads. A two-dimensional five-phase meso-scale level analysis approach was adopted by implementing individual material properties of aggregates, adhered mortar to the aggregates, old interfacial transition zone (ITZ), new ITZ, and cement matrix. Prescribed deformations were applied to the steel concrete interface as corrosion product expansion loads. The physical geometry of each aggregate was mapped from experimental images. The obtained results indicated that the damage first appeared around the ITZs and then propagated into matrix and mortar. It was also found that damage propagated much slower in RAC with ductile matrix compared to that of RAC with brittle matrix. Furthermore, RAC with ductile matrix showed distributed damage pattern while RAC with brittle matrix showed localized damage pattern. Simulation results also showed that non-uniform corrosion product expansion induced faster damage propagation than uniform rust expansion. The new ITZ had higher cracking susceptibility in the RAC composite structure due to its lower tensile strength. With the meso-level modeling technique, aggregate shape and orientation effects on cracking propagation in RAC due to corrosion product expansion were obtained. The uneven damage propagation patterns were observed with the increasing loading level due to the confinement to cement matrix from the aggregate. The less confined area showed more extensive damage areas but lower damage levels. The ITZs experienced less cracking along locations where the aggregates facing the corrosion product expansion load. In contrast, the locations where the faces of aggregates parallel to the load direction had severe damage. It was mainly because of the higher tensile and shear stresses in the parallel direction, even though the distance to the expansion load was greater. The aggregate shape and orientation had significantly less impact on damage pattern and propagation in ductile matrix RAC members than brittle matrix ones. The reason was that ECC material has distributed cracking behavior instead of major cracking in the damage propagation stage even though the studied members have the same tensile strength. The meso-scale numerical simulations of RAC under expansion load around rebar provide insights into the influences of non-uniform corrosion on damage propagation in brittle and ductile RAC members.

1 INTRODUCTION

Concrete made from recycled concrete aggregate has been used to minimize construction industry's impact on climate change due the large greenhouse gas footprint in concrete productions (Winfield & Taylor 2005). However, such efforts were hampered because the major quantity of RAC is used in non-structural works, such as road base, rip rap, and general fill purpose (Du & Jin 2014). The further application of RAC to structural components needs to use recycled concrete aggregate where good aggregate sources are required (Du & Jin 2014).

RAC is a five-phased material with different meso-scale structures, which includes natural aggregate, adhered mortar to the aggregates, old ITZ, new ITZ, and cement paste (Jayasuriya et al. 2018). A heterogeneous material like RAC may be highly susceptible to cracks due to the incompatible material properties of each phase (Jayasuriya et al. 2018). In structural applications, the cement paste can be either brittle or ductile. Ductile cement paste may be useful in cracking suppression. For example, ECC is a ductile construction material characterized by pseudo-strain hardening behavior under tension and exhibits multiple fine cracks prior to localized cracks (Li 2003).

Corrosion of reinforcing bar is one of the main deterioration mechanism that shortens the service life of normal reinforced concrete structures, or even causing catastrophic structural failures (Broomfield 2003).

 DOI 10.1201/9781003316404-15

Steel reinforced RAC could also undergo durability issues. Although a large amount of research has been conducted in studying the corrosion behavior of steel reinforced concrete, the meso-scale investigate is limited (Peng et al. 2019; Zhao et al. 2016). The aggregate geometry may affect the harmful material transport process in the concrete system and consequently influence the corrosion behavior of the steel in reinforced concrete (Yu & Lin 2020). Furthermore, a meso-scale study can provide insights on the cracking initiation and propagation process caused by corrosion product expansion.

The research conducted herein identifies the impact of ductile materials and aggregate geometry on the corrosion initiation and propagation in concrete. Quantitative and qualitative results of chloride concentration, corrosion production expansion, and damage patterns of both brittle and ductile reinforced RAC are explored. The results can used to understand how ductile construction materials can be used to inhibit non-uniform corrosion induced damage.

2 SIMULATION DESCRIPTION AND MODELING PARAMETERS

In this study, one chloride diffusion model, one corrosion development model, and four rust expansion models were completed. The cross section of 100 $mm \times$ 100mm of a known RAC system was selected from literature (Abbas et al. 2009). The geometry of each material was mapped from the image of an experimental study. The reinforcement bar diameter used was 10 mm. As shown in Figure 1, the reinforcing bar was placed in a location to avoid overlapping with the aggregates. The material properties and distribution of ITZs in RAC with ECC were assumed to be the same as ITZs in RAC with normal mortar, so as to avoid variations and focus on the ductility of the cement matrix.

2.1 *Analysis procedures*

First, a diffusion model was set up to study the chloride transport process in RAC system considering the aggregate geometry. Then the steel corrosion model was set up to simulate the corrosion current densities and electrolyte potential distributions. Corrosion product expansion thickness was then calculated based on previous step simulation results. Finally, the rust expansion load was applied to the steel concrete interface and the damage initiation and propagation was studied through mechanical models.

A time-dependent simulation procedure was integrated in the diffusion and corrosion simulations. Chloride concentration at the steel surface that reached critical chloride content increases as chloride exposure continues. At each time step, the anode area was determined according to the chloride concentration at the steel surface. Therefore, the anode/cathode area of the steel surface becomes a time-dependent factor that needs to be updated at each selected time step. A 10 days time interval was selected in this study to ensure analysis accuracy and maintain computing efficiency.

2.2 *Diffusion and corrosion model setup*

The two-dimensional diffusion and corrosion finite element model set up in COMSOL Multiphysics Version 5.4 (COMSOL 2021) is shown in Figure 1. A 0.5 $mm \times 0.5$ mm size quadrilateral element was used to simulate the ECC and normal concrete materials. Oxygen and chloride were assumed to enter only from the bottom side. Blocking of mass transport by reinforcing bars was considered. Natural aggregates were assumed to be impenetrable (Tian et al. 2019). The mass transport properties of adhered mortar, old ITZ, and new ITZ were assumed to be the same as mortar since the volume fraction of them were relatively small compared to cement paste. Diffusion coefficient of oxygen is $D_{O_2} = 3.0E - 9m^2/s$ (Rafiee 2012). The

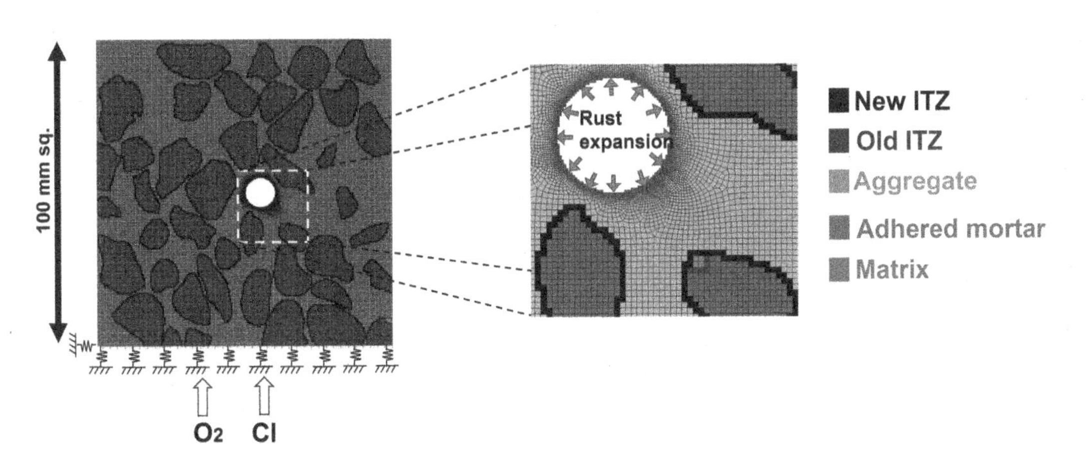

Figure 1. Finite element model set up for RAC system with normal mortar and RAC system with ECC.

diffusion coefficient of chloride in the mortar was calculated from (Zheng & Zhou 2008):

$$D_{Cl}=\frac{2.14\times10^{-10}V_P^{2.75}}{V_P^{1.75}(3-V_P)+14.44(1-V_P)^{2.75}} \quad (1)$$

where V_P is the porosity fo cement paste, and it is given by

$$V_P=\frac{w/c-0.17\alpha}{w/c+0.32\alpha} \quad (2)$$

where w/c is the water to cement ratio and was selected as 0.45 in this study, α denotes the hydration level and was assumed to be 90% (Mouret et al. 1997).

2.3 *Mechanical model setup*

The two-dimensional rust expansion finite element model set up in DIANA FEA Version 10.5 (DIANA 2021) is shown in Figure 1. A total strain based fixed-crack constitute law was implemented in this study. Four-noded 0.5mm $\times$ 0.5 mm size quadrilateral elements were used. All the bottom nodes were restrained in the vertical direction, and the node at the bottom-left of the model was restrained in both vertical and horizontal directions. Rust thickness was calculated from (Böhni 2005)

$$\sigma(t)=\frac{\int_0^t i_{corr}(t)dt\cdot M_s}{Z_{Fe}\cdot F\cdot\rho_s} \quad (3)$$

where t gives the corrosion time (seconds), $M_s=55.85\ g/mol$ is the atomic mass of the iron, $Z_{Fe}=2$ is the valency of anodic reaction, $F=96485\ C/mol$ is the Faraday's constant,$\rho_s=7800\ kg/m^3$ is the steel density. The applied displacement load at the steel cement matrix interface was obtained from

$$u(t)=(n-1)\sigma(t) \quad (4)$$

where n is rust to steel volume expansion ratio and is assumed to be 3 in this study (Cao et al. 2013).

Displacement load in the non-uniform rust expansion simulations was based on previous step study-corrosion simulation. However, an average displacement load around the reinforcement bar circumferential was applied in the uniform rust expansion simulations.

2.4 *Material properties and parameters in simulations*

Diffusion and polarization parameters such as chloride and oxygen diffusion coefficients (D_{Cl},D_{O_2}), Tafel slopes (β_{Fe},β_{O_2}), equilibrium potentials (ϕ^0_{Fe},$\phi^0_{O_2}$), and exchange current densities (i^0_{Fe},$i^0_{O_2}$) are summarized in Table 1 (Rafiee 2012). Surface chloride concentration (Cl_{surf}) was 2% of the concrete mass (Cao 2014). Surface oxygen concentration (O_{2surf}) was 0.268mol/m^3 (Cao 2014). Critical chloride content (Cl_{crit}) was adopted as 0.06% of concrete mass. Concrete conductivity (θ_c) was obtained from experimental results from Rafiee (Rafiee 2012).

Table 1. Mass transport and corrosion polarization parameters.

Input parameters	units	Values
Cl_{surf}	%	2
O_{2surf}	mol/m^3	0.268
D_{Cl}	m^2/s	4.3E-12
D_{O_2}	m^2/s	3.0E-9
θ_c	S/m	0.0063
Cl_{crit}	%	0.06
β_{Fe}	mV/dec	65
β_{O_2}	mV/dec	139
ϕ^0_{Fe}	mV	-600
$\phi^0_{O_2}$	mV	200
i^0_{Fe}	A/m^2	2.75E-4
$i^0_{O_2}$	A/m^2	6E-6

Each of the five-phased materials were defined in the simulations. Mechanical material properties such as elastic modulus, Poisson's ratio of natural aggregates, adhered mortar, and cement paste were obtained from literature (Xiao et al. 2013). Other material properties of the aggregates listed in Table 2 were taken from Winkler (2013). Modulus of elasticity and Poisson's ratio of the old ITZ and new ITZ were determined from Ramesh et al. (1996). The material properties of ECC were obtained from experimental work of Moreno-Luna (2014). The tensile strength of ECC was adjusted from 2.9MPa to 3.0MPa to avoid tensile strength variation and focus on the ductility of the cement matrix. Compressive strength was adjusted from 55MPa to 45.4MPa for the same purpose. The tensile strain at onset of softening of ECC was 0.75% while normal cement paste had an onset softening strain of 0.015% (Moreno-Luna 2014).

3 SIMULATION RESULTS AND DISCUSSION

3.1 *Chloride concentrations*

Figure 2 shows the chloride content contour of the RAC systems after 1200 days of chloride exposure. The natural aggregates and reinforcement bar were assumed to be impenetrable (Tian et al. 2019). Thus there was no chloride in those areas. The aggregate geometry played an important role in chloride transport process. As shown in Figure 2, aggregates that stayed further away from each other provided easier path for chloride to transport. For example, the chloride content in area A (0.89%$wt.$) is higher than that of area B (0.55%$wt.$) at 16 mm distance to the chloride exposure surface (Figure 2).

Figure 3 shows the chloride concentration of the steel surface that reached the critical chloride content. The chloride concentration at the steel surface was developing unsymmetrically due to the impenetrable

Table 2. Mechanical properties.

Material type	Modulus of elasticity	Compressive strength	Tensile strength	Posisson's ratio	Tensile fracture energy
Aggregate	80 GPa	144.0 MPa	9.60 MPa	0.16	0.163 N/mm
Adhered mortar	25 GPa	45.0 MPa	3.00 MPa	0.22	0.051 N/mm
Old ITZ	23 GPa	41.4 MPa	2.76 MPa	0.22	0.041 N/mm
New ITZ	20 GPa	36.0 MPa	2.40 MPa	0.20	0.037 N/mm
Mortar	19.5 GPa	45.4 MPa	3.00 MPa	0.15	0.047 N/mm
ECC	17.2 GPa	45.4 MPa	3.00 MPa	0.15	6.1 N/mm

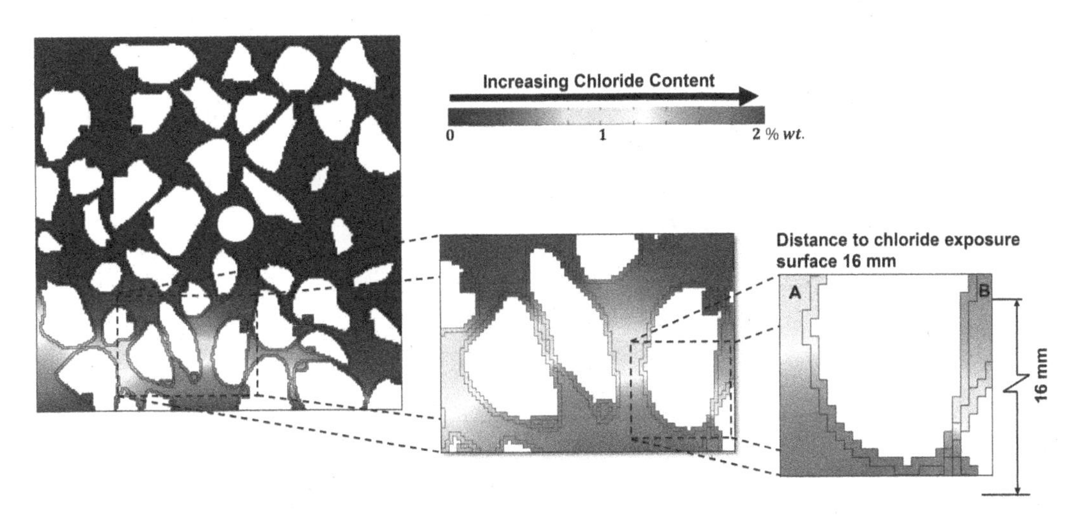

Figure 2. Chloride concentration contour for RAC system with normal mortar and RAC system with ECC.

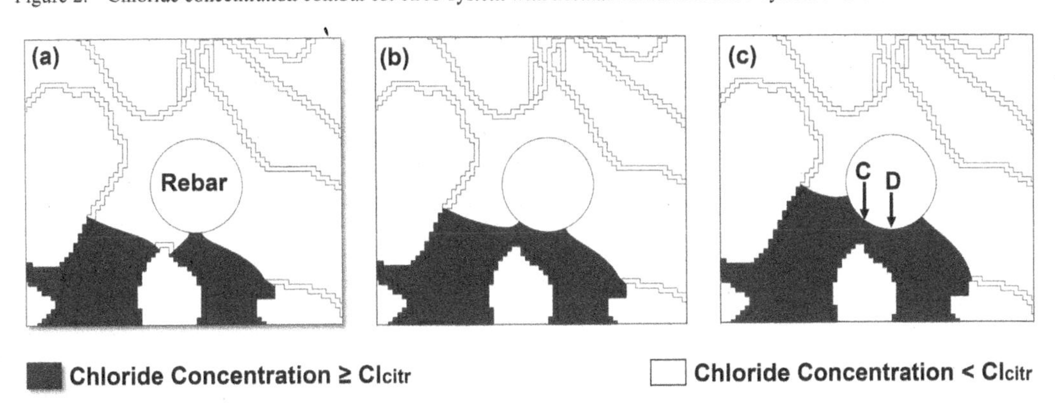

Figure 3. Chloride concentration development for all systems over time at (a) 1170 days, (b) 1200 days, (c) 1290 days of chloride exposure.

feature of aggregates and the unsymmetrical distribution of the aggregates. Consequently, the corrosion front area center was shifted to the left of the reinforcement bar (location C in Figure 3) surface instead of the closest point (location D in Figure 3) to the chloride exposure surface. The result proved the effectiveness of a meso-scale study in terms of mass transport problems including chloride and oxygen diffusion in concrete materials.

3.2 Corrosion behaviors

As a result of unsymmetrical corrosion front area, the electrolyte potential in the RAC system was also unsymmetrical to the reinforcement bar. As shown in Figure 4, the left to the reinforcement bar had higher electrolyte potential of $0.47V$ at location E while the electrolyte potential at location F was $0.48V$.

Figure 5 shows the oxygen level in the RAC system. The area near the reinforcement bar had lower oxygen

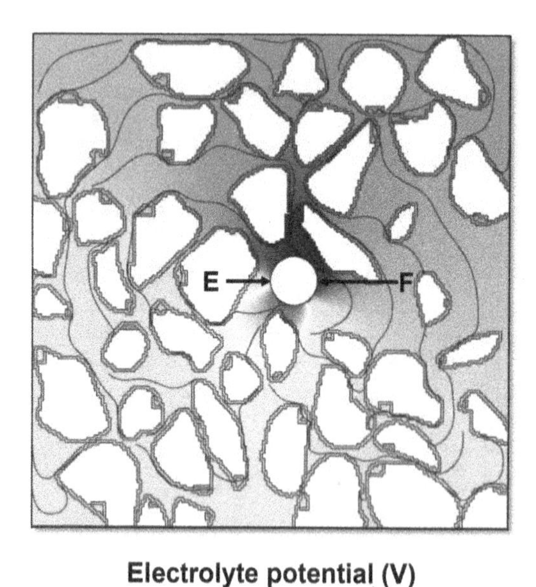

Figure 4. Electrolyte contour.

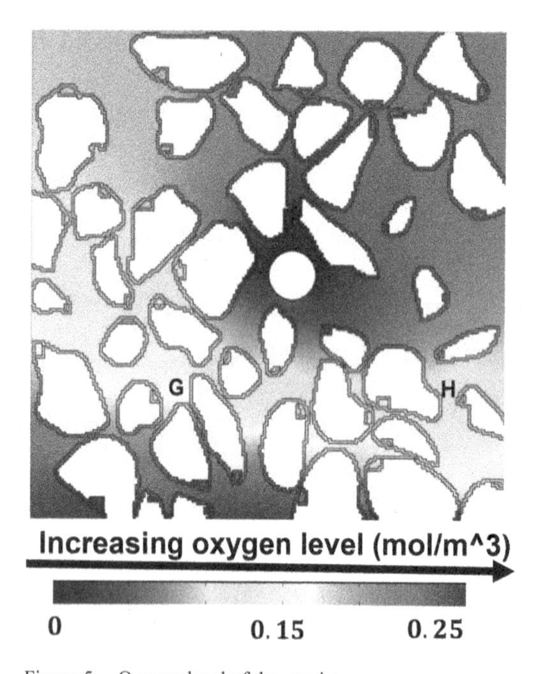

Figure 5. Oxygen level of the specimen.

level compared to areas that had further distance to the bar. It was ascribed to the cathodic reaction at the steel surface, which consumed the initial oxygen in the cement matrix and insufficient oxygen was provided from the specimen surface. The oxygen distribution was also affected by the blocking effect of aggregates since oxygen diffusion in aggregates was negligible. For example, at the same distance to the oxygen exposure surface (bottom side), the oxygen level at location G is 0.16 mol/m^3 while location H had an oxygen level of 0.11 mol/m^3 (in Figure 5).

3.3 Corrosion product distribution

Corrosion product distribution along the reinforcement bar surface is shown in Figure 6. The rust thickness showed an unsymmetrical pattern due to the unsymmetrical development of chloride concentration and oxygen supply. The maximum rust expansion thickness was 45.1 μm while the minimum was zero at 9.5 to 10mm along the reinforcement bar in horizontal direction.

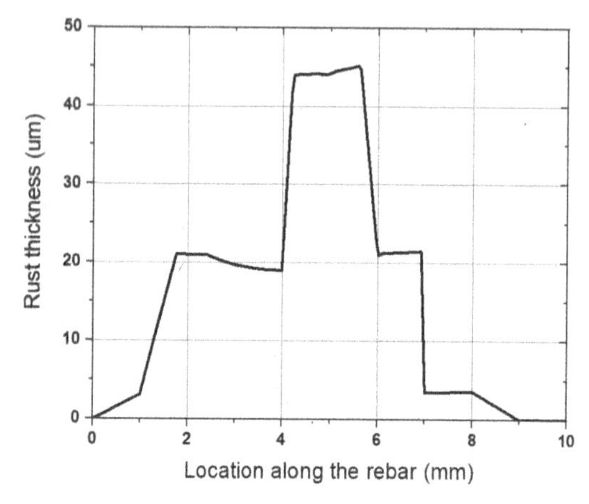

Figure 6. Rust distribution along the reinforcement bar surface.

3.4 Corrosion induced damage

Figure 7 (a) shows the cracking pattern of RAC at high rust expansion level. The damage index scale in Figure 7 indicates the cracking level of the material. The damage index number of zero means the material is still in elastic while number 1 indicates the material is softened. In both uniform and non-uniform cases, RAC with ECC matrix had lower damage level compared to RAC with normal cement matrix. High cracking suppression ability of ECC contributed to the low damage level in RAC with ECC matrix. RAC under uniform loading conditions had more damage compared to non-uniform rust expansion cases. The reason was that only the front of the steel surface was applied with displacement load in non-uniform cases while all the direction of the steel-cement interface was applied with displacement load in uniform cases. Additionally, RAC with normal cement paste under uniform load (13.3 μm) had the same severe damage level compared to non-uniform loading (maximum 45.1 μm). This was because the damage tolerance of the normal cement matrix was low and the applied load in the uniform case was big enough to cause severe damage in all directions. In contrast, RAC with ECC had the same low level of damage even though the damage area in uniform case was bigger. The only

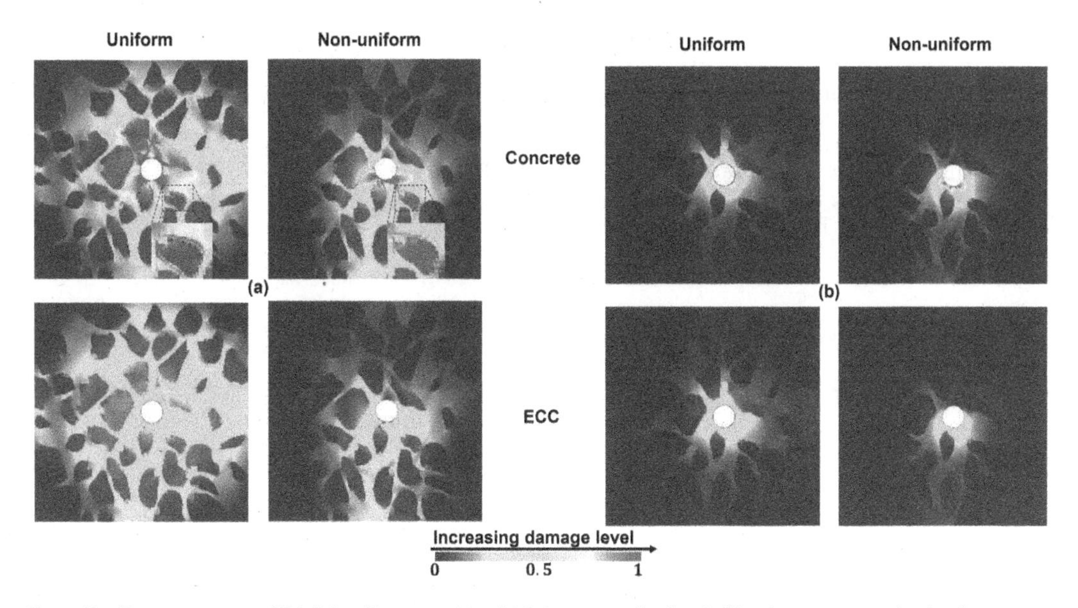

Figure 7. Damage pattern of RAC for all systems (a) at high rust expansion level, (b) at low rust expansion level.

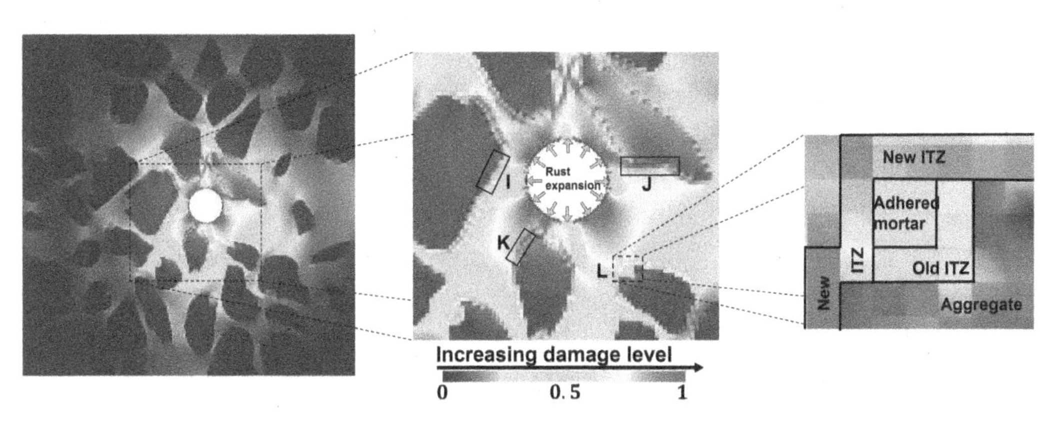

Figure 8. Cracking pattern of RAC with normal cement paste. The damage index number of zero means the material is still in elastic while number 1 indicates the material is softened.

difference of the mechanical properties between normal cement paste and ECC was the ductility, in which ECC had 50 times larger onset softening strain than normal cement paste. Therefore, ductile material such as ECC can provide better cracking resistance than normal concrete material.

Figure 7 (b) shows the cracking pattern of RAC at a relatively lower rust expansion level. Same as the cases in high displacement loading conditions, RAC with ECC showed lower cracking compared to RAC with normal cement paste. However, contradictory to high rust expansion thickness cases, RAC with normal cement paste under uniform loading (2.6 μm) condition had lower damage level than non-uniform (9.0 μm) cases under low loading condition. It was because the uniform load was smaller than the non-uniform load. And more importantly, the uniform load had not exceed the threshold to cause severe damage while the maximum rust expansion load in the non-uniform case was big enough to induce major cracking.

3.5 *Aggregate geometry effect*

Figure 8 shows the cracking at the interfaces and adhered mortar near the reinforcement bar. The new ITZ had higher cracking level in the RAC composite structure due to its lower tensile strength. The uneven damage pattern was ascribed to the randomly distributed aggregates. As shown in Figure 8, the area on the right of the reinforcement bar experienced more extensive damage but lower damage levels. The ITZs had less cracking along locations (e.g. area I in Figure 8) where the ITZs were perpendicular to the corrosion product expansion load. In contrast, the ITZs experienced severe damage at the locations (e.g. area J and

K in Figure 8) where the ITZs were parallel to the loading direction. The main reason was that the tensile and shear stresses in the parallel direction were higher, even though the distance to the expansion load was larger. The aggregate shape and orientation had significantly less impact on damage pattern in ductile matrix RAC than brittle matrix ones. It was beacuse ductile materials such as ECC has distributed cracking characteristics instead of localized cracking pattern even though the studied members have the same tensile strength.

As shown in Figure 8, the adhered mortar in area L had lowest damage level due to its higher tensile strength than the old ITZ and new ITZ materials.

4 CONCLUSIONS

Time-dependent two dimensional five-phased meso-scale nonlinear finite element simulations of chloride diffusion, corrosion, and rust expansion were carried out on RAC members. Both uniform and non-uniform corrosion cases, ductile and brittle cement matrix were considered.

Simulation results showed that aggregate geometry had significant influence on corrosion initiation pattern and lead to unsymmetrical non-uniform corrosion production development. The maximum rust expansion thickness was 45.1 μm while the minimum was zero.

Ductile material such as ECC exhibited higher cracking resistance compared to normal cement paste material. It was because the multiple fine cracking behavior of ECC material.

Non-uniform corrosion at early stage may cause same damage level but less damage to all the systems due to the low and localized internal pressure. However, non-uniform corrosion induced higher damage level at later stage of the corrosion even though the damage area may be smaller.

Aggregate geometry affects the cracking initiation and propagation in RAC systems. The ITZs parallel to the rust expansion direction had higher cracking susceptibility compared to ITZs perpendicular to the rust expansion direction. Adhered mortar had less cracking susceptibility in comparison to old ITZs and new ITZs due to its higher tensile strength.

A complete set of meso-scale models including harmful material ingress, corrosion initiation and propagation, and corrosion product induced damage simulations provide a more reliable prediction to the service life performance of reinforced RAC systems.

ACKNOWLEDGMENT

The authors gratefully acknowledge the support of the John A. Reif, Jr. Department of Civil and Environmental Engineering.

REFERENCES

Abbas, A., G. Fathifazl, B. Fournier, O. B. Isgor, R. Zavadil, A. G. Razaqpur, & S. Foo (2009). Quantification of the residual mortar content in recycled concrete aggregates by image analysis. *Materials characterization 60*(7), 716–728.

Böhni, H. (2005). *Corrosion in reinforced concrete structures*. Elsevier.

Broomfield, J. (2003). *Corrosion of steel in concrete: understanding, investigation and repair*. CRC Press.

Cao, C. (2014). 3d simulation of localized steel corrosion in chloride contaminated reinforced concrete. *Construction and Building Materials 72*, 434–443.

Cao, C., M. M. Cheung, & B. Y. Chan (2013). Modelling of interaction between corrosion-induced concrete cover crack and steel corrosion rate. *Corrosion Science 69*, 97–109.

COMSOL (2021). Comsol multi-physics. *https://www.comsol.com/*.

DIANA (2021). Diana fea. *https://dianafea.com/*.

Du, X. & L. Jin (2014). Meso-scale numerical investigation on cracking of cover concrete induced by corrosion of reinforcing steel. *Engineering Failure Analysis 39*, 21–33.

Jayasuriya, A., M. P. Adams, & M. J. Bandelt (2018). Understanding variability in recycled aggregate concrete mechanical properties through numerical simulation and statistical evaluation. *Construction and Building Materials 178*, 301–312.

Jayasuriya, A., M. Bandelt, & M. Adams (2018). Simulation of cracking susceptibility in recycled concrete aggregate systems. In *Computational Modelling of Concrete Structures*, pp. 421–428. CRC Press.

Li, V. C. (2003). On engineered cementitious composites (ecc). *Journal of Advanced Concrete Technology 1*(3), 215–230.

Moreno-Luna (2014). *ension stiffening in reinforced high performance fiber reinforced cement based composites*. Ph.D., Standford University.

Mouret, M., A. Bascoul, & G. Escadeillas (1997). Study of the degree of hydration of concrete by means of image analysis and chemically bound water. *Advanced Cement Based Materials 6*(3-4), 109–115.

Peng, J., S. Hu, J. Zhang, C. Cai, & L.-y. Li (2019). Influence of cracks on chloride diffusivity in concrete: A five-phase mesoscale model approach. *Construction and Building Materials 197*, 587–596.

Rafiee, A. (2012). *Computer modeling and investigation on the steel corrosion in cracked ultra high performance concrete*, Volume 21. kassel university press GmbH.

Ramesh, G., E. Sotelino, & W. Chen (1996). Effect of transition zone on elastic moduli of concrete materials. *Cement and Concrete Research 26*(4), 611–622.

Tian, Y., C. Chen, N. Jin, X. Jin, Z. Tian, D. Yan, & W. Yu (2019). An investigation on the three-dimensional transport of chloride ions in concrete based on x-ray computed tomography technology. *Construction and Building Materials 221*, 443–455.

Winfield, M. & A. Taylor (2005). *Rebalancing the Load: The need for an aggregates conservation strategy for Ontario*. Pembina Institute for Appropriate Development.

Winkler, E. M. (2013). *Stone: properties, durability in man's environment*, Volume 4. Springer Science & Business Media.

Xiao, J., W. Li, D. J. Corr, & S. P. Shah (2013). Simulation study on the stress distribution in modeled recycled aggregate concrete under uniaxial compression. *Journal of materials in civil engineering 25*(4), 504–518.

Yu, Y. & L. Lin (2020). Modeling and predicting chloride diffusion in recycled aggregate concrete. *Construction and Building Materials 264*, 120620.
Zhao, Y., J. Dong, Y. Wu, & W. Jin (2016). Corrosion-induced concrete cracking model considering corrosion product-filled paste at the concrete/steel interface. *Construction and Building Materials 116*, 273–280.
Zheng, J. & X. Zhou (2008). Analytical solution for the chloride diffusivity of hardened cement paste. *Journal of materials in civil engineering 20*(5), 384–391.

Computational Modelling of Concrete and Concrete Structures – Meschke, Pichler & Rots (Eds)
 ISBN: 978-1-032-32724-2

Simulation of Brazilian tests of ultra-high performance fibre-reinforced concrete

B. Sanz & J. Planas
Departamento de Ciencia de Materiales, Universidad Politécnica de Madrid, Madrid, Spain

J.M. Sancho
Departamento de Estructuras de Edificación, Universidad Politécnica de Madrid, Madrid, Spain

ABSTRACT: Ultra-high performance fibre reinforced concrete (UHPFRC) is an emerging material with a high scientific and technological interest due to its outstanding mechanical behaviour. However there are aspects of its mechanical characterisation that need to be solved. In particular, this work focuses on the determination of the tensile strength, interpreted as the stress at which the crack initiates in pure tension and softening begins. In a recent study, applicability of the Brazilian test or diagonal compression splitting test to determine the tensile strength of UHPFRC was assessed from the results of numerical simulations and experiments conducted on specimens with three contents of fibres and specimens without fibres. From the numerical simulations it was demonstrated that a local maximum in stress may occur in the test for a stress very close to the tensile strength of the material, which can be detected as a pop-in in the curves of results of experiments run under load control with the adequate instrumentation, thus providing an adequate approximation of the tensile strength. In this paper, a numerical study of the Brazilian test is presented which continues the previous work and investigates the size effect and the influence of the material properties. Simulations of the tests have been carried out within the finite element framework COFE (*Continuum Oriented Finite Element*), which implements elements with an embedded adaptable cohesive discrete crack. To reproduce the cracking behaviour, the softening law is assumed to present a steep initial softening, predominantly due to cracking of the matrix, followed by a long tail, due to the contribution of the fibres. In this study, influence of the content of fibres is analysed by modifying the parameters of the softening curve, while the size effect is analysed by modifying the ratio between a fracture length of the matrix, which is defined in detail in the paper, and the diameter of the specimen. The results show a high effect of the size and material properties on the test results, which may guide through the proper design of the experiments in order to ensure appearance of a measurable sufficiently-accurate local maximum.

1 INTRODUCTION

Ultra-high performance fibre-reinforced concrete (UHPFRC) is an emerging material firstly introduced in 1972 (Roy, Gouda, & Bobrowsky 1972; Yudenfreund, Odler, & Brunauer 1972), with a high scientific and technological interest due to its excellent mechanical behaviour (see Yoo & Yoon 2016 for a review of its structural behaviour, design and application). However there are aspects of its mechanical characterisation that still need to be solved. In particular, this work focuses on the determination of the tensile strength, interpreted as the stress at which the crack initiates in pure tension and softening begins. Direct tests are mainly found in the literature (see as examples Habel, Viviani, Denarié, & Brühwiler 2006; Hassan, Jones, & Mahmud 2012; Nguyen, Ryu, Koh, & Kim 2014; Park, Kim, Ryu, & Koh 2012; Toledo Filho, Koenders, Formagini, & Fairbairn 2012; Wille, El-Tawil, & Naaman 2014; Wille & Naaman 2010; Xu & Wille 2015), but they present the inconvenience that failure out of the plane can occur with the subsequent loss of symmetry, as discussed in (Hordijk 1991; Noghabai 1998); hence the interest of using indirect tests.

The diagonal compression splitting test, also known as the *Brazilian test*, has been widely used in the case of ordinary concrete, since it provides a sufficiently accurate approximation of the tensile strength of quasi-brittle materials from the record of the maximum load, as discussed in detail in (Rocco 1996; Rocco, Guinea, Planas, & Elices 1999a; Rocco, Guinea, Planas, & Elices 1999b), provided that the loading rate and the width of the bearing bands are properly limited. However, in the case of fibre-reinforced materials the maximum load occurs for a stress much higher than the tensile strength due to the contribution of the fibres and, thus, performing a *standard* Brazilian test might be not sufficient. Nevertheless, in a recent experimental and numerical study presented in (Sanz, Planas, Rey de Pedraza, Sancho, Sancho, &

DOI 10.1201/9781003316404-16

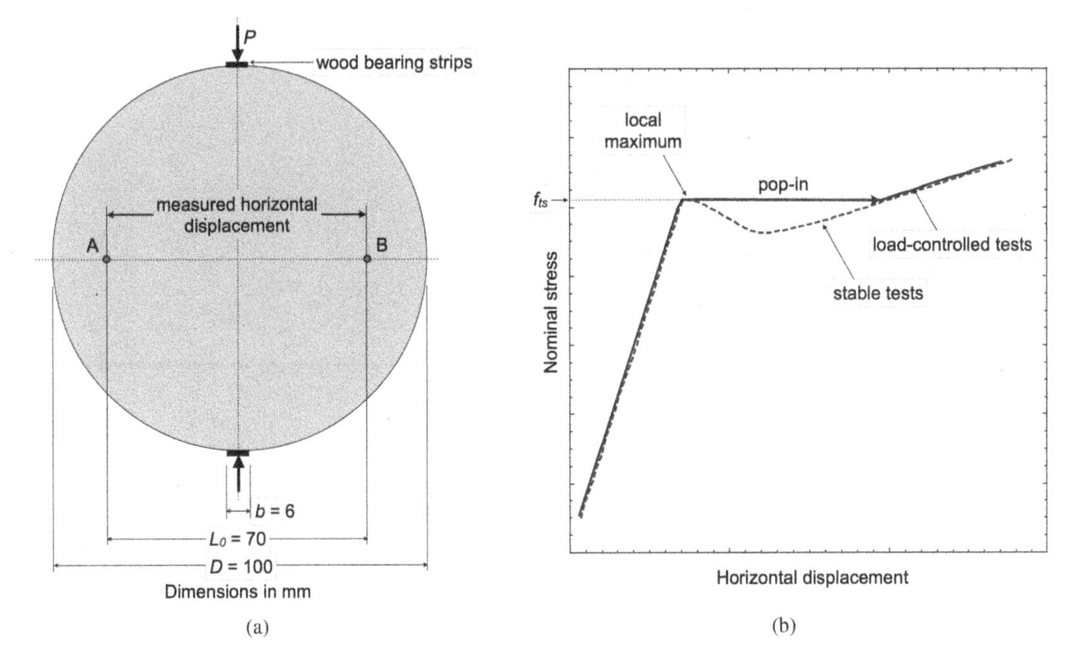

Figure 1. Sketch of the Brazilian test and geometry of the specimen (a), and representative curve of stress versus horizontal displacement, with a local maximum occurring for a stress f_{ts} close to the tensile strength of the material (b).

Gálvez 2022) about the applicability of this test to the case of UHPFRC, it was disclosed that a local maximum in stress occurs at initiation of cracking, which can be detected in experiments with the adequate instrumentation, thus, demonstrating the interest of the Brazilian test in the determination of the tensile strength of UHPFRC.

Figure 1(a) shows a sketch of the version of the test used in (Sanz et al. 2022), in which the horizontal displacement u between the points labeled as A and B was measured perpendicular to the plane of loading. Simulations of the tests were carried out using bi-dimensional models of the specimens by using elements with an embedded adaptable cohesive crack (Sancho, Planas, Cendón, Reyes, & Gálvez 2007; Sancho, Planas, Fathy, Gálvez, & Cendón 2007). From the numerical results, it was verified that a local maximum occurs in the curves of nominal stress versus the horizontal displacement, as illustrated in Figure 1(b), for a stress f_{ts} very close to the tensile strength of the material f_t. The dotted line in the figure shows the result of a representative simulation, which corresponds to the theoretical curve of a stable test. In practise, experiments are typically run under load control, in order to ensure a constant loading rate, since it has a high influence on the measured strength, as discussed in (Rocco et al. 1999b). Consequently, a net pop-in —an abrupt increase of the measured displacement under almost constant load— may occur, resulting the bold line in the figure. Note that, even in that case, the stress f_{ts} corresponding to the local maximum could be measured, provided that the length of the pop-in is sufficient. This was verified in an experimental campaign with specimens with 2%, 3.25% and 4% of fibres by volume and specimens without fibres.

In the current paper a numerical study is presented which continues the previous work and investigates the size effect and the influence of the material properties on the local maximum, since concrete displays a size effect (Bažant & Planas 1998), which may condition the local maximum and length of the pop-in. Dimensionless simulations have been conducted assuming a softening curve which presents a steep initial softening, due to cracking of the matrix, followed by a long tail, due to the contribution of the fibres. Influence of the fibres is analysed by modifying the shape of the softening curve, while the size effect is analysed by modifying the ratio between a fracture length of the material which is described in detail in the corresponding section, and the diameter of the specimen.

In the paper, Section 2 explains the basis of the simulations and the model parameters; Section 3 presents the results of the numerical study and discusses the influence of the material properties and size effect, and, finally, Section 4 presents the main conclusions of this work.

2 NUMERICAL SIMULATIONS

The basis of the simulations are as those reported in (Sanz et al. 2022), which are summarised next for completeness of the text.

2.1 *Fracture behaviour of concrete*

For the fracture behaviour of concrete, a generalisation of the cohesive crack model was assumed, based on the main principles firstly introduced by Hillerborg et al (Hillerborg, Modéer, & Petersson 1976). In that model, it is assumed that when a crack develops in pure opening (Mode I), it still transmits stress σ at its faces depending on the crack width w following a unique relation $f(w)$ which is called the *softening curve*. For general loading, a vectorial traction-separation law is assumed considering the traction vector $\mathbf{t}$ acting on one of the faces of the crack and the crack separation vector $\mathbf{w}$.

In this work, the damage-based vectorial model proposed in (Planas, Sanz, & Sancho 2020) has been used, which is a generalisation of the models found in the literature. This considers a parameter α^2 as the ratio between the fracture energies in modes II and I, and a parameter β^2 as the ratio between the shear and normal stiffnesses, resulting in the following vectorial law:

$$\mathbf{t} = \frac{f(\kappa)}{\kappa}(w_n\mathbf{n} + \beta^2\mathbf{w}_s), \qquad \kappa = \max[w^{eq}(\mathbf{w})] \tag{1}$$

where w_n and $\mathbf{w}_s$ are the normal and shear components of the displacement vector, respectively, which are calculated by considering the unit normal of the reference face of the crack $\mathbf{n}$ as $w_n = \mathbf{w}\cdot\mathbf{n}$ and $\mathbf{w}_s = \mathbf{w} - w_n\mathbf{n}$, and κ is a damage variable which is computed as the maximum value of an equivalente separation w^{eq}. This parameter is defined together with an equivalent traction t^{eq} as

$$w^{eq} := \sqrt{w_n^2 + \frac{\beta^2}{\alpha^2}w_s^2}, \qquad t^{eq} := \sqrt{t_n^2 + \frac{t_s^2}{\alpha^2\beta^2}} \tag{2}$$

where w_s is the modulus of the shear component of the displacement vector, and t_n and t_s the moduli of the normal and the shear components of the traction vector, respectively, which are computed analogously to w_n and w_s. In this work we assume $\alpha = \beta = 1.0$, according to the results reported in (Planas et al. 2020).

A bilinear curve has been used to reproduce the softening curve of concrete, as sketched in Figure 2. This can be defined by the following four parameters: the tensile strength f_t, the stress at the kink point f_1, the horizontal intercept of the first branch with the abscissas axis w_1, and the crack width w_c corresponding to zero stress. To account for the bridging effect of fibres, bilinear curves with a softening tail much longer than that of ordinary concrete have been considered, with a final crack width $w_c = 117w_1$. For convenience, the stress of the kink point is written as $f_1 = \gamma f_t$, where γ is a parameter ranging from 0.0 to 1.0. In this work, five values of γ ranging from 0.5 to 0.95 and $\gamma = 0.0$ have been considered, as indicated in Table 1, in order to study the effect of the material parameters. Note that the curve with $\gamma = 0.0$ corresponds to a linear softening curve as that sketched by the dotted line in the figure, and is interpreted as the behaviour of the material without fibres.

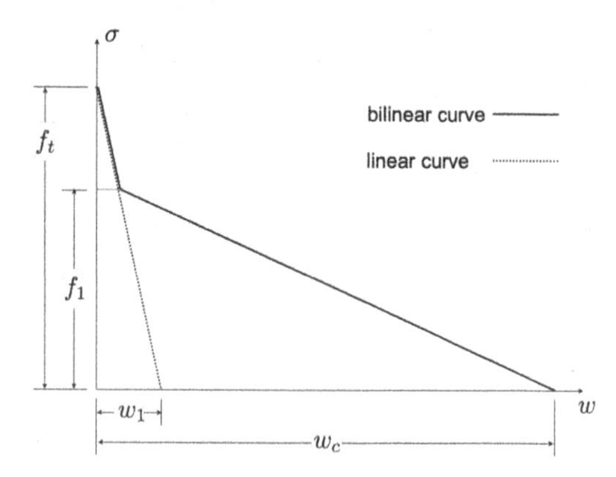

Figure 2. Sketch of the linear and bilinear softening curves of concrete, and parameters defining them.

2.2 *Dimensionless simulations*

Dimensionless simulations have been conducted, following the reasoning in (Planas, Sanz, & Sancho 2021), by considering a dimensionless softening function $\hat{f}$:

$$\sigma = f(w) = f_t\hat{f}\left(\frac{w}{w_1}\right). \tag{3}$$

where f_t and w_1 are the tensile strength and the horizontal intercept of the first segment of the bilinear curve, respectively, as introduced in Figure 2. This implies that all the resulting values of stress and crack traction are divided by f_t, and all the values of displacement and crack separation are divided by w_1. Dimensionless geometrical lengths have been considered as well by dividing all the lengths by the specimen diameter D.

The uncracked material has been modelled as linear elastic, with Poison's ratio $\nu = 0.17$ and an elastic modulus E which is scaled as follows. Let us consider Hooke's law of a unidimensional problem $\sigma = E\varepsilon = E\partial u/\partial x$, and define the unidimensional variables $\sigma^* = \sigma/f_t$, $u^* = u/w_1$ and $x^* = x/D$. Then a dimensionless version of Hooke's law is obtained as

$$\sigma^* = \frac{Ew_1}{f_tD}\frac{\partial u^*}{\partial x^*} = \frac{\ell_2}{D}\frac{\partial u^*}{\partial x^*}, \quad \ell_2 := \frac{Ew_1}{f_t} \tag{4}$$

where ℓ_2 is a brittleness length of the material which we call the *second brittleness length*. From the previous equation, it follows that the dimensionless elastic modulus of the model is calculated as

$$E^* = \frac{\ell_2}{D} \tag{5}$$

It implies that the size effect can be analysed by modifying the size of the specimens, or by modifying any of the parameters of the brittleness length. The option adopted in this work has been to modify the elastic modulus; in particular, ratios ℓ_2/D equal to 1.0, 2.0 and 4.0 have been studied. Note that an increase in the ratio ℓ_2/D is equivalent to diminishing the size of the specimen or to considering a less brittle material with a greater brittleness length.

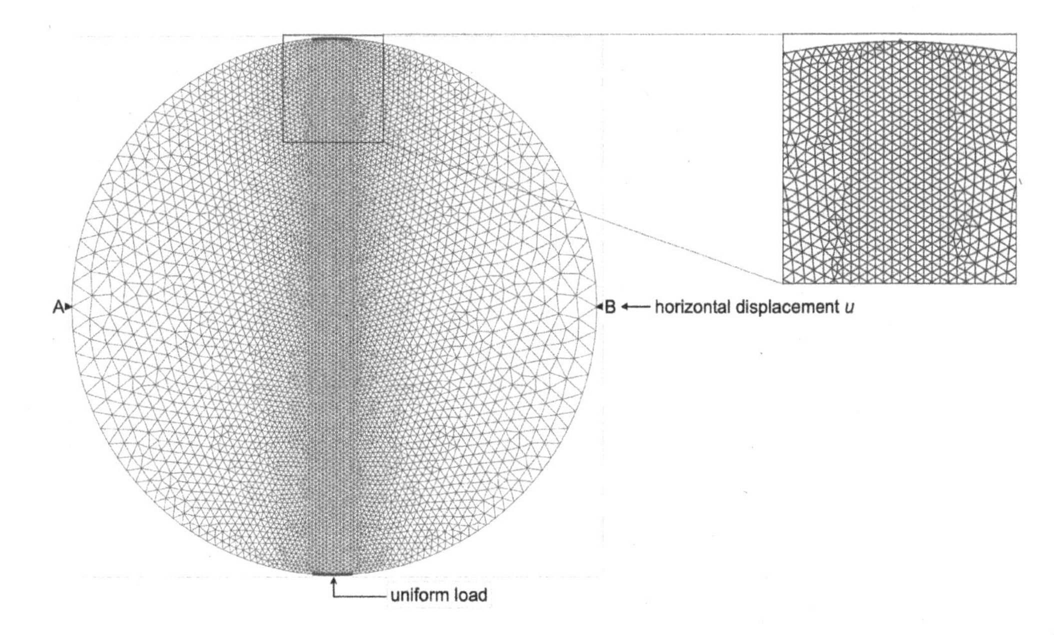

Figure 3. Mesh and boundary conditions in the numerical simulations.

2.3 *Characteristics of the simulations*

Numerical simulations have been carried out within the finite element framework COFE (*Continuum Oriented Finite Element*). It implements elements with an embedded adaptable crack (Sancho et al. 2007; Sancho et al. 2007), in which the crack is allowed to change its direction to adapt to the local stress field, until a given threshold crack width w_{th} is reached. This is calculated as $w_{th} = \alpha' w_1$, where α' is the *adaption factor* of the crack, for which a value of 0.2 was adopted according to the results of (Sancho et al. 2007).

Bi-dimensional models of the specimens were used, as those sketched in Figure 3, in which the mesh was generated by using the program Gmsh (Geuzaine & Remacle 2009), with the meshing algorithm set to "Delaunay". All the elements are constant strain gradient triangles with an embedded cohesive crack. A uniform load was applied at the two lines of nodes marked in blue, simulating the bearing strips of the experiments. The maximum value b/D within the recommended range in (Rocco et al. 1999b) was selected, since it leads to the less brittle behaviour, which is the most unfavourable for the verifications of this work. The simulations were driven by the horizontal displacement u between the two nodes labeled as A and B in the figure, in order to obtain stable calculations. For each family of simulations, a given total displacement was applied in 59 steps with three different magnitudes, in order to capture the local maximum with enough resolution, with a final dimensionless displacement of 1.745 in the case $\ell_2/D = 1.0$, 1.386 for $\ell_2/D = 2.0$ and 1.199 for $\ell_2/D = 4.0$.

Table 1 summarises the range of values of the parameters used to investigate the effect of the material as explained in Section 2.1, and to investigate the size effect as explained in Section 2.2.

Table 1. Parameters of the numerical study, where γ is the ratio of the stress of the kink point of the bilinear curve and the tensile strength, ℓ_2 the selected brittleness length of the material, and D the specimen diameter.

Parameter	Range of values	Effect
γ	0.0, 0.5, 0.6, 0.7, 0.8, 0.9, 0.95	Material effect
ℓ_2/D	1.0, 2.0, 4.0	Size effect

3 RESULTS AND DISCUSSION

In this section, dimensionless plots of results are presented, following the basis described in Section 2.2. From the recorded load P, the nominal stress σ_N is computed, according to the formula of the ASTM-C496 standard as

$$\sigma_N = \frac{2P}{\pi DL} \tag{6}$$

where D is the specimen diameter and L the specimen length. Then dimensionless nominal stress σ_N/f_t is calculated.

Figure 4 shows the curves of dimensionless stress σ_N/f_t versus dimensionless horizontal displacement u/w_1, organised in subfigures by the ratio ℓ_2/D. From the results of the first family, $\ell_2/D = 1.0$, Figure 4(top), it is observed that a local maximum occurs in all the cases for a stress very close to the tensile strength ($\sigma_N/f_t = 1.0$), followed by a softening segment, and a hardening branch in the simulations with $\gamma > 0$, i.e. with a bilinear softening. The effect of the material, which is analysed by modifying the shape of the softening curve through the value of the parameter γ, influences the hardening brach. In particular, as the

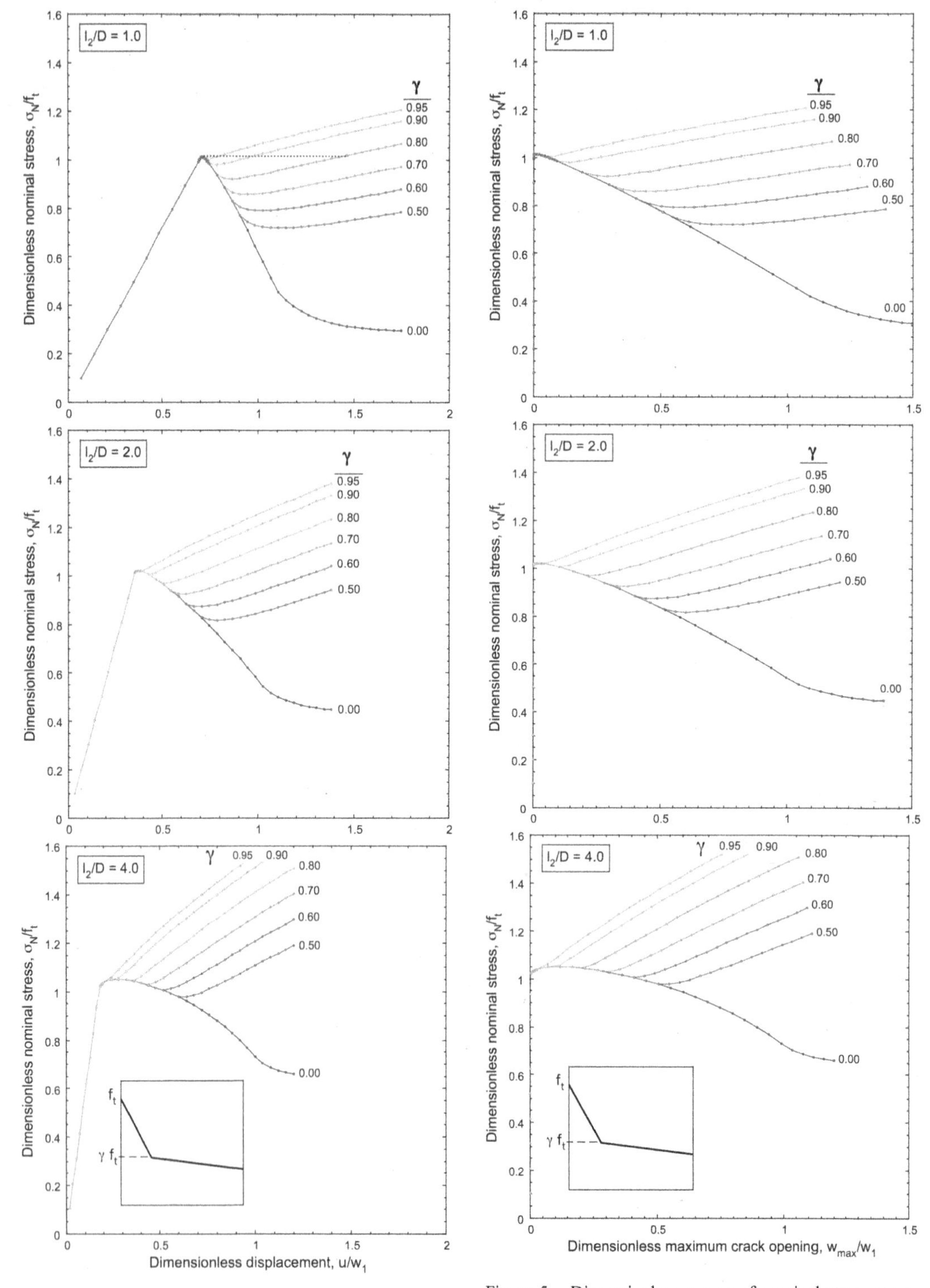

Figure 4. Dimensionless curves of nominal stress versus dimensionless horizontal displacement for ratios of ℓ_2/D equal to 1.0 (top), 2.0 (middle) and 4.0 (bottom) for several values of γ.

Figure 5. Dimensionless curves of nominal stress versus dimensionless maximum crack width for ratios of ℓ_2/D equal to 1.0 (top), 2.0 (middle) and 4.0 (bottom) for several values of γ.

value of γ increases: (1) the stress is greater; (2) as a consequence, the pop-in that would be obtained in load-controlled experiments presents a smaller length, although for this family it is detectable even for the curve corresponding to $\gamma = 0.95$; (3) the stress of the local maximum does not vary.

With regard to the size effect (see the three sub-figures of Figure 4) it is observed that as the ratio ℓ_2/D increases, i.e., as the size of the specimen is smaller or the material is less brittle: (1) the stress of the local maximum slightly grows; (2) the slope of the loading branch of the curves is greater, in accordance with the increase in the elastic modulus of the simulations; (3) the curve corresponding to $\gamma = 0.0$ is less straight; (4) the slope of the hardening branch increases; (5) for a ratio $\ell_2/D = 4.0$ and $\gamma > 0.8$ the local maximum disappears or is very difficult to detect due to the small length of the pop-in, which could be the case of a material exhibiting a high value of the length ℓ_2 and a high content of fibres. This problem could be avoided with a proper test design, by increasing sufficiently the specimen size, in order to obtain an adequate ratio ℓ_2/D which results in a measurable pop-in, precisely taking into account the size effect. Note that using sophisticated techniques could be considered as another solution in order to detect the change in slope and the corresponding stress, but it may entail a greater error, and, whenever possible, the first solution is recommended.

Figure 5 shows the curves of dimensionless stress σ_N/f_t versus dimensionless maximum crack width w/w_1, with the same organisation as the previous figure. Note that similar curves to those in Figure 4 are obtained, except for the elastic contribution of the material. In the new curves, the size effect on the local maximum is more evident. Moreover, completing the list of the previous paragraph, it can be detected that as the ratio ℓ_2/D increases: (6) the crack opening of the local maximum is greater. This effect can be observed more accurately in the zoom views of Figure 6.

Table 2 displays the values of dimensionless nominal stress and dimensionless maximum crack opening corresponding to the local maximum for each ratio ℓ_2/D. Note that the stress of the local maximum differs less than 5.1% from the tensile strength in all the cases, which entails an assumable error. Higher errors might be obtained for higher ratios of ℓ_2/D, but in such cases, it is recommended to modify the size of the specimen, as previously explained, in order to obtain a more brittle behaviour.

In the tests presented in (Sanz et al. 2022) a marked pop-in was detected in all the tests, which means that the size of the specimens was adequate. Further work is in progress in order to determine the main fracture parameter, disclose the ratio ℓ_2/D and the error of the test, and to verify that the local maximum occurs for an opening smaller than that of the kink point of the bilinear curve. Note that in the presented numerical study, this was the case for all the simulations except for $\ell_2/D = 4.0$ and $\gamma > 0.8$.

Table 2. Data of the local maximum, where $l2/D$ is the ratio of the characteristic length and the specimen diameter, σ_N/f_t the dimensionless nominal stress and w_{max}/w_1 the dimensionless maximum crack width.

ℓ_2/D	σ_N/f_t	w_{max}/w_1
1.0	1.014	0.00898
2.0	1.021	0.0230
4.0	1.051	0.120

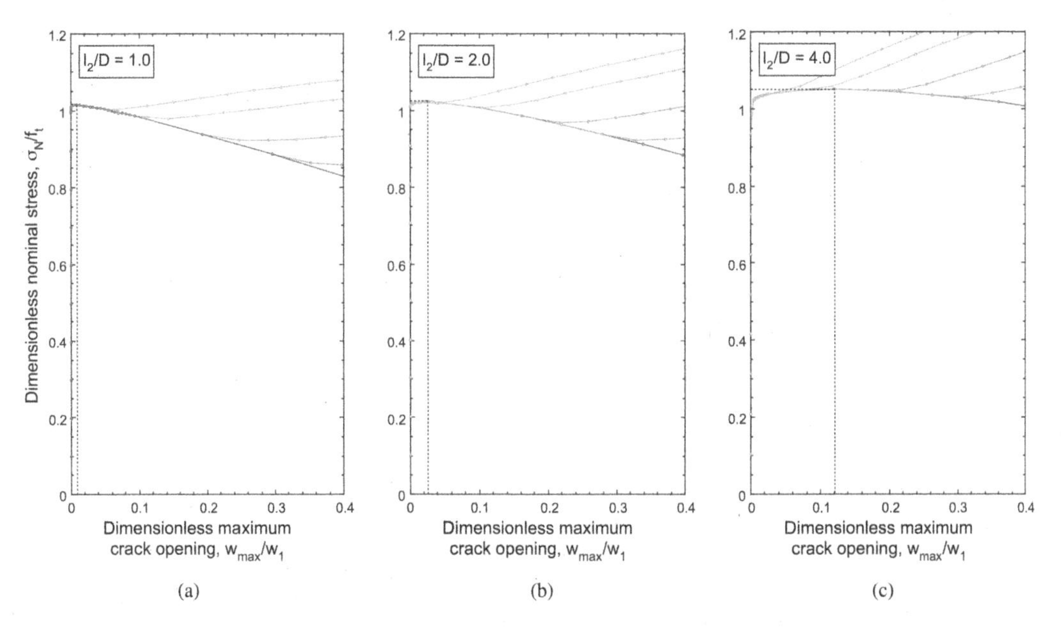

Figure 6. Zoom views of the curves of dimensionless nominal stress versus the dimensionless maximum crack opening.

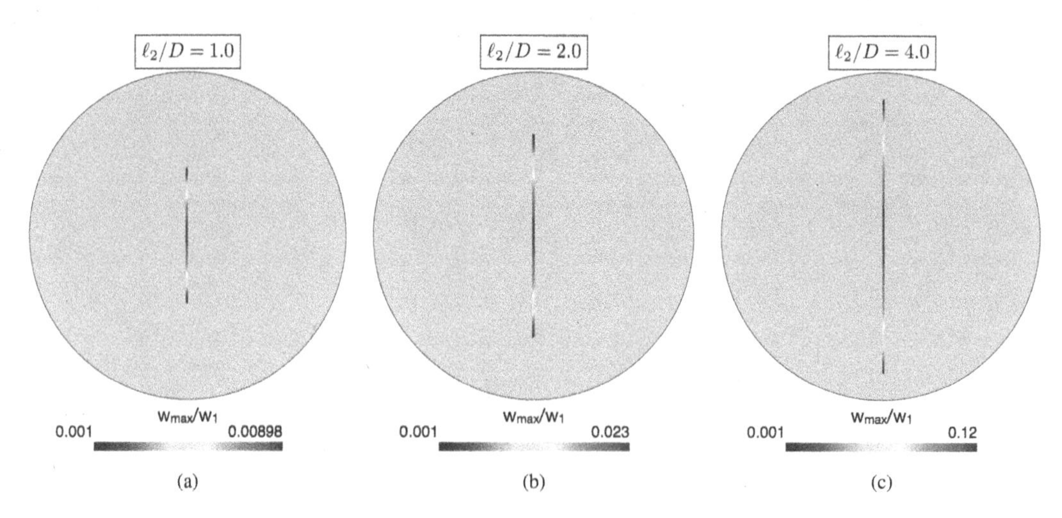

Figure 7. Numerical crack pattern and dimensionless crack width corresponding to the local maximum for $\ell_2/D = 1.0$ (a), $\ell_2/D = 2.0$ (b) and $\ell_2/D = 4.0$ with $\gamma \leq 0.8$ (c).

For completeness of the results, Figure 7 displays the dimensionless crack pattern obtained for the local maximum for $\ell 2/D$ equal to 1.0 and 2.0, and 4.0 with $\gamma \leq 0.8$. In coherence with the results of Table 2, the local maximum of the more brittle behaviour — smaller ratio ℓ_2/D— occurred for a crack less widen and less propagated than that in the other cases.

4 CONCLUSIONS

A numerical analysis of the Brazilian test with application to UHPFRC has been presented, in order to investigate the effect of the material properties, through the modification of the shape of the softening curve, and the size effect, through the modification of the ratio of a brittleness length of the material and the size of the specimen. From the foregoing analysis, the following conclusions are drawn:

- For a wide range of combinations of material properties and specimen size a local maximum exists in a Brazilian test that occurs for a stress very close to the tensile strength. Such maximum can be detected with an adequate instrumentation if the test is properly designed. In stable tests there is a softening branch after the local maximum, followed by a second hardening branch, while in load-controlled tests a pop-in would be obtained.
- An increase in the stress of the kink point of the bilinear softening curve results in a greater stress of the hardening branch and a diminution of the length of the pop-in, but does not affect the value of the local maximum.
- A diminution in the size of the specimen or an increase in the brittleness length results in a slight increment in the stress of the local maximum, a marked one in the crack opening at which it occurs, and a growth in the slope of the hardening branch.
- The study of the influence of the material properties and the size effect may guide in the proper design of the tests, in order to ensure appearance of a measurable local maximum sufficiently proximate to the tensile strength of the material.

ACKNOWLEDGEMENTS

The Authors gratefully acknowledge the *Ministerio de Ciencia, Innovación y Universidades* (MCIU), *Agencia Estatal de Investigación* (AEI) and *Fondo Europeo de Desarrollo Regional* (FEDER) for providing financial support for this work under grant PGC2018-097116-A-I00. The Authors also acknowledge the company *LafargeHolcim, Ductal* for providing the material used in the fabrication of the specimens of this work.

REFERENCES

ASTM-C496 (1990). Standard Test Method for Splitting Tensile Strength of Cylindrical Concrete Specimens. Technical report.

Bažant, Z. P. & J. Planas (1998). *Fracture and Size Effect in Concrete and Other Quasibrittle Materials*. C.R.C. Press, Boca Raton, F.L.

Geuzaine, C. & J.-F. Remacle (2009). Gmsh: A 3-D finite element mesh generator with built-in pre- and post-processing facilities. *International Journal for Numerical Methods in Engineering 79*(11), 1309–1331.

Habel, K., M. Viviani, E. Denarié, & E. Brühwiler (2006). Development of the mechanical properties of an Ultra-High Performance Fiber Reinforced Concrete (UHPFRC). *Cement and Concrete Research 36*(7), 1362–1370.

Hassan, A., S. Jones, & G. Mahmud (2012). Experimental test methods to determine the uniaxial tensile and compressive behaviour of ultra high performance fibre reinforced concrete (UHPFRC). *Construction and Building Materials 37*, 874 – 882. Non Destructive Techniques for Assessment of Concrete.

Hillerborg, A., M. Modéer, & P.-E. Petersson (1976). Analysis of crack formation and crack growth in concrete by means of fracture mechanics and finite elements. *Cement and Concrete Research 6*(6), 773–781.

Hordijk, D. (1991). *Local approach to fatigue of concrete*. Ph. D. thesis, Technical University of Delft.

Nguyen, D. L., G. S. Ryu, K. T. Koh, & D. J. Kim (2014). Size and geometry dependent tensile behavior of ultra-high-performance fiber-reinforced concrete. *Composites Part B: Engineering 58*, 279–292.

Noghabai, K. (1998). *Effect of tension softening on the performance of concrete structures*. Ph. D. thesis.

Park, S. H., D. J. Kim, G. S. Ryu, & K. T. Koh (2012). Tensile behavior of Ultra High Performance Hybrid Fiber Reinforced Concrete. *Cement and Concrete Composites 34*(2), 172–184.

Planas, J., B. Sanz, & J. Sancho (2020). Vectorial stress-separation laws for cohesive cracking: in concrete and other quasibrittle materials. *International Journal of Fracture 223*(77–92).

Planas, J., B. Sanz, & J. Sancho (2021). Numerical analysis of size-effect in UHPFRC beams subjected to wide-span four-point bending. *Revista de Mecánica de la Fractura 1*, 149–154.

Rocco, C. (1996). *Influencia del tamaño y mecanismos de rotura del ensayo de compresión diametral*. Ph. D. thesis, E.T.S.I. Caminos, Canales y Puertos (U.P.M.).

Rocco, C., G. V. Guinea, J. Planas, & M. Elices (1999a). Mechanisms of Rupture in Splitting Test. *ACI Materials Journal 96*(1), 52–60.

Rocco, C., G. V. Guinea, J. Planas, & M. Elices (1999b). Size effect and boundary conditions in the Brazilian test: Experimental verification. *Materials and Structures 32*(3), 210–217.

Roy, D., G. Gouda, & A. Bobrowsky (1972, 05). Very High Strength Cement Pastes Prepared by Hot-Pressing and Other High-Pressure Techniques. *Cement and Concrete Research 2*, 349–366.

Sancho, J., J. Planas, A. M. Fathy, J. C. Gálvez, & D. Cendón (2007). Three-dimensional simulation of concrete fracture using embedded crack elements without enforcing crack path continuity. *International Journal for Numerical and Analytical Methods in Geomechanics 31*(2), 173–187.

Sancho, J. M., J. Planas, D. A. Cendón, E. Reyes, & J. C. Gálvez (2007). An embedded crack model for finite element analysis of concrete fracture. *Engineering Fracture Mechanics 74*(1-2), 75–86.

Sanz, B., J. Planas, V. Rey de Pedraza, R. Sancho, J. M. Sancho, & F. Gálvez (2022). Numerical and experimental study of initiation of cracking of UHPFRC by means of Brazilian tests. *Theoretical and Applied Fracture Mechanics* (accepted for publication).

Toledo Filho, R., E. Koenders, S. Formagini, & E. Fairbairn (2012). Performance assessment of Ultra High Performance Fiber Reinforced Cementitious Composites in view of sustainability. *Materials & Design (1980-2015) 36*, 880–888. Sustainable Materials, Design and Applications.

Wille, K., S. El-Tawil, & A. Naaman (2014). Properties of strain hardening ultra high performance fiber reinforced concrete (UHP-FRC) under direct tensile loading. *Cement and Concrete Composites 48*, 53–66.

Wille, K. & E. Naaman (2010). Fracture energy of UHP-FRC under direct tensile loading. In *Proceedings of the VIII International Conference on Fracture Mechanics of Concrete and Concrete Structures FraMCoS-7, Seoul, Korea*.

Xu, M. & K. Wille (2015). Fracture energy of UHP-FRC under direct tensile loading applied at low strain rates. *Composites Part B: Engineering 80*, 116–125.

Yoo, D.-Y. & Y.-S. Yoon (2016). A Review on Structural Behavior, Design, and Application of Ultra-High-Performance Fiber-Reinforced Concrete. *International Journal of Concrete Structures and Materials 10*(2), 125–142.

Yudenfreund, M., I. Odler, & S. Brunauer (1972). Hardened portland cement pastes of low porosity I. Materials and experimental methods. *Cement and Concrete Research 2*(3), 313 – 330.

Computational Modelling of Concrete and Concrete Structures – Meschke, Pichler & Rots (Eds)
© 2022 Copyright the Author(s), ISBN: 978-1-032-32724-2

Multiscale modeling of the mesotexture of C-S-H and ASR gels

S. Ait-Hamadouche & T. Honorio
Université Paris-Saclay, CentraleSupélec, ENS Paris-Saclay, CNRS, LMPS – Laboratoire de Mécanique Paris-Saclay, Gif-sur-Yvette, France

ABSTRACT: Calcium silicate hydrates (C-S-H) and Alkali-Sillica Reaction (ASR) products exhibit a gel mesostructure in which solid colloidal particles are arranged in the mesoscale forming a phase with mesoporosity. The mesotexture determines the properties of the gel. Modeling the gel scale is challenging since the system sizes required to have a representative volume of the gel can be prohibitive for full atomistic simulations, whereas the approaches based on the continuum mechanics (e.g., homogenization) need to be proven valid at that scale. Coarse-grained (CG) simulations, in which the particles represent chunks of the phase considered interacting *via* potentials of mean force, enable assessing the gel scale. So far, no study has proposed CG simulations to understand ASR gels' mesostructuration. CG simulations have been successfully applied to model C-S-H, but to date, the flexibility of the layers has not been taken into account. Here, we deploy CG simulations to study the mesotexture of ASR and C-S-H gels using the effective interactions identified at the molecular scale. In the case of C-S-H, we propose an original strategy to incorporate the flexibility of the layers. The CG simulations provide configurations of gels at various packing densities, and these configurations are used to assess the structural features and properties at the gel scale.

1 INTRODUCTION

Calcium silicate hydrate (C-S-H) is the main product of cement hydration and is responsible for several important properties of concrete, including setting, hardening, shrinkage, and creep. Alkali-silica reaction (ASR) products are formed from the reaction between the alkali present in the cement paste (generally, sodium or potassium) and the disordered silica present in some of the aggregates (e.g., (Rajabipour et al. 2015)). Both C-S-H and crystalline ASR products are calcium silicates in a hydrated form and present a layered molecular structure (i.e., they are phyllosilicates). C-S-H molecular structures share similarities with defective structures of tobermorite or jennite (Richardson 2004) as a function of the Ca/Si molar ratio, while crystalline ASR products structures are similar to that of shlykovite (Shi et al. 2019). Both C-S-H and ASR products present a gel mesostructure in which solid colloidal particles (i.e., particles with a characteristic size of 1-100 nm (Buckley & Greenblatt 1994)) are arranged in the mesoscale forming a phase with a gel mesoporosity.

Several studies focus on the molecular modeling of the various phases in cement systems. Full atomistic simulations, i.e., simulations in which each atom is explicitly represented, have been deployed to successfully model various properties of the phases relevant to the cement system, including C-S-H (Mishra et al. 2017) and crystalline ASR products (Honorio et al. 2020, 2021; Kirkpatrick et al. 2005). Simulating the gel scale is a more challenging task because the system sizes required to have a representative volume of the gel can be prohibitive for full atomistic simulations (the characteristic size of the gel scale being comprised in the range 100 nm – 10 μm). A strategy to cope with this limitation is employing a *coarse-grained* (CG) representation of the gel. In coarse-grained simulations, the particles represent chunks of the phase considered interacting *via potentials of mean force* (PMF), i.e. potentials that captures the effective interactions from the molecular scale (Ioannidou et al. 2017). Several studies have focused on investigating C-S-H gel behaviour using coarse-grained simulations (Goyal et al. 2020; Ioannidou et al. 2016, 2017; Liu et al. 2019; Masoero et al. 2012, 2013; Masoumi et al. 2020). To date, the majority of approaches for CG simulations of the C-S-H gel relies on a representation of C-S-H grains as mono- or poly-dispersed spheres (Goyal et al. 2020; Ioannidou et al. 2016; Masoero et al. 2013) or by rigid ellipsoidal particles (Masoumi et al. 2020, 2017b; Yu et al. 2016). Experimental evidence shows that C-S-H exists with different mesotextures: including nanometric spherical grains, fibrils, or foils (Richardson 2004), with 2D foils being one of the most prevalent morphologies observed in transmission electron microscopy (TEM) images. Studies on other phyllosilicates (e.g., clays (Honorio et al. 2018)) suggest that the flexibility of the layers can play a significant role in mesostructuration. And experimental

 DOI 10.1201/9781003316404-17

evidence shows that C-S-H layers may appear in bent configuration (Marty et al. 2015). To the best of the author's knowledge, there are no simulations of the gel scale in the literature that take into account the flexibility of the C-S-H layers.

The physical origin of the expansive behavior, the structure, and the properties of the ASR products have not yet been completely understood (Rajabipour et al. 2015). Recently, the effective interactions between the layers of crystalline ASR products have been computed at ambient temperature for liquid water-saturated system (Honorio et al. 2020). This study shows that the effective interactions in crystalline ASR products are depended on the charge balance cation (Na or K). To date, there is no study in the literature dealing with the modeling of ASR products at the gel scale.

The objective of this study is to deploy coarse-grained simulations to study the mesostructuration of (i) *ASR gels*, using the effective interactions identified for both Na- and K-shlykovite (Honorio et al. 2020); (ii) *C-S-H gel*, using the effective interactions (Honorio et al. 2021) identified for a realistic C-S-H model (Kunhi Mohamed et al. 2018), and taking into account the flexibility of C-S-H layers. The CG simulations enable obtaining configurations of gels at a function of the packing density η. These configurations are then used to assess the structural features and properties at the gel scale. We compare our simulation results with experimental data from the literature whenever possible.

2 MATERIALS AND METHODS

2.1 *ASR gels*

We adopt a multi-scale approach by collecting information at the molecular scale, namely the effective interaction of PMF, and using it as input at the gel scale. The PMF was computed using the atomic structure of (K-)shlykovite resolved by Zubkova et al. (Zubkova et al. 2010); the most relevant alkali in ASR, potassium, and sodium, are considered as charge-balancing cations in shlykovite in the simulations (Honorio et al. 2020). The effective interactions between two layers of K- and Na-shlykovite products are obtained using hybrid GCMC-MD simulations considering ambient temperature and liquid saturate conditions (RH = 100%) (Honorio et al. 2020). This PMF represents the interactions between the ASR solid layers in a drained environment (i.e., water is allowed to ingress or leave the pore in systems controlled according to the interlayer distance, as it is generally done for nanolayered materials (Bonnaud et al. 2016; Honorio et al. 2017, 2019; Masoumi et al. 2017a)). Figure 1 shows the effective interactions for both K- and Na-shlykovite.

The use of PMFs computed from crystalline systems to represent the interaction at the gel scale relies on the assumption that the ASR gel is constituted of particles presenting at least some of the structural features of crystalline products. In (Honorio et al. 2020),

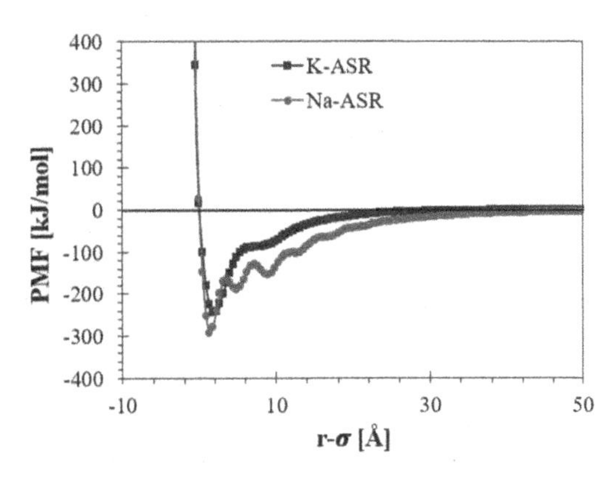

Figure 1. PMF of Na-shlykovite (a) and K-shlykovite (b) as a function of $r - \sigma$.

comparisons of diffraction patterns and pair distribution functions obtained from simulated crystalline systems (based on shlykovite structure) with data from experiments in ASR gels (Benmore & Monteiro 2010) shows that both systems indeed share some structural features.

2.1.1 *Coarse-grained simulations*

Gaboriaud et al. (Gaboriaud et al. 1999) showed that colloidal species in ASR gel could be represented by spheres with the same radius close to 10 Å, whatever the alkaline ions (Li^+, Na^+, or K^+). Therefore, we represent ASR gel as an ensemble of monodisperse spherical grains with a diameter $\sigma = 20$ Å. The use of monodisperse spheres limits the maximum packing density that can be obtained in simulations under zero stress (e.g., (Torquato 2002)): $\eta = 0.64$ for the maximum random packing density and 0.74 for the close-packing for face-centered cubic (fcc) or hexagonal close-packed (hcp) spheres. ASR gels are reported to present a porosity ϕ ranging from 0.4 to 0.8 (Geng et al. 2021), which corresponds to a packing density ($\phi = 1 - \eta$) ranging from 0.2 to 0.6. Monodisperse spheres can fully describe this range of packing density.

The formation of ASR gel via precipitation and aggregation of the spherical grains is simulated here by using hybrid grand canonical Monte Carlo and Molecular Dynamics (GCMC-MD) simulations, as in (Ioannidou et al. 2014). We use LAMMPS software (Plimpton 1995). Starting from an initially empty cubic box of size L = 1000 Å, the simulation box is filled with grains during the GCMC stages following the usual probabilities distributions for insertion (Frenkel & Smit 2002) $P = min\{1, exp(-\frac{(\Delta U - \mu \Delta N)}{k_B T})\}$ where k_B is the Boltzmann constant, T the temperature, ΔU the variation in potential energy caused by the trial insertion/removal, ΔN the variation in the number of ASR grains and μ the chemical potential. For the chemical potential, we have chosen a value that favors the insertion of grains in the simulation box and drives the system towards progressive densification.

Coupling the GCMC insertion/removal events to MD simulation enables investigating specifically the development of the gel properties process under non-equilibrium conditions (Ioannidou et al. 2014). In this context, each GCMC cycle consists of N_{MC} attempts of particle insertion or deletion through MD simulation, which is conducted in the canonical ensemble (NVT) and then the isothermal-isobaric ensemble (NPT). A timestep of 1 fs is adopted. Nosé-Hoover thermostats and barostat are adopted, when relevant, in NVT and NPT simulations. The GCMC process was invoked every $N_{MD} = 100$ MD steps, and in each GCMC stage, $N_{MC} = 100$ attempts for deletion or insertion of the grains are performed. The ratio $R_{prec} = \frac{N_{MC}}{N_{MD}}$ between N_{MD} and N_{MC} is related to the precipitation kinetics rate, i.e. the rate of producing hydrates as determined by the chemical environment (Ioannidou et al. 2016). No experimental value for R_{prec} can be found in the literature for ASR products, therefore we adopt typical R_{prec} used in CG simulation for other silicate gels (Hou et al. 2021; Ioannidou et al. 2016).

2.2 *C-S-H gels*

As for ASR gels, a multi-scale strategy is also adopted to investigate the mesostructuration of C-S-H gel, but here we follow an original approach to take into account the flexibility of C-S-H layers. At the molecular scale, the elastic properties of C-S-H are computed to obtain layers' longitudinal rigidity and flexibility. Then the PMF associated with C-S-H is described. Finally, this information is used as input in the coarse-grained simulation.

At the molecular scale, we adopt the model of Kunhi et al. (Kunhi Mohamed et al. 2018) with molecular formula $C_{1.7}SiO_{3.7} \cdot 1.3H_2O$, consistent with the values of Ca/Si ratio of C-S-H observed in Portland cement (Richardson 2004). The interactions among atoms in C-S-H are described by using ClayFF (Cygan et al. 2004) and SPC/E water model (Berendsen et al. 1987).

2.2.1 *Flexibility of C-S-H layers*

To account for the C-S-H layer's flexibility, we consider the C-S-H solid layer as a thin homogeneous layer. This approximation is valid given the fact that C-S-H layer length L is much larger than its thickness h_s and the characteristic size of atoms (i.e., its atomic granularity), and by assuming that: (i) L is much smaller than the persistence length ξ_p, (ii) C-S-H sheet is symmetric to the midplane and (iii), there are no topological changes under moderate loads. In this context, we can use the thin plate theory in which the energy of shear is neglected, and the free energy associated with bending takes the form (Honorio et al. 2018):

$$U = \frac{1}{2} \sum_{i,j,k,l \in \{x,y\}} D_{ijkl} \frac{\partial^2 u_z}{\partial i \partial j} \frac{\partial^2 u_z}{\partial k \partial l} \tag{1}$$

where D_{ijkl} is the bending modulus, and u_z is the displacement orthogonal to the plane of the plate. The bending modulus is a key property in the characterization of phyllosilicate nanotexture (Honorio et al. 2018). For a thin plate of thickness h the latter parameter is determined by:

$$D_{iiii} = \frac{h^3 E_i}{12(1 - \nu_{ij}\nu_{ji})} \tag{2}$$

with $h = 12.85$ Å being the C-S-H layer thickness, E_i the Young's modulus in the i direction and ν_{ij} the Poisson's ratio in the i direction under a load in the j direction. Hence, we need to determine the elastic properties E and ν to estimate the bending modulus. To do so, we perform a minimization simulation with LAMMPS in which the system is first relaxed under Parinello-Rahman barostat. Then, imposed displacements are performed according to each one of the six directions of interest to compute the full (symmetric) stiffness tensor (xx, yy, zz, xy, xz, yz) . Both positive and negative displacements are performed in each case. The displacement is chosen to keep the same deformation level (on the order of 10^{-5}) in each direction. Voigt-Reuss-Hill (VRH) approximation is used to compute (quasi-)isotropic values for E and ν and compare with the available experimental data from indentation.

Table 1 shows our values of elastic constants and values from the literature for crystalline calcium silicate hydrates (tobermorite 11 Å, tobermorite 14 Å and jennite (Shahsavari et al. 2009)) and another molecular model of C-S-H (cCSH) (Pellenq et al. 2009), based on defective crystalline structures. Our results are generally closer to those of the cCSH model.

The VRH estimate of the elastic properties are gathered in Table 2, along with experimental values of Poisson's ratio ν and indentation modulus $M = 4G\frac{3K+G}{3K+4G}$

Table 1. Comparison between elastic constants of C-S-H (model based on defective tobermorite), and crystalline calcium silicates hydrates as reported in the literature.

Elastic const. [GPa]	Our values	cCSH	Tober. 11 Å	Tober. 14 Å	Jennite
C_{11}	72.74	93.49	116.95	77.6	100.1
C_{22}	81.15	94.87	126.1	104.5	45.7
C_{33}	83.33	68.46	126.35	32.05	59.15
C_{12}	21.24	45.37	45.83	35.9	26.85
C_{13}	41.85	26.07	27.88	20.18	32.03
C_{23}	18.62	30.06	46.2	26.3	4.4
C_{44}	23.33	19.22	30.2	24.5	21.95
C_{55}	17.71	16.11	20.75	14.65	21
C_{66}	23.49	31.23	44.35	38.1	26.55
C_{14}	6.35	0.58	0	0	1.3
C_{15}	−6.84	−0.05	0	0	−6.2
C_{16}	4.15	1.26	0.3	3.08	3.3
C_{24}	1.09	−4.6	0	0	7.35
C_{25}	6.34	1.79	0	0	−6.2
C_{26}	4.67	−3	−14.93	−1.75	−3.18
C_{34}	9.09	−4.6	0	0	−1.3
C_{35}	3.61	1.79	0	0	1.4
C_{36}	0.68	−0.57	−9.35	3.03	0.07
C_{45}	0.72	0.33	−11.1	−9.43	1.73
C_{46}	1.68	1.82	0	0	−1.6
C_{56}	3.61	−0.4	0	0	2.73

Table 2. Elastic properties of C-S-H: comparison between VRH approximation with experiments on cement hydrate samples.

Elastic constants	VRH approx.	Experiment
K [GPa]	43.7	–
G [GPa]	22.5	–
E [GPa]	55.3	–
M [GPa]	60.4	65
ν [–]	0.29	≈0.3

(a parameter to measure material compressive stiffness and relates to bulk and shear moduli (Pellenq et al. 2009)). The indentation modulus and Poisson ratio obtained in our simulation are in agreement with the experiments.

To account for the flexibility of the C-S-H layer, we calculate the rigidities k_{2b} and k_{3b}. We consider the C-S-H layer as a structure discretized in spherical mass points, with 2-body interactions defining the stretch potential between mass points i and j (U_{2b}). Similarly, 3-body interactions define bending potential between mass points i, j and m (U_{3b}). Then this approach is combined with linear elastic beam and plate theory by equating the harmonic potentials used in CG simulations with the free energy expressions of beam theory (Keremides et al. 2018). Therefore k_{2b} and k_{3b} take the forms:

$$k_{2b} = EA/L; \quad k_{3b} = 12EI/L \tag{3}$$

where E is the (in-plan) Young's modulus, $L = 12.85$ Å $(=\sigma)$ is the distance between the center of two mass points (taken as the diameter of each spherical mass), $A = \pi\sigma^2/4$ is section of each mass point, and $I = \frac{1}{4}\pi\,(\sigma/2)^4$ is moment of geometric inertia of each mass point. The bending modulus D and rigidities obtained with this approach are displayed in Table 3.

Table 3. Bending modulus (D) and rigidities (k_{2b} and k_{3b}) of C-S-H layers.

D [N.m]	1.66×10^{-17}
k_{2b} [kcal/mol.Å^2]	80.3
k_{3b} [kcal/mol.rad^2]	9950

Our value of bending modulus is in the same order as that found in a previous study on the flexibility of tobermorite (Honorio & Brochard 2017). To the best of our knowledge, there is no other study where the rigidities of C-S-H layers were calculated. To verify our results, we performed a coarse grained simulation on a single fibril (chain of spherical grains) in which a flexion effort was generated through an imposed displacement $\delta_i = \delta_{i-1} + \Delta\delta$, with $\Delta\delta = 0.02$ Å at the distance of one-quarter from each edge of the chain. Based on the Kirchhoff-Love plate theory, for this configuration the elastic free energy is a quadratic function of the displacement δ: $U = 6\left(\frac{4}{L}\right)^3 D\delta^2$. Our values are in agreement with Kirchhoff-Love plate theory and also the expression proposed by (Prathyusha et al. 2018) linking D and k_{3b}: $D = k_{3b} \times 2\frac{\sigma}{L_{tot}}$ (with L_{tot} being the length of the spherical grain chain)

2.2.2 *Inter- and Intra-particules interactions*

The PMF for C-S-H representing the effective interactions between the layers under ambient temperature and (liquid) saturated conditions is taken from (Honorio et al. 2021)(Figure 2), where hybrid GCMC-NVT simulations were performed on the C-S-H model of Kunhi et al. (Kunhi Mohamed et al. 2018) using as reaction path the interlayer distance. Critical information for the mesoscale simulations can be derived from the PMF: (i) the well depth ε, which is a measure of how strongly two particles attract each other, and (ii) the distance at which the potential crosses zero σ representing the distance between two C-S-H layers when they touch each other.

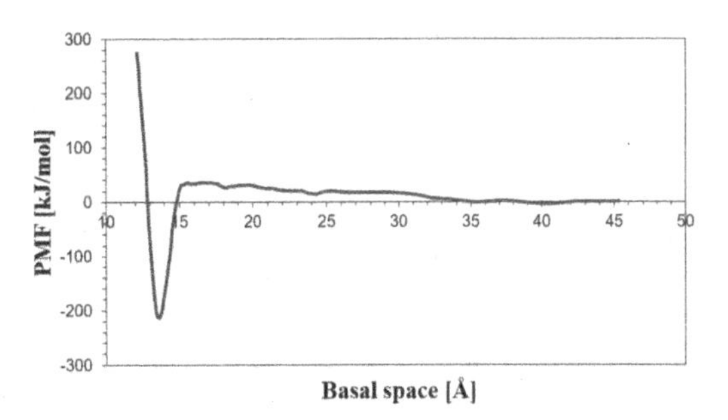

Figure 2. Potential of mean force (PMF) of C-S-H as a function of basal spacing from (Honorio et al. 2021).

Here, we fit the PMF with a Mie (or generalized Lennard-Jones) potential:

$$U_{LJ-G} = 4\varepsilon\left[\left(\frac{\sigma}{r}\right)^{2\alpha} - \left(\frac{\sigma}{r}\right)^{\alpha}\right] \tag{4}$$

with $\sigma = 12.85$ Å, $\varepsilon = 200$ kJ/mol, and $\alpha = 20$ being a coefficient that controls the narrowness of the potential well.

Intra-particle 2- and 3-body interactions are modeled respectively by the harmonic potentials:

$$U_{2b}(r) = k_{2b}\,(r - r_0)^2\,; \quad U_{3b}(\theta) = k_{3b}\,(\theta - \theta_0)^2 \tag{5}$$

where the rigidities are computed from the in-plan Young modulus as detailed above.

2.2.3 *Coarse-grained simulations*

The details of the morphological representation based on the 2D foil-like particles and the CG simulations are presented in this section. The grain size was chosen based on experimental values and in agreement with the PMF (Maruyama et al. 2021). Indeed σ is considered to be the grain's diameter with a value of 12.85 Å, and the disk thickness since the latter is filled with the grains, while the diameter's disk value is 308.4 Å, which is close to the average diameter D $\approx$ 320 Å reported by Maruyama et al. (Maruyama et al. 2021) for a disk form in C-S-H gel.

In order to create a single 2D C-S-H disk, spherical grains are generated in a cubic lattice with a lattice constant equal to σ. Covalent bonds with rigidity k_{2b} are created for all first neighbors in the cubic lattice. Three-body interactions are defined with a rigidity k_{3b} for all relevant triples of grains. Since the resulting structure is hypostatic, we also create covalent bonds with a k_{2b} (computed for a distance $L = \sqrt{2}\sigma$) for all first neighbors in the diagonal of the cubic lattice to brace the resulting layout. This single 2D disk is used as a template to generate the initial mesostructure.

In a simulation box of 1000 nm, 400 disks are randomly placed. Then, we adopt the method proposed by (Masoumi et al. 2020) using high-frequency pressure oscillations and annealing in MD simulations for bringing the system to the most favorable packing density. We use the Nosé-Hoover thermostat and barostat. We begin our simulation with small timesteps to avoid numerical instabilities, then we impose the following cycle of simulation: (i) NVT simulations at a temperature $T = 3000$ K and then at ambient conditions ($T =$ 300 K); and (ii) NPT simulations at $T = 300$ K with at first a pressure $P = 200$ atm, followed by a gradual decrease to 1 atm.

To investigate the effect of layers flexibility on the meso-structuration of C-S-H gel, we launched three simulations named F01, F1, and F10, meaning that the rigidities used as input in coarse-grained simulations were multiplied respectively by 0.1, 1, and 10.

2.3 *Structural features at mesoscale*

The pore size distribution is determined by using the algorithm PSDsolv (Bhattacharya & Gubbins 2006), which is based on a Monte Carlo approach coupled with nonlinear optimization. The isosurfaces and specific surface area were determined using the alpha-shape algorithm described in (Stukowski 2014) as implemented in Ovito program.

2.4 *Mechanical properties at mesoscale*

2.4.1 *Elasticity*

Elastic constants are computed using the same strategy adopted at the molecular scale for C-S-H. We perform a minimization simulation with LAMMPS in which the system is first relaxed; then, to each of the six directions of interest, displacements are imposed, and the pressure is sampled in the system to compute the full (symmetric) stiffness tensor. Due to long-range disorder, the response is expected to be isotropic at the mesoscale.

2.4.2 *Viscosity*

In molecular simulations, the shear and bulk viscosities can be computed using the Green-Kubo formalism with the expressions below. For the shear viscosity (Medina et al. 2011; Allen and Tildesley 1989):

$$\eta_{shear} = \frac{1}{5}\sum_{\alpha,\beta} \lim_{t\to\infty} \frac{V}{k_B T}\int_{t=0}^{\infty} \langle p_{\alpha\beta}(t) p_{\alpha\beta}(0)\rangle dt \qquad (6)$$

where $p_{\alpha\beta}$ are three out-of-diagonal components of stress tensor p_{xy}, p_{xz}, and p_{zy}, and the differences between the diagonal elements $(p_{xx} - p_{yy})/2$ and $(p_{yy} - p_{zz})/2$. For the bulk viscosity (Allen and Tildesley 1989):

$$\eta_{bulk} = \frac{1}{3}\sum_{\alpha,\beta} \lim_{t\to\infty} \frac{V}{k_B T}\int_{t=0}^{\infty} \langle p_{\alpha\beta}(t) p_{\alpha\beta}(0)\rangle dt \qquad (7)$$

where $p_{\alpha\beta}$ are three diagonal components of stress tensor p_{xx}, p_{yy}, and p_{zz}. The stress auto-correlations in these expressions decay fast, allowing adopting an upper integration limit of a few tens of picoseconds.

3 RESULTS AND DISCUSSIONS

3.1 *ASR gels*

As an illustration of the mesostructuration process, Figure 3 displays the network of particle forming clusters as a function of the packing density η. The space occupied by the solid and the void phases is more visible in Figure 4, which shows the isosurfaces of the gel. The maximum packing fraction simulated in this work for both Na-ASR gel and K-ASR gel is $\eta = 40\%$, which means a porosity of 60%. To the best of the author's knowledge, there are no experimental values for the porosity of ASR gel. However, our results are in agreement with those of Geng et al. (Geng et al. 2021) who estimated that the porosity of the ASR gel is between 40 and 80% using the Mori-Tanaka homogenization scheme.

The pore size distributions (PSD), in Figure 5, show that as packing fraction increases, the average pore size decreases for both Na-ASR and K-ASR gel. From $\eta = 30\%$, the emergence of two populations can be noticed: (i) the pores between the grains in a close-packing (which could be associated with the porosity of systems in equilibrium basal spacing) and (ii) the gel pores *per se*. At the maximum packing fraction simulates here, the size of the pores between the grains is around 6 Å for Na-ASR and K-ASR, while the gel pore size is approximately 28 Å for Na-ASR and 24 Å for K-ASR. This observation suggests that the charge balance cation plays a significant role in the pore structure of ASR gels. It is noteworthy that even small changes in pore size can lead to significant changes in effective transport properties such as the permeability (the permeability scale as $K_{perm} \propto \phi R_p^2$, e.g., (Nishiyama & Yokoyama 2017)) even if the total porosity is kept constant.

Figure 6 shows how the SSA varies as a function of η. First, when η increases, the SSA increases, indicating that the grains appear mainly as independent particles (i.e., not a cluster). Then, when the

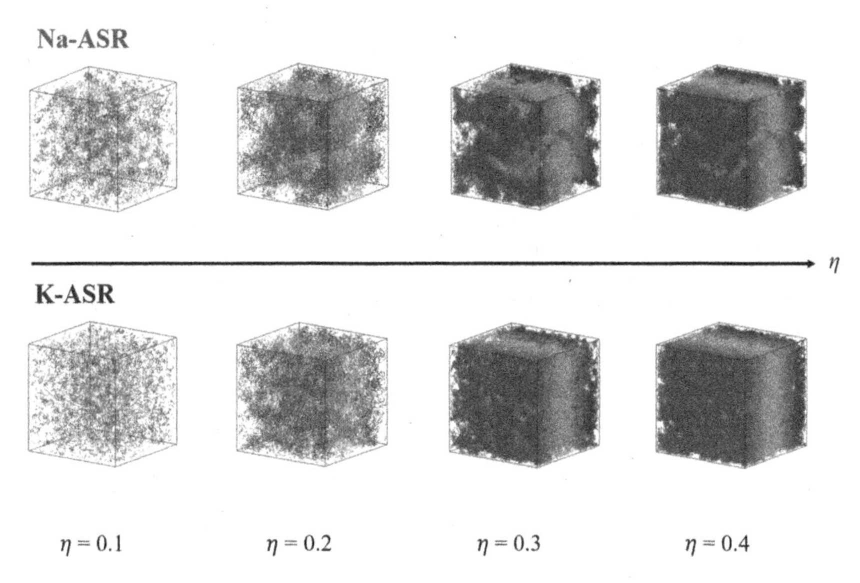

Figure 3. Mesostructuration of Na-ASR and K-ASR gels as a function of packing fraction η.

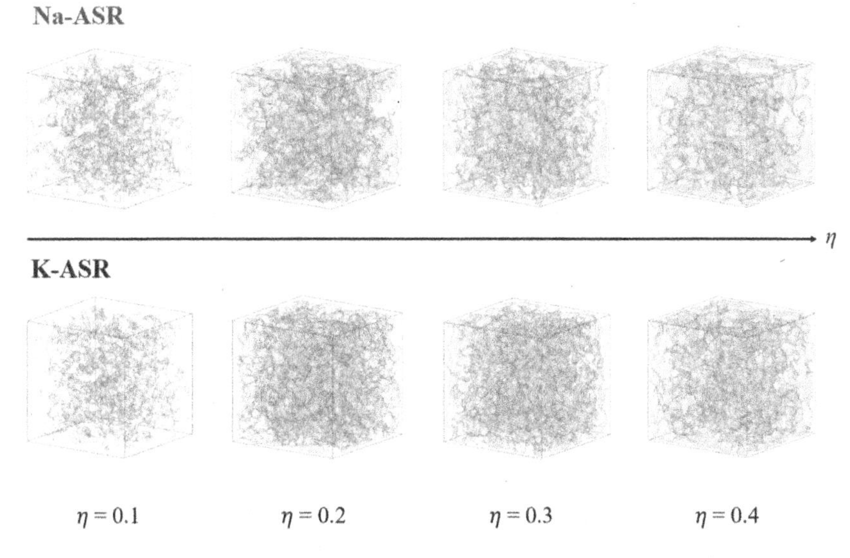

Figure 4. Isosurfaces of Na-ASR and K-ASR gels as a function of packing fraction η.

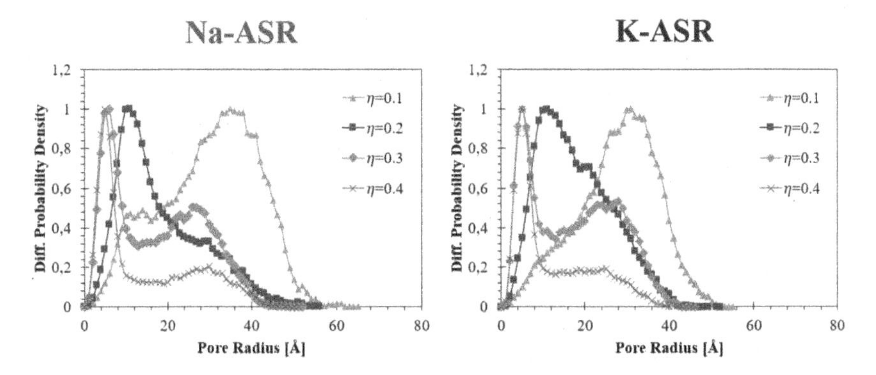

Figure 5. Pore size distribution of Na-ASR and K-ASR gels.

grains begin to aggregate and form clusters, the SSA decreases. We remark that the grains start to aggregate from a packing fraction of 15% for both Na-ASR and K-ASR gel. This process coincides with a significant increase in both bulk and shear viscosity, as can be seen in Figure 7, which suggests a transition from a liquid to an arrested solid-like behavior. In that direction, Del Gado et al. (Del Gado et al. 2002) point out that the bulk viscosity diverges at the percolation. For small values of η, both Na-ASR and K-ASR gel viscosities are low, but their values begin to increase from a packing fraction of roughly 24%, with a higher viscosity for Na-ASR than K-ASR.

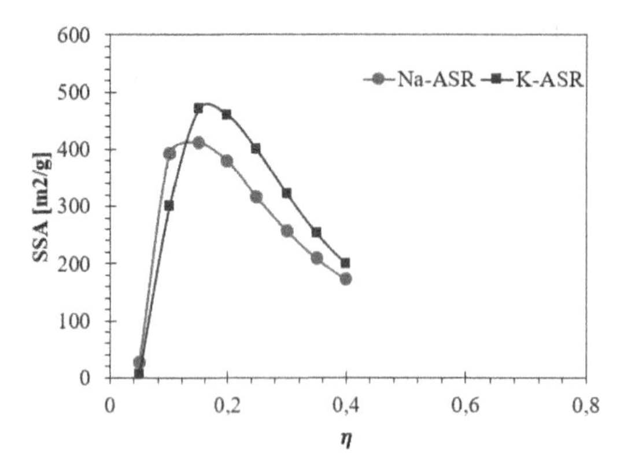

Figure 6. Specific surface area of Na- and K-ASR gels as a function of the packing density.

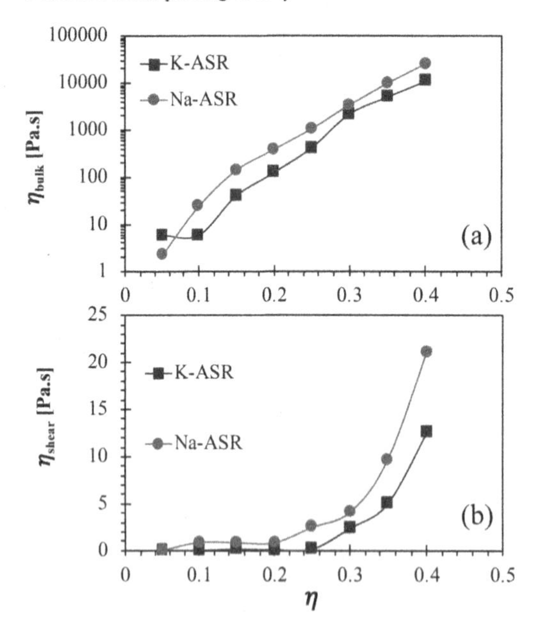

Figure 7. Evolution of (a) the bulk viscosity and (b) the shear viscosity as a function of the packing density.

The evolution of Young modulus as a function of the packing density at zero stress η_0 is shown in Figure 8 for both Na- and K-ASR gels. The results follow the self-consistent (SC) estimated for spherical inclusions (for these estimates, we have used the VRH estimates for the bulk and shear moduli obtained from molecular simulations on Na and K-shlykovite from (Honorio et al. 2020)).

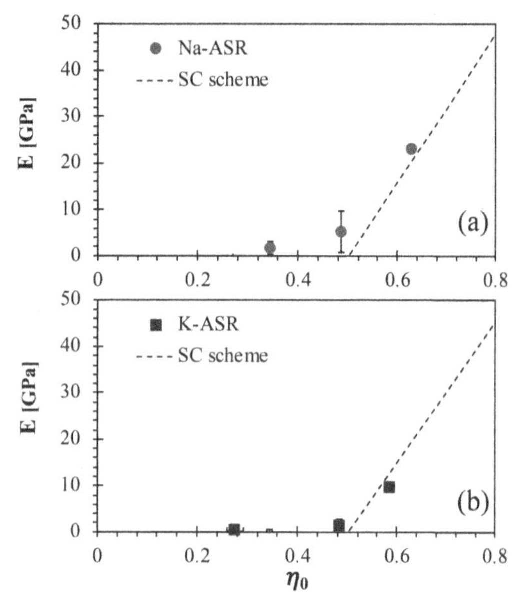

Figure 8. Evolution of Young modulus as a function of the packing density at zero stress η_0 of (a) Na- and (b) K-ASR gels: comparison with the self-consistent estimate. The VRH bulk and shear moduli obtained from molecular simulations on Na and K-shlykovite were used as input in the SC scheme.

3.2 *C-S-H gels*

The aggregation of C-S-H layers forming the C-S-H gel is displayed in Figure 9 (a). The three configurations F01, F1, and F10 composed of stacks of layers have differences that can be visually identified. For example, the layers in F01 being less stiff appears in more pronounced bent/curved configurations than those in F1 and F10. Moreover, for F1, there is predominately face-to-face contact between the layers, while for F10 there is also face to edge contact. The number of layers in stacks had been investigated in some experimental and simulation studies. Chiang et al. (Chiang et al. 2012) experiments found an average number of layers $n = 10.9$ for a C-S-H microstructure at water content of 30%, while Masoumi et al. (Masoumi et al. 2020) simulations found clusters containing up to forty stacked layers at $\eta = 70\%$. In our simulations, it can be visually identified stacks constituted of at least 10 layers. Finally, Figure 9 (b) gives a better view of the space occupied by the solid part and the remaining void.

The pore size distribution is plotted in Figure 10. The population of pores with a radius of about 3 Å represents the voids between the grains inside the disks or neighboring particles in close packing. This population has a similar prevalence in the three configurations. When compared to F1 and F10, the configuration F01 shows a smaller number of intermediary pores with a radius between 9 and 40 Å

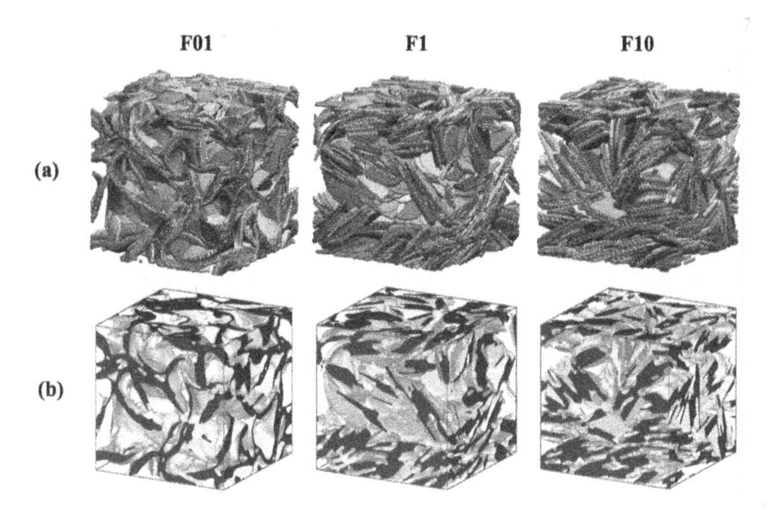

Figure 9. Layers aggregation (a) and isosurface (b) of C-S-H gel according to the flexibility of the layerss for $\eta = 0.25$.

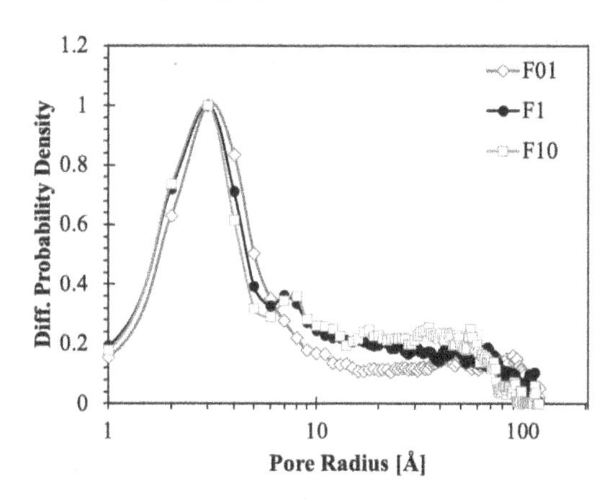

Figure 10. PSD of C-S-H gel according to the flexibility of the layers for $\eta = 0.25$.

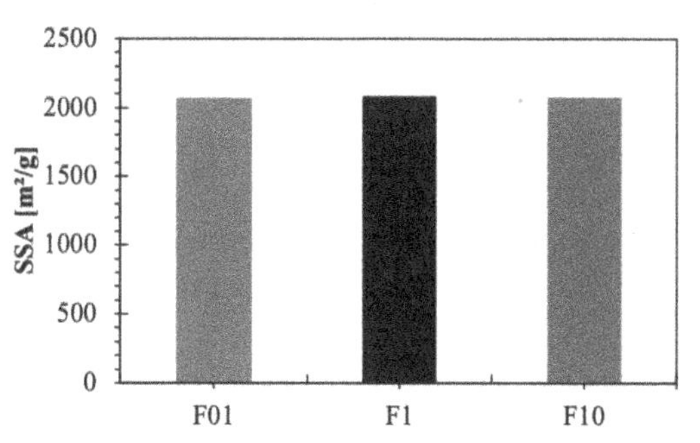

Figure 11. Specific surface area of C-S-H gel according to the flexibility of the layerss for $\eta = 0.25$.

but a slightly larger frequency of large pores with a radius 90–100 Å. As discussed in the case of ASR gels, having a population of larger pores lead to a critical increase in the permeability since this property scales with the square of pore radius.

The SSA shows no significant dependence on the flexibility of the layers (Figure 11).

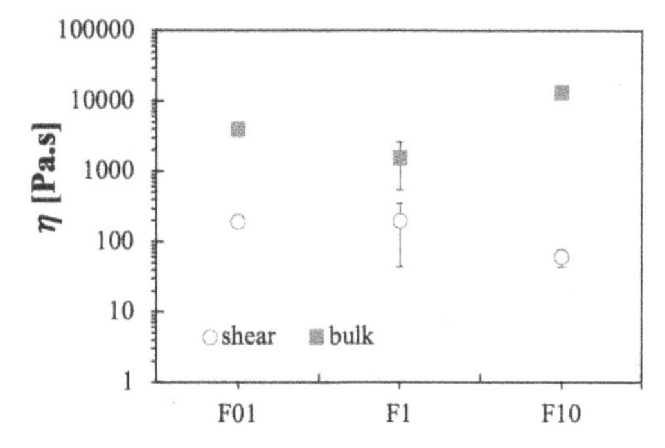

Figure 12. Bulk and shear viscosity of C-S-H according to the flexibility of the layers for $\eta = 0.25$.

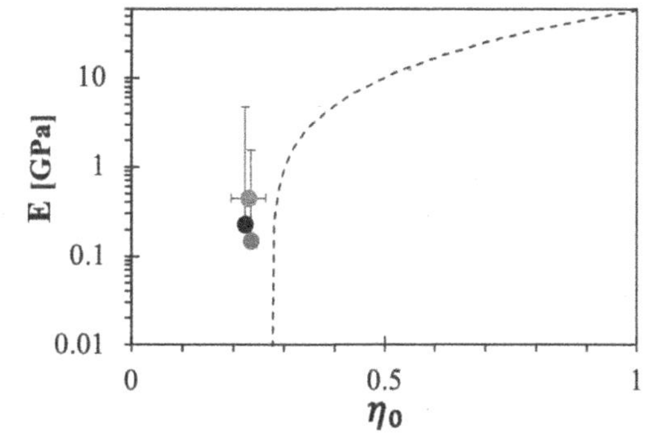

Figure 13. Effective Young modulus of C-S-H a function of the packing density at zero stress η_0 according to the flexibility of the layers. the estimate of self-consistent scheme for ellipsoidal inclusion with an aspect ratio of $a_r = 12.85/154.2 = 0.0833$ is provided for comparison.

We also compute the effective Young modulus of C-S-H as a function of the packing density at zero stress η_0 according to the flexibility of the layers (Figure 13). Higher flexibility leads to a slightly larger Young modulus, but the results are within the standard deviations and are therefore inconclusive. For comparison, we also provide the estimates of the self-consistent scheme for ellipsoidal inclusion with an aspect ratio of $a_r = 12.85/154.2 = 0.0833$. The VRH moduli of C-S-H calculated at the molecular scale (Table 2) are used as input in the homogenization. The percolation threshold of the self-consistent scheme with this oblate inclusion is $\eta_p = 0.28$, which is in agreement with previous work (Sanahuja et al. 2007). CG simulation leads to a slightly earlier percolation. Other packing densities are yet to be assessed. For this, longer simulations (on the order of the μs (Masoumi et al. 2020)) are needed to get configurations at larger packing densities.

4 CONCLUSIONS

Coarse-grained simulations were deployed to understand the mesostructuration and properties of ASR and C-S-H gels.

We simulated the formation of ASR gels for the first time using coarse-grained simulations. We have adopted a mesoscale representation described by mono-disperse spheres. Two different ASR gels were simulated: (i) Na-ASR and (ii) K-ASR gel, respectively, for gel formed from the reaction between the main alkalis present in cement paste and the silica present in aggregates. We simulated systems with packing fraction ranges close to the range of (high) porosity observed in ASR gels (Geng et al. 2021). We have noticed differences between Na-ASR and K-ASR gel structures emerging from the fact that these two gels have different interaction potentials at the molecular level. Na-ASR gels exhibit a coarser gel pore size and higher viscosity than K-ASR.

We proposed an original multi-scale approach to simulate the meso-structuration of C-S-H gel by considering the flexibility of its layers. We investigated the effect of accounting for the layers flexibility by considering three scenarios: one with rigidities as initially determined named F1, another with less rigid layers (F01, with k_{2b} and k_{3b} multiplied by 0.1), and the last one with more rigid layers (F10, k_{2b} and k_{3b} multiplied by 10). Our results showed a difference in the layer aggregation and orientation, with face-to-face contact between the stacked layers being favored for F1 while there is face-to-edge contact for F10. For the same packing density, F01 shows a larger population with large size pores compared to F1 and F10.

Perspectives include using poly-disperse disks to represent the C-S-H gel to reach higher packing fractions reported by (Jennings 2008) for C-S-H gel (between 60 and 70%). Also, future work can deal with the computation of the other properties of ASR gel and take into account the temperature effect on the ASR product morphology.

REFERENCES

Allen, M. P. & D. J. Tildesley (1989). *Computer Simulation of Liquids: Second Edition*. Publication Title: Computer Simulation of Liquids.

Benmore, C. J. & P. J. M. Monteiro (2010). The structure of alkali silicate gel by total scattering methods. *40*(6), 892–897.

Berendsen, H. J. C., J. R. Grigera, & T. P. Straatsma (1987, November). The missing term in effective pair potentials. *The Journal of Physical Chemistry 91*(24), 6269–6271.

Bhattacharya, S. & K. E. Gubbins (2006, August). Fast Method for Computing Pore Size Distributions of Model Materials. *Langmuir 22*(18), 7726–7731.

Bonnaud, P. A., C. Labbez, R. Miura, A. Suzuki, N. Miyamoto, N. Hatakeyama, A. Miyamoto, & K. J. V. Vliet (2016). Interaction grand potential between calciumâŁ"silicateâŁ"hydrate nanoparticles at the molecular level. *8*(7), 4160–4172.

Buckley, A. M. & M. Greenblatt (1994, July). The Sol-Gel Preparation of Silica Gels. *Journal of Chemical Education 71*(7), 599.

Chiang, W.-S., E. Fratini, P. Baglioni, D. Liu, & S.-H. Chen (2012, March). Microstructure Determination of Calcium-Silicate-Hydrate Globules by Small-Angle Neutron Scattering. *The Journal of Physical Chemistry C 116*(8), 5055–5061.

Cygan, R. T., J.-J. Liang, & A. G. Kalinichev (2004, January). Molecular Models of Hydroxide, Oxyhydroxide, and Clay Phases and the Development of a General Force Field. *The Journal of Physical Chemistry B 108*(4), 1255–1266.

Del Gado, E., L. d. Arcangelis, & A. Coniglio (2002). A study of viscoelasticity in gelling systems. *14*(9), 2133–2139.

Frenkel, D. & B. Smit (2002). *Understanding Molecular Simulation* (2nd ed.). Academic Press.

Gaboriaud, F., A. Nonat, D. Chaumont, A. Craievich, & B. Hanquet (1999). 29Si NMR and Small-Angle X-Ray Scattering Studies of the Effect of Alkaline Ions (Li+, Na+, and K+) in Silico-Alkaline Sols.

Geng, G., S. Barbotin, M. Shakoorioskooie, Z. Shi, A. Leemann, D. F. Sanchez, D. Grolimund, E. Wieland, & R. Dähn (2021, March). An in-situ 3D micro-XRD investigation of water uptake by alkali-silica-reaction (ASR) product. *Cement and Concrete Research 141*, 106331.

Goyal, A., K. Ioannidou, C. Tiede, P. Levitz, R. J.-M. Pellenq, & E. Del Gado (2020). Heterogeneous surface growth and gelation of cement hydrates.

Honorio, T. (2019). Monte carlo molecular modeling of temperature and pressure effects on the interactions between crystalline calcium silicate hydrate layers. *35*(11), 3907–3916.

Honorio, T. (2021). Shear deformations in crystalline alkali-silica reaction products at the molecular scale: anisotropy and role of specific ion effects. *54*(2), 86.

Honorio, T. & L. Brochard (2017). Flexibility of C-S-H sheets and stacks from molecular simulations. In *EAC-02 – 2nd International RILEM/COST Conference on Early Age Cracking and Serviceability in Cement-based Materials and Structures*, Brussels, Belgium.

Honorio, T., L. Brochard, & M. Vandamme (2017). Hydration phase diagram of clay particles from molecular simulations. *33*(44), 12766–12776.

Honorio, T., L. Brochard, M. Vandamme, & A. Lebée (2018, September). Flexibility of nanolayers and stacks: implications in the nanostructuration of clays. *Soft Matter 14*(36), 7354–7367.

Honorio, T., O. M. Chemgne Tamouya, & Z. Shi (2020). Specific ion effects control the thermoelastic behavior of nanolayered materials: the case of crystalline alkali-silica reaction products.

Honorio, T., O. M. Chemgne Tamouya, Z. Shi, & A. Bourdot (2020, October). Intermolecular interactions of nanocrystalline alkali-silica reaction products under sorption. *Cement and Concrete Research 136*, 106155.

Honorio, T., F. Masara, & F. Benboudjema (2021, October). Heat capacity, isothermal compressibility, isosteric heat of adsorption and thermal expansion of water confined in C-S-H. *Cement*, 100015.

Hou, D., W. Zhang, P. Wang, M. Wang, & H. Zhang (2021, June). Mesoscale insights on the structure, mechanical performances and the damage process of calcium-silicate-hydrate. *Construction and Building Materials 287*, 123031.

Ioannidou, K., B. Carrier, M. Vandamme, & R. Pellenq (2017). The Potential of Mean Force concept for bridging (length and time) scales in the modeling of complex porous materials. *EPJ Web of Conferences 140*, 01009.

Ioannidou, K., M. KanduÄŽ, L. Li, D. Frenkel, J. Dobnikar, & E. Del Gado (2016). The crucial effect of early-stage gelation on the mechanical properties of cement hydrates. *7*, 12106.

Ioannidou, K., K. J. Krakowiak, M. Bauchy, C. G. Hoover, E. Masoero, S. Yip, F.-J. Ulm, P. Levitz, R. J.-M. Pellenq, & E. D. Gado (2016, February). Mesoscale texture of cement hydrates. *Proceedings of the National Academy of Sciences 113*(8), 2029–2034.

Ioannidou, K., R. J.-M. Pellenq, & E. D. Gado (2014, January). Controlling local packing and growth in calciumâŁ"silicateâŁ"hydrate gels. *Soft Matter 10*(8), 1121–1133.

Jennings, H. M. (2008, March). Refinements to colloid model of C-S-H in cement: CM-II. *Cement and Concrete Research 38*(3), 275–289.

Keremides, K., M. J. Abdolhosseini Qomi, R. J. M. Pellenq, & F.-J. Ulm (2018, August). Potential-of-Mean-Force Approach for Molecular DynamicsâŁ"Based Resilience Assessment of Structures. *Journal of Engineering Mechanics 144*(8), 04018066.

Kirkpatrick, R. J., A. G. Kalinichev, X. Hou, & L. Struble (2005, May). Experimental and molecular dynamics modeling studies of interlayer swelling: water incorporation in kanemite and ASR gel. *Materials and Structures 38*(4), 449–458.

Kunhi Mohamed, A., S. C. Parker, P. Bowen, & S. Galmarini (2018, May). An atomistic building block description of C-S-H – Towards a realistic C-S-H model. *Cement and Concrete Research 107*, 221–235.

Liu, H., S. Dong, L. Tang, N. M. A. Krishnan, G. Sant, & M. Bauchy (2019). Effects of polydispersity and disorder on the mechanical properties of hydrated silicate gels. *122*, 555–565.

Marty, N. C. M., S. Grangeon, F. Warmont, & C. Lerouge (2015). Alteration of nanocrystalline calcium silicate hydrate (c-s-h) at pH 9.2 and room temperature: a combined mineralogical and chemical study. *79*(2), 437–458.

Maruyama, I., G. Igarashi, K. Matsui, & N. Sakamoto (2021, June). Hinderance of C-S-H sheet piling during first drying using a shrinkage reducing agent: A SAXS study. *Cement and Concrete Research 144*, 106429.

Masoero, E., E. Del Gado, R. J.-M. Pellenq, F.-J. Ulm, & S. Yip (2012). Nanostructure and nanomechanics of cement: Polydisperse colloidal packing. *109*(15).

Masoero, E., E. D. Gado, R. J.-M. Pellenq, S. Yip, & F.-J. Ulm (2013, December). Nano-scale mechanics of colloidal CâŁ"SâŁ"H gels. *Soft Matter 10*(3), 491–499.

Masoumi, S., D. Ebrahimi, H. Valipour, & M. J. A. Qomi (2020). Nanolayered attributes of calcium-silicate-hydrate gels. *Journal of the American Ceramic Society 103*(1), 541–557.

Masoumi, S., H. Valipour, & M. J. Abdolhosseini Qomi (2017a). Intermolecular forces between nanolayers of crystalline calcium-silicate-hydrates in aqueous medium. *121*(10), 5565–5572.

Masoumi, S., H. Valipour, & M. J. Abdolhosseini Qomi (2017b). Interparticle interactions in colloidal systems: Toward a comprehensive mesoscale model. *9*(32), 27338–27349.

Medina, J. S., R. Prosmiti, P. Villarreal, G. Delgado-Barrio, G. Winter, B. González, J. V. Alemán, & C. Collado (2011, September). Molecular dynamics simulations of rigid and flexible water models: Temperature dependence of viscosity. *Chemical Physics 388*(1), 9–18.

Mishra, R., A. Kunhi, D. Geissbühler, H. Manzano, T. Jamil, R. Shahsavari, A. G Kalinichev, S. Galmarini, L. Tao, H. Heinz, R. Pellenq, A. van Duin, S. C Parker, R. Flatt, & P. Bowen (2017). cemff: A force field database for cementitious materials including validations, applications and opportunities.

Nishiyama, N. & T. Yokoyama (2017). Permeability of porous media: Role of the critical pore size. *122*(9), 6955–6971.

Pellenq, R. J.-M., A. Kushima, R. Shahsavari, K. J. Van Vliet, M. J. Buehler, S. Yip, & F.-J. Ulm (2009, September). A realistic molecular model of cement hydrates. *Proceedings of the National Academy of Sciences of the United States of America 106*(38), 16102–16107.

Plimpton, S. (1995). Fast parallel algorithms for short-range molecular dynamics. *117*(1), 1–19.

Prathyusha, K. R., S. Henkes, & R. Sknepnek (2018, February). Dynamically generated patterns in dense suspensions of active filaments. *Physical Review E 97*(2), 022606.

Rajabipour, F., E. Giannini, C. Dunant, J. H. Ideker, & M. D. A. Thomas (2015). AlkaliâŁ"silica reaction: Current understanding of the reaction mechanisms and the knowledge gaps. *76*, 130–146.

Richardson, I. G. (2004, September). Tobermorite/jennite- and tobermorite/calcium hydroxide-based models for the structure of C-S-H: applicability to hardened pastes of tricalcium silicate, beta-dicalcium silicate, Portland cement, and blends of Portland cement with blast-furnace slag, metakaolin, or silica fume. *Cement and Concrete Research 34*(9), 1733–1777.

Sanahuja, J., L. Dormieux, & G. Chanvillard (2007). Modelling elasticity of a hydrating cement paste. *Cement and Concrete Research 37*(10), 1427–1439.

Shahsavari, R., M. J. Buehler, R. J.-M. Pellenq, & F.-J. Ulm (2009). First-Principles Study of Elastic Constants and Interlayer Interactions of Complex Hydrated Oxides: Case Study ofTobermorite and Jennite. *Journal of the American Ceramic Society 92*(10), 2323–2330.

Shi, Z., G. Geng, A. Leemann, & B. Lothenbach (2019, July). Synthesis, characterization, and water uptake property of alkali-silica reaction products. *Cement and Concrete Research 121*, 58–71.

Stukowski, A. (2014, March). Computational Analysis Methods in Atomistic Modeling of Crystals. *JOM 66*(3), 399–407.

Torquato, S. (2002). *Random Heterogeneous Materials: Microstructure and Macroscopic Properties*. Springer Science & Business Media.

Yu, Z., A. Zhou, & D. Lau (2016, November). Mesoscopic packing of disk-like building blocks in calcium silicate hydrate. *Scientific Reports 6*(1), 36967.

Zubkova, N. V., Y. E. Filinchuk, I. V. Pekov, D. Y. Pushcharovsky, & E. R. Gobechiya (2010, August). Crystal structures of shlykovite and cryptophyllite: comparative crystal chemistry of phyllosilicate minerals of the mountainite family. *European Journal of Mineralogy 22*(4), 547–555.

Computational Modelling of Concrete and Concrete Structures – Meschke, Pichler & Rots (Eds)
© 2022 Copyright the Author(s), ISBN: 978-1-032-32724-2

Study on detection of internal cracks by inspection of impact echo using CNN

Shunsuke Sano, Toshiaki Mizobuchi & Hiroshi Shimbo
Hosei University, Tokyo, Japan

Tomoko Ozeki
Tokai University, Kanagawa, Japan

Jun-Ichiro Nojima
J-POWER Design Co., Ltd., Tokyo, Japan

ABSTRACT: In this paper, the effectiveness of CNNs trained on the spectrograms of impact echo is examined for micro cracks in concrete caused by corrosion expansion of reinforcing bars. As results of the study, this method is confirmed to be able to detect internal cracks caused by corrosion of reinforcing bars with cover concrete. In addition, the possibility of recognizing the slight change in impact echo by the development of internal cracks and evaluating the progress of deterioration is found.

1 INTRODUCTION

Impact echo method is widely used as an inspection method because it is non-destructive and simple method to conduct. However, impact echo method is not a quantitative judgment because it depends on the skill of the engineer. In addition, the number of skilled engineers is expected to decrease due to the declining birthrate and aging population. Also, the demand for inspection is rapidly increasing with the aging of existing structures. To deal with this situation, detailed maintenance plans should be developed by efficiently judging the initial deterioration. With the above background, quantitative inspection methods to efficiently judge the initial deterioration are required.

To respond to this demand, quantitative detection of defects by impact echo is attempted. For example, if micro cracks appear in the specimen, the sound wave shape will be differed from the normal part because elastic waves by impact are reflected by the cracks. Thus, a method has been proposed to quantitatively judge defects based on the characteristics of the waveform by recording the impact echo with a microphone. Kamata et al. 2002 and Miyoshi et al. 2009 have reported that this method can judge the defective part.

Furthermore, as the characteristics of the waveform are different between the normal part and the defective part, machine learning may be able to judge the defective part automatically and quantitatively. Therefore, in this study, quantitative and automatic judgment of micro cracks in the initial stage of deterioration caused by corrosion expansion of reinforcing bars in concrete is attempted by machine learning of impact echo. Specifically, a method is proposed that transforms the impact echo into a spectrogram image using the Short Time Fourier Transform (STFT) and conducts machine learning using a Convolutional neural network (CNN). CNN is a machine learning method that shows high performance in image recognition by image convolution and pooling. Shimbo et al. (2019; 2020) reported that this method can judge defects with the same level of accuracy as humans on specimen embedded with pseudo defects, and that it can also judge defects in existing structures. Therefore, generalization performance to judge the defects of existing structures from the impact echo of different specimens is aimed to be obtained. However, the impact echo of the specimen is different from that of the existing structures due to its size and simulated defects.

Therefore, obtaining the generalized performance to judge the internal cracks of the real structure by the impact echo of the specimen is considered to be difficult, which is the current issue. So, in this study, reinforced concrete specimen with dimensions of 1800x1800x600mm is placed according to the mixing conditions and shape (thickness, arrangement of reinforcement, diameter of reinforcement and cover concrete) of existing structures, and internal cracks are generated by electrical corrosion. In this way, the impact echo closer to those of existing structures can be obtained. Then, as the first step to obtain the generalization performance, the possibility of detecting micro cracks caused by corrosion of reinforcing bars is examined by this method. Figure 1 shows the appearance of the specimen.

 DOI 10.1201/9781003316404-18

Figure 1. Appearance of the specimen.

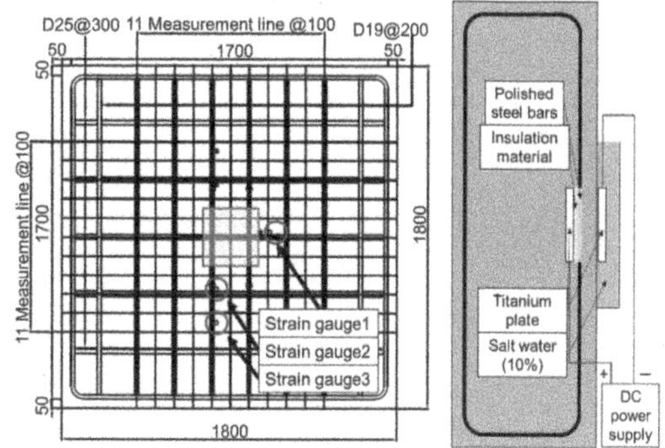

Figure 2. Outline of the specimen and corrosion method of reinforcing bars by electric corrosion.

2 EXPERIMENTAL METHODS

2.1 *Specimen and electrical corrosion experiment*

Reinforced concrete specimen is placed according to the mixing conditions and shape (thickness, arrangement of reinforcement, diameter of reinforcement and cover) of concrete piers.

The dimensions of the specimen are 1800 mm (length) × 1800 mm (width) × 600 mm (height), and the reinforcement was arranged in D25 and D19. The rebar spacing was 300 mm and 200 mm, and the cover concrete was 70 mm. Table 1 shows the mix proportion of concrete. In addition, three strain gauges are embedded in the specimen for the purpose of monitoring the internal conditions of the specimen. The positions of the strain gauges are shown by the red circle in Figure 2. Next, the reinforcing bars in the center of the specimen are corroded and expanded by electric corrosion, and micro cracks are produced. Then, this experiment aims to produce floating in the center of the specimen by the expansion of micro cracks. Figure 2 shows an overview of electrical corrosion. In this experiment, electrophoresis method conducted by Hanaoka et al. 2007 and Toda et al. 2010 is used. This method can produce the actual corrosion of reinforcing bars. Electrical corrosion starts at 23 days after concrete placement and is carried out for 50 days, then stop for 100 days, and resume after that.

Table 1. Mix proportion of concretes.

W/C	S/a	Unit amount				
		W	C	S	G	Ad
%	%	kg/m^3				
56	49	175	313	861	914	3.76

2.2 *Measurement of impact echo*

Measurement of impact echo is carried out once a week after the start of electrical corrosion. Also, the initial measurement is carried out two weeks after the specimen is placed, which is before the start of the electrical corrosion. A hammer with the mass of about 1.5 kg and the length of about 330 mm is used for the measurement. The recordings are carried out at a distance of 10 cm from the hitting point, and the impact echoes are recorded in succession so that one audio file is created for each hitting point. The sampling frequency of the recording is set to 96.0 kHz. The hitting points are 121 intersections of a grid with 100 mm intervals in the center of the specimen to avoid the influence of the edge. This is because the sound wave shape is different from the existing structures due to the reflection of elastic waves at the edge, which may decrease the generalization performance. Figures 2 and 3 show the locations of the measurement lines. In Figure 3, the measurement range of impact echo is surrounded by a square, and the range of electric corrosion is shown by the square in the center. The number of hits is set

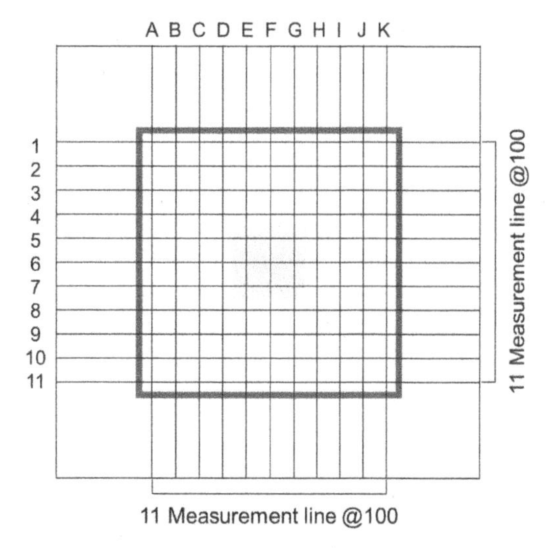

Figure 3. Position of the measurement line.

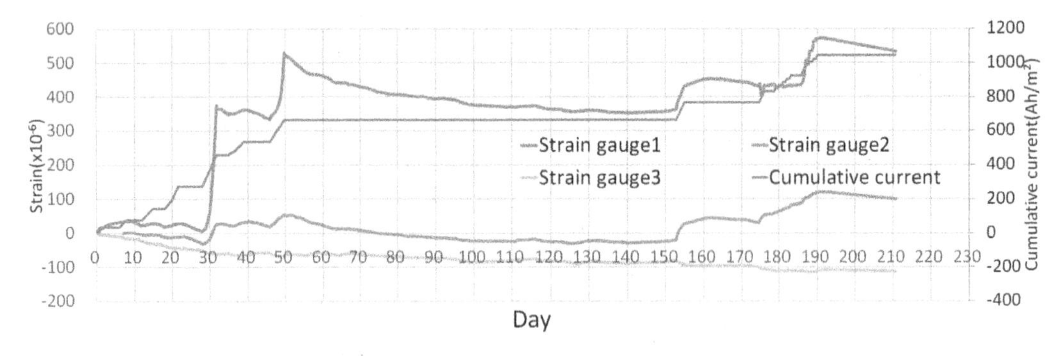

Figure 4. Changes in strain and cumulative current.

at 5 hits per point. As the initial measurements and the measurements on the 60th and 211th day after the start of electrical corrosion are adopted as the training data, 16 hits per point are performed in the measurements on those days.

3 RESULT OF ELECTRICAL CORROSION

Figure 4 shows the strain and the cumulative amount of current per unit surface area of the reinforcing bars from the start of the electrical corrosion until the 211th day. Strain gauge 2 is not measured until the 7th day after the start of electrical corrosion, so the strain is shown from the 7th day. From Figure 4, rapid changes in strain occur around 30, 50, 153, and 188 days. These strain changes are assumed to be due to internal cracks caused by corrosion of the reinforcing bars.

The core is drawn out from the center of the specimen near the reinforcing bars on the 211th day after the start of electrical corrosion. As the results, micro internal cracks (crack width of about 0.05 mm) are observed. Figure 5 shows the internal cracks observed. From the corrosion products observed inside the cracks, these cracks are assumed to be caused by the corrosion of reinforcing bars. These internal cracks may also occur around 30, 50, 153, and 188 days.

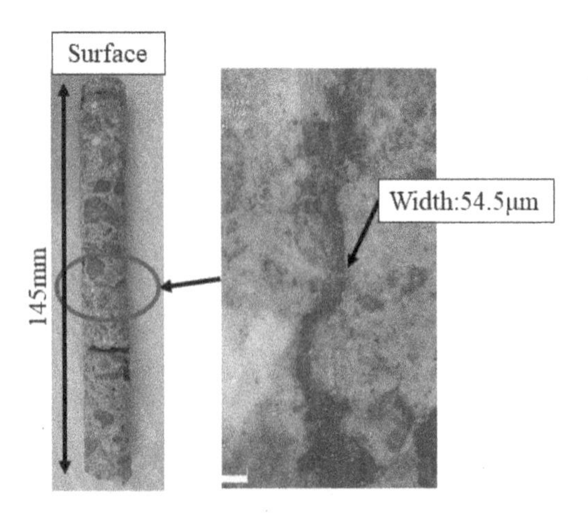

Figure 5. Internal cracks at 211th day.

On the surface of the specimen, several surface cracks are observed in the central part of the specimen on the 81st day after the start of electrical corrosion. The crack width at the 81st day is 0.02 mm. These surface cracks continue to expand, and by the 116th day, all cracks become 0.1 mm wide. Figure 6 shows the sketch of the surface cracks at 116 days.

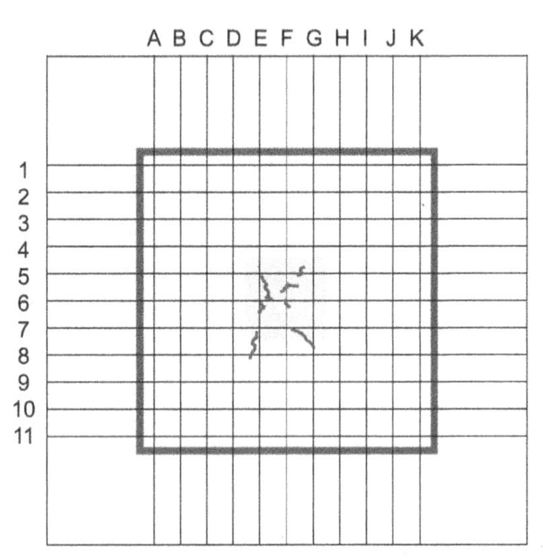

Figure 6. Surface cracks at 116th day.

4 DATA PREPROCESSING

4.1 *Cutting process of impact echo*

The impact echoes are normalized so that the maximum absolute value of the waveform is 1 and then cut to a length of 400 ms using MATLAB ver.R2019b, a numerical analysis software from MathWorks. The timing of the cutout is set to when the absolute value of the waveforms change exceeded 0.1.

4.2 *Converting impact echo into spectrogram*

Next, the waveform cut out to 400ms is converted to a spectrogram using "MATLAB". Short-time Fourier transform (STFT) is used to transform the data into spectrograms. The transformed spectrogram is cropped in the lower left corner and made into a 28x28 pixel spectrogram. Because the accuracy tends to increase as the time axis become longer, the time axis of the spectrogram is limited to 125 ms, and the frequency axis is limited to about 12 kHz. Figure 7 shows the cut-out range of the spectrogram using STFT with a red frame.

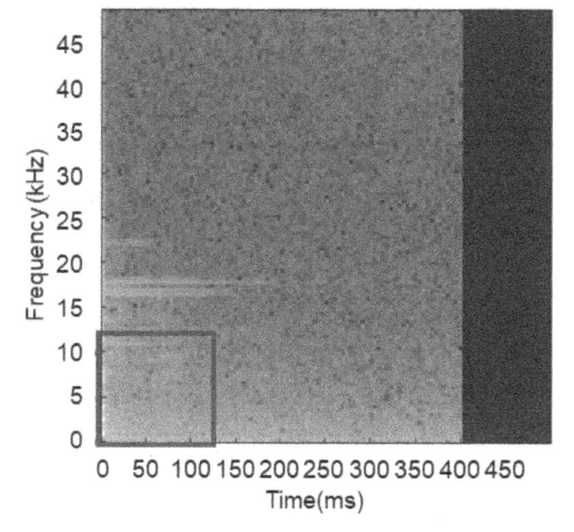

Figure 7. Cut-out range of the spectrogram.

5 MACHINE LEARNING

CNN is trained on the created spectrograms in a supervised training. The network structure of the CNN is 3 convolutional layers. The learning rate is 0.001 and the number of epochs is 150. Two patterns of training data are used. Pattern 1 is the impact echoes before the start of the electrical corrosion and the 60th day after the start of the electrical corrosion, and pattern 2 is the impact echoes before the start of the electrical corrosion and the 211th day after the start of the electrical corrosion. In the labeling of the training data, all the data before the start of electrical corrosion are normal. The data measured after the start of electrical corrosion are labeled as defective at the center and normal at the surroundings, based on the results of impact echo method by a skilled engineer conducted on the 56th day after the start of electrical corrosion. Figure 8 shows the locations of normal and defects. 20% of the training data is used as validation data, for the purpose of checking whether overtraining is occurred.

After the training is completed, the network is evaluated for accuracy of defect detection by judging the test data. The test data are collected during the period of 18–53 days and during the period of 88–193 days. MATLAB is used for machine learning and judging.

Table 2 shows the number of data used for training and the accuracy at the end of training. The accuracy at the end of training is 100% for Pattern 1 and about 98% for Pattern 2. So, both patterns are able to learn with sufficient accuracy.

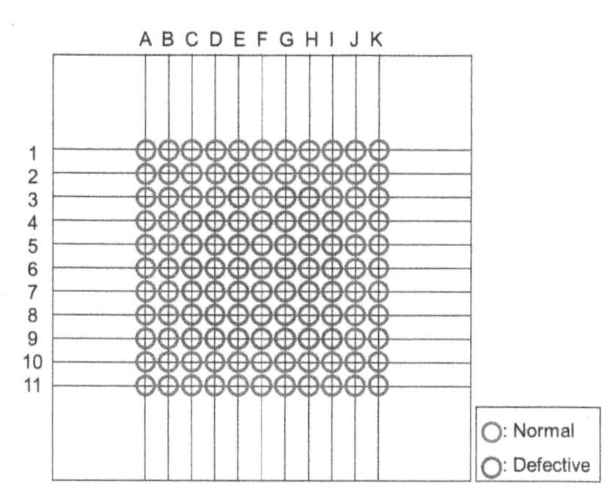

Figure 8. Position of Normal and Defective.

Table 2. Overview of the machine learning conducted.

Pattern	Number of data (normal)	Number of data (defective)	Accuracy %
Pattern 1	6294	1440	100.0
Pattern 2	6294	1440	98.3

6 RESULTS OF JUDGMENT AND DISCUSSION

6.1 *Results and discussion for 18th day to 53rd day*

When the network judges the impact echo, an evaluation score of 1 to 0 is output. 1 means normal, 0 means defective, and 0.5 is the threshold between normal and defective. The scores are arranged in the order of the hitting points on the specimen as shown in (a) of Figure 9. Then, arranged scores are converted into a contour map as shown in (b) of Figure 9. The score is shown in a three-color scale, where 1 is green, 0.5 is yellow, 0 is red, and the color gradually changes in between. Also, the score is the average score in the hitting points because multiple impact echoes are measured for each hitting point.

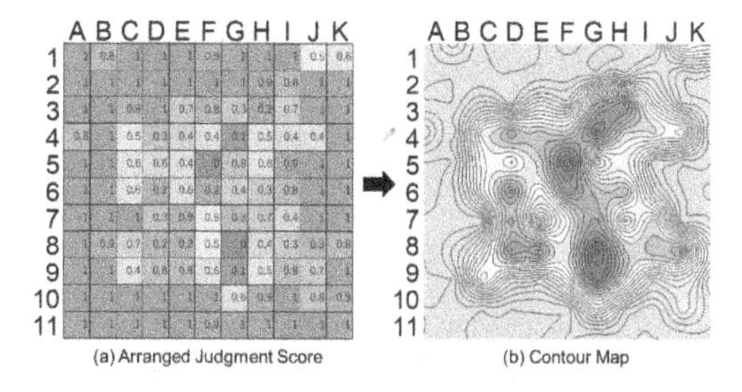

Figure 9. Conversion to contour map.

The results of judgment from 18th to 53rd day are shown in Figure 10. In Figure 10, 18th day from the start of electrical corrosion is represented as Event 1, 27 days as Event 2, and so on. From Figure 10, almost the entire surface is judged to be normal in Events 1 and 2 for both patterns 1 and 2. Also, the central part

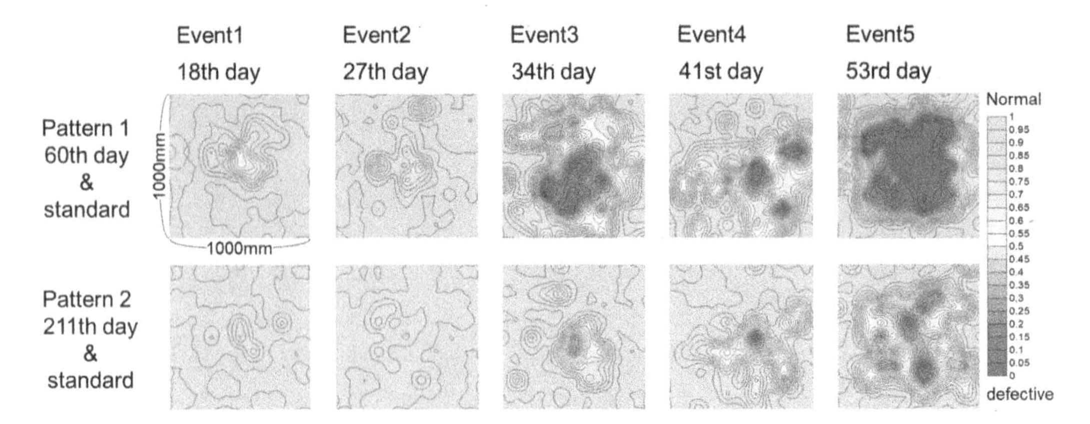

Figure 10. Results of judgment of specimen from 18th day to 53rd day.

of the specimen is judged to be defective in Events 3 through 5. The part judged to be defective is the part undergoing electrical corrosion.

Until Event 2, the corrosion of the reinforcing bars does not progress sufficiently to produce internal cracks. Thus, the judgment of normal from Event 1 to 2 is a reasonable judgment. In addition, as described in Chapter 3, internal cracks are considered to occur after 30 days after the start of the electrical corrosion, when the strain changed rapidly. So, the judgment of defect in the center of the specimen after Event 3 is also reasonable. Therefore, the trained network is considered to be able to judge the internal cracks.

In addition, the range of defects increased significantly in the judgment results at Event 3 and Event 5. Thus, the trained network can sensitively reflect the changes in the specimen due to the growth of internal cracks. Therefore, the trained network may be able to evaluate the progress of degradation. From these results, this method can detect internal cracks caused by corrosion of reinforcing bars with a cover of 70 mm.

However, from Event 3 to Event 4 in Pattern 1, the area judged to be defective became smaller, even though the strain did not change significantly. This may be due to the shrinkage of the concrete and the decrease of micro cracks caused by the decrease in concrete temperature from 30°C to 25°C during this period. The decrease in the concrete temperature is caused by the decrease in the temperature near the test site. Figure 11 shows the concrete temperature, the air temperature near the test site, and the strain values from strain gauge 1.

Next, the judgment results for each pattern are compared with each other. As a result, the range of pattern 2 judged as a defect is smaller than that of pattern 1. In Pattern 1, the data at 60th day is used as the training data, and in Pattern 2, the data at 211th day is used as the defect training data. Figure 4 shows that the values of strain gauge 1 and 2 at 60th day are about 450 μ and 20 μ, while the values of strain gauge 1 and 2 at 210th day are about 530 μ and 100 μ. So, the strain values are larger than that at the 210th day. Therefore, between the 60th day and the 210th day, the deterioration is more severe and the range of deterioration is

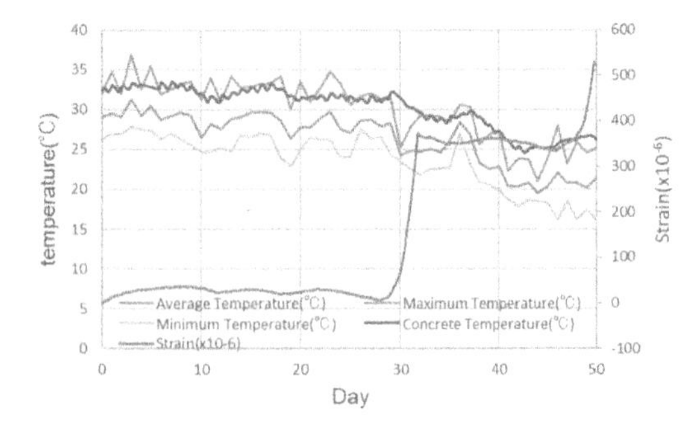

Figure 11. Concrete temperature and temperature near the test site.

larger in the 210th day. Thus, in Pattern 2, the impact echo of the more degraded specimen is learned, which may cause such a change.

In addition, when these two data are used as training data respectively, the judgment results are different, which suggests that CNN using spectrogram may be able to recognize these two data as different data.

6.2 *Results and discussion for 53rd day to 193rd day*

Next, the results of judgment from 53rd to 193rd day are shown in Figure 12. In Figure 12, 53rd day from the start of electrical corrosion is represented as Event 5, 88 days as Event 6, and so on. Figure 12 shows that the range of defects judged in Event 9 is larger than that in Event 5 for both patterns 1 and 2. Also, after Event 6, there was no clear difference in judgments between patterns 1 and 2.

As mentioned in Section 6.1, the strains in Figure 4 indicate that the degradation is more severe on the 210th day than on the 60th day. Therefore, the degradation is considered to be more severe and the range of degradation is larger in Event 9 than in Event 5. So, the judgment result that the range of defects is larger in Event 9 than in Event 5 is reasonable. This result suggests that the trained network is able to detect the progress of internal cracks and degradation even when the specimen is degraded by more electrical corrosion.

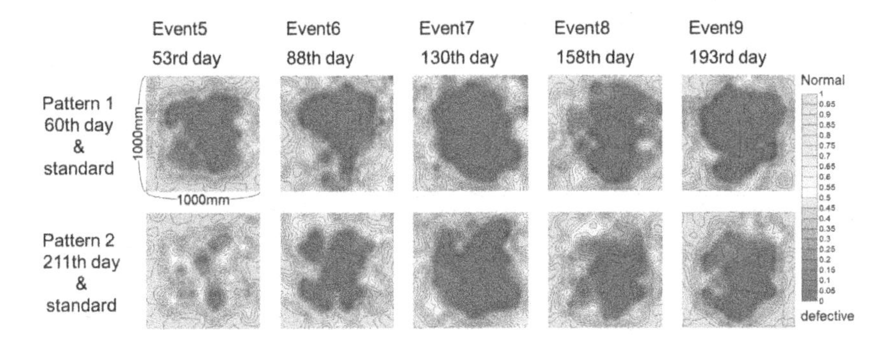

Figure 12. Results of judgment of specimen from 53rd day to 193rd day.

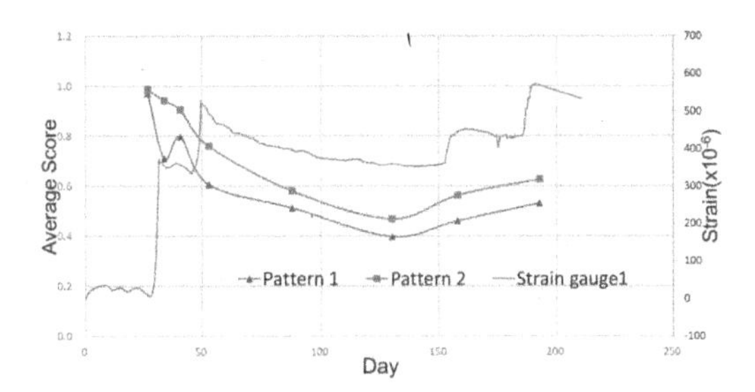

Figure 13. Changes in average score.

Next, the judgment results in Events 6 and 7 are focused on. This is the period when electric corrosion is stopped and the specimens are dried. As a result, the range of defects judged in Event 7 is the largest for both patterns 1 and 2 over the entire period. In order to visualize this, the average of the scores obtained by the judgment for each measurement day is shown in Figure 13.

From Figure 4, the values of strain around the day of Event 7 are the smallest among Event5 to 9. So, the range of defects is expected to be smaller, but the actual result is not smaller. This result is probably caused by the micro cracks that increased due to the drying of the specimens and are judged as defects. When electrical corrosion is resumed, the micro cracks may be closed by the water, and the range of defects may be judged smaller. Also, the characteristics of the impact echo may change by drying of the specimen, which may affect the judgment. The effect of the wet condition of the concrete on the impact echo needs to be studied in the future.

7 CONCLUSION

In this paper, the possibility of the CNN trained on the spectrogram of impact echo to detect the micro cracks in the concrete caused by the corrosion expansion of reinforcing bars is examined. The findings of this study are presented below.

(1) The CNN trained on the spectrogram converted from the impact echo by STFT is confirmed to be able to detect the internal cracks caused by the corrosion of reinforcing bars with deep cover.
(2) This method has potential for recognizing slight changes in impact echo caused by the growth of internal cracks and for judging the progress of deterioration.

If this method can be applied to existing structures, more detailed maintenance management plans can be developed. As a result, the life-cycle cost of structures can be reduced. To apply to existing structures, this method needs to acquire generalization performance. However, in existing structures, internal cracks larger than those in this study may exist. So, generalization performance is difficult to obtain with the current data. Therefore, in the future, the degradation will be advanced by further electrical corrosion to obtain generalized performance.

REFERENCES

Hanaoka Daishin, Yano Sadayosi, Miyazato Shinichi: 2007, Effect of Corrosion Mode and Amount on Flexural Properties of Reinforced Concrete Beams, *Journal of Japan Society of Civil Engineers*, E, Vol63, No. 2, 300–312.

Kamata Toshiro, Asano Masanori, Kunieda Minoru, Rokugo Keitetsu: 2002, Quantitative Nondestructive Evaluation of Defects in Concrete Surface Layer by Impact Acoustic Methods, *Journal of Japan Society of Civil Engineers*, No. 704, V-55, 66–79.

Miyoshi Akane, Sonoda Yoshimi, Kawabata Kenta: 2009, A Fundamental Study on the Degradation Diagnosis using Impact Echo Characteristics, *Proceedings of the Japan Concrete Institute*, Vol. 31, No. 1, 2131–2136.

Shimbo Hiroshi, Mizobuchi Toshiaki, Nojima Jun-ichiro: 2019, A Fundamental Study on the Application of Machine Learning to Impact Echo Detection, *Proceedings of the Japan Concrete Institute*, Vol. 41, No. 1, 1829–1834.

Shimbo Hiroshi, Mizobuchi Toshiaki, Ozeki Tomoko, Nojima Jun-ichiro: 2020, A Study on Quantification of Impact Echo Detection by Machine Learning, *Proceedings of AI and Data Science*, Vol. 1, No. 1, 522–529.

Toda Katsuya, Ueno Atsushi, Uji Kimitaka: 2010, A Study on the Amount of Corrosion of Steel Bars in Concrete during Cracking, *Proceedings of the Japan Concrete Institute, Vol. 21, No. 3, 31–41.*

Computational Modelling of Concrete and Concrete Structures – Meschke, Pichler & Rots (Eds)
© 2022 Copyright the Author(s), ISBN: 978-1-032-32724-2

Effect of lateral confinement on concrete fatigue life under shear loading

M. Aguilar, H. Becks, A. Baktheer, M. Classen & R. Chudoba
Institute of Structural Concrete, RWTH Aachen University, Aachen, Germany

ABSTRACT: In several recent mesoscale and macroscale material model formulations, the authors hypothesized that fatigue evolution in the material structure can be realistically modeled by defining a cumulative measure of inelastic shear strain as the fatigue driving mechanism. The standard method of fatigue characterization using cylinder compression tests induces shear only as a secondary effect within the volume of the specimen. To validate the hypothesis that fatigue at subcritical load levels is determined by a cumulative measure of inelastic shear strain, experimental methods with dominant shear strain appear more appropriate. In the present work, the punch-through shear test (PTST) setup is used to induce shear-dominated strain within the volume of the specimen. Furthermore, the ability to control and measure lateral confinement is utilized. An experimental study of the fatigue behavior of a high-strength concrete is presented, in which the influence of different degrees of confinement on the fatigue life of the concrete at subcritical load levels is evaluated. The study analyzes the accelerating or retarding effect of confinement on the development of fatigue damage that occurs as a result of compressive normal stress. To enable an efficient and realistic representation of the pressure-sensitive, shear dominated fatigue response, an axisymmetric idealization of the PTST test is proposed, modeling the shear ligament using the fatigue microplane model MS1. In this model, the tangential damage at the microplane is linked to a cumulative inelastic strain to reflect the accumulation of fatigue damage owing to internal shear/sliding between aggregates at subcritical load levels. The model aims to capture the basic inelastic mechanisms that are driving the tri-axial stress redistribution within the material zone during the fatigue damage process in concrete.

1 INTRODUCTION

Over the last decade, many endeavors have been made to reduce the amount of concrete, and thus CO2 emissions, necessary to meet the legitimate demands of an increasingly growing population. Therefore, new materials, design techniques and numerical models have been developed. Fatigue response of normal and high strength concrete has been extensively investigated in the literature during the last decades e.g., (Do, Chaallal, & Aïtcin 1993; Kim & Kim 1996; Song, Konietzky, & Cai 2021; Oneschkow, Timmermann, & Löhnert 2022; Schäfer, Gudžulić, Breitenbücher, & Meschke 2021).

While high-cycle fatigue of metals has been thoroughly studied and described so that reliable predictions of fatigue life are available for engineering practice, a complete understanding of the processes of fatigue damage propagation in concrete is still lacking. In spite of a remarkable progress in recent years made in modeling and characterizing the concrete fatigue behavior e.g., (Desmorat, Ragueneau, & Pham 2007; Kirane & Bažant 2015; Rybczynski, Schaan, Dosta, Ritter, & Schmidt-Döhl 2021), many open questions remain that need to be fundamentally addressed in order to develop a deep and general insight into the fatigue phenomenology.

A promising hypothesis in this context postulates that fatigue evolution in concrete material structure is primarily driven by oscillating local shear strains at subcritical load levels within the heterogeneous cement-aggregate material structure. This theory has been investigated in recent publications, where a microplane fatigue material model for concrete showed very promising results under compressive loading (Baktheer, Aguilar, & Chudoba 2021).

To isolate the fundamental fatigue damage mechanisms in high-performance concrete, a new experimental setup "cylindrical punch through shear test (PTST)" has been developed by the authors to characterize the behavior of concrete under combined shear and normal loading. By capturing the fatigue behavior of the PTST for a wide range of loading scenarios and confinement levels, we aim to create an alternative method for characterizing the fatigue behavior of concrete. The controllable degree of confinement extends the scope of validation of both the model and the underlying hypothesis, which is the main goal of the ongoing research.

DOI 10.1201/9781003316404-19

2 EXPERIMENTAL INVESTIGATION

The cylindrical punch thorough shear test (PTST) (Luong 1989) was used to investigate the material behavior of high-strength concrete under combined shear and compression load. Through preliminary studies on similar specimens, the strong influence of compressive stress on fatigue life under mode II fatigue load was investigated at (Becks & Classen 2021). A new test setup was developed to allow simultaneous control of shear and compressive stresses to systematically investigate the influence of the degree of confinement on the fatigue life of high-strength concrete under subcritical fatigue shear loading.

2.1 *Specimen geometry*

With the aim of obtaining an approximately straight shear failure surface, the geometry of the specimen shown in Figure 1 was designed to introduce two different diameters of circular notches and minimize the occurrence of radial tensile stresses on the ligament. In addition, four notches were introduced in the outer concrete ring to prevent the occurrence of radial cracks on the outer cylinder of the specimen and to allow direct transfer of the compressive load to the ligament.

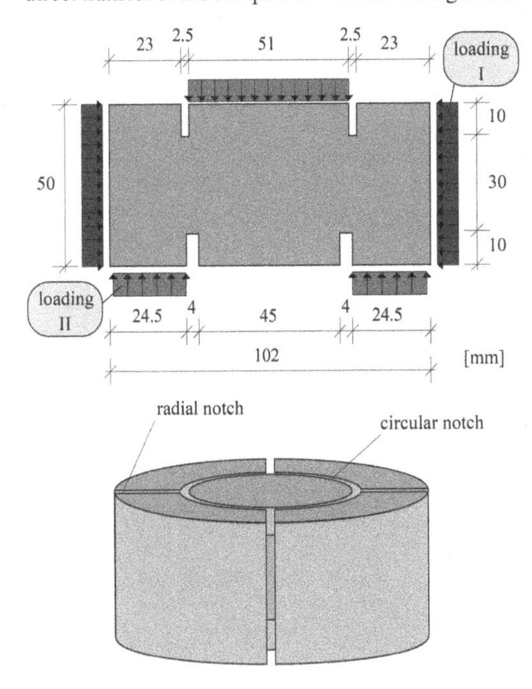

Figure 1. Specimen geometry and loading condition.

2.2 *Materials*

A high-strength concrete with a maximum aggregate size of 8 mm was used for all tests. To evaluate the material properties, concrete specimens were prepared from the same batch and tested after 28 days. The compressive strength was tested on cube specimens (a = 150 mm) and was 96 MPa. The modulus of elasticity and splitting tensile strength were determined on cylinders (h/d = 300 mm/150 mm) and were 39.226 MPa and 4.3 MPa, respectively. The specimens were produced in steel forms, cured for one day and then stored exposed in air until testing.

2.3 *Test setup and instrumentation*

The geometry of the PTST setup is shown in Figure 2. The radial compression stress is controlled by two hydraulic cylinders and applied to the entire surface of the outer ring specimen via four steel jaws. Full-surface support on the outer concrete ring and load application to the entire inner concrete cylinder introduces the mode II load along the test ligament between the upper and lower notches. The relative displacement of the outer and inner rings was measured using six linear variable displacement transducers (LVDTs) attached to the bottom and top of the specimen to capture possible tilting of the inner cylinder.

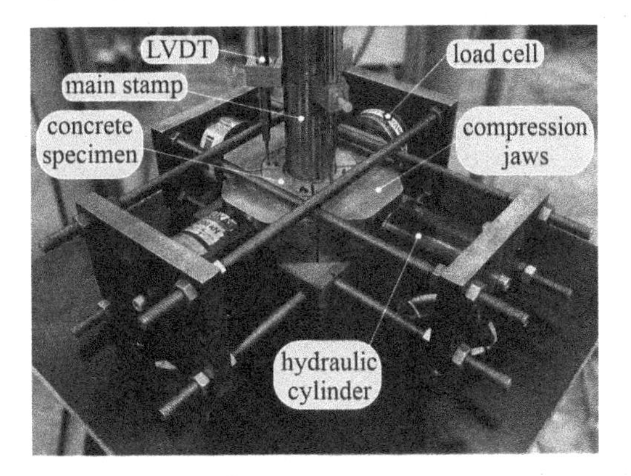

Figure 2. Test setup.

2.4 *Experimental results*

Monotonic tests. To investigate the effect of compression stress on the ultimate shear load, PTSTs were loaded at four different levels of compression σ_c (load I) and subjected to a monotonically increasing displacement-controlled (load II) at a rate of 0.2 mm/min up to a maximum relative displacement of the inner and outer rings of 6 mm. The precompression stress has been evaluated as the radial confinement force normalized w.r.t. surface area of the ligament.

As the experimental results given in Figure 3 show, the precompression load has a clear influence on the pre-peak and post-peak behavior. While the shear force of the specimens with 16 and 32 MPa precompression level steadily decreases after reaching the maximum shear force, the specimen with 4 MPa compression load exhibits a post-peak sudden slippage of the inner concrete core. Specimens without compressive loading fail immediately after reaching the maximum shear load and therefore do not show any post-peak behavior. The peak shear stress and the residual shear force after the peak increase with increasing compressive stress.

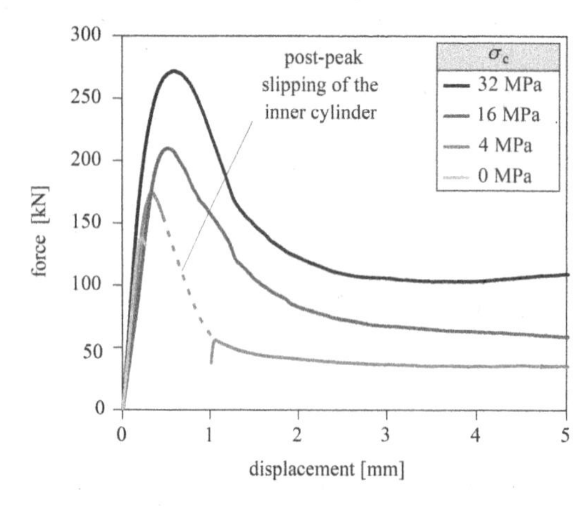

Figure 3. Stress-displacement curve of monotonic shear tests with different compressive loadings.

Fatigue tests. To investigate the material behavior under combined precompression and subcritical shear fatigue loading, eight fatigue tests were performed with three different precompression levels, namely 5, 15, and 30 MPa. The range of the shear fatigue load was set to $S^{max} = 0.85$ and $S^{min} = 0.05$ in relation to the respective monotonic ultimate load. All tests were performed with a cycling frequency of 5 Hz.

Figure 4 shows the force-displacement curve of one selected test with 15 MPa compression load, together with the corresponding monotonic reference test. This test withstood 747 cycles and failed at the displacement of about 0.6 mm. An overview of the results of the experimental fatigue tests is given in Table 1. As can beseen from the experimental studies, the fatigue life under mode II loading increases by two orders of magnitude with each considered increase in compressive loading.

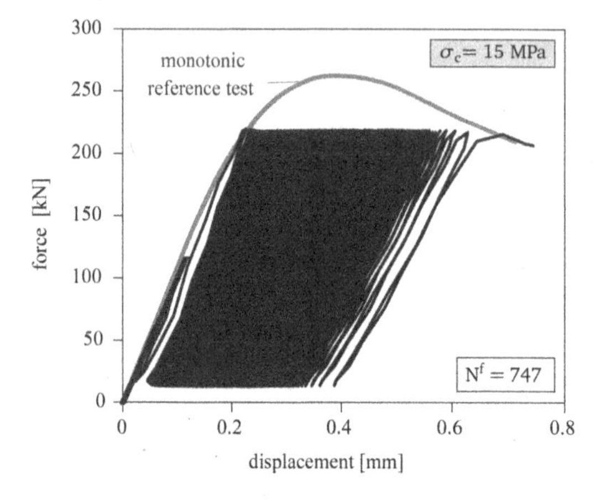

Figure 4. Stress-displacement curve of test F5 with 15 MPa compressive loading.

It should be noted that in the test with the low confinement and the low number of cycles to failure, it was not possible to achieve uniform load amplitudes during the first cycles. Also, the maximum load achieved was

Table 1. Summary of all conducted fatigue tests.

Test	S^{max}	S^{min}	σ_c	Number of cycles N^f
F1	0.80	0.05	5	9
F2	0.80	0.05	5	57
F3	0.80	0.05	5	54
F4	0.85	0.05	15	9454
F5	0.85	0.05	15	747
F6	0.85	0.05	15	3296
F7	0.85	0.05	30	92.349
F8	0.85	0.05	30	201.420

lower than the desired $S^{max} = 0.85$. Despite this deficiency, the qualitative trend shown in Figure 5 can be considered relevant and serves as a basis for the investigations with the microplane material model MS1.

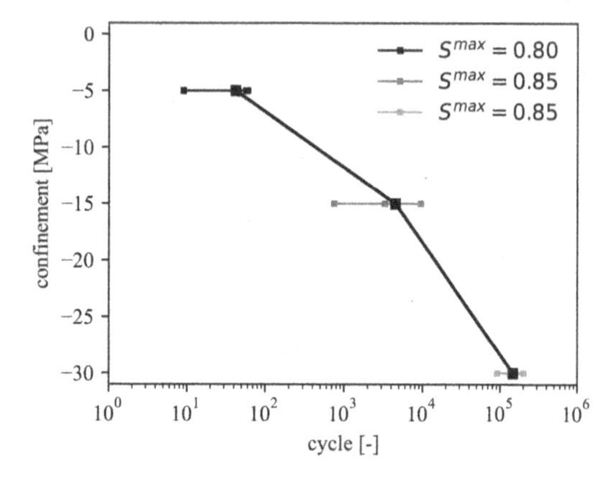

Figure 5. Fatigue life for the studied confinement levels.

3 MICROPLANE MODEL FOR FATIGUE MS1

The recently introduced microplane material model MS1 (Baktheer, Aguilar, & Chudoba 2021) for the fatigue behavior of concrete is used in the current study. The key idea of this material model is to link the evolution of fatigue damage to a measure of cumulative inelastic shear strain (Baktheer & Chudoba 2018a). This hypothesis is based on experimental observations, e.g. by Skarzynski et al. (Skarzynski, Marzec, & Tejchman 2019), indicating that crack initiation and propagation during fatigue loading occurs along the interfaces between the hardened cement paste and aggregates. The model has been formulated within the microplane framework, exploiting the principle of energy equivalence to transform the material state representation on a unit hemisphere to the tensorial stress and stiffness representation. Using this homogenization framework, thermodynamically based constitutive laws that govern the macroscopic behavior can be defined at the level of a microplane, that can be thought of as an oriented plane with an ascribed dissipation behavior within a 3D material structure.

The macroscopic thermodynamic potential is expressed as the sum of the microplane normal and tangential Helmholtz free energies:

$$\psi^{\mathrm{mac}} = \frac{3}{2\pi}\int_{\Omega}\psi^{\mathrm{mic}}d\Omega = \frac{3}{2\pi}\int_{\Omega}\psi_{\mathrm{N}}d\Omega + \frac{3}{2\pi}\int_{\Omega}\psi_{\mathrm{T}}d\Omega. \tag{1}$$

The projection of the thermodynamic potentials onto the normal and tangential direction allows introducing distinguished dissipative mechanisms for each direction, which are summarized in Figure 7.

Normal direction:

The microplane thermodynamic potential of the normal direction was defined as

$$\rho\psi_{\mathrm{N}}^{\mathrm{mic}} = \frac{1}{2}\left[1 - H(\sigma_{\mathrm{N}})\omega_{\mathrm{N}}\right]E_{\mathrm{N}}(\varepsilon_{\mathrm{N}} - \varepsilon_{\mathrm{N}}^{\mathrm{p}})^2 + \frac{1}{2}K_{\mathrm{N}}z_{\mathrm{N}}^2 + \frac{1}{2}\gamma_{\mathrm{N}}\alpha_{\mathrm{N}}^2 + f(r_{\mathrm{N}}), \tag{2}$$

where ρ represents the material density and E_{N} is the normal elastic stiffness given as

$$E_{\mathrm{N}} = \frac{E}{(1-2\nu)}, \tag{3}$$

with E denoting the elastic modulus and ν the Poisson's ratio. To capture the switch of normal behavior between tension and compression, the Heaviside function $H(\sigma_{\mathrm{N}})$ was introduced. If the microplane is subjected to tension in the normal direction, $H(\sigma_{\mathrm{N}}^{+}) = 1$, then normal damage can develop while plastic deformation remains unchanged. In case of compression $H(\sigma_{\mathrm{N}}^{-}) = 0$, the plastic process can take place while the normal damage does not evolve. K_{N} and γ_{N} are the isotropic and the kinematic hardening moduli, respectively. Here the thermodynamic internal variables include: the plastic normal strain $\varepsilon_{\mathrm{N}}^{\mathrm{p}}$, the damage variable ω_{N}, the isotropic and kinematic hardening variables $z_{\mathrm{N}}, \alpha_{\mathrm{N}}$, respectively. The consolidation function $f(r_{\mathrm{N}})$ controls the evolution of tensile damage. The conjugate thermodynamic forces are determined by differentiating the thermodynamic potential (2) with respect to each internal variable.

Tangential direction:

In the tangential direction, cumulative damage is considered as the main source of fatigue damage. This mechanism drives the material deterioration at pulsating subcritical stress levels. The pressure-sensitive interface model presented in (Baktheer & Chudoba 2018b) with fatigue damage due to cumulative inelastic slip is used to describe the tangential constitutive behavior of a microplane. The thermodynamic potential of the microplane in the tangential direction is therefore given as

$$\rho\psi_{\mathrm{T}}^{\mathrm{mic}} = \frac{1}{2}(1-\omega_{\mathrm{T}})E_{\mathrm{T}}(\boldsymbol{\varepsilon}_{\mathrm{T}} - \boldsymbol{\varepsilon}_{\mathrm{T}}^{\pi})\cdot(\boldsymbol{\varepsilon}_{\mathrm{T}} - \boldsymbol{\varepsilon}_{\mathrm{T}}^{\pi}) + \frac{1}{2}K_{\mathrm{T}}z_{\mathrm{T}}^2 + \frac{1}{2}\gamma_{\mathrm{T}}\boldsymbol{\alpha}_{\mathrm{T}}\cdot\boldsymbol{\alpha}_{\mathrm{T}}, \tag{4}$$

where and E_{T} is the tangential elastic stiffness given as

$$E_{\mathrm{T}} = \frac{E(1-4\nu)}{(1+\nu)(1-2\nu)}, \tag{5}$$

while K_{T} and γ_{T} are the isotropic and kinematic strain hardening moduli, respectively. The thermodynamic internal variables are the inelastic tangential strain vector, i.e. the strain vector defining the irreversible strain $\boldsymbol{\varepsilon}_{\mathrm{T}}^{\pi}$, the damage variable ω_{T}, the isotropic hardening internal variable z_{T} and the kinematic hardening vector $\boldsymbol{\alpha}_{\mathrm{T}}$. Similar to the normal direction, the conjugate thermodynamic forces are obtained by differentiating the thermodynamic potential (4) with respect to each internal variable.

The model was calibrated and validated using experimental data for the compressive behavior of concrete under a wide range of loading scenarios as reported in (Baktheer & Chudoba 2021). The results of the validation are exemplified in Figure 6 presenting the ability of the model to reproduce the response under monotonic, cyclic, and fatigue loading with a single set of material parameters. The ability of the macroscale MS1 model to reflect the fatigue induced tri-axial stress redistribution in a material zone using a single point idealization is particularly important in view of high-cycle fatigue modeling when simulating the response cycle by cycle. Further features

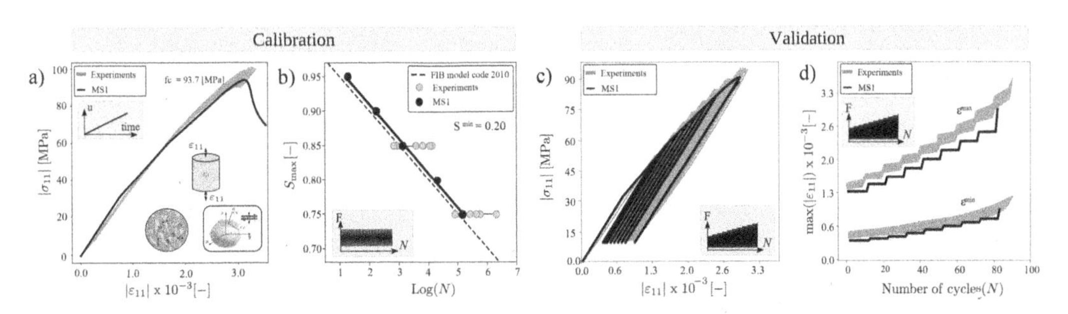

Figure 6. Calibration and validation procedure used for the developed microplane model (MS1) for concrete C80.

and capabilities of the microplane material model MS1 are discussed at (Baktheer, Aguilar, & Chudoba 2021). The breakdown of energy release with fractions ascribed to the included dissipation mechanisms under fatigue loading has been quantified using the MS1 model and discussed in detail in (Aguilar, Baktheer, & Chudoba 2021).

4 PTST NUMERICAL INVESTIGATION

A numerical idealization with a minimal complexity, that would still realistically capture the governing principles of fatigue evolution, is sought with the goal to allow for an efficient cycle-by-cycle simulation of the fatigue response. Moreover, a consistent representation of the fatigue and monotonic degradation for shear dominated loading is required. Therefore, the numerical model only includes the ligament of the PTST, assuming the surrounding bulk material rigid as shown in Figure 7. The axisymmetric representation of the strain and stress state using a FE discretization requires only several degrees of freedom. The microplane fatigue model MS1 is used to model the inelastic phenomena that take place in the ligament. On the inner boundary of the ligament, displacement is constrained in all directions. On the outer ligament boundary, a uniform profile of vertical and horizontal forces/displacements is applied, as visualized in Figure 7.

4.1 Elementary studies

Monotonic behavior. Numerical simulations evaluating the monotonic behavior of the ligament are shown in Figure 8. A displacement-controlled shear load was introduced at three different levels of normal stress as depicted in Figure 8a. The normal stress acting on the ligament was applied first and then held constant while the control slip increased. In Figure 8b, it can be seen that the peak shear force decreases for the tensile normal stress due to normal decohesion. On the other hand, the peak load increases when compressive stress is applied. This result highlight the ability of MS1 to reflect the pressure sensitivity of the shear response observed in the experimental results. The deformation

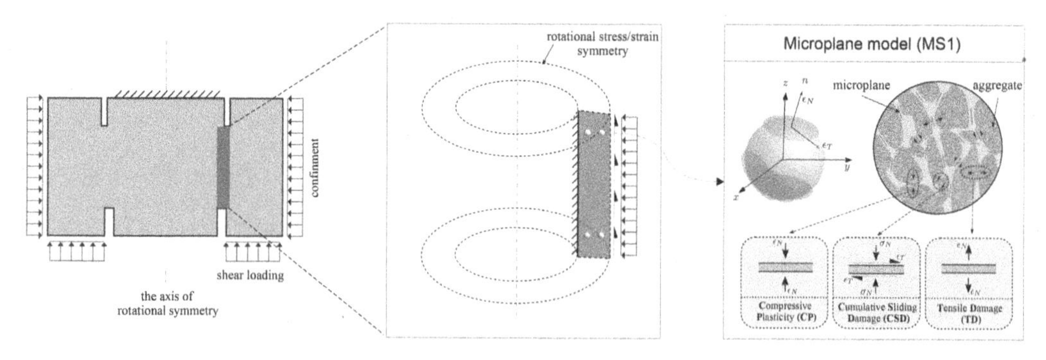

Figure 7. Idealization of the PTST with axi-symmetric representation and microplane material model MS1.

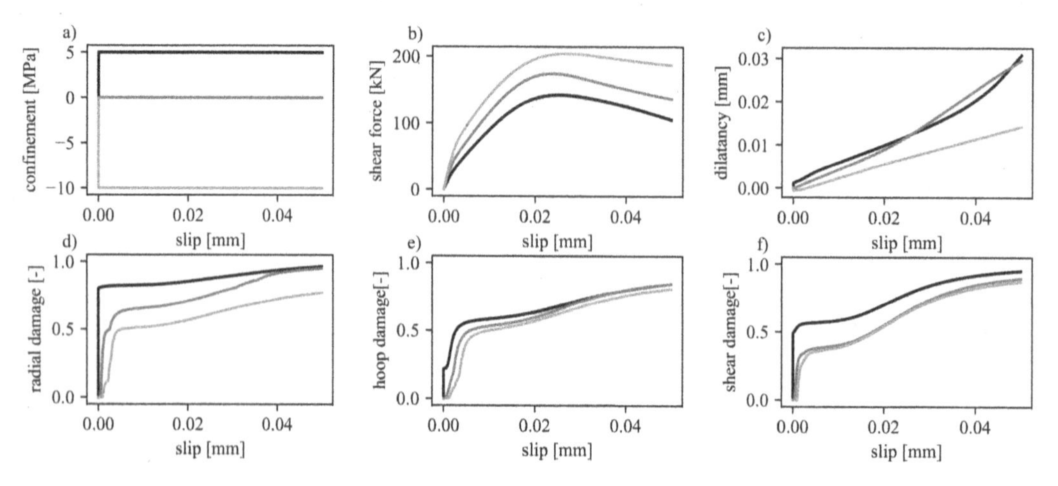

Figure 8. Shear behavior of the ligament under varying normal stresses. Top left panel depicts the applied decohesion/confinement level, top middle panel displays the corresponding force-displacement curve, while the top right panels shows the corresponding evolution of dilatancy displacement. The bottom row displays a measurement of the damage in the radial, hoop and shear directions. Material parameters: $E = 30000.0$, $\nu = 0.18$, $A_d = 5000.0$, $\varepsilon_0 = 0.0001$, $K_N = 0.0$, $\sigma_N^0 = 1000.0$, $\gamma_N = 2000.0$, $\sigma_T^0 = 5.0$, $K_T = 0.0$, $\gamma_T = 2000.0$, $S_T = 0.001$, $c_T = 2.0$, $r_T = 3.0$, $m_T = 0.1$, $p_T = 1.0$.

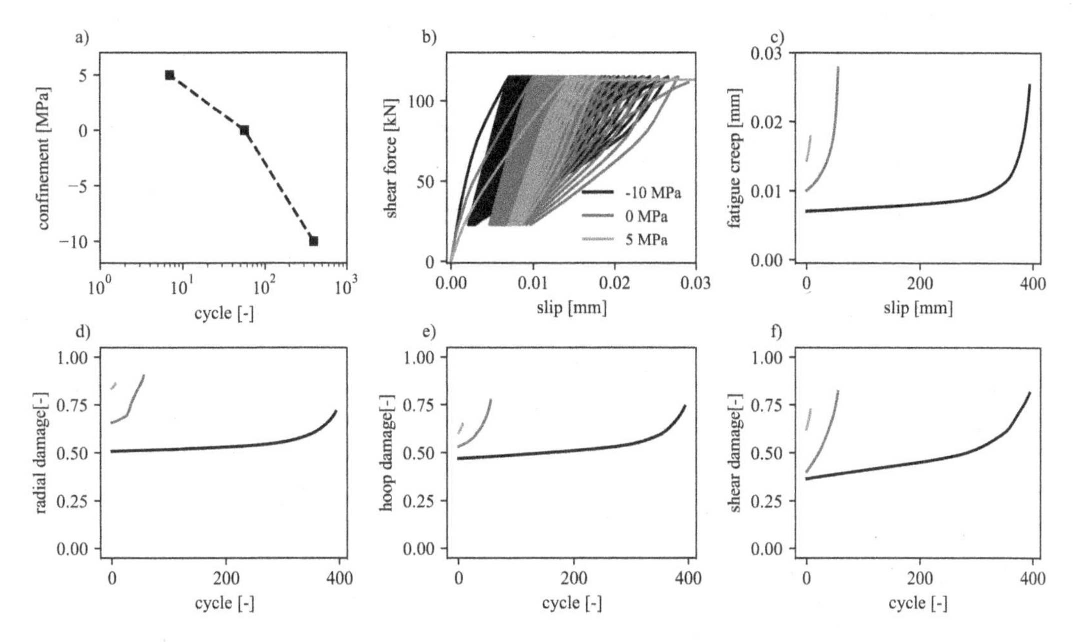

Figure 9. Cyclic shear behavior of the ligament under varying normal stresses. Top left panel depicts the amount of applied cycles for each decohesion/confinement level, top middle panel displays the corresponding force-displacement curve, while the top right panels shows the corresponding evolution of fatigue creep displacements. The bottom row displays a measurement of the damage in the radial, hoop and shear directions. The material parameters are shown in the caption of Figure 8.

of the ligament in the normal direction plotted in Figure 8c reveals a dilatancy effect, which depends on the level of confinement. The obtained dilatancy displacements are of the same order of magnitude as those reported in (Wong, Ma, Wong, & Chau 2007). The lower row in Figure 8 shows the damage evolution in the radial, hoop, and shear directions. In case of the tensile normal loading represented by the black curves, a significant damage develops prior to the application of the sliding displacement.

Cyclic behavior. To examine the ability of the model to reproduce the shear fatigue response under varied normal decohesion/confinement stress, Figure 9 presents a study under subcritical cyclic shear force, with the same material parameters and degree of decohesion/confinement load as in the monotonic studies. The loading range of the cyclic shear force was set between 120 kN for the upper level and 6 kN for the lower level. The number of cycles sustained for the three considered levels of normal stress are shown in Figure 9a. For the tensile normal stress of 5 MPa, the failure occurred after 7 cycles. In the unconfined case the ligament withstood 56 cycles, while for the confined case with precompression of 10 MPa, the material failed after 394 cycles. Figure 9b displays the corresponding force-displacement curves with the opening of the hysteretic loops decreasing for increasing confinement.

The fatigue creep curves showing the slip along the lifetime of the specimen are plotted in Figure 9c for the three levels of lateral stress. These values of slip were recorded in each cycle at the upper load level of the applied shear force. This diagram shows a moderate slope of fatigue creep displacement in its first part for the case associated with the initial compressive level of 10 MPa, which increases drastically during the last cycles until failure. On the other hand, the specimens with normal and zero tensile stress do not exhibit a stable fatigue induced degradation.

This behavior is also reflected in the damage evolution shown in the bottom row of Figure 9 for the radial, hoop and shear damage projections. At a confinement level of 10 MPa, radial and hoop damage dominate during the early cycles. During fatigue loading, damage evolved in a non-proportional manner, reaching higher values of the shear damage right before the ligament failure. The damage accumulation in the radial and circumferential projections results from the triaxial stress redistribution during the fatigue life due to the cumulative shear damage. Note that the material model does not explicitly account for tensile damage accumulation during fatigue.

4.2 *MS1 calibration for monotonic behavior*

In order to quantitatively reproduce the experimental behavior of the considered concrete, the material parameters of the MS1 model were calibrated using a uniaxial stress state to reproduce the strength and stiffness characteristics summarized in Sec. 2.2. The simulated compressive and tensile response plotted Figure 10a and b demonstrate a good fit with the prescribed compressive strength and tensile strength of 96 MPa and 4.3 MPa, respectively.

The calibration of the shear behavior of the material model was performed using the axisymmetric

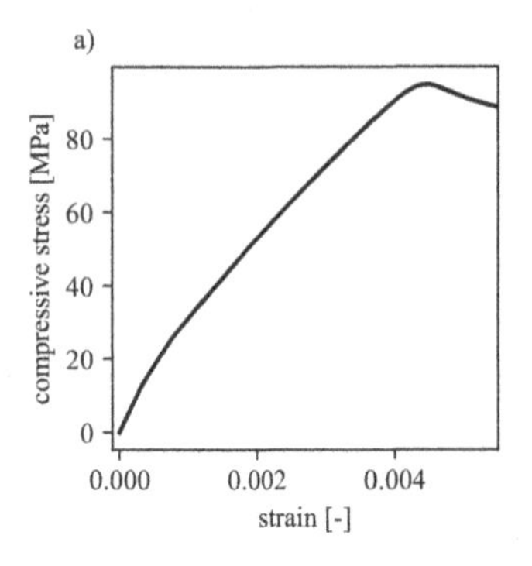

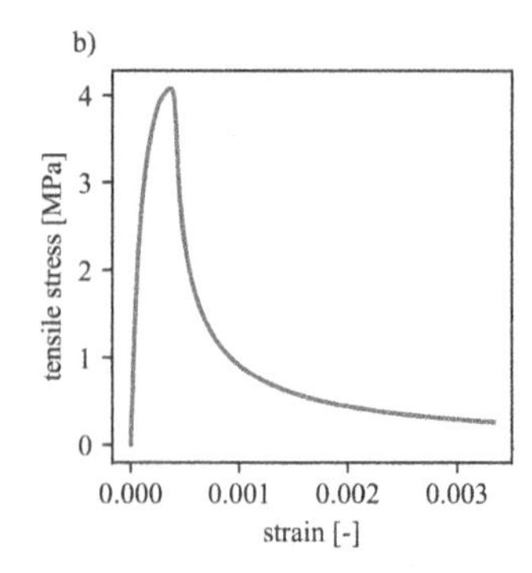

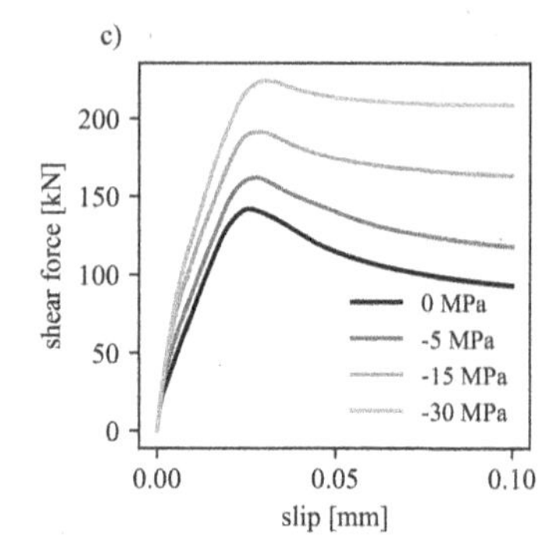

Figure 10. Monotonic compressive, tensile and shear behavior with increasing confinement level. Parameters: $E = 39226.0$, $\nu = 0.18$, $A_d = 7000.0$, $\varepsilon_0 = 1e\text{-}5$, $K_N = 10000.0$, $\sigma_N^0 = 40.0$, $\gamma_N = 60000.0$, $\sigma_T^0 = 2.0$, $K_T = 500.0$, $\gamma_T = 5000.0$, $S_T = 0.003$, $c_T = 10.0$, $r_T = 13.0$, $m_T = 0.1$, $p_T = 6.0$.

idealization shown in Figure 7. The ligament was subjected to 4 precompression levels, namely 0, 5, 15 and 30 MPa. The experimental trend obtained for the monotonic case could be reproduced, as can be seen in Figure 10, with the ultimate loads with the values of 141.99, 162.00, 191.58 and 224.24 kN.

Even though the fatigue behavior, including the tri-axial stress redistribution during subcritical pulsating loading, can already be qualitatively well reproduced, further refinements are required to improve the quantitative representation of the fatigue process.

In particular, the interaction of the normal and tangential dissipative effects at the inter-aggregate level needs to be considered. Integration of the coupled slip-decohesion-compression interface model presented in a companion paper (Chudoba, Vořechovský, Aguilar, & Baktheer 2022), where the behavior of the microplanes in the normal and tangential directions would be coupled in a clear and transparent manner, would allow for a more flexible control of the overall macroscopic response. Further aspects need to be considered at the microplane level to reflect the inter-aggregate consolidation in the initial part of the fatigue creep curve.

5 CONCLUSIONS

The performed studies examine the ability of the fatigue microplane material model MS1 to capture the behavior of concrete subjected to cyclic shear loading under confinement or decohesion normal load. Based on the hypothesis of cumulative shear sliding within the material structure, the study focused on the simulation of the punch-through shear test. This test setup provides the possibility to control the stress configuration along a test ligament in a more flexible way compared to the common cylinder test.

The studies demonstrated the ability of the model to reproduce the qualitative trends observed in the experiments for varied levels of normal decohesion/confinment loads both for monotonic and fatigue types of loading. To enable an efficient simulation of the dissipative mechanisms cycle-by-cycle, the axisymmetric shape of the punch-through shear test was exploited. This modeling concept will be further refined with the aim to effectively support the development of an innovative experimental method for the characterization of concrete fatigue response introducing a flexible control of the normal/shear stress configuration along the test ligament.

AKNOWLEDGMENTS

The authors gratefully acknowledge the support for this research by the German Research Foundation (Deutsche Forschungsgemeinschaft - DFG), in the scope of the Priority Program SPP2020 âŁœ-Cyclic deterioration of high-performance concrete in an experimental virtual lab.âŁž (Project number: 441550460).

REFERENCES

Aguilar, M., A. Baktheer, & R. Chudoba (2021). Numerical investigation of load sequence effect and energy dissipation in concrete due to compressive fatigue loading using the new microplane fatigue model MS1. In *Onate, E., Peric, D., Chiumenti, M., de Souza Neto, E., Eds; COMPLAS 2021; Barcelona, Spain.*

Baktheer, A., M. Aguilar, & R. Chudoba (2021). Microplane fatigue model MS1 for plain concrete under compression with damage evolution driven by cumulative inelastic shear strain. *International Journal of Plasticity*, 102950.

Baktheer, A. & R. Chudoba (2018a). Modeling of bond fatigue in reinforced concrete based on cumulative measure of slip. In *Computational Modelling of Concrete Structures, EURO-C 2018*, pp. 767–776. CRC Press.

Baktheer, A. & R. Chudoba (2018b). Pressure-sensitive bond fatigue model with damage evolution driven by cumulative slip: Thermodynamic formulation and applications to steel- and frp-concrete bond. *International Journal of Fatigue 113*, 277 – 289.

Baktheer, A. & R. Chudoba (2021). Experimental and theoretical evidence for the load sequence effect in the compressive fatigue behavior of concrete. *Materials and Structures 54*(2), 82.

Becks, H. & M. Classen (2021). Mode II behavior of high-strength concrete under monotonic, cyclic and fatigue loading. *Materials 14*(24), 7675.

Chudoba, R., M. Vořechovský, M. Aguilar, & A. Baktheer (2022). Coupled sliding-decohesion-compression model for a consistent description of monotonic and fatigue behavior of material interfaces. *Computer Methods in Applied Mechanics and Engineering*, under review.

Desmorat, R., F. Ragueneau, & H. Pham (2007, February). Continuum damage mechanics for hysteresis and fatigue of quasi-brittle materials and structures. *International Journal for Numerical and Analytical Methods in Geomechanics 31*(2), 307–329.

Do, M.-T., O. Chaallal, & P. Aïtcin (1993). Fatigue behavior of high-performance concrete. *Journal of Materials in Civil Engineering 5*(1), 96–111.

Kim, J.-K. & Y.-Y. Kim (1996). Experimental study of the fatigue behavior of high strength concrete. *Cement and Concrete Research 26*(10), 1513 – 1523.

Kirane, K. & Z. P. Bažant (2015). Microplane damage model for fatigue of quasibrittle materials: Sub-critical crack growth, lifetime and residual strength. *International Journal of Fatigue 70*, 93–105.

Luong, M. P. (1989). Fracture behaviour of concrete and rock under mode ii and mode iii shear loading. *Fracture of Concrete and Rock: Recent Developments, ed by S. Shah, SE Swartz and B. Barr*, 18–26.

Oneschkow, N., T. Timmermann, & S. Löhnert (2022). Compressive fatigue behaviour of high-strength concrete and mortar: Experimental investigations and computational modelling. *Materials 15*(1).

Rybczynski, S., G. Schaan, M. Dosta, M. Ritter, & F. Schmidt-Döhl (2021). Discrete element modeling and electron microscopy investigation of fatigue-induced microstructural changes in ultra-high-performance concrete. *Materials 14*(21).

Schäfer, N., V. Gudžulić, R. Breitenbücher, & G. Meschke (2021). Experimental and numerical investigations on high performance sfrc: Cyclic tensile loading and fatigue. *Materials 14*(24).

Skarzynski, L., I. Marzec, & J. Tejchman (2019). Fracture evolution in concrete compressive fatigue experiments based on x-ray micro-ct images. *International Journal of Fatigue 122*, 256–272.

Song, Z., H. Konietzky, & X. Cai (2021). Modulus degradation of concrete exposed to compressive fatigue loading: Insights from lab testing. *Structural Engineering and Mechanics 78*(3), 000–000.

Wong, R., S. Ma, R. Wong, & K. Chau (2007). Shear strength components of concrete under direct shearing. *Cement and Concrete Research 37*(8), 1248–1256.

Computational Modelling of Concrete and Concrete Structures – Meschke, Pichler & Rots (Eds)
 ISBN: 978-1-032-32724-2

Simulation of self-compacting steel fibre reinforced concrete using an enhanced SPH methodology

Abdulkarim Mimoun & Sivakumar Kulasegaram
School of Engineering, Cardiff University, Cardiff, UK

ABSTRACT: Accurate prediction of self-compacting fibre reinforced concrete (SCFRC) flow, passing and filling behaviour is not a trivial task, particularly in the presence of heavy reinforcement, complex formwork shapes and large size of aggregates. In this regard, complex formwork shapes and large size of aggregate can play an important role in fibre orientation and distribution during the flow of fibre reinforced self-compacting concrete and can thus significantly influence mechanical behaviour of the hardened material. Due to the nature of self-compacting concrete mix and widely varying properties of its constituents, it is hugely challenging to understand the rheological behaviour of the concrete mix. For this reason, it is necessary to thoroughly comprehend fresh property by understanding its rheology. The quality control and accurate prediction of the SCFRC rheology are crucial for the success of its production.

A three-dimensional meshless smoothed particle hydrodynamics (SPH) computational approach, treating the SCFRC mix as a non-Newtonian Bingham fluid constitutive model has been coupled with the Lagrangian momentum and continuity equations to simulate the flow. The aim of this numerical simulation is to investigate the capabilities of the SPH methodology in predicting the flow and passing ability of SCFRC mixes through gaps in reinforcing bars. To confirm that the concrete mixes flow homogeneously, the distribution and orientation of steel fibres in the mixes have been simulated and compared against observations made in the laboratory experiments. It is revealed that the simulated flow behaviour of SCFRC compares well with results obtained in the laboratory tests.

Keywords: Self-compacting steel fibre reinforced concrete, Fibre orientation and distribution, Smooth particle hydrodynamic, 3D simulation.

1 INTRODUCTION

Self-compacting steel fibre reinforced concrete (SCSFRC) may contribute to significant development of high quality complex concrete structures and open new applications for concrete. The addition of fibres makes the fresh concrete stiffer and reduces its workability. This also limits the number of fibres that can be uniformly orientated and distributed in the presence of heavy reinforcement, complex formwork shapes and large aggregates. As the orientation of fibres alter throughout the production of the concrete, it is essential to understand these changes in the fibre orientation. In particular, the orientation and distribution of fibres may be significantly affected, especially when the concrete is cast in the presence of heavy reinforcements. In the past, most of the study focused on visual inspection of fibres in hardened concrete parts cut after casting (Bernasconi et al. 2012; Lee *et al.* 2002; Zak et al. 2001) and the prediction of the typical orientation factor of fibres from the cut sections (Martinie & Roussel 2011).

The prediction of SCSFRC flow and its passing and filling behaviour is very challenging particularly within the congestion of reinforced formwork geometry and in the presence of reinforcing steels. However, an understanding of the flow behaviour and its properties is important for producing high-quality SCC. The most cost-effective way to gain such an understanding is by performing computational simulations, which will enable one to fully characterize the flow behaviour of SCSFRC and to reveal the orientation of fibres inside the complex formwork shapes. The rheological behaviour of fresh concrete mix must be consistent with the formworks of complex shapes to ensure the production of complete and high-quality casting of structural elements. In this work, a three-dimensional Lagrangian smooth particle hydrodynamics (SPH) method is used to simulate and predict the flow of SCSFRC. SPH method is ideal for simulating the flow to analyze the passing ability and filling ability of SCSFRC mixes, irrespective of their characteristic compressive strength. These simulations will provide information about the distribution and orientation of fibres throughout the entire process of casting SCSFRC into the formworks of complex shapes to ensure that the mixture flows as a homogeneous mass without any sign of segregation or blockage. In this

DOI 10.1201/9781003316404-20

paper, a simple method has been established to predict the distribution and orientation of steel fibres in self-compacting concrete mixes during flow-ability (Slump Flow test), pass-ability and fill-ability (L-box test) tests.

2 DEVELOPMENT OF THE SELF COMPACTING NORMAL-STRENGTH STEEL FIBRE REINFORCED CONCRETE MIXES.

A laboratory study was conducted to produce various grades of SCSFRC mixes (with nominal 28 days cube compressive strengths of 30 MPa,40 MPa,50 MPa,60 MPa and 70 MPa). The fundamental materials and secondary materials for the SCC mix are produced following the European Federation of Specialist Construction Chemicals and Concrete Systems (EFNARC) guidelines (EFNARC 2005). These mixes were developed using a mix design method for SCC based on the desired target plastic-viscosity and compressive strength in accordance with mix design method proposed by Abo Dhaheer et al. (2016a, 2016b), which rationalized and simplified the method recommended previously by Karihaloo and Ghanbari (2012), Deeb and Karihaloo (2013). As an example, the amounts and specifications of the components used in the SCC design mix with target compressive strength 40MPa are shown in Table 1. Portland limestone cement (PLC) (CEM II/A-L/32.5R) conforming to (BS EN 197-1 2011) with a specific gravity of 2.95 and Ground granulated blast-furnace slag (GGBS) with a specific gravity of 2.40 were used as the main cement and cement replacement materials respectively. A new generation of polycarboxylic ether-based superplasticiser (SP) with specific gravity of 1.07 was used in all the test mixes. Crushed limestone coarse aggregate with maximum particle size of 20 mm and a specific gravity of 2.80 was used, while the fine aggregate was river sand (less than 2 mm) having a specific gravity of 2.65. Limestone powder (LP) as a filler with maximum particle size of 125 μm (specific gravity 2.40) was used. A part of the river sand was substituted by an equal amount of the coarser fraction of LP in the size range 125 μm – 2 mm. All mixes were tested in the fresh state utilizing slump flow and L-box tests. The plastic viscosity of each mix was calculated using the micro-mechanical procedure described by (Ghanbari & Karihaloo 2009).

Table 1. Constituents and proportions for SCSFRC mixes (kg/m^3).

Mix strength grade 40 MPa	
Cement: kg/m^3	270
GGBS*: kg/m^3	90
Cementitious materials (cement + GGBS): kg/m^3	360
Water: kg/m^3	205
Superplasticiser: kg/m^3	2.3
Water/cementitious materials ratio	0.57
Superplasticiser/cementitious materials ratio	0.64
Limestone powder: kg/m^3	143
steel fibre volume (SF) (0.5%): kg/m^3	40
Fine aggregate (<2 mm)**	740
FA^a: kg/m^3	240
FA^b: kg/m^3	500
Coarse aggregate: kg/m^3 (size of 10 mm)	839
t_{500} in flow: s	1.40
Slump Flow spread : mm	600
Plastic viscosity (PV): Pa. S	27

*Ground granulated blast-furnace slag.
**Fine aggregate <2 mm (Note: a part of the fine aggregate is the coarser fraction of the limestone powder, FA^a 125 μm–2 mm, whereas FA^b refers to natural river sand < 2 mm).

3 MODELLING THE FLOW-ABILITY, PASSING-ABILITY AND FILLING-ABILITY OF STEEL FIBRE SUSPENDED SELF-COMPACTING CONCRETE BASED ON EXPERIMENT

In the rheological studies to characterize fresh concrete, SCSFRC is understood as a suspension of solid particles (coarse aggregate and steel fibre) in a fluid phase (cement paste), in which various of particle size distribution and fluid stage are typically enhanced to meet the three main properties of SCSFRC in the fresh state: filling ability, passing ability, and segregation resistance.

SCSFRC mixes are designed to meet flowability and cohesiveness (i.e., resistance to segregation) standards utilizing the slump cone test. In this test, the time for the SCSFRC mix to spread to a diameter of 500 mm (T_{500}) after the cone filled with the mix has been suddenly lifted is recorded, as well as the diameter of the spread when the flow stops (EFNARC 2005). The resistance to segregation and blockage is checked visually between the coarse aggregate and steel fibre (Figure 1).

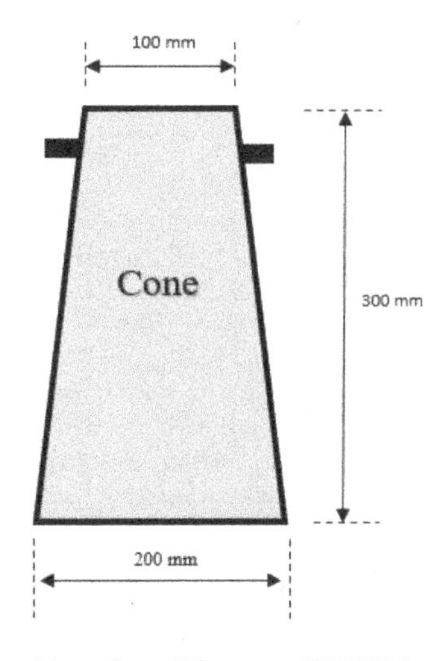

Figure 1. Dimension of Flow test of SCSFRC.

To test the ability of a SCSFRC mix to pass and fill the formwork containing reinforcement under its own weight, the L-box apparatus is used (EFNARC 2005). The vertical leg of the L-box is initially filled with the SCSFRC mix. At the bottom of this leg is a gate with two or three rods in front of it. When the gate is lifted, the mix flows into the horizontal part of the L-box through the gaps between the rods. The times for the mix to reach 200 mm (T_{200}) and 400 mm (T_{400}) from the gate are recorded, as well as the time it takes the mix to level off in the horizontal leg of the L-box. Again, it is required that no coarse large aggregate particles or steel fibres be blocked by the rods (Figure 2).

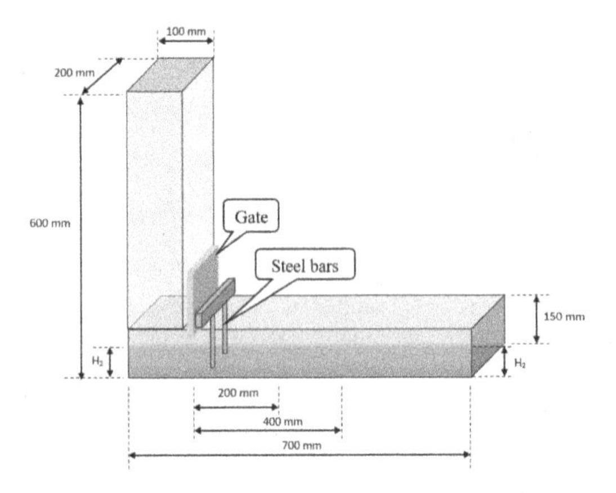

Figure 2. Dimension of L-box test of SCSFRC.

The addition of steel fibres enhances the mechanical characteristics and the ductility of SCC in much the same approach as in vibrated concrete (VC). Nevertheless, the fibres significantly impair the workability of SCC due to their elongated shape and large surface area. The volume of fibre added to a SCC mixture is therefore limited and depends on the fibre type used and the composition of the SCC mix. Therefore, the maximum volume of fibres is decided in such a way to manage the workability, whilst maintaining excellent flowing and passing ability. For the optimum results, the fibres require to be homogeneously distributed in the mixture without clustering or segregation and blockage.

4 MODELLING SIMULATION THE FLOW-ABILITY, PASSING-ABILITY AND FILLING-ABILITY OF SCSFRC

Since fresh SCSFRC flow in slump and L-box test configurations is a gravitational flow with large deformations, a three-dimensional smooth particle hydrodynamic (SPH) mesh-less numerical methodology is chosen here to simulate the fresh state. This section briefly introduces the fundamental governing equations, numerical model and the boundary conditions required for modelling SCSFRC flow in slump flow and L-box of tests with 3D-Lagrangian SPH method.

4.1 *Governing equations*

Fresh SCSFRC mix is a non-Newtonian incompressible fluid which can be described by a bilinear Bingham-type model with relation between the shear stress and shear strain rate which includes two material parameters: the yield stress (τ_y) (Badry et al. 2014) and the plastic viscosity (η) (Ghanbari & Karihaloo 2009) (Papanastasiou 1987).

$$\tau = \eta\dot{\gamma} + \tau_y\left(1 - e^{-m\dot{\gamma}}\right) \tag{1}$$

Where m is a very large number (e.g., m = 100). There are two basic equations to be solved in the SPH method, together with the constitutive relation – the incompressible mass and momentum conservation equations.

$$\frac{1}{\rho}\frac{D\rho}{D_t} + \nabla v = 0 \tag{2}$$

To include the effect of immersed boundary, the momentum conservation equation is modified:

$$\frac{Dv}{Dt} = -\frac{1}{\rho}\nabla P + \frac{1}{\rho}\nabla.\tau + g + f \tag{3}$$

where ρ, t, v, P, g and τ represent the fluid particle density, time, particle velocity, pressure, gravitational acceleration, and shear stress tensor, respectively. Here, f represents the effective reaction force of the fibres on the fluid at any chosen location. However, the reaction force f will not be acting on the fluid particles which do not fall within the radius of influence (or smoothing length) of any boundary particles of a given fibre. In the proposed numerical procedure, fibres are described by immersed boundaries.

4.2 *Numerical implementation*

A projection method based on the predictor–corrector time stepping scheme was adopted to track the Lagrangian non-Newtonian flow. The prediction step is an explicit integration in time without enforcing incompressibility. Only the viscous stress and gravity terms are considered in the momentum equation (Equation 3) and an intermediate particle velocity is obtained as:

$$v_{n+1}^{*} = v_n + \left(g + f + \frac{1}{\rho}\nabla.\tau\right)\Delta t \tag{4}$$

in which v_n and v_{n+1}^{*} are the particle velocity and intermediate particle velocity at time t_n and t_{n+1}, respectively. Then the correction step is performed by considering the pressure term in Equation 3

$$\frac{v_{n+1} - v_{n+1}^{*}}{\Delta t} = -\left(\frac{1}{\rho}\nabla P_{n+1}\right) \tag{5}$$

Rearranging Equation 5 gives

$$\frac{v_{n+1} - v_{n+1}^{*}}{\Delta t} = -\left(\frac{1}{\rho}\nabla P_{n+1}\right) \tag{6}$$

where v_{n+1} is the corrected particle velocity at time step t_{n+1}. By imposing the incompressibility condition in the mass conservation equation (Equation 2), the pressure P_{n+1} in Equation 6 will be obtained. As the particle density remains constant during the flow, the velocity v_{n+1} is divergence-free so that Equation 2 can be simplified as.

$$\nabla \cdot v_{n+1} = 0 \tag{7}$$

Substitution into Equation 6 gives

$$\nabla \left(\frac{1}{\rho} \nabla P_{n+1} \right) = \frac{\nabla \cdot v_{n+1}^*}{\Delta t} \tag{8}$$

which can be rewritten as

$$\nabla^2 P_{n+1} = \frac{\rho}{\Delta t} \nabla v_{n+1}^* \tag{9}$$

where ∇^2 is the Laplacian. Solution of the second-order Poisson equation (Equation 9) gives the pressure from which the particle velocity is updated (see Equation 6). Finally, the instantaneous particle position is updated using the corrected velocity.

$$x_{n+1} = x_n + v_{n+1} \Delta t \tag{10}$$

where x_{n+1} and x_n are the particle positions at tt_{n+1} and t_n, respectively.

The interaction between SPH fluid particles surrounding the rigid fibres can be modelled by treating fibres as immersed boundaries (Figure 3). This would offer a simple and efficient methodology to determine fibre orientations and distribution during the flow.

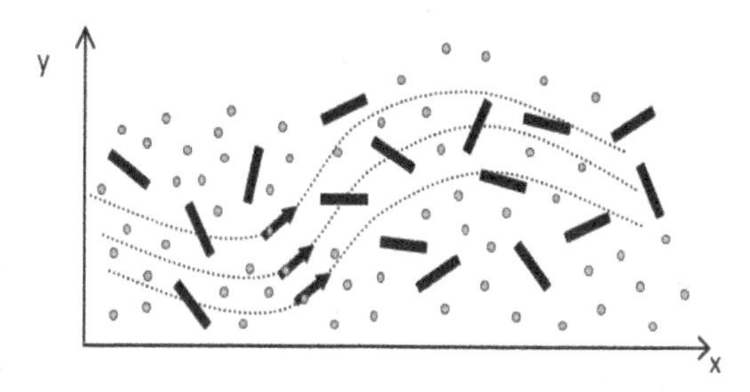

Figure 3. Schematic diagram of the flow of SCSFRC with rigid steel fibres.

5 INITIAL CONFIGURATION AND BOUNDARY CONDITION

When solving the Navier-Stokes and continuity equations, appropriate initial and boundary conditions need to be applied. Three types of boundary conditions need to be considered in the simulation of slump cone test; a zero-pressure condition on the free surface, Dirichlet boundary condition at the wall of the cone, and Neumann conditions on the pressure gradient as illustrated in Figure 4.

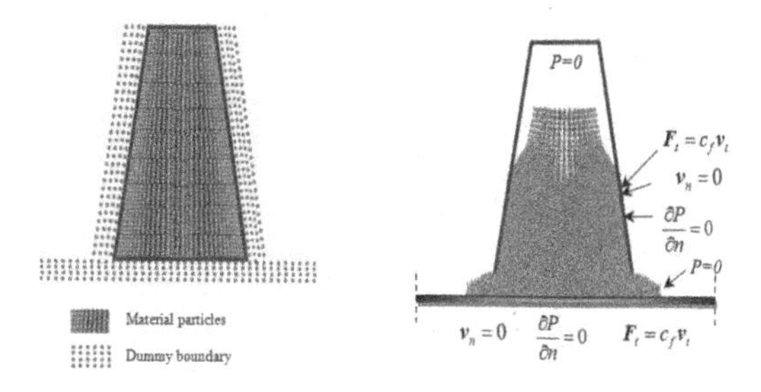

Figure 4. Slump flow and L-box test initial condition.

Four arrays of rigid dummy particles placed outside the wall of the cone were used to implement the wall boundary conditions with space r_o between the arrays, where r_o is the initial particle spacing. To represent the non-slip boundary conditions along the cone wall, the velocity of both the wall and dummy particles must be zero. Friction between SCC flow and boundaries was also considered and imposed on the cone wall and the bottom plate with a dynamic coefficient of friction between the SCC mix and steel equal to 0.55 Ns/m.

6 THREE-DIMENSIONAL SIMULATION RESULTS

To examine how the steel fibres will distribute and orient themselves throughout the filling process, flow (slump flow) and pass/fill (L-box) tests were conducted for SCC mix with steel fibre (Mix 40 MPa, Table 1). The steel fibres were treated as described above. The plastic viscosity (i.e. 27 Pa s) of the mix was estimated analytically using micro-mechanics based formulations. The yield stress and the dynamic coefficient of friction with the steel wall of the cone and the base plate were assumed to be 200 Pa and 0.55 Ns/m respectively.

Figure 5 illustrate the distribution of fibres and their orientation during the numerical simulation of slump flow. During the simulation of slump flow, the time for the mixtures to spread to a diameter of 500 mm ($T_{500} = 1.45$ sec) matches closely with the time measured in the laboratory (Table 2). The surface of the spread is smooth, and the fibres stay homogeneously always distributed during the flow. Similarly, the L-box tests also produced comparable results with experimental observations

The proposed method can be successfully applied in the numerical simulation of SCSFRC flow to analyze the flow, passing and filling behavior of these highly viscous fluids. The numerical results are in excellent agreement with experimental results and validate that

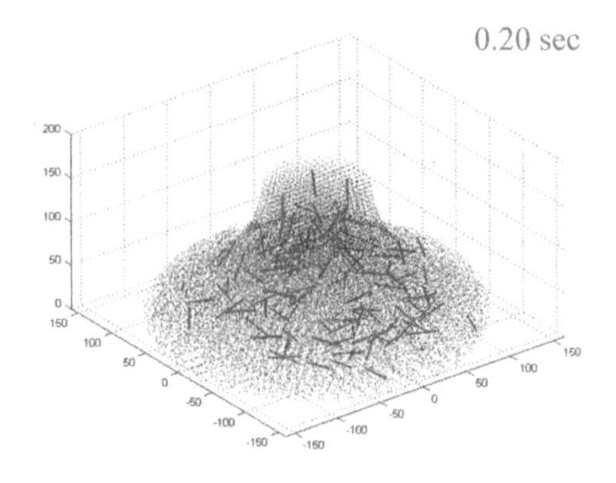

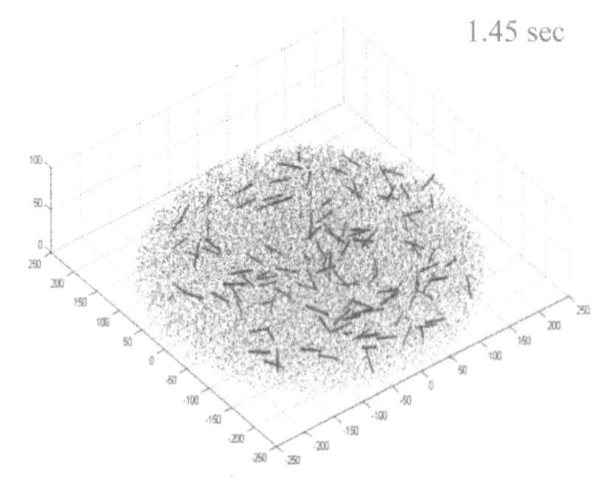

Figure 5. 3D numerical simulation of slump flow test for SCSFRC (0.5%vol fibre).

Table 2. Comparison of experimental and simulations results for slump flow and L-box tests.

	Mix strength grade: 40 MPa	
	Simulation	Experiment
t_{500} mm in Flow: s	1.45	1.40
Flow spread mm	615	600
T_{200} in L-box: s	0.90	0.93
T_{400} mm in L-box: s	1.90	1.90

Figure 6. Experiment of slump flow test for SCSFRC (0.5%vol fibre).

the 3D SPH methodology can effectively predict the flow of fresh SCSFRC mix.

7 CONCLUSIONS

A Lagrangian SPH method has been used to simulate the flow of self-compacting normal performance concrete with steel fibre during the slump flow and L-box tests in 3-dimensional configurations. A appropriate Bingham model (Ghanbari & Karihaloo 2009) has been coupled with the Lagrangian Navier-Stokes and continuity equations to model this flow. The mixture characteristics of the SCSFRC mix have been fully incorporated implicitly through the plastic viscosity, which has been assessed exploiting the micromechanical model described in (Ghanbari & Karihaloo 2009).

The simulation of SCSFRC mixes focused on the orientations of fibres and their distributions during the flow, passing, and filling phases of the slump flow and L-box tests. The established numerical methodology is able to capture the flow, passing, and filling behaviour of SCSFRC mixes and to provide insight into the distribution of fibres and their orientations during these phases. The comparison of the experimental and the simulation results is very encouraging. More flow simulations and validations of SCSFRC mixes (with nominal 28 days cube compressive strengths between 30 to 70 MPa) to be carried out to perform additional parametric studies and to further establish the accuracy of the numerical model.

REFERENCES

Abo Dhaheer, M. S. *et al.* (2016a) 'Proportioning of self-compacting concrete mixes based on target plastic viscosity and compressive strength: Part II – experimental validation', *Journal of Sustainable Cement-Based Materials*. Taylor & Francis, 5(4), pp. 217–232. doi: 10.1080/21650373.2015.1036952.

Abo Dhaheer, M. S. *et al.* (2016b) 'Proportioning of self–compacting concrete mixes based on target plastic viscosity and compressive strength: Part I – mix design procedure', *Journal of Sustainable Cement-Based Materials*. Taylor & Francis, 5(4), pp. 199–216. doi: 10.1080/21650373.2015.1039625.

Badry, F., Kulasegaram, S. and Karihaloo, B. L. (2014) 'Estimation of the yield stress and distribution of large aggregates from slump flow test of self-compacting concrete mixes using smooth particle hydrodynamics simulation', *Journal of Sustainable Cement-Based Materials*. Taylor & Francis, 5(3), pp. 117–134. doi: 10.1080/21650373.2014.979266.

Bernasconi, A., Cosmi, F. and Hine, P. J. (2012) 'Analysis of fibre orientation distribution in short fibre reinforced polymers: A comparison between optical and tomographic methods', *Composites Science and Technology*. Elsevier Ltd, 72(16), pp. 2002–2008. doi: 10.1016/j.compscitech.2012.08.018.

BS EN 197-1 (2011) 'Cemet?: Composition, specifications and conformity criteria for common cements', *BSI*.

Deeb, R. and Karihaloo, B. L. (2013) 'Mix proportioning of self-compacting normal and high-strength concretes', *Magazine of Concrete Research*, 65(9), pp. 546–556. doi: 10.1680/macr.12.00164.

EFNARC (2005) 'The European guidelines for self–compacting concrete–specification, production and use.', *The European Guidelines for Self Compacting Concrete*, (May).

Ghanbari, A. and Karihaloo, B. L. (2009) 'Prediction of the plastic viscosity of self-compacting steel fibre reinforced concrete', *Cement and Concrete Research*. Elsevier Ltd, 39(12), pp. 1209–1216. doi: 10.1016/j.cemconres.2009.08.018.

Karihaloo, B. L. and Ghanbari, A. (2012) 'Mix proportioning of selfcompacting high-and ultrahigh-performance concretes with and without steel fibres', *Magazine of Concrete Research*, 64(12), pp. 1089–1100. doi: 10.1680/macr.11.00190.

Lee, Y. H. *et al.* (2002) 'Characterization of fiber orientation in short fiber reinforced composites with an image processing technique', *Materials Research Innovations*, 6(2), pp. 65–72. doi: 10.1007/s10019-002-0180-8.

Martinie, L. and Roussel, N. (2011) 'Simple tools for fiber orientation prediction in industrial practice', *Cement and Concrete Research*. Elsevier Ltd, 41(10), pp. 993–1000. doi: 10.1016/j.cemconres.2011.05.008.

Papanastasiou, T. (1987) 'Flows of Materials with Yield', *Journal of Rheology*, 31(5), pp. 385–404. doi: 10.1122/1.549926.

Zak, G., Park, C. B. and Benhabib, B. (2001) 'Estimation of three-dimensional fibre-orientation distribution in short-fibre composites by a two-section method', *Journal of Composite Materials*, 35(4), pp. 316–339. doi: 10.1106/65LQ-1UK7-WJ9H-K2FH.

Computational Modelling of Concrete and Concrete Structures – Meschke, Pichler & Rots (Eds)
 ISBN: 978-1-032-32724-2

Decarbonizing concrete with artificial intelligence

Y. Song, B. Ouyang, J. Chen, X. Wang, K. Wang, S. Zhang, Y. Chen, G. Sant & M. Bauchy
University of California, Los Angeles, CA, USA

ABSTRACT: Concrete is a key enabler for modern infrastructures but also a top source of carbon emissions in societal development. Because of the high degree of freedom in concrete mixture design, the optimization of concrete formulas remains broadly empirical and inefficient. Further, the concrete carbon footprints is seldomly considered in the conventional mixture design protocols. Here, we approach the concrete optimization from a novel angle of artificial intelligence, where a machine learning model is trained based on a large dataset of 1,150 representative concrete formulas that are developed in a quality control lab for guiding real concrete production. The results demonstrate that our model achieved an unprecedented accuracy for predicting concrete strength at various ages. By further associating each model input with the corresponding carbon embodiment, the machine learning model is used for designing high-performance concrete mixtures that are optimized for both strength and sustainability.

1 INTRODUCTION

Concrete, a man-made stone with rocks and sands bonded by cement, is the most produced engineering material. The ease of fabrication, low cost, and self-hardening nature make concrete ubiquitous in modern construction (Mehta & Monteiro 2014). With annual production estimated at 33 billion tons over the past years (Bauchy 2017), the consumption of concrete (and cement thereof) surpasses any other alternatives by a large margin and the huge demand for concrete materials will continue surging in the predictable future (CEMBUREAU 2015). Meanwhile, concrete takes a considerable share (approximately 7%) of the global CO_2 emissions (Ali et al. 2011). Given the massive scale of the concrete industry, addressing the high carbon emissions in concrete in the concrete industry is pressing than ever. On the other hand, even a slight drop of CO_2 emissions in concrete production matters for achieving the United Nations' goal of global carbon neutrality by 2050 (Carbon neutrality by 2050: the world's most urgent mission | United Nations Secretary-General 2021). However, despite the profound advances in fundamental concrete science over the past decades, there is still a huge potential to reduce its carbon embodiment.

Among all the influencing factors, the mixture design resides at the core of determining the performance of concrete materials, as well as the carbon embodiment (which is primarily correlated to the cement usage) (Ali et al. 2011). In that regard, continuous efforts have been devoted over the past decades to developing models for predicting the performance of a given concrete design, especially the compressive strength (Breysse & Martínez-Fernández 2014; Chopra et al. 2018; Chou et al. 2011, 2014; Gupta n.d.; Khoury et al. 2002; Moutassem & Chidiac 2016). From a practical perspective, an ideal model should be able to predict the performance of new concrete mixtures, so that it can offer a holistic optimization for multiple performance metrics, such as mechanical properties, constructability, durability, as well as material cost (Biernacki et al. 2018; Provis 2015). With the pressing need for construction sustainability, it is also imperative to extend the scope of the optimization to reduce the carbon footprint in concrete. To this end, conventional studies have achieved different levels of success by building physics/chemistry-based models (Popovics 1998; Powers 1960; Zain & Abd 2009). However, due to the high degree of freedom in concrete mix design (e.g., water-to-cement ratio, dosage of supplementary cementitious materials, aggregate property and gradation, effects of chemical admixtures, etc.) and practical constraints that have to be reconciled (e.g., workability, setting behavior, air content, corrosion potential, etc.), the existing knowledge about concrete materials is often limited to idealized conditions and hard to be systematically scaled to real production (Burris et al. 2015; Wild et al. 1995).

The recent advances in artificial intelligence (AI) provide a promising route for projecting the design of a concrete mixture to its actual engineering performance. In particular, machine learning models are excel at finding the implicit pattern between the input features (e.g., raw material proportions) and output target (e.g., concrete strength), whereby the complex relationship can be established without the need for explicit knowledge (Pedregosa et al. 2011). In recent

DOI 10.1201/9781003316404-21

years, the use of machine learning has become an emerging trend in concrete research. A number of recent studies placed their focuses on applying various machine learning techniques to predict the macroscopic performance of cementitious materials such as strength (Chou et al. 2014; Chou et al. 2011; Chopra et al. 2018; Gupta n.d.; Oey et al. 2020; Ouyang et al. 2021, 2020; Young et al. 2019), durability (Das et al. 2019; Cai et al. 2020; Hoang et al. 2017; Okazaki et al. 2020), and various material qualities (Bangaru et al. 2019; Das et al. 2020; Song et al. 2020). However, limited success is seen in using machine learning models as an innovative tool for the design of concrete mixtures (Choi et al. 2020; Ziolkowski & Niedostatkiewicz 2019), especially for reducing carbon footprints in concrete production.

This study aims to apply state-of-the-art artificial intelligence techniques to model the concrete strength at various ages and further turn the model into a useful tool for guiding the design of sustainable concrete mixtures. To this end, we build a neural network model that can predict concrete strength based on a series of design parameters of a concrete mixture. Our model is trained based on a large concrete database of 1,150 individual mixture formulas that are validated by an industrial lab for guiding the production of more than 20,000 real concrete mixtures. Test results demonstrate that our model achieved unprecedented accuracy in predicting the strength development of these concrete mixtures, where the average prediction error is closed to the intrinsic strength variation of concrete. To the best of our knowledge, this is the first time that AI consistently reaches this level of accuracy on such a large-scale dataset. By using this model to predict a series of mixtures that are never involved in the model training, we further illustrate that AI has a vast potential in advancing the concrete mixture design, wherein the carbon embodiment of many existing concrete mixtures can be slashed without sacrificing concrete strength.

2 METHODOLOGY

2.1 *Concrete dataset*

Data is fundamental to enable any machine learning analysis. In this study, we adopt a concrete dataset comprising 1,150 concrete formulas, as generated by a quality control lab of a major concrete producer in the USA. These lab mixtures are designed and tested for guiding the production of more than 20,000 concrete mixtures in real production, based on the three-point curve method as specified by ACI 318 (ACI CODE-318-19: Building Code Requirements for Structural Concrete and Commentary 2021). Herein, we consider a total number of 20 features as the inputs of the model, which include water-to-cementitious ratio (w/cm), mass fractions between the solid materials (cement, class F fly ash, slag, coarse and fine aggregates), properties of both coarse and fine aggregates (specific gravity, fineness, absorption), dosages of different chemical admixtures involved (high-, mid-, low-range plasticizers, air entrainer, retarder, viscosity modifying admixture, shrinkage reducing admixture) as normalized based on the weight of cementitious materials, along with the age of hydration. On the other side, the labels of those mixtures for training and testing the model are their compressive strengths measured at 3, 7, 28, and 56 days. It should be noted that, due to the actual test schedule from the lab, some of the mixtures do not have the strength measurements at all four ages. The actual numbers of strength labels are 599, 1137, 1183, and 787, chronologically. In this study, since our primary focus of the strength prediction is on the 28-day strength, the fewer labels for 3 and 56 days are not considered to have an impact on our machine learning analysis. As a reference, some correlations between the key input features and 28-day strength from the curated dataset are displayed in Figure 1. It can be seen that the datapoints are distributed widely over the maps, which also indicates the difficulties for achieving an accurate strength prediction.

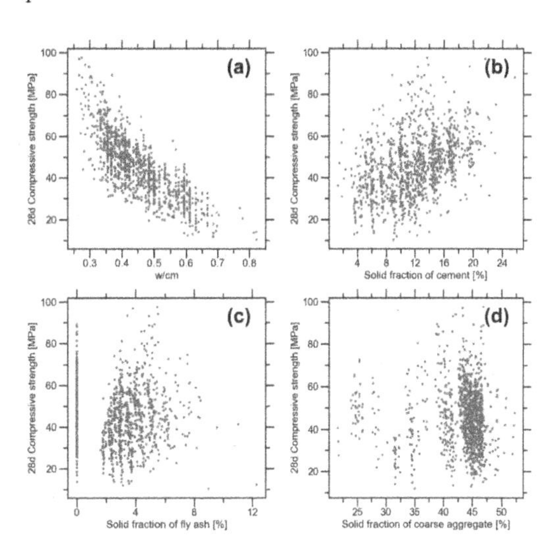

Figure 1. Correlations between the 28-day compressive strength and some key input features: (a) water-to-cementitious ratio and the mass fractions of (b) cement, (c) fly ash, and (d) coarse aggregate over all the solid materials used for each concrete mixture.

2.2 *Machine learning modeling*

In terms of the machine learning model, we build an artificial neural network for the strength prediction. Herein, the neural network is adopted since is one of the most versatile learning algorithms for various regression tasks. Based on the size of the dataset and our previous studies (Ouyang et al. 2020). Using machine learning to predict concrete's strength: learning from small datasets – IOPscience 2021), we design the neural network with two hidden layers, respectively with twelve and three artificial neurons. The modeling work is carried out in PyTorch (Paszke et al. 2019), with the use of Adam optimizer, Rectified Linear Unit (ReLU) activation function, and L2 loss function. Unless specified, the modeling procedures involved in this work broadly follow the machine learning pipeline as detailed in a previous study (Song et al. 2021), and readers are also referred to other studies for additional technical details (Oey et al. 2020; Ouyang et al. 2020).

2.3 *Pipeline of the machine learning analysis*

To build a robust neural network model, we use 85% of the samples to train our model (i.e., training set), while keeping the remaining 15% hidden to assess the model accuracy on generalizing to new samples (i.e., test set)—this is, the test set is reserved to evaluate the model's accuracy for predicting the strength of unknown concrete mixtures that are not involved in the model training. To ensure a fair selection of the testing mixtures, we implement a stratified sampling strategy (Ouyang et al. 2021), so that the potential bias associated with random sampling (e.g., the test mixtures are concentrated at a low/high strength range) can be minimized. This approach has been approved to be an important step for improving the efficiency of machine learning analysis on concrete material datasets with uneven label distribution (Ouyang et al. 2021, Song et al. 2021).

In order to determine the optimal configurations of the neural network (i.e., hyperparameters (Demir-Kavuk et al. 2011)), we split the training set samples with stratified five-fold cross-validation (Pedregosa et al. 2011). In detail, the optimal hyperparameters are identified based on the averaged model performance when using each of the five folds for evaluation. In that regard, we conduct a systematic grid search on four common hyperparameters, which are batch size (16 to 512; selected as 64), learning rate (0.0001 to 1; selected as 0.001), weight decay (0.00001 to 0.1; selected as 0.02), and epoch number (100 to 1000; selected as 300). For both model training and testing, we assess the model performance primary based on the coefficient of determination (R^2) of the strength prediction. Except that, root mean square error (RMSE) and mean absolute percentage error (MAPE) are also adopted as additional accuracy metrics for the model evaluation. Further details about the grid search and hyperparameter optimization are available from previous studies (Ouyang et al. 2020; Song et al. 2021).

3 RESULTS AND DISCUSSION

3.1 *Model accuracy on predicting concrete strength*

After optimizing the hyperparameters for our neural network model (see Sec. 2.3), we use all the training set samples to retrain a final model for predicting the strength of the test set samples at the four ages—3, 7, 28, and 56 days. This is done by fixing the time input of our model at these specific hydration ages. Figure 2 display the comparisons between the actual strength measurements and our model predictions on the concrete strength at the four ages. Note that the accuracy matrices (i.e., R^2, MAPE, RMSE; see Sec. 2.3) reported for each age are calculated only based on test set samples that are not involved in the training of our machine learning model, whereby these metrics provide robust indications of the model performance on predicting never-seen concrete formulas. From the individual accuracy plots, we first observe that, regardless of age, the scatters are all distributed evenly along the line of equality. This suggests that, across the different strength levels, the model prediction does not exhibit an obvious sign of bias (i.e., systematically predict the strength higher or lower). In addition, we note that there are no strong outliers as typically seen in machine-learning-based concrete strength prediction when the database is generated based on the samples collected from the real production (EBOD: An ensemble-based outlier detection algorithm for noisy datasets - ScienceDirect 2021; Young et al. 2019). Although this outlier-free prediction is primarily attributed to the higher quality of lab-generated data, the consistent distribution of each scatters nonetheless demonstrates that our neural network successfully considers the widespread mix designs involved in the adopted concrete dataset.

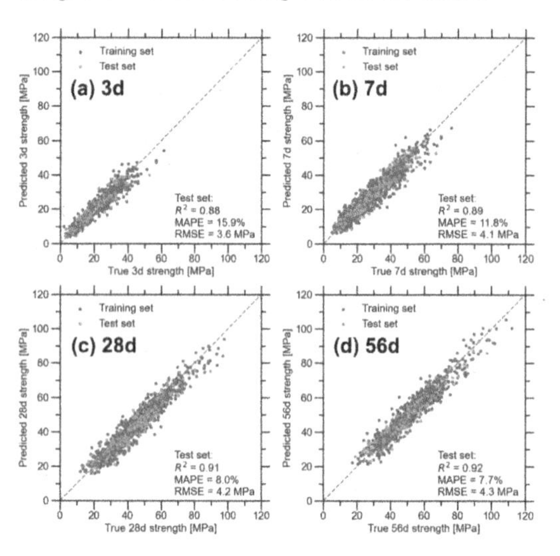

Figure 2. Comparisons between the neural network predicted vs. lab measured compressive strength of the concrete mixtures at (a) 3, (b) 7, (c) 28, and (d) 56 days after fabrication. The orange datapoints correspond to the 15% test set samples that are kept hidden from the model training, and the test set accuracy indicates the true model performance on predicting the strength of new concrete mixtures. The dash in each plot highlights the line of equality, i.e., perfect predictions.

Across the different ages, the changes in R^2 and MAPE metrics indicate a monotonic increase of the model accuracy over time. In comparison, the increase in RMSE is mainly raised from the fact that the averaged sample strength becomes higher as the age extends (whereas this effect is naturally eliminated in the calculation of R^2 and MAPE). The increasing model accuracy can be partially explained by the higher uncertainty involved with measuring the concrete strength at early ages, as concrete properties are more sensitive to the curing and testing conditions at early ages (Juan Luis et al. 2019; Yang 2007; Zhenchao 2020). Among the four ages, the least accurate case is found on predicting 3d strength, but it is actually encouraging to find that the model still gains a fairly high R^2 accuracy of 0.88 (Choi et al. 2020; Ziolkowski & Niedostatkiewicz 2019), especially given that the corresponding strength values vary within a relatively short range between 5 to 50 MPa. In contrast, the model achieves an impressively high test R^2 accuracy of 0.92 on 56-day strength, which varies between 20 and 110 MPa. It should be mentioned here that even an R^2 increase of 0.01 is of significance here, as it becomes exponentially challenging to improve the R^2 accuracy when the baseline accuracy is already close to the maximum (i.e., 1).

To the best of our knowledge, on the same levels of scale (i.e., number of datapoints) and scope (i.e., real production data), our model has surpassed the state-of-the-art accuracy among the studies focusing on using machine learning for concrete strength prediction. As a matter of fact, the prediction accuracy of this model (e.g., RMSE of the 28d test set samples; see Figure 2c) is likely approaching the intrinsic strength variation of concrete under a well-controlled production environment (Zhenchao 2020). This means that any further improvement could be challenging in principle, as it is inevitable to have this kind of intrinsic strength variation involved in any production-based concrete datasets. From a practical viewpoint, this realistic variation also opens up a new possibility of using machine learning to model the strength uncertainty of a given concrete design, which is an important point to be considered in the real concrete production. Hence, the model of this study also represents a key step from simply using machine learning as a predictive model to deploying AI to create new concrete formulas and to guide the routine concrete prediction.

3.2 *Effects of SCMs on the strength development*

The optimized machine learning model can be used as a comprehensive design tool for analyzing the effect of the individual features on the expected concrete strength. As a demonstration, herein we showcase an investigation on the influence of partially replacing cement with two types of supplementary contentious materials (SCMs)—Class F fly ash and slag. To this end, we first select a plain concrete mixture as the baseline mixture, as shown in Table 1. To ensure a fair investigation, this baseline mixture is down-selected from the test set samples, and further narrowed down to the presentative mixtures without complex dosages of chemical and mineral admixtures.

First, we use the optimized model to predict the strength development of the baseline mixture. This is done by using the model to make predictions on strength of a series of assumed inputs, wherein the input of the time is jittered between 1 and 70 days and all the other features related to the mix design remain unchanged (see Table 1). As such, the strength of the baseline mixture is predicted as a function of time, as displayed in Figure 3. Here, we find that our model predicts the strength to increase monotonically over time, while a progressive decline of the rate of strength gain is observed, and the strength gain becomes minuscule after 56 days. In fact, the predicted strength curve also passes through the actual strength of this mixture as measured at 3, 28, and 56 days (7d strength is not available in the dataset), which is also marked in Figure 3. These agreements further confirm that our model is not only able to accurately predict the concrete strength at specific ages, but also offers realistic predictions on the strength evolution of concrete.

Table 1. Comparison of the baseline and modified mix designs investigated in the feature analysis. The solid materials are presented based on their mass fractions.

	w/cm	Cem. %	Fly ash %	Slag %	Coarse agg. %	Fine agg. %	S.P.* ml/kg*
Baseline design	0.41	17	0	0	46.6	36.3	2.6
Modified A	0.41	12	5	0	46.6	36.3	2.6
Modified B	0.41	12	0	5	46.6	36.3	2.6

*Superplasticizer, unit based on the total cementitious

Then, we use the same model to predict two new designs that are modified from the baseline mixture, where we assume 30% of cement is now replaced by the Class F fly ash and slag, respectively (see Table 1). Compared with the baseline, the predicted strength development of the modified mixtures (see Figure 3)

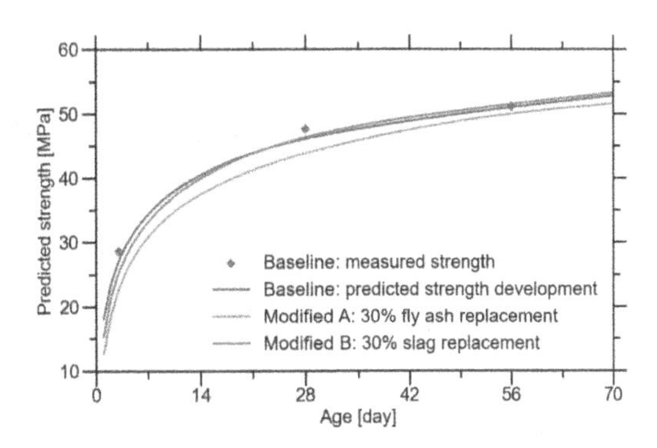

Figure 3. Comparison of the strength developments of three concrete mixtures that are predicted by the optimized neural network machine learning model. Here, the baseline mixture is an actual sample extracted from the test set (see the scatters for its true strength measured at 3, 28, and 56 days). The other two mixtures are modified by replacing 30% of the cement in the baseline mixture with Class F fly ash and slag, respectively.

exhibits very similar trends in terms of strength development. However, the early-age strength of both the assumed mixtures is predicted to be lower than the baseline, though the strength gap between the baseline and modified mixtures is gradually closed up in the long term. In comparison, the analysis also suggests that the 30% fly ash replacement of cement also leads to an even lower early-age strength than that of slag, whereas the slag-replaced mixture reaches the same strength level as the baseline mixture at around 21 days and surpasses it with a margin afterward. These predicted strength behaviors are broadly in line with the characteristic pozzolanic reactions of Class F fly ash and slag, and their featured influence on concrete strength development (Bentz et al. 2013; Chelberg 2019; Durdziński et al. 2017; Menéndez et al. 2003)—both of the materials have less strength contribution at the first few days of hydration (especially for Class F fly ash), but they eventually boost the long-term strength gain of concrete (e.g., after 28 days).

From the modeling perspective, it is encouraging to observe that the machine learning model can predict the unique influence of fly ash and slag replacements on concrete at the different hydration stages. It should be noted again that we did not intentionally encode the effect of any input features to the model. Stated another way, the effects of fly ash and slag, as well as time, are purely learned by our model from the data. On account of the ability to capture the true effects of different input features, our model exhibits a strong potential for predicting the strength performance of new mix designs that are outside of the curated dataset (i.e., extrapolation). This ability provides a foundation for further turning our machine learning model into a tool for discovering new concrete mixtures to pursue many different target performances (e.g., optimal use of a specific material, optimal strength gain at different ages, etc.).

3.3 *AI-guided concrete carbon reduction*

Based on the evidence observed from Sections 3.1 and 3.2, now we extend the investigation to evaluate the potential of using the trained neural network model to guide the optimization of concrete mix design. To this end, we further develop the feature effect analysis in Section 3.2 to investigate the concrete performance under the variation of multiple features. With the aim of improving concrete sustainability in mind, here we focus on reducing the carbon embodiment of the concrete mixture (in terms of the mass of embodied CO_2 in one cubic meter of concrete) while maintaining the 28d strength performance. For calculating the mass of CO_2, we refer to the carbon embodiments of the raw materials as reported in a recent study (Zjup & Adesina 2020), with the exact assumed values provided in Table 2. The concrete optimization is done using a brute force search within the reasonable feature ranges around the baseline design (i.e., the design needs to be optimized). To avoid any implausible assumptions of the design during the searching, we constraint the feature values to satisfy basic physical rules; for instance,

Table 2. Embodied carbon of each concrete ingredient considered in this study (Zjup & Adesina 2020).

Raw material	Embodied Carbon ton CO_2/ton
Cement	0.930
Fly ash	0.010
Slag	0.083
Coarse agg.	0.006
Fine agg.	0.025
Superplasticizer	0.720

the solid fractions are always summed up to 100%. Further, the total mass and volume of the new mixture are recalculated yield one cubic meter of concrete for the correct computation of the embodied carbon.

As an illustration, here we use the example mixture discussed in the last section (see Tab. 1) as the baseline to showcase the AI concrete optimization, and the result of the optimization is displayed in Figure 4. It should be noted that the actual AI optimization as demonstrated in Figure 4 is done automatically and is not broken into steps, since all the design features are simultaneously involved in the optimization process; however, for a rationalized understanding of the optimization outcome, Figure 4 shows the change of both 28d strength and embodied carbon in a discretized fashion, which corresponds to the modification of each feature involved in this optimization.

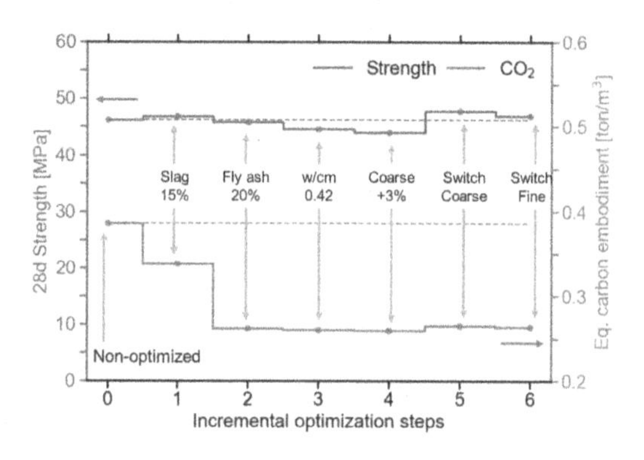

Figure 4. Step-wise decomposition of the AI optimization for the baseline concrete design (see Table 1). The goal of this optimization is to minimize the carbon embodiment of the concrete mixture while preserving the 28d strength at the same level. Here, the changes in carbon embodiment and strength in the individual optimization steps are shown to facilitate the interpretation of the AI optimized design (i.e., Step 6).

In terms of the overall optimization, it can be seen that, between the baseline and the optimized design (Step 6 in Figure 4), our AI-optimization achieves a significant reduction on the carbon embodiment by about 30% (from 0.39 to 0.27 ton/m^3), without sacrificing the 28d strength (which remains at around 46 MPa). The optimized mixture comes with a 35%

cement replacement by both slag and fly ash, slightly increased w/cm (by 0.01), increased coarse-to-fine aggregate ratio, and switched aggregate types (based on searching within a list of available aggregates associated with the raw material database).

Regarding the step-wise AI optimization, we first highlight the contribution of slag and fly ash replacements for reducing the carbon embodiment in concrete, which is attributed to the low embodied carbon in these raw materials (see Tab. 2). When it comes to the other features in Steps 3-to-6, their influence is much smaller yet still notable in sense of the huge volume of concrete production. For example, although switching the coarse aggregate type in Step 5 (to a high-quality rock with a higher density) results in a higher 28d strength, it also leads to an indirect reduction in the volume fraction of the coarse aggregate (as its mass fraction is fixed)—hence, the overall carbon is slightly increased when rebalancing the total volume back to one cubic meter.

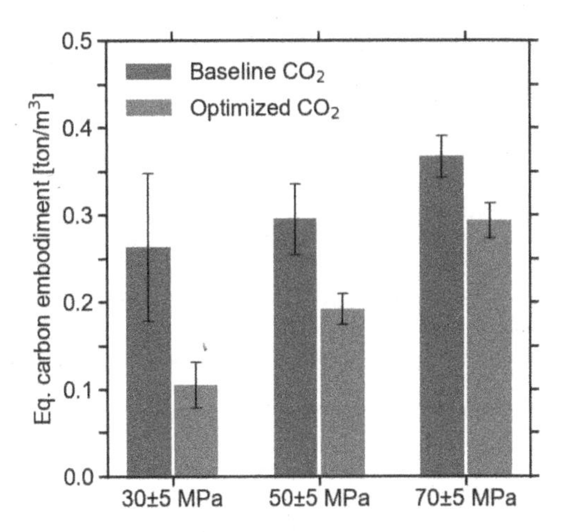

Figure 5. Comparison of the carbon embodiments of three groups of concrete mixtures before and after the AI optimization. The baseline carbon embodiments are calculated directly based on the actual mix designs from the dataset, and reduction on 28d strength of the optimized mixture is not allowed during the AI optimization. The half-length of each error bar in this plot corresponds to one standard deviation.

4 DISCUSSION

In this study, we demonstrate the potential of using an AI-based machine learning model for designing concrete mixtures with significantly reduced carbon embodiment. The pipeline of mixture optimization showcased in Section 3.3 can be generalized for guiding the development of almost any given concrete mixtures to meet different demands of the construction. To gain a further understanding of the plausible ranges of carbon reduction at different strengths, we implement the AI model to optimize the carbon embodiment of three groups of mixtures from the dataset, with their 28-day strength values at 30 ± 5, 30 ± 5, and 70 ± 5 MPa. Similar to the spirit of the optimization in Figure 4, here we still implement the goal of minimizing the carbon embodiment in each mixture under no reduction on the 28d strength. Figure 5 provides a summary of variation of their embodied carbon before and after the optimization.

In general, it can be seen from Figure 5 that the embodied carbon increases with concrete strength, which echoes the trend as reported in several previous studies (Fantilli et al. 2019; Lei et al. 2011; Optimization of the Mixture Design of Low-CO2 High-Strength Concrete Containing Silica Fume 2021; Park et al. 2012). This rise in carbon can be attributed to the improved demand for the minimum cement to achieve the higher strength. Interestingly, the AI model predicts different degrees of potential for trimming the carbon in concrete mixtures at various strengths, where a much higher relative carbon reduction (e.g., >60%) can be fulfilled in the relatively low-strength mixtures at 30 MPa. This should be primarily related to the fact that the low-strength mixtures, such as controlled low strength materials (CLSM) (Brewer 1996; Kaliyavaradhan et al. 2019; Song & Lange 2019), can accommodate a high-volume cement replacement with SCMs that have much lower carbon footprints. In comparison, a 15-to-20% carbon reduction in the high-strength mixtures at 70 MPa is still expected to be achievable based on our model prediction, where the contribution from SCMs should be lighter.

Given the fact that the global production of SCMs like fly ash has been rapidly decreasing (American Coal Ash Association 2018; Benhelal et al. 2013; Schneider et al. 2011), the typical solution of cement replacement for reducing the concrete carbon embodiment may become less efficient in the predictable future. Thus, reducing the embodied carbon from the other materials (e.g., Step 3-to-6 in Figure 4) is expected to become more critical for curbing the carbon demand in concrete production. For optimizing concrete mixtures either across all strength ranges or with the availability of SCMs constrained, we foresee that the AI model should exhibit its advantages over the conventional trial-and-error approach. This is based on the fact that the AI model can maximize the marginal carbon reduction by optimizing each feature simultaneously, so as to yield a holistic optimization on concrete sustainability.

Admittedly, the carbon optimization demonstrated in this paper should be further improved from several practical aspects. Regarding the calculation of the embodied carbon, it should be noted that the embodied carbon of the raw materials considered herein only covers the carbon emissions involved in the production, whereas other sources such as the carbon footprints from the material transportation are not neglectable in real concrete production, which can be sometimes equally influential (CO2 Emissions From Cement Production – CivilDigital – 2013, Environmental impacts and decarbonization strategies in the cement and concrete industries | Nature Reviews Earth & Environment 2021; Lei et al. 2011). With the incorporation of all the sources of the carbon embodiment, it is undoubtful that there is a strong potential to further strengthen the concrete AI optimization in the actual production. Another factor that should be stressed here is that the required concrete performance is barely prescribed by any single criteria in real concrete applications. For instance, the optimization displayed in Figure 4 is very likely involved with a phenomenal strength reduction at an early age (due to the cement replacement; see Figure 3), hence impeding the constructability of the optimized design. For a robust AI concrete optimization, it is of special importance to train the model(s) to optimize the concrete design under multiple constraints, such as early-age strength, slump, air content, etc. Moreover, it is also critical to minimize the material cost, which is of special importance from an operational perspective. In that regard, the cost optimization can be fulfilled in the same spirit of the carbon optimization as discussed in this paper. However, viewing the problem another way, the above requirements further stress the need for a holistic optimization for designing new concrete formulas in the next decades. Therefore, developing the AI-based optimization methodology is found fundamental to fulfill this goal.

5 CONCLUSIONS

In this study, we investigated the use of artificial intelligence (AI) as a tool for designing sustainable concrete mixtures, by specially focusing on reducing its carbon embodiment. To this end, we train a neural network model to predict the concrete strength based on a large lab concrete dataset. The trained AI model is used for finding new concrete formulas that are involved with lower embodied carbon overall the design variables, without compromising strength. Major findings from our investigation are summarized as follows.

- Our model achieves unprecedented accuracy for predicting the 3, 7, 28, and 56-day compressive strength of concrete mixtures that are never involved with the model training. For example, the test R^2 reaches an accuracy of 0.91 for predicting 28-day strength.
- The trained AI model offers continuous and realistic predictions on the strength development for a given concrete design. Our model also accurately captures the influence of raw materials such as fly ash and slag on the time-dependent strength behavior of concrete.
- We demonstrate that the AI model can autonomously discover new high-performance sustainable concrete designs that successfully reduce the concrete carbon embodiment by more than 50%, at no cost of 28-day strength.
- Rather than solely relying on cement replacement, the AI-based concrete optimization reduces the embodied carbon by considering the influence of each design feature, such that a holistic mixture optimization can be actualized.
- With the global reduction of SCMs, we believe that AI-based concrete design optimization represents one of the most promising solution to revolute concrete industry in the next decades.

REFERENCES

ACI CODE-318-19: Building Code Requirements for Structural Concrete and Commentary, 2021.

Ali, M.B., Saidur, R., and Hossain, M.S., 2011. A review on emission analysis in cement industries. *Renewable and Sustainable Energy Reviews*, 15 (5), 2252–2261.

American Coal Ash Association, 2018. *Coal Combustion Product (CCP) production & use survey report.*

Bangaru, S.S., Wang, C., Hassan, M., Jeon, H.W., and Ayiluri, T., 2019. Estimation of the degree of hydration of concrete through automated machine learning based microstructure analysis – A study on effect of image magnification. *Advanced Engineering Informatics*, 42, 100975.

Bauchy, M., 2017. Nanoengineering of concrete via topological constraint theory. *MRS Bulletin*, 42 (1), 50–54.

Benhelal, E., Zahedi, G., Shamsaei, E., and Bahadori, A., 2013. Global strategies and potentials to curb CO2 emissions in cement industry. *Journal of Cleaner Production*, 51, 142–161.

Bentz, D.P., Ferraris, C.F., and Snyder, K.A., 2013. *Best Practices Guide for High-Volume Fly Ash Concretes?: Assuring Properties and Performance.* National Institute of Standards and Technology, No. NIST TN 1812.

Biernacki, J.J., Bullard, J.W., Sant, G., Brown, K., Glasser, F.P., Jones, S., Ley, T., Livingston, R., Nicoleau, L., Olek, J., Sanchez, F., Shahsavari, R., Stutzman, P.E., Sobolev, K., and Prater, T., 2018. Cements in the 21st century: Challenges, perspectives, and opportunities. *Journal of the American Ceramic Society*, 2746–2773.

Brewer, W.E., 1996. CONTROLLED LOW STRENGTH MATERIALS (CLSM). *In*: *Concrete in the Service of Mankind.* CRC Press.

Breysse, D. and Martínez-Fernández, J.L., 2014. Assessing concrete strength with rebound hammer: review of key issues and ideas for more reliable conclusions. *Materials and Structures*, 47 (9), 1589–1604.

Burris, L.E., Alapati, P., Moser, R.D., Ley, M.T., Berke, N., and Kurtis, K.E., 2015. Alternative cementitious materials: Challenges and opportunities. *In*: *International Workshop on Durability and Sustainability of Concrete Structures, Bologna, Italy.*

Cai, R., Han, T., Liao, W., Huang, J., Li, D., Kumar, A., and Ma, H., 2020. Prediction of surface chloride concentration of marine concrete using ensemble machine learning. *Cement and Concrete Research*, 136, 106164.

Carbon neutrality by 2050: the world's most urgent mission | United Nations Secretary-General [online], 2021. Available from: https://www.un.org/sg/en/content/sg/articles/2020-12-11/carbon-neutrality-2050-the-world%E2%80%99s-most-urgent-mission [Accessed 2 Sep 2021].

CEMBUREAU, 2015. *2014 Activity Report.*

Chelberg, M., 2019. The Effect of Fly Ash Chemical Composition on Compressive Strength of Fly Ash Portland Cement Concrete. The Ohio State University.

Choi, H., Venkiteela, G., Gregori, A., and Najm, H., 2020. Advanced Quality Control Models for Concrete Admixtures. *Journal of Materials in Civil Engineering*, 32 (2), 04019349.

Chopra, P., Sharma, R.K., Kumar, M., and Chopra, T., 2018. Comparison of Machine Learning Techniques for the Prediction of Compressive Strength of Concrete. *Advances in Civil Engineering.*

Chou, J.-S., Chiu, C.-K., Farfoura, M., and Al-Taharwa, I., 2011. Optimizing the Prediction Accuracy of Concrete Compressive Strength Based on a Comparison of Data-Mining Techniques. *Journal of Computing in Civil Engineering*, 25 (3), 242–253.

Chou, J.-S., Tsai, C.-F., Pham, A.-D., and Lu, Y.-H., 2014. Machine learning in concrete strength simulations: Multi-nation data analytics. *Construction and Building Materials*, 73, 771–780.

CO2 Emissions From Cement Production, 2013. *CivilDigital*. Available from: https://civildigital.com/co2-emissions-from-cement-production

Das, A., Song, Y., Mantellato, S., Wangler, T., Flatt, R.J., and Lange, D.A., 2020. Influence of Pumping/Extrusion on the Air-Void System of 3D Printed Concrete. *In*: F.P. Bos, S.S. Lucas, R.J.M. Wolfs, and T.A.M. Salet, eds. *Second RILEM International Conference on Concrete and Digital Fabrication*. Cham: Springer International Publishing, 417–427.

Das, A.K., Suthar, D., and Leung, C.K.Y., 2019. Machine learning based crack mode classification from unlabeled acoustic emission waveform features. *Cement and Concrete Research*, 121, 42–57.

Demir-Kavuk, O., Kamada, M., Akutsu, T., and Knapp, E.-W., 2011. Prediction using step-wise L1, L2 regularization and feature selection for small data sets with large number of features. *BMC Bioinformatics*, 12 (1), 412.

Durdziński, P.T., Ben Haha, M., Bernal, S.A., De Belie, N., Gruyaert, E., Lothenbach, B., Menéndez Méndez, E., Provis, J.L., Schöler, A., Stabler, C., Tan, Z., Villagrán Zaccardi, Y., Vollpracht, A., Winnefeld, F., Zajⁱc, M., and Scrivener, K.L., 2017. Outcomes of the RILEM round robin on degree of reaction of slag and fly ash in blended cements. *Materials and Structures*, 50 (2), 135.

Ouyang, B., Song, Y., Li, Y., Sant, G. and Bauchy, M., 2021. EBOD: An ensemble-based outlier detection algorithm for noisy datasets. *Knowledge-Based Systems*, 231, p.107400.

Environmental impacts and decarbonization strategies in the cement and concrete industries | Nature Reviews Earth & Environment [online], 2021. Available from: https://www.nature.com/articles/s43017-020-0093-3?proof=t [Accessed 7 Nov 2021].

Fantilli, A.P., Mancinelli, O., and Chiaia, B., 2019. The carbon footprint of normal and high-strength concrete used in low-rise and high-rise buildings. *Case Studies in Construction Materials*, 11, e00296.

Gupta, S.M., n.d. Support Vector Machines based Modelling of Concrete Strength, 3 (1), 7.

Hoang, N.-D., Chen, C.-T., and Liao, K.-W., 2017. Prediction of chloride diffusion in cement mortar using Multi-Gene Genetic Programming and Multivariate Adaptive Regression Splines. *Measurement*, 112, 141–149.

Juan Luis, F.-M., Zulima, F.-M., and Denys, B., 2019. The uncertainty analysis in linear and nonlinear regression revisited: application to concrete strength estimation. *Inverse Problems in Science and Engineering*, 27 (12), 1740–1764.

Kaliyavaradhan, S.K., Ling, T.-C., Guo, M.-Z., and Mo, K.H., 2019. Waste resources recycling in controlled low-strength material (CLSM): A critical review on plastic properties. *Journal of Environmental Management*, 241, 383–396.

Khoury, G.A., Majorana, C.E., Pesavento, F., and Schrefler, B.A., 2002. Modelling of heated concrete. *Magazine of Concrete Research*, 54 (2), 77–101.

Lei, Y., Zhang, Q., Nielsen, C., and He, K., 2011. An inventory of primary air pollutants and CO2 emissions from cement production in China, 1990–2020. *Atmospheric Environment*, 45 (1), 147–154.

Mehta, P.K. and Monteiro, P.J., 2014. *Concrete: Microstructure, Properties, and Materials*. McGraw-Hill Education.

Menéndez, G., Bonavetti, V., and Irassar, E.F., 2003. Strength development of ternary blended cement with limestone filler and blast-furnace slag. *Cement and Concrete Composites*, 25 (1), 61–67.

Moutassem, F. and Chidiac, S.E., 2016. Assessment of concrete compressive strength prediction models. *KSCE Journal of Civil Engineering*, 20 (1), 343–358.

Oey, T., Jones, S., Bullard, J.W., and Sant, G., 2020. Machine learning can predict setting behavior and strength evolution of hydrating cement systems. *Journal of the American Ceramic Society*, 103 (1), 480–490.

Okazaki, Y., Okazaki, S., Asamoto, S., and Chun, P., 2020. Applicability of machine learning to a crack model in concrete bridges. *Computer-Aided Civil and Infrastructure Engineering*, 35 (8), 775–792.

Optimization of the Mixture Design of Low-CO2 High-Strength Concrete Containing Silica Fume [online], 2021. Available from: https://www.hindawi.com/journals/ace/2019/7168703/ [Accessed 8 Nov 2021].

Ouyang, B., Song, Y., Li, Y., Wu, F., Yu, H., Wang, Y., Sant, G., and Bauchy, M., 2020. Predicting Concrete's Strength by Machine Learning: Balance between Accuracy and Complexity of Algorithms. *Materials Journal*, 117 (6), 125–133.

Ouyang, B., Song, Y., Li, Y., Wu, F., Yu, H., Wang, Y., Yin, Z., Luo, X., Sant, G., and Bauchy, M., 2021. Using machine learning to predict concrete's strength: learning from small datasets. *Engineering Research Express*, 3 (1), 015022.

Park, J., Tae, S., and Kim, T., 2012. Life cycle CO2 assessment of concrete by compressive strength on construction site in Korea. *Renewable and Sustainable Energy Reviews*, 16 (5), 2940–2946.

Paszke, A., Gross, S., Massa, F., Lerer, A., Bradbury, J., Chanan, G., Killeen, T., Lin, Z., Gimelshein, N., Antiga, L., Desmaison, A., Kopf, A., Yang, E., DeVito, Z., Raison, M., Tejani, A., Chilamkurthy, S., Steiner, B., Fang, L., Bai, J., and Chintala, S., 2019. PyTorch: An Imperative Style, High-Performance Deep Learning Library. *In*: *Advances in Neural Information Processing Systems*. Curran Associates, Inc.

Pedregosa, F., Varoquaux, G., Gramfort, A., Michel, V., Thirion, B., Grisel, O., Blondel, M., Prettenhofer, P., Weiss, R., Dubourg, V., Vanderplas, J., Passos, A., and Cournapeau, D., 2011. Scikit-learn: Machine Learning in Python. *Journal of Machine Learning Research*, 12, 2825–2830.

Popovics, S., 1998. History of a Mathematical Model for Strength Development of Portland Cement Concrete. *Materials Journal*, 95 (5), 593–600.

Powers, T.C., 1960. *Physical properties of cement paste.*

Provis, J.L., 2015. Grand Challenges in Structural Materials. *Frontiers in Materials*, 2.

Schneider, M., Romer, M., Tschudin, M., and Bolio, H., 2011. Sustainable cement production—present and future. *Cement and Concrete Research*, 41 (7), 642–650.

Song, Y., Huang, Z., Shen, C., Shi, H., and Lange, D.A., 2020. Deep learning-based automated image segmentation for concrete petrographic analysis. *Cement and Concrete Research*, 135, 106118.

Song, Y. and Lange, D., 2019. Crushing Performance of Ultra-Lightweight Foam Concrete with Fine Particle Inclusions. *Applied Sciences*, 9 (5), 876.

Song, Y., Yang, K., Chen, J., Wang, K., Sant, G., and Bauchy, M., 2021. Machine Learning Enables Rapid Screening of Reactive Fly Ashes Based on Their Network Topology. *ACS Sustainable Chemistry & Engineering*.

Using machine learning to predict concrete's strength: learning from small datasets - IOPscience [online], 2021. Available from: https://iopscience.iop.org/article/10.1088/2631-8695/abe344/meta [Accessed 5 Oct 2021].

Wild, S., Sabir, B.B., and Khatib, J.M., 1995. Factors influencing strength development of concrete containing silica fume. *Cement and Concrete Research*, 25 (7), 1567–1580.

Yang, I.H., 2007. Uncertainty and sensitivity analysis of time-dependent effects in concrete structures. *Engineering Structures*, 29 (7), 1366–1374.

Young, B.A., Hall, A., Pilon, L., Gupta, P., and Sant, G., 2019. Can the compressive strength of concrete be estimated from knowledge of the mixture proportions?: New insights from statistical analysis and machine learning methods. *Cement and Concrete Research*, 115, 379–388.

Zain, M.F.M. and Abd, S.M., 2009. Multiple regression model for compressive strength prediction of high performance concrete. *Journal of applied sciences*, 9 (1), 155–160.

Zhenchao, D., 2020. Discussion on Problem of Standard Deviation of Concrete Strength. *Materials Journal*, 117 (1), 25–35.

Ziolkowski, P. and Niedostatkiewicz, M., 2019. Machine Learning Techniques in Concrete Mix Design. *Materials*, 12 (8), 1256.

Zjup, W. and Adesina, A., 2020. Performance and sustainability overview of alkali-activated self-compacting concrete. *Waste Disposal & Sustainable Energy*, 2.

Computational Modelling of Concrete and Concrete Structures – Meschke, Pichler & Rots (Eds)
© 2022 Copyright the Author(s), ISBN: 978-1-032-32724-2

Comparison of perfect and cohesive adhesion between globules on mechanical properties of C-S-H gel RVE with FEM method

P. Wang, F. Bernard & S. Kamali-Bernard
Laboratory of Civil Engineering and Mechanical Engineering (LGCGM), Rennes University, INSA Rennes, Rennes, France

ABSTRACT: In this paper, we connected both cement research on nanoscale and sub-microscale with FEM method. By adopting parameters transfer and cohesive zone model, C-S-H globule was calibrated from MD results as solid element and interaction between C-S-H globules was represented by cohesive element in The FE software ABAQUS. Then according to the LD/HD C-S-H distribution generated by a particle's placement algorithm, we constructed C-S-H RVE model under 250 nm by assembly C-S-H globule and inserted with cohesive element. Validated by Bolomey formula, this model well exhibits strong coherent results between density and tensile/shearing performances. Finally, we compared mechanical properties under tensile loading for both perfect adhesion and cohesive adhesion.

1 INTRODUCTION

1.1 *Structural elucidation of C-S-H*

Cement concrete is the most widely applied human-made materials in the world (Courland 2011). As the direct blending product of Portland cement hydration, calcium silicate hydrate (hereafter as C-S-H), the near amorphous delicate nanoscale structures which makes up more than 50 vol% among all the hydration phases (Barnes & Bensted 2002; Hewlett & Liska 2019; Olson & Jennings 2001), controls many critical engineering properties (Jennings & Bullard 2011).

As well known, there has been long-lasting difficulties to elucidate the microstructure of C-S-H due to the complexities underlying its composition and structure. Some new insights on cement hydration mechanisms and fundamental explanation towards the nature of C-S-H have been given in (Scrivener & Nonat 2011; Scrivener et al. 2015, 2019). Also, various microstructural characterization methods and nanotechnologies to characterize C-S-H hydrates have been reviewed in (Monteiro et al. 2019; Sanchez & Sobolev 2010).

With the interpretation of experimental results of scattering and water sorption isotherms, Jennings established the colloid model (CM-II) to describe nanostructure of C-S-H, where the most significant feature is the presumption of two types of C-S-H, low density C-S-H and high density C-S-H (hereafter as LD C-S-H and HD C-S-H), in distinct stages (Jennings 2000, 2008; Tennis & Jennings 2000). During the hydration reaction, the formation of HD C-S-H concentrates in the late stage with the packing density as 74%, while LD C-S-H mainly formed in early stage of hydration reaction and the packing density remained around 64% (Jennings et al. 2007).

However, the correlation those exciting advances in nanoscale with practical engineering behaviors in macroscale remains challenging. We must admit that our current understandings are still not adequate, especially on nanostructure under the mesoscale level (Jennings & Bullard 2011; Tennis & Jennings 2000). Therefore, how to seamlessly integrate fundamental data from nanoscale into the coarser model of the microstructure of C-S-H between 1-500 nm precisely has been the long-lasting spot.

1.2 *Upscaling techniques and its challenges*

It has been consensus among the research community that upscaling methods such as bottom-to-up approach and multi-scale modeling would be the essence to the gap between nanoscale and the upper levels (de Souza et al. 2022). Various methods have been proposed to investigate in mesoscale. On the one side, some researchers directly started from MD simulation in nanoscale. Coarse grain molecular dynamic simulation could reach the spatial scale as 500 nm by only keeping the essential information (Qomi et al. 2020). Hou et al. (2021) have applied Grand Canonical Monte Carlo and Peridynamic method to model unilateral tension of C-S-H under the size of 100 nm. Yaphary et al. (2021) have established MD model which incorporated as many as 10000 particles and the size of simulation box could reach as large as 228 nm. The directly upscaling could provide accurate and convincing results, however, it is at the cost of computation capacity and the extension in further is still a doubt.

On the other side, a trendier way is to construct hierarchical multi-scale model to correspond different level. Parameter transfer is the most directly method to established multi-scale model. With transferring essential parameters including Young's modulus, tensile strength, and fracture energy, Hlobil et al. (2016) established four level hierarchical multi-scale model to explore compressive strength. Similar as the hierarchical model, Li et al. (2013) developed contact model within DEM method to study the C-S-H clustering. Parameter transfer is a widely used method, but it highly relies on the calibration of parameter and the model in each scale lack interaction each other.

Voxel-based model integrated in spatial distribution has emerged to introduce porosity to assembly C-S-H RVE. The assembly in spatial distribution could increase the reliability for upscaling from nanoscale to mesoscale. The spatial distribution can be accessed by experimental method such as μCT image or in randomly distribution algorithm (Montero-Chacón et al. 2014).

Combined with parameter transfer and Voxel-based model integrated in spatial distribution, MuMoCC (Multi-scale Modeling of Computational Concrete numerical platform was developed to simulate the mechanical and transport properties of cementitious materials as well as the building of Representative Volume Elements (RVE) from microscale to mesoscale and macro scale (Bernard & Kamali-Bernard 2010a, 2010b; Fu et al. 2018). Multi-scale model could well couple the MD with other numerical simulations such as FEM, DEM, and even Lattice Boltzmann method, which exhibit the better extension degree towards upper scale.

Among the upscaling methods, characterization of interaction between C-S-H globules has always been the one of focus. Recently, Goyal et al. (2021) studied the essential physics mechanism underlying C-S-H cohesion via semi-atomistic simulation. The interlocking of water/ions of C-S-H surfaces consecutively slow down the dielectric screening, which finally resulted in the cohesion between C-S-H layers. Other authors have employed potential of mean force in MD simulation as an effective approach to investigate interaction between C-S-H layers, whose results could be valuable references for the higher scale (Bonnaud et al. 2016; Masoumi et al. 2017, 2019).

Origin from metal & alloy area (Lloyd et al. 2011; Zhou et al. 2009), using cohesive zone models based on MD parameters to simulate adhesion also emerged in cement concrete area recently. There are some papers on developing cohesive zone model at the level of cement paste (Fan & Yang 2018; Trawiński et al. 2018), but currently there have not been reported cohesive zone model for C-S-H RVE at sub-microscale yet.

1.3 *Highlights of this paper*

In this paper, we have constructed a multi-scale model to upscale MD simulation of mechanical properties of C-S-H globule in nanoscale to FEM model of C-S-H RVE in sub-micro scale. The FEM model consists of two types of elements, solid element denotes for C-S-H globule; cohesive elements denote for adhesion between C-S-H globules. Main propose of FEM model is to investigate and compare the influence of perfect adhesion and cohesive adhesion between C-S-H globules on final RVE. With combined three different modelling techniques, the highlights of this paper could be listed as:

A Brittle Cracking model inside the FE software package ABAQUS is used to calibrate the mechanical behavior of solid elements, i.e. C-S-H globules. To this purpose, the parameter transfer has to ensure the solid element could have the same tensile performance as MD model.

Furthermore, a cohesive element inside ABAQUS is considered to model cohesive adhesion based on calibration from MD results. The cohesive zone model could ensure that the adhesion in FEM exhibits the same traction-separation behaviors as in MD model.

For spatial distribution, assembling C-S-H RVE with introducing porosity and spatial distribution generated by a voxel-based model integrated in spatial distribution could ensure the upscaling from 5 nm at nanoscale to 250 nm at sub-microscale.

2 MODELLING

2.1 *Calibration on mechanical properties: solid element*

Brittle Cracking model in ABAQUS is applicable to brittle materials whose pre-cracking tensile behavior can be represented through linear elasticity. It is accurate enough in application where the brittle behavior dominates. As for tension results by molecular dynamic simulation, hereby we adopted Fu's results as reference (Fu 2016). In this work, atomistic simulation of critical component and monolithic structure of C-S-H were conducted to explore the mechanical properties. According to his results such as Young's modulus, the performance of C-S-H globule is brittle enough to apply with *Brittle Cracking model (Fu 2016).

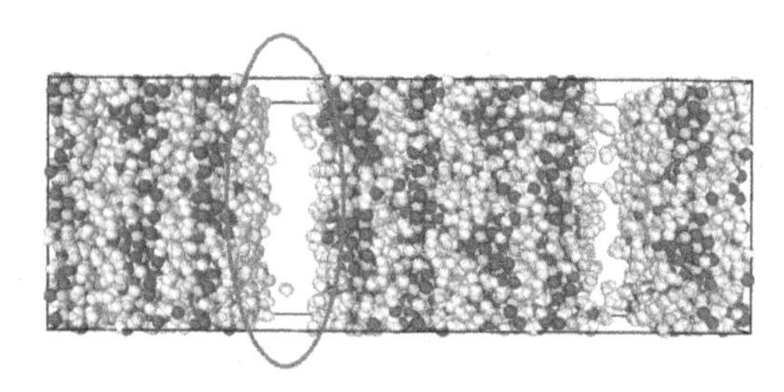

Figure 1. C-S-H globule under tension of MD model (Fu 2016).

The first step is to reproduce C-S-H globule model and impose tension loading with the FE method.

According to Jennings' colloidal model (Jennings 2008), C-S-H globule is as the brick shape. Therefore, we have built cubic solid elements in 3D and 2D with almost the same size of MD model to represent the globules.

Based on tensile performance of MD results, strain-stress curves of C-S-H globule could be calibrated as close as possible to original MD results, as shown in Figure 3.

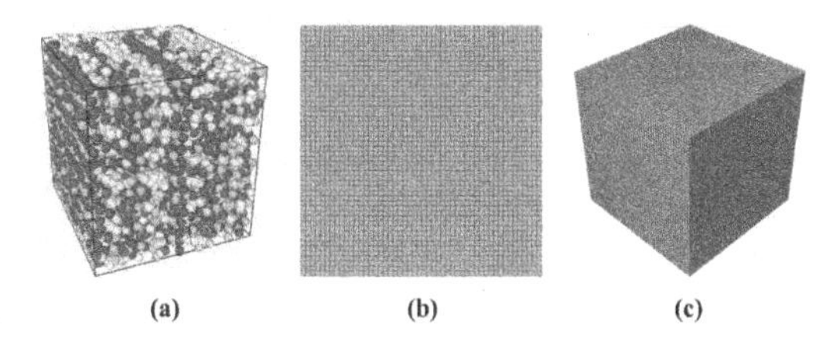

Figure 2. a) C-S-H MD Model, size 5.352 × 4.434 × 4.556 nm^3 (Fu 2016); b) 2D C-S-H globule FEM Model, size 5 nm; c) 3D C-S-H globule FEM Model, size 5 nm.

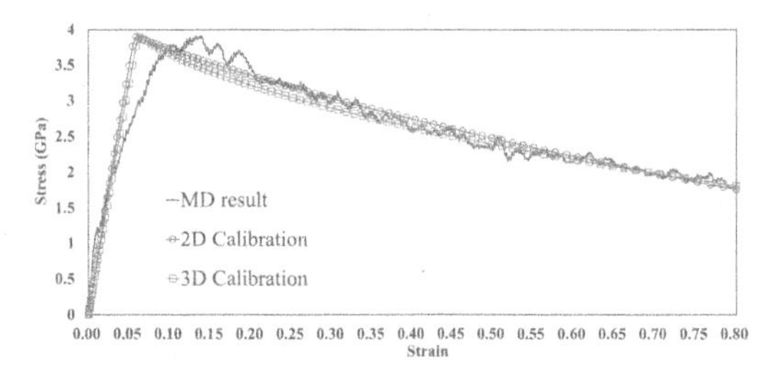

Figure 3. Comparison of strain-stress curve of 2D/3D calibrated model and original MD model.

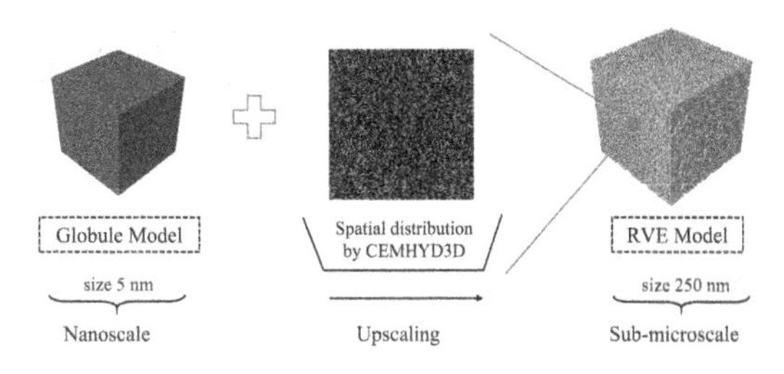

Figure 4. Schematic of building RVE from 2D slice voxels denotes spatial distribution to microstructure of 3D C-S-H LD RVE.

2.2 *Upscaling from globule to RVE: perfect adhesion*

According to literature, there are summarized as three main categories of porosity in hydrated ordinary cement pastes. From the biggest to the smallest, they are compaction/air void, capillary porosity, and gel pores. The size of compaction/air void vary between a few micrometers to a few millimeters. The capillary cavities are reported to be larger than 10 nm to few micrometers. The gel pores can be regarded as the intrinsic porosity of C-S-H hydrate, which know to be nanometer size. Considering the complexity of representing the porosity in C-S-H, we decided not to assign porosity any element but leave it as void. Therefore, C-S-H RVE model would be a scaffold model.

For microstructure digitalization of C-S-H RVE. Bentz and Garboczi in 1989 developed the program entitled as 3-Dimensional CEMent HYDration and microstructure development modeling package (CEMHYD3D), which is a digital-image based computer program to simulate the process of cement hydration (Bentz 2000). There are already plenty of published research on C-S-H RVE meso-structure are based CEMHYD3D (Bernard & Kamali-Bernard 2010b; Hlobil et al. 2016; Montero-Chacón et al. 2014). Besides, recently, some research focused on formation of porosity during cement hydration process and proposed sheet growth model (Etzold et al. 2014; Nguyen-Tuan et al. 2020). We hereby still adopt the program of CEMHYD3D package devoted to generating random spatial distribution of voxels. One vowel represents here a C-S-H globule. The porosity for LD/HD C-S-H RVE was adopted as 37% and 26%, respectively. The procedure of assembly C-S-H RVE could be found in Figure 4.

2.3 *Calibration on interaction properties: Cohesive element*

The definition of cohesive elements between C-S-H and C-S-H were referenced from Bonnaud's work (Bonnaud et al. 2016). Traction modulus of cohesive element of elastic behavior is analyzed as 4 GPa for the normal and the two shear stiffnesses (denoted respectively as E_{nn}, E_{ss}, and E_{tt}), which is quite close to Nemeck's value as 2.5 GPa (Němeček et al. 2018). Tensile strength and cracking displacement are taken equal to respectively 931 MPa and 0.341 nm. With those referenced parameters, we established a corresponding FEM model of a cohesive element layer within two C-S-H globules to validate the elastic behavior in the early stage of traction-separation and calibrate the stiffness degradation stage, as shown in Figure 5. According to the Figure 6, we could see that both the elastic behavior in the early stage and the stiffness degradation stage of FEM results are calibrated as close to MD results.

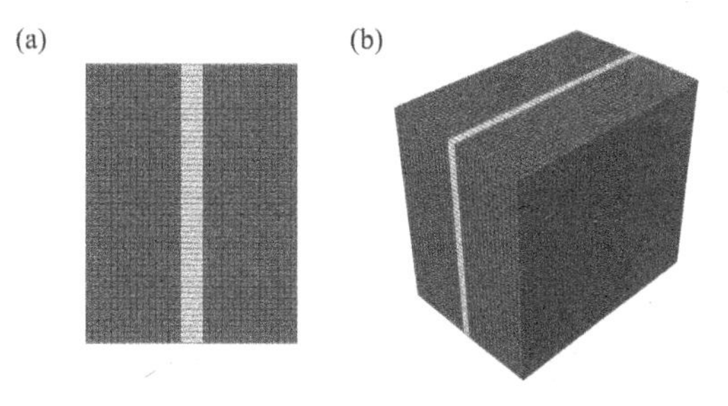

Figure 5. Face-to-face configuration of a) 2D C-S-H globule and b) 3D FEM C-S-H globule simulation.

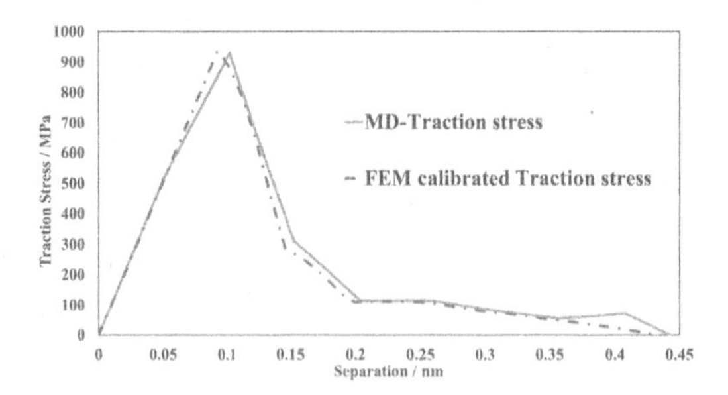

Figure 6. Traction stress - Separation calibration for applying tension between two C-S-H globules.

2.4 *Insertion of cohesive element: cohesive adhesion*

It has been years that researchers applied cohesive element to investigate the fracture (Nguyen et al. 2001; Schwalbe et al. 2012; Zhou & Molinari 2004), and the insertion techniques have gotten developed (Su et al. 2010; Vocialta et al. 2017). In this project, we individually developed python script to fulfill the globally insertion of cohesive element both in 2D and 3D modelling. The schematic of insertion process can be found in Figure 7.

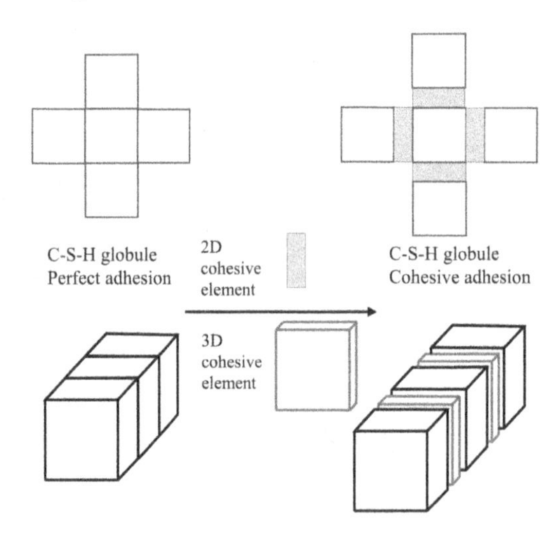

Figure 7. Schematic of insertion process of cohesive element into solid element both in 2D/3D modeling.

The insertion process could be summarized as three stages: break the previous perfection adhesion by construct solid element individually (i.e. not sharing nodes), detect and judge the contacts between solid elements, insert cohesive element and assembly as C-S-H RVE with cohesive adhesion. The final LD C-S-H RVE with cohesive adhesion are display in Figure 8. One thing should be noted, since we have introduced the cohesive element, whose thickness is 0.4 nm which cannot be neglected when comparing with dimension of C-S-H globule or porosity, 5nm, therefore, it is necessary to recalculate the number of C-S-H globules and porosities to ensure the porosity percentage can be kept as 37% and 26%.

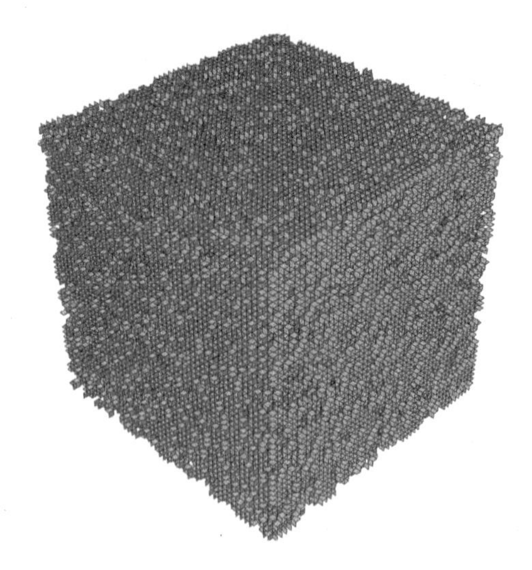

Figure 8. Schematic of adhesion in LD C-S-H RVE model of three dimensions, inferior view of cohesive element.

3 RESULTS AND DISCUSSION

3.1 *Mechanical performance and fitting with Bolomey Formula*

Tensile and pure shearing loading were imposed respectively for all LD, HD, and Pure C-S-H RVE cohesive model to check the mechanical performance. The stress-strain curve is as shown in Figure 9 and 10. Young's moduli E and shearing moduli G can be regressed and the results are summarized in Table 1.

To validate the RVE model with cohesive adhesion, we fitted the revised porosity with moduli by Bolomey formula (Nielsen 1993) in format as $E = E_0$ (1-*Porosity*)^n as shown in Figure 11. Similarly, Figure 12 represents the evolution of the tensile strength.

According to the fitting results, correlation coefficients of both shearing and tension are high enough. It proves that the current C-S-H RVE model with cohesive adhesion could present the coherent relationship between porosity mechanical performances.

Table 1. Tension and shearing moduli and tensile strength of C-S-H RVE.

	Revised Porosity	Young's Moduli E GPa	Shearing Moduli G GPa	Tensile Strength MPa
LD RVE	37.44%	10.64	4.07	362.30
HD RVE	25.81%	19.15	6.44	839.26
Pure C-S-H	13.78%	28.86	12.08	1730.80

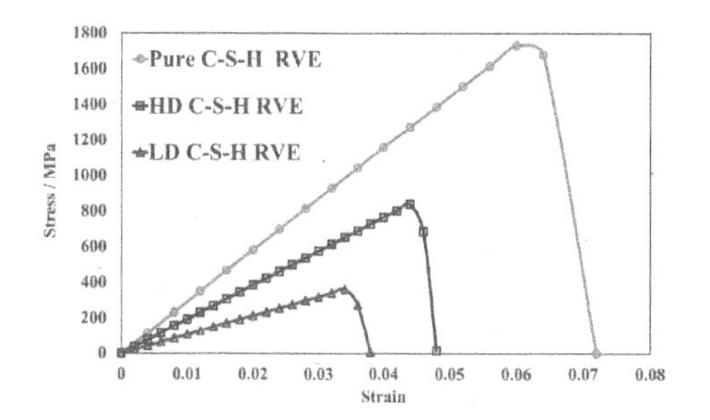

Figure 9. Stress-strain curve of LD/HD/Pure C-S-H RVE model under tensile loading

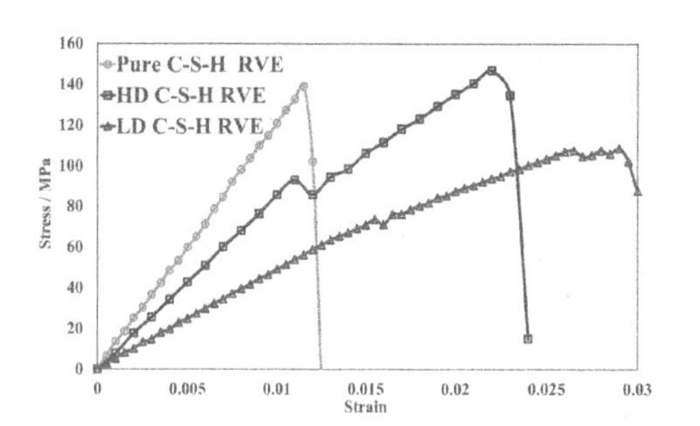

Figure 10. Stress-strain curve of LD/HD/Pure C-S-H RVE model under shearing loading.

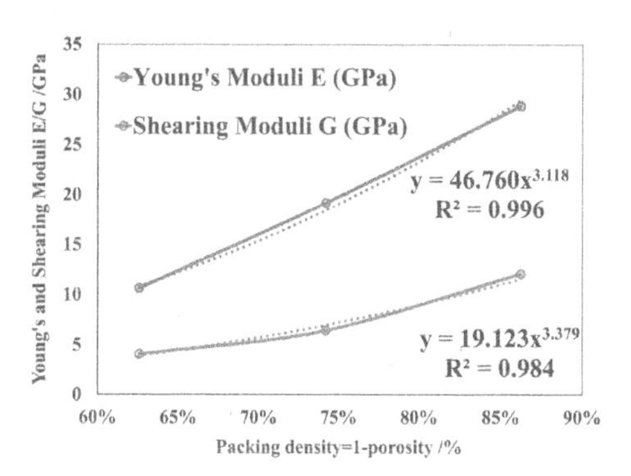

Figure 11. Fitting results by Bolomey's Formula between porosity and moduli.

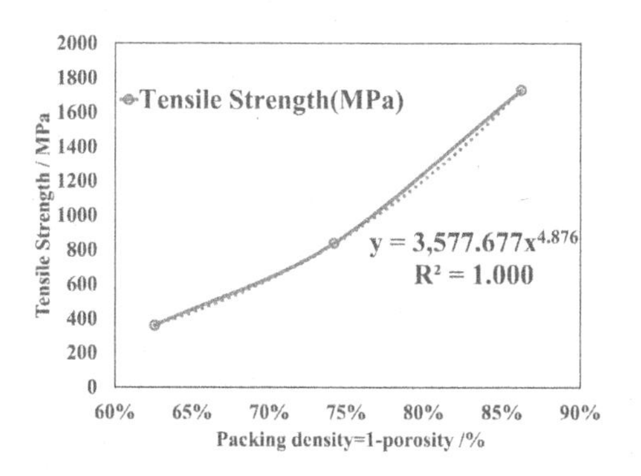

Figure 12. Fitting results by Bolomey's Formula between porosity and tensile strength.

3.2 *Perfect adhesion vs cohesive adhesion*

Then, we compared the tensile stress-strain results of C-S-H with perfect adhesion and cohesive adhesion, respectively, as shown in Figure 13. In addition, we included the microbending experimental value, whose scale is under 20 μm, from reference (Němeček et al. 2016) to compare with results of perfect adhesion and cohesive adhesion in Table 2, where HD C-S-H corresponds to inner product while LD C-S-H corresponds to outer product.

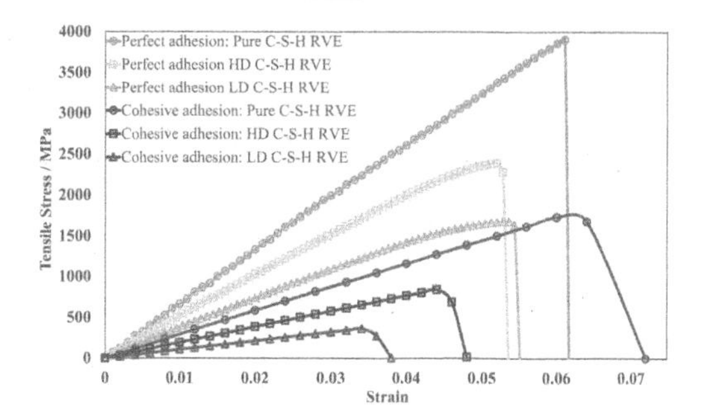

Figure 13. Tensile stress-strain curve for C-S-H RVE: perfect adhesion vs cohesive adhesion.

Table 2. Tensile strength and Young's moduli for C-S-H RVE: perfect adhesion vs cohesive adhesion.

Density Type	Adhesion Type	Tensile Strength MPa	Young's Moduli GPa
Pure C-S-H	Perfect adhesion	3904.87	60.38
	Cohesive adhesion	1730.80	28.86
HD C-S-H	Perfect adhesion	2395.57	47.88
	Reference value	700.2±198.5	34.00
	Cohesive adhesion	839.26	19.15
LD C-S-H	Perfect adhesion	1678.26	32.91
	Reference value	264.1±73.4	23.90
	Cohesive adhesion	362.30	10.64

By comparison, it is clear that packing density and adhesion influence the mechanical properties of C-S-H RVE together. For Young's Modulus, RVE model with perfect adhesion is obviously larger than referenced experimental value as expected while RVE model with cohesive adhesion is less than referenced experimental value. For tensile strength, results of perfect adhesion are still largely higher than referenced experimental value while results of cohesive adhesion almost fall into the range of referenced experimental value. It means that current RVE model with cohesive adhesion has a quite good prediction in tensile strength.

4 CONCLUSION

A multi-scale method was proposed to connect C-S-H globule in nanoscale and C-S-H RVE in microscale. A

FEM model is constructed, where the solid elements denote C-S-H globule were calibrated by *Brittle cracking to simulate mechanical performances of C-S-H globule obtained by MD simulation; cohesive element under traction-separation law were employed to simulate the interaction between C-S-H globules; and the C-S-H RVE was assembly according to a random distribution of voxels.

According to validation of fitting results of Bolomey formula, current C-S-H RVE model can exhibit well coherent relationship between packing density and tensile/shearing properties. By comparing the tensile strength and Young's moduli between perfect adhesion and cohesive adhesion as well as referenced experimental value, preliminary conclusions could be drawn as that: packing density and adhesion type influence tensile strength and Young's moduli mutually. Current RVE model with cohesive adhesion gives quite valuable results and especially favored with experimental value of tensile strength.

REFERENCES

Barnes, P. & Bensted, J. 2002. *Structure and performance of cements,* London, CRC Press.

Bentz, D. P. 2000. *CEMHYD3D: A three-dimensional cement hydration and microstructure development modelling package. Version 2.0,* US Department of Commerce, National Institute of Standards and Technology.

Bernard, F. & Kamali-Bernard, S. Multi-scale modeling to predict ion diffusivity of cracked cement-based materials. *In:* BREUGEL, K. V., YE, G. & YUAN, Y., eds. 2nd International Symposium on Service Life Design for Infrastructures, 2010a. RILEM Publications SARL, 157–166.

Bernard, F. & Kamali-Bernard, S. 2010b. Performance simulation and quantitative analysis of cement-based materials subjected to leaching. *Computational Materials Science,* 50, 218–226.

Bonnaud, P. A., Labbez, C., Miura, R., Suzuki, A., Miyamoto, N., Hatakeyama, N., Miyamoto, A. & Van vliet, K. J. 2016. Interaction grand potential between calcium–silicate–hydrate nanoparticles at the molecular level. *Nanoscale,* 8, 4160–4172.

Courland, R. 2011. *Concrete planet: the strange and fascinating story of the world's most common man-made material,* New York, Prometheus Books.

De souza, F. B., Sagoe–Crentsil, K. & Duan, W. 2022. A century of research on calcium silicate hydrate (C–S–H): leaping from structural characterization to nano - engineering. *Journal of the American Ceramic Society.*

Etzold, M. A., Mcdonald, P. J. & Routh, A. F. 2014. Growth of sheets in 3D confinements — a model for the C–S–H meso structure. *Cement and Concrete Research,* 63, 137–142.

Fan, D. & Yang, S. 2018. Mechanical properties of C-S-H globules and interfaces by molecular dynamics simulation. *Construction and Building Materials,* 176, 573–582.

Fu, J. 2016. *Multiscale modeling and mechanical properties of typical anisotropic crystals structures at nanoscale.* INSA Rennes.

Fu, J., Kamali-Bernard, S., Bernard, F. & Cornen, M. 2018. Comparison of mechanical properties of C-S-H and portlandite between nano-indentation experiments and a modeling approach using various simulation techniques. *Composites Part B: Engineering,* 151, 127–138.

Goyal, A., Palaia, I., Ioannidou, K., Ulm, F.-J., Van Damme, H., Pellenq Roland, J. M., Trizac, E. & Del Gado, E. 2021. The physics of cement cohesion. *Science Advances,* 7, eabg5882.

Hewlett, P. & Liska, M. 2019. *Lea's chemistry of cement and concrete,* Butterworth-Heinemann.

Hlobil, M., ŠMilauer, V. & Chanvillard, G. 2016. Micromechanical multiscale fracture model for compressive strength of blended cement pastes. *Cement and Concrete Research,* 83, 188–202.

Hou, D., Zhang, W., Wang, P., Wang, M. & Zhang, H. 2021. Mesoscale insights on the structure, mechanical performances and the damage process of calcium-silicate-hydrate. *Construction and Building Materials,* 287, 123031.

Jennings, H. M. 2000. A model for the microstructure of calcium silicate hydrate in cement paste. *Cement Concrete Research,* 30, 101–116.

Jennings, H. M. 2008. Refinements to colloid model of CSH in cement: CM-II. *Cement Concrete Research,* 38, 275–289.

Jennings, H. M. & Bullard, J. W. 2011. From electrons to infrastructure: Engineering concrete from the bottom up. *Cement Concrete Research,* 41, 727–735.

Jennings, H. M., Thomas, J. J., Gevrenov, J. S., Constantinides, G. & Ulm, F.-J. 2007. A multi-technique investigation of the nanoporosity of cement paste. *Cement Concrete Research,* 37, 329–336.

Li, K., Stroeven, M., Stroeven, P. & Sluys, L. 2013. CSH globule clustering on nano-scale simulated by the discrete element method for pore structure exploration. *TRANSCEND conference water transport in cementitious materials.* Guildford, UK: RILEM.

Lloyd, J. T., Zimmerman, J. A., Jones, R. E., Zhou, X. W. & Mcdowell, D. L. 2011. Finite element analysis of an atomistically derived cohesive model for brittle fracture. *Modelling and Simulation in Materials Science and Engineering,* 19, 065007.

Masoumi, S., Valipour, H. & Abdolhosseini Qomi, M. J. 2017. Intermolecular Forces between Nanolayers of Crystalline Calcium-Silicate-Hydrates in Aqueous Medium. *The Journal of Physical Chemistry C,* 121, 5565–5572.

Masoumi, S., Zare, S., Valipour, H. & Abdolhosseini Qomi, M. J. 2019. Effective Interactions between Calcium-Silicate-Hydrate Nanolayers. *The Journal of Physical Chemistry C,* 123, 4755–4766.

Monteiro, P. J., Geng, G., Marchon, D., LI, J., Alapati, P., Kurtis, K. E. & Qomi, M. J. A. 2019. Advances in characterizing and understanding the microstructure of cementitious materials. *Cement Concrete Research,* 124, 105806.

Montero-Chacón, F., Marín-Montín, J. & Medina, F. 2014. Mesomechanical characterization of porosity in cementitious composites by means of a voxel-based finite element model. *Computational Materials Science,* 90, 157–170.

Němeček, J., Králík, V., Šmilauer, V., Polívka, L. & Jäger, A. 2016. Tensile strength of hydrated cement paste phases assessed by micro-bending tests and nanoindentation. *Cement and Concrete Composites,* 73, 164–173.

Němeček, J., Šmilauer, V., Kolařík, F. & Maňák, J. 2018. Fracture properties of cement hydrates determined from microbending tests and multiscale modeling. *In:* MESCHKE, G., PICHLER, B. & ROTS, J. G. (eds.) *Computational Modelling of Concrete Structures.* CRC Press.

Nguyen-Tuan, L., Etzold, M. A., Rößler, C. & Ludwig, H.-M. 2020. Growth and porosity of C-S-H phases using the sheet growth model. *Cement and Concrete Research,* 129, 105960.

Nguyen, O., Repetto, E. A., Ortiz, M. & Radovitzky, R. A. 2001. A cohesive model of fatigue crack growth. *International Journal of Fracture,* 110, 351–369.

Nielsen, L. F. 1993. Strength development in hardened cement paste: examination of some empirical equations. *Materials and Structures,* 26, 255–260.

Olson, R. & Jennings, H. 2001. Estimation of CSH content in a blended cement paste using water adsorption. *Cement and Concrete Research,* 31, 351–356.

Qomi, M. J. A., Bauchy, M. & Pellenq, R. J.-M. 2020. Nanoscale Composition-Texture-Property Relation in Calcium-Silicate-Hydrates. *In:* Andreoni, W. & YIP, S. (eds.) *Handbook of Materials Modeling: Applications: Current and Emerging Materials.* Cham: Springer International Publishing.

Sanchez, F. & Sobolev, K. 2010. Nanotechnology in concrete–a review. *Construction Building Materials,* 24, 2060–2071.

Schwalbe, K. H., Scheider, I. & Cornec, A. 2012. *Guidelines for Applying Cohesive Models to the Damage Behaviour of Engineering Materials and Structures*, Springer Berlin Heidelberg.

Scrivener, K., Ouzia, A., Juilland, P. & Mohamed, A. K. 2019. Advances in understanding cement hydration mechanisms. *Cement Concrete Research,* 124, 105823.

Scrivener, K. L., Juilland, P. & Monteiro, P. J. 2015. Advances in understanding hydration of Portland cement. *Cement Concrete Research,* 78, 38–56.

Scrivener, K. L. & Nonat, A. 2011. Hydration of cementitious materials, present and future. *Cement Concrete Research,* 41, 651–665.

Su, X., Yang, Z. & Liu, G. 2010. Finite Element Modelling of Complex 3D Static and Dynamic Crack Propagation by Embedding Cohesive Elements in Abaqus. *Acta Mechanica Solida Sinica,* 23, 271–282.

Tennis, P. D. & Jennings, H. M. 2000. A model for two types of calcium silicate hydrate in the microstructure of Portland cement pastes. *Cement Concrete Research,* 30, 855–863.

Trawiński, W., Tejchman, J. & Bobiński, J. 2018. A three-dimensional meso-scale modelling of concrete fracture, based on cohesive elements and X-ray μCT images. *Engineering Fracture Mechanics,* 189, 27–50.

Vocialta, M., Richart, N. & Molinari, J. F. 2017. 3D dynamic fragmentation with parallel dynamic insertion of cohesive elements. *International Journal for Numerical Methods in Engineering,* 109, 1655–1678.

Yaphary, Y. L., Sanchez, F., Lau, D. & Poon, C. S. 2021. Mechanical properties of colloidal calcium-silicate-hydrate gel with different gel-pore ionic solutions: A mesoscale study. *Microporous and Mesoporous Materials,* 316, 110944.

czhou, F. & Molinari, J. F. 2004. Dynamic crack propagation with cohesive elements: a methodology to address mesh dependency. *International Journal for Numerical Methods in Engineering,* 59, 1–24.

Zhou, X. W., Moody, N. R., Jones, R. E., Zimmerman, J. A. & Reedy, E. D. 2009. Molecular-dynamics-based cohesive zone law for brittle interfacial fracture under mixed loading conditions: Effects of elastic constant mismatch. *Acta Materialia,* 57, 4671–4686.

Computational Modelling of Concrete and Concrete Structures – Meschke, Pichler & Rots (Eds)
© 2022 Copyright the Author(s), ISBN: 978-1-032-32724-2

Predictive approach of the size effect of PFRC simulated by using a softening function

J.C. Gálvez
Departamento de Ingeniería Civil: Construcción, E.T.S de Ingenieros de Caminos, Canales y Puertos Universidad Politécnica de Madrid, Madrid, Spain

F. Suárez
Departamento de Ingeniería Mecánica y Minera, Universidad de Jaén, Jaén, Spain

A. Enfedaque & M.G. Alberti
Departamento de Ingeniería Civil: Construcción, E.T.S de Ingenieros de Caminos, Canales y Puertos Universidad Politécnica de Madrid, Madrid, Spain

ABSTRACT: The size effect on plain concrete specimens is well known and can be correctly captured when performing numerical simulations by using a well characterised softening function. Nevertheless, in the case of polyolefin-fibre-reinforced concrete (PFRC), this is not directly applicable, since using only diagram cannot capture the material behaviour on elements with different sizes due to dependence of the orientation factor of the fibres with the size of the specimen. In previous works, the use of a trilinear softening diagram proved to be very convenient for reproducing fracture of polyolefin-fibre-reinforced concrete elements, but only if it is previously adapted for each specimen size. In this work, a predictive methodology is used to reproduce fracture of polyolefin-fibre-reinforced concrete specimens of different sizes under three-point bending. Fracture is reproduced by means of a well-known embedded cohesive model, with a trilinear softening function that is defined specifically for each specimen size. The fundamental points of these softening functions are defined a priori by using empirical expressions proposed in past works, based on an extensive experimental background. Therefore, the numerical results are obtained in a predictive manner and then compared with a previous experimental campaign in which PFRC notched specimens of different sizes were tested with a three-point bending test setup, showing that this approach properly captures the size effect, although some values of the fundamental points in the trilinear diagram could be defined more accurately.

1 INTRODUCTION

Size effect on plain concrete is well known and is the reason why fracture develops at lower values of the nominal strength when the size of a concrete specimen increases while keeping the same proportions (Bažant 1984). The size effect in fracture of plain concrete is numerically reproduced by means of a cohesive zone formulation that uses a well-characterised softening diagram (Bažant & Planas 1997; Planas, Guinea, & Elices 1999; Jirásek, Rolshoven, & Grassl 2004). The cohesive crack model proposed by Hillerborg can be considered as the most realistic among simple models when quasi-brittle fractures are studied (Bažant & Yu 2009).

Steel fibres as reinforcement in concrete has been used and studied for decades (Di Prisco, Lamperti, Lapolla, & Khurana 2008; Ward & Li 1991), being boosted in recent years, and the range of fibres used for this purpose has increased (Banthia & Gupta 2006; Brandt 2008; Shah & Rangan 1971; Zollo 1997), with polyolefin fibres being one of the most recent types. The use of polyolefin-fibre-reinforced concrete (PFRC) is growing in recent years, due to its good mechanical behaviour and the fact that it reduces and, in some cases, even eliminates some of the problems observed in steel-fibre-reinforced concrete (SFRC) such as corrosion, sensitivity to magnetic fields, or wear and tear of machinery related to its production (concrete pumps and mixers, for example), making PFRC particularly suitable for some uses. The effect of these fibres on the properties of PFRC has been studied in depth during the last years for traditional vibrated concrete (Alberti, Enfedaque, & Gálvez 2015), self-compacting concrete (Alberti, Enfedaque, Gálvez, & Cortez 2020), and in combination with steel fibres (Alberti, Enfedaque, & Gálvez 2017). Many aspects of PFRC are already studied, such as the fibre distribution depending on the production process (Alberti, Enfedaque, Gálvez, & Agrawal 2016) or how it affects

DOI 10.1201/9781003316404-23

fracture in mode I (Alberti, Enfedaque, Gálvez, & Reyes 2017) and mode II (Picazo, Gálvez, Alberti, & Enfedaque 2018). Although this material is starting to count with initial examples of use as a structural material (Alberti, Gálvez, Enfedaque, Carmona, Valverde, & Pardo 2018; Enfedaque, Alberti, Gálvez, Rivera, & Simón-Talero 2018), there is scarce experience with it and the uncertainty on its behaviour in real engineering works under certain situations. One of the key aspects that must be clarified is the size effect; especially to fill the gap if the material properties measured at a laboratory scale are to be used for designing larger structures.

There is not much information about the size effect in fibre-reinforced concrete (FRC), especially in the case of PFRC. In the case of SFRC, some studies can be found (di Prisco, Felicetti, Lamperti, & Menotti 2004; Yoo, Banthia, Yang, & Yoon 2016), and in the case of PFRC, an experimental campaign has been recently carried out (Picazo, Alberti, Gálvez, Enfedaque, & Vega 2019), which has shown that the nominal strength at the limit of proportionality is governed by the matrix (concrete), and the post-cracking residual strength is governed by the fibres.

In previous works, the use of a cohesive zone formulation fed with a trilinear softening curve has proven to be very convenient for reproducing the fracture process in FRC (Enfedaque, Alberti, Gálvez, & Domingo 2017), but it must be adapted depending on several factors such as the fibre length, the fibre proportion (Alberti, Enfedaque, Gálvez, & Reyes 2017), and the specimen size (Suárez, Gálvez, Enfedaque, & Alberti 2019). The adopted trilinear softening diagram describes the contribution of matrix and fibres in the fracture process which, due to the different elastic moduli of both materials, begin to significantly work at different stages of load transmission. Considering the trilinear diagram shown in Figure 1, the initial point t identifies the fracture of the concrete matrix, k the point at which the contribution of fibres starts to predominate over the contribution of the matrix, r the maximum remanent contribution of fibres, and f the eventual failure of the material.

In (Alberti, Enfedaque, Gálvez, & Reyes 2017), some parameters of the PFRC mix were identified, and some expressions were also proposed to define the fundamental points of the trilinear diagram (k and r points). In (Enfedaque, Alberti, & Gálvez 2019) the length and orientation of fibres were observed as key parameters to define the trilinear diagram, also identifying a higher threshold of the PFRC behaviour obtained testing specimens with long fibres oriented in the optimum direction.

There are some approaches and models to simulate fracture. In many cases, these models are calibrated using the experimental results of the test simulated, but this does not guarantee that the parameters represent any other case different from the one under study. From this point of view, the most interesting approach consists of finding models that can reproduce fracture in a predictive way, that is, a model that is fed with parameters obtained by experimental tests that are different from the loading case that wants to be simulated. This type of model is considered more representative of the material than a specific loading case.

The main aim of this contribution is to reproduce fractures on different size specimens of PFRC using a predictive approach. A cohesive model and a softening diagram that corresponds to a trilinear function defined a priori was employed. Using the knowledge obtained in previous works, the coordinates of each of the fundamental points t, k, r and f were identified. To do this, the experimental results of (Picazo, Alberti, Gálvez, Enfedaque, & Vega 2019) were reproduced and compared through a finite element analysis by using an embedded cohesive crack formulation. In the following sections, the experimental work used as a reference of the size effect in PFRC is briefly described, then the main features of the embedded cohesive crack model used to numerically reproduce fracture are presented, and the trilinear softening functions used with each specimen size are obtained by means of the expressions proposed in (Alberti, Enfedaque, Gálvez, & Reyes 2017). Therefore, in the final part of this paper some conclusions are highlighted.

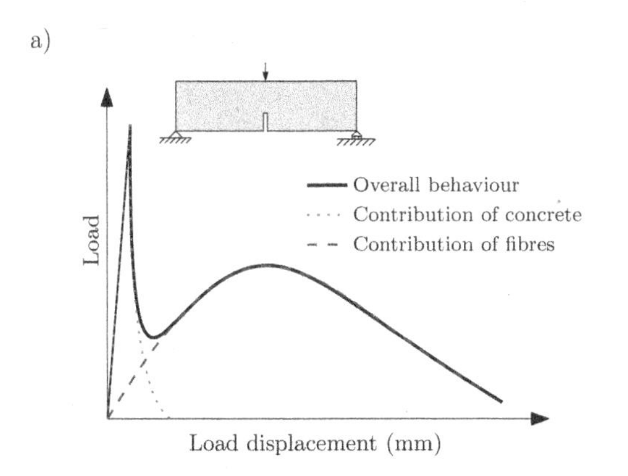

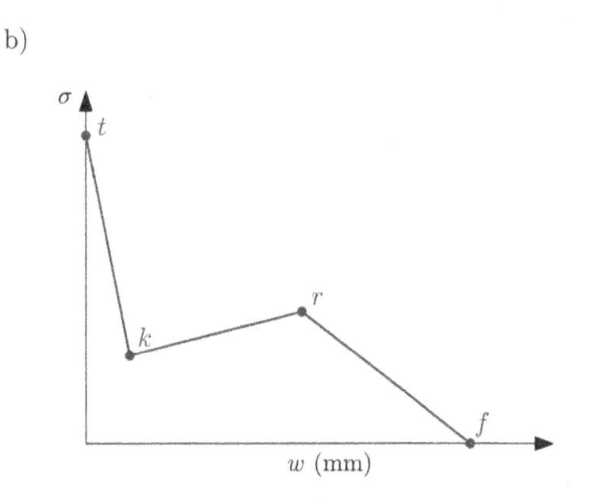

Figure 1. (**a**) Load–displacement diagram obtained in a three-point bending test with a PFRC specimen; (**b**) trilinear softening diagram.

2 EXPERIMENTAL BENCHMARK

The campaign described in (Picazo, Alberti, Gálvez, Enfedaque, & Vega 2019) was used to compare with the nuemrical simulations. For a detailed description of this campaign, the reader is addressed to the referenced work. Table 1 shows the concrete composition, which corresponds to a self-compacting concrete with 10 kg of fibres per m^3 (SCC10).

Table 1. Concrete composition.

Material	SCC10
Cement (kg/m^3)	375
Limestone (kg/m^3)	200
Water (kg/m^3)	188
w/c	0.5
Gravel (kg/m^3)	245
Grit (kg/m^3)	367
Sand (kg/m^3)	918
Superplasticiser (% cement weight)	1.25
PF48 (kg/m^3)	10

Concrete reinforcement consists of 48 mm long polyolefin macrofibres with an embossed surface. The main properties of these fibres can be consulted in Table 2. More information on these fibres can be found in (Alberti, Enfedaque, & Gálvez 2015).

Table 2. Fibres properties.

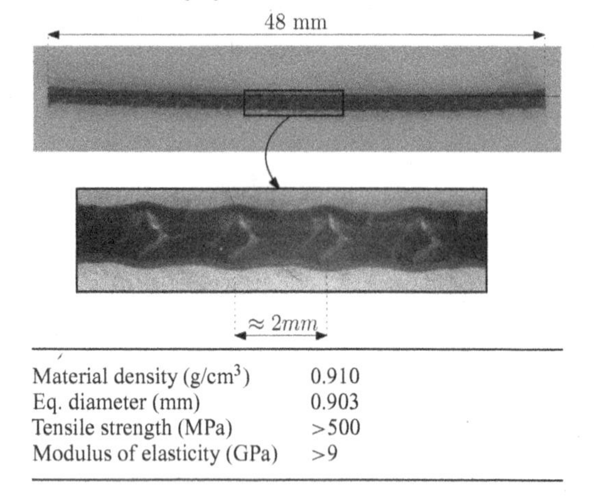

Material density (g/cm^3)	0.910
Eq. diameter (mm)	0.903
Tensile strength (MPa)	>500
Modulus of elasticity (GPa)	>9

The experimental campaign of reference involved three-point bending tests carried out on three samples of each size, following the guidelines of the EN-14651 standard (European Committee for Standardization 2007) (except for the specimen sizes and notch dimensions). Figure 2 shows a schematic drawing of the experimental setup, and Table 3 shows the dimensions of the specimens. In all cases, the concrete composition was the same, and 48 mm long polyolefin fibres were used in a proportion of 10 kg/m^3.

The load and the displacement were recorded by the testing machine, and the evolution of the crack mouth opening displacement (CMOD) was measured by means of a digital image correlation system (DIC). To compare the experimental results, two main diagrams were employed: load versus displacement of the application point of the load and load versus CMOD. Figure 2 shows these values in the scheme of a damaged specimen during the test.

Table 3. Specimens' dimensions.

	Length (mm)	Width (mm)	Height (mm)	Notch (mm)
Large	1350	50	300	150
Medium	675	50	150	75
Small	340	50	75	37.5

3 EMBEDDED COHESIVE CRACK MODEL

The crack process is modelled by using the finite element analysis and adapting a formulation based on the cohesive zone approach developed by Hillerborg (Hillerborg, ModÃ©er, & Petersson 1976), inspired by the work of Dugdale (Dugdale 1960) and Barenblatt (Barenblatt 1962). This formulation simulates fracture inside an element using the strong discontinuity approach and was initially developed for concrete (Sancho, Planas, Cendón, Reyes, & Gálvez 2007; Gálvez, Planas, Sancho, Reyes, Cendón, & Casati 2013) but later adapted to brickwork masonry elements (Reyes, Gálvez, Casati, Cendón, Sancho, & Planas 2009) and fibre-reinforced cementitious materials (Alberti, Enfedaque, Gálvez, & Reyes 2017; Enfedaque, Alberti, Gálvez, & Domingo 2017; Suárez, Gálvez, Enfedaque, & Alberti 2019).

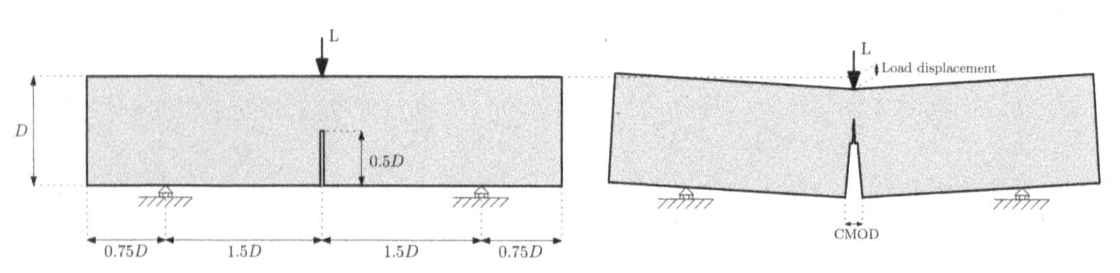

Figure 2. **Left**: scheme of a three-point bending test and specimen geometry; **right**: scheme of crack propagation from the notch tip during the test.

The cohesive zone approach relies on the experimental evidence that fracture usually develops under a predominant local mode I. Thus, this approach assumes that the cohesive stress vector t is perpendicular to the crack opening and parallel to the crack displacement vector w, which is expressed by (1).

$$t = \frac{f(\tilde{w})}{\tilde{w}} w \qquad \text{with } \tilde{w} = \max(|w|) \tag{1}$$

where $f(|\tilde{w}|)$ stands for the material softening function, defined in terms of an equivalent crack opening $\tilde{w}$. This equivalent crack opening stores the maximum historical crack opening to account for possible unloading scenarios. In this case, the softening diagram is defined as trilinear, as shown in Figure 3, and the load–unload branches follow lines towards the origin in all cases. The trilinear diagram is defined by the following expression:

$$\sigma = \begin{cases} f_{ct} + \left(\dfrac{\sigma_k - f_{ct}}{w_k}\right) \cdot w & \text{if } 0 < w \leq w_k \\ \sigma_k + \left(\dfrac{\sigma_r - \sigma_k}{w_r - w_k}\right) \cdot (w - w_k) & \text{if } w_k < w \leq w_r \\ \sigma_r + \left(\dfrac{-\sigma_r}{w_f - w_r}\right) \cdot (w - w_r) & \text{if } w_r < w \leq w_f \\ 0 & \text{if } w > w_f \end{cases} \tag{2}$$

In the finite element models presented later, the embedded cohesive crack formulation is used with constant strain triangular elements. Cracking can only develop in three directions, each parallel to the element sides and at mid height, which guarantees that local and global equilibria are satisfied. Figure 4 shows the only three possible crack paths in an element.

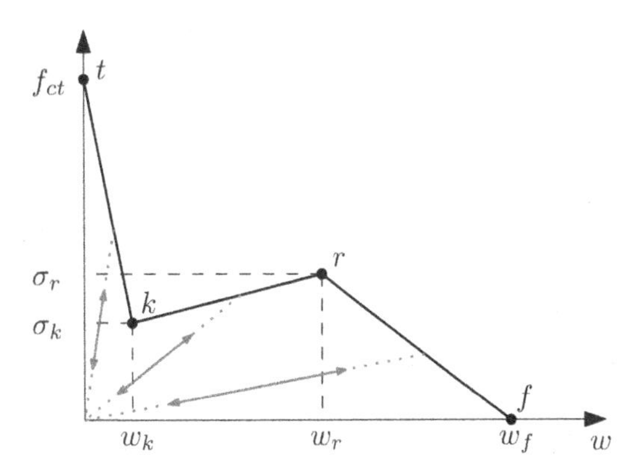

Figure 3. Scheme of a trilinear softening function. Load–unload branches follow a line towards the origin.

Once the crack direction is defined, the element is divided into two parts, A^+ and A^-, and the stress vector t is constant along the crack, expressed by (3).

$$t = \frac{A}{hL} \boldsymbol{\sigma} \cdot n \tag{3}$$

where A stands for the area of the element, h for the height of the triangle over the side opposite to the solitary node, L for the crack length in the element, and n for the unit vector normal to that side and to the crack. Since the crack is parallel to one side of the triangular element and is placed at mid height, Expression (3) turns into $t = \boldsymbol{\sigma} \cdot n$ (the reader can find more details of this and other aspects of the model in (Sancho, Planas, Cendón, Reyes, & Gálvez 2007)).

The material outside the crack is assumed to be elastic, and the crack displacement vector w is solved considering that the stress tensor can be obtained by subtracting an inelastic part, which considers the contribution of the crack displacement to the elastic prediction computed using the apparent strain by means of (4).

$$\boldsymbol{\sigma} = E : \left[\epsilon^a - \left(b^+ \otimes w\right)^S\right] \cdot n \tag{4}$$

where E is the elastic tangent tensor, ϵ^a the apparent strain vector obtained with the nodal displacements, b^+ the gradient vector of the shape function that corresponds to the solitary node, which can be easily obtained in this case by (5), superscript S indicates the symmetric part of the resulting tensor, : the double-dot product ($(A : b)_{ij} = A_{ijkl} b_{kl}$), and $\otimes$ the direct product ($(a \otimes b)_{ij} = a_i b_j$).

$$b^+ = \frac{1}{h} n \tag{5}$$

Since the stress vector t can be obtained as $t = \boldsymbol{\sigma} \cdot n$, using the expression of $\boldsymbol{\sigma}$ obtained with (4) and the expression of t in terms of the crack opening (1), the following expression is defined:

$$\frac{f(\tilde{w})}{\tilde{w}} w = [E : \boldsymbol{\epsilon}^a] \cdot n - \left[E : \left(b^+ \otimes w\right)^S\right] \cdot n$$

which can be rewritten as

$$\left[\frac{f(\tilde{w})}{\tilde{w}} 1 + n \cdot E \cdot b^+\right] \cdot w = [E : \boldsymbol{\epsilon}^a] \cdot n \tag{6}$$

where 1 stands for the second-order identity tensor. Using an iterative process (such as the Newton–Raphson method), the crack displacement w that satisfies (6) can be obtained.

This model is implemented using a UMAT subroutine in ABAQUS and, since vectors n, b^+, crack length L, and the element area A are computed using the nodal coordinates for each element, it reads an external file with this information.

4 DEFINITION OF THE TRILINEAR SOFTENING DIAGRAMS

According to (Alberti, Enfedaque, Gálvez, & Reyes 2017), there are several parameters that can be experimentally measured and help to define the trilinear

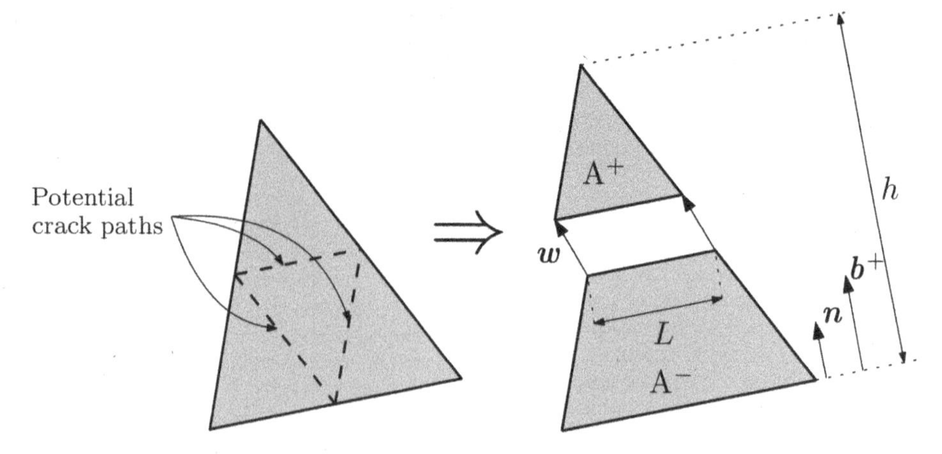

Figure 4. Potential crack paths (**left**) and geometrical definitions of w, n, and b^+ (**right**).

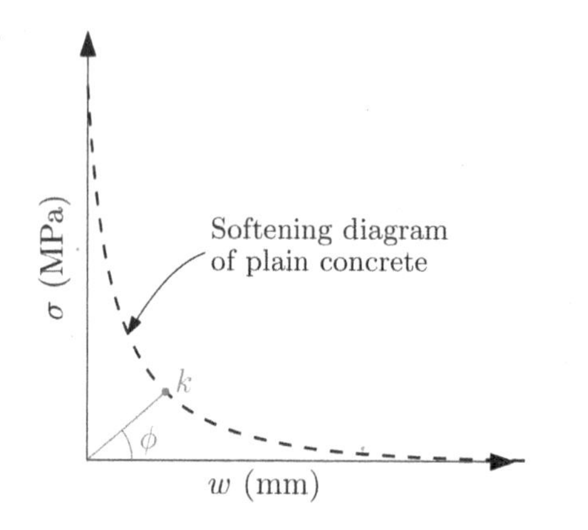

Figure 5. Identification of k point of the trilinear diagram by means of the angle ϕ.

diagram for the PFRC. Apart from the fracture parameters of plain concrete (G_F and f_t), which define the first part of the diagram, these parameters are the volume of fibres (V_f), the orientation factor (θ) and the percentage of pulled out fibres at the fracture surface ($\%Pulled - out$). With the help of V_f, the angle ϕ can be obtained by means of (7).

$$\phi = -3.6046 + 5.0625 \cdot \left(1 - e^{(-6.55 \cdot V_f)}\right) \tag{7}$$

This angle serves to identify the second point of the diagram (point k), which is the intersection of the softening function of plain concrete (here considered as an exponential function: $\sigma = f_t \cdot \exp\left(-\frac{f_t \cdot w}{G_F}\right)$) with a line passing through the origin with a direction defined by ϕ (see Figure 5).

By using the three main parameters mentioned before and the ultimate tensile strength of the fibres (σ_u), the maximum remaining strength (σ_r) can be obtained with (8).

$$\sigma_r = (1 - \%Pulled - out) \cdot V_f \cdot \theta \cdot \sigma_u \tag{8}$$

Considering the scheme of the trilinear diagram shown in Figure 1, the first two points can be identified as follows: point t is identified by f_t, which can be experimentally obtained, while point k depends on the volume fraction of fibres (V_f) by means of the ϕ angle defined with (7) and the softening function of plain concrete. Table 4 shows the intermediate values that result of this calculation.

As regards the remaining two points, r and f, the value of σ_r can be obtained with (8). Table 5 shows the results of this calculation for each size, and σ_f is, obviously, equal to 0, but w_r and w_f must be estimated; they depend on the fibre length, but there are no specific expressions to obtain them. In this case, w_r is estimated as equal to 1.65 mm, since this was the value adopted in (Suárez, Gálvez, Enfedaque, & Alberti 2019) for simulating fracture in specimens made with 48 mm long fibres of the same kind as those used here. As regards w_f, this value is related to the maximum crack opening before completely losing the bonding between the fibres and the matrix; therefore, it is assumed to be proportional to the fibre length. Thus, since in (Alberti, Enfedaque, Gálvez, & Reyes 2017) specimens made with 60 mm long fibres

Table 4. Intermediate values for obtaining point k of the trilinear diagram.

	f_t (MPa)	G_F (N/mm)	ϕ	w_k (mm)	σ_k (MPa)
Small/Medium/Large	3.2	0.13	1.448	0.07143	0.57715

Table 5. Intermediate values for obtaining σ_r for all three considered sizes.

	θ	*% Pulled – Out*	V_f	σ_u (MPa)	σ_r (MPa)
Small	0.63	0.54	0.011	376	1.20
Medium	0.62	0.54	0.011	376	1.18
Large	0.72	0.54	0.011	376	1.37

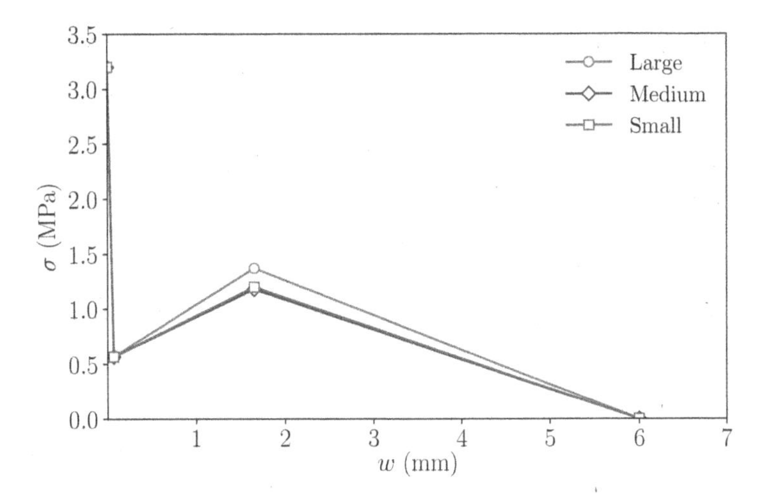

Figure 6. Initial trilinear softening diagrams.

were modelled using $w_f = 7.5$ mm, here, a value of $w_f = \frac{48}{60} \cdot 7.5 = 6.0$ mm is adopted. Figure 6 shows the resulting trilinear softening diagrams for all three sizes.

5 RESULTS AND DISCUSSION

Fracture of the three specimen sizes analysed in (Alberti, Enfedaque, & Gálvez 2015) was carried out using the finite element method, and a displacement control was used to drive the fracture evolution with good convergence. The simulations were computed using ABAQUS (Smith 2009), and the fracture was reproduced by means of a UMAT subroutine that implemented the previously described material behaviour.

Figure 7 shows the three meshes used in this work with the same scale. In all three meshes, the region connecting the notch tip with the load application point was refined in order to better capture the fracture process, while the rest of the specimen was meshed with larger elements, which helped to notably reduce the time of computation. The models were formed by a number of nodes smaller than 800 and a number of triangular finite elements smaller than 1500, thus keeping the model size small enough to have models that perform efficiently. These simulations were run on a computer with an Intel Xeon E5-1620 processor with 4 cores at 3.5 GHz, although only one was used since the user subroutine that reproduces the material behaviour does not allow parallel computing; all the simulations took around 150 min to run. In the case of the large size model (L), the side of minimum element size was around 7 mm, in the case of the medium size model (M), 3.5 mm, and in the case of the small size model (S), 2 mm. The refinement of these meshes was designed based on previous works (see (Suárez, Gálvez, Enfedaque, & Alberti 2019)), in which the mesh dependence was already analysed.

Figure 8 shows the load–load displacement and load–CMOD diagrams for all three sizes and compares them with the experimental results. Each specimen size is identified by a different colour: red for large size, blue for medium size, and green for small size. The shades behind the diagrams correspond to the experimental envelopes, with the same colour code used in the diagrams; therefore, the red shade corresponds to the experimental envelope of the large specimens, the blue shade to the experimental envelope of medium specimens, and the green shade to the experimental envelope of small specimens. Apart from the overestimation of the initial peak, which is a known issue when this type of numerical modelling is used, especially in large-sized specimens (Elices, Guinea, GÃ³mez, & Planas 2002), the models reproduce the experimental results reasonably well. This agreement is particularly good, in the case of the medium size, and presents some differences in the last part of the load–load displacement diagrams, in the cases of large and small sizes, in which the numerical model tends to underestimate the specimen's remaining strength.

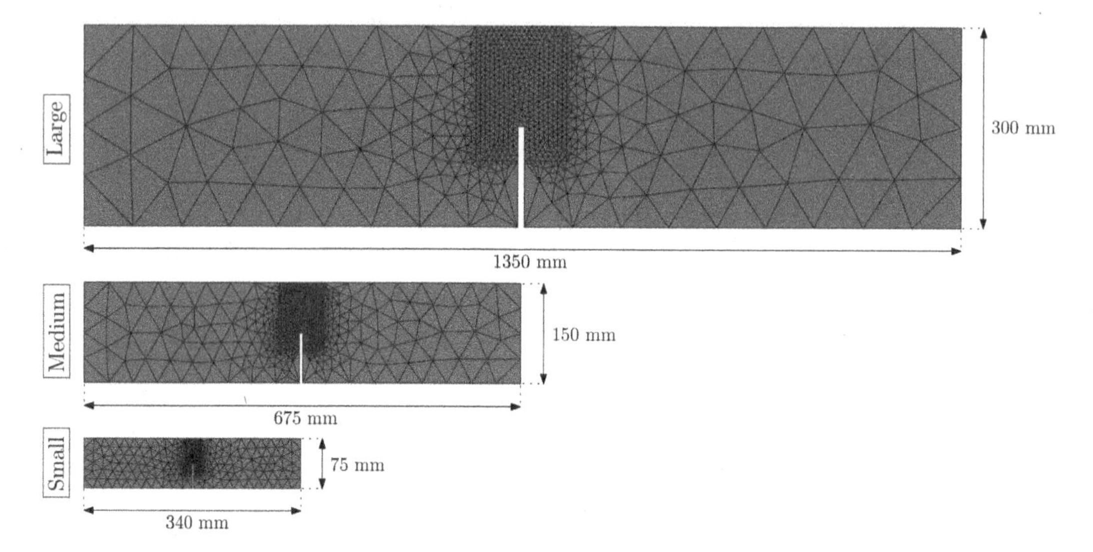

Figure 7. FEM meshes used in the simulations for each specimen size.

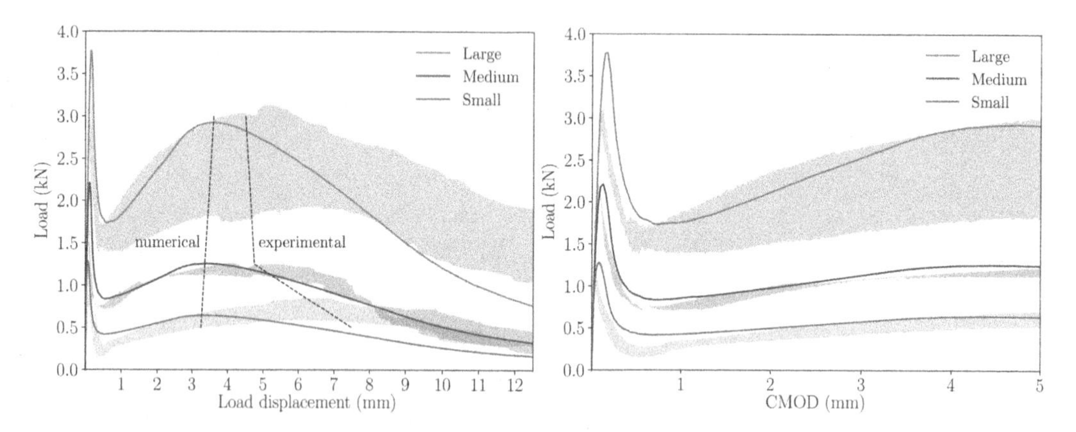

Figure 8. Numerical results compared with the experimental envelopes; each specimen size is identified by a different colour. Experimental envelopes correspond to three specimens tested.

It is also worth noting that in the case of the experimental results, the maximum remanent load occurs at a larger load displacement if compared with the medium and large sizes, while in the case of the numerical results, this maximum load after the first peak occurs approximately at the same load displacement and, in all cases has been retained, following a very linear trend. These trends are depicted by dashed lines on the load–load displacement diagrams of Figure 8.

These results show that expressions (7) and (8), defined in the past by analysing the fracture behaviour of different PFRC mixes, well describe the general behaviour of this material and take into account the main parameters: the volume of fibres in the mix (V_f), the orientation of fibres with respect to the fracture surface (θ), and the quality of bonding between the fibres and concrete, expressed by the fraction of fibres that are pulled out at the fracture surface ($\%Pulled - out$).

Nevertheless, the parameters used to define the trilinear softening diagrams, abscissa values of points k and f, are only estimated based on previous experiences with this type of model, but there are no expressions proposed for them yet. In the following section, the influence of these two values, w_r and w_f, is studied to understand how they modify the diagrams, which can help to propose expressions to quantify them.

6 CONCLUSIONS

In this work, the numerical modelling of the size effect by means of a cohesive model fed with a trilinear softening function was studied using a predictive method. A three-point bending test on specimens of three sizes was numerically reproduced and compared with experimental data from previous works.

The trilinear diagrams for each size were defined by expressions obtained in previous experimental campaigns, resulting in good agreement with the lab observations.

From the work presented above, the following conclusions can be drawn:

1. The complete fracture behaviour of PFRC specimens can be numerically simulated using a predictive trilinear cohesive crack model, which can be defined a priori by means of empirical expressions obtained with lab tests different from those simulated. This diagram is defined by four points, with coordinates that depend on PFRC mechanical characteristics, i.e., the tensile strength of the matrix, the proportion of fibres, and the orientation factor. Abscissa values w_r and w_f (see Figure 3) are fixed based on experimental results obtained in previous literature. It is still an unsolved challenge to obtain expressions to estimate w_r and w_f using the mechanical characteristics of the PFRC.
2. The softening diagrams are not equal for all specimen sizes and should be adjusted for each of them. This is mainly due to a different orientation factor that varies with the size of the specimen.
3. The maximum remanent loads obtained for each size present a linear trend on the load–displacement diagram, which does not agree completely with the experimental observations, although the load–displacement and load–CMOD curves properly agree with the experimental envelopes for the three studied sizes.

ACKNOWLEDGEMENTS

The authors gratefully acknowledge the financial support provided for this research by the Ministry of Science and Innovation of Spain through the Research Fund Project PID2019-108978RB-C31.

REFERENCES

Alberti, M., A. Enfedaque, & J. Gálvez (2015). Comparison between polyolefin fibre reinforced vibrated conventional concrete and self-compacting concrete. *Construction and Building Materials 85*, 182 – 194.

Alberti, M., A. Enfedaque, & J. Gálvez (2017). Fibre reinforced concrete with a combination of polyolefin and steel-hooked fibres. *Composite Structures 171*, 317–325.

Alberti, M., A. Enfedaque, J. Gálvez, & V. Agrawal (2016). Fibre distribution and orientation of macro-synthetic polyolefin fibre reinforced concrete elements. *Construction and Building Materials 122*, 505–517.

Alberti, M., A. Enfedaque, J. Gálvez, & A. Cortez (2020). Optimisation of fibre reinforcement with a combination strategy and through the use of self-compacting concrete. *Construction and Building Materials 235*, 117289.

Alberti, M., A. Enfedaque, J. Gálvez, & E. Reyes (2017). Numerical modelling of the fracture of polyolefin fibre reinforced concrete by using a cohesive fracture approach. *Composites Part B: Engineering 111*, 200 – 210.

Alberti, M. G., J. C. Gálvez, A. Enfedaque, A. Carmona, C. Valverde, & G. Pardo (2018). Use of steel and polyolefin fibres in the la canda tunnels: Applying mives for assessing sustainability evaluation. *Sustainability 10*(12), 4765.

Banthia, N. & R. Gupta (2006). Influence of polypropylene fiber geometry on plastic shrinkage cracking in concrete. *Cement and Concrete Research 36*(7), 1263–1267.

Barenblatt, G. I. (1962). The mathematical theory of equilibrium cracks in brittle fracture. In *Advances in applied mechanics*, Volume 7, pp. 55–129. Elsevier.

Bažant, Z. P. (1984). Size effect in blunt fracture: concrete, rock, metal. *Journal of engineering mechanics 110*(4), 518–535.

Bažant, Z. P. & J. Planas (1997). *Fracture and size effect in concrete and other quasibrittle materials*. CRC press.

Bažant, Z. P. & Q. Yu (2009). Universal size effect law and effect of crack depth on quasi-brittle structure strength. *Journal of engineering mechanics 135*(2), 78–84.

Brandt, A. M. (2008). Fibre reinforced cement-based (frc) composites after over 40 years of development in building and civil engineering. *Composite Structures 86*(1), 3–9. Fourteenth International Conference on Composite Structures.

di Prisco, M., R. Felicetti, M. Lamperti, & G. Menotti (2004). On size effect in tension of sfrc thin plates. *Fracture mechanics of concrete structures 2*, 1075–1082.

Di Prisco, M., M. Lamperti, S. Lapolla, & R. S. Khurana (2008). Hpfrcc thin plates for precast roofing. In *Proceedings of the 2nd international symposium on HPC, Kassel*.

Dugdale, D. S. (1960). Yielding of steel sheets containing slits. *J Mech Phys Solids 8*(2), 100–104.

Elices, M., G. Guinea, J. Gómez, & J. Planas (2002). The cohesive zone model: advantages, limitations and challenges. *Engineering fracture mechanics 69*(2), 137–163.

Enfedaque, A., M. Alberti, J. Gálvez, & J. Domingo (2017). Numerical simulation of the fracture behaviour of glass fibre reinforced cement. *Construction and Building Materials 136*, 108 – 117.

Enfedaque, A., M. G. Alberti, & J. C. Gálvez (2019). Influence of fiber distribution and orientation in the fracture behavior of polyolefin fiber-reinforced concrete. *Materials 12*(2), 220.

Enfedaque, A., M. G. Alberti, J. C. Gálvez, M. Rivera, & J. Simón-Talero (2018). Can polyolefin fibre reinforced concrete improve the sustainability of a flyover bridge? *Sustainability 10*(12), 4583.

European Committee for Standardization (2007). Test method for metallic fibre concrete. measuring the flexural tensile strength (limit of proportionality (lop), residual).

Gálvez, J., J. Planas, J. Sancho, E. Reyes, D. Cendón, & M. Casati (2013). An embedded cohesive crack model for finite element analysis of quasi-brittle materials. *Eng Fract Mech 109*, 369–386.

Hillerborg, A., M. ModÃ©er, & P.-E. Petersson (1976). Analysis of crack formation and crack growth in concrete by means of fracture mechanics and finite elements. *Cem Concr Res 6*(6), 773 – 781.

Jirásek, M., S. Rolshoven, & P. Grassl (2004). Size effect on fracture energy induced by non-locality. *International Journal for Numerical and Analytical Methods in Geomechanics 28*(7-8), 653–670.

Picazo, Á., M. G. Alberti, J. C. Gálvez, A. Enfedaque, & A. C. Vega (2019). The size effect on flexural fracture of polyolefin fibre-reinforced concrete. *Applied Sciences 9*(9), 1762.

Picazo, A., J. Gálvez, M. Alberti, & A. Enfedaque (2018). Assessment of the shear behaviour of polyolefin fibre reinforced concrete and verification by means of digital image correlation. *Construction and Building Materials 181*, 565–578.

Planas, J., G. Guinea, & M. Elices (1999). Size effect and inverse analysis in concrete fracture. *International Journal of Fracture 95*(1-4), 367.

Reyes, E., J. Gálvez, M. Casati, D. Cendón, J. Sancho, & J. Planas (2009). An embedded cohesive crack model for finite element analysis of brickwork masonry fracture. *Engineering Fracture Mechanics 76*(12), 1930 – 1944.

Sancho, J., J. Planas, D. Cendón, E. Reyes, & J. Gálvez (2007). An embedded crack model for finite element analysis of concrete fracture. *Engineering Fracture Mechanics 74*(1), 75 – 86. Fracture of Concrete Materials and Structures.

Shah, S. P. & B. V. Rangan (1971). Fiber reinforced concrete properties. In *Journal Proceedings*, Volume 68, pp. 126–137.

Smith, M. (2009). *ABAQUS/Standard User's Manual, Version 6.9*. United States: Dassault Systèmes Simulia Corp.

Suárez, F., J. Gálvez, A. Enfedaque, & M. Alberti (2019). Modelling fracture on polyolefin fibre reinforced concrete specimens subjected to mixed-mode loading. *Engineering Fracture Mechanics 211*, 244 – 253.

Ward, R. & V. Li (1991). Dependence of flexural behaviour of fibre reinforced mortar on material fracture resistance and beam size. *Construction and Building Materials 5*(3), 151–161.

Yoo, D.-Y., N. Banthia, J.-M. Yang, & Y.-S. Yoon (2016). Size effect in normal- and high-strength amorphous metallic and steel fiber reinforced concrete beams. *Construction and Building Materials 121*, 676 – 685.

Zollo, R. F. (1997). Fiber-reinforced concrete: an overview after 30 years of development. *Cement and Concrete Composites 19*(2), 107–122.

Computational Modelling of Concrete and
Concrete Structures – Meschke, Pichler & Rots (Eds)
© 2022 Copyright the Author(s), ISBN: 978-1-032-32724-2

Estimation of concrete strength by machine learning of AE waves

Y. Shimamoto
Tokyo University of Agriculture and Technology, Tokyo, Japan

S. Tayfur & N. Alver
Ege University, Izmir, Turkey

T. Suzuki
Niigata University, Niigata, Japan

ABSTRACT: Concrete structures are damaged by environmental factors and loads to which they are subjected. To maintain such structures, it is necessary to properly determine the mechanical properties and degree of damage suffered by concrete using core tests. The relationship between the stress level of damaged concrete under compression and acoustic emission (AE) parameters was investigated using clustering and regression analysis with random forests to identify the most important AE parameters. For clusters 1 and 3, R^2 was higher, and RMSE and MAE were lower than for non-clustered cases. Therefore, cluster analysis can be expected to improve the accuracy of AE testing. Finally, the most important parameter was determined to be rise time, the second was the centroid frequency. These two parameters can be used to clarify compressive fracture behavior of damaged concrete.

1 INTRODUCTION

Concrete structures become decrepit and damaged over time. To maintain them, it is necessary to properly assess their condition and damage levels. This is usually done by determining the mechanical properties of the concrete and the degree of damage using core tests. The acoustic emission (AE) technique is recognized as one of many practical ways to quantitatively estimate damage in concrete (Grosse et al. 2021). By means of data collected from AE tests of concrete structures, damage origin (Li et al. 2017; Van Steen et al. 2019; Zhou et al. 2018), type (Alver et al. 2017; JCMS-IIIB5706 2003; Prem et al. 2021; Tayfur et al. 2018; Zhang et al. 2020; location (Boniface et al. 2020; Mirgal et al. 2020; Rodríguez & Celestino 2020; Soltangharaei et al. 2021), and severity (Abouhussien & Hassan 2020; Burud & Kishen 2021) can be determined.

Micro-cracking damage appears predominantly under compression. The increasing number of aging structures and the amount of disastrous damage from recent earthquakes have focused attention on the need to repair damaged concrete structures. Inspection of concrete structures in service is currently done using both destructive and non-destructive tests. Destructive testing uses concrete core samples drilled from structures in use. The cores' chemical and physical properties are then measured. The concrete's uniaxial compressive strength and Young's modulus are genterally specified for testing. The accuracy of non-destructive testing is evaluated by correlating non-destructive monitoring indices with mechanical properties obtained from desructive testing. Thus, the concrete's level of damage is estimated using only properties that correlate with the measured mechanical properties. To date, there are few well established techniques for estimating distributed microscopic damage. Ohtsu proposed measuring AE activity in uniaxial compression testing of core samples because AE-generating behavior is closely associated with the presence of micro-cracks in concrete (Ohtsu 1987). To analyze the occurrence of AE events quantitatively, the rate-process theory was introduced (Ohtsu 1992). Suzuki, one of the co-authors of this article, proposed evaluating concrete damage using AE rate-process analysis and damage mechanics (Suzuki et al. 2007). The procedure is named DeCAT (Damage Estimation of Concrete by Acoustic Emission Technique). It is based on estimating an intact modulus of elasticity in concrete (Suzuki et al. 2007), and suggests quantitatively evaluating concrete damage by applying AE and X-ray CT techniques to core testing (Suzuki et al. 2010; 2014; 2017). Suzuki then improved DeCAT,

DOI 10.1201/9781003316404-24

creating the i-DeCAT system, and proposed that AE energy can be useful for damage evaluation (Suzuki et al. 2019; 2020; Shimamoto & Suzuki 2021). A similar approach has been experimentally tested in studies such as Wu (Wu et al. 2019) and Karcili (Karcili et al. 2016). Other studies have evaluated the damage and fracture behavior of concrete by analyzing various AE parameters such as amplitude and frequency (Shah & Ribakov 2010; Shang et al. 2021; Schiavi et al. 2011. In recent years, AE waves have been classified in detail not only focusing on a single AE parameter but also applying machine learning to multiple AE parameters. By integrating AE with random forests, support vector machines, or neural networks, cracks' locations can be identified for structural inspection (Crivelli et al. 2014; Hübner et al. 2020; Kane & Andhare 2020; Wotzka & Cichoń 2020). AE waves can be classified by frequency characteristics in fracture process (Zhang et al. 2021), damage levels and types in reinforced concrete members can be estimated (Guofeng & Du. 2020; Soltangharaei et al. 2021; Thirumalaiselvi & Sasmal 2021), leakage in tanks can be automatically detected (Rahimi et al. 2020), and AE activities can be localized (Morizet et al. 2016; Suwansin & Phasukkit 2021).

As described above, the usefulness of many AE parameters for evaluating the fracture process is evident, but the effective mix of AE parameters has not been sufficiently identified. In this study, the relationship between the stress level of concrete under compression and AE parameters was investigated by regression analysis with machine learning to identify the most important AE parameters. For this purpose, concrete core samples were drilled out from a headwork which had been subjected to the freeze and thaw process. A random forest regression analysis was used for the machine learning algorithm. In addition, whether prediction accuracy could be improved by clustering AE activities as a preprocessing method was investigated.

2 ANALYTICAL PROCEDURE

2.1 *AE parameter analysis*

According to ASTM E 1316, acoustic emission (AE) is defined as an event producing transient elastic waves by releasing of a number of local sources in materials under stress (ASTM E1316 2002). From a physical viewpoint, fracture in a material takes place as the release of stored strain energy, which is consumed by nucleating new external surfaces (cracks) and emitting elastic waves (Grosse et al. 2021). The elastic waves propagate inside the material and are detected by AE sensors. AE signals are detected as dynamic motions at the surface of a material, and are converted into electric signals. Then the electrical signals are amplified and filtered.

The total AE activity (simply how many "hits" are recorded by any sensor) is indicative of the phenomenon being monitored. When the phenomenon is a fracture, the number of hits is related to the degree of damage and the rate of crack formation and propagation (Grosse et al. 2021). In addition, the shape of the waveform yields important information about the source of the emission. Therefore, many parameters are used to quantify the waveform. The basic parameters are shown in Table 1. Frequency content is also essential for AE analyses.

Table 1. Basic AE parameter.

AE parameter	Meaning
Hit	A signal that exceeds the threshold and causes a system channel to record data.
Energy	The signal strength of the AE waveform
Duration	The time interval between the 1st count and the last descending threshold crossing.
Rise Time	The time interval between the triggering time of AE signal (1st count) and the time of the peak amplitude.
RMS	Root mean square value of AE signal.
Peak frequency	The frequency with the highest magnitude in the FFT.
Centroid frequency	The centroid frequency of the FFT.

In this study, concrete fracture behavior was evaluated using the number of hits, duration, rise time, energy, root mean square value (RMS), centroid frequency and peak frequency of AE signals. AE energy is defined by the following equation:

$$E_{AE} = (a_p)^2. \tag{1}$$

where E_{AE} is detected AE energy (V^2) and a_p is the peak amplitude of the detected AE wave (V). Peak frequency and centroid frequency are additionally determined in real time from the fast Fourier transform (FFT) of the recorded waveforms. Peak frequency is the frequency with the highest magnitude in the FFT and centroid frequency is the centroid of the FFT.

2.2 *k-means clustering algorithm*

k-means clustering is one of the simplest and most commonly used clustering algorithms. It tries to find cluster centers that are representative of certain regions of the data. The algorithm consists of two steps: assigning each data point to the closest cluster center, and setting each cluster center as the mean of the data points that are assigned to it (Andreas & Sarah 2017). Finally, the initial classification is completed when no point is pending. From this point forward, new positions of the centroid must be recalculated until the objective function, J, which sums the distances to the centroids, is minimized. It is calculated using as follows (Kaufman & Rousseeuw 2008):

$$J = \sum_{j=1}^{k}\sum_{i=1}^{n} \|x_i^{(j)} - c_j\|^2. \tag{2}$$

Here, $\|x_i^{(j)} - c_j\|^2$ is the specific length between the point and the cluster center c_j, which is an indicator of the nth point from the respective cluster centers. In other words, the last step is repeated until the centroids no longer move and the objects are separated into classes for which the distances are minimized.

In this study, AE data have been classified into three clusters based on three AE parameters: peak amplitude, peak frequency, and centroid frequency.

2.3 *Regression analysis using random forests*

2.3.1 *Random forest algorithm*

In this study, the relationship between the stress level of concrete in compression and the various AE parameters was examined using random forest regression analysis, which is an ensemble of decision trees.

Decision trees are widely used models for classification and regression tasks (Figure 1). Essentially, they pose a hierarchy of if/else questions, leading to a decision (Beyeler 2018). Each node in the tree either represents a question or terminal node (also called a leaf) that contains the answers. Decision trees have two advantages over many other algorithms: the resulting model can easily be visualized, and the algorithm is completely invariant to scaling of the data. As each feature is processed separately and the possible splits of the data do not depend on scaling, no preprocessing such as normalization or standardization of the features is needed for decision tree algorithms (Beyeler 2018). The main downside of decision trees is that even when pre-pruning is used, they tend to overfit and are not highly generalizable. Therefore, in most applications, ensemble methods are usually used in place of a single decision tree.

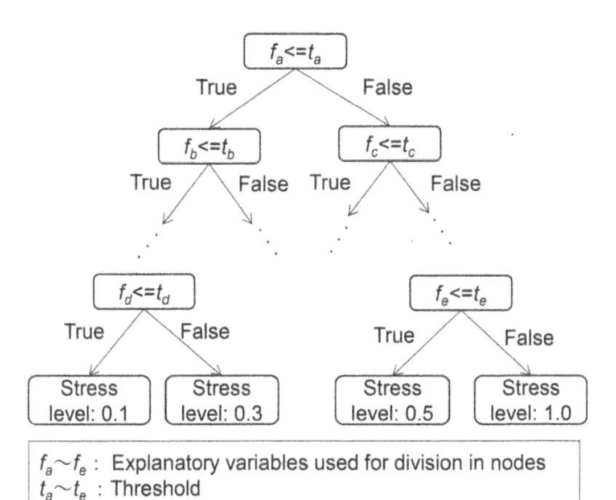

Figure 1. Decision tree method (concept).

Random forests are an ensemble learning algorithm for decision trees. In random forests, a decision tree is constructed from bootstrap data, which are sampled from training data with some overlap (Figure 2). The variables for achieving optimal segmentation are searched for some randomly selected variables among the d variables (instead of all variables in the d-dimensional feature vector) to suppress the correlation among decision trees and to produce more accurate output (Beyeler 2018).

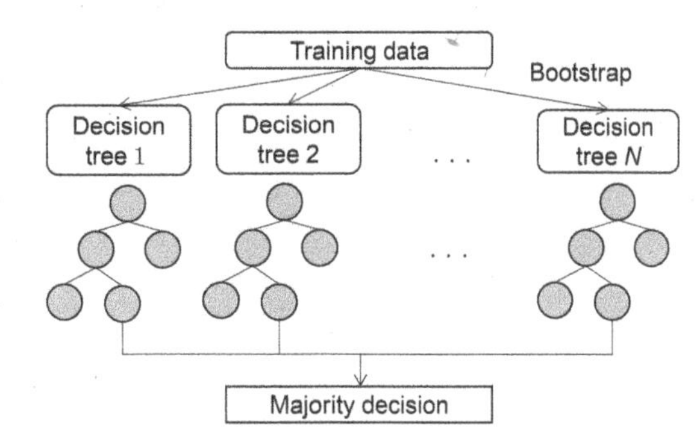

Figure 2. Random forest method (concept).

In this study, the CART (Classification And Regression Tree) method was used to determine the branches of the tree. This method uses the Gini coefficient as a criterion for branching. The Gini coefficient $L(t)$, which represents the impurity at node t, is given by the following equation (Beyeler 2018):

$$L(t) = \sum_{i=1}^{K} p\,(c_i|t)\,(1 - p(c_i|t)) = 1 - \sum_{i=1}^{K} p^2(c_i|t). \quad (3)$$

Here, $p(c_i|t)$ is the probability that the ith class of data is selected at node$t$. In this study, the tree was branched in such a way that the decrease in impurity as measured by the change in the Gini coefficient, $\Delta L(t)$ (shown in Eq. (4)) is maximized.

$$\Delta L(t) = L(t) - (p_L L(t_L) + p_R L(t_R)) \quad (4)$$

In this equation, p_L and p_R are the probabilities of being classified into the left and right branches after splitting, respectively, and t_L and t_R are the nodes at the end of the left and right branches, respectively.

2.3.2 *Cross-validation*

Cross-validation is a statistical method of tuning hyper-parameters. In cross-validation, the data are instead split repeatedly and multiple models are trained. The most commonly used version of cross-validation is k-fold cross-validation. The k-partitioning cross-validation method is to train a model by dividing the training data into k parts, using k-1 datasets as the training set and 1 dataset as the validation set. The method is repeated k times so that every partition becomes a validation set at least once (Beyeler 2018).

Accordingly, in this study, parameter tuning was conducted using a 3-fold cross-validation method. As a result, the tree depth was determined to be 10 in a random forest with 300 trees.

2.3.3 *Verification of accuracy*

Common quantitative performance measures in regression modeling include the coefficient of determination (R^2), root mean square error (RMSE), and mean absolute error (MAE) (Chai & Draxler 2014; Chatur et al. 2013). These metrics, coupled with model diagnostic plots and visualization of predicted versus observed output values, provide a comprehensive picture of a model's performance (Beyeler 2018).

R^2 is a measure of the proportion of the variance in the data that is explained by the model. Accordingly, R^2 is calculated as follows:

$$R^2 = 1 - \frac{\sum_{i=1}^{n} (y_i - \hat{y}_i)^2}{\sum_{i=1}^{n} (y_i - \bar{y})^2}. \quad (5)$$

where y_i is an observed value from the data, $\hat{y}_i$ is the predicted value from the model, and $\bar{y}$ is the average output from the data.

The value of R^2 ranges from zero to one, with higher values indicating a model's better ability to explain the variation in the data. However, R^2 is a measure of correlation, not accuracy, and should be used with other performance measures because it is dependent on the variance of the output variable.

RMSE indicates how closely the data fit around the model. t is measured on the same scale as the output variable, and is always positive due to the squared residuals in its calculation. Using the RMSE accentuates the effect of outliers in the error metric. This means that if median error of the model (usually captured by the mean absolute error) is low, the RMSE of the model can still be large due to the inability to model some outliers in the data. RMSE is calculated as follows:

$$\text{RMSE} = \sqrt{\frac{1}{n}\sum_{i=1}^{n} (y_i - \hat{y}_i)^2}. \quad (6)$$

MAE is a measure of prediction accuracy of a model that uses the absolute value of the errors rather than a squared value. Using the absolute value reduces the influence of very large errors on the measure of performance. Thus, MAE is a measure of the median error of the model and complements the use of R^2 and RMSE. MAE is calculated as follows:

$$\text{MAE} = \frac{1}{n}\sum_{i=1}^{n} |y_i - \hat{y}_i|. \quad (7)$$

3 EXPERIMENTAL PROCEDURE

3.1 *Concrete specimens*

Nine cylindrical specimens which were severely damaged by frost were drilled out from the side walls and bottom slabs of a headwork located in Hokkaido, Japan, which was constructed in 1963 (Figure 3).

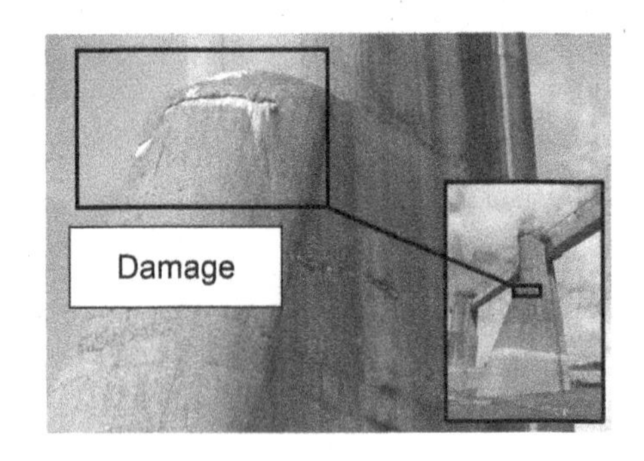

Figure 3. Headwork overview.

3.2 *Compression test with AE*

Each core specimen was monitored for AE in compression testing. The test setup for AE monitoring in a uniaxial compression is shown in Figure 4. Silicon grease was pasted on the top and the bottom of the specimen, and a Teflon sheet was inserted to reduce AE events generated by friction. The SAMOS-AE system (manufactured by PAC) was employed as a measuring device. AE hits were detected and recorded at a threshold level of 42 dB with a 40 dB gain in the pre-amplifier and 20 dB gain in the main amplifier. Hits were counted with six AE sensors of 150 kHz resonance (R15α, PAC). The frequency range and sampling frequency were set from 5 kHz to 400 kHz and 1 MHz, respectively. For event counting, the dead time was set as 2 ms.

Figure 4. AE monitoring of the compression test.

4 DATASETS FOR ESTIMATION OF STRESS LEVEL BY AE PARAMETERS

The flow of the analysis is shown in Figure 5. After compression tests with AE monitoring, the statistics of AE parameters were calculated for every 20×10^{-6} of strain. The total AE hits; the total, mean, standard deviation, and maximum value of AE energy; and the mean, standard deviation, and maximum value of the other five parameters (duration, rise time, RMS, centroid frequency, and peak frequency) were evaluated. These 20 parameters were used as features (explanatory variables) and the stress level of the concrete under compression was used as the target variable. The stress

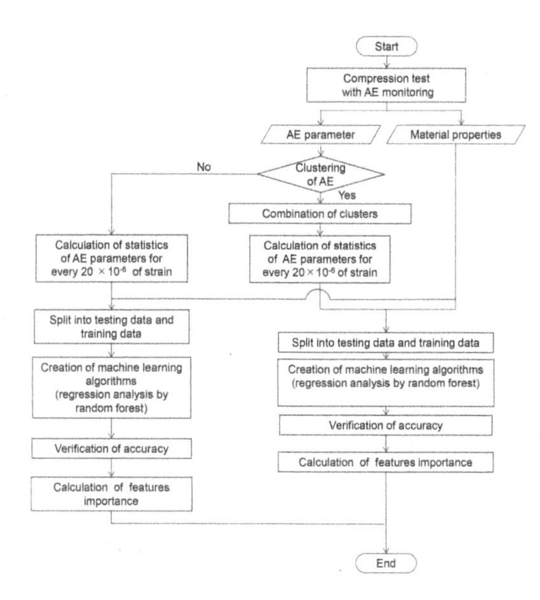

Figure 5. Analytical flowchart.

level was calculated by dividing the stress at every 20×10^{-6} increment of strain by the maximum stress. The accuracy of the regression analysis was compared for seven cases: one for all AE waves (non-clustered) and the others for only AE waves after cluster analysis (six cases). The total data size was 326, where 217 were used for training and 109 for testing.

5 RESULTS AND DISCUSSION

5.1 *Mechanical properties of the concrete core samples*

The mechanical properties of the samples tested under compression are shown in Table 2. The average compressive strength was 15.4 N/mm^2, ranging from 7.2 to 31.7 N/mm^2. The design standard strength of durable agricultural water utilization facilities is 24 N/mm^2. Only one of the nine sample specimens exceeded this value (31.7 N/mm^2).

Table 2. Mechanical properties of the samples.

Statistics		Mean	Standard Deviation
Compression strength	(N/mm^2)	15.4	8.1
Maximum strain	$\times10^{-6}$	715	410
Initial tangent modulus of elasticity	(GPa)	25.0	8.5
Modulus of elasticity	(GPa)	23.2	9.0
Strain energy	(J)	10.2	11.7

The average maximum strain was 726×10^{-6}, which was lower than the standard maximum strain of $2{,}000\times10^{-6}$ for all specimens. The ratio of the initial tangential modulus to the modulus of elasticity averaged 1.1. The average strain energy was 10.7 J, ranging from 2.7 to 38.5 J. These results suggest that the specimens had become brittle due to accumulated frost damage.

5.2 *Characteristics of AE waves classified by cluster analysis*

AE data obtained from the compression tests were classified into three clusters based on three AE parameters: peak amplitude, peak frequency, and centroid frequency. The means of the AE parameters are shown in Table 3. As can be seen, the non-clustered data and cluster 2 are significantly different from cluster 1 or cluster 3 in "rise time" and "centroid frequency" using Tukey's HSD tests ($p < 0.05$), and cluster 3 is significantly different from the other clusters in "duration" ($p < 0.05$). Average peak amplitude is 53 dB, and has no significant differences. On the other hand, there are significant differences in "RMS" and "peak frequency" among all clusters ($p < 0.05$). The RMS of cluster 3 is the highest and cluster 1 is lowest. The peak frequency of cluster 3 is the highest and that of cluster 2 is the lowest. The difference between cluster 2 and cluster 3 is about 50 kHz. Of all the AE parameters, peak frequency is the most effective in distinguishing the clusters. The values of these parameters for each cluster show that activities having lower rise times and peak frequencies (cluster 2) can be attributed to micro-scale damage, while those with higher rise times and peak frequencies (clusters 1 & 3) can be attributed to macro or mezzo-scale damage (Aggelis et al. 2013; De Rousseau et al. 2019). Comparing peak frequencies of clusters 1 and 3, cluster 3 appears to be associated with larger damage than cluster 1, which can be thought of as macro-scale damage.

Table 3. Comparison of AE parameters among clusters.

Cluster	Non-clustering (all data)	Cluster 1	Cluster 2	Cluster 3
Statistics	Mean SE*	Mean SE	Mean SE	Mean SE
Rise time	774	818	693	833
(μs)	6	11	9	16
Duration	3,175	3,013	3,028	3,815
(μs)	37	52	58	108
Peak amplitude	53	53	53	53
(dB)	0.021	0.034	0.034	0.048
RMS	0.0532	0.0480	0.0566	0.0583
(V)	0.0001	0.0002	0.0002	0.0004
Centroid frequency	134	139	126	140
(kHz)	0.04	0.07	0.07	0.07
Peak frequency	93	103	68	118
(kHz)	0.10	0.17	0.13	0.21

*SE: Standard error

The trend for total AE hits in compressive fracturing of No. 4 is shown in Figure 6. No. 4 is the sample

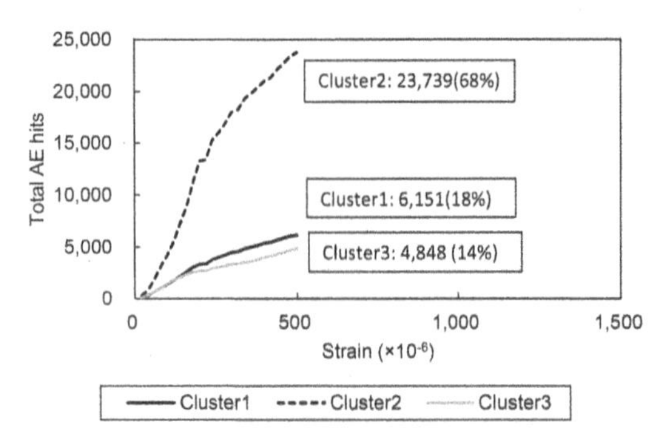

Figure 6. Total AE hit trends in the compressive fracture process (No. 4: 16.1 N/mm^2).

whose compressive strength is closest to the average value. The AE hit trends of clusters 1 & 3 overlap or have similar rates of change. The other samples show the same results as No. 4. These results suggest that the factors for AE occurrence for clusters 1 & 3 differ from those of cluster 2, corresponding with the scale of damage outlined above. The total number of AE hits detected in 9 samples is 258,295. Of them 43% (111,324) fall in cluster 1, 37% (96,492) are in cluster 2, and 20% (50,479) are in cluster 3.

5.3 *Characteristics of AE waves classified by cluster analysis*

5.3.1 *Accuracy of regression analysis*

Because the purpose of this paper is to clarify the relationship between the compressive fracture process and AE parameters in damaged concrete, we attempted to estimate stress levels using the AE parameters. The accuracy of the regression analysis is summarized in Table 4. Estimations for all AEs (non-clustered) and the results for the clustered AEs are shown in Figure 7. When AEs classified as clusters 1 & 3 are used, R^2 increases ($R^2 = 0.720$), and RMSE and MAE decrease compared to the non-clustered case. In Figure 7 the error ranges for clusters 1 & 3 are smaller with a 95% confidence interval. In particular, the error range is smaller and the accuracy improves above a stress level of 0.6 (60%). On the other hand, the accuracy decreases when cluster 2, cluster 3, and clusters 2 & 3 AEs are targeted. In particular, R^2 is below 0.5 for cluster 2 and clusters 2 & 3. These cases have the largest RMSE and MAE values of all the cases. The average peak frequency of cluster 2 is about 30–50 kHz lower than that of clusters 1 & 3 (Table 3). The decrease in accuracy for cluster 2 could be because that cluster comprises micro-scale activities which are seen throughout the test and are not capable of defining damage as well as macro-scale activities. Therefore, identifying AEs due to crack initiation using cluster analysis can be expected to improve accuracy.

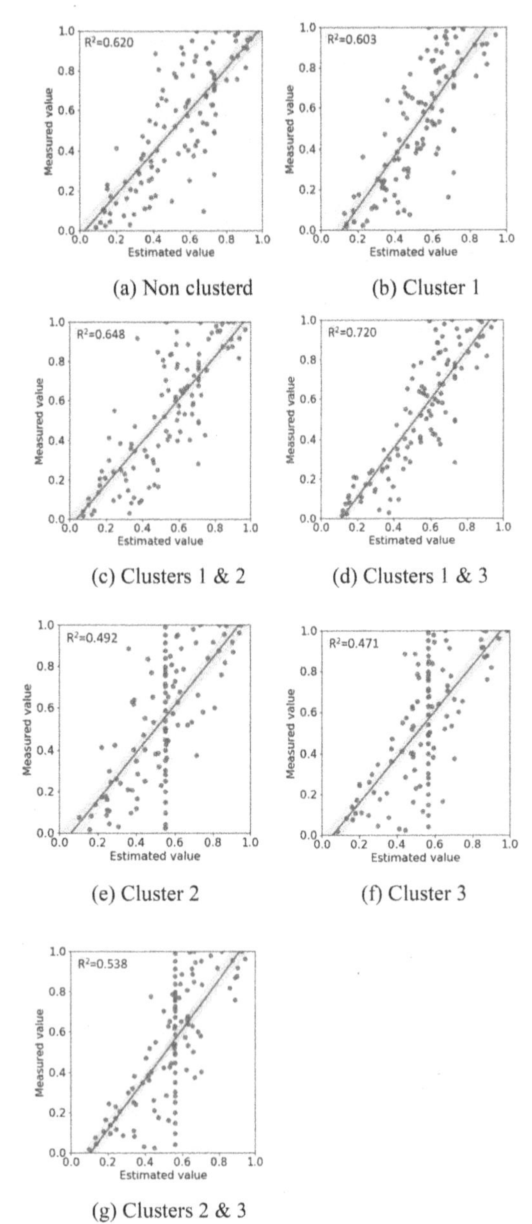

(a) Non clusterd
(b) Cluster 1
(c) Clusters 1 & 2
(d) Clusters 1 & 3
(e) Cluster 2
(f) Cluster 3
(g) Clusters 2 & 3

Figure 7. Stress levels estimated by the random forests.

Table 4. Regression analysis accuracy using random forests.

Index		R^2	RMSE	MAE
Non-clustered		0.62	0.183	0.142
Clustered	Cluster 1	0.603	0.187	0.154
	Clusters 1&2	0.648	0.176	0.135
	Clusters 1&3	0.72	0.157	0.126
	Cluster 2	0.492	0.211	0.173
	Clusters 2&3	0.471	0.216	0.171
	Cluster 3	0.538	0.202	0.161

5.3.2 *Identification of effective AE parameters for evaluation of compressive fracture process*

The most commonly used method of comparing features to rate each one's importance in producing a random forests outcome. These are numbers between 0 and 1 that always sum to 1. There are two methods for calculating importance: one is to use the amount of decrease in the Gini coefficient and the other is to use the variation in the Out-Of-Bag (OOB) error rate (Chai & Draxler 2014). In this study, the Gini coefficient method was adopted. The feature importance was determined by aggregating the amount of decrease in the Gini coefficient for all nodes by dividing them using a chosen variable and averaging the results.

The importance AE parameters are compared between the case of non-clustered and the case of clusters 1 & 3, which had the highest the R^2. Features' importance values for the non-clustered case and for clusters 1 & 3 are shown in Figure 8. The average values of the AE parameters with high importance at each stress level in the non-clustered case and for clusters 1 & 3 are shown in Figures 9–11. In the non-clustered case, the important AE parameters are peak frequency, rise time, and AE hits, while for clusters 1 & 3, the important AE parameters are rise time, centroid frequency, and AE hits. In both cases, the importance of rise time is high, suggesting that it is a useful AE parameter. Peak frequency is the most important parameter in the non-clustered case, while

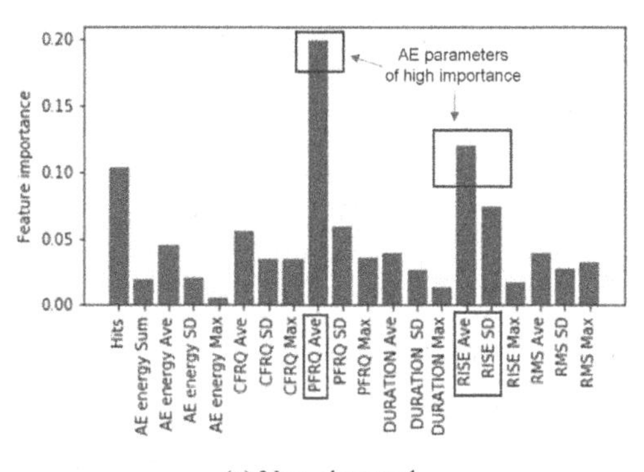

(a) Non-clustered

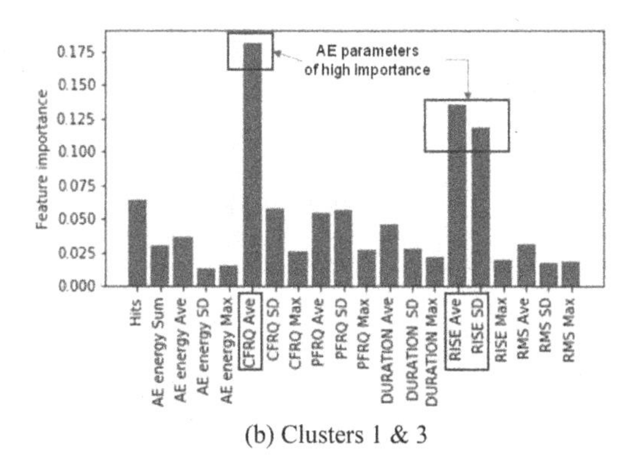

(b) Clusters 1 & 3

Figure 8. Importance of explanatory variables.

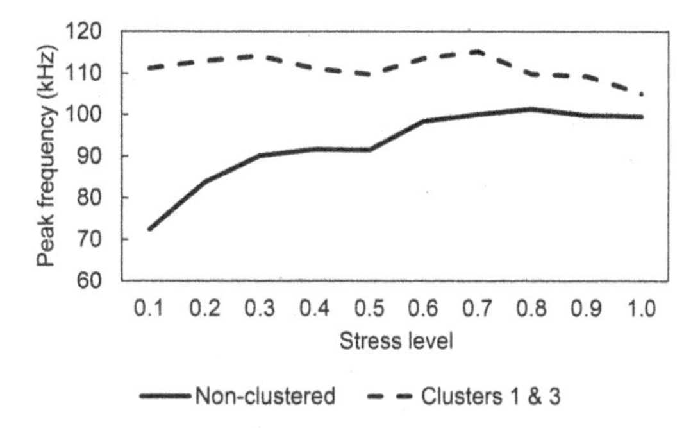

Figure 9. Relation between the average of peak frequency and the stress level (all samples).

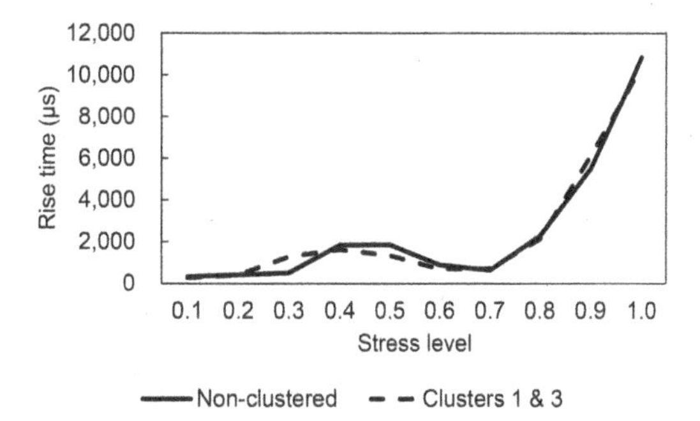

Figure 10. Relation between the average of rise time and the stress level (all samples).

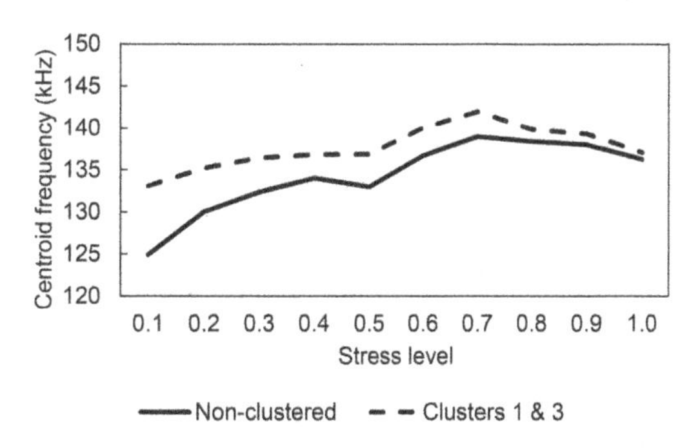

Figure 11. Relation between the average of centroid frequency and the stress level (all samples).

it is the least important for clusters 1 & 3. In this study, the clustering was analyzed based on peak frequency, peak amplitude, and centroid frequency. Peak frequency was shown to be the most effective of all AE parameters in distinguishing the clusters (see 5.2). As a result, the difference in peak frequencies with respect to different stress levels was smaller in the clustered cases (see Figure 9), and the importance of peak frequency decreased. As seen in Figure 10, the rise time increased rapidly above a stress level of 0.9 (90%). This could be due to macro-cracks caused by the consolidation of micro- and mezzo-cracks during the final fracture process. This trend has also been shown by Shahidan et al. (Shahidan et al. 2013). At the centroid frequency, the trends of the two cases are similar. The centroid frequency is high at stress levels above 0.6 (60%) (see Figure 11). For clusters 1 & 3, only AEs with high centroid frequencies were picked up by cluster analysis. For this reason, the difference of centroid frequency does not show as clearly as rise time for the various stress levels.

6 CONCLUSION

This study aimed to identify effective AE parameters for evaluating compressive fracture processes using machine learning. To do so, core samples were drilled out from a damaged concrete structure, and uniaxial compressive strength tests were conducted in which the fracture processes were monitored using the AE method. In order to clarify the trends of AE parameters during the compressive fracture process, regression analysis correlating the stress level of the concrete under compression and AE parameters was performed using random forests. In addition, we have investigated whether the prediction accuracy can be improved by clustering AE activities as a preprocessing method. The conclusions can be summarized as follows:

(1) AE waves were classified into three clusters using a k-mean clustering algorithm based on peak frequency, centroid frequency, and peak amplitude. By evaluating AE features of the clusters, the degree of concrete damage was attributed to the clusters since the characteristics of AE parameters in cluster 2 (micro-scale damage) and in clusters 1 & 3 (mezzo and macro-scale damage) were different. In this process, the most distinctive cluster was identified.

(2) Machine learning is useful for estimating the stress level of the concrete undergoing compressive failure. When the AEs were classified into clusters 1 & 3 (mezzo and macro-scale damage), R^2 increased, and RMSE and MAE decreased compared to the non-clustered case. In other words, the accuracy of the stress level estimation using random forests increased after clustering analyses.

(3) For the clustered AE data, the most important parameter was determined to be rise time, the second was the centroid frequency. These two parameters can be used to clarify the compressive fracture behavior.

REFERENCES

Abouhussien, A.A. and Hassan, A.A.A. 2020. Classification of damage in self-consolidating rubberized concrete using acoustic emission intensity analysis, *Ultrasonics* 100: 105999.

Aggelis, D.G. Dassios, K.G., Kordatos, E.Z. and Matikas, T.E. 2013. Damage accumulation in cyclically-loaded glass-ceramic matrix composites monitored by acoustic emission, *The Scientific World Journal*.

Ai, L., Soltangharaei, V., Bayat, M., Greer, B. and Ziehl, P. 2021. Source localization on large-scale canisters for used nuclear fuel storage using optimal number of acoustic emission sensors, *Nuclear Engineering and Design* 375: 111097.

Alver, N., Tanarslan H.M. and Tayfur S. 2017. Monitoring fracture processes of CFRP-strengthened RC beam by acoustic emission, *Journal of Infrastructure Systems* 23(1): .B4016002.

Andreas, G. and Sarah, C.M. 2017. *Introduction to machine learning with python -a guide of data scientists-*, O'REILLY.

ASTM E1316. 2002. Standard Terminology for NDT.

Beyeler, M. 2018. *Machine learning for OpenCV: Intelligent image processing with python*, Packt Publishing.

Boniface, A., Saliba, J. Sbartaï, Z.M. Ranaivomanana, N. and Balayssac, J.P. 2020. Evaluation of the acoustic emission 3D localisation accuracy for the mechanical damage monitoring in concrete, *Engineering Fracture Mechanics*, 223: 106742.

Burud, N.B. and Kishen, J.M.C. 2021. Response based damage assessment using acoustic emission energy for plain concrete, *Construction and Building Materials*, 269: 121241.

Chai, T. and Draxler R.R. 2014. Root mean square error (RMSE) or mean absolute error (MAE)? – arguments against avoiding RMSE in the literature, *Geosci. Model Dev.*, 7(3): 1247–1250.

Chatur, P.N., Khobragade A.R. and Asudani, D.S. 2013. Effectiveness evaluation of regression models for predictive data-mining, *Int. J. Manag.* IT Eng. 3(3):465–483.

Crivelli, D., Guagliano M. and Monici, A. 2014. Development of an artificial neural network processing technique for the analysis of damage evolution in pultruded composites with acoustic emission, *Composites Part B: Engineering*, 56: 948–959.

De Rousseau, M.A., Laftchiev, E., Kasprzyk, J.R., Rajagopalan, B. and Srubar, W.V.A. 2019. Comparison of machine learning methods for predicting the compressive strength of field-placed concrete, *Construction and Building Materials*, 228:116661.

Grosse, C.U., Ohtsu, M., Dimitrios, A. and Shiotani, T. (Eds) 2021. *Acoustic Emission Testing: Basics for Research – Applications in Engineering*, Springer Tracts in Civil Engineering),

Guofeng, M.A. and Du Q. 2020. Structural health evaluation of the prestressed concrete using advanced acoustic emission (AE) parameters, *Construction and Building Materials*, 250: 118860.

Hübner, H.B., Duarte, M.A.V. and Da Silva R.B. 2020. Automatic grinding burn recognition based on time'|frequency analysis and convolutional neural networks, *The International Journal of Advanced Manufacturing Technology*, 110: 1833–1849.

JCMS-IIIB5706. 2003. Monitoring method for active cracks in concrete by acoustic emission, *Federation of Construction Material Industries, Japan*, 23–28.

Kane, P.V. and Andhare, A.B. 2020. Critical evaluation and comparison of psychoacoustics, acoustics and vibration features for gear Fault correlation and classification, *Measurement*, 154: 107495.

Karcili, M., Alver, N. and Ohtsu, M. 2016. Application of AE rate-process analysis to damaged concrete structures due to earthquake, *Materials and Structures*, 49: 2171–2178.

Kaufman, L. and Rousseeuw P.J. 2008. *Finding groups in data: An introduction to cluster analysis*, John Wiley&Sons.

Li, W., Xu, C., Ho, S.C., Wang, M.B. and Song, G.: Monitoring concrete deterioration due to reinforcement corrosion by integrating acoustic emission and FBG strain measurements, *Sensors (Switzerland)*, 17(3): 657.

Mirgal, P., Pal, J. and Banerjee, S. 2020. Online acoustic emission source localization in concrete structures using iterative and evolutionary algorithms, *Ultrasonics*, 108: 106211.

Morizet, N., Godin, N., Tang, J., Maillet, E., Fregonese, M. and Normand, B. 2016. Classification of acoustic emission signals using wavelets and random forests, Application to localized corrosion, *Mechanical Systems and Signal Processing*, 70–71: 1026–1037.

Ohtsu, M. 1987. Acoustic emission characteristics in concrete and diagnostic applications, *Journal of Acoustic Emission*, 6 (2): 99–108.

Ohtsu, M. 1992. Rate process analysis of acoustic emission activity in core test of concrete, *Concrete Library of JSCE*, 20: 143–153.

Ohtsu, M. and Suzuki, T. 2004. Quantitative damage evaluation of concrete core based on AE rate-process analysis, *Journal of Acoustic Emission*, 22: 30–38.

Prem, P.R., Verma, M. and Ambily, P.S. 2021. Damage characterization of reinforced concrete beams under different failure modes using acoustic emission. *Structures*, 30: 174–187.

Rahimi, M., Alghassi, A. Ahsan, M. and Haider, J. 2020. Deep learning model for industrial leakage detection using acoustic emission signal, Informatics, 7(4):49.

Rodríguez, P. and Celestino, T. B. 2020. Assessment of damage distribution in brittle materials by application of an improved algorithm for three-dimensional localization of acoustic emission sources with P-wave velocity calculation, *Construction and Building Materials*, 23: 117086.

Schiavi, A., Niccolini, G. Tarizzo, P. Lacidogna, G. Manuello, A. and Carpinteri, A. 2011. Analysis of energy released by elastic emission in brittle materials under compression, *Experimental Mechanics on Emerging Energy Systems and Materials*, 5:103–108.

Shah, A.A. and Ribakov, Y. 2010. Effectiveness of nonlinear ultrasonic and acoustic emission evaluation of concrete with distributed damages, *Materials and Design*, 31: 3777–3784.

Shahidan, S., Pulin, R.: Bunnori, N.M. and Holford K.M, Damage classification in reinforced concrete beam by acoustic emission signal analysis, *Construction and Building Materials*, 45: 78–86.

Shang, X., Lu, Y., Li, B. and Peng, K. 2021. A novel method for estimating acoustic emission b value using improved magnitudes. *IEEE Sensors Journal*, 21(15): 16701–16708.

Shimamoto, Y. and Suzuki, T.b2020. Damage evaluation of heavily cracked concrete by initial AE energy parameter, *Advanced Experimental Mechanics*, 5 (2020): 122–127.

Soltangharaei, V., Anay, R., Assi, L., Bayat, M., Rose, J.R., Ziehl, P. 2021. Analyzing acoustic emission data to identify cracking modes in cement paste using an artificial neural network, *Construction and Building Materials*, 267: 121047.

Suwansin, W. and Phasukkit, P. Deep learning-based acoustic emission scheme for nondestructive localization of cracks in train rails under a load, *Sensors*, 21(1): 272.

Suzuki T. and Shimamoto Y. 2019. On-site damage evaluation of cracked irrigation infrastructure by acoustic emission and related non-destructive elastic wave method, *Paddy and Water Environment*, 17(3): 315–321.

Suzuki T., Shiotani T. and Ohtsu, M. 2017. Evaluation of cracking damage in freeze-thawed concrete using acoustic emission and x-ray CT image, *Construction and Building Materials*, 136: 619–626.

Suzuki, T., Nishimura, S., Shimamoto, Y., Shiotani, T. and Ohtsu, M. 2020. Damage estimation of concrete canal due to freeze and thawed effects by acoustic emission and x-ray CT methods, *Construction and Building Materials*, 245(10): 118343.

Suzuki, T., Ogata, H., Takada, R., Aoki, M. and Ohtsu, M.: Use of acoustic emission and x-ray computed tomography for damage evaluation of freeze-thawed concrete, *Construction and Building Materials*, 24: 2347–2352.

Suzuki, T. and Ohtsu M. 2014. Use of acoustic emission for damage evaluation of concrete structure hit by the great east Japan earthquake, *Construction and Building Materials*, 67: 186–191.

Suzuki, T., Shigeishi, M. and Ohtsu, M. 2007. Relative damage evaluation of concrete in a road bridge by AE rate – process analysis, *Materials and Structures*, 40(2): 221–227.

Tayfur S., Alver N., Abdi, S.. Saatcı, S. and Ghiami, A. 2018. Characterization of concrete matrix/steel fiber debonding in an SFRC beam: Principal component analysis and k-mean algorithm for clustering AE data. *Engineering Fracture Mechanics*, 194: 73–85.

Thirumalaiselvi, A., Sasmal, S. 2021. Pattern recognition enabled acoustic emission signatures for crack characterization during damage progression in large concrete structures, Applied Acoustics, 175: 107797.

Van Steen C., Pahlavan, L., Wevers, M., Verstrynge, E. 2019 Localisation and characterisation of corrosion damage in reinforced concrete by means of acoustic emission and X-ray computed tomography. Construction and Building Materials, 197: 21–29.

Wotzka, D. and Cichoń, A. 2019. Study on the influence of measuring AE sensor type on the effectiveness of OLTC defect classification, *Sensors*, 20(11): 3095.

Wu, Y.Q., Li, S.L. and Wang, D.W. 2019. Characteristic analysis of acoustic emission signals of masonry specimens under uniaxial compression test, *Construction and Building Materials*, 196: 637–648.

Zhang, L., Ji, H., Liu, L. and Zhao, J. 2021. Time–frequency domain characteristics of acoustic emission signals and critical fracture precursor signals in the deep granite deformation process, Applied Sciences, 11: 8236.

Zhang, Z.H. and Deng, J.H. 2020. A new method for determining the crack classification criterion in acoustic emission parameter analysis, *International Journal of Rock Mechanics and Mining Sciences*, 130 (2020) 104323.

Computational Modelling of Concrete and Concrete Structures – Meschke, Pichler & Rots (Eds)

Fiber orientation modeling during extrusion-based 3D-concrete-printing

J. Reinold, V. Gudžulić & G. Meschke
Institute for Structural Mechanics, Ruhr University Bochum, Bochum, Germany

ABSTRACT: Fibers tend to align in printing direction during fiber-reinforced 3D-concrete-printing processes, which allows for the orientation of the fibers in desired directions by controlling the printing process. This enables the production of components with advantageous fiber orientation states, which is not possible with conventional casting methods to this extent. To understand correlations between fiber orientation and process (e.g. printing speed or flow rate) and geometric (e.g. extrusion nozzle size and shape) parameters during the printing process, a fiber orientation model is implemented into a framework based on the Particle Finite Element Method (PFEM) to simulate extrusion processes during fiber-reinforced 3D-concrete-printing. The fiber orientation model is based on a representation using a second order orientation tensor, which is combined with an anisotropic Bingham viscsosity constitutive law and upscaling relations for the viscosity and yield stress from literature. A robust PFEM-compatible implementation of the fiber orientation model is proposed and verified using different convergence and parametric studies. Numerical analyses of fiber-reinforced 3D-concrete-printing in 2D revealed that fibers tend to align stronger in printing direction for larger printing speeds, smaller extrusion nozzles and smaller fiber aspect ratios.

1 INTRODUCTION

Automated construction techniques are playing an increasingly important role in the modern construction industry, bringing advantages such as greater efficiency, accuracy and safety. One area of these techniques is additive manufacturing or more specifically extrusion-based 3D-concrete-printing (Mechtcherine et al. 2020). In 3D-concrete-printing, the material is extruded layer-wise via an extrusion nozzle which allows for the construction of novel designs and components. More details related to 3D-concrete-printing can be found in (Buswell et al. 2018; Mechtcherine et al. 2020). Among the many challenges of 3D-concrete-printing, incorporating reinforcement into current production techniques is one of the biggest to solve to increase the product quality and to yield a more effective manufacturing method (Mechtcherine et al. 2021). A possible solution is based on printing technologies using fiber-reinforced fresh concrete. In traditional casting processes of fiber-reinforced concrete the fiber orientation state is greatly influenced by the casting process (e.g. casting direction or flow rate). This can lead to undesired and impractical fiber orientation states in the final product. It was shown that 3D-concrete-printing allows for the orientation of fibers in printing direction (Arunothayan et al. 2021; Huang et al. 2021; Mechtcherine et al. 2021), which can be used to control the printing process and obtain desired fiber orientation states in the final component. To allow for accurate calibration of fiber-reinforced 3D-concrete-printing processes and to understand the correlations between process parameters (e.g. printing speed or flow rate) and the fiber orientation state in printed components, this work focuses on the development and application of a numerical model for simulating fiber-reinforced 3D-concrete-printing extrusion processes.

The most used fiber orientation model is the Folgar-Tucker fiber orientation model (Advani & Tucker 1987; Folgar and Tucker 1984). To get around the huge computational resources needed for discrete fiber modeling approaches, the Folgar-Tucker model translates the evolution law of a single fiber into a probabilistic form representing a fiber bundle to yield a continuity equation for orientation tensors. In case of polymers, such models were already successfully applied to extrusion processes in Fused Deposition Modeling (FDM) for example by using a finite element model in (Heller et al. 2019) or a Smoothed Particle Hydrodynamics (SPH) implementation in (Ouyang et al. 2019). With respect to fiber-reinforced fresh concrete, in (Gudzulic et al. 2018) a Folgar-Tucker fiber orientation model was implemented into a SPH code and applied to different benchmark problems. Beside this approach using a Folgar-Tucker fiber orientation model, mostly discrete approaches were used for modeling of fiber-reinforced fresh concrete, e.g., see (Ŝvec 2013) for a Lattice-Boltzmann based formulation, (Deeb et al. 2014a; 2014b) for a SPH model where fibers have been modeled discretely by linking individual SPH particles or (Ferrara et al. 2012)

DOI 10.1201/9781003316404-25

for a discrete approach based on the Distinct Element Method.

The implementation proposed in this work is based on the combination of two previous works: Due to the successful simulation of extrusion processes during 3D-concrete-printing using the Particle Finite Element Method (PFEM) in (Reinold & Meschke 2019; Reinold et al. 2020), this model is extended by an improved implementation of the Folgar-Tucker fiber orientation model from (Gudzulic et al. 2018). The Folgar-Tucker fiber orientation model is based on the solution of an evolution equation of a second order orientation tensor, which is combined with an anisotropic constitutive model for fiber-reinforced fresh concrete using upscaling relations for the plastic viscosity and yield stress from literature.

The structure of the paper is as follows: Section 2 presents the theoretical and numerical framework of the model, which provides a brief overview of PFEM, the Folgar-Tucker fiber orientation model, the constitutive model for fiber-reinforced fresh concrete and the solution scheme within the PFEM framework. In Section 3 representative numerical studies are discussed to outline characteristics of the proposed implementation and to analyze 3D-concrete-printing extrusion processes in a 2D setup.

2 THEORETICAL AND NUMERICAL FRAMEWORK

In the course of this section, a coherent theoretical and numerical framework for modeling the flow of fresh fiber-reinforced concrete is outlined. First, in Section 2.1 a brief overview about the Particle Finite Element Method (PFEM) and its governing equations for large deformation problems is given. Second, in Section 2.2 the fiber orientation model based on a Folgar-Tucker fiber orientation model is introduced. Section 2.3 introduces a constitutive model for fiber-reinforced fresh concrete and a solution procedure for the PFEM Folgar-Tucker fiber orientation model is presented in Section 2.4. The solution scheme is formulated such that only minor changes to an existing PFEM code are required to couple the fiber orientation model with it.

2.1 Particle Finite Element Method

The Particle Finite Element Method (PFEM) is a numerical method for fluid dynamics and large deformation problems given in an updated Lagrangian description (Idelsohn et al. 2004; Oñate et al. 2004). The spatial discretization is based on linear triangular or tetrahedral finite elements which allow for efficient remeshing algorithms based on the Delaunay triangulation to deal with severe mesh distortions during the simulation. Originally, the free surface of the domain is obtained from the alpha shape method (Edelsbrunner & Mücke 1994). In contrast, for applications with a smooth evolving free surface, constrained Delaunay triangulations are advantageous compared to the alpha shape method with respect to free surface modelling and contact (?). In this paper, PFEM is only summarized in a condensed version. Further information can be found in the original work (Idelsohn et al. 2004; Oñate et al. 2004), in a more recent state of the art report (Cremonesi et al. 2020) and in previous works of the authors (Reinold & Meschke 2019; Reinold et al. 2020; ?).

In all PFEM implementations, at least the balance of momentum and mass must be solved. The discretized residuals of the balance of momentum and mass in a quasi-incompressible framework are summarized as

$$\boldsymbol{R}_v = \boldsymbol{M}_{vv}\dot{\bar{\boldsymbol{v}}} + \boldsymbol{F}_{v,int} - \boldsymbol{F}_{v,ext} = \boldsymbol{0}, \tag{1}$$

$$\boldsymbol{R}_p = \boldsymbol{G}_{vp}^T\bar{\boldsymbol{v}} - \boldsymbol{K}_{pp}\dot{\bar{\boldsymbol{p}}} - \boldsymbol{S}_{pp}\bar{\boldsymbol{p}} = \boldsymbol{0}, \tag{2}$$

in which $\boldsymbol{R}_v$ denotes the residual of the balance of momentum, $\boldsymbol{M}_{vv}$ denotes the lumped mass matrix, $\bar{\boldsymbol{v}}$ denotes the nodal velocities, $\boldsymbol{F}_{v,int}$ denotes the internal force vector, $\boldsymbol{F}_{v,ext}$ denotes the external force vector, $\boldsymbol{R}_p$ denotes the residual of the balance of mass, $\boldsymbol{G}_{vp}$ denotes the velocity-pressure gradient matrix, $\boldsymbol{K}_{pp}$ denotes the compressibility matrix, $\bar{\boldsymbol{p}}$ denotes the nodal pressure and $\boldsymbol{S}_{pp}$ denotes a stabilization matrix in case of incompatible orders of finite element approximations of the velocity and pressure fields. The set of equations (1) and (2) can be solved using any possible solution method and temporal discretization scheme. Here, the set of equations is solved monolithically using an implicit backward Euler discretization in time. This solution procedure is explicitly coupled with the fiber orientation model to yield a robust and efficient algorithm presented in the following sections.

2.2 Folgar-Tucker fiber orientation model

A fiber suspension can be characterized into dilute ($c < 1/r^2$), semi-concentrated ($1/r^2 < c < 1/r$) and highly concentrated regimes ($c > 1/r$) (Doi and Edwards 1978), with the fiber volume fraction c and the aspect ratio r of a single fiber. Due to the low fiber content in case of a dilute suspension, the motion of a fiber is dominated by drag forces from the surrounding fluid. For semi-concentrated regimes also hydrodynamic interactions between fibers may influence the fiber motion. In highly concentrated regimes, the fiber motion is characterized by direct contacts between fibers. Much work has been done in characterization and modeling of fiber-reinforced polymer suspensions, e.g., see (Chung & Kwon 2002; Petrie 1999) for an overview. In contrast, only few fundamental studies have been carried out in case of fresh concrete, e.g., see (Férec et al. 2015; Grünewald 2004; Martinie et al. 2010; Perrot et al. 2013). The fiber orientation model in this work is based on well validated approaches from polymer science and builds upon a previous work, given in (Gudzulic et al. 2018), which was successfully applied to fiber-reinforced fresh concrete flow problems.

Due to the high amount of computational resources needed in simulation approaches using discrete fibers, the majority among the modeling approaches are based on a representation and solution procedure using so-called orientation tensors (Advani & Tucker 1987; Folgar and Tucker 1984). Such models are derived from the continuity equation of a fiber orientation distribution function ψ representing the orientation state of a fiber bundle as (Phelps & Tucker 2009)

$$\dot{\psi} = -\nabla_i^S(\psi \underbrace{(W_{ij}p_j + \lambda(D_{ij}p_j - D_{kl}p_k p_l p_i))}_{=\dot{p}_i^h} \underbrace{- C_I|\dot{\gamma}|\nabla_i^S\psi}_{q_i}), \quad (3)$$

where ∇_i^S is the gradient operator over the unit sphere, $W_{ij} = \frac{1}{2}\left(\frac{\partial v_i}{\partial x_j} - \frac{\partial v_j}{\partial x_i}\right)$ is the skew-symmetric part of the velocity gradient, $D_{ij} = \frac{1}{2}\left(\frac{\partial v_i}{\partial x_j} + \frac{\partial v_j}{\partial x_i}\right)$ is the symmetric part of the velocity gradient, $\lambda = \frac{r^2-1}{r^2+1}$ is a shape factor, $|\dot{\gamma}| = \sqrt{2D_{ij}D_{ij}}$ is the strain rate magnitude, C_I is the isotropic rotary diffusivity factor and $\boldsymbol{p} = [\sin(\theta)\cos(\varphi),\ \sin(\theta)\sin(\varphi),\ \cos(\theta)]^T$ is the direction of a single fiber in a Cartesian frame with the azimuth angle φ and the polar angle θ; see Figure 1.

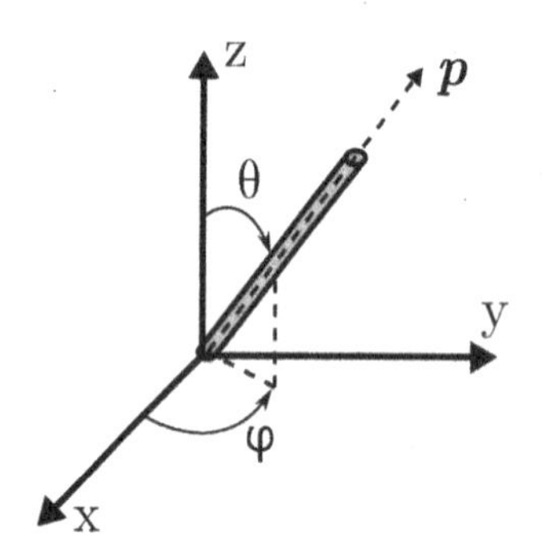

Figure 1. Fiber orientation state in a Cartesian frame.

The hydrodynamic part $\dot{p}_i^h$ of (3) is the well-known fiber orientation evolution law of a single fiber in a Newtonian liquid matrix by Jeffery (Jeffery and Filon 1922). The second part in (3) is the diffusivity term q_i, which accounts for fiber-fiber interactions to reach a steady fiber orientation state during flow (Folgar and Tucker 1984). The isotropic rotary diffusivity factor is obtained from the empirical relation in (Phan-Thien et al. 2002) as $C_I = 0.03(1.0 - e^{-0.224cr})$. Based on the orientation distribution function ψ and the direction of a fiber $\boldsymbol{p}$, the second and fourth order orientation tensors are defined as

$$a_{2,ij} = \int_{\mathbb{S}} p_i p_j \psi \, d\mathbb{S}, \quad (4)$$

$$a_{4,ijkl} = \int_{\mathbb{S}} p_i p_j p_k p_l \psi \, d\mathbb{S}, \quad (5)$$

with the surface of the unit sphere $\mathbb{S}$. The evolution law of the second order orientation tensor is obtained by differentiation of (4) in time, taking into account (3) and integration by parts (Advani & Tucker 1987), yielding

$$\begin{aligned} \dot{a}_{2,ij} = {} & W_{ik}a_{2,kj} - a_{2,ik}W_{kj} + \lambda(D_{ik}a_{2,kj} \\ & + a_{2,ik}D_{kj} - 2(a_{4,ijkl} + (1-\kappa)(L_{ijkl} \\ & - M_{ijmn}a_{4,mnkl}))D_{kl}) + 2C_I|\dot{\gamma}|(\delta_{ij} - 3a_{2,ij}), \end{aligned} \quad (6)$$

where δ_{ij} is the Kronecker delta and $L_{ijkl} = \sum_{m=1}^{3} \lambda_m e_{m,i} e_{m,j} e_{m,k} e_{m,l}$ and $M_{ijkl} = \sum_{m=1}^{3} e_{m,i} e_{m,j} e_{m,k} e_{m,l}$ denote tensors from to the reduced strain closure (RSC) (Wang et al. 2008) with the m-th eigenvalue λ_m and eigenvector $\boldsymbol{e}_m$ of $\boldsymbol{a}_2$ and the RSC tuning factor κ, which takes a value between 0 and 1. The reduced strain closure is introduced to empirically slow down the orientation kinetics, as the unmodified version was shown to overestimate the orientation kinetics compared to experiments (Sepehr et al. 2004). Note that the evolution law of any orientation tensor can be derived in the same way. As seen in (6), the solution of the evolution equation requires a higher order orientation tensor, which also holds for evolution equations of higher order orientation tensors. In order to solve (6) for $\boldsymbol{a}_2$, the fourth order orientation tensor $\boldsymbol{a}_4$ must be known. This contradiction is typically treated by approximating the higher order orientation tensor with the lower one using so-called closure approximations. Among the many possible closure approximations found in literature, the orthotropic fitted closure (ORW3) (Chung & Kwon 2001) is one of the most accurate ones and is adopted in this work. Solution approaches, which are solely based on $\boldsymbol{a}_2$, as in the current case, were found to be sufficiently accurate in most applications (Advani & Tucker 1987). An appropriate solution scheme in a PFEM framework for (6) is given in Section 2.4.

2.3 *Constitutive model for fiber-reinforced fresh concrete*

Fresh concrete consists of water, cement, different types of aggregates and admixtures leading to a rheological behavior, which is dominated by particle interactions depending on the size, shape and roughness of the particles. Due to very different particle sizes and flocculation mechanisms on the nano-scale, fresh concrete rheology cannot be easily modeled in their entirety over all length scales, which is why in most cases empirical macroscopic approaches are chosen. The Bingham fluid model is the most general approach to model the response of materials which exhibit a solid and fluid like behavior, such as fresh concrete. In simple shear flow, the response of a Bingham fluid is summarized as

$$\begin{aligned} \tau_{xy} &= \tau_0 + \mu\dot{\gamma}_{xy} \quad \text{for } |\tau_{xy}| \geq \tau_0, \\ \dot{\gamma}_{xy} &= 0 \qquad\qquad \text{for } |\tau_{xy}| < \tau_0, \end{aligned} \quad (7)$$

with the shear stress τ_{xy}, the shear rate $\dot{\gamma}_{xy}$, the plastic viscosity μ and the yield stress τ_0. In case the stress state lies below the yield point, no deformation is expected. When the stress state exceeds the yield stress, the material response of a Bingham model is similar to that of a Newtonian fluid with a constant plastic viscosity.

2.3.1 *Anisotropic viscosity model for fiber-reinforced fresh concrete*

In case of fiber-reinforced fresh concrete, the material can no longer be described by a standard Bingham model due to fiber reorientation effects during flow, resulting in an anisotropic material behavior, which depends on the current fiber orientation state. An anisotropic constitutive model for the flow of fiber suspensions have been porposed by Sommer et al. (Sommer et al. 2018; Favaloro et al. 2018). Based on the second and fourth order orientation tensors, in this approach a transversely anisotropic fluid model is volume averaged, as described in (Advani & Tucker 1987). Assuming identical inplane shear and transverse shear viscosity, as specified in (Sommer et al. 2018; Favaloro et al. 2018), the fourth order viscosity tensor is defined via

$$\mu_{ijkl} = 4\eta_{23}(R_\eta - 1)\Big[a_{4,ijkl} - \frac{1}{3}\left(a_{2,ij}\delta_{kl} + a_{2,kl}\delta_{ij} - \frac{1}{3}\delta_{ij}\delta_{kl}\right)\Big] + 2\eta_{23}\, I^d_{ijkl} + 2\frac{\tau_0}{|\dot{\gamma}|}\left(1 - e^{-m|\dot{\gamma}|}\right) I^d_{ijkl}, \tag{8}$$

where $I^d_{ijkl} = (\delta_{ik}\delta_{jl} + \delta_{il}\delta_{kj})/2 - \frac{1}{3}\delta_{ij}\delta_{kl}$, m denotes a regularization parameter of the Bingham model, η_{23} denotes the transverse shear viscosity and R_η denotes the anisotropy ratio. The last part of (8) is a modification with respect to a regularized Bingham model (Papanastasiou 1987) which penalizes the rigid response below the yield stress with a large viscosity controlled by m. In this work a value of $m = 1000$ is used.

Compared to the standard Bingham model presented in Section 2.3, the constitutive model given in (8) depends on the second and fourth order orientation tensors and the anisotropy ratio as an additional material parameter. The anisotropy ratio becomes one in case of the absence of fibers and increases to a larger number depending on the aspect ratio and fiber volume fraction, which would increase the anisotropic behavior of the material by several magnitudes. As given in (Favaloro et al. 2018; Pipes et al. 1994) for the case of a highly concentrated Newtonian suspension, the anisotropy ratio is defined as $R_\eta = 1 + \frac{1}{8}c\sqrt{12c/\pi}\,r^2$. Note that no constitutive assumption for R_η is available in case of fiber-reinforced fresh concrete in literature so far. An approximation for the yield stress and the transverse shear viscosity are provided in the next section. Using (8), the Cauchy stress tensor can be given as

$$\sigma_{ij} = \mu_{ijkl}D_{kl} + p\delta_{ij}. \tag{9}$$

Due to major and minor symmetries, the viscosity tensor in (8) can be written in Voigt notation as a 6x6 matrix.

2.3.2 *Rheological properties of fiber-reinforced fresh concrete*

Due to additional contacts and hydrodynamic interactions in a fiber suspension, increasing the fiber volume fraction leads to an increase of the viscosity and yield stress (Grünewald 2004). In case of a Newtonian fluid matrix, various rheological upscaling relations of the viscosity were developed and validated for polymer suspensions, e.g.; see (Chung & Kwon 2002; Petrie 1999). In case of a Bingham fluid matrix or fresh concrete only few theoretical models exist (Férec et al. 2015; 2017), which are not very practical within the adopted fiber orientation modeling framework. Therefore, empirical and experimentally validated upscaling relations for the viscosity and yield stress of fiber-reinforced fresh concrete are taken from the literature. According to (Ghanbari & Karihaloo 2009), a micromechanically motivated approach for the viscosity of a fresh concrete mix including random isotropic oriented rigid fibers is given as

$$\mu = \tilde{\mu}\left((1 - c) + \frac{\pi c r^2}{3\ln(2r)}\right), \tag{10}$$

where $\tilde{\mu}$ is the viscosity of the mixture without fibers. To enforce that (8) yields (10) for a random isotropic fiber orientation state, the transverse shear viscosity must be given as $\eta_{23} = \frac{15}{4R_\eta - 11}\mu$, which can be found by using $\psi = \frac{1}{4\pi}$ to yield an isotropic fiber orientation state.

Similarly, based on (Martinie et al. 2010; Sultangaliyeva et al. 2020) the yield stress of a fresh concrete mix including random isotropic oriented rigid fibers is given as

$$\tau_0 = \tilde{\tau}_0\left(1 - \frac{c}{\phi_{fm}} - \frac{\phi_s}{\phi_{sm}}\right)^{-2}, \tag{11}$$

with the dense packing fraction of fibers $\phi_{fm} = 4/r$, the volume fraction of solid particles ϕ_s and the maximum packing fraction $\phi_{sm} \approx 0.65$. In contradiction to (11), $\tilde{\tau}_0$ must be the yield stress of the cement paste, while $\tilde{\mu}$ may also be the viscosity of the mixture without fibers. As observed in (Férec, Perrot, & Ausias 2015; Martinie, Rossi, & Roussel 2010), the yield stress may also be influenced by the fiber orientation state. However, to the best of the authors knowledge, no experimental data or models are available in literature accounting for such phenomena, which is why the influence of fiber orientation on the yield stress is neglected in this work.

2.4 *Solution scheme*

In PFEM a Lagrangian description of motion is used for the governing equations. The derivative of the orientation tensor can be defined in the same reference frame to yield $\dot{a}_2 = \partial a_2/\partial t$. As a consequence, no spatial derivative of a_2 appears in (6), which can be solved in the strong form directly. To obtain nodal values of the orientation tensor, a node-based smoothed velocity gradient, inspired by the Smoothed Finite Element Method (Liu et al. 2009), is used to calculate a nodal velocity gradient and to solve (6). Note that the smoothed velocity gradient is only used to solve (6). The balance of momentum and mass are still solved using standard linear triangular or tetrahedral elements. For linear triangular and tetrahedral elements the nodal velocity gradient is simply obtained as the volume average of the velocity gradient of all adjacent elements to a certain node (Liu et al. 2009). To improve the robustness of the implementation, an explicit forward Euler time integration scheme is chosen to solve (6). To compensate for the "relatively" large time steps of the implicit PFEM algorithm, a time step n is split into multiple smaller time steps m for the solution of (6) as

$$\begin{aligned}\bar{a}^{m+1}_{2,n+1} = \bar{a}^{m}_{2,n+1} + \frac{\Delta t}{m_{max}}[&\tilde{W}^{m+1}_{n+1}\cdot\bar{a}^{m}_{2,n+1} \\ &- \bar{a}^{m}_{2,n+1}\cdot\tilde{W}^{m+1}_{n+1} + \lambda(\tilde{D}^{m+1}_{n+1}\cdot\bar{a}^{m}_{2,n+1} \\ &+ \bar{a}^{m}_{2,n+1}\cdot\tilde{D}^{m+1}_{n+1} - 2(\bar{a}^{m}_{4,n+1} \\ &+ (1-\kappa)(\bar{L}^{m}_{n+1} - \bar{M}^{m}_{n+1} : \bar{a}^{m}_{4,n+1})) : \tilde{D}^{m+1}_{n+1}) \\ &+ 2C_I|\dot{\bar{\gamma}}|(I - 3\bar{a}^{m}_{2,n+1})]\end{aligned} \tag{12}$$

in which Δt is the time step size, m_{max} is the number of explicit sub-steps and the superimposed bar and tilde denote nodal quantities and quantities obtained from the smoothed nodal velocity gradient, respectively. The deformation rate and spin tensors are defined for an explicit time step as

$$\tilde{D}^{m+1}_{n+1} = \frac{1}{m_{max}}\left[(m_{max} - m)\tilde{D}_n + m\tilde{D}_{n+1}\right], \tag{13}$$

$$\tilde{W}^{m+1}_{n+1} = \frac{1}{m_{max}}\left[(m_{max} - m)\tilde{W}_n + m\tilde{W}_{n+1}\right]. \tag{14}$$

To yield an efficient solution scheme, first the original PFEM equations (1) and (2) are solved based on the orientation tensor from the previous time step. Afterwards, (13) is successively solved m_{max} times for the updated a_2. Here, $m_{max} = 1000$ was chosen, which yielded sufficiently accurate results. The solution scheme within a time step n can be summarized as

1. $t_{n+1} = t_n + \Delta t$. Time step initialization, remeshing and constrained Delaunay triangulation of the domain.
2. Implicit solution of (1) and (2) using $\bar{a}_{2,n}$ and update the velocity, pressure and nodal coordinates.
3. Looped solution of (13) to obtain $\bar{a}^{m+1}_{2,n+1}$. In each solution step of (13) update $m = m + 1$ until $m = m_{max}$.
4. Post-processing of (solution) variables and go back to step 1 until the desired maximum time step $t_{n+1} = t_{max}$ is reached.

3 REPRESENTATIVE NUMERICAL STUDIES

In the following sections some representative numerical studies are discussed to verify and discuss characteristics of the proposed implementation. The first example is the planar Poiseuille flow, which was adopted to analyze fiber orientation in pipe flow problems. Based on these results, plausible inlet fiber orientation states were formulated for 3D-concrete-printing examples studied in a 2D setup. Various convergence and parametric studies were analyzed with respect to the obtained final fiber orientation state in a printed layer.

3.1 Planar Poiseuille flow

The planar Poiseuille flow is a classical benchmark example of a pipe flow problem for which a number of analytical solutions are available for different fluid models (Bird et al. 2002). The numerical simulations were carried out by modeling a part of the pipe using a height of $h = 3$ cm, a mean velocity of $v_{mean} = 0.05$ m/s, $\tilde{\mu} = 10$ Pas, $\tilde{\tau}_0 = 28.4024$ Pa, $\phi_s = 0.45$, $c = 0.01$ and $\kappa = 0.2$. The parameters were chosen to yield values which lie in a range typical for 3D-concrete-printing (Reinold et al. 2020). Simulations with the proposed anisotropic viscosity model and a purely isotropic version (by enforcing $R_\eta = 1$) using two different fiber aspect ratios $r = 30$ and $r = 90$ were performed. Both aspect ratios lie within the range of the semi-concentrated regime according to the definitions given in Section 2.2. The initial fiber orientation state was chosen as random isotropic using $a_{2,ij} = \delta_{ij}/3$ as the initial second order orientation tensor.

The obtained velocity v_x in flow direction x over the height for each numerical model and both aspect ratios are given in Figures 2 a) and b) along with the analytical solution for the isotropic case (Bird et al. 2002). A perfect match between analytical and numerical results can be observed. In addition, only marginal differences between the isotropic and anisotropic model are found, which are more pronounced using a larger aspect ratio due to the larger anisotropy ratio. The shear viscosity during steady state can be compared by neglecting the contribution from the Bingham regularization in the viscosity term. The shear viscosity for $r = 30$ was $\mu_{1212} = 32.18$ Pas considering anisotropy and $\mu_{1212} = 32.92$ Pas considering an isotropic constitutive model using $R_\eta = 1$ showing only a small shear thinning effect due to fiber reorientation. For the larger aspect ratio $r = 90$, the shear viscosity was $\mu_{1212} = 150.4619$ Pas considering anisotropy and $\mu_{1212} = 173.24$ Pas considering an isotropic constitutive model using $R_\eta = 1$, which is already a non-negligible shear thinning effect due to fiber reorientation. By increasing the aspect ratio, viscosity and yield stress also increase. This can lead to a contradictory effect with respect to the height of the plug flow as a larger viscosity decreases and a larger yield stress increases the height of the plug flow. In this example the yield stress effect dominates the influence of the viscosity on the plug flow size due to the larger plug flow size for $r = 90$, which may not necessarily be the case when using different parameters.

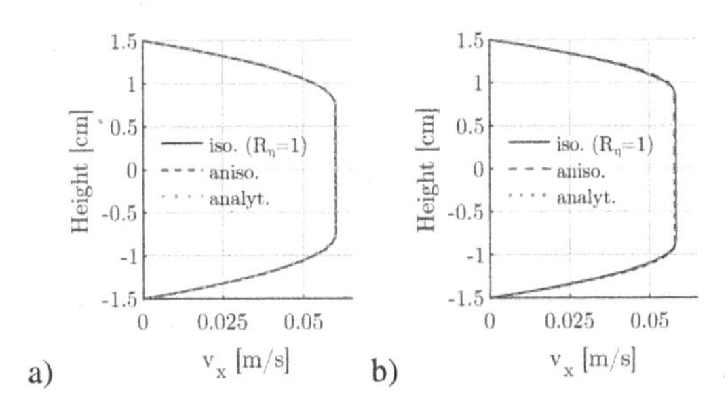

Figure 2. Planar Poiseuille flow: Velocity distribution along the height obtained from the purely isotropic and anisotropic viscosity model and the analytical solution. a) Fiber aspect ratio $r = 30$. b) Fiber aspect ratio $r = 90$.

Figure 3 illustrates how a steady state fiber orientation develops from the random isotropic initial orientation over time across the height. The second order orientation tensor is visualized as an ellipsoid as discussed in (Gudzulic et al. 2018). A long needle-shaped ellipsoid pointing in a certain direction can be interpreted as a high probability of fibers pointing in the same direction. As can be observed, in Figure 3 fibers are aligned in flow direction only in the shear zones near the boundaries. Within the plug flow the fiber distribution remains random isotropic due to the absence of a velocity gradient. Based on these observations, fiber alignment in flow direction can be increased by minimizing the height of the plug flow and increasing the height of the shear zone at the boundary.

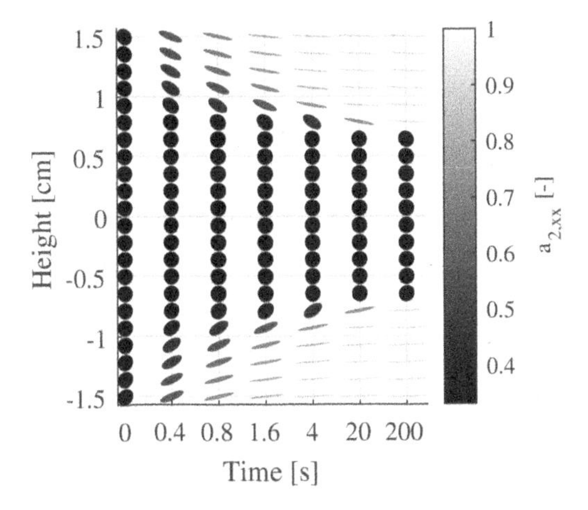

Figure 3. Planar Poiseuille flow: Evolution of the orientation tensor over the height and time showing the component $a_{2,xx}$ of the orientation tensor for $r = 30$.

The time evolution of different orientation tensor entries for both aspect ratios at a height of 1 cm are given in Figures 4 a) and b). The analytical solution is obtained from the analytical velocities of the isotropic viscosity model showing a perfect match with the simulations. In addition, small differences between the isotropic and anisotropic model can be observed in case of $r = 90$. The steady state solution remains almost identical.

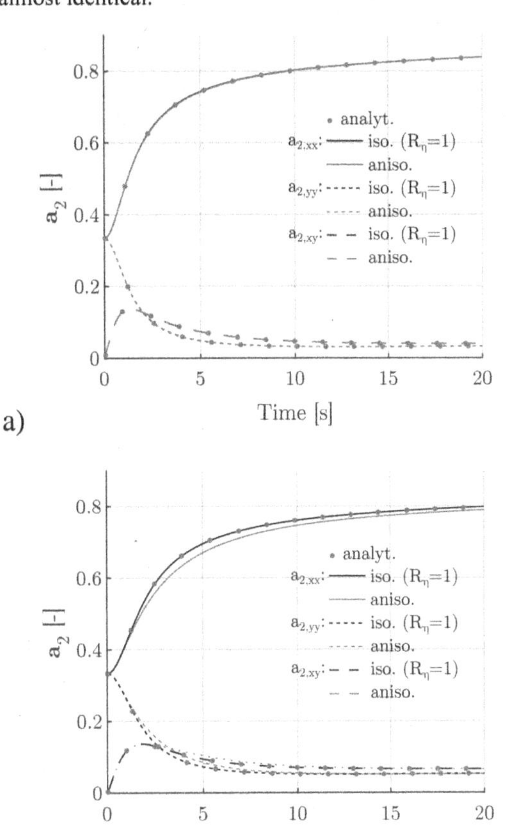

Figure 4. Planar Poiseuille flow: Evolution of orientation tensor components over time at a height of 1 cm. a) Fiber aspect ratio $r = 30$. b) Fiber aspect ratio $r = 90$.

3.2 *Fiber-reinforced 3D-concrete-printing study in 2D*

3.2.1 *General analysis*

In this section the fiber orientation state of rigid fibers in a 3D-concrete-printing extrusion process is modeled in 2D using the proposed model. The material and fiber properties were the same as in Section 3.1 and the geometry of the printing process is depicted in Figure 5 a). The fiber aspect ratio was $r = 30$. The extrusion nozzle width d and separation distance h were $d = h = 2$ cm, the printing speed was $v_p = 5$ cm/s and the flow rate was $Q_{inlet} = 0.98 \cdot v_p \cdot h$. Convergence studies with varying extrusion nozzle modeling heights were conducted using different fiber orientation inlet conditions and the analytical solution of the velocity profile of the planar Poiseuille flow was prescribed at the inlet nodes at the top of the extrusion nozzle. A spatial discretization with a nodal spacing of approximately 1/15 cm was used; see Figure 5 b). No-slip boundary conditions were applied to the ground and the nozzle boundaries. The flow inside the extrusion nozzle can be interpreted as the planar Poiseuille flow problem discussed in the previous section.

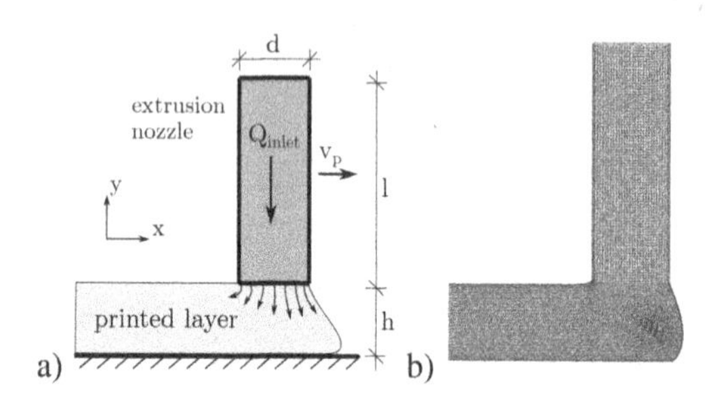

Figure 5. Fiber-reinforced 3D-concrete-printing in 2D: a) Geometry and parameters of the numerical 3D-concrete-printing study. b) Finite element discretization around the extrusion nozzle for the nozzle height $l = 3d$.

In Figure 6 a) the orientation tensor ellipsoids and $a_{2,xx}$ are depicted for $l = 3d$ for the case of a random isotropic fiber orientation inlet condition at the top of the extrusion nozzle. As can be observed, fibers tend to align in printing direction due to increased $a_{2,xx}$ values, especially at the boundaries where the material is subjected to high shear stresses. Indicated by varying orientation states within the extrusion nozzle, the nozzle height l is not large enough to yield a steady fiber orientation state in the nozzle in case of the random isotropic orientation state at the inlet. More realistic results can be obtained by applying the steady fiber orientation state as an inlet condition at the top of the extrusion nozzle, given in Figure 6 b). The steady fiber orientation state was obtained from the analytical solution of the planar Poiseuille flow discussed in Section 3.1. In accordance to observations in experiments (Arunothayan et al. 2021; Huang et al. 2021; Mechtcherine et al. 2021), the final orientation state in the printed layer is dominated by the fiber orientation state developed inside the extrusion nozzle and shearing under the extrusion nozzle is only of secondary importance.

The mean value over the layer height of the orientation tensor component $a_{2,xx}$ and $a_{2,yy}$ are given over different extrusion nozzle modeling heights l for both orientation tensor inlet conditions in Figures 7 a) and b). As can be observed, with an increasing nozzle height, the final orientation tensor values converge to the steady fiber orientation inlet solution, which is plausible. However, very large extrusion nozzle modeling heights l would be necessary to obtain plausible results. The final orientation state remained almost unaffected by the modeling height of the extrusion nozzle in case of the steady state fiber orientation inlet condition, which is supported by observations in Figure 6 b) with respect to the almost constant fiber orientation state within the extrusion nozzle. In

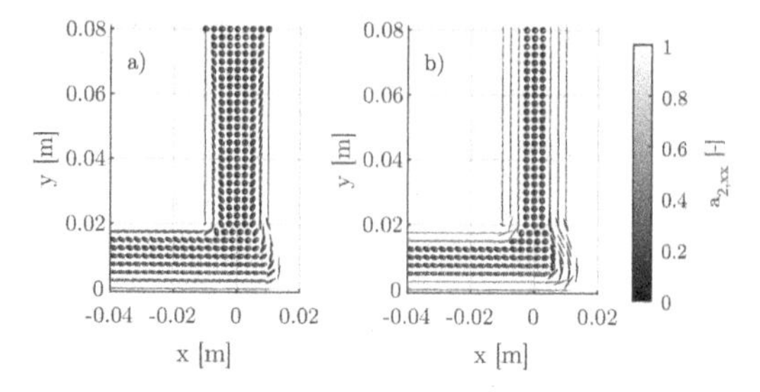

Figure 6. Fiber-reinforced 3D-concrete-printing in 2D: Fiber orientation state and $a_{2,xx}$ around the extrusion nozzle for the nozzle height $l = 3d$. a) Random isotropic fiber orientation inlet condition. b) Steady state fiber orientation inlet condition.

conclusion, the steady state fiber orientation inlet condition yields more plausible and reproducible results, which are independent of the extrusion nozzle modeling height l and should be used in fiber-reinforced 3D-concrete-printing extrusion flow simulations.

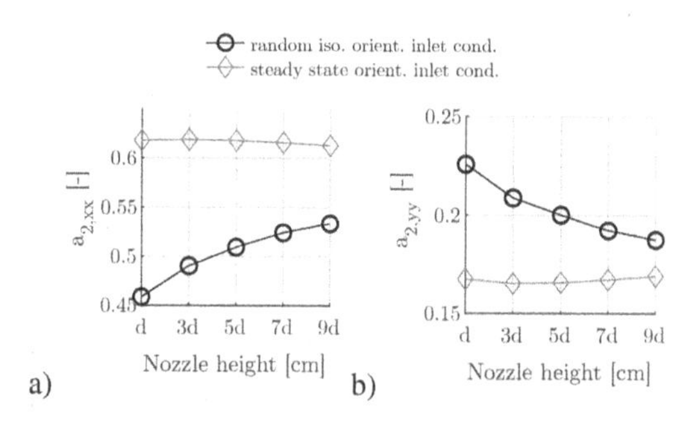

Figure 7. Fiber-reinforced 3D-concrete-printing in 2D: Mean value over the printed layer height of the orientation tensor component a) $a_{2,xx}$ and b) $a_{2,yy}$ over different extrusion nozzle modeling heights l and orientation tensor inlet conditions.

3.2.2 *Parametric studies*

To understand the correlations between process parameters (printing speed, separation distance and extrusion nozzle width) and the fiber orientation state in printed layers, representative parametric studies are discussed in this section. The results may help to tune the printing process for optimizing the printing processes with respect to the desired fiber orientation states in the printed component. Material parameters and geometry of the printing process are taken from Section 3.2.1, which remain unchanged unless otherwise specified.

The influence of different fiber aspect ratios and a purely isotropic ($R_\eta = 1$) and anisotropic viscosity model on the distribution of the second order orientation tensor component $a_{2,xx}$ over the height of the printed layer are given in Figure 8 a) using the steady fiber orientation inlet condition. Similar as for the Poiseuille flow problem in Section 3.1, only marginal differences are observed between the isotropic and anisotropic viscosity modeling approach. Larger $a_{2,xx}$-values over the height, which would lead to a larger mean value of $a_{2,xx}$ over the height, can be interpreted as a higher probability of fiber alignment in printing direction. Fibers with a smaller aspect ratio tend to align more in printing direction due to smaller diffusivity factors for fibers with a smaller aspect ratio (cf. (3)), which leads to a smaller second order orientation tensor component value in flow direction during steady state in the extrusion nozzle. Figure 8 b) shows $a_{2,xx}$ over the height of the printed layer for different printing speeds. As observed, a higher printing speed leads to more fiber alignment in printing direction due to higher shear stresses and a larger shear layer size within the extrusion nozzle. The magnitude of $a_{2,xx}$ in the shear layers (top and bottom of the layer) are the same for different printing speeds due to the same fiber aspect ratio and fiber volume fraction.

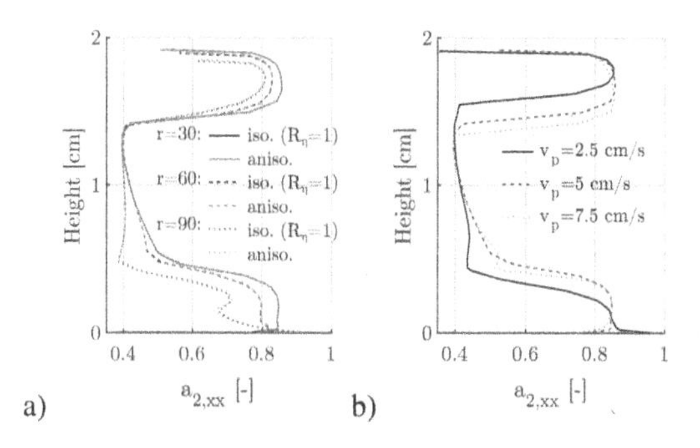

Figure 8. Fiber-reinforced 3D-concrete-printing in 2D: Orientation tensor component $a_{2,xx}$ over the height of the printed layer. a) Different fiber aspect ratios and a purely isotropic and anisotropic viscosity model. b) Different printing speeds v_p.

As depicted in Figure 9 a), a smaller extrusion nozzle width leads to more fiber alignment in printing direction due to higher shear stresses and a larger shear layer size in the extrusion nozzle, which results from larger velocities in the extrusion nozzle when using a constant flow rate Q_{inlet} for different nozzle widths. Figure 9 b) shows the influence of a geometrically scaled printing process by consistently changing the separation distance, nozzle width and flow rate. Note that the height of the printed layer in Figure 9 b) is normalized by the separation distance. As observed, fibers tend to align more in printing direction on a smaller geometric scale due to higher shear stresses and larger shear layer sizes in smaller extrusion nozzles.

4 CONCLUSIONS

In this work, a fiber orientation model was implemented into a numerical framework based on PFEM to model fiber orientation states during extrusion processes of fiber-reinforced 3D-concrete-printing. A brief introduction of the PFEM model was given, which was based on a previous work of the authors

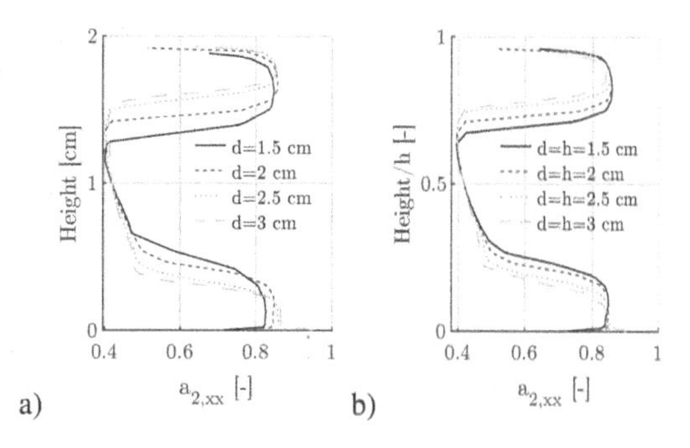

Figure 9. Fiber-reinforced 3D-concrete-printing in 2D: Orientation tensor component $a_{2,xx}$ over the height of the printed layer. a) Different nozzle widths. b) Different nozzle widths and separation distances.

(Reinold & Meschke 2019; Reinold et al. 2020). The fiber orientation state was expressed using a Folgar-Tucker fiber orientation model (Gudzulic et al. 2018). An efficient and robust PFEM compatible solution scheme for the evolution equation of the second order orientation tensor was introduced, which only requires minimal adaptions to an existing PFEM code. The constitutive model for the fiber-reinforced fresh concrete suspension was based on an anisotropic Bingham viscosity model using experimentally validated upscaling relations for the macroscopic plastic viscosity and yield stress from literature.

In representative numerical studies, important features and characteristics of the numerical model were discussed. The planar Poiseuille flow was used to verify the numerical implementation and to discuss the general problem of fiber orientation in Bingham fluids in pipes. It was found, that only in the shear zone at the boundaries fibers tend to align in the flow direction. In the plug flow zone in the channel center, fibers are not influenced due to the absence of a velocity gradient. Steady state fiber orientation solutions from the Poiseuille flow problem were used to model plausible initial flow conditions for 3D-concrete-printing extrusion processes, which were studied in a 2D setup. By applying the steady fiber orientation state as an initial condition at the top of the modeled part of the extrusion nozzle, the final fiber orientation state was unaffected by the modeling height of the extrusion nozzle. Further parametric studies revealed that fibers tend to align more in printing direction for processes with fibers owing a smaller aspect ratio, a higher printing speed and a smaller extrusion nozzle width. These results may help to tune the printing process to yield optimal fiber orientation states in printed components. In all numerical studies the effect of anisotropic viscosity effects were negligible. Such effects may only become important, when higher fiber volume fractions and fibers with a larger aspect ratios are used in the printing process, which is typically not the case for practical reasons.

The numerical examples in this work demonstrated the effectiveness of the proposed model to simulate fiber orientation processed during extrusion based 3D-concrete-printing. Further numerical 3D analyses and experiments are necessary to verify and validate the proposed model.

REFERENCES

Advani, S. & C. Tucker (1987). The use of tensors to describe and predict fiber orientation in short fiber composites. *J. Rheol. 31*, 751–784.

Arunothayan, A., B. Nematollahi, R. Ranade, S. Bong, J. Sanjayan, & K. Khayat (2021). Fiber orientation effects on ultra-high performance concrete formed by 3D printing. *Cem. Concr. Res. 143*, 106384.

Bird, R. B., W. E. Stewart, & E. N. Lightfoot (2002). *Transport Phenomena*. John Wiley & Sons, Ltd.

Buswell, R., W. L. de Silva, S. Jones, & J. Dirrenberger (2018). 3D printing using concrete extrusion: A roadmap for research. *Cem. Concr. Res. 112*, 37–49.

Chung, D. & T. Kwon (2001). Improved model of orthotropic closure approximation for flow induced fiber orientation. *Polym. Compos. 22*(5), 636–649.

Chung, D. & T. Kwon (2002). Fiber orientation in the processing of polymer composites. *Korea-Australia Rheology Journal 14*, 175–188.

Cremonesi, M., A. Franci, S. Idelsohn, & E. Onate (2020). A State of the Art Review of the Particle Finite Element Method (PFEM). *Arch Computat Methods Eng 27*, 1709–1735.

Deeb, R., S. Kulasegaram, & B. Karihaloo (2014a). 3D modelling of the flow of self-compacting concrete with or without steel fibres. Part I: slump flow test. *Comp. Part. Mech. 1*, 373–389.

Deeb, R., S. Kulasegaram, & B. Karihaloo (2014b). 3D modelling of the flow of self-compacting concrete with or without steel fibres. Part II: L-box test and the assessment of fibre reorientation during the flow. *Comp. Part. Mech. 1*, 391–408.

Doi, M. & S. Edwards (1978). Dynamics of rod-like macromolecules in concentrated solution. part 1. *J. Chem. Soc., Faraday Trans. 2 74*, 560–570.

Edelsbrunner, H. & E. P. Mücke (1994). Three-dimensional alpha shapes. *ACM Trans. Graph. 13*(1), 43–72.

Favaloro, A., H.-C. Tseng, & R. Pipes (2018). A new anisotropic viscous constitutive model for composites molding simulation. *Composites Part A: Applied Science and Manufacturing 115*, 112–122.

Férec, J., E. Bertevas, B. Khoo, G. Ausias, & N. Phan-Thien (2017). A rheological constitutive model for semi-concentrated rod suspensions in Bingham fluids. *Phys. Fluids 29*(7), 073103.

Férec, J., A. Perrot, & G. Ausias (2015). Toward modeling anisotropic yield stress and consistency induced by fiber in fiber-reinforced viscoplastic fluids. *J. Non-Newtonian Fluid Mech. 220*, 69–76.

Ferrara, L., S. Shyshko, & V. Mechtcherine (2012). Predicting the flow-induced dispersion and orientation of steel fibers in self-consolidating concrete by distinct element method. In *Fibre Reinforced Concrete: challenges and opportunities, Proceedings BEFIB 2012, 8th International RILEM Symposium*, pp. 1–12.

Folgar, F. & C. Tucker (1984). Orientation behavior of fibers in concentrated suspensions. *J. Reinf. Plast. Compos. 3*, 98–119.

Ghanbari, A. & B. Karihaloo (2009). Prediction of the plastic viscosity of self-compacting steel fibre reinforced concrete. *Cem. Concr. Res. 39*(12), 1209–1216.

Grünewald, S. (2004). *Performance-based design of self-compacting fibre reinforced concrete*. Ph. D. thesis, Delft University of Technology.

Gudzulic, V., T. Dang, & G. Meschke (2018). Computational modeling of fiber flow during casting of fresh concrete. *Comput. Mech. 63(6)*, 1111–1129.

Heller, B., D. Smith, & D. Jack (2019). Planar deposition flow modeling of fiber filled composites in large area additive manufacturing. *Addit. Manuf. 25*, 227–238.

Huang, H., X. Gao, & L. Teng (2021). Fiber alignment and its effect on mechanical properties of UHPC: An overview. *Constr. Build. Mater. 296*, 123741.

Idelsohn, S. R., E. Oñate, & F. Del Pin (2004). The particle finite element method: A powerful tool to solve incompressible flows with free-surfaces and breaking waves. *Int. J. Numer. Methods Eng. 61*(7), 964–989.

Jeffery, G. & L. Filon (1922). The motion of ellipsoidal particles immersed in a viscous fluid. *Proc. R. Soc. London, Ser. A 102*(715), 161–179.

Liu, G. R., T. Nguyen-Thoi, H. Nguyen-Xuan, & K. Y. Lam (2009). A node-based smoothed finite element method (NS-FEM) for upper bound solutions to solid mechanics problems. *Comput. Struct. 87*(1), 14–26.

Martinie, L., P. Rossi, & N. Roussel (2010). Rheology of fiber reinforced cementitious materials: classification and prediction. *Cement and Concrete Research 40*(2), 226–234.

Mechtcherine, V., F. P. Bos, A. Perrot, W. R. L. da Silva, V. N. Nerella, S. Fataei, R. J. M. Wolfs, M. Sonebi, & N. Roussel (2020). Extrusion-based additive manufacturing with cement-based materials – Production steps, processes, and their underlying physics: A review. *Cem. Concr. Res. 132*, 106037.

Mechtcherine, V., R. Buswell, H. Kloft, F. Bos, N. Hack, R. Wolfs, J. Sanjayan, B. Nematollahi, E. Ivaniuk, & T. Neef (2021). Integrating reinforcement in digital fabrication with concrete: A review and classification framework. *Cem. Concr. Compos. 119*, 103964.

Oñate, E., S. R. Idelsohn, F. Del Pin, & R. Aubry (2004). The particle finite element method: An overview. *Int. J. Comput. Methods 1*(02), 267–307.

Ouyang, Z.and Bertevas, E., L. Parc, B. Khoo, N. Phan-Thien, J. Férec, & G. Ausias (2019). A smoothed particle hydrodynamics simulation of fiber-filled composites in a non-isothermal three-dimensional printing process. *Physics of Fluids 31*(12), 123102.

Papanastasiou, T. C. (1987). Flows of Materials with Yield. *J. Rheol. 31*(5), 385–404.

Perrot, A., T. Lecompte, P. Estellé, & S. Amziane (2013). Structural build-up of rigid fiber reinforced cement-based materials. *Mater Struct 46*, 1561–1568.

Petrie, C. (1999). The rheology of fibre suspensions. *J. Non-Newtonian Fluid Mech. 87*(2), 369–402.

Phan-Thien, N., X.-J. Fan, R. Tanner, & R. Zheng (2002). Folgar-tucker constant for a fibre suspension in a newtonian fluid. *J. Non-Newtonian Fluid Mech. 103*(2), 251–260.

Phelps, J. & C. Tucker (2009). An anisotropic rotary diffusion model for fiber orientation in short- and long-fiber thermoplastics. *J. Non-Newtonian Fluid Mech. 156*(3), 165–176.

Pipes, R., D. Coffin, P. Simacek, & S. Shuler (1994). *Flow and Rheology in Polymer Composites Manufacturing*, Chapter Rheological behavior of collimated fiber thermoplastic composite materials, pp. 85–125. Elsevier, Amsterdam.

Reinold, J. & G. Meschke (2019). Particle finite element simulation of fresh cement paste – inspired by additive

manufacturing techniques. *Proc. Appl. Math. Mech. 19*(1), e201900198.
Reinold, J., V. N. Nerella, V. Mechtcherine, & G. Meschke (2020). Extrusion process simulation and layer shape prediction during 3D-concrete-printing using the Particle Finite Element Method. Preprints, 2020070715.
Sepehr, M., G. Ausias, & P. Carreau (2004). Rheological properties of short fiber filled polypropylene in transient shear flow. *J. Non-Newtonian Fluid Mech. 123*(1), 19–32.
Sommer, D., A. Favaloro, & R. Pipes (2018). Coupling anisotropic viscosity and fiber orientation in applications to squeeze flow. *Journal of Rheology 62*(3), 669–679.
Sultangaliyeva, F., H. Carré, C. La Borderie, W. Zuo, E. Keita, & N. Roussel (2020). Influence of flexible fibers on the yield stress of fresh cement pastes and mortars. *Cement and Concrete Research 138*, 106221.
Ŝvec, O. (2013). *Flow modelling of steel fibre reinforced self-compacting concrete*. Ph. D. thesis, Technical University of Denmark.
Wang, J., J. O'Gara, & C. Tucker (2008). An objective model for slow orientation kinetics in concentrated fiber suspensions: Theory and rheological evidence. *J. Rheol. 52*(5), 1179–1200.

Computational Modelling of Concrete and Concrete Structures – Meschke, Pichler & Rots (Eds)
© 2022 Copyright the Author(s), ISBN: 978-1-032-32724-2

Modelling of cracking mechanisms in cementitious materials: The transition from diffuse microcracking to localized macrocracking

I.C. Mihai & A. Bains
Cardiff University, Cardiff, UK

P. Grassl
Glasgow University, Glasgow, UK

ABSTRACT: Cracking in cementitious materials still poses significant and interesting modelling challenges and structural designers need reliable tools for an accurate prediction of crack widths. The paper presents a numerical study into cracking mechanisms in cement based materials using lattice simulations employing the model of Grassl & Antonelli (2019). Furthermore, a micromechanics based constitutive model is proposed that focuses on representing the transition from diffuse microcracking to localized macrocracking. The model includes an Eshelby based two-phase composite solution to represent the aggregate particles embedded in a cementitious matrix, directional microcracking and a criteria for the transition from diffuse microcracking to localised macrocracking. By removing the macrocrack fracture strain component from the strain which drives microcrack growth, the effect of macrocrack development on microcrack growth in various other directions is included. Numerical simulations show that the model captures well the mechanical behaviour as well as key characteristics of the cracking mechanism in cementitious materials.

1 INTRODUCTION

Microcracks are present in concrete before loading is applied and are concentrated at the interfacial transition zone (ITZ) between the cementitious matrix and aggregate particles (Slate& Hover 1984). If an applied tensile load is increased past the initiation threshold, the microcracks propagate and further microcracks are progressively initiated in the ITZ of smaller aggregate particles (Karihaloo 1995). As the load increases further, some microcracks will grow and coalesce to form a macrocrack (Jenq & Shah 1991). For both uniaxial tension and uniaxial compressive loading, these macrocracks tend to form around the peak load and propagate unstably with the material around the zone of macrocracking unloading (Shah et al. 1995; Vonk 1992; i.e., cracking becomes concentrated within a certain zone. The process of cracking becoming concentrated to macrocracks formed by the coalescence of diffuse microcracks is often referred to as crack localisation.

This paper presents the main details of a micromechanics based constitutive model for cementitious materials that simulates crack localization. A series of numerical experiments employing a lattice model were carried out to study the transition from discrete microcracking to localized macrocracking and the results from these studies were used guide the development of the constitutive model.

2 CRACK LOCALIZATION STUDY

2.1 *Lattice model*

A study into the transition from diffuse microcracking to localized macrocracking was carried out with the lattice model of Grassl & Antonelli (2019) which relies on periodic meso-structure generation by employing a representative cell with a periodic lattice network and periodic boundary conditions. Within the computational cell, the meso-structure of concrete was modelled considering three material phases; namely the mortar matrix, the coarse aggregate particles and the ITZ respectively (Figure 1). The aggregate particles are idealized as ellipsoids, the size distribution of which is determined based on Fuller's grading curve.

2.2 *Constitutive relationships for the lattice model*

For this study, the aggregate particles are assumed to have a linear elastic behaviour and the scalar damage relationship in Equation 1 is employed to simulate the mechanical behaviour of both the matrix and the ITZ.

$$\boldsymbol{\sigma} = (1 - \omega_a)\, \mathbf{D}_e \boldsymbol{\varepsilon} \tag{1}$$

where $\boldsymbol{\sigma}$ is the stress vector, $\boldsymbol{\varepsilon}$ is the strain vector, $\mathbf{D}_e$ is the elastic stiffness matrix and ω_a is a scalar

DOI 10.1201/9781003316404-26

damage variable which is 0 at no damage and gradually increases to 1 for complete damage. The damage evolution is given in Equation 2:

$$(1-\omega_a)E\kappa_d = f_t e^{\left(-\frac{\omega_a h \kappa_d}{w_f}\right)} \tag{2}$$

where E is the Young's modulus, f_t is the tensile strength, w_f is a parameter that controls the slope of the softening curve and is related to the fracture energy G_f as follows; $w_f = G_f/f_t$. κ_d is an equivalent strain parameter gouverned by a damage surface based on an ellipsoidal strength envelope in the stress space and standard loading/unloading conditions (Grassl & Bolander 2016).

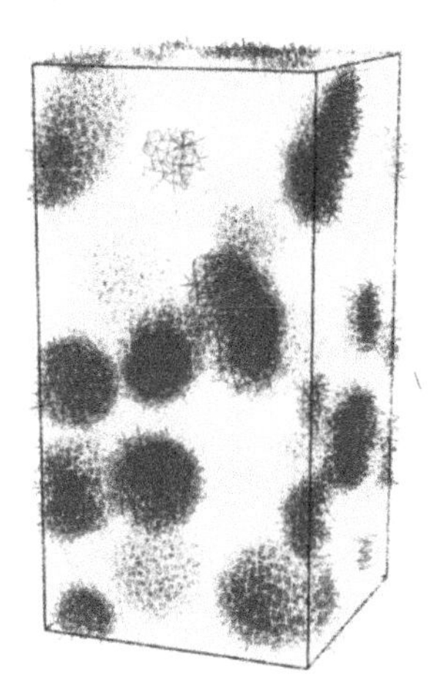

Figure 1. Representative computational cell showing the material phases; mortar matrix (yellow), corase aggregate particles (blue) and the ITZ (red).

2.3 *Crack localization*

A series of lattice simulations using the formulation described above were carried out, employing a 50 × 50 × 100 (mm) periodic cell and the material parameters given in Table 1. Moreover, following a series of convergence studies, a lattice element size of 1.6 mm and aggregate particle diameters ranging from 10 mm to 20 mm were selected respectively.

Table 1. Material parameters - lattice simulations.

E_m(MPa)	30 000
E_{ITZ}(MPa)	45 000
E_Ω(MPa)	90 000
$f_{t,m}$ (MPa)	3
$f_{t,ITZ}$(MPa)	1.5
$G_{f,m}$ (J/m^2)	120
$G_{f,ITZ}$(J/m^2)	60
Volume fraction of aggregate,	V_Ω40%

A typical stress-relative displacement curve from a uniaxial tension simulation is presented in Figure 2 and associated crack patterns at different stages are presented in Figure 3, noting that only the active, growing cracks are shown at each stage.

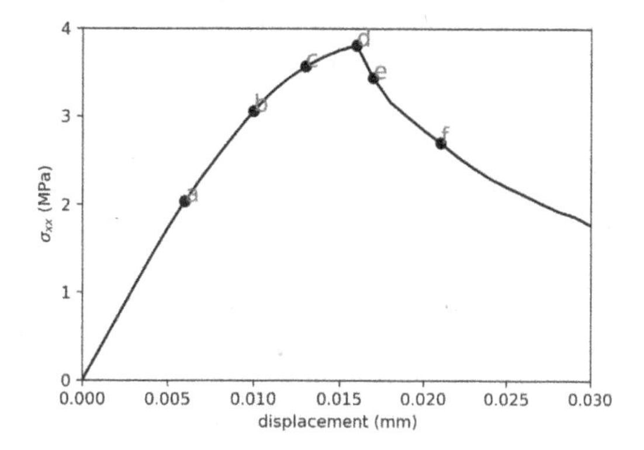

Figure 2. Stress-relative displacement curve from lattice model simulation of uniaxial tension (tension +ve).

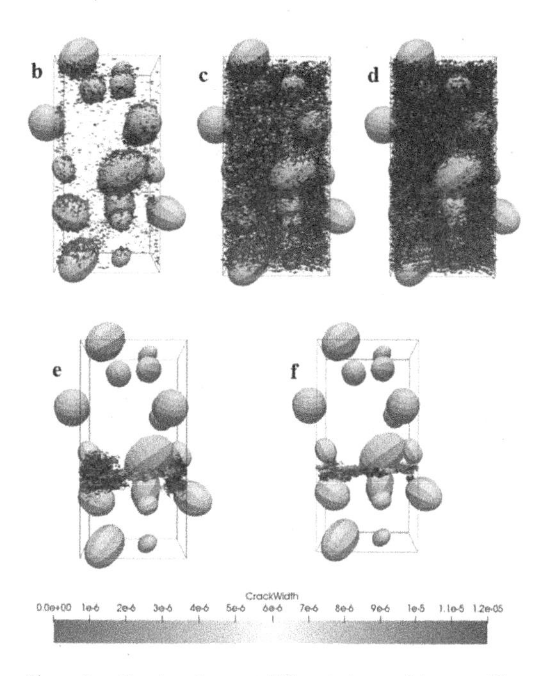

Figure 3. Crack patterns at different stages of damage. The different stages b - f correspond to those marked in Figure 2.

The crack patterns in Figure 3 show a number of cracking mechanisms, captured well by the lattice model. Microcracks are initiated at the matrix-aggregate interface and subsequently propagate in the cementitious matrix to a state of diffuse microcracking associated with pre-peak non-linearity (stages b-d). By contrast, the post-peak response is characterized by a single localized macrocrack (stages e-f).

The representation of these two distinct cracking stages and the transition from diffuse microcracking to localized macrocracking is the focus of the constitutive model presented in Section 3.

3 MICROMECHANICS BASED CONSTITUTIVE MODEL

3.1 *Model concepts*

The constitutive model presented here aims to represent the behaviour at two stages of cracking; (i) the diffuse microcracking stage characteristic of the pre-peak behavior in tension and (ii) the localized macro-crack stage, characteristic of the post-peak behaviour respectively.

In the elastic state, before any damage occurs, the concrete material is modelled as a two-phase composite comprising a matrix representing the mortar and spherical inclusion representing the coarse aggregate particles. The diffuse microcracking stage is represented using a directional microcracking formulation based on the Budiansky & O'Connell (1976) solution. The localised macrocrack stage is then represented by removing a macrocrack fracture strain component from the strain which drives microcrack growth.

3.2 *Two-phase composite*

The elastic constitutive relationship for the two-phase composite is obtained by making use of the micromechanics Eshelby matrix-inclusion solution and the Mori-Tanaka homogenisation scheme (Mura, 1987) for a non-dilute distribution of inclusions:

$$\bar{\boldsymbol{\sigma}} = \mathbf{D}_{\mathrm{m}\Omega} : \bar{\boldsymbol{\varepsilon}} \tag{3}$$

where $\bar{\boldsymbol{\sigma}}$ and $\bar{\boldsymbol{\varepsilon}}$ are the average far-field stress and strain respectively. $\mathbf{D}_{\mathrm{m}\Omega}$ is the elasticity tensor of the composite:

$$\mathbf{D}_{\mathrm{m}\Omega} = (\mathrm{f_m}\mathbf{D}_\mathrm{m} + \mathrm{f}_\Omega \mathbf{D}_\Omega \cdot \mathbf{T}_\Omega) \cdot \left(\mathrm{f_m}\mathbf{I}^{4\mathrm{s}} + \mathrm{f}_\Omega \mathbf{T}_\Omega\right)^{-1} \tag{4}$$

in which $\mathbf{D}_\beta$ represents the elasticity tensor and f_β the volume fraction of β-phase ($\beta = \mathrm{m}$ or Ω), $\mathrm{f}_m + \mathrm{f}_\Omega = 1$. $\mathbf{I}^{4s}$ is the fourth order identity tensor and

$$\mathbf{T}_\Omega = \mathbf{I}^{4s} + \mathbf{S}_\Omega \cdot [(\mathbf{D}_\Omega - \mathbf{D}_\mathrm{m}) \cdot \mathbf{S}_\Omega + \mathbf{D}_\mathrm{m}]^{-1} \cdot (\mathbf{D}_\mathrm{m} - \mathbf{D}_\Omega) \tag{5}$$

$\mathbf{S}_\Omega$ is the Eshelby tensor for spherical inclusions (Nemat-Nasser & Hori, 1993).

3.3 *Directional microcracking*

A solution based on the work of Budiansky & O'Connell (1976) is employed to address microcracking by evaluating the added strain $\boldsymbol{\varepsilon}_a$ from series of penny-shaped microcracks of various orientations distributed according to a crack density function $f(\theta, \psi)$. The added strains resulting from the microcracks are superimposed on the composite such that the constitutive relationship in Equation 3 becomes:

$$\bar{\boldsymbol{\sigma}} = \mathbf{D}_{\mathrm{m}\Omega} : (\bar{\boldsymbol{\varepsilon}} - \boldsymbol{\varepsilon}_\mathrm{a}) \tag{6}$$

The added strain are as follows (Budiansky & O'Connell, 1976):

$$\boldsymbol{\varepsilon}_\mathrm{a} = \left(\frac{1}{2\pi} \int_{2\pi} \int_{\pi/2} \mathbf{N}_\varepsilon : \mathbf{C}_a : \mathbf{N} f(\theta, \psi) sin(\psi) d\psi d\theta \right) : \bar{\boldsymbol{\sigma}} \tag{7}$$

in which $\mathbf{C}_a$ is the local compliance tensor in the local coordinate system of a microcrack (**r,s,t**) and **N** the stress transformation tensor. In each direction, defined by (θ, ψ), the crack density parameter is related to a directional scalar damage parameter ω $(0 \leq \omega \leq 1)$ such that:

$$f(\theta, \psi)\mathbf{C}_a = \frac{\omega(\theta, \psi)}{1 - \omega(\theta, \psi)} \mathbf{C}_\mathrm{L} = \mathbf{C}_\alpha(\theta, \psi) \tag{8}$$

where $\mathbf{C}_\mathrm{L} = \frac{1}{E_m} \begin{bmatrix} 1 & 0 & 0 \\ 0 & \frac{4}{2\text{-}\nu_m} & 0 \\ 0 & 0 & \frac{4}{2\text{-}\nu_m} \end{bmatrix}$ is the local elastic compliance tensor, with ν_m and E_m being Poisson's ratio and Young's modulus of the matrix phase respectively.

The local damage function from Mihai & Jefferson (2011) is employed to govern the evolution of the damage parameter ω and is given by:

$$F_\zeta(\boldsymbol{\varepsilon}_L, \zeta) = \left(\varepsilon_{Lrr} \frac{1 + \alpha_L}{2} + \sqrt{\varepsilon_{Lrr}^2 \left(\frac{1 - \alpha_L}{2} \right)^2 + r_L^2 \left(\varepsilon_{Lrs}^2 + \varepsilon_{Lrt}^2 \right)} \right) - \zeta \tag{9}$$

in which $\alpha_L = \frac{\nu_\mathrm{m}}{1 - \nu_\mathrm{m}}$, $\mathrm{r}_L = \frac{\nu_\mathrm{m} - 1/2}{\nu_\mathrm{m} - 1}$ and noting that the following loading/unloading conditions apply:

$$F_\zeta \leq 0; \ \dot{\zeta} \geq 0; \ F_\zeta \dot{\zeta} = 0 \tag{10}$$

Introducing Equation 7 and Equation 8 into Equation 6 and rearranging gives:

$$\bar{\boldsymbol{\sigma}} = \mathbf{D}_{mc} : \bar{\boldsymbol{\varepsilon}} \tag{11}$$

where;

$$\mathbf{D}_\mathrm{mc} = \left(\mathbf{I}^{4s} + \frac{\mathbf{D}_{\mathrm{m}\Omega}}{2\pi} \int_{2\pi} \int_{\frac{\pi}{2}} \mathbf{N}_\varepsilon : \mathbf{C}_\alpha(\theta, \psi) : \times \mathbf{N} \cdot \sin(\psi) d\psi d\theta \right)^{-1} \cdot \mathbf{D}_{\mathrm{m}\Omega} \tag{12}$$

3.4 *Macrocracking*

The model assumes that macrocracks form when the overall stress reaches its peak value i.e.:

$$\frac{d\sigma_I}{d\varepsilon_I} = \mathbf{0} \tag{13}$$

where σ_I and ε_I are the major principal stress and strain respectively. Under tensile loading, the normal directions of macrocrack plane are based on the orientations of the major principal strains and a maximum of two macrocracks are allowed to form. Under compressive loading, a macrocrack forms with the normal to the crack plane given by the direction which maximises the effective strain parameter at the peak stress.

Macrocrack formation is taken into account in the overall constitutive relationship by removing the macrocrack inelastic strain from the average strain:

$$\bar{s} = \mathbf{D}_{mc} : \left(\bar{\dotplus} - \sum_{i=1}^{n_{sd}} \mathbf{N}_{\varepsilon}(\alpha_i, \beta_i) : \hat{\boldsymbol{\varepsilon}}_i \right) \tag{14}$$

where n_{sd} is the total number of macrocrack planes and $\hat{\boldsymbol{\varepsilon}}$ is the macrocrack inelastic strain. α and β are the orientation angles of the macrocrack plane. The local stress of macrocrack planes $\tilde{\boldsymbol{\sigma}}$ is given by the following local constitutive relationship:

$$\tilde{s}(\alpha, \beta) = (1 - \tilde{\omega}(\alpha, \beta)\mathbf{I}^{4s})\mathbf{C}_L^{-1} : \tilde{\dotplus}(\alpha, \beta) \tag{15}$$

where $\tilde{\dotplus}$ is the macrocrack local strain, $\tilde{\omega}$ is the macrocrack damage parameter. From the above, the inelastic strain of macrocracks can be written in terms of the local strain of macrocracks: $\hat{\dotplus}_i = (\mathbf{I}^{4s} - \tilde{\mathbf{M}}_{s_i}) : \tilde{\dotplus}_i$ where $\tilde{\mathbf{M}}_{s_i} = (1 - \tilde{\omega}_i)\mathbf{I}^{4s}$. The dependencies of $\tilde{\omega}$, including orientation, have been dropped for clarity.

The same damage surface (Eq. 9) employed for microcracks applies for calculating the effective strain parameter of macrocracks $\tilde{\zeta}$ and the evolution of the macrocrack damage parameter $\tilde{\omega}$ respectively.

Once the transition to localised damage has been initiated, inelastic strain $\hat{\dotplus}$ due to macrocracking starts to progress. But it is assumed that microcracks are still present in the band of material outside of the zone of localised cracking. Therefore, to capture the effect of macrocracking on microcrack growth, the inelastic macrocrack strain is removed from the local macrocrack strains:

$$\dotplus_L(\psi_k, \theta_k) = \mathbf{N}_{\dotplus}(\psi_k, \theta_k) : \left(\bar{\dotplus} - \sum_{isd=1}^{n_{sd}} \mathbf{N}_{\varepsilon}(\alpha_{isd}, \beta_{isd}) : \hat{\dotplus}(\alpha_{isd}, \beta_{isd}) \right) \tag{16}$$

A staggered solution is used to calculate the inelastic strain $\hat{\dotplus}$, the full details of which are presented in a forthcoming publication.

4 NUMERICAL SIMULATIONS

Uniaxial tension predictions from the two versions of the model (only microcracking and both microcrack and macrocrack growth) were compared to uniaxial tension lattice simulations of 10 random arrangements of aggregate particles. The intention of the comparisons is to show how a micromechanics based constitutive model for concrete which includes a crack localisation mechanism agrees well with more computationally expensive lattice simulations that discretely model the influence of the heterogeneous material structure of concrete at the meso-scale.

The material parameters employed in the constitutive model for these numerical simulations are given in Table 2. The lattice simulations were carried out using a 40% total volume fraction of aggregate particles and by maintaining the periodic cell and element, dimensions and material parameters described in Section 2

Table 2. Material parameters for the micromechanics based constitutive model.

E_m(MPa)	30 000
E_Ω(MPa)	45 000
ν_m	0.19
ν_Ω	0.21
f_m	0.6
f_Ω	0.4
f_t (MPa)	3
ε_0	0.003

The numerical results are presented in Figure 4. When macrocrack localization is not included the response is overly ductile, whereas the inclusion of the transition to localized cracking leads to more realistic results and a better agreement with the lattice simulations. It can be observed in Figures 4b&c that in the micro-macro transition model, after the peak stress, damage becomes localised to a macrocrack plane and microcrack growth is stalled, much like what has been observed from the lattice experiments. In contrast, in the microcracking only model the microcrack planes continue to become damaged.

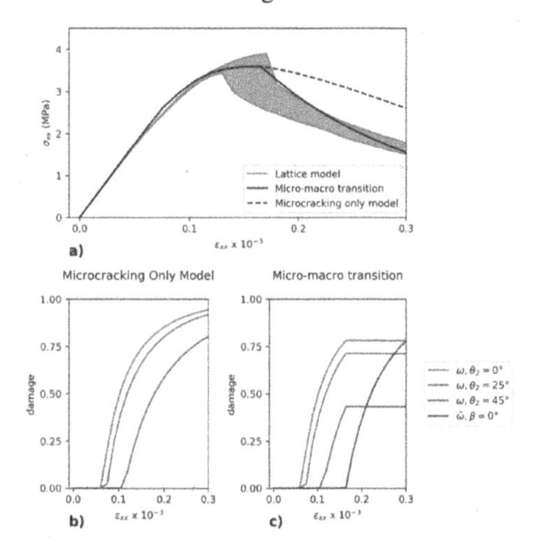

Figure 4. Uniaxial tension predictions. a) Stress-strain response. b) Damage evolution for the microcracking-only model. c) Damage evolution for the micro-macro transition model.

The proposed constitutive model captures well the characteristic behaviour of cementitious materials and associated cracking mechanisms, including the transition from diffuse microcracking to localized macrocracking.

5 CONCLUSIONS

A micromechanics based constitutive model for cementitious materials that addresses the transition from diffuse microcracking to localized macrocracking was presented. The good agreement between the proposed constitutive model and the lattice simulations demonstrated the potential of the constitutive model which captures well the characteristic mechanical behaviour of these materials and associated cracking mechanisms.

REFERENCES

Budiansky, B. & O'Connell, R.J. 1976. Elastic moduli of a cracked solid. *International Journal of Solids and Structures*, 12: 81–97.

Grassl, P. & Antonelli, A., 2019. 3D network modelling of fracture processes in fibre-reinforced geomaterials. *International Journal of Solids and Structures,* Volume 156–157, 234–242.

Grassl, P. & Bolander, J., 2016. Three-dimensional network model for coupling of fracture and mass transport in quasi-brittle geomaterial. *Materials* 9(9): 782–800.

Jenq, Y-S. & Shah, S.P. 1991. Features of mechanics of quasi-brittle crack propagation in concrete. *International Journal of Fracture,* 51:103–120.

Karihaloo, B.L. 1995. *Fracture Mechanics and Structural Concrete.* Essex, England: Longman Scientific & Technical.

Mihai, I.C. & Jefferson, A.D. 2011. A numerical model foe cementitious composite materials with an exterior point Eshelby microcrack initiation criterion. *International Journal of Solids and Structures*, 48: 3312–3325.

Mura T. 1987. *Micromechanics of Defects in Solids. Second, revised edition*, Martinus Nijoff Publishers, The Netherlands.

Nemat-Nasser, S. & Hori, M. 1993. *Micromechanics: Overall Properties of Heterogeneous Materials.* Amsterdam: North-Holland.

Shah, S.P, Swartc, S.E. & C,O. 1995. *Fracture Mechanics of Concrete.* New York: John Wiley & Sons.

Slate, F.O. & Hover, K.C. 1984. Microcracking in concrete. In *Fracture Mechanics of Concrete:Material characterization and testing.* Dordrecht: Springer, 137–159.

Vonk, R.A. 1992. Softening of concrete loaded in compression. *PhD Thesis.* Eindhoven University of Thechnology, The Netherlands.

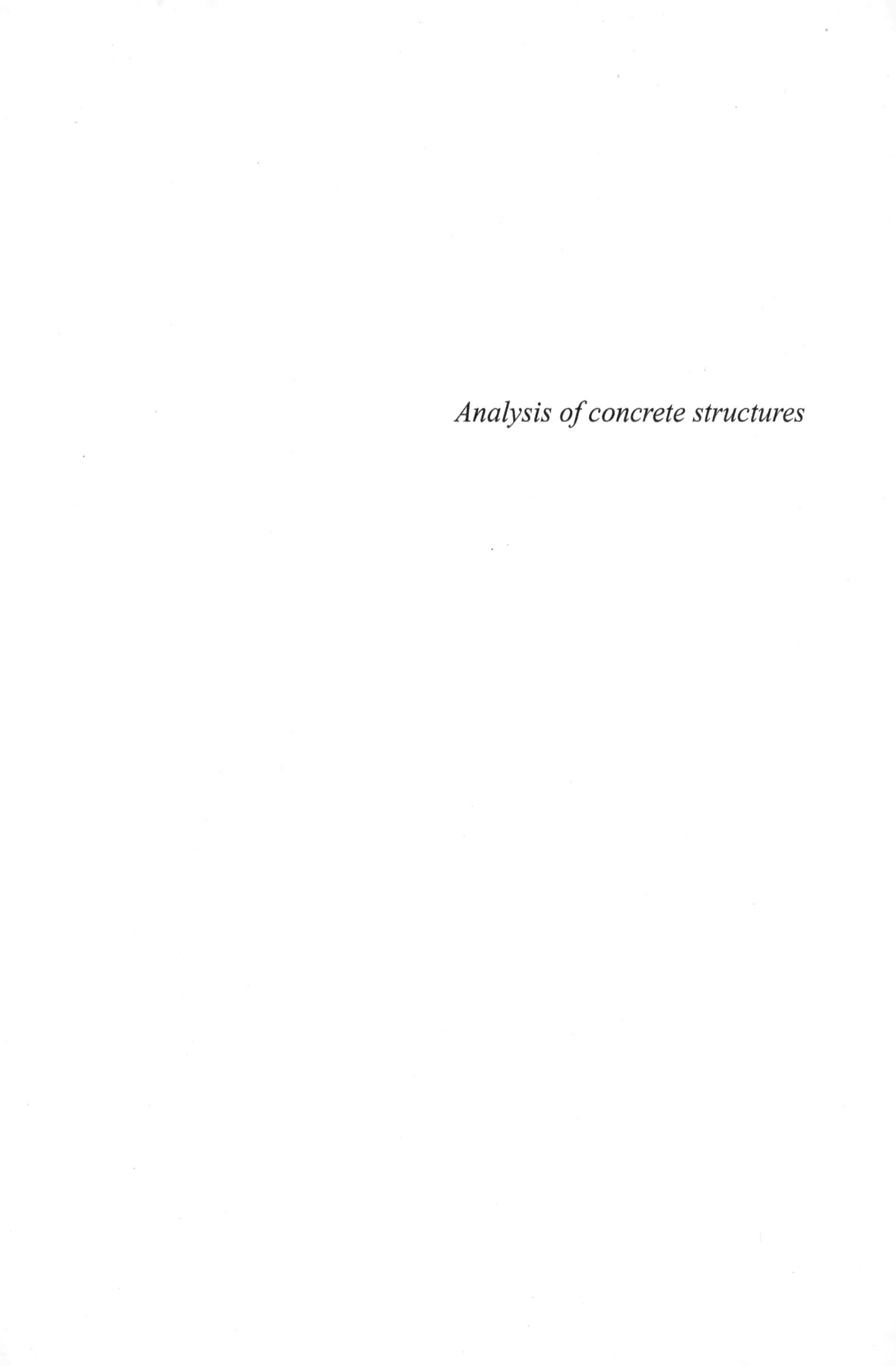

Analysis of concrete structures

Computational Modelling of Concrete and Concrete Structures – Meschke, Pichler & Rots (Eds)
© 2022 Copyright the Author(s), ISBN: 978-1-032-32724-2

Nonlinear FE analysis of fiber reinforced cementitious matrix strengthened RC columns

M. Kyaure & F. Abed
American University of Sharjah, Sharjah, UAE

ABSTRACT: Fiber reinforced cementitious matrix (FRCM) is a noncorrosive two-dimensional high strength fiber reinforced polymer (FRP) mesh saturated with inorganic cementitious mortar. This novel system is evolving as a viable option for retrofitting damaged RC structures. This system is fire resistant, easy to apply and eliminates the toxic hazard of epoxy bonded FRP. While past research investigated the potential of FRCM in shear and flexural applications, limited studies explored the confinement of short columns with different cross sections, particularly using finite element (FE) analysis. In this study, a three-dimensional (3D) nonlinear finite element (FE) model is developed using ABAQUS to investigate the performance of the retrofitting system on corrosion damaged RC columns. Poly-paraphenylene-ben-zobisoxazole (PBO) fibers are modeled in this study. Loading condition is displacement-controlled loading condition and material nonlinearities in concrete, cement mortar and composite are adapted in the FE model. The FE models are validated against experimental studies in published literature. A total of 40 columns are modeled and a parametric study is conducted considering the effects of cross section type (square vs circular), number of FRCM layers (1, 2, 3 and pre-damage severity (mild, moderate and severe). Retrofitting corrosion damaged RC columns with PBO-FRCM effectively resorted and enhanced the original axial capacity and ductility at all damage levels irrespective of cross section shape. Enhancement in axial capacity of 20% was observed in square and 35% in circular columns while axial ductility enhancement of 42% was observed in square and 164% in circular columns. Results also indicated a positive correlation between number of FRCM Layers and axial capacity and ductility enhancement. The performance enhancement is more pronounced in the circular columns. All strengthened specimens failed by matrix damage, indicating effectiveness of the strengthening system. Comparison of column axial capacity computed using ACI 549.4R-13 provisions against FEA revealed that the code provisions underestimate the axial capacity of square and circular short RC columns retrofitted with PBO-FRCM by an average of 20%.

1 INTRODUCTION

1.1 *Background*

Reinforced concrete is the most common choice of construction material in the gulf region due to numerous advantages such as durability, versatility and economic advantages. Concrete structures however suffer from adverse deterioration over their service life. The high temperature, humidity, and chloride content create the perfect condition for corrosion in RC structures. Coastal structures are at higher risk of extensive chloride attack causing reinforcement corrosion, severe cracking and concrete spalling at the youth age of 10 years [1]. The deterioration of RC structures can lead to catastrophic consequences such as the recent collapse of the surfside condominium in Florida. Strengthening of damaged structures is therefore a necessity to avoid disasters during the lifetime of structure.

The cost of infrastructure rehabilitation using traditional approaches such as jacketing, bonded steel plates and load path redistribution using additional elements is typically high and messy. The need for a simpler yet effective solution is imminent. The most practiced approach of strengthening RC columns, the focus of this study is using externally bonded fiber reinforced polymers (FRP) to provide additional confinement. These fibers are corrosion resistant, durable and have a high strength to weight ratio [2].

There is precedence for the potential of externally bonded FRP in strengthening structures however limited research investigates post repair performance. Additionally, epoxy bonded FRP include loss off mechanical, chemical, and bond properties [3] at elevated temperatures leading to delamination [4]. Additional drawbacks to epoxy bonded FRP include the toxic nature of the epoxy and poor compatibility with concrete substrates. These combined drawbacks raise a question on the feasibility of externally bonded FRP in the UAE and regions with high temperatures.

DOI 10.1201/9781003316404-27

Fiber Reinforced Cementitious Matrix (FRCM) reduce the drawbacks of epoxy bonded FRP. FRCM consist of FRP impregnated in cementitious mortar which not only bonds the fibers to the concrete member but also protects the fibers from external environmental factors and is also fire rated. Poly-paraphenylene-ben-zobisoxazole (PBO) are investigated in this paper. This novel system is lightweight, and easy of application in addition to the fire resistance capabilities of the cementitious mortar layers through shielding the embedded fibers and minimizing its vulnerability hazard makes it extremely appealing over epoxy bonded FRP system. The compatibility between the cementitious mortar and the concrete substrate is inherent as both materials have a common cement "base", adding to the various mentioned advantages of FRCM systems. FRCM systems, with their innovative features, ensure the endurance of the rehabilitation process and consequently the sustainability of the repaired structure.

Numerous studies proved the potential of FRCM for structural strengthening. These studies investigated the performance of the novel system in flexural and axial strengthening. Three studies by El Ghazy et al. [5], [6] and [7] investigated different parameters on flexural strengthening such as the effects of FRCM types and bonding schemes on corroded RC beams. The authors concluded that the strengthening system was able to restore original capacity of the corrosion-damaged beams with the level of enhancement in the strength depended on the amount and type of FRCM irrespective of damage severity with enhancement in both ultimate and yield strength as the number of FRCM plies increased. The observed failure modes are slippage and delamination of the fabric within the cementitious mortar. Axial strengthening studies conducted by Colajanni et al. [8], Parretti et al. [9] explored the performance of circular and square column geometries through varying FRCM layers and orientation. The authors reported axial strength enhancement of around 23% with 0.236% confinement ratio for circular columns compared to 0.175% for square columns. A ductile behavior was also observed. The authors also suggested that wrapping the columns with the fibers in the direction of the ties proved optimal. A study by Tello et al. [10] studied the effect of different number of PBO-FRCM layers on square and circular columns. Results revealed higher axial capacity enhancement up to 36% with confinement more pronounced in circular columns than their square counterparts. Ductility enhancement was also observed in the strengthened columns over the control counterparts with activation of the confinement system delayed by the presence of the cementitious mortar. Obaidat [11] developed a nonlinear finite element framework to investigate CFRP on concrete beams using material models from literature and contrasted the results to experimental studies. The author concluded that FEM is capable of predicting the performance and behavior of strengthened beams.

1.2 *Objectives*

The main objective of this study is to develop FE models using material models to investigate the behavior of circular and square corrosion damaged RC columns strengthened with 1, 2, 3 and 4 PBO-FRCM. The potential of PBO-FRCM in restoring the original behavior and axial capacity of the pre-damaged RC columns is investigated and the performance of the system on square and circular geometries is contrasted. Recommendations on modeling FRCM strengthened columns are provided to overcome the drawbacks of epoxy based FRP as a structurally effective system.

2 MATERIALS AND METHODS

Finite element analysis of FRCM strengthened RC columns is a highly nonlinear analysis numerically due to the interaction between the different elements and the plastic behavior of concrete. Simplified numerical material models based on established laws and mechanics are selected and incorporated in the FE model.

2.1 *Concrete*

Concrete compressive behavior is initially linearly elastic proportional to the elastic modulus followed by the onset of micro cracking introducing nonlinear plastic behavior until the ultimate compressive strength is reached. The curve is completed with a descending branch with increasing strain. Concrete tensile behavior is initially linear up to failure stress (or cracking strain) followed by softening with induces a brittle failure.

In this study, concrete damaged plasticity (CDP) model is adapted to model the plastic damage behavior of concrete. This model assumes tensile cracking and compressive crushing as the two main failure mechanisms [11]. Concrete compressive strength is 30MPa and Poisson's ratio is 0.2. Plastic damage parameters area defined as follows: dilation angle is 30 degrees, eccentricity is. 0.1, fb0/fc0 is 1.16, k is 0.667 and viscosity is 0.001 as recommended by the Abaqus manual [12].

2.2 *Steel*

Steel typical behavior is initially linearly elastic up to the yield stress followed by nonlinear branch up to the ultimate tensile stress. Generally, an elastic-plastic behavior (with or without strain hardening) is an adequate representation of steel behavior in finite element modelling. The steel reinforcement in this study is considered to have elastic-perfectly plastic behavior in both tension and compression with Poisson's ratio of 0.3, elastic modulus 200 GPa and yield strength 500 MPa.

2.3 *FRCM composite*

FRCM in this study is composed of poly-paraphenylene-ben-zobisoxazole (PBO) FRP fibers embedded in cementitious mortar. The mortar is compatible with concrete substrates and provides near perfect bond while distributing the load between the fibers while the fibers provide stiffness and load bearing capacity. PBO fibers have a linear elastic behavior until failure. Elastic modulus is 270 GPa and tensile strength is 5800 MPa [10]. The cementitious mortar used has a compressive strength of 30MPa and is assumed to crack under slight loading due to its negligible thickness and therefore provides no additional axial capacity to the columns.

3 FINITE ELEMENT MODEL

3.1 *Element types & constraints*

Three-dimensional FE models are developed on commercial software package ABAQUS [12] to simulate the behavior of FRCM strengthened columns. Nonlinear damage initiation and propagation is monitored to observe the failure mode therefore symmetry is not assumed. Standard 2-node 3D wire truss elements **T3D2** are used to model the embedded reinforcing steel. Three-dimensional hexagonal 8-node linear brick stress elements with reduced integration **C3D8R** are used to model concrete. The FRCM is split into two components. Standard four-node extruded thin shell element with reduced integration **S4R** is selected to model the PBO fibers with and assembled using composite layup of conventional shell with each ply thickness of 0.5mm. A total of 4 layers are modeled with fiber orientation along the direction of the ties. Standard 8-node 3D cohesive element **COH3D8** is used to model the cementitious mortar. Embedded region is used to constraint the steel reinforcement in the concrete. Tangential contact with a penalty formulation and friction coefficient 0.1 is defined for the region between the column and FRCM. The bond between reinforcing steel and concrete is considered to be perfect.

3.2 *Boundary & loading conditions*

Displacements and rotations in the three axes are restrained using the ENCASTRE option at the fixed end of the column. The free end of the column is loaded using a displacement-controlled boundary condition restraining all displacements and rotations except U3. Both boundary conditions are assigned to a reference point on a rigid steel plate to ensure even deformation and stress distribution at the ends of the column. Smooth step linearly increasing static displacement is assigned as the loading condition with a time period of 1 second and 0.001 step increment. The step increment is selected to ensure a smooth step displacement-controlled analysis.

3.3 *Mesh configuration*

A mesh sensitivity conducted different mesh sizes revealed that 15mm to 25mm mesh is optimal therefore 20mm mesh is used which ensured results obtained are within 5% deviation and computational time was between 3 to 4 hours. NLGEOM option is activated to account for large deformations in the nonlinear analysis.

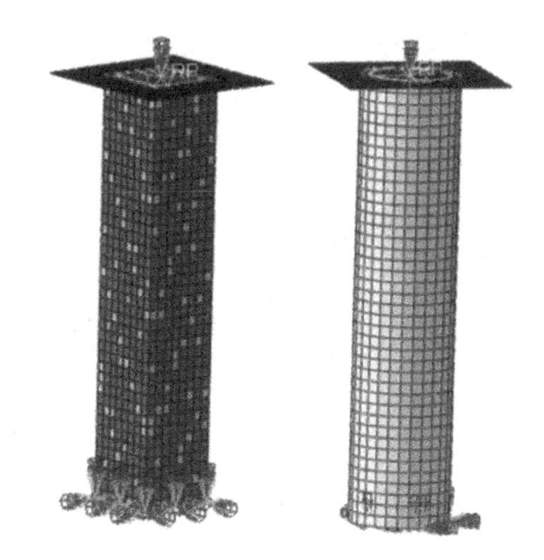

Figure 1. Complete model assembly.

3.4 *Model verification*

Results obtained from the FE model are validated against published literature by Tello et al. [10]. This study presented the behavior of circular and square RC columns strengthened with 1, 2, and 4 layers of PBO-FRCM shown in Figure 2. Results revealed that FE model is capable of predicting the axial capacity to within 4% of the published experimental data. The axial capacity obtained from the FE model is slightly higher than the benchmark experimental results due to imperfections that occur during specimen preparation and testing in the lab. Table 1 presents the comparison.

Table 1. Model verification (FEM vs Tello [10]).

Column ID	Pn FE (kN)	Pn [10] (kN)	% Diff
S0-0-0-SH	740	722	2
S0-0-1-SH	787	759	4
S0-0-2-SH	872	821	6
S0-0-4-SH	896	847	5
C0-0-0-SH	702	687	2
C0-0-2-SH	891	845	5
C0-0-4-SH	964	935	3
		Avg	4

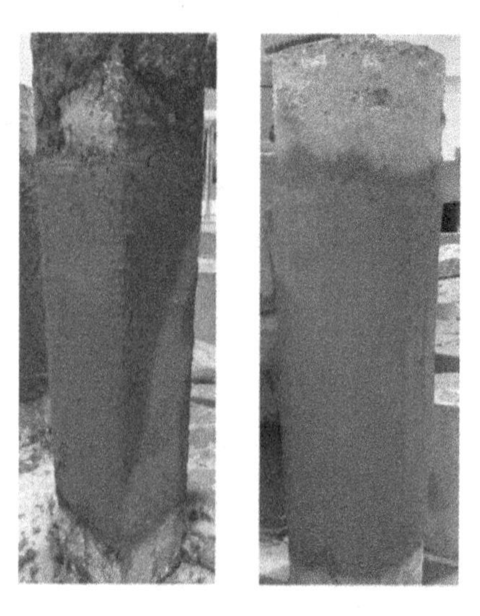

Figure 2. Experimental specimens used in model verification Tello et al. [10].

Furthermore, the axial capacity obtained from the FE models is compared to the recommendations published in ACI 549.4R-13 [13]. The recommendations however are mostly conservative as there are a lot of safety factors embedded in the equations to ensure safe design.

3.5 *Axial ductility*

In this study, axial ductility (η – axial), is determined by recording the ratio of the axial deformation values at 85% of the axial capacity. The ratio of the deformation post the peak to that prior the peak is a dimensionless value given by equation 1 as defined by Kyaure et al. [14] and illustrated in Figure 3.

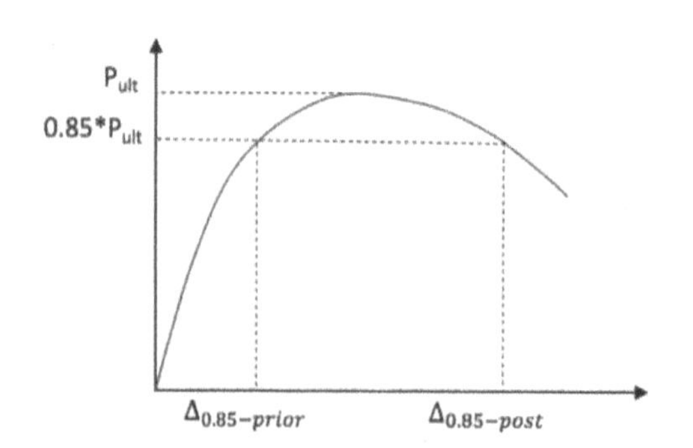

Figure 3. Axial ductility determination (Kyaure et al. [14]).

$$\eta - -\text{axial} = \frac{\Delta_{0.85--post}}{\Delta_{0.85--prior}} \tag{1}$$

4 PARAMETRIC PROGRAM

A total of 40 FE models are developed in this study, 20 square and 20 circular columns. The square columns are identified with a prefix S while the circular columns are designated with a prefix C. Each set of 20 columns contains 4 subgroups of 5 columns which are assigned corrosion pre-damage in form of percentage loss in steel yield stress. The pre-damaged levels are 0%, 30%, 50% and 70% for undamaged, mild, moderate, and severely damaged columns, respectively. The 5 columns are strengthened with 0, 1, 2, 3 and 4 layers of PBO-FRCM.

Table 2. Parametric program.

FRCM Ply	0L, 1L, 2L, 3L, 4L
Corrosion Damage (% loss in yield stress)	0%, 30%, 50%, 70%
Cross Section	Square, Circular

5 RESULTS

Figure 4 below presents a summary of the axial capacity obtained for the 40 specimens. From the chart it is visible that the circular columns demonstrated higher axial capacity enhancement than their square equivalents.

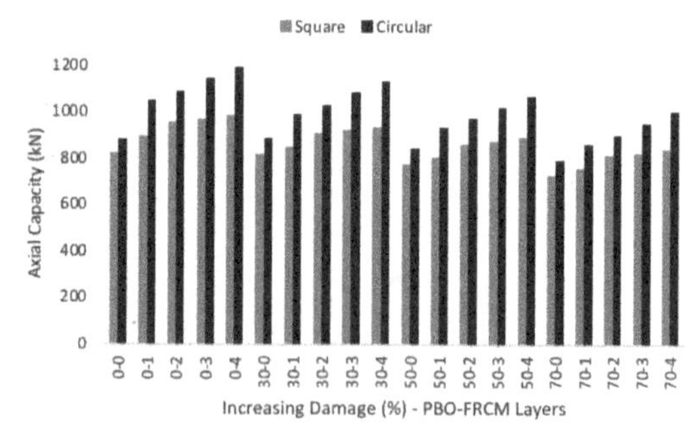

Figure 4. Axial capacity comparison between square and circular columns.

5.1 *Performance enhancement*

Figures 5–8 represent the load versus axial deformation plots of all the specimens. The PBO-FRCM system was successful in restoring the original axial capacity of the corrosion damaged columns. Significant enhancement in the axial capacity is observed with increase in number of PBO-FRCM layers. Mildly damaged specimens required 1 achieve the original capacity of the undamaged column in both square and circular group with axial capacity enhancement of 5%

and 6% respectively. In order to achieve maximum ductility enhancement, 4 layers are required for square to reach enhancement of 34% while only 2 layers are required for maximum ductility enhancement of 154% for circular columns. Similar behavior is observed for the moderately and severely damaged specimens.

In all cases, the axial capacity and ductility enhancement in circular columns is significantly higher their square counterparts indicating better performance for circular cross-sections. This is because the circular columns distribute the stresses evenly to the strengthening system as opposed to square columns which has stress concentrations in the corners. It is recommended to round the edges of square columns before strengthening to reduce the stress concentrations at the corners and improve the distribution of stresses from the column the strengthening system. Axial capacity and ductility of 13% and 35%, respectively was observed based on the ability of the cross-section to be effectively confined with circular columns outperforming their square counterparts. This finding is consistent with published literature [10].

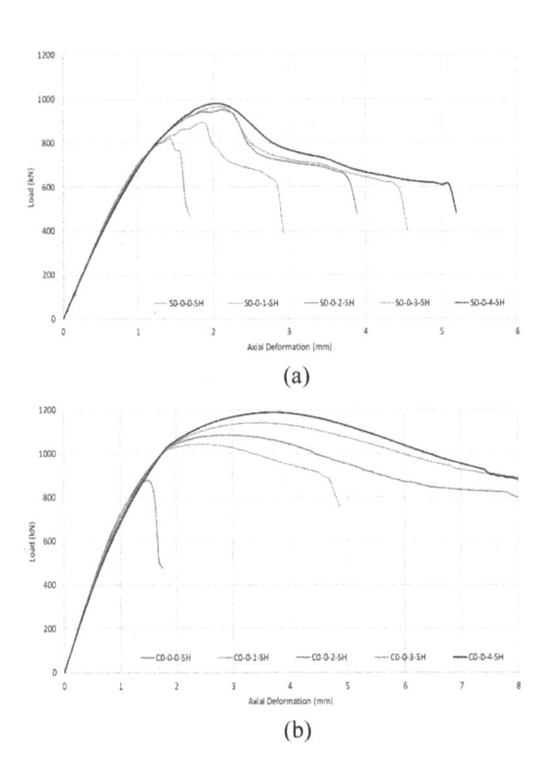

Figure 5. Axial load vs deformation (0% damage) (a) square (b) circular.

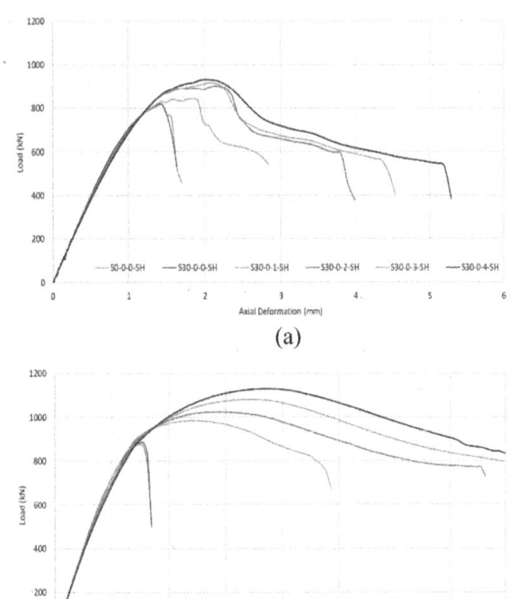

Figure 6. Axial load vs deformation (30% damage) (a) square (b) circular.

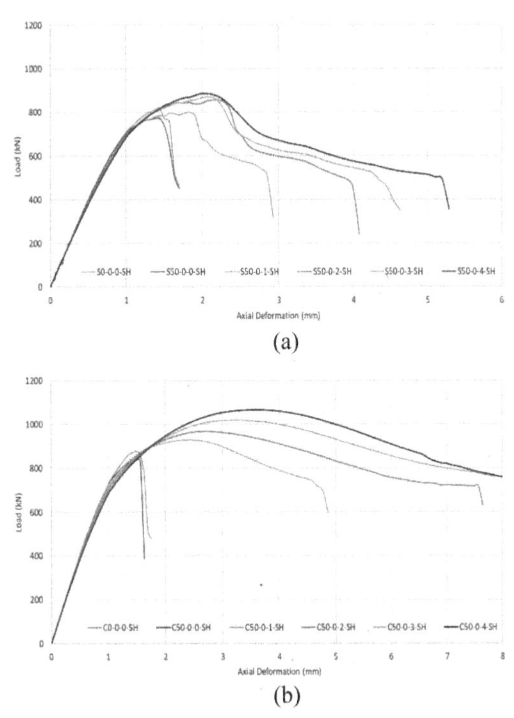

Figure 7. Axial load vs deformation (50% damage) (a) square (b) circular.

5.2 *Failure modes*

The primary failure modes observed are concrete crushing in the control specimens and matrix damage

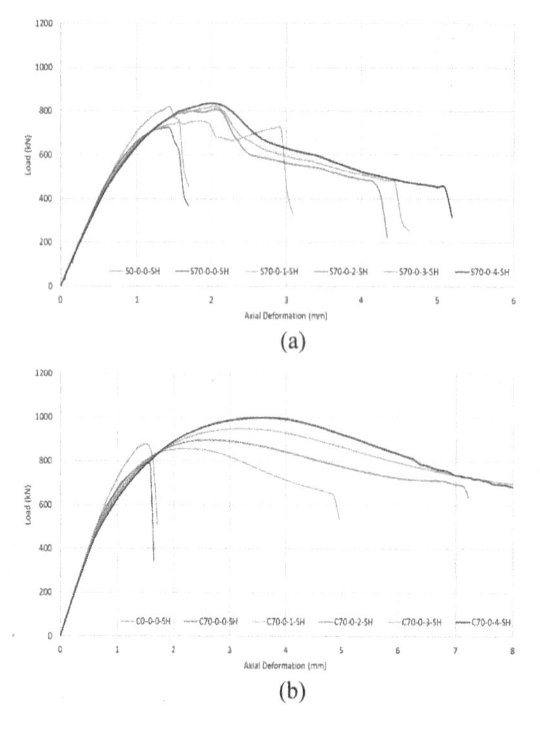

Figure 8. Axial load vs deformation (70% damage) (a) square (b) circular.

in the strengthened columns. Yielding of the steel reinforcement is observed and monitored using the strain levels in the steel. The confinement pressure from the strengthening system delayed the crushing of concrete and as a result the failure mode induced is matrix damage. The circular columns also exhibited a ductile failure mode relative to their square counterparts, visible in the axial load vs deformation plots. Complete damage of the cementitious mortar was achieved during loading proving that the cementitious mortar only contributed to the bond and not the column axial capacity as intended.

5.3 *Code comparison*

Comparison of the results obtained from FE against ACI 549.4R-13 [13] recommended equations for computing axial capacity revealed that the code provisions are over conservative. This is related with the multiple safety factors such as penalizing the axial capacity contribution of the FRP, the environmental safety and the approximation of the confinement contribution incorporated in the code equations. Percentage difference in the prediction of axial capacity of 19% and 22% was observed for square and circular columns respectively. The charts below present a summary of the differences.

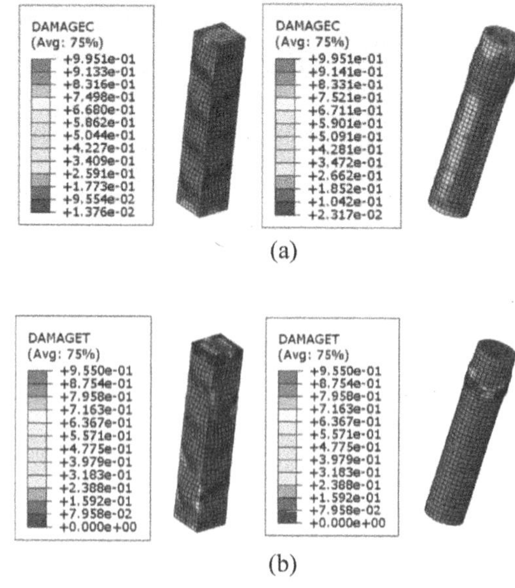

Figure 9. Concrete damage (a) compressive (b) tensile.

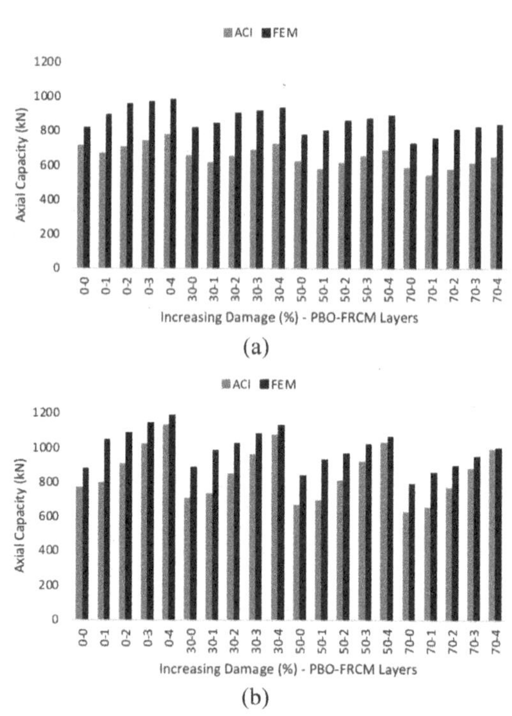

Figure 10. Axial capacity comparison between ACI and FEM (a) square (b) circular.

6 CONCLUSIONS & RECOMMENDATIONS

The potential of strengthening corrosion damage RC columns using PBO-FRCM is investigated and the conclusions and recommendations are presented in this section.

1. PBO-FRCM is capable of restoring and enhancing the axial capacity and ductility of corrosion damaged RC columns.
2. Circular columns outperformed square counterparts than in the square columns. It is recommended to round off the corners of square columns to improve stress distribution in the PBO-FRCM.
3. All columns exhibited similar behaviour irrespective of damage level and number of FRCM layers. This is visible from the load–deformation plots.
4. A positive relationship is established between performance enhancement in axial capacity increasing number of PBO-FRCM layers.
5. A sudden failure mode was observed in the square columns due to stress concentration in the corners of the PBO-FRCM. Primary failure mode observed is concrete crushing in control columns and matrix damage in strengthened columns for both square. And circular cross-sections.
6. The ACI 549.4R-13 provisions are conservative in predicting the actual capacity of PBO-FRCM strengthened RC columns by about 20%. There is room to relax some of the safety factors with. Further research.

REFERENCES

[1] H. M. Shalaby and O. K. Daoud, "Case studies of deterioration of coastal concrete structures in two oil refineries in the Arabian Gulf region," *Cem. Concr. Res.*, vol. 20, no. 6, pp. 975–985, Nov. 1990.

[2] R. Al-Hammoud, K. Soudki, and T. H. Topper, "Fatigue flexural behavior of corroded reinforced concrete beams repaired with CFRP sheets," *J. Compos. Constr.*, vol. 15, no. 1, pp. 42–51, 2011.

[3] J. C. P. H. Gamage, R. Al-Mahaidi, and M. B. Wong, "Bond characteristics of CFRP plated concrete members under elevated temperatures," *Compos. Struct.*, vol. 75, no. 1–4, pp. 199–205, Sep. 2006.

[4] L. A. Bisby, M. F. Green, and V. K. R. Kodur, "Response to fire of concrete structures that incorporate FRP," *Prog. Struct. Eng. Mater.*, vol. 7, no. 3, pp. 136–149, 2005.

[5] M. Elghazy, A. El Refai, U. Ebead, and A. Nanni, "Experimental results and modelling of corrosion-damaged concrete beams strengthened with externally-bonded composites," *Eng. Struct.*, vol. 172, no. June, pp. 172–186, 2018.

[6] M. Elghazy, A. El Refai, U. Ebead, and A. Nanni, "Corrosion-Damaged RC Beams Repaired with Fabric-Reinforced Cementitious Matrix," *J. Compos. Constr.*, vol. 22, no. 5, pp. 1–13, 2018.

[7] M. Elghazy, A. El Refai, U. A. Ebead, and A. Nanni, "Performance of corrosion-aged Reinforced Concrete (RC) beams rehabilitated with Fabric-Reinforced Cementitious Matrix (FRCM)," *Sustain. Constr. Mater. Technol.*, vol. 2016-Augus, pp. 1–9, 2016.

[8] P. Colajanni, F. De Domenico, A. Recupero, and N. Spinella, "Concrete columns confined with fibre reinforced cementitious mortars: Experimentation and modelling," *Constr. Build. Mater.*, vol. 52, pp. 375–384, 2014.

[9] R. Parretti and A. Nanni, "Axial testing of concrete columns confined with carbon FRP: effect of fiber orientation," *Proc. ICCI 2002*, pp. 1–10, 2002.

[10] N. Tello, Y. Alhoubi, F. Abed, A. El Refai, and T. El-Maaddawy, "Circular and square columns strengthened with FRCM under concentric load," *Compos. Struct.*, vol. 255, no. September 2020, p. 113000, 2021.

[11] Y. T. Obaidat, "Structural Retrofitting of Concrete Beams Using FRP -Debonding Issues.," *Thesis (Doctoral in Structural Mechanics) – Departmet of Construction Sciences, Lund University.* Thesis (Doctoral in Structural Mechanics) – Departmet of Construction Sciences, Lund University, p. 185, 2011.

[12] K. & S. I. Pawtucket (America): Hibbitt, "ABAQUS standard user's manual. Version 6.11," vol. IV, 2014.

[13] ACI Committee 549.4R-13, *Guide to Design and Construction of Externally Bonded Fabric-Reinforced Cementitous Matrix (FRCM) Systems for Repair and Strengthening Concrete Structures (ACI 549.4R-13)*, 2013th ed. American Concrete Institute, 2013.

[14] M. Kyaure and F. Abed, "Finite element parametric analysis of RC columns strengthened with FRCM," *Compos. Struct.*, vol. 275, no. July, p. 114498, 2021.

Computational Modelling of Concrete and Concrete Structures – Meschke, Pichler & Rots (Eds)

Interaction diagram for columns with multispiral reinforcement: Experimental data vs. blind prediction using CDPM2

P. Havlásek & Z. Bittnar
Department of Mechanics, Faculty of Civil Engineering, Czech Technical University in Prague, Prague, Czech Republic

B. Li , J.V. Lau & Y.-C. Ou
Department of Civil Engineering, National Taiwan University, Taipei, Taiwan

ABSTRACT: The structural performance of concrete columns with multispiral reinforcement (MSR) developed in Taiwan is significantly superior to conventionally reinforced columns. The increased strength and ductility stem from the passive confinement produced by the partially overlapping spirals which cover almost the entire cross-section. Because of the complex structural behavior and insufficient experimental data, which are limited to monotonic axial compression or cyclic lateral drift with increasing magnitude under constant compression, and due to the novelty of this reinforcement layout, the MSR is not recognized in the design codes. To utilize its full potential, the carrying capacity can be determined computationally for arbitrary loading history via the nonlinear finite element method and the design strength can be obtained by employing the global safety factor approach. The interaction diagram (ID) is a strength envelope surrounding all admissible states of the internal forces and is perfect for assessing the safety and efficiency of the structural design of columns. The ID can be constructed by processing repeatedly run simulations with different loading combinations. This approach is computationally demanding, but the analyses can be defined automatically, run in parallel, and the results for different combinations of material properties and reinforcement layout can be precomputed and stored in a database.

One of the objectives of the current bilateral Czech-Taiwanese project is to develop this approach for the columns with MSR. In summer 2021, within the scope of the present project, the Taiwanese laboratories MOST tested 5 geometrically identical specimens with MSR subject to compression with different values of eccentricity. The aim of this conference contribution is to compare the global behavior of these specimens expressed in the M-N diagram with the blind prediction using FEM and to construct the corresponding ID. In the simulations, concrete is described with CDPM2, the second generation of the well-known Concrete Damage Plastic Model originally proposed by Grassl and Jirásek.

1 INTRODUCTION

The structural performance of concrete columns with multispiral reinforcement (MSR) developed in Taiwan (Yin 2005) is significantly superior to conventionally reinforced columns. The increased strength and ductility (Yin, Wang, & Wang 2012; Yin, Wu, Liu, Sheikh, & Wang 2011) stem from the passive confinement produced by the partially overlapping spirals which cover almost the entire cross-section. This structurally efficient design is also very economical. Highly automated production lines incorporated in the manufacturing process decrease human labor and thus cost. However, more importantly, the MSR reduces the demand on the raw materials; the performance of conventionally reinforced columns can be reached with less reinforcement or possibly with a lower concrete grade.

The most promising configuration of MSR for square columns is the 5-spiral layout (5S4, Figure 1). This setup uses a large spiral in the center and small spirals at the corners. In contrast to the columns with rectilinear transverse reinforcement the 5S4 MSR design offers a superb resistance to the combination of compression with bending and shear, which is salient for seismic active areas. Moreover, on contrary to ordinary circular columns with a single spiral, which also partially benefit from the enhanced behavior of confined concrete, the multispiral concept offers shape variability of the cross-section.

The topology of the transverse reinforcement has a strong influence on both the strength and post-peak response. The topology is fully described by five parameters: pitch of the spirals, H (Figure 2), small spiral outer diameter d_S and the diameter of the large and small spiral rebars, D_L and D_S. The outer diameter

DOI 10.1201/9781003316404-28

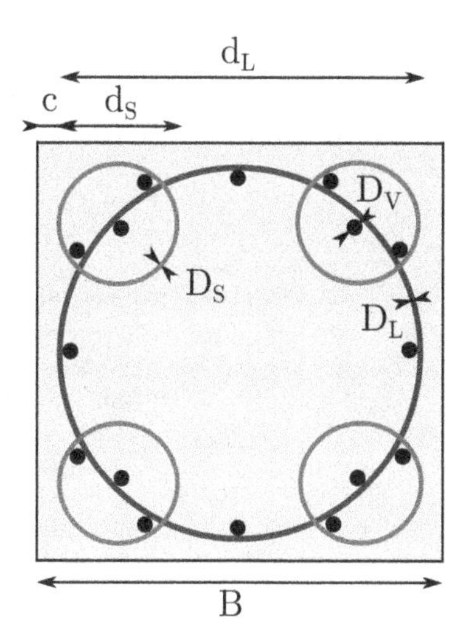

Figure 1. Cross-section in the central part of the specimen.

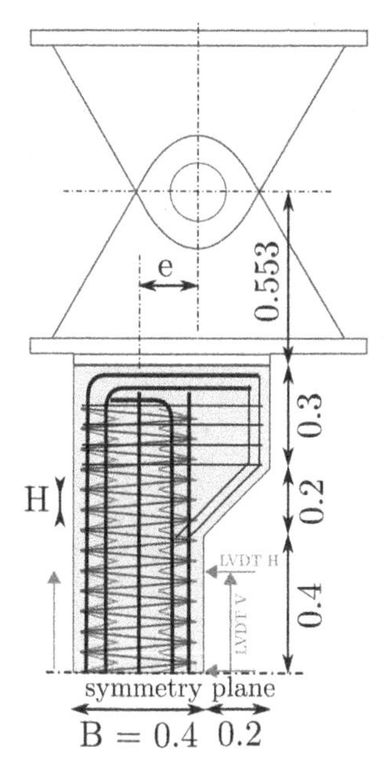

Figure 2. Symmetric half of the column loaded in eccentric compression.

of the large spiral d_L and the position of the small spirals in the plan view are determined by the thickness of the concrete cover, c.

Up till now, no closed-form expressions have been developed to predict the carrying capacity and ductility for an arbitrary combination of these parameters. Therefore, it is not surprising that the current codes for the design of concrete structures do not allow to utilize the full potential of this novel type of reinforcement layout. The updated version of Eurocode 2 (European Committee for Standardization 2018), which stems from the *fib* Model Code 2010 (Fédération Internationale du Béton 2012), recognizes confined strength only in axially compressed structural members with the most typical and conventional reinforcement layouts. Furthermore, the American standard ACI 318 (ACI 2019) defines only criteria on transverse reinforcement but does not permit to further benefit from the confined strength. To overcome these obstacles, the carrying capacity can be determined computationally for an arbitrary loading history via nonlinear finite element method (NLFEM) and the design strength can be obtained by employing the global safety factor approach. Using such an approach, two recent computational studies (Havlásek, Lepš, & Bittnar 2021; Lepš & Havlásek 2021) have demonstrated that the best performance under axial compression is reached when the reinforcement ratios of the small and large spirals are similar.

The interaction diagram (ID) is an envelope surrounding all statically admissible combinations of internal forces and is perfectly suitable for the assessment of safety and efficiency of the structural design of columns. The ID can be computed with different approaches. For simple cross-sections and code-like assumptions on the distribution of stress in concrete, the ID can be determined even from hand calculation. An automated computer routine is necessary for more advanced stress-strain relationships for concrete, such as the Mander's model (Mander, Priestley, & Park 1988) whose response depends on the magnitude of lateral confinement (Ngo & Ou 2021). In such a routine, the cross-section is spatially discretized and is repeatedly subjected to different combinations of prescribed axial strain and curvature. The resulting couples of normal force and bending moment are obtained by integrating the stress response over the cross-section, and the interaction diagram is subsequently constructed as a convex envelope.

A similar numerical technique is presented in this contribution to evaluate the ID of a column with a square cross-section and 5S4 MSR layout. In contrast to the previously outlined procedure, the computational model is loaded by an eccentric force and the resulting couple [M,N] of the ID is obtained as the maximum loading force and the corresponding moment. This is repeated for different values of eccentricity.

2 EXPERIMENTAL PROGRAM

The experimental part examined the behavior of five geometrically identical columns with a square cross-section and MSR subject to compression with different eccentricity and one axially loaded companion specimen with conventional rectilinear reinforcement and a similar reinforcement ratio. The tests were done at NCREE, Taipei, Taiwan on Multi-Axial Testing System (MATS) with the maximum compression capacity of 58.84 MN. The loading was driven by a displacement control and was applied in several subsequent steps with partial unloading to document the evolution of a crack pattern. The initial idea was to utilize the unique capabilities of MATS which can independently prescribe loading for 6 degrees of freedom and to conduct the experimental study on prismatic short columns subject to a combination of vertical displacement and rotation of the bases. However, several technical obstacles appeared and it was decided that a simple eccentric compression is a more robust approach to measure the response to the combination of bending with compression.

The final design of the experimental specimens, their cross-section, and reinforcement is a result of an iterative process; the number of specimens and their dimensions were strictly constrained by the project budget. The typical cross-section of MSR columns 0.6×0.6 m had to be scaled down to 0.4×0.4 m to double the number of specimens from 3 to 6. Yet, it was not possible, in a similar fashion, to rescale the reinforcement. Already the typical design uses the smallest rebars available in Taiwan. Preserving the cross-section diameter of a rebar means significantly higher confining stress, which can be alleviated by increasing the spiral pitch. However, this in turn rapidly decreases the confinement effectiveness and makes the stress distribution less uniform. The resulting reinforcement design presents an acceptable compromise.

A simplified geometry of the specimens is outlined in Figures 1 and 2. The investigated middle prismatic part has a square cross-section with $B = 0.4$ m and is 0.8 m long. Towards the base, the cross-section is widened to 0.6 m to prevent concrete crushing due to eccentric loading. The out-of-plane thickness is constant and equal to 0.4 m. The cross-section is reinforced with 16 longitudinal rebars #5 ($D_V = 16$ mm) spaced approximately uniformly about the circumference of the large spiral #4 ($D_L = 13$ mm) shown in blue color. The outer diameter $d_L = 360$ mm is defined by the concrete cover $c = 20$ mm. The small spirals at the corners drawn in red have outer diameter $d_S = 120$ mm and rebar cross-section #3 ($D_S = 10$ mm). The pitch of all spirals is identical and equal to $H = 60$ mm. In the vertical direction, the reinforcement ratio is 2.0% while the lateral reinforcement ratio (wrt. entire cross-section) is 2.51%. The previous five-spiral designs usually arranged 4 longitudinal rebars in every small spiral to facilitate the construction. The specimens design for this experiment, however, only have 3 longitudinal rebars located in every small spiral to reach a reasonable longitudinal reinforcement ratio.

The magnitude of eccentricity e (Figure 2) was controlled by accurately positioning the specimen between the massive steel hinges. The concrete specimen was connected to a 50 mm thick steel plate by shear studs, afterwards this plate was welded to the hinge. The eccentricity varied from $e = 0$ mm (axial compression) to $e = 200$ mm (resultant at the edge of the cross-section) with 50 mm step. The investigated loading paths should cover the most interesting part of the interaction diagram typical for columns and prove the non-negligible increase in strength under eccentric loading due to the lateral confinement.

The specimens were cast in horizontal position from concrete with cement content 430 kg/m^3, water-to-cement ratio $w/c = 0.44$, and weight proportions among cement, fine and coarse aggregates $c : a_f : a_c = 1 : 1.82 : 2.13$. The specimen is reinforced with steel class SD420W. In the supplementary experiments, the yielding occurs on average around 480 MPa and the peak stress ≈ 700 MPa is reached at approx. 10%.

The compressive strength of concrete was checked on 3 sets of 3 cylinders 120×240 mm at the age of 7, 28, and 100 days. The corresponding values and standard deviations were 31.6 (1.64), 43.0 (4.04), and 44.8 MPa (1.13), all in MPa. The last test was done under displacement control and with externally mounted extensometers to determine the Young modulus.

3 COMPUTATIONAL MODELING

3.1 FEM models

Two different finite element models with different levels of complexity and computational demands were developed to investigate the behavior of columns subjected to eccentric compression and to construct the interaction diagram of its typical cross-section.

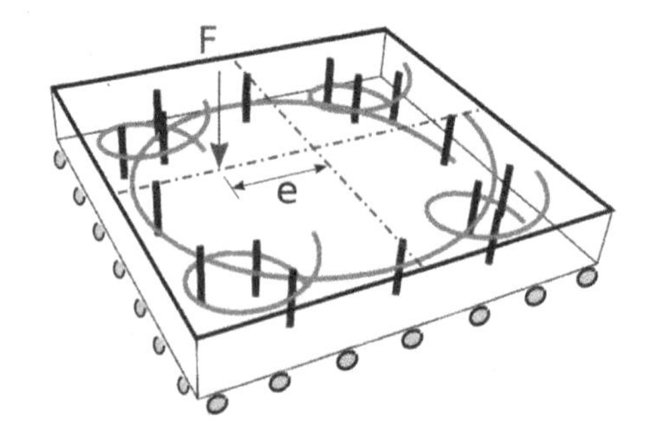

Figure 3. Computational model of a representative section.

The first and computationally more efficient approach uses a model which corresponds to a *representative section* of the central square part of the column with a height equal to the spiral pitch H, see Figure 3. Provided that the failure mode is not localised, this approach (Havlásek, Jirásek, & Bittnar 2019) enables to introduce a significantly denser FE mesh using which the differences among various reinforcement alternatives can be thoroughly identified. The mesh of the finite element model, generated by a preprocessor Malcolm (Havlásek 2019), combines a structured mesh (linear hexahedral elements, reference model $32 \times 32 \times 8$ elements) with a regularly and irregularly discretized longitudinal and spiral reinforcement (truss elements). The two meshes are interconnected using the concept of hanging nodes, and the bond between concrete and steel is treated as rigid. The disabled slip is justified by the assumption that the tensile force in the steel spirals should be almost uniform over the length.

The model uses a generalized master-slave condition to impose periodicity in the axial direction of the column. In the horizontal direction, the nodal displacements on the top horizontal surface are fully linked to the corresponding degrees of freedom on the bottom surface, otherwise the lateral deformation is not restrained as depicted in Figure 3 by the rollers. In the vertical direction, the displacement on the bottom surface is fixed, while on the top surface it obeys a kinematic condition which allows the vertical displacement of the entire surface and its rotation about both horizontal axes.

There are two different approaches to subject the first computational model to the combination of normal force and bending moment. The loading can be defined by means of an eccentric force, in such a case the ratio between the bending moment and normal force remains constant; therefore, in the M-N graph the loading path is a straight line which reaches the interaction diagram at the maximum value of the loading force. In the other approach, the loading can be defined by a fixed ratio of axial deformation and curvature. Then the response becomes highly nonlinear as the loading path approaches the interaction diagram, which might not be reached at the maximum load. This implies that in the latter approach the interaction diagram needs to be extracted as a convex shape of all computed responses which is not favorable.

For this reason, in the present study, the model is loaded by a single force F with eccentricity e. Hereafter, this position is defined by a normalized eccentricity $\hat{e} = e/(B/2)$ so that position $\hat{e} = 0$ corresponds to axial compression and $\hat{e} = 1$ to the outer fibers. The analysis is run under an indirect displacement control. The magnitude of the eccentricity determines whether the controlled displacement corresponds to the overall vertical displacement or to the rotation of the top surface. To construct the interaction diagram the response needs to be investigated for different values of eccentricity $\hat{e}$ between 0 and 1. A direct displacement control is only used to compute the maximum moment under pure bending (zero normal force). In that case, the controlled displacement represents the rotation of the top surface and the axial deformation is left unrestrained.

The second computational model used in this study is a horizontally *symmetric half* of the column. The geometry complies with the actual specimens and includes widening at the ends and a realistic definition of the reinforcement topology. Vertically restrained displacement on the bottom face imposes the symmetry conditions. To properly capture the geometrically nonlinear behavior, the prescribed vertical displacement w with eccentricity e is not defined at the top surface of the concrete specimen but is shifted upwards and coincides with the axis of the steel hinge. The response was evaluated only for the 5 values of eccentricity in the experiment. An in-house mesher T3D (Rypl 2004) was used for spatial discretization. The mesh is structured and predominantly composed of linear hexahedral elements. Due to the computational demands, a coarser mesh had to be adopted in the central part of the column. In the horizontal direction, the difference in the mesh density is insignificant (30×30 instead of 32×32 elements) while in the vertical direction the element size was almost doubled to 13.33 mm (4.5 elements per H). Altogether, the model is composed of 63445 elements and 68001 nodes.

3.2 *Material models and calibration*

In the finite element simulations, the behavior of concrete is described using the Damage-Plastic Model for Concrete Failure (CDPM2) (Grassl, Xenos, Nyström, Rempling, & Gylltoft 2013). This model is an improved and extended version of Concrete damage-plasticity model (CDPM) (Grassl & Jirásek 2006). Both models were implemented by their authors in the finite element package OOFEM (Patzák 2000; Patzák 2012), which is used here in all numerical simulations.

The model CDPM2 is based on plasticity with isotropic hardening and nonassociated flow combined with a scalar damage model with damage driven by plastic flow and by the elastic strain. The yield condition is formulated in the effective stress space and depends on all three stress invariants. The flow rule is derived from a plastic potential that depends only on the hydrostatic stress and the second deviatoric invariant, which improves the efficiency of the implementation and robustness of the model (Grassl & Jirásek 2006).

The model deals with the effective stress

$$\bar{\sigma} = D_e(\varepsilon - \varepsilon_p) \quad (1)$$

which is computed using the plastic part of the model. Here, D_e is the elastic material stiffness matrix, ε is the total strain and ε_p is the plastic strain.

The model uses two independent scalar damage variables ω_t and ω_c for tension and compression, respectively, which enable the transition from the effective to nominal stress. To achieve that, the effective stress is first split into the positive part, $\langle\bar{\sigma}\rangle_+$, and the negative part, $\langle\bar{\sigma}\rangle_-$. The nominal stress is then computed as

$$\sigma = (1 - \omega_t)\langle\bar{\sigma}\rangle_+ + (1 - \omega_c)\langle\bar{\sigma}\rangle_- \quad (2)$$

To prevent mesh-dependent results, the model is regularized using the crack-band approach (Bažant & Oh 1983).

CDPM2 uses a large number of input parameters. In the present case, one of the key parameters is the uniaxial compressive strength at the time of testing (age 100 days), f_{cm}. The compressive strength is used to estimate the tensile strength and fracture energy as recommended in (Fédération Internationale du Béton 2012). The value of tensile strength is here of crucial importance—surprisingly not because of tensile cracking but because it is one of the parameters influencing the shape of the yield surface which in turn defines the response under lateral confinement.

The measured secant modulus of elasticity accurately corresponds with the prediction according to ACI 318 (ACI 2019). As explained in (Bažant & Jirásek 2018) the mean and not the characteristic (specified) value of compressive strength should be used and the formula

$$E_{\mathrm{cm}} = 4733\sqrt{f_{\mathrm{cm}}} \tag{3}$$

gives 31.7 GPa. To reflect the highly compliant nature of the local aggregates, the Taiwanese national standard recommends to reduce this value by 20% which results into 25.3 GPa. This value is consistent with the measured value 26.9 GPa at stress level 40% f_{cm}. To be consistent with the measurements, the parameter q_{0h} which controls the onset of nonlinear behavior is set to 0.4.

The postpeak behavior can be tuned by the parameter A_S whose lower value means faster softening. The uniaxial response computed with a single finite element with $A_S = 10$ and 20 is compared with the experimental data in Figure 4 the first of which will be used hereafter. Even though the experimental scatter of the compressive strength is very small, the postpeak behavior and measurement were not stable which is evident from the grey curves.

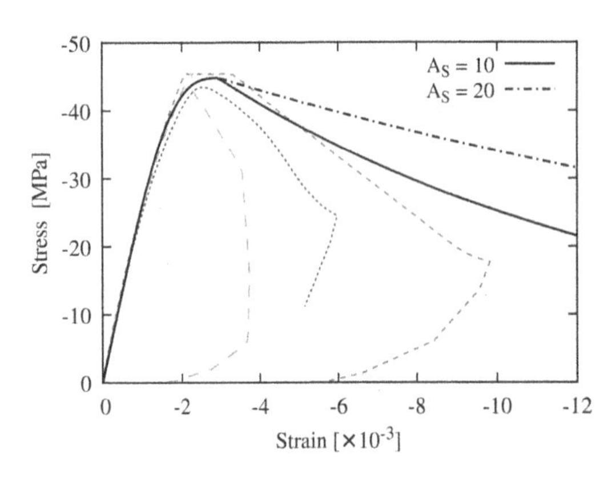

Figure 4. Calibration of CDPM2 on uniaxial compression of concrete cylinders.

The most important material parameters of CDPM2 are summarized in Table 1, the remaining constants were taken with their default recommended (Grassl, Xenos, Nyström, Rempling, & Gylltoft 2013) values.

In a circular column reinforced with hoops or spirals, the average value of passive lateral confinement σ_L (compression treated as positive) can be derived from the condition of equilibrium of stress resultants of confined concrete and yielding reinforcement (in tension).

$$\sigma_L = \frac{2A_\varnothing f_y}{DH} \tag{4}$$

Table 1. Summary of the parameters of CDPM2.

Parameter	Value	Meaning
E	26.9 GPa	Young's modulus
ν	0.2	Poisson's ratio
f_c	44.8 MPa	compressive strength
f_t	3.32 MPa	tensile strength
G_F	144.7 N/m	fracture energy
q_{h0}	0.4	elastic limit in compression
H_p	10^{-2}	hardening modulus
D_f	0.85	dilation factor
A_S	10	softening parameter
ε_{fc}	10^{-4}	softening par. for compression

In the above equation, $A_\varnothing$ is the cross-sectional area of the reinforcing bar, f_y is its yield strength, D is the diameter (center-line) and H is the vertical spacing of the hoops.

CDPM2 is an advanced material model which realistically captures the influence of confinement on strength. Even though the increase in strength $\Delta f_{c,c}$ due to confinement σ_L cannot be determined analytically, at least it can be accurately approximated (Havlásek 2021) as

$$\Delta f_{c,c} = k\sigma_L^p \left[1 + c\,(f_{\mathrm{cm}}/\mathrm{MPa} - 28)^q\right] \tag{5}$$

with constants $k = 7.65, p = 0.80$ and strength correction $c = 0.0085$ and $q = 1.00$. As shown in Figure 5, this resulting dependence is very similar to the formula proposed in *fib* MC2010.

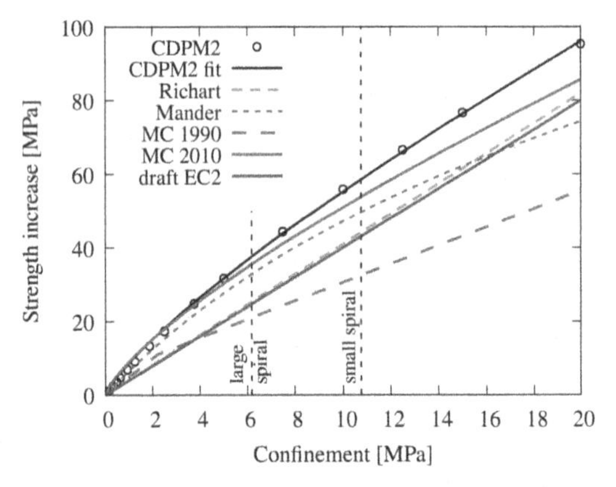

Figure 5. Increase of strength under confined as predicted by the models from the literature, and by the CDPM2 and its approximation given by (5).

To introduce non-uniformity of lateral confinement and to treat different reinforcement layouts, the average confinement is reduced by a multiplicative confinement effectiveness factor. Recently, it has been shown (Havlásek 2021) that for both spiral and circular hoops this factor matches a simple rule $k_e = (1 - 0.5H/D)^2 \approx 1 - H/D$.

In the present case, the average confinement inside the small and large spirals equals 10.76 MPa and

6.18 MPa, respectively, and the effective confinement 7.82 MPa and 5.65 MPa. As shown in Figure 5, the confined strength is expected to be around 70 and 100 MPa in the large and small spirals, respectively, and above 100 MPa in the double-confined region where the spirals overlap.

The response to active and passive confinement is (for the present calibration) depicted in Figure 6. Interestingly, the confined strength is almost independent of the nature of confinement (active or passive), what matters is the magnitude. The strain at peak stress increases with the confinement magnitude. For this reason, in a structurally sound design, the confinement in small and large spirals should not be too much different to reach the peak stress simultaneously (Havlásek, Lepš, & Bittnar 2021).

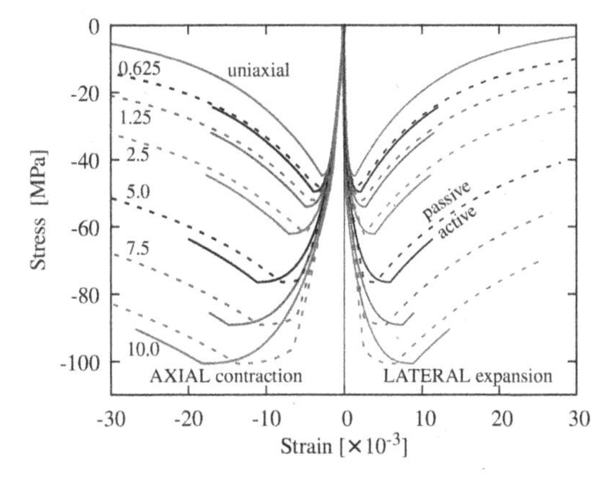

Figure 6. Response of CDPM2 under constant active and passive confinement with magnitude 0.625–10 MPa.

Figure 7 presents experimental data on #3 (10 mm) spiral and #4 (13 mm) spiral and rebar (color lines) compared to the calibrated Mises plasticity material with hardening and damage (black lines with the same pattern). The data for #5 (16 mm) rebar are not available, so the calibration for #4 is adopted. The material is significantly hardening, its strength exceeds the yield stress ($\approx$ 500 MPa) by approx. 150-200 MPa and is attained at approx. 10% elongation. The Young modulus is set to 200 GPa, the constitutive law as well as its parameters are not presented here.

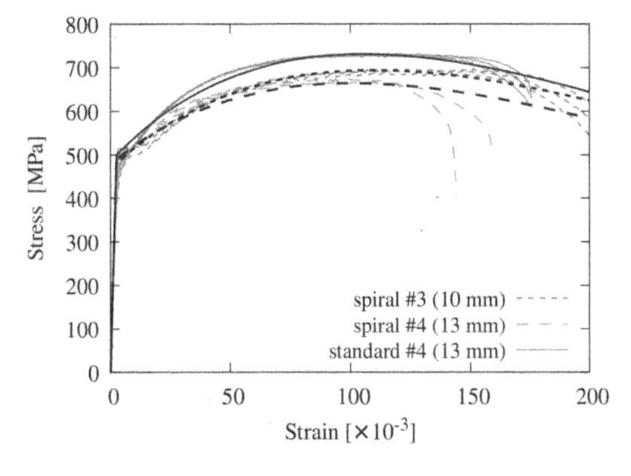

Figure 7. Calibration of Mises plasticity with hardening and damage.

4 RESULTS AND DISCUSSION

4.1 Representative section

The results computed with the *representative section* approach are presented first. In Figure 8 the lines with circles correspond to the interaction diagram obtained from FEM simulations with CDPM2 for different alternatives of reinforcement: no reinforcement (grey color), longitudinal reinforcement (green color), and finally, the MSR (blue color). The significant difference between the last two is the increase originating from lateral confinement. Apparently, this gain is much higher than what would have corresponded to the reinforcement ratio (2.5% laterally, 2.0% longitudinally.) In this particular case, the increase in strength ranges from 25 to 40%, see Figure 9.

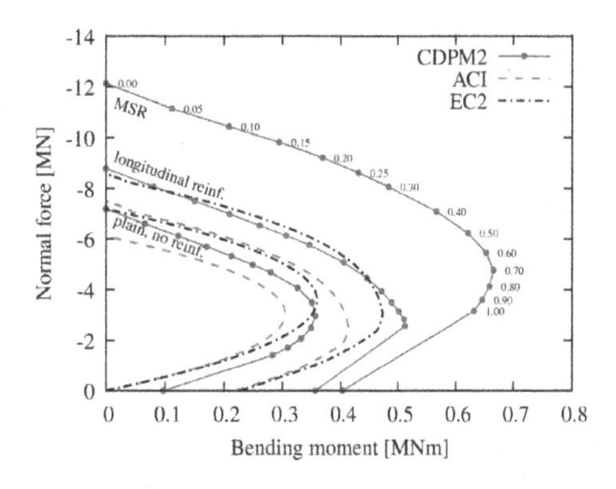

Figure 8. Comparison of ID computed with CDPM2 in OOFEM, and according to ACI-318 and EC2 for plain concrete, CS with longitudinal reinforcement and MSR. The numbers correspond to the normalized eccentricity, $\hat{e}$.

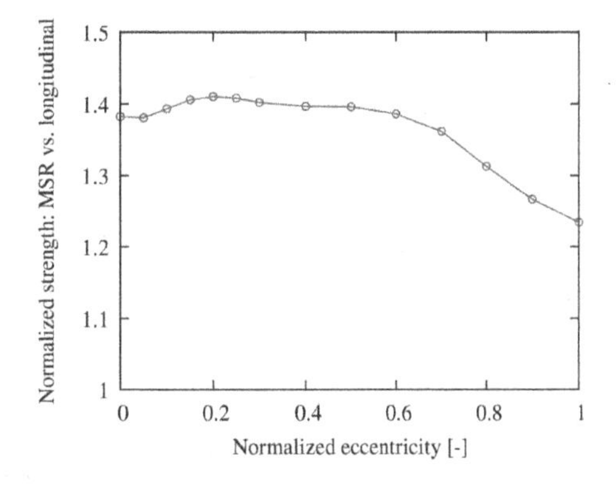

Figure 9. Normalized increase in strength under eccentric compression. Strength of MSR is expressed with respect to cross-section with longitudinal reinforcement only.

A comparison of the numerical results with the code expressions (assuming mean values of the material properties) demonstrates a realistic nature of the simulations of the unreinforced and longitudinally reinforced section. ACI 318 (red dashed line) is more conservative compared to EC2 (black dash-dot lines).

In contrast to ordinary columns with conventional rectilinear reinforcement not designed or detailed to produce confinement, the MSR leads to a highly uneven distribution of normal stress in axial direction. This is illustrated in Figure 10 for uniaxial compression and the peak load. To emphasize the stress non-uniformity, the contour plot is vertically warped.

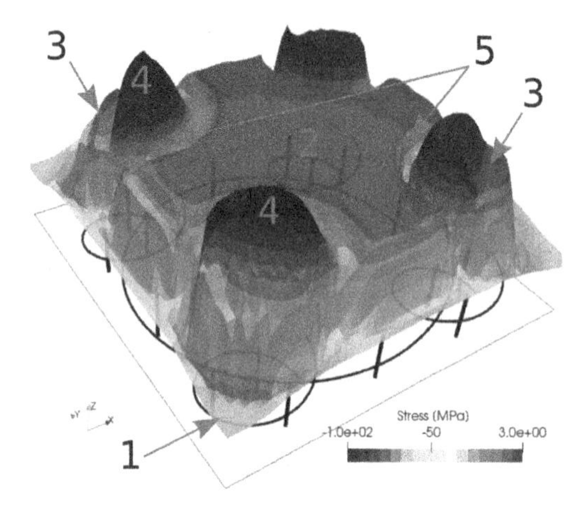

Figure 10. Distribution of vertical stress at peak load under uniaxial compression of a representative section with MSR.

Five regions with different structural responses can be distinguished:

1 *Unconfined concrete* cover is gradually softening after reaching its uniaxial compressive strength. It needs to be noted that in the simulations the concrete cover does not abruptly spall off.

2+3 *Single-confined concrete* inside large (2) and small (3) spirals is hardening until the spiral starts yielding. Similar reinforcement ratios imply similar strain at which the peak stress is attained. Magnitude and distribution of lateral confinement is significantly affected by the confinement effectiveness factor.

4 *Double-confined concrete* whose strength is due to larger confinement higher than that of a single confined concrete. The stress distribution is extremely nonlinear.

5 Concrete just outside the small spiral and inside large spiral, referred to here as ***reduced confinement***, is affected by a small deficiency in lateral confinement. The confinement here is smaller than the average value inside the large spiral and leads to earlier onset of compressive damage and therefore reaches a lower strength. The significance of this phenomenon which is caused by continuity of the displacement field depends on the balance between lateral confinement in the small and large spirals. As shown later, neglecting this phenomenon can cause significant overestimation of the carrying capacity.

The benefit of the MSR concept is apparent from Figure 11 which compares the distribution of normal stress in the vertical (axial) direction for different values of eccentricity and for the case with and without spiral reinforcement. The color scale is the same for all six cases and the state always corresponds to the peak load. With longitudinal reinforcement only, the maximum value of compressive stress in concrete is equal to its uniaxial strength, while with MSR the stresses more than double this value. The increased magnitude of the compressive resultant leads to a slight shift of the neutral axis towards the compression part. The graphical results indicate that with increasing eccentricity, the stresses in double-confined area and small spiral decrease and the confined strength becomes more uniform.

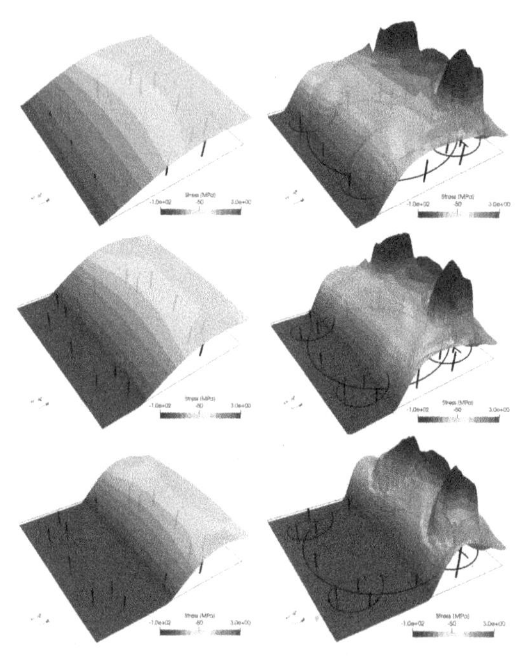

Figure 11. Normalized eccentricity 0.25 (top), 0.5 (middle), and 1.0 (bottom): distribution of vertical stress at peak load without (left) and with multi-spiral reinforcement (right).

4.2 *Mesh sensitivity*

As illustrated in Figure 12, in contrast to plain cross-section (grey color) or cross-section with longitudinal reinforcement only (green color), the mesh discretization has a significant impact on the ID computed for MSR (blue color). Without MSR, the strength increases with mesh refinement, which is varied between $8 \times 8 \times 2$ elements to $46 \times 46 \times 12$ elements; except for the coarsest mesh, the differences are almost not noticeable (different line types in 12). On contrary to this, the results for the MSR differ up to 10% as shown in Figure 13. In this Figure the strength is normalized with respect to the reference mesh 32 × 32 × 8 (the strength under pure bending is not presented). Interestingly, the differences for different mesh densities are almost constant independently of the normalized eccentricity $\hat{e}$.

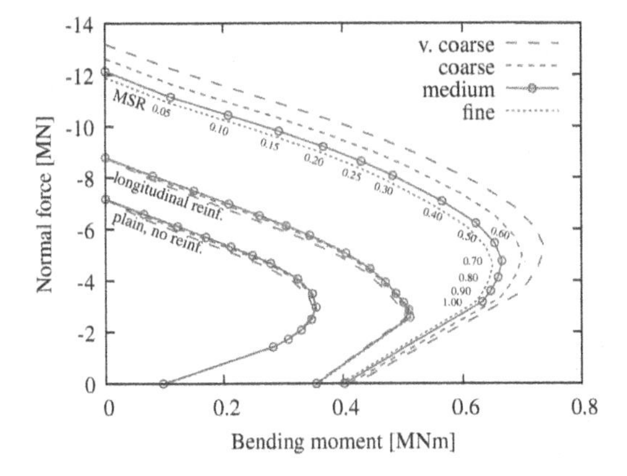

Figure 12. Influence of FE mesh density on the computed ID, v. coarse = 8 × 8 × 2, coarse = 16 × 16 × 4, medium, reference = 32 × 32 × 8, fine = 46 × 46 × 12 elements.

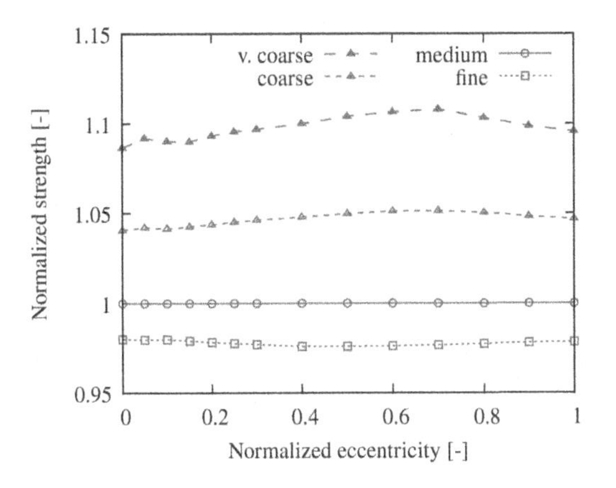

Figure 13. Dependence of the normalized strength of MSR cross-section on the normalized eccentricity evaluated for different FE discretizations. The strength is normalized wrt. medium mesh size, 32 × 32 × 8 elements.

The explanation needs to be sought in the distribution of compressive damage and stress computed on different meshes as shown in Figure 14. Despite very large differences in the mesh density, the overall stress distribution is almost identical. The most significant difference is the response of the region with reduced confinement, labelled "5" in Figure 10. With a fine mesh, the decrease in confinement causing a substantial reduction of strength is captured correctly, but as the mesh becomes coarser, this effect tends to be smeared until it completely disappears.

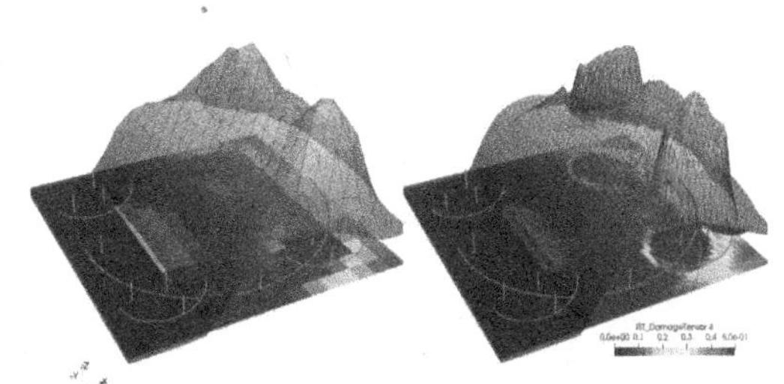

Figure 14. Distribution of compressive damage (contours) and vertical stress (warped grey surface) at peak load computed with coarse (left) and fine (right) finite element mesh and normalized eccentricity 0.5.

The corresponding computational times (1 CPU @ 4.6 GHz) needed to reach the peak load at $\hat{e} = 0.5$ are summarized in Table 2. If the computed constant differences hold also for different reinforcement configurations and concrete strengths, then in practical applications the response can be evaluated on the coarse or very coarse mesh in almost no time and then reduced to ≈ 90%.

Table 2. Computational times needed to reach the peak load of a representative section with MSR reinforcement and $\hat{e} =$ 0.5 and different mesh densities.

Mesh	Computational time [s]
v. coarse, 8 × 8 × 2	21
coarse, 16 × 16 × 4	168
medium, 32 × 32 × 8	1518
fine, 46 × 46 × 12	5284

4.3 *Comparison with experimental data*

The experimental setup was designed to produce as constant response as possible in the central part of the specimen (square cross-section). This would have facilitated a direct comparison with the results computed using the representative section presented in the previous Section. However, the color contours in Figure 15 suggest that this is not be the case as the response computed on the symmetric half is nonuniform over the height. This implies that only the same quantities can be compared with the experiment and that the representative section should be used merely for the ultimate strength while the axial strain and curvature should be interpreted carefully.

A global response of the columns loaded in eccentric compression is presented in Figure 16. Smooth solid lines correspond to FE simulations while the experimental results are shown in noisy lines. Data for the same value of eccentricity are displayed in a similar color. First of all, it is obvious that the strength computed with the symmetric half crosses the boundary of the interaction diagram computed with the representative section. The horizontal discretization was very similar in both cases (32 and 30 elements per edge).

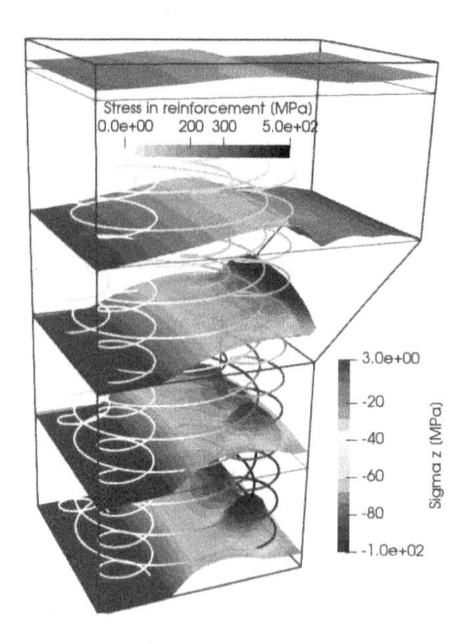

Figure 15. Warped contours of normal stress in vertical direction and stress in the spirals (peak load, normalized eccentricity 0.5).

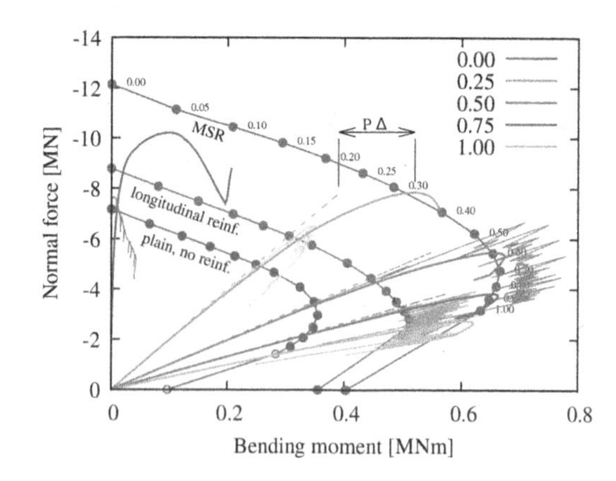

Figure 16. Comparison of the ID evaluated for representative cross-sections (solid lines with circles), FE simulations on a symmetric half (smooth solid lines), and experimental data (noisy thin lines). The numbers correspond to normalized eccentricity. The P-Δ indicates the contribution of geometrical nonlinearity (difference between dashed and solid lines).

The reason is the vertical discretization (8 vs. 4.5 elements per height H). Insufficient number of elements per height spuriously improves confinement effectiveness and as a consequence enhances strength. Next, the P-Δ effect (second-order moment caused by specimen deflection) makes the structural response in the M-N diagram highly nonlinear. This contribution is significant especially in the cases with small eccentricity (e.g. $\hat{e} = 0.25$). The reference first-order moment is shown in thin dashed lines of the same color. The intended position where the M-N curve reaches the strength envelope can eventually become significantly shifted due to the geometrically nonlinear behavior.

The noisy trends in the experimental data have two sources: i) partial unloading during the experiment to document the progress of the crack pattern (approx. 10× in one experiment), ii) tensile cracking in the cases with higher eccentricity. Comparison of the numerical results to the experimental data shows that the overall trend is captured very well if $\hat{e} \geq 0.5$. With $\hat{e} = 0.75$ and $\hat{e} = 1.0$ the strength is slightly overestimated by the numerical model. The sudden drops due to cracking (on the tensile face) cause that the experimental curves fill almost the entire region between the ID with longitudinal reinforcement only (green color) and MSR. With $\hat{e} = 0.5$, the strength exceeds the prediction of the numerical model. Additionally, the experiment demonstrates that a constant moment capacity is maintained even when the normal force decreases to half of its maximum value.

In the cases of low eccentricity, $\hat{e} = 0.0$ and 0.25, the insufficiently designed lateral reinforcement at the ends of the specimen triggered vertical splitting cracks at one of the supports. The primary reason was that the lateral rectilinear reinforcement composed of stirrups was not welded. For this reason, the structural response shown in Figure 16 does not correspond to the behavior of the region of interest because the loading force started decreasing prematurely. In the FEM analysis, the lateral reinforcement was defined without these defects and the bond between the loading plate and the specimen was defined as perfect, therefore the ID was reached with $\hat{e} = 0.25$. Under axial compression ($\hat{e} = 0.0$) the specimen laterally deflects because of the eccentric nonsymmetric widening at the base and fails by a vertical splitting crack, which also appeared in the FEM simulation.

5 CONCLUSIONS

A new experimental study investigated the behavior of geometrically identical concrete columns reinforced with multispiral reinforcement and subjected to eccentric compression. These data will provide a basis for the calibration of the computational models which can be used for the evaluation of the interaction diagrams of MSR columns. The main conclusions are summarized as follows:

- In the case of longitudinal reinforcement only, the numerical results obtained with the *representative section approach* and CDPM2 material model comply with the design codes. The MSR reinforcement enhances strength from ≈25% (high eccentricity) to 40% (low eccentricity).
- The peak load is reached at the onset of spiral yielding, which corresponds to the maximum confinement.
- A considerable mesh dependence of the MSR simulations has been detected. The explanation was found in the behavior of the zone with *reduced confinement*, just outside the small spirals. Coarser

meshes lead to strength overestimation because this effect is not considered properly.
- Second-order moments are important in both experiments and modeling. An exceptional agreement between the experiments and numerical results was obtained, which proves the superior performance of MSR layout stemming from laterally confined concrete.

ACKNOWLEDGMENT

Financial support for this work was provided by the Technology Agency of the Czech Republic (TA ČR), project number TM01000059 (Reducing material demands and enhancing structural capacity of multi-spiral reinforced concrete columns - advanced simulation and experimental validation) and by the Ministry of Science and Technology of Taiwan under Contract No. 109-2923-M-002-006-MY3. The authors acknowledge support from National Center for Research on Earthquake Engineering (NCREE) of Taiwan.

The numerical analyses have been performed with OOFEM, an open-source object-oriented finite element program (Patzák 2000). The finite element meshes have been prepared with the T3D mesh generator (Rypl 2004).

REFERENCES

ACI (2019). *ACI 318-19 Building Code Requirements for Structural Concrete and Commentary*. American Concrete Institute.

Bažant, Z. & M. Jirásek (2018, January). *Creep and hygrothermal effects in concrete structures*, Volume 225 of *Solid Mechanics and its Applications*.

Bažant, Z. P. & B. H. Oh (1983, May). Crack band theory for fracture of concrete. *Matériaux et Construction 16*(3), 155–177.

European Committee for Standardization (2018). *Eurocode 2: Design of concrete structures – Part 1-1: General rules, rules for buildings, bridges and civil engineering structures, Final Version of PT1-draft prEN 1992-1-1 2018 D3.*

Fédération Internationale du Béton (2012). *Model Code 2010*. Number vol. 65 in fib Bulletin. International Federation for Structural Concrete (fib).

Grassl, P. & M. Jirásek (2006). Damage-plastic model for concrete failure. *International Journal of Solids and Structures 43*(22), 7166 – 7196.

Grassl, P., D. Xenos, U. Nyström, R. Rempling, & K. Gylltoft (2013). CDPM2: A damage-plasticity approach to modelling the failure of concrete. *International Journal of Solids and Structures 50*(24), 3805 – 3816.

Havlásek, P. (2019). MaLCoLM, multi-spiral column simulation module version 1.0. http://mech.fsv.cvut.cz/~phavlasek/projects/2018_cestar.

Havlásek, P. (2021). Numerical modeling of axially compressed circular concrete columns. *Engineering Structures 227*, 111445.

Havlásek, P., M. Jirásek, & Z. Bittnar (2019). Modeling of precast columns with innovative multi-spiral reinforcement. fib Proceedings, pp. 2301–2307. FIB - Féd. Int. du Béton.

Havlásek, P., M. Lepš, & Z. Bittnar (2021). Optimum design of axially compressed concrete columns with multi-spiral reinforcement. *AIP Conference Proceedings 2322*(1), 020002.

Lepš, M. & P. Havlásek (2021). Optimum performance of axially compressed concrete columns with multi-spiral reinforcement. *Concrete Structures: New Trends for Eco-Efficiency and Performance, 18th* fib *Symposium Proceedings*, 2175–2184.

Mander, J. B., M. J. N. Priestley, & R. Park (1988). Theoretical stress-strain model for confined concrete. *Journal of Structural Engineering 114*(8), 1804–1826.

Ngo, S.-H. & Y.-C. Ou (2021). Expected maximum moment of multi-spiral columns. *Engineering Structures 249*, 113386.

Patzák, B. (2000). OOFEM home page. http://www.oofem.org.

Patzák, B. (2012). OOFEM - an object-oriented simulation tool for advanced modeling of materials and structures. *Acta Polytechnica 52*(6), 59–66.

Rypl, D. (2004). T3D mesh generator. http://mech.fsv.cvut.cz/~dr/t3d.html.

Yin, S. (2005). Helical rebar structure. US Patent 6,860,077.

Yin, S. Y.-L., J.-C. Wang, & P.-H. Wang (2012). Development of multi-spiral confinements in rectangular columns for construction automation. *Journal of the Chinese Institute of Engineers 35*(3), 309–320.

Yin, S. Y.-L., T.-L. Wu, T. C. Liu, S. A. Sheikh, & R. Wang (2011). Interlocking spiral confinement for rectangular columns. *Concrete International 33*(12).

Computational Modelling of Concrete and Concrete Structures – Meschke, Pichler & Rots (Eds)
 ISBN: 978-1-032-32724-2

FEM modelling of FRP reinforced concrete with a shell element approach

I. De Beuckeleer, T. Molkens & A. Van Gysel
KU Leuven – De Nayer Campus, Sint-Katelijne-Waver, Belgium

E. Gruyaert
KU Leuven – Ghent Technology Campus, Ghent, Belgium

ABSTRACT: This paper outlines a simplified 2D FEM approach to model FRP reinforced concrete members with commercial FEM software. Due to limitations in this type of software, engineers can overcome these design problems with some creative modelling techniques. When linear-elastic calculations are performed, the uncracked behaviour is well captured. As soon as the first crack appears, designers should manually reduce the bending stiffness to investigate the cracked behaviour accurately. In the proposed FEM model, a lumped and smeared cracked stiffness was used to get more realistic simulation results for the cracked phase during the loading process. Moreover, a shell element model was chosen to account for shear deformations of the concrete cross-section, whereas these mechanical effects are neglected in slab or beam models. The reinforcement is included as one single beam element with equivalent stiffness properties in the FEM approach. Finally, a case study for future shear tests will be presented.

1 INTRODUCTION

1.1 *Problem setting*

Fiber Reinforced Polymer (FRP) reinforcement materials, composed of mineral (and sometimes organic) fibers and a polymeric resin matrix, are promising substitutes for conventional carbon steel reinforcement. This non-metallic material is not sensitive to corrosion, even in harsh environmental conditions. A chemical barrier to water, oxides and chlorides is not required anymore, which allows to reduce the concrete cover. Moreover, the overall dimensions of the concrete cross-section and the reinforcement can be diminished owing to the exceptionally high tensile strength of FRPs in comparison with carbon reinforcing steel. As a direct consequence, structures are built with smaller volumes of concrete which implies that the construction industry becomes a more sustainable sector (Bielak et al. 2019). Since FRP reinforcement materials outperform conventional steel reinforcement with regard to their service life, maintenance and reconstruction works are rarely required. These future material savings further reduce the environmental pressure that the concrete industry has caused ever since (El-Sayed et al. 2007).

Despite the ecological and economic benefits of FRPs, engineering practitioners tend to be reluctant to provide this reinforcement type in their design as the underlying mechanisms of the structural behaviour are not fully understood so far. Some preliminary design models for bending have been published in the scientific literature and some experimentally based shear design formulas are at hand (Pilakoutas et al. 2011) As a result, large safety margins are applied in design practice to cover the uncertainties, which restricts an evolution towards more sustainable and economic project designs. (Gudonis et al. 2014) The great potential of FRP reinforcement can only be exploited if all scientific lacunas about the shear behaviour of this brittle matrix composite material are sorted out.

As a starting point, an alternative 2D FEM shell element modelling approach is proposed in this paper, which addresses both the shortcomings of simple 1D beam models and avoids the far-reaching complexity of advanced 3D models, that are time-consuming in terms of input and calculation. The outcomes of this modelling technique will be validated by comparing the numerical results with the experimental results of a dataset of 35 shear tests.

1.2 *Current shear design models*

The background of shear behaviour of longitudinally reinforced members without shear reinforcement has been studied in several theoretical approaches and the interplay of several shear transfer mechanisms has been thoroughly analysed e.g. (Classen 2020). It has turned out (Yang 2014) that the transverse stiffness of the reinforcement bars directly contributes to shear force transfer through dowel action. In addition, the bending stiffness of concrete members reinforced

DOI 10.1201/9781003316404-29

with conventional carbon steel reinforcement is significantly higher than of those reinforced with FRP reinforcement. Hence, their mechanical behaviour is considerably different during the loading process. Larger deflections and crack widths are expected in FRP reinforced elements (Pilakoutas et al. 2011).

The transverse stiffness of the reinforcing materials is not always inserted in design guidelines, even though this variable is of crucial importance. An overview of several shear models, implemented in current design practice, will be provided in the following paragraphs to indicate the absence of this parameter.

1.2.1 *ACI 318-19*

Design formula for the shear capacity (V_n) 1.2.2 prescribed by the American Institute does not take the bending stiffness of the reinforcing materials into account. The main influencing parameters are the size effect factor (λ_s), longitudinal reinforcement ratio (ρ_l), concrete compressive strength (f_c'), width (b_w) and effective depth (d). (Lima et al. 2021).

$$V_n = min\begin{cases} 1.33 * \left(\frac{2}{3} * \lambda_s * \rho_l^{\frac{1}{3}} * \sqrt{f_c'} * b_w * d\right) \\ \frac{5}{6} * \sqrt{f_c'} * b_w * d \end{cases} \quad (1)$$

where

$$\lambda_s = \sqrt{\frac{2}{1 + 0.004 * d}} \quad (2)$$

1.2.2 *Eurocode 2*

The European design standard also does not acknowledge the contribution of the bending stiffness of the reinforcing materials in design formula 1.2.3. The shear capacity ($V_{Rd,c}$) is determined by the size effect factor (k), longitudinal reinforcement ratio (ρ_l), concrete compressive strength (f_{ck}), width (b_w) and effective depth (d). (EN 1992-1-1, 2005)

$$V_{Rd,c} = max\begin{cases} C_{Rd,c} * k * (100 * \rho_l * f_{ck})^{\frac{1}{3}} * b_w * d \\ 0.035 * k^{\frac{3}{2}} * f_{ck}^{\frac{1}{2}} * b_w * d \end{cases} \quad (3)$$

where

$$k = 1 + \sqrt{\frac{200}{d}} \leq 2.0 \quad (4)$$

1.2.3 *Simplified Compression Field Theory*

This shear model by Bentz et al. shows a simplified design approach of the theoretical model by Vecchio & Collins. Shear stresses (v_c) are calculated by means of the tensile stress factor (β) in the cracked concrete and concrete compressive strength (f_c') (Lima et al. 2021). As can be noticed in formula 1.2.4, the effect of the bending stiffness is included in this model.

$$v_c = \beta * \sqrt{f_c'} \quad (5)$$

where

$$\beta = \frac{0.4}{1 + 1500 * \varepsilon_{sl}} * \frac{1300}{1000 + s_{xe}} \quad (6)$$

1.2.4 *AASHTO-LRFD*

The shear model by Bentz et al. is also at the basis of the AASHTO-LRFD formulas. The ultimate shear load results from an analogous design formula (Lima et al. 2021)

$$v_c = \beta * \sqrt{f_c'} \quad (7)$$

where

$$\beta = \frac{0.4}{1 + 750 * \varepsilon_{sl}} * \frac{1300}{1000 + s_{xe}} \quad (8)$$

1.2.5 *Zsutty*

According to Zsutty, ultimate shear stresses (τ_c) can be calculated based on the concrete compressive strength (f_c'), longitudinal reinforcement ratio (ρ_l) and effective depth to shear span ratio (d/a). Two distinct formulas (9) and 1.2.6 are valid for slender elements ($a/d \geq 2.5$) and short elements ($a/d < 2.5$), respectively (Lima et al. 2021). This approach also neglects the effect of the bending stiffness of the reinforcement.

$$\tau_u = 2.17 * \left(f_c' * \rho_l * \frac{d}{a}\right)^{\frac{1}{3}} \quad (9)$$

$$\tau_u = 5.4 * (f_c' * \rho_l)^{\frac{1}{3}} * \left(\frac{d}{a}\right)^{\frac{3}{4}} \quad (10)$$

1.2.6 *Russo et al.*

The maximum diameter of the coarse aggregates (d_a), effective depth (d), longitudinal reinforcement ratio (ρ_l), concrete compressive strength (f_c'), yield stress of the longitudinal reinforcement (f_{yl}) and the shear span to effective depth ratio (a/d) are regarded as principal variables contributing to the ultimate shear capacity (τ_u) (Lima et al. 2021). As many other researchers, they disregard the bending stiffness of the reinforcement.

$$\tau_u = 1.13 * \left[\frac{1 + \sqrt{\frac{5.08}{d_a}}}{\sqrt{1 + \frac{d}{25 * d_a}}}\right] * \left[\rho_l^{0.4} * f_c'^{0.39} + 0.5 * \rho_l^{0.83} * f_{yl}^{0.89} * \left(\frac{a}{d}\right)^{-1.2-0.45*\frac{a}{d}}\right] \quad (11)$$

2 EXPERIMENTAL DATA

In this paper a Finite Element Method (FEM) model is proposed. The results of this numerical model are compared with experimental test data of 35 reinforced concrete slabs without shear reinforcement from the literature to determine the validity of the FEM approach; 16 tests are performed by (Acciai et al. 2016), 2 tests are described in (Noël & Soudki 2014) and 17 tests are carried out by (Abdul-Salam et al. 2016).

2.1 *Acciai et al.*

Acciai et al. (2016) executed a research program to compare the mechanical behaviour of 16 longitudinally reinforced members, see Table 1. In this study, three different influencing parameters were considered: (1) the cross-sectional geometry, (2) the concrete compressive strength and (3) the type of longitudinal reinforcement. These researchers tested a first series with a shallow rectangular cross-section (w = 200 mm; h = 100 mm) and a second series with a deep rectangular cross-section (w = 100 mm; h = 200 mm). All specimens with a total length of 2800 mm spanned 2000 mm and were subjected to a concentrated load at midspan in a three-point bending test.

Table 1. Overview of the specimens out of (Acciai et al. 2016)

N°	Reinforcement	Concrete	Failure
1	1x GFRP (ø = 13 mm)		10.2 kN
	f_u = 690 MPa	f_c = 49 N/mm^2	FC*
	E = 40.8 Gpa	d_c = 20 mm	82 mm#
2	2x GFRP (ø = 13 mm)		15.6 kN
	f_u = 690 Mpa	f_c = 49 N/mm^2	FC*
	E = 40.8 Gpa	d_c = 20 mm	82 mm#
3	1x CFRP (ø = 9 mm)		15.8 kN
	f_u = 2068 Mpa	f_c = 80 N/mm^2	FC*
	E = 124 Gpa	d_c = 20 mm	78 mm#
4	2x CFRP (ø = 9 mm)		32.8 kN
	f_u = 2068 Mpa	f_c = 80 N/mm^2	FC*
	E = 124 Gpa	d_c = 20 mm	82 mm#
5	1x Steel (ø = 8 mm)		4.5 kN**
	f_u = 634 Mpa	f_c = 39 N/mm^2	FT*
	E = 206 Gpa	d_c = 20 mm	10 mm#
6	2x Steel (ø = 8 mm)	9 kN**	
	f_u = 634 Mpa	f_c = 39 N/mm^2	FT*
	E = 206 Gpa	d_c = 20 mm	18 mm#
7	1x Steel (ø = 14 mm)		13 kN**
	f_u = 666 Mpa	f_c = 39 N/mm^2	FT*
	E = 206 Gpa	d_c = 20 mm	21 mm#
8	2x Steel (ø = 14 mm)		25 kN**
	f_u = 666 Mpa	f_c = 39 N/mm^2	FT*
	E = 206 Gpa	d_c = 20 mm	22 mm#
9	1x GFRP (ø = 13 mm)		19.90 kN
	f_u = 690 Mpa	f_c = 49 N/mm^2	S*
	E = 40.8 Gpa	d_c = 20 mm	28 mm#
10	2x GFRP (ø = 13 mm)		24.00 kN
	f_u = 690 Mpa	f_c = 49 N/mm^2	S*
	E = 40.8 Gpa	d_c = 20 mm	18 mm#
11	1x CFRP (ø = 9 mm)		17.70 kN
	f_u = 2068 Mpa	f_c = 80 N/mm^2	S*
	E = 124 Gpa	d_c = 20 mm	13 mm#
12	2x CFRP (ø = 9 mm)		27.90 kN
	f_u = 2068 Mpa	f_c = 80 N/mm^2	S*
	E = 124 Gpa	d_c = 20 mm	12 mm#
13	1x Steel (ø = 8 mm)		12 kN**
	f_u = 634 Mpa	f_c = 39 N/mm^2	B*
	E = 206 Gpa	d_c = 20 mm	3 mm#

(continued)

Table 1. Continued.

N°	Reinforcement	Concrete	Failure
14	2x Steel (ø = 8 mm)		22 kN**
	f_u = 634 Mpa	f_c = 39 N/mm^2	FS*
	E = 206 Gpa	d_c = 20 mm	8 mm#
15	1x Steel (ø = 14 mm)		34 kN**
	f_u = 666 Mpa	f_c = 39 N/mm^2	FS*
	E = 206 Gpa	d_c = 20 mm	10 mm#
16	2x Steel (ø = 14 mm)		49 kN**
	f_u = 666 Mpa	f_c = 39 N/mm^2	S*
	E = 206 Gpa	d_c = 20 mm	12 mm#

(*) FC = flexural (compression excess in concrete)
FT = flexural (tension excess in reinforcement)
S = brittle shear
FS = flexural shear
B = rupture of bar
(**) Exact failure load not mentioned in the article, so the value is inferred from Fig. 3 and Fig. 6 in (Acciai et al., 2016)
(#) Deflection at failure of FRP reinforced elements (or deflection at yielding for steel reinforced elements) collected out of Fig. 3 and Fig. 6 in (Acciai et al. 2016).

2.2 *Noël & Soudki*

Noël and Soudki (2014) performed a four-point bending test with a 1000 mm constant moment region on two GFRP reinforced members with a rectangular cross-section (w = 600 mm; h = 300 mm) to examine the effect of the reinforcement ratio. Both specimens have identical dimensions: a total length of 5000 mm and a total span of 4500 mm. Only data of the slabs without shear reinforcement are presented in Table 2.

Table 2. Overview of the specimens out of (Noël & Soudki, 2014)

N°	Reinforcement	Concrete	Failure
G1	6 GFRP (ø = 15.9 mm)		150 kN*
	f_u = 683 MPa	f_c = C30/37**	/*
	E = 48.2 GPa	d_c = 30 mm	97 mm*
G2	12 GFRP (ø = 15.9 mm)		220 kN*
	f_u = 683 MPa	f_c = C30/37**	/*
	E = 48.2 GPa	d_c = 30 mm	73 mm*

(*) Unknown failure load, failure mode and deflection at failure. Data collected out of Fig. 8 in (Noël & Soudki 2014).
(**) Not mentioned in the article, so this value is based on assumptions.

2.3 *Abdul-Salam et al.*

Abdul-Salam et al. (2016) tested 17 one-way slabs (w = 1000 mm; h = 200 mm) under four-point bending with a constant moment region of 1800 mm. The total length and total span of all specimens is equal to

4000 mm and 3500 mm, respectively. The test specimens are subdivided in six categories according to the separate parameters studied in this research project.

In order to properly investigate the influence of one single parameter, the axial stiffness of the reinforcement was kept similar for each group. The first group reveals the influence of the bar surface texture on the bond and shear behaviour. The second and fifth group are used to explain the effect of the concrete compressive strength on the shear resistance. The third, fourth and sixth group are meant to compare the impact of different reinforcement types. The different groups are indicated by subsequent white or shaded areas in Table 3, remark that test 3 and 8 are identical.

Table 3. Overview of the specimens out of (Abdul-Salam et al. 2016).

N°	Reinforcement	Concrete	Failure
1	5x GFRP (ø = 19.05 mm°) Helically wrapped E = 40.8 GPa	$f_c = 47.9$ N/mm^2 $d_c = 48$ mm**	188 kN DT* 118 mm#
2	4x GFRP (ø = 19.05 mm°) E = 49.8 GPa $d_c = 48$ mm**	$f_c = 48.4$ N/mm^2 93 mm#	211 kN SC_s^*
3	5x GFRP (ø = 19.86 mm°) E = 67.8 GPa $d_c = 48$ mm**	$f_c = 42.9$ N/mm^2 92 mm#	309 kN DT*
4	5x GFRP (ø = 19.86 mm°) E = 67.8 GPa $d_c = 48$ mm**	$f_c = 77.4$ N/mm^2 71 mm#	327 kN SC*
5	5x GFRP (ø = 19.86 mm°) E = 67.8 GPa $d_c = 48$ mm**	$f_c = 82.6$ N/mm^2 91 mm#	290 kN SC*
6	5x CFRP (ø = 13.72 mm) E = 139.2 GPa $d_c = 48$ mm**	$f_c = 49.7$ N/mm^2 78 mm#	237 kN SC_s^*
7	5x GFRP (ø = 25.40 mm) E = 43.9 GPa $d_c = 48$ mm**	$f_c = 47.9$ N/mm^2 67 mm#	242 kN DT_s^*
8	5x GFRP (ø = 19.86 mm°) E = 67.8 GPa $d_c = 48$ mm**	$f_c = 42.9$ N/mm^2 92 mm#	309 kN DT*
9	6x GFRP (ø = 19.86 mm°) E = 67.8 GPa $d_c = 48$ mm**	$f_c = 49.4$ N/mm^2 79 mm#	335 kN DT*
10	6x CFRP (ø = 13.72 mm) E = 147.8 GPa $d_c = 48$ mm**	$f_c = 49.4$ N/mm^2 76 mm#	282 kN SC_s^*
11	7x CFRP (ø = 13.72 mm) E = 144 GPa $d_c = 48$ mm**	$f_c = 52.0$ N/mm^2 77 mm#	317 kN SC*
12	7x CFRP (ø = 13.72 mm) E = 144 GPa $d_c = 48$ mm**	$f_c = 76.0$ N/mm^2 76 mm#	336 kN SC_s^*
13	7x CFRP (ø = 13.72 mm) E = 144 GPa	$f_c = 86.2$ N/mm^2 $d_c = 48$ mm**	274 kN SC_s^* 35 mm#
14	8x CFRP (ø = 16.82 mm°) E = 141 GPa	$f_c = 41.3$ N/mm^2 $d_c = 48$ mm**	385 kN SC_s^* 47 mm#

(*continued*)

Table 3. Continued.

N°	Reinforcement	Concrete	Failure
15	12x GFRP (ø=19.86 mm°) E = 67.8 GPa	$f_c = 48.6$ N/mm^2 $d_c = 48$ mm**	340 kN SB* 45 mm#
16	5x Steel (ø = 19.05 mm°) E = 200 GPa	$f_c = 47.9$ N/mm^2 $d_c = 48$ mm**	252 kN FC* 126 mm#
17	7x GFRP (ø = 30.45 mm°) E = 65.4 GPa	$f_c = 50.3$ N/mm^2 $d_c = 48$ mm**	426 kN SB* 70 mm#

(*) DT = diagonal tension failure
DT_s = diagonal tension failure + cover splitting
SC = shear-compression failure
SC_s = shear-compression failure followed by FRP shearing off
SB = shear-bond failure
FC = flexural-compression failure
(**) Not directly mentioned in the article, this value is based on a derivation out of the reinforcement ratio
(°) Since some contradictions were noticed, the diameter is derived from the reinforcement area mentioned in the reference article and not taken as the specified diameter
(#) Deflection at failure derived from maximum strains

While the authors recognize the effect of the arch action, it is not clear which provisions have been made in the test set-up to avoid this phenomenon.

2.4 *Discussion*

To compare the different test results to a certain extent and to visualize the risk of premature bending failure instead of shear failure, Figure 1 shows the relationship between the normalized shear stress and the reinforcement ratio. The normalized shear stress is in this case the ratio between the shear force due to the applied load and self-weight divided by the product of the width, effective depth and square root of the concrete compressive strength. The reinforcement ratio is the

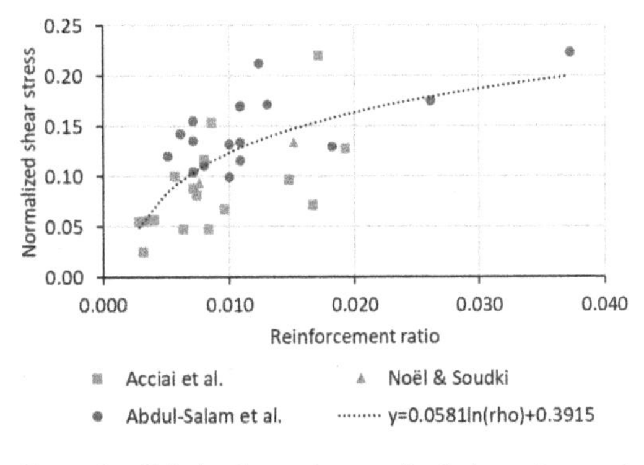

Figure 1. Relation between normalized shear stress and reinforcement ratio for all reference tests.

ratio between the cross-sectional area of the reinforcement (without accounting for the different material properties) and the product of the width and effective depth. No significant differences have been found between the different reinforcement materials, and also the stress levels of the different articles seem to be comparable. The R^2-value of the proposed logarithmic regression by (Abdul-Salam et al. 2016) is relatively weak (0.43). On the other hand, the whole practical application area of reinforcement ratios ($\rho < 0.040$) is covered.

Beside the previous relationship, it is also interesting to have an idea about the sensibility of the test to a shear failure. For that purpose, the ultimate bending capacities (M_u) based on the characteristics of the reinforcement and concrete have been calculated and compared to the applied bending moments (M_a) at failure. Ratios smaller or equal to 1 mean that bending failure can occur, see Figure 2. Only the steel reinforced specimens, which are surrounded by a dashed line in the graph, are prone to bending or combined failure.

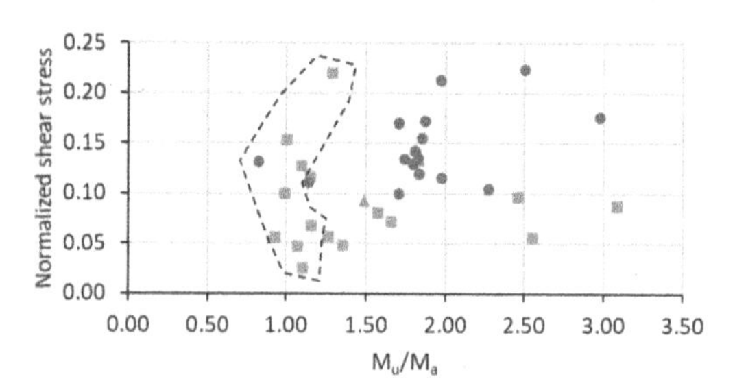

Figure 2. Relation between normalized shear stress and bending resistance level.

As for the design of ordinary reinforced concrete at this stage moment/shear interaction is neglected. Nevertheless, this should be studied in detail.

3 FEM-MODEL

Current design models (see Section 1.2) are mostly based on semi-empirical formulas or a combination between theoretical derivations and regression work out of experimental data. There is however, a tendency to develop models based on mechanics (Xing-lang (2020); Classen (2020)) or kinematic considerations (Mihaylov, (2017). All of them do have a number of advantages but sometimes suffer a limited (practical) application area. To enhance the understanding of different phenomena and for validation purposes also a lot of research is using FEM models. With advanced models, however, it is a challenge to arrive at understandable result processing, reason why a simplified 2D shell model is proposed.

3.1 *Linear elastic material behaviour*

Based on the load-displacement measurements out of the reference articles (Abdul-Salam et al. 2016; Acciai et al. 2016; Noël & Soudki 2014), a bi-linear behaviour can be observed. The first and most stiff reaction is observed when the section is uncracked, and a second linear branch starts when the first crack appears and continues linearly till failure for FRP reinforced elements. Figure 3 out of (Acciai et al. 2016) clearly illustrates the differences in behaviour between FRP (without plastic plateau) and steel reinforcement (including the effect of plasticity).

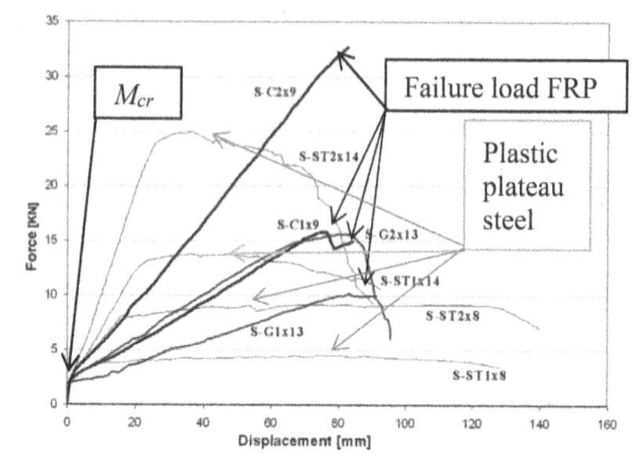

Figure 3. Bi-linear load-displacement graph (Acciai et al., 2016).

To obtain insights in the development of internal forces into the reinforcement, and stresses into the concrete, an approach with shell elements will be worked out. This means that the whole thickness of the element is concentrated into a surface element, which allows for relatively simple postprocessing of data. FEM analysis can be based on linear elastic assumptions as long as the effects of cracking are properly included (see Section 3.4).

An advantage of the use of shell elements is that shear deformations are accounted for in the analysis. This is not the case when slab or beam models are used based on the Bernoulli assumption that straight planes perpendicular to the undeformed axis line stay normal to the deformed axis line of the element. In the proposed FEM approach, reinforcement will be modelled as a beam element with applicable stiffness properties of the provided reinforcement. In that way an overlap of both materials is realized. However, due to the boundary conditions, explained in section 3.4, the influence is minimal.

3.2 *Geometrical and material non-linearity*

As shown in Figure 3, the behaviour of reinforced concrete is defined by the material properties of the reinforcement. The linear behaviour of FRPs leads to an almost perfect elastic response after the appearance of the first cracks. For steel reinforcement, a high degree of ductility is present ensuring large displacements before failure.

In case of fire, material properties are (rapidly) decreasing and the material behaviour becomes non-linear with some softening effects in the concrete and FRP bars. For the latter a critical temperature of 350° C is mentioned by different authors (Kashwani & Al-Tamimi 2014; Wang et al. 2003). Unfortunately, material models accounting for the heating effects on the fibers, the resin matrix and the bond between them are missing so far.

Designers should be aware that even due to the decrease in concrete properties at elevated temperatures, the bearing system and failure mode can change. During a fire, it is already reported (Molkens et al. 2017) that a slab working in compressive membrane action (CMA) may switch to cantilever action due to a loss of compression resistance at the lower heat-affected part of the slab, see Figure 4. In this simulation, the concrete element was modelled as a sequence of layers with shell elements in order to assign different temperature profiles to them varying in time.

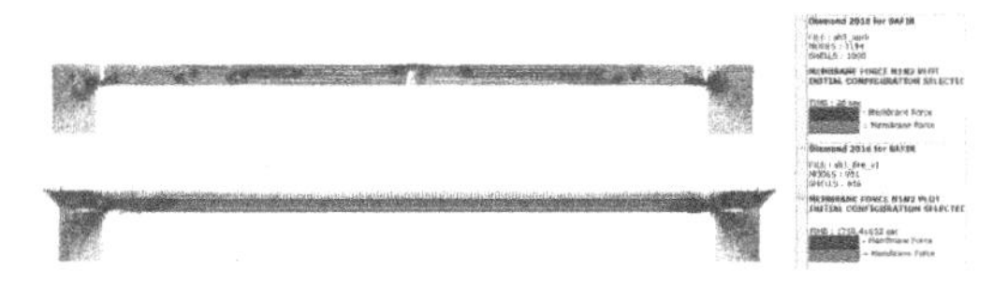

Figure 4. CMA in ULS conditions and cantilever in fire conditions. Principal tensile stresses = red, compressive stresses = blue (Molkens et al. 2017).

3.3 *Equivalent reinforcement bar properties*

Once a shell element approach is used, all bar properties must be concentrated into one beam element. As the axial and bending stiffness will influence respectively the bending and shear behaviour (dowel effect), a rectangular equivalent section with a width b_{eq} and height h_{eq} are used to meet both requirements. By filling in Eq. (12) into Eq. (13), Eq. (14) can be worked out, defining the equivalent height only depending of the bar diameter. By the use of Eq. (12), also b_{eq} is known.

$$A = n \cdot \frac{\pi \O^2}{4} = b_{eq} \cdot h_{eq} \tag{12}$$

$$I = n \cdot \frac{\pi \O^4}{64} = \frac{b_{eq} \cdot h_{eq}^3}{12} \tag{13}$$

$$I = \frac{A \cdot h_{eq}^2}{12} \xrightarrow{yields} h_{eq} = \sqrt{\frac{12 \cdot I}{A}} = \O\sqrt{\frac{3}{4}} \tag{14}$$

3.4 *Lumped and smeared cracked stiffness*

To enhance a correct bending capacity simulation, each model is built up with symmetrical boundary conditions at the middle of the span. At the location of the concrete material, boundary conditions permit horizontal compressive forces but no tensile forces. Only at the location of the reinforcement, compression and tension forces are allowed.

By doing so, a correct tensile force and/or stress is obtained in the reinforcement bars. As the material is calculated in a linear way, this single or lumped numerical initiated crack will lead to a stiffer reaction of the beam to the failure load. Based on the Bernoulli-Euler beam theory, the deformation is the double integral of the the moment-curvature relation ($y=\int\int M/(EI)dx$) over the length. The additional surface of a single crack with a limited width will be minor compared to an uncracked calculation. Reason why, beside the adapted boundary conditions, also the stiffness of the concrete material should be adapted for failure analysis purposes.

As following (EN 1992-1-1 2005), an adequate prediction of a deformation parameter y is given by Eq. (15), for members mainly subjected to flexure. In a simplified way, this deformation parameter may be set equal to the deformation y. While looking to (ACI 440.1R-15 2015), the second moment of area should be adopted, and Eq. (16) becomes valid:

$$y = \left(1 - \beta\left(\frac{M_{cr}}{M_a}\right)^2\right) y_{cr} + \beta\left(\frac{M_{cr}}{M_a}\right)^2 y_{uc} \tag{15}$$

$$I_e = \frac{I_{cr}}{1 - \gamma\left(\frac{M_{cr}}{M_a}\right)^2\left[1 - \frac{I_{cr}}{I_{uc}}\right]} \leq I_{uc} \tag{16}$$

where, M_{cr} and M_a are the cracking and applied moment ($M_a \geq M_{cr}$), respectively, y, y_{cr} and y_{uc} are the deflections, y_{cr} based on a fully cracked section and y_{uc} based on an uncracked section. The coefficient β is taking the influence of the duration of the loading or the repetition on the average strain into account. It is set equal to 1 for a single short-term loading and 0.5 for a sustained or repeated loading. I_e, I_{cr} and I_{uc} are the effective, cracked and uncracked moments of inertia, respectively and γ depends on load and boundary conditions and accounts for the length of the uncracked regions of the member and for the change in stiffness in the cracked regions. The factor can be taken equal to $\gamma = 1.72–0.72(M_{cr}/M_a)$, which is in fact only valid for a simply supported beam with a uniformly distributed load.

At first glance, there are a few similarities between the two approaches of the EC and ACI, but after reworking and using the inverse relationship between deformation and second moment of area, we arrive at the very similar equations (Eqs. (17) and 3.5). Both are referring to the uncracked behaviour, which are the results of a linear elastic analysis. While Eq. (17) is expressing a magnification factor applicable on the deformation, Eq. 3.5 acts as a reduction factor on the moment of inertia or results in an increased deformation.

$$y = \left[\left(1 - \beta\left(\frac{M_{cr}}{M_a}\right)^2\right)\frac{I_{uc}}{I_{cr}} + \beta\left(\frac{M_{cr}}{M_a}\right)^2\right] y_{uc} \tag{17}$$

$$I_e = \left[\frac{1}{\left(1 - \gamma\left(\frac{M_{cr}}{M_a}\right)^2\right)\frac{I_{uc}}{I_{cr}} + \gamma\left(\frac{M_{cr}}{M_a}\right)^2} \right] I_{uc} \tag{18}$$

Note that for Eq. (15), in the original formula (EN 1992-1-1 2005) the ratio of the stress in the reinforcement corresponding to the first cracking and the one with the applied moment is used (σ_{sr}/σ_s) instead of (M_{cr}/M_a). Further on, it can be observed that both equations will lead to identical results for short term loading if $\beta = 1 = \gamma$ which means if $M_{cr}/M_a = 1$.

It is now proposed to account for cracking in the region where moments and shear will interact by a smeared approach, not by reducing the inertia but by adjusting the Young's modulus in an artificial way as the product of EI is determining the flexural behaviour.

3.5 *Results*

To account for the influence of cracking between the middle and the supports, the reduction according to Eq. (17) was used (smeared cracking) in combination with the adapted boundary conditions. The results of the FEM analysis will be discussed per article. In the scope of this contribution, attention is mostly given to the simulations with FRP reinforcement.

3.5.1 *Acciai et al.*

Out of the complete data and simulation set only eight results are presented in Figure 5, showing the influence of the boundary conditions by means of the principal stresses. Compression arch effects become visible when the horizontal displacement at the support is prevented. This finding illustrates the importance of the bearing supports. Unfortunately, none of the articles describe this aspect in detail.

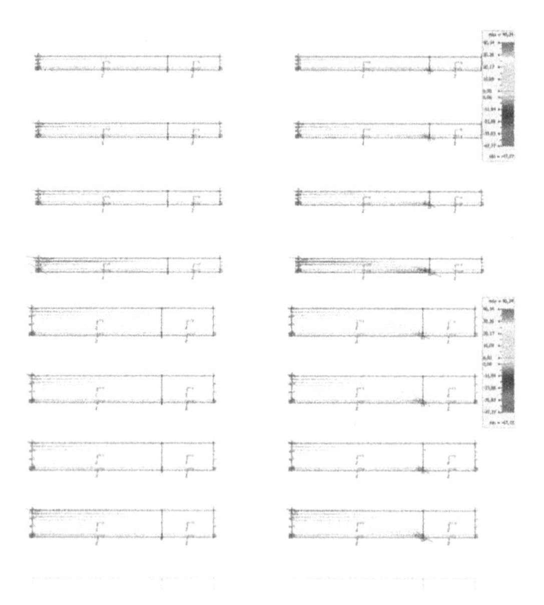

Figure 5. Principal stresses for test 1-4 (top) and 9-12 (bottom), simply supported (left) and with fixed horizontal displacement (right).

Internal normal forces into the reinforcement are developing in a logical way with 0 kN at the supports and maximum values at the symmetrical boundary conditions, see discussion is section 3.4 and Figure 6.

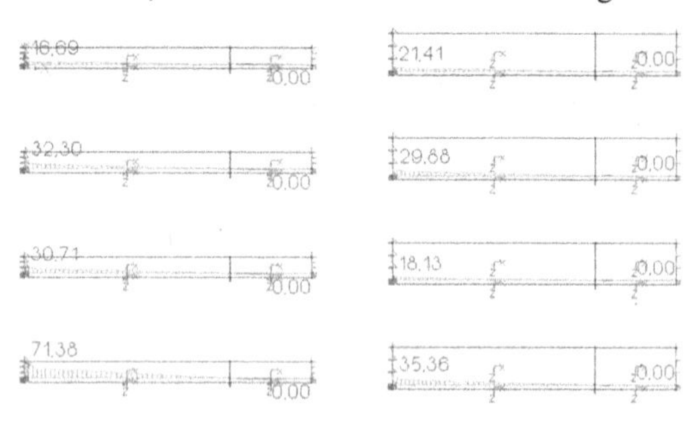

Figure 6. Normal forces into the reinforcement for test 1-4 (left) and 9-12 (right), simply supported conditions.

One of the observations that can be made is that also shear forces are developing in the reinforcement which means that a dowel effect will be activated. The highest shear force and shear deformation region exactly corresponds with the location where shear cracks start. For the slender elements (left hand side of Figure 7) some irregularities will be observed due to the boundary conditions.

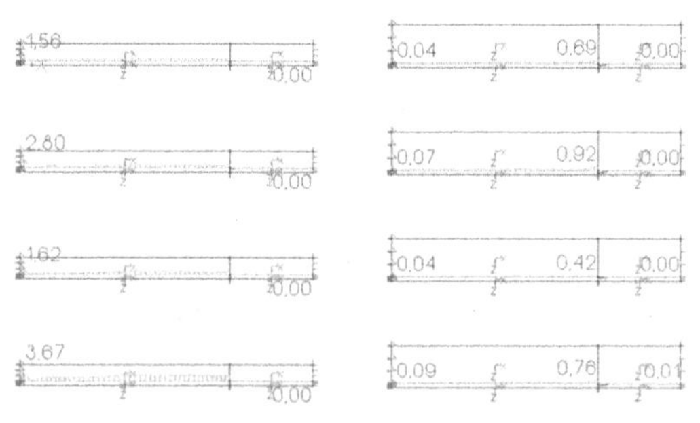

Figure 7. Shear forces into the reinforcement for test 1-4 (left) and 9-12 (right), simply supported conditions.

3.5.2 *Noël & Soudki*

Only two tests have been taken out of this study. Principal stresses are showed in Figure 8 for two types of boundary conditions. The development of normal forces into the reinforcement for the simply supported slab is depicted in Figure 9, and the development of shear forces in Figure 10.

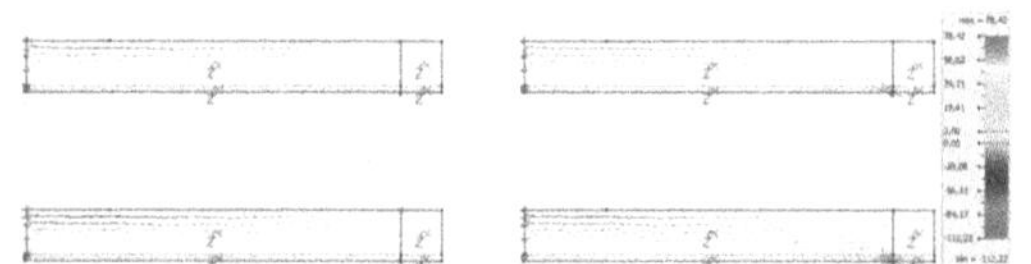

Figure 8. Principal stresses for test G1 (top) and G2 (bottom), simply supported (left) and with fixed horizontal displacement (right).

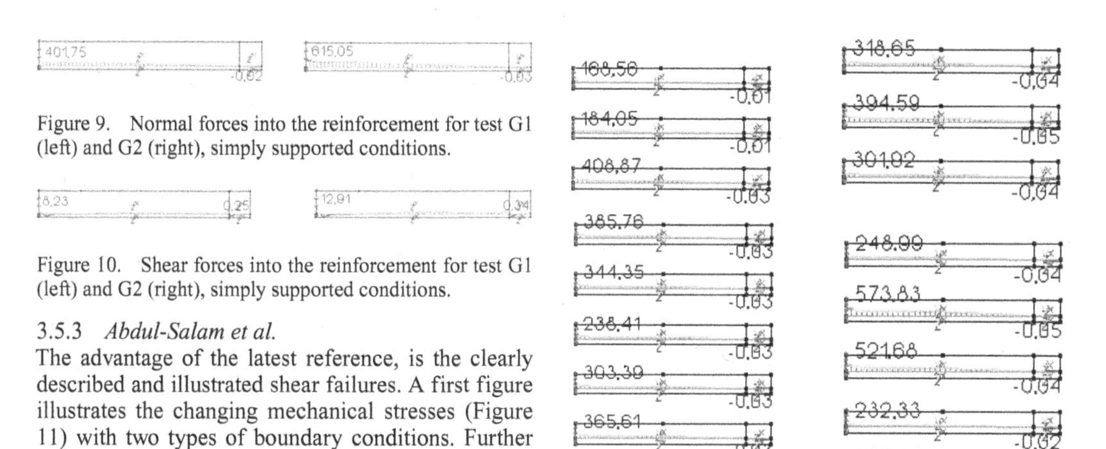

Figure 9. Normal forces into the reinforcement for test G1 (left) and G2 (right), simply supported conditions.

Figure 10. Shear forces into the reinforcement for test G1 (left) and G2 (right), simply supported conditions.

3.5.3 *Abdul-Salam et al.*

The advantage of the latest reference, is the clearly described and illustrated shear failures. A first figure illustrates the changing mechanical stresses (Figure 11) with two types of boundary conditions. Further on, the development of normal (Figure 12) and shear forces (Figure 13) into the reinforcement for the simply supported slab are also shown.

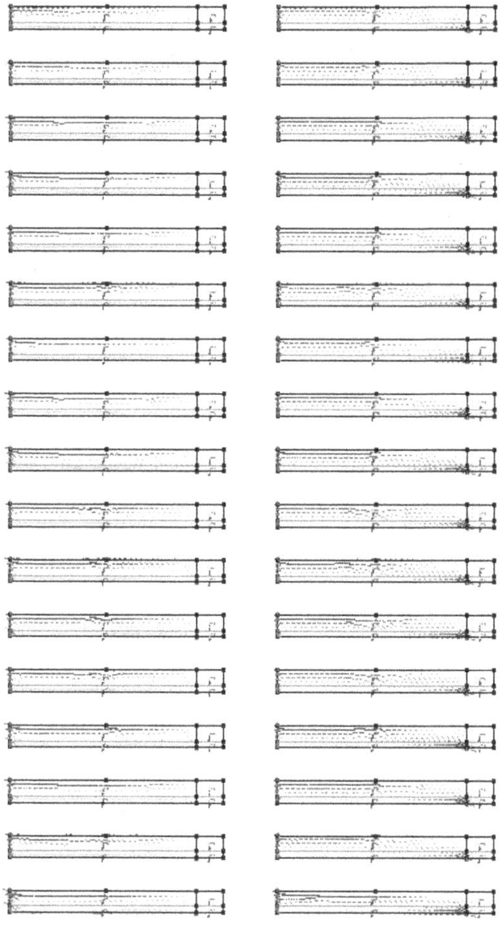

Figure 11. Principal stresses for test 1-17, simply supported (left) and with fixed horizontal displacement (right).

3.5.4 *Validation*

Based on the reduced Young's modulus, following the Eurocode (17), the calculated deformations do show

Figure 12. Normal forces into the reinforcement for test 1-9 (left) and 10-17 (right), simply supported conditions.

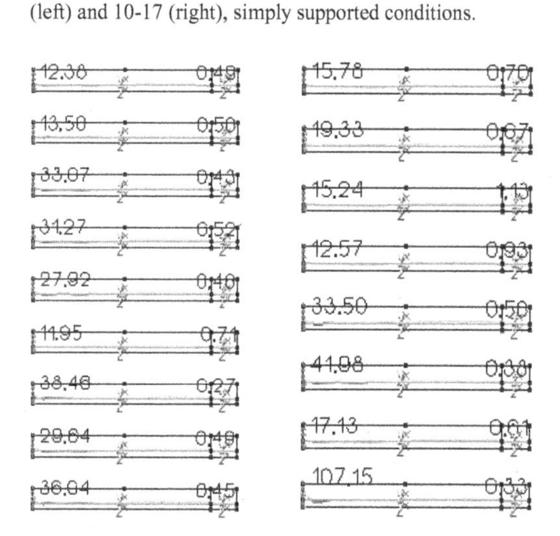

Figure 13. Shear forces into the reinforcement for test 1-9 (left) and 10-17 (right), simply supported conditions.

the same magnitude as the measured ones at failure load, see Figure 14. Results of the tests with steel reinforcement are excluded as they show a plastic plateau which makes the linear elastic approach not appropriate till failure. In general, the calculated values do show an underprediction, so an unsafe result.

Reworking with the ACI approach, 3.5 delivers a slightly better approach but not yet satisfying. This all despite the known and confirmed agreement of both approaches looking to the deformation at the start of yielding (not at failure) when verifying steel reinforced test specimens (Molkens & Van Gysel, 2021). The deviation to the safe side for the tests out of(Noël & Soudki, 2014) can be explained by the relatively low concrete quality which was assumed (C30/37, remember that this information was lacking). With a higher concrete grade, a lower deformation should be calculated resulting in a better agreement.

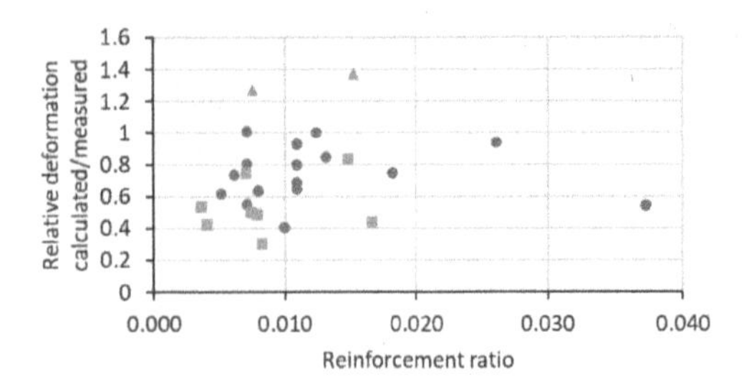

Figure 14. Ratio calculated versus measured deformation (without steel reinforced tests).

4 CASE STUDY (PIER BLANKENBERGE)

4.1 *Description*

The coastal pier in Blankenberge (Belgium) is a carbon steel reinforced concrete construction supporting a walkway of 350 m in the North Sea. Being exposed to a vast amount of chloride ions over the years, the pier has suffered from severe corrosion. Since sustainable concrete repair is not possible anymore, the construction needs to be completely demolished and will be rebuild again according to the original design of 1933. Two alternative reinforcing materials were considered for the new bridge decks: (1) stainless steel rebars and (2) glass fiber rebars. Only the reinforcement configuration with GFRP rebars will be discussed in this paper.

Figure 15. Pier in Blankenberge (*Belgium Pier*, n.d.).

As can be seen from Figure 16, the structure is composed of two repetitive zones. The axis lines of the longitudinal girders and transverse beams create a field pattern, of which only one plate element in zone 1 will be investigated. It spans 2.69 m and 4 m in the pier's longitudinal and transverse direction, respectively.

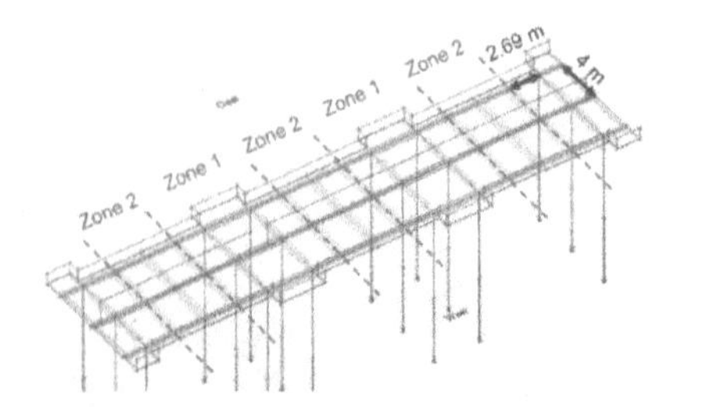

Figure 16. 3D model of the concrete structure.

Table 4 and Table 5 present the geometrical characteristics and mechanical properties of the materials.

Table 4. Geometrical characteristics.

Concrete	
Slab Thickness	180 mm
Concrete cover	25 mm
Glasspree bars of supplier Sireg	
Diameter (top and bottom reinforcement in both directions)	M19
Spacing	80 mm

Table 5. Mechanical properties.

Concrete	
Grade (characteristic/mean compressive strength)	C35/45
Glasspree bars of supplier Sireg	
Modulus of elasticity	46 GPa
Characteristic tensile strength	800 MPa
Transverse shear strength	>150 MPa
Bond strength	>8 MPa

4.2 *Test arrangements*

In the near future, shear tests will be performed on two floor slabs with a length of 3 m and a single span of 2.70 m, which is similar to the distance between the secondary transverse beams in the original design of the pier in Blankenberge.

During the first reference test, a simply supported slab will be loaded up to failure. This specimen only requires 8 GFRP bars (ø = 16 mm) with a spacing of 100 mm at the bottom face, see Figure 17 left hand side.

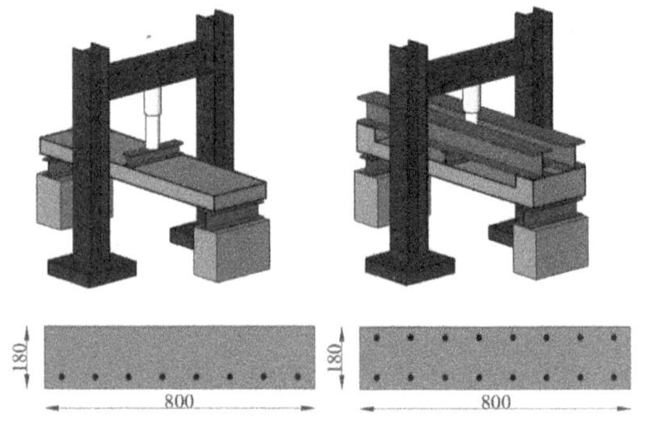

Figure 17. Reference test (left) and test with restrained boundary conditions (right).

During the second test, a slab restrained and clamped at both ends will be tested for the purpose of a robustness check in accidental situations (e.g. collapse of a boat, truck, and even fire). These fixed ends are provided by threaded anchor bar connections through the bottom flange of 2 stiff beams and the concrete slab. In order to check whether membrane action can be formed in this element, it is required to provide 8 GFRP bars (ø = 16 mm) with a spacing of 100 mm at the top face as well. It is generally known that FRP reinforcement only exhibits an elastic behaviour. Due to the lack of a plastic behaviour, FRP bars cannot benefit from any redistribution of forces in accidental conditions. Nevertheless, it might be possible that catenary action develops in such cases thanks to the intact reinforcement at two critical locations. The bottom reinforcement is expected to fail at midspan while failure of the top reinforcement will presumably occur near the supports.

5 FUTURE RESEARCH

As mentioned in chapter 4, the FEM modelling approach will be validated with the experimental results of both shear tests that will take place in the near future. In addition to the shear behaviour, robustness aspects and post-failure behaviour will be investigated.

As earlier discussed, moment/shear interaction should be studied in a more profound way.

Lacking non-linear material models do obstruct further research on the behaviour of FRP-reinforced concrete when subjected to fire.

6 CONCLUSIONS

A wide set of experiments was used to validate the developed FEM approach, in which concrete was modelled as a shell element and the reinforcement was converted to a single rectangular beam element with an effective width and depth. Some lacking information in the data sets of the reference articles hindered an appropriate comparison of the numerical and experimental results.

A closer prediction of the cracked deformation is also needed in order to determine the ultimate deformation capacity of FRP reinforced elements in a better way. With a better understanding of this aspect, failure mechanisms can also be predicted more accurately.

REFERENCES

Abdul-Salam, B., Farghaly, A. S., & Benmokrane, B. (2016). Mechanisms of shear resistance of one-way concrete slabs reinforced with FRP bars. *Construction and Building Materials*, *127*, 959–970. https://doi.org/10.1016/j.conbuildmat.2016.10.015

Acciai, A., D'Ambrisi, A., De Stefano, M., Feo, L., Focacci, F., & Nudo, R. (2016). Experimental response of FRP reinforced members without transverse reinforcement: Failure modes and design issues. *Composites Part B: Engineering*, *89*, 397–407. https://doi.org/10.1016/j.compositesb.2016.01.002

ACI 440.1R-15. (2015). *Guide for the Design and Construction of Structural Concrete Reinforced with Fibre-Reinforced Polymer (FRP) Bars.*

Belgium Pier. (n.d.). https://www.visit-blankenberge.be/belgium-pier

Bielak, J., Adam, V., Hegger, J., & Classen, M. (2019). Shear capacity of textile-reinforced concrete slabs without shear reinforcement. *Applied Sciences (Switzerland)*, *9*(7). https://doi.org/ 10.3390/app9071382

Classen, M. (2020). Shear Crack Propagation Theory (SCPT) – The mechanical solution to the riddle of shear in RC members without shear reinforcement. *Engineering Structures*, *210*(February), 110207. https://doi.org/10.1016/j.engstruct.2020.110207

El-Sayed, A. K., El-Salakawy, E. F., Benmokrane, B., Sherwood, E. G., Bentz, E. C., & Collins, M. P. (2007). Shear strength of FRP-reinforced concrete beams without transverse reinforcement. *ACI Structural Journal*, *104*(1), 113–114. https://doi.org/10.14359/18439

EN 1992-1-1. (2005). *Eurocode 2: Design of concrete structures – Part 1-1: General rules and rules for buildings.*

Gudonis, E., Timinskas, E., Gribniak, V., Kaklauskas, G., Arnautov, A. K., & Tamulėnas, V. (2014). Frp Reinforcement for Concrete Structures: State-of-the-Art Review of Application and Design. *Engineering Structures and Technologies*, *5*(4), 147–158. https://doi.org/10.3846/2029882x.2014.889274

Kashwani, G. A., & Al-Tamimi, A. K. (2014). Evaluation of FRP bars performance under high temperature. *Physics Procedia*, *55*, 296–300. https://doi.org/10.1016/j.phpro.2014.07.043

Lima, J., Reis, L., & Oliveira, D. (2021). A model for shear resistance of reinforced concrete beams. *ACI Structural Journal*, *118*(5), 17–26. https://doi.org/10.14359/51732863

Mihaylov, B.I. (2017). Two-parameter kinematic approach for shear strength of deep concrete beams with internal FRP reinforcement, *Journal of composites for construction, 21,*

Molkens, T., Gernay, T., & Caspeele, R. (2017). *Fire Resistance of Concrete Slabs Acting in Compressive Membrane Action*. 1–8.

Molkens, T., & Van Gysel, A. (2021). Structural behavior of floor systems made by floor plates—mechanical model based on test results. *Applied Sciences (Switzerland)*, *11*(2), 1–25. https://doi.org/10.3390/app11020730

Muttoni, A., Shear crack theory

Noël, M., & Soudki, K. (2014). Estimation of the crack width and deformation of FRP-reinforced concrete flexural members with and without transverse shear reinforcement. *Engineering Structures*, *59*, 393–398. https://doi.org/10.1016/j.engstruct.2013.11.005

Pilakoutas, K., Guadagnini, M., Neocleous, K., & Matthys, S. (2011). Design guidelines for FRP reinforced concrete structures. *Proceedings of the Institution of Civil Engineers: Structures and Buildings*, *164*(4), 255–263. https://doi.org/10.1680/stbu.2011.164.4.255

Wang, Y. C., Wong, P. M. H., & Kodur, V. K. R. (2003). Mechanical properties of FRP reinforcing bars at elevated temperatures. *ASCE/SFPE Specialty Conference of Designing Structures for Fire*, 1–10.

Xing-lang, F., Sheng-jie, G., xi, W., Jiafei, J. (2020). Critical shear crack theory-bassed punching shear model for FRP-reinforced concrete slabs. *Advances in structural engineering*, https://doi.org/10.1177/1369433220978146.

Yang, Y. (2014). Shear behaviour of reinforced concrete members without shear reinforcement: a new look at an old problem. *PhD thesis, TU Delft*, https://doi.org/10.4233/uuid:ac776cf0-4412-4079-968f-9eacb67e8846

Computational Modelling of Concrete and Concrete Structures – Meschke, Pichler & Rots (Eds)

Application of NLFEA for crack width calculations in SLS

O. Terjesen
Department of Structural Engineering, University of Agder, Norway

T. Kanstad & R. Tan
Department of Structural Engineering, Norwegian University of Science and Technology, Norway

ABSTRACT: In this paper, computer-based simulation is carried out using the Finite Element Analysis (FEA) package Abaqus to study crack widths in reinforced concrete beams. A set of experimentally tested beams are investigated, and measured crack widths are compared with crack widths predicted by nonlinear FEA (NLFEA) and relevant design codes. It is shown that Eurocode 2 (EC2), fib Model Code 2010 (MC2010) and the draft for new EC2 underestimates the crack widths at the outermost concrete face to different extents while they are conservative at reinforcement level. Crack widths predicted by NLFEA, on the other hand, provides good crack width predictions at the outermost concrete face for both investigated beams.

1 INTRODUCTION

Crack widths in concrete structures should be limited due to aesthetics, durability, and functional requirements (e.g., tightness). Although research related to this topic has been ongoing since modern time, large uncertainties and large need for further research remains. The large uncertainties are especially due to large scale concrete structures, the large concrete covers applied for structures in harsh environments, and introduction of more eco-friendly modern concretes (Basteskår et al. 2018). Strict crack width limits lead to increased amount of reinforcement and the economic consequences are proven to be large (Basteskår et al. 2019).

The work presented is part of the PhD-project of the first author and are related to the large research activity funded by the large Norwegian infrastructure project "Ferry-free E39" and the PhD work of Reignard Tan (Tan, Reignard 2019).

The main objective of this paper is to investigate how nonlinear finite element analysis (NLFEA) can be applied to predict maximum crackwidths, which furthermore are compared to crack widths predicted by analytical calculation methods in design codes such as Eurocode 2 (EC2) and *fib* Model Code 2010 (MC2010). The study is benchmarked against the experimental results from the comprehensive and well documented beam tests of Hognestad (1962).

2 CONCRETE DAMAGE PLASTICITY

The Concrete Damage Plasticity (CDP) model is a continuum, plasticity-based, damage model for concrete and is in Abaqus based on the models proposed by Lubliner et al. (1989) and by Lee and Fenves (1998). It is assumed that the two main failure mechanisms are tensile cracking and compressive crushing of the concrete material. The evolution of the yield (or failure) surface is controlled by two hardening variables in tension (ε_t^{pl}) and compression (ε_p^{pl}), linked to the respective failure mechanisms.

The experimental behaviour of reinforced concrete beams cannot be captured by elastic damage models or elastic-plastic constitutive laws only. Because in such models irreversible strains cannot be captured. In Figure 1b it can be noticed that a zero stress corresponds to a zero strain which makes the damage value underestimated. On the other hand, when an elastic plastic relation is adopted, the strain will be overestimated since the unloading curve will follow the elastic slope as shown in Figure 1c.

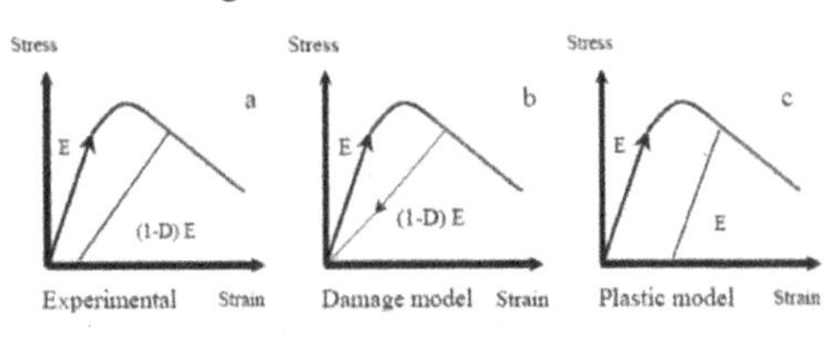

Figure 1. Elastic plastic damage law (Jason et al. 2004).

DOI 10.1201/9781003316404-30

The CDP model is combining the stress-strain curves in Figure 1b and c into Figure 1a so that we can better capture the constitutive behaviour of concrete. In SLS-design, compressive crushing of the concrete is generally not a problem and therefore the damage model for compression is excluded from the analyses described in this paper.

2.1 *Material constitutive behaviours*

The applied numerical models for the constituent material properties are described in this section

2.1.1 *Concrete model*

CDP describes the constitutive behaviour of concrete by introducing scalar damage variables. Both tensile and compressive response of concrete can be characterized by CDP, and the tensile response is depicted in Figure 2. Concrete behaviour in compression are not explained in this section due to investigated beams being within the elastic compression range.

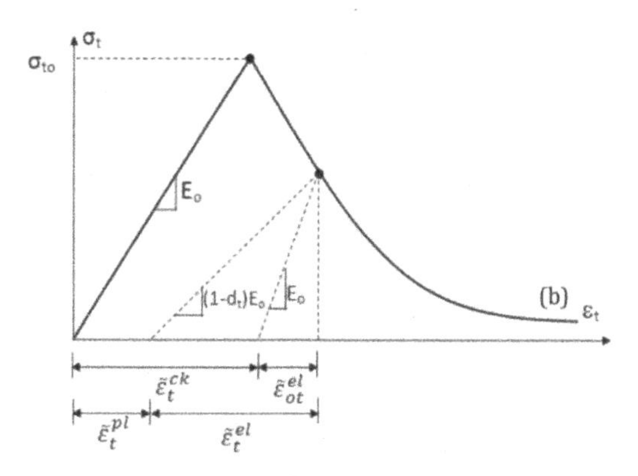

Figure 2. Behaviour of concrete under axial tension according to CDP (Abaqus User Manual 2014).

As shown in Figure 2, the unloading response of concrete specimen is weakened because the elastic stiffness of the material appears to be damaged or degraded. Damage associated with the failure mechanisms of the concrete (cracking and crushing) results in a reduction in the elastic stiffness. The CDP-model characterizes this by a scalar damage variable, d_t which can take values from zero (undamaged material) to one (fully damaged material). (Abaqus User Manual 2014). E_0 is the initial (undamaged) elastic stiffness of the material and $\tilde{\varepsilon}_t^{pl}$ and $\tilde{\varepsilon}_t^{in}$ are tensile plastic strain and inelastic strain respectively. The stress-strain relation under uniaxial tension is taken into account in Eq. (1).

$$\sigma_t = (1 - d_t) \cdot E_0 \cdot (\varepsilon_t - \tilde{\varepsilon}_t^{pl}) \tag{1}$$

A strain softening behaviour at the crack is assumed in the model. Thus, it is necessary to define the behaviour of plain concrete in tension for the CDP-model. ABAQUS allows the user to specify concrete

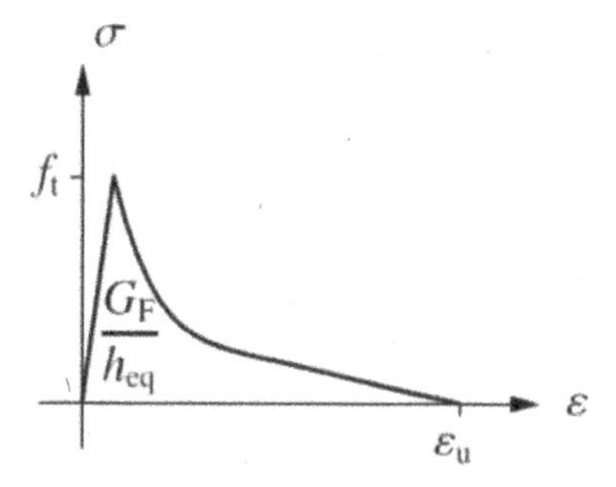

Figure 3. Hordijk softening curve (Hordijk & Dirk Arend 1991).

by post a failure stress-strain relation or by applying a fracture energy cracking criterion (Abaqus User Manual 2014) The former relation is used by the authors.

The stress strain relation for concrete in tension must be given to Abaqus in terms of the cracking strains, $\tilde{\varepsilon}_t^{ck}$, and corresponding yield stresses σ_{t0} which are determined from the nonlinear Hordijk curve (Hordijk, Dirk Arend. 1991). The exponential-type of softening diagram shown in Figure 3 will typically result in localized strains when the concrete in a structural member crack.

The area under the stress-strain curve should be equal to the fracture energy (G_f) divided by the equivalent length (h_{eq}) often called crack bandwidth. After complete softening i.e., when virtually no stresses are transmitted, the crack is said to be "fully open". The ultimate strain parameter in case of the Hordijk softening curve is given by

$$\varepsilon_u = 5.136 \frac{G_F}{h_{eq} f_t} \tag{2}$$

where f_t is the tensile strength of the concrete. The softening curve is given by

$$\sigma = \begin{cases} f_t \left(\begin{array}{l} \left(1 + \left(c_1 \frac{\varepsilon^{cr}}{\varepsilon_u}\right)^3\right) \exp\left(c_2 \frac{\varepsilon^{cr}}{\varepsilon_u}\right) \\ -\frac{\varepsilon^{cr}}{\varepsilon_u}\left(1 + c_1^3\right) \exp\left(-c_2\right) \end{array} \right) & 0 \le \varepsilon^{cr} \le \varepsilon_u \\ 0 & \varepsilon^{cr} > \varepsilon_u \end{cases} \tag{3}$$

where c_1 and c_2 are parameters used to obtain the stress-crack width opening relation for concrete from deformation-controlled uniaxial tensile tests (Hordijk & Dirk Arend 1991). The recommended values are 3 and 6.93 respectively and are also applied in this study. The determination of the fracture energy G_f in tension is more complicated, and the authors have chosen this value to be as recommended by the Dutch guidelines (Hendriks 2017) and *fib* Model Code 2010 (fib 2013).

$$G_F = 0.073 f_{cm}^{0.18} \tag{4}$$

The tension softening data according to the Hordijk curve in Equation 3 are given to Abaqus in terms of cracking strain $\varepsilon_t^{\sim ck}$ and yield stress σ_{t0} as shown in Figure 2. When the unloading data are available, the data are provided to Abaqus in terms of tensile damage curves, $d_t - \varepsilon_t^{\sim ck}$. Abaqus automatically converts the cracking strain values to plastic strain values using the relationship given by:

$$\varepsilon_t^{\sim pl} = \varepsilon_t^{\sim ck} - \frac{d_t}{(1-d_t)}\frac{\sigma_t}{E_0} \tag{5}$$

From this equation the effective tensile cohesion stress ($\bar{\sigma}_t$) determines the size of the yield (or failure) surface as:

$$\bar{\sigma}_t = \frac{\sigma_t}{(1-d_t)} = E_0(\varepsilon_t - \varepsilon_t^{\sim pl}) \tag{6}$$

In Abaqus the parameters required to define the CDP-model consists of four constitutive parameters. First the angle of internal material friction of the concrete 'ψ' measured in the p-q plane at high confining pressure, and in this study, is chosen as recommended default value. The second parameter is the eccentricity ? which defines the rate at which the hyperbolic flow potential flow potential approaches its asymptote and is chosen as default value of 0.1. The third parameter is the ratio of initial biaxial compressive yield stress to initial uniaxial compressive yield stress, 'fb0/fc0', with a default value of 1.16. The fourth parameter is the ratio of the second stress invariant on the tensile meridian to the compressive meridian at initial yield with a default value of 2/3 (Abaqus User Manual 2014).

The parameter 'Kc' should be defined based on the full triaxial tests of concrete, moreover, a biaxial laboratory test is necessary to define the value of 'fb0/fc0'. This paper does not discuss the identification procedure for parameters 'ϵ', 'fb0/fc0', 'Kc' or 'ψ' because the test series that is in this study does not have such information. Therefore, default values have been chosen.

In nonlinear finite element programs, the material models softening behaviour and stiffness degradation can often lead to severe convergence difficulties. A common technique to overcome some of these difficulties is the use of a viscoplastic regularization of the constitutive equations, which causes the consistent tangent stiffness of the softening material to become positive for sufficiently small-time increments. The CDP-model in Abaqus can be regularized by using viscoelasticity to permit stresses to be outside of the yield surface. Using a small value for the viscosity parameter (μ) (small compared to the characteristic time increment) usually helps to improve the rate of convergence of the model in the softening regime, without compromising the results (Abaqus User Manual 2014). The viscosity value used by the authors in this work was chosen as 0 and 0.0001 which is shown to be sufficiently low to give realistically results (Demir et al. 2018). The plasticity damage parameters used by the authors are shown in Table 1.

Tension stiffening is implicitly modelled by the chosen tensile softening law and corresponding chosen mesh, thus causing localization of cracking strains in the tensile zone of the investigated beams for the concrete elements. Distance between localized cracking strains becomes analogous to a crack spacing. This in turn should result in steel strains varying between the crack spacing, having its maximum at a crack and its minimum between two consecutive cracks. This also means that tension stiffening should be accounted for without having to explicitly model the bond between concrete and steel.

Table 1. Plasticity damage parameters.

Ψ	E	fb0/fc0	Kc	μ
35	0.1	1.16	0.667	0 and 0.0001

3 PREDICTION OF CRACK WIDTHS

The crack width calculation methods according to EC2, MC2010 and the drafts for the new versions of EC2 are briefly highlighted in the following. Chosen values for the parameters used in the subsequent crack width calculates are also addressed.

3.1 *Eurocode 2 Part 1-1*

The method for calculation of crack widths applies the following equation:

$$w = S_{r,max}(\varepsilon_{sm} - \varepsilon_{cm}) \tag{7}$$

Where $S_{r,max}$ is the maximum crack spacing for a stabilized cracking stage expressed as:

$$S_{r,max} = k_3 c + k_1 k_2 k_4 \frac{\varphi}{\rho_{s,ef}} \tag{8}$$

Here $k_1 = 0.8$, $k_2 = 0.5$, $k_3 = 3.4$ and $k_4 = 0.425$ are chosen, while φ is the diameter of longitudinal reinforcement and $\rho_{s,ef}$ is the reinforcement ratio in the effective concrete tensile zone. The difference in mean strains is calculated according to:

$$(\varepsilon_{sm} - \varepsilon_{cm}) = \frac{\sigma_s - k_t \frac{f_{ctm}}{\rho_{s,ef}}\left(1 + \alpha_e \rho_{s,ef}\right)}{E_s} \geq 0.6\frac{\sigma_{sr}}{E_s} \tag{9}$$

where σ_s is the reinforcement stress, and k_t is dependent on load duration (short- or long-term loading) and varies from 0.4 to 0.6. The authors have chosen $k_t = 0.6$ due to the probable absence of creep and shrinkage in the experimental results and applies in general as a chosen value for the other codes as well. The ratio between steel and concrete Young's modules is defined as $\alpha_e = E_s/E_{cm}$ (Eurocode 2 Part 1-1, 2004).

3.2 *Model Code 2010*

The maximum calculated crack width at the height of the reinforcement is found by:

$$w = 2l_{s,max}(\varepsilon_{sm} - \varepsilon_{cm}) \quad (10)$$

when the term related to shrinkage strains is neglected. Here, $l_{s,max}$ denotes the length over which slip between concrete and steel is assumed to occur and is expressed by:

$$l_{s,max} = k \cdot c + \frac{1}{4}\frac{f_{ctm}}{\tau_{bms}}\frac{\varphi_s}{p_{s,ef}} \quad (11)$$

where $k = 1$ is an empirical parameter considering the influence of the concrete cover chosen according to the recommended value and c is the concrete cover. The mean bond strength between steel and concrete is chosen as $\tau_{bms} = 1.8f_{ctm}$. The relative mean strain in Equation 10 is the same as chosen in Equation 9 but the lower bound limits between the mean strains are different.

MC2010 allows for extrapolation of the crack width at the reinforcement height given in Equation 10 by a factor (h-x)/(d-x) where, h is cross-section height, x is the height of the compressive sone, and d is the effective height. This extrapolation is valid for cover up to 75mm. For larger covers a more detailed analysis is required and procedures based on fracture mechanics approach would be appropriate.

3.3 *Draft for the new Eurocode 2, 2022 (pr EN 1992-1-1)*

In the draft for the new Eurocode 2 the calculation of crack width is expressed as:

$$w_{k,cal} = k_w S_{rm,cal}(\varepsilon_{sm} - \varepsilon_{cm}) \quad (12)$$

where $k_w = 1.7$ is a factor converting the mean crack width into a calculated crack width and is chosen according to the recommended value. $S_{rm,cal}$ is the calculated mean crack spacing assumed to be valid for both initial cracking and a stabilized crack pattern.

For elements subjected to direct loads or subjected to imposed strains $\varepsilon_{sm} - \varepsilon_{cm}$ can be expressed as:

$$\varepsilon_{sm} - \varepsilon_{cm} = k_{1/r}\frac{\sigma_s - k_t\frac{f_{ctm}}{\rho_{s,ef}}\left(1 + \alpha_e\rho_{s,ef}\right)}{E_s} \geq 0.6\frac{\sigma_{sr}}{E_s} \quad (13)$$

Where $k_{1/r}$ is a coefficient to account for the increase of crack width due to curvature which is expressed as:

$$k_{1/r} = \frac{h - x}{h - a_{y.i} - x} \quad (14)$$

Here x is the distance to the neutral axis, and $a_{y.i}$ is the cover distance plus rebar size. The mean crack spacing is:

$$S_{r,m,cal} = 1.5c + \frac{k_{fl}k_b}{7.2} \cdot \frac{\varphi}{\rho_{p,ef}} \quad (15)$$

where c is cover to the longitudinal reinforcement, φ is bar diameter, $k_b = 0.9$ is a coefficient for bond properties for ordinary reinforcement chosen according to the recommended value and $k_{fl} = (h - h_{c,eff})/h$, where h is cross-section height and $h_{c,eff}$ is the effective tension area.

3.4 *NLFEA and codes*

EC2 and MC2010 both state that SLS verifications using NLFEA can be performed a posteriori. In the case of bending cracks, the crack opening (w) may be calculated according to Dutch guidelines (Hendriks 2017):

$$w = S_{r,max} \cdot \bar{\varepsilon}_s \quad (16)$$

Where $\bar{\varepsilon}_s$ is the mean strain value of the longitudinal reinforcement in the cracked zone obtained in the analysis and $S_{r,max}$ is the maximum crack spacing according to EC2.

4 EXPERIMENTAL TEST AND FEA MODELLING

4.1 *Hognestad beam tests, control of flexural cracking*

From the established database, the investigation carried out by Hognestad (1962) was chosen as appropriate for this paper. This experimental work involved 36 rectangular beams with a length of 3429 mm. Different parameters were chosen as major variables such as bar diameter, bar type, concrete strength, reinforcement ratio, beam width and depth and thickness of cover as shown in Table 2 (Hognestad 1962). All beams were loaded by twin-loads at the third points of the span. To prevent shear failures, the outer thirds were reinforced with ø10 stirrups. The beams examined in this study are No 31 and 32, with respective properties given in Table 3. The different parameter variables shown in Table 2 are included to highlight the extensive work done by Hognestad and are relevant for further work.

Table 2. Parameter variations done by Hognestad.

Beams No.	Major Variable	Description
1–4	Bar diameter	Size and number of rebars
5–7	Bar diameter	Size and number of rebars
8–10	Bar diameter	Size and number of rebars
11–12	Bar diameter	Size and number of rebars
13–16	Bar diameter	Size and number of rebars
17–20	Bar diameter	Size and number of rebars
21–24	Beam width	Size and number of rebars
25–28	Beam depth	Size and number of rebars
29–32	Concrete cover	horizontal cover
33–36	Concrete cover	vertical and horizontal cover

* Both compressive and tensile concrete strength varied for the test series (Hognestad 1962).

Table 3. Geometrical and material properties for Beam No 31 and 32.

Description	mm	Description	MPa
Beam height*	406	fck*	25,1
Beam width*	203	fct*	2,57
Cover vertical B31	63	Es*	200.000
Cover vertical B32	112	Ec*	31.504
Cover horizontal*	25		
Effective depth B31	322		
Effective depth B32	294.5		
Beam length*	3429		
Bar size*	22		
Number of bars*	2		

* Properties shared by both beams No 31 and 32.

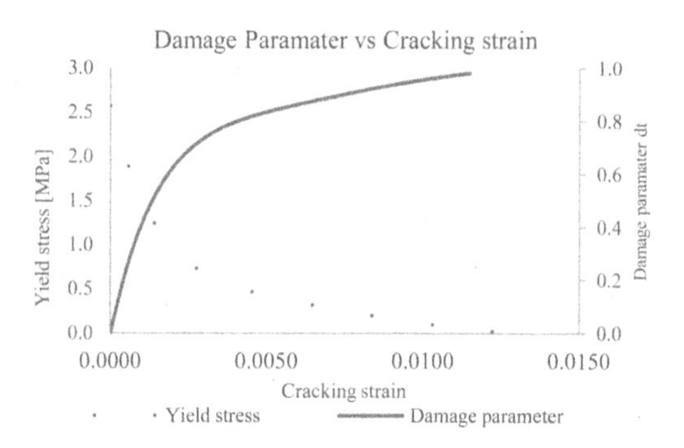

Figure 4. Softening branch of concrete in tension with corresponding damage parameter development applied by Abaqus.

Table 4. Stress-strain values for reinforcement and steel plates.

Yield Stress (σ_t) MPa	Plastic Strain $\varepsilon^{\sim pl}$
Reinforcement:	
575*	0.0
Steel plates:	
275*	0.0

* Both steel plates and reinforcement never reach yielding during the analysis and plastic strains are therefore not calculated

4.2 *Finite element modelling of the RC beams*

To develop the FE models of the RC beams, steel loading- and support plates as well as the concrete cross-section were modelled using 3D brick elements. The FE models thus consist of three types of materials (concrete, steel plate, reinforcement). The embedded reinforcement technique available in ABAQUS is also used. The beams are reinforced with 22 mm rebar diameters with either 84 mm or 122.5mm distance from the outermost surface to the centroid of the reinforcement.

The elements chosen for concrete and steel plates in Abaqus is C3D20R quadratic brick elements with reduced integration (20 nodes and 8 integration points). The element size is approximately 20x20x20 mm and chosen in accordance with Dutch guidelines (Hendriks 2017) maximum element size for NLFEA. For the longitudinal reinforcement wire elements each with a length of 20 mm is used. The loading of both beams are displacement controlled.

There is a mesh sensitivity problem in cases with little or no reinforcement with the specification of a post failure stress-strain relation, in the sense that the finite element predictions do not converge to a unique solution as the mesh is refined because mesh refinement leads to narrower crack bands. In these beam models a post failure material behaviour as explained earlier with tension stiffening derived from Hordijk softening curve is applied and the cracking failure are distributed evenly and results in additional cracks and mesh sensitivity analysis with other element sizes is not performed.

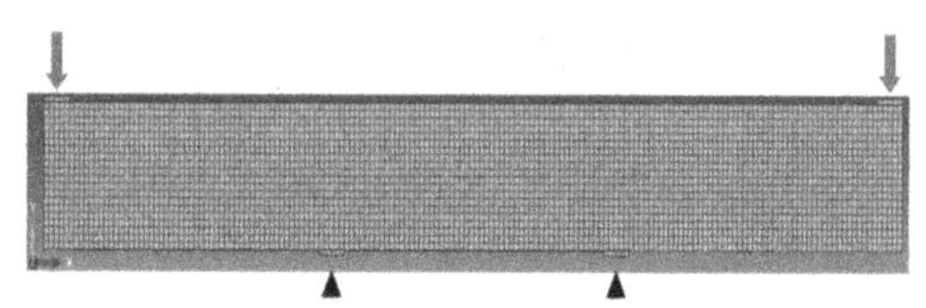

Figure 5. Model of Hognestad Beam in Abaqus.

5 RESULTS

5.1 *Load displacement behaviour*

The load displacement curves were not reported by Hognestad and therefore the FEA load-displacement is used as an indicator for crack development and used to compare when cracking occurs. Also, some sensitivity checks applying various values for the previously discussed viscosity parameter are performed. Viscosity parameters equal to 0 and 0.0001 were used, and from 5.2 we can observe that for beam No. 31 that when initial cracking occurs at approximately 20 kN loading there is a slight difference between the two solutions. This is due to that the viscosity parameter greater than 0 allows for stresses outside the yield surface but provides accurate enough results. For beam No. 32 the Viscosity parameter of 0 are not done due to the iterative process and length of the analysis required.

5.2 *Experimental crack widths*

From the Hognestad beam tests measured surface crack widths at both the height of the steel centroid and concrete top face are reported. The results for the selected beams are given in 5.3. From the measured crack widths, we notice that the crack widths at the height of reinforcement are similar regardless of concrete cover.

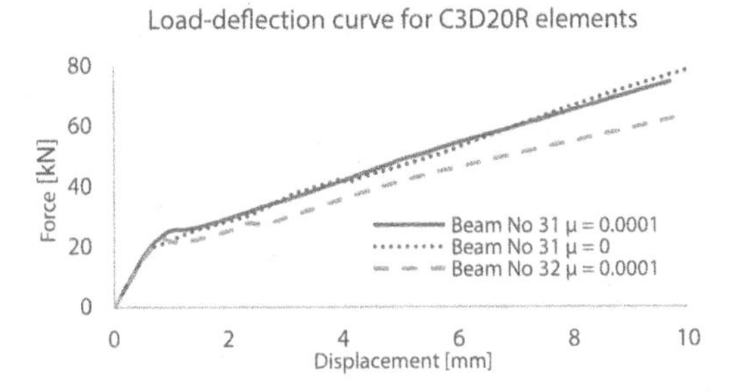

Figure 6. Load deflection curve for different viscosity parameter.

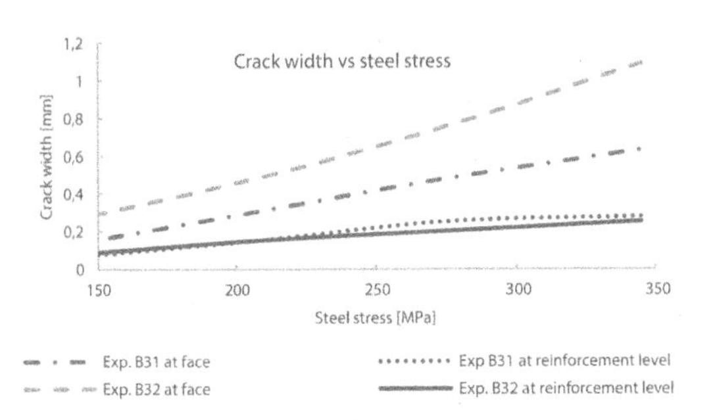

Figure 7. Experimental maximum crack widths vs steel stresses for beam No. 31 and No. 32 (Hognestad 1962).

5.3 *Maximum crack width predicted by design codes*

The predicted maximum crack widths according to EC2, MC2010 and the draft for new EC2 from equations 7,10 and 12 are compared in Figure 8.

It can be noted that for both beams the estimated crack widths are conservative at the height of reinforcement but underestimated at the outermost concrete face for EC2 and the draft for new EC2. MC2010 predict the crack width at the outermost concrete face to a good extent for 62 mm cover but underestimate it for 112 mm cover. The extrapolation of the results to get the crack width at the outermost concrete face are not valid for a larger cover than 75mm but are chosen to be included here.

The new term ($k_{1/r}$) accounting for the curvature in the new EC2 looks to provide a better result for the crack width at increased steel stresses beyond 250 MPa for both beams than the current EC2.

5.4 *Calculations of crack widths combining NLFEA and EC2*

The maximum crack width is calculated from Equation 16. Mean steel strains ($\bar{\varepsilon_s}$) for Beam No. 31 and 32 are extracted from the NLFEA. The maximum crack spacing ($S_{r,max}$) is calculated from equation 8 in accordance with EC2. In addition, the measured maximum spacing between the cracks in the constant moment zone from the Abaqus models at the stabilized cracking stage is also used (steel stress close to 350MPa).

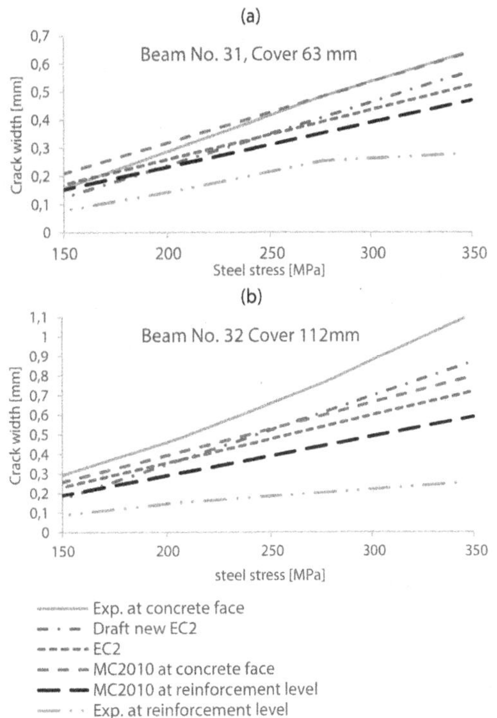

Figure 8. Crack widths predicted by design codes, (a) Beam No. 31, (b) Beam No. 32.

From Figure 9 we can determine the maximum crack spacings from where we have a stabilized cracking pattern at $\sigma_s = 350$ MPa, to (a) $S_{r,max} = 240$ mm and (b) $S_{r,max} = 300$ mm.

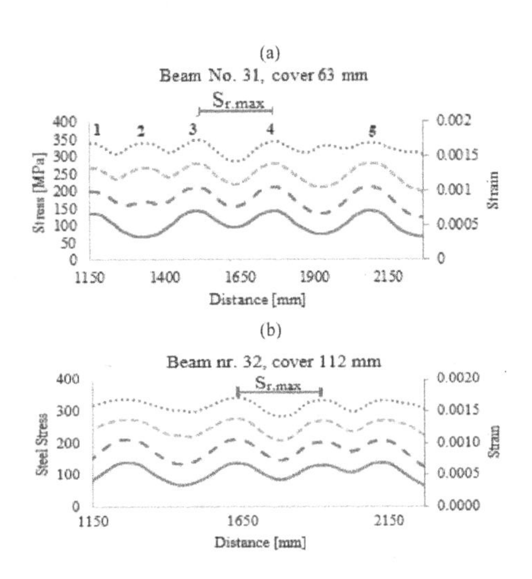

Figure 9. Steel stress levels and corresponding strains along the rebar length in the cracked concrete zone (constant moment), (a) Beam No. 31 numbers 1-5 indicate the localization of cracking strains in Figure 13, (b) Beam No. 32.

From Figure 10 the method based on extracting mean steel strains from the NLFEA and using the EC2 formulation for $S_{r,max}$ and the maximum crack spacing from the analysis shown in Figure 9 to calculate the crack widths at the reinforcement height are conservative. On the other hand, the EC2 formulation for maximum crack spacing fits better at the outmost concrete face than the maximum crack spacing from the analysis.

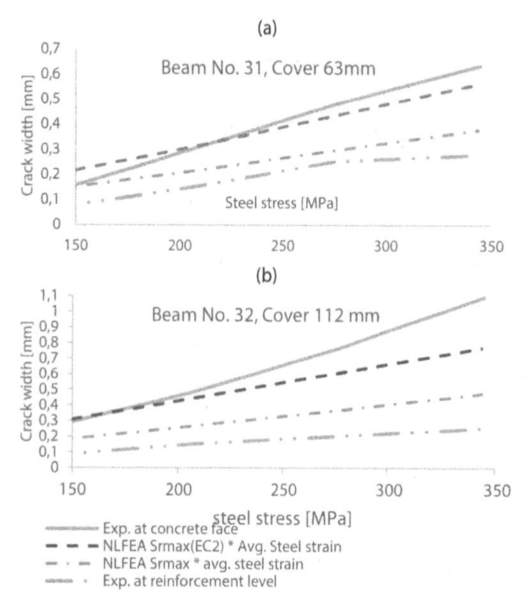

Figure 10. Crack widths estimated by extracting steel strains from NLFEA, (a) Beam No. 31, (b) Beam No. 32.

5.5 *Crack width determined by the Concrete Damage Plasticity model*

From the results in Abaqus the cracking strains are found meaning we can determine the crack width as:

$$w = \varepsilon^{cr} \cdot h_{eq} \quad (17)$$

The cracks localize within the brick elements, and at the top face of the beam the crack widths vary over the width of the beam. The crack widths are calculated by selecting the cracked elements across the beam width and using average cracking strain ε^{cr} multiplied with the crack band width (h_{eq}) which is an essential parameter in constitutive models that describe the softening stress-strain relationship. The preferred method is a method based on the initial direction of the crack and the element dimensions (Hendriks 2017). For both beams the length of the crack band width is 20mm. The development of the crack width using this method is shown in Figure 11. The crack localizations are visualized in 6.

Crack 1 in Figure 11 is selected representing the maximum crack width for both beams and compared

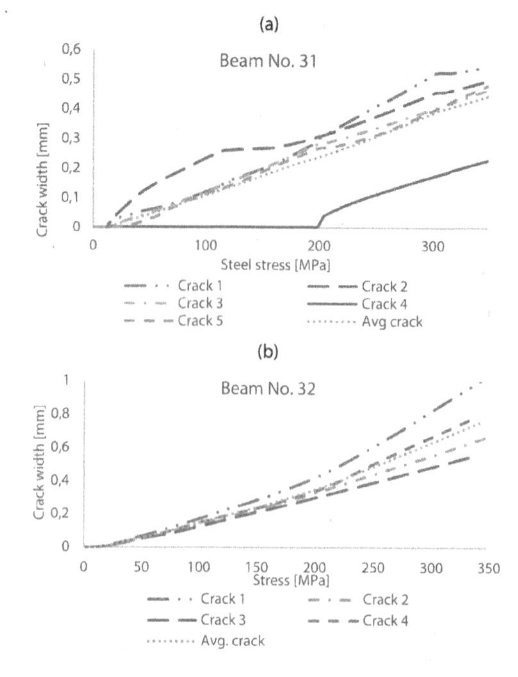

Figure 11. Crack widths of major cracks in the constant moment zone estimated by NLFEA, (a) Beam No. 31, (b) Beam No. 32.

to the reported experimental crack width values in Figure 12.

It is observed that the NLFEA with CDP-model can accurately predict the crack width at the concrete face for the two experimental beams.

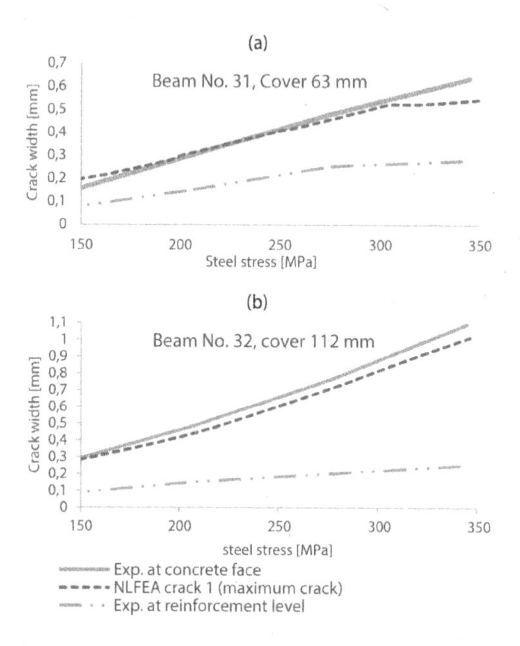

Figure 12. Maximum crack widths estimated by NLFEA CDP-model vs experimental values, (a) Beam No. 31, (b) Beam No. 32.

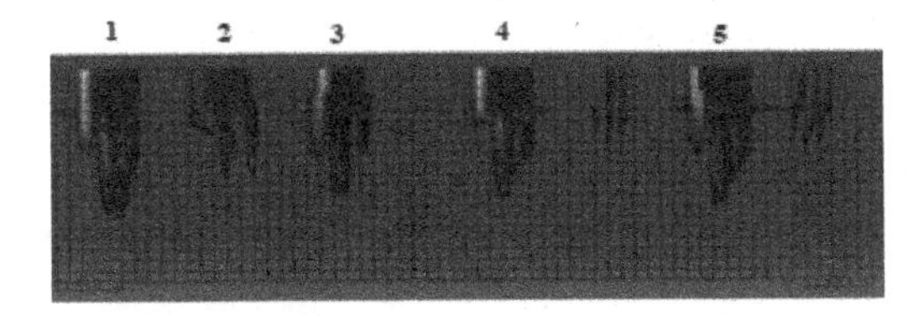

Figure 13. Visualization of localized cracking strains in between the supports for Beam No. 31 at $\sigma_{sr} = 350MPa$.

6 DISCUSSION

By applying the CDP-model with embedded reinforcement (no-slip) and calculating the crack width directly (Equation 18) by the cracking strain and the selected bandwidth as shown in Figure 14, we were able to obtain good crack width predictions of the reported experimental results at the outer most concrete face. Using the Dutch guidelines (Equation 16) with maximum crack spacing ($S_{r,max}$) defined in EC2 (Equation 8) provided also good agreement for beam No. 31 with cover 63 mm, while for beam no. 32 with cover 112 mm the results are to the unconservative side at the outermost concrete face. One reason looks to be that the maximum crack spacing ($S_{r,max}$) in EC2 does not fully consider the curvature effect for beams in bending and the impact of large concrete covers do not seem to be fully accounted for in the current code.

EC2 underestimate the maximum crack width at the outermost concrete face. In fact, it is observed that the underestimation is increasing for larger concrete cover. This seems to be addressed better in the draft for the new EC2 which introduces a coefficient ($k_{1/r}$) to account for increased crack widths due to the curvature from bending. However, it is still underestimating the crack widths at the outermost concrete face, but the results look to be more consistent in comparison with the current EC2. The need for this coefficient for concrete beams subjected to pure bending is supported by the observed results shown in Figure 7 and 14, as it is noticed that both beams have quite similar measured experimental crack widths at the reinforcement level.

MC2010 predict the crack width at the outermost concrete face for beam No. 31 to a very good extent by extrapolating the calculated crack width at reinforcement level, while being conservative at the reinforcement level. The corresponding result for Beam No. 32 by using MC2010 might be considered invalid since the distance from the reinforcement level to the outmost concrete face is larger than 75 mm. It is not clear to the first author how the code accounts for this except stating the following: *"For larger concrete cover a more detailed analysis is required. Procedures based on the fracture mechanics approach would be appropriate"*. However, it seems that methods like the CDP-model are applicable.

From the investigated beams it can be noted that a pivotal question has risen. At which location should the maximum crack width be determined? The term accounting for the curvature in the new EC2 ($k_{1/r}$) is logical, but especially for beams with large concrete cover this gives large crack widths at outermost concrete face. This increase in calculated crack width might have large economic consequences if not the allowed crack limits in the codes are adjusted to this increase. A relevant observation for this discussion is that both beams have quite similar measured experimental crack widths at the reinforcement level that we want to protect with a concrete cover.

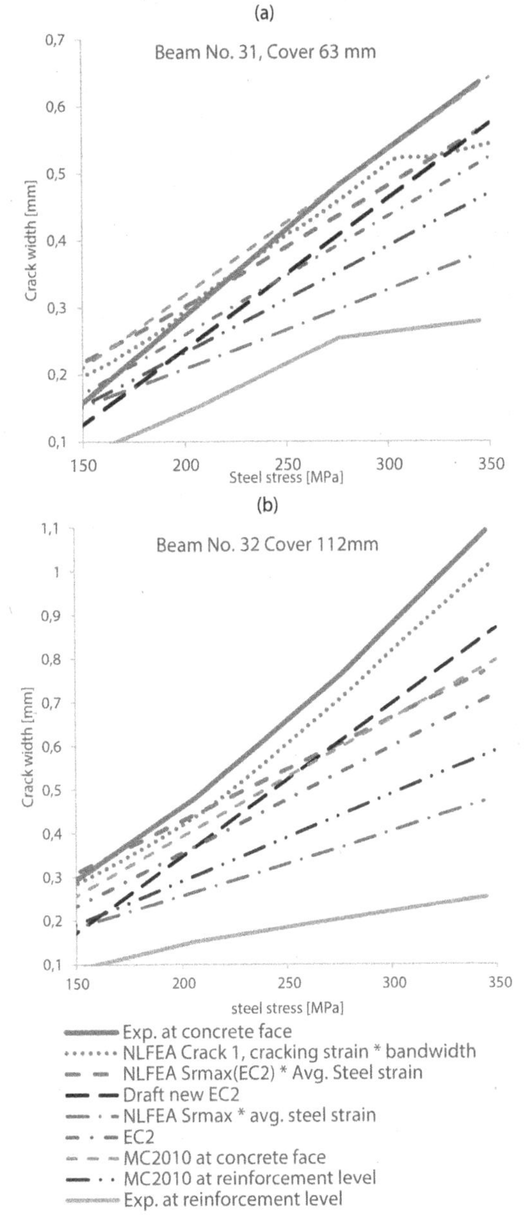

Figure 14. Crack widths vs steel stress for different approaches, (a) Beam No. 31, (b) Beam No. 32.

7 CONCLUSIONS

In this paper NLFEA with the concrete damage plasticity (CDP) model has been used to calculate the maximum crack widths in beams. The results have been compared to experimental values and results from various analytical prediction models. The results suggest that the following conclusion can be drawn:

1. 3D NLFEA analysis with the CDP model and embedded reinforcement is used to calculate the maximum crack width by multiplying the largest average cracking strain at the concrete face through the width of the beam with the selected bandwidth (Equation 18). The resulting crack widths gave predictions in good agreement with the experimental values at the outer most concrete face regardless of the cover size. This suggests that this method take the effect of cover and curvature due to bending into account better than the other NLFEA solutions and the analytical methods in the codes.
2. EC2 gave conservative results for the maximum crack width at the reinforcement level but underestimate the crack width at the outermost concrete face for the investigated beams. This suggest that the current EC2 do not correctly account for the concrete cover and the curvature effect.
3. MC2010 gave conservative results for the maximum crack width at the reinforcement level for both beams. While it gave good predictions at the experimental values at the outermost concrete face for a cover of 63 mm, the prediction was poor for cover size 112 mm. This cover size is greater than the allowed value of 75 mm and thereby clearly shows the limited validity range for beams subjected to bending in MC2010.
4. Calculating the maximum crackwidth from the draft of the new EC2, accounting for the increase in curvature by the factor $k_{1/r}$ gives better agreement than the current EC2 for crackwidth at the outermost concrete surface for increased steel stresses but is still slightly to the unconservative side. This suggests that the introduction of a curvature effect is a more correct solution for beams in bending, but this is based on only two examined beams.
5. Crackwidth calculations based on extracting the average steel strains from the NLFEA with a maximum crack spacing have been performed using two approaches:

   (a) With $S_{r,max}$ from EC2: Good agreement with crack widths at the outermost concrete face was achieved for beam no. 31 but were unconservative for beam no. 32. This suggest that the maximum crack spacing in EC2 do not fully account for the effect of large concrete covers.

   (b) The approach with $S_{r,max}$ extracted directly from the NLFEA is considerably underestimating the crackwidth at the outermost concrete face but is conservative at the reinforcement level.
6. From the conclusions in 1-5 the following can be derived:
   - Predicting crack widths at the outer most concrete face 3D NLFEA with CDP-model using cracking strains and a selected bandwidth (Equation 18) have no visible cover restrictions and gave the best results for the methods involving NLFEA.
   - From the applied codes, the draft for new EC2 seems best suited for a general crack width estimation regardless of concrete cover for beams subjected to bending.

8 FURTHER WORK

The authors are currently establishing a larger crack width database including a large number of experimental studies. Some of these will be investigated further with NLFEA to supply more raw data for recommendations on different solutions for better crack width prediction in beams subjected to bending with large concrete covers.

REFERENCES

Demir, A et al. 2018. "Effect of viscosity parameter on the numerical simulation of reinforced concrete deep beam behavior." *The Online Journal of Science and Technology*

Hordijk, Dirk Arend. 1991. "Local Approach to Fatigue of Concrete." *Delft University of Technology, The Netherlands.*

Hognestad, E. 1961 "High Strength Bars as Concrete Reinforcement Part 1. Introduction to a Series of Experimental Reports" *Journal of the PCA Research and Development Laboratories September 1961*

Hognestad, E. 1962 "High Strength Bars as Concrete Reinforcement Part 2: Control of Flexural Cracking" *Journal of the PCA Research and Development Laboratories September 1962*

M.A.N Hendriks, A. de Boer, B. Belletti, "Guidelines for Nonlinear Finite Element Analysis of Concrete Structures", *Rijkswaterstaat Centre for infrastructure, Report RTD: 1016-1:2017, 2017.*

Basteskår M, Engen M, Kanstad T, Fosså KT. *"A review of literature and code requirements for the crack width limitations for design of concrete structures in serviceability limit states" Structural Concrete. 2019;1.11.*

Basteskår, M., Engen, M., Kanstad, T., Johansen, H., Fosså, K. *"Serviceability limit state design of large concrete structures: Impact on reinforcement amounts and consequences of design code ambiguity." Engineering structures* 2019; Vol. 201.

Lubler, J., J. Oliver, S. Oller, and E. Oate, *"A Plastic-Damage Model for Concrete,"* International Journal of Solids and Structures, vol. 25, no.3, pp. 229–326, 1989.

Lee, J., and G. L. Fenves, *"Plastic-Damage Model for Cyclic Loading of Concrete Structures,"* Journal of Engineering Mechanics, vol. 124, no.8, pp. 892–900, 1998.

Tan, Reignard. 2019 "Consistent crack width calculation methods for reinforced concrete elements to 1D and 2D stress states A mixed experimental, numerical and analytical approach" Doctoral thesis, Norwegian University of Science and Technology.

Computational Modelling of Concrete and Concrete Structures – Meschke, Pichler & Rots (Eds)
 ISBN: 978-1-032-32724-2

Modelling precast concrete structures with equivalent rebar-concrete interaction

Hongning Ye & Yong Lu*
Institute for Infrastructure and Environment, School of Engineering, University of Edinburgh, Edinburgh, UK

ABSTRACT: For the finite element (FE) simulation of reinforced concrete (RC) structures, the concrete-steel "bond" is crucial, especially for the precast structural connection. Despite the wide adoption in practice, there is an insufficient understanding of the "bond" behaviour within the grout sleeve connector, and current design of the connector parameters is largely based on empirical data from the experiment. For seismic applications the ductility of the connection region is a key and this is dependent upon the deformability of the connector; however information about the overall deformation capacity of a sleeve connector is scarce in the existing literature. On the other hand, the ductility of a connection region depends not only on the deformation capacity of the connector itself but also on the interaction between the connector and the surrounding concrete. To cater for these features, a computation model should be capable of representing both the "microscopic" interior rebar-grout-sleeve interaction and the more "macroscopic" exterior sleeve-concrete bond behaviour. This paper presents an overview of an equivalent transitional layer approach which adopts a perfect "bond" at the grout-rebar interface and representing macroscopic "bond strength" "slip" through the strength and deformation of the transition layer. The experimentally observed (macro) bondslip phenomenon is realised through modifying the stressstrain behavior of the solid transitional elements with mesh-objective equivalent properties The proposed bond scheme is verified by FE simulation in ABAQUS for various scenarios included a general pullout test, a grout sleeve connector test, and a precast column test.

1 INTRODUCTION

In a low-carbon oriented era, prefabrication of concrete structures provides a holistic route to reducing environmental impact from construction by enabling more efficient fabrication and increased scope for structural optimisation. While primary components can typically be fabricated at pre-fab plants, the connection is generally done on-site and can significantly affect the assembled structural system's performance.

Various connection methods have been developed, and a representative method is by means of grouted sleeve connectors. Despite the wide adoption in practice, there is an insufficient understanding of the bond behaviour within the sleeve connector, and the current design of the connector parameters is largely based on empirical data from the experiment. For seismic applications, the ductility of the connection region is a key, dependent upon the connector's deformability; however, information about the overall deformation capacity of a sleeve connector is scarce in the existing literature. Furthermore, the ductility of a connection region depends not only on the deformation capacity of the connector itself but also on the interaction between the connector and the surrounding concrete. This means to simulate the inelastic behaviour of a connection reliably, a computation model should be capable of representing both the "microscopic" interior rebar-grout-sleeve interaction and the more "macroscopic" exterior sleeve-concrete bond behaviour.

In this paper, we present an overview of the establishment of the above rebar-concrete interaction model. To cater for the needs for modelling the concrete-rebar interaction (bond) at both levels while minding the computational effort, an equivalent rebar-concrete interaction model has been developed. The model adopts a perfect "bond" at the geometric interface but reflects the "slip" in a macroscopic manner through a transitional layer. A generalised formulation has been proposed to derive the equivalent properties for the transitional layer, preserving the equivalence of the bond strength as well as the macroscopic slip as generally observed from physical experiments.

The model is validated against classical bond tests from the literature, and it is then applied in the simulation of the tensile performance of typical sleeve connectors. Results demonstrate that the model is effective and efficient, and comparison with relevant experiments generally show good agreement. Finally, an example application of the proposed model in analysing the plastic behaviour of a concrete column

*Corresponding Author

DOI 10.1201/9781003316404-31

with a bottom connection via sleeve connectors is given, and the results are discussed.

2 OVERVIEW OF THE EQUIVALENT TRANSITION LAYER SCHEME

In modelling the interface between rebar and concrete with a "perfect bond" scheme, the bond strength is indirectly represented by the shear strength of the transition layer concrete elements in a confined condition. At the same time, the so-called "slip" is also indirectly represented by the shearing deformation of the concrete transition layer. Therefore, the shear behaviour of the concrete transition elements needs to be examined carefully. The macroscopic equivalence bond behaviour needs to be achieved in the following aspects: a) the "bond strength", which is realised through the confined shear strength; b) macroscopic "slip", which is measured as the shear displacement of the transitional layer. General-purpose FE software ABAQUS is used in this study, and concrete is modelled with Concrete Damage Plastic (CDP) material constitutive model.

In the present model, the macroscopic bond strength is realized through the shear strength of transition layer concrete elements. The shear force that is transferred from rebar through an inner concrete element is governed by the equivalent material strength in the transitional layer, and is affected by the hydrostatic pressure, as well as the mesh size of the transition layer. The mesh size aspect is particularly important because the total amount of "bond" force between rebar and concrete at a particular location should be invariant regardless the mesh size of the concrete elements in the inner (transitional) layer surrounding the rebar.

The derivation of the equivalent properties for the transitional layer is demonstrated here using a 2D-axisymmetric numerical scenario with solid elements for both rebar and the concrete, but the same idea can be used when the steel bar is modelled as 1D elements. A simple mesh with square elements is adopted, and for each section the inner layer of concrete surrounding the rebar is the transition layer equivalent properties. The maximum bond force $V_{b,u}$ can be expressed in terms of bond strength τ_u as:

$$V_{b,u} = \pi d_b \times \tau_u dx \tag{1}$$

where τ_u = pullout bond strength, d_b = diameter of the rebar, dx = incremental length in the longitudinal direction.

In the numerical model, the "bond" force $V_{b,num}$ that may be transferred through the inner layer of concrete can be expressed in terms of the shear force developed in the concrete elements surrounding the rebar, $V_{c,in}$, as:

$$V_{b,num} = V_{c,in} = \tau_{c,in} \times A_{c,in} \tag{2}$$

where $\tau_{c,in}$ = shear stress in the inner layer concrete, $A_{c,in}$ = effective shear transfer area in each inner concrete element,

$$A_{c,in} = 2\pi \left(d_b/2 + L_{c,in}/2\right) dx \tag{3}$$

where $L_{c,in}$ = transition layer thickness. The shear strength in the inner layer concrete $\tau_{c,in}$ may be expressed as a function of the compressive strength of concrete f_c' and hydrostatic pressure (confinement) $p_{c,in}$ as:

$$\tau_{c,in} = \widehat{\tau_{c,in}}(f_{c,in}', p_{c,in}) \tag{4}$$

Let the theoretical maximum bond force be equal to bond force in the numerical model, i.e. $V_{b,u} = V_{b,num}$. By substituting Equation 1, 3 & 4 into Equation 2, the equivalent strength for the transitional layer concrete can be determined by:

$$\frac{d_b}{d_b + L_{c,in}} = \frac{\tau_{c,in}(f_{c,in}', p_{c,in})}{\tau_u} \tag{5}$$

Thus,

$$f_{c,in}' = \widehat{f_{c,in}'}(p_{c,in}, \tau_u, d_b, L_{c,in}) \tag{6}$$

It should be noted that the equivalent method applies in the presence of the hydrostatic pressure of the transition element $p_{c,in}$. Thus, beside being mesh-size objective, the equivalent strength properties are affected by other pertinent parameters, including geometric dimensions (e.g. rebar/concrete cover sizes).

As the transition layer element width $L_{c,in}$ directly influences the determination of the equivalent strength, it makes sense to choose a value in a physically meaningful range. Considering the bond damage zone as generally observed from experiments, the size of the equivalent concrete layer $L_{c,in}$ maybe defined on the order of the radius of the rebar. In the situation where 1D truss/beam-element is employed to model the rebar, the $L_{c,in}$ should therefore be defined to be close to the rebar diameter. For a 2D model, the equivalent concrete strength $f_{c,in}'$ in the inner layer elements can then be determined in terms of bond strength τ_u, rebar diameter d_b and mesh-size $L_{c,in}$ as well as a nominal pressure $p_{c,in}$ which may be established from preliminary FE modelling trials.

As can be seen, in the above described steel-concrete interaction modelling scheme, the "bond" strength and "slip" are equivalently represented by the shear strength and deformation of the transition layer concrete element. A single element test, which allows incorporation of varying confining stress, is employed to examine the basic shear behaviour of the concrete constitutive model and the effects of adjusting pertinent model parameters. The CDP constitutive modelin ABAQUS is chosen in this study.

The setup of a single element shear test is indicated in Figure 1. A 10x10mm single element is attached

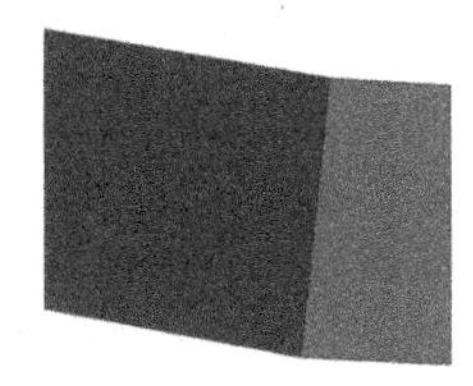

Figure 1. Illustration of the single element test: (a) Singleelement test setup; (b) Deformation of the single element.

on one (right) side to an elastic element which is in turn fixed on the far right side. The load (shear force) is applied on the left side of the main element while the lateral movement is restrained. The involvement of the elastic element is to simulate a flexible constraint to the main element in a way similar to what happens to the transitional layer elements from the outer concrete in an actual bond region. Class C20 class concrete is selected, and the material property parameters in ABAQUS is determined according to the ABAQUS manual (Hibbitt et al. 2011) and existing literature (Jankowiak & Lodygowski 2005), as shown in Table 1.

Table 1. CDP material parameter for C20 class concrete.

Density		Young's modules *MPa*		Poisson's ratio
$2.40E-09$		26200		0.2
Dilation angle ψ	Eccentricity	Ratio of stresses	K	Viscosity parameter
31	0.1	1.16	0.666	0.001

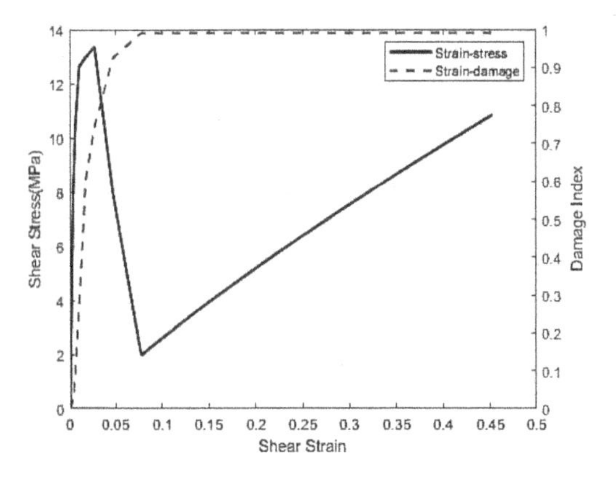

Figure 2. Single element shearing test result of C20 concrete in a flexibly confined condition.

The shear strain-stress test result of the C20 concrete single element is plotted in Figure 2.

As mentioned earlier, the shear strength of the equivalent concrete elements may be expressed in terms of concrete strength $f'_{c,in}$ and hydrostatic pressure $p_{c,in}$ following a general form (Sonnenberg et al. 2003), and after fitting with the CDP concrete material model, the relationship is found as:

$$\tau_{c,in}/f'_{c,in} = 0.75(p_{c,in}/f'_{c,in}) + 0.068 \tag{7}$$

Combining with Equation 5 the equivalent strength $f'_{c,in}$ can be determined to represent the "bond" strength in terms of the expected hydrostatic pressure, the "real" bond strength, steel diameter and transition layer thickness.

Table 2. CDP material parameter for C20 class concrete: uniaxial compressive behavior.

Yield stress - Inelastic strain *MPa*		Damage -Inelastic strain	
6	0	0	0
8	7.47E-005	0	7.47E-05
12	9.88E-005	0	9.88E-05
16	0.000154123	0	0.000154
20	0.000762	0	0.000762
16	0.002558	0.19	0.002558
12	0.005675	0.59	0.005675
2	0.011733	0.89	0.011733
		0.99	0.02

Table 3. CDP material parameter for C20 class concrete: uniaxial tensile behavior.

Yield stress	G_f
MPa	*N/mm*
2.21	133

3 CALIBRATION AND VALIDATION

3.1 *Calibration of CDP model to represent softening branch of bond-slip law*

Following the realisation of the "bond" strength, here we aim to achieve macroscopic "slip", which is indirectly represented by the shear displacement of the transitional layer.

The single element test results in Figure 2 exhibit a late stage shear stress increase in the overall softening phase. This behaviour can be explained by the constant and continued dilation of concrete elements as large shear strain develops in the numerical model. In most situations, we may not be interested in what happens in the late stage shear behavior after the concrete material is severely damaged, but in present situation, it is meaningful to calibrate damage evolution to obtain a smooth decrease of shear stress-strain relationship to represent the softening branch of the bond-slip behaviour.

Extensive calibration analysis has been carried out to evaluate the effect of modifying the damage evolution rule to achieve a more realistic shear behaviour in the descending phase. The modified damage evolution and corresponding numerical test result are shown in Figure 3 Further calibration of the damage evolution with 10mm element size has also been carried out, and the resulting shear strain-stress relationship in the equivalent concrete material is plotted in Figure 4. On the other hand, in order to represent the macroscopic "slip", the mesh sensitivity of the softening branch also needs to be examined. By varying the softening law (damage), it is possible to achieve a meshindependent representation of the bond stress-slip law in the equivalent transition layer.

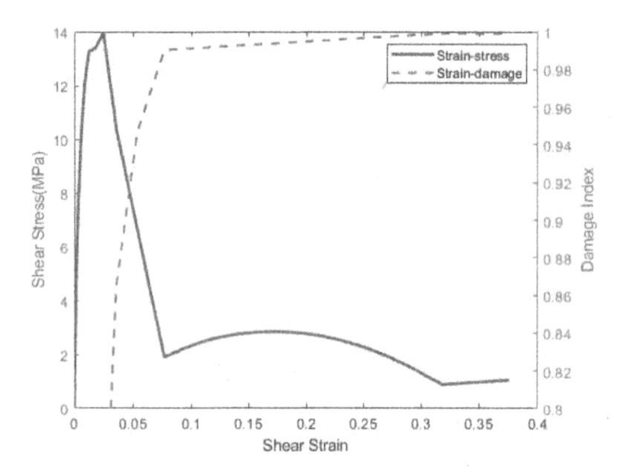

Figure 3. Single element test result of initial calibrated concrete.

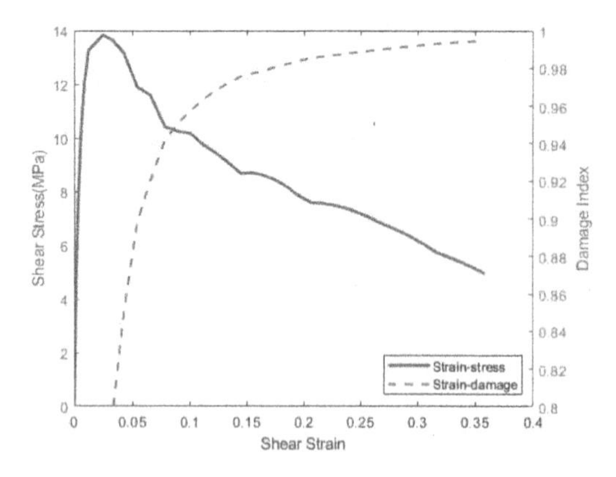

Figure 4. Single element test result of further calibrated concrete model.

3.2 *Verification of proposed equivalent transition layer scheme by pullout experiment*

A typical pullout experiment is firstly modelled to verify the above-proposed bond-slip representation method with the equivalent transition layer for the bond behaviour. The experimental pullout test conducted by Chu and Kwan (2018) schematically shown in Figure 5, is selected. 12-mm rebar is used with 530 MPa yield strength and 663 MPa ultimate strength. The concrete strength is 50.8 MPa, and the embedded length is 50 mm.

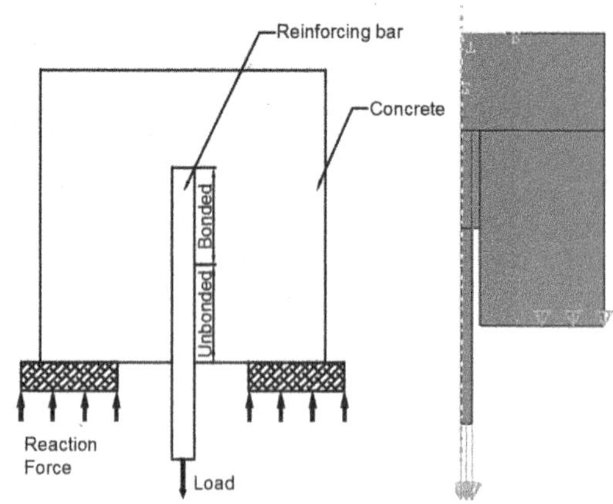

Figure 5. Sketch of pullout experiment (Chu & Kwan (2018)) and 2D axisymmetric model.

The nominal "bond" strength τ_u from the experiment is calculated as the pullout failure load divided by the surface area of the bonded length of the reinforcing bar:

$$\tau_u = \frac{F}{\pi d_b l} \tag{8}$$

where F = pullout failure load, d_b = diameter of the steel bar, and l = embedded length of the steel bar. The experimental "bond strength" is 19.95MPa, which will be the target "bond" strength in the numerical model.

The properties for the equivalent transition layer elements in the numerical model are determined as follows. The average pressure at concrete steel interaction is assumed based on trial analysis to be $p_{c,in} = 20$MPa. From Equation 10, for an element size of 4mm (smaller but close to the rebar radius), the transition element can be assigned with 20MPa equivalent strength. The corresponding numerical model is labelled as M4-C20MD.

The numerical result of specimen M4-C20MD is shown in Figure 6. It can be seen that the numerical "bond" strength matches well with experimental and theoretical macroscopic "bond" strength.

In order to indicate the effectiveness of the mesh-objective equivalent transitional layer, the numerical specimens without adjustment are also compared in Figure 6. The numerical specimen which applied 50MPa normal concrete strength into transitional elements while keeping 4mm mesh size for transitional elements, is labelled as M4-C50MD. The same 20MPa concrete strength, numerical specimens with 7.5mm and 10mm mesh size, labelled as M7.5-C20MD and M10-C20MD, are also plotted in Figure 6. It is obvious that the equivalent "bond"-"slip" is mesh-sensitive and the equivalent properties is mesh-objective. Overall, the numerical result indicates that the equivalent

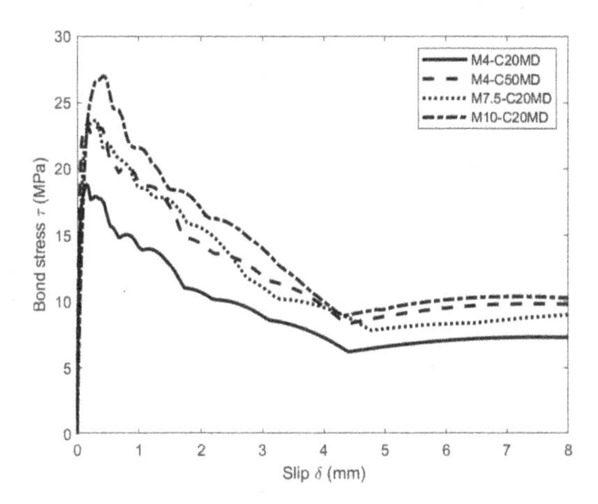

Figure 6. Sketch of pullout experiment and schematic illustration of 2D axisymmetric model.

concrete transition layer scheme effectively represents the macroscopic pullout bond behaviour.

4 APPLICATION OF PROPOSED EQUIVALENT TRANSITION LAYER SCHEME IN PRECAST GROUT SLEEVE CONNECTOR

The proposed equivalent bond modelling scheme can be applied for the general FE analysis of RC structures and in some bond sensitive regions. One motivation of the proposed equivalent bond scheme is to model the grout sleeve connector and the connection as a whole in precast RC members. Compared with the general pullout scenario validated in Section 2, relatively sizeable hydrostatic pressure will be generated at the impending pullout of rebar in grout sleeve connectors. Therefore, parameter identification is employed to calibrate the grouting material property. The grouting thickness in the grout sleeve connector is usually around 0.5 to 2 times of rebar diameter, which means the whole grouting material can be regarded as the transition layer in some cases. Application examples are then given using a sleeve connector axial loading experiment.

4.1 *Calibration for grout sleeve connector*

The confinement in a grout sleeve connector can usually reach a very high level, to around 40 to 60MPa at the grout-steel interface at the stage when the ultimate axial load is reached. In the preliminary simulation, the confinement generated in the transition grouting elements is however much lower than the above experimental observation. In fact, the development of pressure in concrete under a shearing load in a confined condition is very complex and it is difficult to simulate such a phenomenon accurately for a continuum-based material constitutive model. Therefore, it is not unreasonable to employ a parameter identification process to calibrate the material properties for the grouting concrete in a steel tube confined situation, so as to match the hydrostatic pressure as observed from the experiment.

The parameters of grouting materials include the elastic modulus, compressive and tensile strengths, strain-stress relationship, damage evolution (softening) and plastic flow. The confinement in grout sleeve connector is understood to be mainly generated by shear-kind dilation of the grouting material. Therefore, we can expect that the hydrostatic pressure is sensitive to the definition of the dilation angle ψ.

A group of short-anchorage sleeve connectors tested by (Zhang 2020) is selected as the reference benchmark. 12mm, 16mm and 20mm deformed bars are used in the experiments, and specimens with a shorter anchorage length of $3d_b$ to $4d_b$ are selected for benchmarking. High strength 108-MPa grouting material is used for all specimens. To reduce the uncertainty in the calibration, only those specimens with a pullout failure are selected. In the experiment, the hoop strain of the steel sleeve $\varepsilon_{t,sl}$ is measured, so if we assume a uniform distribution of bond stress (due to short bond length), the longitudinal strain of steel sleeve $\varepsilon_{l,sl}$ can be derived as all axial force will be transferred between the (discontinued) rebar and the steel sleeve.

The derivation of pressure at the grouting-rebar interface is shown as follows with the consideration of the Poisson's ratio υ of steel sleeve. The confining stress at the grouting sleeve interface is σ_{sl} and the hydrostatic pressure at the grouting rebar interface σ_0, can be expressed by Equation 12 & 13:

$$\sigma_{sl} = \frac{2E_{sl}(\varepsilon_{t,sl} + \upsilon\varepsilon_{l,sl})t_{sl}}{d_{sl,i}(1-\upsilon^2)} \tag{9}$$

$$\sigma_0 = \frac{d_{sl,i}\sigma_{sl}}{d_b} \tag{10}$$

where $\varepsilon_{l,sl}$ = sleeve longitudinal strain, $\varepsilon_{t,sl}$ = measured hoop strain, E_{sl} = elastic modulus of steel sleeve, t_{sl} = thickness of the sleeve, $d_{sl,i}$ = sleeve inner surface diameter, d_b = rebar diameter.

Table 4 summarises the selected specimens and the corresponding grouting-rebar interface confinement calculated from the strain measurements.

For general applications, the default dilation angle in the concrete damage plastic (CDP) model is given as 31°. Through the parameter identification, a dilation angle around a value of 44° would be required to

Table 4. Confinement stress of selected experimental reference specimens.

	Maximum axial force	Bond strength	Confinement
Specimen	kN	MPa	MPa
S12-0.14-D12-4d	58.5	32.34	28.63
S16-0.14-D16-3d	98.66	40.91	57.84
B12-0.14-D12-4d	61.65	34.09	38.47
B16-0.14-D16-3d	108.44	44.97	42.72
B20-0.14-D16-4d	134.76	41.91	50.12
B20-0.14-D12-4d	63.73	34.68	28.93

result in the level of high confining stress in the grouting material as observed from the sleeve connector test mentioned earlier. An example of pressure distribution using the default and calibrated dilation angles, respectively, is shown in Figure 7. It can be seen that marked difference is made by using the two dilation values and the pressure generated using the calibrated dilation angle value matches well the experimental data given in Table 4.

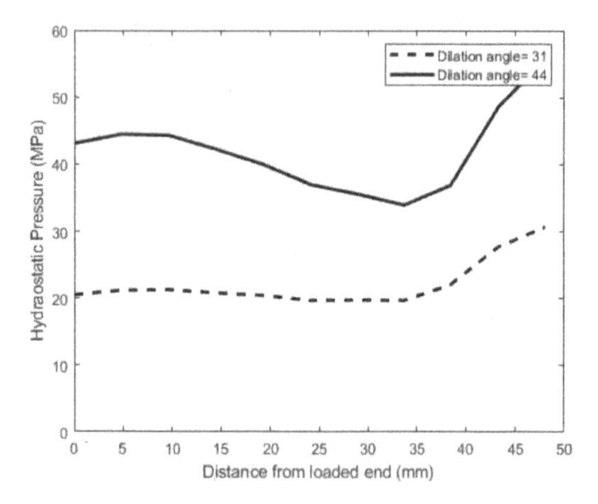

Figure 7. Hydrostatic pressure distribution at the grouting-rebar interface with different dilation angle definition.

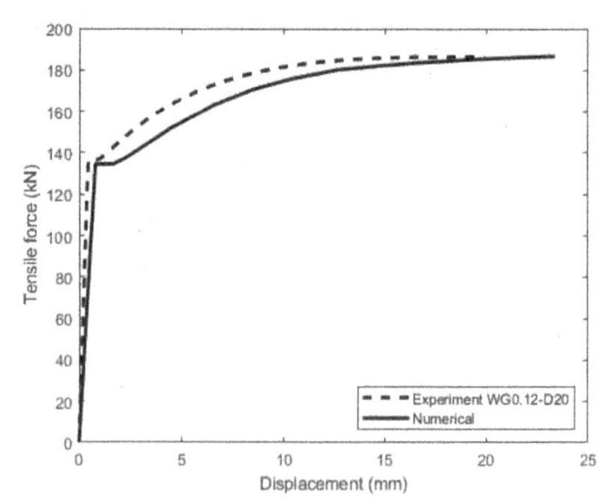

Figure 8. Experimental and numerical displacement-force relationship.

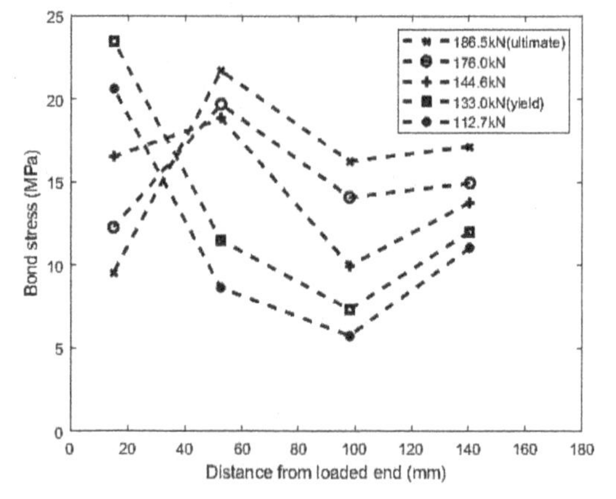

Figure 9. Derived bond stress distribution from strain measurement on the sleeve surface.

4.2 *Application for grout sleeve connector*

In the experiment conducted by Chen et al. (2020) a half grout connector was tested and four groups of longitudinal and hoop strain gauges were mounted on the sleeve surface at four different locations, which allows an estimation of the bond stress distribution.

The experiment was set up with the rebar inserted in the half connector by grouting while the other half was connected by a thread rod to facilitate the test using a universal tension machine. The rebar is 20mm in diameter with a yield strength of 429 MPa, ultimate strength of 578 MPa and 24.08% elongation capacity. The steel sleeve has 396MPa yielding strength and 629MPa ultimate strength. The elastic modulus and Poisson's ratio are 210GPa and 0.269, respectively. The grouting material has a compressive strength is 99 MPa and the rebar anchorage length is 160mm

The measured tension force vs displacement relationship is reproduced with dashed curve in Figure 8. Since the steel rebar is terminated at the middle of the specimen, all the bond force between the rebar and the grout transfers to the steel tube at the middle section of the sleeve connector. We can also consider that the tension force carried by the grouting material is very small and negligible. Therefore, the experimental bond stress distribution can be derived from four groups of hoop/longitudinal strain measurements on the sleeve surface, and the result is plotted in Figure 9.

The equivalent bond model with a transition layer is employed to simulate the tension behavior of the grouted sleeve connector. The deformed shape of the connector from the numerical model is plotted in Figure 10 with axial loading applied at the left end. The grouting material is modelled by a single layer of transition elements. A large "slip" can be observed at the loaded end between the rebar and the sleeve. The specimen fails by steel rebar reaching ultimate strength as expected from the experimental result. Figure 8 plots the displacement-force relstionship from the simulation. The "slip" from the numerical model is generally 20% smaller than that measured from the experiment. This may be explained by the fact that the measured displacement in the experiment was the total displaceemnt at the rebar loading end and it included the elongation of the threaded rod, which is not considered in the numerical model.

The bond stress distribution from the numerical result is shown in Figure 11. A comparison of the numerical and experimental bond stresses at the experimentally measured locations is shown in Figure 12.

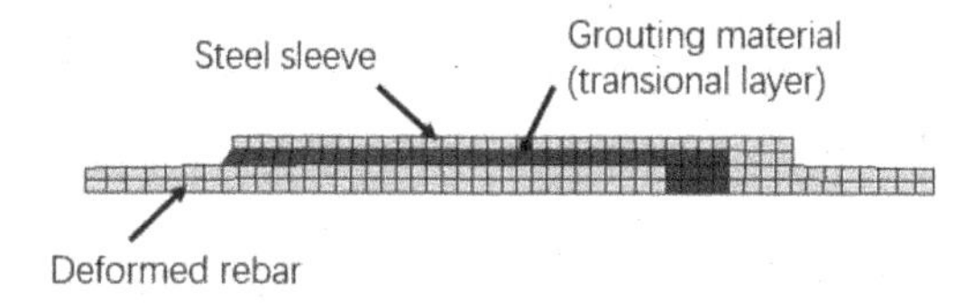

Figure 10. Computed deformed shape of grouted sleeve connector after ultimate axial loading.

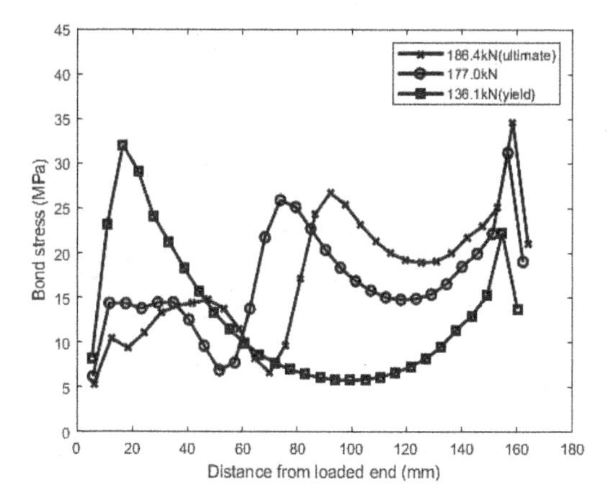

Figure 11. Numerical bond stress distribution.

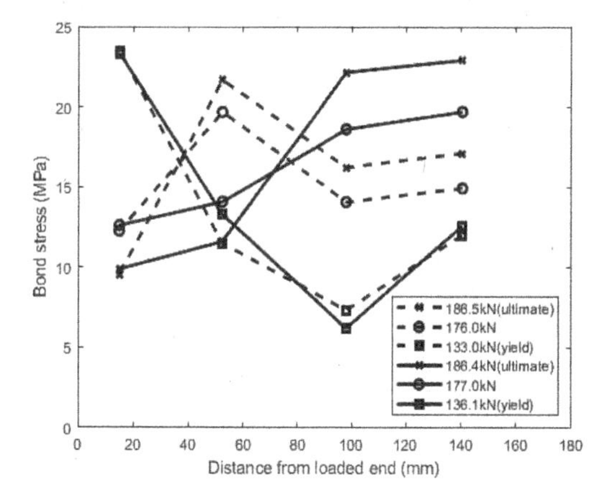

Figure 12. Comparison of numerical and experimental bond stress distribution results (Solid lines = numerical, dashed lines = experimental).

It can be seen that the numerical results match very well the experimental data before rebar yielding. After yielding, the numerical model also undergoes degradation of the "bond" stress near the loading end, similar to the experimental result. However, the bond stress distribution from the numerical model exhibits an increase of the bond stress near the rebar termination (free) end, whereas this phenomenon is not obvious from the experimental data. Neverthless, such behaviour is not unreasonable in a sleeve connection scenario, and could be explained by the interaction between the steel sleeve and the grout concrete around the rebar termination end (middle of the connector), in that the longitudinal stretching (tension) of the sleeve tube at this location will tend to induce an enhanced pressure on the grout material, which in turn enhances the bond stress between the grout and the rebar. In fact, in both numerical and experimental results this passive "bond" stress near the rebar ternmination end can be observed.

Overall, the numerical result indicates that the equivalent concrete transition layer scheme works effectively in representing the macroscopic bond behavior in a grouted sleeve connector

5 MODELLING PRECAST RC MEMBER

In this section the equivalent transitional layer scheme is applied to model an experimental precast column tested by M. J. Ameli and Chris (2016), labelled as GGSS-1 in the original paper. The half-length column has an octagonal cross-section with an outer diameter of 533 mm, and is connected with grouted sleeve connectors at the column base. The height of the test column is 2600 mm, and the horizontal load is applied at an elevation of 2400 mm in a displacement controlled manner The longitudinal steel bars have a diameter of 25.4 mm, yield strength of 469 MPa and ultimate strength of 641 MPa. The transverse reinforcement is in spiral form and has a diameter of 12.7 mm, yield strength of 434 MPa and ultimate strength of 710 MPa. The spiral has a pitch of 64 mm with a double hoop. The concrete has a compressive strength of 40 MPa on the test day, and the high strength grouting material in the steel sleeve has a compressive strength of 99 MPa. The length of the grout connectors is 370 mm, and the length of each dowel bar anchorage in the connector is 180 mm.

In the experiment the column was subjected to cyclic loading. For an illustration purpose, the column is analysed under monotonic loading in the numerical simulation. The equivalent bond model with a transition layer is employed to simulate the bond behaviour of rebar and the grout in the connectors on the tension side of the column, using 3D solid elements. The rest of the steel bars are assigned with 1D beam element and embedded in the surrounding concrete. The setup of the numerical model of the selected grouted sleeve connected precast column is shown in Figure 13.

The concrete vertical strain contour is shown in Figure 14 to illustrate the distribution of damage. It can be seen that the model can successfully capture the concrete damage above and below the connection region at the column base, which indicate the effects of the presence of connector. The rebar axial strain distribution at various lateral displacement levels can also be extracted from the numerical model, plotted in Figure 15 The coordinate origin point of x-axis (x=0) indicates the position at the center of the connector, , and the negative and positive x-axis represent the direction upwards and downwards, respectively. It can be seen from Figure 15 that strain concentrates at both

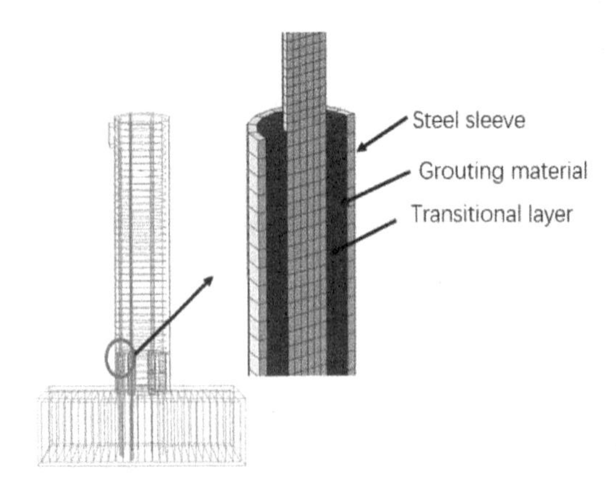

Figure 13. Illustration of grouted sleeve connected precast column numerical model.

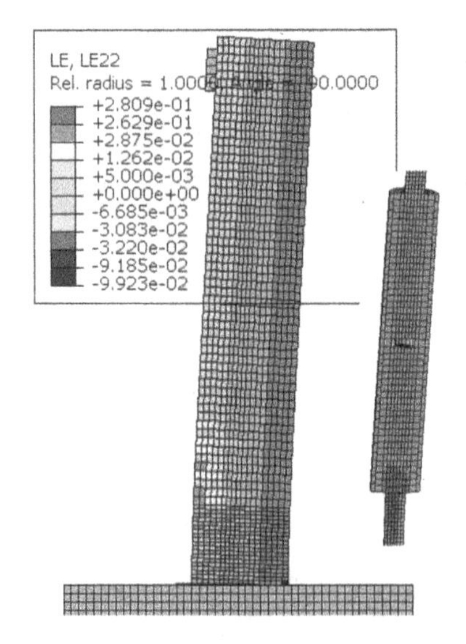

Figure 14. Vertical strain contour of numerical column specimen GGSS-1 and within tensile side connector.

upwards and downwards of connector end, and the yielding rebar is firstly observed at the downwards of connected region with the increase of overall lateral displacement.

The above application of the proposed equivalent bond (transition) layer method shows that the scheme can effectively simulate the structural behaviour in the bond sensitive region. Both the "macroscopic" exterior sleeve-concrete bond behaviour and the "microscopic" interior rebar-grout-sleeve interaction, such as strain and bond stress distribution, can be well captured in the numerical model. The proposed bond scheme provides a computational effective and efficient means for modelling and further investigation into the nonlinear behaviour of RC members and precast structures involving complex plastic deformation and bond-sensitive effects in the plastic regions.

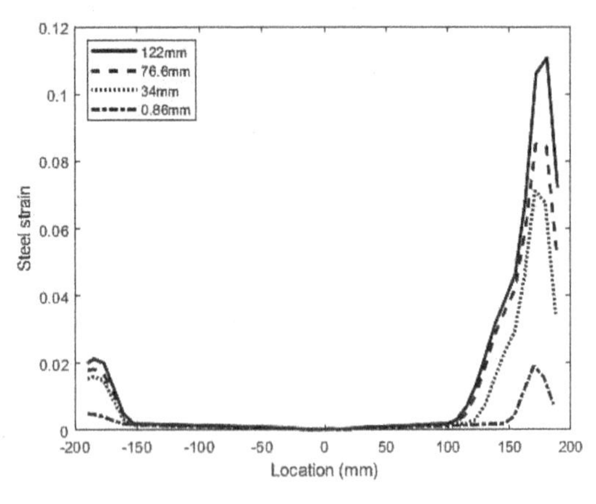

Figure 15. Connector rebar axial strain distribution.

6 CONCLUSIONS

An equivalent bond model with a transitional layer scheme has been developed to cater for the need of modelling both the local bond-related nonlinear behavior in adequate detail and also the global structural response at a large scale. The bond model preserves both the bond strength in a mesh-objective manner and also the macroscopic slip through the shear deformation of the transition layer.

In this paper the proposed bond model is firstly verified by simulating a short anchorage pullout test, showing satisfactory result. The model is then applied in the simulation of grout sleeve connectors, in which parameter identification has been used to calibrate the grouting material property in a consistent manner. Comparison with the relevant experiments shows that the model can capture not only the global force-displacement behaviour in all stages of the response, but also the detailed distribution of the bond stress and strain in the rebar. It is interesting to note that the bond stress distribution exhibits noticeable enhancement towards the rebar termination end (middle of connector), which can be explained by the interaction between the steel tube and the grouting material around the rebar termination point in the middle of the connector.

The model is subsequently applied in the analysis of a precast column with grounted sleeve connectors. The results shows that the model can effectively represent both the local steel-concrete interaction behavior and the nonlinear global structural response for both general and precast RC members in the FE analysis.

REFERENCES

Chen, J. & Wang, Z. & Liu, Z. & Ju, S. 2020. Experimental investigation of mechanical behaviour of rebar in steel half-grouted sleeve connections with defects in water/binder ratio. *Structures,* 26, 487–500.

Chu, S. H. & Kwan, A. K. H. 2018. A new method for pull out test of reinforcing bars in plain and fibre reinforced concrete. *Engineering Structures,* 164, 82–91.

Hibbitt, H. & Karlsson, B. & Sorensen, P. 2011. Abaqus analysis user's manual version 6.10. *Dassault Systèmes Simulia Corp.: Providence, RI, USA.*

Jankowiak, T. & Lodygowski, T. 2005. Identification of parameters of concrete damage plasticity constitutive model. *Foundations of civil and environmental engineering,* 6, 53–69.

M. J. Ameli, D. N. B. J. E. P. & Chris, P. P. 2016. Seismic Column-to-Footing Connections Using Grouted Splice Sleeves. *ACI Structural Journal,* 113, 1021–1030.

Sonnenberg, A. M. & Al-Mahaidi, R. & Taplin, G. 2003. Behaviour of concrete under shear and normal stresses. *Magazine of concrete research,* 55, 367–372.

Zhang, F. 2020. *Experimental study on bond-slip performance of steel bars and sleeve constrained grouting material.* Mater, Hunan University.

Computational Modelling of Concrete and Concrete Structures – Meschke, Pichler & Rots (Eds)
 ISBN: 978-1-032-32724-2

Structural behavior of fiber reinforced concrete foundations

Guomin Ji
Department of Manufacturing and Civil Engineering, Norwegian University of Science and Technology (NTNU), Norway

Terje Kanstad
Department of Structural Engineering, Norwegian University of Science and Technology (NTNU), Norway

Steinar Trygstad
Dr. Ing. Steinar Trygstad, Norway

ABSTRACT: The paper presents a study of application of Fiber Reinforced Concrete (FRC) in foundation slabs through experimental tests and numerical simulations. The fibers are the main reinforcement to take up bending moments and shear stresses in the structural fiber applications. A series of full-scale tests of foundation slabs reinforced with either steel fibers, a composite mineral fiber or ordinary rebars were performed at NTNU. Comparisons between the test results and simplified methods based on recommendations from COIN, NB38, DAfStb, and the recently launched draft of the new Eurocode 2, showed that the calculation methods are highly conservative for the moment capacity. The numerical simulations are performed to investigate the structural behavior under ultimate limit states (ULS). A numerical model closely representing the test setup was established in the finite element software DIANA and the simulation results were compared with test results for fiber reinforced, rebar reinforced and plain concrete foundations. A parameter study has been performed to investigate the effect of shear stiffness of the isolation layer on the structural behavior of the foundations. The research work provide basis for technical approval for application of Structural Fiber Reinforced Concrete (SFRC) in foundation slabs and may contribute to increase use of fibers in load carrying structures.

1 INTRODUCTION

Fiber reinforcement has been used in concrete in many decades, either as a replacement or as a supplement to traditional reinforcement. Over the years, various types of fiber have been researched and developed with different materials, where steel fiber is most widely used. [1] Various studies shows that by using fiber reinforcement material savings, financial savings and reduction of environmental emissions may be achieved [2, 3, 4, 5, 6, 7]. It has been shown that fiber reinforcement has a strong crack-limiting effect, therefore it is widely used in the applications such as industrial floors. In these structures, fiber has mainly been used for crack prevention, and not in so much for load bearing [8]. The reason is mainly lack of guidelines for dimensioning with fiber-reinforced concrete.

In researches from recent years, however, it is shown that fiber reinforced concrete structures may have sufficient load-bearing capacity and ductility, and that fiber reinforcement may be used as replacement for reinforcement bars in certain situations [9, 10, 11, 12, 13, 14, 15]. In March 2020, the Norwegian Concrete Association published NB38: Fiber-reinforced concrete in load-bearing structures to meet the desire for common guidelines for dimensioning and execution of SFRC structures and elements fulfilling a structural function [16]. Such structures need to withstand different types of loads, like wind loads, earthquake loads and various live loads. Foundation slabs are an example of such a structural fiber reinforced concrete application. Foundations must transfer loads from the structure down to the ground, and fibers might be the main or the complementary reinforcement to take up bending moments and shear stresses [9]. The application of fibers in concrete foundations is advantageous to reduce the amount of reinforcement bars and the concrete thickness, but also to improve the structural behavior through the fibers' ability to bridge cracks. Fiber reinforcement might therefore increase both the capacity and durability.

The chemical properties of various composite mineral fiber make it an excellent material for use in particularly corrosion-prone areas, which would normally be challenging for steel reinforcement [6]. Use of such types of fiber will therefore be relevant in parking garages, in structures along the coast and in foundations, which are often placed in salt and humid environment. [3]

 DOI 10.1201/9781003316404-32

In this paper a part of the research activities in a collaborative project between NTNU and several industry partners is presented. In order to achieve technical approval for fiber reinforced concrete in foundations and to increase the use of fibers in load carrying structures in general, full-scale tests of foundation slabs were performed at NTNU [20]. The concrete slabs were reinforced with either steel fibers, a composite mineral fiber or with ordinary rebars. The test results are compared with the analytical results of simplified methods based on recommendations from the comprehensive research project COIN [1], the FRC guidelines published by the Norwegian concrete association NB38 [16], the German rules DAfStb [17], and the recently launched draft of the new Eurocode 2 [18, 19].

Furthermore, numerical simulations are performed in the finite element software DIANA to investigate the structural behavior under ultimate limit states (ULS), and to provide further insight into the failure mechanisms of the fiber reinforced solutions and thereby contribute to new knowledge and improved design of fiber reinforced foundations.

2 EXPERIMENTAL PROGRAM

Four different foundation types and a total of nine full-scale foundation slabs with dimension of $2.0 \times 2.0 \times 0.25$ m were tested under a point load in the center. The foundations were unreinforced, reinforced with rebars, composite mineral fibers or steel fibers respectively. The overview of the test program is shown in Table 1. The foundations that were unreinforced or reinforced with rebars were used as a basis for comparison for the fiber-reinforced foundations.

Table 1. Overview of test program of foundation slabs.

Foundation	Test ID	Reinforcement
1	UA	Unreinforced
2 and 3	SA1, SA2	Steel rebars
4, 5 and 6	BF1, BF2, BF3	$10kg/m^3$ composite fiber
7, 8 and 9	SF1, SF2, SF3	$30kg/m^3$ steel fiber

The setup of the full-scale test represented typical column foundations of buildings and consists of four components: concrete foundation, vapor barrier, isolation layer and ground base combined of concrete blocks and filled sand as shown in Figure 1 a) and c). The insulation layer (Sundolitt XPS700SL) is used to simulate the realistic subgrade of the foundation. The two layers of vapor barrier with a thickness of 0.20 mm allow the gliding between the insulation and the foundation. The floor in the laboratory consists of concrete block and cavities, and the cavities were filled up with sand to create a flat and stable surface which could withstand the slab under loading as shown in Figure 1 a) and b).

2.1 *Material*

A B35 M45 self-compacting concrete (SCC) was used in the foundations because a stable SCC achieves a better distribution of fibers and a more favorable fiber orientation with more fibers in the longitudinal direction [15]. The maximum aggregate size used in the mix design is 22 mm. The fresh concrete has a slump flow of 600-650 mm and air content of 2.5%. The concrete recipe shown in Table 2 is the same for all the foundations apart from the amount of added fibers.

a) The floor in the laboratory

b) The test setup

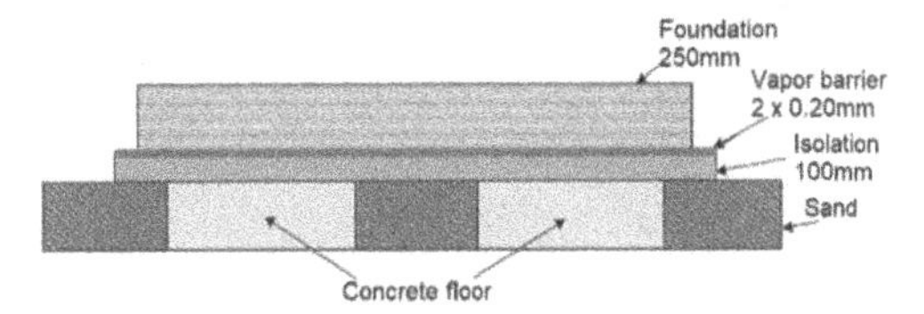

c) Illustration of the different layers in the test setup

Figure 1. Test setup.

Table 2. Concrete mix design.

Material	Content (kg/m^3)
Cemex, Miljøsement (Cem II/B-S 52,5 N)	380
Sand (0–8 mm)	1062.0
Coarse aggregate 8–16 mm	350.0
Coarse aggregate 16–22 mm	400.0
Water	176.55
Super-plasticizer	4.18
Micro silica	15.0
Air Entraining Agent	0.38

Two types of fibers: composite mineral (CM) fibers and steel fibers with hooked ends were used in this research as shown in Table 3. The composite mineral fibers are made of mineral threads twisted and glued together by vinyl resin, and the helical surface structure ensures adhesion of the composite fiber in the concrete matrix in the same way as steel fiber with end hooks, as shown in Figure 2 a) and b). The rebars applied in foundations (SA1 and SA2) are shown in Figure 3.

Table 3. Properties of composite mineral and steel fibers.

Fiber Type	Composite	Steel
Length (mm)	43	50
Diameter (mm)	0.72	0.75
E-modulus (MPa)	44 000	210 000
Tensile strength (MPa)	900	1550
Density (g/cm3)	2.1	7.85
Number of fibers/1 kg	29000	6000

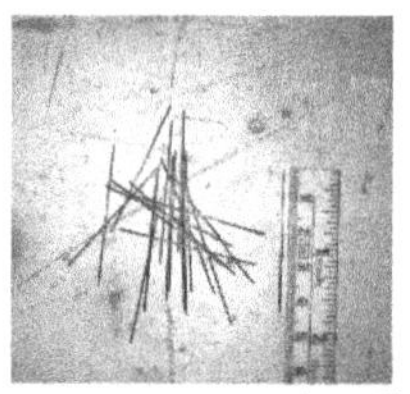

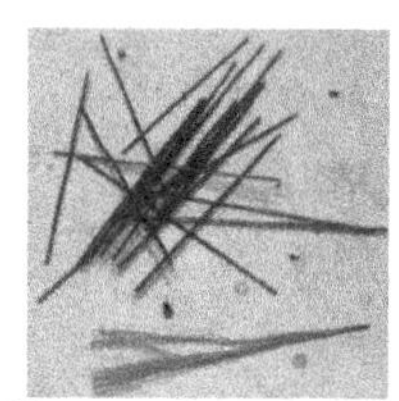

a) Composite fibers

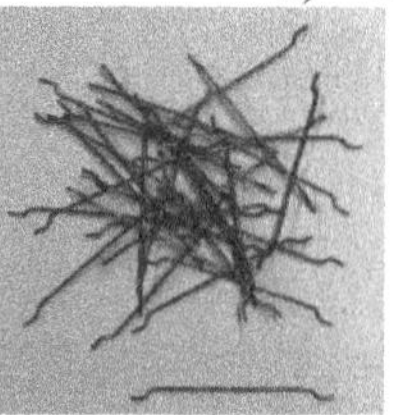

b) Steel fibers

Figure 2. Fibers used in the foundation slabs.

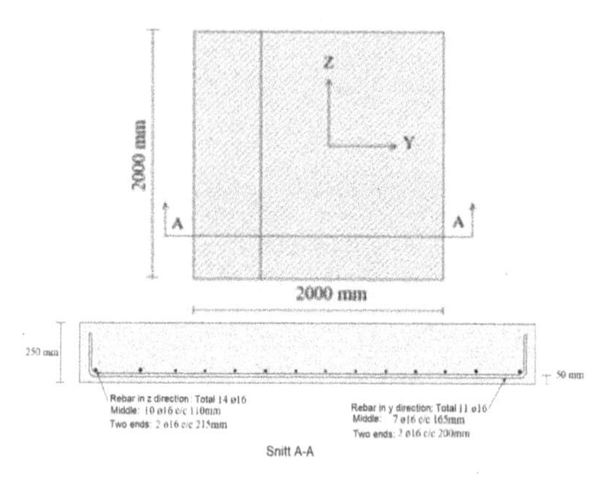

Figure 3. Foundation with steel rebars

2.2 *Experimental tests of material properties*

The compressive strength and E-modulus test (Figure 4 a)) were performed for steel fiber concrete. In addition, the three-point bending test was performed for six beams with 10 kg/m^3 composite fiber and five beams with 30 kg/m^3 steel fiber to determine the residual flexural tensile strength for fiber reinforced beams in accordance with EN14651 (Figure 4b)). The properties of the steel fiber concrete are reported in Table 4.

The nominal flexural stress σ_N-CMOD curves from the three-point bending tests is shown in Figure 5.

a) E-modulus test

b) Three point bending beam test (EN14651)

Figure 4. E-modulus and three-point bending beam test.

Table 4. Properties of fiber concrete.

Properties	Steel fiber concrete
E-modulus	$E_{cm} = 30.9\ GPa$
Compressive strength	$f_{ck} = 54.2\ MPa$

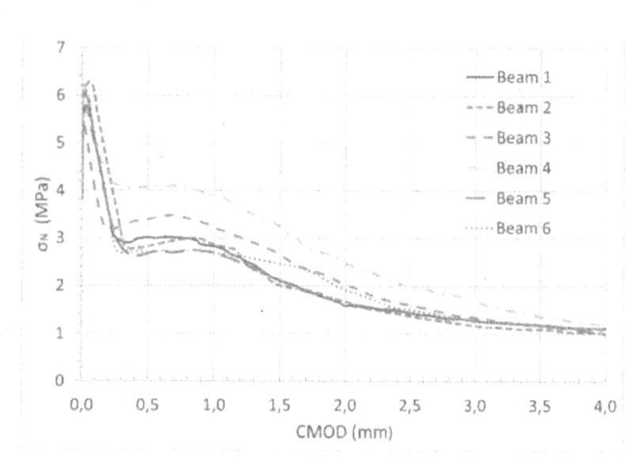

a) Beams with 10 kg/m^3 composite mineral fiber

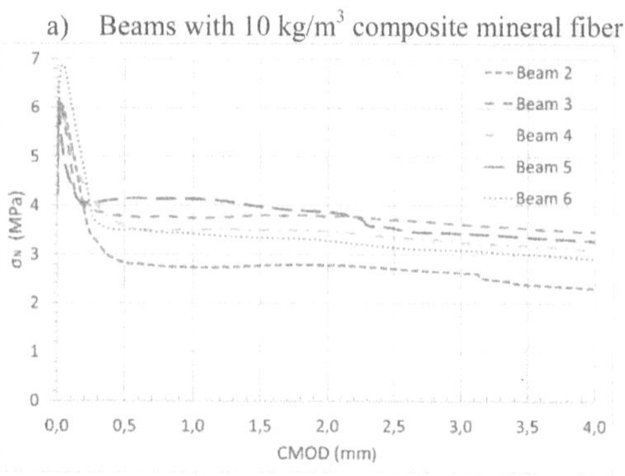

b) Beams with 30 kg/m^3 steel fiber

Figure 5. Nominal stress vs crack mouth-opening displacement (CMOD) from three-point bending tests.

The limit of proportionality f_{LOP}, the residual flexural tensile strengths f_{R1}, f_{R2}, f_{R3} and f_{R4} for a CMOD of 0.5, 1.5, 2.5 and 3.5 mm, respectively, were assessed for all the specimens. The corresponding mean and characteristic values are reported in Table 5 and

Table 6. The beam test results show that both steel and composite fiber concrete have softening post-cracking behavior and the steel fiber gave higher residual flexural tensile strengths than the composite fiber at the large crack widths. The beams reinforced with composite fiber had approximately the same residual tensile strength as steel fiber at smaller crack widths, but at CMOD 2.5 the value was more than halved. This is in accordance with the mechanical properties of composite fibers and the post-cracking strength could therefore be classified as "R2.0a". The post-cracking strength of steel fiber concrete was classified as "R2.5c" according to fib Model Code 2010 specifications [21] with reference to the characteristic values.

In the planned revision of EN 206 and Eurocode 2 an upper bound, $\kappa_{k,max}$, of the characteristic values will be introduced and the adjusted characteristic residual strengths shown in Table 5 and Table 6 could be used as basis for design.

$$f_{R,1kcor} = \min(f_{R,1k}; \kappa_{k,max} \cdot f_{R,1m}) \quad (1)$$

$$f_{R,3kcor} = \min(f_{R,3k}; \kappa_{k,max} \cdot f_{R,3m}) \quad (2)$$

$$\kappa_{k,max} = 0.6$$

Table 5. Three-point bending test results for beams with 10 kg/m^3 composite mineral fibers according to EN 14651 [22].

Reference values (MPa)	f_{LOP}	f_{R1}	f_{R2}	f_{R3}	f_{R4}
Mean $f_{R,im}$	5.94	3.12	2.44	1.56	1.17
St. dev.	0.25	0,50	0,41	0,21	0,32
$f_{R,ik}$	5.51	2.27	1.74	1.20	0.96
$f_{R,ikcor}$		1.87		0.94	

Table 6. Three-point bending test results for beams with 30 kg/m^3 steel fibers according to EN 14651 [22].

Reference values (MPa)	f_{LOP}	f_{R1}	f_{R2}	f_{R3}	f_{R4}
Mean $f_{R,im}$	5.98	3.57	3.48	3.30	3.10
St. dev.	0.65	0,42	0,41	0,34	0,39
$f_{R,ik}$	4.68	2.73	2.66	2.61	2.31
$f_{R,ikcor}$		2.14		1.98	

The E-modulus of the isolation material (XPS700SL) was measured in laboratory as shown in 2.3 and the value is 36.2 MPa.

2.3 *Test results of foundation slabs*

The load was applied in the center of slab in a deformation-controlled manner and a constant speed of 0.4 mm/min was used in the tests. The logging frequency was at 2 Hz throughout the experiment. The

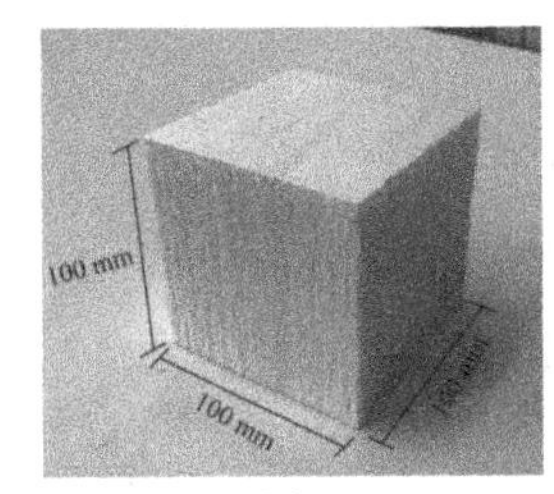

a) Test cubic

b) Test av E-modulus

Figure 6. E-modulus test of isolation material.

vertical displacements at 175, 250 and 950 mm from the center of the jack were measured by Linear Variable Displacement Transducers (LVDTs).

Experimental results are presented in terms of vertical load versus relative displacement between 250 mm and 950 mm from center as shown in Figure 7. The curve of load and deflection is more or less linear

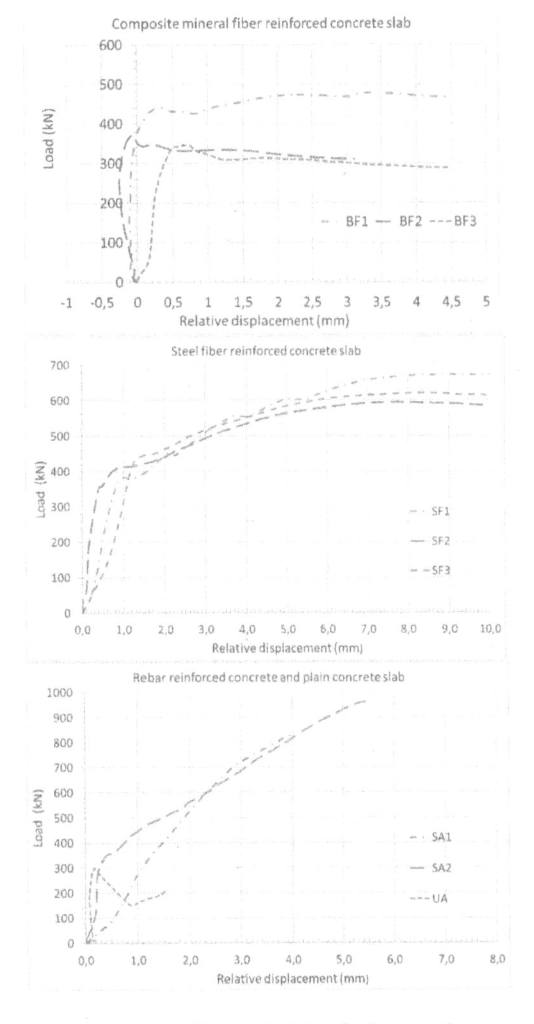

Figure 7. Measured load and relative displacement between 250mm and 950mm from center.

before the first cracking occurs. The load level at initiation of cracking for composite fiber, steel fiber and rebar reinforced slabs is in the similar range, which is between 345 to 445kN as shown in Table 7, and the load level for plain concrete is slightly lower at about 297kN.

The post-cracking behavior of the SFRC slabs is significantly different from the plain concrete slab. The steel fiber effectively enhances the bearing capacity of the slab up to a maximum load of 600kN, and the steel fibers assure a ductile failure while the plain concrete slab showed a brittle failure when the maximum load reached 297kN as shown in Table 7. The composite mineral fiber reinforced slabs reached its capacity at the initiation of crack, but they showed a clear ductility before failure. The test results of the slabs are consistent with the test results of three-point bending beam test. The test results of composite fiber concrete showed larger scatter both for the beam bending and the full-scale test.

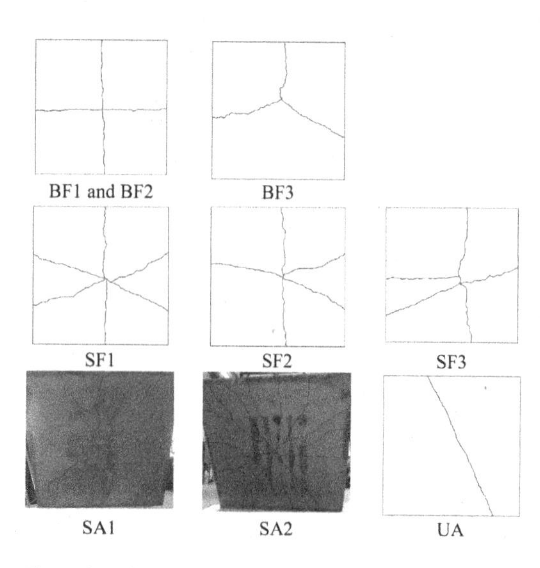

Figure 8. Final crack patterns of slabs.

Table 7. Load and displacement at 250mm from center.

	Initiation of cracking		Maximum loading	
Test ID	Load (kN)	Displacement (mm)	Load (kN)	Displacement (mm)
BF1	440.8	3.97	477.7	6.59
BF2	371.4	3.48	371.4	3.48
BF3	344.4	3.18	344.4	3.18
SF1	384.0	3.41	672.1	9.75
SF2	353.8	3.83	592.8	10.58
SF3	445.7	4.30	618.7	9.37
SA1	424.7	2.75	930.4	5.67
SA2	355.4	3.02	963.5	8.77
UA	297.2	2.82	297.2	2.82

The final crack patterns for the slabs are presented in 3. All the cracks were initialed at the center of the slab and developed further to the edges of slab. The plain concrete slab has only one crack. The final crack patterns for the composite fiber and steel fiber slabs are similar. For composite fiber the crack pattern is characterized by three or four major cracks starting from the slab center, while the steel fiber has five or six major cracks starting from the slab center. The slabs with rebars have more than 10 major cracks starting from the center. It shows that the ductility of the slab is correlated with the crack patterns: more final cracks better ductility, and larger possibilities for load redistribution.

3 ANALYTICAL ANALYSIS

The moment capacity of the slab is calculated by several simplified methods which are based on recommendations from the guidelines published by the Norwegian concrete association NB38 [16], the comprehensive research project COIN [1], the German rules DAfStb [17], and the recently launched draft of the new Eurocode 2 Annex L [19]. The methods are described in following section.

3.1 *NB38 og Eurocode 2 Annex L*

The same approach is adopted for NB38 and Eurocode 2 Annex L. The characteristic residual flexural strength is given in equations 1 and 2:

$$f_{R,1kcor} = \min(f_{R,1k}; 0.6 \cdot f_{R,1m})$$

$$f_{R,3kcor} = \min(f_{R,3k}; 0.6 \cdot f_{R,3m})$$

This approach is included to open up for unfactored design strengths up to 90% of the mean strengths which contributes to economical, safe and robust solutions. [19]

Assuming orientation factor $k_o = 1.0$, which is acceptable for the considered structural members, the uniaxial residual tensile strength is given as:

$$f_{Ftsk} = 0.40 \cdot f_{R,1kcor} \quad \text{for SLS} \tag{3}$$

$$f_{Ftuk} = 0.37 \cdot f_{R,3kcor} \quad \text{for ULS} \tag{4}$$

The design residual tensile strength should be taken as follows:

$$f_{Ftsd} = f_{Ftsk} / \gamma_{SF} \quad \text{for SLS} \tag{5}$$

$$f_{Ftuk} = f_{Ftuk} / \gamma_{SF} \quad \text{for ULS} \tag{6}$$

$$\gamma_{SF} = 1.5$$

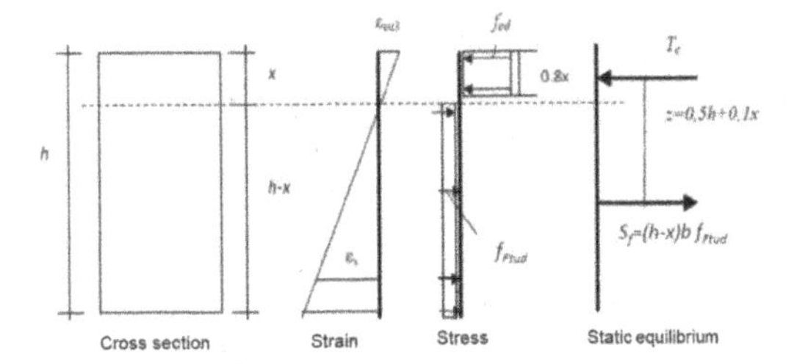

Figure 9. Static equilibrium of the cross section under bending.

The bending moment capacity is calculated based on the static equilibrium shown in *Figure 9*, and the moment capacity for fiber-reinforced cross-sections without conventional longitudinal reinforcement is written as:

$$M_{Rd} = S_f\,(0.5h + 0.1x) \tag{7}$$

$$S_f = (h - x)\,bf_{ftud} \tag{8}$$

$$x = \frac{hf_{Ftud}}{0.8f_{cd} + f_{Ftud}} \tag{9}$$

3.2 *COIN*

The characteristic residual strength is given as:

$$f_{ftk,rec2.5} = 0.37 \cdot f_{R,3k} \quad \text{for ULS} \tag{10}$$

The design residual tensile strength is given as:

$$f_{ftd,rec2.5} = f_{ftk,rec2.5}/\gamma_{SF} \quad \text{for ULS} \tag{11}$$

$$\gamma_{SF} = 1.5$$

By replacement of f_{ftud} with $f_{ftd,rec2.5}$ in equations 7-9 moment capacity for fiber-reinforced cross-sections without conventional longitudinal reinforcement can be calculated.

3.3 *DAfStb*

Two stress-stain relationships (stress block and tri-linear) are specified in DafStb for the tension zone at ULS. The ratio $L2/L1$ determines which one needs to be used. The design residual tensile strength is obtained by multiplying the characteristic residual flexural strength with certain factors and the detailed description may be found in [17].

The bending moment capacity of the cross section without conventional longitudinal reinforcement can be calculated by equations similar to equation 7.

3.4 *Results of analytical analysis*

The moment capacity per meter width of the slab is calculated with the design residual tensile strength with material factor $\gamma_{SF} = 1.0$ and the mean residual tensile strength. The results are presented in Table 8. The moment capacity based on the design residual tensile strength is about 60% to 80% of the moment capacity based on the mean residual tensile strength.

Table 8. Moment capacity of SFRC slabs

Moment capacity (kNm/m)	NB38 and Eurocode 2 Annex L	COIN	DafStb	
BF	Design	10.8	13.7	11.2
	Mean	17.8	17.8	18.6
SF	Design	22.5	29.6	23.5
	Mean	37.3	37.3	38.8

Based on yield line analysis the upper bound solutions for the collapse load of a square spread footing with the edges free to move in vertical direction were presented in [23], and the two failure patterns showed in 4 are similar to the observed final crack pattern of the SFRC slabs in the experiment. The collapse loads for those two failure patterns maybe determined as:

$$P = \frac{8m}{\left(1 - \frac{a_1}{a}\right)^2} \quad \text{Failure pattern+} \tag{12}$$

$$P = \frac{24m}{\left(1 - \frac{a_1}{a}\right)^2\left(2 + \frac{a_1}{a}\right)} \quad \text{Failure pattern x} \tag{13}$$

In which $m = M/a$ the flexural strength of the footing slab per unit width; a_1=the width of the column section; a =the width of the footing slab

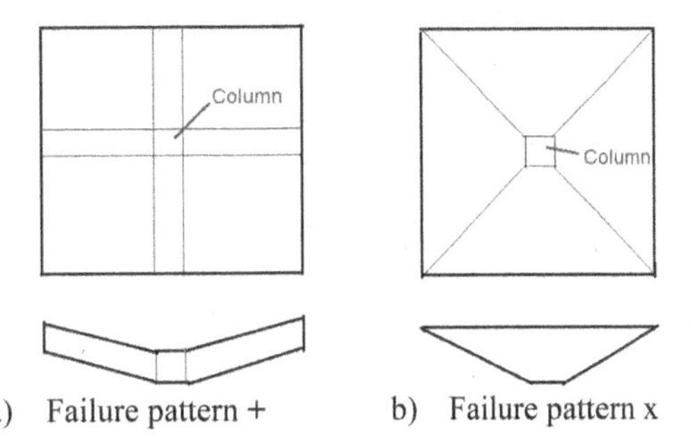

Figure 10. Failure patterns.

The collapse loads are calculated by equations 12 and 13 for composite fiber and steel fiber reinforced slabs. The collapse loads are about 50% higher for failure pattern x compared to the failure pattern +. The collapse loads for both composite fiber and steel fiber reindorsed slabs are significantly lower when they are compared to the test results. It should be noted that the subgrade in the test is not homogeneous and the stiffness properties therefore are different from soil subgrade.

Table 9. Collapse load of SFRC slabs

Collapse load (kN)			NB38	COIN	DafStb
BF	Design	+	100.9	128.5	105.0
		x	145.8	185.8	151.8
	Mean	+	166.9	166.9	173.7
		x	241.3	241.3	251.1
SF	Design	+	210.7	276.4	219.2
		x	304.6	399.6	317.0
	Mean	+	348.9	348.9	362.9
		x	504.4	504.4	524.7

4 NUMERICAL ANALYSIS

A nonlinear finite element model (NLFE) closely representing the test setup was established in the finite element software DIANA 10.4 [26] and the simulation results were compared with test results for composite and steel fiber reinforced, rebar reinforced and plain concrete foundations.

4.1 *Numerical model*

The fracture behavior of concrete was simulated by the total strain rotating crack model. The compressive behavior of concrete was represented by the parabolic stress-strain relationship with compressive strength reduction due to lateral cracking, originally proposed by Vecchio & Collins [24]. The tensile behavior is modelled by tensile failure model for fiber reinforced concrete as defined by the FIB working group [21]. The tensile behavior was considered linear elastic up to the mean tensile strength (f_{ctm}) and the post-peak behavior of the constitutive law is defined based on a linear model with two reference values: serviceability residual tensile strength f_{Fts} and ultimate tensile residual strength f_{Ftu}. They can be determined by residual values of flexural strengths by using the following equations [21]:

$$f_{Fts} = 0.45 f_{R1} \tag{14}$$

$$f_{Ftu} = 0.5 f_{R3} - 0.2 f_{R1} \geq 0 \tag{15}$$

The f_{Fts} and f_{Ftu} of composite and steel fiber reinforced concrete were determined from mean and characteristic flexural residual tensile strengths as presented in Table 10. The tensile constitutive law based on mean ($f_{Fts,m}$ and $f_{Ftu,m}$) and characteristic ($f_{Fts,k} and f_{Ftu,k}$) values are shown in Figure 11. The numerical simulations were performed for both the mean and characteristic curves.

The expression of the tensile constitutive law used in simulation is different from the ones specified in Eurocode 2 (equations 3 and 4) and DAfStb.

Table 10. Properties of basal and steel fiber reinforced concrete used in simulation

Property Designation	Unit	Material CM	Material SF
Classification (according MC2010)		2.0a	2.5c
Mean modulus of elasticity (E_{cm})	MPa		30900
Poisson's ratio (υ)		0.15	0.15
Mean compressive strength (f_{cm})	MPa		57.3
Compressive fracture energy G_f	N/mm		37.8
Mean tensile strength ((f_{ctm})	MPa		3.20
Characteristic tensile strength ((f_{ctk})	MPa		2.20
Mean serviceability residual tensile strength ($f_{Fts,m}$)	MPa	1.40	1.61
Mean ultimate residual tensile strength ($f_{Ftu,m}$)	MPa	0.16	0.94
Characteristic serviceability residual tensile strength ($f_{Fts,k}$)	MPa	1.02	1.23
Characteristic ultimate residual tensile strength ($f_{Ftu,k}$)	MPa	0.15	0.76

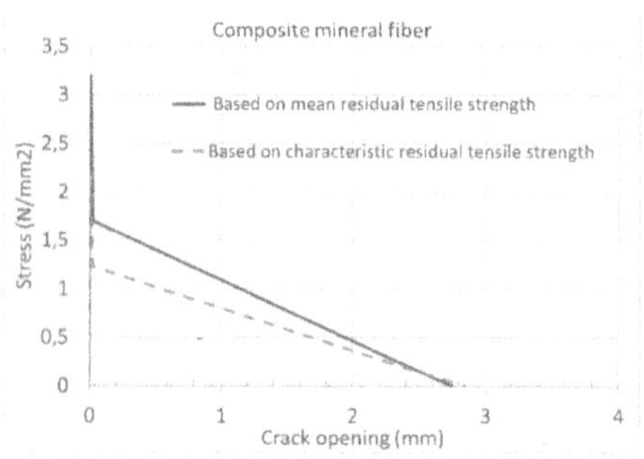

a) Composite mineral fiber reinforced concrete

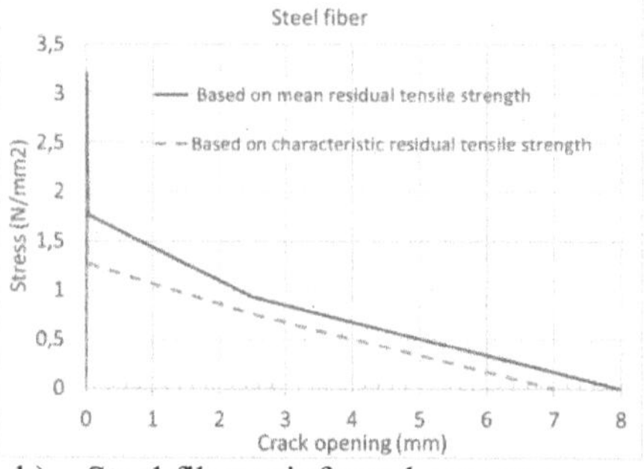

b) Steel fiber reinforced concrete

Figure 11. Constitutive law in tension for SFRC material

The material properties of rebar and concrete used in simulation for rebar reinforced and plain concrete slabs and the material properties of the isolation material XPS700SL are presented in Table 11. The material models used in the FE modelling are presented in Table 12.

The schematic figure of the FE model used in the simulations is shown in Figure 12 for normal mesh size. The slab and isolation layer are model by solid elements with 20 nodes (CHX60) and the vapor barrier is model by 3D surface interface elements with nonlinear elasticity, and the tension and shear force become zero when the slab and isolation layer is separated. The subgrade is also modeled by 3D surface interface elements with nonlinear elasticity to allow separation between isolation layer and subgrade. The subgrade is divided into two areas: concrete and sand subgrade as shown in Figure 12 b). The normal and shear stiffness of surface interface elements used in simulation is presented in Table 13.

Table 11. Properties of rebar and plain concrete used in simulation

Property	Unit	
Rebar		
E-modulus (E_s)	MPa	200000
Poisson's ratio (υ)		0.3
Concrete		
E-modulus (E_{cm})	MPa	30900
Poisson's ratio (υ)		0.2
Compressive strength (f_{cm})	MPa	43.0
Compressive fracture energy G_f	N/mm	35.9
Mean tensile strength ((f_{ctm})	MPa	2.90
Tensile fracture energy	N/mm	0.15
Isolation XPS700SL		
E-modulus	MPa	36.2
Poisson's ratio		0.3

Table 12. The material models used in FE modeling.

	Tensile model	Compressive model	Steel rebar
BF	Mode Code 2010	Parabolic [25]	–
SF	Mode Code 2010	Parabolic	–
SA	Hordijk [25]	Parabolic	Ideal
UA	Hordijk	Parabolic	–

The load is applied as prescribed displacements in the vertical direction and the iterative process used to solve the equations was the Newton-Raphson iteration method. To reach convergency the model was checked to a force norm and energy norm. The mesh sensitivity study is performed for coarse, normal, and fine mesh

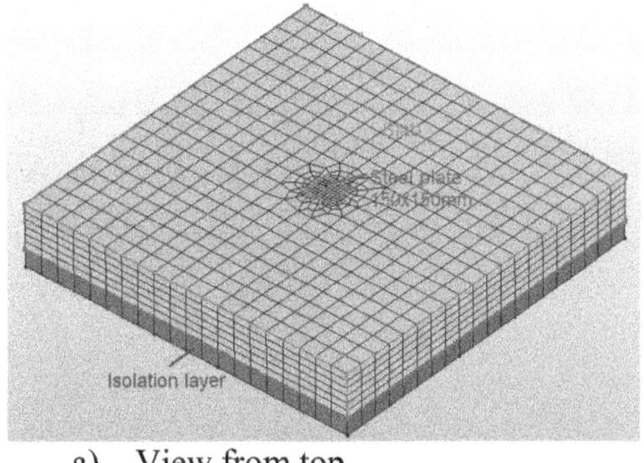

a) View from top

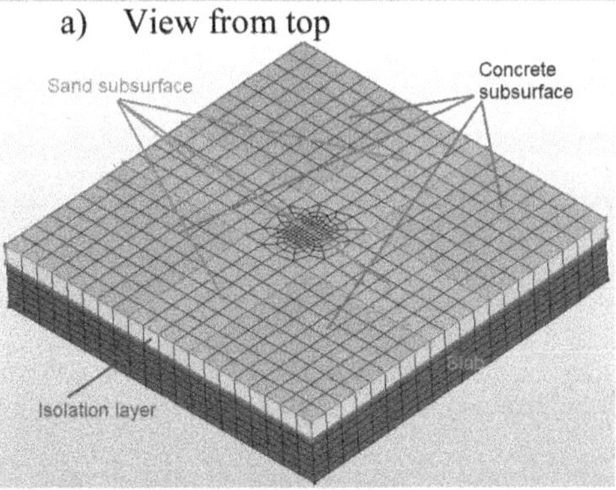

b) View from bottom

Figure 12. Schematic figure of the finite element mesh of slab (Normal mesh).

Table 13. Normal and shear stiffness of 3D surface interface

3D surface interface element	Normal stiffness modulus (N/mm^3)	Shear stiffness modulus (N/mm^3)
Vapor barrier	10	0.1
Concrete subsurface	309	1.0
Sand subsurface	0.005 [25]	0.001

to decide optimal mesh size for the simulations, and the element number of the models with different mesh sizes are shown in Table 14. The simulation results of load-displacement relationship at center of slab are shown in Figure 13, and the three mesh sizes have same results before the maximum load is reached and it is seen that the result with coarse mesh deviates from the other two only at the last stage of loading. The normal mesh size is then selected for the numerical analysis.

4.2 *Simulation results*

The load and relative displacement diagrams in Figure 14 compare the response of the numerical simulations with the results from the full-scale tests.

The numerical response of the steel fiber reinforced slab is in good agreement with the experimental

Table 14. Element number for three mesh sizes

Element number	Coarse	Normal	Fine
Solid element	1320	3843	12510
Contact element	528	1098	2502

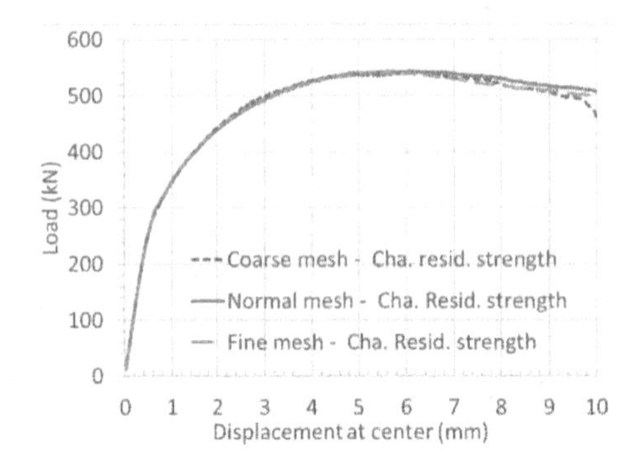

Figure 13. Load-displacement curve at center of slab for three different mesh sizes.

ones in terms of initial stiffness and maximum capacity for the input with mean residual tensile strength. The maximum capacity is about 20% lower for the input with characteristic residual tensile strength. At smaller deflections the results applying the characteristic strengths are in better agreement with the experiment.

Although the full-scale test results of the composite fiber reinforced slabs had relatively large scatter the numerical simulation has a reasonable agreement with the test data. The predicted maximum capacity based on mean residual tensile strength is slightly higher than the maximum capacity of BF1 slab. The maximum capacity based on characteristic residual tensile strength is between BF1 and the other two slabs (BF2 and 3).

There is good agreement between the numerical response and the test results for both the rebar reinforced and the plain concrete slabs. The difference between results based on mean and characteristic residual tensile strength is small for both cases.

The numerical crack patterns at the bottom of the slabs are shown in Figure 15. For the composite and steel fiber slabs the crack initiated at the center of slab and further developed into a cross pattern with one vertical and one horizontal crack path which was located in the area of sand subgrade. For slab with rebars a pattern with multiple crack paths around the center appeared which was also observed in the full-scale test. The numerical and experimental crack pattern was the same for plain concrete, simply a single crack. The good agreement between numerical and experimental crack patterns at failure indicates that the numerical model is reliable.

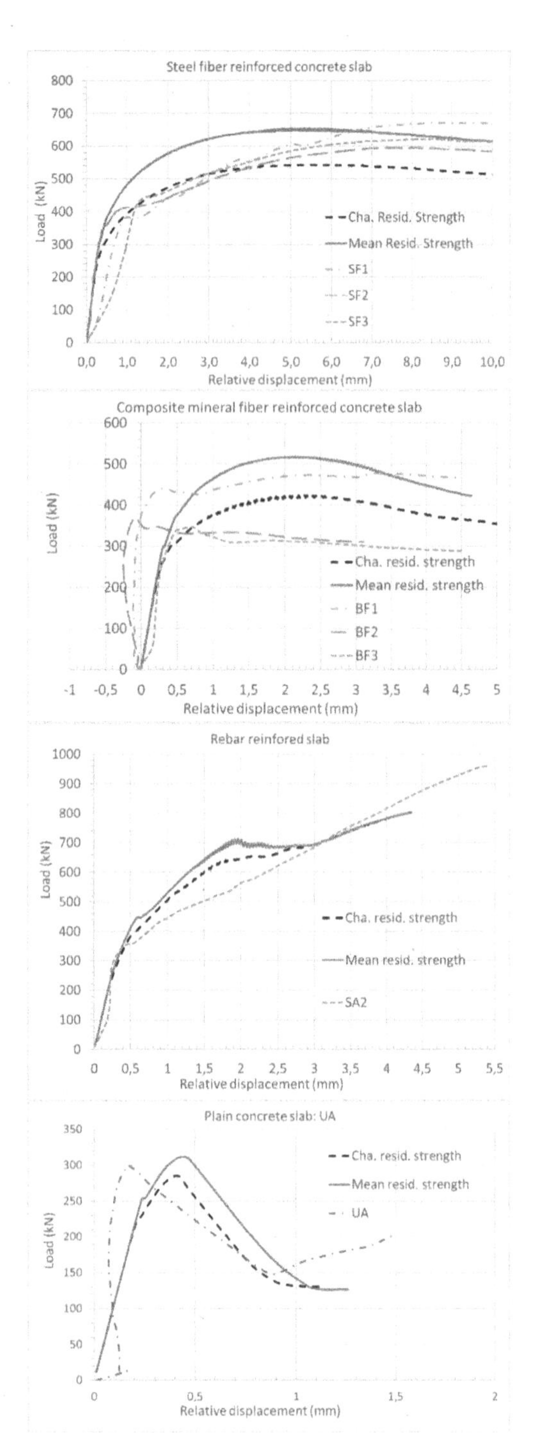

Figure 14. Experimental vs numerical load – relative displacement curve

4.3 *Parameter study*

In the test setup the vapor barrier was used to provide less resistance for lateral movement of the slab when it is under vertical loading. The lateral restraint in the

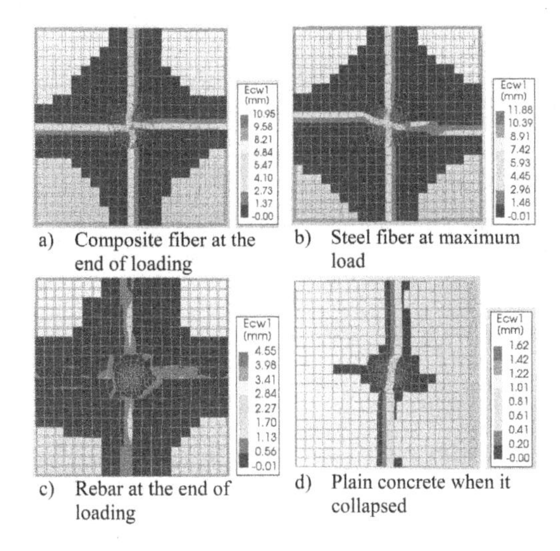

Figure 15. Numerical crack pattern at bottom of slabs.

bottom surface of the slab could be important for the load bearing capacity of slabs. In the parameter study the shear stiffness of the interface between slab and isolation layer is varied from 0.01N/mm^3 to 1.0N/mm^3 which corresponds to loose sand and the concrete surface, and the simulation results are shown in 5. The increase of shear stiffness increases the failure capacity of the slab, but the residual capacity drops sharply at large deformation for the higher shear stiffness.

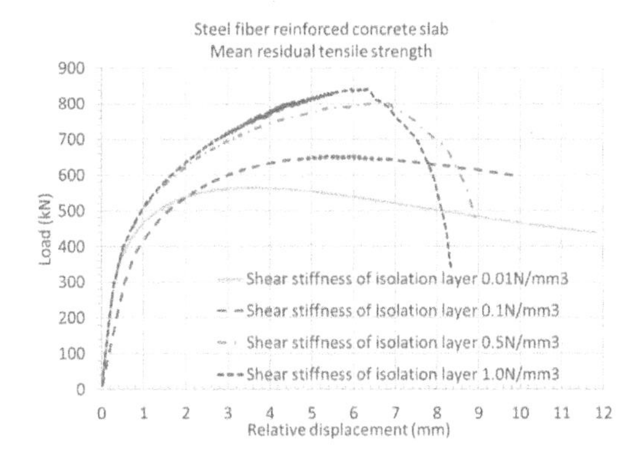

Figure 16. The influence of shear stiffness of interface between slab and isolation layer.

5 CONCLUSION

The results presented and discussed in the manuscript lead to the following remarks:

- The steel fiber effectively enhances the load-carrying capacity of slabs on ground and makes the structural response more ductile; The composite mineral fibers have minor effect on the increase of load-carrying capacity, but the structural response shows certain increased ductility.
- Comparisons between the test results and the simplified methods based on COIN, NB38, DAfStb, and the draft of the new Eurocode 2, show that the calculation methods are conservative for the moment capacity and collapse loads. The main reason for the large deviation between the simplified methods and the experimental results is that a uniform ground pressure over entire foundation slab is assumed in simplified method.
- The analyses of SFRC slabs based on NLFM predict the slab response with sufficient or good accuracy with the input from material testing in the laboratory.
- The lateral constraints provided by the layer under the slabs may have significant effect on the load bearing capacity.
- The results showed that it is safe to use steel fibers or composite mineral fibers as the only reinforcement in column foundations on solid ground.
- The comparison between the experimental results, the simplified calculations and the numerical simulations shows that considerably more economic solutions can be determined if numerical simulation are applied in design.

REFERENCES

[1] Kanstad, T., Juvik, D. A., Vatnar, A., Mathisen, A. E., Sandbakk, S., Vikan, H., Nikolaisen, E., Døssland, Å., Leirud, N., & Overrein, G. O., Forslag til retningslinjer for dimensjonering, utførelse og kontroll av fiberarmerte betongkonstruksjoner. FA 2 Competitive constructions. SP 2.2 High tensile strength all round concrete, SINTEF Building and Infrastructure, 2011.

[2] Alani, A. M. & Aboutalebi, M., Analysis of the subgrade stiffness effect on the behaviour of ground-supported concrete slabs, Structural Concrete, vol. 13, no. 2, pp. 102–108, 2012.

[3] Inman, M., Thorhallson, E. R. & Azrague, K., A mechanical and environmental assessment and comparison of basalt fibre reinforced polymer (BFRP) rebar and steel rebar in concrete beams, Energy Procedia 111 (2017) 31–40.

[4] Marcalikova, Z., Cajka, R., Bilek, V., Bujdos, D., & Sucharda, O., Determination of Mechanical Characteristics for Fiber-Reinforced Concrete with Straight and Hooked Fibers, Crystals, 10(6), 545, 2020.

[5] Marcalikova, Z., Racek, M., Mateckova, P. & Cajka, R., Comparison of tensile strength fiber reinforced concrete with different types of fibers, Procedia structural integrity 28 (2020) 950–956, 2020.

[6] Mohaghegh, A. M., Silfwerbrand, J. & Årskog, V., Properties of Fresh Macro Basalt Fibre (MiniBar) Self-Compacting Concrete (SCC) and Conventional Slump Concrete (CSC) Aimed for Marine Applications, Nordic Concrete Research, ISSN 0800-6377, Vol. 52, nr 1, 2015.

[7] El-Gelani, A. M., High, C. M., Rizkalla, S. H. & Abdalla, E. A., Effects of Basalt Fibres on Mechanical Properties of Concrete, 2018.

[8] Fredvik, T. I., Stemland, H., Kristiansen, B., Vatnar, A., Uppstad, B., Bondestad, O., Eriksen, D., Eikemo, K., Mathisen, A. E. & Cielicki, T. Publikasjon nr.15, Concrete floor – floor on the ground, Norwegian Concrete Association, 2018.

[9] Pouillon, S., Fibre Reinforced Concrete in Structural Applications, 8th International Conference FIBRE CONCRETE 2015, september 10–11, 2015, Prague, Czech Republic.

[10] Janulikova M. & Mateckova, P., Experimental testing of punching shear resistance of concrete foundations, in Proceedings of the 2nd Czech-China Scientific Conference 2016, IntechOpen, 2017.

[11] Sorelli, L. G., Meda, A. & Plizzari G. A., Steel fiber concrete slabs on ground: A structural matter, ACI Structural Journal, V. 103, No. 4, July-August 2006.

[12] Sandbakk, S., Fiber concrete applications in load bearing construction, mur+betong, no. 2, 2020.

[13] Facconi, L., Plizzari G. & Minelli, F., Elevated slabs made of hybrid reinforced concrete: Proposal of a new design approach in flexure, Structural Concrete, 2018.

[14] Cajka, R., Marcalikova, Z., Kozielova, M., Mateckova, P. & Sucharda, O., Experiments on fiber concrete foundation slabs in interaction with the subsoil, Sustainability, vol. 12, no. 9, 2020.

[15] Redaelli, D. & Nseir, J. Y., Structural behavior of prestressed Ultra-High Performance Fibre-Reinforced Concrete beams with and without openings: comparison between experimental results and finite element modelling techniques, FRC2018: Fibre Reinforced Concrete: from Design to Structural Applications, Joint ACI-fib-RILEM International Workshop, 2018.

[16] Kanstad, T., Døssland, Å. L., Vatnar, A., Mathisen, A. E., Brå, H., Hisdal, J. M., Leirud, N., Sandbakk, S., Sandaker, T. K., Bjøntegaard, Ø., & Sæter, Ø., Publikasjon nr.38, Fiber reinforced concrete for load bearing structures, Norwegian Concrete Association, 2020.

[17] Deutscher Ausschuss für Stahlbeton. DafStb Guideline – Steel fibre reinforced concrete, November 2012.

[18] prEN 1992-1-1 Eurocode 2: Design of concrete structures, 2020 – Part 1-1: General rules – Rules for buildings, bridges and civil engineering structures

[19] CEN/TC250/SC2 Project Team T3, Terje Kanstad: Background document to prEN 1992-1-1 D7, rev 7 Annex L, Steel fibre reinforced concrete (SFRC). June 2021.

[20] Salice, V. E., Kråkenes, B. S.-A. & Fauske, B. H., Fiber reinforced concrete for load bearing structures, Universitetet i Agder, 2021

[21] di Prisco, M., Colombo, M. & Dozio, D., Fibre-reinforced concrete in fib Model Code 2010: principles, models and test variation, 2013.

[22] NS-EN 14651:2005+A1:2007, Test method for metallic fibre concrete - Measuring the flexural tensile strength (limit of proportionality (LOP), residual), 2007.

[23] Jiang, D. H., Flexural strength of square spread footing, Journal of Structural Engineering, 1983, 109(8): 1812–1819.

[24] Vecchio, F. J. & Collins, M. P., The modified compression field theory for reinforced concrete elements subjected to shear, ACI Journal 83, 22, 219–231, 1986.

[25] Murthy V.N.S, Soil Mechanics and Foundation Engineering. 1st Edition CBS Publishers and Distributors Pvt. Ltd, India, 2012.

[26] DIANA FEA BV, 2021.

Computational Modelling of Concrete and Concrete Structures – Meschke, Pichler & Rots (Eds)
© 2022 Copyright the Author(s), ISBN: 978-1-032-32724-2

3D FEM analysis of disk shear-key considering the material properties of the existing concrete surface

Y. Ishida, T. Sato & M. Kubota
TOBISHIMA Corporation, Noda-shi, Chiba, Japan

T. Akisawa, H. Sakata & Y. Maida
Tokyo Institute of Technology, Meguro-ku, Tokyo, Japan

Y. Takase
Muroran Institute of Technology, Muroran-shi, Hokkaido, Japan

ABSTRACT: In this study, three-dimensional (3D) finite element modeling (FEM) analysis of the disk shear-key was conducted in consideration of the material properties of the existing concrete surface for the application of the disk shear-key to the seismic retrofitting joint without removal of the finishing material. For this, a 3D FEM model was constructed to reproduce a previous shear element experiment. Subsequently, the analysis was conducted using a model in which the existing concrete surface had a finishing material with a thickness of 10 mm. The results confirmed that the shear resistance performance of the disk shear-key was reduced considerably. Furthermore, the stress state inside the member indicated that the embedded depth of the steel disk, h_d, is an important parameter. As a result of the analysis based on the use of h_d, it was discovered that the shear resistance performance was ensured by extending h_d from 20 to 30 mm.

1 INTRODUCTION

The external seismic retrofitting illustrated in Figure 1 is a seismic retrofitting method used for existing reinforced concrete buildings. External seismic retrofitting is useful as it can be constructed while maintaining the current functions of the building. The joints of seismic retrofitting are required to have sufficient shear strengths and high stiffness values to achieve the reinforcing effect of the reinforcing members.

The disk shear-key depicted in Figure 2(a) is a composite shear resistance system composed of a steel disk and anchor bolt. Because the disk shear-key resists shear force primarily owing to the bearing resistance of concrete, it has a high stiffness value and shear strength compared with the general post-installed anchor illustrated in Figure 2(b). Therefore, the material properties of the concrete surfaces are crucial. In a previous study, the mechanical behavior of the disk shear-key was verified experimentally and analytically. The existing concrete in the test specimens and analysis models used in the experiments and analyses conducted thus far have been in good conditions. However, there are finishing materials on the concrete surfaces in actual concrete buildings. Thus, it is usually necessary to remove all the finishing materials before installing the disk shear-key. However, removing the finishing material leads to an increase in the construction period and cost. In addition, it is necessary to consider the prevention of dust, noise, vibration, and fall accidents of the scraped finishing material that occurs when the finishing material is removed. Hence, it is desirable for the disk shear-key to be installed without removal of the finishing material.

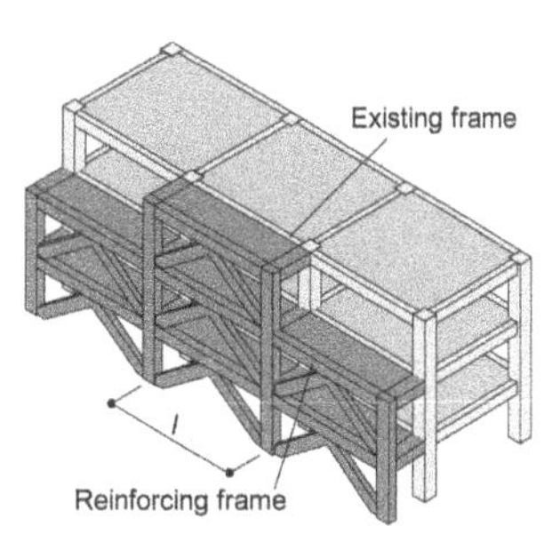

Figure 1. Outline of external seismic retrofitting.

In this study, three-dimensional (3D) finite element modeling (FEM) analysis of the disk shear-key was conducted based on considerations of the material properties of the existing concrete surface. The study aimed at applying the disk shear-key to the seismic retrofitting joint without removal of the finishing material.

DOI 10.1201/9781003316404-33

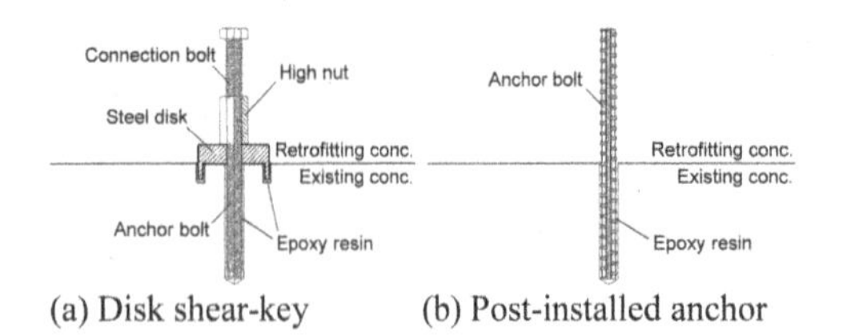

Figure 2. Outline of joint elements.

2 OUTLINE OF STANDARD FEM MODEL

Model 1 is the standard FEM model used in this research. A 3D FEM model of the disk shear-key was constructed by applying the 3D FEM model of the post-installed anchor constructed in a previous study (Ishida et al. 2018). The standard FEM model reproduced the shear element experiment of a seismic retrofitting joint to which a disk shear-key had been applied (Satoh et al. 2017).

2.1 *Outline of target experiment*

2.1.1 *Outline of specimen*

The specimen was composed of the existing concrete, a disk shear-key, and retrofitting concrete, as illustrated in Figure 3. First, the concrete on the existing side was cast, and after demolding, a hole was drilled to install the disk shear-key. After sufficient drying, an epoxy resin adhesive was injected, and the disk shear-key was installed. Reinforcements and a steel plate with welded headed studs were then installed, and the grout on the reinforcing side was cast.

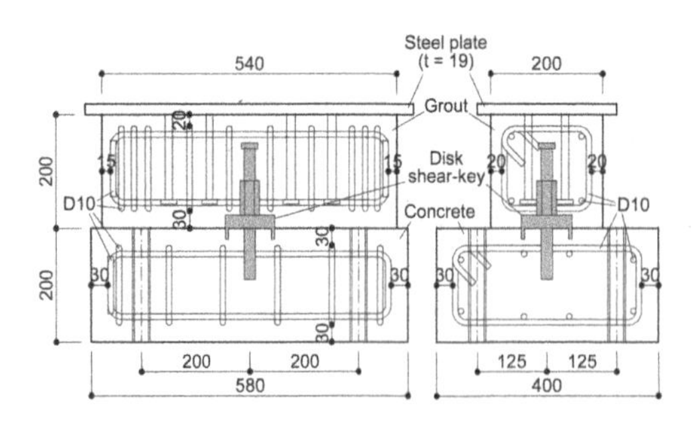

Figure 3. Details of specimen (unit: [mm]).

For the construction of the disk shear-key, a wet core drill was used as the piercing device, and it pierced the concrete in the downward direction.

2.1.2 *Loading device and measurement plan*

The loading device is illustrated in Figure 4. Hydraulic jacks were attached to the left and right sides according to the height of the joint surface, and a shear force was then applied by pressing either of the jacks. In addition, a hydraulic jack was attached in the vertical direction. Subsequently, a constant compressive stress of 0.5 N/mm^2 was applied in accordance with a prior study (Takase et al. 2014). A parallel crank was attached to the top of the U-shaped loading beam to maintain it in a parallel orientation.

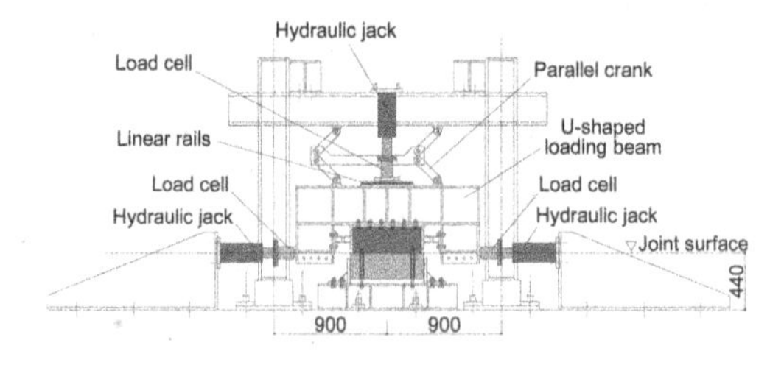

Figure 4. Loading setup (unit: [mm])

The relative horizontal displacement was measured at two positions, as depicted in Figure 5, and their average value was evaluated as δ_h.

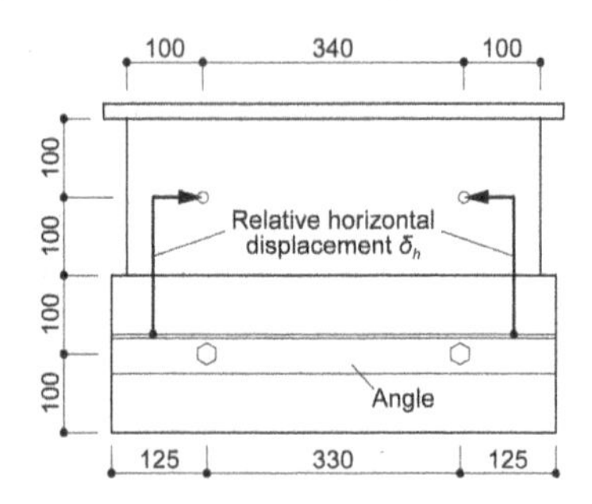

Figure 5. Measurement plan (unit: [mm]).

The loading plan in Figure 6 demonstrates the cyclic loading patterns in the positive and negative directions. Each positive and negative load cycle pattern was applied once up to a relative horizontal displacement of 0.5 mm, and it was then loaded repeatedly twice in each direction up to a relative horizontal displacement of 4 mm. Finally, the specimen was loaded once up to a relative horizontal displacement of 6 mm in the positive and negative directions, and it was then pushed to the positive side until failure.

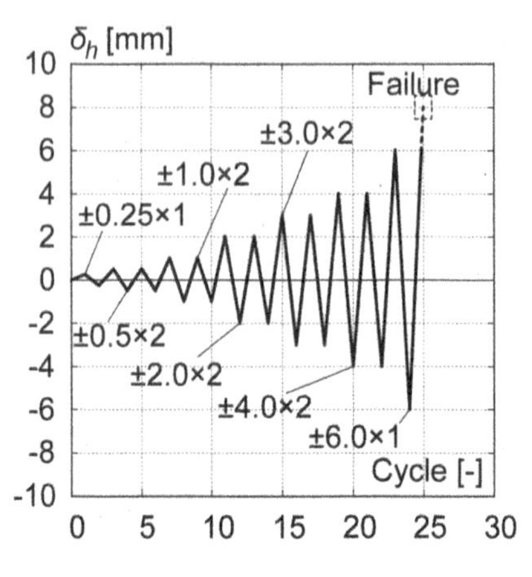

Figure 6. Loading cycle.

2.2 *Analysis conditions*

A general nonlinear FEM analysis program (ITOCHU Techno-Solutions Corporation 2016) was used for the analysis. An outline of the FEM analysis model is depicted in Figure 7. In this analysis model, owing to the symmetry, only half of the construction is modeled against the vertical plane (X-Z plane), which passes through the force axis.

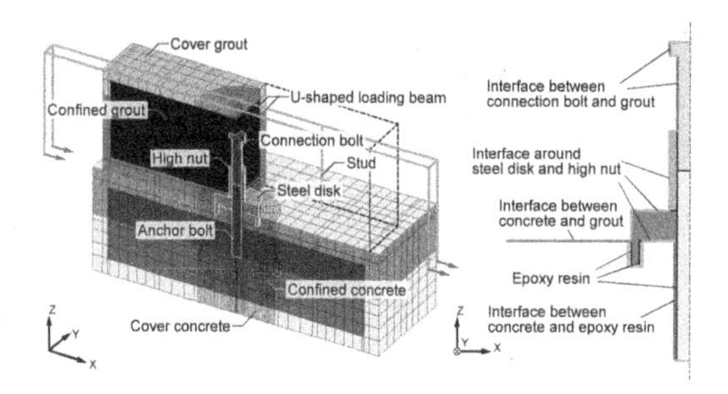

Figure 7. Outline of standard finite element model.

On the sectioned surface, only the Y-translation was restrained. All the degrees-of-freedom were restrained on the lower surface of the concrete, and the upper surface of the grout and U-shaped loading beam were kept parallel to each other. In addition, the X-translational degrees-of-freedom of the concrete both ends in the Y-Z plane were restrained. The total support reaction force in the X-direction was then doubled and used as shear force, Q.

The concrete, grout, and disk shear-key were modeled as hexahedral elements (eight-node isoperimetric elements), and the epoxy resin between the concrete and anchor bolt was modeled as a joint element (eight-node isoperimetric joint elements) with a finite thickness. Finally, depending on the interface, a joint element with zero thickness was inserted to model the bonding properties between the materials, as illustrated in Figure 7.

For loading, a forced displacement was applied to the force application points on the right and left side of the U-shaped loading beam that was modeled with beam elements with large stiffness values. In this analysis, loading was applied in one direction until the relative horizontal displacement reached 4 mm.

2.3 *Material configuration rule*

2.3.1 *Concrete and grout*

The material configuration rules of the concrete and grout are presented in Figure 8. The modified Ahmad model (Naganuma 1995) was used to determine the characteristics of the concrete and grout until the compressive strength was reached. The Nakamura–Higai model (Nakamura & Higai 1999), which is based on fracture energy, was then used to characterize the compression softening zone. In addition, for the failure condition subjected to triaxial stress, the five-parameter model by William–Warnke and the coefficients of Ohnuma and Aoyagi (1981) were used.

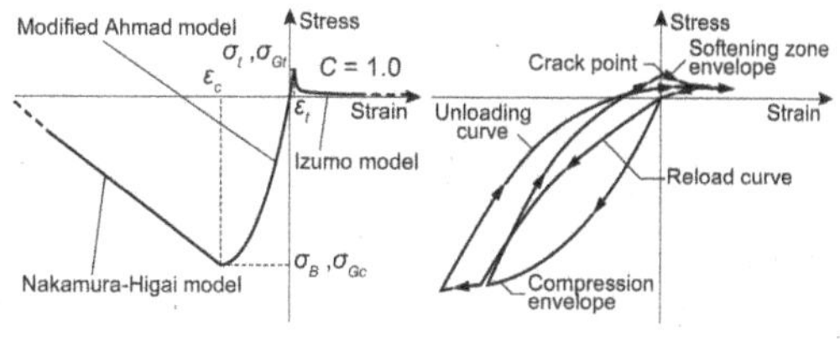

(a) Stress vs. strain (b) Historical characteristics

Figure 8. Material configuration rules (concrete and grout).

In contrast, the tensile side was assumed to be linear until it reached the tensile strength. In this analysis model, the anchor bolt was modeled with hexahedral elements, and the joint elements described below were used. The reinforcements were modeled as embedded reinforcements in concrete and grout elements, and the tension stiffening characteristics were reproduced using the Izumo model (Izumo et al. 1987). Thus, parameter C was set to 1.0 to avoid the doubling of the count of the adhesion property. The Poisson's ratios of the concrete and grout were set to 0.2.

Herein, the material specifications of the concrete and grout used the values of the target experiment, as listed in Table 1(a).

2.3.2 *Disk shear-key and reinforcement*

The material constitution rules for the disk shear-key and reinforcement are illustrated in Figure 9. Both the compression and tensile sides were modeled as bilinear. The stiffness after yielding was 1/100 E_s (where E_s is the Young's modulus).

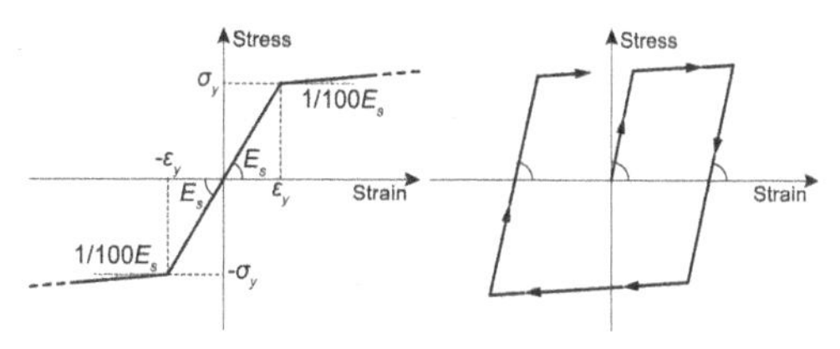

(a) Stress vs. strain (b) Historical characteristics

Figure 9. Material configuration rules (disk shear-key and reinforcement).

In addition, the yielding condition of von Mises was used for the yield condition in the case in which the disk shear-key was exposed to multiaxial stress. The Poisson's ratio of the disk shear-key was set to 0.3.

The numerical values according to the Japanese Industrial Standards were used as the material specifications of the steel material, as indicated in Table 1(b).

Table 1. Materials specifications (unit: [N/mm²]).

Material type	σ_B, σ_{Gc}	E_c, E_G	σ_t, σ_{Gt}
(a) Concrete and grout.			
Concrete	15.8	23,200	1.7
Grout	67.9	22,100	3.2

σ_B,σ_{Gc}: Compressive strength, E_c,E_G: Young's modulus, σ_t, σ_{Gt}: Tensile strength

Material type	Diamter	σ_y	E_s
(b) Steel			
Anchor bolt	M20	325	205,000
Connection bolt	M20	725	205,000
Steel disk	φ90	345	205,000
High nut	M20	400	205,000
Reinforcement	D10	295	205,000

σ_y: Yield strength, E_s: Young's modulus

2.4 *Characteristics of joint elements*

2.4.1 *Interface between concrete and epoxy resin*

The characteristics of the shear direction are illustrated in Figure 10(a). The characteristics were determined based on the adhesion performance experiments of the post-installed adhesive anchors constructed using the same materials and construction method as the test specimen in Setoguchi et al. (2010).

Conversely, the vertical characteristics were modeled to have a large stiffness value during compression and to not transmit tension stress, as depicted in Figure 10(b).

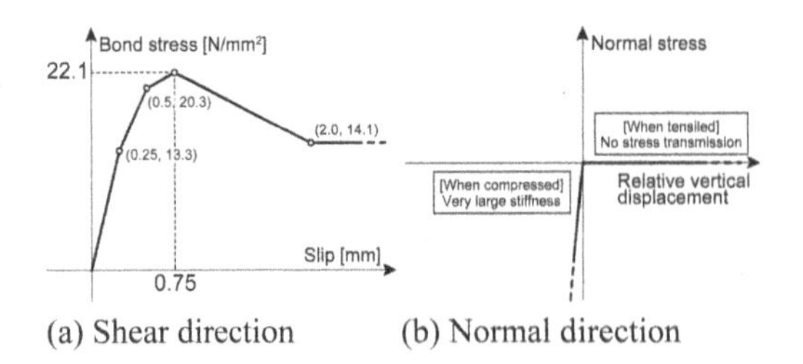

Figure 10. Characteristics of joint element (interface between concrete and epoxy resin).

2.4.2 *Epoxy resin between concrete and anchor bolt*

Previous studies (Nakano et al. 2009; Setoguchi et al. 2010) have reported that adhesion sliding occurs at the interface between the concrete and epoxy resin. Hence, the adhesion property, depicted in Figure 10(a), was inserted between the concrete and epoxy resin, and the epoxy resin and anchor bolt were rigidly connected, as illustrated in Figure 7.

The characteristics of the epoxy resin in the shear direction are already included in the adhesion property, depicted in Figure 10(a). Therefore, in this part, the characteristic in the shear direction has a large stiffness, as illustrated in Figure 11(a).

The characteristic in the vertical direction is assumed to be bilinear, as illustrated in Figure 11(b). The material specifications of the epoxy resin are: specific gravity d_b is 1.2; compressive strength σ_{bc} is 109 N/mm^2; tensile strength σ_{bt} is 75.7 N/mm^2, and the Young's modulus E_b is 2730 N/mm^2 based on the reference (Takase et al. 2016). In this study, the thickness values of the adhesive were 2 mm around the anchor bolt and 2.5 mm around the steel disk. The coordinates of the break points were calculated from the specifications and thickness of the epoxy resin.

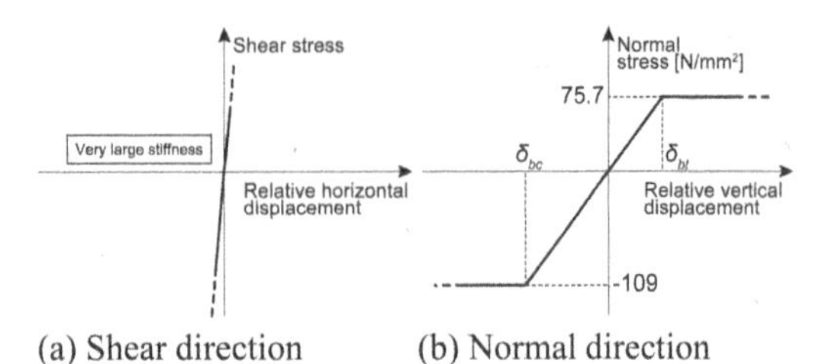

Figure 11. Characteristics of joint element (epoxy resin between concrete and anchor bolt).

2.4.3 *Interface between connection bolt and grout*

The characteristics in the shear direction were determined with reference to the results of the adhesion performance experiment between the concrete and deformed rebar (Nakano et al. 2009), as illustrated in Figure 12(a).

The vertical characteristics were modeled in a way similar to that of the interface between the concrete and epoxy resin, as illustrated in Figure 12(b).

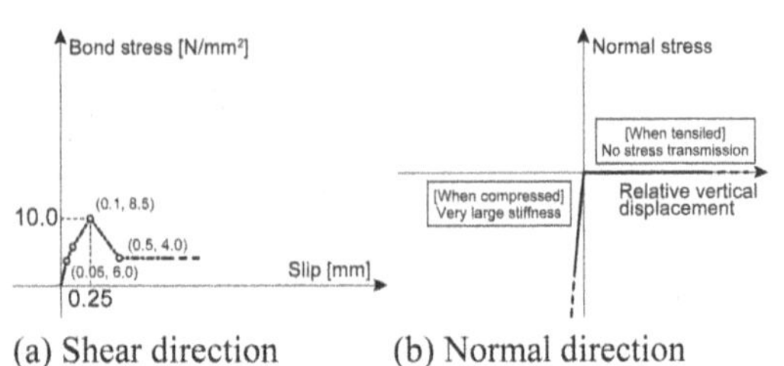

Figure 12. Characteristics of joint element (interface between connection bolt and grout).

2.4.4 *Interface around steel disk and high nut*

The characteristics of the joint element around the steel disk and the high nut are illustrated in Figure 13. In this analysis, because the surfaces of the steel disk and high nut were smooth, adhesion did not occur at the interfaces around them, and the shear stress was set so that it could not to be transmitted.

The characteristics in the vertical direction were modeled in the same manner, as depicted in Figures 10(b) and 12(b).

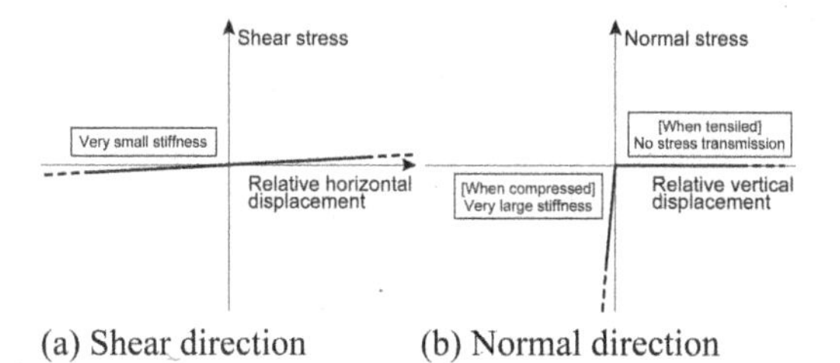

(a) Shear direction (b) Normal direction

Figure 13. Characteristics of joint element (interface around steel disk and high nut).

2.4.5 *Interface between concrete and grout*

The shear direction characteristics are illustrated in Figure 14(a). The shear stress owing to the adhesion between the concrete and grout was set to 0.01 N/mm^2 to ensure that it did not affect the overall shear strength. When compressive stress was applied, the increase in shear stress due to friction was considered. The friction coefficient μ was set to 0.974, which is the apparent shear friction coefficient attributed to mechanical mechanisms, such as the shear friction force of the joint surface and the meshing of the aggregate from a previous study (Katori et al. 1998).

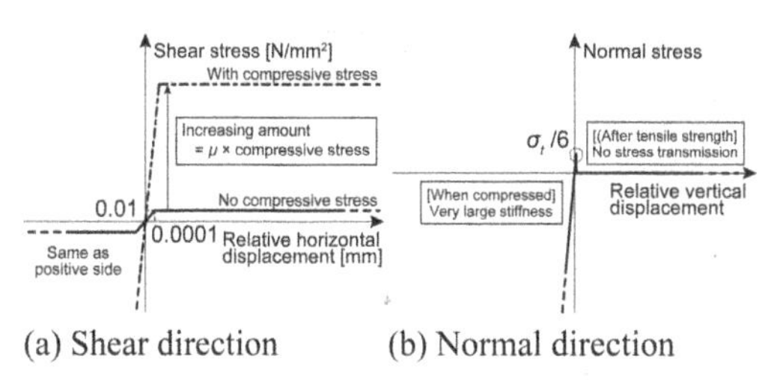

(a) Shear direction (b) Normal direction

Figure 14. Characteristics of joint element (interface between concrete and grout).

The vertical characteristics are illustrated in Figure 14(b). When compressive stress was applied, the interface yielded a large stiffness. The tensile side also had the same stiffness up to the tensile strength. After reaching the tensile strength, the bond was released, and the normal and shear stresses of the joint elements were zero. According to Kimu et al. (2008), the tensile strength was set to one-sixth of the tensile strength of concrete, σ_t.

2.5 *Verification of analysis accuracy*

To verify the accuracy of the analysis, the correspondence between the shear force and relative horizontal displacement is illustrated in Figure 15. The target range of the analysis was up to 4 mm. This value is twice the allowed value of 2 mm for the amount of displacement deformation according to the literature Japan Building Disaster Prevention Association (2002).

While the target shear element experiment was conducted with cyclic loading, this analysis was based on monotonic loading from the viewpoint of limiting

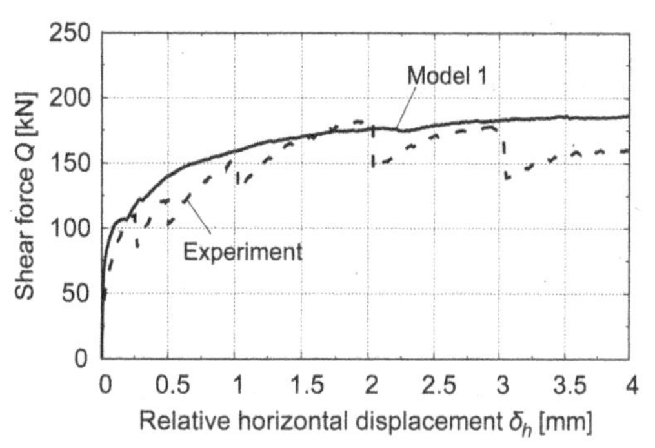

Figure 15. Verification of analysis accuracy (shear force vs. relative horizontal displacement).

the analysis time and ensuring the analysis stability. Although the loading conditions were different, it is judged that the general tendency can be evaluated herein.

3 ANALYSIS USING THE MATERIAL PROPERTIES OF THE EXISTING CONCRETE SURFACE

3.1 *Outline of the analysis*

In this section, the analysis of a model with a finishing material on the surface of existing concrete is presented. Subsequently, the influence of the material properties of the existing concrete surface on the shear resistance performance of the disk shear-key was considered based on comparisons with the analyzed results of Model 1.

3.2 *Analysis parameters*

As illustrated in Figure 16, the parameter used for analysis was the existence of the finishing material on the existing concrete surface, and the model with the finishing material was labelled as Model 2. In this analysis, the thickness of the finishing material was set to 10 mm.

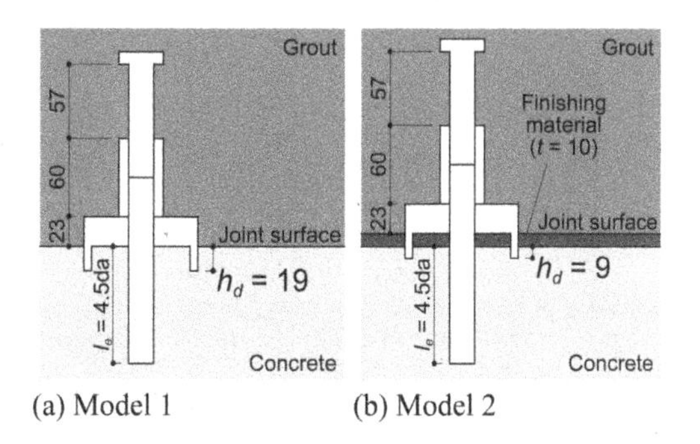

(a) Model 1 (b) Model 2

Figure 16. Analysis parameters (unit: [mm]).

There are various types of finishing materials with different material properties. Therefore, in this analysis, the finishing material was modeled subject to strict conditions. The finishing material was modeled with an elastic body, and the Young's modulus was set to 1.0×10^{-4}, as a sufficiently low value.

The standard embedded depth of the steel disk, h_d, was 19 mm. Because the thickness of the finishing material in Model 2 was set as 10 mm, the embedded depth in concrete became equal to 9 mm. In addition, the embedded length of the anchor bolt, l_e, was modeled such that it did not change at 4.5da (=90 mm), where da is the diameter of the anchor bolt.

3.3 *Results and discussions*

3.3.1 *Shear force vs. relative horizontal displacement*

The relationship between the shear force, Q, and the relative horizontal displacement, δ_h, is illustrated in Figure 17. Figure 17 also depicts the design shear strength, Q_{jd}, calculated using the current design shear strength evaluation formula as follows:

$$Q_{jd} = 0.8 \cdot Q_{disk} \tag{1}$$

$$Q_{disk} = 0.24 \cdot K_1 \cdot K_2' \cdot A_B \sqrt{E_C \cdot \sigma_B} \tag{2}$$

$$A_B = \int_{-\pi/4}^{\pi/4} h_d \cdot \frac{R_d}{2} d\theta = \frac{\pi \cdot R_d \cdot h_d}{4} \tag{3}$$

where Q_{disk} is the ultimate shear strength of the disk shear-key [N]; Q_{jd} is the design shear strength of the disk shear-key [N]; A_B is the effective contact area for obtaining the bearing stress [mm^2]; h_d is the embedded depth of the steel disk [mm]; K_1 is the correction coefficient by edge; K_2' is the correction coefficient by the embedded length of the anchor bolt; E_c is the Young's modulus of concrete [N/mm^2], and σ_B is the compressive strength of the concrete [N/mm^2]. In the analysis model used in this research, $K_1 = K_2' = 1.0$.

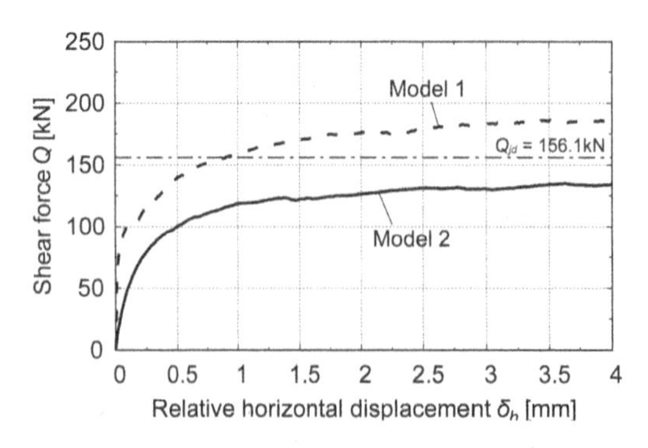

Figure 17. Shear force vs. relative horizontal displacement.

It can be observed that Model 2 has a lower stiffness and shear strength than Model 1, and the shear resistance performance is reduced considerably. The shear strength when the relative horizontal displacement was 2 mm was also lower than the designed shear strength, Q_{jd}. This result suggests that it is not practical to apply a standard disk shear-key to existing frames.

3.3.2 *Mises stress distribution of disk shear-key*

Figure 18 illustrates the Mises stress distribution occurring in the disk shear-key when the relative horizontal displacement, δ_h, is 2 mm. The amount of deformation was displayed five times.

The maximum Mises stress increased by approximately 6% compared with Model 1. In addition, the Mises stress was concentrated and distributed in the anchor bolts embedded in the existing part. This tendency was particularly remarkable at the position of the finishing material. Conversely, the Mises stress generated in the connection bolt was small.

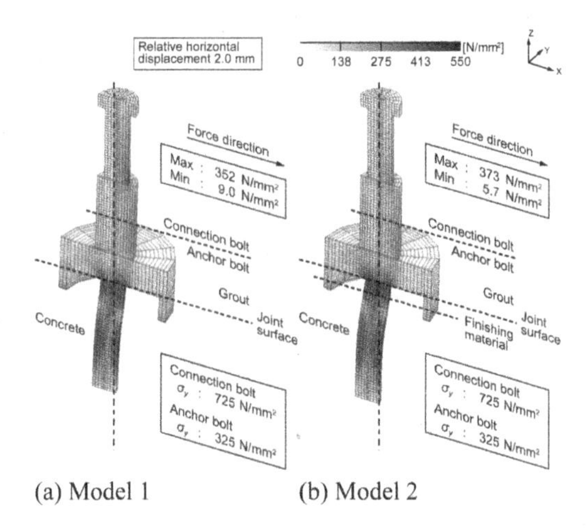

Figure 18. Mises stress distribution of disk shear-key (deformation magnification: five times).

3.3.3 *Curvature distribution of anchor bolt and connection bolt*

The curvature was calculated from the strain in the Z-direction of the outermost element in the cross section, and the curvature distributions at δ_h = 1, 2, and 3 mm are illustrated in Figure 19. The curvature was calculated using Eq. (4).

$$\varphi = (\varepsilon_L - \varepsilon_R) / d_a \tag{4}$$

where φ is the curvature [μ/mm]; ε_L and ε_R are the strains in the Z-direction of the outermost element in the cross-section [μ], and da is the diameter of the anchor bolt and connection bolt [mm].

Moreover, the curvature when ε_L and ε_R reach the yield strain is obtained from Eq. (5), and is also depicted in Figure 19.

$$\varphi_y = 2\varepsilon_y / d_a \tag{5}$$

where φ_y is the curvature when either ε_L or ε_R reaches ε_y [μ/mm], and ε_y is the yield strain of the anchor bolt and the connection bolt [μ].

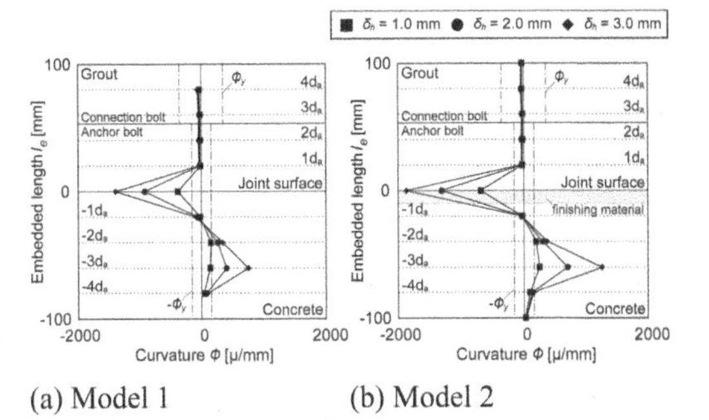

(a) Model 1 (b) Model 2

Figure 19. Curvature distributions of anchor bolt and connection bolt.

As indicated by the Mises stress distribution depicted in Figure 18, it is confirmed that the curvature of the anchor bolt increases at the position of the finishing material. Compared with Model 1, it increased at a rate of +80% at δ_h = 1 mm, +42% at δ_h = 2 mm, and +35% at δ_h = 3 mm.

In addition, the curvature is maximized at a distance of 3da (da: diameter of the anchor bolt) from the joint surface inside the existing concrete. This tendency is the same in both models. The curvature of that position reaches φ_y when the relative horizontal displacement of δ_h = 1 mm, and it is inferred that it yields.

3.3.4 *Minimum principal stress distribution of concrete*

Figure 20 illustrates the minimum principal stress distribution of the concrete when the relative horizontal displacement, δ_h, was 2 mm.

From the distribution of the minimum principal stress, Model 2 yielded values that were smaller than those of Model 1. The maximum value of the minimum principal stress for Model 2 was -33% of that for Model 1. Therefore, it is considered that the shear resistance performance of the disk shear-key decreases owing to the decrease in the bearing resistance of the concrete.

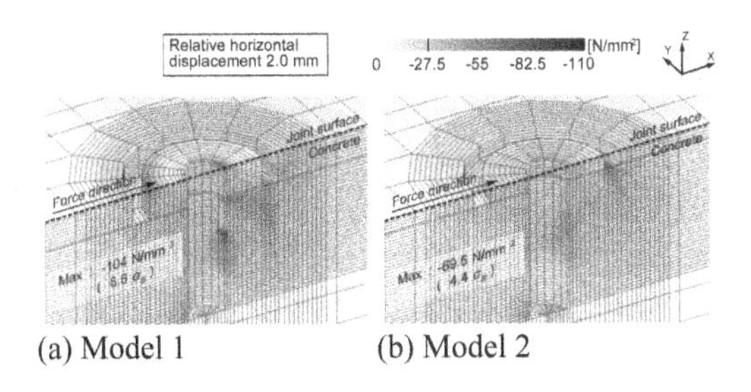

(a) Model 1 (b) Model 2

Figure 20. Minimum principal stress distribution of concrete.

4 ANALYSIS USING THE SHAPE OF THE DISK SHEAR-KEY

4.1 *Outline of the analysis*

The analysis results of Model 2 presented in Section 3 suggest that it would be difficult to apply the standard disk shear-key to the existing structure with the finishing material. This is attributed to the fact that in the case where there is a finishing material on the surface of the existing concrete, h_d—which denotes the depth at which the steel disk is embedded in the concrete—becomes smaller. In other words, the main factor is that the concrete bearing capacity is not fully achieved.

According to the aforementioned listings, it is necessary to improve the shape of the disk shear-key to obtain sufficient bearing pressure resistance of the concrete even at the joint with the finishing material. Therefore, a model with an improved shape of the disk shear-key was constructed and analyzed.

4.2 *Analysis parameters*

Because the shear resistance performance of the disk shear-key was insufficient at h_d = 9 mm in Model 2, further analysis was conducted with a model in which the embedded depth of the steel disk, h_d, was extended, as illustrated in Figure 21.

There are two models, Model 3 with h_d = 20 mm and Model 4 with h_d = 30 mm. Herein, considering the position of the reinforcements inside the concrete, it appears that the maximum h_d is approximately 30 mm.

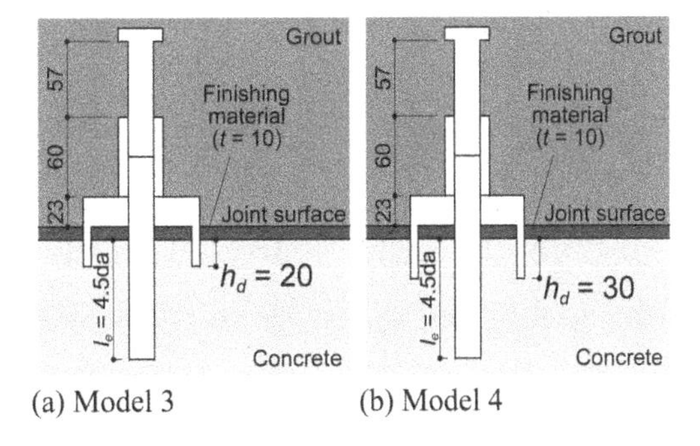

(a) Model 3 (b) Model 4

Figure 21. Analysis parameters (unit: [mm]).

4.3 *Results and discussions*

4.3.1 *Shear force vs. relative horizontal displacement*

The relationship between the shear force, Q, and the relative horizontal displacement, δ_h, is illustrated in Figure 22. Both models exceed the design shear strength of the disk shear-key, Q_{jd}, within a relative horizontal displacement of 2 mm, and the shear resistance performance is improved compared with Model 2.

The response of Model 3 is slightly lower than that of Model 1 but that of Model 4 is above it. However, it should be noted that Model 4 also has a stiffness response up to a relative horizontal displacement of 1 mm, which is less in magnitude than that of Model 1.

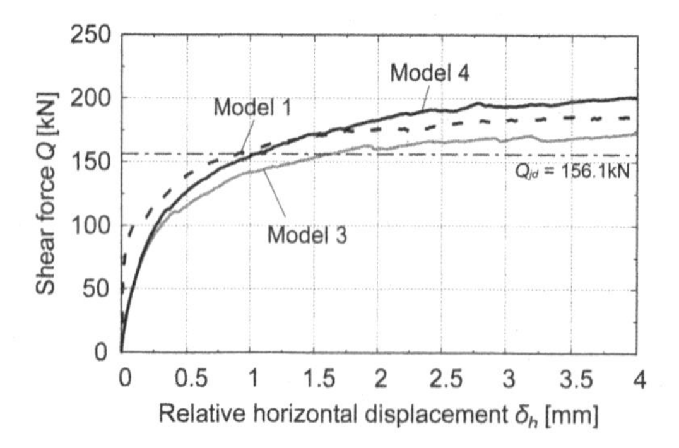

Figure 22. Shear force vs. relative horizontal displacement.

4.3.2 *Mises stress distribution of disk shear-key*

Figure 23 illustrates the Mises stress distribution that occurs in the disk shear-key when the relative horizontal displacement, δ_h, is 2 mm.

Increasing the embedded depth of the steel disk, h_d, tends to reduce the Mises stress of the anchor bolt and tends to increase the Mises stress of the connection bolt. In other words, it is suggested that the bearing pressure of concrete becomes dominant, and the shear force is transmitted sufficiently.

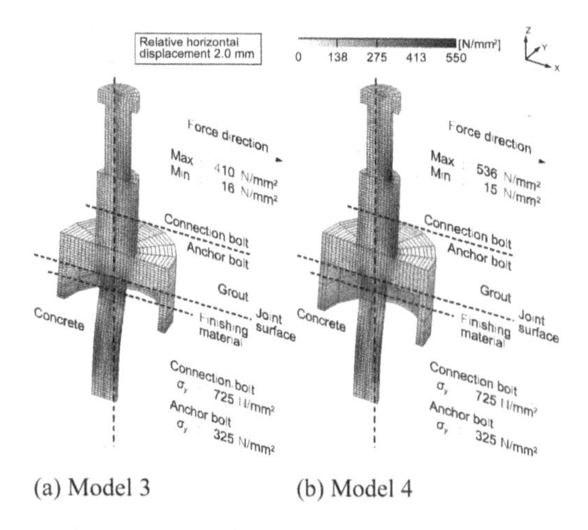

(a) Model 3 (b) Model 4

Figure 23. Mises stress distribution of disk shear-key (deformation magnification: five times).

4.3.3 *Curvature distribution of anchor bolt and connection bolt*

The curvature distributions of the anchor bolt and connection bolt are depicted in Figure 24. Model 3 has almost the same curvature distribution as Model 1. It was also discovered that the bending deformation could be suppressed by increasing the embedded depth of the steel disk, h_d, even in the presence of a finishing material.

Conversely, Model 4 yielded smaller bending deformation than Model 1, thus suggesting that the shear force generated by the anchor bolt is reduced.

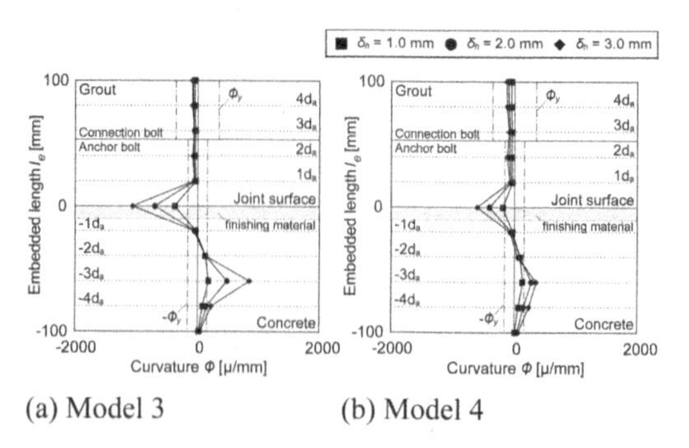

(a) Model 3 (b) Model 4

Figure 24. Curvature distributions of anchor bolt and connection bolt.

4.3.4 *Minimum principal stress distribution of concrete*

Figure 25 illustrates the minimum principal stress distribution of the concrete when the relative horizontal displacement, δ_h, is 2 mm.

In both models, the minimum principal stress increased over a wide range compared with Model 2, thus indicating that the bearing resistance of concrete was achieved. This tendency is particularly remarkable near the bottom of the steel disk. Furthermore, because the highest value of minimum principal stress in Model 4 is higher than Model 3, the embedded depth of the steel disk, h_d, is considered to be one of the most important parameters.

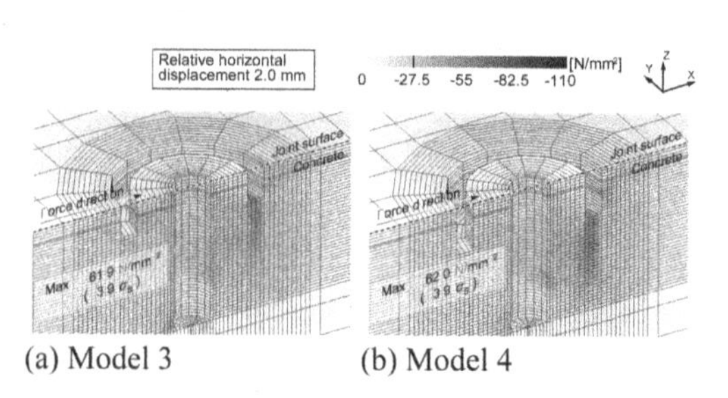

(a) Model 3 (b) Model 4

Figure 25. Minimum principal stress distribution of concrete.

5 CONCLUSIONS

In this study, 3D FEM analysis of the disk shear-key was conducted considering the material properties of the existing concrete surface. Based on these results, the following findings were reported:

1. If the standard disk shear-key is applied when the existing concrete surface has a finishing material with a thickness of 10 mm, the shear resistance performance will decrease, and it will be lower than the design shear strength, Q_{jd}
2. It was clarified—based on the stress state inside the member—that the bearing resistance of the concrete decreased, and the deformation was concentrated on the anchor bolt. Based on this result, it was demonstrated that the embedded depth of the steel disk, h_d, was one of the most important parameters related to this problem
3. When the embedded depth of the steel disk, h_d, was extended to 20 mm, the results indicated that the design shear strength was satisfied within the relative horizontal displacement $\delta_h = 2$ mm. Furthermore, it was suggested that when h_d was extended to 30 mm, the shear resistance performance was almost the same as that of the original one

The results of this study were obtained subject to the condition that the thickness of the finishing material was 10 mm. In addition, the embedding depth of the steel disk, h_d, was treated as an important analysis parameter. In the future, analysis will be conducted with increased number of parameters, such as the thickness of the finishing material and the diameter of the steel disk.

ACKNOWLEDGMENTS

This work was supported by JST Program on Open Innovation Platform with Enterprises (JPMJOP1723), Research Institute and Academia.

REFERENCES

Ishida, Y., Sakata, H., Takase, Y., Maida, Y. & Sato, T. 2018. 3D FEM Analysis of Post-installed Adhesive Anchors under Combined Stress: Stress Transmission Mechanism and Mechanical Behavior of the Joints for External Seismic Retrofitting Part 1. *Journal of Structural and Construction Engineering (Transactions of AIJ)*, Vol.83, No.751: 1307–1317. (in Japanese)

ITOCHU Techno-Solutions Corporation. 2016. Final Help.

Izumo, J., Shima, H. & Okamura, H. 1987. Analysis Model of Reinforced Concrete Board Element Subjected to Inplane Force. *Concrete Journal*, Vol.25, No.87.9–1: 107–120. (in Japanese)

Japan Building Disaster Prevention Association. 2002. External Seismic Retrofitting Manual. (in Japanese)

Katori, K., Hayashi, S., Ushigaki, K. & Norimono, T. 1998. Effects of Surface Roughness on Shear Behavior of Construction Joint Arranging Connecting Bars Perpendicularly: Studies on Shear Behavior of Joint. *Journal of Structural and Construction Engineering (Transactions of AIJ)*, No.508: 101–110. (in Japanese)

Kimu, H., Noguchi, T., Yoneda, N. & Hamasaki, H. 2008. Construction of Adhesion Constitutive Law between Repair Mortar and Frame Concrete by FEM Inverse Analysis. *Proceedings of the Japan Concrete Institute*, Vol.30, No.2: 577–582. (in Japanese)

Naganuma, K. 1995. Stress-strain Relationship for Concrete under Triaxial Compression. *Journal of Structural and Construction Engineering (Transactions of AIJ)*, No.474: 163–170. (in Japanese)

Nakamura, H. & Higai, T. 1999. Compressive Fracture Energy and Fracture Zone Length of Concrete. *Seminar on Post-peak Behavior of RC Structures Subjected to Seismic Load*, JCI-C51E, Vol.2: 259–272.

Nakano, K., Matsuzaki, Y. & Sugiyama, T. 2009. Average Bond Strength of Post-installed Bonded Anchor. *Summaries of Technical Papers of Annual Meeting, Architectural Institute of Japan*, C-2, Structures-IV: 149–150. (in Japanese)

Ohnuma, H. & Aoyagi, Y. 1981. Ultimate Strength Property of Concrete under Triaxial Compressive Stresses, *CRIEPI Research Report*, No.381021 (in Japanese)

Satoh, T., Abe, T., Sakamoto, K., Onaka, A., Yagisawa, Y., Ando, S., Kaneyoshi, T., Tamura, T. & Takase, Y. 2017. Development of Joint Member Using Steel Disk and Anchor Bolt for Seismic Retrofit: Part 19. Outline of Indirect Joints Test with Repair Mortar. *Summaries of Technical Papers of Annual Meeting, Architectural Institute of Japan*, Structures-IV: 367–368. (in Japanese)

Setoguchi, H., Abe, T., Takase, Y., Sato, S., Takahashi, M. & Sato, T. 2010. Experimental Study to Confirm the Performance of Post-installed Anchor that Uses Core Drill Method: Part 2. Test of Bond Strength that Uses Injection Type Anchor. *Summaries of Technical Papers of Annual Meeting, Architectural Institute of Japan*, C-2, Structures-IV: 149–150. (in Japanese)

Takase, Y., Abe, T., Itadani, H., Satoh, T., Onaka, A., Kubota, M. & Ikeda, T. 2014. Estimation Method of Horizontal Capacity of Joint Fracture for Retrofitted Frame Using Disk Shear-key: Study on Shear-key Consisted of Steel Disk and Anchor Bolt for Seismic Retrofitting. *Journal of Structural and Construction Engineering (Transactions of AIJ)*, Vol.79, No.698: 507–515. (in Japanese)

Takase, Y., Wada, T. & Shinohara, Y. 2016. A Study on Mechanical Behavior of Post-installed Adhesive Anchor Receiving Repeated Shear Force under Consist Tensile Force. *Proceedings of the Japan Concrete Institute*, Vol.38, No.2: 1105–1110. (in Japanese)

Computational Modelling of Concrete and Concrete Structures – Meschke, Pichler & Rots (Eds)
 ISBN: 978-1-032-32724-2

Star-shaped Falling Weight Deflectometer (FWD) testing and quantification of the distribution of the modulus of subgrade reaction

R. Díaz Flores, M. Aminbaghai & B.L.A. Pichler
Institute for Mechanics of Materials and Structures, TU Wien, Vienna, Austria

L. Eberhardsteiner & Ronald Blab
Institute for Transportation, TU Wien, Vienna, Austria

M. Buchta
Nievelt Labor GmbH, Höbersdorf, Austria

ABSTRACT: New experimental data obtained from Falling Weight Deflectometer (FWD) tests performed on two concrete slabs is presented. Geophones measured deflections along 8 directions, in a star-shaped configuration. The asymmetry index developed by (Díaz Flores et al. 2021) is used in order to evaluate the new set of data. The results show significant asymmetries in a 22-year-old loaded slab, while a new, freshly installed slab behaved in a virtually double-symmetric manner. A structural analysis of the new slab is performed based on Kirchhoff-Love plate theory with free-edge boundary conditions and a Winkler foundation at the bottom, again following (Díaz Flores et al. 2021). The modulus of subgrade reaction and an auxiliary surface load are used as optimisation variables. This allows for a very satisfactory reproduction of the measured deflections. From this model, a non-linear distribution of the effective modulus of subgrade reaction arises. FE simulations underline the robustness and accuracy of the used method.

1 INTRODUCTION

Falling Weight Deflectometer (FWD) tests are performed worldwide in order to quantify and evaluate the state of pavement structures and their subgrade. For these tests, a standardised mass falls freely from a given height and hits a load plate located on top of the pavement's surface. Displacement sensors, also known as geophones, are used to measure the deflection of the pavement's surface along a specific radial direction. The measured deflections are then used as basis for a backcalculation procedure in which the stiffness of the slab and of the layers underneath may be obtained.

The goal of the backcalculation is to find the elastic properties of the subgrade in such a way that the resulting deflections agree with the ones measured by the geophones. In this context, several types of structural models have been used. These include, e.g., slabs that are lying on top of elastic foundations (Biot 1937; Winkler 1867), multi-layered slabs which are lying on top of a Winkler foundation (Girija Vallabhan et al. 1991), and multi-layered solids (Abd El-Raof et al. 2018; Kausel and Roësset 1981; Pan 1989a, 1989b). However, it has been previously described (Mehta & Roque 2003) that, despite its wide use, the backcalculation procedure is mathematically "ill-posed': different combinations of elastic properties and thickness may result in the same deflection field. In other words, the inverse problem may have multiple solutions. For this reason, the evaluation of FWD tests requires much care, experience and expertise.

When using Winkler foundations, the elastic properties of the whole subgrade are summarised by one value: the modulus of subgrade reaction (Winkler 1867). It has been noted before, however, that the modulus of subgrade reaction, is rather a structural than a material property (Aristorenas & Gómez 2014). Furthermore, if a spatially uniform modulus is used, then unrealistic results may arise (Daloglu and Vallabhan 2000; Eisenberger 1990; Smith 1970). For this reason, non-uniform distributions of the modulus of subgrade reaction have been previously introduced, see e.g. the analysis of concrete slabs (Roesler et al. 2016), and circular plates (Foyouzat et al. 2016).

The present study is based on a method proposed by (Díaz Flores et al. 2021). From the deflections measured by the displacement sensors, they were able to calculate a distribution of the stresses appearing at the bottom of the pavement slab. The resulting distribution of stresses was found to be realistic. Based upon this distribution, a highly non-linear effective modulus of subgrade reaction was also found (Díaz Flores et al. 2021). Furthermore, they developed an index which helps to assess the asymmetry of a structure on the basis of experimental results.

 DOI 10.1201/9781003316404-34

The purpose of this study is to use the methods proposed by (Díaz Flores et al. 2021) for new experimental data, and to find out whether the conclusions drawn are still applicable. In this context, two original slabs are newly investigated: A freshly installed, new slab, and a 22-year-old loaded slab already scheduled for replacement. This is presented in Section 2. The asymmetry index proposed by (Díaz Flores et al. 2021) is used to evaluate the asymmetry of both structures. This is helpful in order to gain further insight into the asymmetry index and assess its informative content. In Section 3, a structural analysis is performed for the virtually double-symmetric new slab. A realistic distribution of the subgrade pressure was found after considering an auxiliary surface load, in the same way as in (Díaz Flores et al. 2021). This model was able to reproduce the measured deflections very satisfactorily. The deflections obtained with the method from (Díaz Flores et al. 2021) are compared with those from an FE simulation under the same conditions. Finally, conclusions are drawn based on the results, see Section 4.

2 STAR-SHAPED FWD TESTING

FWD tests were performed on two concrete slabs on the "A1" highway, in Lower Austria. The concrete slabs were 5.5 m long, 4.2 m wide and 0.22 m thick. The maximum force produced by the falling weight was 202 kN, see Table 1.

Table 1. Properties of the slabs and of the equipment used.

Property	Value
Length of Slab, a	5.50 m
Width of Slab, b	4.20 m
Thickness of Slab, h	0.22 m
Flexural Stiffness of Slab, K	49.5 MNm
Maximum Impact Force	202 kN
Modulus of Elasticity of Concrete, E	36.5 GPa
Poisson's Ratio of Concrete, ν	0.2
Mass Density of Concrete, ρ	2,452 kg/m^3

Both tested slabs were located on the first lane of the highway. They were connected by means of steel bars to their neighbours along all edges: tie bars connected their right and left edges while dowels connected their two edges orthogonal to the driving direction

2.1 *Test protocol*

FWD tests in the centre of the slabs were carried out in eight directions (in a star-shaped manner) described by a local cardinal system, where N refers to the driving direction, see Figure 1 of (Díaz Flores et al. 2021). The angles between neighbouring directions amounted to either 38° or 52°, rather than to 45°, due to constraints from the FWD machine, see Figure 1 and Table 2 of (Díaz Flores et al. 2021).

The tests were performed in the following sequence of directions: N, NE, E, SE, S, SW, W and NW. Three tests were carried out in every direction, except for the N direction, where six tests were carried out (three at the start and three at the end of the protocol). This allows for an assessment of the quality of test repeatability. The six tests in the N direction allow for checking the repeatability of the tests at the start and end of the protocol.

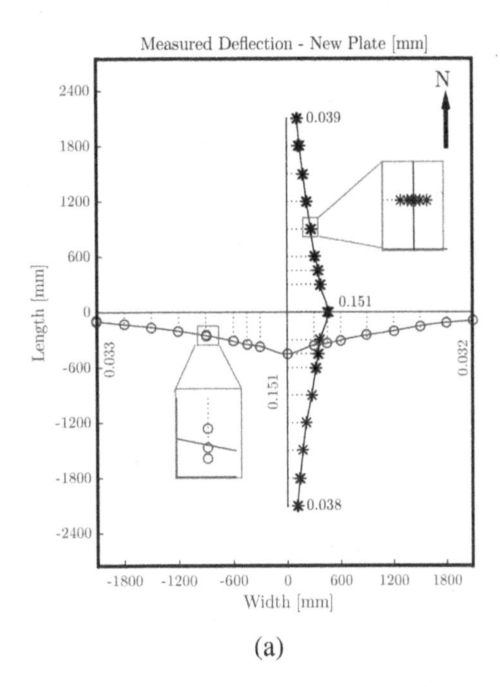

(a)

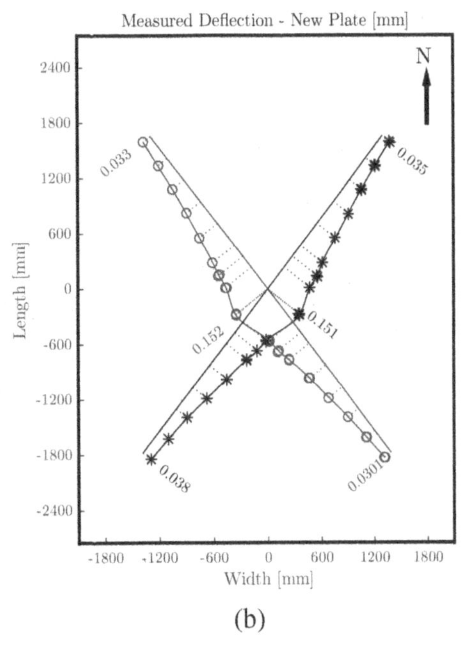

(b)

Figure 1. Results from star-shaped FWD testing on the new slab: 243 deflections measured by the geophones along the (a) N, S, E and W directions, as well as (b) the diagonal directions; the lines refer to splines interpolating between the average deflections measured at each location, see Eq. (1).

Table 2. Coefficients of variation, $CV_{N,g}$, of the deflections measured by each geophone in the N-direction of the new slab, see Eqs. (1)–(3), based on the results of the first and the last three tests ($n_d = 6$), see also Table 6.

Geophone								
$g=1$	$g=2$	$g=3$	$g=4$	$g=5$	$g=6$	$g=7$	$g=8$	$g=9$
2.81%	1.42%	1.54%	1.46%	2.05%	1.70%	1.71%	5.68%	4.28%

Nine geophones recorded the deflections during every single FWD test. Geophone 1 was always located at the centre of the falling weight experiments. The other eight geophones were fixed to a bar, ensuring that the radial distances between them were always the same. In the N, NE, E, S, W and NW directions, the default distances of the machine were used. In the SE and SW directions, further structural constraints of the machine forced the bar to be adjusted. It was thus located 15 cm further away from the centre, see Table 3 of (Díaz Flores et al. 2021).

Table 3. Coefficients of variation, $CV_{N,g}$, of the deflections measured in the N-direction of the old slab, see Eqs. (1)–(3), based on the results of the first and the last three tests ($n_d = 6$), see also Table 8.

Geophone								
$g=1$	$g=2$	$g=3$	$g=4$	$g=5$	$g=6$	$g=7$	$g=8$	$g=9$
1.92%	3.12%	2.83%	2.83%	4.27%	4.05%	3.04%	3.80%	4.73%

2.2 *Experimental data from the new slab*

The first set of tests was performed on a concrete slab which was only a few weeks old at the time of testing. Before that, only site traffic may have passed over the slab.

A total of 27 individual FWD tests were carried out: three tests were performed in every direction and an extra set of three tests was performed in the N direction. A total of 243 individual deflections were thus recorded, see (Table 6 in Appendix A), also illustrated in Figure 1.

The symbols seen in Figure 1 refer to the 243 deflections measured. The symbols that correspond to the same location are barely different from each other, given that the geophones measured almost the same deflections at each location. The avereage values for the deflections are calculated as (Díaz Flores et al. 2021)

$$\overline{w}_{d,g} = \frac{1}{n_d} \sum_{i=1}^{n_d} \max_t w_{d,g,i}(t), \tag{1}$$

where n_d stands for the number of tests performed in direction d. In Figure 1, this results in $n_1 = 6$ for the N direction and $n_d = 3$ for all other measurement directions. In Eq. 1, index d refers to measurement directions (with $d = 1 \Leftrightarrow$ N, $d = 2 \Leftrightarrow$ NE, …, $d = 8 \Leftrightarrow$ NW), index g refers to the geophones, and index i for the ith test in direction d.

Coefficients of variation are used to study how well the tests repeated results. These are calculated as

$$CV_{d,g} = \frac{\sigma_{d,g}}{\overline{w}_{d,g}}. \tag{2}$$

The corresponding standard deviations read as

$$\sigma_{d,g} = \sqrt{\frac{1}{n_d - 1} \sum_{i=1}^{n_d} \left[\max_t w_{d,g,i}(t) - \overline{w}_{d,g}\right]^2}. \tag{3}$$

The quality of repeatability is quantified for all sets of three tests in the same direction. For this purpose, Eqs. (1)–(3) are evaluated for each of the nine directions, and for each of the nine geophones, see Table 7. The obtained 81 coefficients of variation are smaller than 6%, indicating an acceptable level of test repeatability.

The six tests performed in the N direction are evaluated as one sample ($n_d = 6$). Given that the coefficients of variation in this set are smaller than 6%, see Table 2, it is concluded that an acceptable level of test repeatability was ensured.

2.3 *Experimental data from the old slab*

The second set of tests was performed on a 22-year- old concrete slab. It was scheduled to be replaced shortly after the star-shaped FWD testing. It was located on the first lane of the highway.

During the 27 FWD tests, a total of 243 individual deflections were recorded, see (Table 8 in Appendix B), and in Figure 2. The quality of repeatability is quantified for all directions based on coefficients of variation. The obtained 81 coefficients of variation are smaller than 7%, indicating an acceptable level of test repeatability.

2.4 *Asymmetries of the structural behaviour of the tested slabs*

The asymmetries in the structural behavior of the slabs are studied based on the index developed in (Díaz Flores et al. 2021) and on the results from star-shaped FWD experiments (Figures 1 and 2). Such asymmetries may result, for the new slab, from the dowels and tie bars connecting it to its neighbours. Regarding the old slab, asymmetries may also result from long-term service loads. Given that the dowels and tie bars were positioned symmetrically along all edges, it is expected that the deflections of the new slab are virtually the same in the N and S, NE and SE, SW and NW, E and W, NE and NW, SW and SE, NE and SW, as well as the NW and SE directions.

The asymmetry index from (Díaz Flores et al. 2021) reads as

$$A_{d,\delta} = \sqrt{\frac{1}{2.1\,\mathrm{m}} \int_{r=0}^{2.1\,\mathrm{m}} \left[\frac{w_d(r)}{w_d(0)} - \frac{w_\delta(r)}{w_\delta(0)}\right]^2 dr}, \tag{4}$$

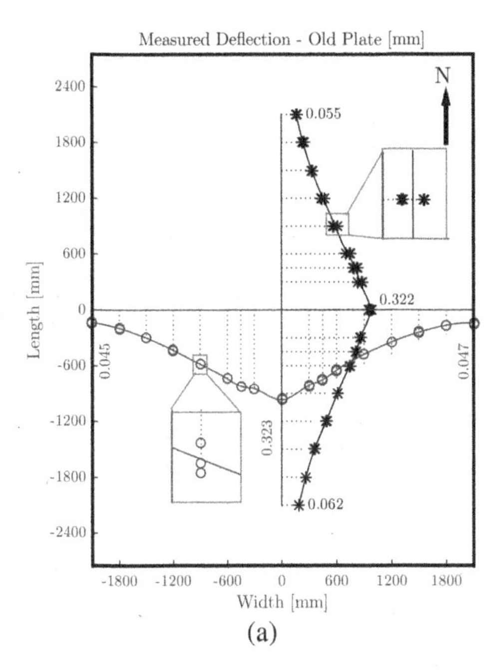

(a)

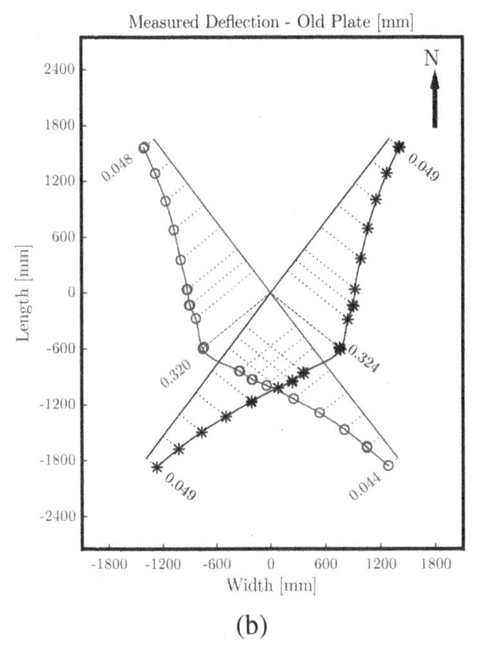

(b)

Figure 2. Results from star-shaped FWD testing on the old slab: 243 deflections measured by the geophones along the (a) N, S, E and W directions, as well as (b) the diagonal directions; the lines refer to splines interpolating between the average deflections measured at each location, see Eq. (1).

where $w_d(r)$ refers to the spline in the d direction, and $r \geq 0$ denotes the radial coordinate. For this index, larger values of $A_{d,\delta}$ refer to a larger asymmetry of the deflections along the d and δ directions. The asymmetry indicators $A_{N,S}$, $A_{NE,SE}$, $A_{SW,NW}$, $A_{E,W}$, $A_{NE,NW}$, $A_{SW,SE}$, $A_{NE,SW}$, and $A_{NW,SE}$ for both the new and the old slab are listed in Table 4.

Table 4. Asymmetry indicators for both slabs, see Eq. (4).

New Slab	Old Slab
$A_{N,S} = 1.65\%$	$A_{N,S} = 2.30\%$
$A_{NE,SE} = 1.96\%$	$A_{NE,SE} = 3.00\%$
$A_{SW,NW} = 3.57\%$	$A_{SW,NW} = 5.03\%$
mean value = 2.39%	mean value = 3.54%
$A_{E,W} = 2.98\%$	$A_{E,W} = 7.56\%$
$A_{NE,NW} = 1.24\%$	$A_{NE,NW} = 2.29\%$
$A_{SW,SE} = 2.62\%$	$A_{SW,SE} = 4.31\%$
$A_{NE,SW} = 3.03\%$	$A_{NE,SW} = 7.17\%$
$A_{NW,SE} = 1.63\%$	$A_{NW,SE} = 1.38\%$
mean value = 2.30%	mean value = 4.54%

Regarding the new slab, the eight asymmetry indicators are on average equal to 2.34%. The mean value of the first three asymmetry indicators is virtually the same as that of the last five indicators, see Table 4. The largest values of the asymmetry indicators are $\leq$4%. This shows that the slab behaved double-symmetrically.

Regarding the old slab, the eight asymmetry indicators are on average equal to 4.17%. This is 1.78 times larger than the average of the new slab. The largest values of the asymmetry indicators of the old slab are $\geq$7% and refer to the E-W and the NE-SW axes. From these results it is concluded that the old slab behaved asymmetrically. This asymmetry may be explained due its long-term service with traffic occasionally turning from and to the second lane along its Western edge. The asymmetry indicators evaluated for the old slab suggest a necessity for replacement.

3 STRUCTURAL ANALYSIS OF THE NEW SLAB

The new slab from Section 2.2 is studied based on Kirchhoff-Love's linear theory of thin plates. A Cartesian coordinate system is used, with the x-axis oriented in the driving direction, see Figure 1 of (Díaz Flores et al. 2021).

The boundary value problem consists of one field equation and boundary conditions. The field equation in the *static* case reads as (Díaz Flores et al. 2021; Vlasov 1966)

$$K\left(\frac{\partial^4 w(x,y)}{\partial x^4} + 2\,\frac{\partial^4 w(x,y)}{\partial x^2 \partial y^2} + \frac{\partial^4 w(x,y)}{\partial y^4}\right) + k\, w(x,y) = p(x,y)\,, \tag{5}$$

where $K = E\,h^3/[12\,(1-\nu^2)]$ denotes the flexural stiffness of the plate, E the modulus of elasticity of concrete and ν its Poisson's ratio, see Table 1. Furthermore, $w(x,y)$ denotes the deflection of the plate, $p(x,y)$ its external load per area and k the modulus of subgrade reaction. Regarding the boundary conditions, all four lateral edges of the rectangular plate are free edges, see (Díaz Flores et al. 2021) as well as (Höller et al. 2019; Vlasov 1966) for the full details. This was shown to be a reasonable assumption in (Díaz Flores et al. 2021), since it was found that the dowels and tie bars had no significant influence on the structural behaviour of the slabs analyzed.

The deflection field $w(x,y)$ consists of a Fourier series of double-symmetric deflection modes, see (Díaz Flores et al. 2021) for more details:

$$w(x,y) = \sum_{m=0}^{N}\sum_{n=0}^{N} C_{m,n} \cos\frac{m\pi x}{a} \cos\frac{n\pi y}{b}\,, \qquad \begin{cases} m = 0,1,3,5,\ldots,N\,,\\ n = 0,1,3,5,\ldots,N\,. \end{cases} \tag{6}$$

The external loading of the plate at the time instant at which the maximum force is produced reads as

$$p(x,y) = \begin{cases} \frac{202\,\text{kN}}{r_c^2\pi} \;\ldots\; \sqrt{x^2+y^2} \le r_c\,,\\ 0 \;\ldots\ldots\; \sqrt{x^2+y^2} > r_c\,, \end{cases} \tag{7}$$

where $r_c = 0.15$ m denotes the radius of the load plate.

3.1 *Identification of the modulus of subgrade reaction*

The modulus of subgrade reaction k is optimised in the interval $[\,0.1\,;\,1.0\,]$ MPa/mm, in order to best reproduce the measured deflections. The differences between measured deflections and simulation results are assessed based on the following square-root of sum of squared errors (SRSSE):

$$SRSSE = \sqrt{\frac{1}{72}\sum_{d=1}^{8}\sum_{g=1}^{m_d}\left[\bar{w}_{d,g} - w(x_{d,g},y_{d,g})\right]^2}\,, \tag{8}$$

with $m_d = 9$.

A value of $k = 0.575$ MPa/mm yields the best reproduction of the measured deflections for the new slab, see Figure 3. The residual error according to Eq. (8) is 33 μm. The agreement between the simulated and the measured deflections is not convincing,

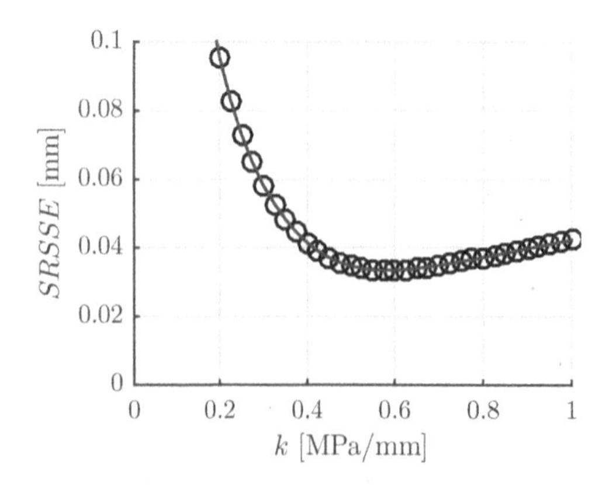

Figure 3. Results of the optimisation of k: SRSSE between measured deflections and simulation results, see Eq. (8).

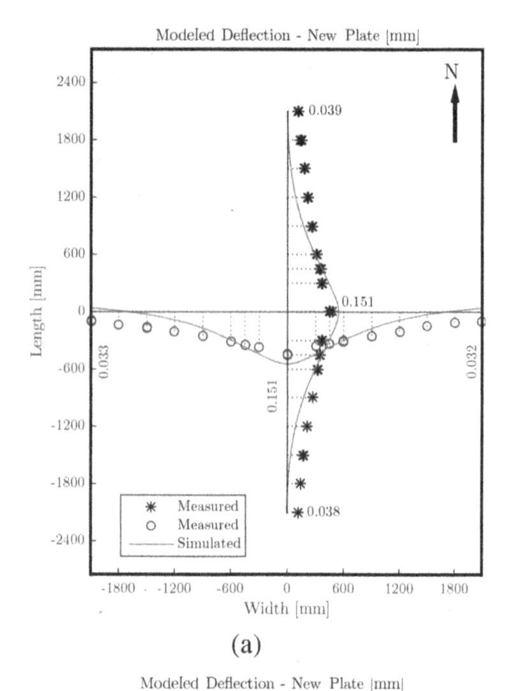

(a)

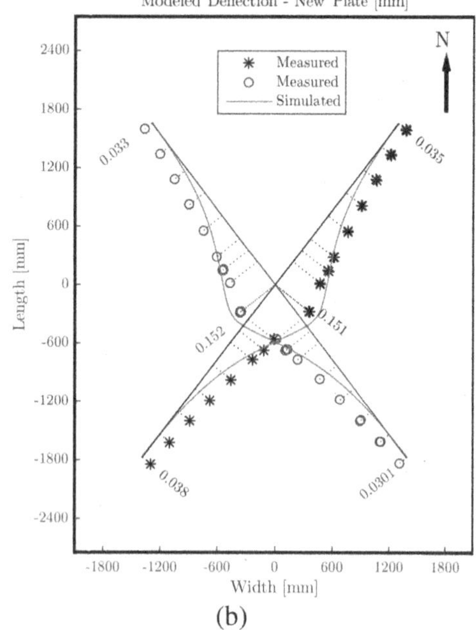

(b)

Figure 4. Results of the optimisation of k: measured deflections (points) and simulation results (lines) obtained with the optimal value of $k = 0.575$ MPa/mm.

see Figure 4. It is concluded that a *uniform* modulus of subgrade reaction cannot explain the measured deflections satisfactorily.

3.2 *Extension towards consideration of an auxiliary surface load*

In order to increase the quality of reproduction of the measured deflections, the same strategy is used as in (Díaz Flores et al. 2021), where a uniform auxiliary load is introduced at the top-surface of the plate. Thus, Eq. (7) is replaced by

$$p(x,y) = \begin{cases} p_{aux} + \frac{202\,\text{kN}}{r_c^2 \pi} \cdots \sqrt{x^2+y^2} \le r_c\,, \\ p_{aux} \cdots\cdots\cdots\cdots \sqrt{x^2+y^2} > r_c\,. \end{cases} \tag{9}$$

The values of the modulus of subgrade reaction and of the auxiliary load are optimised within the intervals $k \in [0.1\,;\,2.0]$ MPa/mm and $p_{aux} \in [-0.01\,;\,+0.1]$ MPa. The values $k = 1.55$ MPa/mm and $p_{aux} = 0.08$ MPa ensure the best reproduction of the measured deflections, see Figure 5. The error according to Eq. (8) is 11 μm. It can thus be concluded that the agreement between the simulated and the measured deflections is satisfactory in a qualitative as well as a quantitative level, see Figure 6.

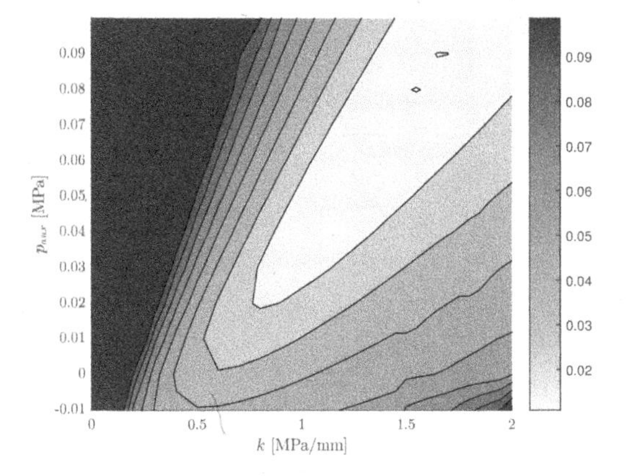

Figure 5. Optimisation of k and p_{aux}: SRSSE between measured deflections and simulation results, see Eq. (8).

A Finite Element (FE) analysis was further performed in order to verify the accuracy of the method proposed. It consisted of a thin plate with free edges with the properties shown in Table 1, on top of an elastic foundation with $k = 1.55$ MPa/mm. As external loads, a point load of 202 kN was placed at the centre of the plate, and a uniform surface load of $p_{aux} = 0.08$ MPa was placed to act downwards. Given that there is virtually no difference between the results from FE simulation and from the proposed method, see the green dashed line in Figure 6, it is concluded that the method does indeed provide accurate results.

The auxiliary load influences the structure in a way that the *effective* modulus of subgrade reaction is spatially *distributed*. In order to show this influence, the plate is conceptually cut free from the Winkler foundation, see also (Díaz Flores et al. 2021) for more details. This way, the auxiliary loading is moved from the top to the bottom of the plate with a changed sign. There, it is superimposed with the stresses resulting from the springs of the Winkler foundation. The resulting distribution at the bottom surface reads as $k\,w(x,y) - p_{aux}$. This is a realistic distribution of the

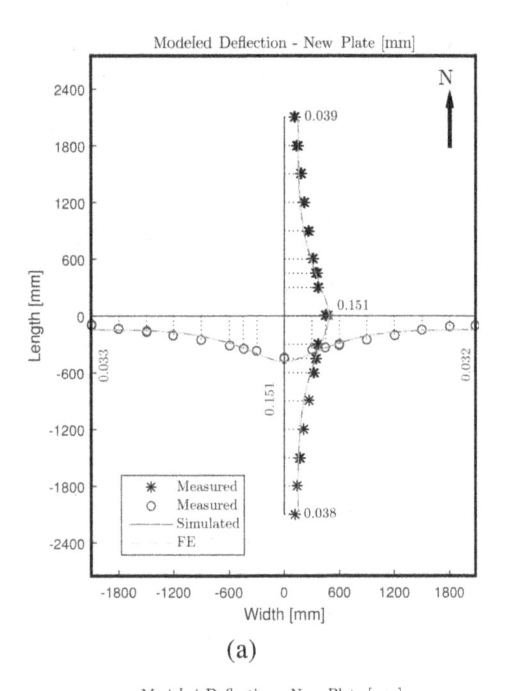

(a)

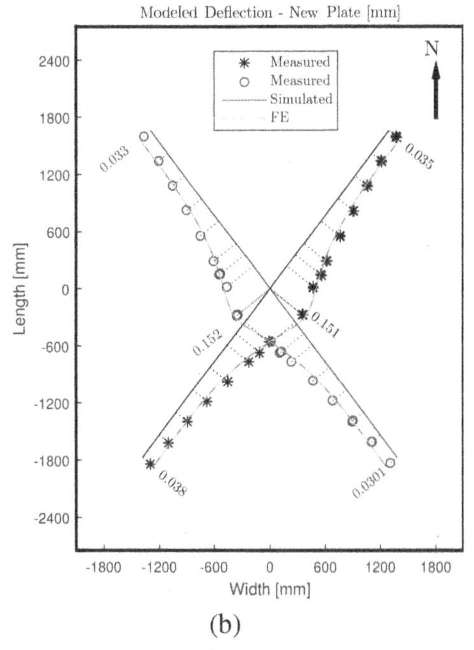

(b)

Figure 6. Results of the optimisation of the values of k and p_{aux}: measured deflections (points) and simulations results (lines) obtained with $k = 1.55$ MPa/mm and $p_{aux} = 0.08$ MPa.

pressure exerted from the subgrade onto the bottom-surface of the plate: It is in equilibrium with the falling weight, while the corresponding deflection field $w(x,y)$ satisfies the plate's field equation and boundary conditions, and it reproduces the measurements accurately.

3.3 *Effective modulus of subgrade reaction*

The effective modulus of subgrade reaction is calculated as

$$k_{eff}(x,y) = \frac{\sigma_{zz,eff}(x,y)}{w_{eff}(x,y)}, \tag{10}$$

where $\sigma_{zz,eff}(x,y)$ refers to the effective pressure at the bottom of the plate, and $w_{eff}(x,y)$ refers to the effective deflection field.

The superposition principle applies to both the pressure (numerator) and to the deflections (numerator) in Eq. (10). However, the superposition principle is not applicable to the effective modulus of subgrade reaction. This happens because k_{eff} is inversely proportional to w_{eff}. Thus, Eq. (10) must be evaluated for the *total* load case. This consists of the dead load of the plate together with the falling weight.

The dead load is given by a uniform load with a value of $\rho gh = 5.29\,\text{kPa}$, where $g = 9.81\,\text{m/s}^2$ refers to gravity. Equilibrium conditions mean that the corresponding subgrade pressure has the same value. The deflection resulting from the dead load, $w_{\rho gh}$, can be calculated as the subgrade pressure divided by the modulus of subgrade reaction:$w_{\rho gh} = \rho gh/k$. Since the value of k for this case is unknown, a sensitivity analysis is performed in the range $k \in [\,0.20\,;\,0.30\,]\,\text{MPa/mm}$, see Table 5. The chosen range was defined in accordance with existing studies (Martin et al. 2016; Murthy 2011; Nielson et al. 1969; Ping and Sheng 2011; Putri et al. 2012).

Table 5. Sensitivity analysis regarding the deflection resulting from the dead load.

ρgh [MPa],	k [MPa/mm]	$w_{\rho gh}$ [mm]
0.00529	0.2	0.026
0.00529	0.3	0.018

Superposition of the load cases "dead load" and "falling weight" at the bottom of the plate, and insertion into Eq. (10), yields

$$k_{eff}(x,y) = \frac{\rho gh + k\,w(x,y) - p_{aux}}{w_{\rho gh} + w(x,y)}. \tag{11}$$

This provides a distribution of the effective modulus of subgrade reaction that is realistic. Interestingly, it is not uniform, but strongly non-linear, see Figure 7.

The distribution of the pressure at the bottom surface of the plate may be calculated as: $\rho gh + k\,w(x,y) - p_{aux}$, see also Figure 8. Given the rectangular geometry of the plate, the subgrade stresses are found to be double-symmetric with respect to the N-S and E-W axes. The maximum pressure amounts to 0.174 MPa.

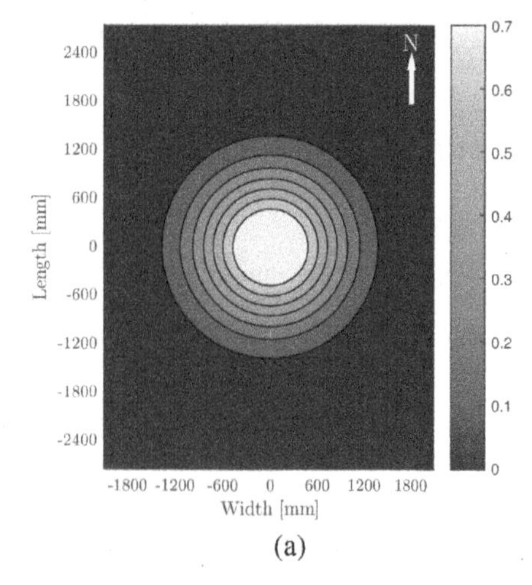

(a)

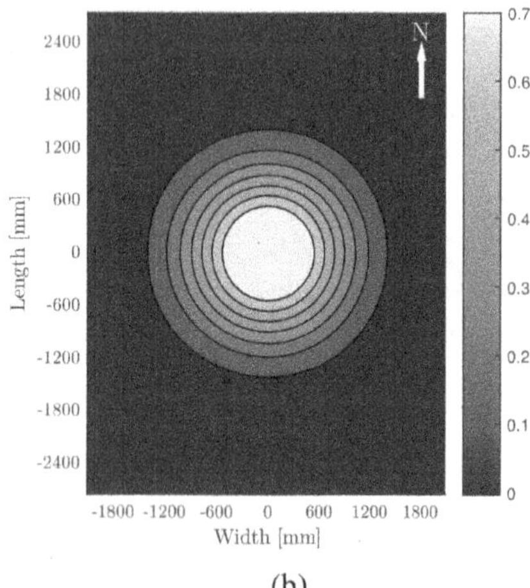

(b)

Figure 7. k_{eff}, in [MPa/mm], according to Eq. (11) with (a) $w_{\rho gh} = 0.026\,\text{mm}$ and (b) $w_{\rho gh} = 0.018\,\text{mm}$, see Table 5.

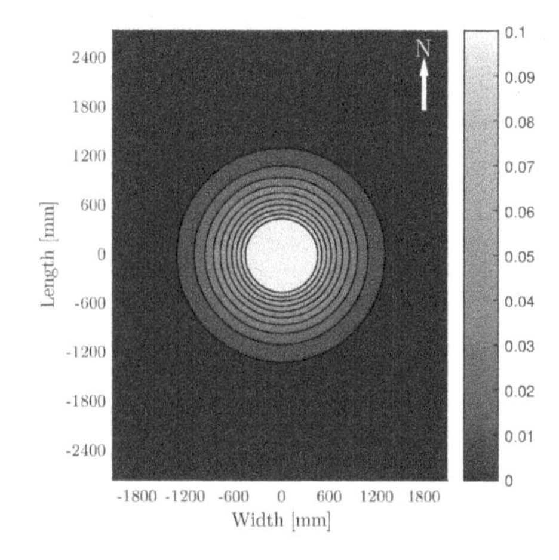

Figure 8. Distribution of pressure exerted from the subgrade onto the bottom-surface of the plate: $\rho gh + k\,w(x,y) - p_{aux}$, in [MPa].

4 CONCLUSIONS

The results of the star-shaped FWD tests allow for the following conclusions to be drawn:

- By confronting the asymmetry index developed by (Díaz Flores et al. 2021) to a new set of original measurement data, it was found that the index was successful in predicting the symmetry or asymmetry present in the behavior of both slabs.
- It can be concluded that values of the asymmetry index amounting to less than 7% refer to virtually double-symmetric structural behaviour. This provides a further tool for decision-making regarding the repair of pavement slabs. This is particularly interesting in the case of slabs exposed to long-term non-symmetric loading.

The structural analysis of the new slab allows for the following conclusions to be drawn:

- A non-linear distribution of the modulus of subgrade reaction is necessary to explain the measurements accurately. The use of a uniform modulus of subgrade reaction was found to be not realistic enough for the purpose of reproducing deflections measured during FWD testing, also for the new experimental data.
- Such a distribution is realistic because it is in equilibrium with the dead load of the plate and the falling weight, the plate's field equation and free-edge boundary conditions are fulfilled, and the deflections obtained from star-shaped FWD testing are reproduced accurately.
- The displacements obtained with the method proposed by (Díaz Flores et al. 2021) and those from FE simulations were virtually the same. This further verifies the robustness and accuracy of the method.
- The structural analysis was limited to the concrete slab and its subgrade. A multi-layered analysis is indispensable if the stresses of the individual layers underneath the slab are of interest.

REFERENCES

Abd El-Raof, H. S., R. T. Abd El-Hakim, S. M. El-Badawy, & H. A. Afify (2018). Simplified closed-form procedure for network-level determination of pavement layer moduli from falling weight deflectometer data. *Journal of Transportation Engineering, Part B: Pavements 144*(4), 04018052.

Aristorenas, G. & J. Gómez (2014). Subgrade modulus—Revisited. *Struct. Mag*, 9–11.

Biot, M. A. (1937). Bending of an infinite beam on an elastic foundation. *Journal of Applied Mechanics 59*(A1–A7).

Daloglu, A. T. & C. G. Vallabhan (2000). Values of k for slab on Winkler foundation. *Journal of Geotechnical and Geoenvironmental Engineering 126*(5), 463–471.

Díaz Flores, R., M. Aminbaghai, L. Eberhardsteiner, R. Blab, M. Buchta, & B. L. A. Pichler (2021). Multi-directional Falling Weight Deflectometer (FWD) testing and quantification of the effective modulus of subgrade reaction for concrete roads. *International Journal of Pavement Engineering*, 1–19.

Eisenberger, M. (1990). Application of symbolic algebra to the analysis of plates on variable elastic foundation. *Journal of Symbolic Computation 9*(2), 207–213.

Foyouzat, M., M. Mofid, & J. Akin (2016). Free vibration of thin circular plates resting on an elastic foundation with a variable modulus. *Journal of Engineering Mechanics 142*(4), 04016007.

Girija Vallabhan, C., W. Thomas Straughan, & Y. Das (1991). Refined model for analysis of plates on elastic foundations. *Journal of Engineering Mechanics 117*(12), 2830–2843.

Höller, R., M. Aminbaghai, L. Eberhardsteiner, J. Eberhardsteiner, R. Blab, B. Pichler, & C. Hellmich (2019). Rigorous amendment of Vlasov's theory for thin elastic plates on elastic Winkler foundations, based on the Principle of Virtual Power. *European Journal of Mechanics-A/Solids 73*, 449–482.

Kausel, E. & J. M. Roësset (1981). Stiffness matrices for layered soils. *Bulletin of the Seismological Society of America 71*(6), 1743–1761.

Martin, U., S. Rapp, D. Camacho, C. Moormann, J. Lehn, & P. Prakaso (2016). Abschätzung der Untergrundverhältnisse am Bahnkörper anhand des Bettungsmoduls. *ETR-Eisenbahntechnische Rundschau 5*, 50–57.

Mehta, Y. & R. Roque (2003). Evaluation of FWD data for determination of layer moduli of pavements. *Journal of Materials in Civil Engineering 15*(1), 25–31.

Murthy, V. (2011). *Textbook of soil mechanics and foundation engineering*. CBS Publishers & Distributors/Alkem Company (S).

Nielson, F. D., C. Bhandhausavee, & K.-S. Yeb (1969). Determination of modulus of soil reaction from standard soil tests. *Highway Research Record 284*, 1–12.

Pan, E. (1989a). Static response of a transversely isotropic and layered half-space to general dislocation sources. *Physics of the Earth and Planetary Interiors 58*(2–3), 103–117.

Pan, E. (1989b). Static response of a transversely isotropic and layered half-space to general surface loads. *Physics of the Earth and Planetary Interiors 54*(3–4), 353–363.

Ping, W. V. & B. Sheng (2011). Developing correlation relationship between modulus of subgrade reaction and resilient modulus for Florida subgrade soils. *Transportation Research Record 2232*(1), 95–107.

Putri, E. E., N. Rao, & M. Mannan (2012). Evaluation of modulus of elasticity and modulus of subgrade reaction of soils using CBR test. *Journal of Civil Engineering Research 2*(1), 34–40.

Roesler, J. R., H. Chavan, D. King, & A. S. Brand (2016). Concrete slab analyses with field-assigned non-uniform support conditions. *International Journal of Pavement Engineering 17*(7), 578–589.

Smith, I. M. (1970). A finite element approach to elastic soil–structure interaction. *Canadian Geotechnical Journal 7*(2), 95–105.

Vlasov, V. Z. (1966). Beams, plates and shells on elastic foundation. *Israel Program for Scientific Translation*.

Winkler, E. (1867). *Die Lehre von der Elasticität und Festigkeit mit besonderer Rücksicht auf ihre Anwendung in der Technik [Lessons on elasticity and strength of materials with special consideration of their application in technology]*.

A RESULTS OF STAR-SHAPED FWD TESTING ON THE NEW SLAB

Table 6. Maximum deflections measured during all FWD tests on the new slab [mm].

Test Direction	Test Number	Geophone $g=1$	$g=2$	$g=3$	$g=4$	$g=5$	$g=6$	$g=7$	$g=8$	$g=9$
$d=1$ (N)	$i=1$	0.159	0.125	0.119	0.107	0.091	0.075	0.061	0.051	0.041
$d=1$ (N)	$i=2$	0.153	0.123	0.117	0.105	0.091	0.074	0.061	0.049	0.041
$d=1$ (N)	$i=3$	0.152	0.123	0.115	0.104	0.088	0.073	0.060	0.048	0.039
$d=2$ (NE)	$i=1$	0.152	0.122	0.119	0.104	0.086	0.071	0.058	0.046	0.036
$d=2$ (NE)	$i=2$	0.150	0.122	0.115	0.102	0.086	0.071	0.056	0.043	0.035
$d=2$ (NE)	$i=3$	0.151	0.122	0.115	0.102	0.086	0.071	0.056	0.044	0.033
$d=3$ (E)	$i=1$	0.152	0.120	0.112	0.102	0.083	0.068	0.051	0.038	0.032
$d=3$ (E)	$i=2$	0.150	0.119	0.112	0.101	0.083	0.068	0.050	0.038	0.032
$d=3$ (E)	$i=3$	0.150	0.118	0.113	0.102	0.083	0.068	0.051	0.038	0.032
$d=4$ (SE)	$i=1$	0.152	0.112	0.111	0.097	0.076	0.063	0.051	0.040	0.030
$d=4$ (SE)	$i=2$	0.151	0.111	0.104	0.094	0.075	0.063	0.049	0.037	0.029
$d=4$ (SE)	$i=3$	0.152	0.112	0.104	0.096	0.078	0.064	0.049	0.039	0.031
$d=5$ (S)	$i=1$	0.150	0.125	0.115	0.106	0.091	0.072	0.056	0.047	0.037
$d=5$ (S)	$i=2$	0.148	0.121	0.117	0.108	0.091	0.071	0.059	0.047	0.037
$d=5$ (S)	$i=3$	0.149	0.123	0.116	0.108	0.090	0.071	0.059	0.047	0.039
$d=6$ (SW)	$i=1$	0.149	0.112	0.106	0.095	0.079	0.065	0.052	0.044	0.038
$d=6$ (SW)	$i=2$	0.147	0.110	0.106	0.095	0.079	0.067	0.052	0.044	0.038
$d=6$ (SW)	$i=3$	0.147	0.110	0.106	0.095	0.079	0.067	0.052	0.044	0.038
$d=7$ (W)	$i=1$	0.148	0.123	0.117	0.105	0.084	0.068	0.055	0.044	0.034
$d=7$ (W)	$i=2$	0.149	0.123	0.115	0.105	0.083	0.068	0.054	0.044	0.031
$d=7$ (W)	$i=3$	0.149	0.124	0.117	0.105	0.085	0.068	0.055	0.044	0.033
$d=8$ (NW)	$i=1$	0.148	0.118	0.113	0.101	0.083	0.068	0.053	0.043	0.033
$d=8$ (NW)	$i=2$	0.148	0.118	0.113	0.101	0.083	0.067	0.053	0.043	0.032
$d=8$ (NW)	$i=3$	0.149	0.121	0.109	0.101	0.084	0.067	0.055	0.043	0.033
$d=1$ (N)	$i=4$	0.148	0.120	0.115	0.104	0.086	0.072	0.059	0.049	0.037
$d=1$ (N)	$i=5$	0.148	0.122	0.115	0.103	0.087	0.073	0.059	0.047	0.038
$d=1$ (N)	$i=6$	0.149	0.122	0.116	0.106	0.088	0.074	0.060	0.043	0.039

Table 7. Coefficients of variation, $CV_{d,g}$, of the maximum deflections measured by each geophone on the new slab, during three subsequent tests in the same direction ($n_d=3$), calculated according to Eqs. (1)–(3), see also Table 6.

Test Direction	Geophone $g=1$	$g=2$	$g=3$	$g=4$	$g=5$	$g=6$	$g=7$	$g=8$	$g=9$
d = 1 (N)	2.24%	1.16%	1.85%	1.24%	1.73%	1.44%	0.78%	2.80%	2.26%
d = 2 (NE)	0.73%	0.31%	2.12%	1.14%	0.23%	0.65%	1.70%	3.45%	5.06%
d = 3 (E)	0.75%	0.48%	0.57%	0.35%	0.44%	0.23%	1.03%	0.61%	0.79%
d = 4 (SE)	0.50%	0.60%	3.64%	1.52%	2.00%	0.40%	2.82%	3.91%	3.41%
d = 5 (S)	0.51%	1.31%	0.65%	1.11%	0.69%	0.72%	3.10%	0.89%	2.74%
d = 6 (SW)	0.80%	0.59%	0.19%	0.11%	0.29%	1.45%	0.59%	0.60%	0.76%
d = 7 (W)	0.37%	0.32%	0.63%	0.29%	1.21%	0.22%	1.32%	1.12%	5.28%
d = 8 (NW)	0.29%	1.24%	1.97%	0.21%	0.77%	0.39%	1.69%	0.59%	2.48%
d = 1 (N)	0.42%	0.67%	0.81%	1.56%	0.98%	1.84%	0.96%	6.99%	2.73%

B RESULTS OF MULTI-DIRECTIONAL FWD TESTING ON THE OLD SLAB

Table 8. Maximum deflections measured during all FWD tests on the old slab [mm].

Test Direction	Test Number	Geophone $g=1$	$g=2$	$g=3$	$g=4$	$g=5$	$g=6$	$g=7$	$g=8$	$g=9$
$d=1$ (N)	$i=1$	0.322	0.295	0.274	0.246	0.203	0.157	0.114	0.081	0.057
$d=1$ (N)	$i=2$	0.329	0.292	0.277	0.250	0.204	0.155	0.114	0.079	0.052
$d=1$ (N)	$i=3$	0.329	0.292	0.275	0.246	0.204	0.159	0.114	0.084	0.059
$d=2$ (NE)	$i=1$	0.321	0.279	0.263	0.234	0.182	0.138	0.099	0.070	0.052
$d=2$ (NE)	$i=2$	0.326	0.280	0.266	0.235	0.183	0.137	0.099	0.070	0.046
$d=2$ (NE)	$i=3$	0.327	0.281	0.262	0.233	0.183	0.135	0.099	0.070	0.048
$d=3$ (E)	$i=1$	0.324	0.274	0.252	0.218	0.160	0.116	0.082	0.056	0.049
$d=3$ (E)	$i=2$	0.323	0.268	0.250	0.216	0.159	0.115	0.080	0.054	0.046
$d=3$ (E)	$i=3$	0.322	0.270	0.250	0.216	0.159	0.114	0.079	0.054	0.046
$d=4$ (SE)	$i=1$	0.320	0.260	0.243	0.217	0.167	0.124	0.090	0.066	0.045
$d=4$ (SE)	$i=2$	0.320	0.265	0.244	0.214	0.168	0.124	0.089	0.065	0.044
$d=4$ (SE)	$i=3$	0.320	0.264	0.245	0.216	0.168	0.124	0.090	0.064	0.044
$d=5$ (S)	$i=1$	0.322	0.288	0.273	0.250	0.205	0.164	0.121	0.088	0.062
$d=5$ (S)	$i=2$	0.319	0.286	0.272	0.248	0.202	0.161	0.119	0.089	0.063
$d=5$ (S)	$i=3$	0.318	0.286	0.269	0.248	0.203	0.160	0.119	0.087	0.060
$d=6$ (SW)	$i=1$	0.318	0.272	0.257	0.232	0.184	0.142	0.104	0.074	0.050
$d=6$ (SW)	$i=2$	0.317	0.269	0.255	0.230	0.182	0.142	0.104	0.073	0.049
$d=6$ (SW)	$i=3$	0.317	0.270	0.255	0.230	0.182	0.142	0.104	0.073	0.049
$d=7$ (W)	$i=1$	0.319	0.284	0.274	0.247	0.194	0.144	0.099	0.067	0.047
$d=7$ (W)	$i=2$	0.318	0.282	0.273	0.245	0.193	0.144	0.099	0.067	0.046
$d=7$ (W)	$i=3$	0.321	0.283	0.275	0.247	0.195	0.145	0.101	0.064	0.042
$d=8$ (NW)	$i=1$	0.314	0.274	0.261	0.236	0.189	0.141	0.102	0.073	0.046
$d=8$ (NW)	$i=2$	0.315	0.274	0.265	0.236	0.189	0.141	0.102	0.073	0.050
$d=8$ (NW)	$i=3$	0.315	0.274	0.261	0.237	0.189	0.141	0.102	0.073	0.049
$d=1$ (N)	$i=4$	0.317	0.276	0.262	0.235	0.188	0.146	0.108	0.078	0.053
$d=1$ (N)	$i=5$	317	0.278	0.260	0.235	0.189	0.146	0.108	0.076	0.053
$d=1$ (N)	$i=6$	0.317	0.277	0.263	0.235	0.188	0.147	0.108	0.077	0.054

Table 9. Coefficients of variation, $CV_{d,g}$, of the maximum deflections measured by each geophone on the old slab, during three subsequent tests in the same direction ($n_d = 3$), calculated according to Eqs. (1)–(3), see also Table 8.

Test Direction	Geophone $g=1$	$g=2$	$g=3$	$g=4$	$g=5$	$g=6$	$g=7$	$g=8$	$g=9$
d = 1 (N)	1.23%	0.56%	0.63%	0.84%	0.27%	1.13%	0.26%	3.27%	6.02%
d = 2 (NE)	0.91%	0.31%	0.83%	0.58%	0.46%	1.10%	0.36%	0.52%	6.72%
d = 3 (E)	0.33%	1.10%	0.45%	0.40%	0.46%	0.87%	2.01%	2.34%	4.11%
d = 4 (SE)	0.10%	0.97%	0.45%	0.74%	0.48%	0.20%	0.22%	1.55%	1.70%
d = 5 (S)	0.55%	0.37%	0.71%	0.46%	0.68%	1.09%	0.79%	1.11%	1.90%
d = 6 (SW)	0.16%	0.55%	0.34%	0.36%	0.47%	0.11%	0.34%	1.07%	1.68%
d = 7 (W)	0.46%	0.27%	0.31%	0.34%	0.50%	0.41%	0.98%	2.87%	5.97%
d = 8 (NW)	0.24%	0.09%	0.85%	0.19%	0.08%	0.04%	0.20%	0.08%	4.44%
d = 1 (N)	0.05%	0.37%	0.48%	0.07%	0.37%	0.34%	0.19%	1.01%	0.66%

Computational Modelling of Concrete and Concrete Structures – Meschke, Pichler & Rots (Eds)
 ISBN: 978-1-032-32724-2

Numerical simulation of evolution of bond strength of GFRP bars with time

M.K. Rahman, M. Fasil & M.M. Al-Zahrani
King Fahd University of Petroleum and Minerals, Dhahran, Saudi Arabia

ABSTRACT: The effect of concrete age on the bond performance of GFRP bars was investigated by performing pullout tests on ribbed GFRP bars partially embedded in a concrete cube with 200 mm edge length. The experiments performed showed that the bond strength increased by 19% after 98 days compared to 7 days. Based on the experimental data, a time dependent analytical model was developed to predict the bond strength and bond-slip curves at different ages. A finite element model of the experimental bond test was developed in commercial software ABAQUS using cohesive interaction between the bonded surfaces and concrete damage plasticity law. The damage parameter is defined from the bond-slip curve obtained analytically or experimentally. Parametric studies were conducted to study the effect of concrete strength and rebar diameter on bond strength of GFRP bars.

1 INTRODUCTION

Extensive research over more than two decades in the application of glass fiber reinforced polymers (GFRPs) has contributed to the recent rapid increase in the acceptance, standardization, and widespread use of GFRP bars in concrete structures. State-of-the-art manufacturing processes and advances in polymer technology have helped improve mechanical properties such as tensile strength and durability. The advantages over steel, such as high strength-to-weight ratio, ease of handling and low maintenance costs, have made FRP bars a compelling substitute for epoxy-coated steel bars.

In reinforced concrete structures, the bond between reinforcing bars and concrete is an important aspect responsible for the transfer of stresses between the two components of the composite material. The bond strength depends on surface finish and strength of the FRP bars, as well as the strength of the concrete (Achillides & Pilakoutas 2004; Maranan et al. 2015; Tekle et al. 2017). The surface texture and tensile strength of GFRP bars are essentially invariant with time. The bond between GFRP bars and concrete, however, evolves with time as the concrete matures and gains strength with time. The time evolution of bond of GFRP bars is of significance, since most RC components are demolded after only 7 days, when only about 65% of the ultimate strength is reached. There is an extensive literature examining the bond strength of different types of GFRP bars, concrete strengths, bar diameters, and the effects of exposure (Parvizi et al. 2020; Rolland et al. 2020; Solyom & Balázs 2020). The authors could not find a study examining the effects of concrete age on the bond performance of GFRP bars embedded in concrete

In order to quantify the evolution of the bond strength of GFRP bars with the age of concrete, an experimental program was conducted to evaluate the bond strength and bond-slip response of GFRP bars in concrete. Pullout tests were performed on GFRP bars embedded in 200 × 200 × 200 mm^3 concrete cubes at 7, 40, and 98 days. Based on the experimental data a new analytical bond-slip model (called as KFM model) capable of predicting the bond-slip response for any age of concrete was developed. A 3D finite element model of the pullout test was developed in ABAQUS with the concrete and GFRP bars modelled with C3D8R brick elements. A concrete damage plasticity (CDP) model was used for the concrete and an elastic-brittle material model was used for GFRP bars. The bond-slip curve obtained experimentally or by the proposed analytical model was used to develop the bond interaction between concrete and GFRP bar and the damage due to slip. The proposed KFM bond-slip analytical model for GFRP bars and the FE model developed in ABAQUS captured the experimental pullout behavior with good accuracy. Finally, parametric studies were performed with the validated finite element model to investigate the effects of variables such as the concrete compressive strength and the diameter of the GFRP bars.

2 EXPERIMENTAL INVESTIGATIONS

2.1 *Materials*

The concrete was produced with a cement content of 320 kg/m^3 and a water-cement ratio of 0.4. The average cube compressive strength of the concrete after 7, 15, 40, and 98 days was found to be 45.2 MPa, 48.9 MPa,

DOI 10.1201/9781003316404-35

51.1 MPa, and 51.5 MPa, respectively. A ribbed GFRP bar (Figure 1) with a nominal diameter of 14 mm and a measured diameter (by dip testing) of 13.7 mm, was used for the present study. Testing in accordance with ASTM D7205M-06 (ASTM 2016), yielded an average tensile strength of 866 MPa and an average modulus of elasticity of 49.5 GPa.

Figure 1. Surface texture characteristics of GFRP bar used.

2.2 *Bond specimen preparation*

For the fabrication of the composite specimens, part of the 980 mm long GFRP bar was embedded in concrete cubes with 200 mm edge length. The bar was inserted into the specimen so that it passed through the center of gravity of the opposite sides of the cube. The part of the GFRP bar in the concrete cube was divided into two zones: 70 mm (5 times the nominal diameter of the rod, d_b) of the bar had full contact with the concrete, while the remaining 130 mm had no contact with the concrete with a bond-breaker formed by a PVC pipe. The loaded end of the specimen was provided with anchors made of galvanized iron pipes (GI) with a diameter of 42 mm and a length of 380 mm according to ASTM D7205M-06 (ASTM 2016). The schematic diagram of the specimen can be seen in Figure 2.

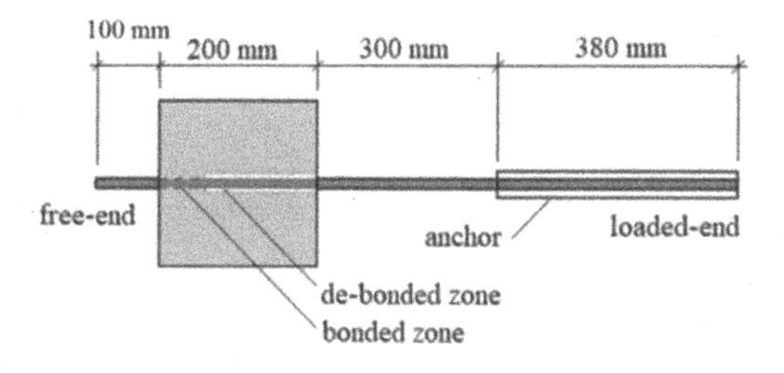

Figure 2. Schematic representation of pullout specimens.

2.3 *Testing setup and instrumentation*

To perform the bond test on a displacement-controlled universal testing machine, a bond test apparatus with dimensions 500 × 300 × 300 mm^3 was fabricated to hold the specimen, as shown in Figure 3. A slot was milled in the base plate so that the bar specimen could be inserted into the fixture. The upper plate of the fixture was attached to the crosshead of the testing machine. The schematic diagram of the setup and an actual loaded specimen are shown in Figures 3a and 3b, respectively. The force was measured using a load cell with a capacity of 250 kN, which was placed between the crosshead and the loading device. The slip of the GFRP bar in the concrete was measured at two locations during the experiment. First, at the free end of the bar with three LVDTs placed between the free

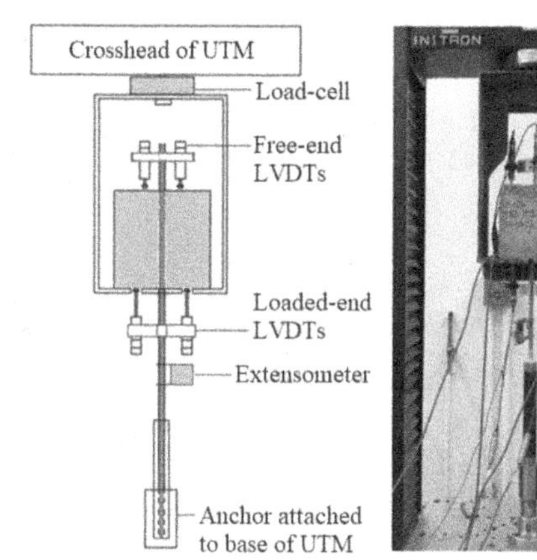

Figure 3. Bond test apparatus: (a) schematic; and (b) experimental setup.

end and the top of the concrete cube to measure the slip at the free end, and second, at the loaded end with two LVDTs placed between the bar and the bottom of the concrete cube to measure the slip at the loaded end. The elongation of the GFRP during the experiment was measured using an extensometer with a gauge length of 50 mm. All sensors were connected to a data logger for data acquisition. Pullout on the GFRP bars was performed by applying a monotonic quasi-static load at a rate of 1.0 mm per minute, according to ASTM D7913M-14 (ASTM 2014).

2.4 *Bond strength and bond-slip response*

Bond samples were tested at 7, 40, and 98 days. All bond specimens tested failed when the bars were pulled out of the concrete, without splitting or cracking the concrete cube. The average bond stress was calculated according to ASTM D7913M-14 (ASTM 2014) and is given by:

$$\tau = \frac{F}{c_b l} \tag{1}$$

where τ is the average bond stress in MPa, F is the tensile force in N, c_b is the effective diameter of the GFRP bar, and $l$is the bonded length in mm. Slip at the free end of the bar during the pullout tests was measured.

The evolution of the average bond strength of the tested specimens is shown in Figure 4. The average bond strengths were 12.4 MPa, 18.8 MPa, and 19.9 MPa at ages of 7, 40 and 98 days, respectively. Figure 5 shows typical bond stress-slip responses of pullout specimens at different ages. The response is characterized by three curves (i) an ascending part of the response, which is nonlinear up to the peak bond stress. The peak stress is the point where the loss of bond between the GFRP bar and the concrete commences, (ii) a linear descending part from the peak value of the bond stress, (iii) a horizontal part showing

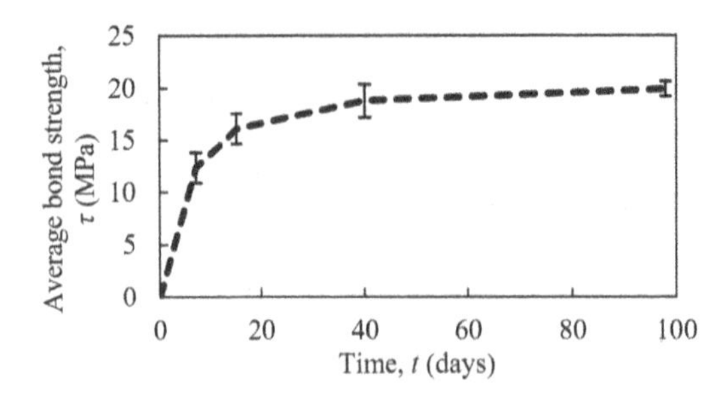

Figure 4. Evolution of average bond strength with time.

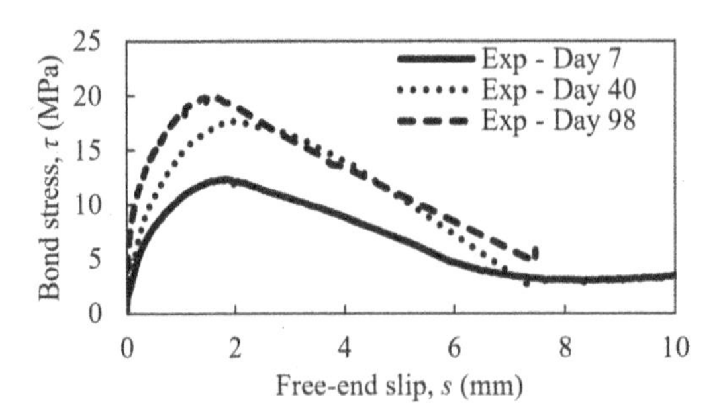

Figure 5. Bond stress, τ (MPa) versus free-end slip, s (mm) at various ages.

the slippage of bars at constant bond stress, indicating complete debonding.

Figure 6 shows the nature of bond failure of the GFRP bar in the embedded zone of the concrete.

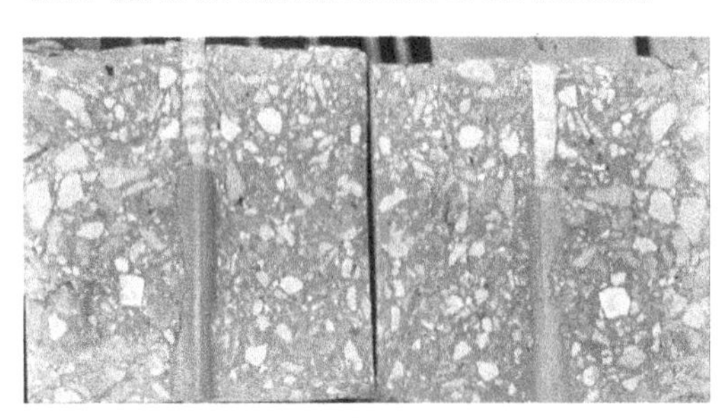

Figure 6. Failed bond specimen split open after test.

It can be seen that bond failure took place due to shearing of the core of the GFRP bars from the ribs, leaving the lugs with the bonded concrete (Figure 7). No damage in the concrete cube was observed in any specimen.

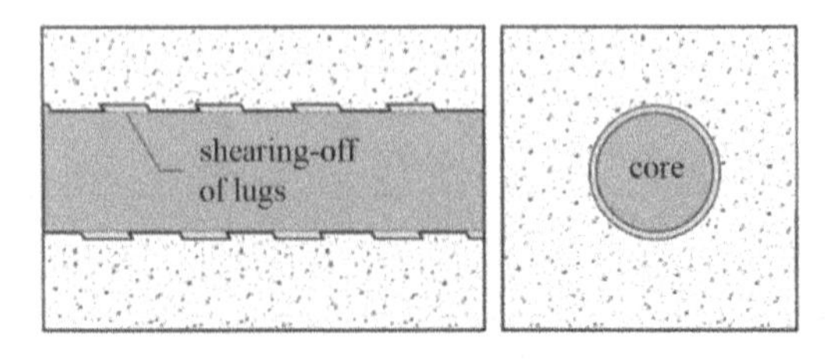

Figure 7. Mechanism of bond loss due to shearing of lugs.

3 ANALYTICAL MODEL

Several analytical models exists in the literatures to predict the bond-slip behavior rebars in concrete, such as the BPE model (Eligehausen et al. 1982) for steel, the modified BPE (Cosenza et al. 1997), the Malvar model (Malvar 1994), the CMR model (Cosenza, E., Manfredi, G. & Realfonzo 1995) for FRP bars. Several international codes have also proposed analytical models to determine the bond strength of FRP bars including the ACI 440.1R-15 (ACI 2015) and CSA S806-02 (CSA 2007). As far as the authors are aware, none of the proposed models in literature considered the evolution of bond strength with the age of concrete and predictions bond strength based on tests carried out at an age of 28-days. At an early age, when the strength and stiffness of the concrete are low, the bond strength is expected to be lower, and it enhances with age. The experimental program conducted in this study focused on evolution of bond strength with time.

Based on the experimental studies conducted, a time dependent analytical model is proposed to predict the bond strength and bond-slip response of a GFRP bar embedded in concrete until failure. The proposed model, referred to as "KFM Model" shown in Figure 8 consists of two parts: a non-linear part increasing up to the peak bond stress τ_b, reached at slip s_b, and a linearly decreasing curve.

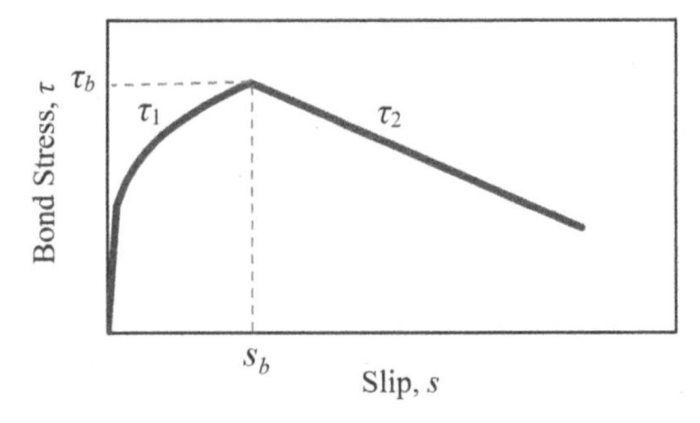

Figure 8. Proposed bond stress-slip model.

The equations proposed for the KFM model are shown in Equations 2 and 3. The equation consists of the parameter α, a shape factor to define the rising curve, and k_1 to define the slope of the falling curve. Equations 4–8 define different parameters used in the proposed equations. k_2 and k_3 define the maximum bond stress (τ_b) and the corresponding slip (s_b). The age of the concrete, t in days, is used to generate the age-dependent behavior of the bond-slip relationship.

$$\tau_1 = \tau_b \left(\frac{s}{s_b}\right)^{\alpha} \tag{2}$$

$$\tau_2 = \tau_b - k_1 \left(\frac{s}{s_b} - 1\right) \tag{3}$$

where

$$\tau_b = 2.6 \left(\frac{f_c}{d_b}\right)^{k_2} \tag{4}$$

$$s_b = k_3 \left(\frac{E_c}{E_{GFRP}} \right) \quad (5)$$

$$k_1 = t^{0.48} \left(1 - 0.07t^{0.45}\right) \quad (6)$$

$$k_2 = t^{0.1} \quad (7)$$

$$k_3 = 2.6 - 0.0044t \quad (8)$$

The peak bond stress, τ_b in the proposed KFM model is a function of concrete strength, f_c, diameter of the GFRP bar d_b, and time-dependent parameter k_2. The slip corresponding to the peak stress, s_b is a function of Young's modulus of concrete, E_c, Young's modulus of GFRP bar, E_{GFRP} and time-dependent factor, k_3. The calibration of the KFM analytical model was performed using experimental results from the present study to include the influence of time on bond strength.

In the absence of concrete strength data, the evolution of concrete compressive strength with time, and the 28-day compressive strength, the concrete compressive strength at any given time t may be predicted using the following simple equation:

$$f_c(t) = f_{c,28}\left(1 - e^{-t\lambda}\right) \quad (9)$$

where λ is the shape factor to fit the curve of concrete strength evolution. For the present experimental data, $\lambda = 0.3$ gave an acceptable fit. By simply varying the age of concrete, the proposed model was used to predict the bond strength-slip response at 7, 40 and 98 days, as shown in Figure 9. The analytical response shown in the figure was obtained by simply varying the age of the concrete.

The KFM model was also validated against the results of an independent study conducted by the authors to test the bond performance of a similar type of bar with a different concrete strength. The 28-days compressive strength of the concrete was reported to be 38.9 MPa. The response from experimental data and the KFM model is shown in Figure 10. Further improvement and validation of the model using other published data is in progress.

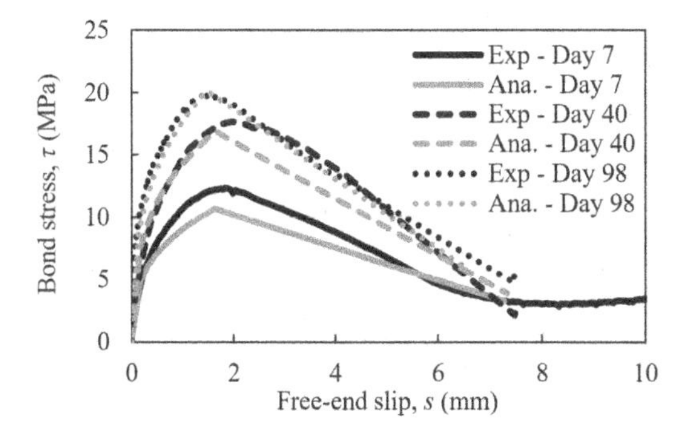

Figure 9. Calibration of model with experimental data.

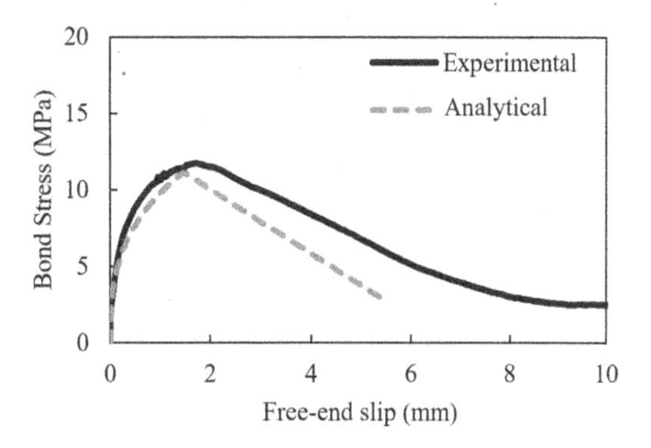

Figure 10. Validation of proposed analytical model.

4 FE MODELING OF PULLOUT TEST

Three-dimensional finite element model of the pullout specimens was developed using the commercial finite element software ABAQUS CAE to simulate the bond-slip behavior of GFRP bars in concrete. The dynamic-explicit module with displacement-controlled loading was used. The geometry of the model simulated the experimental pullout test specimen. The 200 mm concrete cube and the GFRP bars used in pullout test were modeled using 3D 8-noded linear brick elements (C3D8R), as shown in Figure 11. The inelastic mechanical properties of the 50 MPa concrete were modeled using the concrete damage plasticity (CDP) model, which is commonly used for the analysis of quasi-brittle materials such as concrete, rock, and ceramics. The GFRP bars are modelled as elastic-brittle material with an ultimate strength of 866 MPa. The constitutive relationship for concrete in compression was modeled using the Kent and Park model (Kent & Park 1971) and in tension using Wahalathantri et al. (Wahalathantri et al. 2011) model. Concrete damage parameters were determined based on Birtel and

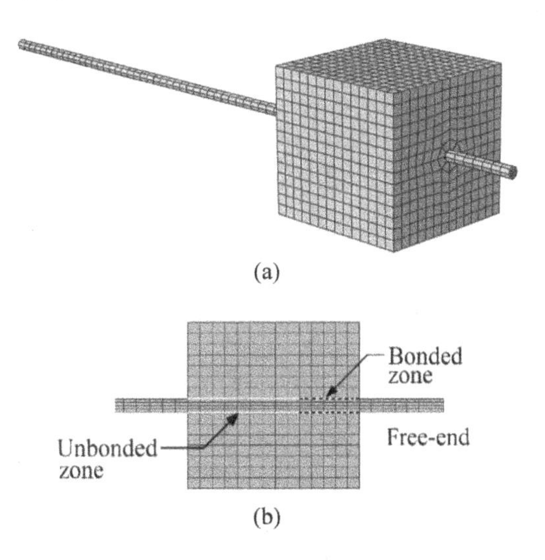

Figure 11. FE model of pull-out specimen developed in ABAQUS: a) 3D geometry; and b) concrete-rebar interface.

Mark model (Birtel & Mark 2006) for tension and the Yu et al. model (Yu et al. 2010) for compression.

A surface-based cohesive interaction was assigned at the concrete- GFRP bar interface, in the bonded part of the sample. A gap of 1 mm was inserted in the unbonded portion of the specimen to simulate the gap between the inner face of PVC and the rebar surface. The cohesive interaction model required three stiffness parameters, namely k_{nn}, k_{ss} and k_{tt}. The elastic bond shear stiffness (k_{tt}) was calculated using $k_{tt} = \tau_b/s_b$ (Rezazadeh et al. 2017), where τ_b is the peak bond stress and s_b is the corresponding slip.

The boundary conditions of the model were assigned according to the experimental setup. The face of the concrete at the loaded-end was assumed to be pinned ($U_1 = U_2 = U_3 = 0$). The displacement-controlled load was applied at the loaded-end of the GFRP bar.

4.1 *Validation of the numerical model*

The proposed KFM analytical model was used to calculate the cohesive stiffness (k_{tt}) and its degradation (damage evolution) at the concrete-rebar interface. The 98-day bond-slip curve obtained from the KFM analytical model was used to validate the numerical model. The elastic bond shear stiffness factor (k_{tt}) was calculated based on the bond stress (τ) at 10% of the slip s_0. The damage evolution curve was developed using the following equation for $D(s)$ (Dugdale 1960; Rolland et al. 2020):

$$D(s) = 1 - \frac{\tau(s)}{k_{tt}s} \tag{10}$$

where $\tau(s)$ is the bond stress corresponding to the slip s. The damage curve for the 98-days analytically developed bond-slip curve is plotted in Figure 12.

The experimental, analytical, and numerically developed bond-slip response of the 98-day bond specimen is compared in Figure 13. The numerical model accurately predicted the bond-slip response of the analytical model and experimental data.

In addition, the bond-slip response was obtained using the validated FE model for the 7-days and 40-days pullout tests from ABAQUS. The response from analytical KFM model and the FE simulation are shown in Figures 14 and 15.

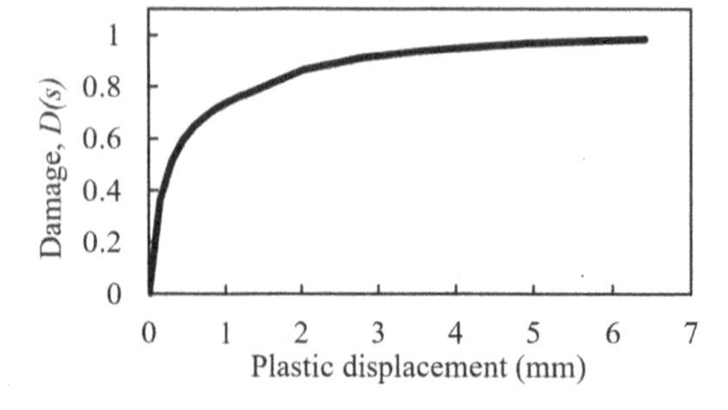

Figure 12. Damage evolution of 98-days bond.

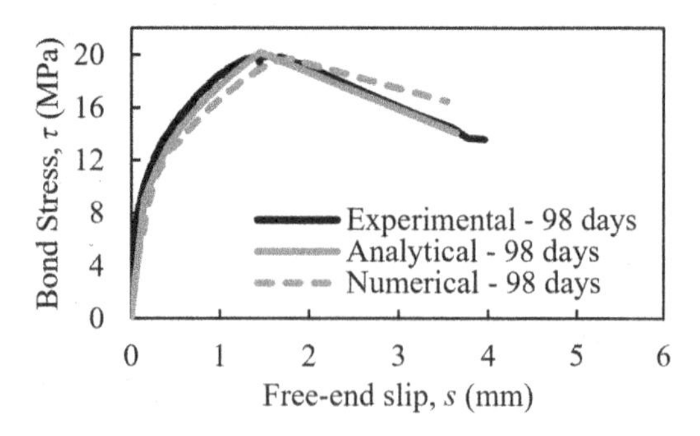

Figure 13. Comparison of experimental, analytical, and numerical 98-days bond-slip responses.

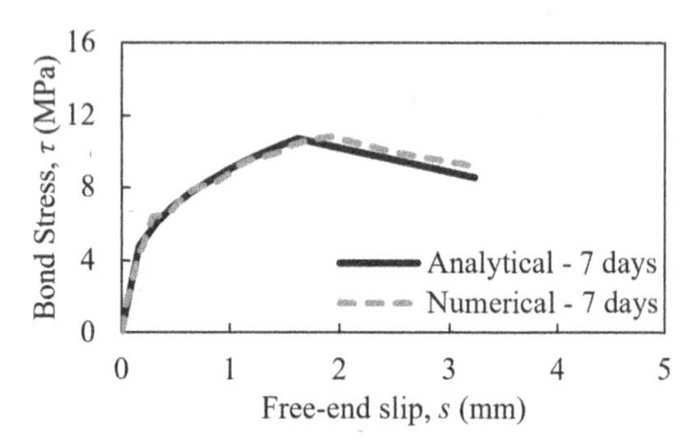

Figure 14. Comparison of analytical and numerical 7-days bond-slip responses.

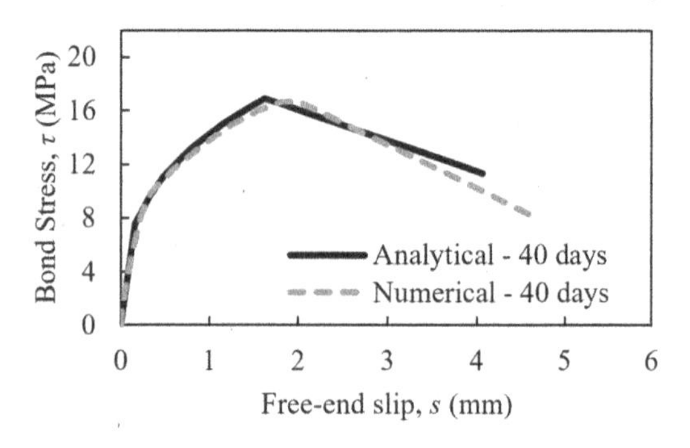

Figure 15. Comparison of analytical and numerical 40-days bond-slip responses.

4.2 *Results from finite element modeling*

The stress contours from the numerical model are shown in Figures 16 and 17. Figure 16a illustrates the gradient of maximum principal stresses developed in the bonded region of the specimen. Figure 16b shows the concrete stress bulb in the vicinity of the bonded region. Figure 17 shows the damage to the concrete bonded to the GFRP bar. The mode of failure achieved from the numerical model is consistent with the failure mode observed in the experimental work, shown in Figure 6.

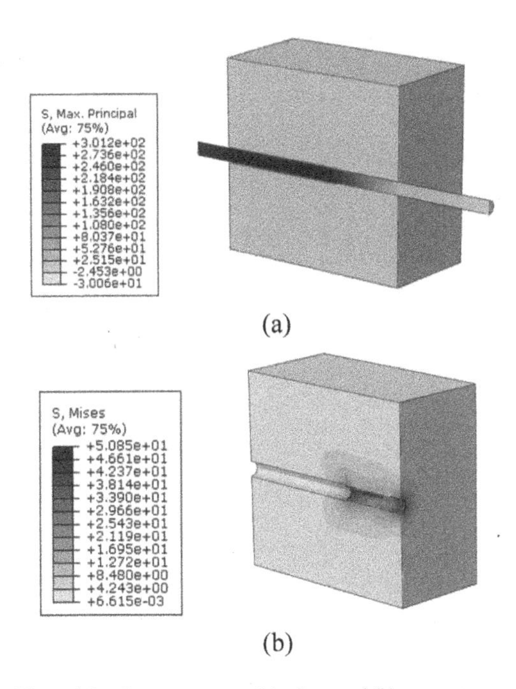

Figure 16. Stress contours: (a) rebar; and (b) concrete.

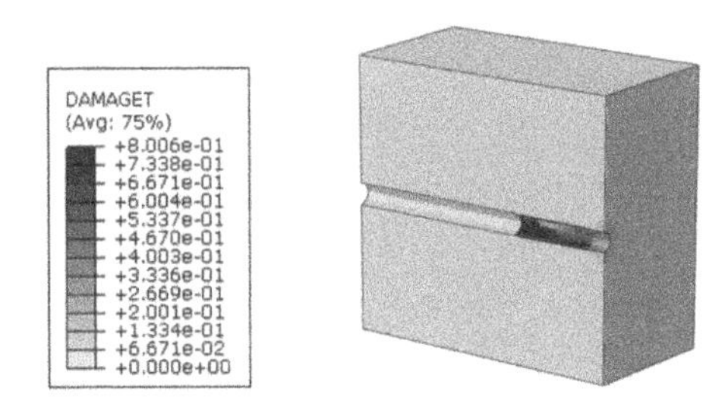

Figure 17. Tension damage.

4.3 *Parametric studies*

Parametric studies were conducted to investigate the effect of parameters such as concrete strength and GFRP bar diameter. First, the analytical model was used to generate the bond-slip curve. Then, the cohesion parameters in the numerical model were calculated and used to analyze the pullout response in ABAQUS.

4.3.1 *Concrete strength*

The bond strength and bond-slip behavior of 30 MPa and 60 MPa were investigated. For the concrete cube, the constitutive law and damage evolution were generated using the KFM analytical bond model employing the bond-slip response obtained at 28 days.

Tensile strengths of 3.4 MPa and 5.2 MPa were assumed for 30 MPa and 50 MPa concrete, to generate the constitutive relationships. From Figure 18, it is evident that the strength of concrete plays a crucial role in the bond-slip response of GFRP bar embedded in concrete.

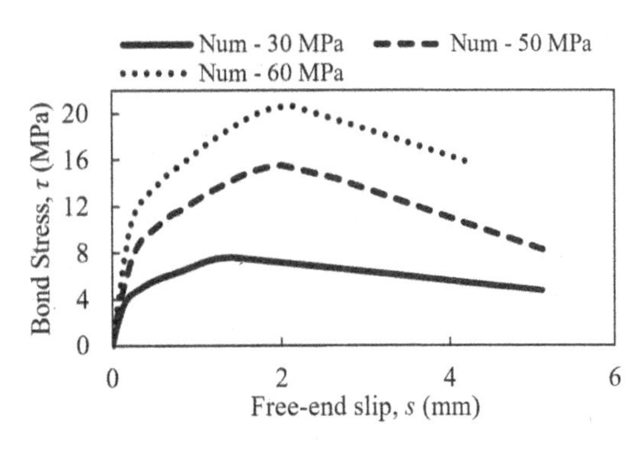

Figure 18. Effect of concrete strength on bond-slip response.

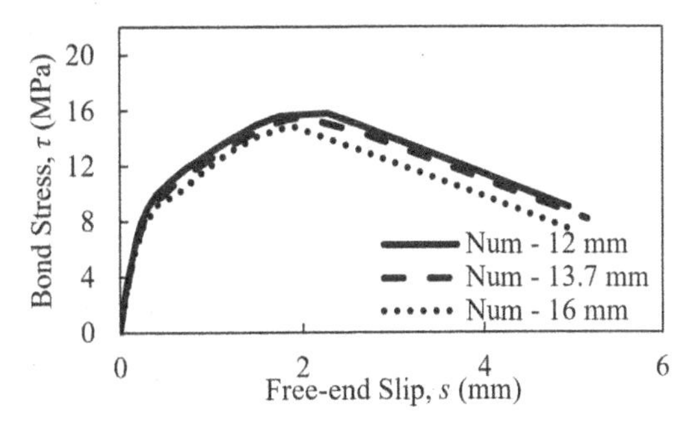

Figure 19. Effect of bar diameter on bond-slip behavior.

4.3.2 *Bar diameter*

GFRP bars with diameters of 12 mm and 16 mm were used to investigate numerically the effect of bar diameter on the bond-slip response of 50 MPa concrete at an age of 28 days. Smaller diameter bars were found to have a slightly larger bond strength as compared to larger diameters (Figure 19). This is consistent with the finding of the experimental work done by Gao et al. (Gao et al. 2019).

5 CONCLUSION

The bond strength of GFRP bars evolves with time as a function of several factors, with concrete strength playing the most dominant role. Experimental investigations have shown that the bond strength increases by about 19% at 98 days compared to the 7-day strength. A new analytical model (KFM Model) is proposed to predict the evolution of bond strength with time. The model captures the experimental response with good accuracy. The nonlinear finite element model of the pullout tests in ABAQUS using cohesive interaction between the bonded surfaces and damage evolution obtained from bond-slip response predicted the experimental response both during the nonlinear response up to bond strength of GFRP bar and subsequent bond degradation. Parametric studies using the calibrated finite element model showed that concrete strength has a significant effect on early age response. Smaller diameter bars have slightly higher bond strength than larger diameter bars.

REFERENCES

Achillides, Z. & Pilakoutas, K. 2004. Bond Behavior of Fiber Reinforced Polymer Bars under Direct Pullout Conditions. *Journal of Composites for Construction*, 8 (2): 173–181.

ACI (American Concrete Institute) 2015. *Guide for the design and construction of structural concrete reinforced with fiber-reinforced polymer (FRP) bars (ACI 440.1R-15)*. Farmington Hills, USA.

ASTM (American Society for Testing Materials) 2014. *Test Method for Bond Strength of Fiber-Reinforced Polymer Matrix Composite Bars to Concrete by Pullout Testing (ASTM D7913/D7913M-14)*. West Conshohocken, USA.

ASTM (American Society for Testing Materials) 2016. *Test Method for Tensile Properties of Fiber Reinforced Polymer Matrix Composite Bars (ASTM D7205/D7205M-06)*. West Conshohocken, USA.

Birtel, V. & Mark, P. 2006. Parameterised Finite Element Modelling of RC Beam Shear Failure. *2006Abaqus User's Conference*: 95–108.

Cosenza, E., Manfredi, G. & Realfonzo, R. 1995. Analytical modelling of bond between frp reinforcing bars and concrete. In Non-Metallic (FRP) Reinforcement for Concrete Structures. *Proceedings of the Second International RILEM Symposium* 29 (CRC Press): 164.

Cosenza, E., Manfredi, G. & Realfonzo, R. 1997. Behavior and modeling of bond of FRP rebars to concrete. *Journal of Composites for Construction* 1(2): 40–51.

CSA (Canadian Standards Association) 2007. *Design and Construction of Building Components with Fibre-Reinforced Polymers (S806-02)*. Ontario, Canada.

Dugdale, D. S. 1960. Yielding of steel sheets containing slits. *Journal of the Mechanics and Physics of Solids* 8 (2): 100–104.

Eligehausen, R., Popov, E. P. & Bertéro, V. V. 1982. Local bond stress-slip relationship of ribbed bars under generalized excitations.

Gao, K., Li, Z., Zhang, J., Tu, J. & Li, X. 2019. Experimental Research on Bond Behavior between GFRP Bars and Stirrups-Confined Concrete. *Applied Sciences* 9 (7): 1340.

Kent, D. C. & Park, R. 1971. Flexural members with confined concrete. *Journal of the Structural Division, Proc. of the American Society of Civil Engineers* 97(ST7): 1969–1990.

Malvar, L. J. 1994. Bond Stress-Slip Characteristics of FRP Rebars. *Technical report, TR-2013-SHR*, February, 50. Virginia: Defense Technical Information Center.

Maranan, G., Manalo, A., Karunasena, K. & Benmokrane, B. 2015. Bond Stress-Slip Behavior: Case of GFRP Bars in Geopolymer Concrete. *Journal of Materials in Civil Engineering* 27 (1): 04014116.

Parvizi, M., Noël, M., Vasquez, J., Rios, A. & González, M. 2020. Assessing the bond strength of Glass Fiber Reinforced Polymer (GFRP) bars in Portland Cement Concrete fabricated with seawater through pullout tests. *Construction and Building Materials* 263: 120952.

Rezazadeh, M., Carvelli, V. & Veljkovic, A. 2017. Modelling bond of GFRP rebar and concrete. *Construction and Building Materials* 153: 102–116.

Rolland, A., Argoul, P., Benzarti, K., Quiertant, M., Chataigner, S. & Khadour, A. 2020. Analytical and numerical modeling of the bond behavior between FRP reinforcing bars and concrete. *Construction and Building Materials* 231: 117160.

Solyom, S. & Balázs, G. L. 2020. Bond of FRP bars with different surface characteristics. *Construction and Building Materials* 264: 119839.

Tekle, B. H., Khennane, A. & Kayali, O. 2017. Bond behaviour of GFRP reinforcement in alkali activated cement concrete. *Construction and Building Materials* 154: 972–982.

Wahalathantri, B. L. L., Thambiratnam, D. P. P., Chan, T. H. T. H. T. & Fawzia, S. 2011. A material model for flexural crack simulation in reinforced concrete elements using ABAQUS. *First International Conference on Engineering, Design and Developing the Built Environment for Sustainable Wellbeing (eddBE2011), 27-29 April 2011*. Brisbane, Australia.

Yu, T., Teng, J. G., Wong, Y. L. & Dong, S. L. 2010. Finite element modeling of confined concrete-II: Plastic-damage model. *Engineering Structures* 32 (3): 680–691.

Computational Modelling of Concrete and Concrete Structures – Meschke, Pichler & Rots (Eds)
 ISBN: 978-1-032-32724-2

Computational modeling of time-dependent shrinkage stresses developed in GFRP bar reinforced slabs-on-ground exposed to ambient environment

M. Fasil, M.K. Rahman & M.M. Al-Zahrani
King Fahd University of Petroleum & Minerals, Dhahran, Saudi Arabia

S. Al-Ghamdi
ARAMCO, Dhahran, Saudi Arabia

ABSTRACT: An experimental program was conducted to monitor the development of strains and cracking in large-scale concrete slabs reinforced with GFRP bars that experience drying shrinkage under ambient conditions. Six slabs reinforced with two types of GFRP bars, namely ribbed and sand-coated with different spacing (200 mm and 300 mm), a slab reinforced with steel bars, and a plain slab were cast on lean concrete layer 100 mm thick laid over compacted soil. This paper presents the development of a numerical model for the slab subjected to drying shrinkage under ambient conditions. The developed finite element model of the slab on grade predicted with reasonable accuracy the evolution of environmentally induced stresses and cracking in the concrete slab. Parametric studies were performed with the validated FE model to investigate the effect of parameters such as concrete strength, bar diameter and slab thickness.

1 INTRODUCTION

After more than two decades of intensive research, concrete structures reinforced with glass fiber-reinforced polymer (GFRP) bars have become increasingly popular worldwide in recent years. Advances in manufacturing processes and resin types, as well as improvements in mechanical properties with advantages in corrosion resistance, ease of construction, and favorable life-cycle costs, have made GFRP bars competitive with epoxy-coated bars. The largest concrete structure in the world reinforced with GFRP bars, a flood control channel, was recently built in Saudi Arabia. While there is an abundance of research literature on the use of GFRP bars in reinforced concrete columns, beams, and shear walls, limited research has been reported in GFRP bars reinforced grade supported structures.

There are several research papers in the literature on the practical application of GFRP bars as embedded reinforcement and dowels in pavements such as continuous reinforced concrete pavements (CRCP), jointed reinforced concrete pavements (JRCP), and jointed plain concrete pavements (JPCP). The first field application of CRCP reinforced with GFRP bars was reported by Benmokrane et al. (Benmokrane et al.) on Highway 40 East (Montréal) in Canada. Analysis of shrinkage and thermal stresses in CRCPs and GFRP reinforcement required in CRCPs was studied by Chen and Choi (Chen & Choi 2011; Choi & Chen 2015).

One of the most common causes of cracking in concrete slabs on ground is the phenomenon of drying shrinkage, in which the concrete loses moisture, causing it to dry out and thus experience a reduction in volume (Carlson 1938). Shrinkage of concrete slabs supported on the ground is resisted by various types of restraints, such as reinforcing bars, friction between the slab base and subgrade, and moisture gradient across the depth. The restraint to free shrinkage leads to the development of tensile stresses in the concrete slab. When tensile stress in concrete exceeds the modulus of rupture, it leads to cracking (ACI 2008). Shrinkage of concrete slabs on grade leads to two other phenomena, namely warping and curling (NP et al. 2002). Warping is developed due to the differential moisture distribution in the concrete and causes the corners of the slab to lift. This leads to loss of support in the slab from the subgrade and failure due to fatigue under impact loads such as traffic loads (Suprenant 2002; Wei & Hansen 2011). Curling is a phenomenon that occurs in concrete grade slabs when they undergo dimensional changes due to uneven moisture/temperature distribution over the depth of the concrete (Shadravan et al. 2015).

An experimental program was carried out to monitor the development of strains and cracks in full-scale GFRP bar reinforced concrete grade slabs with dimensions 6000 × 1100 × 200 mm^3 undergoing drying shrinkage under ambient conditions. Six slabs reinforced with two types of GFRP bars (ribbed and sand-coated types) with different spacing (200 mm and

DOI 10.1201/9781003316404-36

300 mm), one slab reinforced with steel bars and an unreinforced concrete slab cast on lean concrete subbase was monitored for a duration of approximately 180 days.

Computational modeling of the time-dependent evolution of stresses, cracks, and damage in the experimental slabs due to shrinkage was performed using the commercial finite element software ABAQUS. Although the time-dependent behavior of concrete, such as shrinkage, cannot be modeled directly in ABAQUS, the user-defined subroutine feature was used to simulate these phenomena. Two subroutines were developed, USDFLD to define field variables and UEXPAN for calculating the shrinkage strains, which calculate the strains at the beginning of each incremental time step and pass them to the implicit solver of ABAQUS.

The developed finite element model of the slab on ground predicted with reasonable accuracy the evolution of environmentally induced stresses and the cracking of the concrete. After validating the FE model of the slab, numerical parametric studies were performed to investigate the effects of parameters such as the concrete tensile strength, diameter of GFRP bars and slab thickness.

2 EXPERIMENT

2.1 *Materials*

In the present experimental work, concrete with an average 28-day cube compressive strength of 40.2 MPa and split cylinder tensile strength of 3.1 MPa was used.

Three types of reinforcing bars were used in the study, namely ribbed GFRP bars, sand-coated GFRP bars and steel bars. The mechanical properties of these bars are listed in Table 1.

Table 1. Mechanical properties of rebars.

Property	Ribbed Sand-coated GFRP	Conventional GFRP	ribbed steel
Tensile strength (MPa)	1046	1030	575 (yield)
Young's Modulus (GPa)	49	43.5	174
Diameter (mm)	13.7	13.5	12

2.2 *Exposure*

The slab on grade specimens were exposed to ambient environmental conditions in a field station for 180 days. Ambient temperature and relative humidity (RH) were continuously monitored using a weather station installed at the field station (Figure 1). The slabs were cast in January 2021, at an average temperature of

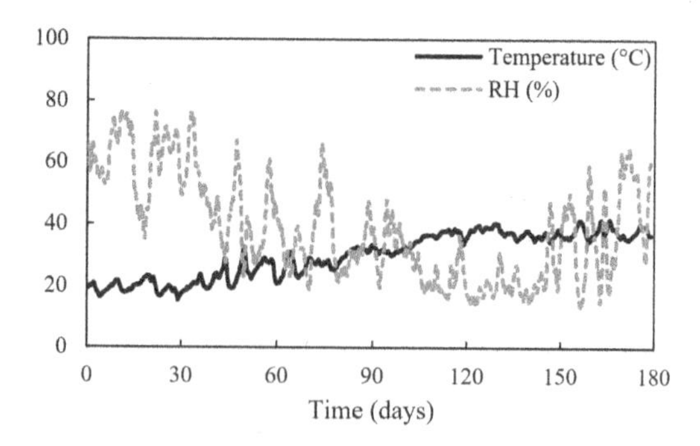

Figure 1. Field exposure conditions: Temperature and relative humidity (1-day moving average).

20°C, which increased gradually to about 40°C in 4 months. The average RH was about 60% at the time of casting and decreased to less than 20% after 4 months.

2.3 *Specimens and instrumentation*

Six slabs on ground specimens were cast in the field to study the response under ambient environment. Specimens measuring 6000 × 1100 × 200 mm^3 (Figure 2) were cast on a 100 mm lean concrete subbase resting on compacted soil subgrade in the field station, as shown in Figure 3. Two slabs were reinforced with ribbed GFRP bars spaced at 200 mm (PG-200) and 300 mm (PG-300), two specimens with sand-coated GFRP bars spaced at 200 mm (GG-200) and 300 mm (GG-300), one slab with conventional steel bars spaced at 200 mm (S-200) and one was unreinforced slab (PL). For all reinforced slabs, the clear cover to the rebars was 50 mm from the top of specimens, as per ACI 360R-10 (ACI 2010). These specimens were reinforced in both longitudinal and transverse directions.

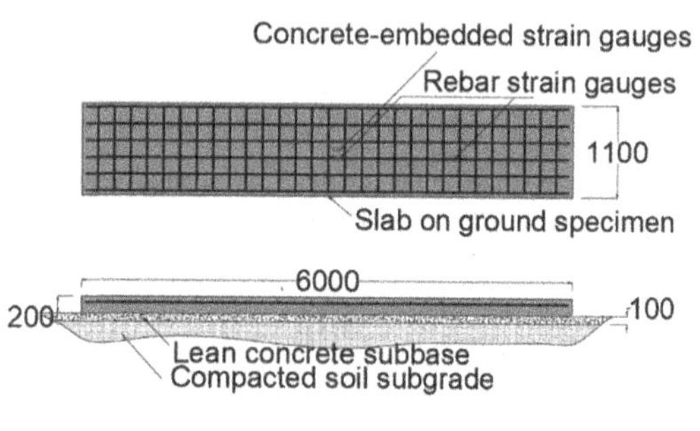

Figure 2. Schematic of slab on ground specimen.

Strain gauges were attached to the reinforcing bars at selected locations of the slab specimens, such as mid-span and quarter-span to monitor the strain under environmental loadings. Strain gauges were also embedded in concrete slab at rebar level at midspan. The cables from the strain gauges were routed to a monitoring room in the field station near the slab specimens. The cables were connected to a data logger to automatically record and monitor the data. Ambient temperature and relative humidity were also recorded using a field weather station (Figure 3).

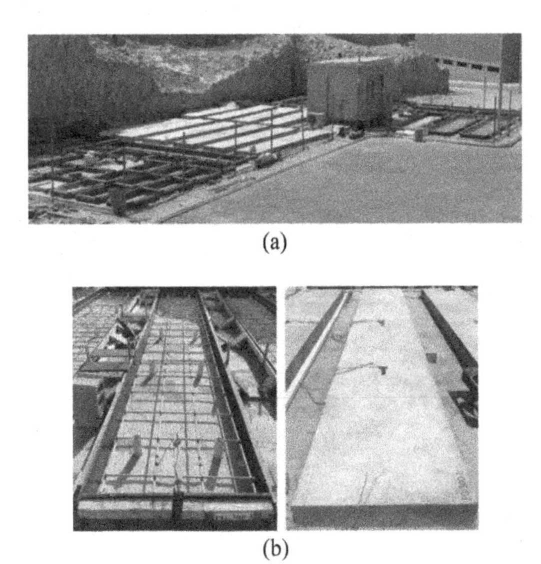

Figure 3. Slab on ground specimens: a) field station; and b) PG-300 specimen: erected reinforcement cage (left) and cast specimen (right).

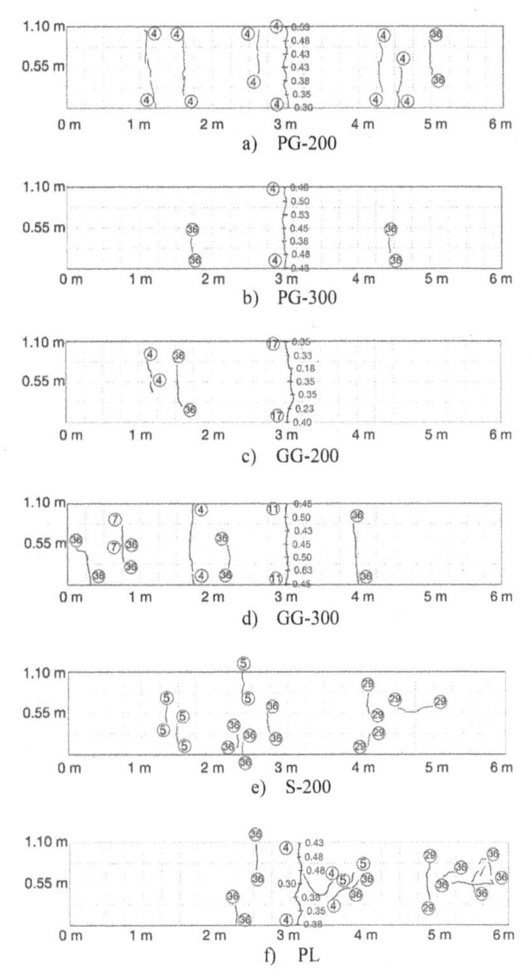

Figure 4. Crack mapping of field specimens.

3 OBSERVATIONS & DISCUSSION

3.1 *Crack mapping*

Figures 4a-4f show the evolution of the cracks in the grade supported slab specimens. The numbers in the circle mark the day the crack occurred. All specimens exhibited cracks varying in locations, lengths, time of occurrence and widths. The four GFRP reinforced slabs and the normal plain concrete slab exhibited a prominent full-depth central drying shrinkage crack. Minor surface cracks, not more than 0.1 mm wide and 0.1 mm deep, were developed at different ages at several other locations along the length of the slab as seen in the figures.

The full-depth transverse central crack developed across the width in the slab with ribbed GFRP bars spaced at 200 mm c/c, had an average crack width of 0.41 mm, while for the slab with bars spaced at 300 mm c/c it was 0.46 mm. The smaller spacing resulted in the development of multiple cracks, while only one crack was observed at 300 mm spacing c/c.

In contrast to the ribbed GFRP bars, the sand-coated GFRP bar reinforced slabs showed an opposite pattern in terms of the number of cracks but the crack width was smaller for the slab with GFRP bars spaced at 200 mm c/c (0.31 mm) than the slab with GFRP bars spaced at 300 mm c/c (0.49 mm).

The steel reinforced slab (Figure 4d), however, did not develop any full-depth central crack. Multiple fine cracks of different lengths are observed in the central portion of the slab. The steel reinforcement effectively distributed the cracks preventing the formation of a wide central crack. The unreinforced concrete slab however, developed a wide full-depth central crack of width 0.46 mm, and had multiple random cracks.

According to ACI Committee 224's guide to reasonable crack width (ACI 2008), the maximum allowable crack width in steel-reinforced concrete slabs for humid, moist air and soil exposure condition is 0.3 mm. ACI-440.1R-15 (ACI 2015) limits the maximum crack width to 0.7 mm in grade-supported GFRP bar reinforced slabs. Higher crack widths are permitted in view of the non-corroding GFRP bars.

3.2 *Strain in the reinforcing bars and concrete*

The total strains recorded for 50 days on the longitudinal reinforcing bar at the middle of the six slabs are shown in Figure 5. The first four days of the plots show expansion due to swelling that occurs in the concrete when the slabs are subjected to water curing (Kucharczyková et al. 2017). When the slabs were exposed to the environment, drying shrinkage ensued

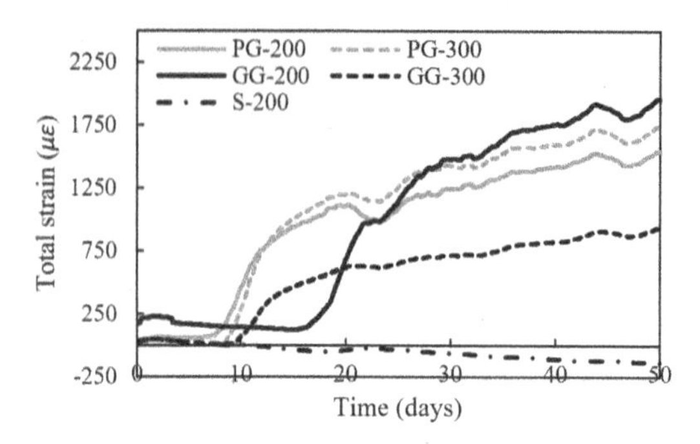

Figure 5. 3-days moving average plot of longitudinal rebar strains up to 50 days.

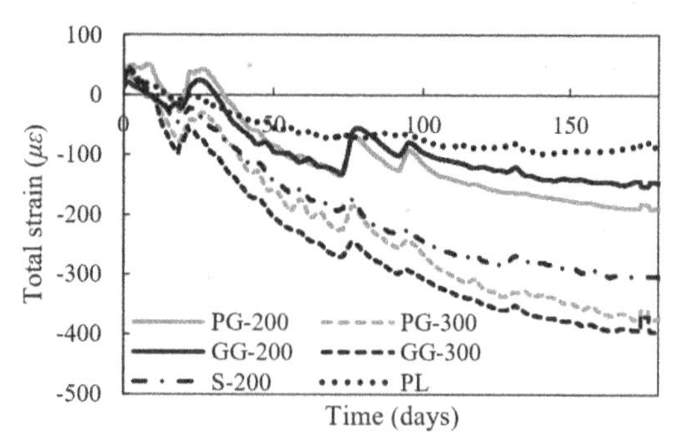

Figure 6. Evolution of concrete stresses at rebar level at mid-span of specimen in the longitudinal direction up to 180 days.

and cracks were developed in the concrete slab and tensile stresses were generated in the reinforcing bars. The steep increase in tensile strain in the rebars occurs when the wide central cracks develop in the slabs. The onset of cracking in the PG-200 and PG-300 slabs occurred at 8 and 10 days, respectively, while cracking in the GG-300 slab occurred at 11 days. On the other hand, for the GG-200 slab cracking occurred at 17 days. The strain in slab reinforced with steel bars, was compressive during the first 60 days and no cracks were observed in the vicinity of the instrumented rebar. After 150 days, a cracked developed near the middle of the slab and tensile stress was measured.

Table 2 shows the maximum strains measured in the rebars of the experimental slabs during the six months and the calculated stresses. Higher strains with a corresponding lower stress developed in GFRP bars compared to the steel rebars due to its lower elastic modulus.

Table 2. Peak strains in rebars within 180 days.

Specimen	Maximum strain ($\mu\varepsilon$)	Day (days)	Calculated stress (MPa)	% of ultimate strength
PG-200	+1816	142	89.0	8.5
PG-300	+2040	142	100.0	9.6
GG-200	+2401	142	104.4	10.1
GG-300	+1374	166	59.8	5.8
S-200	+774	158	134.7	23.4

The measured total strain in the concrete at the center of the slabs in the longitudinal direction, near the rebars at the rebar level, is shown in Figure 6. The maximum total strain in concrete occurs in sand-coated (GG-300) and ribbed (SS-300) GFRP bars spaced at 300 mm c/c (-375 $\mu\varepsilon$ and -395 $\mu\varepsilon$), which decreases significantly for a spacing of 200 mm c/c (-150 $\mu\varepsilon$ and -200 $\mu\varepsilon$). In steel bar reinforced slab, the measured concrete strain is -310 $\mu\varepsilon$. The strain in plain concrete slab is the lowest, which could be due to the misalignment of the embedded strain gauge. The higher concrete strain in slabs with GFRP bars spaced at 300 mm c/c is due to the lower restraint to shrinkage as compared to the slabs with bars spaced at 200 mm c/c.

4 NUMERICAL SIMULATION

4.1 *Model properties*

Numerical finite element (FE) simulations of the slabs monitored in the experimental program were performed with the widely used and powerful commercial software ABAQUS/CAE 2020 (Elchalakani et al. 2018; Abed et al. 2021). The standard module in ABAQUS, typically used for static and low-speed dynamic processes was used. The concrete in the model was assigned as a solid homogeneous material with the concrete damage plasticity model and C3D8R 8-noded linear brick elements with reduced integration. The rebars were modeled as wires with T3D2 2-noded linear displacement truss element assumed to be embedded in concrete using the embedment region constraint. GFRP bar was modelled to be elastic-brittle material with elastic modulus of 49 GPa and ultimate strength of 1046 MPa. The mechanical properties of concrete and GFRP reinforcements adopted in the model are tabulated in Table 3.

ABAQUS does not have the built-in functionality for modeling time-dependent phenomenon such as shrinkage, creep, and thermal loads. For modeling the time-dependent shrinkage in the slabs on grade, two user-defined subroutines were developed in FORTRAN (F95). The subroutine, USDFLD to define field variables and UEXPAN for calculating the shrinkage

Table 3. Mechanical properties of materials for numerical simulation.

Material	Density (kg/m^3)	Poisson's ratio, ν	Young's modulus, E_c (MPa)	Ultimate strength (MPa)
Concrete	2440	0.18	29799	40.2 (comp.) 3.1 (tens.)
GFRP	2200	0.25	49000	1046

strains, which calculate the strains at the beginning of each incremental time step and pass them to the implicit solver of ABAQUS.

Figure 7 shows the constitutive relationships for concrete in compression and tension and the GFRP bars used in the numerical study.

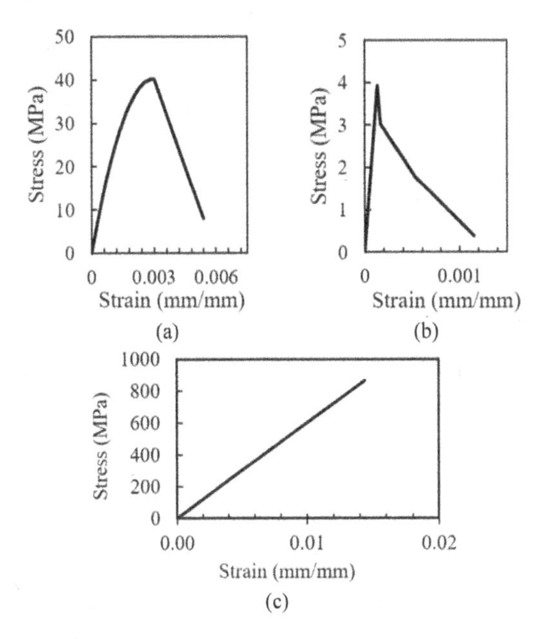

Figure 7. Constitutive relations: a) Concrete in compression; b) concrete in tension; and c) GFRP.

4.2 *Concrete damage plasticity model*

The concrete damage plasticity model in ABAQUS, requires data of compressive stress and inelastic strain. The stress-strain relationship captures the compression behavior starting from the inelastic rising curve, the ultimate stress (peak) and the post-peak response. The present study uses the Kent & Park (1975) model for concrete. The model consists of two parts: a nonlinear curve up to the peak stress at 0.002 mm/mm strain, and beyond this strain, a linear descending path up to failure. In order to simulate the tension behavior of reinforced concrete, the stress-strain relationship proposed by Wahalanthantri et al. (Wahalathantri et al. 2011), which is a modified version of Nayal and Rasheed's model (Nayal & Rasheed 2006) was used. This model efficiently accounts for the interaction between reinforcement and concrete, tension stiffening and strain softening.

4.3 *Concrete slab and subbase interaction*

The contact behavior between the slab and the lean concrete subbase was modelled to have a surface-based cohesive behaviour with the traction-separation behavior defined by the stiffness coefficients k_{nn}, k_{ss} and k_{tt}. The relationship between the uncoupled stiffness coefficients k_{nn}, k_{ss} and k_{tt} given by Henriques et al. (Henriques et al. 2013) as $k_{ss} = k_{tt} = \tau_m / s_m$ and $k_{nn} = 100k_{ss} = 100k_{tt}$, where τ_m is the maximum bond strength and s_m is the corresponding slip.

4.4 *Shrinkage modeling*

The focus of the numerical analysis is to develop a finite element model capable of predicting shrinkage induced stresses, strains, and damage in the concrete slab at any given point of time. The free-shrinkage strain data for the concrete used in the experimental program was obtained from concrete prisms placed at the site instrumented with embedded strain gauges. ACI 209 shrinkage model was then fitted to match the free-shrinkage strain extracted from the experimental model. The ACI 209 shrinkage curve is a function of ultimate free shrinkage strain ε_{shu}, constant f, shape function α and time from the end of initial curing $(t - t_c)$ as shown in Equation 1.

$$\varepsilon_{sh}(t, t_c) = \frac{(t - t_c)^{\alpha}}{f + (t - t_c)^{\alpha}} \varepsilon_{shu} \tag{1}$$

Figure 8 shows the plot of experimental free shrinkage strain curve and the ACI 209 equation fitted to the experimental data. Based on the factors f, α and ε_{shu}, the shrinkage strain function for the UEXPAN subroutine was developed.

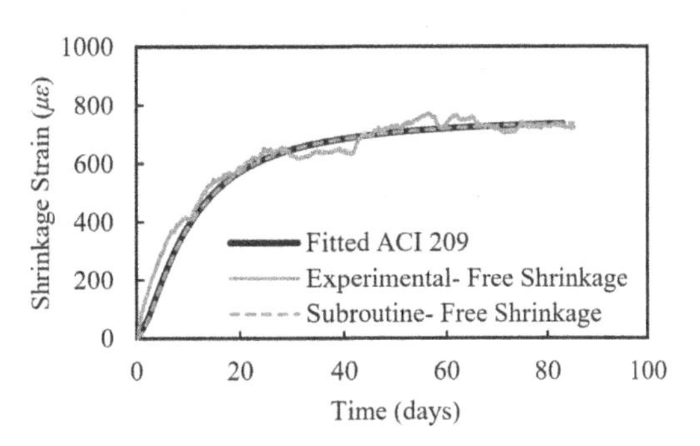

Figure 8. Free shrinkage curves.

4.5 *Calibration of numerical model*

The restraint provided to the concrete slab by the lean concrete subbase has a strong influence on the development of tensile stresses at the top of the slab. The tensile stress, when it exceeds the tensile strength of concrete, leads to the formation of cracks. Various values of the uncoupled stiffness coefficients k_{nn}, k_{ss} and k_{tt} were employed in the finite element model and the cracking patterns in the slab was investigated. DAMAGET (tensile damage, d_t) option in the concrete damage plasticity model was used to define post-cracking stiffness degradation. Based on several iterations performed with several values of k_{nn}, k_{ss} and

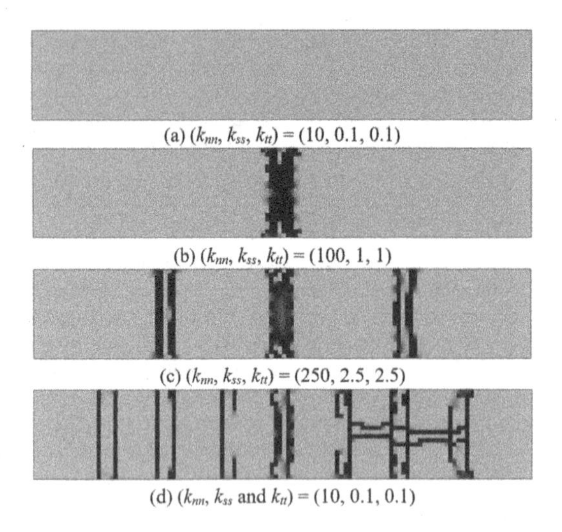

(a) $(k_{nn}, k_{ss}, k_{tt}) = (10, 0.1, 0.1)$

(b) $(k_{nn}, k_{ss}, k_{tt}) = (100, 1, 1)$

(c) $(k_{nn}, k_{ss}, k_{tt}) = (250, 2.5, 2.5)$

(d) (k_{nn}, k_{ss} and k_{tt}) = (10, 0.1, 0.1)

Figure 9. Calibration of model using concrete slab-subbase iteration stiffness parameters.

k_{tt}, as shown in Figure 9, the development of crack patterns was observed.

Based on the above simulations, k_{nn}, k_{ss} and k_{tt} values of (100, 1, 1) closely resembled the experimental observations, in terms of time of appearance of midspan crack and crack pattern. Thus, further numerical parametric studies were done using these values.

4.6 *Parametric studies*

Effects of parameters such as concrete tensile strength, diameter of GFRP bars, and thickness of the slab on cracking and rebar stresses were investigated based on the finite element model developed.

4.6.1 *Tensile strength*

The effect of concrete tensile strength (2 MPa and 4 MPa) are shown in Figure 10. For lower concrete tensile strength, a central crack appeared on the 6th day, followed by cracks at quarter spans on 16th and 19th days. However, for a 4 MPa-tensile strength concrete, only a central crack was developed on the 11th day. Figure 11 shows the stresses developed in a central longitudinal GFRP bar concrete on the 90th day. Although the concrete with a tensile strength of 2 MPa cracked at three locations, the stresses were less than

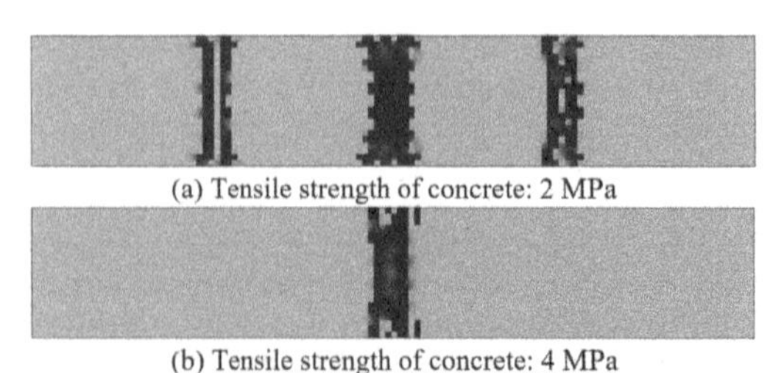

(a) Tensile strength of concrete: 2 MPa

(b) Tensile strength of concrete: 4 MPa

Figure 10. Effect of concrete strength on cracking.

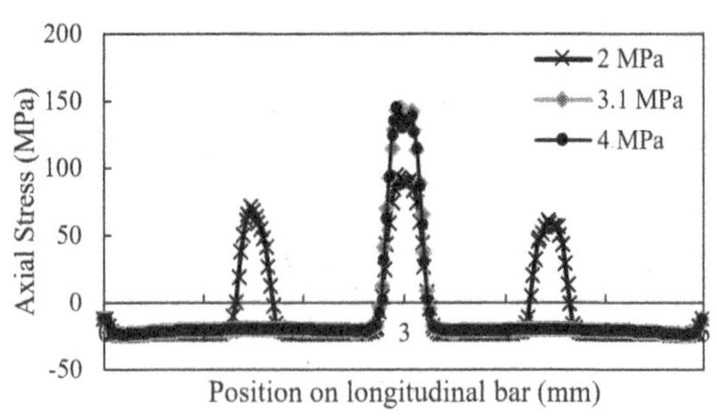

Figure 11. Effect of concrete tensile strength on axial stresses in central longitudinal rebars.

100 MPa, while the stresses were approximately the same at higher concrete strengths (3.1 and 4 MPa). This suggests that the crack width may be wider in higher tensile strength concretes. The stresses developed in higher strength concrete are only up to 14% of the tensile strength of the GFRP bar.

4.6.2 *Bar diameter*

No changes in the crack pattern of the slab were observed when the diameter of the embedded GFRP bars was increased from 12 mm to 16 mm (Figure 12). Cracks in both slabs developed on 8th day from the commencement of shrinkage. The axial stress in the 12 mm dia. GFRP bar was higher compared to the 16 mm diameter bars in the vicinity of the crack (Figure 13). Thus, changing the bar diameter does not have significant effect in terms of crack development.

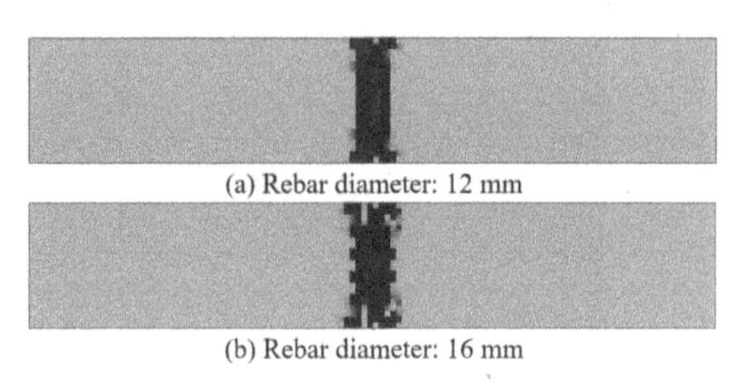

(a) Rebar diameter: 12 mm

(b) Rebar diameter: 16 mm

Figure 12. Effect of rebar diameter on cracking.

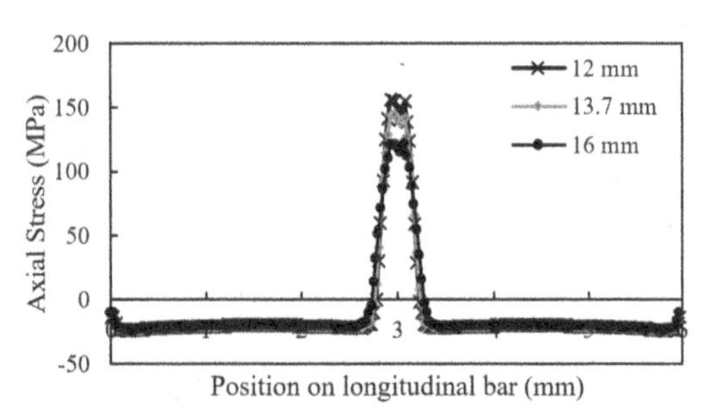

Figure 13. Effect of GFRP rebar diameter on axial stresses in central longitudinal rebars.

4.6.3 *Slab thickness*

The effect of slab thickness on the development of shrinkage cracks was investigated. As shown in Figure 14, the 150 mm thick slab showed cracking in the center on day 7 and in the quarters on days 26 and 53. The 250 mm thick slab, on the other hand, showed only one crack at the center on the day 10. Axial stresses in the central longitudinal bar in 150 mm-thick slab peaks at about 110 MPa at midspan and about 72 MPa at quarter spans, while the 250 mm thick slab developed stresses of approximately 145 MPa at midspan (Figure 15).

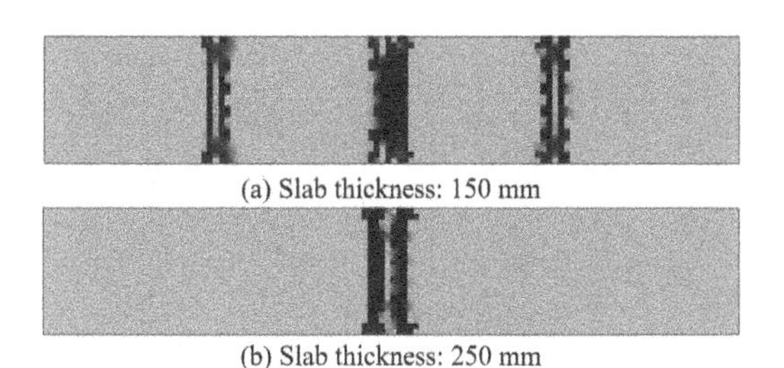

(a) Slab thickness: 150 mm

(b) Slab thickness: 250 mm

Figure 14. Effect of slab thickness on cracking.

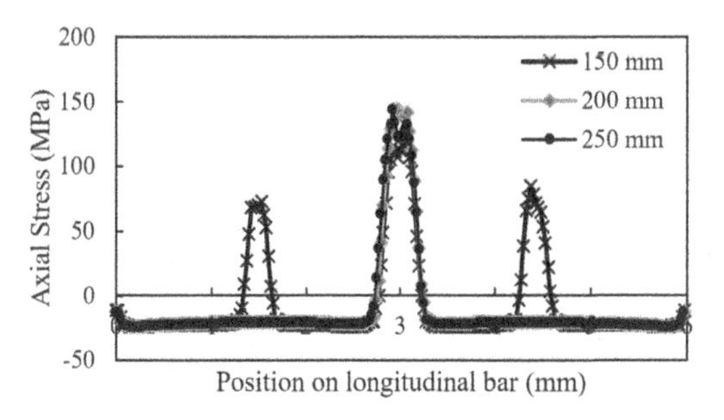

Figure 15. Effect of slab thickness on axial stresses in central longitudinal rebars.

5 CONCLUSIONS

Field monitoring of GFRP bar reinforced slabs on ground provided valuable information on the initiation and evolution of cracks and shrinkage strains in concrete and reinforcements. Slabs with larger bar spacing exhibited larger crack widths. Finite element model of the slab captured the experimental response and predicted the strains and onset of cracking with good accuracy. Parametric investigations showed that the concrete tensile strength and slab thickness influences the number, width, and age at which cracks are developed. Changing bar diameter from 14 to 16 mm did not affect the cracking pattern. Further studies are in progress.

REFERENCES

Abed, F., Oucif, C., Awera, Y., Mhanna, H. H. & Alkhraisha, H. 2021. FE modeling of concrete beams and columns reinforced with FRP composites. *Defence Technology* 17(1): 1–14.

ACI (American Concrete Institute) 2008. *Control of Cracking in Concrete Structures, ACI Manual of Concrete Practice (ACI 224R-01).* Farmington Hills, USA.

ACI (American Concrete Institute) 2015. *Guide for the design and construction of structural concrete reinforced with fiber-reinforced polymer (FRP) bars (ACI 440.1R-15).* Farmington Hills, USA.

ACI (American Concrete Institute) 2010. *Guide to design of slabs-on-ground (ACI 360R-10).* Farmington Hills, USA.

Benmokrane, B., Eisa, M., El-Gamal, S., El-Salakawy, E. & Thebeau, D. 2008. First Use of GFRP Bars as Reinforcement for Continuous Reinforced Concrete Pavement. *Fourth International Conference on FRP Composites in Civil Engineering (CICE 2008), 22–24 July 2008.* Zurich, Switzerland.

Chen, H. R. & Choi, J., 2011. Analysis of Shrinkage and Thermal Stresses in Concrete Slabs Reinforced with GFRP Rebars. *Journal of Materials in Civil Engineering* 23: 612–627.

Choi, J. & Chen, H. R. 2015. Design of GFRP reinforced CRCP and its behavior sensitivity to material property variations. *Construction and Building Materials* 79: 420–432.

Elchalakani, M., Karrech, A., Dong, M., Mohamed Ali, M. S. & Yang, B., 2018. Experiments and Finite Element Analysis of GFRP Reinforced Geopolymer Concrete Rectangular Columns Subjected to Concentric and Eccentric Axial Loading. *Structures* 14: 273–289.

Henriques, J., Simões da Silva, L. & Valente, I. B., 2013. Numerical modeling of composite beam to reinforced concrete wall joints. Part I: Calibration of joint components. *Engineering Structures* 52: 747–761.

Kucharczyková, B., Danik, P., Kocáb, D. & Misák, P., 2017. Experimental Analysis on Shrinkage and Swelling in Ordinary Concrete. *Advances in Materials Science and Engineering* 2017: 3027301.

Nayal, R. & Rasheed, H. A., 2006. Tension stiffening model for concrete beams reinforced with steel and FRP bars. *Journal of Materials in Civil Engineering* 18(6): 831–841.

Wahalathantri, B. L., Thambiratnam, D. P., Chan, T. H. T. & Fawzia, S., 2011. A material model for flexural crack simulation in reinforced concrete elements using ABAQUS. *First International Conference on Engineering, Design and Developing the Built Environment for Sustainable Wellbeing (eddBE2011), 27–29 April 2011.* Brisbane, Australia.

Computational Modelling of Concrete and Concrete Structures – Meschke, Pichler & Rots (Eds)
 ISBN: 978-1-032-32724-2

Validation of reinforced concrete pile caps using non-linear finite element analysis and finite element limit analysis

M.E.M. Andersen & T.W. Jensen
Department of Bridges International, COWI A/S, Kongens Lyngby, Denmark

P.N. Poulsen, J.F. Olesen & L.C. Hoang
Department of Civil Engineering, The Technical University of Denmark, Kongens Lyngby, Denmark

ABSTRACT: Analysis methods based on the rigid-plastic material models, such as the strut-and-tie method (STM), are often used to validate the designs of solid reinforced concrete structures in the ultimate limit state. However, the validation can be quite cumbersome since it involves much manual labor. Another analysis method based on the rigid-plastic material model is Finite Element Limit Analysis (FELA). The workflow when using FELA is easily automatized since the method is fully numerical. The capacity, stress fields, and collapse mode of the structure are the results that can be obtained from a FELA analysis. Another fully numerical method for validating solid reinforced concrete structures is the Non-Linear Finite Element Method (NLFEM). Using advanced material models, NLFEM programs such as DIANA FEA can accurately describe the structural behavior of reinforced concrete structures, even post the peak load. However, this modeling precision comes at the cost of increased complexity, and many material parameters are required for the models. This trade-off between precision and complexity in NLFEM is in contrast to FELA, which requires very few material parameters but only provides information about the structure at peak load. In this paper, the two methods are briefly introduced, whereafter they are compared by analyzing two four-pile cap experiments. Results obtained from the two models are presented and compared both to each other and to the experimental results. At the end, conclusions about the strengths and weaknesses of the two types of analysis are drawn.

1 INTRODUCTION

Design and validation of pile caps for the ultimate limit state are often performed using the Strut-and-Tie Method (STM). This method is based on the lower bound theorem for rigid-plastic materials and thus provides safe designs. However, STM can be inefficient to use, especially when many load cases and different geometries need to be considered. Numerical rigid-plastic limit analysis can be performed using Finite Element Limit Analysis (FELA) (Anderheggen & Knöpfel 1972). The method utilizes a rigid-plastic material model and is, similarly to STM, based on the extremum principles for rigid-plastic materials (Drucker, Prager, & Greenberg 1952; Gvozdev 1960). FELA has been known since the seventies and has been used for geotechnical calculations for several years. Recently, FELA has become popular in reinforced concrete design and has been applied to slab structures (Jensen, Poulsen, & Hoang 2018), wall structures (Poulsen & Damkilde 2000; Herfelt 2017), and solid structures (Vincent, Arquier, Bleyer, & de Buhan 2018; Vincent, Arquier, Bleyer, & de Buhan 2020; Andersen, Poulsen, & Olesen 2021). FELA has reached a mature state for plane reinforced concrete structures, and the first commercial programs are now in use to design and evaluate these structures (Herfelt, Krabbenhøft, & Krabbenhøft 2019).

A requirement for the use of FELA is that the structure should have sufficient deformation capability to enable redistributing of stresses and the development of the predicted collapse mechanism. That is, the structures should exhibit a sufficiently ductile behavior. Ductility is normally achieved by ensuring that yielding of the reinforcement is governing for the capacity, rather than crushing of concrete or rupture of reinforcement. Plane structures that are designed using codes and guidelines possess this ductility provided by minimum reinforcement. For solid reinforced concrete structures, which often do not have minimum reinforcement, the presence of sufficient ductility is not always guaranteed. The presence of the required ductility can be shown by full-scale testing. However, this is not feasible in practice. Instead, advanced non-linear finite element (NLFEM) calculations can be performed, giving a more realistic prediction of the behavior of the structure before the collapse.

This paper studies a series of experiments of reinforced concrete four-pile caps subjected to central

DOI 10.1201/9781003316404-37

compressive loading (Miguel-Tortola, Miguel, & Pallarés 2019). The pile caps are analyzed using FELA and NLFEM. In the FELA model, constant stress elements with normal traction continuity on inter-element surfaces and shear equilibrium in the nodes are used (Andersen, Poulsen, & Olesen 2022). The Modified Mohr-Coulomb yield criterion is used for the concrete with a possible inclusion of the effectiveness factor on the concrete compressive strength and with the inclusion of smeared reinforcement. The NLFEM is modeled in DIANA FEA (DIANA 2017), where a smeared cracking approach is used for concrete in tension, and lateral cracking is considered for reduction of the concrete compressive strength. The material parameters for FELA and NLFEM are based on model codes and guidelines for an objective comparison. The results obtained using the two numerical methods are compared to experimental results, and conclusions are drawn concerning the usefulness of FELA compared to NLFEM for analysis of pile caps.

2 FINITE ELEMENT LIMIT ANALYSIS

Finite Element Limit Analysis (FELA) combines the domain discretization of the finite element method with limit analysis based on a rigid-plastic material model. FELA was first suggested by Anderheggen & Knöpfel (1972) for reinforced concrete membranes and plates and are now used to analyze both geotechnical and reinforced concrete structures in the ultimate limit state.

Limit analysis with a rigid-plastic material model was developed independently by Gvozdev (1960) and Drucker, Prager, & Greenberg (1952) who formulated the extremum principles of rigid-plastic materials. These extremum principles are the lower bound theorem, the upper bound theorem, and the uniqueness theorem. For a thorough review, see (Nielsen & Hoang 2011).

FELA can be based on either the lower bound theorem where a statically admissible stress state is sought or on the upper bound theorem where a kinematically admissible collapse mode is sought. The method used in this paper is based on the lower bound theorem but in a relaxed manner. Consequently, a lower bound on the failure load is not guaranteed. However, the solution will still converge towards the failure load.

In FELA, the structure is divided into several stress-based finite elements. The element which is used is the Normal Traction element (Andersen, Poulsen, & Olesen 2022), which is a partially mixed lower bound element. The element has a constant stress field and thus only one stress point per element. Each stress point corresponds to a full triaxial stress state:

$$\boldsymbol{\sigma}^{\top} = \left[\sigma_{xx}\ \sigma_{yy}\ \sigma_{zz}\ \sigma_{xy}\ \sigma_{xz}\ \sigma_{zy}\right] \tag{1}$$

The stress state of the element needs to be in equilibrium with that of neighboring elements through the force equilibrium of corner nodes and traction equilibrium of the faces.

The shear stress contributions are placed in the corner nodes. The equilibrium for a single node is given by:

$$\sum_{i=1}^{m} Q_{i,n}(\boldsymbol{\sigma}_i) = P_n \tag{2}$$

where $Q_{i,n}$ is the force contribution in the node from element i, which is a function of the stress state of element i, and P_n is the external loading of the node. The subscript n, denotes either the x, y, or z direction. Similarly, the traction equilibrium for an interface between two elements is given by:

$$t_{1,n}(\boldsymbol{\sigma}_1) + t_{2,n}(\boldsymbol{\sigma}_2) = q_n \tag{3}$$

where $t_{1,n}$ and $t_{2,n}$ is the traction for the elements on either side of the interface, and q_n is a traction load on the interface. Here the subscript n denotes the normal direction on the interface since the element only considers strict traction continuity for the normal traction.

The stress point vectors are collected into the system stress vector, $\boldsymbol{\beta}^{\top} = \left[\boldsymbol{\sigma}_1^{\top}\ \dots\ \boldsymbol{\sigma}_n^{\top}\right]$. The stress continuity between the elements and the elements and the loading is ensured via the equilibrium matrix $\mathbf{H}$, which has contributions from each of the elements. The equilibrium matrix, $\mathbf{H}$, multiplied with the system stress vector, $\boldsymbol{\beta}$, should be in equilibrium with the constant loads $\mathbf{R}_0$ and scalable loads $\mathbf{R}\lambda$, where λ is the so-called load-factor, which is sought to be maximized. Furthermore, the stress state in each element should abide by the employed yield criteria. Combined, this gives an optimization problem on the form:

max.	λ	Load	(4a)
s.t.	$\mathbf{H}\boldsymbol{\beta} = \mathbf{R}_0 + \lambda\mathbf{R}$	Stress equilibrium	(4b)
	$f_i(\boldsymbol{\sigma}_i) \leq 0$	Yield conditions	(4c)

The optimization problem above is convex and can to be efficiently solved when certain criteria are met. For the present case the yield conditions (4c) are modeled as semidefinte constraints, and these coupled with an affine objective function (4a) and affine equality constraints (4b) are indeed convex (Boyd & Vandenberghe 2004).

In the following the basis of the stress equilibrium (4a) and the yield conditions (4c) will be briefly elaborated. However, for a thorough examination of the model see Andersen, Poulsen, & Olesen (2022).

2.1 Stress equilibrium

The element used for the FELA analysis of this paper is the Normal Traction element. A sketch of the element can be seen in Figure 1. The element is a so-called partially mixed lower bound element. *Partially* mixed

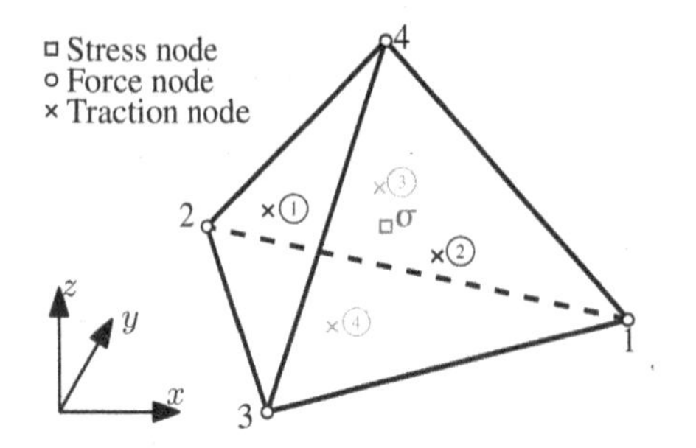

Figure 1. Configuration of the Normal Traction element (Andersen, Poulsen, & Olesen 2022).

because it has strict normal traction continuity on the element faces, but only a relaxed shear traction continuity moved unto the corner nodes as a force equilibrium.

The element contains only one stress node and consequently only one material point, which means that each element has a computational requirement of only one-third of a regular linear stress element. This reduced computational cost per element is desirable when large solid geometries must be modeled, requiring many elements to mesh adequately. Furthermore, the element has proved to have fast convergence (Andersen, Poulsen, & Olesen 2022).

2.2 *Reinforced concrete yield condition*

The stress state of each element needs to abide by a yield condition. In this paper, a yield condition for reinforced concrete based on a separation of stresses into concrete stresses and reinforcement stresses is utilized (here shown as tensors):

$$\boldsymbol{\sigma}_{\square} = \boldsymbol{\sigma}_{\square,c} + \boldsymbol{\rho\sigma}_{\square,s} \tag{5}$$

where $\boldsymbol{\sigma}_{\square}$ is the total stress tensor, $\boldsymbol{\sigma}_{\square,c}$ is the concrete stress tensor given by:

$$\boldsymbol{\sigma}_{\square,c} = \begin{bmatrix} \sigma_{c,xx} & \sigma_{c,xy} & \sigma_{c,xz} \\ \sigma_{c,xy} & \sigma_{c,yy} & \sigma_{c,zy} \\ \sigma_{c,xz} & \sigma_{c,zy} & \sigma_{c,zz} \end{bmatrix} \tag{6}$$

and $\boldsymbol{\rho\sigma}_{\square,s}$ is the smeared reinforcement stress tensor given by:

$$\boldsymbol{\rho\sigma}_{\square,s} = \begin{bmatrix} \rho_x & 0 & 0 \\ 0 & \rho_y & 0 \\ 0 & 0 & \rho_z \end{bmatrix} \begin{bmatrix} \sigma_{s,xx} & 0 & 0 \\ 0 & \sigma_{s,yy} & 0 \\ 0 & 0 & \sigma_{s,zz} \end{bmatrix} \tag{7}$$

where $\boldsymbol{\rho}$ contains the reinforcement degree in each of the three normal directions, and $\boldsymbol{\sigma}_{\square,s}$ is the stress in the reinforcement. As can be seen this assumes that the reinforcement is placed in accordance with the *xyz*-coordinate system, and that only normal stresses can be carried by the reinforcement. This separation of stresses is analogue to the way the Nielsen yield criterion is developed (Nielsen & Hoang 2011).

2.2.1 *Concrete yield criterion*

For the concrete stresses the Modified Mohr-Coulomb yield criterion is used. The yield criterion consist of a friction and a separation criterion and are cast in principal stresses:

$$\sigma_1 \leq \nu_t f_t \tag{8a}$$

$$k\sigma_1 - \sigma_3 \leq \nu f_c \tag{8b}$$

here σ_1 and σ_3 is the largest and smallest principal stress respectively, k is the friction coefficient usually taken as 4, and $\nu_t f_t$ and νf_c are the effective tensile and compressive strength, respectively.

The effective tensile strength is chosen as a small fraction of the effective compressive strength, $\nu_t f_t = \nu f_c / C \approx 0$, where C is a number larger than 1000. The small value of the tensile strength has a negligible influence on the capacity, but improves the quality of the dual variables of the equilibrium equations (4b). The dual variables can be interpreted as the collapse mechanism of the structure (Poulsen & Damkilde 2000).

The conditions of (8a) and (8b) can be cast as a set of two semidefinite constraints with two additional linear constrains and auxiliary variables (Larsen 2010; Martin & Makrodimopoulos 2008; Krabbenhøft, Lyamin, & Sloan 2008; Bisbos & Pardalos 2007):

$$\begin{aligned} \boldsymbol{\sigma}_{\square,c} + (k\alpha_1)\mathbf{I} &\preceq 0 \\ \boldsymbol{\sigma}_{\square,c} - \alpha_2 \mathbf{I} &\succeq 0 \\ \alpha_2 &\leq \nu_t f_t \\ \alpha_1 + \alpha_2 &\leq \nu f_c / k \end{aligned} \tag{9}$$

where $\mathbf{I}$ is the identity matrix of order 3, α_1 and α_2 are auxiliary variables, and the symbols $\succeq 0$ and $\preceq 0$ designate positive and negative semi-definiteness, respectively.

2.2.2 *Reinforcement yield conditions*

The reinforcement stresses are constrained using a set of simple linear constraints on the form:

$$-f_{s,c} \leq \sigma_{s,xx} \leq f_s \tag{10a}$$

$$-f_{s,c} \leq \sigma_{s,yy} \leq f_s \tag{10b}$$

$$-f_{s,c} \leq \sigma_{s,zz} \leq f_s \tag{10c}$$

where $f_{s,c}$ is the reinforcement compressive yield strength, which is set to 0, and f_s is the reinforcement tensile yield strength.

3 NON-LINEAR FINITE ELEMENT ANALYSIS

The commercial program DIANA FEA (DIANA 2017) is used to perform the analysis.

The total strain crack model in DIANA FEA is used, which is based on the modified compression theory

(Vecchio & Collins 1986). The reinforcement bars are modeled with embedded truss elements with bond slip.

The concrete in the model is discretized with solid hexahedron elements with 20 nodes and a quadratic displacement interpolation. The average side length of the element is 50 mm. The reinforcement bars are modeled with embedded truss elements and connected to the solid elements with line interface elements to enable modeling of the bond-slip between the reinforcement and the concrete.

The equilibrium between the external and internal forces is achieved iteratively with the Newton-Raphson method and line-search. For the convergence criterion, the energy norm is chosen with a tolerance of 0.001. The step sizes used in the analyses are between 0:25 mm and 1:00 mm.

3.1 *Material models*

The following material models are used in the NLFEA.

Concrete
The material parameters not presented in the following sections are calculated with the equations stated in fib model code 2010 (Fib 2013), and the guidelines from the Dutch Rijkwaterstaat (Hendriks, de Boer, & Belletti 2017).

The *Total Strain Crack Model* in DIANA FEA is used, which is based on the modified compression theory (Vecchio & Collins 1986). The implementation in 3D is according to Selby (1993). The post-peak tensile behavior is modeled as exponential softening with a crack band based on the individual element size. The damage due to cracking reduces the Poissons ratio at the same pace as the reduction of the secant modulus (see (DIANA 2017)).

The compression behavior is modeled with a parabolic stress-strain relation (Feenstra 1993). The relation is based on Young modulus E_c, the compressive strength f_c, the compressive fracture energy G_c, and the crushing bandwidth h_c (which is equivalent to the crack bandwidth). The compressive fracture energy G_c determine the ductility for the concrete. The guidelines (Hendriks, de Boer, & Belletti 2017) recommend $G_c = 250G_f$, where G_f is the tensile fracture energy. However, this value is based on experiments with concrete strength up to 50 MPa (Nakamura & Higai 2001) and is considerately higher than reported elsewhere (Vonk 1992; Lertsrisakulrat, Watanabe, Matsuo, & Niwa 2001). In (Vonk 1992), the compressive fracture energy is reported as $50G_f - 100G_f$. The influence of the compressive fracture energy on the compressive softening behavior is shown in figure 2. The figure shows that the stress-strain relationship for compressive fracture energy of $50G_f$ results in a significantly more brittle failure when compared to $250G_f$. However, to be consistent with the guidelines, initial compressive fracture energy of $250G_f$ is chosen.

The reduction of the compressive strength due to lateral cracking is considered with model B described in (Vecchio & Collins 1993), as shown in Equation (11).

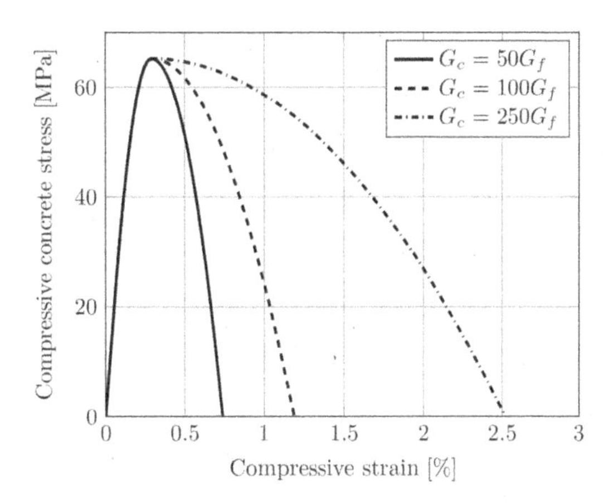

Figure 2. Stress-strain for concrete in compression

In FELA, this effect is taken into account by the effectiveness factor.

$$\beta_{\sigma_{cr}} = \frac{1}{1 + 0.27\left(-\frac{\alpha_{lat}}{\varepsilon_0} - 0.37\right)} \leq 1 \quad (11)$$

where $\alpha_{lat} = \sqrt{\alpha_{l,1}^2 + \alpha_{l,2}^2}$ and $\alpha_{l,1}$ and $\alpha_{l,2}$ are the lateral strains.

Reinforcement
The reinforcement steel is modeled with Von Mises plasticity and a bilinear plastic strain-stress relation for hardning. The simple stress-strain relationship is estimated to have a minor effect on the behaviour.

The bond between the reinforcement and the concrete is modeled with interface elements. The bond-slip material model from fib model code 2010 (Fib 2013) is used where "good bond condition" and pull-out failure are assumed.

4 FOUR-PILE CAP COMPARISON

4.1 *Presentation of specimens*

The two methods are used to analyze some specimens of four-pile caps from an experimental campaign by Miguel-Tortola, Miguel, & Pallarés (2019). The experimental campaign consisted of 21 square four-pile caps with slab width of 1.15 m, height varying from 0:25 m to 0:45 m, varying reinforcement layout, and varying loading setup.

In this paper specimens 4P-N-C2 and 4P-N-C3 are considered. These specimens are loaded by a central normal force and have a height of 0.45 m. The geometry of the four-pile cap can be seen in Figure 3. Both have mesh reinforcement at the bottom face. However, only specimen 4P-N-C3 has shear reinforcement. The reinforcement layouts can be seen in Figure 4. The mesh reinforcement had hooked ends to ensure proper anchorage, which the figure does not show. The number of bars can be seen in Table 1, where A_{sB} is the

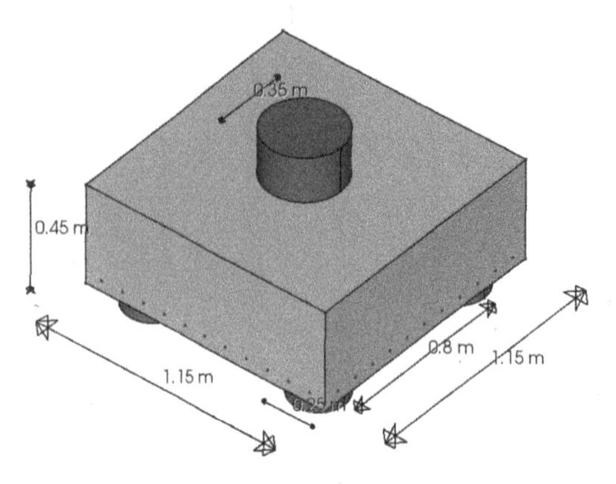

Figure 3. Geometry of the four-pile cap specimens.

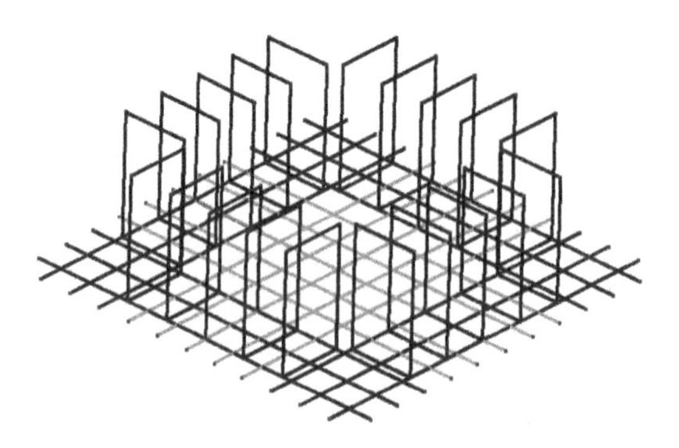

Figure 4. Reinforcement layout of the specimens.

bunched bars over the piles, A_{sH} is the bars between the piles, and A_{sV} is the shear reinforcement. The yield and ultimate strength of the individual bars can be seen in Table 2.

In the experimental setup, the column was fixed, while the load was applied by actuators in each pile to ensure uniform distribution of pile reactions. One actuator was deformation controlled, and the remaining three were synchronized with this and load controlled. The load on the piles was applied through a hinge capable of both rotation and in-plane translation, meaning that only vertical load was applied to the piles.

Table 1 also contains the experimental results of the two specimens with $V_{y,e}$ being the load corresponding to yielding of the bunched reinforcement A_{sB}, $V_{u,e}$ being the ultimate load, and u_z being the vertical deformation at peak load.

Table 2. Material parameters of the reinforcement used in the experiments.

Ø [mm]	f_y [MPa]	f_u [MPa]
8	573.3	650.9
10	519.3	634.7
12	553.8	641.8

4.2 *Description of FELA model*

The FELA calculations are performed using the COWI software package *fela* programmed in Python. The software package uses GMSH as a mesher and Mosek as a solver. The size of the elements in the model can be adjusted by the so-called characteristic length l_c of the elements. The characteristic length is the side-length of the tetrahedral elements, which the mesher aims at providing.

The reinforcement is modeled as smeared reinforcement, as described earlier. The smeared reinforcement approach means that some choices have to be made regarding how detailed the model needs to be. The most detailed model would be to smear out every reinforcement rod independently. However, this creates a model with numerous overlapping solid regions, which would be complicated both to model and mesh. A less detailed way is to separate the reinforcement into groups and model each group as separate smear regions. The less detailed way is often the most practical solution and can adequately capture the effect of the reinforcement in the model. In this paper, the latter method is used. The reinforcement at the bottom face is separated into two groups; the main bending reinforcement over the piles is in one group, and the reinforcement between the piles in another. The reinforcement groups are then further separated for reinforcement in the x- and y- directions, respectively. The stirrups are separated such that each leg of the stirrup, the bottom, and the top part belong to separate groups. The thickness of the smear regions for the bending reinforcement is set to 50 mm, and the thickness of the shear reinforcement regions is set to 25 mm. Figure 5 shows the smeared reinforcement regions of the model for a mesh with $l_c = 50\ mm$.

4.2.1 *Convergence study*

A convergence study is made to make sure that the capacity from the calculations is sufficiently

Table 1. Results from the testing of the two specimens (Miguel-Tortola, Miguel, & Pallarés 2019). * Punching after yielding of bunched reinforcement.

Specimen	f_c [MPa]	f_t [MPa]	A_{sB}	A_{sH}	A_{sV}	$V_{y,e}$ [kN]	$V_{u,e}$ [kN]	u_z [mm]	Failure mode
4P-N-C2	36.3	2.8	4 × (2Ø10 + 1Ø12)	2 × 5Ø8		960.4	1173.9	5.7	Punching*
4P-N-C3	34.0	2.7	4 × (2Ø10 + 1Ø12)	2 × 5Ø8	4 × 5Ø8	1014.1	1317.3	9.7	Flexural

converged. Five computations with characteristic lengths varying between 100 mm and 25 mm is performed. For the convergence study, an effectiveness factor of $\nu = 1.0$ is used. The result of the convergence study can be seen in Figure 6. The solution is converged from the third data point, but the solution precision is adequate from even the first data point. The middle data point of the figure is with a characteristic length of $l_c = 50$ *mm* and has 48610 elements. This mesh is used for the rest of the calculations in this paper.

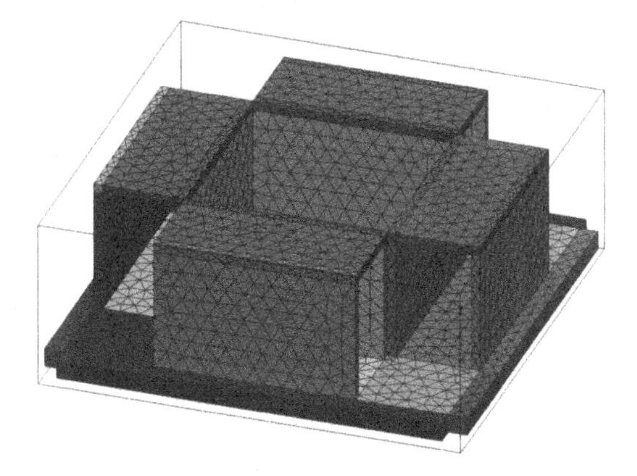

Figure 5. Smeared reinforcement regions for a mesh with characteristic length $l_c = 50$ mm.

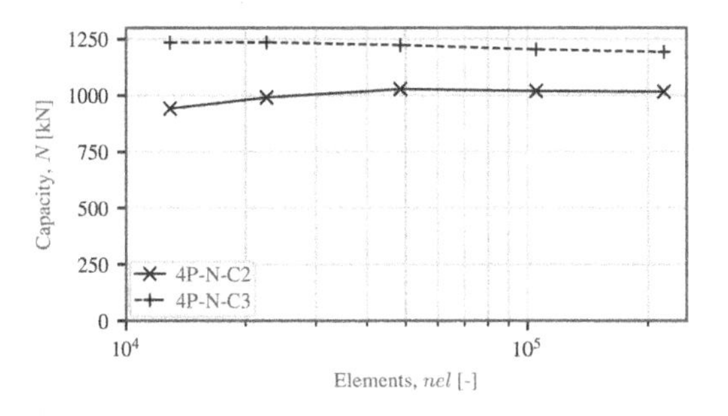

Figure 6. Convergence study for the FELA model.

4.2.2 Influence of effectiveness factor

The influence of the effectiveness factor on the capacity is studied by varying the effectiveness factor between 0.2 and 1.0. A plot of the capacity of the two models as a function of the effectiveness factor can be seen in Figure 7. Interestingly, the effectiveness factor does not significantly influence the capacity unless a very low value below 0.4 is used, which is seldom the case for this type of structure.

This observation indicates that the capacity is mainly limited by yielding of the reinforcement, which would indicate that the failures should be flexural or governed by yielding. This is supported by the failure modes from the experiments, which both had yielding of the primary bending reinforcement before failure (Miguel-Tortola, Miguel, & Pallarés 2019). Consequently, an effectiveness factor of $\nu = 1.0$ is used for the FELA calculations in the remainder of this paper.

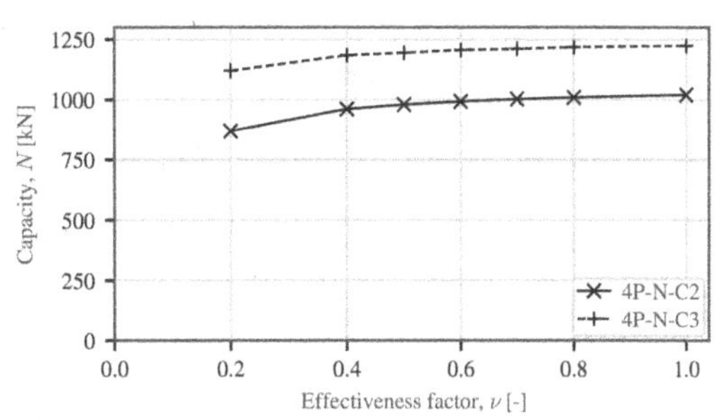

Figure 7. Influence of the effectiveness factor ν on the capacity.

4.3 Description of DIANA model

The material properties, not given in Table 1 and Table 2, are based on (Fib 2013) and the guidelines from the Dutch Rijkwaterstaat (Hendriks, de Boer, & Belletti 2017) to ensure an objective comparison. In Table 2, only f_y and f_u are given as material properties for the reinforcement. Thus, a post-yielding stiffness of $0.01E_s$ is chosen to account for the hardening. After the ultimate capacity, f_u, is reached the stiffness is reduced to zero. The model consist of elements with a size of 0.05 m. To get the post peak, deformation controlled load is applied. The load and support are applied on a load-plate, as shown in Figure 3, which is connected to the concrete in the pile cap with an soft interface.

In the following the results from the two different methods are presented, compared, and discussed. First a comparison of the capacity predicted from the two models are performed, as well as the load displacement plot for the DIANA FEA calculations and the experiments. Thereafter the stress flow from the two models are studied. Finally the predicted collapse from the two models are discussed.

4.4 Results

4.4.1 Capacity and load-deflection

Figure 8 shows the load-displacement plot of the two specimens as well as the same predicted from the DIANA FEA calculations. The capacity found from the FELA calculations is also indicated. The plot shows that the ductility of the two specimens varies quite significantly, with the maximum deflection of the 4P-N-C3 specimen being almost twice that of the 4P-N-C2 specimen. The NLFEA show a more stiff initial behavior compared with the experiments. The material stiffness from the experiments was not measured, so the relation between f_c and E_c was used, which is not an accurate relation. Furthermore, the deformation for the NFLEA is the difference between the top-centre and the point above the supports. After the linear part, both results from the NLFEA shows a decrease of the load. The decrease might be due to the relative low amount of reinforcement which result in small difference between the theoretical uncracked capacity and cracked capacity of the pilecap.

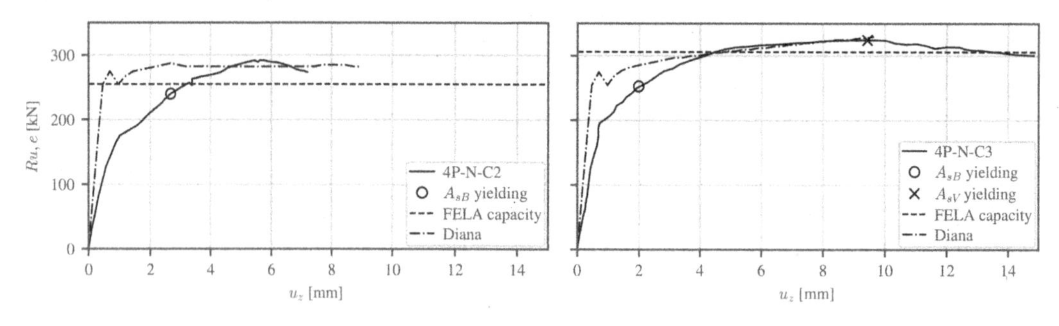

Figure 8. Load-displacement plot for the two specimens. Note that the y-axis shows the load in one pile only.

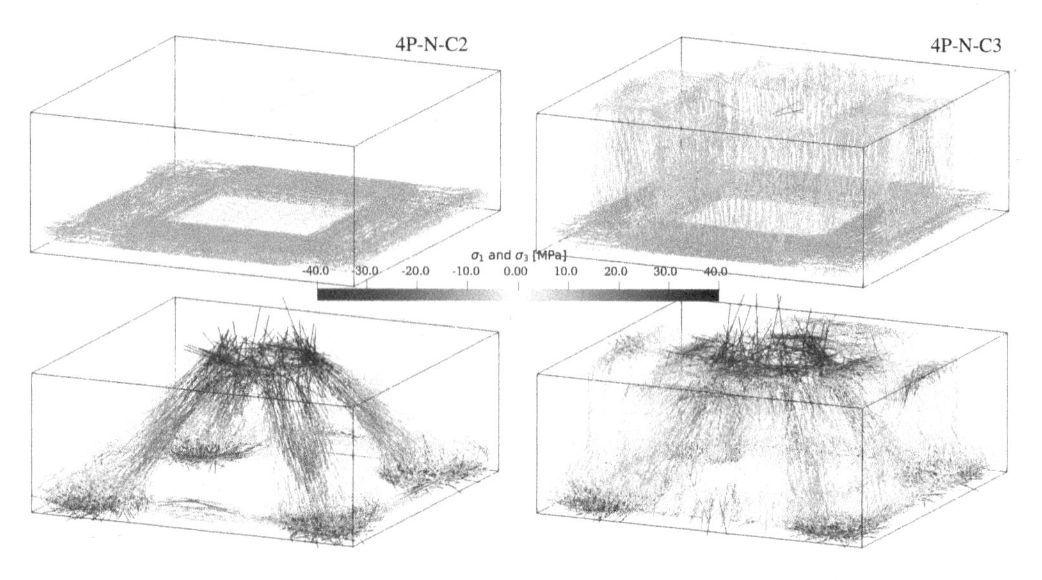

Figure 9. Principal stress flow predicted from by the FELA models for the two specimens. Bottom: Third principal stress, top: first principal stress.

Contrary to the NLFEA calculation, FELA calculations do not give any information about the load-displacement behavior of the model. Instead, the presence of sufficient ductility is a requirement for the safe use of the method. In the present case, even with a brittle failure of specimen 4P-N-C2, the capacities found from FELA are found to be safe.

For both specimens, the capacity found from the FELA calculations is lower than the ultimate load. However, they are in both cases larger than the load, which gives yielding of the bunched reinforcement A_{sB}. From this observation, we can conclude that the FELA calculations can utilize at least some of the reinforcement between the piles. Also, the inclusion of the shear reinforcement makes the FELA model better able to utilize this reinforcement, which can be seen from the larger difference between the yielding of bunched reinforcement and the FELA capacity for specimen 4P-C2-N3 compared to specimen 4P-C2-N2. In other words, the presence of the shear reinforcement allows for new load-paths. Consequently, the capacity is about 200 kN higher when the shear reinforcement is introduced, which is 20% higher than without shear reinforcement. The DIANA FEA calculations are able to very accurately determine the ultimate loads for both specimens. One of the reasons for the accurate calculation, and difference from the FELA model, is that hardening is included in the NLFEA model so the ultimate capacity of the reinforcement can be reached. From 2, it is seen that the ultimate strength of the reinforcement is in average 17% higher than the yield strength.

4.4.2 Stress flow

Figure 9 shows the stress flow found from the FELA calculations by plotting the first and third principal stress vectors of the two specimens. The stress flow for specimen 4P-C-N2 is entirely predictable with a strut originating at each pile and terminating at the column, utilizing the bunched reinforcement and, to a lesser degree, the reinforcement between the piles to carry the tie forces. The stress flow plot is clear and looks similar to a strut-and-tie model. The stress flow for specimen 4P-C-N3 is a bit more complicated since, besides the main strut action, secondary load paths are also present utilizing the shear reinforcement between the piles. For both specimens, it can be seen that the struts terminate on the very edge of the column, which makes sense since this will give the most direct transfer of compression from the piles to the column.

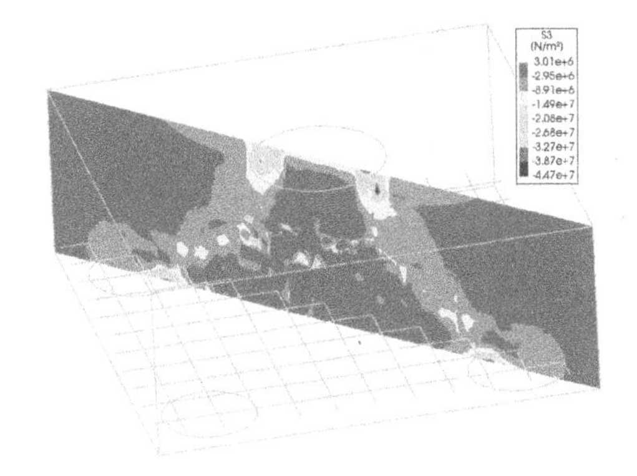

Figure 10. Third principal stress plot for the Diana model of specimen 4P-N-C2 at peak load.

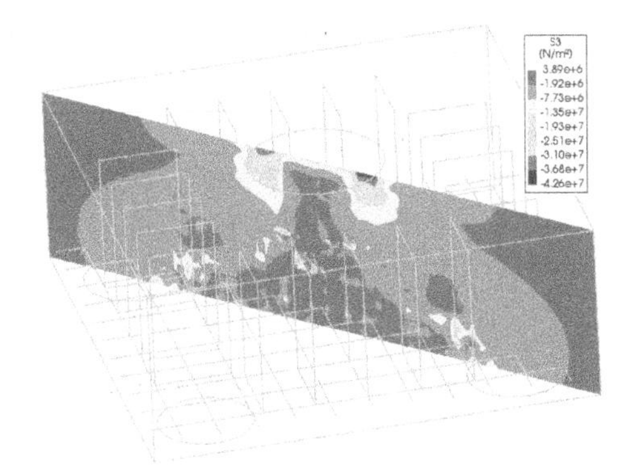

Figure 11. Third principal stress plot for the Diana model of specimen 4P-N-C3 at peak load.

4.4.3 Deformation and collapse mode

The flow of compression at the peak load in the DIANA FEA model can be seen in Figure 10 for specimen 4P-N-C2 and in Figure 11 for specimen 4P-N-C3. The figures show surface plots through the diagonal. The results are quite similar to the FELA results, with a concentrated strut from the pile towards the edge of the column for 4P-N-C2, and a more complicated load path for 4P-N-C3. In both cases the load transferred at the edge of the column, which is also similar to the FELA calculations.

The NLFEA model, which uses a deformation-based model, can show the deformation state at any point in the load history. This is not the case for FELA, which has no information about the deformation due to the rigid-plastic material model. However, the dual variables of the stress equilibrium (4b) can be interpreted as a collapse mode. The collapse mode shows how each part of the model will move with respect to each other at collapse but says nothing about the magnitude of the deformations.

A plot of the collapse mode from the FELA calculations for the two specimens can be seen in Figure 12, and a deformation plot at peak load for the NLFEA calculations can be seen in Figure 13.

Specimen 4P-N-C2 shows a localized collapse with the piles moving upwards. This collapse seems consistent with a punching failure after yielding of the bunched reinforcement, as reported by the experiment. The deformation plot from the NLFEA calculations also shows a localized failure indicative of the aforementioned punching failure mode.

Specimen 4P-N-C3 was reported to have a flexural failure in the experiments, and this also seems consistent with the FELA collapse mode, and the NLFEA deformation plot.

Even if the FELA calculations yield no information about the load-displacement behavior, in this case, the collapse mode still reveals something since a punching failure would generally be assumed to be less ductile than a flexural one.

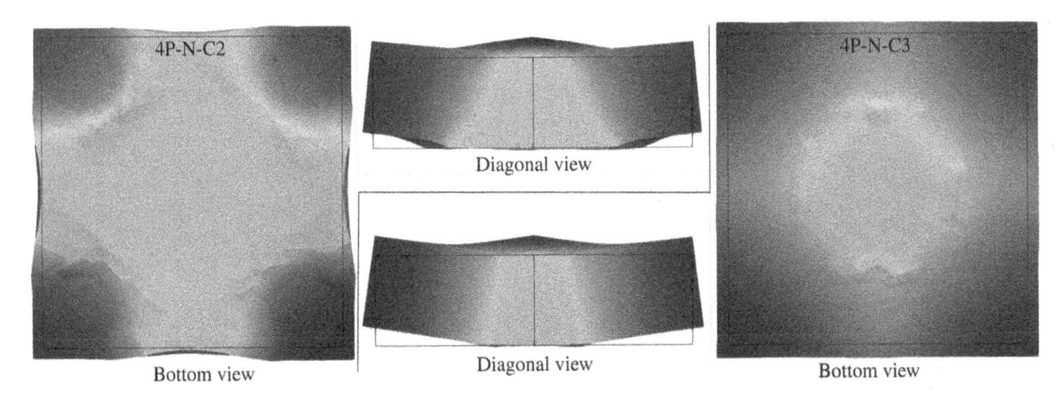

Figure 12. Collapse mode predicted from by the FELA models for the two specimens. The coloring of the faces is the relative vertical displacement.

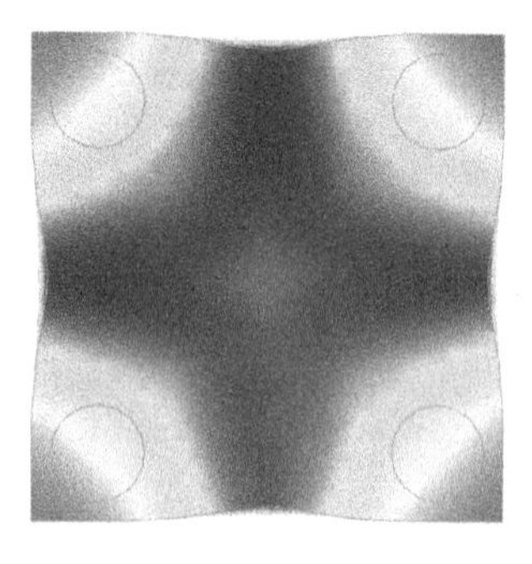
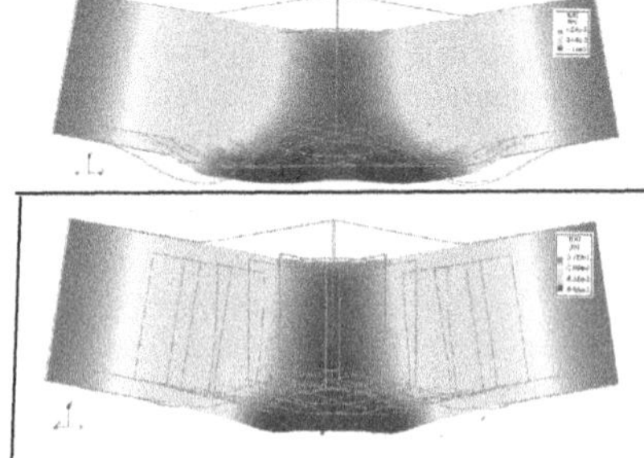
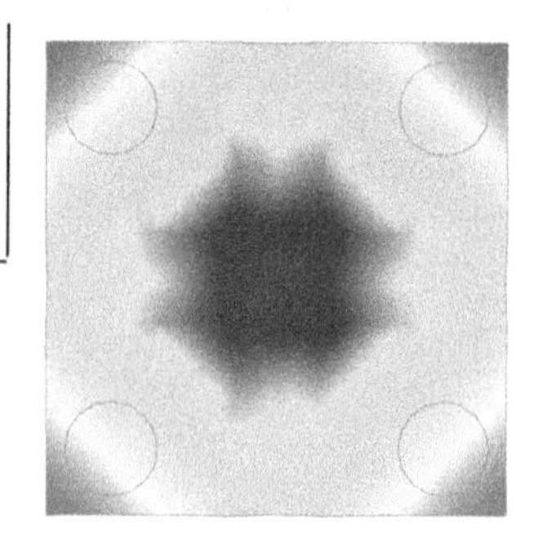

Figure 13. Collapse mode predicted from by the NLFEA models for the two specimens.

5 CONCLUSION

Computational models of two four-pile cap experiments using Finite Element Limit Analysis (FELA) and the Non-Linear Finite Element Method (NLFEM) software DIANA FEA have been presented. Results obtained from the two models have been compared to each other and the experimental findings from the literature. Both models were able to determine the capacity to a satisfactory degree and foresee the mode of collapse. The load-displacement plot obtained from the DIANA FEA model showed good agreement with the experimental results.

While DIANA FEA, and NLFEM in general, can describe the behavior of a structure very accurately, the precision also comes at the cost of increased complexity. On the other hand, FELA can only give information about the ultimate limit state. However, the model is also simpler. The simplicity can mean faster modeling time and calculation time, but the structures must be ductile to use FELA safely. In conclusion, both methods have their strengths and weaknesses, and it is essential to be mindful of these and use the right tool for a given task. Furthermore, the two methods can be used as independent model control for each other.

REFERENCES

Anderheggen, E. & H. Knöpfel (1972, dec). Finite element limit analysis using linear programming. *International Journal of Solids and Structures 8*(12), 1413–1431.

Andersen, M. E. M., P. N. Poulsen, & J. F. Olesen (2021, may). Finite-Element Limit Analysis for Solid Modeling of Reinforced Concrete. *ASCE Journal of Structural Engineering 147*(5), 04021051.

Andersen, M. E. M., P. N. Poulsen, & J. F. Olesen (2022, jan). Partially mixed lower bound constant stress tetrahedral element for Finite Element Limit Analysis. *Computers & Structures 258*, 106672.

Bisbos, C. D. & P. M. Pardalos (2007, aug). Second-Order Cone and Semidefinite Representations of Material Failure Criteria. *Journal of Optimization Theory and Applications 134*(2), 275–301.

Boyd, S. & L. Vandenberghe (2004). *Convex optimization* (1 ed.). New York: Cambridge University Press.

DIANA, F. E. A. (2017). Diana User's Manual, Release 10.2.

Drucker, D. C., W. Prager, & H. J. Greenberg (1952). Extended limit design theorems for continuous media. *Quarterly of Applied Mathematics 9*(4), 381–389.

Feenstra, P. H. (1993). *Computational aspects of biaxial stress in plain and reinforced concrete*. Ph. D. thesis, Delft University of Technology.

Fib (2013, oct). *fib Model Code for Concrete Structures 2010*. Weinheim, Germany: Wiley-VCH Verlag GmbH & Co. KGaA.

Gvozdev, A. (1960). The determination of the value of the collapse load for statically indeterminate systems undergoing plastic deformation. *International Journal of Mechanical Sciences 1*(4), 322–335.

Hendriks, M. A. N., A. de Boer, & B. Belletti (2017). Guidelines for nonlinear finite element analysis of concrete structures. Technical report, Rijkswaterstaat Technisch Document (RTD), Rijkswaterstaat Centre for Infrastructure, RTD.

Herfelt, M. A. (2017). *Numerical limit analysis of precast concrete structures - A framework for efficient design and analysis*. Ph.d. thesis, Technical University of Denmark (DTU), Kgs. Lyngby.

Herfelt, M. A., J. Krabbenhøft, & K. Krabbenhøft (2019). Practical Design and Modelling of Precast Concrete Structures. *Current Trends in Civil and Structural Engineering 3*(2), 8–11.

Jensen, T. W., P. N. Poulsen, & L. C. Hoang (2018, nov). Finite element limit analysis of slabs including limitations on shear forces. *Engineering Structures 174*, 896–905.

Krabbenhøft, K., A. V. Lyamin, & S. W. Sloan (2008). Three-dimensional Mohr-Coulomb limit analysis using semidefinite programming. *Communications in Numerical Methods in Engineering 24*(11), 1107–1119.

Larsen, K. P. (2010). *Numerical Limit Analysis of Reinforced Concrete Structures: Computational Modeling with Finite Elements for Lower Bound Limit Analysis of Reinforced Concrete Structures*. Ph.d. thesis, Technical University of Denmark (DTU), Kgs. Lyngby.

Lertsrisakulrat, T., K. Watanabe, M. Matsuo, & J. Niwa (2001). Experimental study on parameters in localization of concrete subjected to compression. *Doboku Gakkai Ronbunshu 2001*(669), 309–321.

Martin, C. M. & A. Makrodimopoulos (2008). Finite-Element Limit Analysis of Mohr–Coulomb Materials in 3D Using Semidefinite Programming. *Journal of Engineering Mechanics 134*(4), 339–347.

Miguel-Tortola, L., P. F. Miguel, & L. Pallarés (2019, mar). Strength of pile caps under eccentric loads: Experimental study and review of code provisions. *Engineering Structures 182*, 251–267.

Nakamura, H. & T. Higai (2001). Compressive fracture energy and fracture zone length of concrete. In *Modeling of inelastic behavior of RC structures under seismic loads*, pp. 471–487. ASCE Reston, VA.

Nielsen, M. P. & L. C. Hoang (2011). *Limit Analysis and Concrete Plasticity* (3rd ed.). CRC Press.

Poulsen, P. N. & L. Damkilde (2000, oct). Limit state analysis of reinforced concrete plates subjected to in-plane forces. *International Journal of Solids and Structures 37*(42), 6011–6029.

Selby, R. G. (1993). *Three-dimensional Constitutive Relations for Reinforced Concrete.* Ph. D. thesis, University of Toronto. Department of Civil Engineering.

Vecchio, F. J. & M. P. Collins (1986). The modified compression-field theory for reinforced concrete elements subjected to shear. *ACI J. 83*(2), 219–231.

Vecchio, F. J. & M. P. Collins (1993). Compression response of cracked reinforced concrete. *Journal of structural engineering 119*(12), 3590–3610.

Vincent, H., M. Arquier, J. Bleyer, & P. de Buhan (2018, dec). Yield design-based numerical analysis of three-dimensional reinforced concrete structures. *International Journal for Numerical and Analytical Methods in Geomechanics 42*(18), 2177–2192.

Vincent, H., M. Arquier, J. Bleyer, & P. de Buhan (2020). Numerical upper bounds to the ultimate load bearing capacity of three-dimensional reinforced concrete structures. *International Journal for Numerical and Analytical Methods in Geomechanics 44*(16), 2216–2240.

Vonk, R. A. (1992). *Softening of concrete loaded in compression.* Ph. D. thesis, Technische Universiteit Eindhoven.

Computational Modelling of Concrete and Concrete Structures – Meschke, Pichler & Rots (Eds)
 ISBN: 978-1-032-32724-2

Simulation of fracture on PFRC specimens subjected to high temperature using a cohesive model

F. Suárez
Departamento de Ingenieria Mecánica y Minera, Universidad de Jaén, Jaén, Spain

A. Enfedaque, M.G. Alberti & J.C. Gálvez
Departamento de Ingenieria Civil: Construcción, E.T.S de Ingenieros de Caminos, Canales y Puertos, Universidad Politécnica de Madrid, Madrid, Spain

ABSTRACT: Concrete has been traditionally reinforced with steel rebars that confer good tensile properties to this material. Nevertheless, concrete can also be reinforced with fibres, which have been traditionally made of steel, although in the last years new types of fibres have appeared, such as polypropylene fibres, glass fibres or polyolefin fibres. Their use widens the range of application of fibre-reinforced concrete (FRC) and has experienced an significant boost by national and international standards, which now include guidelines for their use in structures. More specifically, textured polyolefin macro-fibres have proved to provide very good tensile properties in concrete. The use of these fibres has significant advantages when compared with traditional steel fibres, since they reduce the tear and wear of devices involved in their production, avoid corrosion problems in concrete and have no influence on magnetic fields, which can be very important in some situations. Concrete properties, both in fresh and hardened states, have been extensively studied in the last years, proving to be a promising alternative to steel fibres. Fracture of FRC, and more specifically of PFRC, has been successfully reproduced using the finite element analysis by means of an embedded cohesive model with a trilinear softening function. On another note, concrete has a good behaviour when subjected to high temperatures and fire, especially when it is compared with other traditional construction materials, such as wood or steel. Nevertheless, concrete reinforcement is usually made of materials that are critically sensitive to these events and the behaviour of the composite material must be assessed to meet the requirements described in the structural standards. With regard to polyolefin-fibre reinforced concrete (PFRC), a recent study has analysed how the fracture properties of this material degrade when subjected to high-temperatures, ranging from 20Â°C to 200Â°C. As temperature increases, fibres modify their geometry and their mechanical properties, which leads to a reduction of their effectiveness. In this work, the fracture behaviour of PFRC specimens subjected to high temperatures is reproduced by using an embedded cohesive model that uses a trilinear softening function. The specific trilinear softening diagram that provides a good numerical simulation of fracture is obtained for each temperature increment. This helps to understand how the trilinear diagram must be adapted when PFRC is subjected to high temperatures and will allow the use of this model to a wider range of situations.

1 INTRODUCTION

The use of reinforced concrete (RC) in the confinement of nuclear reactors boasted the studies regarding the fire resistance of concrete in the last decades of the last century (Bažant & Kaplan 1996). In such studies it was evident that, not only the resistance of concrete was essential, but also the properties of the reinforcing steel that was used to enhance the flexural and tensile strength of RC. Analogously, when fibres are added to concrete forming fibre reinforced concrete (FRC) the response of fibres to high temperatures influences the fire resistance of the material. Such influence might be beneficial as in the case of polypropylene fibres (PF). Using PF fibres in determined dosages has proved to reduce the risk of explosive spalling in concrete (Liu, Ye, De Schutter, Yuan, & Taerwe 2008; Varona, Baeza, Bru, & Ivorra 2018a). PF fibres melt when subjected to high temperatures and generate a capillary network that reduces the high pressures preventing the spalling of the concrete element. However, the contribution of the PF to the structural behaviour of the concrete element cannot be considered due to its reduced mechanical properties.

Steel fibres (SF) have been traditionally employed in structural applications obtaining successful results (Lopez, Serna, Camacho, Coll, & Navarro-Gregori 2014). Moreover, if certain requirements are met, some standards and recommendations enable to reduce, or even substitute, the steel reinforcing bars of the concrete element. However, structural elements manufactured with steel fibre reinforced concrete (SFRC)

 DOI 10.1201/9781003316404-38

exhibits explosive spalling no matter the dosage of steel fibres used. Cocktails of fibres, obtained by mixing SF and PP, have been successfully applied in elements where not only the mechanical behaviour has been improved but also its fire resistance (Varona, Baeza, Bru, & Ivorra 2018a; Yermak, Pliya, Beaucour, Simon, & Noumowé 2017; Varona, Baeza, Bru, & Ivorra 2018b). Moreover, it should be mentioned that several authors have devoted studies to model the constitutive behaviour of SFRC when subjected to high temperatures (Abdallah, Fan, & Cashell 2017; Moradi, Bagherieh, & Esfahani 2020; Ruano, Isla, Luccioni, Zerbino, & Giaccio 2018). The existence of these models is of importance as they might help to estimate the residual load bearing capacity of the material and consequently the reliability of the structural elements after being subjected to high temperatures.

The appearance of polyolefin macro fibres with structural capacities have shown their suitability for their use in structural elements when added to concrete forming polyolefin fibre reinforced concrete (PFRC) (Alberti, Enfedaque, & Gálvez 2015; Alberti, Enfedaque, Gálvez, & Pinillos 2017; Picazo, Gálvez, Alberti, & Enfedaque 2018). In addition, there are recent studies that have analysed the influence of high temperatures in the mechanical behaviour of PFRC (Alberti, Gálvez, Enfedaque, & Castellanos 2021). However, there are not numerous studies concerning the changes that temperature generates in the constitutive models used in the structural design of PFRC structural elements.

Among all the mechanical properties of PFRC, the one that determines the structural character of the material is the fracture behaviour when subjected to flexural stresses. Due to this, significant effort has been recently devoted to find constitutive models that reproduce the flexural fracture behaviour of PFRC. One of the most successful attempts has been carried out by using a cohesive crack approach and an inverse analysis. Merging both concepts, the material behaviour obtained in experimental tests has been reproduced. It has been shown that the trilinear softening functions implemented were capable of considering the dosage of fibres (Alberti, Enfedaque, Gálvez, & Reyes 2017), their orientation (Enfedaque, Alberti, & Gálvez 2019) or even the size effect (Suárez, Gálvez, Alberti, & Enfedaque 2021). Moreover, they were apt also for simulating the behaviour of the material when subjected to tensile streses (Enfedaque, Alberti, Galvez, & Beltran 2018) or a combination of flexural-shear stresses (Suárez, Gálvez, Enfedaque, & Alberti 2019).

Based on the results shown in (Alberti, Gálvez, Enfedaque, & Castellanos 2021) this contribution seeks to determine the changes that should be performed in the constitutive model of PFRC when subjected to a flexural fracture test after being exposed to a range of temperature from 20 Â°C to 200 Â°C. Although in the referred work two PFRC mixes were analysed, with 3 kg of fibres per m^3 (HF3) and with 10 kg of fibres per m^3 (HF10), here only the results of the first of them are numerically reproduced.

2 EXPERIMENTAL BENCHMARCK

The experimental campaign was carried with specimens manufactured by using the concrete formulations and procedures described in (Alberti 2015). Portland cement type EN-197-1 CEM I 52.5 R-SR and Sika Viscocrete 5720, a policarboxylic superplastisicer, were mixed with siliceous aggregates (12.7 mm of maximum size). In addition, a limestone powder with a content of content of 98% calcium carbonate was employed. The mix formulation can be seen in Table 1.

Table 1. Mix proportions used concrete in the experimental campaign (Alberti, Gálvez, Enfedaque, & Castellanos 2021).

	HF	HF3	HF10
Cement (kg/m^3)	375	375	375
Limestone powder (kg/m^3)	100	100	100
Water (kg/m^3)	187.5	187.5	187.5
Sand (kg/m^3)	916	916	916
Gravel (kg/m^3)	300	300	300
Grit (kg/m^3)	450	450	450
Superplasticiser (% cement weight)	0.75	0.75	0.75
Polyolefin fibres (kg/m^3)	–	3	10

In the formulations with fibres 60 mm-long polyolefin fibres were added. Such fibres have a superficial treatment and embossed surface in order to provide a proper response of the fibre-matrix interface. The outlook of the fibres can be seen in Figure 1. Besides, the most representative properties of the fibres can be seen in Table 2.

Figure 1. Appearance of the fibres used (scale in mm).

Table 2. Fibre properties and dimensions.

Density (g/cm^3)	0.91
Length (mm)	60
Eq. diameter (mm)	0.92
Tensile strength (MPa)	>500
Modulus of elasticity (GPa)	>9

The specimens were heated in a convection oven at an approximate heating rate of 2.80 Â°C/min. The specimens remained at the chosen temperature for 3 hours after being cooled progressively during 7 hours inside the oven. The flexural tests were performed after the cooling period. The characterisation of the concrete elements was performed considering three temperatures as reference: room temperature, 150°C and 200°C. The latter two temperatures correspond to a temperature at which the fibres still maintained some integrity and a temperature at which the fibres were remarkably damaged respectively. In order to determine the influence of temperature in a PFRC formulation that was considered as a structural material (HF10) several increments of temperature were chosen between 150°C and 200°C.

The flexural fracture behaviour of the material was determined following the standard RILEM TC-187-SOC (Planas, Guinea, Gálvez, Sanz, & Fathy 2007). The test setup can be seen in Figure 2.

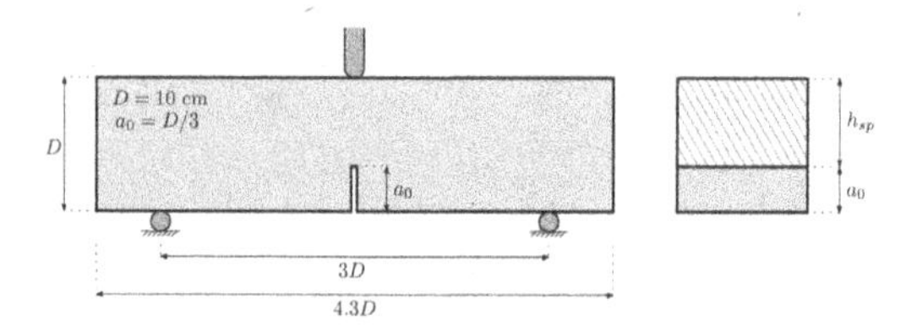

Figure 2. Scheme of the three-point bending test used in (Planas, Guinea, Gálvez, Sanz, & Fathy 2007).

As can be observed in Figure 2, the notch length was equal to $D/3$ and the span between the bearing cylinders was equal to $3D$. In this figure, a_0 corresponds to the notch length, h_{sp} to the ligament length and P to the applied load. The test data was acquired by means of two Linear Variable Differential Transformer (LVDT) placed at each side of the specimens and a Crack Mouth Opening Displacement (CMOD) mounted in the lips of the notch. In addition, the load borne by the sample, the position of the actuator and the elapsed time were also recorded.

In Figure 3 the average load-deflection curves obtained in the tests can be seen.

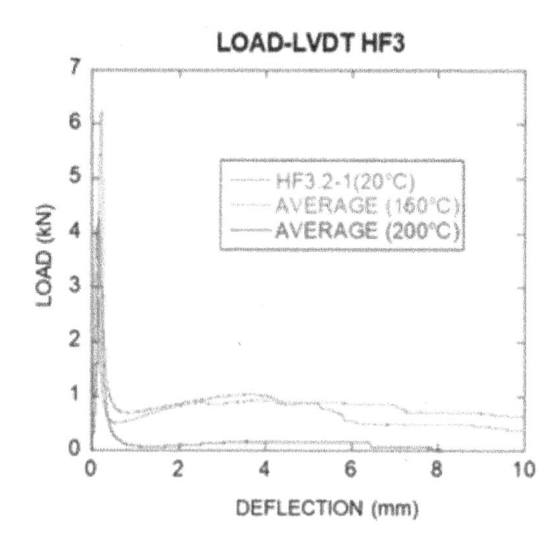

Figure 3. Load-load displacement curves of HF3 at various temperatures (average at each temperature) (Alberti, Gálvez, Enfedaque, & Castellanos 2021).

3 NUMERICAL SIMULATION

In this section the numerical work carried out to reproduce the experimental diagrams obtained for HF3 specimens is described. Firstly, the embedded cohesive crack model used to reproduce fracture is briefly described and, secondly, the finite element models are described. The description of the fracture model is very concise, since its development is not new, and the reader can find it in previous works (Gálvez, Planas, Sancho, Reyes, Cendón, & Casati 2013; Reyes, Gálvez, Casati, Cendón, Sancho, & Planas 2009; Sancho, Planas, Cendón, Reyes, & Gálvez 2007).

3.1 *Embedded cohesive crack model*

The cracking process is reproduced by means of the finite element method (FEM) by using an element formulation that takes advantage of the cohesive zone concept developed by Hillerborg (Hillerborg, Modéer, & Petersson 1976), inspired by the work of Dugdale (Dugdale 1960) and Barenblatt (Barenblatt 1962). This formulation constitutes a strong discontinuity approach proposed by Oliver (Oliver 1996a; Oliver 1996b) that was initially developed for concrete (Gálvez, Planas, Sancho, Reyes, Cendón, & Casati 2013; Sancho, Planas, Cendón, Reyes, & Gálvez 2007), but has also been successfully adapted for brickwork masonry (Reyes, Gálvez, Casati, Cendón, Sancho, & Planas 2009) and fibre-reinforced cementitious materials (Enfedaque, Alberti, Gálvez, & Domingo 2017). Since the cohesive zone approach assumes that fracture develops under mode I conditions, this approach considers that the cohesive stress vector t is perpendicular to the crack opening and parallel to the crack displacement vector w:

$$t = \frac{f(\tilde{w})}{\tilde{w}} w \qquad \text{with } \tilde{w} = \max(|w|) \qquad (1)$$

where $f(|\tilde{w}|)$ stands for the material softening function, defined in terms of an equivalent crack opening $\tilde{w}$, which represents the maximum historical crack opening to account for possible unloading scenarios. In the case of PFRC, past works have proved that trilinear diagrams as shown in Figure 4 properly simulates fracture of this material with varying proportions of fibres, either with vibrated or self-compacting concrete, under mode I or a combination of mode

I and mode II fracture conditions (Suárez, Gálvez, Enfedaque, & Alberti 2019) and correctly capturing fracture on different size specimens (Suárez, Gálvez, Alberti, & Enfedaque 2021). Unloading and reloading branches are aligned with the origin and the softening function is defined by four points (t, k, r and f). The following expression provides the trilinear diagram shown in Figure 4.

$$\sigma = \begin{cases} f_{ct} + \left(\dfrac{\sigma_k - f_{ct}}{w_k}\right) \cdot w & \text{if } 0 < w \leq w_k \\ \sigma_k + \left(\dfrac{\sigma_r - \sigma_k}{w_r - w_k}\right) \cdot (w - w_k) & \text{if } w_k < w \leq w_r \\ \sigma_r + \left(\dfrac{-\sigma_r}{w_f - w_r}\right) \cdot (w - w_r) & \text{if } w_r < w \leq w_f \\ 0 & \text{if } w > w_f \end{cases} \quad (2)$$

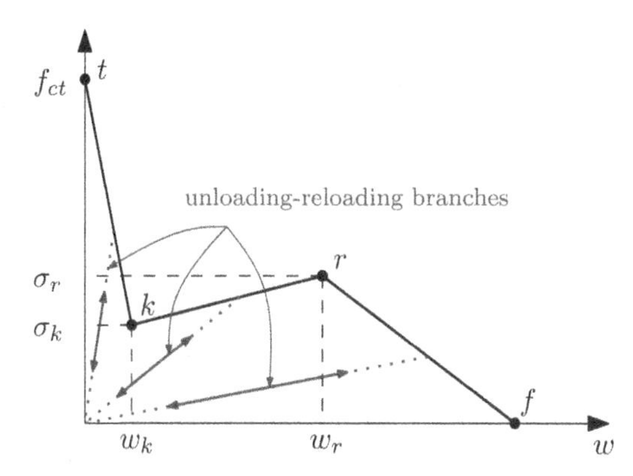

Figure 4. Scheme of the trilinear softening function used for modeling fracture of PFRC specimens.

This element formulation is programmed for constant strain triangular elements, thus accounting for an only integration point. Only three crack directions are considered, each of which are parallel to the triangle sides, and crack is placed at midheight; Figure 5 depicts these characteristics of the fracture model.

Once the crack direction is determined, the element is divided into two parts, A^+ and A^-, and the stress vector t, constant along the crack, can be obtained as:

$$t = \frac{A}{hL}\boldsymbol{\sigma} \cdot n \quad (3)$$

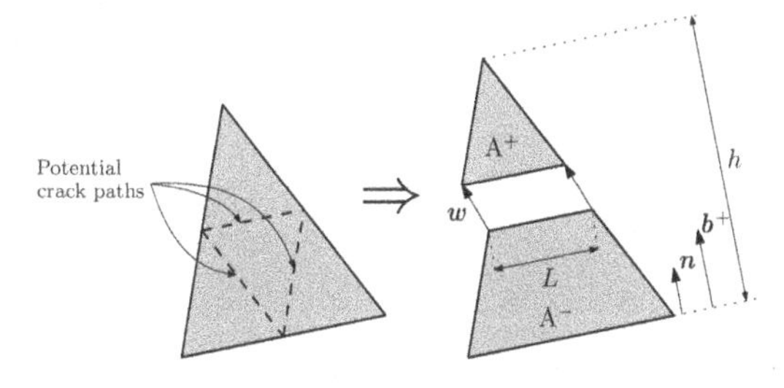

Figure 5. Embedded cohesive crack element.

Where A represents the area of the element, h the triangle height over the side opposite to the solitary node, L the crack length, $\boldsymbol{\sigma}$ the stress tensor and n the unit vector normal to that side and to the crack. Given that the crack is parallel to one side of the triangular element and is placed at midheight, (3) turns into $t = \boldsymbol{\sigma} \cdot n$. The reader can find a more detailed description of the model in (Sancho, Planas, Cendón, Reyes, & Gálvez 2007).

Inside the element, outside the crack the material remains elastic, thus the crack displacement vector w is solved assuming that the stress tensor can be obtained by subtracting an inelastic behaviour, which corrects the elastic prediction of the element by including the effect of the crack displacement, as expressed by (4).

$$\boldsymbol{\sigma} = E : \left[\epsilon^a - \left(b^+ \otimes w\right)^S\right] \cdot n \quad (4)$$

where E stands for the elastic tangent tensor, ϵ^a for the apparent strain vector obtained with the nodal displacements, b^+ for the gradient vector corresponding to the solitary node, which in this case can be obtained with (5). Superscript S denotes the symmetric part of the resulting tensor, : the double-dot product ($(A : b)_{ij} = A_{ijkl} b_{kl}$), and $\otimes$ the direct product ($(a \otimes b)_{ij} = a_i b_j$).

$$b^+ = \frac{1}{h} n \quad (5)$$

Given that $t = \boldsymbol{\sigma} \cdot n$ and by using expression (4) for $\boldsymbol{\sigma}$ and expression (3) for t, the following expression is obtained:

$$\frac{f(\tilde{w})}{\tilde{w}} w = [E : \boldsymbol{\epsilon}^a] \cdot n - \left[E : \left(b^+ \otimes w\right)^S\right] \cdot n$$

which can be rewritten as

$$\left[\frac{f(\tilde{w})}{\tilde{w}} 1 + n \cdot E \cdot b^+\right] \cdot w = [E : \boldsymbol{\epsilon}^a] \cdot n \quad (6)$$

where 1 stands for the identity tensor. By means of an iterative algorithm (such as the Newton-Raphson method), the value of w can be computed to satisfy (6).

This model is implemented for ABAQUS® by means of an UMAT subroutine and, since vector n, b^+, crack length L and the element area A are computed with the nodal coordinates of each element, it reads an external file that stores this information.

3.2 *FEM models*

In order to reproduce the experimental results of (Alberti, Gálvez, Enfedaque, & Castellanos 2021), a bidimensional mesh, shown in Figure 6, has been used and plane stress conditions applied. This mesh is finer in the region where damage develops and coarser out of it, since only the vertical ligament above the notch is relevant in terms of the nonlinear problem to be solved. The adequacy of this refinement has been validated in previous works by the authors (Suárez, Gálvez, Alberti, & Enfedaque 2021; Suárez, Gálvez, Enfedaque, & Alberti 2019). The specimen size is 100 mm × 430 mm, supports are symmetrically placed at 300 mm from each other and loading is applied in the middle point of the upper side of the specimen, aligned with the notch, which is 33.3 mm long.

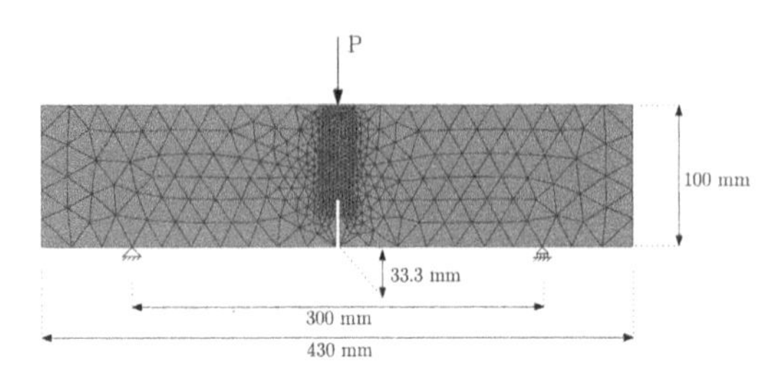

Figure 6. FEM model used in the numerical simulation.

4 RESULTS

In this section the results of the numerical simulations are presented, firstly introducing the trilinear softening diagrams used and, secondly, showing the load-deflection diagrams obtained, which are compared with the experimental results of (Alberti, Gálvez, Enfedaque, & Castellanos 2021).

4.1 *Trilinear softening diagrams*

The trilinear softening diagram is shown in Figure 4, the coordinates of four points must be fixed in the σ-w plane, *t*, *k*, *r* and *f*. In order to use a coherent approach to define the coordinates of these points as temperature increases, the following criteria have been employed:

- **Point *t*.** Due to the different elastic moduli of polyolefin and concrete, together with the negligible proportion of fibres in comparison with concrete, this point represents the tensile strength of concrete. This value is considered to reduce as concrete is subjected to higher temperatures, assuming a gradual degradation of the material.
- **Point *k*.** This point represents the crack opening at which fibres begin to assume tensile stresses across the crack, thus reinforcing the material and producing a load recovery for subsequent load increments. This value has been obtained based on the softening diagram of a plain concrete, that is to say, without fibres. This simplification assumes that fibres contribution to the abscissa of this point is negligible, therefore the material follows the behaviour of a plain concrete.
- **Point *r*.** The abscissa of this point, w_r, has been fixed with a value of 2.25 mm; this assumption is based on previous works by the authors (Suárez, Gálvez, Alberti, & Enfedaque 2021; Suárez, Gálvez, Enfedaque, & Alberti 2019). Regarding the value in the ordinate axis, σ_r, it has been considered to have the highest value at 20°C and only equal or smaller values for higher temperatures have been contemplated. Since this parameter is related to the strength capacity of the fibres, it is assumed that higher temperatures can only degrade the material or the fibres surface, thus leading to equal or smaller values of σ_r.
- **Point *f*.** This value is related to the fibre length and, following past results, a value of w_f=7.5 mm has been considered for all temperatures.

Following these criteria, the results that are presented in the following section are obtained with the trilinear softening diagrams shown in Figure 7 (the coordinates of these points for each trilinear diagram can be consulted in Table 3). It can be observed that σ_k becomes lower as high temperature increases and σ_r is remarkably similar for 20°C and 150°C, unlike in the case of 200°C, for which it drops dramatically.

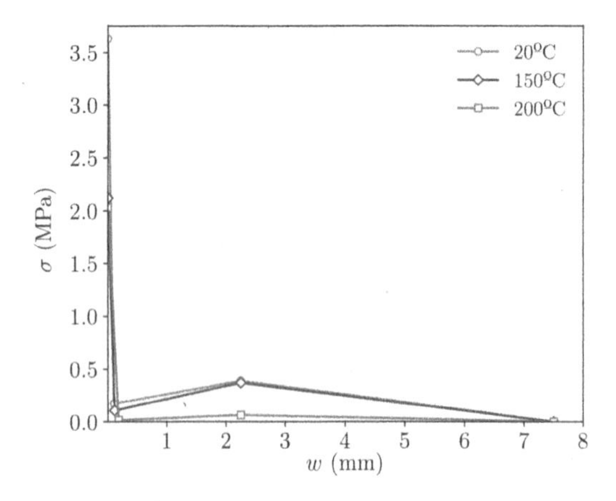

Figure 7. Trilinear softening diagrams used for simulating PFRC fracture specimens with a fibre proportion of 3 kg/m^3 subjected to different high-temperature conditions.

Table 3. Coordinates of *t*, *k*, *r* and *f* points of the trilinear softening diagrams shown in Figure 7.

	20°C	150°C	200°C
w_t	0.000	0.000	0.000
σ_t	3.630	2.120	2.030
w_k	0.102	0.113	0.182
σ_k	0.173	0.108	0.015
w_r	2.250	2.250	2.250
σ_r	0.389	0.372	0.065
w_f	7.500	7.500	7.500
σ_f	0.000	0.000	0.000

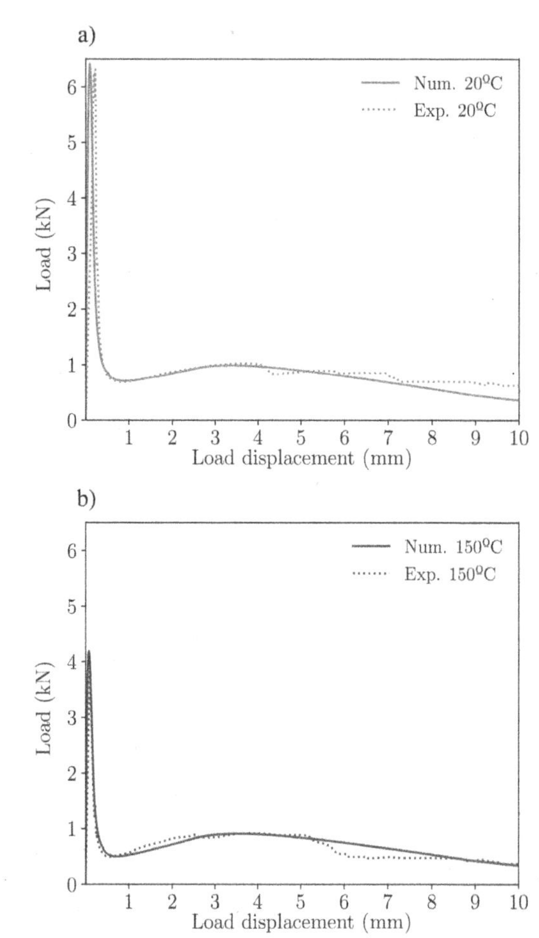

c)

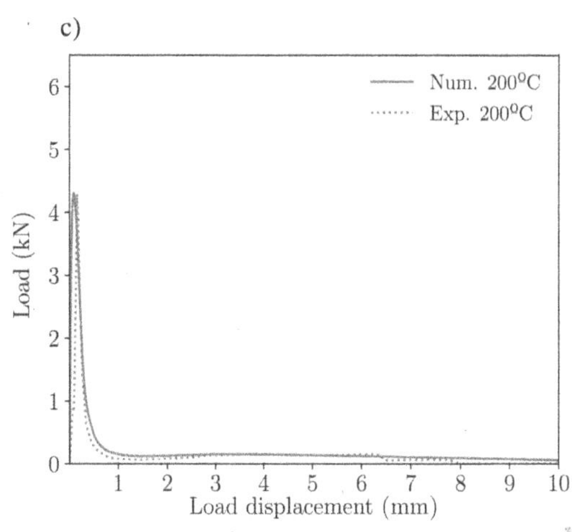

Figure 8. Comparison of load-deflection diagrams obtained numerically and experimentally for specimens of PFRC with 3 kg of fibres per m^3 subjected to different temperature conditions: a) 20°C, b) 150°C, and c) 200°C.

4.2 Load-deflection diagrams

Figure 8 shows the load-deflection diagrams obtained with the trilinear softening diagrams of Figure 7. These results are presented with the same colours used in Figure 7 for each temperature and compare the numerical results, displayed in solid lines, with the experimental curves, in dotted lines.

In general, the initial peak load is well reproduced in all cases considering the inherent experimental scatter in this type of materials. The minimum load after the initial peak load is also very well captured, as well as the remnant peak load that takes place around $w = 4$ mm for all temperatures. In the case of 150°C, the remnant peak load is numerically reproduced at an earlier value of w if compared with the experimental result but in the authorsâĽ™ opinion this can be considered as valid, since only one experimental value for each temperature is available and, based on previous works, if a higher number of specimens had been tested for each case, the experimental envelopes would very likely cover the experimental results in all cases.

These results show that, up to temperatures below 200°C, σ_t and σ_k decrease as temperature increases. This can be a consequence of moderate concrete and fibre degradation. Nevertheless, for temperatures below 200°C, σ_r, which is responsible for the remnant peak load of PFRC, seems to be unaffected and a fixed value of 1.418 provides good numerical results for all temperatures below 200°C. On the contrary, simulation of PFRC subjected to 200°C requires a very different trilinear diagram that, not only requires lower values of σ_t and σ_k, but a very low value of σ_r. This is a consequence of a high degradation of fibres at this temperature, that reduces their size, modifies their shape and their mechanical properties, as reported in (Alberti, Gálvez, Enfedaque, & Castellanos 2021).

5 FINAL COMMENTS

In this work, the numerical simulation of fracture of PFRC specimens exposed to high temperatures has been explored. To do this, some experimental results presented in (Alberti, Gálvez, Enfedaque, & Castellanos 2021) have been considered; the numerical results reproduce the behaviour of PFRC specimens manufactured with a fibre proportion of 3 kg/m^3 (HF3) and exposed to three temperatures: 20°C, 150°C and 200°C. The numerical simulation of fracture has been carried out by means of a cohesive model fed with a trilinear softening function that had been successfully used in the past to reproduce the fracture behaviour of this material.

High temperatures modify the mechanical properties of polyolefin fibres, which results into variations of the softening diagram. The assumptions made to define the modified trilinear diagrams, described in section 4.1, have proved to be reasonable enough to correctly capture the main differences observed in the load-deflection diagrams induced by high temperatures. The softening diagram for a high-temperature exposure of 150°C only presents slight modifications with respect to the ambient temperature of reference,

20°C. On the contrary, the softening diagram corresponding to a high-temperature exposure of 200°C has notable differences, remarkably reducing their reinforcing capacity. This is related to the severe degradation that polyolefin suffers at temperatures over 200°C, which include a change of the fibre shape and modified mechanical properties of the material.

In (Alberti, Enfedaque, Gálvez, & Reyes 2017) the relation of the parameters of points t, k, r and f of the trilinear diagram (see Figure 4) with some measurable values and characteristics of the manufacturing of PFRC were explored and some expressions proposed with the aim of providing a predictive model of fracture for PFRC. The results presented in this work help to better understand how to numerically reproduce fracture of PFRC subjected to high temperatures, how to adapt the trilinear softening diagram when a cohesive model is employed and will hopefully help in providing a predictive fracture model of this material that also includes the effect of exposure to high temperatures. Nevertheless, this work only covers a limited number of temperature and an only fibre proportion (3 kg/m^3), and should be extended to a wider range of cases.

ACKNOWLEDGEMENTS

The authors gratefully acknowledge the financial support provided for this research by the Ministry of Science and Innovation of Spain through the Research Fund Project PID2019-108978RB-C31.

REFERENCES

Abdallah, S., M. Fan, & K. Cashell (2017). Pull-out behaviour of straight and hooked-end steel fibres under elevated temperatures. *Cement and Concrete Research 95*, 132–140.

Alberti, M., A. Enfedaque, J. Gálvez, & E. Reyes (2017). Numerical modelling of the fracture of polyolefin fibre reinforced concrete by using a cohesive fracture approach. *Composites Part B: Engineering 111*, 200–210.

Alberti, M. G. (2015). Polyolefin fibre-reinforced concrete: from material behaviour to numerical and design considerations. *Doctoral Thesis. Universidad Politécnica Madrid*.

Alberti, M. G., A. Enfedaque, & J. C. Gálvez (2015). Improving the reinforcement of polyolefin fiber reinforced concrete for infrastructure applications. *Fibers 3*(4), 504–522.

Alberti, M. G., A. Enfedaque, J. C. Gálvez, & L. Pinillos (2017). Structural cast-in-place application of polyolefin fiber–reinforced concrete in a water pipeline supporting elements. *Journal of Pipeline Systems Engineering and Practice 8*(4), 05017002.

Alberti, M. G., J. C. Gálvez, A. Enfedaque, & R. Castellanos (2021). Influence of high temperature on the fracture properties of polyolefin fibre reinforced concrete. *Materials 14*(3), 601.

Barenblatt, G. I. (1962). The mathematical theory of equilibrium cracks in brittle fracture. In *Advances in applied mechanics*, Volume 7, pp. 55–129. Elsevier.

Bažant, Z. P. & M. F. Kaplan (1996). Concrete at high temperatures: material properties and mathematical models.

Dugdale, D. S. (1960). Yielding of steel sheets containing slits. *J Mech Phys Solids 8*(2), 100–104.

Enfedaque, A., M. Alberti, J. Galvez, & M. Beltran (2018). Constitutive relationship of polyolefin fibre–reinforced concrete: Experimental and numerical approaches to tensile and flexural behaviour. *Fatigue & Fracture of Engineering Materials & Structures 41*(2), 358–373.

Enfedaque, A., M. Alberti, J. Gálvez, & J. Domingo (2017). Numerical simulation of the fracture behaviour of glass fibre reinforced cement. *Construction and Building Materials 136*, 108–117.

Enfedaque, A., M. G. Alberti, & J. C. Gálvez (2019). Influence of fiber distribution and orientation in the fracture behavior of polyolefin fiber-reinforced concrete. *Materials 12*(2), 220.

Gálvez, J., J. Planas, J. Sancho, E. Reyes, D. Cendón, & M. Casati (2013). An embedded cohesive crack model for finite element analysis of quasi-brittle materials. *Eng Fract Mech 109*, 369–386.

Hillerborg, A., M. Modéer, & P.-E. Petersson (1976). Analysis of crack formation and crack growth in concrete by means of fracture mechanics and finite elements. *Cem Concr Res 6*(6), 773 – 781.

Liu, X., G. Ye, G. De Schutter, Y. Yuan, & L. Taerwe (2008). On the mechanism of polypropylene fibres in preventing fire spalling in self-compacting and high-performance cement paste. *Cement and concrete research 38*(4), 487–499.

Lopez, J. A., P. Serna, E. Camacho, H. Coll, & J. Navarro-Gregori (2014). First ultra-high-performance fibre-reinforced concrete footbridge in spain: Design and construction. *Structural Engineering International 24*(1), 101–104.

Moradi, M., A. R. Bagherieh, & M. R. Esfahani (2020). Constitutive modeling of steel fiber-reinforced concrete. *International Journal of Damage Mechanics 29*(3), 388–412.

Oliver, J. (1996a). Modelling strong discontinuities in solid mechanics via strain softening constitutive equations. part 1: Fundamentals. *International journal for numerical methods in engineering 39*(21), 3575–3600.

Oliver, J. (1996b). Modelling strong discontinuities in solid mechanics via strain softening constitutive equations. part 2: Numerical simulation. *International journal for numerical methods in engineering 39*(21), 3601–3623.

Picazo, A., J. Gálvez, M. Alberti, & A. Enfedaque (2018). Assessment of the shear behaviour of polyolefin fibre reinforced concrete and verification by means of digital image correlation. *Construction and Building Materials 181*, 565–578.

Planas, J., G. Guinea, J. Gálvez, B. Sanz, & A. Fathy (2007). Indirect test for stress-crack opening curve, from experimental determination of the stress-crack opening curve for concrete in tension—final report of RILEM Technical Committee TC 187-SOC.

Reyes, E., J. Gálvez, M. Casati, D. Cendón, J. Sancho, & J. Planas (2009). An embedded cohesive crack model for finite element analysis of brickwork masonry fracture. *Engineering Fracture Mechanics 76*(12), 1930 – 1944.

Ruano, G., F. Isla, B. Luccioni, R. Zerbino, & G. Giaccio (2018). Steel fibers pull-out after exposure to high temperatures and its contribution to the residual mechanical behavior of high strength concrete. *Construction and Building Materials 163*, 571–585.

Sancho, J., J. Planas, D. Cendón, E. Reyes, & J. Gálvez (2007). An embedded crack model for finite element analysis of concrete fracture. *Engineering Fracture Mechanics 74*(1), 75 – 86. Fracture of Concrete Materials and Structures.

Suárez, F., J. Gálvez, A. Enfedaque, & M. Alberti (2019). Modelling fracture on polyolefin fibre reinforced concrete

specimens subjected to mixed-mode loading. *Engineering Fracture Mechanics 211*, 244–253.
Suárez, F., J. C. Gálvez, M. G. Alberti, & A. Enfedaque (2021). Fracture and size effect of pfrc specimens simulated by using a trilinear softening diagram: A predictive approach. *Materials 14*(14).
Varona, F., F. J. Baeza, D. Bru, & S. Ivorra (2018a). Evolution of the bond strength between reinforcing steel and fibre reinforced concrete after high temperature exposure. *Construction and Building Materials 176*, 359–370.
Varona, F. B., F. J. Baeza, D. Bru, & S. Ivorra (2018b). Influence of high temperature on the mechanical properties of hybrid fibre reinforced normal and high strength concrete. *Construction and Building Materials 159*, 73–82.
Yermak, N., P. Pliya, A.-L. Beaucour, A. Simon, & A. Noumowé (2017). Influence of steel and/or polypropylene fibres on the behaviour of concrete at high temperature: Spalling, transfer and mechanical properties. *Construction and Building Materials 132*, 240–250.

Computational Modelling of Concrete and Concrete Structures – Meschke, Pichler & Rots (Eds)
 ISBN: 978-1-032-32724-2

Numerical modeling of shear critical T-beam with conventional reinforcement and fibers

A. Kagermanov & I. Markovic
Eastern University of Applied Science (OST), Rapperswil, Switzerland

ABSTRACT: Numerical investigations from a recent blind prediction of T beams with steel-fibers and conventional reinforcement are presented. The blind prediction was organized by the fib Working Group Modelling of Fiber Reinforced Concrete Structures in 2019, where two identical, simply supported T beams were loaded in three point bending up to failure. Both beams failed in shear, reaching an average maximum capacity of 336kN. The submitted prediction resulted in a failure load of 331kN and was ranked 1st among 34 predictions from research and industry. In this paper, first the numerical model used for the blind prediction is presented, which consisted of a two-dimensional FE model with smeared reinforcement calibrated using notched beam tests made available to the participants. Next, analyses addressing objectivity of results with mesh refinement and the choice of constitutive modelling parameters is presented. The influence on the load-displacement response, longitudinal strain underneath the point of load application and crack pattern is investigated.

1 INTRODUCTION

1.1 *The fib blind prediction contest*

A blind prediction contest was organized in March 2020 by the fib Working Group WG 2.4.2 Modelling of Fiber Reinforced Concrete Structures on the capacity of steel fiber reinforced concrete (SFRC) beams with the objective of verifying the performance of current models for numerical simulation. Details on the experimental testing and outcomes of the blind simulations can be found elsewhere (Barros et al. 2020). In the following, a brief description of the test is presented.

Two identical T-beams were tested in March 2020 in the laboratory of Structural Division of the Department of Civil Engineering of Minho University (LEST) under three-point bending (Figure 1). The external load was applied eccentrically at 1.5m from the left support. No shear reinforcement was placed in this region, meaning that most of the shear force has to be transferred through concrete and fiber contributions. The right shear span was well reinforced with Φ6/75 stirrups. 3Φ25 bars were provided for the bottom longitudinal reinforcement, resulting in a reinforcement ratio of $\rho = 1.5\%$.

In addition to conventional reinforcement, hooked-end steel fibers were added to the concrete mix with a content of 60 kg/m^3 (0.76%-Vol). For the characterization of the SFRC, four cylinder tests were performed for the evaluation of compressive strength and Young's modulus, and six three-point notched beam tests for the evaluation of the residual tensile strength parameters according to Model Code 2010.

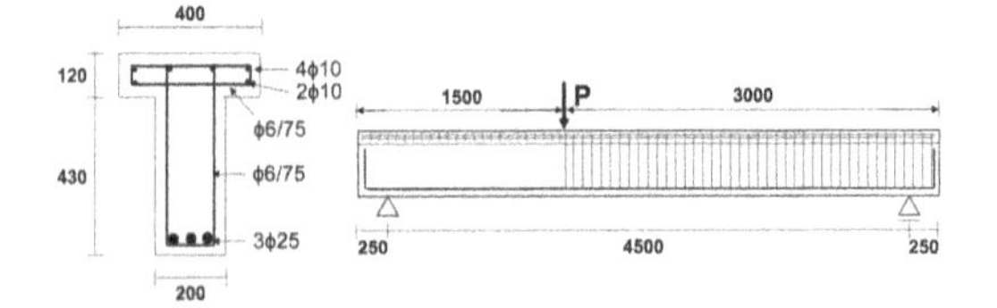

Figure 1. Blind prediction T-Beam: geometry, cross-section and reinforcement (Barros et al. 2020).

The two beams were tested under displacement control at 50 μm/s until failure. The following measurements were taken: (i) deflection below the point of load application, (ii) applied load and (iii) concrete strain at the level of longitudinal reinforcement in the loaded section.

Information regarding material properties (fiber and SFRC properties) as well as load-crack mouth opening displacement (CMOD) curves from 10 notched beam tests at 7 days and 14 days were made available for model calibration. Participants were asked to submit the load-displacement curve and the load-concrete strain curve at the loaded section.

2 BLIND PREDICTION OF THE T-BEAM

2.1 *FE model*

A 2D FE model of the T-beam was created using the in-house FE program IDEEA. The beam was discretized with four-node membrane (plane-stress)

 DOI 10.1201/9781003316404-39

elements with a total of $45 \times 5 = 225$ elements, 5 across the depth and 45 along the span (Figure 2). The thickness of the elements in the flange was set to 400mm and the rest to 200mm. In order to address mesh sensitivity, the element size was refined by a factor of 2, which did not show significant differences in terms of force-displacement response.

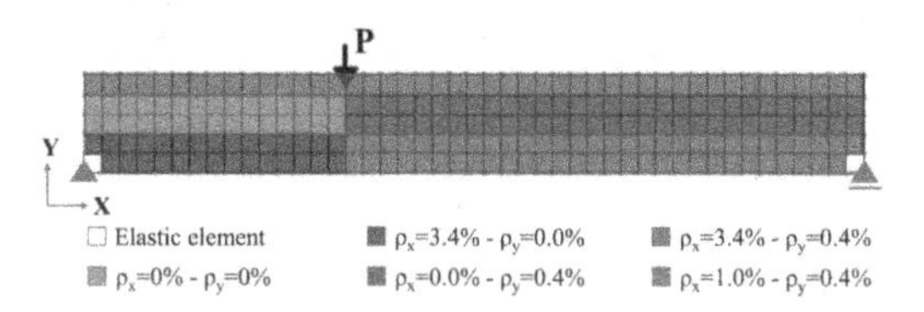

Figure 2. 2D FE model of the T-beam used for the blind prediction.

The nonlinear material model for SFRC was a smeared-crack fixed-crack orthotropic model developed by the author and previously verified for reinforced concrete members (Kagermanov 2019; Kagermanov & Ceresa 2016, 2018) (Figure 4). In these type of models, equivalent constitutive laws in tension and compression are applied in the crack directions. The compressive strength, for example, may be reduced due to orthogonal tensile strains (compression softening) or increased due to biaxial confinement. Shear stresses and strains arising on the crack are related through a shear retention factor, which was assumed constant after cracking but different for the left and right shear span of the beam.

For steel a uniaxial elasto-plastic model with strain hardening of 0.5% was used. Both transverse and longitudinal reinforcement were smeared within the membrane elements according to the provided reinforcement layout. The bottom longitudinal reinforcement was smeared within the bottom two rows of elements, resulting in $\rho_x = 3.4\%$. The top longitudinal reinforcement was assigned to the flange elements only, resulting in $\rho_x = 1.0\%$. A shear reinforcement ratio of $\rho_y = 0.4\%$ was assigned to the flange elements and the web elements with shear reinforcement. As a result of this distribution, a portion of elements in the unreinforced web region had neither longitudinal nor shear reinforcement.

The SFRC constitutive law in tension is shown in Figure 4. For simplicity, the same exponential curve implemented for reinforced concrete was used for SRFC. The curve is defined based on two parameters: the tensile strength, f_t, and the exponential decay parameter Ct, which is a function of the fracture energy and characteristic length. These parameters were calibrated based on FEA of the notched beams using the data provided from flexural tensile tests at 14 days. Figure 3 shows the calculated and experimental load versus CMOD curves. It can be seen that the initial stress drop immediately after cracking and subsequent tension stiffening is not accurately captured. A more reasonable agreement can be observed afterwards in the tension-softening phase up to a CMOD of 4mm. The final values selected for the blind prediction were $f_t = 1.75$ MPa and $C_t = 40$.

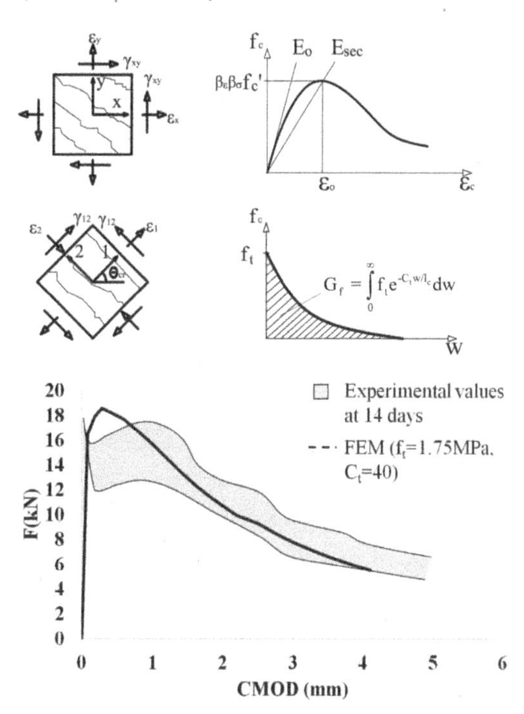

Figure 3. Summary of the orthotropic smeared crack model (top) and tension-softening model calibration using notched beam tests (bottom).

The rest of modeling parameters were relatively straightforward. For the compressive strength of concrete the mean value at 14 days was used, $f_{cm} = 64.2$ MPa, and similarly for the modulus of elasticity with $E_{cm} = 32900$ MPa. The peak strain corresponding to f_{cm} was chosen as $\varepsilon_o = 0.003$. For the shear retention factor a value close to zero ($\beta = 0.001$) was chosen for the region without shear reinforcement, whereas $\beta = 0.15$ was assigned for the region with shear reinforcement. The analysis was performed under displacement control with a step of 0.1mm using the vertical displacement at the point of load application as the controlling degree of freedom.

2.2 *FE results and comparison with experiment*

Results from the numerical simulation of the T-beam and comparison against the average experimental response in terms of load-displacement and strain-displacement curves is shown in Figure 5. A very good agreement was obtained for the initial elastic stiffness, cracking and post-cracking phases as well as maximum capacity. The numerical prediction for the maximum capacity was within 1.5% error on the safe side (331kN vs. 336kN). The corresponding peak displacement was predicted as 16.8mm, whereas the experimentally observed one was 16.1mm. The analysis stops at 19.2mm due to loss of convergence.

Table 1. Material parameters used to model SFRC of the T-beam.

SFRC Parameters				
Compression			Tension	
f_{cm} (MPa)	E_{cm} (MPa)	ε_o	f_{ct} (MPa)	C_t
64.2	32900	0.003	1.75	40
Shear				
G_o (MPa)	β (left shear span)		β (right shear span)	
16450	0.001		0.15	

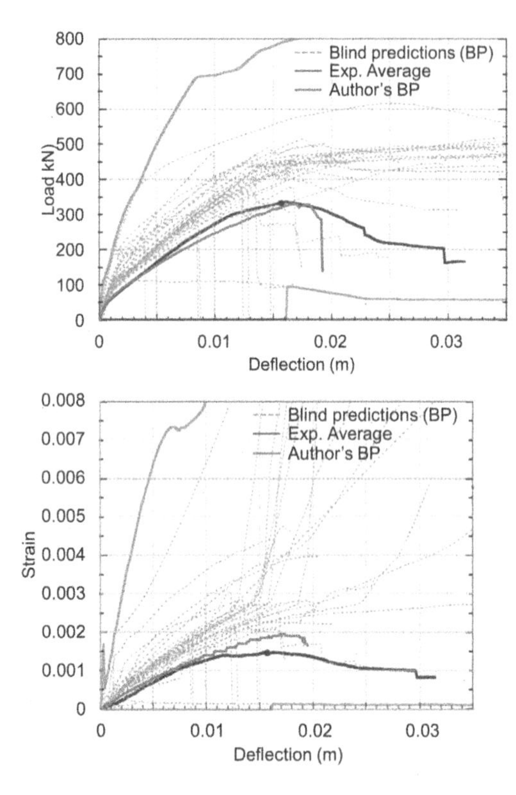

Figure 4. Predicted and experimental load-displacement and strain-displacement responses of the SFRC T-beam.

In the experiment, however, a pronounced softening response was observed which was not captured numerically.

The agreement in terms of longitudinal strain underneath the point of load application was also very good, especially given the complexities associated with capturing (and measuring) local response parameters. The maximum strain remained below the yield strain of 2.3‰, hence shear failure was attained without yielding of longitudinal reinforcement.

It can be seen that several models overestimated the failure load, predicting a ductile flexural failure. Also initiation of cracking was overestimated, which affects the development of the critical shear crack and maximum capacity. Note that the value adopted for the "effective" tensile strength, calibrated from notched beam tests, was 1.75MPa which is lower than typical values.

Tensile strains acting in the crack directions and crack patterns at different displacement levels are shown in Figure 6. Flexural cracks appear at the beginning further developing diagonal shear cracks in the web region. The maximum tensile strains reached 9‰ at failure, which corresponds to crack widths in the order of 5mm. Given the smeared nature of the model and relatively coarse mesh, cracks appear rather distributed in the web region. Note that for graphical representation thin cracks were filtered out at peak load.

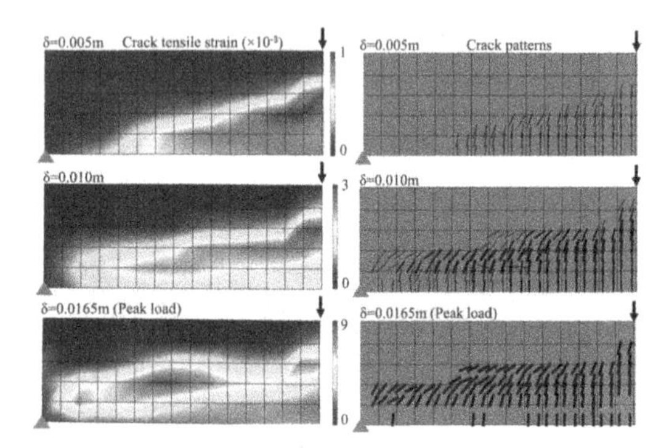

Figure 5. Predicted crack tensile strains and crack patterns at different levels of displacement.

3 INFLUENCE OF MODELLING PARAMETERS

3.1 *Mesh sensitivity*

In the original mesh five elements were used across the depth ("h/5") and forty-five along the span, resulting in an element size of $100\times107\text{mm}^2$. This mesh was refined to an element size of approximately twice the maximum aggregate size ("$2d_{max}$"), with 20 elements across the depth and 180 along the span, resulting in an element size of $25\times27\text{mm}^2$. Two cases of the $2d_{max}$ mesh were investigated: with and without regularization. Regularization was based on keeping the same fracture energy as that of the original coarse mesh. For the exponential tensile-softening law, used to model SFRC, the fracture energy is given as:

$$G_f = \frac{f_{ct} l_{ch}}{C_t} \tag{1}$$

where f_{ct} is the effective tensile strength, l_{ch} is the characteristic length, chosen as the diagonal of the element, and C_t is the exponential decay parameter. For the coarse mesh, equation (1) yields a value of fracture energy equal to 6.43N/mm, which is reasonable for a fiber content of $60 kg/m^3$ (Barros & Sena Cruz 2001). Based on equation (1), the C_t parameter for the regularized $2d_{max}$ mesh is 10. Table 2 summarizes investigated mesh and material parameters and Figure 6 the load-displacement and strain-displacement response. Figure 7 shows the crack patterns at different load levels for the coarse and fine (regularized) mesh.

Table 2. Mesh and material parameters for the investigated meshes.

Mesh	El.Size(mm^2)	l_{ch}(mm)	C_t	G_f(N/mm)
h/5	100 × 107	147	40	6.43
$2d_{max}$	25 × 27	37	40	1.61
$2d_{max}$	25 × 27	37	10	6.43

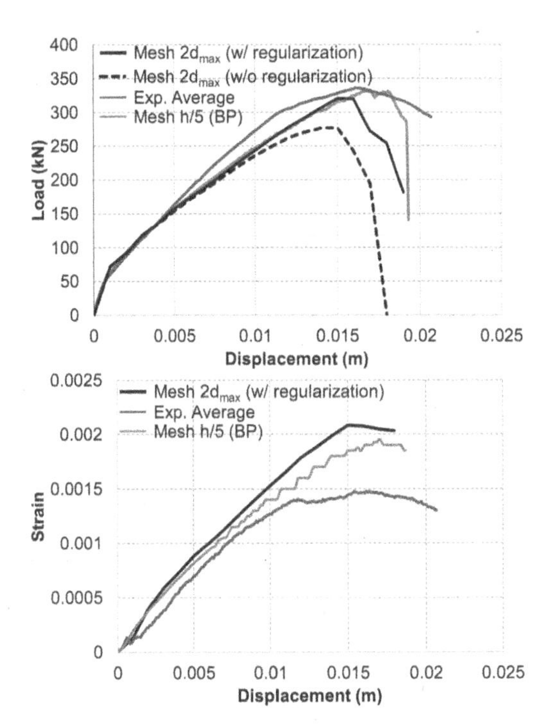

Figure 6. Load-displacement and strain-displacement responses of the T-beam for different meshes and regularization.

The following observations can be made. The applied regularization technique improves objectivity of results. Up to the peak load no significant differences between coarse and fine mesh occur. However, the influence of mesh refinement on the peak load can be still observed. Refining the mesh decreases the peak load about 2% and slightly increases the strain in the bottom reinforcement. Both crack patterns shown the development of a critical shear crack (or cracks) across the web. f_{ct} and C_t parameters of the coarse mesh model were calibrated based on FEA of notched beam tests. These FE analyses used a similar mesh discretization to that of the T-beam (6 elements across the depth). f_{ct} and C_t parameters of the fine mesh model were derived from the coarse mesh model based on the constant fracture energy regularization, using the elements diagonal as the characteristic length.

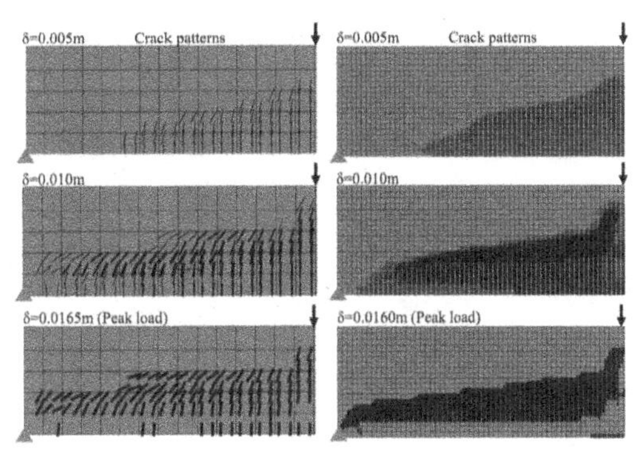

Figure 7. Crack patterns in the left shear span at different load levels for the coarse and fine mesh (regularized).

3.2 *Critical material parameters*

Given the type of failure mechanism (diagonal-tension shear), material parameters related to the tensile and shear response of SFRC were selected for sensitivity analysis. The following parameters were investigated: (i) tensile strength (f_{ct}), (ii) tension-softening parameter (C_t) and (iii) shear retention factor (β) in the left shear span. Sensitivity studies were performed starting from the blind prediction (BP) model and introducing variations to each parameter separately according to Table 3. The coarse mesh (h/5) was used for all investigated models hereafter.

Table 3. Investigated material parameters for each model.

Model	Parameter	f_{ct} (MPa)	C_t	β (left)
BP*	–	1.75	40	0.001
1	f_{ct}	3	40	0.001
2	f_{ct}	4	40	0.001
3	C_t	1.75	30	0.001
4	C_t	1.75	50	0.001
5	B	1.75	40	0.005
6	B	1.75	40	0.01

*BP: blind prediction model

Figures 8 to 10 show the force-displacement and strain displacement responses for each model.

It can be observed that the tensile strength (f_{ct}), besides affecting the cracking point and cracked stiffness, produces a change in the failure mode from brittle shear to flexure-shear, causing yielding of bottom reinforcement and increasing the maximum load by approximately 35%.

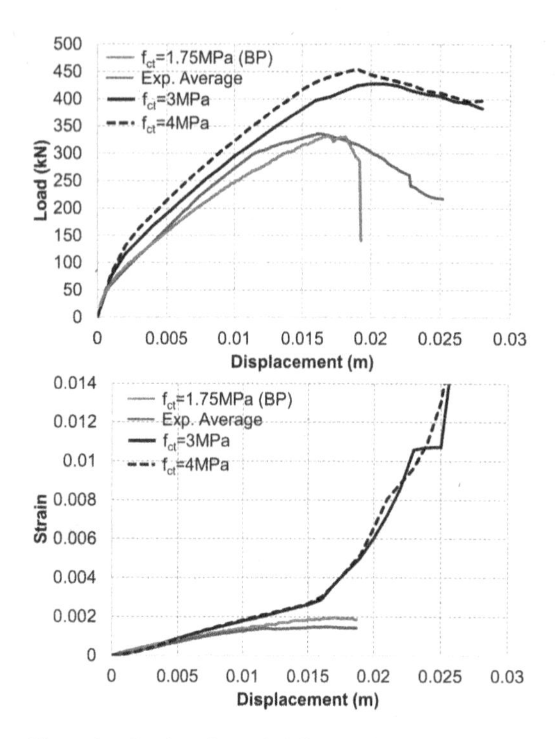

Figure 8. Caption of a typical figure. Photographs will be scanned by the printer. Always supply original photographs.

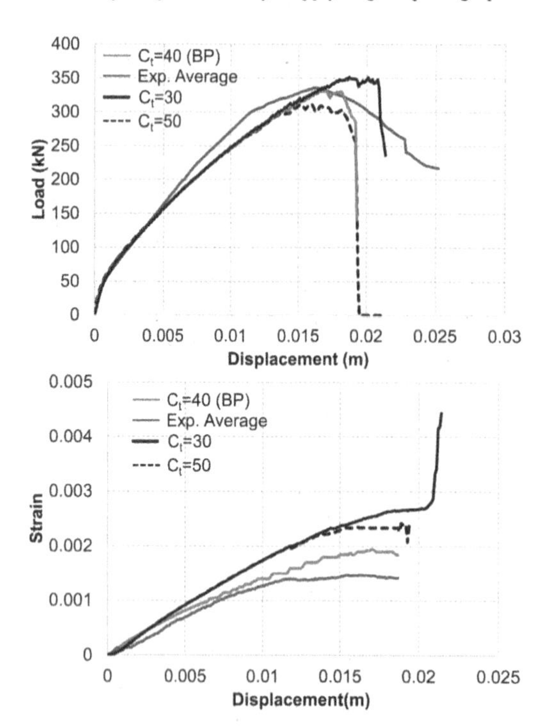

Figure 9. Caption of a typical figure. Photographs will be scanned by the printer. Always supply original photographs.

The tensile-softening parameter (C_t) especially affects the maximum load, increasing or decreasing

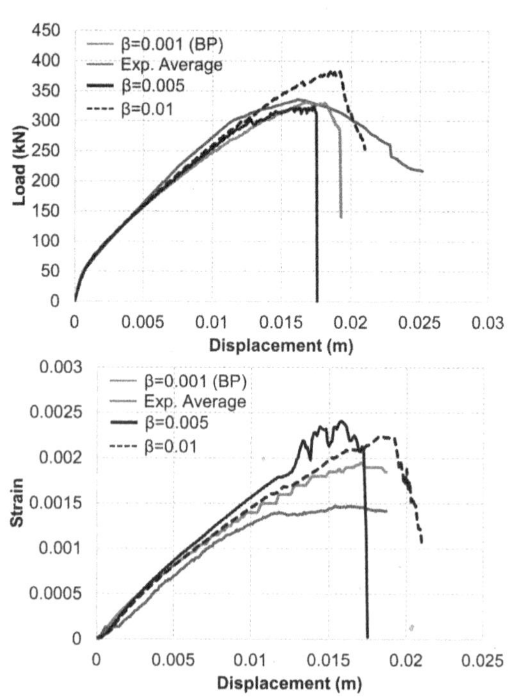

Figure 10. Caption of a typical figure. Photographs will be scanned by the printer. Always supply original photographs.

it by ±10%. Displacement capacity is affected as well but to a lesser extent. Crack initiation and cracked stiffness remains unaffected. Low values of C_t can lead to yielding of bottom reinforcement.

Low values of the shear retention factor (β), below approximately 0.005, do not affect significantly the response. For $\beta > 0.005$ the maximum load increases. The failure mechanism, however, remains the same. No yielding of bottom reinforcement is observed. For $\beta = 0.005$ some numerical fluctuations were observed in the strain response when approaching failure.

Given the coarse mesh and smeared model of reinforcement, all crack patterns were similar to those reported in Figure 5, starting with vertical flexural cracks followed by diagonal shear cracks. For the case of $f_{ct} = 3$MPa more bottom flexural cracks develop compared to diagonal shear cracks in the web.

4 CONCLUSIONS

The blind prediction FE model of a shear critical T-beam and its performance have been presented in detail highlighting the capabilities of numerical simulations for assessment of structural members with conventional reinforcement and steel fibers. Given the relative simplicity of the model, i.e. 2D model, coarse mesh, two-parameter tension-softening model, constant shear retention factor and smeared reinforcement, very good agreement with the experimental failure load, longitudinal strain in the bottom reinforcement and failure mode were achieved. Further investigations concerning mesh sensitivity and material modelling parameters were undertaken. It was shown that reducing the mesh size by a factor of 4 reduced the failure load by 15%, if no regularization was implemented, and by 2% with regularization based on constant fracture energy. The global response was found to be most sensitive to the following material modelling parameters: (i) tensile strength (f_{ct}), (ii) tension-softening parameter (C_t) and (iii) shear retention factor (β). Among these, changes in the failure mode from brittle to ductile were observed when the tensile strength was increased.

REFERENCES

Barros, J., Sanz, B., Kebele, P., Yu, R., Meschke, G., Planas, J., Cunha, V., Caggiano, A., Ozyurt, N., Gouveia, V., Bos., A., Poveda, E., Gal, E., Cervenka, J., Neu, G., Rossi, P., Diasda-Costa, D., Juhasz, P., Cendon, D., Gonzalo, R., Valente, T. 2020. Blind competition on the numerical simulation of steel-fiber-reinforced concrete beams failing in shear. *Structural Concrete*, 1–29.

Barros, J., Sena Cruz, J. 2001. Fracture energy of Steel Fiber-Reinforced Concrete. *Mechanics of Composite Materials and Structures*, 8:1, 29–45.

Kagermanov, A. 2019. Finite element analysis of shear failure of reinforced and prestressed concrete beams. *Hormigón y Acero*, (ACHE), 70(287), 75–84.

Kagermanov A., Ceresa P. 2016. Physically-based cyclic tensile model for RC membrane elements. *Journal of Structural Engineering*, (ASCE), Vol° 142, N° 12, July.

Kagermanov A., Ceresa P. 2018. Fiber-section model with an exact shear strain distribution for RC frame elements. *Journal of Structural Engineering*, (ASCE), Vol° 143, N° 10, October.

Computational Modelling of Concrete and Concrete Structures – Meschke, Pichler & Rots (Eds)
© 2022 Copyright the Author(s), ISBN: 978-1-032-32724-2

Refined and simplified modelling of steel-concrete-steel (SCS) composite beams

R. Calixte
UPL, Univ Paris Nanterre, LEME – Laboratoire Energétique, Mécanique, Electromagnétisme, Ville d'Avray, France
Université Paris-Saclay, CEA, Service d'Études Mécaniques et Thermiques, Gif-sur-Yvette, France

L. Davenne
UPL, Univ Paris Nanterre, LEME – Laboratoire Energétique, Mécanique, Electromagnétisme, Ville d'Avray, France

L. Jason
Université Paris-Saclay, CEA, Service d'Études Mécaniques et Thermiques, Gif-sur-Yvette, France

ABSTRACT: This study presents a general simulation methodology to assess both full and partial composite action of steel-concrete-steel (SCS) structures. This refined methodology, using 3D finite elements, is applied to two three-point bending beams in which a different composite action is provided through variation of the number of studs. The comparison to experimental results validates the methodology and the global and local behavior can be reproduced. However, the important calculation cost reduces the use of this numerical strategy to more complex structures. A simplified methodology is proposed with 1D finite elements to represent the connectors. This modelling choice allows to greatly reduce the computational cost. It also imply a reduction of the reproduced phenomena. Particularly, local damage of the concrete core around the dowels is slightly different. Nevertheless, this strategy allows to accurately reproduce the global behavior as well as the failure modes of SCS beams with full and partial composite action.

1 INTRODUCTION

The Strength, stiffness and durability requirement for civil engineering constructions are steadily increasing. To fulfil these new needs, studies are launched to develop new structural materials with higher specifications. Steel-concrete-steel (SCS) composite structures are one of them (Leekitwattana et al. 2010; Leng & Song 2016; Varma et al. 2015). This composite structure is composed of a concrete core caught between two steel plates. The bond between the components is made thanks to a connection system, generally performed through steel dowels and/or ties (Figure 1). This component is responsible for composite action and ensures the overall behavior of the SCS structure. The structure thus composed has the advantages of reinforced concrete like a good strength and stiffness thanks to the optimal use of the concrete and of the steel. Moreover, the external place of the steel plate increases the stiffness, the sustainability, and the strength under some extreme solicitations (Booth et al. 2015; Bowerman et al. 2002; Oduyemi & Wright 1989; Yan et al. 2015). It also allows their use as lost formwork, which can be prefabricated and SCS are modular structures (Leekitwattana et al. 2010; Schlaseman 2004; Varma et al. 2015). SCS structures have been gradually used in bridge deck (Yan et al. 2015), for the construction of shear walls in high buildings (AISC 2017), for submerged tunnels (Calatrava 2013; Bekarlar 2016) and for blast and impact shield walls or liquid and gas containers (Wright et al. 1991, Liew et al. 2016).

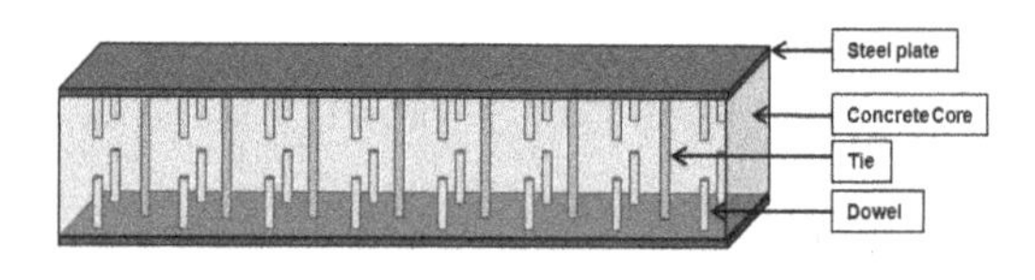

Figure 1. Geometry of a SCS beam with dowels and ties.

The study of bending beams allows to better understand the behavior of this type of structural material. With this aim, the number of research on SCS beams has increased since 1975 (Montague 1975). Several behaviors have been identified depending on the geometric and material characteristics of the structures (Sener et al. 2016; Wright et al. 1991; Yan et al. 2014). Particularly, depending on the degree of composite action, the SCS beam present a full or a partial composite action. Directly affected by the number of connectors, the difference of composite action will

DOI 10.1201/9781003316404-40

impact the global behavior and the failure mode of the structure (Dogan & Roberts 2010; Zhang et al. 2020). The study of both types of behavior is necessary to consider all scenarios, such as design and construction choices, loss of structural integrity or construction difficulties, among others (Lin et al. 2019; Qin et al. 2015; Zhang et al. 2020).

2 MODELING STRATEGY

In this section, a numerical methodology is proposed to represent both full and partial composite actions in SCS structures. It is applied on two representative three-point bending beams.

2.1 *Experiment*

The SP1-1 and SP1-2 beams of the experimental study of Sener et al. (2016) are considered. They have the same geometry (Figure 2 and Table 1) but include a different number of welded headed shear studs. The beams are loaded in three-point bending (simply supported with a load applied at the mid-span). The applied load, the vertical displacements under the loading point, and the cracking evolution are experimentally monitored.

The experimental material properties are given in Tables 2 and 3.

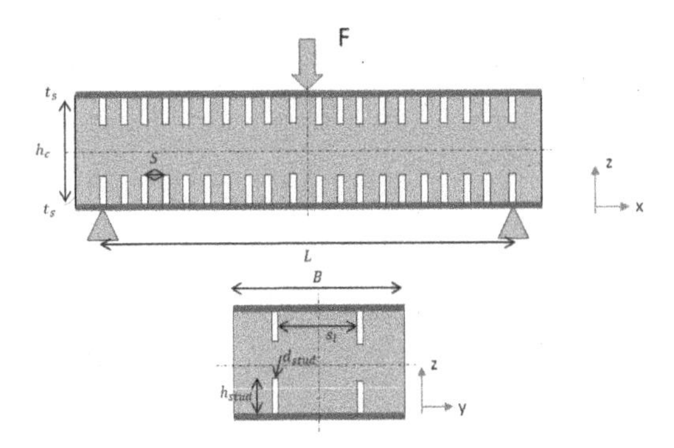

Figure 2. Geometry of SP1 beam.

Table 1. Geometrical parameters of SP1-1 and SP1-2 beams.

	Symbol	SP1-1	SP1-2
Nr of dowels per steel plate	n_{stud}	40	20
Spacing of dowels (length)	S (mm)	152.4	304.8
Length of beam	L (mm)	2896	
Width of beam	B (mm)	305	
Thickness of steel plates	t_s (mm)	6.5	
Height of concrete core	h_c (mm)	445	
Diameter of dowels	d_{stud} (mm)	12.7	
Height of dowels	h_{stud} (mm)	63.5	
Spacing of dowels (width)	sl (mm)	152	

Table 2. Concrete properties.

Compressive strength	f_c (MPa)	42
Tensile strength*	f_{ct} (MPa)	3.15
Young modulus*	E_c (GPa)*	33.85
Poisson's ratio*	ν_c (–)	0.2

* Obtained with Eurocode 2 (CEN 2004a) formulas

Table 3. Steel properties.

		Plates	Dowels
Yield limit	f_y (MPa))	448	489
Young modulus	E_s (GPa)	201	201
Hardening modulus	E_T (GPa)	0.42	0.42
Poisson's ratio	ν_s (–)	0.3	0.3

2.2 *Material behavior modeling*

Concrete behavior is simulated using an isotropic damage model based on Mazars' model (Mazars 1984) with a regularized damage evolution in tension and in compression through the Hillerborg et al. (1976) method. This law introduces a scalar variable D that quantifies the influence of microcracking:

$$\sigma_{ij} = (1 - D)C_{ijkl}\varepsilon_{kl} \tag{1}$$

where σ_{ij} and ε_{kl} are respectively the stress and strain components, respectively, C_{ijkl} is the fourth order elastic tensor and D is the damage variable. For the description of the damage growth, an equivalent strain is introduced from the local strain tensor:

$$\varepsilon_{eq} = \sqrt{\sum_{i=1}^{3}(<\varepsilon_i>_+)^2} \tag{2}$$

where $<\varepsilon_i>_+$ are the positive principal strains.

The loading surface g is defined by:

$$g(\varepsilon, D) = \tilde{d}(\varepsilon) - D \tag{3}$$

where the damage variable D is also the history variable which takes the maximum value reached by $\tilde{d}$ during the history of loading

$$D = \max(\tilde{d}, 0) \tag{4}$$

$\tilde{d}$ is defined by an evolution law which distinguishes the mechanical responses of the material in tension and in compression by introducing two scalars D_t and D_c.

$$\tilde{d}(\varepsilon) = \alpha_t(\varepsilon) D_t(\varepsilon_{eq}) + \alpha_c(\varepsilon) D_c(\varepsilon_{eq})$$

$$D_t = 1 - \frac{\kappa_0}{\varepsilon_{eq}} \exp\left(\frac{l_e.f_{ct}}{G_F}(\kappa_0 - \varepsilon_{eq})\right)$$

$$D_c = 1 - \frac{\kappa_0 (1 - A_c)}{\varepsilon_{eq}} - \frac{A_c}{\exp[B_c(\varepsilon_{eq} - \kappa_0)]} \tag{5}$$

$$\alpha_{t,c} = \left(\sum_{i=1}^{3} \frac{< \varepsilon_i^{t,c} >< \varepsilon_i >_+}{\varepsilon_{eq}^2} \right)^{\beta}$$

D_t and D_c are the tensile and compressive parts of the damage, respectively. The weights α_t and $\alpha_c$are computed from the strain tensor. They are defined as functions of the principal values of the strains $\varepsilon_{ij}^t$and ε_{ij}^c due to positive and negative stresses respectively. The parameter β reduces the effect of damage under shear compared to tension. For the regularization in tension, D_t involves l_e, the average size of the finite element (cubic root of the element volume), G_F the fracture energy and f_{ct} the tensile strength. κ_0 is a parameter (equal to the ratio between the tensile strength and the Young's modulus) and represents the initial threshold from which damage grows. A_c and B_c are two parameters for the compression damage evolution. For the regularization in compression, they are calibrated from uniaxial compression simulations to obtain the same stress – displacement curve for different values of element size l_e. The calibration process is thus based on a constant compressive cracking energy concept, as defined by van Mier 1984.

$$D_c = 1 - \frac{\kappa_0 (1 - A_c)}{\varepsilon_{eq}} - \frac{A_c}{\exp\left[B_c \left(\varepsilon_{eq} - \kappa_0\right)\right]} \tag{6}$$

The Kuhn – Tucker conditions finally determines the evolution of damage:

$$g \leq 0, \quad \dot{d} \geq 0, \quad g\dot{d} = 0 \tag{7}$$

From the experimental data resumed in Table 2 the model parameters in Table 4 are chosen to reproduce the concrete behavior.

Table 4. Concrete model parameters.

A_c	$68l_e^2 + 19l_e$	*
B_c	$26000l_e + 1$	*
G_F ($J{\cdot}m^{-2}$)	150	
κ_0	9.31 10^{-5}	
β	0.6	

* For l_e in m

For the steel plates and the steel dowels, an elastic plastic behavior with an isotropic hardening is chosen.

To be able to capture the shear failure of the dowel, an extremely refined mesh would be required in the plate-stud interface area, leading to very high computational costs. To avoid this, zero-dimension junction elements are used, connecting each stud node to the associated plate node. Their force-displacement law is elastoplastic in the tangential direction, with a very stiff elastic part ($K_s = 10^{12} N/m$). Each junction element has a yield limit proportional to the area attached to the node. The sum of the yield limits of the junction elements is equal to the shear failure of the stud P_{Rd} which is calculated as in the Eurocode 4 (CEN 2004b) without the safety factor:

$$P_{Rd} = 0,8 f_y A_{stud} = 49.5kN \tag{8}$$

2.3 *Refined numerical modelling*

Considering the symmetries, only one fourth of the beams are modelled (Figure 3 and Figure 4). The size of the concrete finite elements ranges from 1.6 mm (near the dowels connector) to 25 mm (far from the connectors).

Given the expected behavior of the structure, a particular attention is paid to the bond between concrete and steel (Figure 5). A one-sided contact relationship is considered between steel plates and concrete, that allows for normal separation and a free slip in the tangential directions (partial bond). The same condition is applied between concrete and dowels. The stud heads are not meshed in detail. A simple perfect bond is imposed at the end of each dowel.

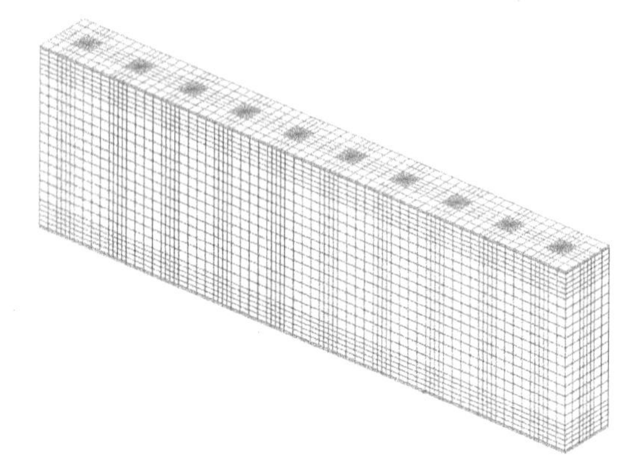

Figure 3. Mesh of SP1-1 beam.

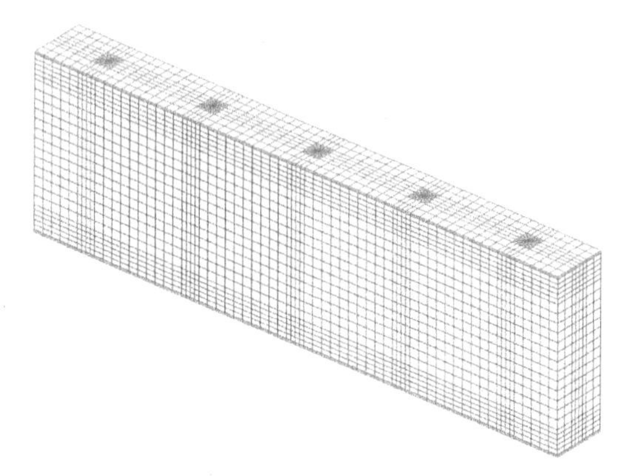

Figure 4. Mesh of SP1-2 beam.

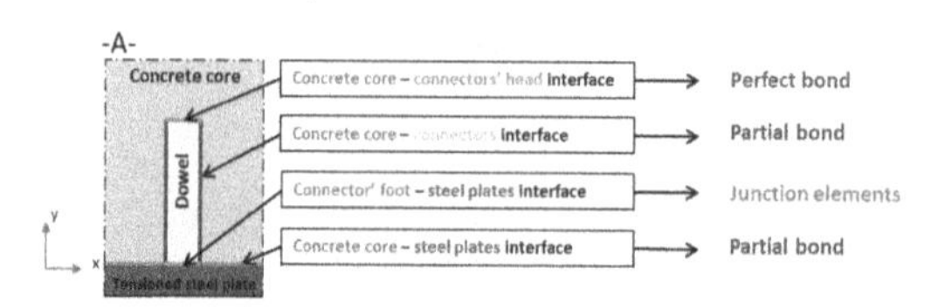

Figure 5. Interfacial bonds between steel and concrete.

The global boundary conditions on the beam are symmetry conditions and displacement in the vertical direction blocked along lines at the position of the experimental support (Figure 6). Finally the loading is imposed through a vertical displacement applied on the upper steel plate at the position of the experimental loading system.

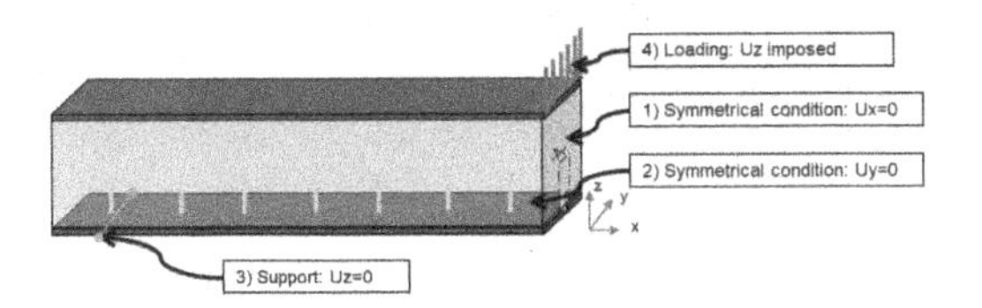

Figure 6. Boundary conditions on the beam.

The simulations are performed using the implicit finite element code Cast3M (CEA 2021).

3 NUMERICAL SIMULATIONS

3.1 *SP1-1 beam*

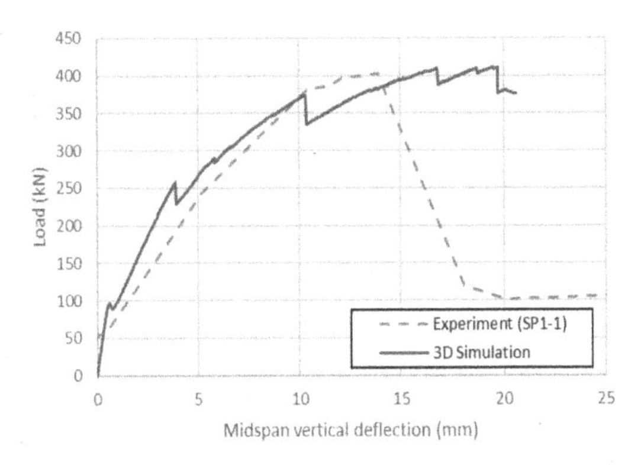

Figure 7. Load – midspan vertical displacement curves for the SP1-1 beam.

Figure 7 presents the global response of SP1-1 beam.

The general mechanical behavior is obtained by the simulation (elastic regime and mechanical degradation). The structural strength is reproduced. For a 13.9 mm deflection (the one of the experimental ruin), the strength obtained with the 3D simulation is 382.02 kN, 6% different of the experimental result. Several partial discharges are observed in the numerical curve. The first one corresponds to the initiation of the vertical flexural crack. As for the test, this crack appears quickly and modifies the stiffness of the structure. The second unloading represents the opening of the 45° inclined concrete shear crack. This crack appears for an applied force of 258 kN. The following discharge corresponds to the opening of concrete cracks in the lower part of the beam, illustrating the propagation of damage parallel to the bottom plate.

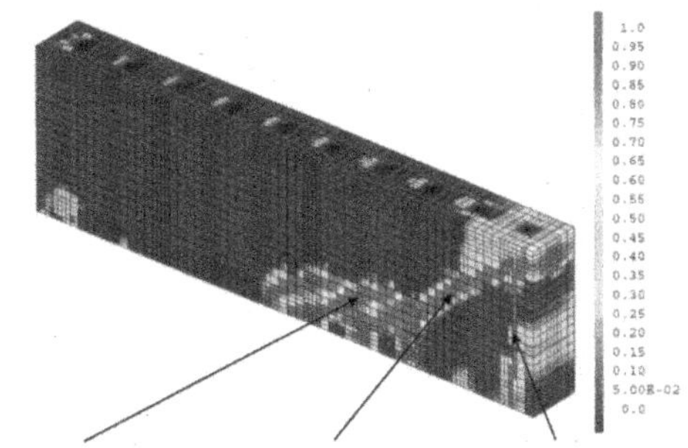

Figure 8. Final damage distribution in concrete for the SP1-1 beam.

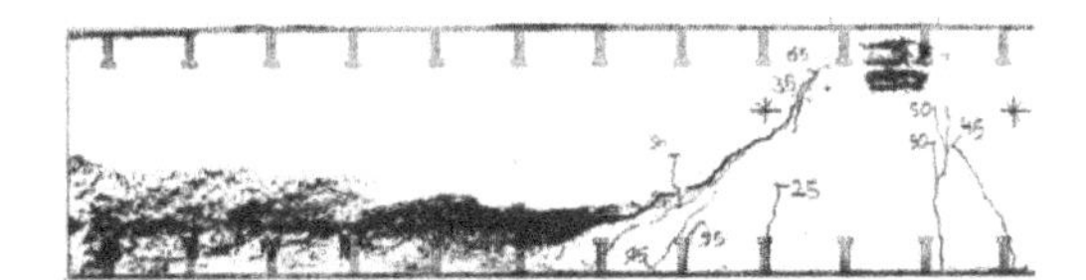

Figure 9. Experimental final crack pattern for SP1-1 beam.

The damage distribution obtained in the simulation (Figure 8) is like the experimental crack pattern (Figure 9). The experimental and numerical longitudinal strains in the steel plates are also in agreement (Figure 10). A local yielding of the bottom steel plate is observed near the position of the shear crack feet for the numerical simulation. For the experimental results, this yielding is visible in the right span of the beam, at the same distance of the midspan. It is to be noted that the pic of strain in the top steel plate at the midspan for the simulation is due to the concentrated applied load (on a line).

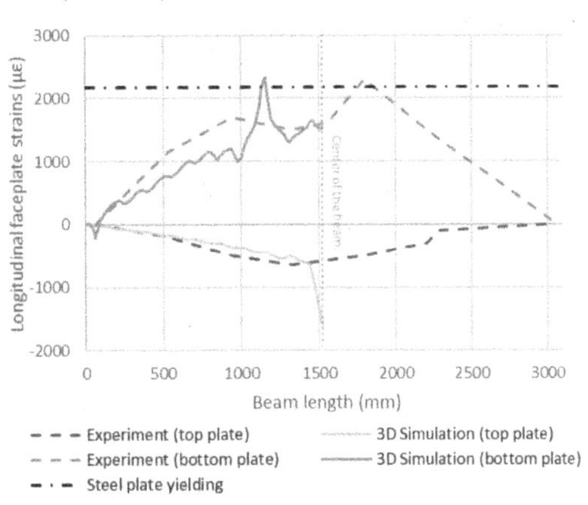

Figure 10. Longitudinal strain in the plates along SP1-1 beam for a deflection of 13.9 mm.

3.2 *SP1-2 beam*

The global mechanical behavior of SP1-2 beam is also correctly captured (Figure 11). The strength is lower (250 kN compared to 400 kN for SP1-1).

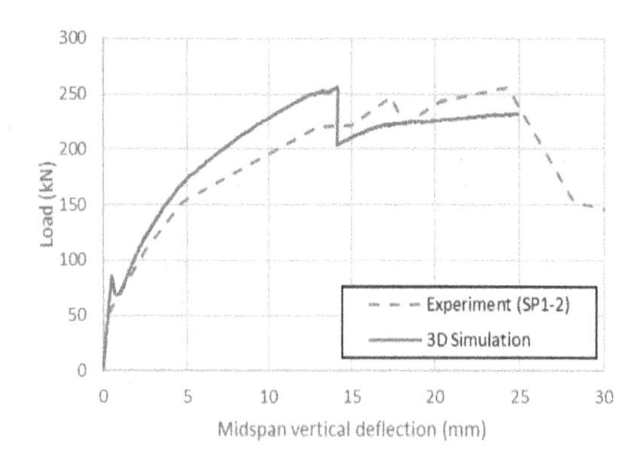

Figure 11. Load – midspan vertical displacement curves for the SP1-2 beam.

As for the SP1-1 beam, the first discharge corresponds to the initiation of a vertical flexural crack at the dowel near the midspan. The second discharge appears at a deflection of 13.9 mm and a load of 256.15 kN. This drop of load corresponds to the apparition of a shear crack inclined at more than 45° (Figure 12) which is not mentioned experiment-ally. However, Zang et al. (2020) obtained this fai-lure mode for beams with a low number of dowels.

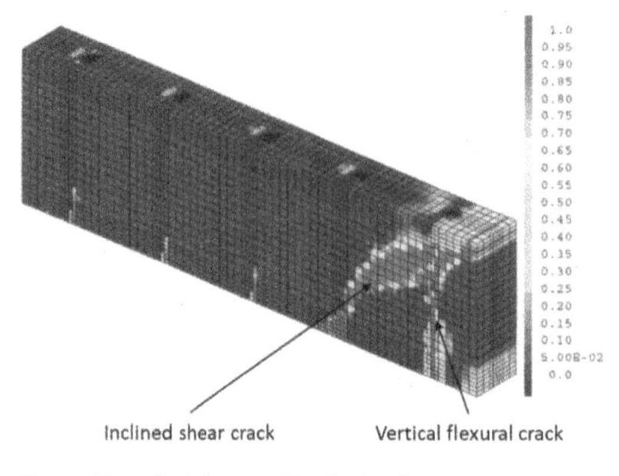

Figure 12. Final damage distribution for the SP1-2 beam.

After the peak load, a constant force is observed for an increasing displacement. Due to the yielding of the dowels, the shear force that can be transferred between the stud and the bottom steel plate has reached its maximum. This is coherent with the experimental failure due to a vertical flexural crack as seen on Figure 13. One can observe a break in the connection of the lower plate due to the failure of the studs.

Finally, the numerical curve shows a strength of the beam of 231.7 kN at a deflection of 25.4 mm (the one of the experimental ruin), less than 8% different of the experimental result.

The simulation of the longitudinal strain of the steel plate shows a slight underestimation of the bottom plate tensile strains compared to the experimental one (Figure 14).

Figure 13. Experimental failure of SP1-2 beam.

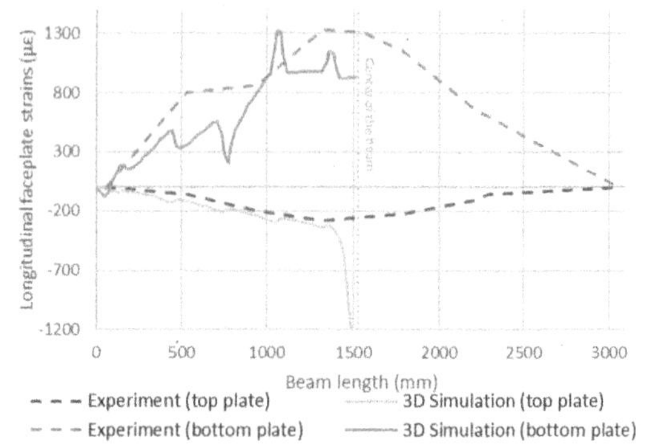

Figure 14. Longitudinal strain in the plates along SP1-2 beam for a deflection of 24.3 mm.

3.3 *Discussion*

The observed differences between experiment and simulation may be explained by some model simplifications: the contact relation without friction between the concrete core and the steel plate, the simplification of the stud heads by a perfect bond to the concrete or the simplification of the behavior of the dowel – steel plates junction elements with a perfect elastic plastic constitutive law. However, the proposed numerical methodology can reproduce the global and local behaviors for both beams. Especially, the differences between a full and a partial composite action beams are obtained: the decrease in the strength and the stiffness, the increase in the ductility and the change in the failure mode.

As the experimental results, the numerical results show a difference in strength and failure behavior between the SP1-1 and SP1-2 beams. In the first one, the number of dowels is sufficient to assure a full composite action. The connection system can support the shear force corresponding to the yielding of the bottom steel plate (Figure 10). This yielding corresponds to the ultimate strength reachable, even with a perfect bond between steel and concrete. It is to be noted that the bottom steel plate is more loaded plate due to the cracking of concrete in tension. On the contrary, in the second one, the connection system is not strong enough and is the weak link. It fails before reaching the yielding of the bottom plate, as it can be seen in Figure 14. In this case there is a partial composite action.

Based on this modeling strategy, one can perform different simulations increasing progressively the number of dowels in the beam and determine the minimum number of studs to reach a full composite action. This work has been done, including a comparison to the provisions of different design codes (Calixte 2021).

4 SIMPLIFIED MODELING

The numerical modelling strategy previously developed allows to represent finely the behavior of the composite structure connection system in shear. However, it can be costly in terms of implementation and calculation time, which reduces its use for SCS industrial structures with larger dimensions and complex geometries. To propose a less expensive numerical simulation, a 1D simplification of the connectors is studied. The representation of the connectors by one-dimensional beam finite elements anchored in the concrete will facilitate the realization of the mesh, reduce the number of nodes and elements, and simplify the interfacial conditions between the components of the system.

4.1 *Simplified modeling*

The implementation of the 1D simplification requires the modification of the interfacial bond conditions (Figure 15). A perfect bond between the concrete and the studs is imposed through kinematic conditions (Lagrange multipliers). A particular attention is paid to the concrete mesh around the studs. The 1D steel beam element nodes should be in the middle of the concrete solid element in which they are embedded, and the size of these solid elements should be near the dowel diameter dimension. For the interface between the stud and the steel plate, a 0D plastic junction element represents the bond between the stud and the steel plate. Finally, the rotation of the studs around their axis are blocked.

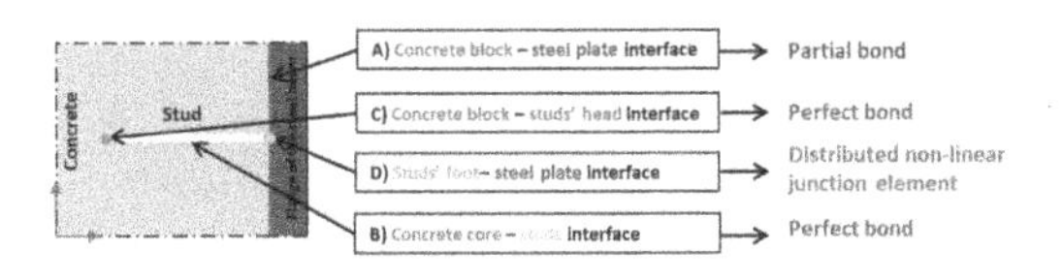

Figure 15. Interfacial bonds of the 1D simplified modelling.

Simulations on push-out tests with 1D elements for the studs showed that, compared to experimental results, the connection system modeled in this way leads to an over-rigidity of the connection behavior (Calixte 2021). The 1D modelling of the studs does not allow to reproduce finely the interactions with the concrete core because the perfect bond reduce the allowable strain of the system. On the other hand, the yield plateau of the stud – steel beam interface junction element is found, thanks to the plastic behavior of the 0D junction element at the stud foot.

To implicitly consider all the phenomena characterizing the shear response of the connection system (crushing and tearing of the concrete, shearing of the connectors, yield of the steel beam), a constitutive law reproducing of the push-out test response is adopted for the 0D junction element at the studs – steel plate interfaces. The chosen law is the one developed by Ollgaard et al. (1971) (Figure 16):

$$P = P_{Rd}\left(1 - \exp\left(-\frac{18}{25.4}\delta\right)\right)^{\frac{2}{5}} \tag{9}$$

where P_{Rd} is the shear strength of the connection in the push-out test, including the shear failure of the stud and the failure of the concrete under the pulling out of the stud:

$$P_{Rd} = \min\left(0.8 f_y A_{stud}; 0.5 A_{stud}\sqrt{f_c E_c}\right) \tag{10}$$

Calixte (2021) showed that with this nonlinear behavior for the 0D element at the dowel foot, the simulations on push-out tests gave a good agreement with the experiment in the global force - displacement response. This is evident since it is the input in the model. But the damage evolution in the concrete around the 1D stud elements is also like the one in the 3D reference simulations, which is not the case if there is no 1D elements (only 0D elements connecting directly steel plate to concrete core).

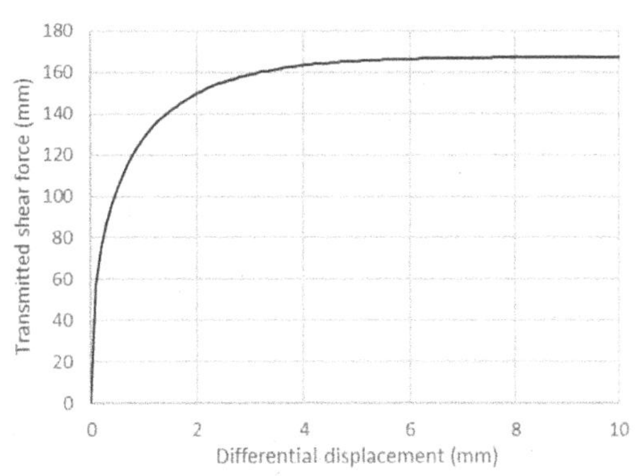

Figure 16. Ollgaard et al. (1971) push-out law.

4.2 *Numerical simulations with simplified model*

The results of the simplified modeling of SP1-1 SCS beam are presented in Figure 17 to Figure 19. The global force-displacement curve of the simplified modeling is like the one of the refined simulation (Figure 17). The difference lies in the forces where the shear crack appears, and where it propagates. The diagonal crack apparition is for a force $F = 338kN$ instead of $F = 258kN$ in the refined simulation with a 3D mesh for the dowels. The mechanical degradation is delayed in the simulation with the 1D beam elements for the dowels.

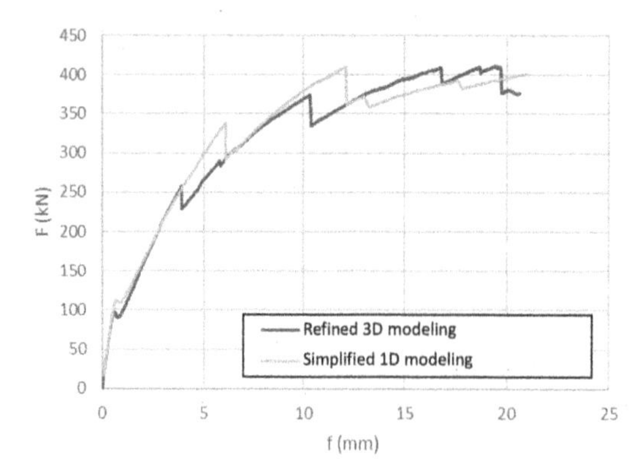

Figure 17. Displacement-force curves for SP1-1 specimen (simplified simulation).

At the end, the damage distribution for the simplified simulation is like the refined simulation one (Figure 18 compared to Figure 8). For the strains in the steel plates (Figure 19), one can see that the yielding in the bottom plate is still not reached in the 1D simulation for a displacement equal to 13.4 mm while it is for the 3D simulation. But the pic near the shear crack foot is here and yielding will soon appear.

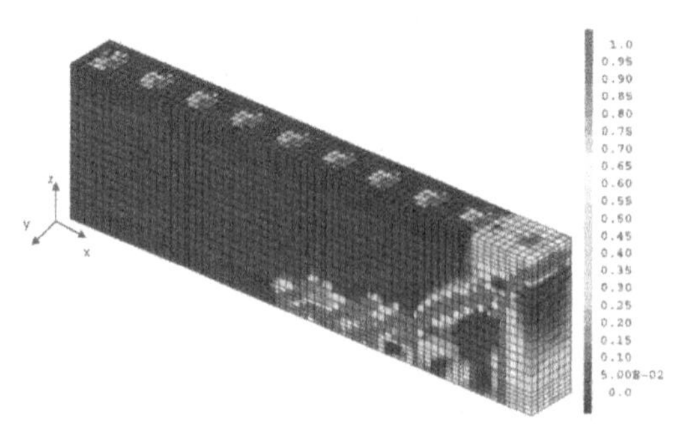

Figure 18. Final concrete damage pattern for the simplified simulation of SP1-1 specimen.

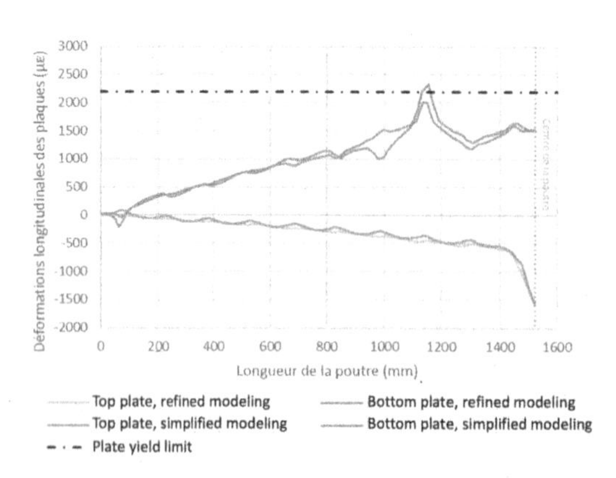

Figure 19. Longitudinal strain in the plates along SP1-1 beam for a deflection of 13.9 mm (simplified modeling simulation).

Equivalent observations are visible for the modeling of the SP1-2 beam (Figures 20 to 22). With this simplified 1D modeling, the crack inclined at more than 45° is not visible displacing the maximum longitudinal strains in the lower plate at the single vertical crack position.

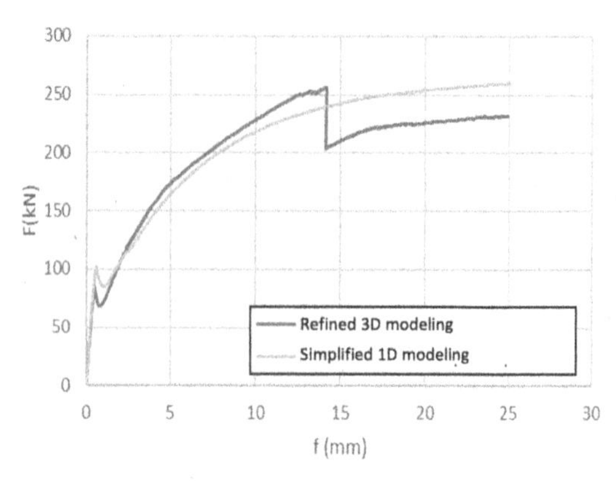

Figure 20. Displacement-force curves for SP1-2 specimen (simplified simulation).

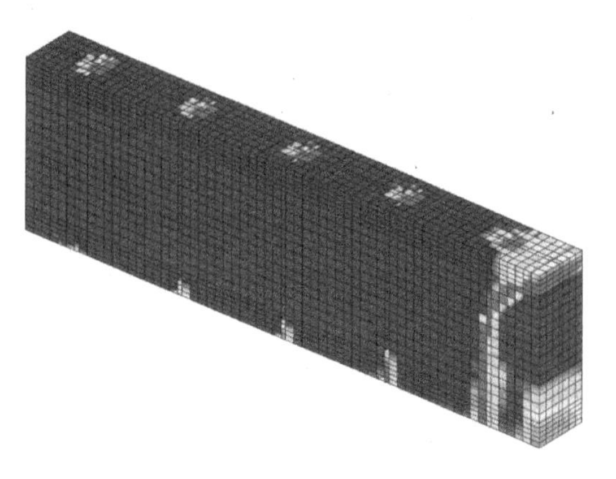

Figure 21. Final concrete damage pattern for the simplified simulation of SP1-2 specimen.

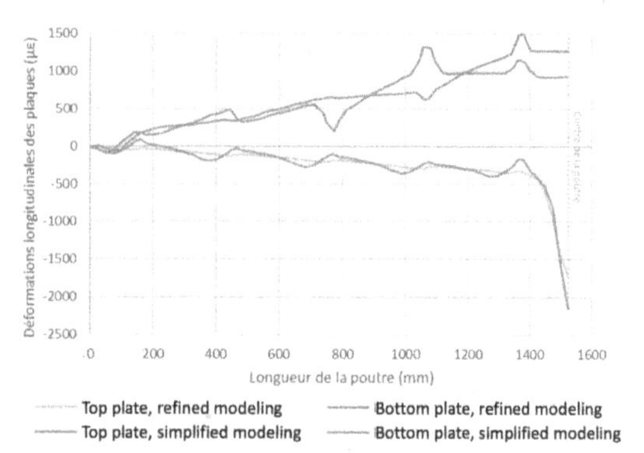

Figure 22. Longitudinal strain in the plates along SP1-2 beam for a deflection of 24.3 mm (simplified modeling simulation).

The main advantage of the simplified simulation is the time to prepare the mesh and the great gain in the computation time (Table 5).

Table 5. Comparison of calculation times.

	SP1-1		SP1-2	
	3D	1D	3D	1D
Number of nodes	51,410	17,160	34,940	15,900
Calculation time (h)	334	78	258	70

5 CONCLUSION

Steel-concrete-steel composite structures are sandwich composite structures combining steel plates and a concrete core through a connection system, which ensures the overall behavior. The structure combines the advantages of reinforced concrete and provides a greater resistance under extreme loading, sustainability, and durability. Moreover, the external position of the steel plates allows their use as formwork and leads to a modular structure, which tends to reduce and ease the construction phase. All these advantages make SCS construction a competitive choice in the construction field.

In this contribution, a general simulation methodlogy was proposed to assess both full and partial composite actions using 3D finite elements. It was validated by comparison to experimental results on three point bending beams. The full composite action was associated to a core concrete shear failure and a local yielding of the bottom steel plate, while the partial composite action was driven by a shear failure of the studs.

Nevertheless, the refined modeling strategy leads to significant computation times. Based on this modeling, a simplified modeling strategy with 1D elements to represent the studs has been develop-ped. The constitutive law of the junction element at the interface stud-plate includes the global law of the push-out test to implicitly consider the concrete – stud interaction not represented by this 1D modelling of the studs. Simulations with this simplified strategy led to significantly reduced computation times and the results, both in terms of global behavior and local degradation, are very similar to those obtained with the refined modeling.

ACKNOWLEDGMENTS

The authors gratefully acknowledge the financial and technical support of EDF R&D for the development and the analysis of the simulation results.

REFERENCES

AISC. 2017. Modern Steel Construction – Steel Core System Revolutionizes High-Rise Construction. *https://www.aisc.org/modernsteel/news/*

Bekarlar, K. 2016. Steel-Concrete-Steel Sandwich Immersed Tunnels For Large Spans, *Master thesis dissertation, Technische Universiteit Delft, Holland.*

Booth, P.N., Varma, A.H., Sener, K.C. & Malushte, S.R. 2015. Flexural behavior and design of steel-plate composite (SC) walls for accident thermal loading. *Nuclear Eengineering and Design* 295: 817–828. https://doi.org/10.1016/j.nucengdes.2015.07.036

Bowerman, H., Coyle, N. & Chapman, J.C. 2002. An innovative steel-concrete construction system. *Structural Engineer* 80(20): 33–38.

Calatrava 2013. Sharq Crossing – Santiago. *Calatrava Architects & Engineers https://calatrava.com/projects/*

Calixte, R. 2021. Simulation of the behavior of steel-concrete-steel structures under mechanical loading. *PhD thesis, University Paris Nanterre, France. (in French)*

CEA Commissariat à l'Energie Atomique et aux energies renouvelables. 2021. Cast3M structural and fluid mechanics calculation code. http://www-cast3m.cea.fr

CEN European Committee for Standardization. 2004a. Eurocode 2: Design of concrete structures – Part 1-1: General rules and rules for buildings. *EN 1992-1-1*

CEN European Committee for Standardization. 2004b. Eurocode 4: Design of composite steel and concrete structures – Part 1-1: General rules and rules for buildings. *EN 1994-1-1*

Dogan, O. & Roberts, T. 2010. Comparing experimental deformations of steel-concrete-steel sandwich beams with full and partial interaction theories. *International Journal of the Physical Sciences,* 5(10): 1544–1557. https://doi.org/10.1016/j.engstruct.2018.12.025

Hillerborg, A., Modéer M. & Peterson, P.E. 1976. Analysis of crack formation and crack growth in concrete by means of fracture mechanics and finite elements. *Cement and Concrete Research* 6: 773–792.

Leekitwattana, M., Boyd, S.W. & Shenoi, R.A. 2010. An alternative design of steel concrete steel sandwich beam, *9th International Conference on Sandwich Structures, Pasadena, United States, 14–16 June 2010.*

Leng, Y.B. & Song, X.B. 2016. Experimental study on shear performance of steel-concrete-steel Sandwich Beams, *Journal of Constructional Steel Research* 38(4): 257–279. *https://doi.org/10.1016/j.jcsr.2015.12.017*

Liew, J.Y.R., Yan J.B. & Huang Z.Y. 2016. Steel-concrete-steel sandwich composite structures – recent innovations, *Journal of constructional Steel research* 130(3): 202–221. *https://doi.org/10.1016/j.jcsr.2016.12.007*

Lin, Y., Yan, J., Wang, Y., Fan, F. & Zou, C. 2019. Shear failure mechanisms of SCS sandwich beams considering bond-slip between steel plates and concrete. *Engineering Structures* 181: 458–475. https://doi.org/10.1016/j.engstruct.2018.12.025

Mazars, J. 1984. Application de la mécanique de l'endommagement au comportement non linéaire et à la rupture du béton de structure. *Ph.D. thesis, University Paris 6, France*

Mier, van, J.G.M. 1984. Strain-softening of concrete under multiaxial loading conditions, *PhD thesis, Technische Hogeschool Eindhoven.* https://doi.org/10.6100/IR145193

Montague, P. 1975. A simple Composite Construction for cylindricl shells subjected to external pressure. *Journal Mechanical Engineering Science* 17(2): 105–113. *https://doi.org/10.1243/JMES_JOUR_1975_017_016_02*

Oduyemi, T. & Wright, H. 1989. An Experimental Investigation into the Behaviour of Double-Skin Sandwich Beams. *J. Construct. Steel Research*, 14(3): 197–220. *https://doi.org/10.1016/0143-974X(89)90073-4*

Ollgaard, J.G., Slutter, R.G. & Fisher, J. 1971. Shear strength of stud connectors in lightweight and normal-weight concrete. *AISC Engineering Journal*, 8: 55–64.

Qin, F., Kong, Q., Li, M., Mo, Y L, Song G. & Fan, F. 2015. Bond slip detection of steel plate and concrete beams using smart aggregates. *Smart Materials and Structures* 24(11): *115039*. https://doi.org/10.1088/0964-1726/24/11/115039

Schlaseman, C. 2004. Application of Advanced Construction technologies to New Nuclear Power Plants. *MPR-2610, prepared for the US Department of Energy under contract for DE-AT01-02NE23476. https://www.nrc.gov/docs/ML0931/ML093160836.pdf*

Sener, K. C., Varma, A. H. & Seo, J. 2016. Experimental and numerical investigation of the shear behavior of steel-plate composite (SC) beams without shear reinforcement. *Engineering Structures.* 127: 495–509. https://doi.org/10.1016/j.engstruct.2016.08.053

Varma, A.H., Malushte, S.R. & Lai Z. 2015. Modularity & Innovation using steel-concrete-steel composite (SC) walls for nuclear and commercial construction, *Advances in Steel-Concrete Composite Structures, Proc 11th International Conference, Beijing, China, 3-5 December 2015. https://doi.org/10.13140/RG.2.1.4665.4804*

Wright, H.D., Oduyemi T.O.S. & Evans H.R. 1991. The experimental behavior of double skin composite elements. *Journal of Constructional Steel Research* 19(2): 97–110. *https://doi.org/10.1016/0143-974X(91)90036-Z*

Yan, J.-B., Liew, J. R., Zhang, M.H. & Wang, J. 2014. Ultimate strength behavior of steel-concrete-steel sandwich beams with ultra-lightweight cement composite, Part 1: Experimental and analytical study. *Steel and Composite Structures* 17(6): 907–927. *http://dx.doi.org/10.12989/scs.2014.17.6.907*

Yan, J.-B., Liew, J. R., Zhang, M.-H. & Sohel, K. 2015. Experimental and analytical study on ultimate strength behavior of steel-concrete-steel sandwich composite beam structures. *Materials and Structures* 48: 1523–1544. *https://doi.org/10.1617/s11527-014-0252-4*

Zhang, W. Huang, Z., Fu, Z., Qian, X., Zhou, Y. & Sui, L. 2020. Shear resistance behavior of partilly composite Steel-Concrete-Steel sandwich beams considering bond-slip effect. *Engineering Structures,* 210: 110394. *https://doi.org/10.1016/j.engstruct.2020.110394*

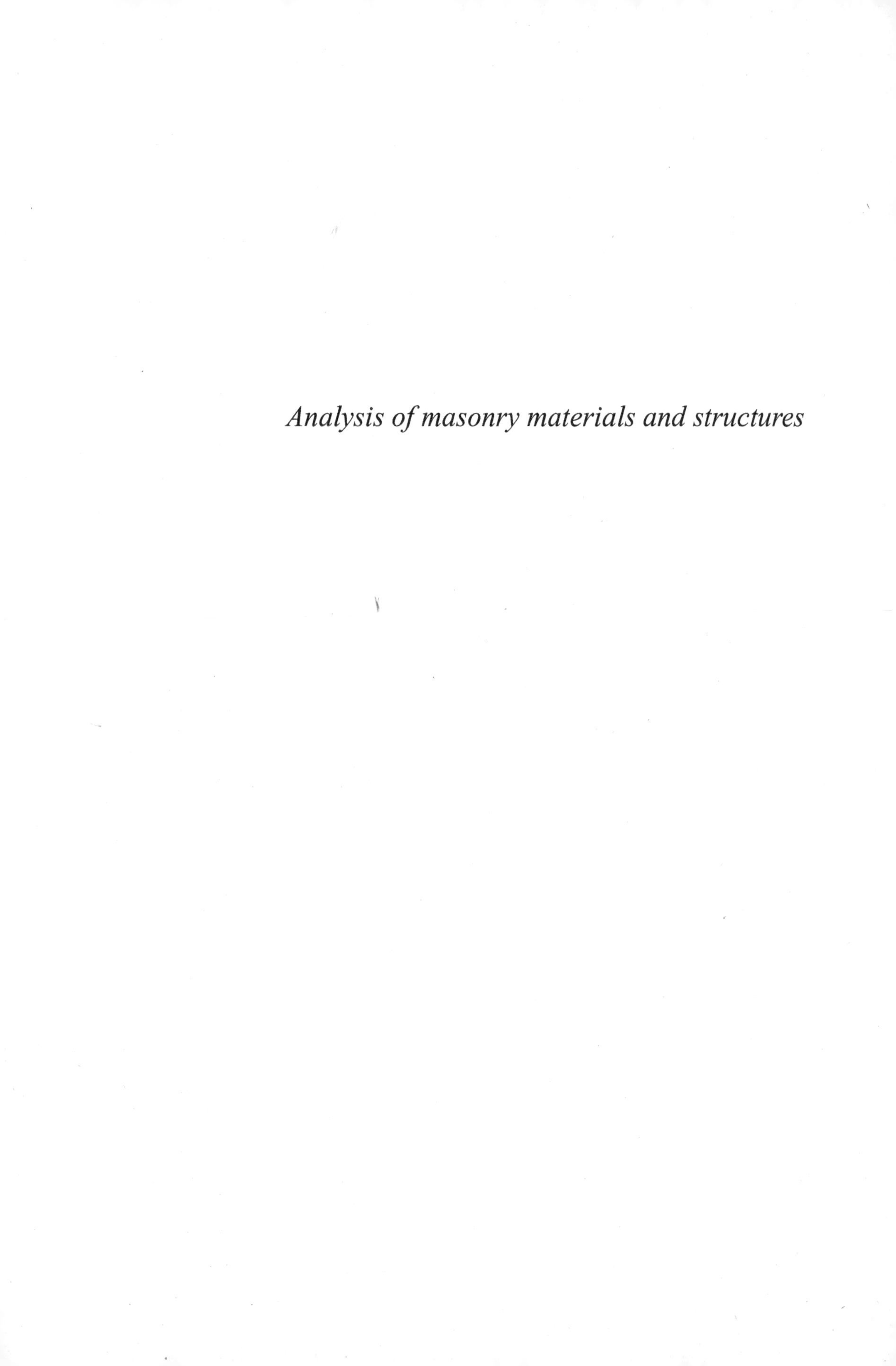

Analysis of masonry materials and structures

Computational Modelling of Concrete and Concrete Structures – Meschke, Pichler & Rots (Eds)
 ISBN: 978-1-032-32724-2

A comparative computational study on the static pushover and dynamic time history response of a masonry building

F. Messali, M. Longo, A. Singla & J.G. Rots
Delft University of Technology, Delft, The Netherlands

ABSTRACT: The paper presents a study on the seismic performance of an unreinforced (URM) two-storey terraced house, a real building located in the north of the Netherlands, which is assessed via both nonlinear pushover (NLPO) and nonlinear time history (NLTH) analyses. The results of both NLPO and NLTH analyses can be considered sufficiently reliable, despite a large number of non-converged steps, mainly due to the use of several different nonlinear constitutive laws. The detailed modelling of connection and floor nonlinearities allows for a more precise definition of the stiffness evolution of the structure at the different storey levels, but it does not affect significantly the ultimate performance of the building at global level. Finally, the assessment of the building via NLPO is highly dependent on the followed assessment procedure, with the original N2 method (recommended by the Eurocode) providing largely unconservative predictions.

1 INTRODUCTION

Nonlinear finite element analyses are often used to assess the seismic vulnerability of unreinforced masonry (URM) structures. Although different modelling strategies may be adopted (D'Altri et al. 2019), the accuracy of the assessment depends in any case on the correctness and on the level of detail of the model created for the analysis, as well as also on the assessment methodology selected (Nakamura et al. 2017). Nonlinear time history (NLTH) analyses are usually considered the most precise tool that may be used to assess the response of a structure to a dynamic input, such as a seismic ground motion. However, NLTH analyses are computationally very demanding, since multiple ground motions should be considered to avoid uncertainties related to the specific characteristics of the imposed shock and for this reason are less frequently used for the assessment of complete buildings. As an alternative, nonlinear pushover (NLPO) analyses have the advantage of decoupling the calculation of the capacity of the structure from the demand. Besides, the loads are imposed monotonically and quasi-statically, resulting in less demanding computations and NLPO analyses are therefore more commonly used for the seismic assessment of complete structures and even building aggregates (e.g. Grillanda et al. 2020; Ramos & Lourenço 2004). On the other hand, due to such simplifications, the analyses disregard the hysteretic behaviour of the structure, which must be therefore included indirectly. Different methods have been proposed in the literature and then adopted by the standards, but they lead to different results of the assessment. For these reasons, few past studies focused on the comparison between the seismic performances of a URM structure computed based on NLTH and NLPO analyses, such as the studies performed by Mendes and Lourenço (2009), Pelà et al. (2013) or Endo et al. (2015). Other works investigated how the use of different methods that were developed to take into account the hysteretic behaviour of the structures in NLPO analyses may lead to different outcomes of the assessment (Guerrini et al. 2017, 2021).

This manuscript presents a study on the seismic performance of a URM two-storey terraced house, considering as case study a real building located in the north of the Netherlands. The paper has a three-fold goal: first, to discuss the modelling assumptions made for the modelling of a URM building, including the modelling of connections, constitutive models and material properties; second, to discuss the robustness and stability of the analyses in connection to the static versus dynamic fashion of the analyses; third, to compare the outcomes of the assessment as computed according to the NLTH and the different NLPO assessment procedures.

The building is modelled via the commercial software Diana FEA 10.4 (Diana 2020) as part of a larger project that aimed to cross-compare the seismic assessment of four URM buildings performed with different modelling approaches and software packages. More details on the performed analyses can be found in

DOI 10.1201/9781003316404-41

Longo et al. (2020), while a description of the outcomes of the whole project is presented in ARUP et al. (2021).

2 METHODOLOGY

2.1 *Building overview*

The assessed building comprises three terraced units. The structure, built in 1973 in the North of the Netherlands, is made of URM cavity walls. The inner leaf and internal walls are made of calcium silicate (CS) bricks while the outer leaf is made of baked clay bricks. Wall ties connects the two masonry leaves. Each unit has two storeys plus an attic level. Also three appendices and an extra one-storey building are connected to the main structure. A picture of the building is shown in Figure 1.

Figure 1. The building assessed: the modelled end unit is identified via a yellow rectangle.

The ground floor is made of prefabricated arched concrete elements (Kwaaitaal floor), while the first and second (attic) floors are made of cast in-situ reinforced concrete (RC) slabs. The roof is composed of timber purlins and trusses with concrete roof tiles. The total mass of each unit is approximately 126 tons, and the height measured at the ridge beam is 8.2 m.

The expected peak ground acceleration (PGA) at the building location is equal to 0.148 g (NEN 2018b).

2.2 *Analysis methods*

Two different methods have been used to assess the vulnerability of the terraced house at the near collapse (NC) limit state, namely the quasi-static NLPO analyses and the dynamic NLTH.

The NLPO bilinear capacity curves and the inelastic acceleration-displacement response spectra (ADRS) are defined in accordance with the procedure described in Annex G of NPR9998 (NEN 2018a). The latter is derived from the elastic spectrum, defined for a return period of 2475 years, by taking into account energy dissipation, ductility and damping of the nonlinear system. The elastic spectrum response is shown in Figure 2 in combination with the accelerograms used for the NLTH analyses. The intersection point of the capacity and the demand curves approximates the seismic response of the structure: if the capacity curve intersects the inelastic ADRS demand curve the structure meets the safety criteria. Both global and local acceptance criteria are considered: the global criteria are applied to the building as a whole, whereas the local criteria are applied to the single structural elements, such as piers and spandrels. The global NC limit state is exceeded at the occurrence of one of the following criteria: (i) exceedance of the drift limit defined at either interstorey level or at the effective height, as reported in Table 1; (ii) drop of the total lateral resistance (base shear) below the 50% of the peak resistance of the structure; (iii) divergence of the analysis related to instability of the structure; (iv) exceedance of the local drift limit for a number of load-bearing elements, whose collapse would lead to an extensive partial collapse of the building. The local acceptance criteria recommended in sections G.9.2.2 and G.9.2.3 of NPR9998 are considered for piers. The masonry spandrels are assumed not to be essential for the stability of the load-bearing system. Therefore, a maximum drift of 2% is assigned to both non-load bearing and load-bearing spandrels, in accordance with the recommendations provided in section G.9.3.1(8) of NPR9998. The effective height is computed by dividing the roof height by the transformation factor (Γ), as defined in B.2 of NEN-EN 1998-1 (NEN 2005) and recommended in Annex G of NPR9998 too (NEN 2018a). This results in a value of about 5 m, slightly below the attic level (5.4 m).

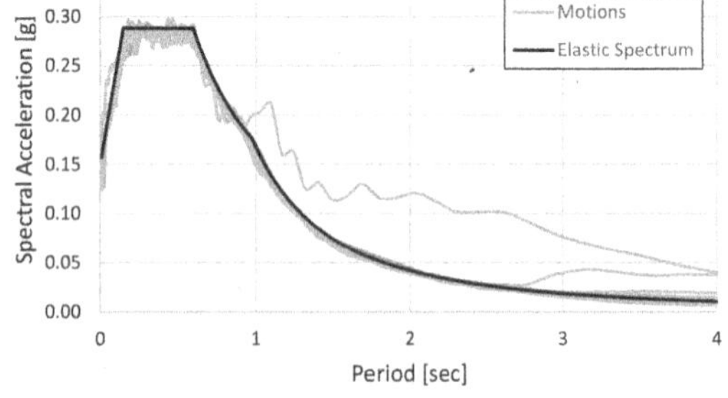

Figure 2. Acceleration-period elastic response spectrum and ground motions used for the assessment of the build.

Table 1. Drift limits defined at near collapse limit state.

	Ductile behaviour	Brittle behaviour
Effective height	0.8%	0.4%
Interstorey	1.5%	0.6%

As regards the NLTH analyses, the structural performance is assessed via an indirect check, for which the deformation of the structure is checked against the same drift limits defined for the NLPO assessment, as recommended in Annex F of NPR9998. A set of 11 tri-directional ground motions (two horizontal and one vertical) is employed: the use of multiple motions provides a more reliable estimate of the expected response of a structure since this latter depends on the characteristics of each single motion. The site-specific ground motions have a PGA in the horizontal directions that ranges from 0.12 g to 0.18 g, with a mean value equal to 0.14 g, in line with the expected PGA at the building location, which is equal to 0.148 g, based on the NEN web tool (NEN 2018b), Ground Motion Model v5 (01-10-2018). The original site-specific motions are scaled up to identify the performance at collapse of the building, which is obtained when the safety criteria are exceeded by the mean of the response of the 11 simulations. The force-displacement curves obtained for each ground motion are also used to create a trilinear backbone curve with the scope to compare the capacity of the building defined via NLTH and NLPO analyses.

2.3 *Modelling approach*

The terraced house is numerically modelled by means of the software package Diana FEA 10.4. Since the three units have same dimensions and nor the loadbearing walls nor the concrete floors of the different units are connected one another, only one building unit is modelled (highlighted in Figure 1). Besides, the three appendices and the extra one-storey building are not modelled since they are not interlocked with the load bearing inner leaves of the structure. A representation of the model is provided in Figure 3.

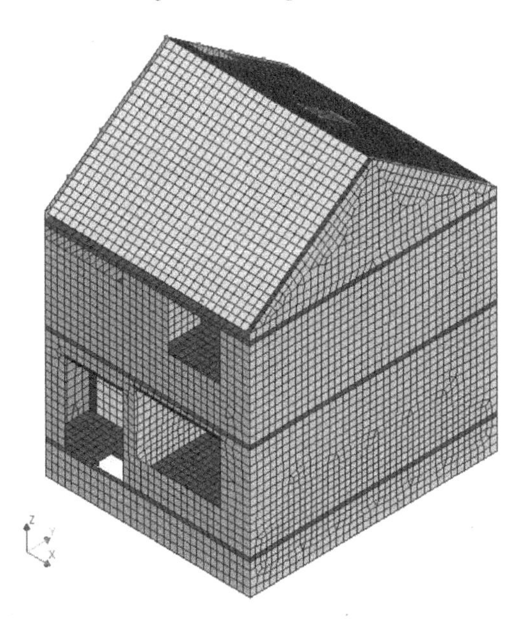

Figure 3. Finite element model of the left unit of the modelled terraced house.

All the masonry walls are modelled via shell elements, and the Engineering Masonry Model (EMM) is used as constitutive model (Schreppers et al. 2016). The EMM is a total-strain based continuum model that accounts for tensile, shear and compression failure of the masonry. The model considers the local axes oriented parallel and perpendicular to the bed joints, respectively. The orthotropy of the masonry is considered by assigning different elastic and inelastic properties for each local direction. The internal walls are explicitly modelled in order to include the extra stiffness that they can provide to the whole building. An overview of the internal walls is shown in Figure 4. The loadbearing internal walls, running transversely in the middle of the building unit, are fully connected to both bottom and top floor and to the longitudinal external façades. The nonloadbearing internal walls are disconnected from the floor above so that no force may be transferred. In addition, the lateral connections with the transversal external façades are modelled with a strip of weak elements that simulates a vertical mortar joints. This is obtained by rotating the local axes of the elements and reducing both the elastic and nonlinear material properties by 30%.

Since the shear stiffness of the wall ties is negligible (Skroumpelou et al. 2018), these ties are assumed to be able to transfer axial loads only. For this reason, the cavity wall system is modelled by explicitly modelling the loadbearing inner leaves and considering the outer leaves as dynamic mass acting in the direction perpendicular to the plane of the walls. The chimney, which runs from the first storey up to the roof, is also included in the model as dynamic mass. The mass density (static and dynamic) assigned to the different walls of the façades is depicted in Figure 5.

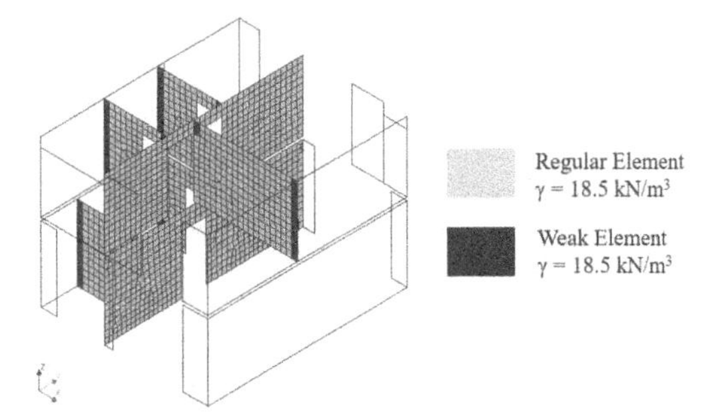

Figure 4. Modelling of the internal walls.

The Kwaaital floor and the RC slabs are modelled via non-linear shell elements, using the Total Rotating Strain Crack Model for the concrete and the Von Mises Plasticity model for the steel reinforcement (Diana 2020). The grid reinforcement at top and bottom of the slab is modelled with an embedded sheet reinforcement, defined via the bar diameter and spacing in the two directions. The reinforcement in the concrete joists is modelled explicitly as line bar reinforcement.

Strips of linear elastic isotropic shell elements, as high as the concrete floor thickness, are modelled between the elements of the masonry façades to simulate the real thickness of the concrete slabs, which cannot otherwise be properly represented by the shell

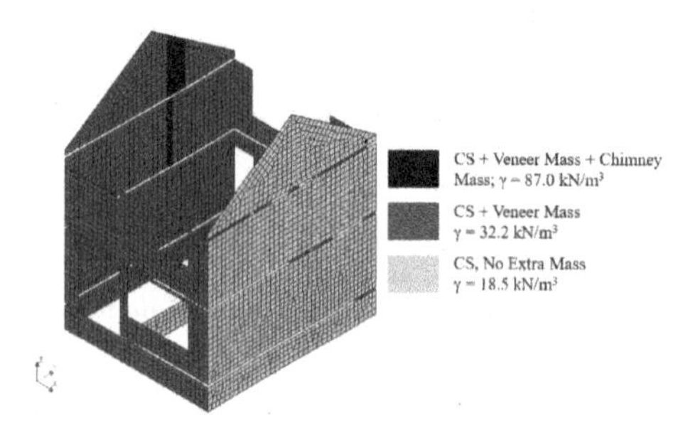

Figure 5. Material density (static and dynamic) assigned to the external loadbearing walls depending on the considered mass.

elements. The concrete lintels above the openings are also modelled via linear elastic shell elements.

The roof purlins, struts, ties and the ridge beam are modelled via beam elements using a linear elastic isotropic material (Figure 6). The connections between the gable and the purlins/ridge beams are modelled with point interface elements using a Coulomb-friction material model to simulate the possible sliding of the beams in the pocket connections. Where the timber beams are connected to the URM walls via a L-shaped steel anchor, no point interface is used and the connection is modelled as rigid. The timber boards are modelled as shell elements using a linear elastic orthotropic material. Dummy beams having negligible stiffness and small cross-section are added on the edges of the roof planks to improve the numerical stability of the analyses (Figure 7).

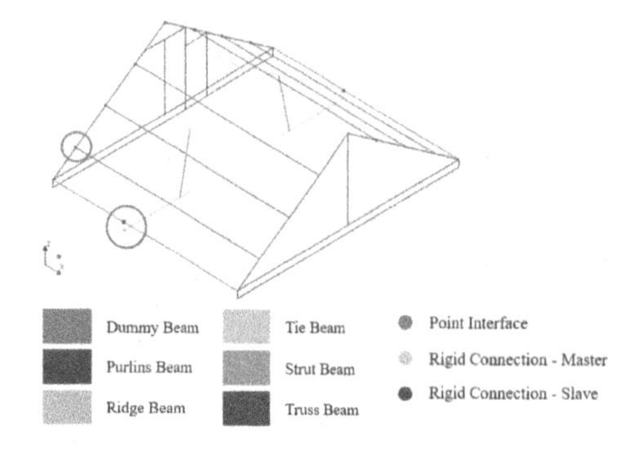

Figure 6. Modelling of beams and connections at the roof level.

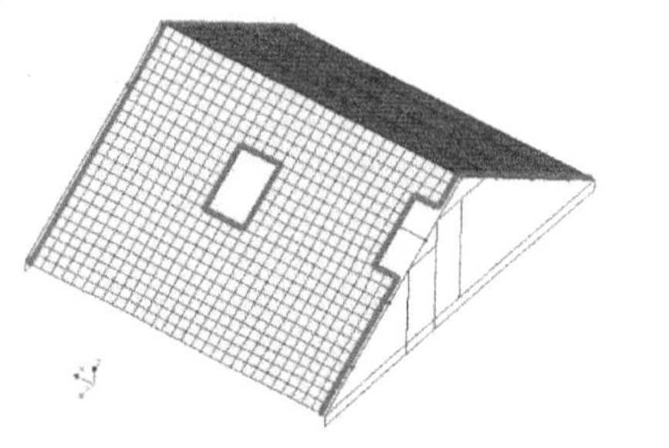

Figure 7. Modelling of roof boards and dummy beams (highlighted in red).

The horizontal components x and y from the NEN web tool are aligned with the respective global x and y axes defined for the numerical models. Since the non-linearities provided by the ground below the foundation are expected to be negligible, a “fixed base” boundary condition is chosen. In this case, the surface level ground motions are applied directly to the base of the building.

Both NLPO and NLTH analyses are performed. For the NLPO analyses, the model is initially subjected to the gravity loads applied in ten equal steps. Then, either mass proportional lateral loads, applied via a uniform lateral acceleration, or modal distributed lateral loads, based on the main eigen-mode of the structure (and the corresponding participating mass), obtained via an eigen-value analysis, are applied so that an average displacement rate of 0.1 mm/step is recorded at floor level. It should be noted that the uniform lateral acceleration does not account for the extra dynamic mass. The Secant BFGS (Quasi-Newton) method is adopted as iterative method in combination with the Arc-Length control. Both displacement and force norms must be satisfied during the iterative procedure within a tolerance of 1%. For the NLTH analyses, the model is first subjected to gravity loads, again applied in ten equal steps. Then, the different acceleration motions are applied in the longitudinal, transversal and vertical direction at the base nodes, using a time step of 2.5 milliseconds. A Rayleigh damping of 2% is accounted in the calculation. The Secant BFGS (Quasi-Newton) method is employed as iterative method. Energy norm must be satisfied during the iterative procedure with a tolerance of 0.01%. For both analyses, the Parallel Direct Sparse method is employed to solve the system of equations. The second order effects are considered via the Total Lagrange geometrical nonlinearity.

2.4 *Material properties*

Since no information was available on the values of the material parameters for the specific modelled building, such values were selected based on the recommendations given in Table F.2 of the Dutch guidelines NPR9998 (NEN 2018a) as well as from experimental tests performed at component or building level and already available in the literature. As regards the masonry properties, the masonry quality was assumed excellent, and the properties provided in the guidelines were used without the application of any reduction factor (CVW 2018). An overview of the parameters employed for the CS masonry is shown in Table 2. For the NLTH calculations, the elastic properties are halved in order to properly capture the cyclic strength degradation, not explicitly described by the EMM.

As regards the modelling of the timber elements, the elastic properties of the timber planks of the roof (tabulated in Table 3) were calibrated based on laboratory experiments performed at TU Delft on similar diaphragms (Mirra et al. 2020). The value of the

Table 2. Material properties assigned to the masonry elements. In parenthesis the values used for the NLTHA.

	CS – Regular	CS – Weak*
E_y [MPa] **	4000 (2000)	2800 (1400)
E_x [MPa]	2667 (1334)	1867 (934)
G [MPa]	1650 (825)	1155 (578)
ρ [kg/m^3]	1850	1850
f_y [MPa] **	0.15	0.10
$f_{x,min}$ [MPa]	0.30	0.20
$G_{f,I}$ [N/m]	10	8.1
α [rad]	0.62	0.62
f_c [MPa]	7.0	7.0
G_c [N/m]	15,000	15,000
ϕ [rad]	0.54	0.54
c [MPa]	0.25	0.175
G_s [N/m]	100	100

* Rotated local axis
** Local y axis is perpendicular to bed joint

Table 3. Elastic properties assigned to the timber planks.

	Timber Planks
E_x [MPa]	1.5
E_y [MPa]	11
E_z [MPa]	400
ρ [Kg/m^3]	380
υ [-]	0.0
G_{xy} [MPa]	1100
G_{yz} [MPa]	1100
G_{xz} [MPa]	500

Young's modulus assigned to the ridge beam, timber purlins, strut and tie beams is 9 GPa, and the Poisson's ratio is 0.35. A smaller value of the Young's modulus (1 GPa) is assigned to the dummy beams. The material properties selected for the Coulomb-Friction model used for the point interfaces between the URM gable walls and the timber purlins are listed in Table 4. The properties attributed to the concrete elements of the floors are defined based on the median values defined for C12/15 in EN 1992-1-1, with a Young's modulus of 27 GPa, and uniaxial tensile and compressive strength of 1.57 MPa and 20 MPa, respectively. A Young's modulus of 200 GPa and a yield strength of 400 MPa is used for the rebars. A plastic behaviour described by Von Mises equations is employed.

Table 4. Properties of the interface elements which simulate the connections between the timber beams and the gable wall.

	Point Interface
k_n [N/mm^3]	1000
k_t [N/mm^3]	100
ϕ [rad]	0.60
Ψ [rad]	0
c [MPa]	0.02

3 ANALYSIS RESULTS

3.1 *NLPO Results*

The NLPO analyses are performed by applying two different distributions of lateral loads: either a uniform or a modal distribution is considered. The former distribution is achieved by using a horizontal equivalent acceleration, increased incrementally. The latter distribution is computed on the basis of the eigen-mode with the highest participating mass. This is the first natural mode (shown in Figure 8), which depicts a global in-plane mechanism of the structure with an almost linear increase of the lateral displacements over the height.

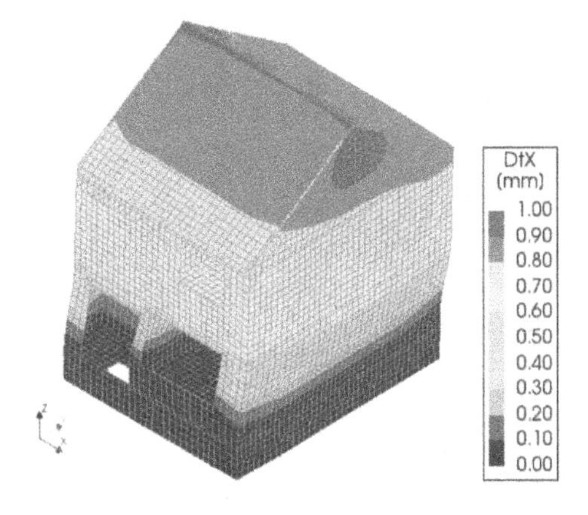

Figure 8. First natural mode of the building model. Plot of the displacements in the longitudinal direction.

For each load distribution, the loads are applied both in the positive and negative directions. The outcomes of the four analyses are qualitatively similar: the failure is governed by a soft-storey mechanism at the ground floor level, characterised by rocking of the piers on both facades and of the internal walls (Figure 9). The horizontal cracks that start from the bottom corners of the windows propagates in the transversal walls too. In case of uniform distributed loads, an additional vertical crack develops between the transversal wall and the internal partition wall, probably due to the larger lateral force applied at the ground location with respect to the modal distribution, but no significant out-of-plane deformations are observed (Figure 10).

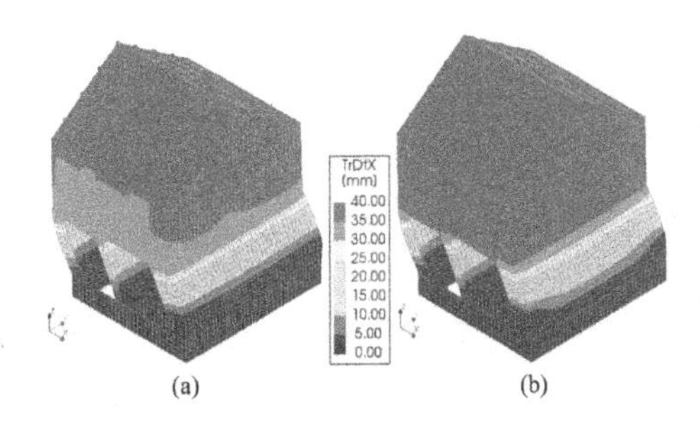

Figure 9. Longitudinal displacements at step 385 of the NLPO analysis for the modal (a) and uniform distribution (b).

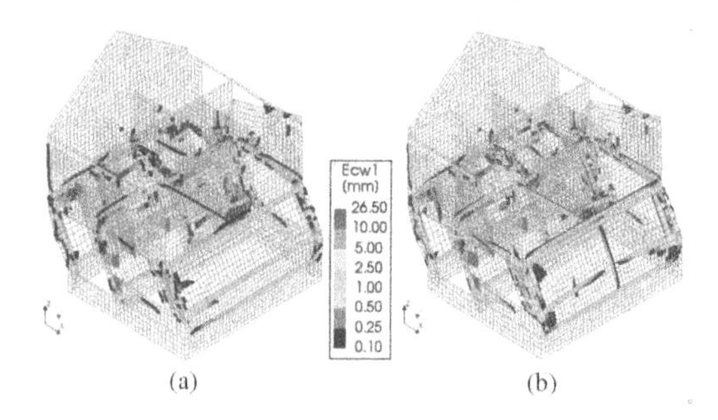

Figure 10. Principal crack width plot at step 385 of the NLPO analysis for the modal (a) and uniform distribution (b).

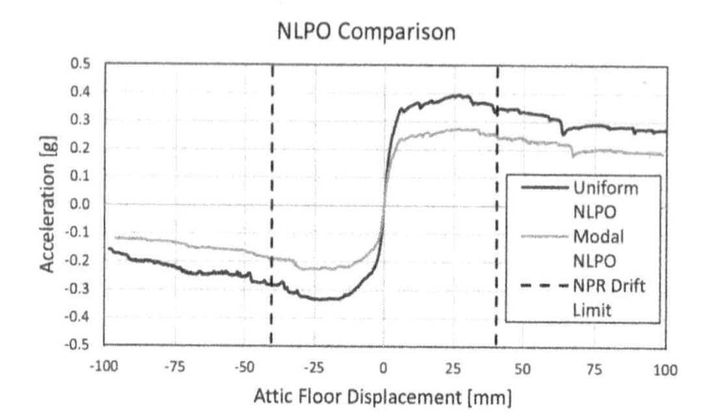

Figure 11. Normalized force-first floor displacement curves of the NLPO analyses. NPR9998 drift limits applied.

The observed global failure mechanism mainly depends on the flexural failure of the piers, so that the global failure may considered ductile. For such mechanisms, the Dutch guidelines recommend a maximum allowable storey drift of 1.5%, a value that is used to cap the capacity curves obtained for each analysis. Figure 11 shows the four base shear-displacement curves, where the base shear is normalized by the effective mass of the building and the displacement is computed at the attic level. The dashed black line represents the NPR9998 limit for the NC displacement, which is equal to 40.4 mm. The relevant displacements and normalized base shear forces of the corresponding bilinearized curves are reported in Table 5.

Table 5. Summary of bilinearized NLPO curves.

	Yield Disp. [mm]	NC Disp. [mm]	Norm. Force [g]
Mod. Positive	4.02	40.35	0.259
Mod. Negative	4.21	40.35	0.207
Unif. Positive	4.00	40.35	0.367
Unif. Negative	4.10	40.35	0.306

3.2 *NLTH results*

The studied building is assessed also via NLTH analyses. As for NLPO analyses, an indirect method recommended in section F.6.3 of NPR9998 (NEN 2018a) is employed to estimate the NC of the building. Both site-specific and scaled ground motions are investigated. The mean value of the 11 performed simulations is adopted to verify the compliance with the failure criteria. For the site-specific analysis (with an expected PGA of 0.148 g) the building complies with the criteria. The average maximum displacement in x-direction recorded at the first floor location is 8.2 mm, equal to 0.31% of the effective height. The average peak force is equal to 449.7 kN. The 11 records are then amplified in order to evaluate the PGA that corresponds to the exceedance of the failure criteria, which is first achieved for a PGA of 0.33 g. A soft-storey at the ground floor along the weak x direction is the global failure mechanism observed for all the applied motions, similar to that observed for the NLPO analyses. An overview of the force-displacement curves computed for the building subjected to all the ground motions is given in Figure 12. The average maximum displacement in x direction recorded at the first floor is 35.5 mm, equal to 1.32% drift, smaller than the drift limit from normative. The displacements in the y direction are relatively small, and at the first floor level equals to 1.54 mm. Extra calculations show that the in-plane interstorey drift limit (1.5%) is exceeded on average for the 11 ground motions having a PGA value comprised between 0.35 g and 0.40 g. The average peak force is equal to 692.8 kN. The overall results are reported in Table 6.

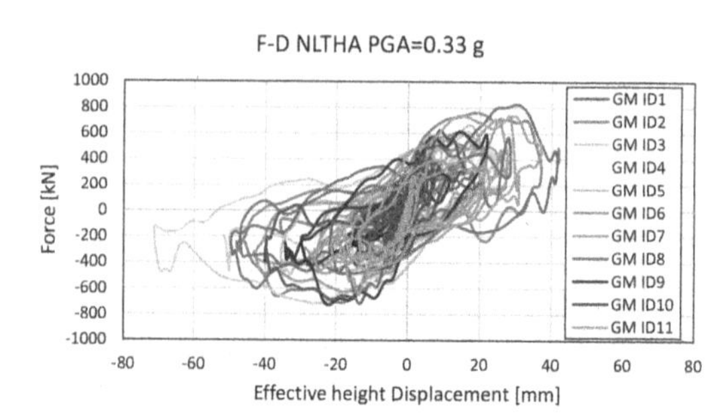

Figure 12. Force-displacement at effective height of the model subjected to scaled motion of 0.33 g.

Table 6. Summary of NLTH data for amplified PGA of 0.33 g.

GM ID	Peak Disp. Floor 1 [mm]	Peak Drift Floor 1 [%]	Peak Base Shear [kN]
1	15.3	0.57	514.9
2	44.5	1.65	760.9
3	66.1	2.46	717.0
4	24.7	0.92	616.5
5	32.1	1.19	665.4
6	39.1	1.45	627.6
7	25.7	0.95	740.7
8	45.2	1.68	826.4
9	31.1	1.16	725.7
10	35.4	1.32	687.2
11	31.2	1.16	738.7
Mean	35.5	1.32	692.8

4 DISCUSSION

4.1 *Convergence of NLPO and NLTH analyses*

The convergence of the calculations is hereinafter discussed in order to assess the reliability of the outcomes. An overview of the converged steps is shown in Figure 13 for the negative direction of the modal NLPO analysis. In total, 69 of the 437 steps (15.8%) do not satisfy the convergence criteria (the orange dots in Figure 13), which are set in terms of residual forces and displacements, with a maximum norm of 1% to be achieved simultaneously in not more than 25 iterations. The residual norms of the steps for which the criteria are not met are plotted in Figure 14. For only 18 of the 69 non-converged steps the criteria are not respected for none of the two residual norms, and the maximum residual norms obtained in those steps are 5.2% for the forces and 2.8% for the displacements. For all the other steps, at least one of the two residual norms is smaller than 1%. The sporadic lack of convergence is mainly caused by the small number of allowed iterations (maintained low to shorten the computational time), the large number of elements cracking simultaneously at the same step, and the use of non-linear constitutive laws for several different materials. Overall, the obtained results are considered sufficiently reliable for an NLPO analysis.

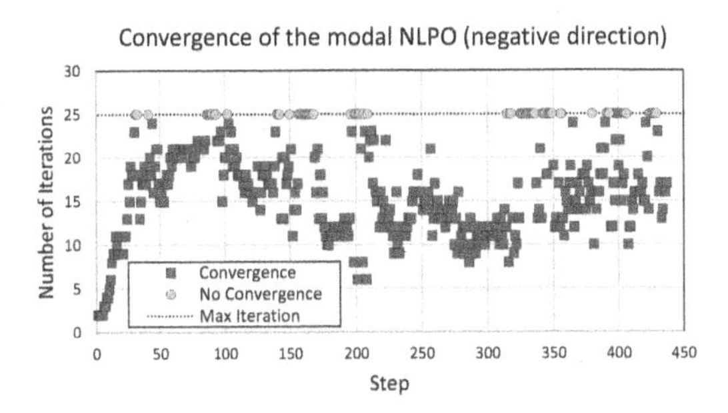

Figure 13. Overview of converged and not converged steps for the modal NLPO analysis with negative loading direction.

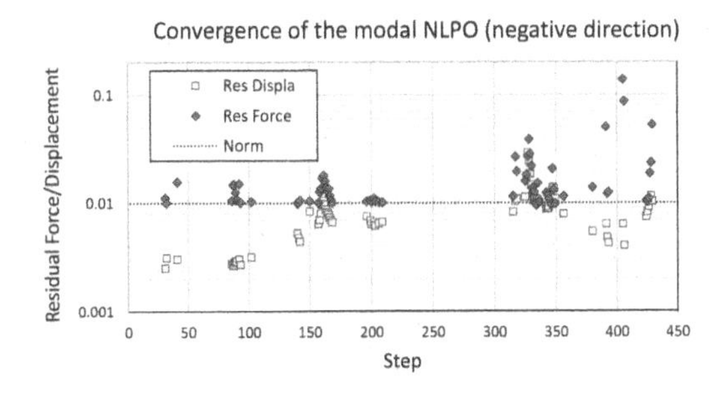

Figure 14. Residual force and displacement of non-converged steps for the modal NLPO with negative loading direction. Selected 1% norm applies for both criteria.

Figure 15 shows the residual energy of both converged and non-converged steps of a NLTH calculation. In total, the 53.6% of the steps do not converge, and the average residual energy evaluated from all steps is equal to 1.9%. The limited number of iterations allowed per step (once more to limit the computational time), the strict tolerance and the multiple nonlinear laws implemented in the model are the cause of such lack of convergence. However, it can be noted that convergence is always reached after few non-converged steps, and the check of the strain contours does not show any unexpected strain localization. For these reasons, also the results of the NLTH analyses are considered sufficiently reliable.

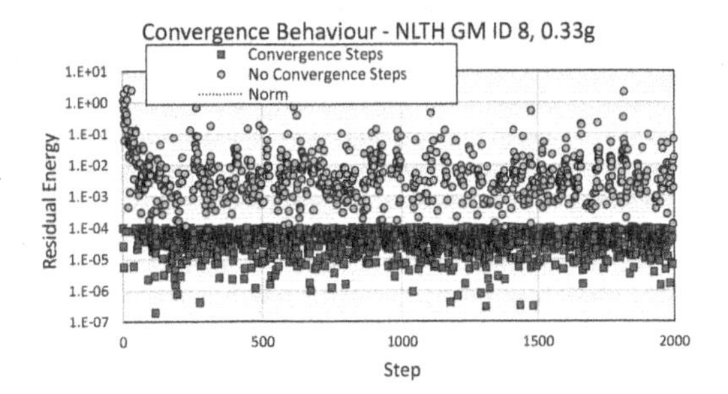

Figure 15. Residual energy of both convergence and non-convergence steps for the GM 8 NLTH with PGA of 0.33 g. Selected 0.01% norm applies.

4.2 *Influence of model assumptions on the outcome*

Both NLPO and NLTH simulations require a number of assumptions, which affect the outcomes of the analyses. In the following, three relevant assumptions are discussed. The NLPO analysis subjected to positive uniform distributed load is considered for this scope.

First, the connection between the internal walls and the transversal external façades is examined: either with or without weak elements that simulates the continuous vertical mortar joints. For these two alternative models, the base shear is plotted against the displacement at the first storey level in Figure 16. The model with weak elements results more flexible and achieves

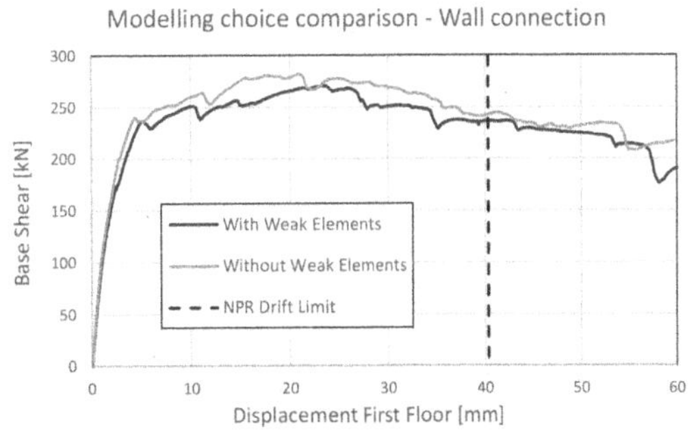

Figure 16. Comparison of the base shear-first floor displacement curves for different modelling of the connection between walls.

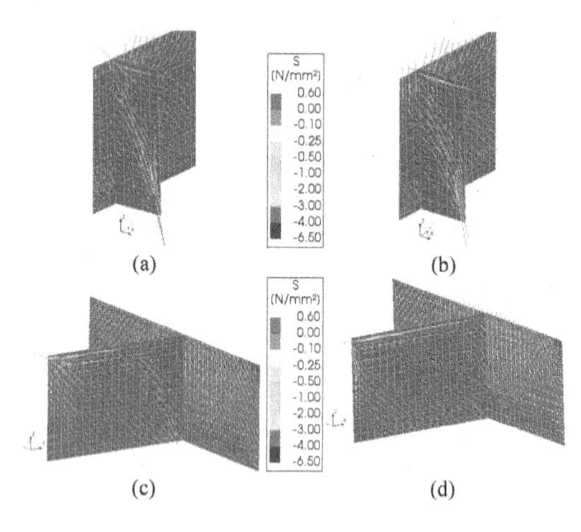

Figure 17. In-plane principal stress components (at step 385): active flange at ground level of model with (a) or without (b) weak elements; passive flange at first level of model with (c) or without (d) weak elements.

a slightly lower force capacity. When the internal wall is connected to the facades, the latter acts as a flange in a T-shape section beam and the internal wall as a web. The different distribution of the in-plane principal stresses in case of an active flange (which is a flange subjected to a reduction of the vertical axial load due to the frame effect determined by the lateral loading) at the ground floor is shown in Figure 17, for both the models with (Figure 17a) and without the row of weak elements (Figure 17b. The comparison shows that the stronger connection at the corner contributes to limit the horizontal rocking crack opening at the base of the internal wall. As a result, the diagonal compressive strut localizes in a smaller area and it results in higher compressive stresses at the toe of the wall. Similarly, the stress distribution for a passive flange (which is a flange subjected to an increase of the vertical axial load due to the frame effect determined by the lateral loading) is shown in Figures 17c and 17d. The latter shows how the presence of a weak row of elements allows a large part of the internal wall to detach from the transversal wall, so that the long internal wall behaves similar to a self-standing wall.

Second, the connection between the timber beams and the masonry walls is modelled by means of point interface elements. This detail allows to provide a more realistic description of the roof stiffness, and to exploit the ductility provided by the local nonlinear mechanism. On the other hand, the elements adopted do not allow for the definition of a maximum limit for the sliding, so that the relative displacement must be checked in the post-processing phase. The in-plane relative displacements of the interface elements are extrapolated and depicted in Figure 18, showing a maximum absolute value of 1.05 mm. It is noted that the nonlinear behaviour is activated for all the connections, but the relative displacements are extremely small compared to the sliding that would lead to the local collapse of the connection (which can be estimated as the length of the supported part of the beam within the pocket, which is typically 60–80 mm for single wythe walls).

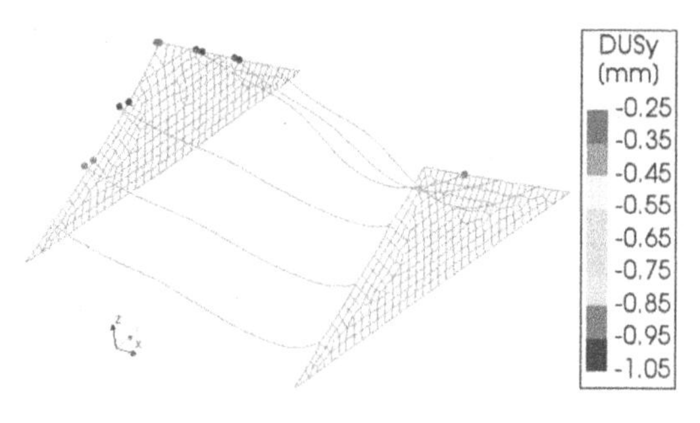

Figure 18. Interface relative in-plane displacement (at step 385).

Third, the nonlinear behaviour of the reinforced concrete floors has been explicitly modelled. Figure 19 shows the cracking in the three concrete floors during the application of the pushover load. Large cracks (up to 0.45 mm) are detected locally at the ground floor, where the rocking of the internal walls largely increases the stresses and causes bending of the floor, but this does not compromise the stability of the supported masonry walls, nor leads to a global failure of the floor system.

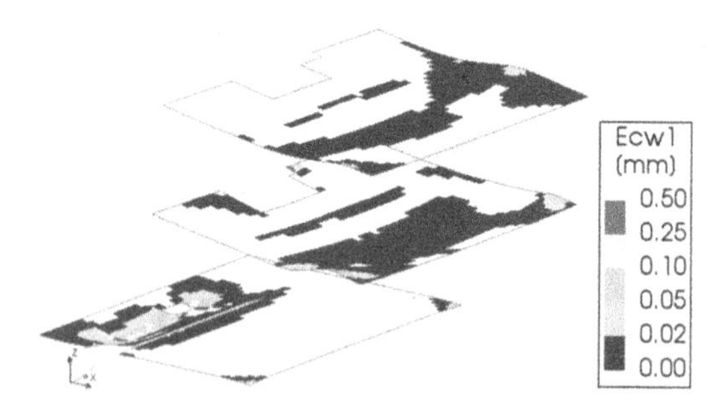

Figure 19. Principal crack width of concrete floors (at step 385).

4.3 *Comparison NLPO-NLTH*

4.3.1 *NLTH backbone curve*

To compare the outcomes of the NLPO and NLTH analyses, first a backbone curve is defined starting from the set of performed NLTH analyses to define an average global behaviour of the building. The maximum (and minimum) base shear force is extrapolated from each analysis and correlated with the corresponding maximum (and minimum) displacement. Both the site-specific hazard and the scaled-up analyses are used to define a trilinear backbone curve, whose significant points (yielding, peak, NC displacement) are obtained from the average of the results of analyses having consistent outcomes (linear behaviour; nonlinear behaviour without exceedance of the failure criteria; exceedance of the failure criteria). The maximum and minimum base shear of the backbone curve are equal to 613.0 kN and −598.6 kN at peak displacements of 22.2 mm and −31.7 mm, respectively. The NC displacement is computed by averaging the maximum displacement of the analyses which show a global drift higher than 1.5%. The computed collapse displacements are 35.0 mm and −56.4 mm for the positive and negative direction, respectively. The NLTH backbone curve is plotted in Figure 20 in comparison with the pushover curves.

Similarities between the curves are observed: namely, the peak base shear of the NLTH backbone curve lies in between the peak forces of the two pushover curves, whereas the NLPO ultimate displacement capacity is comprised between the defined negative and positive values for the NLTH analyses. Finally, the NLPO curves obtained for mass proportional loading match the envelope of the force-displacement curves obtained for the NLTHA. This suggests the fact that the NLPO analysis can be used for reliable predictions of the capacity of a regular URM building such the one considered in this paper.

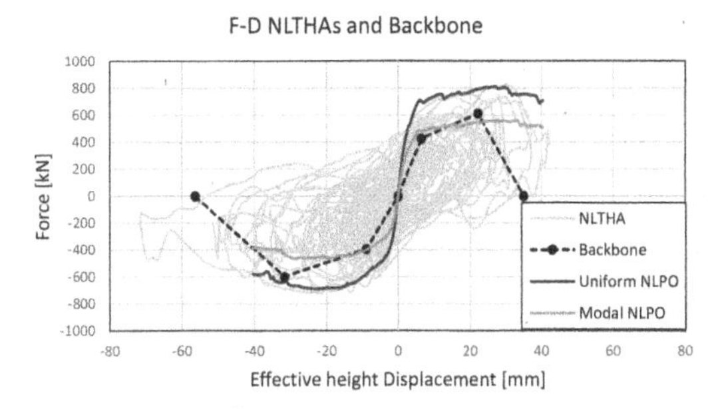

Figure 20. Backbone curve calculated from site-specific and scaled NLTHA analyses.

4.3.2 *Different ADRS methods for NLPO analyses*

The assessment of the building for the NLPO analyses is carried out according to the capacity spectrum method (CSM) as described in Annex G of NPR9998 (NEN 2018a). The CSM, initially proposed by Freeman et al. (1975), is now used in codes and guidelines from New Zealand (NZSEE 2017) and Italy (NTC 2018), besides the Dutch ones, and a modified version has been proposed also in the FEMA 440 guidelines (FEMA 440 2005). When the site-specific hazard is considered (PGA of 0.148 g), all the capacity curves intersect the elastic Acceleration-Displacement Response Spectrum (ADRS) demand. The governing NLPO analysis, i.e. the one with the smallest capacity over demand (C/D) ratio, is the modal NLPO with the loading in the negative direction. The capacity curve of this analysis is used also to define the ADRS demand for which the C/D ratio is equal to one, or (in other words) the maximum PGA for which the building still complies to the failure criteria. It is assumed that all the coordinates of the curve are scaled proportionally as the PGA. So, for instance, at an increase of 50% of the PGA corresponds also an increase of 50% of the displacement of the final vertical branch of the curve. The iterative procedure returns a value of the scaled PGA equal to 0.28 g (Figure 21).

A similar calculation is repeated for two alternative assessment procedures, namely the N2 method, which is recommended by the Eurocode EN 1998-1 (NEN 2005) and an alternative modified version (MN2), proposed by Guerrini et al. (2017) for short-period structures. The PGA corresponding to the capacity of the structure is equal to 0.55 g and 0.37 g for the original and modified N2-methods, respectively. All the scaled ADRS curves (including the inelastic curve for the CSM) and the capacity curve of the building are shown in Figure 21. Both the MN2 and the CSM provide sensible estimates of the PGA at failure compared to that predicted via NLTH analyses (0.33 g): while the MN2 slightly overestimates it (+12%), the CSM underestimates it (−15%). On the opposite, the original N2 is largely unconservative (+66%), showing to be unsuitable for short-period buildings such as the one analyzed in this paper and consistently with the outcomes of the research presented in Guerrini et al. (2021).

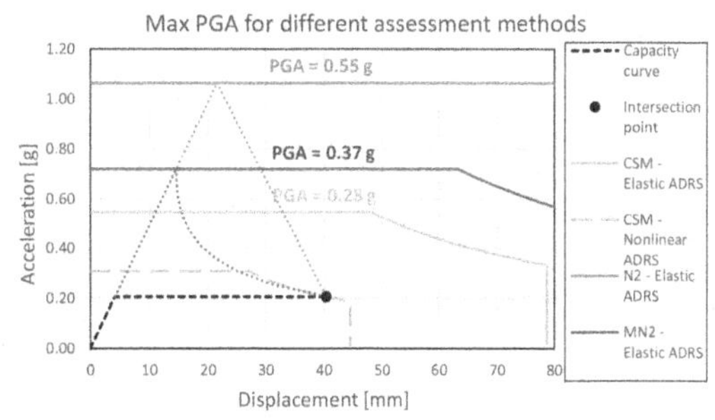

Figure 21. Assessment of the NLPO negative modal pushover for different assessment methods.

5 CONCLUSIONS

This work presents a study on the seismic performance of an unreinforced (URM) two-storey terraced house by considering a real building located in the north of the Netherlands, assessed via both nonlinear pushover (NLPO) and nonlinear time history (NLTH) analyses. The paper discusses the relevance of the modelling assumptions and the stability of the analyses, and provides a comparison between the assessment performed via NLPO and NLTH analyses.

The study highlights that, despite the presence of dummy elements in the roof structure which improve the stability of the solution, the use of non-linear constitutive laws for several materials leads to a large number of non-converged steps, especially for NLTH analyses. However, convergence is always reached after few non-converged steps, and the check of the strain contours does not show any unexpected strain localization. For these reasons, the results of both NLPO and NLTH analyses are considered sufficiently reliable.

Modelling a weak connection between the internal non-loadbearing walls and the loadbearing façades returns a lower peak capacity of the structure, but larger flexibility.

The explicit modelling of the nonlinear behaviour of the timber-masonry connections allows for a more precise definition of the stiffness evolution of the structure at the different storey levels, but it does not affect the ultimate performance of the building. In both cases, the nonlinear behaviour activates, but it does not lead to any failure of the elements which can modify the structural behaviour at the global building level.

The assessment of the building via NLPO is highly dependent on the followed assessment procedure: both a modified version of the N2 method and the capacity spectrum method provide sensible estimates of the PGA at failure compared to that predicted via NLTH analyses, whereas the original N2 is largely unconservative, proving to be unsuitable for buildings with a short-period like the one analysed in this paper.

Overall the paper shows the relevance of the selection of accurate modelling assumptions to properly describe the seismic behaviour of a URM terraced building, which can be adequately described via either NLPO or NLTH analyses.

ACKNOWLEDGEMENT

This research was funded by Stichting Koninklijk Nederlands Normalisatie Instituut (NEN) under the contract number 8505400026-001, which is gratefully acknowledged. The study was developed together with the engineering companies ARUP and Borg, which are thanked for the collaboration.

REFERENCES

ARUP, BORG & TU Delft. 2021. Rapportage resultaten en toepassing verschillende berekeningsmethoden NPR 9998 (module 3)_2021-01. *NEN report.*

CVW. 2018. "Applicatiedocument Beoordeling Seismische Capaciteit (ABSC)". *CVW report no. CVW-ABSC-NPR2018-UK.*

D'Altri, A.M., Sarhosis, V., Milani, G., Rots, J.G., Cattari, S., Lagomarsino, S. et al. 2019. Chapter 1 – A review of numerical models for masonry structures, In Bahman Ghiassi & Gabriele Milani (ed.), *Numerical Modeling of Masonry and Historical Structures. From Theory to Application.* Woodhead Publishing. 3–53. ISBN 9780081024393.

Diana, FEA. 2020. Diana user's manual, release 10.4. *DIANA FEA BV.* https://dianafea.com/manuals/d104/Diana.html

Endo, Y., Pelà, L., Roca, P., Da Porto, F. & Modena, C. 2015. Comparison of seismic analysis methods applied to a historical church struck by 2009 L'Aquila earthquake. *Bulletin of earthquake engineering* 13(12): 3749–3778.

FEMA 440. 2005. Improvement of Nonlinear Static Seismic Analysis Procedures. *Applied Technology Council (ATC-55 Project), California.*

Freeman, S.A., Nicoletti, J.P. & Tyrrell, J.V. 1975. Evaluation of Existing Buildings for Seismic Risk—A Case Study of Puget Sound Naval Shipyard, Bremerton, Washington. In *Proceedings of the 1st US National Conference on Earthquake Engineering*, Oakridge, CA, USA, 18–20 June 1975; pp. 113–122.

Grillanda, N., Valente, M., Milani, G., Chiozzi, A. & Tralli, A. 2020. Advanced numerical strategies for seismic assessment of historical masonry aggregates. *Engineering Structures* 212: 110441.

Guerrini, G., Graziotti, F., Penna, A. & Magenes, G. 2017. Improved evaluation of inelastic displacement demands for short-period masonry structures. *Earthquake Engineering & Structural Dynamics* 46(9): 1411–1430.

Guerrini, G., Kallioras, S., Bracchi, S., Graziotti, F. & Penna, A. 2021. Displacement Demand for Nonlinear Static Analyses of Masonry Structures: Critical Review and Improved Formulations. *Buildings* 11(3): 118.

Longo, M., Singla, A. & Messali, F. 2020. NLTH and NLPO analyses of Building B: Module 3 – Harmonisatie berekeningsmethode. *Delft University of Technology report.*

Mendes, N. & Lourenço, P.B. 2009. Seismic assessment of masonry "Gaioleiro" buildings in Lisbon, Portugal. *Journal of Earthquake Engineering* 14(1): 80–101.

Mirra, M., Ravenshorst, G., & van de Kuilen, J. W. 2020. Experimental and analytical evaluation of the in-plane behaviour of as-built and strengthened traditional wooden floors. *Engineering Structures, 211, 110432.*

Nakamura, Y., Derakhshan, H., Griffith, M.C., Magenes, G. & Sheikh, A.H. 2017. Applicability of nonlinear static procedures for low-rise unreinforced masonry buildings with flexible diaphragms. *Engineering Structures* 137: 1–18.

NEN. 2005. Design of structures for earthquake resistance – Part 1: General rules, seismic actions and rules for buildings. *NEN-EN 1998-1:2005, NEN.*

NEN. 2018a. Assessment of the structural safety of buildings in case of erection, reconstruction, and disapproval – Induced earthquakes – Basis of design, actions and resistances. *NPR9998:2018, NEN.*

NEN. 2018b. Webtool NPR 9998: Bepaling van de seismische belasting. http://seismischekrachten.nen.nl/

NTC. 2018. Norme Tecniche per le Costruzioni. *Ministero delle infrastrutture e dei trasporti, Italy.*

NZSEE. 2017. New Zealand Society for Earthquake Engineering, The seismic assessment of existing buildings, Part C8: Seismic assessment of unreinforced masonry buildings. *Wellington, New Zealand: MBIE, EQC, SESOC, NZSEE and NZGS.*

Pelà, L., Aprile, A. & Benedetti, A. 2013. Comparison of seismic assessment procedures for masonry arch bridges. *Construction and Building Materials* 38: 381–394.

Ramos, L.F. & Lourenço, P.B. 2004. Modeling and vulnerability of historical city centers in seismic areas: a case study in Lisbon. *Engineering structures* 26(9): 1295–1310.

Schreppers, G.M.A., Garofano, A., Messali, F. & Rots, J.G. 2016. DIANA validation report for masonry modelling. *DIANA FEA report 2016-DIANA-R1601 TU Delft Structural Mechanics Report CM-2016-17, 143 pp.*

Skroumpelou, G., Messali, F., Esposito, R., & Rots, J. 2018. Mechanical characterisation of wall tie connection in cavity walls. *In 10th Australasian Masonry Conference: 11 – 14 February, 2018, Sydney, Australia.*

Computational Modelling of Concrete and Concrete Structures – Meschke, Pichler & Rots (Eds)
 ISBN: 978-1-032-32724-2

Numerical evaluation of mortarless interlocking masonry walls under in-plane lateral loading

P.L. Davis & M.J. DeJong
Department of Civil and Environmental Engineering, University of California, Berkeley, USA

ABSTRACT: Interlocking masonry systems have the potential to increase capacity while reducing construction time and cost, particularly in seismic applications. Salient among the challenges faced in advancing this technology is the selection and optimization of the block geometries and interlocking patterns. In this preliminary numerical study, masonry wallets comprised of interlocking units were loaded in-plane until failure using the distinct element program 3DEC. Twenty unique block geometries were used to create wallets of various aspect ratios, and the wallets were simulated under several overburden pressures. To study block splitting, some geometries were altered to include predefined splitting planes and were simulated using several masonry unit strengths. The failure mode and lateral load capacity of each simulation was compared to a control geometry with the same wall aspect ratio and overburden stress. The results (without block splitting) showed that interlocking can force different failure modes and substantially increase lateral load capacity, depending on the wall aspect ratio and the lock location, number and orientation. For slender walls with an aspect ratio of 2, little change versus the control was observed and the lateral load capacity was slightly reduced in most cases; this agrees with expectations as rocking failure governed in all cases including the control. For aspect ratios of 1 and 2/3, the controls and geometries with locks only on the head joints experienced sliding failure along the top course, while samples with locks on the bed joints showed stair-stepping shear failure along the joints which corresponded to a significant increase in lateral load capacity. Lateral load capacity was improved by as much as 130%. However, this substantial increase in lateral load capacity does not reflect block splitting. In the simulations which included predefined splitting planes, the prevalence of block splitting increased with increasing overburden stress and decreased with increasing masonry unit strength. In simulations where little to no splitting was observed (low overburden pressure or high masonry unit strength), results largely agreed with earlier conclusions. On the other hand, samples with low masonry unit strength simulated under high overburden pressure showed little improvement in lateral load capacity. Future investigations will include adding greater complexity to existing models, input parameter sensitivity studies, evaluation of additional interlocking geometries and load patterns, and experimental validation.

1 INTRODUCTION

Masonry is among the oldest forms of construction and continues to be a common building material today. While masonry construction has evolved to include reinforcement in applications where additional strength and ductility are required, the form of masonry units generally remains unchanged. Unreinforced masonry (URM) structures are relatively easy to construct but often perform poorly in seismic events (Lourenço, Leite, & Pereira 2009; Totoev 2015). Despite the danger, URM structures are still constructed today and constitute a substantial portion of existing buildings in regions of seismicity where more advanced construction techniques are not readily available (Ali, Briet, & Chouw 2013; Calvi, Kingsley, & Magenes 1996; Magenes, Kingsley, & Calvi 1995; Magenes & Calvi 1997). Motivated by the need to improve earthquake resistance without increasing construction complexity, interlocking masonry systems have garnered attention for their potential to increase the capacity of masonry structures while reducing or eliminating the need for formwork, reinforcement and mortar. Non-structural advantages of interlocking masonry systems include suitability for use in automated construction systems, the architectural appeal of tessellating block geometries and decreased environmental impact by minimizing construction waste (Loing, Baverel, Caron, & Mesnil 2020; Stinson 2019).

The term interlocking masonry system describes an assemblage of masonry units with locking between adjacent units beyond the contribution of mortar and reinforcement (if present). Interlocking can be produced using one or many masonry unit geometries. Interlocking masonry systems can be comprised of hollow or solid units bonded with or without mortar in a variety of bond patterns. Members can be

DOI 10.1201/9781003316404-42

load-bearing or non-load bearing, reinforced or unreinforced. The method of interlocking creates categories specific to interlocking systems. For example, they can be topologically interlocking (locking by the contact of adjacent surfaces) or mechanically interlocking (locking by protrusions and grooves). The location and orientation of locks plays an important role; e.g., locks on the bed joint orthogonal to the lateral direction will restrict in-plane translation between block courses, while locks on the head joint orthogonal to the vertical direction will restrict translation between adjacent units in the same course. With many possible configurations, the selection of novel block geometries and interlocking patterns for improved structural performance presents a challenge. A limited number of studies have evaluated the structural performance of interlocking masonry systems.

Among the first set of experimental tests on interlocking masonry involved hollow, mechanically interlocking units developed by Thanoon et al. 2004 intended for use in load bearing, mortarless walls (Thanoon, Jaafar, Abdul Kadir, Abang Ali, Trikha, & Najm 2004). Alwathaf et al. 2005 studied the shear characteristics of this interlocking system, Jaafar et al. 2006 investigated the behavior of a prism of similar interlocking blocks in pure compression, Thanoon et al. 2007 tested the same specimen under axial compression with various eccentricities, and Safiee et al. 2011 tested 1m×1.2m wall panels (partially reinforced) under out-of-plane loading at different compressive stresses, (Alwathaf, Thanoon, S., J., & R. 2005; Jaafar, Thanoon, Najm, Abdulkadir, & Abang Ali 2006; Nor, Jaafar, & Alwathaf 2011; Thanoon, Jaafar, Noorzaei, Kadir, & Fares 2007). More recently, Safiee et al. 2018 investigated the lateral load-displacement behavior of masonry wallets under several overburden stresses. The experimental setup used in Safiee et al. 2018 is similar to that chosen for the present numerical study and the results are useful for order of magnitude comparison (Nor, Mohd Nasir, Ashour, & Abu Bakar 2018).

In a separate series of experiments, the behavior of masonry infill panels consisting of units that lock only in the out-of-plane direction was investigated, the system was coined semi-interlocking masonry by (Totev 2010). The semi-interlocking system allows for sliding between block courses in the in-plane direction, dissipating energy through friction (Totoev 2015). Using a select few semi-interlocking geometries, experiments were conducted to study the hysteretic behavior of infill panels with various fillers between neighboring masonry units and between the wall boundaries and frame (Hossain, Totoev, & Masia 2016; Masia, Totoev, & Hossain 2019; Liu, Liu, Lin, & Zhao 2016). Results from the few experimental tests on interlocking masonry systems show promise for improved structural performance; however, experimental methods are infeasible for testing many interlocking geometries. Developing this technology requires optimizing the interlocking geometry and bond pattern, a task which is more easily approached in a numerical setting.

An ongoing series of numerical studies within the framework of 3D limit analysis have been conducted in an effort to create structurally informed interlocking configurations for masonry assemblages of arbitrary shape. Mousavian et al. 2020 investigated the friction coefficient of dry stacked masonry at the limit state for corrugated block geometries (rectangular grooves) with various lock orientations (Mousavian & Casapulla 2020c). Mousavian and Casapulla 2020a/b created a feasibility check, the sliding infeasible measure, to quantify the performance of interlocking structures of arbitrary assemblage shape, again using corrugated block geometries (Mousavian & Casapulla 2020b). An optimization tool in MATLAB was then used to determine the lock orientation which minimizes sliding infeasibility (Mousavian & Casapulla 2020a). This series of studies demonstrates that the numerical setting allows for evaluation of many more interlocking configurations. However, the scope is restricted by solely using the corrugated masonry unit shape. Additionally, while limit analysis provides a powerful tool for optimizing some performance parameter of interlocking assemblages at the limit state, for seismic applications other important aspects of structural response are desired, including the displacement capacity and the hysteretic behavior.

The present study uses 3DEC, a 3-Dimensional Discrete Element Code, to capture the response of interlocking masonry walls under in-plane lateral loading. This study is limited to topologically interlocking mortarless walls with no additional items, focusing on the effect of block geometry. To further confine the problem, each wall was composed of a single interlocking block geometry. Several wall aspect ratios, overburden pressures and masonry unit compressive strengths were investigated. In total, over 300 simulations were completed for 20 masonry unit geometries. The lateral load capacity and deformed shape of each simulation was used to evaluate the relative performance of each configuration. This study is intended as a preliminary stage in a series of numerical and experimental tests. Future investigations will include numerical testing of additional geometries, assemblage shapes and modes of loading, in addition to experimental tests on the best performing geometries.

2 METHODOLOGY

Interlocking block geometries were generated and assembled into masonry wallets using RHINO6, then converted to a geometric input for 3DEC. A standard masonry block of form factor 2 × 1 × 1, (0.4 × 0.2 × 0.2) m, was used as the control. The control geometry can be thought of as a starting point from which point locations are modified to produce other geometries. With this in mind, an equal volume constraint was imposed by moving all points along a line in the direction of change by the same distance. The depth of locks relative to the characteristic dimension was kept constant at 0.2 (i.e. 0.2(0.2) = 0.04 m). Note that holding the depth of locks constant while increasing the number of locks per face increases the lock angle.

18 unique interlocking unit geometries were categorized by three features: (1) location of interlocking, bed joint-only (B), head joint-only (H), or both bed and head joints concurrently (Both); (2) the orientation of locks, preventing translation in-plane (IP), out-of-plane (OP), vertically (V), or a combination referred to as diagonally (D); and (3) the number of locks per face, either 1 or 2. The sample naming system refers to features 1, 2, and 3 in order. For example, a unit with two locks on the bed joints and lock orientation restricting in-plane translation is abbreviated as BIP2, Figure 1a. Figure 1c shows the BothIPOP2 geometry: two locks on each face, in-plane locks on the bed joints and out-of-plane locks on the head joints. For the sample group with locks on both the head and bed joints, bed joint locks restricted in-plane translation exclusively while the head joint locks could take any of the possible orientations (since this study involved only in-plane loading).

Several geometries were selected to include predefined splitting planes: BIP1, BIP2, and Butterfly. In addition, the control geometry was altered to allow splitting vertically through the middle of each unit. In comparison to the BIP1 and BIP2 units where the cross sectional area in the y-z plane is maintained, the Butterfly unit has reduced cross sectional area toward the middle. Figure 1a and 1b show the BIP2 and Butterfly geometries in elevation view with red lines indicating the predefined splitting planes.

A single masonry unit geometry was used in each wall. Units were stacked in a running bond pattern to create walls of three different aspect ratios: H/D=[2/3 1 2], where H=1.6 m. Figure 2 displays the 3DEC test setup. The top and bottom of each masonry wallet ends at a half-course of blocks fixed to the applicator plate and base plate. The base plate was fully fixed while the applicator plate was allowed to translate in the x-direction (in-plane). The applicator plate was constrained against rotation about the x and y axes as well as against translation in the y-direction (into the page). Each wall was loaded by a velocity field in the x-direction uniformly applied to the applicator plate.

Both the applicator plate and base plate were modeled as rigid while masonry units were modeled as deformable. A linearization of the Feenstra-De Borst plasticity model for concrete was adopted to capture crushing failure. The tensile strength of the FE mesh was set artificially high such that tensile failure could occur only along predefined splitting planes. The Coulomb slip model was used for all joints and potential splitting planes. The compressive strength of masonry units, f_{cm}, was used to calculate other parameters. Simulations were conducted with three different values for f_{cm}: 10, 15, and 20 MPa. Table 1 summarizes the values and equations used for the input parameters. Each simulation captured the load-displacement response through a specified number of time steps,

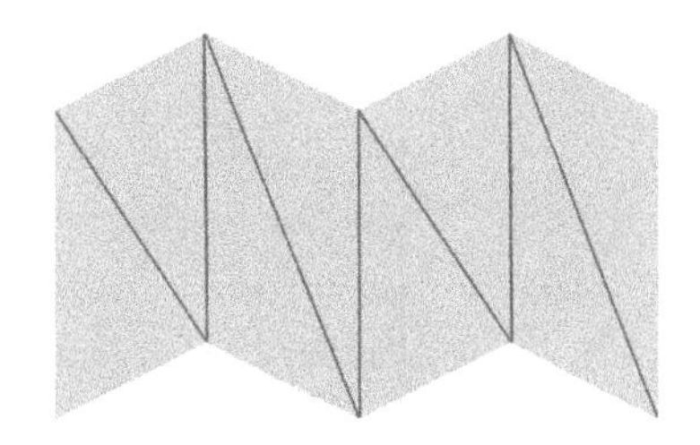

(a) BIP2 w/ Splitting

(b) Butterfly w/ Splitting

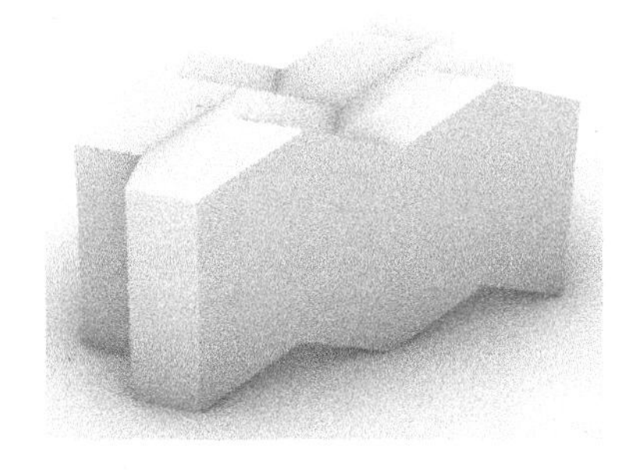

(c) BothIPOP2

Figure 1. Interlocking geometries (red lines indicate potential splitting planes).

$10^6 \leq n \leq 20^6$. The lateral load was taken as the sum of shear forces along the nodes connecting the bottom course of masonry to the base plate, i.e., base shear. Lateral displacement was recorded at the midpoint of the top-right edge.

3 RESULTS

Figures 3 and 4 show two typical lateral load-displacement curves obtained in this study. In Figure 3, the base shear (lateral load) is plotted against the lateral displacement for the Control, BIP2 and BothIPV2 geometries with a wall aspect ratio of 2/3 under each overburden pressure. Notice the control quickly plateaued to pure sliding while the BIP2 and BothIPV2 samples strengthened significantly. Additionally, vertical locks on the head joints increased the lateral load

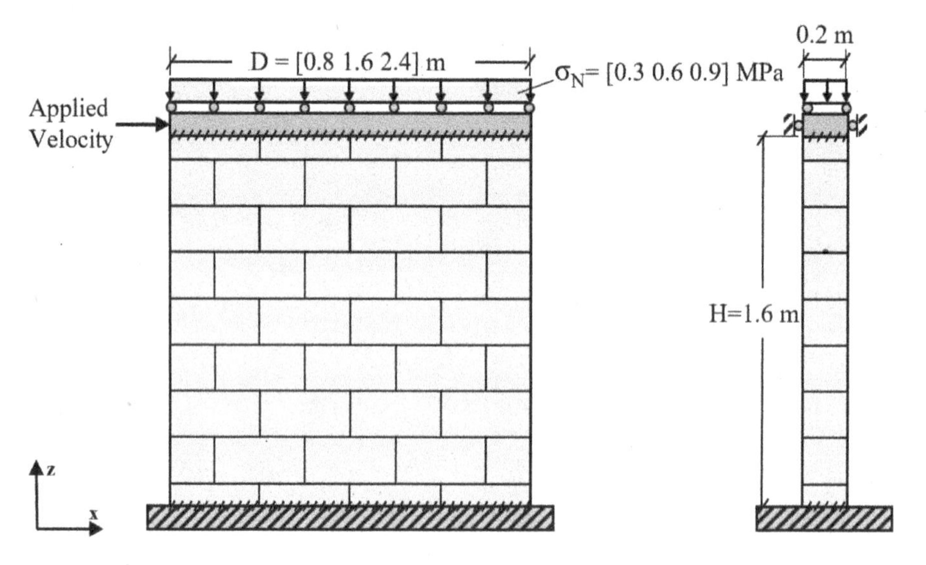

Figure 2. 3DEC setup.

Table 1. General 3DEC, Mohr-Coulomb and Feenstra-De Borst parameters.

E (Pa)	L (m)	ϕ	G	k_n	k_s
10^{10}	0.4	31°	0.4E	E/L	G/L
f_{cm} (MPa)	f_{tb}	f_{tj}	c_j	$f_{t_{sp}}$	c_{sp} *
[10 15 20]	∞	0	0	$0.1f_{cm}$	$\frac{f_{cm}(1-\sin\phi)}{2\cos\phi}$
h (m)	G_c (MPa) †	f_{c3} ‡	f_{c7}	k_e ‡	k_u ‡
0.15	$15+0.43f_{cm}-0.0036f_{cm}^2$	$f_{cm}/3$	$f_{cm}/7$	$\frac{4f_{cm}}{3E}$	$k_e+1.5\frac{G_c}{(.001)hf_{cm}}$

* (Piratheepan, Gnanendran, & Arulrajah 2012; Lelovic & Vasovic 2020; Selimir LELOVIĆ, Dejan VASOVIĆ, & Dragoslav STOJIĆ 2019)
† (Lourenço 2009)
‡ (Feenstra & De Borst 1996)

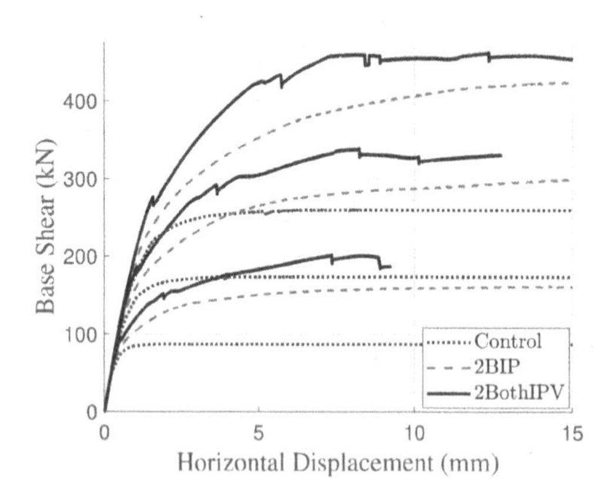

Figure 3. Lateral load-displacement, BIP2, BothIPV2, control without splitting, H/D=2/3.

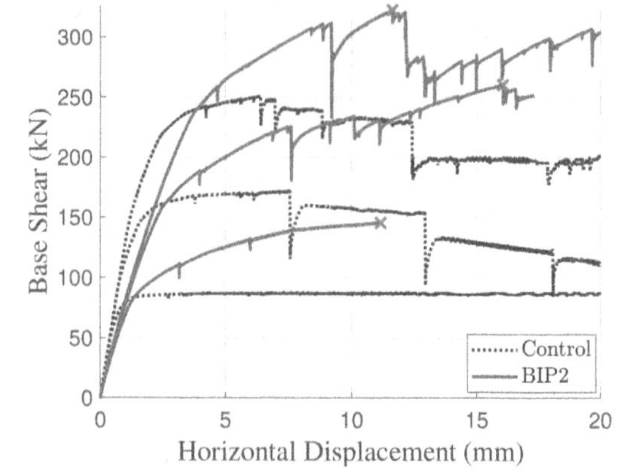

Figure 4. Lateral load-displacement, BIP2, control with splitting, H/D=2/3.

capacity of BothIPV2 beyond that of BIP2 as sliding over the bed joints forces the head joint locks to engage. Figure 4 shows the base shear versus lateral displacement for the control and BIP2 geometries with splitting. At the lowest overburden stress, the response of samples with splitting is nearly identical to the corresponding samples without splitting. However, at higher levels of overburden stress, tensile failure through units begins to occur and the load-displacement curves experience sudden jumps. Strengthening occurs even after the onset of block splitting, typical of squat walls. The lateral load capacity of each simulation was taken as the peak base shear, indicated by green marks in Figure 4.

3.1 Without splitting

Figures 5a-5c show the lateral load capacity of all simulations which did not include block splitting; horizontal lines indicate the control value at each overburden stress. The controls for walls with aspect ratios of 2/3 and 1 experienced sliding failure along the top course as the predominant failure mode, while the controls for walls with an aspect ratio of 2 exhibited rocking failure, as shown in Figure 6a. All walls with an aspect ratio of 2 experienced rocking failure which corresponded to a slight reduction in the lateral load capacity compared with the control, Figure 5c. Across all aspect ratios, interlocking only on the head joints—HV, HOP, HD—had little effect, displaying the same failure modes as the respective controls. For walls with aspect ratios of 2/3 and 1, samples with locks on either the bed joint only or both the bed joint and head joint concurrently showed substantial increases in lateral load capacity. Larger increases in capacity were recorded for geometries with two locks, simulations under lower overburden stress and for walls with an aspect ratio of 2/3. For the geometries with interlocking only on the bed joints, locks normal to in-plane translation resulted in the greatest increases in capacity. For the geometries with interlocking of both the bed and head joints, the addition of locks normal to vertical translation increased capacity the most. In general, the samples with interlocking on both the bed and head joints showed higher lateral load capacity than the bed-only group. Across all simulations, the BothIPV2 geometry in a wall aspect ratio of 2/3 under 0.3 MPa overburden stress produced the highest increase of 130% over the control.

3.2 With splitting

Figures 6c and 7b-d show contours of x-displacement for several samples with predefined splitting planes. It can be observed that the tensile failure of units has a strong effect on the failure mode for samples with the lowest compressive strength of units, $f_{cm} = 10$ MPa. Figure 8 presents the normalized lateral load capacity of simulations which included block splitting with $f_{cm} = 10$ MPa; each data point is normalized by the corresponding control. Walls with an aspect ratio of 2 showed primarily rocking failure with some block splitting at an overburden pressure of 0.9 MPa; a slight reduction in capacity was observed for all cases. For wall aspect ratios of 2/3 and 1, substantial increases in lateral load capacity were recorded, particularly for the BIP2 geometry and for cases with lower overburden pressure. It is important to note that while Figure 8 gives the impression of correlation between normalized capacity and aspect ratio, this trend is only clear

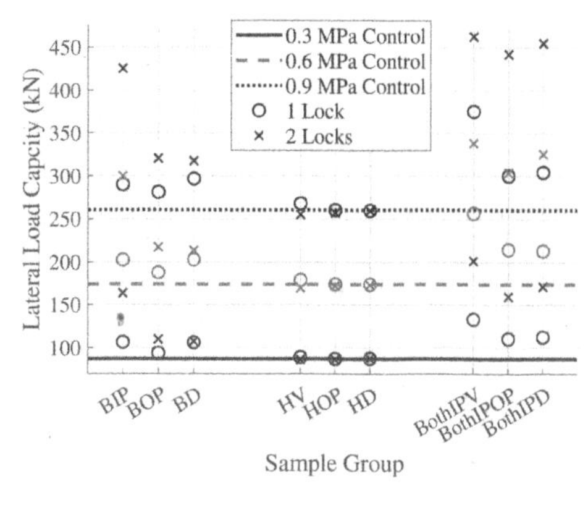

(a) H/D=2/3

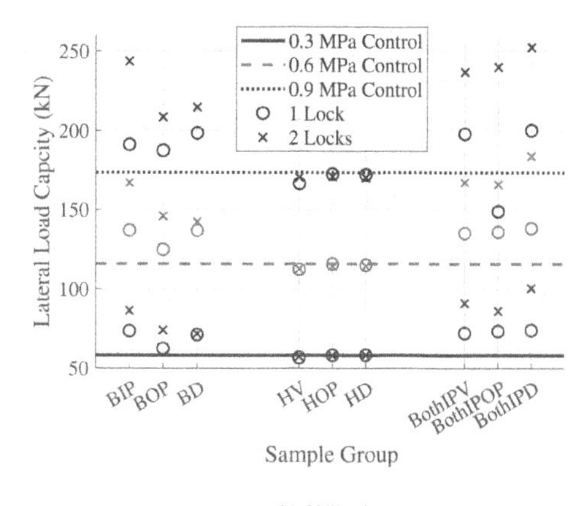

(b) H/D=1

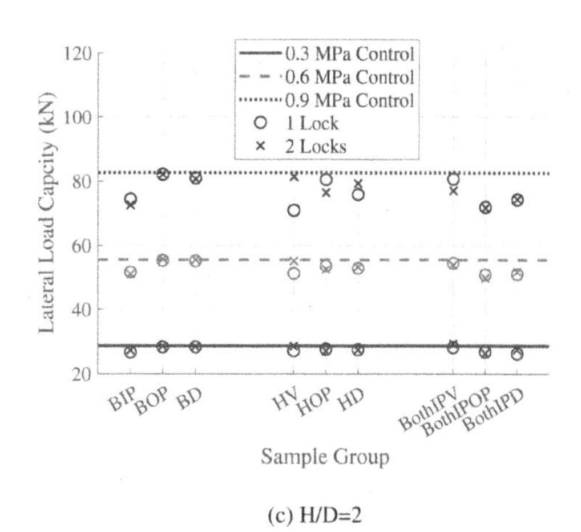

(c) H/D=2

Figure 5. Lateral load capacity, without splitting.

for the BIP2 geometry. Due to the available predefined splitting paths, the BIP2 geometry was forced to split along multiple diagonal bands, Figure 7b, whereas a single band forms for the BIP1 geometry, Figure 6c. Higher overburden pressure correlated to a reduction

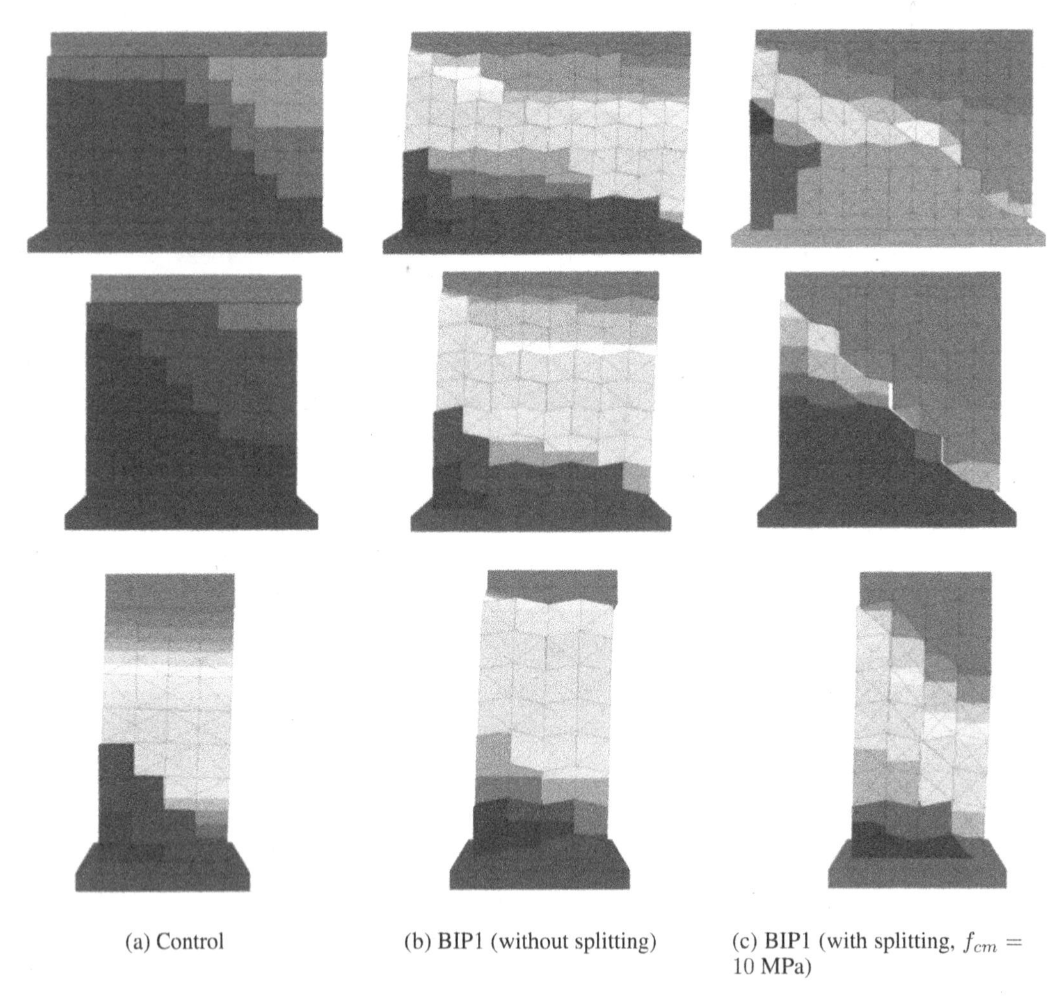

(a) Control (b) BIP1 (without splitting) (c) BIP1 (with splitting, f_{cm} = 10 MPa)

Figure 6. Control and BIP1 x-displacement contour, all aspect ratios.

in lateral load capacity normalized by the control, this correlation was present even for tests without splitting although to a lesser extent.

To more clearly distinguish the relation between overburden pressure and reduction in normalized capacity resulting from block splitting, the lateral load capacity of samples with splitting was normalized by the corresponding interlocking geometries without splitting for walls with an aspect ratio of 1 in Figure 9. Intuitively, higher overburden stress caused an increase in block splitting, thereby reducing the normalized capacity in comparison to simulations without splitting. It can be seen that the reduction in normalized capacity at higher overburden stress for samples with splitting goes beyond that recorded for samples without splitting and a negative correlation exists between overburden pressure and normalized capacity, this is associated with block splitting. The greatest increase in capacity among samples with $f_{cm} = 10$ MPa occurred for BIP2 with a wall aspect ratio of 2/3 under overburden stress 0.3 MPa, a 66% increase.

In passing we mention the results for the butterfly geometry; this geometry was not classified along with the other interlocking shapes because it is unique in the sense that the cross sectional area in the y-z plane is not constant. The Butterfly geometry was found to perform poorly in each case, producing at most a 4% increase in capacity. This poor performance clearly results from the reduced cross-sectional area at the critical section, where splitting occurred.

The compressive strength of masonry units, f_{cm}, had a significant effect on the failure of walls which include splitting. Increasing the compressive strength increased the cohesion, tensile strength of the Coulomb slip model and increased the parameters of the crushing model. As a result, increasing f_{cm} is reflected by an increase in lateral load capacity, particularly at higher overburden stress where splitting becomes more important. Naturally, increasing the strength of masonry units should result in a decrease in block splitting within a sensitive range of overburden pressures, as was the case even for simulations where

(a) Without Splitting

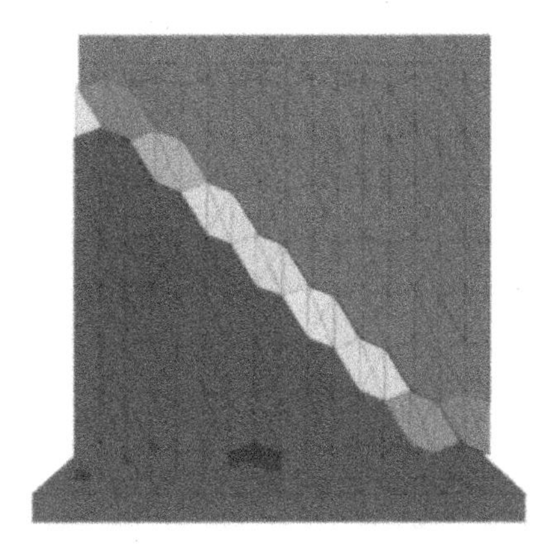

(b) With Splitting ($f_{cm} = 10$ MPa)

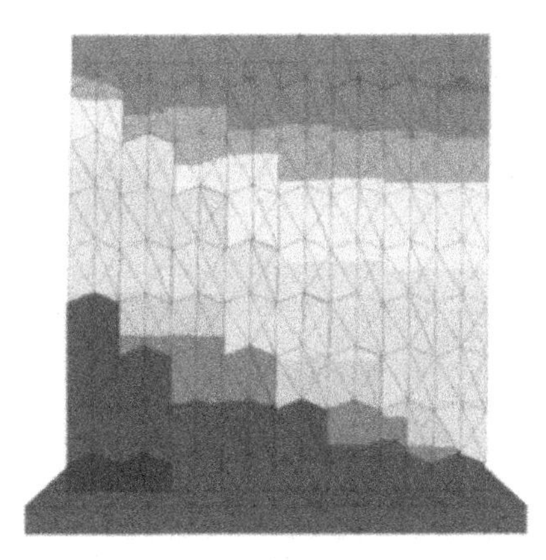

(c) With Splitting ($f_{cm} = 15$ MPa)

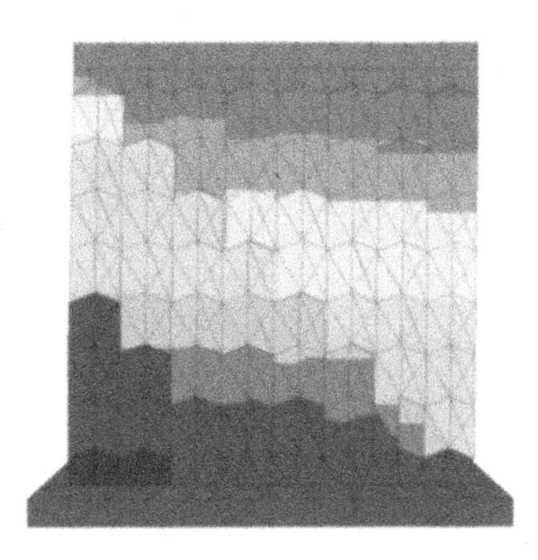

(d) With Splitting ($f_{cm} = 20$ MPa)

Figure 7. BIP2 x-displacement contour, H/D=1, $\sigma_N = 0.9$ MPa.

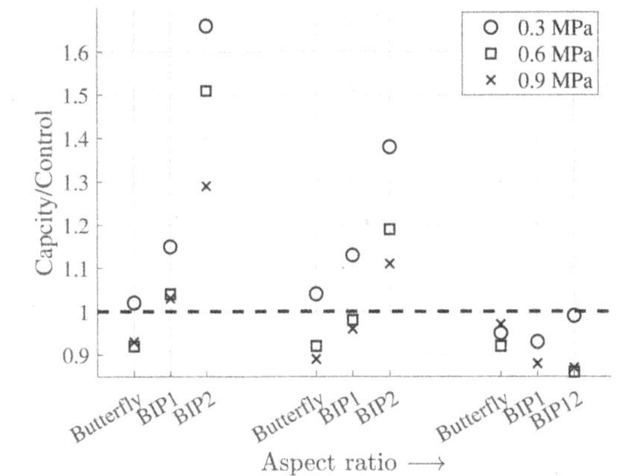

Figure 8. Normalized lateral load capacity (with splitting).

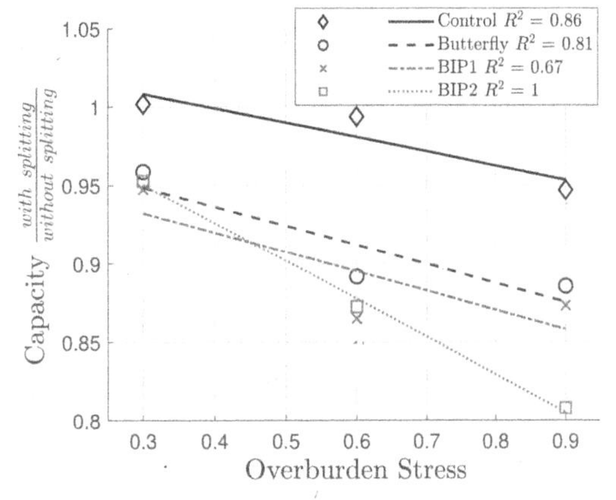

Figure 9. Normalized capacity ($\frac{\text{with}}{\text{without}}$) splitting vs. overburden stress.

overburden pressure was highest, $\sigma_N = 0.9$ MPa, as shown in Figure 7. Rather than opening at the joints as in Figure 7a (without splitting), masonry units split along the predefined planes in Figure 7b (with splitting and $f_{cm} = 10$ MPa). Figure 7c shows that increasing f_{cm} to 15 MPa eliminated the majority of tensile failure through units, with only one splitting plane at the right toe. Upon increasing f_{cm} to 20 MPa only a slight change occurs, preventing virtually all splitting in Figure 7d. Figure 10 shows the lateral load capacity of all simulations which included block splitting at each value for the compressive strength of masonry units: 10, 15, 20 MPa. The change in capacity shown in Figure 9 is noticeable from $f_{cm} = 10$ MPa to $f_{cm} = 15$ MPa and negligible from 15 MPa to 20 MPa. Once the majority of splitting is prevented, further increasing the compressive strength has little effect.

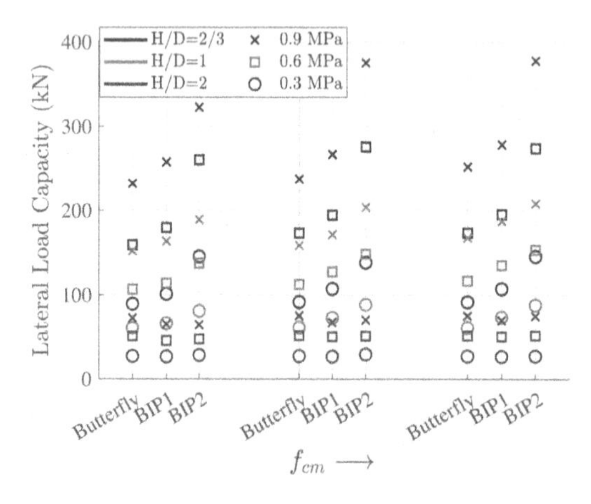

Figure 10. Lateral load capacity (with splitting) varying f_{cm}.

4 DISCUSSION

4.1 *Without splitting*

While ignoring splitting results in an overestimate of the lateral load capacity of interlocking masonry walls, the results (without splitting) confirmed the expected effects of interlocking and provide model validation through agreement with well known concepts of masonry mechanics. For instance, it is well understood that slender walls are susceptible to rocking failure and rocking failure was, in fact, observed for all numerical simulations on walls with an aspect ratio of 2. It follows that if the failure is controlled by rocking, the interlocking of masonry units should have little effect, see Figure 5c.

Another obvious prediction confirmed by analysis in 3DEC is that under the load pattern used in the present study, walls comprised of blocks which interlock only on the head joints should experience sliding failure along the flat bed joints as the head joint locks are never engaged (center column of Figures 5a-c). On the other hand, geometries which included locks on the bed joints perpendicular to the lateral direction should be expected to produce the greatest increase in lateral load capacity of all possible bed joint lock orientations, as was found. Further, the uplifting of block courses resulting from sliding over locks on the bed joints engages locks on the head joints, increasing the lateral load capacity most significantly for geometries with vertical head joint interlocking, see the rightmost columns of Figures 5a/b.

As noted in section two, increasing the number of locks per face while holding the depth of locks constant effectively increases the angle of the locks. It is then expected that samples with two locks should produce a greater increase in lateral load capacity, as confirmed by 3DEC. We also note that the observation that capacity was increased the most for walls with an aspect ratio of 2/3 is in agreement with expectations since a less direct failure path from the top-left to the bottom-right corner is available.

4.2 *With splitting*

The results of simulations which included splitting also agree with expected masonry behavior. For the most slender walls, although some splitting occurred at the higher overburden pressures, rocking failure remained the controlling mechanism and these samples showed no improvement over the control, Figure 8. Secondly, walls with an aspect ratio of 2/3 were able to maintain capacity and even strengthen slightly after the onset of tensile failure through units, a well known characteristic of masonry walls with low aspect ratio, see Figure 4.

The significant change between simulations conducted with and without splitting demonstrates the importance of selecting realistic splitting planes. It should be expected that if little to no splitting occurs, the behavior of walls which included splitting should largely agree with those which did not include splitting. This prediction is verified by simulations at the lowest overburden pressure where results with splitting were closest to the same geometries without splitting; notice the normalized capacity of roughly 1 at $\sigma_N = 0.3$ MPa in Figure 9. Moreover, as splitting becomes increasingly pervasive with increasing overburden pressure, the lateral load capacities recorded for samples with and without splitting diverge; under high overburden pressure, samples which included splitting failed at comparatively lower peak loads.

Of course, the strength of masonry units is essential for accurately modelling the behavior of a masonry assemblage. In this study the compressive strength of units was used to determine several input parameters in the Coulomb-split model (used for tensile failure) and the Feenstra-De Borst model (used for crushing failure). As a result, increasing the compressive strength reduced the prevalence of block splitting. It should be expected that if the strength of units is sufficient to prevent nearly all splitting, the results of simulations with splitting should agree with those without splitting, a prediction confirmed by the observed failure modes of the simulations conducted with $f_{cm} = 15$ MPa and 20 MPa, Figure 7. Additionally, because nearly all splitting was prevented at $f_{cm} = 15$ MPa, little change in capacity occurred beyond this value (see Figure 10).

5 CONCLUSION

Masonry structures have existed for millennia and abound in today's built environment, yet the form of masonry units remains relatively unchanged. The inherent heterogeneity of masonry structures makes their analysis particularly challenging, in large part due to the complex interaction between masonry units. In this paper, the behavior of masonry walls comprised of novel block geometries was studied using the discrete element software 3DEC. Interlocking between masonry blocks was shown to produce a substantial increase in lateral load capacity depending on the wall aspect ratio and the lock location, number and orientation.

The response of masonry wallets simulated in 3DEC agreed well with masonry mechanical phenomena. For instance, slender walls experienced rocking failure and the effect of interlocking was, as a result, negligible. Walls consisting of blocks with flat bed joints experienced sliding failure as with the control. On the contrary, walls with an aspect ratio of 2/3 or 1 showed large increases in lateral load capacity, particularly in cases where interlocking was engaged to the fullest extent, as for sample BOTHIPV2. Samples with two locks on the bed joint (increased lock angle), showed consistently higher lateral load capacity; the relation between lock angle and capacity is the focus of a follow-up study to the present work.

Simulations which included predefined splitting paths provide internal model verification, again confirming predicted behavior. The prevalence of block splitting correlated with overburden pressure and the strength of masonry units. In cases where the overburden pressure was low or the strength of units was high, little splitting occurred and the lateral load capacity and failure mode was consistent with models where splitting was neglected. The converse was also true; simulations conducted under high overburden pressure or with low strength of units showed that block splitting allowed for a more direct failure path, reducing the relative lateral load capacity. This exercise expresses the importance of selecting appropriate predefined splitting paths for a given load case.

Although the scope of this study is modest in comparison to the infinite number of possible interlocking masonry structure configurations, the results demonstrate the potential for interlocking to produce improved structural performance. 3DEC provides a useful environment for studying the many aspects of interlocking masonry behavior that remain to be considered. Future investigations will include additional geometries, load patterns, and the study of interlocking assemblages with mortar, reinforcement, and within framed structures. Eventually, the geometries which perform best in the numerical setting will be experimentally tested.

REFERENCES

Ali, M., R. Briet, & N. Chouw (2013, May). Dynamic response of mortar-free interlocking structures. *Construction and Building Materials 42*, 168–189.

Alwathaf, A., W. Thanoon, J. S., N. J., & A. R. (2005, January). Shear Characteristic of Interlocking Mortarless Block Masonry Joints. *Masonry International 18*, 139–146.

Calvi, G. M., G. R. Kingsley, & G. Magenes (1996, February). Testing of Masonry Structures for Seismic Assessment. *Earthquake Spectra 12*(1), 145–162. Publisher: SAGE Publications Ltd STM.

Feenstra, P. H. & R. De Borst (1996, February). A composite plasticity model for concrete. *International Journal of Solids and Structures 33*(5), 707–730.

Hossain, M. A., Y. Totoev, & M. Masia (2016, August). *Friction on mortar-less joints in semi interlocking masonry.*

Jaafar, M. S., W. A. Thanoon, A. M. Najm, M. R. Abdulkadir, & A. A. Abang Ali (2006, September). Strength correlation between individual block, prism and basic wall panel for load bearing interlocking mortarless hollow block masonry. *Construction and Building Materials 20*(7), 492–498.

Lelovic, S. & D. Vasovic (2020). Determination of Mohr-Coulomb Parameters for Modelling of Concrete. *Crystals 10*(9).

Liu, H., P. Liu, K. Lin, & S. Zhao (2016, March). Cyclic Behavior of Mortarless Brick Joints with Different Interlocking Shapes. *Materials 9*, 166.

Loing, V., O. Baverel, J.-F. Caron, & R. Mesnil (2020, May). Free-form structures from topologically interlocking masonries. *Automation in Construction 113*, 103117.

Lourenço, P., J. Leite, & M. Pereira (2009). Masonry Infills and Earthquakes. Toronto, Canada.

Lourenço, P. B. (2009, September). RECENT ADVANCES IN MASONRY MODELLING: MICROMODELLING AND HOMOGENISATION. In *Multiscale Modeling in Solid Mechanics*, Volume Volume 3 of *Computational and Experimental Methods in Structures*, pp. 251–294. IMPERIAL COLLEGE PRESS.

Magenes, G. & G. M. Calvi (1997, November). In-plane seismic response of brick masonry walls. *Earthquake Engineering & Structural Dynamics 26*(11), 1091–1112. Publisher: John Wiley & Sons, Ltd.

Magenes, G., G. Kingsley, & G. Calvi (1995, January). *Seismic Testing of a Full-Scale, Two-Story Masonry Building: Test Procedure and Measured Experimental Response.*

Masia, M., Y. Totoev, & M. A. Hossain (2019, January). Experimental assessment of large displacement cyclic in-plane shear behaviour of semi-interlocking masonry panels. *International Journal of Masonry Research and Innovation 4*, 378.

Mousavian, E. & C. Casapulla (2020a, November). *Automated Shape Adjustment of Interlocking Joints for Structurally Informed Design of Masonry Block Assemblages.*

Mousavian, E. & C. Casapulla (2020b, November). Quantifiable feasibility check of masonry assemblages composed of interlocking blocks. *Advances in Engineering Software 149*, 102898.

Mousavian, E. & C. Casapulla (2020c, April). Structurally informed design of interlocking block assemblages using limit analysis. *7*, 1–21.

Nor, N. A., M. Jaafar, & A. Alwathaf (2011, December). Structural Behavior of Mortarless Interlocking Load

Bearing Hollow Block Wall Panel under Out-Of-Plane Loading. *Advances in Structural Engineering 14*, 1185.

Nor, Z., N. Mohd Nasir, A. Ashour, & N. Abu Bakar (2018, January). Behaviour of interlocking mortarless hollow block walls under in-plane loading. *Australian Journal of Structural Engineering 19*, 1–9.

Piratheepan, J., C. T. Gnanendran, & A. Arulrajah (2012, September). Determination of c and from IDT and Unconfined Compression Testing and Numerical Analysis. *Journal of Materials in Civil Engineering 24*(9), 1153–1164.

Selimir LELOVIĆ, Dejan VASOVIĆ, & Dragoslav STOJIĆ (2019, April). Determination of the Mohr-Coulomb Material Parameters for Concrete under Indirect Tensile Test. *Tehnicki vjesnik – Technical Gazette 26*(2).

Stinson, L. (2019, Dec). Sleek london building is a marvel of modern masonry.

Thanoon, W. A., M. Jaafar, J. Noorzaei, M. R. A. Kadir, & S. Fares (2007, February). Structural Behaviour of Mortar-Less Interlocking Masonry System under Eccentric Compressive Loads. *Advances in Structural Engineering 10*(1), 11–24.

Thanoon, W. A., M. S. Jaafar, M. R. Abdul Kadir, A. A. Abang Ali, D. Trikha, & A. M. Najm (2004, July). Development of an innovative interlocking load bearing hollow block system in Malaysia. *Construction and Building Materials 18*(6), 445–454.

Totev, Y. Z. (2010). Mortarless Masonry.

Totoev, Y. (2015, May). Design Procedure for Semi Interlocking Masonry. *Journal of Civil Engineering and Architecture 9*.

Computational Modelling of Concrete and
Concrete Structures – Meschke, Pichler & Rots (Eds)
 ISBN: 978-1-032-32724-2

A microporomechanical model to predict nonlinear material behavior of masonry

Yubao Zhou*, Lambertus J. Sluijs & Rita Esposito
Department of Materials, Mechanics, Management and Design (3MD), Faculty of Civil Engineering and Geosciences, Delft University of Technology, Delft, The Netherlands

ABSTRACT: The microporomechanics theory, which combines the mean-field homogenization method and linear fracture mechanics theory, has been successfully adopted to study the nonlinear behavior of composite-like materials, such as alloy, rocks and concretes. The application of such theory is however mainly limited to the isotropic quasi-brittle materials and the study of crack propagation in an initially anisotropic materials, as masonry, has received limited attention. This paper aims to derive the nonlinear material behavior of masonry by adopting microporomechanics theory. In this study, masonry is treated as a composite material, made of bricks, mortar joints and microcracks. At constituents' level, cracks are idealized as three orthotropic families of penny-shaped inclusions, which are then embedded in an undamaged effective masonry matrix formed by bricks and mortar joints. A crack density variable, containing the information of each crack family (e.g., crack radius), is adopted to define the damage state of masonry. The propagation of each crack family is governed by the energy release rate and its critical value. The results shows that the microporomechanics theory can successfully derive the nonlinear behavior of masonry (e.g., the tensile softening). The proposed model allows using limited input parameters mainly related to properties of constituents, and elastic modulus and tensile strength of the composites. However, it should be mentioned that the model developed in this study only considers the cohesive mechanics by modelling the propagation of open cracks, while the friction on the lips of closed microcracks is not taken into consideration and it will be objective of further study.

Keywords: Masonry; Mean-field homogenization; Microporomechanics: Nonlinear behavior; Anisotropic.

1 INTRODUCTION

As one of oldest construction techniques, masonry buildings represent a large number of structures as residential dwellings and heritages throughout the world. Although masonry is not seen as a high-tech material, it is difficult to characterize masonry material and conduct structural analysis accordingly due to its complex mechanical behaviour.

Considering the derivation of the nonlinear response of masonry based on contituentes' behavior, three categories of modelling approaches can be distinguished: discrete models, computational multiscale models and continuum micromechanical approaches (D'Altri et al. 2020). In the discrete models, every brick is separately connected by interfaces (Lourenço 1996) or contact elements (D'Altri et al. 2020). Computational multiscale models (Massart et al. 2007; Petracca et al. 2016) numerically derive the structural response from the behaviour of the homogenized material by means of computational homogenization of a representative element volume (REV). Continuum micromechanical approaches (Addessi et al. 2010; Gambarotta & Lagomarsino 1997; Marfia & Sacco 2012) define the damage evolution starting by postulating the behaviour of a single defect (crack or void) and obtaining the behaviour of the material as a continuum by applying statistical averaging to an ensemble of defects in a REV. Compared with the other two kinds of methods, the continuum micromechanically based approach results in one of the most efficient methods also being user-friendly for their use in practice.

Among various continuum micromechanically based approaches, large numbers of research works mainly focused on phenomenological damage models to describe the induced damage in materials (Chow & Wang 1987; Swoboda & Yang 1999). These phenomenological damage models are easy to be implemented into computer codes and to conduct structural analysis accordingly. However, many assumptions lacking clear physical meaning are made in these models. Additionally, some important physical mechanisms at microscopic scale, such as unilateral effects and frictional sliding, are not properly described in these models (Zhu et al. 2008). The microporomechanics theory provides an efficient solution to overcome these limitations of phenomenological damage models. In this theory, the global (effective) properties of

DOI 10.1201/9781003316404-43

the cracked material, which is deemed as a composite formed by a matrix-inclusion system, can be calculated following a rigorous upscaling procedure (Dormieux et al. 2006).

Following Eshelby's solution (Eshelby 1957) to the elastic matrix-inclusion problem, several mean-field homogenization models have been proposed, creating a new field which can be named as Eshelby-based continuum micromechanics. By imposing that the size of the heterogeneity is one scale lower than the size of a RVE of the continuum material, the strain of the continuum material can be calculated as average strain by considering linear displacements, periodic boundary conditions or uniform traction at the boundary of the volume (Nemat-Nasser & Hori 2013). The global stiffness tensor of the composite can thus be determined by considering the elastic tensors of constituents and the concentration tensor related to the inclusions which contain information regarding their shapes, orientations and volume concentrations.

By combining the mean-field homogenization method and linear elastic fracture mechanics theory, Dourmieux, Kondo and Ulm developed the so-called microporomechanics theory to explain the elastic and strength properties of saturated and unsaturated porous media (Dormieux et al. 2006). The scope was to explain phenomena such as failure of rock, hydration of cementitious materials, degradation processes in concrete in which the combined action of pore pressure and external mechanical load can trigger crack propagation generated by the internal porosity of the material (e.g. Lemarchand et al. 2003). Due to a simple averaging method proposed by the mean-field homogenization technique, the microporomechanics theory has been adopted to study the nonlinear behavior of composite-like materials in different fields ranging from metal composites (Doghri et al. 2016; Pierard et al. 2004), alloys (Pardoen & Hutchinson 2003), rocks (Deude et al. 2002; Pensée et al. 2002), cementitious materials (Pichler et al. 2007; Ulm et al. 2004), geomaterials (Zhu et al. 2009) and bones (Fritsch et al. 2013; Morin et al. 2017). The application of such theory is however mainly limited to isotropic quasi-brittle materials and the study of crack propagation in an initially anisotropic materials, as masonry, has received limited attention.

This paper aims to derive the nonlinear material behaviour of unreinforced brick masonry (URM) by using limited input parameters mainly related to the elastic properties of the constituents (i.e., bricks and mortar) and the elastic properties and tensile strength of the masonry composites. To achieve this target, the microporomechincs theory, which combines the mean-field homogenization method and linear fracture mechanisms theory, is used by following a standard two-step homogenization procedure. In this study, masonry is treated as a composite material, made of bricks, mortar joints and microcracks. In the first step, the orthotropic elastic properties of uncracked masonry are calculated from the isotropic elastic properties of its constituents by an improved mean-field homogenization model. In the second step, microcracks are idealized as three orthotropic families of penny shape inclusions, which are then embedded in the effective uncracked masonry homogenized from the first step. A crack density variable, containing the information of each crack family (e.g., crack radius), is adopted to define the damage state of masonry. The propagation of each crack family is governed by the energy release rate and its critical value. The nonlinear behaviour of masonry is then derived as a result of the evolution and propagation of the microcracks.

2 METHOD

2.1 *Model for orthotropic elastic properties*

In this study, an improved mean-field homogenization scheme is presented to compute the effective elastic properties of masonry. The state-of-the-art for the mean-field homogenization theory and the formation of the proposed model are briefly described.

By considering an inclusion embedded in an infinite elastic matrix subjected to homogeneous boundary conditions (uniform displacement or uniform tractions), the effective (macroscopic) stiffness and compliance tensors, C^* and D^*, for a composite material can be expressed in the following forms (Klusemann & Svendsen 2010):

$$C^* = C_m + \sum_i \varphi_i (C_i - C_m) : A_i \tag{1a}$$

$$D^* = D_m + \sum_i \varphi_i (D_i - D_m) : B_i \tag{1b}$$

where the matrix phase is labeled by m and the inclusion of type-i is labeled by i for a matrix-inclusion system. The colon denotes the tensor operation for the double dot product. $\boldsymbol{C}$ and $\boldsymbol{D}$ are the stiffness and compliance tensors, respectively. φ_i is the volume fraction of type-i inclusion. $\boldsymbol{A}_i$ and $\boldsymbol{B}_i$ are the average strain and stress concentration tensors, respectively.

Various homogenization schemes adopt different expressions for $\boldsymbol{A}_i$ and $\boldsymbol{B}_i$ to evaluate the interaction degree between matrix and inclusions. One of the most used mean-field homogenization schemes is the interaction direct derivative (IDD) scheme. This method assumes that each inclusion (Ω_i) is first embedded into a finite matrix (Ω_m) and then the type-i inclusion-matrix cell denoted by Ω_{Di}, with $\Omega_{Di} = \Omega_i + \Omega_m$, is embedded in the infinite homogenized effective medium denoted by Ω^E with unknown effective (macro) stiffness tensor C^*. The average stress concentration tensor of IDD scheme is determined as follows (Zheng & Du 2001):

$$B_i^{(IDD)} = B_i^{(Dilute)} : \left[I - \sum_i \varphi_i (D_i - D_m) : B_i^{(Dilute)} : C_m : \left(I - S_{Di}^m\right) \right]^{-1}$$

$$\text{with } B_i^{(Dilute)} = C_i : A_i^{(Dilute)} : D_m \tag{2}$$

where I is the fourth-order symmetric identity tensor. S_i^m and S_{Di}^m are the Eshelby tensors for type-i inclusion and type-i inclusion-matrix cell, respectively. $A_i^{(Dilute)}$ is the average strain concentration tensor estimated by the dilute scheme, according to:

$$A_i^{(Dilute)} = [I + S_i^m : C_m^{-1} : (C_i - C_m)]^{-1} \quad (3)$$

By properly choosing an interpolation between the IDD and the inverse IDD models, a new homogenization model is proposed. The IDD method assumes mortar joints as inclusions, while the inverse IDD method assumes the bricks as inclusion phase.

$$C^s = \left[(1 - \zeta(\varphi_i)) D^{*(IDD)} + \zeta(\varphi_i) D^{*(IDD^{-1})}\right]^{-1} \quad (4)$$

where $\boldsymbol{C}^s$ is the stiffness estimate of the proposed model. $\zeta(\varphi_i) = 1/2(\sum \varphi_i)(1 + \sum \varphi_i)$ is a smooth function of interpolation proposed by Lielens (1999).

Figure 1 presents the assumptions of the matrix-inclusion system for the proposed model, where all the inclusions are approximated by elliptical cylindrical inclusions, following the work by (Bati et al. 1999). Each elliptical cylindrical inclusion is first embedded into a matrix-inclusion cell (REV) that is also idealized as an elliptical cylinder. The elliptical cylindrical matrix-inclusion cell is then embedded into the infinite matrix. For the expression of Eshelby's tensor of elliptic cylindrical inclusions, the readers are referred to (Mura 2013).

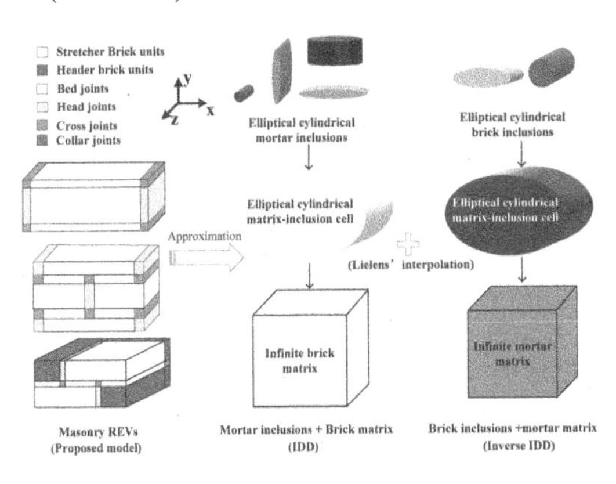

Figure 1. matrix-inclusion assumption for the proposed model.

The proposed model is compared with some classical mean-field homogenization models, including the Reuss, Voigt, dilute, Mori-Tanaka, double inclusion (D-I), self-consistent scheme (SCS), effective self-consistent scheme (ESCS) and interaction direct derivative (IDD) models. For a detailed description of these classical models, the readers are referred to (Klusemann & Svendsen 2010). For each of the classical models, two assumptions regarding the inclusion-matrix system are considered: the assumption with mortar joints as inclusion phase (bricks as matrix) corresponds to the model itself, and the inverse model corresponds to the opposite assumption with bricks as inclusions (mortar as matrix).

Figure 2 shows the prediction results for effective vertical Young's modulus obtained from the proposed model, the classical mean-field homogenization models and a finite element analysis (FEA) for stack bonded masonry. Herein, the dimensions of bricks are $210 \times 52 \times 100\text{mm}^3$ (Length × height × thickness), and a value of 10mm is adopted for the thickness of mortar joints. The Young's modulus and Poisson's ratio of bricks are 20 GPa and 0.15, respectively. The Poisson's ratio of mortar is 0.15. The Young's modulus of mortar is varied to evaluate the influence of the brick-to-mortar stiffness ratio from 1 to 1000 ($1 \leq E_b/E_m \leq 1000$) on the elastic properties of masonry. The results demonstrate that the proposed model outperforms the classical mean-field homogenization models and provides results in very good agreement with the detailed FEA. For a more detailed description of the proposed model, please refer to our recent work in which the proposed model is demonstrated to be able to make accurate evaluation of the three-dimensional orthotropic elastic properties of

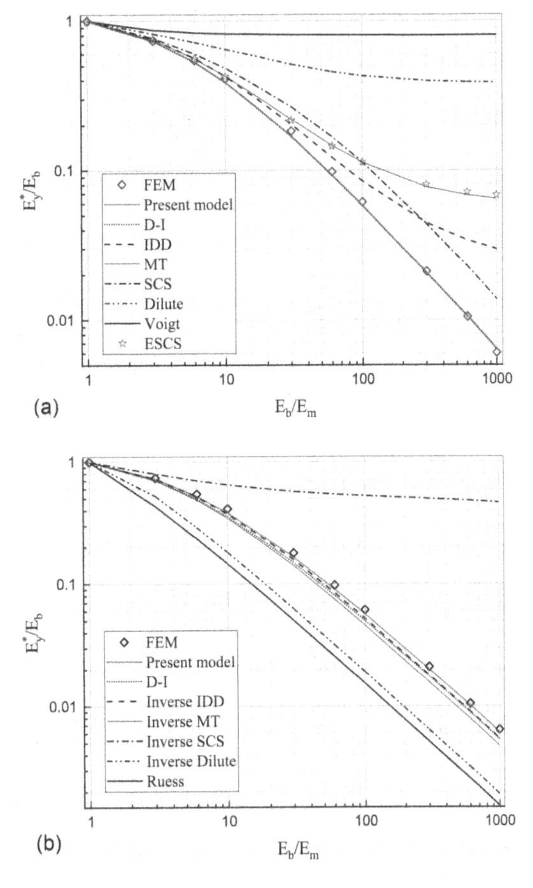

Figure 2. The macroscopic vertical Young's moduli E_y^* calculated by different mean-field homogenization models for stack bonded masonry: (a) the models with mortar joints as inclusions; (b) the inverse models with bricks as inclusions.

stack bonded, running bonded and Flemish bonded masonry (Zhou et al. 2021).

2.2 *Model for nonlinear behaviour*

2.2.1 *Effective elastic properties of cracked masonry*

The homogenized masonry derived from Section 2.1 is hereby named as solid matrix. After deriving the orthotropic elastic properties of masonry from the isotropic properties of its constituents, three orthotropic families of microcracks, which are idealized as penny shaped inclusions, are then embedded in the solid matrix. Description of the two-scale matrix-inclusion system is shown in Figure 3.

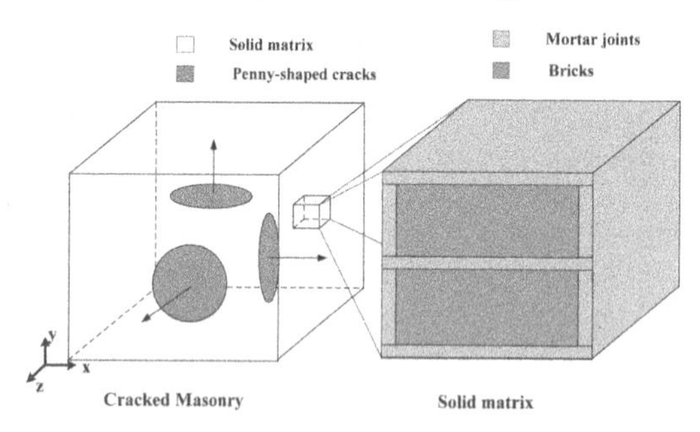

Figure 3. Description for the two-scale matrix-inclusion system.

In the following analysis, the orthotropic stiffness tensor of the solid matrix is denoted by C^s (calculated by Eq. (4)). Each family of penny-shaped cracks is characterized by its normal n, radius a and thickness c. For the i-th crack family ($i = 1, 2$and3), its aspect ratio X_i and volume fraction φ_{ci} are defined as follows:

$$X_i = \frac{c_i}{a_i} \tag{5}$$

$$\varphi_{ci} = \frac{4}{3\pi} c_i n_i a_i^2 \tag{6}$$

where n_i is the number of cracks per unit of volume (crack density).

For a given damage state, the effective stiffness tensor of the cracked masonry C^{hom} is calculated by the Mori-Tanaka method.

$$C^{hom} = C^s : \left(I - \sum_{i=1}^{3} \varphi_{ci} A_{ci} \right)$$

$$\text{with } A_{ci} = A_{ci}^{(Dilute)} : (\varphi_s I + \sum_{i=1}^{3} \varphi_{ci} A_{ci}^{(Dilute)})^{-1} \tag{7}$$

where $\varphi_s = 1 - \sum_{i=1}^{3} \varphi_{ci}$ is the volume fraction of solid matrix. $A_{ci}^{(Dilute)}$ is the average strain concentration tensor estimated by the dilute scheme:

$$A_{ci}^{(Dilute)} = [I - S_{ci}^s]^{-1} \tag{8}$$

where S_{ci}^s is Eshelby's tensor which depends on the orientation and aspect ratio of i-th crack family and the stiffness tensor of the solid matrix (C^s).

The explicit expression for the Eshelby's tensor is however only available for the isotropic or transverse anisotropic matrix. In this study, the stiffness tensor of solid matrix C^s is orthotropic. Therefore, a numerical solution is used. For a generic anisotropic material, the Eshelby tensor is given by the following surface integral (Mura 2013):

$$S_{ci(ijkl)}^s = \frac{1}{8\pi} C_{ijkl}^s \int_{-1}^{1} d\zeta_3 \int_0^{2\pi} \left\{ G_{imjn}\left(\bar{\zeta}\right) + G_{jmin}(\bar{\zeta}) \right\} dw$$

$$\text{with } G_{imjn}\left(\bar{\zeta}\right) = \bar{\zeta}_k \bar{\zeta}_l N_{ij}\left(\bar{\zeta}\right) / D\left(\bar{\zeta}\right)$$

$$\bar{\zeta}_i = \zeta_i / a_i; \quad \zeta_1 = \cos w \sqrt[1/2]{1 - \zeta_3^2};$$

$$\zeta_2 = \sin w \sqrt[1/2]{1 - \zeta_3^2};$$

$$D\left(\bar{\zeta}\right) = \varepsilon_{mnl} K_{m1} K_{n2} K_{l3}; \quad N_{ij}\left(\bar{\zeta}\right) = \frac{1}{2} \varepsilon_{ikl} \varepsilon_{jmn} K_{km} K_{ln};$$

$$K_{ik} = C_{ijkl}^s \bar{\zeta}_j \bar{\zeta}_l \tag{9}$$

where ε_{mnl} is the third-order permutation tensor and C_{ijkl}^s are the components of the orthotropic stiffness tensor of the solid matrix C^s.

The numerical scheme for the evaluation of $S_{ci(ijkl)}^s$ developed by Gavazzi and Lagoudas (1990) is adopted in this study. The double integration in Eq. (9) is transferred into the following form by using Gaussian quadrature formula:

$$S_{ci(ijkl)}^s = \frac{1}{8\pi} \sum_{p=1}^{M} \sum_{q=1}^{N} C_{ijkl}^s \left\{ G_{imjn}\left(w_q, \zeta_{3p}\right) + G_{jmin}\left(w_q, \zeta_{3p}\right) \right\} W_{pq} \tag{10}$$

where M and N refer to the points used for the integration over ζ_{3p} and w, respectively. W_{pq} is the Gaussian weight. In this study, 60×60 Gaussian points are used ($M = N = 60$). For detailed discussion of the determination of the Gaussian points and weight, please refer to the book by Press et al. (2001).

In addition, it should be mentioned that the effective stiffness tensor C^{hom} calculated from Eq. (7) – (10) is asymmetric. Therefore, a simple symmetrization technique is adopted:

$$C_{sym}^{hom} = \frac{1}{2} \left\{ C^{hom} + (C^{hom})^T \right\} \tag{11}$$

where $(C^{hom})^T$ is the transpose of C^{hom}.

2.2.2 *Damage criterion*

The damage criterion adopted in this study is formulated in the framework of linear fracture mechanics theory. When the cracked masonry is subjected to a uniform macroscopic strain E, the state equation can be expressed as follows:

$$\Sigma = C_{sym}^{hom} : E \tag{12}$$

where Σ is the macroscopic stress. The potential energy density (i.e., free enthalpy) W can be determined as follows:

$$W = \frac{1}{2}\Sigma :: E = \frac{1}{2}E : C_{sym}^{hom} : E \tag{13}$$

Considering the crack propagation for the i-th crack family, the damage state is characterized by a crack density variable ϵ_i (Budiansky & O'connell 1976):

$$\epsilon_i = a_i n_i^3 \tag{14}$$

where n_i is the number of cracks per unit of the i-th crack family. Accordingly, the damage criterion is formed in the framework of linear fracture mechanics theory (Dormieux et al., 2006):

$$G_i - G_{ci} \leq 0; \quad \dot{\epsilon}_i \geq 0; \quad (G_i - G_{ci})\dot{\epsilon}_i = 0 \tag{15}$$

where G_i and G_{ci} are the energy release rate and its critical value (threshold value), respectively. The energy release rate G_i acts as the driving force for the damage process, which is determined as the potential energy density conjugate to the crack density variable. For the i-th crack family, the energy release rate G_i reads:

$$G_i\left(E, \dot{\epsilon}_i\right) = \frac{\partial W}{\partial \epsilon_i} = \frac{\partial}{\partial \epsilon_i}\left(\frac{1}{2}E : C_{sym}^{hom}\left(\epsilon_i\right) : E\right) =$$

$$\frac{1}{2}E : \frac{\partial C_{sym}^{hom}}{\partial \epsilon_i} : E \tag{16}$$

where E is positive part of the effective strain E, which is used to exclude the influence of compressive strain (the damage process is only governed by the model-I tension fracture (Dormieux et al. 2006; Zuo et al. 2006).

The critical energy release rate G_{ci} acts as the threshold value for the formation of new crack surface. G_{ci} is associated with the microscopic fracture energy g_f and the damage state (i.e., the crack density variable ϵ_i).

$$G_{ci}\left(\epsilon_i\right) = \frac{2\pi}{3} g_f \left(\frac{n_i}{\epsilon_i}\right)^{\frac{1}{3}} = \frac{2\pi}{3}\frac{g_f}{a_i} \tag{17}$$

3 NUMERICAL ASPECTS

3.1 *Model application*

This section gives a detailed description for the implementation of the model introduced in Section 2. The numerical algorithm used in this study is developed on the basis of the work by Esposito and Hendriks (2016) for the fracture process of concrete. Now, consider the cracked masonry subjected to a uniform macroscopic imposed strain $E = E^{applied}$. The macroscopic strain E can be expressed by Eq. (18).

$$E = \alpha E_{max} \tag{18}$$

where E_{max} is the largest component of the macroscopic strain E. α is the macroscopic strain coefficient tensor, which is determined as follows:

$$\alpha_{max} = 1; \quad \alpha_{ij} = \frac{E_{ij}}{E_{max}} \tag{19}$$

Likewise, the macroscopic stress can also be expressed as follows:

$$\Sigma = \beta \Sigma_{max}$$

$$\text{with } \beta_{max} = 1; \beta_{ij} = \frac{\Sigma_{ij}}{\Sigma_{max}} \tag{20}$$

Accordingly, the state equation in Eq. (12) can be transferred into the following form:

$$\beta \Sigma_{max} = C_{sym}^{hom} : \alpha E_{max} \tag{21}$$

To solve the damage propagation problem, two important issues should be properly addressed: one is to identify the critical crack family that is propagating and the other is to determine the new stress and strain states by applying an (arbitrary) increase of damage variable $\delta_?$ to the critical crack family.

Consider the three orthotropic crack families characterized by their crack density variables ϵ_1, ϵ_2 and ϵ_3. The critical k-th crack family can be identified as the one with the lowest value of the largest component of the macroscopic strain E_{max}. According to the known imposed boundary condition Σ, the values of strain coefficient tensor β can be calculated. The effective stiffness tensor of the cracked masonry C_{sym}^{hom} can also be determined via the volume fractions φ_{ci} and the aspect ratios X_i of the three crack families (related to the crack density variable ϵ_i), and the stiffness of the solid matrix C^s. The macroscopic strain coefficient tensor α can thus be calculated by Eq. (21).

To identify the critical k-th crack family, the critical macroscopic strain tensors $E^{cr,i}$ (i =1, 2 and 3), which leads to the propagation of each crack family, are calculated by imposing the energy release rate (Eq. (16)) is equal to its critical value (Eq. (17)):

$$-\frac{1}{2}\alpha : \frac{\partial C_{sym}^{hom}}{\partial \epsilon_i} : \alpha \left|E_{max}^{cr,i}\right|^2 = \frac{2\pi}{3}\frac{g_f}{a_i} \tag{22}$$

where $E_{max}^{cr,i}$ is the largest components (in absolute sense) of $E^{cr,i}$.

Comparing the calculated values of $E_{max}^{cr,1}$, $E_{max}^{cr,2}$ and $E_{max}^{cr,3}$, the critical k-th crack family can be identified as:

$$E_{max}^{cr,k} = min\left\{\left|E_{max}^{cr,1}\right|, \left|E_{max}^{cr,2}\right|, \left|E_{max}^{cr,3}\right|\right\} \tag{23}$$

The value of macroscopic strain in current damage sate E^* is then determined as follows:

$$E^* = \alpha E_{max}^{cr,k} \tag{24}$$

Using Eq. (12), the macroscopic stress in current damage state Σ^* can be calculated based on the macroscopic strain E^* and the effective stiffness tensor of the cracked masonry C_{sym}^{hom} in current damage state (the constitutive law under current damage state is established). To continue the analysis and derive the constitutive law under new damage state (characterized by the effective stiffness tensor of the cracked masonry C_{sym}^{hom}), an (arbitrary) increase of damage variable δ_ϵ is applied to the critical crack family.

$$\begin{cases} \epsilon_{i+1} = \epsilon_i + \delta_\epsilon & i = k \\ \epsilon_{i+1} = \epsilon_i & i \neq k \end{cases} \tag{25}$$

The loop is continued until the moment at which the sum of volume fractions of the three crack families is equal to 1 ($\sum_{i=1}^{3} \varphi_{ci} = 1$). In the final damage state, the effective stiffness tensor of the cracked masonry $C_{sym(final)}^{hom}$ and the macroscopic stress $\Sigma_{(final)}$ are equal to zero, and the macroscopic strain $E_{(final)}$ reaches its maximum value.

3.2 *Model calibration*

This section provides a description for the calibration procedure of the proposed microporomechanical model.

The proposed model requires a limited number of initial variables, which can be divided into input and calibrated parameters. The input parameters include the volume fracture (φ_b, φ_m), Young's modulus (Y_b, Y_m), Poisson ratio (v_b, v_m), and porosity (θ_b, θ_m) values of bricks and mortar together with the vertical Young's modulus ($Y_{initial}$) and tensile strength ($f_{t,initial}$) of the undamaged masonry and the initial volume fraction and the initial crack thickness ($\varphi_{c,initial}$ and c). It is assumed that initially the three crack families have identical properties. The initial volume fraction of cracks $\varphi_{c,initial}$ is determined as the average of the porosities of bricks and mortar ($\varphi_{c,initial} = (\theta_b\varphi_b + \theta_m\varphi_m)/(\varphi_b + \varphi_m)$, for which a large number of experimental values are available in literature (Cobîrzan et al. 2016; Sassoni et al. 2013). The initial crack thickness c is assumed to be constant and equal to 0.1mm. These input parameters have clear physical meanings and can be experimentally determined.

The calibrated parameters include the initial aspect ratio of the crack $X_{initial}$ and the microscopic fracture energy g_f, which can be determined following a two-step calibration procedure as proposed by Esposito and Hendriks (2016) and shown in Figure 4.

The initial distribution of the three orthotropic cracks is assumed uniform, namely each crack family has the same initial radius ($a_{1,initial} = a_{2,initial} = a_{3,initial} = a_{initial}$) and volume fraction ($\varphi_{c1,initial} = \varphi_{c2,initial} = \varphi_{c3,initial} = \varphi_{c,initial}/3$). The crack thickness c and the number of cracks per unit n are assumed as unchanged constants (i.e., damage is only induced by the growth of crack radius, while the crack nucleation and the opening/closing transformation of cracks are ignored). The cracks are then embedded in the solid matrix which is an orthotropic medium derived from Eq. (4). The stiffness tensor of undamaged masonry $C_{sym(initial)}^{hom}$ calculated from Eq. (11) is orthotropic and its components can be associated to the experimental value of the vertical Young's modulus $Y_{initial}$ obtained by testing masonry with a loading vector perpendicular to bed joints. Consequently, the initial aspect ratios of the cracks $X_{initial}$ can be calibrated. With the known values of $X_{initial}$ and c (crack thickness), the initial crack radius $a_{initial}$, the number of cracks per unit n and the initial crack density variable $\epsilon_{initial}$ can be determined (Eq. (5), (6) and (14)).

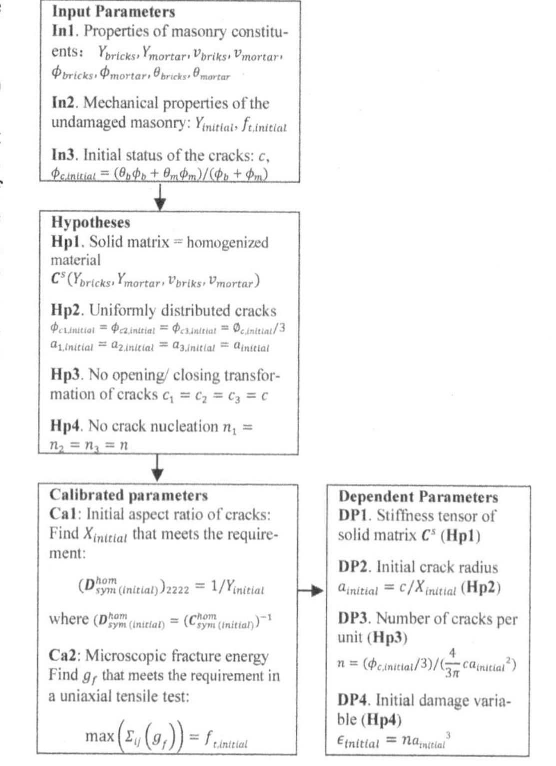

Figure 4. Calibration procedure of the proposed microporomechanical model.

After determining the initial status of the three families of microcracks, the microscopic fracture energy g_f can be calibrated. By imposing the macroscopic stress of the critical crack family equal to the experimental value of the tensile strength of the undamaged masonry $f_{t,initial}$ in a uniaxial tensile test, the microscopic fracture energy g_f can be determined:

$$G_i\left(E^{cr,initial} = D_{sym(initial)}^{hom} : \Sigma^{cr,initial}, \epsilon_i = \epsilon_{initial}\right) = G_{ci}\left(\epsilon_i = \epsilon_{initial}, g_f\right) \tag{26}$$

where $E^{cr,initial}$ and $\Sigma^{cr,initial}$ are the critical strain and stress tensors of the initial crack status (characterized by $\epsilon_{initial}$). $\Sigma^{cr,initial}$ has only one no-zero-component ($\Sigma_{ii}^{cr,initial} = f_{t,initial}$). In addition, g_f is assumed as a constant that is unchanged in the damage process.

4 RESULTS

To evaluate the performance of the proposed microporomechanical model, in this section, the proposed model is validated for the case of uniaxial tension against experiments and its performance regarding compression and pure shear loading is evaluated accordingly. As a case study, the uniaxial tensile tests by Van der Pluijm (1997) are considered. The tests were performed on couplets made of wire cut clay brick, having nominal dimensions 100 × 50 × 100 mm^3(length × height × thickness), and hydrated shell lime mortar with joint thickness of 12.5 mm. Table 1 lists the input and calibrated parameters of the proposed model, where the mechanical properties of masonry and its constituents are from the test data by Van der Pluijm (1997), and the porosities of bricks and mortar are based on the experimental data from the research by Cobîrzan et al. (2016) and Sassoni et al. (2013) where the authors provided the data of the porosities for different types of mortars and bricks. In this section, a right-hand Cartesian frame is defined, with the y-axis aligning with the uniaxial loading direction.

Table 1. Input and calibrated parameters of proposed model.

Properties	Values	Unit
Input parameters		
Brick Young's modulus, Y_{bricks}	16.7	GPa
Mortar Young's modulus, Y_{mortar}	1.22	GPa
Brick Poisson's ratio, ν_{brick}	0.20	-
Mortar Poisson's ratio, ν_{mortar}	0.15	-
Brick volume fraction, φ_{bricks}	0.89	-
Mortar volume fraction, φ_{mortar}	0.11	-
Brick porosity, θ_{bricks}	0.235	-
Mortar porosity, θ_{mortar}	0.127	-
Masonry vertical Young's modulus, $Y_{initial}$	2.37	GPa
Masonry tensile strength, $f_{t,initial}$	0.40	MPa
Crack thickness, c*	0.10	mm
Calibrated parameters		
Initial aspect ratio, $X_{initial}$	14.68	-
Microscopic fracture energy, g_f	6.42e-5	N/mm
Dependent parameters		
Solid matrix Vertical Young's modulus, $Y(D^{hom}_{sym(initial)})$	7.93	GPa
Number of cracks per unit, n	0.0823	mm^{-3}
Initial crack radius, $a_{initial}$	1.468	mm
Initial crack density, $\epsilon_{initial}$	0.260	-
Initial crack volume fraction, $\varphi_{c,initial}$	0.223	-

*Crack thickness, c: assumed value.

Figure 5 shows the prediction results of the proposed model for masonry under uniaxial tension test, which is compared with the experimental data by Van der Pluijm (1997). The results show that the proposed model can successfully predict the tensile softening behaviour of masonry under uniaxial tension. The model results agree well with the experimental results; the Model-I tensile fracture energy predicted by the proposed model is 7.7 N/mm which is close to the average experimental value of 5.5 N/mm. As shown in Figure 6, the evolution of the three orthotropic crack families under tension is consistent with the experimental cracking behaviour in the uniaxial tensile test, namely only the crack family normal to the loading direction (y-axis) propagates.

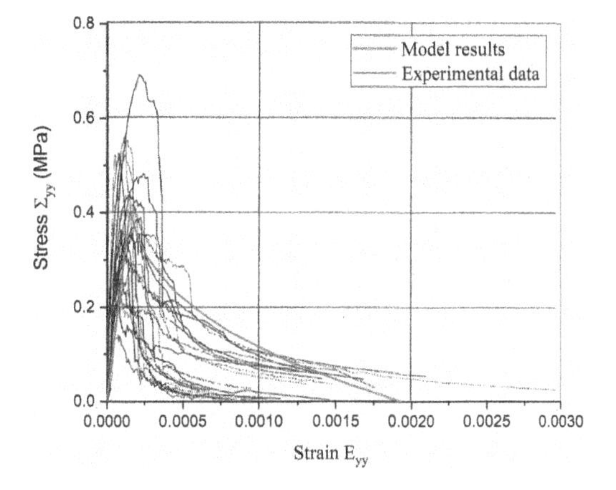

Figure 5. Model performance: uniaxial tension.

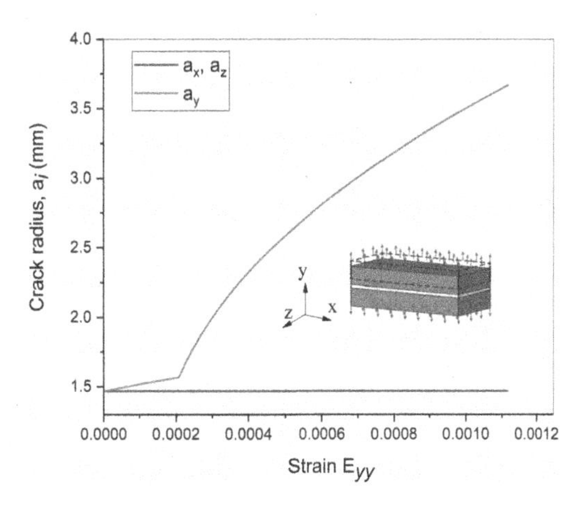

Figure 6. Crack evolution: uniaxial tension.

Figure 7 shows the prediction results of the proposed model for masonry under uniaxial compressive loading. The proposed model is able to capture the complete nonlinear behaviour (including the post-peak

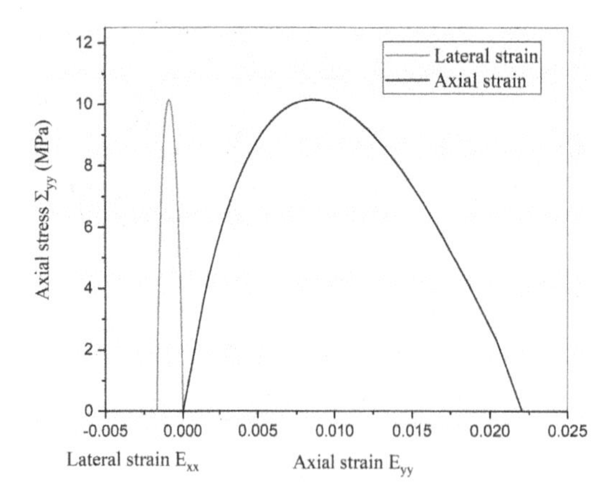

Figure 7. Model performance: uniaxial compression.

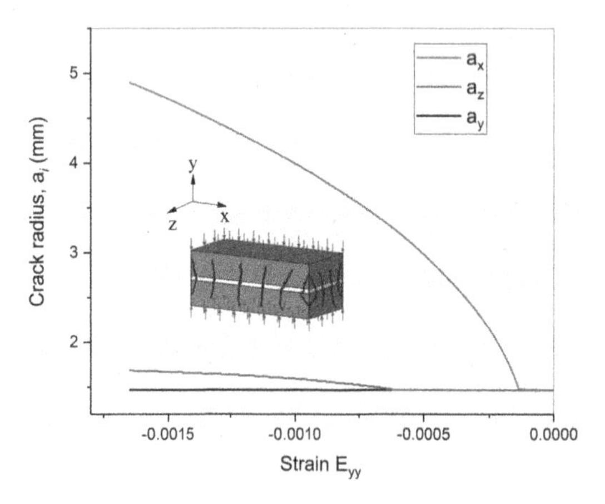

Figure 8. Crack evolution: uniaxial compression.

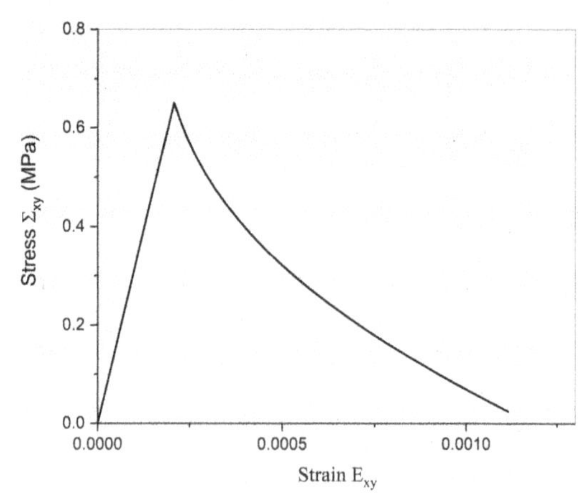

Figure 9. Model performance: pure shear.

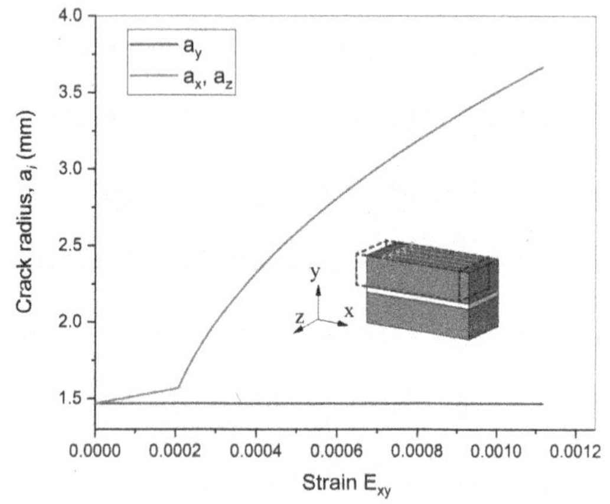

Figure 10. Crack evolution: pure shear.

behaviour) of masonry under compression. The corresponding crack evolution under uniaxial compressive loading is given in Figure 8. The cracking behaviour of the three orthotropic crack families agrees well with the experimental observation in masonry compressive test, e.g. (Jafari et al., 2019). The radius of the crack family normal to the loading direction (y-axis) keeps unchanged, while the other two crack families propagate. Additionally, compared to the crack family normal to the horizontal direction (x-axis), the crack family normal to the thickness direction (z-axis) shows a lower increment in crack radius for the same strain value. This can be explained by the fact that the Young's modulus and Poisson's ratio of masonry in wall thickness direction are larger than those in the horizontal direction, resulting in a weaker resistance for the formation of crack surface in horizontal direction.

The results of the nonlinear behaviour and crack evolution simulated by the proposed model for masonry under pure shear are shown in Figures 9 and 10, respectively. It should be mentioned that only the nonlinear behaviour associated with the de-cohesive mechanism is considered in the model. As a results, in term of stress-strain diagram, the derived material behaviour under shear is similar to that of a tensile test. As shown in Figure 10, for the loading case of pure shear in xy plane, the cracks normal to x- axis and z-axis propagate. This conflicts with the experimental observations in a shear test due to the lack of consideration of the friction mechanism in the proposed model. In the future, the shear-sliding mechanisms should be further considered in the proposed model. By modelling the coupling between Mode-I tensile fracture damage and the friction, the nonlinear behaviour of isotropic rocks under shear loading case has been successfully modelled (Zhu et al., 2009).

5 CONCLUSION

This paper proposes a novel nonlinear constitutive model for masonry in the framework of the microporomechanics theory. The masonry is considered as a composite formed by bricks, mortar joints and penny-shaped microcracks. The orthotropic elastic properties of uncracked masonry are calculated from the isotropic elastic properties of its constituents by an improved mean-field homogenization model. The nonlinear behaviour of masonry is derived as a result of the evolution and propagation of penny-shaped cracks. The results show that the proposed model is able to predict the post-peak nonlinear behaviour of masonry by using a limited number of input parameters mainly related to the elastic properties of the masonry constituents, and the vertical Young's modulus and the tensile strength of masonry.

However, a limitation of the proposed model should be mentioned. According to masonry mechanics, both cohesion and friction mechanisms play a crucial role in the fracture process of quasi-brittle materials, as masonry. Currently, the proposed model only considers the cohesion mechanisms by modelling Mode-I fracture damage. The role of friction between the interface of mortar joints and bricks is ignored. As a results, the shear-sliding behaviour of masonry cannot properly be described by the proposed model. In future work, the coupling between the cohesive damage and the friction mechanism will be investigated.

ACKNOWLEDGEMENT

The financial support of China Scholarship Council (CSC) to the first author is gratefully acknowledged.

REFERENCES

Addessi, D., Sacco, E. & Paolone, A. 2010. Cosserat model for periodic masonry deduced by nonlinear homogenization. *European Journal of Mechanics-A/Solids,* 29, 724–737.

Bati, S. B., Ranocchiai, G. & Rovero, L. 1999. A micromechanical model for linear homogenization of brick masonry. *Materials and Structures,* 32, 22–30.

Budiansky, B. & O'connell, R. J. 1976. Elastic moduli of a cracked solid. *International journal of Solids and structures,* 12, 81–97.

Chow, C. & Wang, J. 1987. An anisotropic theory of elasticity for continuum damage mechanics. *International Journal of fracture,* 33, 3–16.

Cobîrzan, N., Balog, A.-A., Belean, B., Borodi, G., D˘adârlat, D. & STREZA, M. 2016. Thermophysical properties of masonry units: Accurate characterization by means of photothermal techniques and relationship to porosity and mineral composition. *Construction and Building Materials,* 105, 297–306.

D'altri, A. M., Sarhosis, V., Milani, G., Rots, J., Cattari, S., Lagomarsino, S., Sacco, E., Tralli, A., Castellazzi, G. & De Miranda, S. 2020. Modeling strategies for the computational analysis of unreinforced masonry structures: review and classification. *Archives of computational methods in engineering,* 27, 1153–1185.

Deude, V., Dormieux, L., Kondo, D. & Maghous, S. 2002. Micromechanical approach to nonlinear poroelasticity: application to cracked rocks. *Journal of engineering mechanics,* 128, 848–855.

Doghri, I., El Ghezal, M. I. & Adam, L. 2016. Finite strain mean-field homogenization of composite materials with hyperelastic-plastic constituents. *International Journal of Plasticity,* 81, 40–62.

Dormieux, L., Kondo, D. & Ulm, F.-J. 2006. *Microporomechanics*, John Wiley & Sons.

Eshelby, J. D. 1957. The determination of the elastic field of an ellipsoidal inclusion, and related problems. *Proceedings of the royal society of London. Series A. Mathematical and physical sciences,* 241, 376–396.

Esposito, R. & Hendriks, M. A. 2016. A multiscale micromechanical approach to model the deteriorating impact of alkali-silica reaction on concrete. *Cement and Concrete Composites,* 70, 139–152.

Fritsch, A., Hellmich, C. & Young, P. 2013. Micromechanics-derived scaling relations for poroelasticity and strength of brittle porous polycrystals. *Journal of Applied Mechanics,* 80.

Gambarotta, L. & Lagomarsino, S. 1997. Damage models for the seismic response of brick masonry shear walls. Part I: the mortar joint model and its applications. *Earthquake engineering & structural dynamics,* 26, 423–439.

Gavazzi, A. & Lagoudas, D. 1990. On the numerical evaluation of Eshelby's tensor and its application to elastoplastic fibrous composites. *Computational mechanics,* 7, 13–19.

Jafari, S., Rots, J. G. & Esposito, R. 2019. Core testing method to assess nonlinear behavior of brick masonry under compression: A comparative experimental study. *Construction and Building Materials,* 218, 193–205.

Klusemann, B. & Svendsen, B. 2010. Homogenization methods for multi-phase elastic composites. *Technische Mechanik-European Journal of Engineering Mechanics,* 30, 374–386.

Lemarchand, E., Dormieux, L. & Kondo, D. A micromechanical analysis of the observed kinetics of ASR-induced swelling in concrete. Computational Modelling of Concrete Structures (EURO-C), 2003. AA Balkema Publisher, Rotterdam St. Johann im Pongau, Austia, 483–490.

Lielens, G. 1999. Micro-macro modeling of structured materials (PhD thesis). *Universite Catholique de Louvain, Louvain-la-Neuve, Belgium.*

Lourenço, P. B. 1996. A matrix formulation for the elastoplastic homogenisation of layered materials. *Mechanics of Cohesive-frictional Materials: An International Journal on Experiments, Modelling and Computation of Materials and Structures,* 1, 273–294.

Marfia, S. & Sacco, E. 2012. Multiscale damage contact-friction model for periodic masonry walls. *Computer Methods in Applied Mechanics and Engineering,* 205, 189–203.

Massart, T., Peerlings, R. & Geers, M. 2007. An enhanced multi-scale approach for masonry wall computations with localization of damage. *International journal for numerical methods in engineering,* 69, 1022–1059.

Morin, C., Vass, V. & Hellmich, C. 2017. Micromechanics of elastoplastic porous polycrystals: theory, algorithm, and application to osteonal bone. *International Journal of Plasticity,* 91, 238–267.

Mura, T. 2013. *Micromechanics of defects in solids*, Springer Science & Business Media.

Nemat-Nasser, S. & HORI, M. 2013. *Micromechanics: overall properties of heterogeneous materials*, Elsevier.

Pardoen, T. & Hutchinson, J. 2003. Micromechanics-based model for trends in toughness of ductile metals. *Acta Materialia,* 51, 133–148.

Pensée, V., Kondo, D. & Dormieux, L. 2002. Micromechanical analysis of anisotropic damage in brittle materials. *Journal of Engineering Mechanics,* 128, 889–897.

Petracca, M., Pelà, L., Rossi, R., Oller, S., Camata, G. & Spacone, E. 2016. Regularization of first order computational homogenization for multiscale analysis of masonry structures. *Computational mechanics*, 57, 257–276.

Pichler, B., Hellmich, C. & A. Mang, H. 2007. A combined fracture-micromechanics model for tensile strain-softening in brittle materials, based on propagation of interacting microcracks. *International Journal for Numerical and Analytical Methods in Geomechanics*, 31, 111–132.

Pierard, O., Friebel, C. & Doghri, I. 2004. Mean-field homogenization of multi-phase thermo-elastic composites: a general framework and its validation. *Composites Science and Technology*, 64, 1587–1603.

Press, W. H., Vetterling, W. T., Teukolsky, S. A. & Flannery, B. P. 2001. *Numerical recipes in C++ the art of scientific computing*, Cambridge university press.

Sassoni, E., Mazzotti, C., Boriani, M., Gabaglio, R. & Gulotta, D. Assessment of masonry mortar compressive strength by double punch test: The influence of mortar porosity. Proceedings of the International Conference Built Heritage, 2013. Citeseer, 996–1002.

Swoboda, G. & Yang, Q. 1999. An energy-based damage model of geomaterials—II. Deductionof damage evolution laws. *International journal of solids and structures*, 36, 1735–1755.

Ulm, F.-J., Constantinides, G. & Heukamp, F. H. 2004. Is concrete a poromechanics materials?—A multiscale investigation of poroelastic properties. *Materials and structures*, 37, 43–58.

Van Der Pluijm, R. 1997. Non-linear behaviour of masonry under tension. *HERON-ENGLISH EDITION-*, 42, 25–54.

Zheng, Q.-S. & Du, D.-X. 2001. An explicit and universally applicable estimate for the effective properties of multiphase composites which accounts for inclusion distribution. *Journal of the Mechanics and Physics of Solids*, 49, 2765–2788.

Zhou, Y., Sluijs, L.J. & Esposito, R. 2021. An improved mean-field homogenization model for the three-dimensional elastic properties of masonry. *European Journal of Mechanics-A Solids* (under review).

Zhu, Q.-Z., Kondo, D. & Shao, J.-F. 2009. Homogenization-based analysis of anisotropic damage in brittle materials with unilateral effect and interactions between microcracks. *International Journal for Numerical and Analytical Methods in Geomechanics*, 33, 749–772.

Zhu, Q.-Z., Kondo, D. & Shao, J. 2008. Micromechanical analysis of coupling between anisotropic damage and friction in quasi brittle materials: role of the homogenization scheme. *International Journal of Solids and Structures*, 45, 1385–1405.

Zuo, Q., Addessio, F., Dienes, J. & Lewis, M. 2006. A rate-dependent damage model for brittle materials based on the dominant crack. *International journal of solids and structures*, 43, 3350–3380.

Computational Modelling of Concrete and Concrete Structures – Meschke, Pichler & Rots (Eds)
 ISBN: 978-1-032-32724-2

Interpreting size effects on adobe masonry mortar: Experiments and numerical simulations

T. Li Piani
Delft University of Technology, Delft, The Netherlands
Catholic University of Sacred Heart, Milan, Italy

J. Weerheijm
Delft University of Technology, Delft, The Netherlands
TNO-Defense, Safety and Security (DSS), The Hague, The Netherlands

L.J. Sluys
Delft University of Technology, Delft, The Netherlands

ABSTRACT: About one third of the world population lives in earthen dwellings. Adobe connotes an ancient masonry whose bricks and mortar are made out of clay, silt sand and fibres, mixed using water and dried under the sun. The composition of adobe is not standardized yet, namely locally available soil and fibre materials are mixed together, often regardless strict rules on the nature and proportions of these constituents. As a result, adobe mixtures in the field are usually accompanied by a high level of heterogeneity at micro and macro scales. This requires proper assessment of the influence of heterogeneity on the mechanical performance of the finite products at macro-scale. A large portion of the thousands of new and historical adobe buildings in the world are indeed not engineered constructions, often built by house owners themselves with inherent safety issues. Moreover, this building technology is recently gaining significant relevance in light of its good sus-tainability features. This paper investigates the effect of heterogeneity in the mixture on the strength of adobe elements of different sizes. Size dependence is a well-known phenomenon for masonry elements. For indus-trially produced bricks, it is known that larger sizes are generally accompanied by comparatively lower strength values and several theories have been consolidated over the years. In this research, specimens of adobe mortar of different aspect ratios have been statically tested in compression. Nominal values of strength have been calculated and compared. Contrary to initial expectations, lower strength levels appeared to be as-sociated to smaller dimensions. First, this observation has been interpreted as a possible consequence of the effects of the heterogeneity level (sizes and distribution) in the mixture compared to mixture granulometry property. Mixtures which are not standardized and may compromise structural performance of comparatively smaller specimens. Next, this hypothesis has been numerically tested via a series of numerical simulations. A recently developed isotropic damage model called 'Adobe delta damage model' presented at EURO-C 2018, has been used to replicate the observed size effect. This model uses a damage delay framework to obtain mesh-size independent results for both static and dynamic loads in quasi brittle material simulations. Ex-perimental results, physical interpretation and numerical simulations are presented in this paper.

1 INTRODUCTION

Masonry design requires the assessment of its mechanical parameters i.e. compressive strength and critical strain (Kaushik et al. 2007). In building codes, these parameters are often related to the mechanical performance of their constitutive units, namely bricks and mortar (Ingham et al. 2014). Material properties such as compressive strength or the Young's modulus must be determined experimentally on representative samples according to technical standard, which often prescribe strict requirements for the experimental setup or the specimen geometry (Fódi 2011). However, these prescriptions cannot always be met during this material characterization phase, which may happen in the laboratory but also directly in the field, due to limitations of testing machines, site equipment, extracting machineries or simply availability of materials (Bohdan & Tomasz 2013) This often results into the inconsistent evaluation of fundamental material properties on specimens of different proportions or dimensions than the prescribed ones. This is the case for the nominal material strength used for design purposes and calculated as the ratio of the peak load on the sample over its initial cross section area. In fact, it is well known in literature that sample geometry is prone to significantly affect the assessment of the nominal strength for masonry units. Large attention has been devoted to the assessment of the influence of specimen height given the same width. Based on experimental datasets and numerical validations, several functions of correction factors have been proposed to unify nominal strengths of specimens of different slenderness (MacGreggor 1994). Taking into account the specific differences of the various functions proposed in literature, these all show that a smaller slenderness gives a higher apparent nominal strength (HB 195 – The Australian earth building handbook-Standards Australia International, NSW 2001, 2001) In samples with low slenderness, an artificial strengthening is caused by the restraining effects of the steel material platens of the testing apparatus. As a result, lateral expansion is prevented and an increment in the maximum force (thus in nominal strength) is observed (Page & Marshall 1985). This effect appears to steeply reduce above slenderness's of 2 (MacGreggor 1994). If the influence of slenderness on mechanical parameters of units is translated into building standards for masonry design, minor certainty refers to the influence of specimen dimensions given a same slenderness (Krishna et al. 2011) Several theories have been proposed over the past years. When dealing with size effects on masonry materials, the statement "*the smaller, the stronger*" is usually valid; that is the apparent performance in strength decays for larger samples (Bohdan & Tomasz 2013). To this effect, not only restraining mechanisms of steel platens contribute. According to probabilistic principles, the general theory of size dependence explains that larger a volume of material is, the most likely a defect, heterogeneity or void is contained within it, and thus, earlier the specimen is prone to fail (Fódi 2011) The influence of shape and size dependence have been studied for modern construction materials, namely produced according to rigorous product and process standard, which include the phases from raw soil element selection until finite product delivery and certification. In the world, construction technologies which are 'not engineered' still exist, namely these are generally produced using materials and techniques not univocally shared nor standardized in building codes. This is the case of adobe. Adobe is the most ancient masonry technology, whose bricks are made out of soil possibly mixed with fibres and joint together using mud mortar (Austin 1984). Locally available materials are often used and final products are air dried in the field consistently with local building traditions, which of course, also vary over different regions (Varum et al. 2007). Despite not fully standardized yet, almost 1/3 of the world population still lives in earthen dwellings, which are spread in areas of the world prone to earthquakes or involved into military operations (Li Piani 2021). Earthen based architecture is gaining renovated attention also in European urban environments in light of its eco-friendly material properties (Parra-Saldivar & Batty 2006). The seek for product and process standardization is of paramount importance for restoration as well as for new building design purposes all around the world. Still, the nature of the material and building process technology traditions require the proper assessment of the influence of geometry of adobe masonry specimens on the nominal strength parameter.

In the next section, an experimental campaign of material characterization on one mixture of fibre reinforced adobe mortar is reported. Static uniaxial compression tests have been performed on samples of different sizes while keeping the slenderness the same. Resulting values of nominal compressive strengths for the different geometries of the same mortar have been compared. Next, the derived trend has been critically assessed against main theories developed for industrially produced masonry materials. A possible hypothesis prone of justifying the observed trend has been proposed and tested via numerical simulations. A numerical model developed for simulating the dynamic performance of adobe masonry materials has been used. Numerical results have been finally validated against the hypothesis. Numerical simulation robustness is finally investigated against numerical pathology of mesh dependence.

2 THE EXPERIMENTAL PROGRAM

2.1 *Material selection*

Adobe samples were produced at the royal military laboratories of the Netherlands, NLDA. Soil bags and fibres were selected from a German producer of traditional adobe materials. Soil granulometry is shown in Figure 1. The mixture is classified as a "*sandy silt with some clay*" (Li Piani et al. 2018). Maximum aggregate size in the soil mixture is 2mm. Soil bags contained natural fibres up to 3% by weight. Fibre reinforcement consisted of straw and chopped wood. Not an unique geometry of fibre was observed, but an average dimension of 12 mm was measured.

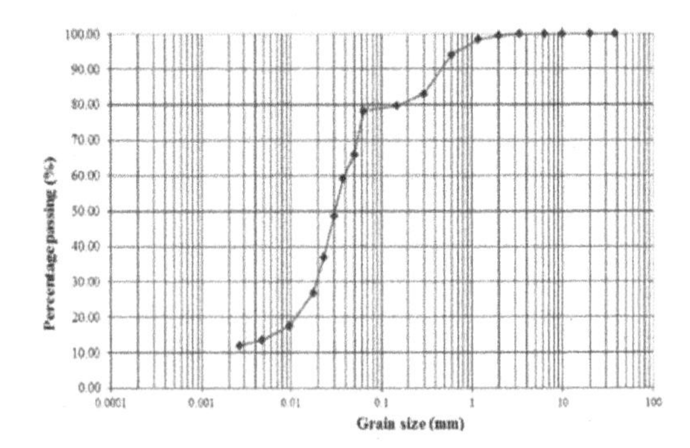

Figure 1. Soil granulometry and fibres in the mixture.

2.2 *Samples production*

Soil, fibre and water were mixed in a concrete mixing machine as prescribed by the company. Next, fresh mortar was poured into prismatic wooden moulds of four different sizes. All sample have the same slenderness equal to two, while square cross section side was progressively increased starting from 40mm with a factor of 1.25 (Figure 2). In the following, the resulting four specimen types are named as Size A, B, C, D. Final dimensions are also reported in Table 1. Mortar specimens were air cured at controlled laboratory conditions for 28 days. After drying, all surfaces were rectified to ensure plan parallelism. The smallest cross section tested was a square of 40x40mm (EN 772-1:2000, Methods of test masonry units Part 1: determination of compressive strength 1999).

Figure 2. Poured mortar (top) and air drying specimens (bottom).

Table 1. Specimens dimensions.

Size	*L x t xH [mmxmmxmm]*
A	40x40x80
B	55x55x110
C	67x67x135
D	80x80x160

2.3 *Test setup*

Compressive tests were performed according to UNI EN 772-1. Six samples per type were subjected to displacement controlled tests at a speed rate of 1 mm/min. For each sample, the apparent stress strain diagram was recorded by dividing the force and displacements by cross sectional areas and specimen height (Figure 3).

2.4 *Results*

From each stress-strain plot, nominal compressive strength and critical strain values were calculated. Mean values for each sample size are plotted in Figure 4. For all types a relatively high scatter in results is observed. Nevertheless, an increasing trend of compressive strength with sample sizes is observed. Type A mortar possesses an average strength of 1.24 MPa, whereas this value increases up to 1.41 MPa for Type D.

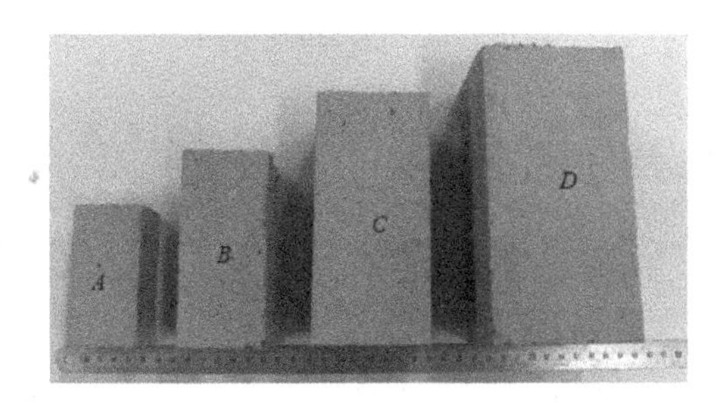

Figure 3. The four different mortar specimens sizes (A-D).

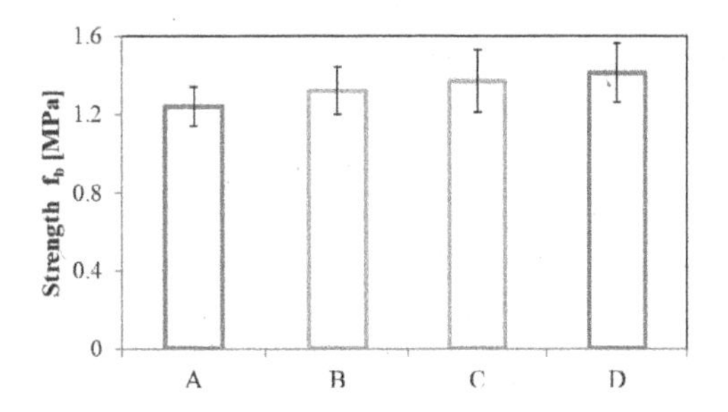

Figure 4. Nominal strength (MPa) for the different mortar sizes (A-D).

Table 2. Mean values (and standard deviations) for compressive strength and strain for sizes A-D.

Type -	Strength MPa	Strain Mm/mm
A	1.24(0.11)	1.23(0.15)
B	1.32(0.12)	0.91(0.11)
C	1.37(0.14)	0.88(0.09)
D	1.41(0.13)	0.91(0.07)

Conversely, smaller samples are more ductile, namely the highest strain at peak is displayed by the sample with 40mm cross section, which is typically characterized by a more diffuse set of cracks at failure (Figure 5).

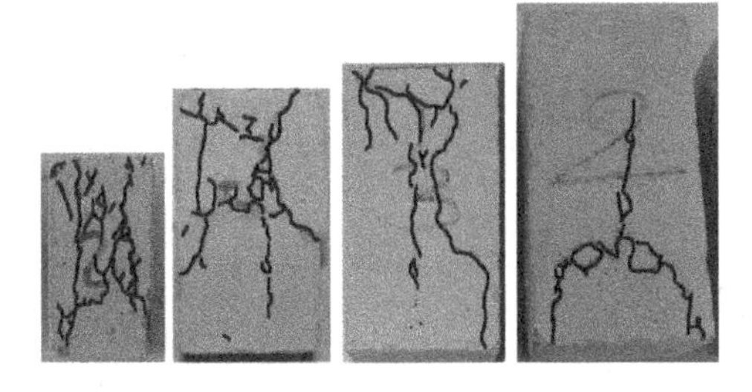

Figure 5. Typical cracking patterns for the different sizes (A-D).

3 INTEPRETATION

3.1 *Physical hypothesis*

In this section, a possible explanation of the observed experimental trend related to the increase of nominal strength levels for larger specimens of adobe is hypothesized. Materials as cement or clay are considered as homogeneous only at a large scale with respect to the dimensions of the largest aggregate. This happens because heterogeneity is intrinsically present in the material. In quasi brittle materials, failure is dictated by the coalescence and propagation of defects and/or voids inside the volume at micro and meso-scales (Phu Nguyen et al. 2010). Probability suggests that the larger the volume, the most probable that a defect is contained in the material is, thus resulting into a reduction of the nominal strength. These theories have been validated for industrially produced materials, where raw elements and inherent proportions, including production processes, are fully standardized to ensure the minimization of the likelihood of the presence and relative dimensions of defects in the final product. In the case of adobe, raw elements in the field are picked up according to local availability, whereas production processes are not defined to ensure the elimination of micro-defects in the final product. As a result, the level of heterogeneity in the mixture of adobe are often significant, both in terms of size and proportions. Namely, in mixtures of adobe currently produced in the field, the presence of defects are more not only probable, but almost sure. This especially counts for fibre reinforced adobes. The presence of fibres in the mixtures of adobe is originally meant to reduce shrinkage rates in the (sun)drying material, limiting the formation of cracks as a consequence of a more efficient drainage system (Li Piani et al. 2018). However, its contribution on the mechanical properties of the resulting brick, especially in terms of the influence of fibre materials is more controversial. Despite still considered as a natural reinforcement in many building guidelines for adobe, the most recent experimental evidence in the field show that the presence of fibres in the mixture usually reduces the mechanical performance of the plain adobe (New Zeland Standards 1998). This can be explained as the result of a lack of adherence between fibres and soil aggregates at micro-scale caused by the non-optimal choices of raw elements and inherent proportions (Li Piani et al. 2020) Fibres are most often chosen based on local availability and different materials, forms and dimensions can be often observed in the same soil mixture. This was evident also in the experimental campaign herein presented. The average geometrical dimensions of the fibre reinforcement of the tested mortar samples was characterized by a considerable scatter. Granulometry analysis on mixtures revealed the presence of large straw and wood elements quite off from the medium declared values for maximum dimensions. This difference could easily correspond i.e. to the 15–18% of the cross section side of "Size A" (Figure 6).

Figure 6. Specimens of adobe with large wood reinforcement visible on the surface (left) and example of fibre geometrical properties revealed in granulometry test (right).

In this setting, not only the likelihood of defects in the mixture, but their relative proportions and dimensions with respect to the specimen size are important matters to consider. The presence of fibres, already associated to a decay of strength with respect to plain adobe in recent literature (Li Piani 2019) is considered in this study the responsible also for the 'opposite' size effect observed. Principles of fracture mechanics can be linked with probability theories considering the effect that one or more large heterogeneity can cause on the capability to withstand external loadings. In particular, it is hypothesized that the sure presence of relatively large areas of de-adherence in the material with respect to specimen size, which are caused by fibre inclusion in the mixture of adobe have a high probability to significantly impact the resistance especially of relatively smaller bricks, namely the likelihood to compromise the overall capability of smaller specimens to redistribute internal stresses. In this setting, this hypothesis explains in the observed opposite trend in nominal strength, where larger specimens appears to be stronger than smaller specimens.

3.2 *Numerical model*

The hypothesis presented in Par. 3.1. on the influence of defect sizes at a meso-scale on the structural mechanical performance of masonry elements is herein numerically tested. To this end, a numerical model recently developed to assess the performance of Adobe bricks and mortar was used (Li Piani et al. 2019). This is an isotropic local damage model (eq.1), which incorporates a smoothed Drucker-Prager failure surface (eq.2). The thermodynamic variables of the material states are expressed as equivalent strains for compression crushing (ε_{eqc}) and tensile cracking (ε_{eqt}). Damage starts when the loading function ψ becomes positive (eq.3). Damage evolution laws dictate the softening process (eq.4). Evolution of damage is directly related to the growth of two monotonic internal variables which account for the maximum equivalent strains reached during loading history in case of non-monotonic loadings (eq.5) Without proper treatment, the local damage model suffers from the well-known numerical pathology called mesh dependence (Sluys & de Borst 1992). In order to solve mesh dependence while keeping physical consistency with a typically rate dependent material, a viscosity based local regularization algorithm has been developed (Allix 2012) The local damage evolution law has been made directly dependent on the loading history based on a decomposition of the Dirichlet boundary condition. Given an arbitrary displacement law evaluated in N points by the Newton-Raphson solver, at each progressive discretized time τ of the analysis after damage initiation, the loading evolution law shows an exponential damage delay to account for a "delta"(δ) od increment based on the prescribed loading history (eq.6).

$$\sigma = (1 - D)\sigma_{eq} \tag{1}$$

$$\epsilon_{eq} = a\epsilon_{oct} + b\gamma_{oct} \tag{2}$$

$$\psi = \epsilon_{eq} - k_0 \tag{3}$$

$$d = 1 - \frac{1}{e^{C_1(k-k_0)}} - \frac{k_0}{C_2 k} \tag{4}$$

$$k = \max\left(\epsilon_{eq}\left(1 - r^{\alpha}\right), k_0\right) \tag{5}$$

$$\delta D^{\tau} = D^{\tau} - D^{\tau-1} = \frac{\Delta}{N} e^{\left(d^{\tau} - d^{\tau-1}\right)} \tag{6}$$

Where D is the damage variable, σ_{eq} is the effective stress vector, $\varepsilon_{oct}$and γ_{oct} are the normal and tangential components of the first and second deviatoric invariants, respectively, k_0 is the damage initiation strain, k the historical maximum equivalent strain reached during loading history, d is the local damage variable before regularization, C_1 and C_2 are material constants, r is derived from the triaxiality factor proposed by Lee and Fenves, α is 0.1, Δ represents a further non-dimensional material parameter and N is needed to make the results independent of discretization of the applied law.

For more detailed information on the model, the reader is referred to (Li Piani, Weerheijm, Koene, et al. 2019) for statics and (Li Piani et al. 2019) for dynamic problems.

3.3 *Numerical simulations of experiments*

Numerical simulations of uniaxial compression tests were performed It is worthy to stress out that the herein presented numerical simulations do not mean to exactly replicate the experimental tests performed. This is an exercise meant to validate a possible hypothesis on a general detected experimental trend. Specimens of 1:2 slenderness were defined as from experiments (Figure 7). Three specimens with the same geometry of Size A, B, D in Par.2 were implemented. Displacement controlled tests were executed consistently with the experimental setup. Static simulations were performed and a constant step displacement history was uniformly applied at the top side of the specimen. A mechanical imperfection was still needed to trigger localization in statics. A small area (in grey in Figure 7) at the left corner of each specimen was connoted by a 30% lower damage initiation strain compared to the one associated to the rest of the volume. Quadrilateral element mesh with four integration points were used. A precision of 10^{-4} was required for the Newton Raphson solver. Values of the parameters of the model were taken from the ones calibrated with respect to another mortar of adobe tested by the authors in (Li Piani et al. 2019). These are: $k_o = 7\%$, $A = 160$, $\Delta = 4.5$, $N = 2000$, $E = 200\text{MPa}$, $\nu = 0.1$. At first, numerical simulations were meant to replicate the theoretical condition of absence of defects in the material. Thus, only the small imperfection needed to trigger localization was present. The results of the numerical simulations are presented in the following for each size. These consist of the normalized stress-strain plots obtained by dividing numerical forces and displacements per the cross section areas and height, respectively and of the damage patterns at failure as numerically obtained (Figure 8).

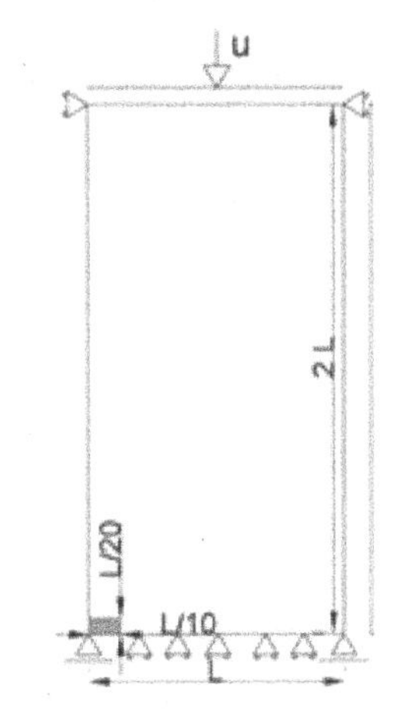

Figure 7. Numerical setup.

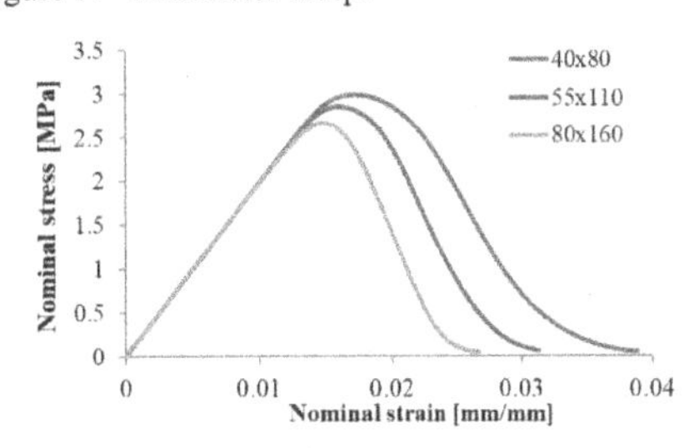

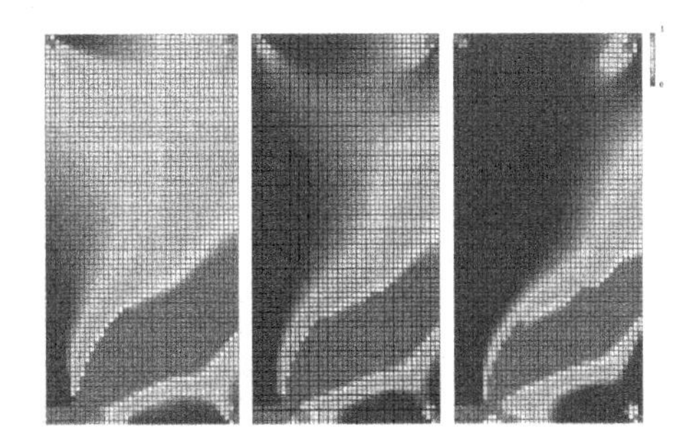

Figure 8. Normalized stress-strain plots and corresponding cracking patterns for Size A, B, D.

The damage pattern is more spread for smaller samples as experimentally exhibited. However, the smaller samples are also the strongest. Next, the same simulations were performed by introducing the effects of possible defects in the mixture. The same weak area approach just presented was used. However, in the following simulations, the value of the weak region was kept the same for all specimen sizes. Three series of simulations were performed by progressively enlarging the area of the weak region from 32mm^2 (corresponding to a 8x4mm defect size) up to 128 mm^2 and 256 mm^2. Results of numerical simulations are presented for Size A and D in terms of compressive strength (Figure 9) and damage spread (Figure 10).

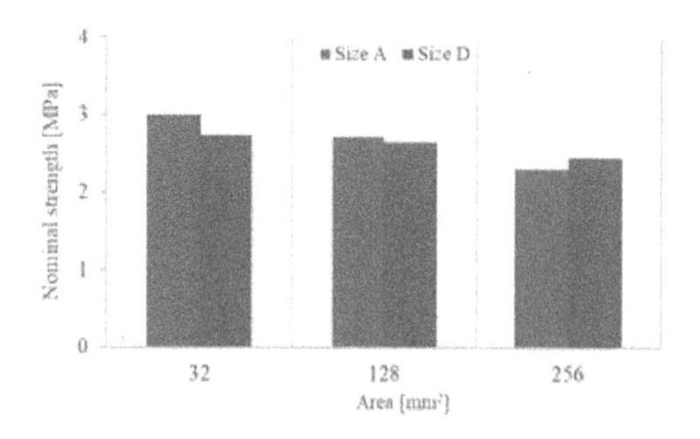

Figure 9. Nominal strength values for progressively larger defect areas in Size A and Size D samples.

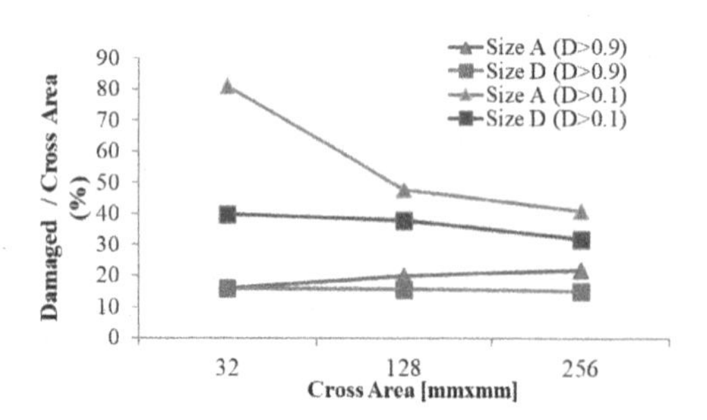

Figure 10. Percentages of damaged areas over cross sections for Size A and Size D for progressively increasing defect areas.

From results, it appears that progressively increasing defect sizes, nominal strength of smaller samples diminishes proportionally faster than for larger samples. At an area of 256 mm^2, Size D possesses a higher strength than Size A, while damage patterns keep being larger for smaller samples (Figure 10). This implies that in the comparison of two samples of given initial geometry with a given initial defect, there is a certain defect size above which smaller samples proportionally lose sooner their mechanical integrity, whereas larger samples still keep the capability of internally redistributing loads (Figure 11).

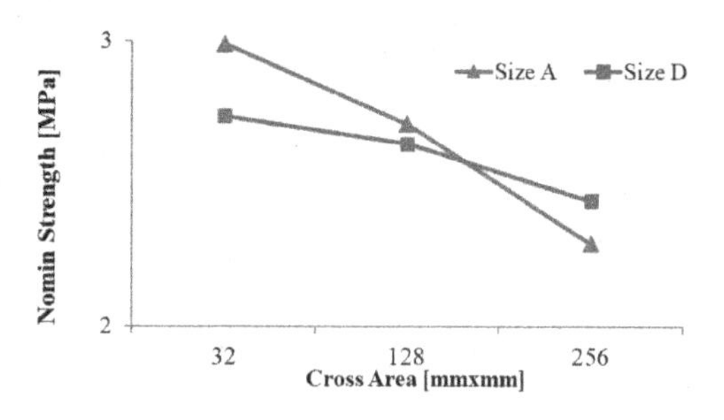

Figure 11. Trends in compressive strength as a function of cross section area.

3.4 *Mesh sensitivity analysis*

In order to guarantee the trustworthiness of the depicted results, a mesh sensitivity analysis has been performed. The same simulations of compression tests as presented in Par.3.3 have been executed using three different mesh sizes from 5mm to 1.25mm. Results are presented in terms of damage patterns and normalized curves. For all simulations, it counts that mesh independence is preserved (Figure 12).

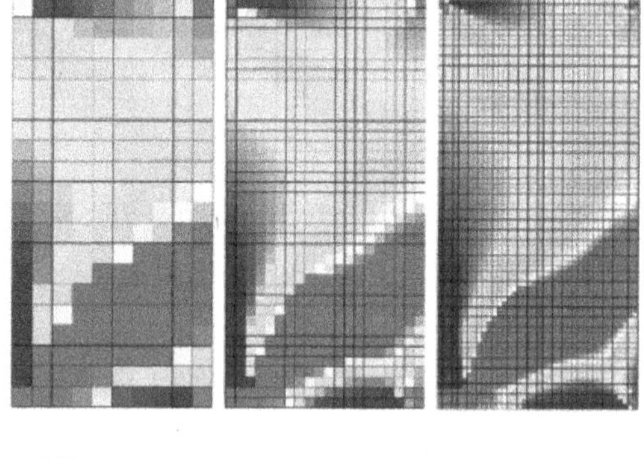

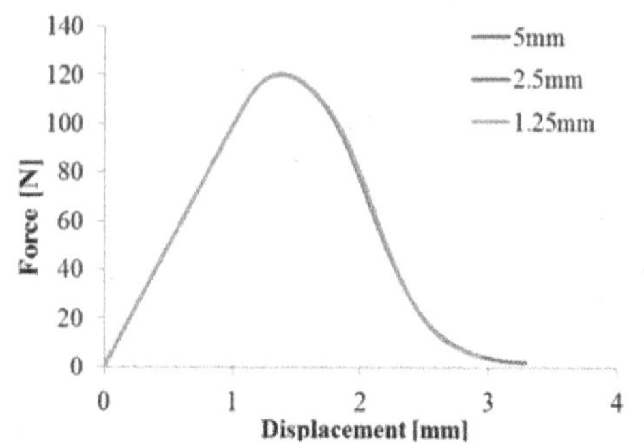

Figure 12. Example of mesh independent results in terms of damage pattern and force-displacement plot.

4 CONCLUSIONS

In this paper, the results of an experimental campaign on adobe mortar has been presented. Size dependency on masonry units has been studied. Static compression tests were executed on specimens with progressively larger sizes and constant slenderness. Results were compared in terms of nominal compressive strength. Contrary to the general trend valid for industrially produced quasi brittle materials, larger samples of adobe comparatively exhibited a higher strength. Probabilistic theories coupled with principles of fracture mechanics have been used to interpret this trend. For not industrialized materials as adobe, the presence of defects inside the mixture is highly probable. These may reach significant dimensions with respect to the dimension of the overall specimen. It is hypothesized that there may exists a certain area of weakness with respect to the overall specimen size which causes the loss of structural integrity of the specimen. Therefore, for a given defect size, larger samples are capable of re-distributing stresses more easily with respect to smaller ones, thus resulting in a higher compressive strength. Overall, for materials as adobe the probability that a defect appears in a given volume must be pondered with the probability that its relative extension is prone to affect or even compromise its load bearing capacity. This especially counts when fibres, usually picked in the field, without proper knowledge on the exact bonding mechanisms with the binder fraction of the mixture are added in the mixture. These may correspond to large areas of de-adherence which compromise the structural integrity of the specimen. In the likelihood of a defect within a given volume of material, the probability that it does affect the mechanical performance of the material depends on its extension with respect to the total volume. This hypothesis has been numerically validated. An isotropic damage model developed for adobe materials has been used to replicate compression tests on specimens with larger sizes. Numerical simulations have shown that the larger the defect size inside the specimen is, the lower its overall capability to withstand the external loads compared to larger specimens. These simulations confirm existing trends of analytical functions of stress intensity factors with respect to initial flaws size (Rooke & Cartwright 1976). If this is sufficiently large, the geometrical correction factor can be larger than 1, thus resulting in a reduced specimen strength. Finally, the reliability of the results of the numerical simulations shown in this paper has been checked. In fact, local damage models are prone to suffer from mesh dependence, thus resulting in inconsistent results with respect to the discretized mesh size. In this study, a mesh sensitivity analysis has been performed. The properties of mesh independence of the model have been confirmed. These are guaranteed by the implementation of a local regularization algorithm which introduces a damage delay consistent with the property of rate dependence experimentally attributed to adobe masonry materials.

REFERENCES

Allix, O. (2012). The bounded rate concept: A framework to deal with objective failure predictions in dynamic within a local constitutive model. *International Journal of Damage Mechanics*, *22*(6), 808–828. https://doi.org/10.1177/1056789512468355

Austin, G. S. (1984). Adobe as a building material. *New Mexico Bureau of Mines and Mineral Resources, Socorro*, 69–71.

Bohdan, S., & Tomasz, K. (2013). Determination of the influence of cylindrical samples dimensions on the evaluation of concrete and wall mortar strength using ultrasound method. *Procedia Engineering*, *57*, 1078–1085. https://doi.org/10.1016/j.proeng.2013.04.136

EN 772-1:2000, Methods of test masonry units Part 1: determination of compressive strength. (1999).

Fódi, A. (2011). Effects influencing the compressive strength of a solid, fired clay brick. *Periodica Polytechnica Civil Engineering*, *55*(2), 117–128. https://doi.org/10.3311/pp.ci.2011-2.04

HB 195 - The Australian earth building handbook-Standards Australia International, NSW 2001, (2001).

Ingham, J., Biggs, D., & R., L. (2014). Uniaxial Compressive Strength and Stiffness of Field-Extracted and Laboratory-Constructed Masonry Prisms. *Journal of Materials in Civil Engineering*, *26*(4), 567–575.

Kaushik, H. B., Rai, D. C., & Jain, S. K. (2007). Uniaxial compressive stress – strain model for clay brick masonry. *Current Science*, *92*(4), 497–501.

Krishna, R., Kumar, R., & Srinivas, B. (2011). Effect of size and shape of specimen on compressive strength of glass fiber reinforced concrete (GFRC). *Facta Universitatis - Series: Architecture and Civil Engineering*, *9*(1), 1–9. https://doi.org/10.2298/FUACE1101001K

Li Piani, T. (2019). *Experimental-numerical material characterization of adobe masonry: tests and simulations on various types of bricks and loading rates*. Delft University of Technology (TU Delft).

Li Piani, T. (2021). Threat assessment and vulnerability mapping for sensitive buildings against terrorism in urban environments. *Security Terrorism Society (STS)*, *1*(13), 60.

Li Piani, T., Krabbenborg, D., Weerheijm, J., Koene, L., & Sluys, L. J. (2018). The Mechanical Performance of Traditional Adobe Masonry Components: An experimental-analytical characterization of soil bricks and mud mortar. *Journal of Green Building*, *13*(3), 17–44.

Li Piani, T., Weerheijm, J., Koene, L., & Sluys, L. J. (2018). Modelling the Mechanical Response of Adobe Components under Uniaxial Loading. *Key Engineering Materials*, 774(Advances in Fracture and Damage Mechanics XVII), 650–657. https://doi.org/10.4028/www.scientific.net/KEM.774.650

Li Piani, T., Weerheijm, J., Koene, L., & Sluys, L. J. (2019). The Adobe delta damage model: A locally regularized rate-dependent model for the static assessment of soil masonry bricks and mortar. *Engineering Fracture Mechanics*, *206*(February), 114–130. https://doi.org/10.1016/j.engfracmech.2018.11.026

Li Piani, T., Weerheijm, J., Peroni, M., Krabbenborg, D., Koene, L., Solomos, G., & Sluys, L. J. (2020). Dynamic behaviour of Adobe bricks in compression: the role of fibres and water content at various loading rates. *Construction & Building Materials*, *230*(October), 117–135.

Li Piani, T., Weerheijm, J., & Sluys, L. J. (2019). Dynamic simulations of traditional masonry materials at different loading rates using an enriched damage delay: Theory and practical applications. *Engineering Fracture Mechanics*, *218*(May). https://doi.org/10.1016/j.engfracmech.2019.106576

MacGregor, F. M. B. and J. G. (1994). Effect of Core Length-to-Diameter Ratio on Concrete Core Strengths. *Materials Journal, 91*(4). https://doi.org/10.14359/4042

New Zeland Standards. (1998). *NZS 4297 (1998): Engineering design of earth buildings. NZS 4297*, 63.

Page, A. W., & Marshall, R. (1985). The Influence of Brick and Brickwork Prism Aspect Ratio on the Evaluation of Compressive Strength. *Proceedings of the Seventh International Brick and Masonry Conference, Melbourne, Australia*, 653–664. https://doi.org/10.1017/CBO9781107415324.004

Parra-Saldivar, M. L., & Batty, W. (2006). Thermal behaviour of adobe constructions. *Building and Environment, 41*(12), 1892–1904. https://doi.org/10.1016/j.buildenv.2005.07.021

Phu Nguyen, V., Lloberas Valls, O., Stroeven, M., & Johannes Sluys, L. (2010). On the existence of representative volumes for softening quasi-brittle materials – A failure zone averaging scheme. *Computer Methods in Applied Mechanics and Engineering, 199*(45–48), 3028–3038. https://doi.org/10.1016/j.cma.2010.06.018

Rooke, D. P., & Cartwright, D. J. (1976). *Compendium of stress intensity factors. HMSO Ministry of Defence. Procurement Executive.*

Sluys, L. J., & de Borst, R. (1992). Wave propagation and localization in a rate-dependent cracked medium-model formulation and one-dimensional examples. *International Journal of Solids and Structures, 29*(23), 2945–2958. https://doi.org/10.1016/0020-7683(92)90151-I

Varum, H., Costa, A., Silveira, D., Pereira, H., Almeida, J., & Martins, T. (2007). Structural Behaviour Assessment and Material Characterization of Traditional Adobe Constructions. *AdobeUSA 2007, El Rito, NM, USA, January*. c:%5CUsers%5CUsuario%5CDesktop%5CPhD%5CPapers%5CMechanical characterization%5CVarum H., Costa A., Silveira D., Pereira H., Almeida J., Martins T. – Structural Behaviour Assessment and Material Characterization of Traditional Adobe Constructions.pdf

Computational Modelling of Concrete and Concrete Structures – Meschke, Pichler & Rots (Eds)
© 2022 Copyright the Author(s), ISBN: 978-1-032-32724-2

Transient shear band and its kinetics around interfaces of cementitious materials and soil/rock foundation

Y. Yamanoi
Central Research Institute of Electric Power Industry, Chiba, Japan

K. Maekawa
Yokohama National University, Kanagawa, Japan

ABSTRACT: The finite sized shear band finally transitions to the assembly of sand and gravel particles owing to meso-scale damaging of cementitious binders. This paper proposes shear band kinetics in consideration of disintegration and/or gravelization of cementitious composites. A model that presents the transient features of confinement from cementitious composites to the rock/soil foundation is proposed. The model is applied to two different structures for verification and validation. One structure is double-beam coupling beams, where the unreinforced concrete strip sandwiched by reinforced concrete beams experiences large shear deformation under seismic load. The other structure is masonry structures, where mortar joints can be a relative weak zone. From the comparison of past experimental results of these structures with local weakness and the analysis results, it was determined that the proposed model may improve and widen the scope of nonlinear analysis in spite of its simplicity.

1 INTRODUCTION

The shear band is a source of nonlinearity and size effect of structural concrete. This paper proposes shear band kinetics in consideration of disintegration and/or gravelization of cementitious composites. The finite sized shear band ultimately transitions to the assembly of sand and gravel particles owing to meso-scale damaging of cementitious binders. It is well known that the characteristics of confinement of concrete differ from those of the soil foundation. The proposed model presents the transient features of confinement from cementitious composites to the rock/soil foundation (Yamanoi & Maekawa 2020).

The main frame of the concrete constitutive model used in this study (Maekawa et al. 2003) applies for shear transfer on the crack surface represented by a contact density model where the ultimate shear transfer hardly depends on the confinement pressure. On the other hand, the mechanical behavior of the assembly of sand and gravel particles can be represented by the elasto-plasticity subjected to the friction law (Towhata 2008). In the proposed model, the two models are combined depending on the degree of disintegration which is expressed as a function of the fracture parameter used in the elasto-plastic and fracture model of concrete.

The proposed model was originally developed for low-strength concrete of lesser cementation. In this case, the localized shear band was experimentally observed being damaged beyond the scope of conventional concrete. Previously, constitutive modeling for soft rocks and cement improved ground has been proposed (Abdulla & Kiousis 1997; Hirai et al. 1989; Namikawa & Mihira 2007; Shen et al. 2019; Sun & Matsuoka 1999) and formulated with respect to the degree of damage (Desai 1996; Yu et al. 1998). The applicable range of these models is limited to uniaxial compressive strength of 2 MPa or less and elasto-plasticity being the basis of the formulation. On the other hand, the constitutive model of concrete has generally targeted concrete with compressive strength of 15 MPa or more. The main scope of the proposed model in this study is situated somewhere between these two cases.

As a new application target of the model, we focus here on two different types of structure. One is double-beam coupling beams (DBCB: Choi et al. 2018) and the other is masonry walls. In the former, the unreinforced concrete strip sandwiched by reinforced concrete beams experiences large shear deformation under seismic load. In the case of masonry structures, mortar joints have been modeled by volume-less joint elements subjected to the friction law. By combining the proposed model with the active crack method, these local weaknesses can be modeled by 3D solid elements.

First, the details of the proposed model are explained. Next, the applicability of this model to the two types of structure is verified by comparison of

DOI 10.1201/9781003316404-45

the analysis results with existing experimental results. Consideration of the changes in material properties associated with shear localization is necessary for structural post-peak analysis to assess the risk of residuals beyond design loads. In addition, the occurrence of crushed areas caused by ground rock faults is an important factor in the risk assessment of underground structures during earthquakes. Opening/exploring a way to comprehensively deal with such aspects to meet future needs is another purpose of this study.

Transition shear band modeling is being applied for engineering practice. One is 3D urban spaces with diversifying underground spaces for the assessment of urban safety and the other is fault attack on underground utility ducts in nuclear power plants (Aoki et al. 2021).

2 TRANSITION MODEL FROM CONCRETE TO GRAVEL IN SHEAR

Concrete and assembly of gravels have different shear resistance mechanisms in the deformed localized bands (Figure 1). The former transmits shear force mainly by the aggregate interlock, and the latter mainly by the contact friction between particles. In the former, the aggregate is fixed by the cement paste binder, while it is rotatable in the latter. Hence, we have different effects of confinement on shear strength and ductility. The contact density model of cracked concrete (Bujadham et al. 1992; Li et al. 1989) was formulated based on this mechanism. Thus, the confinement effect on shear transfer along concrete cracks is comparatively lower than that of sand particle assembly when confinement is higher. For sand-like granules, shear strength is generally proportional to confinement pressure, and elasto-plastic constitutive models that follow the Mohr-Coulomb and Drucker-Prager fracture criteria fit well (Towhata 2008).

The difference between two materials will reduce as the concrete deteriorates to disintegrated graveling by large shear deformation. The transition model is formulated based on this idea (Yamanoi & Maekawa 2020), and we have the total stress σ_{ij} of the localized bands as,

$$\sigma_{ij} = \sigma_{cij}(K) + Z(K)\,\sigma_{sij}, \quad \sigma_{sij} = S_{ij} + \delta_{ij} I_1 \tag{1}$$

where σ_{cij} is the stress tensor yielded by the constitutive model for cracked and uncracked concrete, σ_{sij} is the stress yielded by a perfect elasto-plastic model for sand, S_{ij}, I_1 are the deviator stress tensor and the first invariant of stress tensor, δ_{ij} is Kronecker's delta, and K is a concrete fracture parameter.

The rate of transition to graveling is specified by $Z(K)$, which is the function of the fracture parameter of concrete denoted as K. No damage of the initial state corresponds to $K = 1$, and complete damage is $K = 0$. In this study, $Z(K)$ is tentatively set as Equation 2.

$$Z(K) = 1 - K \tag{2}$$

In the conventional concrete model, which can consider up to 6 cracks, the stress is calculated with regard to the crack state, as shown in Figure 2 (Maekawa & Fukuura 2013). The fracture parameter is calculated based on the elastic strain invariant before cracking. After cracking, it is allocated to each crack as a historical variable. For each crack, the fracture parameter is calculated based on the uniaxial compression fracture in two directions, normal and parallel to the crack surface, and the smaller one represents the fracture of

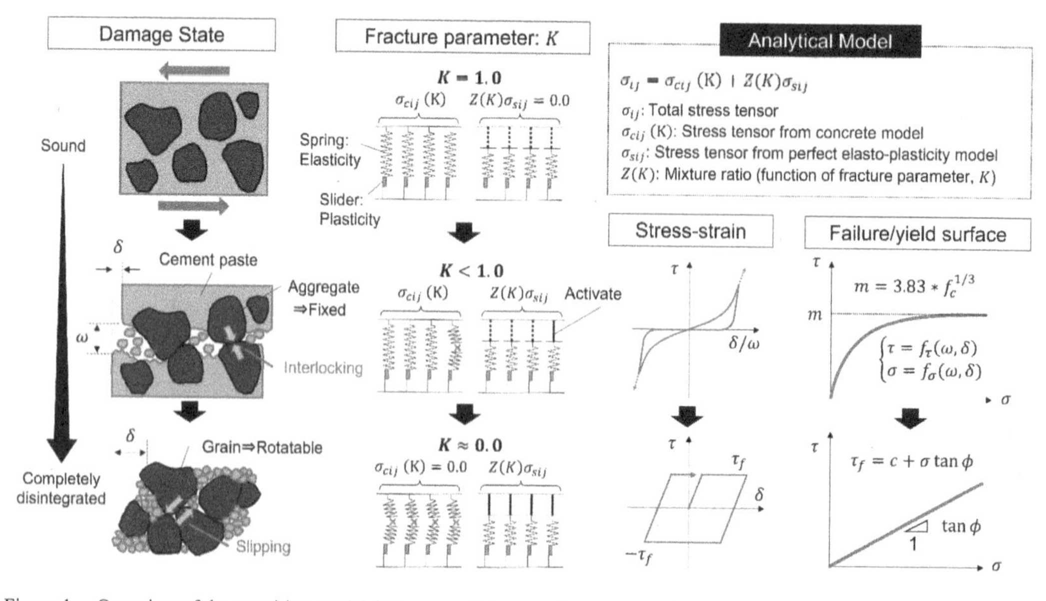

Figure 1. Overview of the transition model from concrete to gravel.

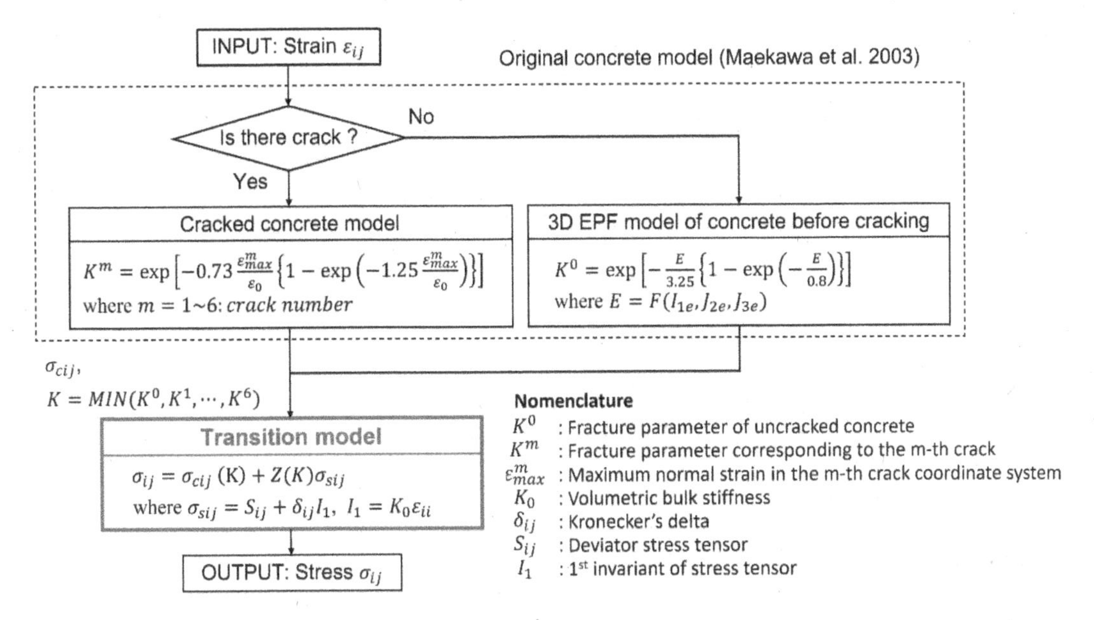

Figure 2. Flow of computing stress in the proposed model.

the crack. As a result, up to 7 fracture parameters can be recorded per Gaussian point. The smallest one is given to the transition model to calculate the ratio of transition.

The stress component of disintegrated graveling (σ_{sij}) is obtained by elasto-plastic modeling for the soil foundation. In this study, we have inelastic flow normal to the deviatoric plane of stresses and Drucker-Prager's yield criterion in Equation 3. Since we have large shear deformation at the time of disintegration, shear dilatancy in progress is assumed to be nil.

$$F = J_2 - S_u = J_2 - (A - BI_1), J_2 = \sqrt{\frac{1}{2}S_{ij}S_{ij}} \qquad (3)$$

where, A, B are constants of the largest yield surface. In the case of dry sand particle assembly, parameters A and B are decided based upon the cohesive strength denoted as c and the internal friction angle denoted as ϕ. Thus, we have the simple model of $A = c = 0$, $B = tan\phi = \mu$ (μ: frictional coefficient).

In previous research, we validated this model for low-strength concrete (Uniaxial compressive strength: $f_c' \approx 8MPa$). A three layers beam with intermediate weak layer was prepared and bending shear force was applied as shown in Figure 3 (Yamanoi & Maekawa 2020). In the experiment, damage was concentrated along the weak layer and the shear band was dispersed and bifurcated. The low-strength concrete observed on the failure surface was pulverized and was like gravel assembly.

Comparing the load-displacement relations and strain distribution between the experiment and analysis, the proposed model successfully reproduced the experimental results (Figure 3). Incidentally, the strain in the experiment was measured by the digital image correlation system (Sutton et al. 2009).

3 APPLICATION OF TRANSITION MODEL TO DBCB (DOUBLE-BEAM COUPLING BEAM)

There are two types of coupling beams: the conventional type consisting of longitudinal main reinforcement and transverse reinforcement, and the diagonally reinforced coupling beam surrounded by shear reinforcement (Lim et al. 2016; Naish 2013). Choi et al. (2018) propose the double-beam coupling beam (DBCB). In DBCB, reinforcement cages are arranged in two stages, and the central part is unreinforced. The unreinforced part may absorb shear deformation, and the upper and lower parts are bent and damaged like double beams. Despite the relatively simple reinforcement arrangement, greater toughness is exhibited throughout the structure. It is expected that the central unreinforced part will undergo large shear deformation locally, and the damage level will exceed the applicable range of the existing concrete constitutive model. In addition, due to repeated loading, graveling will easily proceed. Thus, the authors conducted the experimental verification with the repeated loading experiment of DBCB by Choi and Chao (2020), as shown in Figure 4a.

The analysis model is shown in Figure 4b. The displacement of the lower and side surfaces of the fixed block was confined, and the upper and lower surfaces of the loading block were forcibly displaced. The loading block is always displaced while maintaining parallelism with the fixed block. In the experiment, since the loading block is connected to the fixed block by a steel link, strictly speaking, horizontal displacement occurs in the direction in which the specimen is shortened as the vertical displacement increases. However, the maximum deformation angle loaded in this experiment was about 6 degrees, which is equivalent to 10% of the beam chord rotation. As the displacement

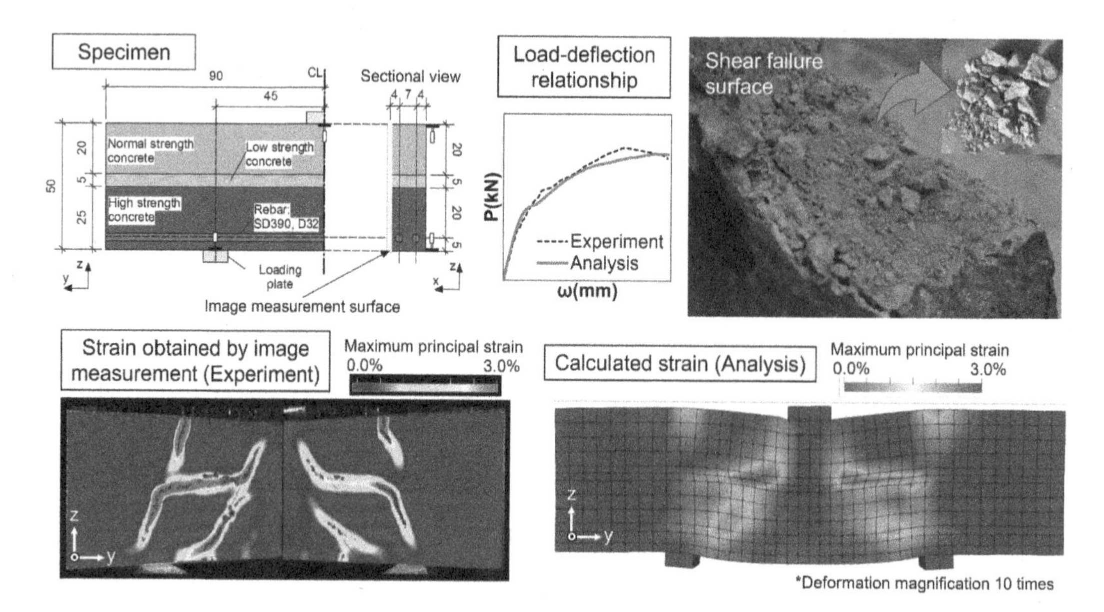

Figure 3. Validation of the proposed model by the experiment of beam with intermediate weak layer (Yamanoi & Maekawa 2020).

error due to the elastic deformation of the steel link and the engagement of the jig is also considered, it was judged that the horizontal displacement can be ignored.

The transition model described in Section 2 was applied to the reinforced and unreinforced concrete parts. The compressive strength was computationally set to the measured value at the test, and Young's modulus and tensile strength were set based on the JSCE code specification (JSCE 2010) from the compressive strength of 30 MPa. Rebar was modeled as smeared reinforcement and the property assuming Grade 60 ($f_y = 420MPa$) was set. The coefficient of friction after graveling was tentatively set 1.0 according to the past experimental fact by Lim and Maekawa (1987). The loading and the fixed blocks were modeled with elastic bodies. The stiffness of those blocks was set to 28 GPa, the same as the initial Young's modulus of concrete.

Figure 5 is a comparison of the experimental results and analysis results. The transition model is shown to be able to capture the experimental results from the viewpoint of both the shear force-beam chord rotation relation (Figure 5a) and failure mode (Figure 5b, c). The DBCB is at its maximum load with damage to the unreinforced parts. As the deformation angle increases, the damage to the unreinforced part increases, but the ductility is large, and it has a residual strength of 60% or more even after peak loading. Focusing on the damage when the deformation angle is 3%, shear cracks occur on the diagonal line of the beam in the experiment. This means that the damaged unreinforced part maintains shear transfer. These behaviors of the experiment can be reproduced accurately with transition of shear localized bands.

We have a tendency to overestimate the energy absorption capacity during repeated loading. A similar tendency is seen when damage to the beam-loading block joint is dominant (Naish 2013, Lim et al. 2016).

Next, we checked the sensitivity of the coefficient of friction in the ultimate state (Figure 6). The maximum shear force does not change significantly even if the coefficient of friction changes from 1.0 to 0.6. On the other hand, the residual strength is greatly affected by the friction angle, and 45 degrees (equivalent to the coefficient of friction of 1.0) appears highly consistent with the experimental result.

It was confirmed that the transition model can be applied not only to low-strength concrete but also normal-strength concrete when the disintegration of the composition becomes significant.

4 APPLICATION OF TRANSITION MODEL TO MASONRY WALL

4.1 *Extended multi-directional fixed crack model*

In the previous sections, large shear deformations occurred in the intended weak zones, which were the application targets for the transition model. In this section, we focus on masonry structures called mortar joints, which have a lot of local weakness.

Aiming for more sophisticated seismic capacity evaluation, various methods for evaluating the seismic performance of existing masonry structures have been developed (Hashimoto et al. 2017; Lourenço & Rots 1997; Pandey & Meguro 2004). There are several types of numerical modelling of masonry structures in order to deal with the high nonlinearity of many mortar joints (Facconi et al. 2014; Lourenço et al. 2007; Maier et al. 1991). However, the authors know of no analysis models capable of handling the state where cracks intersect

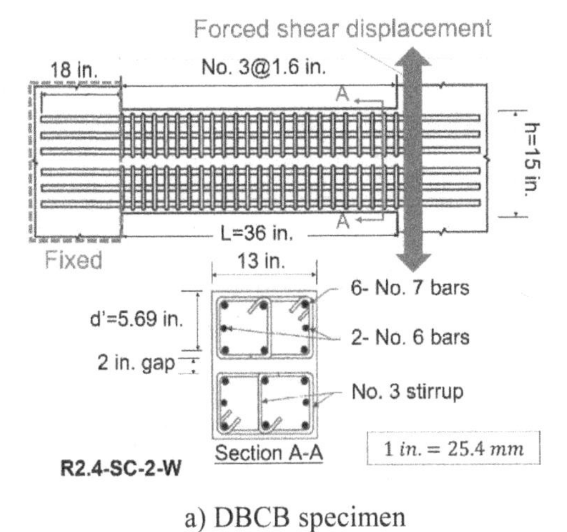

a) DBCB specimen

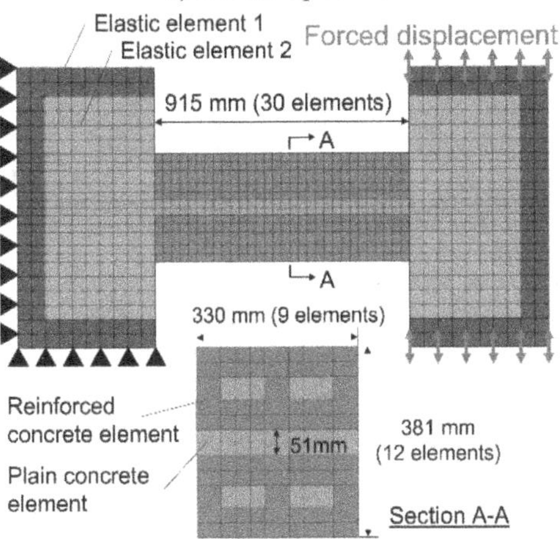

b) Mesh discretization

Figure 4. Target experiment of double-beam coupling beam (DBCB) and mesh discretization for FEM analysis.

in multiple directions within masonry blocks having interaction with multi-directional masonry joints.

Therefore, the authors proposed an extended model of multi-directional fixed crack model as a model of masonry structures (Yamanoi et al. 2021).

Currently, a constitutive model that can consider non-orthogonal cracking in 6 directions is used (Maekawa & Fukuura 2014). For 3-directional quasi-orthogonal crack planes, a non-orthogonal coordinate system is applied. Further, a 3D space averaged constitutive law for a total of six directions of crack groups has been formulated with the addition of a new crack coordinate system.

The authors opted to allocate one of the above two quasi-orthogonal crack coordinate systems to mortar joint planes orthogonal to each other. The remaining quasi-orthogonal coordinate system was allocated to the behavioral analysis of the constituent blocks. This makes it possible to handle kinetics in which multiple cracks occur not only in mortar joints but also in masonry blocks under complex load histories.

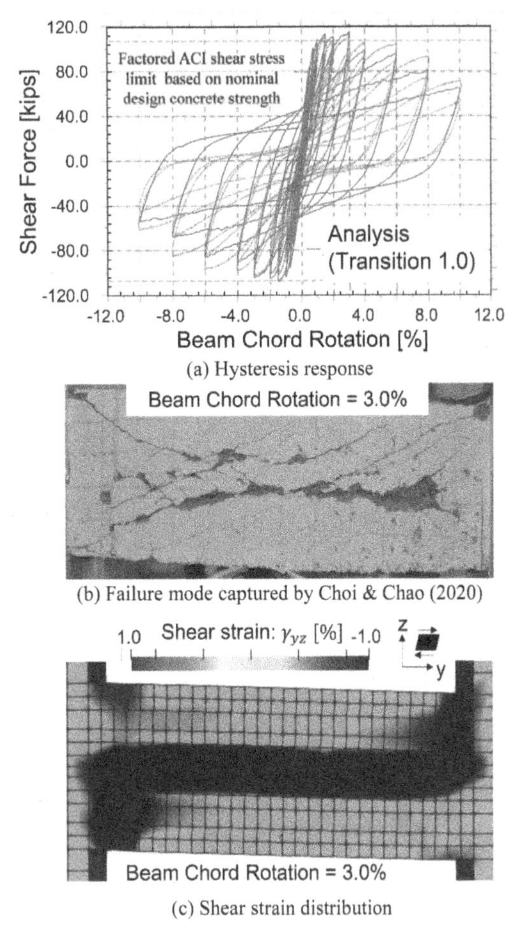

(a) Hysteresis response

(b) Failure mode captured by Choi & Chao (2020)

(c) Shear strain distribution

Figure 5. Comparison of experimental results and analysis results (DBCB).

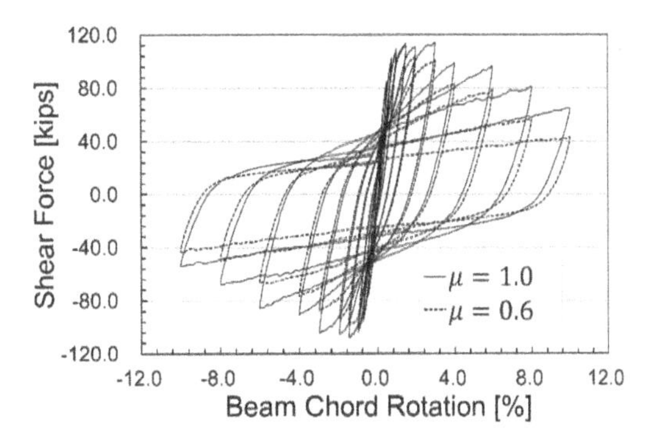

Figure 6. Sensitivity of the ultimate friction in the transition model (DBCB).

The crack criterion and the shear transfer characteristics of cracks can be set separately for the mortar joint and block. The transition model is applied only to the shear of mortar joints. Further details are available in Yamanoi et al. (2021).

4.2 *Validation by the experiment of Ganz's wall*

Ganz and Thurlimann (1984) reported a loading experiment on masonry shear walls. As two types of nonlinearities, slips in the joints and cracks in the masonry blocks, are provided, this experiment was considered appropriate for the verification of the extended multi-directional crack model. Figure 7 shows the dimensions of the specimen in the experiment and the mesh division diagram for FEM analysis.

The specimen consists of hollow clay bricks stacked in 10 layers. The bricks are bonded to each other with 10 mm of mortar. Each brick measures 300 mm × 190 mm × 150 mm. The finite elements placed in the wall were 300 mm × 200 mm × 150 mm, almost the same dimensions as the brick blocks. Thus, one element may contain at most one joint.

An RC constitutive model was applied to the loading beams and specimens (Maekawa et al. 2003). Horizontal displacement was applied to the top plate under a vertical load of 415 kN.

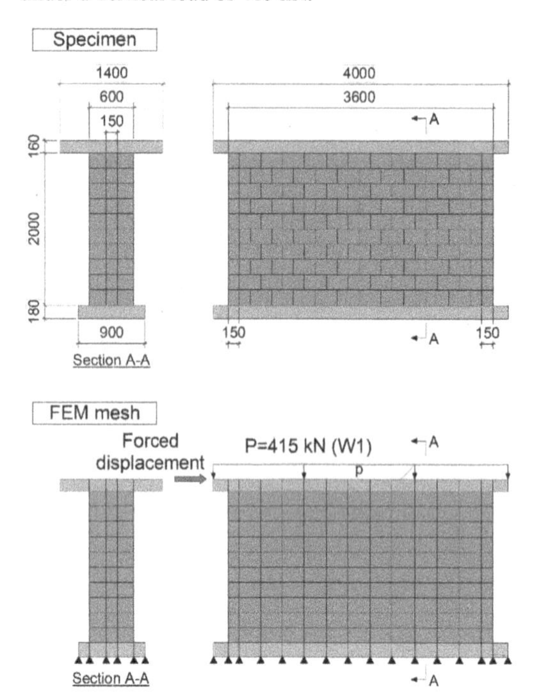

Figure 7. Target experiments of masonry wall and mesh discretization for FEM analysis.

Experimental values were used for the strength of the masonry bricks (Ganz & Thurlimann 1982). Here, the aforementioned transition model was applied to the shear transfer characteristics of the cracks in the mortar joint. From sensitivity analysis and in reference to the commonly known internal friction angle of sands and clays (Rowe 1962; Skempton 1985), the assumed friction coefficient was set to 0.4 for the case when the solidification caused by cement paste disappears.

Figure 8 shows a comparison of the experimental and analysis results obtained by applying the proposed

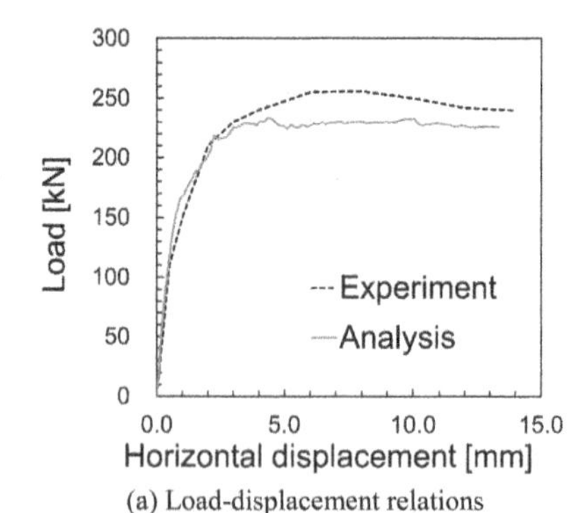

(a) Load-displacement relations

(b) Experimental response at failure (Ganz & Thurlimann 1984)

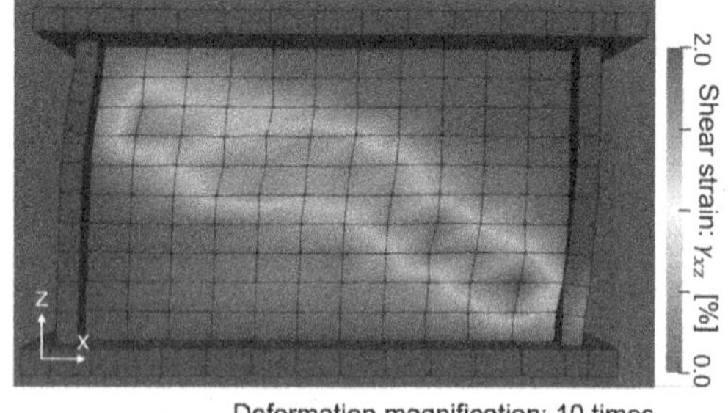

(c) Shear strain distribution of analysis

Figure 8. Comparison of experimental results and analysis results (Masonry wall).

model. In this analysis, the exact position of the joint is not specified within the finite element, but space averaged continuous strain distribution within the element is addressed. The shear strain distribution and the load-displacement relations were reasonably well reproduced by the analysis.

In previous research, the effect of the shear transfer model of mortar joints on the analytical accuracy was investigated (Yamanoi & Maekawa 2021). It was confirmed that the concrete model overestimated the load capacity and failed to reproduce the ultimate failure mode, while the cohesionless friction model underestimated the load capacity. As a result, the validity of the transition model was verified. Further, the sensitivity of the friction angle was newly investigated in this study.

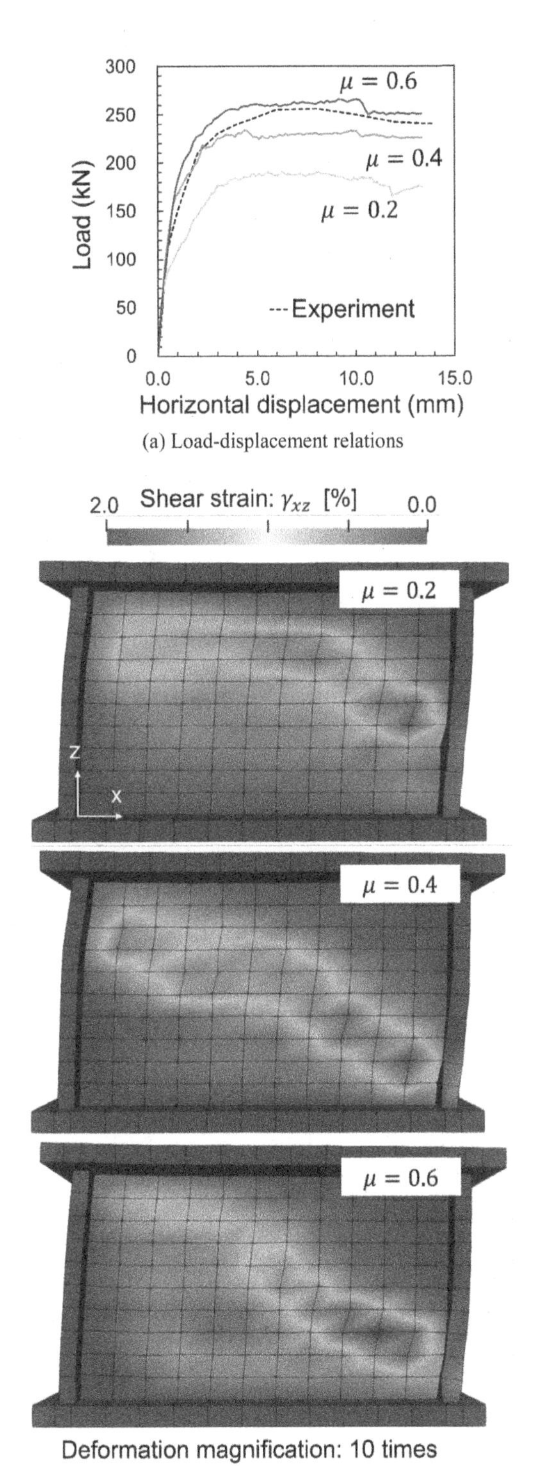

Figure 9. Sensitivity of the ultimate friction angle in the transition model (Masonry wall).

The analysis results when the friction coefficient of the transition model is changed from 0.2 to 0.6 are compared in Figure 9. The higher the coefficient of friction, the higher the shear capacity and stiffness of the wall. When the frictional resistance of the mortar joint is small ($\mu = 0.2$), calculated shear strain distributes horizontally, which indicates that the slip along the horizontal mortar joint is dominant. On the other hand, assuming a higher coefficient of friction ($\mu = 0.6$), the diagonal cracks in the brick appear to be dominant, and the failure mode is determined to be shear compression failure. This is similar to the results of the analysis where shear transfer model on the crack surface of concrete is applied to the mortar joint. This sensitivity analysis shows that the initially assumed coefficient of friction ($\mu = 0.4$) is appropriate.

5 CONCLUSION

For the purpose of evaluating the response of concrete structures beyond the ultimate state, a transition model that can consider the change of shear transfer characteristics of concrete with disintegration has been proposed. It was confirmed that this model originally developed for low-strength concrete can be applied to DBCB made of normal-strength concrete and to mortar joints in masonry structures. The main conclusions are given below.

1. The proposed model can accurately evaluate the capacity and ductility of DBCB.
2. There was room for improvement of the model regarding cyclic behavior.
3. The transition model was preferable for the shear transfer model of mortar joints in masonry.
4. From the transition model, the coefficient of friction after disintegration was inversely identified as 1.0 for normal concrete and as 0.4 for mortar.

It was clarified that the proposed model may improve and widen the scope of nonlinear analysis in spite of its simplicity. As this model can be a mere frictional material model by setting a small concrete strength, it is expected to replace the joint elements with no volume conventionally used in various joints.

ACKNOWLEDGMENTS

This study was financially supported by JSPS KAKENHI Grant Number 20H00260.

REFERENCES

Abdulla, A.A. & Kiousis, P.D. 1997. Behavior of cemented sands - II. Modelling. *International Journal for Numerical and Analytical Methods in Geomechanics* Vol. 21: John Wiley & Sons, Ltd.: 549–568.

Aoki, H., Fan, S., Yamanoi, Y., Mingqian, R., Takahashi, H. & Maekawa, K. 2021. Failure mode of deteriorated concrete tunnel sections under subsidence and localized shear. *Structures and Buildings* [published online ahead of print April 29, 2021], https://doi.org/10.1680/jstbu.20.00150

Bujadham, B., Mishima, T. & Maekawa, K. 1992. Verification of the universal stress transfer model. *Proceeding of JSCE* No. 451/V-17: 289–300.

Choi, Y., Hajyalikhani, P. & Chao, S.H. 2018. Seismic performance of innovative reinforces concrete coupling beam-double-beam coupling beam. *ACI Structural Journal* V. 115 No. 1: 113–125.

Choi, Y. & Chao, S.H. 2020. Analysis and design of double-beam coupling beams. *ACI Structural Journal* Vol. 117 No. 5: 79–96.

Desai, C.S. & Toth, J. 1996. Disturbed state constitutive modeling based on stress-strain and nondestructive behavior. *International Journal of Solids Structures* Vol. 33 No. 11: 1619–1650.

Facconi, L., Plizzari, G. & Vecchio, F. 2014. Disturbed stress field model for unreinforced masonry. *ASCE Journal of Structural Engineering* Vol. 140 No. 4, Article ID 04013085.

Ganz, H.R. & Thurlimann, B. 1982. *Tests on the biaxial strength of masonry* (Report No. 7502-3). Institute of Structural Engineering, Eidgenössische Technische Hochschule Zürich. Zurich, Switzerland (in German).

Ganz, H.R. & Thurlimann, B. 1984. *Versuche an Mauerwerksscheiben unter Normalkraft und Querkraft* (Report No. 7502-4). Institute of Structural Engineering, Eidgenössische Technische Hochschule Zürich. Zurich, Switzerland (in German).

Hirai, H., Takahashi, M. & Yamada, M. 1989. An elastic-plastic constitutive model for the behavior of improved sandy soils. *Soils and Foundations* Vol. 29 No. 2: 69–84.

JSCE. 2010. *Standard specifications for concrete structures (Design)*. Tokyo: Japan Society of Civil Engineers.

Li, B., Maekawa, K. & Okamura, H. 1989. Contact density model for stress transfer across cracks in concrete. *Journal of the Faculty of Engineering*, The University of Tokyo (B), 40(1).

Lim, E., Hwang, S.J., Cheng, C.H. & Lin. P.Y. 2016. Cyclic tests of reinforced concrete coupling beam with intermediate span-depth ratio. ACI Structural Journal Vol. 113 No. 3: 515–524.

Lim, T.B., Li, B. & Maekawa, K. 1987. Mixed-mode strain-softening model for shear fracture band of concrete subjected to in-plane shear and normal compression. *Proc. of Int. Conf. on Computational Plasticity*, June 1987. Barcelona: ECCOMAS: 1431–1443.

Lourenço, P.B. & Rots, J.G. 1997. Multisurface interface model for analysis of masonry structures. *Journal of Engineering Mechanics* Vol. 123 No. 7: 660–668.

Lourenço, P.B., Milani, G., Tralli, A. & Zucchini, A. 2007. Analysis of masonry structures: review of and recent trends in homogenization techniques. *Canadian Journal of Civil Engineering* Vol. 34 No. 11: 1443–1457.

Maekawa, K. & Fukuura, N. 2013. Nonlinear modeling of 3D structural reinforced concrete and seismic performance assessment. In: T. T. C. Hsu, C. Wu and J. Lin Eds. *Infrastructure Systems for Nuclear Energy*. New York: John Wiley & Sons: 153–184.

Maekawa, K., Pimanmas, A. & Okamura, H. 2003. *Nonlinear Mechanics of Reinforced Concrete*. London: CRC Press, Taylor and Francis Group.

Maier, G., Papa, E. & Nappi, A. 1991. On damage and failure of brick masonry. In: *Eurocourse on Experimental and Numerical Methods in Earthquake Engineering, Ispara, Italy, 7–11 October 1991*. Dordrecht, The Netherlands: Kluwer Academic Publishers: 223–245.

Naish, D., Fry, A., Klemencic, R. & Wallace, J. 2013. Reinforced concrete coupling beams-Part I: Testing. *ACI Structural Journal* Vol. 10 No. 6: 1057–1066.

Namikawa, T. & Mihira, S. 2007. Elasto-plastic model for cement-treated sand. *International Journal for Numerical and Analytical Methods in Geomechanics* Vol. 31: John Wiley & Sons, Ltd.: 71–107.

Pandey, B.H. & Meguro, K. 2004. Simulation of brick masonry wall behavior under in-plane lateral loading using applied element method. In: *Proc. 13th World Conference on Earthquake Engineering, Vancouver, Canada 1–6 August 2004*, Paper No. 1664: International Association for Earthquake Engineering.

Shen, Z., Jiang, M. & Wang, S. 2019. Static and kinematic damage characterization in structured sand. Acta Geotechnica Vol. 14: Springer: 1403–1421.

Sun, D. & Matsuoka, H. 1999. An elastoplastic model for frictional and cohesive materials and its application to cemented sands. *Mechanics of Cohesive-Frictional Materials* Vol. 4: 525–543.

Sutton, M.A., Orteu, J.J. & Schreier, H. 2009. *Image Correlation for Shape, Motion and Deformation Measurements: Basic Concepts, Theory and Applications*. New York: Springer US.

Towhata, I. 2008. *Geotechnical earthquake engineering*. Berlin and Heidelberg: Springer-Verlag.

Yamanoi, Y. & Maekawa, K. 2020. Shear bifurcation and gravelization of low-strength concrete. *Journal of Advanced Concrete Technology* Vol. 18: 767–777.

Yamanoi, Y., Miura, T., Soltani, M. & Maekawa, K. 2021. Multi-directional fixed crack model extended to masonry structures. *Journal of Advanced Concrete Technology* Vol. 19: 977–987.

Yu, Y., Pu, J. & Ugai, K. 1998. A damage model for soil-cement mixture. *Soils and Foundations* Vol. 38 No. 3: 1–12.

Computational Modelling of Concrete and Concrete Structures – Meschke, Pichler & Rots (Eds)

3D dissipative mechanisms modelling for masonry-like materials under multiaxial cyclic loads

H. Rostagni, C. Giry & F. Ragueneau
Université Paris-Saclay, ENS Paris-Saclay, CNRS, LMT – Laboratoire de Mécanique et Technologie, Gif-sur-Yvette, France

ABSTRACT: Masonry constructions are a significant part of the existing civil, architectural and cultural heritage. The preservation of their structural integrity requires developing efficient and accurate tools to represent their degradation and assess their safety and vulnerability under complex loading. In this paper, a constitutive model, built within the framework of thermodynamics of irreversible processes, describes the nonlinear mechanical response of masonry by introducing couplings in the free energy and the pseudo-dissipation potential of the material. The damage model decomposes the effects of cracks families that behave independently on the compliance tensor. Plastic flow develops independently along with the orthotropic directions of masonry, which allows friction effects to be decoupled. The unilateral effect related to crack closure during alternating loading is also modelled. A coupling between orthotropic elasticity and damage is introduced for the normal components and an additional coupling between damage and internal friction for the shear components. Using this formalism, the contributions to the overall dissipation of each degradation mechanism can be evaluated. Comparisons between experimental and numerical results performed on masonry shear panels under monotonic and cyclic loading are hereby presented. The model is able to satisfactorily describe the experimental outcomes, reproducing the damage distribution, the hysteresis loops and dissipative processes.

1 INTRODUCTION

Masonry is a composite material that has been widely used in construction and still accounts for about 70% of existing buildings (Wang et al. 2018). Faced with degradation due to human activity and increasingly strong and recurrent environmental hazards (earthquakes, floods), a growing concern for preserving these structures has emerged because of their aesthetic, social, archaeological, cultural, economic and technological values. The risk assessment of these buildings is a challenging task involving developing efficient techniques to represent their degradation and analyse their structural integrity and safety conditions. Recent seismic events such as the *Le Teil* earthquake (France, 2019) have highlighted the need to correctly estimate the vulnerability of unreinforced masonry structures (Taillefer et al. 2021), both locally and globally, under complex loads. By identifying the weakest areas, maintenance and strengthening interventions can be designed to reduce their collapse. To this end, several models have been developed or adapted for masonry in the previous decades (Addessi et al. 2014; D'Altri et al. 2019; Roca et al. 2010; Sacco et al. 2018). However, masonry modelling shows several difficulties related to its composite nature and the specific characteristics of masonry structures. Indeed, its material properties and structural behaviour are difficult to assess.

Masonry is a quasi-brittle material characterised by a heterogeneous nature (bound or unbound blocks), nonlinear and non-symmetric behaviour with the presence of softening branches. Due to the poor resistance of mortar joints to tension stresses, low tensile strength is usually associated with high compressive strength. This led to the 'no-tension material' modelling strategy developed by (Heyman 1966) through the limit analysis method considering simple hypotheses (*i.e.*, no tensile strength, infinite compressive strength and no possible sliding between blocks). This method has been widely used to determine the upper and lower bounds of the collapse load for masonry structures. Although these analytical methods successfully simulate the static response of masonry structures, numerical approaches have been developed to model complex geometries and accurately represent degradation and collapse mechanisms under static and dynamic conditions. These approaches are based on different descriptions – discrete or continuum – and scales of analysis – micro/meso-scale, macro-scale and multiscale. The discrete approach (Cundall 1971; Lemos 2007), which represents masonry as a discrete system of elements (units, joints, interfaces) exhibiting different behaviours (*e.g.* Pina-Henriques & Lourenço

DOI 10.1201/9781003316404-46

2006) and buildings as an assemblage of distinct blocks interacting along the boundaries (*e.g.* Acary & Jean 1998), efficiently describes the cracking and structural failure of masonry structures. Similarly, macro-elements models are used for large buildings where each wall is discretised as an assemblage of piers, spandrels and rigid nodes (*e.g.* Brencich et al. 1998; Lagomarsino et al. 2013). The structural components are connected by special interfaces adopting nonlinear constitutive laws (*e.g.* Marques & Lourenço 2014). The continuum approach modelled masonry as an equivalent continuum media. On the one hand, micro-mechanical methods are based on the distinct modelling of masonry components through different constitutive laws and a detailed description of the interaction between units and joints (*e.g.* Combescure 1996; Lotfi & Shing 1994; Sacco & Toti 2010). They reproduce the masonry micro-structure to provide accurate and reliable results but require high computational efforts, limiting their application to small parts of buildings. On the other hand, macro-mechanical models substitute the masonry material with a fictitious homogeneous continuum medium described by phenomenological constitutive laws (*e.g.* Addessi 2014; Lourenço et al. 1997). These laws rely on damage and/or plasticity model. Macro-models provide high computational efficiency keeping an appropriate numerical accuracy, considering moderate loading that does not lead to complete failure. Macro-models can also be enriched at a lower scale to describe the anisotropic behaviour of masonry structures (*e.g.* computational homogenisation Petracca 2016, Transformation Field Analysis Marfia & Sacco 2015).

Today, macro-mechanical finite element models are the most convenient ones to describe large masonry structures, but some of them present difficulties in representing the nonlinear anisotropic nature of masonry subjected to alternate loading. For this purpose, this paper develops a macro-model based on the Continuum Damage Mechanics by proposing a description of the orthotropic damage through a decomposition of crack families following the natural directions of masonry joints. Unilateral effect and internal sliding coupled with damage are introduced to reproduce the hysteretic behaviour of masonry under cyclic loading.

This paper is organised as follows: first, the theoretical formulation of the proposed damage-friction macro-model within the framework of the thermodynamics of irreversible processes and the adopted regularisation technique are presented. The Gibbs energy of such a model arises the formulation of intrinsic dissipative energy, ensuring the consideration of degradation mechanisms. Then, a study at the local scale on a single finite element is performed to illustrate the response of the model and its related dissipative processes. Finally, comparisons between experimental campaigns and global numerical analyses are carried out using the tools developed at the local scale. Special attention is brought to the description of the dissipation in relation with the nonlinear mechanisms introduced in the model.

2 MACRO-MODEL FOR MASONRY

This section describes the model used to capture the nonlinear behaviour of masonry. It is based on nonlinear macro-scale constitutive equations formulated in the continuous media framework with small perturbation assumption. This model is used in the following to numerically reproduce the structural response of the masonry panels under monotonic and cyclic loading regarding two experimental campaigns. A brief description of the main mechanisms modelled is given, then the intrinsic dissipation arising from the thermodynamic framework of irreversible processes is detailed. Finally, the regularisation technique employed to overcome the mesh dependency of the finite element solution is briefly presented.

2.1 *Damage-friction model*

Masonry is a quasi-fragile material. The constitutive model describes three phases: an orthotropic linear elastic phase, a degradation phase with cracking evolving along three predefined orthogonal planes and a degraded phase considering the unilateral effect as well as the internal shear friction in the cracks. The objective of this continuous modelling is to provide a physical interpretation of the dissipative mechanisms related to crack propagation so that it is possible to reproduce the monotonic and cyclic behaviour of the structure using consistent material parameters, within a 3D framework.

Thermodynamic framework
This model is built within the framework of the thermodynamics of irreversible processes. The nonlinear mechanical response of the masonry is described by introducing couplings in the free energy and dissipation potential. The defined energy takes into account the degradation mechanisms related to damage and friction: a coupling between orthotropic elasticity and damage is introduced for the normal tensile components and coupling between damage and internal friction for the shear components.

Orthotropic elasticity
Due to the architectured nature of the studied masonry, joints correspond to three orthogonal planes whose normal vectors define the orthotropic basis of the material. Its elastic behaviour is therefore characterised by an elastic compliance tensor $\mathbb{S}^0$ which depends on E_i, ν_{ij}, and G_{ij} corresponding respectively to Young's moduli, Poisson's ratios and shear moduli in the masonry natural basis.

Damage model
The damage model is based on the decomposition of the compliance tensor representing the impact of a network of orthogonal cracks. This approach, also used for composite materials (Marcin et al. 2011), considers families of independent cracks that are associated with each of the three orthotropic directions of the masonry

(Kachanov 1993). The damage then affects the compliance tensor by means of an effect tensor $\mathbb{A}^{(i)}$ (Zheng & Betten 1996), which expression can be found in (Tisserand 2020), and a scalar damage variable d_i ranging from 0 for no cracks in direction (i) to infinity for the material completely degraded along this direction. Noting $\mathbb{S}^0$ as the initial compliance tensor, the effective compliance tensor $\mathbb{S}^{eff}$ can be decomposed as

$$\mathbb{S}^{eff} = \mathbb{S}^0 + \Delta\mathbb{S} = \mathbb{S}^0 + \Sigma_i \left(d_i \mathbb{A}^{(i)} : \mathbb{S}^0\right). \tag{1}$$

Damage therefore leads to an increase of the material compliance.

The evolution of cracking is governed by the extension of the material. In the case of shear stress, a coupling has been established since two cracking planes are activated. Thus, a measure of the equivalent strain with respect to (i) direction $(\tilde{\varepsilon}_i)$ is defined as[1]

$$\tilde{\varepsilon}_i = \sqrt{\langle\varepsilon_i\rangle_+^2 + \beta_{ij}\varepsilon_{ij}^2 + \beta_{ik}\varepsilon_{ik}^2} \tag{2}$$

where β_{ij} and β_{ik} are material coefficients modulating the impact of shear strains on the yield strength and damage evolution.

Thus, it is possible to define threshold functions f_i that transcribe the evolution of damage-related softening to an evolution of equivalent strains:

$$f_i = \chi_i \left(\tilde{\varepsilon}_i - k_i\right) - \ln\left[(1 + d_i)\frac{k_i}{\tilde{\varepsilon}_i}\right] \leq 0, \tag{3}$$

where k_i is the equivalent strain threshold that initiates damage and χ_i is a brittleness parameter. Checking the condition $f_i = 0$, the damage is written as

$$d_i = \frac{\tilde{\varepsilon}_i}{k_i}\exp\left[\chi_i\left(\tilde{\varepsilon}_i - k_i\right)\right] - 1. \tag{4}$$

Unilateral effect
Under alternating loading, cracks initially open in tension and then close in compression, restoring largely the stiffness. Thus, to study seismic (cyclic) loading, the unilateral effect related to this crack closure is modelled using a stress partition into a positive and a negative part for the normal components as in (Ladevèze 1983):

$$\text{for } i \in \{1,2,3\}, \quad \sigma_i = \langle\sigma_i\rangle_+ + \langle\sigma_i\rangle_-. \tag{5}$$

This partition is then used directly in the thermodynamic potential expression from which the model constitutive equations are derived.

Friction model
Damage created during the degradation phase induces internal friction modelled by a plastic-type behaviour. Indeed, when cracks appear, the opposing surfaces of these cracks can rub against each other and thus give rise to anelastic frictional stresses and strains.

Non-associated threshold functions are used to introduce the loading-unloading conditions and are constructed by considering frictional resistance and confinement effects. Again, the plastic flow develops independently along with the orthotropic directions of the masonry, thus decoupling the effects of friction. Noting σ_k^π the anelastic frictional stresses, X_k the strain hardening variables and μ_k the frictional coefficients, the threshold function f_k^π is written

$$f_k^\pi = \left|\sigma_k^\pi - X_k\right| + \mu_k\left[\langle\sigma_p\rangle_- + \langle\sigma_q\rangle_-\right] \leq 0 \tag{6}$$

with (p) and (q) corresponding to the normal directions associated with the shear one (k).

Without the application of confinement, kinematic strain hardening induces a slight hysteretic behaviour. This effect increases significantly when confinement is applied due to higher friction inside the cracks.

2.2 *Strain-stress law*

The strain-stress laws can be determined by differentiating the state potential. From the previous description of the dissipative mechanisms, the total normal strains depend on the elastic characteristics and the damage. Thus, taking $(j,l,m) = (1,2,3)$ or any permutation of the set, they can be expressed as

$$\varepsilon_j = \left(1 + d_j\right)\frac{\langle\sigma_j\rangle_+}{E_j} + \frac{\langle\sigma_j\rangle_-}{E_j} - \frac{\nu_{jl}\,\sigma_l}{E_j} - \frac{\nu_{jm}\,\sigma_m}{E_j}. \tag{7}$$

Total shear strains are impacted by the damage and frictional degradation mechanisms. For $k \in [\![4;6]\!]$,

$$\varepsilon_k = \frac{1}{2G_{pq}}\left[\frac{\sigma_k - \sigma_k^\pi}{1 - g_k\left(d_p, d_q\right)}\right], \tag{8}$$

where $(p,q) \in [\![1;3]\!]^2$ are the normal directions associated with the shear direction (k). g_k represents the effect function of the damage variables on shear component and is defined as

$$g_k\left(d_p, d_q\right) = \frac{A_{kk}^{(p)}d_p + A_{kk}^{(q)}d_q}{1 + A_{kk}^{(p)}d_p + A_{kk}^{(q)}d_q} \tag{9}$$

with $A_{kk}^{(p)}$ and $A_{kk}^{(q)}$ the components of the effect tensors $\mathbb{A}^{(p)}$ and $\mathbb{A}^{(q)}$ in the shear direction (k).

2.3 *Dissipation*

The study of energy dissipation ensures that dissipative phenomena are correctly taken into account. Using the Clausius-Duhem inequality, dissipation can be calculated by differentiating the thermodynamic potential

[1] In this paper, the Kelvin notation in the Bechterew basis is considered to decompose stresses and strains by components. In addition, all constitutive equations are written in the natural basis of the masonry.

(Lemaitre et al. 2009). Noting Y_j the thermodynamic forces associated with the damage variable d_j, the intrinsic dissipation is expressed by

$$\mathcal{D} = Y_j \dot{d}_j + \sigma_k^\pi \dot{\varepsilon}_k^\pi - X_k \dot{\alpha}_k \geq 0 \tag{10}$$

and can be divided into two terms: one related to damage $\mathcal{D}_{damage} = Y_j \dot{d}_j$ and the other related to friction $\mathcal{D}_{sliding} = \sigma_k^\pi \dot{\varepsilon}_k^\pi - X_k \dot{\alpha}_k$. Each term has to remain positive to ensure the thermodynamic consistency of the model.

The variables expressions can be determined using state laws or evolution laws related to the dissipation processes. For $j \in [\![1;3]\!]$, damage release rates are calculated as follows

$$Y_j = \frac{1}{2}\frac{\langle \sigma_j \rangle_+^2}{E_j} + \frac{A_{kk}^{(j)}}{4G_{jp}}\left[\left(\sigma_k - \sigma_k^\pi\right)^2 - \frac{\left(\sigma_k^\pi\right)^2}{A_{kk}^{(j)} d_j + A_{kk}^{(p)} d_p}\right] + \frac{A_{ll}^{(j)}}{4G_{jq}}\left[\left(\sigma_l - \sigma_l^\pi\right)^2 - \frac{\left(\sigma_l^\pi\right)^2}{A_{ll}^{(j)} d_j + A_{ll}^{(q)} d_q}\right], \tag{11}$$

where k and l are the shear directions respectively linked to jp and jq considering $(j, p, q) = (1, 2, 3)$.

For $k \in [\![4;6]\!]$, the shear frictional stresses are written as

$$\sigma_k^\pi = 2G_{pq} g_k \left(d_p, d_q\right)\left(\varepsilon_k - \varepsilon_k^\pi\right) \tag{12}$$

and the kinematic back stresses are expressed as

$$X_k = b_k \alpha_k \tag{13}$$

with b_k representing the sliding intensity along (k) direction and α_k the variable driving the energy stored by work hardening during the friction mechanism.

2.4 *Mesh-dependency and regularisation issues*

Different techniques can be adopted to overcome the mesh-dependency of the finite element solution when constitutive laws with strain-softening are developed. In the model presented above, a fracture energy regularisation is implemented to prevent this pathological mesh-sensitivity. This approach relies on the assumption that dissipation within a single finite element has to be independent of its size. It is based on an appropriate adjustment of certain model parameters that control softening (post-peak phase of the stress-strain diagram), taking into account mesh characteristics such as element size (Hillerborg et al. 1976). The constitutive law is then properly modified such that the energy dissipated in one element is equal to the assigned fracture energy $\mathcal{G}_f$. For each element, a characteristic length l_c is defined, and the fracture energy is then written

$$\mathcal{G}_f = \int_0^{+\infty} \sigma \, \mathrm{d}w = l_c \int_0^{+\infty} \sigma \, \mathrm{d}\varepsilon^f = l_c \, g_f, \tag{14}$$

where w and ε^f correspond respectively to the crack opening and the crack opening deformation. The specific fracture energy g_f is then scaled so that $\mathcal{G}_f$ remains constant for any size l_c of the element (i.e. it results $g_f l_c = \mathcal{G}_f$ for each element). This formulation remains local and the algorithmic structure of the finite element code requires only minor adjustments. However, it involves firstly to be able to calculate analytically the expression of g_f by the model, and secondly to be able to express a parameter of the model as a function of the latter. The presented macro-model fulfilled both conditions.

The load-displacement diagram and the energy dissipated by fracture become insensitive to the mesh size but fracture still localises in a single layer of elements. Consequently, the results may suffer by sensitivity to the element shape and orientation (Jirásek & Grassl 2008). Their objectivity with respect to the total dissipated energy is no longer verified when the path or the length of the localisation zone depends on the mesh. Also, formulae to estimate the characteristic length are only approximate and partly empirical. These deficiencies can be alleviated by more sophisticated techniques such as a non-local regularisation method for internal variables (integral Pijaudier-Cabot & Bazant 1987; gradient Peerlings et al. 1996).

3 NUMERICAL APPLICATIONS

Local results are presented for a discriminant shear application. The methods used to calculate the dissipated energies are presented and applied to study the non-linear mechanisms implemented. Then, to validate the proposed damage-friction model and show its effectiveness in describing the nonlinear behaviour on a global scale, responses of masonry shear walls experimentally tested under monotonic and cyclic loading are investigated. The macro-model presented in section 2 is thus used to perform nonlinear static analyses on masonry panels loaded in-plane, investigating their global load-displacement response, damage distribution and energy dissipated. All numerical results presented in this section were obtained using the finite element solver Cast3M (CEA, www-cast3m.cea.fr), in which the model was implemented.

3.1 *Local scale*

To study the relevance and robustness of the model, numerical tests on a linear cubic element (CUB8) were carried out. The mechanical response of this CUB8 element was analysed in tension, compression and shear for monotonic and cyclic load cases. Table 1 gives the material parameters used. The parameters related to shear influence on damage (*i.e.* shear component of fabric tensor $\mathbb{A}^{(i)}$ and coefficient β_{ij}) are taken equal to 0.5 except for $\mathbb{A}_{66}^{(1)}$ and $\mathbb{A}_{66}^{(2)}$, which are assumed equal to one (influence in the (12) shear direction). These values were chosen from literature to represent the orthotropic behaviour of conventional arranged masonry (*e.g.* MADA database Augenti et al. 2012).

Table 1. Local scale: material properties with $i \in \{1, 2, 3\}$ or $ij \in \{23, 13, 12\}$ corresponding to the directions (in that order).

Elastic parameters			
Young's modulus E_i [GPa]	2.0	1.0	2.3
Poisson's ratio ν_{ij} [-]	0.20	0.20	0.20
Shear modulus G_{ij} [GPa]	1.03	1.35	0.95
Damage parameters			
Strain threshold k_i [10^{-4}]	2.0	1.8	1.0
Fracture energy $\mathcal{G}_{fi}$ [J/m^2]	100	80	100
Friction parameters			
Friction coefficient μ_{ij} [-]	0.3	0.3	0.3
Hardening parameter a_{ij} [-]	10^{-4}	10^{-4}	10^{-4}
Sliding intensity b_{ij} [-]	10^8	10^8	10^8

The focus here is on shear loading. Such a loading emphasises the main nonlinear phenomena (damage and internal friction) compared to tension or compression. An imposed displacement in direction (1) is applied on the cube's top face while the bottom face is clamped (see Figure 1). An initial loading to initiate damage in directions (1) and (2) is first applied, followed by an unloading-reloading phase leading to the activation of friction in the cracks. Figure 1 illustrates the stress-strain diagram obtained under shear loading. As described in section 2.1, the linear elastic phase is followed by a softening phase as the material becomes damaged. This then leads to progressive degradation of the shear modulus. Damage is coupled to the friction that develops during the unload-reload phase, as shown by the hysteresis loop. To ensure that these dissipation phenomena have been considered, a study of the dissipation defined by equation (10) has been carried out. In Figure 1, it is possible to estimate graphically the value of the energy dissipated [2] per unit volume ($E_d^{(1)}$) during loading since it corresponds to the area under the stress-strain curve:

$$E_d^{(1)} = \int_{\varepsilon_j} \sigma_j \mathrm{d}\varepsilon_j. \tag{15}$$

It can be compared with $E_d^{(2)}$ calculated from the dissipation $\mathcal{D}$ as

$$E_d^{(2)} = \frac{1}{\mathcal{V}} \int_{\mathcal{V}} \int_t \dot{\mathcal{D}} \mathrm{d}t \mathrm{d}\Omega. \tag{16}$$

By comparing these numerical values with those obtained experimentally on a global scale, the contribution of each degradation phenomenon on energy dissipation during the loading process can be determined.

[2] It corresponds to the total energy after total unloading, which means that the elastic contribution has been restored.

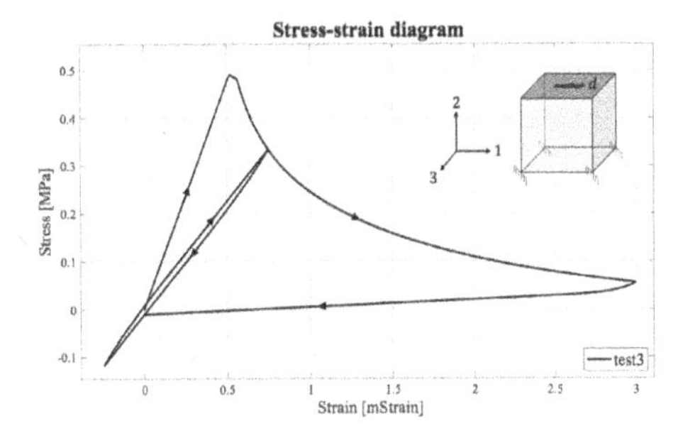

Figure 1. Stress-strain response in 12-shear loading. σ_{12} stress component is represented on the ordinate while $\gamma_{12} = 2\epsilon_{12}$ strain component is displayed on the abscissa

Energy dissipation values obtained for this local test are given in Table 2. Four cases are presented: cases 1 and 2 correspond to a monotonic loading (without unloading-reloading phase), considering respectively the effect of internal friction and not (friction parameters taken equal to zero). Likewise, cases 3 and 4 correspond to the case where the unloading-reloading phase is modelled, considering respectively the effect of internal friction and not. For each case, the element is fully discharged at the end of the test (see Figure 1) to calculate the dissipated energy using equation (15).

Figure 2 shows the numerical dissipation measurements during the loading time for a case without (case 1) and with (case 3) unloading-reloading phase. Arbitrary units are set for time considering the test occurs in 10 units, corresponding to a strain γ_{12} of 3.0e-3.

Table 2. Numerical values of dissipated energy in J/m^3.

Case	1	2	3	4
Method 1 (based on area measurement)				
monotonic	497	445	497	445
hysteretic loop	0	0	16	0
$E_d^{(1)}$	497	445	513	445
Method 2 (based on $\mathcal{D}$ calculation)				
damage	445	445	445	445
sliding	56	0	72	0
$E_d^{(2)}$	501	445	517	445

If no friction is considered (cases 2 and 4), the energy density $E_d^{(1)}$ measured by the area under the stress-strain curve is equal to $E_d^{(2)}$ calculated with the dissipation (445 J/m^3) regardless of the unloading-reloading phase presence. Furthermore, it can be observed that this phase does not influence the damage-related dissipation $\mathcal{D}_{damage}$, which is consistent with the fact that the damage does not evolve during this phase (see Figure 2); the energy is then only dissipated by friction between the crack surfaces. By comparing dissipative energies of cases 1 and 2, it can be noticed that the dissipation due to friction within the cracks occurs also during direct loading (*i.e.* when there is no hysteretic loop). Indeed, even during pure monotonic loading, the energy dissipated $E_d^{(1)}$ (497J/m^3) is greater than 445J/m^3. The difference is related to the frictional dissipation $\mathcal{D}_{sliding}$, which is 52J/m^3 regarding method 1 (it corresponds to 497-445J/m^3) and 56J/m^3 according to method 2. The micro-cracks that develop when the material undergoes damage induce internal friction when loaded in shear. Case 3 shows that the hysteresis loop dissipates 16J/m^3. When no confinement is applied, this value remains low compared to the damage dissipation. $\mathcal{D}_{damage}$ higher than $\mathcal{D}_{sliding}$ is consistent with the fact that the main dissipation mechanism is damage when loading is monotonic (see Figure 2).

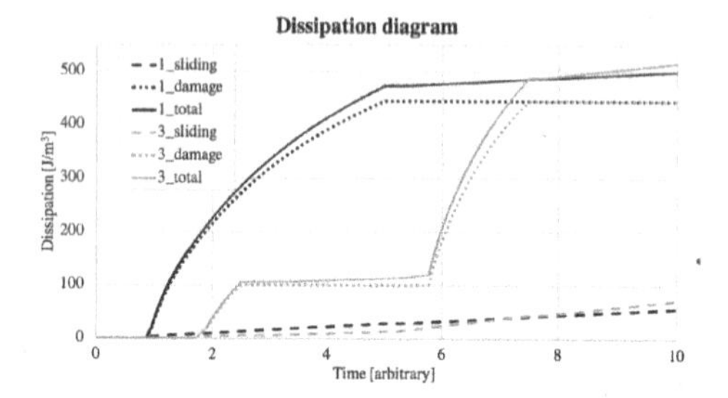

Figure 2. Evolution of dissipation (total, damage, friction) during shear loading (the time step considered is different for the two cases).

This method of calculating dissipated energy is used below at the global scale.

3.2 *Solid wall under monotonic shear loading*

The experimental campaign conducted within the scope of the CUR project (Vermeltfoort et al. 1993) and presented in (Lourenço 1996) is here considered. The geometry of the studied shear wall without opening is shown in Figure 3. The specimen is subjected to shear in a confined way: a uniformly distributed vertical pre-compression load p of 0.3MPa is first applied, followed by a top horizontal displacement d monotonically increased. Bottom and top sides of the wall are kept horizontal and vertical movement is restrained during the second phase (see Figure 3).

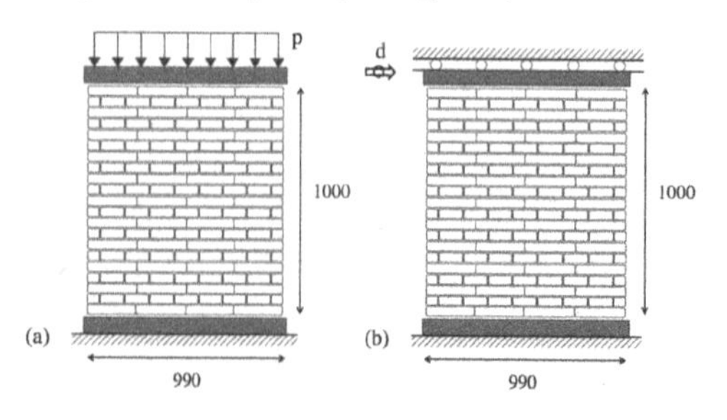

Figure 3. Geometry and load phases for JD shear walls: (a) vertical loading (b) horizontal loading under displacement control (Lourenço & Rots 1997).

Figure 4 compares the numerical (solid line) and experimental (dashed line) in-plane envelope capacity curves representing the total base shear as a function of the top horizontal displacement. The macro-model satisfactorily represents the global behaviour of the masonry shear wall for the elastic and degraded parts.

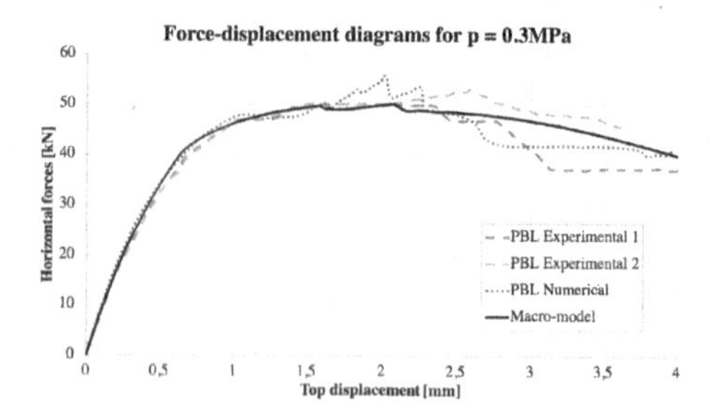

Figure 4. Monotonic response curves of the JD shear wall. *PBL Experimental 1, 2* and *PBL Numerical* correspond respectively to the curves obtained experimentally for walls J4D, J5D and numerically with the micro-model (Lourenço 1996). *Macro-model* (solid line) shows the global response obtained with the presented macro-model.

Figure 5 shows the evolution of the damage profile in direction (2) with increasing imposed displacement. The model rather well captures the shear wall behaviour. As explained in (Lourenço & Rots 1997), its behaviour is characterised by two initial horizontal tension cracks that develop at the bottom and top of the wall. That is indeed the first damage phase observed for a displacement of 2.2mm (Figure 5(2)). A diagonal shear crack immediately follows for a displacement lower than 2.4mm (Figure 5(3)). However, due to the macroscopic nature of the model, its stepped aspect cannot be observed through the head and bed joints. The crack starts in the middle of the wall and, with increasing deformation, progresses towards the supports until it becomes a full diagonal crack. This type of failure is specific to walls loaded in confined shear.

It can be pointed out that the energy regularisation method used is not optimal as damage tends to localise when the mesh is not sufficiently refined. The crack tends to follow the mesh orientation. An analysis of the crack pattern should be carried out using a non-local regularisation method to avoid this directionality problem.

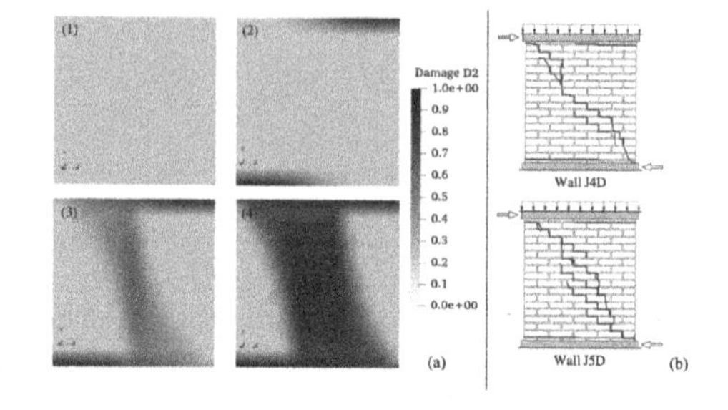

Figure 5. (a) D_2 damage distribution at (1) d = 1.5mm, (2) d = 2.2mm, (3) d = 2.4mm, (4) d = 4mm (b) Experimental crack patterns (Lourenço 1996).

As for the local case (section 3.1), dissipated energies are calculated using the capacity curves (Figure 4) and are given in Table 3. In this way, the contribution of each modelled dissipation mechanism can be estimated. As expected, damage prevails over friction, which accounts for only one fifth of total dissipation.

It can be noticed that calculated dissipated energies are in good agreement with the experiment. Due attention should be given to the slightly higher value of $E_d^{(2)}$ compared to $E_d^{(1)}$ for the numerical model caused by what appears to be an overestimation of slip dissipation. Yet, this satisfactory result allows to estimate the contribution of each mechanism. As expected, damage prevails over friction, which accounts for only one fifth of total dissipation.

Table 3. Numerical values of dissipated energy in J.

Case	Method 1 (area) $E_d^{(1)}$	Method 2 (dissipation $\mathcal{D}$) damage	sliding	$E_d^{(2)}$
PBL Exp. 1	89	–	–	–
PBL Exp. 2	88	–	–	–
PBL Num.	87	–	–	–
Macro-model	93	81	18	99

3.3 *Masonry panels under cyclic loading*

Cyclic tests were carried out and compared with the experimental campaign conducted by (Anthoine et al. 1995) to analyse the model's ability to describe hysteretic response at the structural scale. In this campaign, two solid masonry panels characterised by different height-width ratios are analysed to highlight the effect of geometry on collapsing mechanisms. Specimens geometry, loading and boundary conditions are shown in Figure 6. Experimental boundary conditions have been reproduced: the bottom side of the walls is entirely blocked, while the top side is prevented from rotating. A distributed vertical load p of 0.6MPa is first applied and kept constant during the test, while a cyclic horizontal displacement is applied at the top of the walls (for the applied top displacement evolution with respect to fictitious time variable see Figure 7). Two or three cycles were performed at each amplitude (Anthoine et al. 1995). The main material properties are described in (Nocera et al. 2021).

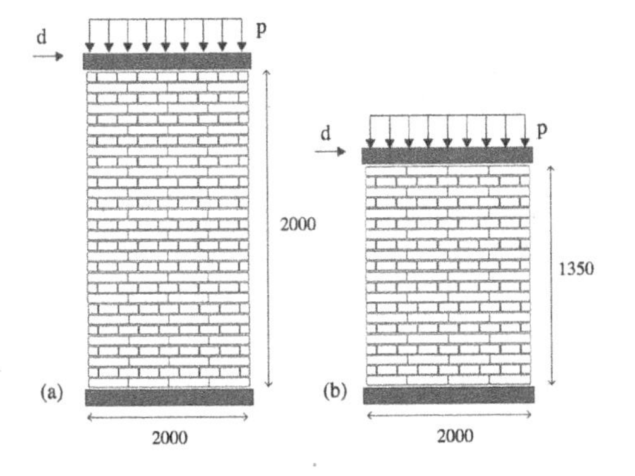

Figure 6. Geometry of the (a) high panel (b) low panel (Gatta et al. 2018).

The cyclic responses of both panels are shown in Figure 7. Under cyclic shear loading, a behaviour change is visible. Indeed, a hysteretic phenomenon appears as the material degrades, indicating friction occurs in the cracks. It can be seen that when the masonry is not yet cracked, the behaviour is elastic and there is no friction. The more cracked the masonry, the more critical the friction, which can be seen from the growth of its related dissipation, estimated by the increased area of the hysteresis loops. As the loading progresses, the shear modulus degrades, causing the loops to tilt horizontally (less rigid behaviour). At the end of the test, the stiffness drops and the last loops are very large for the low panel, indicating the failure of the specimen.

Figure 7 illustrates a different trend for the global response curves of both geometries, which is explained by the different damage mechanisms that arise during the loading. From an envelope point of view, the force-displacement curves obtained numerically are quite satisfactory except for the compression part of the high panel, which does not take a convex shape like the experimental curve but tends to remain straight. At a glance, the numerical model seems to be able to describe the hysteretic dissipation mechanisms, *i.e.* the area under the experimental cyclic curves. This will be developed in more detail in the following.

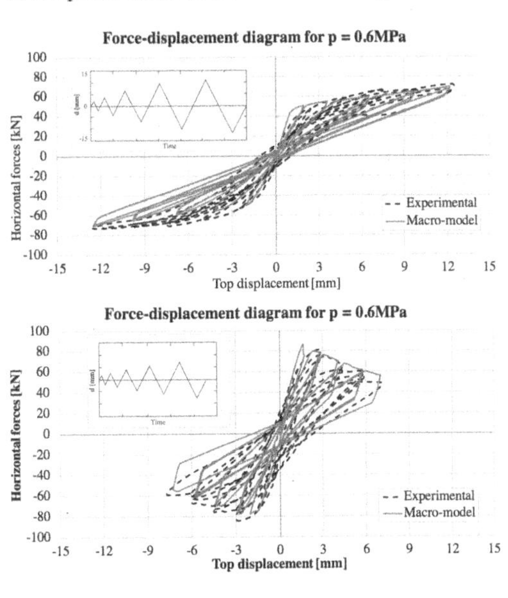

Figure 7. Comparisons between experimental and numerical force-displacement response curve under cyclic loading for the (top) high panel (bottom) low panel.

Dissipations measured on the experimental curve and calculated with the macro-model are presented in Figure 9. In order to compare them, the measurements

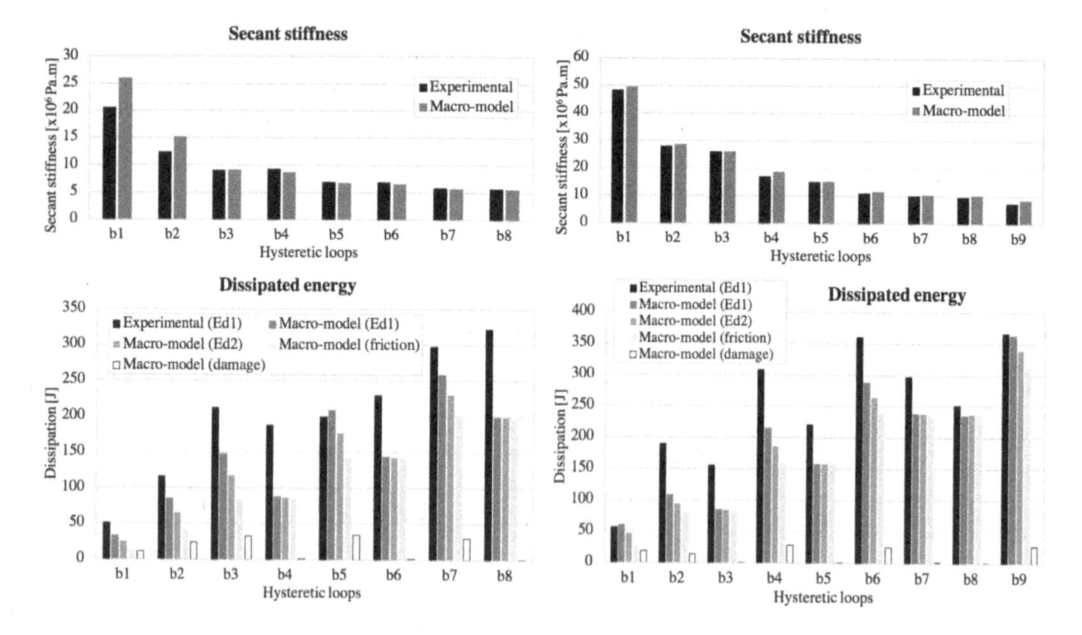

Figure 9. Comparisons of secant stiffness and loop dissipation between experimental and numerical data for the (left) high panel (right) low panel.

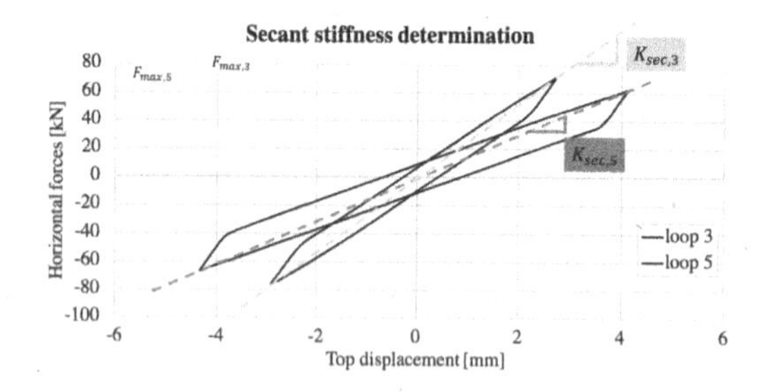

Figure 8. Determination of the secant stiffness of a loop.

are carried out loop by loop. The discrepancy between the experimental and numerical data is significant due to the fact that the loops are not very accurately reproduced as shown in Figure 7. Indeed, the total dissipation obtained for the whole loading is respectively for $E_d^{(1)}$ (experimental), $E_d^{(1)}$ (macro-model) and $E_d^{(2)}$ (macro-model):

- for the low panel: 2205J, 1750J and 1650J;
- for the high panel: 1615J, 1165J and 1040J.

As shown in Figure 9, the general trend in the dissipation evolution is well represented but the numerical values are underestimated by about 25%.

Contrary to the monotonic loading, it can be noticed that in the cyclic case friction is the main degradation mechanism. As the damage does not evolve until the maximum value previously reached is attained, the damage dissipation is low when cycling. Indeed, the histograms show that damage dissipation is almost zero when the imposed strain does not exceed that imposed in the previous cycle. For instance, that is the case for loops b4 and b5 for the low wall. It can be attributed to the fact that damage is not triggered until the strain threshold is exceeded (and its value has not changed between the two loops), resulting in no damage evolution and therefore no dissipation by this mechanism. The friction dissipation is almost identical between the two loops, which leads to a drop in the total dissipation that is found both experimentally and numerically.

Figure 9 also shows the evolution of the secant stiffness K_{sec} of the loops during loading. K_{sec} has been calculated as shown in Figure 8, and is defined as

$$K_{sec,loop} = \frac{F_{\max,loop}}{d_{\max,loop}}. \tag{17}$$

There is a decrease in secant stiffness as the loading rises, which is consistent with the fact that the loops tilt towards the horizontal as the cycles are completed and the shear modulus degrades. The values obtained experimentally are in good agreement with the experimental data, which implies that the model is capable of representing the degradation of the masonry material properties.

4 CONCLUDING REMARKS

A damage-friction model has been proposed with the aim of investigating the nonlinear structural response of unreinforced masonry elements with particular attention to dissipative mechanisms. The nonlinear mechanisms and their interactions have been developed in the framework of thermodynamics of irreversible processes. This macro-model, which considers orthotropic elasticity, compliance growth with damage, unilateral effect and hysteretic mechanisms, has been implemented in the finite element solver Cast3M. For both damage and friction, a direction-driven behaviour has been defined in order to reproduce the masonry orthotropic behaviour and simplify the numerical implementation of the model. A simple fracture energy regularisation technique has been considered to limit the mesh-sensitivity classically observed for softening media. Numerical analyses have been performed on a local scale on a single finite element and a global scale on experimentally tested shear walls under monotonic and cyclic loading.

The local scale study has illustrated the shear response of the model and the associated dissipative mechanisms. The contributions of friction and damage dissipations could be estimated. It has validated that damage is the main degradation mechanism for monotonic loading. However, friction takes over as soon as a cyclic behaviour or a discharge phase is initiated. Tools developed for analysing the dissipated energy were reused at the structural scale.

Structural tests have shown some shortcomings. Indeed, although the numerical analyses showed good agreement in reproducing the force-displacement curves, the damage distributions were not completely satisfactory. The regularisation technique used is not efficient enough as the crack pattern tends to follow the orientation of the mesh, distorting the damage profile. The use of a non-local method that considers as non-local quantities the equivalent strains that determine the damage evolution is an ongoing development.

Regarding the monotonic case, numerical results highlighted the main features of the experimental load-displacement curve and damage distribution. Indeed, maximum loads, energy dissipations and collapse mechanisms are in good agreement with the experimental observations.

For the cyclic case, the general behaviour of the hysteresis mechanism was well represented, notably with the fall in secant stiffness of the loops and the increase in their size linked to rising energy dissipated by friction. Despite a good qualitative distribution of loop dissipation, an underestimation of the total dissipation was observed when compared to experimental results.

In conclusion, the conducted analyses have shown that the proposed model is a suitable and reliable tool to reproduce experimental results and predict the response of masonry walls under monotonic loading. However, some improvements need to be made under cyclic loading, particularly regarding the regularisation method and shear stiffness recovery to represent the crack pattern accurately and not underestimate the dissipated energies.

REFERENCES

Acary, V. & M. Jean (1998). Numerical simulation of monuments by the contact dynamics method. In DGEMN-LNEC-JRC (Ed.), *Monument-98, Workshop on seismic perfomance of monuments*, Lisbon, Portugal, pp. 69–78. Laboratório Nacional de engenharia Civil (LNEC), Lisboa, Portugal.

Addessi, D. (2014). A 2D Cosserat finite element based on a damage-plastic model for brittle materials. *Computers & Structures 135*, 20–31.

Addessi, D., S. Marfia, E. Sacco, & J. Toti (2014). Modeling approaches for masonry structures. *The Open Civil Engineering Journal 8*, 288–300.

Anthoine, A., G. Magonette, & G. Magenes (1995). Shear-compression testing and analysis of brick masonry walls. *Tenth European Conference on Earthquake Engineering 3*, 1657–1662.

Augenti, N., F. Parisi, & E. Acconcia (2012). MADA: online experimental database for mechanical modelling of existing masonry assemblages. In *Proceedings of the 15th World Conference on Earthquake Engineering*.

Brencich, A., L. Gambarotta, & S. Lagomarsino (1998). A macroelement approach to the three-dimensional seismic analysis of masonry buildings. In *11th European Conference on Earthquake Engineering*, pp. 6–11.

Combescure, D. (1996). *Modélisation du comportement sous chargement sismique des structures de bâtiment comportant des murs de remplissage en maçonnerie*. Ph. D. thesis, École Centrale de Paris.

Cundall, P. A. (1971). A computer model for simulating progressive, large-scale movements in blocky rock systems. In *Proceedings of the Symposium on Rock Fracture (ISRM), vol. 1, paper II-8. Nancy, France.*

D'Altri, A., V. Sarhosis, G. Milani, J. Rots, S. Cattari, S. Lagomarsino, E. Sacco, A. Tralli, G. Castellazzi, & S. Miranda (2019). A review of numerical models for masonry structures. In B. Ghiassi and G. Milani (Eds.), *Numerical Modeling of Masonry and Historical Structures*, Woodhead Publishing Series in Civil and Structural Engineering, pp. 3–53. Woodhead Publishing.

Gatta, C., D. Addessi, & F. Vestroni (2018). Static and dynamic nonlinear response of masonry walls. *International Journal of Solids and Structures 155*, 291–303.

Heyman, J. (1966). The stone skeleton. *International Journal of Solids and Structures 2*(2), 249–279.

Hillerborg, A., M. Modéer, & P. E. Petersson (1976). Analysis of crack formation and crack growth in concrete by means of fracture mechanics and finite elements. *Cement and Concrete Research 6*(6), 773–782.

Jirásek, M. & P. Grassl (2008). Evaluation of directional mesh bias in concrete fracture simulations using continuum damage models. *Engineering Fracture Mechanics 75*, 1921–1943.

Kachanov, M. (1993). Elastic solids with many cracks and related problems. *Advances in Applied Mechanics 30*, 259–445.

Ladevèze, P. (1983). *Sur une théorie de l'endommagement anisotrope*. Laboratoire de Mécanique et Technologie.

Lagomarsino, S., A. Penna, A. Galasco, & S. Cattari (2013). TREMURI program: An equivalent frame model for the nonlinear seismic analysis of masonry buildings. *Engineering Structures 56*, 1787–1799.

Lemaitre, J., J. Chaboche, A. Benallal, & R. Desmorat (2009). *Mécanique des matériaux solides - 3ème édition*. Physique. Dunod.

Lemos, J. V. (2007). Discrete element modeling of masonry structures. *International Journal of Architectural Heritage 1*(2), 190–213.

Lotfi, H. R. & P. B. Shing (1994). Interface model applied to fracture of masonry structures. *Journal of Structural Engineering 120*(1), 63–80.

Lourenço, P. B. (1996). *Computational Strategy for Masonry Structures*. Ph. D. thesis, University of Porto.

Lourenço, P. B., R. De Borst, & J. G. Rots (1997). A plane stress softening plasticity model for orthotropic materials. *International Journal for Numerical Methods in Engineering 40*(21), 4033–4057.

Lourenço, P. B. & J. G. Rots (1997). Multisurface interface model for analysis of masonry structures. *Journal of Engineering Mechanics 123*(7), 660–668.

Marcin, L., J.-F. Maire, N. Carrère, & E. Martin (2011). Development of a macroscopic damage model for woven ceramic matrix composites. *International Journal of Damage Mechanics 20*(6), 939–957.

Marfia, S. & E. Sacco (2015). TFA-based homogenization for composites subjected to coupled damage-friction effects. *Procedia Engineering 109*.

Marques, R. & P. B. Lourenço (2014). Unreinforced and confined masonry buildings in seismic regions: Validation of macro-element models and cost analysis. *Engineering Structures 64*, 52–67.

Nocera, M., L. C. Silva, D. Addessi, & P. B. Lourenço (2021). Correlation studies for the in-plane analysis of masonry walls based on macroscopic fe models with damage. In *12th International Conference on Structural Analysis of Historical Constructions SAHC 2020*, pp. 1893–1904. International Centre for Numerical Methods in Engineering (CIMNE).

Peerlings, R., R. Borst, de, W. Brekelmans, & J. Vree, de (1996). Gradient enhanced damage for quasi-brittle materials. *International Journal for Numerical Methods in Engineering 39*(19), 3391–3403.

Petracca, M. (2016). *Computational multiscale analysis of masonry structures*. Ph. D. thesis, University G. d'Annunzio of ChietiPescara (UNICH) - Universitat Politècnica de Catalunya (UPC-BarcelonaTech).

Pijaudier-Cabot, G. & Z. Bazant (1987). Nonlocal damage theory. *Journal of Engineering Mechanics-asce - J ENG MECH-ASCE 113*, 1512–1533.

Pina-Henriques, J. & P. Lourenço (2006). Masonry compression: A numerical investigation at the meso-level. *Engineering Computations 23*, 382–407.

Roca, P., M. Cervera, G. Gariup, & L. Pelà (2010). Structural analysis of masonry historical constructions. classical and advanced approaches. *Archives of Computational Methods in Engineering 17*, 299–325.

Sacco, E., D. Addessi, & K. Sab (2018). New trends in mechanics of masonry. *Meccanica 53*(7), 1565–1569.

Sacco, E. & J. Toti (2010). Interface elements for the analysis of masonry structures. *International Journal for Computational Methods in Engineering Science and Mechanics 11*, 354–373.

Taillefer, N., P. Arroucau, F. Leone, S. Defossez, J. Clément, M.-S. Déroche, C. Giry, M. Lancieri, & P. Quistin (2021). Association française du génie parasismique: rapport de la mission du séisme du teil du 11 novembre 2019 (ardèche).

Tisserand, P.-J. (2020). *Modélisation et calcul de ponts en maçonnerie assisée soumis aux aléas naturels*. Ph. D. thesis, Université Paris-Saclay.

Vermeltfoort, A., T. Raijmakers, & H. Janssen (1993). Shear tests on masonry walls. In A. Hamid and H. Harris (Eds.), *6th North American Masonry Conference, 6-9 June 1993, Philadelphia, Pennsylvania, USA*, pp. 1183–1193. Technomic Publ. Co.

Wang, C., V. Sarhosis, & N. Nikitas (2018). Strengthening/retrofitting techniques on unreinforced masonry structure/element subjected to seismic loads: A literature review. *The Open Construction and Building Technology Journal 12*(1), 251–268.

Zheng, Q. S. & J. Betten (1996). On damage effective stress and equivalence hypothesis. *International Journal of Damage Mechanics 5*(3), 219–240.

Computational Modelling of Concrete and Concrete Structures – Meschke, Pichler & Rots (Eds)
 ISBN: 978-1-032-32724-2

Numerical modeling of compression tests on masonry cores

F. Ferretti & C. Mazzotti
DICAM Department, University of Bologna, Bologna, Italy

ABSTRACT: To assess the safety of existing masonry buildings, one fundamental step is the determination of the mechanical properties of the materials. To evaluate the masonry compressive strength, compression tests on masonry cores can be performed. This is not a standard testing technique: dimensions and bond pattern of the samples may vary together with the mortar cap properties. The objective of the present paper was the study of the behavior of masonry cores subject to uniaxial compression test by performing nonlinear numerical analyses. For this purpose, a 3D numerical model with a brick-to-brick micro-modeling approach was adopted. The results of the nonlinear analyses were compared with the results of a laboratory experimental campaign in which several compression tests on masonry cores were performed, obtaining a good agreement. The numerical results also allowed to better interpret the tests and to investigate the role of confinement, as will be discussed in the paper.

1 INTRODUCTION

In the framework of the vulnerability assessment of existing masonry buildings, one important phase is devoted to the determination of the masonry mechanical properties. To this aim, experimental tests may be performed directly in situ or in laboratory on materials extracted in situ. Generally, the testing techniques can be classified as non-destructive, slightly-destructive or destructive according to their invasiveness on the existing construction (Binda et al. 2000; Ferretti et al. 2019; Jafari et al. 2022).

To determine the masonry compressive strength, one possibility, among the different slightly-destructive testing techniques (ASTM C1197–14, Binda & Tiraboschi 1999), is to extract masonry cores from the load-bearing walls of the building under investigation. After the extraction, high-strength mortar caps are casted at the top and at the bottom of the core samples prior to perform uniaxial compression tests in order to regularize the surfaces, creating horizontal loading planes, and to simulate the confinement of the surrounding masonry on the cores (Brencich & Sterpi 2006; Ispir et al. 2009; Sassoni & Mazzotti 2013).

The compression test on masonry cores is not a standard testing technique: dimensions and bond pattern of the samples, i.e. including horizontal mortar joints only or vertical mortar joints as well, may vary together with the properties of the high-strength mortar cap. In recent years, several research focused on the study of this testing technique by comparing the masonry compressive strength obtained from tests on masonry cores and from standard compression tests on masonry wallets and by establishing correlations between the results obtained with the two tests (Jafari et al. 2019; Pelà et al. 2016; Sassoni et al. 2014). In the cited works, different geometrical and mechanical properties of the masonry cores and of the high-strength mortar cap were considered to cover a variety of possibility, trying to identify the optimal testing methods. In some cases, numerical analyses were also performed to better investigate the reliability of this testing technique in estimating the masonry compressive strength, obtaining interesting results (Pelà et al. 2019). However, the research on this topic is still open and further studies are recommended for the definition of a standard testing method.

The objective of the present paper was the study of the behavior of masonry cores subject to uniaxial compression test by performing nonlinear numerical simulations. For this purpose, a brick-to-brick micro-modeling approach was adopted, and a 3D numerical model was created: the bricks, the mortar joint and the high-strength mortar cap were singularly modelled, and interface elements were included to simulate the interaction between the different materials. A nonlinear behavior was assigned to the materials with the objective of properly describing the crack development and the failure mode of the masonry cores.

In the following, the numerical results, including some parametric analyses, will be presented and compared with the results of a laboratory experimental campaign in which several compression tests on masonry cores were performed.

2 COMPRESSION TEST ON MASONRY CORES

The experimental tests here considered for the comparison with the results of the numerical simulations were

DOI 10.1201/9781003316404-47

conducted on masonry cores extracted from a double-wythe masonry panel, built using fired clay bricks and natural hydraulic lime-based mortar. The cores, having a diameter equal to 100 mm, were characterized by the presence of a horizontal joint only, characterized by a thickness of 10 mm (Figure 1). After the extraction procedure, the cores were cut to obtain single-wythe samples, with a length of 125 mm. To apply a compressive load and provide an adequate confinement to the samples, the masonry cores were capped with a high-strength cement-based mortar. The geometry of the cap, with a width equal to 80 mm and a height of 30 mm, was chosen according to previous studies (Sassoni & Mazzotti 2013; Sassoni et al. 2014).

The compression tests were performed under displacement control, using a servo-hydraulic actuator having a maximum capacity of 100 kN. The adopted displacement rate was equal to 0.02 mm/s. Vertical displacements were monitored during the tests using Linear Variable Differential Transducers (LVDTs), positioned on both sides of the masonry core (Figure 1) with a gage length of approximately 50 mm.

Figure 1. Compression test setup.

Standard laboratory tests, such as three-point bending tests (EN 1015-11, EN 12390-5), and monotonic and cyclic uniaxial compression tests (EN 772-1, EN 1015-11, EN 12390-13) were performed to determine the mechanical properties of the constituent materials, i.e., clay brick, lime-based mortar and high-strength cementitious mortar. The obtained material properties, in terms of compressive strength f_c, flexural strength f_{fl} and elastic modulus E are reported in Table 1. It is worth mentioning that mortar samples were extracted as well from the horizontal joints of the masonry panel, and they were subject to double punch test (Henzel & Karl 1987). The obtained compressive strength f_{dp} was equal to 6.8 MPa, much higher than the value obtained from the standard compression tests. As evidenced by other authors (Pelà et al. 2016), this can be due to the curing conditions of the mortar in the bed joints of the wall panels, significantly different with respect to the curing conditions of the standard prismatic specimens used for the mechanical characterization.

Table 1. Mechanical properties of the materials.

Material	f_c (MPa)	f_{fl} (MPa)	E (MPa)
Brick	18.7	4.6	6846
Mortar	1.4	0.4	2549
Cap	22	5.7	20000*

*given by the producer.

Results in terms of stress at first cracking f_{cr}, masonry compressive strength f_M and masonry elastic modulus E_M are reported in Table 2. Stresses were here obtained considering the entire area of the mortar joint. The elastic modulus was evaluated as the secant modulus between 1/10 and 1/3 of the applied load. It should be mentioned that, in the same experimental program, uniaxial compression tests on masonry wallets were also performed (EN 1052-1), obtaining an average masonry compressive strength equal to 6.4 MPa. It is interesting to observe that the results obtained by testing cores overestimated the masonry compressive strength by 67%, coherently with what observed in previous research on similar masonry typologies (Sassoni & Mazzotti 2013; Sassoni et al. 2014).

Table 2. Results of compression tests on masonry cores.

Sample Code	f_{cr} (MPa)	f_M (MPa)	E_M (MPa)
C1	6.4	11.1	2746
C2	–	11.2	3774
C3	6.1	10.7	3591
C4	8.5	11.3	2326
C5	4.4	9.1	1932
C6	7.7	10.7	2858

The failure mode of the cores was characterized by the presence of several vertical cracks, both in the center of the sample and in correspondence with the extremities of the high-strength mortar cap. Results in terms of stress *vs* strain curves will be shown along with the results of the numerical simulations for sake of comparison.

3 NUMERICAL SIMULATIONS

3.1 *Numerical model*

The modeling strategies which can be adopted to model masonry structural elements differ between one another depending on the way in which the constituents and their interactions are considered (Lourenço et al. 1995, Rots 1997). In general, it is possible to distinguish between micro-modeling and macro-modeling approaches. The former is also denoted as brick-to-brick modeling approach or block-based modeling (D'Altri et al. 2019), in which units and mortar are separately modeled and interface elements are adopted to simulate interactions between them. The latter imply to model masonry as a continuum, by adopting specific homogenization techniques. To investigate the behavior of a masonry core subject to compression, given the small geometry of the sample, a micro-modeling strategy was adopted in this research. The numerical simulations were performed with the software DIANA FEA (v.10.5).

In this framework, a three-dimensional model was considered, in which the bricks, the mortar joint and the high-strength mortar cap were modeled using solid brick elements (cubic, pyramid or tetrahedron), based on quadratic interpolation. To account for the nonlinear behavior of the materials, the rotating total strain crack model (Rots & Blaauwendraad 1989, Selby & Vecchio 1993) was adopted, which follows a smeared approach for the fracture energy. Concerning the tensile behavior of the materials, a linear elastic law was considered until the tensile strength f_t was reached; then, an exponential softening curve was adopted, based on fracture energy G_f^I and related to a crack bandwidth h, as follows:

$$f = f_t e^{-\frac{f_t h}{G_f^I}\alpha_j}, \tag{1}$$

where α_j is a generic strain, bounded between the strain value at peak and the ultimate strain, the latter corresponding to zero residual stress.

In compression, a parabolic behavior was assumed, according to the function (Diana Manual):

$$f = \begin{cases} -f_c \frac{1}{3} \frac{\alpha_j}{\alpha_{c/3}} & \text{if } \alpha_{c/3} < \alpha_j \leq 0 \\ -f_c \frac{1}{3}\left(1 + 4\left(\frac{\alpha_j - \alpha_{c/3}}{\alpha_c - \alpha_{c/3}}\right) - \left(\frac{\alpha_j - \alpha_{c/3}}{\alpha_c - \alpha_{c/3}}\right)^2\right) & \text{if } \alpha_c < \alpha_j \leq \alpha_{c/3} \\ -f_c\left(1 - \left(\frac{\alpha_j - \alpha_c}{\alpha_u - \alpha_c}\right)^2\right) & \text{if } \alpha_u < \alpha_j \leq \alpha_c \\ 0 & \text{if } \alpha_j \leq \alpha_u \end{cases}, \tag{2}$$

where f_c is the compressive strength of the material under consideration, α_j is a generic strain, $\alpha_{c/3}$ is the strain at which one-third of the compressive strength is reached, according to an elastic behavior, α_c is the strain at peak, and α_u is the ultimate strain, defined as:

$$\alpha_c = -\frac{5}{3}\frac{f_c}{E} \tag{3}$$

$$\alpha_u = \min\left(\alpha_c - \frac{3}{2}\frac{G_f^c}{h \cdot f_c}; 2.5\alpha_c\right), \tag{4}$$

where G_f^c is the compressive fracture energy. To properly account for the higher deformability and nonlinear behavior of mortar, the strain at peak (α_c) for the mortar joint was increased 5 times.

Table 3. Input parameters for the numerical model.

	Parameter	Unit	Value
Brick	Elastic modulus (E_b)	MPa	6846
	Poisson's ratio (ν_b)	–	0.10
	Tensile strength ($f_{t,b}$)	MPa	2.95
	Tensile fracture energy ($G^I_{f,b}$)	N/mm	0.036
	Compressive strength ($f_{c,b}$)	MPa	18.7
	Compr. fracture energy ($G^c_{f,b}$)	N/mm	3.62
Mortar	Elastic modulus (E_m)	MPa	2549
	Poisson's ratio (ν_m)	–	0.20
	Tensile strength ($f_{t,m}$)	MPa	0.26
	Tensile fracture energy ($G^I_{f,m}$)	N/mm	8.0e-04
	Compressive strength ($f_{c,m}$)	MPa	4.1
	Compr. fracture energy ($G^c_{f,m}$)	N/mm	0.4
Mortar cap	Elastic modulus (E_{cap})	MPa	20000
	Poisson's ratio (ν_{cap})	–	0.17
	Tensile strength ($f_{t,cap}$)	MPa	3.65
	Tensile fracture energy ($G^I_{f,cap}$)	N/mm	6.1e-03
	Compressive strength ($f_{c,cap}$)	MPa	22
	Compr. fracture energy ($G^c_{f,cap}$)	N/mm	0.613

The effect of lateral confinement, crucial for the correct description of the problem, is directly taken into account in the adopted model (Selby & Vecchio 1993) by modifying the compressive stress-strain relation to incorporate the effects of an increase in the isotropic stress.

Plane quadrilateral 8+8 nodes interface elements were adopted to model the interaction between the bricks and the horizontal mortar joint, and between the bricks and the high-strength mortar cap. Initially, the behavior of these interfaces was assumed elastic and rigid, therefore, the interface elastic stiffness parameters were set to a very high value (10^6 N/mm^3). The choice of assigning a linear behavior to these interface elements was supported by the fact that, experimentally, cracks did not appear in these locations.

The input parameters adopted in the numerical model are reported in Table 3. Some of them were directly obtained from the standard laboratory tests described in Section 2 (Table 1). To be specific, the tensile strength was here determined making use of an empirical formula provided by Eurocode 2, which correlates tensile (f_t) and flexural (f_{fl}) strength as follows:

$$f_t = \frac{f_{fl}}{1.6 - d/1000}, \tag{5}$$

where d is the depth of the tested samples in the standard tests (see Section 2), equal to 40 mm. Regarding the mortar compressive strength, a value of 4.1 MPa was adopted in the numerical simulations, evaluated as the average between the values of compressive strength obtained from the standard laboratory test and the double punch test. Other input parameters, e.g., fracture energy, were determined considering analytical formulations available in literature (Cervera et al. 2010, Diana Manual).

A mesh sensitivity analysis was performed to determine the dimensions of the finite elements, appropriate in terms of a balance between efficiency and accuracy. The adopted mesh is shown in Figure 2. The core sample was considered clamped at the base, while on the top cross section of the high-strength mortar cap a vertical displacement was uniformly applied, with displacement increments equal to 0.02 mm. The nonlinear numerical simulations were performed with the software DIANA FEA (v.10.5). A modified Newton-Raphson method was adopted to control the nonlinear problem solution.

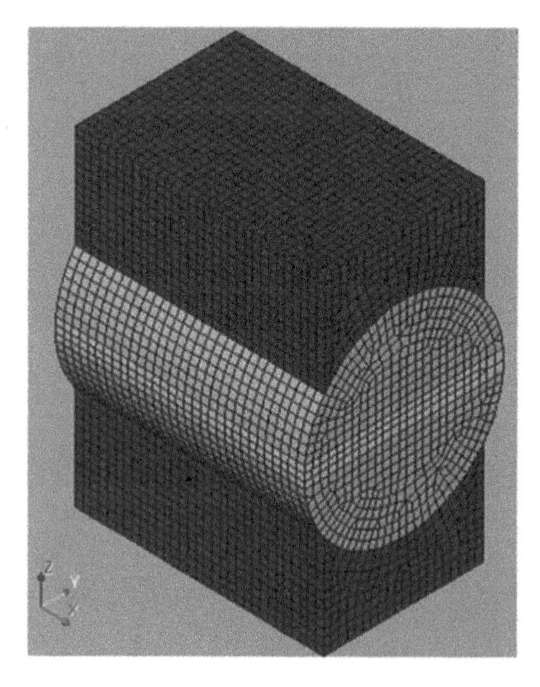

Figure 2. Mesh of the 3D model.

3.2 *Numerical results*

The numerical results are presented in terms of stress *vs* vertical strain diagram in Figure 3, in which the experimental curves are also reported for comparison. As anticipated, stresses were here calculated as the ratio between the applied load and the area of the horizontal mortar joint. To calculate the vertical strain for the numerical simulation, the displacements of two points, positioned in correspondence with the gage points of the LVDTs with a distance of 50 mm between each other, were considered. A good agreement between numerical and experimental results can be noticed, especially in terms of initial stiffness and peak strength. In the post-peak phase, the numerical curve lies below the experimental ones, but it should be taken into account that the behavior of the samples was quite scattered after the development of the cracking process. It is also worth mentioning that, in some cases, deformations were not registered until the end of the experimental tests due to detachment of the LVDTs.

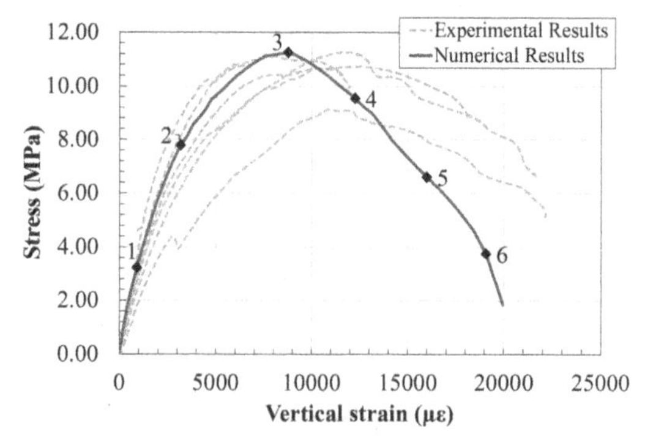

Figure 3. Stress *vs* vertical strain diagram: comparison between numerical and experimental results.

The evolution of the horizontal strain on the external surface of the masonry core is presented in Figure 4. In more detail, specific steps of the numerical simulation are here considered, indicated with numbers along the stress *vs* vertical strain curve of Figure 3. The development of the cracking process can be recognized, with positive horizontal strain progressively involving wider portions of the masonry core and propagating towards the extremities of the high-strength mortar cap. By comparing the horizontal strain contour at the end of the test and the experimental failure mode (Figure 5), a good correspondence was found.

By looking at the horizontal stress evolution reported in Figure 6 at different steps of the analysis, it is possible to notice that, as expected, the mortar joint was confined by bricks, which were subject to tension in the central part of the masonry core. Indeed, the cracking process started within the bricks and it involved later the mortar joint as well. It can also be observed that the lateral compressive stress on the mortar exceeded, in some points, the mortar compressive strength. The effect of confinement, determining a strength increase of the material, was indeed taken into account in the numerical model.

With reference to the interaction between the high-strength mortar cap and the bricks, it can be noticed from Figure 4 and Figure 6 that bricks were laterally restrained by the presence of the cap, i.e., lower horizontal strains were registered in the upper and lower portion of bricks with respect to the portions in the center of the masonry core. Correspondingly, high values of horizontal compressive stress were observed in the bricks in correspondence with the extremities of the cap, i.e. on the left- and right-hand sides of the masonry core, while tensile stresses (positive) were registered in the middle of the cap.

The confinement effect determined by the bricks on the mortar can be considered a local confinement, due to the intrinsic properties of the constituents. The

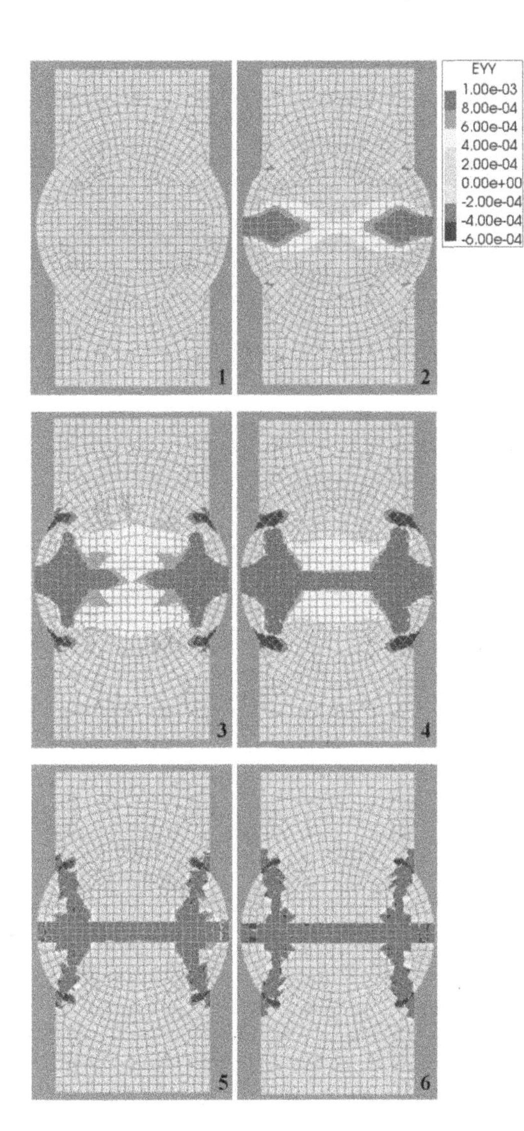

Figure 4. Horizontal strain evolution on the external surface of the masonry core (the reported numbers correspond to the points highlighted in Figure 3).

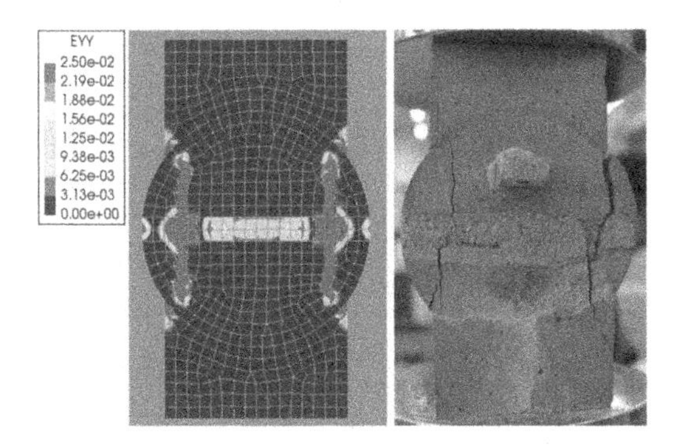

Figure 5. Failure mode: comparison between numerical and experimental results.

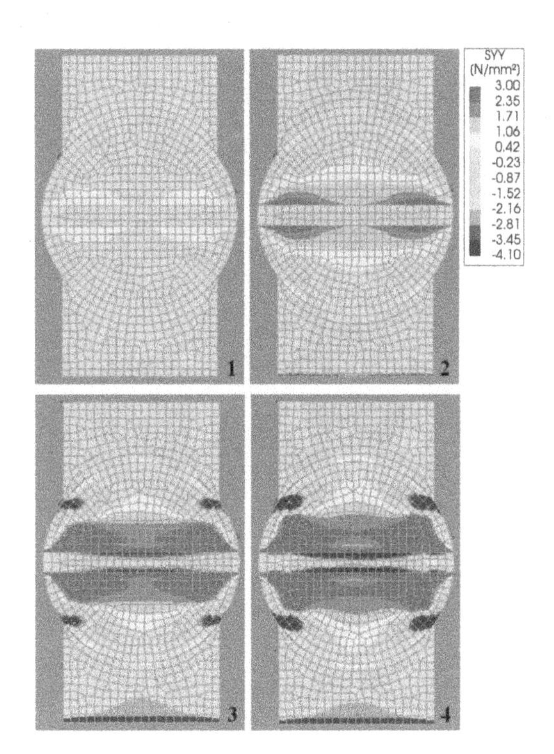

Figure 6. Horizontal stress evolution on the external surface of the masonry core (the reported numbers correspond to the points highlighted in Figure 3).

confinement effect determined by the high-strength mortar cap, instead, can be considered a global phenomenon, since it affects the entire masonry core. Of course, both the effects influenced the results of the compressive test, as discussed.

A focus about the distribution of the vertical stresses is reported in Figure 6. In particular, the vertical stress contour plot in correspondence with an applied displacement equal to 0.1 mm is shown. It is representative of the stress distribution in the first phase of the analysis, i.e. before cracking. The vertical stress distribution on the mortar joint only is also presented. The diffusion of the stresses from the high-strength mortar cap to the masonry core is clearly visible and it can be highlighted that the entire mortar joint was subject to compression, with lower values at the extremities only. Therefore, it is considered reasonable to calculate the average stress on the masonry core by considering the entire area of the mortar joint.

4 PARAMETRIC ANALYSES

In this work, parametric analyses were also performed with the objective of investigating the confinement effect determined, on the one hand, by the high-strength mortar cap on the masonry core and, on the other hand, by the bricks on the mortar joint. To this

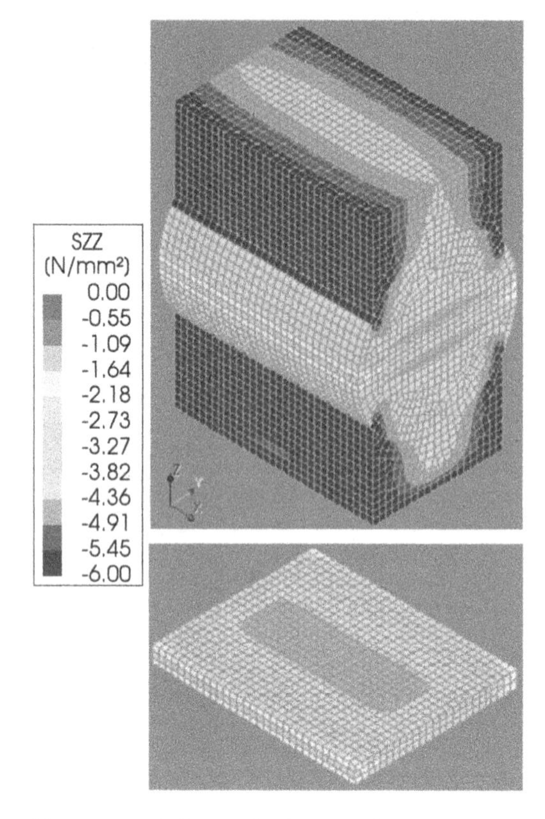

Figure 7. Vertical stress distribution on the entire masonry core and on the mortar joint at a vertical displacement equal to 0.1 mm.

aim, variations in terms of shear stiffness of the interface elements between bricks and mortar and in terms of elastic modulus of the high-strength mortar cap were considered. In more detail, two additional numerical analyses were performed, in which the following modifications with respect to the initial numerical model (reference) were considered:

- PA-01: in the first parametric analysis, the shear stiffness parameters of the brick-to-mortar interfaces were set equal to 40 N/mm^3, while the normal stiffness parameter was maintained equal to 10^6 N/mm^3 as in the previous simulation.
- PA-02: in the second parametric analysis, the elastic modulus of the high-strength mortar cap was reduced to 8250 MPa, a value which is approximately 2.5 times lower than the initial one.

The results of the parametric analyses are here reported in terms of stress *vs* vertical strain diagram (Figure 8). For the first parametric analysis (PA-01), a slight reduction in terms of peak compressive strength was registered with respect to the reference model previously presented. This is coherent with the fact that the horizontal mortar joint was less confined, due to the reduced shear stiffness of the interface elements and, consequently, to the presence of relative displacements between the two materials. In the second parametric analyses (PA-02), a stiffness reduction in the first phase of the analysis was registered, as expected, together with an increase in the vertical strain at peak. Almost no differences were instead observed in terms of peak compressive strength. This can be explained by the fact that, within the variation here considered, the confinement due to the presence of the high-strength mortar cap could depend on the shape of the cap itself rather than on its elastic modulus.

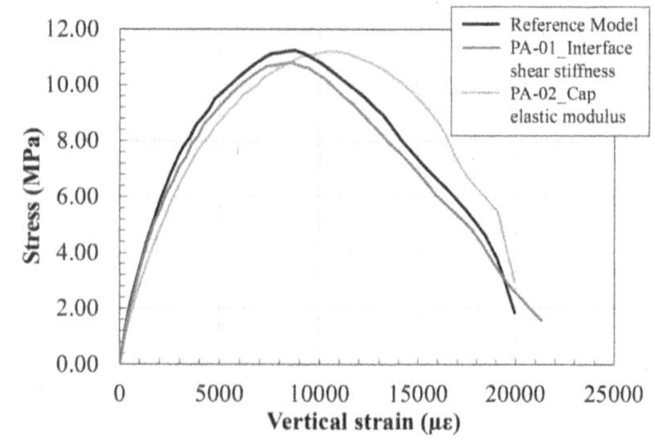

Figure 8. Parametric analyses: stress *vs* vertical strain diagrams.

5 CONCLUSIONS

The objective of the present paper was the study of the behavior of masonry cores subject to uniaxial compression test by performing nonlinear numerical simulations on a 3D numerical model, in which a brick-to-brick micro-modeling approach was adopted. A nonlinear behavior was assigned to the materials with the objective of properly describing the crack development and the failure mode of the masonry cores. The mechanical properties of the materials were obtained from experimental tests or calibrated through available formulations. The results of the nonlinear analyses were then compared with the results of a laboratory experimental campaign in which several compression tests on masonry cores were performed. The comparison was carried out in terms of compressive strength, post-peak behavior and specimen deformability, obtaining a good agreement.

The results of the numerical simulations allowed to better interpret the results of compression tests on masonry cores, such as the interaction between the different materials. Parametric analyses were also performed to investigate the role of confinement, determined both by the bricks on the mortar joint and by the high-strength mortar cap on the masonry core. The effect of confinement was, indeed, quite significant and allowed to explain the differences observed in the experimental campaign between the compressive strength obtained from compression tests on masonry cores and from standard compression tests on masonry wallets.

ACKNOWLEDGEMENTS

This paper was supported by the PRIN 2017 re-search program of the Italian Ministry of Education, University and Research, project DETECT-AGING, grant N. 201747Y73L. The master student Bruno Tavarnesi is gratefully acknowledged for his work on the numerical simulations.

REFERENCES

ASTM C1197. 2014. Standard test method for in situ measurement of masonry deformability properties using the Flatjack method. *American Society for Testing Materials.*

Binda, L., & Tiraboschi, C. 1999. Flat-jack test: a slightly destructive technique for the diagnosis of brick and stone masonry structures. *International Journal for Restoration of Buildings and Monuments* 5(5):449–472.

Binda, L., Saisi, A. & Tiraboschi, C. 2000. Investigation procedures for the diagnosis of historic masonries. *Construction and Building Materials* 14:199–233.

Brencich, A., Sterpi, E. 2006. Compressive strength of solid clay brick masonry: Calibration of experimental tests and theoretical issues. *In Proc. 5th Structural Analysis of Historical Constructions*, 6–8 November 2006, New Delhi, India.

Cervera, M., Pelà, L., Clemente, R. & Roca, P. 2010. A crack-tracking technique for localized damage in quasi-brittle materials. *Engineering Fracture Mechanics* 77(13):2431–2450.

D'Altri, A.M., Sarhosis, V., Milani, G., Rots, J.G., Cattari, S., Lagomarsino, S., et al. 2019. A review of numerical models for masonry structures. *Numerical Modeling of Masonry and Historical Structures*, 3–53.

EN 772-1. 2011. Methods of test for masonry units – Part 1: Determination of compressive strength. *European Committee for Standardization.*

EN 1015-11. 2006. Methods of test for mortar for masonry – Part 11: Determination of flexural and compressive strength of hardened mortar. *European Committee for Standardization.*

EN 1052-1. 1998. Methods of test for masonry – Part 1: Determination of compressive strength. *European Committee for Standardization.*

EN 12390-5. 2009. Testing hardened concrete – Part 5: Flexural strength of test specimens. *European Committee for Standardization.*

EN 12390-13. 2013. Testing hardened concrete – Determination of secant modulus of elasticity in compression. *European Committee for Standardization.*

Ferretti, F., Ferracuti, B., Mazzotti, C. & Savoia M. 2019. Destructive and minor destructive tests on masonry buildings: experimental results and comparison between shear failure criteria. *Construction and Building Materials* 199:12–29.

Henzel, J. & Karl, S. 1987. Determination of strength of mortar in the joints of masonry by compression tests on small specimens. *Darmstadt Concrete* 2:123–136.

Ispir, M., Demir, C., Ilki, A. & Kumbasar, N. 2009. Material characterization of the historical unreinforced masonry Akaretler row houses in Istanbul. *Journal of Materials in Civil Engineering* 22(7):702–713.

Jafari, S., Rots, J.G. & Esposito, R. 2019. Core testing method to assess nonlinear behavior of brick masonry under compression: A comparative experimental study. *Construction and Building Materials* 218:193–205.

Jafari, S., Rots, J.G. & Esposito, R. 2019. A correlation study to support material characterisation of typical Dutch masonry structures. *Journal of Building Engineering* 45:103450.

Lourenco, P.B., Rots, J.G. & Blaauwendraad, J. 1995. Two approaches for the analysis of masonry structures: micro and macro-modeling. *Heron* 40(4): 313–340.

Pelà, L., Canella, E., Aprile, A. & Roca, P. 2016. Compression test of masonry core samples extracted from existing brickwork. *Construction and Building Materials* 119:230–240.

Pelà, L., Saloustros, S. & Roca, P. 2019. Cylindrical samples of brick masonry with aerial lime mortar under compression: Experimental and numerical study. *Construction and Building Materials* 227:116782.

Rots, J.G. & Blaauwendraad, J. 1989. Crack models for concrete: discrete or smeared? Fixed, multi-directional or rotating? *Heron,* 34(1).

Rots, J.G. 1997. *Structural Masonry – An experimental/numerical basis for practical design rules*. Rotterdam: Balkema.

Sassoni, E., Mazzotti, C. & Pagliai, G. 2014. Comparison between experimental methods for evaluating the compressive strength of existing masonry buildings. *Construction and Building Materials* 68: 206–219.

Sassoni, E. & Mazzotti, C. 2013. The use of small diameter cores for assessing the compressive strength of clay brick masonries. *Journal of Cultural Heritage* 14(3): e95–e101.

Selby, R.G., & Vecchio, F.J. 1993. Three-dimensional Constitutive Relations for Reinforced Concrete. *Tech. Rep. 93-02, Univ. Toronto, Dept. Civil Eng., Toronto, Canada.*

Constitutive models and computational frameworks

Computational Modelling of Concrete and Concrete Structures – Meschke, Pichler & Rots (Eds)
 ISBN: 978-1-032-32724-2

How gap tests of ductile and quasibrittle fracture limit applicability of phase-field, XFEM, cohesive, nonlocal and crack-band models?

Z.P. Bažant
Northwestern University, USA

A.A. Dönmez
Northwestern University, USA
Istanbul Technical University, Turkey

H.T. Nguyen
Northwestern University, USA

ABSTRACT: The recently developed gap test exploits the size effect method to determine the effect of crack-parallel compression σ_{xx} on the material fracture energy, G_f, as well as the characteristic size c_f of the fracture process zone (FPZ). The previous gap tests demonstrated that the G_f of concrete can get doubled or reduced to almost zero according to the T-stress (crack-parallel stress) level. A subsequent study of aluminum fracture (Nguyen, Dönmez and Bažant, 2021) concluded that a similar effect exists in ductile fracture of polycrystalline plastic-hardening metals. This paper strengthens this conclusion by presenting and interpreting further gap tests of aluminum. Together with the results of the recent gap tests of crack-parallel stress effect in quasibrittle materials, the experimental evidence shows that the linear elastic fracture mechanics (LEFM), its computational versions XFEM and Phase-Field, and the cohesive crack models are inapplicable in the presence of significant crack-parallel stress—not only for concrete and other quasibrittle materials but also for plastic-hardening polycrystalline metals. On the other hand, the applicability of the crack band model with a realistic tensorial damage law is not limited.

1 INTRODUCTION

A complicating feature of the plastic-hardening metals is that large hardening yielding zone surrounding the fracture front in which the material undergoes softening damage was studied analytically, considering only σ_{xx} in the propagation direction, although the out-of-plane normal and shear components, σ_{zz} and σ_{xz}, of the crack-parallel stress are also expected to play a significant role, as already confirmed for σ_{zz} in concrete (H. Nguyen *et al.* 2020) and shown here for aluminum. Asymptotic matching was used to formulate the general scaling laws of plastic-hardening polycrystalline metals. These laws were related to the material fracture energy G_f as well as the effective radius of the yielding zone, r_p.

In this study, an extension of these gap tests and their theoretical consequences is presented. The changes in the energetic size effect are studied experimentally over a much broader range of crack-parallel compressive stress σ_{xx} and a broader size range. Then the size effect method (Bažant et al. 1991; Bažant & Planas 1997; Nguyen et al. 2021) is used to deduce from these changes the effect of σ_{xx} on the material fracture energy, G_f, and on the effective radius r_p of the yielding zone (YZ) of aluminum. The effect of σ_{xx} on the scaling asymptotes of the small-scale-yielding is also clarified. However, because the range of specimen sizes in this study is much greater than the inhomogeneity size (which is the size of a polycrystalline grain, about 2 to 50 micrometers), the change in the size of the YZ and the fracture process zone (FPZ) can only be distinguished using numerical models.

2 ASYMPTOTIC SCALING REGIMES

The analytical solution of the role of the large yielding zone surrounding the fracture front was studied in recent works (Bažant et al. 2022; Nguyen et al. 2021). A brief review of the hardening plasticity and yielding zone effect on fracture is given here.

The polycrystalline metals have very small FPZ, of micrometer-scale, compared to the millimeter-scale plastic hardening part or the yielding zone. This introduces one more transitional range in the original SEL of Type 2. That additional transition is shown in Figure 1. As shown in Figure 1, three size effects can be considered in the Al alloys and other polycrystalline metals; the transition from the FPZ to large-scale yielding, the transition from the large-scale to small-scale yielding, or LEFM (linear elastic fracture mechanics—the

DOI 10.1201/9781003316404-48

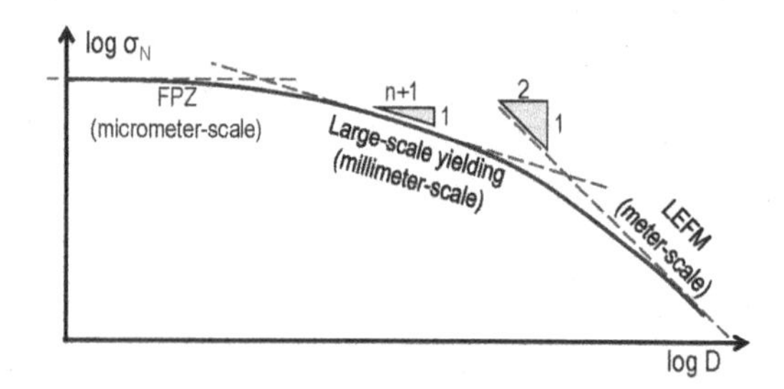

Figure 1. Description of the three transitional zones and the asymptotes in fracture of plastic-hardening polycrystalline metals.

large-scale asymptote for LEFM), and finally the overall transition from the FPZ (micrometer-scale) to small-scale yielding (LEFM). The second transition involves a deviation from the original size effect law. The third transition corresponds to the SEL and is probably the most important one among the three transitional regimes.

3 REVIEW OF STRESS-STRAIN RELATION OF PLASTIC-HARDENNING METALS

The plastic hardening response of metals can be defined by the Ramberg-Osgood model for the uniaxial stress-strain law (Fig. 2) (Ramberg & Osgood 1943).

$$\frac{\varepsilon}{\varepsilon_y} = \frac{\sigma}{\sigma_y} + \alpha_p \left(\frac{\sigma}{\sigma_y}\right)^n \tag{1}$$

where ε_y = initial yield strain, σ_y = initial yield strength; α_p = empirical parameter (usually denoted as α, although α is the standard notation for a dimensionless crack length); and n = plastic hardening exponent, typically 3 to 20. For analysis, it is helpful that the n, the hardening exponent, is so high that the plastic strain dominates and the elastic strain can be ignored. This assumption was the basis of the

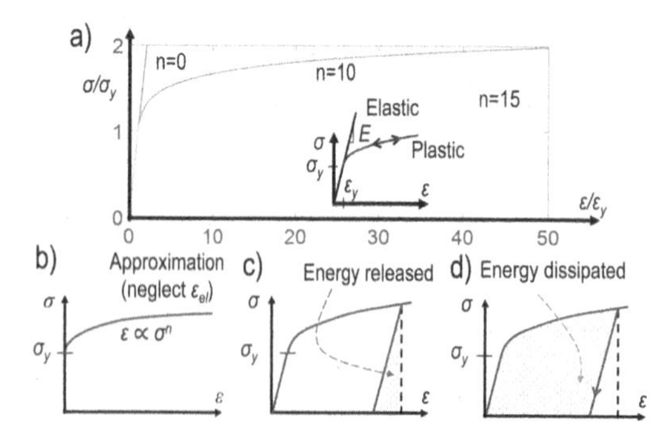

Figure 2. (a) Stress-strain behavior of plastic-hardening metals and response curves for various n(hardening exponent). (b) Approximation when elastic strain is ignored; Elastoplastic constitutive law with various n; (c,d) The partition of strain energy into released and dissipated.

classical Hutchinson-Rice-Rosengren (HRR) theory (Hutchinson 1968; Rice & Rosengren, 1968).

The advantage of the power law in Eq. (1) is that the stress-strain law becomes self-similar for the strain or stress magnitude. Together with the divided form of the power-law singularity, broadly presented as required (Nguyen et al. 2021), the deformation-field at the near-tip asymptote becomes self-similar to radial affine transformations, which makes it feasible for an analytical solution. The uniaxial stress-strain relation is therefore stated as (Hutchinson 1968; Rice & Rosengren 1968):

$$\frac{\varepsilon}{\varepsilon_y} - \alpha_p \left(\frac{\sigma}{\sigma_y}\right)^n \tag{2}$$

$$e_{ij} - \frac{3\alpha_p \varepsilon_y}{2\sigma_y} \left(\frac{\sigma_{ef}}{\sigma_y}\right)^{n-1} \tag{3}$$

$$\sigma_{ef} = \sqrt{\frac{3}{2} s_{kl} s_{kl}} \tag{4}$$

The uniaxial yield stress, σ_y, and the uniaxial yield strain, ε_y, in Eq. (2-4) are the limiting parameters for the effective (or equivalent) yield stress. These parameters point out the distinction of the power-law strain from the previous (largely elastic) regime. The σ_y in Eq. (4) is the scalar effective stress.

Hutchinson and Paris in (Hutchinson & Paris 1979) showed that the deformation theory of plasticity is very accurate in this problem, although its use is a simplifying assumption in the HRR theory and the J-integral.

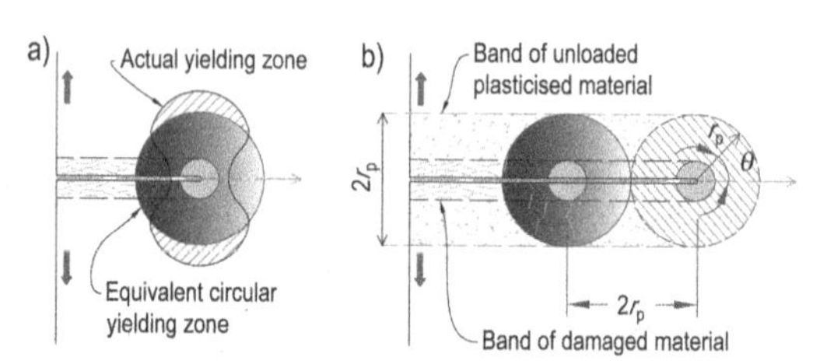

Figure 3. (a) Actual and equivalent (equal area or volume) yielding zones; (b) displacement of the equivalent yielding zone with the crack growth.

Figure 3 illustrates the yielding zone and the approximately equivalent crack growth model. In Figure 3, the (r, θ) are the polar coordinates centered at the tip of the crack. The angle θ is measured from the crack extension line and r_p is the effective size of the hardening part (or the yielding zone, *YZ*).

4 SIZE EFFECT DUE TO ENERGY RELEASE

The energy balance equation can be constructed by using the physical similarities of the ductile and quasibrittle failures with respect to the transitional regimes, as used in (Nguyen et al. 2021). The energy release rates, G_s and G_p arise from two different locations in the structure. The G_s is the rate of energy release from the elastic zone in the structure (or from the undamaged volume of the structure). This energy release is approximately proportional to the characteristic size, D. Second, G_b is the energy release rate from the elastic material traversed by the front of the crack band. It does not depend on D. Note that there is no plastic yielding zone in quasibrittle materials, unlike the plastic hardening metals where the large yielding zone causes a transition between the FPZ and the elastic zone of the structure.

The yielding zone, which is typically of millimeter scale, plays three roles in the failure mechanism. First, it transfers the energy-flux to the FPZ via the yielding zone. The energy is conserved in this transfer. Second, the yielding zone dissipates energy in its wake, with the rate of G_p, as the plastically strained material undergoes unloading. Third, the unloading of the plasticized material in the wake of the advancing yielding zone releases its strain energy, with the rate of G_b. The G_b is defined as the strain energy, $(\gamma_c \sigma_N)^2/2E'$, that was contained in the band of width $2r_p$ prior to the arrival of the yielding zone. Therefore, the energy balance during fracture can be written as:

$$G_s + G_b = G_f + G_p \quad (5)$$

We may use a fixed characteristic length scale for G_b, similar to c_f, defined by the yielding zone width ($2r_p$): $G_b = (\sigma_N^2/E')2r_p$. For G_s we can use the same expression as in quasibrittle materials. Inserting in (5) the energy release rate expressions, we obtain the condition of energy conservation:

$$\frac{\sigma_N^2}{E'} D g_0 + \frac{\sigma_N^2}{E'} 2r_p = G_f + G_p \quad (6)$$

Solving for σ_N, we get the size effect law for fracture of plastic-hardening metals in small-scale yielding:

$$\sigma_N = \frac{\sigma_0}{\sqrt{1 + D/D_0}} \quad (7)$$

Eq. (7) has the same form as the SEL for quasibrittle failures. However, the definitions of its coefficients are not the same:

$$\sigma_0^2 = E' G_f / 2r_p + \sigma_p^2, D_0 = 2r_p/g_0 \quad (8)$$

$$\sigma_p^2 = \frac{1}{2} E' \sigma_y ?_y Q_p \quad (9)$$

The asymptotes of this law have the same slopes as SEL:

$$\sigma_N \underset{D\to 0}{\Rightarrow} \sigma_0 = \text{constant}, \sigma_N \underset{D\to 0}{\Rightarrow} D^{-1/2} \quad (10)$$

The underlying assumption is that r_p is about the same for all specimen sizes. The "triaxiality number" (Anderson 2017) is assumed to remain constant, too.

5 SIZE EFFECT METHOD FOR DUCTILE FRACTURE

The size effect on structural strength is the main consequence of fracture behavior, and the size effect method is the most straightforward and unambiguous procedure to identify the material fracture properties. Eq. (7) can be restated as linear regression:

$$Y = AX + C \text{where} X = D; \ \ Y = 1/\sigma_N^2 \quad (11)$$

$$A = 1/\sigma_0^2 D_0, \ \ C = 1/\sigma_0^2 \quad (12)$$

The fracture energy, G_f, and the effective width of the yielding zone, $2r_p$, can be obtained by fitting these equations with the test results. The required test data consist only of the peak loads (max. loads) of differently sized specimens with a sufficiently broad size range. After getting the dimensionless energy release rate g_0 (and E'), one can find the A and C values by a linear regression of the data in the plane (X, Y). Then one can get $\sigma_0 = 1/\sqrt{C}$ and $D_0 = C/A$ using these values from the regression analysis. Finally, the fracture parameters can be obtained as:

$$G_f = (C^{-\frac{1}{2}} - c_p) g_0 / E' g_0^2 A, \ \ r_p = \frac{g_0}{2\sigma_0^2 A} \quad (13)$$

6 GAP TESTS OF ALUMINUM

In the standard fracture specimens, the stresses $\sigma_{xx}, \sigma_{zz}, \sigma_{xz}$ parallel to the crack plane (x, z) are zero or negligible. It has been implicitly assumed that the cracks are planes with zero thicknesses. If this assumption were correct, then no effect of $\sigma_{xx}, \sigma_{zz}, \sigma_{xz}$ on the crack propagation could be expected. Actually, the FPZ, located in front of the crack tip, has always a finite width, δ_y, measured normal to its plane. This is the fundamental characteristic of the blunt crack (Bažant & Cedolin 1991) and crack band (Bažant 1993; Bažant & Oh 1983) models, which revealed already in 1979 that, if δ_y is finite, the effect of $\sigma_{xx}, \sigma_{zz}, \sigma_{xz}$ must be important and the damage tensor inside the fracture process zone must play a role, and that the scalar stress-displacement law of the cohesive (or fictitious) crack model is inadequate. Some role of the crack-parallel compression in concrete has long ago been suspected by a few researchers (Bažant 1993; Bažant & Cedolin 1979; Bažant & Oh 1983; Tschegg et al. 1995), but a simple unambiguous test had been unavailable until the new gap test was developed, in 2020 (Bažant et al. 2022; H. Nguyen *et al.*, 2020; H. T. Nguyen *et al.*, 2020; Nguyen et al. 2021).

The gap tests conducted here involve notched beam specimens of aluminum, the 6061 series. The specimens are scaled geometrically in compliance with the 2*D* scaling laws, with a fixed width of 10 mm, as shown in Figure 4a, b. Their depths are 12, 24, 48 and, 96 mm. Figs. 4c, d show the results of standard three-point bend tests (no crack-parallel compression). The

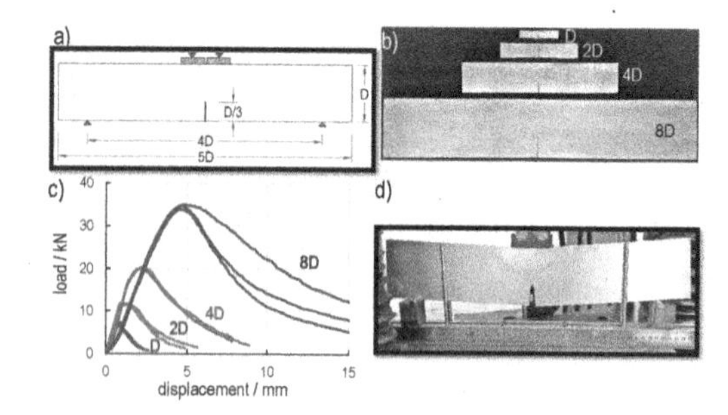

Figure 4. The geometrical properties of the specimens; (b) the scaled specimens of four sizes, with the fixed thickness; (c) Load vs. mid-span deflection curves; (d) the post-failure image showing the crack in the largest specimen.

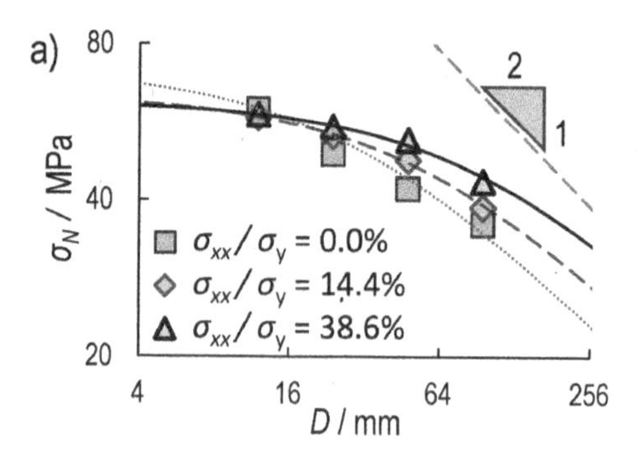

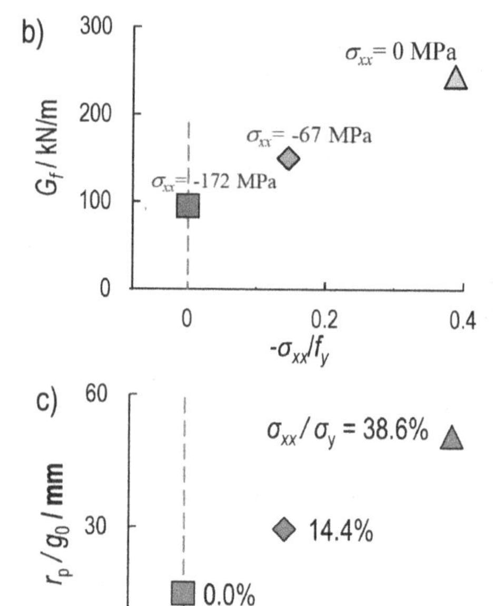

Figure 5. (a) Measured size effect data of aluminum for four different specimen sizes D and three different ratios of crack-parallel stress σ_{xx} to yield strength f_y, in logarithmic scales; Experimentally obtained data on the dependence of (b) fracture energy G_f and (c) half width of yielding zone r_p on the ratio of crack-parallel compressive stress σ_{xx} to yield strength f_y.

deformed state of the largest specimen can be seen in Figure 4d.

The resulting size effect curves are shown in Figure 5. The data points in Figure 5a represent the measured peak values of nominal strength σ_N for scaled gap tests of 4 different sizes $D = 12, 24, 48, 96$ mm, and for three different levels of crack-parallel compression.

In Figure 5a, the crack-parallel stress, σ_{xx}, is presented relative to the yield strength (f_y) of the Al alloy, which is has been measured as 450MPa in the uniaxial compression tests. The three solid curves display the best-fit curves obtained by multivariate nonlinear regression analysis of the data points with the size effect law. The systematic pattern of the curves shows the scatter to be relatively small compared to the overall scatter of all data.

In Figure 5a, the LEFM size effect slope of $-1/2$ is still far from being achieved, even for the largest specimens. This means that a much larger size would be needed to reach the LEFM range (which is the small-scale yielding range). Furthermore, the position of the LEFM asymptote of $-1/2$ determines the fracture energy, G_f (which is equal to J_{cr} and represents a new way of measuring it). The LEFM is displayed as the dashed line for the case of the largest crack-parallel compression. A translation of the LEFM asymptote to the right implies an increase of G_f. These LEFM asymptotes for the three levels of crack-parallel compression (σ_{xx}) are shifted relative to each other (Fig. 5a), which means that the fracture energies of these three cases are different.

The estimates of G_{FPZ}, asapartof the ΔG_f, can be evaluated from the size effect curves and their corresponding fracture parameters. It is known that, for the same crack-parallel stress σ_{xx} (also called the T-stress), the size of the yielding zone in front of the crack tip does not change and stays approximately the same for every size, D. In other words, the dissipation of the energy from the wake of the yielding zone is size-independent for the same T-stress. Thus, the discrepancy in the size effect fits from the gap tests, must be explained by a change of the energy dissipation in the micrometer-scale fracture process zone of the polycrystalline metal.

Consequently, the variation of the fracture energies (G_f) presented in Figure 5b results from a change of both the FPZ (at the micrometer scale) and the yielding zone. Nevertheless, this difference was around 10-30 mm, implying a much more marked contribution from the yielding zone. To differentiate the contributions from these two zones would require either micrometer-scale gap tests or numerical computational results with a realistic damage constitutive model for aluminum. The numerical models for capturing the effects of crack-parallel compression on the strength, size effect, and fracture parameters, require a tensorial damage constitutive law. The phase-field, XFEM, and cohesive crack model cannot capture such effects.

7 CONSEQUENCES FOR APPLICABILITY OF LEFM (WITH XFEM), PHASE-FIELD AND COHESIVE CRACK MODELS

Extension of the gap tests of aluminum to three different levels of the crack-parallel compressive stress σ_{xx} provides clear evidence of the effect of σ_{xx} on the size effect, which is found to be strong. Regression analysis of the gap tests of different sizes for the same σ_{xx} level yields unambiguous evidence of the σ_{xx} effect on the fracture energy G_f of aluminum. Increasing σ_{xx} from 0 to $0.4f_y$, causes G_f to approximately double and r_p to triple (see Fig. 5c). Although no tests were made at $|\sigma_{xx}| > 0.4f_y$, the extension of the curve of G_f versus σ_{xx} is expected not to tend to 0 at $\sigma_{xx} \to -f_y$ because, in contrast to concrete, aluminum yielding under compression in x-direction suffers no softening damages and yields at increasing strength in the y-direction. To reproduce the present experiments mathematically, a fracture process zone of correct finite width, described by a realistic tensorial damage constitutive model for aluminum, will be used in subsequent work. The finite element crack band model serves well for that.

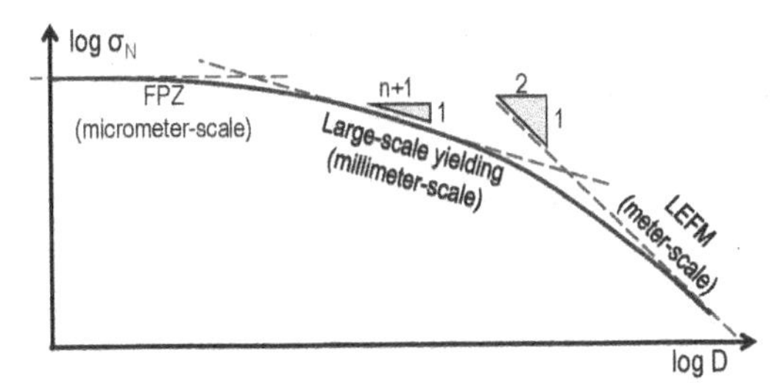

Figure 6. Path dependence of G_f as a function of σ_{xx} (if path dependence were absent, the terminal points encircled by ellipses would have to coincide).

It might be thought that the aforementioned models could be used if the fracture energy, G_f, were considered to be a function of σ_{xx}. However, this is not possible because the dependence of G_f on σ_{xx} is enormously path-dependent (H. Nguyen *et al.*, 2020; H. T. Nguyen *et al.*, 2020), as shown in Fig. 6. There is no way to shrink the FPZ to a line and thus obtain a line crack, except if the crack-parallel stresses are negligible, which is, however, a rare situation in practice.

8 CONCLUSIONS

As a result of the gap tests, the classical fracture mechanics dealing with line cracks is now seen to be severely limited. Nevertheless, this fundamental theory remains necessary for the understanding of brittle and quasibrittle structural failures, for underpinning blunt crack models, and for teaching the fracture behavior of structures.

ACKNOWLEDGMENT

Funding under NSF Grant CMMI-1439960 to Northwestern University is gratefully acknowledged. A.A.D also thanks for funding from Istanbul Technical University, BAP:42833. Conflict of interest: None.

REFERENCES

Anderson, T. (2017) *Fracture Mechanics: Fundamentals and Applications*. 4th edn. Boca Raton, London, New York: CRC Press.

Bažant, Z. P. (1993) 'Scaling Laws in Mechanics of Failure', *Journal of Engineering Mechanics*. American Society of Civil Engineers (ASCE), 119(9), pp. 1828–1844.

Bažant, Z. P. and Cedolin, L. (1979) 'Blunt Crack Band Propagation in Finite Element Analysis', *Journal of the Engineering Mechanics Division*. American Society of Civil Engineers, 105(2), pp. 297–315.

Bažant, Z. P. and Cedolin, L. (1991) *Stability of Structures: Elastic, Inelastic, Fracture and Damage Theories, 2010*. World Scientific Publishing Co. Pte. Lfd., ISBN-13.

Bažant, Z. P., Dönmez, A. A. and Nguyen, H. T. (2022) 'Précis of gap test results requiring reappraisal of line crack and phase-field models of fracture mechanics', *Engineering Structures*, 250, p. 113285.

Bažant, Z. P., Kazemi, M. T. and others (1991) 'Size effect on diagonal shear failure of beams without stirrups', *ACI Structural journal*, 88(3), pp. 268–276.

Bažant, Z. P. and Oh, B. H. (1983) 'Crack band theory for fracture of concrete', *Matériaux et construction*. Springer, 16(3), pp. 155–177.

Bažant, Z. P. and Planas, J. (1997) *Fracture and size effect in concrete and other quasibrittle materials*. Boca Raton, Boston, London: CRC press.

Hutchinson, J. W. (1968) 'Singular behaviour at the end of a tensile crack in a hardening material', *Journal of the Mechanics and Physics of Solids*, 16(1), pp. 13–31.

Hutchinson, J. W. and Paris, P. C. (1979) 'Stability Analysis of J-Controlled Crack Growth', *Elastic-Plastic Fracture*. ASTM International, p. 64.

Nguyen, H. *et al.* (2020) 'New perspective of fracture mechanics inspired by gap test with crack-parallel compression', *Proceedings of the National Academy of Sciences of the United States of America*. National Academy of Sciences, 117(25), pp. 14015–14020.

Nguyen, H. T. *et al.* (2020) 'Gap Test of Crack-Parallel Stress Effect on Quasibrittle Fracture and Its Consequences', *Journal of Applied Mechanics*, 87(7), p. 071012.

Nguyen, H. T., Dönmez, A. A. and Bažant, Z. P. (2021) 'Structural strength scaling law for fracture of plastic-hardening metals and testing of fracture properties', *Extreme Mechanics Letter*, 43, p. 101141.

Ramberg, W. and Osgood, W. R. (1943) *Description of stress-strain curves by three parameters*. Washington.

Rice, J. R. and Rosengren, G. F. G. (1968) 'Plane strain deformation near a crack tip in a power-law hardening material', *Journal of the Mechanics and Physics of Solids*. Pergamon, 16(1), pp. 1–12.

Tschegg, E. K., Elser, M. and Stanzl-Tschegg, S. E. (1995) 'Biaxial fracture tests on concrete — Development and experience', *Cement and Concrete Composites*. Elsevier, 17(1), pp. 57–75.

Computational Modelling of Concrete and Concrete Structures – Meschke, Pichler & Rots (Eds)

Modelling of a PsD hybrid test on a RC column/beam junction combining a multifibre beam model and a POD-ROM approach

B. Bodnar, W. Larbi, M. Titirla & J.-F. Deü
Conservatoire National des Arts et Métiers, Laboratoire de Mécanique des Structures et des Systèmes Couplés, Paris, France

F. Gatuingt & F. Ragueneau
Université Paris-Saclay, Centrale Supélec, ENS Paris-Saclay, CNRS, Laboratoire de Mécanique Paris-Saclay, Gif-sur-Yvette, France

ABSTRACT: In this paper, a pseudo dynamic (PsD) hybrid experimental setup allows for assessing the nonlinear behaviour of a reinforced concrete (RC) column/beam junction under earthquake is proposed. The specimen is linked to a numerical substructure made of multifibre beam elements modelling the other parts of the building. To reduce the CPU time related to the numerical substructure, a proper orthogonal decomposition (POD) projection modal basis is computed from an offline implicit finite element analysis and used to reduce the size of the matrix system. A bilinear elastic-plastic law is used for steel rebars, and a unilateral damage law is used for concrete. Step-by-step calculations are performed using a non-iterative, unconditionally stable and explicit α-OS splitting scheme during the hybrid test (*i.e.* the online phase). A substructuring method is applied to the column-beam junction located at the first-floor level. The reliability of the modelling assumptions as well as the use of POD-modes in the case of quasi-brittle materials are discussed. The analyses are performed by using MATLAB© software. In a first attempt, the column-beam junction response is computed using a 2D nonlinear numerical model defined in Cast3M© software. Results show that using a POD projection modal basis does not significantly reduce the computational cost when the α-OS method is used but improves the response of both numerical and tested substructures thanks to the nonlinearities taken into account into the POD-modes.

1 INTRODUCTION

It is sometimes necessary to perform tests on structural elements to study their behaviour under seismic loading (damage, failure mechanisms, ...) in the civil engineering field. For this purpose, it is possible to carry out quasi-static “push-over” tests (consisting of stressing the specimen by applying step by step the shape of the first vibratory mode) or dynamic tests on a reduced specimen (on a shaking table or in a centrifuge facility). Although, even if these tests are commonly used, they have many limitations. It is impossible to consider the inertial and viscous forces in the first case. In the second one, the similitude theory leads to the addition of masses, sometimes leading to unrealistic collapse mechanisms due to local stresses.

To overcome these limitations, “hybrid tests” have been developed over the last few decades. They allow the assessment of the response of structural elements under seismic loading at full scale by considering the environment in which they are installed. The specimen is loaded at its ends by actuators whose displacements are computed through numerical calculations carried out simultaneously on a complete structure. Displacements are applied at each time step by actuators, and the corresponding measured restoring forces are used as boundary conditions for the numerical substructure. The results give the displacements of actuators for the next time step and so on.

The key idea was introduced by Hakuno et al. (1969), who proposed solving the harmonic oscillator equation by measuring the restoring force of an embedded beam specimen in real-time. However, its study showed that many technical limitations related to the control and delay of actuators do not allow to perform the test in real-time. To overcome these limits, Takanashi et al. (1969) proposed to carry out Pseudo dynamic (PsD) hybrid tests. Actuators thus apply the displacements in deferred time. As a result, only the static restoring forces are measured, while the viscous damping and the inertial forces related to the tested substructure remain unknown. To assess them, Buchet et al. (1994) showed that it is possible to compute them numerically by adding the tested substructure to the numerical model. The specimen can thus be modelled on the common degrees of freedom (DOFs) by a nonlinear oscillator (semi-global approach) or by a complete numerical model (global approach). As a

DOI 10.1201/9781003316404-49

result, the viscous damping and the inertial forces are assessed according to the measured restoring forces but depend on the damping applied numerically.

Performing dynamic finite element analyses in the framework of PsD hybrid tests requires specific non-iterative methods to avoid the risk of overshoot (*i.e.* sudden collapse of the specimen) as well as nonlinear material laws modelling the decrease of stiffness due to damage on the numerical substructure. Nakashima (1992) proposed to implement an Operator Splitting (OS) method to assess the nonlinear restoring forces. The vector is split into a nonlinear term computed from an explicit prediction and a linear term depending on the displacements on the time step. Thus, this scheme remains linearly implicit but becomes nonlinearly explicit, so iterations are unnecessary. The accuracy of this integration scheme, called α-OS, was assessed by Combescure (1997), who showed its reliability when the loss of stiffness does not imply a significant shift in frequency of the high-frequency modes. The α-OS method was later successfully applied by several researchers, including Pegon et al. (2000) and Souid (2009).

Carrying out real-time hybrid tests is still challenging due to many technical limitations related to the delay of the actuators and the computational cost required to solve the nonlinear numerical substructure. However, the CPU time can be reduced by using simplified models, such as macro elements (Moutoussamy 2013), elastic-plastic hinges (Nguyen 2012), and multifibre beam elements (Lebon 2011), as well as reduced-order modelling (ROM) methods (POD-ROM, POD-DEIM, …). To the best of our knowledge, the use of ROM on RC structures made of multifibre beam elements has not been investigated yet in the framework of hybrid tests.

Due to their highly nonlinear behaviour under earthquakes, several researchers performed quasi-static tests on column beam junctions (Iskef 2016). However, in the case of a PsD hybrid test, relevant boundary conditions need to be applied to the specimen to obtain valuable results. Thus, carrying PsD hybrid tests on RC column/beam junctions remains a challenging task.

In this paper, a PsD hybrid experimental setup allows for assessing the nonlinear behaviour of a RC column/beam junction under earthquake is proposed. The modelling of the numerical substructures (made of multifibre beam elements) is first described. The α-OS time integration scheme and the substructuring method are then detailed. A POD projection modal basis (computed from the results of an offline finite element analysis) is also added to the procedure to reduce CPU time. Note that the experimental setup is not yet available in our research. The tested specimen is then replaced by a 2D numerical model defined in Cast3M© software. The reliability of POD projection modal bases in the case of hybrid tests is next assessed by comparison with the offline and full order model (FOM) solutions. The boundary conditions applied to the specimen are finally discussed based on the damage index distribution.

2 NONLINEAR MODELLING OF THE SUBSTRUCTURES

2.1 *Timoshenko multifibre beam elements*

The hybrid test framework requires low time-consuming analyses with numerical models taking account of the loss of stiffness due to damage on the numerical substructure. So, to correctly model the behaviour of RC elements under earthquake, a highly nonlinear "unilateral" damage law needs to be used for concrete. Thus, to ensure a quick convergence of the results and perform real-time or quick PsD hybrid tests, local scale models are usually not used. Semi-global approaches (multifibre beams and multilayer shells) are chosen instead (*cf.* Figure 1). They describe the global kinematic by using a beam (or shell) model whose integration points are linked to a section made of 1D nonlinear fibres (or layers). The deformation of each fibre (or layer) is assessed assuming that the beam cross-sections remain plane. Nonlinear damage laws are then used to update the properties of the fibres at each iteration. Generalized stresses are computed through a double integration: one on the sections and the other on the beam elements. Multifibre beams were previously used by Lebon (2011) to perform PsD hybrid tests on RC frames. In this work, the structure is modelled using the Timoshenko multifibre beam elements developed by Kotronis (2004).

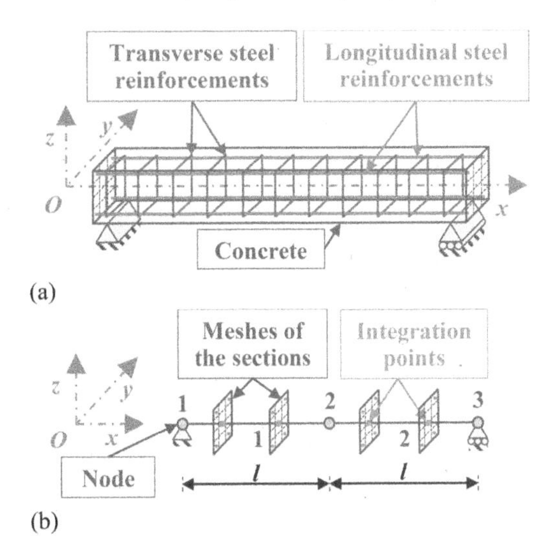

Figure 1. Simply supported RC beam on nodes 1 and 3 (a) and multifibre mesh with two Timoshenko beam elements (b).

2.2 *Nonlinear material laws*

In the case of hybrid tests, nonlinear material laws are required to model the decrease of stiffness due to damage during earthquakes. In addition, cyclic movements generate hardening of plastic steel rebars and opening/closing of cracks in damaged concrete, leading to the appearance of a "unilateral" effect (*i.e.* progressive recovery of stiffness when the cracks are closing). Material laws modelling these phenomena are thus required.

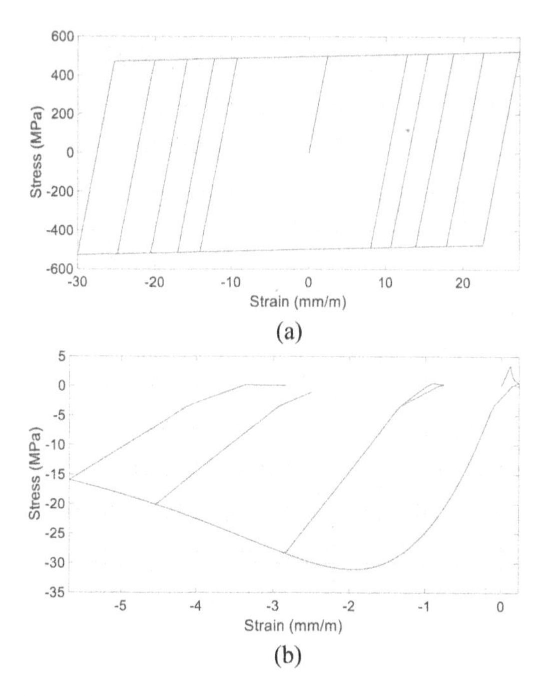

Figure 2. Uniaxial bilinear elastic-plastic law with kinematic hardening for steel rebars under cyclic loading (a) and uniaxial damage law of La Borderie with opening/closing of cracks (b).

The steel rebars are modelled by using a bilinear elastic-plastic law with kinematic hardening (*cf.* Figure 2 (a)). Fe500 steel rebars are used for the reinforcement, with an elastic stiffness of 210 GPa, a yielding stress of 500 MPa and a strain hardening modulus of 1000 MPa. The concrete is modelled with a "unilateral" damage law of La Borderie (1991), commonly used to model quasi-brittle materials under dynamic or cyclic loadings (*cf.* Figure 2 (b)). The parameters of the law of La Borderie are defined by considering an elastic stiffness of 31 GPa, a yielding tensile stress of 3.5 MPa, a yielding compressive stress of -10 MPa and a crack reclosing stress of -3.5 MPa.

3 TIME INTEGRATION SCHEME

Hybrid tests consist in linking a simulated numerical substructure to a tested specimen. An efficient substructuring technique is thus required to introduce the measured restoring forces to the numerical substructure and a non-iterative and unconditionally stable integration scheme dissipating high-frequency content due to the measures on the experimental setup (avoiding the risk of overshoot, *i.e.* collapse of the specimen). Several time-integration schemes were used in the literature to solve the spatially discrete equation of motion (1) during hybrid tests.

$$\begin{aligned} \boldsymbol{M}\cdot\ddot{\boldsymbol{u}}(t)+\boldsymbol{C}\cdot\dot{\boldsymbol{u}}(t)+\boldsymbol{r}(\boldsymbol{u}(t)) &= \boldsymbol{F}(t)\\ \boldsymbol{F}(t) &= -\boldsymbol{M}\cdot\boldsymbol{\Gamma}\cdot\ddot{u}_g(t) \end{aligned} \quad (1)$$

where $\boldsymbol{M}$ is the mass matrix, $\boldsymbol{C}$ the damping matrix, $\boldsymbol{r}(u(t))$ the restoring force vector, $\boldsymbol{F}(t)$ the external force vector, $\boldsymbol{\Gamma}$ the vector used to select the direction of the earthquake at the level of each DOF, $\ddot{u}_g(t)$ the ground acceleration, and $\boldsymbol{u}(t)$, $\dot{\boldsymbol{u}}(t)$ and $\ddot{\boldsymbol{u}}(t)$ the displacement, velocity and acceleration vectors.

Some authors, such as Shing (1991), chose to use an implicit scheme based on the Hilber-Hugues-Taylor (HHT) method (Hilber et al 1977), also called α-method The equation of motion is solved at time $(n+1+\alpha)$ where α is a parameter usually set between $-1/3$ and 0 This scheme is implicit since $\boldsymbol{u}_{n+1}$ depends on $\ddot{\boldsymbol{u}}_{n+1}$. The restoring force vector $\boldsymbol{r}_{n+1}(\boldsymbol{u}_{n+1})$ being a function of $\boldsymbol{u}_{n+1}$, an iterative procedure is thus required to solve (1). This approach was successfully used by Shing (1991) to perform PsD tests. However, in the case of real-time or quick PsD hybrid tests, non-iterative time-integration schemes are used instead to decrease CPU time and the risk of overshoot. To maintain the stability of implicit schemes without iterating, Nakashima (1992) proposed to use an operator splitting (OS) method, based on a linear approximation of the restoring force vector (2).

$$\begin{aligned} \boldsymbol{r}_{n+1}(\boldsymbol{u}_{n+1}) \cong{}& \boldsymbol{K}_I\cdot\boldsymbol{u}_{n+1}\\ &+\left(\tilde{\boldsymbol{r}}_{n+1}(\tilde{\boldsymbol{u}}_{n+1})-\boldsymbol{K}_I\cdot\tilde{\boldsymbol{u}}_{n+1}\right) \end{aligned} \quad (2)$$

where $\boldsymbol{K}_I$ is a secant or tangent stiffness matrix, chosen to be as close as possible to the elastic stiffness matrix $\boldsymbol{K}_E$ (for the sake of stability), and $r_{n+1}(\tilde{\boldsymbol{u}}_{n+1})$ the prediction of the restoring force vector (Combescure 1997). The system of linear equations to solve in order to compute $\ddot{\boldsymbol{u}}_{n+1}$ is thus given in (3).

$$\widehat{\boldsymbol{M}}\cdot\ddot{\boldsymbol{u}}_{n+1}=\widehat{\boldsymbol{F}}_{n+1+\alpha} \quad (3)$$

where $\widehat{\boldsymbol{M}}$ is the pseudo mass matrix (4), and $\widehat{\boldsymbol{F}}_{n+1+\alpha}$ the pseudo force vector (5).

$$\widehat{\boldsymbol{M}}=\boldsymbol{M}+\gamma\cdot\Delta t\cdot(1+\alpha)\cdot\boldsymbol{C}+\beta\cdot\Delta t^2\cdot(1+\alpha)\cdot\boldsymbol{K}_I \quad (4)$$

$$\begin{aligned} \widehat{\boldsymbol{F}}_{n+1+\alpha} ={}& (1+\alpha)\cdot\boldsymbol{F}_{n+1}-\alpha\cdot\boldsymbol{F}_n\\ &+\alpha\cdot\boldsymbol{r}_n-(1+\alpha)\cdot\tilde{\boldsymbol{r}}_{n+1}\\ &+\alpha\cdot\boldsymbol{C}\cdot\tilde{\dot{\boldsymbol{u}}}_n-(1+\alpha)\cdot\boldsymbol{C}\cdot\tilde{\dot{\boldsymbol{u}}}_{n+1}\\ &+\alpha\cdot\left(\gamma\cdot\Delta t\cdot\boldsymbol{C}+\beta\cdot\Delta t^2\cdot\boldsymbol{K}_I\right)\cdot\ddot{\boldsymbol{u}}_n \end{aligned} \quad (5)$$

Note that $\tilde{\boldsymbol{u}}$ and $\tilde{\dot{\boldsymbol{u}}}$ are the explicit predictions of the displacement and velocity vectors (6), and β and γ

are the parameters of the time-integration scheme of Newmark, defined according to the α parameter (7).

$$\tilde{u}_{n+1} = u_n + \Delta t \cdot \dot{u}_n + \Delta t^2 \cdot \left(\frac{1}{2} - \beta\right) \cdot \ddot{u}_n$$
$$\tilde{\dot{u}}_{n+1} = \dot{u}_n + \Delta t \cdot (1 - \gamma) \cdot \ddot{u}_n \qquad (6)$$

α is used to dampen the high-frequency content, mainly introduced by the measures in the case of hybrid tests. Its value of commonly set at -0.05 (Hilber et al. 1977).

$$\beta = \frac{(1-\alpha)^2}{4} \& \gamma = \frac{(1 - 2 \cdot \alpha)}{2} \qquad (7)$$

The α-OS method is implicit in the linear phase and explicit in the nonlinear phase. As demonstrated by Combescure et al. (1995) in practical cases, the α-OS method competes very well in terms of accuracy with iterative implementations of the α-method, even if a residual error appears due to the approximation in (2). Note that the predictive restoring force vector $\tilde{r}_{n+1}$ ($\tilde{u}_{n+1}$) is assessed once per time step: it is thus not necessary to solve (6) in increments. In addition, since $K_I = K_E$, the matrix $\widehat{M}^{-1}$ is computed before entering the time step loop, decreasing CPU time.

4 SUBSTRUCTURING METHOD

In the case of hybrid tests, numerical and experimental substructures are split to introduce the measured restoring forces as external loads on common DOFs. The complete structure is thus substructured, as described in the example in Figure 3.

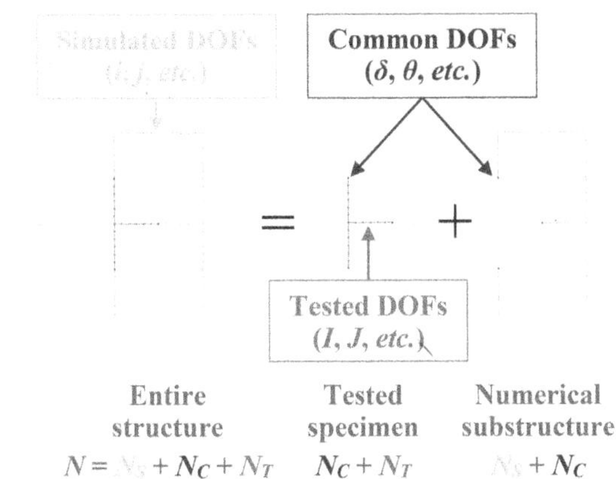

Figure 3. Substructuring of an in-plane two-storey frame: complete structure with 72 nodes, the tested specimen with 19 nodes, and numerical substructure with 56 nodes.

Among the N DOFs in the matrix system of (1), N_S DOFs only belong to the modelled substructure (subscript i, j, *etc.*), N_C belong to both the modelled substructure and the tested specimen (subscript δ, θ, *etc.*) and N_T only belong to the tested specimen (subscript I, J, *etc.*).

By distinguishing in (3) the systems of equations coming from the numerical substructure (subscripted S) and the tested specimen (subscripted T), it is possible to reorganize the matrix $\widehat{M}$ and the related terms as described in (8).

$$\begin{bmatrix} {}^S\widehat{M}_{ij} & {}^S\widehat{M}_{i\theta} & 0 \\ {}^S\widehat{M}_{\delta j} & {}^S\widehat{M}_{\delta\theta} + {}^T\widehat{M}_{\delta\theta} & {}^T\widehat{M}_{\delta J} \\ 0 & {}^T\widehat{M}_{I\theta} & {}^T\widehat{M}_{IJ} \end{bmatrix} \cdot \begin{bmatrix} \ddot{u}_{j,n+1} \\ \ddot{u}_{\theta,n+1} \\ \ddot{u}_{J,n+1} \end{bmatrix} = \begin{bmatrix} {}^S\widehat{F}_{i,n+1+\alpha} \\ {}^S\widehat{F}_{\delta,n+1+\alpha} + {}^T\widehat{F}_{\delta,n+1+\alpha} \\ {}^T\widehat{F}_{I,n+1+\alpha} \end{bmatrix} \qquad (8)$$

where $\ddot{u}_{j,n+1}$, $\ddot{u}_{\theta,n+1}$ and $\ddot{u}_{J,n+1}$ are the acceleration vectors respectively related to the simulated, common, and tested DOFs. So, by condensing the components of $\ddot{u}_{j,n+1}$, (8) can be rewritten on the DOFs related to the tested specimen (9).

$$\begin{bmatrix} {}^T\widehat{M}_{\delta\theta} + {}^S\widehat{M}^*_{\delta\theta} & {}^T\widehat{M}_{\delta J} \\ {}^T\widehat{M}_{I\theta} & {}^T\widehat{M}_{IJ} \end{bmatrix} \cdot \begin{bmatrix} \ddot{u}_{\theta,n+1} \\ \ddot{u}_{J,n+1} \end{bmatrix} = \begin{bmatrix} {}^T\widehat{F}_{\delta,n+1+\alpha} + {}^S\widehat{F}^*_{\delta,n+1+\alpha} \\ {}^T\widehat{F}_{I,n+1+\alpha} \end{bmatrix} \qquad (9)$$

where ${}^S\widehat{M}^*_{\delta\theta}$ and ${}^S\widehat{F}^*_{\delta,n+1+\alpha}$ are the condensed pseudo mass matrix and pseudo force vector defined in (10) & (11).

$$ {}^S\widehat{M}^*_{\delta\theta} = {}^S\widehat{M}_{\delta\theta} - {}^S\widehat{M}_{\delta j} \cdot {}^S\widehat{M}^{-1}_{ij} \cdot {}^S\widehat{M}_{i\theta} \qquad (10)$$

$$ {}^S\widehat{F}^*_{\delta,n+1+\alpha} = {}^S\widehat{F}_{\delta,n+1+\alpha} - {}^S\widehat{M}_{\delta j} \cdot {}^S\widehat{M}^{-1}_{ij} \cdot {}^S\widehat{F}_{i,n+1+\alpha} \qquad (11)$$

The measured restoring force vector ${}^T\tilde{r}_{\delta,n+1}$ is introduced in the pseudo force vector ${}^T\widehat{F}_{\delta,n+1+\alpha}$ (5), whereas the restoring force vectors computed on the numerical substructure ${}^S\tilde{r}_{i,n+1}$ and ${}^S\tilde{r}_{\delta,n+1}$ are introduced in ${}^S\widehat{F}_{i,n+1+\alpha}$ and ${}^S\widehat{F}_{\delta,n+1+\alpha}$.

Note that the components related to the internal tested DOFs (indexed I, J in Figure 3) are computed by modelling the tested specimen. The same finite elements and nonlinear material laws are usually used on both the numerical and tested substructures. The elastic stiffness matrix of the tested specimen can be initially set based on measurements performed on the experimental setup. The displacement of the tested DOFs (stored in ${}^T\tilde{u}_{J,n+1}$) can either be predicted or measured at the level of the neutral axis (by using field measurements or interpolation methods). Knowing the values of the reactions applied to the common DOFs, it is thus possible to approximate the restoring force vector ${}^{T}\tilde{\boldsymbol{r}}_{I,n+1}$ as well as the displacements of the tested specimen under earthquake (stored in ${}^{T}\boldsymbol{u}_{J,n+1}$).

At the time step $(n+1)$, (9) is firstly solved to compute the acceleration vector on the common DOFs. Once $\ddot{\boldsymbol{u}}_{\theta,n+1}$ is known, the acceleration vector related to the simulated DOFs (named $\ddot{\boldsymbol{u}}_{j,n+1}$) is then assessed by solving (12).

$$ {}^{S}\widehat{\boldsymbol{M}}_{ij} \cdot \ddot{\boldsymbol{u}}_{j,n+1} = {}^{SC}\widehat{\boldsymbol{F}}_{n+1+\alpha} \tag{12} $$

where ${}^{SC}\widehat{\boldsymbol{F}}_{n+1+\alpha}$ is the condensed pseudo force vector defined in (13).

$$ {}^{SC}\widehat{\boldsymbol{F}}_{n+1+\alpha} = {}^{S}\widehat{\boldsymbol{F}}_{i,n+1+\alpha} - {}^{S}\widehat{\boldsymbol{M}}_{i\theta} \cdot \ddot{u}_{\theta,n+1} \tag{13} $$

During hybrid tests, (9) and (12) are solved on two computers exchanging data, decreasing CPU time. The first one (called master PsD computer) is responsible for the tested specimen. It sends instructions to the experimental setup and receives measures. It ensures the analogue to digital (A/D) conversion of data with the help of an acquisition card and computes the acceleration vector $\ddot{\boldsymbol{u}}_{\theta,n+1}$ by solving (9). The second computer is responsible for the modelled substructure and computes the acceleration vector $\ddot{\boldsymbol{u}}_{j,n+1}$, according to (12).

5 REDUCED ORDER MODELLING BY USING A POD-ROM METHOD

In the case of hybrid tests, solving nonlinear substructures at each time step increases CPU time, even if a non-iterative time-integration scheme is used. Carrying these tests in real-time can thus be difficult due to the additional delay of the actuators, especially when the numerical substructure is modelled with a high number of DOFs. In the literature, most of the researchers use either a linear model (Bonnet et al. 2008) or nonlinear macroelements (Moutoussamy 2013). Other methods need to be used to reduce the CPU time with many DOFs and nonlinear material laws during the online phase (*i.e.* during hybrid tests). Among them, the POD-ROM method allows reducing the size of matrix systems by projecting equations on a basis made of few nonlinear POD-modes.

The key idea is to perform first a full offline step-by-step nonlinear analysis on the complete structure (including both numerical and tested substructures). Snapshots are then extracted from the results to compute N nonlinear POD-modes by using a Singular Value Decomposition (SVD) procedure (with N the number of DOFs). m POD-modes are then selected to build a modal projection basis. This reduces the number of DOFs and decreases CPU time during the online phase. The displacement vectors $\boldsymbol{u}_j(t)$ (related to the simulated DOFs) can thus be expressed in a new basis $\boldsymbol{\Phi} = [\boldsymbol{\varphi}_1 \quad \cdots \quad \boldsymbol{\varphi}_m]$ of dimension $m \ll N$ as described in (14).

$$ \boldsymbol{u}_j(t) \cong \boldsymbol{\Phi} \cdot \boldsymbol{q}(t) \tag{14} $$

where $\boldsymbol{q}(t)$ is the vector of size $m \times 1$ containing the coordinates of displacements in the new basis $\boldsymbol{\Phi}$ and $\boldsymbol{\varphi}_{i=1,\ldots,m}$ the POD-modes computed from a SVD procedure. So, by substituting $\ddot{\boldsymbol{u}}_{j,n+1}$ with $\ddot{\boldsymbol{q}}_{n+1}$ in (12), it comes:

$$ \boldsymbol{\Phi}^{T} \cdot {}^{S}\widehat{\boldsymbol{M}}_{ij} \cdot \Phi \cdot \ddot{\boldsymbol{q}}_{j,n+1} = \boldsymbol{\Phi}^{T} \cdot {}^{SC}\widehat{\boldsymbol{F}}_{n+1+\alpha} \tag{15} $$

Note that when the α-OS time-integration scheme is used, operator $\boldsymbol{\Phi}^{T} \cdot {}^{S}\widehat{\boldsymbol{M}}_{ij} \cdot \boldsymbol{\Phi}$ is computed once and set as a constant during the online phase. However, the nonlinear restoring force vectors ${}^{S}\boldsymbol{r}_{n+1}$ always needs to be computed in the full coordinates, making this operation the most time-consuming part of the entire process. The use of a non-iterative α-OS method is thus relevant to avoid multiple reassessments of ${}^{S}\boldsymbol{r}_{n+1}$ at each time step.

6 APPLICATION

6.1 *Case study*

One of the aims of this paper is to propose a PsD hybrid experimental set that allows for assessing the behaviour of a RC column/beam junction under earthquake. In the following, all the simulations are performed using the ground x-acceleration drawn in Figure 4. It is an artificial signal typical of a French average (close to strong) seismic hazard area. Its peak ground acceleration (PGA) equals 2.32 m/s^2 and is reached at time 3.16 s.

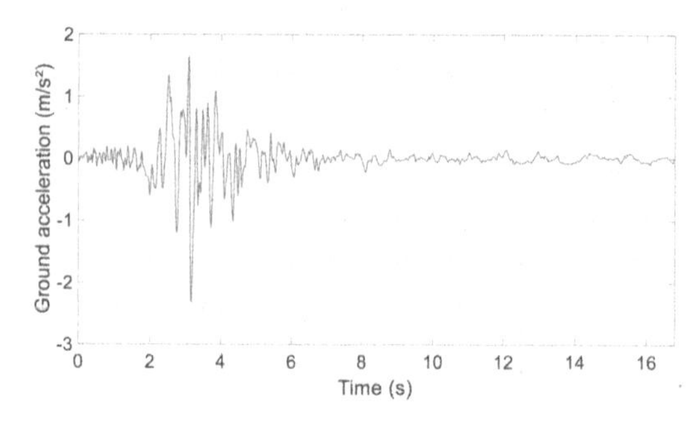

Figure 4. Ground acceleration versus time.

The case study is a three-storey RC frame of 3 m long spans and 3 m high storeys (*cf.* Figure 5). All the columns are fixed to the foundation level and have a 15×15 cm square cross-section, while the beams have a 15×25 cm rectangular one. The diameter of each longitudinal steel rebar is set at 12 mm, and the steel coating is equal to 20 mm (*cf.* Figure 6 (a)). A mass per unit of length equal to 900 kg/m is applied to each floor *via* the longitudinal beams (live loads), in addition to the dead loads. The last storey is two-span long, while the others are made of four spans: a rooftop is thus located at the 2nd-floor level. As a result, the masses and the dead loads are the highest on the 1st-floor column/beam junction located on the western side of the building. This structural element is thus assumed as the tested specimen in the following.

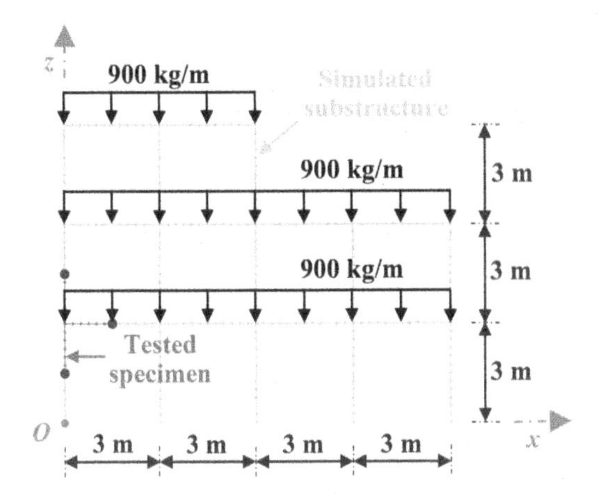

Figure 5. Elevation view of the three-storey RC frame.

A viscous damping ratio set at 2 % at $f_1 = 1.19$ Hz (*i.e.* the main eigenfrequency) is applied to dampen the high-frequency content. The damping matrix is thus defined such as $\boldsymbol{C} = \beta_M \cdot \boldsymbol{K}_E$, with $\beta_M = 0.02 /(\pi \cdot f_1)$, *i.e.* $\beta_M = 5.3 \times 10^{-3}$ s/rad. Several researchers showed from experiments that this damping matrix is well suited to model damaging reinforced concrete structures, knowing that the damping cannot depend on the mass matrix when the section is fully broken (Faria 2002). The damping is thus managed by the concrete damage in the low-frequency range.

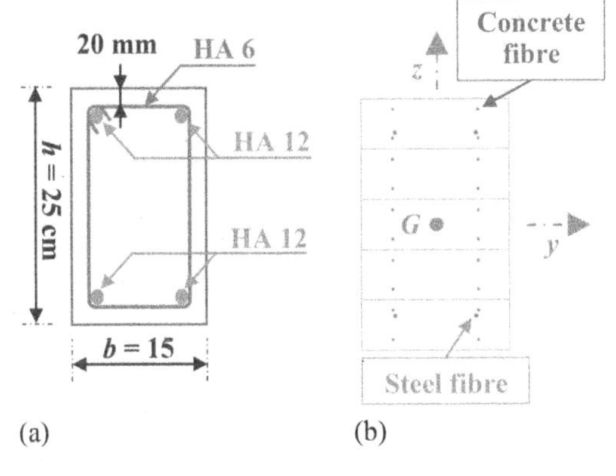

Figure 6. Cross-section of the beams (a) and mesh of the cross-section of the beams (b).

The complete structure comprises 798 free DOFs (*i.e.* 271 nodes), while all the cross-sections are divided into 1×5 surface elements. The concrete fibres are located at the integration points of the surface elements (grey dots), while the steel fibres (blue dots) are located at 32 mm from the edges of the cross-sections (*cf.* Figure 6 (b)).

During the hybrid test, the restoring forces applied to the common DOFs (see blue dots in Figure 5) are measured on a specimen of the columns/beam junction. Thus, relevant boundary conditions need to be applied to the experimental setup to achieve viable results, as described in Section 6.2.

6.2 *Virtual experimental setup*

The column/beam junction includes the mid-length of the right beam and the mid-heights of the upper and lower columns. Even if all ends are embedded in an actual structure, several simplifications can be assumed on the experimental setup based on the properties of the building as well as the loading applied on it.

Firstly, it is commonly assumed in earthquake engineering that the mass of the building is mainly located at floor level. As a result, the bending moment evolves linearly along the columns when horizontal forces are applied to the floors, as it is the case during earthquakes. So, the bending moment can be considered as close to zero at mid-heigh of the storeys (as shown in Figure 7 (b)). Pin connections are thus applied at the ends of the half-columns on the tested specimen, allowing them to rotate freely.

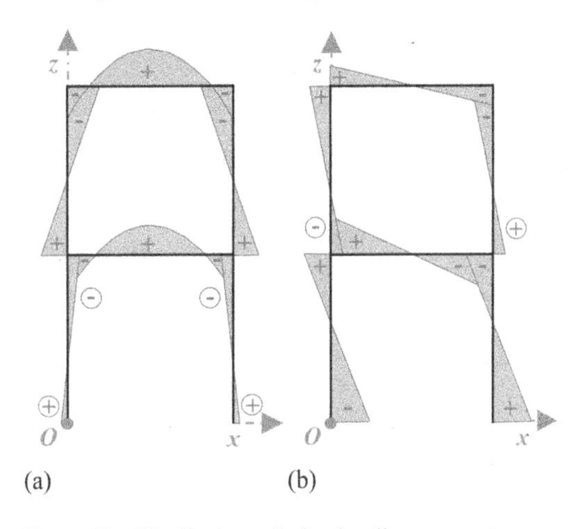

Figure 7. Distribution of the bending moment on a two-storey frame under vertical live loads (a) and horizontal earthquake (b).

Secondly, by considering that the PGA of the ground acceleration in Figure 4 (equal to 2.16 m/s^2) is more than four times lower than the gravitational acceleration (equal to 9.81 m/s^2), it is reasonable to consider that the horizontal earthquake has a low influence on the vertical displacements of the spans (compared to the dead and live loads). As a result, the beams are mainly loaded by the vertical live loads, so the bending moment reaches an extremum close to the mid-length of the spans (*cf.* Figure 7 (a)). The end of the half-beam is thus not able to rotate on the experimental setup.

Vertical and horizontal displacements U_{x1}, U_{x2} and U_{z2} are applied at the ends of the upper half column and half beam by using three actuators (*cf.* Figure 8). F_{z1} is applied at the top of the upper half column by using pre-stressed steel rebars. Its value is set as a constant and equal to 27.8 kN according to the dead and live loads (the earthquake is thus neglected).

The live loads are applied to the half beam by using an additional static actuator. Knowing that the mass per unit length is 900 kg/m and that the half beam is

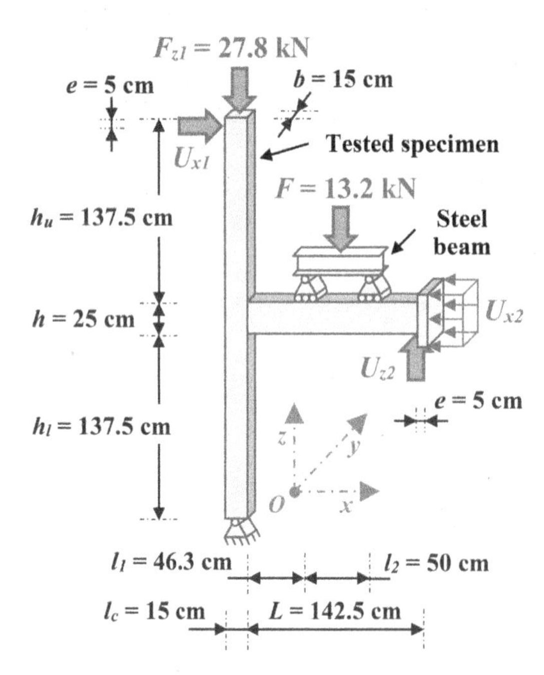

Figure 8. Elevation view of the three-storey RC frame.

1.5 m long, the force applied by this actuator is equal to 13.2 kN. This vertical load is transmitted to the half beam *via* a 50 cm long simply supported steel beam. The forces F_{x3} and F_{z3} (applied to the pin connection at the end of the lower half column) are measured, whereas the moment M_{y2} at the end of the half beam can be assessed by writing the equilibrium of the tested specimen.

As a result, the restoring forces vector ${}^T\dot{r}_\delta$ related to the common DOFs has seven non-zero components: F_{x1}, F_{z1},F_{x2}, F_{z2}, M_{y2}, F_{x3}, F_{z3}. Note that this experimental setup is complex but realistic. If necessary, the vertical live load F = 13.2 kN can be reasonably neglected on the span linked to the column/beam junction (on both substructures).

6.3 *Numerical modelling of the experimental setup*

Before performing the PsD hybrid test, a detailed finite elements analysis is firstly required to set the properties of the actuators (strength, stroke, …) as well as to check the reliability of the boundary conditions detailed in Section 6.2. To do so, the tested specimen is replaced by a 2D numerical model of the column/beam junction defined in Cast3M© software.

Here, the concrete is modelled by using 980 quadratic surface elements, while the steel rebars are made of 624 uniaxial rods whose properties are defined per unit of length (*cf.* Figure 9). The transverse steel rebars are explicitly modelled, contrary to the multifibre beam elements. Their spacing varies between 5 and 15 cm. They are mainly placed at the ends of the specimen and at the level of the connection between the beam and the columns (to avoid the appearance of shearing collapse mechanisms in case of earthquake). The area of the longitudinal steel rebars is equal to 15.1 cm^2/m, while it is set at 8.3 cm^2/m for the transverse ones.

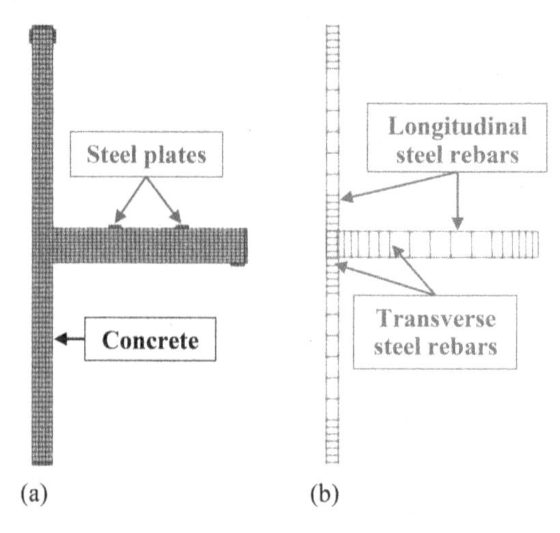

Figure 9. Elevation view of the 2D mesh of the tested specimen (a) and steel rebars (b) defined in Cast3M© software.

Note that 2.5 cm thick steel plates are located at the ends of the specimen and at the location of the actuators (as it is the case during actual experiments on RC structures). This avoids the appearance of local stress concentrations and prevents the concrete from tearing off.

The bilinear elastic-plastic law with kinematic hardening defined in Section 2.2 is used with the same parameters to model the behaviour of the steel rebars. The concrete is modelled by an accurate quasi-brittle material law available in Cast3M© software. The stiffness recovery, inelastic strains and frictional sliding are all considered (Richard et al. 2010). The law is defined by considering an elastic stiffness of 31 GPa, a yielding tensile stress of 3.5 MPa, a Poisson ratio of 0.2, a tension brittleness of 1.10^{-2}, a compression brittleness of 4.710^{-4}, a kinematic hardening of 7.10^9 Pa and a nonlinear hardening of 7.10^{-7} Pa^{-1}. Contrary to the La Borderie damage law, the energy dissipation due to frictional sliding is modelled, although the "unilateral" effect is partial (*cf.* Figure 10).

The finite element analysis is next performed on the simulated substructure (made of multifibre beam elements) by using a solver defined in MATLAB©, while the displacements of the actuators are sent to a console running Cast3M© software in parallel. Restoring forces are then computed and sent to MATLAB© for the next time step and so on. The analysis is first performed using the full order model (FOM). A POD projection modal basis is then added to the α-OS solver, as described in Section 5. The results are compared and discussed in Section 6.4.

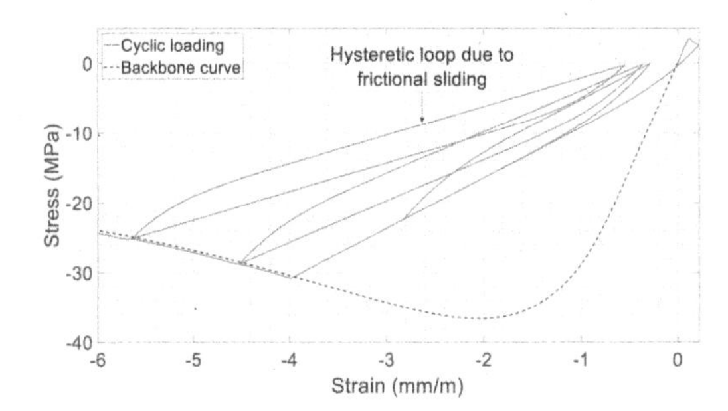

Figure 10. Uniaxial damage law of concrete under cyclic loading with partial "unilateral" effect and frictional sliding.

6.4 *Results with the full and reduced models*

The numerical substructure is reduced by using a modal projection basis made of POD-modes computed from the results of an offline implicit finite element analysis performed on the entire structure.

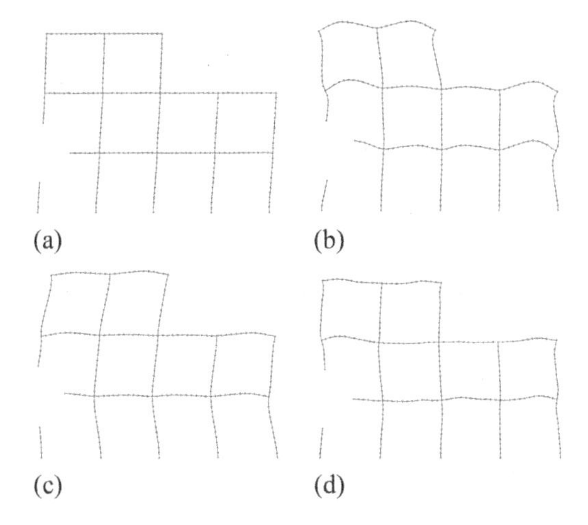

Figure 11. POD-modes of the numerical substructure: 1st mode (a), 2nd mode (b), 3rd mode (c) and 4th mode (d).

The use of POD-modes introduces additional information on the nonlinearities (*i.e.* location of damage and plasticity, as it is clearly visible in the shape of the 2nd POD-mode drawn in Figure 11 (b)) as well as the global response of the building, despite cutting part of the high-frequency content. As a result, their use to perform hybrid tests can change the displacements applied to the tested specimen and the response of the numerical substructure. To assess the reliability of the reduced-order model (ROM) in the framework of PsD hybrid tests, a comparison with the full order model (FOM) is thus necessary.

According to Ayoub (2021), the number of POD-modes can reasonably be assessed by guaranteeing that at least 99 % of the total system energy is considered for the ROM. Knowing that the singular value Λ_i indicates the amount of energy brought by the ith POD-mode, the energy criterion used to assess m can thus be written as described in (16).

$$\frac{\sum_{i=1}^{m}\Lambda_i}{\sum_{j=1}^{N}\Lambda_j} \geq 0.99 \tag{16}$$

Figure 12 shows that this criterion is fully reached with m = 10 POD-modes: this value is thus used to build the POD projection modal base related to the numerical substructure.

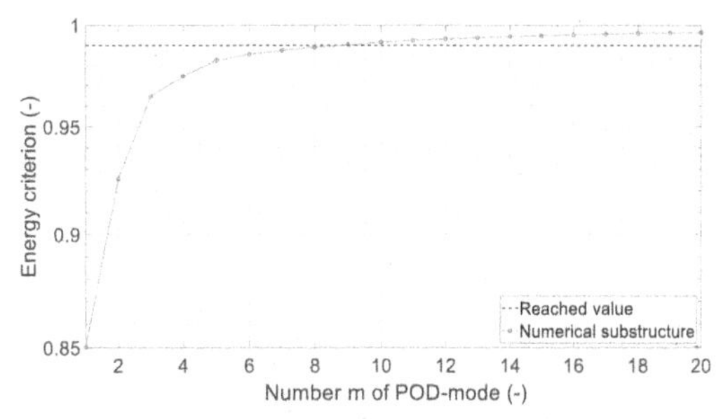

Figure 12. Energy criterion based on the singular values.

The displacement response is plotted in Figure 13 at times $t = 4.78$ s and $t = 5.38$ s (*i.e.* when the horizontal displacements reach their extrema during the strong motion phase). Results related to the implicit offline phase, hybrid test with the FOM and hybrid test with the ROM are compared. Note that the displacements of the tested specimen are plotted by post-processing, and that deformations are amplified by a factor 50.

The global responses of the RC frame computed in hybrid test conditions fit well with the implicit Newmark reference, despite the simplifications made on the boundary conditions applied to the tested specimen. However, with the FOM, note that the numerical model of the tested specimen undergoes lower deformations than the other RC column/beam junctions, leading to lower horizontal displacements on the complete structure, especially at the top of the building. On the contrary, the dynamic response is similar on all junctions with the ROM, leading to more consistent and "realistic" results, each POD-modes carries information about the local nonlinearities and the global response of the RC frame.

Simplifications being most of the times necessary on PsD experimental setups, these results show that using POD-modes computed from a fully numerical implicit finite element analysis carried out on the complete structure partially corrects the induced error and improves the consistency of the dynamic responses of both the numerical substructure and the tested specimen. However, it should be noted that the CPU time related to the numerical substructure (modelled on an Intel™ Core™ i9-10900K CPU @ and 64 GB RAM personal computer using MATLAB© software) is approximately equal to 42 s with the FOM and 40 s with the ROM. As a result, the number of DOFs is not high enough to save significant CPU time when

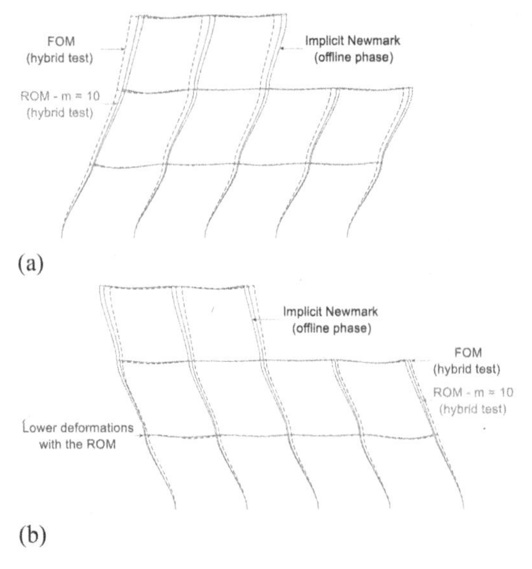

Figure 13. Response in displacements of the entire RC frame at times $t = 4.78$ s (a) and $t = 5.38$ s (b).

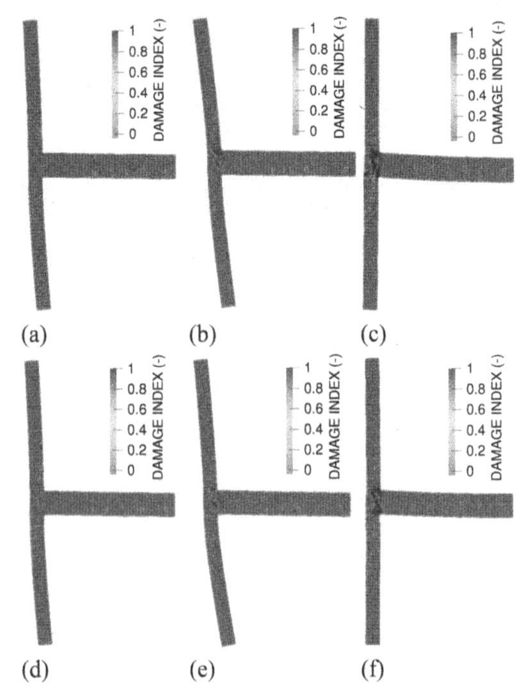

Figure 14. Damage index of specimen: FOM at times $t = 2.60$ s (a), $t = 2.73$ s (b) and $t = 16.79$ s (c), and ROM (m = 10) at times $t = 2.60$ s (d), $t = 2.73$ s (e) and $t = 16.79$ s (f).

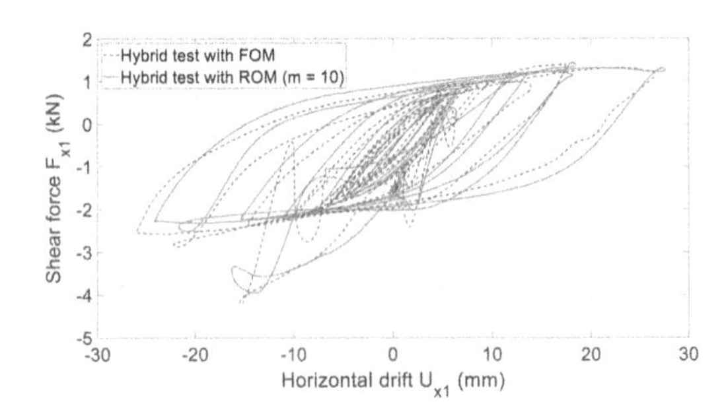

Figure 15. Shear/drift response of the tested specimen.

a POD projection modal base is used, the α-OS time integration scheme is non-iterative and all operators are pre-computed. To save more CPU time and allow for faster testing, it is thus necessary to reduce the computational cost due to assessing the restoring force vector ${}^{S}\tilde{\boldsymbol{r}}\left(u\right)$ at each time step. This can be achieved by using a POD Discrete Empirical Interpolation Method (DEIM) approach.

To ensure that the boundary conditions applied to the tested specimen lead to actual damage mechanisms, the distributions of the damage index computed with the FOM and the ROM are compared in Figure 14.

Contrary to the dynamic response of the full structure, the static response of the tested specimen is almost the same with the FOM and the ROM (*cf.* Figure 15). As a result, the same damage mechanisms appear. Damage index at time $t = 2.60$ s shows that cracks due to bending first appear at the transverse steel rebars (where the reinforced concrete is locally stiffer). Then, at time $t = 2.73$ s, shearing led to the appearance of a 45° inclined crack on the node connecting the columns to the beam. At the end of the hybrid test (*i.e.* $t = 16.79$ s), damage is thus mainly located around these areas. These results are in accordance with the experiments performed by Masi et al. (2013), which highlighted similar damage mechanisms and shear/drift behaviour (*cf.* Figure 15). As a result, the boundary conditions applied to the specimen can be considered as well suited to perform hybrid tests on column/beam junctions.

The minimum requirements to consider for the actuators applying the displacements U_{x1}, U_{x2} and U_{z2} are finally assessed based on the previous results. Note that the data given in Table 1 will be soon used to perform an actual PsD hybrid test on a column/beam junction.

Table 1. Minimum requirements to consider for the actuators.

Displacement	Strength kN	Stroke mm	Type
U_{x1}	4.2	52.6	Double acting
U_{x2}	4.6	28.9	Double acting
U_{z2}	9.0	2.7	Simple acting

7 CONCLUSIONS

In this paper, a PsD hybrid experimental setup allows for assessing the nonlinear behaviour of a RC column/beam junction under earthquake is proposed.

Pins connections are applied to the ends of the half-columns, while the rotations are not allowed at the end of the half beam. The static loads applied to the upper storeys are modelled by pre-stressing the half-columns with steel rebars, while the displacements at the ends of the tested specimen are applied by using three actuators. The numerical substructure is modelled by using nonlinear multifibre beam elements, and the use of a POD projection modal basis to reduce its computational cost is investigated. FEM analyses are carried in hybrid test conditions by substituting the tested specimen with a numerical model defined in Cast3M© software.

Results showed that using a POD projection modal basis computed from an offline implicit finite element analysis improves the consistency of the response of both numerical and tested substructures (additional information on the global response of the structure as well as nonlinearities being added) but does not significantly reduce the CPU time. In addition, the simplified boundary conditions applied to the specimen led to actual damage mechanisms, showing their relevancy.

To reduce the computational cost due to the assessment of the restoring force vector ${}^{S}\tilde{\boldsymbol{r}}_i\left(\tilde{\boldsymbol{u}}_j\right)$, further investigations are led to assess the reliability of the POD-DEIM hyper reduction method in the framework of PsD hybrid tests.

REFERENCES

Ayoub, N., Larbi, W., Pais, J., Rouleau, L., & Deü, J.-F. 2021. An application of the proper orthogonal decomposition method or nonlinear dynamic analysis of reinforced concrete structures subjected to earthquakes. *COMPDYN 2021, 8th ECCOMAS Thematic Conference on Computational Methods and Structural Dynamics and Earthquake Engineering.* Athens, Greece.

Bonnet, P., Williams, M., & Blakeborough, A. 2008. Evaluation of numerical time-integration schemes for real-time hybrid testing. *Earthquake Engineering and Structural Dynamics*, 37:1467–1490.

Buchet, P., & Pegon, P. 1994. PsD testing with substructuring: Implementation and use. *Special publication, ISPRA, I.94.25.*

Combescure, D., & Pegon, P. 1997. α-operator splitting time integration technique for pseudodynamic testing. error propagation analysis. *Soil Dynamic end Earthquake Engineering*, 16:427–443.

Combescure, D., Pegon, P., & Magonette, G. 1995. Numerical investigation of the impact of experimental errors on various pseudo-dynamic integration algorithms. *Proceeding of the 10th European Conference on Earthquake Engineering* (pp. 2479–2484). Rotterdam, The Netherlands: Duma G. (ed.) Balkema.

Faria, R., Vila Pouca, N., & Delgado, R. 2002. Seismic behavior of a r/c wall: numerical simulation and experimental validation. *Journal of Earthquake Engineering*, 6(4):473–498.

Hakuno, M., Shidawara, M., & Hara, T. 1969. Dynamic destructive of a cantilever beam controlled by an analog-computer. *In Procedings of JSCE*, n°171.

Hilber, H., Hugues, T., & Taylor, R. 1977. Improved numerical dissipation for time integration algorithms in structural dynamics. *Earthquake Engineering and Structural Dynamics*, 5(3):282–292.

Iskef, A. 2016. Technologies informatiques pour l'étude du comportement expérimental et numérique d'un assemblage poteau-poutre en béton armé. *PhD Thesis. ENS Cachan. (In French).*

Kotronis, P., Davenne, L., & Mazars, J. 2004. Poutre multifibre de Timoshenko pour la modélisation de structures en béton armé. Théorie et applications numériques. *Revue Française de Génie Civil, Taylor & Francis*, 8(2–3):329–343. *(In French).*

La Borderie, C. 1991. Phenomènes unilatéraux dans un matériau endommageable: modélisation et application à l'analyse de structures en béton. *PhD Thesis. Paris VI University. (In French).*

Lebon, G. 2011. Analyse de l'endommagement des structures de Génie Civil : techniques de sous-structuration hybride couplées à un modèle d'endommagement anisotrope. *PhD Thesis, ENS Cachan. (In French).*

Masi, A., Santarsiero, G., Lignola, G., & Verderame, G. M. 2013. Study of the seismic behavior of external RC beam-column joints through experimental tests and numerical simulations. *Engineering Structures*, 52:207–219.

Moutoussamy, L. 2013. Essais hybrides en temps réels sur structures de Génie Civil. *PhD Thesis. ENS Cachan. (In French).*

Nakashima, M., Kato, H., & Takaoka, E. 1992. Development of real-time pseudo dynamic testing. Earthquake Engineering and Structural Dynamics, 21(1):79–92.

Nguyen, T. (2012). Analyses du comportement de rupteurs thermiques sous sollicitations sismiques. *PhD Thesis, ENS Cachan. (In French).*

Pegon, P., & Pinto, V. 2000. Pseudo-dynamic testing with substructing at the ELSA Laboratory. *Earthquake Engineering and Structural Dynamics*, 29:905–925.

Richard, B., Ragueneau, F., Cremona, C., & Adelaide, L. 2010. Isotropic continuum damage mechanics for concrete under cyclic loading: Stiffness recovery, inelastic strains, and frictional sliding. *Engineering Fracture Mechanics*, 1203–1223.

Shing, P., Vannan, M., & Carter, E. (1991). Implicit time integration for pseudodynamic tests. *Earthquake Engineering and Structural Dynamics*, 20:551–576.

Souid, A., Delaplace, A., Ragueneau, F., & Desmorat, R. 2009. Pseudo-dynamic testing and nonlinear substructuring of damaging structures under earthquake loading. *Engineering Structures*, 31(5):1102–1110.

Takanashi, K., & Nakashima, M. 1987. Japanese activities on online testing. *Journal of Engineering Mechanics*, 113(7):1014–1032.

Computational Modelling of Concrete and Concrete Structures – Meschke, Pichler & Rots (Eds)
© 2022 Copyright the Author(s), ISBN: 978-1-032-32724-2

Integration of the principle of mesh refinement in the Adaptive Static Condensation (ASC) method

A. Mezher*
Université Paris-Saclay, CEA, Service d'Études Mécaniques et Thermiques, Gif-sur-Yvette, France
Paris Nanterre University, LEME Laboratory, Ville-d'Avray, France

L. Jason & G. Folzan
Université Paris-Saclay, CEA, Service d'Études Mécaniques et Thermiques, Gif-sur-Yvette, France

L. Davenne
Paris Nanterre University, LEME Laboratory, Ville-d'Avray, France

ABSTRACT: In order to evaluate the cracking process in large reinforced and pre-stressed concrete structures, a predictive model of concrete damage with refined mesh and a nonlinear law can be required. Because of the computational load, such modelling is not applicable directly on large-scale structures whose characteristic dimensions are over several meters. To overcome this difficulty, a method based on the static condensation [1], called "ASC" has been proposed. It concentrates the computational effort on the damaged area, which can evolve due to crack propagation or initiation [2].

In order to reach a higher level of representativeness, mesh refinement is integrated into the ASC method in the proposed contribution. The method first uses a coarse initial mesh and refines it only in the domain of interest (DI) to reach a density of mesh adapted to a non-local model (order of the centimeter). This principle of refinement, integrated in the ASC method, is described. It is then applied to a notched bending beam. The use of mesh refinement combined to ASC method allow to keep results similar to a calculation with a fine mesh on the entire structure. However, it also allows a saving in computational time and in memory occupation that increases with the increase in the mesh density of the refined zones.

1 INTRODUCTION

When designing civil engineering structures, it is necessary to take into account all the factors that can lead to the ruin and to predict their effects. As the experiments at the scale of the large structures are very limited, it is often necessary to use the numerical modeling. These models must be able to predict the behavior of these structures for the different loading cases that are likely to occur during the life of the structure.

In this contribution, cracking in concrete for large structures (e.g. containment vessel) is studied, which is a localized phenomenon. Its modeling requires a nonlinear constitutive law with a fine mesh of the order of one centimeter in order to achieve a good accuracy and robust simulations. This type of modeling can be expensive and sometimes unattainable on large structures. Several solutions for solving this problem exist in the literature.

One of the possible solutions is the adaptive mesh refinement method (AMR) which consists in locally increasing the fineness of the mesh in the areas where it is essential to calculate the solution with greater precision than in the rest of the structure. In [3], the concept of mesh refinement was introduced first for the resolution of hyperbolic partial differential equations in 1984. In [4], it was then proposed an adaptive version of 2D mesh refinement for conservation laws .The efficiency of AMR for gas dynamics was shown in 1989. More recently, the AMR method has been adapted to several fields of physics such as compressible and incompressible fluid dynamics, solid mechanics and combustion [5,6]. Two families of adaptive mesh refinement techniques exist: remeshing techniques and refinement techniques. The first consists in completely reconstructing the mesh where it is necessary as in [7–9] or to move the mesh to generate adaptive meshes as in [10,11]. The second is based on the elements of the initial mesh to generate a new one by locally enriching the spatial discretization [12]. The advantage of the AMR method lies mainly in its performance in terms of memory size and CPU time [13]. However, with this approach, the whole structure is modelled with a nonlinear behavior even in the part where the materials

*Corresponding Author

DOI 10.1201/9781003316404-50

stay in the linear part of the behavior. In the case of localized damaged zones, this step can become very costly and the solution is not very satisfying. The decomposition of the system combined to the use of parallel computation to exploit several computation units simultaneously is another possible solution. It can significantly reduce the calculation time. Several approaches have been developed for parallel computation analysis. A review of the fundamental concepts and issues of parallel processing is made in [14]. The parallel processing can be applied in fluid mechanics [15], for nonlinear analysis of reinforced concrete three-dimensional frame [16], among others domains. Parallelism brings a computing power and a storage capacity, which increases with the increase of the number of processors. This makes it possible to reach higher simulation levels, within a reasonable timeframe [17]. However, with this type of approach, the overall computational load increases because of the cost of communication between the different calculation units. Therefore, for a given computing power, this method does not improve the computation and even with this approach. In computational mechanics, static condensation is a model reduction method that reduces the number of degrees of freedom by eliminating variables from the linear system in the stiffness matrix. Condensed substructures are thus created. "Superelements" are generally defined by eliminating the internal unknowns in the condensed zones. Complex problems, whose complexity is related to their size, can be calculated at a more reasonable cost. The condensation method was first introduced by Guyan in 1965 [9] and has been widely used in mechanics. The method is called "static condensation" since it is only exact for static problems, even if it has been widely applied in structural dynamics. In this case, the condensation of mass is approximated [1]. The initial formulation have been gradually improved in dynamic calculations [6,18,19] and/or associated with substructuring [20] in the analysis of structures with localized nonlinearities. In [13], the static condensation was used for efficient conceptual-level aerospace structural design.

Regarding nonlinear cracking in reinforced concrete structures, a so-called adaptive static condensation method has been developed [2]. It consists in concentrating the computational effort of the nonlinear calculation on preliminary defined "zones of interest" (zones with expected cracking behavior), by "eliminating" the zones with a linear elastic behavior. This approach uses Guyan's static condensation method [1] to replace the elastic zones by a set of boundary conditions applied at the boundaries of the zones of interest. As the system evolves (evolution of a given crack or apparition of a new one), criteria are used to detect if damage is likely to appear and to make evolve the geometry of the zones of interest. This method enables to reduce the dimension of the nonlinear problem without altering the quality of the results compared to a complete reference computation. The method was applied to several structures: a notched bending beam, a prestressed beam and a simplified containment building in reinforced concrete subjected to an internal pressure. Results were identical to a complete reference computation (without condensation) with a very variable time saving compared to the complete computation. This gain varied between 3 and reached 15 for the case of prestressing. However, even with the ASC method, elements size of 1 cm still seems impossible to reach with current machines for the simulation of very large structures and in particular containment vessels. In order to overcome this difficulty, the principle of mesh refinement is integrated in the ASC method. The main idea is to refine only the domain of interest by keeping a coarse mesh for the rest of the structure.

After a brief description of the ASC method in part 2, the integration of the mesh refinement in the ASC method is presented in section 3.1. In section 3.2, this approach is applied on a notched bending beam to test the efficiency of the method.

2 ASC METHOD

The first section presents Guyan's static condensation principle on which the ASC method is based. Then, the second one details the algorithm of the method.

2.1 *Guyan's static condensation [1]*

Static condensation, also known as Guyan reduction simplifies the resolution of linear systems of large dimensions by eliminating part of the degrees of freedom (DoF). For a given mechanical problem discretized on n dofs, the static equilibrium can be expressed by the equation (1):

$$K\,u = f \tag{1}$$

where K represents the stiffness matrix of the structure, u is the displacement vector and f the nodal force vector. We broke down the structure into two subdomains Ω_C (slave) and Ω_M (master):

$$\begin{pmatrix} K_{C,C} & K_{C,M} \\ K_{M,C} & K_{M,M} \end{pmatrix} \begin{pmatrix} U_C \\ U_M \end{pmatrix} = \begin{pmatrix} f_C \\ f_M \end{pmatrix} \tag{2}$$

with $K_{C,C}$ ($\in\mathbb{R}^{p,p}$); $K_{M,M}$ ($\in\mathbb{R}^{q,q}$); $K_{M,C}$ $\in\mathbb{R}^{p,q}$ ($\in\mathbb{R}^{p,q}$) $K_{C,M}$ ($\in\mathbb{R}^{q,p}$) f_C, U_c ($\in\mathbb{R}^{p}$); f_M, U_M ($\in\mathbb{R}^{q}$); $n = p + q$.

By developing equation (2), a reduced (condensed) problem can be defined by:

$$\widehat{K}.U_M = \widehat{F} \tag{3}$$

The condensed force $\widehat{F}$ and the reduced stiffness matrix $\widehat{K}$ are given by the system of two equations:

$$\begin{cases} \widehat{K} = K_{M,M} - K_{M,c}K_{c,c}^{-1}K_{c,M} \\ \widehat{F} = f_M - K_{M,c}K_{c,c}^{-1}f_c \end{cases} \tag{4}$$

The reduced system given by equation (3) is of dimension q that is smaller than n and its resolution gives the solution on the master domain Ω_M. However, it is always possible to obtain the vector U_c by applying a "decondensation" given by equation (5):

$$U_C = K_{c,c}^{-1}(f_c - K_{c,M}.U_M) \tag{5}$$

It is noted that if the domain Ω_C is large, the computation of the inverse term $K_{c,c}^{-1}$ can become an expensive task. It is also noted that the static condensation is only exact if the condensed domain has a linear elastic behavior.

2.2 *Detailed algorithm of the ASC method*

This ASC method was developed to deal with the problems of large dimension structures with localized damage. It uses Guyan's static condensation method to transform the elastic zones to a set of boundary conditions applied to the boundaries of the DI. If damage evolves, criteria allow evolving the DI with the evolution of damage and maintain the accuracy of the calculation. Figure 1 presents the algorithm of the ASC method.

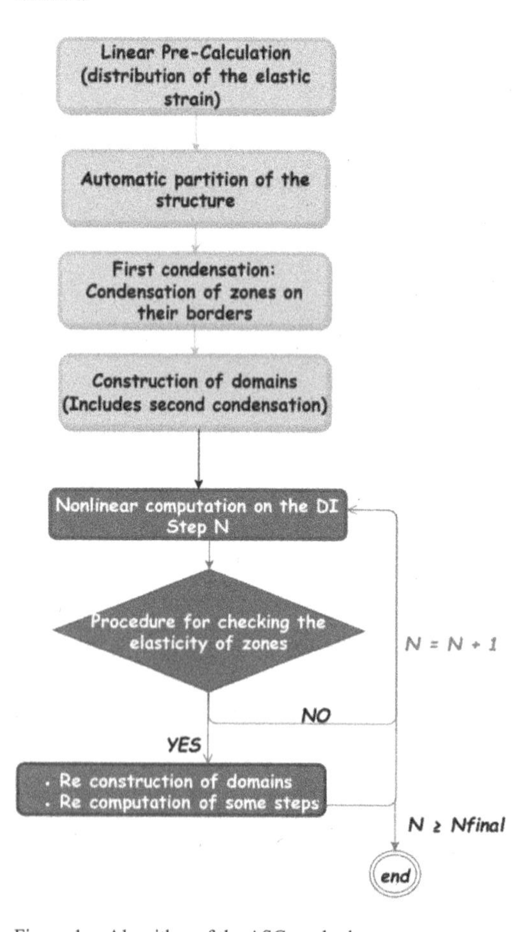

Figure 1. Algorithm of the ASC method.

The first step consists in carrying out a linear precomputation on the whole structure to obtain a distribution of the elastic strain. A quantity of interest is then calculated. In this contribution, related to damage mechanics, this quantity is Mazars's equivalent strain, ε_{eq}, which is computed using the following equation [21]:

$$\varepsilon_{eq} = \sqrt{\langle\varepsilon_1\rangle_+^2 + \langle\varepsilon_2\rangle_+^2 + \langle\varepsilon_3\rangle_+^2} \tag{6}$$

where $\langle\varepsilon_i\rangle_+$ represents the principal positive value of the strain. The next step is to partition the structure into zones. This partitioning is carried out using an automatic cutting procedure adapted to the ASC method [22]. This partitioning is based on the distribution of the elastic strain and takes into account the initiation and propagation of the damage over time. Then we carry out the "first condensation" in which each zone is condensed and replaced by equivalent boundary conditions on its borders. This step is done only once before starting the computation and the condensed matrices are saved and used when needed during all the simulation. Then, the next step called "construction of domains" consists in building the two domains: the DI that will be fully represented and the elastic domain (ED) which is represented by a set of boundary conditions on the borders of the DI (second condensation). Our starting DI is defined from the zone, which is most likely to be damaged based on the elastic strain distribution that has been already calculated (the zone associated to max (ε_{eq})). All these steps, colored in orange on the figure, forms what is called the preparation phase. Now, the system on which we carry out the nonlinear calculation is built from the domain of interest and the boundary conditions which represent the rest of the structure and which has a reduced size compared to the global system.

As we have already mentioned, the condensation is only exact if the condensed part has an elastic behavior. To keep the accuracy of the method, we test the elasticity of the elastic domain during the calculation. This is done using a verification procedure, which has two criteria: the propagation and the initiation criteria. The first one evaluates the potential propagation of damage from the existing DI to neighboring condensed zones. It detects if the damage is approaching the border of the DI. Propagation bands are defined over a width L around the border of the DI (in orange on the Figure 3). If damage reaches this band, the neighboring zone is added to the DI. This criterion ensures that the interface between the DI and the elastic domain has an elastic behavior during computation. As this test is only geometric, the associated computational cost is not expensive. This criterion is thus checked at the end of each time step. The second one is the initiation criterion that evaluates the potential apparition of damage on all the condensed zones. It supposes to check the elasticity of the condensed zones. To do so, a double "decondensation" is performed to obtain the values of the displacements in the overall structure then we test if we are still in the elastic domain. Otherwise, if the

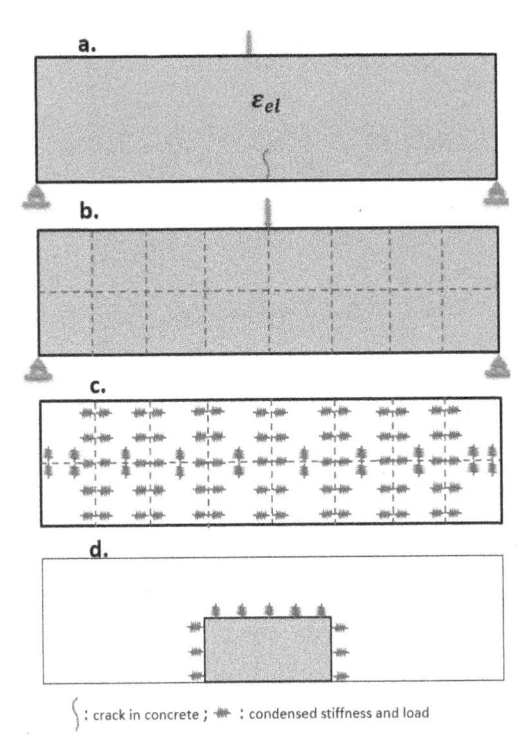

Figure 2. The preparation phase, a: linear pre-calculation, b: partitioning of the structure, c: first condensation, d: the starting DI.

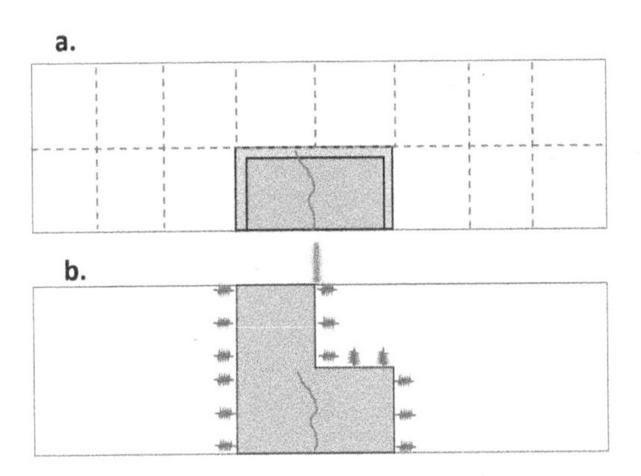

Figure 3. a: Detection of the propagation criterion; b: evolution of the DI.

elasticity threshold is exceeded in a given zone outside the DI, this zone has to be included in the new DI. Because of the cost of the double "decondensation", which has a significant cost and especially if the ED is large and finely meshed, this criterion is only checked after each p loading steps.

In the case of an evolution of the DI (evolution of damage to the condensed domain), a domain reconstruction procedure is used in which only the second condensation is carried out. It especially shows the advantage of the double condensation as a re-condensation of the whole structure is not needed. It is to be noted that in both cases, if the DI is changed, previous loading steps need to be recalculated (p steps in case of initiation, one-step for propagation). In the case of detection of initiation of the damage in a new zone, this zone was condensed between the step N and the step$N - p$. To avoid any loss of information in this case, we return to the step number$N - p$, the step during which this zone was elastic and does not present any damage. To ensure in the case of propagation at the step N that the borders between the two domains always have elastic behaviors, we recalculate the step N by adding the new detected zone. The implementation of the method is done on the finite element computation software cast3m [23].

3 MESH REFINEMENT

In order to improve the representativeness of the modelling strategy and to overcome the problem of mesh dependence, the use of a regularization technique is necessary. The use of these techniques and in particular the integral method [24] requires a mesh fineness of the order of one centimeter. However, on a large structure, the use of a fine mesh over the entire structure may not be applicable because of the calculation cost and the memory occupation. On the other hand, in the case where the damage is located in a small part of the structure, the use of a fine mesh in the areas outside the areas of interest is not necessary.

For this, we have introduced the principle of mesh refinement with the ASC method. The objective is to apply the regularization (a nonlocal model) in the DI only, while keeping the initial coarse mesh in the condensed elastic domain. The main idea in this approach is not to use, in any step of the method, a fine mesh on the entire structure but only on the DI while maintaining the accuracy of the ASC method. The reason is that even a linear operation may be inapplicable on a large structure with a very fine mesh.

In this part, we first present the steps necessary for the integration of the approach of mesh refinement in the ASC method then one application of this approach on a notched bending beam is presented.

3.1 *Method*

The principle is to use a fine mesh only where it is necessary. Let us take the algorithm of the method presented in Figure 1. The linear precomputation, the partitioning of mesh into zones, which is done using an automatic partitioning procedure independent of mesh size, the first and the second condensation do not require a fine mesh. The entire preparation phase that contains the four first steps (in orange) is thus carried out on the initial coarse mesh.

The first intervention for the application of mesh refinement in the ASC method takes place after this phase and just before starting the nonlinear computation. Once we detect the starting DI, we refine it by subdividing the elements until we reach the required mesh density using a hierarchical h-refinement method [25]. We chose this method because of its hierarchical nature which makes it easy to implement and can ease the projection of the fields of the initial mesh on the new fine one (all the nodes of the initial mesh exist in the new mesh). The target density is related to the application of a non-local damage model (order of 1 cm).

As explained in part 2, with the propagation criteria and the recalculation steps, it is ensured that the first row of elements cannot be damaged during computation. That is why, in order to optimize the refined part of the domain of interest, we refine only the inner part of the DI without modifying the boundary elements. Figure 4 shows an application of this refinement on simplified containment vessel (Figure 4-a), the initial coarse mesh of the structure (Figure 4-b) and the domain of interest refined and without boundary refinement (Figure 4-c).

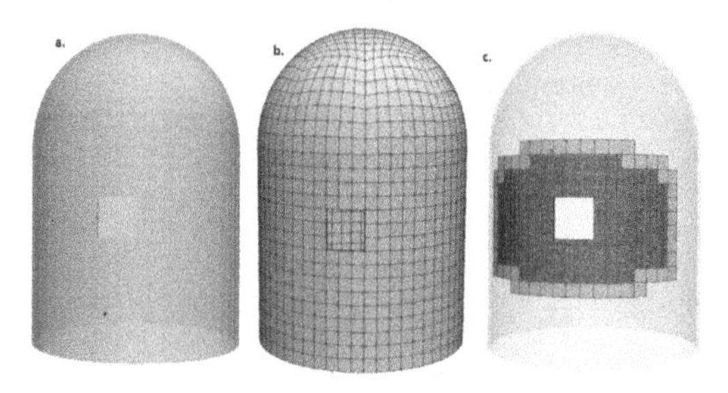

Figure 4. Example of refined DI on a cylindrical reinforced concrete structure; a: the structure; b: the initial coarse mesh; c: the mesh of the DI after refinement.

With the process of mesh refinement and in particular when using several mesh sizes, a nonconformity in the mesh occurs. This results in the creation of hanging nodes, which are generated after each refinement if the adjacent elements do not have the same size. Figure 6 shows on the right an example of a conformal mesh and on the left a non-conforming mesh that contains one "hanging node". Usually, with a finite element mesh, the vertices are shared with their other neighboring elements, but the node circled in red in the figure (node C) does not belong to the top triangle [12]. The hanging nodes are either removed, by connecting it to another vertex and thus creating two new elements as in [26], or by imposing relations on the hanging node, what is called compatibility relations by imposing for example in this case $U_c = 0.5(U_A + U_B)$ [27]. The second option is chosen here. These relations are imposed via Lagrange multipliers [28] in the same way that the boundary conditions of Dirichlet . This refinement step is performed each time the DI evolves.

As already explained in part 2.2, some computation steps must be recalculated when the DI evolves to ensure the accuracy of the calculation. To initialize the results in the new activated zones, a decondensation is carried out to obtain the results in these zones because these zones were condensed and therefore the results did not exist. However, this decondensation gives the

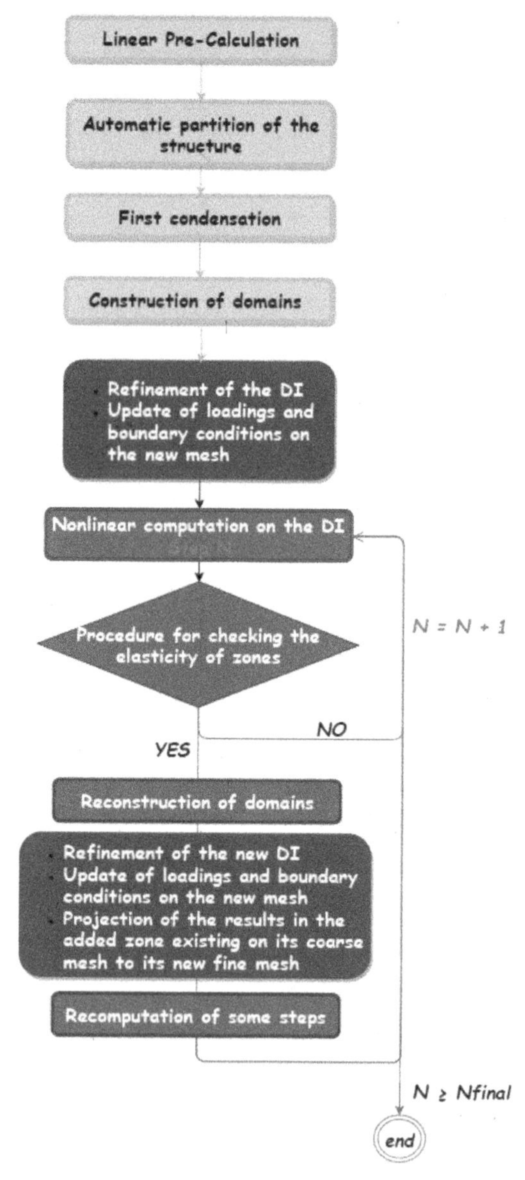

Figure 5. The algorithm of the ASC method after the integration of the mesh refinement into it.

results on the coarse mesh because the condensation is carried out on the initial coarse mesh during the preparation phase. To go to the fine mesh, remappings are performed for all the required fields on the new fine mesh. It is noted that with the hierarchical refinement, this remap step is only exact in the case of linear elastic behavior and the remap problem is one of the main reasons why many studies based on automatic mesh refinement are limited to the elastic frame [12]. In the ASC method and with the recalculations that we perform (see paragraph 2.2), we ensure that all the zones that we activate have linear behaviors at the time of the projection and therefore this projection is exact.

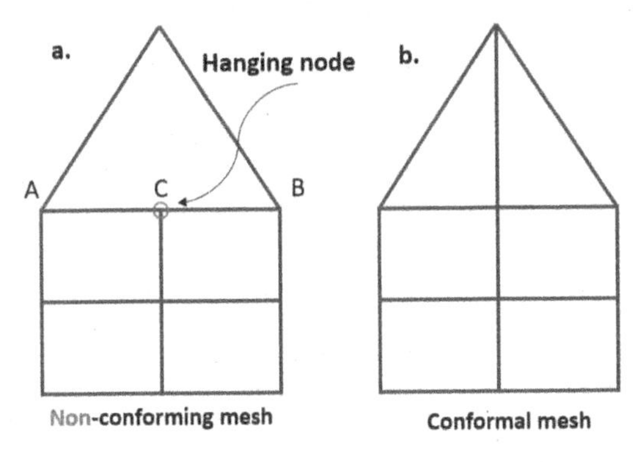

Figure 6. Illustrative example of the difference between conformal and non-conforming mesh.

The fields defined at the nodes of the initial mesh are transferred to the new refined one by using the interpolation of the form functions [29] [30]. In the case of the fields defined at the Gauss points (like the constraints and internal variables), no continuous approximation of this fields overall the space but only in discrete points. The method used for these fields is that presented in [29]. This method consists in building a continuous approximation of these fields while passing by a field at the nodes of the initial mesh then to interpolate it on the Gauss points of the new mesh.

Loadings like the internal pressure, the prestressed and the self-weight must be projected on the new mesh when the DI evolves. The kinematic relations (steel-concrete bonding links for example) and the boundary conditions must also be updated with every DI evolution. Figure 5 presents the updated algorithm of the ASC method after the integration of the principle of mesh refinement into it. The new required tasks are those in blue on this algorithm.

3.2 *Application: notched bending beam*

To test the applicability of the method, we apply the ASC method with the approach of mesh refinement on the example of the notched bending beam. It is noted that in [2], the ASC method was validated on this structure and on a simplified containment vessel and results similar to a complete reference computation (without condensation) in the DI were obtained.

The beam is 160 cm long and 40 cm high and notched at mid span. The notch is 80 mm high and 8 mm large. It is modeled in two dimensions (plane stress with thickness of 20 cm), although the method allows the use of 3D elements. The initial (coarse) mesh is made with 610 quadrilateral elements of 4 cm length (Figure 8).

Concrete is modeled using Mazars's damage model [31] with the same parameters as in [32]. It gives a compressive strength of 41.4 MPa and a tensile strength of 3.0 MPa. The load is applied on a 8 cm zone on the top of the beam, through an imposed vertical displacement downwards increasing from 0 up to 20 mm. Vertical displacement is blocked at the support points and horizontal displacement is blocked at the top middle point only and free elsewhere. For the application of the condensation, the structure is partitioned into 35 zones using the automatic partitioning procedure that we have developed (see Figure 7).

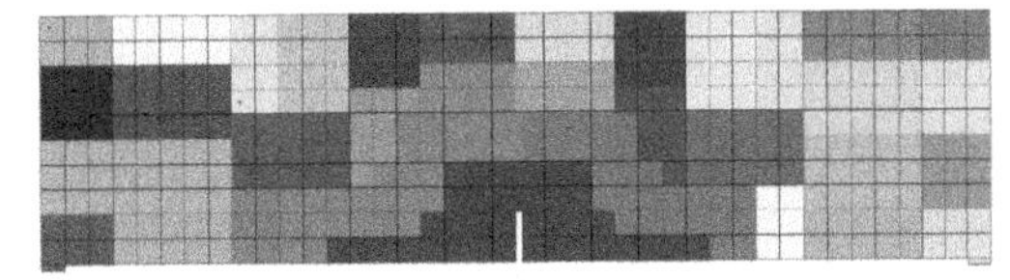

Figure 7. The notched beam partitioned into 35 zones.

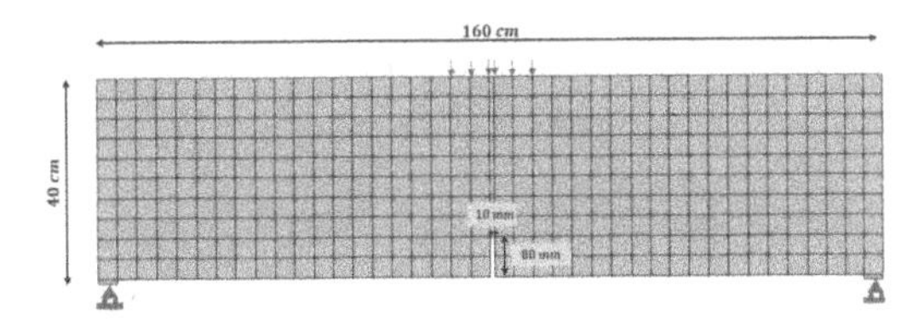

Figure 8. Geometry, mesh and load of the notched in bending.

Once the DI is detected, we refine it until we reach a final density of 1cm. This density allows us to apply a nonlocal integral method [24] to limit the damage mesh dependency in concrete with an internal length $L_c = 3cm$.

The calculation results using the ASC method with the DI refinement approach are presented in Figure 9. In this figure, we see the evolution of the DI (in blue on the figure) and the damage over time. The damage

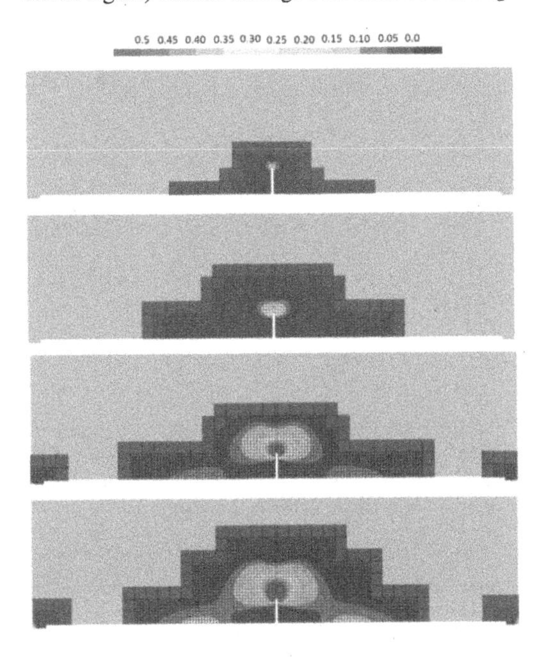

Figure 9. Evolution of the DI (in blue) and the damage.

starts around the notch, then propagates, and initiates at the support zones. On these profiles, we can very well notice that all the damaged part is refined and that all the part, which has a coarse mesh, does not show any damage.

To compare the results with a reference computation in order to see the effect of the mesh refinement on the accuracy of computation, we carried out a complete computation (without condensation) on the same beam starting from a fine initial mesh on the entire structure (of 1 cm of length). Results similar to the condensed calculation are found at all times during the simulation. Figure 10 presents the damage distribution at the end of computation using the ASC method and this same profile resulting from the complete computation.

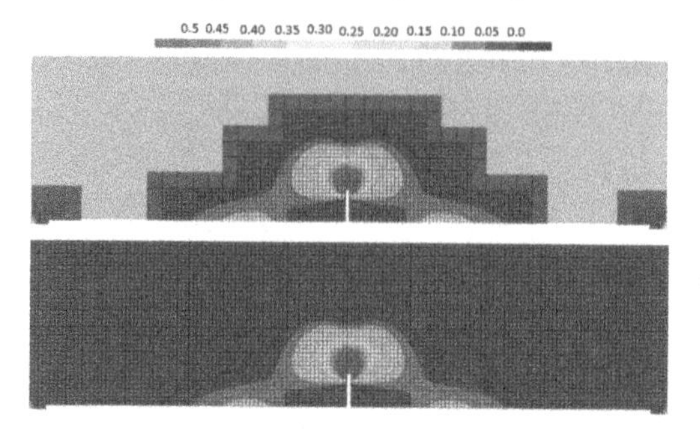

Figure 10. Damage profile of the condensed calculation (top) and the complete calculation (bottom) at the end.

A comparison was made and showed that these two profiles are similar in the DI.

Finally, to test the efficiency in terms of computational performance, we performed a third computation with the ASC method but this time starting from a fine mesh (1cm) on the entire structure. It is noted that the three computations were performed on the same computation node (32 cores). The computation time results of the three computations are shown in Table 1.

Table 1. Numerical efficiency results.

	Complete computation	ASC computation (1 cm)	ASC computation (4 cm – 1cm)
Time	549,57 s	87,52s	57,70 s
Gain Factor	1	6,2	9,6

A time saving factor of 9.6 instead of 6.2 with respect to the complete computation is obtained by integrating the mesh refinement approach in this example while keeping the same results in the DI;

4 CONCLUSION AND PERSPECTIVES

The adaptive static condensation (ASC) method consists in reducing the size of the system on only the damaged parts of the structure by condensing the elastic part. It also consists in testing the elasticity of the condensed zones during computation to allow the promotion of condensed zones into zones of interest if necessary. This method initially proposed in [2] was developed and became totally automatic and its field of application was extended (in particular on prestressing).

In order to improve the representativeness of computation with the ASC method, the principle of automatic mesh refinement of the DI has been integrated. The main objective is not to make operations using a fine mesh on the entire structure but only on the DI (where it is necessary). This approach was applied to an example of a notched bending. Results similar to a fine calculation on the totality of the structure were obtained and this principle of refinement brought a saving of time of the order of 1.5 for the ASC method.

In the next future, applications to more complex and representative structures are expected to fully evaluate the efficiency of the proposed developments.

ACKNOWLEDGMENTS

The authors gratefully acknowledge the financial and technical support of EDF R&D for the development and the analysis of the simulation results.

REFERENCES

[1] R. J. Guyan, "Reduction of stiffness and mass matrices", AIAA J.3, 1965.

[2] A. Llau, L. Jason, F. Dufour, and J. Baroth, "Adaptive zooming method for the analysis of large structures with localized nonlinearities," *Finite Elem. Anal. Des.*, vol. 106, pp. 73–84, 2015.

[3] M. J. Berger and J. Oliger, "Adaptive mesh refinement for hyperbolic partial differential equations," *J. Computational Physics*, vol. 53, no. 3, pp. 484–512, 1984.

[4] M. J. BERGER, "Local Adaptive Mesh Refinement for Shock Hydrodynamics," *J. Computational Physics*, vol. 82, p. 184, 1989.

[5] I . Babuska , A . Craig , J . Mandel and J . Pitkaranta. "Efficient Preconditioning for the p-Version Finite Element Method in Two Dimensions", SIAM Journal on Numerical Analysis, vol. 28, no. 3, pp. 624–661, 1991.

[6] R. Craig *et al.*, "Coupling of Substructures for Dynamic Analyses To cite this version: HAL Id: hal-01537654 Coupling of Substructures for Dynamic Analyses," vol. 6, no. 7, pp. 1313–1319, 2017.

[7] R. Zhang, L. Li, L. Zhao, and G. Tang, "An adaptive remeshing procedure for discontinuous finite element limit analysis", *Int. J. Numer. Methods Eng.*, no. October 2017, pp. 287–307, 2018.

[8] P. Yue, C. Zhou, J. J. Feng, C. F. Ollivier-Gooch, and H. H. Hu, "Phase-field simulations of interfacial dynamics in viscoelastic fluids using finite elements with adaptive meshing," *J. Computational Physics*, vol. 219, no. 1, pp. 47–67, 2006

[9] N. S. Lee and K. J. Bathe, "Error indicators and adaptive remeshing in large deformation finite element analysis,"

Finite Elements in Analysis and Design, vol. 16, no. 2, pp. 99–139, 1994.

[10] W. A. Wood and W. L. Kleb, "On Multi-dimensional Unstructured Mesh Adaption", 14th AIAA Computational Fluid Dynamics Conference On Multi-dimensional Unstructured Mesh Adaption, 1999.

[11] G. Bono and A. M. Awruch, "An adaptive mesh strategy for high compressible flows based on nodal Reallocation", *J. of the Brazilian Soc. Mech. Sci. Eng.*, vol. 30, no. 3, pp. 189–196, 2008,

[12] G. Gibert, "Propagation de fissures en fatigue par une approche X-FEM avec raffinement automatique de maillage", Thesis. Université de Lyon, 2019.

[13] S S. Delage Santacreu, "Méthode de raffinement de maillage adaptatif hybride pour le suivi de fronts dans les écoulements incompressibles",Thesis. Université Bordeaux I, 2006.

[14] A. K. Noor, "Parallel processsing in finite flement structural analysis," *Eng. Comput.*, vol. 3, no. 4, pp. 225–241, 1988.

[15] C. Bartels, M. Breuer, K. Wechsler, and F. Durst, "Computational fluid dynamics applications on parallel-vector computers: Computations of stirred vessel flows", *Comput. Fluids*, vol. 31, no. 1, pp. 69–97, 2002.

[16] M. L. Romero, P. F. Miguel, and J. J. Cano, "A parallel procedure for nonlinear analysis of reinforced concrete three-dimensional frames", *Computers and Structures*, vol. 80, no. 16–17, pp. 1337–1350, 2002.

[17] P. Bassomo, "Contribution à la parallélisation de méthodes numériques à matrices creuses skyline. Application à un module de calcul de modes et fréquences propres de Systus". Thesis, Université jean monnet de Saint-Etienne et de l'école nationale supérieure des mines de Saint-Etienne,1999.

[18] J. H. Ong, "Automatic masters for eigenvalue economization", J. Earthquake Engineering and Structural Dynamics, vol. 3, pp. 375–383, 1975.

[19] A. Y. Leung, "An accurate method of dynamic condensation in structural analysis", International Journal for Numerical Methods in Engineering, vol. 12, pp. 1705–1715, 1978.

[20] M. A. Lee, "an approach for efficient , conceptual-level aerospace structural design using the static condensation reduced basis element method", Thesis, Georgia Institute of Technology , 2018.

[21] J. Mazars, F. Hamon, and S. Grange, "A new 3D damage model for concrete under monotonic, cyclic and dynamic loadings" , Material and Structures , vol. 48, pp. 3779–3793, 2015.

[22] A. Mezher, L. Jason, G. Folzan, and L. Davenne, "New adaptative static condensation method for the simulation of large dimensions prestressed reinforced concrete structures", XVI International Conference on Computational Plasticity, Fundamentals and Applications, Sep 2021, Barcelona, Spain.

[23] CEA, Description of the finite element code Cast3M, 2021.

[24] G. Pijaudier-Cabot and Z. P. Bazant, "Nonlocal damage theory", J. Engineering Mechanics, vol. 113 (10), pp. 1512–1533,1987.

[25] K. Carlberg, "Adaptive h-refinement for reduced-order models", *Int. J. Numer. Methods Eng.*, vol. 102, pp. 1102–1119, 2015.

[26] G. Nicolas, T.Fouqueut, "EDF-Logiciel HOMARD", 2014.

[27] R. Bank, A. H. Sherman, and A. Weiser, "Some refinement algorithms and data structures for regular local mesh refinement", *Scientific Computing*, vol. 1, no. 48, pp. 3–17, 1983.

[28] P. Verpeaux and T. Charras, "Multiplicateur de Lagrange, Condensation Statique et Conditions Unilatérales", CSMA conference, 2011.

[29] D. Peric, C. Hochard, M. Dutko, and D. R. J. Owen, "Transfer operators for evolving meshes in small strain elasto-placticity", J. *Comput. Methods Appl. Mech. Eng.*, vol. 137, pp. 331–344, 1996.

[30] A. Bérard, "Transferts De Champs Entre Maillages De Type Éléments Finis Et Applications Numériques En Mécanique Non Linéaire Des Structures", Thesis, Université de Franche-Comté, 2011.

[31] C. Giry, "Modélisation objective de la localisation des déformations et de la fissuration des structures en béton soumises à des chargements statiques ou dynamiques", Thesis, Université De Grenoble, 2011.

[32] F. Dufour, G. Legrain, G. Pijaudier-Cabot, A. Huerta, "Estimation of crack opening from a two-dimensional continuum-based finite element computation", J. International Journal for Numerical and Analytical Methods in Geomechanics, vol. 36 (16), pp. 1813–1830, 2012.

Computational Modelling of Concrete and Concrete Structures – Meschke, Pichler & Rots (Eds)
© 2022 Copyright the Author(s), ISBN: 978-1-032-32724-2

Recent advancements in Sequentially Linear Analysis (SLA) type solution procedures

M. Pari & J.G. Rots
Faculty of Civil Engineering and Geosciences, Delft University of Technology, Delft, The Netherlands

M.A.N. Hendriks
Faculty of Civil Engineering and Geosciences, Delft University of Technology, Delft, The Netherlands
Norwegian University of Science and Technology, Trondheim, Norway

ABSTRACT: Sequentially Linear Analysis (SLA) type non-incremental non-iterative solution procedures; wherein a sequence of scaled linear analyses is performed with decreasing secant stiffness of one integration point per analysis, representing local damage increments; are proven robust alternatives to traditional incremental-iterative simulations of quasi-brittle fracture. Although several enhancements have been made over the last two decades, this class of methods is still not a practical alternative to the traditional incremental-iterative class of solution procedures. To this end, more recently, SLA has been extended to 3D simulations, and under non-proportional loading conditions, using total strain based 3D smeared cracking models – both fixed and rotating, and a composite interface constitutive model allowing for cracking-shearing-crushing failures, typical of masonry damage. Furthermore, the approach has been made relatively efficient using tailor-made solvers which efficiently use the favourable event-by-event approach of SLA. The global stiffness matrix is factorised intermittently at only a certain number of linear analyses, and the solution for the remaining intermediate linear analyses is found for low-rank corrections to the factorised stiffness matrix, which is possible using additional matrix-vector manipulations. Moreoever, in a first of its kind, several experimental benchmarks that exhibit structural collapse were simulated both using SLA and an incremental sequentially linear approach, the Force-Release method to gain further insight into the topic of non-proportional loading in such approaches. This article presents an overview of all these recent developments which pushes the SLA type of solution procedures towards being a practical alternative to NLFEA in engineering practice. Additionally, simulations of a reinforced concrete slab subject to one-way brittle shear failure, a skew-notched beam subject to prestress and bending load, and a squat shear masonry wall exhibiting a brittle diagonal shear failure are shown to illustrate the developments.

1 INTRODUCTION

Sequentially Linear Analysis (SLA) is a proven alternative to incremental-iterative solution methods in nonlinear finite element analysis (NLFEA) of quasi-brittle specimen. The core of the method is in its departure from a load, displacement or arc-length driven incremental approach, aided by internal iterations to establish equilibrium, to a damage driven event-by-event approach that approximates the nonlinear response by a sequence of scaled linear analyses. The constitutive relations are discretised into secant-stiffness–based *saw-tooth* laws, with successively reducing strengths and stiffnesses. In each linear analysis, the global load is scaled such that the *critical* integration point, with the largest stress, jumps to its next saw-tooth representing local damage increments.

The approach has been under development from the early 2000s and is a proven alternative for applications in masonry (Giardina et al. 2013), reinforced concrete (Van de Graaf 2017) and glass (Invernizzi et al. 2011). Advancements in SLA include contributions to make the procedure mesh-objective (Van de Graaf 2017; Rots et al. 2008); saw-tooth laws for extremely brittle materials like glass (with snap-back at constitutive level) (Invernizzi et al. 2011); approaches to non-proportional loading situations (DeJong et al. 2008; Eliáš et al. 2010; Eliáš 2015; Graça-E-Costa et al. 2013; Pari et al. 2018; Pari et al. 2019); incremental sequentially linear approach (Yu et al. 2018); extensions to interface elements with discrete cracking (Van de Graaf et al. 2010), bond-slip (Ensink et al. 2012), and step-wise secant Coulomb friction laws (Van de Graaf 2017); creep induced cracking (Hendriks and Rots 2009); combined incremental-total approaches like Non-Iterative Energy based Method

DOI 10.1201/9781003316404-51

(NIEM) and the automatic method (Graça-E-Costa et al. 2013); efficient linear solvers to improve the speed of SLA (Pari et al. 2020); SLA in a stochastic setting (Georgioudakis et al. 2014); combining SLA with crack tracking technique (Slobbe 2014); and mesh free SLA (Al-Sabah & Laefer 2016).

SLA is a feature, as a part or whole, of several state-of-the-art solution methods which are hereon referred to as *Sequentially Linear Methods* (SLM), although alternatively referred to as non-iterative methods in literature (Graça-E-Costa et al. 2013). These methods can be classified into three categories: purely *total* approaches (Alfaiate and Sluys 2018; DeJong et al. 2008; Van de Graaf 2017) wherein unloading and reloading are done non-proportionally, purely *incremental* approaches (Eliáš et al. 2010; Eliáš 2015; Graça-E-Costa et al. 2013) wherein the stress and loading history is explicitly tracked, and finally, a class of *combined incremental-total* approaches (Graça-E-Costa et al. 2013).

Despite the advantages of simplicity and numerical robustness in comparison to NLFEA, SLM as a class of solution procedures still needed significant developments to be used in engineering practice as a numerical tool for structural applications, such as the pushover analysis of a masonry structure or the capacity assessment of a shear-critical reinforced concrete slab. To this end, the following scientific contributions were made recently. Firstly, there was lack of complete understanding on the non-proportional loading problem in the sequentially linear class of solution procedures and the associated redistributed mechanisms. An extensive qualitative review was made on the different SLM highlighting the primary differences in load modification, and on how they address the multiple failures. Simultaneously, in a first of its kind, case studies involving real structural collapse were analysed to exemplify the differences between a *total* approach: SLA with the double load multiplier strategy, and an *incremental* approach: the Force-Release method. The latter is an incremental sequentially linear method that allows for gradual stress redistribution after each damage increment, while simultaneously keeping track of the loading history. The findings of these studies are summarised in Section 3. Secondly, most of the existing constitutive formulations used in SLA were rather simplistic. Structural level simulations motivate the need for elaborate constitutive formulations with view to non-proportional loading conditions. Accordingly, 3D Orthogonal smeared fixed and rotating cracking models (Bresser 2019; Pari et al. 2018) (with additionally possibility of crushing failures) under non-proportional loading conditions were revisited in the SLA set-up. Furthermore, composite interface constitutive models suitable for 2D and 3D sequentially linear simulations were also proposed. These are summarised in Section 4 with illustrations. Finally, to improve on solution times which have been a bottleneck for the SLM, two tailor-made solvers capable of re-using the factorised stiffness matrix were proposed and this is touched upon in Section 5.

2 SLA: METHODOLOGY

Sequentially Linear Analysis (SLA) is a non-incremental (*total*) secant stiffness-based event-by-event approach, wherein one linear analysis is performed at a time to identify and damage the *critical* integration point in the FE model. Therefore, it approximates the nonlinear response as a sequence of linear analyses with gradually increasing damage (damage-driven). The definition of the load multiplier per analysis step j for each integration point i, over all elements in the FE model, is shown below in a general sense, where f_i^j and $\sigma_{\text{gov},i}$ are the corresponding allowable strengths and the governing stresses respectively. The *critical* integration point is identified as the one with the minimum of all such positive load multipliers: the critical load multiplier λ_{crit}.

$$\lambda_i^j = \frac{f_i^j}{\sigma_{\text{gov},i}^j}, \quad \lambda_{\text{crit}}^j = \min_i \left(\lambda_{\text{crit},i}^j\right) \quad \forall \quad \lambda_{\text{crit},i}^j > 0. \tag{1}$$

The linear analysis results i.e. displacements, forces, stresses and strains are then scaled using the critical load multiplier λ_{crit}. Subsequently, the strength and stiffness of this integration point are reduced in a step-wise manner based on a discretised constitutive relation, with successively reducing secant stiffnesses and allowable strengths, called the *saw-tooth law* (Figure 1). This process of identifying critical events and load scaling is repeated until a user-defined stop criteria is reached or when the FE model is completely damaged. In summary, the method avoids multiple

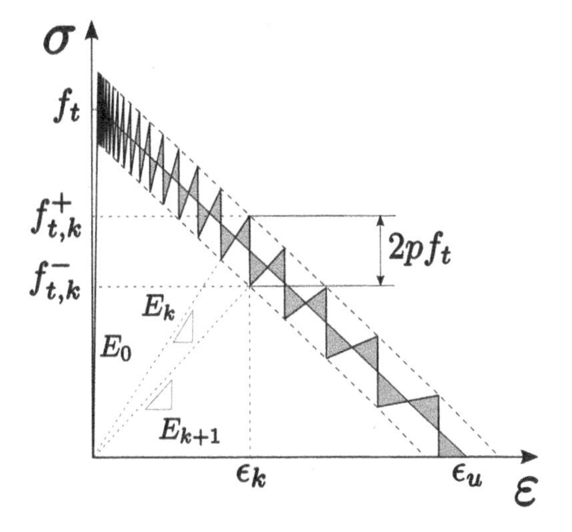

Figure 1. Linear tension softening saw tooth law, with p the saw-teeth discretisation factor, based on the band width ripple approach to ensure mesh objectivity. In this approach, a strength range p is defined as a percentage of the undamaged material strength f_t and a band is introduced into the softening part of the base curve, enclosing it such that the upper and lower triangles cancel each other out to eventually yield the same fracture energy.

integration points being pushed simultaneously into failure, as in an incremental-iterative approach, and is therefore robust. In other words, SLA traces through every event, i.e. a jump or snap back, that may occur in the response of the structure. The combination of a *total* (load-unload) approach and the saw-tooth laws forms the crux of the method.

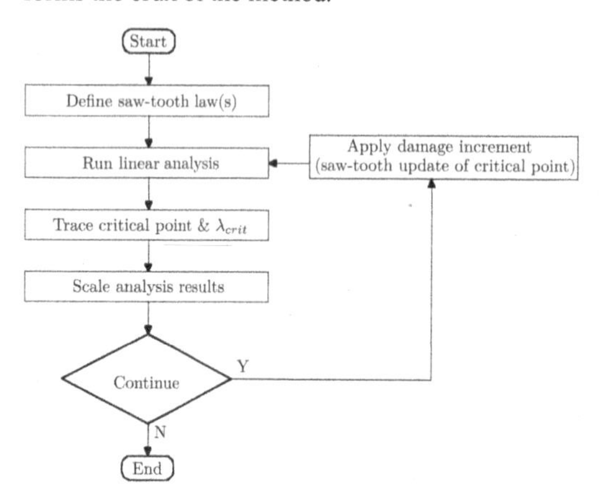

Figure 2. SLA workflow for proportional loading conditions.

The aforementioned methodology was initially conceived for proportional loading conditions, and subsequently, modified for real loading schemes which very often have multiple loads. The simplest and most common case is when there are constant loads on the structure like dead loads, precompression, overburden etc., and the structure is subsequently subject to variable loads like earthquake or wind or vehicle loads. Under such situations, the loading is considered to be non-proportional and in the *total* approach of SLA, the system is loaded by constant loads (L_{con}) and a unit variable load (L_{var}). The stresses are expressed as the superposition of the stresses due to constant and the scaled variable loads as shown in Eq. 2 for each integration point i. The governing stress is then limited by the allowable strengths f, corresponding to the failure criterion, as shown in Eq. 3, such that only the critical integration point i lies on the failure surface while all non-critical points lie below it. These equations apply for orientations depending on the failure criterion and the type of element (continuum or interfaces). As long as Eq. 3 holds, Eq. 4 applies at the global level. Contrarily, when Eq. 3 fails in a certain analysis step j, the simulation runs into a so-called *limit point* and the procedure is steered into the *Intermittent Proportional Loading* (IPL) (Van de Graaf 2017), while implicitly reducing the constant load, as shown in Eq. 5 to reinstate Eq. 3. Such regions indicate the need for multiple failures representing a sudden propagation of damage.

$$\sigma_i = \sigma_{i,\text{con}} + \lambda\ \sigma_{i,\text{var}} \tag{2}$$

$$(\sigma_{i,\text{con}} + \lambda\Delta\sigma_{i,\text{var}}) = f \qquad \forall i \neq k : (\sigma_{k,\text{con}} + \lambda\Delta\sigma_{k,\text{var}}) < f \tag{3}$$

$$L^j_{\text{crit}} = \lambda_{\text{con}} L_{\text{con}} + \lambda_{\text{var}} L_{\text{var}} \tag{4}$$

where $\lambda_{\text{con}} = 1$ and $\lambda_{\text{var}} = \lambda_{\text{crit}}$

$$L_{\text{ipl}} = L_{\text{con}} + \lambda^{j-1}_{\text{crit}}\ L_{\text{var}} \tag{5a}$$

$$L^j_{\text{ipl}} = \lambda^j_{\text{crit}}\ L_{\text{ipl}} \tag{5b}$$

Alternatively, in an *incremental* version like the Force-Release method (Eliáš et al. 2010), the non-proportional load path is discretised into a series of piece-wise proportional loading paths. Each prescribed load is discretised into a series of load vectors with magnitudes ensured to be non-decreasing, so that the proper loading/stress history is taken care of. Linear analyses are performed with load increments of a certain load vector, each of which may or may not lead to damage at a critical integration point i according to Eq. 6, wherein all quantities with Δ are the corresponding *incremental* values caused by the load increment. Upon damage, the stress from a damaged element is released gradually through a sequentially linear redistribution loop wherein the unbalanced forces due to the previous damage are the applied as loads on the FE model, while all previously applied loads are kept constant, and other elements may be damaged. When the redistribution loop does not lead to further damage, the response stays in equilibrium. Otherwise, it evolves through states of disequilibrium and eventually returns to equilibrium. A comprehensive overview on the workflow of SLA and the Force-Release methods and on the differences between such total and incremental sequentially linear methods for continuum applications can be found in Reference (Pari et al. 2020), and for lattice applications in References (Eliáš et al. 2010; Liu & Sayed 2012).

$$(\sigma_i + \lambda\Delta\sigma_i) = f \ \wedge\ \forall i \neq k : (\sigma_k + \lambda\Delta\sigma_k) < f \tag{6}$$

Unlike NLFEA which is considered as *one analysis* containing several *steps*, SLA comprises several linear analysis which are referred interchangeably as *'analysis steps'* or *'steps'* as such.

3 NON-PROPORTIONAL LOADING

The redistribution mechanism involved in SLA under non-proportional loading problems is more prominent in real-life quasi-brittle problems/experiments, at component or structural level, rather than in simple experimental benchmarks. Such redistributions, for instance previously observed with a masonry facade settlement example (Van de Graaf 2017), were attributed to the lack of crack closure algorithm in SLA but there was still a lack of complete understanding on this topic. The need for redistribution stems from the the fundamental problem of using a static approach to model an intrinsically dynamic phenomenon like cracking or crushing; and in an attempt to understand this, investigations were made at structural level and under non-proportional loading conditions. In a first of

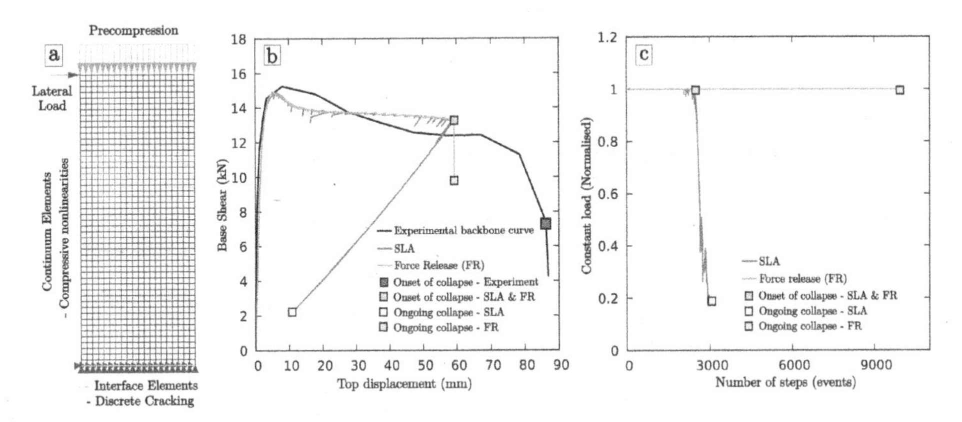

Figure 3. (a) FE model of the TUDCOMP20 test, (b) the base shear vs top displacement response for SLA & Force-Release methods compared against the experimental backbone curve - including the reference points during the collapse mechanism, and (c) evolution of the precompression load (normalised) during the simulations.

its kind, several experimental benchmarks that exhibit structural collapse were simulated both using SLA and an incremental sequentially linear approach, the Force-Release method; thereby comparing the inherent differences in their redistribution mechanism and a qualitaive overview on several methods was made (Pari 2020). One of these studies is also presented here for illustrative purposes.

The explosive failure in a non-proportionally loaded pushover experiment on a Calcium silicate masonry wall (Esposito & Ravenshorst 2017) was simulated using the SLA and Force-Release methods. The explosive failure led to instability of the wall and its eventual collapse (refer Figure 4). Information about the simulation can be found in Pari (2020). The explosive failure is captured adequately by both methods, but the mechanism of redistribution differs. The point of onset of the mechanism is denoted by a green mark in the Figure 3(b), and it occurs around the same imposed displacement for both simulations. The damage patterns are also identical as seen in Figure 5(a) where DmSS is the amount of damage accumulated in the discrete cracking interface and the crushing continuum. DmSS ranges from 0 to 1, which corresponds to undamaged and fully damaged conditions. It is clear that the two continuum elements at the bottom right corner of the wall are fully crushed and that all interface elements to the left of this region are completely cracked, leaving a tiny portion which effectively supports the wall. The ensuing mechanism is described by both approaches

differently. On the one hand, SLA runs into the limit point situation described previously in Section 2, where there is no constitutively admissible critical load multiplier. The intermittent proportional loading (IPL) commences and the last successful load combination is scaled proportionally. Firstly, the IPL occurs a little before the onset of collapse as well but recovers back to the conventional non-proportional loading. However, once the collapse begins, the IPL never recovers which is evident from the amount of precompression, the first load applied (constant), that remains on the structure in the rest of the simulation, refer Figure 3(c). The IPL implicitly reduces the constant load, thereby describing the entire dynamic brittle collapse mechanism while maintaining equilibrium. On the other hand, the Force-Release method runs into an avalanche of ruptures while going through disequilibrium states. Since the previously applied load can not be altered, for the same imposed displacement and the full value of precompression, the Force-Release method attempts to allow for redistribution due to successive failure events by gradually releasing the stresses. The ongoing failure is therefore captured differently by both approaches as seen in Figure 5(b).

Figure 4. Failure pattern of masonry wall in the TUDCOMP20 experiment before collapse (left) and after collapse (right).

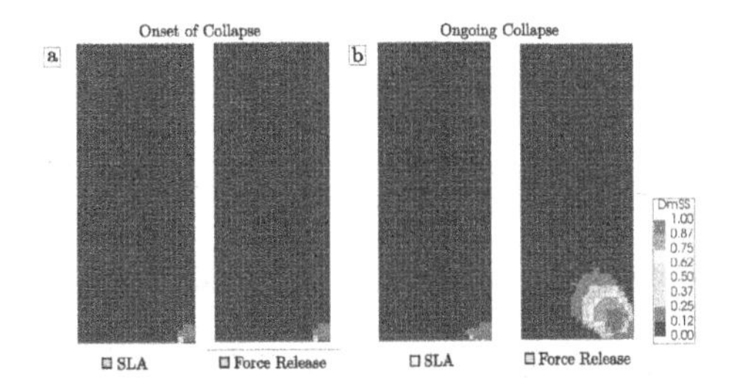

Figure 5. Failure patterns of SLA & Force-Release simulations of the TUDCOMP20 experiment: (a) at the onset of collapse, and (b) during collapse, in relation to Figure 3.

The difference between the approaches in describing collapse may be interpreted as two extremes of the time scales for a dynamic redistribution process (Eliáš 2015). SLA in this situation is essentially assuming that the loading equipment is fast enough to react to the collapse mechanism, alter the load and consequently, release the stresses quickly to avoid further failures. This is clear as the eventual failure pattern additionally only involves the crushing of the tiny effective portion supporting the wall before the onset of collapse. The crush zone appears to be more realistic, in a quasi-static sense, wherein SLA gives room for damage propagation while quasi-statically releasing the loads. The Force-Release method, on the other hand, stays true to the displacement controlled experiment, and realises the full collapse by gradually releasing the stresses in a sequentially linear redistribution loop. Although the process is dynamic, the Force-Release method effectively neglects all inertial forces. Since vertical equilibrium is not possible anymore, the simulation could be interpreted to have been completed, and the wider crush-zone is indicative of this instability. In summary, both approaches adequately describe a real non-proportionally loaded experiment involving true brittle collapse, in terms of the failure patterns and the eventual mechanism, differing only in their respective approaches to the latter.

The conclusions on suitability of either methods drawn from the studies are summarised below:

1. Comparative studies between the SLA and Force-Release approaches (Pari 2020) substantiate the difference in their approaches to the dynamic propagation of damage, which is addressed statically by both approaches through a sequence of failures referred to as the *avalanche of ruptures*. On the one hand, the SLA approach does so by temporarily releasing existing loads and maintaining equilibrium, while the Force-Release traverses through disequilibrium states for a constant imposed displacement. These main points of attention are as follows:

   - The differences in the Force-Release and SLA force-displacement curves are due to their inherent load modification approaches. Since every damaged element's stress is released instantaneously in SLA, the neighbouring elements whose stresses are close to their respective allowable strengths, subsequently, become critical at a considerably lower load. This is possible in SLA only by the temporary release of the load, which essentially explains the snap-backs. The Force-Release, on the other hand, releases the stresses gradually through disequilibrium states while maintaining all previously applied loads (displacement history), and therefore shows drops of load for constant displacements.
   - In general, it is observed across the case studies that the non-proportional loading strategy in a *total* approach like SLA, and the *incremental* solution obtained using a Force-Release method result in the qualitatively similar results i.e. damage patterns. The contrasting differences observed between the approaches in lattice modelling applications, for e.g. in the elemental failure sequence as in the work of Eliáš et al. (2010), are not observed in structural case studies since the change in stiffness due to a single damage event is not so abrupt and large. Therefore, the redistribution of the energy into the vicinity as is done in the Force-Release method does not cause further failure before attaining equilibrium. In principle, it could also be extended that a very fine saw-teeth formulation would result in near-equivalent responses using the SLA and Force-Release methods.
   - The suitability of the two methods depends on the type of problem/experiment being simulated. Force-Release method is suitable for typical displacement controlled experiments which actually exhibit instabilities. On the other hand, it may not be suitable for physical processes which exhibit snap backs or for truly quasi-static experiments. SLA is more preferable when the damage process zone is unique and controlled for quasi-static evolution in an experiment (Rots et al. 2006). However, for a CMOD controlled experiment with multiple cracking zones, SLA may not be appropriate. Force-Release method, in this case, may increase the CMOD due to the redistribution. In a quasi-static sequentially linear setup, a truly CMOD controlled experiment with multiple evolving damage zones can be appropriately simulated by the *general* method (Eliáš 2015).

2. It was clear from several case studies (Pari 2020) that the limit point situation, and the associated need for intermittent proportional loading in SLA, is not an artefact of the stress locking problem in a typical smeared fixed model (also verified for smeared rotating cracking model (Bresser 2019)), Coulomb friction model, or even the discrete cracking model. It means that irrespective of the constitutive model used, there is a need for multiple failures at certain points in an SLA simulation (as previously concluded (Van de Graaf 2017)). In such a scenario in SLA, a problem arises owing to the inherent *non-proportional unloading & reloading* on a damaged state of the structure, and therefore intermittent proportional loading follows.

3. It was also clear that the oncoming dynamic failure processes at limit points in an SLA type response for structural level examples can be distinguished into *intermediate local instabilities* or the eventual *collapse* mechanism. In case of *intermediate local instabilities*, if the intermittent proportional loading allows for a redistribution which helps recover the full value of constant load, the redistribution is deemed acceptable. However, if the redistribution results in gradual loss of constant loads to extremely low values, much ahead of the actual structural

collapse, this could either correspond to alternate equilibrium paths of damage propagation that do not culminate in the actual expected collapse mechanism or be interpreted as premature structural failure. On the contrary, in case of the eventual *collapse*, the intermittent proportional loading forces a relaxed mechanism maintaining equilibrium all through the simulation. In other words, SLA lets the damage progress quasi-statically by releasing previously applied loads thereby allowing the structure to relax during a dynamic collapse. Herein, as against overall unloading of the structure, only the elastic parts on either side of active damage zones unload. This is acceptable under non-proportional loading conditions *only* if the experiments are controlled quasi-statically. Since the system as whole is allowed for an *overall* quasi-static damage propagation, it may be interpreted to be equivalent to CMOD controlled experiments as in Reference (Rots et al. 2006) which involve a *unique* damage process zone. In case of *multiple* cracks developing in the system, SLA does not control a *unique* damage process zone as in a CMOD experiment, and therefore may incorrectly decrease it.

4 CONSTITUTIVE MODELLING DEVELOPMENTS

4.1 *Total strain based smeared cracking models*

4.1.1 *Fixed crack model*

The Fixed Crack Model (FCM), a type of total strain based smeared cracking model allows for orthogonal cracking (e.g. Reference (Feenstra et al. 1998)) and describes the cracking/crushing that arises in the fracture zone to be smeared over the continuum. It is rather straightforward to use since it describes the tensile and compressive behaviour of a material along orthogonal directions, that are fixed upon crack/crush initiation governed by the principal stress criterion, with uniaxial tensile and compressive saw-tooth laws. They are suitable for the sequentially linear framework due to the simplicity, and similar to the 2D plane stress version, in a 3D stress state as soon as the principal stress violates the allowable strength at an integration point in tension or compression, the isotropic stress-strain relation $\boldsymbol{\sigma} = \mathbf{D}\boldsymbol{\varepsilon}$ transforms into a 3D orthotropic relation as $\boldsymbol{\sigma}_{nst} = \mathbf{D}_{nst}\boldsymbol{\varepsilon}_{nst}$, and the cracked coordinate system denoted by *nst* is fixed along the directions of the principal stresses. A simple Rankine type failure surface is used to initiate damage for tension or compression failures in the fixed crack set-up. The primary principal stress direction's Young's modulus and strength are damaged according to the uniaxial saw-tooth law of the appropriate failure mode. In the event that normal stresses in the orthogonal direction (secondary) violates the corresponding allowable strength, caused by stress rotations or redistribution of stresses or application of another load non-proportionally, damage is introduced in that direction similarly. So every integration point essentially requires three uniaxial saw-tooth laws each for tension and compression.

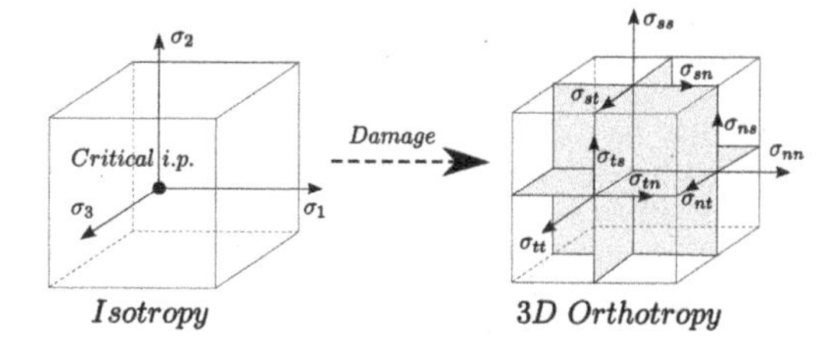

Figure 6. The change in isotropic to orthotropic formulation (*nst* crack coordinate system) upon damage initiation in a fixed crack framework for 3D stress states.

Furthermore, the model uses the crack band approach proposed by Bazant and Oh (Bažant and Oh 1983), which states that fracture energy is spread over the cracked area characterised by a certain crack band length h, to ensure that the constitutive curve depends on the size of the crack band. This, therefore, triggers the energy consumed due to smeared cracking in the FCM to be mesh independent. Alternative projection based crack band approaches are also available (Oliver 1989; Slobbe et al. 2013; Volokh 2013) but have been sparsely used in the SLM set-up. The shear behaviour is represented using a variable shear retention function that reduces with increasing damage in normal directions of the cracked/crushed plane (Slobbe et al. 2012). Also, the Poisson's ratio is reduced at the same rate as the associated Young's modulus. The orthotropic degradation i.e. the crux of the fixed crack set up necessitates the simultaneous reduction of Poisson behavior during damage to avoid spurious lateral cracking/crushing. Furthermore, this yields a favourable symmetric reduced stress-strain relationship in which the orthotropic degradation is solely dependent on the reduced moduli of elasticity. In the current framework, immediately upon violation of the allowable strengths either along the maximum or minimum principal stress direction, the transition from an isotropic to orthotropic formulation is made, and the 3D orthogonal fixed crack system (*nst*) is established. Alternatively, the fixing of the secondary and tertiary directions of the crack system could be postponed until failure in the secondary direction, due to violation of the allowable strength by the principal stress (computed anew) in that direction. This is a more realistic representation of the multi-directional cracking phenomenon in comparison to the simplified former approach.

4.1.2 *Elastic-brittle fraction model*

Within a Rotating Crack Model (RCM), the crack directions co-rotate with the principal stress directions such that the development of spurious stresses due to the fixed cracking planes is avoided. In the sequentially linear set-up several researchers tried to envisage a RCM compatible to the event-by-event approach. Since only update of one integration point at a time is allowed, whilst the RCM requires the crack directions of all cracked integration points to be updated during each event; and furthermore the rotation of the principal stresses and strains is non-linearly related to the load multiplier (Vorel & Boshoff 2015); several *partial* rotating crack models were proposed (Cook et al. 2018; Slobbe 2010; Vorel & Boshoff 2015).

Hendriks and Rots (Rots & Hendriks 2015) developed the Elastic-brittle fraction approach in order to simulate the effects of rotational cracking within sequentially linear analysis; and is also essentially a *partial* rotating crack model. This model divides each element in a set of N parallel fractions, each of which is elastic-perfectly brittle with tensile strength, stiffness and thickness chosen appropriately to represent the continuum's constitutive law. The total behaviour is found by the superposition of the stresses and strains of the perfectly brittle fractions, all of whose strains are the same (refer Figure 7).

Each fraction, denoted by the subscript k, essentially is isotropic to begin with and as soon as the principal stress violates the allowable strength at an integration point, the isotropic stress-strain relation $\boldsymbol{\sigma}_k = \mathbf{D}_k\boldsymbol{\varepsilon}$ transforms into an orthotropic relation as $\boldsymbol{\sigma}_{k,nt} = \mathbf{D}_{k,nt}\boldsymbol{\varepsilon}_{k,nt}$ with nt denoting the fixed cracked coordinate system. The fraction is fully cracked but still retains strength in the tangential direction which can also be damaged upon rotation of stresses. Once the fraction is lost, the next layers are damaged and the model thereby describes softening as a gradual reduction of the cross-sectional area, which is physically in accordance with the micro-cracking coalescing to form a macro-crack as in the fictitious crack model (Hillerborg et al. 1976). The superposition of fixed crack fractions results in a rotating effect of the crack as shown in Figure 7. Since the fractions may also be seen as sublayers, the model is alternatively referred to as the *sublayer model*. A recent study (Bresser 2019) shows that the fraction model consistently exhibits sharper crack localisation, lesser

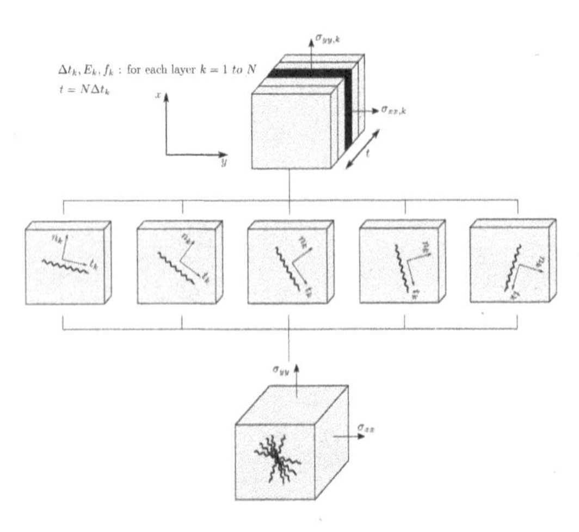

Figure 7. Schematic representation of the elastic perfectly-brittle fraction model.

stress-locking, generally more flexible behaviour for several experimental plain-concrete benchmarks, and sometimes overcomes bifurcations that may be missed with SLA using the FCM. The approach was also extended to 3D fractions in the same study (Bresser 2019) with appreciable results.

The fraction model is similar to the approaches of Slobbe (Slobbe 2010) and Vorel (Vorel & Boshoff 2015) in the sense that only the critical point undergoes crack rotation but there are differences. Slobbe's approach differentiates damage and crack rotation as two different events, while the fraction model introduces crack rotation with each damage. In comparison to Vorel's approach which explicitly allows for redistribution by gradual release of forces due to damage (avalanche of ruptures using the force-release approach), the redistribution in the fraction model is implicit with each event. The fraction model also shows similarities with the approach of Cook et al. (Cook et al. 2018) in the sense that both approaches allow for multiple cracking planes per integration point. The difference lies in the fact that the orientation of the cracking planes are predefined in Cook's approach while each of the fractions in the fraction model can have their own cracking direction depending on the principal stress. More recently, Liu (Liu 2018) proposed the *sub-element* method which is fundamentally similar to the fraction model but is applied for lattice problems.

4.1.3 Illustrations

The fixed-crack and elastic brittle fraction models are exemplified here using a 3D non-planar curved crack propagation problem with the aid of the skew-notched beam in a three-point bending test. The skew-notched beam has been used by others as a benchmark test to verify 3D (often XFEM related) numerical algorithms (Ferté et al. 2016; Jäger et al. 2008). The geometry, FE model and its associated parameters and other details can be found in the studies of Pari et al. (2018), Bresser (2019).

5 types of simulations are run. the first two are with the fixed crack and fraction models using SLA, and the rest are all nonlinear finite element analyses (NLFEA) based on the choice of variable or constant shear and Poisson reductions. The force-displacement curves of the simulations are shown in Figure 8. NLFEA with the FCM and damage based Poisson's ratio and shear reduction shows very good agreement with SLA with FCM. The constant Poisson's ratio NLFEA is not in line with the theoretical framework of fixed crack framework of SLA and therefore clear differences are found between both NLFEA and SLA. Since Poisson's ratio is not reduced during damage increments, larger spurious stresses develop within the crack plane, ensuring quicker application of damage increments and therefore, the post-peak load reduces relatively fast as well. In the NLFEA with the RCM, a more flexible response is found compared to SLA with FCM. The load displacement curve is qualitatively similar to the curve of the fraction model: the same peak load

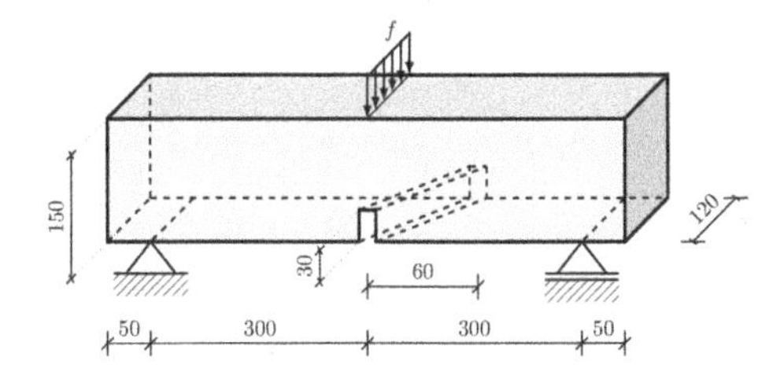

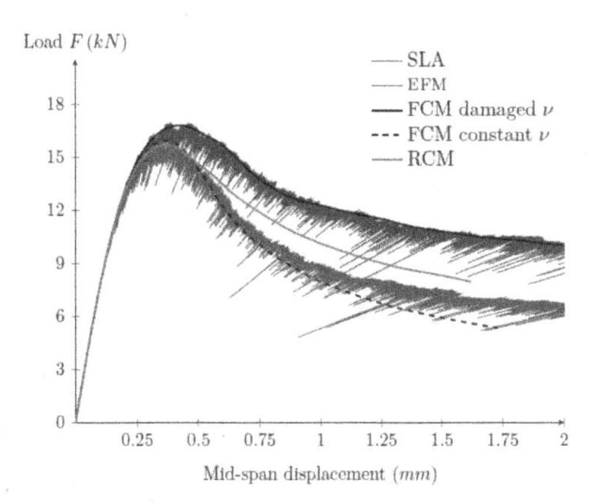

Figure 8. Geometry of the skew-notched beam in a three-point bending test with all dimensions in mm (left), and Load F versus mid-span displacement for the skew-notched beam for SLA with the FCM, the fraction model and NLFEA with FCM and RCM.

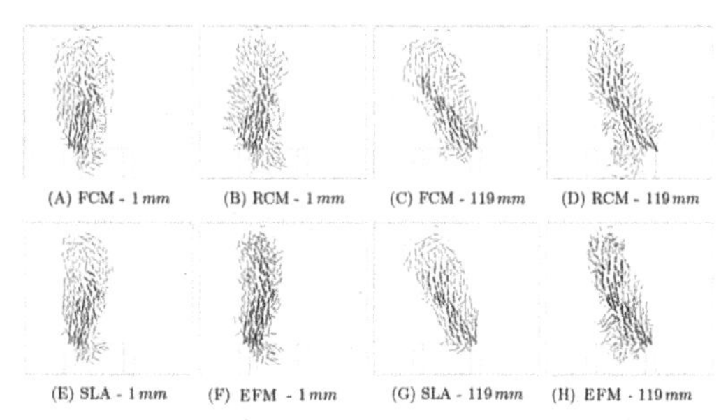

Figure 9. Comparison of crack paths FCM and RCM of NLFEA with regular SLA and the elastic-brittle fraction model in vertical slicing planes at 1 and 119 mm from front side for a mid-span displacement of 1.2 mm.

is predicted and both analyses seem to converge to approximately the same load. However, as has already been discussed in previous case studies, the fraction model is not able to exactly match with the RCM due to previously cracked fractions (with outdated crack angles) contributing to the total behaviour. To that end, it is a remarkable finding that the fraction model for this case results in an even more flexible response than the RCM, indicating the presence of even less spurious stresses, where one would contrarily expect a somewhat stiffer response; which in turn is due to pronounced crack rotations at the top of the crack path (Bresser 2019). It can however be concluded that better agreement with the RCM is encountered by invoking the 3D implementation of the fraction model.

A curved three-dimensional crack pattern is observed for all methods: Starting from the front side, the straight vertically directed crack gradually transfers to an inclined curved crack at the rear side of the beam, see Figure 9. Furthermore, starting from the bottom side of the beam, the crack gradually straightens towards the top, rotating from the notch direction towards the direction of the line load. In this way, a non-planar 3-dimensional crack path is obtained above the inclined notch. Furthermore, the crack strain plots of NLFEA FCM and SLA FCM are very similar as shown in Figure 9. Although FCM results in a slightly wider localization band, both exhibit the same main crack path and a U-turn type of behaviour at the top, troubling further crack propagation. The similarity also explains the excellent agreement between the load-displacement curves of NLFEA FCM and SLA FCM. The crack strain plots of NLFEA RCM and the fraction model also show similarities. The fully developed crack paths show an almost one-to-one agreement. Also, the additional crack zone next to the notch is similarly captured by both analyses. However, as becomes very clear from subfigures (B) and (F) in Figure 9, RCM results in a wider band of spurious stresses compared to the fraction model. Apparently, the SLA-type of procedure restricts the development of spurious crack paths, explaining the even more flexible behaviour that is obtained by the fraction model in the load-displacement curve. A possible cause of this remarkable observation might be that for SLA-type of procedures, only a single damage increment is performed at a time, potentially allowing for a certain degree of self-correction in the next steps, while for NLFEA, damage increments are performed in any step anywhere throughout the structure, such that a complete zone of integration points can enter the spurious regime simultaneously. To this end, it is less likely for SLA-type of procedures that large zones of spurious stresses develop. A more thorough study on the generation of spurious stresses in RCM and FCM is necessary, especially compared to SLA-type of analyses.

4.2 *Composite interface model*

A discretised tension-shear-compression criterion for 2D line interfaces and 3D planar interfaces was also recently proposed. This discretised composite interface model makes it possible to analyse quasi-brittle structures in a sequentially linear framework using predefined interfaces as potential discrete cracks, shear or crush planes. It is especially suitable for masonry, with the simplified micro-modelling strategy separating the continuum as linear bricks and nonlinear interfaces, to simulate cracking-shearing-crushing failures typical of masonry damage until structural collapse. The model uses a tension gap criterion coupled with a uniaxial tensile softening law, a compression

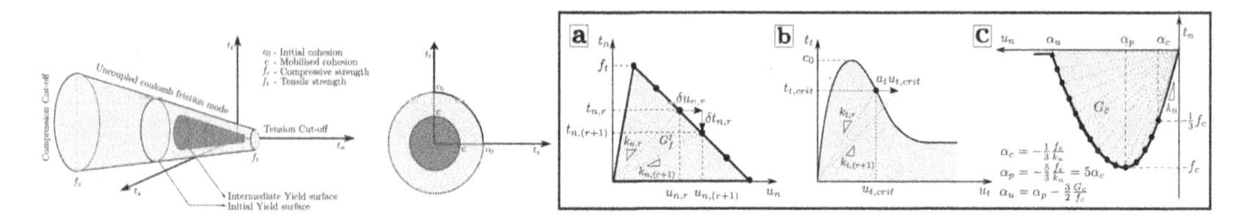

Figure 10. Failure surface for the 3D planar interfaces; and the saw-tooth laws from (a) tension, (b) cohesion and (c) compression.

cut-off criterion coupled with a compressive parabolic hardening-softening law, and an uncoupled step-wise Coulomb friction formulation with cohesion softening and without dilatancy effects (refer Figure 10). The model (Pari et al. 2021) was validated using a pushover experiment of a masonry wall subject to precompression followed by an in-plane lateral load, and is shown to ably reproduce the force-displacement curves and the brittle diagonal shear failure followed by toe-crushing due to the compressive strut action (refer Figure 11). Furthermore, the formulation was also extended to the 3D case of planar interfaces, and validated using the same case study thereby enabling 3D masonry applications with SLA (Pari et al. 2021). The extension to include dilatancy and a more advanced cap-type model for compression are features to be investigated in the future.

5 SOLVER IMPROVEMENTS

Structural simulations of SLA were extremely computationally intensive. Computational intensity has been pointed out previously to be one of SLA's major bottlenecks (Alnaas 2016; Al-Sabah & Laefer 2016; Van de Graaf 2017; Vorel & Boshoff 2015). For instance, considering an SLA simulation to predict tensile failure in an FE model with x truss elements, there can be a maximum of $x \times y \times z$ linear analyses or damage events, where y & z correspond to the number of integration points per truss element, and the number of saw-teeth in tension per integration point respectively. This would become 2 or 3 times larger in case of 2D or 3D elements considering the appropriate directions of the orthogonal smeared cracking model, and possibly even more if compression nonlinearities were to be considered. This indicates the need for an extremely high number of linear analyses (each corresponding to a unique damage location) to bring about an equivalent nonlinear response as in traditional NLFEA (damaging multiple locations). A departure from the event-by-event nature of SLA into multiple failures per analysis could be considered, but the robustness may be lost in the process in attempting to establish equilibrium using internal iterations.

Under such a premise where the event-by-event nature of SLA is not compromised, the time taken to solve the system of linear equations using direct solvers was targeted, which was the most dominant part of the computing time in each SLA step. Since only one element is effectively damaged at a time, the system of linear equations to be solved actually changes locally between these analyses. Traditional direct solution techniques do not exploit this property and calculate a rather expensive stiffness matrix factorisation every

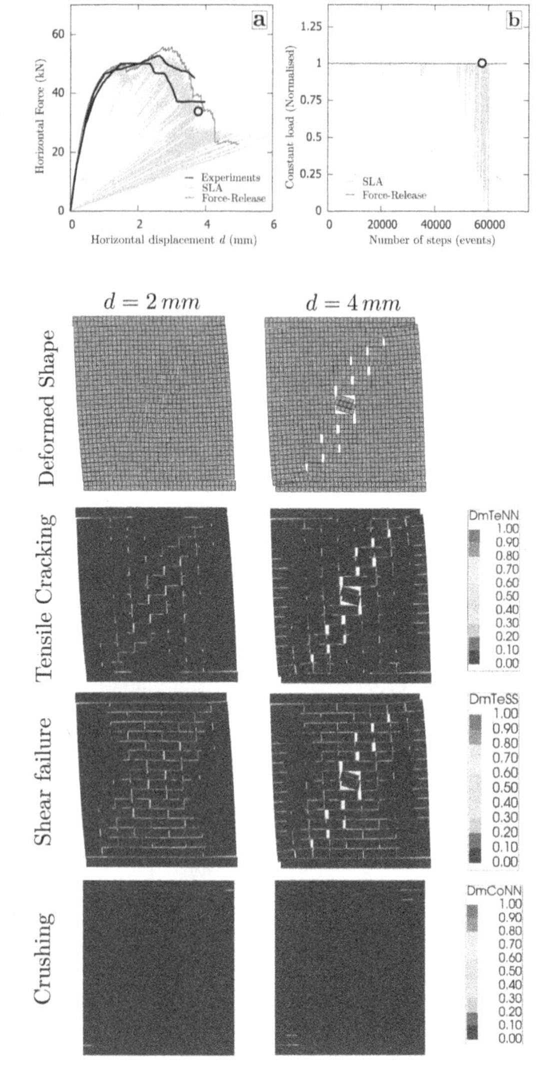

Figure 11. Force-displacement curves from SLA and Force-Release method compared to the experiment; and the damage plots of SLA simulation.

step, resulting in high computational times. This led to the proposal of two tailor-made solvers (Pari et al. 2020) which efficiently use the favourable event-by-event approach of SLA. Both solvers factorise the global stiffness matrix intermittently at only a certain number of linear analyses. The solution for the remaining intermediate linear analyses is found for low-rank corrections to the factorised stiffness matrix, which is possible using additional matrix-vector manipulations. The first is a direct solver based on the Woodbury-Identity matrix to find the inverse of an arbitrary rank-r corrected matrix. The second is a Preconditioned Conjugate Gradient (PCG) solver that uses the factorised stiffness matrix as the preconditioner for the remaining analyses. When the elapsed time in these intermediate analyses grows, a restart step is prescribed wherein a new factorisation is calculated. These points of restart are deduced such that the total analysis times are minimised. The performance of the solvers are analysed using a 2D and 3D case study, including additional saw-teeth and mesh sensitivity studies. Both solvers perform better than a traditional direct solver like Intel's Parallel Direct Sparse Solver (Pari et al. 2020), especially for large 3D problems, and the Woodbury-Identity based direct solver is more efficient among the two. Additionally, some branches of the work-flow of SLA are now computed in parallel, using multi-threading.

6 CONCLUSIONS

This article presents the recent advancements made in the realm of Sequentially Linear Methods, towards the topics of non-proportional loading, constitutive modelling and computational efficiency. The class of methods has been extended to 3D structural applications involving cracking, crushing, and shear failures, both in a smeared and discrete manner. The method in general has also been made relatively efficient. Nevertheless, approach still needs to be extended on important topics such as crack-closure effects, which requires a dedicated event & an algorithm and is possibly difficult to incorporate in the *total* frame-work. The computational intensity could be further addressed using smart damage tracking algorithms that distinguish the potential elements to be damaged; or a departure to the incremental SLA which needs investigation on constitutive and performance aspects. Furthermore, topics including influence of tension-compression interactions for damage initiation and propagation in 2D and 3D stress states, and the extension to anisotropic failure surfaces are important from a constitutive modelling aspect.

REFERENCES

Al-Sabah, A. S. & D. F. Laefer (2016). Meshfree sequentially linear analysis of concrete. *Journal of Computing in Civil Engineering 30*(2), 04015009.

Alfaiate, J. & L. J. Sluys (2018). On the use of non-iterative methods in cohesive fracture. *International Journal of Fracture 210*(1-2), 167–186.

Alnaas, W. (2016). *Nonlinear finite element analysis of quasi-brittle materials*. Ph. D. thesis, Cardiff University.

Bažant, Z. P. & B. H. Oh (1983). Crack band theory for fracture of concrete. *Matériaux et Constructions 16*(3), 155–177.

Bresser, D. (2019). Mimicking a rotating crack model within sequentially linear analysis using an elastic-perfectly brittle sublayer model. Master's thesis, Delft University of Technology.

Cook, A. C., S. S. Vel, & S. E. Johnson (2018). Pervasive cracking of heterogeneous brittle materials using a multi-directional smeared crack band model. *International Journal of Mechanical Sciences 149*, 459–474.

DeJong, M. J., M. A. N. Hendriks, & J. G. Rots (2008). Sequentially linear analysis of fracture under non-proportional loading. *Engineering Fracture Mechanics 75*(18), 5042–5056.

Eliáš, J. (2015). Generalization of load–unload and force-release sequentially linear methods. *International Journal of Damage Mechanics 24*(2), 279–293.

Eliáš, J., P. Frantík, & M. Vořechovský (2010). Improved sequentially linear solution procedure. *Engineering fracture mechanics 77*(12), 2263–2276.

Ensink, S. W. H., A. V. van de Graaf, A. T. Slobbe, M. A. N. Hendriks, J. A. den Uijl, & J. G. Rots (2012). Modelling of bond behaviour by means of sequentially linear analysis and concrete-to-steel interface elements. In *Proceedings of the Fourth Bond In Concrete Conference.*

Esposito, R. & G. J. P. Ravenshorst (2017). Quasi-static cyclic in-plane tests on masonry components 2016/2017. Technical Report C31B67WP3-4,(1), Delft University of Technology.

Feenstra, P. H., J. G. Rots, A. Arnesen, J. G. Teigen, & K. V. Hoiseth (1998). A 3D constitutive model for concrete based on a co-rotational concept. In *Computational Modelling of Concrete Structures, Bad Gastein, Austria; Editors : R. de Borst et al.*, pp. 13–22.

Ferté, G., P. Massin, & N. Moës (2016). 3D crack propagation with cohesive elements in the extended finite element method. *Computer Methods in Applied Mechanics and Engineering 300*, 347 – 374.

Georgioudakis, M., G. Stefanou, & M. Papadrakakis (2014). Stochastic failure analysis of structures with softening materials. *Engineering Structures 61*, 13–21.

Giardina, G., A. V. Van de Graaf, M. A. N. Hendriks, J. G. Rots, & A. Marini (2013). Numerical analysis of a masonry facade subject to tunnelling-induced settlements. *Engineering Structures 54*, 234–247.

Graça-E-Costa, R., J. Alfaiate, D. Dias-Da-Costa, P. Neto, & L. J. Sluys (2013). Generalisation of non-iterative methods for the modelling of structures under non-proportional loading. *International Journal of Fracture 182*(1), 21–38.

Hendriks, M. A. N. & J. G. Rots (2009). Simulation of creep induced cracking based on sequentially linear analysis. In *Proc. of the 8th International Conference on Creep, Shrinkage and Durability Mechanics of Concrete and Concrete Structures; Editors: T. Tanabe et al.*, pp. 579–585.

Hillerborg, A., M. Modéer, & P. E. Petersson (1976). Analysis of crack formation and crack growth in concrete by means of fracture mechanics and finite elements. *Cement and Concrete Research 6*(6), 773–781.

Invernizzi, S., D. Trovato, M. A. N. Hendriks, & A. V. Van de Graaf (2011). Sequentially linear modelling of local snap-back in extremely brittle structures. *Engineering Structures 33*(5), 1617–1625.

Jäger, P., P. Steinmann, & E. Kuhl (2008). Modeling three-dimensional crack propagation—a comparison of crack path tracking strategies. *International Journal for Numerical Methods in Engineering 76*(9), 1328–1352.

Liu, J. X. (2018). Simulating quasi-brittle failures including damage-induced softening based on the mechanism of stress redistribution. *Applied Mathematical Modelling 55*, 685–697.

Liu, J. X. & T. E. Sayed (2012). On the load–unload (l–u) and force–release (f–r) algorithms for simulating brittle fracture processes via lattice models. *International Journal of Damage Mechanics 21*(7), 960–988.

Oliver, J. (1989). A consistent characteristic length for smeared cracking models. *International Journal for Numerical Methods in Engineering 28*(2), 461–474.

Pari, M. (2020). *Simulating quasi-brittle failure in structures using Sequentially Linear Methods: Studies on non-proportional loading, constitutive modelling, and computational efficiency*. Ph. D. thesis, Delft University of Technology.

Pari, M., M. A. N. Hendriks, & J. G. Rots (2019). Non-proportional loading in sequentially linear analysis for 3D stress states. *International Journal for Numerical Methods in Engineering 119*(6), 506–531.

Pari, M., M. A. N. Hendriks, & J. G. Rots (2020). Non-proportional loading in sequentially linear solution procedures for quasi-brittle fracture: A comparison and perspective on the mechanism of stress redistribution. *Engineering Fracture Mechanics*, 106960.

Pari, M., J. G. Rots, & M. A. N. Hendriks (2018). Non-proportional loading for 3-D stress situations in Sequentially Linear Analysis. In *Computational Modelling of Concrete Structures; Editors: G. Meschke et al.*, pp. 931–940. CRC Press.

Pari, M., W. Swart, M. B. Van Gijzen, M. A. N. Hendriks, & J. G. Rots (2020). Two solution strategies to improve the computational performance of sequentially linear analysis for quasi-brittle structures. *International Journal for Numerical Methods in Engineering 121*(10), 2128–2146.

Pari, M., A. Van de Graaf, M. Hendriks, & J. Rots (2021). A multi-surface interface model for sequentially linear methods to analyse masonry structures. *Engineering Structures 238*, 112123.

Rots, J. G., B. Belletti, & S. Invernizzi (2008). Robust modeling of RC structures with an "event-by-event"ž strategy. *Engineering Fracture Mechanics 75*(3-4), 590 – 614.

Rots, J. G. & M. A. N. Hendriks (2015). Elastic-brittle fraction model for robust post-peak analysis of masonry structures. *Key Engineering Materials 624*, 27–39.

Rots, J. G., S. Invernizzi, & B. Belletti (2006). Saw-tooth softening/stiffening-a stable computational procedure for rc structures. *Computers and Concrete 3*(4), 213–233.

Slobbe, A. T. (2010). Sequentially linear analysis of shear critical reinforced concrete beams. Master's thesis, Delft University of Technology.

Slobbe, A. T. (2014). *Propagation and band width of smeared cracks*. Ph. D. thesis, Delft University of Technology.

Slobbe, A. T., M. A. N. Hendriks, & J. G. Rots (2012). Sequentially linear analysis of shear critical reinforced concrete beams without shear reinforcement. *Finite Elements in Analysis and Design 50*, 108 – 124.

Slobbe, A. T., M. A. N. Hendriks, & J. G. Rots (2013). Systematic assessment of directional mesh bias with periodic boundary conditions: Applied to the crack band model. *Engineering Fracture Mechanics 109*, 186–208.

Van de Graaf, A. V. (2017). *Sequentially linear analysis for simulating brittle failure*. Ph. D. thesis, Delft University of Technology.

Van de Graaf, A. V., M. A. N. Hendriks, & J. G. Rots (2010). A discrete cracking model for sequentially linear analysis. *Computational Modelling of Concrete Structures; Editors: Nenad Bicanic et al., EURO-C*, 409–418.

Volokh, K. Y. (2013). Characteristic length of damage localization in concrete. *Mechanics Research Communications 51*, 29–31.

Vorel, J. & W. P. Boshoff (2015). Computational modelling of real structures made of strain-hardening cement-based composites. *Applied Mathematics and Computation 267*, 562 – 570.

Yu, C., P. C. J. Hoogenboom, & J. G. Rots (2018). Incremental sequentially linear analysis to control failure for quasi-brittle materials and structures including non-proportional loading. *Engineering Fracture Mechanics 202*, 332 – 349.

Computational Modelling of Concrete and Concrete Structures – Meschke, Pichler & Rots (Eds)

Investigation of an extended damage-plasticity model for concrete considering nonlinear creep behavior

A. Dummer, M. Neuner & G. Hofstetter
Unit of Strength of Materials and Structural Analysis, Institute of Basic Sciences in Engineering Sciences, University of Innsbruck, Innsbruck, Austria

ABSTRACT: For concrete, the dependence of the creep strain rate on the acting stress under high stress levels is nonlinear. Furthermore, very high stress levels may lead to a growth of microcracks resulting in material failure. For the realistic life time assessment of creep sensitive concrete structures by means of numerical simulations, the appropriate description of nonlinear creep is crucial. However, many material models developed for concrete focus on either time-dependent behavior, restricted to linear creep, or nonlinear short-term material behavior. This is the motivation for an extended damage-plasticity model, aiming at a unified and computational efficient approach for representing the highly nonlinear time-dependent behavior in large-scale finite element simulations. Well established approaches for modeling the evolution of material properties, inelastic deformation, damage, creep and shrinkage serve as basis for considering nonlinear creep and material failure due to creep.

1 INTRODUCTION

For the detailed study of the mechanical long-term behavior of concrete structures by means of numerical simulations the appropriate representation of the creep behavior is of vital importance: Whereas for moderate stress levels the creep strain rate is approximately proportional to the acting stress, for higher stress levels the relation between the creep strain rate and the acting stress becomes nonlinear. Furthermore, very high sustained stress levels lead to an increase of the creep strain rate due to growth of microcracks, possibly resulting in material failure. Even though many material models have been developed for describing either linear creep behavior or the evolution of material damage of concrete under short-term loading, only few focus on nonlinear creep and the coupling of creep and material damage in the literature, cf. e.g. Mazzotti & Savoia (2003); Boumakis, Di Luzio, Marcon, Vorel, & Wan-Wendner (2018); Ren, Wang, Ballarini, & Gao (2020).

Currently, a material model for concrete, focusing on the appropriate representation of nonlinear creep and damage due to high sustained degrees of material utilization, is developed (Dummer, Neuner, & Hofstetter 2022). In this contribution some features of the model are investigated and its time-dependent response is compared to experimental results of the creep tests in moderate uniaxial and multiaxial compression by Kim, Kwon, Kim, & Kim (2005) and the creep tests in moderate and high uniaxial compression up to failure by Rüsch (1968).

2 CONSTITUTIVE MODEL

The model is based on the well-known damage-plasticity model by Grassl & Jirásek (2006), the Solidification Theory by Bažant & Prasannan (1989) and the B4 model by Bažant, Jirásek, Hubler, & Carol (2015). The nonlinear stress-strain relation is expressed in rate form as

$$\begin{aligned}\dot{\boldsymbol{\sigma}} &= (1-\omega)\,\dot{\bar{\boldsymbol{\sigma}}} = (1-\omega)\,\mathbb{C}:\dot{\boldsymbol{\varepsilon}}^{\mathrm{el}} \\ &= (1-\omega)\,\mathbb{C}:\left(\dot{\boldsymbol{\varepsilon}} - \dot{\boldsymbol{\varepsilon}}^{\mathrm{p}} - \dot{\boldsymbol{\varepsilon}}^{\mathrm{ve}} - \dot{\boldsymbol{\varepsilon}}^{\mathrm{f}} - \dot{\boldsymbol{\varepsilon}}^{\mathrm{dc}} - \dot{\boldsymbol{\varepsilon}}^{\mathrm{shr}}\right),\end{aligned} \tag{1}$$

in which $\boldsymbol{\sigma}$ denotes the nominal stress tensor, i.e. force per total area, $\bar{\boldsymbol{\sigma}}$ the effective stress tensor, i.e., force per undamaged area, ω the isotropic damage variable and $\mathbb{C}$ the fourth order instantaneous elastic stiffness tensor. The total strain tensor $\boldsymbol{\varepsilon}$ is decomposed additively into the instantaneous elastic strain $\boldsymbol{\varepsilon}^{\mathrm{el}}$, the plastic strain $\boldsymbol{\varepsilon}^{\mathrm{p}}$, the viscoelastic strain $\boldsymbol{\varepsilon}^{\mathrm{ve}}$, the viscous strain $\boldsymbol{\varepsilon}^{\mathrm{f}}$, the drying creep strain $\boldsymbol{\varepsilon}^{\mathrm{dc}}$ and the total shrinkage strain $\boldsymbol{\varepsilon}^{\mathrm{shr}}$.

The evolution of the shrinkage strain $\boldsymbol{\varepsilon}^{\mathrm{shr}}$ is described by means of the B4 model proposed by Bažant, Jirásek, Hubler, & Carol (2015). Therein, the evolution of the total shrinkage strain $\boldsymbol{\varepsilon}^{\mathrm{shr}}$ is assumed to be the sum of the autogenous shrinkage strain $\boldsymbol{\varepsilon}^{\mathrm{shr,au}}$ and the drying shrinkage strain $\boldsymbol{\varepsilon}^{\mathrm{shr,d}}$. The autogenous shrinkage strain $\boldsymbol{\varepsilon}^{\mathrm{shr,au}}$ is described as

$$\boldsymbol{\varepsilon}^{\mathrm{shr,au}}(t) = \mathbf{I}\,\varepsilon_{\infty}^{\mathrm{shr,au}}\left[1+\left(\frac{\tau_{\mathrm{shr,au}}}{t}\right)^{\alpha}\right]^{r_t} \tag{2}$$

DOI 10.1201/9781003316404-52

in which $\varepsilon_{\infty}^{\text{shr,au}}$ is the ultimate autogenous shrinkage strain, $\tau_{\text{shr,au}}$ is the autogenous shrinkage halftime, and α and r_t are parameters depending on the concrete composition and on the cement type.

The evolution of the drying shrinkage strain $\boldsymbol{\varepsilon}^{\text{shr,d}}$ is described as

$$\boldsymbol{\varepsilon}^{\text{shr,d}}(t)=\mathbf{I}\,\varepsilon_{\infty}^{\text{shr,d}}\,k_h\,\tanh\sqrt{\frac{\langle t-t_0\rangle}{\tau_{\text{shr,d}}}}, \tag{3}$$

in which $\langle\bullet\rangle$ are the MACAULY brackets. In Eq. 3, $\varepsilon_{\infty}^{\text{shr,d}}$ denotes the ultimate drying shrinkage strain, k_h a parameter depending on the ambient relative humidity h_{env}, t_0 the start of drying and $\tau_{\text{shr,d}}$ the drying shrinkage halftime.

The evolution of the instantaneous elastic strain and the basic creep strain are represented by the Solidification Theory proposed by Bažant & Prasannan (1989), generalized to three dimensional stress states. Thereby, the evolution of the instantaneous elastic strain $\boldsymbol{\varepsilon}^{\text{el}}$, the viscoelastic strain $\boldsymbol{\varepsilon}^{\text{ve}}$ and the viscous (flow) strain $\boldsymbol{\varepsilon}^{\text{f}}$ are defined as

$$\dot{\boldsymbol{\varepsilon}}^{\text{el}}=q_1\,\mathbb{D}_\nu:\dot{\bar{\boldsymbol{\sigma}}}(t), \tag{4}$$

$$\dot{\boldsymbol{\varepsilon}}^{\text{ve}}=\frac{F(\bar{\boldsymbol{\sigma}},t)}{v(t)}\int_0^t\dot{\Phi}(t-t')\,\mathbb{D}_\nu:\mathrm{d}\bar{\boldsymbol{\sigma}}(t'), \tag{5}$$

$$\dot{\boldsymbol{\varepsilon}}^{\text{f}}=q_4\,\frac{F(\bar{\boldsymbol{\sigma}},t)}{t}\,\mathbb{D}_\nu:\bar{\boldsymbol{\sigma}}(t), \tag{6}$$

in which $\mathbb{D}_\nu$ denotes the unit compliance tensor (Jirásek & Bažant 2002). The evolution of the viscoelastic strain, cf. Eq. (5), is formulated in terms of the the solidified volume function $v(t)$ and the compliance rate of the hydrated constituent $\dot{\Phi}$. The solidified volume function

$$v(t)=\left[\left(\frac{\lambda_0}{t}\right)^m+\frac{q_3}{q_2}\right]^{-1}, \tag{7}$$

controls the time-dependent aging of the material. Therein, m and λ_0 are material parameters with default values $m=0.5$ and $\lambda_0=1$ d according to Bažant & Prasannan (1989). The anti-derivative of the compliance rate $\dot{\Phi}$ in Eq. (5) is defined as

$$\Phi(t-t')=q_2\,\ln\left(1+\left(\frac{t-t'}{\lambda_0}\right)^n\right), \tag{8}$$

in which n is the exponent of the log-power law. In Eqs. (4) and (6) to (8), q_1, q_2, q_3 and q_4 are the compliance parameters.

The evolution of the drying creep strain $\boldsymbol{\varepsilon}^{\text{dc}}$ is defined as

$$\dot{\boldsymbol{\varepsilon}}^{\text{dc}}=q_5\,\frac{F(\bar{\boldsymbol{\sigma}},t)}{\exp(p_{5H}/2)}\int_0^t\dot{\Phi}_d(t,t',t_0)\,\mathbb{D}_\nu:\mathrm{d}\bar{\boldsymbol{\sigma}}(t'), \tag{9}$$

in which q_5 is the drying creep compliance parameter, p_{5H} a parameter depending on the cement type and $\dot{\Phi}_d$ the drying creep compliance rate. The anti-derivative of the drying creep compliance rate is defined as

$$\Phi_d(t,t',t_0)=\sqrt{\left\langle\exp\left(b\,\tanh\sqrt{\xi-\xi_0}\right)-\exp\left(b\,\tanh\sqrt{-\xi_0}\right)\right\rangle} \tag{10}$$

considering the dimensionless times $\xi=(t-t')/\tau_{\text{shr,d}}$ and $\xi_0=(t_0-t')/\tau_{\text{shr,d}}$, and $b=p_{5H}(1-h_{\text{env}})$.

In order to account for a nonlinear dependence of the creep strain rates $\dot{\boldsymbol{\varepsilon}}^{\text{ve}}$, $\dot{\boldsymbol{\varepsilon}}^{\text{f}}$ and $\dot{\boldsymbol{\varepsilon}}^{\text{dc}}$ on the acting stress, the creep amplification function $F(\bar{\boldsymbol{\sigma}},t)$, originally introduced by Bažant & Prasannan (1989) for basic creep, is reformulated as

$$F(\bar{\boldsymbol{\sigma}},t)=1+a_{\text{cr}}\,\mathcal{U}(\bar{\boldsymbol{\sigma}},t)^2, \tag{11}$$

in which, $\mathcal{U}(\bar{\boldsymbol{\sigma}},t)$ represents the degree of material utilization, and a_{cr} is a parameter for scaling the influence of the stress level.

The elastic domain is delimited by the time-dependent generalization of the yield function of the CDP model by Grassl & Jirásek (2006) as introduced by Neuner, Gamnitzer, & Hofstetter (2017). It is formulated in terms of three invariants of the effective stress tensor, i.e., the effective mean stress $\bar{\sigma}_m$, the effective deviatoric radius $\bar{\rho}$ and the LODE angle θ, as

$$f_p(\bar{\boldsymbol{\sigma}},q_h(\alpha_p),t)=\left((1-q_h(\alpha_p))\left(\frac{\bar{\rho}}{\sqrt{6}f_{cu}(t)}+\frac{\bar{\sigma}_m}{f_{cu}(t)}\right)^2+\sqrt{\frac{3}{2}}\frac{\bar{\rho}}{f_{cu}(t)}\right)^2 + m_0\,q_h^2(\alpha_p)\left(\frac{\bar{\rho}}{\sqrt{6}f_{cu}(t)}r(\theta)+\frac{\bar{\sigma}_m}{f_{cu}(t)}\right)-q_h^2(\alpha_p). \tag{12}$$

In Eq. (12), $r(\theta)$ is a function proposed by Willam & Warnke (1975) defining the shape of the yield function in the deviatoric sections, q_h denotes the normalized stress-like internal hardening variable, and m_0 is the friction parameter, defined in terms of the current material strength parameters, i.e., the uniaxial compressive strength $f_{cu}(t)$, the uniaxial tensile strength $f_{tu}(t)$ and the equi-biaxial compressive strength $f_{cb}(t)$. The evolution of the plastic strain $\boldsymbol{\varepsilon}^{\text{p}}$ is described by means of a non-associated flow rule as

$$\dot{\boldsymbol{\varepsilon}}^{\text{p}}=\dot{\lambda}\,\frac{\partial g_p(\bar{\boldsymbol{\sigma}},q_h(\alpha_p),t)}{\partial\bar{\boldsymbol{\sigma}}}, \tag{13}$$

with the plastic multiplier $\dot{\lambda}$ and the plastic potential function g_p according to Neuner, Gamnitzer, & Hofstetter (2017).

The hydration-dependent evolution of the uniaxial compressive strength is assumed to follow the relation by CEB-FIP (2013) as

$$f_{cu}(t)=f_{cu}^{(28)}\,\exp\left(\left[1-\sqrt{\frac{28}{t}}\right]s_f\right), \tag{14}$$

in which $f_{\mathrm{cu}}^{(28)}$ denotes the uniaxial compressive strength at the material age of 28 days, and s_{f} is a parameter controlling the time-dependent evolution of the material strength.

The evolution of the scalar damage variable ω is formulated in the framework of the gradient-enhanced continuum theory by means of the exponential softening law

$$\omega = 1 - \exp\left(-\bar{\beta}_{\mathrm{d}}\right), \tag{15}$$

in which $\bar{\beta}_{\mathrm{d}}$ is the nonlocal damage driving variable. The nonlocal damage driving field of $\bar{\beta}_{\mathrm{d}}$ is defined implicitly by the screened Poisson equation in terms of its local counterpart β_{d} as

$$\bar{\beta}_{\mathrm{d}} - l^2 \nabla \bar{\beta}_{\mathrm{d}} = \beta_{\mathrm{d}}. \tag{16}$$

The evolution of the local damage driving variable β_{d} is driven by (i) the accumulation of plastic deformation, and (ii) high sustained degrees of material utilization for modeling tertiary creep as

$$\dot{\beta}_{\mathrm{d}} = \frac{\dot{\alpha}_{\mathrm{d}}^{(\mathrm{p})}}{\epsilon_{\mathrm{f}}} + \dot{\alpha}_{\mathrm{d}}^{(\mathrm{cr})}, \tag{17}$$

in which $\dot{\alpha}_{\mathrm{d}}^{(\mathrm{p})}$ is an equivalent plastic strain rate measure, ϵ_{f} is the softening modulus, and $\dot{\alpha}_{\mathrm{d}}^{(\mathrm{cr})}$ is an equivalent creep strain rate measure related to high sustained acting stress. For describing the evolution of material damage due to inelastic deformation by means of the first term in (17), the equivalent plastic strain rate is defined according to Grassl & Jirásek (2006). For describing the evolution of material damage due to high sustained acting stress by means of the second term in (17), for $\dot{\alpha}_{\mathrm{d}}^{(\mathrm{cr})}$ a constitutive law inspired by the 1D damage model proposed by Bažant & Jirásek (2018) is employed as

$$\dot{\alpha}_{\mathrm{d}}^{(\mathrm{cr})} = \frac{1}{\tau_{\mathrm{f}}} \mathcal{U}(\bar{\boldsymbol{\sigma}}, t)^{n_{\mathrm{f}}}, \tag{18}$$

depending on the degree of material utilization $\mathcal{U}(\bar{\boldsymbol{\sigma}}, t)$ and the material parameters τ_{f} and n_{f}.

The degree of material utilization for the constitutive model is expressed in terms of the fully hardened yield function, cf. Eq. (12), as

$$\mathcal{U}(\bar{\boldsymbol{\sigma}}, t) = \left\langle 1 + f_{\mathrm{p}}(\bar{\boldsymbol{\sigma}}, q_{\mathrm{h}}(\alpha_{\mathrm{p}} = 1), t) \right\rangle = \left\langle \frac{3}{2}\left(\frac{\bar{\rho}}{f_{\mathrm{cu}}(t)}\right)^2 + m_0 \left(\frac{\bar{\rho}}{\sqrt{6} f_{\mathrm{cu}}(t)} r(\theta) + \frac{\bar{\sigma}_{\mathrm{m}}}{f_{\mathrm{cu}}(t)}\right) \right\rangle. \tag{19}$$

Figure 1 shows the degree of material utilization $\mathcal{U}$ for common stress paths and variable stress intensity, i.e., the ratio of nominal stress σ to the respective material strength f. While equality of the degree of material utilization and the stress intensity holds for uniaxial tension, a nonlinear relation is predicted for uniaxial and equi-biaxial compression.

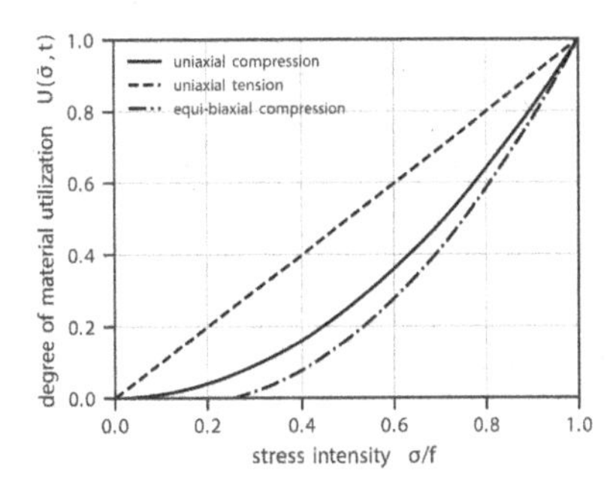

Figure 1. Illustration of the degree of material utilization for different stress intensities and stress paths.

3 INVESTIGATION OF THE MODEL EXTENSIONS FOR NONLINEAR CREEP

The formulations for (i) the nonlinear dependence of the creep strain rate on the acting stress, cf. Eq. (11), and (ii) the evolution of material damage due to creep, cf. Eq. (18), are investigated using a fictive uniaxial compressive creep test. For all examples, a uniaxial compressive stress intensity, i.e., the ratio of the nominal compressive stress to the uniaxial compressive strength $\sigma/f_{\mathrm{cu}}(t')$, is assumed at $t' = 28$ days and kept constant throughout the test. The evolution of uniaxial compressive strength is taken into account with $s_{\mathrm{f}} = 0.25$ for normal hardening cement according to CEB-FIP (2013).

Figure 2 shows the evolution of the creep amplification factor $F(\bar{\boldsymbol{\sigma}}, t)$ considering $a_{\mathrm{cr}} = 1$ (solid lines), $a_{\mathrm{cr}} = 2$ (dashed lines) and $a_{\mathrm{cr}} = 3$ (dash-dotted lines). Therein, for an isolated investigation of the nonlinear creep amplification factor the evolution of damage due to creep is neglected. It can be seen that for moderate loading levels, i.e., $\sigma/f_{\mathrm{cu}}^{(28)} \leq 0.4$, the creep amplification factor equals approximately 1, resulting in an approximately linear dependence of the creep strain rate on the acting stress. For higher loading levels $F(\bar{\boldsymbol{\sigma}}, t)$ increases disproportionately. As the load duration increases, the material strength evolves, and thus, the degree of material utilization decreases, resulting in a decreasing creep amplification factor $F(\bar{\boldsymbol{\sigma}}, t)$.

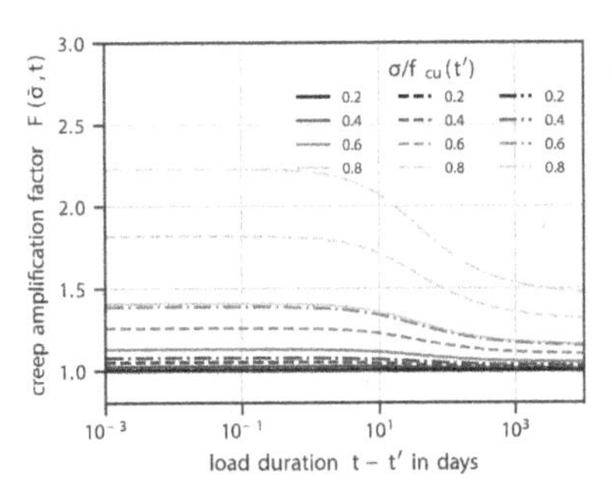

Figure 2. Evolution of the creep amplification factor $F(\bar{\boldsymbol{\sigma}}, t)$ in a uniaxial compressive creep test loaded at $t' = 28$ days considering $a_{\mathrm{cr}} = 1$ (solid lines), $a_{\mathrm{cr}} = 2$ (dashed lines) and $a_{\mathrm{cr}} = 3$ (dash-dotted lines).

Figure 3 shows the time-dependent evolution of material damage considering an evolving material strength due to hydration, i.e., aging material behavior with $s_f = 0.25$ (solid lines), and constant material strength, i.e., non-aging behavior with $s_f = 0$ (dashed lines). The parameters for the creep damage law, cf. Eq. (18) are chosen as $\tau_f = 0.4$ h and $n_f = 18$. It can be seen that for the non-aging material the damage variable increases continuously and failure occurs for all investigated stress intensities after a certain time. In contrast, if the evolution of material strength is taken into account, the evolution of material damage is slowed down as the material strength evolves. For the chosen parameters the model predicts failure only for sustained loads higher than $\sigma/f_{cu}^{28} = 0.7$ within the usual life time of concrete structures i.e., approximately 4×10^4 days. This is in line with experimental observations, cf. e.g. Rüsch (1968).

Figures. 4 and 5 illustrate the influence of the material parameters τ_f and n_f on the predicted failure time,

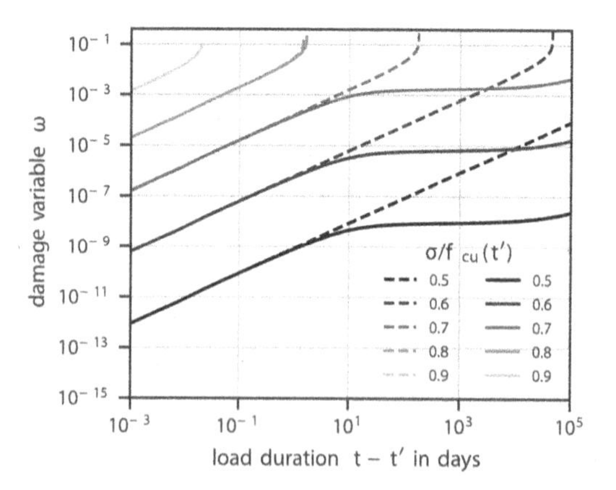

Figure 3. Evolution of the damage variable ω in a uniaxial compressive creep test loaded at $t' = 28$ days considering $s_f = 0.25$ (solid lines) and $s_f = 0$ (dashed lines).

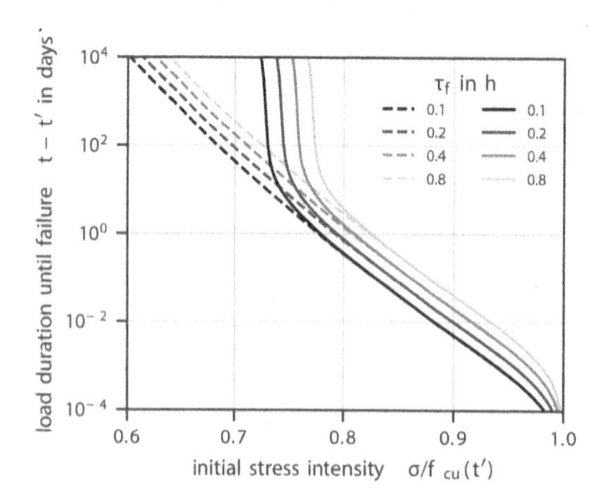

Figure 4. Influence of parameter τ_f on the predicted failure time in a uniaxial compressive creep test loaded at $t' = 28$ days considering $s_f = 0.25$ (solid lines) and $s_f = 0$ (dashed lines).

considering a constant material strength, i.e., non-aging behavior, and an evolving material strength, i.e., aging behavior, respectively.

For investigating parameter τ_f in Figure 4, a constant creep damage exponent $n_f = 18$ is assumed. It can be seen that an increase of τ_f yields an increase of the failure time.

For investigating parameter n_f in Figure 5, a constant creep damage time parameter $\tau_f = 0.4$ h is assumed. In contrast to the influence of the creep damage time parameter τ_f, a change in n_f yields a disproportionately higher increase of the predicted failure time with a decreasing initial stress intensity.

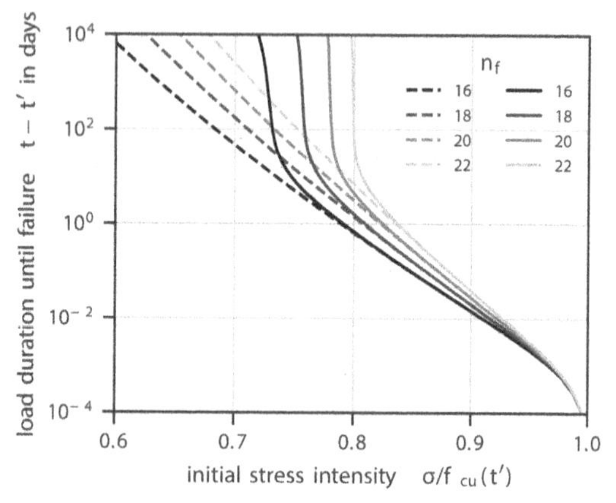

Figure 5. Influence of parameter n_f on the predicted failure time in a uniaxial compressive creep test loaded at $t' = 28$ days considering $s_f = 0.25$ (solid lines) and $s_f = 0$ (dashed lines).

4 NUMERICAL IMPLEMENTATION FOR FINITE ELEMENT SIMULATIONS

For the present constitutive model the screened Poisson equation Eq. (16), which defines the nonlocal damage driving field, and the quasi-static equilibrium equation

$$\nabla \sigma + f = 0, \tag{20}$$

with f denoting the body forces, form a fully coupled system of partial differential equations.

The weak forms of the governing equations, cf. Eqs. (16 20, are discretized in space by finite elements and subsequently discretized in time. The solution of the fully coupled, time-dependent, quasi-static problem is then computed by means of an incremental-iterative Newton-Raphson scheme employing full algorithmic tangent operators. This way, quadratic convergence is achieved in the global iterative solution procedure.

At integration point level, the evolution of the plastic strain is computed by means of the fully implicit return mapping algorithm. In order to achieve quadratic convergence and stability within the return mapping algorithm, the Jacobian of the local problem is computed using automatic differentiation (Leal (2018)). The evolution of the viscoelastic strain and drying creep strain is computed by means of step-by step integration of two Kelvin chains which correspond to the Dirichlet series approximations of the respective compliance functions, cf. Eqs. (8 and 10. The determination of the coefficients in the Dirichlet series is based on an approximation of the continuous retardation spectrum of the compliance function. In this context, for a k-th order approximation of the continuous retardation spectrum, the k-th derivative of the compliance function is required. Especially for the approximation of the drying creep compliance function higher order derivatives are required for sufficient accuracy. Due to the fact that the manual derivation of Eq. (10) is laborious and error-prone, automatic differentiation is employed for computing the higher order derivatives.

The constitutive model is implemented within the framework of the Marmot material modeling toolbox (Dummer, Mader, Neuner, & Schreter 2018) which is based on the `C++` programming language. In order to achieve high computational efficiency, the `Eigen` library (Guennebaud, Jacob, et al. 2010) is used for linear algebra computations. Employing the Marmot framework, the constitutive model can be used in in-house finite element codes directly as well as in the finite element software Abaqus (2015) as a user-defined subroutine.

5 CALIBRATION AND VALIDATION

The constitutive model is calibrated and validated by means of numerical simulations at material point level using experimental data provided by (i) Kim, Kwon, Kim, & Kim (2005) for moderate uniaxial and multiaxial compression, and (ii) Rüsch (1968) for moderate and high uniaxial compression up to failure.

Basic Creep in Moderate Multiaxial Compression

Kim, Kwon, Kim, & Kim (2005) studied the creep behavior of concrete in multiaxial compression on sealed cubic specimens with an edge length of 200 mm considering three different concrete compositions. For the present numerical study, the mixture CI is chosen. The specimens were loaded at the material age of 28 days with the loading scheme outlined in Table 1. The strain was measured using embedded gauges with a measurement length of 100 mm.

Since the uniaxial test U2 exhibits the highest loading, corresponding to only about 37% of the uniaxial compressive strength at loading of $f_{cu}^{28} =$ 26 MPa, nonlinear creep behavior is not expected in this experiments.

Hence, the capability of the constitutive model to represent approximately linear creep behavior for moderate uniaxial and multiaxial stresses will be demonstrated. In lack of experimental data for

Table 1. Loading conditions of the creep tests carried out on the concrete mixture CI by Kim, Kwon, Kim, & Kim (2005).

designation	acting stress σ_1 (MPa)	σ_2 (MPa)	σ_3 (MPa)
U1	-4.90	–	–
U2	-9.80	–	–
B1	-4.90	-0.98	–
B2	-4.90	-1.96	–
B3	-9.80	-1.96	–
T1	-4.90	-0.49	-0.49
T2	-4.90	-0.98	-0.98
T3	-4.90	-1.96	-1.96
T4	-4.90	-1.96	-0.98

calibrating the creep amplification factor for nonlinear creep a_{cr}, it is assumed $a_{cr} = 7.6$ as identified from the creep experiments by Rüsch (1968) later on. For the evolution of material strength $s_f = 0.25$ is assumed for normal hardening cement according to the recommendation by CEB-FIP (2013). As reported in Kim, Kwon, Kim, & Kim (2005), the effective Poisson's ratio is chosen as $\nu = 0.17$. The compliance parameters q_1 and q_4 are estimated according to Bažant, Jirásek, Hubler, & Carol (2015). The parameters n, q_2 and q_3, which control the aging viscoelastic behavior, are calibrated by means of a nonlinear least square optimization procedure using experimental data for the axial strain of the uniaxial creep test U2. The parameters obtained by this calibration procedure are summarized in Table 3.

Figure 6 shows a comparison of the experimental results and the numerical predictions by the constitutive model. Except for the uniaxial test U2, which was used for model calibration, all numerical simulations can be viewed as prediction. Accordingly, it is concluded that the predictions of the model are in very good agreement with the experimental data reported by Kim, Kwon, Kim, & Kim (2005).

Creep at drying conditions up to failure

Rüsch (1968) studied the evolution of deformations of dog-bone like unsealed concrete specimens in moderate and high sustained uniaxial compression for different concrete compositions. For the present investigation, the uniaxial creep tests of the set denoted as S3 are chosen.

Table 2. Applied compressive stress in the uniaxial creep tests of set S3 by Rüsch (1968). The * and † indicate tests used for model calibration.

t'	applied compressive stress in MPa						
20 d	-16.9*	-18.5	-20.2	-21.9	-24.6		
56 d	-16.5	-19.8	-23.1	-26.5†	-29.8	-37.8	
170 d	-23.1	-26.5*	-29.8	-33.1*†	-37.2	-39.4	
600 d	-29.7	-31	-32.1	-33.1	-33.5	-34.7	-40.9

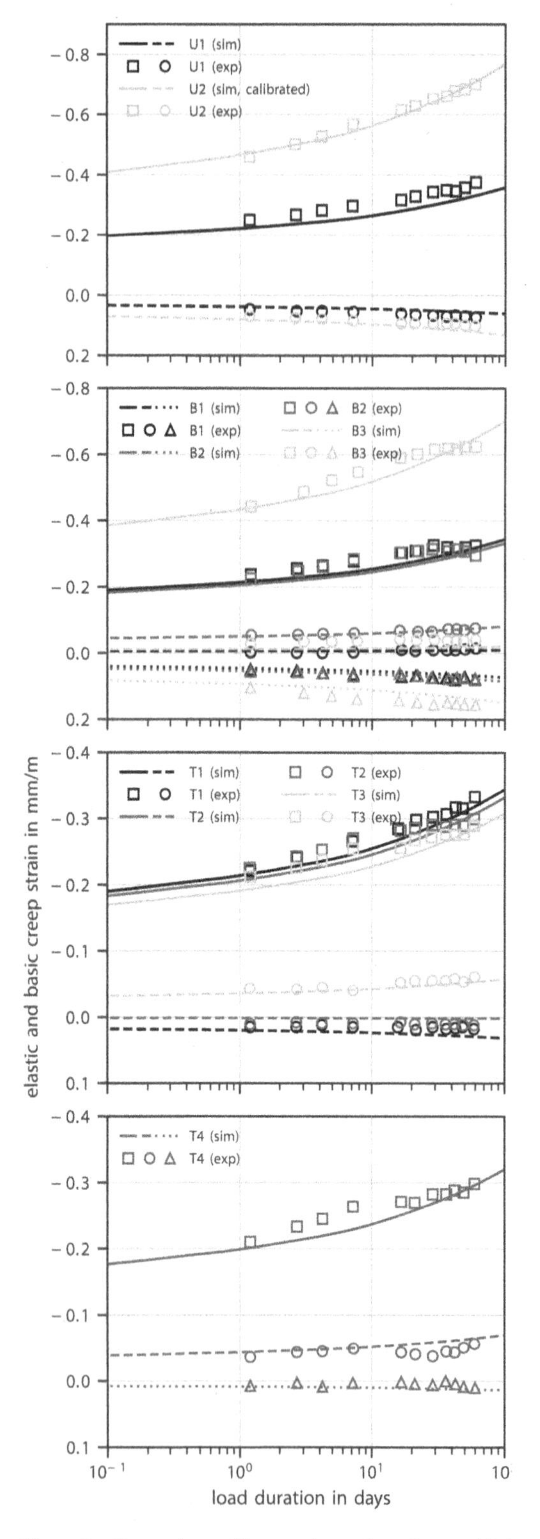

Figure 6. Comparison of the experimental results, provided by Kim, Kwon, Kim, & Kim (2005), with the respective numerical predictions for the evolution of the principal strain components ε_1 (rectangles and solid lines), ε_2 (circles and dashed lines) and ε_3 (triangles and dotted lines).

The tests loaded at the material age of 56 days and 170 days are used for evaluating the capabilities of the constitutive model for representing nonlinear creep and failure due to creep by means of numerical simulations at material point level.

The calibration of the constitutive model is performed employing the following strategy:

(i) The evolution of the uniaxial compressive strength is calibrated by means of the uniaxial compression tests on cylinders at 14 days and 28 days material age.
(ii) The parameters for autogenous shrinkage and drying shrinkage are estimated based on the concrete composition and the experimental data from the accompanying shrinkage tests on unsealed specimens according to Bažant, Jirásek, Hubler, & Carol (2015).
(iii) The parameters q_1 and q_4 are estimated according to Bažant, Jirásek, Hubler, & Carol (2015) based on the provided Young's modulus at the material age of 28 days of $E^{28} = 29.5$ MPa and the concrete composition.
(iv) The parameters n, q_2, q_3, q_5 and a_{cr} are calibrated by means of a nonlinear least-square optimization using the tests marked with * in Table 4.
(v) The creep damage parameters in Eq. (18) are chosen as $n_f = 16.5$ and $\tau_f = 0.77$ h, for representing the failure times of the creep tests marked with † in Table 4.

The parameters resulting from this calibration procedure are summarized in Table 4.

Figures. 7 and 8 depict the experimental results for the tests loaded at 56 days and 170 days respectively, together with the response of the calibrated constitutive model. Excellent agreement of the numerical simulations with the experimental results can be seen for all tests.

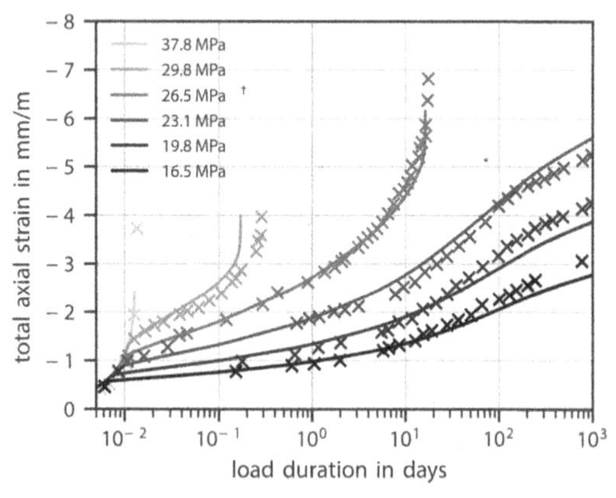

Figure 7. Evolution of the total axial strain in uniaxial compressive creep tests loaded at the concrete age of 56 days: experimental data provided by Rüsch (1968) together with the predictions by the calibrated material model. The † indicates the test used for model calibration.

Table 3. Material parameters used for the uniaxial and multiaxial creep tests by Kim, Kwon, Kim, & Kim (2005).

ν (-)	n (-)	q_1 (1/MPa)	q_2 (1/MPa)	q_3 (1/MPa)	q_4 (1/MPa)	a_{cr} (-)	s_f (-)	$f_{cu}^{(28)}$ (MPa)
0.17	0.21	29.2×10^{-6}	28.5×10^{-6}	16.4×10^{-6}	10.2×10^{-6}	7.6	0.25	26

Table 4. Material parameters for the uniaxial creep tests by Rüsch (1968).

q_1 (1/MPa)	q_2 (1/MPa)	q_3 (1/MPa)	q_4 (1/MPa)	q_5 (1/MPa)	a_{cr} (-)	$\varepsilon_{\infty}^{shr,d}$ ($‰^{\infty}$)	$\tau_{shr,d}$ (h)	$\varepsilon_{\infty}^{shr,au}$ ($‰^{\infty}$)	$\tau_{shr,au}$ (h)	s_f (-)	$f_{cu}^{(28)}$ (MPa)
23.7×10^{-6}	118×10^{-6}	19.4×10^{-6}	9.3×10^{-6}	420×10^{-6}	7.6	–0.627	2215	–0.065	72.7	0.55	29.6

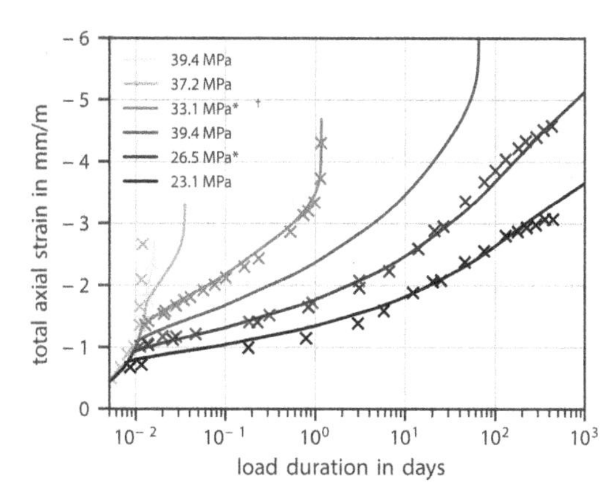

Figure 8. Evolution of the total axial strain in uniaxial compressive creep tests loaded at the concrete age of 170 days: experimental data by Rüsch (1968) together with the predictions by the calibrated material model. The * and † indicates tests used for model calibration.

6 SUMMARY AND CONCLUSIONS

An extended material model for concrete with a novel formulation for modeling nonlinear creep and the evolution of damage due to high sustained degrees of material utilization was investigated. In particular the determination of the degree of material utilization, the influence of evolution of the material strength, and the influence of the material parameters on the nonlinear creep behavior and damage due to creep were discussed. Finally, the experimental results provided by Kim, Kwon, Kim, & Kim (2005) and Rüsch (1968) were compared to the predictions by the calibrated constitutive model. It was shown that excellent agreement of the model prediction and the experimental results is obtained for various loading conditions.

ACKNOWLEDGEMENTS

Partial financial support for A. D. by the Tyrolean Science Fund (TWF, Project No. F.18719) is gratefully acknowledged.

REFERENCES

Abaqus (2015). *ABAQUS v6.14 Documentation*. Providence, RI, USA: Dassault SystÃ¨mes.

Bažant, Z. P. & S. Prasannan (1989). Solidification theory for concrete creep. I: Formulation. *J Eng Mech 115*(8), 1691–1703.

Bažant, Z. P., M. Jirásek, M. Hubler, & I. Carol (2015). RILEM draft recommendation: TC-242-MDC multi-decade creep and shrinkage of concrete: material model and structural analysis. Model B4 for creep, drying shrinkage and autogenous shrinkage of normal and high-strength concretes with multi-decade applicability. *Mater Struct 48*(4), 753–770.

Bažant, Z. P. & M. Jirásek (2018). *Creep and hygrothermal effects in concrete structures*, Volume 38. Springer.

Boumakis, I., G. Di Luzio, M. Marcon, J. Vorel, & R. Wan-Wendner (2018). Discrete element framework for modeling tertiary creep of concrete in tension and compression. *Eng Fract Mech 200*, 263–282.

CEB-FIP (2013). *Fib model code for concrete structures 2010* (1 ed.). Ernst & Sohn GmbH & Co. KG.

Dummer, A., T. Mader, M. Neuner, & M. Schreter (2018). Marmot library. https://github.com/MAteRialMOdeling Toolbox/marmot.

Dummer, A., M. Neuner, & G. Hofstetter (2022). An extended gradient-enhanced damage-plasticity model for concrete considering nonlinear creep and failure due to creep. *submitted*.

Grassl, P. & M. Jirásek (2006). Damage-plastic model for concrete failure. *Int J Solids Struct 43*(22–23), 7166–7196.

Guennebaud, G., B. Jacob, et al. (2010). Eigen v3. http://eigen.tuxfamily.org.

Jirásek, M. & Z. P. Bažant (2002). *Inelastic Analysis of Structures*. John Wiley & Sons, Ltd.

Kim, J., S. Kwon, S. Kim, & Y. Kim (2005). Experimental studies on creep of sealed concrete under multiaxial stresses. *Mag Concr Res 57*(10), 623–634.

Leal, A. M. M. (2018). autodiff, a modern, fast and expressive C++ library for automatic differentiation. https://autodiff.github.io.
Mazzotti, C. & M. Savoia (2003). Nonlinear creep damage model for concrete under uniaxial compression. *J Eng Mech 129*(9), 1065–1075.
Neuner, M., P. Gamnitzer, & G. Hofstetter (2017). An Extended Damage Plasticity Model for Shotcrete: Formulation and Comparison with Other Shotcrete Models. *Materials 10*(1), 82.
Ren, X., Q. Wang, R. Ballarini, & X. Gao (2020). Coupled creep-damage-plasticity model for concrete under long-term loading. *J Eng Mech 146*(5), 04020027.
Rüsch, H. (1968). Festigkeit und Verformung von unbewehrtem Beton unter konstanter Dauerlast. *Schriftenreihe DafStb 198*.
Willam, K. & E. Warnke (1975). Constitutive model for the triaxial behaviour of concrete. In *Seminar on Concrete Structure Subjected to Triaxial Stresses*, Volume 19, Bergamo, Italy, pp. 1–30.

Computational Modelling of Concrete and Concrete Structures – Meschke, Pichler & Rots (Eds)
 ISBN: 978-1-032-32724-2

3D DEM simulations of fracture in reinforced concrete beams

M. Nitka & J. Tejchman
Faculty of Civil and Environmental Engineering, Gdańsk University of Technolog, Gdańsk, Poland

ABSTRACT: The paper deals with the behaviour of a reinforced concrete beam without vertical reinforcement under three-point bending. The beam failed in shear due to over-reinforcement. The experiments were performed at the laboratory scale using a micro-CT system. They were next reproduced in numerical analyses using the 3D discrete element method (DEM). The 4-phase model of concrete was used with meso-strucure, based directly on micro-CT images. A satisfactory agreement between experimental and numerical outcomes was achieved with respect to the location and direction of a critical macro-crack.

1 INTRODUCTION

Quasi-brittle multiphase materials, such as concrete, are largely used in engineering structures. Concrete is generally referred to as a strongly heterogeneous and discontinuous material (Königsberger et al. 2018; Pichler & Hellmich 2011; Skarżyński & Tejchman 2013). It may be considered at the meso-scale as a composite material wherein four key constituents (phases) may be isolated: aggregate, cement matrix, interfacial transition zones (ITZs) between aggregates and cement matrix and macro-pores. The mesoscopic material heterogeneity has a pronounced influence on complex crack growth trajectory paths at the macro-scale, composed of various macro-crack branches, complementary cracks and micro-cracks. The mechanical concrete performance depends on material properties of all its constituents and their mutual interaction which makes the modelling of the crack formation and development at the meso-scale a real challenge in terms of efficiency and accuracy. The optimization and safety assessment of structures composed of quasi-brittle materials (like concrete) requires, however, a comprehensive understanding of the initiation, formation and propagation of micro- and macro-cracks. Recently, great efforts were made to accurately and efficiently capture the failure behaviour (damage and fracture) of concrete structures at the aggregate level by meso-scale models. The meso-scale behaviour of concretes may be modelled with continuous (e.g. Gitman et al. 2008; Kim & Abu Al-Rub 2011; Shahbeyk et al. 2011; Skarżyński & Tejchman 2010; Zhou & Chen 2019) and discontinuous models (Yang et al. 2009, Su et al. 2010, Wang et al. 2016, Trawinski et al. 2016, 2018) within continuum mechanics and discrete models, including lattice models (e.g. Cu-satis et al. 2011; Herrmann et al. 1989; Karavelić et al. 2019; Kozicki & Tejchman 2007, 2008; Lilliu & van Mier 2003; Pan et al. 2018; Schlangen & Gar-boczi 1996; Šavija et al. 2019), interface models based on fracture mechanics (Carol et al. 2001; López et al. 2008) and particulate discrete models (e.g. Donze et al. 1999; Dupray et al. 2009; Groh et al. 2011; Hentz et al. 2004; Krenzer et al. 2019; Nguyen et al. 2019; Nitka & Tejchman 2018, 2020; Rangari et al. 2018; Skarżyński et al. 2015).

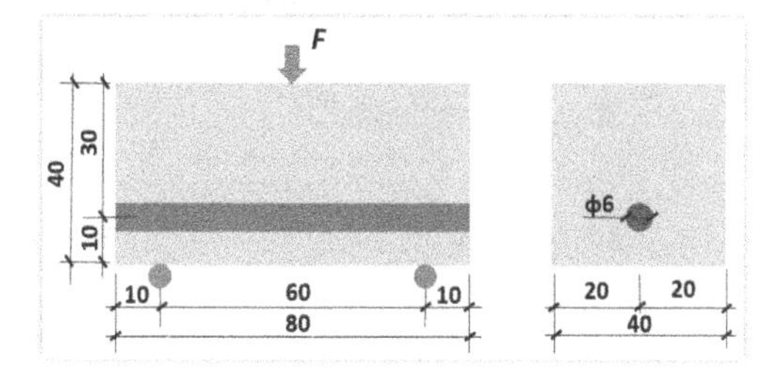

Figure 1. Geometry of RC beam (Skarżyński & Tejchman 2021).

The main objective of the current paper is to clarify the usefulness of the discrete element method (DEM) for studying a fracture pattern in reinforced concretes under 3D conditions and 2) to achieve a better insight regarding micro- and macro-cracks at the mesoscopic level. The DEM calculations were carried out for one RC beam under three-point bending without vertical reinforcement that was tested in experiments (Skarżyński & Tejchman 2021). The beam-meso-structure was assumed, based on micro-CT images. The 3D particulate discrete element model YADE was employed that was developed at the University of Grenoble (Kozicki & Donze 2008; Šmilauer & Chareyre 2011). As compared to mesoscopic continuum calculations, DEM is able to simulate fracture from the beginning of deformation since it possesses strongly diverse local failure criteria. It does not also need material softening to be imposed.

DOI 10.1201/9781003316404-53

2 OWN EXPERIMENTS

The experiments were carried out on a RC beam under three-point bending without vertical reinforcement (length of 80 mm, span length of 60 mm, cross-section of 40×40 mm^2 and shear span ratio $a/D = 0.75$, Figure 1), composed of round aggregate particles (the maximum diameter $d_{max} = 16$ mm and mean diameter $d_{50} = 2$ mm) (Skarżyński & Tejchman 2021). The effective height was $D = 30$ mm. The average uniaxial compressive strength of concrete was $fc = 49.75$ MPa, mean tensile strength during bending was $f_t = 3.96$ MPa, average Young's modulus $E = 34.8$ GPa and mean Poisson's ratio $\nu = 0.21$. The initial concrete porosity was 2.7%. The reinforcement ratio was high to avoid reinforcement yielding ($\rho = 1.8\%$). As reinforcement, a steel bar with a diameter of $d = 6$ mm) was used. The laboratory tests were performed with a displacement-controlled option using the rate of 0.002 mm/min. The beam was continuously scanned using the micro-CT system SkyScan 1173, mounted on the Instron 5569 loading machine (Figure 2). It was three times scanned by micro-CT for different beam deflections: close to the peak load, after the peak load in a softening regime and close to the failure. The scanning process lasted 45 minutes. The voxel size was 46 microns and the exposure time was 3000 ms. The beam was scanned at 180° with a single rotation step of 0.6°.

The beam failure took place in a rapid brittle due to a diagonal shear crack moving from the support region through a beam compressive zone towards the loading point. (Figure 3). The critical shear crack propagated as the outermost crack. Some secondary cracks also occurred on the lateral end sides of the beam just before the beam failure. The maximum vertical force of the RC beam was $F = 10.46$ kN, shear strength was $V = F/bD = 8.72$ MPa and flexural strength was 14.71 MPa. The beam deflection corresponding to the maximum force was 0.26 mm. The force-deflection diagram indicated a pre-peak hardening region, softening after the peak load, re-hardening and re-softening The mean inclination of the critical shear crack to the bottom was 59° (front side) and 57° (rear side). The critical shear crack width in the tensile region non-linearly changed with the beam height from $w_{cs} = 0.19$ mm to $w_{cs} = 1.25$ mm (the average value was 0.61 mm). The relationship between the crack volume and beam deflection (based on micro-CT measurements) was bilinear. The change of the curve inclination occurred at the peak load region where the crack volume was about 0.2–0.5%. The final crack volume was about 3.3%.

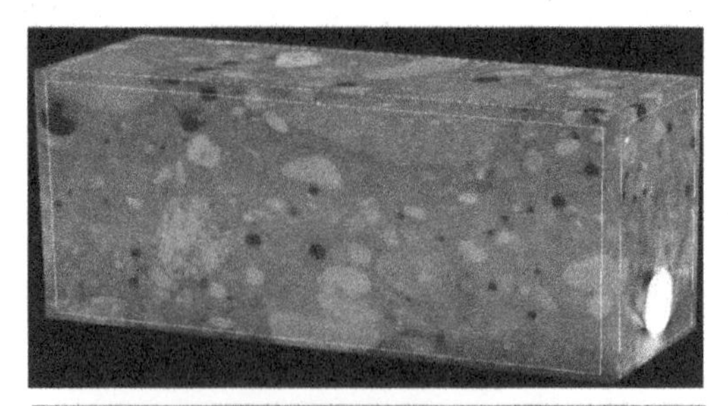

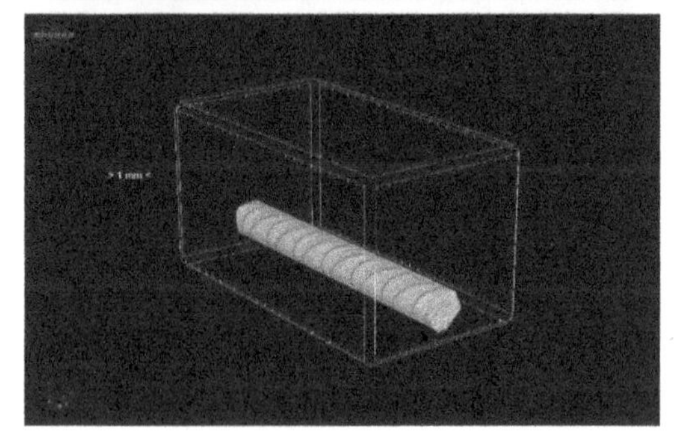

Figure 2. General view on non-cracked RC beam and distribution of aggregates and steel bar in 3D micro-CT images before loading (Skarżyński & Tejchman 2021).

3 DEM MODEL

DEM directly simulates meso-structure and thus it may be used to comprehensively study the mechanism of the initiation, growth and formation of localized zones, cracks and fractures that greatly affect the macroscopic behaviour of frictional-cohesive materials (Skarżyński et al. 2015, Suchorzewski et al. 20118a, 2018b, Nitka & Tejchman 2018, 2020). It easily represents discontinuities caused by cracking or fragmentation. The disadvantages are the huge computational cost. For normal contacts, a linear relationship between forces and displacements in compression and tension was assumed with a limit tensile force F^n_{min}. For tangential contacts, the bi-linear cohesive-frictional law was chosen with the initial cohesive force F^s_{max}. The bond breakage between elements appeared if the tensile force/shear force reached their limits. If any contact between spheres after the failure re-appeared, the cohesion was not considered. The critical cohesive F^s_{max} and tensile forces F^n_{min} were assumed as a function of the cohesive stress C, tensile normal stress T and element radius R. The following five main local material parameters were needed for our discrete simulations: E_c (modulus of elasticity of the grain contact), υ_c (Poisson's ratio of the grain contact), μ_c (inter-particle friction angle)C (cohesive stress) and T (tensile normal stress). In addition, the particle radius R, particle

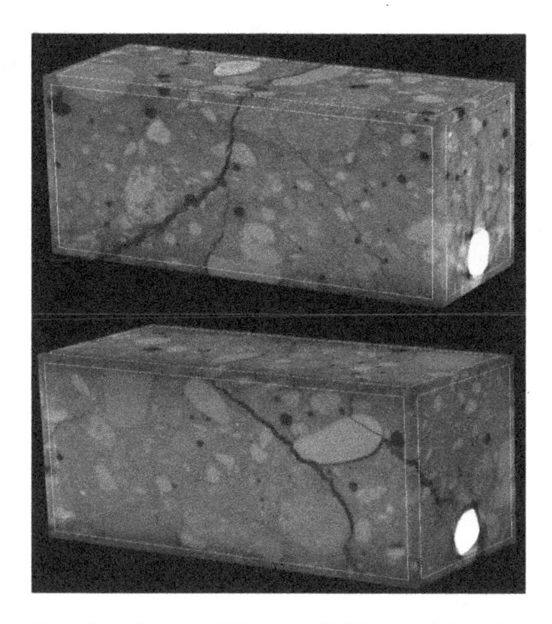

Figure 3. 3D micro-CT images of RC beam reinforced on both sides at failure.

mass density ρ and damping parameters α_d were required.

The concrete was numerically modelled as a 4-phase material composed of aggregate, cement matrix, ITZs (interfacial transitional zones) and pores. The aggregate range was assumed to be between 2 mm and 16 mm. The aggregate particles consisted of clusters of spheres (each sphere had a diameter varying from 0.9 mm to 2.40 mm). The size, shape and position of each aggregate were taken directly from the real specimen, based on micro-CT images. The macro-voids (modelled as empty spaces) were also determined with the aid of micro-CT images (Skarżyński & Tejchman 2021). The cement matrix in the beam was modelled with spheres of the diameter between 0.50 mm and 2 mm, with an initial porosity of 2.7% (as in the experiment). ITZs were modelled as weaker contact between aggregates and mortar particles (they had no physical width). The geometry of the steel bar with ribs was again transferred from the micro-CT image. The bar was fulfilled with spherical elements with a diameter 0.25–2 mm that were very tightly packed. The rib height was 0.85 mm. The contact model between steel spheres was linear both in compression and tension (without a plastic region). The contact stiffness was matched with the global elastic modulus of steel (about 200 GPa). The concrete parameters were calibrated with uniaxial compression and tension tests (Nitka & Tejchman 2018, 2020) (Table 1). The modulus elasticity of contacts spheres in the steel bar was small due to their high density (coordination number was about 100).

Table 1. DEM parameters for different phases in simulations.

Parameters	Cement	ITZs	Bar
Contact stiffness E [GPa]	11.2	11.2	1.4
k_n/k_s [−]	0.2	0.2	0.1
friction angle μ [°]	18	18	7
cohesion C [GPa]	22.5	22.5	100
tensile normal stress T [GPa]	22.5	22.5	100

4 NUMERICAL RESULTS

The numerical beam had the same geometry as in the experiment. The total number of DEM elements was about 200 000 (15 000 in the steel bar and 15 600 in aggregates) (Figure 4).

First, the calculated global force-deflection curve ($F = \mathrm{f}(u)$) was compared with the experimental one (Figure 5). The maximum value of the force was similar as in the experiment, about $F = 10{,}5$ kN. The initial

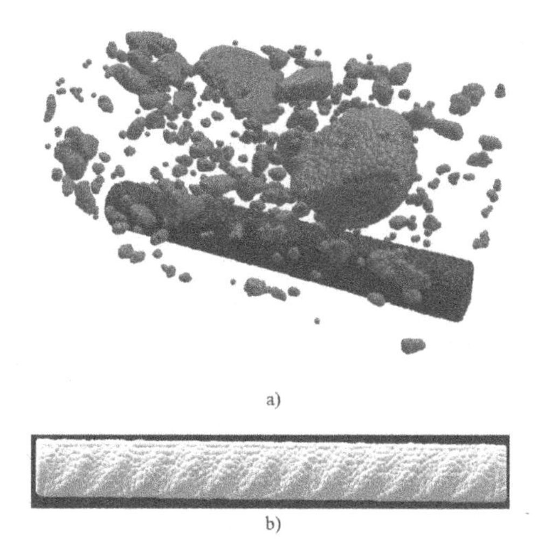

Figure 4. 3D DEM model: a) view on aggregates and steel bar (based on Figure 2) and b) view on steel bar with ribs.

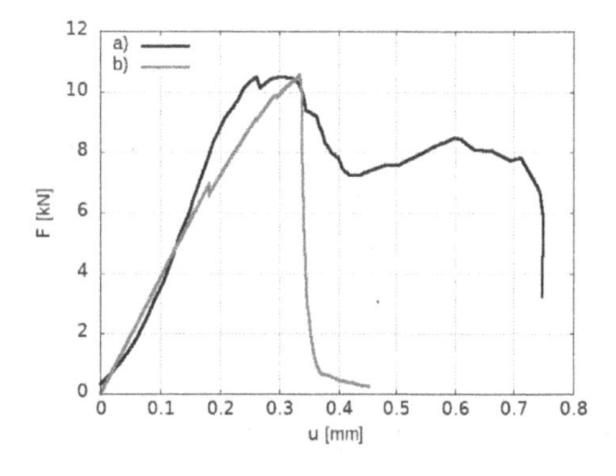

Figure 5. Macroscopic force-deflection curve: a) experiment (Skarżyński & Tejchman 2021) and b) 3D DEM calculations.

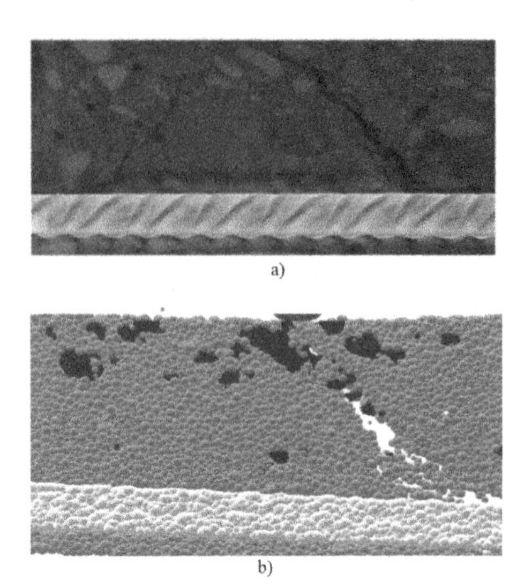

Figure 6. Final critical shear macro-crack path in beam mid-part: a) experiment (Skarżyński & Tejchman 2021) and b) 3D DEM simulation (white colour corresponds to steel bar, black colour to aggregates and light grey colour denotes cement matrix, macro-crack was enlarged 10 times).

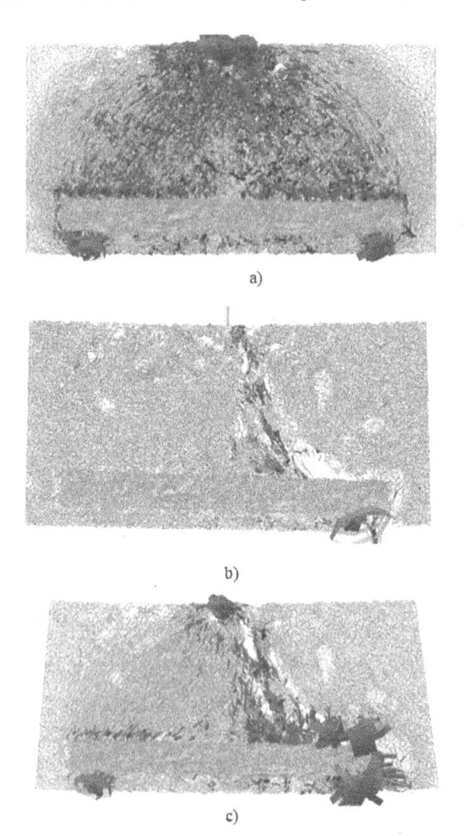

Figure 7. Contact forces from 3D DEM calculations in beam mid- cross-section: a) before peak, b) after peak and c) at test end (red colour denotes compressive normal forces and blue colour tensile normal forces (green colour shows small force values).

global stiffness was also similar. However, the post-peak behaviour was different. In the DEM analysis, rapid damage appeared for the deflection $u > 0.35$ mm. In contrast, in the experiment, the re-hardening response occurred for $0.4 \text{ mm} < u < 0.6$ mm. The maximum normal stress in the bar was about 85 MPa.

The calculated final critical shear macro-crack in the beam mid-part is shown in Figure 6. Its location is similar to that in the experiment. The crack propagated along the steel bar and later up to the loading point. The inclination of the macro-crack was about 45–50°. The experimental secondary cracks were not obtained in calculations. In Figure 7, the contact normal forces between DEM elements are presented in three steps: before the peak, just after peak and at the test end (the red colour denotes compressive and blue colour tensile forces, and the green colour corresponds to small force values). The strong compressive stresses appeared below the loading point and at two supports. Some high compressive forces were also observed along the beam. Strong interlocking appeared in the macro-crack (Figures 7b and 7c).

The evolution of broken contacts is shown in Figure 8. The cracking process already started for $u = 0.15$ mm. It was well visible after $u = 0.2$ mm (far before the peak force). A pronounced increase

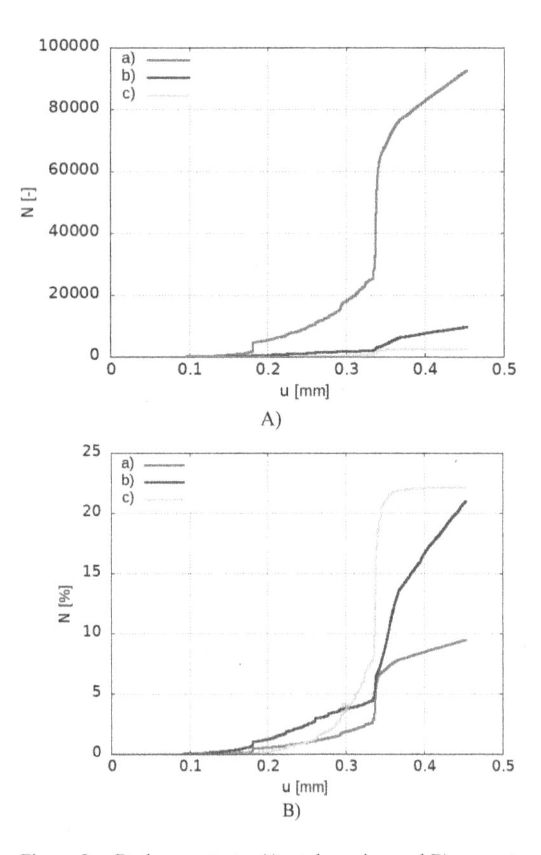

Figure 8. Broken contacts: A) total number and B) percent content with respect to initial one for beam deflection u (a) inside cement matrix, b) in ITZs and c) at steel bar surface.

of broken contact took place after the force peak $u = 0.33$ mm. The total number of broken contacts was more than 100 000. Most contacts were damaged in the cement matrix (Figure 8Aa), then in ITZs (about 10 000, Figure 8Ab) and the least at the steel bar interface (about 5 000, Figure 8Ac). However, if the percent of broken contacts was plotted with respect to initial contacts (Figure 8B), the most damaged region was the steel bar interface (Figure 8Bc) wherein almost 23% of all initial contacts were broken. In ITZs 20% of initial contacts were damaged, and in the cement matrix, 9% were broken.

5 CONCLUSIONS

The DEM model proved to be realistic for describing fracture in a reinforced concrete beam failing in shear. The calculated final macro-crack path was found to be similar to the experimental one. The micro-cracking process started far before the peak force. Most contacts were damaged in the cement matrix, then in ITZs and at the steel bar interface. If the percent volume of broken contacts was considered with respect to the initial one, the most damaged region was the steel bar interface. The calculated maximum vertical force matched the experimental value. However, a different post-peak behaviour was obtained in DEM simulations (material softening instead of re-hardening/re-softening material).

REFERENCES

Carol I, López C.M., Roa O. 2001. Micromechanical analysis of quasi-brittle materials using fracture-based interface elements. *International Journal for Numerical Methods in Engineering* 52: 193–215.

Cusatis G., Pelessone D., Mencarelli A. 2011. Lattice discrete particle model (LDPM) for failure behavior of concrete. I: theory. *Cem. Concr. Compos*. 33: 881–890.

Donze F.V., Magnier S.A., Daudeville L., Mariotti C. 1999. Numerical study of compressive behaviour of concrete at high strain rates. *Journal for Engineering Mechanics* 122 (80): 1154–1163.

Dupray. F, Malecot Y., Daudeville L., Buzaud E.A. 2009. Mesoscopic model for the behaviour of concrete under high confinement. *International Journal for Numerical and Analytical Methods in Geomechanics* 33: 1407–23.

Gitman I.M., Askes H., Sluys L.J. 2008. Coupled-volume multi-scale modelling of quasi-brittle material. *European Journal of Mechanics A/Solids* 27: 302–327.

Groh U., Konietzky H., Walter K. et al. 2011. Damage simulation of brittle heterogeneous materials at the grain size level, *Theoretical and Applied Fracture Mechanics* 55: 31–38.

Hentz S., Daudeville L., Donze F. 2004. Identification and validation of a Discrete Element Model for concrete. *Journal of Engineering Mechanics* ASCE 130 (6): 709–719.

Herrmann H.J., Hansen A., Roux S. 1989. Fracture of disordered, elastic lattices in two dimensions. *Physical Rev B* 39: 637–647.

Karaveliæ E., Nikoliæ M., Ibrahimbegovic A., Kurtoviæ A. 2019. Concrete meso-scale model with full set of 3D failure modes with random distribution of aggregate and cement phase. Part I: Formulation and numerical implementation. *Comput. Methods Appl. Mech. Eng.* 344: 1051–1072.

Kim S.M., Abu Al-Rub R.K. 2011. Meso-scale computational modeling of the plastic-damage response of cementitious composites. *Cement and Concrete Research* 41: 339–358.

Kozicki J., Tejchman J. 2007. Effect of aggregate structure on fracture process in concrete using 2D lattice model. *Archives of Mechanics* 59 (4–5): 1–20.

Kozicki J., Tejchman J. 2008. Modelling of fracture processes in concrete using a novel lattice model. *Granular Matter* 10: 377–388

Königsberger, M., Hlobil, M., Delsaute, B., Staquet, S., Hellmich, C., Pichler, B. 2018. Hydrate failure in ITZ governs concrete strength: A micro-to-macro validated engineering mechanics model, *Cem. Concr. Res. 103*: 77–94.

Krenzer K., V. Mechtcherine V., Palzer U. 2019. Simulating mixing processes of fresh concrete using the discrete element method (DEM) under consideration of water addition and changes in moisture distribution. *Cement and Concrete Research* 115: 274–282.

Lilliu G., van Mier J.G.M. 2003. 3D lattice type fracture model for concrete. *Engineering Fracture Mechanics* 70: 927–941.

López C.M., Carol I., Aguado, A. 2008. Meso-structural study of concrete fracture using interface elements. I: Numerical model and tensile behavior. *Mater. Struct*. 41: 583–599.

Nguyen T.T., Bui H.H., Ngo T.D., Nguyen G.D., Kreher M.U., Darve F. 2019. A micromechanical investigation for the effects of pore size and its distribution on geopolymer foam concrete under uniaxial compression. *Eng. Fract. Mech* 209: 228–244.

Nitka M., Tejchman J. 2018. A three-dimensional meso scale approach to concrete fracture based on combined DEM with X-ray μCT images. *Cement and Concrete Research* 107: 11–29

Nitka M., Tejchman J. 2020. Meso-mechanical modelling of damage in concrete using discrete element method with porous ITZs of defined width around aggregates. *Engineering Fracture Mechanics* 231, 107029.

Pan Z., Ma R., Wang D., Chen A. 2018. A review of lattice type model in fracture mechanics: theory, applications, and perspectives. *Engineering Fracture Mechanics* 190: 382–409.

Pichler, B., Hellmich, C. 2011. Upscaling quasi-brittle strength of cement paste and mortar: A multi-scale engineering mechanics model. *Cem. Concr. Res. 41*: 467–476.

Rangari S., Murali K., Deb A. 2018. Effect of meso-structure on strength and size effect in concrete under compression. *Eng. Fract. Mech.* 195: 162–185.

Schlangen E., Garboczi E.J. 1996. New method for simulating fracture using an elastically uniform random geometry lattice. *Internat. J. Engrg. Sci.* 34: 1131–1144.

Shahbeyk S., Hosseini M., Yaghoobi M. 2011. Mesoscale finite element prediction of concrete failure. *Computational Materials Science* 50 (7): 1973–1990.

Skarżyński L., Nitka M., Tejchman J. 2015. Modelling of concrete fracture at aggregate level using FEM and DEM based on x-ray μCT images of internal structure. *Engineering Fracture Mechanics* 10 (147): 13–35.

Skarżyński Ł., Tejchman, J. 2010. Calculations of fracture process zones on meso-scale in notched concrete beams

subjected to three-point bending. *European Journal of Mechanics* A/Solids 29 (4): 746–760.
Skarżyński, Ł., Tejchman, J. 2013. Modeling the effect of material composition on the tensile properties of concrete. Understanding the tensile properties of concrete *(edited by Jaap Weerheijm), Woodhead Publishing Limited 48*: 52–97.
Skarżyński, Ł., Tejchman, J. 2021. Investigations on fracture in reinforced concrete beams in 3-point bending using continuous micro-CT scanning. *Construction and Building Materials* 284.
Su X.T. Yang Z.J., Liu G.H. 2010. Monte Carlo simulation of complex cohesive fracture in random heterogeneous quasi-brittle materials: A 3D study. *Int. J. Solids Struct.* 47: 2336–2345.
Suchorzewski, J., Tejchman, J., Nitka, M.2018a. Experimental and numerical investigations of concrete behaviour at meso-level during quasi-static splitting tension. *Theor. Appl. Fract. Mech.* 96: 720–739.
Suchorzewski, J., Tejchman, J., Nitka, M. 2018b. DEM simulations of fracture in concrete under uniaxial compression based on its real internal structure. *Int. J. Damage Mech.* 27 (4): 578–607.
Šavija B., Smith G.E., Liub D., Schlangen E., Flewitt P.E.J. 2019. Modelling of deformation and fracture for a model quasi-brittle material with controlled porosity: Synthetic versus real microstructure. *Engineering Fracture Mechanics* 205: 399–417.
Trawinski W., Bobinski J., Tejchman J. 2016. Two-dimensional simulations of concrete fracture at aggregate level with cohesive elements based on X-ray micro-CT images. *Engineering Fracture Mechanics* 168: 201–226.
Trawiński W., Tejchman J., Bobiński J. 2018. A three-dimensional meso-scale approach with cohesive elements to concrete fracture based on X-ray μCT images. Engineering Fracture Mechanics 189: 27–50.
Wang X.; Zhang M. Jivkov A.P. 2016. Computational technology for analysis of 3D meso-structure effects on damage and failure of concrete. *Int. J. Solids Struct.* 80: 310–333.
Yang Z.J. Su, X.T. Chen J.F. Liu G.H. 2009. Monte Carlo simulation of complex cohesive fracture in random heterogeneous quasi-brittle materials. *Int. J. Solids Struct.* 46: 3222–3234.
Zhou R., Chen H.-H. 2019. Mesoscopic investigation of size effect in fracture process zone. *Engineering Fracture Mechanics* 212: 136–152.

Computational Modelling of Concrete and Concrete Structures – Meschke, Pichler & Rots (Eds)

 ISBN: 978-1-032-32724-2

Numerical reproduction of three plain concrete tests

M. Szczecina
Faculty of Civil Engineering and Architecture, Kielce University of Technology, Kielce, Poland

A. Winnicki
Faculty of Civil Engineering, Cracow University of Technology, Cracow, Poland

ABSTRACT: There are many material models dedicated to concrete, but they have their limitations and drawbacks and FEM software users should be aware of them. The models would give very similar results for simple loading tests, e.g. uniaxial tension, but they can produce completely different results for the mixed-mode fracture. The authors of this paper have chosen the following tests (at the structural level): Nooru-Mohamed test, Schlangen test and L-specimen test under quasi-static loads. The tests are performed with two concrete models implemented in the Abaqus software: Concrete Damaged Plasticity (CDP) and Concrete Smeared Cracking (CSC). Results of the tests allow to compare the behavior of both models in the mode-I and mode-II of crack growth. Additionally some important practical issues concerning the FEM modeling are analyzed in the paper: influence of a type of a FEM mesh – structural vs. random and influence of a finite element type – 4-node vs. 3-node.

1 INTRODUCTION

Proper numerical modeling of concrete behavior demands use of a carefully checked, robust material model. The Abaqus (2012) software offers a few models, of which two were selected by the authors of this paper for further analysis. These two models are: Concrete Damaged Plasticity (CDP) and Concrete Smeared Cracking (CSC). The authors have already gained some experience using the CDP model, especially in analysis of concrete frame corners (Szczecina & Winnicki 2018 2021). The next stage of the ongoing work is the comparison of behavior of the CDP and CSC models in three selected numerical tests for plain concrete. The main goal of the research was a verification which of the two selected models "passed" the tests. Furthermore, some important practical issues were also analyzed, namely influence of a type of a FEM mesh and influence of a finite element type.

The three selected tests are: the Nooru-Mohamed test (Bobiński & Tejchman 2016; Nooru-Mohamed 1992), the Schlangen test (Gontarz & Podgórski 2020; Schlangen 1993) and the L-specimen test (Winkler et al. 2001), all described in details in the next section. The first andsecond tests allow to observe mixed-mode fracture (mode-I and mode-II – see the Figure 1). Results obtained by the authors were compared with laboratory and numerical results presented in the cited works.

This paper is the continuation of research presented at the CFRAC 2019 conference (Szczecina & Winnicki 2019). The authors have also some experience with the Willam test performedusing a few selected material models (Wosatko et al. 2020).

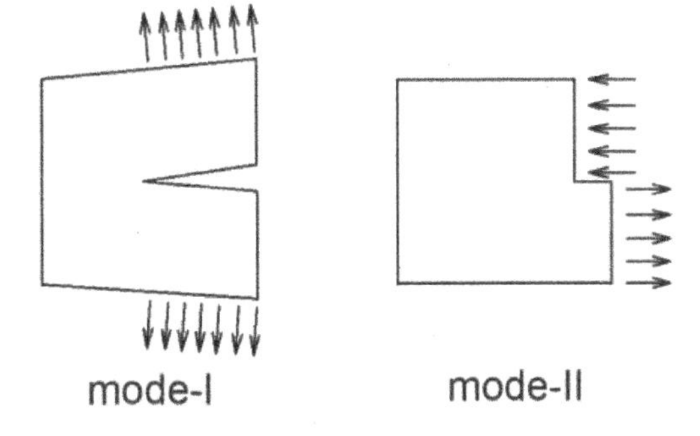

Figure 1. A general view of two modes of fracture: mode-I (opening mode) and mode-II (shearing mode).

2 OVERVIEW OF REPRODUCED TESTS

2.1 *Nooru-Mohamed test*

In the Nooru-Mohamed test rectangular concrete specimens of different sizes with notches under different in-plane load paths were examined. The authors decided to reproduce numerically the load-path 4, i.e. axial tension at a constant shear force $P_s = 5$ kN. In the first step of the test, a compressive shear force P_s is

DOI 10.1201/9781003316404-54

applied to the specimen, which is then accompanied in the following step by an axial tensile force P till failure. Geometry and loading of the specimen are shown in the Figure 2. The main goal of laboratory tests was to obtain reliable mixed-mode results.

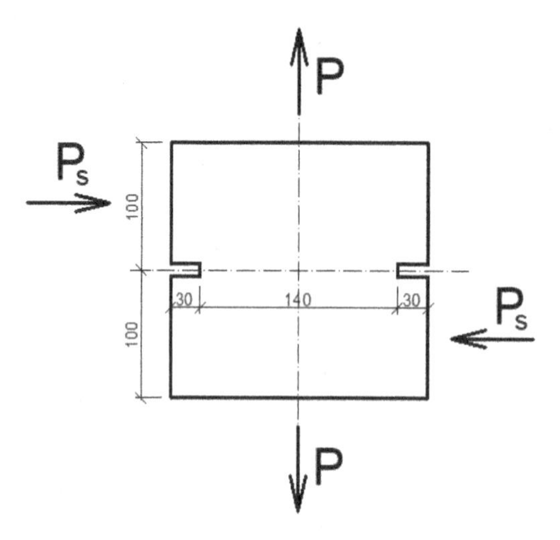

Figure 2. Geometry (dimensions in [mm]) and loading of the specimen in the Nooru-Mohamed test.

2.2 *Schlangen test*

In the Schlangen test, single edge notched (SEN) and double edge notched (DEN) beams with different cross sections and lengths were tested in a four-point shear. This test is useful for assessing the performance of concrete models in the simulation of the mixed-mode cracking (the mode I and mode II). Schlangen investigated experimentally the influence of boundary conditions and different materials (ordinary concrete, lightweight concrete, fiber reinforced concrete) on the fracture process. He also examined the experimentally obtained fracture energy in the mode I and mode II tests. The authors decided to reproduce numerically the SEN specimen. Geometry of the SEN specimen tested by Schlangen is shown in the Figure 3.

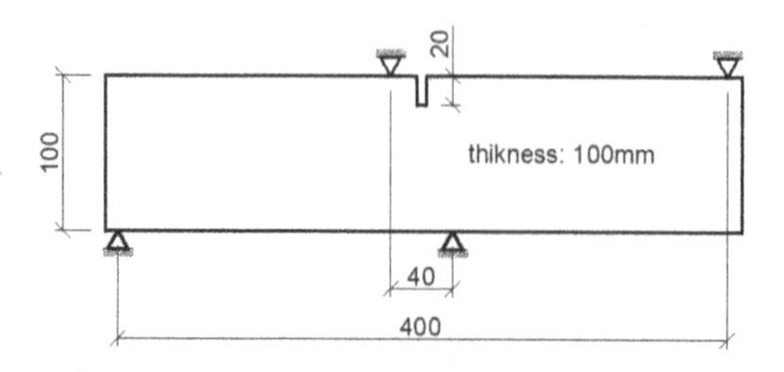

Figure 3. Geometry (dimensions in [mm]) of the SEN specimen in the Schlangen test.

2.3 *L-specimen test*

The L-specimen test had been performed by Winkler et al. (2001) and then reproduced numerically by Ožbolt et al. (2002), Ožbolt & Sharma (2012) and also tested in laboratory by Ožbolt et al. (2015). The latter authors considered a few different loading rates and a quasi-static load as well. The L-specimen test is often used to demonstrate the correctness and capabilities of material models. Moreover, it is also used to show the problem of mesh sensitivity. Dimensions and loading of the L-specimen are shown in the Figure 4.

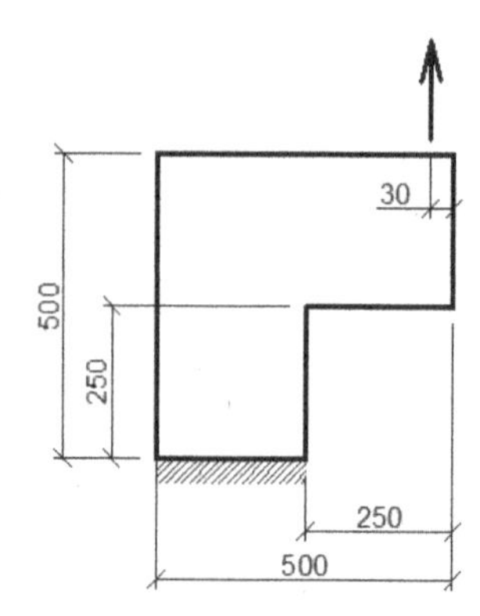

Figure 4. Geometry (dimensions in [mm]) of the L-specimen.

3 OVERVIEW OF APPLIED MATERIAL MODELS

Numerical models for concrete can be based on different theoretical formulations, e.g. plasticity, damage, smeared crack approach, discrete cracks (Hillerborg et al. 1976; Lubliner et al. 1989; Willam et al. 1987). There are many material models dedicated to concrete, e.g. Cichoñ & Winnicki 1998, band numerical reproduction of laboratory tests for plain concrete using different models is a current scientific issue, especially using X-FEM approach (Gontarz & Podgórski 2020) or smeared-cracking modeling(Chen & de Borst 2019). As mentioned before, the authors decided to use two material models implemented in Abaqus software. Especially the CDP model (also called "the Barcelona model") is very popular and widely applied in numerical simulations of RC structures, e.g. Kossakowski & Uzarska (2019), Szczecina & Winnicki (2021).

3.1 *Concrete Damaged Plasticity (CDP)*

The CDP model was theoretically described by Lubliner et al. (1989, 1990) and developed by Lee (1996) and Lee & Fenves (1998).

The yield function and the flow potential function in the CDP model are presented below, respectively:

$$\sigma = (1 - -d)\mathbf{D}_0^{el} : (\varepsilon - -\varepsilon_{pl}) \quad (1)$$

$$F = \frac{1}{1 - -\alpha}\left(\overline{q} - -3\alpha\overline{p} + \beta(\tilde{\varepsilon}_{pl})\left\langle\hat{\overline{\sigma}}_{\max}\right\rangle - \gamma\left\langle-\hat{\overline{\sigma}}_{\max}\right\rangle\right) - -\overline{\sigma_c}(\tilde{\varepsilon}_{pl}) = 0 \tag{2}$$

$$G = \sqrt{(\varepsilon\sigma_{t0}\tan\psi)^2 + \overline{q}^2} - -\overline{p}\tan\psi \tag{3}$$

where $\boldsymbol{\sigma}$ is a stress tensor, $\mathbf{D}_0^{\mathrm{el}}$ denotes an initial elasticity matrix, d is a damage parameter, α, β and γ are parameters of the yield surface, p is a hydrostatic equivalent pressure stress, q is von Mises equivalent effective stress, ε is a flow potential eccentricity and ψ is a dilatation angle. The yield function is presented in the plane stress state in Figure 5.

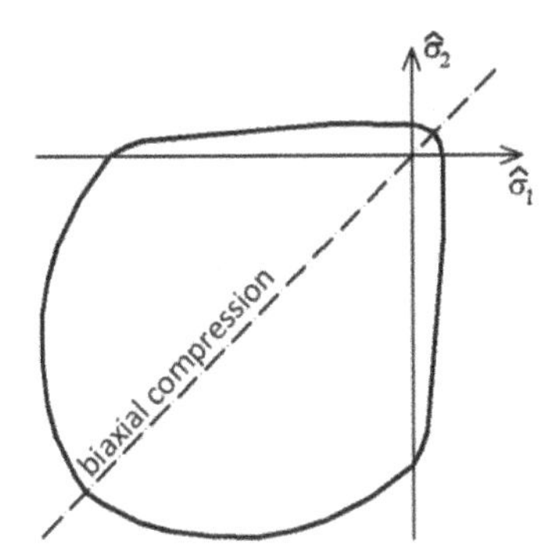

Figure 5. Yield function in the plane stress state.

The viscoplastic regularization in the CDP model can be introduced according to Duvaut-Lions (Duvaut & Lions 1976) approach. A plastic viscous strain rate is obtained from the formula (4):

$$\dot{\varepsilon}_v^{pl} = \frac{1}{\mu}\left(\varepsilon^{pl} - \varepsilon_v^{pl}\right) \tag{4}$$

where μ denotesthe relaxation time (the so-called viscosity parameter in Abaqus software).

The tension behavior of concrete in the post-critical range in the CDP model can be defined in three different ways: by defining the $\sigma - \varepsilon_{in}$ or σ-u_{cr} curves, or by inputting the fracture energy G_f. The compressive behavior is defined with the $\sigma - \varepsilon$ relation for the compression of concrete. The proper choice of the two crucial CDP model parameters, i.e. the dilatation angle and the relaxation time was discussed in previous papers (Szczecina & Winnicki, 2016, 2021).

3.2 *Concrete Smeared Cracking (CSC)*

The CSC model in Abaqus is based on the classical theory of plasticity in compression and the fixed smeared crack model in tension. It is intended for monotonic loadings under low confining pressures. A crack occurs when stresses reach the so-called "crack detection surface". The post-failure behavior of cracked concrete is described with damaged elasticity and the compressive behavior of concrete is ruled by the compressive yield surface. The plastic flow is associated and an isotropic hardening is used in the CSC model.

In the CSC model, the strain rate in compression and in tension is decomposed into elastic and plastic strain rate. A compression yield surface is described by the Coulomb-Mohr criterion (5):

$$F_c^{pl} = q - -\sqrt{3}a_o p - -\sqrt{3}\tau_c = 0 \tag{5}$$

where a_o is a constant defining the slope of the surface and τ_c is the yield stress in pure shear. The associated flow equation in compression is expressed in the form (6):

$$d\boldsymbol{\varepsilon}_c^{pl} = d\lambda_c\left(1 + c_o\left(\frac{p}{\sigma_c}\right)^2\right)\cdot\frac{\partial F_c^{pl}}{\partial\boldsymbol{\sigma}} \tag{6}$$

where c_o is a constant and λ_c is the hardening parameter in compression.

The uniaxial behavior assumed in the CSC model is presented in the form of a graphs in the Figures 6a and 6b:

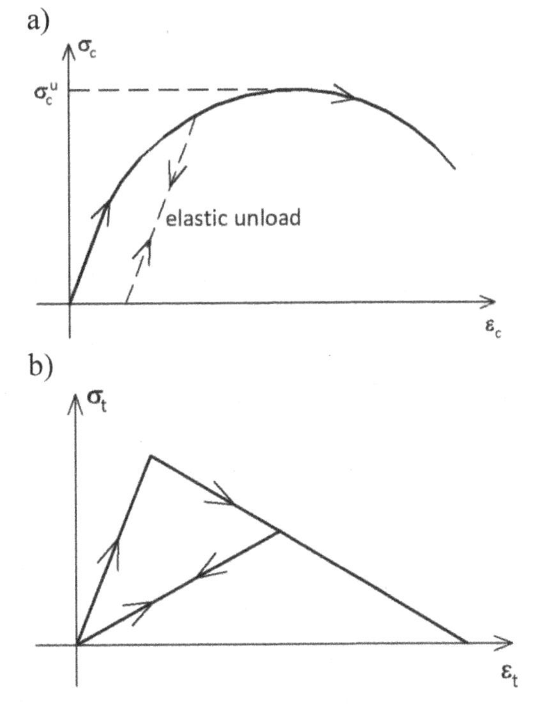

Figure 6. Uniaxial behavior of concrete in the CSC model: a) in compression, b) in tension.

The crack detection surface is described with the formula (7):

$$F_t^{pl} = \hat{q} - \left(3 - -b_o\cdot\frac{\sigma_t}{\sigma_t^u}\right)\hat{p} - \left(3 - \frac{b_o}{3}\cdot\frac{\sigma_t}{\sigma_t^u}\right)\sigma_t = 0 \tag{7}$$

and the associated flow equation with the formula (8):

$$d\varepsilon_t^{pl} = d\lambda_t \cdot \frac{\partial F_t^{pl}}{\partial \sigma} \tag{8}$$

where b_o is a material parameter and λ_t is the hardening parameter in tension.

The cracking behavior in the CSC model is based on the fracture energy. After cracking, components of stress are rotated to remain in the local coordinates system defined by the crack orientation vector normal to the crack faces. The CSC is therefore a fixed crack model.

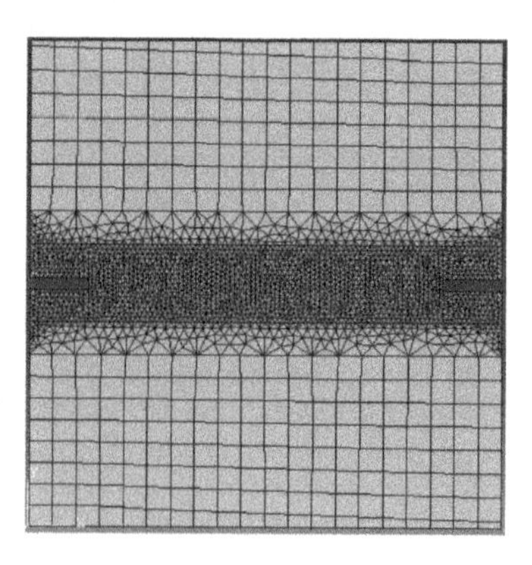

Figure 8. Refined mesh in the Nooru-Mohamed test.

4 SETUP OF NUMERICAL TESTS

For all the performed tests, the compressive behavior of concrete was defined as non-linear according to the Eurocode 2 (2004) and the tensile behavior of concrete was defined by inputting the fracture energy G_f. Calculations were performed in plane stress state using displacement control.

4.1 *Nooru-Mohamed test*

The Nooru-Mohamed test was performed using 4-node or 3-node finite elements and a structural or random mesh. The finite element size varied from 2.5 to 5.0 mm. For the sake of brevity, only a coarse mesh for all the cases is presented in the Figure 7.

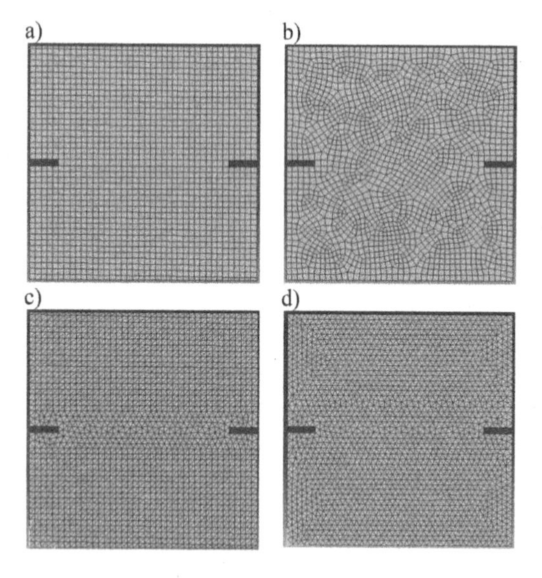

Figure 7. Meshing of the specimen in the Nooru-Mohamed test: a) 5 mm structural mesh, b) 5 mm random mesh, c) 5 mm structural mesh, d) 5 mm random mesh.

The authors performed also calculations using a refined mesh (mesh size: 2–5 mm), presented in the Figure 8.

The properties of concrete and input variables of the CDP and CSC models are presented in the Tables 1–3.

Table 1. Properties of concrete in the numerical reproduction of the Nooru-Mohamed test.

Input variable [unit]	Value
Compressive strength [MPa]	46.24
Tensile strength [MPa]	3.67
Tangent modulus of elasticity [GPa]	35.00
Poisson's ratio	0.167

Table 2. Parameters of the CDP model.

Input variable [unit]	Value
Fracture energy [Nm^{-1}]	145.55
Dilatation angle [deg]	5.00
Eccentricity [–]	0.10
fb0/fc0 [–]	1.16
K parameter [–]	0.67
Relaxation time [s]	10^{-6}

Table 3. Parameters of the CSC model.

Input variable [unit]	Value
Failure ratio 1	1.160
Failure ratio 2	0.079
Failure ratio 3	1.280
Failure ratio 4	0.333
Tension stiffening defined by displacement [mm]	0.079

4.2 *Schlangen test*

In the Schlangen test also 4-node or 3-node finite elements with structural or random mesh were used. The finite element size varied from 2.5 to 5.0 mm. Calculations were also performed for a refined mesh. Some sample meshing is presented in the Figures 9 and 10. The properties of concrete and input variables of the CDP and CSC models are presented in the Tables 4–6.

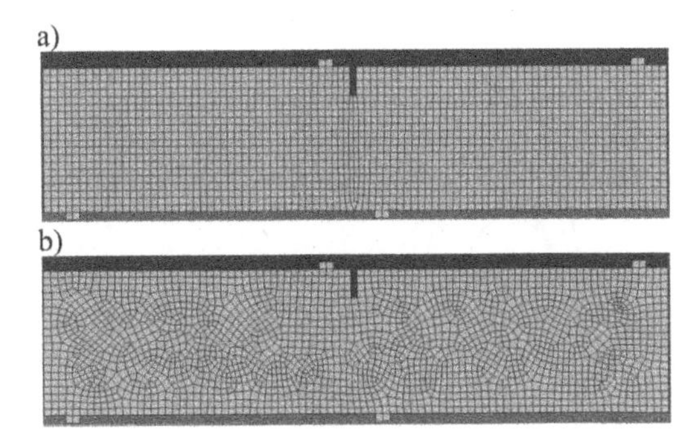

Figure 9. Meshing of the specimen in the Schlangen test: a) 5 mm structural mesh, b) 5 mm random mesh.

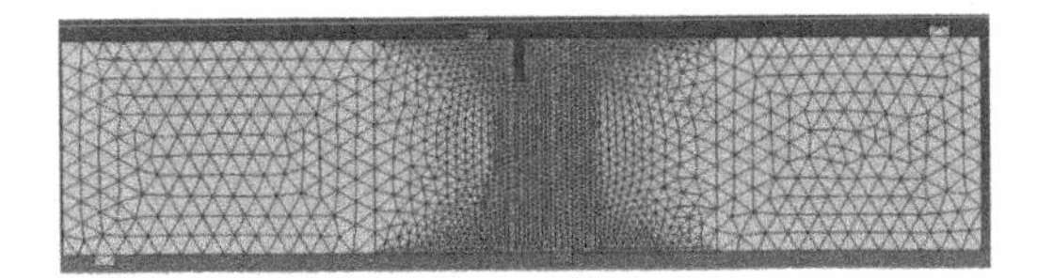

Figure 10. Refined mesh in the Schlangen test.

Table 4. Properties of concrete in the numerical reproduction of the Schlangen test.

Input variable [unit]	Value
Compressive strength [MPa]	46.60
Tensile strength [MPa]	3.44
Tangent modulus of elasticity [GPa]	35.00
Poisson's ratio	0.20

Table 5. Parameters of the CDP model.

Input variable [unit]	Value
Fracture energy [Nm^{-1}]	145.76
Dilatation angle [deg]	5.00
Eccentricity [–]	0.10
fb0/fc0 [–]	1.16
K parameter [–]	0.67
Relaxation time [s]	10^{-4}

Table 6. Parameters of the CSC model.

Input variable [unit]	Value
Failure ratio 1	1.160
Failure ratio 2	0.074
Failure ratio 3	1.280
Failure ratio 4	0.333
Tension stiffening defined by displacement [mm]	0.085

4.3 *L-specimen test*

In the latter test a random, refined mesh, presented in the Figure 11, was applied. Four variants of specimen were taken into consideration, namely:

- a specimen without reinforcement, called "series A" in the Winklertest et al. (2001) work,
- a specimen reinforced with four $\phi 6$ bars (see the Figure 12a) – "series B",
- a specimen reinforced with an orthogonal mesh consisted of $\phi 6$ bars (the Figure 12b) – "series C",
- a specimen reinforced with a diagonal mesh consisted of $\phi 6$ bars (the Figure 12c) – "series D".

Boundary conditions, identical for each series, are presented in the Figure 12d.

Figure 11. Meshing of the L-specimen test.

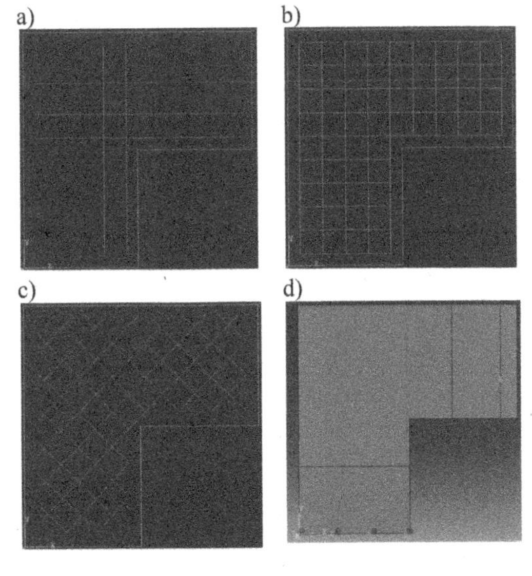

Figure 12. Reinforcement of the a) series B, b) series C, c) series D and d) boundary conditions of the L-specimen test.

Properties of concrete and reinforcing steel in the L-specimen test are presented in the Tables 7 and 8. For all the series, Poisson's ratio was assumed as 0.18.

Table 7. Properties of concrete in the L-specimen test.

Parameter [unit]	Series A	Series B	Series C	Series D
Compressive strength [MPa]	31.00	34.85	38.90	29.45
Tangent modulus of elasticity [GPa]	25.85	27.98	29.30	26.08
Tensile strength [MPa]	2.66	2.39	3.28	2.88
Fracture energy [Nm^{-1}]	60.00	60.00	60.00	60.00

Table 8. Properties of reinforcing steel in the L-specimen test.

Parameter [unit]	Series B	Series C	Series D
Yield strength [MPa]	544.8	533.2	526.3
Ultimate strength [MPa]	603.7	597.2	584.5
Ultimate strain [%]	19.19	20.30	23.23
Tangent mod. of elast. [GPa]	193.2	201.6	197.1

5 RESULTS OF FEM CALCULATIONS

For all the performed tests, a few common results and output variables are presented in this section, namely:

- a graph of a force-displacement relationship, created for a node where the displacement is imposed,
- a map of the equivalent plastic strain in tension (PEEQT) when using the CDP model,
- a map of the plastic strain (PE) in the CSC model.

The results are compared with those obtained in laboratory tests and/or numerical tests of the above mentioned authors.

5.1 *Nooru-Mohamed test*

Results obtained in the Nooru-Mohamed test are presented in the Figures 13-19. A general conclusion is that the CDP model returns quite correct results while the CSC model does not cope well with the test. A proof of this statement are: a post-peak behavior (see the Figure 14) and a very limited crack propagation (e.g. the Figure 15c) and d)).

An impact of a mesh size and a mesh type can be particularly visible when considering the crack propagation. For the regular mesh, the crack propagates in one row of finite elements while for the random mesh it resembles that from the Nooru-Mohamed laboratory test (see the Figure 19c)).

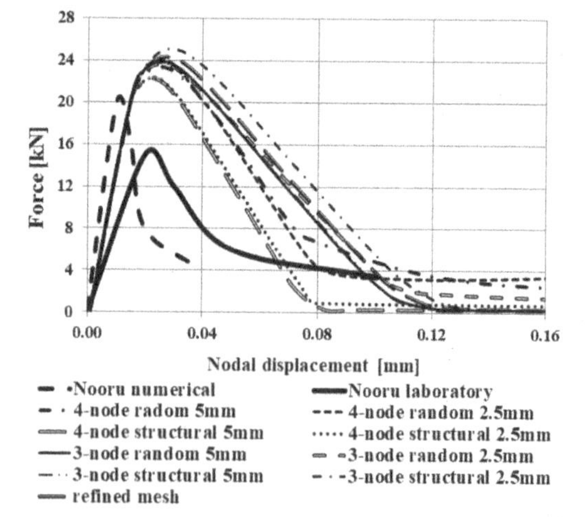

Figure 13. Force-displacement curves obtained in the Nooru-Mohamed test using the CDP model.

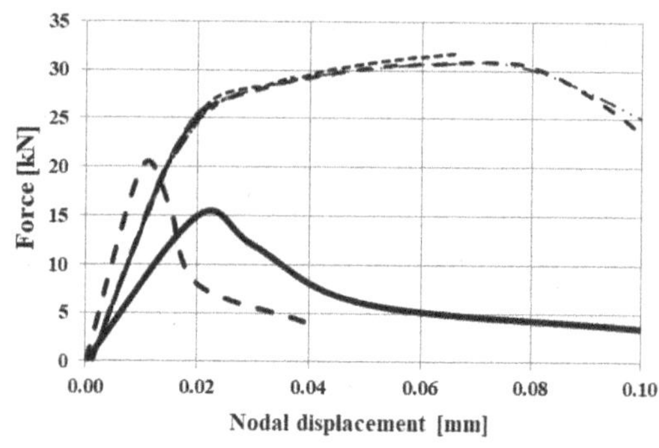

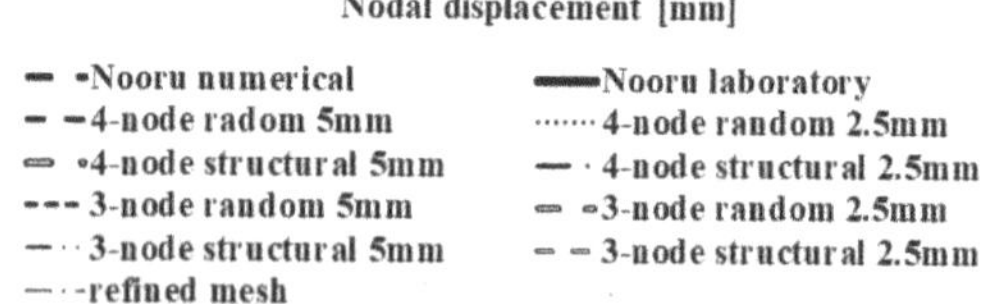

Figure 14. Force-displacement curves obtained in the Nooru-Mohamed test using the CSC model.

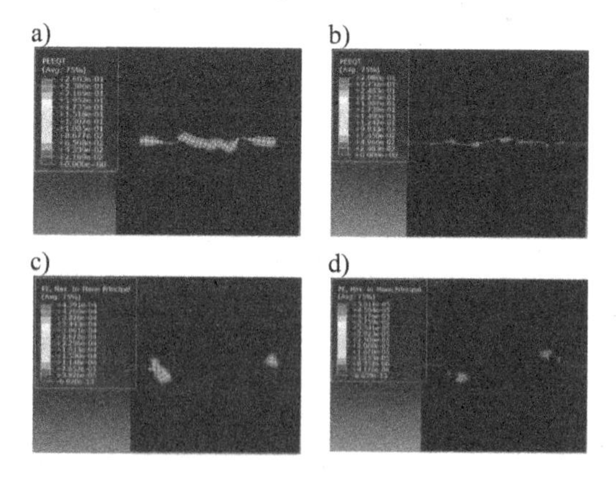

Figure 15. PEEQT or PE output variable in case of a) CDP, 5 mm 4-node random mesh, b) CDP, 2.5mm 4-node random mesh, c) CSC, 5 mm 4-node random mesh, d) CSC, 2.5 mm 4-node random mesh.

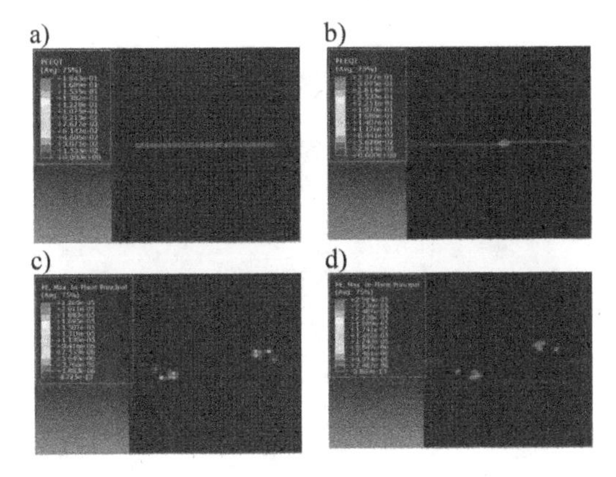

Figure 16. PEEQT or PE output variable in case of a) CDP, 5 mm 4-node structural mesh, b) CDP, 2.5 mm 4-node structural mesh, c) CSC, 5 mm 4-node structural mesh, d) CSC, 2.5 mm 4-node structural mesh.

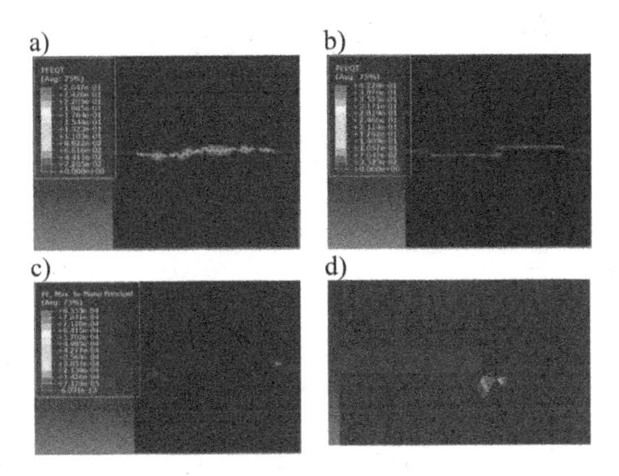

Figure 17. PEEQT or PE output variable in case of a) CDP, 5 mm 3-node random mesh, b) CDP, 2.5mm 3-node random mesh, c) CSC, 5 mm 3-node random mesh, d) CSC, 2.5 mm 3-node random mesh (magnified notched area).

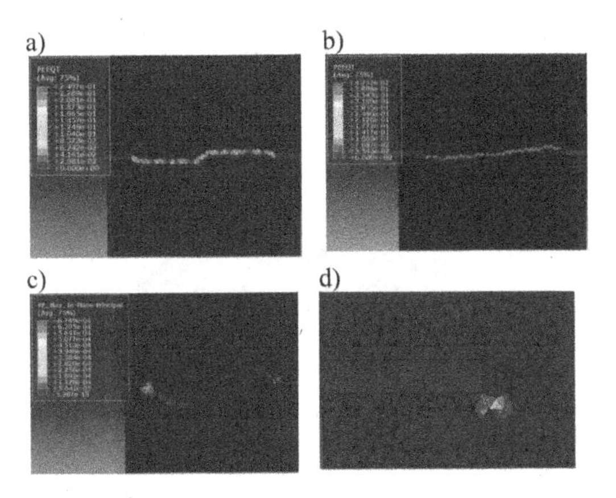

Figure 18. PEEQT or PE output variable in case of a) CDP, 5 mm 3-node structural mesh, b) CDP, 2.5 mm 3-node structural mesh, c) CSC, 5 mm 3-node structural mesh, d) CSC, 2.5 mm 3-node structural mesh (magnified notched area).

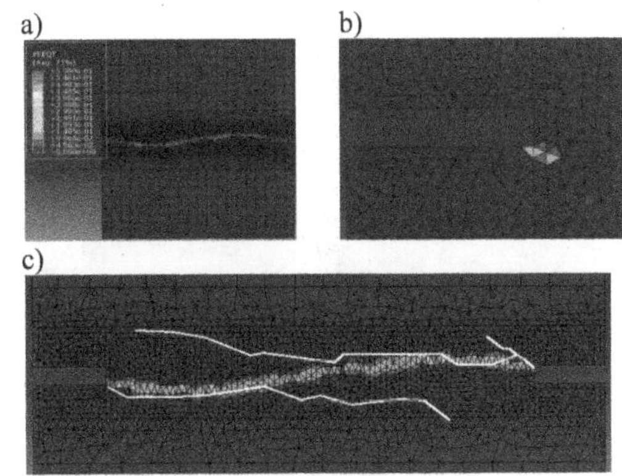

Figure 19. PEEQT or PE output variable in case of a) CDP, refined mesh, b) CSC, refined mesh (magnified notched area) compared with c) comparison of a crack pattern of the specimen in the Nooru-Mohamed laboratory test (in white) and in Abaqus using CDP model (colored map).

5.2 *Schlangen test*

The results of the numerically reproduced Schlangen test (the Figures 20-26) allow to draw very similar judgment as for the Nooru-Mohamed test. Once again

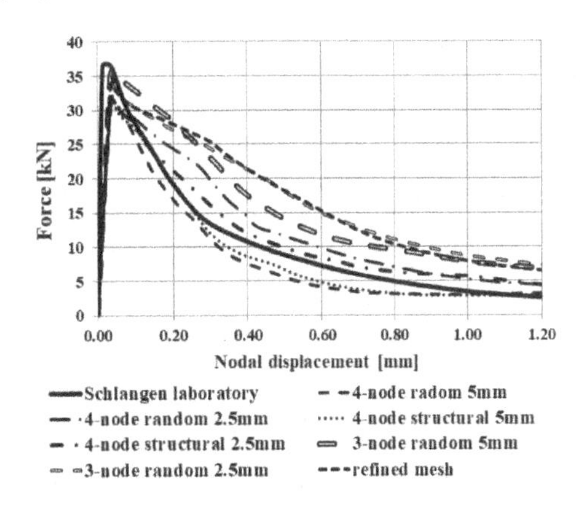

Figure 20. Force-displacement curves obtained in the Schlangen test using the CDP model.

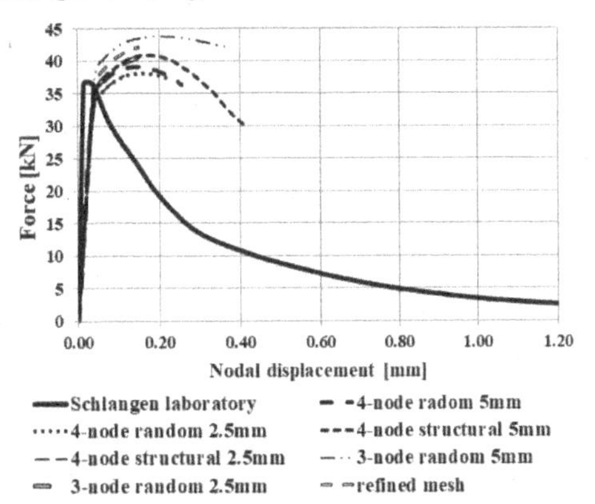

Figure 21. Force-displacement curves obtained in the Schlangen test using the CSC model.

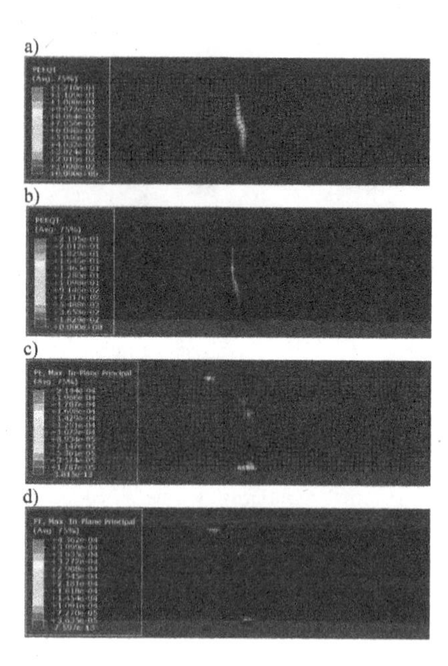

Figure 22. PEEQT or PE output variable in case of a) CDP, 5 mm 4-node random mesh, b) CDP, 2.5 mm 4-node random mesh, c) CSC, 5 mm 4-node random mesh, d) CSC, 2.5 mm 4-node random mesh.

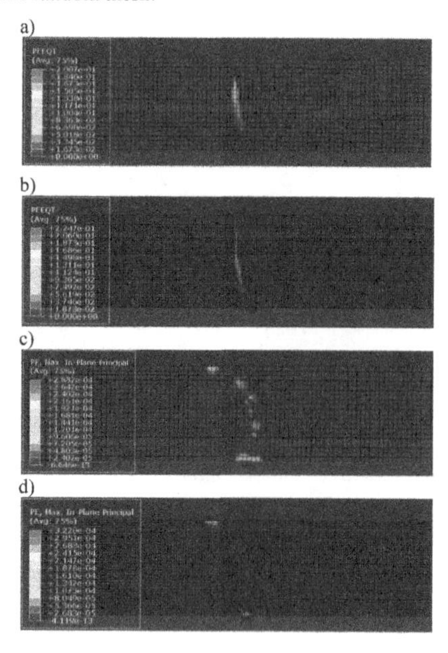

Figure 23. PEEQT or PE output variable in case of a) CDP, 5 mm 4-node structural mesh, b) CDP, 2.5 mm 4-node structural mesh, c) CSC, 5 mm 4-node structural mesh, d) CSC, 2.5 mm 4-node structural mesh.

the CSC model returns unsatisfactory results, although there is also one optimistic crack pattern (see the Figure 23c)). On the other hand, the CDP model behaves correctly regardless of a mesh size or a mesh type. The only problem is that the crack does not reach the opposite side of the specimen, as it happened in the laboratory test (see the Figure 26).

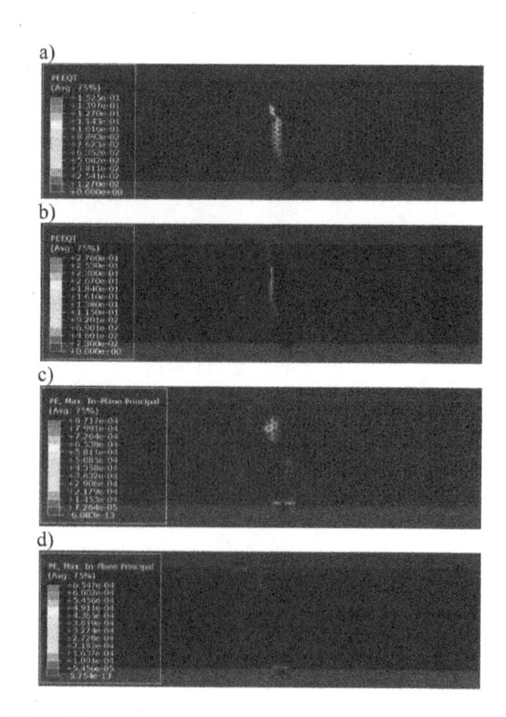

Figure 24. PEEQT or PE output variable in case of a) CDP, 5 mm 3-node random mesh, b) CDP, 2.5 mm 3-node random mesh, c) CSC, 5 mm 3-node random mesh, d) CSC, 2.5 mm 3-node random mesh.

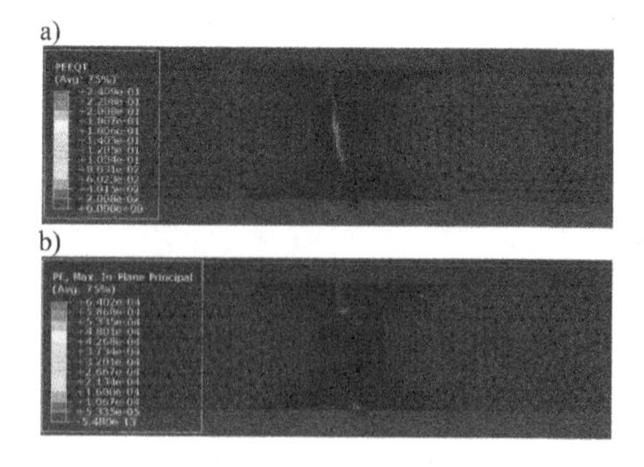

Figure 25. PEEQT or PE output variable in case of a) CDP, refined mesh, b) CSC, refined mesh.

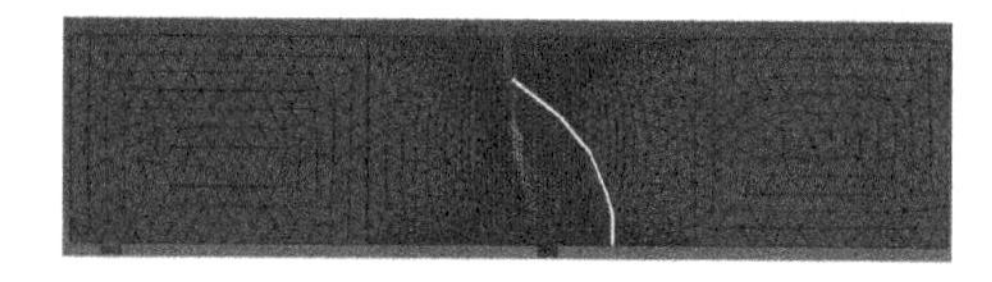

Figure 26. Comparison of a crack pattern of the specimen obtained in the Schlangenlaboratory test (in white) and in Abaqus using CDP model (colored map).

5.3 *L-specimen test*

The above mentioned problems with the CSC model are still present in the L-specimen test (the Figures 27–29), even for the specimens with reinforcement.

Still the crack propagation is limited (the Figure 29) and the force-displacement curves seem to terminate prematurely.

The CDP model once again reproduces the laboratory test pretty well. The numerically obtained force-displacement curves are comparable with those plotted in the experiment. The crack propagation depends on

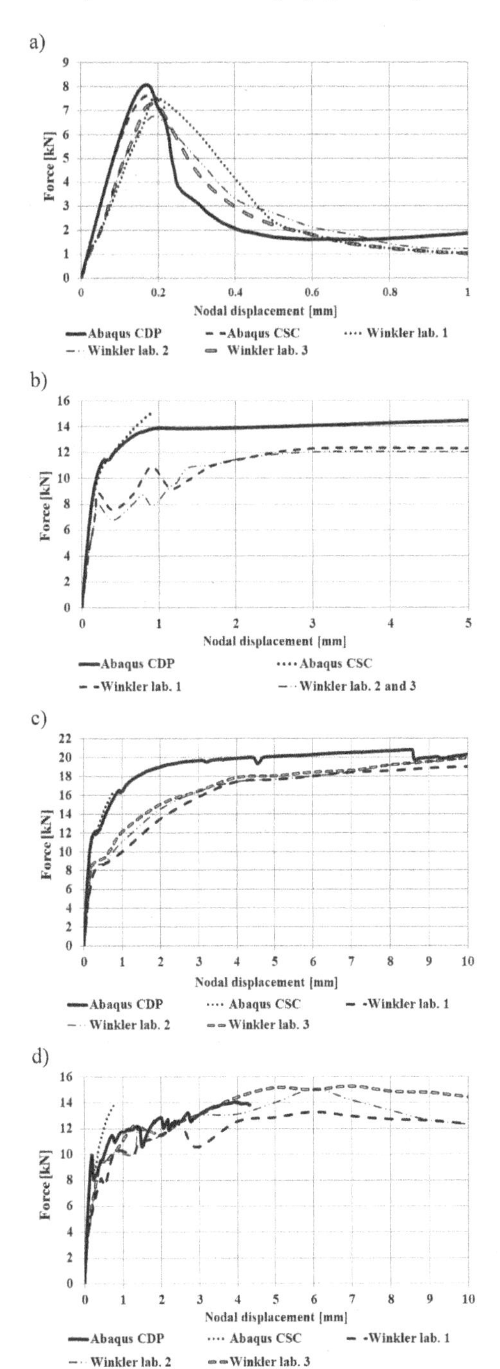

Figure 27. Force-displacement curves obtained numerically for the a) series A, b) series B, c) series C, d) series D.

a kind of the provided reinforcement (the Figure 28). For the series B and C a crack occurs outside the corner zone and for the series D there are three concurrent cracks.

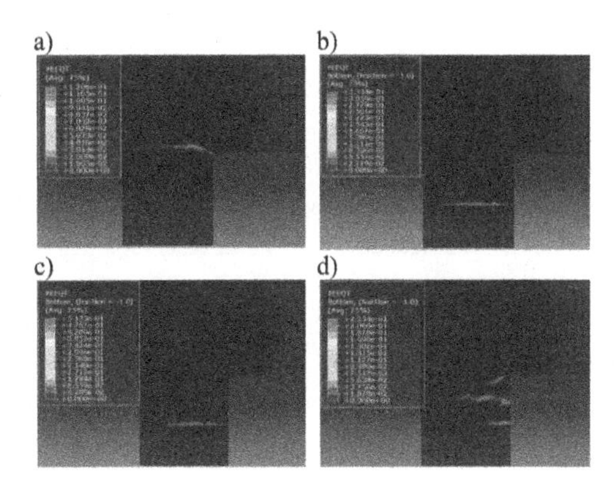

Figure 28. PEEQT variable in the L-specimen test using the CDP model: a) series A, b) series B, c) series C, d) series D.

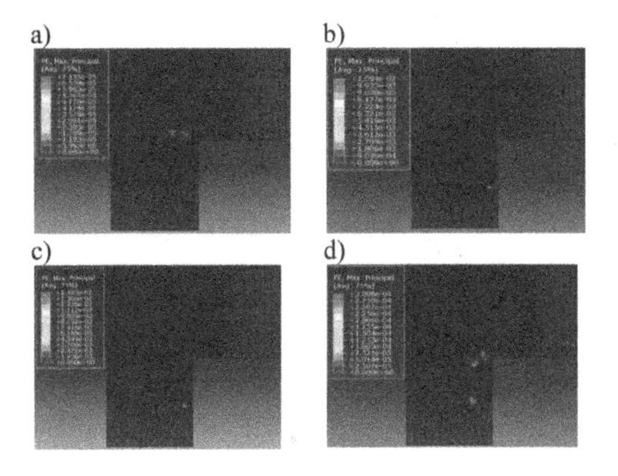

Figure 29. PE output variable in the L-specimen test using the CSC model: a) series A, b) series B, c) series C, d) series D.

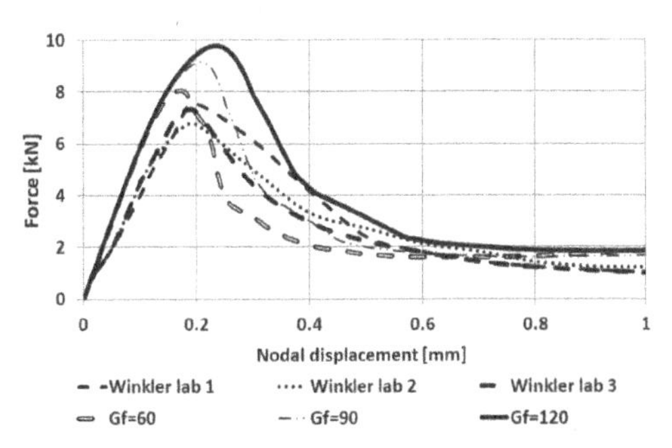

Figure 30. Force-displacement curves for different values of the fracture energy G_f.

The authors performed also a parametric study analyzing how does the force-displacement relationship depend on the fracture energy value in the CDP model. Three values of G_f were assumed: 120, 90 and 60 Nm^{-1}. Results of the study are presented in the Figure 30. An increase of the fracture energy value (with the unchanged value of the tensile strength) leads to a significant increase of the load peak. Such behavior indicates the stable crack propagation in the pre-peak regime typical for elements with notches or singular points.

6 CONCLUSIONS

The results of the calculations presented in the paper allow to draw the following conclusions:

- the CDP model deals with all the tests satisfactorily well,
- on the other hand, the results obtained using the CSC model do not match the experimental observations; the crack propagation is limited and the post-peak behavior is improper,
- the influence of the finite element type (3-node vs. 4-node) is not that significant, but the choice of the structural mesh can force a crack to be localizedin one row of finite elements, which is not consistent with the analyzed laboratory tests,
- the CDP model is recommendable for FEM calculations both for plain and reinforced concrete,
- some effort should yet be done to explain the improper behavior of the CSC model.

ACKNOWLEDGEMENTS

This paper is supported by the programme of the Polish Minister of Science and Higher Education under the name: Regional Initiative of Excellence (Regionalna Inicjatywa Doskonałości) in 2019–2022, project number 025/RID/2018/19, financing amount PLN 12.000.000.

REFERENCES

Abaqus/CAE ver. 6-12.2 documentation, Dassault Systemes Simulia Corp., 2012.

Bobiński, J., Tejchman, J. 2016. Comparison of continuous and discontinuous constitutive models to simulate concrete behaviour under mixed-mode failure conditions, *International Journal for Numerical and Analytical Methods in Geomechanics* 40: 406–435.

Cichoń, C., Winnicki, A. 1998. Plastic Model for Concrete in Plane Stress State. Part I: Theory, *Journal of Engineering Mechanics – ASCE* 124: 591–602.

Chen, L., de Borst, R. 2019. Cohesive fracture analysis using Powell-Sabin B-splines, *International Journal for Numerical and Analytical Methods in Geomechanics* 43: 625–640.

Duvaut, G., Lions, J. L. 1976. *Inequalities in Mechanics and Physics*. Berlin-Heidelberg-New York: Springer Verlag.

EN1992-1-1 (2004) Eurocode 2 – Concrete structures – Part 1-1: General rules and rules for buildings.

Gontarz, J., Podgórski, J. 2020. Simulation of four-point beam bending test using the X-FEM method, *Budownictwo i Architektura* 19(3): 53–62.

Hillerborg, A., Modeer, M., Petersson, P. E. 1976. Analysis of Crack Formation and Crack Growth in Concrete by Means of Fracture and Finite Elements, *Cement and Concrete Research* 6: 773–782.

Kossakowski, P., Uzarska, I. 2019. Numerical modeling of an orthotropic RC slab band system using the Barcelona model, *Advances in Computational Design* 4(3): 211–221.

Lee, J., 1996. *Theory and implementation of plastic-damage model for concrete structures under cyclic and dynamic loading*. PhD thesis, Berkeley: University of Berkeley.

Lee, J., Fenves, G. L. 1998. Plastic-Damage Model for Cyclic Loading of Concrete Structures, *Journal of Engineering Mechanics – ASCE* 124: 892–900.

Lubliner, J., Oliver, J., Oller, S., Oñate, E. 1989. A plastic-damage model for concrete, *International Journal of Solids Structures* 25: 229–326.

Lubliner, J., Oliver, J., Oller, S., Oñate, E. 1990. Finite element nonlinear analysis of concrete structures using a "plastic-damage model", *Engineering Fracture Mechanics* 35: 219–321.

Nooru-Mohamed, M. B. 1992. *Mixed-mode fracture of concrete: an experimental approach*. PhD thesis, Delft: Delft University of Technology.

Ožbolt, J., Pivonka, P., Lackner, R. Three dimensional FE analyses of fracture of concrete – material models and mesh sensitivity, *Proc. of the 5th World Congress on Computational Mechanics (WCCM V), Vienna, Austria*.

Ožbolt, J., Sharma, A. 2012. Numerical simulation of dynamic fracture of concrete through uniaxial tension and L-specimen, *Engineering Fracture Mechanics* 85: 88–102.

Ožbolt, J., Bede, N., Sharma, A., Mayer, U. 2015. Dynamic fracture of concrete L-specimen: Experimental and numerical study, *Engineering Fracture Mechanics* 148: 27–41.

Schlangen, E. 1993. Experimental and numerical analysis of fracture processes in concrete, *Heron* 38 (2): 1–117.

Szczecina, M., Winnicki, A. 2016. Selected aspects of computer modeling of reinforced concrete structures, *Archives of Civil Engineering* 52: 51–64.

Szczecina, M., Winnicki, A. 2018. Analysis of RC frame corners using CDP Model, *Computational Modelling of Concrete Structures: Proc. of the Conference on Computational Modelling of Concrete and Concrete Structures (EURO-C 2018), February 26 – March 1, 2018, Bad Hofgastein, Austria*, 1: 569–578.

Szczecina, M., Winnicki, A. 2019. Comparison of different constitutive models for concrete in simple numerical tests, *Proc. of the VI International Conference on Computational Modeling of Fracture and Failure of Materials and Structures (CFRAC 2019), June 12–14, 2019, Braunschweig, Germany*.

Szczecina, M., Winnicki, A. 2021. Rational Choice of Reinforcement of Reinforced Concrete Frame Corners Subjected to Opening Bending Moment, *Materials* 14(12): e1-33.

Willam, K., Pramono, E., Sture, S. 1987. Fundamental issues of smeared crack models, *Proc. of the SEM/RILEM International Conference on Fracture of Concrete and Rock, Houston, USA*.

Winkler, B., Hofstetter, G., Niederwanger G. 2001. Experimental verification of a constitutive model for concrete cracking, *Proceedings of the Institution of Mechanical Engineers, Part L: Journal of Materials: Design and Applications*, 215: 75–86.

Wosatko, A., Szczecina, M., Winnicki, A. 2020. Selected Concrete Models Studied Using Willam's Test, *Materials* 13(21): e1-23.

Computational Modelling of Concrete and Concrete Structures – Meschke, Pichler & Rots (Eds)
 ISBN: 978-1-032-32724-2

Comparison of classical and higher order continuum models for shear failure of concrete

P. Hofer, M. Neuner & G. Hofstetter
Institute of Basic Sciences in Engineering Sciences, University of Innsbruck, Austria

ABSTRACT: Classical and higher order continuum formulations of the widely used Concrete Damage Plasticity model by Grassl & Jirásek (2006) are assessed with regard to their ability to represent shear failure of plain concrete in an objective manner. To this end, a transverse shear test for anchor channels serves as benchmark. The experimental results are compared with the results of complex 3D numerical simulations, performed on the basis of (i) a classical local formulation with mesh-adjusted softening modulus, (ii) a higher order gradient-enhanced formulation by Poh & Swaddiwudhipong (2009) and (iii) a recently proposed gradient-enhanced micropolar formulation by Neuner et al. (2020). The comparison highlights the limitations of the classical local formulation, whereas the higher order continuum approaches show great potential, in particular for modeling the structural post-peak response of the anchoring system. Moreover, the load-displacement behavior, which is associated with different failure modes observed for varying values of the edge distance of the anchor channel is accurately predicted by means of the higher order continuum formulations, further highlighting their predictive capabilities.

1 INTRODUCTION

Predicting the mechanical behavior of concrete structures in the pre- and post-peak domain of the structural response has been the subject of extensive research endeavors for many decades. In this context, damage-plasticity continuum models allow describing the inelastic material behavior in combination with the degradation of elastic properties due to damage. Accordingly, they represent a popular class of material models for the highly nonlinear mechanical behavior of concrete. One prominent member of this group is the Concrete Damage Plasticity (CDP) model proposed by Grassl & Jirásek (2006). The model serves as the core of the formulations used in this contribution, and encompasses a smooth, pressure-dependent yield surface formulated in invariants of the effective stress tensor, as well as hardening and softening behavior driven by the evolution of the plastic strain. It is suitable for describing concrete subjected to a broad range of stress states, and it has been successfully used for large scale simulations of concrete structures in the past (Neuner et al. 2022; Poh & Swaddiwudhipong 2009; Valentini & Hofstetter 2013).

Failure in cohesive-frictional materials like concrete is typically accompanied by the emergence of highly localized deformation. For classical continuum models, the onset of localization results in the loss of ellipticity of the governing boundary value problem entailing loss of uniqueness of the solution and a pathological dependence of numerical results on the employed discretization. This is especially true for plasticity models with pressure-dependent yield surfaces exhibiting softening behavior and/or non-associated plastic flow Rudnicki & Rice (1975). Without remedy, localized failure will emerge in the smallest possible bandwidth, causing the results to be dependent on the numerical discretization. Consequently, objectivity of the results with respect to the discretization may not be expected.

Various approaches for objectively describing the material behavior of concrete have been developed in the past. On the one hand, the classical remedy of the mesh-adjusted softening modulus is available, which serves as a numerical regularization technique. It is, however, characterized by deficiencies as highlighted in Jirásek & Bauer (2012). On the other hand, higher order continuum approaches, like nonlocal or gradient-enhanced formulations offer superior alternatives by introducing intrinsic material length scales, which naturally provide a remedy for the pathological mesh-sensitive behavior. Special attention has to be paid to the fact that localization may arise also in modes deviating from pure mode I failure. In particular, the realistic representation of shear dominated failure is an important aspect, which is often paid little attention compared to mode I failure.

Due to the local character of the resulting formulation, the mesh-adjusted softening modulus allows for an efficient and simultaneously simple numerical implementation. The numerical regularization of

DOI 10.1201/9781003316404-55

the problem is approximately achieved by adjusting the softening behavior of the underlying material based on characteristic length scales of the numerical discretization.

In contrast to classical local models, gradient-enhanced continua account for the nonlocal character of material damage. This is achieved by additional balance equations concerning the gradients of internal variables, which describe the spatial interaction of micro cracks during the evolution of material damage. The localization of deformation is thereby limited and a finite size of the fracture process zone is ensured, especially for mode I failure.

The micropolar continuum, which goes back to the work of the Cosserat brothers Cosserat & Cosserat (1909), accounts for averaged rotations within the material microstructure, and naturally introduces an intrinsic length scale for shear failure. Accordingly, it provides a remedy for pathological behavior related to the localization of shear dominated deformation in cohesive-frictional materials. A remedy for the pathological behavior related to the localization of deformation in mode I failure is, however, not provided by the micropolar continuum Iordache & Willam (1998).

This work is motivated by an apparent lack of large scale 3D numerical studies investigating the predictive capabilities of classical and higher-order continuum material models for concrete, particularly for complex shear dominated failure modes.

Accordingly, the performance of (i) a classical continuum version of the CDP model Grassl & Jirásek (2006) using the mesh-adjusted softening modulus, herein simply denoted as the CDP model, is compared to (ii) a gradient-enhanced continuum formulation of the CDP model by Poh & Swaddiwudhipong (2009), denoted as the GCDP model and (iii) a gradient-enhanced micropolar continuum formulation of the CDP model by Neuner et al. (2020), denoted as the GMCDP model.

The comparison is performed by means of 3D finite element simulations of a transverse shear test for anchor channels. During this displacement controlled test, a cast-in anchor channel, which is placed in close proximity to the edge of a concrete slab, is transversely loaded up to failure, closely resembling a typical loading scenario of such anchor channels in real structures. Experimental data indicates that the expected failure mode depends on the edge distance of the anchor channel, making clear that the chosen benchmark example is a perfect candidate for assessing the classical and higher order continuum formulations of the CDP model.

In particular, the aim of the present contribution is to provide additional results of recent work performed by the authors, which were not presented in Neuner et al. (2022).

By comparing the predicted results of the three formulations with experimental data, it will be shown that their performance varies considerably. For obtaining meaningful results, proper knowledge of both their capabilities and limitations are thereby deemed imperative.

2 CLASSICAL AND HIGHER ORDER CONTINUUM FORMULATIONS OF THE CDP MODEL

The Concrete Damage Plasticity (CDP) Model by Grassl & Jirásek (2006) is based on a combination of the flow theory of plasticity and the isotropic continuum damage theory. Hence, the nominal stress tensor σ_{ij}, which in combination with the vector of body forces f_j satisfies the equilibrium conditions

$$\sigma_{ij,i} + f_j = 0, \tag{1}$$

is related to the effective stress tensor $\bar{\sigma}_{ij}$ using a scalar damage parameter ω as

$$\sigma_{ij} = (1 - \omega)\bar{\sigma}_{ij}. \tag{2}$$

Using the fourth order elastic stiffness tensor $\mathbb{C}_{ijkl}$ and performing an elastic-plastic split of the linearized strain tensor ε_{ij} into an elastic part $\varepsilon^{\mathrm{e}}_{ij}$ and a plastic part $\varepsilon^{\mathrm{p}}_{ij}$, the effective stress tensor is related to the elastic strain as

$$\bar{\sigma}_{ij} = \mathbb{C}_{ijkl}(\varepsilon_{kl} - \varepsilon^{\mathrm{p}}_{kl}) = \mathbb{C}_{ijkl}\varepsilon^{\mathrm{e}}_{kl}. \tag{3}$$

The yield function of the model

$$f_{\mathrm{p}}(\bar{\sigma}_{\mathrm{m}}, \bar{\rho}, \theta, q_{\mathrm{h}}) = \\ = \left((1 - q_{\mathrm{h}})\left(\frac{\bar{\rho}}{\sqrt{6}f_{\mathrm{cu}}} + \frac{\bar{\sigma}_{\mathrm{m}}}{f_{\mathrm{cu}}}\right)^2 + \sqrt{\frac{3}{2}}\frac{\bar{\rho}}{f_{\mathrm{cu}}}\right)^2 \\ + m_0 q_{\mathrm{h}}^2\left(\frac{\bar{\rho}}{\sqrt{6}f_{\mathrm{cu}}}r(\theta) + \frac{\bar{\sigma}_{\mathrm{m}}}{f_{\mathrm{cu}}}\right) - q_{\mathrm{h}}^2 \tag{4}$$

and the corresponding yield surface $f_{\mathrm{p}} = 0$ are formulated in terms of the three invariants of the effective stress tensor, i.e., the mean effective stress $\bar{\sigma}_{\mathrm{m}}$, the effective deviatoric radius $\bar{\rho}$ and the Lode angle θ. The material strength parameters, i.e., the uniaxial compressive strength f_{cu}, the biaxial compressive strength f_{cb} and the uniaxial tensile strength f_{tu} enter the yield function either directly or via the friction-like parameter m_0. The evolution of the yield surface due to hardening material behavior is controlled by a single scalar stress-like internal variable q_{h}. The shape of the yield surface in deviatoric sections is controlled by the Willam-Warnke polar radius function $r(\theta)$ (Willam & Warnke 1975). Its influence is, however, depending on the levels of hardening and confinement.

The non-associated plastic flow rule

$$\dot{\varepsilon}^{\mathrm{p}}_{ij} = \dot{\lambda}\frac{\partial g_{\mathrm{p}}}{\partial \bar{\sigma}_{ij}} = \dot{\lambda}\frac{\partial g_{\mathrm{p}}(\bar{\sigma}_{\mathrm{m}}, \bar{\rho}, q_{\mathrm{h}})}{\partial \bar{\sigma}_{ij}} \tag{5}$$

with the plastic multiplier $\dot{\lambda}$ makes use of a plastic potential function g_p.

The exponential evolution of the scalar damage variable

$$\omega = 1 - \exp\left(-\frac{\alpha_\mathrm{d}}{\varepsilon_\mathrm{f}}\right) \tag{6}$$

is driven by the internal variable α_d, which is driven by the rate of volumetric plastic strain $\dot{\varepsilon}^\mathrm{p}$. The initial slope of the evolution of the damage variable is controlled by the softening modulus ε_f.

2.1 *CDP model with mesh-adjusted softening modulus*

For the local version of the CDP model, the softening behavior of each finite element is adjusted by modifying the softening modulus ε_f in (6) based on the characteristic element length l_char, the specific mode I fracture energy G_I^f, the uniaxial tensile strength f_tu and the Young's modulus E acc. to Valentini (2011):

$$\varepsilon_\mathrm{f} = \frac{G_\mathrm{I}^\mathrm{f}}{f_\mathrm{tu} l_\mathrm{char}} - \frac{f_\mathrm{tu}}{2E}. \tag{7}$$

2.2 *Gradient-Enhanced Concrete Damage-Plasticity (GCDP) model*

Poh & Swaddiwudhipong (2009) formulated an implicit gradient-enhanced continuum version of the CDP model. To this end, the damage driving variable α_d in (6) is replaced by its nonlocal counterpart $\tilde{\alpha}_\mathrm{d}$, which is implicitly defined as the solution of the screened Poisson equation

$$\tilde{\alpha}_\mathrm{d} - l_\mathrm{d}^2 \frac{\partial^2}{\partial x_i \partial x_i} \tilde{\alpha}_\mathrm{d} = \alpha_\mathrm{d}. \tag{8}$$

Therein, α_d acts as the driving variable on the right hand side, and l_d is a length parameter.

2.3 *Gradient-Enhanced Micropolar Concrete Damage-Plasticity (GMCDP) model*

In a recent publication (Neuner et al. 2020), both the benefits of the gradient-enhanced continuum for describing mode I failure and the micropolar continuum for describing shear failure were incorporated into a new framework for cohesive-frictional materials, and applied to the CDP model resulting in a gradient-enhanced micropolar version.

For the micropolar continuum, the displacement field u_i is complemented by a field of independent micro-rotations w_i. Making use of the Levi-Civita symbol ϵ_{ijk}, the linearized strain tensor ε_{ij} is defined as

$$\varepsilon_{ij} = u_{j,i} - \epsilon_{ijk} w_k. \tag{9}$$

In addition to the translational equilibrium conditions (1) the rotational equilibrium conditions

$$m_{ij,i} + \epsilon_{jkl}\sigma_{kl} + c_j = 0 \tag{10}$$

serve as the governing balance equations. Therein, m_{ij} denotes the nominal couple-stress tensor, and c_j is the body couple vector.

The gradient of the micro-rotations is denoted as the linearized micro-curvature measure κ_{ij}

$$\kappa_{ij} = w_{j,i}, \tag{11}$$

for which an additive split into an elastic part κ_{ij}^e and a plastic part κ_{ij}^p is assumed.

By analogy to the definition of the effective force-stress tensor (2), the effective couple-stress tensor $\bar{m}_{ij}$ is related to the nominal couple-stress tensor m_{ij}:

$$m_{ij} = (1-\omega)\bar{m}_{ij}. \tag{12}$$

The effective couple-stress tensor $\bar{m}_{ij}$ is related to the elastic part of the micro-curvature κ_{ij}^e using the fourth order elastic stiffness tensor $\mathbb{C}_{ijkl}^\mathrm{m}$ as

$$\bar{m}_{ij} = \mathbb{C}_{ijkl}^\mathrm{m}(\kappa_{ij} - \kappa_{kl}^\mathrm{p}) = \mathbb{C}_{ijkl}^\mathrm{m}\kappa_{kl}^\mathrm{e}. \tag{13}$$

The fourth order stiffness tensors $\mathbb{C}_{ijkl}$ and $\mathbb{C}_{ijkl}^\mathrm{m}$ which relate the linearized strain tensor and the linearized micro-curvature measure to the respective effective stress measures

$$\bar{\sigma}_{ij} = (G+G_c)\varepsilon_{ij}^\mathrm{e} + (G-G_c)\varepsilon_{ji}^\mathrm{e} + \lambda\varepsilon_{kk}^\mathrm{e}\delta_{ij} = \mathbb{C}_{ijkl}\varepsilon_{kl}^\mathrm{e}, \tag{14}$$

$$\bar{m}_{ij} = (\gamma+\beta)\kappa_{ij}^\mathrm{e} + (\gamma-\beta)\kappa_{ji}^\mathrm{e} + \alpha\kappa_{kk}^\mathrm{e}\delta_{ij} = \mathbb{C}_{ijkl}^\mathrm{m}\kappa_{kl}^\mathrm{e}, \tag{15}$$

are defined by 6 material parameters, i.e., the LamÃ© parameters λ and G, the coupling modulus G_c, and 3 micropolar elastic parameters γ, β, and α, which are commonly expressed in terms of the polar ratio ψ, the characteristic length for bending l_b and the characteristic length for torsion l_t:

$$\psi = \frac{2\gamma}{2\gamma+\alpha}, \quad l_\mathrm{b} = \sqrt{\frac{\gamma+\beta}{4G}}, \quad l_\mathrm{t} = \sqrt{\frac{\gamma}{G}}. \tag{16}$$

The plasticity part of the micropolar GMCDP model is formulated by replacing the invariants of the effective stress tensor $\bar{\sigma}_\mathrm{m}$, $\bar{\rho}$, and θ in the yield function (4) and the plastic potential function (5) by generalized invariants $\tilde{\bar{\sigma}}_\mathrm{m}$, $\tilde{\bar{\rho}}$ and $\tilde{\theta}$, formulated in terms of the states of the effective force-stress and the effective couple-stress. The evolution of $\varepsilon_{ij}^\mathrm{p}$ and κ_{ij}^p is then given by the generalized flow rule

$$\dot{\varepsilon}_{ij}^\mathrm{p} = \dot{\lambda}\frac{\partial g_\mathrm{p}}{\partial \bar{\sigma}_{ij}}, \quad \dot{\kappa}_{ij}^\mathrm{p} = \dot{\lambda}\frac{\partial g_\mathrm{p}}{\partial \bar{m}_{ij}}. \tag{17}$$

The damage part of the GMCDP model is identical to the one of the gradient-enhanced GCDP model for all intents and purposes of this contribution.

In summary, three independent fields are considered for the GMCDP model: (i) the displacement field u_i, (ii) the field of micro-rotations w_i describing averaged rotations of the microstructure, and (iii) the field of the nonlocal damage-driving variable $\tilde{\alpha}_d$.

3 FINITE ELEMENT MODEL OF THE TRANSVERSE SHEAR TEST

Anchor channels are used in the construction of superstructures to transfer forces between load bearing reinforced concrete structures and other components. Typical applications include the anchorage of curtain-wall facades or elevator and tunnel equipment. One important component test for obtaining the technical approval for anchor channels is the transverse shear test. In this test a cast-in anchor channel in close proximity to the edge of a concrete slab is subjected to tensile forces acting in the direction normal to the edge of the slab. The resulting failure mode depends on the edge distance of the anchor channel, and it is dominated by concrete failure. Hence, this test is a well-suited benchmark example for investigating the performance of different material models for concrete.

The model setup of the transverse shear test is illustrated in Figure 1. It consists of (i) a concrete body representing a part of the concrete slab, (ii) a cast-in anchor channel, (iii) a load transfer plate which is connected to the anchor channel via a T-bolt as well as (iv) a steel support. The anchor channel consists of a C-shaped steel channel and two headed anchors. Symmetry of the test setup is exploited in the numerical simulations. The test procedure is represented in the numerical simulations by increasing the horizontal displacement at the outer end of the load transfer plate as indicated in Figure 1.

A penalty based node to surface contact formulation is used for modeling contact between the steel and concrete components. Frictional contact with coefficients of $\mu_{s/s} = 0.2$ (steel to steel) and $\mu_{s/c} = 0.35$ (steel to concrete) is assumed. The individual contact pairs are illustrated in Figure 2.

Experimental results for the transverse shear test for two values of the edge distance of the anchor channel in the form of normalized load-displacement curves serve for validating the predicted numerical results.

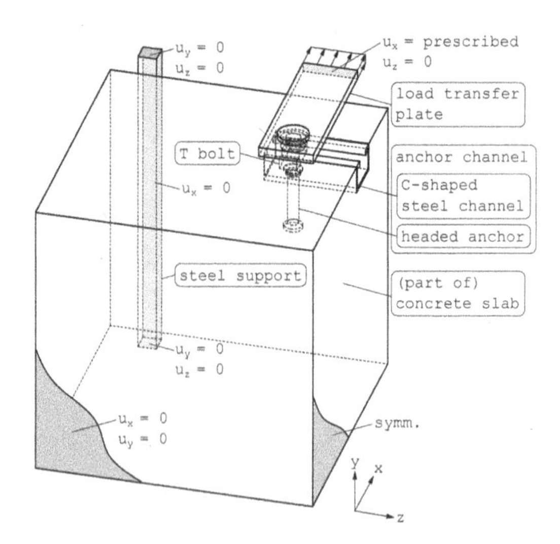

Figure 1. Illustration of the transverse shear test, showing the considered part of the concrete slab and the steel parts, boundary conditions and exploited symmetry. Modified reprint from Neuner et al. (2022).

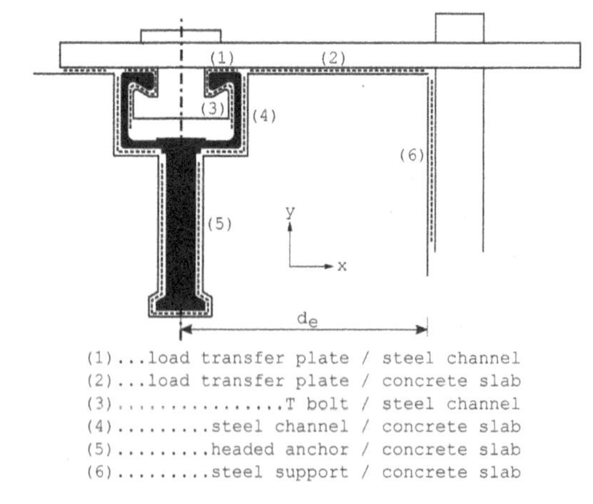

Figure 2. Section through the test setup with d_e indicating the edge distance. The considered contact interfaces are indicated as dotted lines. Modified reprint from Neuner et al. (2022).

3.1 *Material parameters*

For calibrating the core of the CDP model a set of standard material parameters for concrete is used, which can be determined from laboratory tests. These material parameters are the Young's modulus E, the Poisson's ratio ν, the uniaxial compressive strength f_{cu}, the uniaxial yield stress f_{cy}, the biaxial compressive strength f_{cb} and the uniaxial tensile strength f_{tu}.

The choice of material parameters for the investigated test setup is based on a set of standard parameters for the specified grade C20 of the investigated concrete according to Model Code fib (2013). The material parameters used in the numerical model are listed in Table 1.

An additional set of 6 model parameters, i.e., A_h, B_h, C_h, D_h, D_f, and A_s is used, which cannot be directly related to standard lab test results. The default values proposed in Grassl & Jirásek (2006) are used for the present study. They are summarized in Table 2.

The remaining parameters are the softening modulus ε_f, as well as the additional parameters for the higher order continuum GCDP and GMCDP models. Softening material behavior is calibrated using a

Table 1. Material parameters for the CDP, GCDP and GMCDP model for the concrete grade C20 acc. to Model Code fib (2013).

E (GPa)	ν (−)	f_{cu} (MPa)	f_{cy} (MPa)	f_{tu} (MPa)	f_{cb} (MPa)
30.3	0.2	28	9.33	2.2	31.4

Table 2. Model parameters according to Grassl & Jirásek (2006) for the CDP, GCDP and GMCDP model.

A_h	B_h	C_h	D_h	D_f	A_s
0.08	0.003	2.0	1e-6	0.85	15

specific mode I fracture energy of $G_I^f = 133$ N/mm according to Model Code fib (2013) for the given concrete grade.

For the CDP model with mesh-adjusted softening modulus, the softening modulus ε_f is computed from the specific mode I fracture energy and the characteristic element length l_{char} according to (7). Assuming that localization of deformation occurs in a crack band with a width of a single finite element, the characteristic element length is computed directly from the volume of a finite element V_e as

$$l_{char} = \sqrt[3]{V_e}. \quad (18)$$

For perfectly cubic elements with linear shape functions this approach yields accurate results when the direction of crack opening is aligned with the edges of the finite element. However, for other directions the determination of the characteristic element length becomes inaccurate leading to an overestimation of the dissipated energy during crack opening, and for higher order elements localization of deformation may arise in sub-domains of the element Jirásek & Bauer (2012). While various improved approaches for computing the characteristic element length have been proposed e.g., (Oliver 1989), such methods have to be implemented at finite element level losing the simplicity of the mesh-adjusted softening modulus approach. Since the practice of relating the characteristic element length to the volume of the finite element (18) enjoys some degree of popularity in engineering practice, and due to its forementioned simplicity, this method is employed in the present contribution.

For the GCDP model, the softening modulus ε_f is chosen in such a manner that the specific mode I fracture energy is reproduced in a direct uniaxial tension test with the characteristic length parameter l_d being specified in advance. While both the physical interpretation of l_d and its determination are still the subject of research, it is commonly related to the characteristic size of the microstructure of the material, as it determines the size of the damaged zone. By choosing a value of $l_d = 4$ mm the thickness of the damaged zone obtained in the numerical analyses is approximately equal to the experimentally observed thickness of the fracture process zone, which is two to three times the maximum aggregate size (Bažant & Pijaudier-Cabot 1989). The same value of the length parameter has been successfully employed in the past for the numerical analysis of experimental tests on concrete specimens (Neuner et al. 2020). For the length parameter of $l_d = 4$ mm and the given fracture energy of $G_I^f = 133$ N/mm calibration of the softening modulus resulted in a value of $\varepsilon_f = 0.0064$.

In addition to the parameters required for the GCDP model, for the GMCDP model the micropolar constants G_c, l_b, l_{J2}, l_t and ψ need to be specified. The coupling modulus G_c is commonly chosen in a range between $0.1G$ and $1.0G$, with G denoting the shear modulus. In this contribution, a value of $0.1G$ is assumed, and the characteristic length for bending is chosen as $l_b = 2$ mm. The length scale parameter l_{J2} controls the resistance against plastic shear deformation of the microstructure. It is chosen as $l_{J2} = 2$ mm. The remaining parameters are chosen as $l_t = 2l_b$ and $\psi = \frac{3}{2}$, satisfying the principle of bounded stiffness (Neff, Jeong, & Fischle 2010). For an in depth discussion of the choice of material parameters the reader is referred to (Neuner et al. 2020).

A summary of the model-specific parameters for the three investigated models is presented in Table 3.

Table 3. Model-specific parameters for the investigated material models.

	ε_f (−)	l_d (mm)	G_c/G (−)	l_b (mm)	l_t (mm)	ψ (−)	l_{J2} (mm)
CDP	acc. to (7)						
GCDP	0.0064	4					
GMCDP	0.0064	4	0.1	2	4	$\frac{3}{2}$	2

3.2 *Numerical implementation of the material models and employed software*

The three investigated models were implemented in Abaqus/Standard (Abaqus 2015) using the respective UMAT (CDP model) and UEL (GCDP and GMCDP models) interfaces. For the GCDP and GMCDP models, user defined finite elements are employed, as fully coupled problems are being solved. All models were implemented in the Marmot material modeling toolbox library (Dummer, Mader, Neuner, & Schreter 2021) using C++ programming language. For achieving a quadratic rate of convergence in nonlinear simulations, the consistent tangent operators are computed for all models. To this end, the Eigen template library (Guennebaud et al. 2010) is used for linear algebra computations for the CDP and the GCDP models, while for the GMCDP model tensor contraction operations are performed using the Fastor library (Poya, Gil, & Ortigosa 2017) at material and finite element level.

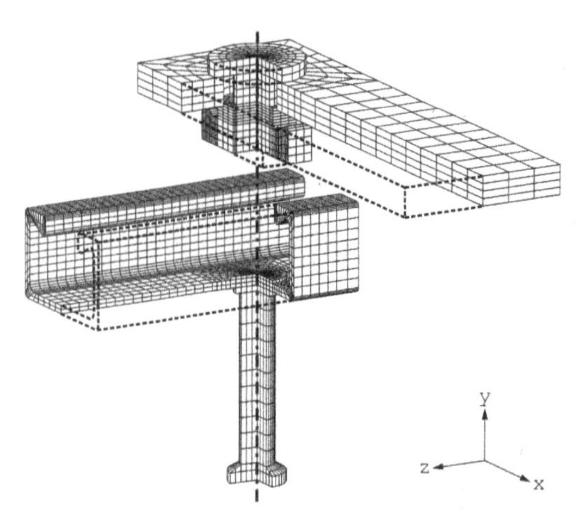

Figure 3. Finite element mesh for the load transfer plate with mounted T-bolt and the anchor channel. Modified reprint from (Neuner et al. 2022).

The Fastor library provides a convenient and performant way of implementing tensor operations using the Einstein notation and utilizing SIMD vectorization.

The visualization capabilities of Abaqus/CAE for simulation results produced with user defined elements (`UEL`) are severely limited. Therefore, after exporting all results using the `fil` file format, they are converted to the open `Ensight Gold` format employing the AbaqusFilFile-Translator software (Neuner 2021). Using ParaView (Ahrens, Geveci, & Law 2005), the converted results in `Ensight Gold` format are read and visualized.

3.3 *Finite element discretization*

For the respective edge distances of the anchor channel, finite element meshes employing hexahedral elements are used to discretize the concrete slab. The finite element mesh for the small edge distance is illustrated in Figure 4. For an illustration of the very similar finite element mesh employed for the large edge distance the reader is referred to (Neuner et al. 2022). For the steel parts, finite element meshes employing mainly hexahedral and a few pentahedral elements for easing the meshing procedure are used (Figure 3).

Since the use of elements with quadratic shape functions is not advised when the mesh-adjusted softening modulus is utilized (Jirásek & Bauer 2012), for use with the CDP model fully integrated 8-node elements with linear shape functions are employed, whereas for the higher order continuum GCDP and GMCDP models 20-node hexahedral elements with quadratic shape functions and reduced integration are used in order to represent the steep gradient of the nonlocal damage-driving field $\tilde{\alpha}_d$ with sufficient accuracy.

For all steel parts, 8-node hexahedral elements with an enhanced assumed strain formulation (Simo & Rifai 1990; ?) with 13 incompatible modes (`C3D8I` in Abaqus/Standard) are used to prevent numerical locking.

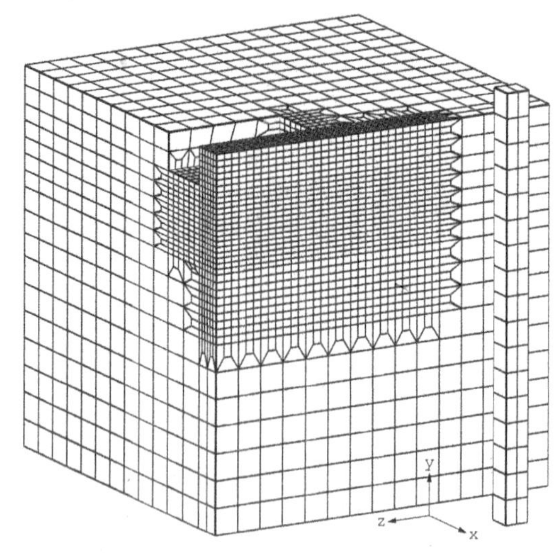

Figure 4. Finite element mesh for the concrete slab for the small edge distance of the anchor channel and the steel support.

4 RESULTS

The predicted load-displacement behavior on the basis of the three material models for both edge distances of the anchor channel is depicted along with results from experimental tests in Figures 5 and 7. The respective failure modes predicted by the CDP and GMCDP models are illustrated in Figures 6, 8 and by means of contour plots of the first invariant of plastic strain I_1^p as a measure for accumulated plastic deformation. The illustration of the results for the GCDP model is omitted here, since they are very similar to the results of the GMCDP model (Neuner et al. 2022).

4.1 *Large edge distance of the anchor channel*

In the experimental tests for the large edge distance of the anchor channel, a virtually linear load-displacement relation is observed until the peak load is attained. The load subsequently drops in a quasi-brittle fashion until a plateau of residual load is attained. A further increase of displacement leads to no substantial increase of the attained load level.

After an initially too soft mechanical response, consistent results closely matching the experimentally observed stiffness are obtained with all of the three material models up to the initiation of concrete damage. The initiation of concrete damage occurs near the lower front edge of the steel channel. For the higher order GCDP and GMCDP models the subsequent propagation of damage and the accompanying development of plastic deformation mark the attainment of the peak load and the onset of softening in the

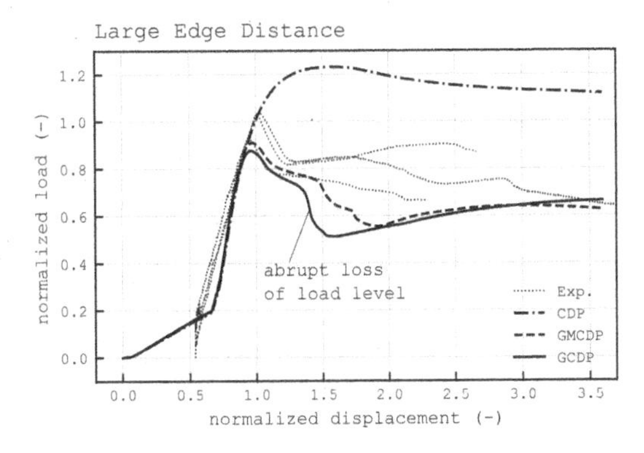

Figure 5. Normalized predicted load-displacement curves for the large edge distance of the anchor channel: Results obtained with the three material models, and the results from three experimental tests. Modified reprint from (Neuner et al. 2022).

load-displacement curves. This is in sharp contrast to the behavior predicted by the CDP model: While the initiation of damage is predicted approximately at the same time as for the higher order models, the damage progression is associated with continuing hardening behavior, leading to a considerable overestimation of the peak load.

While the peak load is only slightly underestimated by the higher order models, the immediate post-peak response is in very good agreement with the experimental data. However, as indicated in Figure 5 an abrupt loss of the load level in the post peak regime is predicted by the GCDP model, contrasting the experimental data. This loss of load level is attributed to pathological localization behavior not being completely ruled out by the employed gradient enhancement. The micropolar GMCDP model serves as a remedy for this behavior, which is further counteracted by higher values of the parameter l_{J2}, as demonstrated in (Neuner et al. 2022).

In contrast to the experimental data and the results obtained with the higher order models, no significant softening behavior is predicted with the CDP model, despite the fact that the plastic zone is nearly fully developed. This behavior is explained by two contributing factors: Firstly, due to the spatial orientation of the crack the employed simple method of determining the characteristic element length (18) leads to an overestimation of the fracture energy. Secondly, the employed 8-node hexahedral elements are prone to locking, especially in situations when volumetric and deviatoric plastic flow are coupled, as is here the case (de Borst & Groen 1995; ?).

Although the illustrations of the predicted failure modes on the basis of the CDP and GMCDP models in Figure 6 show some similarities, by means of the CDP model branching of the plastic zone with a mesh bias in close proximity to the concrete edge can be observed, while the higher order models do not show this spurious behavior.

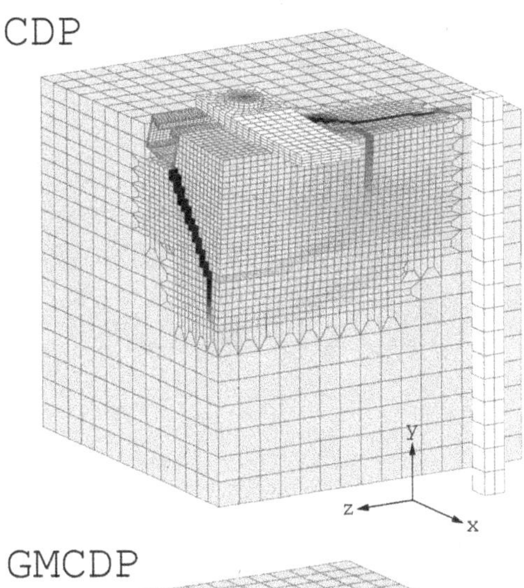

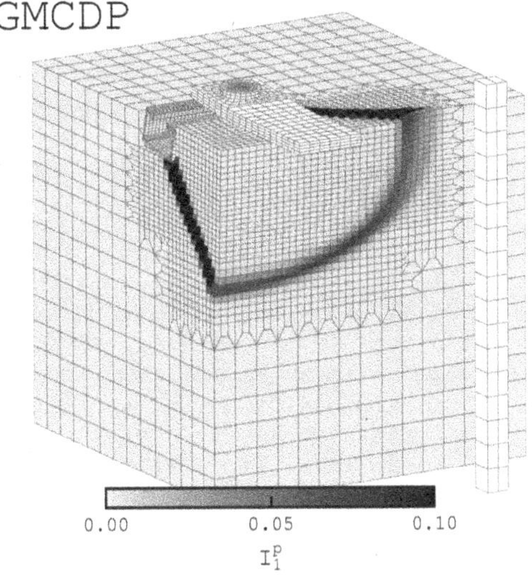

Figure 6. Predicted failure modes for the large edge distance of the anchor channel on the basis of the CDP and GMCDP material models. First invariant of plastic strain I_1^p. Displacements are scaled by a factor of 3. Results from (Neuner et al. 2022).

4.2 Small edge distance of the anchor channel

In contrast to the large edge distance of the anchor channel and despite some experimental scatter, experimental results for the small edge distance of the anchor channel unambiguously show distinct nonlinear pre-peak hardening behavior with gradually decreasing slope followed by post-peak softening persisting up to the termination of the experimental procedure. No residual plateau is observed for the small edge distance of the anchor channel.

The pre-peak hardening behavior, which is associated with gradual failure of the concrete edge is predicted by all of the three investigated material models. Its onset coincides with the initiation of concrete damage, and marks the point where the predicted load displacement curves start deviating from each other.

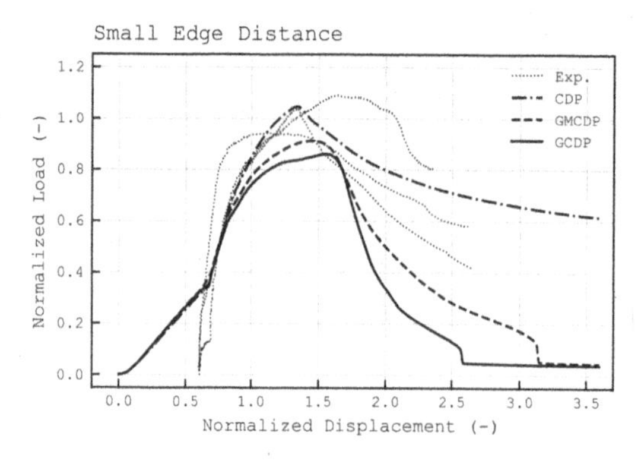

Figure 7. Normalized predicted load-displacement curves for the small edge distance of the anchor channel along with the results from three experimental tests.

As for the large edge distance of the anchor channel, the ultimate load bearing capacity is slightly underestimated by the higher order GCDP and GMCDP models. The post-peak regime predicted by these models is characterized by pronounced softening behavior, which ultimately leads to a total loss of load bearing capacity. By means of the micropolar GMCDP model, a slightly higher peak load and a less brittle post-peak response are predicted compared to the GCDP model. These differences are due to the mobilization of the micropolar part of the model as a result of localized shear deformation of the concrete.

Although the load-displacement behavior predicted by the CDP model is in very good agreement with the experimental data, this should not disguise the fact that the same deficiencies as mentioned above are still present. Mesh bias of the results is clearly apparent in Figure 8, and despite the fully developed plastic zone a significant residual load is predicted.

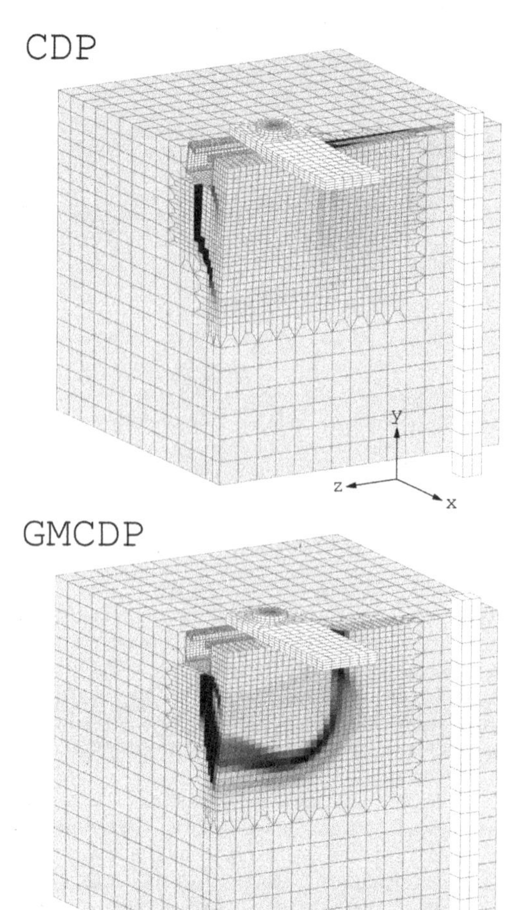

Figure 8. Failure modes for the small edge distance of the anchor channel on the basis of the CDP and GMCDP material models. First invariant of plastic strain I_1^p. Displacements are scaled by a factor of 3.

5 CONCLUSION

Three formulations of a popular damage plasticity model for concrete, i.e., (i) a classical local formulation with mesh-adjusted softening modulus denoted as the CDP model, (ii) a higher order gradient-enhanced formulation denoted as the GCDP model and (iii) a recently proposed higher order gradient-enhanced micropolar formulation denoted as the GMCDP model were outlined. In the subsequent finite element analyses of transverse shear tests for anchor channels with varying edge distances, their predictive capabilities for complex structural applications were investigated. The following conclusions are drawn:

- The experimentally observed nonlinear pre-peak hardening behavior for the small edge distance of the anchor channel is predicted by all of the investigated material models. While a significant amount of residual load is retained with the CDP model, the higher order models accurately predict distinct softening behavior and the eventual total loss of the load bearing capacity associated with the break out of the whole anchor channel.
- For the large edge distance of the anchor channel, apart from a slight underestimation of the peak load, the structural behavior predicted by means of the higher order GCDP and GMCDP models is in good agreement with the experimental results. An abrupt loss of load level in the post-peak domain observed for the GCDP model is attributed to pathological localization behavior. The GMCDP model serves as a remedy for this behavior.
- The predicted failure modes for the CDP model exhibit severe mesh bias, which is not the case for the investigated higher order models. Moreover, the peak load for the large edge distance of the anchor channel is considerably overestimated by means of

the CDP model and post-peak softening is under-predicted for both edge distances. This is due to the employed method for determining the characteristic element length, and due to suspected locking behavior of the employed 8-node hexahedral finite elements.

- For both edge distances of the anchor channel, the higher order GCDP and GMCDP models represent very well the experimental results, which is apparent from the load-displacement curves.
- The edge distance of the anchor channel affects the failure pattern and the size of the broken out part of the concrete edge. This circumstance is quantitatively and qualitatively reflected by means of the higher order GCDP and GMCDP models. In the present study, the CDP model employing the mesh adjusted softening modulus is not able to compete in this regard.

ACKNOWLEDGEMENT

The computational results presented here have been achieved (in part) using the LEO HPC infrastructure of the University of Innsbruck.

REFERENCES

Abaqus (2015). *ABAQUS v6.14 Documentation*. Providence, RI, USA: Dassault Systémes.

Ahrens, J., B. Geveci, & C. Law (2005). Paraview: An end-user tool for large data visualization. *The visualization handbook 717*(8).

Bažant, Z. P. & G. Pijaudier-Cabot (1989). Measurement of characteristic length of nonlocal continuum. *Journal of Engineering Mechanics 115*(4), 755–767.

Cosserat, E. & F. Cosserat (1909). *Théorie des corps déformables*. A. Hermann et fils.

de Borst, R. & A. E. Groen (1995). Some observations on element performance in isochoric and dilatant plastic flow. *International Journal for Numerical Methods in Engineering 38*(17), 2887–2906.

Dummer, A., T. Mader, M. Neuner, & M. Schreter (2021). Marmot library.

fib (2013). *Model Code for Concrete Structures 2010*. Wilhelm Ernst & Sohn, Verlag fÃ¼r Architektur und technische Wissenschaften GmbH & Co. KG.

Grassl, P. & M. Jirásek (2006). Damage-plastic model for concrete failure. *International Journal of Solids and Structures 43*(22–23), 7166–7196.

Guennebaud, G., B. Jacob, et al. (2010). Eigen v3. http://eigen.tuxfamily.org.

Iordache, M.-M. & K. Willam (1998). Localized failure analysis in elastoplastic Cosserat continua. *Computer Methods in Applied Mechanics and Engineering 151*(3–4), 559–586.

Jirásek, M. & M. Bauer (2012). Numerical aspects of the crack band approach. *Computers & Structures 110*, 60–78.

Neff, P., J. Jeong, & A. Fischle (2010). Stable identification of linear isotropic cosserat parameters: bounded stiffness in bending and torsion implies conformal invariance of curvature. *Acta Mechanica 211*(3), 237–249.

Neuner, M. (2021). AbaqusFilFile-Translator.

Neuner, M., P. Gamnitzer, & G. Hofstetter (2020). A 3d gradient-enhanced micropolar damage-plasticity approach for modeling quasi-brittle failure of cohesive-frictional materials. *Computers & Structures 239*, 106332.

Neuner, M., P. Hofer, & G. Hofstetter (2022, January). On the prediction of complex shear dominated concrete failure by means of classical and higher order damage-plasticity continuum models. *Engineering Structures 251*.

Oliver, J. (1989). A consistent characteristic length for smeared cracking models. *International Journal for Numerical Methods in Engineering 28*(2), 461–474.

Poh, L. H. & S. Swaddiwudhipong (2009). Over-nonlocal gradient enhanced plastic-damage model for concrete. *International Journal of Solids and Structures 46*(25–26), 4369–4378.

Poya, R., A. J. Gil, & R. Ortigosa (2017). A high performance data parallel tensor contraction framework: Application to coupled electro-mechanics. *Computer Physics Communications*.

Rudnicki, J. W. & J. R. Rice (1975, December). Conditions for the localization of deformation in pressure-sensitive dilatant materials. *Journal of the Mechanics and Physics of Solids 23*(6), 371–394.

Simo, J. C. & M. Rifai (1990). A class of mixed assumed strain methods and the method of incompatible modes. *International journal for numerical methods in engineering 29*(8), 1595–1638.

Valentini, B. (2011). *A three-dimensional constitutive model for concrete and its application to large scale finite element analyses*. Ph. D. thesis, Leopold-Franzens-UniversitÃ¤t Innsbruck.

Willam, K. & E. Warnke (1975). Constitutive models for the triaxial behavior of concrete. In *Proceedings of the International Association for Bridge and Structural Engineering*, Volume 19, Bergamo, Italy, pp. 1–30.

Computational Modelling of Concrete and Concrete Structures – Meschke, Pichler & Rots (Eds)
 ISBN: 978-1-032-32724-2

A combined VEM and interface element-based approach for stochastic failure analysis of concrete

G. Chacón & F.L. Rivarola
Faculty of Engineering, University of Buenos Aires, Argentina

D. van Huyssteen & P. Steinmann
Institute of Applied Mechanics, Friedrich-Alexander-Universität Erlangen-Nürnberg, Germany

G. Etse
Faculty of Exact Sciences and Technology, University of Tucuman, Argentina

ABSTRACT: In this work, the efficacy and efficiency of a discretization technique based on virtual elements (VEs) and non-linear interface elements (IEs) are assessed for the modelling of representative boundary value problems (BVPs) of fracture processes in concrete components characterized by tortuous crack propagations. In particular, the efficiency of randomness in mesh refinement is evaluated for meshes comprising polyhedral VEs.

The results demonstrate the limitations of structured meshes and the effectiveness of random polyhedral virtual meshes in modelling the tortuous propagation of cracks in concrete subjected to mode II fracture, or shear. It is concluded that random h refinement applied to polyhedral meshes allows a statistically accurate prediction of the crack path. In particular, the influence of random mesh generation is studied. The results demonstrate that the average predictions generated by a series of random coarse mesh perturbations of polyhedral elements are very accurate, particularly in terms of peak load and crack path. Finally, an adaptative refinement technique is discussed as a possible technique for efficient identification of the localization zones.

1 INTRODUCTION

Failure analysis of concrete components requires efficient and effective numerical procedures to reproduce the complex fracture mechanisms that develop in this composite material. This is due to its severe brittleness under tension and shear states as a consequence of unstable and abrupt processes of coalescence of microdefects during loading beyond the elastic limit. During failure only a small portion of the relevant boundary value problem (BVP) is involved in governing the failure behavior while in other, considerably larger, subdomains homogeneous and elastic stress states develop. This dichotomy motivates the consideration of discretization strategies that allow for abrupt jumps between the passive, or elastic, subdomains and the active, and strongly inelastic, subdomain where the failure processes are localized. At the same time, appropriate discretization densification strategies need to be considered to allow strong concentration of degrees of freedom in the active subdomains through non-regular arrangements and densifications that favor the tortuous and intricate propagation of cracks during the evolution of the failure processes.

By exploiting the great versatility of virtual elements (VEs) with respect to meshing it is possible to strongly modify the concentration of elements along the boundary between regions. In this work VEs are used to model the abrupt jumps between elastic and inelastic regions of concrete components involved in failure processes. In the zones of the BVP where the processes of cracks and inelastic dissipation are concentrated VEs are combined with inelastic interface elements (IEs) that define potential lines of cohesive-frictional crack propagations.

Regarding the VE and IE strategy considered in the inelastic subdomains, the need arises to evaluate the influence of the discretization density on the failure mechanics and crack paths predictions. In that sense, and based on the geometric versatility of the VEM, it is convenient to analyze comparatively the efficacy of meshes based on rectangular and polyhedral elements, as well as the sensitivity of the results with respect to discretization density and randomness. In addition, the non-trivial question that arises is the identification of the active zone before running numerical analysis. This is particularly important in the case of highly hyperstatic and complex concrete components.

To evaluate the efficacy of the proposed strategy based on VEs and IEs, and provide answers to the aforementioned questions, in this work the failure

DOI 10.1201/9781003316404-56

predictions obtained with discretizations of the active zones composed of rectangular and polyhedral virtual elements are comparatively evaluated. Additionally, random h densification strategies are considered for the polyhedral VEs located in the failure concentration zones. Finally, an elastic analysis of the considered BVP using the VE refinement technique proposed in van Huyssteen et al. (2022) is performed to assess its capability to predict the active zones of failure localization.

From the results obtained it is clear that polyhedral VEs provide the necessary mesh irregularity to reproduce tortuous crack propagations such as those that occur during concrete fracture. Additionally, the mean values of peak load, ductility at peak load, and crack path, obtained from random mesh perturbations of coarse meshes with polyhedral VEs, lead to similar predictions to those obtained with highly refined meshes. This suggests that a strategy based on the application of the discrete crack approach with a series of perturbed coarse discretizations of polyhedral VEs and IEs can efficiently and effectively predict the failure behavior of quasi-brittle materials such as concrete. Finally, it is found through an elastic analysis that the adaptive refinement procedure proposed in van Huyssteen et al. (2022) for VEs is able to provide an accurate prediction of the active/localization zone of the BVP.

2 FRAMEWORK AND FORMULATION

2.1 *Fundamental equations of the mixed variational form for fracture equilibrium problems.*

The approach for the fracture problem consists of a cohesive zone model implemented in a mixed formulation based on an augmented Lagrangian functional. A summary of the basic elements of this approach is provided in this section. For a complete description refer to Rivarola et al. (2020). Further details can be found in Labanda et al. (2018a) and Labanda et al. (2018b) for implementation issues and Doyen et al. (2010) for theoretical aspects.

The symbols L^2 and H^1 denote the classical Sobolev spaces of square integrable functions and functions with square integrable weak first derivatives whose trace vanishes on the Dirichlet boundary $\partial^u\Omega$. The energetic behaviour of an infinitesimally deformable body in $\mathbb{R}^d$ with a discontinuity in the displacement field, as schematically represented in Figure 1, can be expressed in its weak form as follows:

Given $f \in L^2(\Omega), p \in L^2(\partial^t\Omega)$ and a penalty parameter $\gamma \in \mathbb{R}^+$, find $(u, \boldsymbol{\lambda}) \in \mathscr{V} \times \mathscr{X}$ such that

$$a(u, \boldsymbol{\delta}, \boldsymbol{\lambda}, v) = l(v), \ \forall v \in \mathscr{V} \tag{1}$$

$$b(u, \boldsymbol{\delta}, \boldsymbol{\lambda}, \boldsymbol{\mu}) = 0, \quad \forall \boldsymbol{\mu} \in \mathscr{X} \tag{2}$$

$$c(u, \boldsymbol{\delta}, \boldsymbol{\lambda}) = 0, \quad \boldsymbol{\delta} \in \mathscr{W} \tag{3}$$

where u and $\boldsymbol{\lambda}$ are fixed in collocation points of the interface in equation (3) and the functional operators are given by

$$a(u, \boldsymbol{\delta}, \boldsymbol{\lambda}, v) = \int_\Omega \boldsymbol{\varepsilon}(u) : \mathcal{C} : \boldsymbol{\varepsilon}(v) \, d\Omega \tag{4}$$

$$+ \int_\Gamma [\boldsymbol{\lambda} + \gamma \left([\![u]\!] - \boldsymbol{\delta} \right)] \cdot [\![v]\!] \, d\Gamma \, , \tag{5}$$

$$l(v) = \int_{\partial^t\Omega} p \cdot v \, d\partial^t\Omega + \int_\Omega f \cdot v \, d\Omega \, , \tag{6}$$

$$b(u, \boldsymbol{\delta}, \boldsymbol{\lambda}, \boldsymbol{\mu}) = \int_\Gamma [[\![u]\!] - \boldsymbol{\delta}(u, \boldsymbol{\lambda})] \cdot \boldsymbol{\mu} \, d\Gamma \, , \tag{7}$$

$$c(u, \boldsymbol{\delta}, \boldsymbol{\lambda}) = \partial_{\boldsymbol{\delta}} \Psi \left(\boldsymbol{\delta}, \kappa \right) - \boldsymbol{\lambda} - \gamma \left([\![u]\!] - \boldsymbol{\delta}(u, \boldsymbol{\lambda}) \right) \tag{8}$$

with

$$\begin{aligned} \mathscr{V}\left(\Omega \setminus \Gamma\right) &= \left\{ v \in \mathbf{H}^1 : v|_{\partial^u_h\Omega} = 0 \right\} , \\ \mathscr{W}\left(\Gamma\right) &= \left\{ w \in \mathbf{H}^{\frac{1}{2}} : w \cdot n \geq 0, \forall w \in \Gamma \right\} , \\ \mathscr{X}\left(\Gamma\right) &= \left\{ x \in \mathbf{H}^{-\frac{1}{2}} \right\} . \end{aligned} \tag{9}$$

The definition of $\mathscr{V}\left(\Omega \setminus \Gamma\right)$ indicates that the functions $v \in \mathscr{V}$ belong to $\mathbf{H}^1$ in the open set defined by the continuous body minus the set of points that define the interface.

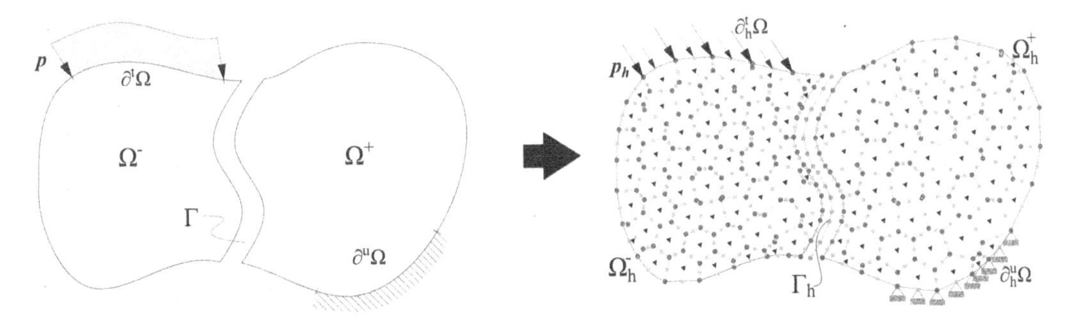

Figure 1. Domain decomposition of Ω considering a cohesive fracture Γ. Continuum and discrete notation with DOFs corresponding to a 2nd order VEM discretization.

In the above equations $\partial^t\Omega$ represents the Neumann boundary, u is the displacement field, v is the test displacement field, $p \in L^2(\partial^t\Omega)$ is a prescribed boundary traction, C the elastic tensor, $\boldsymbol{\varepsilon}$ is the strain tensor, γ is a penalty parameter, $\boldsymbol{\lambda}$ represents the Lagrange multipliers, $\boldsymbol{\mu}$ the test Lagrange multipliers, $f \in L^2(\Omega)$ are the volumetric loads, $[\![\bullet]\!] = (\bullet)|_{\Gamma^+} - (\bullet)|_{\Gamma^-}$ represents the jump of field $(\bullet)$ over domain Γ. Finally, $\boldsymbol{\delta}$ is a supplementary variable and Ψ is a pseudo-potential over the discontinuity Γ (Lorentz 2008) defined as

$$\Psi(\boldsymbol{\delta}, \kappa) = I_{\mathbb{R}^+}(\delta_n) + \psi(\boldsymbol{\delta}, \kappa), \tag{10}$$

with $I_{\mathbb{R}^+}$ an indicator function to prevent penetration in the normal direction n, and ψ representing the cohesive free energy function. The internal variable κ controls the irreversible nature of the process.

Note that outside of the discontinuity Γ, the body remains elastic. Equation (1) represents the balance between internal and external forces. Equation (2) weakly enforces the equality $[\![u]\!] = \boldsymbol{\delta}$. The supplementary variable $\boldsymbol{\delta}$ is solved in a staggered scheme motivated by the coordination decomposition method (Fortin and Glowinski 1983) in the collocation points along the interface. The traction-separation law is solved in an inverse way for the supplementary variable, starting with a predicted Lagrange multiplier and displacement field. To this end $c(u, \boldsymbol{\delta}, \boldsymbol{\lambda})$ in equation (3), is defined for each collocation point, and solved independently from the rest of the system, as u, $\boldsymbol{\delta}$ and $\boldsymbol{\lambda}$ are fixed in each iteration step and for each collocation point.

We note that in this paper we are not considering a fully mixed VEM formulation. Rather, the standard VEM formulation is used to discretize the displacement field on each element and a mixed formulation is only present on the element boundaries. Here, interface elements are added to account for the involved function spaces of displacements, interface stresses, and interface gaps.

2.2 *Traction-separation law*

The free energy on the interfaces is defined as follows:

$$\psi(\delta_{eq}) = \begin{cases} G_c \frac{\delta_{eq}}{\delta_c}\left[2 - \frac{\delta_{eq}}{\delta_c}\right], & \text{if } \delta_{eq} \leq \delta_c \\ G_c, & \text{if } \delta_{eq} > \delta_c \end{cases} \tag{11}$$

where $G_c = \frac{1}{2}\sigma_c\delta_c$ is the fracture energy, σ_c the critical tension and δ_c the critical displacement. The potential from which the free energy is derived is defined by only two of the three parameters $(G_c, \sigma_c, \delta_c)$. This gives rise to a simple traction-separation law consisting of a linear cohesive zone model, presented in Figure 2 (a) and (b) for the cases of pure Mode II and Mode I separation, respectively. The formulation results in an extrinsic law, as the interface remains completely closed until the critical tension is reached. Contact and adhesion regimes are considered, with linear unloading. An equivalent displacement scalar δ_{eq} drives the cohesive forces in the crack and is defined as

$$\delta_{eq} = \|\boldsymbol{\delta}\| = \sqrt{\boldsymbol{\delta} \cdot \boldsymbol{\delta}}. \tag{12}$$

This choice of δ_{eq} is symmetric in the normal and tangential components and results in the same critical tension for both directions, but other choices are possible. Furthermore, non linear damage curves can be obtained by varying the exponent of δ_{eq} in (11). An irreversibility variable κ is introduced to control the crack opening/closure:

$$\kappa(t) = \sup_{t' < t^*} \delta_{eq}(t'), \tag{13}$$

where t^* is the current time. The initiation criteria used to detect the cracking onset is shown in Figure 2 (c). Note that coupling between responses in the normal and tangential directions is present in eq. (11) through the computation of δ_{eq} in eq. (12). For a more comprehensive description and for the final algorithm of the constitutive model integration the reader is referred to Labanda et al. (2018b).

2.3 *The virtual element method*

The VEM is a generalization of the standard finite element method (FEM) to meshes comprising arbitrary polyhedra. It was first introduced in Beirão da Veiga et al. (2013) and the basic ideas are recalled here for the case of the general second order elliptic equations (as described in Beirão da Veiga et al. (2016)).

A summary of the VEM numerical procedure will be given in this section. For more details see Artioli et al. (2017).

Discretization Given a domain Ω divided into a mesh τ_h, for a desired order of accuracy k and with the space $\mathbb{P}_k$ of polynomials of maximum degree k, let us define the local space $V_{k,h}^{El}$ as

$$\begin{aligned} V_{k,h}^{El} &= \{v_h \in \mathrm{H}^1(El) : v_{h|\partial El} \in C^0(\partial El), \\ v_{h|e} &\in \mathbb{P}_k(e)\ \forall e \subset \partial El,\ \Delta v_h \in \mathbb{P}_{k-2}(El)\}, \end{aligned} \tag{14}$$

where h is a mesh parameter, El is an element of the mesh, ∂El is its boundary and e an edge. From the definition it can be seen that the base functions in the VEM space are not explicitly known for the entire domain, *i.e.* hey are only known on the boundary of an element.

The global virtual element space is then

$$\begin{aligned} \mathcal{V}_{k,h} &= \{v_h \in \mathrm{H}_D^1(\Omega) \cap C^0(\Omega) : v_{h|El} \\ &\in \mathcal{V}_{k,h}^{El}, \forall El \in \mathcal{T}_h\}. \end{aligned} \tag{15}$$

As in the FEM the discrete solutions of the variational problem are required: Find $u_h \in \mathcal{V}_{k,h}$ such that

$$a_h(u_h, v_h) = l_h(v_h) \qquad \forall v_h \in \mathcal{V}_{k,h}. \tag{16}$$

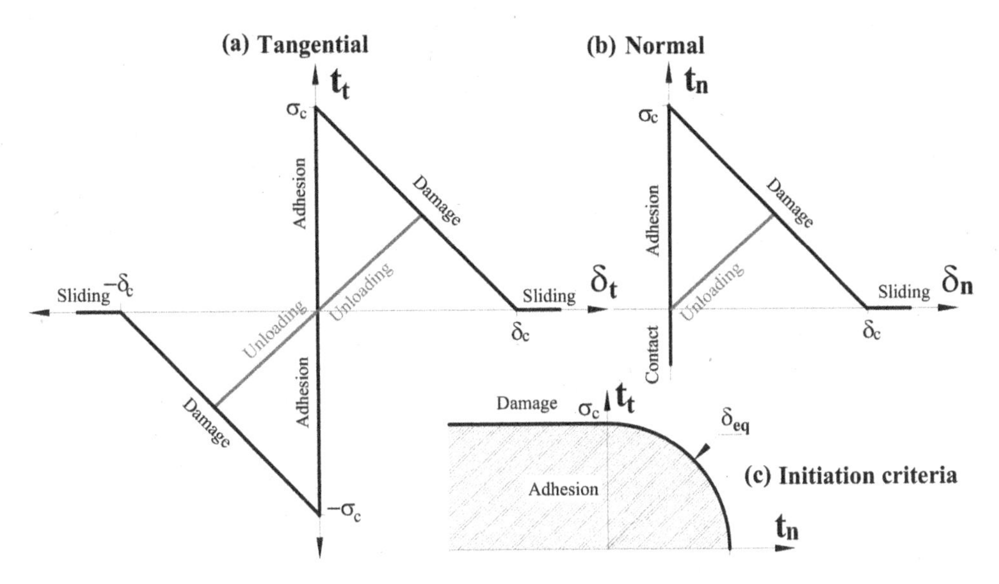

Figure 2. Proposed traction separation law with linear unloading. (a) Tangential traction-separation law (pure Mode II). (b) Normal traction-separation law (pure Mode I). (c) Initiation criteria.

The discrete version of the bilinear form is defined element-wise as

$$a_h(u_h, v_h) = \sum_{El \in \mathcal{T}_h} a_h^{El}(u_h, v_h) \tag{17}$$

$$= \sum_{El \in \mathcal{T}_h} \int_{El} \boldsymbol{\varepsilon}(u_h) : \mathcal{C} : \boldsymbol{\varepsilon}(v_h)\, dx\,. \tag{18}$$

As the base functions in the local spaces are not explicitly known inside an element the introduction of a projection operator is required.

Projection operator The local projector operator $\Pi_k^{El} : \boldsymbol{\mathcal{V}}_{k,h}^{El} \rightarrow [\mathcal{P}_k(El)]^2$ acting on a function $v_h \in \boldsymbol{\mathcal{V}}_{k,h}^{El}$ is defined by

$$\underline{a}_h^{El}(\Pi_k^{El}(v_h), p) = \underline{a}_h^{El}(v_h, p) \quad \forall p \in [\mathcal{P}_k(El)]^2\,. \tag{19}$$

In this work it is assumed that the coefficients in $\mathcal{C}$ are constant within each element. Although the base functions are not known in the interior of the elements, the projector can be exactly computed for functions in the local space in terms of the DOFs using integration by parts. The definition of the projection operator guarantees exact results when tested against polynomials of degree up to k.

Stiffness matrix The local bilinear form needs to be decomposed into a consistency and a stability term

$$a_h^{El}(u_h, v_h) = \underbrace{a_h^{El}(\Pi_k^{E}(u_h), \Pi_k^{El}(v_h))}_{consistency} + \underbrace{s^{El}(u_h - \Pi_k^{El}(u_h)), v_h - \Pi_k^{El}(v_h))}_{stabilization}\,. \tag{20}$$

The consistency term approximates the bilinear form using the projection operator, while the stabilization term is applied to the higher order terms ($> k$) whose contribution is not accounted for by the projection. The latter is taken simply as the scalar product of the values at the DOFs of the difference between the VEM function and its projection,

$$s^{El}(u_h - \Pi_k^{El}(u_h), v_h - \Pi_k^{El}(v_h)) = \tau \sum_{l=1}^{2n_{k,D}^{El}} \mathrm{dof}_l(u_h - \Pi_k^{El}(u_h))\, \mathrm{dof}_l(v_h - \Pi_k^{El}(v_h)), \tag{21}$$

where dof_l is the value at the l-th DOF, and τ is a material parameter which for linear elasticity is constant and depends on Young's modulus and Poisson's ratio (see Artioli et al. (2017)).

By defining a base for the local space $\{\varphi_i\}_{i=1,\ldots,2n_{k,D}^{El}}$, where each function takes the value 1 at its associated DOF and 0 otherwise, the stiffness matrix is computed as

$$\left[k^{El}\right]_{ij} = a_h^{El}(\varphi_i, \varphi_j) \qquad i,j = 1, \ldots, 2n_{k,D}^{El}\,. \tag{22}$$

The assembly of the global matrices system is done as in standard FEM.

Loading terms As the base functions are known on the element boundaries, load terms are treated just as in standard FEM. Volumetric terms will not be considered in this work.

3 SUBDOMAIN DISCRETIZATION

In addition to the benefit of the VEM allowing for elements with complex polygonal geometries, the method also allows for discrete jumps in mesh size between selected subdomains of a given geometry.

This feature is particularly useful in BVP undergoing localized failure processes. Here, the transition between the active zone, requiring fine meshes, and the passive zone, where coarse meshes are sufficient, can be easily created.

The main assumptions made in this work are: (i) the continuum VEs remain elastic throughout the entire deformation history, (ii) the entire inelastic behaviour and energy dissipation is concentrated in the interfaces. For this purpose, once the specimen discretization with continuum VEs is completed, IEs are introduced only within the pre-established active zone. For the special case of an interface between active and passive zones a perfectly bonded interface is considered, *i.e.* a null displacement jump is strongly enforced using Lagrange multipliers. This is necessary since the interfaces generated by the meshing algorithm entirely surround the virtual elements.

3.1 *Zero-thickness interface elements*

Interface elements are added on edges of the VE mesh. Since the functions in the local VEM spaces are polynomials on the boundaries of the element, the insertion of the interface elements is the same as with standard FEM. These interfaces must form a closed path on the mesh, or else a physically inconsistent non-propagating crack would occur.

In this work the interfaces surround any solid element but only in the active zone, and therefore, a closed loop is always ensured. This is particularly beneficial for remeshing as the process is performed locally for each element. However, the effectiveness and accuracy of this procedure strongly depends on the discretization to avoid introducing constraints on the crack evolution during the deformation history of the concrete component.

4 NUMERICAL EXAMPLES

4.1 *Three point concrete beam problem*

In this work the proposed procedure is used to model a notched concrete specimen tested under a three point bending problem set-up.

The geometry and boundary conditions of the problem are shown in Figure 3, where the values are in [mm]. The specimen comprises a vertical notch of 30mm height and 4mm in width. A prescribed vertical displacement is applied at the mid point of the top edge. All numerical simulations of this problem assume plane stress conditions and load results are reported as per unit of width (in [mm]). Material parameters are shown in Table 1.

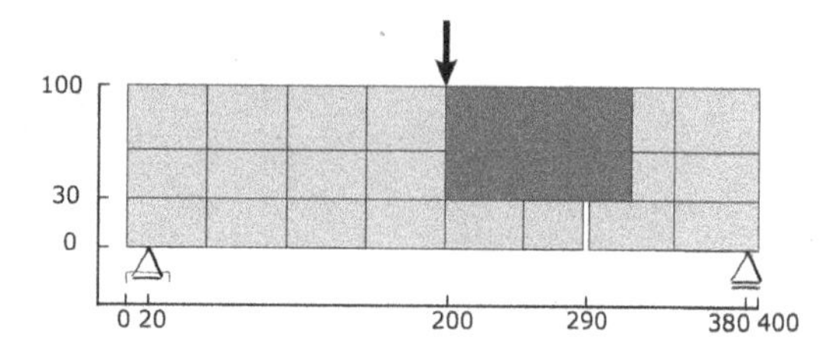

Figure 3. Three-point beam geometry.

Table 1. Material properties.

	Continuum
Young's modulus [MPa]	22750
Poisson's ratio	0.19
	Interface
Critical stress σ_c [MPa]	1.3
Fracture energy G_c [N/mm]	0.07

The method proves very stable during crack initiation and propagation, and is able to reach the post-peak behaviour without the need of any special solution algorithms other than classical Newton iterations to reduce residual forces and achieve equilibrium on every load step.

4.1.1 *Discretization strategy*

Based on prior knowledge of the problem an active zone was pre-established. This region is where the inelastic behaviour and cracking process localizes, and is shown in Figure 3 in dark grey. In this zone finer meshes with interface elements along all element edges are used to capture the fracture process. Details of the refinement are given in the next sections. A coarser mesh of elastic elements is used for the rest of the domain, shown in light grey.

The influence of the discretization of the elastic zone has been analyzed in Rivarola et al. (2020). Several different mesh refinements were tested to evaluate their influence on the global response behaviour and failure prediction. It was found that the crack path did not change and the peak load and residual strength were very similar for all meshes. The discretization of the elastic zone mainly modifies the elastic stiffness, which is overestimated in coarser meshes. This indicates that discretization of the active zone is of significantly greater importance than that of the elastic zone. For this reason a coarse discretization was used for the elastic zone and all meshes considered have the same discretization in this region, as depicted in Figure 3.

4.2 *Structured mesh refinement*

In this section we evaluate the failure predictions generated using structured rectangular VEs and IEs in the active zone. Six different meshes were considered, where the active zone was divided into 48, 96, 192, 384, 768 and 1536 elements. A diagram comparing total applied force vs. displacement of the top central node for all cases can be found in Figure 4. The different meshes resulted in almost identical force vs. displacement curves.

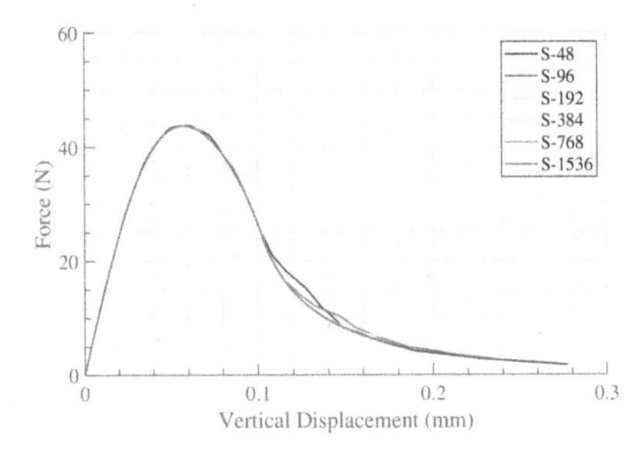

Figure 4. Force vs. displacement curve for structured meshes.

In Figure 5 the considered mesh discretizations and resulting crack paths can be seen. In all cases the crack propagates in a straight line, which explains why in Figure 4 the curves were almost identical. Thus, refining using a structured mesh does not improve the prediction of either the peak load or crack path.

4.3 *Unstructured mesh refinement*

In this section we evaluate the failure predictions generated using polygonal meshes. These polygons can be numerically generated through the so-called Voronoi/Delaunay tessellation and slightly perturbing a structured set of seed points before the tessellation procedure.

Six different mesh refinement levels were considered, with the active zone divided into 50, 100, 200, 400, 800 and 1600 polygonal elements. For each refinement level an ensemble of 20 different randomized mesh perturbations was considered.

Figure 6 shows a sample mesh for each refinement level with the crack path indicated in red. The final deformed configuration, magnified 200 times, for the case with 1600 elements in the active zone is shown in Figure 7. Here it is clear that the failure is driven by both mode I and mode II fracture. showing clearly the failure mode. In all cases the failure process starts at the top of the notch and propagates upwards and slightly towards the applied load. These crack paths show good agreement with experimental results, as can be seen in Carpinteri & Brighenti (2010).

Figure 8 shows the force-displacement curves for the 20 different cases at each refinement level. Additionally, the mean peak load and mean displacement are reported along with their maximum absolute deviations. A summary of the results is presented in Table 2, where the mean load and displacement values at the peak load are detailed, as well as the standard deviation (σ) for each case. Additionally, post-peak energy dissipation up to a displacement of 0.2mm was calculated from the area under the force-displacement curve between the peak value and the specified displacement.

It is clear that as the mesh is refined, the predictions of the peak load and ductility at peak load improve. A relatively wide spread can be found in the results of the coarser meshes, which becomes progressively narrower at higher mesh refinement levels. The deviation decreases by approximately 25% for each refinement level, which doubles the amount of elements in the active zone.

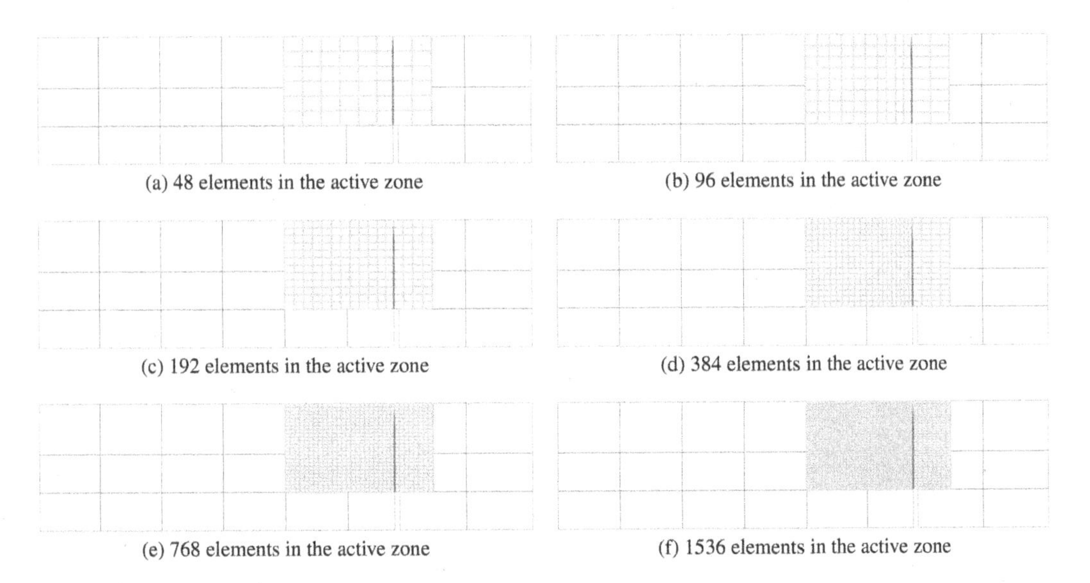

Figure 5. Mesh and crack paths for the structured meshes.

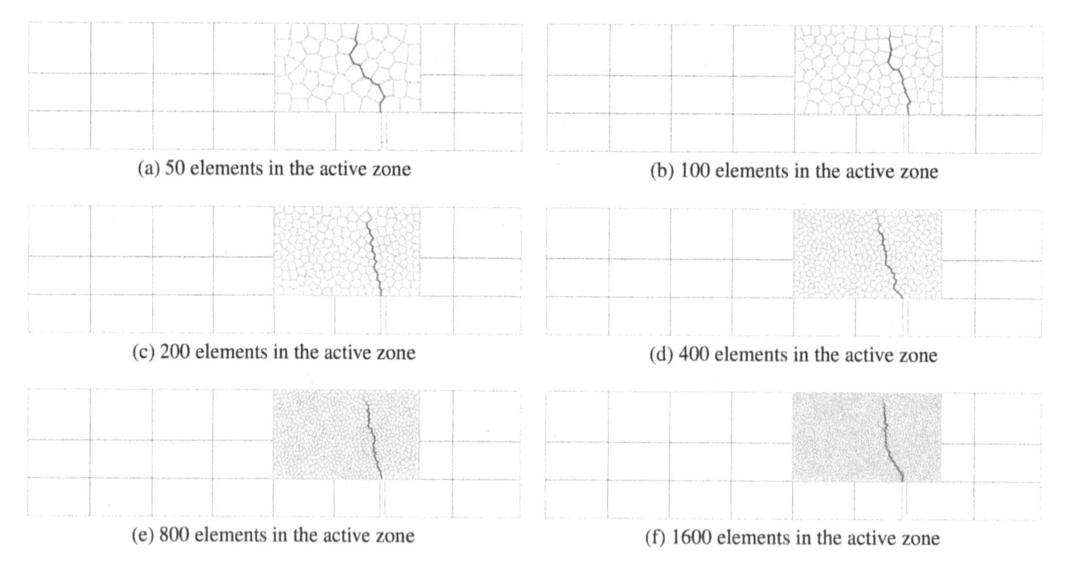

(a) 50 elements in the active zone

(b) 100 elements in the active zone

(c) 200 elements in the active zone

(d) 400 elements in the active zone

(e) 800 elements in the active zone

(f) 1600 elements in the active zone

Figure 6. Mesh and crack paths for one case of each refinement level for unstructured meshes.

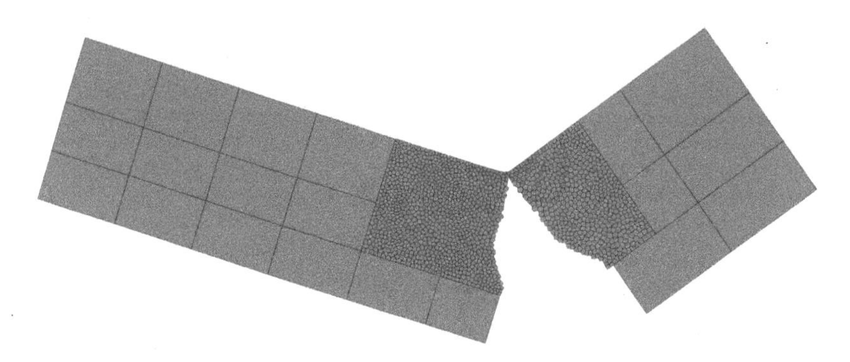

Figure 7. Final deformed configuration (magnified x200) for one case with 1600 elements in the active zone.

Additionally, it is note that a reduction in peak load occurs with increasing mesh refinement, although the difference is small. This decrease in peak load can be attributed to the fact that a finer discretization of the crack zone introduces more favorable crack paths as more elements/interfaces are added. Interestingly, the mean results generated using meshes of 100 elements in the active zone could be considered sufficiently accurate as they differ by less than 2% from those generated using 1600 elements.

The crack paths that resulted from the ensemble of 20 meshes at each refinement level are shown in Figure 9 in blue, and the mean crack path for each case in red. Once again, a wider spread of results is observed in the case of coarser meshes. However, the spread of crack paths becomes progressively narrower as the mesh is refined. The average crack path for all refinement levels resulted in an almost straight line with a similar inclination angle, given in Table 4.3. This shows, once again, that the mean results generated using an ensemble of random coarse meshes could prove to be sufficiently accurate. Running several coarse meshes and calculating the average can not only be computationally more efficient than running just one very fine mesh but could also be more precise.

4.4 *Future work – adaptative refinement*

The presented results have shown that unstructured refinement of the active zone improved the accuracy of the results and led to better crack path precision when compared to experimental results. It can, however, be computationally expensive to use very fine meshes in non-linear problems, as is the case of failure processes in concrete components. This is particularly important when using interface elements, as the amount of degrees of freedom more than doubles in the subdomains where they are used. The non-trivial question that arises is the identification of the active zone. If this zone is not known in advance IEs need to be added everywhere in the domain.

An adaptive refinement procedure could be used to a priori identify the active zones of localized stresses/strains where failure processes are likely to take place. The zone where IEs are added can then be

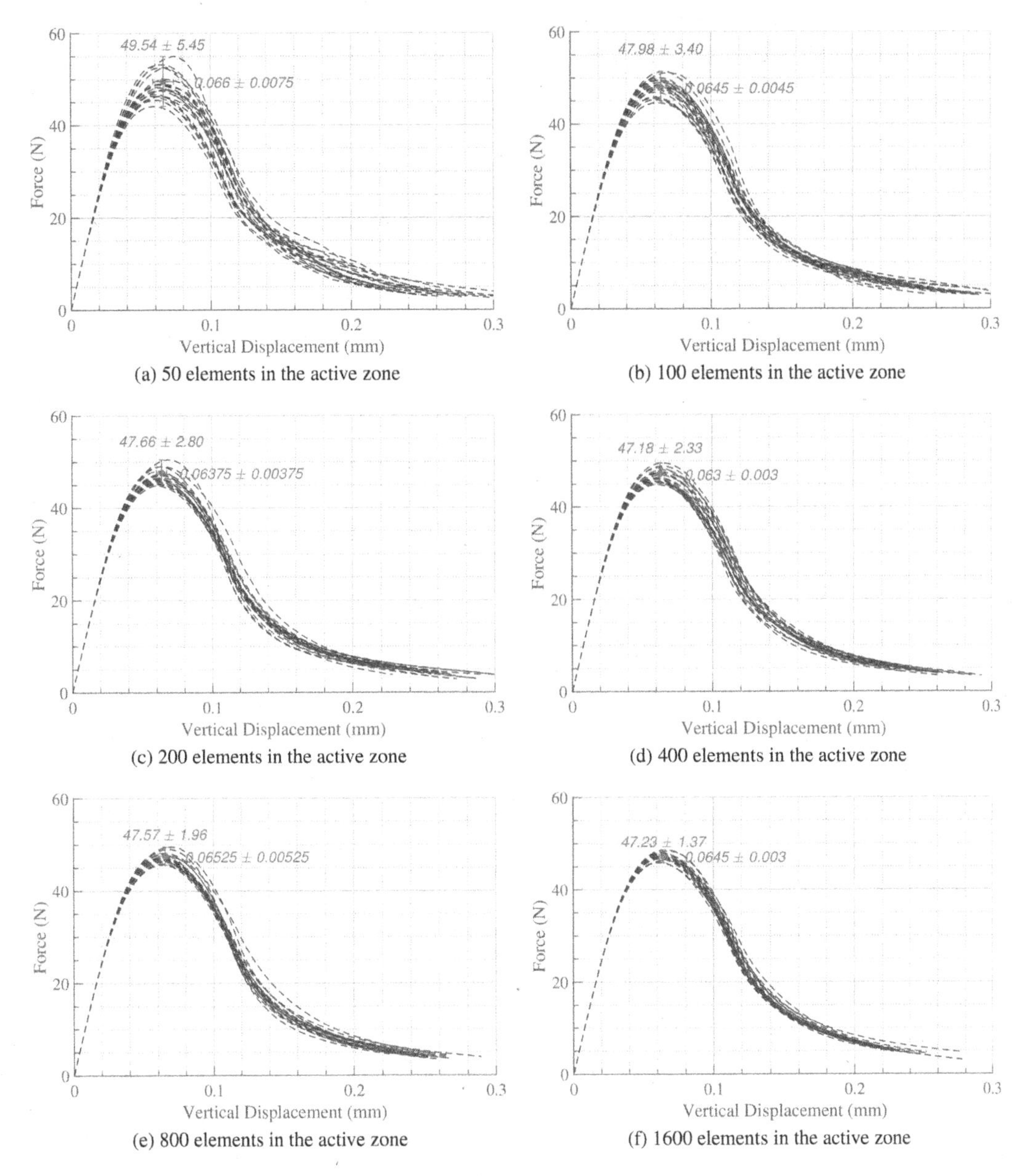

Figure 8. Force vs. displacement curve for all cases of each refinement level.

Table 2. Summary of results from unstructured meshes.

Elements	Mean Load [N]	σ Load [N]	Mean Disp. [mm]	σ Disp. [mm]	Mean Energy [Nmm]	σ Energy [Nmm]	Mean crack angle [°]
50	48.57	2.858	0.06345	0.003771	3.368	0.2129	70.31
100	48.13	1.845	0.06292	0.002642	3.308	0.1221	65.24
200	46.73	1.478	0.06247	0.001773	3.187	0.1406	70.57
400	46.60	1.335	0.06217	0.001975	3.232	0.09642	70.12
800	47.22	0.9682	0.06345	0.002580	3.254	0.1411	69.69
1600	47.19	0.7372	0.06322	0.001705	3.261	0.07644	69.39

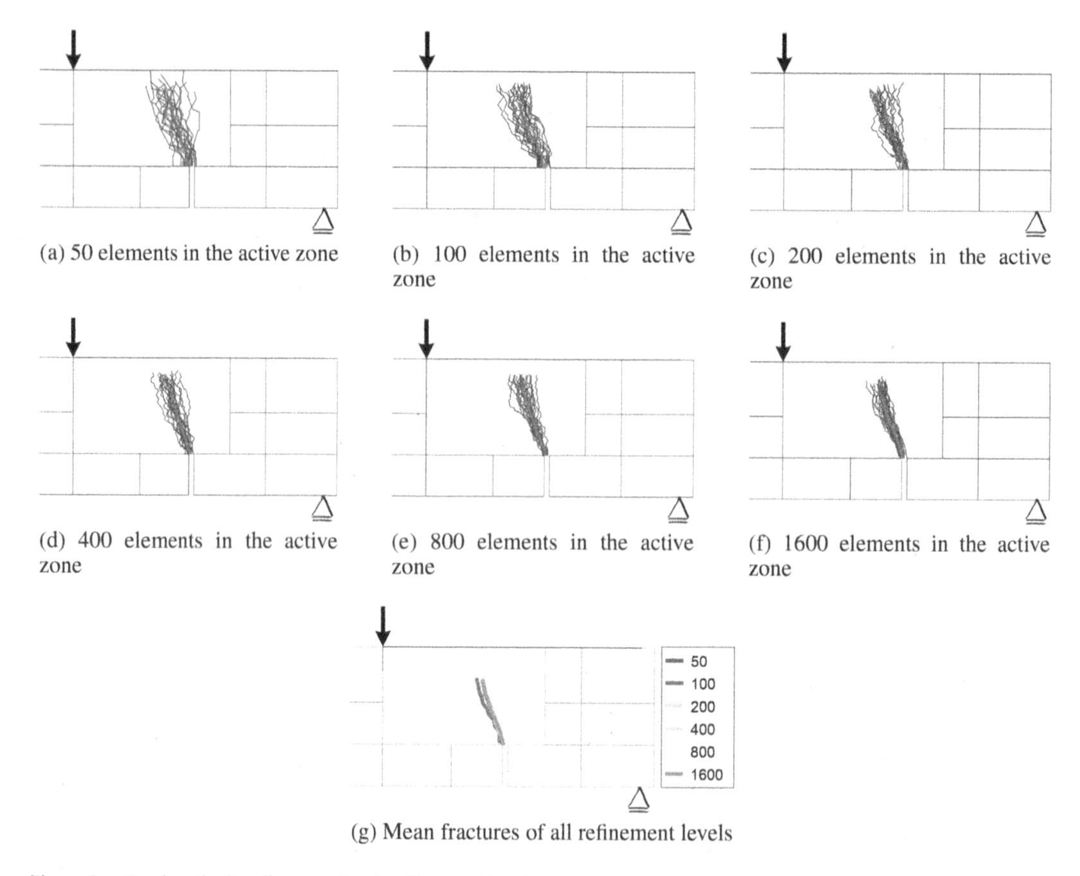

(a) 50 elements in the active zone (b) 100 elements in the active zone (c) 200 elements in the active zone

(d) 400 elements in the active zone (e) 800 elements in the active zone (f) 1600 elements in the active zone

(g) Mean fractures of all refinement levels

Figure 9. Crack paths for all cases of each refinement level.

restricted to where the mesh most refined. This could lead to a computationally efficient procedure, as an adaptive procedure usually takes much less run time than using IEs in the entire domain. In the context of adaptive remeshing, the VEM provides significant advantages over the FEM as additional nodes may be inserted arbitrarily along element edges with no consideration or treatment of hanging nodes required.

In van Huyssteen et al. (2022) a variety of novel approaches for the computation of isotropic and anisotropic mesh refinement indicators suited for the VEM were presented and comparatively assessed through a range of numerical examples for the case of two-dimensional linear elasticity. In particular, the refinement technique based on displacement and strain indicators demonstrated the best performance in terms of efficacy and efficiency.

Figure 10 depicts the resulting refined mesh obtained with the refinement procedure of van Huyssteen et al. (2022) for the three points beam problem considered in this work. The initial discretization comprised a coarse mesh of structured rectangular elements.

The refined mesh shows a greater refinement in the subdomains close to the top of the notch and the supports. Figure 11 shows the elastic von Mises stresses obtained with the refined mesh, in a logarithmic scale. The compressive arch that characterizes the beam behavior can be clearly recognized as well as the localization of stresses above the notch, indicating a possible location of fracture initialization.

In the framework of this research program the refinement procedure will be extended to polyhedral VEs. This could be an efficient procedure for identifying in advance the zones of localized failure processes requiring the inclusion of IEs between the elastic VEs.

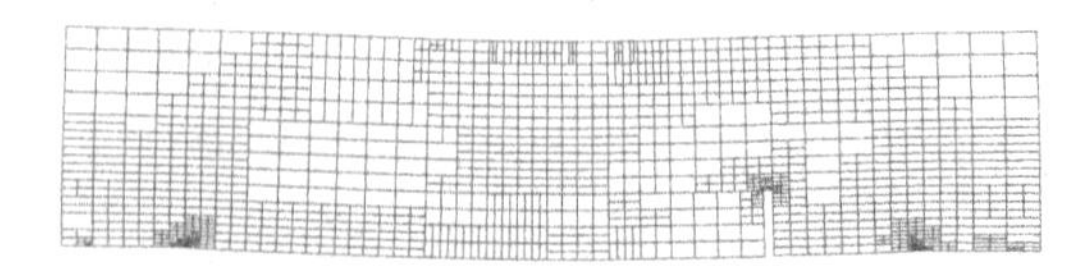

Figure 10. Adaptive mesh refinement of the three-point beam.

Figure 11. Stress field for the adaptive mesh refinement.

5 CONCLUSIONS

In this work, the efficiency of structured and unstructured discretizations of virtual elements and non-linear interface elements was assessed for modelling localized failure processes of BVPs representing concrete components subject to tortuous crack propagations. In addition, the efficiency of random mesh refinement for meshes comprising rectangular and polyhedral VEs was analyzed.

Refinement of structured meshes did not improve the prediction of either the peak load or crack path. However, refining unstructured polygonal meshes did improved the results. The obtained results demonstrate the effectiveness of ensembles of random polyhedral VE meshes for modelling the tortuous propagation of cracks in active concrete zones subject to mode II type of fracture. The crack paths also show good agreement with the experimental results. Additionally, it was found that the average values of peak load, ductility, and crack path, generated from a series of perturbations of coarse polygonal meshes, provide similar predictions to those obtained with fine meshes. Running several coarse meshes and calculating the average could be computationally more efficient and give more precise results than running just one very fine mesh. Thus, the discrete crack approach based on the use of polyhedral VEs and non-linear interfaces in the active failure zone is efficient and effective for the prediction of the failure behavior of quasi-brittle materials such as concrete.

Finally, the results also indicate that the extension of the adaptive refinement method proposed in van Huyssteen et al. (2022) to elements could provide an effective method for the a priori determination of the active zones in which failure processes occur.

REFERENCES

Artioli, E., L. Beirão da Veiga, C. Lovadina, & E. Sacco (2017, Sep). Arbitrary order 2D virtual elements for polygonal meshes: part I, elastic problem. *Computational Mechanics 60*(3), 355–377.

Beirão da Veiga, L., F. Brezzi, A. Cangiani, G. Manzini, L. D. Marini, & A. Russo (2013). Basic principles of Virtual Element methods. *Math. Models and Methods in Applied Sciences*.

Beirão da Veiga, L., F. Brezzi, L. D. Marini, & A. Russo (2016). Virtual element method for general second-order elliptic problems on polygonal meshes. *Mathematical Models and Methods in Applied Sciences 26*(04), 729–750.

Carpinteri, A. & R. Brighenti (2010). Fracture behaviour of plain and fiber-reinforced concrete with different water content under mixed mode loading. *Materials and Design 31*(4), 2032 – 2042. Design of Nanomaterials and Nanostructures.

Doyen, D., A. Ern, & S. Piperno (2010). A three-field augmented lagrangian formulation of unilateral contact problems with cohesive forces. *ESAIM: M2AN 44*(2), 323–346.

Fortin, M. & R. Glowinski (1983). *Augmented lagrangian methods: application to the numerical solution of boundary-value problems*. North-Holland – Studies in mathematics and its applications.

Labanda, N. A., S. M. Giusti, & B. M. Luccioni (2018a). Meso-scale fracture simulation using an augmented lagrangian approach. *International Journal of Damage Mechanics 27*(1), 138–175.

Labanda, N. A., S. M. Giusti, & B. M. Luccioni (2018b). A path-following technique implemented in a lagrangian formulation to model quasi-brittle fracture. *Engineering Fracture Mechanics 194*, 319 – 336.

Lorentz, E. (2008). A mixed interface finite element for cohesive zone models. *198*, 302–317.

Rivarola, F. L., N. Labanda, M. F. Benedetto, & G. Etse (2020). A virtual element and interface based concurrent multiscale method for failure analysis of quasi brittle heterogeneous composites. *Computers and Structures 239*(106338).

van Huyssteen, D., F. L. Rivarola, G. Etse, & P. Steinmann (2022). On anisotropic mesh refinement procedures for the virtual element method for two-dimensional elastic problems. *Under review*.

Computational Modelling of Concrete and Concrete Structures – Meschke, Pichler & Rots (Eds)
 ISBN: 978-1-032-32724-2

Examination of advanced isotropic constitutive laws under complex stress states in plain and reinforced concrete specimens

J. Bobiński, P. Chodkowski & J. Schönnagel
Faculty of Civil and Environmental Engineering, Gdańsk University of Technology, Gdańsk, Poland

ABSTRACT: The performance of advanced isotropic constitutive laws under complex stress states in plain and reinforced concrete specimens is investigated. Three different formulations are chosen: original Mazars model, Mazars μ model and model proposed by Pereira and coworkers. The degradation of the material in all formulations is described via a single variable, but a strain/stress state is taken into account via quite sophisticated relationships. In order to better reproduce experimentally observed stress-strain curves in uniaxial tension and compression, some modifications and extensions are proposed. An integral non-local approach is used to ensure FE mesh insensitive results. Two benchmarks are simulated. Nooru-Mohamed test is chosen to analyse numerically the growth of cracks in plain concrete under mixed-mode stress conditions. A geometrically scaled set of longitudinally reinforced beams under four-point bending load serves to assess the ability of reproducing different failure mechanisms. All results obtained from calculations are compared with experimental outcomes.

1 INTRODUCTION

Cracks in quasi-brittle materials like concrete are the primary source of their complex behaviour observed as nonlinear stress-strain relationship with hardening and softening phases, material orthogonality and different strengths under different loading conditions. The proper capture of mentioned phenomena in the formulation of a constitutive law for concrete is a highly demanding task. Usually some simplifying assumptions are made and only selected subset of experimentally observed properties is included in the model. In numerical simulations of concrete and reinforced concrete (RC) members at the macro-scale there is a huge number of alternative formulations defined within elasto-plasticity, continuum damage mechanics (CDM) and also based on coupling two aforementioned approaches. Constitutive laws defined within CDM are the most popular among them, especially their isotropic versions. They are widely used not only to simulate cracks in a smeared sense but also to describe the behaviour of discrete cracks in connection with interface elements. The proper choice of basic ingredients (like equivalent strain definitions, evolution laws) and the adequate combination of terms describing the behaviour under simple loading cases (e.g. uniaxial tension, uniaxial compression) defines the group of problems, to which a selected constitutive law can be applied properly.

The simplest formulation of the constitutive law defined within continuum damage mechanics requires two variables to be defined: an equivalent strain measure and the damage (degradation) evolution law. Usually one equivalent strain measure and one evolution law is declared explicitly defined from the known strain state. The influence of the stress (strain) state is taken into account in the definition of the equivalent strain only. There are several examples using this approach. The most popular definitions of the equivalent strain to describe concrete were given by Mazars and Pijaudier-Cabot (1989), de Vree et al. (1995), Jirásek (2004) or Haüsler-Combe & Pröchtel (2005). They are used with the evolution laws adopting linear, exponential or polynomial relationships.

More advanced constitutive laws still defined as isotropic damage models introduce e.g. two independent state variables or/and two independent damage evolution curves to describe selected degradation processes independently, e.g. in tension and compression. Such split enables also to simulate stiffness recovery. The resultant damage variable may be averaged by using appropriate weights, which take into account the stress/strain state in a material point. Alternative formulations do not follow the explicit dependence of the strains; degradation of a material is obtained implicitly based e.g. on the formulation analogous to elasto-plasticity (Comi & Perego 2001; Qi et al. 2020).

Proposed formulations are then confronted with selected experimental outcomes or with other constitutive laws. Plain concrete specimens subjected to uniaxial tension tests or/and beams under bending are usually simulated first. More sophisticated benchmarks create more complex stress state in material (mixed mode failure mode). The replicability of the observed failure mechanism and crack patterns in RC

 DOI 10.1201/9781003316404-57

elements is also examined. The most popular specimens are beams with different reinforcement schemes, but other elements can be also tested, e.g. columns, corbels.

In the paper the performance of three different isotropic damage constitutive laws under complex stress state is examined. All models share several common features and the origin from the same root, a classical formulation by Mazars (1986). His model evolved later into so called μ model (Mazars et al. 2014, 2015). On that basis Pereira et al. (2017, 2018) proposed another constitutive law.

The intention of the paper is to examine the behaviour of all three above constitutive laws under complex stress state. Two test simulations are be performed. The first benchmark is to verify the ability to simulate the crack growth in plane concrete specimens in dominated tensile loading. In the second benchmark the behaviour of a reinforced concrete beam under bending is reproduced. It should be emphasized that not only tension-like, but also failure mechanism governed by compression will be simulated.

In order to simulate the concrete behaviour in plane specimens classical Nooru-Mohamed (1992) test is applied. It has been simulated by many researchers and some of them used isotropic and anisotropic damage models. Grassl & Jirásek (2004) used anisotropic micro-plane model equipped with the external algorithm for crack propagation tracking. Patzák & Jirásek (2004) applied micro-plane based anisotropic damage model regularized by a non-local integral theory with adaptive remeshing technique. Later Jirásek & Grassl (2008) adopted orthogonal fixed-crack approach with crack band regularisation. Similar idea was employed by Cervera & Chiumenti (2006). Based on Mazar's definition of the equivalent strain Desmorat et al. (2007) created anisotropic damage model enhanced by integral non-local theory with characteristic length. Recently the performance of different isotropic formulations was investigated by Bobiński & Tejchman (2016).

As a second benchmark experimental results from Suchorzewski et al. (2018) are chosen. In the test a reinforced concrete beam with different dimensions under four-point bending was examined and different failure mechanisms were observed. Such topic has been investigated experimentally by many authors. Kim & Park (1994) investigated the behaviour and ultimate shear capacity of strongly reinforced, high strength concrete beams without stirrups. Belgin & Şener (2008) analysed the size effect phenomenon in beams under four-point bending. Tan et al. (2005) investigated the effects of shear span-to-depth ratio l/d and effective depth d on shear strength and on the behaviour of large beams. Ghahremannejad & Abolmaali (2018) used concrete damaged plasticity model from Abaqus to calculate the ultimate shear strength of RC beams. Marzec et al. (2019), Marzec & Bobiński (2020), and Marzec & Tejchman (2021) carried out a series of finite element analysis of RC beams using different continuum constitutive laws. Sanabria Diaz et al. (2020) combined nonlinear finite element analysis with reliability theory approaches for advanced safety assessment of deep beam design. In numerical calculations concrete fracture and plasticity theory with crack band approach was employed.

2 CONSTITUTIVE LAWS

2.1 *General relationship*

All constitutive laws analysed here fall within isotropic continuum damage mechanics (CDM) where the degradation of the material is described via a single scalar variable D. The following general constitutive relationship holds:

$$\boldsymbol{\sigma} = (1 - D)\, \boldsymbol{C}^e \boldsymbol{\varepsilon} \tag{1}$$

where $\boldsymbol{\sigma}$ = stress vector; $\boldsymbol{C}^e$ = elasticity matrix; and $\boldsymbol{\varepsilon}$ = strain vector. The variable D changes from 0 (undamaged state) to 1 (fully damaged state), so it can be interpreted as a stiffness reduction factor. The evolution of the variable D is governed by a state variable κ defined as:

$$\kappa\,(t) = \max_{\tau \leq t} \tilde{\varepsilon}\,(\tau) \tag{2}$$

where so called equivalent strain measure $\tilde{\varepsilon}$ transforms strain vector into a scalar value. There are several different proposals presented in literature how to define the equivalent strain measure and how to calculate the degradation variable D.

2.2 *Original Mazars model*

The first constitutive law chosen for comparison was formulated by Mazars (1986). The equivalent strain measure $\tilde{\varepsilon}$ is defined here as:

$$\tilde{\varepsilon} = \sqrt{\sum \langle \varepsilon_i \rangle^2} \tag{3}$$

where ε_i = i-th principal value of the strain vector and a symbol $\langle \blacksquare \rangle$ stands for a Macaulay bracket (it returns given number for positive values and zero for negative values). Degradation variable D is defined via the following formulae:

$$D = \alpha_t D_t + \alpha_c D_c \tag{4}$$

Damage variable D_t describes the degradation in the tension as an exponential softening with the parameters A_t and B_t:

$$D_t = 1 - \frac{\kappa_0}{\kappa}(1 - A_t) - A_t \exp(-B_t(\kappa - \kappa_0)) \tag{5}$$

while the damage variable D_c reflects the softening in compression (again using an exponential curve with the parameters A_c and B_c):

$$D_c = 1 - \frac{\kappa_0}{\kappa}(1 - A_c) - A_c \exp(-B_c(\kappa - \kappa_0)) \tag{6}$$

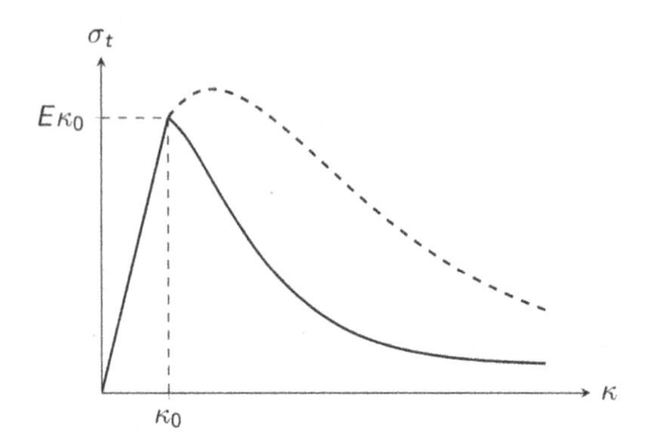

Figure 1. Exponential softening curve by Equation (5).

In Equations (5) and (6) κ_0 is a threshold value of the κ state variable when the damage starts. Note that the same value of κ_0 is applied to both damage mechanisms, i.e. in tension and compression. Figure 1 presents schematically the shape of this curve (a hardening – dashed line – is obtained if $B_t < 1/\kappa_0$).

Coefficients α_t and α_c reflect the stress state (formally strain state) in a point under consideration and they are defined as:

$$\alpha_t = \left(\sum_{i=1}^{3} \frac{\varepsilon_{ti}\varepsilon_i}{\tilde{\varepsilon}^2}\right)^{\beta}, \alpha_c = \left(1 - \sum_{i=1}^{3} \frac{\varepsilon_{ti}\varepsilon_i}{\tilde{\varepsilon}^2}\right)^{\beta} \tag{7}$$

where $\varepsilon_{ti} = i$-th principal strain of positive strains and $\beta =$ coefficient that slows the degradation process under shear. In the original paper $\beta = 1.0$ was assumed, but later a slightly larger value was taken as $\beta = 1.06$.

2.3 *Mazars μ model*

The second constitutive law analysed in this paper was proposed again by Mazars and his coworkers (Mazars et al. 2014, 2015). Two equivalent strain measures are defined here to describe the behaviour in tension:

$$\tilde{\varepsilon}_t = \frac{1}{2}\frac{I_1}{1-2\nu} + \frac{1}{2}\frac{\sqrt{3J_2}}{1+\nu} \tag{8}$$

and in compression:

$$\tilde{\varepsilon}_c = \frac{1}{5}\frac{I_1}{1-2\nu} + \frac{6}{5}\frac{\sqrt{3J_2}}{1+\nu} \tag{9}$$

In two above equations I_1 is the first invariant of the strain tensor, while J_2 stands for the second invariant of the deviatoric strain tensor. Similarly, two independent state variables are also introduced; in tension:

$$\kappa_t = \max(\kappa_{t0}, \max(\tilde{\varepsilon}_t)) \tag{10}$$

and in compression:

$$\kappa_c = \max(\kappa_{c0}, \max(\tilde{\varepsilon}_c)) \tag{11}$$

where $\kappa_{t0} =$ initial threshold value of κ_t (in tension) and $\kappa_{c0} =$ initial threshold value of κ_c (in compression). The weighted state variable κ is used to calculate the value of the degradation variable D:

$$\kappa = r\kappa_t + (1-r)\kappa_c \tag{12}$$

with the weighted initial threshold value κ_0:

$$\kappa_0 = r\kappa_{t0} + (1-r)\kappa_{c0} \tag{13}$$

The weight factor r in Equations (12) and (13) reflects the stress state and it is defined as (Lee & Fenves 1998):

$$r = \frac{\sum_i \bar{\sigma}_i}{|\bar{\sigma}_i|} \tag{14}$$

where $=i$-th principal effective stress (i.e. calculated with Eqn. (1) and D=0). The weight factor r changes from 0 (only negative i.e. compressive principal stresses) to 1 (only positive, i.e. tensile principal stresses).

The evolution of the degradation variable D is governed (as in original model by Mazars) by the exponential law:

$$D = 1 - \frac{\kappa_0}{\kappa}(1-A) - A\exp(-B(\kappa - \kappa_0)) \tag{15}$$

with the parameters A and B declared as:

$$A = A_t\left[2r^2(1-2k) - r(1-4k)\right] + A_c[A_r] \tag{16}$$

$$B = B_t r^{(r^2-2r+2)} + B_c\left[1 - r^{(r^2-2r+2)}\right] \tag{17}$$

and auxiliary variable A_r:

$$A_r = 2r^2 - 3r + 1 \tag{18}$$

A parameter k in Equation (16) is used to tune the behaviour of the model in pure shear. By default the value $k = 0.7$ is applied (Mazars & Grange 2015).

2.4 *Pereira model*

The third constitutive law (called here 'Pereira model') was formulated by Pereira and coworkers (Pereira et al. 2017, 2018). It is based on the Mazars μ model described in Section 2.3 with some modifications. As in the Mazars μ model two equivalent strain measures are defined using Equations (8) and (9). Two state variables are also defined, but using alternate formulas:

$$\kappa_t = \max(\kappa_{t0}, \kappa_t(\tau), r^{\alpha}\max(\tilde{\varepsilon}_t)) \tag{19}$$

$$\kappa_c = \max(\kappa_{c0}, \kappa_c(\tau), (1-r)^{\alpha}\max(\tilde{\varepsilon}_c)) \tag{20}$$

where α=coefficient. The calculation of the damage variable D is changed; the following relationship is assumed here:

$$D = 1 - (1-D_t)(1-D_c) \tag{21}$$

with damage variables D_t and D_c calculated as in the original Mazars model using Equations (5) and (6). The only improvement in Pereira model is fact that different threshold values κ_{t0} and κ_{c0} are substituted as κ_0 in Equations (5) and (6), respectively.

2.5 *Non-local regularization*

Finite element calculations with conventional constitutive laws (i.e. with stress-strain relationships) require the special treatment of the softening phase in order to ensure the mesh sensitivity of the results. Here an integral non-local approach is used as a regularisation technique.

In original Mazars model (Section 2.2) the equivalent strain measure in Equation (2) is replaced by its non-local counterpart defined as:

$$\bar{\varepsilon}(x) = \frac{\int \alpha_0 (x-y)\, \tilde{\varepsilon}(y)\, dy}{\int \alpha_0 (x-y)\, dy} \tag{22}$$

where $\boldsymbol{x}$ = a considered point; $\boldsymbol{y}$ = a neighbour point; and α_0 = weight function declared for 2D case as:

$$\alpha_0 (r) = \frac{1}{l\sqrt{\pi}} \exp\left(-\left(\frac{r}{l}\right)^2\right) \tag{23}$$

where r = distance between two points and l = a characteristic length of a microstructure.

In Mazars μ model (Section 2.3) and Pereira model (Section 2.4) equivalent strains $\tilde{\varepsilon}_t$ and $\tilde{\varepsilon}_c$ are averaged independently using the Equation (22). Then the values $\bar{\varepsilon}_t$ and $\bar{\varepsilon}_c$ replace their local counterparts in Equations (10) and (11) in Mazars μ model and in Equations (19) and (20) in Pereira model, respectively.

2.6 *Softening curves*

All three models roughly use the same exponential relationship originally proposed by Mazars (1986). Despite its apparent attractiveness, simplicity and ability to produce (under specific circumstances) stress-strain curves in uniaxial tension and compression close to realistic outcomes (Pijaudier-Cabot et al. 1991) this relationship has two serious drawbacks. It does not allow for independent definition of the peak point (strain and stress values) and the fracture energy density (it is extremely important in compression). Moreover, it does not allow for easy scaling the softening regime (fracture energy density) in order to obtained physically sound total fracture energy when small widths of the localisation zones are assumed. The fracture energy density g_f for the 1D case and damage evolution curve defined by Equations (5), (6) or (15) can be derived as (assuming $\alpha = 1$, otherwise this energy is infinite):

$$g_f = \int_0^\infty \sigma(\varepsilon)\, d\varepsilon = \frac{1}{2} E\kappa_0^2 + \frac{E}{\beta^2}(\beta\kappa_0 + 1) \tag{24}$$

Application of the curve by Mazars with suggested parameter values (Pijaudier-Cabot et al. 1991) with small localization widths (small characteristic lengths l) results in unrealistically small values of the total fracture/compression energy, i.e. almost perfectly brittle behaviour. For instance, taking values after Pijaudier-Cabot et al. (1991): $E = 40$ GPa, $\kappa_0 = 1.1 \cdot 10^{-4}$, $\beta = 2 \cdot 10^4$ the fracture energy density is equal to $g_f = 562$ N/m^2. Pereira et al. (2017) assumed the following set of parameters: $E = 36$ GPa, $\kappa_0 = 1.06 \cdot 10^{-4}$, $\beta = 10^4$, which gives $g_f = 944$ N/m^2. Multiplying the second value by the width of the localisation zone (here it can be assumed as $w = 3.5l$) for l = 5 mm the total fracture energy is calculated as $G_F = 16.5$ N/m. This result is much lower than a realistic value for concrete (about 100 N/m).

Therefore slightly modified exponential relationship is used thorough the paper (Figure 2):

$$D = 1 - \frac{\kappa_0}{\kappa}[(1-A) - A\exp(-B(\kappa - \kappa_0))] \tag{25}$$

as a replacement (with required index adjustments) for Equations (5), (6) and (15). This relationship allows for easy scaling of the softening regime, but still does not include a nonlinear hardening phase (compression). In order to eliminate this deficiency another curve in compression has been proposed with a hardening region described by a parabola and a softening phase with exponential curve (Figure 3). In the hardening regime stress (uniaxial case) is calculated as:

$$\sigma = -\left(f_c - E\kappa_p\right)\left(\frac{\kappa - \kappa_p}{\kappa_p - \kappa_0}\right)^2 + E\kappa \tag{26}$$

while the softening part is described as:

$$\sigma = f_c\left[(1 - A_c) - A_c\exp(-B_c(\kappa - \kappa_p))\right] \tag{27}$$

Then the damage variable D can be retrieved as:

$$D = 1 - \frac{\sigma}{E\kappa} \tag{28}$$

In Equations (26) and (27) f_c is the compressive strength, κ_p is the value of the state variable κ (uniaxial

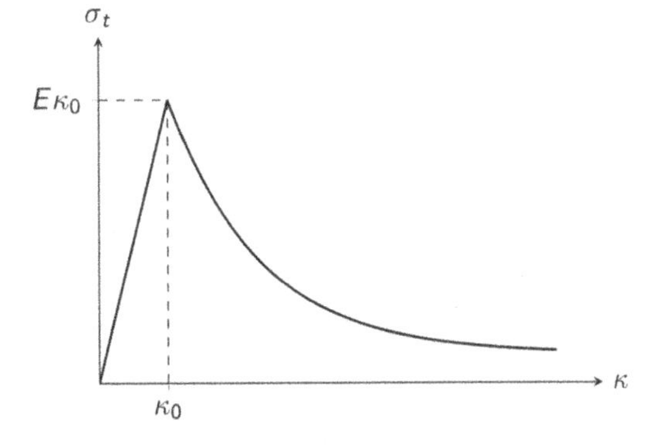

Figure 2. Exponential softening curve by Equation (25).

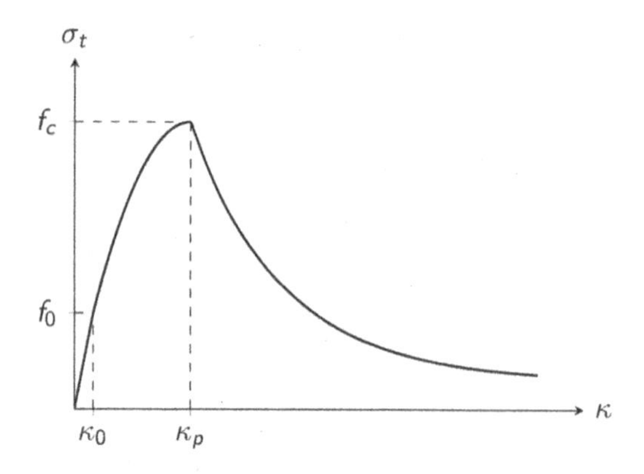

Figure 3. Softening curve by Equations (26) and (27).

compression strain) at the peak (when stress is equal to f_c), κ_0 stands here for the value of the state variable κ when a nonlinear hardening starts, and E = Young modulus. Note this curve cannot be used for Mazars μ model (Section 2.3) where only one evolution curve is defined (Equation (15)). It means a hardening phase is not defined. In original Mazars model (Section 2.2) in uniaxial compression the following relationship holds between the strain ε and the state variable κ (in plane stress):

$$\kappa = \nu\sqrt{2}|\varepsilon| \tag{29}$$

Therefore the parameters describing the stress-strain relationship in compression, namely κ_p, f_c and β_c, have to modified in the model definition in order to retrieve assumed curve based on experimental outcomes.

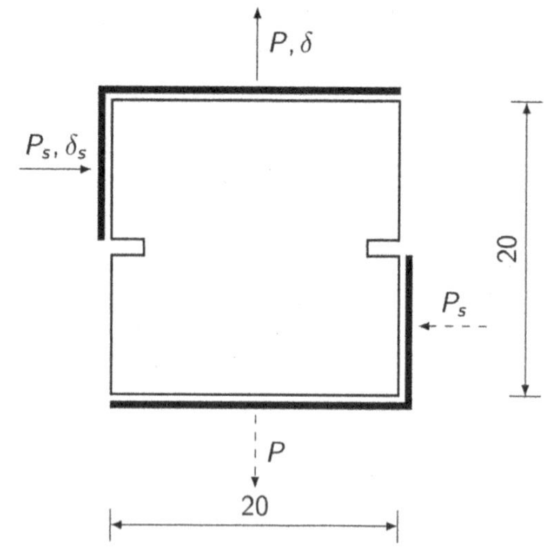

Figure 4. Nooru-Mohamed test: geometry and schematic boundary conditions (dimensions in mm).

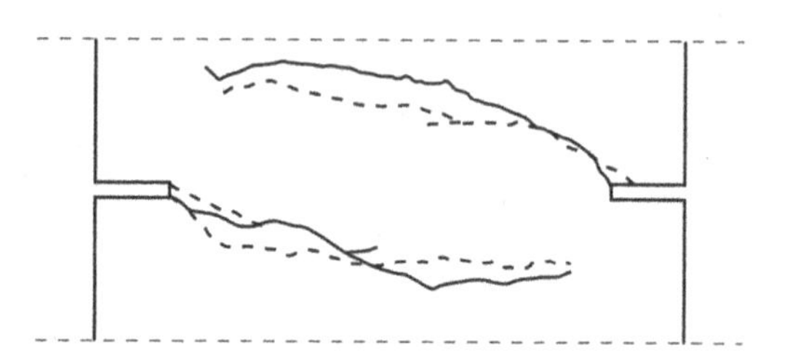

Figure 5. Nooru-Mohamed test: experimental crack pattern for $P_s = 10$ kN.

3 NOORU-MOHAMED TEST

3.1 *Experiment*

In order to examine the performance of all constitutive laws under complex stress state (but with dominating tension load) in plain concrete as a first benchmark a double edge notched (DEN) specimen under combined tension and shear was chosen. This problem was formulated and experimentally examined by Nooru-Mohamed (1992). In his PhD thesis he analysed specimens with different dimensions and different loading scenarios. In the paper the most popular configuration is numerically reproduced. The concrete specimen had a length and height of 200 mm (with thickness equal to 50 mm) and two notches 25x5 mm^2 at the middle of the vertical edges, see Figure 4. The test consists of two phases. First a horizontal shear force P_s is applied until a specific value is reached, while the vertical edge if force free. Then the force P_s is held constant and the vertical displacement δ is imposed. As a consequence two cracks are formed starting from the notches, see Figure 5. Their curvature and "distance" between them depend on the prescribed value of the shear force P_s. The obtained maximum force in vertical direction decreases with increasing the level of the shear force P_s. However, in the paper simulations with only one value of the shear force, namely P_s=10 kN, will be presented.

3.2 *FE calculations*

The following elastic parameters are taken in FE calculations: Young modulus $E = 32.8$ GPa and Poisson's ratio $\nu = 0.2$. Plane stress conditions are assumed. Three node triangle elements and four node quad elements are defined in FE mesh. The characteristic length is equal to $l = 2$ mm.

3.2.1 *Original Mazars model*

In the simulations the in tension the softening curve given by Equation (25) with parameters $\kappa_0 = 7 \cdot 10^{-5}$, $A_t = 0.98$ and $B_t = 220$. It corresponds approximately to the fracture energy $G_F = 100$ N/m in uniaxial tension. In compression parabolic-exponential relationship is assumed (Equations (26) and (27)) with the parameters $\kappa_0 = 7 \cdot 10^{-5}$, $\kappa_p = 0.622 \cdot 10^{-3}$, $f_c =$ 10.86 MPa, $A_c = 0.98$ and $B_c = 141.4$. All above parameters (except A_c) are modified to take into account the performance of the equivalent strain

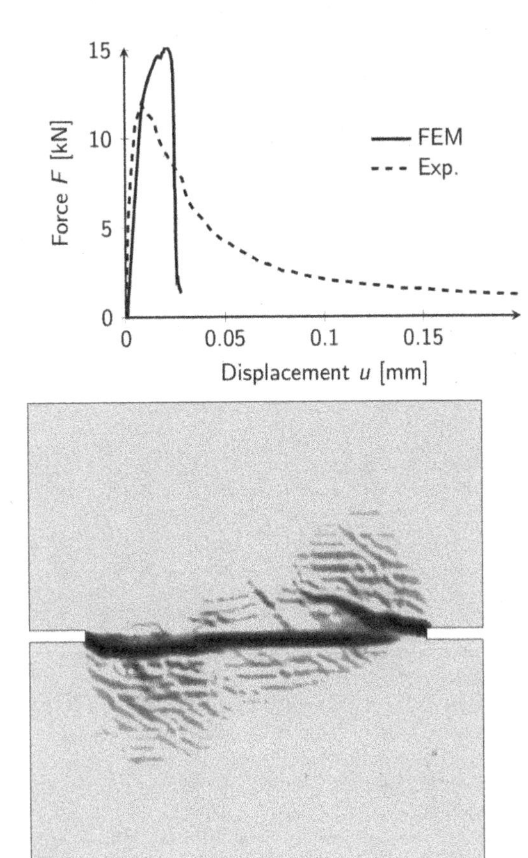

Figure 6. Nooru-Mohamed test and original Mazars model: force-displacement diagram (top) and cracks pattern (damage variable D bottom).

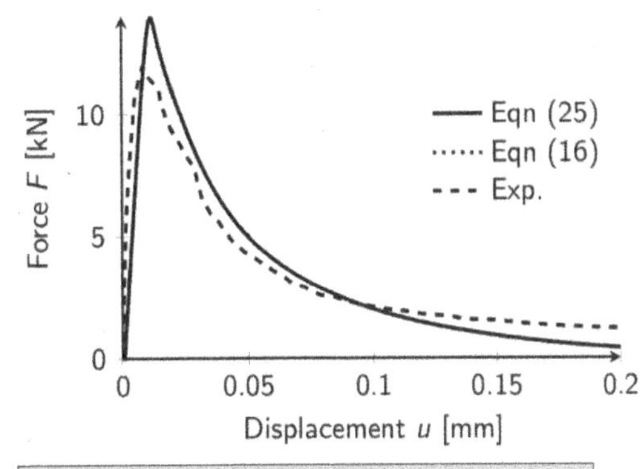

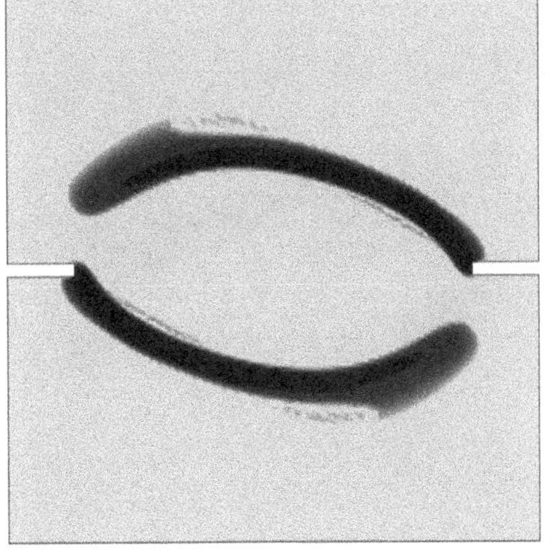

Figure 7. Nooru-Mohamed test and Mazars μ model: force-displacement diagrams (top) and cracks pattern obtained with model A (damage variable D, bottom).

measure (Equation (3)) in uniaxial compression (Equation (29)) and to define physically sound relationship in uniaxial compression with the strength 38.4 MPa and strain 0.22%. The crushing energy G_c is approximately equal to 8000 N/m. The β exponent is set to its default value 1.06. Figure 6 presents force-displacement curve and obtained cracks pattern. Sudden drop of the force is observed after the peak and only one horizontal crack is formed. Definitely original Mazars model is not capable to simulate properly Nooru-Mohamed test. This behaviour is similar to results obtained with isotropic damage constitutive laws and 'Rankine' equivalent strain (Bobiński & Tejchman 2016).

3.2.2 *Mazars μ model*

The threshold values are set as $\kappa_{t0} = 7 \cdot 10^{-5}$ and $\kappa_{c0} = 1.17 \cdot 10^{-3}$ (f_c/E) in tension and compression, respectively. Two cases are investigated. First, Equation (25) with parameters $A_t = 0.98$ and $B_t = 220$ (as in Section 3.2.1) is used (model A). Second, original evolution law (using Equations (16) and (17)) but with Equation (25) instead of Equation (15) is tested (model B). The following parameters are assumed then: $A_t = 0.98$ and $B_t = 220$, $A_c = 0.98.0$ and $B_c = 40$ and $k = 0.7$. In both cases the same fracture energy is set $G_F = 75$ N/m. Figure 7 shows obtained results. It clear that despite some minor discrepancies, obtained results are in good agreement with experiment.

3.2.3 *Pereira model*

Finally, a model by proposed Pereira et al. (2018) is verified. In tension a softening curve given by Equation (25) with parameters $\kappa_{t0} = 7 \cdot 10^{-5}$, $A_t = 0.98$ and $B_t = 220$ is set. In compression a parabolic-exponential relationship (Equations (26) and (27)) with the parameters $\kappa_{c0} = 3.5 \cdot 10^{-4}$, $\kappa_p = 2.2 \cdot 10^{-3}$, $f_c = 38.4$ MPa, $A_c = 0.98$ and $B_c = 40$ is set. It can be easily checked that this model returns the same stress-strain curve in uniaxial tension in compression as original Mazars model from Section 3.2.1 (despite different parameters in compression). Such uniqueness with the respect to Mazars μ model holds only for uniaxial tension case. Coefficient α in Equations (19) and (20) is assumed to its default value 0.1.

Figure 8 presents force-displacement diagram and cracks patterns. Results are very close to those

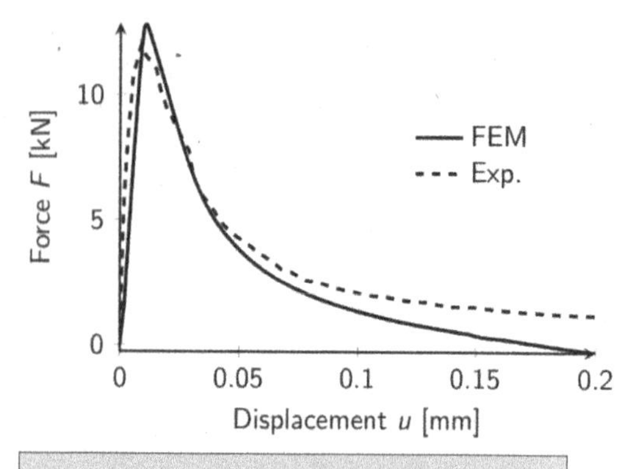

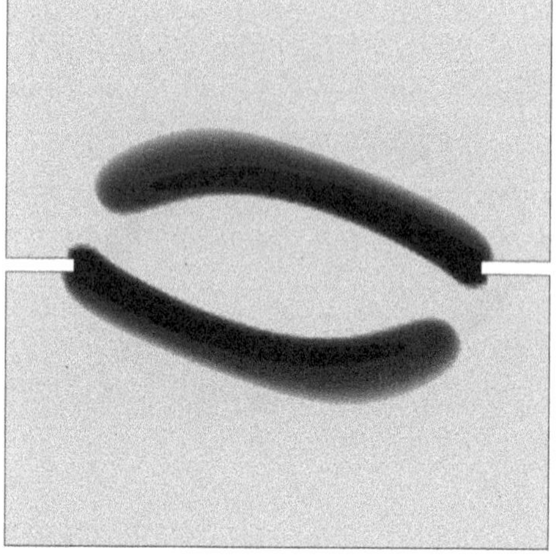

Figure 8. Nooru-Mohamed test and Pereira model: force-displacement diagram (top) and cracks pattern (damage variable D, bottom).

obtained with Mazars μ model. A good correlation with experimental outcomes is achieved. The only differences occur at the late phase of the loading (numerically too small vertical force is reproduced). Also slightly too curved curves are obtained (similarly as for isotropic damage model with modified von Mises definition as the equivalent strain measure, see Bobiński & Tejchman 2016). It is worth to mention that α coefficient a has a minimal impact on results. Simulations with $\alpha = 1.0$ give almost identical results comparing with outcomes with the basic set of parameters.

3.2.4 *Failure envelopes*

Additionally, in order to understand the obtained results, limit failure envelopes are created for local versions (the characteristic length is not taken into account) of all three models with parameters defined in previous sections. Figure 9 shows obtained failure envelopes for plane stress case. All formulations correctly reproduce tensile and compressive assumed uniaxial strengths. The original Mazars constitutive law gives unrealistic output, especially in biaxial compression. The Mazars μ model significantly overestimates

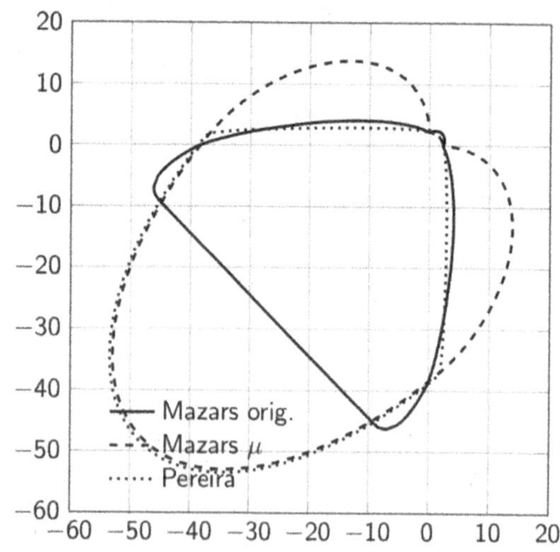

Figure 9. Nooru-Mohamed test: Failure envelopes.

the material strength in tension-compression regime. Its performance in biaxial compression is quite realistic. The best envelope is produced with Pereira model (in biaxial compression it coincides with Mazars μ curve). The detailed analysis shows, however, that in tension compression regime an increase of the tensile strength is found (from 2.3 MPa to 3.5 MPa). The similar trend was observed in FE simulations with simple isotropic laws and modified von Mises definition of the equivalent strain measure (Bobiński & Tejchman 2016). This phenomenon was not observed in the model with original damage evolution formulation (Pereira & et al. 2017). The differences between all envelopes in biaxial tension and tension-compression regimes (biaxial compression is not active in this test) does not allow to draw ultimate conclusions. Further study is required on this topic.

4 REINFORCED CONCRETE BEAM TEST

4.1 *Experiment*

As a second benchmark reinforced concrete beams under four-point bending are numerically reproduced based on experimental campaign run at Gdańsk University of Technology (Suchorzewski et al. 2018). The goal of this research was to investigate a size effect phenomenon in RC beams with independently scaled their height or length. In total four series were executed: S1 and S2 with specimens with longitudinal reinforcement only and S3 and S4 with added stirrups. As a consequence different failure mechanism were obtained.

In the paper only two selected geometries from the series S1 and S2 (without stirrups) will be analysed. The geometry of the beams is shown in Figure 10. In the first series S1 the effective depth D was taken as 18 cm, 36 cm and 72 (ratio 1:2:4) for a specimen

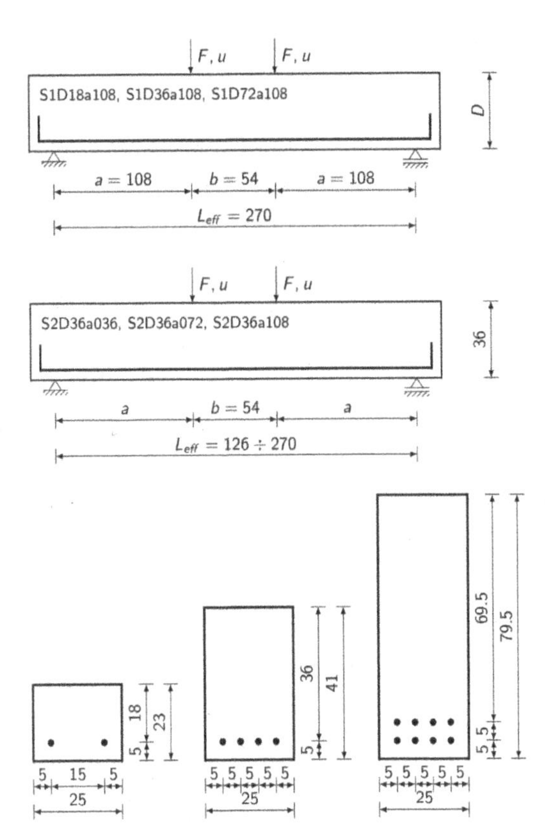

Figure 10. Geometry and boundary conditions for the series S1 (top), series S2 (middle) and cross sections (bottom).

labelled as S1D18a108, S1D36a108 and S1D72a108, respectively. In this series the span length was equal $l_{eff} = 270$ cm and the distance between the support and a loading point (shear span length) was taken as $a = 108$ cm for all geometries. The distance between two loads was fixed as $b = 54$ cm. In the second series S2 the shear span length a was defined as 36 cm, 72 cm and 108 cm (ratio 1:2:3) for a specimen named as S2D36a036, S2D36a072 and S2D36a108, respectively. The effective height was $D = 36$ cm and the distance $b = 54$ cm were fixed, so the span length l_{eff} was determined as 126 cm, 198 cm and 270 cm for the beam S2D36a036, S2D36a072 and S2D36a108, respectively. The thickness of all beams in both series was constant $t = 25$ cm.

Longitudinal reinforcement consisted of bars with diameter of 20 mm. The reinforcement ratio was fixed as $\rho = 1.4\%$ in all beams in both series. The beam S1D18a108 had two bars, beams S1D36a108, S2D36a36 and S236a72 – four bars, and the beam S1D72a108 – eight bars (located in two rows).

Different failure mechanisms were observed. In the beam S1D18a108 yielding of the reinforcement occurred. The beam S1D36a108 was destroyed in so called shear-tension failure mode with normal crack displacements. In the beam S1d72a108 significant

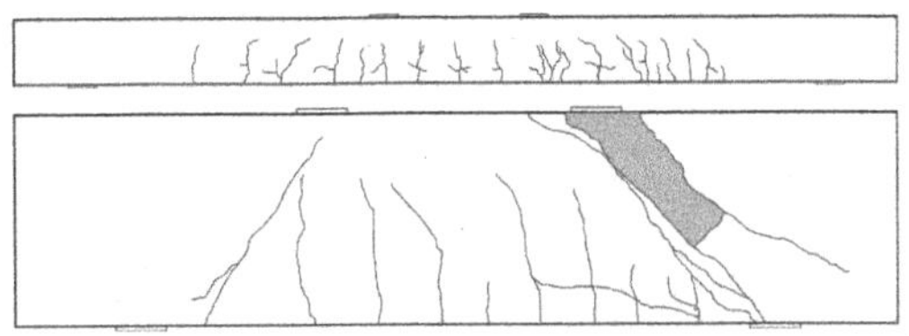

Figure 11. Experimental crack pattern for the beam S1D18a108 (top) and S2D36a36 (bottom).

normal and tangential displacements in the critical crack were observed (diagonal shear-compression mode). Combined shear and compression caused the failure of the beam S2D36a36. Finally two failure mechanisms were observed in the specimens S2D36a72: diagonal shear-tension or diagonal shear-compression mode. Figure 11 presents experimental cracks patterns for the beams S1D18a108 and S2D36a36 (only these beams are simulated in this paper).

4.2 *FE calculations*

In all simulations of both beams elastic parameters are taken as: Young modulus $E = 34.2$ GPa and Poisson's ratio $\nu = 0.2$. Plane stress conditions are assumed. Three node triangle elements and four node quad elements are defined in FE mesh. The characteristic length is equal to $l = 5$ mm (in order to speed up calculations). In the calculations with original Mazars model the following parameters are set: $\kappa_0 = 9.4 \cdot 10^{-5}$, $A_t = 0.98$ and $B_t = 550$ for exponential softening in tension and $\kappa_0 = 9.4 \cdot 10^{-5}$, $\kappa_p = 0.622 \cdot 10^{-3}$, $f_c = 17.36$ MPa, $A_c = 0.98$ and $B_c = 441.9$ for parabolic-exponential relationship in compression. In simulations with Mazars μ law the threshold values are set as $\kappa_{t0} = 9.4 \cdot 10^{-5}$ in tension and $\kappa_{c0} = 1.81 \cdot 10^{-3}$ compression. The softening is described via the exponential curve with $A_t = 0.98$ and $B_t = 550$. In Pereira model the following set of parameters is assumed: $\kappa_{t0} = 9.4 \cdot 10^{-5}$, $A_t = 0.98$ and $B_t = 550$ in tension and $\kappa_{c0} = 3.5 \cdot 10^{-4}$, $\kappa_p = 2.2 \cdot 10^{-3}$, $f_c = 61.5$ MPa, $A_c = 0.98$ and $B_c = 125$ in compression. Reinforced bars are modelled as 1D truss elements with an elasto-perfectly plastic material law and the following parameters: Young modulus $E_s = 200$ GPa and yield strength $\sigma_y = 650$ MPa. Perfect bond (no slip) between concrete and bars is set.

4.2.1 *Beam S1D18a108*

Figure 12 shows obtained force-displacement curves while computed crack patterns are presented in Figure 13. It can be clearly seen no of all three formulations is capable to correctly reproduce the experimental failure mechanism i.e. longitudinal reinforcement yielding. Obtained maximum forces are smaller than experimental values. In simulations the premature shear crack is responsible for failure in all simulations.

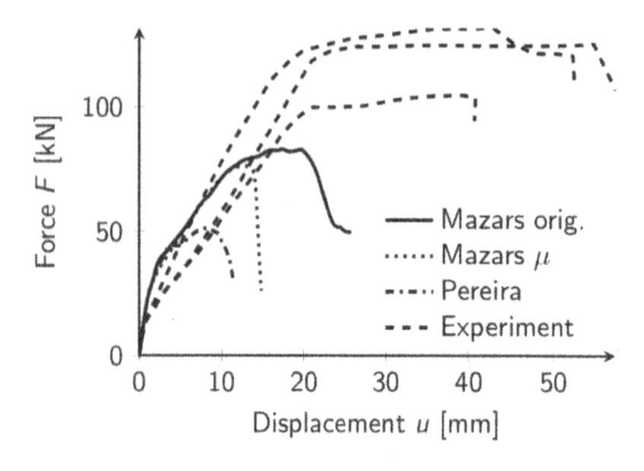

Figure 12. Beam S1D18a108: force-displacement curves.

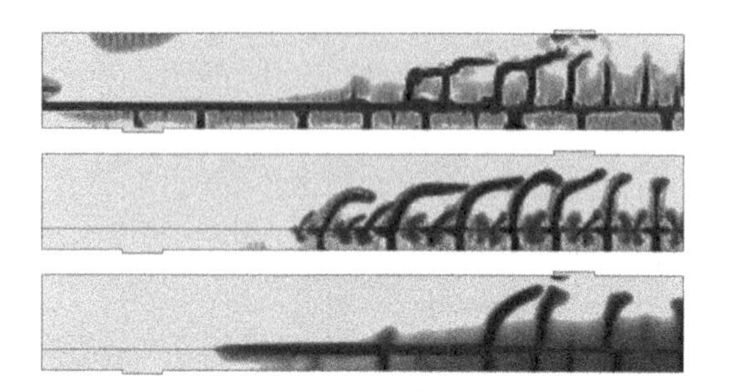

Figure 13. Beam S1D18a108: crack pattern (damage variable D) obtained with original Mazars model (top), Mazars μ model (middle) and Pereira model (bottom).

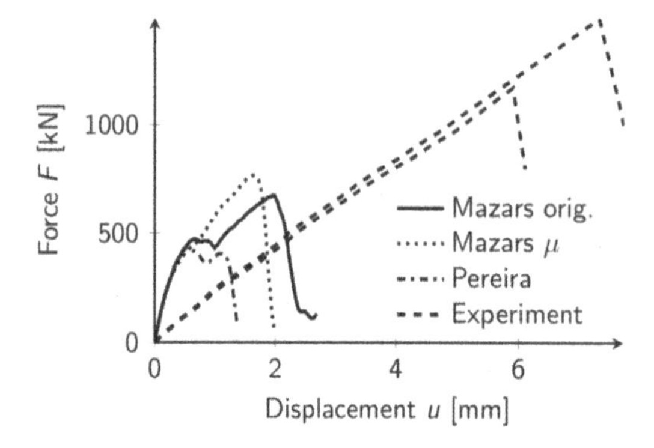

Figure 14. Beam S2D36a108: force-displacement curves.

4.2.2 *Beam S2D36a108*

The results of analysis of the beam S2D36a108 are depicted on Figures 14 (force-displacement curves) and 15 (crack patterns). The experimental failure mechanism is moderately reproduced. A critical shear crack is not fully developed, but it starts to dominate (with a small exception for original Mazars model). However obtained maximum loads are heavily underestimated. Also obtained stiffness is too large comparing to experiments.

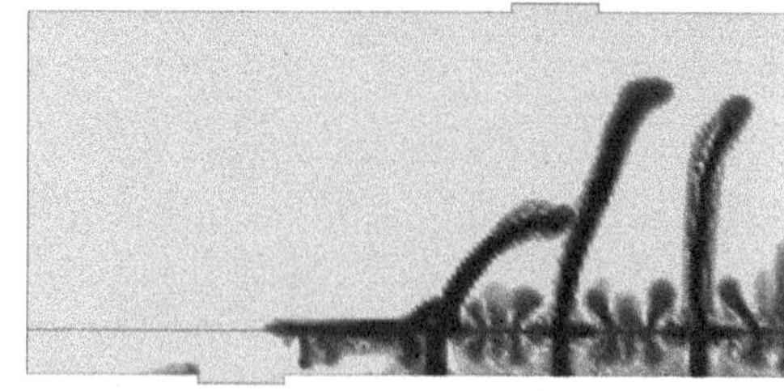

Figure 15. Beam S2D36a108: crack pattern (damage variable D) obtained with original Mazars model (top), Mazars μ model (middle) and Pereira model (bottom).

5 CONCLUSIONS

In the paper three advanced isotropic damage constitutive laws were discussed and their ability to reproduce concrete's behaviour under complex stress state was examined. Two benchmarks were simulated: Nooru-Mohamed test and RC beam with different failure mechanisms.

FE calculations have shown that none of all three damage material laws is able to follow properly experimental outcomes. While Mazars μ and Pereira models were able to simulate Nooru-Mohamed test, they failed in simulations of both cases of the RC beam due to premature shear. Original Mazars constitutive law did not succeed in any analysed benchmarks.

ACKNOWLEDGMENTS

Calculations were carried out at the Centre of Informatics Tricity Academic Supercomputer and Network.

REFERENCES

Belgin, Ç.M. & Şener, S. 2008. Size effect on failure of overreinforced concrete beams. *Engineering Fracture Mechanics* 75:2308–2319.

Bobiński, J. & Tejchman, J. 2016. Comparison of continuous and discontinuous models to simulate concrete behaviour under mixed-mode failure conditions. *International Journal for Numerical and Analytical Methods in Geomechanics* 40:406–435.

Cervera, M. & Chiumenti, M. 2006. Smeared crack approach: back to the original track. *International Journal for Numerical and Analytical Methods in Geomechanics* 30(12): 1173–1199.

Comi, C. & Perego, U. 2001. Fracture energy based bi-dissipative damage model for concrete. *International Journal of Solids and Structures* 38: 6427–6454.

Desmorat, R., Gatuingt, F. & Ragueneau, F. 2007. Nonlocal anisotropic damage model and related computational aspects for quasi-brittle materials. *Enginnering Fracture Mechanics* 74(10): 1539–1560.

Ghahremannejad, M. & Abolmaali, A. 2018. Prediction of shear strength of reinforced concrete beams using displacement control finite element analysis. *Engineering Structures* 169: 226–237.

Grassl, P. & Jirásek M. 2004. On mesh bias of local damage models for concrete. *Proceedings of 5th International Conference on Fracture Mechanics of Concrete and Concrete Structures*, FraMCoS-5, Vail, Colorado, USA.

Haüßler-Combe, U. & Pröchtel, P. 2005. Ein dreiaxiales Stoffgesetz für Betone mith normaler und hoher Festigkeit. *Beton- und Stahlbetonbau* 100(1):52 – 62.

Jirásek, M. (2004). Non-local damage mechanics with application to concrete. *Revue française de génie civil* 8(5–6): 683–707.

Jirásek, M. & Grassl, P. 2008. Evaluation of directional mesh bias in concrete fracture simulations using continuum damage models. *Engineering Fracture Mechanics* 75(8): 1921–1943.

Kim, J.K. & Park, Y.D. 1994. Shear strength of reinforced high strength concrete beams without web reinforcement. *Magazine of Concrete Research* 46(166): 7–16.

Lee, J. & Fenves, G. L. 1998, Plastic-damage model for cyclic loading of concrete structures. *Journal of Engineering Mechanics ASCE* 124(8): 892–900.

Marzec I., Tejchman J. & Mróz Z. 2019. Numerical analysis of size effect in RC beams scaled along height or length using elasto-plastic-damage model enhanced by non-local softening. *Finite Elements in Analysis and Design* 157:1–20.

Marzec, I. & Bobiński, J. 2020. Performance of isotropic constitutive laws in simulating failure mechanisms in scaled RC beams. *Archives of Mechanics* 72(3): 193–215.

Marzec, I. & Tejchman, J. 2021. Experimental and numerical investigations on RC beams with stirrups scaled along height or length. *Engineering Structures* 252.

Mazars, J. 1986. A description of micro- and macroscale damage of concrete structures. *Engineering Fracture Mechanics* 25 (5–6): 729–737.

Mazars, J. & Grange, S. 2015. Modeling of reinforced concrete structural members for engineering purposes. *Computers and Concrete* 16(5): 683–701.

Mazars, J., Hamon, F. & Grange, S. 2014. A model to forecast the response of concrete under severe loadings the μ damage model. *Procedia Materials Science* 3: 979–984.

Mazars, J., Hamon, F. & Grange, S. 2015. A new 3d damage model for concrete under monotonic, cyclic and dynamic load. *Materials and Structures* 48: 3779–3793.

Mazars, J. & Pijaudier-Cabot, G. 1989. Continuum damage theory – application to concrete. *Journal of Engineering Mechanics ASCE* 115(2): 345–365.

Nooru-Mohamed, M.B. 1922. *Mixed mode fracture of concrete: an experimental research.* PhD Thesis, TU Delft.

Patzák, B. & Jirásek, M. 2004. Adaptive resolution of localized damage in quasi-brittle materials. *Journal of Engineering Mechanics ASCE* 130(6): 720–723.

Pereira, L., Weerheijm, J. & Sluys, L. 2017. A numerical study on crack branching in quasi-brittle materials with a new effective rate-dependent nonlocal damage model. *Engineering Fracture Mechanics* 182: 689–707.

Pereira, L., Weerheijm, J. & Sluys, L. 2018. Simulation of compaction and crushing of concrete in ballistic impact with a new damage model. *International Journal of Impact Engineering* 111: 208–221.

Pijaudier-Cabot, G., Mazars, J. & Pulikowski, J. 1991. Steel-concrete analysis with nonlocal continuous damage. *Journal of Structural Engineering ASCE* 117(3): 862–882.

Qi, H., Li, T., Liu, X., Zhao, L., Lin, C. & Fan, S. 2020. A variable parameters damage model for concrete. *Engineering Fracture Mechanics* 228: 106898.

Sanabria Diaz, R.A., Sarmiento Nova, S.J., Teixeria da Silva, M.C.A., Trautwein, L.M. & de Almeida, L.C. 2020. Reliability analysis of shear strength of reinforced concrete deep beams using NLFEA. *Engineering Structures* 203.

Suchorzewski, J., Korol, E., Tejchman, J. & Mróz, 2018. Z. Experimental study of shear strength and failure mechanisms in RC beams scaled along height or length. *Engineering Structures* 157: 203–233.

Tan, K.H., Cheng, G.H. & Cheong, H.K. 2005. Size effect in shear strength of large beams – behavior and finite element modelling. *Magazine of Concrete Research* 57(8): 497–509.

de Vree, J.H.P., Brekelmans, W.A.M. & van Gils, M.A.J. 1995. Comparison of nonlocal approaches in continuum damage mechanics. *Computers and Structures* 55(4): 581–588.

Computational Modelling of Concrete and Concrete Structures – Meschke, Pichler & Rots (Eds)
 ISBN: 978-1-032-32724-2

Numerical modeling of concrete fracturing and size-effect of notched beams

M. Pathirage
Universite de Pau et des Pays de l'Adour, E2S UPPA, CNRS, TotalEnergies, LFCR, Anglet, France
Department of Civil and Environmental Engineering, Northwestern University, Evanston, IL, USA

D. Tong
Department of Civil and Environmental Engineering, Northwestern University, Evanston, IL, USA

F. Thierry
Universite de Pau et des Pays de l'Adour, E2S UPPA, CNRS, TotalEnergies, LFCR, Anglet, France

G. Cusatis
Department of Civil and Environmental Engineering, Northwestern University, Evanston, IL, USA

D. Grégoire & G. Pijaudier-Cabot
Universite de Pau et des Pays de l'Adour, E2S UPPA, CNRS, TotalEnergies, LFCR, Anglet, France

ABSTRACT: Size-effect of quasi-brittle materials such as concrete defines the relation between nominal strength and structural size when material fractures. One main type of size-effect, which is the focus of this manuscript, is the so-called energetic size-effect and is due to the release of stored energy of the structure into the fracture front. In contrast to brittle materials, the fracture process zone size has a non-negligible size in concrete, which makes the size-effect law non-linear. In order to simulate size-effect, a numerical model must be able to describe accurately the development and propagation of the fracture process zone. Over the years, a number of models have been proposed to describe the fracturing process in concrete. Nevertheless, it appears challenging to obtain a correct description of fracture and size-effect when the structural dimension and shape are varying. In this study, the Lattice Discrete Particle Model (LDPM) was proposed to overcome this lack of accurate models. The use of mesoscale discrete models such as LDPM, which describes concrete at the aggregate level, is especially adequate in simulating complex cracking mechanisms. In order to investigate the effect of structural dimension and geometry on the fracturing process and the nominal strength, one of the most comprehensive experimental data set available in the literature was considered, which includes three-point bending tests of notched and unnotched beams. First, the relevant material parameters in LDPM were calibrated on a single size notched beam on the corresponding entire load-Crack Mouth Opening Displacement (CMOD) curve. The model was then used to predict the load-CMOD curves of different beam sizes with the same notch length. Predictions on one unnotched beam were also made to test the model's capability to simulate crack initiation from a smooth surface. Preliminary results show very a good agreement with the experimental data, which suggests that LDPM is an efficient model in predicting concrete size-effect.

1 INTRODUCTION

The effect of structural size on the nominal strength of concrete and other quasi-brittle materials has been explored numerous times in the literature, experimentally, theoretically, and numerically. In quasi-brittle materials, an increase in structural size is accompanied by a reduction in strength for geometrically similar structures. This phenomenon is the so-called size-effect and was confirmed experimentally for a wide variety of materials and in particular for plain concrete (see for instance the work of Bažant & Pfeiffer 1987; Grégoire et al. 2013; Hoover et al. 2013). Size-effect is mainly caused by two distinct phenomena: (i) the stress redistribution due to stable crack propagation and release of stored energy into the fracture front, and (ii) the randomness in material strength. The reader is referred to the seminal work of Bažant and coworkers (Bažant 2002; Bazant & Le 2017; Bažant & Planas 1997) for extensive details. This study focuses solely on the energetic size-effect which is by the way a purely deterministic process. The release of stored energy in the structure combined with the finite nature of the fracture process zone size in concrete make the

 DOI 10.1201/9781003316404-58

size-effect behavior deviate from classical linear elastic fracture mechanics predictions. In order to describe the non-linear scaling in concrete and the development and propagation of the fracture front for different specimen sizes and shapes, one often needs to carry out numerical modeling.

Over the years, a number of models have been proposed to describe the fracturing process and size-effect in concrete. The major ones are the cohesive (Elices et al. 2002) and crack-band model (Bažant & Oh 1983), non-local continuum damage models (Bažant & Jirásek 2002; Pijaudier-Cabot & Bažant 1987), and discrete models (Bolander et al. 2021). In the aforementioned models, the two ingredients necessary to capture strain softening and size-effect in concrete are taken into account: (i) crack localization and (ii) existence of an internal characteristic length related to the size of the heterogeneity. In this regard, random lattice or particle models are especially appealing as crack formation, localized and oriented events such as frictional slip can be naturally captured. Moreover, realistic features can be taken into account such as the actual particle size distribution used in the mix design and the randomness in the spatial distribution of particles which reproduces the statistically isotropic nature of the material and avoids directional mesh bias during the fracturing process. A recent argument in favor of lattice particle models is the necessity to capture the effect of stress parallel to cracks on the size of the fractured zone, which was shown to be significant in concrete (Nguyen et al. 2020a, 2020b).

In this study, the Lattice Discrete Particle Model (Cusatis et al. 2011a, 2011b) is adopted. This model has been used extensively to simulate the behavior of concrete and other granular materials at the mesoscale, i.e. at the coarse aggregate level, by modeling their interaction. Size and geometry effects are here investigated by considering the experimental data set generated by Grégoire and coworkers (Grégoire et al. 2013) which includes three-point bending tests of notched and unnotched concrete beams. This data set is among one of the very few available in the literature that encompasses a large range of beam depths, two notch sizes and unnotched beams. It is worth noting that two previous studies attempted to simulate this data set. In the first study, an integral-type non-local model was used but was unsuccessful in capturing size and geometry effects (Grégoire et al. 2013). Whereas the second study was able to simulate the data with a good accuracy (Grassl et al. 2012), the discrete lattice model used in that work falls in the miniscale category, where each particle is discretized along with the matrix and the aggregate-matrix interface, making any simulations quickly computationally prohibitive as the structural size increases. Another limitation is the use of a two-dimensional model, which might however not affect the fracture test results of similar specimens.

In the following, LDPM is first explained with a focus on the constitutive equations describing the fracturing behavior. The relevant model parameters were calibrated on the load-CMOD curve corresponding to one beam configuration and on the compressive strength. Size-effect predictions were then performed on different sizes of beams. Preliminary results on one unnotched beam are also provided.

2 LATTICE DISCRETE PARTICLE MODEL

The Lattice Discrete Particle Model was proposed by Cusatis and coworkers to simulate the mechanical behavior of concrete (Cusatis et al. 2011a, 2011b). It has also been used to simulate a wide range of granular quasi-brittle materials such as mortar (Han et al. 2020; Pathirage et al. 2019), fiber reinforced concrete and engineered cementitious composites (Feng et al. 2022; Rezakhani et al. 2021; Schauffert & Cusatis 2011), irregular stone masonry (Angiolilli et al. 2020, 2021; Mercuri et al. 2020, 2021a, 2021b, 2022), shale (Li et al. 2017) and to reproduce multi-physics phenomena such as hygro-thermo-chemical processes, alkali-silica reaction, aging, and self-healing in concrete (Alnaggar et al. 2013; Cibelli et al. 2019, 2022; Pathirage et al. 2018, 2019; Yang et al. 2021, 2022).

In this model, spherical particles are placed in the considered volume of material following a specific particle size distribution given the cement content c, the water-to-cement ratio w/c, the density ρ, and the maximum and minimum aggregate size d_a and d_0, respectively. The geometry of the interaction between aggregates is described as follows: first, a lattice system is defined to describe the interaction between particles by means of a Delaunay tetrahedralization performed with the centers of the particles; next a domain tessallation is performed to define the potential failure locations, which finally generates a system of polyhedral cells. The surface of each polyhedral cell is composed of triangular facets where the LDPM constitutive equations, facet stresses and strains are formulated in a vectorial form. The interaction between polyhedral cells is governed by specific constitutive equations describing tensile fracturing with strain-softening, cohesive and frictional shearing, and compressive response with strain-hardening. In particular, the fracturing behavior in LDPM incorporates effective strains and stresses through a damage-type constitutive equation. The strain-dependent limiting boundary is characterized by an exponential decay which starts when the maximum effective strain reaches its elastic limit. The relevant constitutive equations are given next. For the complete set of constitutive equations, as well as the compatibility and equilibrium equations, the reader is referred to the original work of Cusatis et al. 2011.

Let us denote by $\mathbf{x}_i$ and $\mathbf{x}_j$ the positions of nodes i and j, adjacent to the facet k, the facet strains are defined as $\mathbf{e}_k = [e_N \; e_M \; e_L]^T = [\mathbf{n}_k^T[\![\mathbf{u}_k]\!]/l \; \mathbf{m}_k^T[\![\mathbf{u}_k]\!]/l \; \mathbf{l}_k^T[\![\mathbf{u}_k]\!]/l]^T$, where e_N is the normal strain component, and e_M, e_L are the tangential strain components, $[\![\mathbf{u}_k]\!] = \mathbf{u}_j - \mathbf{u}_i$ is the displacement jump at the centroid of the facet k, $l = \|\mathbf{x}_j - \mathbf{x}_i\|_2$ is the distance between the two nodes, $\mathbf{n}_k = (\mathbf{x}_j - \mathbf{x}_i)/l$ and $\mathbf{m}_k$, $\mathbf{l}_k$ are two unit vectors mutually orthogonal in the facet plane projected orthogonally to the line connecting the adjacent nodes. One can define the traction vector as $\mathbf{t}_k = [t_N \; t_M \; t_L]^T$, where t_N is the normal component, t_M and t_L are the shear components. For the sake of readability, the subscript k that designates the facet is dropped in the following equations.

2.1 *Elastic behavior*

The elastic behavior is defined through linear relations between the normal and shear stresses, and the corresponding strains as

$$t_N = E_N e_N \ , \ t_M = E_T e_M \ , \ t_L = E_T e_L \tag{1}$$

where $E_N = E_0$ and $E_T = \alpha_0 E_0$, $E_0 \approx E/(1-2\nu)$ and $\alpha_0 \approx (1-4\nu)/(1+\nu)$ are the effective normal modulus and the shear-normal coupling parameter, respectively, and E is the macroscopic Young's modulus and ν is the macroscopic Poisson's ratio.

2.2 *Fracture behavior*

Although three-point bending tests are designed to generate pure mode I opening, some facets in LDPM might be under a mixed-mode tension-shear because of the irregular shape of the polyhedral cells, very much like in real fracture taking place at the interface of aggregates. It is therefore important to explain the fracturing and cohesive behavior under tension but also tension/shear. It occurs for $e_N > 0$. One can define an effective strain as $e = (e_N^2 + \alpha_0(e_M^2 + e_L^2))^{\frac{1}{2}}$, and an effective stress as $t = (t_N^2 + (t_M^2 + t_L^2)/\alpha_0)^{\frac{1}{2}}$ and write the relationship between stresses and strains through $t_N = te_N/e$, $t_M = \alpha_0 te_M/e$ and $t_L = \alpha_0 te_L/e$. The effective stress t is defined incrementally as $\dot{t} = E_N\dot{e}$ and its magnitude is limited by a strain-dependent boundary $0 \leqslant t \leqslant \sigma_{bt}(e,\omega)$ in which

$$\sigma_{bt}(e,\omega) = \sigma_0(\omega)\exp\left[-H_0(\omega)\frac{\langle e_{\max} - e_0(\omega)\rangle}{\sigma_0(\omega)}\right] \tag{2}$$

$\langle x\rangle = \max(x,0)$, ω is a variable defining the degree of interaction between shear and normal loading defined as $\tan(\omega) = (e_N)/(\sqrt{\alpha_0}e_T) = (t_N\sqrt{\alpha_0})/(t_T)$; e_T is the total shear strain defined as $e_T = (e_M^2 + e_L^2)^{\frac{1}{2}}$, and t_T is the total shear stress defined as $t_T = (t_M^2 + t_L^2)^{\frac{1}{2}}$. The maximum effective strain is time dependent and is defined as $e_{\max}(\tau) = (e_{N,\max}^2(\tau) + \alpha_0 e_{T,\max}^2(\tau))^{\frac{1}{2}}$, where $e_{N,\max}(\tau) = \max_{\tau'<\tau}[e_N(\tau')]$ and $e_{T,\max}(\tau) = \max_{\tau'<\tau}[e_T(\tau')]$. The strength limit of the effective stress that defines the transition between pure tension and pure shear is

$$\sigma_0(\omega) = \sigma_t \frac{-\sin(\omega) + \sqrt{\sin^2(\omega) + 4\alpha_0\cos^2(\omega)/r_{st}^2}}{2\alpha_0\cos^2(\omega)/r_{st}^2} \tag{3}$$

where $r_{st} = \sigma_s/\sigma_t$ is the shear to tensile strength ratio, σ_s is the shear strength and σ_t is the tensile strength. The post-peak softening modulus is controlled by the effective softening modulus $H_0(\omega) = H_s/\alpha_0 + (H_t - H_s/\alpha_0)(2\omega/\pi)^{n_t}$, in which $H_t = 2E_0/(l_t/l - 1)$, $H_s = r_s E_0$ and n_t is the softening exponent; l_t is the tensile characteristic length defined as $l_t = 2E_0 G_t/\sigma_t^2$, G_t is the mesoscale fracture energy. The model was recently implemented by the authors in the finite element solver CAST3M (Verpeaux et al. 1988) (2020 version) within a dynamic implicit framework (Pathirage et al. 2022). All the simulations presented in this study were performed using CAST3M.

3 MODEL CALIBRATION

3.1 *Experimental data*

The experiments performed by Grégoire et al. 2013 included three-point bending tests of four different sizes of geometrically similar specimens, with depths D varying between 50 mm and 400 mm, span-to-depth ratio of 2.5, and out-of-plane thickness of 50 mm. Three different notch-to-depth ratios α were tested in the experimental work. However only one ratio, namely $\alpha = 0.2$, was considered in this paper. The tests were performed under CMOD control in order to obtain a stable post-peak. For unnotched beams, the plates of the extensometer were glued at a distance from mid-span of half the beam depth to ensure that crack initiates between the plates. In addition, unconfined compression tests on cylinders were conducted and elastic parameters were measured. For more details on the experimental program and the concrete mix design, the reader is referred to Grégoire et al. 2013.

3.2 *Calibration process*

First, the mix design parameters used to define the LDPM geometry were identified based on the actual mix used in the experiments. The particle size distribution given in Grégoire et al. 2013 was matched exactly with a cut-off size at $d_0 = 4$ mm and values of $d_a = 10$ mm, $c = 260$ kg/m^3, $w/c = 0.626$, and $\rho = 2121$ kg/m^3 were taken.

Next, the parameters related to the elastic behavior were identified. More specifically, the elastic modulus and Poisson's ratio were taken as $E = 37000$ MPa and $\nu = 0.176$ based on the mean values obtained by measurements on cylinders. Values of $E_0 = 57180$ MPa and $\alpha_0 = 0.25$ were then deduced from the equations listed in section 2.1. Finally, notched three-point bending and unconfined compression simulations were performed to calibrate the model parameters in the inelastic regime. The mesoscale tensile strength $\sigma_t = 2.9$ MPa, characteristic length $l_t = 400$ mm, and shear-to-tensile strength ratio $\sigma_s/\sigma_t = 3.276$ were obtained simultaneously based on: (i) the load-CMOD response corresponding to the medium size beam of depth

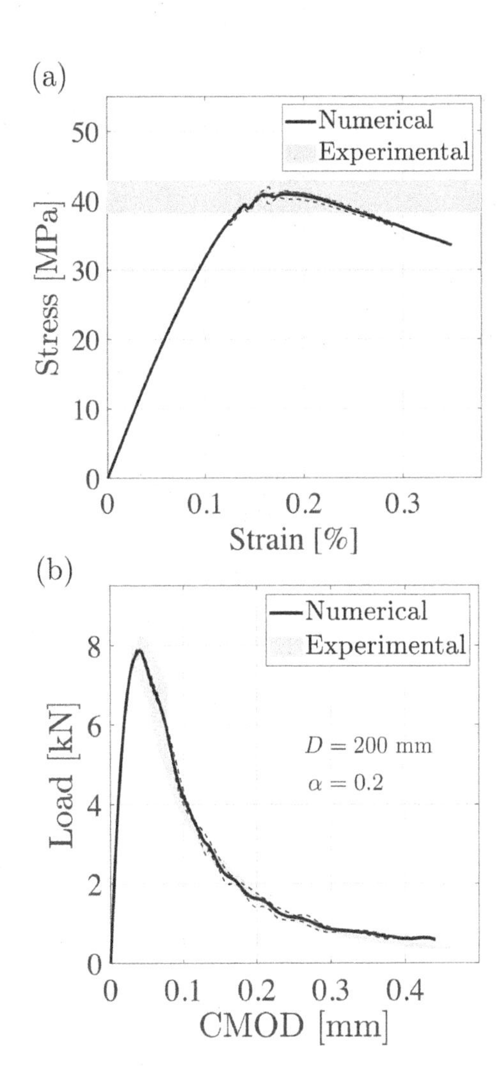

Figure 1. Calibration results: (a) stress-strain curve of unconfined compression test on cylinder of diameter $D_c = 74$ mm and height $H_c = 142$ mm, (b) load-CMOD curve of three-point bending test with $D = 200$ mm and $\alpha = 0.2$.

$D = 200$ mm with $\alpha = 0.2$ and (ii) the compressive strength obtained from cylinders of diameter $D_c = 74$ mm and height $H_c = 142$ mm. For the compression tests, rigid plates at the top and bottom of the specimens were used. High friction between steel plates and concrete was simulated through a classical Coulomb friction law with a friction coefficient of $\mu = 0.13$. Both fracture and compression tests were conducted under displacement control and with a constant velocity of 0.01 mm s^{-1} to ensure quasi-static conditions. For all the simulations, three different random particle placements were used to take into account the spatial variability of aggregate size and distribution. In addition, values of $n_t = 0.2$ and $r_s = 0$ were assumed fixed. The remaining parameters, not related to unconfined compression or fracture, were assumed based on section 5.3 in Cusatis et al. 2011 and are listed here for the sake of completeness: $\sigma_{c0} = 120$ MPa, $H_{c0}/E_0 = 0.4$, $k_{c0} = 2$, $k_{c1} = 1$, $k_{c2} = 5$, $\mu_0 = 0.2$, $\mu_\infty = 0$, $\sigma_{N0} = 600$ MPa, and $E_d/E_0 = 1$. Last but not least, the simulations were performed in dynamic implicit with a constant time step of $\Delta t = 0.1$ s and a criterion on the residual of 10^{-3}.

Figure 1(a) shows the simulated stress-strain curve and the compressive strength obtained experimentally. Figure 1(b) shows the load-CMOD curves for the bending test. On each figure, three dashed lines are displayed: they correspond to three simulations performed with different particle distributions. The solid line is the mean of the latter three curves. The grey area represents the experimental scatter where the bounds corresponds to the maximum and minimum values. One can see that the LDPM responses match well with experiments within the scatter of the data.

4 PREDICTION RESULTS

4.1 *Size-effect*

In order to assess the capability of the model to predict size-effect, blind predictions were carried out without adjusting the model parameters on the three remaining sizes of the bending tests while keeping the same notch-to-depth ratio. Figure 2(a) shows the full predictions of the load-CMOD curves for sizes $D = 50$ mm, $D = 100$ mm, and $D = 400$ mm. One can see that the predictions are in good agreement with the experimental curves since all the simulations lie within the scatter of the experimental data. Figure 2(b) shows the mesoscale crack opening in the four samples at some displacement in the post-peak, more specifically at a displacement 0.05 mm for $D = 50$ mm, 0.05 mm for $D = 100$ mm, 0.10 mm for $D = 200$ mm, and 0.15 mm for $D = 400$ mm. As expected, cracks are localized and propagate through the sample almost vertically: the slight deviations are due to the heterogeneity of the material.

4.2 *Crack initiation from smooth surface*

Preliminary work was also done on evaluating the capability of LDPM to predict fracture propagation in samples without notch, i.e. to simulate crack initiation from a smooth surface. Figure 3(a) shows the full prediction of the load-CMOD curve for the beam of depth $D = 50$ mm. One can see again a good agreement experiments. The predicted mean peak value is slightly higher than the experimental one. In addition, there is a relatively small difference between experiments and simulation in the mid-far post-peak. The reason for these slight deviations in predictions might be the size of the beam, i.e. $D = 50$ mm. Indeed, this depth is very close to the smallest beam one could cast or simulate with a maximum aggregate size of 10 mm. Moreover, one might amplify additional sources of errors when considering such small specimen sizes, for instance the correct application of

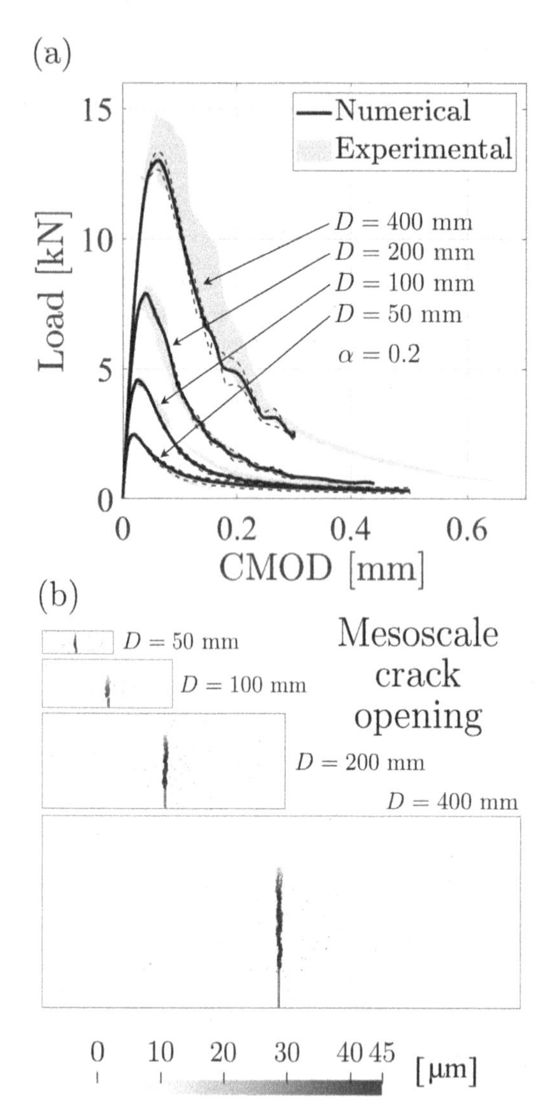

Figure 2. Size-effect predictions: (a) load-CMOD curves for the four beam depths ($D = 200$ mm is the calibration, the remaining curves are full predictions) and (b) mesoscale crack opening.

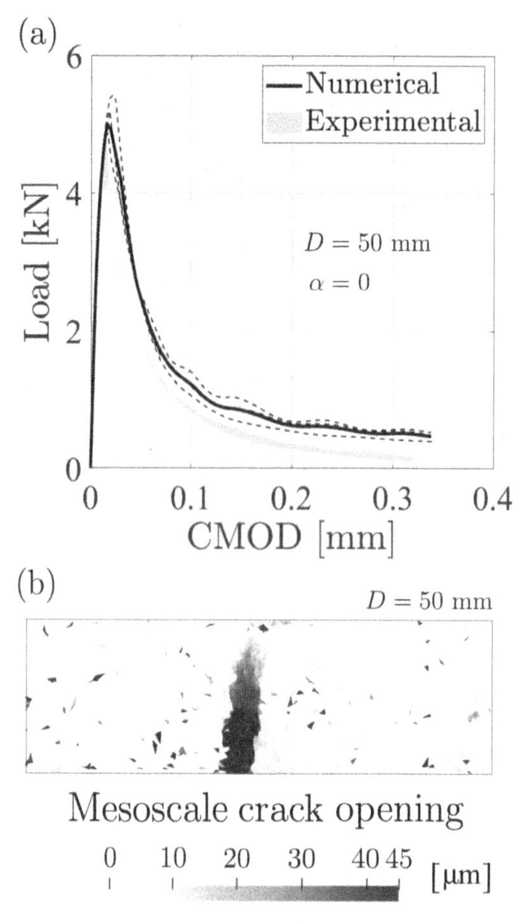

Figure 3. Unnotched beam: (a) load-CMOD curve for beam depth $D = 50$ mm and (b) mesoscale crack opening.

boundary conditions. Figure 3(b) shows the mesoscale crack opening for a displacement of 0.05 mm. The main crack does not initiate exactly at midspan, which is of course expected and observed in experiments. This is well captured by the model since the heterogeneity of the material is reproduced in the LDPM internal geometry. The presence of individual cracks, sometimes located far from the main propagating crack, is due to the large time step taken for the simulation. This spurious noise effect does not affect the accuracy of the load-displacement curve as almost all the energy is dissipated in the main crack. As a matter of fact, these individual cracks can be eliminated by reducing the time step or by running a static simulation (these results are not shown here).

5 CONCLUSIONS

A set of numerical simulations were performed in order to study fracture and size-effect in concrete. For this purpose, the Lattice Discrete Particle Model was used to simulate experimental data of four different sizes notched three-point bending specimens reported in Grégoire et al. 2013. The model was first calibrated on one size bending beam and on a compression test. Predictions of load-CMOD curves were finally made on the remaining three sizes and on one unnotched beam. Results show that:

- LDPM is able to predict the fracture propagation and size-effect for a wide range of beam depths.
- Results on the unnotched beam confirms that the model is able to capture crack initiation and propagation from a smooth surface.

Ongoing work focuses on extending this study by predicting size-effect on beams with different notch lengths including unnotched specimens.

6 ACKNOWLEDGMENTS

Partial financial support from the investissement d'avenir French programme (ANR-16-IDEX-0002) under the framework of the E2S UPPA hub Newpores is gratefully acknowledged. This research was supported in part through the computational resources provided for the Quest high performance computing facility at Northwestern University which is jointly supported by the Office of the Provost, the Office for Research, and Northwestern University Information Technology.

REFERENCES

Alnaggar, M., G. Cusatis, & G. Di Luzio (2013). Lattice discrete particle modeling (ldpm) of alkali silica reaction (asr) deterioration of concrete structures. *Cement and Concrete Composites 41*, 45–59.

Angiolilli, M., A. Gregori, M. Pathirage, & G. Cusatis (2020). Fiber reinforced cementitious matrix (frcm) for strengthening historical stone masonry structures: Experiments and computations. *Engineering Structures 224*, 111102.

Angiolilli, M., M. Pathirage, A. Gregori, & G. Cusatis (2021). Lattice discrete particle model for the simulation of irregular stone masonry. *Journal of Structural Engineering 147*(9), 04021123.

Bažant, Z. P. (2002). *Scaling of structural strength*. CRC Press.

Bažant, Z. P. & M. Jirásek (2002). Nonlocal integral formulations of plasticity and damage: survey of progress. *Journal of engineering mechanics 128*(11), 1119–1149.

Bazant, Z. P. & J.-L. Le (2017). *Probabilistic mechanics of quasibrittle structures: strength, lifetime, and size effect*. Cambridge University Press.

Bažant, Z. P. & B. H. Oh (1983). Crack band theory for fracture of concrete. *Matériaux et construction 16*(3), 155–177.

Bažant, Z. P. & P. A. Pfeiffer (1987). Determination of fracture energy from size effect and brittleness number. *ACI Materials Journal 84*(6), 463–480.

Bažant, Z. P. & J. Planas (1997). *Fracture and size effect in concrete and other quasibrittle materials*. CRC press.

Bolander, J. E., J. Eliáš, G. Cusatis, & K. Nagai (2021). Discrete mechanical models of concrete fracture. *Engineering Fracture Mechanics 257*, 108030.

Cibelli, A., G. Di Luzio, L. Ferrara, G. Cusatis, M. Pathirage, et al. (2019). Modelling of autogenous healing for regular concrete via a discrete model. In *10th International Conference on Fracture Mechanics of Concrete and Concrete Structures, FraMCoS-X*, pp. 1–12. IA-FraMCoS.

Cibelli, A., M. Pathirage, G. Cusatis, L. Ferrara, & G. Di Luzio (2022). A discrete numerical model for the effects of crack healing on the behaviour of ordinary plain concrete: Implementation, calibration, and validation. *Engineering Fracture Mechanics 263*, 108266.

Cusatis, G., A. Mencarelli, D. Pelessone, & J. Baylot (2011). Lattice discrete particle model (ldpm) for failure behavior of concrete. ii: Calibration and validation. *Cement and Concrete composites 33*(9), 891–905.

Cusatis, G., D. Pelessone, & A. Mencarelli (2011). Lattice discrete particle model (ldpm) for failure behavior of concrete. i: Theory. *Cement and Concrete Composites 33*(9), 881–890.

Elices, M., G. Guinea, J. Gomez, & J. Planas (2002). The cohesive zone model: advantages, limitations and challenges. *Engineering fracture mechanics 69*(2), 137–163.

Feng, J., W. Sun, L. Chen, B. Chen, E. Arkin, L. Du, & M. Pathirage (2022). Engineered cementitious composites using chinese local ingredients: Material preparation and numerical investigation. *Case Studies in Construction Materials 16*, e00852.

Grassl, P., D. Grégoire, L. R. Solano, & G. Pijaudier-Cabot (2012). Meso-scale modelling of the size effect on the fracture process zone of concrete. *International Journal of Solids and Structures 49*(13), 1818–1827.

Grégoire, D., L. B. Rojas-Solano, & G. Pijaudier-Cabot (2013). Failure and size effect for notched and unnotched concrete beams. *International Journal for Numerical and Analytical Methods in Geómechanics 37*(10), 1434–1452.

Han, L., M. Pathirage, A.-T. Akono, & G. Cusatis (2020, 11). Lattice Discrete Particle Modeling of Size Effect in Slab Scratch Tests. *Journal of Applied Mechanics 88*(2). 021009.

Hoover, C. G., Z. P. Bažant, J. Vorel, R. Wendner, & M. H. Hubler (2013). Comprehensive concrete fracture tests: description and results. *Engineering fracture mechanics 114*, 92–103.

Li, W., R. Rezakhani, C. Jin, X. Zhou, & G. Cusatis (2017). A multiscale framework for the simulation of the anisotropic mechanical behavior of shale. *International Journal for Numerical and Analytical Methods in Geomechanics 41*(14), 1494–1522.

Mercuri, M., M. Pathirage, A. Gregori, & G. Cusatis (2020). Computational modeling of the out-of-plane behavior of unreinforced irregular masonry. *Engineering Structures 223*, 111181.

Mercuri, M., M. Pathirage, A. Gregori, & G. Cusatis (2021a). Lattice discrete modeling of out-of-plane behavior of irregular masonry. In *Proceedings of the 8th ECCOMAS Thematic Conference on Computational Methods in Structural Dynamics and Earthquake Engineering*, pp. 546–562.

Mercuri, M., M. Pathirage, A. Gregori, & G. Cusatis (2021b). On the collapse of the masonry medici tower: An integrated discrete-analytical approach. *Engineering Structures 246*, 113046.

Mercuri, M., M. Pathirage, A. Gregori, & G. Cusatis (2022). Masonry vaulted structures under spreading supports: Analyses of fracturing behavior and size effect. *Journal of Building Engineering 45*, 103396.

Nguyen, H., M. Pathirage, M. Rezaei, M. Issa, G. Cusatis, & Z. P. Bažant (2020). New perspective of fracture mechanics inspired by gap test with crack-parallel compression. *Proceedings of the National Academy of Sciences 117*(25), 14015–14020.

Nguyen, H. T., M. Pathirage, G. Cusatis, & Z. P. BaÅ¾ant (2020, 05). Gap Test of Crack-Parallel Stress Effect on Quasibrittle Fracture and Its Consequences. *Journal of Applied Mechanics 87*(7). 071012.

Pathirage, M., D. Bentz, G. Di Luzio, E. Masoero, & G. Cusatis (2019). The onix model: a parameter-free multiscale framework for the prediction of self-desiccation in concrete. *Cement and Concrete Composites 103*, 36–48.

Pathirage, M., D. P. Bentz, G. Di Luzio, E. Masoero, G. Cusatis, et al. (2018). A multiscale framework for the prediction of concrete self-desiccation. In *Computational modelling of concrete structures: proceedings of the conference on computational modelling of concrete and concrete structures (EURO-C 2018)*.

Pathirage, M., F. Bousikhane, M. D'Ambrosia, M. Alnaggar, & G. Cusatis (2019). Effect of alkali silica reaction on the mechanical properties of aging mortar bars: Experiments and numerical modeling. *International Journal of Damage Mechanics 28*(2), 291–322.

Pathirage, M., F. Thierry, D. Tong, G. Cusatis, D. Grégoire, & G. Pijaudier-Cabot (2022). Comparative investigation of dynamic implicit and explicit methods for the lattice discrete particle model. In *Proceeding Euro-C*, Volume (this volume). CRC press.

Pijaudier-Cabot, G. & Z. P. Bažant (1987). Nonlocal damage theory. *Journal of engineering mechanics 113*(10), 1512–1533.

Rezakhani, R., D. A. Scott, F. Bousikhane, M. Pathirage, R. D. Moser, B. H. Green, & G. Cusatis (2021). Influence of steel fiber size, shape, and strength on the quasi-static properties of ultra-high performance concrete: Experimental investigation and numerical modeling. *Construction and Building Materials 296*, 123532.

Schauffert, E. A. & G. Cusatis (2011). Lattice discrete particle model for fiber-reinforced concrete. i: Theory. *Journal of Engineering Mechanics 138*(7), 826–833.

Verpeaux, P., T. Charras, & A. Millard (1988). Castem 2000: une approche moderne du calcul des structures. *Calcul des structures et intelligence artificielle 2*, 261–271.

Yang, L., M. Pathirage, H. Su, M. Alnaggar, G. Di Luzio, & G. Cusatis (2021). Computational modeling of temperature and relative humidity effects on concrete expansion due to alkali—silica reaction. *Cement and Concrete Composites 124*, 104237.

Yang, L., M. Pathirage, H. Su, M. Alnaggar, G. Di Luzio, & G. Cusatis (2022). Computational modeling of expansion and deterioration due to alkali—silica reaction: Effects of size range, size distribution, and content of reactive aggregate. *International Journal of Solids and Structures 234–235*, 111220.

Computational Modelling of Concrete and Concrete Structures – Meschke, Pichler & Rots (Eds)
 ISBN: 978-1-032-32724-2

Comparative investigation of dynamic implicit and explicit methods for the Lattice Discrete Particle Model

M. Pathirage
Universite de Pau et des Pays de l'Adour, E2S UPPA, CNRS, TotalEnergies, LFCR, Anglet, France
Department of Civil and Environmental Engineering, Northwestern University, Evanston, IL, USA

F. Thierry
Universite de Pau et des Pays de l'Adour, E2S UPPA, CNRS, TotalEnergies, LFCR, Anglet, France

D. Tong & G. Cusatis
Department of Civil and Environmental Engineering, Northwestern University, Evanston, IL, USA

D. Grégoire & G. Pijaudier-Cabot
Universite de Pau et des Pays de l'Adour, E2S UPPA, CNRS, TotalEnergies, LFCR, Anglet, France

ABSTRACT: There are in general two classes of time integration algorithms for dynamics problems, namely implicit and explicit methods. While explicit methods are conditionally stable and often require a very small time step, implicit algorithms are unconditionally stable and larger time steps can be used. In terms of memory usage, implicit algorithms require more memory as a system of equations needs to be solved one or several times per step for the solution to advance. Implicit methods are usually used to solved problems in which low frequency modes dominate. Nevertheless, one often faces convergence issues when the material behavior is highly nonlinear and explicit methods seem more appropriate in that case. In this study, the performance of the Lattice Discrete Particle Model (LDPM), newly implemented in the implicit solver CAST3M was investigated. LDPM is a mesoscale model developed to simulate concrete and other granular quasi-brittle materials. It incorporates complex nonlinear constitutive equations and for this reason, it is currently used within the dynamic explicit framework ABAQUS Explicit. This manuscript presents preliminary results on the comparison between explicit central difference algorithm and implicit average acceleration scheme for LDPM. For this purpose, a classical three-point bending test under quasi-static conditions was considered. Three different integration methods were used: dynamic explicit, dynamic implicit and static implicit. Load-displacement responses were obtained and discussed, along with the force imbalance at loading points to assess static equilibrium. For each simulation, the computational cost was also obtained. Results show that quasi-static simulations can be performed using dynamic explicit integration method if the effect of inertia is small enough. In addition, the computational cost for static and dynamic implicit simulations is much lower than for dynamic explicit calculations. Last but not least, the time step size in the dynamic implicit method needs to be chosen carefully to avoid spurious energy growth or higher modes due to time discretization.

1 INTRODUCTION

In computational mechanics, there are in general two classes of time integration algorithms designed to solve dynamic problems, namely implicit and explicit methods. In implicit algorithms, a system of equations needs to be solved one or several times at each step. Equilibrium conditions are considered at the same time step for which the solution is sought. In contrast, explicit algorithms do not require to store in memory a system of equations if the mass matrix is diagonal or lumped, which is most often the case in practice. Implicit methods are either conditionally or unconditionally stable. In the Newmark family, the average acceleration method is for instance unconditionally stable. It is one of the reason why this method has become popular over the years in the field of structural dynamics. The time step in such methods can be large and its maximum size is only governed by the solution accuracy and not the stability of the integration scheme. On the other hand, explicit methods are always conditionally stable (at least for linear problems) and the maximum time step is inversely proportional to the highest frequency of the discrete system, which in many cases is much smaller than what is needed to obtain a reasonably accurate solution for engineering problems. In terms of computational cost, implicit methods are expensive since

DOI 10.1201/9781003316404-59

a system of equation needs to be solved at least one time per step and a large amount of memory can be required. However depending on the size of the problem, i.e. the number of degrees of freedom, the time step can be set large enough to compensate the aforementioned cost. In explicit algorithms, the cost and the required memory per step are low but the size of the critical time step prohibits simulations of moderate and long-term dynamic problems. For more details on the different integration schemes, the reader is for example referred to Zienkiewicz et al. 2005, Hughes 2012 or Belytschko et al. 2014. In general, implicit methods are preferred over explicit algorithms for low-velocity dynamic problems where the response is mostly governed by a few low-frequency modes (Subbaraj & Dokainish 1989). Wave propagation problems and high-velocity dynamics are usually better handled with explicit methods (Dokainish & Subbaraj 1989). However, the choice of one algorithm over the other also depends strongly on the behavior of the material model. For highly non-linear material models, low frequency mode simulations are often performed with explicit algorithms because of convergence issues observed in implicit schemes. Although nonlinear computations are usually guided by linear stability estimates (Belytschko et al. 2014), there is no general rule and the type of time integration scheme needs to be chosen carefully and is problem-dependent.

In this study, the Lattice Discrete Particle Model (LDPM) (Cusatis et al. 2011a, 2011b) is considered. LDPM is a mesoscale model that simulates concrete at the aggregate particle level. In this model, the spatial discretization is intrinsically performed when the LDPM geometry is defined (see Section 2) and only time discretization should be considered. This material model incorporates complex and highly non-linear constitutive equations. Over the years, all the calculations with LDPM to simulate the mechanical behavior of concrete have been performed using the dynamic explicit central difference method in order to avoid convergence issues (due to the nonlinear nature of the model) that one would typically face if an implicit method was used. LDPM is currently used within the ABAQUS Explicit solver.

For the first time, LDPM was implemented within an implicit solver, in the general finite element program CAST3M (Verpeaux et al. 1988) (2020 version) which incorporates a robust implicit average acceleration time-integration scheme. This manuscript explains the details of this implementation and shows preliminary results on the comparison between dynamic explicit and implicit methods for LDPM. For this purpose, a three-point bending test was performed under displacement control and the load-displacement curves were obtained for dynamic explicit, dynamic implicit and static implicit methods. The simulations were performed under quasi-static conditions and the force imbalance from equilibrium was evaluated for each case. Last but not least, the computational cost was obtained and is discussed.

2 LATTICE DISCRETE PARTICLE MODEL

The Lattice Discrete Particle Model was proposed by Cusatis and coworkers to simulate the mechanical behavior of concrete (Cusatis et al. 2011a, 2011b). It has also been used to simulate a wide range of granular quasi-brittle materials and complex failure mechanisms. For instance, LDPM was used to simulate mortar (Pathirage et al. 2019) and fracture propagation in scratch testing (Han et al. 2020). The effect of fibers in concrete was also studied (Feng et al. 2022; Rezakhani et al. 2021; Schauffert & Cusatis 2011). Other materials such as irregular masonry were simulated and realistic and complex damage patterns were successfully reproduced (Angiolilli et al. 2020, 2021; Mercuri et al. 2020, 2021a, 2021b, 2022). In addition, long-term behavior of concrete was also considered, such as hygro-thermo-chemical processes, alkali-silica reaction, aging, and self-healing in concrete (Alnaggar et al. 2013; Cibelli et al. 2019, 2022; Pathirage et al. 2018, 2019; Yang et al. 2021, 2022). In LDPM, spherical particles are placed in the considered volume of material following a specific particle size distribution given the cement content c, the water-to-cement ratio w/c, the density ρ, and the maximum and minimum aggregate size d_a and d_0, respectively. The geometry of the interaction between aggregates is described as follows: first, a lattice system is defined to describe the interaction between particles by means of a Delaunay tetrahedralization performed with the centers of the particles; next a domain tessallation is performed to define the potential failure locations, which finally generates a system of polyhedral cells. The surface of each polyhedral cell is composed of triangular facets where the LDPM constitutive equations, facet stresses and strains are formulated in a vectorial form.

The relevant constitutive equations for the description of fracture in three-point bending configuration are given next. Let us denote by $\mathbf{x}_i$ and $\mathbf{x}_j$ the positions of nodes i and j, adjacent to the facet k, the facet strains are defined as $\mathbf{e}_k = [e_N \; e_M \; e_L]^T = [\mathbf{n}_k^T [\![\mathbf{u}_k]\!]/l_k \; \mathbf{m}_k^T [\![\mathbf{u}_k]\!]/l_k \; \mathbf{l}_k^T [\![\mathbf{u}_k]\!]/l_k]^T$, where e_N is the normal strain component, and e_M, e_L are the tangential strain components, $[\![\mathbf{u}_k]\!] = \mathbf{u}_j - \mathbf{u}_i$ is the displacement jump at the centroid of the facet k, $l_k = \|\mathbf{x}_j - \mathbf{x}_i\|_2$ is the distance between the two nodes, $\mathbf{n}_k = (\mathbf{x}_j - \mathbf{x}_i)/l_k$ and $\mathbf{m}_k$, $\mathbf{l}_k$ are two unit vectors mutually orthogonal in the facet plane projected orthogonally to the line connecting the adjacent nodes. One can define the traction vector as $\mathbf{t}_k = [t_N \; t_M \; t_L]^T$, where t_N is the normal component, t_M and t_L are the shear components. For the sake of readability, the subscript k that designates the facet is dropped in the following equations. The elastic behavior is defined through linear relations between the normal and shear stresses, and the corresponding strains as $t_N = E_N e_N$, $t_M = E_T e_M$, $t_L = E_T e_L$, where $E_N = E_0$ and $E_T = \alpha_0 E_0$, $E_0 \approx E/(1-2\nu)$ and $\alpha_0 \approx (1-4\nu)/(1+\nu)$ are the effective normal modulus and the shear-normal coupling parameter, respectively, and E is the macroscopic

Young's modulus and ν is the macroscopic Poisson's ratio. Although three-point bending tests are designed to generate pure mode I opening, some facets in LDPM might be under a mixed-mode tension-shear because of the irregular shape of the polyhedral cells, very much like in real fracture taking place at the interface of aggregates. It is therefore important to explain the fracturing and cohesive behavior under tension but also tension/shear. It occurs for $e_N > 0$. One can define an effective strain as $e = (e_N^2 + \alpha_0(e_M^2 + e_L^2))^{\frac{1}{2}}$, and an effective stress as $t = (t_N^2 + (t_M^2 + t_L^2)/\alpha_0)^{\frac{1}{2}}$ and write the relationship between stresses and strains through $t_N = te_N/e$, $t_M = \alpha_0 te_M/e$ and $t_L = \alpha_0 te_L/e$. The effective stress t is defined incrementally as $\dot{t} = E_N\dot{e}$ and its magnitude is limited by a strain-dependent boundary $0 \leqslant t \leqslant \sigma_{bt}(e, \omega)$ in which $\sigma_{bt}(e, \omega) = \sigma_0(\omega)\exp[-H_0(\omega)\langle e_{\max} - e_0(\omega)\rangle/\sigma_0(\omega)]$ $\langle x \rangle = \max(x, 0)$, ω is a variable defining the degree of interaction between shear and normal loading defined as $\tan(\omega) = (e_N)/(\sqrt{\alpha_0}e_T) = (t_N\sqrt{\alpha_0})/(t_T)$; e_T is the total shear strain defined as $e_T = (e_M^2 + e_L^2)^{\frac{1}{2}}$, and t_T is the total shear stress defined as $t_T = (t_M^2 + t_L^2)^{\frac{1}{2}}$. The maximum effective strain is time dependent and is defined as $e_{\max}(\tau) = (e_{N,\max}^2(\tau) + \alpha_0 e_{T,\max}^2(\tau))^{\frac{1}{2}}$, where $e_{N,\max}(\tau) = \max_{\tau'<\tau}[e_N(\tau')]$ and $e_{T,\max}(\tau) = \max_{\tau'<\tau}[e_T(\tau')]$. The strength limit of the effective stress that defines the transition between pure tension and pure shear is $\sigma_0(\omega) = \sigma_t(-\sin(\omega) + \sqrt{\sin^2(\omega) + 4\alpha_0\cos^2(\omega)/r_{st}^2})/(2\alpha_0\cos^2(\omega)/r_{st}^2)$ where $r_{st} = \sigma_s/\sigma_t$ is the shear to tensile strength ratio, σ_s is the shear strength and σ_t is the tensile strength. The post-peak softening modulus is controlled by the effective softening modulus $H_0(\omega) = H_s/\alpha_0 + (H_t - H_s/\alpha_0)(2\omega/\pi)^{n_t}$, in which $H_t = 2E_0/(l_t/l_k - 1)$, $H_s = r_s E_0$ and n_t is the softening exponent; l_t is the tensile characteristic length defined as $l_t = 2E_0G_t/\sigma_t^2$, G_t is the mesoscale fracture energy. For the complete set of constitutive equations, compatibility and equilibrium equations, definitions of mass and strain-displacement matrices, the reader is referred to the original work of Cusatis et al. 2011.

3 INTEGRATION ALGORITHMS

The general dynamic nonlinear structural equation without damping writes as follows:

$$\mathbf{M}\ddot{\mathbf{X}} + \mathbf{P} = \mathbf{F} \tag{1}$$

where $\mathbf{M}$ is the mass matrix (lumped in this study), $\mathbf{P}$ is the vector of nodal points containing nonlinear internal forces, and $\mathbf{F}$ is the vector of nodal points containing externally applied forces. $\ddot{\mathbf{X}}$ is the acceleration vector which is the second derivative in time of the displacement vector $\mathbf{X}$. Equation (1) is an initial-value problem described by a system of differential equations in which the initial conditions are $\mathbf{X}(0) = \mathbf{X}_0$ and $\dot{\mathbf{X}}(0) = \dot{\mathbf{X}}_0$ where $\dot{\mathbf{X}}$ is the velocity vector.

Each tetrahedral element in LDPM is composed of 12 facets (equivalent of integration points in classical finite elements). For each facet k, the traction vector $\mathbf{t}_k$ defined in Section 2 can be formally written as $\mathbf{t}_k = \mathbf{t}_k(\mathbf{e}_k)$, where $\mathbf{e}_k$ is the facet strain vector. The internal force vector can then be calculated as:

$$\mathbf{P} = \sum_e \sum_{k=1}^{12} l_k A_k \mathbf{B}_k^t \mathbf{t}_k(\mathbf{e}_k) \tag{2}$$

where $\mathbf{B}_k^t$ is the transpose of the strain-displacement matrix for facet k, A_k is the projected facet area and l_k is the length of the edge shared by facet k (see Section 2). The sum is performed over the 12 facets of each LDPM element and over all the elements of the considered volume.

The Newmark family of methods are based on the following two time-discretizations:

$$\dot{\mathbf{X}}_{t+\Delta t} = \dot{\mathbf{X}}_t + \Delta t[(1-\gamma)\ddot{\mathbf{X}}_t + \gamma\ddot{\mathbf{X}}_{t+\Delta t}] \tag{3}$$

and

$$\mathbf{X}_{t+\Delta t} = \mathbf{X}_t + \Delta t\dot{\mathbf{X}}_t + (\Delta t)^2[(1/2-\beta)\ddot{\mathbf{X}}_t + \beta\ddot{\mathbf{X}}_{t+\Delta t}] \tag{4}$$

where $\mathbf{X}_t$ (and $\mathbf{X}_{t+\Delta t}$), $\dot{\mathbf{X}}_t$ (and $\dot{\mathbf{X}}_{t+\Delta t}$), $\ddot{\mathbf{X}}_t$ (and $\ddot{\mathbf{X}}_{t+\Delta t}$) are the displacement, velocity, and acceleration vectors at time t and $t + \Delta t$, respectively. Δt is the time step. γ and β are constants defining the stability and accuracy of the algorithm.

For $\gamma = 1/2$ and $\beta = 0$, the integration scheme is the so-called central difference method (if M is diagonal, which is the case here). This method is the one used in ABAQUS Explicit and is conditionally stable. The central difference method is known to have the highest accuracy and maximum stability limit among explicit methods of second order in time (Dokainish & Subbaraj 1989). It is worth noting that the algorithm does not possess numerical dissipation. The semi-discrete dynamic equation to be solved for LDPM at time t then becomes:

$$\mathbf{M}\ddot{\mathbf{X}}_t = \mathbf{F}_t - \mathbf{P}_t \tag{5}$$

where $\mathbf{F}_t$ and $\mathbf{P}_t$ are the external and internal force vectors at time t, respectively.

When $\gamma = 1/2$ and $\beta = 1/4$, the algorithm is called average acceleration method and is the one used in CAST3M. This scheme is unconditionally stable and the global rate of convergence for displacements and velocities is second-order in time. There is no numerical dissipation associated with this method, similarly to the central difference method. In this case, the semi-discrete dynamic equation writes as:

$$\frac{4\mathbf{M}}{(\Delta t)^2}(\mathbf{X}_{t+\Delta t} - \mathbf{X}_t) = \mathbf{F}_{t+\Delta t} + \mathbf{F}_t - \mathbf{P}_{t+\Delta t} - \mathbf{P}_t + \frac{4\mathbf{M}\dot{\mathbf{X}}_t}{\Delta t} \tag{6}$$

This equation cannot be solved directly since $\mathbf{P}_{t+\Delta t}$ is not known. In CAST3M, a global residual is defined and minimized. For static simulations, the mass matrix term drops, i.e. the inertia is not considered. It is worth noting that the calculation of acceleration is not required to make the scheme progress in the average acceleration method.

4 COMPARISON IMPLICIT/EXPLICIT

4.1 Three-point bending as case study

In order to compare the performance of the new implicit implementation, a classical three-point bending simulation was considered. To provide realistic numerical data, the mix design and dimensions of the beam were chosen based on a real set of experimental data (Grégoire et al. 2013). The depth of the beam was $D = 100$ mm, with a span-to-depth ratio of 2.5, and out-of-plane thickness of 50 mm. The specimen was notched and the notch-to-depth ratio was $\alpha = 0.2$. The LDPM geometry were identified based on the actual mix design used in the experiments. The particle size distribution given in Grégoire et al. 2013 was matched exactly with a cut-off size at $d_0 = 4$ mm and values of $d_a = 10$ mm, $c = 260$ kg/m^3, $w/c = 0.626$, and $\rho = 2121$ kg/m^3 were taken. The model parameters were taken as follows: $E_0 = 57180$ MPa, $\alpha_0 = 0.25$, $\sigma_t = 2.9$ MPa, $l_t = 400$ mm, $\sigma_s/\sigma_t = 3.276$, $n_t = 0.2$, $r_s = 0$. The reader is referred to Pathirage et al. 2022 where the calibration process to obtained these parameters is discussed in detail. The remaining parameters not related to fracture were assumed based on section 5.3 in Cusatis et al. 2011 and are listed here for the sake of completeness: $\sigma_{c0} = 120$ MPa, $H_{c0}/E_0 = 0.4$, $k_{c0} = 2$, $k_{c1} = 1$, $k_{c2} = 5$, $\mu_0 = 0.2$, $\mu_\infty = 0$, $\sigma_{N0} = 600$ MPa, and $E_d/E_0 = 1$.

Three types of simulations were performed: dynamic explicit, dynamic implicit, and static implicit. All the simulations were performed with one single mesh, i.e. one random particle placement to avoid the scatter in the numerical data due to the spatial variability of aggregate size and distribution. The specimen was loaded under displacement control. For both explicit and implicit methods, the same velocity of 1 mm s^{-1} was used. For the implicit calculation, the criterion on the global residual was set equal to 10^{-3} and three different times steps were used: $\Delta t = 10^{-2}$ s, 10^{-3} s, and 10^{-4} s. The data output frequency was set equal to the time step. For the explicit calculation, the time step required for stability was $\Delta t_{cr} = 2.32 \times 10^{-7}$ s, and results were outputted every 10^{-4} s.

4.2 Load-displacement

Figure 1 shows the load-displacement curves for the three types of time integration and for the three different time steps in the implicit calculations. Two important points should be here highlighted. First, a velocity of 1 mm s^{-1} in combination with the density and

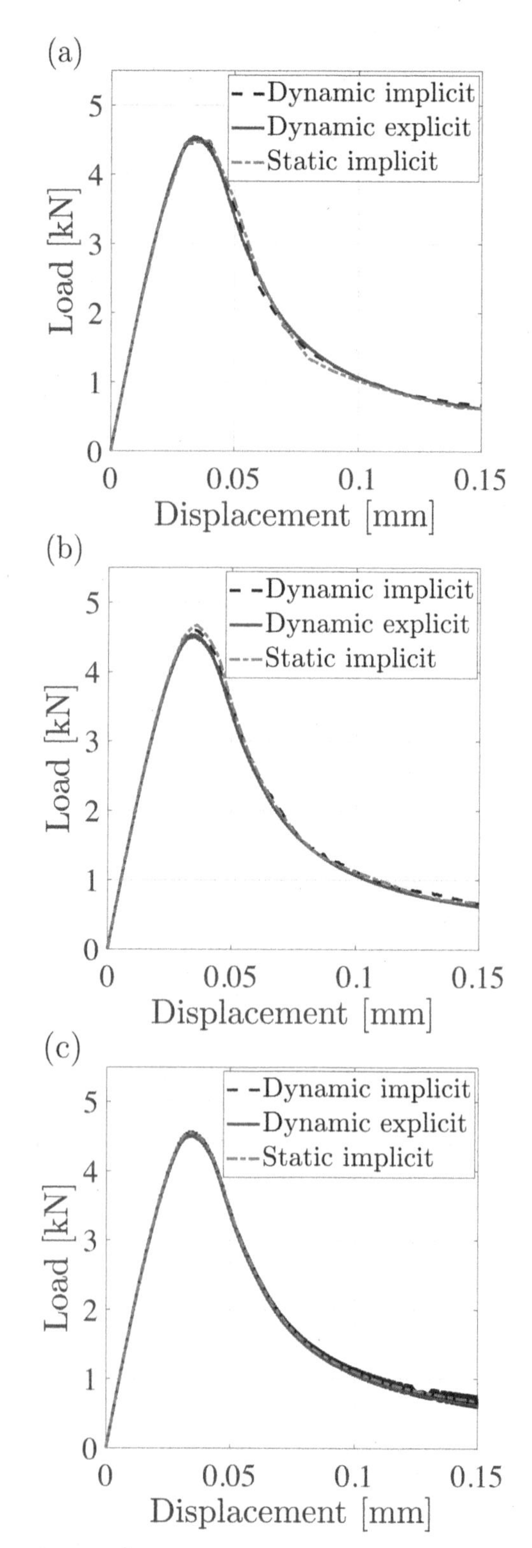

Figure 1. Load-displacement curves: time step in implicit calculations of (a) $\Delta t = 10^{-2}$ s, (b) $\Delta t = 10^{-3}$ s, and (c) $\Delta t = 10^{-4}$ s.

the model parameters defined earlier ensures a quasi-static loading condition. In addition, this result proves that one can perform quasi-static tests in dynamic explicit, as long as the effects of inertia are negligible, which is the case here since the dynamic explicit and static implicit curves overlap. In the implicit calculations, when the time step becomes smaller, i.e. from Figure 1(a) to (c), one can make two additional observations: (i) the static implicit result gets closer to the explicit one, not so much because the results gets more accurate but because there is an increase in the number of data points appearing in the figure, and (ii) high frequency oscillations appear in the post-peak for the smallest time step $\Delta t = 10^{-4}$ s and the source of these oscillations is explained next.

4.3 *Force imbalance from static equilibrium*

As shown above, the simulations were performed under quasi-static conditions. Therefore, a useful indicator, aside from comparing the load-displacement curves obtained with static implicit and dynamic explicit/implicit simulations, is the force imbalance from the static equilibrium at the loading points. This imbalance in percentage is here defined as $\Delta = 100(F_l - F_s)/F_s$ where F_l is the force at the loading points at mid-span and F_s is the vertical resultant reaction from the left and right supports. Figure 2 shows the force imbalance as a function of displacement for the three different time steps. At the very beginning of the simulation, there is a large force imbalance which increases if smaller time steps are considered, simply because the specimen is loaded from the top with a non-zero force whereas the reactions at the supports are initially zero. For the static implicit case, results showed that the imbalance was in the order of 10^{-11}% whatever the value of time step and for any value of displacement, i.e. in the elastic regime and in the post-peak. For the explicit simulation (see for example Figure 2(a)), the deviation Δ decreases up to the displacement corresponding to the peak load (about 0.37 mm).

This is due to the fact that the reactions at the supports slowly compensate the top force. However, before a steady state is reached, e.g. where the amplitude of the imbalance remains constant since there is a minimal or no energy dissipation, a small jump appears at peak load. Next, one can observe a continuous decrease up to 0.15 mm. This latter decrease is due to the energy dissipation in the fracture process and correspond the post-peak of the load-displacement curve. For the dynamic implicit simulations, the imbalance decays up to the displacement corresponding to the peak. As a matter of fact, the dynamic explicit and dynamic implicit curves shown in Figure 2(b) and (c) overlap up to 0.37 mm. In the post-peak regime, one can observe three different behaviors for the three time steps. When $\Delta t = 10^{-2}$ s (Figure 2(a)), the imbalance increases in time, even though the energy should be dissipated in damage. The reason might be that the chosen time step is too large and spurious energy growth (larger than the amount dissipated in fracture)

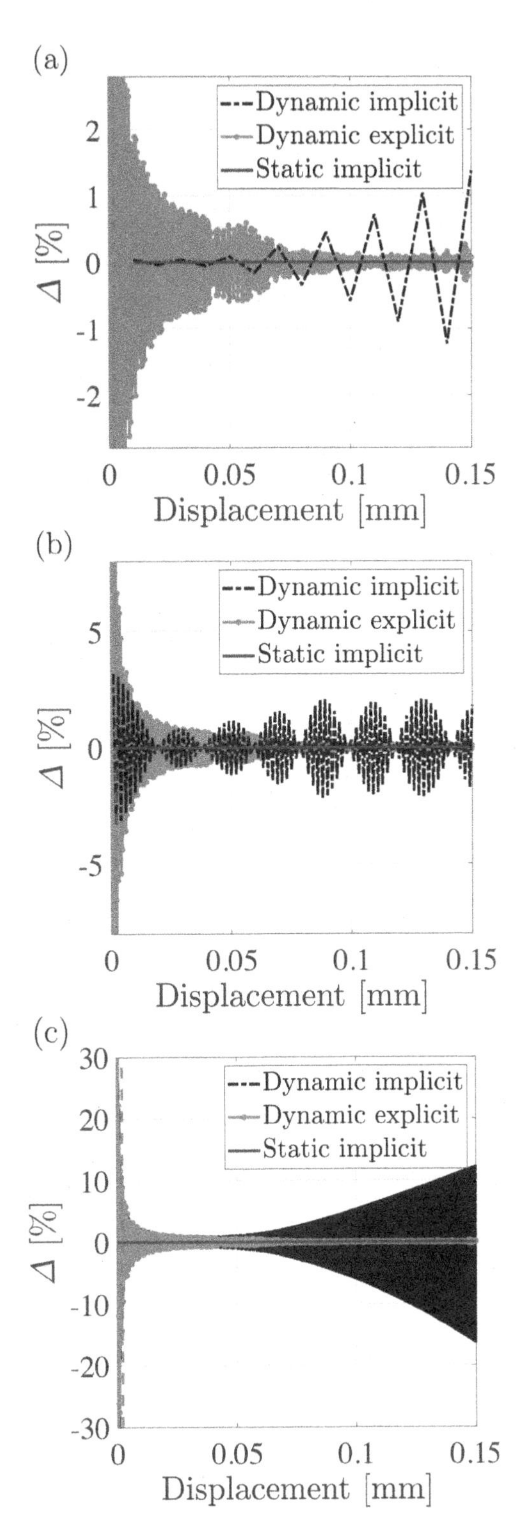

Figure 2. Force imbalance from static equilibrium: time step in implicit calculations of (a) $\Delta t = 10^{-2}$ s, (b) $\Delta t = 10^{-3}$ s, and (c) $\Delta t = 10^{-4}$ s. The red dashed vertical line is the limit after which the first crack appears, i.e. start of inelastic behavior.

Table 1. Computational cost in implicit calculations.

Integration scheme	Time step [s]	Memory [Gb]	Real time [hh:mm]	Time step [s]	Memory [Gb]	Real time [hh:mm]	Time step [s]	Memory [Gb]	Real time [hh:mm]
Dynamic implicit	10^{-2}	14.16	00:17	10^{-3}	16.71	00:35	10^{-4}	46.86	10:22
Static implicit	10^{-2}	13.82	00:15	10^{-3}	13.12	00:22	10^{-4}	46.99	10:08

occurs when integrating the internal forces in presence of softening behavior (Hughes 1976; Verpeaux & Charras 2013). This is a well-known phenomenon in the average acceleration method: the energy is not conserved when the material is nonlinear. When $\Delta t = 10^{-3}$ s (Figure 2(b)), it seems that the energy dissipated in cracks compensates the aforementioned spurious energy growth: the average amplitude remains the same but with periodic growth and decay events which might be related to the spurious energy introduced in the system and dissipated in damage periodically. Last but not least, when $\Delta t = 10^{-4}$ s (Figure 2(c)), one can observe an almost exponential increase of the force imbalance as displacement increases. For such a small time step, the spurious energy growth due to the integration of the internal force becomes negligible. The increase of Δ can be here explained by the presence of high-frequency modes that do not attenuate and is larger than the dissipation of energy in the fracturing process. These modes are artefacts of the discretization in time and does not originate from the actual model behavior (Hughes 2012). These higher-modes appear in the tail of the load-displacement curve in Figure1(c). It is worth noting that for $\Delta t = 10^{-2}$ s and $\Delta t = 10^{-3}$ s, low-modes of frequency dominate and the simulations are not plagued by these numerically introduced high-frequency modes. Finally, it is interesting to note that the criterion on the global residual was set equal to 10^{-3} for the dynamic implicit simulations. The reason why the force imbalance could exceed this value is that the global residual is computed by taking into account the overall specimen volume whereas the measure of imbalance at the loading points is much more conservative as it is a local measurement to assess equilibrium.

4.4 *Computational cost*

All the simulations were performed in Quest high performance computing facility at Northwestern University. For each simulation, 16 cores were used. Tables 1 and 2 provides the memory (RAM) and the wall-clock time used for each integration method. One can observe that the performance of the static implicit and dynamic implicit methods are equivalent. With respect to the accuracy reached using $\Delta t = 10^{-2}$ s and $\Delta t = 10^{-3}$ s in the implicit calculations (see Figure 1(a) and (b)), one can conclude that using the static or dynamic implicit solver is much faster than the dynamic explicit calculation in terms of running time, at least for this case study. In terms of memory, the implicit method requires more resources than for the explicit algorithm, which is expected. It is worth noting

Table 2. Computational cost in explicit calculation.

Integration scheme	Stable time step [s]	Memory [Gb]	Real time [hh:mm]
Dynamic explicit	2.32×10^{-7}	2.79	06:30

that increasing the output frequency in the explicit calculation increases the running time, which might be due to disk access time and other processes. A fair comparison would require to run the simulations without any output. However, the difference in running time would remain consequent and the conclusions in terms of computational cost would be unchanged.

5 CONCLUSIONS

In this study, preliminary results on the newly implemented Lattice Discrete Particle Model within the implicit solver CAST3M are presented. In order to compare it with the current implementation in Abaqus Explicit, a classical three-point bending test under quasi-static conditions was considered. Three different integration methods were used: dynamic explicit, dynamic implicit and static implicit. The results of the simulations were discussed in terms of load-displacement response and force imbalance at loading points to assess static equilibrium. The computational costs for each simulation were also obtained. Results show that:

- Quasi-static simulations can be performed using dynamic explicit integration method, with a higher velocity as compared to the actual velocity used in the test, if the effect of inertia is small enough.
- In the studied example, the computational cost for static and dynamic implicit simulations is much lower than for the dynamic explicit case.
- If the dynamic implicit method is employed, the time step size is to be chosen appropriately to avoid spurious energy growth or higher modes due to time discretization.

ACKNOWLEDGMENTS

Partial financial support from the investissement d'avenir French programme (ANR-16-IDEX-0002) under the framework of the E2S UPPA hub Newpores is gratefully acknowledged. This research was supported in part through the computational resources provided for the Quest high performance computing facility at Northwestern University which is jointly supported by the Office of the Provost, the Office for Research, and Northwestern University Information Technology.

REFERENCES

Alnaggar, M., G. Cusatis, & G. Di Luzio (2013). Lattice discrete particle modeling (ldpm) of alkali silica reaction (asr) deterioration of concrete structures. *Cement and Concrete Composites 41*, 45–59.

Angiolilli, M., A. Gregori, M. Pathirage, & G. Cusatis (2020). Fiber reinforced cementitious matrix (frcm) for strengthening historical stone masonry structures: Experiments and computations. *Engineering Structures 224*, 111102.

Angiolilli, M., M. Pathirage, A. Gregori, & G. Cusatis (2021). Lattice discrete particle model for the simulation of irregular stone masonry. *Journal of Structural Engineering 147*(9), 04021123.

Belytschko, T., W. K. Liu, B. Moran, & K. Elkhodary (2014). *Nonlinear finite elements for continua and structures*. John wiley & sons.

Cibelli, A., G. Di Luzio, L. Ferrara, G. Cusatis, M. Pathirage, et al. (2019). Modelling of autogenous healing for regular concrete via a discrete model. In *10th International Conference on Fracture Mechanics of Concrete and Concrete Structures, FraMCoS-X*, pp. 1–12. IA-FraMCoS.

Cibelli, A., M. Pathirage, G. Cusatis, L. Ferrara, & G. Di Luzio (2022). A discrete numerical model for the effects of crack healing on the behaviour of ordinary plain concrete: Implementation, calibration, and validation. *Engineering Fracture Mechanics 263*, 108266.

Cusatis, G., A. Mencarelli, D. Pelessone, & J. Baylot (2011a). Lattice discrete particle model (ldpm) for failure behavior of concrete. ii: Calibration and validation. *Cement and Concrete composites 33*(9), 891–905.

Cusatis, G., D. Pelessone, & A. Mencarelli (2011b). Lattice discrete particle model (ldpm) for failure behavior of concrete. i: Theory. *Cement and Concrete Composites 33*(9), 881–890.

Dokainish, M. & K. Subbaraj (1989). A survey of direct time-integration methods in computational structural dynamicsâŁ"i. explicit methods. *Computers & Structures 32*(6), 1371–1386.

Feng, J., W. Sun, L. Chen, B. Chen, E. Arkin, L. Du, & M. Pathirage (2022). Engineered cementitious composites using chinese local ingredients: Material preparation and numerical investigation. *Case Studies in Construction Materials 16*, e00852.

Grégoire, D., L. B. Rojas-Solano, & G. Pijaudier-Cabot (2013). Failure and size effect for notched and unnotched concrete beams. *International Journal for Numerical and Analytical Methods in Geomechanics 37*(10), 1434–1452.

Han, L., M. Pathirage, A.-T. Akono, & G. Cusatis (2020, 11). Lattice Discrete Particle Modeling of Size Effect in Slab Scratch Tests. *Journal of Applied Mechanics 88*(2). 021009.

Hughes, T. J. (1976). Stability, convergence and growth and decay of energy of the average acceleration method in nonlinear structural dynamics. *Computers & Structures 6*(4-5), 313–324.

Hughes, T. J. (2012). *The finite element method: linear static and dynamic finite element analysis*. Courier Corporation.

Mercuri, M., M. Pathirage, A. Gregori, & G. Cusatis (2020). Computational modeling of the out-of-plane behavior of unreinforced irregular masonry. *Engineering Structures 223*, 111181.

Mercuri, M., M. Pathirage, A. Gregori, & G. Cusatis (2021a). Lattice discrete modeling of out-of-plane behavior of irregular masonry. In *Proceedings of the 8th ECCOMAS Thematic Conference on Computational Methods in Structural Dynamics and Earthquake Engineering*, pp. 546–562.

Mercuri, M., M. Pathirage, A. Gregori, & G. Cusatis (2021b). On the collapse of the masonry medici tower: An integrated discrete-analytical approach. *Engineering Structures 246*, 113046.

Mercuri, M., M. Pathirage, A. Gregori, & G. Cusatis (2022). Masonry vaulted structures under spreading supports: Analyses of fracturing behavior and size effect. *Journal of Building Engineering 45*, 103396.

Pathirage, M., D. Bentz, G. Di Luzio, E. Masoero, & G. Cusatis (2019). The onix model: a parameter-free multiscale framework for the prediction of self-desiccation in concrete. *Cement and Concrete Composites 103*, 36–48.

Pathirage, M., D. P. Bentz, G. Di Luzio, E. Masoero, G. Cusatis, et al. (2018). A multiscale framework for the prediction of concrete self-desiccation. In *Computational modelling of concrete structures: proceedings of the conference on computational modelling of concrete and concrete structures (EURO-C 2018)*.

Pathirage, M., F. Bousikhane, M. D'Ambrosia, M. Alnaggar, & G. Cusatis (2019). Effect of alkali silica reaction on the mechanical properties of aging mortar bars: Experiments and numerical modeling. *International Journal of Damage Mechanics 28*(2), 291–322.

Pathirage, M., D. Tong, F. Thierry, G. Cusatis, D. Grégoire, & G. Pijaudier-Cabot (2022). Numerical modeling of concrete fracturing and size-effect of notched beams. In *Proceeding Euro-C*, Volume (this volume). CRC press.

Rezakhani, R., D. A. Scott, F. Bousikhane, M. Pathirage, R. D. Moser, B. H. Green, & G. Cusatis (2021). Influence of steel fiber size, shape, and strength on the quasi-static properties of ultra-high performance concrete: Experimental investigation and numerical modeling. *Construction and Building Materials 296*, 123532.

Schauffert, E. A. & G. Cusatis (2011). Lattice discrete particle model for fiber-reinforced concrete. i: Theory. *Journal of Engineering Mechanics 138*(7), 826–833.

Subbaraj, K. & M. Dokainish (1989). A survey of direct time-integration methods in computational structural dynamicsâŁ"ii. implicit methods. *Computers & structures 32*(6), 1387–1401.

Verpeaux, P. & T. Charras (2013). Dynamique du solide: modification du schéma de newmark aux cas non-linaires. *CSMA, Giens*.

Verpeaux, P., T. Charras, & A. Millard (1988). Castem 2000: une approche moderne du calcul des structures. *Calcul des structures et intelligence artificielle 2*, 261–271.

Yang, L., M. Pathirage, H. Su, M. Alnaggar, G. Di Luzio, & G. Cusatis (2021). Computational modeling of temperature and relative humidity effects on concrete expansion due to alkali–silica reaction. *Cement and Concrete Composites 124*, 104237.

Yang, L., M. Pathirage, H. Su, M. Alnaggar, G. Di Luzio, & G. Cusatis (2022). Computational modeling of expansion and deterioration due to alkali–silica reaction: Effects of size range, size distribution, and content of reactive aggregate. *International Journal of Solids and Structures 234-235*, 111220.

Zienkiewicz, O. C., R. L. Taylor, & J. Z. Zhu (2005). *The finite element method: its basis and fundamentals*. Elsevier.

Computational Modelling of Concrete and Concrete Structures – Meschke, Pichler & Rots (Eds)
© 2022 Copyright the Author(s), ISBN: 978-1-032-32724-2

Monotonic and fatigue behavior of cementitious composites modeled via a coupled sliding-decohesion-compression interface model

A. Baktheer, M. Aguilar & R. Chudoba
Institute of Structural Concrete, RWTH Aachen University, Aachen, Germany

M. Vořechovský
Institute of Structural Mechanics, Brno University of Technology, Brno, Czech Republic

ABSTRACT: In this paper we introduce a consistent constitutive model capturing the behavior of a 3D interface under both monotonic and cyclic loading. The model accounts for the interaction of dissipative effects during a combined decohesion-compression and sliding loading introduced through a smooth cap threshold function and non-associative flow potential. The proposed flow potential provides the possibility to couple the damage evolution of decohesion and sliding and to control the degree of this coupling. High computational efficiency is achieved by requiring a constant gradient of the threshold function in the normal direction with respect to the yield locus. As a result, a single-step return-mapping procedure without the need for iteration increases the computational efficiency. To capture the fatigue behavior in heterogeneous structures, the model consistently reflects the dissipative mechanisms of fatigue damage evolution at subcritical load levels using a cumulative measure of deformation as a damage-driving variable. Further effects of the interface response, such as shear dilatancy and vertex effect, are captured by the proposed model as demonstrated using elementary studies. A systematic calibration and validation procedure is included for selected applications showing a pull-out response of concrete-steel interface under monotonic and cyclic loading captured with a consistent set of material parameters.

1 INTRODUCTION

A deep understanding of the interfacial behavior is an essential prerequisite for realistic modeling and valid prediction of material behavior in various applications. To capture the behavior of an interface, it is necessary to construct the relationship between the main kinematic components such as decohesion-compression and sliding along the interface, which play the main role in the failure process of the interface. The crucial importance of a general and realistic description of the interfacial behavior lies in the fact that interacting effects of decohesion-compression and sliding determine the overall behavior of heterogeneous material structures and components.

The development of interface models has been one of the major research topics over the last decades. Several cohesive zone interface models have been developed by many authors, e.g., (Högberg 2006; McGarry, Máirtín, Parry, & Beltz 2014) to capture the mode I and mode II fracture, as well as the mixed mode loading condition (Dimitri, Trullo, Zavarise, & De Lorenzis 2014). An example of a thermodynamic consistent mixed-mode cohesive interface model with a single scalar damage variable can be found in (Serpieri, Sacco, & Alfano 2015). All these cohesive models focused on the monotonic behavior of the material interfaces.

The use of interface models to simulate the cyclic behavior of interface materials has been widely reported in the literature. General cohesive contact models have been introduced to study the response of material interfaces to cyclic loading histories in (Roe & Siegmund 2003; Harper & Hallett 2010). A slip based thermodynamic formulation of zero thickness interface model capable of capturing pressure sensitive bond behavior, interaction of damage and sliding at the interface was presented in (Ragueneau, Dominguez, & Ibrahimbegović 2006). Another bond model with an independent slip field has been used to study the bond behavior for cyclic loading histories in (Kwak & Kim 2006). Further model proposed recently in (Huang, Chi, Xu, & Deng 2019) has been applied for simulation of the bond in RC members under monotonic and cyclic loading. It should be noted that in all these models, however, no damage accumulation mechanism upon loading and reloading was considered, an effect that the authors consider essential for modeling of the fatigue behavior (Baktheer & Chudoba 2019; Baktheer, Aguilar, & Chudoba 2021).

To consider fatigue degradation, a thermodynamically consistent model coupling damage and plasticity

 DOI 10.1201/9781003316404-60

to describe the fatigue behavior of interfaces between FRP sheets and concrete surface under shear cyclic loading conditions has been introduced in (Carrara & De Lorenzis 2015). Another fracture mechanics based model used to simulate the monotonic and fatigue behavior of FRP/concrete interface proposed in (Martinelli & Caggiano 2014). A one dimensional thermodynamic based numerical model for bond fatigue behavior recently proposed by the authors (Baktheer & Chudoba 2018a, 2018b) is based on a coupled damage and inelastic slip within the bond interface with sensitivity to the lateral pressure/tension. The fatigue damage in this model is governed by a cumulative measure of the inelastic slip.

The objective of the present work is to promote a recent publication by the authors (Chudoba, Vořechovský, Aguilar, & Baktheer 2022) which introduces a general thermodynamically based 3D interface model that can consistently capture the monotonic and fatigue responses under all possible modes, i.e., mode I, mode II, and mixed mode, with a coupling of damage evolution in normal and tangential directions. The proposed constitutive model can be applied at different structural scales: 1) fatigue pull-out behavior of metallic and non-metallic fibers embedded in concrete, reflecting the effect of lateral pressure/tension, 2) microplane model for concrete with the microplane response governed by the described interface model, 3) discrete lattice or particle mesoscale model with the inter-aggregate interaction represented by the described interface model, 4) discrete crack models, including XFEM and embedded crack models, cohesive-zone models, semi-analytical models for shear crack propagation. Furthermore, the thermodynamic formulation of the model provides the possibility to evaluate the individual fractions of energy dissipated due to damage or plasticity in either tension, shear or compression. This feature will serve as a basis for a sound specification of regularized fatigue propagation criteria applicable in a broad range of applications.

2 THERMODYNAMIC BASED FORMULATION

2.1 *Free energy potential, state variables and thermodynamic forces*

The relative displacement of two points connected via the interface is represented by a normal (out-of-plane) component $u_\mathrm{N} \equiv u_z$ and an (in-plane) sliding vector $u_\mathrm{T} = \{u_x, u_y\}^\mathrm{T}$. The vector of kinematic variables defining the irreversible state of the interface is introduced as follows

$$\boldsymbol{\mathcal{E}} := \left[u_\mathrm{N}^\mathrm{p},\, \omega_\mathrm{N},\, u_\mathrm{T}^\mathrm{p},\, \omega_\mathrm{T},\, z,\, \boldsymbol{\alpha}\right]. \tag{1}$$

To provide a transparent representation of dissipative mechanisms in the normal and tangential directions, the free energy is introduced as a sum of out-of-plane opening (N) and in-plane sliding (T) contributions

$$\rho\psi(\boldsymbol{\mathcal{E}}) := \rho\psi_\mathrm{N}(\boldsymbol{\mathcal{E}}) + \rho\psi_\mathrm{T}(\boldsymbol{\mathcal{E}}). \tag{2}$$

Free energy associated with interface opening and closing (N) is defined as a function of total displacement u_N, plastic displacement u_N^p, and damage ω_N as

$$\rho\psi_\mathrm{N}(u_\mathrm{N}, u_\mathrm{N}^\mathrm{p}, \omega_\mathrm{N}) := \frac{1}{2}\left(1 - H(\sigma_\mathrm{N})\,\omega_\mathrm{N}\right) E_\mathrm{N}(u_\mathrm{N} - u_\mathrm{N}^\mathrm{p})^2, \tag{3}$$

where E_N denotes the stiffness. The Heaviside step function $H(.)$ is used to introduce the unilateral effect by activating the damage only for positive values of the traction stress σ_N. The free energy associated with the interface sliding is defined as a function of total sliding vector u_T, plastic sliding vector u_T^p, tangential damage ω_T, and the displacement variables corresponding to isotropic and kinematic hardening, z and $\boldsymbol{\alpha} = [\alpha_x, \alpha_y]$, respectively, as

$$\rho\psi_\mathrm{T}(u_\mathrm{T}, u_\mathrm{T}^\mathrm{p}, \omega_\mathrm{T}, z, \boldsymbol{\alpha}) := \frac{1}{2}(1 - \omega_\mathrm{T})E_\mathrm{T} \tag{4}$$
$$\left[(u_\mathrm{T} - u_\mathrm{T}^\mathrm{p})^\mathrm{T} \cdot (u_\mathrm{T} - u_\mathrm{T}^\mathrm{p})\right] + \frac{1}{2}Kz^2 + \frac{1}{2}\gamma(\boldsymbol{\alpha}^\mathrm{T} \cdot \boldsymbol{\alpha}),$$

where E_T denotes the tangential stiffness, K the isotropic and γ the kinematic hardening moduli. The thermodynamic forces are obtained by differentiating the free energy with respect to the kinematic state variables

$$\boldsymbol{\mathcal{S}} = \boldsymbol{\Upsilon}\,\frac{\partial\rho\psi(\boldsymbol{\mathcal{E}})}{\partial\boldsymbol{\mathcal{E}}}. \tag{5}$$

The sign vector operator $\boldsymbol{\Upsilon}$ is introduced to render positive thermodynamic force for positive state variable. To distinguish the thermodynamic forces based on the in correspondence with the definition of the state vector in Eq. (1), let us introduce the generalized vector of thermodynamic forces as

$$\boldsymbol{\mathcal{S}} := \left[\sigma_\mathrm{N}^\mathrm{p},\, Y_\mathrm{N},\, \boldsymbol{\sigma}_\mathrm{T}^\mathrm{p},\, Y_\mathrm{T},\, Z,\, X\right]. \tag{6}$$

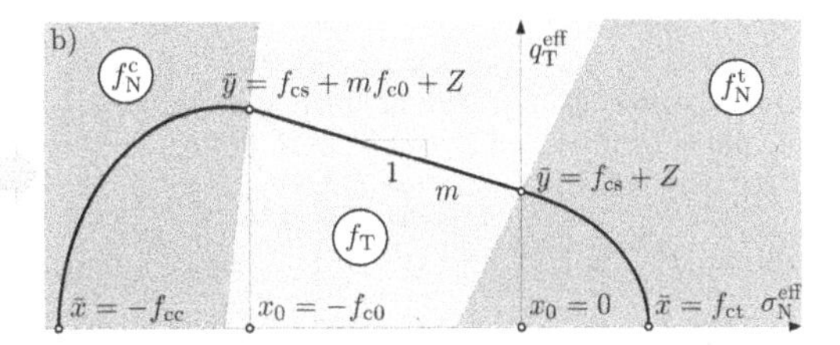

Figure 1. Transition between elliptic and linear domains of the introduced threshold function.

The individual components of this vector can be obtained using Eq. (5) The individual components of this vector are obtained using Eq. (5). The normal plastic stress σ_N^p and normal energy release rate Y_N read

$$\sigma_N^p = -\rho \frac{\partial \psi(\mathcal{E})}{\partial u_N^p} = \left(1 - H(\sigma_N^p)\,\omega_N\right) E_N(u_N - u_N^p), \quad (7)$$

$$Y_N = -\rho \frac{\partial \psi(\mathcal{E})}{\partial \omega_N} = \frac{1}{2} E_N (u_N - u_N^p)^2. \quad (8)$$

The Heaviside switches off the damage in compression to represent the stiffness recovery upon interface closure. The tangential plastic stress vector $\boldsymbol{\sigma}_T^p$, energy release rate Y_T, isotropic hardening stress Z and back stress $X = \{X_x, X_y\}^T$ are obtained as follows

$$\boldsymbol{\sigma}_T^p = -\rho \frac{\partial \psi(\mathcal{E})}{\partial u_T^p} = (1 - \omega_T) E_T (u_T - u_T^p) \quad (9)$$

$$Y_T = -\rho \frac{\partial \psi(\mathcal{E})}{\partial \omega_T} = \frac{1}{2} E_T \left[(u_T - u_T^p)^T \cdot (u_T - u_T^p)\right] \quad (10)$$

$$Z = \rho \frac{\partial \psi(\mathcal{E})}{\partial z} = Kz, \quad X = \rho \frac{\partial \psi(\mathcal{E})}{\partial \boldsymbol{\alpha}} = \gamma\, \boldsymbol{\alpha}. \quad (11)$$

2.2 *Threshold function*

The threshold function defines the elastic domain in terms of the effective normal and tangential stresses σ_N^{eff}, $\boldsymbol{\sigma}_T^{eff}$, which represent the stress level in the undamaged skeleton of material. Therefore, the elastic threshold function expressed in effective stress space. The envelope represents an intrinsic material property prescribing the character of interaction between normal and shear response of the material interface. The relationship between the effective stresses σ_N^{eff}, $\boldsymbol{\sigma}_T^{eff}$ and apparent stresses $\sigma_N, \boldsymbol{\sigma}_T$ can be established by realizing that they are identical with their plastic counterparts $\sigma_N^p, \boldsymbol{\sigma}_T^p$. With reference to Eqs. (7) and (9) we can write

$$\sigma_N^{eff}(\sigma_N^p, \omega_N) = \frac{\sigma_N^p}{1 - H(\sigma_N^p)\omega_N} = E_N(u_N - u_N^p) \quad (12)$$

$$\boldsymbol{\sigma}_T^{eff}(\boldsymbol{\sigma}_T^p, \omega_T) = \frac{\boldsymbol{\sigma}_T^p}{1 - \omega_T} = E_T(u_T - u_T^p). \quad (13)$$

The shape of the elastic domain in the effective stress space displayed in Figure 1b. It is convenient to introduce the norm q_T^{eff} representing a stress-like variable related to the elastic-plastic behavior of the material skeleton. It is defined as the norm of the difference between the effective tangential elastic $\boldsymbol{\sigma}_T^{eff}$ and the tangential back stress X representing the shift of the origin of the elastic domain, i.e.

$$q_T^{eff} = \|\boldsymbol{\sigma}_T^{eff} - X\| \quad (14)$$

The shear limit, f_T, follows the Mohr-Coulomb criterion, which represents the frictional, pressure-sensitive interface enhanced with kinematic and isotropic hardening as

$$f_T(\sigma_N^{eff}, q_T^{eff}, Z; f_{cs}, m) = q_T^{eff} - (f_{cs} + Z) + m\,\sigma_N^{eff} \quad (15)$$

where f_{cs} denotes the shear stress limit and m is the pressure sensitivity factor. This form of threshold function has been used by the authors in (Baktheer & Chudoba 2018b) to simulate the fatigue behavior of the bond loaded in pull-out condition. Since the shear stress dominates in the pull-out problem, the threshold function given in Eq. (15) was sufficient to deliver realistic results. However, for structural configurations affected by an interaction of normal and tangential dissipative mechanisms, the threshold function must be extended to account for the tensile and compressive strength limits. The mathematical formulation of the multi-domain level set function satisfies the requirements of convexity, continuity and smoothness at an arbitrary level $f(\mathcal{E}, \mathcal{S}) = \ell$. It consists of three parts corresponding to the tension-, compression- and shear-dominated subdomains within the effective stress half-space shown in Figure 1b. The subdomains associated with the tensile and compressive limits in normal direction are introduced using elliptical functions f_N^t and f_N^c, respectively, which are connected by the linear, pressure-sensitive shear threshold f_T given in Eq. (15) as illustrated in Figure 1b.

The elliptical parts of the threshold function f_N^t and f_N^c with smooth transition to the linear part f_T are derived by imposing the continuity and smoothness conditions displayed in Figure 1a. The derivation starts with the parameterized form of the linear and the elliptic part within the (x, y) domain:

$$f_{lin}(x, y; \bar{y}, x_0, m) := |y| - \bar{y} + m(x - x_0), \quad (16)$$

$$f_{ell}(x, y; \bar{x}, \bar{y}, x_0, m) := \sqrt{\frac{y^2}{b^2} + \frac{(x - x_0 - x_c)^2}{a^2}} - c,$$

where the parameters $\bar{x}, x_0, \bar{y}, m$ prescribe the shape of the envelope and a, b, c, x_c are unknown variables that are solved to comply the following compatibility and smoothness conditions:

Table 1. Threshold compatibility and smoothness conditions.

Normal limit:	$f_{ell}\|_{x=\bar{x}, y=0} = 0$
Tangential limit:	$f_{ell}\|_{x=x_0, y=\bar{y}} = 0$
Compatible N-T transition:	$f_{lin}(x_c, 0) = f_{ell}(x_c, 0)$
Smooth N-T transition:	$\left.\frac{f_{ell,x}}{f_{ell,y}}\right\|_{x=x_0, y=\bar{y}} = -m$

By introducing the distance between the normal limit $\bar{x}$ and the transition point x_0 as $\hat{x} = x_0 - \bar{x}$ the parameters a, b, c and x_c satisfying these conditions are obtained as follows

$$c=\bar{y}+\frac{m^2\hat{x}^2}{2m\hat{x}+\bar{y}},\quad a=-\frac{\bar{y}-m\hat{x}}{\bar{y}-2m\hat{x}}\cdot\frac{\hat{x}}{c},\qquad (17)$$
$$b=\frac{\bar{y}-m\hat{x}}{\sqrt{\bar{y}-2m\hat{x}}}\cdot\frac{\sqrt{\bar{y}}}{c},\quad x_c=-\frac{m\hat{x}^2}{2m\hat{x}+\bar{y}}$$

After substituting these values in f_{lin} and f_{ell} in Eqs. (16), the solution of the smooth transition between the linear and elliptic parts shown in Figure 1a can be instantiated to the tensile and compression domains as illustrated in Figure 1b using the substitutions

$$\begin{aligned}\mathcal{N}^{\text{t}} &:= \{x=\sigma_{\text{N}}^{\text{eff}}, y=q_{\text{T}}^{\text{eff}}, \bar{x}=f_{\text{ct}}, \bar{y}=f_{\text{cs}}+Z, x_0=0\}\\ \mathcal{N}^{\text{c}} &:= \{x=-\sigma_{\text{N}}^{\text{eff}}, y=q_{\text{T}}^{\text{eff}}, \bar{x}=f_{\text{cc}},\\ &\qquad \bar{y}=f_{\text{cs}}+mf_{\text{c0}}+Z, x_0=f_{\text{c0}}\}\end{aligned}\qquad (18)$$

To compose the multi-domain threshold function, the transition line between the linear and elliptical domains, plotted as a dashed line in Figure 1a is defined as the line connecting the points $[x_c, 0]$ and $[x_0, \bar{y}]$. Then, the subdomains of the elliptic part can be readily expressed using a level set function

$$\mathcal{T}(x, y; \bar{x}, \bar{y}, x_0) := \frac{\bar{y}}{x_0 - x_c}(x - x_c) - |y| > 0, \qquad (19)$$

which is instantiated for the tensile or compressive subdomains in Figure 1b using the substitutions $\mathcal{T}(\mathcal{N}^{\text{t}})$ and $\mathcal{T}(\mathcal{N}^{\text{c}})$ given in Eq. (18). The threshold function within the effective stress space $\{\sigma_{\text{N}}^{\text{eff}}, \boldsymbol{\sigma}_{\text{T}}^{\text{eff}}\} \subset \boldsymbol{\mathcal{S}}^{\text{eff}}$ consisting of the tensile, shear and compressive subdomains can now be defined as a piecewise function

$$f(\boldsymbol{\mathcal{S}}^{\text{eff}}) := \begin{cases} f_{\text{N}}^{\text{t}} = f_{\text{ell}}(\mathcal{N}^{\text{t}}) & \text{if } \mathcal{T}(\mathcal{N}^{\text{t}}) > 0 \\ f_{\text{N}}^{\text{c}} = f_{\text{ell}}(\mathcal{N}^{\text{c}}) & \text{if } \mathcal{T}(\mathcal{N}^{\text{c}}) > 0 \\ f_{\text{T}} & \text{otherwise} \end{cases} \qquad (20)$$

Note that the parameters defining the shape of the yield locus $f_{\text{ct}}, f_{\text{cs}}, f_{\text{cc}}$, f_{c0} are substituted for the parameters $\bar{x}, \bar{y}, x_0$ appropriately in the two instantiations of f_{ell} for the tensile and compressive subdomains. Finally, to transform the threshold function $f(\boldsymbol{\mathcal{S}}^{\text{eff}})$ into the apparent stress space $\{\sigma_{\text{N}}, \boldsymbol{\sigma}_{\text{T}}\} \subset \boldsymbol{\mathcal{S}}$, the effective stresses in Eq. (20) must be substituted using Eqs. (12) and (13).

$$f(\boldsymbol{\mathcal{E}}, \boldsymbol{\mathcal{S}}) = f(\boldsymbol{\mathcal{S}}^{\text{eff}})\Big|_{\sigma_{\text{N}}^{\text{eff}}=\frac{\sigma_{\text{N}}}{1-\omega_{\text{N}}}, \boldsymbol{\sigma}_{\text{T}}^{\text{eff}}=\frac{\boldsymbol{\sigma}_{\text{T}}}{1-\omega_{\text{T}}}} \qquad (21)$$

After this substitution, the threshold function becomes dependent on the damage variables $\omega_{\text{N}}, \omega_{\text{T}}$. By including the internal variables $\boldsymbol{\mathcal{E}}$ explicitly in the argument list of $f(\boldsymbol{\mathcal{E}}, \boldsymbol{\mathcal{S}})$, we emphasize the fact that f depends on the internal variables $\boldsymbol{\mathcal{E}}$ not only indirectly, via the constitutive laws $\boldsymbol{\mathcal{S}}(\boldsymbol{\mathcal{E}})$ given in Eqs. (7)-(11), but also directly, via $\{\omega_{\text{N}}, \omega_{\text{T}}\} \subset \boldsymbol{\mathcal{E}}$.

2.3 *Non-associative flow potential accounting for damage interaction*

The flow potential extends the threshold function $f(\boldsymbol{\mathcal{E}}, \boldsymbol{\mathcal{S}})$ with additional terms controlling the evolution of damage. The goal of the flow potential definition is to account for all possible interactions between the normal damage and shear damage (ω_{N} and ω_{T}) in a general and transparent way.

To define the level of interaction between damage in normal direction (compression-decohesion) and shear direction (sliding), we introduce a material parameter $\eta \in \langle 0, 1 \rangle$, which provides a smooth transition between an uncoupled damage potential ($\eta = 0$) and a fully coupled potential ($\eta = 1$). In particular, we propose a linear transition between the limiting cases of fully uncoupled and fully coupled damage potentials. The transition is controlled via η parameter by reweighting the uncoupled and fully coupled contributions to the dissipation potential

$$\varphi(\boldsymbol{\mathcal{E}}, \boldsymbol{\mathcal{S}}) := f(\boldsymbol{\mathcal{E}}, \boldsymbol{\mathcal{S}}) + (1-\eta)\underbrace{(\varphi_{\text{N}} + \varphi_{\text{T}})}_{\varphi_{\text{u}}\ldots\text{uncoupled}} + \eta\underbrace{\varphi_{\text{NT}}}_{\text{coupled}} \qquad (22)$$

in which the term φ_{N} depends solely on the normal (out-of-plane) displacement and the related material parameters for the normal direction and, analogously, φ_{T} depends solely on the shear (in-plane) displacement and the related material parameters. A particular form of the φ_{N} and φ_{T} can be inferred by realizing that the damage evolution law is obtained by differentiating the damage potential terms w.r.t. energy density release rates Y_{N} and Y_{T}, respectively.

The potential functions corresponding to the normal and shear directions are defined as follows

$$\varphi_{\text{N}} := (1-\omega_{\text{N}})^{c_{\text{N}}} \frac{S_{\text{N}}}{r+1}\left(\frac{Y_{\text{N}}}{S_{\text{N}}}\right)^{r+1} H\left(\sigma_{\text{N}}^{\text{p}}\right) \qquad (23)$$

$$\varphi_{\text{T}} := (1-\omega_{\text{T}})^{c_{\text{T}}} \frac{S_{\text{T}}}{r+1}\left(\frac{Y_{\text{T}}}{S_{\text{T}}}\right)^{r+1} \qquad (24)$$

These potential functions allow to obtain a damage evolution equation in the form of the first ordinary differential equation which results in a damage law that asymptotically approach the value of 1. The Heaviside function $H\left(\sigma_{\text{N}}^{\text{p}}\right)$ applied to the normal direction in Eq. (23) secures that damage is accumulated only when the stress $\sigma_{\text{N}}^{\text{p}}$ is positive (tension).

Analogously to the formulations for normal and shear directions above, the *fully coupled* (mixed) flow potential term is proposed to read

$$\varphi_{\text{NT}} := (1-\omega_{\text{NT}})^{c_{\text{NT}}} \frac{S_{\text{NT}}}{r+1}\left(\frac{Y_{\text{N}} + Y_{\text{T}}}{S_{\text{NT}}}\right)^{r+1} \qquad (25)$$

where we introduce the geometric mean values for the two material parameters: the exponent, c_{NT}, and the S_{NT} parameter related to material ductility as

$$S_{\text{NT}} := \sqrt{S_{\text{N}} S_{\text{T}}}, \qquad c_{\text{NT}} := \sqrt{c_{\text{N}} c_{\text{T}}} \qquad (26)$$

Apart from the averaged *material parameters*, the coupled term introduces a single *damage parameter*, ω_{NT}. We propose to derive the damage of the fully coupled model from the two damage parameters for normal and shear directions. We again introduce a kind of averaging between the two damage parameters ω_{N} and ω_{T}. However, instead of averaging the damage variables, we average their complementary values, i.e. the *integrity* parameters

$$1-\omega_{\mathrm{NT}} := \sqrt{(1-\omega_{\mathrm{N}})\,(1-\omega_{\mathrm{T}})} \tag{27}$$

The reason for this choice that for small values of ω_{N} and ω_{T}, this formulation behaves similar to the arithmetic average $\omega_{\mathrm{NT}} \approx \frac{1}{2}(\omega_{\mathrm{N}}+\omega_{\mathrm{T}})$ which is intuitively acceptable. However, when any of the two damages becomes large, Eq. (27) sets ω_{NT} closer to the maximum of the two: $\max(\omega_{\mathrm{N}}, \omega_{\mathrm{T}})$ while the arithmetic average would simply take the average integrity.

2.4 *Evolution equations*

The corresponding directions of flow is given as the product of the sign operator Υ and of the derivatives with respect to state variables $\boldsymbol{\Upsilon}\,\nabla_{\boldsymbol{S}}\,\varphi$. This renders the following flow direction vector

$$\boldsymbol{\Phi}(\boldsymbol{\mathcal{E}},\boldsymbol{\mathcal{S}}) = -\Upsilon\,\frac{\partial\varphi(\boldsymbol{\mathcal{E}},\boldsymbol{\mathcal{S}})}{\partial\boldsymbol{\mathcal{S}}(\boldsymbol{\mathcal{E}})} \tag{28}$$

The evolution equations (normality rule) of the state variables can then be written as follows

$$\dot{\boldsymbol{\mathcal{E}}} = \lambda\,\boldsymbol{\Phi}(\boldsymbol{\mathcal{E}},\boldsymbol{\mathcal{S}}) = -\lambda\,\Upsilon\,\frac{\partial\varphi(\boldsymbol{\mathcal{E}},\boldsymbol{\mathcal{S}})}{\partial\boldsymbol{\mathcal{S}}(\boldsymbol{\mathcal{E}})} \tag{29}$$

2.5 *Energy dissipation*

The thermodynamic admissibility of the proposed interface model for any kind of loading scenario can be evaluated with the help of the Clausius-Duhem inequality ($\mathcal{D} \geq 0$). Where $\mathcal{D}$ is the dissipated energy given as the sum of the normal and tangential dissipated energy

$$\begin{aligned}\mathcal{D} = \mathcal{D}_{\mathrm{N}} + \mathcal{D}_{\mathrm{T}} &= -\frac{\partial\rho\psi(\boldsymbol{\mathcal{E}})}{\partial\boldsymbol{\mathcal{E}}}\cdot\dot{\boldsymbol{\mathcal{E}}} \\ &= -\left[\frac{\partial\rho\psi_{\mathrm{N}}(\boldsymbol{\mathcal{E}})}{\partial\boldsymbol{\mathcal{E}}}\cdot\dot{\boldsymbol{\mathcal{E}}} + \frac{\partial\rho\psi_{\mathrm{T}}(\boldsymbol{\mathcal{E}})}{\partial\boldsymbol{\mathcal{E}}}\cdot\dot{\boldsymbol{\mathcal{E}}}\right],\end{aligned} \tag{30}$$

where the normal dissipated energy $\mathcal{D}_{\mathrm{N}}$ is given as

$$\mathcal{D}_{\mathrm{N}} = -\frac{\partial\rho\psi_{\mathrm{N}}(\boldsymbol{\mathcal{E}})}{\partial\boldsymbol{\mathcal{E}}}\cdot\dot{\boldsymbol{\mathcal{E}}} = \sigma_{\mathrm{N}}^{\mathrm{p}}\,\dot{u}_{\mathrm{N}}^{\mathrm{p}} + Y_{\mathrm{N}}\,\dot{\omega}_{\mathrm{N}}, \tag{31}$$

and the tangential dissipated energy $\mathcal{D}_{\mathrm{T}}$ is obtained as

$$\begin{aligned}\mathcal{D}_{\mathrm{T}} = -\frac{\partial\rho\psi_{\mathrm{T}}(\boldsymbol{\mathcal{E}})}{\partial\boldsymbol{\mathcal{E}}}\cdot\dot{\boldsymbol{\mathcal{E}}} &= \boldsymbol{\sigma}_{\mathrm{T}}^{\mathrm{p}}\cdot\dot{\boldsymbol{u}}_{\mathrm{T}}^{\mathrm{p}} \\ &\quad -(Z\,\dot{z} + \boldsymbol{X}\cdot\dot{\boldsymbol{\alpha}}) + Y_{\mathrm{T}}\,\dot{\omega}_{\mathrm{T}}.\end{aligned} \tag{32}$$

The energy dissipation owing to plasticity i.e. plastic dissipation for both normal and tangential directions can be written as

$$\begin{aligned}\mathcal{D}^{\mathrm{P}} = \mathcal{D}_{\mathrm{N}}^{\mathrm{P}} + \mathcal{D}_{\mathrm{T}}^{\mathrm{P}} &= \sigma_{\mathrm{N}}^{\mathrm{p}}\,\dot{u}_{\mathrm{N}}^{\mathrm{p}} + \boldsymbol{\sigma}_{\mathrm{T}}^{\mathrm{p}}\cdot\dot{\boldsymbol{u}}_{\mathrm{T}}^{\mathrm{p}} \\ &\quad -(Z\,\dot{z} + \boldsymbol{X}\cdot\dot{\boldsymbol{\alpha}}),\end{aligned} \tag{33}$$

which represents the difference between the plastic work $\mathcal{W}^{\mathrm{P}}$ and the plastic free energy related to the isotropic and kinematic hardening, respectively $\mathcal{U}^{\mathrm{iso}}, \mathcal{U}^{\mathrm{kin}}$ (Yang, Sinha, Feng, McCallen, & Jeremić 2018).

$$\mathcal{D}^{\mathrm{P}} = \mathcal{W}^{\mathrm{P}} - (\mathcal{U}^{\mathrm{iso}} + \mathcal{U}^{\mathrm{kin}}) \tag{34}$$

The plastic work for both normal and tangential direction reads

$$\mathcal{W}^{\mathrm{P}} = \sigma_{\mathrm{N}}^{\mathrm{p}}\,\dot{u}_{\mathrm{N}}^{\mathrm{p}} + \boldsymbol{\sigma}_{\mathrm{T}}^{\mathrm{p}}\cdot\dot{\boldsymbol{u}}_{\mathrm{T}}^{\mathrm{p}}, \tag{35}$$

and the free isotropic and kinematic energy can be written as

$$\mathcal{U}^{\mathrm{iso}} = Z\,\dot{z}, \qquad \mathcal{U}^{\mathrm{kin}} = \boldsymbol{X}\cdot\dot{\boldsymbol{\alpha}}. \tag{36}$$

In similar way to the plastic dissipation Eq. (33), the damage dissipation can be written as

$$\mathcal{D}^{\omega} = \mathcal{D}_{\mathrm{N}}^{\omega} + \mathcal{D}_{\mathrm{T}}^{\omega} = Y_{\mathrm{N}}\,\dot{\omega}_{\mathrm{N}} + Y_{\mathrm{T}}\,\dot{\omega}_{\mathrm{T}} \tag{37}$$

The total input work $\mathcal{W}^{\mathrm{tot}}$ can be defined as a sum of the elastic strain energy i.e stored energy $\mathcal{W}^{\mathrm{el}}$ and the inelastic work $\mathcal{W}^{\mathrm{in}}$ of both the damage and the plastic mechanism

$$\mathcal{W}^{\mathrm{tot}} = \mathcal{W}^{\mathrm{el}} + \mathcal{W}^{\mathrm{in}} \tag{38}$$

where the inelastic work is the sum of the plastic work and the damage dissipation

$$\mathcal{W}^{\mathrm{in}} = \mathcal{W}^{\mathrm{P}} + \mathcal{D}^{\omega} \tag{39}$$

To demonstrate the described different parts of the energy, elementary example of the model behavior under monotonic loading is presented in Sec. 3.4 showing the described portions of the energy.

3 ELEMENTARY STUDIES

3.1 *Coupled sliding-decohesion-compression behavior under proportional loading*

An elementary example of the coupled sliding-decohesion-compression behavior of a single material point under proportional loading is summarized in Figure 2. Five different loading scenarios are considered, including pure tension, pure compression, pure sliding, and sliding with tension or compression, as depicted in Figure 2a. The decohesion-compression response of the normal direction for the studied proportional loading is shown in Figure 2b. This response demonstrates the decrease of

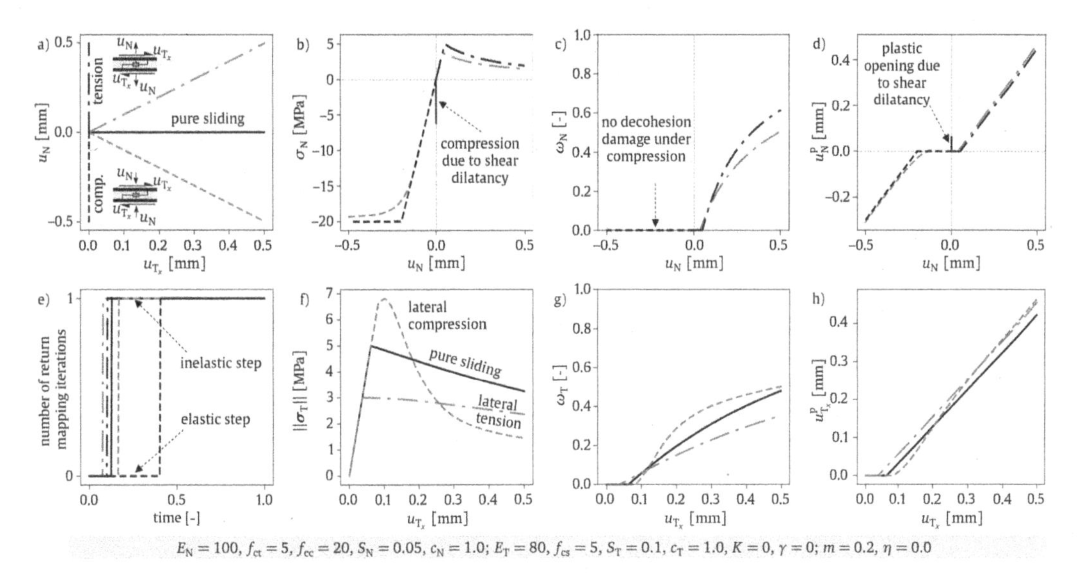

Figure 2. Elementary study showing the coupled sliding-decohesion-compression behavior under simultaneous loading.

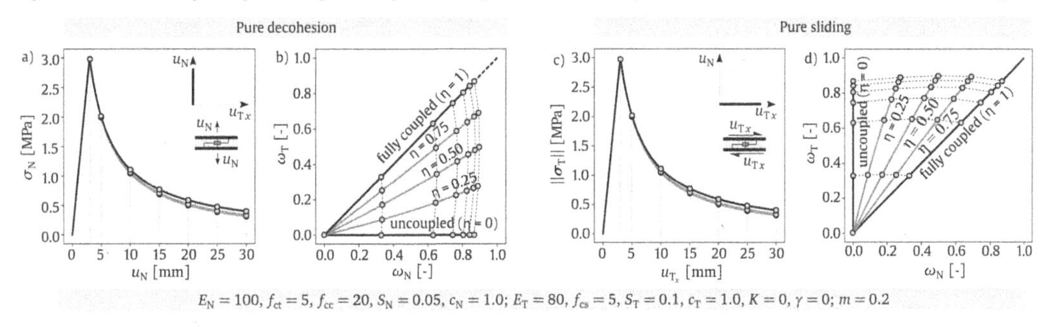

Figure 3. Elementary study showing the damage interaction feature between decohesion and sliding.

the decohesion/compression strength in case of combined normal-tangential loading in comparison to the pure decohesion/compression behavior. This mutual interaction on the achieved strength is governed by the introduced smooth threshold cap function, where the maximum tensile/compression strength f_{ct},f_{cc} can be only achieved in case of pure decohesion/compression loading. The corresponding damage evolution curves are shown in Figure 2c. This shows that the normal damage develops only under tension, while no normal damage can develop in the cases of compression. This feature of the proposed model is introduced through the Heaviside function in Eq. (3). On the other hand, the model considers that plastic deformation can develop under both tensile and compressive normal loading, as shown in Figure 2d. Although the tensile behavior of cementitious materials is usually described in terms of pure damage behavior, experimental observations of the tensile behavior of various materials such as concrete inherently exhibit inelastic deformation e.g. (Hordijk 1992; Horii, Shin, & Pallewatta 1992). The effect of decohesion/compression on the sliding behavior is summarized in Figures. 2f, g, h. The response to the studied loading scenarios, i.e. pure sliding, sliding under lateral compression and sliding under lateral tension, is illustrated in Figure 2f. The response shows that the achieved shear strength increases under lateral compression and decreases under lateral tension. This response represents a plausible trend that has been observed experimentally for many types of interfaces, such as the bond behavior between concrete and steel reinforcement e.g. (Eligehausen, Popov, & Bertero 1982; Lindorf, Lemnitzer, & Curbach 2009). The corresponding evolutions of damage and plastic deformation for the studied cases are shown in Figure 2g, f, respectively. This qualitative study highlights that the sensitivity of the sliding behavior with respect to lateral compression/tension is reflected by the model, thus providing a solid representation of the underlying physical phenomenon. Further more, the recorded number of iterations needed for the return mapping is shown in Figure 2e which demonstrates the single step return mapping needed in most of the cases.

3.2 *Decohesion-sliding damage interaction*

The aim of the study shown in Figure 3 is to highlight the *damage interaction* feature of the proposed model. Besides the coupling between the decohesion and

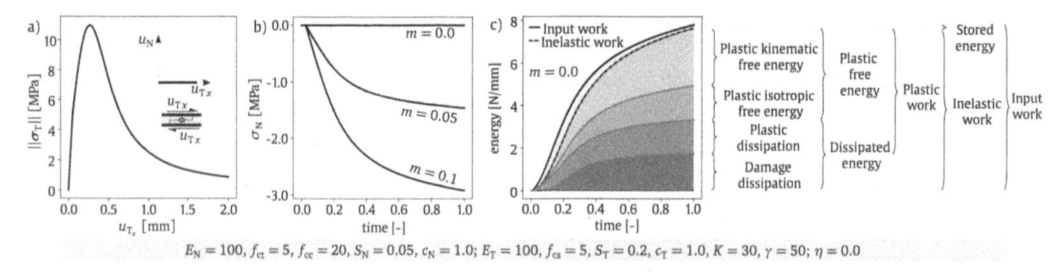

Figure 4. Elementary study showing the captured shear dilatancy and the evaluation of energy dissipation.

sliding through the threshold function Eq. (20) which governs the strength both of decohesion-compression and sliding, another level of coupling has been introduced through the non-associative flow potential Eq. (22). Due to this coupling, the sliding damage can develop under purely decohesion loading, as well as the decohesion damage can develop under purely sliding loading, which indicates the mutual interaction of decohesion and sliding damage controlled by the material parameter η. The left part of Figure 3 represents the pure decohesion case, while the pure sliding case is shown in the right part of Figure 3. The relationship between the evolution of decohesion and sliding damage for the pure decohesion and pure sliding responses is illustrated in Figures. 3b, d, respectively, for varied interaction parameter η, ranging from the uncoupled case ($\eta = 0$) to the fully coupled case ($\eta = 1$).

3.3 *Shear dilatancy*

The aim of the study shown in Figure 4 is to highlight the shear dilatancy behavior captured by the proposed model. In this study, the interface is subjected to pure sliding loading, where the shear stress-displacement response is depicted in Figure 4a. However, during the shear loading especially after the elastic range the decohesion behavior exhibits a growth of compressive stress when the pressure sensitivity parameter $m > 0$ as shown in Figure 4b. This phenomenon is known as *shear dilatancy*, where a compression stress in the normal direction are induced by the shearing of the interface (Bažant & Gambarova 1984). The perpendicular return mapping to the yield surface for the cases with pressure sensitivity parameter $m > 0$ leads to the occurrence of compressive stresses in the normal direction. Indeed, such a phenomenon has been observed experimentally in the behavior of some interfaces, e.g., the frictional behavior of the crack surface in concrete, as documented in (Bažant & Gambarova 1980; Paulay & Loeber 1974).

3.4 *Analysis of energy dissipation*

Energy dissipation in a mechanical system represents the irreversible process, e.g., plastic deformation and damage in which energy is transformed from one form to another. The determination of the individual proportions of energy dissipation is of great importance, as it can be considered as an effective indicator of the main dissipative processes. Such an indicator can serve as a basis to capture the thermo-mechanical interaction effects within the material structure, especially under cyclic and fatigue loading. A comprehensive analysis of the energy dissipation of an elasto-plastic material behavior within the framework of thermodynamics was presented in (Yang, Sinha, Feng, McCallen, & Jeremić 2018). In this analysis, the separation of plastic work into two parts, namely plastic free energy and plastic dissipation, was introduced.

To highlight the possibility to evaluate the individual portion of energy with the proposed model, the example of pure sliding behavior presented in Figure 4a is accompanied with the evaluation of energy dissipation as shown in Figure 4c. With the thermodynamic base formulation of the proposed model, the energy fractions of the total input work, i.e. stored energies as well as dissipated energies, can be clearly evaluated and distinguished for any type of loading scenario as depicted in Figure 4c based on the description in Sec. 2.5.

3.5 *Cumulative damage for fatigue simulation*

In the classical damage models, the damage usually evolves once the control state variable, e.g. displacement, exceeds the last maximum displacement obtained so far during the loading history e.g., (Ragueneau, Dominguez, & Ibrahimbegović 2006), which can be only used to simulate the monotonic behavior. However, to obtain a unified model for monotonic, cyclic and fatigue behavior, it is essential that the damage evolution is governed by cumulative measure of strain/displacement allowing the damage to evolve during the unloading and reloading conditions (Baktheer, Spartali, Hegger, & Chudoba 2021; Desmorat, Ragueneau, & Pham 2007; Kirane & Bažant 2015; Lemaitre & Desmorat 2005), which is establishing a mechanism driving the material deterioration under cyclic and fatigue loading. This feature has been introduced through the non-associative flow potential Eq. (22), resulting that the damage evolution is linked with the plastic multiplier similar to the potential proposed in (Lemaitre & Desmorat 2005).

The study shown in Figure 5 presents an example of monotonic and cyclic behavior of the proposed model under shear/sliding displacement. The degradation of the sliding behavior under monotonic and cyclic loading with constant range of slip applied for

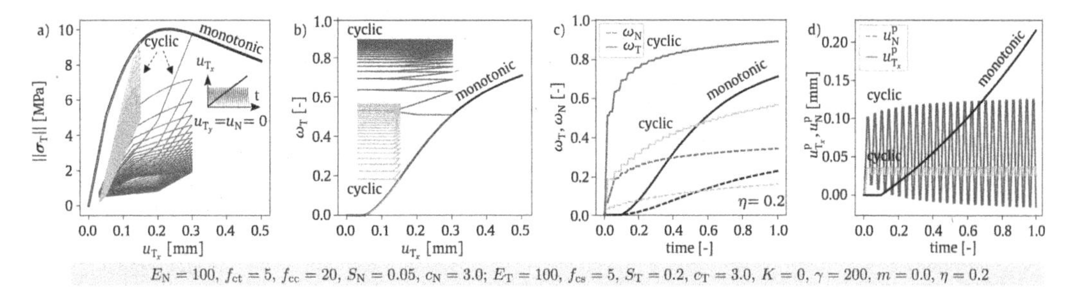

Figure 5. Elementary study showing the feature of cumulative damage for fatigue simulation.

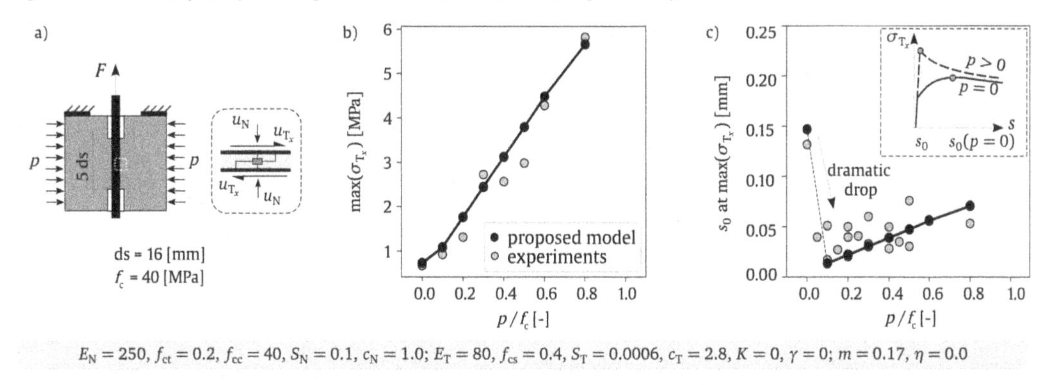

Figure 6. Bond between concrete and plain rebar steel reinforcement under lateral compression: comparison with experimental data.

50 cycles is presented in Figure 5a. The corresponding sliding damage evolution for monotonic and cyclic loading is depicted in Figure 5b, which show the feature of damage accumulation during unloading and reloading stages. As demonstrated in (Kirane & Bažant 2015) when linking the damage to a cumulative measure of strain/displacement with the goal to cover the high cycle fatigue behavior, damage must be accumulated slowly, within a large range of cumulative strain approaching $\omega = 1.0$ asymptotically. Another view of the damage accumulation is depicted in Figure 5c, showing the sliding and decohesion damage grow during the loading history for both monotonic and cyclic cases. The shape of the damage accumulation during the cyclic loading resulting in asymptotic manner. This feature has been covered by introducing a modified function of the Lemaitre's damage potential (Lemaitre & Desmorat 2005). The corresponding evolution of the plastic slip and irreversible opening displacement are plotted in Figure 5d for both monotonic and cyclic cases.

4 NUMERICAL APPLICATIONS

4.1 Bond between concrete and steel under lateral compression

In this example we study the ability of the model to reproduce the pullout behavior of plain steel reinforcement from concrete block and subjected to lateral pressure Figure 6. The results of the test program performed by (Xu, Wu, Zheng, Hu, & Li 2014) have been used in this study. Plain steel reinforcement with bar diameter of $d_s = 16$ mm in combination with concrete matrix C40 were used. The bond length was set to $L_b = 5ds$. The experimental results show no difference between the pullout response of the loaded and unloaded ends. Therefore, the assumption of constant bond stress distribution along the bond length cab be considered valid in this experimental study. Therefore, a single material point simulation of the interface is sufficient to reproduce the observed experimental behavior Figure 6a. The shear behavior has been studied for different levels of lateral pressure. The obtained results show an increase of the maximum bond stress with the increase of the level lateral pressure. The fit of the model response with the experimental data is shown in Figure 6b. The horizontal axis is normalized with respect to the concrete compressive strength. As a result of the maximum bond stress fit, the obtained slip values at the maximum bond stress are compared with the values recorded during the tests as shown in Figure 6c. The obtained numerical results show similar trend to the experimental results, especially the interesting dramatic drop of the slip value for the cases under lateral pressure in comparison to the case without lateral pressure shown in Figure 6c as discussed in (Xu, Wu, Zheng, Hu, & Li 2014).

4.2 Bond between concrete and steel under cyclic loading

To demonstrate the applicability of the model to the monotonic and cyclic behavior, another numerical

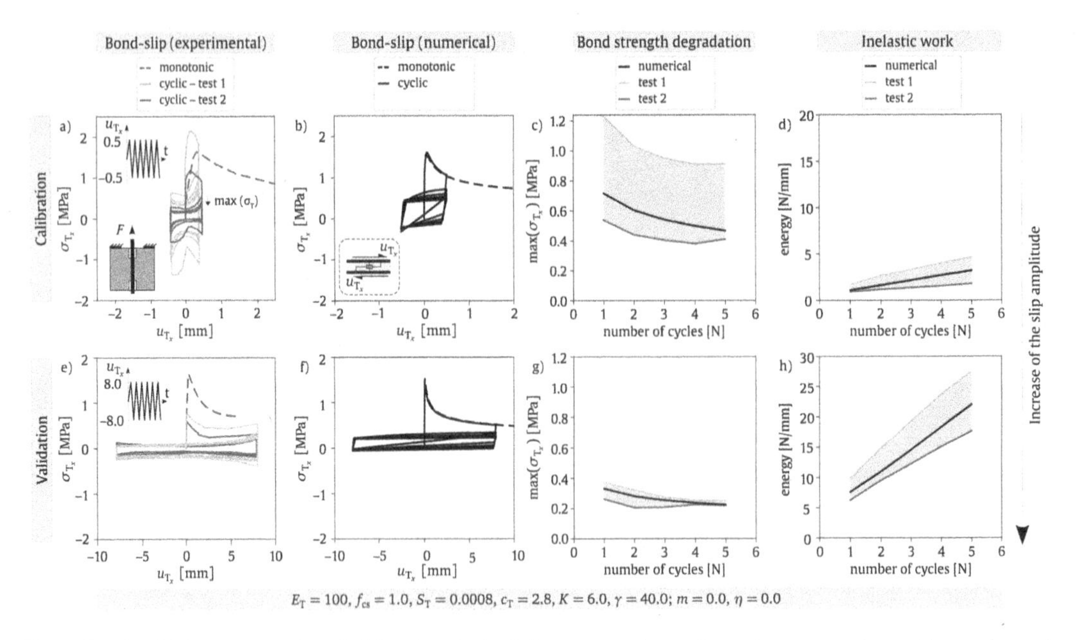

Figure 7. Bond between concrete and plain rebar steel reinforcement under cyclic loading: comparison with experimental data.

example is presented in Figure 7, where the bond behavior between concrete and plain steel reinforcement under monotonic and cyclic loading are studied and compared with the results of the test program presented in (Verderame, Ricci, Carlo, & Manfredi 2009). Due to the short bond length used in the test program, single material point simulations of the interface have been used in this study as well. In the experimental program, the bond behavior has been studied under monotonic and reversed slip control cyclic loading with constant amplitude i.e. pull-out and push-in loading. Two different amplitudes have been used i.e. 1.0 mm and 16.0 mm with two repetitions for each case. As a constant bond stress distribution along the bond length has been observed in the tests, the experimentally obtained pull-out/push-in curves have been normalized with respect to the contact area and plotted as bond stress vs. slip as shown in Figures. 7a, e. The obtained numerical curves of bond stress vs. slip are depicted in Figures. 7b, f. The bond stress degradation under cyclic loading is shown in Figures. 7c, g, where experimental and numerical results are compared. The evaluated energy dissipation from the experimental results are compared with the numerically evaluated energy dissipation obtained from the thermodynamic state variables as explained in Sec. 2.5 as depicted in Figures. 7d, h. It should be noted that the material model parameters have been identified to obtain a reasonable fit of the monotonic response, as well as the cyclic response with the slip amplitude equal to 1.0 mm see Figures. 7a-d. The model has been used to predict the cyclic response under the larger slip amplitude as depicted in Figures. 7e-h. The proposed model show the ability to reproduce the monotonic and cyclic behavior of the concrete-steel interface with consistent set of material parameters.

5 CONCLUSIONS

The introduced model was shown capable of simulating both monotonic and cyclic behavior of a material interface, e.g., between steel and concrete, using a consistent set of material parameters. The introduced hypothesis of damage accumulation using a cumulative deformation measure as the damage driving variable allows the model to consistently reflect the dissipative mechanisms of fatigue damage development at subcritical load levels. In addition, the model accounts for the interaction of dissipative effects under combined decohesion-compression and sliding loading through a smooth cap threshold function and a non-associative flow potential that couple damage evolution in the normal and tangential directions. All these features makes the proposed constitutive model applicable to different structural scales of representations.

ACKNOWLEDGMENTS

The work was supported by the German Research Foundation (DFG), project no. 412131890 and by the Czech Science Foundation, project no. GC19-06684J in the framework of the joint research project CumFatiCon "Fatigue of structural concrete driven by cumulative measure of shear strain".

REFERENCES

Bažant, Z. & P. Gambarova (1980, January). Rough cracks in reinforced concrete. *Journal of Structural Engineering 106*(4), 819–842.

Bažant, Z. P. & P. G. Gambarova (1984). Crack shear in concrete: Crack band microflane model. *Journal of Structural Engineering 110*(9), 2015–2035.

Baktheer, A., M. Aguilar, & R. Chudoba (2021). Microplane fatigue model MS1 for plain concrete under compression with damage evolution driven by cumulative inelastic shear strain. *International Journal of Plasticity*, 102950.

Baktheer, A. & R. Chudoba (2018a). Modeling of bond fatigue in reinforced concrete based on cumulative measure of slip. In *Computational Modelling of Concrete Structures, EURO-C 2018*, pp. 767–776. CRC Press.

Baktheer, A. & R. Chudoba (2018b). Pressure-sensitive bond fatigue model with damage evolution driven by cumulative slip: Thermodynamic formulation and applications to steel- and frp-concrete bond. *International Journal of Fatigue 113*, 277–289.

Baktheer, A. & R. Chudoba (2019). Classification and evaluation of phenomenological numerical models for concrete fatigue behavior under compression. *Construction and Building Materials 221*, 661–677.

Baktheer, A., H. Spartali, J. Hegger, & R. Chudoba (2021). High-cycle fatigue of bond in reinforced high-strength concrete under push-in loading characterized using the modified beam-end test. *Cement and Concrete Composites 118*, 103978.

Carrara, P. & L. De Lorenzis (2015, December). A coupled damage-plasticity model for the cyclic behavior of shear-loaded interfaces. *Journal of the Mechanics and Physics of Solids 85*, 33–53.

Chudoba, R., M. Vořechovský, M. Aguilar, & A. Baktheer (2022). Coupled sliding-decohesion-compression model for a consistent description of monotonic and fatigue behavior of material interfaces. *Computer Methods in Applied Mechanics and Engineering*, under review.

Desmorat, R., F. Ragueneau, & H. Pham (2007). Continuum damage mechanics for hysteresis and fatigue of quasi-brittle materials and structures. *International Journal for Numerical and Analytical Methods in Geomechanics 31*(2), 307–329.

Dimitri, R., M. Trullo, G. Zavarise, & L. De Lorenzis (2014). A consistency assessment of coupled cohesive zone models for mixed-mode debonding problems. *Frattura ed IntegritÃ Strutturale 8*(29), 266–283.

Eligehausen, R., E. P. Popov, & V. V. Bertero (1982). Local bond stress-slip relationships of deformed bars under generalized excitations. In *Proceedings of the 7th European Conference on Earthquake Engineering*, pp. 69–80.

Harper, P. W. & S. R. Hallett (2010). A fatigue degradation law for cohesive interface elements–development and application to composite materials. *International Journal of Fatigue 32*(11), 1774–1787.

Högberg, J. (2006). Mixed mode cohesive law. *International Journal of Fracture 141*(9), 549–559.

Hordijk, D. A. (1992). Tensile and tensile fatigue behaviour of concrete; experiments, modelling and analyses. *Heron 37*(1).

Horii, H., H. C. Shin, & T. M. Pallewatta (1992). Mechanism of fatigue crack growth in concrete. *Cement and Concrete Composites 14*(2), 83–89. Special Issue on Micromechanics of Failure in Cementitious Composites.

Huang, L., Y. Chi, L. Xu, & F. Deng (2019). A thermodynamics-based damage-plasticity model for bond stress-slip relationship of steel reinforcement embedded in fiber reinforced concrete. *Engineering Structures 180*, 762–778.

Kirane, K. & Z. P. Bažant (2015). Microplane damage model for fatigue of quasibrittle materials: Sub-critical crack growth, lifetime and residual strength. *International Journal of Fatigue 70*, 93–105.

Kwak, H.-G. & J.-K. Kim (2006). Implementation of bond-slip effect in analyses of rc frames under cyclic loads using layered section method. *Engineering structures 28*(12), 1715–1727.

Lemaitre, J. & R. Desmorat (2005). *Engineering damage mechanics: ductile, creep, fatigue and brittle failures*. Springer Science & Business Media.

Lindorf, A., L. Lemnitzer, & M. Curbach (2009). Experimental investigations on bond behaviour of reinforced concrete under transverse tension and repeated loading. *Engineering Structures 31*(7), 1469–1476.

Martinelli, E. & A. Caggiano (2014). A unified theoretical model for the monotonic and cyclic response of frp strips glued to concrete. *Polymers 6*(2), 370–381.

McGarry, J. P., É. Ó. Máirtín, G. Parry, & G. E. Beltz (2014). Potential-based and non-potential-based cohesive zone formulations under mixed-mode separation and over-closure. part I: Theoretical analysis. *Journal of the Mechanics and Physics of Solids 63*, 336–62.

Paulay, T. & P. Loeber (1974). Shear transfer by aggregate interlock. *Special Publication 42*, 1–16.

Ragueneau, F., N. Dominguez, & A. Ibrahimbegović (2006, November). Thermodynamic-based interface model for cohesive brittle materials: Application to bond slip in RC structures. *Computer Methods in Applied Mechanics and Engineering 195*(52), 7249–7263.

Roe, K. & T. Siegmund (2003). An irreversible cohesive zone model for interface fatigue crack growth simulation. *Engineering fracture mechanics 70*(2), 209–232.

Serpieri, R., E. Sacco, & G. Alfano (2015). A thermodynamically consistent derivation of a frictional-damage cohesive-zone model with different mode I and mode II fracture energies. *European Journal of Mechanics - A/Solids 49*, 13–25.

Verderame, G. M., P. Ricci, G. D. Carlo, & G. Manfredi (2009). Cyclic bond behaviour of plain bars. part i: Experimental investigation. *Construction and Building Materials 23*(12), 3499 – 3511.

Xu, F., Z.-m. Wu, J.-j. Zheng, Y. Hu, & Q.-b. Li (2014). Bond behavior of plain round bars in concrete under complex lateral pressures. *ACI Structural Journal 111*(1), 15.

Yang, H., S. K. Sinha, Y. Feng, D. B. McCallen, & B. Jeremić (2018). Energy dissipation analysis of elastic–plastic materials. *Computer Methods in Applied Mechanics and Engineering 331*, 309–326.

Computational Modelling of Concrete and Concrete Structures – Meschke, Pichler & Rots (Eds)
 ISBN: 978-1-032-32724-2

On the modelling of the rate dependence of strength using a crack-band based damage model for concrete

X. Liu, C.H. Lee & P. Grassl
James Watt School of Engineering, University of Glasgow, Scotland, UK

ABSTRACT: Concrete strength is reported in literature to increase with increasing strain rate, which needs to be considered in concrete constitutive models used for finite element analyses of concrete structures. One attractive group of constitutive models for concrete failure are scalar damage models using the crack band approach. These models are computationally efficient for rate independent loading because they produce mesh independent results with coarse discretisations as long as the strain field localises in mesh-dependent zones. The aim of this study is to incorporate the strain rate dependence in crack band based damage models while maintaining their ability to produce mesh-independent results. First the proposed model is described. Then, it used for direct tensile analysis in which strain softening occurs. It is demonstrated that a combination of strain rate dependent model for the undamaged response combined with a relative displacement rate dependent model for the damaged response provides stress-displacement curves which converge with mesh refinement.

1 INTRODUCTION

Critical infrastructure made of reinforced concrete are required to be designed to resist fast dynamic loading in the form of impact, sudden ground acceleration and blast, which can arise due to collisions, earthquakes and explosions, respectively. Concrete structures subjected to dynamic loading exhibit a complex nonlinear failure response which differs significantly from the one due to quasi-static loading. For instance, structures subjected to impact and blast can exhibit localised shear failure for loads, which, if they were applied slowly, would result in bending failure. Furthermore, dynamic loading produces compressive shock waves which can cause tensile spalling of the concrete cover, if reflected at free boundaries. Fast dynamic loading produces higher stiffness and strength than quasi-static loading. Strength of concrete is reported in the literature to be significantly increased for high strain rates (Bischoff & Perry 1991; Malvar & Ross 1998), whereas for fracture energy the results are less conclusive (Doormaal et al. 1994; de Pedraza et al. 2018).

For finite element modelling of failure of concrete structures, constitutive models for concrete are required, which can model localised tensile and diffuse compressive fracture. Approaches suitable for this are regularised continuum approaches which describe tensile fracture by means of mesh-independent zones of localised strains. One group of these regularised models for dynamic loading are based on the concept of damage delay (Häussler-Combe and Kühn 2012; Piani et al. 2019). One of the limitations of these approaches is that a very fine mesh is needed so that the zones of localised strain are modelled mesh-independently. Therefore, this approach less suitable for modelling structural components of reinforced concrete. Alternatively, discrete element approaches, in the form of lattice or particle models are used to model fracture in concrete. In these approaches, rate dependence of strength is modelled by cohesive laws which depend on the rate of the crack opening (Cusatis 2011). The third group of constitutive models are hybrid approaches in which the continuum and discrete models are combined. One numerically efficient hybrid approach are crack band models (Bažant & Oh 1983; Pietruszczak & Mróz 1981; Willam et al. 1986). In these approaches, cracks are described by using strain softening, which results in mesh-dependent localised zones. Mesh dependence of load displacement curves is avoided by adjusting the softening part with respect to the element size. These crack band models are popular choice for concrete, because they produce mesh-independent results for coarse discretisations. Therefore, they can be used for analysing the failure response of reinforced concrete components. The challenge for formulating crack band approaches for dynamic loading as a function of the strain rate is that the strain profiles obtained are mesh-dependent. As a consequence, the strain rate in the localised zones are also mesh-dependent. As the mesh is refined, the strain rate in the localised zones increases, which results in an artificial strengthening of the material.

DOI 10.1201/9781003316404-61

The aim of the present study is to propose a technique to provide mesh-independent load-displacement curve for strain rate dependent material responses used within the crack band approach. We use a scalar damage model to the illustrate the modelling concept. However, crack band models can be applied to a range of constitutive models, such as plasticity and damage-plasticity models (Grassl et al. 2013). The first part of the paper is used to introduce the rate dependent damage model. This is done in three steps. First, the standard rate independent model is presented. Next, the approach in which the damage evolution is made a function of the strain rate. Finally, the newly proposed formulation which produces mesh-independent results is presented. All three approaches are applied to an one-dimensional bar subjected to uniaxial tension to illustrate the differences of the formulations.

2 SCALAR DAMAGE MODEL

The approach to model the rate dependence of concrete is demonstrated here by a scalar damage model. First, the standard rate independent model is presented. This is followed by a description of the extension of the model that is modified to take into account rate dependence.

2.1 *Rate independent model*

The constitutive model used here is a strain based damage model for which the nominal stress $\boldsymbol{\sigma}$ is

$$\boldsymbol{\sigma} = (1 - \omega)\, \bar{\boldsymbol{\sigma}} = (1 - \omega)\, \mathbf{D}_{\mathrm{e}} : \boldsymbol{\varepsilon} \tag{1}$$

where $\mathbf{D}_{\mathrm{e}}$ is the elastic stiffness based on the Young's modulus E and Poisson's ratio ν, $\bar{\boldsymbol{\sigma}}$ is the effective stress, tensor $\boldsymbol{\varepsilon}$ is the strain tensor and ω is the damage variable ranging from 0 (undamaged) to 1 (fully damaged).

The damage variable ω is determined from a history variable κ. The history variable κ is obtained by a damage loading function of the form

$$f = \tilde{\varepsilon} - \kappa \tag{2}$$

with the loading and unloading conditions

$$f \leq 0 \;\; \dot{\kappa} > 0 \;\; \dot{\kappa} f = 0 \tag{3}$$

Here, the equivalent strain $\tilde{\varepsilon}$ is

$$\tilde{\varepsilon} = \frac{1}{E} \sqrt{\sum_{I=1}^{3} \langle \bar{\sigma}_{\mathrm{I}} \rangle^2} \tag{4}$$

where $\bar{\sigma}_{\mathrm{I}}$ are the principal values of the effective stress and $\langle \bar{\sigma}_{\mathrm{I}} \rangle$ are their positive parts. This equivalent strain definition gives a modified Rankine strength envelope at the onset of damage as shown in Figure 1. For $\tilde{\varepsilon} > f_{\mathrm{t}}/E$, damage occurs. The evolution of the damage variable is formulated so that in uniaxial tension

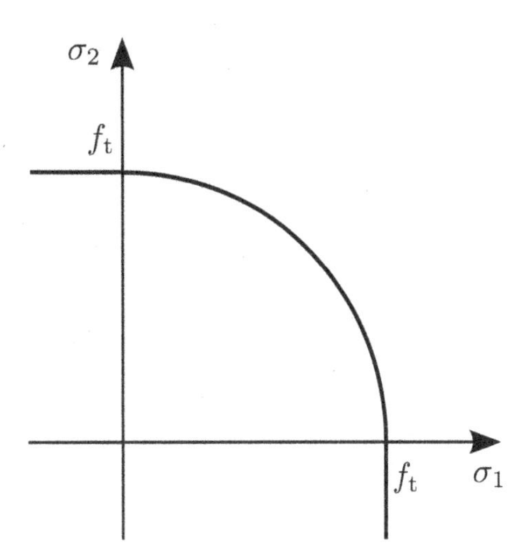

Figure 1. Strength envelope.

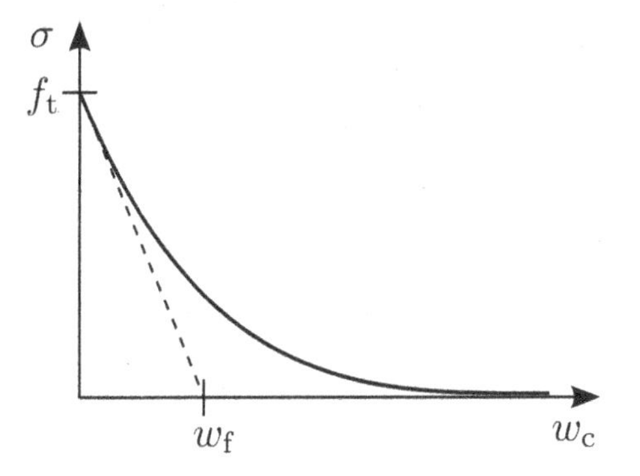

Figure 2. Stress crack-opening curve in uniaxial tension.

an exponential stress-crack opening curve as shown in Figure 2 is obtained. This is achieved by solving

$$(1 - \omega)\, E\kappa = f_{\mathrm{t}} \exp\left(-\omega\kappa h_{\mathrm{e}}/w_f\right) \tag{5}$$

for ω using the standard Newton-Raphson method. Here, h_e is the characteristic element length, f_t represents the tensile strength and w_{f} is the crack opening threshold in Figure 2. The left hand side of (5) is equal to the expression for $\boldsymbol{\sigma}$ in (1) for the case of monotonic uniaxial tension with the uniaxial strain replaced by κ. The right hand side shows the exponential softening law (Figure 2) whereby the crack opening is expressed by $\omega\kappa h_{\mathrm{e}}$.

2.2 *Extension to strain rate dependence*

The scalar damage model presented in the previous section is for quasi-static loading in which the strength is independent of the strain rate. The main purpose of this work is to introduce a modification of the damage model so that strain rate dependence of strength can be taken into account in the constitutive model. One approach would be to delay the evolution of the equivalent strain so that damage is initiated at a higher stress.

This can be achieved by reformulating expression (5) to be

$$(1-\omega)\,E\kappa_1 = f_t \exp\left(-\omega\kappa_2 h_e / w_f\right) \tag{6}$$

where κ_1 and κ_2 are determined in rate form as

$$\dot{\kappa}_1 = \dot{\kappa}/\alpha \text{ and } \dot{\kappa}_2 = \dot{\kappa}\alpha \tag{7}$$

Here, α is a (scalar) strain rate dependent factor, which, in this work, is based on *fib* Model Code (CEB-FIP 2012). It has the form

$$\alpha = \begin{cases} 1 & \text{for } \dot{\tilde{\varepsilon}} \leq \dot{\varepsilon}_1 \\ \left(\dfrac{\dot{\tilde{\varepsilon}}}{\dot{\varepsilon}_1}\right)^{0.018} & \text{for } \dot{\varepsilon}_1 \leq \dot{\tilde{\varepsilon}} \leq \dot{\varepsilon}_2 \\ 0.0062\left(\dfrac{\dot{\tilde{\varepsilon}}}{\dot{\varepsilon}_1}\right)^{1/3} & \text{for } \dot{\varepsilon}_2 \leq \dot{\tilde{\varepsilon}} \end{cases} \tag{8}$$

where $\dot{\varepsilon}_1 = 1 \times 10^{-6}\ \text{s}^{-1}$ and $\dot{\varepsilon}_2 = 10\ \text{s}^{-1}$. It should be noted that the specific expression of α is not important for this study, since it is not aimed to reproduce experimental results, but to propose a formulation that is mesh-independent. Comparing expression (6) with (5), the term κ on the left hand side of (6) is replaced by κ_1 so that for strain rates which result in $\alpha > 1$, the stress at which damage starts is greater than the tensile strength. Keeping the same spirit, the term κ_2 is used on the right hand side of (6) for the crack opening in order to ensure that the fracture energy remains constant with increasing strain rate. Both κ_1 and κ_2 are given in rate form as described in (7).

The problem with this formulation is that the rate factor α is a function of the strain rate, which for the crack band model is mesh-dependent once damage is initiated. Therefore, finer meshes will produce greater rate factors and therefore, higher strengths. To overcome this mesh dependence, the rate factor α is made a function of the displacement rate once damage is induced. To achieve a continuous evolution of the rate factor for the transition from undamaged to damage state, the strain rate before the onset of damage is linked to the displacement rate after the onset of damage by means of the incremental form

$$\frac{\tilde{\varepsilon}^n - \tilde{\varepsilon}^{n-1}}{t^n - t^{n-1}} = \beta h_e \frac{\tilde{\varepsilon}^{n+1} - \tilde{\varepsilon}^n}{t^{n+1} - t^n} \tag{9}$$

where $n+1$ is the first step where damage is nonzero. The parameter β is determined once at the start of damage and then kept constant. The material models described above were implemented in the open source finite element program OOFEM (Patzák 2012).

3 MESH DEPENDENCE STUDY

The response of the three crack band damage models above, namely rate-independent, strain-rate dependent and displacement rate dependent is investigated for possible mesh-dependence by means of a direct tensile analysis. For being able to investigate the material response independent of wave propagation, the problem is solved assuming zero mass for the material. Consequently, the results presented here are based on force equilibrium only without the inertia term. The geometry of the bar is shown in Figure 3. One element in the centre of the bar is strongly weakened to trigger the onset of failure. The length of the bar is $L = 0.1$ m and the cross-sectional area is $A = 0.01$ m^2. Four meshes with 1, 5, 10 and 20 equally sized elements are used. For the weakened element, the properties are $f_t = 3$ MPa, Young's modulus $E = 30$ GPa and fracture energy $G_F = 100$ N/m. The adjacent elements have the same Young's modulus, but a much higher strength of 15 MPa so that damage is limited to the weakened element for all the analyses. The analysis is displacement controlled at the end of the specimen with a displacement rate of 5 m/s. For the elastic stage during which the strain is uniformly distributed, this corresponds to a strain rate of 50 1/s and rate factor of $\alpha = 2.28$ according to (8). The first set of analyses were carried out with the strain-rate independent model. The load displacement curve and strain profiles are shown in Figures 4 and 5, respectively. The load-displacement curves are mesh-independent. However, the strain profiles depend on the number of elements. This is a typical result for crack band models with rate independent material models in which the cracks are represented by mesh-dependent zones of high strain values. In the present setup, the high strain occurs in the weakened element, whereas in the other elements unloading occurs. For quasi-static simulation,

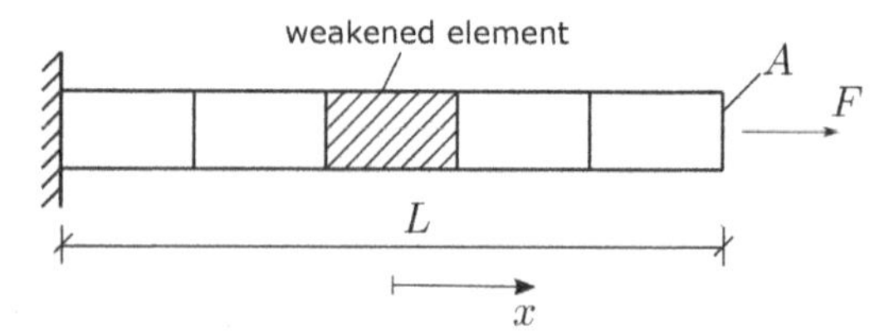

Figure 3. Geometry of specimen for mesh dependence study.

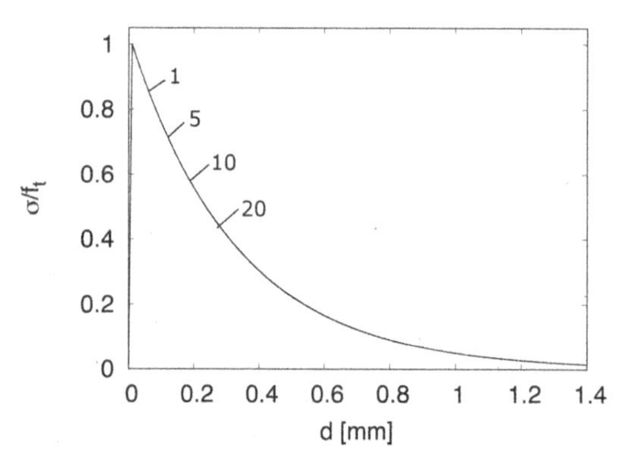

Figure 4. Normalised stress versus displacement for four meshes for the rate independent damage model.

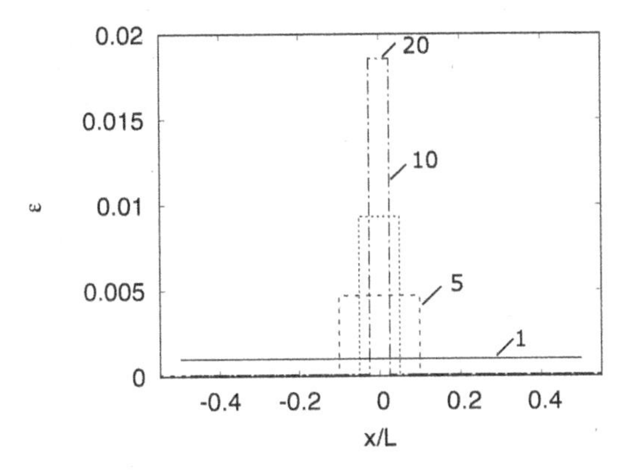

Figure 5. Strain versus x-coordinate for four meshes for the rate independent damage model at a displacement of 0.1 mm.

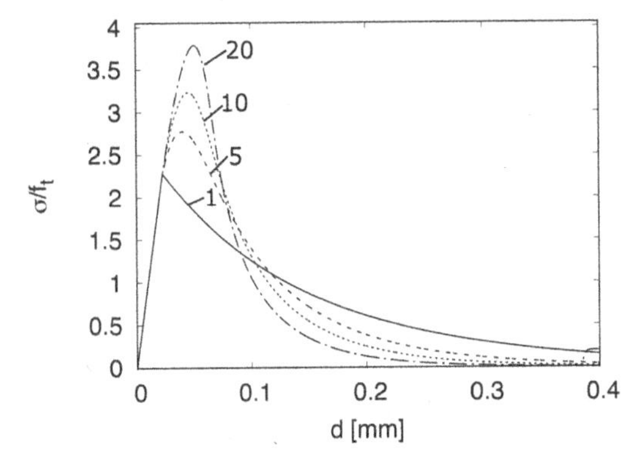

Figure 6. Normalised stress versus displacement for four meshes for the strain rate dependent damage model.

the mesh-dependent zones are not problematic, since in general the main aim is to obtain mesh-independent load-displacement curves. However, this is not the case for strain rate dependent models as the mesh-dependence of the strain causes problems which will be illustrated in the next part.

In the second part, the results for the strain rate dependent scalar damage model are presented. In Figures 6 and 7, the load-displacement and rate factor are shown, respectively. The load-displacement response in Figure 6 and the rate factor in Figure 7 are strongly mesh-dependent. The finer the mesh is, the greater is maximum peak load and rate factor. This mesh-dependence of the load-displacement response and rate factor is explained by Figure 5. The strain localises in mesh-dependent region. Therefore, for the same displacement, the strain and also strain rate in smaller zones is greater. Therefore, the finer the mesh, the greater is the strain rate and the rate factor. Note that an increase in the rate factor does not result in a jump in the stress, because the history variables κ_1 and κ_2 are formulated in rate form in 7.

In the next part, the results of the modified rate-dependent damage model is presented. In this model, the rate factor is determined from the deformation rate in the element once damage has started. The

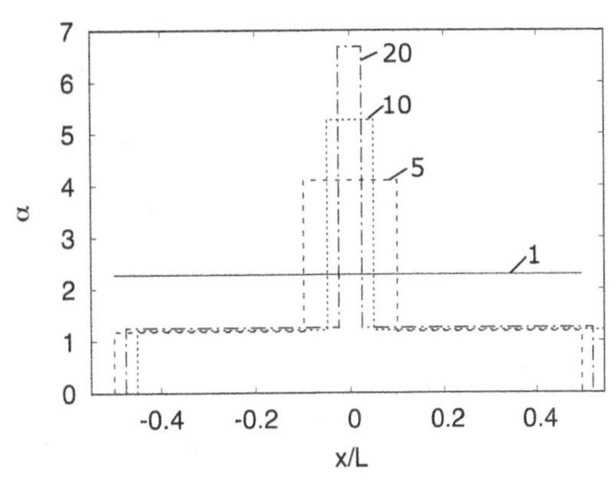

Figure 7. Rate factor versus x-coordinate for four meshes for the strain rate dependent damage model at a displacement of 0.1 mm.

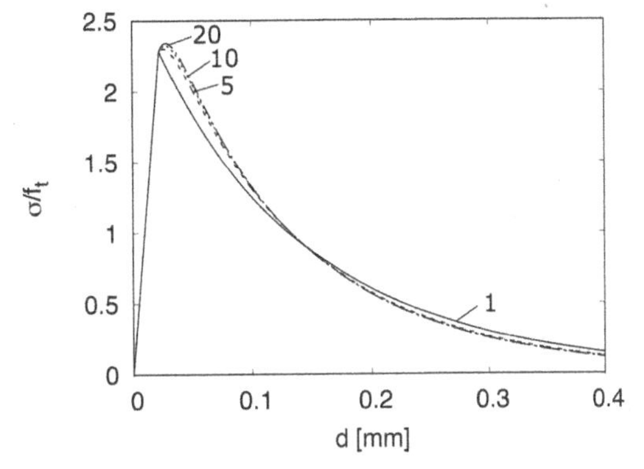

Figure 8. Load-displacement curves for four meshes for the deformation rate dependent damage model.

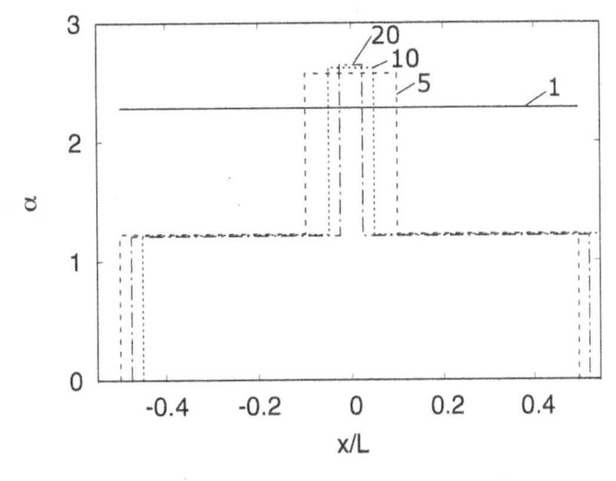

Figure 9. Rate factor profile for four meshes for the deformation rate dependent damage model at a displacement of 0.1 mm.

normalised stress versus displacement is shown in Figure 8. Furthermore, the rate factor versus the x-coordinate is shown in Figure 9. The normalised stress displacement curves are converging as the mesh is refined. Also, the rate factor in the damaged element converges with mesh refinement. Still, there is some mesh dependence visible, because the rate factor in the damaged element is obtained from the total deformation rate in the element and not the crack opening rate. The crack band model describes crack openings mesh-independently, but not the deformation of the damaged element. Nevertheless, the finer the mesh, the smaller is the difference between deformation and crack opening rate.

4 CONCLUSIONS

We proposed a new approach to make a scalar damage model based on the crack band approach rate-dependent by switching at the onset of damage from a strain rate based to a deformation rate based formulation. It is shown that this formulation provides load-displacement curves which converge with mesh refinement. In the next step, the approach presented here will be applied to more comprehensive damage-plasticity models reported in (Grassl, Xenos, Nyström, Rempling, & Gylltoft 2013) and then used to investigate problems with wave propagation such as the spalling experiments reported in (Schuler, Mayrhofer, & Thoma 2006).

REFERENCES

Bažant, Z. P. & B.-H. Oh (1983). Crack band theory for fracture of concrete. *Materials and Structures 16*, 155–177.

Bischoff, P. H. & S. H. Perry (1991). Compressive behaviour of concrete at high strain rates. *Materials and structures 24*(6), 425–450.

CEB-FIP (2012). *CEB-FIP Model Code 2010, Design Code.*

Cusatis, G. (2011). Strain-rate effects on concrete behavior. *International Journal of Impact Engineering 38*, 162–170.

de Pedraza, V. R., F. Galvez, & D. C. Franco (2018). Measurement of fracture energy of concrete at high strain rates. In *EPJ Web of Conferences*, Volume 183, pp. 02065.

Doormaal, J. C. A. M. V., J. Weerheijm, & L. J. Sluys (1994). Experimental and numerical determination of the dynamic fracture energy of concrete. *Le Journal de Physique IV 4*(C8), C8–501.

Grassl, P., D. Xenos, U. Nyström, R. Rempling, & K. Gylltoft (2013). CDPM2: A damage-plasticity approach to modelling the failure of concrete. *International Journal of Solids and Structures 50*(24), 3805–3816.

Häussler-Combe, U. & T. Kühn (2012). Modeling of strain rate effects for concrete with viscoelasticity and retarded damage. *International Journal of Impact Engineering 50*, 17–28.

Malvar, L. J. & C. A. Ross (1998). Review of strain rate effects for concrete in tension. *ACI Materials Journal 95*, 735–739.

Patzák, B. (2012). OOFEM – An object-oriented simulation tool for advanced modeling of materials and structure. *Acta Polytechnica 52*, 59–66.

Piani, T. L., J. Weerheijm, & L. J. Sluys (2019). Dynamic simulations of traditional masonry materials at different loading rates using an enriched damage delay: Theory and practical applications. *Engineering Fracture Mechanics 218*, 106576.

Pietruszczak, S. T. & Z. Mróz (1981). Finite element analysis of deformation of strain-softening materials. *International Journal for Numerical Methods in Engineering 17*(3), 327–334.

Schuler, H., C. Mayrhofer, & K. Thoma (2006). Spall experiments for the measurement of the tensile strength and fracture energy of concrete at high strain rates. *International Journal of Impact Engineering 32*(10), 1635–1650.

Willam, K., N. Bićanić, & S. Sture (1986). Composite fracture model for strain-softening and localised failure of concrete. In E. Hinton and D. R. J. Owen (Eds.), *Computational Modelling of Reinforced Concrete Structures*, Swansea, pp. 122–153. Pineridge Press.

Computational Modelling of Concrete and Concrete Structures – Meschke, Pichler & Rots (Eds)
© 2022 Copyright the Author(s), ISBN: 978-1-032-32724-2

Parametric study of the Lattice Discrete Particle Model (LDPM) constitutive law for fiber reinforced concretes (FRCs)

C. Del Prete
University of Bologna, Bologna, Italy

R. Wan-Wendner
Ghent University, Ghent, Belgium

N. Buratti & C. Mazzotti
University of Bologna, Bologna, Italy

ABSTRACT: The growing adoption of fiber reinforced concrete (FRC) as a structural material is motivating plenty of research in this field, especially those regarding the experimental aspects and the development of numerical models. The composite nature of the material suggests that the final mechanical performance is due to the contribution of both components and to their interaction. In this respect, the experimental research has a fundamental importance for the identification of the mechanical response scatter, as the natural heterogeneity of concrete is further incremented by the randomness of fiber distribution.

When the mechanical behavior of FRCs is simulated numerically, this aspect needs to be properly reproduced to get a reliable response of the fiber reinforced concretes. In this framework, the present paper illustrates a numerical model describing the behavior of a FRC concrete reinforced with polymeric fibers, developed with the Lattice Discrete Particle Model (LDPM). This theory is able to reproduce the behavior of only concrete (LDPM), by describing the mechanical interaction between the aggregates, and also the interaction with fibers (LDPM-F). The model has been already validated for the plain concrete short- and long-term behavior (M-LPDM); in the recent years, the fiber-bridging action due to the reinforcement has been introduced.

Many numerical parameters concerning the fiber geometry and its mechanics determine the whole response: the discretization of each fiber, the definition of its shape, its elastic modulus and also the orientation of the fibrous reinforcement in the concrete matrix. Furthermore, polymeric fibers may be characterized by a crimped profile to improve the matrix-to-fiber bond and, so, it is fundamental to consider their actual shape also numerically. Their geometry is defined by the number of segments in which each fiber is divided and its tortuosity.

This paper performs a parametric analysis of these specific aspects showing how they affect the flexural behavior of macro-synthetic fiber reinforced concrete beams. The fiber elastic modulus handles the force transferring when concrete is cracked so defines the post-peak strength in the flexural behavior: a value between the tangent and secant elastic modulus has to be considered in the calibration. The orientation of the reinforcement, especially in the crack surroundings, drives the crack development: the randomness is what influences more the scatter in the response that, in turn, depends on the number of fibers connecting the crack. In this numerical approach the counting of fibers has been also performed and, at a given fibers dosage, the orientation is a parameter calibrated to make the numerical count close to the experimental. Finally, regarding the concrete composition, the aggregates are here generated according to the minimum and maximum size: the minimum value given is shown to influence the post peak behavior especially in terms of cracking evolution under flexural load.

1 INTRODUCTION

The increasing number of structural applications for fiber reinforced concretes (FRCs) are fostering research on the experimental investigation of their performance and on the development of numerical advanced numerical models (Di Prisco et al. 2013). The durability improvement given by the adoption of plastic fibers (Xu et al. 2021), is promoting their use instead of those made of steel (Camille et al. 2021; Nana et al. 2021). On the other side, the identification of the creep deformations of FRCs with MS fibers is widely studied in the perspective of the test method standardization and the quantification in terms of deformations (Llano-Torre et al. 2021).

In this scenario, many advances have been done in the numerical simulation of the FRCs performance: the formulation of material models (Blanco et al. 2013; Thai et al. 2020), and the calibration of predicting models using different approaches, parametric studies (Liu et al. 2022), multi-scale models (Nonato Da Silva et al. 2020) and meso-scale analysis (Gal & Kryvoruk 2011). One of the main physical phenomena to be reproduced in FRC models is the effective dispersion and orientation of the fibers (Ferrara et al. 2012; Leporace-Guimil et al. 2021; Nonato Da Silva et al. 2020); fundamental in defining the post-cracking performance and the variability of results, proper of FRC. The relation between the number of fibers on the cracked surface on FRCs specimens in tension or bending, and the residual strength is considered in the studies (Del Prete et al. 2017), also for the long term deformations (C. Del Prete et al. 2021; Vrijdaghs et al. 2020).

The study here presented concerns a parametric analysis of the Lattice Discrete Particle Model (LDPM-F) applied to FRCs (Jin et al. 2016; Schauffert et al. 2011). This model has been also extended to the viscoelasticity of concrete (Abdellatef et al. 2019) and of macro-synthetic fibers (C. Del Prete et al. 2021). The formulation depends on two sets of parameters, those of concrete and fibers: here the effect of the geometrical and mechanical fiber parameters and the aggregate size is discussed.

When dealing with the geometry of the fibers for the numerical model, they must be in compliance with the geometry is intended to reproduce. For example, a crimped shape improves the fiber-to-matrix bond strength so, beside the geometry, also the mechanical parameters should be calibrated, bond strength and debonding fracture energy. The total framework must be reliable with the mechanics.

2 LATTICE DISCRETE PARTICLE MODEL FOR FIBER REINFORCED CONCRETE (LDPM-F)

The Lattice Discrete Particle Model (LDPM) approach is a mesoscale theory used to describe the concrete particles interaction by means of constitutive laws applied on the internal facets in which the matrix structure is organized (Cusatis, Mencarelli, et al. 2011; Cusatis, Pelessone, et al. 2011). The aggregates are connected and merged into a unique matrix with the Delaunay tetrahedralization that connects four particles center, so producing the tetrahedron in Figure 1, creating a lattice system.

So, the LDPM formulation consists of the solution of the Principle of Virtual Work (PVW), or equilibrium of the internal and external work, and the compatibility of displacement, applied on the facets of each tetrahedron (Figure 1).

The concrete structure is generated according to the mix design properties, i.e. w/c ratio, cement content and aggregate minimum and maximum size.

The mechanics of concrete depend on a set of parameters, describing the elastic, E_0 effective normal

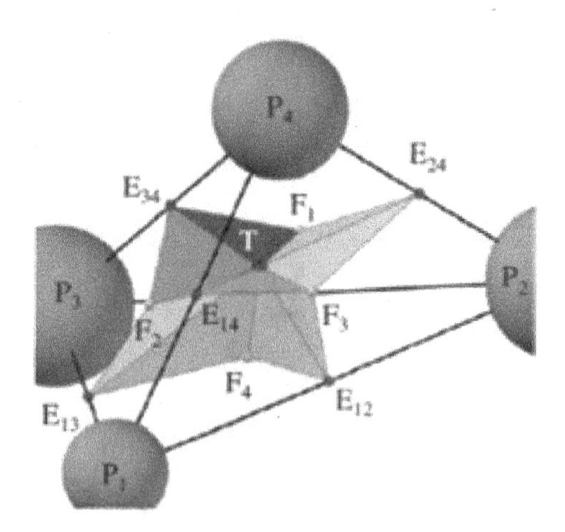

Figure 1. LDPM tethraedron connecting four particles (Cusatis, Pelessone, et al. 2011).

modulus and α the shear-normal coupling parameter, and inelastic phase. This stage includes different mechanisms, each of them regulated by a specific group of parameters: those calibrated for the present study have been σ_t, tensile strength, l_t, characteristic length, n_t, softening exponent, σ_s/σ_t the shear strength ratio. The other parameters, listed and detailed described in (Cusatis, Mencarelli, et al. 2011), have been assumed from literature.

The LDPM approach has been extended to include the effect if fibers and describe their crack-bridging mechanism (Jin et al. 2016; Schauffert et al. 2011). They are quasi-randomly generated in the concrete volume and, so, intersect the facets of the system: their orientation is defined by the intersection of each element with a facet (Figure 2).

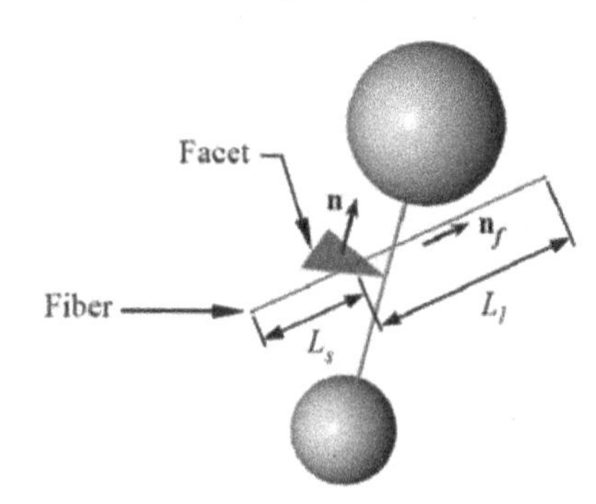

Figure 2. Fiber intersecting the facet (Jin et al. 2016).

Fibers are geometrically and mechanically described by a wide set of parameters. The diameter d_f, length l_f, edges per fiber and tortuosity. Their dosage is set according to a volume fraction. It is straightforward to get the key role of the fiber orientation in the failure mechanism, i.e. fiber debonding and pull-out (Del Prete et al. 2019). These mechanisms are dependent on a set of parameters including, the debonding fracture energy G_d, the frictional stress τ_0, a dimensionless factor β. The mechanical behaviour of the fibers is described by their tensile strength σ_{fu} and elastic modulus E_f.

Furthermore, the snubbing, spalling and cook-Gordon effects starting at the fiber-matrix interface are also simulated through this formulation, by means of additional parameters (Jin et al. 2016; Schauffert et al. 2011).

The LDPM-F formulation considered in the present paper is implemented in the MARS (Modeling and analysis of the Response of Structures) software.

3 EXPERIMENTAL TESTS SIMULATED

In this study the influence of different parameters on the mechanical response of MSFRCs, is analyzed by examining the flexural response of prismatic specimens in bending (Figure 3).

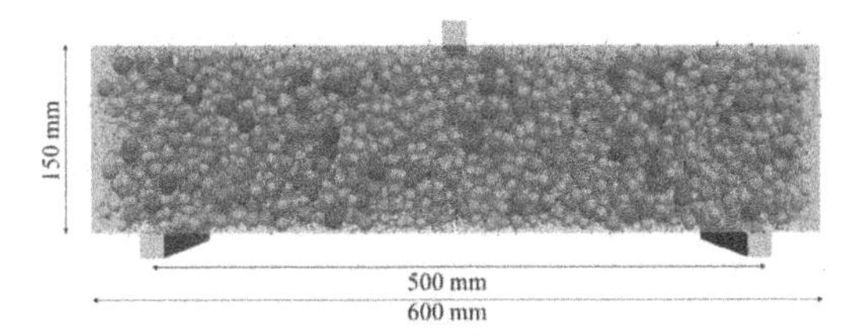

Figure 3. Geometry of the samples used for the simulations.

In a first phase, experimental data from bending and compression tests are considered for calibrating the LDPM-F parameters (Tables 1 and 2). In Figure 4 the black dashed line represents the mean experimental curve whose scatter is identified by the grey area, while the black solid line represents the mean numerical curve. Numerically four different arrangements of fiber and aggregates, named seeds, are simulated to describe the actual variability inside the concrete particles and fibers distribution (grey dashed lines).The procedure of calibration parameter is detailed described in (C. Del Prete et al. 2019; C. Del Prete, Boumakis, et al. 2021).

Table 1. LDPM parameters.

E_0 [GPa]	α [-]	σ_t [MPa]	l_t [mm]	$\sigma_s/\sigma_t n_t$ [-]	σ_{c0} [-]	H_{c0}/E_0 [MPa]	k_{c0} [-]	n_F [-]	[-]
40	0.25	2.0	800	4.0	1.5	190	0.4	2	0.5

k_{c1} [-]	k_{c2} [-]	μ_0 [-]	μ_∞ [-]	σ_{N0} [MPa]	d_0 [mm]	d_a [mm]	c [kg/m^3]	w/c [-]	ρ [kg/m^3]	a/c [-]
1	5	0.35	0	600	8*	15	400	0.46	2400	4.34

Table 2. LDPM-F parameters.

G_d [N/m]	τ_0 [MPa]	β [-]	k_{sp} [-]	k_{sn} [-]	σ_{uf} [MPa]	k_{rup} [-]	E_f [GPa]	tortuosity [-]	Edg [-]
1.0	4.0	0.5	6.2	1.0	473	0.0	3.3*	0.6*	8*

l_f [mm]	d_f [mm]	V_f [%]	ρ [kg/m^3]	shape [-]	orientation(x) [-]
54	0.81	0.85	946	crimped	10*

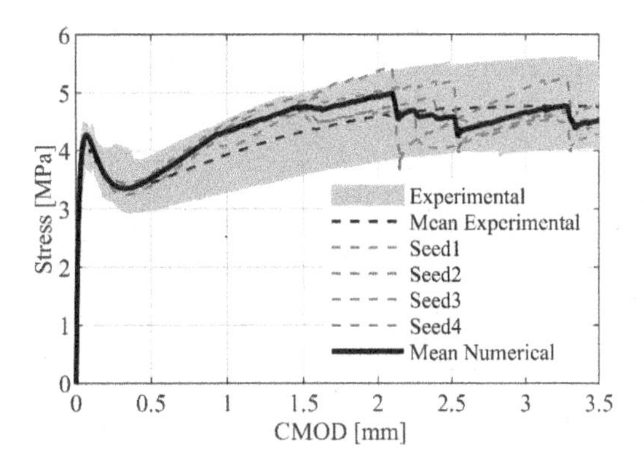

Figure 4. MSFRC bending tests calibration.

4 INFLUENCE OF AGGREGATE SIZE AND FIBER GEOMETRY PARAMETERS ON MSFRCS FLEXURAL BEHAVIOUR

The parameters investigated, highlighted with a (*) in Tables 1 and 2, are: minimum concrete aggregate size, fiber tortuosity, orientation, edges number and elastic modulus. The parameters are varied respect to the reference set in Tables 1 and 2 and four seeds are generated for each variation.

4.1 *Minimum aggregate size*

In LDPM models, the aggregate size range is typically included between the real maximum aggregate diameter and half of it (for computational time reason), so the lower bound of the range is set according to the maximum size. The influence of this parameter is investigated by considering three different sizes: 5 mm, 6 mm, 8 mm (this last is used as a reference). For each dimension, four seeds are simulated, i.e. four different distributions of aggregates and fibers. For each value of the minimum aggregate size Figure 5 reports

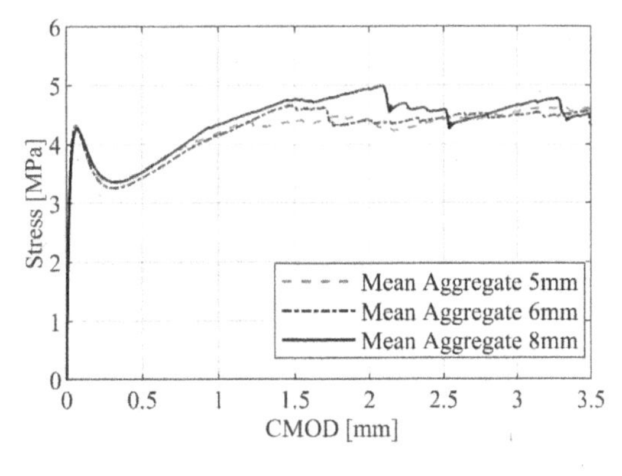

Figure 5. MSFRC bending tests – influence of the minimum aggregate size.

the mean curve obtained combining the results of the four seeds.

Looking at Figure 5, it can be clearly seen that a lower aggregate size implies a slightly lower post-peak strength in the range of 1.5 mm < CMOD < 2.5 mm.

After this CMOD value, the responses appear to be more similar. Furthermore, the curve for a minimum aggregate size of 8 mm presents stronger discontinuities than the other curves. This can be explained considering that the neutral axis of the cracked cross section is very close to its top, at a distance that can become lower than the aggregate size in experimental tests. Therefore, the jumps in the post-peak curve are attributed to progressive cracking of concrete. Figures 6 and 7 show examples of the particle distribution in the cracked section for 5 mm and 8 mm, respectively.

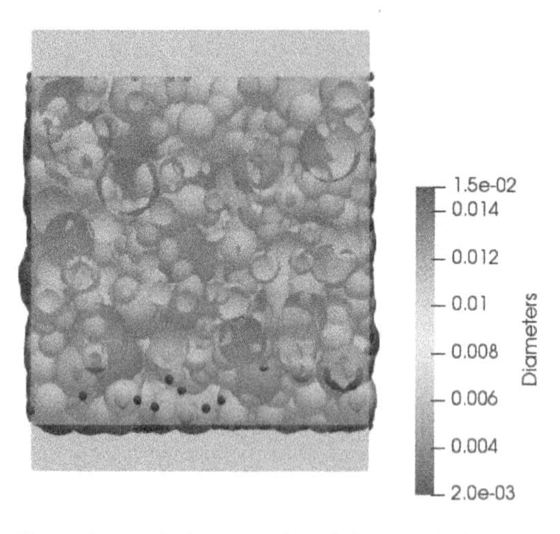

Figure 6. Notched cross section of the numerical model: distribution of aggregates with 5 mm minimum diameter.

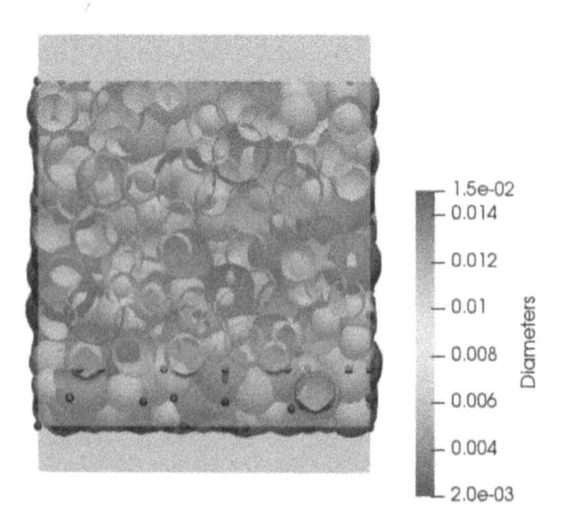

Figure 7. Notched cross section of the numerical model: distribution of aggregates with 8 mm minimum diameter.

4.2 *Fiber elastic modulus* (E_f)

The elastic modulus of the fibers influences the force transfer between the concrete matrix and the fibers. Three different values of this parameter are considered here, i.e. 2600 MPa, 3300 MPa and 3900 MPa.

In Figure 8 each curve represents the mean value of four seeds and, at a value of CMOD 0.5 mm, the residual strength is 3.3 MPa for E_f 2600 MPa, 3.5 MPa for E_f 3300 MPa and 3.7 MPa for E_f 3900 MPa. Furthermore, the valley of the curves after the first peak shifts towards larger CMOD values as the elastic modulus decreases. In fact, a fiber with lower elastic modulus requires a larger elongation to take the same force taken by a stiffer fiber.

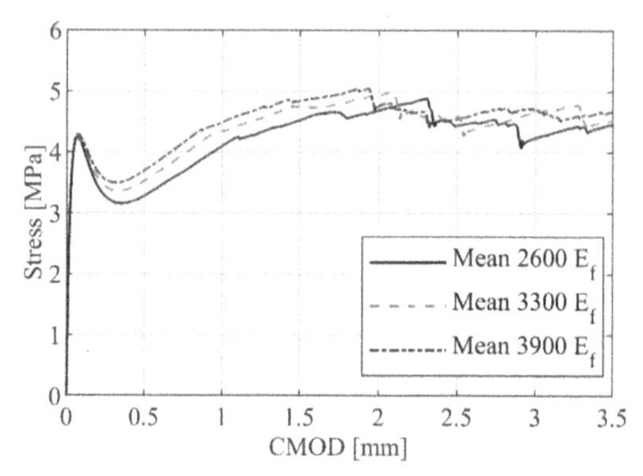

Figure 8. MSFRC bending tests – influence of fiber elastic modulus.

4.3 *Fiber discretization: edges per fiber*

In the LDPM-F formulation considered here, each fiber is divided into a number of segments, named edges. Defining one edge (Figure 9), the fiber appears straight, while setting a number of edges higher than one, a curved shape is obtained. In the present study 1, 4 and 8 edges are considered. Referring to Figure 10, it is possible to observe that when the number of edges per fiber is increased, the residual strength decreases: examining the strength at CMOD 0.5 mm (f_{R1} as in

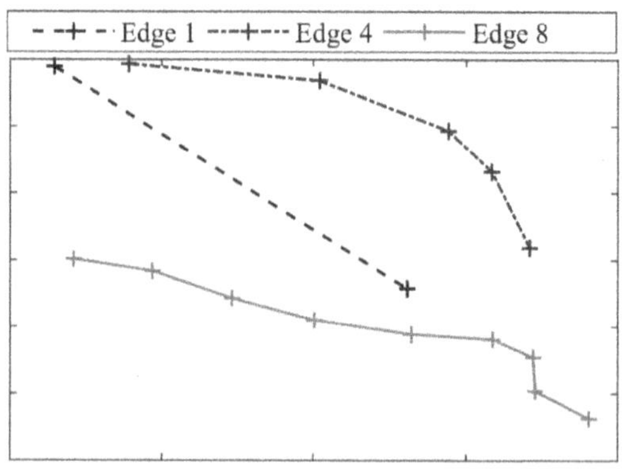

Figure 9. Fiber profile at different edges.

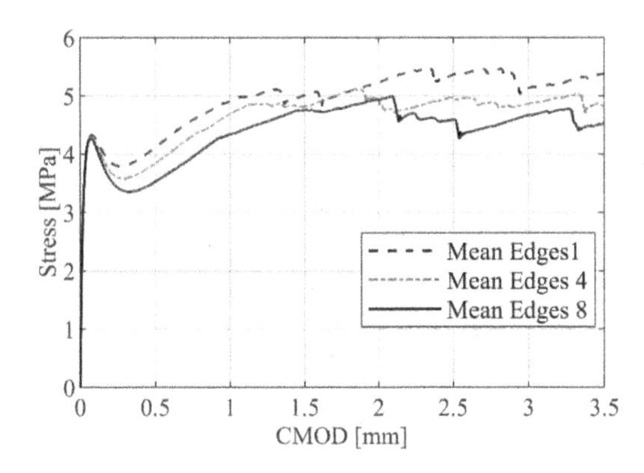

Figure 10. MSFRC bending tests – influence of the number of edges per fiber.

Table 3. Number of fibers counted on the notched section for fibers with 1, 4 and 8 edges.

Edges per fiber		
1	4	8
Number of fibers		
354	272	218

Model Code 2010 (Di Prisco et al. 2013)), we have a residual strength of 3.5 MPa, 3.8 MPa and 4.1 MPa passing from 8 to 1 edges per fiber. The number of edges affects the distribution of fibers in the specimen. If they are straight (one edge), they will tend to orient along only one axis of the beams, while, if curved, they will have a more random orientation. This can be revealed by counting the fibers crossing mid-span section of surfaces of the prismatic model (Table 3). The number of fibers drastically decreases passing from 1 to 4 edges, while closer values are obtained for 4 and 8 edges. Figure 11 shows histograms with the inclination of the fibers crossing the mid span section with respect to the cross-section plane (the inclination is defined considering a straight segment connecting the ends of the fiber). Clearly with 1 edge there are more fibers with an inclination higher that 85°.

4.4 *Fiber orientation*

The orientation of the fibers is a parameter that can be specified: it is possible to indicate a value of orientation between 0 and 10 (a sort of weight) along the three axis, x, y and z. Indicating this value is possible to enforce the orientation of the reinforcement along a specified direction: in these simulations this parameter – orientation – along the x-axis is varied, with values of 1, 4 and 10. When they are more distributed along the x-axis, corresponding to the longitudinal direction of the prism, the residual strength increases because the number of fibers on the middle cross section (notched section) is higher (Figure 12).

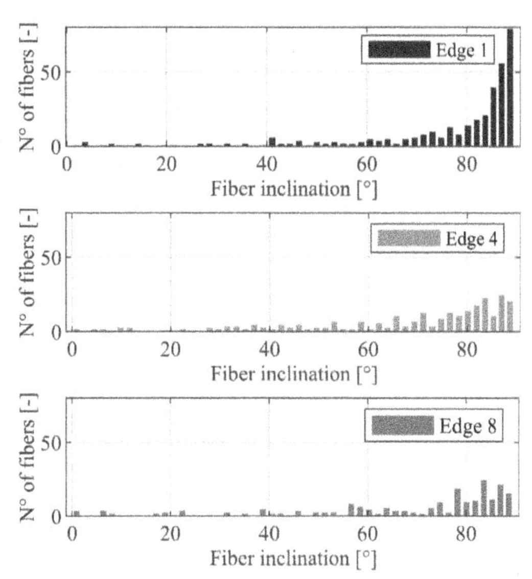

Figure 11. N° of fibers *vs* fibers inclination at different edges specification.

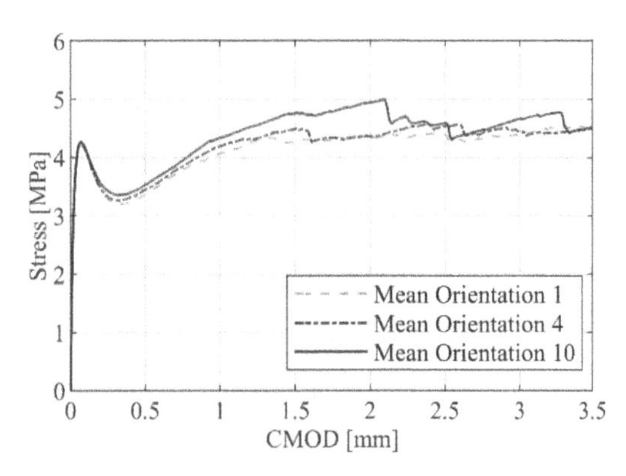

Figure 12. MSFRC bending tests – influence of fiber orientation.

Table 4 reports the number of fibers at the three values of the fiber orientation parameter, as a mean value of four seeds.

Table 4. Number of fibers counted on the notched section at fibers orientation of 1, 4 and 10 along x-axis.

Orientation of fibers x-axis		
1	4	10
Number of fibers		
161	192	218

Moreover, in Figures 13 and 14 the fiber distribution on the notched surface and their inclination is represented, given the same volume fraction; the histograms suggest that the number of fibers increases especially at higher inclination (after 80°).

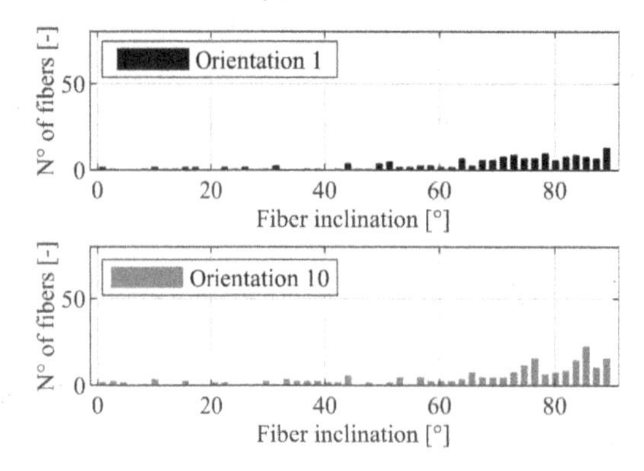

Figure 13. N° of fibers *vs* fibers inclination at different orientation.

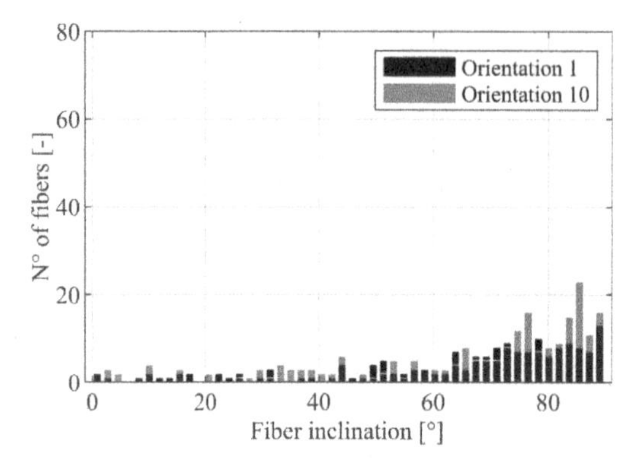

Figure 14. N° of fibers *vs* fibers inclination at different orientation.

4.5 *Fiber tortuosity*

The shape of the fiber can be also defined by the tortuosity parameter, calibrating a value between 0 and 1. The profile changes as in Figure 15 when a tortuosity equal to 0 or 1 (the lower and upper bound of the range) is set. The curves in Figure 16 suggest that a lower value of the parameter produces a higher toughness in the response, increasing the residual strength. Moreover, comparing the behavior of a straight fiber

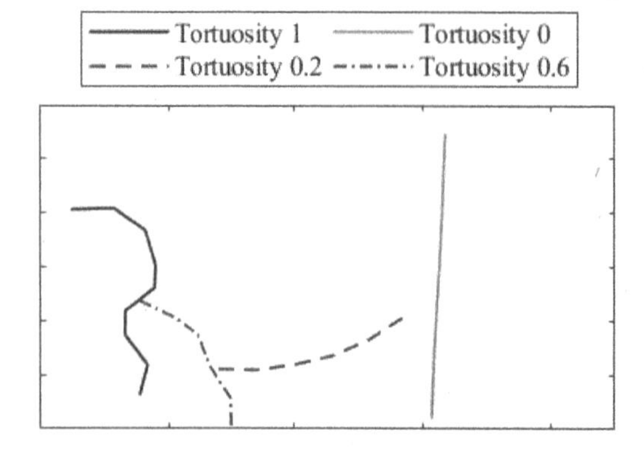

Figure 15. Fiber profile at tortuosity 0, 0.2, 0.6 and 1.

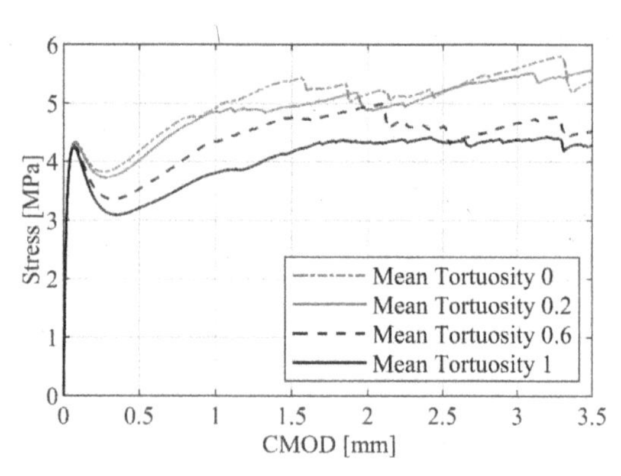

Figure 16. MSFRC bending tests – influence of fiber tortuosity.

Table 5. Number of fibers counted on the notched section at fibers tortuosity of 0, 0.2, 0.6, 1.

Tortuosity			
0	0.2	0.6	1
Number of fibers			
336	321	218	150

(solid black line) with a crimped one (dashed grey line) the shape of the curve becomes more irregular. This can be due to the different interlock at interface that produces a different slip mechanism between a crimped fiber, rather than a straight one.

A different value of the tortuosity, generates a different number of fibers that crosses the prism section: in particular, counting the fibers on the notched section (Figure 17), the total amount changes a lot between 0 and 1. The increment of the fibers is more pronounced when passing from 0.6 to 0, also the orientation moves from higher values of the angle inclination, from 60° to 80°.

5 CONCLUSIONS

The research presented deals with the calibration of the parameters for the LDPM-F model used to simulate the behaviour of macro-synthetic fiber reinforced concretes.

The results shown make possible to draw the following conclusions:

- The minimum aggregate size specified to generate the aggregates influences the cracking behaviour, especially under flexural loads, since the height of the compression zone is similar to the aggregate size;
- The orientation of the fibers along one direction determines the effective number of fibers involved in the cracking process;

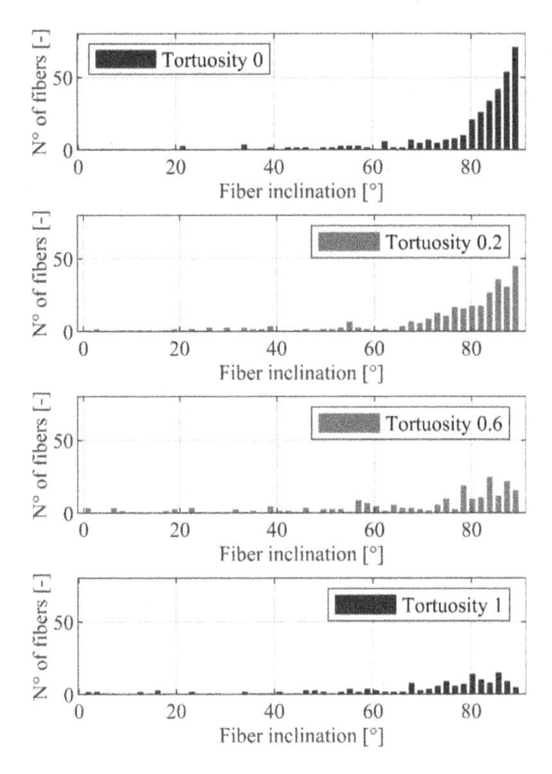

Figure 17. N° of fibers *vs* fibers inclination at different tortuosity.

- The specification of the tortuosity and the number of edges per fiber strongly influences the fibers shape and their contribution to the mechanical response after the crack formation: the number and inclination of the fibers crossing the middle section strongly depends on this variable;
- The elastic modulus of the fibers influences the stress transfer for the concrete matrix to fibers, in particular right after crack formation.

Thus, the geometrical parameters can be strongly defined by the effective number of fibers experimentally counted (Bernard 2017).

REFERENCES

Abdellatef, M., Boumakis, I., Wan-Wendner, R., & Alnaggar, M. (2019). Lattice Discrete Particle Modeling of concrete coupled creep and shrinkage behavior: A comprehensive calibration and validation study. *Construction and Building Materials*, 211, 629–645.

Bernard, E. (2017). *Influence of fibre count on variability in post- crack performance of fibre reinforced concrete. June*. https://doi.org/10.1617/s11527-017-1035-5

Blanco, A., Pujadas, P., De La Fuente, A., Cavalaro, S., & Aguado, A. (2013). Application of constitutive models in European codes to RC-FRC. *Construction and Building Materials*, 40, 246–259.

Camille, C., Kahagala Hewage, D., Mirza, O., Mashiri, F., Kirkland, B., & Clarke, T. (2021). Performance behaviour of macro-synthetic fibre reinforced concrete subjected to static and dynamic loadings for sleeper applications. *Construction and Building Materials*, 270, 121469. https://doi.org/10.1016/j.conbuildmat.2020.121469

Cusatis, G., Mencarelli, A., Pelessone, D., & Baylot, J. (2011). Lattice Discrete Particle Model (LDPM) for failure behavior of concrete. II: Calibration and validation. *Cement and Concrete Composites*, 33(9), 891–905.

Cusatis, G., Pelessone, D., & Mencarelli, A. (2011). Lattice Discrete Particle Model (LDPM) for failure behavior of concrete. I: Theory. *Cement and Concrete Composites*, 33(9), 881–890.

Del Prete, C., Boumakis, I., Wan-Wendner, R., Vorel, J., Buratti, N., & Mazzotti, C. (2021). A lattice discrete particle model to simulate the viscoelastic behaviour of macro – synthetic fibre reinforced concrete. *Construction and Building Materials*, 295, 123630.

Del Prete, C., Buratti, N., & Mazzotti, C. (2021). Experimental Investigation on the Influence of Temperature Variations on Macro-synthetic Fibre Reinforced Concrete Short and Long Term Behaviour. In *RILEM Bookseries* (pp. 331–341).

Del Prete, C., Tilocca, A., Buratti, N., & Mazzotti, C. (2017). *EFFECT OF FIBER DOSAGE AND MATRIX COMPRESSIVE STRENGHT ON MSFRC PERFORMANCE*. 1–10.

Del Prete, C., Wan-Wendner, R., Buratti, N., & Mazzotti, C. (2019). Lattice Discrete Particle Modeling of MSFRC. *SSCS19 Numerical Modeling Strategies for Sustainable Concrete Structures*, 27–36.

Del Prete, Clementina, Buratti, N., Manzi, S., & Mazzotti, C. (2019). Macro-synthetic fibre reinforced concrete: influence of the matrix mix design on interfacial bond behaviour. *IOP Conference Series: Materials Science and Engineering*.

Di Prisco, M., Colombo, M., & Dozio, D. (2013). Fibre-reinforced concrete in fib Model Code 2010: Principles, models and test validation. *Structural Concrete*, 14(4), 342–361.

Ferrara, L., Bamonte, P., Caverzan, A., Musa, A., & Sanal, I. (2012). *Testing the Fresh and Hardened State Performance of Steel Fibre Reinforced Self-Compacting Concrete*. 1–12.

Gal, E., & Kryvoruk, R. (2011). Meso-scale analysis of FRC using a two-step homogenization approach. *Computers and Structures*, 89(11–12), 921–929. https://doi.org/10.1016/j.compstruc.2011.02.006

Jin, C., Buratti, N., Stacchini, M., Savoia, M., & Cusatis, G. (2016). Lattice discrete particle modeling of fiber reinforced concrete: Experiments and simulations. *European Journal of Mechanics, A/Solids*, 57, 85–107.

Leporace-Guimil, B., Mudadu, A., Conforti, A., & Plizzari, G. A. (2021). Influence of fiber orientation and structural-integrity reinforcement on the flexural behavior of elevated slabs. *Engineering Structures*, 252(June 2021), 113583. https://doi.org/10.1016/j.engstruct.2021.113583

Liu, X., Sun, Q., Song, W., & Bao, Y. (2022). Numerical modeling and parametric study of hybrid fiber-rebar reinforced concrete tunnel linings. *Engineering Structures*, 251(PB), 113565. https://doi.org/10.1016/j.engstruct.2021.113565

Llano-Torre, A., Serna, P., Garcia-Taengua, E., Vrijdaghs, R., Pauwels, H., Del Prete, C., &..., & Bernard, E. S. (2021). Analysis of the RRT Results. In Springer (Ed.), *Round-Robin Test on Creep Behaviour in Cracked Sections of FRC: Experimental Program, Results and Database Analysis* (pp. 147–270).

Nana, W. S. A., Tran, H. V., Goubin, T., Kubisztal, G., Bennani, A., Bui, T. T., Cardia, G., & Limam, A. (2021).

Behaviour of macro-synthetic fibers reinforced concrete: Experimental, numerical and design code investigations. *Structures*, 32(March), 1271–1286. https://doi.org/10.1016/j.istruc.2021.03.080

Nonato Da Silva, C. A., Ciambella, J., Barros, J. A. O., dos Santos Valente, T. D., & Costa, I. G. (2020). A multiscale model for optimizing the flexural capacity of FRC structural elements. *Composites Part B: Engineering*, 200(August), 108325. https://doi.org/10.1016/j.compositesb.2020.108325

Schauffert, E. A., Cusatis, G., Pelessone, D., O'Daniel, J. L., & Baylot, J. T. (2011). Lattice Discrete Particle Model for Fiber-Reinforced Concrete. II: Tensile Fracture and Multiaxial Loading Behavior. *Journal of Engineering Mechanics*, 138(7), 834–841.

Thai, D. K., Nguyen, D. L., & Nguyen, D. D. (2020). A calibration of the material model for FRC. *Construction and Building Materials*, 254, 119293. https://doi.org/10.1016/j.conbuildmat.2020.119293

Vrijdaghs, R., di Prisco, M., & Vandewalle, L. (2020). Creep of polymeric fiber reinforced concrete: A numerical model with discrete fiber treatment. *Computers and Structures*, 233, 106233. https://doi.org/10.1016/j.compstruc.2020.106233

Xu, H., Wang, Z., Shao, Z., Cai, L., Jin, H., Zhang, Z., Qiu, Z., Rui, X., & Chen, T. (2021). Experimental study on durability of fiber reinforced concrete: Effect of cellulose fiber, polyvinyl alcohol fiber and polyolefin fiber. *Construction and Building Materials*, 306(June), 124867. https://doi.org/10.1016/j.conbuildmat.2021.124867

Computational Modelling of Concrete and Concrete Structures – Meschke, Pichler & Rots (Eds)

Convex and effective yield surfaces for numerical rigid plastic limit analysis of reinforced concrete structures with in-plane forces

M.E.M. Andersen
Department of Bridges International, COWI A/S, Kongens Lyngby, Denmark

P.N. Poulsen, J.F. Olesen & L.C. Hoang
Department of Civil Engineering, The Technical University of Denmark, Kongens Lyngby, Denmark

ABSTRACT: Many reinforced concrete structures are validated in the ultimate limit state (ULS) using analysis methods based on the theorems of plasticity and the rigid-plastic material model. The rigid-plastic material model significantly simplifies the actual stress-strain relationship of reinforced concrete. However, good agreement with capacities found from experiments has been shown when a reduced or so-called effective concrete compressive strength is used. The effective strength is mainly dependent on the transverse tensile strain when a single material point is considered, and well-accepted expressions are given in the codes. The Modified Mohr-Coulomb yield criterion with an effective strength is combined with the elasto-plastic behavior of the reinforcement to create an effective yield surface for reinforced concrete for plane stress states. Based on this, the paper presents an approximate convex effective yield surface, which can be used for Finite Element Limit Analysis (FELA) calculations. The convex effective yield surface is based on auxiliary strains linked to the reinforcement stresses on a material point level. The effective yield surface is tested on a material point level using an experimental database for reinforced concrete panels and on a structural level with an example of a reinforced concrete deep beam with holes. Both tests yield satisfactory results.

1 INTRODUCTION

Concrete is a material with a highly non-linear material behavior in both compression and tension. Advanced Non-Linear Finite Element Analysis (NLFEA) programs such as Diana (Ferreira 2020), and Atena (Červenka & Červenka 2017) can account for the non-linearity using expressions from, for instance, the fib Model Code (fib 2013). By using these non-linear expressions, detailed modeling of structures is possible. However, the analysis can also be cumbersome and requires expert knowledge to alleviate convergence problems in the loading of the structures. Furthermore, many material parameters are needed to describe the non-linear relationship, which can be challenging to determine.

For these reasons, many designs are validated in the ultimate limit state (ULS) using limit analysis methods based on the theorems of plasticity and the rigid-plastic material model (Drucker, Prager, & Greenberg 1952; Gvozdev 1960). Finite Element Limit Analysis (FELA) applies the theorems of plasticity and is a numerical method based on optimization, and since the problem can be posed as a convex problem, it can be solved efficiently (Anderheggen & Knöpfel 1972).

In a FELA framework based on the lower bound theorem, the structure is divided into stress-based finite elements. Scalable load is applied to the structure, and equilibrium is ensured in elements and on boundaries, while a yield surface constrains the stress state of the elements. The largest possible load which the structure can sustain is then sought. For reinforced concrete, the yield surface is often based on the Modified Mohr-Coulomb yield criterion with the possible inclusion of smeared reinforcement using additional linear constraints.

Using a rigid-plastic material model is an extreme simplification compared to the actual stress-strain relationship. However, in combination with a reduced concrete compressive strength, the load-bearing capacities obtained using these methods have shown good agreement with those obtained from experiments on beams, plates, and other structural elements. The reduced concrete compressive strength is obtained by multiplying the cylinder compression strength with a so-called effectiveness factor, ν. Historically, the effectiveness factor has been obtained empirically for individual problem types, such as beams in bending and beams in shear, through the fitting of experimental results with results from exact rigid plastic solutions. Large test databases exist to make these fits for many

DOI 10.1201/9781003316404-63

different structure types. However, all situations cannot be tested, and a general method is needed. For FELA this would mean the development of an effective yield surface, which is the topic of this paper. Previously the topic has been treated in a purely stress-based approach (Herfelt, Poulsen, & Hoang 2018).

To determine how an effective yield surface would look a deformation-based model is considered since several authors have suggested that the effective compressive strength of concrete is linked to the transverse tensile strain ε_1 (Collins & Vecchio 1982; Hoang, Jacobsen, & Larsen 2012). Expressions to determine the effectiveness factor based on ε_1 exist in, for example, the fib model code. Using these expressions combined with a linear elastic perfectly plastic constitutive law for concrete and reinforcement, the stress state for a given strain state can be found. By repeating this calculation for many different strain states, effective yield surfaces are found in the stress space, depending on the degree of allowed strain. The effective elasto-plastic yield surfaces found in this manner are clearly reduced compared to yield surfaces where the effective strength of concrete is not considered.

The next step is to develop a yield surface that can be used in FELA, approximating the yield surfaces from the deformation-based model. The effective elasto-plastic yield surfaces are not convex due to the expression for the effectiveness factor. Therefore, linearization is performed. Furthermore, the elasto-plastic yield surfaces require knowledge of the strains. However, strains are not available on a structural level in FELA due to the rigid-plastic material model. This challenge is overcome by introducing strains as an auxiliary variable on a material point level. The auxiliary strains are constrained and linked to the stresses of the reinforcement by assuming an elasto-plastic behavior. In this way, an effective and convex yield surface is established.

The effective rigid-plastic and convex yield surface for plane reinforced concrete is tested in two examples. The first example is of a reinforced concrete panel loaded in shear, with and without biaxial compression or tension, to test the performance of the yield surface on a material point level. The second example is of a reinforced concrete deep beam with holes, this example is made to see the performance of the yield surface on a structural level.

2 FINITE ELEMENT LIMIT ANALYSIS

Finite Element Limit Analysis (FELA) is a combination of the domain discretization of the Finite Element Method and limit analysis based on the extremum principles of plasticity as postulated by Gvozdev (1960), and Drucker, Prager, & Greenberg (1952). The method was first proposed by Anderheggen & Knöpfel (1972). This paper will only give a brief explanation of the method. For further information, refer to, e.g., Andersen, Poulsen, & Olesen (2022).

The FELA method of this paper is based on the lower bound theorem and is posed as a constrained optimization problem in the following way:

$$\begin{aligned} &\text{max.} \quad \lambda && \text{Load parameter} && \text{(1a)}\\ &\text{s.t.} \quad \mathbf{H}\boldsymbol{\beta} = \mathbf{R}_0 + \lambda\mathbf{R} && \text{Stress equilibrium} && \text{(1b)}\\ &\qquad f_i(\boldsymbol{\sigma}_i) \leq 0 && \text{Yield conditions} && \text{(1c)} \end{aligned}$$

The parameter λ scales the load, and is maximized via the objective function (1a). Equation (1b) ensures the stress equilibrium between internal and external forces. The stress continuity is ensured by $\mathbf{H}\boldsymbol{\beta}$ where $\mathbf{H}$ is the so-called equilibrium matrix, which consists of contributions from each of the elements, and $\boldsymbol{\beta}$ which is a vector collection of the stress variables. The element used for the calculations is a mixed linear lower bound triangle (Herfelt 2017; Krabbenhøft 2016), which is a relaxed version of the lower bound element by Poulsen & Damkilde (2000). The external forces are given by the constant loads $\mathbf{R}_0$ and scalable loads $\lambda\mathbf{R}$.

The last part of the optimization problem is the yield conditions (1c). The elements have a number of material points which contain stress variables. For a plane model the stress variables will be described by the vector:

$$\boldsymbol{\sigma} = \begin{bmatrix} \sigma_{xx} \\ \sigma_{yy} \\ \sigma_{xy} \end{bmatrix} \tag{2}$$

Equation (1c) states that the stresses of the material points should be on or inside the yield surfaces defined by f_i. These yield surfaces are the subject of this paper.

3 MODELING OF PLANE REINFORCED CONCRETE

The models of this paper all use the so-called *smeared reinforcement* approach, whereby the reinforcement bars are assumed to be placed sufficiently close for this to be a reasonable simplification. Furthermore, the reinforcement is assumed to be orthogonally placed coinciding with the x- and y-axis of the Cartesian coordinate system. The amount of reinforcement is described as the reinforcement ratios $\rho_{s,x}$ and $\rho_{s,y}$, see Figure 1. The yield strength of the reinforcement is f_s and the reinforcement is assumed to carry normal tensile stresses only.

The concrete is assumed to be a material with compressive strength, f_c, and negligible tensile strength. Consequently, the reinforced concrete is considered a composite material, where the compressive capacity comes from the concrete and the tensile capacity from the reinforcement.

The elasto-plastic models also use the modulus of elasticity of concrete and steel, E_c and E_s, as well as the crushing strain of concrete ε_{cu} and the rupture strain of the reinforcement ε_{su}. This paper considers a fixed set of parameters which can be seen in Table 1.

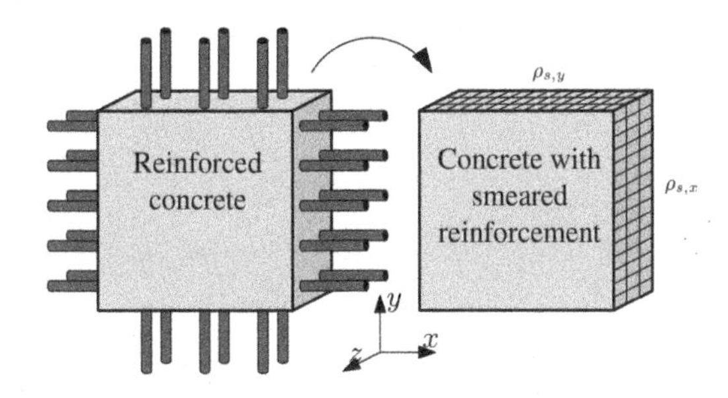

Figure 1. Representative reinforced concrete membrane.

Table 1. Material parameters used to generate the yield surfaces.

f_c	[MPa]	30
E_c	[GPa]	33
ε_{cu}	[‰]	3.5
f_s	[MPa]	500
E_s	[GPa]	210
ε_{su}	[‰]	50
$\rho_{s,x} = \rho_{s,y}$	[%]	0.6

3.1 *Separation of stresses*

Separation of the total stress into concrete and reinforcement stresses is performed to enable the modeling of the yield surfaces:

$$\boldsymbol{\sigma}_{\square} = \boldsymbol{\sigma}_{\square,c} + \boldsymbol{\rho\sigma}_{\square,s} \tag{3}$$

where $\boldsymbol{\sigma}_{\square}$ is the total stress tensor given by:

$$\boldsymbol{\sigma}_{\square} = \begin{bmatrix} \sigma_{xx} & \sigma_{xy} \\ \sigma_{xy} & \sigma_{yy} \end{bmatrix} \tag{4}$$

and $\boldsymbol{\sigma}_{\square,c}$ is the concrete stress tensor given by:

$$\boldsymbol{\sigma}_{\square,c} = \begin{bmatrix} \sigma_{c,xx} & \sigma_{c,xy} \\ \sigma_{c,xy} & \sigma_{c,yy} \end{bmatrix} \tag{5}$$

and $\boldsymbol{\rho\sigma}_{\square,s}$ is the reinforcement stress tensor given by:

$$\boldsymbol{\rho\sigma}_{\square,s} = \begin{bmatrix} \rho_x & 0 \\ 0 & \rho_y \end{bmatrix} \begin{bmatrix} \sigma_{s,xx} & 0 \\ 0 & \sigma_{s,yy} \end{bmatrix} \tag{6}$$

This separation of stresses is analogue to the way the Nielsen yield criteria is developed (Nielsen & Hoang 2011).

3.2 *The effectiveness factor*

The effectiveness factor, ν, is a parameter introduced to enable the usage of limit analysis methods based on the theory of rigid-plastic materials for reinforced concrete structures, even though the actual material behavior is not rigid-plastic. However, the limit analysis methods can still be used to provide failure loads in good agreement with tests when a reduction of the concrete compressive strength via the effectiveness factor is applied:

$$f_{c,\text{eff}} = \nu f_c \tag{7}$$

where $f_{c,\text{eff}}$ is the effective concrete compressive strength. The effectiveness factor accounts for several different strength reduction effects related to softening, micro-, and macro-cracking (Nielsen & Hoang 2011). Several authors have suggested formulas for determining the effectiveness factor based on different geometrical and material properties. See Ref. (Hoang, Jacobsen, & Larsen 2012) for an overview of different works. Several of these authors suggest that the effectiveness factor should be a function of the transverse tensile strain, ε_1, and this has also been adopted in the fib model code 2010 (fib 2013) and in the new enquiry version of Eurocode 2 (pr EN1992-1-1 2021). The effectiveness factor for structures that meet the demand for minimum reinforcement may be written in the following way:

$$\nu(\varepsilon_1) = \eta_{fc}\eta_\varepsilon(\varepsilon_1) \tag{8}$$

The first factor, η_{fc}, accounts for the brittleness of the concrete and can according to (pr EN1992-1-1 2021) be taken as:

$$\eta_{fc} = \sqrt[3]{f_{c0}/f_c} \leq 1.0, \quad f_c \text{ in MPa} \tag{9}$$

where f_{c0} is a reference strength in the order of 30–40 MPa. In this paper, the value is taken as 30 MPa. The second factor, η_ε, is dependent on the transverse tensile strain and can be formulated as:

$$\eta_\varepsilon(\varepsilon_1) = \frac{1}{c_1 + c_2\varepsilon_1} \leq c_3 \tag{10}$$

where c_1, c_2, and c_3 are some calibration constants. Herfelt, Poulsen, & Hoang (2018) chose values of $c_1 = 1$, $c_2 = 80$, and $c_3 = 1$, which have also been adopted here.

The left hand side of equation (10) is non-convex and thus also equation (8), making the formula unusable in a convex optimization framework. For his reason, a simple linear relation is adopted in the convex approximations:

$$\eta_\varepsilon(\varepsilon_1) = 1 - a\varepsilon_1 \leq 1 \tag{11}$$

where a is the proportionality factor. Figure 2 shows the graph of the left hand side of Expression (10) and the simple linear expression with different values of the a-parameter. The a-parameters in the figure correspond to the slope required to get a reduction similar to Expression (10) for different maximal strains $\varepsilon_{1,\max}$, at a transverse strain corresponding to the yielding strain of the reinforcement.

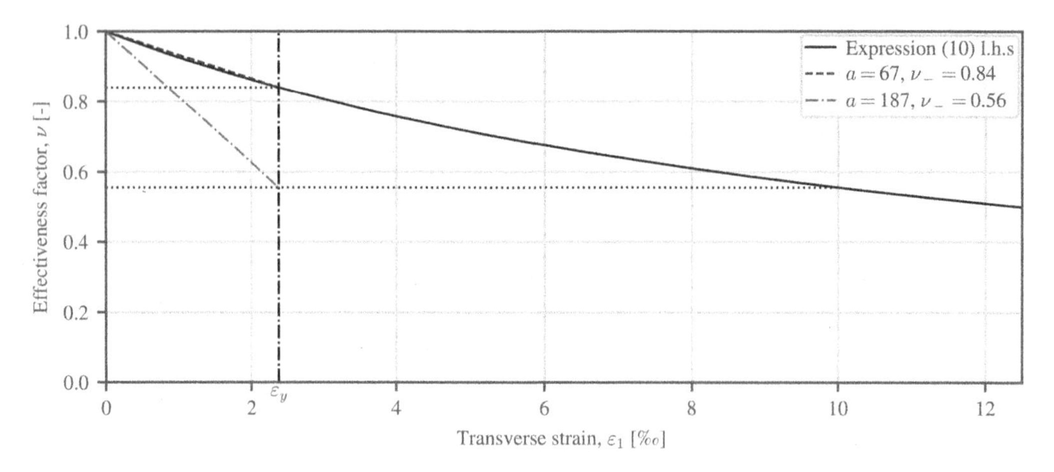

Figure 2. Effectiveness factor as a function of the transverse strain with linear approximations yielding the same reduction at ε_y, as Expression (10) yields for $\varepsilon_{1,\max}$.

4 YIELD SURFACES

Four implementations of yield surfaces for plane reinforced concrete are shown in the following. The yield surfaces are plotted in the $(\sigma_{xx}, \sigma_{yy}, \sigma_{xy})$-coordinate system. Only the positive values of the shear stress are plotted since the yield surfaces are symmetrical with respect to the $(\sigma_{xx}, \sigma_{yy})$-plane.

The rigid-plastic yield surface for plane stress states, is presented as a reference. The rigid-plastic yield surface can only consider a constant reduction of the compressive strength. Therefore, it is a helpful comparison, to see the effect of the reductions due to the transverse strain. Thereafter, two effective elasto-plastic yield surfaces are developed, one as a lower envelope and one as an upper envelope of the effective yield surface. Lastly, a convex effective rigid-plastic yield surface is developed.

4.1 *Rigid-plastic reinforced concrete yield surface*

If only a fixed value of the effectiveness factor is considered, a rigid-plastic yield surface can be developed based only on the concrete compressive strength f_c, the reinforcement yield strength f_s, and the reinforcement ratios $\rho_{s,x}$ and $\rho_{s,y}$.

The Rigid-plastic reinforced concrete yield surface separates stresses into concrete and reinforcement stresses as described above. The concrete should then abide by the Modified Mohr-Coulomb yield criterion with a tensile cutoff of zero and the reinforcement by a simple uni-axial relation. The mathematics of the yield surface is described in Nielsen & Hoang (2011), and a convex implementation can be found, e.g., in Herfelt (2017). A plot of the yield surface can be seen in Figure 3 using the material parameters of Table 1.

It should be noted that Nielsen proposed introducing the effectiveness factor by an additional constraint $|\sigma_{xy}| \leq 0.5\,\nu f_c$ on the shear stress. This additional constraint is omitted for the comparisons in this paper.

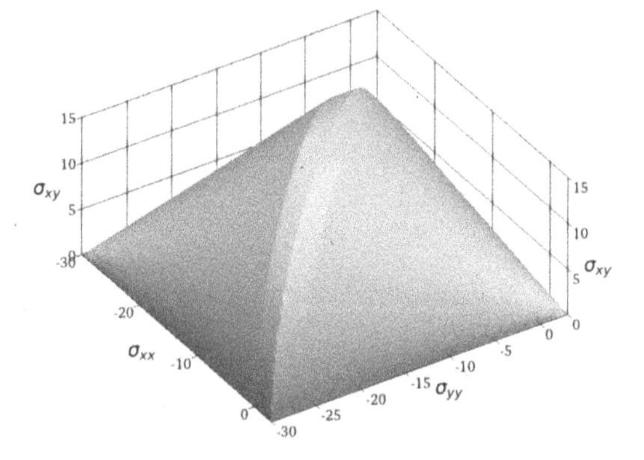

Figure 3. The rigid-plastic reinforced concrete yield for parameters in Table 1.

4.2 *Effective elasto-plastic reinforced concrete yield surface*

The following shows the methodology used to generate two different effective elasto-plastic yield surfaces. These two yield surfaces will represent an upper and a lower bound envelope, of which the significance will be explained later.

The model takes a strain tensor in the form:

$$\boldsymbol{\varepsilon}_{\square} = \begin{bmatrix} \varepsilon_{xx} & \varepsilon_{xy} \\ \varepsilon_{xy} & \varepsilon_{yy} \end{bmatrix} \tag{12}$$

and based on the constitutive equations of the reinforcement and concrete determines a stress state. This process is repeated many times for different strain tensors. The strain tensors are generated in a step-wise process. A unit strain tensor is generated, which is equivalent to a direction in the strain space. The unit strain tensor is then multiplied by a linearly increasing factor to control the magnitude. By repeating this for many different unit strain tensors the strain space is covered. Applying the non-linear constitutive relation will result in many different stress tensors, and thus a so-called point cloud will be generated in the plane stress space. The two yield surfaces can then be determined from the point cloud by certain criteria.

4.2.1 Constitutive relation of the reinforcement
The constitutive relation for the reinforcement is a simple uniaxial relation in each of the two directions, since the reinforcement is assumed only to carry normal stresses and to be placed according to the (x, y)-coordinate system. The linear elastic perfectly plastic material model is applied:

$$\sigma_{s,n}(\varepsilon_n) = \begin{cases} 0, & \varepsilon_n \leq 0 \\ \varepsilon_n E_s, & 0 \leq \varepsilon_n \leq \varepsilon_s \\ f_y, & \varepsilon_s \leq \varepsilon_n \end{cases} \tag{13}$$

where the subscript n denotes either the x- or y-normal. A graph of the relation can be seen in Figure 4.

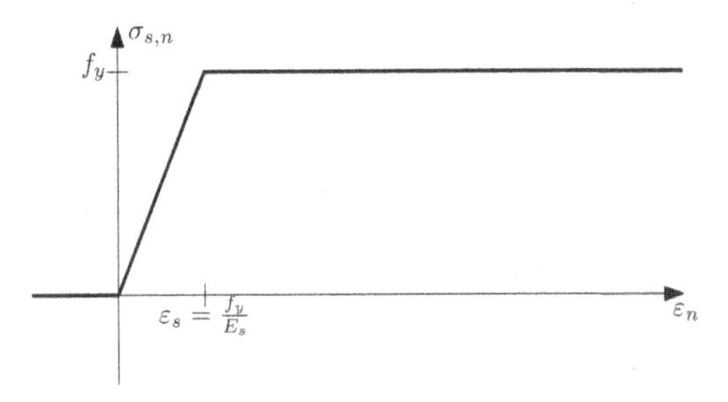

Figure 4. Constitutive relation of the reinforcement.

4.2.2 Constitutive relation of the concrete
The constitutive relation of the concrete is based on a linear elastic perfectly plastic relation, same as for the reinforcement. However, the effective uniaxial compressive strength depends on the transverse strain via the effectiveness factor. For these reasons, the constitutive relation of the concrete is based on principal stresses and principal strains. Due to the effect of the transverse strain and the potentially complicated expression for the effectiveness factor, the equations are not easily posed with limits. However, they can be posed in the following way:

$$\sigma_{c,1}(\varepsilon_1, \varepsilon_2) = \min\{\max\{E_c \varepsilon_1, \nu(\varepsilon_2) f_c\}, 0\} \tag{14}$$

$$\sigma_{c,2}(\varepsilon_1, \varepsilon_2) = \min\{\max\{E_c \varepsilon_2, \nu(\varepsilon_1) f_c\}, 0\} \tag{15}$$

With the usual ordering of the principal strains and stresses, that is, $\varepsilon_1 \geq \varepsilon_2$ and $\sigma_{c,1} \geq \sigma_{c,2}$, only the second principal concrete stress can be influenced by the transverse strain, since if ε_2 is positive ε_1 must also be positive, which implies that $\sigma_{c,1}$ is zero.

A plot of the constitutive relation of the second principal concrete stress as a function of the principal strains can be seen in Figure 5, where material parameters from Table 1 are used. With tensile strain in both principal directions, no concrete stress is present, whereas a linear relation is seen with increasing negative principal strains. The effective compressive strength limits the maximum principal stress, and the increasing transverse strain makes the strength decrease.

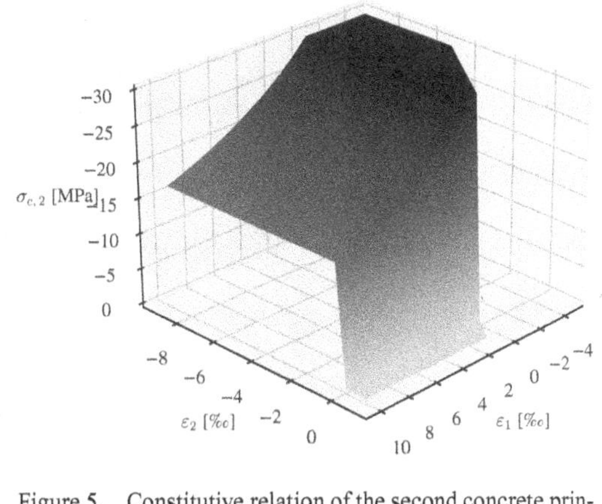

Figure 5. Constitutive relation of the second concrete principal stress in principal strain space.

4.2.3 Generation of yield surfaces
With the constitutive relation of the concrete and the reinforcement established, it is possible to determine the corresponding stress state of the composite material for a given strain. The calculation procedure is as follows:

1. Given a strain tensor in the form of equation (12).
2. Compute reinforcement stresses by equation (13) using ε_{xx} and ε_{yy}.
3. Compute principal strains and then compute principal concrete stresses from equations (14) and (15).
4. Transform concrete principal stresses back into directions of original coordinate system.
5. Compute the composite stress state from equation (3).

The above algorithm is used to generate the point cloud of possible stress states from which the yield surfaces are generated. The first yield surface will be called the *upper envelope* (UE) yield surface, which will be generated from the concave envelope of the entire point cloud. The second yield surface will be called the *lower envelope* (LE) yield surface, which will be generated from the concave envelope of the points where there is either no tension and the concrete has reached the crushing strain ε_{cu}, or for points with tension, where the maximum normal tensile strain reaches $\varepsilon_{1,\max} = 10‰$. With these criteria, very similar stress states can exist with varying shear capacity. In these situations, the point with the least shear capacity is shown. The choice of the value 10‰ is arbitrary, and it could be argued that a larger value should be chosen. For instance, the ductility requirement according to the Eurocode for type B reinforcement is required to be 50‰ (Eurocode 2 2008). However, a transverse strain of 50‰ would correspond to a prohibitively large reduction of the concrete strength, and therefore a lower value is used for these examples.

A plot of the UE yield surface with material parameters from Table 1 can be seen in Figure 6. The coloring on the surface is the perpendicular distance from the current yield surface to the rigid-plastic reinforced concrete yield surface, which can be used to distinguish what has been cut away by introducing the effectiveness factor.

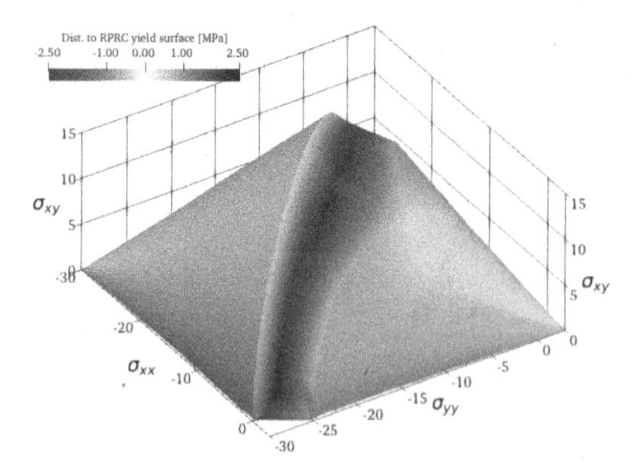

Figure 6. Upper envelope (UE) yield surface.

Firstly, it can be seen that a cone in the compression side of the plot remains unaltered, which is the part corresponding to concrete in biaxial compression. These stresses can be carried without activating the reinforcement in tension, and therefore are not influenced by the effectiveness factor.

Notably, "the right-hand side" of the plot is no longer shaped like a cone. From the coloring of the figure, it can be seen that the reduction is most pronounced in a band around the middle of the yield surface. These are stress states with either predominately shear stress, or shear stress with normal stresses of opposite signs.

From the unaltered part, a decrease in the shear capacity and the maximum compression with transverse tension is seen. Looking at the $\sigma_{xx}\sigma_{yy}$-plane, for maximum transverse tension, it can be seen that the effective compressive strength is reduced from 30 MPa to about 25 MPa. This reduction is equivalent to the effectiveness factor for a transverse strain of ε_y, which is also the required transverse tension to activate the reinforcement fully and thus as expected for the upper envelope.

A plot of the lower envelope yield surface with material parameters from Table 1 can be seen in Figure 7. The cone in the compression part of the yield surface corresponding to biaxial compression is still unaltered. However, the rest of the yield surface is much more reduced due to the larger strains meaning an additionally reduced compressive strength. Looking at the $\sigma_{xx}\sigma_{yy}$-plane, the effective concrete compressive strength is reduced from 30 MPa to about 17 MPa, which is consistent with a transverse strain of $\varepsilon_{1,\max} = 10‰$.

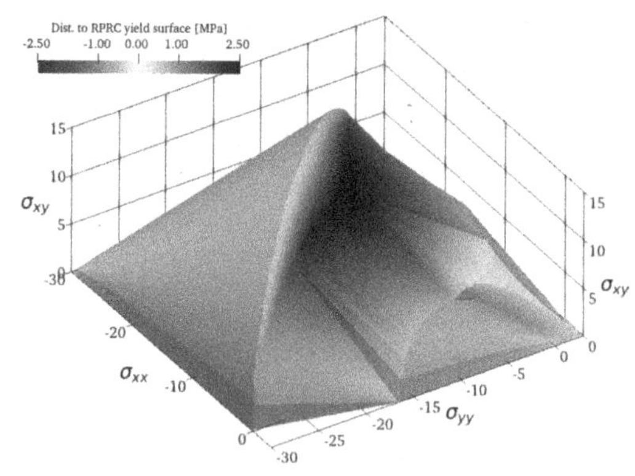

Figure 7. Lower envelope (LE) yield surface.

4.3 *Effective rigid-plastic reinforced concrete yield surface*

To develop a yield surface that can be used in a FELA context, it must be convex and based on the available variables, which are stresses. However, auxiliary strains can be introduced on a material point level by assuming a restriction between the stresses of the model and the auxiliary strains. The strains are introduced as variables: $[\varepsilon_{xx}, \varepsilon_{yy}, \varepsilon_{xy}]$. The strains are linked to the material point and not to a structural deformation-based model, and therefore the strains of the material point are only indirectly influenced by the rest of the structure via the stress equilibrium.

The strains are introduced in relation to the reinforcement stresses in the following way:

$$\sigma_{s,xx} - \varepsilon_{xx}E_s \leq 0 \quad (16a)$$

$$\sigma_{s,yy} - \varepsilon_{yy}E_s \leq 0 \quad (16b)$$

The relations above create a link between the strains and the reinforcement stresses. So in order for the reinforcement to be activated, positive strains are required.

From the plane strains, principal strains can be found in the following way:

$$C_\varepsilon = 1/2(\varepsilon_{xx} + \varepsilon_{yy}) \quad (17a)$$

$$R_\varepsilon = \sqrt{1/2(\varepsilon_{xx} - \varepsilon_{yy})^2 + \varepsilon_{xy}^2} \quad (17b)$$

$$\varepsilon_1 = C_\varepsilon + R_\varepsilon \quad (17c)$$

where ε_1 is the transverse strain. Equation (17b) is equivalent to a second-order cone and can therefore be cast in a convex form. Hereby the transverse strain is available for the implementation.

The concrete stresses should abide by the Modified Mohr-Coulomb yield criterion:

$$\sigma_1 \leq 0 \quad (18a)$$

$$k\sigma_1 - \sigma_3 \leq \nu(\varepsilon_1)f_c \quad (18b)$$

where σ_1, and σ_3 are the largest and smallest principal stress, respectively, and k is the frictional parameter usually taken as 4. Equation (18a) describes the separation criterion meaning that no concrete tensile strength is considered, and equation (18b) describes the friction criterion. The compressive strength is now a function of the transverse strain via the effectiveness factor. For the proposed yield surface to be convex, the linear approximation of the strain-dependent part of the effectiveness factor, equation (11), is used. The Modified Mohr-Coulomb yield criterion is implemented in the usual manner.

The reinforcement stresses are restricted by simple uni-axial bounds:

$$0 \leq \sigma_{s,xx} \leq f_s \tag{19a}$$

$$0 \leq \sigma_{s,yy} \leq f_s \tag{19b}$$

where f_s is the strength of the reinforcement. With this the effective rigid-plastic reinforced concrete yield surface is presented.

Figure 8 shows the yield surface generated for the material parameters of Table 1 and with the slope parameter a in equation (11) of 67, which is equivalent to a straight line rendering the same value as expression (10) at a transverse strain equal to the yield strain of the reinforcement. The surface is colored after the distance to the UE yield surface shown in Figure 7. The rigid-plastic yield surface generally has the same shape as the UE yield surface, however, as can be seen from the red coloring, the rigid-plastic yield surface is generally less conservative. Figure 9 shows the yield surface with a slope parameter of 187, which is equivalent to a reduction from Expression (10) of 10‰, but at the yielding strain of the reinforcement. This figure is equivalent to the lower envelope and is colored by the distance to the LE yield surface. Again the approximation is quite good. However, there are still areas that are non-conservative with respect to the elasto-plastic yield surface. Nevertheless, this is expected since the non-convex parts can not be accurately captured in a convex approximation.

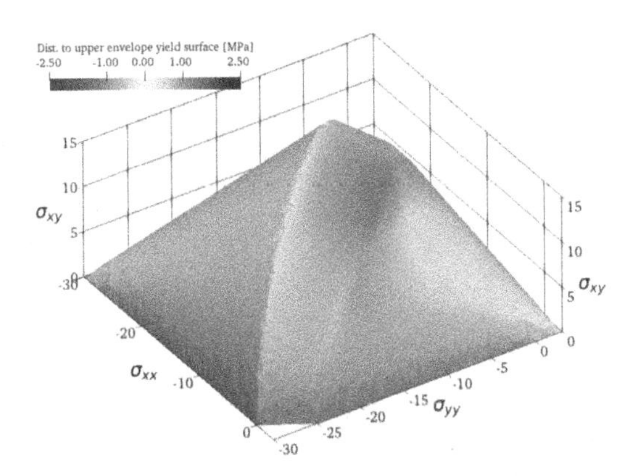

Figure 8. Effective rigid-plastic yield surface with $a = 67$.

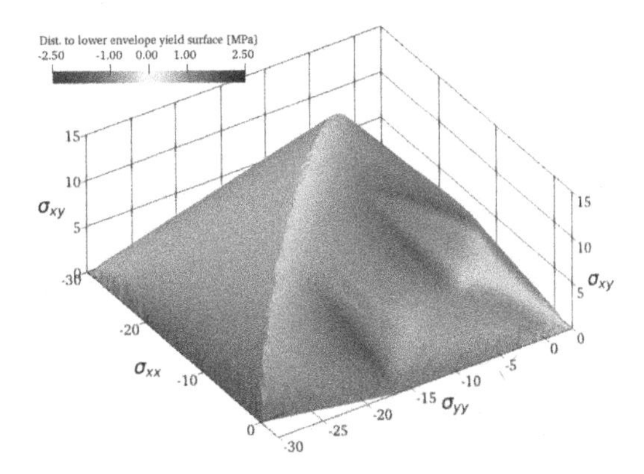

Figure 9. Effective rigid-plastic yield surface with $a = 187$.

5 EXAMPLE: REINFORCED CONCRETE PANEL IN SHEAR WITH AND WITHOUT NORMAL FORCE

The effective yield surface is tested on some experiments of reinforced concrete panels. The reinforced concrete panel experiments have been collected by Hoang, Jacobsen, & Larsen (2012). However, a modified version of the database by Brask & Xuan (2019) is used. The modified database omitted panels that experienced local failure or failure in the experimental setup.

The setup of the panel experiments varies. However, they all seek to emulate a reinforced concrete panel with a concrete stress state in pure shear or shear with biaxial compression or tension. The idealized model in FELA can be seen in Figure 10. The database consists of 72 specimens, with 60 panels in pure shear, 5 in shear with biaxial tension, and 7 with shear and biaxial compression. The biaxial compression and tension are included as a fraction κ of the shear. Of the 72-panels, roughly half ($N = 34$) is isotropically reinforced. The

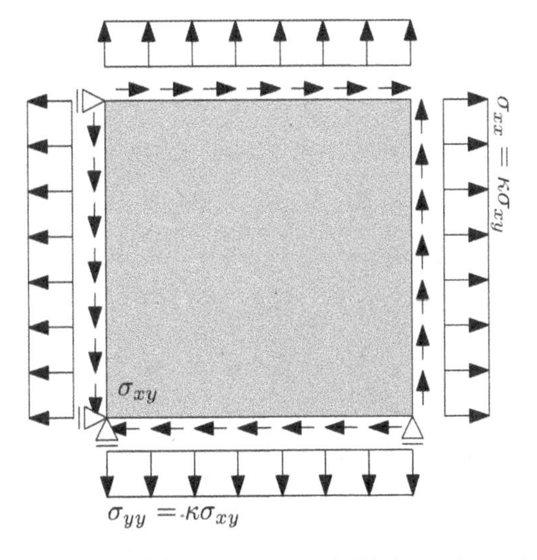

Figure 10. Reinforced concrete panel with shear and normal load.

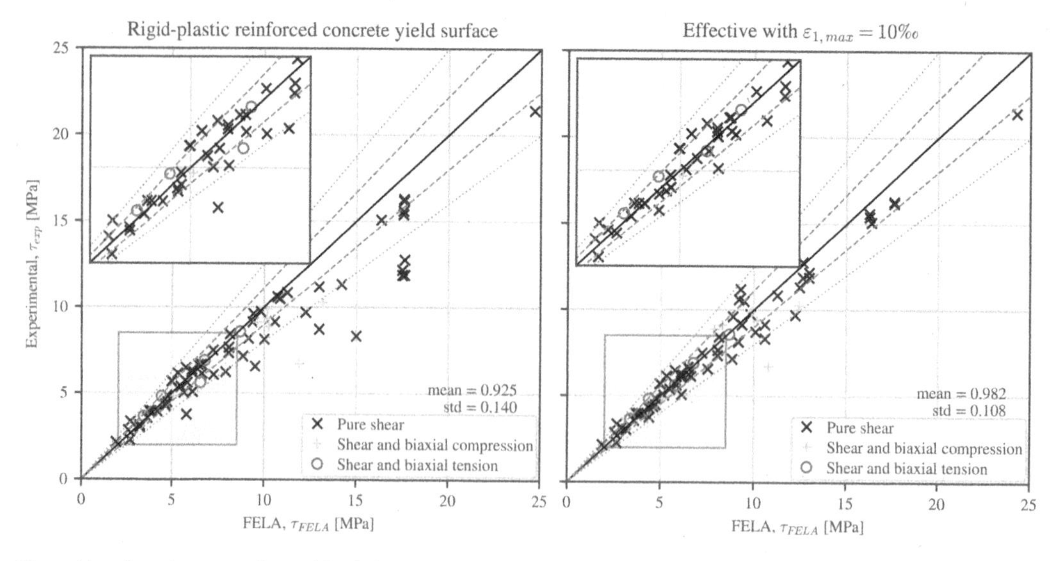

Figure 11. Capacity comparison with reinforced concrete panel experiments.

material parameters for the tests vary, and all the details will not be given here, but they can be found in Brask & Xuan (2019).

The FELA calculations are performed using two yield surfaces. First the rigid-plastic reinforced concrete yield yield surface, where the effectiveness factor is simply 1.0 everywhere (Figure 3), and secondly the effective yield surface with a maximum transverse tensile strain of $\varepsilon_{1,\max} = 10‰$, where the effectiveness factor can vary from 1.0 (Figure 9). The yield surfaces use the material parameters of the specimen and will therefore not be exactly equal to the ones presented so far.

Figure 11 shows a comparison between the experimental capacity τ_{exp} on the ordinate and the capacity found from the FELA calculations τ_{FELA} on the abscissa. The specimens have different markers depending on the loading scenario. The plots also shows a thick line corresponding to $\tau_{\text{exp}} = \tau_{\text{FELA}}$, and two additional lines on either side corresponding to a 5% and 10% deviation. Observations to the right of the thick line will have an overestimated capacity and opposite for points to the left. Furthermore, the plot also shows the basic statistics of the capacity ratio $\tau_{\text{exp}}/\tau_{\text{FELA}}$, where a mean value close to 1 and a low standard deviation would indicate a good fit between the experimental and calculated capacity.

The rigid-plastic reinforced concrete yield generally overestimates the capacity with several data points way outside the 10% deviation line. The result using the effective yield surface is much improved. Almost all the worst outliers are now within the 10% deviation line, and the mean value of the capacity ratio went from 0.925 to 0.982, while the standard deviation has gone down, which indicates that the effective yield surface works well on a material point level. However, one thing to consider is which values of $\varepsilon_{1,\max}$ and the calibration parameters c_1, c_2, and c_3 from equation (10), are used to find the slope parameter a. Since the slope parameter is what defines how much the yield surface is reduced.

6 EXAMPLE: DEEP BEAM WITH HOLES

The previous example showed the behavior of the effective yield surface when compared to experiments performed on reinforced concrete panels. Here the effective yield surface improved the scatter of the results. However, the FELA calculations of those experiments yield a constant stress state over the entire model, and therefore it is also desirable to see the effect on an example with a complicated stress distribution. Therefore, an example for a reinforced concrete deep beam with holes is considered.

A sketch of the beam can be seen in Figure 12. The beam is thicker at the top and the bottom, with the

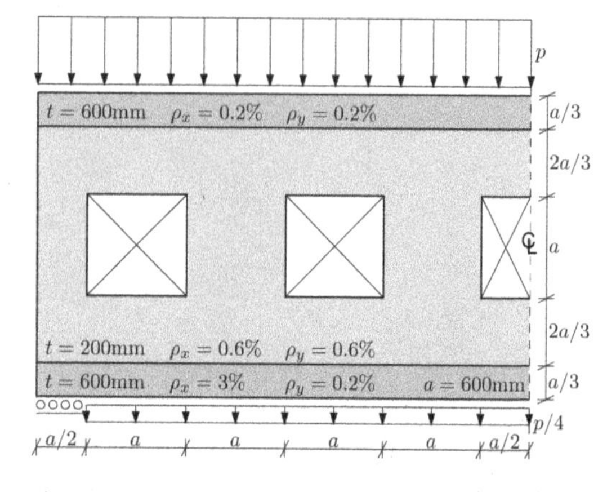

Figure 12. Reinforced concrete deep beam with holes.

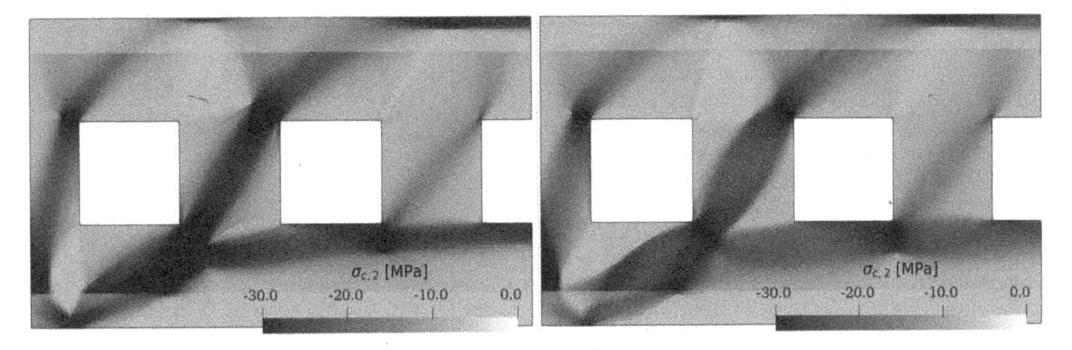

Figure 13. Smallest principal concrete stress for the rigid-plastic reinforced concrete yield surface (left), and the effective yield surface with $a = 187$ (right).

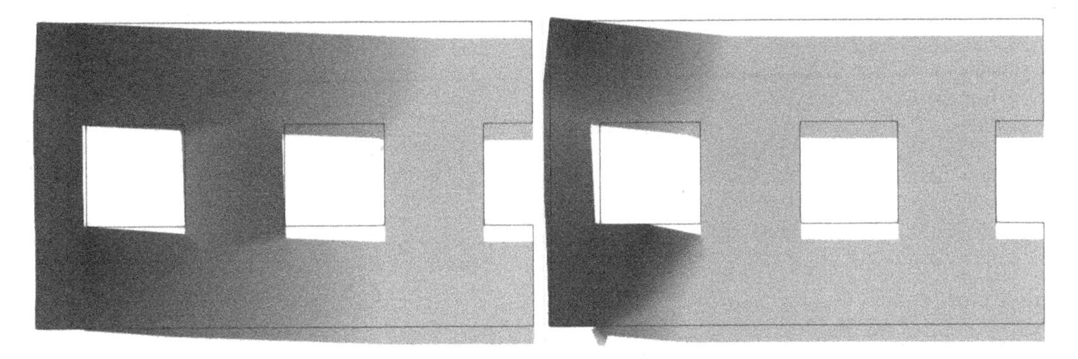

Figure 14. failure mode of the model for the rigid-plastic reinforced concrete yield surface (left), and the effective yield surface with $a = 187$ (right).

top and bottom flanges being three times the thickness of the rest of the beam. The material parameters are the same as listed in Table 1, except for the thickness and the reinforcement ratio. These vary between the web and the flanges. The vertical reinforcement in the flanges is chosen to correspond to the ratio between the thickness of the flanges and the web. Furthermore, horizontal bending reinforcement is added to the bottom flange.

The model is supported vertically at the left end with a support width of 300 mm, and with a symmetry boundary condition on the section at the right-hand side. The loading consists of a distributed load of λp on the top face and $\lambda p/4$ on the bottom face.

For the calculations, an unstructured mesh with an element side length of 25 mm is used, which corresponds to 17160 elements.

The resulting load factor λ is 0.455 using the Rigid-plastic reinforced concrete yield surface and 0.415 using the effective yield surface, which corresponds to $p = 273$ kN/m and $p = 249$ kN/m, respectively. The capacity is thus reduced by 9% when the effective yield surface is used.

Figure 13 shows the value of the smallest principal concrete stress for the model using the rigid-plastic reinforced concrete and effective yield surfaces, respectively. A clear difference between the layout of the compressive stresses is visible. For the rigid-plastic reinforced concrete yield surface, the compression is carried through struts with more or less constant spread and stresses close to f_c, whereas the struts are more diffused in the example with the effective yield surface. This effect is especially visible between the first and the second window when counting from the right, where the strut for the effective yield surface looks like a typical bulging strut, and as such, also has a decreased effective compressive strength.

A comparison of the failure mode using the two different yield surfaces can be seen in Figure 14. The failure mode for the model using the rigid-plastic reinforced concrete yield surface is a combination of a bending and shear failure, whereas the failure mode is much more localized when using the effective yield surface.

7 CONCLUSION

The yield surface of a plane reinforced concrete material point considering the effect of transverse tension on the effective compressive strength has been examined with the goal of developing a convex yield surface for use in Finite Element Limit Analysis (FELA). First, a strain-based elasto-plastic model was developed utilizing an expression for the effectiveness factor similar to the one given in the fib model code. Secondly, a stress-based convex effective yield surface was developed. The yield surface limits the effective concrete compressive strength by introducing strains linked to the reinforcement stresses on a material point level. The convex yield surface applied a linearized approximation of the effectiveness factor expression. The convex effective yield surface was compared to the elasto-plastic yield surface and was found to be a good approximation. After that, two examples were shown utilizing the effective yield surface compared to the rigid-plastic reinforced concrete yield surface where the effective compressive strength is not considered. The first example used a test database of reinforced concrete panels. The panels were subjected to shear stresses with and without biaxial compression or tension. The effective yield surface improved the predicted failure load compared to the experimental failure load, which indicates that the effective yield surface works well on a material point level. The second example was of a reinforced concrete deep beam with holes. Here the use of the effective yield surface reduced the capacity of the beam by 9%, and a difference in the stress flow and failure mode was seen, which indicates that the effective yield surface also works well on a structural level.

REFERENCES

Anderheggen, E. & H. Knöpfel (1972, dec). Finite element limit analysis using linear programming. *International Journal of Solids and Structures 8*(12), 1413–1431.

Andersen, M. E. M., P. N. Poulsen, & J. F. Olesen (2022, jan). Partially mixed lower bound constant stress tetrahedral element for Finite Element Limit Analysis. *Computers & Structures 258*, 106672.

Brask, S. L. & W. Xuan (2019). *Effective compressive strength and ductility of reinforced concrete panels*. Ph. D. thesis, The Technical University of Denmark (DTU).

Červenka, V. & J. Červenka (2017). *ATENA Program Documentation Part 2-2 User's manual for ATENA 3D*. Cervenka Consulting.

Collins, M. & F. Vecchio (1982). *The Response of Reinforced Concrete to In-plane Shear and Normal Stresses*. University of Toronto, Department of Civil Engineering.

Drucker, D. C., W. Prager, & H. J. Greenberg (1952). Extended limit design theorems for continuous media. *Quarterly of Applied Mathematics 9*(4), 381–389.

Eurocode 2 (2008). *Eurocode 2: Design of concrete structures – Part 1-1: General rules and rules for buildings*. European Committee for Standardization.

Ferreira, D. (2020). *Diana User Manual, Release 10.4*. Delft: DIANA FEA.

fib (2013, oct). *fib Model Code for Concrete Structures 2010*. Weinheim, Germany: Wiley-VCH Verlag GmbH & Co. KGaA.

Gvozdev, A. (1960). The determination of the value of the collapse load for statically indeterminate systems undergoing plastic deformation. *International Journal of Mechanical Sciences 1*(4), 322–335.

Herfelt, M. A. (2017). *Numerical limit analysis of precast concrete structures – A framework for efficient design and analysis*. Ph.d. thesis, Technical University of Denmark (DTU), Kgs. Lyngby.

Herfelt, M. A., P. N. Poulsen, & L. C. Hoang (2018). Closed form adaptive effectiveness factor for numerical models. In *The International Federation for Structural Concrete 5th International fib Congress*.

Hoang, L. C., H. J. Jacobsen, & B. Larsen (2012). Compressive Strength of Reinforced Concrete Disks with Transverse Tension. *Bygningsstatiske Meddelelser (Proceedings of the Danish Society for Structural Science and Engineering) 83*(2-3), 23–61.

Krabbenhøft, K. (2016). Shell finite element. Technical report, Optum Computational Engineering.

Nielsen, M. P. & L. C. Hoang (2011). *Limit Analysis and Concrete Plasticity* (3rd ed.). CRC Press.

Poulsen, P. N. & L. Damkilde (2000, oct). Limit state analysis of reinforced concrete plates subjected to in-plane forces. *International Journal of Solids and Structures 37*(42), 6011–6029.

pr EN1992-1-1 (2021). prEN 1992-1-1 ver. 2021-01 Eurocode 2, Design of concrete structures – Part 1-1: General rules – Rules for buildings, bridges and civil engineering structures.

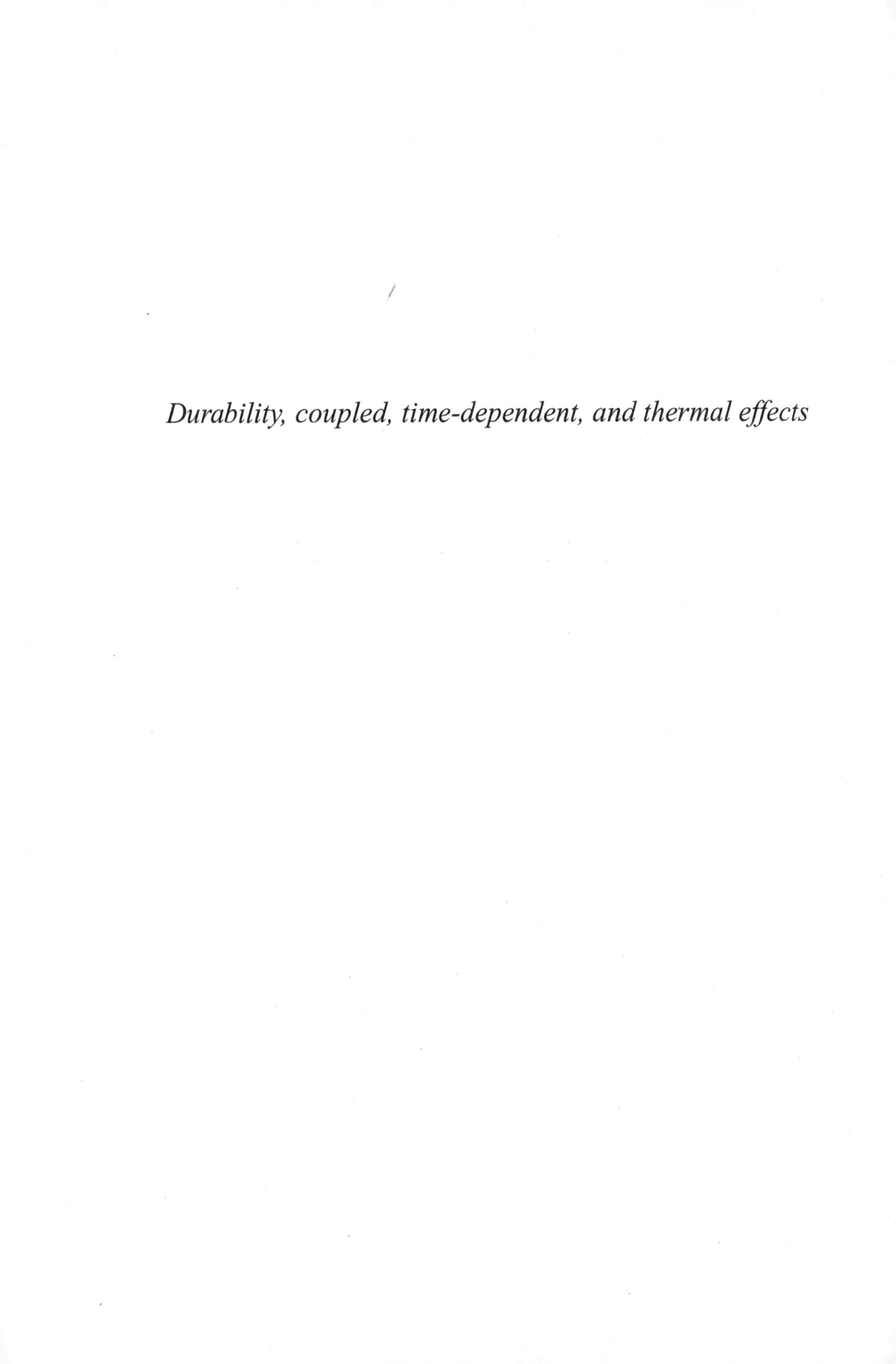

Durability, coupled, time-dependent, and thermal effects

Computational Modelling of Concrete and Concrete Structures – Meschke, Pichler & Rots (Eds)

Optimization of the corrosion initiation time of RC structures considering uncertainties

S. Schoen, P. Edler, V. Gudzulic & G. Meschke
Institute for Structural Mechanics, Ruhr University Bochum, Germany

S. Freitag
Institute for Structural Analysis, Karlsruhe Institute of Technology, Germany

ABSTRACT: According to the design codes, the concrete cover is simply selected based on the predefined exposure classes. This implies a thicker concrete cover if moisture or chloride is expected to attack the concrete. However, a thicker concrete cover will result in larger cracks assuming the same area of reinforcement and dimensions of the reinforced concrete (RC) structure, or the area of reinforcement has to be increased. Considering chloride ingress as observed in marine environment, the ion transport is significantly affected by the crack pattern, which is characterized by a remarkable degree of uncertainty. In the scope of this work, by considering aleatory and epistemic uncertainties, an optimal concrete cover is determined using a multiphysics finite element model (FE). Since usually a large number of realization is needed, which will result in high computation time of a fully coupled multiphysical FE model, two subproblems are solved. The first FE model simulates tensile cracking of a concrete beam under mechanical loading, and the second FE model computes the coupled moisture and chloride transport within the cracked unsaturated concrete. Additionally, the enhanced diffusion of cracked concrete is accounted for by adjusting the diffusion coefficients of moisture and chloride. In this concept, the uncertainty of the design parameter (concrete cover) is quantified by an interval. This uncertain parameter and uncertain structural actions lead to polymorphic uncertain prognoses of cracks and subsequently to a polymorphic uncertain prediction of corrosion initiation. In conclusion, within the framework of this work, the crack induced corrosion initiation time will be maximized by optimizing the concrete cover.

1 INTRODUCTION

Corrosion of reinforcing steel is one of the most common damage mechanisms of reinforced concrete and can significantly affect the safety and serviceability of the structure. Therefore, predicting chloride ingress into concrete is critical to accurately determine the service life of structures. Over the past few decades, extensive research has been conducted to investigate the transport properties of concrete. A major drawback of some models is the formulation based on perfect laboratory condition for uncracked concretes (eg. Pack et al. 2010; Song et al. 2008; Wang et al. 2005; Zhang & Gjorv O.E. 1996). In reality, however, cracks in concrete are unavoidable. For example, the choice of reinforcement position can have a decisive influence on concrete cracking. According to the design rules, the concrete cover is selected based on predefined exposure classes. Respectively, for a structure in marine environment, it is recommended to place the reinforcement deeper into the concrete, thus increasing the transport path to the reinforcement. However, greater concrete cover will lead to cracks, which may act as flow channels for chlorides, thus accelerating chloride penetration and the onset of corrosion (Djerbi et al. 2008 and Rodriguez 2001).

In this paper, considering aleatory and epistemic uncertainties, an optimal concrete cover is determined using finite element (FE) models. For durability-oriented design the consideration of aleatoric and epistemic uncertainties enables probabilistic lifetime prognoses, substituting classical safety factors. Aleatory uncertainty is characterized by a known variability and can be modeled by stochastic distributions, while epistemic uncertainty is quantified by a lack of knowledge and therefore can be modeled by unvervals or fuzzy numbers (Möller & Beer 2008). In the framework of this paper, the uncertainty of the design parameter (concrete cover) is quantified by an interval. Additionally, the uncertainties of the mesh topology are considered by running the model with different meshes. The analysis of a model with uncertainty parameters usually requires a large number of realizations. Since this leads to a high computation time for a fully coupled FE model, two subproblems are solved sequentially. In the first FE simulation concrete tensile cracking is simulated with the help of cohesive zero thickness interface elements

DOI 10.1201/9781003316404-64

(Snozzi & Molinari 2012), while for steel reinforcement a discrete representation is applied. The interaction between concrete and steel is taken into account by a contact-based tying algorithm using bond-slip characteristics (Gall et al. 2018). The second FE simulation computes the moisture transport within the cracked concrete, whereby coupled transport of moisture and chloride in unsaturated concrete is considered Samson & Marchand (2007). Additionally, the enhanced diffusion of cracked concrete is accounted for by adjusting the diffusion coefficients of moisture and chloride Zhang et al. (2017). To further reduce the computation time for solving the optimization problem, the FE simulation models are approximated by artificial neural network surrogate models.

2 MATERIAL MODEL FOR REINFORCED CONCRETE

In order to assess the influence of cracks in reinforced concrete on the corrosion initiation of reinforcement, the crack pattern should first be determined as accurately as possible using finite element simulation. To model propagating cracks, a discrete crack finite element model (Carol et al. 2001; Ortiz & Pandolfi 1999 and Snozzi & Molinari 2012) is utilized. This model is characterized by cohesive zero-thickness interface elements, which are inserted between small-strain linear-elastic triangular finite elements (i.e. bulk elements). The fracture behavior of plain concrete is modeled by a nonlinear traction-separation law of the interface elements. The cohesive traction acting on the interface element is evaluated as follows (Gudzulic & Meschke 2021, Snozzi et al. 2012)

$$t = \frac{t(\alpha)}{\tilde{u}_{max}} K_i \begin{bmatrix} ||u||_n \\ \frac{\beta^2}{\kappa} ||u||_t \\ \frac{\beta^2}{\kappa} ||u||_s \end{bmatrix} = (1-d) K_i \begin{bmatrix} ||u||_n \\ \frac{\beta^2}{\kappa} ||u||_t \\ \frac{\beta^2}{\kappa} ||u||_s \end{bmatrix}, \quad (1)$$

where d is the scalar damage variable and K_i is the initial stiffness. The loading criterion is defined as a function of the displacement

$$f(\tilde{u}, \alpha) = \tilde{u} - u_0 - \alpha \leq 0, \quad (2)$$

where u_0 corresponds to the limit state of the elastic interface with a tensile strength f_{ct} and is defined as

$$u_0 = \frac{f_{ct}}{K_i}, \quad (3)$$

and $\tilde{u}$ is the effective separation and defined as

$$\tilde{u} = \sqrt{||u||_n^2 + \frac{\beta^2}{\kappa^2} ||u||_t^2 + \frac{\beta^2}{\kappa^2} ||u||_s^2}. \quad (4)$$

In Eq. (4), $||u||_n$, $||u||_t$, $||u||_s$ are the normal and tangential components (local x-, y- and z- direction respectively) of the displacement jump $||\boldsymbol{u}||$ across the element. The two mixed mode parameters β and κ represent the ratio between shear and tensile strength (Eq. (5)) and the ratio between mode II and mode I fracture energy as in Eq. (6), respectively (Gudzulic & Meschke 2021, Snozzi et al. 2012).

$$\beta = \frac{f_{cs}}{f_{ct}} \quad (5)$$

$$\kappa = \frac{G_{f,\mathrm{II}}}{G_{f,\mathrm{I}}} \quad (6)$$

In Eq. (2), α is the internal parameter defining the maximum value of the effective separation appeared during the loading history

$$\alpha = \langle \tilde{u}_{max} - u_0 \rangle, \quad (7)$$

where $\tilde{u}$ is always positive, regardless of the sign of u (see Eq. (4)).
Considering exponential softening for plain concrete according to Karihaloo (1995), the traction-separation relationship is defined as

$$t(\alpha) = f_{ct} \exp\left(-\frac{\alpha f_{ct}}{G_f}\right), \quad (8)$$

from which the scalar damage variable, as used in Eq. (1) can be derived as

$$d(\alpha) = 1 - \frac{t(\alpha)}{K_i \alpha}. \quad (9)$$

The Reinforcement is accounted for by a discretization-independent embedded rebar model with geometrically linear truss elements. The steel properties are considered with an elasto-plastic material behavior using the v. Mises yield criterion with linear hardening. The interaction between rebar and concrete is established with a bond-slip-law according to the *fib* Model Code 1990 (International Federation for Structural Concrete). The internal force contribution of the rebar slip with respect to the bulk matrix is penalized to enforce a displacement constraint between the rebar and the concrete bulk (Gall et al. 2018). In the *fib* Model Code, the bond stress τ_b between concrete and rebar is formulated for monotonic loading as a function of the relative displacement s (see Figure 1 and Eq. (10)).

$$\tau_b = \begin{cases} \tau_{max} \left(\frac{s}{s_1}\right)^{\alpha} & 0 \leq s \leq s_1 \\ \tau_{max} & s_1 \leq s \leq s_2 \\ \tau_{max} - (\tau_{max} - \tau_f) \left(\frac{s - s_2}{s_3 - s_2}\right) & s_2 \leq s \leq s_3 \\ \tau_f & s_3 < s \end{cases} \quad (10)$$

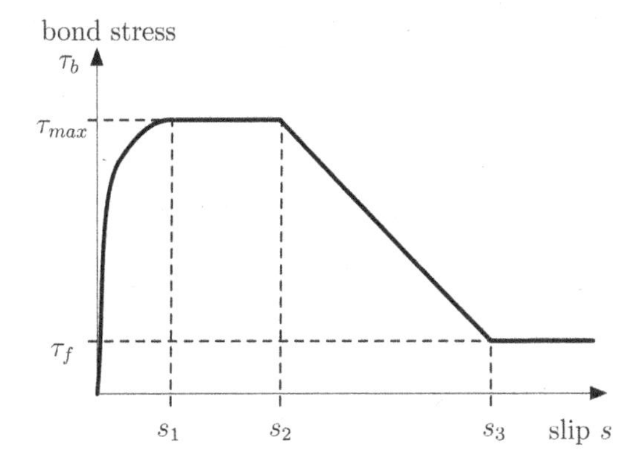

Figure 1. Schematic representation of the analytical bond stress-slip relation according to *fib* Model Code.

3 MASS TRANSPORT MODEL

The development of moisture content and ion concentration within cracks should be taken into account when estimating the service life of concrete structures. Using a second FE simulation, the coupled moisture and chloride transport within an unsaturated cracked concrete is modeled.

3.1 *Moisture transport model for unsaturated, uncracked concrete*

Assuming that the transport process takes place in a fully hardened concrete, the moisture transport can be described by the volume fraction of pore water in the concrete using Richards' equation (Ozbolt et al. 2010)

$$\frac{\partial \theta}{\partial t} - \frac{\partial}{\partial x}\left(D_\theta \frac{\partial \theta}{\partial x}\right) = 0, \tag{11}$$

where t is the time [s] and $\theta(t,x)$ is the fractional volume of water content [m^3 water per m^3 concrete] whose maximum possible value is the concrete porosity, i.e. $\theta \in [0,\phi]$. D_θ in Eq. (11) represents the moisture diffusion coefficient [m^2/s]

$$D_\theta = \frac{K(\theta)}{\frac{\partial \theta}{\partial h}}, \tag{12}$$

where h is the pressure head and $K(\theta)$ is the hydraulic conductivity written as

$$K(\theta) = K_r(\theta)\, K_s. \tag{13}$$

In Eq. (13) K_r, represents the relative hydraulic conductivity, while K_s is the hydraulic conductivity for a fully saturated and undamaged porous medium. According to Van Genuchten (1980) the relative hydraulic conductivity is computed as

$$K_r(\theta) = S_e^{\,l}\,[1 - (1 - \theta^{\frac{1}{m}})^m], \tag{14}$$

where S_e is the normalized moisture content

$$S_e = \frac{\theta - \theta_r}{\theta_s - \theta_r}, \tag{15}$$

and θ_s and θ_r are the moisture content corresponding to the fully saturated and the dry state, respectively. In Eq. 14, l, α and m are material dependent parameters.

3.2 *Modification of moisture transport model for cracked concrete*

Cracks in concrete accelerate the moisture transport. Therefore, the crack pattern should be included in the transport equation. Ozbolt et al. (2010) argued that the relative increase in water diffusivity corresponds to the relative increase in water conductivity of the concrete, and therefore the hydraulic conductivity is written as

$$K(\theta) = K_r(\theta)\, K_s\, K_{cr}(w_{cr}), \tag{16}$$

where $K_{cr}(w_{cr})$ is the normalized conductivity coefficient, which is obtained by dividing the conductivity coefficient corresponding to cracked concrete by that of the uncracked.

3.3 *Transport of chloride ions in unsaturated uncracked concrete*

In unsaturated porous materials, the transport of chloride ions occurs as a combination of diffusion and advection. As explained in Bear & Bachmat (1991), the transport process of ions is modeled by averaging the extended Nernst-Planck equation with an advection term over a Representative Elementary Volume (REV) and leads to the following transport equation

$$\frac{\partial(\theta\, c)}{\partial t} - \frac{\partial}{\partial x}\left(\underbrace{\theta\, D_c \frac{\partial c}{\partial x}}_{\text{diffusion}} + \underbrace{c\, D_\theta \frac{\partial \theta}{\partial x}}_{\text{advection}}\right) = 0, \tag{17}$$

where c is the chloride concentration of the species in solution [mmol/L], and D_c the chloride diffusion coefficient [m^2/s]. According to Samson & Marchand (2007), D_c is expressed as

$$D_c = \tau D^0 \left(\frac{\theta^{7/3}}{\phi^{7/3}}\right), \tag{18}$$

where τ [–] and ϕ [m^3/m^3] are the tortuosity and porosity, respectively. D^0 describes the diffusion coefficient in freewater for chloride ions [m^2/s]. The term τD^0 in Eq. (18) represents the diffusion coefficient for a saturated material. However, to take the reduction of diffusion properties due to the volume decrease of the aqueous phase into account, the term in parenthesis in Eq. (18) is needed and based on the power relationship derived by Millington & Quirk (1961).

3.4 *Modification of chloride transport for cracked concrete*

Similar to moisture transport, the chloride transport is accelerated through cracks in concrete. For this reason, the chloride diffusion coefficient needs to be adjusted for the cracked concrete. Accoring to Jang et al. (2011), the relationship between crack width and diffusivity of concrete is similar with an analogy to the relationship between diffusion and pore structure of hardened cement paste. Therefore, the diffusion coefficient in a single crack is defined as

$$D_{c,cr} = \beta_{cr} D^0, \tag{19}$$

where β_{cr} is the "crack geometry factor"ž accounting for tortuousness, connectivity and constrictivity of the crack path perpendicular to the flow direction. To account for crack width, Zhang et al. (2017) expanded Eq. (19) as:

$$D_{c,cr} = \begin{cases} D_c & w_{cr} \leq w_1 \\ \dfrac{D_c + D^0}{2} + \dfrac{D^0 - D_c}{2} \sin(\alpha) & w_1 < w_{cr} \leq w_2 \\ D^0 & w_{cr} \geq w_2 \end{cases} \tag{20}$$

$$\text{with } \alpha = \frac{\pi}{w_2 - w_1}\left(w_{cr} - \frac{w_2 + w_1}{2}\right),$$

where w_1 and w_2 are threshold values of the crack width in the concrete. According to Zhang et al. (2017), Akhavan (2012) and Djerbi et al. (2008), it can be assumed that up to a minimal crack width of w_1, the transport is equal to that in the uncracked state, and above a maximal crack width of w_2, the transport is equal to that in the fully cracked concrete (see Figure 2)

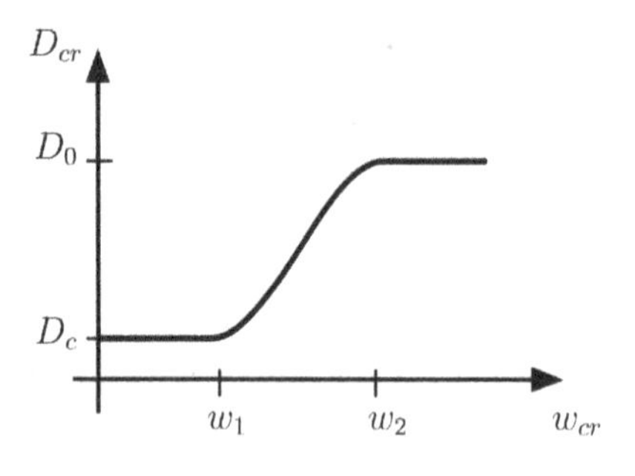

Figure 2. Schematic representation of the relationship between crack width (w_{cr}) and chloride diffusion coefficient within a crack (D_{cr}).

4 VALIDATION OF THE PROPOSED MODELING CONCEPT

The two FE models described in the previous sections are validated separately. The two FE models described in the previous sections are first validated separately from each other. Afterwards, the coupling of the two models, which is influenced by the mesh topology of the mechanical model, is explained.

4.1 *Validation of the proposed material model*

In order to investigate the admissibility of the fracture process of the described reinforced concrete model, a flexural test was simulated as performed experimentally by Suchorzewski et al. (2018). The test setup is presented in Figure 3 and the material properties are listed in Tables 1 and 2. In the experiment as well as in the finite element simulation the reinforced concrete beam failed in bending by reinforcement yielding (Figure 4). Moreover, it can be observed that the proposed finite element model determines the crack pattern with sufficient accuracy (see Figure 5).

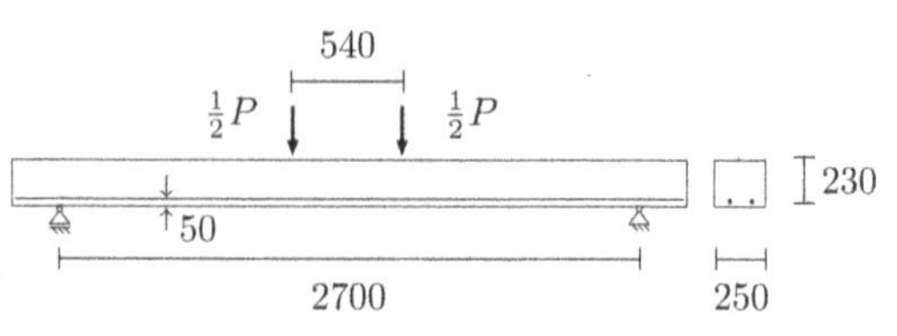

Figure 3. Experimental setup of the reinforced concrete beam under four-point bending according to Suchorzewski et al. (2018).

Table 1. Material properties of concrete C 45/55 for the mechanical model.

Property		Value	Unit
Young's modulus	E	34000	N/mm^2
Poisson's ratio	ν	0.2	-
Tensile strength	f_{ct}	3.2	N/mm^2
Compressive strength	f_{cm}	60	N/mm^2
Fracture energy	G_f	0.12	N/mm

Table 2. Material properties of steel B500 for the mechanical model.

Property		Value	Unit
Young's modulus	E	205000	N/mm^2
Yield stress	σ_{ys}	560	N/mm^2
Area	A	2ø20	mm^2

4.2 *Validation of the proposed transport model*

To verify the accuracy of the proposed approach to model mass transport in uncracked and cracked concrete, a chloride ion transport as performed by Sahmaran (2007) has been simulated. The experimental setup is illustrated in Figure 6 and the diffusion properties of the concrete beam are provided in Table 3

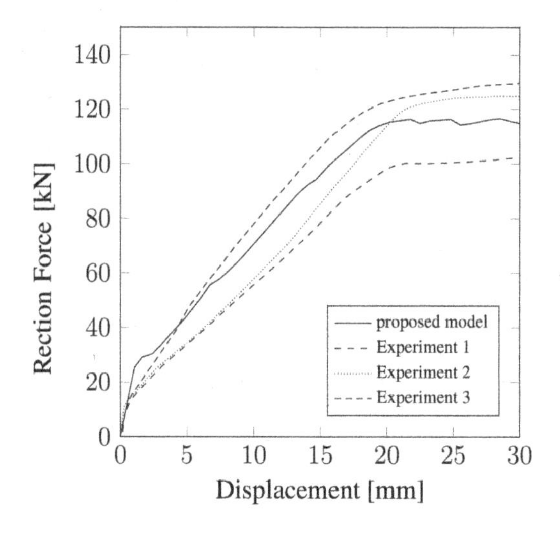

Figure 4. Comparison of experimental data reported by Suchorzewski et al. (2018) and the results of the proposed numerical model.

according to Zhang et al. (2017). In the experiment performed by Sahmaran (2007), the cracked surface of the concrete beam was exposed to a chloride concentration of 0.51% by weight (wt.) of cement (CMT) solution for 30 days. Subsequently, the chloride concentration was measured over the depth of the beam for different crack widths. Figure 7 presents the penetration depth of chloride concentration after 30 days in the experiment and in the numerical model for different crack width. It can be observed that the proposed numerical model gives quite satisfying results for the uncracked concrete sample as well as for the two cracked concrete samples.

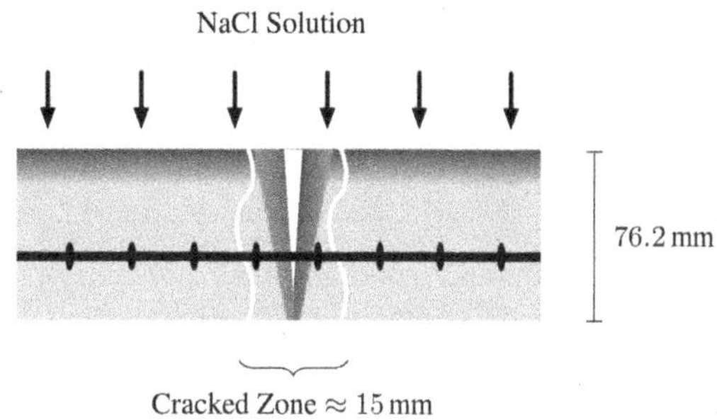

Figure 6. Locations of the sampling for chloride profiling Sahmaran (2007).

Table 3. Chloride diffusion properties of the concrete beam.

Property		Value	Unit
Diff. coef. in freewater	D^0	$2 \cdot 10^{-9}$	m^2/s
Diff. coef. in concrete	D_c	$2.34 \cdot 10^{-11}$	m^2/s
Threshold values	w_1	0.03	mm
	w_2	0.12	mm

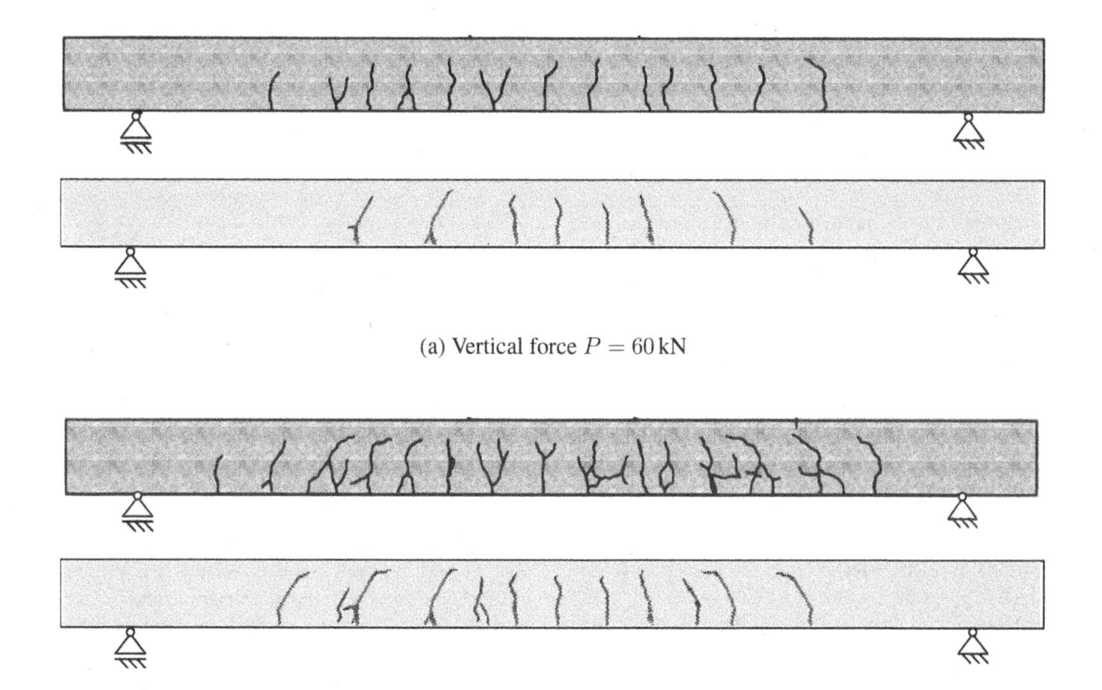

(a) Vertical force $P = 60\,\text{kN}$

(b) Vertical force $P = 90\,\text{kN}$

Figure 5. Comparison of the crack pattern evolution of the experiment according to Suchorzewski et al. (2018) (top) and the proposed numerical model (bottom) for the vertical force of a) $P = 60$ kN and b) $P = 90$ kN.

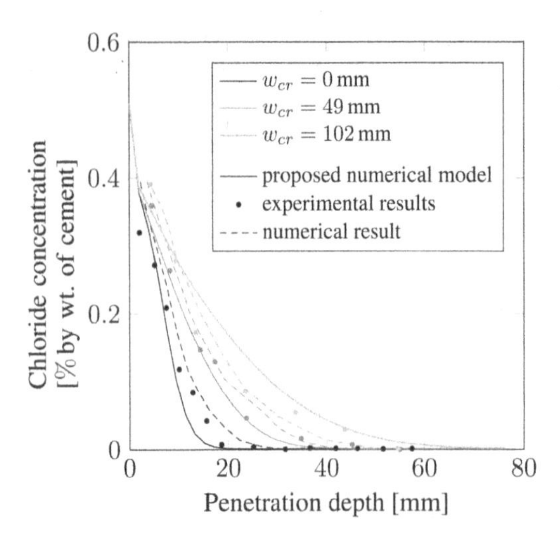

Figure 7. Comparison of experimental data reported by Sahmaran (2007), the numerical data presented by Zhang et al. (2017) and the results of the proposed numerical model for a crack width of 0 mm, 49 mm and 102 mm.

4.3 *Combining the mechanical model and the mass transport model*

To ensure that the concrete properties from the mechanical model correspond to those within the transport model, a direct correlation between the mechanical and the transport model is necessary. According to experiments (eg. Chindaprasirt et al. 2009, Lian et al. 2011 and Stroeven 2000) the relationship between porosity and concrete strength can be formulated as

$$\phi = \frac{\ln\left(\frac{f_{cm}}{231.44}\right)}{-9}, \tag{21}$$

and additionally,the porosity correlation to the concrete tortuosity as

$$\tau = \frac{1}{\frac{3}{2} - \frac{1}{2} \cdot \phi}. \tag{22}$$

Furthermore, to determine the initiation time, the decisive crack is determined from the mechanical model and its crack geometry (crack width and crack depth) is passed to the transport model as input parameters. However, the use of cohesive interface elements allows cracks to propagate only across the boundaries between solid finite elements. In other words, the topology of the mesh forces the cracks to follow paths that generally require more energy per unit crack extension (larger driving forces) than the paths they would follow in the original continuum. This means that the crack geometry is mesh dependent and by passing it to the transport model, the initiation time will also dependent on the FE mesh. To determine the relationship between mesh topology and initiation time, a mesh analysis was performed based on the example from the previous section with a concrete cover of $c_{nom} = 50$ mm, using ten different FE meshes with the same element size (15 mm). For the mesh analysis, the amount of reinforcement and the material properties of the reinforced concrete are taken from Table 1 and 2. In addition, Table 4 and 5 summarize the parameters utilized within the transport model. These parameters are based on purely experimental boundary conditions and imply that the dry reinforced concrete beam is constantly exposed to very high humidity and an extremely salty atmosphere. At this point, it should be mentioned, that for the determination of the Initiation time only the range of the service load ($P = 30 \pm 10$ kN according to the design rules) is relevant. Moreover, a maximum value of 0.6 % by weight of cement at the reinforcement was defined as the end of the initiation time (Gehlen 2000).

Table 4. Properties of concrete for the moisture transport.

Property		Value	Unit
Degree of saturation	S_e	0.2 (dry)	%
Porosity	ϕ	0.15	–
Tortuosity	τ	0.70	–
Relative humidity	RH	82	%
Initial moisture content	$\theta(0)$	0.077	mm^3/mm^3

Table 5. Properties of concrete for the chloride ion transport.

Property		Value	Unit
Diff. coef. in freewater	D^0	$2 \cdot 10^{-9}$	m^2/s
Initial chloride content	$c(0)$	0.0	% by wt. of CMT
Chloride exposure	$c(t)$	1.88 [1]	% by wt. of CMT
Threshold values	w_1	0.03 [2]	mm
	w_2	0.08 [2]	mm

[1] (Gehlen 2000), [2] (Djerbi et al. 2008)

For a better illustration of the results, only the bandwidth obtained by the ten meshes, i.e. the upper (solid line) and lower bounds (dotted line) are shown in Figure 8 to 10. Figure 8 presents the load-displacement curve for the range of the expected service load. Here, when the first crack propagates, the curve of the upper bound mesh is only slightly above the lower bound one. Thus, if one considers the overall mechanical model, based on the load-displacement curve, one may assume that the mesh geometry has almost no influence. However, if the macroscopic level is examined, the crack geometry strongly dependence on the mesh topology as it can be seen in Figure 9. This graph depicts the evolution of the maximum crack width w_{cr} of the beam for an increasing load P. Due to the later cracking of the upper bound mesh, the curve is shifted to the right with regard to the lower limit mesh. This leads to the fact that, for instance, the lower bound mesh reaches a maximal crack width

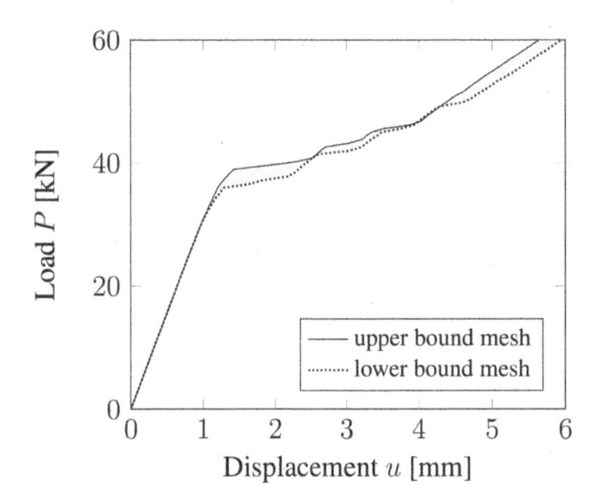

Figure 8. Load-displacement curve for results of different FE-meshes for a concrete cover of $c_{nom} = 50$ mm.

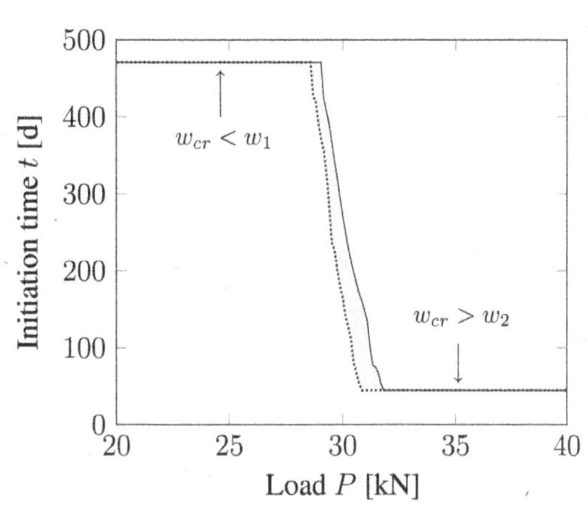

Figure 10. Band width of initiation time computed by different FE-meshes for a concrete cover of $c_{nom} = 50$ mm.

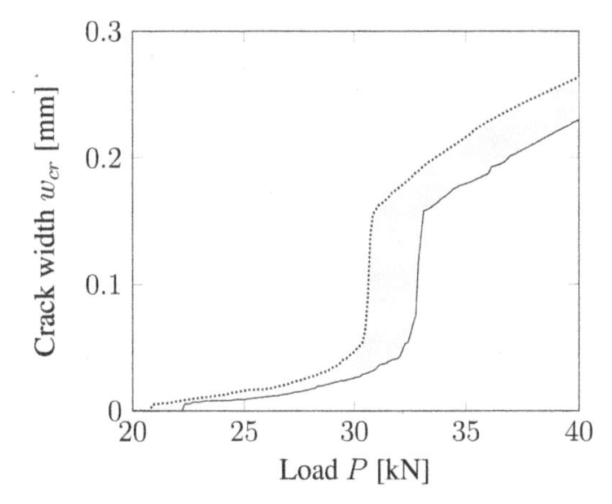

Figure 9. Crack evolution computed by different FE-meshes for a concrete cover of $c_{nom} = 50$ mm.

of $w_{cr,l} = 0.16$ mm at a load of $P = 31$ kN, while the maximal crack width for the upper bound is barely $w_{cr,u} = 0.03$ mm for the same load.

In Figure 10, the initiation time is plotted as a function of the load. Until the first crack ($w_{cr} < w_1$) develops, the initiation time is constant and equal for all meshes, because here the initiation time only depends on the transport model. Then, the curve of the initiation time has a sharp drop until the cracks are large enough ($w_{cr} > w_2$ up to the reinforcement) that freewater transport in the crack can be assumed. Here again, for large cracks, the initiation time only depends on the transport model. It is also worth mentioning once again that, since the cracks in the upper bound mesh develop at a higher load, the drop of the initiation time occurs at a higher load compared to the lower bound mesh.

In Figure 11 the initiation time t is plotted as a function of the concrete cover c_{nom} for different values of the load P. It can be noted that, regardless of the mesh topology and the load P, the curve of the initiation time follows the path of the uncracked concrete until the first crack ($w_{cr} < w_1$) appears at a certain concrete cover. Then the curve drops and follows the path of the completely cracked concrete ($w_{cr} > w_2$ assumption of free-water transport up to the reinforcement). In between this extrema the ignition time is depending on the mesh topology which can be seen as ranges in Figure 11.

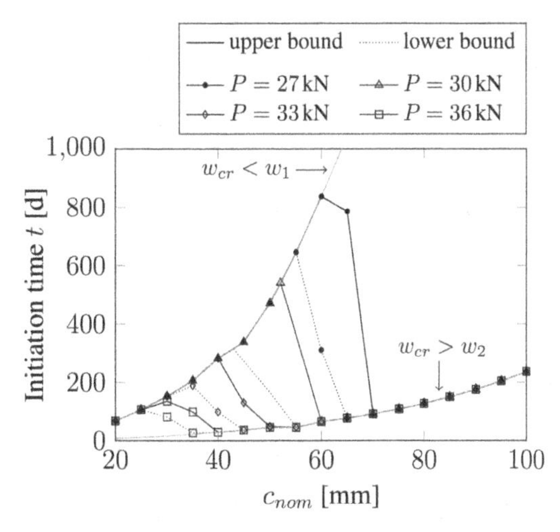

Figure 11. Initiation time over the concrete cover for different loads.

At this point it can be summarized, that the mesh topology has an influence to the initiation time. However, it should be noted that for all investigated load levels, one out of the ten meshes yields the lower bound results and one out of the ten meshes yields the upper bound results. The eight other tested meshes are within the bandwidth. This means, that the mesh topology uncertainty is considered as an interval by evaluating the corresponding lower bound mesh and the upper bound mesh in further computations.

The FE simulations performed by the two lower and upper bound meshes are replaced by two artificial neural networks with feed forward architecture to speed up the computation time for further investigations, i.e. uncertainty quantification and durability optimization. For each mesh, an ANN with two inputs (c_{nom}) and P), 10 hidden neurons and one output neuron t is sufficient to approximate the FE simulations with an accuracy of $R = 0.998$.

5 DESIGN MODEL WITH CONSIDERATION OF POLYMORPHIC UNCERTAINTY

After the models have been validated in the previous section, now the optimal concrete cover of a reinforced concrete structure with respect to a maximal corrosion initiation time is investigated. The maximal initialization time is determined for the previously validated reinforced concrete beam (see Figure 3) with the same material parameters (see Table 1 and 2 for the material model and Table 4 and 5 for the transport model, respectively) by optimizing the concrete cover c_{nom}. In addition to the mesh uncertainty modeled as an interval, the load is assumed as a normal distributed random variable (stochastic a priori parameter) with different mean values investigated in a range of [20; 40] kN and a fixed standard deviation $\sigma\ (P) = 1000$ kN. Two optimization problems are formulated, where the concrete cover has to be optimized. For Problem 1, the concrete cover is modeled as a deterministic design variable and for Problem 2 as an interval design parameter with an interval radius of ${}_r c_{nom} = 5$mm and interval midpoint to be optimized. The objective is to maximize the mean value of corrosion initiation time and can be formulated as

$$\begin{aligned} \max: &\quad \hat{z} = \mu\left(t\left(c_{nom}, P\right)\right) \\ \text{s.t.}: &\quad c_{nom} \in [20; 100] \end{aligned}, \tag{23}$$

for Problem 1 and

$$\begin{aligned} \max: &\quad \hat{z} = {}_l\mu\left(t\left(c_{nom}, P\right)\right) \\ \text{s.t.}: &\quad {}_m c_{nom} \in [25; 95] \end{aligned}, \tag{24}$$

for Problem 2. For the optimization, the Particle Swarm Optimization (PSO) Kennedy & Eberhart. (1995) is applied with an extension to consider aleatory and epistemic uncertainty Edler et al. (2019).

In Figure 12 and Figure 13, the optimized concrete cover c^*_{nom} and the corresponding mean value of the initiation time $\mu(t)^*$ are shown for different mean values of the load are shown for Problem 1. With increasing mean value of the load, the optimized value of the concrete cover for both, upper and lower bound mesh, decreases until a value of $c^*_{nom} = 41.5$ mm is reached at $\mu(P) = 30.75$ kN for the lower bound and $c^*_{nom} = 41.6$ mm at $\mu(P) = 32.50$ kN for the upper bound, respectively. With further increase of the mean value of the load, the optimized concrete cover increases again. The jump from small optimal concrete covers at loads around $\mu(P) = 30$ kN to the maximal possible concrete cover of $c_{nom} = 100$ mm can be explained by the fact, that for higher loads, the crack width w_{cr} exceeds the threshold value of $w_2 = 0.08$ mm, for all concrete covers within the design space. This means that in this case, the optimal concrete cover is equal to the longest distance from the concrete surface to the reinforcement.

It can be observed, that the mesh topology has an influence on the optimized concrete cover as well. For instance, due to the uncertain mesh topology, for a mean value of the load of $\mu(P) = 30$ kN the optimized concrete cover is in the range of $c^*_{nom} = [44.1, 50.9]$ mm and the mean value of the corresponding initiation time is in the range of $\mu(t)^* = [0.8, 1.2]$ years, as can be seen in Figure 13. This small range is acceptable and can be taken as the interval width for the optimum.

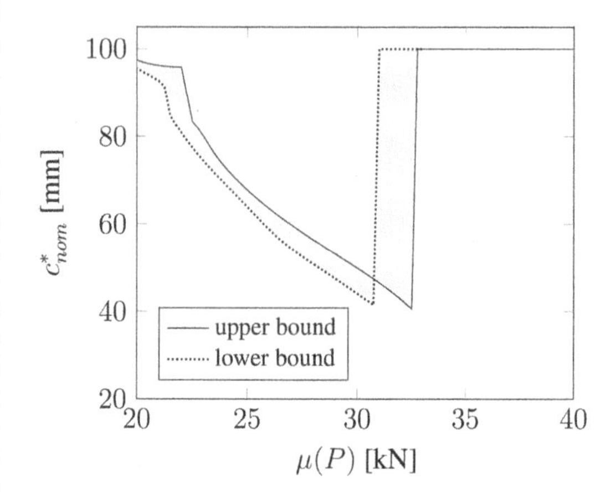

Figure 12. Optimized concrete cover c^*_{nom} for different mean values of the load $\mu(P)$ using the upper and lower bound meshes.

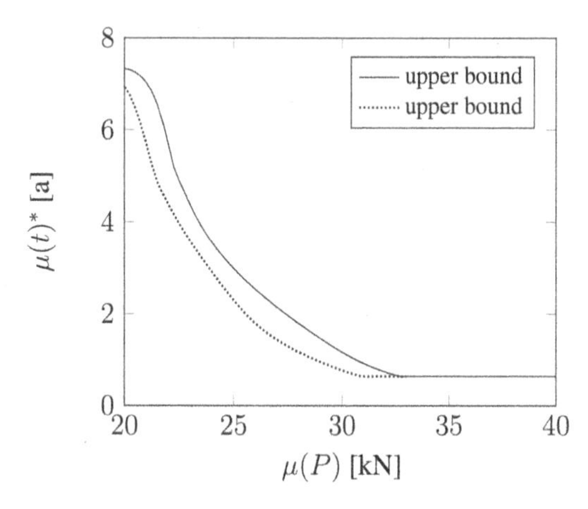

Figure 13. Mean value of the initiation time $\mu(t)^*$ for different the mean values of the load $\mu(P)$ using the upper and lower bound meshes.

In Figure 14 and Figure 15, the optimized midpoint of the concrete cover ${}_m c^*_{nom}$ and the corresponding worst case mean value of the initiation time ${}_l\mu(t)^*$ are shown for different mean values of the load, for

Problem 2. It can bee seen, that the optimized midpoint of the concrete cover does not significantly differ from the optimized concrete cover of Problem 1 (cf. Figure 12 and Figure 14), while the worst case of the mean value of initiation time is smaller compared to Problem 1 for all mean values of the load (cf. Figure 13 and Figure 15). For instance, $\mu(P) = 30\,\mathrm{kN}$, the optimized concrete cover is in the range of ${}_{m}c_{nom} = [43.2, 50.1]\,\mathrm{mm}$ and the mean value of the corresponding initiation time is in the range of ${}_{I}\mu(t)^{*} = [0.65, 0.93]$ years.

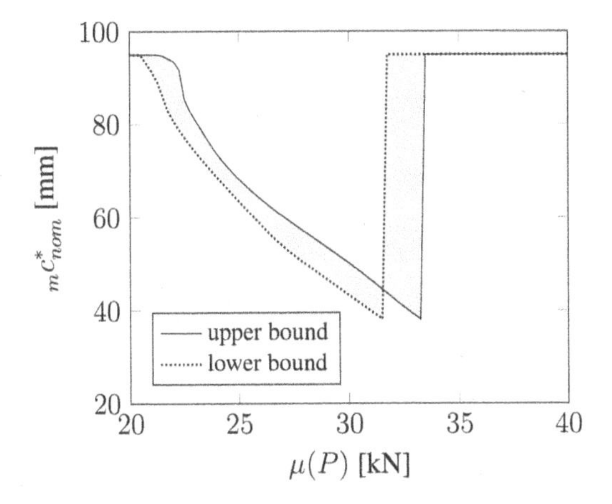

Figure 14. Optimized midpoint of the concrete cover ${}_{m}c^{*}_{nom}$ over the mean value of the load $\mu(P)$ for an interval radius of 5 mm.

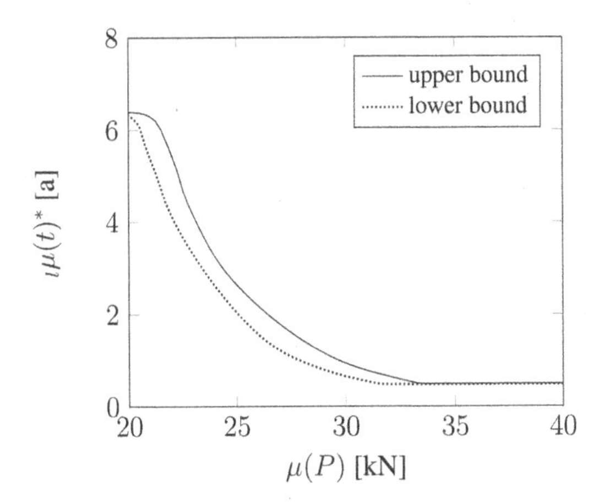

Figure 15. Worst case mean value of the initiation time ${}_{I}\mu(t)^{*}$ over the mean value of the load $\mu(P)$.

6 CONCLUSION

Within this paper, two FE models, one simulating the tensile cracking of concrete structures and the other one computing the coupled moisture and chloride transport within the cracked unsaturated concrete, have been introduced and validated using experimental data. Within a mesh study of the structural model it has been figured out, that the crack pattern and consequently also corrosion initiation time of the reinforcement is influenced by the mesh topology. The uncertainty resulting from the mesh topology has been taken into account as an interval by evaluating the corresponding lower bound mesh and the upper bound mesh for the durability optimization of a simple reinforced concrete beam. For maximizing the corrosion initiation time of the reinforcement, the concrete cover has been optimized in two different optimization problems considering a stochastic load.

The results of both optimization problems have shown, that the uncertain mesh topology has an influence on the optimization by resulting in a range of the optimal concrete cover and corresponding initiation time. For small as well as for high mean values of the load, the optimized concrete cover is the maximal value of the design space, because in these cases, the optimum is driven only by the transport model. More precisely, if the transport distance is larger, the initiation time is longer in case of very small cracks ($w_{cr} < w_1$) or very large cracks ($w_{cr} > w_2$). For mean values of the load between $\mu(P) = 22\,\mathrm{kN}$ and $\mu(P) = 34\,\mathrm{kN}$, the initiation time is driven by two counteracting mechanisms. On the one hand, it makes sense to choose a higher concrete cover to maximize the transport distance for chlorides to the reinforcement, but on the other hand, a higher concrete cover leads to higher crack width, which increases the transport velocity of the chloride ions. Therefore, a balance between these two mechanisms is found as a result of the optimization.

In future works, the distribution of the results is investigated using random fields. It will be analyzed, if the uncertainty resulting from different mesh topology can be reduced by applying random fields.

ACKNOWLEDGMENT

The authors gratefully acknowledge the financial support of the German Research Foundation (DFG) within Subproject 6 (Project number 312921814) of the Priority Programme "Polymorphic uncertainty modelling for the numerical design of structures - SPP 1886".

REFERENCES

Akhavan, A. (2012). *Characterizing saturated mass transport in fractured cementitious materials*. Ph. D. thesis, The Pennsylvania State University.

Bear, J. & Y. Bachmat (1991). *Introduction to Modeling of Transport Phenomena in Porous Media*. Springer Netherlands.

Carol, I., C. M. López, & O. Roa (2001). Micromechanical analysis of quasi-brittle materials using fracture-based interface elements. *International Journal for Numerical Methods in Engineering 52*(1-2), 193–215.

Chindaprasirt, P., S. Hatanaka, N. Mishima, Y. Yuasa, & T. Chareerat (2009). Effects of binder strength and aggregate size on the compressive strength and void ratio of porous concrete. *International Journal of Minerals, Metallurgy and Materials 16*(6), 714–719.

Djerbi, A., S. Bonnet, A. Khelidj, & V. Baroghel (2008). Influence of traversing crack on chloride diffusion into concrete To cite this version :. *Hal.Archives-Ouvertes.Fr*.

Edler, P., S. Freitag, K. Kremer, & G. Meschke (2019). Optimization Approaches for the Numerical Design of Structures Under Consideration of Polymorphic Uncertain Data. *ASCE-ASME Journal of Risk and Uncertainty in Engineering Systems, Part B: Mechanical Engineering 5*(4), 1–12.

Gall, V. E., S. N. Butt, G. E. Neu, & G. Meschke (2018). An embedded rebar model for computational analysis of reinforced concrete structures with applications to longitudinal joints in precast tunnel lining segments. In *Computational Modelling of Concrete Structures – Proceedings of the conference on Computational Modelling of Concrete Structures, EURO-C 2018*, pp. 705–714.

Gehlen, C. (2000). Probabilistische Lebensdauerbemessung von Stahlbeton- bauwerken – Zuverlässigkeitsbetrachtungen zur wirksamen Vermeidung von Bewehrungskorrosion. *Deutscher Ausschuss für Stahlbeton* (September), 106.

Gudzulic, V. & G. Meschke (2021). Multi-level approach for modelling the post-cracking response of steel fibre reinforced concrete under monotonic and cyclic loading. In *Pamm*.

International Federation for Structural Concrete (1990). CEB-FIP model code 1990.

Jang, S. Y., B. S. Kim, & B. H. Oh (2011). Effect of crack width on chloride diffusion coefficients of concrete by steady-state migration tests. *Cement and Concrete Research 41*(1), 9–19.

Karihaloo, B. (1995). *Fracture mechanics and structural concrete*. Harlow, Essex, England : Longman Scientific & Technical ; New York : Wiley.

Kennedy, J. & R. Eberhart. (1995). Particle Swarm Optimization. pp. 1942–1948. IEEE Xplore.

Lian, C., Y. Zhuge, & S. Beecham (2011). The relationship between porosity and strength for porous concrete. *Construction and Building Materials 25*(11), 4294–4298.

Millington, R. J. & J. P. Quirk (1961). Permeability of porous solids. *Transactions of the Faraday Society 57*, 1200–1207.

Möller, B. & M. Beer (2008). Engineering computation under uncertainty – Capabilities of non-traditional models. *Computers and Structures 86*(10), 1024–1041.

Ortiz, M. & A. Pandolfi (1999). Finite-łdeformation irreversible cohesive elements for three-łdimensiona crack-łpropagation analysi. *International Journal for Numerical Methods in Engineering 44*(9), 1267–1282.

Ozbolt, J., G. Balabanić, G. Periškić, & M. Kušter (2010, sep). Modelling the effect of damage on transport processes in concrete. *Construction and Building Materials 24*(9), 1638–1648.

Pack, S. W., M. S. Jung, H. W. Song, S. H. Kim, & K. Y. Ann (2010). Prediction of time dependent chloride transport in concrete structures exposed to a marine environment. *Cement and Concrete Research 40*(2), 302–312.

Rodriguez, O. (2001). *Influence of cracks on chloride ingress into concrete*. Ph. D. thesis, University of Toronto.

Sahmaran, M. (2007). Effect of flexure induced transverse crack and self-healing on chloride diffusivity of reinforced mortar. *Journal of Materials Science 42*(22), 9131–9136.

Samson, E. & J. Marchand (2007). Modeling the transport of ions in unsaturated cement-based materials. *Computers and Structures 85*(23-24), 1740–1756.

Snozzi, L., F. Gatuingt, & J. F. Molinari (2012). A meso-mechanical model for concrete under dynamic tensile and compressive loading. *International Journal of Fracture 178*(1-2), 179–194.

Snozzi, L. & J.-F. Molinari (2012). A cohesive element model for mixed mode loading with frictional contact capability. *International Journal for Numerical Methods in Engineering 89*(9), 1102–1119.

Song, H. W., C. H. Lee, & K. Y. Ann (2008). Factors influencing chloride transport in concrete structures exposed to marine environments. *Cement and Concrete Composites 30*(2), 113–121.

Stroeven, P. (2000). Stereological approach to roughness of fracture surfaces and tortuosity of transport paths in concrete. *Cement and Concrete Composites 22*(5), 331–341.

Suchorzewski, J., E. Korol, J. Tejchman, & Z. Mróz (2018). Experimental study of shear strength and failure mechanisms in RC beams scaled along height or length. *Engineering Structures 157*(March 2017), 203–223.

Van Genuchten, M. T. (1980). A closed-form equation for predicting the hydraulic conductivity of unsaturated soils. *Soil Science Society of America Journal 44*, 892–898.

Wang, Y., L. Y. Li, & C. L. Page (2005). Modelling of chloride ingress into concrete from a saline environment. *Building and Environment 40*(12), 1573–1582.

Zhang, R., L. Jin, M. Liu, X. L. Du, & Y. Li (2017). Numerical investigation of chloride diffusivity in cracked concrete. *Magazine of Concrete Research 69*(16), 850–864.

Zhang, T. & Gjorv O.E. (1996). Diffusion behavior of chloride ions in concrete. *Modern at large: Cultural dimensions of globalization 26*(6), 907–917.

Computational Modelling of Concrete and Concrete Structures – Meschke, Pichler & Rots (Eds)
 ISBN: 978-1-032-32724-2

Engineering mechanics analysis of a moderate fire inside a segment of a subway station

M. Sorgner, R. Díaz Flores & B.L.A. Pichler
Institute for Mechanics of Materials and Structures, TU Wien – Vienna University of Technology, Vienna, Austria

H. Wang
School of Naval Architecture, Ocean and Civil Engineering, Shanghai Jiao Tong University, Shanghai, China

ABSTRACT: Reinforced concrete structures must be designed to withstand extreme-case scenarios such as fires. Structural engineers are interested to analyze the behavior of reinforced concrete structures subjected to a combination of mechanical loads *and* elevated temperatures. In the present study, an engineering mechanics approach is used to describe the structural behavior of a segment of a subway station subjected to regular service loads and a moderate fire. This approach combines fundamental concepts of thermo-elasto-mechanics with beam analysis software. The three-dimensional reinforced concrete structure is idealized as a frame consisting of straight beams. The rectangular columns are transformed into cylindrical ones with equivalent extensional stiffness. The obtained temperature changes of the structural elements are used to quantify thermal eigenstrains. The latter are decomposed into three parts: an eigenstretch and an eigencurvature of the axis of the structural element, as well as an eigenwarping of its cross-sections. Beam analysis software is used to study the load carrying behavior of the frame structure subjected to mechanical loads as well as to thermal eigenstretches and eigencurvatures of all structural elements. The obtained axial forces and bending moments result in axial stresses which are linear across the cross-sections. The latter remain plane even under combined mechanical and thermal loading. This activates self-equilibrated thermal eigenstresses which are spatially nonlinear across the cross-sections. Total axial stresses are obtained from adding the thermal eigenstresses to the axial stresses quantified based on the axial forces and bending moments. The total stresses agree well with the results of a three-dimensional thermo-elastic Finite Element simulation. It is concluded that the subdivision of the developed engineering mechanics approach into a sequence of several smaller problems allows for relating causes to effects in a clear and insightful fashion, such as appreciated by structural engineers.

1 INTRODUCTION

Reinforced concrete is the most commonly used and one of the most intensively investigated construction materials in the world. Structural engineers are interested in the load-carrying behavior of reinforced concrete structures. The development of appropriate models that reliably and efficiently predict their behavior when subjected to a combination of elevated temperatures and mechanical loads, however, remains a challenge.

Investigations in this area can be separated into two main fields of research, which are very often found in combination with each other: experimental investigations based on large-scale tests, and numerical simulations. The latter address either the full complexity of a full-scale test, or a simplified subproblem referring to a specific element of a structure. As regards experimental campaigns, Vecchio & Sato (1990) performed three large scale tests to reinforced concrete frames subjected to a combination of mechanical loads and controlled thermal loads specified at the inner surface of the frame. Ring et al. (2014a) performed a large-scale fire test on an underground concrete frame structure, the results of which were used for the elaboration and validation of numerical Finite Element simulations in (Ring et al. 2014b). Kamath et al. (2015) conducted a full-scale fire test on a reinforced concrete frame, which was first subjected to simulated seismic damage to investigate the material behavior of reinforced concrete due to fire following an earthquake. Recently, Lu et al. (2018) performed a large-scale fire test on a segment of a subway station.

Regarding purely numerical simulations, thermo-mechanical three-dimensional Finite Element (FE) simulations have been performed in order to study reinforced concrete beams (Albrifkani & Wang 2016; Gao et al. 2013; Ozbolt et al. 2014; Sun et al. 2018; Zha 2003) or columns (Bratina et al. 2005; Zha 2003) subjected to elevated temperatures. Diaz et al. (2018) modeled, by means of three-dimensional FE simulations, the structural behavior of an underground

DOI 10.1201/9781003316404-65

frame structure subjected to the first 30 minutes of the fire test presented in (Lu et al. 2018), representative of a moderate fire scenario (Diaz et al. 2018). Simplified methods have also been developed in order to reduce the computational power required by complex FE simulations that attempt to reproduce fire scenarios. In this context, simple numerical models have been developed to analyze the response of reinforced concrete frames (Vecchio 1987), the resistance of reinforced beams (Choi and Shin 2011; Dwaikat & Kodur 2008; Kodur & Dwaikat 2008) and of cylindrical columns (Franssen & Dotreppe 2003). El-Fitiany & Youssef (2017) presented a simplified method to calculate internal compression forces and corresponding moments for heated concrete structures that can be easily applied using available commercial linear structural analysis software to predict the fire performance of reinforced concrete structures. El-Tayeb et al. (2017) gave a more intuitive insight into the effect of thermal loads on reinforced concrete structures by separating the temperature distribution into a uniform, a linear and a non-linear contribution, and described the importance of the non-linear part, exclusive of non-stationary thermal conditions. Wang et al. (2019) further advanced this method by introducing a semi-analytical solution which determines the non-linear thermal stresses developed by concrete pavements due to temperature changes on the top surface in non-stationary thermal conditions. Finally, Schmid (2020) extended this solution for concrete pavements with temperature changes on both, the top and bottom surface.

The present study attempts to describe the structural behavior of the segment of a subway station presented in (Lu et al. 2018) subjected to a moderate fire scenario, as described in (Diaz et al. 2018), by means of an inexpensive, practice-oriented model that combines the use of inexpensive beam models and thermo-elasto-mechanics fundamentals. This simulation is organized in five steps. The first one refers to the idealization of the originally three-dimensional structure into a frame structure with one-dimensional beam elements, where the Euler-Bernoulli beam theory is applicable. In order to consider the reinforcement of the structure and its mechanical properties, the initially non-uniform cross section is transformed into an equivalent cross-section with uniform properties. The second step refers to the solution of the non-stationary heat conduction problem. The third step refers to the quantification of the thermal stress contributions, based on a thermo-elasto-mechanics analysis and semi-analytical solutions of the heat conduction problem. The fourth step refers to the use of beam analysis software in order to predict the internal forces resulting by the combination of the mechanical and the thermal loading. The fifth and last step refers to the superposition of the stresses resulting from both load cases and the self-equilibrated, nonlinear part of the thermal stresses, resulting from prevented eigenwarping of the cross-section.

The present paper is organized as follows: Section 2 refers to the prerequisites for this study, see (Diaz et al. 2018) for details. Section 3 refers to the engineering mechanics analysis pushed forward in this study. In Section 4, the results of the engineering mechanics analysis are compared with those from the FE simulation. Section 5 contains the conclusions drawn from the present study.

2 PREREQUISITES FOR THE PRESENT STUDY

A scaled fire test was performed (Lu et al. 2018) on a structure inspired by the upper floor of a three-span, two-floor reinforced concrete frame, commonly used in underground stations in China. The motivation to perform the test, was to identify the temperature at the inner surface as well as the temperature histories and strains inside the structure. The tested structure was placed sidelong on top of a furnace and closed with a fire-resistant cover. A frame of steel with hydraulic presses and supports was located around the model to simulate service conditions. The real structure described in (Diaz et al. 2018) was tested at a scale of 1:4. Only the columns were scaled by 1:5, see (Diaz et al. 2018) for details. The structure was produced with normal concrete "C40", with a mass density of 2373 kg/m^3 and a concrete cover of 30 mm. Before the concrete was cast, the temperature and strain sensors at the slabs and walls were installed, see (Diaz et al. 2018) for details. During the first 1800 s of heating the inner surface, the temperature of the outer surface was constant. The temperature sensors recorded readings every 20 s.

Based on the scaled fire test, the structure was simulated by means of three-dimensional, non-stationary Finite Element (FE) simulations using the commercial software Abaqus FEA 2016, for details see (Diaz et al. 2018). Figure 1 illustrates the idealized geometric boundary conditions and their locations as well as the locations of the point loads $P_1 = 192.0$ kN, $P_2 = 151.2$ kN, and $P_3 = 120.0$ kN, that represent the hydraulic presses. Given that the thermal properties of the tested structure were unknown, the values of the specific heat capacity and the thermal conductivity were estimated in accordance with building codes

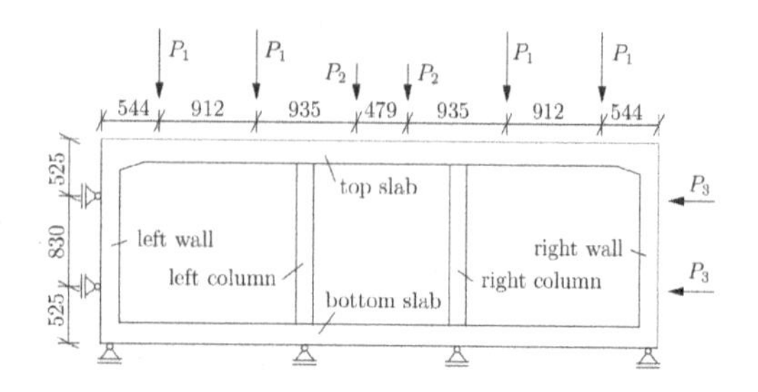

Figure 1. Support and loading conditions of the tested structure (Diaz et al. 2018).

and scientific studies, see Table 1. According to the insights in (Diaz et al. 2018), the temperature change at the steel bars may be assumed to be equal to that of the concrete at its immediate vicinity. Thus, this thermal simulation of the structure depends on the thermal properties of concrete only.

Table 1. Thermal properties of concrete (Diaz et al. 2018).

Property	Value
Specific heat capacity [J/kgK]	900
Thermal conductivity [W/mK]	1.6

For the thermo-mechanical analysis, the steel needs to be taken into account. The reinforcement were modeled as one-dimensional truss elements. The values of specific properties of concrete and steel, defined in (Diaz et al. 2018), were assigned to the corresponding elements, see Table 2.

Table 2. Mechanical properties of concrete and steel at room temperature (Diaz et al. 2018).

Property		concrete	steel
Modulus of elasticity	[GPa]	33.4	195
Poisson's ratio	[-]	0.2	0.3
Thermal expansion coefficient	[10^{-6} °C^{-1}]	9.03	12.2

For simulation of non-stationary heat conduction with ABAQUS CAE (Dassault Systemes Simulia Corp 2019) the temperature data of the scaled fire test was used to approximate the temperature histories at the inner surface of the structure, see (Diaz et al. 2018) for details. The thermo-mechanical results of the FE simulation were evaluated at selected sections of the structure, representing the stress component σ_{xx} of Cauchy's stress tensor at the mid-plane in axial direction. The linear shape-functions of the chosen elements yield element-wise constant stresses.

3 ENGINEERING MECHANICS ANALYSIS

In the following, a simplified engineering mechanics analysis of the scaled fire test in (Lu et al. 2018) is performed using a beam analysis software under the assumption of the first-order beam theory. For this analysis, the slabs and walls of the structure will be simulated as prismatic beams with double-symmetric cross-sections in a Cartesian coordinate system with origin at the axis of the beam. Because of the geometry of the plates and the constant temperature at the outer surface, the generally three-dimensional heat conduction problem may be approximated as a one-dimensional heat conduction in thickness direction. The prismatic columns, located in the mid-plane of the structure, are thermally loaded on all four lateral surfaces. To reduce the two-dimensional heat conduction problem to a one-dimensional one, the prismatic columns will be transformed into cylindrical ones with axisymmetric cross-sections in a Polar coordinate system with origin at the axis of the column. In order to consider the reinforcement of concrete, a non-uniform modulus of elasticity $E(y,z)$ and thermal expansion coefficient $\alpha_T(y,z)$ are considered at a cross-sectional scale.

3.1 *Semi-analytic solutions of the heat conduction problem*

The isotropic form of the heat equation for the case of three-dimensional heat conduction problem reads as

$$\dot{T} - a\nabla \cdot (\boldsymbol{I} \cdot \nabla T) = 0\,, \tag{1}$$

where a denotes the thermal diffusivity, ∇ denotes the nabla operator and $\boldsymbol{I} = \sum_{i=1}^{3} \underline{e}_i \otimes \underline{e}_i$ denotes the second order identity tensor. The partial differential equation (1) will be solved for one-dimensional heat conduction in thickness direction of prismatic beams and radial heat conduction of cylindrical columns with time-dependent boundary conditions at the surfaces according to the temperature histories documented in (Diaz et al. 2018). At the beginning, the temperature at the surface is equal to a reference temperature $T(z,t=0) = T_{ref}$ (initial condition). This yields a constant initial condition for solving the heat conduction problem. Because of the linearity of the partial differential equation (1), the superposition principle applies and the time-dependent boundary conditions in temperature can be discretized in N_T temperature increments ΔT_k at the surfaces, where $k = 1, 2, ..., N_T$. Elementary solutions referring to each temperature increment ΔT_k are superimposed to a semi-analytical solution for the given boundary conditions. In this context, the constant boundary condition at the surface for one increment reads as:

$$\Delta T_k = T(t_k) - T(t_{k-1}) \tag{2}$$

where t_k and t_{k-1} are time instants.

3.1.1 *One-dimensional heat conduction in thickness direction of prismatic beams*

Specification of the heat equation (1) for a Cartesian coordinate system and one-dimensional heat conduction in thickness direction z with temperature $T = T(z,t)$ and thermal diffusivity a, yields:

$$\frac{\partial T}{\partial t} - a\frac{\partial^2 T}{\partial z^2} = 0\,. \tag{3}$$

After each time step t_k with $k = 1, 2, ...N_T$, where N_T is the total number of temperature increments, the temperature of the bottom surface of the beam T_k^{bot} is constant and depends on the temperature histories as documented in (Diaz et al. 2018). It is assumed that the heat flux across the lateral surfaces is equal to zero and the one-dimensional heat equation (3) is applicable. The temperature of the top surface is equal to T_{ref}. In this case the initial condition and the boundary conditions can be written as:

$$\left.\begin{aligned} T(z=-\tfrac{h}{2},t) &= T_{ref} \\ T(z=+\tfrac{h}{2},t) &= T_k^{bot} \end{aligned}\right\} \text{boundary conditions.} \tag{4}$$

The solution of the one-dimensional heat equation (3) for one temperature increment is documented in the literature, e.g., (Ausweger 2016). Summation of the elementary solutions of N_T temperature increments at the bottom surface of prismatic beams reads as

$$\begin{aligned} \Delta T(z,t) = \sum_{k=1}^{N_T} \Delta T_k^{bot} \Big(\frac{1}{2} + \frac{z}{h}\Big) \\ + \sum_{n=1}^{\infty} \exp\Big(-\frac{(2n-1)^2\pi^2 a\langle t-t_k\rangle}{h^2}\Big) \\ \Big[\frac{2\Delta T_k^{bot}(-1)^n}{(2n-1)\pi} \cos\Big(-(2n-1)\pi\frac{z}{h}\Big)\Big] \\ - \sum_{n=1}^{\infty} \exp\Big(-\frac{(2n\pi)^2 a\langle t-t_k\rangle}{h^2}\Big)\Big[\frac{\Delta T_k^{bot}(-1)^n}{n\pi} \sin\Big(-2n\pi\frac{z}{h}\Big)\Big], \end{aligned} \tag{5}$$

where ΔT_k^{bot} denotes the temperature step at the inner surface at time t_k, h denotes the thickness of the cross-section and the angled brackets denote the Macaulay operator, see, e.g., (Wang et al. 2019).

3.1.2 *Radial heat conduction in axisymmetric cylindrical columns*

Specification of the heat equation (1) for a cylindrical coordinate system with spatially uniform boundary conditions at the lateral surface delivers an axisymmetric heat conduction problem where the temperature $T = T(r,t)$ is a function of the radial coordinate r and the time t. The resulting partial differential equation with variable coefficients is also known as Bessel differential equation. It reads as

$$\frac{\partial T}{\partial t} - a\Big(\frac{\partial^2 T}{\partial r^2} + \frac{1}{r}\frac{\partial T}{\partial r}\Big) = 0. \tag{6}$$

After each time step t_k with $k = 1, 2, \ldots N_T$, where N_T is the total number of temperature steps, the temperature of the lateral surface of the column T^{lat} is constant and depends on the temperature history as documented in (Diaz et al. 2018). The boundary condition can be written as:

$$T(r=R,t) = T^{lat}. \tag{7}$$

The solution of the radial heat conduction problem in Eq. (6) for one temperature increment prescribed on the lateral surface ΔT_k^{lat} of the structure as defined in Eq. (2) is derived analogous to the one-dimensional solution of prismatic beams. Summation of the elementary solutions of N_T temperature increments on the lateral surface of cylindrical columns and subtracting the reference temperature T_{ref} reads as

$$\begin{aligned} \Delta T(r,t) = \Delta T^{lat} - \sum_{k=1}^{N_T} \frac{2\,\Delta T_k^{lat}}{R} \\ \sum_{n=1}^{\infty} \frac{J_0(\lambda_n r)}{J_1(\lambda_n R)\lambda_n} \exp\big(-\lambda_n^2 a\langle t-t_k\rangle\big), \end{aligned} \tag{8}$$

where ΔT^{lat} denotes the total temperature change on the lateral surface, defined as the summation of all temperature increments ΔT_k^{lat} in Eq. (2). J_0 and J_1 denote the Bessel functions of the first kind with their eigenvalues λ_n, R denotes the radius of the column and the angled brackets denote the Macauley operator.

3.2 *Engineering mechanics modeling at cross-sectional scale*

The thermal eigenstrains, ε^e, developed at the points y and z inside the cross-section of a beam, see Figure 2, are equal to the thermal expansion coefficient evaluated at those points, $\alpha_T = \alpha_T(y,z)$, multiplied with the change of temperature, measured relative to the reference configuration, $\Delta T(z,t) = T(z,t) - T_{ref}$, as

$$\varepsilon_{xx}^e = \varepsilon_{yy}^e = \varepsilon_{zz}^e = \alpha_T \Delta T. \tag{9}$$

In a non-stationary heat conduction problem, the thermal eigenstrains are spatially nonlinear along the thickness direction, see e.g., (Hasenbichler 2019; Schmid 2020; Wang et al. 2019). When it comes to the quantification of thermal stresses, the question must be answered whether the eigenstrains are free to develop, constrained, or prevented. This question is answered partly at the larger, structural level, and partly at the smaller, cross-sectional level. To this end, the spatially nonlinear eigenstrains are subdivided into three parts,

$$\alpha_T \Delta T = \varepsilon_0^e + \kappa_0^e z + \varepsilon_w^e, \tag{10}$$

where ε_0^e refers to an eigenstretch of the beam, κ_0^e refers to an eigencurvature of the beam, and ε_w^e refers to an eigenwarping of the cross-section (?). The eigenstretch and the eigencurvature of the axis of the beam cause axial stresses depending on the boundary conditions that constrain the deformation of the structure. On the other hand, the assumption that plane sections remain plane means that the eigenwarping of the cross-section of the beam is prevented, thus always resulting in nonlinear thermal stresses. Herein, we focus on double-symmetric cross-sections with non-uniform modulus of elasticity and thermal expansion coefficient, $E(y,z)$ and $\alpha_T(y,z)$, respectively, assuming a coordinate system with origin at the axis of the beam, see Figure 2.

The kinematics of the Euler-Bernoulli theory for slender beams, that essentially describes that cross-sections remain plane and normal to the deformed axis

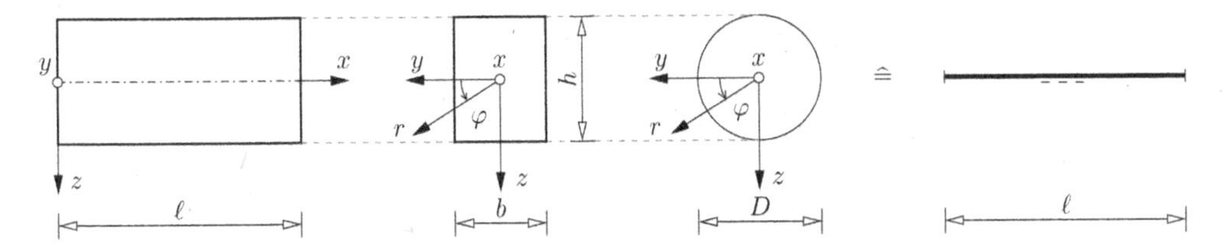

Figure 2. Local coordinate systems describing positions inside and at the boundary of the structural elements: x denotes the longitudinal axis; y and z denote Cartesian coordinates of the cross-section; r and φ denote cylindrical coordinates.

of the beam, also in the deformed configuration (= Euler-Bernoulli hypothesis), reads as

$$u = u_0 - \frac{\partial w_0}{\partial x} z, \tag{11}$$

where u denotes the displacement components in x-direction, at any point within the volume of the beam and u_0 and w_0 denote the displacement components at the axis of the beam. Inserting Eq. (11) in the ("total") axial strain component ε_{xx} of the linearized strain tensor and in Hooke's law for thermoelasticity, yields the axial stress component σ_{xx} of Cauchy's stress tensor. This reads as

$$\sigma_{xx} = E\,(\varepsilon_0 + \kappa_0\, z - \alpha_T \Delta T), \tag{12}$$

where $\varepsilon_0 = \partial u_0/\partial x$ is the stretch of the axis of the beam, and $\kappa_0 = -\partial^2 w_0/\partial x^2$ its curvature. Herein, the modulus of elasticity is a function of the y- and z-coordinates which describe points inside the cross-sections of the beam, i.e. $E = E(y, z)$. Thus, one can conclude from inserting Eq. (12) in the definition of the axial force N, which is energetically conjugate to the displacements u_0, that the eigenstretch of the axis of the beam is calculated as

$$\varepsilon_0^e = \frac{1}{\overline{EA}} \int_A E\, \alpha_T \Delta T \,\mathrm{d}A, \tag{13}$$

where $\overline{EA} = \int_A E \,\mathrm{d}A$ is the effective extensional stiffness of the beam. In the case of a reinforced concrete beam, that presents a constant temperature in each reinforcement bar, Eq. (13) reads as

$$\varepsilon_0^e = \frac{1}{A_{tr}} \left[\int_{A_c} \alpha_{T,c} \Delta T \,\mathrm{d}A + n_E \sum_{j=1}^{L} \alpha_{T,s} \Delta T_j A_{s,j} \right], \tag{14}$$

where n_E refers to the ratio between the modulus of elasticity of steel, E_s, and concrete, E_c, $\alpha_{T,c}$ is the coefficient of thermal expansion of concrete and $\alpha_{T,s}$ is that of steel, j refers to each one of the L individual reinforcement bars of the cross-section, and $A_{tr} = A_c + n_E \sum_{j=1}^{L} A_{s,j}$ refers to the total area of the âŁœtransformedâŁž section with A_c denotes the area of concrete and A_s denotes the area of steel. In Eq. (10), ε_0^e denotes a spatially constant contribution according to Eq. (13), representing an eigenstretch of the beam.

Inserting Eq. (12) in the definition of the bending moment M, that is energetically conjugate to the cross-sectional rotation $\partial w_0/\partial x$, leads to the expression for the eigencurvature of the beam as

$$\kappa_0^e = \frac{1}{\overline{EI}} \int_A E\, \alpha_T \Delta T\, z \,\mathrm{d}A, \tag{15}$$

with $\overline{EI} = \int_A E z^2 \,\mathrm{d}A$ is the effective bending stiffness of the beam. In the case of a reinforced concrete beam that presents a constant temperature within each reinforcement bar, Eq. (15) reads as

$$\kappa_0^e = \frac{1}{I_{tr}} \left[\int_{A_c} \alpha_{T,c} \Delta T z \,\mathrm{d}A + n_E \sum_{j=1}^{L} \alpha_{T,s} \Delta T_j A_{s,j} z_{s,j} \right]. \tag{16}$$

where $I_{tr} = I_c + n_E \sum_{j=1}^{L} A_{s,j}\, z_{s,j}^2$ refers to the second moment of inertia of the âŁœtransformedâŁž cross section, and $z_{s,j}$ refers to the distance between each individual reinforcement bar and the axis of the beam. In Eq. (10), $\kappa_0^e\, z$ denotes a spatially linear contribution with vanishing mean value, see Eq. (15), representing an eigencurvature of the beam. Rearranging the definitions of the stress resultants, M and N, considering the axial stress σ_{xx}, for the stretch of the axis of the beam, ε_0, and its curvature, κ_0, and inserting this expressions in Eq. (12) yields

$$\sigma_{xx}(y,z) = \frac{N\, E(y,z)}{\overline{EA}} + \frac{M\, E(y,z)}{\overline{EI}} z - E(y,z)\big[\alpha_T(y,z)\Delta T(z) - \varepsilon_0^e - \kappa_0^e z\big]. \tag{17}$$

The expression in the square brackets of Eq. (17) is equal to the nonlinear part of the eigenstrains, ε_w^e, representing an eigenwarping of the cross-section of the beam, see Eq. (10).

3.3 *Engineering mechanics modeling at the structural scale*

In thermoelasticity the internal forces M and N depend on the eigenstretch and eigencurvature ε_0^e and κ_0^e:

$$N = \overline{EA}\left(\varepsilon_0 - \varepsilon_0^e\right), \qquad M = \overline{EI}\left(\kappa_0 - \kappa_0^e\right). \tag{18}$$

In statically determinate structures, the thermal eigenstretches ε_0^e and eigencurvatures κ_0^e are free to develop.

Thus, there are no internal forces. As a consequence, the total internal forces depend on the mechanical loading only. In case of a statically *in*determinate structure, the boundary conditions constrain the thermal eigenstretches and eigencurvatures. Corresponding stresses need to be quantified based on a simulation of the behavior of the whole structure, accounting for its boundary conditions. The tested structure (Diaz et al. 2018) will be modeled with the beam analysis software RStab (Dlubal Software GmbH 2020) using first-order beam theory.

In a first step, the three-dimensional FE model (Diaz et al. 2018) will be used as a basis in order to build a simplified one-dimensional beam model. For this purpose, the structure will be idealized. Rigid connections between the structural elements are assumed. In order to perform the simulation as simple as possible and as complex as necessary, the tapered part of the top slab will not be part of the model. This yields an idealized structure for the simulation with beam analysis software which is statically indeterminate to the twelfth degree ($n = 12$), see Figure 3. The local coordinate system of the frame is defined by the dashed line, see Fig 2. x_1 denotes the axial location at which results will be discussed.

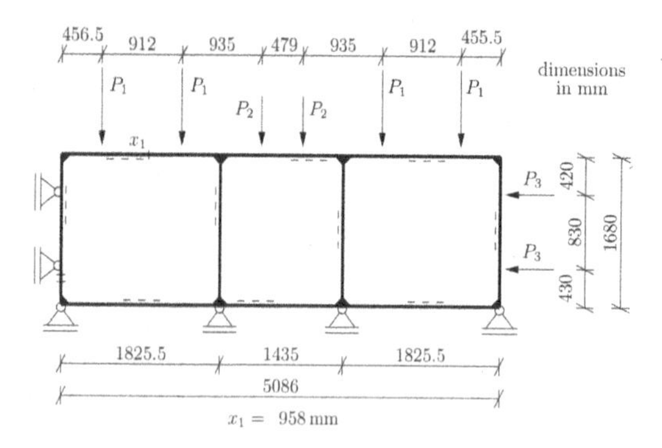

Figure 3. Idealized representation of the tested segment of a subway station, as the basis for structural analysis using beam analysis software: for the numerical values of the point loads P_1, P_2, and P_3.

In a second step, the cross-section of the columns is transformed from its prismatic shape with dimension 160/240 mm to a circular shape with diameter $D = 221.1$ mm, equivalent in extensional stiffness. The expectation of a predominance of axial force at the columns provides the motivation for performing this transformation. In a third step, the reinforced concrete cross-sections of the structure will be idealized as transformed cross-sections depending on the material behavior of concrete. The ratio between the modulus of elasticity of steel and concrete reads as $n_E = 5.838$, see Table 2. The "transformed" cross-sectional properties are input for the beam analysis software RStab (Dlubal Software GmbH 2020). This requires the quantification of ratio factors defined as

$$\eta_A = \frac{A_{tr}}{A}, \qquad \eta_I = \frac{I_{tr}}{I}, \tag{19}$$

where A denotes the real area defined as $A = A_c + A_s$ and I denotes the real second moment of inertia defined as $I = I_c + I_s$. The ratio factors of each part of the structure as defined in Eq. (19) are documented in Table 3.

Material properties that are required for simulation with beam analysis software RStab (Dlubal Software GmbH 2020) based on the simulation with FE software ABAQUS CAE in (Diaz et al. 2018) are documented in Tables 1 and 2.

4 RESULTS AND DISCUSSION

4.1 *Results from the simulation of heat conduction*

The non-stationary heat conduction problem described in Section 3.1 is solved by inserting the thermal properties of concrete from Table 1 and the surface temperature histories (Diaz et al. 2018) into the derived semi-analytical solutions, see Eqs. (5) and (8). The infinite sums are approximated based on the first 1000 terms. This is more than sufficient to obtain a well-converged solution. The temperature change obtained half an hour after the start of the fire is illustrated,

Table 3. Cross-sectional properties of the structural elements: "transformed properties" refer to a cross-section consisting of concrete only, but being equivalent to the actual reinforced concrete cross-section, "real properties" refer to the actual concrete part of the actual cross-section, "ratio factors" are defined in Eq. (19).

cross-section	transformed property A_{tr} [mm^2], I_{tr} [mm^4]	real property A [mm^2], I [mm^4]	ratio factors [–], see Eq. (19)
Top Slab	$A_{tr} = 2.684 \times 10^5$	$A = 2.520 \times 10^5$	$\eta_A = 1.065$
	$I_{tr} = 1.037 \times 10^9$	$I = 9.261 \times 10^8$	$\eta_I = 1.120$
Bottom Slab	$A_{tr} = 2.411 \times 10^5$	$A = 2.280 \times 10^5$	$\eta_A = 1.058$
	$I_{tr} = 7.529 \times 10^8$	$I = 6.859 \times 10^8$	$\eta_I = 1.098$
Lateral Wall	$A_{tr} = 2.279 \times 10^5$	$A = 2.100 \times 10^5$	$\eta_A = 1.085$
	$I_{tr} = 6.073 \times 10^8$	$I = 5.359 \times 10^8$	$\eta_I = 1.133$
Columns	$A_{tr} = 4.387 \times 10^4$	$A = 3.840 \times 10^4$	$\eta_A = 1.143$
	$I_{tr} = 1.388 \times 10^8$	$I = 1.173 \times 10^8$	$\eta_I = 1.183$

for the top slab, as a solid line in Figure 4. It is compared with results from the FE simulation by Diaz et al. (2018), see the dashed line in Figure 4.

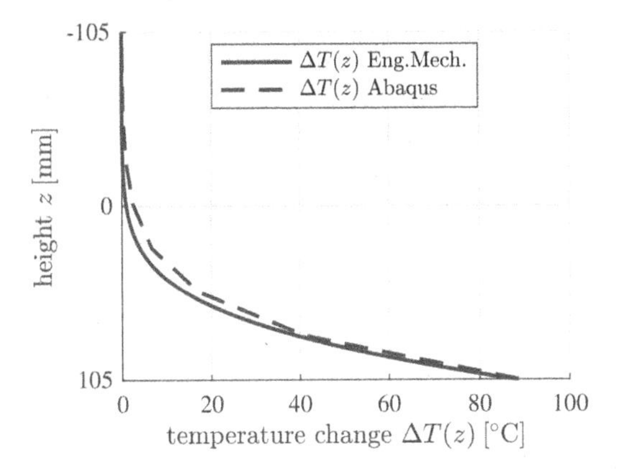

Figure 4. Temperature changes at the top slab obtained half an hour after the start of the fire test.

4.2 *Decomposition of thermal eigenstrains*

In order to compute thermal eigenstrains for all structural elements, the temperature changes (e.g., of the top slab, see Figure 4) are multiplied with the coefficients of thermal expansion of steel and concrete; e.g., see the dotted red line in Figure 5 for the eigenstrain distribution of the top slab experienced by concrete. This spatially nonlinear distribution is subdivided into (i) a constant part, related to the eigenstretch, ε_0^e, of the axis of the structural element, see Eq. (13) and (14) as well as the blue solid line in Figure 5, (ii) a linear part, related to the eigencurvature, κ_0^e, of the axis of the structural element, see Eqs. (15) and (16) as well as the blue dashed line in Figure 5, and (iii) the

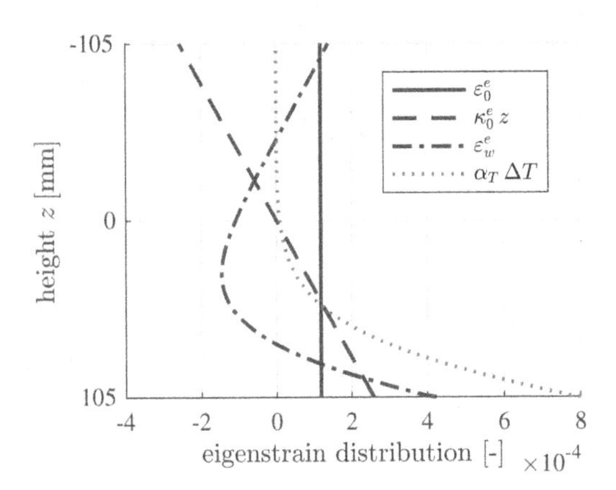

Figure 5. Thermal eigenstrains of the concrete at the top slab half an hour after the start of the fire, obtained with the engineering mechanics analysis: the dotted graph refer to total eigenstrains, the solid graph to the eigenstretch, the dashed graph to the eigencurvature, and the dash-dotted graph to the eigenwarping of the cross-section of the structural element.

nonlinear rest, representing the thermal eigenwarping of the cross-section of the structural element:

$$\varepsilon_w^e = \alpha_T \Delta T - \varepsilon_0^e - \kappa_0^e \, z \,, \tag{20}$$

and the dash-dotted blue line in Figure 5. The numerical values of the eigenstretches and eigencurvatures of all structural elements are listed in Table 4.

Table 4. Numerical values of the thermal eigenstretches and eigencurvatures of the axes of the structural elements, half an hour after the start of the fire, obtained with the engineering mechanics approach, see Eqs. (14) and (16).

Cross-section	Eigenstretch $[10^{-4}]$	Eigencurvature $[10^{-6}\,\text{mm}^{-1}]$
Top Slab	$\varepsilon_0^e = 1.1848$	$\kappa_0^e = 2.4632$
Bottom Slab	$\varepsilon_0^e = 1.2860$	$\kappa_0^e = 2.8946$
Lateral Wall	$\varepsilon_0^e = 1.1362$	$\kappa_0^e = 2.7662$
Columns	$\varepsilon_0^e = 12.5200$	$\kappa_0^e = 0.0000$

4.3 *Structural analysis using beam analysis software*

Two simulations are carried out. The first one refers to the point loads representing ground pressure. This simulation provides insight into the structural behavior before the start of the fire. The second simulation refers to the point loads representing ground pressure *and* eigenstretches as well as eigencurvatures representing the thermal loading half an hour after the start of the fire.

4.4 *Stress distribution at the top slab*

Under combined mechanical and non-stationary thermal loading, axial stresses result, in every cross-section, from three contributions: the axial force, the bending moment, and eigenstresses. In the present context of reinforced concrete members, we focus on the axial stresses of the top slab experienced by the concrete. The axial force refers to spatially constant axial stresses, the bending moment to spatially linear stresses, and the eigenstresses are nonlinearly distributed across the cross-section. The axial forces and the bending moments depend on the mechanical loading as well as the thermal eigenstretches and the eigencurvatures of all structural members. These contributions are accounted for by means of the simulation with the beam analysis software, see Section 4.3. The eigenstresses account for the *non-stationary* nature of the heat conduction problem. They are equal to the eigenwarping-part of the thermal eigenstrains, multiplied with the modulus of elasticity and -1. The eigenwarping-parts of the thermal eigenstrains were determined in Section 4.2.

Axial stresses at position x_1 of the top slab are discussed in two diagrams. The first diagram displays, with blue graphs, the stress distribution $\sigma_{xx,c}$ of concrete resulting from the point loads only, referring to the service condition *before* the fire, see Figure 6a. The second diagram displays, with red graphs, the stresses of concrete computed for the time instant half an hour after the start of the fire, see Figure 6b. In both cases, results from the simulation with beam analysis software, from the engineering mechanics analysis, and from FE simulation are compared.

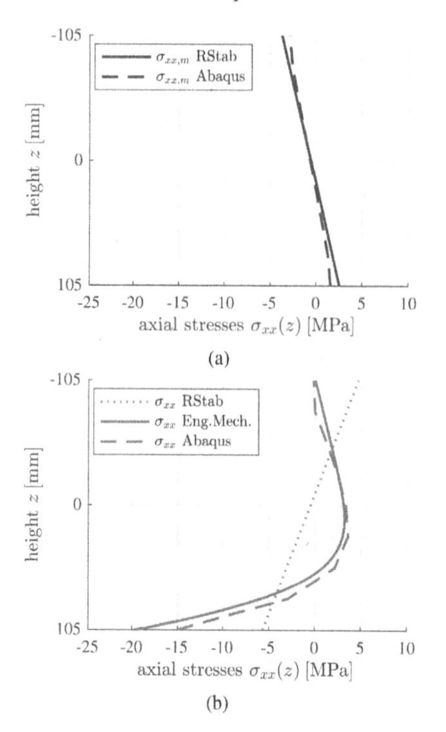

Figure 6. Top slab location $x_1 = 0.958$ m: axial stresses of concrete **(a)** before the fire, and **(b)** half an hour after the start of the fire.

As for the situation before the fire, the stress distributions obtained from the FE simulation and from the engineering mechanics analysis are virtually linear and perfectly linear, respectively. The engineering mechanics analysis is reproducing the overall structural behavior very well.

As for the time instant half an hour after the start of the fire, the total stresses are nonlinearly distributed over the height of the cross-section. It is appealing that the engineering mechanics analysis allows for a decomposition of the total stresses into (i) spatially linear stress contributions resulting from the axial forces and the bending moments, as well as (ii) spatially nonlinear stress contributions resulting from the prevented eigenwarping of the cross-sections. Both simulation approaches suggest that the maximum tensile stresses are activated inside the volume of the top slab rather than at the upper or lower surface. These tensile stresses amount to some 3 MPa. It is to be expected that the tensile strength of concrete would be reached during the third quarter of an hour after the start of the fire, and that cracking of concrete will occur visually unnoticeable *inside* the bulk of the top slab.

5 CONCLUSIONS

A reinforced concrete segment of a subway station subjected, in a large-scale laboratory test (Lu et al. 2018), to ground pressure and a moderate fire, was analyzed based on an engineering mechanics model, in order to check whether or not it can reproduce the results from a linear-elastic three-dimensional Finite Element analysis (Diaz et al. 2018). From this analysis, several conclusions are drawn.

The first set of conclusions refers to the question whether or not spatially nonlinear temperature fields, which are a characteristic of *non-stationary* heat conduction problems, can be translated into statically equivalent linear temperature fields. Non-stationary heat conduction orthogonal to the axis of a reinforced concrete beam is associated with spatially nonlinear distributions of thermal eigenstrains inside individual cross-sections. The *nonlinear* part of the eigenstrains represents an eigenwarping of the cross-sections. The latter remain plane in slender beams, even in case of non-stationary heat conduction. This planarity is related to spatially *linear* total strains. Thus, the spatially nonlinear part of the thermal eigenstrains is prevented at the scale of the cross-sections. Because the resulting thermal eigenstresses have a vanishing mean value and a vanishing first moment, they are "self-equilibrated", i.e. they do neither contribute to the axial force nor to the bending moment. Subtracting from the total thermal eigenstrains the nonlinear part, results in eigenstrains which are related to the thermal eigenstretch and the thermal eigencurvature of the axis of the beam analyzed. Whether eigenstretch and eigencurvature are free to develop, constrained, or prevented must be answered at the scale of the entire structure. They are free to develop in statically determinate structures. They are at least constrained in statically indeterminate structures. Because eigenstretches and eigencurvatures are constrained at the *larger* scale of a statically indeterminate reinforced concrete *structure*, and because the eigenwarping is prevented at the *smaller* scale of the *cross-sections*, it is impossible to translate spatially nonlinear temperature fields into statically equivalent linear temperature fields.

The second set of conclusions refers to the potential of the engineering-mechanics approach, regarding the reproduction of results obtained with a computationally much more expensive three-dimensional Finite Element model. Commercially available beam analysis software is typically capable of accounting for eigenstretches and eigencurvatures. This allows for computing axial forces and bending moments as well as the related axial stresses which are linear functions within the individual cross-sections. Adding to these stress fields the spatially nonlinear thermal eigenstresses resulting from the prevented eigenwarping of the cross-sections delivers total stresses which are in good agreement with the results from the elaborate Finite Element model. FE analyses intrinsically suffer from a discretization error. Its quantification requires convergence analyses, i.e. the same problem must be solved based on different Finite Element meshes, and important output quantities are illustrated as a function of the discretization effort, in order to find a trade-off between discretization effort and reliability of the results obtained. Such convergence analyses require a significant amount of time, given that pre-processing of FE simulations frequently represents a large (if not the dominating) part of the time investments required for the overall analysis. The remaining discretization error manifests itself in frequently kinky rather than smooth distributions of output quantities.

Overall, it is concluded that the developed mode of thermo-elastic analysis provides interesting insight into nontrivial aspects of the structural behavior. Still, the current limitations of the presented engineering mechanics analysis shall also be addressed. Because a moderate fire was analyzed, mechanical properties of concrete were treated as constants and set equal to values at room temperature. In the future, these constants can be replaced by functions of temperature. Such mathematical relationships are provided by many codes for the design of reinforced concrete structures. The simulation with beam analysis software is based on the Euler-Bernoulli hypothesis. The latter is questionable, e.g. in the immediate vicinity of connections of different structural elements and of point loads. In order to gain detailed insight into stress distributions in such regions, Finite Element simulations appear to be indispensable.

ACKNOWLEDGEMENT

The work of Dr. Hui Wang is sponsored by Shanghai Pujiang Program (Grant No. 20PJ1406100).

REFERENCES

Albrifkani, S. & Y. C. Wang (2016). Explicit modelling of large deflection behaviour of restrained reinforced concrete beams in fire. *Engineering Structures 121*, 97–119.

Ausweger, M. (2016). Spannungen und Verformungen gerader EinzelstÃ¤be zufolge Temperaturbeanspruchung. Bachelorarbeit, Technische UniversitÃ¤t Wien.

Bratina, S., B. Cas, M. Saje, & I. Planinc (2005). Numerical modelling of behaviour of reinforced concrete columns in fire and comparison with Eurocode 2. *Solids and Structures 42*, 5715–5733.

Choi, E. & Y. Shin (2011). The structural behavior and simplified thermal analysis of normal-strength and high-strength concrete beams under fire. *Engineering Structures 33*, 1123–1132.

Dassault Systemes Simulia Corp (2019). ABAQUS CAE.

Diaz, R., H. Wang, H. Mang, Y. Yuan, & B. Pichler (2018). Numerical Analysis of a Moderate Fire inside a Segment of a Subway Station. *applied sciences 8*, 1–34.

Dlubal Software GmbH (2020). RStab.

Dwaikat, M. & V. Kodur (2008). A numerical approach for modeling the fire induced restraint effects in reinforced concrete beams. *Fire Safety 43*, 291–307.

El-Fitiany, S. & M. Youssef (2017). Fire performance of reinforced concrete frames using sectional analysis. *Engineering Structures 142*, 165–181.

El-Tayeb, E. H., S. E. El-Metwally, H. S. Askar, & A. M. Yousef (2017). Thermal analysis of reinforced concrete beams and frames. *HBRC Journal 13*, 8–24.

Franssen, J.-M. & J.-C. Dotreppe (2003). Fire Tests and Calculation Methods for Circular Concrete Columns. *Fire Technology 39*, 89–97.

Gao, W., J.-G. Dai, J. Teng, & G. Chen (2013). Finite element modeling of reinforced concrete beams exposed to fire. *Engineering Structures 52*, 488–501.

Hasenbichler, T. (2019). Semi-analytische Berechnung von WÃ¤rmespannungen in StÃ¤ben. Bachelorarbeit, Technische UniversitÃ¤t Wien.

Kamath, P., U. Kumar Sharma, V. Kumar, P. Bhargava, A. Usmani, B. Singh, Y. Singh, J. Torero, M. Gillie, & P. Pankaj (2015). Full-scale fire test on an earthquake-damaged reinforced concrete frame. *Fire Safety Journal 73*, 1–19.

Kodur, V. & M. Dwaikat (2008). A numerical model for predicting the fire resistance of reinforced concrete beams. *Cement & Concrete Composites 30*, 431–443.

Lu, L., J. Qiu, Y. Yuan, J. Tao, H. Yu, H. Wang, & H. Mang (2018). Large-scale test as the basis of investigating the fire-resistance of underground RC substructures. *Engineering Structures 178*, 12–23.

Ozbolt, J., J. Bošnjak, G. Periškic, & A. Sharma (2014). 3D numerical analysis of reinforced concrete beams exposed to elevated temperature. *Engineering Structures 58*, 166–174.

Ring, T., M. Zeiml, & R. Lackner (2014a). Underground concrete frame structures subjected to fire loading: Part I – Large-scale fire tests. *Engineering Structures 58*, 175–187.

Ring, T., M. Zeiml, & R. Lackner (2014b). Underground concrete frame structures subjected to fire loading: Part II – Re-analysis of large-scale fire tests. *Engineering Structures 58*, 188–196.

Schmid, S. (2020). Stresses in a concrete pavement resulting from transient heat conduction: engineering analysis of in situ temperature measurements. Diplomarbeit, Technische UniversitÃ¤t Wien.

Sun, R., B. Xie, R. Perera, & Y. Pan (2018). Modeling of Reinforced Concrete Beams Exposed to Fire by Using a Spectral Approach. *Advances in Materials Science and Engineering*, 1–12.

Vecchio, F. J. (1987). Nonlinear Analysis of Reinforced Concrete Frames Subjected. *ACI Structural Journal 84*, 492–501.

Vecchio, F. J. & J. A. Sato (1990). Thermal gradient effects in reinforced concrete frame structures. *ACI Structural Journal 87*, 262–275.

Wang, H., R. Höller, M. Aminbaghai, C. Hellmich, Y. Yuan, H. A. Mang, & B. L. A. Pichler (2019). Concrete pavements subjected to hail showers: A semi-analytical thermoelastic multiscale analysis. *Engineering Structures 200*, 1–19.

Wang, H., Y. Yuan, H. A. Mang, Q. Ai, X. Huang, & B. L. Pichler (2022). Thermal stresses in rectangular concrete beams, resulting from constraints at microstructure, cross-section, and supports. *European Journal of Mechanics-A/Solids*, 104495.

Zha, X. (2003). Three-dimensional non-linear analysis of reinforced concrete members in fire. *Building and Environment 38*, 297–307.

Computational Modelling of Concrete and Concrete Structures – Meschke, Pichler & Rots (Eds)
© 2022 Copyright the Author(s), ISBN: 978-1-032-32724-2

Curling stresses and thermal eigenstresses in a concrete pavement slab

S.J. Schmid, R. Díaz Flores, M. Aminbaghai & B.L.A. Pichler
Institute for Mechanics of Materials and Structures, TU Wien, Vienna, Austria

L. Eberhardsteiner & R. Blab
Institute of Transportation, TU Wien, Vienna, Austria

H. Wang
School of Naval Architecture, Ocean and Civil Engineering, Shanghai Jiao Tong University, Shanghai, China

ABSTRACT: Concrete pavements are mostly subjected to unsteady heat conduction, resulting in spatially nonlinear temperature distributions in thickness direction, and leading to thermal stresses. Thermal stresses of a concrete slab are analyzed based on in-situ temperature measurements and the evaluation method proposed in (Schmid et al. 2022). Temperature data are taken from the same field testing site as in (Schmid et al. 2022), but from a different monitoring period covering 7-days in spring. Unsteady heat conduction in vertical direction is simulated using histories of surface temperatures as boundary conditions. Their numerical values are obtained by spatial extrapolation of temperatures measured in four depths of the slab. Temperature profiles are computed and translated into thermal eigenstrains. The latter are split into three parts: eigenstretches of the slab, which are unconstrained, eigencurvatures of the slab, which are constrained by its support conditions such that curling stresses are activated, and eigendistortions of plate-generators, which are virtually prevented such that self-equilibrated eigenstresses are activated. The latter are evaluated analytically. Curling stresses are quantified numerically, prescribing a modulus of subgrade reaction of 100 MPa/m. Disregarding thermal eigenstresses overestimates tensile stresses in the afternoon, at the bottom of the slab, in its central region, by 14 %. Disregarding thermal eigenstresses underestimates tensile stresses in the morning, at the top of the slab, in its corner regions, by 59 %.

1 INTRODUCTION

Daily and seasonal temperature fluctuations are responsible for mostly nonlinear temperature profiles in thickness direction of a pavement slab and the activation of thermal stresses (Thomlinson 1940; Westergaard 1927). Total thermal stresses quantified in concrete slabs consist of spatially linear thermal stresses (= curling stresses) and spatially nonlinear thermal stresses (= eigenstresses) (Choubane & Tia 1992). Although the existence of self-equilibrated eigenstresses is known since Thomlinson (1940) decomposed nonlinear temperature distributions into a constant, a linear, and a nonlinear part, they are not accounted for in current codes of practice for the design of concrete pavements. Instead, concrete pavements are designed using the linear part of the temperature distribution, which corresponds to the temperature gradient, and thus to the effective difference between the top and bottom surface temperature (Janssen & Snyder 2000; Westergaard 1927). However, thermal eigenstresses have a significant impact on total thermal stresses in extreme weather situations (Wang et al. 2019) as well as in regular climatic conditions (Choubane & Tia 1992; Choubane & Tia 1995; Schmid et al. 2022). Consideration of thermal stresses, and especially of eigenstresses, requires (nonlinear) temperature profiles in thickness direction of the slab. They can either be simulated from climatic conditions, e.g. air temperature, wind velocity and solar radiation (Barber 1957; Bentz 2000), or quantified using directly measured data (Bayraktarova et al. 2021; Choubane & Tia 1992; Teller & Sutherland 1935). The latter requires an approach to spatially interpolate and extrapolate the pointwisely recorded temperature measurements. The simplest way to account for the nonlinearity of the temperature field is to approximate measurements using a low degree polynomial function, e.g. a quadratic or cubic polynomial (Choubane & Tia 1992; Ioannides & Khazanovich 1998; Mohamed & Hansen 1997). Schmid et al. (2022) used a semi-analytical series-solution for simulating one-dimensional unsteady heat conduction in order to obtain nonlinear temperature profiles in thickness direction of the slab. For this, surface temperatures are prescribed as boundary conditions. They are obtained

DOI 10.1201/9781003316404-66

Table 1. Material properties of the slab and of concrete

Property and source	Numerical value
length of the slab	$\ell_x = 5.50\,\text{m}$
width of the slab	$\ell_y = 2.70\,\text{m}$
thickness of the slab	$h = 0.25\,\text{m}$
Thermal diffusivity (Schmid et al. 2022)	$a = 1.4 \times 10^{-6}\,\text{m}^2/\text{s}$
Coefficient of thermal expansion (Wang et al. 2019)	$\alpha_T = 1.153 \times 10^{-5}\,/^\circ\text{C}$
Modulus of elasticity (Wang et al. 2019)	$E = 31.76\,\text{GPa}$
Poisson's ratio (Wang et al. 2019)	$\nu = 0.203$
Mass density	$\rho = 2{,}400\,\text{kg/m}^3$

by fitting quadratic polynomials to the temperature measurements taken in four specific depths of a concrete pavement slab and extrapolating these functions to its top and bottom surface. The obtained temperature profiles are used to quantify thermal eigencurvatures of the plate and eigendistortions of the generators of the plate. Eigencurvatures and eigendistortions are translated into curling stresses and into eigenstresses, respectively (Schmid et al. 2022).

Herein, this procedure is taken over and applied to a set of temperature measurements monitored in the concrete pavement slab, which is part of the highway "A2 – Süd Autobahn" in Lower Austria (Schmid et al. 2022). The temperature was recorded during one week in April 2015 in 5 cm, 9 cm, 14 cm, and 19 cm under the top surface of the slab using PT100A temperature sensors. Material properties, such as the thermal diffusivity, the coefficient of thermal expansion, the modulus of elasticity and Poisson's ratio are taken from (Schmid et al. 2022). Curling stresses are quantified numerically in nonlinear Finite Element analyses, whereby the elastic subgrade of the slab is modeled by means of an elastic Winkler foundation (Winkler 1867). Kinematic restrictions of the plate referring to Kirchhoff's hypothesis, prevent the development of thermal eigendistortions (Wang et al. 2019; Höller et al. 2019). They are nullified by mechanical strains which activate thermal eigenstresses.

The study is organized as follows. Section 2 is devoted to the application of the thermal stress evaluation method proposed in (Schmid et al. 2022) to a set of temperature data measured in spring. In Section 3, curling stresses and eigenstresses are quantified and discussed in the context of total thermal stresses prescribing a modulus of elasticity of 100 MPa/m. This part is followed by the conclusions drawn from the presented analysis.

2 TEMPERATURE MONITORING AND THERMO-MECHANICAL ANALYSIS OF A CONCRETE SLAB

2.1 *Temperature monitoring in a concrete slab*

Schmid et al. (2022) proposed an evaluation method for quantification of thermal stresses based on in situ temperature measurements recorded in a concrete pavement slab during a 23 days period in autumn 2015. This method is applied to temperature data obtained from a 7-days measurement period in the same slab in spring 2015.

Temperature data were recorded in the concrete pavement slab, located in Lower Austria at kilometer 21 of the highway "A2". The length, width, and thickness of the slab and the material properties of concrete are provided in Table 1.

From 9 Apr. 00:00 to 15 Apr. 24:00, four sensors of type PT100A logged temperature data every 30 minutes in depths amounting to 5 cm, 9 cm, 14 cm, and 19 cm measured from the top surface of the slab. The vertical positions of the sensors are described by the z-coordinate of a Cartesian coordinate system. Its origin corresponds to the center of gravity of the slab. The positive z-axis runs downwards in the direction of thickness. Thus, the z-coordinates of the sensors follow to $z_1 = -0.075\,\text{m}$, $z_2 = -0.035\,\text{m}$, $z_3 = +0.015\,\text{m}$ and $z_4 = +0.065\,\text{m}$. Spline functions are used to temporally interpolate the measured data. They are evaluated every three minutes, see Figure 1.

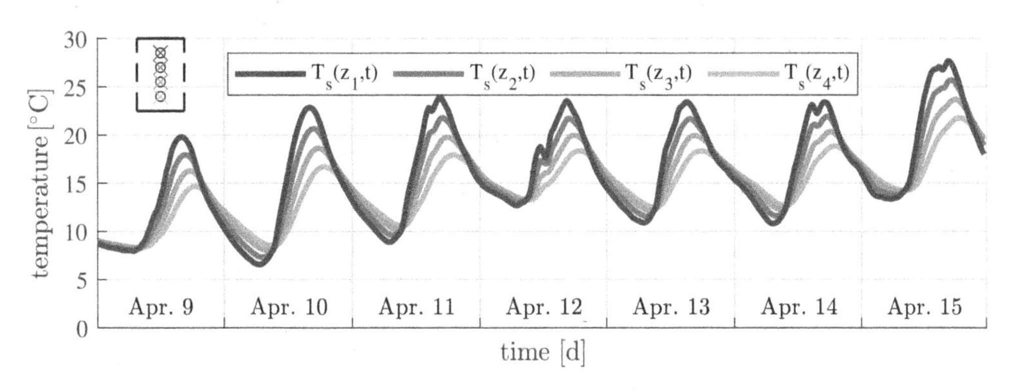

Figure 1. Recorded and temporally interpolated temperature data within the volume of the slab, referring to depths of 5 cm, 9 cm, 14 cm and 19 cm measured from the top surface.

For every time instant, a set of four temperature values, represented by $T_s(z_1, t)$, $T_s(z_2, t)$, $T_s(z_3, t)$, and $T_s(z_4, t)$, is obtained.

2.2 *Unsteady heat conduction in thickness direction of the slab*

Daily temperature variations go along with (one-dimensional) unsteady heat-conduction in thickness direction of the slab during most of the time. Heat-conduction is a boundary value problem. Solutions of such problems require a field equation, an initial condition and boundary conditions, for details see (Schmid et al. 2022; Wang et al. 2019).

The field equation is the heat equation $\partial T(z,t)/\partial t = a \cdot \partial^2 T(z,t)/\partial z^2$. Thereby, T stands for the temperature, t for the time variable and a denotes the thermal diffusivity, see Table 1.

The initial condition corresponds to a linear temperature distribution:

$$T(z, t{=}0) = \frac{T_{ini}^{bot} + T_{ini}^{top}}{2} + \left[T_{ini}^{bot} - T_{ini}^{top}\right]\frac{z}{h}\,. \qquad (1)$$

T_{ini}^{top} and T_{ini}^{bot} denote the temperature at the top and bottom surface, respectively, at the starting point of the simulation on 9 Apr. at 00:00.

Boundary conditions correspond to the histories of surface temperatures. As to obtain them, every set of four temperature values is best-fitted using a quadratic polynomial which is spatially extrapolated to the top and bottom surface (Schmid et al. 2022). The surface temperatures follow as:

$$\begin{aligned} T_{top}(t) = {} & +2.3701\; T_s(z_1,t) - 0.6527\; T_s(z_2,t) \\ & -1.5309\; T_s(z_3,t) + 0.8135\; T_s(z_4,t)\,, \end{aligned} \qquad (2)$$

$$\begin{aligned} T_{bot}(t) = {} & +1.1223\; T_s(z_1,t) - 1.5082\; T_s(z_2,t) \\ & -1.3261\; T_s(z_3,t) + 2.7119\; T_s(z_4,t)\,, \end{aligned} \qquad (3)$$

and are exemplarily illustrated for 10 Apr., see Figure 2. The temperature histories $T^{top}(t)$ and $T^{bot}(t)$

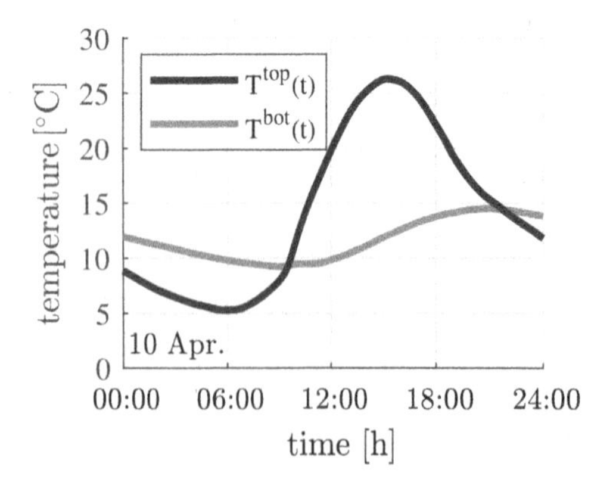

Figure 2. Surface temperatures obtained for 10 Apr.

are approximated by a superposition of temperature steps ΔT_i^{loc} with $loc = \{top, bot\}$, using the Heaviside function $H(t - t_i)$:

$$T(z^{loc}, t) = T_{ini}^{loc} + \sum_{i=1}^{N_i} \Delta T_i^{loc}\, H(t - t_i)\,, \qquad (4)$$

where N_i denotes the number of considered temperature steps. They are prescribed every three minutes.

The semi-analytical solution of the unsteady heat conduction problem is the temperature solution and reads as:

$$\begin{aligned} T(z,t) = {} & \frac{T^{bot}(t) + T^{top}(t)}{2} + \left[T^{bot}(t) - T^{top}(t)\right]\frac{z}{h} \\ & - \sum_{i=1}^{N_i}\left(\Delta T_i^{top} - \Delta T_i^{bot}\right)\sum_{n=1}^{\infty}\frac{(-1)^n}{n\pi}\,\sin\left(\frac{2n\pi z}{h}\right) \\ & \qquad \exp\left(-(2n\pi)^2\frac{a\langle t - t_i\rangle}{h^2}\right) \\ & + \sum_{i=1}^{N_i}\left(\Delta T_i^{top} + \Delta T_i^{bot}\right)\sum_{n=1}^{\infty}\frac{2(-1)^n}{(2n-1)\pi}\,\cos\left(\frac{(2n-1)\pi z}{h}\right) \\ & \qquad \exp\left(-(2n-1)^2\pi^2\frac{a\langle t - t_i\rangle}{h^2}\right), \end{aligned} \qquad (5)$$

where the angled brackets denote the Macaulay operator. Once the surface temperatures are quantified, Eq. (5) enables the reproduction of spatially nonlinear temperature profiles for slabs subjected to a series of temperature steps at its top and bottom surface (Schmid et al. 2022). Reliable temperature results are obtained for time instants more than 12 hours after the starting time of the simulation. The infinite sums are truncated after 9 summands. The numerical value of the thermal diffusivity is taken from (Schmid et al. 2022), see Table 1.

The simulation of unsteady heat conduction starts at 00:00 on 9 Apr. This results in reliable temperature profiles from 9 Apr. 12:00 onwards. Such temperature profiles are exemplarily illustrated for representative time instants on 10 Apr. in Figure 3. For the sake of simplicity, the quality of reproduction of temperature measurements is evaluated only for 10 Apr. to 15 Apr., referring to 288 time instants times four temperature measurements in four depths. The quality of reproduction is quantified by means of the square-root of the mean of squared errors:

$$SRMSE = \sqrt{\frac{1}{4 \times 288}\sum_{s=1}^{4}\sum_{m=1}^{288}\left[T(z_s, t_m) - T_m(z_s, t_m)\right]^2}\,, \qquad (6)$$

and amounts to 0.3°C. This corresponds to the quality reached for the autumn-period analyzed in (Schmid et al. 2022).

2.3 *Decomposition of thermal eigenstrains*

The temperature solution, see Eq. (5), result in spatially nonlinearly distributed temperature profiles. They are translated into total thermal eigenstrains ε^e_{xx} ($= \varepsilon^e_{yy} = \varepsilon^e_{zz}$) by multiplying the coefficient of thermal expansion α_T by the change of temperature, measured relative to the reference temperature T_{ref}:

$$\varepsilon^e_{xx} = \alpha_T \left[T(z,t) - T_{ref} \right]. \tag{7}$$

The input values amount to $\alpha_T = 1.153 \times 10^{-5}/°C$ and $T_{ref} = 17°C$ (Wang et al. 2019). Total thermal eigenstrains are distributed nonlinearly in thickness direction, compare Eqs. (5) and (7). The decomposition rules of eigenstrains for quantifying eigenstretches $\varepsilon^e(t)$ and eigencurvatures $\kappa^e(t)$ as well as eigendistortions $\varepsilon^e_{dist}(z,t)$ follow from the structural behavior of the slab according to the Kirchhoff-Love hypothesis, for details see (Schmid et al. 2022):

$$\varepsilon^e_{xx}(z,t) = \varepsilon^e(t) + \kappa^e(t)\, z + \varepsilon^e_{dist}(z,t). \tag{8}$$

The decomposition is exemplarily shown for 11 Apr., 13:00 in Figure 4.

Thermal stresses are activated if thermal eigenstrains are constrained or prevented (Wang et al. 2019). Schmid et al. discuss in detail that predominantely eigencurvatures of the slab and eigendistortions of the generators of the slab contribute to thermal stresses:

- Eigencurvatures of the slab result in convex or concave curling. This is constrained by the pavement layers on which the slab rests, and activates linearly distributed curling stresses. They can be either quantified semi-analytically (Höller et al. 2019) or by means of a nonlinear Finite Element simulation.
- Eigendistortions of the generators of the slab are essentially prevented, because the generators of the plate remain straight according to the Kirchhoff-Love hypothesis. Thus, the eigendistortions are

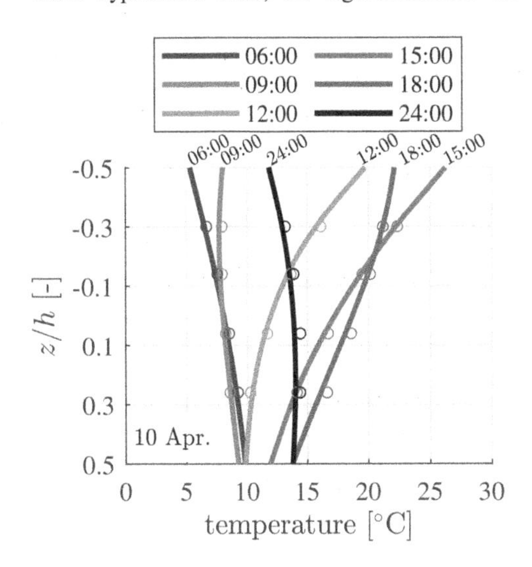

Figure 3. Temperature profiles referring to 10 Apr.: the circles label the temperature measurements, the solid lines refer to the computed distributions of temperature (= solution of the heat conduction problem).

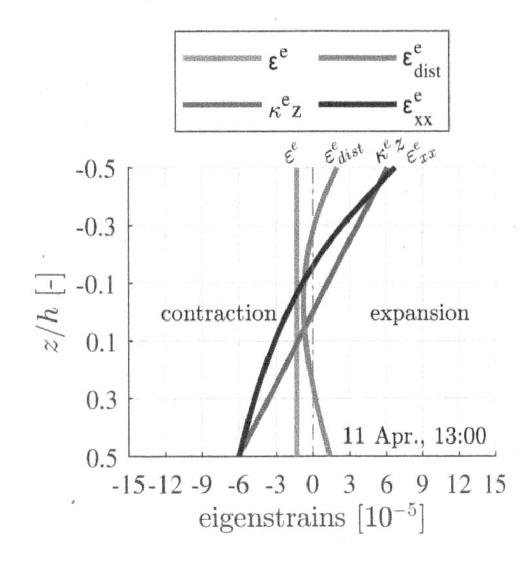

Figure 4. Decomposition of total thermal eigenstrains into a constant, linear and nonlinear part, shown exemplarily for 11 Apr., 13:00.

nulliï¬'ed by mechanical strains of the same size and distribution, but of opposite sign. Multiplying them with $E/(1-\nu)$, where E and ν denote the modulus of elasticity and the Poisson's ratio, compare Table 1, leads to thermal eigenstresses:

$$\sigma_T(\varepsilon^e_{dist}) = \frac{E}{1-\nu} \times (-\varepsilon^e_{dist}). \tag{9}$$

They are distributed nonlinearly in thickness direction.

3 THERMAL STRESSES RESULTING FROM TEMPERATURE VARIATIONS

3.1 *Curling stresses*

Curling stresses result from constrained eigencurvatures and can be computed in a Finite Element analysis according to e.g. (Schmid et al. 2022). The geometric dimensions of the simulated plate are listed in Table 1. The plate has free edges and rests on an elastic Winkler foundation (Winkler 1867), see Figure 5. The modulus of subgrade reaction is set equal to $k_s = 100$ MPa/m. The plate is subjected to eigencurvatures computed in Section 2.3, and its dead load, amounting to $p = \rho g h =$

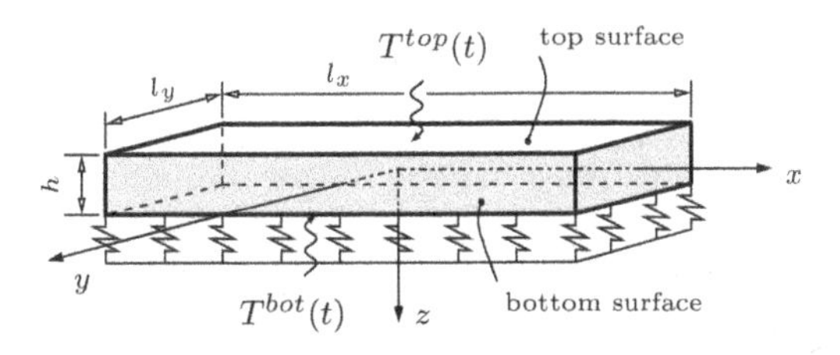

Figure 5. One-dimensional heat conduction in thickness-direction of a rectangular plate resting on a Winkler foundation.

5.89 kN/m², where ρ and h are given in Table 1 and g stands for the gravitational acceleration. The eigencurvatures vary between -6.61×10^{-4}/m and $+2.13 \times 10^{-4}$/m. The negative extreme value lies outside the interval of $[-5 \times 10^{-4}/\text{m};+5 \times 10^{-4}/\text{m}]$, for which numerical results of the nonlinear FE simulation are available (Schmid et al. 2022). To cover the total range of computed eigencurvatures, the FE simulation is performed analogously to (Schmid et al. 2022), prescribing eigencurvatures amounting to -6×10^{-4}/m and -7×10^{-4}/m, respectively. The FE software RFEM, version 5.27.01 (Dlubal Software GmbH 2020), is used, considering $110 \times 54 = 5{,}940$ quadratic, 5 cm long Finite Elements of type "Kirchhoff bending theory". The result of interest is the largest principal tensile curling stress, for numerical results see Table 2 and for computational details see (Schmid et al. 2022).

Table 2. Results from nonlinear FE analyses, partly taken over from (Schmid et al. 2022): largest principal tensile curling stress as a function of the eigencurvature κ^e, prescribing the modulus of subgrade reaction as $k_s = 100$ MPa/m.

	$k_s = 100$ MPa/m
$\kappa^e = -7 \times 10^{-4}\,\text{m}^{-1}$	1.800 MPa
$\kappa^e = -6 \times 10^{-4}\,\text{m}^{-1}$	1.739 MPa
$\kappa^e = -5 \times 10^{-4}\,\text{m}^{-1}$	1.627 MPa
$\kappa^e = -4 \times 10^{-4}\,\text{m}^{-1}$	1.404 MPa
$\kappa^e = -3 \times 10^{-4}\,\text{m}^{-1}$	1.171 MPa
$\kappa^e = -2 \times 10^{-4}\,\text{m}^{-1}$	0.801 MPa
$\kappa^e = -1 \times 10^{-4}\,\text{m}^{-1}$	0.422 MPa
$\kappa^e = +1 \times 10^{-4}\,\text{m}^{-1}$	0.140 MPa
$\kappa^e = +2 \times 10^{-4}\,\text{m}^{-1}$	0.253 MPa
$\kappa^e = +3 \times 10^{-4}\,\text{m}^{-1}$	0.333 MPa

The obtained stress values are interpolated using spline functions. Thus, the maximum tensile curling stresses can be now quantified for all values of κ^e from Section 2.3. As the curling stresses are distributed linearly in thickness direction, these values are sufficient to determine the total curling stress profile, as it is exemplarily shown for 10 Apr. in Figure 6. For the correct assignment of tensile stresses to the top or bottom surface of the slab, two types of structural behavior are distinguished:

- Positive eigencurvatures correspond to concave curling. The top surface of the plate is cooler than its bottom. The corners of the plate lose contact with the foundation, whereas the central region is pressed downwards into the base layer. This leads to tensile stresses at the top surface. This behavior occurs primarily during nighttime and the early morning, see the blue and black curves in Figure 6.
- Negative eigencurvatures correspond to convex curling. The top surface of the plate is warmer than its bottom. The corners of the plate are pressed into the foundation, whereas the central region loses contact with the base layer. This leads to tensile stresses at the bottom surface. This behavior occurs primarily during daytime, see the orange, red, and green curve in Figure 6.

For the analyzed monitoring period, the maximum values of the tensile curling stresses amount to 0.27 MPa and 1.77 MPa at the top and the bottom of the slab, respectively.

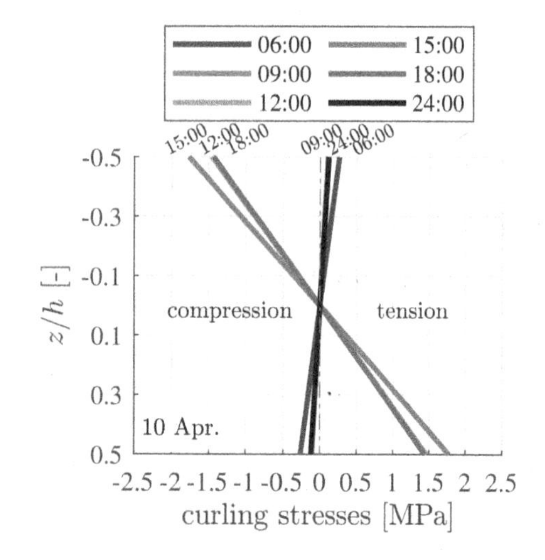

Figure 6. Exemplary results of nonlinear FE analyses: thermal stresses resulting from constrained eigencurvature of the slab; the results refer to 10 Apr.

3.2 *Thermal eigenstresses*

Thermal eigenstresses are quantified according to Eq. (9). They are nonlinear functions of the vertical coordinate z, but independent of the modulus of subgrade reaction. For specific values of z and arbitrary values of $|x| \leq \ell_x/2$ and $|y| \leq \ell_y/2$, eigenstresses are constant throughout the entire observed plane of the slab. The obtained stress distributions have a vanishing mean value and a vanishing first moment, see Figure 7. Heating of the slab results in compressive stresses at the top and bottom surface, and tensile stresses in the region around the midplane, see the magenta, orange and red curves in Figure 7. The maximum tensile stresses are reached around noon. Vice versa, cooling of the plate yields tensile stresses at the surfaces, and compressive stresses around the midplane, see the blue, green and black curve in Figure 7. For the analyzed monitoring period, the maximum values of the tensile eigenstresses amounted to 0.64 MPa and 0.41 MPa at the top and the bottom of the slab, respectively. In the region around the midplane, the maximum value was 0.43 MPa.

3.3 *Total thermal stresses*

The summation of the thermal eigenstresses and curling stresses result in total thermal stresses. They are distributed nonlinearly in thickness direction of the slab, see Figure 8. This underlines that eigenstresses

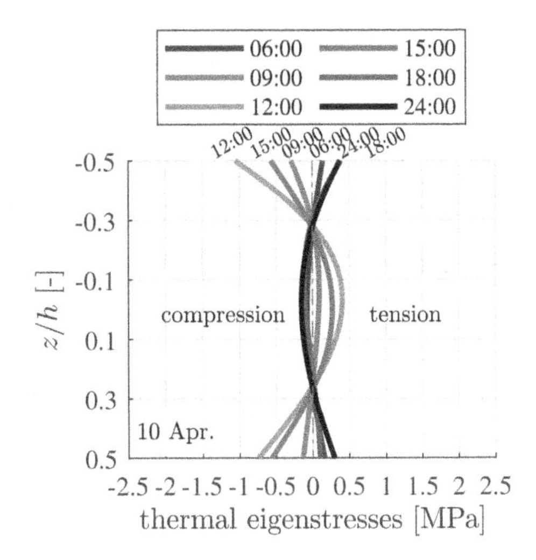

Figure 7. Exemplary results of engineering thermo-mechanical analysis: thermal stresses resulting from virtually prevented eigendistortions of the generators of the slab; the results refer to 10 Apr.

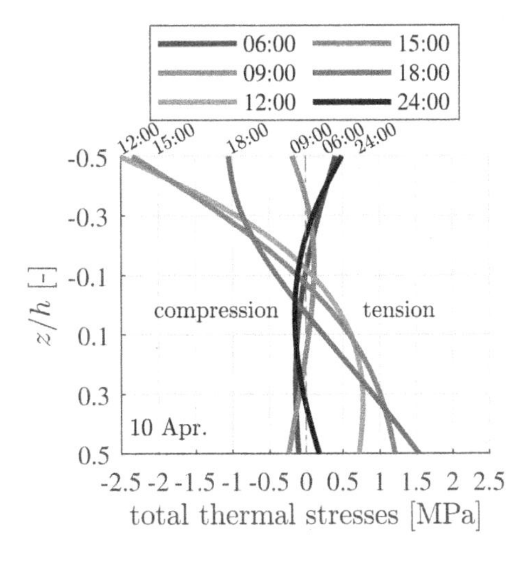

Figure 8. Exemplary results of engineering thermo-mechanical analysis: total thermal stresses; the results refer to 10 Apr.

have a significant impact on the total thermal stresses (Choubane & Tia 1992; Schmid et al. 2022). The design of concrete pavements focuses on primarily tensile thermal stresses. Thus, they are here discussed in more detail with respect to six specific time instants on 10 Apr. by comparing Figs. 6 and 7 with Figure 8, and with respect to the quasi-continuous temporal stress evolution illustrated in Figure 9 for 10 Apr.

At the *midplane*, non-zero eigenstresses are superimposed with permanently vanishing curling stresses. Thus, the total thermal stresses are equal to the thermal eigenstresses. The largest tensile stresses occur around noon and amounts to +0.43 MPa, see the red diamond at the mid-gray curve in Figure 9.

At the *top surface* of the slab, the largest tensile stresses occur during nighttime in corner regions, see e.g. the black stress profile in Figure 8 and the red diamond on the dark-gray solid line in Figure 9. At that time, the curling stresses and eigenstresses are both tensile and amplify each other, see the dark-gray dashed and dotted curves in Figure 9 around 01:30. During the entire monitoring period, the largest tensile thermal stress amounted to +0.62 MPa. The largest compressive stresses occurred around noon in the central region, see the dark-gray solid line in Figure 9. Curling stresses and eigenstresses again amplify each other. The compressive total stresses were smaller than or equal to −2.78 MPa during the entire observation period.

At the *bottom surface* of the slab, the largest tensile stresses occur in the early evening in the central region, see the green curves of Figure 8 and the red diamond on the bright-gray solid line in Figure 9. At the same time, curling stresses are tensile, but eigenstresses are compressive, compare the dashed and dotted bright-gray curves in Figure 9 around 18:00. Thus, the curling stresses and eigenstresses diminish each other, see also the orange and red curves in Figure 8. During the entire monitoring period, the largest tensile thermal stress amounted to +1.59 MPa. The largest compressive stresses occurred in the morning, see the blue and magenta curves of Figure 8 and the bright-gray solid line in Figure 9. The curling stresses vanish for this time instant and the total thermal stresses are exceptionally equal to the eigenstresses. For the entire observation period, the total compressive stresses were always smaller than or equal to −0.31 MPa.

The maximum value of tensile curling stresses occur with a time lag to the maximum value of tensile eigenstresses, compare the red stars and circles, respectively, at the dashed and dotted curves of Figure 9, labeling the maximum tensile stresses of curling stresses and eigenstresses, respectively. As a consequence, the sum of the largest curling stresses and eigenstresses is larger than the largest total thermal stress. The latter occurs between the extreme values of curling stresses and eigenstresses, compare the red symbols in Figure 9, referring to three curves of the same color.

For the quantification of the significance of eigenstresses and curling stresses on total thermal stresses, the daily maxima of tensile total stresses and tensile curling stresses are computed for the top and bottom surface, respectively, for day 2 onwards to the end of the monitoring period, see Table 3.

At the top surface, the mean value of the daily maximum of total thermal stresses amounts to 0.54 MPa, whereas the mean value of the daily maximum of curling stresses reads as 0.22 MPa. Disregarding the (nonlinear) eigenstresses would underestimate the largest tensile stresses in the slab by 59 %. At the bottom of the slab, the mean value of the daily maximum of total thermal stresses amounts to 1.51 MPa, whereas the

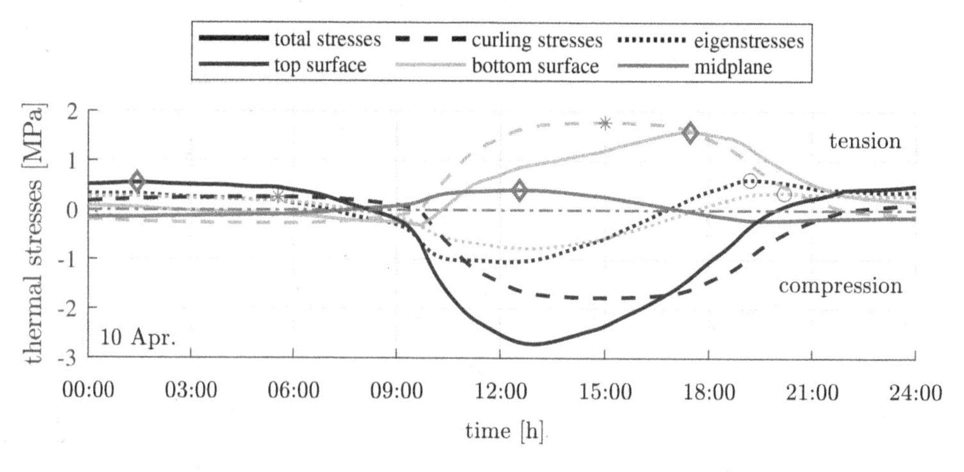

Figure 9. Exemplary results of engineering thermo- mechanical analysis: temporal evolution of thermal stresses on 10 Apr. The red markers label the maximum values of tensile stresses at the top, the midplane, and the bottom of the slab, where the diamond refers to total stresses, the stars to curling stresses and the circles to eigenstresses.

Table 3. Mean values of daily tensile stress maxima for curling stresses and total thermal stresses at the top and the bottom of the slab for the time period of 10 Apr. to 15 Apr., and quantification of misestimations.

$k_s = 100$ [MPa/m]		top surface	bottom surface
total thermal stresses	[MPa]	0.54	1.51
curling stresses	[MPa]	0.22	1.72
misestimations	[-]	−59%	+14%

mean value of the daily maximum of curling stresses reads as 1.72 MPa. Disregarding the (nonlinear) eigenstresses would here overestimate the largest tensile stresses in the slab by 14%.

4 CONCLUSIONS

Temperature measurements from a 7-days monitoring period were analyzed based on the evaluation method proposed in (Schmid et al. 2022). The analysis has led to the following conclusions:

- The value of the thermal diffusivity quantified in (Schmid et al. 2022) for 2 Oct. also holds for the same slab in spring.
- Curling stresses, resulting from constrained eigencurvatures of the plate, were computed by means of nonlinear Finite Element analyses. The modulus of subgrade reaction of the simulated Winkler foundation was set equal to 100 MPa/m. The obtained results extend the curling stress data provided in (Schmid et al. 2022).
- Thermal eigenstresses follow simply from the spatially nonlinear eigendistortions of the plate-generators, which are virtually prevented, because the generators remain straight according to the Kirchhoff-Love hypothesis.
- The slab exhibits concave curling during night-time and the early morning. Curling stresses and eigenstresses are both tensile at the top surface and amplify each other. At the bottom surface, compressive curling stresses are partly balanced by tensile eigenstresses.
- The slab exhibits convex curling during daytime and the early evening. Curling stresses and eigenstresses are both compressive at the top surface and amplify each other. At the bottom surface, tensile curling stresses are partly balanced by compressive eigenstresses.
- The temporal evolution and the magnitude of the curling stresses, the eigenstresses, and the total thermal stresses are similar to those obtained in (Schmid et al. 2022).
- Disregarding the nonlinear eigenstresses underestimates tensile stresses at the top surface of the slab during the early morning in the corner regions by 59%. This misestimation is by 14 percentage points larger than the one quantified for the observation period in autumn.
- Disregarding the nonlinear eigenstresses overestimates tensile stresses at the bottom surface of the slab during the early evening in the central region by 14%. This misestimation is by 15 percentage points smaller than the one quantified for the observation period in autumn.

REFERENCES

Barber, E. (1957). Calculation of maximum pavement temperatures from weather reports. *Highway Research Board Bulletin 168*, 1–8.

Bayraktarova, K., L. Eberhardsteiner, D. Zhou, & R. Blab (2021). Characterisation of the climatic temperature variations in the design of rigid pavements. *International Journal of Pavement Engineering* (https://doi.org/10.1080/10298436.2021.1887486), 1–14.

Bentz, D. P. (2000). A computer model to predict the surface temperature and time-of-wetness of concrete pavements and bridge decks. *National Institute of Standards and Technology – Technology Administration, U.S. Department of Commerce 6551*, 1–19.

Choubane, B. & M. Tia (1992). Nonlinear temperature gradient effect on maximum warping stresses in rigid pavements. *Transportation Research Record 1370*, 11–19.

Choubane, B. & M. Tia (1995). Analysis and verification of thermal-gradient effects on concrete pavement. *Journal of Transportation Engineering 121*(1), 75–81.

Dlubal Software GmbH (2020). RFEM – FEM structural analysis software.

Höller, R., M. Aminbaghai, L. Eberhardsteiner, J. Eberhardsteiner, R. Blab, B. Pichler, & C. Hellmich (2019). Rigorous amendment of Vlasov's theory for thin elastic plates on elastic Winkler foundations, based on the Principle of Virtual Power. *European Journal of Mechanics–A/Solids 73*, 449–482.

Ioannides, A. M. & L. Khazanovich (1998). Nonlinear temperature effects on multilayered concrete pavements. *Journal of Transportation Engineering 124*, 128–136.

Janssen, D. J. & M. B. Snyder (2000). Temperature-moment concept for evaluating pavement temperature data. *Journal of Infrastructure Systems 6*, 81–83.

Mohamed, A. R. & W. Hansen (1997). Effect of nonlinear temperature gradient on curling stress in concrete pavements. *Transportation Research Record 1568*, 65–71.

Schmid, S. J., R. Díaz Flores, M. Aminbaghai, L. Eberhardsteiner, H. Wang, R. Blab, & B. L. A. Pichler (2022). Signiï¬'cance of eigenstresses and curling stresses for total thermal stresses in a concrete slab, as a function of subgrade stiffness. *Submitted in January 2022 to the International Journal of Pavement Engineering*, 1–28.

Teller, L. & C. Sutherland (1935). The structural design of concrete pavements; parts 1+2. *Division of tests, Bureau of Public Roads 16(8-9)*, 145–189.

Thomlinson, J. (1940). Temperature variations and consequent stresses produced by daily and seasonal temperature cycles in concrete slabs. *Concrete Constructional Engineering 36(6)*, 298–307.

Wang, H., R. Höller, M. Aminbaghai, C. Hellmich, Y. Yuan, H. Mang, & B. L. Pichler (2019). Concrete pavements subjected to hail showers: A semi-analytical thermoelastic multiscale analysis. *Engineering Structures 200 (109677)*, 1–19.

Westergaard, H. (1927). Analysis of stresses in concrete pavements due to variations of temperature. In *Highway Research Board Proceedings*, Volume 6, pp. 201–215.

Winkler, E. (1867). *Die Lehre von der ElastizitÃ¤t und Festigkeit: Mit besonderer Rücksicht auf ihre Anwendung in der Technik [Lessons on elasticity and strength of materials: with special consideration of their application in technology]*. Dominicus.

Computational Modelling of Concrete and Concrete Structures – Meschke, Pichler & Rots (Eds)
© 2022 Copyright the Author(s), ISBN: 978-1-032-32724-2

Modeling of capillary fluid flow in concrete using a DEM-CFD approach

M. Krzaczek, M. Nitka & J. Tejchman
Gdansk University of Technology, Gdańsk, Poland

ABSTRACT: A novel coupled approach to modelling hydraulic- and capillary-driven two-phase water flow in unsaturated concrete was formulated. The flow process was numerically analyzed at meso-scale in two-dimensional (2D) conditions by combining the discrete element method (DEM) with computational fluid dynamics (CFD) under isothermal conditions. Fully coupled hydro-mechanical simulation tests were carried out on small concrete specimens of a simplified particulate meso-structure. The pure DEM represented by bonded spheres was calibrated with the aid of uniaxial compression while the pure CFD was calibrated with the aid of a permeability and sorptivity test for a sphere assembly. DEM/CFD calculations were successively performed for specimens of the pure cement matrix, cement matrix including aggregate and cement matrix including aggregate and interfacial transition zone (ITZ) of a defined thickness. The major goal of investigations was to show the effect of ITZs on fluid flow in unsaturated concrete driven by hydraulic/capillary pressure.

1 INTRODUCTION

Concrete is a strongly heterogeneous, discontinuous and porous composite material with four crucial phases at the meso-scale: aggregates, cement matrix, interfacial transition zones (ITZs) between aggregates and cement matrix and macro-pores. ITZs are adjacent to aggregates and indicate pronounced compositional differences against the cement matrix (Bentz 1992). They possess e.g. more and larger pores, smaller particles and less anhydrous cement and C-S-H (calcium silicate hydrate) gel which causes higher transport properties (i.e. permeability, diffusivity and conductivity) than the cement matrix (Delagrave et al. 1997; Schwartz et al. 1995; Stroeven et al. 2017a,b). They facilitate penetration of external aggressive agents into concrete that deteriorates both concrete and reinforcement. They promote also humidity transport through concrete (Stroeven et al. 2017a). If the moisture content inside concrete is less than its saturation level, water may be absorbed into the concrete by capillary forces arising from a contact of very small pores with a liquid phase. This is an important mechanism of water flow into concrete that is often observed in field applications subjected to wetting and drying cycles.

Water acts as a transport means for aggressive agents (e.g. chlorides, sulfates, carbon dioxides). Therefore, the durability of concrete structures largely depends on their resistance to water transport and dissolved aggressive species. Penetration, diffusion and absorption are three main transport mechanisms in concrete materials (Ababneh et al. 2003; Šavija et al. 2013). The spatial distribution of water in concrete depends on applied external loading (Wyrzykowski & Lura 2014) that deforms the concrete structure by re-sizing pores and capillaries, promotes a crack generation and development that increases a water penetration rate. For investigating fluid problems in concrete, it is important to take porous ITZs around aggregates into account since they usually allow for the faster penetration of external aggressive agents (Stroeven et al. 2017a).

The current paper demonstrates a novel mathematical mesoscopic approach to modelling viscous and capillary-driven two-phase fluid flow in unsaturated uncracked concrete at the meso-scale under isothermal conditions. In the model, aggregates and ITZs were explicitly taken into account. Since an understanding of the pore-scale behaviour is essential to a successful interpretation and prediction of the macroscopic behaviour, DEM was applied to capture the mechanical behaviour of concrete and CFD was used to describe the laminar viscous two-phase liquid/gas flow in pores between the discrete elements by employing fluid flow network.

The main goal of our simulations was to show the effect of ITZs on fluid flow in unsaturated concretes driven by both hydraulic and capillary pressure that cannot be easily evaluated experimentally. Fully coupled hydro-mechanical simulation tests were carried out on unsaturated specimens of a simplified meso-structure imitating the pure cement matrix, cement matrix including aggregate and cement matrix including aggregate with ITZ of a defined thickness.

The physics of capillary-driven fluid flow at the mesoscale is complex (Chatzis & Dullien 1983; Mason & Morrow 1991; Tsakiroglou et al. 2007). In general, the capillary pressure constitutes a difference between partial pressures of liquid (wetting) and gas phases.

DOI 10.1201/9781003316404-67

Consequently, capillary pressure exists wherever a liquid phase and a gas phase coexist. Three types of pore-level physics exist that are relevant to dynamic pore-scale modelling: 1) piston-like displacement, 2) pore cooperative filling and 3) snap-off. For a piston-like displacement, one phase completely displaces the other from the pore space. The capillary pressure is defined by a local geometry and radius of curvature of the interface. For the pore co-operative filling, the displacement of one fluid by another depends not only on a local pore geometry but also on whether there exist other adjacent pinned interfaces (Hughes & Blunt 2000). For snap-off, the interface movement occurs in the form of films through corners and rough surfaces. The wetting phase maintains hydraulic continuity, leading to pore filling ahead of the connected front (Rossen 2003).

Capillary-driven fluid flow in porous materials at the meso-scale was studied using a variety of approaches and models. In general, three types of models to determine the liquid conductivity function can be distinguished (empirical, tube and network models). The empirical models are the most simplified ones. They provide a set of analytical functions to be adjusted to measured data (Galbraith 1992; Pedersen 1990). However, due to their limited flexibility and adjustability as well as the implied restrictions to a specific moisture range such models are not applicable for an entire moisture range description. By contrast, network models are the most extensive ones (Carmeliet & Roels 2001; Descamps 1997; Dullien 1979; Fatt 1956; Roels et al. 2003; Scheffler & Plagge 2010; Xu et al. 1997). Their basic idea is to approximate the pore structure by a lattice of tubes and bonds at the meso- or micro-scale. The tubes are randomly distributed. The tube radii follow the measured pore structure data of the material. Simulating the penetration of the network by fluid, storage and transport functions based on the measured pore volume distribution data can be derived. The main advantage of the network approaches is the ability to approximate the pore structure and to account for structural effects by percolation rules. An extensive review of the existing capillary-driven fluid flow models based on the pore-network approach is presented by Sheng and Thompson 2016. However, they did not couple DEM with CFD.

The innovative element of the numerical mesoscopic coupled DEM/CFD approach developed by the authors for modelling hydraulic- and capillary-driven fluid flow in unsaturated concrete at the meso-scale subjected to the external load as compared to other existing models in the literature is the detailed tracking of water/gas fractions in pores regarding their varying geometry, size and location. There are no available in the literature coupled hydro-mechanical DEM/CFD models of multi-phase fluid flow that might be employed to simulate capillary-driven multiphase flow in unsaturated concrete subjected to external loads. The ability of the approach to faithfully reproduce the material meso-structure (based e.g. on micro-computed tomography (Skarżyński et al. 2015; Zeng et al. 2021)) allows for realistic studying fluid flow in ITZs. The approach enables also to study of capillary-driven multiphase flow in unsaturated concrete during initiation and propagation of cracks.

The arrangement of the current paper is as follows. After the introductory Section 1, the discrete element method (DEM) is summarized in Section 2. Section 3 describes a coupled DEM/CFD approach for two-phase flow. A discussion on the model calibration is in Section 4. Section 5 includes some numerical simulation results on capillary-driven water flow in unsaturated cement matrix/concrete specimens. Finally, some concluding remarks are offered in Section 6.

2 DEM FOR COHESIVE-FRICTIONAL MATERIALS

DEM calculations were performed with the 3D spherical explicit discrete element open code YADE (Kozicki & Donzé 2011; Šmilauer & Chareyre 2011). The method allows for a small overlap between two contacted bodies (the so-called soft-particle model). Thus, an arbitrary micro-porosity can be obtained in DEM. In DEM, particles interact with each other during translational and rotational motions through a contact law and Newton's 2nd law of motion using an explicit time-stepping scheme (Cundall & Strack 1979). In the model, a cohesive bond is assumed at the grain contact exhibiting brittle failure under the critical normal tensile load. The shear cohesion failure initiates contact slip and sliding obeying the Coulomb friction law under normal compression. Damage occurs if a cohesive joint between spheres disappears after reaching a critical threshold. If any contact between spheres after failure re-appears, the cohesion does not appear more. A simple local non-viscous damping is used (Cundall & Strack 1979) to accelerate convergence in quasi-static analyses. Note that material softening is not considered in the DEM model.

In general, the material constants are identified in DEM with the aid of simple laboratory tests on the material (uniaxial compression, uniaxial tension, shear and biaxial compression). The detailed calibration procedure for frictional-cohesive materials was described by Nitka and Tejchman (2015) and Suchorzewski et al. (2018).

The DEM model demonstrated its usefulness for both local and global simulations of macro- and micro-cracks in concretes under bending (2D and 3D analyses) (Kozicki et al. 2014; Nitka & Tejchman 2015) uniaxial compression (2D and 3D simulations) (Caggiano et al. 2018) and splitting tension (2D analyses) (Kozicki et al. 2014; Skarżyński et al. 2015). The combined DEM/x-ray μ-CT images mesoscopic approach proved to be an extremely appealing computational tool for investigating fracture in concrete. In those calculations, ITZs had no either physical width and were simulated by weaker contacts between aggregates and cement matrix (Nitka & Tejchman 2015; Skarżyński et al. 2015) or had a defined width with higher porosity (Nitka & Tejchman 2020). To study coupled mechanical-hydro-thermal problems in concrete, the second approach should be used. The coupled problems must, namely, take porous ITZs into account since their porosity favours heat and fluid transport (Stroeven et al. 2017a).

3 LAMINAR TWO-PHASE FLUID FLOW MODEL

The 2D laminar two-phase fluid flow model was described in detail by Krzaczek et al. (2021) and Abdi et al. (2021). Here, the most important information is provided for the sake of clarity. The original system consists of two coexisting domains: 3D discrete domain (one layer of spheres) and 2D fluid domain. The gravity centers of spheres are located on the XOY specimen mid-plane. To create a two-dimensional fluid domain, spherical 3D particles are projected onto the mid-plane of the specimen. Consequently, a remeshing procedure interprets the set of 3D spheres as a set of 2D disks (circles). The remeshing procedure discretizes the overlapping circles (projected spheres), determines the contact segments and deletes the overlapping areas (Krzaczek et al. 2020). The algorithm of discretization is based on the Alfa Shapes theory (Bernardini & Bajaj 1997). The displacements of spheres in the perpendicular direction OZ and the rotations around the axes OX and OY are fixed. The remeshing procedure results in a triangular mesh representing the fluid domain. The gravity centers of the mesh triangles (VP) between the discrete elements are connected by channels composed of two parallel plates that form a virtual network of pores (VPN) to accurately reproduce their changing geometry (shape, surface and position) (Figure 1). The isolated pores in 2D are not isolated in 3D. The virtual (artificial) channels are introduced in the 2D fluid flow network to reproduce real flow in 3D. They are located between spheres in contact and connect the isolated pores.

Two types of channels are introduced (Krzaczek et al. 2020, 2021) (Figure 1): 1) the channels between spheres in contact (called here the virtual 'S2S' channels) and 2) the channels connecting grid triangles in pores (called here the 'T2T' channels). The channel length is equal to the distance between the gravity centers of adjacent grid triangles. The hydraulic aperture h (height) of virtual channels 'S2S' is related to the normal stress by a modified empirical formula of (Hökmark et al. 2010) and calculated as

$$h=\beta(h_{inf}+(h_0--h_{inf})e^{-1.5\cdot10^{-7}\sigma_n}) \qquad (1)$$

where – the hydraulic aperture for the infinite normal stress, – the hydraulic aperture for the zero normal stress, – the effective normal stress at the particle contact and β – the aperture coefficient.

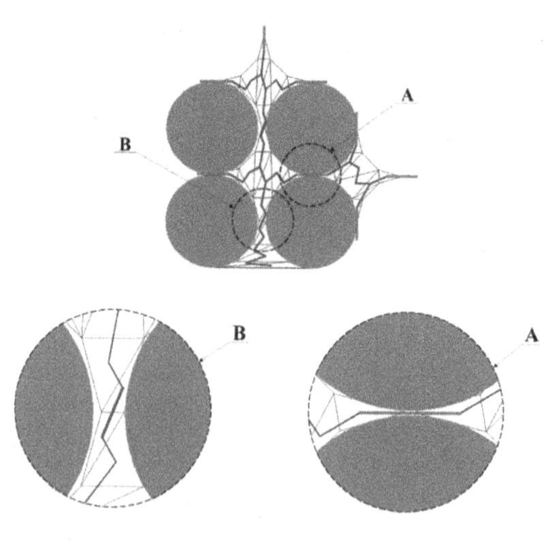

Figure 1. Fluid flow network in non-homogeneous granular specimen with triangular discretization of pores: types of channels (A) virtual channel type 'S2S' (red colour) and B) channel type 'T2T' (blue colour)).

The hydraulic aperture of the channel type 'T2T' is directly related to the geometry of the adjacent triangles as

$$h=\gamma e\cos(90^\circ-\omega) \qquad (2)$$

where e – the edge length between two adjacent triangles, ω – the angle between the edge with the length e and the centerline of the channel that connects two adjacent triangles and γ – the reduction factor, necessary to fit the fluid flow intensity to real complex fluid flow conditions in concrete. The reduction factor γ is determined to keep the maximum Reynolds number Re below 2100.

There is no fluid flow in triangles by assumption. VPs accumulate pressure and store fluid phase fractions and densities. The mass change in VPs is related to the density change in a fluid phase that results in pressure variations. The equation of momentum conservation is, thus, neglected in triangles but the mass is still conserved in the entire volume of triangles. The equations of state and continuity are employed to compute the density of fluid phases stored in VPs. The fluid phase fractions in VPs are calculated by applying the continuity equation for each phase assuming that fluid phases share the same pressure. The fluid flow in channels is estimated by solving continuity and momentum equations for the laminar flow of the incompressible fluid. As a result of discretization, each pore is discretized into several triangles (VP). The channels connect the gravity centers of triangles and create a fluid flow network (VPN). The liquid and gas may initially exist in virtual pores.

In capillary-driven water flow calculations, piston displacement, snap-off physics and viscous flow are taken as the primary fluid flow mechanisms. Due to the discretization of a single pore, the cooperative pore filling effect is indirectly taken into account. The liquid phase is assumed to be a wetting (invading) fluid and the gas phase is a non-wetting (defending) fluid. Three flow regimes are distinguished: a) single-phase flow of gas with α_p=1 (α_p is the volume fraction of the gas-phase), b) single-phase flow of liquid with α_q=1 (α_q is the volume fraction of the liquid phase) and c) two-phase flow (liquid and gas) with $0<\alpha_q<1$. Capillary pressure is solely considered in flow regime (b) when VP completely pre-filled with a liquid (wetting) phase is adjacent to VP partially filled with a liquid-phase (or only a gas-phase). In this flow regime, the fluid flows in channels through a thin film region separated by two closely spaced parallel plates according to a classical lubrication theory (Reynolds 1883), based on the Poiseuille flow law (Batchelor 2000).

To link viscous forces and capillary forces, the Washburn (Washburn 1921) equation is combined with the Poiseuille equation. Hence, the mass flow rate of the capillary-driven flow along channels for flow regime (b) (single-phase flow) in channels (capillaries) is

$$M_x = \rho \frac{h^3}{12\mu} \frac{P_i - P_j - P_c}{L}, \tag{3}$$

where M_x = the mass fluid flow rate (per unit length) across the film thickness in the x-direction [kg/(m s)], h = the hydraulic channel aperture (its perpendicular width) [m], ρ = the fluid density [kg/m3], t = the time [s], μ = the dynamic fluid (liquid or gas) viscosity [Pa s] and P = the fluid pressure [Pa], P_i and P_j are the pressure in adjacent VPs, P_c is the capillary pressure and L is the length of the channel connecting VPs. The capillary pressure P_c due to the interface between the two phases (a meniscus) is given by Young-Laplace law

$$P_c = \frac{2\sigma \cos\Theta}{r_t} \tag{4}$$

where σ is the interfacial tension [N/m], Θ is contact angle [deg] and r_t is the throat radius equal to half the channel aperture [m]. However, the capillary sorption and water/gas permeability of cement-based materials are physically determined by their current pore structure and current water distribution which change during a pore filling process with water. Hence, the capillary pressure in individual capillaries or pores varies with time and depends on the local current concentration of water. The problem was investigated in (Aker et al. 2000) for the drainage-dominated fluid displacement. They proposed the dependence of P_c on the position of the meniscus in the tube (capillary). This concept is adopted to relate the capillary pressure in the channel to the liquid phase fraction in VP. In the piston displacement scenario, VP_i is filled in with water while the adjacent pore VP_j is partially filled in with water. VPs are connected by the 'S2S' or 'T2T' channel that can perform a capillary function depending on their dimensions (height) only. It is assumed that the capillary pressure is greater than zero in the channels 1×10^{-9} m to 1×10^{-6} in height. Moreover, it is assumed that the relative position of the meniscus in the channel is equal to the liquid phase fraction α_q in the partially filled VP. The corrected capillary pressure is computed as

$$\tilde{P}_c = e_{cor} P_c, \tag{5}$$

where e_{cor} is the correction coefficient

$$e_{cor} = \frac{1}{2}(1 - \cos(2\pi\alpha_q))(1 - k) + k \tag{6}$$

with α_q as the liquid phase fraction in the partially filled VP and as is the calibration parameter. It can be noted that the correction coefficient e_{cor} in Eq. 5 varies in the range $(k, 1)$ and reaches the maximum value for α_q=0.5.

The maximum corrected capillary pressure is equal to the capillary pressure in Eq. 4. For the snap-off filling, the movement of the interface occurs in the form of films through corners of the pore space (the corners are formed by edges of triangles that make up the throat). The wetting phase maintains hydraulic continuity, leading to the pore filling ahead of the wetting front. Adopting a concept of Hughes and Blunt (2000), a criterion for the snap-off threshold capillary pressure is implemented:

$$P_{cr} = \frac{\sigma}{r_t}(\cos\Theta - \sin) \tag{7}$$

where ω is the half-angle of the pore corner. If the hydraulic pressure difference ΔP_{ij} in adjacent virtual pores is greater than P_{cr} and both virtual pores are not fully saturated, the liquid phase fraction $\alpha_{q,i}$ is set to 1.0 (full saturation). When the adjacent virtual pores are partially filled with the liquid phase, a two-phase flow (flow regime (c)) of two immiscible and incompressible fluids in a horizontal channel is assumed to simulate two-phase isothermal fluid flow, driven by a pressure gradient in adjacent VPs. The liquid/gas-phase surface is assumed to be horizontal and constant along the channel. A detailed description of the two-phase fluid flow was presented in (Krzaczek et al. 2021). The fluid flow model is coupled with DEM and is implemented into the open-source code YADE (Kozicki & Donzé 2011).

When the external load is applied, the material structure becomes significantly deformed. This can result in significant variations in the pore-capillary system topology. The topology of the fluid flow network may change as well as the dimensions of pores and capillaries. As a consequence, the pressure and fluid phases fractions in pores are affected. The numerical algorithm is divided into 4 main stages (Krzaczek et al. 2020, 2021):

a) computing the mass flow rate for fluid flowing in channels surrounding VPi,
b) calculating liquid and gas phase fractions in VPi by employing the continuity equations,
c) computing density of phases in VPi by employing the equation of state,

d) calculating pressure in VPi by employing the equation of state.

This algorithm is repeated for each VP in VPN using an explicit formulation. The discretization algorithm is based on the alpha-shape theory and the Delaunay triangulation. The grid remeshing is automatically performed when the topological properties of the grid geometry change (Krzaczek et al. 2020). The computational results (e.g. pressures) are accurately transformed from the old grid to the new one by assuming that the mass is a topological invariant. The coupling scheme of DEM with CFD is described in detail in (Krzaczek et al. 2020).

The results of high-pressure fluid flow in densely packed granules with the coupled DEM/CFD approach were successfully compared with the full 3D CFD model (Abdi et al. 2021).

4 MODEL CALIBRATION

The pure DEM represented by spheres was calibrated with the aid of a simple uniaxial compression test while the pure CFD was calibrated with the aid of a permeability and sorptivity test for an assembly of spheres.

In the first step, an extremely simple 2D DEM mesoscopic structure was chosen to imitate the cement matrix/ concrete in contrast to our previous detailed 3D simulations (Nitka & Tejchman 2018). Spheres as discrete elements were chosen only. A small bonded granular specimen of 10×10 mm^2 included 400 spheres. One layer of spheres was applied along with the specimen depth. The sphere diameter was in the range 0.25-0.75 mm (with the mean sphere diameter of d_{50}=0.5 mm). The macro-pores were neglected for the sake of simplicity. The same simplified meso-structure of concrete was assumed in simulations (Section 5). The coupled DEM/CFD calculations were carried out on 3 different bonded specimens: pure cement matrix specimen (the so-called 1-phase concrete material), cement matrix specimen with one aggregate (the so-called 2-phase concrete material) and cement matrix specimen with one aggregate and ITZ around (the so-called 3-phase concrete material) (Figure 2).

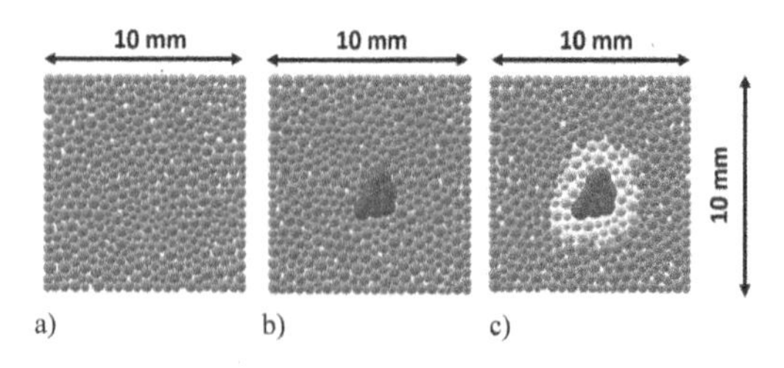

Figure 2. Numerical specimens in DEM/CFD calculations: a) pure cement matrix (grey colour) with initial porosity p=5%, b) cement matrix including aggregate (dark grey colour) with initial porosity p=5% and c) cement matrix including aggregate with initial porosity p=5% and ITZ layer around it (light grey colour) with initial porosity p=20%.

The micro-pores corresponded to the free space between spheres. A non-spherical aggregate was modelled as a rigid non-breakable cluster composed of 12 spheres. ITZ was assumed in the form of 3 layers of spheres of d=0.25 mm around the aggregate with the initial porosity of p=20% (the cement matrix had the initial porosity of p=5%).

Both the initial porosities of the cement matrix (5%) and ITZ (20%) were assumed from experiments (Nitka & Tejchman 2020). The width of ITZ was about t_{ITZ}=0.75 mm, i.e. was 10 times higher than this in real concrete, measured with the scanning electron microscope (SEM) (Nitka & Tejchman 2020).

4.1 *Calibration of CFD*

The fluid flow model was calibrated on an already calibrated DEM specimen. Calibration constituted a two-step procedure: 1) permeability tests (Darcy test) and 2) sorptivity test.

Permeability test

A simple permeability Darcy test with two cement matrix specimens of Figure 2a was performed for calibration purposes of the CFD model (Figure 3). Single-phase flow was assumed. Two 2D DEM specimens with different initial porosity were prepared. Their size was again 10×10 mm^2 (Figure 3).

The first specimen '1' simulated the cement matrix and had the initial porosity of p=5% and the second specimen '2' simulated ITZ around aggregates and had the initial porosity of p=20%. The number of spheres and the number of contacts were equal to 470 and 1280 (specimen '1') and 370 and 830 (specimen '2'). The constant water pressure of 4 MPa was applied to the bottom edge and the constant water pressure of 1 MPa was applied to the top edge. At the left and right edges, the zero-flux conditions were imposed (Figure 3c). The following material constants were assumed. The dynamic viscosity of water was $\mu=10.02\cdot10^{-4}$ Pa·s, its compressibility $C=4.0\bullet10^{-10}$ Pa^{-1} and density ρ_0=998.321 kg/m^3 for the reference pressure P_0=0.1 MPa. The virtual S2S channel apertures in Eq.1 were equal to $h_{inf}=4.5\cdot10^{-7}$ m and $h_0=3.25\cdot10^{-6}$ m. The reduction factor in T2T channels was γ=0.012 in Eq. 2 and the aperture coefficient was β=1.0 in Eq. 1 (in virtual S2S channels).

Assuming that the volumetric flow rate at horizontal walls was the same at the equilibrium state, the macroscopic permeability coefficient κ was calculated using Darcy's law:

$$\kappa = \frac{Q}{A}\mu_q\frac{L}{\Delta P} \tag{8}$$

where Q is the volumetric flow rate at the equilibrium state [m^3/s], A is the specimen cross-section [m^2], L denotes the specimen height and ΔP is the pressure difference between the bottom and top edges [Pa].

For the further numerical tests, the permeability coefficient of κ=4e-16 m^2 was chosen for the cement

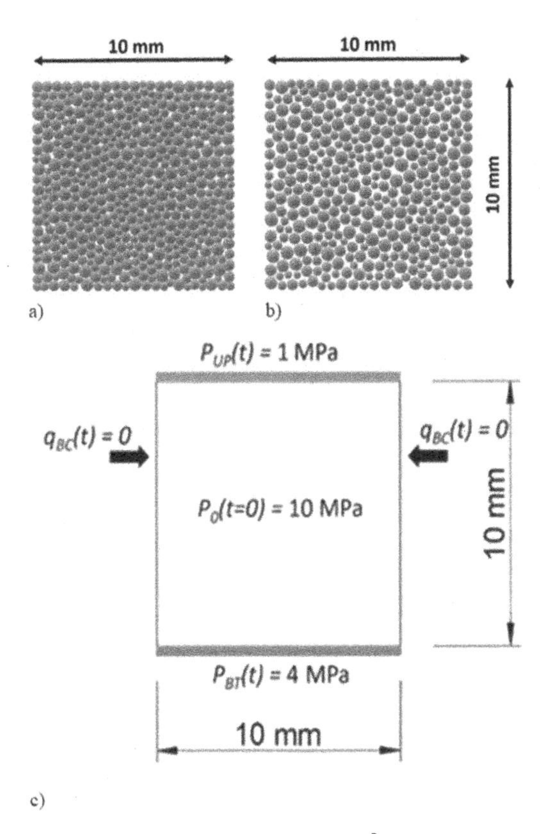

Figure 3. Cement matrix specimen in permeability test: a) and b) DEM granular specimens composed of spheres with different initial porosity p: 5% (a) and 20% (b), c) boundary and initial conditions.

matrix that agreed well with laboratory test results for ordinary cement matrices with the water/cement ratio of 0.40 and degree of hydration of 0.95 (Stroeven et al. 2017a). For ITZ with p=20%, the permeability coefficient was 10 times higher κ=4e-15 m^2 (in agreement with test results for high saturation degrees (Stroeven et al. 2017a)).

Sorptivity test

The dry specimen of porosity 5% (Figure 9a) was used. The initial fluid (gas phase) pressure was 0.1 MPa. To fill in the pores in contact with water, constant pressure of 0.14 MPa was adopted as a boundary condition at the lower specimen edge. The boundary pressure maintains a small pressure gradient on the boundary simulating a slight fluid flow, replenishing the water in the specimen. At the top edge of the specimen, constant pressure of 0.10 MPa was defined. No mass flow rate was defined along the vertical edges of the specimen. The simulations were carried out in isothermal conditions at the temperature of 293.16 K. The permeability κ=4e-16 m2 was assumed.

The calculated sorptivity S_0 of the cement matrix specimen was 0.405 mm/min$^{1/2}$, being in agreement with laboratory tests results (Hall 1989).

5 NUMERICAL RESULTS OF CAPILLARY-DRIVEN FLUID FLOW IN CEMENT MATRIX SPECIMENS

Three bonded granular specimens of Figure 2 were again selected for capillary-driven fluid flow testing. The cement mortar with the initial porosity of 5% had a permeability coefficient of κ_{mort}=4•e^{-16} m^2, and ITZ with the initial porosity of 20% had a permeability coefficient of κ_{ITZ}=4•e^{-15} m^2 (Section 4.1). The basic material constants for the cement matrix and fluid were adopted. The initial conditions and boundary conditions were the same for all specimens. The initial pressure of the fluid (gas and liquid phase) was set at 0.1 MPa (close to the atmospheric pressure). The initial saturation degree was 0.0 (dry specimen). To simulate the pore filling with water, constant pressure of 0.14 MPa was chosen. The boundary pressure maintained a small pressure gradient on the boundary simulating a slight fluid flow, replenishing the water in the specimen. It was defined in pores in contact with the specimen bottom. This pressure was slightly higher than the initial pressure in the fluid domain (0.1 MPa) to fill in the pores with the water before the capillary pressure became effective. Along the remaining specimen boundaries, a mass flux rate of zero was defined (the so-called sealed surfaces). Hence, the dominant factor driving the fluid flow was the capillary pressure. The transient process took place under isothermal conditions. A constant and uniform temperature of 293.16 K was defined in the solid-fluid domain. The external pressure equal to the atmospheric pressure (0.1 MPa) was also applied to sealed surfaces. The fluid flow in the specimens was observed at three stages: 1. the first stage - after the zone of full saturation reached the height of 0.15H (where H is the specimen height), 2. the second stage – after the zone of full saturation reached the height of 0.33H (corresponds to the lower surface of the aggregate grain) and 3. the third stage – after the zone of full saturation reached the height of 0.60H (corresponds to the upper surface of the aggregate).

5.1 *Pure cement matrix*

Once the pores in contact with their surroundings were filled in with water, capillary pressure appeared in the virtual S2S channels and fluid began to flow. The capillary pressure depended on the channel aperture and its maximum value reached 3.52 MPa. The capillary pressure started to drive the fluid from one virtual pore to the adjacent virtual pore. It solely affected the mass flow rate through the common edge of two adjacent virtual pores. The pressure in virtual pores depended on the amount of mass flowing to or from virtual pores, properties of compressible fluids in virtual pores, virtual pore volume changes (due to the mechanical interaction with spheres) and pressures in surrounding pores. Therefore, the pressure in virtual pores differed from the capillary pressure. In addition, the fluid flow rate depended on the gas-phase fraction in virtual pores and channels. The fluid flow rate between two adjacent virtual pores that were not fully water-filled was strongly affected by permeability while the fluid flow rate between the fully saturated virtual pore and adjacent partially water-filled virtual pore was affected by the capillary pressure. Figure 4 shows the capillary-driven fluid flow at the third stage of fluid flow. The capillary pressure solely appeared along the surface that separated fully saturated virtual pores from partially saturated virtual pores, hereinafter referred to as the wetting front.

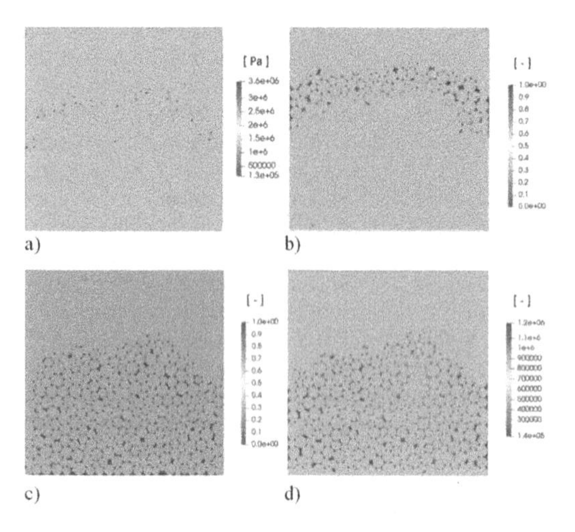

Figure 4. Calculation results of capillary-driven fluid flow in a specimen of the cement matrix at third stage of fluid flow: (a) capillary pressure, b) water phase fraction in range from 0.0 to 0.999 and c) full saturation state of water content and d) high hydraulic pressure zone).

At the first stage of fluid flow, the virtual pores were only partially filled in with water above the wetting front and under the wetting front, the virtual pores were completely filled in with water (fully saturated). At this stage, the progress of the wetting front was relatively fast. The fluid flow above the wetting front was driven solely by the hydraulic pressure.

At the second flow stage, the fluid slightly slowed down due to an increasing distance from the specimen bottom, resulting in a hydraulic pressure reduction below the capillary pressure in the wetting front. The mean capillary pressure remained the same as in the first flow stage. The influence of vertical boundary conditions on the shape of the wetting front increased. It was lower in partially saturated virtual pores. The increased hydraulic pressure zone was more dispersed than in the first flow stage.

At the third flow stage (Figure 4), the fluid continuously continued to slow down. The mean capillary pressure remained the same (Figure 4a) as in the previous flow stages The influence of vertical boundary conditions on the shape of the wetting front remained the same (Figure 4c) – it was again lower in partially saturated virtual pores. The increased hydraulic pressure zone was not dispersed as in the second stage. The area with the increased hydraulic pressure was much larger than in the previous flow stages. The calculated sorptivity S_0 was 0.405 mm/min$^{1/2}$ and corresponded to that of the mortar (cement/sand ratio 1:7) (Hall 1989).

5.2 *Cement matrix with aggregate*

After filling in boundary pores with water to full saturation, the fluid started to flow into the fluid domain (capillaries and pores) under the capillary pressure. Figure 5 presents the third flow stage of the capillary-driven flow.

At the first stage of fluid flow, the capillary pressure was present in the cement matrix only. Thus, fewer capillaries were involved in driving the fluid flow. Unexpectedly, the increased hydraulic pressure zone concentrated just below the aggregate. The area of fully saturated virtual pores coincided with the increased hydraulic pressure zone. The fluid flow, driven solely by the hydraulic pressure (apparent pores not filled in with water) was the strongest outside the area of interaction between the aggregate and vertical specimen surface boundaries. The fluid flow in this area was highly dependent upon permeability.

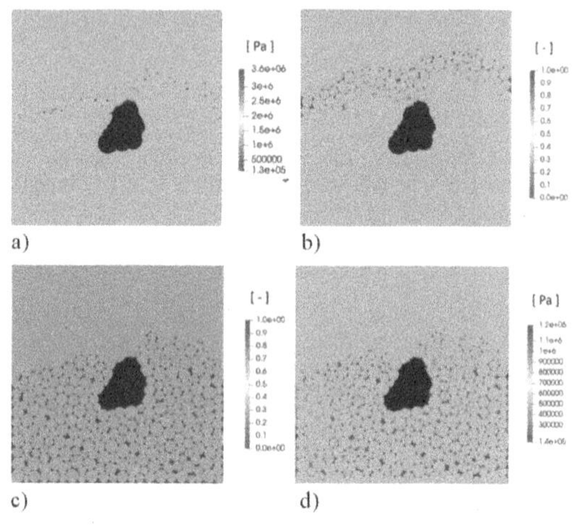

Figure 5. Computation results of capillary-driven fluid flow in cement matrix with aggregate at third fluid flow stage: (a) capillary pressure, b) water content (water phase fraction) in range from 0.0 to 0.999 and c) full saturation state of water content and d) high hydraulic pressure zone).

In contrast to the cement matrix specimen (Section 5.1), the fluid slowed down at the second flow stage due to the presence of the aggregate, being an obstacle to the flowing fluid. A significant influence of the aggregate on the shape of the wetting front occurred. The largest area of the increased hydraulic pressure zone was between the aggregate and vertical specimen boundaries. The highest hydraulic pressure was in this zone, reaching the value of 1.18×10^6 Pa.

At the third flow stage (Figure 5), the fluid continued to flow more slowly than at the first flow stage. The presence of the aggregate still affected the shape of the wetting front (Figure 5a). Above the wetting front, the fluid flow was driven by the hydraulic pressure only (Figure 5b). Due to the presence of the aggregate and its shape, the increased hydraulic pressure zone was asymmetric and its right part between the aggregate and the right vertical specimen boundary was closer to the upper specimen boundary (Figures 5d). As a result, the fluid moved slightly faster along the right vertical specimen boundary surface than along its left vertical specimen boundary (Figure 5c). The calculated sorptivity S_0 was 0.401 mm/min$^{1/2}$ and corresponded to that of the concrete cement/sand/aggregate ratio 1:2:4 and water/cement ratio 0.4) (Hall 1989).

5.3 *Cement matrix with aggregate and ITZ*

A strong influence of the presence of the aggregate and ITZ on the wetting front took place at the first fluid flow stage. The maximum capillary pressure was 3.54 MPa and the maximum hydraulic pressure was 1.16 MPa. At the first stage, the mass flow rate of the fluid above the wetting front depended on the permeability only. Hence, the highest flow rate occurred in ITZ. Thus, the water content and hydraulic pressure increased most rapidly in ITZ and not in the cement matrix.

In contrast to the cement matrix with the aggregate (Section 5.2), the lowest height of the full saturation zone was just below the aggregate and ITZ since the water moved much faster through ITZ than through the cement matrix. As a result, the time was long to fill in the pores below the aggregate and ITZ.

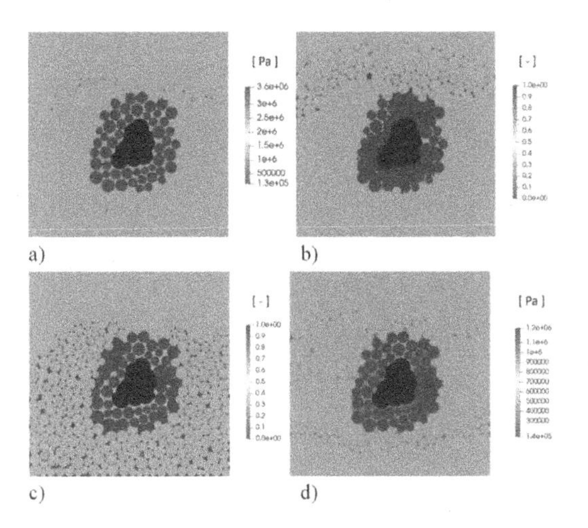

Figure 6. Computation results of capillary-driven fluid flow in cement matrix with aggregate and ITZ at third fluid flow stage: (a) capillary pressure, b) water phase fraction in range from 0.0 to 0.999 and c) full saturation state of water content and d) high hydraulic pressure zone).

At the second fluid flow stage, the water completely filled in the pores in the cement matrix almost to half the specimen height but not in ITZ. At this stage, the fluid mass flow rate above the wetting front still depended on the permeability only. Thus, the highest mass flow rate of the fluid took place in and above ITZ. Along the surface of the wetting front, the capillary-driven fluid flow dominated. The wetting front was not present in ITZ. The highest hydraulic pressure was not thus in ITZ but in the cement matrix. Due to the high porosity of ITZ (porosity 20% in contrast to the cement matrix porosity of 5%), the pores and capillaries in ITZ were too large for developing capillary pressure. Despite the greater permeability of ITZ, the pores fully filled in with water at a slower rate than in the cement matrix. As in the first stage, the maximum capillary pressure was 3.54 MPa and the maximum hydraulic pressure was 1.07 MPa.

At the third stage of the fluid flow, the wetting front passed ITZ (Figure 6a). The process of full filling of pores with water accelerated caused by a significant growth of the number of pores and capillaries in which the capillary pressure developed. Even if the pores outside ITZ were fully saturated (Figure 6b), the pores in ITZ were not filled in with water (Figure 6c). At this stage, the highest hydraulic pressure was observed in ITZ (Figure 6d). The calculated sorptivity S_0 was 0.276 mm/min$^{1/2}$ and corresponded to that of the concrete (cement/sand/aggregate ratio 1:2:4 and water/cement ratio 0.6 (Hall 1989)). There existed a significant difference in sorptivity between the specimen of the cement matrix with the aggregate (S_0=0.67 mm/min$^{1/2}$) and the specimen of the cement matrix with the aggregate and ITZ (S_0=0.48 mm/min$^{1/2}$)).

It can be concluded that ITZs in concrete decelerate the capillary fluid flow and consequently reduce sorptivity. However, for the sufficiently high hydrostatic pressure of water on the outer surface of a structural element made of concrete, the hydraulic pressure and not the capillary pressure becomes the dominant factor driving the fluid flow in unsaturated concretes. In this case, ITZs accelerate the fluid flow and the process of filling the pores with water due to the higher permeability of ITZ than the cement matrix. This conclusion is consistent with the results in (Stroeven et al. 2017a). This is of particular importance when designing water-tight concretes as the presence of ITZs with high permeability accelerates the penetration of water through the concrete. It should be noted that the DEM-CFD model in the current paper solely reproduces the free water flow under isothermal conditions (the mass and heat transfer in porous materials under non-isothermal conditions may be more complex).

6 SUMMARY AND CONCLUSIONS

The paper proposes a novel hydro-mechanical DEM/CFD model of multi-phase fluid flow in unsaturated concretes under isothermal conditions. The model enables the detailed tracking of liquid/gas fractions in pores and fractures concerning their varying geometry and topology, size and location. The model results provide a deeper understanding of the effects of ITZ on water transport through cementitious materials. The following main conclusions can be offered from our numerical analyses:

- Porous ITZs in concretes decelerate the capillary fluid flow and consequently reduce its sorptivity.
- Sufficiently high hydraulic water pressures become the dominant factor driving the fluid flow in unsaturated concretes. In this case, porous ITZs accelerate the full saturation of pores. For low hydraulic pressures, the capillary pressure becomes the dominant factor driving the fluid flow in unsaturated concretes. In this case, porous ITZs slow down the full saturation of pores. The aggregates without ITZs increase the fluid flow time as compared to the pure cement matrix under the capillary pressure and do not have an influence on fluid flow under the hydraulic pressure.
- The direction of the applied external pressure affects the fluid flow velocity in cement matrix specimens. The fastest fluid flow takes place for the horizontal external pressure and the slowest one for the vertical external pressure.

ACKNOWLEDGEMENTS

The present study was supported by the research project "Fracture propagation in rocks during hydro-fracking – experiments and discrete element method coupled with fluid flow and heat transport" (years 2019-2022) financed by the National Science Centre (NCN) (UMO-2018/29/B/ST8/00255).

REFERENCES

Ababneh, A., Benboudjema, F., Xi, Y. 2003. Chloride penetration in nonsaturated concrete, *J. Mater. Civ. Eng.* 15(2): 183–191.

Abdi, R., Krzaczek, M., Tejchman, J. 2021. Comparative study of high-pressure fluid flow in densely packed granules using a 3D CFD model in a continuous medium and a simplified 2D DEM-CFD approach, *Granular Matter*, accepted.

Aker, E., Måløy, K.J., Hansen, A. Batrouni, G.G. 1998. A Two-Dimensional Network Simulator for Two-Phase Flow in Porous Media, *Transport in Porous Media* 32: 163–186.

Batchelor, G. 2000. An Introduction to Fluid Dynamics, *Cambridge University Press*, Cambridge.

Bentz, D.P., Stutzman, P.E., Garboczi, E.J. 1992. Experimental and simulation studies of the interfacial zone in concrete, *Cem. Conc. Res.* 22(5): 891–902.

Bernardini, F., Bajaj, C. 1997. Sampling and reconstructing manifolds using alpha-shapes, *Technical Report CSD-TR-97-013*, Dept. Comput. Sci., Purdue Univ., West Lafayette, IN.

Caggiano, A., Schicchi, D.S., Mankel, C., Ukrainczyk, N., Koenders, E.A.B. 2018. A mesoscale approach for modeling capillary water absorption and transport phenomena in cementitious materials, *Computers and Structures* 200: 1–10.

Carmeliet, J., Roels, S. 2001. Determination of isothermal moisture transport properties of porous building material, *J. Therm. Environ. Build. Sci.* 24: 83–210.

Chatzis, I., Dullien, F.A.L. 1983. Dynamic immiscible displacement mechanisms in pore doublets: theory versus experiment, *J. Colloid. Interface Sci.* 91(1): 199–222.

Delagrave, A., Bigas, J.P., Olivier, J., Marchand, M., Pigeon, M. 1997. Influence of the interfacial zone on the chloride diffusivity of mortars, *Adv. Cem. Based Mater* 5: 86–92.

Descamps, F. 1979. Continuum and discrete modelling of isothermal water and air transfer in porous media, *Ph.D. Thesis*, Catholic University, Leuven.

Dullien, F.A.L. 1979. Porous Media – Fluid Transport and Pore Structure, *Academic Press*, New York.

Fatt, I. 1956. The network model of porous media: I–III, *Trans. AIME Petrol. Div.* 207: 144–177.

Galbraith, G.H. 1992. Heat and mass transfer within porous building materials, *Ph.D. Thesis*, University of Strathclyde, Glasgow.

Hall, C. 1989. Water sorptivity of mortars and concretes: a review, *Magazine of Concrete Research* 41(147): 1–61

Hökmark, H., Lönnqvist, M., Fälth, B. 2010. THM-issues in repository rock – thermal, mechanical, thermo-mechanical and hydro-mechanical evolution of the rock at the Forsmark and Laxemar sites, *Technical Report TR-10-23*, SKB-Swedish Nuclear Fuel and Waste Management Co., pp. 26–27.

Hughes, R.G., Blunt, M.J. 2000. Pore scale modeling of rate effects in imbibition, *Transp. Porous Media* 40(3):295–322.

Kozicki, J., Donzé, F.V. 2011. A new open-source software developer for numerical simulations using discrete modeling methods, *Computer Methods in Applied Mechanics and Engineering* 197: 4429–4443.

Krzaczek, M., Kozicki, J., Nitka, M., Tejchman, J. 2020. Simulations of hydro-fracking in rock mass at meso-scale using fully coupled DEM/CFD approach, *Acta Geotechnica* 15(2): 297–324.

Krzaczek, M., Nitka, M., Tejchman, J. 2021. Effect of gas content in macro-pores on hydraulic fracturing in rocks using a fully coupled DEM/CFD approach, *International Journal for Numerical and Analytical Methods in Geomechanics* 45(2): 234–264.

Li, K., Stroeven, P., Stroeven, M., Sluys, L.J. 2017a. A numerical investigation into the influence of the interfacial transition zone on the permeability of partially saturated cement paste between aggregate surfaces, *Cem. Concr. Res.* 102: 99–108.

Li, K., Stroeven, P., Stroeven, M. Sluys, L.B. 2017b. Effects of technological parameters on permeability estimation of partially saturated cement paste by a DEM approach, *Cem. Concr. Compos.* 84: 222–31.

Mason, G., Morrow, N.R. 1991. Capillary behavior of a perfectly wetting liquid in irregular triangular tubes, J. Colloid. Interface Sci. 141(1): 262–74.

Nitka, M., Tejchman, J. 2015. Modelling of concrete behaviour in uniaxial compression and tension with DEM, *Granular Matter* 17(1): 145–164.

Nitka, M., Tejchman, J. 2018. A three-dimensional meso scale approach to concrete fracture based on combined DEM with X-ray μCT images, *Cement and Concrete Research* 107: 11–29.

Nitka, M., Tejchman, J. 2020. Meso-mechanical modelling of damage in concrete using discrete element method with porous ITZs of defined width around aggregates, *Engineering Fracture Mechanics* 231, 107029.

Pedersen, C.R., 1990. Combined moisture and heat transfer in building constructions, *Ph.D. Thesis*, Technical University of Denmark.

Reynolds, O. 1883. An experimental investigation of the circumstances which determine whether the motion of water

shall be direct or sinous, and of the law of resistances in parallel channels, *Phil. Trans. Roy. Soc.* 174: 935–982.
Roels, S., Carmeliet, J., Hens, H. 2003. Modelling unsaturated moisture transport in heterogeneous limestone, Part 1: a mesoscopic approach, *Transport Porous Med.* 52: 333–350.
Šavija, B., Pacheco, J., Schlangen, E. 2013. Lattice modeling of chloride diffusion in sound and cracked concrete, *Cem. Concr. Compos.* 42: 30–40.
Scheffler, G.A., Plagge, R. 2010. A whole range hygric material model: Modelling liquid and vapour transport properties in porous media, *International Journal of Heat and Mass Transfer* 53: 286–296.
Sheng, Q., Thompson, K. 2016. A unified pore-network algorithm for dynamic two-phase flow, *Advances in Water Resources* 95: 92–108.
Skarżyński, L., Nitka, M., Tejchman, J. 2015. Modelling of concrete fracture at aggregate level using FEM and DEM based on x-ray μCT images of internal structure, *Engineering Fracture Mechanics* 10(147): 13–35.
Šmilauer, V., Chareyre, B. 2011. Yade DEM Formulation, *Manual*.
Suchorzewski, J., Tejchman, J., Nitka, M. 2018. Discrete element method simulations of fracture in concrete under uniaxial compression based on its real internal structure, *International Journal of Damage Mechanics* 27 (4): 578–607.
Tsakiroglou, C.D., Avraam, D.G., Payatakes, A.C. 2007. Transient and steady-state relative permeabilities from two-phase flow experiments in planar pore networks, *Adv. Water Resour.* 30(9): 1981–1992.
Washburn, E.W. 1921. The Dynamics of Capillary Flow, *Physical Review* 17: 273.
Wyrzykowski, M., Lura, P. 2014. The effect of external load on internal relative humidity in concrete. *Cement and Concrete Research* 65: 58–63.
Xu, K., Daian, J.-F., Quenard, D. 1997. Multiscale structures to describe porous media. Part I: theoretical background and invasion by fluids, *Transport Porous Med.* 26: 51–73.
Zeng, Q., Wang, X., Yang, R., Jike, N., Peng, Y., Wang, J., Tian, Y., Zhou, C., Ruan, S., Yan, D. 2021. Transmission microfocus X-ray radiographic measurements towards in-situ tracing capillary imbibition fronts and paths in ultra-thin concrete slices. *Measurement* 175, 109141.

Computational Modelling of Concrete and Concrete Structures – Meschke, Pichler & Rots (Eds)

Modeling of carbonation, de-carbonation and re-carbonation processes of structural concrete subjected to high temperature

K. Iwama, T. Nagayasu & K. Maekawa
Yokohama National University, Yokohama, Japan

K. Higuchi
Maeda Corporation, Tokyo, Japan

ABSTRACT: This paper proposes a multi-chemo-physics model to incorporate carbonation, de-carbonation and re-carbonation processes under high temperature. Experimental validation of the proposed integrated model is conducted by using the thermo-gravimetry experiments of cement paste and the strength of mortar composites immediately after high-temperature heating and after post-fire-curing. The CO_2 concentration and the humidity are experimentally changed as the thermodynamic boundary conditions for wide-range verification and validation. The compressive strength is treated not as the material property but the computed structural capacity of a cylindrical solid in which the temperature, hydration degree and carbonation develop non-uniformly. The proposed model allows practically reasonable assessment of fire-damaged and moist-cured concrete as a multi-scale composite.

1 INTRODUCTION

In fire-exposed structural concrete, various changes in chemical and mechanical properties proceed because of the complex combination of thermal, hygral, chemical and mechanical phenomena (e.g. Hertz 2003; Liu et al. 2018; Phan et al. 2001). The authors have proposed some models (Higuchi et al. 2021; Iwama et al. 2020, 2021) aiming at performance evaluation during and after fire of concrete structures (see Figure 1) by integrating the characteristics of concrete and reinforcing bars at high temperature into a multi-scale platform (Maekawa et al. 2003, 2008), as shown in Figure 1. The meso-scale modeling of cementitious composite is linked with the macroscopic model of large referential volume including multi-directional cracks. Then, reinforced concrete that was exposed to fire can be rebuilt inside the computer system and its capacity and ductility can be simulated as shown in Figure 2.

This paper is part of the upgrading multi-scale modeling (Higuchi et al. 2021; Iwama et al. 2020, 2021), and the authors focus on the carbonation of cement hydrates during and after heating as well as carbonation under normal climate conditions where neutralization is the major factor affecting durability. There are studies investigating the carbonation depth of fired concrete and the rate of carbonation after heating (Li et al. 2013, 2014; Oliveira et al. 2019; Yatsushiro et al. 2019). On the other hand, carbonation has the effect of densifying the micro-pore structure of concrete and increasing its strength. Thus, these conflicting effects on solid mechanics may be prominent especially at elevated temperature.

According to the analysis of combustion gas generated at fire sites in Japan (Suzuki et al. 1989), the carbon dioxide (CO_2) concentration often rises to 5%, but there are also sites where it rises to 15% depending on the fire situation. Recent years have seen the emergence of studies focusing on carbonation during heating (e.g. Yatsushiro et al. 2019). However, strength gain by carbonation during fire has received comparatively little attention. It is surmised that increase in strength related to changed pore structure is a non-negligible aspect that ought to be considered in the performance assessment of structural concrete.

In this paper, the authors introduce carbonation and de- and re-carbonation models of cement hydrates in all processes of fire heating. The proposed models are computationally verified and validated by multifaceted experiments, and further refinement of the existing multi-scale modeling for fire conditions (Higuchi et al. 2021; Iwama et al. 2020, 2021) is envisaged to extend its applicability and versatility.

2 CARBONATION, DE-CARBONATION AND RE-CARBONATION MODELS DURING HIGH TEMPERATURE AND POST-FIRE-CURING

Three models are proposed and described, namely a carbonation model of quicklime (CaO), which

 DOI 10.1201/9781003316404-68

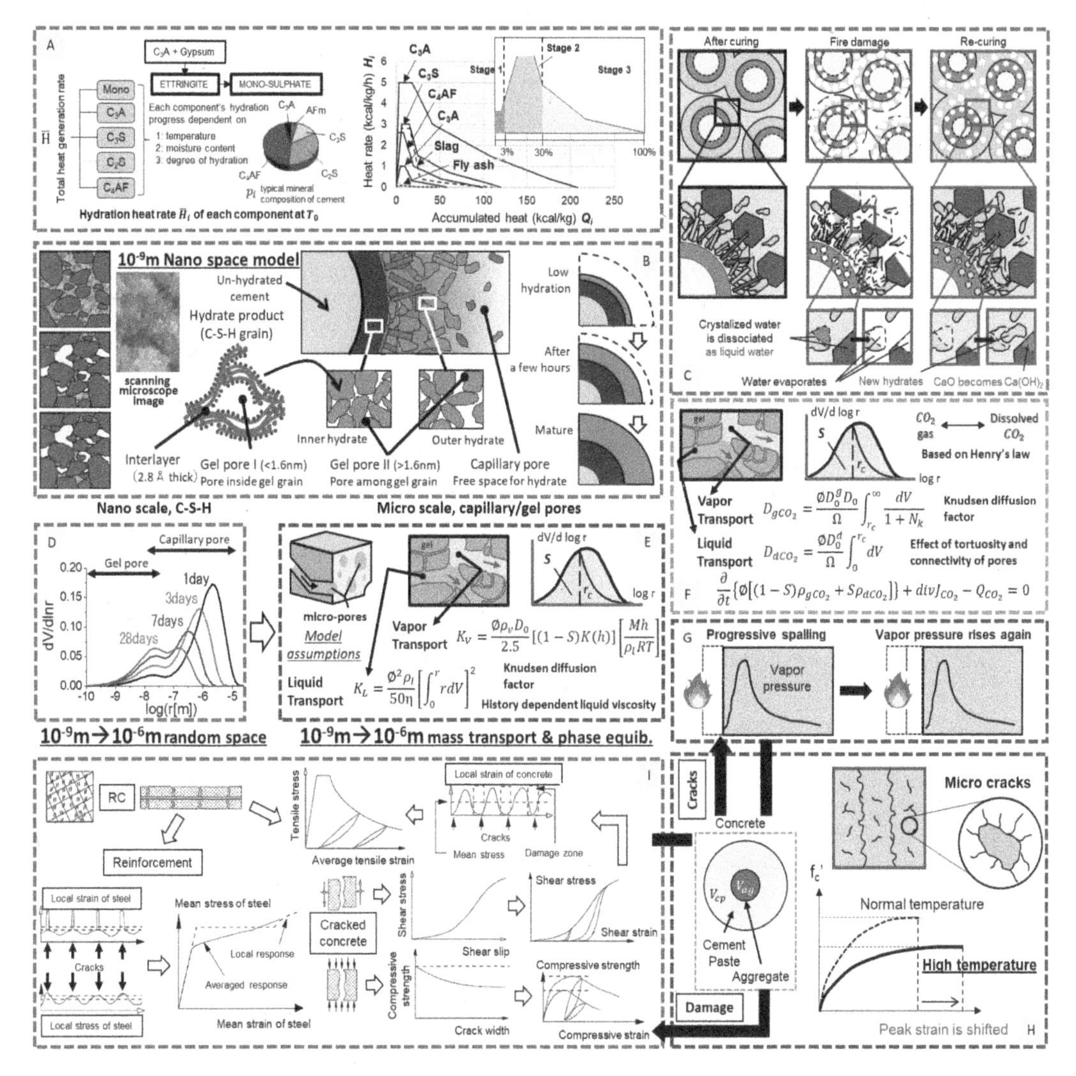

Figure 1. Outline of multi-scale modeling including upgrading models in fire conditions (Iwama et al. 2020, Higuchi et al. 2021). A) Multi-component cement heat hydration model, B) Solidified micro-pore structure formation model according to the hydra-tion and carbonation, C) Deterioration and re-hydration model of CSH and CH solids with moisture and carbon dioxide, D) Sto-chastic model of micro-pores, E) Moisture equilibrium and kinetics inside the micro-pores, F) Coupled migration and equilibri-um of pore solutions, G) Spalling model and transient boundary condition, H) Aggregate-cement binder interaction and solidification, I) Smeared crack modeling for structural concrete elements.

can be created by dehydration of calcium hydroxide ($Ca(OH)_2$) at high temperature, a de-carbonation model of calcium carbonate ($CaCO_3$), which can be generated by carbonation of calcium silicate hydrate (C-S-H), $Ca(OH)_2$, and CaO, and a re-carbonation model of $Ca(OH)_2$, which is created by CaO rehydration at post-fire-curing. These models are integrated in a multi-scale platform (Maekawa et al. 2003, 2008), allowing automatic prediction of the properties of solid concrete as accurate as the previous models (Higuchi et al. 2021; Iwama et al. 2020, 2021).

The base multi-scale platform (Maekawa et al. 2003, 2008, Figure 3A) incorporates the carbonation models of C-S-H and $Ca(OH)_2$, and its applicability has been confirmed at normal ambient room temperature (Ishida & Li 2008; Ishida et al. 2004). This system also computes the migration of carbon dioxide and moisture and their thermodynamic equilibrium. As chemo-physics events are not uniform in space, dense clusters of carbonated solid are computationally generated near the surface of structural concrete.

In this study, the existing carbonation model at ambient states is kept unchanged and the characteristic events at high temperature as stated above are further focused on, as shown in Figure 3, to extend the scope of the platform. This is expected to allow the assessment of structural concrete performance under accidental events to upgrade the resilience.

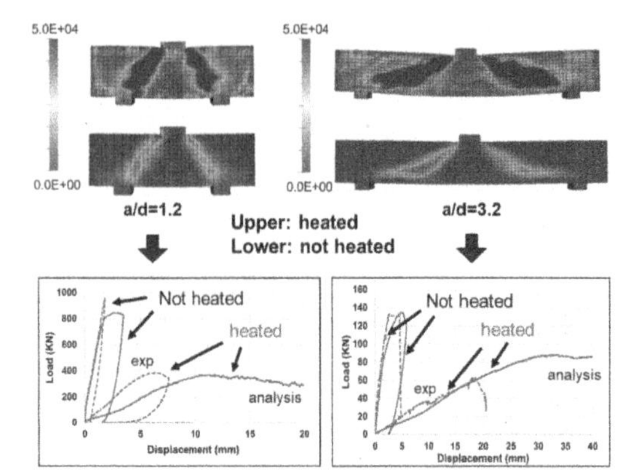

Figure 2. Shear capacity of RC beams under fire actions (Higuchi et al. 2021).

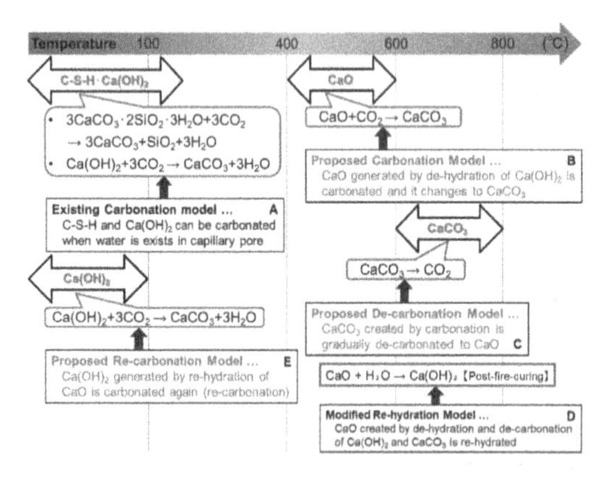

Figure 3. Outline of carbonation, de-carbonation and re-carbonation models during and after heating.

2.1 *Carbonation of CaO over 400°C*

The model of CaO quicklime carbonation generated by the dehydration of $Ca(OH)_2$ is introduced at high temperature heating exceeding the boiling point of 100°C. According to a previous study (Yatsushiro et al. 2019), it has been confirmed that CaO generated by the decomposition of $Ca(OH)_2$ instantly reacts with CO_2 gas to generate $CaCO_3$ in the temperature range of 400°C to 500°C. It is generally held that in this temperature range, the strength of concrete decreases sharply (Schneider 1988, Li et al. 2018) due to the increase in porosity of cement paste caused by dehydration of $Ca(OH)_2$.

However, according to Yatsushiro et al. (2019), in the presence of high concentrations of CO_2, carbonation may densify the micro-pore structure and increase its strength. In the present study, it is assumed that CaO generated by the decomposition of $Ca(OH)_2$ directly reacts with CO_2 gas in micro-pores to generate $CaCO_3$ (see Figure 3B). Thus, we simply propose the CaO carbonation model as,

$$R_{CaO} = \int C_{CaO} \cdot M_{CO2} \tag{1}$$

where R_{CaO} = the amount of reacted CaO with CO_2 gas (mol/m^3), which does not exceed the amount of CaO (mol/m^3) generated by dehydration of $Ca(OH)_2$ (Iwama et al. 2020); C_{CaO} = the reaction rate of CaO at high temperature, which is tentatively assumed as 1.0; and M_{CO2} = the amount of CO_2 gas in the micro-pores (mol/m^3), which is calculated by the base original carbonation model (Ishida & Maekawa 1999; Maekawa et al. 2003, 2008). This base model is a linear rate model of the amount of CO_2 gas whose migration and diffusion are computed in parallel and linked with the updated micro pore structures (Ishida et al. 2004; Maekawa et al. 2003).

2.2 *De-carbonation of $CaCO_3$ at 600°C to 800°C*

As one of the decomposition models of cement hydrates at high temperature, the de-carbonation model of $CaCO_3$ calcium carbonate, which is generated by carbonation of C-S-H, $Ca(OH)_2$, and CaO, is introduced. According to previous studies (Alonso & Fernandez 2004; Sabeur et al. 2016), $CaCO_3$ de-carbonation is reported to take place in the temperature range of 600°C to 800°C. In the micro-pore structure previously densified by carbonation, it is considered that porosity increases rapidly due to the decomposition of $CaCO_3$, and that this de-carbonation of $CaCO_3$ has a great effect on the strength of concrete exposed to high temperature.

In this study, $CaCO_3$ generated by carbonation of C-S-H, $Ca(OH)_2$, and CaO is assumed to be gradually decomposed from 600°C to 800°C (Figure 3C). Thus, we have the simplified de-carbonation model as,

$$W_{d(CO2),CaCO3} = W_{CO2,CaCO3}\left(\frac{T_{max}-600}{800-600}\right) \tag{2}$$
$$(600° \leq T_{max} \leq 800°)$$

$$W_{d(CO2),CaCO3} = W_{CO2,CaCO3} \tag{3}$$
$$(800° < T_{max}$$

where $W_{d(CO2),CaCO3}$ = the amount of CO_2 generated by de-carbonation of $CaCO_3$ (mol/m^3); $W_{CO2,CaCO3}$ = the amount of CO_2 chemically bound in $CaCO_3$ (mol/m^3); and T_{max} = the past maximum temperature. The progression in the quantity of these minerals is used to identify porosity and the distribution of pore sizes based upon a statistical function (Maekawa et al. 2003).

2.3 *Re-carbonation of $Ca(OH)_2$ during post-fire-curing*

An event that occurs in concrete after a fire is the rehydration of CaO during post-fire-curing (Park et al. 2015; Poon et al. 2001; Suh et al. 2020). This is attracting attention as a self-healing process of concrete exposed to high temperatures, and in the previous upgrading of the multi-scale platform (Higuchi et al. 2021; Iwama et al. 2020), it has been introduced and verified.

In this study, this rehydration model is integrated with the carbonation model. Specifically, not only CaO generated by dehydration of $Ca(OH)_2$ but also CaO generated by de-carbonation of $CaCO_3$ are assumed to rehydrate during post-fire-curing and return to $Ca(OH)_2$, as shown in Figure 3D.

It is assumed that in the presence of CO_2 in the atmosphere, $Ca(OH)_2$ produced by rehydration of CaO is carbonated again (re-carbonation) by the same process as the existing carbonation model (Ishida & Maekawa 1999; Maekawa et al. 2003, 2008), as shown in Figure 3. The upgraded rehydration model is succinctly expressed as,

$$W_{r(CH),CaCO3} = \int C_{rh,CaCO3} \cdot \left(W_{d(CO2),CaCO3} \cdot \frac{18.0}{44.0} \right) \cdot W_{free} \cdot dt \quad (4)$$

$$W_{r(CH)} = W_{r(CH),Ca(OH)2} + W_{r(CH),CaCO3} \quad (5)$$

where $W_{r(CH),CaCO3}$ = the weight of rehydrated crystallized water of $Ca(OH)_2$ created by rehydration of CaO, which is generated by de-carbonation of $CaCO_3$ (kg/m^3); $C_{rh,CaCO3}$ = the reaction rate of rehydration of $CaCO_3$-based CaO, which is tentatively assumed as 0.5; W_{free} = the weight of free water that reacts with CaO; $W_{r(CH)}$ = the weight of all rehydrated crystallized water of $Ca(OH)_2$ (kg/m^3); and $W_{r(CH),Ca(OH)2}$ = the weight of rehydrated crystallized water of $Ca(OH)_2$ created by rehydration of $Ca(OH)_2$-based CaO, which is proposed in Iwama et al. (2020) (kg/m^3).

The rehydrated $Ca(OH)_2$ can be carbonated again (re-carbonated) based on the existing original carbonation model (Ishida & Maekawa 1999; Maekawa et al. 2003, 2008), as shown in Figure 3E. It was also reported that expansion of concrete composites and adhesion appear when $Ca(OH)_2$ is generated by rehydration of CaO (Higuchi et al. 2021). It is simply assumed in this study that this property does not change, even in the rehydration model that is incorporated in the carbonation model.

On the contrary, according to previous studies (e.g. Kangni-Foli et al. 2021; Powers 1962), it has been also reported that shrinkage occurs in the process of carbonation of the hardened cement paste. In this study, the model includes both events, i.e., volumetric shrinkage is introduced at carbonation of $Ca(OH)_2$ in addition to the expansion at rehydration of CaO. Thus, the overall volumetricchange can be positive or negative according to the balance of both chemical events related to the mechanistic actions.

Table 1. Mix proportion of mortar specimens.

W/C (%)	s/c	Unit weight (kg/m^3 W	C	S	SP
25	1.2	242	968	1162	14.5

W: water, C: cement, S: fine aggregate*1, SP: super plasticizer, s/c: fine aggregate to cement ratio
*1 river sand, surface-dry density=2.58g/c^{m3}, rate of water absorption=2.21%

3 EXPERIMENTAL STUDY OF COMPRESSIVE STRENGTH WITH CO_2 CONCENTRATION DURING HEATING

The varying compressive strength was presented in the presence of high-concentration CO_2 especially during heating. Here, a high CO_2 concentration environment was experimentally created by placing dry ice (CO_2 solidified under low temperature) in an electric furnace at high temperature, as illustrated in Figure 4. As a result, the chamber of the furnace was mostly occupied by carbon dioxide when heating began.

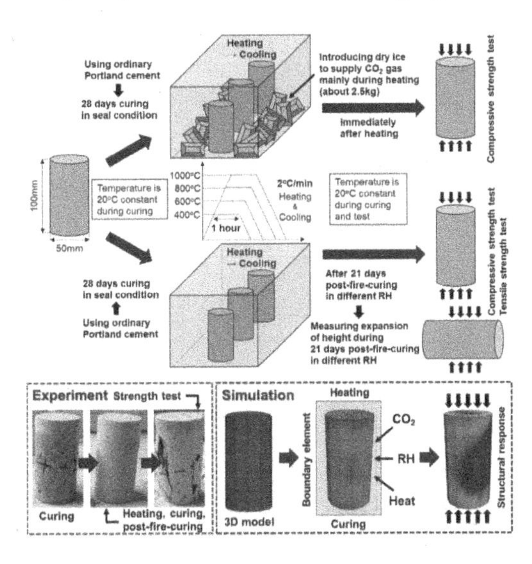

Figure 4. Outline of experiments related in this study.

Table 1 shows the mix proportion of the high strength mortar with water to cement ratio of 25% using ordinary Portland cement. The specimen was cured in sealed condition under constant temperature of 20°C for 28 days after casting. In this experiment, the compressive strength was compared with four levels of heating temperatures of 400, 600, 800 and 1,000°C with and without dry ice (source of CO_2 in the furnace) during heating. The heating rate was 2°C/min for heating and cooling, and the temperature was maintained at the specified level for one hour. When the surface temperature of the test pieces was about the same as room temperature, the compression test was promptly performed. The compressive strength of all cases is shown in Figure 5.

An increase in compressive strength is clearly seen around the temperature of 400°C, when a high concentration of CO_2 was achieved in the furnace through the introduction of dry ice. We can say that a high CO_2 environment is effective for strength gain. It is

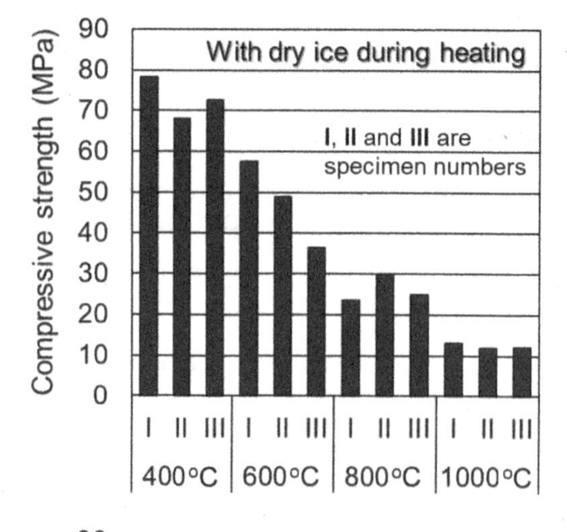

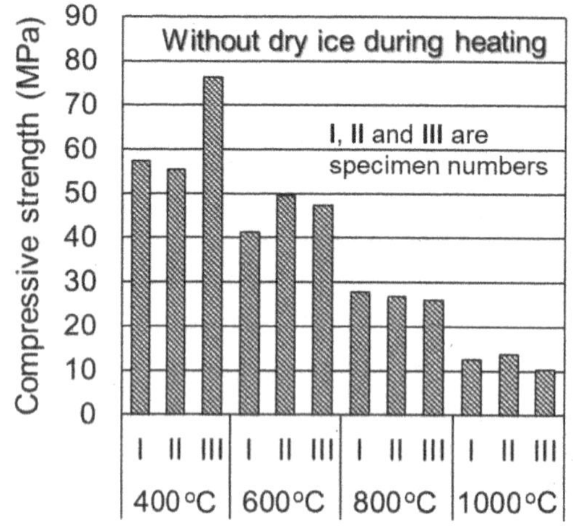

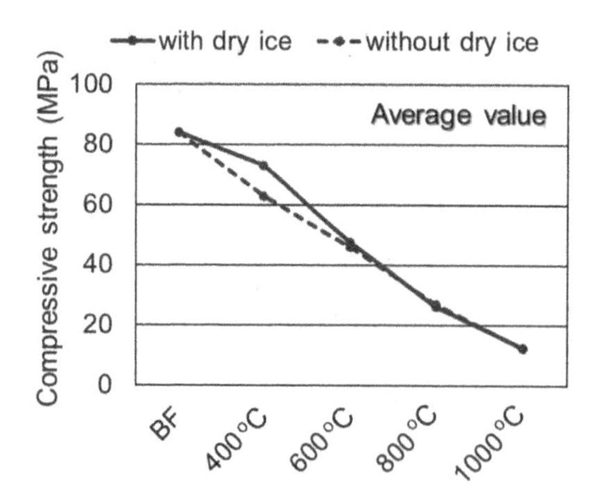

Figure 5. Experimental results of effect of CO_2 during heat-ing for change in compressive strength.

thought that the micro-pore structure would become denser and the compressive strength would increase based on the carbonation of CaO in the temperature range of 400°C to 500°C as reported in Yatsushiro et al. (2019). In the multi-chemo-physics analysis, the diffusion of CO_2 gas is considered through the media with cracking. Thus, gas penetration reaches deeper inside the specimens compared with the case of normal temperature, in which the densified carbonation layers are limited to around the surface.

However, the impact of CO_2 was not shown in the case of 600°C. There are two possible causes: a) CO_2 gas was emitted quickly from the heating furnace, and the CO_2 concentration had already dropped in the temperature range of 400°C to 500°C, b) de-carbonation that occurs usually at about 600°C to 800°C occurred from a slightly lower temperature. At any rate, it can be said that the CO_2 concentration during heating may have an impact on the compressive strength of cementitious composites with micro-pores.

4 EXPERIMENTAL VALIDATION OF PROPOSED MODEL

4.1 *Thermo-gravimetric analysis*

The weight change of the hardened cement paste during high temperature heating was confirmed by using thermo-gravimetric analysis (TGA). The confirmed weight change due to dehydration and carbonation of cement paste during high-temperature heating may lead to validation of the proposed model for micro-pore structures associated with the CO_2 related chemical reactions.

According to the previous study by Yatsushiro et al. (2019), the test specimen used for TGA was made of cement paste with water to cement ratio of 40%, using ordinary Portland cement. After placing, the specimen was sealed and cured for 28 days in a constant temperature environment of 20°C. Then, the specimen was crushed into a powder sample. The mass change when the temperature was raised to 1,000°C at 10°C/min under N_2 flow and CO_2-5% flow was measured.

The comparison of the experiment and simulation is shown in Figure 6. As for the N_2 flow, although mass loss due to dehydration of $Ca(OH)_2$ in the temperature range of 400°C to 450°C was less overestimated in the proposed model, the proposed model could reproduce the experimental facts. Looking at the case of CO_2-5% flow, the proposed model can capture the tendency of mass gain due to CaO car bonation around 400°C to 500°C and mass loss due to de-carbonation of $CaCO_3$ around 600°C to 800°C. Thus, the proposed model can be said to be capable of reproducing the trend of experimental results.

4.2 *Compressive strength after heating with and without dry ice*

The proposed model is further validated by using the strength immediately after heating with and without dry ice during heating described in Section 3.1. The compressive strength is the computed structural capacity of the cylinder in which the local temperature

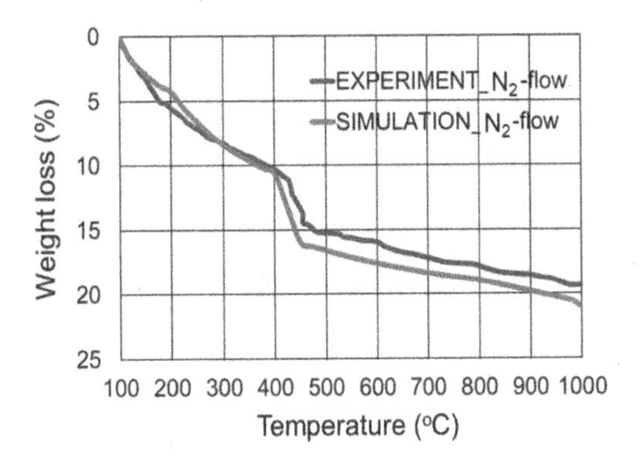

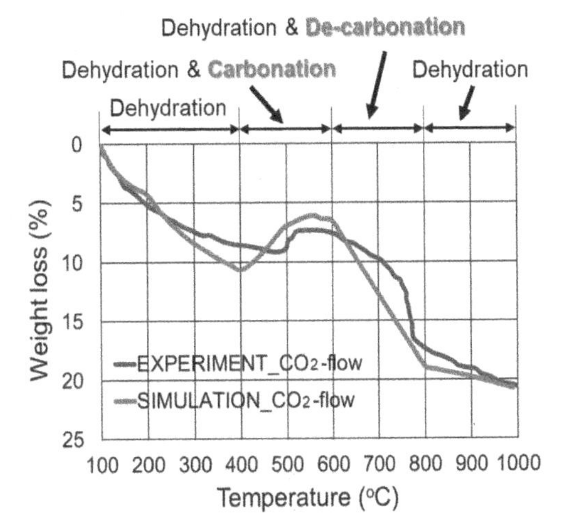

Figure 6. Validation: Thermogravimetric analysis (the cases of N_2 flow and CO_2-5% flow).

and strains are not uniform place-by-place (see Figure 4). The local strength of meso-scale cement paste is computed based upon the micro-pore structures and examined with a wide range of water to cement ratio, curing, and material age by Otabe and Kishi (2005) as,

$$f_c = f_\infty \left\{1 - exp\left(-\alpha D_{hyd.out}^{\beta}\right)\right\} \tag{6}$$

where f_c = compressive strength of concrete (MPa); f_∞ = ultimate compressive strength (MPa), which is determined in Otabe and Kishi (2005); $D_{hyd.out}$ = ratio of the volume of cement hydrates created outside of cement particles to the amount of capillary porosity when hydration is started (Otabe & Kishi 2005); α, β = constant value (3.0 and 4.0, respectively).

Micro-cracking and its impact on local stresses are taken into account (Maekawa et al. 2003, 2008). The mix proportion, curing conditions, heating conditions and other items are described in Section 3.1.

The measured concentration of CO_2 was 95% in the furnace at the start of heating and decreased to 0.03% of general atmospheric CO_2 concentration at the end of the experiments after 3 hours. The concentration could not be measured when the temperature inside the furnace was high. Thus, linear change of CO_2 concentration is assumed for the boundary conditions of the multi-chemo-physics simulation.

Figure 7 shows the validation of compressive strength. The average value of the compressive strength yielded by the experiment and the calculated value obtained by the proposed model were compared for all 5 levels. The proposed model can be said to roughly grasp the experimental results. Although the compressive strength was underestimated at 400°C, the proposed model can reproduce the difference due to the presence of dry ice. According to the TGA result in the CO_2 flow of Section 4.1, there are some differences in mass loss due to the amount of C-S-H and $Ca(OH)_2$ produced and the degree of carbonation. Large mass loss indicates greater decomposition of the cement hydrates. Thus, it is considered that the porosity of the micro-structure increases owing to decomposition of cement hydrates, and this leads to reduced compressive strength.

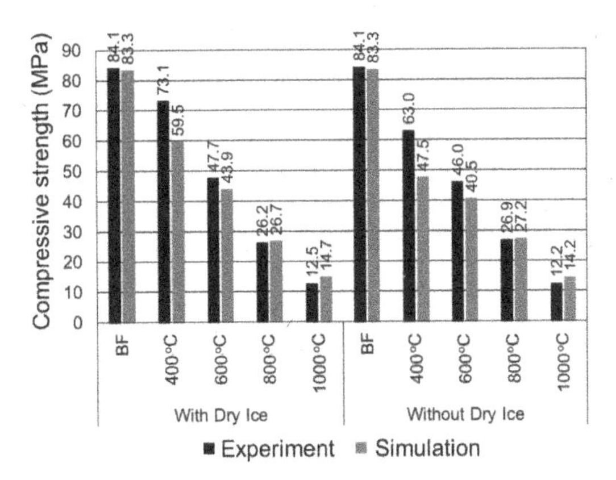

Figure 7. Compressive strength just after heating with and without high CO_2 concentration due to introduction of dry ice.

4.3 *Deformation during post-fire-curing*

The proposed model was examined by using the experimentally measured size change of the specimen during post-fire-curing. The mix proportion of the mortar was the same as that listed in Table 1. In this experiment, the sole heating temperature case was 1,000°C, but four levels of relative humidity (RH) environment at the post-fire-curing were prepared: 30%, 60% and RH-90% curing, and wet curing in water. The heating rate was 2°C/min for raising and lowering the temperature, and the temperature was maintained at 1,000°C for 1 hour. After heating and cooling, the specimens were cured for 21 days in each RH environment, and the changes in specimen height and strength were measured. A detailed explanation of this experiment can be found in Higuchi et al. (2021). Figure 8 shows the validation of the deformation after the heating. The proposed model was found to be able to capture the specimen size change trend.

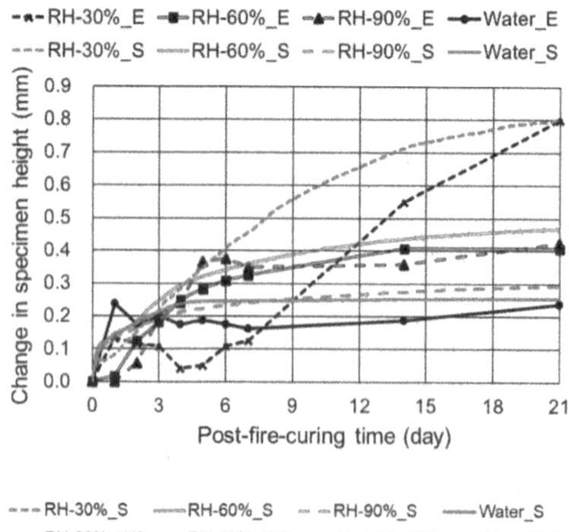

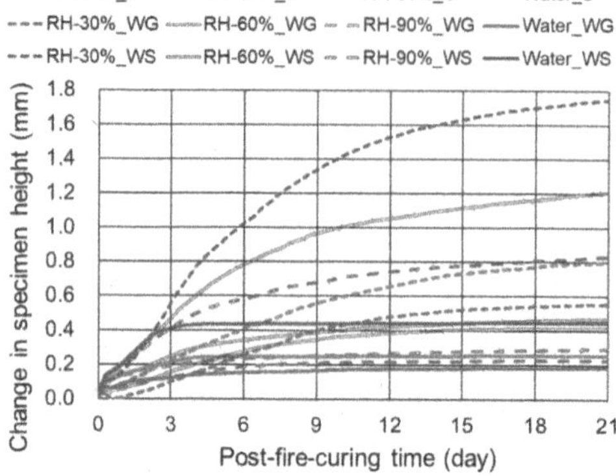

Figure 8. alidation: Size change of specimen height dur-ing post-fire-curing: Left: E=experiment, S=simulation, right: S=simulation, WG=model without carbonation, WS=model without considering shrinkage by carbonation.

Figure 8 shows the results of the parametric study. Expansion by absorption of atmospheric moisture can be seen to be small in the model that does not consider carbonation itself (the previous model by the authors). This is because the amount of CaO generated with consideration of carbonation is larger than that without carbonation. It can also be seen that the expansion after 21 days is more than twice that of the proposed model when the shrinkage due to carbonation is not considered. It was confirmed that the effect of shrinkage due to carbonation of $Ca(OH)_2$ has a large effect on the calculation. The authors plan to carry out more detailed experiments related to the volumetric change due to carbonation.

4.4 *Compressive and tensile strength after post-fire-curing*

The proposed model was examined by using the experimentally obtained compressive and tensile strength after post-fire-curing in various RH environments. After 21 days of post-fire-curing under various RH conditions as shown in Section 4.3, the compression test and the split tensile test were analytically conducted and the strengths were calculated as shown in Figure 9.

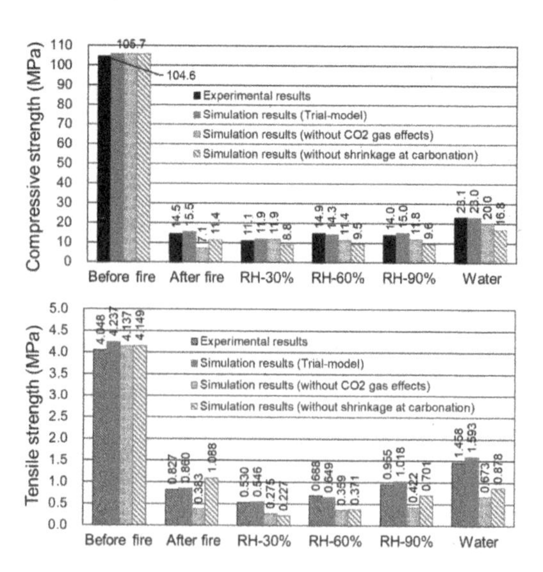

Figure 9. Validation: Compressive and tensile strength after post-fire-curing.

The results of the proposed model show that the experimental results can be reproduced with high accuracy. At RH-30% curing, both compressive and tensile strength are lower than those immediately after heating. The proposed model can express that the negative effect of expansion during post-fire-curing is predominant at low humidity. In the model that does not consider the effect of shrinkage due to carbonation, the strength after post-fire-curing is also small. This is because the negative effect of expansion due to rehydration exceeds the positive effect of self-healing behavior since the shrinkage due to carbonation is not taken into account.

5 CONCLUSIONS

The authors undertook this study with the aim to extend the applicability of multi-scale modeling at carbonation during and after high temperature heating. The conclusions reached in this study are summarized as follows.

I. Three models related to carbonation are proposed: 1) carbonation model of CaO at high temperature, 2) de-carbonation model of $CaCO_3$ at the temperature range 600°C to 800°C and 3) re-carbonation model of $Ca(OH)_2$ at post-fire-curing, incorporated in the multi-scale platform for structural concrete. By multifaceted experiments in terms of weight change, moisture migration, compressive strength and CO_2 environment, the proposed models were validated with sufficient accuracy.
II. Compressive strength was experimentally confirmed to clearly increase around the temperature of 400°C with CO_2 gas release from dry ice and the strength gain was quantitatively reproduced by

simulating the micro-structures resulting from de- and re-carbonation.

III. Based on the experimental and analytical investigations, it was clarified that there are positive and negative effects on the strength of the solidified media during the post-fire-curing, to wit, the positive effect of the self-healing process of rehydration and re-carbonation, and the negative effect of the self-destruction process due to structural expansion.

ACKNOWLEDGEMENTS

This study was financially supported by JSPS KAKENHI Grant Number 20H00260.

REFERENCES

Alonso, C. & Fernandez, L. 2004. Dehydration and rehydration processes of cement paste exposed to high temperature environments. *Journal of Materials Science* 39(9): 3015–3024.

Hertz, K.D. 2003. Limits of spalling of fire-exposed concrete. *Fire Safety Journal* 38(2): 103–116.

Higuchi, K., Iwama, K. & Maekawa, K. 2021. Remaining shear capacity of fire-damaged high strength RC beams after moist curing. *Journal of Advanced Concrete Technology* 19(8): 897–912.

Ishida, T. & Li, C.H. 2008. Modeling of carbonation based on thermo-hygro physics with strong coupling of mass transport and equilibrium in micro-pore structure of concrete. *Journal of Advanced Concrete Technology* 6(2): 303–316.

Ishida, T. & Maekawa, K. 1999. An integrated computational system of mass/energy generation, transport and mechanics of materials and structures. *Journal of JSCE* 627(44): 13–25 (in Japanese).

Ishida, T., Maekawa, K. & Soltani, M. 2004. Theoretically identified strong coupling of carbonation rate and thermodynamic moisture states in micropores of concrete. *Journal of Advanced Concrete Technology* 2(2): 213–222.

Iwama, K., Higuchi, K. & Maekawa, K. 2020. Model-based assessment of long-term serviceability and fire resistance for underground reinforced concrete ducts. *Structural Engineering International* 30(4): 506–514.

Iwama, K., Higuchi, K. & Maekawa, K. 2020. Thermo-mechanistic multi-scale modeling of structural concrete at high temperature. *Journal of Advanced Concrete Technology* 18(5): 272–293.

Iwama, K., Kato, Y., Baba, S., Higuchi, K. & Maekawa, K. 2021. Accelerated moisture transport through local weakness of high-strength concrete exposed to high temperature. *Journal of Advanced Concrete Technology* 19(2): 106–117.

Kangni-Foli, E., Poyet, S., Le Bescop, P., Charpentier, T., Bernachy-Barbé F., Dauzères, A., L'Hôpital, E. & d'Espinose de Lacaillerie, J.B. 2021. Carbonation of model cement pastes: The mineralogical origin of microstructural changes and shrinkage. *Cement and Concrete Research* 144: Article ID 106446.

Li, Q., Li, Z., Yuan, G. & Shu, Q. 2013. The effect of a proprietary inorganic coating on compressive strength and carbonation depth of simulated fire-damaged concrete. *Magazine of Concrete Research* 65(11): 651–659.

Li, Q., Yuan, G. & Shu, Q. 2014. Effects of heating/cooling on recovery of strength and carbonation resistance of fire-damaged concrete. *Magazine of Concrete Research* 66(18): 925–936.

Li, Z., Li, L., Wang, J. & Wu, X. 2018. Effect of elevated temperature on meso- and micro-structure and compressive strength of high-strength concrete and mortar containing blast-furnace slag. *Journal of Advanced Concrete Technology* 16(10): 498–511.

Liu, J.C., Tan, K.H. & Yao, Y. 2018. A new perspective on nature of fire-induced spalling in concrete. *Construction and Building Materials* 184: 581–590.

Maekawa, K., Ishida, T. & Kishi, T. 2008. *Multi-scale Modeling of Structural Concrete*. London: Taylor & Francis.

Maekawa, K., Ishida, T. & Kishi, T. 2003. Multi-scale modeling of concrete performance—Integrated material and structural mechanics (Invited). *Journal of Advanced Concrete Technology* 1(2): 91–126.

Maekawa, K., Pimanmas, A. & Okamura, H. 2003. *Nonlinear Mechanics of Reinforced Concrete*. London and New York: Spon Press.

Oliveira, J.A., Ribeiro, J.C.L., Pedroti, L.G., Faria, C.S., Nalon, G.H. & Oliveira, A.L. 2019. Durability of concrete after fire through accelerated carbonation tests. Materials Research 22(1): 1–7.

Otabe, Y. & Kishi, T. 2005. Development of hydration and strength model for quality evaluation of concrete. *Industrial Science (University of Tokyo)* 57(2): 37–42 (in Japanese).

Park, S.J., Yim, H.J. & Kwak, H.G. 2015. Effects of post-fire curing conditions on the restoration of material properties of fire-damaged concrete. *Construction and Building Materials* 99: 90–98.

Phan, L.T., Lawson, J.R. & Davis, F.L. 2001. Effects of elevated temperature exposure on heating characteristics, spalling, and residual properties of high performance concrete. *Materials and Structures* 34(2): 83–91.

Poon, C.S., Azhar, S., Anson, M. & Wong, Y.L. 2001. Strength and durability recovery of fire damaged concrete after post-fire-curing. *Cement and Concrete Research* 31(9): 1307–1318.

Powers, T.C. 1962. A hypothesis on carbonation shrinkage. *Journal of the PCA Research and Development Laboratories* 4(2): 40–50.

Sabeur, H., Platret, G. & Vincent, J. 2016. Composition and microstructural changes in an aged cement pastes upon two heating–cooling regimes, as studied by thermal analysis and X-ray diffraction. *Journal of Thermal Analysis and Calorimetry* 126: 1023–1043.

Schneider, U. 1988. Concrete at high temperatures–A general review. *Fire Safety Journal* 13(1): 55–68.

Suh, H., Jee, H., Kim, J., Kitagaki, R., Ohki, S., Woo, S., Jeong, K. & Bae, S. 2020. Influences of rehydration conditions on the mechanical and atomic structural recovery characteristics of Portland cement paste exposed to elevated temperatures. *Construction and Building Materials* 235: Article ID 117453.

Suzuki, Y., Takeda, M., Inamura, T. & Tanaka, Y. 1989. Analyzed gases of burning products collected at fire scenes. *Report of Fire Science Laboratory* 26: 45–52.

Yatsushiro, D., Atarashi, D., Yoshida, N. & Okumura, Y. 2019. Study on carbonation mechanism of hardened cement paste due to fire damage. In *The 41th JCI Technical Conference, Sapporo, 10–12 July 2019*. Tokyo: Japan Concrete Institute.

Computational Modelling of Concrete and Concrete Structures – Meschke, Pichler & Rots (Eds)
 ISBN: 978-1-032-32724-2

Multi-physics simulation of steel corrosion in reinforced UHPC beams under coupled sustained loading and chloride attack

J. Fan, M.P. Adams & M.J. Bandelt
John A. Reif, Jr. Department of Civil and Environmental Engineering, New Jersey Institute of Technology, USA

ABSTRACT: Ultra-high performance concrete (UHPC) is a novel construction material associated with enhanced mechanical and durability characteristics compared to traditional concrete, such as very high compressive and tensile strengths, high ductility, and low permeability to aggressive materials. In reinforced concrete structures, the development of cracks accelerates the corrosion of steel as water and aggressive ions are able to infiltrate the concrete more easily. Accelerated steel corrosion leads to increased corrosion products which subsequently induces more cracks. Consequently, the corrosion of steel in cracked concrete is a time-dependent coupled process. Additionally, reinforced concrete structures are often under sustained loading, which can affect the initial and time-dependent damage state of the structure. In this paper, the steel corrosion process and flexural behavior of both reinforced concrete and reinforced UHPC beams subjected to sustained loading and chloride attack are investigated through multi-physics simulation techniques. Numerical simulations are conducted on members considering the time-dependent deterioration process. First, a service load acting on the beams was selected to induce an initial damage state. Second, the mass transport of chloride in normal reinforced concrete and reinforced UHPC was modeled considering the initial cracking due to the service load. Next, the active area of the rebar was chosen based on the critical chloride value to initiate corrosion. The corrosion of the reinforcement was then simulated and the resulting rust expansion thickness was calculated. Finally, the combined effects of corrosion product expansion and mechanical loading was simulated. In the next time step, the process was repeated until severe damage was observed. The simulation results show that the reinforced UHPC beams exhibit distributed cracking patterns and smaller crack widths while the reinforced concrete beams had localized cracking behavior and major cracks under initial mechanical loading. Unlike the pitting corrosion mode in concrete without initial damage reported from many experimental results, both reinforced concrete and reinforced UHPC beams have more uniform corroding areas of the reinforcing bar when initial cracks from mechanical loading are simulated. Reinforced UHPC beams experienced significantly slower chloride ingress than reinforced concrete beams due to smaller crack widths and higher material density. Furthermore, the reinforced UHPC beam showed smaller corrosion current densities along the reinforcing bar resulting in less rust expansion. With lower rust expansion and higher damage tolerance, the reinforced UHPC beam shows better damage control after corrosion. The simulation results show that the reinforced concrete beam experiences much faster deterioration. In contrast, the reinforced UHPC beam provides excellent resistance to chloride attack and negligible increase of damage area under sustained loading even after a much longer time in a harmful environment.

1 INTRODUCTION

Across a range of durability challenges, corrosion is the most common source of deterioration mechanism in normal reinforced concrete structures that causes loss to the economic life of a structure (Broomfield 2003; Zhao & Jin 2016). With widespread interest in mitigating corrosion induced service life disruption issues, innovative alternative material systems with superior durability properties have been used to achieve improved life-cycle performance (Alkaysi et al. 2016). Ultra-high-performance concrete (UHPC) is one such material that may slow down chloride ingress and corrosion initiation and restrain the damage that is caused or promoted by corrosion (Sohail et al. 2021).

UHPC is a ductile high strength material associated with fine cracking characteristics. Furthermore, UHPC is distinguished from conventional concrete because of its improved durability performance. Due to its low permeability and distributed cracking pattern, UHPC has shown outstanding resistance to harmful materials (Alkaysi et al. 2016).

A substantial number of studies have been conducted to investigate the mass transport and corrosion features of UHPC members. The oxygen diffusivity, and chloride permeability are reported to be at least one order of magnitude smaller than normal

DOI 10.1201/9781003316404-69

reinforced concrete (Voort 2008). The corrosion rate of a UHPC member was approximately two times lower than that of reinforced high performance concrete according to Ghafari et al. (2015). However, the previously mentioned research did not include cracking of UHPC. Although the cracks in UHPC are well distributed, the material's durability performance could be adversely influenced by the multiple cracks induced by the combined mechanical loading and environmental conditions(Lv et al. 2021).

Among the limited studies considered cracking effects in UHPC corrosion, most are focused on the material transport phase of cracked UHPC (Lv et al. 2021). However, the cracking in the reinforced UHPC system has an impact on material transport and corrosion propagation stage. A fully coupled time-dependent multi-physics analysis method considering the damage effects throughout the corrosion process is needed to better predict the service life of reinforced UHPC system.

The steel's corrosion under coupled sustained loading and chloride ingress is a dynamic process. The cracks accelerate the corrosion of the steel, and in return, the accelerated corrosion of the steel causes more damage to the reinforced system. A time-dependent model that updates the corrosion initiation status and corrosion propagation conditions in each time step has rarely been implemented in reinforced concrete (Firouzi et al. 2020; Li et al. 2021), while no such fully coupled studies have been conducted on UHPC system to the authors' knowledge to date.

This paper presents a fully coupled time-dependent multi-physics numerical study on the steel corrosion of reinforced concrete and reinforced UHPC beams under sustained loading and chloride ingress. The comparison study of reinforced concrete and reinforced UHPC systems accounts for differences in electrochemical process and mechanical properties. This study allows the complex process of structural deterioration under combined mechanical and environmental loading conditions to be investigated in an efficient and reliable way.

2 SIMULATION DESCRIPTION AND MODELING PARAMETERS

In this study, mechanical models were coupled to chloride diffusion models and steel corrosion models. A time-dependent simulation framework incorporating different simulation platforms was integrated. Multiple phenomena were simulated in the study, such as mass transport in porous media, electrochemical process, and structural responses.

2.1 *Analysis procedures*

A time-dependent simulation procedure was integrated in the diffusion and corrosion simulations. First, a structural analysis was conducted for simply supported beams under three point bending. The transport properties of the cementitious systems were then assigned based on the simulated crack widths under a service loading, set to 80% of the yielding load level. A diffusion model was set up to study the chloride transport process in the reinforced concrete and reinforced UHPC systems considering the damage condition and cracking from the initial structural analysis. At each time step, the anode area was determined according to the chloride concentration at the steel surface. The steel corrosion behavior was modeled based on this information, and a rust expansion load was applied to the structural model at the steel-concrete interface. The process was repeated until next time step. The time intervals for reinforced concrete and reinforced UHPC systems were 180 days and 720 days, respectively.

2.2 *Mechanical model setup*

The nonlinear finite element model set up in DIANA FEA Version 10.5 (DIANA 2021) is shown in Figure 1. The longitudinal reinforcement was symmetrically placed on top and bottom with areas of 258 mm^2, respectively. The transverse reinforcement was 16 mm^2 in area and was placed with spacing of 75 mm. Element size was 10 $mm \times 10\ mm$. The beam was simply supported and the span was 2160 mm. Rust thickness was calculated from (Böhni 2005):

$$\sigma(t) = \frac{\int_0^t i_{corr}(t)dt \cdot M_s}{Z_{Fe} \cdot F \cdot \rho_s} \tag{1}$$

where t is the corrosion time (seconds), $M_s = 55.85\ g/mol$ is the atomic mass of the iron, $Z_{Fe} = 2$ is the valency of anodic reaction, $F = 96485\ C/mol$ is the Faraday's constant, $\rho_s = 7800\ kg/m^3$ is the steel density. The rust expansion thickness was applied at the steel concrete interface as a displacement load and was obtained from:

$$u(t) = (n - 1)\sigma(t) \tag{2}$$

where n is rust to steel volume expansion ratio and is assumed to be 3 in this study (Cao et al. 2013).

2.3 *Diffusion and corrosion model setup*

The two-dimensional diffusion and corrosion modeling was conducted in COMSOL Multiphysics Version 5.4 (COMSOL 2021). A 25mm size triangular element was used. Oxygen were assumed to enter from four sides of the beam while chloride was assumed to ingress only from the bottom side of the beams. The diffusion coefficient of chloride in the cracked concrete was calculated from (Djerbi et al. 2008):

$$D_{Cl\_concrete} = \begin{cases} 2 \times 10^{-11} w - 4 \times 10^{-10}, \\ \quad 30\mu m \leq w \leq 80\mu m \\ 14 \times 10^{-10}, w \geq 80\mu m \end{cases} \tag{3}$$

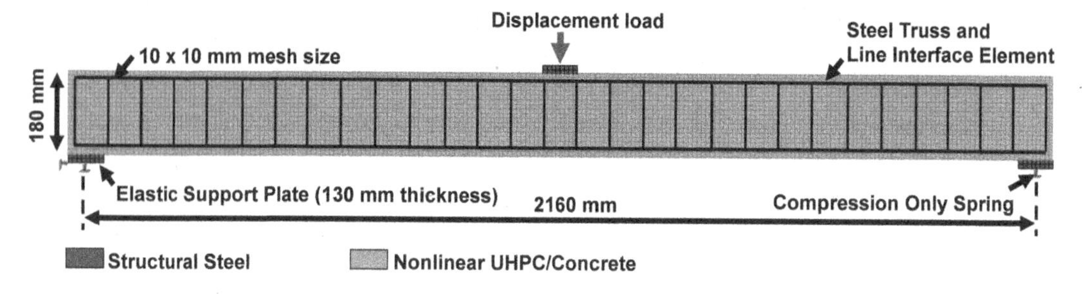

Figure 1. Mechanical model set up and geometry configuration.

where w is crack width (μm). A modified equation for diffusion coefficient in cracked UHPC based on data in literature and inverse analysis is proposed by the authors of this study (e.g., equation 4).

$$D_{Cl\_UHPC} = \begin{cases} 4 \times 10^{-12}w - 3 \times 10^{-11}, \\ \quad 10\mu m \leq w \leq 80\mu m \\ 3 \times 10^{-10}, w \geq 80\mu m \end{cases} \quad (4)$$

Oxygen transport in cracks is described as follows:

$$D_{O_2}^{crack} = \begin{cases} D_{O_2}^{sound}, w \leq w_{cr} \\ D_{O_2}^{sound} \times (w/w_{cr})^3, w \geq w_{cr} \end{cases} \quad (5)$$

where w_{cr} is the critical crack width and is adopted as $100\mu m$. The oxygen transport in the cracked concrete and UHPC was assumed to be the same due to insufficient test data on air permeability of the materials (Beglarigale et al. 2021; Guo et al. 2019).

2.4 *Parameters in the simulations*

Surface chloride concentration (Cl_{surf}) was assumed to be 2% of the concrete/UHPC mass (Cao 2014). Surface oxygen concentration (O_{2surf}) was 0.268 mol/m^3 (Cao 2014). Mass transport properties and electrochemical reaction parameters such as chloride and oxygen diffusion coefficients (D_{Cl}, D_{O_2}), Tafel slopes (β_{Fe}, β_{O_2}), equilibrium potentials (ϕ_{Fe}^0, $\phi_{O_2}^0$), and exchange current densities (i_{Fe}^0, $i_{O_2}^0$) are summarized in Table 1 (Rafiee 2012). The critical chloride content (Cl_{crit}) of concrete was adopted as 0.06% of concrete mass (Isgor & Razaqpur 2006), which is one magnitude of order smaller than that of UHPC (0.65% of UHPC mass) (Dauberschmidt 2006). The mechanical properties such as compressive strength, modulus of elasticity, tensile and compressive fracture energy, PoissonâŁ™s ratio strain at crack initiation, and strain at onset of softening were collected from literature (Moreno-Luna 2014; Shafieifar et al. 2017). Modulus of elasticity and yield strength of the reinforcement bar were assumed from experiments (Bandelt & Billington 2016). The mechanical properties are summarized in Table 2.

Table 1. Mass transport and corrosion polarization parameters.

Input parameters	Units	Values	Resources
Cl_{surf}	%	2	1
O_{2surf}	mol/m^3	0.268	1
$D_{Cl\_concrete}$	m^2/s	1.3E-11	3
$D_{O_2\_concrete}$	m^2/s	3.02E-9	3
$\theta_{\_concrete}$	S/m	0.0063	3
$Cl_{crit\_concrete}$	%	0.06	2
$\beta_{Fe\_concrete}$	mV/dec	65	3
$\beta_{O_2\_concrete}$	mV/dec	-138.6	3
$\phi^0_{Fe\_concrete}$	mV	-600	3
$\phi^0_{O_2\_concrete}$	mV	200	3
$i^0_{Fe\_concrete}$	A/m^2	2.75E-4	3
$i^0_{O_2\_concrete}$	A/m^2	6E-6	3
D_{Cl_UHPC}	m^2/s	4.5E-13	3
$D_{O_2\_UHPC}$	m^2/s	4.2E-10	3
$\theta_{\_UHPC}$	S/m	4.33E-5	3
Cl_{crit_UHPC}	%	0.65	4
β_{Fe_UHPC}	mV/dec	61	3
$\beta_{O_2\_UHPC}$	mV/dec	-130.9	3
$\phi^0_{Fe\_UHPC}$	mV	-600	3
$\phi^0_{O_2\_UHPC}$	mV	200	3
$i^0_{Fe\_UHPC}$	A/m^2	2.75E-4	3
$i^0_{O_2\_UHPC}$	A/m^2	6E-6	3

[1] Cao et al. (2013),[2] Isgor and Razaqpur (2006),[3] Rafiee (2012),] [4] Dauberschmidt (2006).]

3 SIMULATION RESULTS AND DISCUSSION

3.1 *Damage patterns*

Figure 2 shows the cracking strain contour of the reinforced concrete and reinforced UHPC system before and after corrosion. The reinforced UHPC beam showed excellent crack suppression during the corrosion propagation phase. As illustrated in Figure 2,

Table 2. Mechanical properties.

Properties	Unit	UHPC	Normal Concrete	Longitudinal Steel	Transverse Steel
Tensile strength	MPa	9.7	3.98	–	–
Strainat crack initiation	%	0.019	0.01	–	–
Strain at onset of softening	%	0.2	0.01	–	–
Tensile fracture energy	MPa-mm	19.1	0.154	–	–
Compressive strength	MPa	138	63.4	–	–
Compressive fracture energy	MPa-mm	184	38.5	–	–
Modulus of elasticity	GPa	54.7	39.8	200	205
Poisson's ratio	mm/mm	0.18	0.18	0.30	0.30
Yield strength	MPa	–	–	455	690
Ultimate strength	MPa	–	–	675	–
Strain at hardening	%	–	–	1.36	–
Strain at ultimate strength	%	–	–	16	–
Resources		1,2	3	4	4

[1] Shafieifar et al. (2017), [2] Moreno-Luna (2014), [3] de Putter (2020), [4] Bandelt and Billington (2016).]

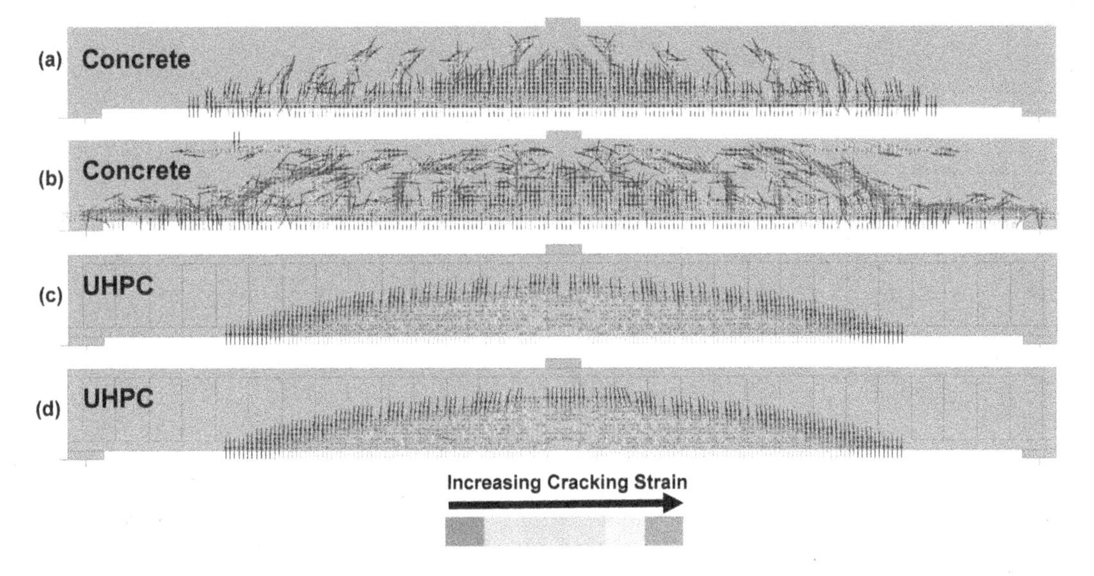

Figure 2. Cracking strain contour of reinforced concrete and reinforced UHPC (a) and (c) before corrosion, (b) 3 years after corrosion, (d) 12 years after corrosion.

the damage area of reinforced concrete beam experienced an increase of 77.0% while the UHPC beam increased 3.1%.

3.2 Chloride concentrations

The chloride accumulated quickly in reinforced concrete beams and reached 2% of concrete mass within half a year. However, the chloride concentration at the same location in the reinforced UHPC beam did not exceed 1% of UHPC mass. Chloride penetration depth at the end of three years of chloride exposure was shown in Figure 3. The chloride concentration of the reinforced concrete beam remained steady and was close the surface concentration within a 50*mm* range from the chloride ponding surface. However, the chloride concentration of the reinforced UHPC beam

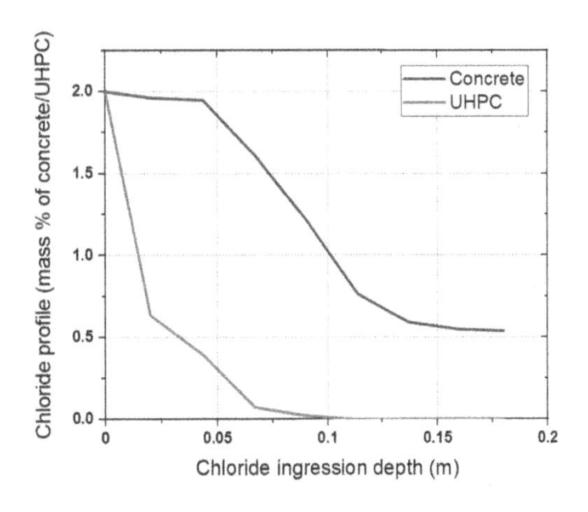

Figure 3. Chloride profile along beam depth at the midspan.

dropped sharply within the same depth. It was because of the dense material properties and distributed fine cracks of UHPC, which suppressed the diffusion of chloride process.

3.3 *Corrosion product distributions*

From the simulation results, the rust thickness in reinforced concrete developed much faster than that of reinforced UHPC beam. The reinforcing bar closer to chloride exposed soffit of the beams (bottom reinforcing bar) had a rust thickness of 189μm after 3 years of chloride exposure. In contrast, the bottom reinforcing bar of the reinforced UHPC beam had a rust thickness of 80.8μm after 12 years.

The corrosion product was distributed uniformly along the bottom reinforcing bar in reinforced concrete beams while the corrosion product was localized in the center of the beam in reinforced UHPC beams. The uniform corrosion pattern in reinforced concrete beams was due to the rapid chloride ingress in the cracked zone of the specimen. However, most of the cracks in reinforced UHPC that were further away from the midspan were too small to allow for accelerated chloride diffusion.

4 CONCLUSIONS

The paper presents a time-dependent multi-physics simulation of concrete structural durability. The framework proposed here is able to capture the corrosion and structural response of conventional and ductile concrete systems in a time efficient manner.

The simulation results show that the UHPC material exhibited excellent resistance to chloride intrusion and corrosion propagation. This behavior is attributed to the smaller crack widths and higher material density in reinforced UHPC system. The corrosion product development in reinforced concrete was twelve times faster than that of reinforced UHPC. The reinforced concrete beam had a rust layer thickness of 79μm after one year while the reinforced UHPC beam developed a similar level of rust layer thickness (80.8μm) after twelve years.

The reinforced concrete beam deteriorated faster than the reinforced UHPC beam under sustained mechanical loading and rust expansion. The damaged area and cracking in the reinforced concrete beams provided a path for chloride and oxygen, which promoted the corrosion initiation and propagation, respectively.

A coupled mechanical-electrochemical model considering time-dependent loading and environmental conditions provides an approach to predict the service life performance of reinforced concrete structures. Furthermore, the time-dependent multi-physics models can be extended to various loading conditions and material systems.

ACKNOWLEDGMENT

The authors gratefully acknowledge the support of John A. Reif, Jr. Department of Civil and Environmental Engineering at New Jersey Institute of Technology and by the New Jersey Department of Transportation (NJDOT) through Contract ID# 19-60155.

REFERENCES

Alkaysi, M., S. El-Tawil, Z. Liu, & W. Hansen (2016). Effects of silica powder and cement type on durability of ultra high performance concrete (uhpc). *Cement and Concrete Composites 66*, 47–56.

Bandelt, M. J. & S. L. Billington (2016). Impact of reinforcement ratio and loading type on the deformation capacity of high-performance fiber-reinforced cementitious composites reinforced with mild steel. *Journal of Structural Engineering 142*(10), 04016084.

Beglarigale, A., D. Eyice, B. Tutkun, & H. Yazıcı (2021). Evaluation of enhanced autogenous self-healing ability of UHPC mixtures. *Construction and Building Materials 280*, 122524.

Böhni, H. (2005). *Corrosion in reinforced concrete structures*. Elsevier.

Broomfield, J. (2003). *Corrosion of steel in concrete: understanding, investigation and repair*. CRC Press.

Cao, C. (2014). 3D simulation of localized steel corrosion in chloride contaminated reinforced concrete. *Construction and Building Materials 72*, 434–443.

Cao, C., M. M. Cheung, & B. Y. Chan (2013). Modelling of interaction between corrosion-induced concrete cover crack and steel corrosion rate. *Corrosion Science 69*, 97–109.

COMSOL (2021). COMSOL Multi-physics. *https://www.comsol.com/*.

Dauberschmidt, C. (2006). Untersuchungen zu den korrosionsmechanismen von stahlfasern in chloridhaltigem beton. *Technischen Hochschule Aachen*.

de Putter, A. (2020). Towards a uniform and optimal approach for safe NLFEA of reinforced concrete beams: quantification of the accuracy of multiple solution strategies using a large number of samples.

DIANA (2021). DIANA FEA. *https://dianafea.com/*.

Djerbi, A., S. Bonnet, A. Khelidj, & V. Baroghel-Bouny (2008). Influence of traversing crack on chloride diffusion into concrete. *Cement and Concrete Research 38*, 877–883.

Firouzi, A., M. Abdolhosseini, & R. Ayazian (2020). Service life prediction of corrosion-affected reinforced concrete columns based on time-dependent reliability analysis. *Engineering Failure Analysis 117*, 104944.

Ghafari, E., M. Arezoumandi, H. Costa, & E. Julio (2015). Influence of nano-silica addition on durability of UHPC. *Construction and Building Materials 94*, 181–188.

Guo, J.-Y., J.-Y. Wang, & K. Wu (2019). Effects of self-healing on tensile behavior and air permeability of high strain hardening UHPC. *Construction and Building Materials 204*, 342–356.

Isgor, O. B. & A. G. Razaqpur (2006). Modelling steel corrosion in concrete structures. *Materials and Structures 39*(3), 291–302.

Li, C.-z., X.-b. Song, & L. Jiang (2021). A time-dependent chloride diffusion model for predicting initial corrosion time of reinforced concrete with slag addition. *Cement and Concrete Research 145*, 106455.

Lv, L.-S., J.-Y. Wang, R.-C. Xiao, M.-S. Fang, & Y. Tan (2021). Chloride ion transport properties in microcracked ultra-high performance concrete in the marine environment. *Construction and Building Materials 291*, 123310.

Moreno-Luna, D. M. (2014). *Tension stiffening in reinforced high performance fiber reinforced cement based composites*. Stanford University.

Rafiee, A. (2012). *Computer modeling and investigation on the steel corrosion in cracked ultra high performance concrete*, Volume 21. kassel university press GmbH.

Shafieifar, M., M. Farzad, & A. Azizinamini (2017). Experimental and numerical study on mechanical properties of Ultra High Performance Concrete (UHPC). *Construction and Building Materials 156*, 402–411.

Sohail, M. G., R. Kahraman, N. Al Nuaimi, B. Gencturk, & W. Alnahhal (2021). Durability characteristics of high and ultra-high performance concretes. *Journal of Building Engineering 33*, 101669.

Voort, T. L. V. (2008). *Design and field testing of tapered H-shaped Ultra High Performance Concrete piles*. Iowa State University.

Zhao, Y. & W. Jin (2016). *Steel corrosion-induced concrete cracking*. Butterworth-Heinemann.

Computational Modelling of Concrete and Concrete Structures – Meschke, Pichler & Rots (Eds)
© 2022 Copyright the Author(s), ISBN: 978-1-032-32724-2

A novel DEM based pore-scale thermo-hydro-mechanical model

M. Krzaczek, M. Nitka & J. Tejchman
Gdansk University of Technology, Gdańsk, Poland

ABSTRACT: Concrete is a strongly heterogeneous, discontinuous and porous material. Under non-isothermal conditions, the movement of fluid in the pore and capillary system is strongly coupled with heat transfer. In such conditions, the pores and cracks facilitate the penetration of external aggressive agents into concrete that degrades both concrete and reinforcement. An innovative DEM-based thermo-hydro-mechanical model was developed to track in detail the liquid/gas fractions in pores and cracks with respect to their different geometry, size, location and temperature. A coarse 2D mesh was generated to create a fluid flow network and to solve the energy conservation equation. The thermo-hydro-mechanical model was verified by comparing the results with the analytical solution to the problem of one-dimensional heat transfer in a solid. Finally, the relevance of a fully coupled thermo-hydro-mechanical model is illustrated by the simulation of an experiment in which a saturated porous specimen is subjected to a cooling process.

1 INTRODUCTION

Most of physical phenomena in engineering problems occur under non-isothermal conditions. Moreover, even if the physical system is initially in a state of thermodynamic equilibrium, the physical phenomena or chemical reactions that occur may lead to local temperature changes and, consequently, to heat transfer. Therefore, understanding heat transfer in particulate systems is of great interest to many scientific disciplines and engineering applications such as environmental science, chemical and food processing, powder metallurgy, energy management, geotechnics, or geological engineering. The need to take into account the effect of heat transfer becomes critical in the analysis of many multi-field problems in porous and fractured materials. Complex thermo-hydro-mechanical (THM) processes, including heat transfer, fluid flow, and material deformations occur simultaneously and are affected by many non-linear processes.

The most common approach in THM models is the continuous medium phenomena approach, which is based on a mathematical framework linking sets of equations to describe the laws of thermodynamics, solid mechanics, and hydraulics, e.g. the finite element implementations of such concepts (Kolditz et al. 2012; Olivella et al. 1996; Rühaak and Sass 2013; Selvadurai et al. 2015; Tang et al. 2021; Zarei-darmiyan et al. 2020) or the finite difference scheme (Rutqvist et al. 2002). Nonetheless, even though attractive for macro-scale applications, continuum modeling approaches based on the finite element method (FEM) or the finite volume method (FVM) suffer critical computational and continuity limitations when applied to discontinuous and highly deformable media such as packed or fluidized beds and granular or fractured porous materials.

In porous media with low porosity (less than about 15%), such as concrete or rocks, classical methods lead to huge problems with generating sufficiently fine mesh (Abdi et al. 2022). The problem increases when simulating the crack initiation and propagation process in porous materials with low porosity (e.g. concrete).

On the other hand, discrete approaches like, for instance, the discrete element method (Cundall & Strack 1979) or the finite-discrete element method (FDEM) prove successful at modeling the behavior of these discrete systems. FDEM method was used by Yan and Zheng 2017 to formulate a thermo-mechanical model for simulating thermal cracking of rock and by Yan et al. (2022) to develop a 2D coupled thermal-hydro-mechanical model for describing rock hydraulic fracturing.

The strength of DEM in modeling particulate systems opened up recent efforts to extend its predictive capabilities to meso- and micro-scale THM processes. Different approaches have been used to couple DEM with fluid flow and heat transfer models. Direct numerical simulations (DNS) can be used to couple TH processes with DEM. To solve governing equations, DNS models can use different numerical methods (e.g. FEM, FVM). Deen et al. (2012) proposed immersed boundary implementation that does not require using any effective diameter. The method was dedicated to THM processes in dense fluid–particle systems. However, the proposed method was limited to invariant geometries, their topologies and relatively high porosity (porosity greater than that of concrete or rock). In practice, DNS-DEM models are restricted to systems comprised of a smaller number of particles than CFD.

DOI 10.1201/9781003316404-70

Another approach is based on the lattice Boltzmann method (LBM) (Chen et al. 2018; Jiao et al. 2021; Yang et al. 2017). LBM relies strongly on the accurate representation of solid-fluid boundaries which can be difficult both numerically and computationally and can lead to the same limitations as with DNS-DEM models.

The use of DEM to study THM processes in very dense fluid-particle systems with very low porosity (even below 5%), such as concrete or rocks, fluid flow and heat transfer models should be simplified in order to reduce computational costs and avoid numerical limitations. Most of the recently available DEM-based THM models separate the fluid flow in the reservoirs (pores, macropores, pre-existing cracks, etc.) and flow between the reservoirs. The assumption is that the fluid in the reservoirs is compressible, while the fluid flowing between the reservoirs is incompressible. This concept of simplification was first introduced and developed by (Al-Busaidi et al. 2005; Cundall 2000; Hazzard et al. 2002). In addition to highlighting the properties of fluids, fluid flow regimes are also distinguished. The fluid flow regime is stagnant or close to stagnation in the reservoirs, while the fluid flow between the reservoirs is laminar to estimate the mass flow rates at the reservoirs' boundaries. In most cases, a Poiseuille flow model in pipes or between two parallel plates is adopted to estimate mass flow rates. The pressure in the pore is computed directly from assumed the equation of state (Al-Busaidi et al. 2005; Hazzard et al. 2002) or solving Stocks equation (Catalano et al. 2014; Papachristos et al. 2017). All models assume a single-phase fluid flow of a pure liquid or mixture. However, in the case of a mixture, the phase fractions are not tracked. The fluid is barotropic according to the assumption. In this approach, different heat transfer models are coupled with DEM-CFD approaches. Tomac and Gutierrez 2017 solved the energy conservation equation for each cell (reservoir) volume. The adopted energy conservation equation corresponded to the energy transport in the laminar flow of an incompressible fluid. Caulk et al. (2020) proposed more advanced DEM-based THM model. They proposed a 3D model based on the framework of the pore-scale finite volume (PFV) scheme initially proposed by Chareyre et al. (2012) and extended by Scholtès et al. (2015) for up-scaling compressible viscous flow and oriented toward dense grain packing applications.

The goal of the current paper is to demonstrate the DEM-based pore-scale thermal-hydro-mechanical model of two-phase fluid flow coupled with heat transfer in porous materials of very low porosity (e.g. concretes). Calculations were carried out with a 3D DEM model coupled with a 2D CFD and 2D heat transfer model that combined solid mechanics with fluid mechanics and heat transfer at the meso-scale. Previously, our coupled DEM/CFD model was successfully used to describe a hydraulic fracturing process in rocks with one- or two-phase laminar viscous two-phase fluid flow composed of a liquid and gas (Krzaczek et al. 2020, 2021).

The innovative elements of our DEM-based THM mesoscopic approach for modelling fluid flow and heat transfer as compared to other existing models in the literature are: the detailed tracking of water/gas fractions in pores regarding their varying geometry, size and location; an algorithm for automatic meshing and remeshing domains of solids and fluids to capture changes in geometry and their topology; the use of a coarse mesh of solid and liquid domains to generate a virtual fluid flow network (VPN) and solve energy conservation equation; adoption of the corrected Peng-Robinson equation of state for both fluid phases to study supercritical fluids flow (necessary e.g. for the study of THM processes in the hydrofracturing process); FVM was used to solve the energy conservation equation on a very coarse mesh of cells in both domains.

The current paper is structured as follows. After the introductory Section 1, a mathematical model of the DEM based coupled thermal-hydro-mechanical approach is presented. The model validation is presented in section 3. Section 4 analyzes the effect of advection on the cooling of a cohesive granular bar specimen. Section 5 discusses the mechanism of damage of the cohesive granular bar due to cooling. Finally, some concluding remarks are offered in Section 6.

2 THERMO-HYDRO-MECHANICAL MODEL

The novel concept of the model is based on the assumption that two different domains coexist in a physical system: the 3D discrete (solid) domain and the 2D continuous (fluid) domain. Originally, the solid domain is consisted of one layer of 3D spherical elements while the fluid domain is two-dimensional (Figure 1a).

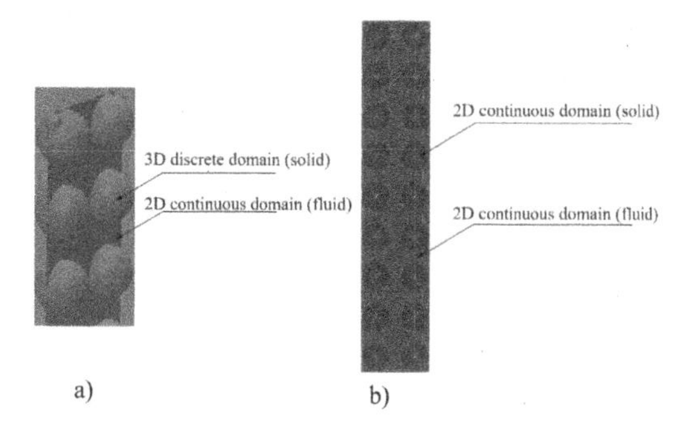

Figure 1. Two domains coexisting in one physical system: a) co-existing domains before projection and discretization, b) solid and fluid domains after discrete elements projection and discretisation (fluid domain in red colour and solid domain in black colour).

Spheres are arranged in such a way that their gravity centres are located on a mid-plane (2D surface). The spheres are projected onto the plane to form circles (Figure 1b). After projection, both the domains are discretized into a very coarse grid of cells (triangles)

(Figure 1b). Consequently, the equations of motion of discrete elements are solved in the 3D discrete domain and the equations of fluid flow are solved in the 2D fluid continuous domain (red colour in Figure 1b) and heat transfer equations are solved in the 2D fluid and solid continuous domains (red and black colours in Figure 1b).

2.1 *DEM for cohesive-frictional materials*

DEM calculations were performed with the 3D spherical explicit discrete element open code YADE (Kozicki & Donzé 2008; Šmilauer & Chareyre 2011). The method allows for a small overlap between two contacted bodies (the so-called soft-particle model). Thus, an arbitrary micro-porosity can be obtained in DEM wherein particles interact with each other during translational and rotational motions through a contact law and Newton's 2nd law of motion using an explicit time-stepping scheme (Cundall & Strack 1979). In the model, a cohesive bond is assumed at the grain contact exhibiting brittle failure under the critical normal tensile load. The shear cohesion failure initiates contact slip and sliding obeying the Coulomb friction law under normal compression. Damage occurs if a cohesive joint between spheres disappears after reaching a critical threshold. If any contact between spheres after failure re-appears, the cohesion does not appear more. A simple local non-viscous damping is used (Cundall & Strack 1979) to accelerate convergence in quasi-static analyses. The material softening is not considered in the DEM model.

In general, the material constants are identified in DEM with the aid of simple laboratory tests on the material (uniaxial compression, uniaxial tension, shear, biaxial compression). The detailed calibration procedure for frictional-cohesive materials was described in (Nitka & Tejchman 2015; Suchorzewski et al. 2018).

The DEM model demonstrated its usefulness for both local and global simulations of macro- and micro-cracks in concretes under bending (2D and 3D analyses) (Nitka & Tejchman 2015) uniaxial compression (2D and 3D simulations) (Caggiano et al. 2018) and splitting tension (2D analyses) (Skar¿yñski et al. 2015). The combined DEM/x-ray μ-CT images mesoscopic approach proved to be an extremely appealing computational tool for investigating fracture in concrete. In those calculations, ITZs had no either physical width and were simulated by weaker contacts between aggregates and cement matrix (Nitka & Tejchman 2015; Skar¿yñski et al. 2015) or had a defined width with higher porosity (Nitka & Tejchman 2020). To study combined mechanical-hydro-thermal problems in concrete, the second approach should be used.

2.2 *Fluid flow model*

The general concept of a fluid flow algorithm using DEM was adopted from (Al-Busaidi et al. 2005; Cundall 2000; Hazzard et al. 2005). The model in the current paper significantly differs from this general concept. The reservoirs (pores, cracks, pre-existing cracks, etc.) store now not only pressures but also phase fractions, fluids densities, energy and temperature. The continuity equation is employed to compute the density of fluid phases stored in reservoirs. The fluid phase fractions in reservoirs are computed by applying the equation of state for each phase assuming that fluid phases share the same pressure (as in the Euler model of multi-phase flow). The mass flow rate in artificial channels of fluid flow network is now estimated by solving continuity and momentum equations for laminar flow of incompressible fluid.

The gravity centres of the 3D spheres are located on the XOY plane. The 3D spherical particles are projected onto the 2D midplane and next discretized into the 2D polygons (Krzaczek et al. 2020). To get a more realistic distribution of the unknown variables (pressure, fluid-phase fractions and densities), a remeshing procedure discretizes the overlapping circles, determined the contact lines and deletes the overlapping areas (Krzaczek et al. 2020). As a result, each reservoir is discretized into a number of triangles (in 2D problem). Each triangle in the fluid domain is called the Virtual Pore (VP) (Figure 3). The artificial channels connect the gravity centres of triangles (VPs) to create a fluid flow network called the Virtual Pore Network (VPN). VPs accumulate pressure, store both fluid-phase fractions and densities, energy and temperature. The mass change in VPs is related to the density change in a fluid phase that results in pressure variations. There is no fluid flow in triangles by assumption (flow regime is close to stagnant). The equation of momentum conservation is thus neglected in triangles but the mass is to be still conserved in the entire volume of triangles. The numerical algorithm can be divided into 5 main stages:

a) estimating the mass flow rate for each phase of fluid flowing through the cell faces (in channels surrounding VP) by employing continuity and momentum equations,
b) computing the phase fractions and their densities in VP by employing equations of state and continuity,
c) computing pressure in VP by employing the equation of state,
d) solving energy conservation equation in fluid and solids,
e) updating material properties.

This algorithm is repeated for each VP in VPN and each solid cell (stage (d)) using an explicit formulation. According to the above algorithm, incompressible laminar two-phase fluid (liquid/gas) flow under non-isothermal conditions is assumed in the channels of the fluid flow network. The liquid and gas initially exist in the matrix and pre-existing discontinuities. Two channel types in the Virtual Pore Network are introduced (Krzaczek et al. 2020): (A) the channels between discrete elements of the material matrix in contact (called the 'S2S' channels) and (B) the channels connecting grid triangles in pores that touch each other by a common edge (called the 'T2T' channels).

The channel length is assumed to be equal to the distance between the gravity centres of adjacent grid triangles. In real 3D problems, the fluid flows around the spheres in contact. However, in 2D problems, there is no free space for fluid flow. Therefore, the concept of virtual S2S channels is introduced (Krzaczek et al. 2020).

2.2.1 *Mass flow rate estimation in channels*

Three flow regimes in the VPN channels are distinguished: a) single gas-phase flow with gas phase fraction $\alpha_p = 1$, b) single liquid-phase flow with liquid phase fraction $\alpha_q = 1$ and c) two-phase flow (liquid and gas) with $0 < \alpha_q < 1$. For single-phase flow in channels (flow regime 'a' and 'b'), the fluid moves in channels through a thin film region separated by two closely spaced parallel plates.

A two-phase flow of two immiscible and incompressible fluids in a channel is assumed to simulate two-phase fluid flow (flow regime 'c'), driven by a pressure gradient in adjacent VPs. The liquid-gas interface is parallel to channel plates and constant along the channel. Gravity forces are neglected. The interface between the fluids, labelled as $j = q,p$ (q – the lower liquid-phase, p – the upper gas-phase), is assumed to be flat in the undisturbed flow state. Under this assumption, the model allows for a plane-parallel solution. The interface position is known and is related to fractions of fluid phases in adjacent VPs while the volumetric flow rates of fluid phases are unknown.

The method of the mass flow rate estimation in channels was described in detail by Krzaczek et al. 2021.

2.2.2 *Fluid flow in virtual pores*

Contrary to the model of fluid flow in the channels, the VPs assume that the fluid is compressible. In some problems, incl. during the hydraulic fracturing process, the fluid pressure exceeds 70 MPa. Under these conditions, the gas phase exceeds the critical point and becomes a supercritical fluid. To describe the behaviour of the fluid above the critical point, the Peng-Robinson equation of state is adopted for both the fluid phases in VP:

$$P = \frac{RT}{\left(V_{q/p} - b_{q/p}\right)} - \frac{a_{q/p}}{\left(V_{q/p}^2 + 2b_{q/p}V_q - b_{q/p}^2\right)}, \quad (1)$$

where P is the pressure [Pa], R = 8314,4598 J/(kmol K) is the gas constant, $V_{q/p}$ the molar volume of liquid (q) and gas (p) fraction [m^3/kmol] and T is the temperature [K]. Equations 1 provide a good fit of the vapor pressure for most substances but the prediction of molar volumes can be seriously in error. In particular, the prediction of saturated liquid molar volumes can be in error by 10–40% (Mathias et al. 1989). An effective correction term was already suggested by Peneloux and Rauzy 1982:

$$V_q^{corr} = V_q + s, \quad (2)$$

where s is the small molar volume correction term that is component dependent, V_q is the molar volume predicted by Eqn. 1 and V_q^{corr} refers to the corrected molar volume.

By solving the mass conservation equation for both the phases, the density of the liquid and gas phases can be computed. Since the fluid phases share the same pressure, the fluid phase fractions can be computed.

2.3 *Heat transfer in fluid*

A homogeneous heat transfer model in multiphase fluid flow is assumed. For simplicity, incompressible and homogeneous fluid is assumed. The viscous dissipation of energy is neglected. The energy conservation equation is shared among the phases in homogeneous model, and is expressed in integral form:

$$\int_V \frac{\partial}{\partial t}\left(\rho_{eff} E\right) \cdot dV + \oint \nabla \cdot \left(\rho_{eff} \vec{v} E\right) \cdot d\vec{A} = \oint \left(\lambda_{eff} \nabla T\right) \cdot d\vec{A} + \int_V S_h, \quad (3)$$

where ρ_{eff} is the effective fluid density [kg/m^3], E is the total energy [J], t is time [s], v is velocity vector [m/s], T is the temperature [K], λ_{eff} is the effective thermal conductivity of fluid [W/(mK)] and S_h is a source term that includes energy sources. Assuming an incompressible and laminar flow of a homogeneous fluid, the enthalpy h equation of state is:

$$h = \int_{T_{ref}}^{T} c_p dT, \quad (4)$$

where T_{ref} is the reference temperature [K] and c_p is specific heat in constant pressure [J/(kg·K)]. The effective fluid properties and velocity are computed by volume averaging over the phases. If the time derivative is discretized using backward differences and assuming that the total energy E is equal to the enthalpy h and applying the enthalpy h equation of state to Eq.3, the energy conservation equation can be expressed in terms of temperature T:

$$T^{n+1} = T_{ref} + \frac{c_{p,eff}^{n}\left(T^n - T_{ref}\right)}{c_{p,eff}^{n+1}} + \frac{\Delta t}{V\rho^n c_{p,eff}^{n+1}} \sum_f^{N_{faces}} \lambda_{eff} \nabla T_f^n \cdot \vec{A}_f - \frac{\Delta t}{V\rho^n c_{p,eff}^{n+1}} \times \sum_f^{N_{faces}} \rho_f^n \vec{v}_f^{\,n} c_{p,eff}^n \left(T_f^n - T_{ref}\right) \cdot \vec{A}_f + \frac{\Delta t}{\rho^{n+1} c_{p,eff}^{n+1}} S_h, \quad (5)$$

where N_{faces} is the number of faces enclosing the cell, T_f is the value of T on the face f, $\rho_f \quad v_f \cdot A_f$ is the mass flux through the face f, A_f is the area vector of the face f, ∇T_f is the gradient of T at the face f, V is the cell volume and S_h is related to the internal enthalpy source of diffusive energy [W/m^3] of heat transferred by diffusion along the channel S2S.

The FVM method is used to solve the energy conservation equation (Eqn. 5).

2.4 *Heat transfer in solids*

Assuming no convective energy transfer, no internal heat sources and constant density ρ_s, in solid regions, the energy conservation equation has the following integral form:

$$\rho_s \int_V \frac{\partial E}{\partial t} \cdot dV = \oint (\lambda_s \nabla T) \cdot d\vec{A} \tag{6}$$

where E is the total energy and is equal to enthalpy h, ρ_s is the density of solid [kg/m^3], λ_s is the solid thermal conductivity [W/(mK)], T_{ref} is the reference temperature and c_p is the specific heat in constant pressure.

Equation 6 is applied to each cell (triangle) in the solid domain. The discretization of Eq. 6 yieldsfor a given cell

$$T^{n+1} = T^n + \frac{\Delta t}{V \rho_s c_p} \sum_f^{N_{faces}} \lambda_s \nabla T_f^n \cdot \vec{A}_f \tag{7}$$

The FVM method is used to solve Eq.7.

3 VALIDATION OF THM MODEL

First, the THM model was calibrated. The calibration procedure was presented by Krzaczek et al. 2021. The model validation was carried out by comparing the numerical results with the analytical solution of the 1D heat transfer problem

$$\frac{\partial T}{\partial t} = \alpha_{eqv} \frac{\partial^2 T}{\partial x^2}, \tag{8}$$

where α_{eqv} is the effective value of thermal diffusivity [m^2/s] and t is time [s]. The analytical solution of the 1D heat equation is constrained by the following initial and boundary conditions: $T(x,t) = 323.16$ K for $x \in <0, L>$ and $T(0,t) = T(L,t) = 293.16$ K for $t \geq 0$, where L is the length of the bar specimen. The unsteady solution to Eq.8 was obtained using the Fourier series. For model validation, a specimen with a random distribution of spheres (Figure 2a) and porosity $p = 10.70\%$ was selected.

The assumed effective thermal diffusivity and boundary conditions imitated heat transfer only by diffusion in an equivalent solid bar, made of a fictitious homogeneous material with effective thermal properties (volume-averaged over the phases).The effective material properties of the equivalent solid were: $\lambda_{eff} = 3.357$ W/(m·K), $c_{p,eff} = 929.51$ J/(kg·K), $\rho_{eff} = 2422.74$ kg/m^3. The comparison is shown in Figure 3 for two different time steps (100 s and 400 s) and the corresponding Fourier numbers.

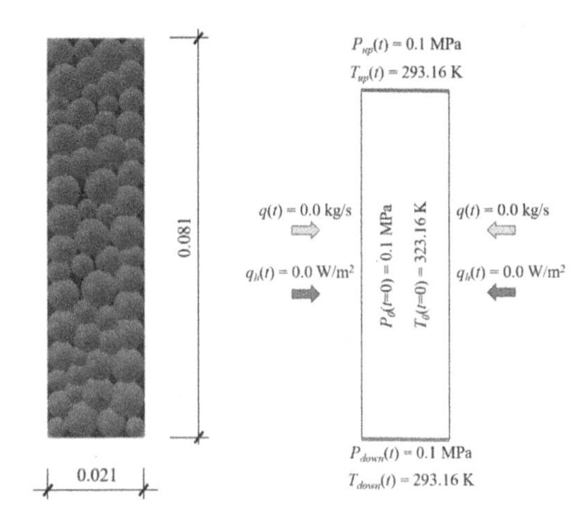

Figure 2. Cohesive granular bar specimen used to validating purposes: a) random distribution of spheres and b) initial and boundary conditions (q – fluid mass flow rate and q_h – heat transfer rate).

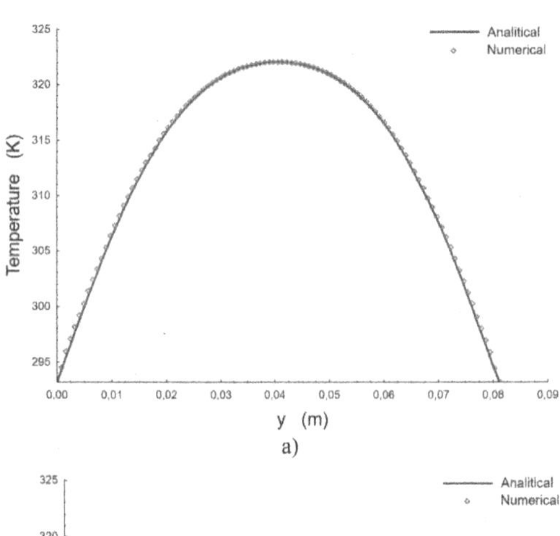

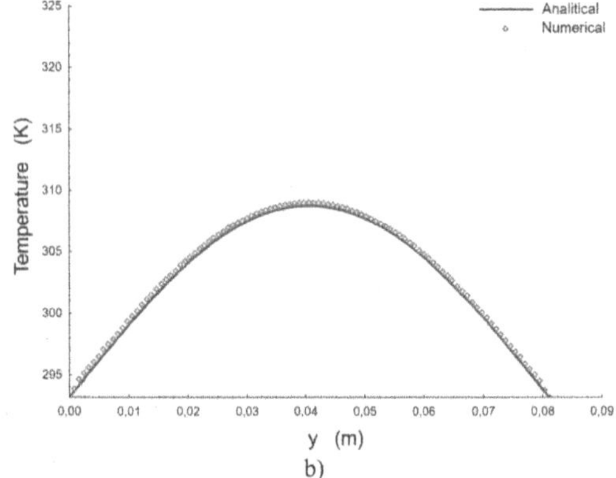

Figure 3. Temperature along vertical center line of bar specimen in Figure 2a: a) after 100 s ($F_o = 0.0227$), b) after 400 s ($F_o = 0.0907$).

The numerical results are in good agreement with the results of the analytical solution. The maximum discrepancy between numerical and analytical results was 0.52 K after 400 s of cooling. Figure 4 shows the temperature distribution and liquid density in the bar specimen after 400 s of cooling. The density varied from 1000.02 kg/m^3 to 1014.28 kg/m^3. The Peng-Robinson equation of state (Eq. 1) with correction (Eq. 2) introduced a very small error (less than 1.3%) in the estimation of the water density. The density of the fluid was slightly overestimated.

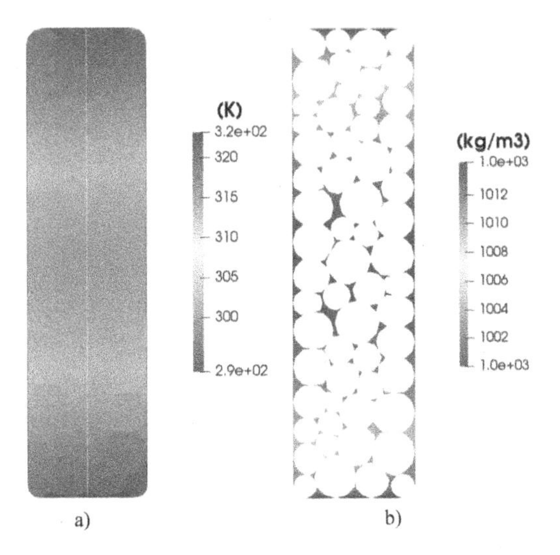

Figure 4. Cohesive granular bar specimen with random distribution of spheres after 400 s of cooling: a) temperature distribution and b) liquid density distribution.

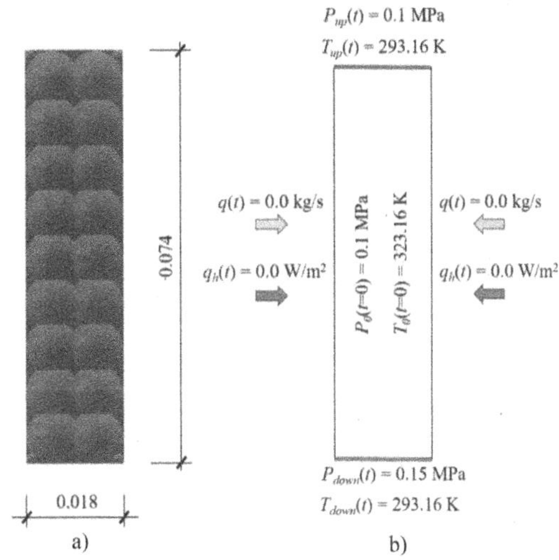

Figure 5. Specimen used for validating purposes (diffusion and advection): a) structured distribution of spheres and b) initial and boundary conditions (q – fluid mass flow rate and q_h – heat transfer rate).

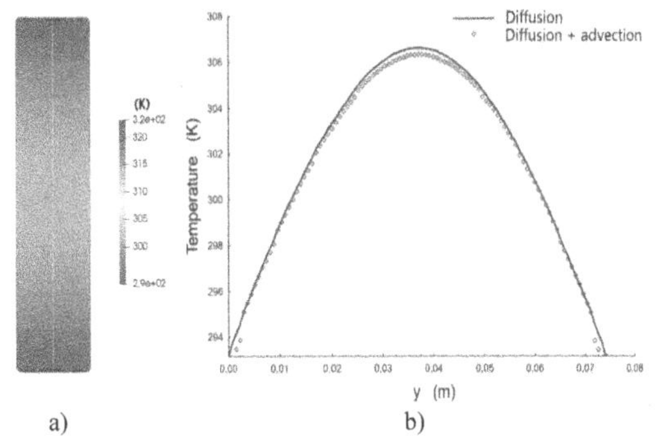

Figure 6. Temperature in cohesive granular bar specimen after 400 s of cooling (diffusion and advection): a) in entire specimen and b) along vertical center line.

4 ADVECTION INFLUENCE ON COHESIVE GRANULAR BAR COOLING

The influence of advection on the cooling of the bar specimen was investigated. The specimen of a structured distribution of spheres (Figure 5a) was chosen. The simulation results were compared with the simulation results of the diffusion test (not presented here) carried out on a specimen with a structured distribution of spheres. The single-phase flow of water was assumed. The adopted initial and boundary conditions (Figure 5b) simulated heat transfer by diffusion and advection in the bar.

The maximum temperature difference between cooling by diffusion and diffusion with advection was 2.26 K after 400 s of cooling (Figures 6a and 6b). Advection slightly speeded up the cooling process. However, it should be noted that the pressure difference between the lower and upper boundary was very small, 0.05 MPa (Figure 7a). This resulted in very low fluid velocity. The maximum fluid velocity did not exceed 1.8·10-5 m/s (Figure 7c). Generally, the velocity vectors were parallel to the vertical boundaries (Figure 7c) which confirmed a 1D fluid flow in the specimen. The fluid pressure varied almost linearly along the bar (Figure 7a) from the lower boundary to the upper boundary. The fluid density ranged from 1002.3 kg/m^3 to 1010.5 kg/m^3 (Figure 7b) after 400 s of cooling. The Peng-Robinson equation of state (Eqn. 1) with correction (Eqn. 2) introduced a very small error (less than 1.3%) in the estimation of the density. The fluid density was slightly overestimated.

5 THERMAL EXPANSION TEST OF COHESIVE GRANULAR BAR DURING COOLING

The bar specimen of random sphere distribution (Figure 2a) was used to perform a thermal expansion

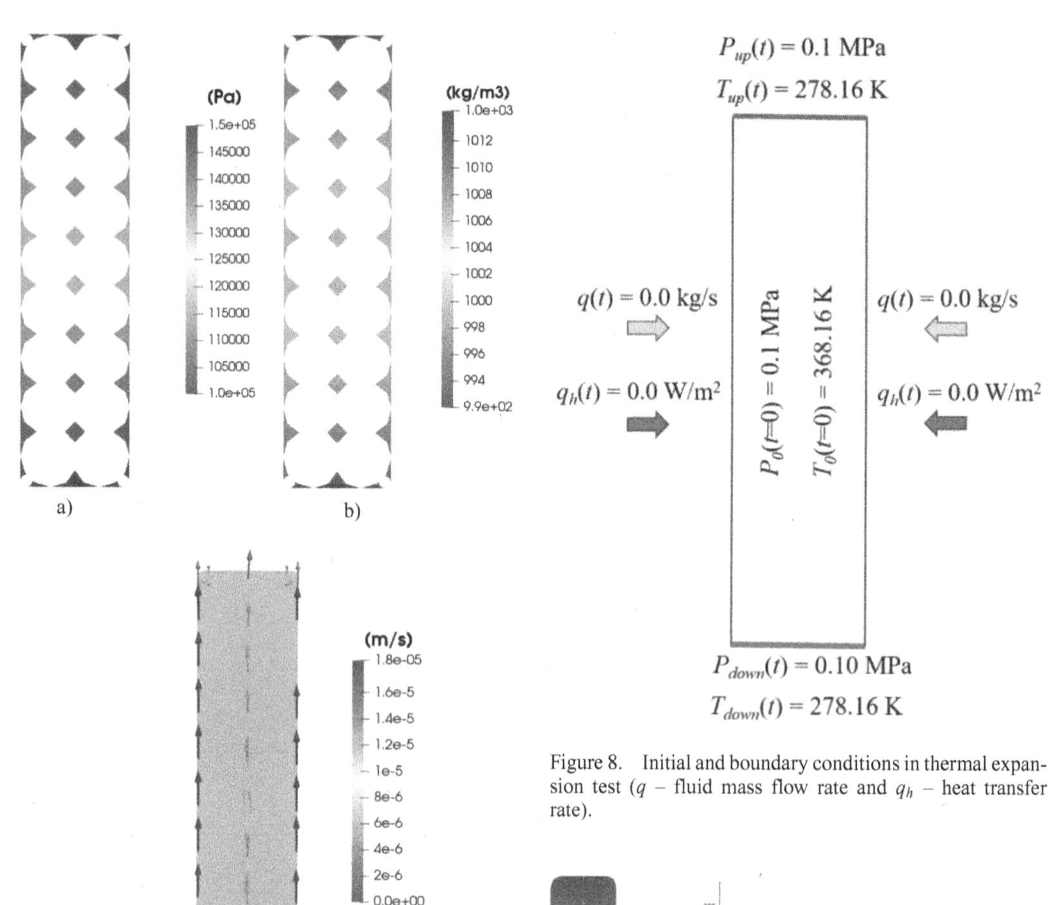

Figure 7. Cohesive granular bar specimen after 400 s of cooling (diffusion and advection): a) pressure distribution, b) density distribution and c) velocity vectors.

Figure 8. Initial and boundary conditions in thermal expansion test (q – fluid mass flow rate and q_h – heat transfer rate).

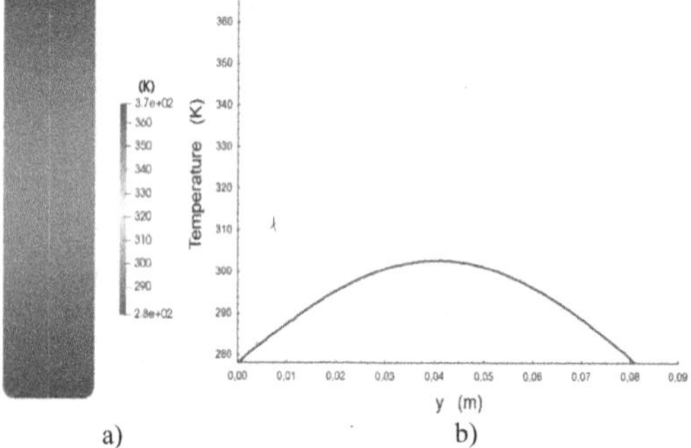

Figure 9. Temperature in bar after 700 s of cooling (thermal expansion test): a) in entire specimen and b) along vertical center line.

test during bar cooling. The boundary and initial conditions were adopted to eliminate advection (Figure 8) and keep the fluid out of phase change conditions. The single-phase flow of water was investigated. Displacements and rotations of the discrete elements at lower and upper boundaries were fixed. In the test, the influence of temperature changes on the mechanical properties was obtained. The initial temperature of the bar (solids and fluid) was relatively high (368.16 K) but still below the boiling point. The bar was then cooled at the lower and upper boundaries by defining a constant temperature of 278.16 K at the boundaries. The simulation was stopped after 700 s. Figure 9a shows the temperature distribution in the bar after 700 s. The maximum temperature was 302.74 K (Figure 9b).

During the test, the high vertical tensile forces appeared due to temperature changes (Figure 10). The forces grew in the specimen quite homogeneously (Figures 10a and 10b). The mean tensile force increased up to 42.5 N (with the maximum force equal to 132 kN). Simultaneously, the mean radius of the element decreased from 3.016 mm to 3.014 mm (~0.07%). After reaching the maximum tensile stress, the 5 contacts were broken (Figure 10c). The mean tensile force decreased down next to 15 N.

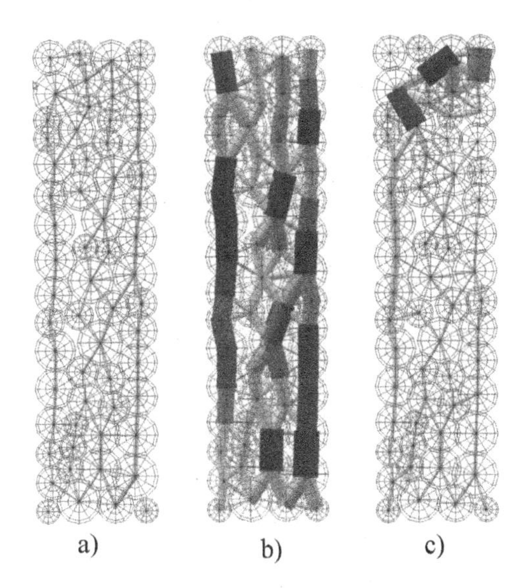

Figure 10. Internal contact forces (tensile) in bar specimen during cooling test after time : a) $t = 41$ s, b) $t = 413$ s and c) $t = 447$ s.

The final macro-crack after 700 s is presented in Figure 11. It appeared exactly at the same place as during pure uniaxial tension.

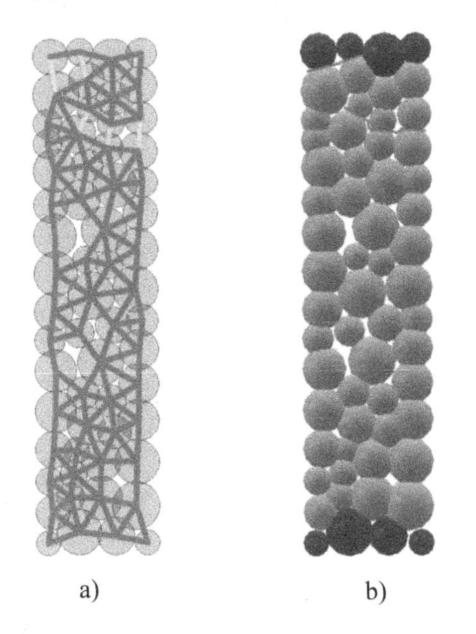

Figure 11. Final macro-crack in specimen during cooling test: a) tensile normal contact forces (green – broken contacts) and b) deformed specimen (macro-crack is in red).

6 SUMMARY AND CONCLUSIONS

The paper presents a novel DEM-based pore-scale thermal-hydro-mechanical model of two-phase fluid flow coupled with heat transfer in porous materials of very low porosity (e.g. concrete or rocks). The validation of the model was performed by comparing the numerical results with the analytical solution of the 1D heat transfer problem in an equivalent bar specimen. Two different cohesive granular bar specimens with initial porosity of 10–13% were tested.

The maximum discrepancy between the numerical and analytical results during the 1D heat transfer problem were solely 0.52–0.64 K.

Advection speeded a cooling process of the bar specimen. The maximum temperature difference between cooling by diffusion and cooling by diffusion with advection was 2.26 K after 400 s of cooling for a small pressure drop (0.05 MPa) along the specimen height.

The thermal shrinkage of the bar specimen during cooling showed the same tensile failure mechanism as in the purely mechanical uniaxial tension.

ACKNOWLEDGEMENTS

The present study was supported by the research project "Fracture propagation in rocks during hydro-fracking – experiments and discrete element method coupled with fluid flow and heat transport" (years 2019–2022) financed by the National Science Centre (NCN) (UMO-2018/29/B/ST8/00255).

REFERENCES

Abdi, R., Krzaczek, M., Tejchman, J. 2022. Comparative study of high-pressure fluid flow in densely packed granules using a 3D CFD model in a continuous medium and a simplified 2D DEM-CFD approach. *Granular Matter* 24, 15, https://doi.org/10.1007/s10035-021-01179-2.

Al-Busaidi, A., Hazzard, J.F., Young, R.P. 2005. Distinct element modeling of hydraulically fractured Lac du Bonnet granite. *J. Geophys. Res.*, 110, B06302, doi:10.1029/2004JB003297.

Catalano, E., Chareyre, B., Barthélémy, E. 2014. Pore-scale modeling of fluid-particles interaction and emerging poromechanical effects. *Int. J. Numer. Anal. Meth. Geomech.*, 38:51–71.

Caulk, R., Sholtès, L., Krzaczek, M., Chareyre, B. 2020. A pore-scale thermo–hydro-mechanical model for particulate systems. *Comput. Methods Appl. Mech. Engrg.*, 372 113292.

Caggiano, A., Schicchi, D.S., Mankel, C., Ukrainczyk, N., Koenders, E.A.B. 2018. A mesoscale approach for modeling capillary water absorption and transport phenomena in cementitious materials, *Computers and Structures* 200: 1–10.

Chareyre, B., Cortis, A., Catalano, E., Barthélemy, E. 2012. Pore-scale modeling of viscous flow and induced forces in dense sphere packings, *Transp. Porous Media* 94(2): 595–615.

Chen, Z., Jin, X., Wang, M. 2018. A new thermo-mechanical coupled DEM model with non-spherical grains for thermally induced damage of rocks. *Journal of the Mechanics and Physics of Solids*, 116: 54–69.

Cundall, P., Strack, O. 1979. A discrete numerical model for granular assemblies, *Géotechnique* 29(1): 47–65.

Cundall P. 2000. Fluid formulation for PFC2D. *Itasca Consulting Group: Minneapolis*, Minnesota.
Deen, N.G., Kriebitzsch, S.H., van der Hoef, M.A., Kuipers, J. 2012. Direct numerical simulation of flow and heat transfer in dense fluid–particle systems, *Chem. Eng. Sci.* 81: 329–344.
Hazzard, JF, Young, RP, Oates, SJ. 2002. Numerical modeling of seismicity induced by fluid injection in a fractured reservoir. *Proceedings of the 5th North American Rock Mechanics Symposium*, Miningand Tunnel Innovation and Opportunity, Toronto, Canada, 7-10 July 2002, pp.1023–1030.
Jiao, K., Han, D., Li, J., Bai, B., Gong, L., Yu, B. 2021. A novel LBM-DEM based pore-scale thermal-hydro-mechanical model for the fracture propagation process. *Computers and Geotechnics*, 139 104418.
Kolditz, O., Bauer, S., Bilke, L., Böttcher, N., Delfs, J.O., Fischer, T., Görke, U.J., Kalbacher, T., Kosakowski, G., McDermott, C.I., Park, C.H., Radu, F., Rink, K., Shao, H., Shao, H.B., Sun, F., Sun, Y.Y., Singh, A.K., Taron, J., Walther, M., Wang, W., Watanabe, N., Wu, Y., Xie, M., Xu, W., Zehner, B. 2012. Opengeosys: an open-source initiative for numerical simulation of thermo-hydro-mechanical/chemical (thm/c) processes in porous media, *Environ. Earth Sci.* 67(2): 589–599.
Kozicki, J., Donzé, F.V. 2008. A new open-source software developer for numerical simulations using discrete modeling methods, *Computer Methods in Applied Mechanics and Engineering* 197: 4429–4443.
Krzaczek, M., Kozicki, J., Nitka, M., Tejchman, J. 2020. Simulations of hydro-fracking in rock mass at meso-scale using fully coupled DEM/CFD approach, *Acta Geotechnica* 15(2): 297–324.
Krzaczek, M., Nitka, M., Tejchman, J. 2021. Effect of gas content in macro-pores on hydraulic fracturing in rocks using a fully coupled DEM/CFD approach, *International Journal for Numerical and Analytical Methods in Geomechanics* 45(2): 234–264.
Li, T., Tang, C., Rutqvist, J., Hu, M. 2021. TOUGH-RFPA: Coupled thermal-hydraulic-mechanical Rock Failure Process Analysis with application to deep geothermal wells. *International Journal of Rock Mechanics & Mining Sciences*, 142 104726.
Mathias P.M., Naheiri M., Oh E.M. 1989. A Density Correction for the Peng-Robinson Equation of State. *Elsevier Science Publishers B.V.*, Amsterdam, Fluid Phase Equilibria, 47:77–87.
Nitka, M., Tejchman, J. 2015. Modelling of concrete behaviour in uniaxial compression and tension with DEM, *Granular Matter* 17(1): 145–164.
Nitka, M., Tejchman, J. 2020. Meso-mechanical modelling of damage in concrete using discrete element method with porous ITZs of defined width around aggregates, *Engineering Fracture Mechanics* 231, 107029.
Olivella, S., Gens, A., Carrera, J., Alonso, E. 1996. Numerical formulation for a simulator (code bright) for the coupled analysis of saline media, *Eng. Comput.* 13(7): 87–112.
Papachristos, E., Scholtès, L., Donzé, F.V., Chareyre, B. 2017. Intensity and volumetric characterizations of hydraulically driven fractures by hydro-mechanical simulations. *International Journal of Rock Mechanics & Mining Sciences*, 93: 163–178.
Peneloux, A., Rauzy, E., 1982. *Fluid Phase Equilibria*, 8: 7–23.
Rutqvist, J., Wu, Y.-S., Tsang, C.-F., Bodvarsson, G. 2002. A modeling approach for analysis of coupled multiphase fluid flow, heat transfer, and deformation in fractured porous rock, *Int. J. Rock Mech. Min. Sci.* 39 (4): 429–442, Numerical Methods in Rock Mechanics.
Rühaak, W., Sass, I. 2013. Applied thermo-hydro-mechanical coupled modeling of geothermal prospection in the Northern Upper Rhine Graben. *Proceedings, Thirty-Eighth Workshop on Geothermal Reservoir Engineering*, Stanford University, Stanford, California, 2013.
Scholtès, L., Chareyre, B., Michallet, H., Catalano, E., Marzougui, D. 2015. Modeling wave-induced pore pressure and effective stress in a granular seabed, *Contin. Mech. Thermodyn.* 27(1): 305–323.
Selvadurai, A. P. S., Suvorov, A. P., Selvadurai, P. A. 2015. Thermo-hydro-mechanical processes in fractured rock formations during a glacial advance. *Geosci. Model Dev.*, 8: 2167–2185.
Skar¿yñski, L., Nitka, M., Tejchman, J. 2015. Modelling of concrete fracture at aggregate level using FEM and DEM based on x-ray μCT images of internal structure, *Engineering Fracture Mechanics* 10(147): 13–35.
Suchorzewski, J., Tejchman, J., Nitka, M. 2018. Discrete element method simulations of fracture in concrete under uniaxial compression based on its real internal structure, *International Journal of Damage Mechanics* 27(4): 578–607.
Šmilauer, V., Chareyre, B. 2011. Yade DEM Formulation, *Manual*.
Tomac, I., Gutierrez, M. 2017. Coupled hydro-thermo-mechanical modeling of hydraulic fracturing in quasi-brittle rocks using BPM-DEM. *Journal of Rock Mechanics and Geotechnical Engineering*, 9: 92–104.
Yan, C., Zheng, H. 2017. A coupled thermo-mechanical model based on the combined finite-discrete element method for simulating thermal cracking of rock. *International Journal of Rock Mechanics & Mining Sciences*, 91: 170–178.
Yan, C., Xie, X., Ren, Y., Ke, W., Wang, G. 2022. A FDEM-based 2D coupled thermal-hydro-mechanical model for multiphysical simulation of rock fracturing. *International Journal of Rock Mechanics & Mining Sciences 149*, 104964.
Yang, B., Chen, S., Liu, K. 2017. Direct numerical simulations of particle sedimentation with heat transfer using the lattice boltzmann method, *Int. J. Heat Mass Transfer* 104: 419–437.
Zareidarmiyan, A., Salarirad, H., Vilarrasa, V., Kim, K.-I., Lee, J., Min, K.-B. 2020. Comparison of numerical codes for coupled thermo-hydro-mechanical simulations of fractured media. *Journal of Rock Mechanics and Geotechnical Engineering* 12: 850–865.

Computational Modelling of Concrete and Concrete Structures – Meschke, Pichler & Rots (Eds)

Modeling inundation flooding in urban environments using density functional theory

E.D. Vartziotis, F.-J. Ulm & K. Boukin
Massachusetts Institute of Technology, Cambridge, MA, USA

R.J.-M. Pellenq
EpiDaPo Lab - CNRS / George Washington University, Children's National Medical Center, Children's Research Institute, Washington, DC, USA

Y. Magnin
Consultant, TotalEnergies, OneTech, Upstream R&D, CSTJF, Pau Cedex, France

K. Ioannidou
Laboratoire de Mécanique et Génie Civil, CNRS, Université de Montpellier, Montpellier, France

ABSTRACT: Evaluating the risk of inundation flooding and its deleterious effects in urban environments is key, considering that such natural disasters are poorly predictable, costly, and are expected to increase with global warming. To insight and evaluate flooding impact in cities and the role played by city textures, we propose a statistical physics computational approach called on-lattice density functional theory. Originally developed in Materials Science, the model is applied to the city scale, considered here as a porous media. We thus show that the strength of such an equilibrium-based approach stems from the combination of three aspects: *i.* the model has a minimum of inputs and an efficient computational time, *ii.* the model comes with an ease of modeling a variety of city elements that are critical for inundation flooding (*e.g.*, buildings, pavements, permeable soils, and drainage systems), *iii.* the model has physically meaningful output parameters, such as adsorption and desorption isotherms, which can be linked to a city's drainage capacity and steady-state gage heights. We found that isotherms exhibit a pronounced hysteresis, indicating that flooding and draining properties can be blocked in metastable microstates. Such behavior is key since it provides a fundamental means to qualitatively identify the risk of inundation flooding.

1 INTRODUCTION

Floods are one of the most common and costliest natural disasters in the United States and worldwide (FEMA 2021). Urban flooding is one of the greatest challenges to human safety, and it can cause severe damage to the economy and lead to high devastation (Tsubaki & Fujita 2010). The main factor that affects floodwater inundation and movement is the development of impervious surfaces, which do not allow natural drainage and inhibit infiltration of stormwater (Shuster et al. 2005). In addition to this, urban flooding may also occur due to non-mandatory building ordinances and non-consideration of the impact the city texture has. Specifically, the development of very dense building areas is prone to floods (National Academies of Sciences, Medicine, et al. 2019). Flood problems are expected to continue to rise and worsen due to climate change, as well as the expansion of cities and the increase of urban population (Pielke et al. 2002; Velasco et al. 2016). Studies have also shown that high atmospheric CO_2 concentrations lead to increased frequency of heavy daily precipitation events (Schreider et al. 2000). So, it is predicted that by 2100 the precipitation rates are expected to increase by 10–30% (Group 2021), and at the same time, the urban population is projected to increase from 55% to 68% by 2050 (UN 2018). These projections emphasize the importance of urban flood modeling and highlight the need for further development of existing flood models. In particular, they call for new reliable tools, that go beyond coarse grained empirical urban planning approaches based on mean land occupation density values (buildings, streets, drainage, green areas) (Özgen et al. 2016); empirical flood maps (FEMA 2018); and high-resolution empirical, hydrodynamic (2D shallow water equations (Teng et al. 2017)), and simplified conceptual models appropriate for building-project scale evaluation of flood inundation (Özgen et al. 2016). More specifically, to enhance the resilience of cities, there is a need to ascertain both a qualitative and quantitative link between city

DOI 10.1201/9781003316404-71

texture parameters, characteristic of the neighborhood scale in urban environment, and the risk of inundation flooding.

We approach this problem by means of a 'physics-by-analogy' simulation using on-lattice Density Functional Theory (DFT) (Kierlik et al. 2002; Zhou et al. 2019). This method, [1] was originally developed in statistical physics for pore-size characterization, gas adsorption-desorption, and capillary condensation phenomena in micro- and mesoporous materials (Lowell et al. 2006; Monfared et al. 2020; Rigby & Chigada 2009). In the DFT simulations, a virtual reservoir of particles is connected to the simulation box (here a portion of city as shown in Figure 1(a)), and particles are exchanged depending on the water particles chemical potential μ, which mimics the prescribed precipitation. This is being done to establish adsorption–desorption isotherms [Figure 1(b)], which can be used to qualitatively pin down the risk of inundation. The DFT-method hence minimizes the Grand potential [Eq.1], to determine the local equilibrium density on each lattice site i, with $\rho_i \in [0, 1]$ ($\rho = 0$ dry site, $\rho = 1$ water saturated site). In the system, each site interacts with its nearest neighbors (Kierlik et al. 2002), while those located at system edges interact through boundary periodic conditions. In addition, for the system we consider an unstructured lattice, allowing to locally increase the number of lattice sites, and so the accuracy of calculations in those regions. Thus, in this work, the lattice size in the elevation direction (z-direction) has been fixed smaller than the in-plane one in (x,y) directions.

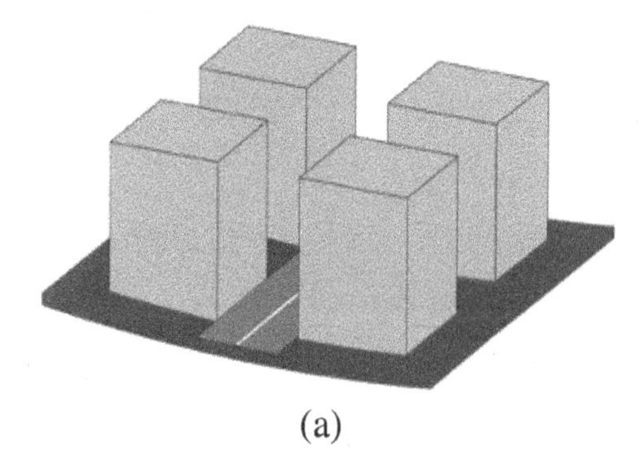

(a)

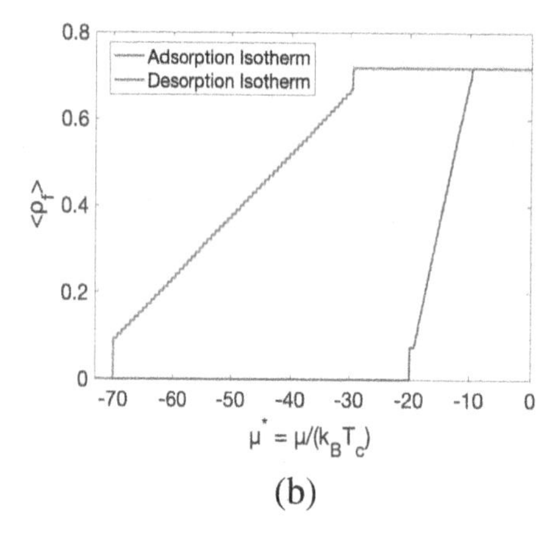

(b)

Figure 1. City model and isotherms. (a) Schematic of urban topography of side-length $L_x = L_y = 36$ m and height $H = 10$ m, consisting of buildings, and impermeable surface; (b) Adsorption - desorption density isotherms as a function of the dimensionless chemical potential, $\mu^* = \mu/(k_B T_c)$ and simulation parameters corresponding to: $a_{x,y}/a_z = 10$, $T^* = 0.8$, $\overline{w}_{sf}^0 = 3$, $\overline{w}_{ff}^0 = 6$.

[1] The DFT method herein employed should not be confused with the solid-state physics computational approach by the same name employed for electronic structure investigations of atoms and molecules (Roy 2019).

2 METHODOLOGY

For a given urban configuration, the DFT method minimizes the dimensionless Grand potential, by adjusting iteratively the local densities $\rho_i \in [0, 1]$ (Kierlik et al. 2001):

$$\begin{aligned}\frac{\Omega}{k_B T_c} &= \min_{\rho_i} \Bigg(T^* \sum_i [\rho_i \ln \rho_i + (\eta_i - \rho_i) \ln (\eta_i - \rho_i)] \\ &\quad - \sum_{<i,j>} (\overline{w}_{ff}^i \rho_i \rho_j + \overline{w}_{sf}^i [\rho_i (1 - \rho_j) + \rho_j (1 - \eta_i)]) \\ &\quad - \mu^* \sum_i \rho_i \Bigg) \\ &= T^* \sum_i \eta_i \ln (1 - \frac{\rho_i}{\eta_i}) + \sum_{<i,j>} \overline{w}_{ff}^i \rho_i \rho_j\end{aligned} \tag{1}$$

where the local density of sites i read:

$$\rho_i = \eta_i [1 + \exp \left(- \frac{\overline{v}_i^{eff}}{T^*} \right)]^{-1}. \tag{2}$$

Here, η_i is the occupation number of site i; $T^* = T/T_c$ is the (kinetic) temperature, normalized by the critical temperature, T_c; $\mu^* = \mu/(k_B T_c)$ is the dimensionless chemical potential; $\overline{\omega}_{ff}^i = w_{ff}/(k_B T_c)$ and $\overline{\omega}_{sf}^i = w_{sf}/(k_B T_c)$ stand for dimensionless fluid-fluid and solid-fluid interaction parameters [in units of $k_B T_c$, with k_B the Boltzmann constant]; and $\overline{v}_i^{eff}$ stands for the dimensionless effective potential of site i,

$$\overline{v}_i^{eff} = \frac{v_i^{eff}}{k_B T_c} = \mu^* + \sum_{j/i} [\overline{w}_{ff}^i \rho_j + \overline{w}_{sf}^i (1 - \eta_j)]. \tag{3}$$

The sum over j is limited to the nearest neighbors of site i.

The Grand potential is minimized in a range of chemical potentials chosen to cover a full isotherm. To do so, we incrementally increase the chemical potential μ^* during the adsorption steps, from a large negative value, where no water adsorption occurs, up to a maximum value corresponding to the water saturation.

Once the adsorption phase is fully covered, we progressively decrease μ^* to simulate the desorption. For each subsequent value of $\mu_n^* = \mu_{n-1}^* \pm \Delta\mu^*$, we use the converged values of ρ_i at μ_{n-1}^* as initial condition for the next adsorption or desorption step (μ_n^*). For each μ^*, the convergence is considered to be reached when the density of two subsequent iterations is smaller than a threshold defined as, $(1/N)\sum_i(\rho_i^{(m+1)} - \rho_i^{(m)})^2 < 10^{-8}$ (Kierlik et al. 2002). Moreover, $\Delta\mu^*$ step sizes ranging from 10^{-1} to -10^{-3} have been chosen. The determination of isotherms, is finally obtained by averaging the full density as the sum of local densities as, $< \rho_f > = \frac{1}{N}\sum_i \rho_i$.

In the resulting isotherms, the water level or gage height, and the hysteresis loop qualitatively pin down the risk of inundation, and the city drainage efficiency, [Figure 1(b)] depending on the specific role played by city elements in the system. It is worth mentioning that for a linear filling process in a flat terrain, the average density $< \rho_f >$, coincides with the normalized gage height; whereas it is an integrated value of gage height distribution for complex terrains and city elements, that define the inundation risk of urban environments.

The entire system is composed of N lattice sites, occupied by either a fluid or a solid to delineate city elements (building, road, etc). This is defined by an occupation number η with fixed values 0 and 1, for solid and liquid sites, respectively. Hence, buildings are considered as repulsive surfaces with $\eta = 0$. Furthermore, $\eta = 0$ describes impermeable (*e.g.*, streets) surfaces; whereas permeable surfaces (*e.g.*, green areas) are identified by a mixture of occupation numbers of both ones and zeros, representative of the soil's porosity, $\phi = \overline{\eta}^{\text{soil}}$, where an overbar stands for averaging. Depending on the surface density of zero occupation numbers, we can identify soils with different porosities. Finally, water retention or the drainage system is identified as areas underground with $\eta = 1$, indicating sites where water can be stored before accumulating on the surface. Table 2 summarizes city elements with corresponding occupation numbers.

Table 1. Occupation number of city elements.

City elements	Occupation number η
Building	0
Street	0
Impermeable Soil	0
Permeable soil	0 & 1
Drainage System	1

One original addition to the DFT method for inundation modeling in urban environments is the elevation (z-direction) dependent fluid-fluid and solid-fluid interaction parameters, which were in the original model constant (Ioannidou et al. 2014; Pellenq et al. 2009). This allows us to consider the gravity-driven water filling process during precipitation. Thus, the z-direction dependency ensures a bottom-up filling during the adsorption process, and a top-down receding of the fluid phase during the desorption process. Following a constant hydrostatic gradient, we then use a linear relationship of the form:

$$\overline{w}_{kk}^i = \overline{w}_{kk}^0 \left(1 - \frac{z_i - z_0}{\alpha_1 \cdot H_{ref}}\right); (kk = ff, sf), \tag{4}$$

where $1/(\alpha_1 \cdot H_{ref})$ defines the hydrostatic gradient, with α_1 a constant, and H_{ref} a reference height; z_i is the lattice site elevation; z_0 is the elevation of the lattice at the origin; and $\overline{w}_{kk}^0 (kk = ff, sf)$ stand for reference values of the interaction parameters, with $\overline{\omega} = w_{sf}^0 / w_{ff}^0$ in a dimensionless form. In the original DFT model, the interaction parameters ratio for water adsorption in cement has been reported as $w_{sf}/w_{ff} = 2.5$ (Bonnaud et al. 2012). However, such a parameter has been determined for capillary condensation in microscale structures, irrelevant at the city scale where such an effect is meaningless. For city modelization, dimensionless interaction parameters are city-specific, require a dedicated calibration, and will be discussed bellow. They need to be chosen with care in order to avoid occurrence of microscale phenomena that are irrelevant at city scale, such as capillary bridges (Zhou et al. 2019).

3 ISOTHERMS AS A MEANS TO ASSESS RISK OF INUNDATION FLOODING

To illustrate our purpose, we present below some features of the isotherms in section 3.1 in function of the most important DFT model parameters, and in section 3.2 for different drainage systems. As model system, we consider the city block displayed in Figure 1(a).

3.1 *Influence of model parameters*

To insight effects of DFT parameters on the water isotherm, we evaluate changes induced by the lattice sizes and their geometries, the effect of the temperature $T^* = T/T_c$, and the ratio of solid-fluid and fluid-fluid interactions, $\overline{\omega} = w_{sf}/w_{ff}$.

In Figure 2, we evidenced a negligible effect of the lattice sizes in the isotherm. A negligible effect is also found when varying the lattice geometry (keeping constant lattice sizes, $a_{x,y} = 0.5$ m and $a_{x,y} = 1$ m), from an isotropic case where $\alpha = a_{x,y}/a_z = 1$, to highly anisotropic ones, up to $\alpha = 100$. The lattice distortion, with a finer lattice discretization in the elevation direction, is found to be an important parameter affecting isotherm accuracy. More specifically, a fine enough elevation discretization reduces the size of finite jumps in the isotherms, while it does not modify the isotherm shape, value, or hysteresis loop. Thus, for determining relevant information about inundation gage heights, a fine enough elevation discretization is needed (typically, $a_z = 0.05$ m), whereas the in-plane discretization

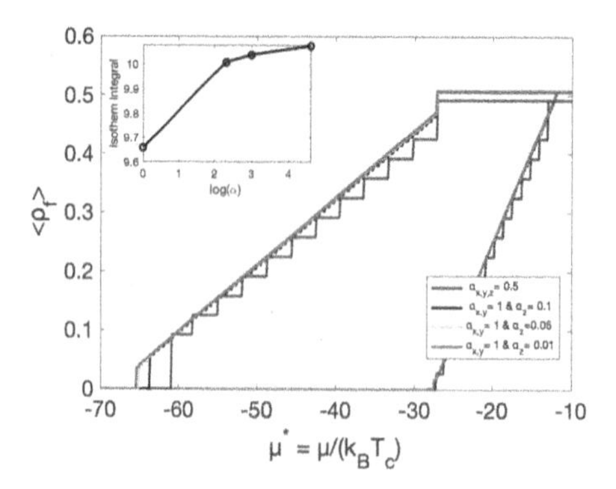

Figure 2. Impact of lattice size on isotherms: $a_{x,y,z} = 0.5$m (blue line), $a_{x,y} = 1$m and $a_z = 0.1$m (black line), $a_{x,y} = 1$m and $a_z = 0.05$m (green line), $a_{x,y} = 1$m and $a_z = 0.01$m (red line). The inset corresponds to the isotherm integral, plotted as the log of the lattice distortion α. Other simulation parameters correspond to $T^* = 0.8$, $\overline{\omega} = 0.5$.

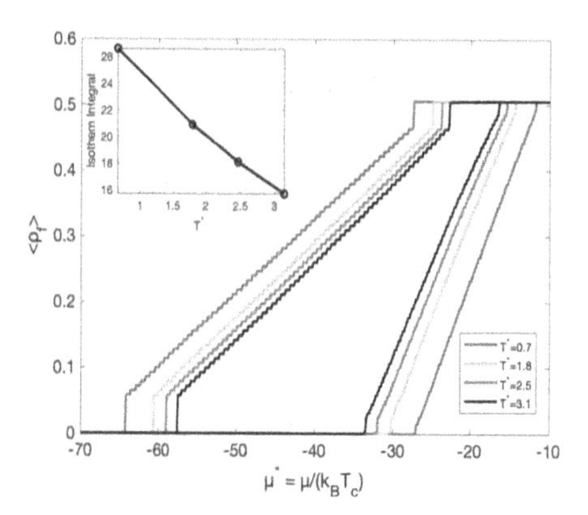

Figure 3. Impact of dimensionless temperature T^* on isotherms: $T^* = 0.7$ (red); $T^* = 1.8$ (green); $T^* = 2.5$ (pink); and $T^* = 3.1$ (black). The inset corresponds to the isotherm integral plotted vs. the dimensionless temperature T^*. Other simulation parameters correspond to $\alpha = 10$, $\overline{\omega} = 0.5$.

is fixed from the size of city elements (buildings, streets, green areas, etc.).

We now look at the effect of the temperature $T^* = T/T_c$ in isotherms. We recall that T^* is a dimensionless parameter, where the temperature T is normalized by the critical temperature T_c. In the simulations, we observe that values $T^* > 1$ entail a diffuse boundary between the gage height (water level) and the air, with density values ρ varying between 0 (gas) and 1 (liquid) over a finite thickness, whereas the (free) water boundary is demarcated for $T^* < 1$. This diffuse boundary for $T^* > 1$ affects the adsorption-desorption isotherms, both in size and absolute values, as shown in Figure 3.

In contrast to the overall role of T^* in the system's response, the fluid-fluid and solid-fluid interactions, or more specifically the interaction parameter ratio $\overline{\omega} = w_{sf}/w_{ff}$, play a prominent role in capturing the physics of adsorption and desorption of liquid into the city texture. This prominent role is shown in form of the adsorption-desorption isotherm in Figure 4 (a), and the snapshots shown in Figures. 4 (b) and (c) for two different values, $\overline{\omega} = 0.5$ and $\overline{\omega} = 5$. Figure 4 (a) shows that the onset of adsorption occurs at lower chemical potentials with increasing value of $\overline{\omega}$; and the same shift occurs at the end of the desorption process. This shift is due to the increase in the solid-fluid interaction, which favors the adsorption of water onto surfaces leading eventually to menisci formation [compare Figures. 4 (b) and (c)]. Such menisci are highly relevant for wetting phenomena at nano- and mesoscale of porous materials, but they do not represent the physics of inundation at the city scale, and need to be avoided.

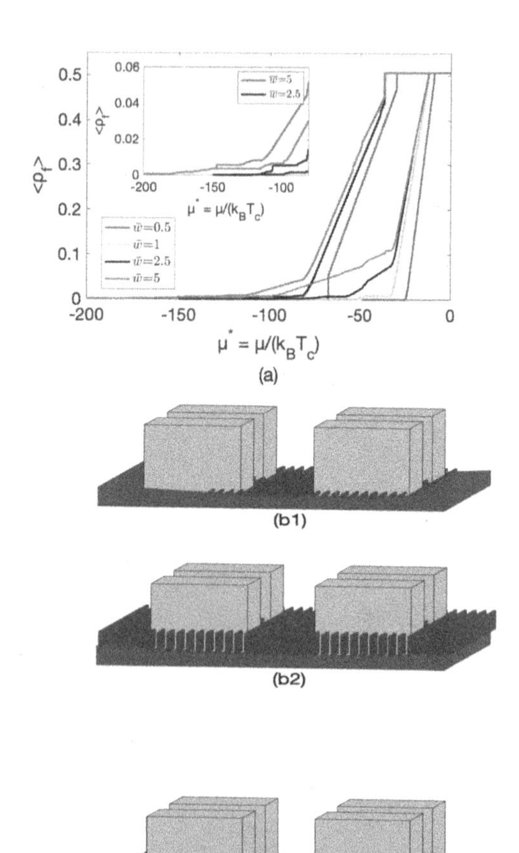

Figure 4. (a) Impact of interaction parameter ratio $\overline{\omega} = w_{sf}/w_{ff}$ on isotherms: $\overline{\omega} = 0.5$ (red line); $\overline{\omega} = 1$ (green line); $\overline{\omega} = 2.5$ (black line); $\overline{\omega} = 5$ (pink line). (b1-c2) Snapshots of surface filling for different values of $\overline{\omega}$ and chemical potential μ^*: (b1) $\overline{\omega} = 0.5$ & $\mu^* = -24$; (b2) $\overline{\omega} = 0.5$ & $\mu^* = -19$; (c1) $\overline{\omega} = 5$ & $\mu^* = -98$; (c2) $\overline{\omega} = 5$ & $\mu^* = -69$. Other simulation parameters correspond to $\alpha = 10$, $T^* = 0.8$.

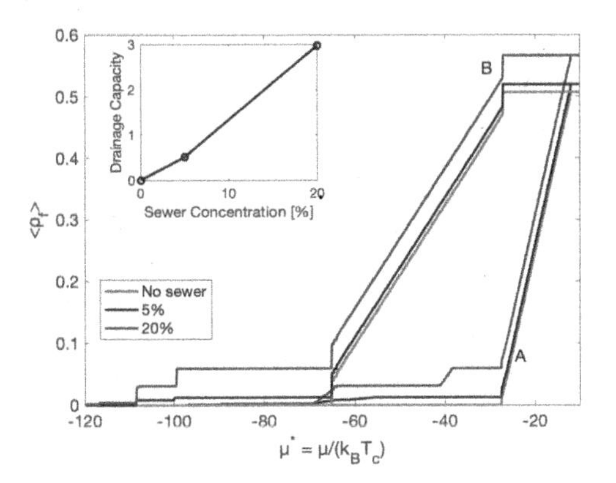

Figure 5. Impact of the city's drainage system on isotherms: No sewer (red line); 5% sewer (black line); 20% drainage (blue line); the inset corresponds to the area of hysteresis loop for the adsorption - desorption loop of the sewer vs. sewer concentration. Other simulation parameters correspond to $\alpha = 10$, $T^* = 0.8$.

3.2 *Impact of drainage and soil porosity on risk of inundation flooding*

Inundation flooding of urban environments results from the imbalance between precipitation and drainage. While precipitation is incorporated in the DFT approach by the chemical potential, μ^*, drainage in our DFT approach is captured in form of either a pipe system or a distributed soil drainage system, considering a spatial distribution of the occupation number in the soil (see Table 2). We compare the impact of pipe and soil drainage on isotherms.

Simulations are carried out for three drainage system concentrations, 0, 5, and 20% representative of the volume fraction of the underground where water can accumulate. The adsorption-desorption isotherms for the three systems are shown in Figure 5. These isotherms are characterized by a change in slope, from the almost horizontal branch during pipe drainage, followed by an almost vertical branch once the pipes are filled. The necking point, $<\rho_f>_d$, between these two regimes [denoted by (A) in Figure (5)] increases with the drainage concentration, from $<\rho_f>_d = 0$ for the system without drainage, to $<\rho_f>_d = 0.06$ for the 20%-drainage system. This density shift is constant over the entire adsorption-desorption isotherm [point (B) in Figure (5)], and can thus be considered as a qualitative measure of the impact of drainage at the scale of the simulated city block. If one considers an adsorption-desorption loop to the chemical potential μ_d at the necking point, the area below the curve can be viewed as a measure of building block's drainage capacity [see inset of Figure 5]. This means that the effective gage height above ground is reduced by this drainage (water storage) capacity.

Similar drainage capacity measures are obtained when considering a permeable soil, i.e. a soil which can retain water in our equilibrium-based approach. Such a soil is modeled in the DFT approach through an occupation number distribution in the soil, with a porosity of $\phi = \overline{\eta}^{\text{soil}}$ and surface solid-fluid interactions, $\omega = 1$. A comparison for three different porosity values, $\phi = 0, 25, 50\%$, is shown in Figure 6 in terms of the adsorption-desorption isotherms. Similar to pipe drainage [Figure 5], a porous soil entails a vertical shift of the isotherms, defined by a necking point. However, the drainage capacity calculated from the integral over the adsorption-desorption hysteresis loop [inset of Figure 6], exhibits smaller values when compared to pipe drainage [inset of Figure 5]. This observation can be attributed to the distributed nature of the porosity in the ground when compared to the concentrated nature of drainage pipes. While difference in different drainage systems merits further investigation, it shows the potential of the DFT approach to qualitatively discern the role of city-scale drainage systems in inundation flooding.

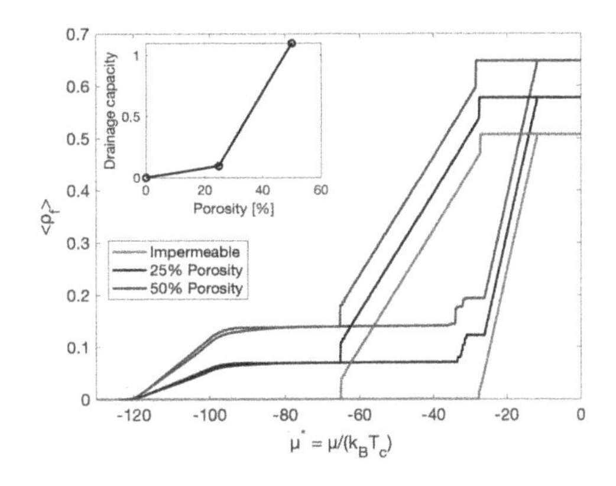

Figure 6. Impact of the soil's permeability on isotherms: Impermeable soil $\phi = 0\%$ (red line); permeability $\phi = 25\%$ (black line); permeability $\phi = 50\%$ (blue line); the inset corresponds to the area of hysteresis loop for the adsorption - desorption loop of the soil vs. soil porosity. Other simulation parameters correspond to $\alpha = 10$, $T^* = 0.8$.

4 CALIBRATION

This on-lattice DFT approach requires calibration of the α_1 and the $\overline{w}^0_{kk}$ parameters of the interaction parameters, and the chemical potential μ^*, to gain quantitative capabilities for urban flood risk evaluation. For that purpose, we use an optimization algorithm, the genetic algorithm (GA) (MathWorks 2021), and a reference model developed by the MIT Office of Sustainability (MITOS), against which we calibrate the model parameters. MITOS is using the InfoWorks ICM (Integrated Catchment Modelling) model (Innovyze 2021), a software platform for integrated 1D /2D hydrodynamic modeling (Ltd 2021). The applied optimization algorithm finds the best fit for the three model parameters by minimizing the mean of the height difference between the DFT model and the reference predictions:

$$fl = \min_{\alpha_1, \overline{w}^0_{kk}, \mu^*} \left(mean\left[ICM_{Height} - DFT_{Height} \right]^2 \right), \quad (5)$$

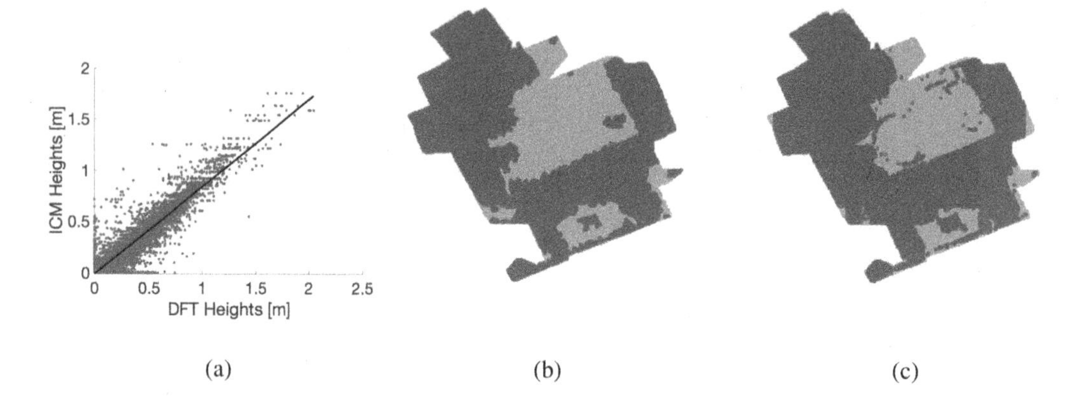

Figure 7. Results for Danforth St. sub-catchment for the 100 year storm event under current climate conditions (a) ICM vs. DFT height values; water distribution for (b) the ICM model; (c) the DFT model at $\mu^* = -0.577$.

where fl is the minimized function; ICM_{Height} are the flood heights obtained by the ICM model for six precipitation events (see Table 5); and DFT_{Height} are the equivalent flood heights obtained by the DFT model.

For the model calibration, we have chosen as our study area the MIT campus, and more specifically, the sub-catchment on Danforth Street. This sub-catchment is considered for the calibration with impermeable surfaces and no city elements. In Table 5 we see the results after the calibration for the average flood heights for the ICM and the DFT model and the corresponding chemical potential μ^*. The quality of the calibration is illustrated by comparing the ICM and DFT height values for the same (x,y) points. Figure 7 (a) displays the spread of height values obtained from the two models for the 100-year storm event under current climate conditions in the form of a cross plot. The R^2-value for the simulation is 0.9154 showing a close to linear correlation. The corresponding water distribution on the sub-catchment for the 100-year precipitation event, shown in Figure 7 (b) and (c), indicate that both models are mainly flooding the same areas. The correctness of this distribution must be further studied and validated.

5 DISCUSSION

The proposed DFT-model for inundation flooding departs from the premise that an equilibrium-based model built around the minimization of the Grand potential [i.e., Eq. (1)] can capture essential features of the physics of inundation of urban environments. It is thus of interest to test this conjecture by comparing the DFT approach with the most common approach for inundation modeling, which is based upon the shallow-water equations (SWE). The SWE-model condenses the laws of mass and momentum conservation into a continuum 2D-boundary layer in between terrain height $b(x,y)$ and free surface height $b+h$, considering the in-plane flow velocity vector $\vec{u}(x,y,t) = (u,v)$. Mass conservation reads (Vreugdenhil 1994):

$$\frac{\partial h}{\partial t} + \nabla \cdot (h\vec{u}) = p_z - d_z, \tag{6}$$

and momentum conservation:

$$\frac{\partial (h\vec{u})}{\partial t} + \nabla \cdot (h\vec{u} \otimes \vec{u}) = -gh\nabla(b+h), \tag{7}$$

where p_z is the precipitation rate; d_z the drainage rate; g the earth acceleration, b the terrain height (measured e.g. w.r.t. seal level). The underlying assumptions of the SWE-model are (Teng et al. 2017; Vreugdenhil 1994): (1) horizontal variations in velocity are much greater than vertical variations, so that the velocity $\vec{u}(x,y,t)$ can be viewed as depth-averaged in-plane velocities; (2) vertical accelerations can be neglected (boundary layer assumption); and (3) the pressure gradient over the height is negligible compared to in-plane pressure gradients. The focus of this discussion is to identify commonalities and differences with the proposed DFT approach.

The first point in common is the outcome of both approaches, namely spatially resolved flood height estimates, coined gage height in the case of classical approaches (Ali et al. 2016), and adsorption-desorption isotherms in the DFT approach. In the SWE approach, this gage height at a point (x,y) is attained once mass and momentum conservation has balanced in time the imbalance between precipitation and drainage rate. In the DFT approach the system is put in contact with an (outside) virtual bath of particles at a chemical potential μ, and so in first order, the precipitation rate, p_z, can be captured by a Newton-type kinetics law of the form $p_z \sim \frac{\mu_0^* - \mu^*}{k_p}$, with k_p an exchange coefficient. Similarly, it is tempting to formulate the drainage rate in terms of DFT modeling quantities. For instance, for drainage by a permeable soil, the drainage rate, d_z, is driven by the soil saturation rate S, meaning $d_z \sim \frac{dS}{dt}$, where $S \sim \frac{\overline{\eta\rho}^{soil}}{\overline{\eta}^{soil}}$.

Herein, $\overline{\eta}^{soil}$ is the soil porosity, whereas $\overline{\eta\rho}^{soil}$ is the part of this porosity saturated by the fluid phase. The determination of the drainage rate in the SWE model thus requires additional information about water retention in order to become operational (Massari et al. 2014). In contrast, in the DFT model, the soil and its water retention capacity are explicitly included in the minimization of the Grand potential [see Eq. 1 and Table 2].

Table 2. Calibration results for Danforth Street.

Storm	ICM Height	μ^*	DFT Height
10Y 24H today	0.1255	–0.5824	0.1235
10Y 24H 2030	0.1439	–0.58	0.1789
10Y 24H 2070	0.163	–0.5789	0.2121
100Y 24H today	0.2267	–0.577	0.2444
100Y 24H 2030	0.2598	–0.5743	0.2783
100Y 24H 2070	0.2982	–0.571	0.3148

In the equilibrium-based DFT approach, all relations are time-independent. It excludes flash flood phenomena (Vreugdenhil 1994), the complex interplay between soil saturation and drainage (Teng et al. 2017), and the non-stationarity of gage height distribution in urban environments (Group 2021). In return, it can be viewed as an asymptotic state of inundation flooding for urban environments, and hence an appropriate means for a first-order evaluation of the risk of inundation flooding. Finally, it should be noted that the assumption of stationarity does not necessary imply a zero-flow velocity. It only means that the first term on the l.h.s. of respectively Eq. (6) and Eq. (7) is zero, but not the second one.

6 CONCLUSIONS AND PERSPECTIVES

The method herein proposed provides a simulation-based adsorption technique dedicated to urban environments at the city scale. To the best of our knowledge, such an approach is original in its attempt to transpose by analogy the physics of coarse-grain adsorption techniques commonly employed in Materials Science to city scale. While the equilibrium-based approach discards the time-dependence of flooding, the strength of the DFT method stems from a combination of a minimum of input quantities, the ease of modeling city elements (buildings, pavements, drainage systems, soil saturation, etc.), and physically meaningful output parameters which can be linked to a city's drainage capacity, risk of inundation flooding and so on. A further strength of the approach is the well-posed and computationally efficient minimization problem.

While the results so far obtained illustrate the potential to qualitatively capture inundation features in an urban environment, there are several steps required for the approach to gain quantitative capabilities for risk evaluation. This includes (1) the further development of the calibration by adding more elements into the calibration, such as permeable soils, drainage systems, etc.; (2) the water distribution validation via the so-called two-point correlation function; (3) the systematic investigation of the role of city texture parameters; and (4) based on a clear understanding of the governing city texture parameters, it should be possible to identify means of mitigating the impact of inundation flooding. It is expected that the proposed approach will contribute to the emerging field of 'urban physics' (Sobstyl et al. 2018).

7 ACKNOWLEDGEMENT

This research was carried out by the Concrete Sustainability Hub (CSHub@MIT), with funding provided by the Portland Cement Association (PCA) and the Ready Mixed Concrete Research & Education Foundation (RMC EF). The CSHub@ MIT is solely responsible for content. Additional support was provided by the MIT Office of Sustainability (MITOS@MIT).

REFERENCES

Ali, H., P.-Y. Lagrée, & J.-M. Fullana (2016). Application of the shallow water equations to real flooding case.

Bonnaud, P., Q. Ji, B. Coasne, R.-M. Pellenq, & K. Van Vliet (2012). Thermodynamics of water confined in porous calcium-silicate-hydrates. *Langmuir 28*(31), 11422–11432.

FEMA, F. E. M. A. (2018). Guidance for Flood Risk Analysis and Mapping - Flood Risk Assessments. https://www.fema.gov/sites/default/files/2020-02/Flood_Risk_Assessment_Guidance_Feb_2018.pd. [Online; accessed January-2021].

FEMA, F. E. M. A. (2021). Floods. https://www.ready.gov/floods. [Online; accessed June-2021].

Group, W. B. (2021). Climate Change Knowledge Portal, Laos County Historical Climateb Data. https://climateknowledgeportal.worldbank.org/country/laos/climate-data-projections. [Online; accessed January-2021].

Innovyze (2021). ICM. https://www.innovyze.com/en-us/products/infoworks-icm. [Online; accessed June-2021].

Ioannidou, K., R. J.-M. Pellenq, & E. Del Gado (2014). Controlling local packing and growth in calcium–silicate–hydrate gels. *Soft Matter* 10(8), 1121–1133.

Kierlik, E., P. Monson, M. Rosinberg, & G. Tarjus (2002). Adsorption hysteresis and capillary condensation in disordered porous solids: a density functional study. *Journal of Physics: Condensed Matter 14*(40), 9295.

Kierlik, E., M. Rosinberg, G. Tarjus, & P. Viot (2001). Equilibrium and out-of-equilibrium (hysteretic) behavior of fluids in disordered porous materials: Theoretical predictions. *Physical Chemistry Chemical Physics 3*(7), 1201–1206.

Lowell, S., J. E. Shields, M. A. Thomas, & M. Thommes (2006). Characterization of porous solids and powders: surface area, pore size and density. *16*.

Ltd, A. (2021). InfoWorks ICM – The Most Powerful 1D/2D Integrated Catchment Modeling Solution. https://www.aquamod.eu/en/index.php/software/infoworks-icm?showall=1. [Online; accessed June-2021].

Massari, C., L. Brocca, T. Moramarco, Y. Tramblay, & J.-F. D. Lescot (2014). Potential of soil moisture observations in flood modelling: Estimating initial conditions and correcting rainfall. *Advances in Water Resources 74*, 44–53.

MathWorks (2021). What Is the Genetic Algorithm? https://www.mathworks.com/help/gads/what-is - the - genetic - algorithm.html. [Online; accessed July-2021].

Monfared, S., T. Zhou, J. E. Andrade, K. Ioannidou, F. Radjaï, F.-J. Ulm, & R. J.-M. Pellenq (2020). Effect of confinement on capillary phase transition in granular aggregates. *Physical Review Letters 125*(25), 255501.

National Academies of Sciences, E., Medicine, et al. (2019). Framing the challenge of urban flooding in the united states.

Özgen, I., J. Zhao, D. Liang, & R. Hinkelmann (2016). Urban flood modeling using shallow water equations with depth-dependent anisotropic porosity. *Journal of Hydrology 541*, 1165–1184.

Pellenq, R. J.-M., A. Kushima, R. Shahsavari, K. J. Van Vliet, M. J. Buehler, S. Yip, & F.-J. Ulm (2009). A realistic molecular model of cement hydrates. Proceedings of the National Academy of Sciences 106(38), 16102–16107.

Pielke, R. A., M. W. Downton, & J. B. Miller (2002). Flood damage in the united states, 1926-2000: a reanalysis of national weather service estimates.

Rigby, S. P. & P. I. Chigada (2009). Interpretation of integrated gas sorption and mercury porosimetry studies of adsorption in disordered networks using mean-field dft. *Adsorption 15*(1), 31–41.

Roy, A. K. (2019). A new density functional method for electronic structure calculation of atoms and molecules. *arXiv preprint arXiv:1904.08806*.

Schreider, S. Y., D. Smith, & A. Jakeman (2000). Climate change impacts on urban flooding. *Climatic Change 47*(1), 91–115.

Shuster, W. D., J. Bonta, H. Thurston, E. Warnemuende, & D. Smith (2005). Impacts of impervious surface on watershed hydrology: A review. *Urban Water Journal 2*(4), 263–275.

Sobstyl, J., T. Emig, M. A. Qomi, F.-J. Ulm, & R.-M. Pellenq (2018). Role of city texture in urban heat islands at nighttime. *Physical review letters 120*(10), 108701.

Teng, J., A. J. Jakeman, J. Vaze, B. F. Croke, D. Dutta, & S. Kim (2017). Flood inundation modelling: A review of methods, recent advances and uncertainty analysis. *Environmental modelling & software 90*, 201–216.

Tsubaki, R. & I. Fujita (2010). Unstructured grid generation using lidar data for urban flood inundation modelling. *Hydrological Processes: An International Journal 24*(11), 1404–1420.

UN (2018). 2018 Revision of World Urbanization Prospects Web-based resource (New York: The Population Division of the Department of Economic and Social Affairs of the United Nations). https://www.un.org/development/desa/publications/2018 - revision - of - world - urbanization - prospects.html. [Online; accessed January-2021].

Velasco, M., À. Cabello, & B. Russo (2016). Flood damage assessment in urban areas. application to the raval district of barcelona using synthetic depth damage curves. *Urban Water Journal 13*(4), 426–440.

Vreugdenhil, C. B. (1994). Numerical methods for shallow-water flow. *13*.

Zhou, T., K. Ioannidou, E. Masoero, M. Mirzadeh, R. J.-M. Pellenq, & M. Z. Bazant (2019). Capillary stress and structural relaxation in moist granular materials. *Langmuir 35*(12), 4397–4402.

Zhou, T., K. Ioannidou, F.-J. Ulm, M. Z. Bazant, & R.-M. Pellenq (2019). Multiscale poromechanics of wet cement paste. *Proceedings of the National Academy of Sciences 116*(22), 10652–10657.

Computational Modelling of Concrete and Concrete Structures – Meschke, Pichler & Rots (Eds)
 ISBN: 978-1-032-32724-2

Weak finite-discrete element coupling for the simulation of drying shrinkage cracking in concrete

C. Oliver-Leblond, N. Chan, F. Benboudjema & F. Ragueneau
Université Paris-Saclay, ENS Paris-Saclay, CentraleSupélec, CNRS, LMPS – Laboratoire de Mécanique Paris-Saclay, Gif-sur-Yvette, France

ABSTRACT: Micro-cracks can appear at the surface of concrete specimens due to differential drying along their depth. The influence of those micro-cracks on the mechanical behavior is studied. The proposed sequential analysis is composed of two steps: hydric modeling via a finite element model and mechanical modeling via a beam-particle model. This latter model provides an explicit description of cracking and the associated mechanisms: initiation, propagation and contact during closing. The approach allows a weak coupling between the hydric and mechanical phenomena achieved in a non-intrusive way in both models. The study of cracking under drying conditions followed by three-points bending test is realized. Numerical results are compared with an available experimental campaign.

1 INTRODUCTION

The drying of concrete structures modifies their mechanical and transport properties, and especially because of the appearance of cracks on the surface. This cracking is due to the difference in humidity between the core of the material and the external environment. Indeed, a hydraulic moisture gradient induces local tension and compression stresses, which can lead to surface cracking (Bisschop & van Mier 2002; Hossain & Weiss 2004).

Drying shrinkage is a strongly coupled problem. Indeed, moisture transport induces shrinkage which can lead to the development of cracks. These cracks will then interfere with moisture transport since they constitute preferential pathways for water flow. Thus, the study of drying shrinkage should theoretically require the simultaneous analysis of water transport and mechanical failure. However, these two can be solved separately, by first performing the water transport analysis, then calculating the shrinkage strain, and finally deducing the mechanical strain and crack propagation. This weak coupling method is justified by the fact that, experimentally, the drying rate of mortars and concretes has been shown to be scarcely influenced by the presence of surface cracks.

In this study, the moisture field in the concrete specimen during drying is obtained with a water transport model developed in the finite element code Cast3m. The fracture modeling is then performed with a beam-particle model. The coupling variable is the shrinkage strain field which is calculated from the moisture field and applied through an external force in the mechanical model.

The beam-particle model is used here because it offers an explicit description of cracking. It is composed of randomly distributed rigid particles connected by Euler-Bernoulli brittle beams. The development of cracking is a consequence of the breaking and the removal of those cohesive beams. In addition, interactions between particles, such as contact and friction, are considered to manage crack closure.

To correctly reproduce the impact of drying on the mechanical behavior, the positive contribution of the capillary pressure is modeled. Indeed, drying in a porous network – such as the one found in concrete – generates a pore pressure which leads to the prestressing of the solid part of the microstructure. Here, the influence of this capillary pressure is added to the failure criterion of discrete beams.

The formation and influence of surface drying cracks on the mechanical behavior are studied. For this purpose, numerical analyses are performed on three-point bending tests of specimens kept under autogenous drying conditions. The results are compared with experimental results obtained on concrete specimens by Soleilhet et al. (2021). Simulations are performed on notched and unnotched specimens, as these specimens are not affected by surface cracks in the same way.

2 FINITE-DISCRETE ELEMENT MODELING

2.1 *Weak coupling principle*

Drying shrinkage is a strongly coupled problem that theoretically requires the simultaneous analysis of moisture transport and mechanical strains. Indeed, the

DOI 10.1201/9781003316404-72

moisture transport induces shrinkage which might lead to cracking. Those cracks will then affect the moisture transport as they are ideal paths for the water flow. Models exist to study this strong coupling between transport and cracking (Grassl & Bolander 2016; Roth et al. 2020). However, this problem is classically solved by analyzing water transport and fracture separately and successively.

This weak coupling method is justified by experimental observations. Indeed, the drying rate of mortars and concretes is scarcely influenced by the presence of micro-cracks (Bisschop & van Mier 2008).

In this study, the moisture field due to drying is obtained with a hydric transport model implemented in the finite element code Cast3m. The shrinkage strain field is then computed for this moisture field and applied through an external force in the discrete beam-particle model. The cracking patterns due to drying followed by mechanical loadings can be thus obtained.

2.2 *Hydric transport model*

Although drying is related to the combined transport of liquid water, water vapor and dry air, it can be studied by considering only the mass balance equation for the liquid water:

$$\frac{dS}{dP_c}\frac{dP_c}{dt} = div\left(\frac{Kk_r}{\mu_l\varphi}grad\left(P_c\right)\right) \tag{1}$$

where S is the degree of saturation, P_c is the capillary pressure, K is the intrinsic permeability, k_r is the relative permeability, μ_l is the viscosity of the liquid water and ϕ is the porosity.

To solve equation (1), the relations proposed by van Genutchen (1980) are used which linked together the degree of saturation, the capillary pressure and the relative permeability:

$$k_r = \sqrt{S}\left(1-\left(1-S^{\frac{1}{\beta}}\right)^{\beta}\right)^2 \tag{2}$$

$$S = \left(1+\left(\frac{|P_c|}{P_0}\right)^{\frac{1}{1-\gamma}}\right)^{-\gamma} \tag{3}$$

where β, γ and P_0 are material parameters.

A linear relationship is then used to obtain the drying shrinkage rate from the variation of the degree of saturation:

$$\dot{\boldsymbol{\varepsilon}}^{sh} = \kappa^{sh}\dot{S}\mathbf{1} \tag{4}$$

where κ^{sh} is a hydrous compressibility factor and $\mathbf{1}$ is the unit matrix.

A more complete description of this model as well as the procedure used to identify the material parameters are presented in Soleilhet et al. (2020).

2.3 *Beam-particle model*

The beam-particle model has been developed to explicitly simulate cracking in concrete (Delaplace 2008; Vassaux et al. 2016). This discrete model is constituted by a set of polygonal rigid particles. A center, randomly placed in a grid dividing the specimen into regions of constant size, is associated with each of these particles (see Figure 1a). Those particles are connected together with Euler-Bernoulli beams to represent the cohesion between particles (see Figure 1b-c).

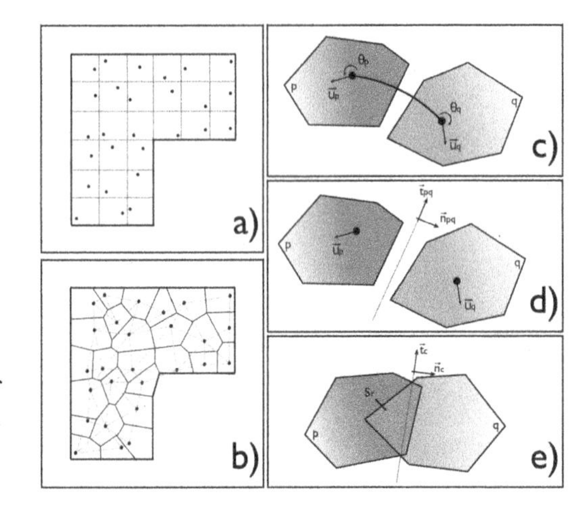

Figure 1. Generation of the random mesh (a-b) and mechanical interactions between the particles (c-e) for the beam-particle model.

The Euler-Bernoulli beams constitute a lattice of stiffness matrix $\boldsymbol{K}$. To obtain the displacement vector due to the external forces, the global equilibrium of the lattice system must be solved at each time step:

$$\boldsymbol{K}\vec{u} = \vec{f}^{ext} + \vec{f}^{sh} \tag{5}$$

To compute the shrinkage forces vector, it is supposed that each beam pq undergoes a shrinkage strain in its axis only that is constant along its length and corresponds to the average of the values at its two extremities. Those values are obtained from the finite element hydric simulation and projected on the discrete beam-particle mesh following the method proposed by Oliver-Leblond et al. (2013). The vector of shrinkage forces for the particle p is thus given by:

$$\vec{f}_p^{sh} = E\sum_{q=1}^{N_p} A_{pq}\frac{\varepsilon_p^{sh}+\varepsilon_q^{sh}}{2}\vec{n}_{pq} \tag{6}$$

where N_p is the number of particles q connected to p by a beam pq,E is the Young modulus of the beams, A_{pq} is the section of the beam pq, $\vec{n}_{pq}$ is the axial vector of this beam and ε_p^{sh} (resp. ε_q^{sh}) is the value of the finite element shrinkage strain interpolated on the node p (resp. q). The forces acting on the particles at both ends of the beam compensate each other, so the resulting shrinkage force on all particles is zero.

A brittle behavior is imposed to the beams in order to introduce cracks. The beams are removed one by one once they reach their failure criteria (see Figure 1d). The failure criterion for the beam linking the particles p and q is written as:

$$\frac{\varepsilon_{pq} - \varepsilon_{pq}^{sh} + \alpha\left|\kappa_{pq}\right|}{\varepsilon_{pq}^{cr} - b_{\varepsilon}\varepsilon_{pq}^{sh}} > 1 \tag{7}$$

The failure of the beam pq thus depends on:

- the axial strain of the beam ε_{pq}computed from the displacement vectors of the centroids of the particles p and q;
- the curvature of the beam κ_{pq} computed from the rotation of the particles p and q around their centroids multiplied here by the material parameter α to reproduce the dissymmetric tension/compression behavior;
- the shrinkage strain of the beam defined as $\varepsilon_{pq}^{sh} = \left(\varepsilon_{p}^{sh} + \varepsilon_{q}^{sh}\right)/2$;
- the critical strain of the beam ε_{pq}^{cr} randomly generated following a Weibull distribution in order to introduce material heterogeneities allowing to obtain the quasi-brittle behavior of concrete;
- the capillary pressure calculated in a simplified way here as proportional to the drying strain with a factor b_{ε}, linked to the Biot coefficient, which must be positive to ensure the positive effect of the capillary pressure during shrinkage.

Once a beam is removed between two particles as a result of its failure, the crack can close which leads to the appearance of contact between these two particles (see Figure 1e). In this case, geometric interpenetration is allowed and a repulsive contact force proportional to the interpenetration area is computed.

3 APPLICATION

The test cases studied here are derived from the experimental campaign of Soleilhet et al. (2021). Concrete beams of section 10cm x 10cm and length 84cm were subjected to drying conditions at 30% relative humidity for 70 days or kept in water for the same duration. Half of those beams were then sawn to obtain a notch of 2cm. Those beams are then submitted to a three-points bending test. The experimental Force-Displacement responses of the beams under three-points bending test are presented on Figure 2 for the unnotched beams and Figure 3 for the notched beams.

First, the three-point bending tests on wet specimens are simulated using the beam-particle model. The 2D discrete mesh of the concrete specimens consists of 21000 particles. Thus, the average length of the lattice beams is 2 mm. The parameters of this discrete model are identified in order to find the average Force-Displacement responses for notched and unnotched specimens (see Figures 4 and 5). Since the model introduces heterogeneity, it is necessary to perform several

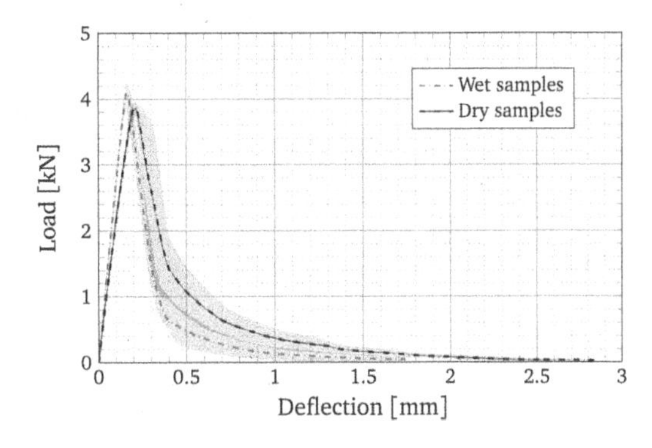

Figure 2. Force-Displacement responses of the unnotched beams under three-point bending tests from the experimental campaign of Soleilhet et al. (2021).

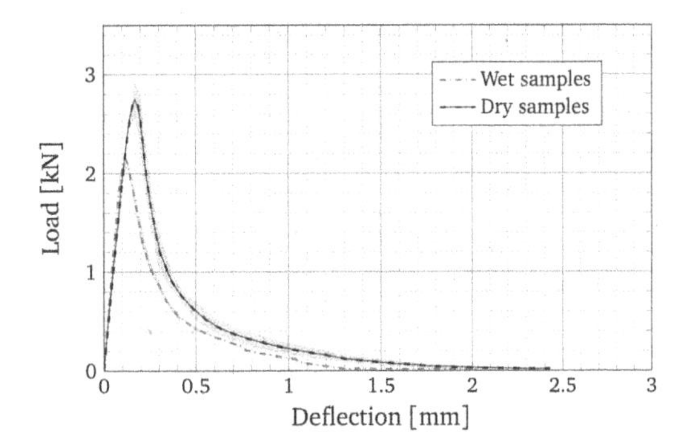

Figure 3. Force-Displacement responses of the notched beams under three-point bending tests from the experimental campaign of Soleilhet et al. (2021).

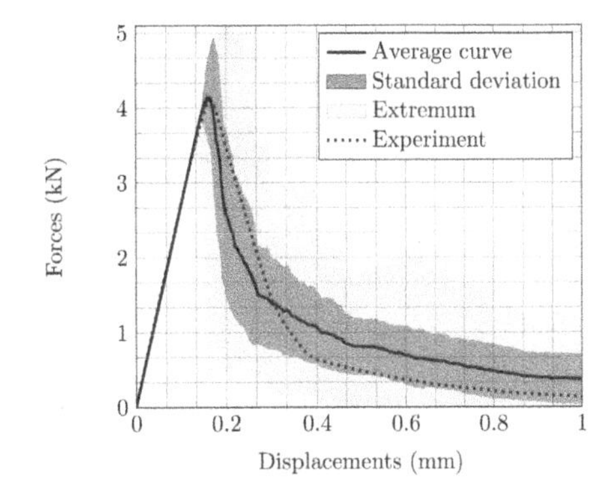

Figure 4. Numerical (average curve, standard deviation and extremum) and Experimental Force-Displacement responses of the wet unnotched beams under three-point bending tests.

simulations. The average results are derived from 100 Monte-Carlo runs.

Firstly, 3D finite element hydric simulations are performed to obtain the drying shrinkage field. Then, 2D discrete simulations allow to capture the displacement field and the cracking pattern. As observed on Figure 4, the cracks are distributed on the whole periphery of the beam in a perpendicular way. Their size

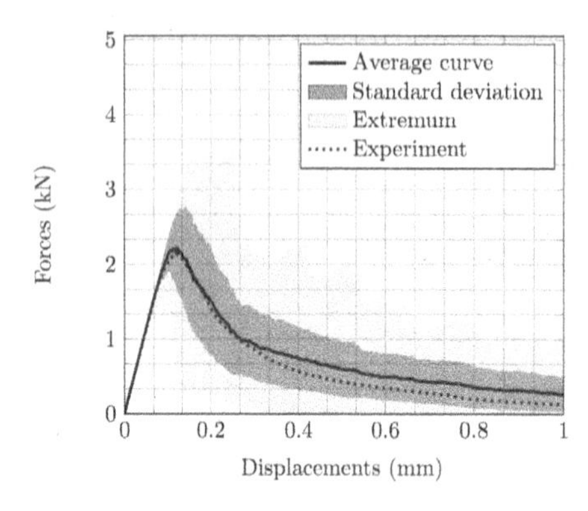

Figure 5. Numerical (average curve, standard deviation and extremum) and Experimental Force-Displacement responses of the wet notched beams under three-point bending tests.

and opening are not constant since the beam-particle model introduces randomness in the mesh and fracture properties.

Figure 6. Numerical concrete beam after 70 days of drying.

Figure 7 shows the distribution of drying micro-crack lengths obtained for all 100 Monte Carlo runs. The majority of the cracks are millimeter long and very few are centimeter long. However, these longer cracks will create preferential paths for cracking during the three-point bending test.

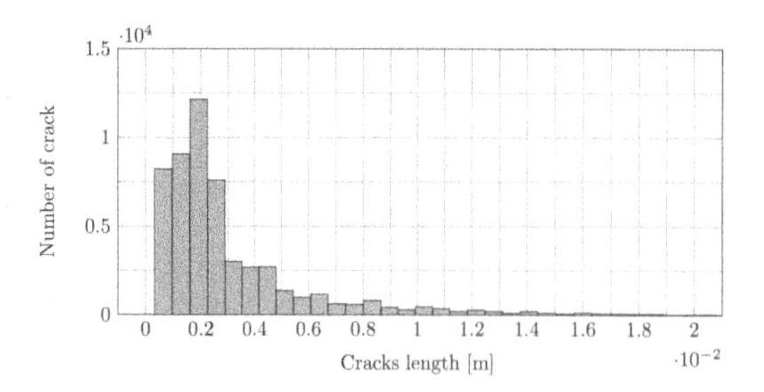

Figure 7. Distribution of microcrack lengths on the 100 numerical specimens subjected to 70 days of drying.

Three-point bending tests on specimens subjected to 70-day drying conditions were simulated for two values of the capillary pressure coefficient: b_ε= 0,0 and b_ε= 0,1. On Figures 8 and 9, the initial bending stiffness is reproduced regardless of the value of this coefficient. On the other hand, its influence is important for the maximum bending strength. The capillary pressure acts here as a pre-stress that will prevent cracking (Bažant et al. 1997).

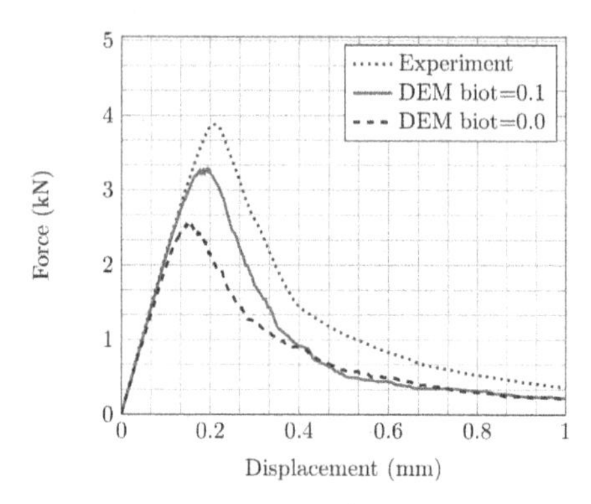

Figure 8. Numerical and Experimental Force-Displacement responses of the dried unnotched beams under three-point bending tests.

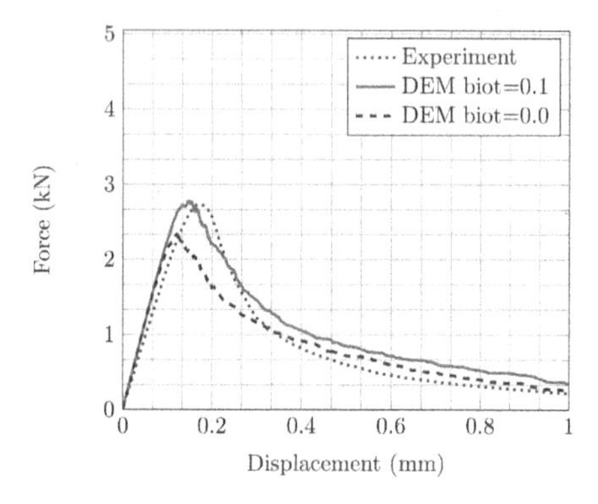

Figure 9. Numerical and Experimental Force-Displacement responses of the dried notched beams under three-point bending tests.

4 DISCUSSION AND CONCLUSIONS

The following remarks can be made on the use of a weak finite-discrete element coupling approach to study differential drying shrinkage cracking in concrete:

- This approach is simple to implement since the coupling only requires the interpolation of the shrinkage field on the discrete mesh to perform the coupling. The method is therefore non-intrusive.
- The use of a beam-particle model allows to obtain a fine description of the cracking. It is therefore possible to numerically quantify the penetration depth of micro-cracks following drying shrinkage as well as their openings.
- It is possible to perform mechanical test simulations on the cracked specimens under drying. These studies allow to quantify the impact of these drying micro-cracking on the mechanical properties.

A strong assumption is made on the choice to consider the positive effect of capillary pressure on the mechanical strength of concrete after drying. This allows us to correctly approximate the peak loads during three-point bending tests on notched and unnotched specimens. Indeed, the predicted bending strength would be lower without considering the capillary pressure. A next step is to consider creep, both basic and drying creep, which would reduce drying-related micro-cracking and thus also have a positive impact on post-drying mechanical strength.

REFERENCES

Bažant, Z.P., Hauggaard, A.B., Baweja, S. & Ulm, F.-J. (1997). Microprestress-solidification theory for concrete creep. i: aging and drying effects. *Journal of Engineering Mechanics* 123, 1188–1194.

Bisschop, J. & van Mier, J.G. (2002). How to study drying shrinkage microcracking in cement-based materials using optical and scanning electron microscopy. *Cement and Concrete Research* 32, 279–287.

Bisschop, J. & van Mier, J.G. (2008). Effect of aggregates and microcracks on the drying rate of cementitious composites. *Cement and Concrete Research* 38(10), 1190–1196.

Delaplace, A. (2008). Modélisation discrete appliquée au comportemet des matériaux et structures. *Mémoire d'habilitation à diriger des recherches de l'ENS Cachan.*

van Genutchen (1980). A closed-form equation for predicting the hydraulic conductivity of unsaturated soils. *Soil Science Society of America Journal* 44, 892–898.

Grassl, P. & Bolander, J. (2016). Three-dimensional network model for coupling of fracture and mass transport in quasi-brittle geomaterials. *Materials* 9(9), 782.

Hossain, A.B. & Weiss, J. (2004). Assessing residual stress development and stress relaxation in restrained concrete ring specimens. *Cement and Concrete Composites* 26(5), 531–540.

Oliver-Leblond, C., Delaplace, A., Ragueneau, F. & Richard, B. (2013). Non-intrusive global/local analysis for the study of fine cracking. *International Journal for Numerical and Analytical Methods in Geomechanics 37(8), 973–992.*

Roth, S.N., Léger, P. & Soulaïmani, A. (2020). Fully-coupled hydro-mechanical cracking using xfem in 3d for application to complex flow in discontinuities including drainage system. *Computer Methods in Applied Mechanics and Engineering* 370, 113282.

Soleilhet, F., Benboudjema, F., Jourdain, F. & Gatuingt, F. (2020). Role of pore pressure on cracking and mechanical performance of concrete subjected to drying. *Cement and Concrete Composites* 114, 103727.

Soleilhet, F., Benboudjema, F., Jourdain, F. & Gatuingt, F. (2021). Effect of transient drying on mechanical properties of concrete specimens. *European Journal of Environmental and Civil Engineering* 6, 1–20.

Vassaux, M., Oliver-Leblond, C., Richard, B. & Ragueneau, F. (2016). Beam-particle approach to model cracking and energy dissipation in concrete: Identification strategy and validation. *Cement and Concrete Composites* 70, 1–14.

Computational Modelling of Concrete and Concrete Structures – Meschke, Pichler & Rots (Eds)
© 2022 Copyright the Author(s), ISBN: 978-1-032-32724-2

Upscaling of coupled mechanical and mass transport discrete model

J. Eliáš
Institute of Structural Mechanics, Faculty of Civil Engineering, Brno University of Technology, Brno, Czechia

G. Cusatis
Department of Civil and Environmental Engineering, Northwestern University, Evanston, IL USA

ABSTRACT: A computational homogenization approach for mesoscale discrete models of coupled mechanics and mass transport in concrete is developed via asymptotic expansion. Primary fields of the model (pressure, displacements and rotations) are decomposed into macroscopic and microscopic components. Taylor expansion is then applied to relate field values between the neighboring nodes. The expanded primary fields propagate through geometric, constitutive and balance equations to provide a two-level model. At the microscale, heterogeneous and discrete Representative Volume Element (RVE) problem is obtained. The homogenization renders the RVE to be steady state even for transient tasks. Periodic boundary conditions are applied and the load is imposed in a form of eigen pressure gradient, eigen strains and eigen curvatures, which are computed as projections of macroscopic tensors of pressure gradient, strain and curvature. The mechanical RVE is solved first as it is independent on the transport part. The transport RVE is then solved taking into account crack openings from the mechanical RVE. At the macrocale, homogeneous and continuous coupled transient equations emerge. These equations are solved with a help of the finite element method with a mechanical and transport RVE couple attached to each integration point. Biot's coupling terms between transport and mechanics appear at the macroscale only. Simple examples are presented verifying the homogenization technique.

1 INTRODUCTION

Transport of a fluid in heterogeneous quasi brittle materials is strongly coupled with their mechanical behavior. The pressure affects the stress state in the solid phase according to the Biot's theory (Detournay & Cheng 1993), while volumetric changes and cracking have an effect on pressure and material conductivity. Recently, fully coupled models were developed in a discrete framework at mesoscale. Their advantage in the coupled analysis is a detailed representation of the cracking pattern, including transition from diffused to localized cracking state and anisotropy of the cracking phenomenon (Bolander, Eliáš, Cusatis, & Nagai 2021). The mechanical model at hand uses *physical* discretization, i.e., the mechanical rigid bodies correspond to the heterogeneous units of a material internal structure. It is similar to the LDPM approach (Cusatis, Mencarelli, Pelessone, & Baylot 2011; Cusatis, Pelessone, & Mencarelli 2011) but the domain tessellation is treated differently. Geometry of the transport part of the model is conveniently built as a dual diagram to the mechanical part (Grassl 2009; Grassl & Bolander 2016). Each transport element is then aligned with some potential crack direction and can therefore easily accommodate an effect of cracking on its permeability coefficient. Consequently, realistic anisotropic transport behavior in the inelastic regime is provided.

Since the model geometry is generated according to material heterogeneities, there is a large computational cost associated with it. Various techniques are available to reduce this cost including the classical computational homogenization technique. Such homogenization was already derived for mechanical discrete models (Rezakhani, Alnaggar, & Cusatis 2019; Rezakhani & Cusatis 2016; Rezakhani, Zhou, & Cusatis 2017) via an asymptotic expansion. This contribution extends the homogenization further considering the mechanical part being fully coupled with the mass transport part. There are two coupling mechanisms considered: (i) the Biot's theory and (ii) an effect of cracking on the conductivity as the open cracks create channels for the fluid.

Two verification examples are included: flow through a compresses cylinder and hydraulic fracturing of a hollow cylinder. Reasonable correspondence between results from the full and homogenized model is obtained, time savings in a form of speed-up factors are reported.

 DOI 10.1201/9781003316404-73

2 FORMULATION OF THE DISCRETE MODEL

The discrete model at hand is generated by random placing of spherical aggregates into the domain. Aggregate diameters are drawn according to the Fuller curve and they are placed without any overlapping. Weighted Delaunay triangulation is performed on centers of the spheres to define mechanical connectivity and tetrahedrons serving as control volumes for flow. Dual power/Laguerre tessellation serves to find connectivity of the transport part of the model and polyhedral shapes of the mechanical rigid bodies. Each rigid body therefore contains one mineral grain and surrounding matrix.

Centers of the spherical grains bear six degrees of freedom (three translations, $\mathbf{u}$, and three rotations, $\boldsymbol{\theta}$), nodes of the power tessellation bear one degree of freedom with meaning of fluid pressure, p. Contact between rigid bodies I and J is provided by mechanical elements, e, with length $\mathbf{x}^{IJ} = l$, normal $\mathbf{n}$ and projected area $A^{\star}$. One also defines a local coordinate system at each mechanical contact by orthogonal unit vectors $\mathbf{e}_N = \mathbf{n}$, $\mathbf{e}_M$ and $\mathbf{e}_L$, or shortly $\mathbf{e}_\alpha$, $\alpha \in \{N, M, L\}$. Conduit elements, d, connecting transport nodes P and Q have length $\mathbf{x}^{PQ} = h$, normal $\mathbf{e}_\lambda$ and projected area $S^{\star}$. The volume of a polyhedral mechanical particle is denoted V, the volume of a transport tetrahedron is W.

The first set of equations provides pressure gradient, q, between transport nodes P and Q and strain, $\boldsymbol{\varepsilon}$, and curvature, $\boldsymbol{\chi}$, at the contact between rigid bodies I and J

$$g = \frac{p^Q - p^P}{h} \tag{1}$$

$$\varepsilon_\alpha = \frac{1}{l}\left(\mathbf{u}^J - \mathbf{u}^I + \boldsymbol{\mathcal{E}} : (\boldsymbol{\theta}^J \otimes \mathbf{c}_J - \boldsymbol{\theta}^I \otimes \mathbf{c}_I)\right) \cdot \mathbf{e}_\alpha \tag{2}$$

$$\chi_\alpha = \frac{1}{l}\left(\boldsymbol{\theta}^J - \boldsymbol{\theta}^I\right) \cdot \mathbf{e}_\alpha \tag{3}$$

where vectors $\mathbf{c}_I$ and $\mathbf{c}_J$ vectors point from the nodes I or J to the centroid of the mechanical contact face, $\boldsymbol{\mathcal{E}}$ is the third order Levi-Civita tensor.

The second set of equations, called constitutive, gives the flux scalar, j, the total traction vector, $\mathbf{t}$, and the couple traction vector, $\mathbf{m}$. The form of these equations can be arbitrary (providing they are thermodynamically admissible) but one needs to know at this point all the variables which are entering the constitutive functions. Examples in the last section of this contribution use simplified LDPM mechanical constitutive equations (Cusatis & Cedolin 2007) and fluxes dependent on crack openings, δ_λ, and the average pressure, p_λ, in the conduit element

$$j = f_j(p_\lambda, g, \delta_\lambda) = -\lambda(p_\lambda, \delta_\lambda) g \tag{4}$$

$$\mathbf{t} = f_s(\boldsymbol{\varepsilon}) - b p_a \mathbf{e}_N \tag{5}$$

$$\mathbf{m} = f_m(\boldsymbol{\chi}) \tag{6}$$

b is Biot coefficient and p_a is an average pressure in the mechanical element.

The final set of equations describe balance of fluid mass in the control volume (Delaunay tetraheron) of a fully saturated medium and balance of forces and moments at each rigid polyhedral body.

$$\sum_{Q \in W}\left[S^{\star} j - \rho_{w0} W \dot{v}_c \left(1 + b + \frac{p_\lambda - p_0}{K_w}\right) - \rho_{w0} W v_c \frac{\dot{p}_\lambda}{K_w}\right] - \rho_{w0}\left(3b\dot{\varepsilon}_V + \frac{\dot{p}_\lambda}{M_b}\right) W - Wq = 0 \tag{7}$$

$$V\rho \ddot{\mathbf{u}}^I + \mathbf{M}_{u\theta} \cdot \ddot{\boldsymbol{\theta}}^I - V\mathbf{b} = \sum_J A^{\star} t_\alpha \mathbf{e}_\alpha \tag{8}$$

$$\mathbf{M}_\theta \cdot \ddot{\boldsymbol{\theta}}^I + \mathbf{M}_{u\theta}^T \cdot \ddot{\mathbf{u}}^I = \sum_J A^{\star}\left[\mathbf{w} + m_\alpha \mathbf{e}_\alpha\right] \tag{9}$$

ρ is the density of the solid, $\mathbf{M}_{u\theta}$ and $\mathbf{M}_\theta$ are moment of inertia tensors, $\mathbf{b}$ is the volumetric load, $\mathbf{w} = \boldsymbol{\mathcal{E}} : (\mathbf{c}_I \otimes \mathbf{t}) = t_\alpha \boldsymbol{\mathcal{E}} : (\mathbf{c}_I \otimes \mathbf{e}_\alpha)$ is the moment of traction with respect to the mechanical node I, ρ_{w0} is the initial fluid density, v_c is the crack density (volume of cracks over the total volume), p_0 is the reference pressure, ε_V is the volumetric strain, K_w is the fluid bulk modulus and M_b is Biot modulus.

3 ASYMPTOTIC EXPANSION

The homogenization assumes one global macroscopic reference system X and infinite number of local systems y defined at every macroscopic material point. It is assumed that the material appears continuous and homogeneous from the viewpoint of the global system, while discrete and heterogeneous from the viewpoint of the local reference system. The scale factor that relates X to y is η. It is called separation of scales constant and expected to be positive but much lower than one.

The primary fields are now expanded into macroscopic terms, $\bullet^{(0)}$, fluctuating terms, $\bullet^{(1)}$, and higher order terms that are eventually omitted. Such expansion assumes that the material internal structure is composed of some period units called Representative Volume Elements (RVE) and that the fluctuating terms are periodic and zero on average over the RVEs .

$$p(\mathbf{X}, \mathbf{y}) = p^{(0)}(\mathbf{X}, \mathbf{y}) + \eta p^{(1)}(\mathbf{X}, \mathbf{y}) + \dots \tag{10}$$

$$\mathbf{u}(\mathbf{X}, \mathbf{y}) = \mathbf{u}^{(0)}(\mathbf{X}, \mathbf{y}) + \eta \mathbf{u}^{(1)}(\mathbf{X}, \mathbf{y}) + \dots \tag{11}$$

$$\boldsymbol{\theta}(\mathbf{X}, \mathbf{y}) = \eta^{-1}\boldsymbol{\omega}^{(-1)}(\mathbf{X}, \mathbf{y}) + \boldsymbol{\omega}^{(0)}(\mathbf{X}, \mathbf{y}) + \boldsymbol{\varphi}^{(0)}(\mathbf{X}, \mathbf{y}) + \eta \boldsymbol{\varphi}^{(1)}(\mathbf{X}, \mathbf{y}) + \dots \tag{12}$$

The special type of expansion of rotations is adopted from Rezakhani & Cusatis (2016).

The second fundamental component is the Taylor series expansion of the primary fields around nodes I

or P to estimate values at neighboring nodes J or Q for mechanical or transport fields, respectively. This expansion is developed according to Fish, Chen, & Li (2007)

$$p(\mathbf{X}_Q, \mathbf{y}_Q) = p(\mathbf{X}_P, \mathbf{y}_Q) + \frac{\partial p(\mathbf{X}_P, \mathbf{y}_Q)}{\partial X_i} x_i^{PQ} + \frac{1}{2}\frac{\partial^2 p(\mathbf{X}_P, \mathbf{y}_Q)}{\partial X_i X_j} x_i^{PQ} x_j^{PQ} + \mathcal{O}(h^3) \quad (13)$$

$$\mathbf{u}(\mathbf{X}_J, \mathbf{y}_J) = \mathbf{u}(\mathbf{X}_I, \mathbf{y}_J) + \frac{\partial \mathbf{u}(\mathbf{X}_I, \mathbf{y}_J)}{\partial X_j} x_j^{IJ} + \frac{1}{2}\frac{\partial^2 \mathbf{u}(\mathbf{X}_I, \mathbf{y}_J)}{\partial X_j X_k} x_j^{IJ} x_k^{IJ} + \mathcal{O}(l^3) \quad (14)$$

$$\boldsymbol{\theta}(\mathbf{X}_J, \mathbf{y}_J) = \boldsymbol{\theta}(\mathbf{X}_I, \mathbf{y}_J) + \frac{\partial \boldsymbol{\theta}(\mathbf{X}_I, \mathbf{y}_J)}{\partial X_j} x_j^{IJ} + \frac{1}{2}\frac{\partial^2 \boldsymbol{\theta}(\mathbf{X}_I, \mathbf{y}_J)}{\partial X_j X_k} x_j^{IJ} x_k^{IJ} + \mathcal{O}(l^3) \quad (15)$$

Expansions (10)-(15) are now substituted into the equations for pressure gradient (1), strain (2) and curvature (3) to obtain expansion of these variables. Then, an expansion of flux, traction and couple traction is developed using the Taylor series expansion of functions f_j, f_s and f_m (Eqs. 4-6) with respect to all its input variables. Finally, all these expansions are substituted into the balance equations of the discrete model (7)-(9). The set of three balance equations falls apart into several such sets collecting expressions with the same power of the scale separation constant η. Each of the sets is then solved independently resulting in (i) mechanical and transport RVE problems at the mesoscale and (ii) a continuous differential equations of the macroscale problem.

The macroscale problem is defined as

$$\nabla_X \cdot \mathbf{f} = \rho_{w0}\left[\dot{v}_{c0}\left(1 + b + \frac{p^{(0)} - p_0}{K_w}\right) + v_{c0}\frac{\dot{p}^{(0)}}{K_w} + 3b\dot{\varepsilon}_V^{(0)} + \frac{\dot{p}^{(0)}}{M_b}\right] + q \quad (16)$$

$$\nabla_X \cdot \boldsymbol{\sigma}_{\mathrm{s}} - \nabla_X p^{(0)} \cdot \boldsymbol{\xi} = \langle \rho \rangle \ddot{\mathbf{v}}^{(0)} - \mathbf{b} \quad (17)$$

$$\nabla_X \cdot \boldsymbol{\mu}_{\mathrm{s}} - \nabla_X p^{(0)} \cdot \boldsymbol{\zeta} + \boldsymbol{\mathcal{E}} : \boldsymbol{\sigma}_{\mathrm{s}} - p^{(0)}\boldsymbol{\mathcal{E}} : \boldsymbol{\xi} = = \eta^{-1}\mathbf{I}_0 \cdot \ddot{\boldsymbol{\omega}}^{(-1)} \quad (18)$$

with degrees of freedom being macroscopic pressure $p^{(0)}$, macroscopic displacement $\mathbf{v}^{(0)}$ and macroscopic rotation $\eta^{-1}\boldsymbol{\omega}^{(-1)}$. Tensor $\mathbf{I}_0$ denotes inertia tensor of the whole RVE with entries

$$I_{ij}^0 = \frac{1}{V_0}\sum_{I \in V_0}\left[V\rho\mathcal{E}_{ikl}\mathcal{E}_{ljm}\left(\eta r_k^0\right)\left(\eta r_m^0\right) + M_{ij}^\theta\right] \quad (19)$$

Vectors $\mathbf{r}$ are positions of the centroid of rigid bodies in the reference system y and V_0 is the volume of the whole RVE. Symbol v_{c0} in Eq. (16) is the average crack volume density in the RVE

$$v_{c0} = \frac{1}{V_0}\sum_{P \in V_0}\sum_{Q \in W} v_c^{(0)} W \quad (20)$$

Tensors $\mathbf{f}$, $\boldsymbol{\sigma}_{\mathrm{s}}$, $\boldsymbol{\mu}$ are macroscopic flux, stress in the solid and couple stress in the solid, respectively. Along with the two auxiliary tensors $\boldsymbol{\xi}$ and $\boldsymbol{\zeta}$ they read

$$\mathbf{f} = \frac{1}{V_0}\sum_{d \in V_0} hS^\star j^{(0)}\mathbf{e}_\lambda \quad (21)$$

$$\boldsymbol{\sigma}_{\mathrm{s}} = \frac{1}{V_0}\sum_{e \in V_0} lA^\star s_\alpha^{(0)}\mathbf{e}_N \otimes \mathbf{e}_\alpha \quad (22)$$

$$\boldsymbol{\mu}_{\mathrm{s}} = \frac{1}{V_0}\sum_{e \in V_0} lA^\star \mathbf{e}_N \otimes \left[\eta m_\alpha^{(1)}\mathbf{e}_\alpha + s_\alpha^{(0)}\boldsymbol{\mathcal{E}} : (\mathbf{x}_c \otimes \mathbf{e}_\alpha)\right] \quad (23)$$

$$\boldsymbol{\xi} = \frac{1}{V_0}\sum_{e \in V_0} lA^\star b\mathbf{e}_N \otimes \mathbf{e}_N \quad (24)$$

$$\boldsymbol{\zeta} = \frac{1}{V_0}\sum_{e \in V_0} lA^\star b\mathbf{e}_N \otimes [\boldsymbol{\mathcal{E}} : (\mathbf{x}_c \otimes \mathbf{e}_N)] \quad (25)$$

The flux scalar $j^{(0)}$, the traction vector in the solid $\mathbf{s}^{(0)}$ and the couple traction vector $\eta\mathbf{m}^{(1)}$ are computed at each integration point of the macroscale problem by solving a couple of RVE problems (one for mechanics and one for mass transport). RVEs are loaded by projection of macroscopic tensors of pressure gradient, Cosserat strain and curvature that comes from macroscale kinematics. The projection generates eigen components of the pressure gradient, strain and curvature in the discrete elements. The mechanical RVE is completely independent on the transport RVE, while the transport RVE requires on its input information about crack openings in the mechanical RVE. The Biot's coupling effects appears only at the macroscale. Both mechanical and transport RVEs are steady state, transient terms appear only at the macroscale.

4 VERIFICATION

Two examples are presented to verify the derived homogenization scheme. Results of the full and homogenized model are compared by simulating (i) a compressed cylinder where fluid flows parallel to the loading direction and (ii) a hydraulic fracturing of a hollow cylinder. Cosserat trilinear isoparametric brick elements are used at the macroscale to solve weak form of Eqs. (16)-(18). Mechanical constitutive equations of the discrete model are implemented according to on the older LDPM version (Cusatis & Cedolin 2007) and simplified, see Eliáš (2016). Definition of the permeability coefficient, λ, is adopted from Grassl & Bolander (2016). The maximum diameter of the aggregates in the discrete model is 10 mm, the minimum diameter directly represented in the model is 4 mm. Smaller aggregates are left to be phenomenologically included in the constitutive model. RVE size considered in the examples is 50 × 50 × 50 mm^3. An adaptive switching from linear pre-computed RVE states to full nonlinear states is implemented according to Rezakhani, Zhou, & Cusatis (2017).

4.1 *Flow through compressed cylinder*

The first test features cylinder of diameter 0.1 m and depth 0.1 m compressed vertically, see Figure 1. There is also prescribed fluid pressure difference 1 MPa between the top and the bottom specimen surface. The steady state solution is calculated considering Biot coefficient $b = 0$. The test is performed to asses the effect of cracks on the permeability coefficient, λ. The full model has 32, 500 degrees of freedom and took 68 minutes to run.

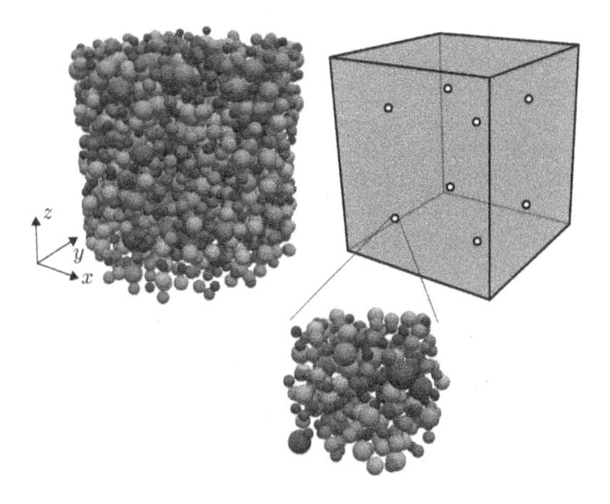

Figure 1. Cylindrical specimen: the full model and the homogenized model composed of a single brick element along with an RVE at one integration point.

The homogenized model is represented by a single Cosserat brick element (of equivalent cross section area) with eight integration points and therefore eight RVE couples. The total number of degrees of freedom (RVEs plus the macroscale) is 33, 416 and computational time is 25 minutes. The speed-up factor is therefore about 2.7.

A comparison of results obtained by the two models is presented in Figure 2. Both stresses and fluxes show good correspondence, therefore one can conclude that the homogenization correctly preserves the effect of cracks on material conductivity. The strain localization, taking place at the end of the simulation, is in this case analyzed correctly by both (full & homogenized) models, because the size of the RVEs approximately corresponds to the volume associated with integration points (Rezakhani & Cusatis 2016).

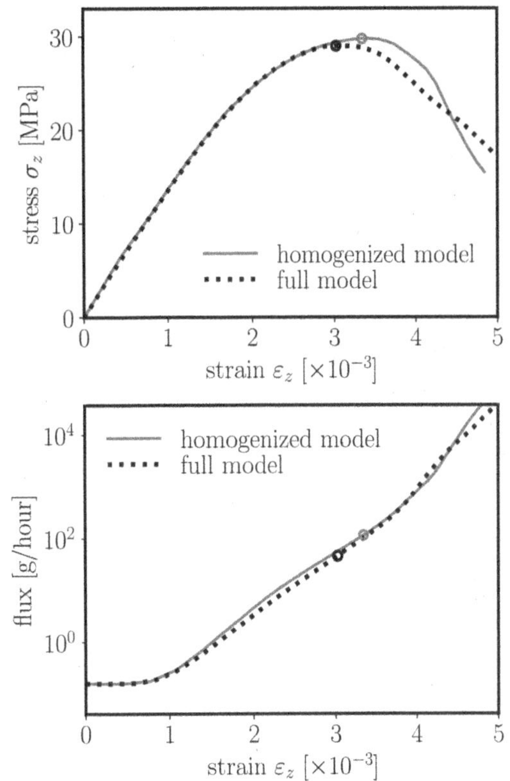

Figure 2. Mechanical responses and fluxes obtained from the full and homogenized model during the virtual compression test.

4.2 *Hydraulic fracturing of hollow cylinder*

The second verification example applies the homogenization to hydraulic fracturing. A hollow cylinder of outer diameter 0.4 m, inner diameter 0.1 m and depth 0.05 m is fractured by an increasing fluid pressure in the hole. The outer fluid pressure is set to zero. The simulation is transient, the inner pressure increases by 200 Pa per second. Different Biot coefficients are considered, $b \in \{0, 0.5, 1\}$. The full model contains approximately 150, 000 degrees of freedom and takes about 500 ($b = 0$), 315 ($b = 0.5$) and 200 ($b = 1$) minutes to run. It is terminated by divergence of the implicit solver when the fluid pressure becomes too large and macroscopic cracks over the whole thickness of the cylinder develop.

The homogenized model is composed of 40 Cosserat brick elements, total number of degrees of freedom is approximately 1.3 × 10^6. Speed-up factors obtained by the homogenization are 1.2, 5.2 and 5 for Biot coefficient 0, 0.5 and 1, respectively. In this case, the homogenized models have many more degrees of freedom at the end of the simulation when all the RVEs are switched from the linear pre-computed state to the full nonlinear state.

The RVE volume does not correspond to the material volume associated with the integration point. The

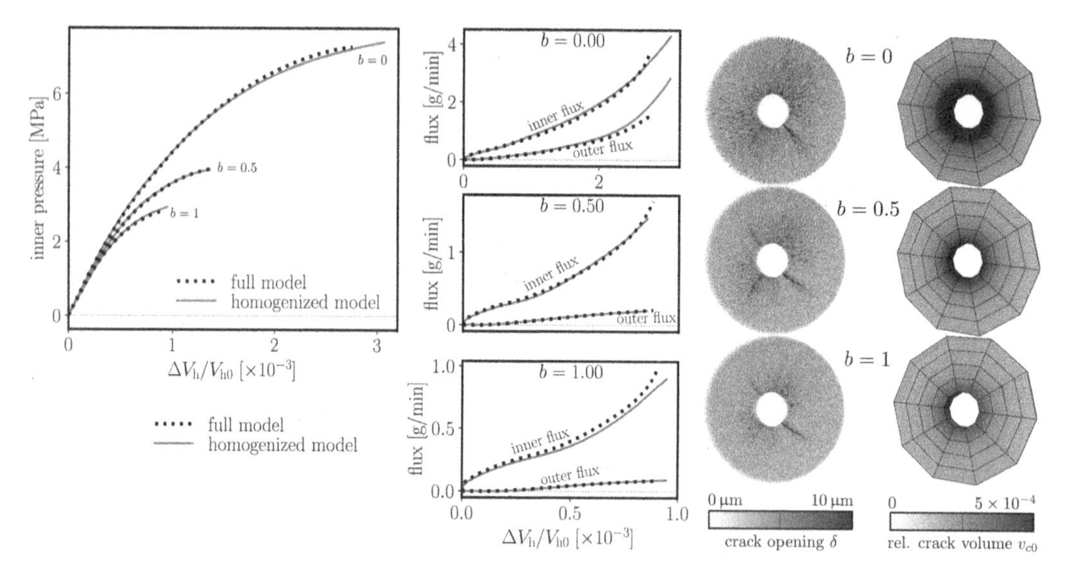

Figure 3. Left: fluid pressure inside the hole and fluxes through the cylinder for different Biot coefficients; right: cracks at the final step of the full and homogenized model simulation. ΔV_h is change of volume of the hole and V_{h0} is the original volume of the hole.

strain localization phenomenon, which is generally considered unfeasible for homogenization (Gitman, Askes, & Sluys 2007), plays an important role and causes discrepancies between responses from the full and homogenized model in the terminal stages of the simulation, see Figure 3 left. Another issue is that the periodic boundary conditions of the RVE do not allow the localized crack to propagate under arbitrary angle (Stránský & Jirásek 2011; Coenen, Kouznetsova, & Geers 2012). For that reason, all the RVEs are rotated so that their y reference system i aligned with the cylindrical coordinate system at the integration point. Unfortunately, such a rotation constitutes perfect symmetry in the homogenized model and the strain localization does not take place before the simulation is terminated, Figure 3 right.

5 CONCLUSIONS

An asymptotic expansion homogenization of discrete models for coupled mechanics and mass transport in a fully saturated media is developed. The homogenization derives Cosserat type equations for mechanics and a standard differential equation of mass transport at the macroscale, there are however several additional coupling terms. At the mesoscale, two periodic steady-state RVE problems emerge. The mechanical RVE is independent on the transport one, while the transport RVE depends on crack openings from the mechanical RVE problem, they are therefore solved in a sequence.

Two verification examples showed reasonably small errors of the homogenized solution. The errors are mostly attributed to (i) a poor approximation of the primary fields at the macroscale, (ii) the strain localization phenomenon present in late stages of simulations and (iii) a boundary layer or wall effect (Eliáš 2017). The first error source can be easily reduced by improving the mesh or type of elements, but such improvement would also increase computational time. The second error source is not easy to eliminate as the homogenization actually does not apply in the case of the strain localized into a macroscopic crack. There are few remedies suggested in the literature (Unger 2013; Coenen, Kouznetsova, Bosco, & Geers 2012), but this issue is left for further research.

The speed-up factors achieved by the homogenization are not astonishing. There is, however, possibility to distribute RVE computations over several computational cores. Since they are independent, distributed computing should be trivial and extremely effective.

ACKNOWLEDGEMENTS

Jan Eliáš gratefully acknowledges financial support from the Czech Science Foundation under project no. GA19-12197S.

REFERENCES

Bolander, J. E., J. Eliáš, G. Cusatis, & K. Nagai (2021). Discrete mechanical models of concrete fracture. *Eng Fract Mech 257*, 108030.

Coenen, E., V. Kouznetsova, E. Bosco, & M. Geers (2012). A multi-scale approach to bridge microscale damage and macroscale failure: a nested computational homogenization-localization framework. *Int J Fracture 178*, 157–178.

Coenen, E., V. Kouznetsova, & M. Geers (2012). Novel boundary conditions for strain localization analyses in microstructural volume elements. *Int J Numer Meth Eng 90*(1), 1–21.

Cusatis, G. & L. Cedolin (2007). Two-scale study of concrete fracturing behavior. *Eng Fract Mech 74*(1), 3–17. Fracture of Concrete Materials and Structures.

Cusatis, G., A. Mencarelli, D. Pelessone, & J. Baylot (2011). Lattice discrete particle model (LDPM) for failure behavior of concrete. II: Calibration and validation. *Cement Concrete Composites 33*(9), 891–905.

Cusatis, G., D. Pelessone, & A. Mencarelli (2011). Lattice discrete particle model (LDPM) for failure behavior of concrete. I: Theory. *Cement Concrete Composites 33*(9), 881–890.

Detournay, E. & A. H.-D. Cheng (1993). Fundamentals of poroelasticity. In C. Fairhusrt (Ed.), *Analysis and Design Methods*, pp. 113–171. Oxford: Pergamon.

Eliáš, J. (2016). Adaptive technique for discrete models of fracture. *Int J Solids Struct 100–101*, 376–387.

Eliáš, J. (2017). Boundary layer effect on behavior of discrete models. *Materials 10*, 157.

Fish, J., W. Chen, & R. Li (2007). Generalized mathematical homogenization of atomistic media at finite temperatures in three dimensions. *Comput Method Appl M 196*(4), 908–922.

Gitman, I., H. Askes, & L. Sluys (2007). Representative volume: Existence and size determination. *Eng Fract Mech 74*(16), 2518–2534.

Grassl, P. (2009). A lattice approach to model flow in cracked concrete. *Cement Concrete Composites 31*(7), 454–460.

Grassl, P. & J. Bolander (2016). Three-dimensional network model for coupling of fracture and mass transport in quasi-brittle geomaterials. *Materials 9*(9), 782.

Rezakhani, R., M. Alnaggar, & G. Cusatis (2019). Multiscale homogenization analysis of alkali–silica reaction (ASR) effect in concrete. *Engineering 5*(6), 1139–1154.

Rezakhani, R. & G. Cusatis (2016). Asymptotic expansion homogenization of discrete fine-scale models with rotational degrees of freedom for the simulation of quasi-brittle materials. *J Mech Phys Solids 88*, 320–345.

Rezakhani, R., X. Zhou, & G. Cusatis (2017). Adaptive multi-scale homogenization of the lattice discrete particle model for the analysis of damage and fracture in concrete. *Int J Solids Struct 125*, 50–67.

Stránský, J. & M. Jirásek (2011). Calibration of particle-based models using cells with periodic boundary conditions. In *II International Conference on Particle-based Methods - Fundamentals and Applications*, pp. 1–12.

Unger, J. F. (2013). An FE2-X1 approach for multiscale localization phenomena. *J Mech Phys Solids 61*(4), 928–948.

Computational Modelling of Concrete and Concrete Structures – Meschke, Pichler & Rots (Eds)
© 2022 Copyright the Author(s), ISBN: 978-1-032-32724-2

PARC_CL 2.1: Modelling of the time-dependent behaviour of reinforced concrete slabs

F. Vecchi, L. Franceschini & B. Belletti
Department of Engineering and Architecture, University of Parma, Parma, Italy

ABSTRACT: During their lifetime, reinforced concrete (RC) structures may suffer a combination of time-dependent effects that are directly associated to properties of concrete, loading, humidity, temperature, and other environmental actions. For the prediction of long-term strains and cracking, creep and shrinkage should be accurately considered. In particular, the underestimation of cracks width and deflections of RC elements can be identified as primary consequences of the inadequate evaluation of time-dependent effects.

The present study focuses on the modelling of a RC slab by considering the time-dependent effects of concrete. To this aim, a non-linear finite element approach, based on multi-layer shell elements, named PARC_CL 2.1, is adopted. Finally, considering deemed-to-satisfy approach previsions, preliminary considerations on cracks width induced by corrosion deterioration are carried out.

As a conclusive remark, the PARC_CL 2.1 crack model can be considered as a useful tool for the time-dependent modelling of RC elements.

1 INTRODUCTION

The service life design of new reinforced concrete (RC) structures and the residual life estimation of existing RC structures, require an appropriate description of the mechanical properties of the materials, including the time-dependent strains of the concrete (induced by creep and shrinkage effects) and the deterioration phenomena (induced by environmental exposure).

Creep and shrinkage have a considerable impact upon the performance of RC structures by inducing unexpected excessive cracking and deformations over time. Therefore, to properly determine the structural performances at serviceability limit state (SLS), the detrimental effects of shrinkage and creep must be accurately considered with the aim to ensure adequate safety and durability throughout the life of the structure.

In this framework, several experimental research has been carried out to quantify the effects of creep and shrinkage, Bazant & Li (2008). On the other hand, scientific committee such as ACI Committee 209 (2008) and Model Code 2010 (fib 2013) proposed deterministic models that correlate analytical formulations based on sectional analysis and experimental data tests.

Nevertheless, Standard Codes and Guidelines are lacking provisions for the assessment of the rheological phenomena, coupled with the deterioration induced by corrosion.

Nowadays, existing structures and infrastructures – such as road, motorway, rail network – show evident signs of deterioration and requires a huge number of investments for repairing and maintenance. For this reason, the forthcoming challenges on the assessment of existing structures affected by aging and deterioration are addressing the scientific debate. An example, is the international workshop on the capacity assessment of corroded structures (https://www.cte-it.org/attivita-e-programmi/cacrcs-days-2021/) that since 2019 has seen the increasing participation of international experts coming from universities, public institutions, and private studios. Several cases study of structures requiring proactive or reactive interventions are reported in the proceedings of CACRCS DAYS (2020, 2021). In particular, the case of RC slabs in underground car parking are frequently cited as members affected by durability problems. Durability problems are caused at the extrados by excessive crack opening combined with deficiencies in the water drainage system, and use of water, salt, and chemical agents for the de-icing solutions, while at the intrados by relative humidity, temperature and carbonatation, as highlighted in Figure 1.

In the absence of specific requirements, crack control at SLS is imposed by Eurocode 2 (2004). According to the deemed-to-satisfy approach, crack opening widths must be limited, under the quasi-permanent combination of loads and during the entire service live of members, to defined values of w_{max}, that depend on the class of exposure.

The present paper deals with the evaluation of the crack opening and the deflection of RC slabs over time to appreciate the effects of shrinkage and creep.

DOI 10.1201/9781003316404-74

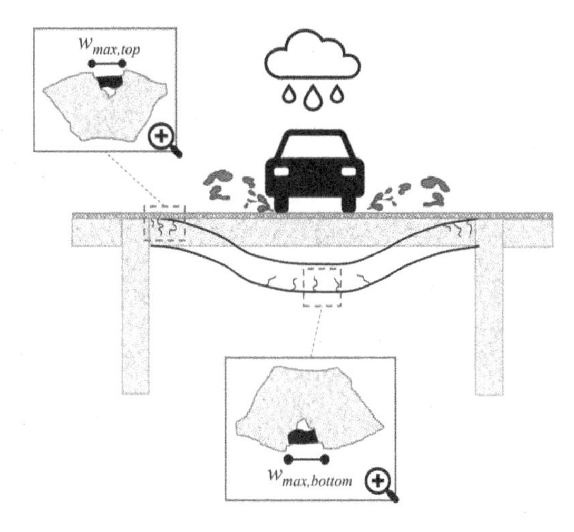

Figure 1. Durability problems and excessive cracking of RC slab.

The calculation of the deflection of RC slabs over time had been deeply investigated by Kilpatrick & Gilbert (2017) and by Bertero & Bertero (2018).

Referring to scientific literature – among the others – Motter et al. (2018) proposed a single function for the combination of creep and shrinkage effects, while a non-linear shrinkage profile along both depth and width of RC elements was applied by Sirico et al. (2017) and Hasan et al. (2018). Moreover, a finite element analysis method based on CEB-FIB MC 90 was suggested by Hossain & Vollum (2002). However, it is worth to mention that most of the available numerical models consider creep and shrinkage as separated phenomena, even if, their action is coupled.

The paper focuses on the response prediction over time of continuous RC slabs by adopting non-linear finite element analyses (NLFEA), based on multi-layer shell elements and PARC_CL 2.1 crack model, (Belletti at al. 2017; Shu et al. 2017). The presented numerical approach is validated by comparing NLFEA results with experimental outcomes on RC slabs tested by Gilbert & Guo (2005).

The PARC_CL 2.1 model, that stands for *Physical Approach of Reinforced Concrete for Cyclic Loading*, is a model developed at the University of Parma that incorporates different cyclic constitutive laws of materials and the evolution of corrosion deterioration, shrinkage, and creep over time (Belletti et al. 2019; 2020).

In the present work, the time-dependent effects of creep and shrinkage are evaluated by adopting the Model Code 2010 formulations (fib 2013). However, to properly calibrate creep and shrinkage strains, experimental measurements on the variation of the mechanical properties of concrete are considered and directly used as an input data for the analysis.

Based on the obtained results, the new updated version of PARC_CL 2.1 crack model can be considered as a useful tool for the time-dependent modelling of RC and prestressed concrete (PC) members.

Finally, some remarks on the evolution of the cracks over time with regards to the crack control verification imposed in the framework of the deemed-to-satisfy approach for service life design will be pointed out.

2 MAIN FEATURES OF PARC_CL 2.1

The PARC_CL 2.1 is a smeared and fixed crack model in which the reinforcement is assumed smeared in the hosting concrete element.

As widely discussed in previous works carried out by the Authors (Belletti et al. 2017; Franceschini et al. 2021; Vecchi & Belletti 2021), the PARC_CL 2.1 crack model is implemented in a user subroutine (UMAT.for) implemented in the software Abaqus.

The PARC_CL 2.1 crack model was previously validated against the capability to predict the response of RC elements subjected to static, cyclic and dynamic loading conditions (Franceschini et al. 2021; Vecchi & Belletti 2021). The paper presents, as main novelty of the new release, the implementation of the time-dependent effects of creep and shrinkage.

Figure 2 shows that the PARC_CL 2.1 crack model is characterized by two reference systems at each integration point, denoted as (i) the local x, y-coordinate system and (ii) the 1,2-coordinate system, related to the fixed crack orientation. In detail, the parameter ψ stands for the angle between the 1-direction, perpendicular to cracks, and the x-direction, while the parameter θ_i corresponds to the angle between the direction of the ith order of rebars and the x-direction.

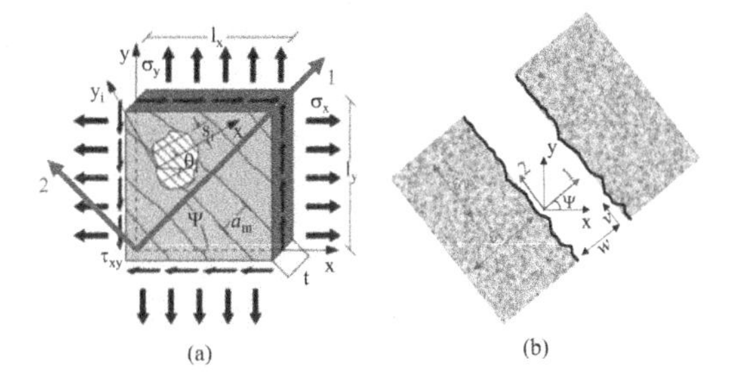

Figure 2. (a) Plane stress state of a RC element and (b) crack pattern (Belletti et al. 2017).

For a clear understanding, the main steps of the implemented algorithm are given in the flowchart reported in Figure 3.

2.1 *Strain field for concrete*

The PARC_CL 2.1 crack model is suitable for membrane elements of multi-layered shell elements subjected to plane state of stresses. Given the strain field of the element, $\{\varepsilon_{x,y}\}$, at each integration point in the global system (x, y-coordinate system), the strain field

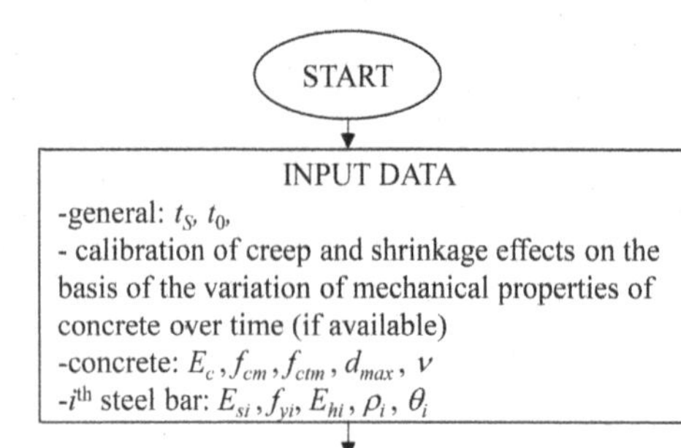

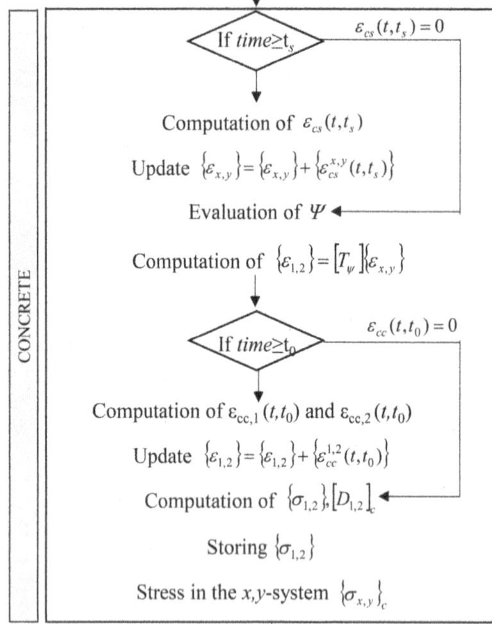

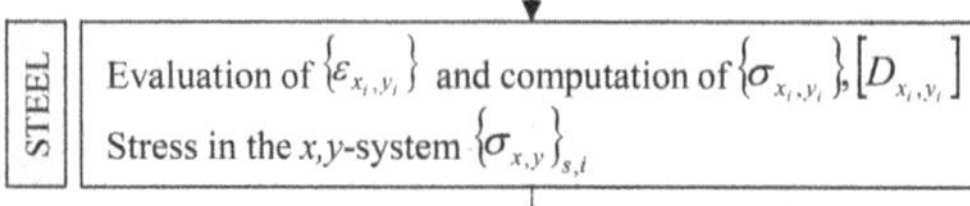

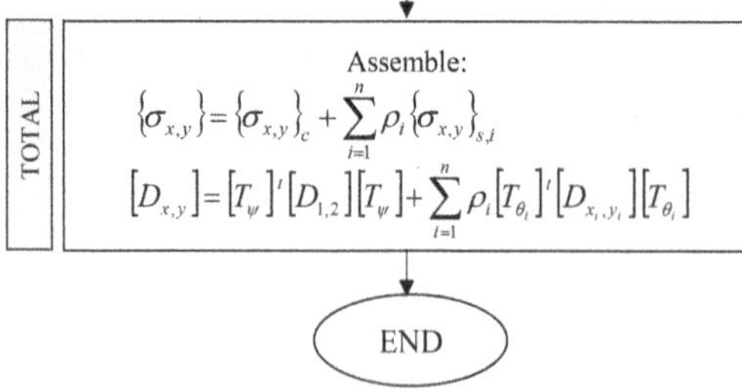

Figure 3. Flowchart of the PARC_CL 2.1 crack model for creep and shrinkage analysis.

for concrete in the 1,2-coordinate system, $\{\varepsilon_{1,2}\}$ can be obtained, as expressed by Eq.(1):

$$\{\varepsilon_{1,2}\} = [T_\psi]\{\varepsilon_{x,y}\} \tag{1}$$

where (T_ψ) is the transformation matrix.

In particular, the concrete behaviour is assumed orthotropic both before and after cracking.

However, the total strain field of concrete results to be strictly affected by the long-term strains induced by shrinkage and creep effects.

In this context, the total strain of a RC element at a time t, $\varepsilon_c(t)$, – that has been subjected to a uniaxial load at time t_0 – can be expressed as the sum of three different components, as highlighted in Eq.(2):

$$\varepsilon_c(t) = \varepsilon_{ci}(t_0) + \varepsilon_{cc}(t, t_0) + \varepsilon_{cs}(t, t_s) \tag{2}$$

where $\varepsilon_{ci}(t_0)$ is the instantaneous deformation, $\varepsilon_{cc}(t, t_0)$ is the strain due to creep, and $\varepsilon_{cs}(t, t_s)$ is the strain due to shrinkage. In particular, as shown in Figure 4, $\varepsilon_{cc}(t, t_0)$ is related to the age of the loading application, t_0; whereas $\varepsilon_{cs}(t, t_s)$ is expressed as a function of the concrete age at the beginning of drying, t_s.

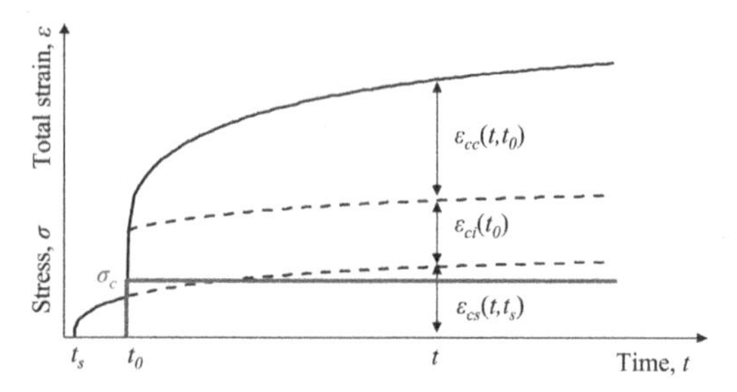

Figure 4. Example of sustained load and strain components development in a concrete specimen over time.

2.1.1 *Shrinkage effect*

Shrinkage is a load-independent strain due to the volume reduction of concrete at constant temperature, which is primarily caused by drying and hydration. In the PARC_CL 2.1 crack model, shrinkage is considered to be direction independent, while the shrinkage shear strain is assumed equal to zero. The shrinkage strain vector is applied along the x, y-coordinate system of the element, Figure 2(a), according to Eq.(3).

$$\{\varepsilon_{cs}^{x,y}(t, t_s)\} = \{\varepsilon_{cs}(t, t_s) \quad \varepsilon_{cs}(t, t_s) \quad 0\}^t \tag{3}$$

The axial shrinkage strain, $\varepsilon_{cs}(t, t_s)$, is negative and its magnitude is calculated by Model Code 2010 (*fib* 2013). In detail, the axial shrinkage strain $\varepsilon_{cs}(t, t_s)$, which occurs between times t_s at the beginning of shrinkage and time t (as shown in Figure 4), can be evaluated as the sum of the basic shrinkage $\varepsilon_{cbs}(t)$ and the drying shrinkage $\varepsilon_{cds}(t, t_s)$. Eq.(4) reports the entire formulation:

$$\varepsilon_{cs}(t, t_s) = \varepsilon_{cbs}(t) + \varepsilon_{cds}(t, t_s) \tag{4}$$

where, the basic shrinkage component, $\varepsilon_{cbs}(t)$, evaluated through Eq.(5), depends on the notional shrinkage coefficient, $\varepsilon_{cbs0}(f_{cm})$, which is evaluated as a function of the mean compressive strength at the age of 28 days, f_{cm}, and on the time function β_{bs}, expressed by Eq.(6).

$$\varepsilon_{cbs}(t) = \varepsilon_{cbs0}(f_{cm}) \cdot \beta_{bs}(t) \tag{5}$$

$$\beta_{bs}(t) = 1 - \exp(-0.2\sqrt{t}) \tag{6}$$

On the other hand, the evolution of the drying shrinkage strain, $\varepsilon_{cds}(t, t_s)$, is given in Eq.(7) by multiplying the notional drying shrinkage coefficient $\varepsilon_{cds0}(f_{cm})$ by the coefficient β_{RH} for taking into account the effect of ambient relative humidity RH (in %) and by the time-dependent function β_{ds}, expressed by Eq.(8):

$$\varepsilon_{cds}\,(t, t_s) = \varepsilon_{cds0}(f_{cm}) \cdot \beta_{RH}(RH) \cdot \beta_{ds}(t - t_s) \tag{7}$$

$$\beta_{ds}\,(t - t_s) = \left(\frac{(t - t_s)}{0.035h^2 + (t - t_s)}\right)^{0.5} \tag{8}$$

where h stands for the notional size of the member, which is calculated as the ratio between two times the cross-section area and the perimeter of the cross-section in contact with the atmosphere.

2.1.2 *Creep effect*

Creep is a time-dependent deformation that develops at a decreasing rate under a sustained loading. In the PARC_CL 2.1 crack model the creep strain vector is formulated in the local 1, 2-coordinate system of the element (Figure 2), assuming the shear strain component equal to zero, as reported in Eq.(9).

$$\left\{\varepsilon_{cc}^{1,2}(t, t_0)\right\} = \left\{\varepsilon_{cc,1}(t, t_0) \quad \varepsilon_{cc,2}(t, t_0) \quad 0\right\}^t \tag{9}$$

The creep strain in the 1-direction (perpendicular to the crack direction), $\varepsilon_{cc,1}(t, t_0)$, is evaluated in function of the stress, σ_1, while the stress, σ_2, determinates the creep strain in the 2-direction (parallel to the crack direction), $\varepsilon_{cc,2}(t, t_0)$. The axial creep strains are evaluated according to Model Code 2010 (*fib* 2013).

In particular, the creep is assumed to be linearly related to stress within the range of service stresses, $\sigma_c^{1,2} \leq 0.4 f_{cm}$. The creep strains in the 1, 2-coordinate system, $\varepsilon_{cc}(t, t_0)$, are defined in Eq.(10):

$$\varepsilon_{cc}(t, t_0) = \sigma_{1,2}(t_0) / E_c \cdot \varphi(t, t_0) \tag{10}$$

where E_c is the modulus of elasticity at the age of 28 days, and $\varphi\,(t, t_0)$ is the creep coefficient calculated through Eq.(11).

$$\phi(t, t_0) = [\phi_{bc}(t, t_0) + \phi_{dc}(t, t_0)] \tag{11}$$

As highlighted by Eq.(11), the total creep coefficient is given by the sum of two components: the basic creep coefficient, $\varphi_{bc}(t, t_0)$, estimated through Eq.(12)-(14), and the drying creep coefficient, $\varphi_{dc}(t, t_0)$, calculated by means of Eq.(15)–(20).

$$\phi_{bc}(t, t_0) = \beta_{bc}(f_{cm}) \cdot \beta_{bc}(t, t_0) \tag{12}$$

being:

$$\beta_{bc}(f_{cm}) = 1.8 / (f_{cm})^{0.7} \tag{13}$$

$$\beta_{bc}(t, t_0) = \ln\left(\left(\frac{30}{t_0} + 0.035\right)^2 (t - t_0) + 1\right) \tag{14}$$

$$\phi_{dc}(t, t_0) = \beta_{dc}(f_{cm}) \cdot \beta(RH) \cdot \beta_{dc}(t_0) \cdot \beta_{dc}(t, t_0) \tag{15}$$

where:

$$\beta_{dc}(f_{cm}) = \frac{412}{(f_{cm})^{1.4}} \tag{16}$$

$$\beta(RH) = \frac{1 - \frac{RH}{100}}{\sqrt[3]{0.1 \frac{h}{100}}} \tag{17}$$

$$\beta_{dc}(t_0) = \frac{1}{0.1 + t_0^{0.2}} \tag{18}$$

$$\beta_{dc}(t, t_0) = \left[\frac{(t - t_0)}{\beta_h + (t - t_0)}\right]^{\gamma(t_0)} \tag{19}$$

being:

$$\gamma(t_0) = \frac{1}{2.3 + \frac{3.5}{\sqrt{t_0}}} \tag{20}$$

It is worth noting that the previous formulations are established for concrete in compression. Therefore, as suggested by Li et al. (2002) and Kristiawan (2006), the magnitude of tensile creep – considering the same stress magnitude – is assumed equal to three times the magnitude of creep in compression.

2.2 *Stress field for concrete*

As highlighted in Figure 3, once the shrinkage and creep strains are evaluated, the concrete stress field, Eq.(21), in the 1,2-coordinate system $\{\sigma_{1,2}\}$ is determined by adopting the cyclic constitutive relationship proposed by He-Wu (2008) and implemented in the PARC_CL 2.1 crack model, Figure 5. The bi-axial state of concrete in compression is modelled by adopting the relationship proposed by Vecchio & Collins (1993) – extended to cyclic loading. The aggregate interlock law proposed by Gambarova (1983) is implemented in the PARC_CL 2.1 crack model.

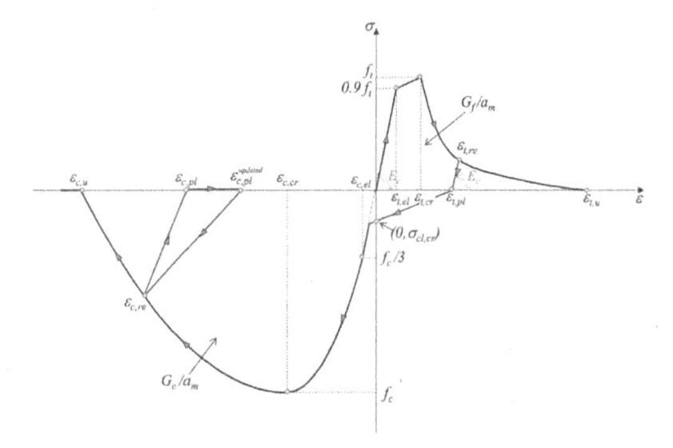

Figure 5. Cyclic behaviour of concrete, Belletti et al. (2017).

The stress field for concrete is given by Eq.(21):

$$\{\sigma_{1,2}\} = \begin{Bmatrix} \sigma_1 \\ \sigma_2 \\ \tau_{1,2} \end{Bmatrix} \tag{21}$$

while $[D_{1,2}]$ the stiffness matrix for concrete given in Eq.(22):

$$[D_{1,2}] = \begin{bmatrix} \frac{\partial\sigma_1}{\partial\varepsilon_1}\frac{1}{(1-\upsilon^2)} & \frac{\partial\sigma_1}{\partial\varepsilon_2}\frac{\upsilon}{(1-\upsilon^2)} & 0 \\ \frac{\partial\sigma_2}{\partial\varepsilon_1}\frac{\upsilon}{(1-\upsilon^2)} & \frac{\partial\sigma_2}{\partial\varepsilon_2}\frac{1}{(1-\upsilon^2)} & 0 \\ 0 & 0 & \frac{\partial\tau_{1,2}}{\partial\gamma_{1,2}} \end{bmatrix} \quad (22)$$

For the sake of brevity, refer to Belletti et al. (2017) for more details.

2.3 *Strain and stress field for steel*

Similarly to concrete, the strain field for steel along the reference system of each bar, $\{\varepsilon_{xi,yi}\}$, is evaluated by rotating the strain field in x,y-coordinate system, $\{\varepsilon_{x,y}\}$, by means of the transformation matrix (T_{Oi}), according to Eq.(23):

$$\{\varepsilon_{xi,yi}\} = [T_{\theta i}]\{\varepsilon_{x,y}\} \quad (23)$$

Thereafter, the stress field, $\{\sigma_{xi,yi}\}$, for each ith order of bars in the x_i,y_i-coordinate system, reported in Eq.(24), can be evaluated by adopting three different stress-strain relationships: (i) the uniaxial model introduced by Menegotto & Pinto (1973), the model proposed by Monti & Nuti (1992), and (iii) the relation suggested by Kashani et al. (2015). The implemented models are graphically reported in Figure 6. In particular, the last two models are useful to reproduce the buckling phenomenon of reinforcements subjected to cyclic loading.

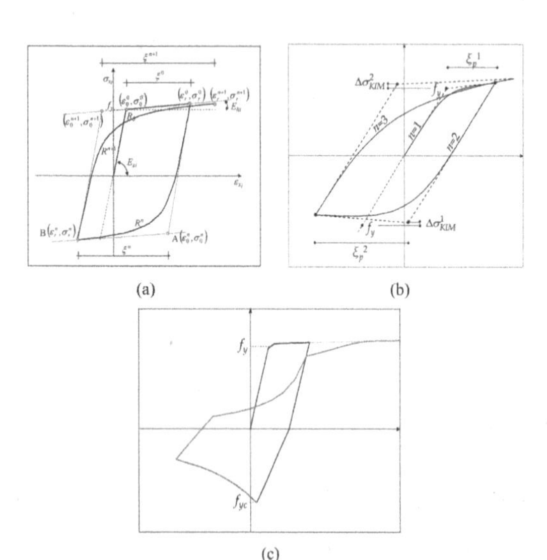

Figure 6. Stress-strain relationships implemented in the PARC_CL 2.1: (a) Menegotto-Pinto law, (b) Monti-Nuti law, and (c) Kashani et al. law (Vecchi & Belletti (2021).

The stress field for steel is given by Eq.(24):

$$\{\sigma_{x_i y_i}\} = \begin{Bmatrix} \sigma_{x_i} \\ \sigma_{y_i} \\ \tau_{x_i y_i} \end{Bmatrix} = \begin{Bmatrix} \sigma_{x_i} \\ 0 \\ 0 \end{Bmatrix} \quad (24)$$

while $[D_{xi,yi}]$ the stiffness matrix for steel given in Eq.(25):

$$[D_{x_i y_i}] = \begin{bmatrix} \frac{\partial\sigma_{x_i}}{\partial\varepsilon_{x_i}} & 0 & 0 \\ 0 & 0 & 0 \\ 0 & 0 & 0 \end{bmatrix} \quad (25)$$

For more details regarding the implemented models and the effect of corrosion on the mechanical properties of steel refer to Vecchi & Belletti (2021).

2.4 *Total stress field*

Finally, the total stress field in the x,y-coordinate system, $\{\sigma_{x,y}\}$, is obtained by coupling the concrete and the reinforcement behaviour as two springs working in parallel, according to Eq.(26) – Figure 3.

$$\{\sigma_{x,y}\} = \{\sigma_{x,y}\}_c + \sum_{i=1}^{n} \rho_i \{\sigma_{x,y}\}_{s,i} \quad (26)$$

where the two components of the stress field in x,y-coordinate system are calculated according to Eq.(27) and Eq.(26), respectively.

$$\{\sigma_{x,y}\}_c = [T_\psi]^t \cdot \{\sigma_{1,2}\} \quad (27)$$

$$\{\sigma_{x,y}\}_{s,i} = [T_{\vartheta_i}]^t \cdot \{\sigma_{x_i y_i}\} \quad (28)$$

3 VALIDATION OF PARC_CL 2.1

In the present work, the updated version of the PARC_CL 2.1 crack model – characterized by the implementation of creep and shrinkage effects – is validated through the comparison of NLFE results and the experimental outcomes of a continuous RC flat slab investigated by Gilbert & Guo (2005), named S7, subjected to sustained uniformly distributed service loads.

3.1 *Details of RC flat slab*

The slab S7 consisted of four identical square panels with dimensions 3000 × 3000 mm and a cantilevered region 600 mm long over the external columns at the eastern and western edge that led to an overall plan dimension of 7200 × 6200 mm in longitudinal and transversal direction, respectively, see Figure 7(a). The thickness of slab S7 was equal to 90 mm and the clear concrete cover was equal to 8 mm.

The analysed slab was supported by a total of nine square RC columns – indicated with the letter C in Figure 7 – with dimensions of the transversal cross

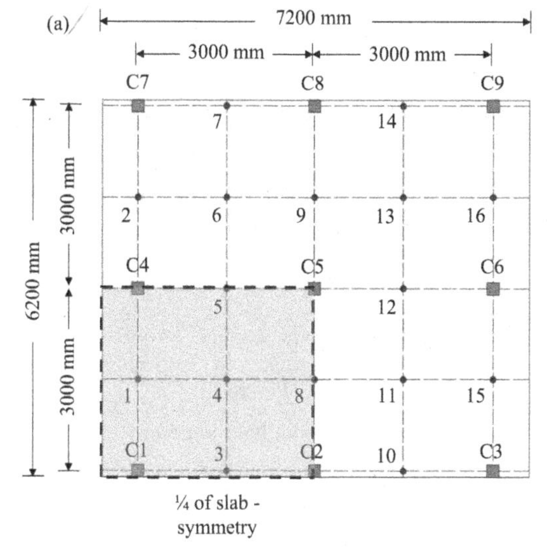

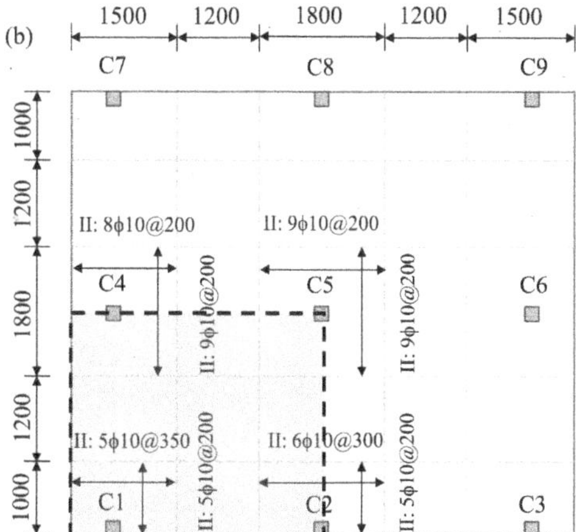

Figure 7. (a) Plan view of the investigated continuous RC flat slab, and (b) top reinforcement layout of slab S7.

section equal to 200 × 200 mm and height of about 1250 mm. All columns were fixed and clamped at their base.

Considering mechanical properties of concrete – measured at 14 days –, the compressive, f_c, and the tensile, f_{ct}, strength were equal to 15.4 MPa and 2.37 MPa, respectively, while the elastic modulus of concrete, E_c, was equal to 19030 MPa. The relative humidity (RH), the time at the beginning of shrinkage (t_s), and the notional size of the member (h) were assumed equal to 70%, 9 days, and 88.89, respectively.

Referring to reinforcements, the average yield strength, f_y, was equal to 650 MPa with an elastic modulus, E_s, of 219 GPa. The bottom reinforcement consisted of $\varphi 10$ with a spacing equal to 300 mm in both directions, while the top reinforcement layout is reported in Figure 7(b).

Figure 8 shows the loading history for slab S7. In detail, fourteen days after casting, the constraints were removed, and the slab started to deflect under

Figure 8. Loading history for slab S7.

the combination of superimposed loads (3.41 kPa) and self-weight (2.16 kPa). Indeed, the total load equal to 5.57 kPa was applied until reaching 508 days.

For more details on the RC flat slab and loading setup refer to Gilbert & Guo (2005).

3.1.1 *Experimental evaluation of creep coefficient and shrinkage strain*

Throughout the period of testing, the variation of concrete mechanical properties was experimentally measured by referring to cylindrical or prismatic samples with dimensions of 100 mm of diameter and 100 × 100 × 150 mm, respectively. These latter samples were casted together with the reference slab under the same curing and drying conditions.

It is worth noting that the creep coefficient was evaluated by taking into account samples subjected to a constant sustained stress of 5 MPa applied at the same age as the corresponding slab; whereas the shrinkage strain was measured on unloaded shrinkage samples.

On the basis of the available data, the creep strains were determined by subtracting the free shrinkage

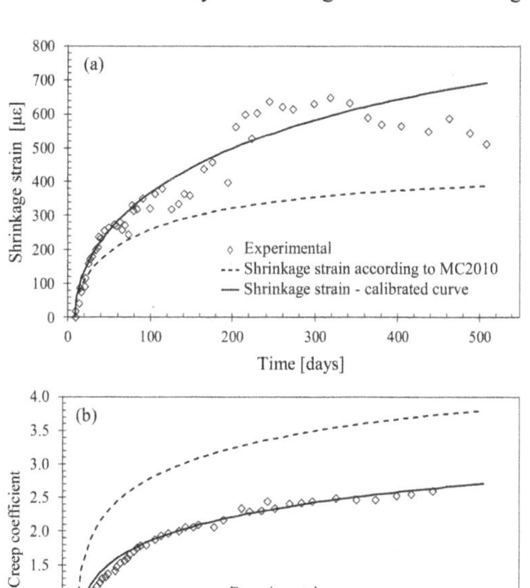

Figure 9. Mechanical properties variation over time: (a) shrinkage strain, and (b) creep coefficient.

strain and the instantaneous strain from the total creep strain measured on the creep specimens, while the creep coefficient was defined as the measured creep strain divided by the instantaneous strain.

Referring to Figure 9(a) and Figure 9(b), the square dots show the experimental trend of the shrinkage strain and the creep coefficient over time, respectively. In particular, the experimental outcomes are compared with the dashed lines obtained by adopting the Model Code 2010 formulations (MC 2010) (fib 2013). From the comparison of the two trends, it is possible to observe how the experimental measurements over time of shrinkage strain and creep coefficient were respectively underestimated and overestimated by adopting the MC 2010 formulations. To better reproduce the creep and shrinkage effects, further trends, identified by a continuous line in Figure 9(a) and Figure 9(b), were calibrated on the basis of experimental measurements and directly used as an input data for the analysis carried out by adopting the PARC_CL 2.1 crack model.

3.2 *NLFE model*

The mesh of the slab S7 has been modelled on the basis of the following assumptions:

- the symmetry of geometry and loading conditions is considered by modelling only a quarter of the RC flat slab S7 and by imposing appropriate boundary conditions;
- four node multi-layered shell elements with reduced integration schemes are adopted;
- seven layers are used to describe the thickness of the slab: (i) two layers of concrete cover, (ii) four layers for reinforcements, and (iii) an intermediate layer between reinforcements. Three Simpson integration points for each layer are used resulting in a total of twenty-one integration points along the thickness, Figure 10.
- brick elements are used for modelling columns and a linear elastic material is assigned to columns;
- uniformly distributed pressure is applied on the slab to reproduce the loading time history previously reported in Figure 8;
- the Regular Newton-Raphson method is used to reach the solution at each load increment;
- since the slab is thin, the profile of drying shrinkage strain is assumed to be uniform.

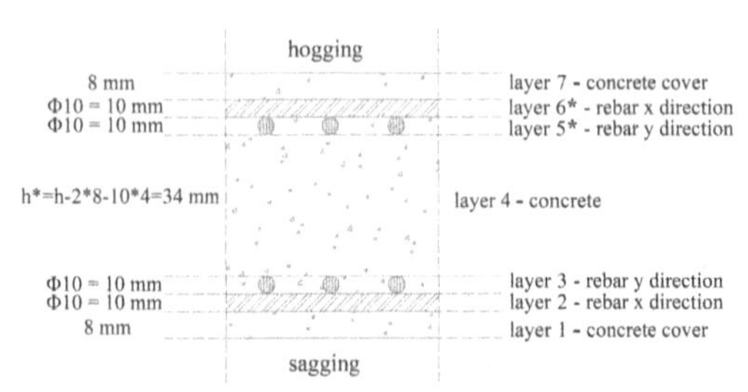

Figure 10. Subdivision in layers to describe the thickness of the slab.

4 RESULTS

4.1 *Deflections*

In this paragraph the comparison between experimental results and NLFE outcomes in terms of deflections recorded at various position over time is illustrated. In detail, Figure 11(a) and Figure 11(b) report the deflection trends at position 4 and at position 8 – identified in Figure 7(a) – that correspond to the mid-panel and the mid-span between adjacent columns, respectively.

Several NLFE analyses were carried out to properly estimate the effects associated to creep and shrinkage over time, as reported in Figure 11:

- the black dashed line in bold represents the NLFE analysis performed by adopting the MC 2010 formulations for creep and shrinkage without calibration;
- the black continuous line in bold represents the NLFE analysis obtained by considering the MC 2010 formulations calibrated on the basis of experimental outcomes;
- the grey dashed line stands for the NLFE analysis carried out by neglecting both creep and shrinkage effects;

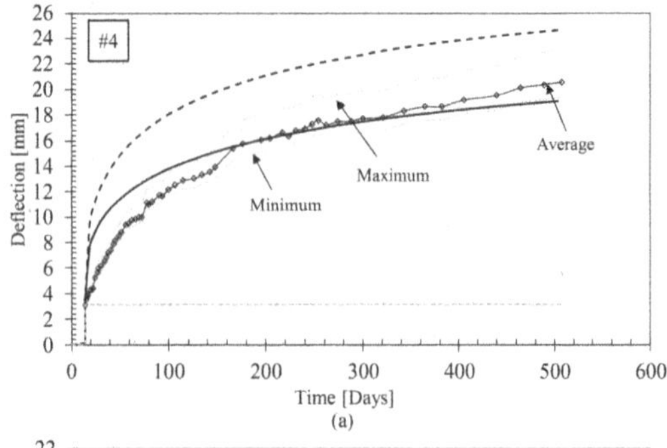

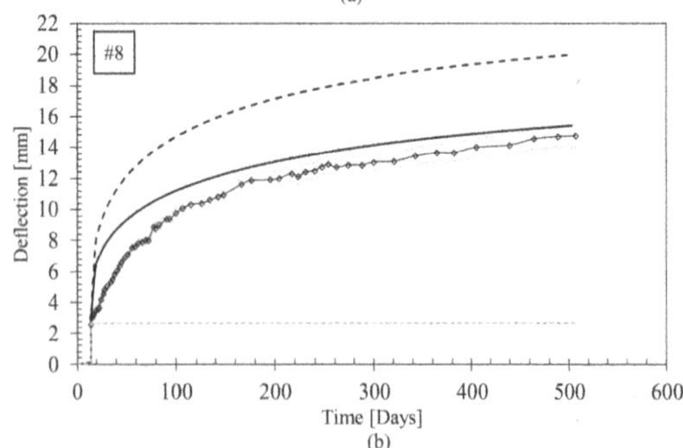

Figure 11. Deflection measurements of slab S7 over time: (a) at mid-panel, and (b) at mid-span between adjacent columns.

- the dotted line represents the experimental results in terms of average, maximum, and minimum deflections measured – in correspondence of each panel of the slab – by means of dial gauges throughout the period of testing.

Based on the obtained results, the PARC_CL 2.1 crack model is able to reproduce the experimental trend of the investigated continuous RC flat slab in terms of long-term deflections, as shown in Figure 11.

Firstly, a strong underestimation of the experimental trends is observed by neglecting the time-dependent effects of creep and shrinkage, as shown by the comparison between the black dashed line in bold and the grey dashed line.

Secondly, even if the use of MC 2010 formulations can generally catch the experimental trend in terms of deflections, a not negligible overestimation of experimental results is obtained.

Finally, a good approximation of experimental outcomes is achieved by adopting the PARC_CL 2.1 crack model when the creep and shrinkage effects are calibrated based on experimental measurements and directly used as an input data for the analysis.

4.2 *Crack pattern and crack opening width*

Figure 12 show the position where the crack opening widths are measured over time, together with the experimental and numerical crack pattern at the extrados of the slab at the end of the test (508 days).

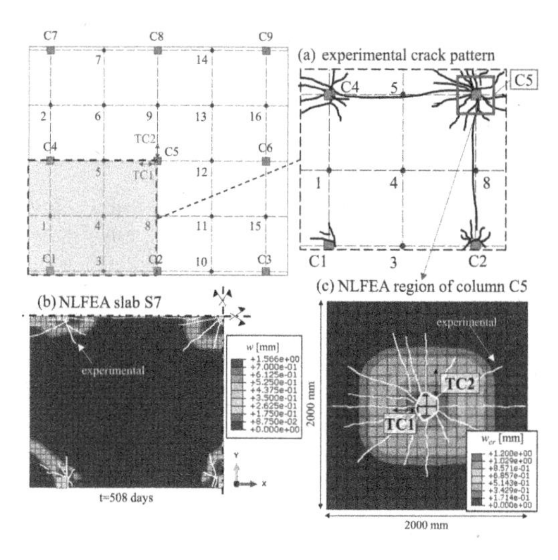

Figure 12. (a) Experimental crack pattern at the extrados of the slab, Gilbert & Guo (2005), (b) predicted crack pattern from NLFE analysis – at the extrados of slab S7 after 508 days, and (c) focus on the NLFEA crack pattern in proximity to column C5.

As shown in Figure 12(a), the most relevant cracks occurred at the extrados of the slab over each column, while – as pointed out by Gilbert & Guo (2005) – no significant cracks were observed at the intrados of slab S7. After 508 days, the maximum crack width was measured in proximity to columns C5 and was equal to 1.00 mm. On the contrary, the maximum crack width at the intrados was significantly smaller and resulted equal to 0.175 mm, Gilbert & Guo (2005).

Referring to Figure 12(b), similarly to the experimental evidence, the cracks develop in both the tangential and radial directions and are properly reproduced by the NLFE analysis crack pattern. On the other hand, the longitudinal cracks – that were observed between columns C4–C5 and C5–C2 at the end of the test Figure 12(a) – are not well predicted in the NLFE analysis.

Figure 12 shows the value of tangential crack opening width measured at the extrados of the slab in positions TC1 and TC2 over time, see Figure 12.

As reported in Table 1 the crack control limit value defined by Eurocode 2 (2004) is irrespective of crack opening forming at the extrados or intrados of slabs and exposure classes, except for exposure class X0 and XC1. It is interesting to observe that the crack opening width evaluated by neglecting the creep and shrinkage effects results lower that the limit value while if rheological phenomena are considered the crack control results not satisfied.

Table 1. Limit value of crack opening, w_{max}, for different exposure classes, Eurocode 2 (2004).

Exposure Class	w_{max} for RC members [mm]
X0, XC1	0.4
XC2, XC3, XC4	0.3
XD1, XD2	0.3
XS1, XS2, XS3	0.3

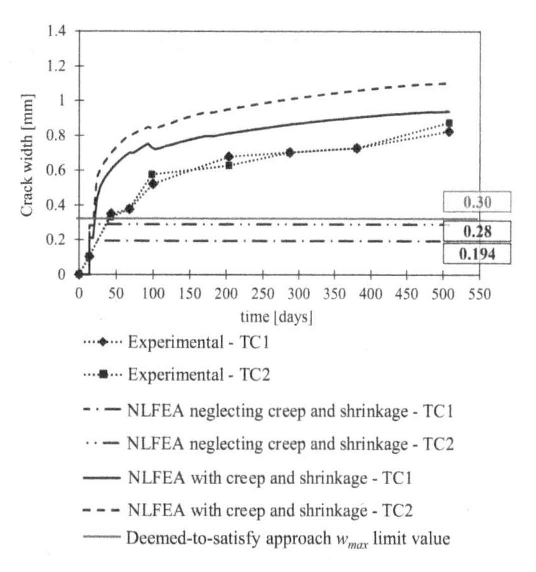

Figure 13. Evolution over time of tangential crack opening in proximity to column C5 in position TC1 and TC2, by considering: (i) Experimental results, (ii) NLFEA outcomes, and (iii) deemed-to-satisfy limitations.

The dependency of the corrosion rate on crack opening width will be better evaluated in future works to appreciate the reliability of deemed-to-satisfy approach for RC slabs.

5 CONCLUSIONS

In this paper the crack model PARC_CL 2.1, for the response prediction of two-dimensional RC element subjected to time-dependent effects (i.e. creep and shrinkage) is presented. The proposed model incorporates the Model Code 2010 formulations of the time-dependent values of creep and shrinkage with the possibility to consider – as input parameters – the experimental outcomes resulting from the variation of the mechanical properties of concrete over time. Secondly, the applicability and accuracy of the PARC_CL 2.1 crack model are validated through the analysis of a continuous RC flat slab, named S7, tested by Gilbert & Guo (2005). Finally, some remarks on the crack opening width assessment for the crack control in the framework of the deemed-to-satisfy approach are given.

Based on the obtained results, the following conclusions can be drawn:

- creep and shrinkage effects play a fundamental role in the durability assessment and serviceability limit state verification of RC flat slab;
- the PARC_CL 2.1 crack model is able to accurately predict the long-term effects induced by creep and shrinkage; in detail, accurate predictions in terms of long-term deflections and crack pattern are obtained by adopting the proposed model;
- the use of MC 2010 formulations for time dependent properties of concrete generally leads to the overestimation of long-term deflections of the analysed slab; whereas the same formulations calibrated by means of experimental outcomes allows a better approximation of the experimental behaviour;
- the crack opening width is strongly influenced by rheological phenomena and the coupled effects of environmental exposure will be deeply investigated in future works.

REFERENCES

ACI Committee 209. 2008. Guide for Modeling and Calculating Shrinkage and Creep in Hardened Concrete; ACI: Farmington Hills, MI, USA.

Bazant, Z.P. & Li, G.H. 2008. Comprehensive Database on Concrete Creep and Shrinkage. *Materials Journal* 105(6): 635-637.

Belletti, B., Muttoni, A., Vecchi, F. & Halimi, M. 2020. Modelling of Time-Dependent Effects of Reinforced Concrete Flat Slabs. Italian Concrete Days 2020. On-line, Italia; 14–17 April 2021.

Belletti, B., Muttoni, A., Ravasini, S., & Vecchi, F. 2019. Parametric analysis on punching shear resistance of reinforced concrete continuous slabs. *Magazine of Concrete Research* 71(20): 1083-1096.

Belletti, B., Scolari, M. & Vecchi, F. 2017. PARC_CL 2.0 crack model for NLFEA of reinforced concrete structures under cycling loadings. *Computers and Structures* 86(11): 1305-1317.

Bertero, R. & Bertero, A. Statistical Evaluation of Minimum Thickness Provisions for Slab Deflection Control. *ACI Structural Journal* 115(6): 1659- 1670.

Eurocode 2: Design of concrete structures – Part 1-1: General rules and rules for buildings (EN 1992-1-1). 2004. CEN, Bruxelles, Belgium.

fib – International federation for structural concrete. fib model code for concrete structures 2010. 2013. 431 Hernst & Sohn: 434 p.

Franceschini, L., Vecchi, F. & Belletti, B. 2021. The PARC_CL 2.1 Crack Model for NLFEA of Reinforced Concrete Elements Subjected to Corrosion Deterioration. *Corrosion and Materials Degradation* 2: 474-492.

Gambarova, P. 1983. Sulla trasmissione del taglio in elementi bidimensionali piani di c.a. fessurati. *In the Proceedings of the Giornate AICAP, Bari, Italy, 26-29 May*: 141-156. (in Italian)

Gilbert, R.I. & Guo, X.H. 2005. Time-Dependent Deflection and Deformation of Reinforced Concrete Flat Slabs – An Experimental Study. *ACI Structural Journal* 102(3): 363-373.

Hasan, N.M.S., Visitin, P., Oehlers, D.J., Bennett, T. & Sobuz, H.R. 2019. Time dependent deflection of RC beams allowing for partial interaction and non-linear shrinkage. *Mater Struct* 52(3).

He, W., Wu, J.F. & Liew, K.M. 2008. A fracture energy based constitutive model for the analysis of reinforced concrete structures under cycling loading. *Comput. Methods Appl. Mech. Eng.* 197: 4745-4762.

Hoissan, T.R. & Vollum, R.L. 2002. Prediction of slab deflections and validation against Cardington data. *Proceedings of the Institution of Civil Engineers Structures & Buildings* 152(3): 235-248.

Kashani, M.M., Lowes, L.N., Crewe, A.J. & Alexander, N.A. 2015. Phenomenological hysteretic model for corroded reinforcing bars including inelastic buckling and low-cycle fatigue degradation. *Computers and Structures* 156: 58-71.

Kilpatrick, A. E. & Gilbert, R. I. 2017. Simplified calculation of the long-term deflection of reinforced concrete flexural members. *Australian Journal of Structural Engineering.*

Kristiawan, S. 2006. Strength, shrinkage and creep of concrete in tension and compression. *Civil Engineering Dimension* 8(2): 73-80.

Li, H., Wee, T.H., & Wong, S.F. 2002. Early-Age Creep and Shrinkage of Blended Cement Concrete. *ACI Material Journal,* Jan-Feb 2002: 3-10.

Menegotto, M. & Pinto, P.E. 1973. Method of Analysis for Cyclically Loaded R.C. Plane Frames Including Changes in Geometry and Non-Elastic Behavior of Elements under Combined Normal Force and Bending. *In Symposium on Resistance and Ultimate Deformability of Structures Acted on by Well Defined Repeated Load; IASBSE: Lisbon, Portugal*: 15-22.

Monti, G. & Nuti, C. 1992. Nonlinear Cyclic Behavior of Reinforcing Bars Including Buckling. *Journal of Structural Engineering* 118: 3268-3284.

Motter, C.S. & Scanlon, A. 2018. Modeling of Reinforced Concrete Two-Way Floor Slab Deflections due to Construction Loading. *Journal of Structural Engineering* 144(6).

Proceedings of the fib CACRCS DAYS 2020. 2020. On-line, 1-4 December.

Proceedings of the fib CACRCS DAYS 2021. 2021. On-line, 30 November-3 December.
Shu, J., Belletti, B., Muttoni, A., Scolari, M., & Plos M. 2017. Internal force distribution in RC slabs subjected to punching shear. *Engineering Structures* 153: 766–781.
Sirico, A., Michelini, E., Bernardi, P. & Cerioni, R. 2017. Simulation of the response of shrunk reinforced concrete elements subjected to short-term loading: a bi-dimensional numerical approach. *Engineering Fracture Mechanics* 174(2017): 64-79.
Vecchi, F. & Belletti, B. 2021. Capacity Assessment of Existing RC Columns. *Buildings* 11(61): 1-19.
Vecchio, F.J. & Collins, M.P. 1993. Compression Response of Cracked Reinforced Concrete. *Journal of Structural Engineering* 119: 3590-3610.

Computational Modelling of Concrete and Concrete Structures – Meschke, Pichler & Rots (Eds)
 ISBN: 978-1-032-32724-2

Investigation of drying shrinkage effects on sloped concrete-concrete composites

D. Daneshvar, K. Deix & A. Robisson
Research Group of Building Materials and Technology, Faculty of Civil Engineering, Vienna University of Technology (TU Wien), Vienna, Austria

Behrouz Shafei
Department of Civil, Construction, and Environmental Engineering, Iowa State University, Ames, IA, United States

ABSTRACT: Multi-layer concrete systems have been widely used in bridge decks, rigid pavements, and floors. The restrained drying shrinkage of the overlay is of great concern as it can lead to overlay cracking and/or interfacial debonding. In such constructions, a transverse slope is typically considered to drain off the surface water. Despite advances made in understanding the drying shrinkage of non-sloped concrete-concrete composites, there are still standing questions regarding the overlay cracking, as well as bond failure, in sloped concrete-concrete composites. The current study establishes a high-fidelity computational model validated with experimental tests to evaluate the structural performance of sloped, double-layer overlay systems under drying shrinkage. The simulation scenarios systematically cover the effects of key overlay properties and interface conditions. The numerical analysis results reveal the critical role of overlay thickness and mechanical properties in the time of overlay cracking and interfacial debonding failures. Based on the obtained results, the implementation of a cross slope may delay the failures, depending on the initial thickness. Higher interfacial stiffness also induces stronger restraint against overlay shrinkage strain, leading to a faster overlay cracking.

1 INTRODUCTION

Concrete overlays have been used to protect concrete structures exposed to the excessive loading and harsh environmental conditions (Emmons 1994). Not limited to repair applications, concrete overlays have also been employed in large concrete structures where several concrete pours are required at various stages of construction (Loo et al. 1995). Thus, the use of multi-layer concrete systems covers a wide range of applications such as bridge decks, rigid pavements, dams, slabs and floors. According to the American Concrete Pavement Association (ACPA), the use of concrete overlays has rapidly increased from 2% in 2000 to 12% of the entire concrete paving applications in 2017 (Gross & Harrington 2018). As a durable, cost-effective, and sustainable solution, concrete overlays provide additional strength and protect the underlying reinforced concrete layer from deleterious agents (Harrington et al., 2007). Therefore, it is essential for the overlays to retain their strength under a wide variety of mechanical and environmental stressors (Çolak et al. 2009; Daneshvar et al. 2021; Emmons 1994).

Drying shrinkage represents a major issue upon constructing overlays. In multi-layer concrete systems, the water loss from the overlay is attributed to the environmental drying, ongoing hydration, and moisture absorption by the concrete substrate. As the drying shrinkage-induced strains of the concrete overlay are restrained by the concrete substrate, they cause tensile stresses in the overlay, and additional shear (friction) and normal (delamination) stresses at the interface of the overlay and the concrete substrate (Li & Li 2006). These induced stresses are influenced by the magnitude of overlay's shrinkage such that larger shrinkage of the overlay results in higher stresses and hence a higher risk of failure. Overlay cracking and/or interface debonding are the most common types of failures observed in concrete-concrete composites (Emmons & Vaysburd 1994). The outlined failure modes impact the load transfer mechanism between concrete layers, causing a non-monolithic behavior that eventually jeopardizes the load-bearing capacity and overall structural performance of the concrete-concrete composites. Such defects can also facilitate the ingress of water and deleterious agents into the composite system, resulting in long-term durability issues and extensive repair needs.

Moreover, during the lifetime of concrete overlays, standing surface water is a major concern. Ponding of water on the overlay surface can result in slippage, hydroplaning, and icing in winter, while facilitating the gravitational ingress of standing water into the reinforced concrete substrate (Smith et al. 2014). A

 DOI 10.1201/9781003316404-75

sufficient transverse slope, however, can ensure the proper drainage of surface water from the overlay and minimize its associate issues. To drain off the surface water, a transverse cross slope is typically considered for the overlay. The American Association of State Highway and Transportation Officials (AASHTO) policy on geometric design of highways and streets recommends a cross slope of 2% for usual conditions to mitigate the risk of hydroplaning. Depending on the rainfall and application, a lower or higher cross slope can also be implemented (AASHTO 2011).

Previous studies investigated the effects of overlay properties, interface conditions, external environmental conditions, and boundary conditions on the structural performance and durability of concrete-concrete composites subjected to drying shrinkage (Li & Li 2006; Santos & Julio 2011). Despite advances made in understanding the drying shrinkage of non-sloped concrete-concrete composites, there are still standing questions regarding the crack formation, as well as bond failure, in sloped concrete-concrete composites. Furthermore, numerous shrinkage and creep prediction models have been developed based on extensive experimental datasets. Although these models have been widely and successfully used, most of the finite-element (FE)-based studies used an analytical or semi-analytical approach to take shrinkage and creep behavior of concrete into consideration. These approaches, in most cases, employed a heat transfer analysis, missing factors contributing to shrinkage and creep.

Given the outlined research gaps and questions, the current study established a high-fidelity computational model validated with experimental tests to evaluate sloped, double-layer concrete systems under drying shrinkage. For this purpose, three-dimensional FE simulations were performed. The shrinkage strain and creep coefficient of concrete were incorporated as a function of time based on the criteria of the empirical American Concrete Institute model (ACI 209.2R-08 2008). The simulation scenarios under consideration systematically cover the effects of key overlay properties and interface conditions, such as overlay geometry (thickness and slope), overlay material properties, and interfacial degree of restraint. The outcome sheds light on the optimized combinations of overlay geometries and interfacial conditions required to enhance the short- and long-term performance of concrete-concrete composites.

2 MODELLING AND VALIDATION

2.1 *Numerical model*

The Abaqus software package (2021) was used in this study. The concrete overlay was modeled with a length of 5 m and a width of 2 m. The overlay length was selected long enough (greater than 2 times of the width) to account for the continuity of the overlay. The thickness and slope of the overlay were varied between 20 and 200 mm, and 0 and 10% slope, respectively. The overlay crowned surface slopes from both sides of the centerline. The substrate length was 8 m with a width of 4 m and a constant thickness of 200 mm. Due to the symmetry of the models, a quarter of the described concrete-concrete composites were modeled. Figure 1 shows the generated 3D FE model.

Figure 1. 3D FE model generated for numerical simulations.

As for boundary conditions, the bottom surface of the substrate was fixed against translation in all three orthogonal directions. Due to the restraint provided by adjacent layers, the movement of the lateral faces of the substrate were fixed via a roller support. 20-node quadratic brick elements (C3D20) were used for the model. Considering the mesh sensitivity analysis and cracking occurrence in the overlay, a finer mesh was employed for the overlay while the substrate was modeled with a coarser mesh. The interface between concrete layers was modeled by including a cohesive contact behavior. This contact allows to capture possible delamination at the interface. The bi-linear separation-traction and force-slip constitutive curves were employed to formulate the cohesive contact behavior. They are typically characterized by penalty stiffness defined as the slope of linear elastic part (pre-damage response) and peak strength (damage initiation). The normal and shear bond strength was assumed as 1.5 and 3.0 MPa, respectively, according to the experimental values reported in the literature (Momayez et al. 2005). Depending on the degree of restraint between concrete layers, a wide range of interfacial normal and shear stiffnesses is proposed (Tsioulou & Dritsos 2011). To account for this, and based on the experimental results reported in the literature, the interfacial stiffness between concrete layers was modeled with values in the range of 0.5 to 100 N/mm^3. The concrete damage plasticity model (CDP) was utilized to represent the behavior of the concrete layer in both elastic and plastic domains. Depending on the application and field requirements, various types of overlays have been utilized, among which the normal-strength concrete (NC) and ultra high-performance concrete (UHPC) are the most common types (Gross & Harrington 2018; Haber et al. 2017). Therefore, to compare their structural performance, these two concrete types (i.e., NC and UHPC) were considered for the overlay while the substrate was modeled with NC. For this purpose, experimental tests including uniaxial tensile, compressive and elastic modulus tests were carried out to capture their behavior at 28 days. Table 1 presents the experimental results used for the properties of concrete overlay and substrate in the simulations.

Table 1. Experimental results of concrete properties at 28 days.

Properties	Value	
	NC	UHPC
Tensile strength (MPa)	3.6	8.3
Compressive strength (MPa)	63.4	134.3
Elastic modulus (GPa)	26.3	36.2
Density (kg/m^3)	2400	2500
Poisson's ratio	0.2	0.2

2.2 *Drying shrinkage simulation*

The empirical ACI 209 shrinkage and creep model was employed to simulate the shrinkage strain and creep of concrete as a function of time (ACI 209.2R-08 2008). Thermal loading was artificially used to input shrinkage and creep. For this purpose, the UEXPAN and USDFLD user subroutines were developed. The UEXPAN subroutine applied a thermal strain increment, which represented the defined time-dependent strain caused by the shrinkage and creep of concrete. The USDFLD subroutine stored the elastic strains in each increment as state variables, which were used in creep strain calculations. In ACI 209, the time-dependent shrinkage and creep strain are described based on the age of the concrete, curing method, time of drying, relative humidity, volume to surface ratio, slump, fine aggregate to total aggregate ratio, cement content, and air content (ACI 209.2R-08 2008). Table 2 lists the coefficients and correction factors used in the FE simulations. The age of concrete at the start of drying and loading was set to 2 days. In this study, the development of drying shrinkage was considered during the first three months of casting so that the time period of analysis was set at 9×10^6 seconds (104 days).

Table 2. Correction factors for shrinkage strain and creep coefficient based on ACI 209 model.

Correction factor	Value	
	Shrinkage	Creep
Initial moist curing	1.13	–
Age of loading	–	1.15
Ambient relative humidity	0.69	0.8
Member size factor	0.33	0.67
Slump factor	1.03	1.05
Fine aggregate factor	1	1
Cement content factor	0.97	1
Air content factor	0.99	6
Time ratio (α, ψ)	0.85	0.6
Time ratio (f, d)	35	15
Ultimate shrinkage strain	1.3	–
Ultimate tensile creep	–	1.8
Ultimate compressive creep	–	1.4

2.3 *Validation*

To validate the developed FE models, the structural performance of a concrete-concrete composite model was compared to the experimental results reported by (Li & Li 2006). Similar material properties, geometry and boundary conditions were employed in the model. Figure 2 compares the end corner delamination height of the overlay in the model with the data measured experimentally. As can be seen, the results extracted from the FE simulation are consistent with those experimental measurements. Specifically, a similar evolution of delamination height tends to prove consistent drying shrinkage strain and stress distribution in both cases. Moreover, the overlay cracking occurred within the first 7 days in both cases. It must be pointed out that the Abaqus standard solver was used to capture the failure occurrence and its corresponding time in each simulation. Therefore, the numerical analysis continued only until cracking occurrence. Thus, large plastic deformations are not part of this study.

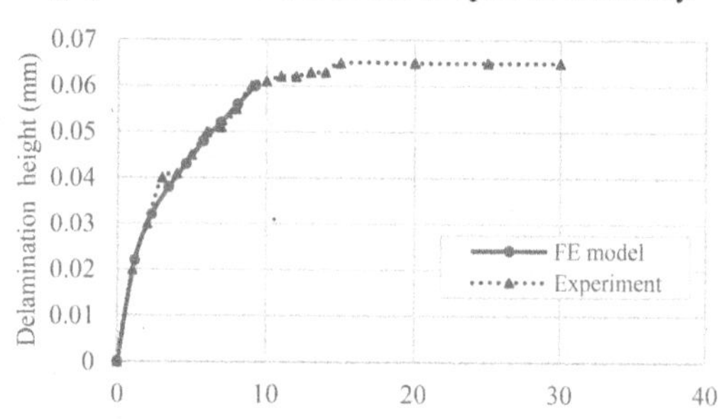

Figure 2. Comparison of the FE model results with the experimental test data recorded for the corner delamination height in concrete-concrete composites over time.

3 NUMERICAL ANALYSIS AND RESULTS

Overlay cracking and interface debonding make up a majority of failures in concrete-concrete composites. Therefore, the emphasis of this study is on evaluating the occurrence of these two types of failure, as a function of overlay geometry (thickness and slope), overlay material properties, and interfacial degree of restraint (normal and shear stiffness). Specifically, the onset of the development of maximum principal plastic strain, based on the CDP model, was considered as the overlay cracking occurrence. To determine the interfacial failure, based on the defined bi-linear traction-separation behavior, the corresponding time to the state in which the separation and/or slip between layers exceeds the maximum allowable strength (damage initiation) was taken as the interfacial debonding/slippage failure. Before damage initiation, the behavior of the cohesive contact is ensured. This was defined as a linear elastic behavior so the interfacial debonding and slippage thresholds are calculated by dividing the defined bond strength to the corresponding interfacial stiffness. It must be highlighted that debonding was the only type of interfacial failure that occurred in the models that experienced failure. Figure 3 show the typical 3D stress distribution and deformed shaped of the concrete overlay subjected to the restrained drying shrinkage.

Figure 3. Typical 3D deformed shape of the overlay in the FE model subjected to restrained drying shrinkage. The contours represent the maximal principal stress before damage initiation. Deformations are amplified by a factor of 3000 for visualization purposes.

3.1 *Thickness*

The concrete overlay thickness varies depending on the purpose of the application, overlay material properties, and field limitations. Thus, based on the reported values in the literature, a broad range of thicknesses, between 20 and 200 mm, was considered in this study. Figure 4 shows the overlay cracking and interfacial debonding corresponding times. To capture the effect of thickness on concrete-concrete composites failure, the thresholds for cracking of NC overlay and debonding of the interface were set at 3.6 MPa and 0.03 mm, respectively. As can be seen, increasing the overlay thickness from 20 mm to 100 mm leads to a negligible increase (2 days) in overlay cracking time. However, further increasing the thickness results in a drop of overlay cracking time by 10 days. Moreover, the interface debonding occurred earlier than overlay cracking in composites containing thick overlays (above 100 mm).

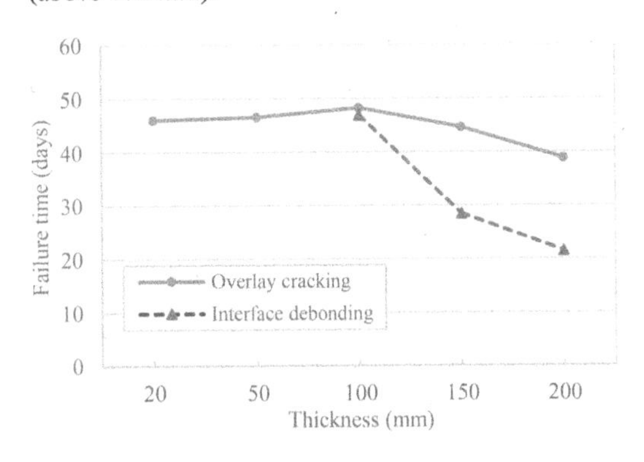

Figure 4. Overlay drying shrinkage cracking and interface debonding corresponding times as a function of overlay thickness.

The drying shrinkage cracking in overlays with a thickness up to 100 mm was initiated from the center part of the overlays where the induced tensile stress was highest. However, further increasing the thickness resulted in the shift of cracking location to the end corners of the overlay. To compare the stress evolution in overlays with different thicknesses, the stresses induced by the restrained drying shrinkage are plotted in Figure 5 and Figure 6. It must be highlighted that the results are plotted at time of 1.75 × 106 seconds (20 days). This time was the latest possible time at which all the composites were still intact (without any overlay cracking and interface debonding) so that the development of the stresses was not influenced by any stress release through failure occurrence. As shown in Figure 5 and Figure 6, the interfacial shear and normal stresses are predominant at the outer edge and gradually increase in thicker overlays. This is mainly attributed to the significantly higher differential displacement between the top free surface and the bottom restrained surface in thicker overlays. Specifically, this was found to increase from 0.004 to 0.042 mm upon increasing the overlay thickness from 20 to 200 mm. In case of having interfacial restraint, the higher tendency of the overlay to move, the greater the induced interfacial normal and shear stresses. In thick overlays,

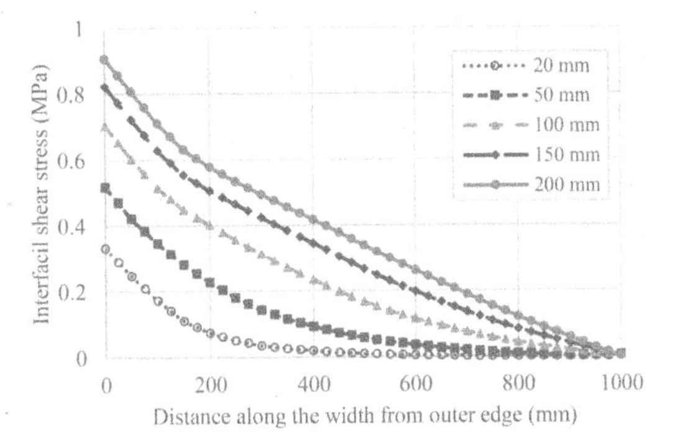

Figure 5. Distribution of interfacial shear stress along the width of concrete overlays with thicknesses from 20 to 200 mm.

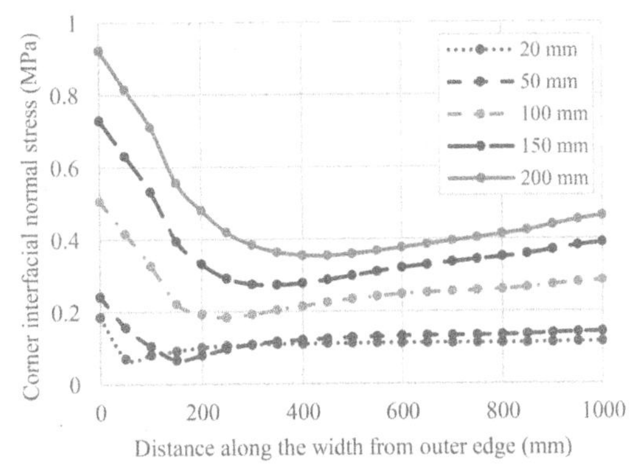

Figure 6. Distribution of corner interfacial normal stress along the width of concrete overlays with thicknesses from 20 to 200 mm.

the interfacial normal stresses in end corners exceeded the thresholds earlier than the center part, resulting in a faster occurrence of debonding and development of plastic strains. On the other hand, in thin overlays where the differential shrinkage deformation between the bottom and the top overlay surfaces was not remarkable, the induced tensile stress was concentrated in the center part of the overlay, and therefore, surface cracking was the predominant type of failure.

3.2 *Slope*

To evaluate the effect of slope in various applications, the overlay cross slope up to 10% was considered with crowned surfaces such that the overlay surface slopes from either side of the centerline. Figure 7 compares the corresponding overlay cracking times. In thin overlays (20 and 50 mm in the outer side, i.e., thinnest part), implementation of a cross slope up to 5% provides better resistance against induced tensile stress and hence extends the overlay cracking time. However, employing a cross slope of 10% in thin overlays and all cross slopes in thick overlays (100 and 150 mm) results in a faster overlay cracking. Furthermore, in each specific cross slope, increasing the thickness from 20 to 50 mm leads to an increase in overlay cracking time. It must be pointed out that overlay cracking location was observed at the end corners in case of thick sloped overlays. These results are consistent with the effect of thickness described in the previous section, i.e., increasing the cross slope leads to a transverse increase in the overlay thickness and intensifies end corner delamination.

Interfacial debonding occurred prior to the overlay cracking in thick overlays (100 and 150 mm) including all implemented slopes. For thinner overlay, the interface failure was never observed within the studied time.

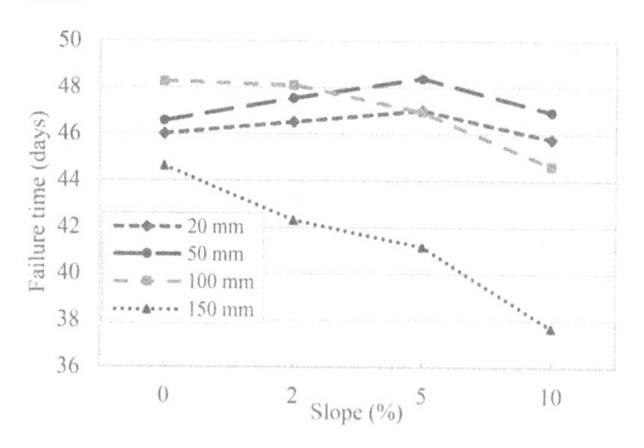

Figure 7. Overlay drying shrinkage cracking times as a function of overlay cross slope and thickness.

3.3 *Interfacial stiffness*

The non-sloped overlay with a thickness of 100 mm was used in this set of simulations. Greater interfacial stiffness typically induces stronger restraint against overlay displacement along the interface, inducing higher shear and tensile stress. This led to an overlay cracking in composites with an interfacial stiffness greater than 1 N/mm^3 (Figure 8). Specifically, there is a sharp drop of 30 days in overlay cracking time upon increasing the interfacial stiffness from 2 to 20 N/mm^3. In composites with a smoother interface and lower interfacial stiffness (0.5 and 1 N/mm^3), the drying shrinkage of the overlay is less restrained and hence the induced tensile and shear stress stayed below the overlay strength, and thus, cracking was not observed within the studied time period (104 days).

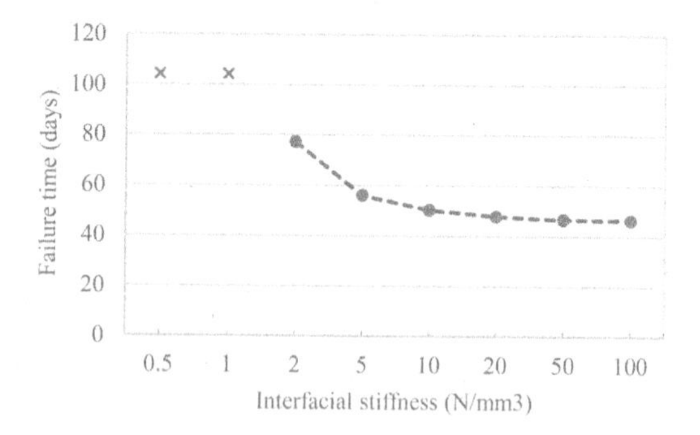

Figure 8. Overlay drying shrinkage cracking times as a function of interfacial stiffness. The cross marks represent points where failure did not occur during analysis time period.

3.4 *Overlay strength*

The effect of overlay characteristics was assessed by comparing the structural performance of UHPC with conventional NC. The non-sloped overlay with a thickness of 100 mm was used in the simulation. Figure 9 compares the structural performance of normal and UHPC overlays in concrete-concrete composites subjected to the restrained drying shrinkage. In all the studied cases, the UHPC resisted the induced tensile and shear stress, and hence the drying shrinkage cracking did not occur within the studied period time (104 days). This is mainly due to the superior mechanical properties of UHPC and specifically its high tensile strength, which was 2.3 times greater than that of NC. It must be highlighted that the main part of induced stress was released thorough end corner interface

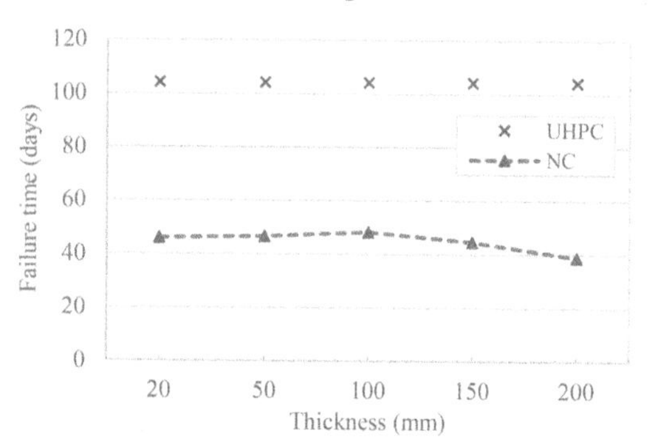

Figure 9. Comparison of NC and UHPC overlay drying shrink-age cracking times. The cross marks represent points where failure did not occur during the analysis time period.

delamination. Increasing the overlay thickness intensified this phenomenon and resulted in early-stage interfacial debonding failure.

Figure 10 shows the interfacial performance of composites including UHPC and NC overlay. As described before, the greater resistance of UHPC against induced tensile and shear stresses leads to the predominant mechanism of stress release through end corner interface delamination. Therefore, the debonding failure occurs (3–10 days) earlier in UHPC overlaid composites than NC ones.

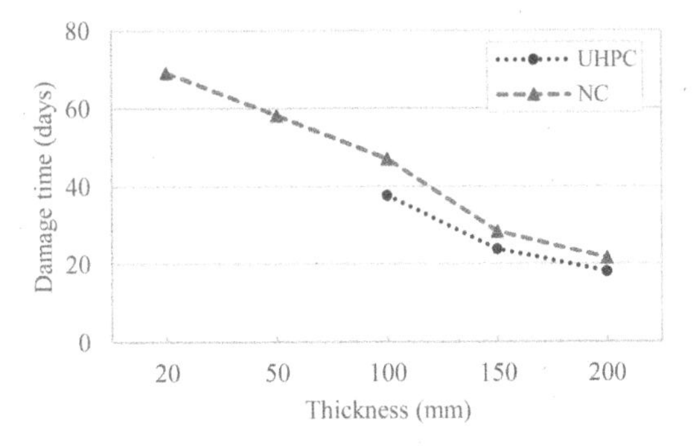

Figure 10. Comparison of NC and UHPC overlay drying shrinkage cracking times. The cross marks represent points where failure did not occur during the analysis time period.

4 CONCLUSIONS

The effects of overlay geometry (thickness and slope), overlay material properties, and interface restraint conditions were investigated in concrete-concrete composites subjected to drying shrinkage. The numerical analysis results showed that the overlay thickness plays a pivotal role in the structural performance of the composites. It was found that there is an optimum point for the overlay thickness (in this study, 100 mm), below which the overlays are more vulnerable to restrained shrinkage and demonstrate earlier cracking. Employing thicker overlays, on the other hand, was found to result in a significant end corner delamination and hence early-stage interfacial debonding. Consistent with this, the effect of overlay transverse slope was determined to depend on the initial thickness of the overlay such that for the overlays up to 50 mm thick, the overlay crack can be delayed by employing a cross slope up to 5%.

Further increasing the cross slope and employing higher initial thickness would adversely impact the structural performance of concrete-concrete composites through early-age overlay cracking and/or interfacial debonding. Furthermore, the interfacial degree of restraint was found to significantly affect the performance of the double-layer concrete systems such that a sharp drop of 30 days in overlay cracking time was observed upon increasing the interfacial stiffness from 2 to 20 N/mm^3. In the case of smoother interfaces, i.e., with an interfacial stiffness less than 2 N/mm^3, the overlay restrained shrinkage strain was insignificant to the extent that overlay drying shrinkage cracking did not occur. The type and mechanical properties of the overlay materials also influenced the behavior of the concrete-concrete composites under restrained drying shrinkage conditions. It was found that application of UHPC overlay can delay the overlay shrinkage cracking at least 60 days compared to the NC overlays. However, the end corner interface delamination occurred earlier in UHPC overlays.

REFERENCES

ACI 209.2R-08. (2008). *Guide for Modeling and Calculating Shrinkage and Creep in Hardened Concrete*. American Concrete Institute, USA.

American Association of State Highway and Transportation Officials. (2011). *A Policy on Geometric Design of Highways and Streets*. Washington, DC, USA.

Çolak, A., Çoşgun, T., & Bakirci, A. E. (2009). Effects of environmental factors on the adhesion and durability characteristics of epoxy-bonded concrete prisms. *Construction and Building Materials*, *23*(2), 758–767. https://doi.org/10.1016/j.conbuildmat.2008.02.013

Daneshvar, D., Deix, K., & Robisson, A. (2021). Effect of casting and curing temperature on the interfacial bond strength of epoxy bonded concretes. *Construction and Building Materials*, *307*, 124328. https://doi.org/10.1016/j.conbuildmat.2021.124328

Emmons, P. H., & Vaysburd, A. M. (1994). Factors affecting the durability of concrete repair: the contractor's viewpoint. *Construction and Building Materials*, *8*(1), 5–16. https://doi.org/10.1016/0950-0618(94)90003-5

Emmons, Peter H. (1994). *Concrete Repair and Maintenance Illustrated*. R. S. Means Company.

Gross, J., & Harrington, D. (2018). *Guide for the Development of Concrete Overlay Construction Documents* (Issue August).

Haber, Z. B., Munoz, J. F., & Graybeal, B. A. (2017). *Field Testing of an Ultra-High Performance Concrete Overlay*. https://www.fhwa.dot.gov/publications/research/infrastructure/structures/bridge/17096/index.cfm

Harrington, D., DeGraaf, D., Riley, R., Rasmussen, R. O., Grove, J., & Mack, J. (2007). *Guide to Concrete Overlay Solutions* (Issue January).

Li, M., & Li, V. C. (2006). Behavior of ECC-concrete layered repair system under drying shrinkage conditions. In *Restoration of Buildings and Monuments* (pp. 143–160).

Loo, Y. H., Peterson, J. S., Swaddiwudhipong, S., & Tam, C. T. (1995). Application of the layering method on large concrete pours. *Magazine of Concrete Research*, *47*(172), 209–217. https://doi.org/10.1680/macr.1995.47.172.209

Momayez, A., Ehsani, M. R., Ramezanianpour, A. A., & Rajaie, H. (2005). Comparison of methods for evaluating bond strength between concrete substrate and repair materials. *Cement and Concrete Research*, *35*(4), 748–757. https://doi.org/10.1016/j.cemconres.2004.05.027

Santos, P. M. D., & Julio, E. N. B. S. (2011). Factors affecting bond between new and old concrete. *ACI Materials Journal*, *108*(4), 449–456. https://doi.org/10.14359/51683118

Smith, K., Harrington, D., Pierce, L., Ram, P., & Smith, K. (2014). *Concrete Pavement Preservation Guide, Second Edition* (Vol. 32, Issue 5).

Tsioulou, O. T., & Dritsos, S. E. (2011). A theoretical model to predict interface slip due to bending. *Materials and Structures/Materiaux et Constructions*, *44*(4), 825–843. https://doi.org/10.1617/s11527-010-9669-6

Computational Modelling of Concrete and Concrete Structures – Meschke, Pichler & Rots (Eds)
 ISBN: 978-1-032-32724-2

Time-dependent behavior of the twisted columns of New Marina Casablanca Tower

S.A. Brown & G. Cusatis
Department of Civil and Environmental Engineering, Northwestern University, Evanston IL, USA

U. Folco & D. Masera
Advanced and Special Structures Department, Masera Engineering Group Srl, Turin, Italy

ABSTRACT: Predicting the behavior of reinforced concrete structures beyond the construction phase is a critical aspect of structural design. In particular, the differential shortening of columns due to varying stress conditions over time can significantly impact the long-term health of a structure. This effect is not always considered at low building heights as the comparative effect might be minimal – e.g. 1% shortening in a single story building is on the order of millimeters. However, as the number of stories increases, this effect compounds, such that column shortening for a 60 story structure might be on the order of 30 mm. At this scale the shortening can cause problems not only within the structural members but in the nonstructural components as well, such as partitions and pipe lines. It is therefore necessary to design a structure such that it is able to mitigate the shortening which occurs in the years and decades following construction.

Such is the case for the twisted columns of the New Marina Casablanca Tower in Morocco. This 160 m tall tower (currently in conception) was designed as part of a new convention center along the coast in Casablanca. The primary architectural feature is exhibited in the form of a spiraling tower where each successive floor is rotated as the building ascends, amounting to a total of 135° twist. The structural system consists of an inner core and inclined columns which follow the angle of rotation. Although a system of core and columns is common, the introduction of the inclined columns leads to unique behavior of the structure, and the time-dependent shortening effects cannot be assumed based on previous work intended for straight columns.

In this study the authors analyze the time-dependent behavior of these columns at two scales. At the building scale, the construction procedure is simulated to determine the load history of the columns and initial elastic deformations. Multiple construction analysis methods are considered. Then, based on the results from this global simulation, a single-story column is modeled to the material level, providing more nuanced analysis. First the hygro-thermal-chemical (HTC) theory is used to determine the temperature and relative humidity in the column under actual environmental conditions. This is then coupled with the mechanical loads determined from the global analysis to model the mechanical behavior using the solidification-microprestress-microplane (SMM) model. This analysis is then repeated at multiple angles to determine the effects of inclination on the shortening behavior.

1 INTRODUCTION

1.1 *Time dependent behavior of concrete*

The time-dependent material behavior of concrete is well studied (Bažant & Jirasek 2018), and its application to structural scale deformations has been consequently investigated by many authors (ACI 2008; Khan, Cook, & Mitchell 1997). The two primary mechanisms of time-dependent deformations in concrete, and the two considered in this study, are creep and shrinkage.

Shrinkage in concrete arises as a result of both mechanical and chemical behaviors. These behaviors may be classified as: a) plastic (due to water loss), b) drying (due to reduction in capillary water by cement reaction), c) autogenous (due to change in particle volume after hydration), and d) carbonation (due to reaction of water with CO_2). All of these factors depend on the initial water-to-cement ratio, the volume-to-surface area ratio, and the environmental conditions at the surface of the concrete specimen (Bažant & Wittmann 1982). Also related is the thermal expansion due to the heat of hydration (Aitcin, Neville, & Acker 1997). Generally, shrinkage is here treated load-independent strain.

Creep is generally defined as a continued increase in strain under a sustained constant load. In application to concrete these deformations are caused by both 'basic'

DOI 10.1201/9781003316404-76

creep and drying creep (Bažant & Yunping 1994). Basic creep is so-called as it is an intrinsic property not dependent on the specimen size or environmental conditions. Drying creep, conversely, is an additional deformation due to changes in relative humidity, similar to (but separate from) shrinkage. It is therefore a factor of both size and environmental conditions. In addition, temperature changes can result in transient thermal creep (Bažant, Cusatis, & Cedolin 2004).

The modern approach to comprehensively describe concrete creep is taken by a combination of the solidification theory (Bažant & Prasannan 1989a, 1989b) and the microprestress theory (Bažant, Hauggaard, & Baweja 1997; Bažant, Hauggaard, Baweja, & Ulm 1997). In this framework, the hardening cement gel is age-independent, the chemical aging is simulated by a reaction degree, and the sum of creep behavior (basic, drying, and transient) are described by the same theory, where the relaxation of self-equilibriated stresses occurs due to hydration and chemical potential imbalance of water phases. This approach has shown to match well with experimental data (Di Luzio & Cusatis 2009a, 2009b).

Both creep and shrinkage depend heavily on surrounding environmental conditions. In particular, the relative humidity, temperature, and their corresponding rates and time histories, are necessary for accurate material modeling, even at a structure scale.

It has been shown that both drying and wetting induce higher creep than that of concrete at constant moisture content (Gamble & Parrott 1978; Pickett 1942). Similar behavior is found during temperature changes (Chern, Bažant, & Marchertas 1985; Fahmi, Polivka, & Bresler 1972). From this it can be determined that long-term deformations of concrete structures must take into account the hygrothermal phenomenon to accurately predict total creep.

1.2 Column shortening

Any material which creeps will lead to the shortening of vertical structure members. This is a common concern even in "traditional" concrete structures with columns under pure vertical loading and even spacing. Figure 1 illustrates how such vertical deformations can cause, for example, a differential shortening effect in a structure where the columns are under higher vertical stress than a lateral-force-resisting core. For unique structural systems as the one considered here, the vertical shortening behavior may cause even more of a concern. Furthermore, the shortening of columns due to varying stress conditions over time can significantly impact the long-term health of a structure.

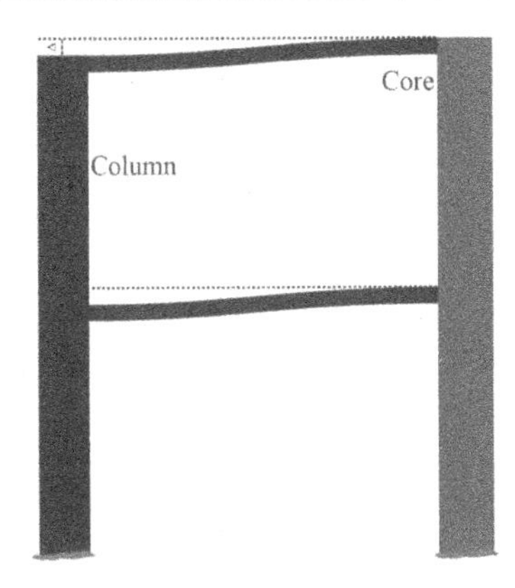

Figure 1. Vertical differential shortening schematic.

Though the effect may vary depending on stress, floor plan, and mitigating design, column shortening for a 60 story structure might be on the order of 30 mm (Pan, Liu, & Bakoss 1993). At this scale the shortening can cause problems not only within the structural members, but the nonstructural components as well, such as partitions and pipe lines. Furthermore, as construction technologies advance, engineers and architects are able to design increasingly complex and unique structural elements. These result in unusual movements not necessarily designed for or observed during construction, but rather occur due to the stress state during service, over the lifespan of the building. Unique structural systems must thus be given additional consideration, which they often do not receive.

2 CASE STUDY: NEW MARINA CASABLANCA TOWER

The New Marina Casablanca Tower (Figure 2) was proposed as part of new development in the Casablanca

Figure 2. Rendering of New Marina Casablanca Tower (Wimberly Allison Tong & Goo).

Marina, Morocco. Although construction was not completed according to schedule, it remains a unique structural case-study. The award-winning design was intended to revitalize the skyline and deploy innovative building techniques and components (Folco 2017). The designers achieved these goals primarily through the unique column structure, which presents a striking twisting profile. Each floor (see Figure 3) has been rotated slightly, amounting to a total 135° twist over 43 stories.

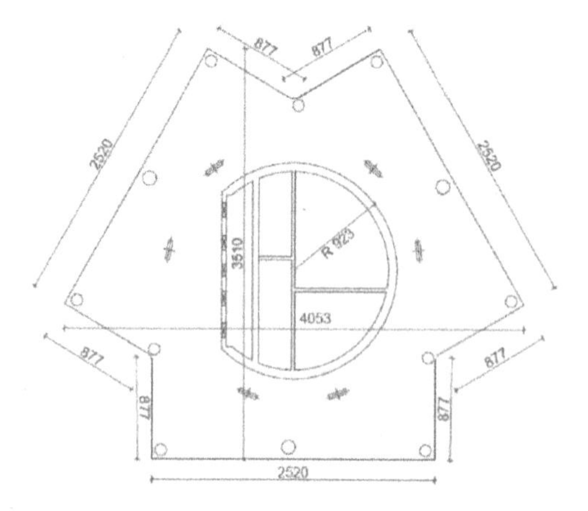

Figure 3. Typical floor plan.

The focus of this study will be on the external circular columns, which undergo the most dramatic change in geometry. The diameter varies between 1 to 0.6 meters, with a corresponding variation in concrete design strength from 50 MPa to 30 MPa. It is important to note the difference in design specification, as mix design has a significant impact on the long-term behaviors which contribute to vertical shortening, including both stress variation as well as moisture and temperature diffusion.

3 ANALYSIS METHOD

A two-scale approach is taken for this structure. First, a global model of the tower is analyzed in a structural finite element program (MIDAS). Typical loads were applied, and built-in code equations were used for material behavior, including creep. The global model also considers include construction stage analysis.

Following this, a single-story base column is simulated in a general finite element program (ABAQUS) with boundary conditions according to the global model. Here both moisture diffusion and heat transfer are considered, and creep behavior was modeled according to the well-established SMM theory (Di Luzio & Cusatis 2013). This provides more nuanced results which would be too computationally intensive to calculate for the entire structure.

The combination of these results are then reviewed and conclusions regarding expected structural behavior are discussed.

3.1 Global

The global model is composed of beam, plate, and wall elements defined by MIDAS. Beam elements have 12 degrees of freedom (DOFs), and are formulated according to Timoshenko beam theory. The plate elements are analogous to shell elements in typical finite element programs, formulated with two in-plane translation DOFs , one out-of-plane rotational DOFs and two out-of-plane rotational DOFs. Wall elements are a MIDAS specific formulation intended for shear wall use in structural application, and is composed of a plate element with beam elements at the perimeter nodes. The model is built as seen in Figure 4.

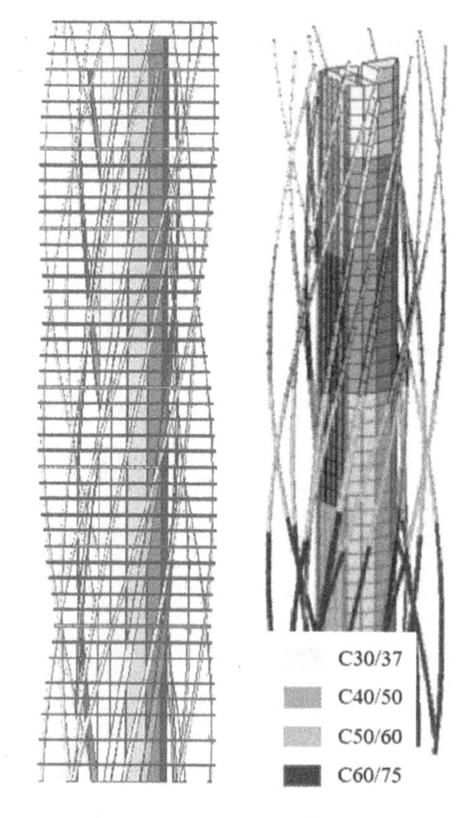

Figure 4. Global model.

The global model was analyzed for the duration of construction, using a multi-step construction analysis in-built in MIDAS. In the conventional analysis, floors are loaded simultaneously, while in multi-phase analysis each floor is "built" and then sequentially loaded, leading to a varying stress history within the members. This is relevant for time-dependent behaviors such as creep, particularly when the construction process is prolonged, as in tall structures.

To define material behavior, MIDAS has a number of built-in creep and shrinkage models. For the purposes of this analysis, the *fib* Model Code 2010 was used, where the coefficients were calculated by the software based on the provided material properties – standard EN04(RC) was adopted. Section properties were also assigned as given.

During construction analysis, only gravitational loads are considered: self-weight, dead load, and reduced live loads (30%). Dead and live loads vary by floor depending on use, but are typically 2 kPa and 3 kPa (1 kPa reduced) respectively. For modeling purposes, each floor was built in 8 days, and at mid-construction (192 days) dead loads are applied starting at the base, with sequential floor dead loads added after 4 days.

3.2 Local

Following the global analysis, a single-story column from the base floor (Figure 5) is modeled using a more refined approach. First, the three dimensional temperature and humidity field histories were computed. These field histories were passed to the mechanical model through ABAQUS, where the displacements over 400 days are modeled.

Figure 5. Local column model.

The temperature and humidity fields were calculated using the hygrothermochemical model (HTC), as formulated by Di Luzio & Cusatis (2009a) and (2009b). Boundary conditions were assumed to 23°C and sealed (i.e., no change in humidity across the boundary).

The temperature and humidity results were passed to ABAQUS for use in the mechanical analysis. The SMM formulation as described by Di Luzio & Cusatis (2013) was used. The base of the column was constrained and the three-dimensional column load history was calculated from the global analysis.

4 RESULTS

An elastic analysis was first performed to quantify the important of utilizing a multi-phase construction model. Figure 6 shows the results, which indicate that simultaneous loading would produce significant

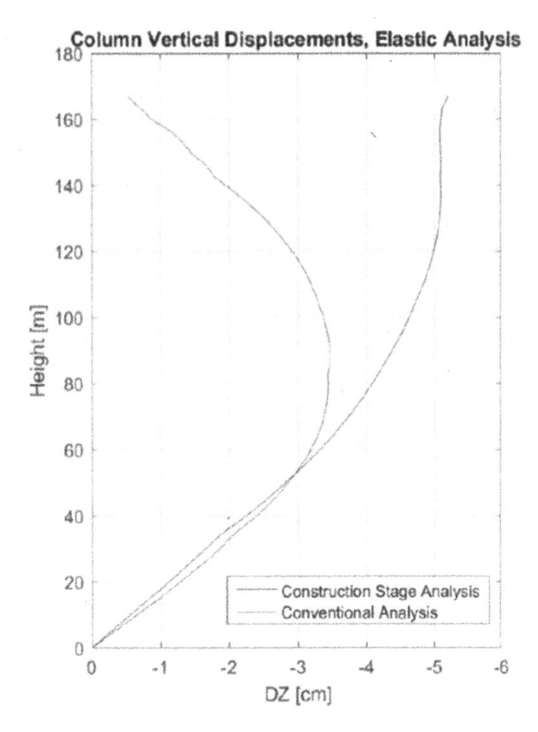

Figure 6. Comparison of elastic vertical displacement using conventional vs multi-phase constructing analysis.

unnecessary deformation at the height of the structure. Though this may be considered a more conservative approach, in the case of such unique geometry it limits a realistic understanding of the column behavior.

Vertical displacements and rotations of the structure are calculated for 30000 days (roughly 100 years – the expected service life). Results are shown in Figure 7. Here we see the importance of a viscoelastic analysis, as the displacement curve through the height is clearly different after 100 years, compared with an elastic calculation.

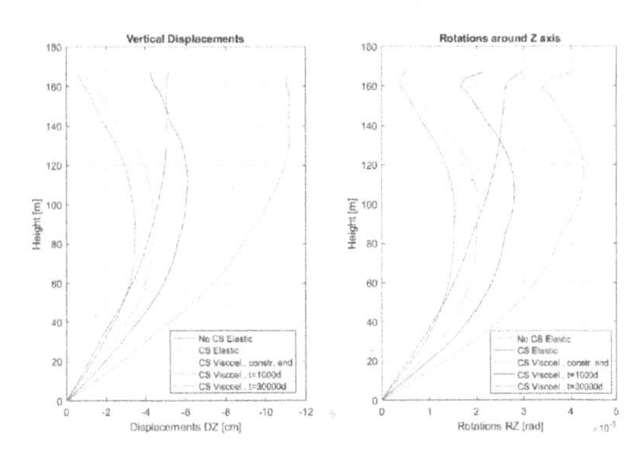

Figure 7. Vertical viscoelastic displacements.

When looking at the lateral displacements, the asymmetric nature of the structure is clear. Although in the elastic case it may be assumed to have similar displacements in the two directions, this is clearly no the

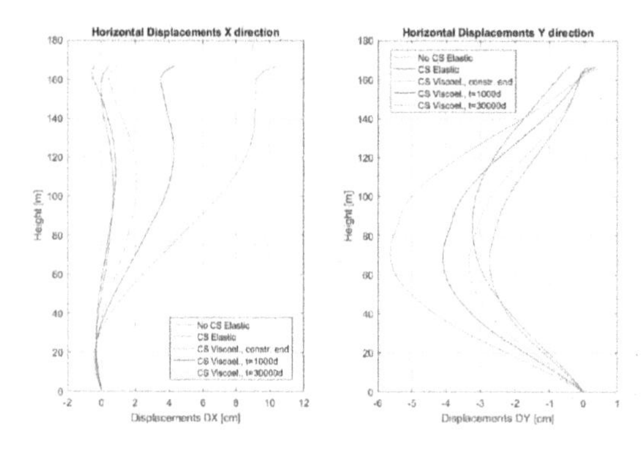

Figure 8. Lateral viscoelastic displacements.

case in the viscoelastic analysis. Furthermore the absolute lateral displacements are quite large, 5–10 cm. This must clearly be considered during the design process, and would not be expected for a building with vertical columns.

In addition to the overall displacements, there is also a local variation in vertical displacements which results in differential shortening. Figure 9 shows a clear difference between the core and exterior columns, and furthermore this difference increases over the service life of the structure from < 1 cm up to 4 cm difference.

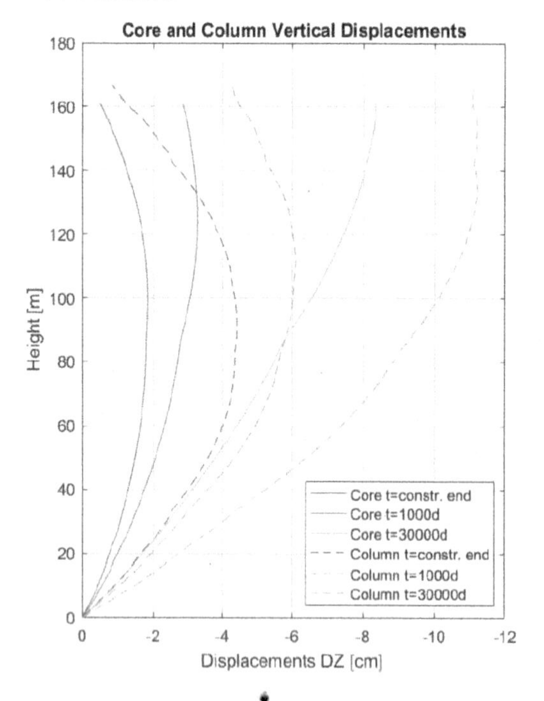

Figure 9. Differential viscoelastic displacements.

When looking at a more nuanced analysis for a single column, we can see there is actually an initial upwards displacement due to thermal expansion as a result of hydration. This takes course over a couple days, but the trend back to neutral displacement (and then compression) takes multiple weeks. This is

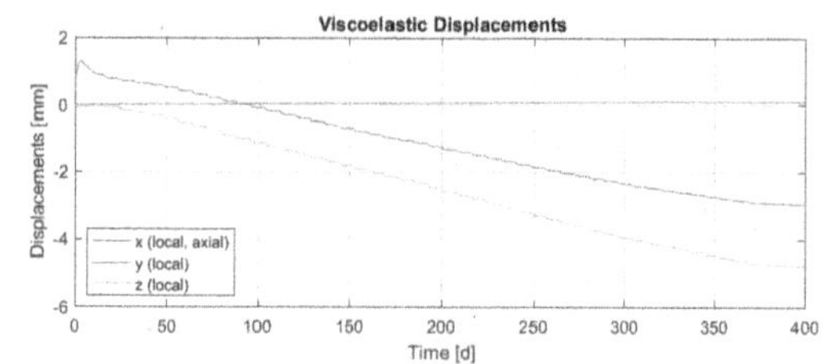

Figure 10. Viscoelastic displacements for a single column.

a critical consideration for construction as long-term mitigation techniques for creep may be exacerbated by this effect initially.

Figure 11 shows an analysis of the column under varying angles. Here it is clear that while the vertical deformations are not significantly impacted by the geometry, the lateral displacements can more than double from a single change in degree. Although it was assumed a constant angle through the structure, the reality is that not all the floors where the same in this respect, and thus the "twisting" effect could be severely exacerbated over time if the angle is not correctly considered.

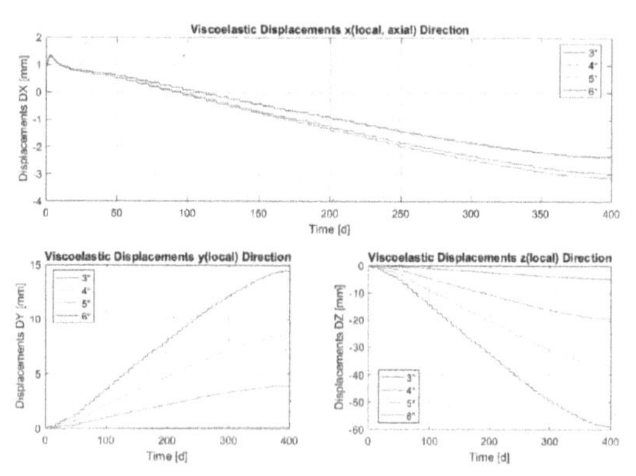

Figure 11. Viscoelastic displacements as a function of column angle.

5 CONCLUSIONS

The time-dependent behavior of twisted concrete columns was analyzed at two scales to explore the long-term effects of a unique structural geometry. Results demonstrate that the three dimensional displacements of the overall structure are notably asymmetric, and furthermore the use of multi-phase construction loading is necessary to accurately understand this behavior. The effect of column angle is also pronounced, as expected, and show how dependent long-term effects are on geometry. Both models showed a trend of delayed creep and shrinkage effects, with the local model also producing an early-stage thermal dilation effect, indicating the need for a comprehensive design approach to properly manage the resulting deformations.

REFERENCES

Aitcin, P.-C., A. Neville, & P. Acker (1997). Integrated view of shrinkage deformation. *Concrete International 19*(9), 35–41.

American Concrete Institute (2008). *209.2R-08: Guide for Modeling and Calculating Shrinkage and Creep in Hardened Concrete*. American Concrete Institute.

Bažant, Z. P. & M. Jirasek (2018). *Creep and Hygrothermal Effects in Concrete Structures*, Volume 225 of *Solid Mechanics and Its Application*. Springer.

Bažant, Z. P. & F. H. Wittmann (1982). *Creep and shrinkage in concrete structures*. Wiley New York.

Bažant, Z. P., G. Cusatis, & L. Cedolin (2004). Temperature effect on concrete creep modeled by microprestress-solidification theory. *Journal of engineering mechanics 130*(6), 691–699.

Bažant, Z. P., A. B. Hauggaard, & S. Baweja (1997). Microprestress-solidification theory for concrete creep. ii: Algorithm and verification. *Journal of Engineering Mechanics 123*(11), 1195–1201.

Bažant, Z. P., A. B. Hauggaard, S. Baweja, & F.-J. Ulm (1997). Microprestress-solidification theory for concrete creep. i: Aging and drying effects. *Journal of Engineering Mechanics 123*(11), 1188–1194.

Bažant, Z. P. & S. Prasannan (1989a). Solidification theory for concrete creep. i: Formulation. *Journal of engineering mechanics 115*(8), 1691–1703.

Bažant, Z. P. & S. Prasannan (1989b). Solidification theory for concrete creep. ii: Verification and application. *Journal of Engineering mechanics 115*(8), 1704–1725.

Bažant, Z. P. & X. Yunping (1994). Drying creep of concrete: constitutive model and new experiments separating its mechanisms. *Materials and structures 27*(1), 3–14.

Chern, J., Z. Bažant, & A. Marchertas (1985). Concrete creep at transient temperature: Constitutive law and mechanism. Technical report, Argonne National Lab.

Di Luzio, G. & G. Cusatis (2009a). Hygro-thermo-chemical modeling of high performance concrete. i: Theory. *Cement and Concrete composites 31*(5), 301–308.

Di Luzio, G. & G. Cusatis (2009b). Hygro-thermo-chemical modeling of high-performance concrete. ii: Numerical implementation, calibration, and validation. *Cement and Concrete composites 31*(5), 309–324.

Di Luzio, G. & G. Cusatis (2013). Solidification–microprestress–microplane (smm) theory for concrete at early age: Theory, validation and application. *International Journal of Solids and Structures 50*(6), 957–975.

Fahmi, H. M., M. Polivka, & B. Bresler (1972). Effects of sustained and cyclic elevated temperature on creep of concrete. *Cement and Concrete Research 2*(5), 591–606.

Folco, U. (2017). Time-dependent behavior of twisted-shape tall buildings with inclided columns. Master's thesis, Politecnico di Torino.

Gamble, B. & L. Parrott (1978). Creep of concrete in compression during drying and wetting. *Magazine of concrete research 30*(104), 129–138.

Khan, A. A., W. D. Cook, & D. Mitchell (1997). Creep, shrinkage, and thermal strains in normal, medium, and high-strength concretes during hydration. *ACI Materials Journal 94*(2).

Pan, L., P. Liu, & S. Bakoss (1993). Long-term shortening of concrete columns in tall buildings. *Journal of Structural Engineering 119*.

Pickett, G. (1942). The effect of change in moisture-content on the crepe of concrete under a sustained load. In *Journal Proceedings*, Volume 38, pp. 333–356.

Computational Modelling of Concrete and Concrete Structures – Meschke, Pichler & Rots (Eds)
 ISBN: 978-1-032-32724-2

Portlandite dissolution: Part 1. Mechanistic insight by Molecular Dynamics (MD)

K.M. Salah Uddin & Bernhard Middendorf
Institute of Structural Engineering, University of Kassel, Kassel, Germany

Mohammadreza Izadifar, Neven Ukrainczyk & Eduardus Koenders
Institute of Construction and Building Materials, Technical University of Darmstadt, Darmstadt, Germany

ABSTRACT: The dissolution of portlandite leads to carbonation, which affects the sustainability of the concrete. Therefore, it is important to understand the dissolution behavior at the portlandite-water interface. The current contribution aims at the development of a multi-scale bridging modeling approach that connects the atomistic scale to the (sub-) micro scale. In this work, first, the biased molecular dynamics, metadynamics coupled with ReaxFF is employed to calculate the reaction path as a free energy profile of calcium dissolution at 298K in diluted water from the different surfaces of portlandite. The reason for the reactivity of (010) crystal plane is higher compared to the (001) surface is explained. In addition the influence of neighboring Ca on the dissolution rate is also investigated. The calculated rate constant of most important atomistic reaction steps provided an input for developing the upscaled model using a kinetic Monte Carlo (KMC) method.

1 INTRODUCTION

The portlandite ($Ca(OH)_2$) is a major by-product of the cement hydration reaction, which results in the passivation of steel reinforcement. The dissolution of portlandite leads to the carbonation that accelerates the corrosion of the reinforcement by dropping the pH value of the pore solution by approximately three units. Which plays a vital role in the reduction of the service life of the concrete (Taylor 1997). The carbonation of portlandite is also played a critical role during the setting of the concrete mixture, since the calcite formed by carbonation is comparatively less soluble than the portlandite itself. During the hydration of cement, portlandite is not only precipitated in the hardened cement paste but also forms a thin crystalline layer between the steel reinforcement and aggregate. Those interfacial layers of portlandite influence the resistance of the reinforced concrete (Lea 2004).

Therefore, it is important to understand the dissolution behavior at the portlandite-water interface. Moreover, due to portlandite's relatively simple crystal structure, therefore, it has proposed benchmarking minerals in developing atomistic modeling approaches for the dissolution/precipitation process of other (more complex) cementitious minerals in general. The dissolution of cementitious minerals especially at an atomistic scale is not fully understood yet due to the lack of experimental techniques available to reach this resolution. Within the last decade, computational methods have been expanded to the atomistic description of cementitious materials. Atomistic simulation using Reactive Force Fields (ReaxFF) parameterized by quantum mechanical calculation, in combination with metadynamics (metaD) can be an effective solution to study the chemical reactions pathways with sufficient accuracy and reasonable computing times.

ReaxFF has been developed to investigate the reaction mechanism at the material interface. It has already been implemented successfully in many materials, i.e. hydrocarbons (Chenoweth et al. 2008), polymer chemistry (Senftle et al. 2016), metal oxides (Si/SiO_2) (Fogarty et al. 2010), metal hydrides (Cheung et al. 2005). ReaxFF usually calculates molecular dynamics in femtosecond (i.e. 10^{-15} seconds) time steps. Therefore, it could be computationally expensive especially during the Transition state calculation despite its efficiencies compared to classical force field theory. Metadynamics (metaD) is integrated into ReaxFF to solve these timescale issues. Metadynamics is an efficient algorithm to accelerate observing the rare events by adding biased potential on a selected number of collective variables (CVs) (Barducci et al. 2011). During the MD simulation, the bias potential is applied as a sum of Gaussian acting directly on the microscopic coordinates of the system (Aktulga et al. 2012).

The combined approach has already been implemented successfully to calculate the dissolution mechanism and the reactivity of different surfaces of Portland cement clinkers at the atomistic scale (Salah Uddin et al. 2019). The current contribution aims to develop the links between the atomistic scale to the

*Corresponding Author

 DOI 10.1201/9781003316404-77

(sub-)micro-scale. It will explain new insights into the dissolution mechanism of portlandite.

In this work (Part-1), a multistep modeling approach has been taken to get depth information of the dissolution and reactivity of different surfaces of Portlandite at room temperature (298K). At first, the hydration of (001), (100), (010) surfaces of portlandite were allowed for 600 picoseconds. Later, all pre-hydrated surfaces were used as an input to study the dissolution mechanism of calcium by using ReaxFF coupled with metaD. Afterward, the reactivity of surfaces was compared with the dissolution profile (free energy surface) of calcium. Besides, the orientation of calcium and the number of neighbors is different on each surface. Therefore, the influence of crystal site neighbors was also be calculated. Finally, all the calculated microscopic rate constant using transition state theory (TST) will be provided as an input for upscaling (in Part-2 contribution presented within this conference proceedings) using a Kinetic Monte Carlo (KMC) approach and calculating the overall rate of the dissolution (Martin et al. 2021).

2 COMPUTATIONAL DETAILS

Many simulation methods have been developed over the decade. Density functional theory (DFT) using quantum mechanics is accurate compared to the force field in general (Izadifar et al. 2021). However, due to its higher computational cost, this method becomes practically not applicable for a large system.

In contrast, classical force field theory is unable to explain the chemical reaction (bond formation and bond breaking). The ReaxFF parameterized by DFT has been considered a comparatively efficient method of calculating the reaction mechanism with reasonable simulation cost. It has been implemented to the cementitious system by combining two parameter sets (Si-O-H and Ca-O-H) developed individually by Fogarty et al. (Fogarty et al. 2010) and Manzano et al. (Manzano et al. 2012), respectively. This parameter set already has explained the absorption of water to C_3S clinker phases (Manzano et al. 2015). The computations were carried out by using ReaxFF in LAAMPS (Large-scale Atomic/Molecular Massively Parallel Simulator) platform (Plimpton 1995).

In addition, the metaD, (history-dependent bias potential) is applied as a sum of Gaussians directly on the microscopic coordinates of the system. Those small hills in energy representation are placed on top of the underlying free energy landscape. MetaD is not only able to explore the new pathways but also reconstruct the free energy surfaces (FES).

2.1 *Model construction*

The fresh cleaved (001) surface of portlandite (Busing & Levy 1957) orthogonal periodic $(17.80 \times 19.64 \times 38.65) \times 10^{-30}$ m^3 simulation cell composed of 1302 atoms was constructed by virtual nano lab (VNL) (Schneider et al. 2017) and Avogadro (Hanwell et al. 2012). Then the geometry has been optimized using with Hessian-free truncated Newton algorithm (htfn) (Tuckerman et al.2006) with the cutoff tolerances (energy and force) of 4.18×10^{-4} and 4.18×10^{-8} kJ mol^{-1}. Maximum iterations for the minimizer were 100. Later on, an additional 6.99×10^{-27} m^{3}periodic cell filled with water was added to the optimized (001) portlandite surface using packmole (Martínez et al. 2009). The number of water molecules matched a density of 1000 kg m^{-3} with a random distribution.

The simulation cells were equilibrated to 298K for 150 picoseconds with 0.5 femtoseconds time steps using canonical ensemble (nvt) with a Nose–Hoover thermostat, integrating the non-Hamiltonian equations of motion (Tuckerman et al. 2006). Subsequently, they were hydrated for 600 picoseconds using Nose-Hoover barostat (npt) with all three diagonal components of the pressure tensor to be coupled together (iso). A periodic boundary condition was applied during the simulation.

The last geometry after 600 picoseconds of hydration for (001) surface was considered as an initial geometry to calculate the dissolution mechanism of calcium using the combined approach of ReaxFF and metaD. The PLUMED package (Bonomi et al. 2009; Laio & Gervasio 2008; Tribello et al. 2014) was used as an extension of LAMMPS for metaD simulation.

The central calcium (Ca-588) of hexagonally oriented neighboring Ca of portlandite surface from (001) surface was selected. A well-tempered metaD scheme was applied to remove calcium from the surface to pore solution. The distance between the center of mass (COM) and the selected calcium atom is selected as CVs and computed by adding biased potential in every 40 time steps. Besides, Gaussian hills height of 6.28 kJ /mol and a full width at half-maximum of 0.2×10^{-10} m are added every 0.02 picoseconds. The simulation was performed for 500 picoseconds (till converged) using NPT ensemble at temperature 298K.

Table 1. Crystallographic data for the orthogonal simulation cells of Portlandite consisting of different crystalline planes at 298K.

Crystal plane of portlandite	Orthogonal Cell dimension (Å^3 / 10^{-30} m^3)	No. of atoms in the simulation cell
(100)	20.75, 19.94, 38.42	1431
(010)	17.67, 20.75, 34.94	1155

A similar approach was applied and calculated the FES central Ca in absence of 1,2,3,4,5,6 hexagonally oriented neighbors (total 7 scenarios) removed clockwise in order to understand the effect of neighbors.

The same modeling approach was used for other simulation cells of portlandite containing (100) and (010) surfaces (Tab. 1). Unlike (001), the Ca are arranged differently (in a rowing manner) on both (100) and (010) surfaces. Therefore, a total of three scenarios was considered to calculate the dissolution profile of the central Ca before and after removing 1 and 2 neighboring Ca located on both sides in the same row. The reactivities were calculated from the dissolution profile (FES) of the calcium of the particular surface.

Finally, the free energy of activation of all selected events for three different surfaces of portlandite was analyzed to calculate the individual rate constant of all individual events using transition state theory (TST) and provided the results as an input of KMC simulation for calculating the overall rate of the dissolution.

3 RESULTS AND DISCUSSION

3.1 *Hydration of portlandite*

To investigate the interfacial interaction between portlandite surfaces and water bulk, the hydration was run for 600 picoseconds at room temperature (298K) and pressure applying periodic boundary conditions.

In order to capture the movement of the lighter element, hydrogen 0.5 fs (femtosecond) time steps were used for the entire simulation. Since lighter atom usually vibrates so fast and if we increase the time steps the distance between two atoms increases. As ReaxFF considers bond order at a higher distance these atoms are considered as non-bonded which provides wrong results.

According to our observation, the water molecules interact with the surfaces of portlandite initially and dissociate to hydroxyl pair by protonating the oxygen on the (010) surface. Afterward, the proton transfer (hopping process) from the hydroxyl to inner oxygen by leaving the first oxygen-free for further reaction as described by Manzano et al (Manzano et al. 2015). Hydration is carried out only for 600 picoseconds.

The (010) surface of portlandite (Figure 1c, II) shows higher reactivity during hydration compared to the (100) (Figure 1c, I) and (001) (Figure 1b) surfaces due to water tessellation at the (001) surface which prevents water from penetration into the crystal and the dissolution of calcium ions from the surface. However, further study is required to get a clear overview of the different reactivity of different crystal surfaces. The dissolution profile of calcium from three surfaces of portlandite can deliver a proper explanation about the reactivity by comparing the total free energy changes during dissolution.

3.2 *Dissolution of calcium from (001) surface of Portlandite*

Free energy calculations have received significant importance in molecular dynamics (MD) simulation for proper understanding of reaction mechanisms including transition state. A straightforward traditional sampling approach is often not possible in order to get FES due to the higher barrier.

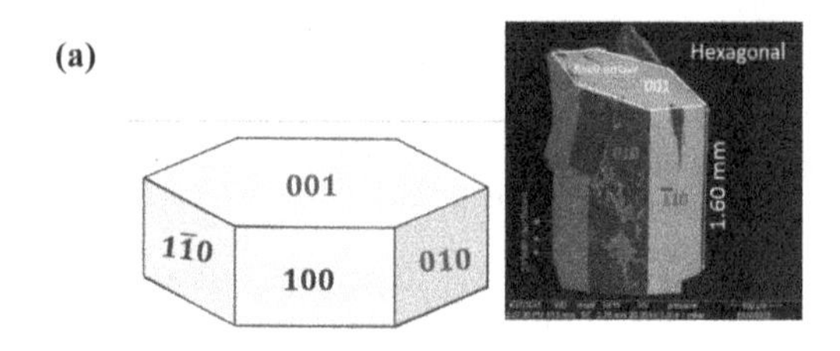

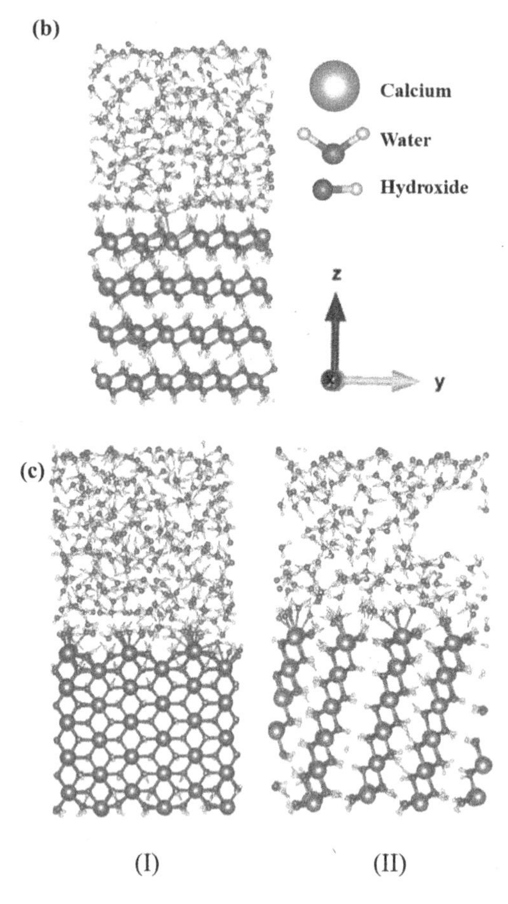

Figure 1. (a) SEM image of the hexagonal Single crystal of portlandite Comparison of different reactivity of (a) basal (001) surface and (b) prismatic (100), (010) surface of portlandite during hydration for 600 picoseconds at 298K.

Well-tempered metaD is able to force the system to overcome the free energy barriers by selecting the correct CVs. It offers to control and compute the region of FES that we are interested in. MetaD simulation was performed to investigate the dissolution mechanism of calcium from the hydrated (001) surface of portlandite.

Figure 2a represents the free energy surface for the dissolution of calcium (red Ca) from (001) crystal surface of portlandite at 298K. Where the distance between the central Ca-588 surrounded by six Ca neighbors and the center of mass of the crystal was selected as CVs. The x-axis represents the reaction coordinate in terms of distance in Å (10^{-10} m).

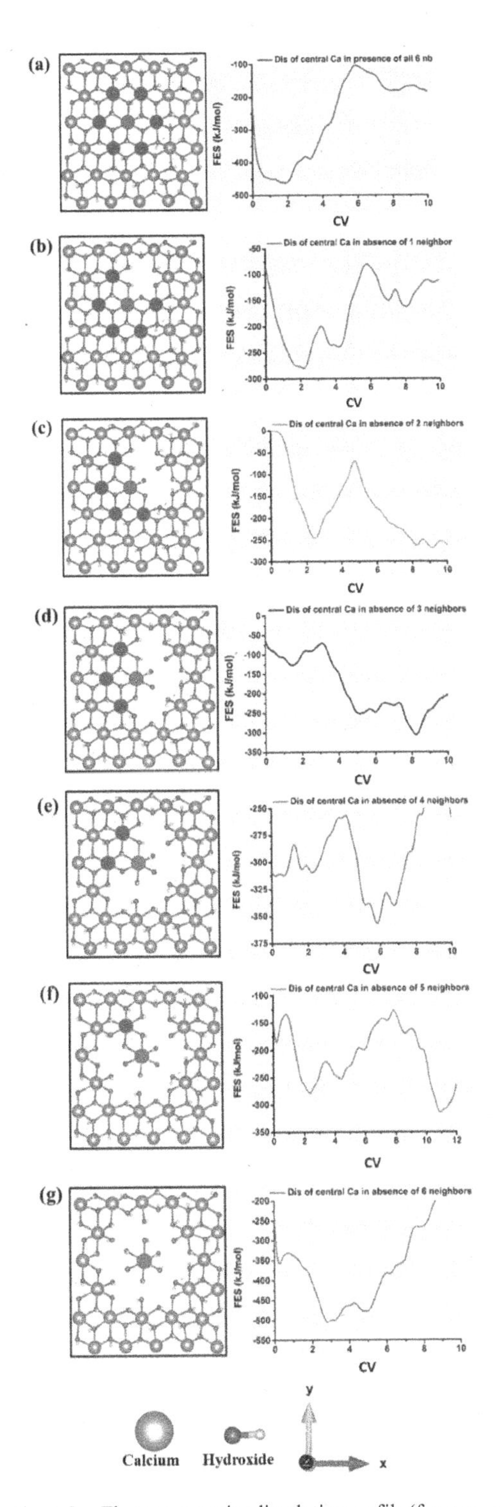

Figure 2. The representative dissolution profile (free energy surface) of the hexagonally oriented central Ca (red Ca) from (001) surface of Portlandite in different scenarios: the presence of all six Ca neighbors (a-g) and absence of the different number of 1,2,3,4,5,6 hexagonally oriented neighbors at 298K. CV represents the distance in Angstrom (Å) (10^{-10} m).

The free energy surface represents the movement of central Ca from the surface to the pore solution by overcoming the huge barrier of 352.00 kJ/mol (Figure 2a). The total free energy change of (ΔG) of +280.80 kJ/mol indicated the endergonic, thermodynamically unfavorable, and less reactive surface (Table 2). However, in absence of the first and second neighbors, the activation energy was reduced to 175.40 kJ/mol and the dissolution process becomes thermodynamically favorable (Figure 2b, c). Further removal of neighbors one by one anti-clockwise decreases the activation energy and reaches the minimum value of 25.90 kJ/mol when all six neighbors are missing (Figure 2g). The results were in good agreement with the surface interaction with water during pre-hydration, where (001) surface was found less reactive due to water tessellation (Manzano et al. 2015). Therefore, the surface is not favorable for Ca dissolution, however, the crystal defect and missing neighbors increase its reactivity.

3.3 *Dissolution of calcium from (100) and (010) surface of portlandite*

The FES for the dissolution of central Ca (red marked) from (010) surface of portlandite shows higher reactivity as observed after the 600 picoseconds of hydration (Figure 1c II). The complete dissolution of Ca from (010) surface required a comparatively lower energy barrier of 29.90 kJ/mol at 298K (Figure 3a). Besides, the free energy change (ΔG) of −299.69 kJ/mol indicates the exergonic and thermodynamically favorable process which, explains the higher interaction

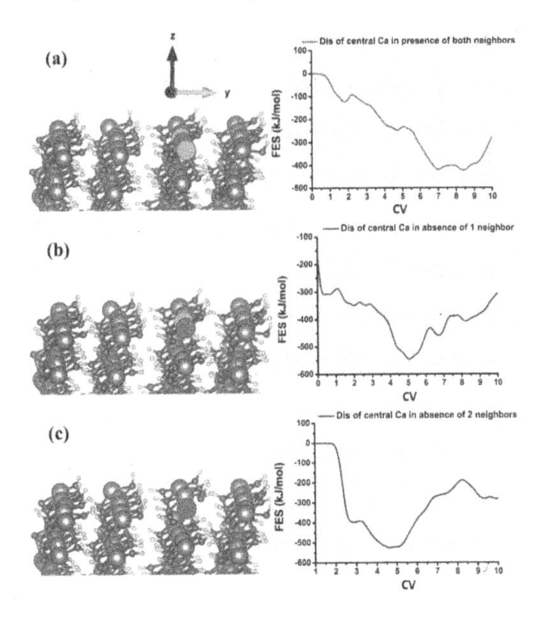

Figure 3. (a–c) Representative snapshot of the dissolution profile (represents the activation barrier) of the central calcium (red Ca) from (010) crystalline plane of portlandite in different scenarios (in absence of the different number of neighbors) at 298K. CV represents the distance in Angstrom (Å) (10^{-10} m).

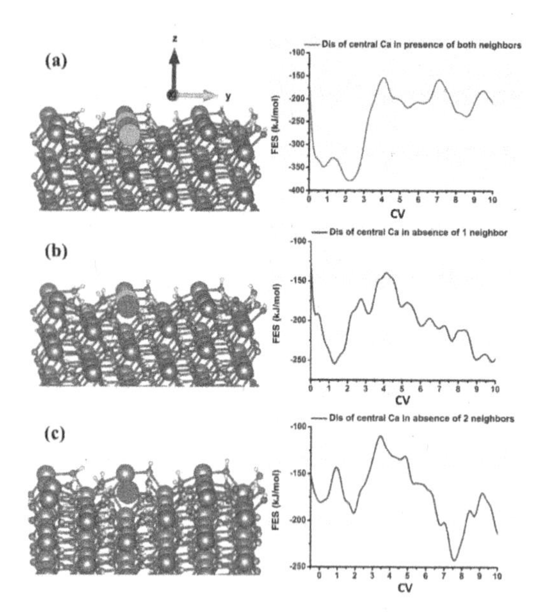

Figure 4. (a-c) Representative snapshot of the dissolution profile (represents the activation barrier) of the central calcium (red Ca) from (100) crystalline plane of portlandite in different scenarios (in absence of the different number of neighbors) at 298K. CV represents the distance in Angstrom (Å) (10^{-10} m).

with water. In addition, Further reduction of activation barrier (same red Ca) to 20.55 kJ/mol and 7.10 kJ/mol in absence of first and second neighboring Ca indicate the increase of reactivity toward dissolution (Figure 3b, c)

Among the prismatic surfaces of portlandite, the reactivity and effect of neighbors for the dissolution of (100) surface follows the same trend as the basal (001) surface. The dissolution of central calcium (Ca-542) in presence of all neighbors needs to overcome 195.30 kJ/mol which is lower than the (001) surface in a similar scenario (Figure 4a). The ΔG of +111.15 kJ/mol at 298 K represents the endergonic and thermodynamically unfavorable process. After removing the first neighbor the activation barrier was reduced to 114.60 kJ/mol but remained unfavorable. Nevertheless, in absence of both neighbors (Figure 4b, c) results in the further reduction of the activation barrier (70.00 kJ/mol) as well as the dissolution process becomes favorable (ΔG = −62.33 kJ/mol) (Table 2).

Table 2. Free energy change of the different surfaces of Portlandite during the dissolution of calcium.

	Scenarios of dissolution of central Ca	Free Energy Activation of (ΔG^*) kJ/mol	Free Energy Change (ΔG) kJ/mol
(001)	In presence of 6 neighbors	352.00	+280.80
	After removing 1 neighbor	199.10	+117.70
	After removing 2 neighbors	175.40	−23.60
	After removing 3 neighbors	56.14	−180.57
	After removing 4 neighbors	55.80	−44.30
	After removing 5 neighbors	54.90	−130.00
	After removing 6 neighbors	25.90	−147.60
(100)	In presence of 2 neighbors	195.30	+111.15
	After removing 1 neighbor	114.60	+5.60
	After removing 2 neighbors	70.00	−62.33
(010)	In presence of 2 neighbors	29.90	−299.69
	After removing 1 neighbor	20.55	−237.96
	After removing 2 neighbors	7.10	−126.00

4 CONCLUSION

The objective of this contribution was to elucidate the dissolution mechanism of portlandite and developed the upscaling approach which connects the atomistic simulation using ReaxFF coupled with metaD and sub-micro KMC method.

The free energy calculation at different scenarios was indicated the variation of reactivity of different surfaces of portlandite. Both (001) and (100) surfaces were found less reactive initially, however, the reactivity increases with increasing the number of missing neighbors. Among them, (010) surface was found most reactive. The reactivity of the surfaces of portlandite is in following order:

(010) > (100) > (001)

The calculated activation barrier obtained in the current contribution by MD simulations for most important scenarios provided input data for KMC simulations in Part 2 proceedings, where the rate of individual scenarios for calcium dissolution was calculated using transition state theory. Furthermore, the MD-KMC simulation enabled to calculate the overall mesoscopic rate of portlandite dissolution in far from equilibrium conditions.

ACKNOWLEDGMENTS

The authors grateful to Nikolas Luke, M. Sc. for his technical support in running MD simulation successfully in the high-performance Linux cluster at the University of Kassel. The authors also acknowledge support by the German Research Foundation (DFG).

REFERENCES

Aktulga, H.M., Fogarty, J.C., Pandit, S.A., & Grama, A.Y. (2012). Parallel reactive molecular dynamics: Numerical methods and algorithmic techniques. *Parallel Computing, 38*(4–5), 245–259. doi:10.1016/j.parco.2011.08.005

Barducci, A., Bonomi, M., & Parrinello, M. (2011). Metadynamics. *Wiley Interdisciplinary Reviews: Computational Molecular Science, 1*(5), 826–843. doi:10.1002/wcms.31

Bonomi, M., Branduardi, D., Bussi, G., Camilloni, C., Provasi, D., Raiteri, P., … Parrinello, M. (2009). PLUMED: A portable plugin for free-energy calculations with molecular dynamics. *Computer Physics Communications, 180*(10), 1961–1972. doi:10.1016/j.cpc.2009.05.011

Busing, W.R., & Levy, H.A. (1957). Neutron Diffraction Study of Calcium Hydroxide. *The Journal of chemical physics, 26*(3), 563–568. doi:10.1063/1.1743345

Chenoweth, K., van Duin, A.C.T., & Goddard, W.A. (2008). ReaxFF reactive force field for molecular dynamics simulations of hydrocarbon oxidation. *The journal of physical chemistry. A, 112*(5), 1040–1053. doi:10.1021/jp709896w

Cheung, S., Deng, W.-Q., van Duin, Adri C T, & Goddard, W.A. (2005). ReaxFF(MgH) reactive force field for magnesium hydride systems. *The journal of physical chemistry. A, 109*(5), 851–859. doi:10.1021/jp0460184

Fogarty, J.C., Aktulga, H.M., Grama, A.Y., van Duin, Adri C T, & Pandit, S.A. (2010). A reactive molecular dynamics simulation of the silica-water interface. *The Journal of chemical physics, 132*(17), 174704. doi:10.1063/1.3407433

Hanwell, M.D., Curtis, D.E., Lonie, D.C., Vandermeersch, T., Zurek, E., & Hutchison, G.R. (2012). Avogadro: an advanced semantic chemical editor, visualization, and analysis platform. *Journal of cheminformatics, 4*(1), 17. doi:10.1186/1758-2946-4-17

Izadifar, M., Dolado, J.S., Thissen, P., & Ayuela, A. (2021). Interactions between Reduced Graphene Oxide with Monomers of (Calcium) Silicate Hydrates: A First-Principles Study. *Nanomaterials, 11*(9), 2248. doi:10.3390/nano11092248

Laio, A., & Gervasio, F.L. (2008). Metadynamics: a method to simulate rare events and reconstruct the free energy in biophysics, chemistry and material science. *The Journal of Physical Chemistry A, 71*(12), 126601. doi:10.1088/0034-4885/71/12/126601

Lea, F.M. (2004). *Lea's chemistry of cement and concrete* (4. ed.) (P.C. Hewlett, Ed.). Oxford: Elsevier Butterworth-Heinemann.

Manzano, H., Durgun, E., López-Arbeloa, I., & Grossman, J.C. (2015). Insight on Tricalcium Silicate Hydration and Dissolution Mechanism from Molecular Simulations. *ACS applied materials & interfaces, 7*(27), 14726–14733. doi:10.1021/acsami.5b02505

Manzano, H., Pellenq, R.J.M., Ulm, F.-J., Buehler, M.J., & van Duin, Adri C T. (2012). Hydration of calcium oxide surface predicted by reactive force field molecular dynamics. *Langmuir : the ACS journal of surfaces and colloids, 28*(9), 4187–4197. doi:10.1021/la204338m

Martin, P., Gaitero, J.J., Dolado, J.S., & Manzano, H. (2021). New Kinetic Monte Carlo Model to Study the Dissolution of Quartz. *ACS Earth and Space Chemistry, 5*(3), 516–524. doi:10.1021/acsearthspacechem.0c00303

Martínez, L., Andrade, R., Birgin, E.G., & Martínez, J.M. (2009). PACKMOL: a package for building initial configurations for molecular dynamics simulations. *Journal of computational chemistry, 30*(13), 2157–2164. doi:10.1002/jcc.21224

Plimpton, S. (1995). Fast Parallel Algorithms for Short-Range Molecular Dynamics. *Journal of Computational Physics, 117*(1), 1–19. doi:10.1006/jcph.1995.1039

Salah Uddin, K.M. Elucidation of Chemical Reaction Pathways in Cementitious Materials. doi:10.17170/kobra-202009211835

Salah Uddin, K.M., & Middendorf, B. (2019). Reactivity of Different Crystalline Surfaces of C3S During Early Hydration by the Atomistic Approach. *Materials (Basel, Switzerland), 12*(9). doi:10.3390/ma12091514

Schneider, J., Hamaekers, J., Chill, S.T., Smidstrup, S., Bulin, J., Thesen, R., … Stokbro, K. (2017). ATK-ForceField: A new generation molecular dynamics software package. *Modelling and Simulation in Materials Science and Engineering, 25*(8), 85007. doi:10.1088/1361-651X/aa8ff0

Senftle, T.P., Hong, S., Islam, M.M., Kylasa, S.B., Zheng, Y., Shin, Y.K., … van Duin, A.C.T. (2016). The ReaxFF reactive force-field: development, applications and future directions. *npj Computational Materials, 2*(1), 9396. doi:10.1038/npjcompumats.2015.11

Taylor, H.F.W. (1997). *Cement chemistry* (Vol. 2). Thomas Telford London.

Tribello, G.A., Bonomi, M., Branduardi, D., Camilloni, C., & Bussi, G. (2014). PLUMED 2: New feathers for an old bird. *Computer Physics Communications, 185*(2), 604–613. doi:10.1016/j.cpc.2013.09.018

Tuckerman, M.E., Alejandre, J., López-Rendón, R., Jochim, A.L., & Martyna, G.J. (2006). A Liouville-operator derived measure-preserving integrator for molecular dynamics simulations in the isothermal–isobaric ensemble. *Journal of Physics A: Mathematical and General, 39*(19), 5629–5651. doi:10.1088/0305-4470/39/19/S18

Computational Modelling of Concrete and Concrete Structures – Meschke, Pichler & Rots (Eds)
© 2022 Copyright the Author(s), ISBN: 978-1-032-32724-2

Portlandite dissolution: Part 2. Forward rates by Kinetic Monte Carlo (KMC)

Mohammadreza Izadifar, Neven Ukrainczyk & Eduardus Koenders
Institute of Construction and Building Materials, Technical University of Darmstadt, Darmstadt, Germany

K.M. Salah Uddin & Bernhard Middendorf
Institute of Structural Engineering, University of Kassel, Kassel, Germany

ABSTRACT: Portlandite or calcium hydroxide ($Ca(OH)_2$) is considered as a most soluble hydration product, which is formed through the hydration reaction of tricalcium silicate (alite) and dicalcium silicate (belite) with water during curing of concrete. In the present work, an atomistic kinetic Monte Carlo (KMC) upscaling approach is implemented in MATLAB code in order to investigate the dissolution time of portlandite crystal. First simulations demonstrate far-from-equilibrium dissolution behavior, which encompass 119323 atoms and 26011 initial surface sites. First, the atomistic rate constants of individual Ca dissolution events are computed for three different morphologies of 100 or $\bar{1}00$, 010 or $0\bar{1}0$, and 001 or $00\bar{1}$ crystal planes, resulting in a total of 13 different scenarios. We observed that the dissolution process preferentially takes place from edges, sides, and surfaces of 010 or $0\bar{1}0$ crystal plane. Those sites have a significantly higher event probability to be selected, as the event probability is proportional to the atomistic rate constant. On the one hand, the dissolution time of sites follow a liner trend up to 23000, and then the time of site dissolution increases due to the reduction of the surface sites, i.e., the active surface area of the crystal, for the computation of the total rate constant. The steady-state dissolution rates are 0.706 mol/ (s cm^2) for 010 or $0\bar{1}0$, 1.548×10^{-7} mol/ (s cm^2) for 001 or $00\bar{1}$, and 1.58×10^{-17} mol/ (s cm^2) for 100 or $\bar{1}00$ surfaces.

1 INTRODUCTION

Concrete is the most used man-made construction material and extensively exerted for the construction of railways, dams, roads, skyscrapers, bridges, and public infrastructure globally. The mixture of cement as a major material with water, resulting in the production of calcium silicate hydrate (CSH) gel to bind coarse and fine aggregates in concrete. On the other hand, portlandite or calcium hydroxide ($Ca(OH)_2$) is also a significant mineral precipitated through the hydration of tricalcium silicate (alite) and dicalcium silicate (belite) with water during. It plays a substantial role in mechanical and durability properties of cement paste (Perko et al. 2020). Portlandite in other words is widely exerted in multiple aspects, such as dewatering sludge (Czechowski & Marcinkowski 2006; Ma et al. 2019), improving the mechanical properties of fly ash cement (Wang 2014; Yu & Ye 2013), delaying steel corrosion (Glass et al. 2000), water (Perko et al. 2020; Sato et al. 2007; Sun et al. 2012) and acid (Drugă et al. 2018; Ukrainczyk et al. 2019) resistance.

The main disadvantage of cement clinker is its enormous environmental CO_2 footprint. Producing cement clinker not only requires high amounts of energy, but it also releases approximately 5% of the global anthropogenic CO_2 (Worrell et al. 2001) and retreats 1.7% of total global freshwater (Miller et al. 2018). Portlandite presents the most soluble hydration phase (Perko et al. 2020), and an intense reactivity to the CO_2 resulting entire re-crystallization of portlandite into the calcite in a short while (Gu et al. 2006; Regnault et al. 2005). In fact, the process of degradation of concrete can be ascribed through the interaction of portlandite with CO_2 from the ambient atmosphere in the presence of water, producing calcium carbonate ($CaCO_3$) via the following reaction ($Ca(OH)_2 + CO_2 + H_2O \rightarrow CaCO_3 + 2H_2O$), which is oriented on the portlandite substrate. Although, the pH of the Portland cement (PC) in the absence of carbonation or any other attack can almost reach to the value of 14.1, carbonation of portlandite causes a low-pH of 12.5 aqueous solution due to the dissolution of hydrates; and therefore, making concrete mass porous, resulting in lower strength, and durability due to the subsequent corrosion of reinforcement steel (Boualleg et al. 2017).

Hence, the main objective of this study is to develop an elementary physical/chemical bridging model for the initial dissolution of portlandite. It is proposed here, as the simplest benchmark representative of cementitious minerals, for a long-sought goal of connecting the nanoscale to the upscaled microscale level. To understand the effects of equilibrium crystal morphology of portlandite (Galmarini et al. 2011) during the dissolution process, different surface orientations of 100 or $\bar{1}00$, 001 or $00\bar{1}$, and 010 or

DOI 10.1201/9781003316404-78

$0\bar{1}0$ according to Wulff construction were selected to upscale the atomistic dissolution rates by employment of far-from-equilibrium kinetic Monte Carlo (KMC) approach, representing forward reaction rate. Moreover, to implement the KMC approach in a proper and accurate way, Uddin et al. in parallel study (Part 2 contribution of this conference proceedings) provided input information about the reaction energy barriers (E_D) of Ca dissolution for different (depending on the neighbors) scenarios of different surface orientations (crystal planes) at the room temperature by a molecular dynamic (MD) coupled with metadynamics simulation method in order to compute the dissolution rates (r_D). In fact, in the KMC upscaling approach, the dissolution and precipitation energy barriers for a given site are sometimes written as the sum of the contribution of the *n* bonded neighbors. Then, by application of a MATLAB code, the time of dissolution of Ca is computed dependent on the dissolution rates of Ca already computed for different scenarios through the KMC approach.

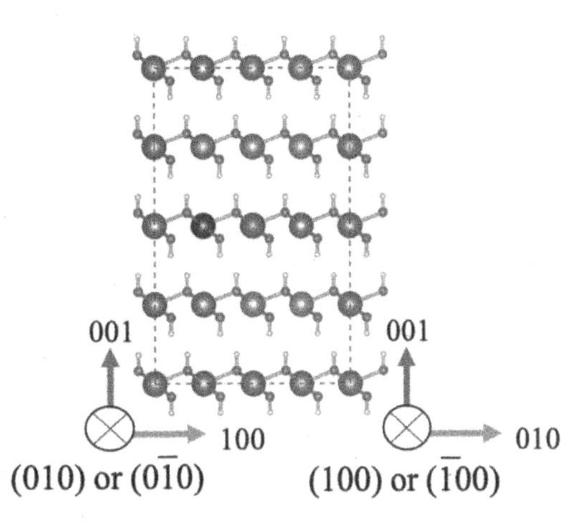

Figure 2. Illustration of three different scenarios for red Ca dissolution depending on the existing neighbors on the 100 or $\bar{1}00$, and 010 or $0\bar{1}0$ surface orientations of portlandite employing KMC upscaling approach.

2 METHODOLOGY

In order to implement kinetic Monte Carlo (KMC) upscaling approach for dissolution of portlandite in the aqueous ambient atmosphere, a MATLAB code was developed to compute the time of dissolution of portlandite for a supercell consisting of 119323 atoms and 26011 sites. To execute the MATLAB code, initially was needed to compute the dissolution rate constant of Ca for seven various scenarios depending on the existing neighbors for the surface orientation of 001 or $00\bar{1}$; moreover, three different scenarios for the surface orientations of 100 or $\bar{1}00$, and 010 or $0\bar{1}0$ as shown in Figures 1 and 2, respectively. The reason of choosing different scenarios (also called atomistic events) for computation of each atomistic Ca dissolution rate is due to the effects of the neighbors on the computation of activation energy barrier during the dissolution of a particular Ca. In this way, reaction energy barrier (activation energy) of Ca for each scenario has been calculated in order to compute the dissolution rate constant according to the equation reported by Martin (Martin et al. 2021).

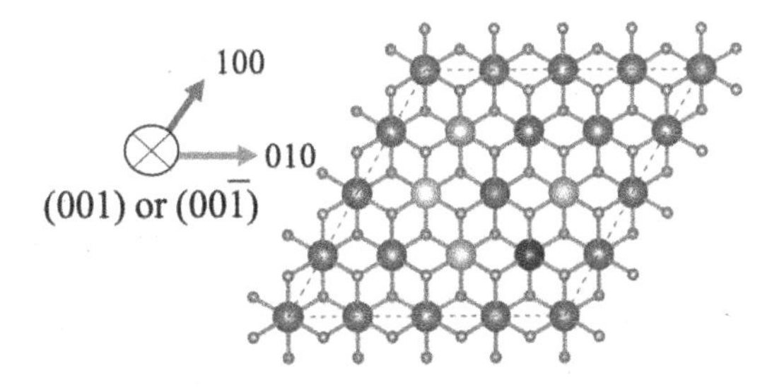

Figure 1. Illustration of seven different scenarios for red Ca dissolution depending on the existing neighbors on the 001 or $00\bar{1}$ surface orientations of portlandite employing KMC upscaling approach.

3 RESULTS AND DISCUSSIONS

According to the reported energy barrier (ΔG) by Uddin et al. regarding Ca dissolution for the three different surface orientations, we initially computed the rate constants for all possible scenarios on the three different surface orientations of 001 or $00\bar{1}$, 100 or $\bar{1}00$, and 010 or $0\bar{1}0$.

Regarding Figure 1, (001 or $00\bar{1}$ surface orientations), the rate constant of the red Ca dissolution for seven different scenarios depending on the existing all 6 neighbors, and 1, 2, 3, 4, 5 ,6 missed neighbors were computed to be $1.2432*10^{-49}$, $7.8493*10^{-23}$, $1.1193*10^{-18}$, $0.8077*10^{3}$, $1.0290*10^{3}$, $1.4797*10^{3}$, and $1.7918*10^{8}s^{-1}$, respectively. The rate constant of the red Ca dissolution for three different scenarios as shown in Figure 2 (100 or $\bar{1}00$ surface orientations) depending on the present all 2 neighbors, and 1, 2, missed neighbors were computed to be $3.6382*10^{-22}$, $5.0813*10^{-8}$, and $3.3377\ s^{-1}$, respectively. Moreover, the rate constants of $3.5659*10^{7}$, $1.5525*10^{9}$, and $3.5362*10^{11}\ s^{-1}$ were computed for three different scenarios of red Ca dissolution on the 010 or $0\bar{1}0$ surface orientations as illustrated in Figure 2.

After computation of the rate constants for different scenarios of all surface orientations, we implemented our MATLAB code to compute the time of dissolution for the system consisting of 26011 sites as shown in Figure 3 from two different perspectives. The following process briefly explains how to implement KMC algorithm for dissolution time computation of each site. Each time iteration, it is initially needed to update all surface sites of crystal after dissolution of each site to compute the total rate constant (k_{tot}). Tabulate probability of each event between 0 to 1 (13 possible events for all different scenarios) is computed by normalizing the rate of each event, which is multiplied by the number of sites and then dividing to

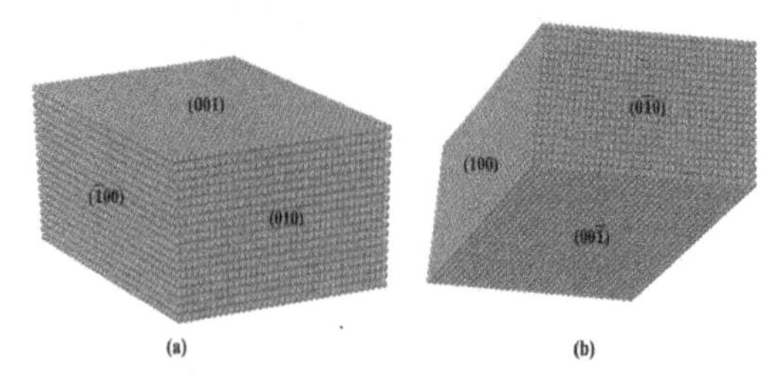

Figure 3. Snapshots of the initial simplified portlandite crystal system consisting of 26011 sites from two various perspectives.

the k_{tot}. A random number between 0 to 1 is then generated to select the probability of occurring event; and consequently, random selection of the site from that event. Finally, the time of selected site for dissolution is then computed by division of another random number between 0 to 1 with k_{tot}.

Figure 4 shows site-by-site dissolution model along 6000 (a,d), 18000 (b,e), and 23000 (c,f) steps. Each step is representative of one site dissolution. In fact, the dissolution process of the crystal is performed for scenarios of 010 or $0\bar{1}0$ surface orientations, the common sides with 010 or $0\bar{1}0$, and the common edges with 001 or $00\bar{1}$ due to the greater value for event probability according to the greater computed rate constants.

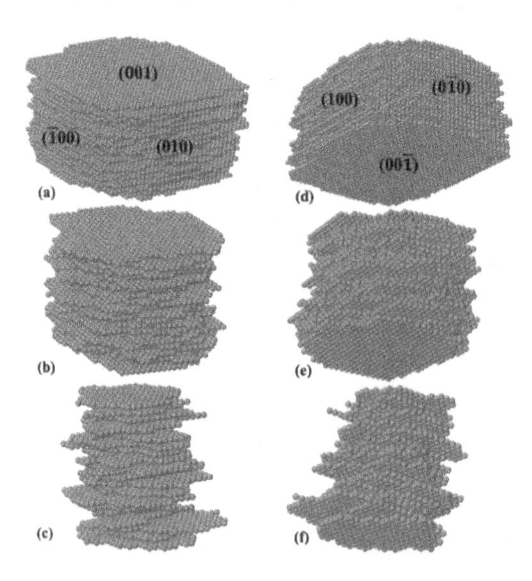

Figure 4. Process of site-by-site dissolution model of 26011 sites for 6000 (a,d), 18000 (b,e), and 23000 (c,f) dissolution steps. Each step is representative of one site dissolution.

In other words, all different events for 100 or $\bar{1}00$, and 001 or $00\bar{1}$ have very small chance for event probability to be selected by the initial random number due to their small rate constants. Figure 5 has also been plotted to illustrate the time evolution of sites dissolution after 6000 (a), 18000 (b), and 23000 (c) steps. The slope of the line stays constant up to almost 23000 sites dissolution. This is due to the almost identical total

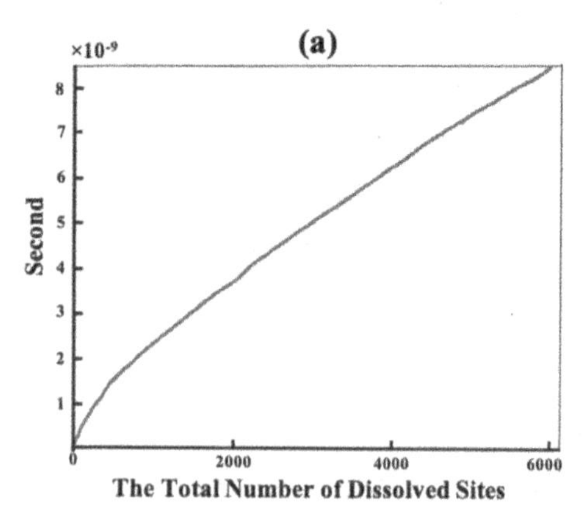

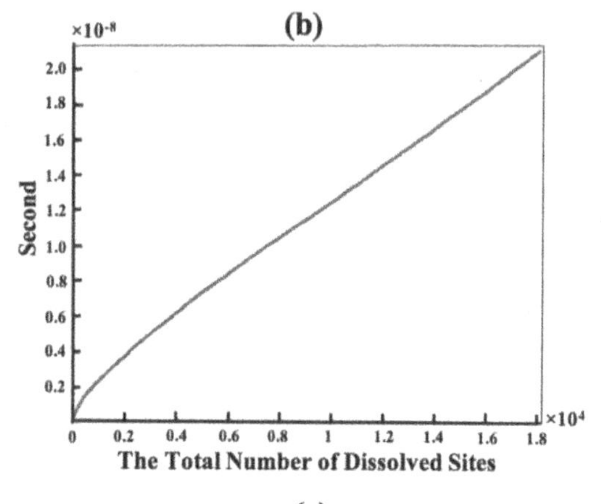

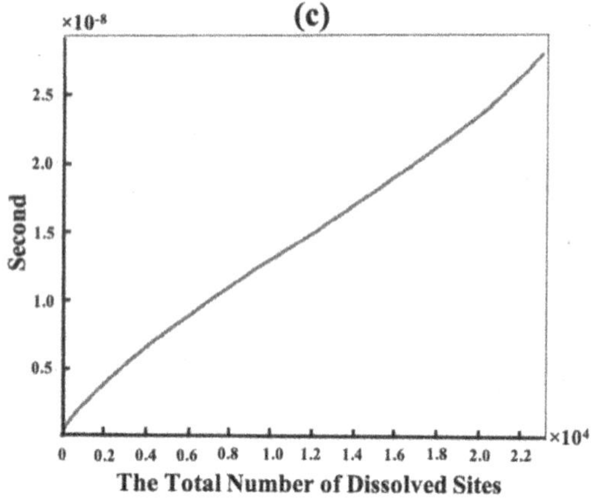

Figure 5. Time evolution of sites dissolution for the crystal consisting of 26011 sites after 6000 (a), 18000 (b), and 23000 (c) dissolution steps. Each step is representative of one site dissolution.

rate constant for computation of dissolution time. On the one hand in Figure 6, we showed the dissolution time of each individual sites along 6000 (a), 18000 (b), and 23000 (c) steps. It is clear that the majority of sites have been dissolved between 10^{-11} to 10^{-14} seconds for 23000 steps. Those sites which have been dissolved for the time less than 10^{-14}second is concerning to the larger second random number selection

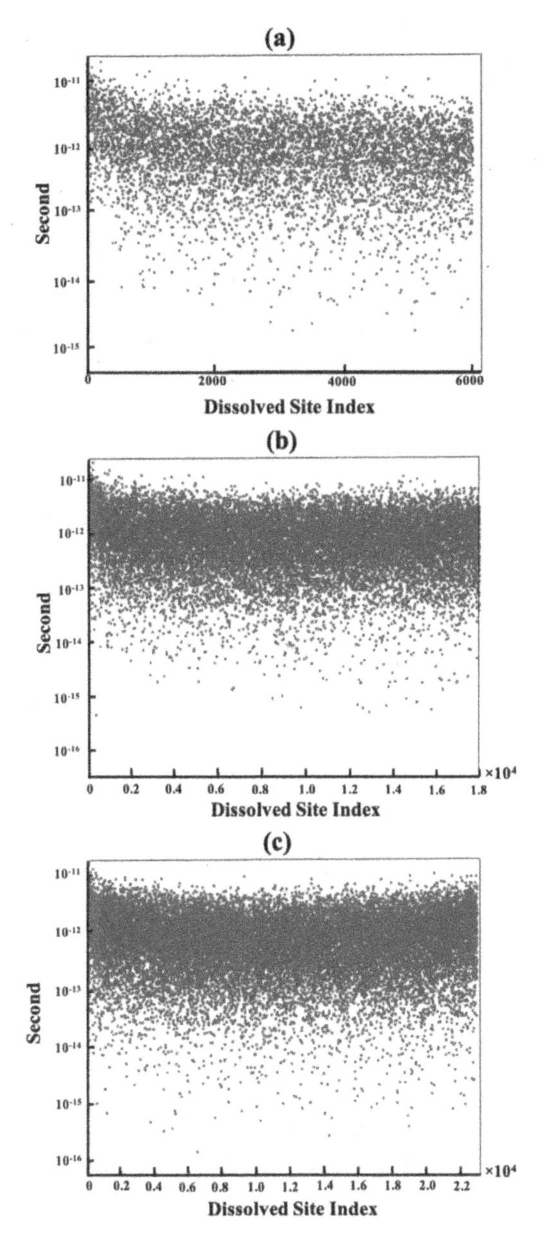

Figure 6. Individual Point represents the dissolution time for each site for 6000 (a), 18000 (b), and 23000 (c) steps. Each step is representative of one site dissolution.

close to 1 for dissolution time computation. However, the trend of random number selection is uniform and it is impossible to avoid those larger number selections close to 1. Finally, from the total number of dissolved sites for the crystal as shown in Figure 7 (a), can be observed that the dissolutions time of remaining sites between 23000 to 26011 increase resulting a decreasing trend in the average dissolution rate due to the smaller k_{tot} since the number of updated surface sites declined dramatically. The dissolutions time of all individual site of crystal have also been shown in

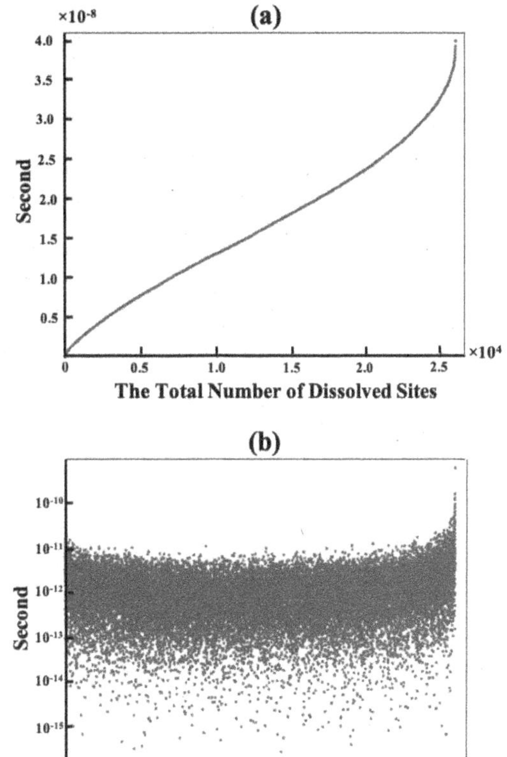

Figure 7. (a) shows the total time evolution of all dissolved sites of crystal. (b) illustrates the dissolution time of each individual site for the whole crystal.

Figure 7(b), and those between 10^{-10} to 10^{-11} second belong to the remaining sites between 23000 to 2601. From the slope of the curve in Figure 7a, the steady-state dissolution rate for the most reactive facets (010 or $0\bar{1}0$) was computed to be 0.706 mol/ (s cm^2) by considering the Avogadro's constant and that the initial facet area of the most reactive (010 and $0\bar{1}0$) facets is in total 226.319 nm^2 (= 8.768 nm × 12.906 nm × 2 facets). Obtained steady-state dissolution rates are in decreasing order for the rest of less reactive surface orientations as follows: 1.548×10^{-7} mol/ (s cm^2) for 001 or $00\bar{1}$, and 1.58×10^{-17} mol/ (s cm^2) for 100 or $\bar{1}00$.

4 CONCLUSIONS

MATLAB code was developed to employ a far-from-equilibrium kinetic Monte Carlo (KMC) upscaling approach in order to investigate the dissolution time of portlandite crystal consisting of 119323 atoms and 26011 sites. To perform the KMC approach, Uddin et al. in parallel study provided necessary input information about the reaction energy barriers (E_D) of Ca dissolution. This allowed to compute KMC for upscaling of the atomistic rate constants of the different scenarios into mesoscale rate and enabled to visualize the evolution of crystal morphologies during the dissolution process. In fact, seven different atomistic scenarios for Ca dissolution were considered depending on the existing neighbors for surface orientation of 001 or $00\bar{1}$; moreover, three different scenarios for surface orientations of 100 or $\bar{1}00$, and 010 or $0\bar{1}0$. The results showed that 001 or $00\bar{1}$, and 100 or $\bar{1}00$ surface orientations represented very small dissolution rate constant, which allowed scenarios of 010 or $0\bar{1}0$ surface orientations, adjoining edge with 001 or $00\bar{1}$, and adjoining side with 100 or $\bar{1}00$ to be dissolved. Moreover, the dissolution time of sites follow almost a liner trend up to 23000, and then the upscaled time of site dissolution increases due to the reduction of the surface sites for the computation of the total rate constant resulting a decreasing trend in the average dissolution rate. The steady-state dissolution rate for the most reactive facets (010 or $0\bar{1}0$) was also reported to be 0.706 mol/ (s cm^2), followed by less reactive surface orientations: 1.548×10^{-7} mol/ (s cm^2) for 001 or $00\bar{1}$, and 1.58×10^{-17} mol/ (s cm^2) for 100 or $\bar{1}00$.

ACKNOWLEDGMENTS

The financial support by German Research Foundation (DFG) is gratefully acknowledged.

REFERENCES

Boualleg, S., Bencheikh, M., Belagraa, L., Daoudi, A., & Chikouche, M. A. (2017). The Combined Effect of the Initial Cure and the Type of Cement on the Natural Carbonation, the Portlandite Content, and Nonevaporable Water in Blended Cement. Advances in Materials Science and Engineering, 2017, 5634713. https://doi.org/10.1155/2017/5634713

Czechowski, F., & Marcinkowski, T. (2006). Primary sludge stabilisation with calcium hydroxide. Environmental Chemistry Letters, 4(1), 11–14. https://doi.org/10.1007/s10311-005-0012-3

Drugă, B., Ukrainczyk, N., Weise, K., Koenders, E., & Lackner, S. (2018). Interaction between wastewater microorganisms and geopolymer or cementitious materials: Biofilm characterization and deterioration characteristics of mortars. International Biodeterioration & Biodegradation, 134, 58–67. https://doi.org/https://doi.org/10.1016/j.ibiod.2018.08.005

Galmarini, S., Aimable, A., Ruffray, N., & Bowen, P. (2011). Changes in portlandite morphology with solvent composition: Atomistic simulations and experiment. Cement and Concrete Research, 41(12), 1330–1338. https://doi.org/https://doi.org/10.1016/j.cemconres.2011.04.009

Glass, G. K., Reddy, B., & Buenfeld, N. R. (2000). Corrosion inhibition in concrete arising from its acid neutralisation capacity. Corrosion Science, 42(9), 1587–1598. https://doi.org/https://doi.org/10.1016/S0010-938X(00)00008-1

Gu, W., Bousfield, D. W., & Tripp, C. P. (2006). Formation of calcium carbonate particles by direct contact of Ca(OH)2 powders with supercritical CO2. J. Mater. Chem., 16(32), 3312–3317. https://doi.org/10.1039/B607184H

Ma, X., Ye, J., Jiang, L., Sheng, L., Liu, J., Li, Y.-Y., & Xu, Z. P. (2019). Alkaline fermentation of waste activated sludge with calcium hydroxide to improve short-chain fatty acids production and extraction efficiency via layered double hydroxides. Bioresource Technology, 279, 117–123. https://doi.org/10.1016/j.biortech.2019.01.128

Martin, P., Gaitero, J. J., Dolado, J. S., & Manzano, H. (2021). New Kinetic Monte Carlo Model to Study the Dissolution of Quartz. ACS Earth and Space Chemistry, 5(3), 516–524. https://doi.org/10.1021/acsearthspacechem.0c00303

Miller, S. A., Horvath, A., & Monteiro, P. J. M. (2018). Impacts of booming concrete production on water resources worldwide. Nature Sustainability, 1(1), 69–76. https://doi.org/10.1038/s41893-017-0009-5

Perko, J., Ukrainczyk, N., Šavija, B., Phung, Q. T., & Koenders, E. A. B. (2020). Influence of Micro-Pore Connectivity and Micro-Fractures on Calcium Leaching of Cement Pastes-A Coupled Simulation Approach. Materials (Basel, Switzerland), 13(12), 2697. https://doi.org/10.3390/ma13122697

Regnault, O., Lagneau, V., Catalette, H., & Schneider, H. (2005). Experimental study of pure mineral phases/supercritical CO2 reactivity. Implications for geological CO2 sequestration.

Sato, T., Beaudoin, J. J., Ramachandran, V. S., Mitchell, L. D., & Tumidajski, P. J. (2007). Thermal decomposition of nanoparticulate Ca(OH)2-anomalous effects. Advances in Cement Research, 19(1), 1–7. https://doi.org/10.1680/adcr.2007.19.1.1

Sun, Z., Chi, H., & Fan, L.-S. (2012). Physical and Chemical Mechanism for Increased Surface Area and Pore Volume of CaO in Water Hydration. Industrial & Engineering Chemistry Research, 51(33), 10793–10799. https://doi.org/10.1021/ie300596x

Ukrainczyk, N., Muthu, M., Vogt, O., & Koenders, E. (2019). Geopolymer, Calcium Aluminate, and Portland Cement-Based Mortars: Comparing Degradation Using Acetic Acid. Materials, 12(19). https://doi.org/10.3390/ma12193115

Wang, X.-Y. (2014). Effect of fly ash on properties evolution of cement based materials. Construction and Building Materials, 69, 32–40. https://doi.org/https://doi.org/10.1016/j.conbuildmat.2014.07.029

Worrell, E., Price, L., Martin, N., Hendriks, C., & Meida, L. O. (2001). CARBON DIOXIDE EMISSIONS FROM THE GLOBAL CEMENT INDUSTRY. Annual Review of Energy and the Environment, 26(1), 303–329. https://doi.org/10.1146/annurev.energy.26.1.303

Yu, Z., & Ye, G. (2013). New perspective of service life prediction of fly ash concrete. Construction and Building Materials, 48, 764–771. https://doi.org/https://doi.org/10.1016/j.conbuildmat.2013.07.035

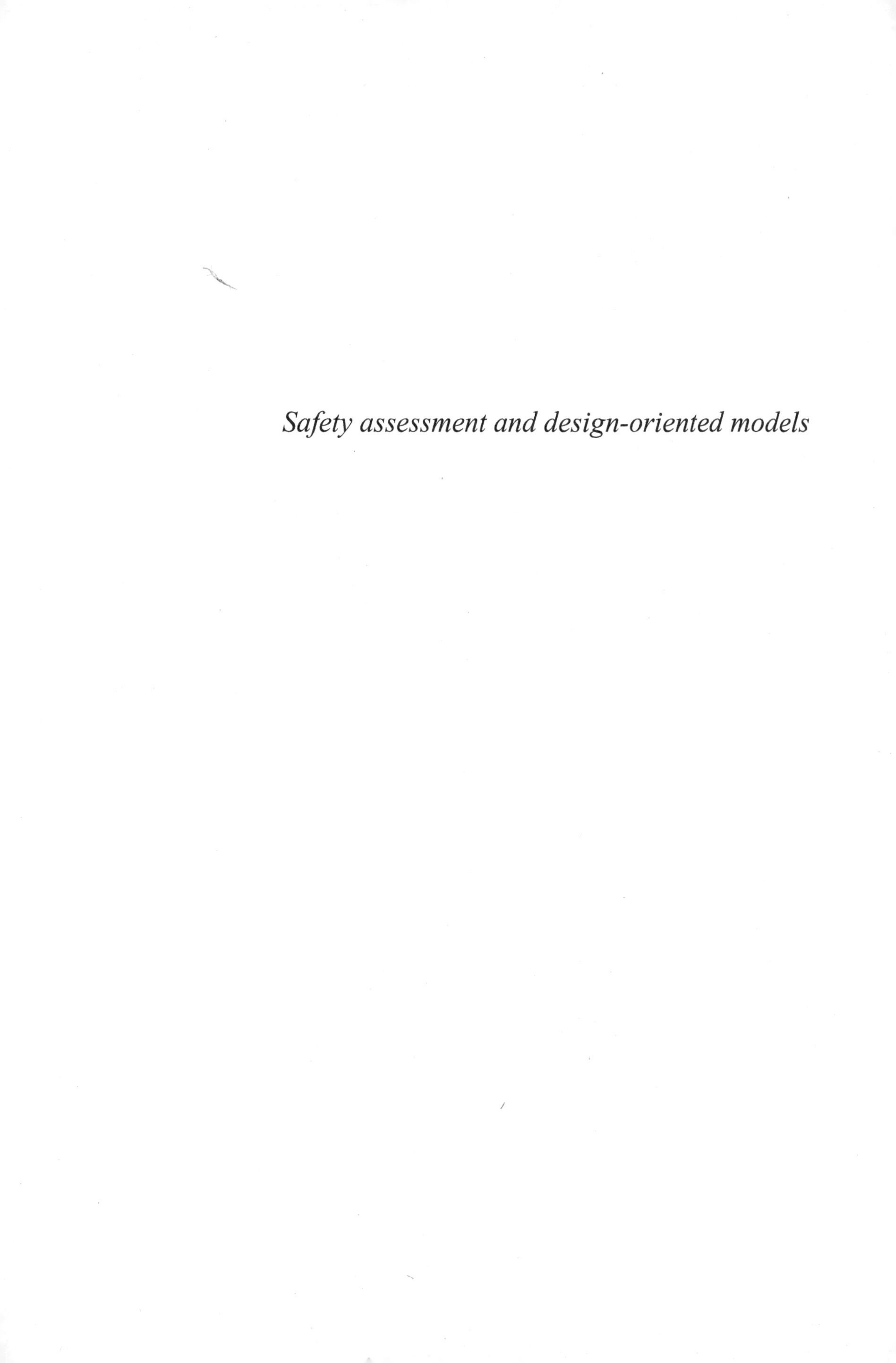

Safety assessment and design-oriented models

Computational Modelling of Concrete and Concrete Structures – Meschke, Pichler & Rots (Eds)

Retrofitting of existing structures by advanced analysis

S. Van Hout
Sweco Belgium bv, Berchem, Belgium
Department of Civil Engineering, KU Leuven, campus De Nayer, Sint-Katelijne Waver, Belgium

T. Molkens
Department of Civil Engineering, KU Leuven, campus De Nayer, Sint-Katelijne Waver, Belgium

M. Classen
Institute of Structural Concrete, RWTH University, Aachen, Germany

E. Verstrynge
Department of Civil Engineering, KU Leuven, Leuven, Belgium

ABSTRACT: To meet the climate objectives, there is a clear need to deal differently with our building patrimony. In the future, building structures must be re-used as much as possible, if necessary, in combination with changes in their intended function. In this contribution, an overview is given of mostly known but often not yet fully exploited techniques that can be applied for reliability assessment of existing structures. Obviously, the starting point is the load history of the building, although this is not a conclusive verification. An often-underestimated aspect here is the modelling of the load arrangements. Despite that this is the starting point of further (advanced) analyses, disproportionally little research attention has been given to this to date. Furthermore, advanced structural analysis methods as well as testing and monitoring of structures are also discussed, followed by some sub-aspects. To conclude, three practical examples are given, underlining the importance of this work.

1 INTRODUCTION

1.1 *Global challenges of the construction sector*

The impact of the construction sector on climate change can be called massive; about 8% of the world-wide greenhouse gasses is dedicated to the production of cement, again 8% to the production of steel. The share is even greater when looking to the whole live-time consumption for buildings: 40% of the green-house gasses, 33% of the water consumption, 35% of the waste generation, 50% of the raw (primary) materials and 50% of the energy consumption (Level(s) 2019; #EUGreendeal 2020).

Before discussing the use of "green" cements or recycling of materials, the first step should be to assess whether the re-use of a structure is possible or not, facing the direct impact on the items listed above. To achieve the intended goals of the climate objectives (UN 2015; COP26 2021), the re-use of structures with admittedly different functions will become more common or even compulsory. It is even stated that greater material efficiency can save up to 80% of the emissions (#EUGreendeal 2020).

Besides the needed change towards more sustainable and responsible use of existing building structures, a market change is also starting to develop. The origin of this market evolution can be found in the more stringent urban requirements for the design of new buildings. Existing building permits allow for denser land use and sometimes also a higher number of storeys. The authors performed a small survey by several real estate developers active in city centres on the Belgian market. Based on historical numbers and their prediction for the near future, the reinstatement market of existing structures, Figure 1 (solid lines), will be as important as the development of new buildings (dashed lines). Ratios are given for a mix of dwellings

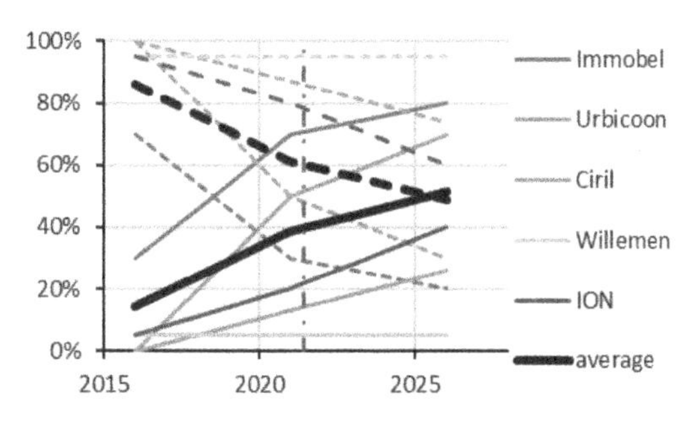

Figure 1. Market evolution following real estate developers active on the city market in Belgium.

DOI 10.1201/9781003316404-79

and offices, and are based on the balance weighted average, around 2025.

1.2 *Assessment of existing structures*

Building owners and authorities should be aware that the outcome of a cost-benefit analysis of such more time-consuming engineering work is totally different for the assessment of existing buildings compared to the design of new buildings. A more thorough analysis can lead to an upgrade of the calculated load-bearing capacity depending on the structural system. For the design of new buildings, the study cost will be governed by the market and the floor area (as a first approximation), so an easy comparison of offers is possible. The opposite is true for assessing an existing building; technical means and a high skill environment can lead to significant savings or influence even the feasibility of a project.

More advanced calculation models are then used to define more realistic load arrangements and structural response. The latter is based on the classical bending theory and membrane effects, either in compression or tension. In this case, the assessment is no longer made at the element level but at the system level since the system dictates the boundary conditions for making additional or alternative load-bearing paths possible. Typically, the use of non-linear finite element analysis (NLFEA) is performed, including geometrical and material non-linear behaviour

The principles of such advanced load and structural models are well known and often applied in accidental situations such as fire or robustness checks. However, its full potential for ordinary renovation and re-use design situations has not yet been fully exploited so far. Research to date has focused on numerical simulations of simple idealised models (Genikomsou & Polak 2017; Thoma & Malisia 2018; Kang et al. 2020) or the consequences of a column loss scenario in the context of a robustness check (Botte et al. 2014; Belletti et al. 2016; Xiao & Hedegaard 2018).

1.3 *Design codes*

While for the design of new buildings, codes are available with several requirements for resistance, serviceability, durability and fire, this type of information is mostly lacking for existing buildings.

At a European level, a technical specification is recently released (TS17440, 2020) dealing with assessing and retrofitting procedures in a general way. It is worth mentioning the work of the fédération international du béton (fib bulletin 80, 2016), which facilitates insights into the composition of the partial safety factors and the desired reliability level for existing concrete structures. In the United States, interesting work is published by the American Concrete Institute (ACI), which describes some aspects more in detail. The (ACI364.1, 2007) permits the use of secondary elements as structural elements in existing concrete structures but requires, on the other hand, that the rules for new buildings should be followed. Furthermore, (ACI562.13, 2016) states in section 6.5, *analysis shall consider the load path from the load applied through the structure to the foundation. Three-dimensional distribution of loads and forces in the complete structural system shall be considered unless a two-dimensional analysis represents the part of the structure*". The guidelines also stipulate that time-dependent effects should be considered to account for material degradation and the load history.

From these general statements, it follows that, as stated before, more advanced analyses can and should be performed at the level of load arrangements and structural models. Methodologies to execute such analysis are surprisingly still missing and are therefore the focus of this contribution. Nevertheless, the problem of model uncertainties of advanced structural analysis shows scientific interest with several publications (Gino et al. 2021; Engen et al. 2021; Castaldo et al. 2018).

1.4 *Assessment procedure*

At the start of the assessment of an existing building a first verification can be done by checking the load history of the building. A warehouse for example, could have been heavily loaded in the past, more than what can be derived from the available reinforcement. In this contribution, an assessment methodology for existing buildings is proposed, as shown in Figure 2; the full detailed explanation of each step is elaborated in the following sections.

The load arrangement is determined based on linear elastic models, using 3D finite element analysis (3D-FEA). The reason for the elastic approach can be found in the validity of the superposition principle. It allows a straightforward combination of load cases. Next, the

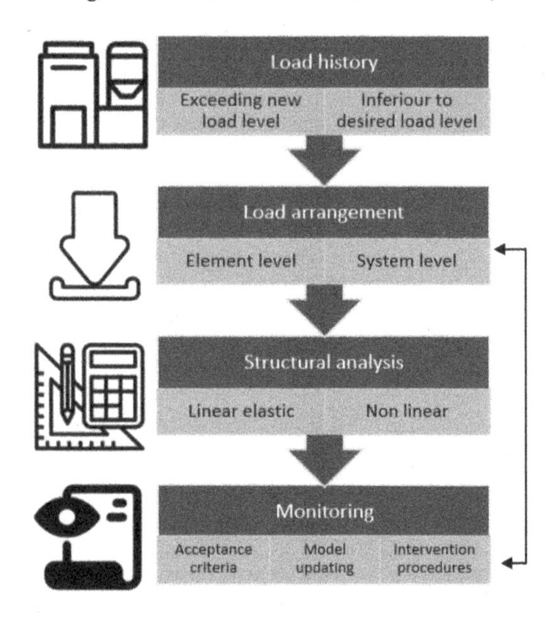

Figure 2. Assessment procedures for existing buildings.

decisive load combination is imposed on a numerical model that can handle nonlinearities in the material and structural behaviour (NLFEA). For example, shell elements are used here instead of plate elements for the floors.

With this proposed weak-coupling strategy, an iteration is needed whether the deformation from the NLFEA model is compatible with the assumptions used to calculate the load arrangement (3D-EFEA). If necessary, several iterations should be executed. In the end, the conditions for a monitoring campaign shall be pointed out.

To conclude this paper, three case studies will be presented to validate the proposed methodology. The advanced theoretical analysis results are compared with two in-situ large-scale load tests on existing structures in Brussels and Paris. A transformation of an office and a parking garage to flats and offices at those locations became possible.

2 LOAD HISTORY

In the first respect, a quick estimation of the bearing capacity could be established by evaluating the historical load. If this historical load level exceeds the projected load level, there is a tendency, out of practice, to claim that the structure is safe. Especially when the foundation system of an existing building is to be assessed, there is often a lack of information, and no other option than using the load history for its capacity assessment is available.

Unfortunately, such a statement is not justified from the point of view of structural reliability. There is lacking knowledge about uncertainties and quantification of risks with such a practical approach.

On the other hand, also the reversal is true; if the historical load level was below the future desired level, it does not give a verdict on the actual bearing capacity. It is commonly recognised that only a large number of tests, which conflicts with practical issues in the case of real buildings, or calculation can justify the reliability level of a structure.

Nevertheless, the building can and will tell the observer how it reacted to previous loads. Such insights are crucial for determining the boundary conditions of the structural system, at least during service conditions assuming that no failure occurred in the past. In this way, this is the first important step.

3 LOAD ARRANGEMENT

3.1 *Manual methods – element level*

Manual methods, which are in practice till today frequently if not primarily used, start from the assumption that gravity loads just descend vertically without any redistribution depending on the axial stiffness of vertical elements and bending stiffness of horizontal ones. In other words, the stiffness of the surrounding structure is neglected; for that reason, it can be called an element level approach. Loads from slabs are only transferred in the shortest span direction or in a slightly more advanced way following the yield line theory, but again without any redistribution.

This historical way of working deviates from the natural load arrangement due to simplifications. However, this type of elementary models is used here to illustrate the possible benefits that can be realised and underline this work's research relevance. Assume a supporting beam with a wall on it and a maximum load p (kN/m), a span of L and, a bending stiffness EI wherein E represents the Young modulus and I the second moment of area. Internal shear forces, bending moment and deflections can be expressed as given by Eqs. (1) to (3).

$$V_{Max} = \alpha_V pL \tag{1}$$

$$M_{Max} = \alpha_M pL^2 \tag{2}$$

$$\delta_{Min} = \alpha_D \frac{pL^4}{EI} \tag{3}$$

Dimensionless parameters α_V, α_M, and α_D are used, respectively valid for shear, moments and displacements. Those factors can be found in Table 1 and depends on the load arrangements type going from a rectangular to a triangular shape. With relative α-factors next to the absolute values compared to the values obtained for a uniform load case. Separate columns are used for the forces while the relative α_D-factor can be found behind the bar (/).

Table 1. Internal force and deformation parameters for a simply supported beam.

Load	Internal forces		Relative forces		Deformation
arrangement	α_V	α_M	Shear	Moment	α_D/rel.
Uniform	0.50	0.125	1.00	1.00	0.013/1.00
Parabolic	0.33	0.104	0.67	0.83	0.011/0.81
Sinusoidal	0.32	0.101	0.64	0.81	0.010/0.79
Bloc (L/2)	0.25	0.094	0.50	0.75	0.009/0.71
Triangle	0.25	0.083	0.50	0.67	0.008/0.64

Table 1 shows that a significant reduction in internal forces can be obtained if some compression arch effect can be accounted for in the masonry wall above (parabolic load). This reduction becomes more important with a strut and tie model (triangular load). Savings up to 50% and 33% of the uniform load case are possible. Also, for the deformations, which will influence the serviceability, a saving of 36% is possible in the most favourable case.

For the design of new buildings, it is a common practice to keep the distribution of shear forces, moments, and displacements constant for service limit state (SLS) and ultimate limit state (ULS). The distinction between both limit states in analysis is only reflected by the applied safety factors. However, the action on the bearing element can be completely different depending on the stiffness ratio between the bearing and loading element. Presume a beam that will transfer from bending (in SLS) to catenary (in ULS) action. Table 1 shows that even a smooth transition of load arrangement can result in meaningful reductions of loads on bearing elements. So, differentiation should be made between models suitable for SLS and ULS assessment.

3.2 *System approach by 3D-FEA*

The effect of implementing a system approach instead of a calculation on element level will be pointed out using a case study. Project Haagbeuk was assigned to Sweco (engineering office) to assess the fire resistance of the transfer slab above a parking garage with apartments on top of it; see also section 7.3.

After a revision of the original design from another party, a serious mistake was discovered, which undermined the structural reliability of the building. The building was fully equipped and almost in service. However, no signs of damage or extensive deformations could be observed while the construction was already finished.

A major structural issue was pointed out in a transfer slab that was foreseen supported by columns to realise in the basement an open parking space: a flat slab of 40 cm thickness in concrete class C30/37 with mainly one-directional reinforcement. Figure 3 shows the bearing elements in the basement with a grey shade; columns, beams and concrete walls are indicated. The superstructure, made out of glued lime blocks, can be recognised; a hallway and square-shaped apartments. A reinforcement ratio of 3664 mm^2/m BE500 was available from W05 to W07, while perpendicular to this direction (transverse), only 524 mm^2/m was available. Upper reinforcement was also foreseen but not discussed. Note that the wide beam ($2\times0.6\,m^3$) does not add much stiffness and is eccentrically loaded. On top of that, the slab next to the beam is thinner, namely only 20 cm. Based on a linear elastic approach, it can be expected that the reinforcement ratios should be comparable in both directions.

The transfer slab was reassessed using a 3D-FEA in the software package Diamonds of Buildsoft to define the load arrangement in combination with software package SAFIR® (Franssen & Gernay 2017) for the structural analysis. The latter is a temperature-dependent, geometrical and material non-linear software. Simulations had been started up with this software due to the in origin fire-related questionary.

By using 3D-FEA, the differences in stiffness of surrounding elements could be taken into account. Figure 3 shows the differences in load arrangements applied on the transfer slab. At the top are the results of a manual calculation, and at the bottom, the results of the 3D FEA can be found. This analysis excluded the compression arch effect in the bearing walls on top of the transfer slab.

Firstly, a difference in the shape of the line load can be found. While the manual load arrangement considers a uniform line load, the 3D FEA indicates a triangular distribution in combination with concentrated loads. It should be considered in the comparison that a double value of a triangular load corresponds to the same total load as a uniformly distributed load.

The other relevant difference is the distribution of the loads. While the manual load arrangement results in high loads in the upper wall (W06, Figure 3) next to the hallway, the 3D FEA reduces this load. On the other hand, the load increases in the directly supported W07. The opposite is observed for some partition walls (W02_2, W02_3) between apartments. In the manual calculation, only the own weight of the walls was taken into account, while by the 3D-FEA, those walls are heavily loaded.

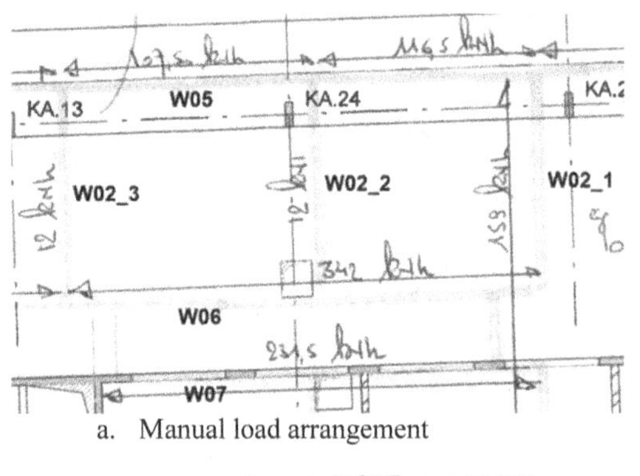

a. Manual load arrangement

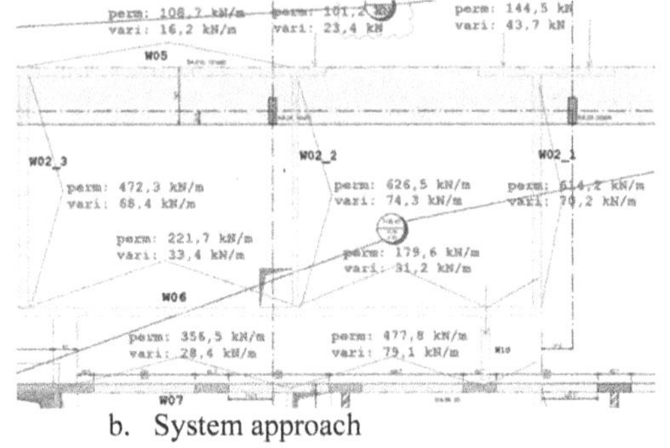

b. System approach

Figure 3. Comparison of loads using a manual calculation on an element level (a) and a system approach with a linear elastic FEA (b).

Numerical values have been set next to each other in Table 2, to facilitate a comparison of the results. Loads are guided towards the parts of the slab directly supported by bearing elements and in the direction of the reinforcement. This redistribution causes a decrease in bending moment in the transfer slab and, therefore, decreases the need for longitudinal and transverse reinforcement. Activation of the upper structure makes such favourable load flow possible.

Table 2. Numerical comparison of loads (service limit state) using a manual calculation on element level and a system approach with a linear elastic FEA for the same building.

Wall	R_{eq} [kN] Manual	3D-FEA	Relative FEA/Man
W05	107.5	124.9	1.17
W06	342.0	255.1/210.8	0.75/0.62
W07	231.5	384.9/556.9	2.41/1.66
W02_1	159	684.4	4.30
W02_2	12.0	700.8	58.4
W02_3	12.0	540.7	45.1

The system approaches using software package Diamonds does not explicitly consider non-linear behaviour. Effects of cracking ask for iterative simulations, and creep is introduced by reducing the Young modulus. In reality, additional compressive and tensile membrane action will lead to a more rigid slab. Therefore, in a next step, the line loads, derived from the linear elastic FEA, are implemented in the non-linear software to consider the previously mentioned membrane effects.

4 ADVANCED ANALYSIS METHODS

4.1 *Linear and non-linear effects*

Due to a lack of numerical tools, older existing buildings are designed mainly by tabulated data, based on an elastic structural response (i.e., the strip method starts from a Poisson coefficient = 0) and idealised supports. Nowadays, with FEA as a numerical tool, it is possible to analyse the exact shape of the slab, while this is not always the case using tabulated data. Time-dependent effects as creep, nonlinearities due to cracking and even orthotropy caused by a difference in reinforcement can be accounted for.

Table 3 makes the comparison between the maximum reaction forces and internal moments (pro unit width) of a slab with a short span of 5 m, a total load of 10 kN/m, a thickness of 0.20 m and a Young modulus of 33 GPa. An analysis is performed with the aid of a commercial software tool (Diamonds, by Buildsoft) that allows for linear elastic calculations with cracked sections, including creep effects based on triangular Kirchhoff elements.

It is seen that for linear elastic approaches, the influence of the method of analysis is minimal on the internal forces. Only one exception can be made if the slab is calculated using a crude single strip model without any interaction of the secondary direction. The reaction force would be in $10 \cdot 5/2 = 25$ kN/m and the moment $10 \cdot 5^2/8 = 31.25$ kNm/m. However, important differences can be found in the calculated deformations once cracking, and creep are accounted for. This phenomenon is widely known. Note that including the practical reinforcement, using a higher minimum (mesh) reinforcement only slightly decreases the calculated theoretical deformations.

Table 3. Comparison of the outcomes following different slab analysis methods for a simply supported slab ratio $L_1/L_2 = 2$.

Method	Assessment parameters R(kN/m)	M(kNm/m)	δ (mm)
Tabulated data*	18.75	23.58	−2.89
Linear elastic°	18.16	25.14	−2.75
Cracked and min. reinf.	18.21	24.53	−7.87
With practical reinforcement	18.20	24.45	−7.71

* Based on the strip method of Marcus (Girkmann 1986).
° No influence was found by the presence of min. reinf.

A much more pronounced difference can be found by adapting the boundary conditions. Assume that after a survey of the local situation, one long side of the previous slab reacts as fixed for rotational moments parallel along the long edge of the slab. A significant change in reaction forces, moments and deformation is observed, see Table 4. For the moments not only the positive values are given (with tension at the bottom side of the slab = span) but also the negative ones (with tension at the upper side = at the support).

Table 4. Comparison of the outcomes following different slab analysis methods for a simply supported slab (except one long side L_1 = fixed) ratio $L_1/L_2 = 2$.

Method	Assessment parameters R(kN/m)	M(kNm/m)	δ (mm)
Tabulated data*	25,85	15.2/−30,49	−1.38
Linear elastic°	25,68	16.7/−30,44	−1.37
Cracked and min. reinf.	25,39	16.9/−26,50	−4.23
With practical reinforcement	25,62	16.4/−30,43	−3.77

* Based on the strip method of Marcus.
° No influence was found by the presence of minimum reinf.

So far, only linear elastic results are shown, but in some way already accounting for creep and cracking. In the previous examples, the line support blocks any vertical and horizontal displacements. With the plate elements used, no normal stresses will be generated even with the calculated vertical deformations. Once shell elements are involved in the analysis, also membrane effects can develop, discussed in the following two sections. This requires that the software used can handle geometrical nonlinearities. Mostly these tools include also material nonlinearity like the software SAFIR® (Franssen & Gernay 2017), used in section 3.2.

4.2 *CMA*

When horizontal displacements and rotations are (partially) blocked, two types of successive membrane effects can develop, see Figure 4, called compressive (CMA) and tensile membrane action (TMA). CMA can develop as long as the section height allows for the development of moderate compression stresses without entering too far in the plastic region (to limit deformations). Accounting for this, also called arch effect, has the advantage that deformations stay limited, and deformations are almost reversible, making it suitable to combine with service limit states.

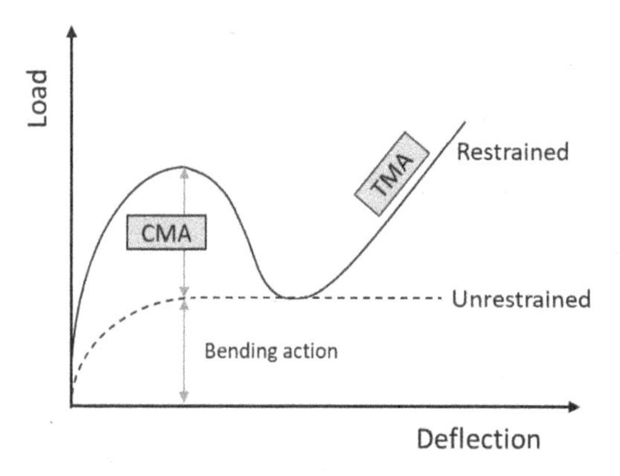

Figure 4. Successive development of CMA and TMA, adopted from (Botte et al. 2014).

As shown in Figure 4, CMA is in some way an unstable situation; a slightly higher deformation will end in lower load-bearing capacity, followed by even higher deformations and the development of tensile membrane action.

4.3 *TMA*

While for CMA, the presence of reinforcement is of lesser importance (except to limit compression stresses), TMA depends mainly on the reinforcement, its yield strength and deformation properties. It is necessary to investigate the reinforcement's properties (strength and ductility) when using this type of action. Reflecting on the needed substantial deformations, this type of action can only validate an ultimate limit state and/or accidental situation.

While the difference in bending moments are minor (although membrane forces are developing), a significant difference can be found in reaction forces and the calculated deformations. The decrease in deformations is due to the membrane effect, introducing CMA into the slab. The maximum reaction forces are in line

Table 5. Comparison of NLFEA (including CMA) results for a simply supported slab and one long edge fixed, ratio $L_1/L_2 = 2$.

	Assessment parameters		
Method	R(kN/m)	M(kNm/m)	δ (mm)
SAFIR® – simply supported	25.16	−24.90	−3.57
SAFIR® – one long edge fixed	31.13	−16.7/25.77	−1.74

with the values obtained using a simple single strip model. As vertical equilibrium must be respected, the increase in maximum values can only be compensated by another shape of the load arrangement.

5 MONITORING

As mentioned before in Section 2, understanding the system behaviour of an existing structure is extremely important to assess the reliability of the structure for its different limit states. Unfortunately, not all design assumptions and execution details are available, which makes a possible conflict of unknowns.

Monitoring of the structure can fill, or at least reduce, this gap of unknowns and deliver the needed proof of evidence.

5.1 *Acceptance criteria*

Acceptance criteria can be formulated at two levels. In a preliminary phase, before assessing the structure and after finishing refurbishment works.

Load tests can be executed on the existing structure until service and ultimate load level to understand the structure and its system behaviour better. A comparison of measured and calculated deformations will confirm the model assumptions or will create the need for an update of the model. This way of working is supported by the (TS17440, 2020); which allows testing and monitoring of existing structures to verify and improve structural analysis assumptions. In Figure 6 the configuration of a swimming pool test can be seen to validate CMA in an existing 16 cm thin, 4.58 m span concrete slab.

After the reinstatement of an existing building in its new configuration, it can be useful to continue or set up a monitoring campaign to verify if the time-dependent behaviour is consistent with what was predicted by numerical models. In general, an asymptote should be reached, which makes this type of measurement more frequently repeated in the beginning and will end after some time (can be years).

5.2 *Model updating*

According to (TS17440, 2020) the measurement of static or dynamic properties should be compared with those predicted by structural analysis models based on the actual conditions of the existing structure. When a significant deviation from the prediction is observed, the reasons should be investigated and explained, requiring additional tests or model updating. The latter can be expressed in modifications of the load arrangement, boundary conditions, type of analysis or the reliability level. Threshold values should be agreed upon for a specific project by the relevant parties to start up interventions.

5.3 *Intervention procedures*

To limit risks or achieve an adequate reliability level, it should be considered to prepare some intervention procedures. When passing some threshold values, a scenario should be available about possible maintenance, replacing, repair, or strengthening works.

Typical threshold values are based on measurements of deformations at well positioned and indicated survey points. On the other hand, dynamic measuring techniques have also been developed to monitor the state of health of a structure. It is always the purpose to verify if the exposed existing construction is reacting as assumed in the assessment documents. When a specified deviation from the prediction is observed, the reason should be investigated and explained.

In a post-intervention file belonging to the building, it can be determined i) which possible damage should be considered as normal, ii) which should lead to a higher inspection frequency and/or additional investigations, iii) from which limit strengthening works are necessary and iv) in the worst case from which extreme limit even evacuation should be proceeded with. Sometimes it is stated that a relative rotation of 1/150 would cause an ultimate limit state.

6 ADDITIONAL ASPECTS

6.1 *Reliability level*

For buildings categorised in standard Consequence Class 2 (EN 1990, 2015), the reliability index β corresponds to a probability of failure of $7.23 \cdot 10^{-5}$ and equals 3.8 (Gulvanessian et al. 2012). It has previously been shown by some researchers (Sykora et al. 2017), (Caspeele et al. 2013) that there should be a distinction between new, existing and even temporary structures regarding their reliability indexes.

The reliability index depends on the structural safety level as well as cost optimizations. As the cost for increasing "beta" is generally larger in case of existing structures compared to an adaptation of a still to build design, "beta" may be reduced as a result of the effect of cost implications on the optimization (fib bulletin 80, 2016).

Following ISO 2394 (ISO 2394, 2015) it is proposed to limit β to a lower limit of 3.3 for an existing building in the post-fire condition, for which societal and human risks are still satisfactory (Sykora et al. 2017). Another reason to adopt this reliability factor is the reduced projected lifetime (Holicky et al. 2013).

6.2 *Model uncertainties*

Model uncertainties arise at the load level (γ_{Sd}) and resistance (γ_{Rd}) side. The *fib* recommendation (Modelcode 2010, 2012) allows a reduced value of 1 instead of 1.06 when evidence of the model validation of the design conditions is available. However, it was proven that this γ_{Rd}-factor for NLFEA should be slightly higher than for more conventional methods (1.15 instead of 1.06) (Castaldo et al. 2018).

In line with these findings, upcoming documents such as Modelcode 2020 (Engen et al. 2021) and the new version of Eurocode 2 (in annex F) show the tendency to increase the safety factor for the model uncertainty if it was not derived from probabilistic calibrations.

6.3 *Fire*

One of the most challenging and rare events for a building is a fire. Seldom, a full compartment fire can develop, and mostly the area affected by a fire is relatively limited. Assessment procedures should be adapted when the re-use of a building after a well-documented or forgotten fire will be considered. Different formats make such assessment possible, reference is made to literature handling concrete (Van Coile 2015; Molkens 2022) and steel buildings (Molkens et al. 2021; Molkens et al. 2021).

During a fire, it is already observed by the authors (Molkens et al. 2017) that a slab working in CMA may switch to cantilever action due to a loss of compression resistance at the lower heat-affected part of the slab, see Figure 5 which is taken from the example discussed in section 7.1.

Figure 5. CMA in ULS conditions and cantilever in fire conditions. Principal tensile stresses = red, compressive = blue.

7 CASE STUDIES

Validation of solution strategies and economic relevance should be based on several benchmark analyses relevant for the type of retrofitting works at hand. So far, no codified minimum number of benchmark studies is available. Referring to (Engen et al. 2021), a minimum of 2 to 3 should be available.

7.1 *Brussels, Leopold II-building*

In the '70s there was a great need for offices buildings in Brussels in the area between the city centre and the national airport. With the known evolution of HVAC techniques and other office equipment, the Leopold II building was outdated and needed a serious upgrade. However, the relatively low ceiling height was seen as an obstacle, and on the market, there was a more important (and beneficial) demand for dwellings instead of offices.

Questions arose about the bearing capacity of the slabs in ambient under fire conditions. The building was constructed before the actual fire regulation, but the real estate developer wanted to meet the latest requirements. By means of the thermo-plastic FEA software SAFIR® developed at the University of Liège (Franssen & Gernay 2017), mechanical response in service, ultimate, and fire limit state has been assessed. For model validation, load tests (up to 80 cm of water) have been performed with the aid of swimming pools, see Figure 6.

7.2 *Paris*

In the early years when cars entered the city centres, parking buildings were constructed that offered a parking place and other services. Maintenance and cleaning works were offered so that the driver was assured about a safe journey back homewards out the city. From our present point of view, this is relatively strange, but those buildings are still present in city centres nowadays.

The location of the envisaged building in Paris is uniquely located close to the city centre and public transport but hidden in a densely populated area. Due to the actual urban regulations, it would never be allowed to rebuild the building at the same spot and with the same number of levels or building height. This is the reason why the real estate developer opted for retrofitting, while the passage for fire brigade should be increased, the free room between the columns should be respected and the fire resistance guaranteed.

A first assessment claimed that these goals could not be met based on simple design models and current codes. However, the beams have been fixed (without rotational freedom) to the columns, the slab was reinforced in a continuous way, and the second bearing direction was fully disregarded. With the aid of a simple LE-FEA it could be shown that the structure met all targets. This model was validated by a load test (executed up to 60 cm water load), see Figure 7. In the end, NLFEA has been applied to assess fire resistance.

7.3 *Haagbeuk*

An apartment building in Belgium, Figure 8, was designed with a structural system of load bearing masonry starting from the ground floor, as discussed in section 3.2. Remember a design error was found while the building was almost in service.

Figure 6. Building before, during load test and after rehabilitation.

Figure 7. Load test existing parking building.

Extensive strengthening seemed to be needed. To reduce the influence on the free height in the basement and avoid conflicts with the technical equipment suspended on the ceiling, this would be executed with CFRP (carbon fibre reinforced polymer) laminates. Unfortunately, the glues applied in these strengthening systems are very sensitive to high temperatures and should be protected for a critical temperature of around 75°C. Benefits in height realised with those laminates would be lost in combination with a 10 cm thick fire protection, where only 15 mm space was available.

As previously discussed (section 3.2), many efforts have been made from the engineering side to define a

Figure 8. Views on the transfer slab with 10 mm of fire protection added.

more realistic load arrangement on the transfer slab. After analysis of the transfer slab by NLFEA in ambient conditions, it was found that by doing so, the reinforcement was capable of guaranteeing the system's structural reliability. Redistribution of internal forces and activation of membrane forces made it possible.

However, in the case of fire, the loss of compressive strength at the lower part of the slab would be disastrous. Finally, it was decided that only a 10 mm thick fire protection was needed and no strengthening at all (Van Hout & Molkens 2021).

8 CONCLUSIONS

There is a clear need for a more uniform methodology for assessing existing buildings. A consistent proposal was worked out in this contribution.

Besides historical load data, most of the attention is given to load arrangements. Based on simple models, savings up to 50% can be obtained by proper load arrangements. This was illustrated by a case study of a real-life application.

Advanced structural analysis has been subsequently discussed, and it is found that including membrane effects can significantly impact the calculated structure's bearing capacity. NLFEA opens many possibilities but also creates a need for knowledge about the system behaviour, which can be validated by monitoring and load tests.

Additional aspects have been briefly discussed, such as the reliability level of existing structures, model uncertainties, and fire assessment.

Three benchmark case studies have been presented, which also underlined the possible unexpected outcomes in view of fire protection to guarantee the compressive resistance, for example, when counting for membrane effects.

9 FUTURE RESEARCH

In this publication, the importance of a correct load arrangement was underlined. Unfortunately, accurate data from tests on scaled models or in real scale are lacking. The influence of system parameters such as axial stiffness of load bearing walls and bending stiffness of slabs is missing in scientific literature and publications.

Future research will aim at contributing to this not yet fully recognised field, focussing on validation of presumed load arrangements, results of monitoring, model uncertainties, applicability, and valorisation.

REFERENCES

#EUGreendeal, 2020. New circular economy action plan, For a cleaner and more competitive Europe, Brussels: European Union.

ACI364.1, 2007. Guide for evaluation of concrete structures before rehabilitation. s.l.:ACI.

ACI562.13, 2016. Code requirements for evaluation, repair and rehabilitation of concrete structures. s.l.:ACI.

Belletti, B., Damoni, C., Cervenka, V. & Hendriks, M. A., 2016. Catenary action effects on the structural robustness assessment of RC slab strips subjected to shear and tensile forces. Structural concrete, pp. 1003–1016.

Botte, W., Caspeele, R., Gouverneur, D. & Taerwe, L., 2014. Influence of membrane action on robustness indicators and a global resistance factor design. Shangai, IABMAS.

Caspeele, R., Sykora, M., Allaix, D. L. & Steenbergen, R., 2013. The design value method and adjusted partial factor approach for existing structures. Structural Engineering International, 23(4), pp. 386–393.

Castaldo, P., Gino, D., Bertagnoli, G. & Mancini, G., 2018. Partial safety factor for resistance model uncertainties in 2D non-linear finite element analysis of reinforced concrete structures. Engineering structures, pp. 746–762.

COP26, 2021. UN Climat Change Conference in Glasgow, Glasgow: UN.

EN 1990, 2015. Eurocode 0 Basis of structural design (consolidated version including A1:2005 and AC:2010). Brussels, Belgium: CEN.

Engen, M., Hendriks, M. A., Monti, G. & Allaix, D. L., 2021. Treatment of modelling uncertainty of NLFEA in fib Model Code 2020. Structural Concrete, pp. 3202–3212.

fib bulletin 80, 2016. Partial factor methods for existing structures. Lausanne: Fédération intenationale bu béton.

Franssen, J. & Gernay, T., 2017. Modelling structures in fire with SAFIR. Journal of Structural Fire Engineering, vol. 8, nr. (3), pp. 300–323.

Genikomsou, A. & Polak, M., 2017. 3D finite element investigation of the compressive membrane action effect in concrete flat slabs. Engineering Structures, pp. 233–244.

Gino, D., Castaldo, P., Giordono, L. & Mancini, G., 2021. Model uncertainty in non-linear numerical analyses of slender reinforced concrete members. Structural concrete, pp. 845–870.

Girkmann, K., 1986. Flächentragwerke. Sechste Auflage ed. Wien: Springer Verlag.

Gulvanessian, H. C., Calgaro, J. & Holicky, M., 2012. Designers' Guide to Eurocode: Basis of Structural Design – 2nd Ed.. Croydon, UK: Thomas Telford Ltd.

Holicky, M. et al., 2013. Basics for assessment of existing structures. Prague: Klockner institute.

ISO 2394, 2015. General principles on reliability for structures. Geneva: International standard organization.
Kang, S.-B., Wang, S. & Gao, S., 2020. Analytical study on one-way reinforced concrete beam-slab sub-structures under compressive arch action and catenary action. Engineering structures, p. 110032.
Level(s), 2019. Taking action on the total impact of the construction sector, Luxembourg: European union.
Modelcode 2010, 2012. Lausanne: fib.
Molkens, T., 2022. The cooling phase, a key factor in the post-fire performance of RC columns. Fire Safety Journal, p. accepted for publication.
Molkens, T. et al., 2021. Performance of structural stainless steel following a fire. Engineering structures, p. 112001.
Molkens, T., Cashell, K. & Rossi, B., 2021. Post-fire mechanical properties of carbon steel and safety factors for the reinstatement of steel structures. Engineering structures, p. 111975.
Molkens, T., Gernay, T. & Caspeele, R., 2017. Fire Resistance of Concrete Slabs Acting in Compressive Membrane Action. Napels, IFireSS 2017.
Sykora, M., Diamantidis, D., Holicky, M. & Jung, K., 2017. Target reliability levels for existing structures considering economic and societal aspects. Structure and Infrastructure Engineering, Issue 13, pp. 181–194.
Thoma, K. & Malisia, F., 2018. Compressive membrane action in RC one-way slabs. Engineering structures, pp. 395–404.
TS17440, 2020. Assessment and retrofitting of existing structures. Brussels: CEN250.
UN, 2015. Paris agreement, Paris: United Nations.
Van Coile, R., 2015. Reliability-based decision making for concrete elements exposed to fire. Doctoral dissertation ed. Ghent: Ghent University.
Van Hout, S. & Molkens, T., 2021. Rekennota: Haagbeuk Willemen Real Estate, Berchem, Belgium: Sweco Belgium nv.
Xiao, H. & Hedegaard, B., 2018. Flexural, compressive arch, and catenary mechanisms in pseudo static progressive collapse analysis. Journal of Performance of Constructed Facilities, p. 04017115.

Computational Modelling of Concrete and Concrete Structures – Meschke, Pichler & Rots (Eds)
 ISBN: 978-1-032-32724-2

Modelling aspects of non-linear FE analyses of RC beams and slabs failing in shear

M. Harter, V. Jauk & G.A. Rombach
Institute for Structural Concrete, Hamburg University of Technology, Hamburg, Germany

ABSTRACT: Complex 3-dimensional FE models are increasingly used in practice for the design and evaluation of new and existing concrete structures, respectively. In this paper, the difficulties of realistic finite element modelling of shear failure in simple reinforced concrete beams and slabs, for which experimental results are available, are presented. The results show that a comprehensive validation of the used software for each field of application is required. The paper aims to discuss challenges in the user's decision of modelling parameters for concrete and their influence in the calculated load bearing capacity.

1 INTRODUCTION

It is well-known that the utilization of non-linear finite element analyses (NLFEA) for practical applications is accompanied by uncertainties and requires experience in this field (Rombach 2011). This applies to the evaluation of existing reinforced concrete structures as well as the design of new structures. As a result of increased computation performance and the desire to more accurately describe structural behavior of RC structures, the application of NLFEA has become increasingly common.

Finite element analyses are oftentimes classified by structural elements or by the failure mode of the experiments, especially when model uncertainties are regarded. In Hendriks et al. (2017a, 2017b) recommendations for NLFEA of concrete structures for ultimate limit state considerations are made. Therefore, numerous tests were simulated, treating bending failure, bending shear in beams and shear in slabs separately. It was found that simulating shear failure in slab specimen is rather challenging (Hendriks et al. 2017a). The failure mechanisms in slabs subjected to single loads are more diverse compared to beams due to load transfer actions in transversal direction, which can additionally influence the load bearing capacity (Hendriks et al. 2017a, b).

At Hamburg University of Technology (TUHH) test series investigating shear failure in beams without stirrups (Rombach & Jauk 2022) and in slabs without stirrups subjected to single loads (Rombach & Henze 2017) have been conducted. To gain more insight into the structural behavior of shear failure in both structural members, the test specimens are analyzed by means of NLFEA. Simulating shear failure in beams has the advantage of a significantly lower computational effort compared to slab structures. Therefore, one RC beam of the test series from Rombach and Jauk (2022) is selected to validate the finite element (FE) model. The resulting FE-model decisions are checked by the shear failure analysis of a second beam and compared to two test results of slabs subjected to single (block) loads. It is discussed to what extent the modelling decisions regarding shear failure in beams can be applied to describe the structural behavior of slabs subjected to single loads. Various modelling aspects and challenges, that arise during the definition of a NLFEA are emphasized. NLFEA for ultimate limit state by means of mean values of material properties for two different structural members failing in shear are evaluated and compared to test results without regard to reliability considerations.

The commercial software ATENA Studio (Version 5.6.1) and GiD (Version 14.0.5), the fracture-plastic constitutive model CC3DNonLinCementitious2 for concrete (Cervenka & Papanikolaou 2008, Cervenka et al. 2020) and three-dimensional (3D) modelling were used for all non-linear simulations.

2 TESTS ON CONCRETE MATERIAL VERSUS CODE PROVISIONS

When modelling concrete structures by means of NLFEA the designer may use material properties based on material tests conducted additionally for the regarded test specimen. In practice such measurements, like e.g. for the elastic modulus E_c or the tensile strength f_{ct} of concrete, are not available and code provisions for material properties are used. These formulations are oftentimes based on the concrete compressive strength f_c. In ATENA-GiD (Cervenka

DOI 10.1201/9781003316404-80

et al. 2020), one can choose to use recommendations for additional parameters for the concrete model CC3DNonLinCementitious2 (Cervenka & Papanikolaou 2008; Cervenka et al. 2020) based on f_c or adjust individual values.

The necessity to include predefined material values for a certain software increases, as the application of NLFEA for common practice is used more frequently. This applies especially in the case of modelling concrete. The material itself scatters and testing procedures induce a certain variance in such a way that resulting uncertainties can be specified as aleatory or epistemic depending on the particular case (Gino 2019).

Two material parameters that have a great influence on the results of a FEA – the Young's modulus E_c and the fracture energy G_f – are presented in the following section to illustrate this. Material tests conducted at the Institute for Structural Concrete at TUHH on both variables are taken and compared to code provisions.

2.1 *Young's modulus*

The Young's modulus describes the relation between stress and strain in uncracked concrete structures for uniaxial compression and by approximation for uniaxial tension (fib Bul. No. 70 2013). It can be defined as tangent or secant modulus. Fib Model Code 2010 (2013) ('MC10') defines the Young's modulus E_{ci} as the tangent modulus at the origin of the stress-strain diagram according to Equation 1.

$$E_{ci} = 21.500 \cdot \alpha_E \cdot (f_{cm}/10)^{1/3} \cdot (E_{ci} \text{ and } f_{cm} \text{ in MPa}) \quad (1)$$

To account for initial plastic deformations, E_{ci} is reduced to E_c by a factor α_i according to Equation 2:

$$E_c = \alpha_i \cdot E_{ci} \quad \text{(MC10 2013)} \quad (2)$$

with: $\alpha_i = 0.8 + 0.2\ (f_{cm}/88) \leq 1.0$

The modulus of elasticity predominantly depends on the properties of its components (fib Bul. No. 70 2013). The factor α_E in Equation 1 varies between −30% and +20% depending on the type of aggregate and is set to $\alpha_E = 1.0$ for quartzite aggregates (MC10 2013).

According to DIN EN 1992-1-1 (2011) ('EC2') the tangent modulus of elasticity for concrete with quartzite aggregates is defined as E_c. E_{cm} is specified as the secant modulus of elasticity (Eq. 3, 4).

$$E_c = 1.05 \cdot E_{cm} \quad \text{(tangent modulus)} \quad (3)$$

$$E_{cm} = 22.000 \cdot (f_{cm}/10)^{0.3} \quad \text{(secant modulus)} \quad (4)$$

Figure 1 shows the results of 75 material tests which were conducted in the laboratory of the Institute for Structural Concrete (TUHH) in 2015 – 2021 (Schütte 2018; Rombach & Henze 2017; Rombach & Jauk 2022) following the testing procedure (static tests) given in DIN EN 12390-13 (2014). $E_{c,0}$ and $E_{c,s}$ are values from material tests explained in the following.

For comparison, the Young's modulus according to EC2 (2011) and MC10 (2013) are displayed in Figure 1. The secant modulus of elasticity E_{cm} according to EC2 (2011) shows higher values compared to E_c according to MC10 (2013).

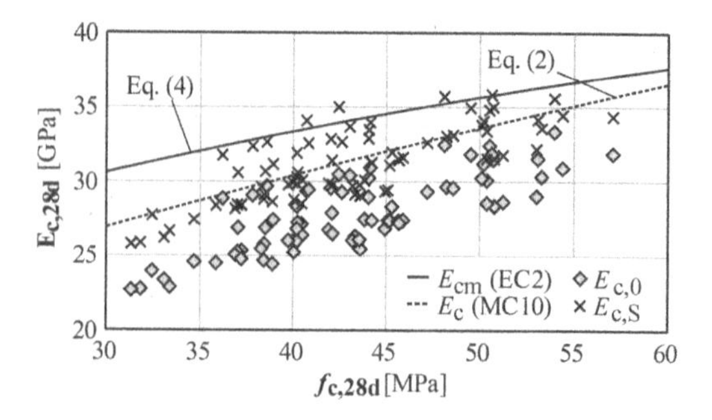

Figure 1. Young's modulus E_c – material tests versus code provisions.

The Young's modulus E_c does not only depend on the composition of concrete but also on the used testing method and the storage conditions. To identify realistic material properties for the structural members, the material tests were conducted on cylindric specimens with 300 mm height and a diameter of 150 mm on the same day as the large-scale specimen were tested. Further, the cylinders were stored under the same ambient conditions as the large-scale specimen, not in water as specified in DIN EN 12390-13 (2014). The composition of aggregates differs and the aggregates were not clearly specified. For the comparison with code provisions the factor $\alpha_E = 1.0$ for quartzite aggregates was assumed. To identify the Young's modulus for one structural member the mean value E_{cm} out of three tested dry cylinders was calculated. Figure 1 shows the results of the single cylinders. According to DIN EN 12390-13 (2014) two testing methods A and B exist. For Method A, three preloading cycles and three main loading cycles are applied. An initial Young's modulus $E_{c,0}$ is then calculated between the end of the preloading and the first loading cycle of the upper proof stress σ_a^m (Figure 2).

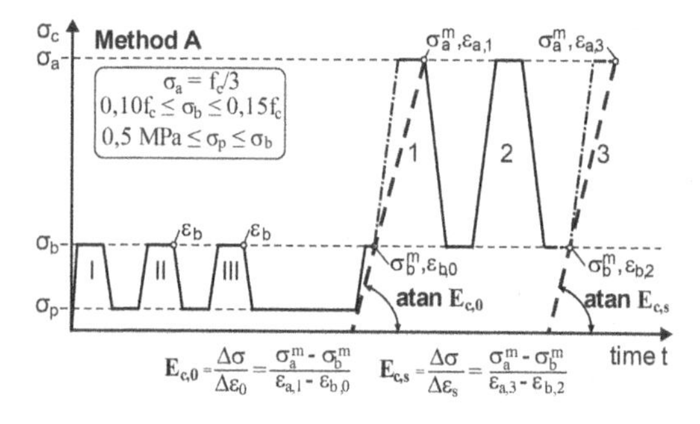

Figure 2. Loading sequence acc. to DIN EN 12390-13 (2014).

The stabilized modulus of elasticity $E_{c,S}$ is determined between the last lower proof stress σ_b^m and the upper proof stress σ_a^m of the last loading cycle. For Method B solely three main loading cycles are provided and the stabilized modulus of elasticity $E_{c,S}$ is calculated.

Method A was applied for the material tests displayed in Figure 1. Since the equations according to MC10 (2013) and EC2 (2011) account for the concrete compressive strength f_c and Young's modulus E_c tested 28 days after concreting, the measured properties were calculated back following EC2 (2011). The calculated values for f_{cm} (28d), that were determined from dry cylinders tested at $t > 28$ days, fit well to the concrete compressive strength measured f_{cwm} after 28 days on wet cylinders acc. EC2 (2011) ($-5\% < f_{cm,calc}/f_{cwm}(28d) < +15\%$). Therefore, it is assumed that cylinders stored and tested under the same ambient conditions as the structural members can be used to calculate the values for 28 days and can then be compared to code provisions.

The tested values are slightly smaller compared to the code provisions, which may be caused by the inherent aggregates. Further, when the ratio $\gamma = E_{c,28d}/E_{c,MC10}$ is regarded, the initial modulus of elasticity $E_{c,0}$ (mean $m_\gamma = 0.89$) is generally smaller compared to the stabilized modulus $E_{c,S}$ (mean $m_\gamma = 0.99$). Despite that, the variance of the test results is comparably small since concrete suppliers presumably use similar materials of the same origin and all tests were executed in the same laboratory ($CoV = s_\gamma/m_\gamma$; $CoV_{(Ec,0)} = 0.057$; $CoV_{(Ec,S)} = 0.051$). In comparison, for in-situ concrete deviations of the exemplary tested Young's modulus compared to the values in a structural member of about 10% cannot be controlled according to DAfStb (2020). Faber & Vrouwenvelder (2001) recommend to use a lognormal distribution with a coefficient of variation of $CoV_{(Ec)} = 0.15$ and mean $m = 1.0$ for the Young's Modulus of concrete used for statistical finite element analyses. The influence of E_c on the results of NLFEA are presented in Section 4.2. Since material tests for both test series were available, particularly the application of the measured initial and stabilized Young's modulus is examined.

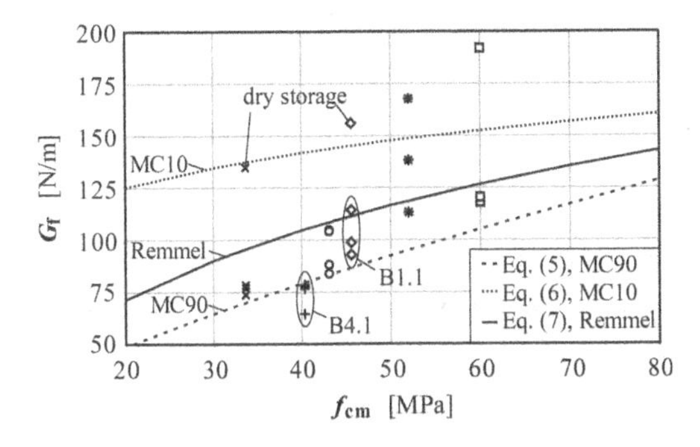

Figure 3. Fracture energy G_F – measured data (Rombach & Jauk, 2022) versus different approaches.

2.2 *Fracture energy*

The fracture energy G_F is a material parameter that describes the energy required to cause a surface to tear up such that stresses can no longer be transferred, thus causing a crack to form.

A pure tensile test is the most accurate method to determine the fracture energy. For simplification, Hillerborg (1985) developed an easy-to-implement experimental setup, namely a notched three-point bending beam. This was the basis for the Rilem Draft Recommendation (1985) for determining the fracture energy for plain concrete.

The latter procedure was used for the test series from Rombach and Jauk (2022). For maximum aggregates of 8 mm and 16 mm, the beam has a depth and width of 10 cm each and a length of 84 cm with a span width of 80 cm. The notch is located in the formwork. After the concrete specimens have cured, they are stored under water until about 30 min before the start of the experiment. The tests were conducted 28 days after concreting.

However, if this parameter is not determined by tests, formulas can be used. Equation 5 is given in CEB-FIB Model Code 1990 (1993) ('MC90'), whereby G_F depends on the type of aggregates and f_{cm}.

$$G_F = G_{F0} \cdot (f_{cm}/10\ \text{MPa})^{0.7} \quad \text{where: } f_{cm} < 80\ \text{MPa} \tag{5}$$

In the current MC10 (2013), another formula (Eq. 6) is given because, according to the state of the art, the influence of the aggregate size is negligible. Therefore, G_F depends only on f_{cm}.

$$G_F = 73 \cdot f_{cm}^{0.18} \tag{6}$$

Remmel (1994) proposed a formula (Eq. 7), where the fracture energy depends on the mean tensile strength f_{ctm} only.

$$G_F = 0.0307 \cdot f_{ctm} \quad \text{for } f_{cm} \leq 80\ \text{MPa} \tag{7}$$

with: $f_{ctm} = 2.12 \cdot \ln\{1 + (f_{cm}/10\ \text{MPa})\}$

In Figure 3, the results of the before mentioned approaches and the values from the tests according to the RILEM Draft Recommendation (1985) are plotted. An important factor in determining the fracture energy is the ambient condition during storage and testing of the notched beams. Mi et al. (2020) conducted a test series, where the beams were not stored under water. The fracture energies determined are in the same order of magnitude as the dry stored ones of the present test series.

3 NUMERICAL SIMULATION

Mean material values, which were derived from material tests for every specimen, are used as input parameters for concrete and reinforcement for the FE analyses in ATENA – GiD (Cervenka et al. 2000). Where no material parameters were measured by means of material tests, recommendations of the CC3DNonLinCementitious2 constitutive model (Cervenka & Papanikolaou 2008, Cervenka et al. 2020) based on the tested mean compressive strength f_c are applied. The concrete is modeled by means of 20 noded brick elements with identical edge lengths and quadratic form functions. The reinforcement is inserted discretely into the concrete elements. Perfect bond is assumed.

Steel plates, used in the tests for spreading the load from the jack and at the supports, are modeled as linear-elastic material. Fixed contact is assumed.

In the first step, the dead load of the concrete volume is applied. Subsequently, the steel plate is deflected incrementally. Newton-Raphson method with a maximum of 50 iterations is used for the simulations. Errors for force equilibrium are set to 0.01 and the relative error in energy is limited to 0.0001.

For the evaluation, load-deflection-curves and crack patterns are compared to the test results.

4 NUMERICAL SIMULATION OF BEAMS

4.1 *Test series on beams*

A series of beam tests was conducted at the TUHH to study the influence of crack kinematics on the shear or torsional strength (Rombach & Jauk 2022). In this paper experimental and numerical results of concrete test beam no. B1.1 and B4.1 without stirrups will be presented. All details of the beams are given in Figure 4 and Table 1. The FE-simulations and the necessary parameters are calibrated on the RC beam B1.1.

First, the load on the test beam B1.1 was increased in 3 steps of 25 kN. Then a displacement-controlled load with 0.012 mm/sec was applied until the failure. The first bending cracks start to open at a load of approx. 10 kN. The failure shear crack starts to develop at a load of more than 100 kN. After the maximum load of $F_{max} = 106.7$ kN and a deformation of $w_{max} = 6.35$ mm was reached, the shear crack gets unstable and the beam fails.

Table 1. Material test results for RC beams B1.1 and B4.1 (Rombach & Jauk 2022).

		B1.1	B4.1
aggregate size a_g	[mm]	16	8
mean compressive strength f_{cm}	[MPa]	45.6	40.4
splitting tensile strength $f_{ct,sp}$	[MPa]	3.55	3.53
initial Young's modulus $E_{c,0}$	[MPa]	28,290	26,127
stabilized Young's modulus $E_{c,S}$	[MPa]	32,940	29,350
fracture energy G_f	[N/m]	102.1	75,3
yield strength of rebar R_{eL}	[MPa]	528.7	528.7
tensile strength of rebar $R_{m,sl}$	[MPa]	648.1	648.1
failure load F_{max}	[kN]	106.7	154.2
deflection at failure w_{max}	[mm]	6.35	12.22

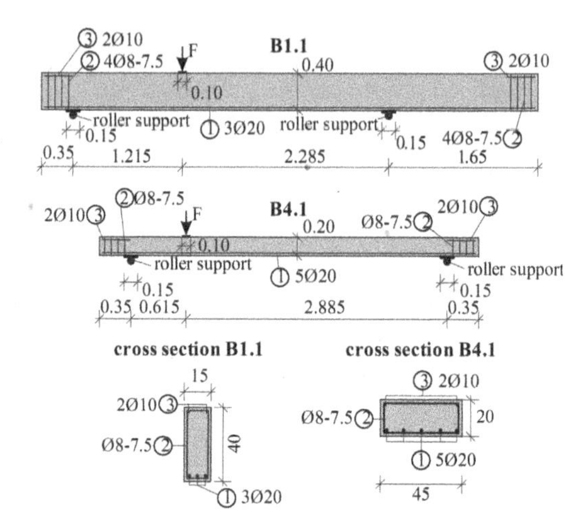

Figure 4. RC beams B1.1 and B4.1 (Rombach & Jauk 2022).

4.2 *Parameter study on RC beam B1.1*

The FE-model of the beam is shown Figure 5. To reduce the calculation time, symmetry constraints are considered by cutting the beam along the longitudinal plane and providing the nodes with appropriate boundary conditions. For the FE analysis, the material parameters from Table 1 were used. The analysis proceeds with an incremental deformation by 0.08 mm/step due to convergence criteria.

First, the influence of the element size on the results is studied. For this purpose, the beam is analyzed with different edge lengths of 20 mm, 31 mm, 40 mm, 50 mm and accordingly 20, 13, 10 and 8 finite elements over the height of the beam ($h = 0.40$ m). The finite elements were supposed to have a quadratic shape. Therefore, 4, 3, 2, 1 elements are arranged over the width of 0.075 m. The load-deformation curve, the crack pattern and the calculation time are evaluated. Crack patterns showed a best fit compared to the test results when a finite element size of 20 mm was used. The results with an element size of 40 mm was found to be sufficient regarding the crack pattern as well as load-deflection curves.

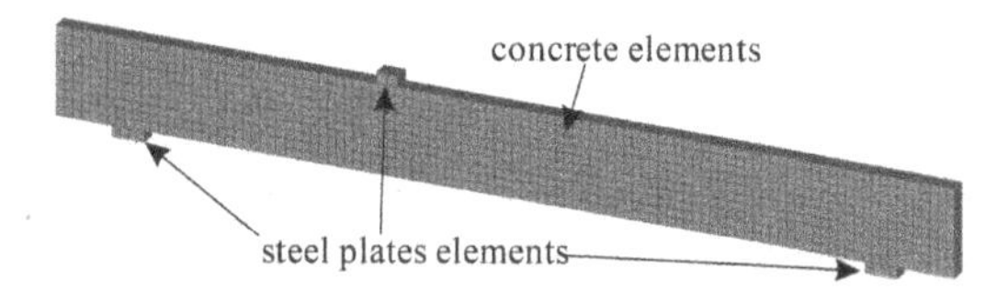

Figure 5. FE-Model for beam B1.1.

Next, the influence of the Young's modulus is studied. While the stabilized Young's modulus $E_{c,s}$ was found to be in good accordance with the code provisions according to MC10 (2013) (see Sec. 2.1), the analysis with the initial Young's modulus $E_{c,0}$ shows a slightly better agreement with the load-deflection curve of the tested beam B1.1 (see Figure 6). The initial

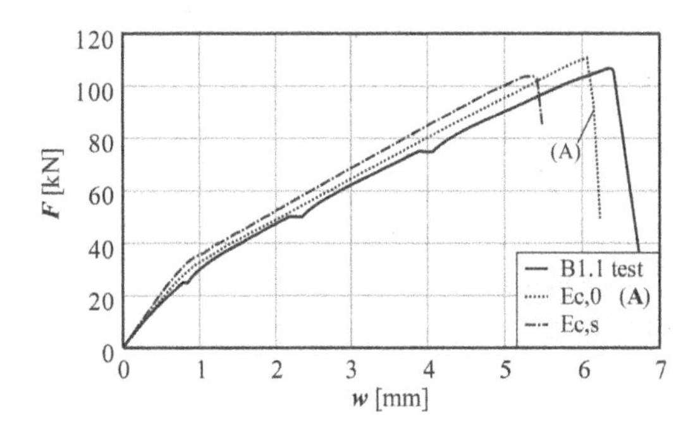

Figure 6. Load-deflection curves for beam B1.1 and FE-simulations with $E_{c0} = 28,290$ MPa and $E_{c,s} = 32,940$ MPa (measured material parameters, FIX = 1.0, $s_F = 30$).

Young's modulus was utilized for NLFEA simulations of RC beams in Schuette (2018) as well.

In Cervenka et al. (2016) the shear factor s_F was described as a parameter which has a great influence on the results of the non-linear simulations of RC beams. The shear factor is defined as the ratio between the shear stiffness and the normal stiffness in the crack (Cervenka et al. 2020). The parameter s_F was varied for the analysis of beam B1.1 as a mechanical based value is not available. Best results were obtained with a shear factor $s_F = 30$. That means that the shear stiffness of a crack is 30 times greater than the normal stiffness of the crack. It was found that if a shear factor is applied, the crack pattern of the simulation can be improved.

The comparison of measured fracture energy G_f and code provisions in Section 2.2 showed significant differences. Thus, analyses with different G_f-values were conducted. The load deflection curves are shown in Figure 7. The calculated failure load varies between $F_u = 101$ kN (95 %) with $G_F = 87$ N/m (MC 90 1993) and $F_u = 129$ kN (121 %) with $G_F = 145$ N/m (MC10 2013). The analysis with the measured fracture energy $G_F = 102.1$ N/m is in good agreement with the results of the beam test ($F_u = 111$ kN (104 %), $w_{max} = 6.1$ mm (96 %)). Usually, tests to determine the fracture energy are not conducted.

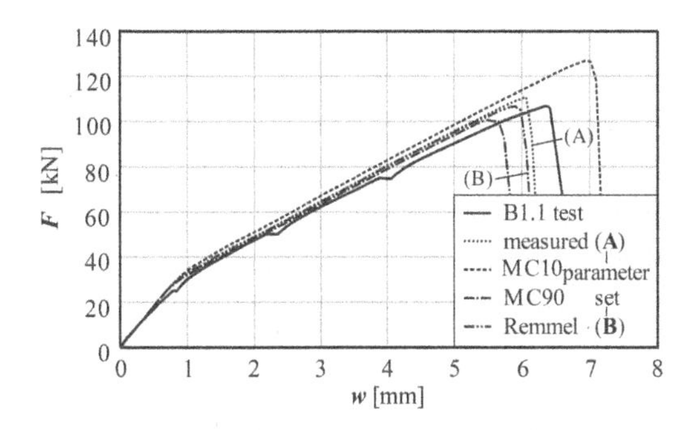

Figure 7. Load-deflection curves for B1.1 – variation of fracture energy G_f (FIX = 1.0, $s_F = 30$).

In such a case, the approach of Remmel (1994) based on the concrete tensile strength f_{ctm} appears to

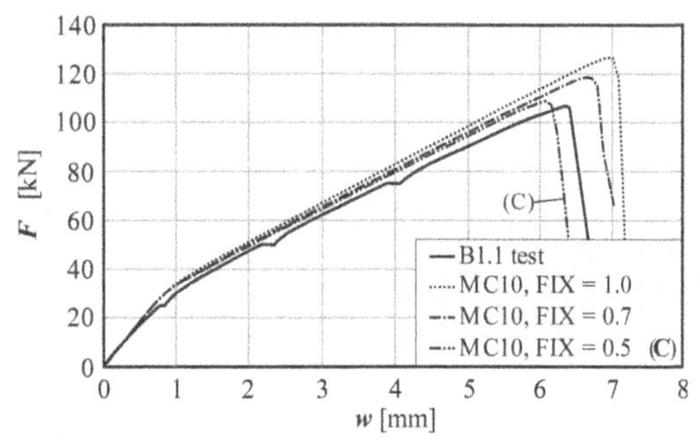

Figure 8. Load-deflection curves for B1.1 – fixed and rotated cracks.

be most appropriate. The experimental value $f_{ctm} = 3.20$ MPa gives a fracture energy of $G_f = 98.2$ N/mm. As expected, good results for the load-deflection curve in the range of the simulation with mean measured values was achieved (load bearing capacity $F_u = 106.6$ kN (100 %) and deflection of $w_{max} = 5.90$ mm (93%)).

Simulations have been conducted with the fixed crack model for smeared cracks. The fixed crack model is applied in other publications when the regarded constitutive model and software ATENA are used for nonlinear simulations (Cervenka et al. (2016), Cervenka et al. 2018). For the fixed crack model (Cervenka 1985) material axes for cracked elements are defined according to a fixed crack direction while the orientation of principal strains rotates independent from the material axes. This results in shear stresses parallel to the crack surface (Cervenka et al. 2020). A rotated crack model for smeared crack simulations is also provided in ATENA (Cervenka et al. 2000). This model is based on the fact that the principal stresses and strains have the same direction at each load level. The CC3DNonLinCementitious2 material model (Cervenka & Papanikolaou 2008; Cervenka et al. 2020) offers the option to apply a combination of the fixed crack and the rotated crack model. In this case, the direction of the cracks is fixed, when the residual tensile stress is below a defined level of tensile strength (Cervenka et al. 2020). It was found, that this parameter can be used to regulate the load bearing behavior in the FE-simulation when fracture energy according to MC10 (2013) was applied. Figure 8 shows the load-deflection curves for different fixed crack factors (FIX). When the fracture energy according to MC10 (2013) was combined with a fixed crack factor of FIX = 0.5, a maximum load of $F_u = 108.7$ kN (102 %) and a deflection of $w_u = 6.08$ mm (96 %) was calculated leading to a very good agreement with the test results.

The combination of both, the rotated and fixed crack model for a simulation, was a new aspect of the parameter study and little literature on the topic was found. In Cervenka & Bergmeister (1999), the simulation of a reinforcement bar embedded into a concrete prism subjected to tension with the concrete model SBETA (Cervenka et al. 2020) for two-dimensional simulations in ATENA is presented. It was found, that the rotated crack model and the fixed crack model deliver a lower and upper boundary when the formed cracks in the simulations are compared to the test results (Cervenka & Bergmeister 1999). Since the option was available as a modelling aspect for the constitutive model CC3DNonLinCementitious2 (Cervenka et al. 2020) and delivered good convergence behavior, it was additionally regarded. The modelling aspects according to Model A, B and C (see Tab. 2) should be transferred to the other structural members due to a good agreement with the load-deflection curves and crack pattern.

In Figure 9, crack patterns for the simulations with parameter sets A, B, and C are compared to the test specimen B1.1. A good agreement between the beam tests and the numerical analysis can be seen.

Table 2. Parameter sets for simulation A – C.

	A	B	C	
f_{ct}	[MPa]	$0.9 \cdot f_{ct,sp}$	$0.9 \cdot f_{ct,sp}$	$0.9 \cdot f_{ct,sp}$
E_c	[MPa]	$E_{c,0}$	$E_{c,0}$	$E_{c,0}$
G_f	[N/m]	from tests	Remmel (1994)	MC10 (2013)
s_F	[-]	30	30	30
fix	[-]	1.0	1.0	0.5

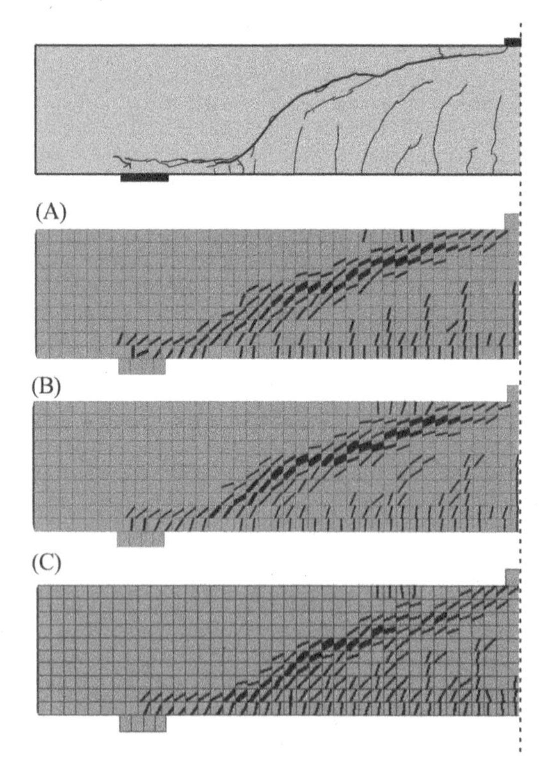

Figure 9. Crack patterns at failure for B1.1 – test results (Rombach & Jauk 2022) versus analysis to model set A, B, C.

4.3 *Simulation of RC beam B4.1*

According to the previous simulation of concrete beam B1.1, good agreement between test and simulations can be achieved by setting the element size to 40 mm and use the parameter sets A – C (see Table 2). To verify these parameter sets, the slab-like beam B4.1 from the test series (Rombach & Jauk 2022) is analyzed.

The results of the FE analyses are shown in Table 3 and Figure 10. The gradients of the load-deflection curves are identical for all parameter sets, but a clear difference in ultimate loads of models A – C can been seen. The main parameter that was varied is the fracture energy. Table 3 shows that the ultimate load also increases with increasing fracture energy. In the case of B4.1, the FE analysis of set A with a mean measured fracture energy gives significantly underestimated values compared to the ultimate loads of the test (71.9%).

Table 3. B4.1 – Fracture energy and results of simulation A – C.

	A	B	C
G_f [N/m]	75.3	97.5	142.1
$F_{max,FE}$ [kN]	110.7	137.3	163.3
$F_{max,FE}/F_{max}$	71.7%	89.0%	105.8%

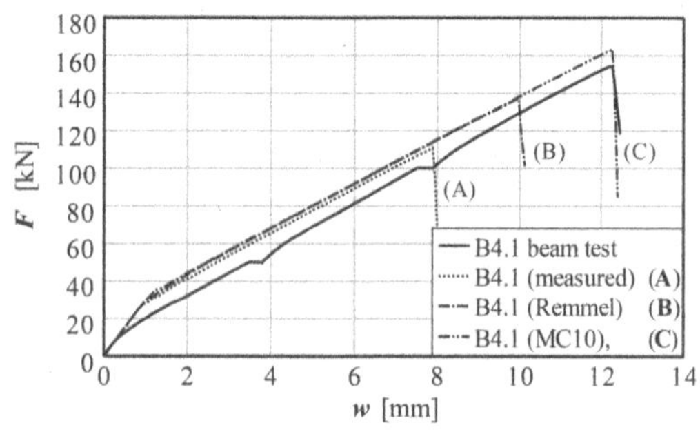

Figure 10. Load-deflection curves for B4.1.

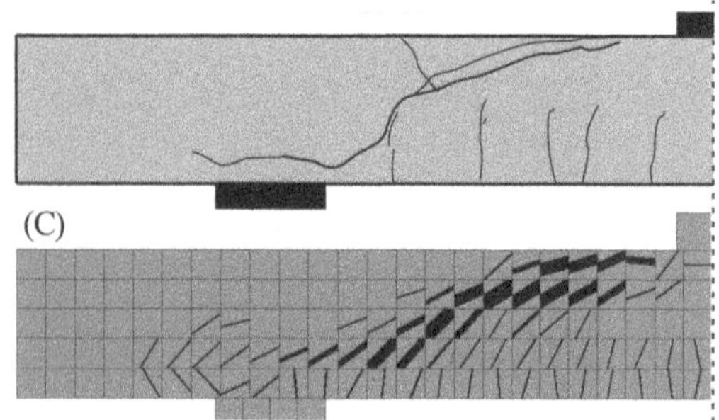

Figure 11. Crack patterns at failure for B4.1 – test results (Rombach & Jauk 2022) versus analysis to model set C.

Set B with the fracture energy according to Remmel (1994) provides sufficient results with respect to the ultimate load (89.0%), but an early failure occurs, so that the deformation is significantly underestimated. Set C provides the best results of the three parameter sets. Both the ultimate load (105.8%) and the deformation agree well. Furthermore, the comparison of the crack patterns (set C) is shown in Figure 11. Here there is also a high degree of agreement.

Based on the performed FE-analyses, it can be shown that the experimental determination of the fracture energy can lead to significant deviations in the models. The approach of the fracture energy according to MC10 (2013) with FIX = 0.5 provides very acceptable results for B1.1 as well as for B4.1.

5 NUMERICAL SIMULATION OF SLABS

5.1 *Test series on slabs*

In the test series on slabs (Rombach & Henze 2017), 14 identical large-scale cantilever slabs without stirrups were loaded by a concentrated force up to failure. The length of the cantilever was $l_c = 1.9$ m, the slab had a total width of $b = 4.5$ m and an effective depth $d = 0.215$ m. Block loads were applied by means of steel plates with a contact area of 0.4 m × 0.4 m.

For one specimen with a load distance to the edge of the support of $a_v = 4{\cdot}d$ the longitudinal reinforcement was reduced. These two slabs P3/4d-1 with longitudinal reinforcement of Ø16/80 mm ($\rho_l = 1.16\%$) and the test P13/4d-2 with a reduced longitudinal reinforcement of Ø14/100 mm ($\rho_l = 0.71\%$) are modelled according to the before mentioned modelling aspects. The geometry and support conditions of the slab tests are shown in Figure 12. Mean values for the tested material properties are given in Table 4.

In Rombach and Henze (2017) the behavior of the slab during failure was described as follows: Since the slab thickness was measured during testing, cracks developing inside the slab could be detected.

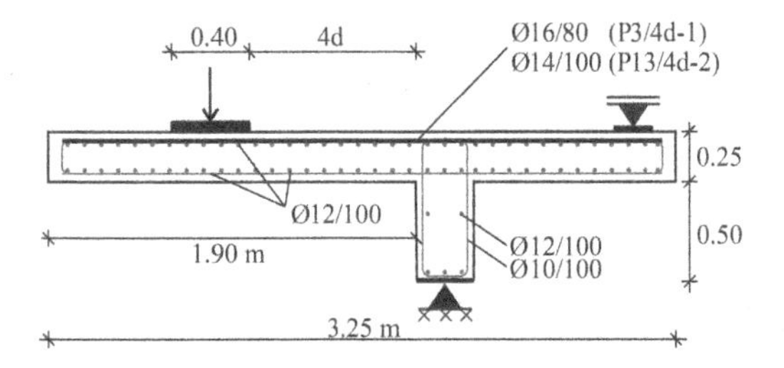

Figure 12. Test setup for P3/4d-1 and P13/4d-2 – section view (Rombach & Henze 2017).

During testing, the loading plate was pressed into the slab and failure occurred in the region of the steel plate, when one block load was applied. For both test specimen inclined cracks close to the loading plates were obtained when the intersections were analyzed after testing (Rombach & Henze 2017).

Table 4. Material test results for slabs P3/4d-1 and P13/4d-2 (Rombach & Henze 2017).

		P3/4d-1	P13/4d-2
mean compressive strength f_{cm}	[MPa]	40.0	47.1
splitting tensile strength $f_{ct,sp}$	[MPa]	2.78	3.86
initial Young's modulus $E_{c,0}$	[MPa]	29,660	26,733
stabilized Young's modulus $E_{c,S}$	[MPa]	32,753	30,473
yield strength of long. rebar R_{el}	[MPa]	522	559
tensile strength of long. rebar R_m	[MPa]	625	682
failure load F_u	[kN]	677.4	725.5

In Henze (2019) it was discussed, that test P13/4d-2 had a higher risk of bending failure due to the low reinforcement ratio but the specimen showed a diagonal crack. Further, the load-deflection curve for slab P13/4d-2 shows an increase of deformation when a load of 600 kN was reached (Henze 2019).

Since this specimen is more in the threshold region between shear and bending failure it is investigated in the following to what extent the modelling aspects chosen for shear failure can be applied here as well.

5.2 *Simulation of slab P3/4d-1*

For the simulation of slab P3/4d-1 the general assumptions for numerical simulations according to Section 3 were applied. Symmetry conditions were considered and half of the slab with a width of 2.25 m was modelled to minimize the computational effort. Fixed contact conditions in transverse direction were applied for all surfaces in the cross section where the slab was cut in half. Since the symmetry axis is in the middle of the load application the development of cracks in this section were visualized during calculation. The dead weight was applied in a first step before the load was incrementally increased by means of displacements steps of 0.16 mm. Other than for the test execution, no preloading up to 200 kN and unloading is applied for the simulations. Monitors for resulting deflection were set to a point on the bottom side of the load plate and the applied force was obtained from reactions in the middle of the top surface of the loading plate where the displacements were applied.

The parameter set C (Table 2) showed a best fit for both analyzed beams B1.1 and B4.1 and a FE-size of 40 mm. Further, no material tests for the fracture energy according to simulation A were available for the slab specimen. Therefore, parameter set C is analyzed in the following. Figure 13 shows the load-deflection curves of the regarded FE-models compared to the test results for P3/4d-1.

For the simulations of P3/4d-1, single steps at a load of about 145 kN fail to fulfill the convergence criteria for the relative error of residual forces. Here, flexural cracks in the support region occur. It is assumed that the resulting error has little influence on the ongoing analysis. Further, when the load bearing capacity of the model is reached, convergence criteria usually can't be achieved for several steps. The simulation is not interrupted but it can't be assured that the non-converged steps have no influence on the further analysis. Therefore, the next steps are displayed in grey color (Figures 13 and 15). Due to the extensive calculation time the applied step size is not reduced to overcome convergence difficulties in this case.

In simulation C, the diagonal crack forms at a load level which is in good agreement with the test ($F_u = 687{,}5$ kN (101,5%)). The flexural crack width in the support region matches the test results quite well. Cracks with a minimum width of 0.1 mm for simulation C and the crack pattern of the test specimen are displayed in Figure 14.

In Figure 13, FE-models with a fracture energy according to MC10 (2013) and a fixed crack model FIX = 0.7 and FIX = 1.0 are displayed. The calculated load bearing capacity of these additional simulations exceed the test results more clearly.

5.3 *Simulation of slab P13/4d-2*

The finite element model for slab P13/4d-2 is, disregarding the lower bending reinforcement ratio, identical to P3/4d-1. Due to the lower longitudinal reinforcement ratio, more flexural cracks appeared in the simulations of slab P13/4d-2. Therefore, the displacements were applied with an increment of 0.054 mm/step for simulation C, which led to a comparably high number of load steps. In Figure 15, the load-deflection curve of simulation C is compared to the test specimen P13/4d-2. Additionally, simulations with a fracture energy according to MC10 (2013) and a factor FIX = 0.7 and FIX = 1.0 are displayed (displacement increment for block load: 0.107 mm/step).

For the simulation with parameter set C, first diagonal cracks in the cutting plane appear for a load of $F \approx 570$ kN. These cracks are visualized by a step change in the load-deflection curve, where several steps in a row can't find convergence criteria for residual forces.

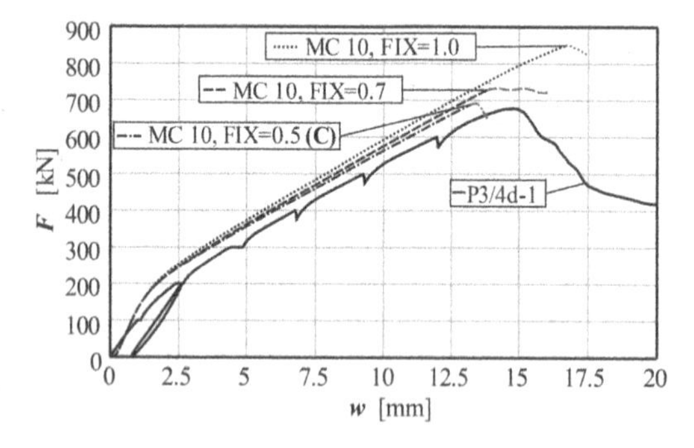

Figure 13. Load-deflection curves for slab P3/4d-1 – test results versus numerical simulations.

Thereafter, the load is further increased until another step change occurs for a load $F = 678$ kN (93,5%). To gain reliable results for loading $F > 550$ kN, the step size would have to be reduced significantly to meet convergence criteria for residual forces. Despite this consideration, the further redistribution and course of the load-deflection curve of the simulation decreased slightly compared to the test results and it can be seen, that cracks around the loading plate appear in the simulation as in the test where it was pushed into the slab.

Unlike for the previously considered specimen, when the fracture energy according to MC10 (2013) with a fixed crack model (FIX = 1.0) is used, the load-deflection curve matches the test specimen better, but no step change occurs. The curve is almost horizontal

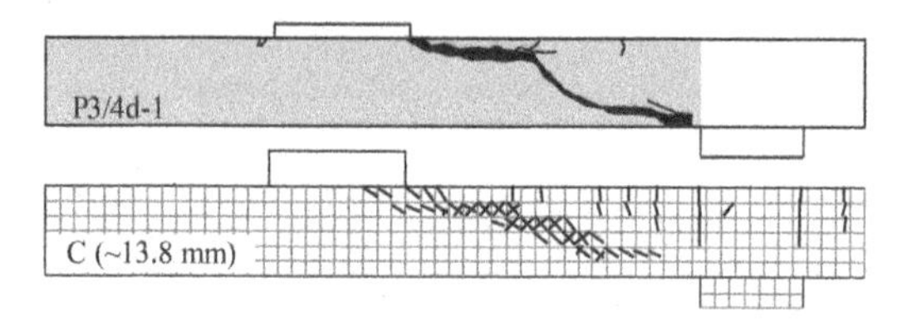

Figure 14. Crack patterns for slab P3/4d-1 and deflection at midspan (Rombach & Henze 2017).

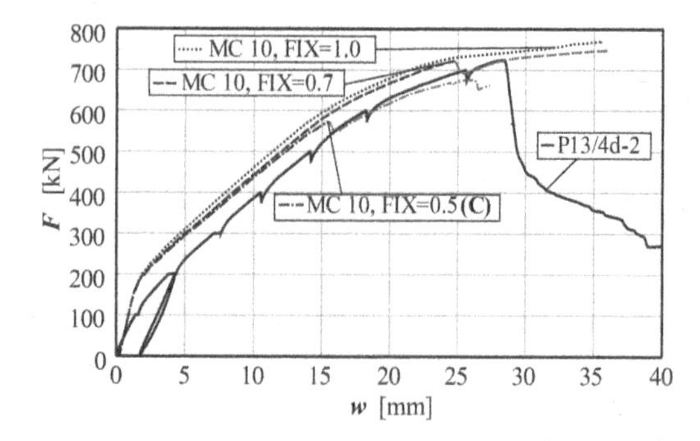

Figure 15. Load-deflection curves for slab P13/4d-2 – test results versus numerical simulations.

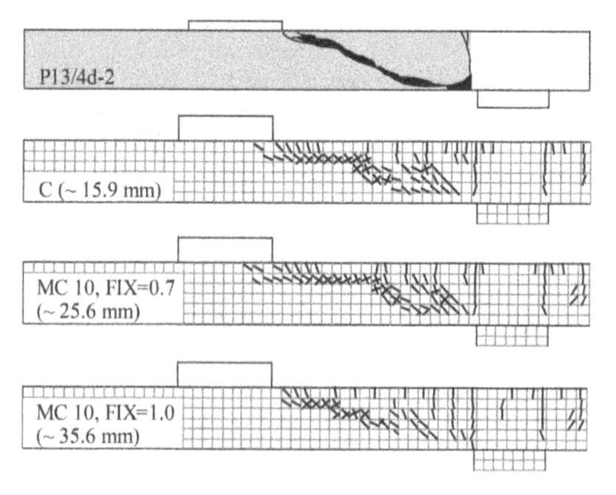

Figure 16. Crack patterns for slab P13/4d-2 and deflection at midspan (Rombach & Henze 2017).

for deflections greater than 25 mm. This refers to bending failure and the longitudinal reinforcement shows plastic strains near the support. Here, a first diagonal crack appears for a deflection of about 35 mm. Plastic strains in the longitudinal reinforcement in the support region occur at a load $F > 500$ kN while yielding was measured for $F > 550$ kN during slab test. Where the fracture energy according to MC10 (2013) with FIX = 0.7 is regarded, the first diagonal crack appears for $F =$ 722 kN (99,5%) and 24 mm maximum deflection. Plastic strains in the reinforcement occurred for $F > 550$ kN. For P13/4d-2 the combination of a fixed and rotated crack model with a higher factor FIX > 0.5 seems to meet the test results better compared to P3/4d-1 or the regarded beams.

The description of the load bearing behavior for P13/4d-2 is clearly more complex, since failure is more in the threshold between shear and bending failure as expected by Henze (2019). The crack patterns after diagonal cracks were formed are compared to the cut out of the test specimen in Figure 16 (min. crack width 0.1 mm).

6 CONCLUSIONS

The following conclusions can be drawn from the presented investigations on beams and slabs without stirrups failing in shear.

Generally, NLFEM models for shear failure in RC beams can be modeled by means of different modelling aspects with good accordance to the test results. The fracture energy had a high influence on the results of the regarded FE-models. Measured values for the fracture energy should be handled with care, since testing procedures may induce an additional uncertainty due to the difficult testing procedure. The combination of a rotated and fixed crack model with a factor of FIX = 0.5 resulted in good accordance with the beam tests, when a fracture energy according to MC10 (2013) and a relatively large FE size of 40 mm was applied.

The transfer of modeling decisions for shear failure in beams to shear failure in slabs is possible with limitations. Generally, the failure mechanisms in slabs are more versatile and therefore it is more difficult to achieve reliable NLFEM models. While the failure in P3/4d-1 can be displayed well when modelling aspects from beams are transferred, the load bearing behavior of P13/4d-2 diverges more significantly and a fracture energy according to MC10 (2013) with a fixed crack model (FIX = 1.0) or a factor FIX = 0.7 delivers significantly better results compared to the before regarded specimen.

The conducted numerical simulations demonstrate that a software user should have an in-depth knowledge of the software, especially the used material model. Validation for each structural type and failure mode (bending, shear, ..) is required. Material parameters used in NLFEM should always be considered in relation to the basics of the material model. Values from codes, like EC2 (2011) or MC10 (2013), may be better than the values from material tests.

ACKNOWLEDGEMENTS

The authors acknowledge the German Research Foundation (DFG) for the financial support of the project (project no. RO 793/14-1).

REFERENCES

ATENA Studio (Version 5.6.1) [Computer Software]. Na Hrebenkach 55, 150 00 Prague, Czech Republic: Cervenka Consulting s.r.o.

GiD (Version 14.0.5) [Computer Software]. Edificio C1, Campus Norte UPC, Gran Capitàn s/n, 08034 Barcelona, Spain: CIMNE.

Červenka, V., Jendele, L., & Červenka, J. (2020). ATENA Program Documentation–Part 1. *Cervenka Consulting sro.* https://www.cervenka.cz/assets/files/atena-pdf/ATENA_Theory.pdf

Cervenka, J. & Papanikolaou, V.K. (2008). Three-dimensional combined fracture-plastic material model for concrete. *International Journal of Plasticity* 24(12): 2192–2220. DOI: 10.1016/j.ijplas.2008.01.004

Cervenka, V. (1985). Constitutive model for cracked reinforced concrete. *Journal Proceedings ACI* 82(6): 877–882.

Cervenka, V. & Bergmeister, K. (1999). Nichtlineare Berechnung von Stahlbetonkonstruktionen – Finite-Elemente-Simulationen unter Bemessungsbedingungen. *Beton- und Stahlbetonbau* 94(10): 413–419. DOI: 10.1002/best.199901440

Cervenka, V, Cervenka, J, Rukl, R. & Sajdlova, T. (2016) Prediction of shear failure of large beam based on fracture mechanics. *9th International Conference on Fracture Mechanics of Concrete and Concrete Structures (FraMCoS-9).* V. Saouma, V., Bolander, J. & Landis, E. (Eds). Berkeley, California USA. DOI: 10.21012/FC9.029

Cervenka, V., Cervenka, J. & Kadlec, L. (2018). Model uncertainties in numerical simulations of reinforced concrete structures. *Structural Concrete* 19(6): 2004–2016. DOI: 10.1002/suco.201700287

Cervenka, V., Jendele, L. & Cervenka, J. (2020). ATENA Program Documentation Part 1 Theory. Na Hrebenkach 55, 150 00 Prague, Czech Republic: Cervenka Consulting s.r.o.

Comite Euro-International du Beton (CEB-F) (1993). *CEB-FIB Model Code 1990.* CEB Bulletin No. 213/214. London: Thomas Telford Ltd.

Deutscher Ausschuss für Stahlbeton (DAfStb)(ed.). (2020). *Erläuterungen zu DIN EN 1992-1-1 und DIN EN 1992-1-1/NA* (2nd. Ed.) (600). Berlin, Wien, Zürich: Beuth Verlag GmbH.

DIN EN 1992-1-1:2004 + AC:2010 (2011). *Eurocode 2: Design of concrete structures – Part 1-1: General rules and rules for buildings; German version.* (EN 1992-1-1:2011)

DIN EN 12390-13: 2013 (2014). *Testing hardened concrete – Part 13: Determination of secant modulus of elasticity in compression; German version.* (EN 12390-13:2013)

Faber, M.H. & Vrouwenvelder, T. (2001). JCSS Probabilistic Model Code Part 3: Material properties. Joint Committee on Structural Safety (ed.). ISBN 978-3-909386-79-6

Fédération internationale du béton (fib). (2013). *Model code for concrete structures 2010.* Lausanne: Ernst & Sohn. DOI: 10.1002/9783433604090

Fédération internationale du béton (fib) Bulletin No 70. (2013). *Code-type models for structural behaviour of concrete.* State-of-art report. Lausanne: International Federation for Structural Concrete (fib).

Gino, D. (2019). *Advances in Reliability Methods for reinforced concrete structures.* PhD thesis. Turin: Politecnico di Torino.

Hendriks, M.A.N., de Boer, A., Belletti, B., Damoni, C. (2017a). Guidelines for Nonlinear Finite Element Analysis of Concrete Structures – Part: Overview of results. Rijkswaterstaat Centre for Infrastructure, (Report RTD:1016-2:2017, version 1.0)

Hendriks, M.A.N., de Boer, A., Belletti, B., Damoni, C. (2017b). Guidelines for Nonlinear Finite Element Analysis of Concrete Structures – Part: slabs. Rijkswaterstaat Centre for Infrastructure, (Report RTD:1016-3C:2017, Version 1.0)

Henze, L. (2019). *Querkrafttragverhalten von Stahlbeton-Fahrbahnplatten.* PhD thesis (in German). Hamburg: Hamburg University of Technology. DOI: 10.15480/882.2270

Hillerborg, A. (1985). The theoretical basis of a method to determine the fracture energy G_F of concrete. *Materials and structures* 18(4), 291-296. DOI: 10.1007/BF02472919

Mi, Z., Li, Q., Hu, Y., Liu, C., Qiao, Y. (2020). Fracture Properties of Concrete in Dry Environments with Different Curing Temperatures. *Applied Sciences* 10(14): 4734. DOI: 10.3390/app10144734

Remmel, G. (1994). Zum Zug- und Schubtragverhalten von Bauteilen aus hochfestem Beton. *Deutscher Ausschuss für Stahlbeton (DAfStb)(ed.) (444).* Beuth Verlag GmbH. DOI: 10.2366/3715966

Rilem Draft Recommendation (1985). Determination of the fracture energy of mortar and concrete by means of three-point bend tests on notched beams. *Materials and structures, 18*(106), 285–290.

Rombach, G.A. (2011). *Finite-element design of concrete structures: practical problems and their solutions* (2nd Ed.). London: ICE Publishing.

Rombach, G.A. & Henze, L. (2017). *Experimental investigations on shear capacity of concrete slabs with concentrated loads close to the support.* Technical report (in German). Hamburg: Hamburg University of Technology, Institute for Structural Concrete. DOI: 10.15480/882.1443

Rombach, G.A. & Jauk, V. (2022). *Investigation of force transmission in cracks of concrete beams due to shear force and torsion.* Technical Report (in German). Hamburg: Hamburg University of Technology, Institute for Structural Concrete.

Schütte, Björn (2018). *Zum Tragwiderstand querkraftbewehrter Stahlbetonbalken.* PhD Thesis (in German). Hamburg:
Hamburg University of Technology.

Computational Modelling of Concrete and Concrete Structures – Meschke, Pichler & Rots (Eds)
 ISBN: 978-1-032-32724-2

Non-linear finite element analysis affected by ill-defined concrete parameters

F. Sattler & A. Strauss
Department of Civil Engineering and Natural Hazards, University of Natural Resources and Life Sciences, Vienna, Austria

ABSTRACT: Performance and lifetime assessment of new and existing concrete structures is a complex multidisciplinary topic that has been developed and intensively discussed for many years. The non-linear finite element analysis (NLFEA) can support significantly in this issue because NLFEA is used to develop safety formats that allow to guaranty the safety demands of our society. The provisions associated with NLFEA in codes enable a range of interpretation which values of material input parameters should be or must be used in order to derive reliability-consistent design values of the structural resistance. The objectives of this contribution are as follows: Non-linear analysis safety formats and deterministic equations for the derivation of the fracture energy are examined and subsequently it is investigated how uncertainties in the input variables of the resistance model and the different equations for the fracture energy affect the system responses and the design values of the structural resistance.

1 INTRODUCTION

The assessment of the performance and reliability of new and existing structures is a complex multidisciplinary topic and becomes more important due to aging infrastructure buildings across Europe. This topic has been discussed and developed for many years and non-linear numerical modelling or non-linear finite element analysis (NLFEA) can support in this issue. There are different levels of numerical finite element modelling – from linear to non-linear, from deterministic to probabilistic, from holistic system modelling to detail modelling. This assessment of the reliability and safety can be performed with different methods and approaches, which represent the resistance side with different accuracy, and differ in accounting for the uncertainties.

NLFEA in praxis has become quite common and a widely used tool in the design of new and in the assessment of existing structures. In particular, these methods are the basis of advanced safety assessment methods like the ECOV method or the full probabilistic reliability assessment methods, as proposed in the *fib* Model Code 2010 (2013) and EN 1992 (CEN 2008) and in literature by several authors (Castaldo et al. 2019; Červenka 2013; Strauss et al. 2018). Nonetheless, the NLFEA associated provisions in such codes enable a more or less wide range of interpretation. For instance, in some of the proposed procedures it is not clearly defined which values of material input parameters should be or must be used in order to match the level of safety that exists in reality or to derive reliability-consistent design values of the structural resistance. Since the input material parameters, which are continuously adjusted during a non-linear numerical analysis, have a significant influence on the calculation results and in consequence on the safety and reliability computed by e.g. classical and advanced reliability-based methods, the objectives of this contribution are as follows:

Non-linear analysis safety formats and the determination of the material parameters are examined and subsequently investigated how uncertainties in the input variables of the resistance model affect the system responses, especially the influence of different equations for the deterministic determination of the fracture energy are a matter of interest. These analyses are carried out on two critical details of reinforced concrete structures for which size effects and second-dependent effects are dominant.

2 SAFETY FORMATS IN FIB MODEL CODE 2010 (MC2010)

In *fib* Model Code 2010 (2013) the design principles for the analysis and assessment of non-linear analysis is based on the global resistance design method and the design condition can be written as

$$F_d \leq R_d, \text{ with } R_d = \frac{R_m}{\gamma_R \cdot \gamma_{Rd}}, \tag{1}$$

DOI 10.1201/9781003316404-81

where F_d is the design value of external actions, R_d is the design value of the structural load bearing capacity, R_m is the mean value of the structural load bearing capacity, γ_R is the global resistance factor and γ_{RD} is the model uncertainty factor.

There are three different methods proposed in *fib* Model Code 2010 (2013) how to derive the design structural resistance R_d

(1) the probabilistic method for a required reliability index β
(2) global resistance methods
(3) partial safety factor method.

In this document only the global resistance methods and the partial safety factor method will be discussed and presented in more detail, as they were applied to the case studies.

2.1 *Global resistance methods*

Global resistance methods and safety formats were initiated by the increasing use of non-linear analysis for safety assessment since it is based on a global structural model. The uncertainties in these types of formats are described on the level of structural resistance (fib 2013). As these methods use a more general and global approach than the partial factor method, it is more similar to the probabilistic safety concept. To derive the design resistance R_d using a global resistance approach the design resistance computed with a NLFEA with chosen representative values f_{rep} is divided by a global safety factor γ_R^*. (Castaldo et al. 2019; fib 2013)

The global safety factor γ_R^* is accounting for both the uncertainties of the materials γ_R and the model formulation γ_{RD}. The uncertainties related to the randomness of the model parameters especially of material properties (aleatory uncertainties) are accounted for using the global resistance factor γ_R (Castaldo et al. 2019; fib 2013). To account for uncertainties due to the model formulation, a separate safety factor for model uncertainty γ_{RD} needs to be applied. Typical values for the model uncertainty can be found in *fib* Model Code 2010 (2013). The global design resistance can be derived as follows:

$$R_d = \frac{R\left(f_{rep,\ldots}\right)}{\gamma_R \cdot \gamma_{Rd}}. \tag{2}$$

In *fib* Model Code 2010 (2013) there are two alternative safety formats proposed: the method of estimating a coefficient of variation of the resistance (ECOV method) and the global resistance factor method.

2.1.1 *Global resistance factor method*

For the calculation according to the global resistance factor method (GRF), *mean values* shall be used for the material properties for the NLFEA (fib 2013). Equation 3 and equation 4 are used to determine the *mean values* for concrete and reinforcing steel, respectively, where f_{ck} and f_{yk} are the corresponding characteristic material parameters. The mean compressive strength of the concrete as well as all resulting parameters are assumed to be 85% of the characteristic values, the mean value of the tensile strength of the reinforcing steel is assumed to be 110% of the corresponding characteristic value.

$$f_{\mathrm{cm,\,GRF}} = 0.85 \cdot f_{\mathrm{ck}} \tag{3}$$

$$f_{\mathrm{ym,GRF}} = 1.1 \cdot f_{\mathrm{yk}} \tag{4}$$

The global safety factor γ_R^* is calculated from the product of the model uncertainty factor $\gamma_{RD} = 1.06$ and the global resistance safety factor or partial factor for resistance $\gamma_R = 1.20$ and results in $\gamma_R^* = 1.27$ (fib 2013; Hendriks et al. 2017). It should be noted that the ratio 1.27/0.85 corresponds to the partial factor for resistance for concrete with 1.5 and the ratio 1.27/1.1 corresponds to the partial factor for steel with 1.15 (Castaldo et al., 2019; Hendriks et al. 2017).

2.1.2 *Method of estimation of coefficient of variation of resistance (ECoV)*

Using the ECoV method (Červenka 2013) the safety factor for material uncertainties (the global safety factor) can be estimated assuming a log-normal distribution for the load bearing capacity of the resistance structure according to Eq. (4):

$$\gamma_R = e^{\alpha_r \cdot \beta \cdot V_R}, \tag{5}$$

where α R is the FORM sensitivity factor, β the reliability index and V_R is the coefficient of variation (CoV) of the resistance side. For ultimate limit states, a target lifetime of 50 years and moderate consequences of failure the values $\alpha R = 0.8$ and $\beta = 3.8$ are to be used according to Schneider and Schlatter (1994), EN 1990:2002 (CEN, 2013) and *fib* Model Code 2010 (2013). Using this approach just two simulations of NLFEA are needed and CoV can be calculated, assuming the resistance follows a log-normal distribution, by the means of two nonlinear models:

$$V_R = \frac{1}{1.65} ln \cdot \left(\frac{R_m}{R_k}\right), \tag{6}$$

where R_m and R_k are the structural resistance determined by performing an NLFEA with mean values and with characteristic values of the input random variables, respectively (Červenka 2013). Taking into account the model uncertainty $\gamma_{RD} = 1.06$, according to MC2010 (2013), the design resistance can be derived, according to Equation (1).

2.2 *Partial safety factor method*

Using the partial factor method (PFM) according to (fib, 2013), the design resistance R_d is determined by means of a single non-linear FE analysis (NLFEA), where the material parameters are selected with the design values of the material resistances f_d. By dividing the design resistance $R_{NLFEA}(f_d)$ obtained from this NLFEA by the safety factor accounting for model uncertainties γ_{RD}, the design resistance R_d is estimated (Castaldo et al. 2019; fib 2013). However, it should be mentioned that the use of design values of the material properties when performing an NLFEA can lead to an incorrect assessment of the structural load bearing capacity, especially in the case of slender columns (Allaix et al. 2013; CEB 1995).

3 FRACTURE ENERGY

To realistically describe the non-linear behavior of concrete structure, the non-linear effects of the concrete material properties need to be considered (Červenka 2013; Slowik et al. 2021; Sucharda 2020). According to Zimmermann et al. (2014) the non-linear material properties of concrete can be characterized by its variable modulus of elasticity (Youngs Module) E_c, its tensile strength f_t and its specific fracture energy G_f as well as the material and geometric uncertainties (i.e. aleatory uncertainties).

As stated in the introduction, one key parameter for reliable non-linear modelling is the fracture energy GF and its stochastic parameters and it has been a research subject for various authors (Bažant & Becq-Giraudon 2002; Bažant & Planas 2019; Červenka 2013; Sucharda 2020). The *fib* Model Code 2010 (2013) defines the specific fracture energy of concrete G_f as the energy that is required to propagate a tensile crack of unit area. The fracture energy G_f is defined by Bažant and Becq-Giraudon (2002) as the area under the complete stress-separation curve of the cohesive crack model and it can be obtained by dividing the area under the measured load-deflection curve by the ligament area using the work-of-fracture method. (Bažant & Becq-Giraudon 2002; Bažant & Planas 2019; Hillerborg 1985; Zimmermann et al. 2014).

In *fib* Model Code 2010 (2013) and in publications (Czernuschka et al. 2018; Zimmermann et al. 2014) it is suggested that the determination of the fracture parameters should be done by related tests, e.g. the uniaxial tension test, the three-point bending test, the wedge-splitting test, the Brazilian splitting test or the compact tension test.

In the absence of experimental data there are several different deterministic equations in publications (CEB-FIP 1993; CEB-FIP 2008; fib 2013; Marí et al. 2015; Vos 1983) to estimate the fracture energy for ordinary concretes of normal weight. The different deterministic equations to determine the fracture energy used for the NLFEA as input parameter leaves a wide range of uncertainty for the designer of the non-linear finite element model (NLFEM). Especially in structures where the failure mode is shear failure, as in the case studies presented later in this contribution, the choice of fracture energy can lead to an over- and underestimation regarding the assessment of the maximal load bearing capacity and the safety level of the structure. Figure 1 shows the relation of fracture energy of concrete versus the characteristic compressive strength for Equations 7–10.

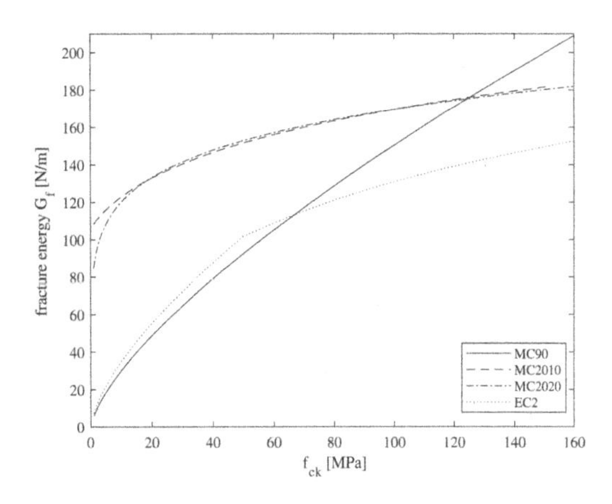

Figure 1. Fracture energy of concrete versus characteristic compressive strength – comparison of deterministic equations (7–10).

3.1 *Formulation VOS/EC2/ATENA-GID*

The NLFEAs presented in this paper are performed in ATENA Studio and ATENA-GID (Červenka et al. 2013). When generating concrete material properties using the EC2 code generator implanted within ATENA-GID software (Červenka et al. 2013), the following equation is used to derive the fracture energy from the tensile strength f_{ct} (Sucharda 2020).

$$G_f = 25 f_{ct}. \tag{7}$$

The formulation is based on the relationship according to Vos (1983) and Červenka et al. (2013) and results in much lower values for the fracture energy (see Figure 1) compared to *fib* Model Code 2010 (2013) or compared to the proposed formulation in the draft for *fib* Model Code 2020 (2022).

3.2 *Formulation in CEB-FIP Model Code 1990 (MC90)*

The following deterministic equation is recommended in CEB-FIP Model Code 1990 (1993) in the absence of experimental data:

$$G_f = G_{F0}\left(\frac{f_{cm}}{f_{cmo}}\right)^{0.7}, \tag{8}$$

where f_{cmo} = 10 MPa and G_{F0} is the base value of the fracture energy that depends on the maximum aggregate size d_{max} (see Table 1). According to CEB-FIP Model Code 1990 (1993) the equation does not take into account the size of structural members and other concrete properties resulting in deviations of ± 30%. This equation (8) also shows a too pronounced effect of the compressive strength on the fracture energy G_f when compared to experimental data (CEB-FIP 2008).

Table 1. Base values of fracture energy G_{F0}.

d_{max} [mm]	8	16	32
G_{Fo} [N/mm]	0.025	0.03	0.058

3.3 *Formulation in fib Model Code 2010 (MC2010)*

In absence of experimental data, *fib* Model Code 2010 (2013) following deterministic equation for normal weight concrete is given to estimate the value of the fracture energy G_f from the mean compressive strength f_{cm}:

$$G_f = 73 f_{cm}^{0.18}. \tag{9}$$

A similar formulation has already been mentioned in *fib* Bulletin 42 (CEB-FIP 2008), where the underestimation of the value of the fracture energy using Eq. 8 for low-strength concretes and the overestimation of the fracture energy for high-strength concretes is shown.

3.4 *Formulation in fib Model Code 2020 (MC2020)*

In the proposal for *fib* Model Code 2020 (2022) the formulation for the deterministic determination for the estimation of the value for the fracture energy is modified in comparison to *fib* Model Code 2010 (2013). The equation is based on the characteristic compressive strength f_{ck} in contrary to the mean compressive strength f_{cm} in equation (9). This formulation results in slightly lower values for the fracture energy G_f for concrete with a characteristic strength lower than 18 MPa, and slightly higher values for a compressive strength between 20 to 100 MPa (see Figure 1).

$$G_F = 85 f_{ck}^{0.15}. \tag{10}$$

4 CASE STUDIES AND METHOLOGY

In order to show the impact of the safety formats and the choice of material parameters in the applications of these on assessment of the load bearing capacity and subsequently on the safety level and the reliability index, two different structures are selected. These structures were investigated within European INTERREG AUSTRIA-CZECH REPUBLIC "ATCZ190 SAFEBRIDGE" project. The project focused on advanced numerical analysis of existing bridges and their safety formats. In particular, this paper presents an in-situ produced reinforced concrete bridge in Austria and a fictive reinforced T-beam both failing in shear:

(a) reinforced concrete bridge (detail v2-1): detailed model of a reinforced T-beam concrete bridge
(b) set 8: a simple reinforced T-beam

For both case studies the same methodology has been applied. To study the influence of the deterministic equations of the fracture energy on the assessment of structures, NLFEMs for the 4 formulations described in Section 3 have been created. For the computation of the parameters for Model Code 90 (CEB-FIP,1993) the maximum aggregate size was assumed with $d_{\max} = 8$ mm resulting in a base value of the fracture energy $G_{F0} = 0.03$ N/mm.

This approach has been applied for the three safety formats briefly presented in Section 2. As mentioned in the introduction, in the proposed procedures and safety formats it is not clearly defined which values of material input parameters should be or must be used in order to derive reliability-consistent design values of the structural resistance. There is more or less conflicting information and guidelines, which input parameters need to be changed for a specific safety format and how to derive these values (Belletti et al., 2011, Belletti et al., 2017, fib, 2013, Hendriks et al., 2017). In order to take into account the scope of interpretation left by the codes, the ECoV method and the PFM is computed with 2 variants of input parameters of the material properties and the GRF method with 3 variants, resulting in 8 NLFEM per code and 27 design values R_d for the structural resistance per case study.

4.1 *Reinforced concrecte bridge*

The first investigations focused on the Freudenauer Hafenbrücke B0209 with a span length of 352.6 m and a width of 15 m. The bridge was built in the years 1957/1958. The superstructure of the bridge relevant to the analysis and discussed in this paper is a four-span, four-girder T-beam bridge with cross girders at the abutments and in the center of the span. The analysed section has a total length of 105.25 m and a span length of 26.20 m. The width is about 15.50 m and the width for traffic lanes is 12 meters. The bridge structure was designed for the load models according to the standards at the time of the construction (Ing. Mayreder et al., 1957) and the structure was also recalculated in 1984 according to relevant standards of the time (Fritsch and Chiari, 1984). As part of the research Project ATCZ190 SAFEBRIDGE, the bridge was recalculated to modern load models according to Eurocode (CEN 2010) with the relevant traffic load model LM 1. With a span of 12 m, the bridge is designed to accommodate four traffic lanes (Sattler et al. 2022).

An initial linear FE assessment has been performed for the whole bridge structure using SOFiSTiK software (SOFiSTiK 2020). The goal of the linear assessment was to find the areas of the structures with the highest bending moments and highest shear forces - the areas that are likely to cause failure of the structure - and the corresponding load cases (Sattler et al. 2022).

The non-linear finite element analyses are performed in ATENA Studio developed by Cervenka Consulting (Červenka et al. 2013). In the NLFEM, only the edge beam is modelled as shown in Figure 2.

This model is a cut-out of the bridge including parts of field one, support G, and parts of field two. These are the parts of the edge beam with the highest bending moments and highest shear forces and the areas where the verification of for shear force capacity according to Eurocode 2 (CEN 2004) in the initial linear assessment could not be achieved. The beginning and the end of the detail model are respectively the points of the zero crossing of the bending moment, i.e., the points where the curvature of the deformation line changes (Sattler et al. 2022).

A preliminary study comparing the edge beam model presented here with larger mode in the transverse directions and comparing it to different load cases showed that the model presented here has the same failure mechanism and achieves similar ultimate loads (Sattler & Strauss 2022). Due to the reduced size of the model, the calculation time is significantly reduced modelling just the edge beam. The cross section for the NLFEM is created as a three-dimensional system using volumes/brick elements (Sattler et al. 2022).

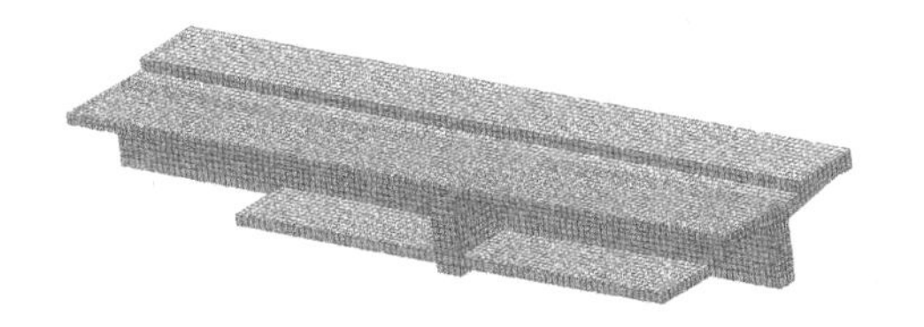

Figure 2. Generated mesh and the cross section of model v2-1.

The generated mesh for the NELFM consists of 15,422 hexahedral finite elements shown in Figure 2. The concrete type B300 is modelled using the advanced material model *CC3DNonLinCementious2* in ATENA-GID (Červenka et al. 2013, Strauss et al. 2018). The longitudinal reinforcement of the beam and the slab as well as the stirrup reinforcement is modelled as discrete 1D reinforcement according to the original drawings. To reduce the model size and increase the computation time, the transverse reinforcement of the slab is modeled as smeared reinforcement (Sattler et al. 2022).

As shown in Figure 3, the loading has been applied stepwise to the yellow, pink, blue and green colored areas, whereas the loads are applied uneven in order to take into account the different length on the left and right side of the support, respectively. To take into the load model LM1 according to EN 1991 (CEN 2010), configuration of the different traffic lanes, the loads are applied uneven in transverse direction, also shown in Figure 3 (bottom). Regions of 10 to 20 cm of linear elastic materials at the areas of the boundary conditions and the loading have been applied to avoid unrealistic non-linear effects resulting from singularities, see Figure 3 (top). The horizontal displacements are restricted along the concrete slab as well as along the crossbeam at the level of the support in order to simulate the subsequent parts of the remaining bridge. To prevent tipping and instability a boundary condition hindering the vertical and longitudinal displacement at the bottom area of the support has been applied (Sattler et al. 2022).

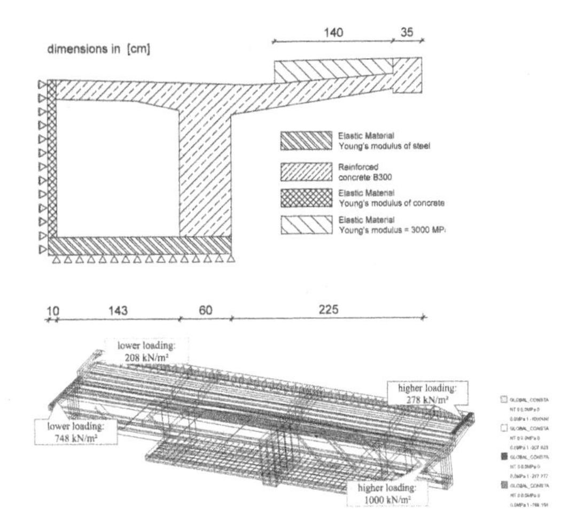

Figure 3. Boundary conditions of model v2-1 (top), load configuration and application to account for eccentricity (bottom).

The material parameters from the bridge's original design – concrete B300 and reinforcement steel, type Torstahl 40 (RT40) were transformed to mean values according to the applied codes (ASI 2018; CEB-FIP 1993; fib 2013; fib 2022). For the stochastic and semi-probabilistic evaluation, 6 random variables were chosen for the concrete B300 and 2 random variables for the reinforcement RT40 and these are as follows: Young's modulus of concrete E_c, compressive strength of concrete f_c, tensile strength of concrete f_{ct}, fracture energy G_f, Plastic Strain and Onset of crushing, Young's modulus of reinforcement E_y and yield strength f_y. The mean values for the concrete corresponding to the code formulation are displayed in Table 2 (Sattler et al. 2022).

The random variables for the NLFEA using characteristic material parameters for ECoV format, *mean* material parameters for the GRF format and design material parameters for the PFM as input variables were derived according to different interpretations of proposals in the literature (Belletti et al., 2011, 2017; fib 2013; Hendriks et al. 2017). The input values of *concrete* for the corresponding NLFEM are displayed in Table 2. For the fracture energy G_f, the formulations according to Section 2 were used.

4.1.1 *Results*

For the non-linear finite element models, the design values of resistance R_d are determined under the assumption of a log-normal distribution with the target reliability factor $\beta_{ULS} = 3.8$ and the FORM sensitivity factor α_R set equal to 0.8. To account for model uncertainties, the design values are reduced by the model uncertainty factor $\gamma_{Rd} = 1.06$, as recommended in EN

Table 2. Material parameters for concrete B300 for the different safety formats used for the semi-probabilistic assessments.

Model Code 1990, Model Code 2010, Model Code 2020

	unit	mean values	characteristic (k)	characteristic 2 (k2)	GRF	GRF2	GRF3	PFM	PFM2
E_c	N/mm²	29488	29488	26056	29488	24682	25065	29488	22762
f_{cs}	N/mm²	25.8	17.8	17.8	15.13	15.13	15.13	11.87	11.87
f_t	N/mm²	2.045	1.432	1.432	1.217	1.111	1.217	0.954	0.954
G_f (MC90)	N/m	58.24	44.92	44.92	38.18	40.09	38.18	33.82	33.82
G_f (MC2010)	N/m	131.04	122.57	122.57	104.19	119.04	104.19	113.95	113.95
G_f (MC2020)	N/m	130.91	130.91	130.91	111.28	114.13	111.28	130.91	130.91

Eurocode 2

	unit	mean values	characteristic (k)	characteristic 2 (k2)	GRF	GRF2	GRF3	PFM	PFM2
E_c	N/mm²	29200	29200	26056	29200	24682	24820	29200	23159
f_c	N/mm²	25.8	17.8	17.8	15.13	15.13	15.13	11.87	11.87
f_t	N/mm²	2.045	1.432	1.432	1.217	1.111	1.217	0.954	0.954
G_f (VOS)	N/m	51.13	35.79	35.79	30.42	27.78	30.42	23.86	23.86

1992-2 (CEN 2008) and in line with the proposals of the *fib* Model Code 2010 (2013). The maximum load capacity was selected for each NLFEM by means of a chosen failure criterion with respect to the delta of the displacement from one calculation step to the next calculation step. Figure 4 shows an example of the load versus delta-displacement curves of the NLFEAs performed with mean material parameters (see Table 2). Please note that the reference to the specific code refers to the corresponding formulation of the fracture energy, where EC2 corresponds to the formulation according to VOS (1983). In this plot it can be clearly seen that the different formulations of the fracture energy have an influence on the maximum load bearing capacity of the structure.

The design values for the resistance R_d have been determined using the ECoV approach, the global resistance factor method (GRF) and the partial factor method (PFM) as described in Section 2.

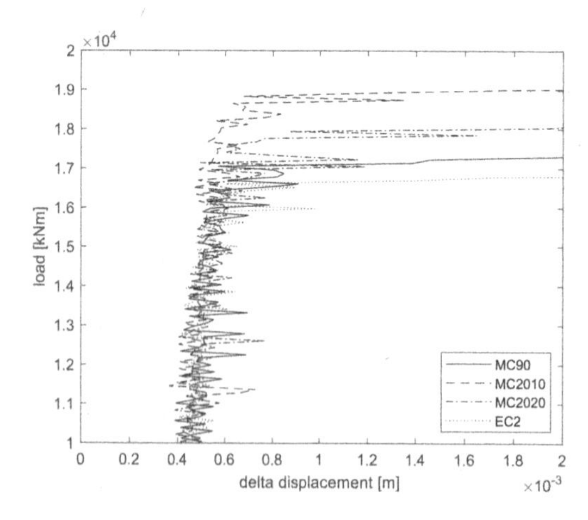

Figure 4. Load versus delta of the displacement curves of model v2-1 with mean material parameters.

The mean value of the results of the design value R_d, the derived mean value R_m as well as the CoV and the global resistance factor safety factor γ_R are shown in Table 3 grouped by the applied code for determining the value of the fracture energy. In Figure 5 the derived design values R_d (Figure 5 left) and the CoV (Figure 5 right) of the design values are displayed as a bar chart. For the sake of comparison, the mean design value of all derived design values and the corresponding CoV is also displayed. The design values obtained using the draft of *fib* Model Code 2020 (2022) lead to the highest design loads and are roughly 16% higher compared to the values obtained from EC2 input parameters. However, compared to the formulations of Model Code 2010 (2013), the difference with just 2% is negligible. Figure 6 shows the obtained design load of model v2-1 in relation to the input value of the fracture energy, again grouped by the 4 different codes as scatter plot. One can see that there is a correlation between the value of fracture energy and the corresponding design load.

Table 3. Derived design value R_d, mean value R_m and CoV of the resistance *R* for model v2-1 grouped by applied code.

Model v2-1

Code	R_d [kNm]	R_m [kNm]	CoV [-]	γ_R [-]	γ_{rd} [-]
MC 90	10692	14762	0.076	1.13	1.06
MC 2010	11308	14856	0.064	1.11	1.06
MC 2020	11560	16443	0.084	1.16	1.06
EC2	9975	12730	0.056	1.09	1.06
mean of all NLFEA	10884	15786	0.089	1.17	1.06

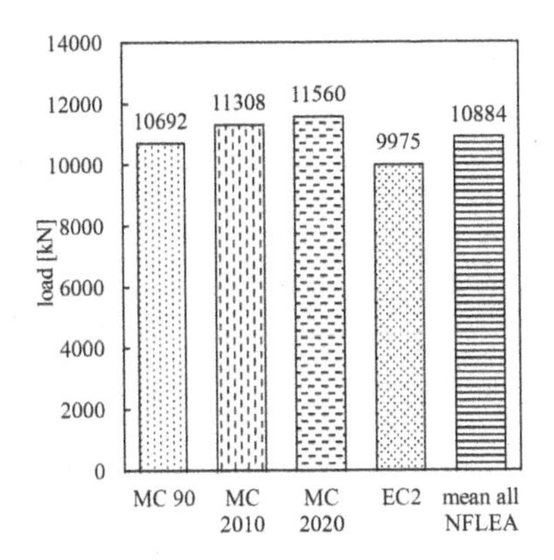

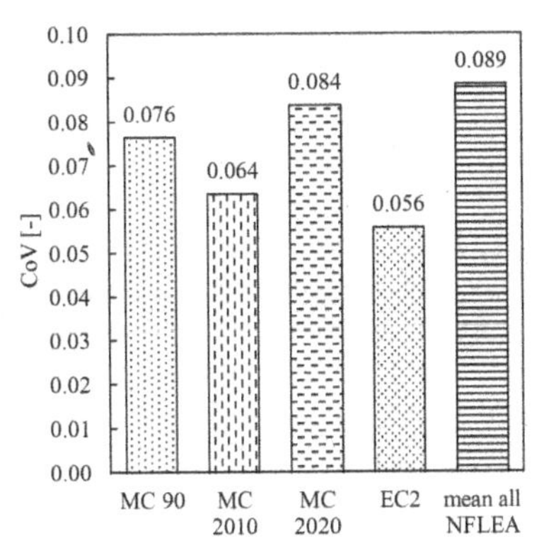

Figure 5. Bar chart of the derived design values for model v2-1(left), derived CoV for the design values of model v2-1 (right).

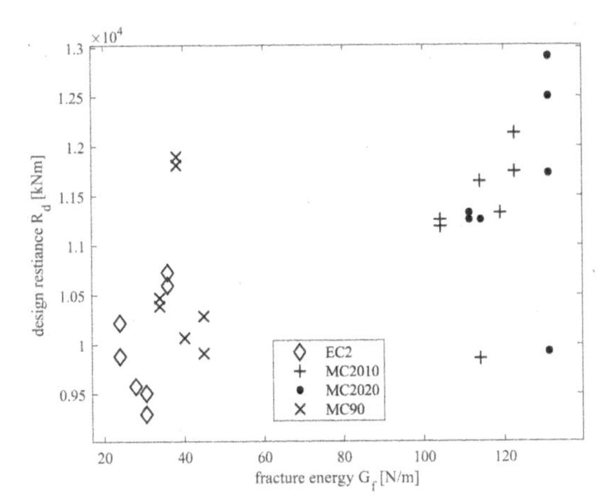

Figure 6. Scatter plot model v2-1: derived design values R_d versus fracture energy G_f.

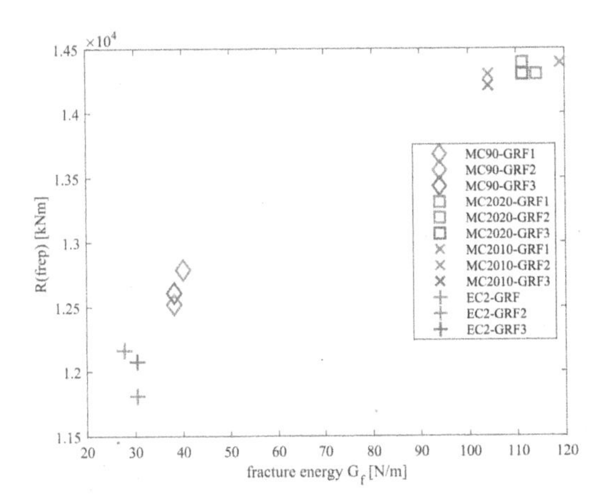

Figure 7. Scatter plot model v2-1: maximum load capacity versus fracture energy using GRF input parameters.

The calculation of the correlation coefficient ρ delivers a value of $\rho = 0.6236$. However, there are 4 NLFEM simulations that do not correspond to that trend. An overestimation compared to the expected design load R_d for a certain value of fracture energy G_f occurs for variants MC90-GRF and MC90-GRF3 (north-west side of the plot) and an underestimation occurs for variant MC2010-PFM and MC2020-PFM (south-east side of the plot).

Figure 7 shows the ultimate loads derived from the modelled variants of the Global Resistance Factor Method (GRF1-3) with *mean* material parameters (see Table 2) as input variables. They are grouped by specific markers regarding the used code and differentiated by color regarding the used variant (GRF1-3). A notable change in the maximum load capacity can only be determined for the variants of Model Code 1990 (diamond marker) and EC2 (+ Marker).

4.2 *T-beam set 8*

In the context of the ATCZ project, an analysis was carried out to determine the optimum between computation time and computational accuracy of a specific non-linear finite element model. This study was performed on a structural system with similar geometry and reinforcement layout as detail v2-1, i.e. based on the detail of the Freudenauer harbor bridge. Thirteen sets of mesh sizes and reinforcement layouts were modelled and the resulting crack widths and crack developments, steel stresses and reaction forces were calculated and analyzed. In general, a distinction was made between two mesh sizes, two variants of beam geometry and 1-span beams and 2-span beams. The T-beam described in this section (set 8) is a result of this NLFEA study and Figure 8 shows the dimensions of the modelled T-beam and the position of the loading force. For set 8 a fictitious load position was selected and applied as point-like displacement to the system. The loading force is distributed via a 60 × 60 × 15 mm steel plate (Figure 8). A solid Elastic 3D material model implanted within ATENA-GID (Červenka et al. 2013) using a Young Modulus of $2*10^5$ MPa was utilized for the plate. The NLFEM consists of 19,650 elements of hexahedra type, whereby the ratio between edge sizes of a single element never exceeds 3:1. For general numerical stability and for mesh compatibility

Figure 8. Geometry and loading configuration of set8, dimensions in [cm].

Table 4. Material parameters for concrete C30/37 for the different safety formats used for the semi-probabilistic assessments.

Model Code 1990, Model Code 2010, Model Code 2020

	unit	mean values	characteristic (k)	characteristic 2 (k2)	GRF	GRF2	GRF3	PFM	PFM2
E_c	N/mm^2	33551	33551	31008	33551	29373	28518	33551	27088
f_{cs}	N/mm^2	38	30	30	25.50	25.50	25.50	20	20
f_t	N/mm^2	2.896	2.028	2.028	1.723	2.022	1.723	1.35	1.35
G_f (MC90)	N/m	76.38	64.73	64.73	55.02	57.77	55.02	48.74	48.74
G_f (MC2010)	N/m	140.50	134.65	134.65	114.45	130.77	114.45	125.17	125.17
G_f (MC2020)	N/m	141.58	141.58	141.58	120.34	130.58	120.34	141.58	141.58

Eurocode 2

	unit	mean values	characteristic (k)	characteristic 2 (k2)	GRF	GRF2	GRF3	PFM	PFM2
E_c	N/mm^2	32837	32837	30589	32837	29133	27911	32837	27085
f_c	N/mm^2	38	30	30	25.50	25.50	25.50	20	20
f_t	N/mm^2	2.896	2.028	2.028	1.723	2.022	1.723	1.35	1.35
G_f (VOS/EC2)	N/m	72.41	50.69	50.69	43.09	50.55	43.09	33.79	33.79

between two connected volumes, brick elements were used (Sattler et al., 2022).

Moreover, brick elements allow an easy definition and application of a structured mesh using hexahedral elements. The *concrete* of the beam type C30/37 is modelled using the advanced material model CC3DNonLinCementious2 in ATENA-GID (Červenka et al. 2013; Strauss et al. 2018). The longitudinal reinforcement and the stirrup reinforcement is modelled as discrete 1D reinforcement material, the reinforcement in the slab is modelled as smeared reinforcement (Červenka et al. 2013).

4.2.1 *Results*

The maximum load capacity was obtained for each NLFEM, Figure 9 shows an example of the load versus displacement curves of the NFLEMs obtained with mean values for material parameters. Please note that the reference to the specific code refers to the corresponding formulation of the fracture energy.

As described in section 4.1.1, the design values of resistance R_d of the performed NLFEM are determined under the assumption of a log-normal distribution with the target reliability factor $\beta_{ULS} = 3.8$ and with α_R set equal to 0.8. To account for model uncertainties the design values are reduced by the model uncertainty factor $\gamma_{Rd} = 1.06$. The design values were obtained using the safety formats described in section 2. Figure 10 shows the derived *mean* design values R_d (Figure 10 left) as well as the CoV (Figure 10 right) of the design values for the corresponding code as bar chart. For comparison the *mean* design value and CoV of all four codes is also displayed. One can see that the differences between the design values R_d of MC2020 and MC2010 are negligible (1.5%), but

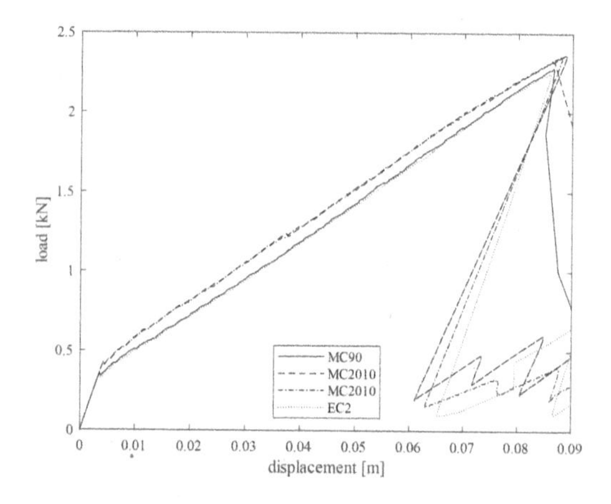

Figure 9. Load versus displacement curves for set 8 using mean material parameters

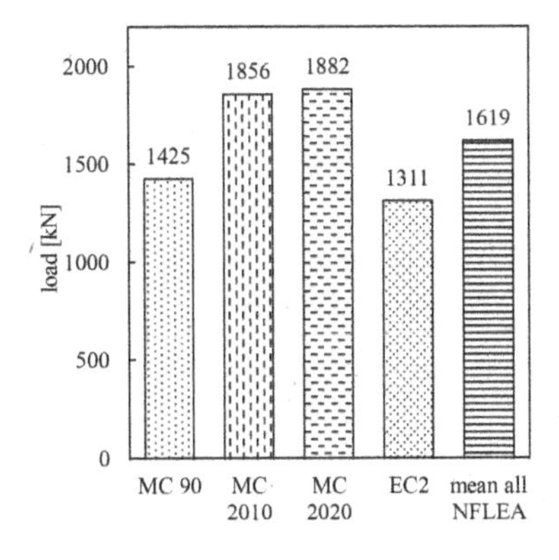

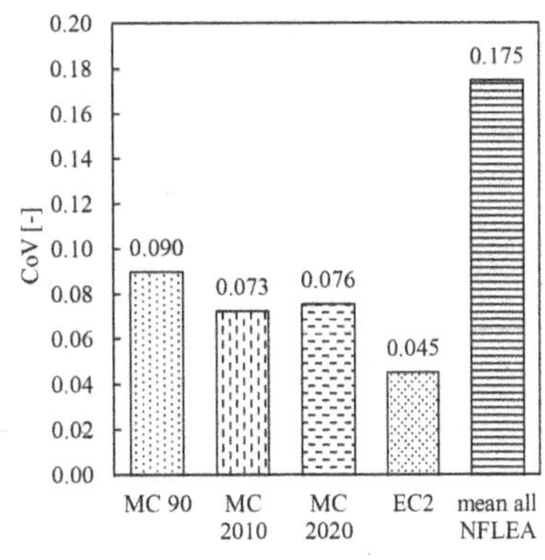

Figure 10. Bar chart of the derived design values for set 8 (left), derived CoV for the design values of set 8 (right).

the differences compared to MC90 and EC2 are 32% and 43% respectively. This results in a high coefficient of variation (CoV) for the *mean* of all design values. Figure 11 shows the obtained design values of the structural resistance to the corresponding value of fracture energy used in the NLFEA. In comparison to Figure 6 the correlation between the two values R_d and G_f is more pronounced and the correlation coefficient $\rho = 0.9329$ can be computed. Figure 12 displays the ultimate loads derived from the 3 modelled variants of the Global Resistance Factor Method (GRF1-3) with *mean* material parameters as input variables.

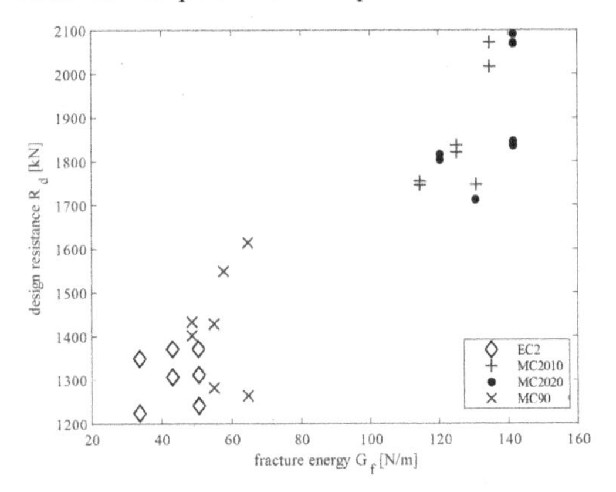

Figure 11. Scatter plot set 8: derived design values Rd versus fracture energy G_f.

Table 5. Derived design value R_d, mean value R_m and CoV of the resistance R for model set 8 grouped by applied code.

set 8					
Code	R_d [kN]	R_m [kN]	CoV [-]	γ_R [-]	γ_{rd} [-]
MC 90	1425	2077	0.090	1.24	1.06
MC 2010	1856	2524	0.073	1.20	1.06
MC 2020	1882	2591	0.076	1.20	1.06
EC2	1311	1614	0.045	1.14	1.06
mean of all NLFEA	1619	3658	0.175	1.61	1.06

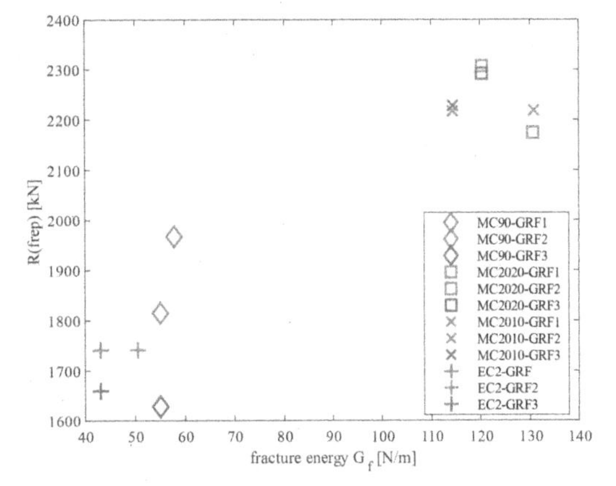

Figure 12. Scatter plot set 8: maximum load capacity versus fracture energy using GRF input parameters.

They are grouped by specific markers regarding the used code and differentiated by color regarding the used variant (GRF1-3). Please note that the same evaluations were performed for the other safety formats but are not presented due to space limitations.

5 CONCLUSIONS

On the basis of the results of the two case studies presented in Section 4, it can be concluded that the deterministic formulations for the fracture energy (7-8) according to VOS/EC2 (1983) and according to *fib* Model Code 1990 (CEB-FIP, 1993) lead to similar design values R_d. On the other hand, the NLFEM with these two formulations of the fracture energy (7-8) leads to significantly lower design values of the resistance side compared to the formulations (9–10) according to *fib* Model Code 2010 (2013) and *fib* Model Code 2020 (2022), respectively. Consequently, the safety index Beta and the safety margin also increase with these newer deterministic equations. Furthermore, the analyses show that with the new formulations of the fracture energy, the CoV also increases.

Further literature research and calculations are needed to show the influence of fracture energy on different structures and to give a recommendation for the choice of the deterministic fracture energy equation. It is necessary to investigate which of the deterministic fracture energy equations are compatible with the models implemented in the NLFEA to best reproduce the results compared to laboratory experiments. Nevertheless, it would be necessary for users to include clear formulations in this regard in future codes.

REFERENCES

Allaix, D. L., Cabrone, V. I. & Mancini, G. (2013) Global safety format for non-linear analysis of reinforced concrete structures. *Structural Concrete,* 14, 29–72.

ASI (2019) *ÖNORM B 4008-2 Bewertung der Tragfähigkeit bestehender Tragwerke - Teil 2: Brü-ckenbau,* Wien, Austrian Standards Institute.

Bažant, Z. P. & Becq-Giraudon, E. (2002) Statistical prediction of fracture parameters of concrete and implications for choice of testing standard. *Cement and Concrete Research,* 32, 529–556.

Bažant, Z. P. & Planas, J. (2019) *Fracture and Size Effect in Concrete and Other Quasibrittle Materials,* Boca Raton, Routledge.

Belletti, B., Damoni, C. & Hendriks, M. A. N. (2011) Development of guidelines for nonlinear finite element analyses of existing reinforced and pre-stressed beams. *European Journal of Environmental and Civil Engineering,* 15, 1361–1384.

Belletti, B., Damoni, C., Hendriks, M. A. N. & Boer, A. D. (2017) Validation of the Guidelines for Nonlinear Finite Element Analysis of Concrete Structures - Part: Reinforced beams. Delft University of Technology, Netherlands, Rijkswaterstaat Centre for Infrastructure.

Castaldo, P., Gino, D. & Mancini, G. (2019) Safety formats for non-linear finite element analysis of reinforced concrete structures: discussion, comparison and proposals. *Engineering Structures*, 136–153.

CEB-FIP (1993) *CEB-FIP Model Code 1990,* Lausanne, Thomas Telford Services Ltd.

CEB-FIP (2008) *Constitutive modelling of high strength/high performance concrete, fib Bulletin 42,* Lausanne, fib Fédération Internationale du Béton.

CEB (1995) *New Developments in Non-linear Analysis Method CEB Bulletin No. 229,* Lausanne, fib Fédération Internationale du Béton.

CEN (2004) *BS EN 1992-1-1:2004 Eurocode 2: Design of concrete structures - Part 1-1: General rules and rules for buildings,* Brussels, European Committee for Standardization.

CEN (2008) *BS EN 1992-2:2005 Eurocode 2: Design of concrete structures - Part 2: Concrete bridges - Design and detailing rules,* Brussels, European Committee for Standardization.

CEN (2010) *BS EN 1991-2:2003 Eurocode 1: Actions on structures - Part 2: Traffic loads on bridges,* Brussels, European Committee for Standardization.

CEN (2013) *BS EN 1990:2002 Eurocode - Basis of structural design,* Brussels, European Committee for Standardization.

Červenka, V. (2013) Reliability-based non-linear analysis according to fib Model Code 2010. *Structural Concrete,* 14, 19–28.

Červenka, V., Jendele, L. & Červenka, J. (2013) ATENA Program Documentation, Part I, Theory, Czechia: Cervenka Consulting.

Czernuschka, L.-M., Wan-Wendner, R. & Vorel, J. (2018) Investigation of fracture based on sequentially linear analysis. *Engineering Fracture Mechanics,* 202, 75–86.

FIB (2013) *fib Model Code for Concrete Structures 2010,* Wilhelm Ernst & Sohn.

FIB (2022) *fib Model Code 2020,* fib - Fédération Internationale du Béton, Draft.

Fritsch, P. & Chiari, G. (1984) Freudenauer Hafenbrücke Objekt 0209 Brücke über die Hafeneinfahrt und Donaukanal - Standberechnung - Kurzfassung. Wien, MA 29 Brückenbau und Grundbau.

Hendriks, M. A. N., De Boer, A. & Belleti, B. (2017) Guidelines for Nonlinear Finite Element Analysis of Concrete Structures. Delft University of Technology, Netherlands, Rijkswaterstaat Ministerie van Infrastructure en Milieu.

Hillerborg, A. (1985) The theoretical basis of a method to determine the fracture energy GF of concrete. *Materials and Structures,* 18, 291–296.

Ing. Mayreder, Kraus & Co. (1957) Originalstatik Freudenauer Hafenbrücke.

Marí, A., Bairán, J., Cladera, A., Oller, E. & Ribas, C. (2015) Shear-flexural strength mechanical model for the design and assessment of reinforced concrete beams. *Structure and Infrastructure Engineering,* 11, 1399–1419.

Sattler, F. & Strauss, A. (2022) Probabilistic and Semi-probabilistic Analyses of Bridge Structures - Multi-level Modelling Based Assessment of Existing Structures. *Proceedings of the 1st Conference of the European Association on Quality Control of Bridges and Structures.*

Sattler, F., Strauss, A., Novák, D. & Novák, L. (2022) Application of a novel safety format technique on concrete bridges. *1th International Conference on Bridge Maintenance, Safety and Management (Barcelona IABMAS 2022),* Unpublished.

Schneider, J. & Schlatter, H. P. (1994) *Sicherheit und Zuverlässigkeit im Bauwesen,* ETH Zürich, v/d/f Hochschulverlag AG an der ETH Zürich.

Slowik, O., Novák, D., Novák, L. & Strauss, A. (2021) Stochastic modelling and assessment of long-span precast prestressed concrete elements failing in shear. *Engineering Structures,* 228, 111500.

Sofistik (2020) SOFiSTiK 2020 ASE Handbuch.

Strauss, A., Zimmermann, T., Lehký, D., Novák, D. & Keršner, Z. (2014) Stochastic fracture-mechanical parameters for the performance-based design of concrete structures. *Structural Concrete,* 15, 380–394.

Strauss, A., Krug, B., Slowik, O. & Novak, D. (2018) Combined shear and flexure performance of prestressing concrete T-shaped beams: Experiment and deterministic modeling. *Structural Concrete,* 19, 16–35.

Sucharda, O. (2020) Identification of Fracture Mechanic Properties of Concrete and Analysis of Shear Capacity of Reinforced Concrete Beams without Transverse Reinforcement. *Materials,* 13, 2788.

Vos, E. (1983) Influence of Loading Rate and Radial Pressure on Bond in Reinforced Concrete. Delft University, TU Delft.

Zimmermann, T., Strauss, A., Lehký, D., Novák, D. & Keršner, Z. (2014) Stochastic fracture-mechanical characteristics of concrete based on experiments and inverse analysis. *Construction and Building Materials,* 73, 535–543.

Computational Modelling of Concrete and Concrete Structures – Meschke, Pichler & Rots (Eds)
© 2022 Copyright the Author(s), ISBN: 978-1-032-32724-2

Using submodels for a probabilistic nonlinear analysis of corroded RC-structures

M. Kwapisz, M. Ralbovsky & A. Vorwagner
AIT, Austrian Institute of Technology, Vienna, Austria

M. Rebhan
TU Graz, Austria

ABSTRACT: Probabilistic analysis is best suited to cover a wide range of concrete and steel properties including corrosion effects and its influence on structural behaviour. If a reduction of the computational time to calculate sufficient samples is needed, a use of submodels might be very useful. The general idea is to compute very accurately only a relatively small section that is of interest to capture the nonlinear steel properties including any localized corrosion and its bond with the concrete. The rest of the structure is considered either with simple FE elements or with an analytical solution. The procedure was verified by large-scale tests on cantilever walls with artificially induced corrosion and with a conventional nonlinear FE analysis. The presented method combines the advantages of a detailed and complex nonlinear FE analysis with the applicability and performance of an analytical solution or simple FE calculation. The advantages, accuracy and limitations of the method are broadly discussed, and the field of application is described.

1 MOTIVATION

Although nonlinear modeling of a reinforced concrete has become state-of-the-art, managed by most of FE software, it is still a time-consuming process. A major problem is that the obtained results are highly dependent on the assumed material parameters. With careful model updating or material testing, it is possible to obtain correct parameters that lead to a satisfactory result. In the case of corrosion modeling of reinforcement bars, which was a main topic of the presented studies, there is yet another uncertainty in the estimation or prediction. Therefore, it is important to acquire the full range of possible solutions in order to obtain the reliability and the possible structural behavior. In such a case, probabilistic analysis is best suited to cover a wide range of concrete and steel properties including corrosion effects and its influence on structural behavior. Reduction of the computational time needed to calculate a sufficient number of samples is achieved by using submodels which are described in detail.

2 METHODOLOGY

2.1 *Submodeling technique*

If only a few parts of a structure require detailed analysis whereas the rest could be greatly simplified, it is a common practice in the FE computation to use a submodel. It can be divided into two categories: one where a general behavior is independent on the local stress – strain correlation and second where the local behavior is crucial for the global deformation state. The first one is frequently used for example in fatigue assessment of weld joints, where the stress level at given point can be obtained with a detailed model by applying the deformation resulting from a coarse model of the whole structure as a boundary condition (FKM-Richtlinie 2020). The latter, which is in focus of this paper, works the other way around. Firstly, a very precise model of a small detail is created, from which the dependency between force and deformation or bending moment and rotation is extracted. Once it is done, it can be simplified as nonlinear spring element or described with an analytical equation and used for assessing the behavior of the whole structure.

2.2 *Detailed model and parameters*

Presented study focuses on the pitting corrosion of reinforcing bars in RC-structures and its influence on the deflection and inclination changes of the structure. The submodel consists of concrete, reinforcing bars and nonlinear springs representing the bond connection (Figure 1). Material properties of concrete, steel and bond were distributed using values taken from the model code (fib model code 2010). The corrosion itself was included by decreasing of the cross section over the chosen length and by reducing the bond to the concrete according to (Bhargava 2007).

DOI 10.1201/9781003316404-82

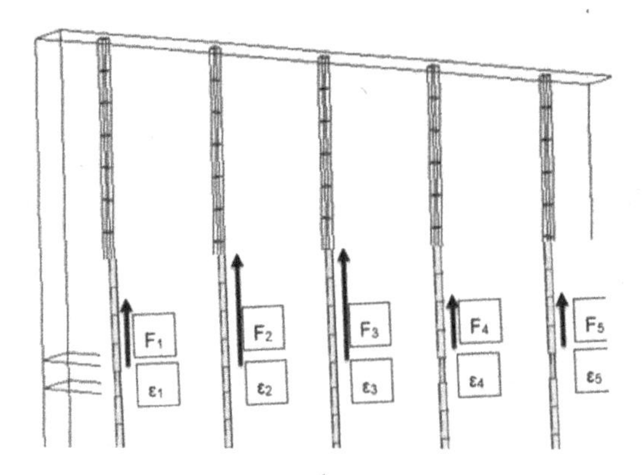

Figure 1. Submodel consisting of reinforcing bars and concrete.

Corrosion extend was also a probabilistic parameter, with separately generated values for each bar. It was an important issue, as it is in compliance with repeated observations during the inspections of the corroded structures, mostly retaining walls and with validation tests in the lab. It was noticed, that corrosion ratio often varies strongly within a single element which leads to strongly corroded bars neighboring almost healthy ones. Providing accurate corrosion distributions was definitely a difficult and uncertain task. An example of 500 generated samples is shown on Figure 2. The average values were based on inspections data (Vollenweider 2014) where also the indication on the distribution could be found.

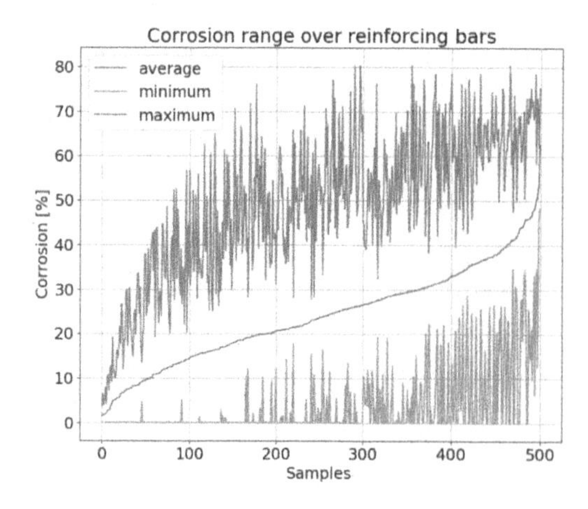

Figure 2. Range of ccorrosion rates of reinforcement bars

The assumption of an existing crack where a corrosion damage can occur was made. An example of a detail is shown on Figure 1. Submodel is cut from a tension side of the wall having the dimension of one meter length and a thickness equals double of the concrete cover. Height was chosen as anchor length in each direction starting from the induced crack. An appropriate number of bars was included and fixed with the concrete in transversal directions, whereas in longitudinal a nonlinear springs representing bond were used. Reduction of bond and bar cross section was undertaken only at the given corrosion length, whereas the rest was considered healthy.

Averaged length change for calculating the spring characteristic of all corroded reinforcing bars was determined with equation 1:

$$\Delta l_m = \frac{\sum_i^n \varepsilon_i}{n} * l_k \tag{1}$$

where Δl_m = Averaged length change of bars; ε_i= strain of single reinforcement bar in corroded section; n = bar amount and l_k = notch length.

3 VALIDATION

The simulation procedure, chosen material parameters and their distributions were carefully validated using full-scale tests and conventional FE analysis. The latter was also compared to the test results. Detailed information about the test procedure and results are described (Rebhan 2020).

3.1 *Test setup*

The corrosion of the reinforcement bars in the zone of the construction joint between the foundation and the wall segment was measured on several test specimens. For this purpose, U-shaped test specimens (Figure 3) were made of reinforced concrete. The test planning, test layout and execution were accompanied by numerical calculation models.

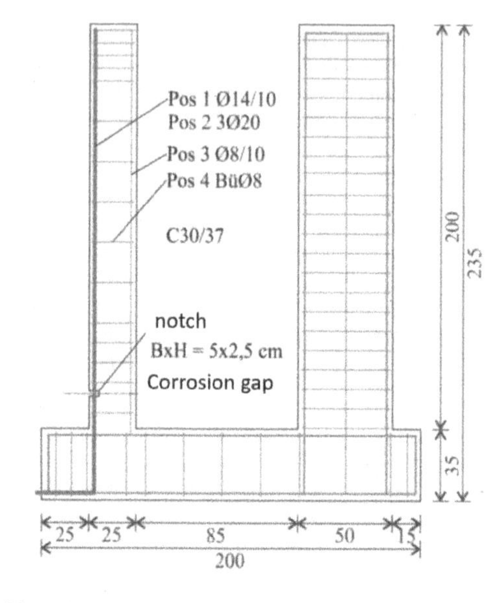

Figure 3. Geometry of a retaining test wall; bars subjected to corrosion marked in red.

The more massive of the two vertical legs was used as an abutment for the load application, the second (thinner) one formed the vertical wall leg of an angle retaining wall. The horizontal foundation bar of the structure served on the one hand as a force short-circuit between the two vertical segments, and on the other hand as a foundation. This was necessary to represent not only the vertical wall but also the connection area to the foundation and thus the area of the construction joint. The stressing of the structure was achieved by a concentrated load at the head, resulting in a linear moment load (earth pressure on an angle retaining wall) that increases towards the bottom.

By using the process of electrochemical erosion, it is possible to reproduce artificial corrosion in the form of cross-section reduction. The process of electrochemical ablation makes it possible to achieve a continuous reduction in diameter. On the one hand, this can be achieved by arranging a continuous cathode for the corrosion of all bars; on the other hand, local cathodes can also be used to weaken only individual reinforcing bars.

3.2 *Comparison between test results and conventional FE method*

The design of the test geometry and its execution was accompanied by nonlinear numerical simulations. The computations were a very helpful tool for a successful test program but could not obtain the exact values that were than measured during corrosion and ultimate loading process even if some of the material parameters were known beforehand. The agreement could be achieved by a model update once the measurement was fully analyzed (Figure 4).

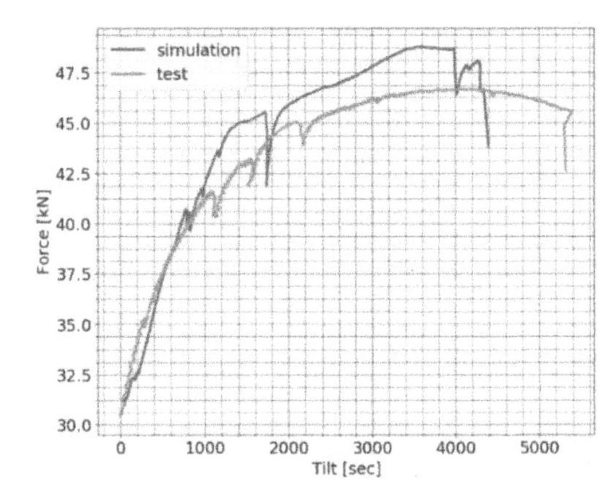

Figure 4. Comparison of the force – inclination diagrams showing test results (orange) and full FE simulations (blue).

This leads to an obvious conclusion that even very sophisticated analysis does not ensure obtaining accurate results while using a deterministic approach. In most cases it is thus more useful to make probabilistic analysis and be able to work with probabilities of obtaining certain values and with a range of possible solutions. This is true even if some simplifications must be done in order to reduce the computational effort.

3.3 *Comparison with the deterministic solution*

One of the test scenarios described in 3.1 was, that a wall with around 50% corrosion ratio of reinforcement was subjected to an additional load until the bars plastified and fracture. Resulting inclination was measured and after substruction of the part resulting from elastic wall movement an increase due to corrosion was estimated. These values are marked in the Figure 5 with dots, showing the increase of force with 2kN step, starting with 36kN until the ultimate force of 44kN. Solid lines with the same colors represent the solution based on FE submodel and analytical rigid body rotation. Although a perfect match could not be reached, the method can be evaluated as thoroughly satisfactory. It is also to be noted, that the corrosion level obtained in the lab was measured manually which might lead to minor uncertainty.

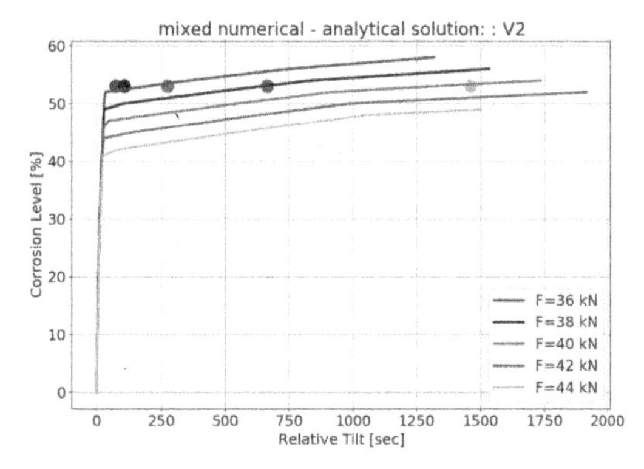

Figure 5. Comparison of a corrosion dependent inclination between test (dots) and mixed FE and analytical solution.

3.4 *Comparison with a probabilistic study*

The main advantage of using submodels was a simplicity and quickness of performing probabilistic analysis. To verify its accuracy, it was confronted with the results of all five test wall setups. It is visualized on Figure 6 with solid lines showing test results and each dot representing generated sample.

To compare different wall configurations the diagram was generated using force to the wall collapse on the vertical axis. It results from the substruction of the current force from the ultimate force and therefore must always be negative. It was introduced to represent the remaining capacity of a retaining wall and is used in further assessment of existing and generic walls described in 4.1. Horizontal axis represents a corrosion caused increase of inclination on the top of the wall.

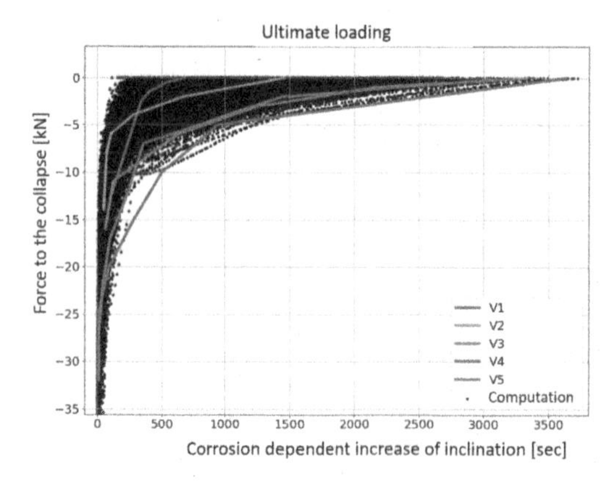

Figure 6. Comparison of a corrosion dependent inclination between five tests (solid lines) and mixed FE and analytical solution.

By generating over 10,000 samples with different material properties and corrosion extend a wide range of results was calculated. Test results seam to behave slightly more ductile in comparison to the computation, but a match was satisfactory.

4 CASE STUDIES

A nonlinear probabilistic analysis was carried out for two different tasks, both of which should estimate the influence of the corrosion on the deformation of the structure. One was a broad population of retaining walls with varying thickness and reinforcement ratio being subjected to a pitting corrosion located at the construction joint. The other was a rail bridge with a corrosion problem in the mid span. Exactly the same procedure to divide the calculations into two parts could be used: detailed submodel and a simplified evaluation of the whole structure.

4.1 *Retaining walls*

The most common location of the corrosion problems of the retaining walls is clearly defined as the construction joint. Thus, it can be solely modeled with a detailed submodel, whereas the rest of the wall is considered as rigid body with induced rotation over the construction joint. It must be stated that only an increase of deformation due to corrosion is of interest, so the elastic load-dependent deformation was not calculated. No additional cracks ware expected, as the load (earth pressure) remains constant and introducing a week section amplified the existing crack at this point.

For each wall configuration a submodel was created and for each sample a force – deflection dependence was extracted. The recalculation into bending moment – rotation allowed a direct output as a corrosion dependent increase of inclination/deformation. Those were used as a basis for an analytical dependency between earth pressure and wall inclination change (Figure 7).

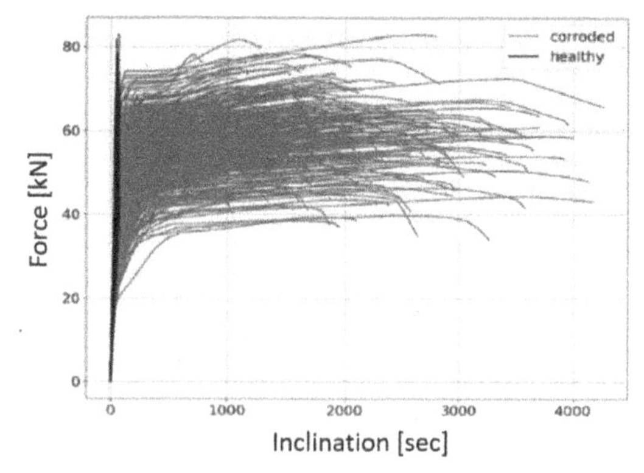

Figure 7. Results from a probabilistic analysis of the corroded retaining walls.

The most interesting was to assess the probability of damage detection with a given measurement accuracy. In order to do it, another parameter has to be involved, namely a possible change of the earth pressure due to for example increase of water saturation of the soil. A detailed outcome can be found in (Final Report on SIBS 2019) one example is shown in Figure 8. It shows a dependency of the probability of detection and the relative change of the earth pressure. The results are clustered into three categories based on diversity of corrosion: mainly uniform (blue), nonuniform (red) and purple in between. According to expectation, the chances of damage detection are raising the less uniform the corrosion is.

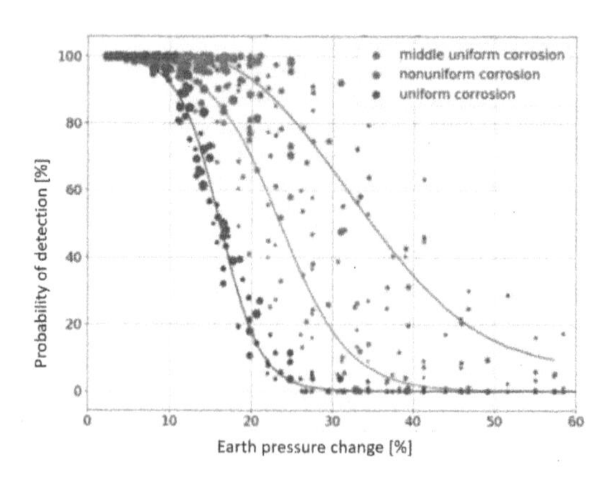

Figure 8. Probability of detection dependent on the change of earth pressure and uniformity of corrosion.

4.2 *Generic bridge*

The second use case is a rail bridge with an anticipated corrosion problem in the midspan. It consists of three spans, whereas the middle one is with 13m the longest. It was assumed, that the corrosion would take place locally which implies it could be represented with a rotational spring elements evenly distributed over the bridge width. Over this line representing corrosion, shell elements used for the bridge deck are split with a hinge so that only translational degree of freedom are coupled. Rotation over the axis in a transversal direction is overtaken by the implemented additional spring elements, having properties extracted from the submodel (Figure 9).

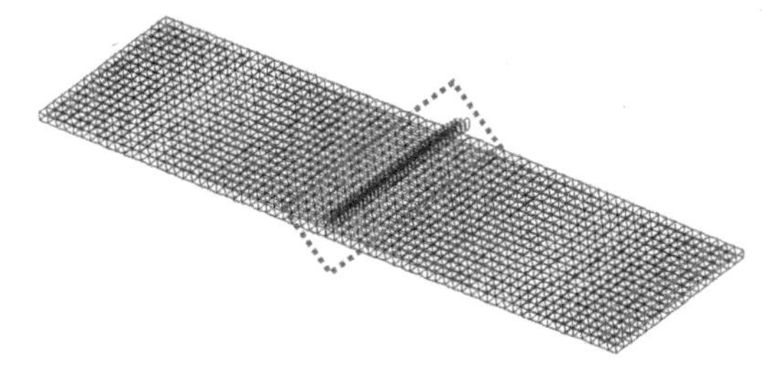

Figure 9. FE shell bridge model with marked section where nonlinear spring elements were implemented.

The submodel was created in the same manner as in 4.1 but instead of an analytical equation describing the deflection a simple FE model was used. Between each corrosion step the bridge was exposed to a norm train load, but the resulting deflection was read after the unloading, with the dead load only. It was consistent with the usage of the result, as the deformation should be investigated in an unloaded state.

Probabilistic analysis involved one thousand samples generated using Latin hypercube which indicated the correlation between additional deformation and corrosion as can be seen in figure 10. The increase number of observation measurements would reduce the uncertainty due to unknown parameters.

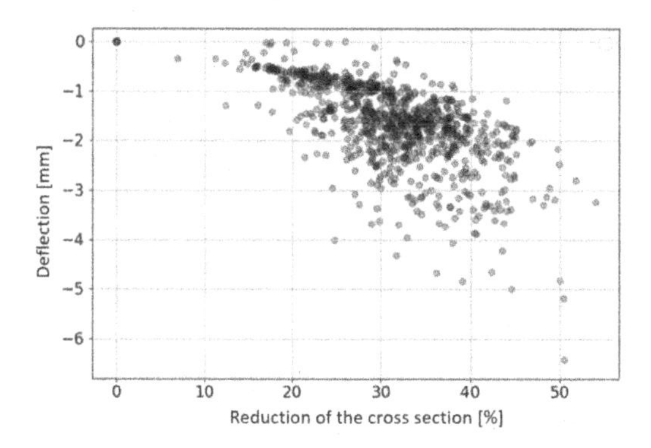

Figure 10. Influence of the corrosion on the bridge deflection.

5 CONCLUSIONS

The presented method combines the advantages of a detailed and complex nonlinear FE analysis with the applicability and performance of an analytical solution or simple FE calculation. The advantages, accuracy and limitations of the method are discussed, and the field of application is described. The method and its application have been described in detail, validated and the application presented on several case studies. A natural limitation is a relatively small region of interest i.e., corrosion or other nonlinear behaviour. Otherwise, if the corrosion extend is large, the advantage of splitting of the model becomes insignificant.

ACKNOWLEDGEMENT

The authors would like to acknowledge the Austrian Research Promotion Agency FFG for financial support through the projects SIBS: "Safety assessment of existing retaining structures" and VerBew-Ing: "Deformation-based Assessment of Engineering Structures".

REFERENCES

Rechnerischer Festigkeitsnachweis für Maschinenbauteile au Stahl, Eisenguss und Aluminiumwerkstoffen.

FKM-Richtlinie, Frankfurt, VDMA-Verlag, ISBN 978-3-81630-605-4.

fib Model Code for Concrete Structures, Wilhelm Ernst & Sohn, 2010.

Bhargava K., Ghosh A., Mori Y., Ramanujam S.: Corrosion-induced bond strength degradation in reinforced concrete – Analytical and empirical models, Nuclear Engineering and Design 237(2007) 1140–1157

VÖBU and SIBS Project partners; Final Report on SIBS – Saftey asssessment of exiting retaioningwalls; FFG Project #861163; VÖBU; Wien; 2019.

Vollenweider AG, Gefährdung von Winkelstützmauern durch Korrosion – Technischer Bericht Untersuchung des Bruchverhaltens – Phase 2, Schweizereidgenossenschaft Bundesamt für Strassen ASTRA, 2014.

Rebhan, M. J., Vorwagner, A., Burtscher S.L., Marte R., Kwapisz, M., Versuchstechnische Untersuchungen zu Korrosionsschäden an Winkelstützmauern. Beton Stahlbetonbau 2020.

Computational Modelling of Concrete and Concrete Structures – Meschke, Pichler & Rots (Eds)
 ISBN: 978-1-032-32724-2

Numerical analysis of experimentally tested frame corners with opening moments using the Compatible Stress Field Method (CSFM)

M.A. Kraus, M. Weber, J. Mata-Falcón & W. Kaufmann
Institute of Structural Engineering (IBK), Concrete Structures and Bridge Design, ETH Zurich, Switzerland

L. Bobek
IDEA StatiCa s.r.o, Brno, Czech Republic

ABSTRACT: The critical parts of structures are typically discontinuity regions, where abrupt changes in geometry occur or large concentrated loads are applied. In engineering practice the verification of the ultimate limit stage of such discontinuity regions employs strut-and-tie models or stress fields based on the lower bound theorem of plasticity theory. These models are mechanically consistent but they can be prohibitively time consuming and are not directly applicable for serviceability limit state analysis as they do not consider strain compatibility. To overcome these limitations the Compatible Stress Field Method (CSFM) was developed for the design and assessment of discontinuity regions in concrete structures. The CSFM consists of a simplified nonlinear finite element-based stress field analysis procedure. Considering compatibility and equilibrium conditions at stress-free cracks, uniaxial constitutive laws as provided in concrete standards are used. While the concrete tensile strength neglected in terms of strength, the CSFM accounts for tension stiffening to obtain realistic predictions of deflections and crack widths, and cover the deformation capacity aspects. The effective compressive strength of concrete is automatically evaluated based on the transverse strain state. The present work validates the ability of the CSFM to reproduce the observed behaviour of experimentally tested frame corners with an opening moment. A quantitative comparison between the outcomes from the numerical analyses and reported experimental results proves the CSFM to be a reliable tool for assessing the structural behaviour of discontinuity regions. In addition, a numerical study is conducted to investigate the sensitivity of the CSFM to several input and model parameters.

1 INTRODUCTION

Reinforced concrete structures can be divided into "B" (Bernoulli) and "D" (Discontinuity) regions (see Figure 1). In "B" regions, the hypothesis of plane strain distribution is satisfied and conventional design approaches are applicable (e.g. cross-section analysis). In "D" regions the strain distributions are nonlinear (due to abrupt changes in the geometry or concentrated loads), and sectional design is not applicable. Common examples of "D" regions include corbels, dapped-ends, deep beams, anchorage zone, walls with openings or frame corners. Strut-and-tie models and stress fields (Marti 1985; Schlaich et al. 1987) are common methods in engineering practice for the ultimate limit design of "D" regions. These models are mechanically consistent, powerful tools that yield direct insight into the load-carrying behaviour and give the engineer a high level of control over the design. Yet, being based on the lower bound theorem of plasticity theory, they do not consider compatibility and hence, they are not directly applicable for verification of serviceability criteria (e.g. deformations, crack widths).

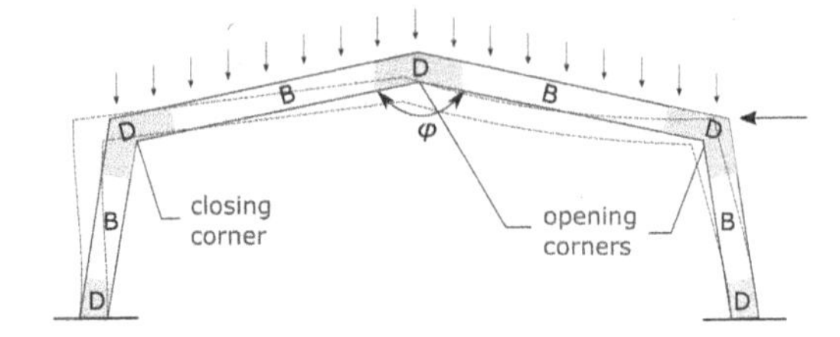

Figure 1. Examples of Bernoulli (B) and discontinuity (D) regions in a concrete frame structure.

Furthermore, the design of real-life structures using strut-and-tie models or stress fields is often tedious due to the iterative nature and typically hand calculations, particularly where different load arrangements and combinations need to be considered.

Several attempts were made in the past to automate the development of stress fields in "D" regions by using computer-aided models. Some applications were developed explicitly for the design of "D" regions such as e.g. CAST (Tjhin & Kuchma 2002) and AStrutTie

 DOI 10.1201/9781003316404-83

(AStrutTie 2020). However, the user has to manually propose a truss model beforehand and assign the effective concrete compressive strength to each truss member and node, which makes the design process inefficient and user unfriendly. To overcome these drawbacks, (Fernández Ruiz & Muttoni 2007) developed the elastic-plastic stress field method (EPSF), where a nonlinear finite element procedure is used to automatically generate suitable solutions for the design and considering the influence of cracking on the concrete strength. The EPSF method has shown to be robust to predict the failure load and associated mechanisms, but as its use requires Java coding, user-friendliness is limited. Furthermore, the EPSF method is not suitable for realistic serviceability checks since tension stiffening is neglected. A more refined modelling of the material behaviour is possible by using more sophisticated nonlinear finite element tools such as e.g. Abaqus, Atena, Ansys or MASA. Although these tools may give a deeper insight in the "real" structural behaviour, they have not found widespread application in engineering practice, mainly due to the large number of required input parameters, typically including non-standard material properties unknown in the design stage, causing a certain arbitrariness of the results; in addition, the modelling effort is typically excessive for common design tasks.

The Compatible Stress Field Method (CSFM) (Kaufmann et al. 2020; Mata-Falcón et al. 2018) was developed to overcome the limitations of classical design tools and existing computer-aided models. This papers recapitulates the basic principles of the CSFM and validates its capability to capture the strength and deformation capacity of reinforced frame corners, a common type of discontinuity region, by the analysis of selected experiments from literature. Furthermore, based on the CSFM analyses, a numerical study is conducted to investigate the sensitivity of the CSFM results to selected input and model parameters (e.g. tensions stiffening, finite element mesh size). Finally, the main findings are summarised, allowing to identify remaining open points of the method. A detailed description of the CSFM and an extensive validation against test results can be found in (Kaufmann et al. 2020; Mata-Falcón et al. 2018).

2 COMPATIBLE STRESS FIELD METHOD

The CSFM consists of a nonlinear finite element-based stress field analysis procedure suitable for the code-compliant design and assessment of discontinuity regions subjected to in-plane loadings. Common uniaxial constitutive laws provided in concrete codes and specified in the design stage are used for concrete and reinforcement, explicitly accounting for concrete and reinforcement strain limitations. Hence the designers do not have to provide additional, often uncertain and/or highly stochastic material properties as typically required for nonlinear FE analysis, making the method perfectly suitable for engineering practice. Since the state of strain is evaluated through the analysis of the structure, the effective compressive strength of concrete can be automatically computed based on the state of transverse strain. Moreover, the CSFM considers tension stiffening, providing realistic stiffness and covering all design code prescriptions including serviceability and deformation capacity aspects, which are not consistently addressed by previous strut-and-tie or stress field approaches. The CSFM has been implemented into the finite element software IDEA StatiCa Detail, a user-friendly software developed jointly by ETH Zurich and the software company IDEA StatiCa s.r.o.

2.1 *Assumptions and idealisations*

The CSFM considers fictitious rotating cracks, which open perpendicularly (see Figure 2a). Furthermore, a discrete crack spacing is assumed, and any variation in the stresses due to bond stresses between the cracks is neglected. Hence, the orientation of the cracks depends on the stress state, and the cracks are stress-free. Equilibrium is formulated at the stress-free cracks, while compatibility is expressed in terms of average strains of the reinforcement ϵ_m. Hence, the model considers maximum concrete stresses σ_{c3r} and reinforcement stresses σ_{sr} at the cracks while neglecting the concrete tensile strength ($\sigma_{c1r} = 0$), expect for its tension stiffening effect on the reinforcement (see Figure 2b). According to the assumptions of the model, the directions of the principal strains and stresses coincide (compatibility condition: $\theta_r = \theta_\sigma = \theta_\epsilon$), and the principal directions are decoupled expect for the compression softening effect.

2.2 *Constitutive models*

2.2.1 *Concrete*

The CSFM considers the uniaxial compression constitutive laws provided by EN 1992-1-1 (EN 1992-1-1 1 01). As seen from Figure 2c the stress-strain relationship is defined by the compressive strength f_c, the modulus of elasticity E_c and the strain ϵ_{c0} at the compressive strength as well as the ultimate strain ϵ_{cu} (which all are given by the strength class of the concrete). To account for compression softening, the compressive strength f_c and the strain ϵ_{c0} are reduced by the factor k_c, which is calculated based on the transverse tensile strain ϵ_1 (see Figure 2e). The implemented relationship is a generalisation of the fib Model Code 2010 (Taerwe, Matthys, et al. 2013) proposal for shear verification, but without the limitation $k_c \leq 0.65$ that is incompatible with other actions than shear. This is consistent with the main assumptions of the CSFM (see Sec. 2.1), since it is also derived in terms of maximum stresses at the cracks: As previously mentioned, the CSFM neglects the tensile strength of concrete in terms of strength just as in standard structural concrete design, which justifies higher values

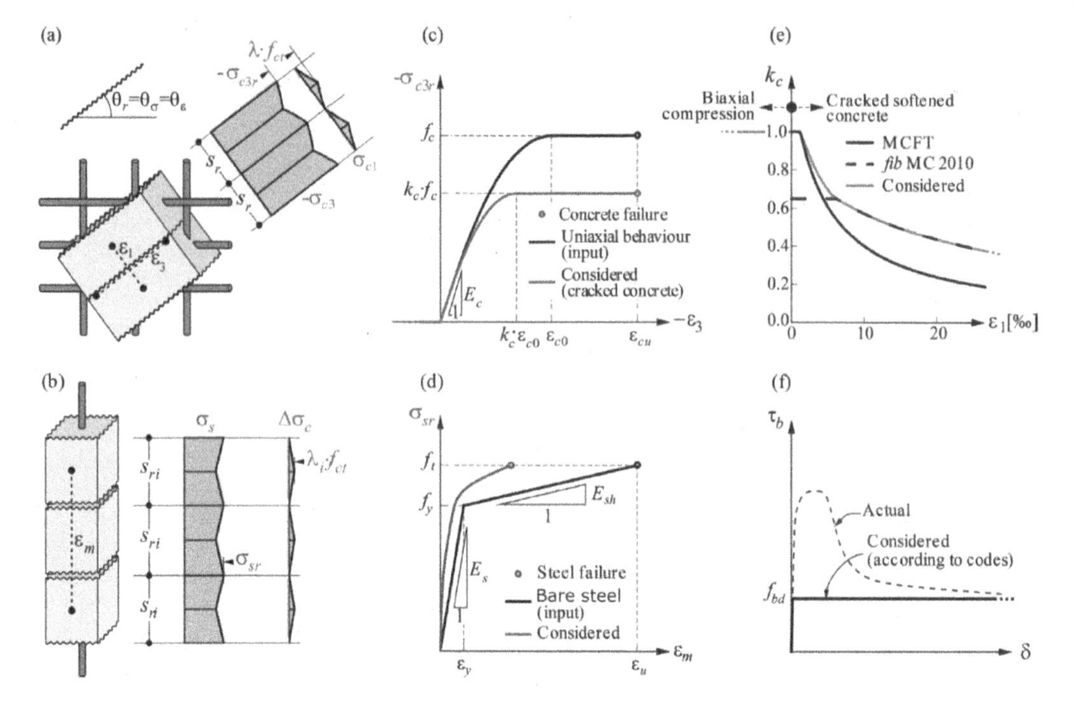

Figure 2. Compatible stress field model: (a) principal stresses in concrete and reinforcement; (b) stresses in the reinforcement direction; (c) constitutive law for concrete ; (d) constitutive law for reinforcement; (e) compression softening law; (f) bond shear stress-slip relationship (adapted from Mata-Falcón et al. 2018).

of k_c than according to models accounting for average stresses, such as the Modified Compression Field Theory.

2.2.2 Reinforcement

As defined in design codes, a bilinear idealisation of the stress-strain relationships of the bare reinforcing bars is used in the CSFM by default (see Figure 2d). The definition of this stress-strain curve only requires basic properties of the reinforcement, such as the yield strength f_y, ultimate strength f_t as well as the ultimate strain ϵ_u and the modulus of elasticity E_s (which all are well-known during the design phase by the specified strength and ductility class). Where known, the actual stress-strain relationship of the reinforcement (hot-rolled, cold-worked) can be considered instead. While an elastic-perfectly plastic characteristic (as implemented in EPSF) is also possible, it would not allow verifying deformation capacity due to the lack of explicit failure criteria.

2.2.3 Tension stiffening

The effect of bond on the load-deformation behaviour of reinforced concrete members loaded in tension is known as tension stiffening. In the CSFM, tension stiffening is captured by modifying the input stress-strain relationship of the bare reinforcing bar in order to capture the overall stiffer response of cracked reinforced concrete compared to bare reinforcing bars of equal cross-section. Basically, the CSFM differentiates between stabilised and non non-stabilised cracking.

Stabilised cracking exists in regions where rupture of the reinforcement at cracking is avoided by a geometric reinforcement ratio ρ higher than the minimum reinforcement
amount ρ_{cr}. Formulating equilibrium on a tension chord subjected to its cracking load, one gets the required minimum reinforcement amount p_{cr}:

$$\rho_{cr} = \frac{1}{f_y/f_{ct} - n + 1} \tag{1}$$

where f_{ct} and $n = \frac{E_s}{E_c}$ correspond to the concrete tensile strength and modular ratio, respectively.

In case of stabilised cracking, tension stiffening is implemented by means of the Tension Chord Model (TCM) (Alvarez 1998). The TCM assumes a stepped, rigid-perfectly plastic bond shear stress-slip relationship along a cracked element (i.e. a tension chord element bounded by two cracks according to Figure 3a). Hence, the bond shear stresses τ_b are independent of the slip δ and are fully determined at a specific location with $\tau_b = \tau_{b0} = 2f_{ct}$ for $\sigma_s \leq f_y$ and $\tau_b = \tau_{b1} = f_{ct}$ for $\sigma_s \geq f_y$. Therefore, the steel stresses σ_{sr} at the cracks follow directly from equilibrium in a closed analytical form as a function of the crack spacing s_{rm}, the bond stresses τ_b and the average reinforcement strain ϵ_m. The average concrete strains in the crack element due to composite action are small compared to the average steel strains and are neglected in the determination of crack widths. The crack spacing s_{rm} is given by λs_{rm0}, where the maximum crack spacing s_{rm0} is determined

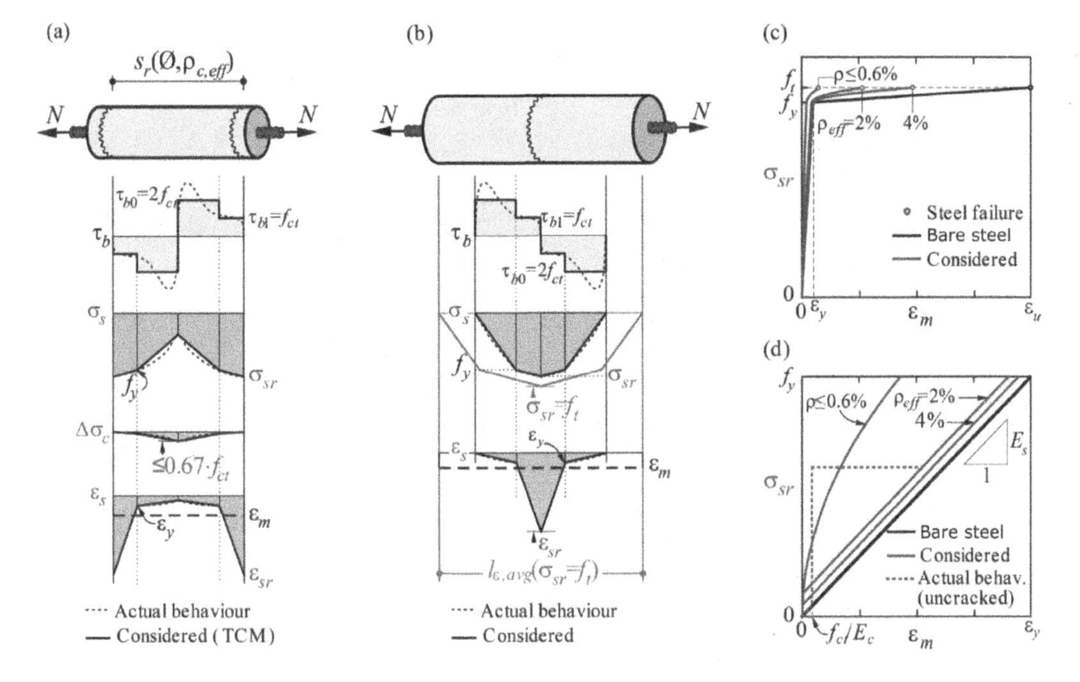

Figure 3. Tension-stiffening models: (a) tension chord element for stabilised cracking with distribution of bond shear, steel and concrete stresses, and steel strains between cracks; (b) pull-out assumption for non-stabilised cracking with distribution of bond shear and steel stresses and strains around the crack (c) resulting tension chord behaviour in terms of reinforcement stresses at the cracks and average strains for B500B steel; (d) detail of the elastic part of the tension chord response (adapted from Mata-Falcón et al. 2018).

from the stress distribution along the crack element according to Figure 2, provided that σ_{c1} is limited to the concrete maximum tensile strength at the centre between two cracks. The crack spacing coefficient λ is assumed to be 2/3 by default in CSFM, based on statistical considerations and experimental evidence (Beeby 1979).

Since the crack spacing s_{rm} according to the TCM depends on the reinforcement ratio, an appropriate concrete area acting in tension between the cracks needs to been assigned to each reinforcing bar. To this end, an automatic procedure is used to determine the corresponding geometric reinforcement ratio ρ for any configuration of the reinforcing bars. A detailed description about this algorithm is found in (Kaufmann et al. 2020; Mata-Falcón et al. 2018).

In regions where ρ is smaller than ρ_{cr} cracking is non-stabilised and tension stiffening is considered using the Pull-Out Model (POM), described in Figure 3b. The POM analyses the behaviour of a single crack (i) considering no mechanical interaction between cracks, (ii) neglecting tensile concrete strains and (iii) assuming the same bond shear stress-slip relationship as used in the TCM. Given the fact that the crack spacing is unknown for a non-fully developed crack pattern, the average strain is computed for any load level over the distance between points with zero slip, when the reinforcement bar reaches its tensile strength f_t at the crack. Similar as in the TCM, this allows determining the average reinforcement strain ϵ_m for any steel stress at the crack σ_{sr} directly from equilibrium.

Figure 3 c) and d) illustrate the behaviour including tension stiffening for the most common European reinforcing steel (B500B, with $f_t/f_y = 1.08$ and $\epsilon_u = 5\%$). Obviously, the consideration of tension stiffening increases the stiffness and reduces the ductility depending on the reinforcement ratio ρ, but does not affect the ultimate strength.

The use of a stepped, rigid-perfectly plastic bond shear stress-slip relationship in the TCM and POM, simplifies the formulation of the load-deformation behaviour of tension chords significantly since no integration of the differential equation of bond is necessary as used in other approaches. Hence, tedious iterations within the nonlinear numerical procedure are reduced, which improves the stability and robustness of the analyses.

2.2.4 *Bond and anchorage*

To verify the anchorage prescriptions according to design codes, the simplified, perfectly plastic constitutive relationship presented in Figure 2d is implemented, with f_{bd} being the design value of the ultimate bond stress for anchorage specified by the design codes. Since none of the analysed experiments in Sec. 3 exhibited an bond or anchorage failure, this verification was switched off for the analyses described in this paper.

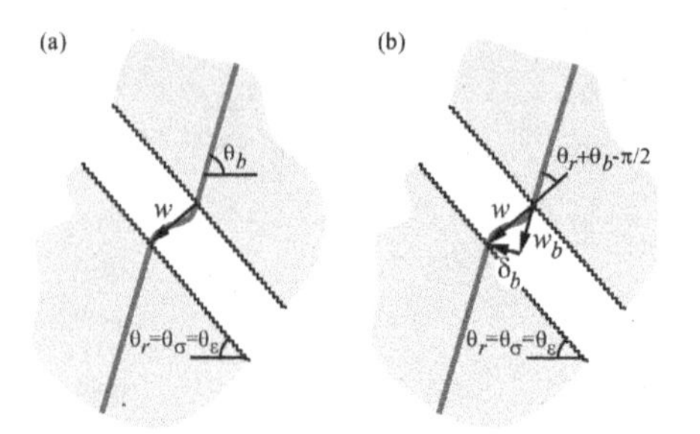

Figure 4. Crack width calculation: (a) considered crack kinematics; (b) projection of crack width into principal directions of stresses and strains (adapted from Mata-Falcón et al. 2018).

2.3 *Crack width calculation*

Based on the presented tension stiffening models, the crack width w_b in the direction of the reinforcement θ_b (see Figure 4a) is consistently calculated by integrating the reinforcement strains. For those regions with stabilised cracking, the average strains ϵ_m along the reinforcement are integrated along the crack spacing s_{rm}. For the case of non-stabilised cracking the crack width calculation based on the maximum strains at the cracks, which in this case are more reliable than the average strains (see (Mata-Falcón et al. 2018)).

The crack width w perpendicular to the crack opening is given by the projection of the crack width w_b in the direction of the principal stresses and strains (see Figure 4b). This is consistent with the main model assumptions, since the inclinations of the cracks θ_r are coincide with the principal directions of stresses θ_σ and strains θ_ϵ (compatibility condition, see Sec. 2.1).

2.4 *Numerical implementation*

The CSFM introduced in the previous sections is implemented in the user-friendly finite element software environment IDEA StatiCa Detail. The structure to be analysed is discretised using 1D bar elements for the reinforcement and 2D plane elements for the concrete. The reinforcement and concrete elements are connected using rigid multi-point constraint elements to ensure the relative position of the reinforcement in relation to the concrete. Full Newton-Raphson algorithm is used to solve the set of nonlinear equations.

3 EXPERIMENTAL VALIDATION

This section validates the capability of the CSFM to capture the observed behaviour of the frame corners, which were experimentally tested by (Campana et al. 2013; Muttoni et al. 2011)). This paper compares the ultimate loads and failure modes predicted by the CSFM against the experimental results. Furthermore, the CSFM predicted load-deformation responses are compared to these observed during the experiments in order to verify the applicability of the CSFM for serviceability checks.

If not stated otherwise, the CSFM analyses use the default material and analysis parameters as described in previous sections. However, in the present study, concrete crushing was defined as soon as the principal stress σ_{c3} reaches the effective compressive strength. Hence, the horizontal plateau of the constitutive law for concrete according to Figure 2c is neglected (ϵ_{c0}=ϵ_{cu}).

3.1 *Specimens*

(Campana et al. 2013; Muttoni et al. 2011) studied the load-deformation response of 16 simply supported frame corners with a corner angle of 125° subjected to opening moments through four-point bending. The corresponding test setup is shown in Figure 5. For

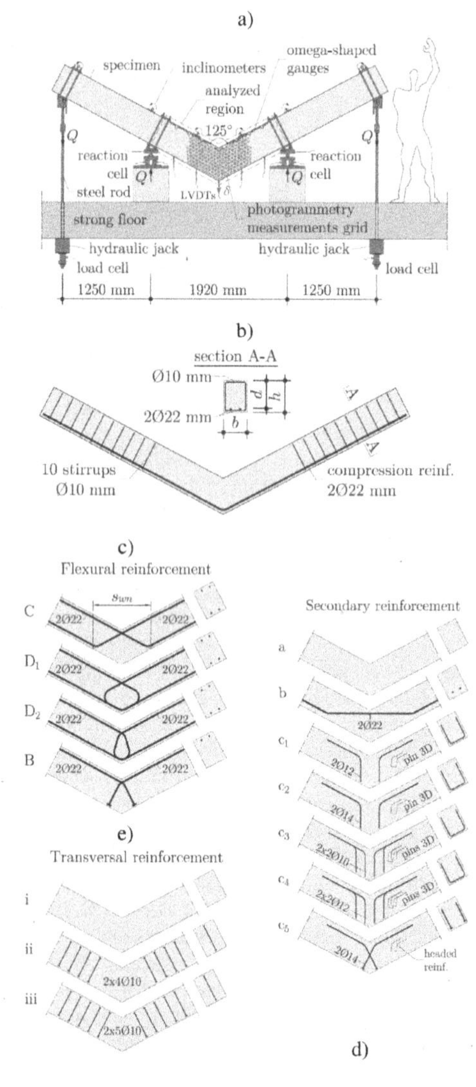

Figure 5. Experimental programme: (a) test setup; (b) compression reinforcement and stirrups; (c) flexural reinforcement layout; (d) secondary reinforcement; and (e) transverse reinforcement (adapted from Campana et al. 2013).

Table 1. Designation of the test specimens and combinations of the reinforcement layout.

Test	Flexural Reinforcement					Secondary Reinforcement					Transversal Reinforcement					Concrete
	Layout	D [mm]	f_y [MPa]	f_t [MPa]	ϵ_u [%]	Layout	D [mm]	f_y [MPa]	f_t [MPa]	ϵ_u [%]	Layout	D [mm]	f_y [MPa]	f_t [MPa]	ϵ_u [%]	f_c [MPa]
SC 26	D1	22	515	630	11.1	a	-	-	-	-	i	-	-	-	-	41.9
SC 27	D1	22	515	630	11.1	b	22	515	630	11.1	i	-	-	-	-	41.6
SC 30	D2	22	515	630	11.1	a	-	-	-	-	i	-	-	-	-	42.0
SC 31	D2	22	515	630	11.1	b	22	515	630	11.1	i	-	-	-	-	41.7
SC 34	B	22	515	652	11.6	a	-	-	-	-	i	-	-	-	-	41.4
SC 35	B	22	515	652	11.6	b	22	515	630	11.1	i	-	-	-	-	42.1
SC 38	C	22	500	596	11.4	c1	12	555	610	4.70	ii	10	568	641	6.20	31.3
SC 39	C	22	500	596	11.4	c1	12	555	610	4.70	iii	10	568	641	6.20	31.1
SC 40	C	22	500	596	11.4	c2	14	560	600	4.10	ii	10	568	641	6.20	30.9
SC 41	C	22	500	596	11.4	c2	14	560	600	4.10	iii	10	568	641	6.20	30.9
SC 42	C	22	500	596	11.4	c3	10	575	620	3.60	iii	10	568	641	6.20	31.0
SC 43	C	22	500	596	11.4	c4	12	555	610	4.70	iii	10	568	641	6.20	31.0
SC 44	C	22	500	596	11.4	c5	14	560	614	4.30	ii	10	568	641	6.20	30.9
SC 45	C	22	500	596	11.4	c5	14	560	614	4.30	iii	10	568	641	6.20	30.8

all members the cross section (b=300 mm width and h=400 mm height) was kept constant. The main test parameter was the layout of the flexural, transverse as well as the secondary reinforcement. Four different layouts for the flexural reinforcement (details C, D1, D2 and B in Figure 5c) were investigated, which were provided either alone or in combination with different arrangements of secondary reinforcement (details a, b and c1 to c5 in Figure 5d) and/or transverse reinforcement (details i, ii and iii in Figure 5e). The designations of the specimens and the corresponding combination of the reinforcement details are summarised in Table 1.

It should be noted that the specimens SC 22 and SC 23 reported in (Campana et al. 2013) were not analysed with the CSFM, since they exhibited a premature failure due to lack of reinforcement for crack control. In general such failures are difficult to be captured with approaches such as the CSFM method, since these presume structural continuity and are not capable of predicting such local failure modes caused by inadequate detailing.

3.2 *Material properties*

Table 1 summarises the material properties of the concrete and reinforcement used in the CSFM analyses. All material properties listed were available in the test report. For all analyses the Youngs' moduli of concrete and reinforcement were assumed as E_c=35 GPa and E_s=200 GPa, respectively. The concrete ultimate strain was approximately the same for all experiments and therefore fixed at $\epsilon_{c0} = 2.1‰$ for all analyses.

3.3 *CSFM modelling*

The geometry, reinforcement, support and loading conditions were modelled in CSFM according to the experimental setup. Figure 6 shows the modelling of specimen SC45 as an example. By default, the mesh was generated automatically in Idea StatiCa Detail, which leads to 4 elements over the frame depth for all analyses. As described in Sec. 2.2.3, tension stiffening is included by means of the TCM for the flexural reinforcement ($\rho_{flex} = 2.1\% > \rho_{cr} \approx 0.6\%$ with $f_{ct} \approx 2.9MPa$ and $f_y \approx 500MPa$) and the POM for the transverse reinforcement ($\rho_{trans} = 0.16\% < \rho_{cr} \approx 0.5\%$ with $f_{ct} \approx 2.9MPa$ and $f_y \approx 570MPa$).

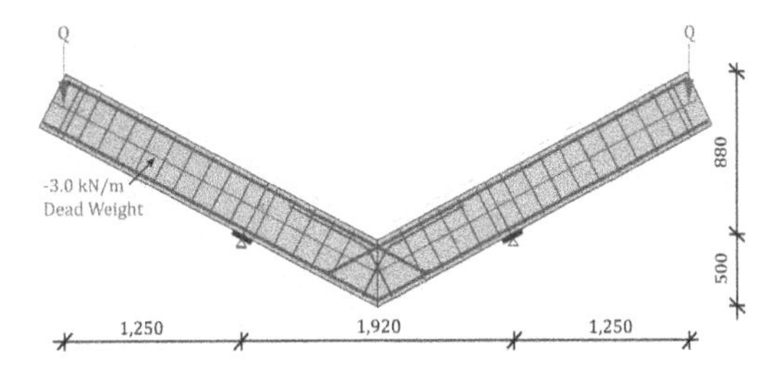

Figure 6. CSFM modelling of specimen SC 45: geometry, loading and supports.

3.4 *Results of the numerical investigation*

Figure 7 summarises the ratios of experimental and predicted peak loads ($Q_{u,exp}/Q_{u,CSFM}$) for the investigated experiments, including the mean value and coefficient of variation (CoV) of this ratio. Q denotes the applied force at each side of the specimens (see Figure 5a). Ratios $Q_{u,exp}/Q_{u,CSFM} > 1$ denote conservative predictions of the ultimate load. The predictions of the ultimate loads are very satisfactory, yielding slightly conservative results on average ($Q_{u,exp}/Q_{u,CSFM}$=1.02 for all experiments), with reasonable scatter among the analysed frame corners (CoV=16.3%). Note that the latter would be significantly reduced by excluding Specimen SC 30, whose strength was clearly below the predicted value. As discussed in the test report, this might have been a consequence of a an assembly error, affecting the strength by causing a premature failure in the experiment, cf. (Campana et al. 2013; Muttoni et al. 2011).

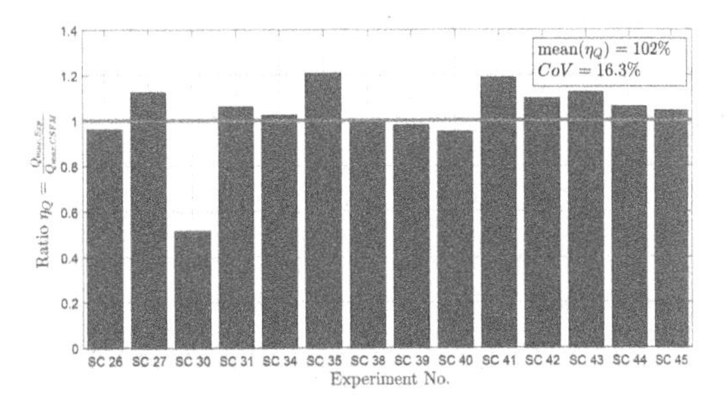

Figure 7. Ratio of experimentally observed to predicted (CSFM) ultimate load.

In all of the numerical analyses, failure was triggered by concrete crushing, preceded by yielding of the flexural reinforcement. This agrees with the experimental observations except for some specimens with low transverse reinforcement, which exhibited a premature brittle failure without yielding of the reinforcement cf. (Muttoni et al. 2011).

Figure 8a shows the calculated stress fields (principal compressive stresses σ_c and steel stresses σ_{sr} at the cracks) for specimen SC 27 with marks for the predicted failure mode and location as an example for the analyses conducted in this study. In addition, the computation of cracked regions and the magnitudes of the crack widths (represented by the length of the lines) at the onset of yielding are plotted in Figure 8b. The predicted cracked regions and crack orientations agree well with the experimental observations at failure illustrated in Figure 8c.

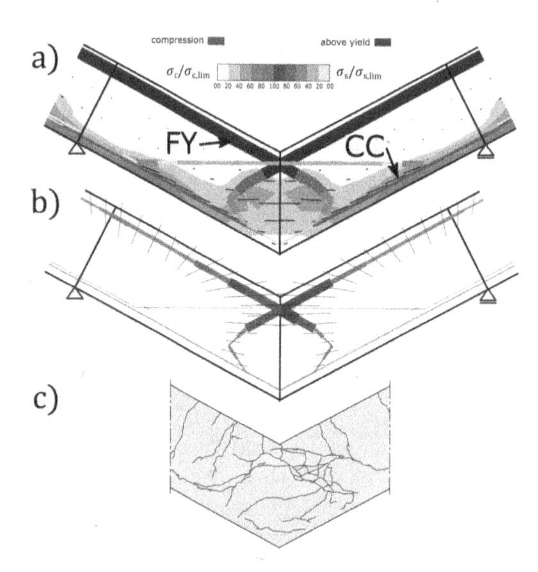

Figure 8. Results for SC 27: (a) CSFM stress fields at ultimate load, (b) CSFM crack pattern at reinforcement yielding, (c) experimentally observed crack pattern at ultimate load (from Campana et al. 2013)

Figure 9 compares the load-deformation responses (Q-δ_{ext}) predicted by CSFM with the measured load-deformation responses of the experiments. δ_{ext} corresponds to the vertical deflection at the loading point of the frame (see Figure 5a). Up to yielding of the reinforcement, the calculated load-deformation responses agree very well with the experimental results. After the one-set of yielding, the numerical analyses tend to overestimate the deformation capacity particularly for tests with low transverse reinforcement ratios. For those tests, the deformation behaviour were not properly recaptured by the numerical analyses, as these predict a pronounced yielding of the reinforcement at failure as opposed to the experimentally observed sudden loss of strength due to brittle failure of the concrete (vertical separation of the compression zone from the remaining part of the specimen). These differences highlight, that the CSFM, similarly to classical strut-and-tie models and stress fields, cannot account well for such brittle failures caused by strain localisation in regions without any (transverse) minimum reinforcement. Regarding specimen SC 30, the same observations as made above in relation to the predicted and observed strength apply.

4 PARAMETRIC STUDY

In this section a discussion of a sensitivity study towards the default parameters of the CSFM used in Sec. 3 is presented. The sensitivity analysis investigates the dependence of the CSFM results on the mesh size (default: 4 elements over the depth of the frame) and on the consideration of tension stiffening (default: tension stiffening is included by means of the TCM and POM). Based on these analyses, the suitability of the default values can be judged. The corresponding variations of the CSFM computational parameters are outlined within the subsequent subsections accordingly.

4.1 *Influence of mesh size*

The influence of the mesh size is studied for the two test specimens SC 26 (brittle failure in experiment) and SC 45 (ductile failure in experiment). The mesh size is varied from 4 elements (default element size in Idea StatiCa Detail used in the previous section) to 8 and 16 elements over the cross-sectional depth. Figure 10 shows the ratio of observed to predicted ultimate loads $Q_{u,exp}/Q_{u,CSFM}$ for the different mesh sizes. It can be seen that the smaller the finite element mesh, the lower and more conservative are the strength predictions by the CSFM. This dependency is mainly due to the more pronounced localisation of the transverse strains for smaller finite element mesh size, which reduces the compressive strength via the compression softening factor and causes failure at a lower load. The failure mode remains insensitive to the considered mesh size, except that yielding of the reinforcement was prevented using a very fine mesh (16 elements over frame's depth) for SC 26 (as observed in the experiment).

4.2 *Influence of tension stiffening effects*

The influence of considering the tension stiffening effect during CSFM computations is studied for Test Specimen SC 45 (ductile failure during experimentation) only. The corresponding load-deformation response is shown in Figure 11 on the example of SC 45. Tension stiffening has neither a significant influence on the stiffness and ductility, nor on the ultimate load and failure mechanism. This is due to the fact that the reinforcement ratio of the flexural reinforcement (which governs the structural behaviour, see Sec. 2.2.3) is very high ($\rho_{flex} = 2.1\%$, see Sec. 3.1)

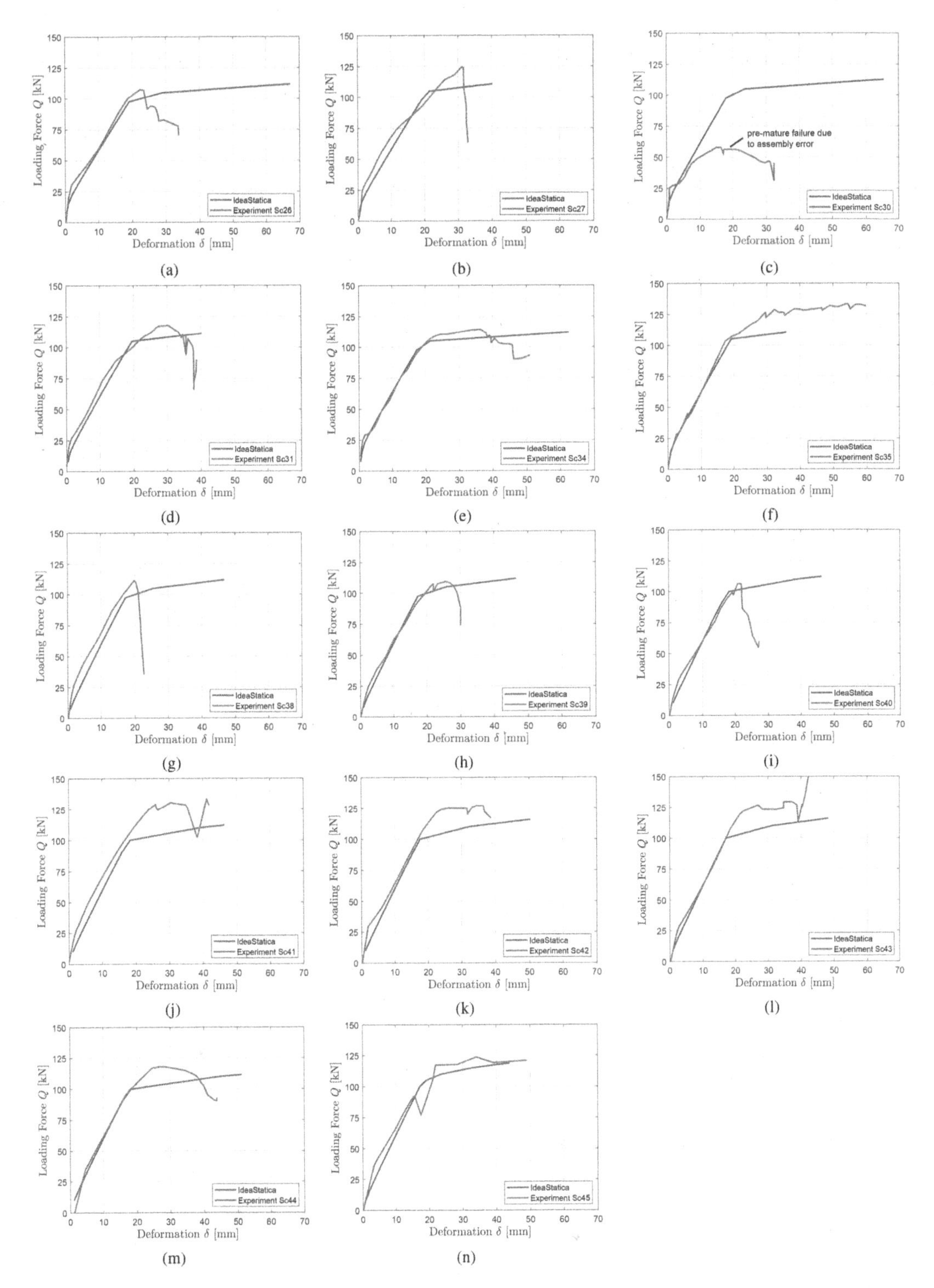

Figure 9. Load-deformation responses: (a) SC 26 (b) SC 27 (c) SC 30 (d) SC 31 (e) SC 34 (f) SC 35 (g) SC 38 (h) SC 39 (i) SC 40 (j) SC 41 (k) SC 42 (l): SC 43 (m) SC 44 (n) SC 45.

and therefore, tension stiffening does not significantly reduce the deformation capacity of the reinforcement. This behaviour applies to all analysed experiments in Sec. 3, since the flexural reinforcement amount was the same for all specimens. It should be noted however, that tension stiffening could have a relevant impact,

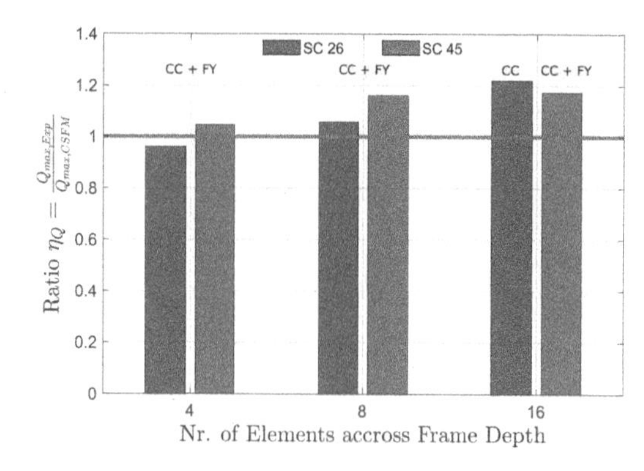

Figure 10. Comparison of results for the mesh sensitivity study for specimen SC 26 and SC 45.

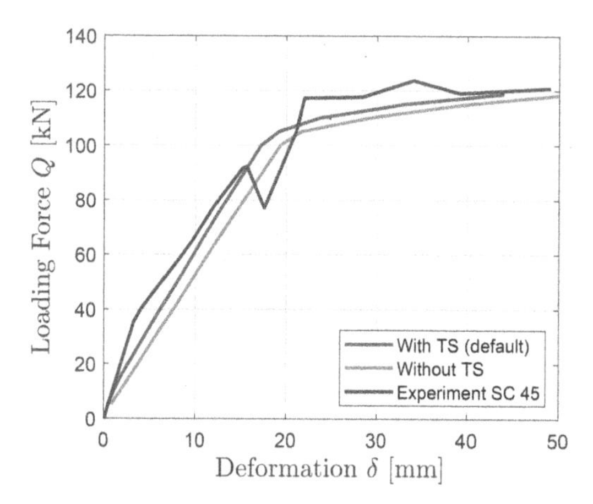

Figure 11. Comparison of results for the sensitivity study for specimen SC 45 (TS=tension stiffening).

if the amount of reinforcement would be low (see Sec. 2.2.3). Neglecting tension stiffening, in spite of not affecting the strength of the reinforcement, could lead to a pronounced overestimation of the ultimate load in such cases, and therefore to an unsafe design (see (Mata-Falcón et al. 2018)). On the other hand, if concrete crushing is governing the failure, neglecting tension stiffening of the transverse reinforcement could result in overly conservative predictions of the failure load due to the overestimation of transverse strains and hence, compression softening.

5 SUMMARY AND CONCLUSIONS

The paper recapitulates the basic principles of the Compatible Stress Field Model (CSFM), a consistent nonlinear finite element-based stress field analysis procedure suitable for the design and assessment of discontinuity regions subjected to in-plane loadings. To validate its capability towards capturing ultimate strength and deformation capacity of reinforced frame corners, all experiments of the campaign reported in (Campana et al. 2013; Muttoni et al. 2011) were analysed. A sensitivity study investigated the influence of the mesh size as well as accounting for the tension stiffening effect. Summarising, the following conclusions can be drawn from the numerical analyses:

- The comparison of the CSFM results with experimental results in terms of ultimate load and failure modes shows a good agreement.
- The use of the default parameters in Idea StatiCa on average leads to slightly conservative estimates of ultimate loads.
- The results show that the deformation capacity for brittle failure cannot be predicted with the same accuracy as failures modes with a certain deformation capacity by the CSFM approach. This could be expected since the CSFM, similar to classic stress fields, is not intended for such types of failure. In design this is compensated by the higher safety coefficient for concrete in compression than for reinforcement yielding.
- The analysis of the sensitivity of the model to parameters differing from the default ones in Idea StatiCa Detail shows that the most relevant parameter in this case is the mesh size: The smaller the finite element mesh size, the lower the strength predicted by the CSFM. To guarantee a safe design, it is thus recommended to use a finer mesh than the default one in the CSFM, in particular if a brittle failure is to be expected. Furthermore, it is highly recommended that the sensitivity of the model to changes in the mesh size is always investigated.

REFERENCES

Alvarez, M. (1998). *Einfluss des Verbundverhaltens auf das Verformungsvermögen von Stahlbeton*, Volume 236. ETH Zurich.

AStrutTie (2020).

Beeby, A. (1979). The prediction of crack widths in hardened concrete. *The Structural Engineer 57*(1), 9–17.

Campana, S., M. F. Ruiz, & A. Muttoni (2013). Behaviour of nodal regions of reinforced concrete frames subjected to opening moments and proposals for their reinforcement. *Engineering Structures 51*, 200–210.

EN 1992-1-1, (2011-01). *Bemessung und Konstruktion von Stahlbeton- und Spannbetontragwerken - Teil 1-1: Allgemeine Bemessungsregeln und Regeln fÃ¼r den Hochbau.*

Fernández Ruiz, M. & A. Muttoni (2007). On development of suitable stress fields for structural concrete. *ACI, Structural Journal 104*(ARTICLE), 495–502.

Kaufmann, W., J. Mata-FalcÃ³n, M. Weber, T. Galkovski, D. T. Tran, J. Kabelac, M. Konecny, J. Navratil, M. Cihal, & P. Komarkova (2020). *Compatible Stress Field Design of Structural Concrete.* ETH Zurich, Institute of Structural Engineering and IDEA StatiCa s.r.o., Brno, Czech Republic.

Marti, P. (1985). Truss models in detailing. *Concrete International 7*(12), 66–73.

Mata-Falcón, J., D. T. Tran, W. Kaufmann, & J. Navrátil (2018). Computer-aided stress field analysis of

discontinuity concrete regions. In *Computational Modelling of Concrete Structures*, pp. 641–650. CRC Press.
Muttoni, A., S. Campana, O. Burdet, M. Fernandez-Ruiz, G. Guignet, N. Kostic, & G. Messi (2011, December 6th). Essais d'ouverture d'angles de cadre d'une tranchee couverte a section polygonale. Rapport d'essai IBETON 08.03 - RE02, EPFL - ENAC - IIC, EPFL Lausanne.
Schlaich, J., K. Schäfer, & M. Jennewein (1987). Toward a consistent design of structural concrete. *PCI journal 32*(3), 74–150.
Taerwe, L., S. Matthys, et al. (2013). *Fib model code for concrete structures 2010*. Ernst & Sohn, Wiley.
Tjhin, T. N. & D. A. Kuchma (2002). Computer-based tools for design by strut-and-tie method: Advances and challenges. *Structural Journal 99*(5), 586–594.

Computational Modelling of Concrete and Concrete Structures – Meschke, Pichler & Rots (Eds)
© 2022 Copyright the Author(s), ISBN: 978-1-032-32724-2

Efficiently determining the structural reliability of a corroding concrete bridge girder using finite element modelling for multiple limit states

Rutger Vrijdaghs & Els Verstrynge
Department of Civil Engineering, KU Leuven, Leuven, Belgium

ABSTRACT: In this paper, an combined approach based on response surface sampling and importance sampling is developed to efficiently assess the structural reliability of a corroding concrete bridge girder. The approach employs cubic surface sampling with smart parameter updating and directed importance sampling to limit the computational cost of the reliability assessment. The case study considers a post-tensioned concrete bridge girder undergoing severe corrosion. Multiple limit states and various stochastic parameters related to the material and damage properties are considered in a finite element based reliability assessment. The combined approach limits the required number of simulations as much as possible and yields relatively quick convergence. The results show that the corrosion level and compressive strength play a significant role in the structural behavior. Given the high post-tensioning loads, the girder is limited by its ultimate capacity rather than serviceability constraints.

1 INTRODUCTION AND STATE OF THE ART

Corrosion of the reinforcement is a major problem that can seriously endanger the durability and structural safety of reinforced concrete elements as they age. Recently, corrosion led to structural failures with fatalities in Genoa (Ponte Morandi bridge) and Miami (Champlain Towers North). Despite decades of research, the structural effects of corrosion induced damage are not easily assessed due to the number of variables involved, the loading and damage history and the complex interaction of the various materials. To predict the structural behavior of corroding elements, complex non-linear models are needed. Assessing then the reliability of these elements requires an efficient computational scheme.

Indeed, the assessment of the structural reliability is an important task in civil engineering (Melchers 1999). The reliability is often expressed through the reliability index β, which is related to the failure probability p_f according to the well-known Eq. 1, where Φ^{-1} is the inverse cumulative standard normal distribution.

$$\beta = \Phi^{-1}(1 - p_f) \tag{1}$$

The Eurocode for structural design imposes limits for β based on the consequence class (European Committee for Standardization 2002). For structures with great consequences of failure, the reliability index $\beta = 4.3$ in the ultimate limit state (ULS) for a 50 year reference design period. The failure probability can be determined by integrating the joint probability density function (PDF) $f_{\mathbf{X}}(\mathbf{x})$ of the m variables $\mathbf{x}$ in the failure domain D (where the limit state function (LSF) $g(\mathbf{x}) < 0$). Mathematically, the failure probability is defined in Eq. 2.

$$p_f = \int_D f_{\mathbf{X}}(\mathbf{x}) d\mathbf{x} \tag{2}$$

While Eqs. 1–2 can be used to determine the structural safety theoretically, in practice, a number of problems arise: an analytical expression for the joint PDF is often not available, nor can it be easily integrated over the failure domain. Therefore, a number of different methods are developed to approximate p_f or β. The most commonly used methods are the first (or second) order reliability method (Hasofer & Lind 1974) and Monte-Carlo sampling (Zio 2013). Other methods include subset sampling (Au & Beck 2001), importance sampling (Melchers 1989), response surface sampling (Bucher & Bourgund 1990) or surrogate modelling (Hu, Nannapaneni, & Mahadevan 2017). For a comprehensive comparison between these (and other) methods, reference is made to literature (Ditlevsen & Madsen 1996; Aldosary, Wang, & Li 2018; Shittu, Kolios, & Mehmanparast 2021). The first and second order reliability methods (FORM/SORM) remain the preferred method for reliability assessment thanks to their ease of use but require an analytical expression of the limit state function, which might hamper general usage. For realistic structural engineering cases, the limit state functions are non-linear, time-dependent or implicit (or all three simultaneously). Similarly, the high computational requirements

DOI 10.1201/9781003316404-84

of solving complex, non-linear systems often impedes the use of Monte-Carlo sampling.

Probabilistic studies and the associated reliability assessment of corroding (reinforced) concrete structures are published in literature (Bastidas-Arteaga 2018; Gu, Guo, Zhou, Zhang, & Jiang 2018; Kioumarsi, Hendriks, Kohler, & Geiker 2016; Lim, Akiyama, & Frangopol 2016; Zhang, Song, Lim, Akiyama, & Frangopol 2019). In most cases, the reliability assessment and probabilistic studies are based on sectional design approaches which cannot take into account the change in failure mode and/or the change in bond strength. However, pairing a complete finite element model with a probabilistic analysis is very computationally intensive.

In this paper, a combined approach based on subsequent response surface sampling and importance sampling is developed. A cubic response surface is proposed, where a novel coefficient selection scheme is implemented to take only the most important parameters into account, rather than updating all parameters consecutively. In this way, significant computational gains can be achieved. The combined methodology is presented and applied to a corroding post-tensioned concrete bridge girder for assessing the ULS and SLS. Reference is made to the literature for a comprehensive discussion and analysis of the structural behavior of the corroded concrete bridge girder (Vrijdaghs & Verstrynge 2022).

2 PROBABILISTIC FRAMEWORK

As shown in the state of the art, correctly assessing the reliability of a complex or degrading structure is a difficult task due to the high number of variables involved. In this paper, a so-called response surface sampling + importance sampling (RSS+IS) approach is developed to be compatible with implicit (numerical model-based output) and multiple limit state functions.

A short summary of the implemented RSS+IS method is presented below; for more details reference is made to a previous publication (Vrijdaghs, Van Steen, Nasser, & Verstrynge 2020). RSS+IS combines the quick convergence to a solution and the accuracy and robustness of Monte Carlo based sampling methods. In this two-step approach, the response surface sampling aims to locate the most probable point (MPP) in the **U**-space, which is the space composed of independent standard normal variables, which are related to the variables through a Rosenblatt transformation. Here, a full cubic description of the response surface (RS) is employed, Eq. 3.

$$G(\mathbf{u}) \approx \tilde{G}(\mathbf{u}) = a + \sum_{i=1}^{m} b_i u_i + \sum_{i=1}^{m}\sum_{j=i}^{m} c_{ij} u_i u_j + \sum_{i=1}^{m}\sum_{j=i}^{m}\sum_{k=j}^{m} d_{ijk} u_i u_j u_k \quad (3)$$

Herein, m is the dimensionality and u_i is the independent standard normal variable i in the **U**-space. Starting with a linear hyperplane, the coefficients a and b_i are calculated to best fit (in a least-square sense) the first $m+1$ function evaluations $G(\mathbf{u}_i)$. A FORM analysis is then performed on the linear hyperplane to estimate the location of the MPP $\tilde{\mathbf{u}}_{MPP}$. An exact LSF evaluation is performed at $\tilde{\mathbf{u}}_{MPP}$, yielding $g(\tilde{\mathbf{x}}_{MPP})$. The new information is used to take an additional coefficient of the cubic hyperplane into account. The coefficient that is taken into account has the highest associated weight w_i as determined by the normalized weight vector $\mathbf{w} = [1\ \mathbf{w}_1\ \mathbf{w}_2\ \mathbf{w}_3]$, which is based on the linear coefficients b_i, see Eq. 4.

$$\begin{aligned} \mathbf{w}_1 &= |\mathbf{b}_1| / \|\mathbf{b}_1\| \\ &= \left[|b_1|\ \ldots\ |b_m|\right] / \|\mathbf{b}_1\| \\ \mathbf{w}_2 &= |\mathbf{b}_2| / \|\mathbf{b}_2\| \\ &= \left[|b_1^2|\ |b_1 b_2|\ \ldots\ |b_m b_{m-1}|\ |b_m^2|\right] / \|\mathbf{b}_2\| \\ \mathbf{w}_3 &= |\mathbf{b}_3| / \|\mathbf{b}_3\| \\ &= \left[|b_1^3|\ |b_1^2 b_2|\ \ldots\ |b_m^2 b_{m-1}|\ |b_m^3|\right] / \|\mathbf{b}_3\| \end{aligned} \quad (4)$$

Then, in subsequent function evaluations, additional coefficients are taken into account. The main advantage of the proposed method is that the various higher-order coefficients c_{ij} and d_{ijk} are taken into account and updated in order of importance. Subsequent iterations in the RSS algorithm further refine the response surface and converge to the MPP (Başağa, Bayraktar, & Kaymaz 2012; Guan & Melchers 2001). RSS convergence is reached if at least m simulations predict failure. The RSS output is used as input for the IS algorithm. In this second step, random samples are taken around the MPP. Here, convergence is reached if the number of failed simulations is at least 60 (a user-defined parameter based on previous experience). Once the IS method is converged, an estimation of the failure probability can be calculated, and consequently, the reliability index β is obtained.

In (Vrijdaghs, Van Steen, Nasser, & Verstrynge 2020), the combined RSS+IS approach is calibrated with 52 benchmark examples, chosen for their (highly) non-linear behavior, non-Gaussian distributions or mathematical complexity. The predicted failure probability of the RSS+IS approach differs 0.69% (median value) compared to Monte-Carlo sampling at a relative computational cost of $7.5 \cdot 10^{-4}$, boasting a very good accuracy with minimal computational effort.

3 CASE STUDY: POST-TENSIONED HIGHWAY BRIDGE GIRDER

The combined approach is applied to a concrete bridge girder (built in 1957) with a total length of 40.9 m which carries highway traffic in and out of Brussels, Belgium. It consists of three spans. The outer span girders (9 m span) are made of reinforced concrete, while 13 I-shaped post-tensioned reinforced concrete girders with a center-to-center distance of 1.3 m cross the central span of 22.9 m. The girders are experiencing severe corrosion damage and are subject of the numerical study. The girders are monolithically connected with an in-situ cast concrete slab. In Figure 1, a photo of the bridge is shown.

Figure 1. Global photo of the bridge.

3.1 *Geometry and reinforcement of the bridge girders*

The post-tensioned girders have a height and (maximum) width of 0.72 m and 0.7 m, respectively with a web thickness of 0.25 m. On top of the girder, a 0.18 m thick concrete slab is cast which brings the total height to 0.9 m. A schematic cross-section of one girder is shown in Figure 2.

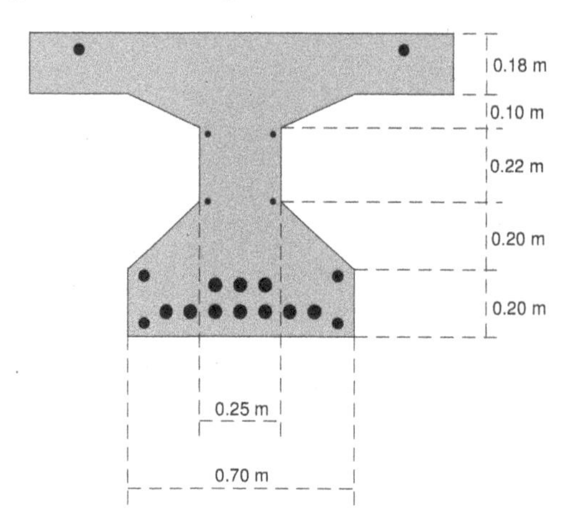

Figure 2. Cross-section of the post-tensioned bridge girder with longitudinal reinforcement (diameters not to scale, stirrups not shown).

All mid-span girders are post-tensioned with 10 strands which are routed upwards near the beams' ends. The strands consist of 12 wires with diameter of 7 mm and are divided over two layers, with a minimum concrete cover at the bottom layer of 50 mm. Additional tensile reinforcement (6×16 mm $+ 4 \times 8$ mm) is placed in the flanges and web of the girder. Stirrups (8mm) are used for shear reinforcement with a center-to-center distance of 200 mm. It should be noted that all dimensional data are deduced from the original design plans.

The central span girders have experienced significant corrosion damage. A visual inspection revealed longitudinal cracks at the bottom face of the girders which run along the post-tensioning strands. Severe concrete spalling leads to exposed rebars and strands in some places, where the ducts of the post-tensioning strands are completely corroded away (Figure 3). A material characterization and long-term monitoring is planned in the near future.

Figure 3. Corrosion damage of the reinforcement in the central bridge girders.

3.2 *Modelling approach and mesh size*

A 2D plane stress analysis is performed on a mid-span girder in a finite element model (built in DIANA). The mesh size for the entire model is 0.1 m, such that the model consists of 14106 nodes and 2310 elements. Cracks are modelled through a smeared crack approach, and reinforcement elements that represent the corroding rebars are defined independently from the mesh of concrete elements.

3.3 *Stochastic parameters and limit states*

In total, 6 stochastic parameters are taken into account: (1) the rebar yield strength $x_1 = f_{yk}$, (2) the strand yield strength $x_2 = f_{pk}$, (3) the concrete compressive strength $x_3 = f_{ck}$, (4) the length of the corrosion zone $x_4 = \ell_{CL}$, (5) the corrosion level $x_5 = CL$, which is defined as the mass loss fraction or fraction of broken wires, and (6) the post-tensioning load ratio $x_6 = LR$. The various parameters and their distributions are shown in Eq. 5–10 and Figure 4, and are discussed in more details in the following sections.

$$x_1 = f_{yk} \sim LN(400, 20) \text{ [MPa]} \tag{5}$$

$$x_2 = f_{pk} \sim LN(1400, 100) \text{ [MPa]} \tag{6}$$

$$x_3 = f_{ck} \sim LN(50, 5) \text{ [MPa]} \tag{7}$$

$$x_4 = \ell_{CL} \sim LN(0.1, 0.04) \text{ [m]} \tag{8}$$

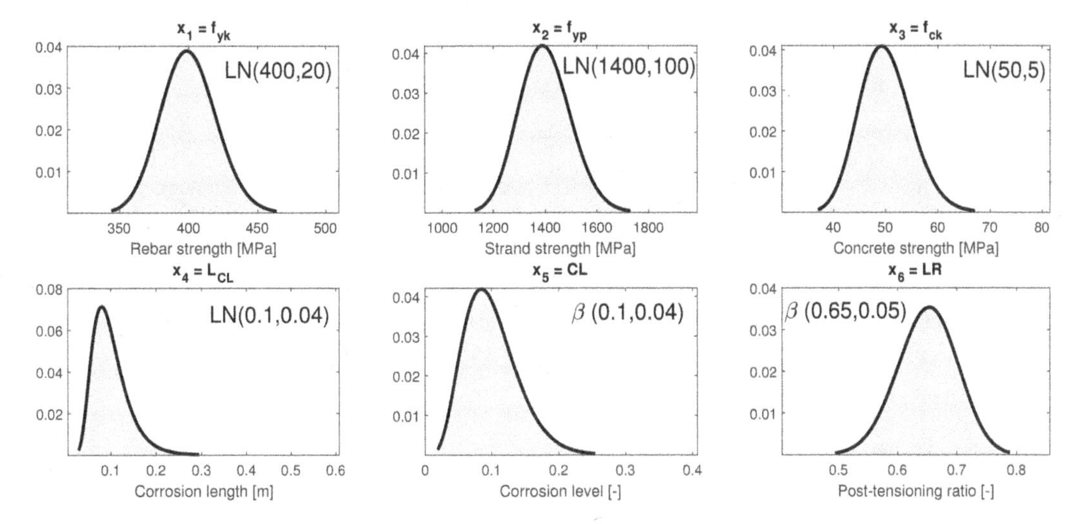

Figure 4. Distribution of all stochastic parameters.

$$x_5 = CL \sim \beta(0.1, 0.04) \text{ [-]} \quad (9)$$

$$x_6 = LR \sim \beta(0.65, 0.05) \text{ [-]} \quad (10)$$

Two limit states are implemented in the model. The first limit state $g_1(\mathbf{x})$ is related to the ULS, while $g_2(\mathbf{x})$ refers to the allowable crack width in the SLS. The global limit state $g(\mathbf{x})$ is then the minimum of $g_1(\mathbf{x})$ and $g_2(\mathbf{x})$, as expressed mathematically in Eq. 11–13.

$$g_1(\mathbf{x}) = 1 - \frac{E_d}{R_d(\mathbf{x})} \quad (11)$$

$$g_2(\mathbf{x}) = 1 - \frac{w(\mathbf{x})}{w_{lim}} \quad (12)$$

$$g(\mathbf{x}) = min[g_1(\mathbf{x}), g_2(\mathbf{x})] \quad (13)$$

Herein, E_d is the applied load of the LM1 load model from Eurocode in ULS (discussed later) and w_{lim} is the allowable bending crack width in SLS, here set equal to 0.3 mm (European Committee for Standardization 2003). It should be noted that the model allows for the easy definition of additional limit states in this form, e.g. for limiting deflections or vibrations if needed.

3.4 *Material properties*

Different material models are incorporated in the numerical model, namely a non-linear concrete model, and a Von Mises plasticity model for both the traditional as well as the post-tensioning reinforcement. A bond-slip interface is also defined for the reinforcement elements, as it may affect the crack widths.

3.4.1 *Non-linear concrete*

The characteristic compressive strength of the concrete is lognormally distributed $f_{ck} \sim LN(50, 5)$ MPa. This value is based on commonly used concrete strength classes for bridge girders, and the standard deviation corresponds to a coefficient of variation of 10%. All other (elastic) parameters are based on the value of f_{ck} as defined in Eurocode 2. The compressive behavior is modelled as a Thorenfeldt curve determined by $f_{cd} = f_{ck}/\gamma_c$, with $\gamma_c = 1$ or 1.5, in the SLS or ULS, respectively. In SLS tension, a Hordijk curve is implemented. In ULS, the tensile behavior of concrete is neglected.

3.4.2 *Non-linear steel*

The steel reinforcement is described by a Von Mises plasticity yield criterion and a elastoplastic stress-strain curve. The yield strength is a lognormal parameter $f_{yk} \sim LN(400, 20)$ and $f_{pk} \sim LN(1400, 100)$ for the rebars and strands. In ULS, the design yield strengths are used $f_{yd} = f_{yk}/\gamma_s$ and $f_{pd} = f_{pk}/\gamma_s$ with $\gamma_s = 1.15$. The stiffness of the strands and rebars is set to 190 and 200 GPa, respectively.

3.4.3 *Bond-slip interface*

The rebars and strands are allowed to undergo relative deformations with respect to the surrounding concrete. For the ribbed rebars, the bond-slip interface during pull-out is described for good bond conditions by the relative deformation-shear stress relation proposed in Model Code 2010 (fédération internationale du béton (fib) 2010). The shear stiffness (unit: N/mm^3) is equal to the slope of of the shear-slip curve at the origin. The normal stiffness is set to 10 times the shear stiffness (DIANA FEA). For the strands, a Dörr interfacial behavior is chosen with the shear stiffness K_s being equal to the slope at the origin $K_s = 5 \cdot f_{ctm}/s_0$, and the normal stiffness $K_n = 10 \cdot K_s$.

3.4.4 *Corrosion damage*

The focus of this paper lies on the structural effects of corrosion damage and its influence on the reliability. Here, the corrosion damage is considered over a length $x_4 = \ell_{CL} \sim LN(0.1, 0.04)$ m with a corrosion level $x_5 = CL \sim \beta(0.1, 0.04)$.

The length of the corrosion zone ℓ_{CL} is the length – at mid-span – over which significantly degraded material properties are implemented for all bottom 7 strands. This length is lognormally distributed with a mean of 0.1 m. The corrosion level in this zone represents the sectional loss of the steel due to corrosion and wire rupture and it is Beta distributed with a mean of 0.1, i.e. 10% corrosion level. The Beta distribution ensures that the corrosion level is bound between 0 and 1.

The effects of corrosion are taken into account on three levels: (1) a linear sectional reduction of the strand, (2) an exponential change of the bond-slip relation and (3) exponential embrittlement of the rebars and strands.

Sectional reduction
The reduced section of the strand $A_{corr} = A_0 \cdot (1 - CL)$ is only implemented in the corroded length ℓ_{CL}. While this is a simplification, preliminary analyses and engineering judgement show that the effect of corrosion is mostly of importance at mid-span, i.e. between the axle loads of the LM1 load model (discussed later).

Bond-slip relation
Due to corrosion, the tension stiffening effect can be degraded and a shift in failure mode can be observed from bending to a shear-splitting failure (Nasser, Van Steen, Vandewalle, & Verstrynge 2021). This degradation of the bond-slip relation applied here is based on experimental data reported in literature (Koulouris & Apostolopoulos 2021) (rebars) and (Wang, Zhang, Zhang, Yi, & Liu 2017) (strands), refer to Figure 5.

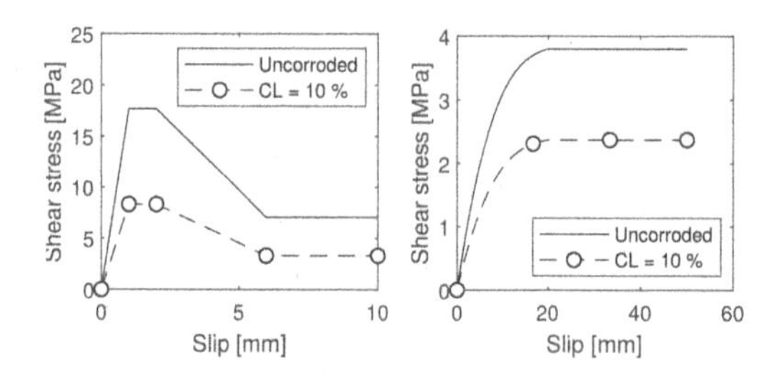

Figure 5. Bond-slip relation of corroded (left) rebars and (right) strands.

Embrittlement of the steel
Research has shown that corrosion decreases the ductility, i.e. ultimate strain, of steel reinforcement. The exponential relations in literature are implemented here (Fernandez, Bairán, & Marí 2015).

3.5 *Boundary conditions and loading*

The girder is simply supported in the vertical direction, and supported in the horizontal direction with a spring with a stiffness corresponding to the bending stiffness of the column.

The initial load on the model is the self weight of the girder and slab (SW) and the post-tensioning loads. As stated before, the post-tensioning load ratio (LR) is a stochastic parameter, with the post-tensioning load P per strand calculated as $P = LR \cdot A_p \cdot f_{pk}$. The variable load on the girder is the Eurocode EN 1991-2 Load Model 1 (LM1), refer to Figure 6 (European Committee for Standardization 2003). LM1 consists of a distributed load of $\alpha_{q1} \cdot 9$ kN/m^2 on the heaviest loaded lane of the bridge deck together with 2 concentrated axle loads of $\alpha_{Q1} \cdot 300$ kN, consisting of two wheel loads (center-to-center distance of 2 m) of 150 kN each.

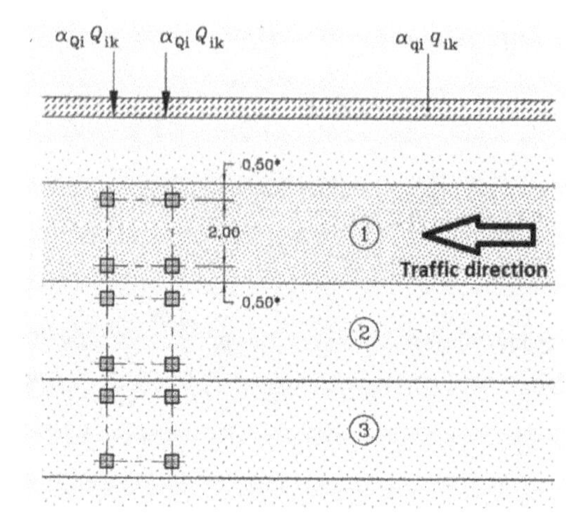

Figure 6. Load Model 1 (LM1) according to EN 1991-2. Lane number 1-3: $Q_{1k} = 300$ kN, $q_{1k} = 9$ kN/m^2, $Q_{2k} = 200$ kN, $q_{2k} = 2.5$ kN/m^2, $Q_{3k} = 100$ kN, $q_{3k} = 2.5$ kN/m^2.

In ULS, the loads are multiplied by the safety factors $\gamma_g = 1.35$ and $\gamma_q = 1.5$ for the permanent and variable loads. In the SLS, both safety factors are set equal to 1. In Table 1, a comparison between the SLS and ULS models is given.

3.6 *Loading sequence, analysis parameters and outputs*

Each FE analysis consists of two phases. In the first phase, the SW+post-tensioning load is applied at the anchor ends. In the second phase, the bond between the post-tensioning cables and concrete is developed and the LM1 load model is applied.

The iterative method in both phases is a Quasi-Newton method (BFGS) with a relative displacement convergence criterion equal to 5% and 1% in the first and second phase, respectively.

In an SLS analysis, the output is the crack width under full load, i.e. under self-weight and the LM1 load model, while in a ULS analysis, the primary output is the load ratio $E_d/R_d(\mathbf{x})$.

Table 1. Comparison between the model parameters in SLS and ULS.

	SLS	ULS
Concrete strength	f_{ck}	$f_{cd}=f_{cm}/\gamma_c$
Tensile strength	f_{ctm}	0
Rebar yield strength	f_{yk}	$f_{yd}=f_{yk}/\gamma_s$
Strand yield strength	f_{pk}	$f_{pd}=f_{pk}/\gamma_s$
Self-weight	SW	$\gamma_g \cdot$ SW
LM1	11.7 kN/m + 2 × 150 kN	$\gamma_q \cdot 11.7$ kN/m + 2 × $\gamma_q \cdot 150$ kN
Output	Crack width $w(\mathbf{x})$	Load/resistance $E_d/R_d(\mathbf{x})$

From national annex to Eurocode 2: $\gamma_c = 1.5, \gamma_s = 1.15, \gamma_g = 1.35, \gamma_q = 1.5$

4 RESULTS AND DISCUSSION

4.1 *Calculation of the reliability index*

The RSS+IS algorithm required 8 limit state evaluations (referred to as g-evals), each consisting of an SLS and ULS simulation, to converge to an estimated MPP. Of these 8 simulations, 6 predicted failure ($g(\mathbf{x}) < 0$). In the RSS, the reliability index β_{RSS} is defined as the Euclidean distance from the origin to the MPP which is equal to $\beta_{RSS} = 2.12$.

In the second step, importance sampling is set up around the MPP. The importance sampling required 492 g-evals. The reliability index is calculated and is equal to $\beta_{IS} = 1.40$. The most probable failure point in the $\mathbf{U}$-space is shown in Eq. 14.

$$\mathbf{u}_{MPP} = \begin{bmatrix} 0.21 \\ -0.47 \\ -0.85 \\ -0.84 \\ 1.20 \\ 0.73 \end{bmatrix} \tag{14}$$

This reliability index is quite low. Often, a range is given for the target reliability index of existing structures between 1 and 3, depending on the consequence class and the age of the structure (Sýkora, Holický, & Marková 2012). Furthermore, since IS is a random process and the number of (failed) simulations is relatively low, variation on the reliability index can be expected. Nevertheless, the high corrosion level and its associated damage would decrease the reliability index.

4.2 *Analysis of the importance factors*

The importance factor α_i is defined as the normalized partial derivative of a linear response surface around the MPP, as shown in Eq. 15.

$$\alpha_i = \frac{\left| \dfrac{\partial \tilde{G}_{lin}(\mathbf{u}_{MPP})}{\partial u_i} \right|}{||\nabla \tilde{G}_{lin}(\mathbf{u}_{MPP})||} \tag{15}$$

The importance factors indicate how sensitive the result is to changes in a certain parameter. Higher values indicate that that parameter has a large effect on the response surface, and thus on the failure probability and reliability.

For the bridge girder, the resulting importance factors are shown in Figure 7. It is clear that the reliability is mainly determined by (changes of) the corrosion level, accounting for over 60 percent. Next, the compressive strength of the concrete plays an important role as well, where lower concrete strengths cause earlier failure due to concrete crushing in the compressive zone between the axle loads. Together, these two parameters account for 80% of the sensitivity of the model to changes in its parameters. Then in order of decreasing importance, the following importance cadence is determined: strand yield strength > corrosion length > post-tensioning load ratio > rebar yield strength.

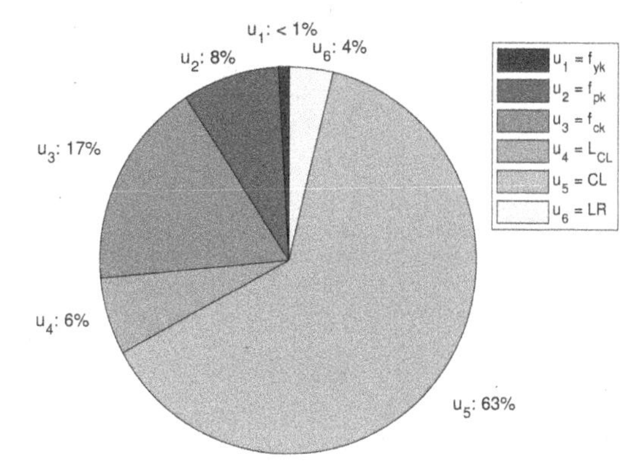

Figure 7. Importance factors for the 6 stochastic variables of the bridge girder.

Figure 7 can be used to rank the variables in order of importance when performing an on-site inspection. Indeed, from this analysis, correctly identifying the corrosion level and the concrete compressive strength are very important. On-site investigations can then be used to update the stochastic distribution for those parameters through Bayesian updating.

4.3 *Analysis of the limit states*

The minimum, median, mean and maximum value for the load factor LF at ULS and the crack width and deflection at full SLS load is shown in Table 2.

Table 2. Results of all iterations at ULS and SLS (LF is the load factor, w the crack width and δ the midspan deflection.

	LF ULS [%]	w_{max} [mm]	δ_{max} [mm]
Minimum	8.8	0	28.8
Median	109	0.023	37.6
Mean	102	0.029	38.4
Maximum	119	0.21	51.7

The histograms across all simulations for the three key performance indicators (LF, w, L/δ) are shown in Figure 8. The results show that all simulations remain below the SLS requirement by a large margin. All simulations that predict failure, i.e. $g(\mathbf{x}) < 0$, are due to ULS load exceedance. The ULS thus seems the limiting factor across all simulations for the LSF considered. This result is not unexpected, as the considered girders are heavily post-tensioned, thereby largely reducing the occurrence of bending cracks. However, it should be noted that corrosion induced cracking cannot be considered in this analysis and would require a very different type of model compared to the structural approach presented here. In Figure 9, the LF-δ curves in ULS are shown on the left, the right figure shows the crack width vs the LF in SLS. In this figure, some curves do not reach $LF = 1$ in the ULS, constituting a failure mode while all simulations predict that $w_{max} < 0.3$ mm at the SLS load. Moreover, the crack width limit is only reached (and exceeded) at an

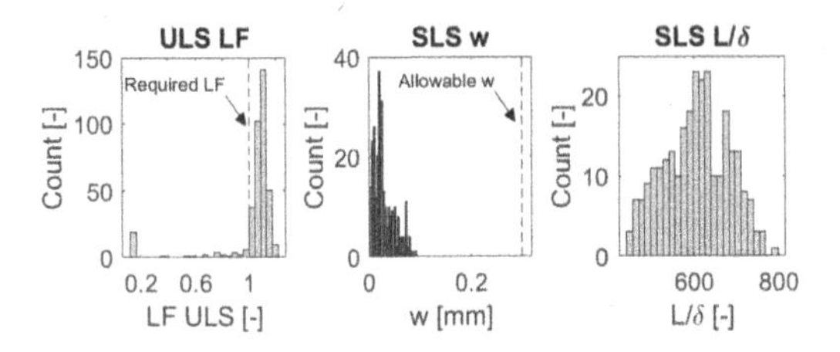

Figure 8. Histograms of the (left) LF in ULS, (middle) w in SLS, (right) L/δ in SLS.

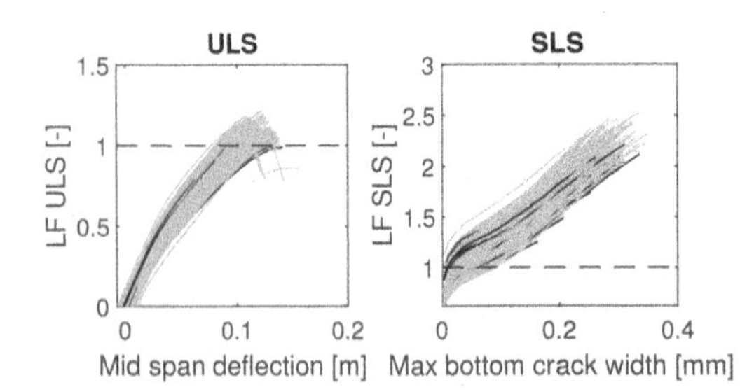

Figure 9. (left) LF-δ curves in ULS (right) w-LF curves in SLS.

SLS load factor equal to 2.25 on average (minimum = 1.72, maximum = 2.58). This means that the SLS load can – on average – be 2.25 higher for the crack width to be the relevant failure mode. Note that the deflections are not included in the reliability assessment as there is no generally accepted limit on the L/δ range. Here, the calculated values lie between 442 and 796 (mean: 596), which is close to the span-to-deflection limit (600) for train bridges (case 1) (European Committee for Standardization 2003), but does not satisfy the limit according to AASHTO for new bridges (800, case 2) (American Association of State Highway and Transportation Officials (AASHTO) 2020). Indeed, should the SLS deflection be taken into account, then nearly 45% and 100% of the simulations would become deflection limited based on the $\delta < L/600$ and $\delta < L/800$ limit respectively. This means that the SLS becomes nearly as dominant (case 1) or the dominant limit state (case 2) in the simulations.

5 CONCLUSION

The correct assessment of the structural reliability of existing structures is often a difficult task, due to the multitude of stochastic variables, the determination of a correct and adequate limit state function and the required computational effort. In this paper, a combined approach based on subsequent response surface sampling (RSS) and importance sampling is presented and applied to a real-life case study, where a corroding post-tensioned concrete bridge girder is modelled in a 2D finite element analysis.

It is found that the developed RSS+IS algorithm is very efficient for structural reliability assessment of a post-tensioned corroding concrete bridge girder, with stochastic material and damage properties. From the (computationally free) analysis of the importance factors, it is shown that the corrosion level and concrete compressive strength influence the reliability the most. The bridge girder is ULS limited in all simulations, owing to the high post-tensioning load and the absence of generally accepted deflection criteria.

The research has shown the potential of the combined RSS+IS approach in a practical application. The limited number of required simulations allows the approach to be used in a variety of different real-life use cases such as existing structures, and the importance factors can provide valuable input for on-site investigations to further improve the reliability assessment.

REFERENCES

Aldosary, M., J. Wang, & C. Li (2018). Structural reliability and stochastic finite element methods. *Engineering Computations 35*(6), 2165–2214.

American Association of State Highway and Transportation Officials (AASHTO) (2020). *AASHTO LRFD Bridge Design Specifications (9th Edition)*. Washington, D.C.: American Association of State Highway and Transportation Officials,.

Au, S.-K. & J. L. Beck (2001). Estimation of small failure probabilities in high dimensions by subset simulation. *Probabilistic Engineering Mechanics 16*(4), 263–277.

Başağa, H., A. Bayraktar, & I. Kaymaz (2012). An improved response surface method for reliability analysis of structures. *Structural Engineering and Mechanics 42*.

Bastidas-Arteaga, E. (2018). Reliability of reinforced concrete structures subjected to corrosion-fatigue and climate change. *International Journal of Concrete Structures and Materials 12*(1), 10.

Bucher, C. G. & U. Bourgund (1990). A fast and efficient response surface approach for structural reliability problems. *Structural Safety 7*(1), 57–66.

DIANA FEA. What is the guideline for reasonable choice values of tangent and normal stiffness of interface elements? accessed on 1/6/2021.

Ditlevsen, O. & H. Madsen (1996). *Structural Reliability Methods*. John Wiley and Sons Ltd.

European Committee for Standardization (2002). En 1990 eurocode 0 – basis of structural design.

European Committee for Standardization (2003). Eurocode 1: Actions on structures – part 2: Traffic loads on bridges (+ac:2010).

fédération internationale du béton (fib) (2010). Model code 2010 first complete draft.

Fernandez, I., J. M. Bairán, & A. R. Marí (2015). Corrosion effects on the mechanical properties of reinforcing steel bars. fatigue and stressâŁ"strain behavior. *Construction and Building Materials 101*, 772–783.

Gu, X., H. Guo, B. Zhou, W. Zhang, & C. Jiang (2018). Corrosion non-uniformity of steel bars and reliability of corroded rc beams. *Engineering Structures 167*, 188–202.

Guan, X. L. & R. E. Melchers (2001). Effect of response surface parameter variation on structural reliability estimates. *Structural Safety 23*(4), 429–444.

Hasofer, A. M. & N. Lind (1974). Exact and invariant second-moment code format. *Journal of the Engineering Mechanics Division 100*(1), 111–121.

Hu, Z., S. Nannapaneni, & S. Mahadevan (2017). Efficient kriging surrogate modeling approach for system reliability analysis. *Artificial Intelligence for Engineering Design, Analysis and Manufacturing 31*(2), 143–160.

Kioumarsi, M. M., M. A. N. Hendriks, J. Kohler, & M. R. Geiker (2016). The effect of interference of corrosion pits on the failure probability of a reinforced concrete beam. *Engineering Structures 114*, 113–121.

Koulouris, K. & C. Apostolopoulos (2021). Study of the residual bond strength between corroded steel bars and concreteâŁ"a comparison with the recommendations of fib model code 2010. *Metals 11*(5), 757.

Lim, S., M. Akiyama, & D. M. Frangopol (2016). Assessment of the structural performance of corrosion-affected rc members based on experimental study and probabilistic modeling. *Engineering Structures 127*, 189–205.

Melchers, R. (1999). *Structural Reliability Analysis and Prediction*. Wiley.

Melchers, R. E. (1989). Importance sampling in structural systems. *Structural Safety 6*(1), 3–10.

Nasser, H., C. Van Steen, L. Vandewalle, & E. Verstrynge (2021). An experimental assessment of corrosion damage and bending capacity reduction of singly reinforced concrete beams subjected to accelerated corrosion. *Construction and Building Materials 286*, 122773.

Shittu, A. A., A. Kolios, & A. Mehmanparast (2021). A systematic review of structural reliability methods for deformation and fatigue analysis of offshore jacket structures. *Metals 11*(1), 50.

Sýkora, M., M. Holický, & J. Marková (2012). Target reliability levels for assessment of existing structures. In Strauss, Frangopol, and Bergmeister (Eds.), *Life-Cycle and Sustainability of Civil Infrastructure Systems*, pp. 813–820. Taylor And Francis Group.

Vrijdaghs, R., C. Van Steen, H. Nasser, & E. Verstrynge (2020). Efficiently assessing the structural reliability of corroded reinforced concrete bridge girders. In B. Belletti and D. Coronelli (Eds.), *CACRCS 2020*, pp. 457–464.

Vrijdaghs, R. & E. Verstrynge (2022). Probabilistic structural analysis of a real-life corroding concrete bridge girder incorporating stochastic material and damage variables in a finite element approach. *Engineering Structures 254*, 113831.

Wang, L., X. Zhang, J. Zhang, J. Yi, & Y. Liu (2017). Simplified model for corrosion-induced bond degradation between steel strand and concrete. *Journal of Materials in Civil Engineering 29*(4), 04016257.

Zhang, M., H. Song, S. Lim, M. Akiyama, & D. M. Frangopol (2019). Reliability estimation of corroded rc structures based on spatial variability using experimental evidence, probabilistic analysis and finite element method. *Engineering Structures 192*, 30–52.

Zio, E. (2013). *System Reliability and Risk Analysis by Monte Carlo Simulation*, pp. 59–81. London: Springer London.

Computational Modelling of Concrete and Concrete Structures – Meschke, Pichler & Rots (Eds)
© 2022 Copyright the Author(s), ISBN: 978-1-032-32724-2

Assessment by in situ load tests of historical steel-concrete bridge decks

Ane de Boer
Ane de Boer Consultancy, Arnhem, The Netherlands
Municipality of Amsterdam, Amsterdam, The Netherlands

Long Ha & Andrew Quansah
Engineering Office, Municipality of Amsterdam, Amsterdam, The Netherlands

ABSTRACT: In old inner cities of the Netherlands a lot of steel-concrete bridge decks are built in the period between 1880 till 1960. Some of these bridges are renovated while the bearing capacity had to be upgraded by introducing the tram as public transport possibility. The assessment of those existing structures aren't covered by the current Eurocode of steel-concrete bridge decks. A minimum of interaction between the steel part and the concrete part should be present in the current design. However the old designed steel-concrete bridge decks show proven bearing capacity, so the feeling is that there is a lot of hidden bearing capacity in this type of historical bridge decks. Renovation can be avoided, which saves a lot of money and hindrance for the inhabitants of Amsterdam. To get more insight information of the bearing capacity of the historical bridge decks some in-situ load tests combined with nonlinear analysis were planned to setup an additional part for the recommendation for existing steel-concrete bridge decks.

1 INTRODUCTION

The current design of steel-concrete bridge decks show a lot of interaction behaviour between the steel part and the concrete part. Of course this interaction can be split into no interaction, partial interaction and full interaction. No interaction isn't allowed anymore in the current Eurocode design code for steel-concrete decks. Upon the steel girders are dowels installed to get a distributed shear force interaction behaviour between the two material parts. Full interaction beha-viour can be assumed when a fixed connection is assumed between concrete and steel parts. Partial behaviour is assumed when there is a type of interaction present which doesn't fit to a full interaction behaviour, but still transfers shear forces from the concrete to the steel girders.

At the end of the 19th century Melan [1], bridge designer from Austria, published analytical formulas to examine the strain behaviour over the bridge deck height including the type of interaction between the concrete part and the steel part.

Later on in between 1940 and1956 the interaction behaviour was discussed by Utescher [2] and resulted in to an update of these analytical formulas. However the historical steel-concrete bridge decks installed in a lot of European countries, based on the Melan design theory, were still in operation and didn't got automatically a renovation. Later on a lot of research is done to include dowels to the old bridge decks. A nice overview is given by Goralski[3] in 2006.

The last 20 years the structural safety of existing infra structures got more interest and became an important issue in asset management of the infra network of countries, provinces and towns. Now, by law the government has to prove that the structures have enough bearing capacity to ensure the structural safety of the recommendations. Beside the Eurocode set of recommendations there is a need for recommendations for existing structures. In relation to the Eurocode set of new structures, a start is made with de basic principles and the loads. Later on the set will be extended with the material related Eurocode set, like concrete, steel etc.

The steel-concrete design checks are mainly covered by the separated design rules of concrete and steel recommendations for the existing structures. However the historical steel-concrete bridge decks are already not covered by the current Eurocode set by the missing link to no interaction between the steel and concrete structural parts. Also the quality of the material properties on concrete and steel were on a lower level then today, so there could be some discrepancies between the used material in the past and the material of today. The interaction between a plane steel profile girder is also different from a steel reinforcement bar with ribs. So the first stage is to setup an inventory about all those aspects.

DOI 10.1201/9781003316404-85

2 INVENTORY

The inventory of the steel-concrete bridge decks shows the following aspects:

- the number of bridges with a steel-concrete bridge deck
- the number of lanes
- the span lengths and the type of those bridges
- the variation of the dimensions of the cross section
- the two phase in constructing the bridge
- information of inspection reports
- support conditions

2.1 *Number of steel-concrete bridge decks*

The estimated number of bridges with a steel-concrete bridge deck is 150 bridges. More then 30 of these bridge decks donn't have dowels placed on the steel girders. There is also only shrinkage reinforcement present on the top side of the bridge deck. That means that the structural safety isn't covered by the current Eurocode. More then 50 bridge decks have reinforcement on the top, bottom and edge side of the concrete. Sometimes there are small L-profiles installed at the start and end of the steel girders installed. These L-profiles doesn't fulfil the requirements of the current Eurocode related to a fixed interaction between the steel girders and the belonging concrete part. The first amount of 30 bridges means an renovation budget of 60 million euros, which gives the relevance of further research to keep the bridges in operation.

2.2 *Number of lanes*

The number of lanes is important for the traffic load of the bridges. Almost all bridges have 2 lanes for the road traffic. Additional there are mostly always lanes for bicycles and pedestrian on both edges of the bridge deck. That means that the traffic load can be spread over more than the lane width. On the other hand it is also possible that a vehicle takes a parking area on the bike or pedestrian lane, while the difference in height of the lane levels is mostly 10 cm, which is less then the Eurocode recommendation of 20 cm.

2.3 *Span length and bridge type*

A lot of bridges in the inner city show one or two side spans, where the span length varies from 5 till 8 meter. The main span of the bridges are 8.5 till 14 meter long. Mainly the spans are statically determinate. Sometimes the bridge type is coming from a renovation of a bridge type with steel girders and a timber deck. Here the steel girders are stabilized with truss cables or small steel girders to keep the deck in the right shape. When the timber deck is removed, the concrete deck is casted on a formwork with was temporarily placed on the bottom flange of the steel girders. Another renovation was the introduction of the tram as public transport in the city. In this period the steel girders were replaced by higher and wider steel girders, to increase the bearing capacity of this part of the bridge deck.

Another aspect is the pre cambering of the steel girders. To get a arching effect over the span length of the bridge the steel girders were cambered at the midspan cross section of each girder.

2.4 *Variation of the dimensions of the cross section*

The used steel girders are very different and related to the use of the bridge deck by traffic or tram. The steel girders are starting with DIN type steel profiles with heights form 260 mm to 700 mm. Also DIR, NP, BP, INP or HE profiles are found in the technical archive of the city with heights in the same order like the already mentioned DIN profile.

The head to head distance between the steel girders is stark related to the height of the steel girder and differs from 600 mm till 800 mm. The bridge decks with lanes for trams have mostly head to head distances of 700 till 750 mm.

The belonging concrete part have mostly 150 mm till 200 mm height. This means that there is also a spreading length for the wheel print load with an Eurocode are of 400 by 400 mm^2 of 300 mm till 400 mm. The height of the asphalt layer upon the concrete part is 70 mm, so that means another spreading length of 140 mm. Between the bottom flanges there is always a cut out with a height of 140 mm. The length of the cut out is related to the head to head distance of the steel girders. Figure 1 shows an idea of the cross section of a steel girder with the belonging concrete part.

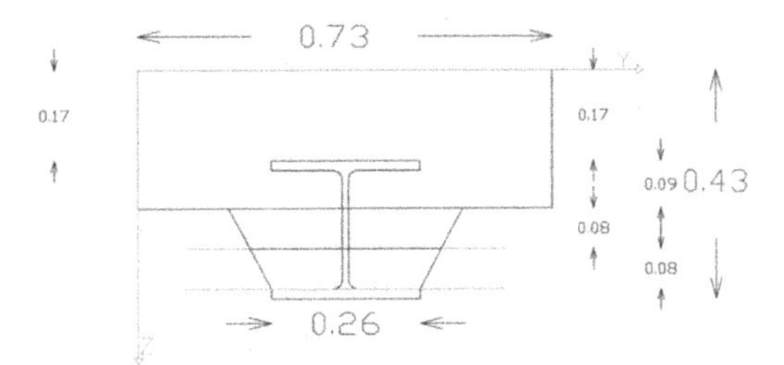

Figure 1. Cross section of a historical steel-concrete bridge deck.

Figure 1 shows the cross section of a steel girder (DIN26) and the concrete part. Here the overall construction height is 0.43 meter and the head to head dimension of the steel girders is 0.72 meter.

Between the surfaces of the steel and concrete part are interface assumed, which properties can be related to the partial interaction of the composite cross section. This interaction can be divided into slip with a belonging slip limit and friction.

The center of the cross-sectional area of the uncracked cross section lies always under the top flange of the steel girder, so it is assumed that the concrete part lying under the top flange of the steel girder is cracked. The crack width in this area are dependent of the (overloaded) traffic loads passing the bridge decks during the lifetime of the structure.

2.5 *Constructing the bridge deck*

The technical archive shows mostly three construction stages. The first stage is the installation of the steel girders crossing the side and/or main span. The second stage shows placing the formwork on top oof the bottom flanges of the steel girder. The third stage is casting the green concrete, what implies only an additional load to the steel girder coming from the dead weight of the concrete. When the concrete got his stiffness the formwork is removed and other loads are added to the structure. These stages are important for the yield limit of steel girders, while the assumed yield limit of the used steel is probably 230 MPa, where the starting stress after stage three is already 25 MPa. The concrete stress is after that stage three still 0.0 MPa.

2.6 *Information from the inspection reports*

The periodic inspection reports show some corrosion attacks of the top surface of the bottom flange of the steel girder. Sometimes the coating on the bottom of the bottom flange of the steel girder is gone. Also delamination over the thickness of the bottom flange is observed in those case. Beside that there are sometimes cracks found at the bottom fibre of the bottom surface of the concrete starting at the voute of the cut-out of the concrete part, see Figure 2. The depth of the crack is never measured and there are also no cores drilled to get a better idea of the crack depth.

Figure 2. Concrete cracks located at the voute.

The surfaces of the bottom flanges of the steel girders are green coated, where the concrete is showing a grey surface.

2.7 *Support conditions*

The analytic formulas show a statically dependent system with the support lines at the start and end of the bridge decks. However almost all historical steel-concrete bridge decks show a support length of 0.45 meter and a width similar to the width of flange of the steel girder. Above this support area the concrete doesn't have a cut-off anymore, so a rectangle cross beam is casted, where reinforcement is included. This part of the structure is supported by masonry, which can be seen as a structural part with its own normal stiffness. A rather thick steel plate between the masonry and the bottom flange of steel girder will distribute the reaction forces over the masonry area, see Figure 3.

Figure 3. Additional support plate.

Again the green coating is the steel girder, where the support plate is crumbling and corroded.

3 THE IN-SITU TESTS

3.1 *Preparation*

During the period of the inventory of the historical steel-concrete bridge decks, a renovation of a heavy traffic corridor to the inner city of Amsterdam was in a construction phase. Some bridge decks should be demolished, where a period of 2 weeks was open for testing for the West side. The East side of the bridge is staying open for the traffic. One of these corridor bridges was a 3 span bridge with a bridge deck without any dowels inside.

Figure 4. Side view of the tested 3 span bridge deck.

This was an opportunity to setup some load tests in-situ on a historic steel-concrete. Only the side spans could be used for testing because of the fact that the main span has to be open for 24 hours a day for the business and tourists boats.

The tested part are the 6 girders left of the pedestrian path, where the height of the bridge deck is given in Figure 5. This means that there is an uplift support at the 6th girder, which is also indicated in Figure 5.

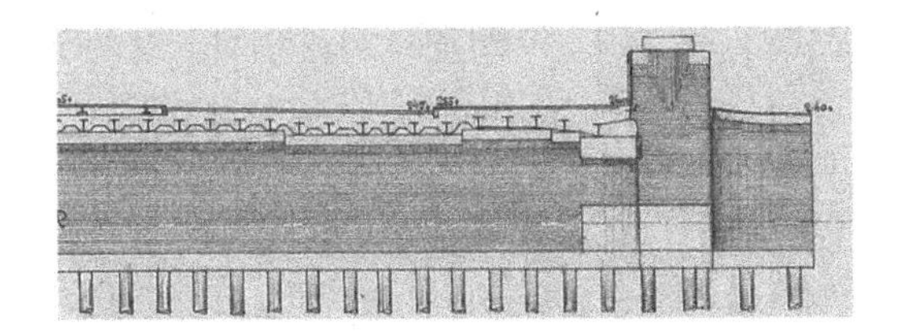

Figure 5. Cross sectional abutment pier view.

Before testing the bridge deck, material tests were done by drilling concrete cylinders from the bottom side of the bridge deck to minimize hindrance for the traffic, which is passing the bridge on the East side.

Of course there are made some FE simulations to get an idea of the deflections of the deck and the strains of the steel girders. The results of the FE simulations show that the longitudinal bearing capacity should be sufficient, however the transversal bearing capacity is not sufficient. The FE shell model of the orthotropic bridge deck doesn't give the right results, where the more advanced FE solid models give realistic results.

The focus of the tests was to get insight information of the punching behaviour of the concrete part between the girders. The construction documents are signed in 1934, where there was only a concept design recommendation for traffic loads in the Netherlands. The expected lifetime of the bridge is already passed, but the condition is rather good. However the current heavy traffic flow for a corridor to the inner city is insufficient.

3.2 *In-situ tests*

In total there has been 5 wheel print load locations over the length of the spans. The tested bridge deck part counts 6 girders with a total width of 3.95 meter over a length of 6.5 meter. This part is sawn over the length and the width, so there should be no connection anymore with the rest of the bridge deck. The support lines should be similar to the original bridge deck. Therefor the wheel print load locations are chosen to be symmetric. The north side span is tested at the 1/4L and 3/4L location, the south side span is tested first at the midspan location and later on at 1/6L and 5/6L location. Only the midspan load location test is described more in detail in this contribution. Figure 6 gives the setup of the load and measurement locations of the midspan load test.

Figure 6 shows the blue rectangles as the locations of the strain sensors on the bottom flange of the steel girders, the orange circles as the vertical displacement lasers locations, the green lines as the horizontal displacements between the bottom flanges of the steel girders and at least the red rectangle the load cell location. On the top the 6 girders are numbered, at the left the rows where the measurements rows which are related to the ratio of the span length on the right side of the figure.

The load counts some stepwise load levels, where every level will have 3 cycles. The load levels are 0,

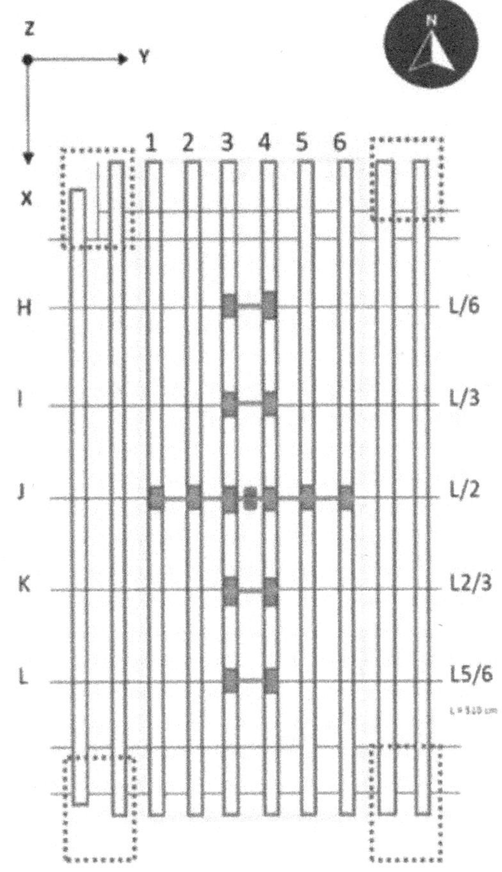

Figure 6. Load and sensor location midspan test.

50, 100, 200, 300, 400 and 475 kN. The capacity of the foundation, which has to be intact after these test, has a limit of 500 kN. For safety reasons, regarding the foundation capacity and the used load cell, the in-situ an overall load of 475 kN. Figure 7 shows the load scheme during all five tests. The load speed is set to 2.5 kN/sec. A realistic modified wheel print area of 230×300 mm^2 is chosen as load area between girder 3 and 4.

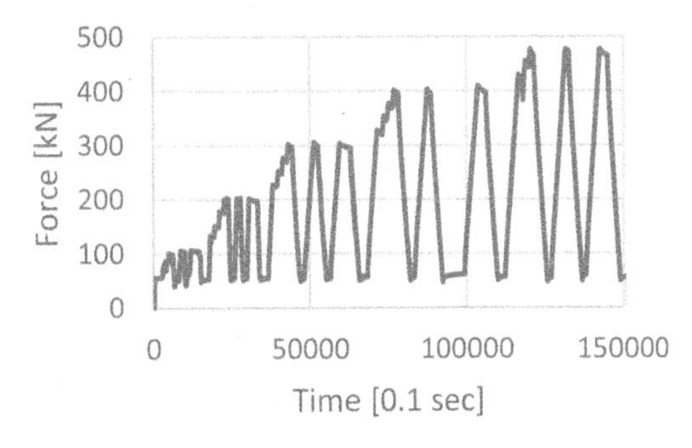

Figure 7. Load scheme 5 performed tests.

Figure 7 shows also that there has been intermediate load steps of 25 kN, when the first cycle is started to a new mail load level. Every load test takes 3.5 hours, which results in 2 weeks of testing in total.

After these 2 weeks each bridge deck part were splitted into 2 parts, which are stored for additional laboratory tests later on. Also a 3 girder width bridge deck part from the main span of the bridges is stored and will be tested also in the laboratory in the future.

4 EVALUATION OF THE MEASUREMENTS

The most important aspects of the measurement was behaviour of the concrete, which was cracking only on the micro level at a wheel print load level of 475 kN. From the strain measurements of the bottom flange of the steel girders and the vertical displacements of the steel girders the conclusion can be set to a force distribution over four neighbouring steel girders. However the expected stiffness of the bridge deck is higher than assumed and there is not always symmetry in the measurements when there is symmetry at the symmetric loading tests. Unfortunely we didn't measure the vertical displacements of the support lines.

Another aspect is the slipping behaviour over the width at the midspan load test. The other four tests doesn't show this slip behaviour. The slipping behaviour isn't only measured at the steel girder next to the load cell, but later on it is also measured at the next cycle at the other two girders left and right of the mid girders, so all girders shows this slip behaviour in the transversal direction of the bridge deck. Figure 8 shows the transversal displacement of cycles 1-3 for girders 1 and 3. The first cycle of in the load levels from 50 till 475 kN shows no transversal displacement but the following cycles show at a lower load force level. Girder 1 which is laying on the edge of the bridge deck part shows also a similar transversal displacement. Then an extra girder with belonging concrete is demolished complete at the store location. This operation results in a very smooth surface of the outline of the 85 year old steel girder, so the stick/slip behaviour should be an additional aspect for the nonlinear FE analysis.

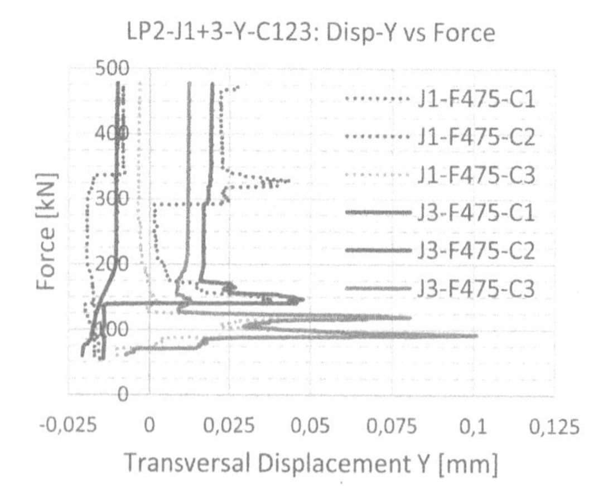

Figure 8. Displacement in transversal direction.

Later on when the bridge deck parts were lifted out of the bridge not all parts were free fully sawn over the height of the bridge deck from the original bridge deck.

Also an additional steel plate was part of the support area between the bottom flange of the steel girder and the masonry pier and abutment pier. An additional observation was that some steel girders seemed to be partly fixed to the top surface of the abutment pier. That causes a hidden internal bending moment at the support line during the tests of the bridge deck.

5 INPUT FE SIMULATION

The FE model of the six steel girders and the concrete is a complete quadratic solid element model, with interface elements between the outline of the steel girder and the concrete part, see figure 9 and 10.

Figure 9. FE solid model bridge deck.

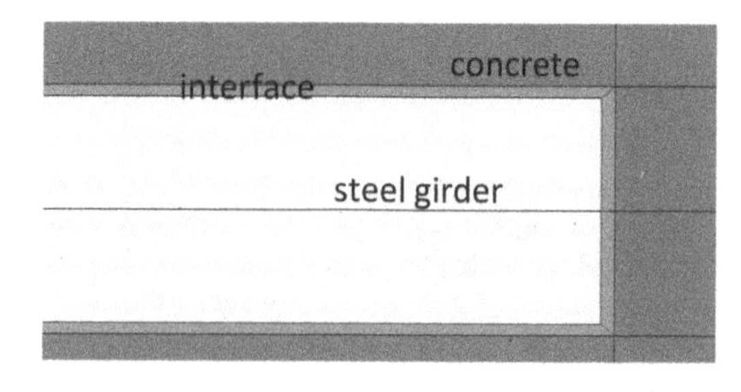

Figure 10. Top flange of a steel girder with interface elements.

The support areas contain also interface elements to simulate the observed internal bending moment behaviour.

The interface elements are divided in separate sets, while the web and the flanges can get different material properties. The different colors in Figure 10 indicate this difference in material properties. At the start stage of simulations the material properties of the interface elements are similar.

The size of the elements of the steel girder and the interface elements can also be seen in Figure 10. The width of the steel girder counts 4 large elements and 2 small elements. With steel girders DIN26 and DIN28 the element size over the width various in ge-neral from 65 till 70 mm. The height of the flange counts 2 elements, so this element size is 10 mm. Over the length of the bridge deck the element size is 100 mm. All interface elements on the outline of the steel girders have an height of 1 mm, where the support interface elements have an height of 10 mm. The cross sectional area of one steel girder with its belonging concrete part is shown in Figure 11.

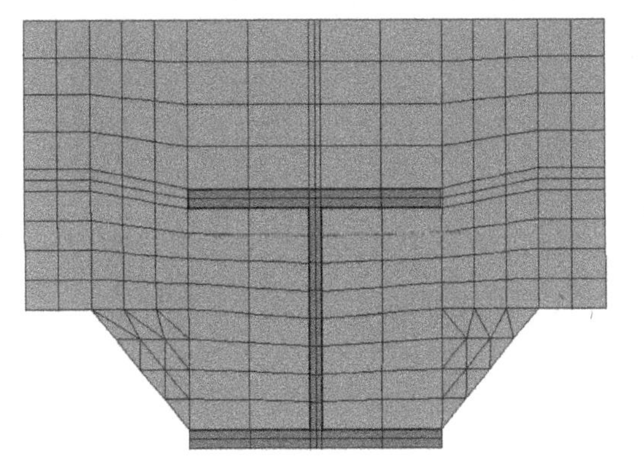

Figure 11. Cross section one steel girder + concrete.

The mean value of the concrete cylinders show a value of 50.9 MPa, with a standard deviation of 6.2 MPa. Later on there are also concrete cylinders bored from the side of the bridge deck and the skew cut-off side. The concrete strength in the horizontal direction show a mean value of 42.1 MPa with a standard deviation of 3.3 MPa. The concrete strength perpendicular drilled in the cut-off skew direction have a value of 49.1 MPa with a standard deviation of 6.9 MPa.

Four samples were taken from one steel girder to get an idea of the yield stress value and the Young's modulus. A mean peak stress value of 261.8 MPa was observed at these tests, followed by a yield level of 233.8 MPa with a standard deviation of 9.0 MPa. One of these sample results is shown in Figure 12.

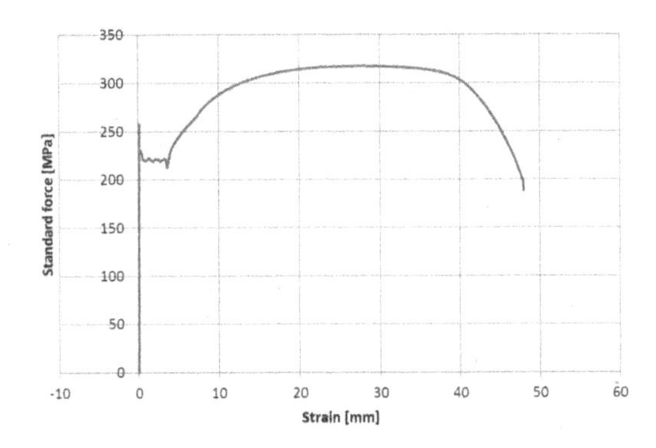

Figure 12. Stress strain relation steel sample.

Also Ultrasonic Pulse Velocity(UPV) research has been done to the drilled concrete cylinders, which results in a mean value for the Young's modulus of 52461 MPa and a standard deviation of 5923 MPa. The UPV mean value is rather high related to the ModelCode2010 [4]. For the FE simulations the common relations of the MC2010 are used based on the compression strength.

The ultimate mean stress of the four steel samples was 317.5 MPa, where the standard deviation counts 2.6 MPa.

The most unknown material properties are the slip behaviour of the different surfaces between the steel girders and the belonging concrete and the friction angle between steel and concrete after 80 years lifetime. Different authors like Leskela(2000), Takami(2005) and Hegger(2005) and others did in the past research to slip behaviour between concrete and steel. The maximum stress value can be set to 1.0 MPa with a belonging slip limit value of 0.1 mm. Of course the worst case is a slip limit value of 0.0 mm and a belonging stress of 0.0 MPa. The influence of these value variations can be found by running different FE simulations. Table 1 shows the different material properties for the FE simulations.

Table 1. Overview material properties.

Material property	Symbol	Value
Concrete		
Young's modulus	E	37 900 N/mm^2
Poisson ratio	ν	0.15
Density	ρ	2 400 kg/m^3
Type curve softening	Hordijk	
Mean tensile strength	fctm	3.88 N/mm^2
Fracture energy	Gf	105 N/m
Crack bandwidth method	Govindjee	
Reduction Poisson ratio	Damage based	
Type curve compression	Parabolic	
Mean compression strength	fcm	54.6 N/mm^2
Compressive fracture energy	Gc	32 000 N/m
Reduction lateral cracks	Yes	
Minimum reduction factor	0.4	
Influence lateral locking	Yes	
Steel		
Young's modulus	E	210 000 N/mm^2
Poisson ratio	ν	0.3
Density	ρ	7 850 kg/m^3
Yield stress	σ	235 N/mm^2
Hardening method	Strain, EN1993-1-5	

6 FE SIMULATION PUNCHING BEHAVIOUR

6.1 *Analysis type*

Following the Guidelines for Nonlinear Finite Element Analysis for Concrete Structures(2020) shows table 1 already the nonlinear parameters so the FE simulation will be a nonlinear analysis. The used method is Newton Raphson Regula with convergence tolerances based on energy with a basic tolerance of 1.E-3. The load increments are similar to the in-situ tests with values of 100 kN and 25 kN between after the load level of 300 kN. Cracking of concrete is expected after a load level of 300 kN.

Related to the construction stages a phased analysis is added. In the first phase only the steel girders are active with the dead load of the steel girders and concrete. In this first stage the dead weight of the concrete is simulated as an distributed load to the top surface of the top flange of the steel girder. In the second phase the dead weight of steel and concrete parts will be the starting load and will be added with the punching load, the wheel print load between the steel girders three and four on the midspan location.

6.2 *Output nonlinear analysis*

Several aspects could be concluded from the nonlinear analysis like:

- Crack pattern punching behaviour
- Slip behaviour
- Load displacement diagram till 475 kN
- Distribution wheel print load over a row of girders

The first aspect was the punching behaviour of the bridge deck with the belonging crack pattern. The crack pattern at a load level of 475 kN can be shown in Figure 13a and 13b.

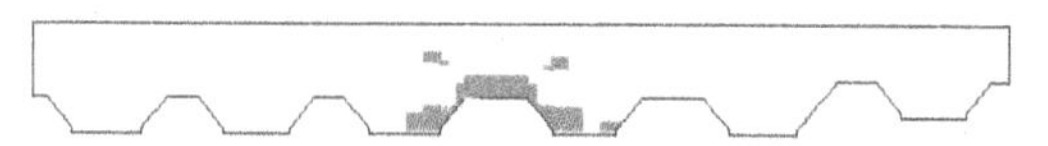

a. Cross sectional view crack pattern at midspan

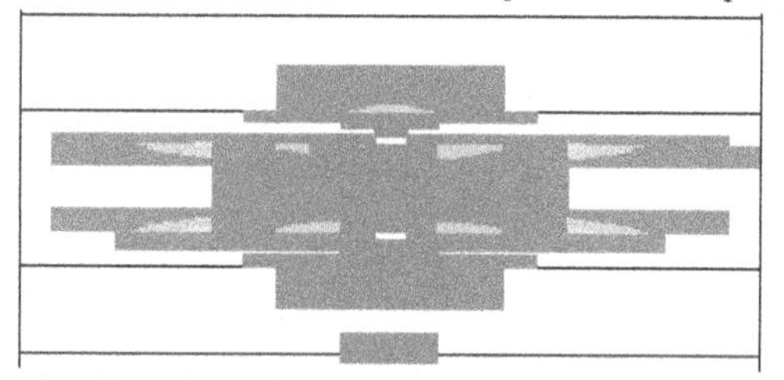

b. Top view of the crack pattern

Figure 13. Crack patterns load level 475 kN.

As expected the most cracks are developed in the area of the wheel print load in de concrete of girder 3 and 4. There is almost no cracking pattern at the other girders. The maximum crack width is 0.02 mm, so the observation after the in-situ load tests was right that only micro cracks could be observed. The depth of the crack was 60 mm, starting from the bottom fibre of the cut-out of the concrete bridge deck. This is shown in Figure 14.

The length of the relevant crack pattern in longitudinal direction is 1100 mm. The axle load distance of the Eurocode tandem axle configuration is head to head 1200 mm, so both wheel print loads together will influence the overall crack pattern of an axle load. However the netto wheel print load in the Eurocode is 150 kN, so still there is a factor of 3.0 present. . Over the width of the bridge deck the Eurocode distance is 2 meters, so that belongs to two other concrete parts of the bridge deck.

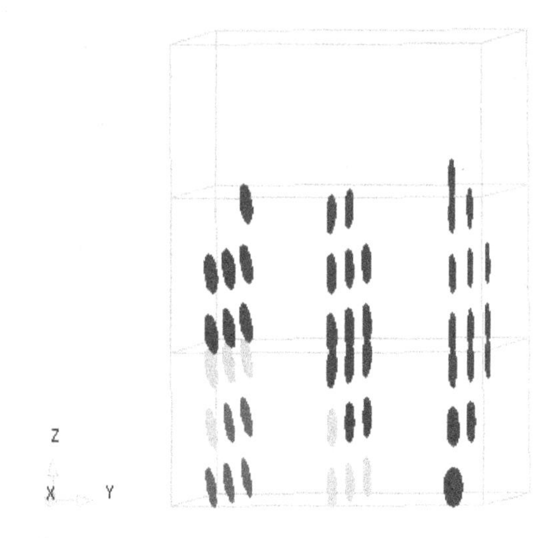

Figure 14. Crack strain including the orientation.

To get more information about the influence fields of the crack pattern in longitudinal direction the FE simulation is continued till an unrealistic ultimate limit load, like 1900 kN. Similar pictures, like Figure 14, can be shown on different load levels in Figure 15, but now over the height of the bridge deck.

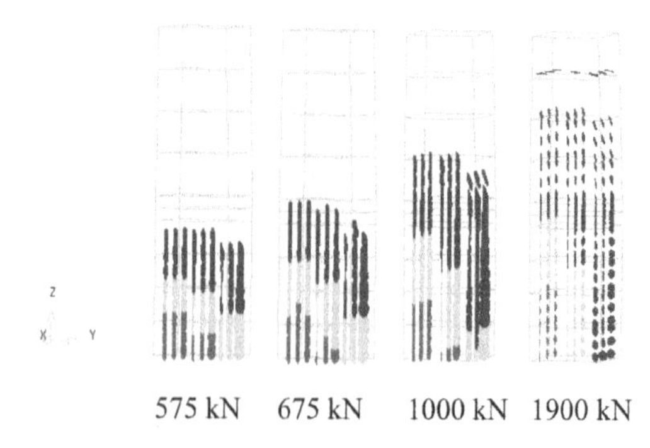

Figure 15. Crack strain development.

Figure 15 shows that the cracks is reaching at a load level of 675 kN the top surface to the top flange of the steel girder. To ensure that there is always an uncracked concrete part above the top flange the extra load capacity should not come above the 675 kN for two wheel print loads. The partial factor in the Eurocode for axles is 1.5, so there is enough space between the calculated 675/300 = 2.25 and the EC factor 1.5. The netto concrete length of the girders 3 and 4 is 730 − 270 = 560 mm, where the wheel print width was 230 mm. In this case the punching cone under the wheel print could occur.

Meanwhile the main goal of the in-situ tests was successful. A second goal was the slip behaviour between the steel girders and the belonging concrete. Variations of the slip limit from 0.1 mm till a value of 0.0 mm were done with the FE model. The slip values at a load level of 475 in both directions are shown in Figure 16.

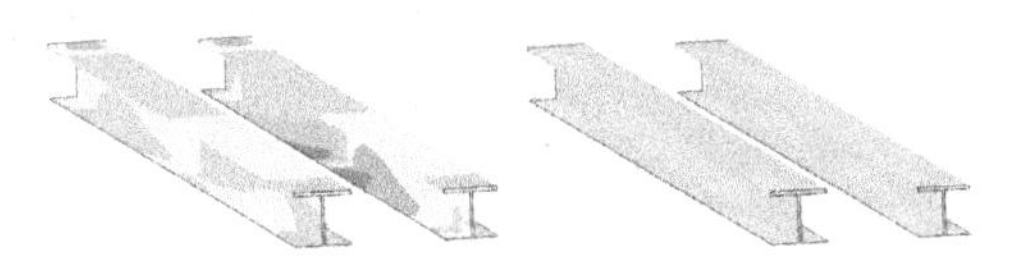

a. X-direction at 475 kN b. Y-direction at 475 kN

Figure 16. Slip results at girders 3 and 4 in longitudinal(X) and transversal direction(Y).

Figure 16 shows very clear that the slip behaviour at the load level of 475 kN is still almost uniform distributed in transversal direction, where the slip beha-viour in longitudinal direction is already varying. Nearby the wheel print load location at midspan the dark blue and dark red colored parts are also beyond the chosen sliplimit of 0.1 mm. The interface elements on the bottom flange of the steel girders show already slip behaviour beyond the chosen sliplimit of 0.1 mm. When the sliplimit is decreasing to a value of 0.05 or 0.025 mm these areas will be reached at a lower load level of the wheel print load.

Further research has to be done to tackle this phenomena more in detail.

The load displacement diagram of the in-situ test is still very lineair, so it is not so important to include a figure of this results. However it is good to know that an engineer can simulate this behaviour with a linear elastic analysis.

The wheel print load is equally distributed over 4 girders as result from the measurements of all in-situ tests which is also simulated by the described FE model.

7 FUTURE RESEARCH

As mentioned before three parts of the bridge deck were shipped to the laboratorium in Delft to get better measurement conditions then at the in-situ test location. Nevertheless the in-situ tests are madee to create some 'virgin' tests, while the negative effects of lifting and transporting of bridge deck samples to the lab were not predictable on forehand.

The results of the foreseen lab tests can be compared with the results of the in-situ tests. The prediction of the lab tests can be done based on the results of the in-situ tests. In that way the measurement plan of sensor locations etc. can be planned more in detail.

8 CONCLUSIONS

The following conclusions can be drawn from the in-situ tests and FE simulations sofar:

- The in-situ tests show a linear behaviour till a wheel print loa of 475 kN
- The maximum acceptable wheel print load is probably 675 kN
- Four girders are bearing one wheel print load equally distributed
- Slip behaviour was assumed and is measured in one of the in-situ tests
- Slip behaviour will be tested further on in a more conditioned environment like the laboratorium
- The phase analysis showed an maximum intital stress in the steel girders, which has an influence of 10% to the yield stress of steel
- The FE ULS simulation has reached a load level of 1900 kN which shows that there should be a lot of extra bearing capacity for these type of bridge decks
- Additional lab tests are needed to prove more in detail the above mentioned conclusions

Most important conclusion is that the authors have the feeling, that design recommendations, like the Eurocode need additional articles for re-examination of historical steel-concrete bridge decks without dowels.

ACKNOWLEDGEMENT

The authors wish to express their gratitude and sincere appreciation to the Programme Bridges and Quay Walls of the Engineering Office of the municipality of Amsterdam for their support of the pilot project "Steel-Concrete Bridge Decks(Verbundträger)". Beside these thanks we like to thank also our other project team members and external partners, which were involved in the in-situ tests for their contribution.

REFERENCES

[1] Eggerman, H., Kurrer, K-E., Zur Internationalen Verbreitung des Systems Melan seit 1892: Konstruktion und Brückenbau, BETON- UND STAHLBAU, 07 NOVEMBER 2006

[2] Utescher, J., Bemsssungsverfahren für Verbundträger, Springer Verlag, 1956

[3] Goralski, C., Zusammenwirken von Beton und Stahlprofil bei kammerbetonierten Verbundträgern, PhD thesis RWTH, Germany, 2006

[4] Takami, K., Ishii, N., Hamada, S., Shear properties of bonded T-shaped shear connector, Proceedings of the 4th European Conference on Steel and Composite Structures Eurosteel, Vol. B, pp. 4.2-1 - 4.2-6, Maastricht, 2005

[5] Hegger J, Rauscher S, Goralski C, Russell H (2005). Push-out tests on headed studs in high-strength concrete. Special Publication, 228: 769-786

[6] CEN 2005. Eurocode 2: Design of Concrete Structures - Part 1-1 General Rules and Rules for Buildings. NEN-EN 1992-1-1:2005. Brussels, Belgium: Comité Européen de Normalisation.

[7] fib 2012. ModelCode 2010 Final Draft, Lausanne, International Federation for Structural Concrete (fib)

[8] M.A.N. Hendriks & M. Roosen, "Guidelines for Non-linear Finite Element Analysis of Concrete Structures", Rijkswaterstaat Centre for Infrastructure, Report RTD:1016-1:2020, version 2.2, Rijkswaterstaat, the Netherlands, 2020

Computational Modelling of Concrete and Concrete Structures – Meschke, Pichler & Rots (Eds)
 ISBN: 978-1-032-32724-2

Joint free pavements made with HPFRC

Alessandro P. Fantilli, Nicholas S. Burello & Masood Khan
Politecnico di Torino, Torino, Italy

Giovanni Volpatti
CEMEX Innovation Holding AG – Brügg Branch, Brügg, Switzerland

Jorge C. Diaz Garcia
CEMEX LATAM Holding AG – Swiss Branch, Brügg, Switzerland

Davide Zampini
CEMEX Innovation Holding AG, Zug, Switzerland

ABSTRACT: Concrete has been widely used in the construction of roads, highways, industrial floors and pavements since early twentieth century. Construction methods generally include placement of joints at specific distances to control the cracking phenomenon. The latter is due to the development of tensile strains caused by the shrinkage of concrete and by environmental factors, such as temperature gradient. However, joints result in reduced load carrying capacity, local failure, and pavement damage. To reduce the number of joints, the fracture toughness of concrete can be enhanced by adding fibers. As the models available for conventional fiber-reinforced concrete (FRC) cannot be extended to high-performance fiber-reinforced concrete (HPFRC), the aim of this work is to describe a new model to design HPFRC joint free slabs. Specifically, a composite cross section made of soil and concrete, which is subjected to imposed strains, is modelled through the Colonnetti's theory of elastic coactions. In this way, not only the effect produced by concrete shrinkage but also the nonlinear response of HPFRC in the strain hardening stage are taken into account. For given maturity curves, crack does not appear if the maximum tensile strain provided by the model is lower than the strain that produces localization in HPFRC.

1 INTRODUCTION

Nowadays, reinforced concrete is the most used building material, especially in structures and infrastructures. Focusing on pavements, concrete is almost the only alternative to asphalt, and the first guidelines for the design of concrete pavements were introduced about 30 years ago (AASHTO 1993). Afterwards, several guidelines were also proposed (Choi et al. 2005; Rasmussen et al. 2009; Roesler et al. 2016; Söderqvist 2006).

Although concrete slabs guarantee longer term service life and can be easily built, a specific attention is required during the first days of curing. Indeed, due to the shrinkage phenomenon and to the low strength in tension, concrete is prone to cracking at early age. More precisely, when applied stress overcomes the tensile strength, concrete fails and crack growths. It can occur just after casting, when strength increases at a slower rate than the constrained stresses induced by the reduction of volume.

As cracks can compromise both durability and functionality of pavement, contraction joints are frequently used to control the cracking phenomenon (FHA 2019). In practice, the upper side of the slabs is sawn at regular intervals during the first hours after casting.

Nevertheless, joints generate several problems to the pavement in service. Traffic movements can damage the joint (joint-edge chipping and cracking) and facilitate the penetration of aggressive chemicals, which in turn affect the durability of pavements. For these reasons, jointless pavements, also made with fiber reinforced concrete, have been proposed (ChunPing et al. 2015; Larrard et al. 2011; Zhang et al. 2013). In fact, several studies have shown that the addition of fibers, either steel or polymeric, improves the mechanical properties in tension, and controls the widening of cracks (Tehmina et al. 2014; Yoo et al. 2018). At early age, experimental results show that the presence of $0.2 \div 0.3\%$ of nylon fibers may reduce the effect of drying shrinkage (up to 75%) in a cementitious matrix (Choi et al. 2011). Furthermore, the performances improve when high performance fiber reinforced concrete (HPFRC) is used to cast thin concrete pavements, even in the case of instant repairs or quick renewal of roads (Burger 2010; Hachiya et al.

 DOI 10.1201/9781003316404-86

2006, R. 2009). As HPFRC shows a strain-hardening behavior (Graybeal 2016) before reaching the localization of tensile strain (Ramadoss 2008; Savino et al. 2018), the fibers may completely substitute the conventional reinforcement, and reduce/eliminate the presence of joints.

For instance, Destrée et al. (2016) discussed the main parameters of concrete shrinkage and provided a model for the analysis of slab made of FRC (see Figure 1). It is based on the classical tension stiffening equations of RC structures, and includes the bond slip mechanisms between soil and FRC and between fiber and matrix, and the fracture mechanics of the matrix in tension.

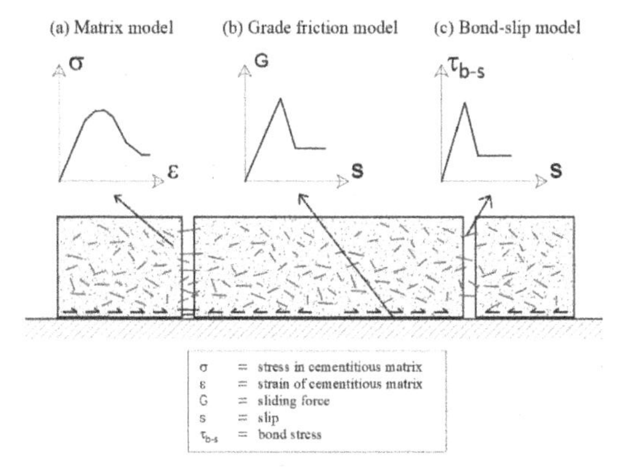

Figure 1. Mechanical behavior of cracked fiber-reinforced cementitious composite (Destrée et al. 2016): (a) stress-strain relationship of the cement-based material; (b) bond-slip model between FRC and soil; (c) bond-slip model of the fiber within the matrix.

Although the model can be used to design jointless slabs using steel fibers, the application cannot be extended to HPFRC, because, for this cement-based material, the strain localization in tension, $\varepsilon_{c,cp}$, does not occur at the first cracking, $\varepsilon_{c,cr}$ (see Figure 2b). Accordingly, a new model, able to predict the mechanical behavior of slabs on grade is proposed. More precisely, a composite cross-section made of concrete and elastic soil (see Figure 3), and subjected to the imposed strain ε_{sh} (sh = shrinkage), is analysed. The proposed model calculates the internal states of stress and strain by means of the Colonnetti's theory of elastic coactions (Colonnetti 1950), when material properties are known at every stage of curing. The model works within Stage I (i.e., in absence of strain localization) and can be applied also in the cases of conventional concrete and FRC, in which $\varepsilon_{c,cr} = \varepsilon_{c,cp}$ (Figure 2a).

2 PROPERTIES OF MATERIALS

The model analyzes slabs on ground during the curing stage, just after casting. To evaluate the state of stress and strain in this scenario, it is necessary to

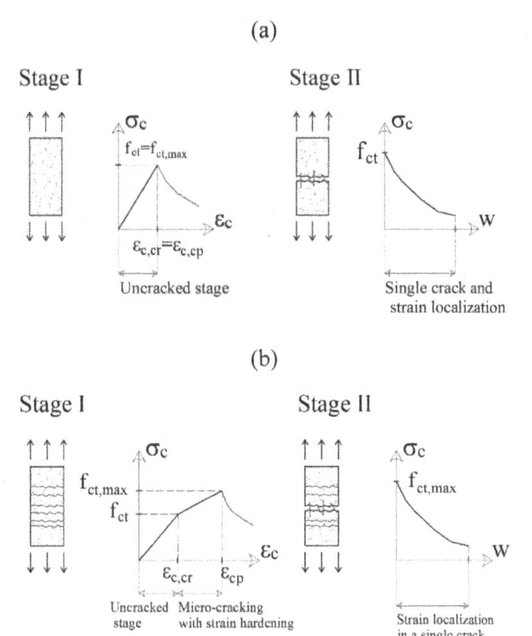

Figure 2. Tensile behavior of (a) conventional FRC and (b) HPFRC.

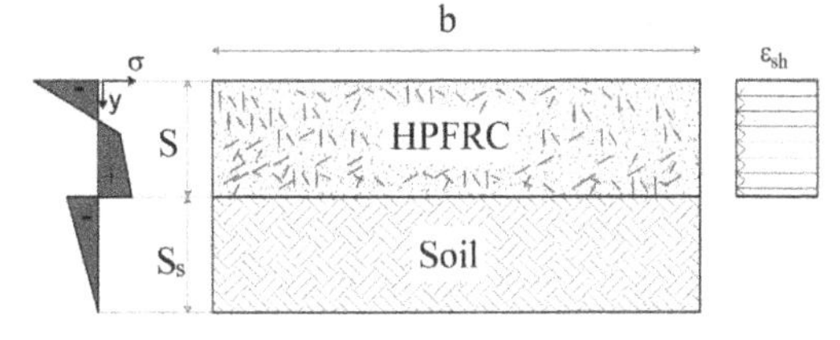

Figure 3. Composite cross-section analyzed by the proposed model, in the case of uniform shrinkage strain ε_{sh}.

know the mechanical properties of concrete, including shrinkage, at any time. However, each parameter (e.g., shrinkage strain, tensile strength, compressive strength, etc.) shows different rate of development after casting.

2.1 *Mechanical performances of concrete and soil*

HPFRC has a strain hardening response in tension (as depicted in Figure 2b) and a linear elastic behavior in compression. The parameters of a possible stress-strain relationship are described by exponential equations (Eurocode 2 1-1 1992; fib Model Code 2010; ACI 209R-92 1997). However, tensile strength tends to increase more rapidly than the compressive strength (Bentur 2003).

Some studies (Boshoff 2012; Combrinck et al. 2019; Hammer et al. 2007; Roziere et al. 2015) defined the tensile strain capacity of concrete at early age, which reaches the minimum during the setting time (up to 10 hours) and before early hardening (see Figure 4). This is due the fact that a significant increase of the elastic modulus occurs earlier than the increment of the tensile strength. For this reason, two types of analyses are carried out after casting:

- short term analysis (STA), from 2 hours to 96 hours;
- long term analysis (LTA), from 4 days to 28 days.

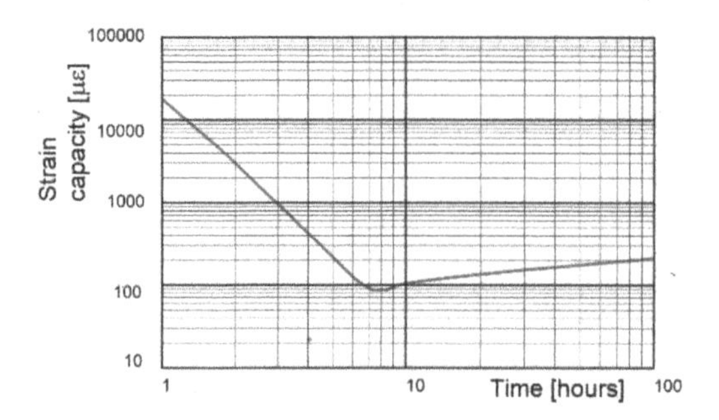

Figure 4. Tensile strain capacity *vs*. time (Boshoff 2012).

In both the cases, the equations provided by Eurocode 2 are taken into consideration. Specifically, compressive strength of concrete, at various ages, may be estimated as follows:

$$f_{cm}(\tau) = \beta_{cc}(\tau) \cdot f_{cm} \tag{1}$$

$$\beta_{cc}(\tau) = exp\left\{s\left[1 - \left(\frac{28}{\tau}\right)^{1/2}\right]\right\} \tag{2}$$

where $f_{cm}(\tau)$ = mean concrete compressive strength at time τ; f_{cm} = mean compressive strength at 28 days; $\beta_{cc}(\tau)$ = coefficient which depends on τ; and s = coefficient that depends on the type of cement.

As a first approximation, the value of the tensile strength with time, $f_{ctm}(\tau)$, is given by:

$$f_{ct}(\tau) = (\beta_{cc}(\tau))^{\alpha} \cdot f_{ctm} \tag{3}$$

where f_{ctm} = mean tensile strength at 28 days of curing; and $\alpha = 1$ when $\tau < 28$ days.

Focusing on the stiffness, the variation of the modulus of elasticity with time, $E_{cm}(\tau)$, is:

$$E_{cm}(\tau) = \beta_{cc}(\tau)^{0.3} E_{cm} \tag{4}$$

where E_{cm} = modulus of elasticity at 28 days.

Tensile strain at cracking can be calculated by using the modulus of elasticity in compression, $E_{cm}(\tau)$, and the tensile strength $f_{ctm}(\tau)$:

$$\varepsilon_{c,cr}(\tau) = f_{ct}(\tau) / E_{cm}(\tau) \tag{5}$$

To take into account the results of previous studies (Boshoff 2012; Combrinck et al. 2019; Hammer et al. 2007; Roziere et al. 2015), in the case of STA, the tensile strain capacity is calculated by means of a suitable correction:

$$\varepsilon_{c,cr}(\tau) = \varepsilon(\tau) \cdot \frac{\varepsilon_{c,cr}(\tau = 96h)}{\varepsilon(\tau = 96h)} \tag{6}$$

where $\varepsilon(\tau) = \varepsilon(\tau = 96\,h)$ = tensile strain at first cracking, calculated at $\tau = 96$ h, respectively on the function drawn in Figure 4; and $\varepsilon_{c,cr}$ ($\tau = 96$ h) = strain at first cracking calculated through Eq.(5).

In other words, the tensile strain at first cracking is calculated by scaling the curve reported in Figure 4 (Boshoff, 2012), with respect to the value computed at 96h (Eurocode 2 1-1 1992). Young's modulus is consequently updated with respect to the new value of tensile strain capacity by means of Eq. (5). Ultimate tensile strain (i.e., $\varepsilon_{c,cp}$ in Figure 2b) and the slope of the hardening branch can be similarly calculated.

The soil, assumed as an aged material, is characterized by an elastic modulus, E_{to}, at the interface with the slab. It linearly increases with the depth according to the coefficient K_t, as shown in Figure 5. Hence, for a given thickness, S_S, of the soil, the average value of the elastic modulus is given by:

$$E_t\left(z = \frac{S_S}{2}\right) = E_{t0} + \frac{K_t \cdot S_s}{2} \tag{7}$$

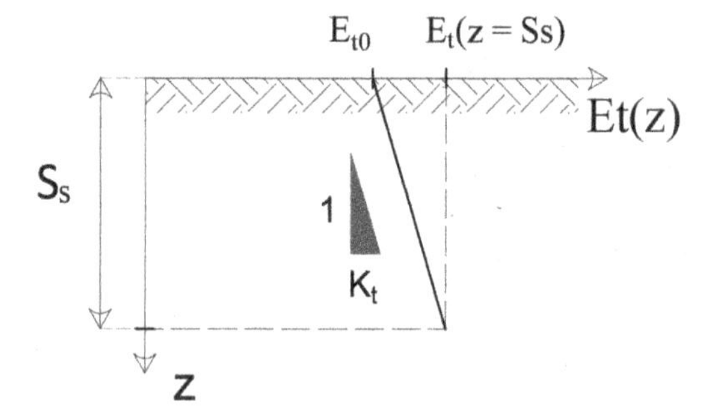

Figure 5. The Young's modulus of soil.

2.2 *Shrinkage model*

Focusing on the curing of concrete, several studies have been carried out in the last years. The shrinkage strain ε_{cs} is composed by the drying, ε_{cd}, and the autogenous, ε_{ca}, contributions:

$$\varepsilon_{cs}(\tau) = \varepsilon_{cd}(\tau) + \varepsilon_{ca}(\tau) \tag{8}$$

Drying shrinkage strain develops slowly, since it is a function of the migration of the water through the hardened concrete. As autogenous shrinkage increases during the hardening of concrete, a major part of ε_{ca} develops in the early days, as illustrated in Figure 6a, where the two components of shrinkage are plotted as a function of concrete aging (Gribniak et al. 2011).

Zhang et al. (2003) noted that most of the total shrinkage in high-strength concrete can be attributed to autogenous shrinkage (see Figure 6b), rather than drying shrinkage. Whereas, due to the higher water-binder ratio, drying shrinkage is dominant in normal concrete (Yoo et al. 2018).

(a)

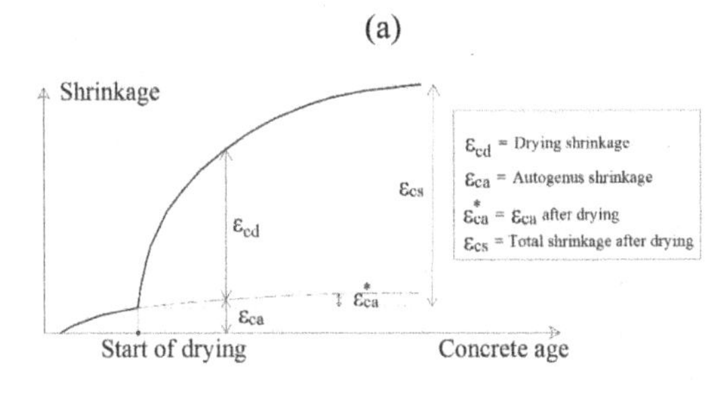

(b)

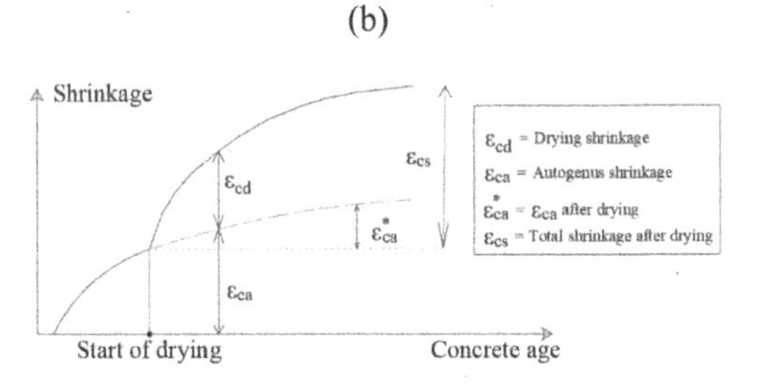

Figure 6. Shrinkage strain components in (a) normal and (b) high-strength concrete.

Gardner & Lockman (2001) provided the GL-2000 model, which can be applied to conventional concrete having the water-cement ratio within the range 0.4–0.6, and a compressive strength lower than 82 MPa (at 28 days). Another model was proposed by Bazant & Baweja (2001). It is called B3 Model and shows very low coefficient of variations, if compared with the results provided by ACI209R (1997) and Eurocode 2 1-1 (1992). In the latter, used herein for the structural analysis, the drying shrinkage strain with the time can be calculated as follows:

$$\varepsilon_{cd}(t) = \beta_{ds}(\tau, \tau_s) \cdot k_h \cdot \varepsilon_{cd,0} \tag{9}$$

where k_h = coefficient depending on the notional size, and:

$$\varepsilon_{cd,o} = 0.85\left[(220 + 110 * \alpha_{ds1}) * exp\left(-\alpha_{ds2}\frac{f_{cm}}{f_{cmo}}\right)\right] * 10^{-6} * \beta_{RH} \tag{10}$$

$$\beta_{RH} = 1.55\left[1 - \left(\frac{RH}{RH_o}\right)^3\right] \tag{11}$$

$$\beta_{ds}(\tau, \tau_s) = \frac{(\tau - \tau_s)}{(\tau - \tau_s) + 0.04\sqrt{h_o^3}} \tag{12}$$

In these equations, $f_{cmo} = 10$ MPa; α_{ds1} and α_{ds2} = coefficients depending on the type of cement; RH = ambient relative humidity (%); RHo = 100%; τ_s = age of concrete (days) at the beginning of drying shrinkage (or swelling); and h_0 = notional size of the cross-section.

The autogenous strain can be calculated as:

$$\varepsilon_{ca}(t) = \beta_{as}(t)\,\varepsilon_{ca}(\infty) \tag{13}$$

where:

$$\varepsilon_{ca}(\infty) = 2.5\,(f_{ck} - 10)\,10^{-6} \tag{14}$$

$$\beta_{as}(t) = 1 - exp\left(-0.2t^{0.5}\right) \tag{15}$$

The maximum shrinkage strains computed with the previous formulae can vary between 300 and 900 $\mu\varepsilon$. However, some researchers (Al-Saleh 2014; Gűneyisi et al. 2014; Yoo et al. 2014; Zhang et al. 2013) suggested an increment of the upper bound, in order to maximize the structural effect of shrinkage on the slabs on ground. Moreover, also the distribution within the thickness of a slab needs to be better investigated. For instance, Rasmussen & McCullough (1998) assumed a linear distribution, in which the full shrinkage strain appears on the surface of the concrete pavement, whereas zero shrinkage is at the mid-depth of the slab. Heath & Roesler (1999) measured the distribution of drying shrinkage by installing strain gauges at different depths of a full-scale slabs on grade. They observed a remarkable difference of shrinkage strain on the top and the bottom of the slab. Thus, shrinkage strain cannot be neglected also on the bottom surface, but, at the same time, it cannot be equal to that on the top. More recently, a new model (Tiberti et al. 2018) assumes 100% of total shrinkage, calculated through Eq. (8), on the free surface. Then, a linear decrement of the drying shrinkage strain is assumed in the rest of the cross-section (it is 50% or 75% lower on the bottom surface).

3 A NEW MODEL FOR SLABS ON GROUND

Referring to the composite cross-section depicted in Figure 3, and made by HPFRC and soil, a new algorithm for measuring the effect of shrinkage of the cement-based material can be developed. To be on the safe side, the stress in the HPFRC is maximized by assuming the existence of the perfect bond between the two layers. Indeed, if a slip between HPFRC and soil exists, both strain and stress reduce in the upper layer.

According to Figure 3, the geometrical input data of the problem are S (= thickness of the slab), S_S (= thickness of the soil), and b (= width of both slab and underlying soil). If the whole cross-section resists to the external actions, the soil can be considered as the steel reinforcement in a reinforced concrete cross-section. After casting, the only load applied on the slab is the shrinkage, which can be considered as imposed strain acting in the concrete layer. Consequently, both stresses and strains in the composite cross-section of Figure 3 can be calculated by using the Colonnetti's theory of elastic coactions (Colonnetti, 1950).

Obviously, the elastic properties of the HPFRC layer vary with the time, therefore, at a fixed time $\tau > 0$, the homogenized geometrical parameters of the composite cross-section are calculated:

$$E_o = E_c(\tau) \tag{16}$$

$$A_o = \frac{E_c}{E_o} \cdot S \cdot b + \frac{E_t}{E_o} \cdot S_s \cdot b \quad (17)$$

$$Y_G = \frac{S_{x0}}{A_0} = \frac{b \cdot s\left(\frac{S}{2} + S_s\right) + \frac{E_t}{E_c} \cdot b \cdot \frac{S_s^2}{2}}{A_0} \quad (18)$$

$$Ix_0 = \left[\frac{bS^3}{12} + b \cdot S\left(\frac{S}{2} + S_s - Y_G\right)^2\right] + \left[\frac{bS_s^3}{12} + b \cdot S_s\left(\frac{S_s}{2} - Y_G\right)^2\right]\frac{E_t}{E_c} \quad (19)$$

where, E_o = Young's modulus of elasticity used to homogenize the cross-section; A_o = homogenized area of the cross-section; Y_g = ordinate of the centroid; S_{xo} = homogenized static moment; and I_{x0} = homogenized moment of inertia.

Assuming that plane section remains plane, the state of stress in concrete ($\sigma_{z,c}$) and soil ($\sigma_{z,s}$) are orthogonal to the Z direction (see Figure 3). They can be calculated as:

$$\begin{cases} \sigma_{z,c} = E\,(\lambda + \mu_x \cdot y - \varepsilon_{im}) & if\ 0 \leq y \leq S \\ \sigma_{z,s} = E\,(\lambda + \mu_x \cdot y) & if\ S \leq y \leq S_s \end{cases} \quad (20)$$

where, y = ordinate of the point with respect to the intrados; ε_{im} = imposed strain (in this case, t is due to shrinkage); λ = total axial deformation; and μ_x = in plane total curvature.

The strain parameters (i.e., λ and μ_x) are the sum of the elastic contribution, due to the external explicit actions (i.e., λ_{el} and μ_{el}), and of the effect produced by imposed strain (λ_{pl} and μ_{pl}). As no external loads are applied (i.e., $\lambda_{el} = \mu_{el} = 0$) the total strain parameters are calculated as follows:

$$\lambda = \lambda_{tot} = \lambda_{el} + \lambda_{pl} = \frac{1}{A_0} \int_{Ac} \frac{E_c}{E_c} \cdot \varepsilon_{im} dA \quad (21)$$

$$\mu_x = \mu_{tot,x} = \mu_{el} + \mu_{pl} = \frac{1}{A_0} \int_{Ac} \frac{E_c}{E_c} \cdot \varepsilon_{im} \cdot y dA \quad (22)$$

However, when the non-linear stage of the cementitious matrix is reached (i.e., $\varepsilon_{c,cr} \leq \varepsilon \leq \varepsilon_{cp}$ in Figure 2b), Eqs.(18)–(19) are not valid.

Nevertheless, according the Colonnetti's theorem (Colonnetti 1950), nonlinear contributions can be taken into account by introducing suitable imposed strain, ε_{nl}, as shown in Figure 7 (where the subscripts "E" = "elastic" and "R" = "real law" indicate the type of stress calculation, respectively). In practice, for a given ε, the linear stress-strain relationship is translated up to the real relationship through ε_{nl}:

$$\varepsilon_{nl} = \frac{E \cdot \varepsilon - \sigma(\varepsilon)}{E} \quad (23)$$

Finally, the state of stress can be calculated with Eqs.(21)–(22) by assuming $\varepsilon_{im} = \varepsilon_{sh} + \varepsilon_{nl}$.

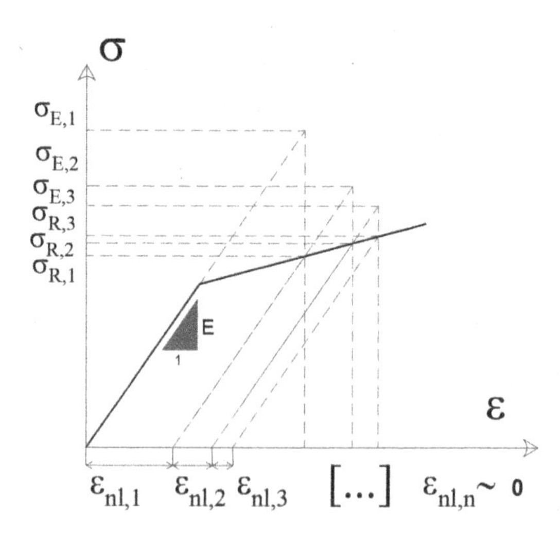

Figure 7. Effect of nonlinear behavior of the constitutive law and calculation of the ε_{nl}.

As ε_{nl} has to be continuously updated, an iterative procedure for the calculation of the states of stress and strain is introduced. The iterations end when the difference between two consecutive values of ε_{nl} is negligible (i.e., when two consecutive states of stress are mathematically coincident).

If the strain in each point of the cross-section is lower than the strain at beginning of strain localization:

$$\varepsilon(y,\tau) \leq \varepsilon_{c,cp}(\tau) \quad 0 \leq y \leq S \quad (24)$$

macrocracks do not appear (or crack width is much lower than the admissible values). In these cases, joints are not necessary.

As tensile stresses in the soil are not allowed, the thickness S_S must be iteratively changed until only compressive stresses are present in the soil (see Figure 3). In other words, the thickness of the soil substrate is not input of the problem. In this way, the whole procedure is composed by the two encapsulated iterative parts shown in Figure 8.

4 NUMERICAL RESULTS

The procedure previously described is herein applied in two different slabs: slab_1 and slab_2. In both the cases, the thickness is the same (S = 100 mm), whereas the mechanical parameters of soil are summarized in Table 1.

The slabs are made with different types of concrete, normal concrete (C30/37) in the slab_1, and HPFRC in slab_2, whose mechanical properties are illustrated in Table 2. The properties of C30/37 are those suggested by Eurocode 2 1-1 (1992) and measured at 28 days. On the contrary, the properties of HPFRC have been provided by a building material supplier, and concern a product available on the market (Esser et al. 2015).

The model computes the states of stress and strain in both the slabs under the hypothesis of a curing at 20°C

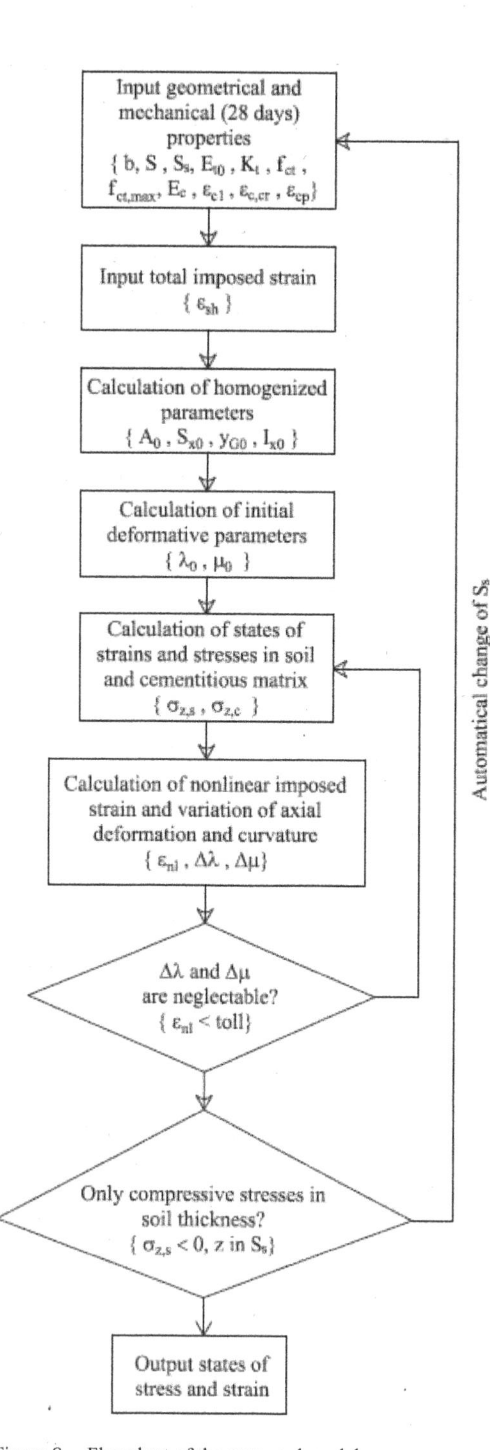

Figure 8. Flowchart of the proposed model.

Table 1. Mechanical properties of the soil.

E_{t0} (MPa)	5000
K_t (MPa/mm)	5

Table 2. Mechanical properties of concrete and HPFRC.

	slab_1	slab_2
b (mm)	5000	5000
S (mm)	100	100
Cement class	32.5N	42.5R
Rck (MPa)	37	50
f_{ck} (MPa)	30	40
E_{cm} (MPa)	33000	36300
f_{ct} (MPa)	2.00	2.50
$f_{ct,max}$ (MPa)	2.00	3.00
$\varepsilon_{c,cr}$ (‰)	0.06	0.07
ε_{cp} (‰)	0.06	2.00
ε_{cu} (‰)	3.5	3.5

(RH = 50%) for 28 days. In this period, the shrinkage is assumed to be the only external action on the slab.

The dashed curves reported in Figure 9a and 9b show the maximum tensile strains reached in the slab_1 and slab_2, respectively. More precisely, these strains are calculated after considering three different distributions of shrinkage (Tiberti et al. 2018). In the same figures, the maximum strain capacity of the cement-based materials is also reported.

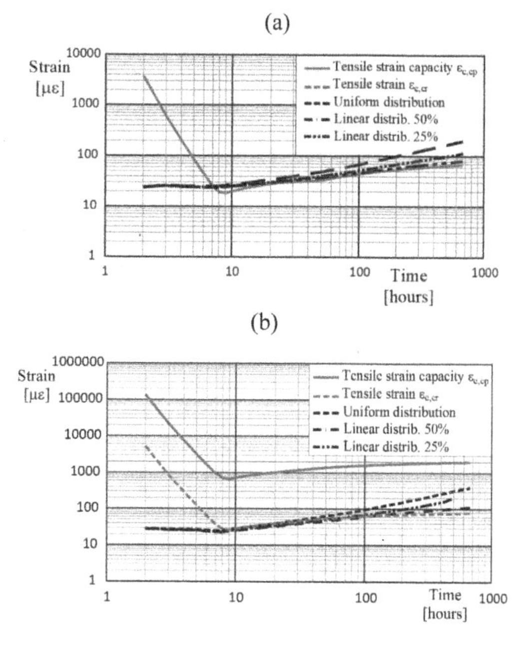

Figure 9. Numerical outcomes in case of (a) slab_1 (C30/37) and (b) slab_2 (HPFRC).

Figure 9a is representative of a concrete that produces a crack 7 hours after casting (i.e., when the strain capacity of concrete is the lowest). In this situation the presence of joints is necessary. In the case of HPFRC (see Figure 9b), due to the absence of crack localization, contraction joints may be avoided.

Nevertheless, in both the slabs, the bi-logarithmic diagrams do not show remarkable variation of tensile strains with the shrinkage distribution. On the contrary, numerical results show that the most important parameters to obtain jointless slabs on ground are the tensile properties, especially the strain capacity, and shrinkage evolutions (both autogenous and drying).

5 CONCLUSION

Tensile strain capacity of the cementitious matrix is important for designing jointless slabs, and controlling crack widths after strain localization, as well. Current methods for the characterization of tensile properties in concrete and FRC cannot be extended to HPFRC, because in the latter strain localization in tension does not occur at cracking (enhanced capacity). At the same time, mechanical models devoted to the analysis of conventional reinforced concrete (or FRC) slabs on ground behave differently than those made with HPFRC. Hence, a new model for designing jointless slab on ground, made with HPFRC (i.e., a strain hardening material) has been proposed. Based on the results of the analyses previously described, it can be observed that to avoid shrinkage cracking few hours after casting, it is of fundamental importance to know the mechanical properties of cement-based materials at very early age. Specifically, the evolutions of tensile strengths and strains with time are the most significant parameters, like the shrinkage actions. If the tensile strain capacity increases, as in the case of HPFRC, the jointless slabs can be built.

REFERENCES

AASHTO, 1993. AASHTO Guide for design of pavement structures". America.

ACI 209R-92: Prediction of Creep, Shrinkage, and Temperature Effects in Concrete Structures, American Concrete Institute, Farmington Hills, 1997.

Al-Saleh, S. A. 2014. Comparison of theoretical and experimental shrinkage in concrete. Constr. Build. Mater., volume 72: 326–332. The Netherlands: Elsevier.

Bazant, Z.P. & Baweja, S. 2001. Creep and Shrinkage Prediction Model for Analysis and Design of Concrete Structures: Model B3," ACI Special Publications, volume 194: 1-84. USA: ACI Material Journal.

Bentur, A. 2003. Early-age Cracking in Cementitious Systems. RILEM Technical Committee 181-EAS: Early-age Shrinkage Induced Stresses and Cracking in Cementitious Systems. France: RILEM.

Boshoff, W.P. 2012. Plastic Shrinkage Cracking of Concrete, Part 2: Commentary. Report number: ISI2012-17. Germany: Insitute of Structural Engineering.

Burger, A.F 2010. Experience with the construction of an ultrathin continuously reinforced concrete pavement. 11th International Symposium on concrete roads.

Choi, J. & Chen, R. H. L. 2005. Design of Continuously Reinforced Concrete Pavements Using Glass Fiber Reinforced Polymer Rebars. U.S. Department of transportation FHA. America.

ChunPing, G. & Guang, Y. & Wei, S. 2015.Ultrahigh performance concrete–properties, applications and perspectives". Volume 58: 587-599. China: Technological Sciences.

Colonnetti, G. 1950. Elastic equilibrium in the presence of permanent set. Quarterly of applied mathematics Vol. VII No.4: 353-362. https://doi.org/10.1090/qam/33732

Combrinck, R. & Boshoff, W. P. 2019. Tensile properties of plastic concrete and the influence of temperature and cyclic loading. Cement and Concrete Composites, volume 97: 300-311. The Netherlands: Elsevier.

Destrée, X. & Yao, Y. & Mobasher, B. & Asce, M. 2016. Sequential Cracking and Their Openings in Steel-Fiber-Reinforced Joint-Free Concrete Slabs. J. Mater. Civ. Eng., volume 28: (04015158) 1-11. USA: ASCE.

Esser J., Guerini A, Volpatti G, Zampini D. 2015. Advanced fiber reinforced concrete mix designs and admixtures systems, Patent EP3307692 B1 [priority 2015-06-11]. Commercially available as CEMEX®Resilia®.

Eurocode 2: Design of concrete structures – Part 1-1 : General rules and rules for buildings. 2004.

FHA – U.S. Department of transportation, 2019. Technical Advisory: Concrete Pavement Joints. America.

fib Model Code for Concrete Structures 2010 –.

Gardner, N.J. & Lockman, M.J. 2001. Design provisions for drying shrinkage and creep of normal strength concrete. ACI, volume 98: 159-167. USA: ACI Material Journal.

Graybeal, B. F.2013. Development of a Direct Tension Test Method for UHPFRC. ACI Mater J, volume 110: 177-186. USA: 2013.

Gribniak, V. & Kaklauska, G. & Bacinskas, D. & Sung, W. & Sokolov, A. & Ulbinas, D 2011. Investigation of shrinkage of concrete mixtures used for bridge construction in Lithuania. Volume 6: 7-83. Lithuania: The Bàltic Journal of Road and Bridge Engineering.

Güneyisi, E. & Gesoglu, M. & Mohammedameen, A. 2014. Enhancement of shrinkage behavior of lightweight aggregate concretes by shrinkage reducing admixture and fiber reinforcement. Construction and Building Materials, volume 54: 91-98. The Netherlands: Elsevier.

Hachiya, N. J. K. S. 2006. Applicability of concrete with high flexural strength to airport pavements. 10th International Symposium on Concrete Roads.

Hammer, T. & Bjontegaard, O. & Fossa, K. 2007. Cracking tendency of HSC :Tensile strength and self-generated stress in the period of setting and early hardening. Materials and Structures Journal, volume 40: 319-324. The Netherlands: Springer.

Heath, A.C. & Roesler, J.R. 1999. Shrinkage and Thermal Cracking of Fast Setting Hydraulic Cement Concrete Pavements in Palmdale. California.

Larrard, F. & Sedran, T. 2011. High and ultra-High performance concrete in pavement :tools for the road eternity. France: Hal.

R., A.-T. 2009. Renewal of concrete slabs using high performance concrete. 9th International Symposium on concrete roads.

Ramadoss, K. N. P. 2008. Tensile strength and durability characteristics of high-performance fiber reinforced concrete. Arabian J Sci Eng, vol. 33 : 307-319. Saudi Arabia: Springer.

Rasmussen, R.O. & McCullough, B.F. 1998. A Foundation for High Performance Jointed Concrete Pavement Design and Construction Guidelines; Transtec Consultants: Austin, TX. USA.

Rasmussen, R.O. & Rogers, R. & Ferragut, T. R. 2009. Continuously reinforced concrete pavements – Design & Construction guidelines. U.S. Department of transportation FHA. America.

Roesler, J. R. & Hiller, J. E. & Brand, A. S. 2016. Continuously Reinforced Concrete Pavement Manual – Guidelines for Design, Construction, Maintenance, and

Rehabilitation. U.S. Department of transportation FHA. America.
Roziere, E. R. & Cortas, R. & Loukili, A. 2015. Tensile behaviour of early age concrete: New methods of investigation", Cement & Concrete Composites, volume 55: 153-161. The Netherlands: Elsevier.
S. J. CHOI, S. Y. & Park, J. S. & Jung, W. T. 2011. A Study on the Shrinkage Control of Fiber Reinforced Concrete Pavement. Procedia Engineering, Volume 14:2815–2822. The Netherlands: Elsevier.
Savino, V. & Lanzoni, M. & Tarantino A. M. & Viviani, M. 2018. Simple and effective models to predict the compressive and tensile strength of HPFRC as the steel fiber content and type changes. Composites Part B: Engineering, volume 137: 153-162. The Netherlands: Elsevier.
Söderqvist, J. 2006. Design of Concrete Pavements – Design Criteria for Plain and Lean Concrete. Stockholm: licentiate thesis.
Tehmina, A. & Shafiq, N. & Nuruddin, M. 2014. Mechanical Properties of High-Performance Concrete Reinforced with Basalt Fibers. Procedia Engineering, volume 77: 31-139. The Netherlands: Elsevier.
Tiberti, G. & Mudadu, A. & Barragan, B. & Plizzari, G. 2018. Shrinkage Cracking of Concrete Slabs-On-Grade: A Numerical Parametric Study. MDPI Journal Fibers, volume 6:, 64., 2018.
Yoo, D. & Kim, M. & Kim, S. & Ryu, G & Koh, K. 2018. Effects of mix proportion and curing condition on shrinkage behavior of HPFRCCs with silica fume and blast furnace slag. Construction and Building Materials, volume 166: 241-256. The Netherlands: Elsevier.
Yoo, D. & Min, K. & Yoon, Y. 2014. Shrinkage and cracking of restrained ultra-high-performance fiber-reinforced concrete slabs at early age. Construction and Building Materials, volume 73: 357-365. The Netherlands: Elsevier.
Zhang, J. & Wang, Z. & Ju, X. 2013. Application of ductile fiber reinforced cementitious composite in jointless concrete pavements. Composites Part B: Engineering, Volume 50: 224-231. The Netherlands: Elsevier.
Zhang, M. & Leow, M.P. 2003. Effect of water-to-cementitious materials ratio and silica fume on the autogenous shrinkage of concrete. Cement and Concrete Research, volume 10: 687–1694. The Netherlands: Elsevier.
Zhang, M. & Zakaria, M. & Hama, Y. 2013. Influence of aggregate materials characteristics on the drying shrinkage properties of mortar and concrete. Construction and Building Materials, volume 49:500-510. The Netherlands: Elsevier.

Computational Modelling of Concrete and Concrete Structures – Meschke, Pichler & Rots (Eds)
 ISBN: 978-1-032-32724-2

Modelling stability of reinforced concrete walls applying convex optimization

D. Vestergaard
Rambøll Denmark A/S, Copenhagen, Denmark
Department of Civil Engineering, Technical University of Denmark, Kgs. Lyngby, Denmark

P.N. Poulsen & L.C. Hoang
Department of Civil Engineering, Technical University of Denmark, Kgs. Lyngby, Denmark

K.P. Larsen & B. Feddersen
Rambøll Denmark A/S, Copenhagen, Denmark

ABSTRACT: Due to time constraints on structural design processes, modelling and computational complexity is often a key concern in limit state analysis of reinforced concrete (RC) structures, causing practitioners to choose efficient but inaccurate methods of analysis over more advanced ones. Recently, a framework using convex optimization for elasto-plastic, geometrically linear analysis of RC walls was proposed, enabling analysis of models with more than 10,000 finite elements within minutes on a standard PC. In order to improve the applicability and relevance of the framework as a design tool, this paper proposes an extension that enables the determination of the critical buckling load. Based on the nonlinear solution obtained from the elasto-plastic optimization problem, the cracked tangent stiffness of the RC sections is determined, and a linearized buckling problem is posed and solved as a linear eigenvalue problem. This allows the actual critical buckling load to be determined by solving a sequence of optimization and eigenvalue problems. The accuracy of the proposed method is assessed, and its applicability to practical design scenarios is demonstrated by an analysis of an RC wall with a door hole, showing an average solution time of approximately 30 seconds per load step.

1 INTRODUCTION

When designing reinforced concrete (RC) structures, the attractiveness of a given method for structural analysis is highly affected by its computational efficiency and robustness. Consequently, one of the most popular methods of analysis among practitioners of structural engineering is still the linear-elastic, displacement-based finite element method (FEM). Being based upon a linear-elastic and isotropic material model, however, linear-elastic FEM produces somewhat inaccurate results when applied to structures with nonlinear material behaviour, e.g. reinforced concrete stressed beyond the cracking limit for the concrete. Consequently, for reinforced concrete structures, the applicability of linear-elastic FEM is mainly limited to serviceability limit state (SLS) analysis.

In recent years, the concept of finite element limit analysis (FELA) has proven to be an efficient and robust method for ultimate limit state (ULS) analysis of reinforced concrete structures. By use of a rigid-plastic material model and stress-based finite elements, a load-maximization problem based on the lower bound theorem can be cast as a convex optimization problem that can be solved using commercial solvers. Currently, numerical frameworks for limit analysis of reinforced concrete structures have been developed for a variety of stress-based finite element types, e.g., membranes, solids, plates, and shells (Herfelt 2017; Jensen 2019; Larsen 2011; Poulsen & Damkilde 2000). Due to the rigid-plastic material models, however, the solutions do not include any finite deformations, meaning that they cannot be used to assess structural ductility, crack widths, or structural displacements.

To overcome this shortcoming, a framework for efficient elasto-plastic analysis of reinforced concrete walls subjected to in-plane loading was recently proposed (Vestergaard et al. 2021). Using stress-based finite elements and a hyper-elastic material model, the framework poses the principle of minimum complementary energy as a convex optimization problem. Since this approach does not rely on incremental load-stepping, it enables the analysis of models with more than 10,000 finite elements within a few minutes on a standard PC. An extension of the framework was proposed, enabling the analysis of thin walls subjected to combined in-plane and transverse loading (Vestergaard et al. 2022). This extension, which uses a layer-based submodel to represent the nonlinear stress

DOI 10.1201/9781003316404-87

variation over the wall thickness, increases the number of variables per finite element by a factor of approx. 10.

In order to improve the applicability and relevance of the framework as a design tool, it should be able to assess structural stability. To this end, this paper proposes an extension of the framework that enables the determination of the critical buckling load. Based on the nonlinear solution obtained from the complementary energy minimization problem, the stress and strain state are known, enabling the determination of the tangent stiffness of the RC sections. Using the well-established displacement-based CST and Specht (Specht 1988) finite elements, and by applying a nonlinear strain measure, the linearized buckling problem is formulated as an eigenvalue problem, which is solved using the tangent stiffness matrix. With this approach, the actual critical buckling load can be determined more accurately as the load is increased by solving a sequence of optimization and eigenvalue problems.

Initially, a description is given of the proposed procedure of analysis, and the applied constitutive model is presented. Subsequently, expressions for the linearized (i.e., tangent) sectional stiffness are derived, followed by a description of the stress-based and displacement-based finite elements and the corresponding discretized expressions. Based on these expressions and the principle of virtual work, the linearized buckling problem is derived. Finally, the accuracy of the method is assessed using a simple validation example, and its applicability to practical design scenarios is demonstrated using an example involving a wall with a door hole.

2 ANALYSIS PROCEDURE

As described in the Introduction, the framework presented by (Vestergaard et al. 2021) has proven extremely efficient for the analysis of fully cracked reinforced concrete structures with nonlinear material behaviour. By posing the principle of minimum complementary energy as a convex optimization problem (more specifically, a *Second-Order Cone Programme*), this framework uses stress-based finite elements and state-of-the-art commercial convex optimization algorithms to solve the finite element (*FE*) problem without the need of incremental application of loads. This is the case even for nonlinear material behaviour as long as the stiffness is positive and non-increasing, e.g., reinforcement with yielding, and concrete with crushing and zero tensile strength (cracking). Thus, the framework provides a method for structural analysis which is efficient both in terms of modelling effort and computational complexity while being substantially more accurate than the traditional linear-elastic FE methods often applied in practice.

Since the principle of minimum complementary energy is based upon a linear strain-displacement relation, the framework described above is not directly applicable to structures where geometrical nonlinearity is of importance. However, due to its efficiency for solving geometrically linear problems, it can be used to predict the through-thickness strain field, and thereby the (cracked) tangent section stiffness in a pre-buckling state where geometrical nonlinearity is negligible.

The approach proposed in this paper is based on posing and solving the linearized buckling problem as a linear eigenvalue problem. In its essence, the method is closely linked to the modification to Euler's critical load originally proposed by (Engesser 1889) to account for material nonlinearity by simply substituting the elastic modulus with the tangent modulus; this concept is illustrated in Figure 1.

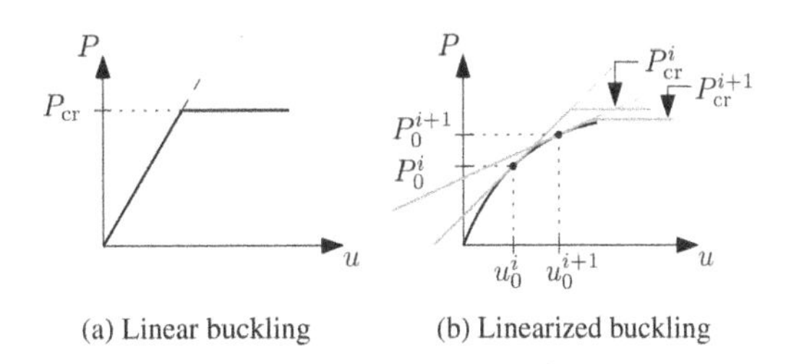

(a) Linear buckling (b) Linearized buckling

Figure 1. Load-displacement curves with bifurcations.

Since this approach presupposes knowledge of the stiffness for a given static/kinematic configuration, the method proposed in this paper consists of a two-step procedure in which the existing framework based on the principle of minimum complementary energy is used to predict the state of the sections of the structure in a geometrically linear analysis, based upon which the linearized buckling problem is posed and solved as a linear eigenvalue problem. As Figure 1 illustrates, the accuracy of the linearization increases as the applied load approaches the critical buckling load, meaning that the two-step procedure is to be repeated for a range of load factors to estimate the critical buckling load. In order to take into account the adverse effects of geometrical imperfections, the analysis is performed on a slightly modified geometry which is constructed by imposing upon the original geometry a scaled version of the critical buckling mode for the first load step.

3 CONSTITUTIVE MODEL

3.1 Material models

The material considered in this paper is reinforced concrete with orthogonal reinforcement. The concrete is assumed to be fully cracked, i.e., with no tensile stresses, and with a Poisson's ratio of $\nu_c = 0$. The reinforcement is considered smeared in the in-plane directions with full strain compatibility with the concrete, and it is assumed to carry axial stresses only, i.e., dowel action is neglected. The concrete and the reinforcement are modelled independently as piecewise-linear elastic materials with stress-strain curves as illustrated in Figure 2. These models apply to the axial reinforcement stress components $\{\sigma_{sx}, \sigma_{sy}\}$ and the (in-plane) principal concrete stress components $\{\sigma_{cI}, \sigma_{cII}\}$, respectively. Note that the slope of the concrete hardening branch may be chosen as close to zero as numerical stability allows.

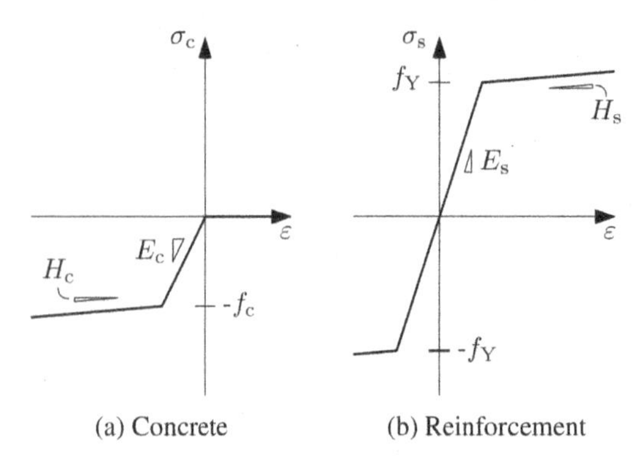

Figure 2. Material stress-strain curves.

3.2 *Section model*

Given that buckling generally concerns slender structures, the section model is based on Kirchhoff shell theory for thin plates, i.e., only the in-plane stress and strain components, $\boldsymbol{\sigma} = [\,\sigma_{xx},\ \sigma_{yy},\ \tau_{xy}\,]^T$ and $\boldsymbol{\varepsilon} = [\varepsilon_{xx},\ \varepsilon_{yy},\ 2\varepsilon_{xy}]^T$, are considered. The in-plane section forces $\mathbf{n} = [\,n_{xx},\ n_{yy},\ n_{xy}\,]^T$ and the section moments $\mathbf{m} = [\,m_{xx},\ m_{yy},\ m_{xy}\,]^T$ are related directly to the in-plane stresses as

$$\mathbf{n} = \int_{-\frac{t}{2}}^{\frac{t}{2}} \boldsymbol{\sigma}\,\mathrm{d}z \quad , \quad \mathbf{m} = \int_{-\frac{t}{2}}^{\frac{t}{2}} z\boldsymbol{\sigma}\,\mathrm{d}z \tag{1}$$

where z is the thickness coordinate starting at the center plane, and component-wise integration is implied. The in-plane strain components are assumed to vary linearly over the shell thickness, allowing them to be stated as

$$\boldsymbol{\varepsilon} = \boldsymbol{\varepsilon}_0 + z\boldsymbol{\kappa} \tag{2}$$

where $\boldsymbol{\varepsilon}_0 = [\varepsilon^0_{xx},\ \varepsilon^0_{yy},\ 2\varepsilon^0_{xy}]^T$ are the center-plane strain components, and $\boldsymbol{\kappa} = [\kappa_{xx},\ \kappa_{yy},\ 2\kappa_{xy}]^T$ are the curvature components.

Assuming moderate displacements, the center-plane strain is related to the displacement field $\mathbf{u} = [\,u_x,\ u_y,\ u_z\,]^T$ by the Green strain measure, while the curvature $\boldsymbol{\kappa}$ is described with sufficient accuracy using the small-strain measure. Using comma derivative notation, i.e., $\mathbf{u}_{,x} = \partial\mathbf{u}/\partial x$, these relations are

$$\boldsymbol{\varepsilon}_0 = \boldsymbol{\partial}_1\mathbf{u} + \frac{1}{2}\begin{bmatrix} \mathbf{u}_{,x}^T\mathbf{u}_{,x} \\ \mathbf{u}_{,y}^T\mathbf{u}_{,y} \\ \mathbf{u}_{,x}^T\mathbf{u}_{,y} + \mathbf{u}_{,y}^T\mathbf{u}_{,x} \end{bmatrix} \quad , \quad \boldsymbol{\kappa} = \boldsymbol{\partial}_2\mathbf{u} \tag{3}$$

where the operators

$$\boldsymbol{\partial}_1 = \begin{bmatrix} \frac{\partial}{\partial x} & 0 & 0 \\ 0 & \frac{\partial}{\partial y} & 0 \\ \frac{\partial}{\partial y} & \frac{\partial}{\partial x} & 0 \end{bmatrix}, \quad \boldsymbol{\partial}_2 = \begin{bmatrix} 0 & 0 & \frac{\partial^2}{\partial x^2} \\ 0 & 0 & \frac{\partial^2}{\partial y^2} \\ 0 & 0 & \frac{\partial^2}{\partial x \partial y} \end{bmatrix} \tag{4}$$

have been introduced.

4 STIFFNESS LINEARIZATION

To simplify notation, the section forces $\mathbf{n}$ and section moments $\mathbf{m}$, and the center-plane strain $\boldsymbol{\varepsilon}_0$ and curvature $\boldsymbol{\kappa}$, are collected in the generalized stress and strain vector, respectively:

$$\bar{\boldsymbol{\sigma}} = \begin{bmatrix} \mathbf{n} \\ \mathbf{m} \end{bmatrix}, \quad \bar{\boldsymbol{\varepsilon}} = \begin{bmatrix} \boldsymbol{\varepsilon}_0 \\ \boldsymbol{\kappa} \end{bmatrix} \tag{5}$$

In the neighborhood of a given state of generalized strain and stress, $\{\bar{\boldsymbol{\varepsilon}}_0, \bar{\boldsymbol{\sigma}}_0\}$, the generalized stress-strain relation is assumed to be well-approximated by the first-order Taylor expansion:

$$\bar{\boldsymbol{\sigma}} = \bar{\boldsymbol{\sigma}}_0 + \left.\frac{\partial\bar{\boldsymbol{\sigma}}}{\partial\bar{\boldsymbol{\varepsilon}}}\right|_0 (\bar{\boldsymbol{\varepsilon}} - \bar{\boldsymbol{\varepsilon}}_0) := \bar{\boldsymbol{\sigma}}_0 + \mathbf{D}_\mathrm{T}(\bar{\boldsymbol{\varepsilon}} - \bar{\boldsymbol{\varepsilon}}_0) \tag{6}$$

where

$$\mathbf{D}_\mathrm{T} = \int_{-\frac{t}{2}}^{\frac{t}{2}} \begin{bmatrix} \left(\frac{\partial\boldsymbol{\sigma}}{\partial\boldsymbol{\varepsilon}}\right) & z\left(\frac{\partial\boldsymbol{\sigma}}{\partial\boldsymbol{\varepsilon}}\right) \\ z\left(\frac{\partial\boldsymbol{\sigma}}{\partial\boldsymbol{\varepsilon}}\right) & z^2\left(\frac{\partial\boldsymbol{\sigma}}{\partial\boldsymbol{\varepsilon}}\right) \end{bmatrix}_0 \mathrm{d}z \tag{7}$$

is the tangent constitutive matrix. As seen, $\mathbf{D}_\mathrm{T}$ is fully determined by the stress-strain gradient,

$$\frac{\partial\boldsymbol{\sigma}}{\partial\boldsymbol{\varepsilon}} = \frac{\partial\boldsymbol{\sigma}_\mathrm{s}}{\partial\boldsymbol{\varepsilon}} + \frac{\partial\boldsymbol{\sigma}_\mathrm{c}}{\partial\boldsymbol{\varepsilon}} \tag{8}$$

where the subscript "s" and "c" refer the steel reinforcement and the concrete, respectively. The first term of (8) is given directly in terms of the tangent reinforcement stiffness $E_{\mathrm{sT}}(\varepsilon)$:

$$\frac{\partial\boldsymbol{\sigma}_\mathrm{s}}{\partial\boldsymbol{\varepsilon}} = \begin{bmatrix} E_{\mathrm{sT}}(\varepsilon_{xx}) & & \\ & E_{\mathrm{sT}}(\varepsilon_{yy}) & \\ & & 0 \end{bmatrix} \tag{9}$$

Since the model defines the tangent stiffness of concrete $E_{\mathrm{cT}}(\varepsilon)$ in the principal directions (which are co-aligned for both stresses and strains),

$$\frac{\partial\boldsymbol{\sigma}_\mathrm{cp}}{\partial\boldsymbol{\varepsilon}_\mathrm{p}} = \begin{bmatrix} E_{\mathrm{cT}}(\varepsilon_\mathrm{I}) & \\ & E_{\mathrm{cT}}(\varepsilon_\mathrm{II}) \end{bmatrix} \tag{10}$$

the second term of (8) should be expressed using the chain rule. Viewing $\boldsymbol{\sigma}_\mathrm{c}$ as a function of the principle stress magnitudes $\boldsymbol{\sigma}_\mathrm{cp} = [\,\sigma_\mathrm{cI}, \sigma_\mathrm{cII}\,]^T$ and orientation θ,

$$\boldsymbol{\sigma}_\mathrm{c} = \frac{1}{2}\begin{bmatrix} (1+\cos 2\theta)\,\sigma_\mathrm{cI} + (1-\cos 2\theta)\,\sigma_\mathrm{cII} \\ (1-\cos 2\theta)\,\sigma_\mathrm{cI} + (1+\cos 2\theta)\,\sigma_\mathrm{cII} \\ -\sin 2\theta\sigma_\mathrm{cI} + \sin 2\theta\sigma_\mathrm{cII} \end{bmatrix} \tag{11}$$

the chain rule takes the form

$$\frac{\partial \boldsymbol{\sigma}_{\mathrm{c}}}{\partial \boldsymbol{\varepsilon}}=\frac{\partial \boldsymbol{\sigma}_{\mathrm{c}}}{\partial \boldsymbol{\sigma}_{\mathrm{cp}}} \frac{\partial \boldsymbol{\sigma}_{\mathrm{cp}}}{\partial \boldsymbol{\varepsilon}_{\mathrm{p}}} \frac{\partial \boldsymbol{\varepsilon}_{\mathrm{p}}}{\partial \boldsymbol{\varepsilon}}+\frac{\partial \boldsymbol{\sigma}_{\mathrm{c}}}{\partial \cos 2\theta} \frac{\partial \cos 2\theta}{\partial \boldsymbol{\varepsilon}} \tag{12}$$

where $\cos 2\theta$ has been chosen as the principal orientation angle variable for convenience. By introducing $\varepsilon_{\mathrm{m}}=(\varepsilon_{xx}+\varepsilon_{yy})/2$, $\varepsilon_{\mathrm{d}}=(\varepsilon_{xx}-\varepsilon_{yy})/2$ and $\varepsilon_{\mathrm{r}}=\sqrt{\varepsilon_{\mathrm{d}}^2+\varepsilon_{xy}^2}$, the double-angle sine and cosine can be expressed simply as

$$\sin 2\theta=-\frac{\varepsilon_{xy}}{\varepsilon_{\mathrm{r}}}, \quad \cos 2\theta=\frac{\varepsilon_{\mathrm{d}}}{\varepsilon_{\mathrm{r}}} \tag{13}$$

and the derivatives of $\varepsilon_{\mathrm{I}}=\varepsilon_{\mathrm{m}}+\varepsilon_{\mathrm{r}}$ and $\varepsilon_{\mathrm{II}}=\varepsilon_{\mathrm{m}}-\varepsilon_{\mathrm{r}}$ can be found directly as

$$\frac{\partial \boldsymbol{\varepsilon}_{\mathrm{p}}}{\partial \boldsymbol{\varepsilon}}=\frac{1}{2\varepsilon_{\mathrm{r}}}\begin{bmatrix}(\varepsilon_{\mathrm{r}}+\varepsilon_{\mathrm{d}}) & (\varepsilon_{\mathrm{r}}-\varepsilon_{\mathrm{d}}) & \varepsilon_{xy} \\ (\varepsilon_{\mathrm{r}}-\varepsilon_{\mathrm{d}}) & (\varepsilon_{\mathrm{r}}+\varepsilon_{\mathrm{d}}) & -\varepsilon_{xy}\end{bmatrix} \tag{14}$$

Based on (11) and the relations in (13), the remaining components of (12) can be established as

$$\frac{\partial \boldsymbol{\sigma}_{\mathrm{c}}}{\partial \boldsymbol{\sigma}_{\mathrm{cp}}}=\frac{1}{2\varepsilon_{\mathrm{r}}}\begin{bmatrix}(\varepsilon_{\mathrm{r}}+\varepsilon_{\mathrm{d}}) & (\varepsilon_{\mathrm{r}}-\varepsilon_{\mathrm{d}}) & \varepsilon_{xy} \\ (\varepsilon_{\mathrm{r}}-\varepsilon_{\mathrm{d}}) & (\varepsilon_{\mathrm{r}}+\varepsilon_{\mathrm{d}}) & -\varepsilon_{xy}\end{bmatrix}^T \tag{15}$$

and

$$\frac{\partial \boldsymbol{\sigma}_{\mathrm{c}}}{\partial \cos 2\theta}=\frac{\sigma_{\mathrm{cI}}-\sigma_{\mathrm{cII}}}{2\varepsilon_{xy}}\left[\varepsilon_{xy},\,-\varepsilon_{xy},\,-\varepsilon_{\mathrm{d}}\right]^T$$

$$\frac{\partial \cos 2\theta}{\partial \boldsymbol{\varepsilon}}=\frac{\varepsilon_{xy}}{2\varepsilon_{\mathrm{r}}^3}\left[\varepsilon_{xy},\,-\varepsilon_{xy},\,-\varepsilon_{\mathrm{d}}\right] \tag{16}$$

where it is utilized that $\mathrm{d}(\sin 2\theta)/\mathrm{d}(\cos 2\theta)=-\cot 2\theta=-\varepsilon_{\mathrm{d}}/\varepsilon_{xy}$. In effect, (12) is seen to yield a positive semidefinite matrix for $\varepsilon_{\mathrm{r}}>0$. Since $\varepsilon_{\mathrm{r}}=0$ corresponds to an un-strained material, the concrete stiffness can be taken as uncracked and linear-elastic in this special case. Due to the complexity of the expression for $\partial \boldsymbol{\sigma}_{\mathrm{c}}/\partial \boldsymbol{\varepsilon}$, the integrals in (7) are evaluated numerically.

5 FINITE ELEMENT MODEL

5.1 Stress-based, geometrically linear model

For the geometrically linear analysis, a stress-based finite element introduced by (Vestergaard et al. 2022) is used within a convex optimization framework based on the principle of minimum complementary energy. Using the in-plane section forces and moments $\mathbf{n}$ and $\mathbf{m}$ as the primary degrees-of-freedom (dofs), this type of finite element rigorously satisfies equilibrium within and between elements. The dofs are chosen such that they define a linear and quadratic variation of $\mathbf{n}$ and $\mathbf{m}$, respectively, and in 10 points within the element, these generalized stresses are coupled to the section stress variation using a discrete layer model. Based on these stress variations, the complementary energy is given for each submodel point, which is interpolated and integrated assuming a cubic variation within each element. The element and the layer submodel are illustrated in Figures 3 and 4, respectively.

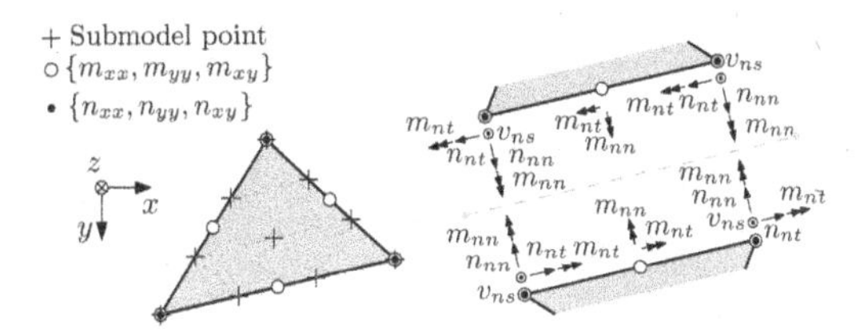

Figure 3. Stress-based element. Left: Location of dofs and submodel points. Right: Equilibrium between element sides.

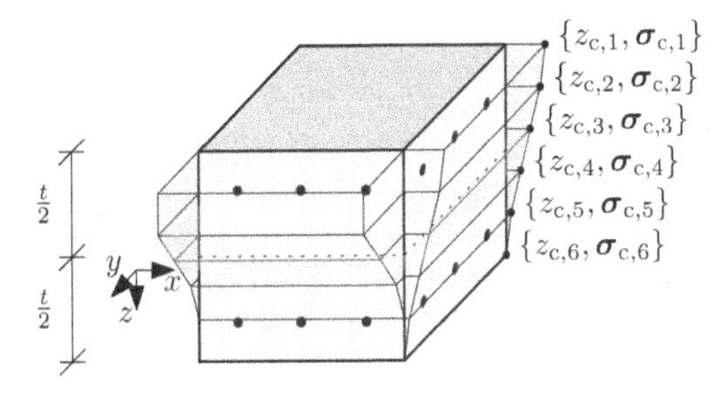

Figure 4. Layer submodel with five discrete layers.

By stating the principle of minimum complementary energy as a convex optimization problem, the problem can be solved efficiently and robustly by commercial convex optimization algorithms.

5.2 Displacement-based, geometrically nonlinear model

In the geometrically nonlinear analysis, the Constant Strain Triangle (CST) and Specht's element (Specht 1988) are used to model the in-plane and out-of-plane (bending) behavior, respectively. The elements and their dofs are illustrated in local coordinates (denoted by a superscript "ℓ") in Figure 5.

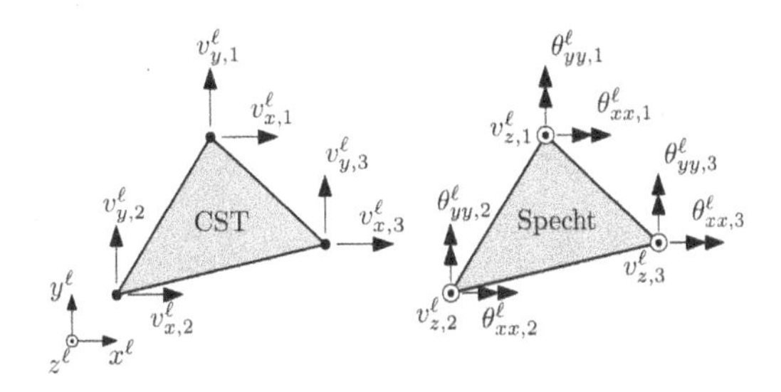

Figure 5. CST and Specht element dofs.

Collecting the CST and Specht's element dofs in $\mathbf{v}_\varepsilon^\ell$ and $\mathbf{v}_\kappa^\ell$, respectively, the three-dimensional element displacement field can be represented as

$$\underbrace{\begin{bmatrix} \mathbf{u}_\varepsilon(\boldsymbol{\xi}) \\ \mathbf{u}_\kappa(\boldsymbol{\xi}) \end{bmatrix}}_{\mathbf{u}(\boldsymbol{\xi})} = \underbrace{\begin{bmatrix} \mathbf{N}_\varepsilon(\boldsymbol{\xi}) & \\ & \mathbf{N}_\kappa(\boldsymbol{\xi}) \end{bmatrix}}_{\mathbf{N}(\boldsymbol{\xi})} \underbrace{\begin{bmatrix} \mathbf{v}_\varepsilon^\ell \\ \mathbf{v}_\kappa^\ell \end{bmatrix}}_{\mathbf{v}^\ell} \tag{17}$$

where $\mathbf{u}_\varepsilon(\boldsymbol{\xi})$ and $\mathbf{u}_\kappa(\boldsymbol{\xi})$ are the in-plane and out-of-plane displacement fields, and $\mathbf{N}_\varepsilon(\boldsymbol{\xi})$ and $\mathbf{N}_\kappa(\boldsymbol{\xi})$ are the CST and Specht interpolation matrices. In the following, the dependence on the spatial coordinates $\boldsymbol{\xi}$ is made implicit to simplify notation.

Introducing the linear and nonlinear strain interpolation matrices $\mathbf{B}$ and $\{\mathbf{G}_{xx}, \mathbf{G}_{yy}, \mathbf{G}_{xy}\}$

$$\mathbf{B} = \begin{bmatrix} \partial_1 \\ \partial_2 \end{bmatrix} \mathbf{N} \quad , \quad \mathbf{G}_{ij} = \mathbf{N}_{,i}^T \mathbf{N}_{,j} + \mathbf{N}_{,j}^T \mathbf{N}_{,i} \tag{18}$$

as well as the index vectors $\mathbf{I}_1 = [\,1, 0, 0, 0, 0, 0\,]^T$, $\mathbf{I}_2 = [\,0, 1, 0, 0, 0, 0\,]^T$, and $\mathbf{I}_3 = [\,0, 0, 1, 0, 0, 0\,]^T$, the generalized strain-displacement relation (3) can be expressed as

$$\begin{aligned} \bar{\boldsymbol{\varepsilon}} &= \mathbf{B}\mathbf{v}^\ell + \frac{1}{2}\mathbf{I}_1 \left(\mathbf{v}^\ell\right)^T \mathbf{G}_{xx}\mathbf{v}^\ell \ldots \\ &\ldots + \frac{1}{2}\mathbf{I}_2 \left(\mathbf{v}^\ell\right)^T \mathbf{G}_{yy}\mathbf{v}^\ell + \frac{1}{2}\mathbf{I}_3 \left(\mathbf{v}^\ell\right)^T \mathbf{G}_{xy}\mathbf{v}^\ell \\ &:= \bar{\boldsymbol{\varepsilon}}_{\mathrm{L}}(\mathbf{v}^\ell) + \bar{\boldsymbol{\varepsilon}}_{\mathrm{NL}}(\mathbf{v}^\ell) \end{aligned} \tag{19}$$

where $\bar{\boldsymbol{\varepsilon}}_{\mathrm{L}}$ and $\bar{\boldsymbol{\varepsilon}}_{\mathrm{L}}$ are introduced as the linear and nonlinear generalized strain functions, respectively.

Any virtual displacement field $\delta\mathbf{u}$ is represented analogously to the physical displacement field, i.e., as $\delta\mathbf{u} = \mathbf{N}\delta\mathbf{v}^\ell$, giving rise to a virtual generalized strain field of the form

$$\begin{aligned} \delta\bar{\boldsymbol{\varepsilon}} &= \frac{\partial\bar{\boldsymbol{\varepsilon}}}{\partial\mathbf{v}^\ell}\delta\mathbf{v}^\ell = \mathbf{B}\delta\mathbf{v}^\ell + \mathbf{I}_1 \left(\mathbf{v}^\ell\right)^T \mathbf{G}_{xx}\delta\mathbf{v}^\ell \ldots \\ &\ldots + \mathbf{I}_2 \left(\mathbf{v}^\ell\right)^T \mathbf{G}_{yy}\delta\mathbf{v}^\ell + \mathbf{I}_3 \left(\mathbf{v}^\ell\right)^T \mathbf{G}_{xy}\delta\mathbf{v}^\ell \\ &:= \delta\bar{\boldsymbol{\varepsilon}}_{\mathrm{L}}(\delta\mathbf{v}^\ell) + \delta\bar{\boldsymbol{\varepsilon}}_{\mathrm{NL}}(\mathbf{v}^\ell, \delta\mathbf{v}^\ell) \end{aligned} \tag{20}$$

where $\delta\bar{\boldsymbol{\varepsilon}}_{\mathrm{L}}$ and $\delta\bar{\boldsymbol{\varepsilon}}_{\mathrm{L}}$ are introduced as the linear and nonlinear generalized virtual strain functions, respectively.

6 STABILITY ANALYSIS BY THE PRINCIPLE OF VIRTUAL WORK

The displacement field within an element is decomposed into a component related to the element nodal forces, $\mathbf{u}_1(\mathbf{q}^\ell)$, and an independent perturbation, $\epsilon\mathbf{u}_2$ where $\epsilon \ll 1$:

$$\mathbf{u} = \mathbf{u}_1(\mathbf{q}^\ell) + \epsilon\mathbf{u}_2 = \mathbf{N}\left(\mathbf{v}_1^\ell(\mathbf{q}^\ell) + \epsilon\mathbf{v}_2^\ell\right) \tag{21}$$

In the neighborhood of a given state $\{\mathbf{u}_0, \mathbf{q}_0^\ell\}$, the relation between the nodal load vector $\mathbf{q}^\ell = \lambda\mathbf{q}_0^\ell$ and the load-dependent nodal displacement vector $\mathbf{v}_1^\ell$ is assumed to be well-approximated by the first-order Taylor expansion:

$$\begin{aligned} &\lambda\mathbf{q}_0^\ell = \mathbf{q}_0^\ell + \left.\frac{\partial\mathbf{q}^\ell}{\partial\mathbf{v}_1^\ell}\right|_0 \left(\mathbf{v}_1^\ell - \mathbf{v}_{1,0}^\ell\right) \\ &\Updownarrow \mathbf{v}_1^\ell = (\lambda - 1)\hat{\mathbf{v}}^\ell + \mathbf{v}_{1,0}^\ell \quad , \quad \hat{\mathbf{v}}^\ell = \left(\mathbf{k}_{\mathrm{T}}^\ell\right)^{-1}\mathbf{q}_0^\ell \end{aligned} \tag{22}$$

where $\mathbf{k}_{\mathrm{T}}^\ell = (\partial\mathbf{q}^\ell/\partial\mathbf{v}_1^\ell)|_0$ is identified as the element tangent stiffness matrix. Assuming that the strain field produced by $\mathbf{u}_1$ is dominated by the linear term, the strain field can be approximated as

$$\begin{aligned} \bar{\boldsymbol{\varepsilon}} &\simeq \bar{\boldsymbol{\varepsilon}}_{\mathrm{L}}(\mathbf{v}_1^\ell + \epsilon\mathbf{v}_2^\ell) + \bar{\boldsymbol{\varepsilon}}_{\mathrm{NL}}(\epsilon\mathbf{v}_2^\ell) \\ &= \bar{\boldsymbol{\varepsilon}}_{\mathrm{L}}(\mathbf{v}_1^\ell + \epsilon\mathbf{v}_2^\ell) + \mathcal{O}(\epsilon^2) \\ &= (\lambda - 1)\mathbf{B}\hat{\mathbf{v}}^\ell + \mathbf{B}\mathbf{v}_{1,0}^\ell + \epsilon\mathbf{B}\mathbf{v}_2^\ell + \mathcal{O}(\epsilon^2) \end{aligned} \tag{23}$$

where $\mathcal{O}(\epsilon^2)$ is negligible since $\epsilon \ll 1$. The virtual strains are assumed to be load-independent, i.e.,

$$\begin{aligned} \delta\bar{\boldsymbol{\varepsilon}} &= \delta\bar{\boldsymbol{\varepsilon}}_{\mathrm{L}}(\delta\mathbf{v}^\ell) + \delta\bar{\boldsymbol{\varepsilon}}_{\mathrm{NL}}(\mathbf{v}_2^\ell, \delta\mathbf{v}^\ell) \\ &= \Big(\mathbf{B} + \epsilon\mathbf{I}_1 \left(\mathbf{v}_2^\ell\right)^T \mathbf{G}_{xx} + \epsilon\mathbf{I}_2 \left(\mathbf{v}_2^\ell\right)^T \mathbf{G}_{yy} \ldots \\ &\quad \ldots + \epsilon\mathbf{I}_3 \left(\mathbf{v}_2^\ell\right)^T \mathbf{G}_{xy}\Big)\,\delta\mathbf{v}^\ell \end{aligned} \tag{24}$$

Using the linearized relation between the generalized stresses and strains (6), the virtual internal work in a finite element can be approximated as

$$\begin{aligned} \delta W_{\mathrm{int}}^e &= \int_A \delta\bar{\boldsymbol{\varepsilon}}^T\bar{\boldsymbol{\sigma}}\,\mathrm{d}x \simeq \int_A (\delta\mathbf{v}^\ell)^T \big(\mathbf{B}^T \ldots \\ &\ldots + \epsilon\mathbf{G}_{xx}\mathbf{v}_2^\ell\mathbf{I}_1^T + \epsilon\mathbf{G}_{yy}\mathbf{v}_2^\ell\mathbf{I}_2^T \ldots \\ &\ldots + \epsilon\mathbf{G}_{xy}\mathbf{v}_2^\ell\mathbf{I}_3^T\big)\left(\bar{\boldsymbol{\sigma}}_0 + (\lambda - 1)\mathbf{D}_{\mathrm{T}}\mathbf{B}\hat{\mathbf{v}} \ldots\right. \\ &\ldots \left.+ \mathbf{D}_{\mathrm{T}}\left(\mathbf{B}\mathbf{v}_{1,0}^\ell - \bar{\boldsymbol{\varepsilon}}_0\right) + \epsilon\mathbf{D}_{\mathrm{T}}\mathbf{B}\mathbf{v}_2^\ell\right)\mathrm{d}A \end{aligned} \tag{25}$$

where, by definition, $\mathbf{B}\mathbf{v}_{1,0}^\ell - \bar{\boldsymbol{\varepsilon}}_0 = \mathbf{0}$ and $\hat{\boldsymbol{\sigma}} = \mathbf{D}_{\mathrm{T}}\mathbf{B}\hat{\mathbf{v}} \simeq \bar{\boldsymbol{\sigma}}_0$ for a well-interpolated strain field. By expanding the expression and neglecting terms containing ϵ^2, the virtual internal work can be approximated as

$$\begin{aligned} \delta W_{\mathrm{int}}^e &\simeq (\delta\mathbf{v}^\ell)^T \big(\mathbf{q}_0^\ell + (\lambda - 1)\mathbf{k}_{\mathrm{T}}^\ell\hat{\mathbf{v}} \ldots \\ &\ldots + \epsilon\left[\mathbf{k}_{\mathrm{T}}^\ell + \mathbf{k}_{\mathrm{g}}^\ell + (\lambda - 1)\hat{\mathbf{k}}_{\mathrm{g}}^\ell\right]\mathbf{v}_2^\ell\big) \end{aligned} \tag{26}$$

where

$$\begin{aligned} &\mathbf{q}_0^\ell = \int_A \mathbf{B}^T\bar{\boldsymbol{\sigma}}_0\,\mathrm{d}A \quad , \quad \mathbf{k}_{\mathrm{T}}^\ell = \int_A \mathbf{B}^T\mathbf{D}_{\mathrm{T}}\mathbf{B}\,\mathrm{d}A \\ &\mathbf{k}_{\mathrm{g}}^\ell = \int_A \mathbf{I}_1^T\bar{\boldsymbol{\sigma}}_0\mathbf{G}_{xx} + \mathbf{I}_2^T\bar{\boldsymbol{\sigma}}_0\mathbf{G}_{yy} + \mathbf{I}_3^T\bar{\boldsymbol{\sigma}}_0\mathbf{G}_{xy}\,\mathrm{d}A \\ &\hat{\mathbf{k}}_{\mathrm{g}}^\ell = \int_A \mathbf{I}_1^T\hat{\boldsymbol{\sigma}}\mathbf{G}_{xx} + \mathbf{I}_2^T\hat{\boldsymbol{\sigma}}\mathbf{G}_{yy} + \mathbf{I}_3^T\hat{\boldsymbol{\sigma}}\mathbf{G}_{xy}\,\mathrm{d}A \end{aligned} \tag{27}$$

To allow evaluation of the integrals containing $\mathbf{D}_T$, its components are assumed to vary linearly between the submodel points.

By expanding (26) to system level,

$$\delta W_{int} = \delta\mathbf{V}^T\left(\mathbf{Q}_0 + (\lambda - 1)\mathbf{K}_T\hat{\mathbf{V}} \ldots \right.$$
$$\left. \ldots + \epsilon\left[\mathbf{K}_T + \mathbf{K}_g + (\lambda - 1)\hat{\mathbf{K}}_g\right]\mathbf{V}_2\right) \tag{28}$$

and stating the external virtual work in terms of the current load vector $\mathbf{R}_0$,

$$\delta W_{ext} = \lambda\delta\mathbf{V}^T\mathbf{R}_0 \tag{29}$$

the principle of virtual works leads to the following equation:

$$(\lambda - 1)\left(\mathbf{K}_T\hat{\mathbf{V}} - \mathbf{R}_0\right)\ldots$$
$$\ldots + \epsilon\left[\mathbf{K}_T + \mathbf{K}_g + (\lambda - 1)\hat{\mathbf{K}}_g\right]\mathbf{V}_2 = \mathbf{0} \tag{30}$$

It is seen that the first term is simply a linear system of equations requiring equilibrium of the current state, while the second term is a linear eigenvalue problem which must be satisfied for any nonzero perturbation ϵ. Note that for a well-interpolated strain field where $\mathbf{K}_g = \hat{\mathbf{K}}_g$, this closely resembles the well-known linear buckling problem from linear elasticity.

7 EXAMPLES

7.1 Validation: Euler column

To assess the accuracy of the proposed method, a reinforced concrete concrete wall with simple end-point supports is analyzed and compared to the solution for an Euler column. The wall has the height $h = 3$ m, the width $b = 1$ m and the thickness $t = 0.2$ m, and it is reinforced in the longitudinal direction with two layers of Ã˜8 mm bars per 150 mm ($A_s = 335$ mm/m) positioned as $z_s = \pm 71$ mm. The reinforcement has the stiffness $E_s = 200$ GPa for absolute stress values smaller than $f_Y = 500$ MPa, and the reduced stiffness $H_s = 0.08 f_Y/(\varepsilon_u - f_y/E_s) = E_s/257.5$ for stresses beyond this limit where $\varepsilon_u = 0.05$ is the ultimate reinforcement strain. Note that this corresponds to the requirements for Class B reinforcement as specified in (EN1992 2004). The concrete has the compressive stiffness $E_c = 33$ GPa for compressive stresses not exceeding $f_c = 30$ MPa, and the reduced stiffness $H_c = E_c/2575$ for compressive stresses exceeding this value. The maximum initial eccentricity of the wall is taken as $e_1 = 20$ mm.

A simple approach for estimating the critical buckling load of an Euler column consists of analyzing the stress/strain state of the critical section only (the mid-section), and including the second-order effect in terms of an initial imperfection e_1 and an assumed curvature variation along the column. Thus, by assuming a triangular curvature distribution at the critical load (which is more accurate than a sine distribution when localization can occur), a baseline result for the critical buckling load can be determined using the following approach:

1. Choose the applied axial compressive load $p > 0$
2. Assume ε_0
3. Find $\kappa \geq 0$ such that $n(\varepsilon_0, \kappa) = -p$ and compute $m(\varepsilon_0, \kappa)$
4. Compute the mid-section displacement $e_2 = \left(\frac{L^2}{12}\right)\kappa$
5. Compute the residual moment capacity $m_0 = m(\varepsilon_0, \kappa) - p\,(e_1 + e_2)$

The steps 2–5 are repeated for different values of ε_0 as to find the maximum of m_0. By performing this process for increasing values of p, the critical load p_{cr} can be found as the axial load for which the residual moment capacity m_0 vanishes. Using this approach, the critical load is estimated as $p_{cr}^{est} = 4.37 \cdot 10^3$ kN/m.

The wall cross-section is modelled using 10 concrete layers in the stress-based analysis and a regular mesh of 3 × 10 rectangular tiles of 4 finite elements each. The predicted critical load p_{cr} is shown as a function of the applied load p in Figure 6.

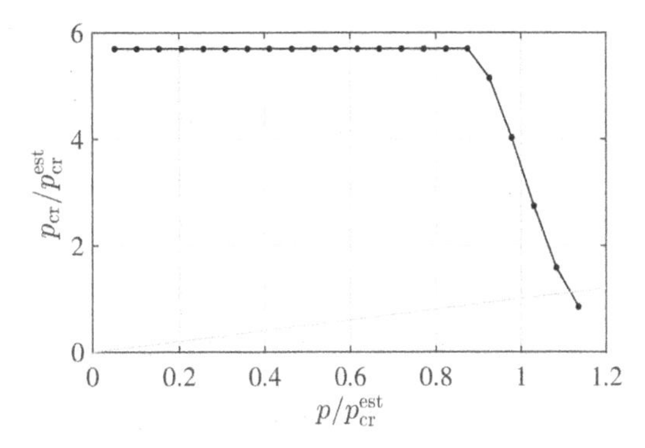

Figure 6. Predicted critical load for Euler wall as a function of the applied load.

For low load levels, the predicted critical load corresponds to the Euler load for an uncracked section, $p_E = 24.9 \cdot 10^3$ kN/m $= 5.70 p_{cr}^{est}$, since the whole section is in compression. For higher load levels, the predicted critical load decreases until crossing the line $p_{cr} = p$ at $p = 1.11 p_{cr}^{est}$, i.e., 11% higher than the estimated value. This corresponds to the critical buckling load as predicted by the proposed method. Note that the p_{cr}^{est} is expected to be lower than the actual critical buckling load due to the simplified calculation procedure, e.g. the assumed curvature variation. Thus, a part of the 11% deviation is presumably due this, while another part is due to the assumption of a linear strain-displacement relation for loads lower than the critical buckling load in (23).

7.2 Demonstration: wall with door hole

As a demonstration of the applicability of the method to practical design scenarios, a rectangular reinforced concrete wall with a door hole is considered; see Figure 7. The wall is $b = b_1 + b_2 + b_3 = 5$ m wide and $h = h_1 + h_2 = 3$ m high, and the door hole with dimensions $b_2 \times h_1 = 1\,\text{m} \times 2\,\text{m}$ is positioned $b_1 = 3$ m from the leftmost edge. The wall is supported along all directions along the bottom and against transverse movement along the top, and it is subjected to a distributed vertical load p with an out-of-plane eccentricity $e_p = 20$ mm causing also the bending moment $m_p = pe_p$. As in the previous example, the wall is $t = 0.2$ m thick, it is orthogonally reinforced with two layers of Ã˜8 mm steel bars per 150 mm positioned at $z_{s,i} = \pm 71$ mm, and the material strength and stiffness parameters are taken as $f_Y = 500$ MPa, $E_s = 200$ GPa and $H_s = E_s/257.5$ for the reinforcement, and $f_c = 30$ MPa, $E_c = 33$ GPa and $H_c = E_c/2575$ for the concrete.

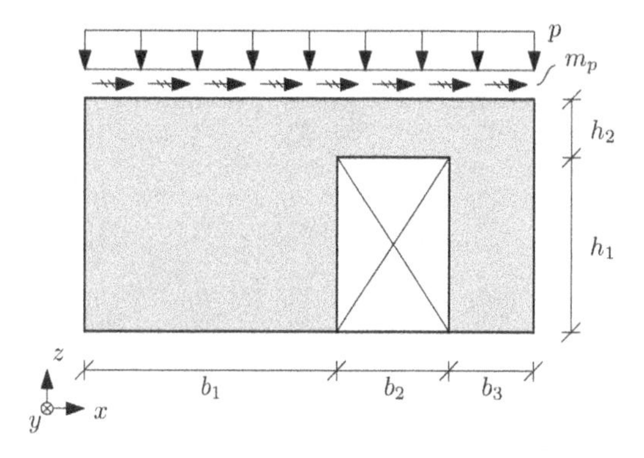

Figure 7. Wall with door hole.

In this case, a simple estimate for the critical buckling load can be obtained by considering the segment to the right of the door hole as an Euler column subjected to the axial load $(1 + b_2/2b_3)p = 1.5p$. Using the approach described in Section 7.1, this results in the estimate $p_{cr}^{est} = 2.92 \cdot 10^3$ kN/m.

The wall cross-section is modelled using 5 concrete layers in the stress-based analysis, and an unstructured mesh of 345 elements produced by *MESH2D* (Engwirda 2014). The predicted critical load p_{cr} is shown as a function of the applied load p in Figure 8 where the axes have been normalized with respect to p_{cr}^{est}.

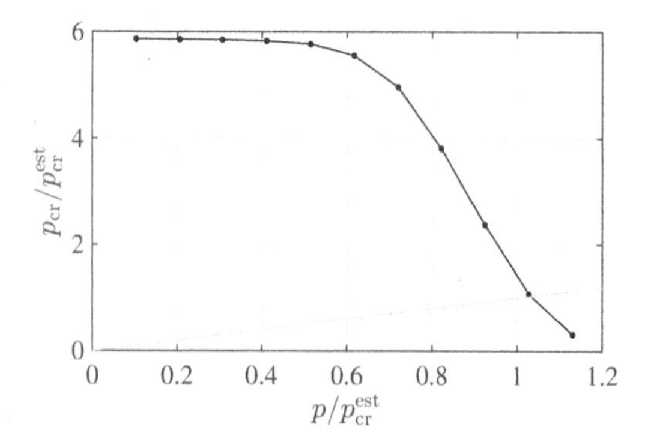

Figure 8. Predicted critical load for wall with door hole as a function of the applied load.

It is seen that the range of the initial and final normalized critical buckling loads are similar to those of the previous example; however, the decrease in the critical buckling load is more gradual than in the previous example due to the progression of the tensile zone in the lintel. The buckling modes obtained from the eigenvectors in load steps 1 and 10, respectively, are shown in Figure 9. From these, it is clear that in load step 1 the part to the right of the door hole is stiffened by the lintel, whereas in load step 10 the buckling mode has localized due cracking and yielding in the lintel.

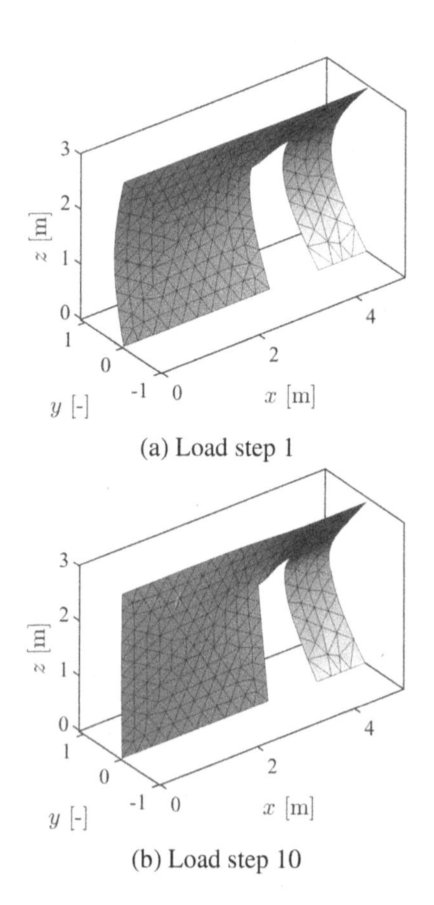

Figure 9. Buckling modes for wall with door hole.

The solution time for the geometrically linear and geometrically nonlinear problem is shown for each load step in Figure 10. It is seen the solution time is approximately 30 s per load step, and that approx. 75% of the solution time goes to solving the geometrically linear problem. Note that each load step is independent, i.e., in the context of design verification, a single load step is, in principle, sufficient to check if the design load is higher or lower than the critical buckling load.

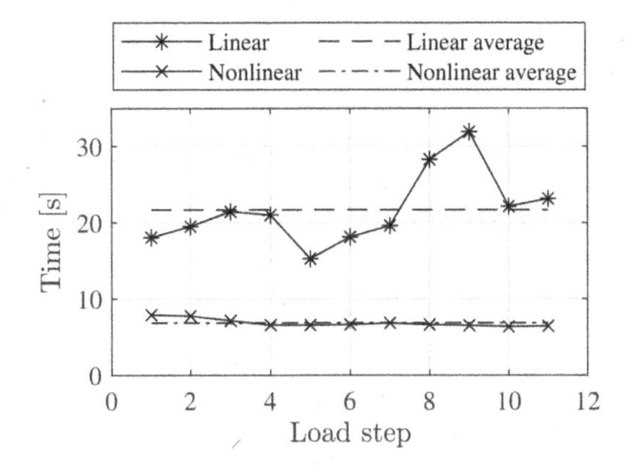

Figure 10. Solution time for wall with door hole.

8 CONCLUSIONS

A framework for stability analysis of cracked reinforced concrete walls has been presented. Consisting of a two-step procedure, the framework finds the cracked tangent stiffness of the reinforced concrete sections in a stress-based, geometrically linear finite element analysis using convex optimization, and subsequently poses a linearized buckling problem which is solved as a linear eigenvalue problem. By performing the two-step procedure for a sequence of loads, the critical buckling load is estimated with increasing accuracy as the applied load approaches the critical buckling load. The method was validated by comparing the solution for a wall with simple end-point supports to that of an Euler column, showing a slight overestimation of the critical buckling load. Finally, the applicability of the method to practical design scenarios was demonstrated on a reinforced concrete wall with a door hole, which produced results in the expected range and solution times of approximately 30 seconds per load step.

ACKNOWLEDGEMENTS

The work presented in this paper is partly funded by the Innovation Fund Denmark under File No. 9065-00170B, and the Ramboll Foundation under File No. 2018-103.

REFERENCES

EN1992 (2004). Eurocode 2: Design of concrete structures - part 1-1: General rules and rules for buildings. Standard, European Committee for Standardization, Brussels, BE.

Engesser, F. (1889). Über die knickfestigkeit gerader stäbe. *Zeitschrift des Architekten- und Ingenieur-Vereins zu Hannover 35*, 455–462.

Engwirda, D. (2014). *Locally-optimal Delaunay-refinement and optimisation-based mesh generation*. Phd thesis, The University of Sydney, School of Mathematics and Statistics.

Herfelt, M. A. (2017). *Numerical Limit Analysis of Precast Concrete Structures: A framework for efficient design and analysis*. Ph. D. thesis, Technical University of Denmark, Department of Civil Engineering.

Jensen, T. W. (2019). *Modelling the load-carrying capacity of reinforced concrete slab bridge: With a focus on slabs constructed with inverted T-beams*. Ph. D. thesis, Technical University of Denmark, Department of Civil Engineering.

Larsen, K. P. (2011). *Numerical Limit Analysis of Reinforced Concrete Structures: Computational Modeling with Finite Elements for Lower Bound Limit Analysis of Reinforced Concrete Structures*. Ph. D. thesis, Technical University of Denmark, Department of Civil Engineering.

Poulsen, P. N. & L. Damkilde (2000). Limit state analysis of reinforced concrete plates subjected to in-plane forces. *International Journal of Solids and Structures 37*(42), 6011–6029.

Specht, B. (1988). Modified shape functions for the three-node plate bending element passing the patch test. *International Journal for Numerical Methods in Engineering 26*(3), 705–15.

Vestergaard, D., K. P. Larsen, L. C. Hoang, P. N. Poulsen, & B. Feddersen (2021). Design-oriented elasto-plastic analysis of reinforced concrete structures with in-plane forces applying convex optimization. *Structural Concrete 22*, 3272–3287.

Vestergaard, D., K. P. Larsen, L. C. Hoang, P. N. Poulsen, & B. Feddersen (2022). A shell element for design-oriented elasto-plastic analysis of reinforced concrete walls using convex optimization. Manuscript to be submitted for publication.

Computational Modelling of Concrete and Concrete Structures – Meschke, Pichler & Rots (Eds)
 ISBN: 978-1-032-32724-2

Design of steel fiber reinforced concrete tunnel lining segments by nonlinear finite-element analysis with different safety formats

G.E. Neu, V. Gudžulić & G. Meschke
Institute for Structural Mechanics, Ruhr University Bochum, Germany

ABSTRACT: The use of Nonlinear Finite Element Analysis (NLFEA) for the assessment of structures requires special considerations regarding the employed safety format. While some guidelines are available, the provided information is focused on Reinforced Concrete (RC) and the application for Steel Fiber Reinforced Concrete (SFRC) structures is not addressed. In order to evaluate the application of NLFEA and the use of SFRC as reinforcement scheme for segmental tunnel lining design, a multi-level model for the analysis of SFRC structures is used. This multi-level SFRC model allows to directly assess the influence of a chosen fiber type, content and fiber orientation on the structural response. The post-cracking behavior is captured by a discrete crack model based on cohesive interface elements. After a brief introduction of the numerical model and the use of safety formats in NLFEA, the influence of the applied safety format on the local structural response of a SFRC beam subjected to 3-point bending is investigated. The proposed multi-level SFRC model is used to carry out a full probabilistic analysis due to its capability to capture the influence and scatter of important SFRC parameters. Finally, a segmental lining ring subjected to earth & water pressure is designed using different safety formats. In order to discuss the influence of the reinforcement scheme, the design by NLFEA is carried out for SFRC as well as RC segments. It is shown that the use of NLFEA can result in a greatly increased design resistance and that SFRC segments offer a better performance compared to RC segments.

1 INTRODUCTION

The circular lining in mechanized tunneling consists of concrete segments, which are exposed to different loading cases during tunnel construction. Reinforced concrete linings were typically designed using traditional steel reinforcement bars, but SFRC is used in more and more mechanized tunneling construction projects due to their economic benefits and serviceability performance (Bakshi & Nasri 2017; Carlo, Meda, & Rinaldi 2016). Difficulties arise for the design of SFRC segmental linings, because the current versions of many design codes do not explicitly account for the dimensioning and design of SFRC structures and therefore a specific set of guidelines determining the SFRC design process must contractually be agreed upon. In general, the post-cracking response of SFRC is affected by the size, shape, concentration as well as orientation of the fibers. Within the framework of available guidelines, the residual strength of an SFRC is characterized by bending tests and therefore the above listed fiber properties are not explicitly accounted for in the design. In this contribution, a multi-level model for the analysis of SFRC structures is used, which allows to directly assess the influence of a chosen fiber type, content and fiber orientation on the structural response (Zhan & Meschke 2016). Based on the explicit fiber geometry, the interface conditions, the material properties and an assumed fiber orientation, the model generates an equivalent traction-separation law. In order to evaluate the post-cracking response of SFRC structures by Finite-Element (FE) analysis, a discrete crack approach using interface-elements (Ortiz & Pandolfi 1999), whose post-cracking behavior is governed by the derived traction-separation law, is used.

The focus of this work is to use non-linear FE analysis (NLFEA) to design a SFRC tunnel lining segment. The use of NLFEA requires special considerations regarding the employed safety concept (Castaldo, Gino, Bertagnoli, & Mancini 2018) and heavily influences the resulting structural design. In order to ensure the required probability of failure, (fib Model Code for Concrete Structures 2010 2013) proposes different methods for the verification of a structural system. While (fib Model Code for Concrete Structures 2010 2013) provides some guidance for the use of these safety formats in conjunction with NLFEA, only with (Hendriks & Roosen 2020) a specific guideline for the application of NLFEA for the design of concrete structures is available. Furthermore, the provided information is focused on RC and the application on SFRC structures is not addressed.

 DOI 10.1201/9781003316404-88

The structure of the paper is as follows: After a brief introduction of the numerical model (Section 2) and the use of safety formats in NLFEA (Section 3), the influence of the applied safety format on the local structural response of a SFRC member is investigated (Section 4). For this purpose, 3-point bending tests of a previously carried out systematic validation campaign (Gudzulic, Neu, Gebuhr, Anders, & Meschke 2020) are re-analyzed by adopting the different safety formats. For a fully probabilistic analysis the proposed multi-level SFRC model is employed so that the influence and scatter of important material parameters can be assessed. In the last section (Section 5), a segmental lining ring subjected to earth & water pressure is designed using different safety formats. In order to discuss the influence of the reinforcement scheme, the design by NLFEA is carried out for SFRC as well as RC segments.

2 MODELING OF STEEL FIBER AND CONVENTIONAL REINFORCED CONCRETE

Steel fibers provide a residual strength after onset of cracking depending on the type, content and orientation of the fibers. Available guidelines (i.e. (fib Model Code for Concrete Structures 2010 2013)) characterize the residual strength of SFRC based on bending tests and derive uniaxial stress-strain relationships for Ultimate Limit State (ULS) and Service Limit State (SLS) design. As an alternative, a multi-level SFRC model has been developed in (Zhan & Meschke 2016), which allows to directly assess the influence of the individual fiber type and the fiber cocktail on the structural behavior. As illustrated in Figure 1, the proposed multi-level model consists of submodels related to three different scales involved in the numerical analyses of SFRC structures.

2.1 *Fiber scale – single fiber pullout behavior*

At the level of the individual fibers and the matrix, the pull-out behavior of a single fiber is controlled by the interface conditions, the fiber shape and the fiber inclination with respect to a crack. A semi-analytical model predicting the pullout force-displacement relation of single fibers $F(w, \theta, \tilde{x})$, which depends on the position of the centroid $\tilde{x}$ and the inclination θ of the fiber with respect to the crack plane (Figure 1, left) has been developed in (Zhan & Meschke 2014). The model is capable of capturing the major mechanisms (straightening of the hooked-end, concrete spalling and fiber rupture) activated during the pullout of a single steel fiber embedded in a concrete matrix, accounting for different configurations of fiber type and strength, concrete strength, fiber inclination and embedment length (see Figure 1, left). The basis of the semi-analytical model is the specification of a stress-slip relationship $\tau(s)$ along the interface between fiber and concrete (Figure 1, left), which reflects the three stages of the single fiber pull-out process (bonded state, debonding stage and sliding phase). The plastification

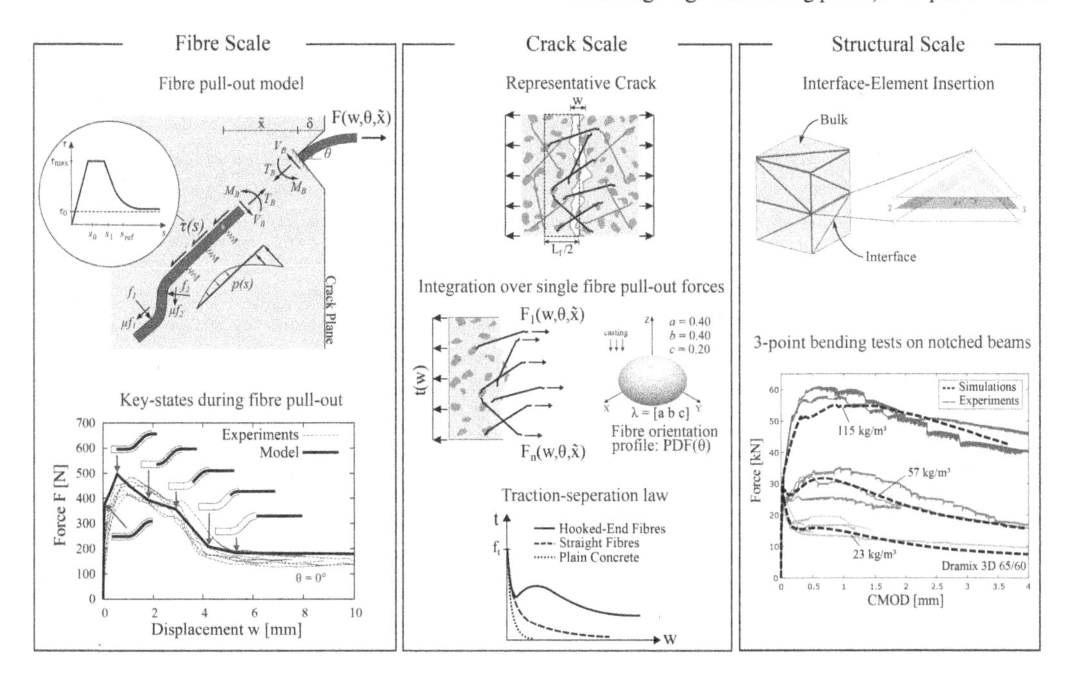

Figure 1. Multi-level modelling of SFRC: Semi-analytical model for single fiber pull-out considering various 'key states' during elongation, plastification and concrete spalling of hooked-end fibers (fiber Scale); Integration of the pull-out response of all fibers crossing a representative crack and considering their orientation to calculate the traction-separation law (Crack Scale); Modeling of discrete cracks via zero-thickness interface elements inserted between the regular bulk elements and validation for bending tests on notched beams (Structural Scale).

of the hooked-end and thereby its straightening during pull-out is captured by multiple characteristic key-states (Figure 1, left bottom) based on (Laranjeira, Molins, & Aguado 2010). The contribution of the hooked-end is calculated by taking the explicit fiber hooked-end geometry and the interaction with the surrounding concrete matrix into account. Concrete spalling and fiber rupture are considered based on (Laranjeira, Molins, & Aguado 2010).

2.2 *Crack scale – crack bridging stress*

At the level of an opening crack within the fiber-concrete composite, the fibers crossing the crack are activated and ensure a residual post-cracking strength depending on the fiber content and the fiber orientation. In the multi-level SFRC model, the post cracking response is approximated by a traction-separation law which is derived via the integration of the pullout force-displacement relations $F(w, \theta, \tilde{x})$ of all single fibers intercepting the crack (Figure 1, center) and taking an anisotropic orientation of fibers into consideration (Zhan & Meschke 2016). According to (Wang, Backer, & Li 1989), the bridging stress of SFRC $t_{fibers}(w)$ is estimated by

$$t_{\text{fibers}}(w) = \int_{\tilde{x}=0}^{L_f/2} \left[\int_{\theta=0}^{\operatorname{acos}(2\tilde{x}/L_f)} F(\tilde{x}, \theta, w)\, p(\theta)\, \mathrm{d}\theta \right] \dots$$

$$\dots p(\tilde{x})\, \mathrm{d}\tilde{x} \cdot \frac{c_f}{A_f}, \quad (1)$$

where c_f is the volume fraction of the fibers and A_f and L_f are the cross-section area and length of one fiber, respectively. The spatial dispersion characteristics of the fibers in the composite is represented by the probability densities $p(\theta)$ as functions of the inclination angle θ and embedment length $\tilde{x}$ of the fiber. The spatial orientation of fibers is depending on the dimensions of the structural member and the casting direction. In principle, the complete and general description of the fiber orientation requires a distribution function w.r.t. all spatial directions, which can be obtained by simulations of the casting procedure (Gudzulic, Dang, & Meschke 2018). In order to capture the anisotropic fiber orientation as a consequence of the casting procedure in a straightforward and practical manner, (Zhan & Meschke 2016) proposed a method to compute the probability density $p(\theta)$ based on a given fiber orientation profile $\lambda_{cast} = [a, b, c]$. The spatial preference of the fibers in the global coordinate system is represented by means of an ellipsoid, with the semi-axes a, b, and c representing the assumed fiber orientation profile (Figure 1, center). An isotropic fiber orientation can be graphically represented by a sphere with the fiber orientation profile $\lambda_{cast} = [0.33, 0.33, 0.33]$). The distribution of fibers is considered to be homogeneous, so that $p(\tilde{x})$ is reduced to the constant $\frac{2}{L_f}$ (Wang, Backer, & Li 1989).

The resulting post cracking response of the fiber-concrete composite $t(w)$ is composed of the fiber bridging stresses $t_{fibers}(w)$ and the cohesive traction of the plain concrete $t_{coh}(w)$

$$t(w) = t_{coh}(w) + t_{fibers}(w), \quad (2)$$

where the cohesive traction of the plain concrete $t_{coh}(w)$ is described by an exponential softening law taking the uniaxial tensile strength f_t of the SFRC and the mode I fracture energy $G_{F,I}$ into account.

2.3 *Structural scale – discrete crack model with interface elements*

At the structural level, the post-cracking behavior is captured by a discrete crack model based on cohesive interface elements (Ortiz & Pandolfi 1999). Between the regular finite elements (bulk elements), zero-thickness interface elements are inserted (Figure 1, right), which allow a discrete mapping of cracks and provide direct information on crack widths. Therefore, a sufficiently fine mesh is required to accurately predict the crack pattern and not restrict the crack initiation and propagation. The behavior of the zero-thickness interface elements is governed by the traction separation law derived on the crack scale with the multi-level SFRC model, but it should be noted that also arbitrary stress-crack opening relationships can be used (i.e. the stress-crack opening relationships proposed in the fib model code 2010 for modeling the tensile behavior of SFRC at SLS and ULS). The bulk material is assumed to behave as a linearly elastic material.

For the convenient use of the traction-separation law in FE simulations, the integral in Eq. (1) is numerically evaluated and replaced by an analytical surrogate function

$$t(\alpha) = (f_t - t_1)\exp(-\frac{\alpha}{G_f/f_t}) + \dots \quad (3)$$

$$\dots + t_1 \frac{w_u - \alpha}{w_u} + t_2\, \alpha \exp(c_1 - c_2 \alpha),$$

where t_1, t_2, c_1 and c_2 are coefficients, which are determined by fitting the surrogate function to the numerical evaluation of Eq. (1). The parameter w_u represents the ultimate crack opening and can also be used for obtaining a better fit. The crack opening w is replaced by the effective separation α to account for multi-axial stress states:

$$\alpha = \sqrt{u_N^2 + \frac{\beta_\tau^2}{\kappa^2} u_T^2}. \quad (4)$$

In Eq. (4), u_N is the normal and u_T the tangential separation of the interface element. Based on (Snozzi & Molinari 2013), the parameter β_τ controls the shear strength ($\tau_{SFRC} = \beta_\tau f_t$) and the parameter κ defines the ratio between mode II and mode I fracture energy ($\kappa = \frac{G_{F,II}}{G_{F,I}}$). In this study values of $\beta_\tau = 5$ and $\kappa = 20$ are used. The integration of the constitutive rate equations is performed following the IMPL-EX algorithm (Oliver, Huespe, & Cante 2008) to enhance the robustness of the solution procedure.

In order to account for reinforcement bars if present, the rebars are modeled as linear trusses and coupled with the concrete matrix using a constraint condition between control points located on the rebar elements and their respective projection points within the solid elements in which they are embedded (Gall, Butt, Neu, & Meschke 2018). The constraint condition includes the bond-slip mechanism as provided in (fib Model Code for Concrete Structures 2010 2013). The steel behavior itself is considered with an elastoplastic v. Mises yield surface with linear hardening.

3 SAFETY CONCEPTS

A structure should be designed that the action E is not exceeding the resistance R as

$$E(e) \leq R(r). \tag{5}$$

The action as well as the resistance are in general random variables and may vary throughout the lifetime of a structure. Loads can be acting persistently, temporarily or on rare occasions and therefore result in different states of actions. Those situations have to be taken into account during the design and their respective frequency should be considered accordingly. The resistance is depended on the structural properties, which can be time dependent as well as scattering due to material uncertainties. In order to verify entire structures, structural elements or local regions, limit state design principles are adopted which using a limit state function g to separate acceptable from unacceptable states of the structure. An unacceptable state or failure is represented as

$$g(E, R) = R(r) - E(e) \leq 0, \tag{6}$$

when the limit state function can be separated into a resistance and a loading function. It is assumed that the action E(e) and the resistance R(r) can be described by log-normal distributions (fib Model Code for Concrete Structures 2010 2013). The area between the probability density functions, where E(e) is greater than R(r) represents the failure probability P_f (Figure 2, shaded area). The purpose of structural design is to reduce the failure probability to an acceptable value. Often the reliability index β is used to ensure a sufficient distance between the limit state $g(E, R) = 0$ and the two probability density functions (Figure 2). The reliability index β is related to the failure probability by the distribution function of standardized normal distribution Φ as

$$P_f = \Phi(-\beta). \tag{7}$$

The safety formats given in (fib Model Code for Concrete Structures 2010 2013) as well as the (European Comittee for Standardisation 2002) are formulated to ensure a certain failure probability P_f. The

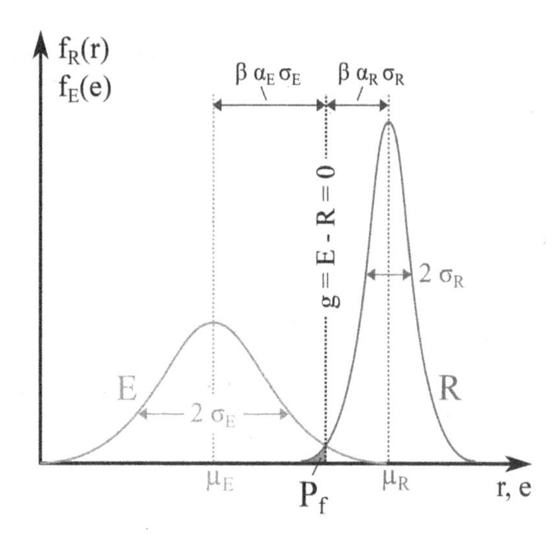

Figure 2. Definition of the limit state regarding log-normal distributions of actions E and the resistance R. The failure probability P_f corresponds to the area where the resistance is smaller than the action.

definition of the failure probability is a compromise between the economic feasibility and the safety of the structure. The variability of the loads and the material parameters is taken into account so that a certain failure is accepted under rare conditions (i.e. strongest loading and worst possible material strength). The consequence and type of failure is also considered by different failure probabilities in the Ultimate Limit State (ULS) as well as the Service Limit State (SLS). For example, a failure probability of $P_f = 7.2 \cdot 10^{-5}$ for the load bearing capacity in the ULS and a failure probability of $P_f = 0.067$ in the SLS are required for a reference period of 50 years assuming consequence class CC2 (i.e. reliability class RC2) according to (European Comittee for Standardisation 2002). Expressed in terms of the reliability index, a value of $\beta = 1.5$ for the SLS and $\beta = 3.8$ for the ULS are required.

3.1 Safety formats for NLFEA

In order to assess the structural reliability by NLFEA, the design criterion according to (fib Model Code for Concrete Structures 2010 2013) can be expressed as

$$F_d \leq R_d = \frac{R_{NLFEA}}{\gamma_R \cdot \gamma_{Rd}}, \tag{8}$$

where F_d is the design value of actions (i.e. formulated in terms of external loads or displacements) and R_{NLFEA} is the global resistance of a structure evaluated by NLFEA. The global resistance safety factor γ_R accounts for the uncertainty of the material properties and is dependent on the chosen safety format. The global safety factor γ_{Rd} accounts for the model uncertainties and its value has not been conclusively clarified. In (fib Model Code for Concrete Structures 2010 2013), values from $\gamma_{Rd} = 1.06$ for non-linear models with low uncertainty up to $\gamma_{Rd} = 1.1$ for non-linear models with high uncertainty are proposed. Explicit definitions to classify the uncertainty of the model are not provided. In case of bending failure, (Hendriks & Roosen 2020) recommends a value of $\gamma_{Rd} = 1.06$. If other or mixed types of failure occurs, a value of $\gamma_{Rd} = 1.15$ should be used.

In contrast to the assessment of structural members by local cross-sectional verification used in engineering practice, a NLFEA of a structural member can capture the redistribution of stresses and therefore offers a higher level of approximation of the global behavior. This also requires a global definition of failure and a evaluation if the model can capture all possible failure modes. In the following different safety formats proposed in (fib Model Code for Concrete Structures 2010 2013) are described.

Partial Safety Factor method (PSF)
The most common approach for the verification of structures is the partial safety factor method and is commonly applied for the evaluation of local material points. It is a simplified verification concept where actions and material properties are modified by partial safety factors to achieve the required level of safety. The partial safety factors given in (European Comittee for Standardisation 2002) are based on reliability indexes $\beta = 1.5$ for the SLS and $\beta = 3.8$ for the ULS for a period of 50 years. If a different failure probability is required, the partial safety factors has to be adjusted. The design resistance is obtained by using the design values of the material properties in a single NLFEA:

$$R_{d,PSF} = \frac{R(f_{cd}, f_{ctd}, \ldots)}{\gamma_{Rd}}. \tag{9}$$

No global resistance safety factor γ_R is applied, because the material properties are already modified by partial safety factors (i.e. $f_{cd} = f_{ck}/1.5$). A concern, especially when using the PSF in NLFEA, is that the application of safety factors on the material properties can influence the global failure mode.

Global Resistance Factor method (GRF)
Using the global resistance factor method, the structural resistance can be expressed by

$$R_{d,GRF} = \frac{R_{NLFEA}(f_{cmd}, f_{ctmd}, \ldots)}{1.2 \cdot \gamma_{Rd}}. \tag{10}$$

The global resistance safety factor is chosen to $\gamma_R = 1.2$ according to (fib Model Code for Concrete Structures 2010 2013). The material properties for the GRF are derived from the characteristic mechanical properties. The characteristic material properties of the concrete (fracture energy, compressive and tensile strength) are multiplied by a factor of 0.85 (i.e. $f_{cmd} = 0.85 f_{ck}$) while the yield strength of the reinforcement is calculated by $f_{ym} = 1.1 f_{yk}$. In this way, the GRF is approximately consistent with the PSF and therefore based on the same reliability index $\beta = 3.8$ (Castaldo, Gino, Bertagnoli, & Mancini 2018).

Method of estimating the coefficient of variation of the structural resistance (ECoV)
The global resistance using the ECoV method can be calculated by

$$R_{d,ECoV} = \frac{R_{NLFEA}(f_{cm}, f_{ctm}, \ldots)}{\gamma_R \cdot \gamma_{Rd}} \tag{11}$$

using the structural resistance predicted by a NLFEA considering the mean values of the material properties. A log-normal distribution for the global load bearing capacity is assumed (see Figure 2) and therefore the global resistance safety factor can be written as:

$$\gamma_R = exp(\alpha_R \cdot \beta \cdot V_R), \tag{12}$$

where α_R is the FORM sensitivity factor ($\alpha_R = 0.8$ in accordance to (fib Model Code for Concrete Structures 2010 2013)), β is the required reliability index (can be explicitly defined) and V_R is the coefficient of variation of the the global structural resistance. The coefficient of variation can be estimated as

$$V_R = \frac{1}{1.65}\, ln\left(\frac{R_{NLFEA}(f_{cm}, f_{ctm}, \ldots)}{R_{NLFEA}(f_{ck}, f_{ctk}, \ldots)}\right), \tag{13}$$

where the variance between the structural resistance predicted by two NLFEA, considering the mean as well as the characteristic values of the material properties, are used.

Probabilistic Method (PM)
The structural reliability can directly be calculated by the probabilistic method. In order to do so, the material parameters have to be modeled by a suitable uncertain variable. In general, a parameter is uncertain, if it is concerned with randomness or a lack of knowledge, and it can be classified into aleatory or epistemic uncertainty. Aleatory uncertainty is characterized by variability and can be represented by random variables, which are quantified by probability density functions. For example, the concrete strength can be modeled by a log-normal distribution (distribution parameters has to be chosen on the corresponding strength class) and material parameters with a strong correlation (i.e. the tensile strength and the fracture energy) can derived from the concrete strength by the formulas given in (fib Model Code for Concrete Structures 2010 2013). In case of epistemic uncertainty, which is characterized by limited data and lack of knowledge, intervals can be used to quantify the range of a parameter (i.e. the influence of the fiber orientation on the post-cracking behavior of SFRC).

By adopting a sampling technique, i.e Latin Hypercube Sampling, corresponding input data for the specified number of samples is generated. In order to calculate the failure probability explicitly, the following Equation has to be evaluated:

$$P_f(R_{NLFEA} - F_d \leq 0) \leq P_{f,requiered}. \tag{14}$$

A number of NLFEA in the magnitude of the failure probability ($\approx 10^6$ for ULS) has to be carried out for an accurate estimate of the failure probability. To overcome the related computational costs, surrogate models can be used to replace the simulation model (Neu, Edler, Freitag, Gudzulic, & Meschke 2022) or the numerical results of several NLFEA can be fitted by an appropriate probabilistic model (Castaldo, Gino, Bertagnoli, & Mancini 2018).

4 INFLUENCE OF THE SAFETY FORMAT ON THE SFRC STRENGTH

In order to investigate the influence of the applied safety concept on the local structural response, the results of a previously carried out validation campaign of the Multi-Level FRC model (Sec. 2) for Dramix 3D 65/60 hooked-end steel fibers are taken as a basis (Gudzulic, Neu, Gebuhr, Anders, & Meschke 2020). The experimental results (light grey) and the numerical prediction (straight black line, donated as 'Validation') of the 3-point bending tests on notched beams (150x150x600 mm) with a fiber content of 57 kg/m³ are shown in Figure 3.

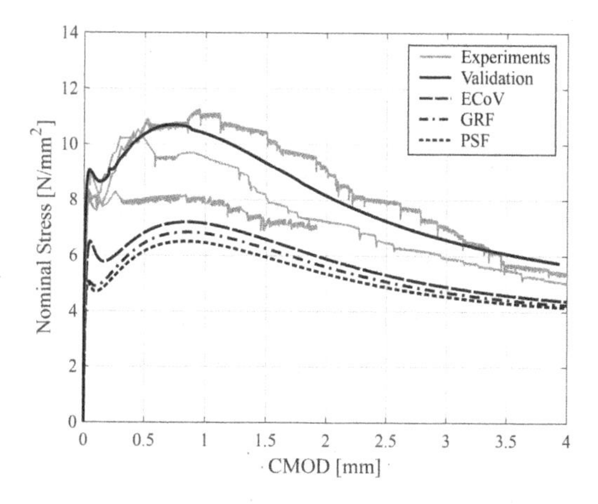

Figure 3. Experimentally and numerically obtained nominal Stress vs. CMOD curves of 3-point bending tests on notched SFRC beams. A comparison between the validation results and the different investigated safety concepts.

4.1 *Safety formats – PSF, GRF & ECoV*

Within the framework of the experimental campaign, only the mean value of the compressive strength was determined ($f_c = 112\,\mathrm{N/mm^2}$) and therefore a C100/115 concrete is assumed for the following evaluation of the different safety formats. The material properties are calculated in accordance to (fib Model Code for Concrete Structures 2010 2013). For the post-cracking response of the SFRC, the traction-separation law $t_{fibers}(\alpha)$ including the values for the coefficients (see Eq. 3) derived in (Gudzulic, Neu, Gebuhr, Anders, & Meschke 2020) is used as a basis. The material parameters are given in Table 1. None of the available guidelines for the structural design by NLFEA contains information regarding the use of SFRC. Therefore, the same reduction factors for the post cracking response of SFRC as for the concrete properties are used (Tab. 1, right column). As no characteristic values for the residual strength of SFRC are defined, the fiber response $t_{fibers}(\alpha)$ is reduced through a multiplication of the traction by a value of 0.85.

Table 1. Material properties used in the NLFEA.

	f_t [N/mm²]	$G_{F,I}$ [N/mm]	$\gamma_{FRC} \cdot t_{FRC}(\alpha)$
PM	Log-Normal	$0.073 \cdot f_c^{0.18}$	1.0
PSF	2.47 (f_{cd})	0.0797	1/1.5
GRF	3.16 (0.85 f_{ctk})	0.1016	0.85
ECoV_m	5.30 (f_{ctm})	0.1707	1.0
ECoV_k	3.71 (f_{ctk})	0.1195	0.85

The resulting nominal stress – Crack Mouth Opening Displacement (CMOD) curves are shown in Figure 3. In terms of the three-point bending test all safety concepts show a similar response. The ECoV method utilize a higher tensile strength, which results in a higher first peak in comparison to the GRF and PSF method. The assumption for the characteristic values of the residual SFRC strength seems reasonable due to the good agreement between the PSF and the GRF/ECoV approaches.

4.2 *Probabilistic method*

The choice of the model used to capture the uncertainty of a material property and the considered range have a great influence on the calculated failure probability by the PM. In order to quantify the influence of the different material parameters on the structural response, a parametric study is carried out first. Three different sets of input parameters are identified:

- Concrete strength: The tensile strength is modeled as log-normal distribution corresponding to a C100/115 concrete ($\mu = 5.3$, $\sigma = 0.94$). The compressive strength, the elasticity modulus and the fracture energy are calculated according to (fib Model Code for Concrete Structures 2010 2013) based on the tensile strength.
- Fiber orientation and distribution: An isotropic fiber orientation is considered by a spherical distribution of the probability density $p(\theta)$ (see Eq.1), with the semi-axis $\lambda_{cast} = [\lambda_F, \frac{1-\lambda_F}{2}, \frac{1-\lambda_F}{2}]$ in which the fiber orientation parameter λ_F is modeled as an interval $\tilde{\lambda}_F = [0.25, 0.5]$. This assumption is based on suggestions in (Tiberti, Germano, Mudadu, & Plizzarri 2018), where the average fiber inclination in cross-sections was investigated for more than 500 bending tests on notched beams. The inclination was measured to be in a range of 34.9° to 53.1°, which correspond to a fiber orientation parameter λ_F between 0.3 and 0.45. Here, this range is slightly

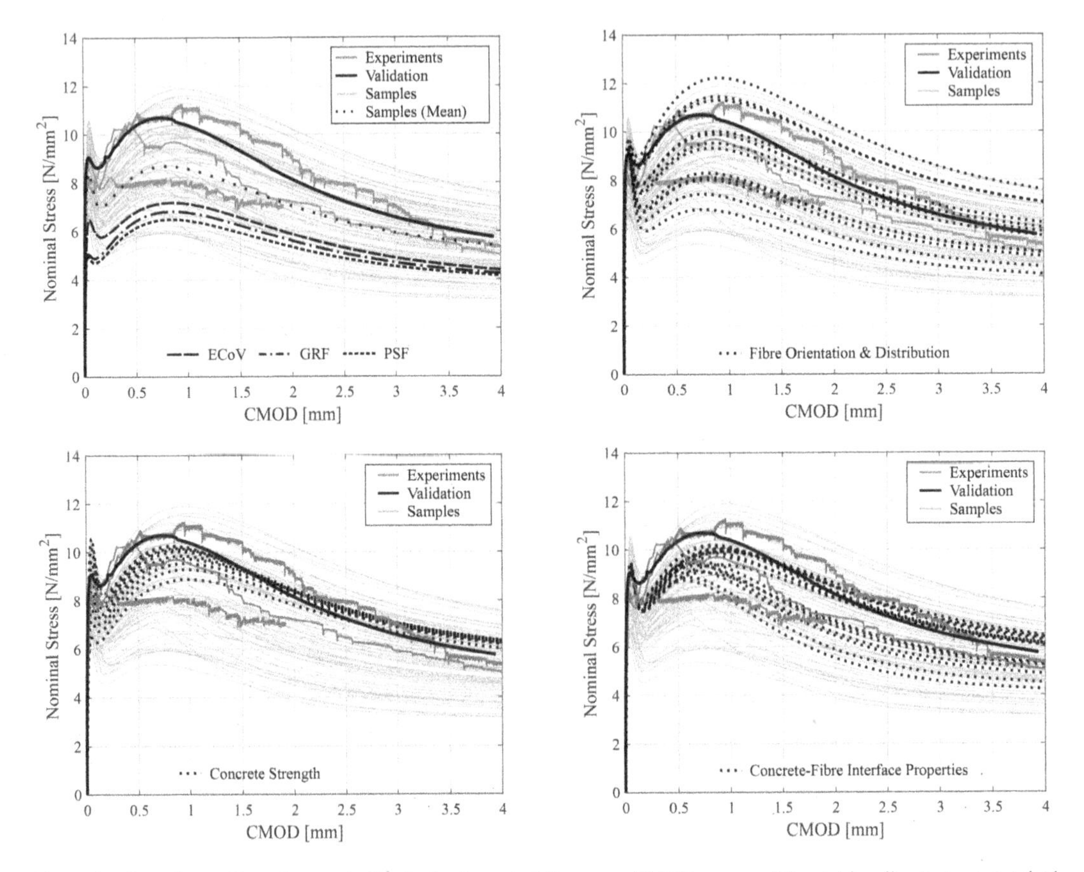

Figure 4. Experimentally and numerically obtained nominal Stress vs. CMOD curves of 3-point bending tests on notched SFRC beams. A comparison between the validation results, the different investigated safety concepts and samples covering the complete material input space is provided (top, left). In addition, different sets of material parameters were independently sampled to investigate their influence on the structural response (rest).

extended in order to be on the save side. A random variation of the fiber content of +/-10% is assumed.

- Concrete-Fiber interface: Based on the investigations of fiber pull-out tests in (Gudzulic, Neu, Gebuhr, Anders, & Meschke 2020), the bond strength τ_{max} is modeled as an interval $\bar{\tau}_{max} = [1.5, 5.0]$ N/mm^2 as well as the residual bond strength $\bar{\tau}_0 = [0.5, 4.0]$ N/mm^2 and the reference slip $\bar{s}_{ref} = [0.02, 0.25]$ mm (see Figure 1, left).

A total number of 128 samples is generated by LHS and the numerical results of the corresponding simulations are show in Figure 4 (top, left). It can be observed that the samples can capture the experimental scatter and also that the lower bound samples predict a smaller residual strength than predicted by the different safety concepts. This indicates that the assumed ranges for the material uncertainty, especially regarding the fiber orientation, are too conservative. In order to evaluate the influence of each set of input parameters on the post-cracking response of SFRC, only samples in the according set are generated while all other material parameters are fixed to their values used in (Gudzulic, Neu, Gebuhr, Anders, & Meschke 2020). The overall greatest influence on the post cracking behavior of SFRC has the fiber orientation and distribution (Figure 4 – top, right). The concrete strength parameters are onlyn affecting the structural response for CMOD values smaller than 1.0mm while the concrete-fiber interface parameters have a great influence at higher CMOD values (Figure 4).

For the calculation of the design resistance by the PM, a feed-forward Artificial Neural Network (ANN) with two hidden layers is used as a surrogate model to replace the FE model of the beam. The ANN is trained by 700 samples generated in the material parameter space and provide the maximum nominal stress f_{max} as an output. The interval to account for the fiber orientation is adjusted to $\bar{\lambda}_F = [0.3, 0.45]$ due to the previous findings (assumed range for the fiber orientation too conservative). Figure 5 shows the histograms based on 10 000 samples evaluated by the ANN (light grey, 'Refined Samples') and based on the 128 samples used in the parametric study (dark grey, 'Random Samples').

In order to calculate the design resistance corresponding to a probability index of $\beta = 3.8$, a log-normal distribution is fitted to both histograms. The 'refined' samples lead to a $\approx$ 13% higher design resistance in comparison to the 'random' samples (6.88 vs

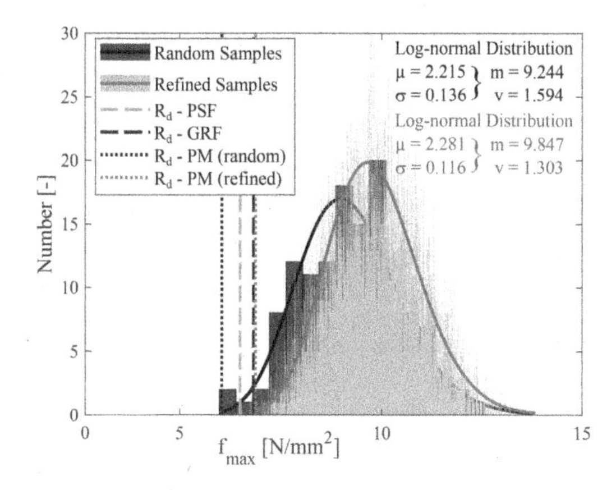

Figure 5. Histograms of the two sample sets for applying the Probability Method (PM) and comparison of the calculated design resistances R_d.

6.06 N/mm^2). When compared to the other safety formats, the PM offers a similar design resistance (PSF: 6.51 N/mm^2, GRF: 6.83 N/mm^2).

The calculated failure probability P_f is dependent on the chosen distribution fitted to the generated samples. To quantify the influence, the failure probability regarding a assumed failure load of $f_{max,fail} = 7.0$ N/mm^2 is calculated directly by using 10^6 samples in conjunction with the ANN and by fitting the samples to different distribution functions. While a failure probability of 0.0013 is calculated by the direct evaluation of samples, the fit to a log normal distribution lead to a value of 0.002. Using only 128 samples for fitting a log normal distribution results in a calculated failure probability of 0.0026. This emphasize how sensitive the estimation of the failure probability is to the fitted probability density distribution function.

5 DESIGN OF A SEGMENTAL LINING RING

In order to investigate the potential use of SFRC as a reinforcement scheme for segmental linings and show the potentials of numerical design, a comparison between conventional reinforced and steel fiber reinforced segmental linings for a reference tunnel project was carried out. The overburden of the tunnel is 21.5 m and the lowest ground water level is assumed at 13.7 m. The surrounding sandy soil has an elasticity modulus of 120 N/mm^2 with a saturated and unsaturated weight of 11 and 21 kN/m^3, respectively. For the lateral earth pressure coefficient K_0 a value of 0.4 is assumed. The dimensions and the resulting stress resultants from the earth & water pressure loading are given in Figure 6. A conventionally reinforced segment (ϕ10-10 + additional rebars in high stressed regions) is compared to a SFRC segment reinforced by 57 kg/m^3 of steel fibers. The overall steel content of the investigated segments is comparable (RC $\approx$ 260 kg/segment and SFRC $\approx$ 258 kg/segment). The material properties used in Section 4 are used. For the rebars an elasticity modulus of 200 000 N/mm^2 and a yield strength according to B500 rebars (European Comittee for Standardisation 2005) is chosen (Validation: 550 N/mm^2, GRF: 550 N/mm^2, ECoV$_k$: = 500 N/mm^2,

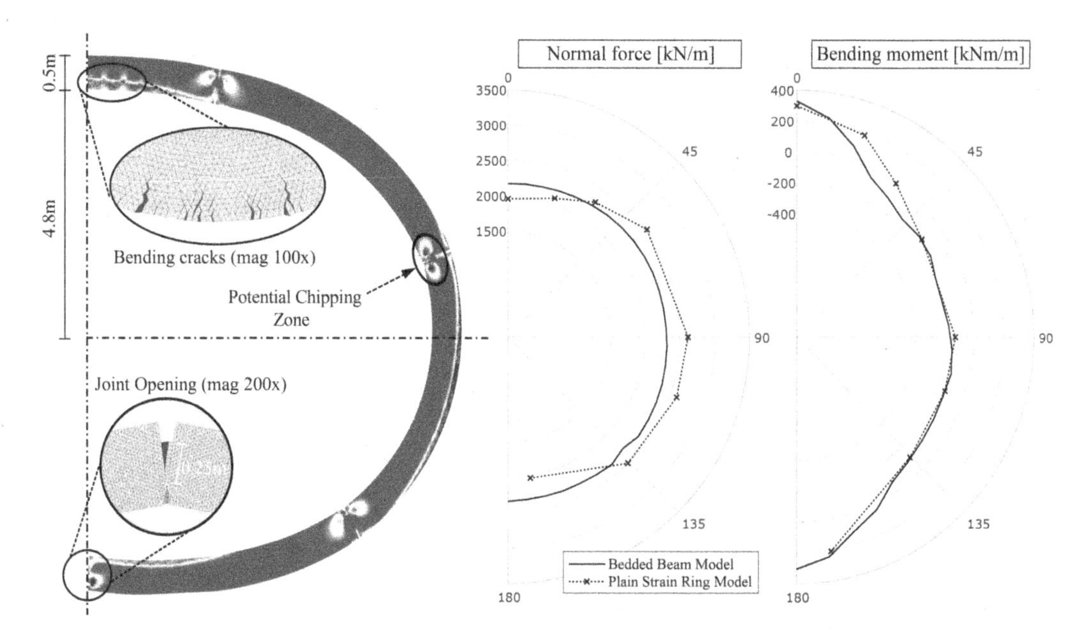

Figure 6. Left – Dimension of the ring and the longitudinal joint, the maximum principal stresses in the segmental lining model and an indication of potential failure mechanisms. Right – Characteristic normal force and bending moment distribution from a linear elastic analysis compared to the stress resultants obtained by a bedded beam model.

ECoV$_m$: = 550 N/mm^2, PM: Gaussian – $\mu = 500.0$, $\sigma = 32.6$). The bond slip behavior is assumed according to (fib Model Code for Concrete Structures 2010 2013), where the bond strength is based on the compressive strength. A plane strain numerical model is used with an element size of 25mm (155 825 elements) and the external load is increased by a factor λ. A load increment of $\Delta\lambda = 0.0005$ is used.

5.1 SFRC vs. RC segments – SLS

In order to evaluate the model response and to compare the conventional with the fiber reinforced segment, their serviceability performance is analyzed. The SLS verifications are carried out by mean values of the material properties without the application of safety factors. Therefore, only the results using the 'Validation' material parameters (see Sec. 4) are evaluated. The corresponding load factor – maximum crack width curves are shown in Figure 7 (black lines). Three stages can be identified: Up to a maximum crack width of 0.06 mm bending cracks are initiated and propagate at the crown segment (Figure 7, A). Then a splitting crack at the bottom longitudinal joint is initiated and propagates together with the bending cracks in the crown segment (Figure 7, B). In Stage C, splitting cracks at the remaining longitudinal joints are initiated and propagating until failure occurs due to a chipping of the longitudinal joint at the bottom.

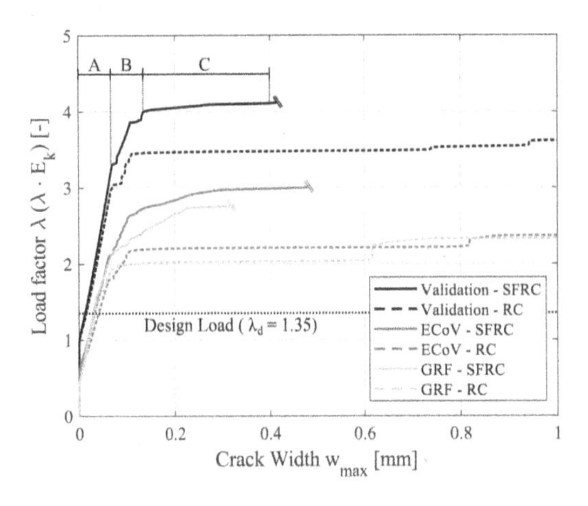

Figure 7. Load-crack width curves for the lining ring containing RC as well as SFRC segments.

In order to ensure a sufficient serviceability performance, the maximum crack width is restricted to 0.2 mm. Both reinforcement schemes show a similar performance up to a crack width of 0.07 mm (Figure 7). After this point, the SFRC segments provide a better crack width control as well as a ≈18% higher peak load ($\lambda_{d,SFRC} = 4.1$ vs. $\lambda_{d,RC} = 3.47$). While the SFRC segment achieves a higher peak load, a less ductile response (failure occurs at smaller crack widths and is not announced by strongly increasing crack widths) in comparison to the RC segment can be observed.

5.2 Application of safety concepts in nonlinear FEA – ULS

The assessment of the ULS by nonlinear FEA requires a definition of the segmental lining ring failure. As structural failure criteria the development of a failure strain or crack width can be considered. In case of RC, (European Comittee for Standardisation 2005) suggests for ductile rebars an ultimate strain of $\epsilon_{ud} = 25$ ‰. For SFRC, an ultimate crack opening of $w_u = 2.5$ mm (or expressed as strain $\epsilon_{Fu} = 2\%$) is suggested by (fib Model Code for Concrete Structures 2010 2013) as failure criterion. This large crack width cannot develop in a confined system such as the investigated segmental lining ring (see Figure 7). Therefore, the shear opening is restricted to 0.2 mm, because such large shear deformations indicate the initiation of a brittle chipping failure. In addition, the principal compressive stresses can be limited (a compressive stress in the magnitude of the compressive strength $\sigma_{comp} = f_{ck} = 100$ N/mm^2 is reached at at approximately $\lambda = 5$) to prevent brittle failure.

The design resistance is calculated by using the PM and the GRF method. For the PM, 128 samples are generated for the SFRC as well as RC segmental lining and fitted to a log normal distribution (analog to Section 4). The corresponding histograms and log-normal distributions are shown in Figure 8. It can be observed that the SFRC segment provides a ≈16% greater mean resistance ($\lambda_{mean,SFRC} = 3.96$ vs. $\lambda_{mean,RC} = 3.4$) and ≈19% greater design resistance ($\lambda_{mean,SFRC} = 3.0$ vs. $\lambda_{mean,RC} = 2.55$). In comparison to the design resistance obtained by the GRF method, the PM provides a ≈8% greater design resistance for SFRC segments and a ≈18% greater design resistance for RC segments. If the design resistance of the bottom joint is calculated by a strut and tie model (German Tunnelling Committee (DAUB) 2013) based on the existing reinforcement area, the design by NLFEA using the GRF can provide a ≈36% higher design resistance.

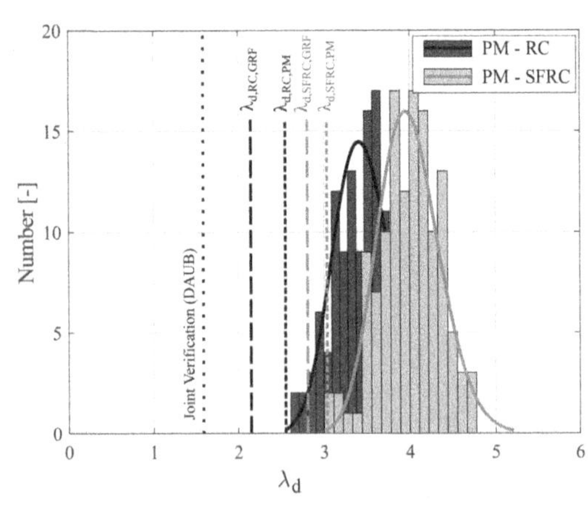

Figure 8. Histograms of 128 samples related to SFRC as well as RC segments. The calculated design resistances R_d by the PM are compared to the ones obtained by the GRF method.

6 CONCLUSION

In this work, the application of NLFEA and potential benefits for the verification of segmental tunnel linings were addressed. The proposed multilevel modeling approach for SFRC is suitable for the model based full probabilistic design of SFRC structural members. Investigations on notched beams subjected to 3-point bending showed that the model is capable to predict the scatter caused by the variance of the concrete properties and the fiber orientation due to the explicit consideration of these important parameters. The chosen uncertainty models for the material properties has a noticeable influence on the predicted design resistance and therefore the material input space was verified by comparing the design resistance obtained by a full probabilistic analysis with the one obtained by simplified safety formats proposed in available guidelines for the assessment of structures by NLFEA. Finally, a design of segmental lining ring subjected to earth & water pressure loading by NLFEA was carried out. It is shown that SFRC segments provide a better crack width control as well as a $\approx 18\%$ higher peak load in comparison to a conventional RC segment while a less ductile response is observed. The design resistance obtained by NLFEA is increased by up to 36% when compared to a conventional design carried out by a strut and tie model. The chosen safety format for the nonlinear design has also a noticeable effect on the calculated design resistance. This contribution highlights, how advanced numerical models used in the nonlinear design of structures could reduce material consumption and therefore provide more cost-effective segmental lining designs with a reduced environmental impact.

ACKNOWLEDGMENTS

Financial support was provided by the German Research Foundation (DFG) in the framework of project B2 of the Collaborative Research Center SFB 837 *Interaction modeling in mechanized tunnelling* (Project number: 77309832) and the Priority Programme SPP 2020 *Cyclic deterioration of High-Performance Concrete in an experimental-virtual lab* (Project number: 353819637). This support is gratefully acknowledged.

REFERENCES

Bakshi, M. & V. Nasri (2017, 1–4 June 2016). Design of fiber-reinforced tunnel segmental lining according to new aci report. In *Canadian Society for Civil Engineering Annual Conference 2016: Resilient Infrastructure*. Canadian Society for Civil Engineering (CSCE).

Carlo, F. D., A. Meda, & Z. Rinaldi (2016). Design procedure for precast fibre-reinforced concrete segments in tunnel lining construction. *Structural Concrete 17*, 747–759.

Castaldo, P., D. Gino, G. Bertagnoli, & G. Mancini (2018). Partial safety factor for resistance model uncertainties in 2d non-linear finite element analysis of reinforced concrete structures. *Engineering Structures 176*, 746–762.

European Comittee for Standardisation (2002). *EN 1990 – Eurocode: Basis of structural design*. European Comittee for Standardisation.

European Comittee for Standardisation (2005). *EN 1992 – Eurocode 2: Design of concrete structures*. European Comittee for Standardisation.

fib Model Code for Concrete Structures 2010 (2013). *Fédération internationale du béton (fib)*. Ernst & Sohn.

Gall, V. E., S. Butt, G. Neu, & G. Meschke (2018). An embedded rebar model for computational analysis of reinforced concrete structures with applications to longitudinal joints in precast tunnel lining segments. In G. Meschke, B. Pichler, and J. G. Rots (Eds.), *Computational Modelling of Concrete Structures (EURO-C 2018)*, pp. 705–714. CRC press.

German Tunnelling Committee (DAUB) (2013). Recommendations for the design, production and installation of segmental rings. Technical report, Deutscher Ausschuss für unterirdisches Bauen e. V. (DAUB).

Gudzulic, V., T. Dang, & G. Meschke (2018). Computational modeling of fiber flow during casting of fresh concrete. *Computational Mechanics 63(6)*, 1111–1129.

Gudzulic, V., G. Neu, G. Gebuhr, S. Anders, & G. Meschke (2020). Numerisches Mehrebenen-Modell für Stahlfaserbeton: Von der Faser- zur Strukturebene. *Beton und Stahlbetonbau 115*, 146–157.

Hendriks, M. & M. Roosen (2020). Guidelines for nonlinear finite element analysis of concrete structures – rtd 1016-1:2020. Technical report, Rijkswaterstaat Centre for Infrastructure.

Laranjeira, F., C. Molins, & A. Aguado (2010). Predicting the pullout response of inclined hooked steel fibers. *Cement and Concrete Research 40*, 1471–1487.

Neu, G., P. Edler, S. Freitag, V. Gudzulic, & G. Meschke (2022). Reliability based optimization of steel-fibre segmental tunnel linings subjected to thrust jack loading (accepted). *Engineering Structures*.

Oliver, J., A. Huespe, & J. Cante (2008). An implicit/explicit integration scheme to increase computability of non-linear material and contact/friction problems. *Computational Methods in Applied Mechanics and Engineering 197*(4), 1865–1889.

Ortiz, M. & A. Pandolfi (1999). Finite-deformation irreversible cohesive elements for three-dimensional crack-propagation analysis. *International Journal for Numerical Methods in Engineering 44*(9), 1267–1282.

Snozzi, L. & J. F. Molinari (2013). A cohesive element model for mixed mode loading with frictional contact capability. *International Journal for Numerical Methods in Engineering 93*(5), 510–526.

Tiberti, G., F. Germano, A. Mudadu, & G. Plizzarri (2018). An overview of the flexural post-cracking behavior of steel fiber reinforced concrete. *Structural Concrete 19*(3), 695–718.

Wang, Y., S. Backer, & V. Li (1989). A statistical tensile model of fibre reinforced cementitious composites. *Composites 20*(3), 265–274.

Zhan, Y. & G. Meschke (2014). Analytical model for the pullout behavior of straight and hooked-end steel fibers. *Journal of Engineering Mechanics (ASCE) 140(12)*, 04014091(1–13).

Zhan, Y. & G. Meschke (2016). Multilevel computational model for failure analysis of steel-fiber - reinforced concrete structures. *Journal of Engineering Mechanics (ASCE) 142*(11), 04016090(1–14).

Computational Modelling of Concrete and Concrete Structures – Meschke, Pichler & Rots (Eds)
 ISBN: 978-1-032-32724-2

Optimised strut and tie model for integrated ULS- and SLS design of RC structures

J. Larsen, P.N. Poulsen, J.F. Olesen & L.C. Hoang
Department of Civil Engineering, Technical University of Denmark, Kgs. Lyngby, Denmark

ABSTRACT: The structural designer has to take serviceability - and ultimate- limit states into account when choosing layout and materials as there exists no universally used tool, which can consider both simultaneously. A tool for the ultimate limit state is the Finite Element Limit Analysis, which can perform material optimisation on a ground structure, which has been proven effective. Here the serviceability limit state is adapted to fit the optimisation in order to handle both limit states simultaneously in a convex solver.

As a first step towards a general tool, the method is set up for bar elements and applied to concrete structures yielding a so-called strut and tie model. For simple ground structures, the tool reproduces known solutions. For large scale structures subjected to multiple load cases the tool shows good results but did suffer from minor numerical instabilities.

1 INTRODUCTION

When designing structures, the designer has to ensure that the structure adheres to requirements in both the Serviceability Limit State (SLS) and the Ultimate Limit State (ULS). Today many design tools exist that can help the designer. These tools can help by visualising the force distribution in the structure or help minimise the material usage through material optimisation. However, in general, these tools do not consider both SLS and ULS requirements simultaneously when doing the material optimisation. In ULS, very effective methods for solid concrete structures exist, which can find optimal designs of large scale structures with reasonable computational cost, e.g. Finite Element Limit Analysis (FELA) (Andersen, Poulsen, & Olesen 2022). These design tools utilise material optimisation on a ground structure as the method of finding optimised designs. The feasible set of FELA is convex, from which the global minimum is directly found by means of a convex solver.

For SLS structural optimisation, no universally used tool exists. One of the most commonly used elasticity based methods is Topology Optimisation (Bendsoe & Sigmund 2003). However, this method is based upon linear material models and is originally not suitable for modelling the non-linear behaviour of cracking in concrete. The cracking of concrete can be modelled with numerical tools, as seen in (Vestergaard, Larsen, Hoang, Poulsen, & Feddersen 2021), from which the current papers has drawn its inspiration. However, the referenced paper, does not performer material optimisation. In general, there is a lack of methods for material optimisation while considering the cracking of concrete. Thus a design tool taking into account the cracking of concrete, as well as both SLS and ULS requirements, is needed. The effectiveness of the methods for material optimisation in ULS applying FELA is due to the convex solver that effectively solves large-scale problems (Boyd & Vandenberghe 2004). To solve the combined problem of SLS and ULS requirements the aim is to also include the SLS requirements in the convex solver. However, the SLS requirements are non-convex.

For the convex solver to be used on a non-convex problem, the problem must be reformulated. Approximations can be made in several ways, such as linearisation of the model or second-order Taylor expansion. For the second order Taylor expansion, the hessian has to be positive-definite for a convex solver. If the hessian is not positive-definite, it can be approximated by removing the negative eigenvalues of the hessian (Duchi 2018). However, in this paper, only a first-order Taylor expansion is applied.

A first step in developing a general tool for solids is presented in this paper, where bar elements are used instead of solids. For reinforced concrete structures, the use of bar elements is often called a strut and tie model (Schlaich, Shafer, Jennewein, & Kotsovos 1987).

2 CONSTITUTIVE MODEL

To simulate the behaviour of reinforced concrete, a constitutive model is needed. The behaviour is approximated by a bi-linear model in both compression and tension, such that plastic and elastic behaviour can be represented. The difference in stiffness of concrete and reinforcement is furthermore included in the model.

DOI 10.1201/9781003316404-89

This leads to a quad-linear stress-strain curve which, as seen in Figure 1.

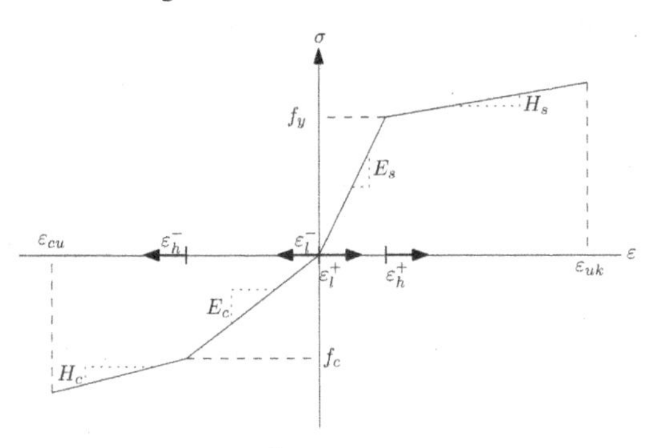

Figure 1. Constitutive model.

Where E_c and E_s are the young's moduli and H_c and H_s is the hardening modulus of concrete and reinforcement, respectively. f_c and f_s are the yield strengths, and ε_{cu} and ε_{uk} are the ultimate strain of concrete and reinforcement, respectively.

Table 1 shows the parameters used in the model. The values correspond to those found in (for Standardization 2005). The first set of parameters in Table 1 represents an SLS condition, i.e. a pure elastic behaviour, where no hardening is allowed, and the stresses are constrained to be within the pure elastic domain of reinforced concrete. The limit of the elastic regime is chosen to be half of the yield strength. The second set of parameters represents ULS conditions, where strains can reach the ultimate strain of the respective material.

Table 1. Material parameters.

SLS			
$\sigma_{c,max}$	$\varepsilon_{c,max}$	E_c	H_c
17.5 MPa	0.000515	34 GPa	6.63 Mpa
$\sigma_{s,max}$	$\varepsilon_{s,max}$	E_s	H_s
250 MPa	0.00125	200 Gpa	410 MPa
ULS			
f_c	ε_{cu}	E_c	H_c
35 MPa	0.0035	34 GPa	6.63 Mpa
f_y	ε_{uk}	E_s	H_s
500 MPa	0.05	200 Gpa	410 MPa

For modelling purposes, the strains in the model have to be split into different parts. The strains in an element are given as the sum of the concrete strains and the reinforcement strains. As the reinforcement is in tension it is given that $\varepsilon_l^- \geq 0$ and $\varepsilon_h^- \geq 0$. While for the concrete in compression, and the strains are thus negative, a minus is introduced such that $\varepsilon_l^+ \geq 0$ and $\varepsilon_h^+ \geq 0$. This is expressed as:

$$\varepsilon_e = \varepsilon_{l,e}^+ + \varepsilon_{h,e}^+ - \varepsilon_{l,e}^- - \varepsilon_{h,e}^-$$

This formulation does not ensure compatible solutions. However, as minimum potential energy is obtained, compatibility is ensured at the optimal point. To use this model in a FEM formulation the strains in the element, ε_e, is calculated as:

$$\varepsilon_e = \mathbf{B}_e \mathbf{v}_e \tag{1}$$

where $\mathbf{v}_e$ is the nodal displacement of element e and $\mathbf{B}_e$ is the global strain-displacement matrix as defined in e.g. (Kuna 2013).

3 MINIMUM POTENTIAL ENERGY

The principle of minimum potential energy is utilised to ensure an admissible displacement field, which states: The actual displacement, which satisfies stable equilibrium, renders the potential energy minimum.

The total potential energy can be found as the sum of the strain energy and the potential energy associated with applied forces:

$$E_{pot}^{tot} = E_{pot}^{in} + E_{pot}^{ext}$$

This method can find the displacement field for the quad-linear model without using a stiffness matrix.

The internal strain energy can be determined as the integration of the potential energy density.

$$E_{pot}^{in} = \int_\Omega P' \mathbf{d}\Omega$$

The strain energy is given as the absolute sum of areas under the stress-strain curve. This can be expressed through two triangular contributions from the concrete and reinforcement, along with a rectangular contribution from each material. This is expressed in Equation 2 for element e.

$$\begin{aligned}\int_{\Omega_e} P' \mathbf{d}\Omega_e = {} & \tfrac{1}{2} E_c A_{c,e} L_e (\varepsilon_{l,e}^-)^2 + \tfrac{1}{2} H_c A_{c,e} L_e (\varepsilon_{h,e}^-)^2 \\ & + \tfrac{1}{2} E_s A_{s,e} L_e (\varepsilon_{l,e}^+)^2 + \tfrac{1}{2} H_s A_{s,e} L_e (\varepsilon_{h,e}^+)^2 \\ & + A_{c,e} L_e f_c (\varepsilon_{h,e}^-) + A_{s,e} L_e f_y (\varepsilon_{h,e}^+)\end{aligned} \tag{2}$$

where $A_{s,e}$ and $A_{c,e}$ are the reinforcement and concrete areas of element e respectively, with L_e being the length of the element. P' is the potential energy density which is integrated over the domain Ω. This can be written as a second order equation:

$$\int_{\Omega_e} P' \mathbf{d}\Omega_e = \tfrac{1}{2} \mathbf{x}_e^{\mathrm{T}} \mathbf{Q}_e \mathbf{x}_e + \mathbf{c}_e^{\mathrm{T}} \mathbf{x} \tag{3}$$

Where $\mathbf{Q}_e$, $\mathbf{x}_e$ and $\mathbf{c}_e$ are given by

$$\mathbf{Q}_e = \begin{bmatrix} E_c A_{c,e} L_e & 0 & 0 & 0 \\ 0 & H_c A_{c,e} L_e & 0 & 0 \\ 0 & 0 & E_s A_{s,e} L_e & 0 \\ 0 & 0 & 0 & H_s A_{s,e} L_e \end{bmatrix}$$

$$\mathbf{x}_e = \begin{bmatrix} \varepsilon_{c,l,e} \\ \varepsilon_{c,h,e} \\ \varepsilon_{s,l,e} \\ \varepsilon_{s,h,e} \end{bmatrix}, \qquad \mathbf{c}_e = \begin{bmatrix} 0 \\ A_{c,e} L_e f_c \\ 0 \\ A_{s,e} L_e f_y \end{bmatrix}$$

The external potential energy can be found by the product of external forces and displacements, which can be expressed in a vectorised form, as seen below.

$$E_{pot}^{ext} = -\sum_{i} F_i u_i = -\mathbf{R}^{\mathrm{T}}\mathbf{V}$$

Where F is a force and u is the displacement. $\mathbf{R}$ is the nodal load vector, and $\mathbf{V}$ is the nodal displacement vector.

To find the displacement field that leads to minimum potential energy, a convex optimisation problem is presented. The objective is to minimise the potential energy while adhering to equilibrium and stress constraints, which is presented in Equation 4.

$$\text{Min.} \sum_{e=1}^{n_{el}} \gamma_e - \mathbf{R}^{\mathrm{T}}\mathbf{V} \tag{4a}$$

$$\text{S.t. } \mathbf{H}(\mathbf{E}\odot\mathbf{A}\odot\boldsymbol{\varepsilon}) = \mathbf{R} \tag{4b}$$

$$\mathbf{B}_e\mathbf{v}_e = \varepsilon_{l,e}^{+} + \varepsilon_{h,e}^{+} - \varepsilon_{l,e}^{-} - \varepsilon_{h,e}^{-}, \quad \forall e \tag{4c}$$

$$\tfrac{1}{2}\mathbf{x}_e^{\mathrm{T}}\mathbf{Q}_e\mathbf{x}_e + \mathbf{c}_e^{\mathrm{T}}\mathbf{x}_e - \gamma_e \leq 0, \quad \forall e \tag{4d}$$

Where $\boldsymbol{\varepsilon}$ is a strain vector, and $\mathbf{H}$ is the equilibrium matrix (see e.g. (Damkilde 1991)), which is assembled from contributions of each element. $\mathbf{A}$ and $\mathbf{E}$ are the generalised areas and Young's moduli of the elements, consisting of contributions from the reinforcement and concrete for both $\mathbf{A}$ and $\mathbf{E}$, while contributions from both the linear and hardening stiffness are also given in $\mathbf{E}$. The symbol $\odot$ is the Hadamard product signifying entrywise product.

In Equation 4 the objective Equation 4a is to minimise the potential energy, Equation 4b is the equilibrium constraint, Equation 4c is the strain split constraint and Equation 4d is the constraint, defining the potential energy at the optimal point. It should be noted that equilibrium is ensured at the point of minimum potential energy. However, for the use with material optimisation, the equilibrium is also formulated explicitly.

This minimisation problem can be solved directly as the problem is convex, which it is since the objective along with the equilibrium, and yielding constraints are linear and thus convex. The potential energy constraint is, however quadratic, and is only convex if $\mathbf{Q}$ is symmetric and positive semidefinite, which is true as $\mathbf{Q}$ is diagonal with non-negative diagonal entries. This rotated quadratic cone can be assembled for all elements, such that a single cone can be used to represent the total potential energy of the structure. The inequality in Equation 4d is needed for convexity. However, in the case of minimum potential energy, the inequality becomes an equality.

4 MATERIAL OPTIMISATION

The theory of minimum potential energy will lead to a stable deformation field and thus an admissible strut and tie model. However, this is not a optimal model as redundant material might be used. Thus the use of material optimisation is needed. The areas of reinforcement and concrete are introduced as variables. This is added as an objective function, along with the potential energy, which leads to the following multi-criterion optimisation problem.

$$\text{Min.} \left[\sum_{e=1}^{n_{el}} \gamma_e - \mathbf{R}^{\mathrm{T}}\mathbf{V} \;\; \sum_{e=1}^{n_{el}} \left(\tfrac{f_y}{f_c} A_{s,e} + A_{c,e}\right)\right]^{\mathrm{T}}$$

$$\text{S.t. } \mathbf{H}(\mathbf{E}\odot\mathbf{A}\odot\boldsymbol{\varepsilon}) = \mathbf{R}$$

$$\mathbf{B}_e\mathbf{v}_e = \varepsilon_{l,e}^{+} + \varepsilon_{h,e}^{+} - \varepsilon_{l,e}^{-} - \varepsilon_{h,e}^{-}, \quad \forall e$$

$$\int_{\Omega_i} P' \mathrm{d}\Omega_e \leq \gamma_e, \quad \forall e$$

The problem is, however, non-convex, which can be seen by investigating the potential energy density given by the first equation in Equation 2. It is noted that this is given as a third-order polynomial, which can never be convex. To solve the problem, a sequential convex program is formulated, where a series of convex approximations of the problem is solved.

4.1 *Local convex approximation*

The potential energy density defined by Equation 2, where the reinforcement and concrete areas and strains are split into an increment and a value given from the former iteration. This is illustrated for the reinforcement area in Equation 5.

$$A_{s,e} = A_{s,0,e} + \Delta A_{s,e} \tag{5}$$

The same is introduced for the concrete area, as well as all strains. When introduced into Equation 2 a third order equation arises, which is approximated by a first-order Taylor expansion, leading to a purely linear model, as seen in Equation 6.

$$\int_{\Omega_e} P'\mathbf{d}\Omega_e \approx \mathbf{c}_{lin,e}^{\mathrm{T}}\Delta\mathbf{x}_e, \quad \Delta\mathbf{x}_e = \begin{bmatrix} \Delta\varepsilon_{l,e}^{-} \\ \Delta\varepsilon_{h,e}^{-} \\ \Delta\varepsilon_{l,e}^{+} \\ \Delta\varepsilon_{h,e}^{+} \\ \Delta A_c \\ \Delta A_s \end{bmatrix}$$

$$\mathbf{c}_{lin,e} = \begin{bmatrix} L_e E_c A_{c,0,e}\varepsilon_{l,0,e}^{-} \\ L_e H_c A_{c,0,e}\varepsilon_{h,0,e}^{-} + L_e f_{ck} H_c \varepsilon_{h,0,e}^{-} \\ L_e E_s A_{s,0,e}\varepsilon_{l,0,e}^{+} \\ L_e H_s A_{s,0,e}\varepsilon_{h,0,e}^{+} + L_e f_y L H_s \varepsilon_{h,0,e}^{+} \\ \frac{L_e}{2}\left(E_c(\varepsilon_{l,0,e}^{-})^2 + H_c(\varepsilon_{h,0,e}^{-})^2\right) + f_{ck}\varepsilon_{h,0,e}^{-} \\ \frac{L_e}{2}\left(E_s(\varepsilon_{l,0,e}^{+})^2 + H_s(\varepsilon_{h,0,e}^{+})^2\right) + f_y\varepsilon_{h,0,e}^{+} \end{bmatrix} \tag{6}$$

This can be formulated as an equality constraint, as it is linear and convex.

The equilibrium constraint can be proven to be non-convex, by writing it in a second order form, and proving that the matrix coefficient of the second order term is not positive-semidefinite. Thus a approximation based upon a first order Taylor expansion is used:

$$\mathbf{H}(\mathbf{E}\odot\mathbf{A}_0\odot\Delta\boldsymbol{\varepsilon} + \mathbf{E}\odot\Delta\mathbf{A}\odot\boldsymbol{\varepsilon}_0) \approx \mathbf{R} - \mathbf{H}_0 \tag{7}$$

where $\mathbf{H}_0 = \mathbf{H}(\mathbf{E} \odot \mathbf{A}_0 \odot \boldsymbol{\varepsilon}_0)$. This approximation is obviously linear and thus convex. The yielding constraints are still found to be linear and are thus convex as is.

The convex approximations are only viable when the changes are small. A limit on the design variables is thus implemented, through a box constraint, to ensure adequately good approximations. These approximations also turn the second-order cone program described in Equation 4 into a linear program.

4.2 *Weighted sum method*

The weighted sum method is used to solve the multi-criterion optimisation problem, where the two objectives are combined in a weighting sum. If the weighting on the potential energy is high, the algorithm will increase the material to minimise the potential energy. If the weight is low, minimum potential energy will not be guaranteed for the observed structure. The weighting is chosen as unity on the material optimisation and a weighting of α_{ws} on the potential energy. This leads to the following problem.

$$
\begin{aligned}
&\text{Min. } \alpha_{ws}\left(\sum_{e=1}^{n_{el}+1} \gamma_e - \mathbf{R}^{\mathrm{T}}\mathbf{V}\right) + \sum_{e=1}^{n_{el}} \frac{f_y}{f_c} A_{s,e} + A_{c,e} \\
&\text{S.t. } \mathbf{H}(\mathbf{E} \odot \mathbf{A}_0 \odot \Delta\boldsymbol{\varepsilon} + \mathbf{E} \odot \Delta\mathbf{A} \odot \boldsymbol{\varepsilon}_0) = \mathbf{R} - \mathbf{H}_0 \quad (8) \\
&\quad \mathbf{B}_e \mathbf{v}_e = \varepsilon_{l,e}^{+} + \varepsilon_{h,e}^{+} - \varepsilon_{l,e}^{-} - \varepsilon_{h,e}^{-}, \quad \forall e \\
&\quad \mathbf{c}_{lin,e} \mathbf{x}_{lin,e} = \gamma_e, \quad \forall e
\end{aligned}
$$

With the volume given in m^3 and the potential energy given in MJ, it was empirically found that a weighting of $\alpha_{ws} = 10^{-2}$ lead to reliable results.

A pseudo-code that represents the algorithm can be seen in Algorithm 1.

Algorithm 1: Pseudo-code for the sequential convex program for strut and tie.

```
Initialisation of Topology;
FELA;
Update Areas;
Solve Equation 4;
Define Strains;
for i = 1 To n do
    Solve Equation 8;
    Update Areas;
    Solve Equation 4;
    Define Strains;
end
```

Note that Equation 8 utilise changes of design variables and is solved through approximations, while Equation 4 is solved with the total design variables and is solved precisely in each iteration.

4.3 *Multiple load cases for different limit states*

The method is expanded to contain several load cases where each load case can be of a different limit state. This is done by allocating the respective material properties to each load case, depending on if the calculations belong to SLS or ULS. Each new load case will introduce new strain variables, each independent of the other load cases. However, the cross-sectional areas are shared between all load cases.

5 RESULTS

The method will be applied to two examples, where the solution is presented.

5.1 *Example 1 – deep beam*

The first example is a deep beam loaded at the quarter-point. The structure is illustrated in Figure 2.

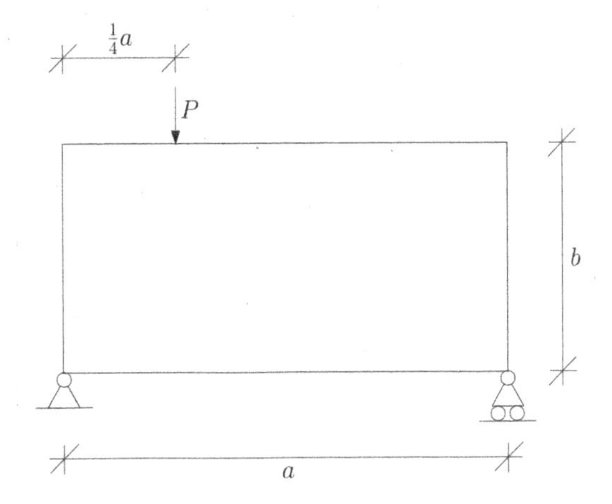

Figure 2. First Example - deep beam.

The example will be investigated for two different ground structures. The first one is given as coarse mesh, while the second will be for a finer mesh. The parameters for the model are summarised in Table 2.

Table 2. Parameters for example 1.

a	b	P	Limit State
5 m	10 m	200 kN	ULS

First, the coarse ground structure is chosen, with 15 nodes, where each node is connected to other nodes by bar elements, leading to a total of 105 elements. The ground structure, as well as the optimised structure, can be seen in Figure 3 and Figure 4 respectively.

The optimised strut and tie is seen to be relatively simple and is verifiable by hand calculation.

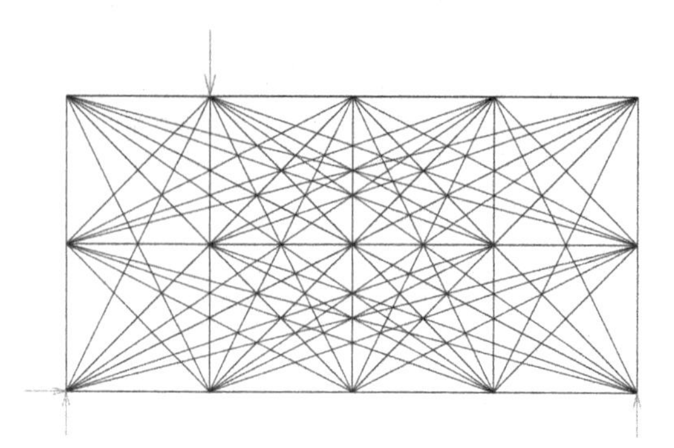

Figure 3. Course ground structure of first example.

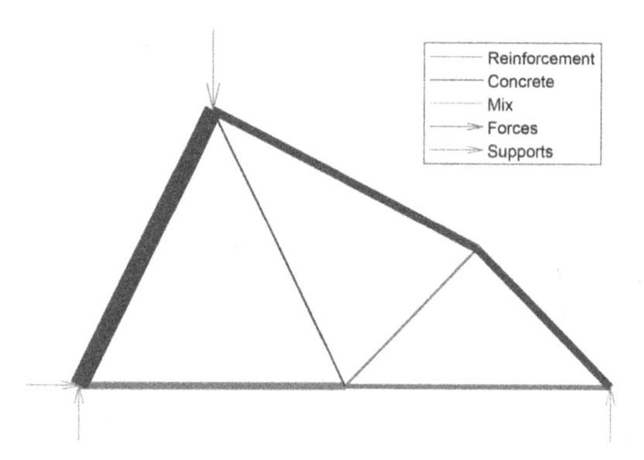

Figure 4. Optimised structure of first example with course ground structure. Only elements with a capacity of more than 1% of the highest capacity is plotted.

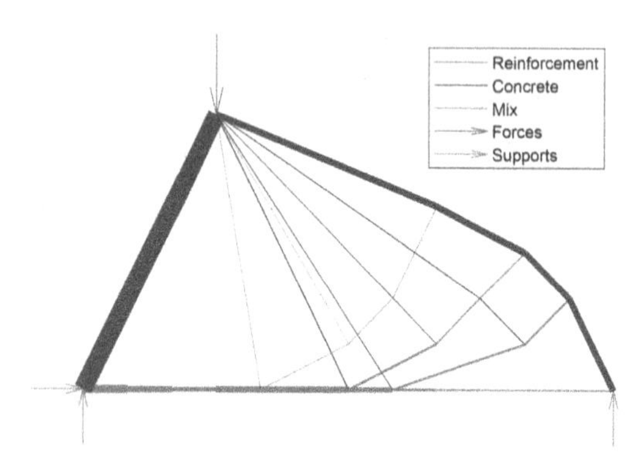

Figure 5. Optimised structure of first example with fine ground structure. Only elements with a capacity of more than 1% of the highest capacity is plotted.

As stated, the example is also investigated with a much finer mesh, given by 153 nodes, which are connected with elements, leading to 11628 elements.

The results in Figure 5 is seen to be similar to a well-known Michell structure (Michell 1904). Notably, some bars seem to change thickness, even though there is no joint with other elements. This is due to the number of elements connecting to each node along with the elements. These elements are tiny in area and are thus below the plotting threshold. However, the sum of the normal forces in these elements is enough to accumulate a significant force to facilitate these changes in thickness.

An example is also run to investigate if the Michell structure is also present for two load cases, where a second load case is added, with an equal force in the 3/4 point.

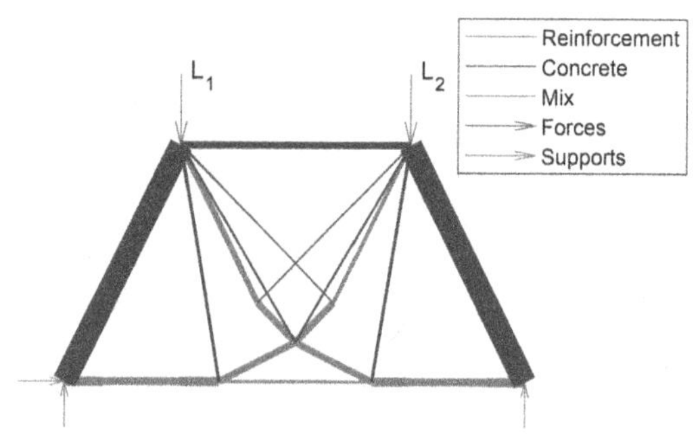

Figure 6. First example with fine ground structure and two load cases. Only elements with a capacity of more than 1% of the highest capacity is plotted. L_1 and L_2 indicates the loads for the first and second load case respectively.

Here the Michell-like structure disappears, and a more straightforward structure is achieved.

To investigate the robustness of the solution, the coarse mesh is also optimised based upon a homogeneous initial guess of material instead of a FELA optimised initial guess. When doing this, the solutions converged slowly to the same solution as the initial guess of FELA, indicating that even though the method is non-convex and globally optimal solutions are not guaranteed, the method seems to find solution solutions close to the initial guess of FELA.

5.2 Example 2 – multi-storey shear wall

The second example is given as a multi-storey shear wall with holes, which can be seen in Figure 7.

Each floor of the wall is 5 m wide and 5 m tall. Furthermore, the door opening is 1 m wide and 3 m tall. The structure is loaded by four different load cases, which are summarised in Table 3.

The ground structure, along with the optimised strut and tie, can be seen in Figure 8 and Figure 9 respectively.

It is noted that the optimised strut and tie is too complex to be found by hand and probably also too complex ever to be built in reality. However, this model could be used for an engineer to understand the stress distribution and create a simpler strut and tie model based upon the material distribution.

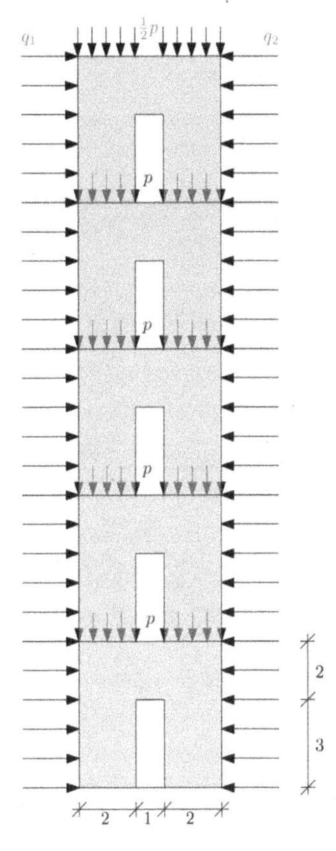

Figure 7. Mesh of structure for second example.

Table 3. Load cases for example 2.

Load Case	q_1 kN/m	q_2 kN/m	p kN/m	Limit state
1	20	0	30	ULS
2	0	20	30	ULS
3	13.3	0	30	SLS
4	0	13.3	30	SLS

6 DISCUSSION

The strut and tie models found by the method are seen to be very complex, and some post-processing is needed. The strut and tie models often produce Michell-like structures and would thus lead to designs where reinforcement bars need to be bent into curves, which is often not practical. Thus the designs should be used as an initial model for understanding the stress-distribution in the structure, and a simpler and reasonable strut and tie model could be created based upon the knowledge acquired. Furthermore, the local stress state in the nodes have to be investigated, however this is not considered in this paper.

When investigating example 2, the method was unstable, and the optimisation could become infeasible before convergence. This behaviour was especially

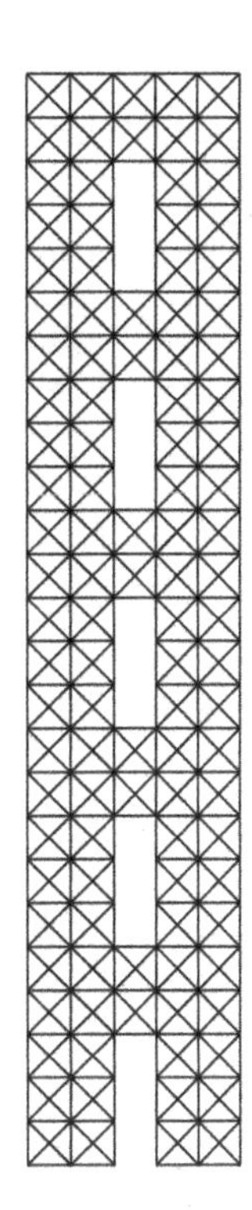

Figure 8. Ground structure of second example.

Figure 9. Optimised structure of Second example with course ground structure. Only elements with a capacity of more than 1% of the highest capacity is plotted.

present when optimising for many load cases simultaneously. This is a problem that can occur when performing sequential convex programming (Duchi 2018).

This is thought to be caused by the narrow solution space of the elastic solutions. Widening of the solution space might lead to better stability, which could be done through relaxation of the constraints. To ensure that the method will converge to a feasible solution, when the relaxation is applied, a penalty function could be introduced, where the constraints are introduced into the objective function, such that relaxation of the constraints are allowed at a price.

Along with this an obvious next step is to expand the method to 3-dimensional solid elements, such that realistic structures can be represented.

7 CONCLUSION

In this paper, a method for finding Strut and Tie models for reinforced concrete structures, subjected to both SLS and ULS load cases is presented.

The method suffers from instability and effectiveness issues, which was mostly seen for large scale structures, especially when subjected to many load cases, where the sequential convex programming could be infeasible after some iterations, before convergence was achieved. The convergence of the methods is also quite slow as the step-size has to be small to mitigate this instability. However, for simple structures, well-known solutions were reproduced, with Michell-like structures being achieved for fine meshed ground structures. The method also produced reliable results for relative complex structures, which were loaded with multiple load cases where both SLS and ULS were considered, despite the instability and effectiveness issues stated earlier.

REFERENCES

Andersen, M., P. N. Poulsen, & J. F. Olesen (2022). Partially mixed lower bound constant stress tetrahedral element for finite element limit analysis. *Computers & Structures 258*.

Bendsoe, M. & O. Sigmund (2003). *Topology Optimization: Theory, Methods, and Applications*. Springer Berlin Heidelberg.

Boyd, S. & L. Vandenberghe (2004). *Convex Optimization*. Cambridge University Press.

Damkilde, L. (1991). An efficient implementation of limit state calculations based on lower-bound solutions. In *Proceedings of Fourth Nordic Seminar on Computational Mechanics*.

Duchi, J. (2018). *Sequential Convex Programming*. Stanford Universit.

for Standardization, E. C. (2005). *EN 1992-1-1 Eurocode 2: Design of concrete structures - Part 1-1: General ruels and rules for buildings*. CEN.

Kuna, M. (2013, 01). *Finite elements in fracture mechanics: Theory – Numerics – Applications*, Volume 201. Springer Berlin Heidelberg.

Michell, A. (1904). The limits of economy of material in frame-structures. *The London, Edinburgh, and Dublin Philosophical Magazine and Journal of Science 8*(47), 589–597.

Schlaich, J., K. Shafer, M. Jennewein, & M. Kotsovos (1987, 05). Toward a consistent design of structural concrete. *PCI Journal 32*, 74–150.

Vestergaard, D., K. Larsen, L. Hoang, P. Poulsen, & B. Feddersen (2021). Design-oriented elasto-plastic analysis of reinforced concrete structures with in-plane forces applying convex optimization. *Structural Concrete 22*(6), 3272–3287.

Computational Modelling of Concrete and
Concrete Structures – Meschke, Pichler & Rots (Eds)
© 2022 Copyright the Editor(s), ISBN: 978-1-032-32724-2

Author index